Some Physical Constants

Quantity	Symbol	Value[a]
Atomic mass unit	u	$1.660\,538\,86\,(28) \times 10^{-27}$ kg $931.494\,043\,(80)$ MeV/c^2
Avogadro's number	N_A	$6.022\,141\,5\,(10) \times 10^{23}$ particles/mol
Bohr magneton	$\mu_B = \dfrac{e\hbar}{2m_e}$	$9.274\,009\,49\,(80) \times 10^{-24}$ J/T
Bohr radius	$a_0 = \dfrac{\hbar^2}{m_e e^2 k_e}$	$5.291\,772\,108\,(18) \times 10^{-11}$ m
Boltzmann's constant	$k_B = \dfrac{R}{N_A}$	$1.380\,650\,5\,(24) \times 10^{-23}$ J/K
Compton wavelength	$\lambda_C = \dfrac{h}{m_e c}$	$2.426\,310\,238\,(16) \times 10^{-12}$ m
Coulomb constant	$k_e = \dfrac{1}{4\pi\epsilon_0}$	$8.987\,551\,788\ldots \times 10^9\ \mathrm{N \cdot m^2/C^2}$ (exact)
Deuteron mass	m_d	$3.343\,583\,35\,(57) \times 10^{-27}$ kg $2.013\,553\,212\,70\,(35)$ u
Electron mass	m_e	$9.109\,382\,6\,(16) \times 10^{-31}$ kg $5.485\,799\,094\,5\,(24) \times 10^{-4}$ u $0.510\,998\,918\,(44)$ MeV/c^2
Electron volt	eV	$1.602\,176\,53\,(14) \times 10^{-19}$ J
Elementary charge	e	$1.602\,176\,53\,(14) \times 10^{-19}$ C
Gas constant	R	$8.314\,472\,(15)$ J/mol·K
Gravitational constant	G	$6.674\,2\,(10) \times 10^{-11}\ \mathrm{N \cdot m^2/kg^2}$
Josephson frequency–voltage ratio	$\dfrac{2e}{h}$	$4.835\,978\,79\,(41) \times 10^{14}$ Hz/V
Magnetic flux quantum	$\Phi_0 = \dfrac{h}{2e}$	$2.067\,833\,72\,(18) \times 10^{-15}\ \mathrm{T \cdot m^2}$
Neutron mass	m_n	$1.674\,927\,28\,(29) \times 10^{-27}$ kg $1.008\,664\,915\,60\,(55)$ u $939.565\,360\,(81)$ MeV/c^2
Nuclear magneton	$\mu_n = \dfrac{e\hbar}{2m_p}$	$5.050\,783\,43\,(43) \times 10^{-27}$ J/T
Permeability of free space	μ_0	$4\pi \times 10^{-7}$ T·m/A (exact)
Permittivity of free space	$\epsilon_0 = \dfrac{1}{\mu_0 c^2}$	$8.854\,187\,817\ldots \times 10^{-12}\ \mathrm{C^2/N \cdot m^2}$ (exact)
Planck's constant	h $\hbar = \dfrac{h}{2\pi}$	$6.626\,069\,3\,(11) \times 10^{-34}$ J·s $1.054\,571\,68\,(18) \times 10^{-34}$ J·s
Proton mass	m_p	$1.672\,621\,71\,(29) \times 10^{-27}$ kg $1.007\,276\,466\,88\,(13)$ u $938.272\,029\,(80)$ MeV/c^2
Rydberg constant	R_H	$1.097\,373\,156\,852\,5\,(73) \times 10^7\ \mathrm{m^{-1}}$
Speed of light in vacuum	c	$2.997\,924\,58 \times 10^8$ m/s (exact)

Note: These constants are the values recommended in 2002 by CODATA, based on a least-squares adjustment of data from different measurements. For a more complete list, see P. J. Mohr and B. N. Taylor, "CODATA Recommended Values of the Fundamental Physical Constants: 2002." *Rev. Mod. Phys.* **77**:1, 2005.

[a] The numbers in parentheses for the values represent the uncertainties of the last two digits.

Solar System Data

Body	Mass (kg)	Mean Radius (m)	Period (s)	Distance from the Sun (m)
Mercury	3.18×10^{23}	2.43×10^{6}	7.60×10^{6}	5.79×10^{10}
Venus	4.88×10^{24}	6.06×10^{6}	1.94×10^{7}	1.08×10^{11}
Earth	5.98×10^{24}	6.37×10^{6}	3.156×10^{7}	1.496×10^{11}
Mars	6.42×10^{23}	3.37×10^{6}	5.94×10^{7}	2.28×10^{11}
Jupiter	1.90×10^{27}	6.99×10^{7}	3.74×10^{8}	7.78×10^{11}
Saturn	5.68×10^{26}	5.85×10^{7}	9.35×10^{8}	1.43×10^{12}
Uranus	8.68×10^{25}	2.33×10^{7}	2.64×10^{9}	2.87×10^{12}
Neptune	1.03×10^{26}	2.21×10^{7}	5.22×10^{9}	4.50×10^{12}
Pluto[a]	$\approx 1.4 \times 10^{22}$	$\approx 1.5 \times 10^{6}$	7.82×10^{9}	5.91×10^{12}
Moon	7.36×10^{22}	1.74×10^{6}	—	—
Sun	1.991×10^{30}	6.96×10^{8}	—	—

[a] In August 2006, the International Astronomical Union adopted a definition of a planet that separates Pluto from the other eight planets. Pluto is now defined as a "dwarf planet" (like the asteroid Ceres).

Physical Data Often Used

Average Earth–Moon distance	3.84×10^{8} m
Average Earth–Sun distance	1.496×10^{11} m
Average radius of the Earth	6.37×10^{6} m
Density of air (20°C and 1 atm)	1.20 kg/m^3
Density of water (20°C and 1 atm)	1.00×10^{3} kg/m^3
Free-fall acceleration	9.80 m/s^2
Mass of the Earth	5.98×10^{24} kg
Mass of the Moon	7.36×10^{22} kg
Mass of the Sun	1.99×10^{30} kg
Standard atmospheric pressure	1.013×10^{5} Pa

Note: These values are the ones used in the text.

Some Prefixes for Powers of Ten

Power	Prefix	Abbreviation	Power	Prefix	Abbreviation
10^{-24}	yocto	y	10^{1}	deka	da
10^{-21}	zepto	z	10^{2}	hecto	h
10^{-18}	atto	a	10^{3}	kilo	k
10^{-15}	femto	f	10^{6}	mega	M
10^{-12}	pico	p	10^{9}	giga	G
10^{-9}	nano	n	10^{12}	tera	T
10^{-6}	micro	μ	10^{15}	peta	P
10^{-3}	milli	m	10^{18}	exa	E
10^{-2}	centi	c	10^{21}	zetta	Z
10^{-1}	deci	d	10^{24}	yotta	Y

PHYSICS
for Scientists and Engineers

PHYSICS
for Scientists and Engineers

Seventh Edition

Raymond A. Serway
Emeritus, James Madison University

John W. Jewett, Jr.
California State Polytechnic University, Pomona

THOMSON
BROOKS/COLE

Australia • Brazil • Canada • Mexico • Singapore • Spain • United Kingdom • United States

THOMSON
BROOKS/COLE

Physics for Scientists and Engineers, Seventh Edition
Raymond A. Serway and John W. Jewett, Jr.

Physics Acquisition Editor: Chris Hall
Publisher: David Harris
Vice President, Editor-in-Chief, Sciences: Michelle Julet
Development Editor: Ed Dodd
Assistant Editor: Brandi Kirksey
Editorial Assistant: Shawn Vasquez
Technology Project Manager: Sam Subity
Marketing Manager: Mark Santee
Marketing Assistant: Melissa Wong
Managing Marketing Communications Manager: Bryan Vann
Project Manager, Editorial Production: Teri Hyde
Creative Director: Rob Hugel
Art Director: Lee Friedman
Print Buyers: Barbara Britton, Karen Hunt
Permissions Editors: Joohee Lee, Bob Kauser
Production Service: Lachina Publishing Services
Text Designer: Patrick Devine Design
Photo Researcher: Jane Sanders Miller
Copy Editor: Kathleen Lafferty
Illustrator: Rolin Graphics, Progressive Information Technologies, Lachina Publishing Services
Cover Designer: Patrick Devine Design
Cover Image: Front: © 2005 Tony Dunn; Back: © 2005 Kurt Hoffmann, Abra Marketing
Cover Printer: R.R. Donnelley/Willard
Compositor: Lachina Publishing Services
Printer: R.R. Donnelley/Willard

Printed in the United States of America
1 2 3 4 5 6 7 11 10 09 08 07

Library of Congress Control Number: 2006936870

Student Edition:
ISBN-13: 978-0-495-01312-9
ISBN-10: 0-495-01312-9

Instructor Edition:
ISBN-13: 978-0-495-11375-1
ISBN-10: 0-495-11375-1

Thomson Higher Education
10 Davis Drive
Belmont, CA 94002-3098
USA

We dedicate this book to our wives Elizabeth and Lisa and all our children and grandchildren for their loving understanding when we spent time on writing instead of being with them.

John W. Jewett, Jr.

Courtesy of NASA

© Thomson Learning/Charles D. Winters

Courtesy of Henry Leap and Jim Lehman

Contents

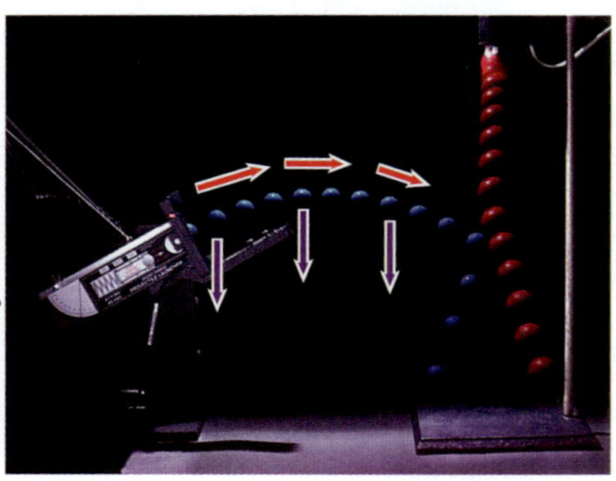

© Thomson Learning/Charles D. Winters

NASA

© Thomson Learning/Charles D. Winters

PART 6 MODERN PHYSICS 1111

© Thomson Learning/Charles D. Winters

Raymond A. Serway received his doctorate at Illinois Institute of Technology and is Professor Emeritus at James Madison University. In 1990, he received the Madison Scholar Award at James Madison University, where he taught for 17 years. Dr. Serway began his teaching career at Clarkson University, where he conducted research and taught from 1967 to 1980. He was the recipient of the Distinguished Teaching Award at Clarkson University in 1977 and of the Alumni Achievement Award from Utica College in 1985. As Guest Scientist at the IBM Research Laboratory in Zurich, Switzerland, he worked with K. Alex Müller, 1987 Nobel Prize recipient. Dr. Serway also was a visiting scientist at Argonne National Laboratory, where he collaborated with his mentor and friend, Sam Marshall. In addition to earlier editions of this textbook, Dr. Serway is the coauthor of *Principles of Physics,* fourth edition; *College Physics,* seventh edition; *Essentials of College Physics;* and *Modern Physics,* third edition. He also is the coauthor of the high school textbook *Physics,* published by Holt, Rinehart, & Winston. In addition, Dr. Serway has published more than 40 research papers in the field of condensed matter physics and has given more than 70 presentations at professional meetings. Dr. Serway and his wife, Elizabeth, enjoy traveling, golf, singing in a church choir, and spending quality time with their four children and eight grandchildren.

John W. Jewett, Jr., earned his doctorate at Ohio State University, specializing in optical and magnetic properties of condensed matter. Dr. Jewett began his academic career at Richard Stockton College of New Jersey, where he taught from 1974 to 1984. He is currently Professor of Physics at California State Polytechnic University, Pomona. Throughout his teaching career, Dr. Jewett has been active in promoting science education. In addition to receiving four National Science Foundation grants, he helped found and direct the Southern California Area Modern Physics Institute. He also directed Science IMPACT (Institute for Modern Pedagogy and Creative Teaching), which works with teachers and schools to develop effective science curricula. Dr. Jewett's honors include the Stockton Merit Award at Richard Stockton College in 1980, the Outstanding Professor Award at California State Polytechnic University for 1991–1992, and the Excellence in Undergraduate Physics Teaching Award from the American Association of Physics Teachers in 1998. He has given more than 80 presentations at professional meetings, including presentations at international conferences in China and Japan. In addition to his work on this textbook, he is coauthor of *Principles of Physics,* fourth edition, with Dr. Serway and author of *The World of Physics . . . Mysteries, Magic, and Myth.* Dr. Jewett enjoys playing keyboard with his all-physicist band, traveling, and collecting antiques that can be used as demonstration apparatus in physics lectures. Most importantly, he relishes spending time with his wife, Lisa, and their children and grandchildren.

In writing this seventh edition of *Physics for Scientists and Engineers*, we continue our ongoing efforts to improve the clarity of presentation and include new pedagogical features that help support the learning and teaching processes. Drawing on positive feedback from users of the sixth edition and reviewers' suggestions, we have refined the text to better meet the needs of students and teachers.

This textbook is intended for a course in introductory physics for students majoring in science or engineering. The entire contents of the book in its extended version could be covered in a three-semester course, but it is possible to use the material in shorter sequences with the omission of selected chapters and sections. The mathematical background of the student taking this course should ideally include one semester of calculus. If that is not possible, the student should be enrolled in a concurrent course in introductory calculus.

Objectives

This introductory physics textbook has two main objectives: to provide the student with a clear and logical presentation of the basic concepts and principles of physics and to strengthen an understanding of the concepts and principles through a broad range of interesting applications to the real world. To meet these objectives, we have placed emphasis on sound physical arguments and problem-solving methodology. At the same time, we have attempted to motivate the student through practical examples that demonstrate the role of physics in other disciplines, including engineering, chemistry, and medicine.

Changes in the Seventh Edition

A large number of changes and improvements have been made in preparing the seventh edition of this text. Some of the new features are based on our experiences and on current trends in science education. Other changes have been incorporated in response to comments and suggestions offered by users of the sixth edition and by reviewers of the manuscript. The features listed here represent the major changes in the seventh edition.

QUESTIONS AND PROBLEMS A substantial revision to the end-of-chapter questions and problems was made in an effort to improve their variety, interest, and pedagogical value, while maintaining their clarity and quality. Approximately 23% of the questions and problems are new or substantially changed. Several of the questions for each chapter are in objective format. Several problems in each chapter explicitly ask for qualitative reasoning in some parts as well as for quantitative answers in other parts:

> **19.** ● Assume a parcel of air in a straight tube moves with a constant acceleration of $-4.00\ \text{m/s}^2$ and has a velocity of 13.0 m/s at 10:05:00 a.m. on a certain date. (a) What is its velocity at 10:05:01 a.m.? (b) At 10:05:02 a.m.? (c) At 10:05:02.5 a.m.? (d) At 10:05:04 a.m.? (e) At 10:04:59 a.m.? (f) Describe the shape of a graph of velocity versus time for this parcel of air. (g) Argue for or against the statement, "Knowing the single value of an object's constant acceleration is like knowing a whole list of values for its velocity."

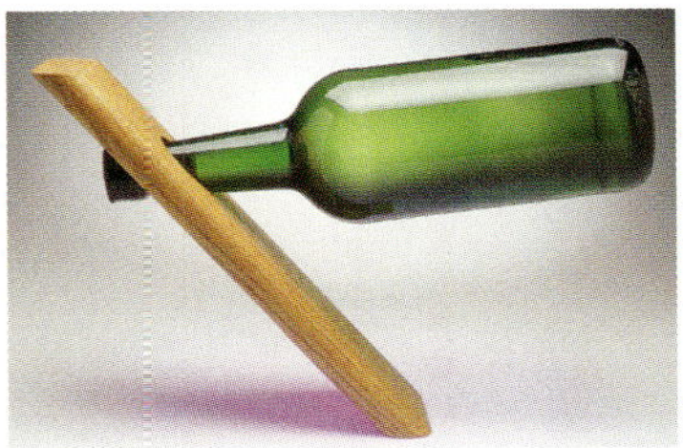

© Thomson Learning/Charles D. Winters

WORKED EXAMPLES All in-text worked examples have been recast and are now presented in a two-column format to better reinforce physical concepts. The left column shows textual information that describes the steps for solving the problem. The right column shows the mathematical manipulations and results of taking these steps. This layout facilitates matching the concept with its mathematical execution and helps students organize their work. These reconstituted examples closely follow a General Problem-Solving Strategy introduced in Chapter 2 to reinforce effective problem-solving habits. A sample of a worked example can be found on the next page.

Each solution has been reconstituted to more closely follow the General Problem-Solving Strategy as outlined in Chapter 2, to reinforce good problem-solving habits.

EXAMPLE 3.2 A Vacation Trip

A car travels 20.0 km due north and then 35.0 km in a direction 60.0° west of north as shown in Figure 3.11a. Find the magnitude and direction of the car's resultant displacement.

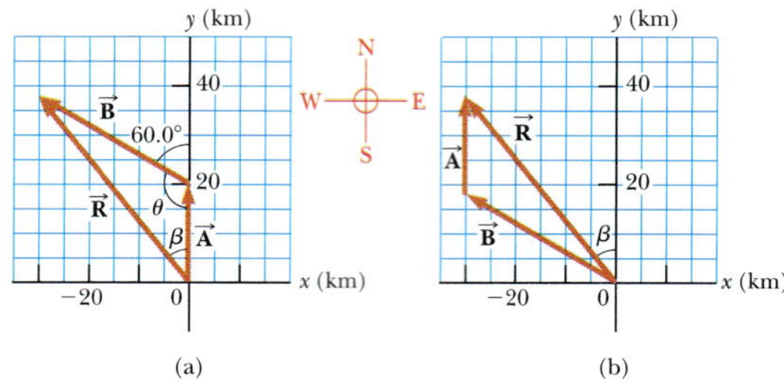

Figure 3.11 (Example 3.2) (a) Graphical method for finding the resultant displacement vector $\vec{\mathbf{R}} = \vec{\mathbf{A}} + \vec{\mathbf{B}}$. (b) Adding the vectors in reverse order $(\vec{\mathbf{B}} + \vec{\mathbf{A}})$ gives the same result for $\vec{\mathbf{R}}$.

SOLUTION

Conceptualize The vectors $\vec{\mathbf{A}}$ and $\vec{\mathbf{B}}$ drawn in Figure 3.11a help us conceptualize the problem.

Categorize We can categorize this example as a simple analysis problem in vector addition. The displacement $\vec{\mathbf{R}}$ is the resultant when the two individual displacements $\vec{\mathbf{A}}$ and $\vec{\mathbf{B}}$ are added. We can further categorize it as a problem about the analysis of triangles, so we appeal to our expertise in geometry and trigonometry.

Each step of the solution is detailed in a two-column format. The left column provides an explanation for each mathematical step in the right column, to better reinforce the physical concepts.

Analyze In this example, we show two ways to analyze the problem of finding the resultant of two vectors. The first way is to solve the problem geometrically, using graph paper and a protractor to measure the magnitude of $\vec{\mathbf{R}}$ and its direction in Figure 3.11a. (In fact, even when you know you are going to be carrying out a calculation, you should sketch the vectors to check your results.) With an ordinary ruler and protractor, a large diagram typically gives answers to two-digit but not to three-digit precision.

The second way to solve the problem is to analyze it algebraically. The magnitude of $\vec{\mathbf{R}}$ can be obtained from the law of cosines as applied to the triangle (see Appendix B.4).

Use $R^2 = A^2 + B^2 - 2AB\cos\theta$ from the law of cosines to find R:

$$R = \sqrt{A^2 + B^2 - 2AB\cos\theta}$$

Substitute numerical values, noting that $\theta = 180° - 60° = 120°$:

$$R = \sqrt{(20.0\text{ km})^2 + (35.0\text{ km})^2 - 2(20.0\text{ km})(35.0\text{ km})\cos 120°}$$

$$= \boxed{48.2\text{ km}}$$

Use the law of sines (Appendix B.4) to find the direction of $\vec{\mathbf{R}}$ measured from the northerly direction:

$$\frac{\sin\beta}{B} = \frac{\sin\theta}{R}$$

$$\sin\beta = \frac{B}{R}\sin\theta = \frac{35.0\text{ km}}{48.2\text{ km}}\sin 120° = 0.629$$

$$\beta = \boxed{38.9°}$$

The resultant displacement of the car is 48.2 km in a direction 38.9° west of north.

Finalize Does the angle β that we calculated agree with an estimate made by looking at Figure 3.11a or with an actual angle measured from the diagram using the graphical method? Is it reasonable that the magnitude of $\vec{\mathbf{R}}$ is larger than that of both $\vec{\mathbf{A}}$ and $\vec{\mathbf{B}}$? Are the units of $\vec{\mathbf{R}}$ correct?

Although the graphical method of adding vectors works well, it suffers from two disadvantages. First, some people find using the laws of cosines and sines to be awkward. Second, a triangle only results if you are adding two vectors. If you are adding three or more vectors, the resulting geometric shape is usually not a triangle. In Section 3.4, we explore a new method of adding vectors that will address both of these disadvantages.

What If? Suppose the trip were taken with the two vectors in reverse order: 35.0 km at 60.0° west of north first and then 20.0 km due north. How would the magnitude and the direction of the resultant vector change?

Answer They would not change. The commutative law for vector addition tells us that the order of vectors in an addition is irrelevant. Graphically, Figure 3.11b shows that the vectors added in the reverse order give us the same resultant vector.

What If? statements appear in about 1/3 of the worked examples and offer a variation on the situation posed in the text of the example. For instance, this feature might explore the effects of changing the conditions of the situation, determine what happens when a quantity is taken to a particular limiting value, or question whether additional information can be determined about the problem situation. This feature encourages students to think about the results of the example and assists in conceptual understanding of the principles.

All worked examples are also available to be assigned as interactive examples in the Enhanced WebAssign homework management system (visit **www.pse7.com** for more details).

ONLINE HOMEWORK It is now easier to assign online homework with Serway and Jewett and Enhanced WebAssign. All worked examples, end-of-chapter problems, active figures, quick quizzes, and most questions are available in WebAssign. Most problems include hints and feedback to provide instantaneous reinforcement or direction for that problem. In addition to the text content, we have also added math remediation tools to help students get up to speed in algebra, trigonometry, and calculus.

SUMMARIES Each chapter contains a summary that reviews the important concepts and equations discussed in that chapter. A marginal note next to each chapter summary directs students to additional quizzes, animations, and interactive exercises for that chapter on the book's companion Web site. The format of the end-of-chapter summary has been completely revised for this edition. The summary is divided into three sections: Definitions, Concepts and Principles, and Analysis Models for Problem-Solving. In each section, flashcard-type boxes focus on each separate definition, concept, principle, or analysis model.

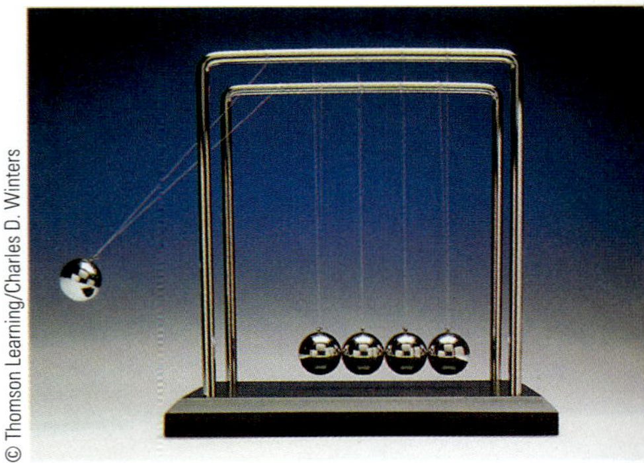

MATH APPENDIX The math appendix, a valuable tool for students, has been updated to show the math tools in a physics context. This resource is ideal for students who need a quick review on topics such as algebra, trigonometry, and calculus.

CONTENT CHANGES The content and organization of the textbook are essentially the same as in the sixth edition. Many sections in various chapters have been streamlined, deleted, or combined with other sections to allow for a more balanced presentation. Vectors are now denoted in boldface with an arrow over them (for example, $\vec{\mathbf{v}}$), making them easier to recognize. Chapters 7 and 8 have been completely reorganized to prepare students for a unified approach to energy that is used throughout the text. A new section in Chapter 9 teaches students how to analyze deformable systems with the conservation of energy equation and the impulse-momentum theorem. Chapter 34 is longer than in the sixth edition because of the movement into that chapter of the material on displacement current from Chapter 30 and Maxwell's equations from Chapter 31. A more detailed list of content changes can be found on the instructor's companion Web site.

Content

The material in this book covers fundamental topics in classical physics and provides an introduction to modern physics. The book is divided into six parts. Part 1 (Chapters 1 to 14) deals with the fundamentals of Newtonian mechanics and the physics of fluids; Part 2 (Chapters 15 to 18) covers oscillations, mechanical waves, and sound; Part 3 (Chapters 19 to 22) addresses heat and thermodynamics; Part 4 (Chapters 23 to 34) treats electricity and magnetism; Part 5 (Chapters 35 to 38) covers light and optics; and Part 6 (Chapters 39 to 46) deals with relativity and modern physics.

Text Features

Most instructors believe that the textbook selected for a course should be the student's primary guide for understanding and learning the subject matter. Furthermore, the textbook should be easily accessible and should be styled and written to facilitate instruction and learning. With these points in mind, we have included many pedagogical features, listed below, that are intended to enhance its usefulness to both students and instructors.

Problem Solving and Conceptual Understanding

GENERAL PROBLEM-SOLVING STRATEGY A general strategy outlined at the end of Chapter 2 provides students with a structured process for solving problems. In all remaining chapters, the strategy is employed explicitly in every example so that students learn how it is applied. Students are encouraged to follow this strategy when working end-of-chapter problems.

MODELING Although students are faced with hundreds of problems during their physics courses, instructors realize that a relatively small number of physical situations form the basis of these problems. When faced with a new problem, a physicist forms a *model* of the problem that can be solved in a simple way by identifying the common physical situation that occurs in the problem. For example, many problems involve particles under constant acceleration, isolated systems, or waves under refraction. Because the physicist has studied these situations extensively and understands the associated behavior, he or she can apply this knowledge as a model for solving a new problem. In certain chapters, this edition identifies Analysis Models, which are physical situations (such as the particle under constant acceleration, the isolated system, or the wave under refraction) that occur so often that they can be used as a model for solving an unfamiliar problem. These models are discussed in the chapter text, and the student is reminded of them in the end-of-chapter summary under the heading "Analysis Models for Problem-Solving."

PROBLEMS An extensive set of problems is included at the end of each chapter; in all, the text contains approximately three thousand problems. Answers to odd-numbered problems are provided at the end of the book. For the convenience of both the student and the instructor, about two-thirds of the problems are keyed to specific sections of the chapter. The remaining problems, labeled "Additional Problems," are not keyed to specific sections. The problem numbers for straightforward problems are printed in black, intermediate-level problems are in blue, and challenging problems are in magenta.

- **"Not-just-a-number" problems** Each chapter includes several marked problems that require students to think qualitatively in some parts and quantitatively in others. Instructors can assign such problems to guide students to display deeper understanding, practice good problem-solving techniques, and prepare for exams.
- **Problems for developing symbolic reasoning** Each chapter contains problems that ask for solutions in symbolic form as well as many problems asking for numerical answers. To help students develop skill in symbolic reasoning, each chapter contains a pair of otherwise identical problems, one asking for a numerical solution and one asking for a symbolic derivation. In this edition, each chapter also contains a problem giving a numerical value for every datum but one so that the answer displays how the unknown depends on the datum represented symbolically. The answer to such a problem has the form of a function of one variable. Reasoning about the behavior of this function puts emphasis on the *Finalize* step of the General Problem-Solving Strategy. All problems developing symbolic reasoning are identified by a tan background screen:

> **53.** ● A light spring has an unstressed length of 15.5 cm. It is described by Hooke's law with spring constant 4.30 N/m. One end of the horizontal spring is held on a fixed vertical axle, and the other end is attached to a puck of mass m that can move without friction over a horizontal surface. The puck is set into motion in a circle with a period of 1.30 s. (a) Find the extension of the spring x as it depends on m. Evaluate x for (b) $m = 0.070\ 0$ kg, (c) $m = 0.140$ kg, (d) $m = 0.180$ kg, and (e) $m = 0.190$ kg. (f) Describe the pattern of variation of x as it depends on m.

- **Review problems** Many chapters include review problems requiring the student to combine concepts covered in the chapter with those discussed in previous chapters. These problems reflect the cohesive nature of the principles in the text and verify that physics is not a scattered set of ideas. When facing a real-world issue such as global warming or nuclear weapons, it may be necessary to call on ideas in physics from several parts of a textbook such as this one.
- **"Fermi problems"** As in previous editions, at least one problem in each chapter asks the student to reason in order-of-magnitude terms.

- **Design problems** Several chapters contain problems that ask the student to determine design parameters for a practical device so that it can function as required.
- **"*Jeopardy!*" problems** Some chapters give students practice in changing between different representations by stating equations and asking for a description of a situation to which they apply as well as for a numerical answer.
- **Calculus-based problems** Every chapter contains at least one problem applying ideas and methods from differential calculus and one problem using integral calculus.

The instructor's Web site, **www.thomsonedu.com/physics/serway,** provides lists of problems using calculus, problems encouraging or requiring computer use, problems with "**What If?**" parts, problems referred to in the chapter text, problems based on experimental data, order-of-magnitude problems, problems about biological applications, design problems, *Jeopardy!* problems, review problems, problems reflecting historical reasoning about confusing ideas, problems developing symbolic reasoning skill, problems with qualitative parts, ranking questions, and other objective questions.

QUESTIONS The questions section at the end of each chapter has been significantly revised. Multiple-choice, ranking, and true–false questions have been added. The instructor may select items to assign as homework or use in the classroom, possibly with "peer instruction" methods and possibly with "clicker" systems. More than eight hundred questions are included in this edition. Answers to selected questions are included in the *Student Solutions Manual/Study Guide,* and answers to all questions are found in the *Instructor's Solutions Manual.*

19. O (i) Rank the gravitational accelerations you would measure for (a) a 2-kg object 5 cm above the floor, (b) a 2-kg object 120 cm above the floor, (c) a 3-kg object 120 cm above the floor, and (d) a 3-kg object 80 cm above the floor. List the one with the largest-magnitude acceleration first. If two are equal, show their equality in your list. **(ii)** Rank the gravitational forces on the same four objects, largest magnitude first. **(iii)** Rank the gravitational potential energies (of the object–Earth system) for the same four objects, largest first, taking $y = 0$ at the floor.

23. O An ice cube has been given a push and slides without friction on a level table. Which is correct? (a) It is in stable equilibrium. (b) It is in unstable equilibrium. (c) It is in neutral equilibrium (d) It is not in equilibrium.

WORKED EXAMPLES Two types of worked examples are presented to aid student comprehension. All worked examples in the text may be assigned for homework in WebAssign.

The first example type presents a problem and numerical answer. As discussed earlier, solutions to these examples have been altered in this edition to feature a two-column layout to explain the physical concepts and the mathematical steps side by side. Every example follows the explicit steps of the General Problem-Solving Strategy outlined in Chapter 2.

The second type of example is conceptual in nature. To accommodate increased emphasis on understanding physical concepts, the many conceptual examples are labeled as such, set off in boxes, and designed to focus students on the physical situation in the problem.

WHAT IF? Approximately one-third of the worked examples in the text contain a **What If?** feature. At the completion of the example solution, a **What If?** question offers a variation on the situation posed in the text of the example. For instance, this feature might explore the effects of changing the conditions of the situation, determine what happens when a quantity is taken to a particular limiting value, or question whether additional information can be determined about the situation. This feature encourages students to think about the results of the example, and it also assists in conceptual understanding of the principles. **What If?** questions also prepare students to encounter novel problems that may be included on exams. Some of the end-of-chapter problems also include this feature.

QUICK QUIZZES Quick Quizzes provide students an opportunity to test their understanding of the physical concepts presented. The questions require students to make decisions on the basis of sound reasoning, and some of the questions have been written to help students overcome common misconceptions. Quick Quizzes have been cast in an objective format, including multiple-choice, true–false, and ranking. Answers to all Quick Quiz questions are found at the end of each chapter. Additional Quick Quizzes that can be used in classroom teaching are available on the instructor's companion Web site. Many instructors choose to use such questions in a "peer instruction" teaching style or with the use of personal response system "clickers," but they can be used in standard quiz format as well. Quick Quizzes are set off from the text by horizontal lines:

Quick Quiz 7.5 A dart is loaded into a spring-loaded toy dart gun by pushing the spring in by a distance x. For the next loading, the spring is compressed a distance $2x$. How much faster does the second dart leave the gun compared with the first? (a) four times as fast (b) two times as fast (c) the same (d) half as fast (e) one-fourth as fast

---

PITFALL PREVENTION 16.2
Two Kinds of Speed/Velocity

Do not confuse v, the speed of the wave as it propagates along the string, with v_y, the transverse velocity of a point on the string. The speed v is constant for a uniform medium, whereas v_y varies sinusoidally.

PITFALL PREVENTIONS More than two hundred Pitfall Preventions (such as the one to the left) are provided to help students avoid common mistakes and misunderstandings. These features, which are placed in the margins of the text, address both common student misconceptions and situations in which students often follow unproductive paths.

Helpful Features

STYLE To facilitate rapid comprehension, we have written the book in a clear, logical, and engaging style. We have chosen a writing style that is somewhat informal and relaxed so that students will find the text appealing and enjoyable to read. New terms are carefully defined, and we have avoided the use of jargon.

IMPORTANT STATEMENTS AND EQUATIONS Most important statements and definitions are set in **boldface** or are highlighted with a background screen for added emphasis and ease of review. Similarly, important equations are highlighted with a background screen to facilitate location.

MARGINAL NOTES Comments and notes appearing in the margin with a ▶ icon can be used to locate important statements, equations, and concepts in the text.

PEDAGOGICAL USE OF COLOR Readers should consult the **pedagogical color chart** (inside the front cover) for a listing of the color-coded symbols used in the text diagrams. This system is followed consistently throughout the text.

MATHEMATICAL LEVEL We have introduced calculus gradually, keeping in mind that students often take introductory courses in calculus and physics concurrently. Most steps are shown when basic equations are developed, and reference is often made to mathematical appendices near the end of the textbook. Vector products are introduced later in the text, where they are needed in physical applications. The dot product is introduced in Chapter 7, which addresses energy of a system; the cross product is introduced in Chapter 11, which deals with angular momentum.

SIGNIFICANT FIGURES Significant figures in both worked examples and end-of-chapter problems have been handled with care. Most numerical examples are worked to either two or three significant figures, depending on the precision of the data provided. End-of-chapter problems regularly state data and answers to three-digit precision.

UNITS The international system of units (SI) is used throughout the text. The U.S. customary system of units is used only to a limited extent in the chapters on mechanics and thermodynamics.

APPENDICES AND ENDPAPERS Several appendices are provided near the end of the textbook. Most of the appendix material represents a review of mathematical concepts and techniques used in the text, including scientific notation, algebra, geometry, trigonometry, differential calculus, and integral calculus. Reference to these appendices is made throughout the text. Most mathematical review sections in the appendices include worked examples and exercises with answers. In addition to the mathematical reviews, the appendices contain tables of physical data, conversion factors, and the SI units of physical quantities as well as a periodic table of the elements. Other useful information—fundamental constants and physical data, planetary data, a list of standard prefixes, mathematical symbols, the Greek alphabet, and standard abbreviations of units of measure—appears on the endpapers.

Course Solutions That Fit Your Teaching Goals and Your Students' Learning Needs

Recent advances in educational technology have made homework management systems and audience response systems powerful and affordable tools to enhance the way you teach your course. Whether you offer a more traditional text-based course, are interested in using or are currently using an online homework management system such as WebAssign, or are ready to turn your lecture into an interactive learning environment with JoinIn on TurningPoint, you can be confident that the text's proven content provides the foundation for each and every component of our technology and ancillary package.

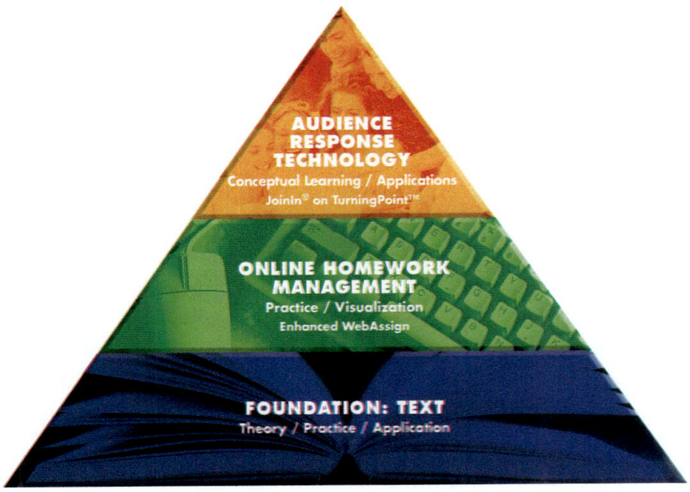

Homework Management Systems

Enhanced WebAssign Whether you're an experienced veteran or a beginner, Enhanced WebAssign is the perfect solution to fit your homework management needs. Designed by physicists for physicists, this system is a reliable and user-friendly teaching companion. Enhanced WebAssign is available for *Physics for Scientists and Engineers*, giving you the freedom to assign

- every end-of-chapter Problem and Question, enhanced with hints and feedback
- every worked example, enhanced with hints and feedback, to help strengthen students' problem-solving skills
- every Quick Quiz, giving your students ample opportunity to test their conceptual understanding.

- animated Active Figures, enhanced with hints and feedback, to help students develop their visualization skills
- a math review to help students brush up on key quantitative concepts

Please visit **www.thomsonedu.com/physics/serway** to view a live demonstration of Enhanced WebAssign.

The text also supports the following Homework Management Systems:

LON-CAPA: A Computer-Assisted Personalized Approach
http://www.lon-capa.org/

The University of Texas Homework Service
contact **moore@physics.utexas.edu**

Personal Response Systems

JoinIn on TurningPoint Pose book-specific questions and display students' answers seamlessly within the Microsoft® PowerPoint slides of your own lecture in conjunction with the "clicker" hardware of your choice. JoinIn on TurningPoint works with most infrared or radio frequency keypad systems, including Responsecard, EduCue, H-ITT, and even laptops. Contact your local sales representative to learn more about our personal response software and hardware.

Personal Response System Content Regardless of the response system you are using, we provide the tested content to support it. Our ready-to-go content includes all the questions from the Quick Quizzes, test questions, and a selection of end-of-chapter questions to provide helpful conceptual checkpoints to drop into your lecture. Our series of Active Figure animations have also been enhanced with multiple-choice questions to help test students' observational skills.

We also feature the Assessing to Learn in the Classroom content from the University of Massachusetts at Amherst. This collection of 250 advanced conceptual questions has been tested in the classroom for more than ten years and takes peer learning to a new level.

Visit **www.thomsonedu.com/physics/serway** to download samples of our personal response system content.

Lecture Presentation Resources

The following resources provide support for your presentations in lecture.

MULTIMEDIA MANAGER INSTRUCTOR'S RESOURCE CD An easy-to-use multimedia lecture tool, the Multimedia Manager Instructor's Resource CD allows you to quickly assemble art, animations, digital video, and database files with notes to create fluid lectures. The two-volume set (Volume 1: Chapters 1–22; Volume 2: Chapters 23–46) includes prebuilt PowerPoint lectures, a database of animations, video clips, and digital art from the text as well as editable electronic files of the *Instructor's Solutions Manual* and *Test Bank.*

TRANSPARENCY ACETATES Each volume contains approximately one hundred transparency acetates featuring art from the text. Volume 1 contains Chapters 1 through 22, and Volume 2 contains Chapters 23 through 46.

Assessment and Course Preparation Resources

A number of resources listed below will assist with your assessment and preparation processes.

INSTRUCTOR'S SOLUTIONS MANUAL by Ralph McGrew. This two-volume manual contains complete worked solutions to all end-of-chapter problems in the textbook as well as answers to the even-numbered problems and all the questions. The solutions to problems new to the seventh edition are marked for easy identification. Volume 1 contains

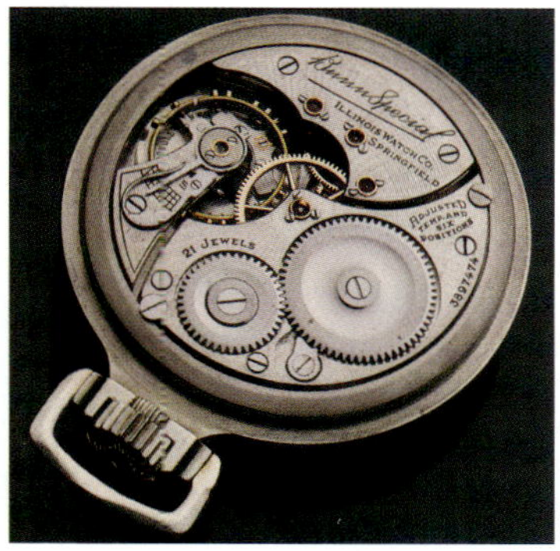

© Thomson Learning/George Semple

Chapters 1 through 22, and Volume 2 contains Chapters 23 through 46. Electronic files of the Instructor's Solutions are available on the Multimedia Manager CD as well.

PRINTED TEST BANK by Edward Adelson. This two-volume test bank contains approximately 2 200 multiple-choice questions. These questions are also available in electronic format with complete answers and solutions in the ExamView test software and as editable Word® files on the Multimedia Manager CD. Volume 1 contains Chapters 1 through 22, and Volume 2 contains Chapters 23 through 46.

EXAMVIEW This easy-to-use test generator CD features all of the questions from the printed test bank in an editable format.

WEBCT AND BLACKBOARD CONTENT For users of either course management system, we provide our test bank questions in the proper format for easy upload into your online course. In addition, you can integrate the ThomsonNOW for Physics student tutorial content into your WebCT or Blackboard course, providing your students a single sign-on to all their Web-based learning resources. Contact your local sales representative to learn more about our WebCT and Blackboard resources.

INSTRUCTOR'S COMPANION WEB SITE Consult the instructor's site by pointing your browser to **www.thomsonedu.com/physics/serway** for additional Quick Quiz questions, a detailed list of content changes since the sixth edition, a problem correlation guide, images from the text, and sample PowerPoint lectures. Instructors adopting the seventh edition of *Physics for Scientists and Engineers* may download these materials after securing the appropriate password from their local Thomson•Brooks/Cole sales representative.

Student Resources

STUDENT SOLUTIONS MANUAL/STUDY GUIDE by John R. Gordon, Ralph McGrew, Raymond Serway, and John W. Jewett, Jr. This two-volume manual features detailed solutions to 20% of the end-of-chapter problems from the text. The manual also features a list of important equations, concepts, and notes from key sections of the text in addition to answers to selected end-of-chapter questions. Volume 1 contains Chapters 1 through 22, and Volume 2 contains Chapters 23 through 46.

THOMSONNOW PERSONAL STUDY This assessment-based student tutorial system provides students with a personalized learning plan based on their performance on a series of diagnostic pre-tests. Rich interactive content, including Active Figures, Coached Problems, and Interactive Examples, helps students prepare for tests and exams.

Teaching Options

The topics in this textbook are presented in the following sequence: classical mechanics, oscillations and mechanical waves, and heat and thermodynamics followed by electricity and magnetism, electromagnetic waves, optics, relativity, and modern physics. This presentation represents a traditional sequence, with the subject of mechanical waves being presented before electricity and magnetism. Some instructors may prefer to discuss both mechanical and electromagnetic waves together after completing electricity and magnetism. In this case, Chapters 16 through 18 could be covered along with Chapter 34. The chapter on relativity is placed near the end of the text because this topic often is treated as an introduction to the era of "modern physics." If time permits, instructors may choose to cover Chapter 39 after completing Chapter 13 as a conclusion to the material on Newtonian mechanics.

For those instructors teaching a two-semester sequence, some sections and chapters could be deleted without any loss of continuity. The following sections can be considered optional for this purpose:

2.8 Kinematic Equations Derived from Calculus
4.6 Relative Velocity and Relative Acceleration
6.3 Motion in Accelerated Frames
6.4 Motion in the Presence of Resistive Forces
7.9 Energy Diagrams and Equilibrium of a System
9.8 Rocket Propulsion
11.5 The Motion of Gyroscopes and Tops
14.7 Other Applications of Fluid Dynamics
15.6 Damped Oscillations
15.7 Forced Oscillations
17.5 Digital Sound Recording
17.6 Motion Picture Sound
18.6 Standing Waves in Rods and Membranes
18.8 Nonsinusoidal Wave Patterns
22.8 Entropy on a Microscopic Scale
25.7 The Millikan Oil-Drop Experiment
25.8 Applications of Electrostatics
26.7 An Atomic Description of Dielectrics
27.5 Superconductors
28.5 Electrical Meters
28.6 Household Wiring and Electrical Safety
29.3 Applications Involving Charged Particles Moving in a Magnetic Field
29.6 The Hall Effect
30.6 Magnetism in Matter
30.7 The Magnetic Field of the Earth
31.6 Eddy Currents
33.9 Rectifiers and Filters
34.6 Production of Electromagnetic Waves by an Antenna
36.5 Lens Aberrations
36.6 The Camera
36.7 The Eye
36.8 The Simple Magnifier
36.9 The Compound Microscope
36.10 The Telescope
38.5 Diffraction of X-Rays by Crystals
39.10 The General Theory of Relativity

Acknowledgments

This seventh edition of *Physics for Scientists and Engineers* was prepared with the guidance and assistance of many professors who reviewed selections of the manuscript, the prerevision text, or both. We wish to acknowledge the following scholars and express our sincere appreciation for their suggestions, criticisms, and encouragement:

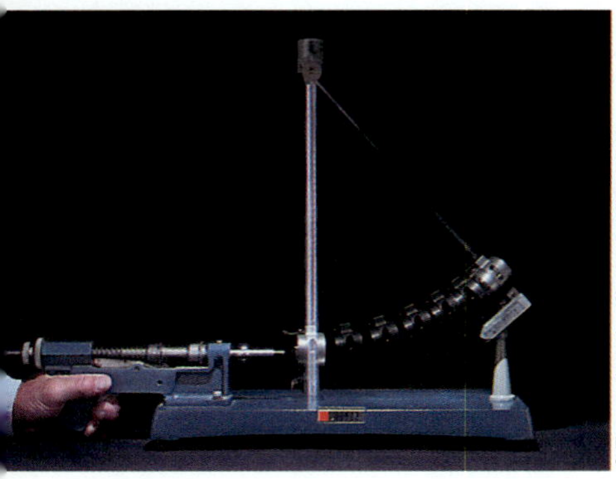

David P. Balogh, *Fresno City College*
Leonard X. Finegold, *Drexel University*
Raymond Hall, *California State University, Fresno*
Bob Jacobsen, *University of California, Berkeley*
Robin Jordan, *Florida Atlantic University*
Rafael Lopez-Mobilia, *University of Texas at San Antonio*
Diana Lininger Markham, *City College of San Francisco*
Steven Morris, *Los Angeles Harbor City College*
Taha Mzoughi, *Kennesaw State University*
Nobel Sanjay Rebello, *Kansas State University*
John Rosendahl, *University of California, Irvine*
Mikolaj Sawicki, *John A. Logan College*

Glenn B. Stracher, *East Georgia College*
Som Tyagi, *Drexel University*
Robert Weidman, *Michigan Technological University*
Edward A. Whittaker, *Stevens Institute of Technology*

This title was carefully checked for accuracy by Zinoviy Akkerman, *City College of New York;* Grant Hart, *Brigham Young University;* Michael Kotlarchyk, *Rochester Institute of Technology;* Andres LaRosa, *Portland State University;* Bruce Mason, *University of Oklahoma at Norman;* Peter Moeck, *Portland State University;* Brian A. Raue, *Florida International University;* James E. Rutledge, *University of California at Irvine;* Bjoern Seipel, *Portland State University;* Z. M. Stadnick, *University of Ottowa;* and Harry W. K. Tom, *University of California at Riverside.* We thank them for their diligent efforts under schedule pressure.

We are grateful to Ralph McGrew for organizing the end-of-chapter problems, writing many new problems, and suggesting improvements in the content of the textbook. Problems and questions new to this edition were written by Duane Deardorff, Thomas Grace, Francisco Izaguirre, John Jewett, Robert Forsythe, Randall Jones, Ralph McGrew, Kurt Vandervoort, and Jerzy Wrobel. Help was very kindly given by Dwight Neuenschwander, Michael Kinney, Amy Smith, Will Mackin, and the Sewer Department of Grand Forks, North Dakota. Daniel Kim, Jennifer Hoffman, Ed Oberhofer, Richard Webb, Wesley Smith, Kevin Kilty, Zinoviy Akkerman, Michael Rudmin, Paul Cox, Robert LaMontagne, Ken Menningen, and Chris Church made corrections to problems taken from previous editions. We are grateful to authors John R. Gordon and Ralph McGrew for preparing the *Student Solutions Manual/Study Guide.* Author Ralph McGrew has prepared an excellent *Instructor's Solutions Manual.* Edward Adelson has carefully edited and improved the test bank. Kurt Vandervoort prepared extra Quick Quiz questions for the instructor's companion Web site.

Special thanks and recognition go to the professional staff at the Brooks/Cole Publishing Company—in particular, Ed Dodd, Brandi Kirksey (who managed the ancillary program and so much more), Shawn Vasquez, Sam Subity, Teri Hyde, Michelle Julet, David Harris, and Chris Hall—for their fine work during the development and production of this textbook. Mark Santee is our seasoned marketing manager, and Bryan Vann coordinates our marketing communications. We recognize the skilled production service and excellent artwork provided by the staff at Lachina Publishing Services, and the dedicated photo research efforts of Jane Sanders Miller.

Finally, we are deeply indebted to our wives, children, and grandchildren for their love, support, and long-term sacrifices.

Raymond A. Serway
St. Petersburg, Florida

John W. Jewett, Jr.
Pomona, California

It is appropriate to offer some words of advice that should be of benefit to you, the student. Before doing so, we assume you have read the Preface, which describes the various features of the text and support materials that will help you through the course.

How to Study

Instructors are often asked, "How should I study physics and prepare for examinations?" There is no simple answer to this question, but we can offer some suggestions based on our own experiences in learning and teaching over the years.

First and foremost, maintain a positive attitude toward the subject matter, keeping in mind that physics is the most fundamental of all natural sciences. Other science courses that follow will use the same physical principles, so it is important that you understand and are able to apply the various concepts and theories discussed in the text.

Concepts and Principles

It is essential that you understand the basic concepts and principles before attempting to solve assigned problems. You can best accomplish this goal by carefully reading the textbook before you attend your lecture on the covered material. When reading the text, you should jot down those points that are not clear to you. Also be sure to make a diligent attempt at answering the questions in the Quick Quizzes as you come to them in your reading. We have worked hard to prepare questions that help you judge for yourself how well you understand the material. Study the **What If?** features that appear in many of the worked examples carefully. They will help you extend your understanding beyond the simple act of arriving at a numerical result. The Pitfall Preventions will also help guide you away from common misunderstandings about physics. During class, take careful notes and ask questions about those ideas that are unclear to you. Keep in mind that few people are able to absorb the full meaning of scientific material after only one reading; several readings of the text and your notes may be necessary. Your lectures and laboratory work supplement the textbook and should clarify some of the more difficult material. You should minimize your memorization of material. Successful memorization of passages from the text, equations, and derivations does not necessarily indicate that you understand the material. Your understanding of the material will be enhanced through a combination of efficient study habits, discussions with other students and with instructors, and your ability to solve the problems presented in the textbook. Ask questions whenever you believe that clarification of a concept is necessary.

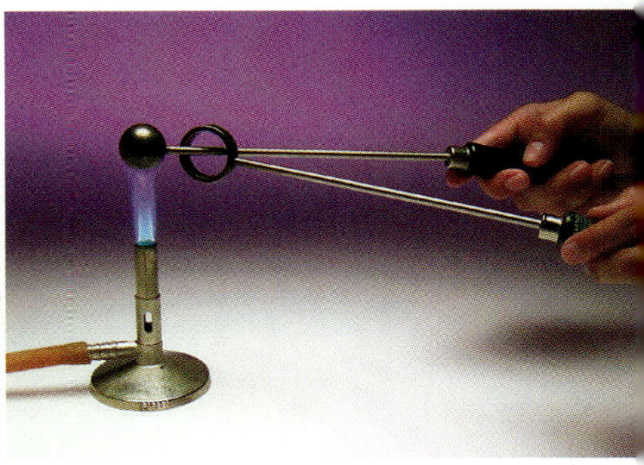

© Thomson Learning/Charles D. Winters

Study Schedule

It is important that you set up a regular study schedule, preferably a daily one. Make sure that you read the syllabus for the course and adhere to the schedule set by your instructor. The lectures will make much more sense if you read the corresponding text material *before* attending them. As a general rule, you should devote about two hours of study time for each hour you are in class. If you are having trouble with the course, seek the advice of the instructor or other students who have taken the course. You may find it necessary to seek further instruction from experienced students. Very often, instructors offer review sessions in addition to regular class periods. Avoid the practice of delaying study until a day or two before an exam. More often than not, this approach has disastrous results. Rather than undertake an all-night study session before a test, briefly review the basic concepts and equations, and then get a good night's rest. If you believe that you need additional help in understanding the concepts, in preparing for exams, or in problem solving, we suggest that you acquire a copy of the *Student Solutions Manual/Study Guide* that accompanies this textbook; this manual should be available at your college bookstore or through the publisher.

Use the Features

You should make full use of the various features of the text discussed in the Preface. For example, marginal notes are useful for locating and describing important equations and concepts, and **boldface** indicates important statements and definitions. Many useful tables are contained in the appendices, but most are incorporated in the text where they are most often referenced. Appendix B is a convenient review of mathematical tools used in the text.

Answers to odd-numbered problems are given at the end of the textbook, answers to Quick Quizzes are located at the end of each chapter, and solutions to selected end-of-chapter questions and problems are provided in the *Student Solutions Manual/Study Guide*. The table of contents provides an overview of the entire text, and the index enables you to locate specific material quickly. Footnotes are sometimes used to supplement the text or to cite other references on the subject discussed.

After reading a chapter, you should be able to define any new quantities introduced in that chapter and discuss the principles and assumptions that were used to arrive at certain key relations. The chapter summaries and the review sections of the *Student Solutions Manual/Study Guide* should help you in this regard. In some cases, you may find it necessary to refer to the textbook's index to locate certain topics. You should be able to associate with each physical quantity the correct symbol used to represent that quantity and the unit in which the quantity is specified. Furthermore, you should be able to express each important equation in concise and accurate prose.

Problem Solving

R. P. Feynman, Nobel laureate in physics, once said, "You do not know anything until you have practiced." In keeping with this statement, we strongly advise you to develop the skills necessary to solve a wide range of problems. Your ability to solve problems will be one of the main tests of your knowledge of physics; therefore, you should try to solve as many problems as possible. It is essential that you understand basic concepts and principles before attempting to solve problems. It is good practice to try to find alternate solutions to the same problem. For example, you can solve problems in mechanics using Newton's laws, but very often an alternative method that draws on energy considerations is more direct. You should not deceive yourself into thinking that you understand a problem merely because you have seen it solved in class. You must be able to solve the problem and similar problems on your own.

The approach to solving problems should be carefully planned. A systematic plan is especially important when a problem involves several concepts. First, read the problem several times until you are confident you understand what is being asked. Look for any key words that will help you interpret the problem and perhaps allow you to make certain assumptions. Your ability to interpret a question properly is an integral part of problem solving. Second, you should acquire the habit of writing down the information given in a problem and those quantities that need to be found; for example, you might construct a table listing both the quantities given and the quantities to be found. This procedure is sometimes used in the worked examples of the textbook. Finally, after you have decided on the method you believe is appropriate for a given problem, proceed with your solution. The General Problem-Solving Strategy will guide you through complex problems. If you follow the steps of this procedure *(Conceptualize, Categorize, Analyze, Finalize)*, you will find it easier to come up with a solution and gain more from your efforts. This Strategy, located at the end of Chapter 2, is used in all worked examples in the remaining chapters so that you can learn how to apply it. Specific problem-solving strategies for certain types of situations are included in the text and appear with a blue heading. These specific strategies follow the outline of the General Problem-Solving Strategy.

Often, students fail to recognize the limitations of certain equations or physical laws in a particular situation. It is very important that you understand and remember the assumptions that underlie a particular theory or formalism. For example, certain equations in kinematics apply only to a particle moving with constant acceleration. These equations are not valid for describing motion whose acceleration is not constant such as the motion of an object connected to a spring or the motion of an object through a fluid. Study the Analysis Models for Problem-Solving in the chapter summaries carefully so that you know how each model can be applied to a specific situation.

© Thomson Learning/Charles D. Winters

Experiments

Physics is a science based on experimental observations. Therefore, we recommend that you try to supplement the text by performing various types of "hands-on" experiments either at home or in the laboratory. These experiments can be used to test ideas and models discussed in class or in the textbook. For example, the common Slinky toy is excellent for studying traveling waves, a ball swinging on the end of a long string can be used to investigate pendulum motion, various masses attached to the end of a vertical spring or rubber band can be used to determine their elastic nature, an old pair of Polaroid sunglasses and some discarded lenses and a magnifying glass are the components of various experiments in optics, and an approximate measure of the free-fall acceleration can be determined simply by measuring with a stopwatch the time it takes for a ball to drop from a known height. The list of such experiments is endless. When physical models are not available, be imaginative and try to develop models of your own.

New Media

We strongly encourage you to use the **ThomsonNOW** Web-based learning system that accompanies this textbook. It is far easier to understand physics if you see it in action, and these new materials will enable you to become a part of that action. **ThomsonNOW** media described in the Preface and accessed at **www.thomsonedu.com/physics/serway** feature a three-step learning process consisting of a pre-test, a personalized learning plan, and a post-test.

It is our sincere hope that you will find physics an exciting and enjoyable experience and that you will benefit from this experience, regardless of your chosen profession. Welcome to the exciting world of physics!

> *The scientist does not study nature because it is useful; he studies it because he delights in it, and he delights in it because it is beautiful. If nature were not beautiful, it would not be worth knowing, and if nature were not worth knowing, life would not be worth living.*
>
> **—Henri Poincaré**

PART 1

Mechanics

Physics, the most fundamental physical science, is concerned with the fundamental principles of the Universe. It is the foundation upon which the other sciences—astronomy, biology, chemistry, and geology—are based. The beauty of physics lies in the simplicity of its fundamental principles and in the manner in which just a small number of concepts and models can alter and expand our view of the world around us.

The study of physics can be divided into six main areas:

1. *classical mechanics,* concerning the motion of objects that are large relative to atoms and move at speeds much slower than the speed of light;
2. *relativity,* a theory describing objects moving at any speed, even speeds approaching the speed of light;
3. *thermodynamics,* dealing with heat, work, temperature, and the statistical behavior of systems with large numbers of particles;
4. *electromagnetism,* concerned with electricity, magnetism, and electromagnetic fields;
5. *optics,* the study of the behavior of light and its interaction with materials;
6. *quantum mechanics,* a collection of theories connecting the behavior of matter at the submicroscopic level to macroscopic observations.

The disciplines of mechanics and electromagnetism are basic to all other branches of classical physics (developed before 1900) and modern physics (c. 1900–present). The first part of this textbook deals with classical mechanics, sometimes referred to as *Newtonian mechanics* or simply *mechanics.* Many principles and models used to understand mechanical systems retain their importance in the theories of other areas of physics and can later be used to describe many natural phenomena. Therefore, classical mechanics is of vital importance to students from all disciplines.

An electric vehicle recharges from a public charging station. Electric cars, as well as gasoline-powered vehicles and hybrid vehicles, use many of the concepts and principles of mechanics that we will study in this first part of the book. Quantities that we can use to describe the operation of vehicles include position, velocity, acceleration, force, energy, and momentum. (© Mehau Kulyk/Photo Researchers, Inc.)

A close-up of the gears inside a mechanical clock. Complicated timepieces have been built for centuries in an effort to measure time accurately. Time is one of the basic quantities that we use in studying the motion of objects. (© Photographer's Choice/Getty Images)

1 Physics and Measurement

ThomsonNOW™ Throughout this chapter and others, there are opportunities for online self-study, linking you to interactive tutorials based on your level of understanding. Sign in at **www.thomsonedu.com** to view tutorials and simulations, develop problem-solving skills, and test your knowledge with these interactive resources.

WebAssign Interactive content from this chapter and others may be assigned online in WebAssign.

Like all other sciences, physics is based on experimental observations and quantitative measurements. The main objectives of physics are to identify a limited number of fundamental laws that govern natural phenomena and use them to develop theories that can predict the results of future experiments. The fundamental laws used in developing theories are expressed in the language of mathematics, the tool that provides a bridge between theory and experiment.

When there is a discrepancy between the prediction of a theory and experimental results, new or modified theories must be formulated to remove the discrepancy. Many times a theory is satisfactory only under limited conditions; a more general theory might be satisfactory without such limitations. For example, the laws of motion discovered by Isaac Newton (1642–1727) accurately describe the motion of objects moving at normal speeds but do not apply to objects moving at speeds comparable with the speed of light. In contrast, the special theory of relativity developed later by Albert Einstein (1879–1955) gives the same results as Newton's laws at low speeds but also correctly describes the motion of objects at speeds approaching the speed of light. Hence, Einstein's special theory of relativity is a more general theory of motion than that formed from Newton's laws.

Classical physics includes the principles of classical mechanics, thermodynamics, optics, and electromagnetism developed before 1900. Important contributions to classical physics were provided by Newton, who was also one of the originators of calculus as a mathematical tool. Major developments in mechanics continued in the 18th century, but the fields of thermodynamics and electromagnetism were not developed until the latter part of the 19th century, principally because before that time the apparatus for controlled experiments in these disciplines was either too crude or unavailable.

A major revolution in physics, usually referred to as *modern physics,* began near the end of the 19th century. Modern physics developed mainly because many physical phenomena could not be explained by classical physics. The two most important developments in this modern era were the theories of relativity and quantum mechanics. Einstein's special theory of relativity not only correctly describes the motion of objects moving at speeds comparable to the speed of light; it also completely modifies the traditional concepts of space, time, and energy. The theory also shows that the speed of light is the upper limit of the speed of an object and that mass and energy are related. Quantum mechanics was formulated by a number of distinguished scientists to provide descriptions of physical phenomena at the atomic level. Many practical devices have been developed using the principles of quantum mechanics.

Scientists continually work at improving our understanding of fundamental laws. Numerous technological advances in recent times are the result of the efforts of many scientists, engineers, and technicians, such as unmanned planetary explorations and manned moon landings, microcircuitry and high-speed computers, sophisticated imaging techniques used in scientific research and medicine, and several remarkable results in genetic engineering. The impacts of such developments and discoveries on our society have indeed been great, and it is very likely that future discoveries and developments will be exciting, challenging, and of great benefit to humanity.

1.1 Standards of Length, Mass, and Time

To describe natural phenomena, we must make measurements of various aspects of nature. Each measurement is associated with a physical quantity, such as the length of an object.

If we are to report the results of a measurement to someone who wishes to reproduce this measurement, a *standard* must be defined. It would be meaningless if a visitor from another planet were to talk to us about a length of 8 "glitches" if we do not know the meaning of the unit glitch. On the other hand, if someone familiar with our system of measurement reports that a wall is 2 meters high and our unit of length is defined to be 1 meter, we know that the height of the wall is twice our basic length unit. Whatever is chosen as a standard must be readily accessible and must possess some property that can be measured reliably. Measurement standards used by different people in different places—throughout the Universe—must yield the same result. In addition, standards used for measurements must not change with time.

In 1960, an international committee established a set of standards for the fundamental quantities of science. It is called the **SI** (Système International), and its fundamental units of length, mass, and time are the *meter, kilogram,* and *second,* respectively. Other standards for SI fundamental units established by the committee are those for temperature (the *kelvin*), electric current (the *ampere*), luminous intensity (the *candela*), and the amount of substance (the *mole*).

The laws of physics are expressed as mathematical relationships among physical quantities that we will introduce and discuss throughout the book. In mechanics, the three fundamental quantities are length, mass, and time. All other quantities in mechanics can be expressed in terms of these three.

Length

We can identify **length** as the distance between two points in space. In 1120, the king of England decreed that the standard of length in his country would be named the *yard* and would be precisely equal to the distance from the tip of his nose to the end of his outstretched arm. Similarly, the original standard for the foot adopted by the French was the length of the royal foot of King Louis XIV. Neither of these standards is constant in time; when a new king took the throne, length measurements changed! The French standard prevailed until 1799, when the legal standard of length in France became the **meter** (m), defined as one ten-millionth of the distance from the equator to the North Pole along one particular longitudinal line that passes through Paris. Notice that this value is an Earth-based standard that does not satisfy the requirement that it can be used throughout the universe.

As recently as 1960, the length of the meter was defined as the distance between two lines on a specific platinum–iridium bar stored under controlled conditions in France. Current requirements of science and technology, however, necessitate more accuracy than that with which the separation between the lines on the bar can be determined. In the 1960s and 1970s, the meter was defined as 1 650 763.73 wavelengths[1] of orange-red light emitted from a krypton-86 lamp. In October 1983, however, the meter was redefined as **the distance traveled by light in vacuum during a time of 1/299 792 458 second.** In effect, this latest definition establishes that the speed of light in vacuum is precisely 299 792 458 meters per second. This definition of the meter is valid throughout the Universe based on our assumption that light is the same everywhere.

Table 1.1 lists approximate values of some measured lengths. You should study this table as well as the next two tables and begin to generate an intuition for what is meant by, for example, a length of 20 centimeters, a mass of 100 kilograms, or a time interval of 3.2×10^7 seconds.

PITFALL PREVENTION 1.1
Reasonable Values

Generating intuition about typical values of quantities when solving problems is important because you must think about your end result and determine if it seems reasonable. If you are calculating the mass of a housefly and arrive at a value of 100 kg, this answer is *unreasonable* and there is an error somewhere.

TABLE 1.1
Approximate Values of Some Measured Lengths

	Length (m)
Distance from the Earth to the most remote known quasar	1.4×10^{26}
Distance from the Earth to the most remote normal galaxies	9×10^{25}
Distance from the Earth to the nearest large galaxy (Andromeda)	2×10^{22}
Distance from the Sun to the nearest star (Proxima Centauri)	4×10^{16}
One light-year	9.46×10^{15}
Mean orbit radius of the Earth about the Sun	1.50×10^{11}
Mean distance from the Earth to the Moon	3.84×10^{8}
Distance from the equator to the North Pole	1.00×10^{7}
Mean radius of the Earth	6.37×10^{6}
Typical altitude (above the surface) of a satellite orbiting the Earth	2×10^{5}
Length of a football field	9.1×10^{1}
Length of a housefly	5×10^{-3}
Size of smallest dust particles	$\sim 10^{-4}$
Size of cells of most living organisms	$\sim 10^{-5}$
Diameter of a hydrogen atom	$\sim 10^{-10}$
Diameter of an atomic nucleus	$\sim 10^{-14}$
Diameter of a proton	$\sim 10^{-15}$

[1] We will use the standard international notation for numbers with more than three digits, in which groups of three digits are separated by spaces rather than commas. Therefore, 10 000 is the same as the common American notation of 10,000. Similarly, $\pi = 3.14159265$ is written as 3.141 592 65.

(a) (b)

Courtesy of National Institute of Standards and Technology. U.S. Department of Commerce

Figure 1.1 (a) The National Standard Kilogram No. 20, an accurate copy of the International Standard Kilogram kept at Sèvres, France, is housed under a double bell jar in a vault at the National Institute of Standards and Technology. (b) The primary time standard in the United States is a cesium fountain atomic clock developed at the National Institute of Standards and Technology laboratories in Boulder, Colorado. The clock will neither gain nor lose a second in 20 million years.

Mass

The SI fundamental unit of **mass,** the **kilogram** (kg), is defined as **the mass of a specific platinum–iridium alloy cylinder kept at the International Bureau of Weights and Measures at Sèvres, France**. This mass standard was established in 1887 and has not been changed since that time because platinum–iridium is an unusually stable alloy. A duplicate of the Sèvres cylinder is kept at the National Institute of Standards and Technology (NIST) in Gaithersburg, Maryland (Fig. 1.1a). Table 1.2 lists approximate values of the masses of various objects.

TABLE 1.2

Approximate Masses of Various Objects

	Mass (kg)
Observable Universe	$\sim 10^{52}$
Milky Way galaxy	$\sim 10^{42}$
Sun	1.99×10^{30}
Earth	5.98×10^{24}
Moon	7.36×10^{22}
Shark	$\sim 10^{3}$
Human	$\sim 10^{2}$
Frog	$\sim 10^{-1}$
Mosquito	$\sim 10^{-5}$
Bacterium	$\sim 1 \times 10^{-15}$
Hydrogen atom	1.67×10^{-27}
Electron	9.11×10^{-31}

Time

Before 1960, the standard of **time** was defined in terms of the *mean solar day* for the year 1900. (A solar day is the time interval between successive appearances of the Sun at the highest point it reaches in the sky each day.) The fundamental unit of a **second** (s) was defined as $(\frac{1}{60})(\frac{1}{60})(\frac{1}{24})$ of a mean solar day. The rotation of the Earth is now known to vary slightly with time. Therefore, this motion does not provide a time standard that is constant.

In 1967, the second was redefined to take advantage of the high precision attainable in a device known as an *atomic clock* (Fig. 1.1b), which measures vibrations of cesium atoms. One second is now defined as **9 192 631 770 times the period of vibration of radiation from the cesium-133 atom.**[2] Approximate values of time intervals are presented in Table 1.3.

TABLE 1.3

Approximate Values of Some Time Intervals

	Time Interval (s)
Age of the Universe	5×10^{17}
Age of the Earth	1.3×10^{17}
Average age of a college student	6.3×10^{8}
One year	3.2×10^{7}
One day	8.6×10^{4}
One class period	3.0×10^{3}
Time interval between normal heartbeats	8×10^{-1}
Period of audible sound waves	$\sim 10^{-3}$
Period of typical radio waves	$\sim 10^{-6}$
Period of vibration of an atom in a solid	$\sim 10^{-13}$
Period of visible light waves	$\sim 10^{-15}$
Duration of a nuclear collision	$\sim 10^{-22}$
Time interval for light to cross a proton	$\sim 10^{-24}$

[2] *Period* is defined as the time interval needed for one complete vibration.

TABLE 1.4

Prefixes for Powers of Ten

Power	Prefix	Abbreviation	Power	Prefix	Abbreviation
10^{-24}	yocto	y	10^{3}	kilo	k
10^{-21}	zepto	z	10^{6}	mega	M
10^{-18}	atto	a	10^{9}	giga	G
10^{-15}	femto	f	10^{12}	tera	T
10^{-12}	pico	p	10^{15}	peta	P
10^{-9}	nano	n	10^{18}	exa	E
10^{-6}	micro	μ	10^{21}	zetta	Z
10^{-3}	milli	m	10^{24}	yotta	Y
10^{-2}	centi	c			
10^{-1}	deci	d			

In addition to SI, another system of units, the *U.S. customary system,* is still used in the United States despite acceptance of SI by the rest of the world. In this system, the units of length, mass, and time are the foot (ft), slug, and second, respectively. In this book, we shall use SI units because they are almost universally accepted in science and industry. We shall make some limited use of U.S. customary units in the study of classical mechanics.

In addition to the fundamental SI units of meter, kilogram, and second, we can also use other units, such as millimeters and nanoseconds, where the prefixes *milli-* and *nano-* denote multipliers of the basic units based on various powers of ten. Prefixes for the various powers of ten and their abbreviations are listed in Table 1.4. For example, 10^{-3} m is equivalent to 1 millimeter (mm), and 10^{3} m corresponds to 1 kilometer (km). Likewise, 1 kilogram (kg) is 10^{3} grams (g), and 1 megavolt (MV) is 10^{6} volts (V).

The variables length, time, and mass are examples of *fundamental quantities.* Most other variables are *derived quantities,* those that can be expressed as a mathematical combination of fundamental quantities. Common examples are *area* (a product of two lengths) and *speed* (a ratio of a length to a time interval).

A table of the letters in the Greek alphabet is provided on the back endpaper of this book. ▶

Another example of a derived quantity is **density.** The density ρ (Greek letter rho) of any substance is defined as its *mass per unit volume:*

$$\rho \equiv \frac{m}{V} \tag{1.1}$$

In terms of fundamental quantities, density is a ratio of a mass to a product of three lengths. Aluminum, for example, has a density of 2.70×10^{3} kg/m^{3}, and iron has a density of 7.86×10^{3} kg/m^{3}. An extreme difference in density can be imagined by thinking about holding a 10-centimeter (cm) cube of Styrofoam in one hand and a 10-cm cube of lead in the other. See Table 14.1 in Chapter 14 for densities of several materials.

Quick Quiz 1.1 In a machine shop, two cams are produced, one of aluminum and one of iron. Both cams have the same mass. Which cam is larger? (a) The aluminum cam is larger. (b) The iron cam is larger. (c) Both cams are the same size.

1.2 Matter and Model Building

If physicists cannot interact with some phenomenon directly, they often imagine a **model** for a physical system that is related to the phenomenon. For example, we cannot interact directly with atoms because they are too small. Therefore, we build a mental model of an atom based on a system of a nucleus and one or more electrons outside the nucleus. Once we have identified the physical components of the model, we make predictions about its behavior based on the interactions among the components of the system or the interaction between the system and the environment outside the system.

As an example, consider the behavior of *matter.* A 1-kg cube of solid gold, such as that at the top of Figure 1.2, has a length of 3.73 cm on a side. Is this cube nothing but wall-to-wall gold, with no empty space? If the cube is cut in half, the two pieces still retain their chemical identity as solid gold. What if the pieces are cut again and again, indefinitely? Will the smaller and smaller pieces always be gold? Such questions can be traced to early Greek philosophers. Two of them—Leucippus and his student Democritus—could not accept the idea that such cuttings could go on forever. They developed a model for matter by speculating that the process ultimately must end when it produces a particle that can no longer be cut. In Greek, *atomos* means "not sliceable." From this Greek term comes our English word *atom.*

The Greek model of the structure of matter was that all ordinary matter consists of atoms, as suggested in the middle of Figure 1.2. Beyond that, no additional structure was specified in the model; atoms acted as small particles that interacted with one another, but internal structure of the atom was not a part of the model.

In 1897, J. J. Thomson identified the electron as a charged particle and as a constituent of the atom. This led to the first atomic model that contained internal structure. We shall discuss this model in Chapter 42.

Following the discovery of the nucleus in 1911, an atomic model was developed in which each atom is made up of electrons surrounding a central nucleus. A nucleus of gold is shown in Figure 1.2. This model leads, however, to a new question: Does the nucleus have structure? That is, is the nucleus a single particle or a collection of particles? By the early 1930s, a model evolved that described two basic entities in the nucleus: protons and neutrons. The proton carries a positive electric charge, and a specific chemical element is identified by the number of protons in its nucleus. This number is called the **atomic number** of the element. For instance, the nucleus of a hydrogen atom contains one proton (so the atomic number of hydrogen is 1), the nucleus of a helium atom contains two protons (atomic number 2), and the nucleus of a uranium atom contains 92 protons (atomic number 92). In addition to atomic number, a second number—**mass number,** defined as the number of protons plus neutrons in a nucleus—characterizes atoms. The atomic number of a specific element never varies (i.e., the number of protons does not vary) but the mass number can vary (i.e., the number of neutrons varies).

Is that, however, where the process of breaking down stops? Protons, neutrons, and a host of other exotic particles are now known to be composed of six different varieties of particles called **quarks,** which have been given the names of *up, down, strange, charmed, bottom,* and *top.* The up, charmed, and top quarks have electric charges of $+\frac{2}{3}$ that of the proton, whereas the down, strange, and bottom quarks have charges of $-\frac{1}{3}$ that of the proton. The proton consists of two up quarks and one down quark, as shown at the bottom of Figure 1.2 and labeled u and d. This structure predicts the correct charge for the proton. Likewise, the neutron consists of two down quarks and one up quark, giving a net charge of zero.

You should develop a process of building models as you study physics. In this study, you will be challenged with many mathematical problems to solve. One of the most important problem-solving techniques is to build a model for the problem: identify a system of physical components for the problem and make predictions of the behavior of the system based on the interactions among its components or the interaction between the system and its surrounding environment.

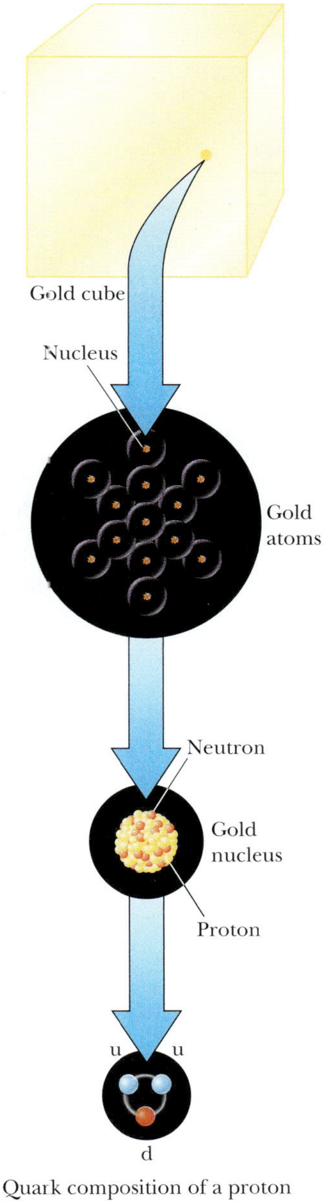

Figure 1.2 Levels of organization in matter. Ordinary matter consists of atoms, and at the center of each atom is a compact nucleus consisting of protons and neutrons. Protons and neutrons are composed of quarks. The quark composition of a proton is shown.

1.3 Dimensional Analysis

The word *dimension* has a special meaning in physics. It denotes the physical nature of a quantity. Whether a distance is measured in units of feet or meters or fathoms, it is still a distance. We say its dimension is *length.*

TABLE 1.5

Dimensions and Units of Four Derived Quantities

Quantity	Area	Volume	Speed	Acceleration
Dimensions	L^2	L^3	L/T	L/T^2
SI units	m^2	m^3	m/s	m/s^2
U.S. customary units	ft^2	ft^3	ft/s	ft/s^2

PITFALL PREVENTION 1.2

Symbols for Quantities

Some quantities have a small number of symbols that represent them. For example, the symbol for time is almost always *t*. Others quantities might have various symbols depending on the usage. Length may be described with symbols such as *x*, *y*, and *z* (for position); *r* (for radius); *a*, *b*, and *c* (for the legs of a right triangle); ℓ (for the length of an object); *d* (for a distance); *h* (for a height); and so forth.

The symbols we use in this book to specify the dimensions of length, mass, and time are L, M, and T, respectively.[3] We shall often use brackets [ ] to denote the dimensions of a physical quantity. For example, the symbol we use for speed in this book is v, and in our notation, the dimensions of speed are written $[v] = L/T$. As another example, the dimensions of area A are $[A] = L^2$. The dimensions and units of area, volume, speed, and acceleration are listed in Table 1.5. The dimensions of other quantities, such as force and energy, will be described as they are introduced in the text.

In many situations, you may have to check a specific equation to see if it matches your expectations. A useful and powerful procedure called **dimensional analysis** can assist in this check because **dimensions can be treated as algebraic quantities.** For example, quantities can be added or subtracted only if they have the same dimensions. Furthermore, the terms on both sides of an equation must have the same dimensions. By following these simple rules, you can use dimensional analysis to determine whether an expression has the correct form. Any relationship can be correct only if the dimensions on both sides of the equation are the same.

To illustrate this procedure, suppose you are interested in an equation for the position x of a car at a time t if the car starts from rest at $x = 0$ and moves with constant acceleration a. The correct expression for this situation is $x = \frac{1}{2}at^2$. Let us use dimensional analysis to check the validity of this expression. The quantity x on the left side has the dimension of length. For the equation to be dimensionally correct, the quantity on the right side must also have the dimension of length. We can perform a dimensional check by substituting the dimensions for acceleration, L/T^2 (Table 1.5), and time, T, into the equation. That is, the dimensional form of the equation $x = \frac{1}{2}at^2$ is

$$L = \frac{L}{\cancel{T^2}} \cdot \cancel{T^2} = L$$

The dimensions of time cancel as shown, leaving the dimension of length on the right-hand side to match that on the left.

A more general procedure using dimensional analysis is to set up an expression of the form

$$x \propto a^n t^m$$

where n and m are exponents that must be determined and the symbol $\propto$ indicates a proportionality. This relationship is correct only if the dimensions of both sides are the same. Because the dimension of the left side is length, the dimension of the right side must also be length. That is,

$$[a^n t^m] = L = L^1T^0$$

Because the dimensions of acceleration are L/T^2 and the dimension of time is T, we have

$$(L/T^2)^n T^m = L^1T^0 \quad \rightarrow \quad (L^nT^{m-2n}) = L^1T^0$$

[3] The *dimensions* of a quantity will be symbolized by a capitalized, nonitalic letter, such as L or T. The *algebraic symbol* for the quantity itself will be italicized, such as *L* for the length of an object or *t* for time.

The exponents of L and T must be the same on both sides of the equation. From the exponents of L, we see immediately that $n = 1$. From the exponents of T, we see that $m - 2n = 0$, which, once we substitute for n, gives us $m = 2$. Returning to our original expression $x \propto a^n t^m$, we conclude that $x \propto at^2$.

Quick Quiz 1.2 True or False: Dimensional analysis can give you the numerical value of constants of proportionality that may appear in an algebraic expression.

EXAMPLE 1.1 Analysis of an Equation

Show that the expression $v = at$, where v represents speed, a acceleration, and t an instant of time, is dimensionally correct.

SOLUTION

Identify the dimensions of v from Table 1.5:

$$[v] = \frac{\text{L}}{\text{T}}$$

Identify the dimensions of a from Table 1.5 and multiply by the dimensions of t:

$$[at] = \frac{\text{L}}{\text{T}^{\cancel{2}}}\cancel{\text{T}} = \frac{\text{L}}{\text{T}}$$

Therefore, $v = at$ is dimensionally correct because we have the same dimensions on both sides. (If the expression were given as $v = at^2$, it would be dimensionally *incorrect*. Try it and see!)

EXAMPLE 1.2 Analysis of a Power Law

Suppose we are told that the acceleration a of a particle moving with uniform speed v in a circle of radius r is proportional to some power of r, say r^n, and some power of v, say v^m. Determine the values of n and m and write the simplest form of an equation for the acceleration.

SOLUTION

Write an expression for a with a dimensionless constant of proportionality k:

$$a = kr^n v^m$$

Substitute the dimensions of a, r, and v:

$$\frac{\text{L}}{\text{T}^2} = \text{L}^n\left(\frac{\text{L}}{\text{T}}\right)^m = \frac{\text{L}^{n+m}}{\text{T}^m}$$

Equate the exponents of L and T so that the dimensional equation is balanced:

$$n + m = 1 \quad \text{and} \quad m = 2$$

Solve the two equations for n:

$$n = -1$$

Write the acceleration expression:

$$a = kr^{-1}v^2 = k\frac{v^2}{r}$$

In Section 4.4 on uniform circular motion, we show that $k = 1$ if a consistent set of units is used. The constant k would not equal 1 if, for example, v were in km/h and you wanted a in m/s^2.

PITFALL PREVENTION 1.3

Always Include Units

When performing calculations, include the units for every quantity and carry the units through the entire calculation. Avoid the temptation to drop the units early and then attach the expected units once you have an answer. By including the units in every step, you can detect errors if the units for the answer turn out to be incorrect.

1.4 Conversion of Units

Sometimes you must convert units from one measurement system to another or convert within a system (for example, from kilometers to meters). Equalities between SI and U.S. customary units of length are as follows:

$$1 \text{ mile} = 1\,609 \text{ m} = 1.609 \text{ km} \qquad 1 \text{ ft} = 0.304\,8 \text{ m} = 30.48 \text{ cm}$$

$$1 \text{ m} = 39.37 \text{ in.} = 3.281 \text{ ft} \qquad 1 \text{ in.} = 0.025\,4 \text{ m} = 2.54 \text{ cm (exactly)}$$

A more complete list of conversion factors can be found in Appendix A.

Like dimensions, units can be treated as algebraic quantities that can cancel each other. For example, suppose we wish to convert 15.0 in. to centimeters. Because 1 in. is defined as exactly 2.54 cm, we find that

$$15.0 \text{ in.} = (15.0 \cancel{\text{in.}})\left(\frac{2.54 \text{ cm}}{1 \cancel{\text{in.}}}\right) = 38.1 \text{ cm}$$

where the ratio in parentheses is equal to 1. We must place the unit "inch" in the denominator so that it cancels with the unit in the original quantity. The remaining unit is the centimeter, our desired result.

Quick Quiz 1.3 The distance between two cities is 100 mi. What is the number of kilometers between the two cities? (a) smaller than 100 (b) larger than 100 (c) equal to 100

EXAMPLE 1.3 Is He Speeding?

On an interstate highway in a rural region of Wyoming, a car is traveling at a speed of 38.0 m/s. Is the driver exceeding the speed limit of 75.0 mi/h?

SOLUTION

Convert meters in the speed to miles:

$$(38.0 \cancel{\text{m}}/\text{s})\left(\frac{1 \text{ mi}}{1\,609 \cancel{\text{m}}}\right) = 2.36 \times 10^{-2} \text{ mi/s}$$

Convert seconds to hours:

$$(2.36 \times 10^{-2} \text{ mi}/\cancel{\text{s}})\left(\frac{60 \cancel{\text{s}}}{1 \cancel{\text{min}}}\right)\left(\frac{60 \cancel{\text{min}}}{1 \text{ h}}\right) = 85.0 \text{ mi/h}$$

The driver is indeed exceeding the speed limit and should slow down.

What If? What if the driver were from outside the United States and is familiar with speeds measured in km/h? What is the speed of the car in km/h?

Answer We can convert our final answer to the appropriate units:

$$(85.0 \cancel{\text{mi}}/\text{h})\left(\frac{1.609 \text{ km}}{1 \cancel{\text{mi}}}\right) = 137 \text{ km/h}$$

Figure 1.3 shows an automobile speedometer displaying speeds in both mi/h and km/h. Can you check the conversion we just performed using this photograph?

Phil Boorman/Getty Images

Figure 1.3 The speedometer of a vehicle that shows speeds in both miles per hour and kilometers per hour.

1.5 Estimates and Order-of-Magnitude Calculations

Suppose someone asks you the number of bits of data on a typical musical compact disc. In response, it is not generally expected that you would provide the exact number but rather an estimate, which may be expressed in scientific notation. An *order of magnitude* of a number is determined as follows:

1. Express the number in scientific notation, with the multiplier of the power of ten between 1 and 10 and a unit.
2. If the multiplier is less than 3.162 (the square root of ten), the order of magnitude of the number is the power of ten in the scientific notation. If the multiplier is greater than 3.162, the order of magnitude is one larger than the power of ten in the scientific notation.

We use the symbol ~ for "is on the order of." Use the procedure above to verify the orders of magnitude for the following lengths:

$$0.008\,6 \text{ m} \sim 10^{-2} \text{ m} \qquad 0.002\,1 \text{ m} \sim 10^{-3} \text{ m} \qquad 720 \text{ m} \sim 10^{3} \text{ m}$$

Usually, when an order-of-magnitude estimate is made, the results are reliable to within about a factor of 10. If a quantity increases in value by three orders of magnitude, its value increases by a factor of about $10^3 = 1\,000$.

Inaccuracies caused by guessing too low for one number are often canceled by other guesses that are too high. You will find that with practice your guesstimates become better and better. Estimation problems can be fun to work because you freely drop digits, venture reasonable approximations for unknown numbers, make simplifying assumptions, and turn the question around into something you can answer in your head or with minimal mathematical manipulation on paper. Because of the simplicity of these types of calculations, they can be performed on a *small* scrap of paper and are often called "back-of-the-envelope calculations."

EXAMPLE 1.4 Breaths in a Lifetime

Estimate the number of breaths taken during an average human life span.

SOLUTION

We start by guessing that the typical human life span is about 70 years. Think about the average number of breaths that a person takes in 1 min. This number varies depending on whether the person is exercising, sleeping, angry, serene, and so forth. To the nearest order of magnitude, we shall choose 10 breaths per minute as our estimate. (This estimate is certainly closer to the true average value than 1 breath per minute or 100 breaths per minute.)

Find the approximate number of minutes in a year:

$$1 \cancel{\text{yr}} \left(\frac{400 \cancel{\text{days}}}{1 \cancel{\text{yr}}}\right)\left(\frac{25 \cancel{\text{h}}}{1 \cancel{\text{day}}}\right)\left(\frac{60 \text{ min}}{1 \cancel{\text{h}}}\right) = 6 \times 10^5 \text{ min}$$

Find the approximate number of minutes in a 70-year lifetime:

$$\text{number of minutes} = (70 \text{ yr})(6 \times 10^5 \text{ min/yr}) = 4 \times 10^7 \text{ min}$$

Find the approximate number of breaths in a lifetime:

$$\text{number of breaths} = (10 \text{ breaths/min})(4 \times 10^7 \text{ min}) = 4 \times 10^8 \text{ breaths}$$

Therefore, a person takes on the order of 10^9 breaths in a lifetime. Notice how much simpler it is in the first calculation above to multiply 400×25 than it is to work with the more accurate 365×24.

What If? What if the average life span were estimated as 80 years instead of 70? Would that change our final estimate?

Answer We could claim that $(80 \text{ yr})(6 \times 10^5 \text{ min/yr}) = 5 \times 10^7$ min, so our final estimate should be 5×10^8 breaths. This answer is still on the order of 10^9 breaths, so an order-of-magnitude estimate would be unchanged.

1.6 Significant Figures

When certain quantities are measured, the measured values are known only to within the limits of the experimental uncertainty. The value of this uncertainty can depend on various factors, such as the quality of the apparatus, the skill of the experimenter, and the number of measurements performed. The number of **significant figures** in a measurement can be used to express something about the uncertainty.

As an example of significant figures, suppose we are asked to measure the area of a compact disc using a meter stick as a measuring instrument. Let us assume the accuracy to which we can measure the radius of the disc is ± 0.1 cm. Because of the uncertainty of ± 0.1 cm, if the radius is measured to be 6.0 cm, we can claim only that its radius lies somewhere between 5.9 cm and 6.1 cm. In this case, we say that the measured value of 6.0 cm has two significant figures. Note that **the significant figures include the first estimated digit.** Therefore, we could write the radius as (6.0 ± 0.1) cm.

Now we find the area of the disc by using the equation for the area of a circle. If we were to claim the area is $A = \pi r^2 = \pi(6.0\text{ cm})^2 = 113\text{ cm}^2$, our answer would be unjustifiable because it contains three significant figures, which is greater than the number of significant figures in the radius. A good rule of thumb to use in determining the number of significant figures that can be claimed in a multiplication or a division is as follows:

> When multiplying several quantities, the number of significant figures in the final answer is the same as the number of significant figures in the quantity having the smallest number of significant figures. The same rule applies to division.

Applying this rule to the area of the compact disc, we see that the answer for the area can have only two significant figures because our measured radius has only two significant figures. Therefore, all we can claim is that the area is $1.1 \times 10^2\text{ cm}^2$.

Zeros may or may not be significant figures. Those used to position the decimal point in such numbers as 0.03 and 0.007 5 are not significant. Therefore, there are one and two significant figures, respectively, in these two values. When the zeros come after other digits, however, there is the possibility of misinterpretation. For example, suppose the mass of an object is given as 1 500 g. This value is ambiguous because we do not know whether the last two zeros are being used to locate the decimal point or whether they represent significant figures in the measurement. To remove this ambiguity, it is common to use scientific notation to indicate the number of significant figures. In this case, we would express the mass as 1.5×10^3 g if there are two significant figures in the measured value, 1.50×10^3 g if there are three significant figures, and 1.500×10^3 g if there are four. The same rule holds for numbers less than 1, so 2.3×10^{-4} has two significant figures (and therefore could be written 0.000 23) and 2.30×10^{-4} has three significant figures (also written 0.000 230).

For addition and subtraction, you must consider the number of decimal places when you are determining how many significant figures to report:

> When numbers are added or subtracted, the number of decimal places in the result should equal the smallest number of decimal places of any term in the sum.

PITFALL PREVENTION 1.4
Read Carefully

Notice that the rule for addition and subtraction is different from that for multiplication and division. For addition and subtraction, the important consideration is the number of *decimal places,* not the number of *significant figures.*

For example, if we wish to compute $123 + 5.35$, the answer is 128 and not 128.35. If we compute the sum $1.000\,1 + 0.000\,3 = 1.000\,4$, the result has five significant figures even though one of the terms in the sum, 0.000 3, has only one significant figure. Likewise, if we perform the subtraction $1.002 - 0.998 = 0.004$, the result has only one significant figure even though one term has four significant figures and the other has three.

In this book, most of the numerical examples and end-of-chapter problems will yield answers having three significant figures. When carrying out order-of-magnitude calculations, we shall typically work with a single significant figure.

If the number of significant figures in the result of an addition or subtraction must be reduced, there is a general rule for rounding numbers: the last digit retained is increased by 1 if the last digit dropped is greater than 5. If the last digit dropped is less than 5, the last digit retained remains as it is. If the last digit dropped is equal to 5, the remaining digit should be rounded to the nearest even number. (This rule helps avoid accumulation of errors in long arithmetic processes.)

A technique for avoiding error accumulation is to delay rounding of numbers in a long calculation until you have the final result. Wait until you are ready to copy the final answer from your calculator before rounding to the correct number of significant figures.

EXAMPLE 1.5 Installing a Carpet

A carpet is to be installed in a room whose length is measured to be 12.71 m and whose width is measured to be 3.46 m. Find the area of the room.

SOLUTION

If you multiply 12.71 m by 3.46 m on your calculator, you will see an answer of 43.976 6 m^2. How many of these numbers should you claim? Our rule of thumb for multiplication tells us that you can claim only the number of significant figures in your answer as are present in the measured quantity having the lowest number of significant figures. In this example, the lowest number of significant figures is three in 3.46 m, so we should express our final answer as 44.0 m^2.

Summary

ThomsonNOW™ Sign in at **www.thomsonedu.com** and go to ThomsonNOW to take a practice test for this chapter.

DEFINITIONS

The three fundamental physical quantities of mechanics are **length, mass,** and **time,** which in the SI system have the units **meter** (m), **kilogram** (kg), and **second** (s), respectively. These fundamental quantities cannot be defined in terms of more basic quantities.

The **density** of a substance is defined as its *mass per unit volume:*

$$\rho \equiv \frac{m}{V} \tag{1.1}$$

CONCEPTS AND PRINCIPLES

The method of **dimensional analysis** is very powerful in solving physics problems. Dimensions can be treated as algebraic quantities. By making estimates and performing order-of-magnitude calculations, you should be able to approximate the answer to a problem when there is not enough information available to specify an exact solution completely.

When you compute a result from several measured numbers, each of which has a certain accuracy, you should give the result with the correct number of **significant figures.** When multiplying several quantities, the number of significant figures in the final answer is the same as the number of significant figures in the quantity having the smallest number of significant figures. The same rule applies to division. When numbers are added or subtracted, the number of decimal places in the result should equal the smallest number of decimal places of any term in the sum.

Questions

☐ denotes answer available in *Student Solutions Manual/Study Guide;* **O** denotes objective question

1. Suppose the three fundamental standards of the metric system were length, *density,* and time rather than length, *mass,* and time. The standard of density in this system is to be defined as that of water. What considerations about water would you need to address to make sure the standard of density is as accurate as possible?
2. Express the following quantities using the prefixes given in Table 1.4: (a) 3×10^{-4} m (b) 5×10^{-5} s (c) 72×10^{2} g
3. **O** Rank the following five quantities in order from the largest to the smallest: (a) 0.032 kg (b) 15 g (c) 2.7×10^{5} mg (d) 4.1×10^{-8} Gg (e) 2.7×10^{8} μg. If two of the masses are equal, give them equal rank in your list.
4. **O** If an equation is dimensionally correct, does that mean that the equation must be true? If an equation is not dimensionally correct, does that mean that the equation cannot be true?
5. **O** Answer each question yes or no. Must two quantities have the same dimensions (a) if you are adding them? (b) If you are multiplying them? (c) If you are subtracting them? (d) If you are dividing them? (e) If you are using one quantity as an exponent in raising the other to a power? (f) If you are equating them?
6. **O** The price of gasoline at a particular station is 1.3 euros per liter. An American student can use 41 euros to buy gasoline. Knowing that 4 quarts make a gallon and that 1 liter is close to 1 quart, she quickly reasons that she can buy (choose one) (a) less than 1 gallon of gasoline, (b) about 5 gallons of gasoline, (c) about 8 gallons of gasoline, (d) more than 10 gallons of gasoline.
7. **O** One student uses a meterstick to measure the thickness of a textbook and finds it to be 4.3 cm ± 0.1 cm. Other students measure the thickness with vernier calipers and obtain (a) 4.32 cm ± 0.01 cm, (b) 4.31 cm ± 0.01 cm, (c) 4.24 cm ± 0.01 cm, and (d) 4.43 cm ± 0.01 cm. Which of these four measurements, if any, agree with that obtained by the first student?
8. **O** A calculator displays a result as $1.365\,248\,0 \times 10^{7}$ kg. The estimated uncertainty in the result is ±2%. How many digits should be included as significant when the result is written down? Choose one: (a) zero (b) one (c) two (d) three (e) four (f) five (g) the number cannot be determined

Problems

WebAssign The Problems from this chapter may be assigned online in WebAssign.

ThomsonNOW™ Sign in at **www.thomsonedu.com** and go to ThomsonNOW to assess your understanding of this chapter's topics with additional quizzing and conceptual questions.

1, 2, 3 denotes straightforward, intermediate, challenging; ☐ denotes full solution available in *Student Solutions Manual/Study Guide;* ▲ denotes coached solution with hints available at **www.thomsonedu.com;** ▭ denotes developing symbolic reasoning; ● denotes asking for qualitative reasoning; ▪ denotes computer useful in solving problem

Section 1.1 Standards of Length, Mass, and Time

Note: Consult the endpapers, appendices, and tables in the text whenever necessary in solving problems. For this chapter, Table 14.1 and Appendix B.3 may be particularly useful. Answers to odd-numbered problems appear in the back of the book.

1. ● Use information on the endpapers of this book to calculate the average density of the Earth. Where does the value fit among those listed in Table 14.1? Look up the density of a typical surface rock, such as granite, in another source and compare the density of the Earth to it.
2. The standard kilogram is a platinum-iridium cylinder 39.0 mm in height and 39.0 mm in diameter. What is the density of the material?
3. A major motor company displays a die-cast model of its first automobile, made from 9.35 kg of iron. To celebrate its one-hundredth year in business, a worker will recast the model in gold from the original dies. What mass of gold is needed to make the new model?
4. ● A proton, which is the nucleus of a hydrogen atom, can be modeled as a sphere with a diameter of 2.4 fm and a mass of 1.67×10^{-27} kg. Determine the density of the proton and state how it compares with the density of lead, which is given in Table 14.1.
5. Two spheres are cut from a certain uniform rock. One has radius 4.50 cm. The mass of the second sphere is five times greater. Find the radius of the second sphere.

Section 1.2 Matter and Model Building

6. A crystalline solid consists of atoms stacked up in a repeating lattice structure. Consider a crystal as shown in Figure P1.6a. The atoms reside at the corners of cubes of side L = 0.200 nm. One piece of evidence for the regular

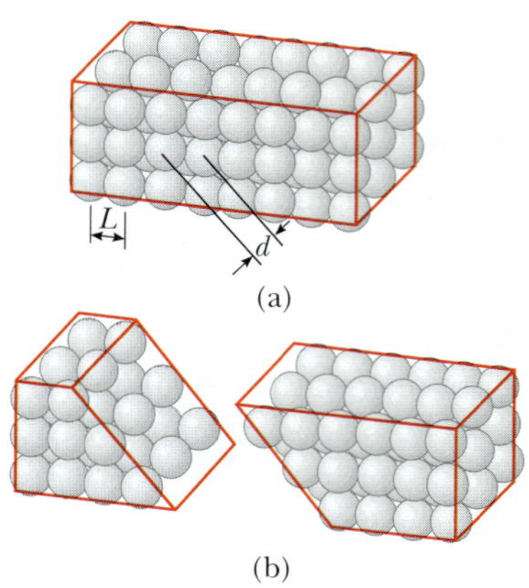

Figure P1.6

arrangement of atoms comes from the flat surfaces along which a crystal separates, or cleaves, when it is broken. Suppose this crystal cleaves along a face diagonal as shown in Figure P1.6b. Calculate the spacing d between two adjacent atomic planes that separate when the crystal cleaves.

Section 1.3 Dimensional Analysis

7. Which of the following equations are dimensionally correct? (a) $v_f = v_i + ax$ (b) $y = (2 \text{ m}) \cos(kx)$, where $k = 2 \text{ m}^{-1}$

8. Figure P1.8 shows a *frustum of a cone.* Of the following mensuration (geometrical) expressions, which describes **(i)** the total circumference of the flat circular faces, **(ii)** the volume, and **(iii)** the area of the curved surface? (a) $\pi(r_1 + r_2)[h^2 + (r_2 - r_1)^2]^{1/2}$, (b) $2\pi(r_1 + r_2)$ (c) $\pi h(r_1^2 + r_1 r_2 + r_2^2)/3$

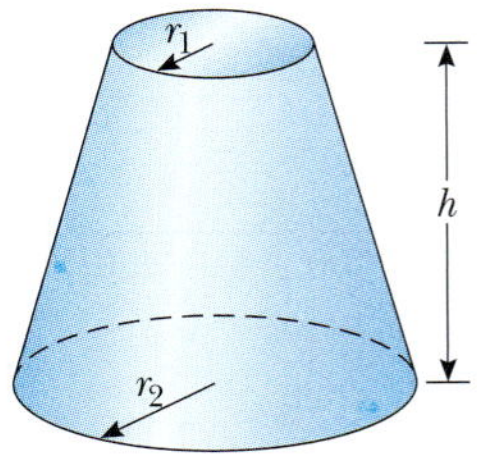

Figure P1.8

9. Newton's law of universal gravitation is represented by

$$F = \frac{GMm}{r^2}$$

Here F is the magnitude of the gravitational force exerted by one small object on another, M and m are the masses of the objects, and r is a distance. Force has the SI units $\text{kg} \cdot \text{m/s}^2$. What are the SI units of the proportionality constant G?

Section 1.4 Conversion of Units

10. Suppose your hair grows at the rate 1/32 in. per day. Find the rate at which it grows in nanometers per second. Because the distance between atoms in a molecule is on the order of 0.1 nm, your answer suggests how rapidly layers of atoms are assembled in this protein synthesis.

11. A rectangular building lot is 100 ft by 150 ft. Determine the area of this lot in square meters.

12. An auditorium measures 40.0 m × 20.0 m × 12.0 m. The density of air is 1.20 kg/m^3. What are (a) the volume of the room in cubic feet and (b) the weight of air in the room in pounds?

13. ● A room measures 3.8 m by 3.6 m, and its ceiling is 2.5 m high. Is it possible to completely wallpaper the walls of this room with the pages of this book? Explain your answer.

14. Assume it takes 7.00 min to fill a 30.0-gal gasoline tank. (a) Calculate the rate at which the tank is filled in gallons per second. (b) Calculate the rate at which the tank is filled in cubic meters per second. (c) Determine the time interval, in hours, required to fill a 1.00-m^3 volume at the same rate. (1 U.S. gal = 231 in.^{3})

15. A solid piece of lead has a mass of 23.94 g and a volume of 2.10 cm^3. From these data, calculate the density of lead in SI units (kg/m^3).

16. An ore loader moves 1 200 tons/h from a mine to the surface. Convert this rate to pounds per second, using 1 ton = 2 000 lb.

17. At the time of this book's printing, the U.S. national debt is about \$8 trillion. (a) If payments were made at the rate of \$1 000 per second, how many years would it take to pay off the debt, assuming no interest were charged? (b) A dollar bill is about 15.5 cm long. If eight trillion dollar bills were laid end to end around the Earth's equator, how many times would they encircle the planet? Take the radius of the Earth at the equator to be 6 378 km. *Note:* Before doing any of these calculations, try to guess at the answers. You may be very surprised.

18. A pyramid has a height of 481 ft, and its base covers an area of 13.0 acres (Fig. P1.18). The volume of a pyramid is given by the expression $V = \frac{1}{3}Bh$, where B is the area of the base and h is the height. Find the volume of this pyramid in cubic meters. (1 acre = 43 560 ft^2)

Sylvain Grandadam/Photo Researchers, Inc.

Figure P1.18 Problems 18 and 19.

19. The pyramid described in Problem 18 contains approximately 2 million stone blocks that average 2.50 tons each. Find the weight of this pyramid in pounds.

20. A hydrogen atom has a diameter of 1.06×10^{-10} m as defined by the diameter of the spherical electron cloud around the nucleus. The hydrogen nucleus has a diameter of approximately 2.40×10^{-15} m. (a) For a scale model, represent the diameter of the hydrogen atom by the playing length of an American football field (100 yards = 300 ft) and determine the diameter of the nucleus in millimeters. (b) The atom is how many times larger in volume than its nucleus?

21. ▲ One gallon of paint (volume = 3.78×10^{-3} m^3) covers an area of 25.0 m^2. What is the thickness of the fresh paint on the wall?

22. The mean radius of the Earth is 6.37×10^6 m and that of the Moon is 1.74×10^8 cm. From these data calculate (a) the ratio of the Earth's surface area to that of the Moon and (b) the ratio of the Earth's volume to that of the Moon. Recall that the surface area of a sphere is $4\pi r^2$ and the volume of a sphere is $\frac{4}{3}\pi r^3$.

23. ▲ One cubic meter (1.00 m^3) of aluminum has a mass of 2.70×10^3 kg, and the same volume of iron has a mass of 7.86×10^3 kg. Find the radius of a solid aluminum sphere that will balance a solid iron sphere of radius 2.00 cm on an equal-arm balance.

24. Let ρ_{Al} represent the density of aluminum and ρ_{Fe} that of iron. Find the radius of a solid aluminum sphere that balances a solid iron sphere of radius r_{Fe} on an equal-arm balance.

Section 1.5 Estimates and Order-of-Magnitude Calculations

25. ▲ Find the order of magnitude of the number of table-tennis balls that would fit into a typical-size room (without being crushed). In your solution, state the quantities you measure or estimate and the values you take for them.

26. An automobile tire is rated to last for 50 000 miles. To an order of magnitude, through how many revolutions will it turn? In your solution, state the quantities you measure or estimate and the values you take for them.

27. Compute the order of magnitude of the mass of a bathtub half full of water. Compute the order of magnitude of the mass of a bathtub half full of pennies. In your solution, list the quantities you take as data and the value you measure or estimate for each.

28. ● Suppose Bill Gates offers to give you $1 billion if you can finish counting it out using only one-dollar bills. Should you accept his offer? Explain your answer. Assume you can count one bill every second, and note that you need at least 8 hours a day for sleeping and eating.

29. To an order of magnitude, how many piano tuners are in New York City? Physicist Enrico Fermi was famous for asking questions like this one on oral doctorate qualifying examinations. His own facility in making order-of-magnitude calculations is exemplified in Problem 48 of Chapter 45.

Section 1.6 Significant Figures

Note: Appendix B.8 on propagation of uncertainty may be useful in solving some problems in this section.

30. A rectangular plate has a length of (21.3 ± 0.2) cm and a width of (9.8 ± 0.1) cm. Calculate the area of the plate, including its uncertainty.

31. How many significant figures are in the following numbers: (a) 78.9 ± 0.2 (b) 3.788×10^9 (c) 2.46×10^{-6} (d) 0.005 3?

32. The radius of a uniform solid sphere is measured to be (6.50 ± 0.20) cm, and its mass is measured to be (1.85 ± 0.02) kg. Determine the density of the sphere in kilograms per cubic meter and the uncertainty in the density.

33. Carry out the following arithmetic operations: (a) the sum of the measured values 756, 37.2, 0.83, and 2 (b) the product $0.003\,2 \times 356.3$ (c) the product $5.620 \times \pi$

34. The *tropical year,* the time interval from one vernal equinox to the next vernal equinox, is the basis for our calendar. It contains 365.242 199 days. Find the number of seconds in a tropical year.

Note: The next 11 problems call on mathematical skills that will be useful throughout the course.

35. **Review problem.** A child is surprised that she must pay $1.36 for a toy marked $1.25 because of sales tax. What is the effective tax rate on this purchase, expressed as a percentage?

36. ● **Review problem.** A student is supplied with a stack of copy paper, ruler, compass, scissors, and a sensitive balance. He cuts out various shapes in various sizes, calculates their areas, measures their masses, and prepares the graph of Figure P1.36. Consider the fourth experimental point from the top. How far is it from the best-fit straight line? (a) Express your answer as a difference in vertical-axis coordinate. (b) Express your answer as a difference in horizontal-axis coordinate. (c) Express both of the answers to parts (a) and (b) as a percentage. (d) Calculate the slope of the line. (e) State what the graph demonstrates, referring to the shape of the graph and the results of parts (c) and (d). (f) Describe whether this result should be expected theoretically. Describe the physical meaning of the slope.

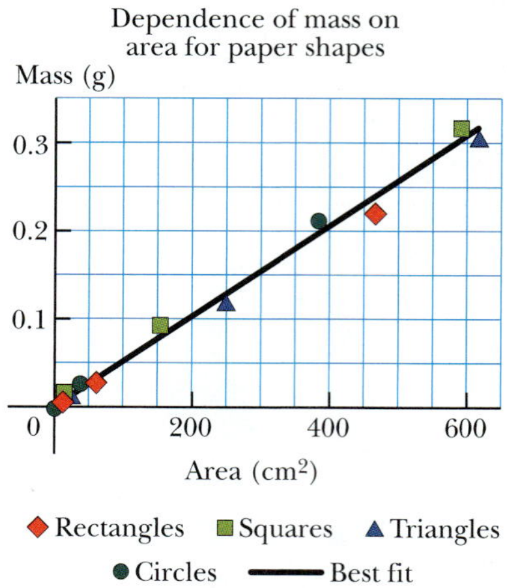

Figure P1.36

37. **Review problem.** A young immigrant works overtime, earning money to buy portable MP3 players to send home as gifts for family members. For each extra shift he works, he has figured out that he can buy one player and two-thirds of another one. An e-mail from his mother informs him that the players are so popular that each of 15 young neighborhood friends wants one. How many more shifts will he have to work?

38. **Review problem.** In a college parking lot, the number of ordinary cars is larger than the number of sport utility vehicles by 94.7%. The difference between the number of cars and the number of SUVs is 18. Find the number of SUVs in the lot.

39. **Review problem.** The ratio of the number of sparrows visiting a bird feeder to the number of more interesting birds is 2.25. On a morning when altogether 91 birds visit the feeder, what is the number of sparrows?

40. **Review problem.** Prove that one solution of the equation

$$2.00x^4 - 3.00x^3 + 5.00x = 70.0$$

is $x = -2.22$.

41. **Review problem.** Find every angle θ between 0 and 360° for which the ratio of $\sin\theta$ to $\cos\theta$ is -3.00.

42. **Review problem.** A highway curve forms a section of a circle. A car goes around the curve. Its dashboard compass shows that the car is initially heading due east. After it travels 840 m, it is heading 35.0° south of east. Find the radius of curvature of its path. *Suggestion:* You may find it useful to learn a geometric theorem stated in Appendix B.3.

43. **Review problem.** For a period of time as an alligator grows, its mass is proportional to the cube of its length. When the alligator's length changes by 15.8%, its mass increases by 17.3 kg. Find its mass at the end of this process.

44. **Review problem.** From the set of equations

$$p = 3q$$
$$pr = qs$$
$$\tfrac{1}{2}pr^2 + \tfrac{1}{2}qs^2 = \tfrac{1}{2}qt^2$$

involving the unknowns p, q, r, s, and t, find the value of the ratio of t to r.

45. ● **Review problem.** In a particular set of experimental trials, students examine a system described by the equation

$$\frac{Q}{\Delta t} = \frac{k\pi\, d^2\,(T_h - T_c)}{4L}$$

We will see this equation and the various quantities in it in Chapter 20. For experimental control, in these trials all quantities except d and Δt are constant. (a) If d is made three times larger, does the equation predict that Δt will get larger or smaller? By what factor? (b) What pattern of proportionality of Δt to d does the equation predict? (c) To display this proportionality as a straight line on a graph, what quantities should you plot on the horizontal and vertical axes? (d) What expression represents the theoretical slope of this graph?

Additional Problems

46. In a situation in which data are known to three significant digits, we write 6.379 m = 6.38 m and 6.374 m = 6.37 m. When a number ends in 5, we arbitrarily choose to write 6.375 m = 6.38 m. We could equally well write 6.375 m = 6.37 m, "rounding down" instead of "rounding up," because we would change the number 6.375 by equal increments in both cases. Now consider an order-of-magnitude estimate, in which factors of change rather than increments are important. We write 500 m ~ 10^3 m because 500 differs from 100 by a factor of 5, whereas it differs from 1 000 by only a factor of 2. We write 437 m ~ 10^3 m and 305 m ~ 10^2 m. What distance differs from 100 m and from 1 000 m by equal factors so that we could equally well choose to represent its order of magnitude either as ~10^2 m or as ~10^3 m?

47. ● A spherical shell has an outside radius of 2.60 cm and an inside radius of a. The shell wall has uniform thickness and is made of a material with density 4.70 g/cm^3. The space inside the shell is filled with a liquid having a density of 1.23 g/cm^3. (a) Find the mass m of the sphere, including its contents, as a function of a. (b) In the answer to part (a), if a is regarded as a variable, for what value of a does m have its maximum possible value? (c) What is this maximum mass? (d) Does the value from part (b) agree with the result of a direct calculation of the mass of a sphere of uniform density? (e) For what value of a does the answer to part (a) have its minimum possible value? (f) What is this minimum mass? (g) Does the value from part (f) agree with the result of a direct calculation of the mass of a uniform sphere? (h) What value of m is halfway between the maximum and minimum possible values? (i) Does this mass agree with the result of part (a) evaluated for a = 2.60 cm/2 = 1.30 cm? (j) Explain whether you should expect agreement in each of parts (d), (g), and (i). (k) **What If?** In part (a), would the answer change if the inner wall of the shell were not concentric with the outer wall?

48. A rod extending between $x = 0$ and $x = 14.0$ cm has uniform cross-sectional area $A = 9.00$ cm^2. It is made from a continuously changing alloy of metals so that along its length its density changes steadily from 2.70 g/cm^3 to 19.3 g/cm^3. (a) Identify the constants B and C required in the expression $\rho = B + Cx$ to describe the variable density. (b) The mass of the rod is given by

$$m = \int_{\text{all material}} \rho dV = \int_{\text{all } x} \rho A dx = \int_0^{14\text{ cm}} (B + Cx)\ (9.00\text{ cm}^2)\ dx$$

Carry out the integration to find the mass of the rod.

49. The diameter of our disk-shaped galaxy, the Milky Way, is about 1.0×10^5 light-years (ly). The distance to Andromeda, which is the spiral galaxy nearest to the Milky Way, is about 2.0 million ly. If a scale model represents the Milky Way and Andromeda galaxies as dinner plates 25 cm in diameter, determine the distance between the centers of the two plates.

50. ● Air is blown into a spherical balloon so that, when its radius is 6.50 cm, its radius is increasing at the rate 0.900 cm/s. (a) Find the rate at which the volume of the balloon is increasing. (b) If this volume flow rate of air entering the balloon is constant, at what rate will the radius be increasing when the radius is 13.0 cm? (c) Explain physically why the answer to part (b) is larger or smaller than 0.9 cm/s, if it is different.

51. ▲ The consumption of natural gas by a company satisfies the empirical equation $V = 1.50t + 0.008\,00t^2$, where V is the volume in millions of cubic feet and t is the time in months. Express this equation in units of cubic feet and seconds. Assign proper units to the coefficients. Assume a month is 30.0 days.

52. In physics it is important to use mathematical approximations. Demonstrate that for small angles (< 20°),

$$\tan \alpha \approx \sin \alpha \approx \alpha = \frac{\pi \alpha'}{180°}$$

where α is in radians and α' is in degrees. Use a calculator to find the largest angle for which $\tan \alpha$ may be approximated by α with an error less than 10.0%.

53. A high fountain of water is located at the center of a circular pool as shown in Figure P1.53. Not wishing to get his feet wet, a student walks around the pool and measures its circumference to be 15.0 m. Next, the student stands at the edge of the pool and uses a protractor to gauge the angle of elevation of the top of the fountain to be 55.0°. How high is the fountain?

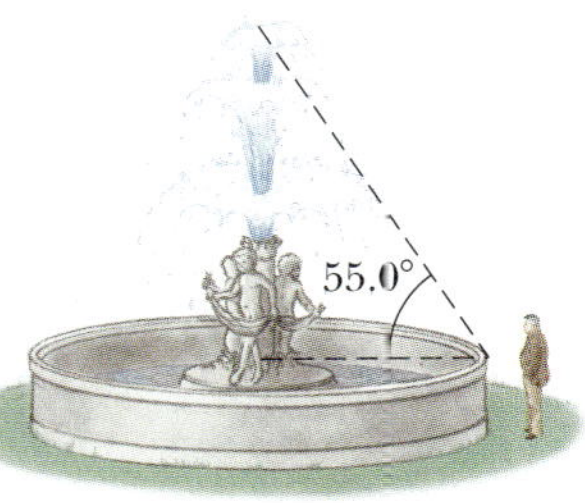

Figure P1.53

54. ● Collectible coins are sometimes plated with gold to enhance their beauty and value. Consider a commemorative quarter-dollar advertised for sale at $4.98. It has a

diameter of 24.1 mm and a thickness of 1.78 mm, and it is completely covered with a layer of pure gold 0.180 μm thick. The volume of the plating is equal to the thickness of the layer times the area to which it is applied. The patterns on the faces of the coin and the grooves on its edge have a negligible effect on its area. Assume the price of gold is $10.0 per gram. Find the cost of the gold added to the coin. Does the cost of the gold significantly enhance the value of the coin? Explain your answer.

55. One year is nearly $\pi \times 10^7$ s. Find the percentage error in this approximation, where "percentage error" is defined as

$$\text{Percentage error} = \frac{|\text{assumed value} - \text{true value}|}{\text{true value}} \times 100\%$$

56. ● A creature moves at a speed of 5.00 furlongs per fortnight (not a very common unit of speed). Given that 1 furlong = 220 yards and 1 fortnight = 14 days, determine the speed of the creature in meters per second. Explain what kind of creature you think it might be.

57. A child loves to watch as you fill a transparent plastic bottle with shampoo. Horizontal cross sections of the bottle are circles with varying diameters because the bottle is much wider in some places than others. You pour in bright green shampoo with constant volume flow rate 16.5 cm^3/s. At what rate is its level in the bottle rising (a) at a point where the diameter of the bottle is 6.30 cm and (b) at a point where the diameter is 1.35 cm?

58. ● The data in the following table represent measurements of the masses and dimensions of solid cylinders of aluminum, copper, brass, tin, and iron. Use these data to calculate the densities of these substances. State how your results for aluminum, copper, and iron compare with those given in Table 14.1.

Substance	Mass (g)	Diameter (cm)	Length (cm)
Aluminum	51.5	2.52	3.75
Copper	56.3	1.23	5.06
Brass	94.4	1.54	5.69
Tin	69.1	1.75	3.74
Iron	216.1	1.89	9.77

59. Assume there are 100 million passenger cars in the United States and the average fuel consumption is 20 mi/gal of gasoline. If the average distance traveled by each car is 10 000 mi/yr, how much gasoline would be saved per year if average fuel consumption could be increased to 25 mi/gal?

60. The distance from the Sun to the nearest star is about 4×10^{16} m. The Milky Way galaxy is roughly a disk of diameter $\sim 10^{21}$ m and thickness $\sim 10^{19}$ m. Find the order of magnitude of the number of stars in the Milky Way. Assume the distance between the Sun and our nearest neighbor is typical.

Answers to Quick Quizzes

1.1 (a). Because the density of aluminum is smaller than that of iron, a larger volume of aluminum than iron is required for a given mass.

1.2 False. Dimensional analysis gives the units of the proportionality constant but provides no information about its numerical value. To determine its numerical value requires either experimental data or geometrical reasoning. For example, in the generation of the equation $x = \frac{1}{2}at^2$, because the factor $\frac{1}{2}$ is dimensionless there is no way to determine it using dimensional analysis.

1.3 (b). Because there are 1.609 km in 1 mi, a larger number of kilometers than miles is required for a given distance.

In drag racing, a driver wants as large an acceleration as possible. In a distance of one-quarter mile, a vehicle reaches speeds of more than 320 mi/h, covering the entire distance in under 5 s. (George Lepp/Stone/Getty)

2 Motion in One Dimension

As a first step in studying classical mechanics, we describe the motion of an object while ignoring the interactions with external agents that might be causing or modifying that motion. This portion of classical mechanics is called *kinematics.* (The word *kinematics* has the same root as *cinema.* Can you see why?) In this chapter, we consider only motion in one dimension, that is, motion of an object along a straight line.

From everyday experience we recognize that motion of an object represents a continuous change in the object's position. In physics, we can categorize motion into three types: translational, rotational, and vibrational. A car traveling on a highway is an example of translational motion, the Earth's spin on its axis is an example of rotational motion, and the back-and-forth movement of a pendulum is an example of vibrational motion. In this and the next few chapters, we are concerned only with translational motion. (Later in the book we shall discuss rotational and vibrational motions.)

In our study of translational motion, we use what is called the **particle model** and describe the moving object as a *particle* regardless of its size. In general, **a particle is a point-like object, that is, an object that has mass but is of infinitesimal size.** For example, if we wish to describe the motion of the Earth around the Sun, we can treat the Earth as a particle and obtain reasonably accurate data about its orbit. This approximation is justified because the radius of the Earth's orbit is large compared with the dimensions of the Earth and the Sun. As an example on a much smaller scale, it is possible to explain the pressure exerted by a gas on the walls of a container by treating the gas molecules as particles, without regard for the internal structure of the molecules.

2.1 Position, Velocity, and Speed

Position ▶

The motion of a particle is completely known if the particle's position in space is known at all times. A particle's **position** is the location of the particle with respect to a chosen reference point that we can consider to be the origin of a coordinate system.

Consider a car moving back and forth along the x axis as in Active Figure 2.1a. When we begin collecting position data, the car is 30 m to the right of a road sign, which we will use to identify the reference position $x = 0$. We will use the particle model by identifying some point on the car, perhaps the front door handle, as a particle representing the entire car.

We start our clock, and once every 10 s we note the car's position relative to the sign at $x = 0$. As you can see from Table 2.1, the car moves to the right (which we have defined as the positive direction) during the first 10 s of motion, from position Ⓐ to position Ⓑ. After Ⓑ, the position values begin to decrease, suggesting the car is backing up from position Ⓑ through position Ⓕ. In fact, at Ⓓ, 30 s after we start measuring, the car is alongside the road sign that we are using to mark our origin of coordinates (see Active Figure 2.1a). It continues moving to the left and is more than 50 m to the left of the sign when we stop recording information after our sixth data point. A graphical representation of this information is presented in Active Figure 2.1b. Such a plot is called a *position–time graph.*

Notice the *alternative representations* of information that we have used for the motion of the car. Active Figure 2.1a is a *pictorial representation*, whereas Active Figure 2.1b is a *graphical representation*. Table 2.1 is a *tabular representation* of the same information. Using an alternative representation is often an excellent strategy for understanding the situation in a given problem. The ultimate goal in many problems is a *mathematical representation*, which can be analyzed to solve for some requested piece of information.

TABLE 2.1

Position of the Car at Various Times

Position	t (s)	x (m)
Ⓐ	0	30
Ⓑ	10	52
Ⓒ	20	38
Ⓓ	30	0
Ⓔ	40	−37
Ⓕ	50	−53

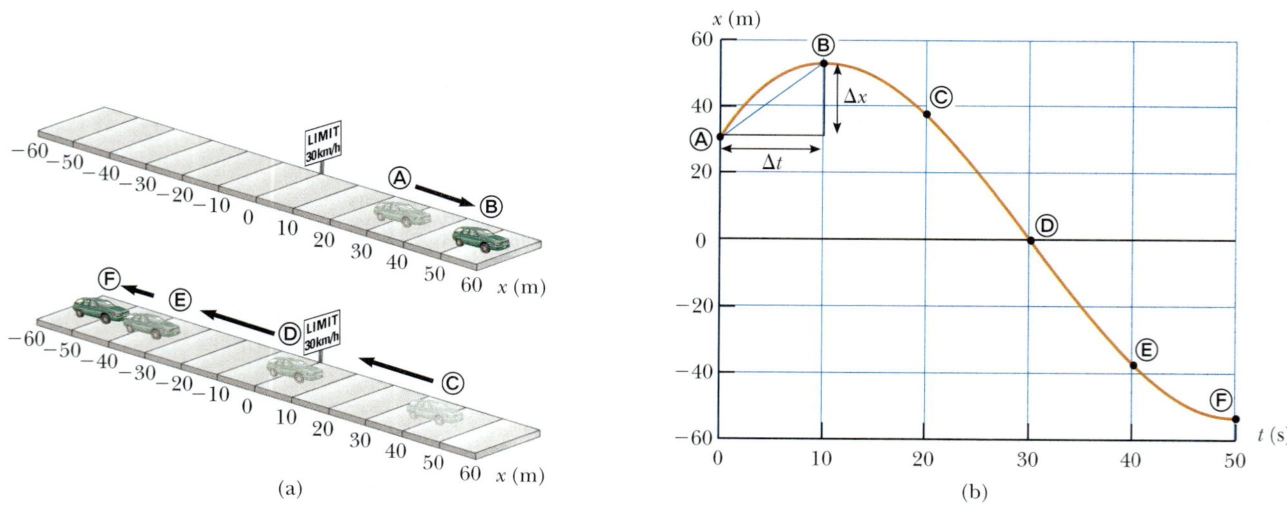

ACTIVE FIGURE 2.1

A car moves back and forth along a straight line. Because we are interested only in the car's translational motion, we can model it as a particle. Several representations of the information about the motion of the car can be used. Table 2.1 is a tabular representation of the information. (a) A pictorial representation of the motion of the car. (b) A graphical representation (position–time graph) of the motion of the car.

Sign in at www.thomsonedu.com and go to ThomsonNOW to move each of the six points Ⓐ through Ⓕ and observe the motion of the car in both a pictorial and a graphical representation as it follows a smooth path through the six points.

Given the data in Table 2.1, we can easily determine the change in position of the car for various time intervals. The **displacement** of a particle is defined as its change in position in some time interval. As the particle moves from an initial position x_i to a final position x_f, its displacement is given by

$$\Delta x \equiv x_f - x_i \tag{2.1}$$

◀ Displacement

We use the capital Greek letter delta (Δ) to denote the *change* in a quantity. From this definition we see that Δx is positive if x_f is greater than x_i and negative if x_f is less than x_i.

It is very important to recognize the difference between displacement and distance traveled. **Distance** is the length of a path followed by a particle. Consider, for example, the basketball players in Figure 2.2. If a player runs from his own team's basket down the court to the other team's basket and then returns to his own basket, the *displacement* of the player during this time interval is zero because he ended up at the same point as he started: $x_f = x_i$, so $\Delta x = 0$. During this time interval, however, he moved through a *distance* of twice the length of the basketball court. Distance is always represented as a positive number, whereas displacement can be either positive or negative.

Ken White/Allsport/Getty Images

Figure 2.2 On this basketball court, players run back and forth for the entire game. The distance that the players run over the duration of the game is nonzero. The displacement of the players over the duration of the game is approximately zero because they keep returning to the same point over and over again.

Displacement is an example of a vector quantity. Many other physical quantities, including position, velocity, and acceleration, also are vectors. In general, **a vector quantity requires the specification of both direction and magnitude**. By contrast, **a scalar quantity has a numerical value and no direction**. In this chapter, we use positive (+) and negative (−) signs to indicate vector direction. For example, for horizontal motion let us arbitrarily specify to the right as being the positive direction. It follows that any object always moving to the right undergoes a positive displacement $\Delta x > 0$, and any object moving to the left undergoes a negative displacement so that $\Delta x < 0$. We shall treat vector quantities in greater detail in Chapter 3.

One very important point has not yet been mentioned. Notice that the data in Table 2.1 result only in the six data points in the graph in Active Figure 2.1b. The smooth curve drawn through the six points in the graph is only a *possibility* of the actual motion of the car. We only have information about six instants of time; we have no idea what happened in between the data points. The smooth curve is a *guess* as to what happened, but keep in mind that it is *only* a guess.

If the smooth curve does represent the actual motion of the car, the graph contains information about the entire 50-s interval during which we watch the car move. It is much easier to see changes in position from the graph than from a verbal description or even a table of numbers. For example, it is clear that the car covers more ground during the middle of the 50-s interval than at the end. Between positions Ⓒ and Ⓓ, the car travels almost 40 m, but during the last 10 s, between positions Ⓔ and Ⓕ, it moves less than half that far. A common way of comparing these different motions is to divide the displacement Δx that occurs between two clock readings by the value of that particular time interval Δt. The result turns out to be a very useful ratio, one that we shall use many times. This ratio has been given a special name: the *average velocity*. **The average velocity $v_{x,\,\text{avg}}$ of a particle is defined as the particle's displacement Δx divided by the time interval Δt during which that displacement occurs:**

$$v_{x,\,\text{avg}} \equiv \frac{\Delta x}{\Delta t} \tag{2.2}$$

◀ Average velocity

where the subscript x indicates motion along the x axis. From this definition we see that average velocity has dimensions of length divided by time (L/T), or meters per second in SI units.

The average velocity of a particle moving in one dimension can be positive or negative, depending on the sign of the displacement. (The time interval Δt is always positive.) If the coordinate of the particle increases in time (that is, if $x_f > x_i$), Δx is positive and $v_{x,\,\text{avg}} = \Delta x/\Delta t$ is positive. This case corresponds to a particle moving in the positive x direction, that is, toward larger values of x. If the coordinate decreases in time (that is, if $x_f < x_i$), Δx is negative and hence $v_{x,\,\text{avg}}$ is negative. This case corresponds to a particle moving in the negative x direction.

We can interpret average velocity geometrically by drawing a straight line between any two points on the position–time graph in Active Figure 2.1b. This line forms the hypotenuse of a right triangle of height Δx and base Δt. The slope of this line is the ratio $\Delta x/\Delta t$, which is what we have defined as average velocity in Equation 2.2. For example, the line between positions Ⓐ and Ⓑ in Active Figure 2.1b has a slope equal to the average velocity of the car between those two times, (52 m − 30 m)/(10 s − 0) = 2.2 m/s.

In everyday usage, the terms *speed* and *velocity* are interchangeable. In physics, however, there is a clear distinction between these two quantities. Consider a marathon runner who runs a distance d of more than 40 km and yet ends up at her starting point. Her total displacement is zero, so her average velocity is zero! Nonetheless, we need to be able to quantify how fast she was running. A slightly different ratio accomplishes that for us. The **average speed** v_{avg} of a particle, a scalar quantity, is defined as **the total distance traveled divided by the total time interval required to travel that distance:**

Average speed ▶

$$v_{\text{avg}} \equiv \frac{d}{\Delta t} \tag{2.3}$$

PITFALL PREVENTION 2.1

Average Speed and Average Velocity

The magnitude of the average velocity is *not* the average speed. For example, consider the marathon runner discussed before Equation 2.3. The magnitude of her average velocity is zero, but her average speed is clearly not zero.

The SI unit of average speed is the same as the unit of average velocity: meters per second. Unlike average velocity, however, average speed has no direction and is always expressed as a positive number. Notice the clear distinction between the definitions of average velocity and average speed: average velocity (Eq. 2.2) is the *displacement* divided by the time interval, whereas average speed (Eq. 2.3) is the *distance* divided by the time interval.

Knowledge of the average velocity or average speed of a particle does not provide information about the details of the trip. For example, suppose it takes you 45.0 s to travel 100 m down a long, straight hallway toward your departure gate at an airport. At the 100-m mark, you realize you missed the restroom, and you return back 25.0 m along the same hallway, taking 10.0 s to make the return trip. The magnitude of your average *velocity* is +75.0 m/55.0 s = +1.36 m/s. The average *speed* for your trip is 125 m/55.0 s = 2.27 m/s. You may have traveled at various speeds during the walk. Neither average velocity nor average speed provides information about these details.

Quick Quiz 2.1 Under which of the following conditions is the magnitude of the average velocity of a particle moving in one dimension smaller than the average speed over some time interval? (a) a particle moves in the $+x$ direction without reversing (b) a particle moves in the $-x$ direction without reversing (c) a particle moves in the $+x$ direction and then reverses the direction of its motion (d) there are no conditions for which this is true

EXAMPLE 2.1 Calculating the Average Velocity and Speed

Find the displacement, average velocity, and average speed of the car in Active Figure 2.1a between positions Ⓐ and Ⓕ.

SOLUTION

Consult Active Figure 2.1 to form a mental image of the car and its motion. We model the car as a particle. From the position–time graph given in Active Figure 2.1b, notice that $x_Ⓐ = 30$ m at $t_Ⓐ = 0$ s and that $x_Ⓕ = -53$ m at $t_Ⓕ = 50$ s.

Use Equation 2.1 to find the displacement of the car: $\Delta x = x_Ⓕ - x_Ⓐ = -53\text{ m} - 30\text{ m} = -83\text{ m}$

This result means that the car ends up 83 m in the negative direction (to the left, in this case) from where it started. This number has the correct units and is of the same order of magnitude as the supplied data. A quick look at Active Figure 2.1a indicates that it is the correct answer.

Use Equation 2.2 to find the average velocity:

$$v_{x,\,\text{avg}} = \frac{x_{Ⓕ} - x_{Ⓐ}}{t_{Ⓕ} - t_{Ⓐ}}$$

$$= \frac{-53\text{ m} - 30\text{ m}}{50\text{ s} - 0\text{ s}} = \frac{-83\text{ m}}{50\text{ s}} = -1.7\text{ m/s}$$

We cannot unambiguously find the average speed of the car from the data in Table 2.1 because we do not have information about the positions of the car between the data points. If we adopt the assumption that the details of the car's position are described by the curve in Active Figure 2.1b, the distance traveled is 22 m (from Ⓐ to Ⓑ) plus 105 m (from Ⓑ to Ⓕ), for a total of 127 m.

Use Equation 2.3 to find the car's average speed:

$$v_{\text{avg}} = \frac{127\text{ m}}{50\text{ s}} = 2.5\text{ m/s}$$

Notice that the average speed is positive, as it must be. Suppose the brown curve in Active Figure 2.1b were different so that between 0 s and 10 s it went from Ⓐ up to 100 m and then came back down to Ⓑ. The average speed of the car would change because the distance is different, but the average velocity would not change.

2.2 Instantaneous Velocity and Speed

Often we need to know the velocity of a particle at a particular instant in time rather than the average velocity over a finite time interval. In other words, you would like to be able to specify your velocity just as precisely as you can specify your position by noting what is happening at a specific clock reading—that is, at some specific instant. What does it mean to talk about how quickly something is moving if we "freeze time" and talk only about an individual instant? In the late 1600s, with the invention of calculus, scientists began to understand how to describe an object's motion at any moment in time.

To see how that is done, consider Active Figure 2.3a, which is a reproduction of the graph in Active Figure 2.1b. We have already discussed the average velocity for the interval during which the car moved from position Ⓐ to position Ⓑ (given by the slope of the blue line) and for the interval during which it moved from Ⓐ to Ⓕ (represented by the slope of the longer blue line and calculated in Example 2.1). The car starts out by moving to the right, which we defined to be the positive direction. Therefore, being positive, the value of the average velocity during the interval from Ⓐ to Ⓑ is more representative of the initial velocity than is the value

PITFALL PREVENTION 2.2

Slopes of Graphs

In any graph of physical data, the *slope* represents the ratio of the change in the quantity represented on the vertical axis to the change in the quantity represented on the horizontal axis. Remember that *a slope has units* (unless both axes have the same units). The units of slope in Active Figure 2.1b and Active Figure 2.3 are meters per second, the units of velocity.

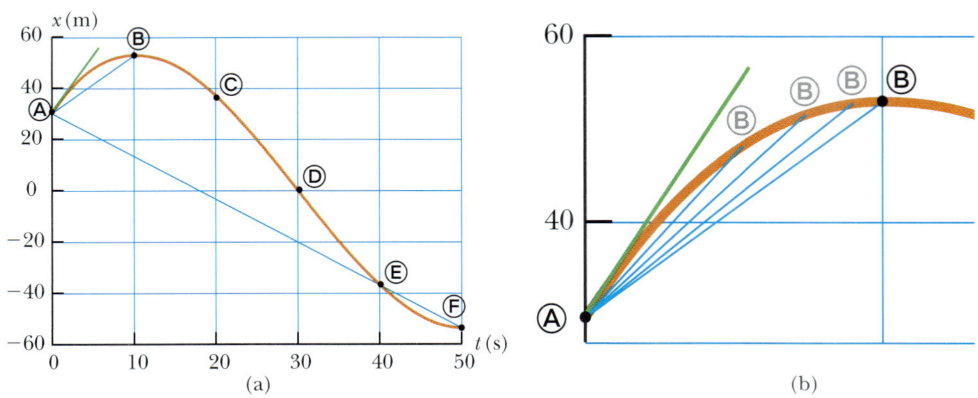

ACTIVE FIGURE 2.3

(a) Graph representing the motion of the car in Active Figure 2.1. (b) An enlargement of the upper-left-hand corner of the graph shows how the blue line between positions Ⓐ and Ⓑ approaches the green tangent line as point Ⓑ is moved closer to point Ⓐ.

Sign in at www.thomsonedu.com and go to ThomsonNOW to move point Ⓑ as suggested in part (b) and observe the blue line approaching the green tangent line.

of the average velocity during the interval from Ⓐ to Ⓕ, which we determined to be negative in Example 2.1. Now let us focus on the short blue line and slide point Ⓑ to the left along the curve, toward point Ⓐ, as in Active Figure 2.3b. The line between the points becomes steeper and steeper, and as the two points become extremely close together, the line becomes a tangent line to the curve, indicated by the green line in Active Figure 2.3b. The slope of this tangent line represents the velocity of the car at point Ⓐ. What we have done is determine the *instantaneous velocity* at that moment. In other words, **the instantaneous velocity v_x equals the limiting value of the ratio $\Delta x/\Delta t$ as Δt approaches zero:**[1]

$$v_x \equiv \lim_{\Delta t \to 0} \frac{\Delta x}{\Delta t} \tag{2.4}$$

In calculus notation, this limit is called the *derivative* of x with respect to t, written dx/dt:

Instantaneous velocity ▶

$$v_x \equiv \lim_{\Delta t \to 0} \frac{\Delta x}{\Delta t} = \frac{dx}{dt} \tag{2.5}$$

PITFALL PREVENTION 2.3

Instantaneous Speed and Instantaneous Velocity

In Pitfall Prevention 2.1, we argued that the magnitude of the average velocity is not the average speed. The magnitude of the instantaneous velocity, however, *is* the instantaneous speed. In an infinitesimal time interval, the magnitude of the displacement is equal to the distance traveled by the particle.

The instantaneous velocity can be positive, negative, or zero. When the slope of the position–time graph is positive, such as at any time during the first 10 s in Active Figure 2.3, v_x is positive and the car is moving toward larger values of x. After point Ⓑ, v_x is negative because the slope is negative and the car is moving toward smaller values of x. At point Ⓑ, the slope and the instantaneous velocity are zero and the car is momentarily at rest.

From here on, we use the word *velocity* to designate instantaneous velocity. When we are interested in *average velocity*, we shall always use the adjective *average*.

The **instantaneous speed** of a particle is defined as the magnitude of its instantaneous velocity. As with average speed, instantaneous speed has no direction associated with it. For example, if one particle has an instantaneous velocity of $+25$ m/s along a given line and another particle has an instantaneous velocity of -25 m/s along the same line, both have a speed[2] of 25 m/s.

Quick Quiz 2.2 Are members of the highway patrol more interested in (a) your average speed or (b) your instantaneous speed as you drive?

CONCEPTUAL EXAMPLE 2.2 The Velocity of Different Objects

Consider the following one-dimensional motions: **(A)** a ball thrown directly upward rises to a highest point and falls back into the thrower's hand; **(B)** a race car starts from rest and speeds up to 100 m/s; and **(C)** a spacecraft drifts through space at constant velocity. Are there any points in the motion of these objects at which the instantaneous velocity has the same value as the average velocity over the entire motion? If so, identify the point(s).

SOLUTION

(A) The average velocity for the thrown ball is zero because the ball returns to the starting point; therefore, its displacement is zero. There is one point at which the instantaneous velocity is zero: at the top of the motion.

(B) The car's average velocity cannot be evaluated unambiguously with the information given, but it must have some value between 0 and 100 m/s. Because the car will have every instantaneous velocity between 0 and 100 m/s at some time during the interval, there must be some instant at which the instantaneous velocity is equal to the average velocity over the entire motion.

(C) Because the spacecraft's instantaneous velocity is constant, its instantaneous velocity at *any* time and its average velocity over *any* time interval are the same.

[1] Notice that the displacement Δx also approaches zero as Δt approaches zero, so the ratio looks like 0/0. As Δx and Δt become smaller and smaller, the ratio $\Delta x/\Delta t$ approaches a value equal to the slope of the line tangent to the x-versus-t curve.

[2] As with velocity, we drop the adjective for instantaneous speed. "Speed" means instantaneous speed.

EXAMPLE 2.3 Average and Instantaneous Velocity

A particle moves along the x axis. Its position varies with time according to the expression $x = -4t + 2t^2$, where x is in meters and t is in seconds.[3] The position–time graph for this motion is shown in Figure 2.4. Notice that the particle moves in the negative x direction for the first second of motion, is momentarily at rest at the moment $t = 1$ s, and moves in the positive x direction at times $t > 1$ s.

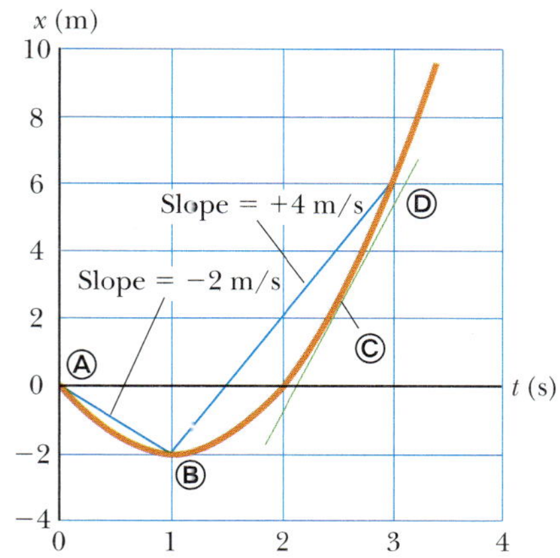

Figure 2.4 (Example 2.3) Position–time graph for a particle having an x coordinate that varies in time according to the expression $x = -4t + 2t^2$.

(A) Determine the displacement of the particle in the time intervals $t = 0$ to $t = 1$ s and $t = 1$ s to $t = 3$ s.

SOLUTION

From the graph in Figure 2.4, form a mental representation of the motion of the particle. Keep in mind that the particle does not move in a curved path in space such as that shown by the brown curve in the graphical representation. The particle moves only along the x axis in one dimension. At $t = 0$, is it moving to the right or to the left?

During the first time interval, the slope is negative and hence the average velocity is negative. Therefore, we know that the displacement between Ⓐ and Ⓑ must be a negative number having units of meters. Similarly, we expect the displacement between Ⓑ and Ⓓ to be positive.

In the first time interval, set $t_i = t_Ⓐ = 0$ and $t_f = t_Ⓑ = 1$ s and use Equation 2.1 to find the displacement:

$$\Delta x_{Ⓐ\to Ⓑ} = x_f - x_i = x_Ⓑ - x_Ⓐ$$
$$= [-4(1) + 2(1)^2] - [-4(0) + 2(0)^2] = -2 \text{ m}$$

For the second time interval ($t = 1$ s to $t = 3$ s), set $t_i = t_Ⓑ = 1$ s and $t_f = t_Ⓓ = 3$ s:

$$\Delta x_{Ⓑ\to Ⓓ} = x_f - x_i = x_Ⓓ - x_Ⓑ$$
$$= [-4(3) + 2(3)^2] - [-4(1) + 2(1)^2] = +8 \text{ m}$$

These displacements can also be read directly from the position–time graph.

(B) Calculate the average velocity during these two time intervals.

SOLUTION

In the first time interval, use Equation 2.2 with $\Delta t = t_f - t_i = t_Ⓑ - t_Ⓐ = 1$ s:

$$v_{x,\text{ avg }(Ⓐ\to Ⓑ)} = \frac{\Delta x_{Ⓐ\to Ⓑ}}{\Delta t} = \frac{-2 \text{ m}}{1 \text{ s}} = -2 \text{ m/s}$$

In the second time interval, $\Delta t = 2$ s:

$$v_{x,\text{ avg }(Ⓑ\to Ⓓ)} = \frac{\Delta x_{Ⓑ\to Ⓓ}}{\Delta t} = \frac{8 \text{ m}}{2 \text{ s}} = +4 \text{ m/s}$$

These values are the same as the slopes of the lines joining these points in Figure 2.4.

(C) Find the instantaneous velocity of the particle at $t = 2.5$ s.

SOLUTION

Measure the slope of the green line at $t = 2.5$ s (point Ⓒ) in Figure 2.4:

$$v_x = +6 \text{ m/s}$$

Notice that this instantaneous velocity is on the same order of magnitude as our previous results, that is, a few meters per second. Is that what you would have expected?

[3] Simply to make it easier to read, we write the expression as $x = -4t + 2t^2$ rather than as $x = (-4.00 \text{ m/s})t + (2.00 \text{ m/s}^2)t^{2.00}$. When an equation summarizes measurements, consider its coefficients to have as many significant digits as other data quoted in a problem. Consider its coefficients to have the units required for dimensional consistency. When we start our clocks at $t = 0$, we usually do not mean to limit the precision to a single digit. Consider any zero value in this book to have as many significant figures as you need.

2.3 Analysis Models: The Particle Under Constant Velocity

An important technique in the solution to physics problems is the use of *analysis models.* Such models help us analyze common situations in physics problems and guide us toward a solution. An **analysis model** is a problem we have solved before. It is a description of either (1) the behavior of some physical entity or (2) the interaction between that entity and the environment. When you encounter a new problem, you should identify the fundamental details of the problem and attempt to recognize which of the types of problems you have already solved might be used as a model for the new problem. For example, suppose an automobile is moving along a straight freeway at a constant speed. Is it important that it is an automobile? Is it important that it is a freeway? If the answers to both questions are no, we model the automobile as a *particle under constant velocity*, which we will discuss in this section.

This method is somewhat similar to the common practice in the legal profession of finding "legal precedents." If a previously resolved case can be found that is very similar legally to the current one, it is offered as a model and an argument is made in court to link them logically. The finding in the previous case can then be used to sway the finding in the current case. We will do something similar in physics. For a given problem, we search for a "physics precedent," a model with which we are already familiar and that can be applied to the current problem.

We shall generate analysis models based on four fundamental simplification models. The first is the particle model discussed in the introduction to this chapter. We will look at a particle under various behaviors and environmental interactions. Further analysis models are introduced in later chapters based on simplification models of a *system*, a *rigid object*, and a *wave*. Once we have introduced these analysis models, we shall see that they appear again and again in different problem situations.

Let us use Equation 2.2 to build our first analysis model for solving problems. We imagine a particle moving with a constant velocity. The **particle under constant velocity** model can be applied in *any* situation in which an entity that can be modeled as a particle is moving with constant velocity. This situation occurs frequently, so this model is important.

If the velocity of a particle is constant, its instantaneous velocity at any instant during a time interval is the same as the average velocity over the interval. That is, $v_x = v_{x,\text{ avg}}$. Therefore, Equation 2.2 gives us an equation to be used in the mathematical representation of this situation:

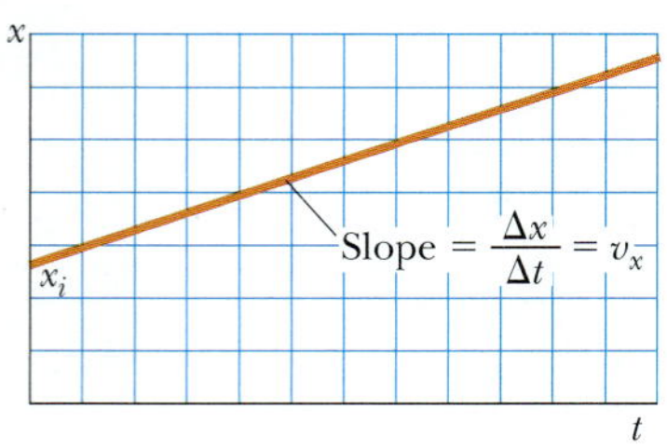

Figure 2.5 Position–time graph for a particle under constant velocity. The value of the constant velocity is the slope of the line.

$$v_x = \frac{\Delta x}{\Delta t} \tag{2.6}$$

Remembering that $\Delta x = x_f - x_i$, we see that $v_x = (x_f - x_i)/\Delta t$, or

$$x_f = x_i + v_x\,\Delta t$$

This equation tells us that the position of the particle is given by the sum of its original position x_i at time $t = 0$ plus the displacement $v_x\,\Delta t$ that occurs during the time interval Δt. In practice, we usually choose the time at the beginning of the interval to be $t_i = 0$ and the time at the end of the interval to be $t_f = t$, so our equation becomes

Position as a function of time ▶

$$x_f = x_i + v_x t \quad \text{(for constant } v_x\text{)} \tag{2.7}$$

Equations 2.6 and 2.7 are the primary equations used in the model of a particle under constant velocity. They can be applied to particles or objects that can be modeled as particles.

Figure 2.5 is a graphical representation of the particle under constant velocity. On this position–time graph, the slope of the line representing the motion is constant and equal to the magnitude of the velocity. Equation 2.7, which is the equation of a straight line, is the mathematical representation of the particle under constant velocity model. The slope of the straight line is v_x and the y intercept is x_i in both representations.

EXAMPLE 2.4 Modeling a Runner as a Particle

A scientist is studying the biomechanics of the human body. She determines the velocity of an experimental subject while he runs along a straight line at a constant rate. The scientist starts the stopwatch at the moment the runner passes a given point and stops it after the runner has passed another point 20 m away. The time interval indicated on the stopwatch is 4.0 s.

(A) What is the runner's velocity?

SOLUTION

Think about the moving runner. We model the runner as a particle because the size of the runner and the movement of arms and legs are unnecessary details. Because the problem states that the subject runs at a constant rate, we can model him as a particle under constant velocity.

Use Equation 2.6 to find the constant velocity of the runner:

$$v_x = \frac{\Delta x}{\Delta t} = \frac{x_f - x_i}{\Delta t} = \frac{20\text{ m} - 0}{4.0\text{ s}} = \boxed{5.0\text{ m/s}}$$

(B) If the runner continues his motion after the stopwatch is stopped, what is his position after 10 s has passed?

SOLUTION

Use Equation 2.7 and the velocity found in part (A) to find the position of the particle at time $t = 10$ s:

$$x_f = x_i + v_x t = 0 + (5.0\text{ m/s})(10\text{ s}) = \boxed{50\text{ m}}$$

Notice that this value is more than twice that of the 20-m position at which the stopwatch was stopped. Is this value consistent with the time of 10 s being more than twice the time of 4.0 s?

The mathematical manipulations for the particle under constant velocity stem from Equation 2.6 and its descendent, Equation 2.7. These equations can be used to solve for any variable in the equations that happens to be unknown if the other variables are known. For example, in part (B) of Example 2.4, we find the position when the velocity and the time are known. Similarly, if we know the velocity and the final position, we could use Equation 2.7 to find the time at which the runner is at this position.

A particle under constant velocity moves with a constant speed along a straight line. Now consider a particle moving with a constant speed along a curved path. This situation can be represented with the **particle under constant speed** model. The primary equation for this model is Equation 2.3, with the average speed v_{avg} replaced by the constant speed v:

$$v = \frac{d}{\Delta t} \qquad \textbf{(2.8)}$$

As an example, imagine a particle moving at a constant speed in a circular path. If the speed is 5.00 m/s and the radius of the path is 10.0 m, we can calculate the time interval required to complete one trip around the circle:

$$v = \frac{d}{\Delta t} \quad \rightarrow \quad \Delta t = \frac{d}{v} = \frac{2\pi r}{v} = \frac{2\pi\ (10.0\text{ m})}{5.00\text{ m/s}} = 12.6\text{ s}$$

2.4 Acceleration

In Example 2.3, we worked with a common situation in which the velocity of a particle changes while the particle is moving. When the velocity of a particle changes with time, the particle is said to be *accelerating*. For example, the magnitude of the velocity of a car increases when you step on the gas and decreases when you apply the brakes. Let us see how to quantify acceleration.

Suppose an object that can be modeled as a particle moving along the x axis has an initial velocity v_{xi} at time t_i and a final velocity v_{xf} at time t_f, as in Figure 2.6a. The **average acceleration** $a_{x,\,\text{avg}}$ of the particle is defined as the *change* in velocity Δv_x divided by the time interval Δt during which that change occurs:

Average acceleration ▶

$$a_{x,\,\text{avg}} \equiv \frac{\Delta v_x}{\Delta t} = \frac{v_{xf} - v_{xi}}{t_f - t_i} \qquad \textbf{(2.9)}$$

As with velocity, when the motion being analyzed is one dimensional, we can use positive and negative signs to indicate the direction of the acceleration. Because the dimensions of velocity are L/T and the dimension of time is T, acceleration has dimensions of length divided by time squared, or L/T^2. The SI unit of acceleration is meters per second squared (m/s^2). It might be easier to interpret these units if you think of them as meters per second per second. For example, suppose an object has an acceleration of $+2\ m/s^2$. You should form a mental image of the object having a velocity that is along a straight line and is increasing by 2 m/s during every interval of 1 s. If the object starts from rest, you should be able to picture it moving at a velocity of +2 m/s after 1 s, at +4 m/s after 2 s, and so on.

In some situations, the value of the average acceleration may be different over different time intervals. It is therefore useful to define the **instantaneous acceleration** as the limit of the average acceleration as Δt approaches zero. This concept is analogous to the definition of instantaneous velocity discussed in Section 2.2. If we imagine that point Ⓐ is brought closer and closer to point Ⓑ in Figure 2.6a and we take the limit of $\Delta v_x/\Delta t$ as Δt approaches zero, we obtain the instantaneous acceleration at point Ⓑ:

Instantaneous acceleration ▶

$$a_x \equiv \lim_{\Delta t \to 0} \frac{\Delta v_x}{\Delta t} = \frac{dv_x}{dt} \qquad \textbf{(2.10)}$$

PITFALL PREVENTION 2.4

Negative Acceleration

Keep in mind that *negative acceleration does not necessarily mean that an object is slowing down.* If the acceleration is negative and the velocity is negative, the object is speeding up!

PITFALL PREVENTION 2.5

Deceleration

The word *deceleration* has the common popular connotation of *slowing down.* We will not use this word in this book because it confuses the definition we have given for negative acceleration.

That is, **the instantaneous acceleration equals the derivative of the velocity with respect to time**, which by definition is the slope of the velocity–time graph. The slope of the green line in Figure 2.6b is equal to the instantaneous acceleration at point Ⓑ. Therefore, we see that just as the velocity of a moving particle is the slope at a point on the particle's x–t graph, the acceleration of a particle is the slope at a point on the particle's v_x–t graph. One can interpret the derivative of the velocity with respect to time as the time rate of change of velocity. If a_x is positive, the acceleration is in the positive x direction; if a_x is negative, the acceleration is in the negative x direction.

For the case of motion in a straight line, the direction of the velocity of an object and the direction of its acceleration are related as follows. **When the object's velocity and acceleration are in the same direction, the object is speeding up. On the other hand, when the object's velocity and acceleration are in opposite directions, the object is slowing down.**

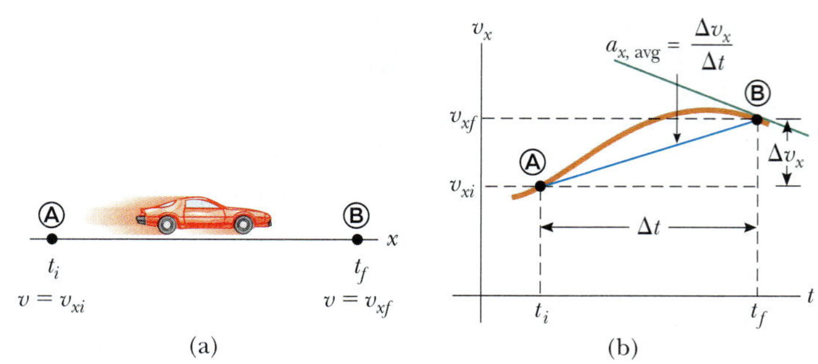

Figure 2.6 (a) A car, modeled as a particle, moving along the x axis from Ⓐ to Ⓑ, has velocity v_{xi} at $t = t_i$ and velocity v_{xf} at $t = t_f$. (b) Velocity–time graph (brown) for the particle moving in a straight line. The slope of the blue straight line connecting Ⓐ and Ⓑ is the average acceleration of the car during the time interval $\Delta t = t_f - t_i$. The slope of the green line is the instantaneous acceleration of the car at point Ⓑ.

To help with this discussion of the signs of velocity and acceleration, we can relate the acceleration of an object to the total *force* exerted on the object. In Chapter 5, we formally establish that **force is proportional to acceleration:**

$$F_x \propto a_x \tag{2.11}$$

This proportionality indicates that acceleration is caused by force. Furthermore, force and acceleration are both vectors and the vectors act in the same direction. Therefore, let us think about the signs of velocity and acceleration by imagining a force applied to an object and causing it to accelerate. Let us assume the velocity and acceleration are in the same direction. This situation corresponds to an object that experiences a force acting in the same direction as its velocity. In this case, the object speeds up! Now suppose the velocity and acceleration are in opposite directions. In this situation, the object moves in some direction and experiences a force acting in the opposite direction. Therefore, the object slows down! It is very useful to equate the direction of the acceleration to the direction of a force, because it is easier from our everyday experience to think about what effect a force will have on an object than to think only in terms of the direction of the acceleration.

Quick Quiz 2.3 If a car is traveling eastward and slowing down, what is the direction of the force on the car that causes it to slow down? (a) eastward (b) westward (c) neither eastward nor westward

From now on we shall use the term *acceleration* to mean instantaneous acceleration. When we mean average acceleration, we shall always use the adjective *average.* Because $v_x = dx/dt$, the acceleration can also be written as

$$a_x = \frac{dv_x}{dt} = \frac{d}{dt}\left(\frac{dx}{dt}\right) = \frac{d^2x}{dt^2} \tag{2.12}$$

That is, in one-dimensional motion, the acceleration equals the *second derivative* of x with respect to time.

Figure 2.7 illustrates how an acceleration–time graph is related to a velocity–time graph. The acceleration at any time is the slope of the velocity–time graph at that time. Positive values of acceleration correspond to those points in Figure 2.7a where the velocity is increasing in the positive x direction. The acceleration reaches a maximum at time $t_{Ⓐ}$, when the slope of the velocity–time graph is a maximum. The acceleration then goes to zero at time $t_{Ⓑ}$, when the velocity is a maximum (that is, when the slope of the v_x–t graph is zero). The acceleration is negative when the velocity is decreasing in the positive x direction, and it reaches its most negative value at time $t_{Ⓒ}$.

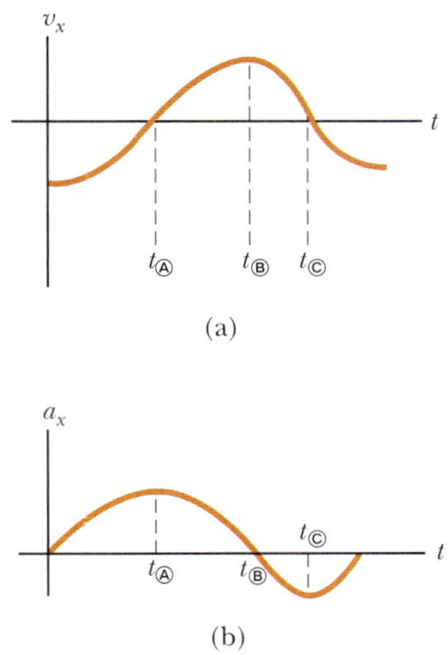

Figure 2.7 The instantaneous acceleration can be obtained from the velocity–time graph (a). At each instant, the acceleration in the graph of a_x versus t (b) equals the slope of the line tangent to the curve of v_x versus t (a).

Quick Quiz 2.4 Make a velocity–time graph for the car in Active Figure 2.1a. The speed limit posted on the road sign is 30 km/h. True or False? The car exceeds the speed limit at some time within the time interval 0 – 50 s.

CONCEPTUAL EXAMPLE 2.5 Graphical Relationships Between x, v_x, and a_x

The position of an object moving along the x axis varies with time as in Figure 2.8a (page 30). Graph the velocity versus time and the acceleration versus time for the object.

SOLUTION

The velocity at any instant is the slope of the tangent to the x–t graph at that instant. Between $t = 0$ and $t = t_{Ⓐ}$, the slope of the x–t graph increases uniformly, so the velocity increases linearly as shown in Figure 2.8b. Between $t_{Ⓐ}$ and $t_{Ⓑ}$, the slope of the x–t graph is constant, so the velocity remains constant. Between $t_{Ⓑ}$ and $t_{Ⓓ}$, the slope of the x–t graph decreases, so the value of the velocity in the v_x–t graph decreases. At $t_{Ⓓ}$, the slope of the x–t graph is zero, so the velocity is zero at that instant. Between $t_{Ⓓ}$ and $t_{Ⓔ}$, the slope of the x–t graph and therefore the velocity are negative and decrease uniformly in this interval. In the interval $t_{Ⓔ}$ to $t_{Ⓕ}$, the slope of the x–t graph is still negative, and at $t_{Ⓕ}$ it goes to zero. Finally, after $t_{Ⓕ}$, the slope of the x–t graph is zero, meaning that the object is at rest for $t > t_{Ⓕ}$.

The acceleration at any instant is the slope of the tangent to the v_x–t graph at that instant. The graph of acceleration versus time for this object is shown in Figure 2.8c. The acceleration is constant and positive between 0 and $t_Ⓐ$, where the slope of the v_x–t graph is positive. It is zero between $t_Ⓐ$ and $t_Ⓑ$ and for $t > t_Ⓕ$ because the slope of the v_x–t graph is zero at these times. It is negative between $t_Ⓑ$ and $t_Ⓔ$ because the slope of the v_x–t graph is negative during this interval. Between $t_Ⓔ$ and $t_Ⓕ$, the acceleration is positive like it is between 0 and $t_Ⓐ$, but higher in value because the slope of the v_x–t graph is steeper.

Notice that the sudden changes in acceleration shown in Figure 2.8c are unphysical. Such instantaneous changes cannot occur in reality.

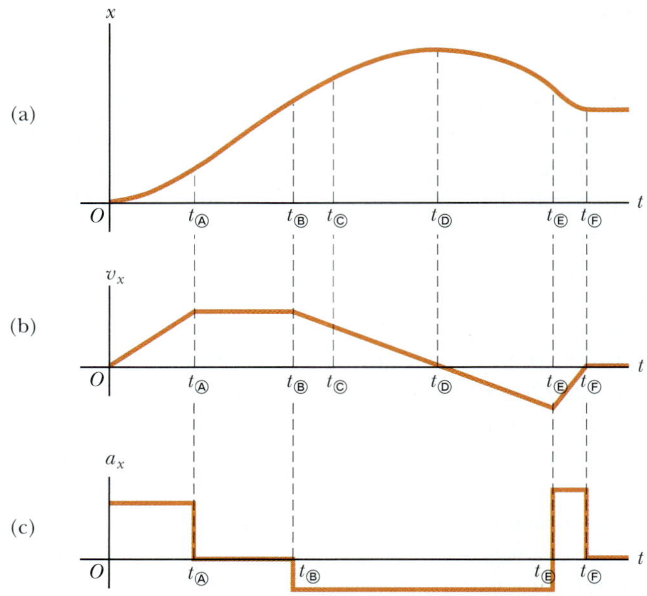

Figure 2.8 (Example 2.5) (a) Position–time graph for an object moving along the x axis. (b) The velocity–time graph for the object is obtained by measuring the slope of the position–time graph at each instant. (c) The acceleration–time graph for the object is obtained by measuring the slope of the velocity–time graph at each instant.

EXAMPLE 2.6 Average and Instantaneous Acceleration

The velocity of a particle moving along the x axis varies according to the expression $v_x = (40 - 5t^2)$ m/s, where t is in seconds.

(A) Find the average acceleration in the time interval $t = 0$ to $t = 2.0$ s.

SOLUTION

Think about what the particle is doing from the mathematical representation. Is it moving at $t = 0$? In which direction? Does it speed up or slow down? Figure 2.9 is a v_x–t graph that was created from the velocity versus time expression given in the problem statement. Because the slope of the entire v_x–t curve is negative, we expect the acceleration to be negative.

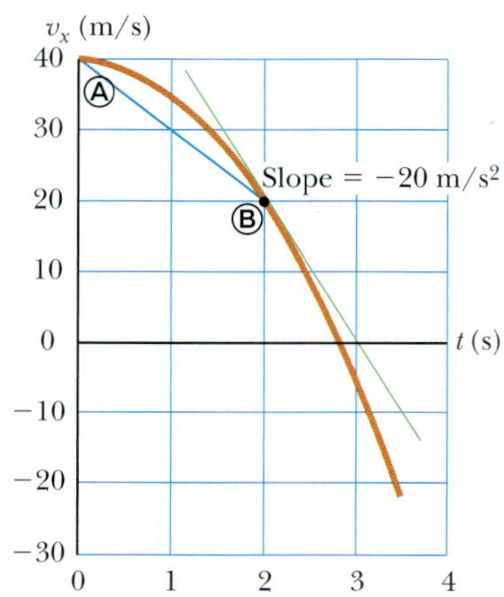

Figure 2.9 (Example 2.6) The velocity–time graph for a particle moving along the x axis according to the expression $v_x = (40 - 5t^2)$ m/s. The acceleration at $t = 2$ s is equal to the slope of the green tangent line at that time.

Find the velocities at $t_i = t_Ⓐ = 0$ and $t_f = t_Ⓑ = 2.0$ s by substituting these values of t into the expression for the velocity:

$$v_{xⒶ} = (40 - 5t_Ⓐ^{\,2})\text{ m/s} = [40 - 5(0)^2]\text{ m/s} = +40\text{ m/s}$$

$$v_{xⒷ} = (40 - 5t_Ⓑ^{\,2})\text{ m/s} = [40 - 5(2.0)^2]\text{ m/s} = +20\text{ m/s}$$

Find the average acceleration in the specified time interval $\Delta t = t_Ⓑ - t_Ⓐ = 2.0$ s:

$$a_{x,\text{ avg}} = \frac{v_{xf} - v_{xi}}{t_f - t_i} = \frac{v_{xⒷ} - v_{xⒶ}}{t_Ⓑ - t_Ⓐ} = \frac{(20 - 40)\text{ m/s}}{(2.0 - 0)\text{ s}}$$

$$= -10\text{ m/s}^2$$

The negative sign is consistent with our expectations—namely, that the average acceleration, represented by the slope of the line joining the initial and final points on the velocity–time graph, is negative.

(B) Determine the acceleration at $t = 2.0$ s.

SOLUTION

Knowing that the initial velocity at any time t is $v_{xi} = (40 - 5t^2)$ m/s, find the velocity at any later time $t + \Delta t$:

$$v_{xf} = 40 - 5(t + \Delta t)^2 = 40 - 5t^2 - 10t\,\Delta t - 5(\Delta t)^2$$

Find the change in velocity over the time interval Δt:

$$\Delta v_x = v_{xf} - v_{xi} = [-10t\,\Delta t - 5(\Delta t)^2] \text{ m/s}$$

To find the acceleration at any time t, divide this expression by Δt and take the limit of the result as Δt approaches zero:

$$a_x = \lim_{\Delta t \to 0} \frac{\Delta v_x}{\Delta t} = \lim_{\Delta t \to 0} (-10t - 5\Delta t) = -10t \text{ m/s}^2$$

Substitute $t = 2.0$ s:

$$a_x = (-10)(2.0) \text{ m/s}^2 = \boxed{-20 \text{ m/s}^2}$$

Because the velocity of the particle is positive and the acceleration is negative at this instant, the particle is slowing down.

Notice that the answers to parts (A) and (B) are different. The average acceleration in (A) is the slope of the blue line in Figure 2.9 connecting points Ⓐ and Ⓑ. The instantaneous acceleration in (B) is the slope of the green line tangent to the curve at point Ⓑ. Notice also that the acceleration is *not* constant in this example. Situations involving constant acceleration are treated in Section 2.6.

So far we have evaluated the derivatives of a function by starting with the definition of the function and then taking the limit of a specific ratio. If you are familiar with calculus, you should recognize that there are specific rules for taking derivatives. These rules, which are listed in Appendix B.6, enable us to evaluate derivatives quickly. For instance, one rule tells us that the derivative of any constant is zero. As another example, suppose x is proportional to some power of t, such as in the expression

$$x = At^n$$

where A and n are constants. (This expression is a very common functional form.) The derivative of x with respect to t is

$$\frac{dx}{dt} = nAt^{n-1}$$

Applying this rule to Example 2.5, in which $v_x = 40 - 5t^2$, we quickly find that the acceleration is $a_x = dv_x/dt = -10t$.

2.5 Motion Diagrams

The concepts of velocity and acceleration are often confused with each other, but in fact they are quite different quantities. In forming a mental representation of a moving object, it is sometimes useful to use a pictorial representation called a *motion diagram* to describe the velocity and acceleration while an object is in motion.

A motion diagram can be formed by imagining a *stroboscopic* photograph of a moving object, which shows several images of the object taken as the strobe light flashes at a constant rate. Active Figure 2.10 (page 32) represents three sets of strobe photographs of cars moving along a straight roadway in a single direction, from left to right. The time intervals between flashes of the stroboscope are equal in each part of the diagram. So as to not confuse the two vector quantities, we use red for velocity vectors and violet for acceleration vectors in Active Figure 2.10. The vectors are shown at several instants during the motion of the object. Let us describe the motion of the car in each diagram.

In Active Figure 2.10a, the images of the car are equally spaced, showing us that the car moves through the same displacement in each time interval. This equal spacing is consistent with the car moving with *constant positive velocity* and *zero acceleration.*

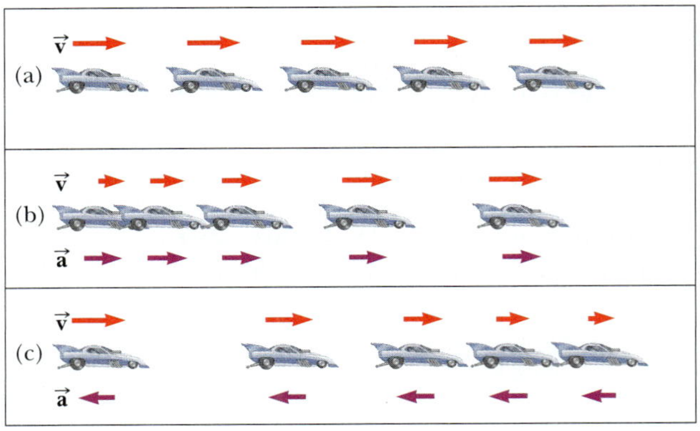

ACTIVE FIGURE 2.10

(a) Motion diagram for a car moving at constant velocity (zero acceleration). (b) Motion diagram for a car whose constant acceleration is in the direction of its velocity. The velocity vector at each instant is indicated by a red arrow, and the constant acceleration is indicated by a violet arrow. (c) Motion diagram for a car whose constant acceleration is in the direction *opposite* the velocity at each instant.

Sign in at www.thomsonedu.com and go to ThomsonNOW to select the constant acceleration and initial velocity of the car and observe pictorial and graphical representations of its motion.

We could model the car as a particle and describe it with the particle under constant velocity model.

In Active Figure 2.10b, the images become farther apart as time progresses. In this case, the velocity vector increases in length with time because the car's displacement between adjacent positions increases in time. These features suggest that the car is moving with a *positive velocity* and a *positive acceleration.* The velocity and acceleration are in the same direction. In terms of our earlier force discussion, imagine a force pulling on the car in the same direction it is moving: it speeds up.

In Active Figure 2.10c, we can tell that the car slows as it moves to the right because its displacement between adjacent images decreases with time. This case suggests that the car moves to the right with a negative acceleration. The length of the velocity vector decreases in time and eventually reaches zero. From this diagram we see that the acceleration and velocity vectors are *not* in the same direction. The car is moving with a *positive velocity,* but with a *negative acceleration.* (This type of motion is exhibited by a car that skids to a stop after applying its brakes.) The velocity and acceleration are in opposite directions. In terms of our earlier force discussion, imagine a force pulling on the car opposite to the direction it is moving: it slows down.

The violet acceleration vectors in parts (b) and (c) of Figure 2.10 are all of the same length. Therefore, these diagrams represent motion of a *particle under constant acceleration.* This important analysis model will be discussed in the next section.

Quick Quiz 2.5 Which one of the following statements is true? (a) If a car is traveling eastward, its acceleration must be eastward. (b) If a car is slowing down, its acceleration must be negative. (c) A particle with constant acceleration can never stop and stay stopped.

2.6 The Particle Under Constant Acceleration

If the acceleration of a particle varies in time, its motion can be complex and difficult to analyze. A very common and simple type of one-dimensional motion, however, is that in which the acceleration is constant. In such a case, the average acceleration $a_{x,\text{ avg}}$ over any time interval is numerically equal to the instantaneous acceleration a_x at any instant within the interval, and the velocity changes at the same rate throughout the motion. This situation occurs often enough that we identify it as an analysis model: the **particle under constant acceleration.** In the discussion that follows, we generate several equations that describe the motion of a particle for this model.

If we replace $a_{x,\text{ avg}}$ by a_x in Equation 2.9 and take $t_i = 0$ and t_f to be any later time t, we find that

$$a_x = \frac{v_{xf} - v_{xi}}{t - 0}$$

or

$$v_{xf} = v_{xi} + a_x t \quad \text{(for constant } a_x\text{)} \tag{2.13}$$

This powerful expression enables us to determine an object's velocity at *any* time t if we know the object's initial velocity v_{xi} and its (constant) acceleration a_x. A velocity–time graph for this constant-acceleration motion is shown in Active Figure 2.11b. The graph is a straight line, the slope of which is the acceleration a_x; the (constant) slope is consistent with $a_x = dv_x/dt$ being a constant. Notice that the slope is positive, which indicates a positive acceleration. If the acceleration were negative, the slope of the line in Active Figure 2.11b would be negative. When the acceleration is constant, the graph of acceleration versus time (Active Fig. 2.11c) is a straight line having a slope of zero.

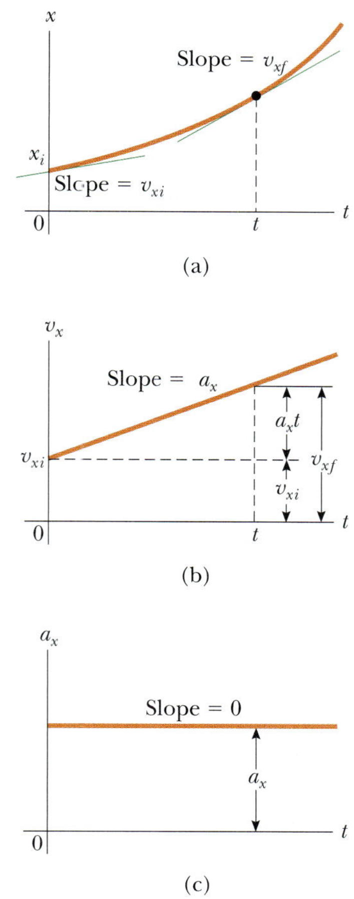

ACTIVE FIGURE 2.11

A particle under constant acceleration a_x moving along the x axis: (a) the position–time graph, (b) the velocity–time graph, and (c) the acceleration–time graph.

Sign in at www.thomsonedu.com and go to ThomsonNOW to adjust the constant acceleration and observe the effect on the position and velocity graphs.

Because velocity at constant acceleration varies linearly in time according to Equation 2.13, we can express the average velocity in any time interval as the arithmetic mean of the initial velocity v_{xi} and the final velocity v_{xf}:

$$v_{x,\text{ avg}} = \frac{v_{xi} + v_{xf}}{2} \quad \text{(for constant } a_x\text{)} \tag{2.14}$$

Notice that this expression for average velocity applies *only* in situations in which the acceleration is constant.

We can now use Equations 2.1, 2.2, and 2.14 to obtain the position of an object as a function of time. Recalling that Δx in Equation 2.2 represents $x_f - x_i$ and recognizing that $\Delta t = t_f - t_i = t - 0 = t$, we find that

$$x_f - x_i = v_{x,\text{ avg}}\, t = \tfrac{1}{2}(v_{xi} + v_{xf})t$$

◀ Position as a function of velocity and time

$$x_f = x_i + \tfrac{1}{2}(v_{xi} + v_{xf})t \quad \text{(for constant } a_x\text{)} \tag{2.15}$$

This equation provides the final position of the particle at time t in terms of the initial and final velocities.

We can obtain another useful expression for the position of a particle under constant acceleration by substituting Equation 2.13 into Equation 2.15:

$$x_f = x_i + \tfrac{1}{2}[v_{xi} + (v_{xi} + a_x t)]t$$

◀ Position as a function of time

$$x_f = x_i + v_{xi}t + \tfrac{1}{2}a_x t^2 \quad \text{(for constant } a_x\text{)} \tag{2.16}$$

This equation provides the final position of the particle at time t in terms of the initial velocity and the constant acceleration.

The position–time graph for motion at constant (positive) acceleration shown in Active Figure 2.11a is obtained from Equation 2.16. Notice that the curve is a parabola. The slope of the tangent line to this curve at $t = 0$ equals the initial velocity v_{xi}, and the slope of the tangent line at any later time t equals the velocity v_{xf} at that time.

Finally, we can obtain an expression for the final velocity that does not contain time as a variable by substituting the value of t from Equation 2.13 into Equation 2.15:

$$x_f = x_i + \tfrac{1}{2}(v_{xi} + v_{xf})\left(\frac{v_{xf} - v_{xi}}{a_x}\right) = x_i + \frac{v_{xf}^{\,2} - v_{xi}^{\,2}}{2a_x}$$

Velocity as a function of position

$$v_{xf}^{\,2} = v_{xi}^{\,2} + 2a_x(x_f - x_i) \quad \text{(for constant } a_x\text{)} \tag{2.17}$$

This equation provides the final velocity in terms of the initial velocity, the constant acceleration, and the position of the particle.

For motion at *zero* acceleration, we see from Equations 2.13 and 2.16 that

$$\left.\begin{aligned} v_{xf} &= v_{xi} = v_x \\ x_f &= x_i + v_x t \end{aligned}\right\} \quad \text{when } a_x = 0$$

That is, when the acceleration of a particle is zero, its velocity is constant and its position changes linearly with time. In terms of models, when the acceleration of a particle is zero, the particle under constant acceleration model reduces to the particle under constant velocity model (Section 2.3).

Quick Quiz 2.6 In Active Figure 2.12, match each v_x–t graph on the top with the a_x–t graph on the bottom that best describes the motion.

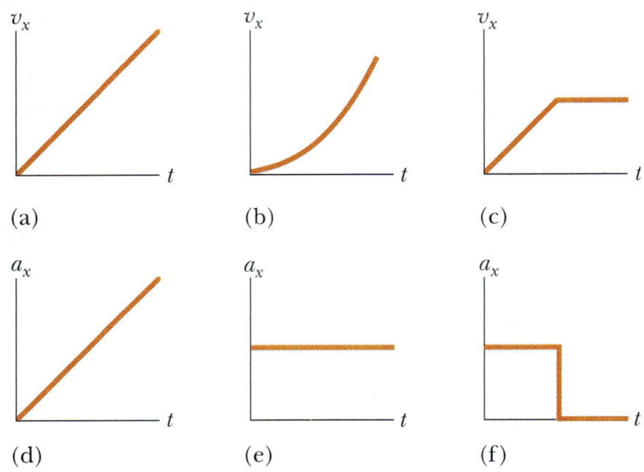

ACTIVE FIGURE 2.12

(Quick Quiz 2.6) Parts (a), (b), and (c) are v_x–t graphs of objects in one-dimensional motion. The possible accelerations of each object as a function of time are shown in scrambled order in (d), (e), and (f).

Sign in at www.thomsonedu.com and go to ThomsonNOW to practice matching appropriate velocity versus time graphs and acceleration versus time graphs.

Equations 2.13 through 2.17 are **kinematic equations** that may be used to solve any problem involving a particle under constant acceleration in one dimension. The four kinematic equations used most often are listed for convenience in Table 2.2. The choice of which equation you use in a given situation depends on what you know beforehand. Sometimes it is necessary to use two of these equations to solve for two unknowns. You should recognize that the quantities that vary during the motion are position x_f, velocity v_{xf}, and time t.

You will gain a great deal of experience in the use of these equations by solving a number of exercises and problems. Many times you will discover that more than

TABLE 2.2

Kinematic Equations for Motion of a Particle Under Constant Acceleration

Equation Number	Equation	Information Given by Equation
2.13	$v_{xf} = v_{xi} + a_x t$	Velocity as a function of time
2.15	$x_f = x_i + \frac{1}{2}(v_{xi} + v_{xf})t$	Position as a function of velocity and time
2.16	$x_f = x_i + v_{xi}t + \frac{1}{2}a_x t^2$	Position as a function of time
2.17	$v_{xf}^{\,2} = v_{xi}^{\,2} + 2a_x(x_f - x_i)$	Velocity as a function of position

Note: Motion is along the x axis.

one method can be used to obtain a solution. Remember that these equations of kinematics *cannot* be used in a situation in which the acceleration varies with time. They can be used only when the acceleration is constant.

EXAMPLE 2.7 Carrier Landing

A jet lands on an aircraft carrier at 140 mi/h ($\approx$ 63 m/s).

(A) What is its acceleration (assumed constant) if it stops in 2.0 s due to an arresting cable that snags the jet and brings it to a stop?

SOLUTION

You might have seen movies or television shows in which a jet lands on an aircraft carrier and is brought to rest surprisingly fast by an arresting cable. Because the acceleration of the jet is assumed constant, we model it as a particle under constant acceleration. We define our x axis as the direction of motion of the jet. A careful reading of the problem reveals that in addition to being given the initial speed of 63 m/s, we also know that the final speed is zero. We also notice that we have no information about the change in position of the jet while it is slowing down.

Equation 2.13 is the only equation in Table 2.2 that does not involve position, so we use it to find the acceleration of the jet, modeled as a particle:

$$a_x = \frac{v_{xf} - v_{xi}}{t} \approx \frac{0 - 63 \text{ m/s}}{2.0 \text{ s}}$$

$$= \boxed{-32 \text{ m/s}^2}$$

(B) If the jet touches down at position $x_i = 0$, what is its final position?

SOLUTION

Use Equation 2.15 to solve for the final position:

$$x_f = x_i + \tfrac{1}{2}(v_{xi} + v_{xf})t = 0 + \tfrac{1}{2}(63 \text{ m/s} + 0)(2.0 \text{ s}) = \boxed{63 \text{ m}}$$

If the jet travels much farther than 63 m, it might fall into the ocean. The idea of using arresting cables to slow down landing aircraft and enable them to land safely on ships originated at about the time of World War I. The cables are still a vital part of the operation of modern aircraft carriers.

What If? Suppose the jet lands on the deck of the aircraft carrier with a speed higher than 63 m/s but has the same acceleration due to the cable as that calculated in part (A). How will that change the answer to part (B)?

Answer If the jet is traveling faster at the beginning, it will stop farther away from its starting point, so the answer to part (B) should be larger. Mathematically, we see in Equation 2.15 that if v_{xi} is larger, x_f will be larger.

EXAMPLE 2.8 Watch Out for the Speed Limit!

A car traveling at a constant speed of 45.0 m/s passes a trooper on a motorcycle hidden behind a billboard. One second after the speeding car passes the billboard, the trooper sets out from the billboard to catch the car, accelerating at a constant rate of 3.00 m/s^2. How long does it take her to overtake the car?

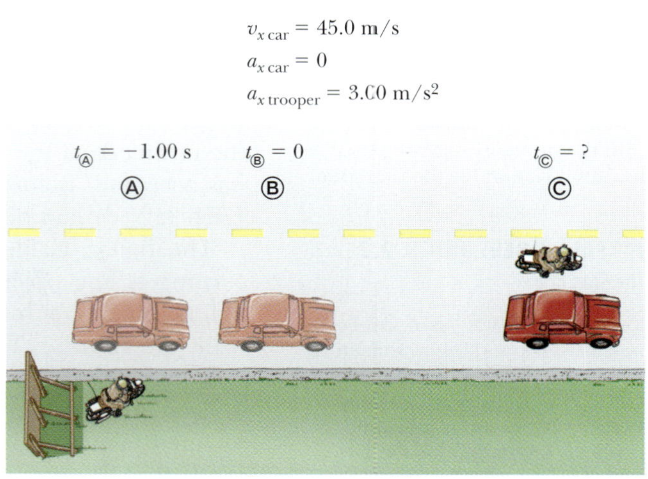

Figure 2.13 (Example 2.8) A speeding car passes a hidden trooper.

SOLUTION

A pictorial representation (Fig. 2.13) helps clarify the sequence of events. The car is modeled as a particle under constant velocity, and the trooper is modeled as a particle under constant acceleration.

First, we write expressions for the position of each vehicle as a function of time. It is convenient to choose the position of the billboard as the origin and to set $t_Ⓑ = 0$ as the time the trooper begins moving. At that instant, the car has already traveled a distance of 45.0 m from the billboard because it has traveled at a constant speed of $v_x = 45.0$ m/s for 1 s. Therefore, the initial position of the speeding car is $x_{Ⓑ} = 45.0$ m.

Apply Equation 2.7 to give the car's position at any time t:

$$x_{car} = x_{Ⓑ} + v_{x\,car}t = 45.0 \text{ m} + (45.0 \text{ m/s})t$$

A quick check shows that at $t = 0$, this expression gives the car's correct initial position when the trooper begins to move: $x_{car} = x_{Ⓑ} = 45.0$ m.

The trooper starts from rest at $t_{Ⓑ} = 0$ and accelerates at 3.00 m/s² away from the origin. Use Equation 2.16 to give her position at any time t:

$$x_f = x_i + v_{xi}t + \tfrac{1}{2}a_x t^2$$

$$x_{trooper} = 0 + (0)t + \tfrac{1}{2}a_x t^2 = \tfrac{1}{2}(3.00 \text{ m/s}^2)t^2$$

Set the two positions equal to represent the trooper overtaking the car at position Ⓒ:

$$x_{trooper} = x_{car}$$

$$\tfrac{1}{2}(3.00 \text{ m/s}^2)t^2 = 45.0 \text{ m} + (45.0 \text{ m/s})t$$

Simplify to give a quadratic equation:

$$1.50t^2 - 45.0t - 45.0 = 0$$

The positive solution of this equation is $t = 31.0$ s.

(For help in solving quadratic equations, see Appendix B.2.)

What If? What if the trooper has a more powerful motorcycle with a larger acceleration? How would that change the time at which the trooper catches the car?

Answer If the motorcycle has a larger acceleration, the trooper will catch up to the car sooner, so the answer for the time will be less than 31 s.

Cast the final quadratic equation above in terms of the parameters in the problem:

$$\tfrac{1}{2}a_x t^2 - v_{x\,car}t - x_{Ⓑ} = 0$$

Solve the quadratic equation:

$$t = \frac{v_{x\,car} \pm \sqrt{v_{x\,car}^2 + 2a_x x_{Ⓑ}}}{a_x} = \frac{v_{x\,car}}{a_x} + \sqrt{\frac{v_{x\,car}^2}{a_x^2} + \frac{2x_{Ⓑ}}{a_x}}$$

where we have chosen the positive sign because that is the only choice consistent with a time $t > 0$. Because all terms on the right side of the equation have the acceleration a_x in the denominator, increasing the acceleration will decrease the time at which the trooper catches the car.

PITFALL PREVENTION 2.6

g and g

Be sure not to confuse the italic symbol g for free-fall acceleration with the nonitalic symbol g used as the abbreviation for the unit gram.

PITFALL PREVENTION 2.7

The Sign of g

Keep in mind that g is a *positive number*. It is tempting to substitute -9.80 m/s² for g, but resist the temptation. Downward gravitational acceleration is indicated explicitly by stating the acceleration as $a_y = -g$.

2.7 Freely Falling Objects

It is well known that, in the absence of air resistance, all objects dropped near the Earth's surface fall toward the Earth with the same constant acceleration under the influence of the Earth's gravity. It was not until about 1600 that this conclusion was accepted. Before that time, the teachings of the Greek philosopher Aristotle (384–322 BC) had held that heavier objects fall faster than lighter ones.

The Italian Galileo Galilei (1564–1642) originated our present-day ideas concerning falling objects. There is a legend that he demonstrated the behavior of falling objects by observing that two different weights dropped simultaneously from the Leaning Tower of Pisa hit the ground at approximately the same time. Although there is some doubt that he carried out this particular experiment, it is well established that Galileo performed many experiments on objects moving on inclined planes. In his experiments, he rolled balls down a slight incline and measured the distances they covered in successive time intervals. The purpose of the incline was to reduce the acceleration, which made it possible for him to make accurate measurements of the time intervals. By gradually increasing the slope of the incline, he was finally able to draw conclusions about freely falling objects because a freely falling ball is equivalent to a ball moving down a vertical incline.

You might want to try the following experiment. Simultaneously drop a coin and a crumpled-up piece of paper from the same height. If the effects of air resistance are negligible, both will have the same motion and will hit the floor at the same time. In the idealized case, in which air resistance is absent, such motion is referred to as *free-fall* motion. If this same experiment could be conducted in a vacuum, in which air resistance is truly negligible, the paper and coin would fall with the same acceleration even when the paper is not crumpled. On August 2, 1971, astronaut David Scott conducted such a demonstration on the Moon. He simultaneously released a hammer and a feather, and the two objects fell together to the lunar surface. This simple demonstration surely would have pleased Galileo!

When we use the expression *freely falling object,* we do not necessarily refer to an object dropped from rest. **A freely falling object is any object moving freely under the influence of gravity alone, regardless of its initial motion. Objects thrown upward or downward and those released from rest are all falling freely once they are released. Any freely falling object experiences an acceleration directed *downward*, regardless of its initial motion.**

We shall denote the magnitude of the *free-fall acceleration* by the symbol g. The value of g near the Earth's surface decreases with increasing altitude. Furthermore, slight variations in g occur with changes in latitude. At the Earth's surface, the value of g is approximately 9.80 m/s^2. Unless stated otherwise, we shall use this value for g when performing calculations. For making quick estimates, use $g = 10$ m/s^2.

If we neglect air resistance and assume the free-fall acceleration does not vary with altitude over short vertical distances, the motion of a freely falling object moving vertically is equivalent to motion of a particle under constant acceleration in one dimension. Therefore, the equations developed in Section 2.6 for objects moving with constant acceleration can be applied. The only modification for freely falling objects that we need to make in these equations is to note that the motion is in the vertical direction (the y direction) rather than in the horizontal direction (x) and that the acceleration is downward and has a magnitude of 9.80 m/s^2. Therefore, we always choose $a_y = -g = -9.80$ m/s^2, where the negative sign means that the acceleration of a freely falling object is downward. In Chapter 13, we shall study how to deal with variations in g with altitude.

North Wind Picture Archives

GALILEO GALILEI
Italian physicist and astronomer
(1564–1642)
Galileo formulated the laws that govern the motion of objects in free fall and made many other significant discoveries in physics and astronomy. Galileo publicly defended Nicolaus Copernicus's assertion that the Sun is at the center of the Universe (the heliocentric system). He published *Dialogue Concerning Two New World Systems* to support the Copernican model, a view that the Catholic Church declared to be heretical.

PITFALL PREVENTION 2.8
Acceleration at the Top of the Motion

A common misconception is that the acceleration of a projectile at the top of its trajectory is zero. Although the velocity at the top of the motion of an object thrown upward momentarily goes to zero, *the acceleration is still that due to gravity* at this point. If the velocity and acceleration were both zero, the projectile would stay at the top.

Quick Quiz 2.7 Consider the following choices: (a) increases, (b) decreases, (c) increases and then decreases, (d) decreases and then increases, (e) remains the same. From these choices, select what happens to **(i)** the acceleration and **(ii)** the speed of a ball after it is thrown upward into the air.

CONCEPTUAL EXAMPLE 2.9 The Daring Skydivers

A skydiver jumps out of a hovering helicopter. A few seconds later, another skydiver jumps out, and they both fall along the same vertical line. Ignore air resistance, so that both skydivers fall with the same acceleration. Does the difference in their speeds stay the same throughout the fall? Does the vertical distance between them stay the same throughout the fall?

SOLUTION

At any given instant, the speeds of the skydivers are different because one had a head start. In any time interval Δt after this instant, however, the two skydivers increase their speeds by the same amount because they have the same acceleration. Therefore, the difference in their speeds remains the same throughout the fall.

The first jumper always has a greater speed than the second. Therefore, in a given time interval, the first skydiver covers a greater distance than the second. Consequently, the separation distance between them increases.

EXAMPLE 2.10 Not a Bad Throw for a Rookie!

A stone thrown from the top of a building is given an initial velocity of 20.0 m/s straight upward. The building is 50.0 m high, and the stone just misses the edge of the roof on its way down, as shown in Figure 2.14.

(A) Using $t_{\text{Ⓐ}} = 0$ as the time the stone leaves the thrower's hand at position Ⓐ, determine the time at which the stone reaches its maximum height.

SOLUTION

You most likely have experience with dropping objects or throwing them upward and watching them fall, so this problem should describe a familiar experience. Because the stone is in free fall, it is modeled as a particle under constant acceleration due to gravity.

Use Equation 2.13 to calculate the time at which the stone reaches its maximum height:

$$v_{yf} = v_{yi} + a_y t \quad \rightarrow \quad t = \frac{v_{yf} - v_{yi}}{a_y}$$

Substitute numerical values:

$$t = t_{\text{Ⓑ}} = \frac{0 - 20.0 \text{ m/s}}{-9.80 \text{ m/s}^2} = \boxed{2.04 \text{ s}}$$

(B) Find the maximum height of the stone.

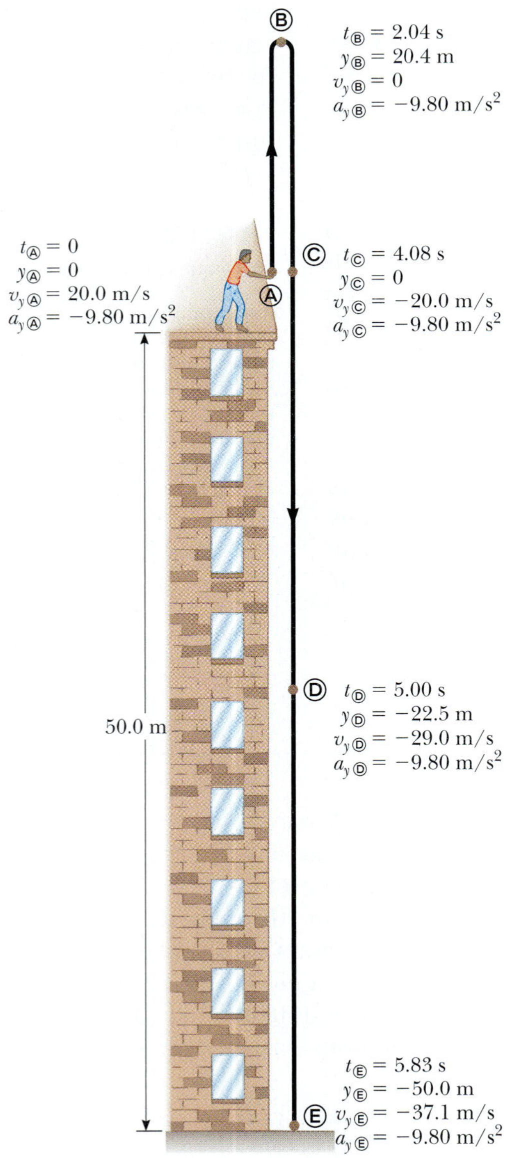

Figure 2.14 (Example 2.10) Position and velocity versus time for a freely falling stone thrown initially upward with a velocity $v_{yi} = 20.0$ m/s. Many of the quantities in the labels for points in the motion of the stone are calculated in the example. Can you verify the other values that are not?

SOLUTION

Set $y_Ⓐ = 0$ and substitute the time from part (A) into Equation 2.16 to find the maximum height:

$$y_{max} = y_Ⓑ = y_Ⓐ + v_{xⒶ}t + \frac{1}{2}a_y t^2$$

$$y_Ⓑ = 0 + (20.0 \text{ m/s})(2.04 \text{ s}) + \frac{1}{2}(-9.80 \text{ m/s}^2)(2.04 \text{ s})^2 = \boxed{20.4 \text{ m}}$$

(C) Determine the velocity of the stone when it returns to the height from which it was thrown.

Substitute known values into Equation 2.17:

$$v_{yⒸ}^2 = v_{yⒶ}^2 + 2a_y(y_Ⓒ - y_Ⓐ)$$

$$v_{yⒸ}^2 = (20.0 \text{ m/s})^2 + 2(-9.80 \text{ m/s}^2)(0 - 0) = 400 \text{ m}^2/\text{s}^2$$

$$v_{yⒸ} = \boxed{-20.0 \text{ m/s}}$$

When taking the square root, we could choose either a positive or a negative root. We choose the negative root because we know that the stone is moving downward at point Ⓒ. The velocity of the stone when it arrives back at its original height is equal in magnitude to its initial velocity but is opposite in direction.

(D) Find the velocity and position of the stone at $t = 5.00$ s.

Calculate the velocity at Ⓓ from Equation 2.13:

$$v_{yⒹ} = v_{yⒶ} + a_y t = 20.0 \text{ m/s} + (-9.80 \text{ m/s}^2)(5.00 \text{ s}) = \boxed{-29.0 \text{ m/s}}$$

Use Equation 2.16 to find the position of the stone at $t_Ⓓ = 5.00$ s:

$$y_Ⓓ = y_Ⓐ + v_{yⒶ}t + \frac{1}{2}a_y t^2$$

$$= 0 + (20.0 \text{ m/s})(5.00 \text{ s}) + \frac{1}{2}(-9.80 \text{ m/s}^2)(5.00 \text{ s})^2$$

$$= \boxed{-22.5 \text{ m}}$$

The choice of the time defined as $t = 0$ is arbitrary and up to you to select as the problem-solver. As an example of this arbitrariness, choose $t = 0$ as the time at which the stone is at the highest point in its motion. Then solve parts (C) and (D) again using this new initial instant and note that your answers are the same as those above.

What If? What if the building were 30.0 m tall instead of 50.0 m tall? Which answers in parts (A) to (D) would change?

Answer None of the answers would change. All the motion takes place in the air during the first 5.00 s. (Notice that even for a 30.0-m tall building, the stone is above the ground at $t = 5.00$ s.) Therefore, the height of the building is not an issue. Mathematically, if we look back over our calculations, we see that we never entered the height of the building into any equation.

2.8 Kinematic Equations Derived from Calculus

This section assumes the reader is familiar with the techniques of integral calculus. If you have not yet studied integration in your calculus course, you should skip this section or cover it after you become familiar with integration.

The velocity of a particle moving in a straight line can be obtained if its position as a function of time is known. Mathematically, the velocity equals the derivative of the position with respect to time. It is also possible to find the position of a particle if its velocity is known as a function of time. In calculus, the procedure used to perform this task is referred to either as *integration* or as finding the *antiderivative*. Graphically, it is equivalent to finding the area under a curve.

Suppose the v_x–t graph for a particle moving along the x axis is as shown in Figure 2.15. Let us divide the time interval $t_f - t_i$ into many small intervals, each of duration Δt_n. From the definition of average velocity we see that the displacement of the particle during any small interval, such as the one shaded in Figure 2.15, is

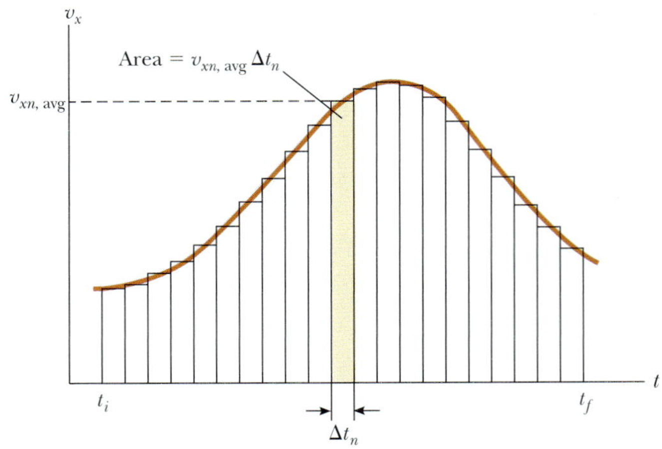

Figure 2.15 Velocity versus time for a particle moving along the x axis. The area of the shaded rectangle is equal to the displacement Δx in the time interval Δt_n, whereas the total area under the curve is the total displacement of the particle.

given by $\Delta x_n = v_{xn,\,avg}\,\Delta t_n$, where $v_{xn,\,avg}$ is the average velocity in that interval. Therefore, the displacement during this small interval is simply the area of the shaded rectangle. The total displacement for the interval $t_f - t_i$ is the sum of the areas of all the rectangles from t_i to t_f:

$$\Delta x = \sum_n v_{xn,\,avg}\,\Delta t_n$$

where the symbol Σ (uppercase Greek sigma) signifies a sum over all terms, that is, over all values of n. Now, as the intervals are made smaller and smaller, the number of terms in the sum increases and the sum approaches a value equal to the area under the velocity–time graph. Therefore, in the limit $n \to \infty$, or $\Delta t_n \to 0$, the displacement is

$$\Delta x = \lim_{\Delta t_n \to 0} \sum_n v_{xn}\,\Delta t_n \qquad \textbf{(2.18)}$$

Notice that we have replaced the average velocity $v_{xn,\,avg}$ with the instantaneous velocity v_{xn} in the sum. As you can see from Figure 2.15, this approximation is valid in the limit of very small intervals. Therefore, if we know the v_x–t graph for motion along a straight line, we can obtain the displacement during any time interval by measuring the area under the curve corresponding to that time interval.

The limit of the sum shown in Equation 2.18 is called a **definite integral** and is written

Definite integral ▶

$$\lim_{\Delta t_n \to 0} \sum_n v_{xn}\,\Delta t_n = \int_{t_i}^{t_f} v_x(t)\,dt \qquad \textbf{(2.19)}$$

where $v_x(t)$ denotes the velocity at any time t. If the explicit functional form of $v_x(t)$ is known and the limits are given, the integral can be evaluated. Sometimes the v_x–t graph for a moving particle has a shape much simpler than that shown in Figure 2.15. For example, suppose a particle moves at a constant velocity v_{xi}. In this case, the v_x–t graph is a horizontal line, as in Figure 2.16, and the displacement of the particle during the time interval Δt is simply the area of the shaded rectangle:

$$\Delta x = v_{xi}\,\Delta t \quad \text{(when } v_x = v_{xi} = \text{constant)}$$

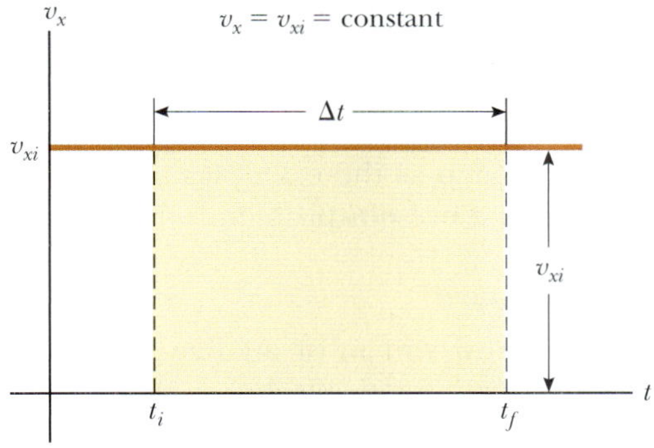

Figure 2.16 The velocity–time curve for a particle moving with constant velocity v_{xi}. The displacement of the particle during the time interval $t_f - t_i$ is equal to the area of the shaded rectangle.

Kinematic Equations

We now use the defining equations for acceleration and velocity to derive two of our kinematic equations, Equations 2.13 and 2.16.

The defining equation for acceleration (Eq. 2.10),

$$a_x = \frac{dv_x}{dt}$$

may be written as $dv_x = a_x\,dt$ or, in terms of an integral (or antiderivative), as

$$v_{xf} - v_{xi} = \int_0^t a_x\,dt$$

For the special case in which the acceleration is constant, a_x can be removed from the integral to give

$$v_{xf} - v_{xi} = a_x \int_0^t dt = a_x(t - 0) = a_x t \qquad \textbf{(2.20)}$$

which is Equation 2.13.

Now let us consider the defining equation for velocity (Eq. 2.5):

$$v_x = \frac{dx}{dt}$$

We can write this equation as $dx = v_x\,dt$, or in integral form as

$$x_f - x_i = \int_0^t v_x\,dt$$

Because $v_x = v_{xf} = v_{xi} + a_x t$, this expression becomes

$$x_f - x_i = \int_0^t (v_{xi} + a_x t)\,dt = \int_0^t v_{xi}\,dt + a_x \int_0^t t\,dt = v_{xi}(t - 0) + a_x\left(\frac{t^2}{2} - 0\right)$$

$$x_f - x_i = v_{xi}t + \tfrac{1}{2}a_x t^2$$

which is Equation 2.16.

Besides what you might expect to learn about physics concepts, a very valuable skill you should hope to take away from your physics course is the ability to solve complicated problems. The way physicists approach complex situations and break them into manageable pieces is extremely useful. The following is a general problem-solving strategy to guide you through the steps. To help you remember the steps of the strategy, they are *Conceptualize*, *Categorize*, *Analyze*, and *Finalize*.

GENERAL PROBLEM-SOLVING STRATEGY

Conceptualize

- The first things to do when approaching a problem are to *think about* and *understand* the situation. Study carefully any representations of the information (e.g., diagrams, graphs, tables, or photographs) that accompany the problem. Imagine a movie, running in your mind, of what happens in the problem.
- If a pictorial representation is not provided, you should almost always make a quick drawing of the situation. Indicate any known values, perhaps in a table or directly on your sketch.
- Now focus on what algebraic or numerical information is given in the problem. Carefully read the problem statement, looking for key phrases such as "starts from rest" ($v_i = 0$), "stops" ($v_f = 0$), or "falls freely" ($a_y = -g = -9.80 \text{ m/s}^2$).
- Now focus on the expected result of solving the problem. Exactly what is the question asking? Will the final result be numerical or algebraic? Do you know what units to expect?
- Don't forget to incorporate information from your own experiences and common sense. What should a reasonable answer look like? For example, you wouldn't expect to calculate the speed of an automobile to be 5×10^6 m/s.

Categorize

- Once you have a good idea of what the problem is about, you need to *simplify* the problem. Remove the details that are not important to the solution. For example, model a moving object as a particle. If appropriate, ignore air resistance or friction between a sliding object and a surface.
- Once the problem is simplified, it is important to *categorize* the problem. Is it a simple *substitution problem* such that numbers can be substituted into an equation? If so, the problem is likely to be finished when this substitution is done. If not, you face what we call an *analysis problem:* the situation must be analyzed more deeply to reach a solution.
- If it is an analysis problem, it needs to be categorized further. Have you seen this type of problem before? Does it fall into the growing list of types of problems that you have solved previously? If so, identify any analysis model(s) appropriate for the problem to prepare for the Analyze step below. We saw the first three analysis models in this chapter: the particle under constant velocity, the particle under constant speed, and the particle under constant acceleration. Being able to classify a problem with an analysis model can make it much easier to lay out a plan to solve it. For example, if your simplification shows that the problem can be treated as a particle under constant acceleration and you have already solved such a problem (such as the examples in Section 2.6), the solution to the present problem follows a similar pattern.

Analyze

- Now you must analyze the problem and strive for a mathematical solution. Because you have already categorized the problem and identified an analysis model, it should not be too difficult to select relevant equations that apply to the type of situation in the problem. For example, if the problem involves a particle under constant acceleration, Equations 2.13 to 2.17 are relevant.
- Use algebra (and calculus, if necessary) to solve symbolically for the unknown variable in terms of what is given. Substitute in the appropriate numbers, calculate the result, and round it to the proper number of significant figures.

Finalize

- Examine your numerical answer. Does it have the correct units? Does it meet your expectations from your conceptualization of the problem? What about the algebraic form of the result? Does it make sense? Examine the variables in the problem to see whether the answer would change in a physically meaningful way if the variables were drastically increased or decreased or even became zero. Looking at limiting cases to see whether they yield expected values is a very useful way to make sure that you are obtaining reasonable results.
- Think about how this problem compared with others you have solved. How was it similar? In what critical ways did it differ? Why was this problem assigned? Can you figure out what you have learned by doing it? If it is a new category of problem, be sure you understand it so that you can use it as a model for solving similar problems in the future.

When solving complex problems, you may need to identify a series of subproblems and apply the problem-solving strategy to each. For simple problems, you probably don't need this strategy. When you are trying to solve a problem and you don't know what to do next, however, remember the steps in the strategy and use them as a guide.

For practice, it would be useful for you to revisit the worked examples in this chapter and identify the *Conceptualize, Categorize, Analyze,* and *Finalize* steps. In the rest of this book, we will label these steps explicitly in the worked examples. Many chapters in this book include a section labeled Problem-Solving Strategy that should help you through the rough spots. These sections are organized according to the General Problem-Solving Strategy outlined above and are tailored to the specific types of problems addressed in that chapter.

Summary

ThomsonNOW™ Sign in at **www.thomsonedu.com** and go to ThomsonNOW to take a practice test for this chapter.

DEFINITIONS

When a particle moves along the x axis from some initial position x_i to some final position x_f, its **displacement** is

$$\Delta x \equiv x_f - x_i \quad (2.1)$$

The **average velocity** of a particle during some time interval is the displacement Δx divided by the time interval Δt during which that displacement occurs:

$$v_{x,\text{ avg}} \equiv \frac{\Delta x}{\Delta t} \quad (2.2)$$

The **average speed** of a particle is equal to the ratio of the total distance it travels to the total time interval during which it travels that distance:

$$v_{\text{avg}} \equiv \frac{d}{\Delta t} \quad (2.3)$$

The **instantaneous velocity** of a particle is defined as the limit of the ratio $\Delta x/\Delta t$ as Δt approaches zero. By definition, this limit equals the derivative of x with respect to t, or the time rate of change of the position:

$$v_x \equiv \lim_{\Delta t \to 0} \frac{\Delta x}{\Delta t} = \frac{dx}{dt} \quad (2.5)$$

The **instantaneous speed** of a particle is equal to the magnitude of its instantaneous velocity.

The **average acceleration** of a particle is defined as the ratio of the change in its velocity Δv_x divided by the time interval Δt during which that change occurs:

$$a_{x,\text{ avg}} \equiv \frac{\Delta v_x}{\Delta t} = \frac{v_{xf} - v_{xi}}{t_f - t_i} \quad (2.9)$$

The **instantaneous acceleration** is equal to the limit of the ratio $\Delta v_x/\Delta t$ as Δt approaches 0. By definition, this limit equals the derivative of v_x with respect to t, or the time rate of change of the velocity:

$$a_x \equiv \lim_{\Delta t \to 0} \frac{\Delta v_x}{\Delta t} = \frac{dv_x}{dt} \quad (2.10)$$

CONCEPTS AND PRINCIPLES

When an object's velocity and acceleration are in the same direction, the object is speeding up. On the other hand, when the object's velocity and acceleration are in opposite directions, the object is slowing down. Remembering that $F_x \propto a_x$ is a useful way to identify the direction of the acceleration by associating it with a force.

An object falling freely in the presence of the Earth's gravity experiences free-fall acceleration directed toward the center of the Earth. If air resistance is neglected, if the motion occurs near the surface of the Earth, and if the range of the motion is small compared with the Earth's radius, the free-fall acceleration g is constant over the range of motion, where g is equal to 9.80 m/s^2.

Complicated problems are best approached in an organized manner. Recall and apply the *Conceptualize, Categorize, Analyze,* and *Finalize* steps of the General Problem-Solving Strategy when you need them.

(*continued*)

ANALYSIS MODELS FOR PROBLEM-SOLVING

Particle Under Constant Velocity. If a particle moves in a straight line with a constant speed v_x, its constant velocity is given by

$$v_x = \frac{\Delta x}{\Delta t} \tag{2.6}$$

and its position is given by

$$x_f = x_i + v_x t \tag{2.7}$$

Particle Under Constant Speed. If a particle moves a distance d along a curved or straight path with a constant speed, its constant speed is given by

$$v = \frac{d}{\Delta t} \tag{2.8}$$

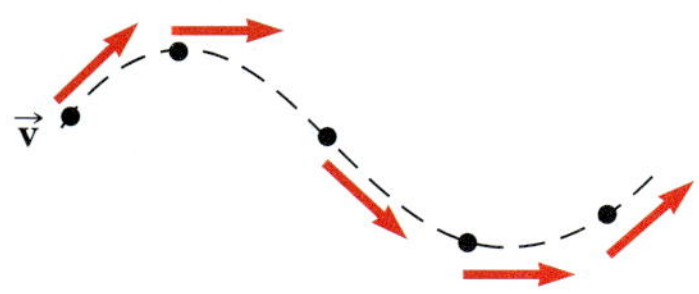

Particle Under Constant Acceleration. If a particle moves in a straight line with a constant acceleration a_x, its motion is described by the kinematic equations:

$$v_{xf} = v_{xi} + a_x t \tag{2.13}$$

$$v_{x,\,\text{avg}} = \frac{v_{xi} + v_{xf}}{2} \tag{2.14}$$

$$x_f = x_i + \tfrac{1}{2}(v_{xi} + v_{xf})t \tag{2.15}$$

$$x_f = x_i + v_{xi}t + \tfrac{1}{2}a_x t^2 \tag{2.16}$$

$$v_{xf}^{\,2} = v_{xi}^{\,2} + 2a_x(x_f - x_i) \tag{2.17}$$

Questions

☐ denotes answer available in *Student Solutions Manual/Study Guide;* **O** denotes objective question

1. **O** One drop of oil falls straight down onto the road from the engine of a moving car every 5 s. Figure Q2.1 shows the pattern of the drops left behind on the pavement. What is the average speed of the car over this section of its motion? (a) 20 m/s (b) 24 m/s (c) 30 m/s (d) 100 m/s (e) 120 m/s

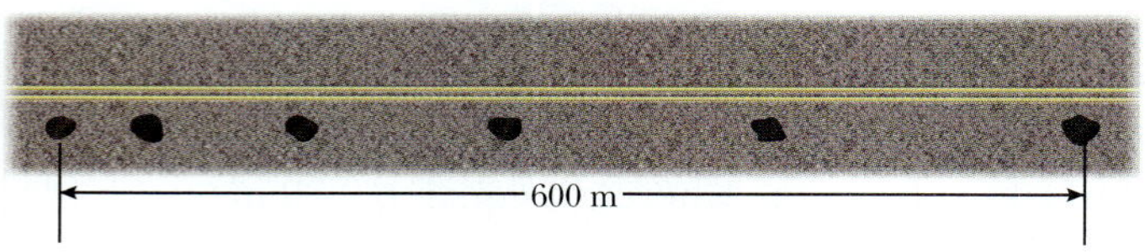

Figure Q2.1

2. ☐ If the average velocity of an object is zero in some time interval, what can you say about the displacement of the object for that interval?

3. **O** Can the instantaneous velocity of an object at an instant of time ever be greater in magnitude than the average velocity over a time interval containing the instant? Can it ever be less?

4. **O** A cart is pushed along a straight horizontal track. (a) In a certain section of its motion, its original velocity is $v_{xi} = +3$ m/s and it undergoes a change in velocity of $\Delta v_x = +4$ m/s. Does it speed up or slow down in this section of its motion? Is its acceleration positive or negative? (b) In another part of its motion, $v_{xi} = -3$ m/s and $\Delta v_x = +4$ m/s. Does it undergo a net increase or decrease in speed? Is its acceleration positive or negative? (c) In a third segment of its motion, $v_{xi} = +3$ m/s and $\Delta v_x = -4$ m/s. Does it have a net gain or loss in speed? Is its acceleration positive or negative? (d) In a fourth time interval, $v_{xi} = -3$ m/s and $\Delta v_x = -4$ m/s. Does the cart gain or lose speed? Is its acceleration positive or negative?

5. ☐ Two cars are moving in the same direction in parallel lanes along a highway. At some instant, the velocity of car A exceeds the velocity of car B. Does that mean that the acceleration of A is greater than that of B? Explain.

6. **O** When the pilot reverses the propeller in a boat moving north, the boat moves with an acceleration directed south. If the acceleration of the boat remains constant in magnitude and direction, what would happen to the boat (choose one)? (a) It would eventually stop and then remain stopped. (b) It would eventually stop and then start to speed up in the forward direction. (c) It would eventually stop and then start to speed up in the reverse direction. (d) It would never quite stop but lose speed more and more slowly forever. (e) It would never stop but continue to speed up in the forward direction.

7. **O** Each of the strobe photographs (a), (b), and (c) in Figure Q2.7 was taken of a single disk moving toward the right, which we take as the positive direction. Within each photograph, the time interval between images is constant. **(i)** Which photograph(s), if any, shows constant zero velocity? **(ii)** Which photograph(s), if any, shows constant zero acceleration? **(iii)** Which photograph(s), if any, shows constant positive velocity? **(iv)** Which photograph(s), if any, shows constant positive acceleration? **(v)** Which photograph(s), if any, shows some motion with negative acceleration?

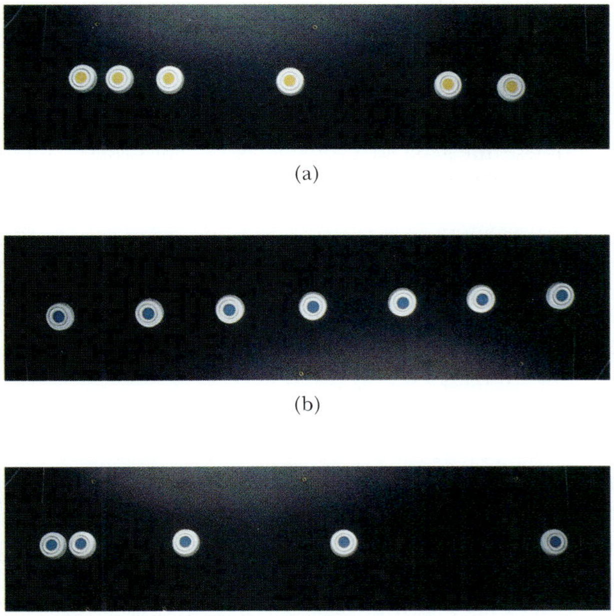

Figure Q2.7 Question 7 and Problem 17.

8. Try the following experiment away from traffic where you can do it safely. With the car you are driving moving slowly on a straight, level road, shift the transmission into neutral and let the car coast. At the moment the car comes to a complete stop, step hard on the brake and notice what you feel. Now repeat the same experiment on a fairly gentle uphill slope. Explain the difference in what a person riding in the car feels in the two cases. (Brian Popp suggested the idea for this question.)

9. O A skateboarder coasts down a long hill, starting from rest and moving with constant acceleration to cover a certain distance in 6 s. In a second trial, he starts from rest and moves with the same acceleration for only 2 s. How is his displacement different in this second trial compared with the first trial? (a) one-third as large (b) three times larger (c) one-ninth as large (d) nine times larger (e) $1/\sqrt{3}$ times as large (f) $\sqrt{3}$ times larger (g) none of these answers

10. O Can the equations of kinematics (Eqs. 2.13–2.17) be used in a situation in which the acceleration varies in time? Can they be used when the acceleration is zero?

11. A student at the top of a building of height h throws one ball upward with a speed of v_i and then throws a second ball downward with the same initial speed $|v_i|$. How do the final velocities of the balls compare when they reach the ground?

12. O A pebble is released from rest at a certain height and falls freely, reaching an impact speed of 4 m/s at the floor. **(i)** Next, the pebble is thrown down with an initial speed of 3 m/s from the same height. In this trial, what is its speed at the floor? (a) less than 4 m/s (b) 4 m/s (c) between 4 m/s and 5 m/s (d) $\sqrt{3^2 + 4^2}$ m/s = 5 m/s (e) between 5 m/s and 7 m/s (f) (3 + 4) m/s = 7 m/s (g) greater than 7 m/s **(ii)** In a third trial, the pebble is tossed upward with an initial speed of 3 m/s from the same height. What is its speed at the floor in this trial? Choose your answer from the same list (a) through (g).

13. O A hard rubber ball, not affected by air resistance in its motion, is tossed upward from shoulder height, falls to the sidewalk, rebounds to a somewhat smaller maximum height, and is caught on its way down again. This motion is represented in Figure Q2.13, where the successive positions of the ball Ⓐ through Ⓖ are not equally spaced in time. At point Ⓔ the center of the ball is at its lowest point in the motion. The motion of the ball is along a straight line, but the diagram shows successive positions offset to the right to avoid overlapping. Choose the positive y direction to be upward. **(i)** Rank the situations Ⓐ through Ⓖ according to the speed of the ball $|v_y|$ at each point, with the largest speed first. **(ii)** Rank the same situations according to the velocity of the ball at each point. **(iii)** Rank the same situations according to the acceleration a_y of the ball at each point. In each ranking, remember that zero is greater than a negative value. If two values are equal, show that they are equal in your ranking.

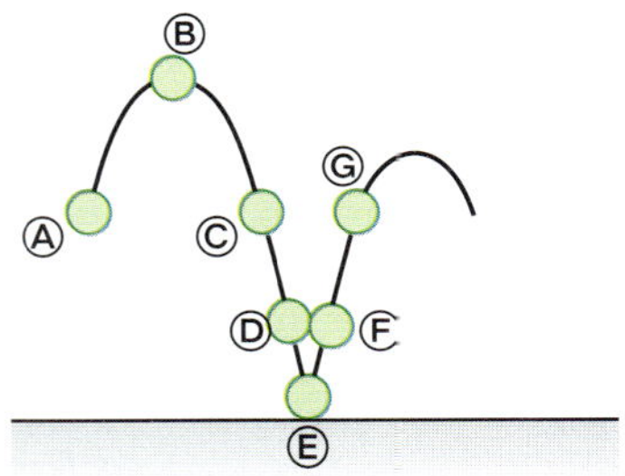

Figure Q2.13

14. O You drop a ball from a window located on an upper floor of a building. It strikes the ground with speed v. You now repeat the drop, but you ask a friend down on the ground to throw another ball upward at speed v. Your friend throws the ball upward at the same moment that you drop yours from the window. At some location, the balls pass each other. Is this location (a) *at* the halfway point between window and ground, (b) *above* this point, or (c) *below* this point?

Problems

WebAssign The Problems from this chapter may be assigned online in WebAssign.

ThomsonNOW™ Sign in at **www.thomsonedu.com** and go to ThomsonNOW to assess your understanding of this chapter's topics with additional quizzing and conceptual questions.

1, 2, 3 denotes straightforward, intermediate, challenging; □ denotes full solution available in *Student Solutions Manual/Study Guide*; ▲ denotes coached solution with hints available at **www.thomsonedu.com;** denotes developing symbolic reasoning; ● denotes asking for qualitative reasoning; denotes computer useful in solving problem

Section 2.1 Position, Velocity, and Speed

1. The position versus time for a certain particle moving along the x axis is shown in Figure P2.1. Find the average velocity in the following time intervals. (a) 0 to 2 s (b) 0 to 4 s (c) 2 s to 4 s (d) 4 s to 7 s (e) 0 to 8 s

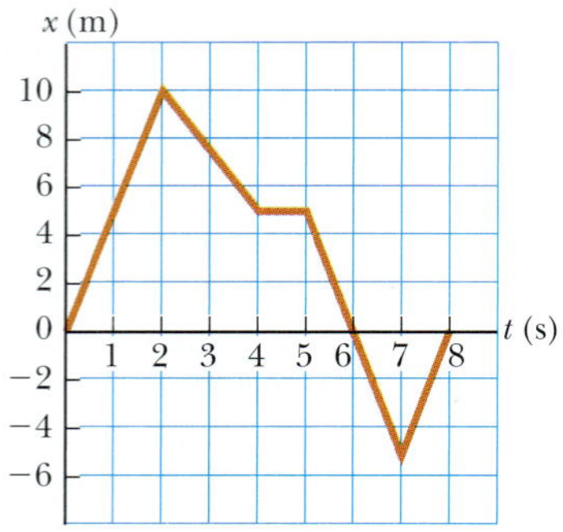

Figure P2.1 Problems 1 and 8.

2. The position of a pinewood derby car was observed at various moments; the results are summarized in the following table. Find the average velocity of the car for (a) the first 1-s time interval, (b) the last 3 s, and (c) the entire period of observation.

***t* (s)**	0	1.0	2.0	3.0	4.0	5.0
***x* (m)**	0	2.3	9.2	20.7	36.8	57.5

3. A person walks first at a constant speed of 5.00 m/s along a straight line from point A to point B and then back along the line from B to A at a constant speed of 3.00 m/s. (a) What is her average speed over the entire trip? (b) What is her average velocity over the entire trip?

4. A particle moves according to the equation $x = 10t^2$, where x is in meters and t is in seconds. (a) Find the average velocity for the time interval from 2.00 s to 3.00 s. (b) Find the average velocity for the time interval from 2.00 s to 2.10 s.

Section 2.2 Instantaneous Velocity and Speed

5. ▲ A position–time graph for a particle moving along the x axis is shown in Figure P2.5. (a) Find the average velocity in the time interval $t = 1.50$ s to $t = 4.00$ s. (b) Determine the instantaneous velocity at $t = 2.00$ s by measuring the slope of the tangent line shown in the graph. (c) At what value of t is the velocity zero?

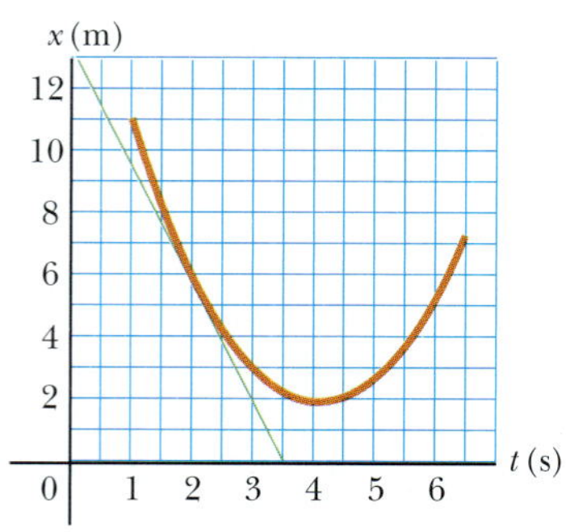

Figure P2.5

6. The position of a particle moving along the x axis varies in time according to the expression $x = 3t^2$, where x is in meters and t is in seconds. Evaluate its position (a) at $t =$ 3.00 s and (b) at 3.00 s + Δt. (c) Evaluate the limit of $\Delta x/\Delta t$ as Δt approaches zero to find the velocity at $t =$ 3.00 s.

7. (a) Use the data in Problem 2.2 to construct a smooth graph of position versus time. (b) By constructing tangents to the $x(t)$ curve, find the instantaneous velocity of the car at several instants. (c) Plot the instantaneous velocity versus time and, from the graph, determine the average acceleration of the car. (d) What was the initial velocity of the car?

8. Find the instantaneous velocity of the particle described in Figure P2.1 at the following times: (a) $t = 1.0$ s (b) $t =$ 3.0 s (c) $t = 4.5$ s (d) $t = 7.5$ s

Section 2.3 Analysis Models: The Particle Under Constant Velocity

9. A hare and a tortoise compete in a race over a course 1.00 km long. The tortoise crawls straight and steadily at its maximum speed of 0.200 m/s toward the finish line. The hare runs at its maximum speed of 8.00 m/s toward the goal for 0.800 km and then stops to tease the tortoise. How close to the goal can the hare let the tortoise approach before resuming the race, which the tortoise wins in a photo finish? Assume both animals, when moving, move steadily at their respective maximum speeds.

Section 2.4 Acceleration

10. A 50.0-g Super Ball traveling at 25.0 m/s bounces off a brick wall and rebounds at 22.0 m/s. A high-speed camera records this event. If the ball is in contact with the wall for 3.50 ms, what is the magnitude of the average acceleration of the ball during this time interval? *Note:* 1 ms = 10^{-3} s.

11. A particle starts from rest and accelerates as shown in Figure P2.11. Determine (a) the particle's speed at $t = 10.0$ s and at $t = 20.0$ s and (b) the distance traveled in the first 20.0 s.

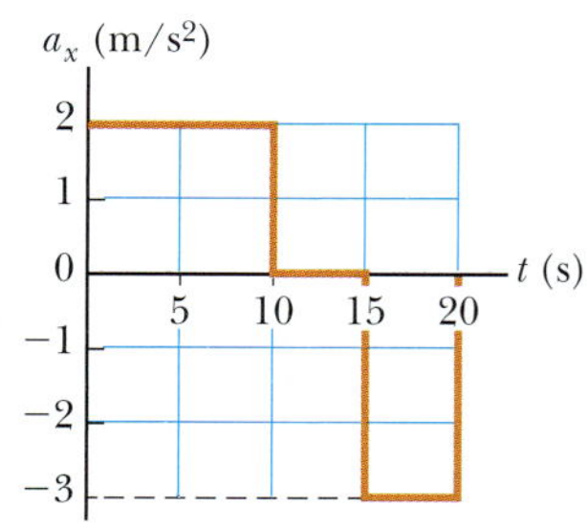

Figure P2.11

12. A velocity–time graph for an object moving along the x axis is shown in Figure P2.12. (a) Plot a graph of the acceleration versus time. (b) Determine the average acceleration of the object in the time intervals $t = 5.00$ s to $t =$ 15.0 s and $t = 0$ to $t = 20.0$ s.

13. ▲ A particle moves along the x axis according to the equation $x = 2.00 + 3.00t - 1.00t^2$, where x is in meters and t is in seconds. At $t = 3.00$ s, find (a) the position of the particle, (b) its velocity, and (c) its acceleration.

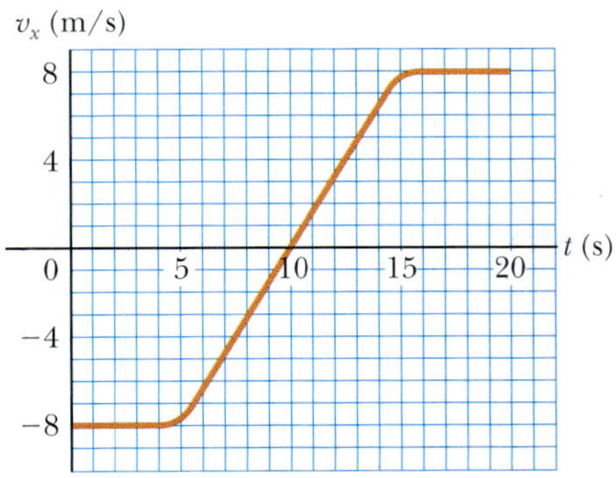

Figure P2.12

14. A child rolls a marble on a bent track that is 100 cm long as shown in Figure P2.14. We use x to represent the position of the marble along the track. On the horizontal sections from $x = 0$ to $x = 20$ cm and from $x = 40$ cm to $x = 60$ cm, the marble rolls with constant speed. On the sloping sections, the speed of the marble changes steadily. At the places where the slope changes, the marble stays on the track and does not undergo any sudden changes in speed. The child gives the marble some initial speed at $x = 0$ and $t = 0$ and then watches it roll to $x = 90$ cm, where it turns around, eventually returning to $x = 0$ with the same speed with which the child initially released it. Prepare graphs of x versus t, v_x versus t, and a_x versus t, vertically aligned with their time axes identical, to show the motion of the marble. You will not be able to place numbers other than zero on the horizontal axis or on the velocity or acceleration axes, but show the correct relative sizes on the graphs.

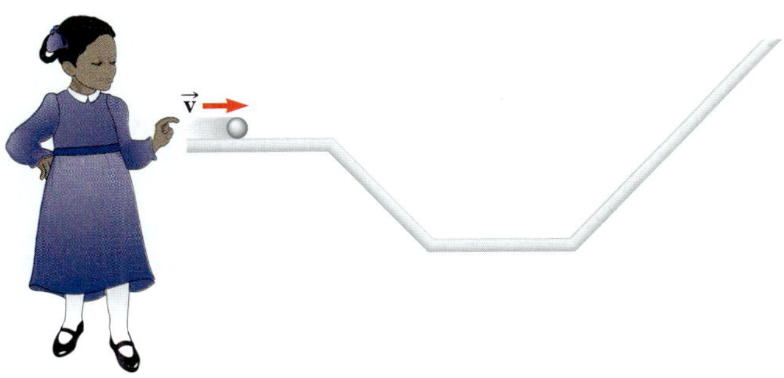

Figure P2.14

15. An object moves along the x axis according to the equation $x(t) = (3.00t^2 - 2.00t + 3.00)$ m, where t is in seconds. Determine (a) the average speed between $t = 2.00$ s and $t = 3.00$ s, (b) the instantaneous speed at $t = 2.00$ s and at $t = 3.00$ s, (c) the average acceleration between $t = 2.00$ s and $t = 3.00$ s, and (d) the instantaneous acceleration at $t = 2.00$ s and $t = 3.00$ s.

16. Figure P2.16 shows a graph of v_x versus t for the motion of a motorcyclist as he starts from rest and moves along

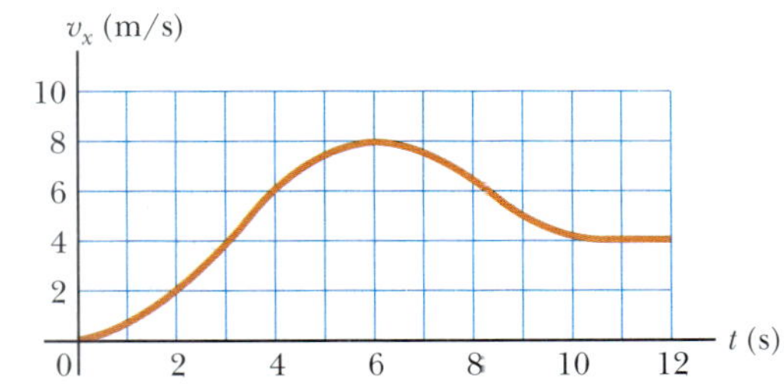

Figure P2.16

the road in a straight line. (a) Find the average acceleration for the time interval $t = 0$ to $t = 6.00$ s. (b) Estimate the time at which the acceleration has its greatest positive value and the value of the acceleration at that instant. (c) When is the acceleration zero? (d) Estimate the maximum negative value of the acceleration and the time at which it occurs.

Section 2.5 Motion Diagrams

17. ● Each of the strobe photographs (a), (b), and (c) in Figure Q2.7 was taken of a single disk moving toward the right, which we take as the positive direction. Within each photograph the time interval between images is constant. For each photograph, prepare graphs of x versus t, v_x versus t, and a_x versus t, vertically aligned with their time axes identical, to show the motion of the disk. You will not be able to place numbers other than zero on the axes, but show the correct relative sizes on the graphs.

18. Draw motion diagrams for (a) an object moving to the right at constant speed, (b) an object moving to the right and speeding up at a constant rate, (c) an object moving to the right and slowing down at a constant rate, (d) an object moving to the left and speeding up at a constant rate, and (e) an object moving to the left and slowing down at a constant rate. (f) How would your drawings change if the changes in speed were not uniform; that is, if the speed were not changing at a constant rate?

Section 2.6 The Particle Under Constant Acceleration

19. ● Assume a parcel of air in a straight tube moves with a constant acceleration of -4.00 m/s^2 and has a velocity of 13.0 m/s at 10:05:00 a.m. on a certain date. (a) What is its velocity at 10:05:01 a.m.? (b) At 10:05:02 a.m.? (c) At 10:05:02.5 a.m.? (d) At 10:05:04 a.m.? (e) At 10:04:59 a.m.? (f) Describe the shape of a graph of velocity versus time for this parcel of air. (g) Argue for or against the statement, "Knowing the single value of an object's constant acceleration is like knowing a whole list of values for its velocity."

20. A truck covers 40.0 m in 8.50 s while smoothly slowing down to a final speed of 2.80 m/s. (a) Find its original speed. (b) Find its acceleration.

21. ▲ An object moving with uniform acceleration has a velocity of 12.0 cm/s in the positive x direction when its x coordinate is 3.00 cm. If its x coordinate 2.00 s later is -5.00 cm, what is its acceleration?

2 = intermediate; 3 = challenging; □ = SSM/SG; ▲ = ThomsonNOW; = symbolic reasoning; ● = qualitative reasoning

22. Figure P2.22 represents part of the performance data of a car owned by a proud physics student. (a) Calculate the total distance traveled by computing the area under the graph line. (b) What distance does the car travel between the times $t = 10$ s and $t = 40$ s? (c) Draw a graph of its acceleration versus time between $t = 0$ and $t = 50$ s. (d) Write an equation for x as a function of time for each phase of the motion, represented by (i) $0a$, (ii) ab, and (iii) bc. (e) What is the average velocity of the car between $t = 0$ and $t = 50$ s?

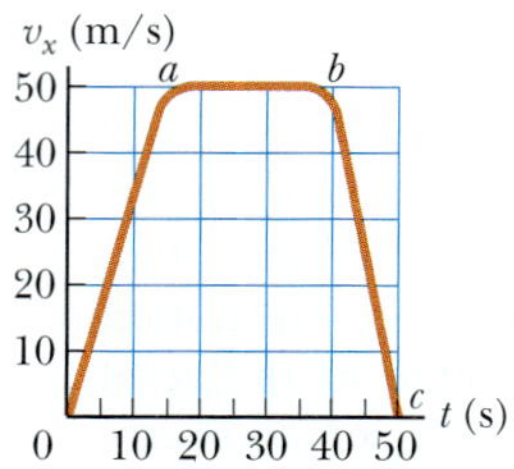

Figure P2.22

23. ● A jet plane comes in for a landing with a speed of 100 m/s, and its acceleration can have a maximum magnitude of $5.00\ \text{m/s}^2$ as it comes to rest. (a) From the instant the plane touches the runway, what is the minimum time interval needed before it can come to rest? (b) Can this plane land on a small tropical island airport where the runway is 0.800 km long? Explain your answer.

24. ● At $t = 0$, one toy car is set rolling on a straight track with initial position 15.0 cm, initial velocity -3.50 cm/s, and constant acceleration $2.40\ \text{cm/s}^2$. At the same moment, another toy car is set rolling on an adjacent track with initial position 10.0 cm, an initial velocity of $+5.50$ cm/s, and constant acceleration zero. (a) At what time, if any, do the two cars have equal speeds? (b) What are their speeds at that time? (c) At what time(s), if any, do the cars pass each other? (d) What are their locations at that time? (e) Explain the difference between question (a) and question (c) as clearly as possible. Write (or draw) for a target audience of students who do not immediately understand the conditions are different.

25. The driver of a car slams on the brakes when he sees a tree blocking the road. The car slows uniformly with an acceleration of $-5.60\ \text{m/s}^2$ for 4.20 s, making straight skid marks 62.4 m long ending at the tree. With what speed does the car then strike the tree?

26. *Help! One of our equations is missing!* We describe constant-acceleration motion with the variables and parameters v_{xi}, v_{xf}, a_x, t, and $x_f - x_i$. Of the equations in Table 2.2, the first does not involve $x_f - x_i$, the second does not contain a_x, the third omits v_{xf}, and the last leaves out t. So, to complete the set there should be an equation *not* involving v_{xi}. Derive it from the others. Use it to solve Problem 25 in one step.

27. For many years Colonel John P. Stapp, USAF, held the world's land speed record. He participated in studying whether a jet pilot could survive emergency ejection. On March 19, 1954, he rode a rocket-propelled sled that moved down a track at a speed of 632 mi/h. He and the sled were safely brought to rest in 1.40 s (Fig. P2.27). Determine (a) the negative acceleration he experienced and (b) the distance he traveled during this negative acceleration.

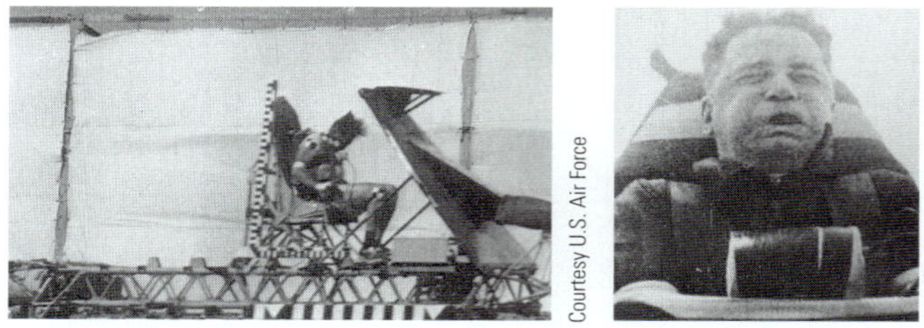

Figure P2.27 (*Left*) Col. John Stapp on rocket sled. (*Right*) Stapp's face is contorted by the stress of rapid negative acceleration.

28. A particle moves along the x axis. Its position is given by the equation $x = 2 + 3t - 4t^2$, with x in meters and t in seconds. Determine (a) its position when it changes direction and (b) its velocity when it returns to the position it had at $t = 0$.

29. An electron in a cathode-ray tube accelerates from a speed of 2.00×10^4 m/s to 6.00×10^6 m/s over 1.50 cm. (a) In what time interval does the electron travel this 1.50 cm? (b) What is its acceleration?

30. ● Within a complex machine such as a robotic assembly line, suppose one particular part glides along a straight track. A control system measures the average velocity of the part during each successive time interval $\Delta t_0 = t_0 - 0$, compares it with the value v_c it should be, and switches a servo motor on and off to give the part a correcting pulse of acceleration. The pulse consists of a constant acceleration a_m applied for time interval $\Delta t_m = t_m - 0$ within the next control time interval Δt_0. As shown in Figure P2.30, the part may be modeled as having zero acceleration when the motor is off (between t_m and t_0). A computer in the control system chooses the size of the acceleration so that the final velocity of the part will have the correct value v_c. Assume the part is initially at rest and is to have instantaneous velocity v_c at time t_0. (a) Find the required value of a_m in terms of v_c and t_m. (b) Show that the displacement Δx of the part during the time interval Δt_0 is given by $\Delta x = v_c\,(t_0 - 0.5t_m)$. For specified values of v_c and t_0, (c) what is the minimum displacement of the part? (d) What is the maximum displacement of the part? (e) Are both the minimum and maximum displacements physically attainable?

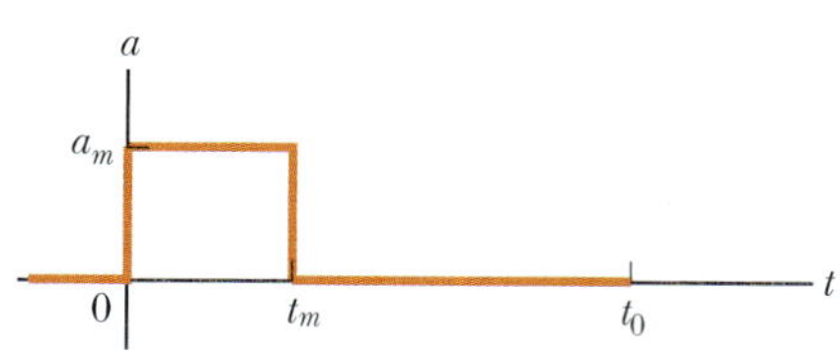

Figure P2.30

31. ● A glider on an air track carries a flag of length ℓ through a stationary photogate, which measures the time

2 = intermediate; 3 = challenging; □ = SSM/SG; ▲ = ThomsonNOW; ▭ = symbolic reasoning; ● = qualitative reasoning

interval Δt_d during which the flag blocks a beam of infrared light passing across the photogate. The ratio $v_d = \ell/\Delta t_d$ is the average velocity of the glider over this part of its motion. Suppose the glider moves with constant acceleration. (a) Argue for or against the idea that v_d is equal to the instantaneous velocity of the glider when it is halfway through the photogate in space. (b) Argue for or against the idea that v_d is equal to the instantaneous velocity of the glider when it is halfway through the photogate in time.

32. ● Speedy Sue, driving at 30.0 m/s, enters a one-lane tunnel. She then observes a slow-moving van 155 m ahead traveling at 5.00 m/s. Sue applies her brakes but can accelerate only at $-2.00\ \text{m/s}^2$ because the road is wet. Will there be a collision? State how you decide. If yes, determine how far into the tunnel and at what time the collision occurs. If no, determine the distance of closest approach between Sue's car and the van.

33. *Vroom, vroom!* As soon as a traffic light turns green, a car speeds up from rest to 50.0 mi/h with constant acceleration 9.00 mi/h · s. In the adjoining bike lane, a cyclist speeds up from rest to 20.0 mi/h with constant acceleration 13.0 mi/h · s. Each vehicle maintains constant velocity after reaching its cruising speed. (a) For what time interval is the bicycle ahead of the car? (b) By what maximum distance does the bicycle lead the car?

34. Solve Example 2.8 (Watch Out for the Speed Limit!) by a graphical method. On the same graph plot position versus time for the car and the police officer. From the intersection of the two curves read the time at which the trooper overtakes the car.

35. ● A glider of length 12.4 cm moves on an air track with constant acceleration. A time interval of 0.628 s elapses between the moment when its front end passes a fixed point Ⓐ along the track and the moment when its back end passes this point. Next, a time interval of 1.39 s elapses between the moment when the back end of the glider passes point Ⓐ and the moment when the front end of the glider passes a second point Ⓑ farther down the track. After that, an additional 0.431 s elapses until the back end of the glider passes point Ⓑ. (a) Find the average speed of the glider as it passes point Ⓐ. (b) Find the acceleration of the glider. (c) Explain how you can compute the acceleration without knowing the distance between points Ⓐ and Ⓑ.

Section 2.7 Freely Falling Objects

Note: In all problems in this section, ignore the effects of air resistance.

36. In a classic clip on *America's Funniest Home Videos,* a sleeping cat rolls gently off the top of a warm TV set. Ignoring air resistance, calculate (a) the position and (b) the velocity of the cat after 0.100 s, 0.200 s, and 0.300 s.

37. ● *Every morning at seven o'clock*
There's twenty terriers drilling on the rock.
The boss comes around and he says, "Keep still
And bear down heavy on the cast-iron drill

And drill, ye terriers, drill." And drill, ye terriers, drill.
It's work all day for sugar in your tea
Down beyond the railway. And drill, ye terriers, drill.

The foreman's name was John McAnn.
By God, he was a blamed mean man.
One day a premature blast went off
And a mile in the air went big Jim Goff. And drill . . .

Then when next payday came around
Jim Goff a dollar short was found.
When he asked what for, came this reply:
"You were docked for the time you were up in the sky."
And drill . . .

—American folksong

What was Goff's hourly wage? State the assumptions you make in computing it.

38. A ball is thrown directly downward, with an initial speed of 8.00 m/s, from a height of 30.0 m. After what time interval does the ball strike the ground?

39. ▲ A student throws a set of keys vertically upward to her sorority sister, who is in a window 4.00 m above. The keys are caught 1.50 s later by the sister's outstretched hand. (a) With what initial velocity were the keys thrown? (b) What was the velocity of the keys just before they were caught?

40. ● Emily challenges her friend David to catch a dollar bill as follows. She holds the bill vertically, as shown in Figure P2.40, with the center of the bill between David's index finger and thumb. David must catch the bill after Emily releases it without moving his hand downward. If his reaction time is 0.2 s, will he succeed? Explain your reasoning.

George Semple

Figure P2.40

41. A baseball is hit so that it travels straight upward after being struck by the bat. A fan observes that it takes 3.00 s for the ball to reach its maximum height. Find (a) the ball's initial velocity and (b) the height it reaches.

42. ● An attacker at the base of a castle wall 3.65 m high throws a rock straight up with speed 7.40 m/s at a height of 1.55 m above the ground. (a) Will the rock reach the top of the wall? (b) If so, what is its speed at the top? If not, what initial speed must it have to reach the top? (c) Find the change in speed of a rock thrown straight down from the top of the wall at an initial speed of 7.40 m/s and moving between the same two points. (d) Does the change in speed of the downward-moving rock agree with the magnitude of the speed change of the rock moving upward between the same elevations? Explain physically why it does or does not agree.

43. ▲ A daring ranch hand sitting on a tree limb wishes to drop vertically onto a horse galloping under the tree. The constant speed of the horse is 10.0 m/s, and the distance

from the limb to the level of the saddle is 3.00 m. (a) What must the horizontal distance between the saddle and limb be when the ranch hand makes his move? (b) For what time interval is he in the air?

44. The height of a helicopter above the ground is given by $h = 3.00t^3$, where h is in meters and t is in seconds. After 2.00 s, the helicopter releases a small mailbag. How long after its release does the mailbag reach the ground?

45. A freely falling object requires 1.50 s to travel the last 30.0 m before it hits the ground. From what height above the ground did it fall?

Section 2.8 Kinematic Equations Derived from Calculus

46. A student drives a moped along a straight road as described by the velocity-versus-time graph in Figure P2.46. Sketch this graph in the middle of a sheet of graph paper. (a) Directly above your graph, sketch a graph of the position versus time, aligning the time coordinates of the two graphs. (b) Sketch a graph of the acceleration versus time directly below the v_x–t graph, again aligning the time coordinates. On each graph, show the numerical values of x and a_x for all points of inflection. (c) What is the acceleration at $t = 6$ s? (d) Find the position (relative to the starting point) at $t = 6$ s. (e) What is the moped's final position at $t = 9$ s?

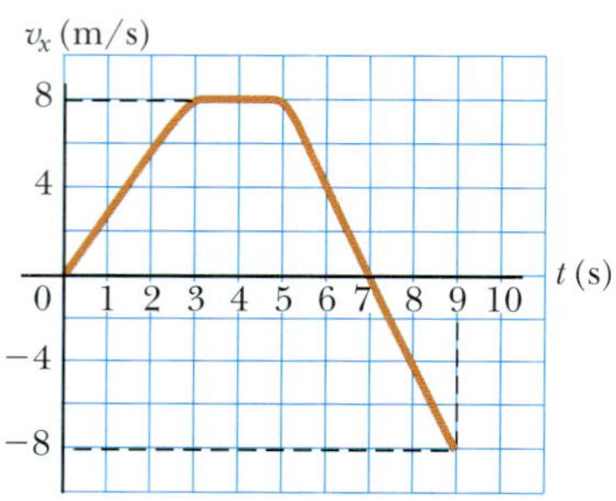

Figure P2.46

47. Automotive engineers refer to the time rate of change of acceleration as the "jerk." Assume an object moves in one dimension such that its jerk J is constant. (a) Determine expressions for its acceleration $a_x(t)$, velocity $v_x(t)$, and position $x(t)$, given that its initial acceleration, velocity, and position are a_{xi}, v_{xi}, and x_i, respectively. (b) Show that $a_x^2 = a_{xi}^2 + 2J(v_x - v_{xi})$.

48. The speed of a bullet as it travels down the barrel of a rifle toward the opening is given by $v = (-5.00 \times 10^7)t^2 + (3.00 \times 10^5)t$, where v is in meters per second and t is in seconds. The acceleration of the bullet just as it leaves the barrel is zero. (a) Determine the acceleration and position of the bullet as a function of time when the bullet is in the barrel. (b) Determine the time interval over which the bullet is accelerated. (c) Find the speed at which the bullet leaves the barrel. (d) What is the length of the barrel?

Additional Problems

49. An object is at $x = 0$ at $t = 0$ and moves along the x axis according to the velocity–time graph in Figure P2.49. (a) What is the acceleration of the object between 0 and 4 s? (b) What is the acceleration of the object between 4 s and 9 s? (c) What is the acceleration of the object between 13 s and 18 s? (d) At what time(s) is the object moving with the lowest speed? (e) At what time is the object farthest from $x = 0$? (f) What is the final position x of the object at $t = 18$ s? (g) Through what total distance has the object moved between $t = 0$ and $t = 18$ s?

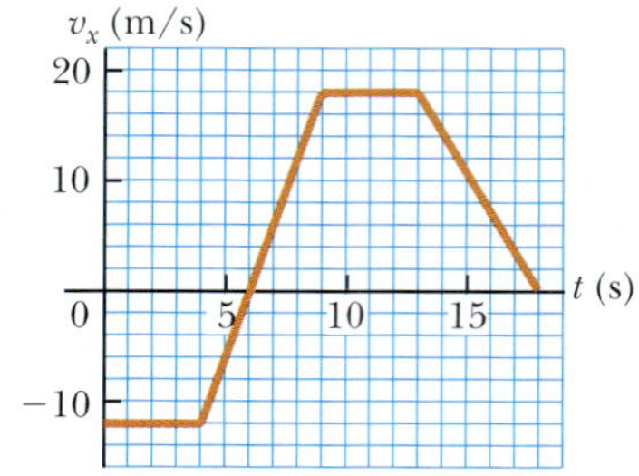

Figure P2.49

50. ● The Acela (pronounced ah-SELL-ah and shown in Fig. P2.50a) is an electric train on the Washington–New York–Boston run, carrying passengers at 170 mi/h. The carriages tilt as much as 6° from the vertical to prevent passengers from feeling pushed to the side as they go around curves. A velocity–time graph for the Acela is shown in Figure P2.50b. (a) Describe the motion of the train in each successive time interval. (b) Find the peak positive acceleration of the train in the motion graphed. (c) Find the train's displacement in miles between $t = 0$ and $t = 200$ s.

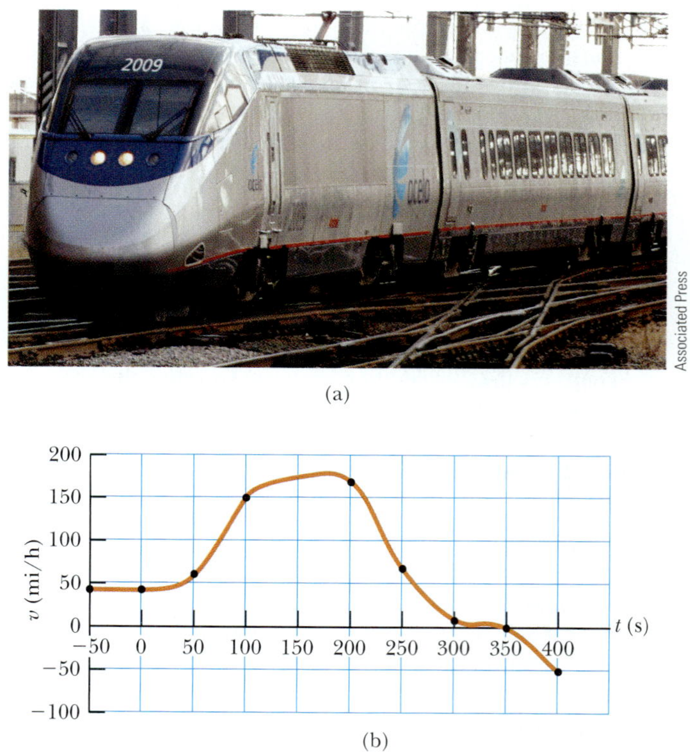

Figure P2.50 (a) The Acela: 1 171 000 lb of cold steel thundering along with 304 passengers. (b) Velocity-versus-time graph for the Acela.

51. A test rocket is fired vertically upward from a well. A catapult gives it an initial speed of 80.0 m/s at ground level.

Its engines then fire and it accelerates upward at 4.00 m/s^2 until it reaches an altitude of 1 000 m. At that point its engines fail and the rocket goes into free fall, with an acceleration of -9.80 m/s^2. (a) For what time interval is the rocket in motion above the ground? (b) What is its maximum altitude? (c) What is its velocity just before it collides with the Earth? (You will need to consider the motion while the engine is operating separate from the free-fall motion.)

52. ● In Active Figure 2.11b, the area under the velocity versus time curve and between the vertical axis and time t (vertical dashed line) represents the displacement. As shown, this area consists of a rectangle and a triangle. Compute their areas and state how the sum of the two areas compares with the expression on the right-hand side of Equation 2.16.

53. Setting a world record in a 100-m race, Maggie and Judy cross the finish line in a dead heat, both taking 10.2 s. Accelerating uniformly, Maggie took 2.00 s and Judy took 3.00 s to attain maximum speed, which they maintained for the rest of the race. (a) What was the acceleration of each sprinter? (b) What were their respective maximum speeds? (c) Which sprinter was ahead at the 6.00-s mark, and by how much?

54. ● *How long should a traffic light stay yellow?* Assume you are driving at the speed limit v_0. As you approach an intersection 22.0 m wide, you see the light turn yellow. During your reaction time of 0.600 s, you travel at constant speed as you recognize the warning, decide whether to stop or to go through the intersection, and move your foot to the brake if you must stop. Your car has good brakes and can accelerate at -2.40 m/s^2. Before it turns red, the light should stay yellow long enough for you to be able to get to the other side of the intersection without speeding up, if you are too close to the intersection to stop before entering it. (a) Find the required time interval Δt_y that the light should stay yellow in terms of v_0. Evaluate your answer for (b) v_0 = 8.00 m/s = 28.8 km/h, (c) v_0 = 11.0 m/s = 40.2 km/h, (d) v_0 = 18.0 m/s = 64.8 km/h, and (e) v_0 = 25.0 m/s = 90.0 km/h. **What If?** Evaluate your answer for (f) v_0 approaching zero, and (g) v_0 approaching infinity. (h) Describe the pattern of variation of Δt_y with v_0. You may wish also to sketch a graph of it. Account for the answers to parts (f) and (g) physically. (i) For what value of v_0 would Δt_y be minimal, and (j) what is this minimum time interval? *Suggestion:* You may find it easier to do part (a) after first doing part (b).

55. A commuter train travels between two downtown stations. Because the stations are only 1.00 km apart, the train never reaches its maximum possible cruising speed. During rush hour the engineer minimizes the time interval Δt between two stations by accelerating for a time interval Δt_1 at a rate a_1 = 0.100 m/s^2 and then immediately braking with acceleration a_2 = -0.500 m/s^2 for a time interval Δt_2. Find the minimum time interval of travel Δt and the time interval Δt_1.

56. A Ferrari F50 of length 4.52 m is moving north on a roadway that intersects another perpendicular roadway. The width of the intersection from near edge to far edge is 28.0 m. The Ferrari has a constant acceleration of magnitude 2.10 m/s^2 directed south. The time interval required for the nose of the Ferrari to move from the near (south) edge of the intersection to the north edge of the intersection is 3.10 s. (a) How far is the nose of the Ferrari from the south edge of the intersection when it stops? (b) For what time interval is *any* part of the Ferrari within the boundaries of the intersection? (c) A Corvette is at rest on the perpendicular intersecting roadway. As the nose of the Ferrari enters the intersection, the Corvette starts from rest and accelerates east at 5.60 m/s^2. What is the minimum distance from the near (west) edge of the intersection at which the nose of the Corvette can begin its motion if the Corvette is to enter the intersection after the Ferrari has entirely left the intersection? (d) If the Corvette begins its motion at the position given by your answer to part (c), with what speed does it enter the intersection?

57. An inquisitive physics student and mountain climber climbs a 50.0-m cliff that overhangs a calm pool of water. He throws two stones vertically downward, 1.00 s apart, and observes that they cause a single splash. The first stone has an initial speed of 2.00 m/s. (a) How long after release of the first stone do the two stones hit the water? (b) What initial velocity must the second stone have if they are to hit simultaneously? (c) What is the speed of each stone at the instant the two hit the water?

58. ● A hard rubber ball, released at chest height, falls to the pavement and bounces back to nearly the same height. When it is in contact with the pavement, the lower side of the ball is temporarily flattened. Suppose the maximum depth of the dent is on the order of 1 cm. Compute an order-of-magnitude estimate for the maximum acceleration of the ball while it is in contact with the pavement. State your assumptions, the quantities you estimate, and the values you estimate for them.

59. Kathy Kool buys a sports car that can accelerate at the rate of 4.90 m/s^2. She decides to test the car by racing with another speedster, Stan Speedy. Both start from rest, but experienced Stan leaves the starting line 1.00 s before Kathy. Stan moves with a constant acceleration of 3.50 m/s^2 and Kathy maintains an acceleration of 4.90 m/s^2. Find (a) the time at which Kathy overtakes Stan, (b) the distance she travels before she catches him, and (c) the speeds of both cars at the instant she overtakes him.

60. A rock is dropped from rest into a well. (a) The sound of the splash is heard 2.40 s after the rock is released from rest. How far below the top of the well is the surface of the water? The speed of sound in air (at the ambient temperature) is 336 m/s. (b) **What If?** If the travel time for the sound is ignored, what percentage error is introduced when the depth of the well is calculated?

61. ● In a California driver's handbook, the following data were given about the minimum distance a typical car travels in stopping from various original speeds. The "thinking distance" represents how far the car travels during the driver's reaction time, after a reason to stop can be seen but before the driver can apply the brakes. The "braking distance" is the displacement of the car after the brakes are applied. (a) Is the thinking-distance data

consistent with the assumption that the car travels with constant speed? Explain. (b) Determine the best value of the reaction time suggested by the data. (c) Is the braking-distance data consistent with the assumption that the car travels with constant acceleration? Explain. (d) Determine the best value for the acceleration suggested by the data.

Speed (mi/h)	Thinking Distance (ft)	Braking Distance (ft)	Total Stopping Distance (ft)
25	27	34	61
35	38	67	105
45	49	110	159
55	60	165	225
65	71	231	302

62. ● Astronauts on a distant planet toss a rock into the air. With the aid of a camera that takes pictures at a steady rate, they record the height of the rock as a function of time as given in the table in the next column. (a) Find the average velocity of the rock in the time interval between each measurement and the next. (b) Using these average velocities to approximate instantaneous velocities at the midpoints of the time intervals, make a graph of velocity as a function of time. Does the rock move with constant acceleration? If so, plot a straight line of best fit on the graph and calculate its slope to find the acceleration.

Time (s)	Height (m)	Time (s)	Height (m)
0.00	5.00	2.75	7.62
0.25	5.75	3.00	7.25
0.50	6.40	3.25	6.77
0.75	6.94	3.50	6.20
1.00	7.38	3.75	5.52
1.25	7.72	4.00	4.73
1.50	7.96	4.25	3.85
1.75	8.10	4.50	2.86
2.00	8.13	4.75	1.77
2.25	8.07	5.00	0.58
2.50	7.90		

63. Two objects, A and B, are connected by a rigid rod that has length L. The objects slide along perpendicular guide rails as shown in Figure P2.63. Assume A slides to the left with a constant speed v. Find the velocity of B when $\theta = 60.0°$.

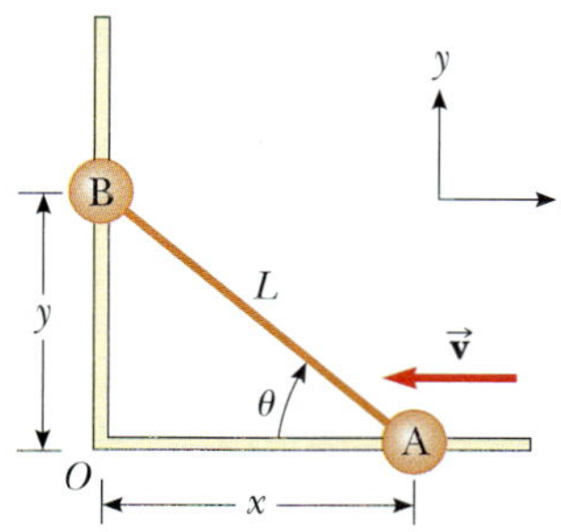

Figure P2.63

Answers to Quick Quizzes

2.1 (c). If the particle moves along a line without changing direction, the displacement and distance traveled over any time interval will be the same. As a result, the magnitude of the average velocity and the average speed will be the same. If the particle reverses direction, however, the displacement will be less than the distance traveled. In turn, the magnitude of the average velocity will be smaller than the average speed.

2.2 (b). Regardless of your speeds at all other times, if your instantaneous speed at the instant it is measured is higher than the speed limit, you may receive a speeding ticket.

2.3 (b). If the car is slowing down, a force must be pulling in the direction opposite to its velocity.

2.4 False. Your graph should look something like the following.

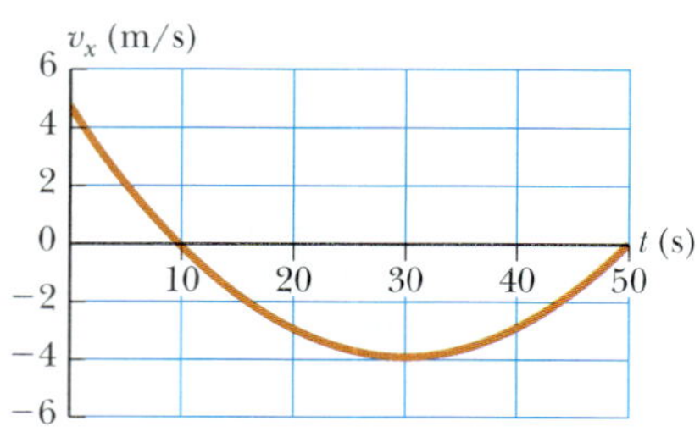

This v_x–t graph shows that the maximum speed is about 5.0 m/s, which is 18 km/h (= 11 mi/h), so the driver was not speeding.

2.5 (c). If a particle with constant acceleration stops and its acceleration remains constant, it must begin to move again in the opposite direction. If it did not, the acceleration would change from its original constant value to zero. Choice (a) is not correct because the direction of acceleration is not specified by the direction of the velocity. Choice (b) is also not correct by counterexample; a car moving in the $-x$ direction and slowing down has a positive acceleration.

2.6 Graph (a) has a constant slope, indicating a constant acceleration; it is represented by graph (e).

Graph (b) represents a speed that is increasing constantly but not at a uniform rate. Therefore, the acceleration must be increasing, and the graph that best indicates that is (d).

Graph (c) depicts a velocity that first increases at a constant rate, indicating constant acceleration. Then the velocity stops increasing and becomes constant, indicating zero acceleration. The best match to this situation is graph (f).

2.7 **(i)**, (e). For the entire time interval that the ball is in free fall, the acceleration is that due to gravity. **(ii)**, (d). While the ball is rising, it is slowing down. After reaching the highest point, the ball begins to fall and its speed increases.

These controls in the cockpit of a commercial aircraft assist the pilot in maintaining control over the velocity of the aircraft—how fast it is traveling and in what direction it is traveling—allowing it to land safely. Quantities that are defined by both a magnitude and a direction, such as velocity, are called *vector quantities.* (Mark Wagner/Getty Images)

3 Vectors

In our study of physics, we often need to work with physical quantities that have both numerical and directional properties. As noted in Section 2.1, quantities of this nature are vector quantities. This chapter is primarily concerned with general properties of vector quantities. We discuss the addition and subtraction of vector quantities, together with some common applications to physical situations.

Vector quantities are used throughout this text. Therefore, it is imperative that you master the techniques discussed in this chapter.

3.1 Coordinate Systems

Many aspects of physics involve a description of a location in space. In Chapter 2, for example, we saw that the mathematical description of an object's motion requires a method for describing the object's position at various times. In two dimensions, this description is accomplished with the use of the Cartesian coordinate system, in which perpendicular axes intersect at a point defined as the origin (Fig. 3.1). Cartesian coordinates are also called *rectangular coordinates.*

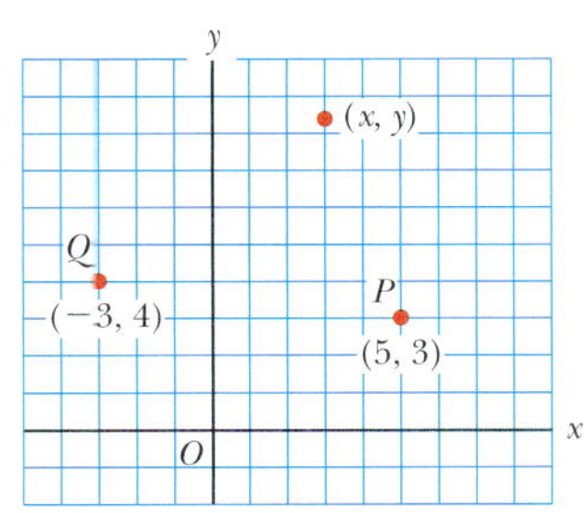

Figure 3.1 Designation of points in a Cartesian coordinate system. Every point is labeled with coordinates (x, y).

Sometimes it is more convenient to represent a point in a plane by its *plane polar coordinates* (r, θ) as shown in Active Figure 3.2a (see page 54). In this *polar coordinate system,* r is the distance from the origin to the point having Cartesian coordinates (x, y) and θ is the angle between a fixed axis and a line drawn from the origin to the point. The fixed axis is often the positive x axis, and θ is usually measured counterclockwise from it. From the right triangle in Active Figure 3.2b,

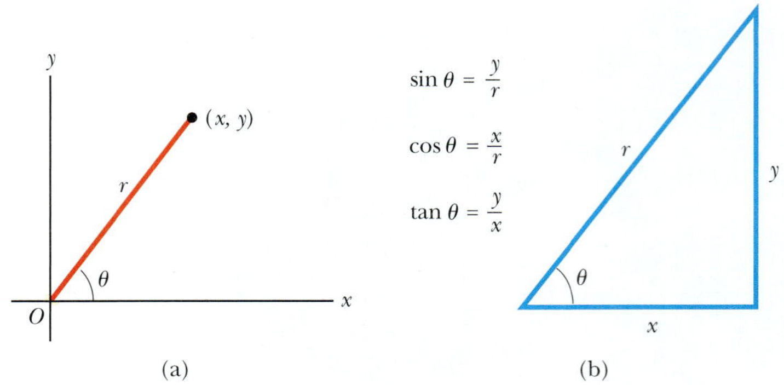

ACTIVE FIGURE 3.2

(a) The plane polar coordinates of a point are represented by the distance r and the angle θ, where θ is measured counterclockwise from the positive x axis. (b) The right triangle used to relate (x, y) to (r, θ).

Sign in at www.thomsonedu.com and go to ThomsonNOW to move the point and see the changes to the rectangular and polar coordinates as well as to the sine, cosine, and tangent of angle θ.

we find that $\sin\theta = y/r$ and that $\cos\theta = x/r$. (A review of trigonometric functions is given in Appendix B.4.) Therefore, starting with the plane polar coordinates of any point, we can obtain the Cartesian coordinates by using the equations

$$x = r\cos\theta \quad \textbf{(3.1)}$$

$$y = r\sin\theta \quad \textbf{(3.2)}$$

Furthermore, the definitions of trigonometry tell us that

$$\tan\theta = \frac{y}{x} \quad \textbf{(3.3)}$$

$$r = \sqrt{x^2 + y^2} \quad \textbf{(3.4)}$$

Equation 3.4 is the familiar Pythagorean theorem.

These four expressions relating the coordinates (x, y) to the coordinates (r, θ) apply only when θ is defined as shown in Active Figure 3.2a—in other words, when positive θ is an angle measured counterclockwise from the positive x axis. (Some scientific calculators perform conversions between Cartesian and polar coordinates based on these standard conventions.) If the reference axis for the polar angle θ is chosen to be one other than the positive x axis or if the sense of increasing θ is chosen differently, the expressions relating the two sets of coordinates will change.

EXAMPLE 3.1 Polar Coordinates

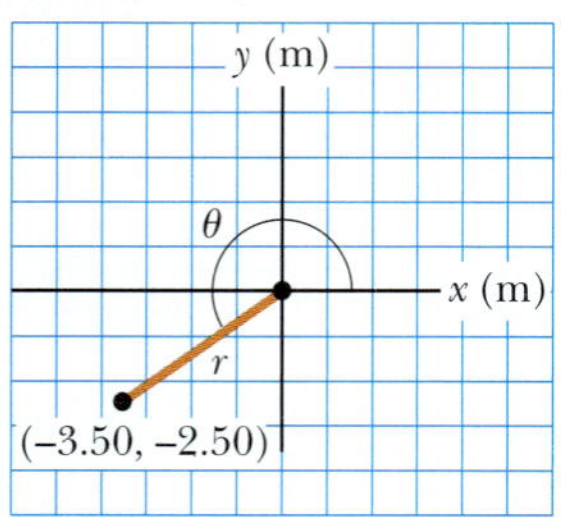

ACTIVE FIGURE 3.3

(Example 3.1) Finding polar coordinates when Cartesian coordinates are given.

Sign in at www.thomsonedu.com and go to ThomsonNOW to move the point in the xy plane and see how its Cartesian and polar coordinates change.

The Cartesian coordinates of a point in the xy plane are $(x, y) = (-3.50, -2.50)$ m as shown in Active Figure 3.3. Find the polar coordinates of this point.

SOLUTION

Conceptualize The drawing in Active Figure 3.3 helps us conceptualize the problem.

Categorize Based on the statement of the problem and the Conceptualize step, we recognize that we are simply converting from Cartesian coordinates to polar coordinates. We therefore categorize this example as a substitution problem. Substitution problems generally do not have an extensive Analyze step other than the substitution of numbers into a given equation. Similarly, the Finalize step consists primarily of checking the units and making sure that the answer is reasonable. Therefore, for substitution problems, we will not label Analyze or Finalize steps.

Use Equation 3.4 to find r: $r = \sqrt{x^2 + y^2} = \sqrt{(-3.50 \text{ m})^2 + (-2.50 \text{ m})^2} = 4.30 \text{ m}$

Use Equation 3.3 to find θ: $\tan\theta = \frac{y}{x} = \frac{-2.50 \text{ m}}{-3.50 \text{ m}} = 0.714$

$$\theta = 216°$$

Notice that you must use the signs of x and y to find that the point lies in the third quadrant of the coordinate system. That is, $\theta = 216°$, not 35.5°.

3.2 Vector and Scalar Quantities

We now formally describe the difference between scalar quantities and vector quantities. When you want to know the temperature outside so that you will know how to dress, the only information you need is a number and the unit "degrees C" or "degrees F." Temperature is therefore an example of a *scalar quantity*:

A **scalar quantity** is completely specified by a single value with an appropriate unit and has no direction.

Other examples of scalar quantities are volume, mass, speed, and time intervals. The rules of ordinary arithmetic are used to manipulate scalar quantities.

If you are preparing to pilot a small plane and need to know the wind velocity, you must know both the speed of the wind and its direction. Because direction is important for its complete specification, velocity is a *vector quantity*:

A **vector quantity** is completely specified by a number and appropriate units plus a direction.

Another example of a vector quantity is displacement, as you know from Chapter 2. Suppose a particle moves from some point Ⓐ to some point Ⓑ along a straight path as shown in Figure 3.4. We represent this displacement by drawing an arrow from Ⓐ to Ⓑ, with the tip of the arrow pointing away from the starting point. The direction of the arrowhead represents the direction of the displacement, and the length of the arrow represents the magnitude of the displacement. If the particle travels along some other path from Ⓐ to Ⓑ, such as shown by the broken line in Figure 3.4, its displacement is still the arrow drawn from Ⓐ to Ⓑ. Displacement depends only on the initial and final positions, so the displacement vector is independent of the path taken by the particle between these two points.

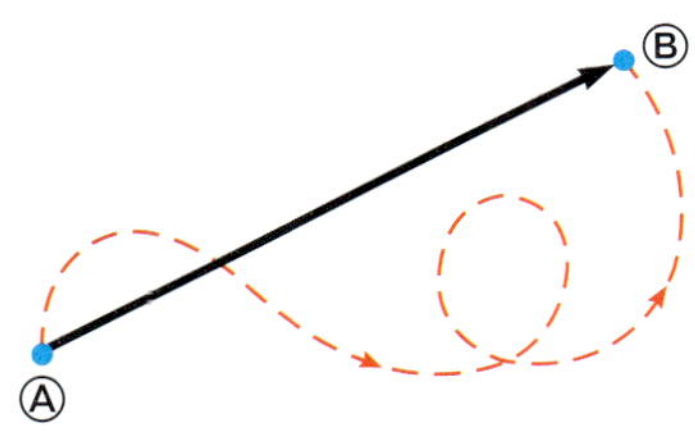

Figure 3.4 As a particle moves from Ⓐ to Ⓑ along an arbitrary path represented by the broken line, its displacement is a vector quantity shown by the arrow drawn from Ⓐ to Ⓑ.

In this text, we use a boldface letter with an arrow over the letter, such as $\vec{\mathbf{A}}$, to represent a vector. Another common notation for vectors with which you should be familiar is a simple boldface character: **A**. The magnitude of the vector $\vec{\mathbf{A}}$ is written either A or $|\vec{\mathbf{A}}|$. The magnitude of a vector has physical units, such as meters for displacement or meters per second for velocity. The magnitude of a vector is *always* a positive number.

Quick Quiz 3.1 Which of the following are vector quantities and which are scalar quantities? (a) your age (b) acceleration (c) velocity (d) speed (e) mass

3.3 Some Properties of Vectors

In this section, we shall investigate general properties of vectors representing physical quantities. We also discuss how to add and subtract vectors using both algebraic and geometric methods.

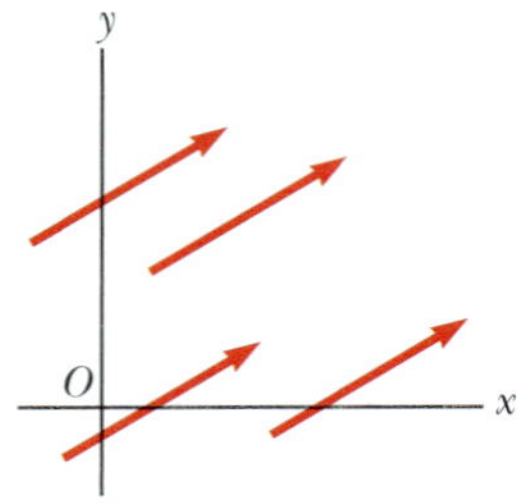

Figure 3.5 These four vectors are equal because they have equal lengths and point in the same direction.

Equality of Two Vectors

For many purposes, two vectors $\vec{\mathbf{A}}$ and $\vec{\mathbf{B}}$ may be defined to be equal if they have the same magnitude and if they point in the same direction. That is, $\vec{\mathbf{A}} = \vec{\mathbf{B}}$ only if $A = B$ and if $\vec{\mathbf{A}}$ and $\vec{\mathbf{B}}$ point in the same direction along parallel lines. For example, all the vectors in Figure 3.5 are equal even though they have different starting points. This property allows us to move a vector to a position parallel to itself in a diagram without affecting the vector.

Adding Vectors

The rules for adding vectors are conveniently described by a graphical method. To add vector $\vec{\mathbf{B}}$ to vector $\vec{\mathbf{A}}$, first draw vector $\vec{\mathbf{A}}$ on graph paper, with its magnitude represented by a convenient length scale, and then draw vector $\vec{\mathbf{B}}$ to the same scale, with its tail starting from the tip of $\vec{\mathbf{A}}$, as shown in Active Figure 3.6. The **resultant vector** $\vec{\mathbf{R}} = \vec{\mathbf{A}} + \vec{\mathbf{B}}$ is the vector drawn from the tail of $\vec{\mathbf{A}}$ to the tip of $\vec{\mathbf{B}}$.

A geometric construction can also be used to add more than two vectors as is shown in Figure 3.7 for the case of four vectors. The resultant vector $\vec{\mathbf{R}} = \vec{\mathbf{A}} + \vec{\mathbf{B}} + \vec{\mathbf{C}} + \vec{\mathbf{D}}$ is the vector that completes the polygon. In other words, **$\vec{\mathbf{R}}$ is the vector drawn from the tail of the first vector to the tip of the last vector**. This technique for adding vectors is often called the "head to tail method."

When two vectors are added, the sum is independent of the order of the addition. (This fact may seem trivial, but as you will see in Chapter 11, the order is important when vectors are multiplied. Procedures for multiplying vectors are discussed in Chapters 7 and 11). This property, which can be seen from the geometric construction in Figure 3.8, is known as the **commutative law of addition:**

$$\vec{\mathbf{A}} + \vec{\mathbf{B}} = \vec{\mathbf{B}} + \vec{\mathbf{A}} \tag{3.5}$$

When three or more vectors are added, their sum is independent of the way in which the individual vectors are grouped together. A geometric proof of this rule for three vectors is given in Figure 3.9. This property is called the **associative law of addition:**

$$\vec{\mathbf{A}} + (\vec{\mathbf{B}} + \vec{\mathbf{C}}) = (\vec{\mathbf{A}} + \vec{\mathbf{B}}) + \vec{\mathbf{C}} \tag{3.6}$$

In summary, **a vector quantity has both magnitude and direction and also obeys the laws of vector addition** as described in Figures 3.6 to 3.9. When two or more vectors are added together, they must all have the same units and they must all be the same type of quantity. It would be meaningless to add a velocity vector (for example, 60 km/h to the east) to a displacement vector (for example, 200 km to the north) because these vectors represent different physical quantities. The same

PITFALL PREVENTION 3.1

Vector Addition versus Scalar Addition

Notice that $\vec{\mathbf{A}} + \vec{\mathbf{B}} = \vec{\mathbf{C}}$ is very different from $A + B = C$. The first equation is a vector sum, which must be handled carefully, such as with the graphical method. The second equation is a simple algebraic addition of numbers that is handled with the normal rules of arithmetic.

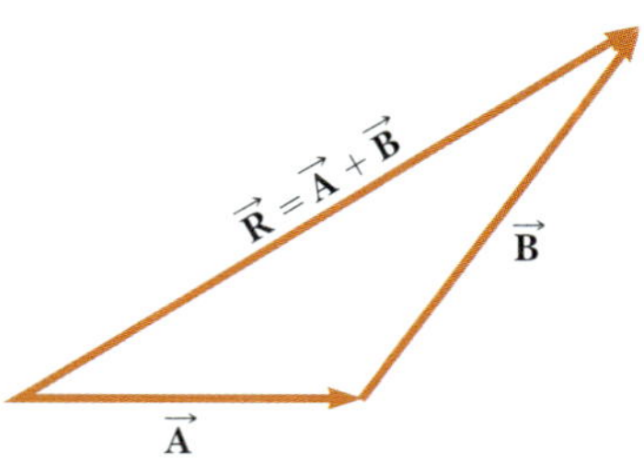

ACTIVE FIGURE 3.6

When vector $\vec{\mathbf{B}}$ is added to vector $\vec{\mathbf{A}}$, the resultant $\vec{\mathbf{R}}$ is the vector that runs from the tail of $\vec{\mathbf{A}}$ to the tip of $\vec{\mathbf{B}}$.

Sign in at www.thomsonedu.com and go to ThomsonNOW to explore the addition of two vectors.

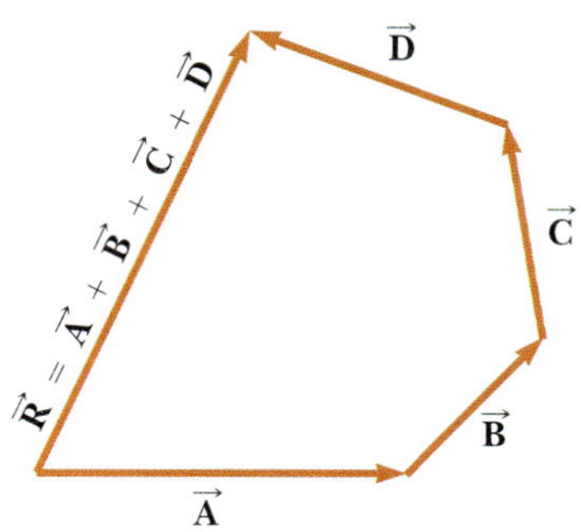

Figure 3.7 Geometric construction for summing four vectors. The resultant vector $\vec{\mathbf{R}}$ is by definition the one that completes the polygon.

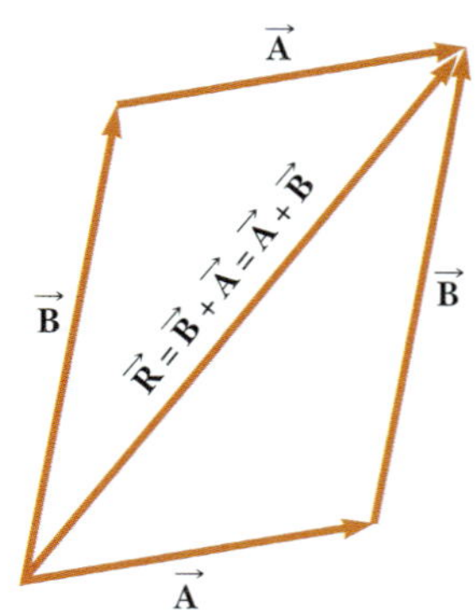

Figure 3.8 This construction shows that $\vec{\mathbf{A}} + \vec{\mathbf{B}} = \vec{\mathbf{B}} + \vec{\mathbf{A}}$ or, in other words, that vector addition is commutative.

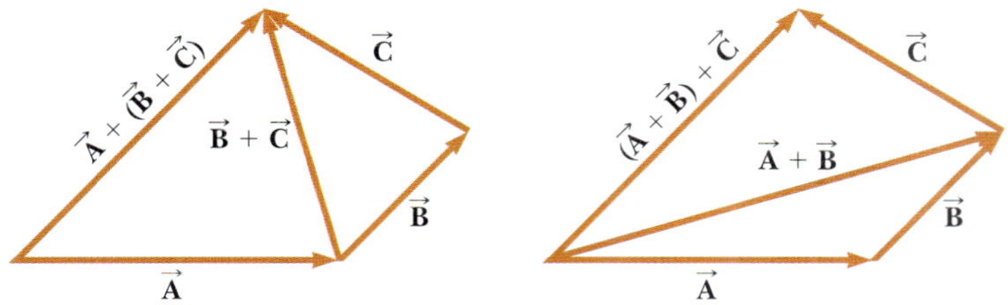

Figure 3.9 Geometric constructions for verifying the associative law of addition.

rule also applies to scalars. For example, it would be meaningless to add time intervals to temperatures.

Negative of a Vector

The negative of the vector $\vec{\mathbf{A}}$ is defined as the vector that when added to $\vec{\mathbf{A}}$ gives zero for the vector sum. That is, $\vec{\mathbf{A}} + (-\vec{\mathbf{A}}) = 0$. The vectors $\vec{\mathbf{A}}$ and $-\vec{\mathbf{A}}$ have the same magnitude but point in opposite directions.

Subtracting Vectors

The operation of vector subtraction makes use of the definition of the negative of a vector. We define the operation $\vec{\mathbf{A}} - \vec{\mathbf{B}}$ as vector $-\vec{\mathbf{B}}$ added to vector $\vec{\mathbf{A}}$:

$$\vec{\mathbf{A}} - \vec{\mathbf{B}} = \vec{\mathbf{A}} + (-\vec{\mathbf{B}}) \tag{3.7}$$

The geometric construction for subtracting two vectors in this way is illustrated in Figure 3.10a.

Another way of looking at vector subtraction is to notice that the difference $\vec{\mathbf{A}} - \vec{\mathbf{B}}$ between two vectors $\vec{\mathbf{A}}$ and $\vec{\mathbf{B}}$ is what you have to add to the second vector to obtain the first. In this case, as Figure 3.10b shows, the vector $\vec{\mathbf{A}} - \vec{\mathbf{B}}$ points from the tip of the second vector to the tip of the first.

Multiplying a Vector by a Scalar

If vector $\vec{\mathbf{A}}$ is multiplied by a positive scalar quantity m, the product $m\vec{\mathbf{A}}$ is a vector that has the same direction as $\vec{\mathbf{A}}$ and magnitude mA. If vector $\vec{\mathbf{A}}$ is multiplied by a negative scalar quantity $-m$, the product $-m\vec{\mathbf{A}}$ is directed opposite $\vec{\mathbf{A}}$. For example, the vector $5\vec{\mathbf{A}}$ is five times as long as $\vec{\mathbf{A}}$ and points in the same direction as $\vec{\mathbf{A}}$; the vector $-\frac{1}{3}\vec{\mathbf{A}}$ is one-third the length of $\vec{\mathbf{A}}$ and points in the direction opposite $\vec{\mathbf{A}}$.

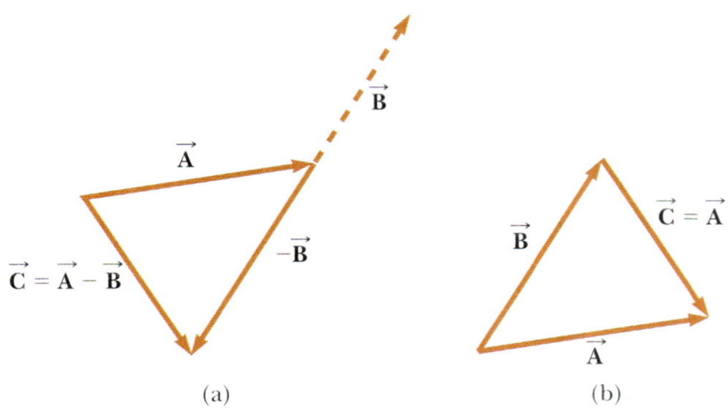

Figure 3.10 (a) This construction shows how to subtract vector $\vec{\mathbf{B}}$ from vector $\vec{\mathbf{A}}$. The vector $-\vec{\mathbf{B}}$ is equal in magnitude to vector $\vec{\mathbf{B}}$ and points in the opposite direction. To subtract $\vec{\mathbf{B}}$ from $\vec{\mathbf{A}}$, apply the rule of vector addition to the combination of $\vec{\mathbf{A}}$ and $-\vec{\mathbf{B}}$: first draw $\vec{\mathbf{A}}$ along some convenient axis and then place the tail of $-\vec{\mathbf{B}}$ at the tip of $\vec{\mathbf{A}}$, and $\vec{\mathbf{C}}$ is the difference $\vec{\mathbf{A}} - \vec{\mathbf{B}}$. (b) A second way of looking at vector subtraction. The difference vector $\vec{\mathbf{C}} = \vec{\mathbf{A}} - \vec{\mathbf{B}}$ is the vector that we must add to $\vec{\mathbf{B}}$ to obtain $\vec{\mathbf{A}}$.

Quick Quiz 3.2 The magnitudes of two vectors $\vec{\mathbf{A}}$ and $\vec{\mathbf{B}}$ are $A = 12$ units and $B = 8$ units. Which of the following pairs of numbers represents the *largest* and *smallest* possible values for the magnitude of the resultant vector $\vec{\mathbf{R}} = \vec{\mathbf{A}} + \vec{\mathbf{B}}$? (a) 14.4 units, 4 units (b) 12 units, 8 units (c) 20 units, 4 units (d) none of these answers

Quick Quiz 3.3 If vector $\vec{\mathbf{B}}$ is added to vector $\vec{\mathbf{A}}$, which *two* of the following choices must be true for the resultant vector to be equal to zero? (a) $\vec{\mathbf{A}}$ and $\vec{\mathbf{B}}$ are parallel and in the same direction. (b) $\vec{\mathbf{A}}$ and $\vec{\mathbf{B}}$ are parallel and in opposite directions. (c) $\vec{\mathbf{A}}$ and $\vec{\mathbf{B}}$ have the same magnitude. (d) $\vec{\mathbf{A}}$ and $\vec{\mathbf{B}}$ are perpendicular.

EXAMPLE 3.2 A Vacation Trip

A car travels 20.0 km due north and then 35.0 km in a direction 60.0° west of north as shown in Figure 3.11a. Find the magnitude and direction of the car's resultant displacement.

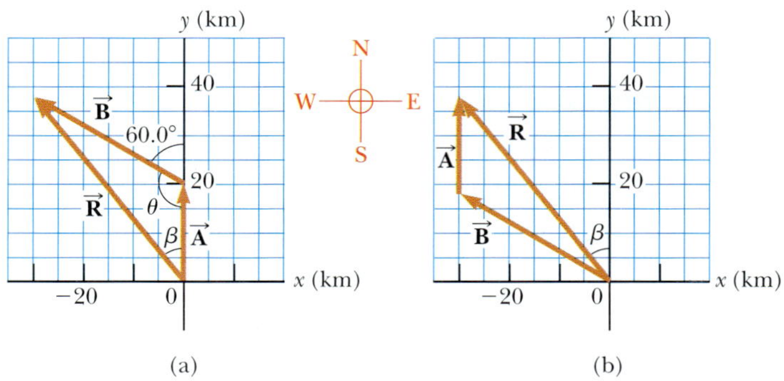

Figure 3.11 (Example 3.2) (a) Graphical method for finding the resultant displacement vector $\vec{\mathbf{R}} = \vec{\mathbf{A}} + \vec{\mathbf{B}}$. (b) Adding the vectors in reverse order $(\vec{\mathbf{B}} + \vec{\mathbf{A}})$ gives the same result for $\vec{\mathbf{R}}$.

SOLUTION

Conceptualize The vectors $\vec{\mathbf{A}}$ and $\vec{\mathbf{B}}$ drawn in Figure 3.11a help us conceptualize the problem.

Categorize We can categorize this example as a simple analysis problem in vector addition. The displacement $\vec{\mathbf{R}}$ is the resultant when the two individual displacements $\vec{\mathbf{A}}$ and $\vec{\mathbf{B}}$ are added. We can further categorize it as a problem about the analysis of triangles, so we appeal to our expertise in geometry and trigonometry.

Analyze In this example, we show two ways to analyze the problem of finding the resultant of two vectors. The first way is to solve the problem geometrically, using graph paper and a protractor to measure the magnitude of $\vec{\mathbf{R}}$ and its direction in Figure 3.11a. (In fact, even when you know you are going to be carrying out a calculation, you should sketch the vectors to check your results.) With an ordinary ruler and protractor, a large diagram typically gives answers to two-digit but not to three-digit precision.

The second way to solve the problem is to analyze it algebraically. The magnitude of $\vec{\mathbf{R}}$ can be obtained from the law of cosines as applied to the triangle (see Appendix B.4).

Use $R^2 = A^2 + B^2 - 2AB\cos\theta$ from the law of cosines to find R:

$$R = \sqrt{A^2 + B^2 - 2AB\cos\theta}$$

Substitute numerical values, noting that $\theta = 180° - 60° = 120°$:

$$R = \sqrt{(20.0\text{ km})^2 + (35.0\text{ km})^2 - 2(20.0\text{ km})(35.0\text{ km})\cos 120°}$$

$$= 48.2\text{ km}$$

Use the law of sines (Appendix B.4) to find the direction of $\vec{\mathbf{R}}$ measured from the northerly direction:

$$\frac{\sin\beta}{B} = \frac{\sin\theta}{R}$$

$$\sin\beta = \frac{B}{R}\sin\theta = \frac{35.0\text{ km}}{48.2\text{ km}}\sin 120° = 0.629$$

$$\beta = 38.9°$$

The resultant displacement of the car is 48.2 km in a direction 38.9° west of north.

Finalize Does the angle β that we calculated agree with an estimate made by looking at Figure 3.11a or with an actual angle measured from the diagram using the graphical method? Is it reasonable that the magnitude of $\vec{\mathbf{R}}$ is larger than that of both $\vec{\mathbf{A}}$ and $\vec{\mathbf{B}}$? Are the units of $\vec{\mathbf{R}}$ correct?

Although the graphical method of adding vectors works well, it suffers from two disadvantages. First, some people find using the laws of cosines and sines to be awkward. Second, a triangle only results if you are adding two vectors. If you are adding three or more vectors, the resulting geometric shape is usually not a triangle. In Section 3.4, we explore a new method of adding vectors that will address both of these disadvantages.

What If? Suppose the trip were taken with the two vectors in reverse order: 35.0 km at 60.0° west of north first and then 20.0 km due north. How would the magnitude and the direction of the resultant vector change?

Answer They would not change. The commutative law for vector addition tells us that the order of vectors in an addition is irrelevant. Graphically, Figure 3.11b shows that the vectors added in the reverse order give us the same resultant vector.

3.4 Components of a Vector and Unit Vectors

The graphical method of adding vectors is not recommended whenever high accuracy is required or in three-dimensional problems. In this section, we describe a method of adding vectors that makes use of the projections of vectors along coordinate axes. These projections are called the **components** of the vector or its **rectangular components**. Any vector can be completely described by its components.

Consider a vector $\vec{\mathbf{A}}$ lying in the xy plane and making an arbitrary angle θ with the positive x axis as shown in Figure 3.12a. This vector can be expressed as the sum of two other *component vectors* $\vec{\mathbf{A}}_x$, which is parallel to the x axis, and $\vec{\mathbf{A}}_y$, which is parallel to the y axis. From Figure 3.12b, we see that the three vectors form a right triangle and that $\vec{\mathbf{A}} = \vec{\mathbf{A}}_x + \vec{\mathbf{A}}_y$. We shall often refer to the "components of a vector $\vec{\mathbf{A}}$," written A_x and A_y (without the boldface notation). The component A_x represents the projection of $\vec{\mathbf{A}}$ along the x axis, and the component A_y represents the projection of $\vec{\mathbf{A}}$ along the y axis. These components can be positive or negative. The component A_x is positive if the component vector $\vec{\mathbf{A}}_x$ points in the positive x direction and is negative if $\vec{\mathbf{A}}_x$ points in the negative x direction. The same is true for the component A_y.

PITFALL PREVENTION 3.2
Component Vectors versus Components

The vectors $\vec{\mathbf{A}}_x$ and $\vec{\mathbf{A}}_y$ are the *component vectors* of $\vec{\mathbf{A}}$. They should not be confused with the quantities A_x and A_y, which we shall always refer to as the *components* of $\vec{\mathbf{A}}$.

From Figure 3.12 and the definition of sine and cosine, we see that $\cos\theta = A_x/A$ and that $\sin\theta = A_y/A$. Hence, the components of $\vec{\mathbf{A}}$ are

$$A_x = A\cos\theta \qquad \textbf{(3.8)}$$

$$A_y = A\sin\theta \qquad \textbf{(3.9)}$$

◀ Components of the vector $\vec{\mathbf{A}}$

PITFALL PREVENTION 3.3
***x* and *y* Components**

Equations 3.8 and 3.9 associate the cosine of the angle with the x component and the sine of the angle with the y component. This association is true *only* because we measured the angle θ with respect to the x axis, so do not memorize these equations. If θ is measured with respect to the y axis (as in some problems), these equations will be incorrect. Think about which side of the triangle containing the components is adjacent to the angle and which side is opposite and then assign the cosine and sine accordingly.

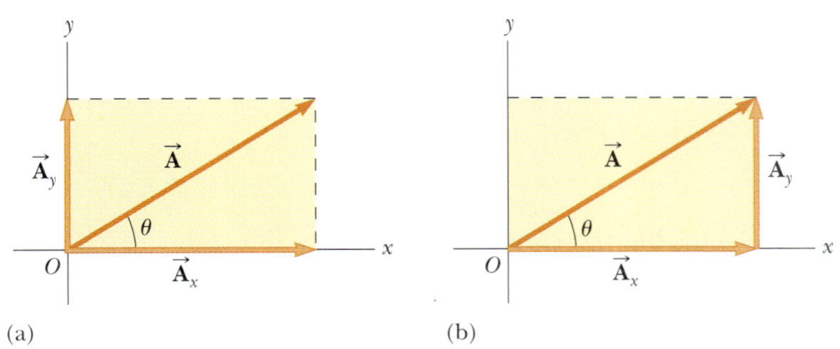

Figure 3.12 (a) A vector $\vec{\mathbf{A}}$ lying in the xy plane can be represented by its component vectors $\vec{\mathbf{A}}_x$ and $\vec{\mathbf{A}}_y$. (b) The y component vector $\vec{\mathbf{A}}_y$ can be moved to the right so that it adds to $\vec{\mathbf{A}}_x$. The vector sum of the component vectors is $\vec{\mathbf{A}}$. These three vectors form a right triangle.

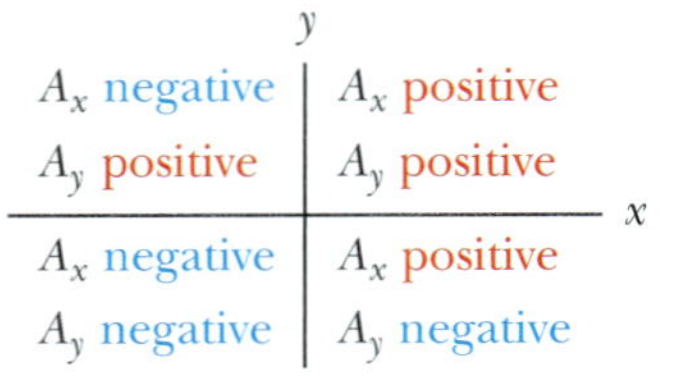

Figure 3.13 The signs of the components of a vector $\vec{\mathbf{A}}$ depend on the quadrant in which the vector is located.

The magnitudes of these components are the lengths of the two sides of a right triangle with a hypotenuse of length A. Therefore, the magnitude and direction of $\vec{\mathbf{A}}$ are related to its components through the expressions

$$A = \sqrt{A_x^2 + A_y^2} \quad \textbf{(3.10)}$$

$$\theta = \tan^{-1}\left(\frac{A_y}{A_x}\right) \quad \textbf{(3.11)}$$

Notice that **the signs of the components A_x and A_y depend on the angle θ.** For example, if $\theta = 120°$, A_x is negative and A_y is positive. If $\theta = 225°$, both A_x and A_y are negative. Figure 3.13 summarizes the signs of the components when $\vec{\mathbf{A}}$ lies in the various quadrants.

When solving problems, you can specify a vector $\vec{\mathbf{A}}$ either with its components A_x and A_y or with its magnitude and direction A and θ.

Suppose you are working a physics problem that requires resolving a vector into its components. In many applications, it is convenient to express the components in a coordinate system having axes that are not horizontal and vertical but that are still perpendicular to each other. For example, we will consider the motion of objects sliding down inclined planes. For these examples, it is often convenient to orient the x axis parallel to the plane and the y axis perpendicular to the plane.

Quick Quiz 3.4 Choose the correct response to make the sentence true: A component of a vector is (a) always, (b) never, or (c) sometimes larger than the magnitude of the vector.

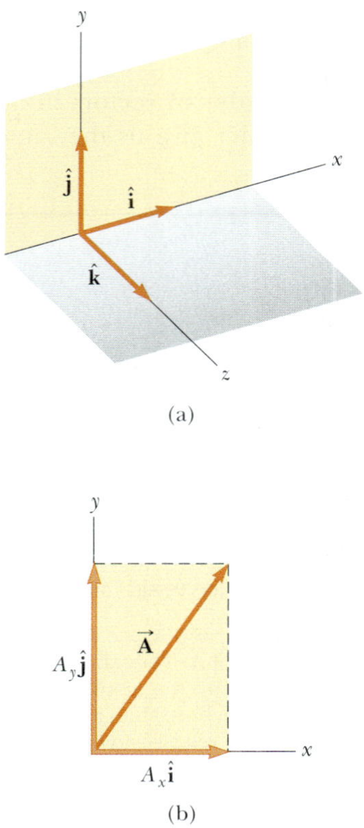

ACTIVE FIGURE 3.14

(a) The unit vectors $\hat{\mathbf{i}}$, $\hat{\mathbf{j}}$, and $\hat{\mathbf{k}}$ are directed along the x, y, and z axes, respectively. (b) Vector $\vec{\mathbf{A}} = A_x\hat{\mathbf{i}} + A_y\hat{\mathbf{j}}$ lying in the xy plane has components A_x and A_y.

Sign in at www.thomsonedu.com and go to ThomsonNOW to rotate the coordinate axes in three-dimensional space and view a representation of vector $\vec{\mathbf{A}}$ in three dimensions.

Unit Vectors

Vector quantities often are expressed in terms of unit vectors. **A unit vector is a dimensionless vector having a magnitude of exactly 1.** Unit vectors are used to specify a given direction and have no other physical significance. They are used solely as a bookkeeping convenience in describing a direction in space. We shall use the symbols $\hat{\mathbf{i}}$, $\hat{\mathbf{j}}$, and $\hat{\mathbf{k}}$ to represent unit vectors pointing in the positive x, y, and z directions, respectively. (The "hats," or circumflexes, on the symbols are a standard notation for unit vectors.) The unit vectors $\hat{\mathbf{i}}$, $\hat{\mathbf{j}}$, and $\hat{\mathbf{k}}$ form a set of mutually perpendicular vectors in a right-handed coordinate system as shown in Active Figure 3.14a. The magnitude of each unit vector equals 1; that is, $|\hat{\mathbf{i}}| = |\hat{\mathbf{j}}| = |\hat{\mathbf{k}}| = 1$.

Consider a vector $\vec{\mathbf{A}}$ lying in the xy plane as shown in Active Figure 3.14b. The product of the component A_x and the unit vector $\hat{\mathbf{i}}$ is the component vector $\vec{\mathbf{A}}_x = A_x\hat{\mathbf{i}}$, which lies on the x axis and has magnitude $|A_x|$. Likewise, $\vec{\mathbf{A}}_y = A_y\hat{\mathbf{j}}$ is the component vector of magnitude $|A_y|$ lying on the y axis. Therefore, the unit–vector notation for the vector $\vec{\mathbf{A}}$ is

$$\vec{\mathbf{A}} = A_x\hat{\mathbf{i}} + A_y\hat{\mathbf{j}} \quad \textbf{(3.12)}$$

For example, consider a point lying in the xy plane and having Cartesian coordinates (x, y) as in Figure 3.15. The point can be specified by the **position vector** $\vec{\mathbf{r}}$, which in unit–vector form is given by

$$\vec{\mathbf{r}} = x\hat{\mathbf{i}} + y\hat{\mathbf{j}} \quad \textbf{(3.13)}$$

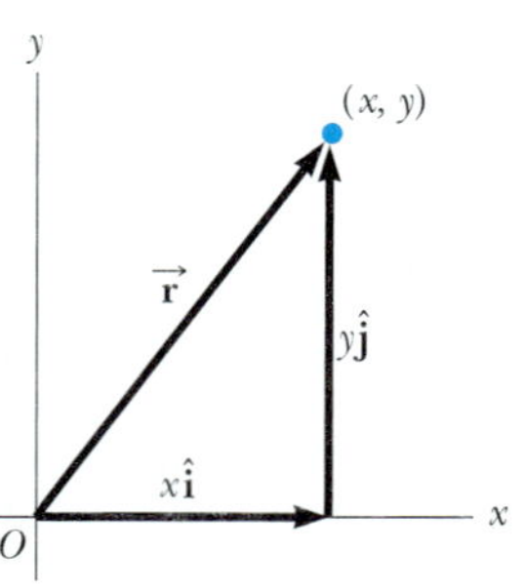

Figure 3.15 The point whose Cartesian coordinates are (x, y) can be represented by the position vector $\vec{\mathbf{r}} = x\hat{\mathbf{i}} + y\hat{\mathbf{j}}$.

This notation tells us that the components of $\vec{\mathbf{r}}$ are the coordinates x and y.

Now let us see how to use components to add vectors when the graphical method is not sufficiently accurate. Suppose we wish to add vector $\vec{\mathbf{B}}$ to vector $\vec{\mathbf{A}}$ in Equation 3.12, where vector $\vec{\mathbf{B}}$ has components B_x and B_y. Because of the bookkeeping convenience of the unit vectors, all we do is add the x and y components separately. The resultant vector $\vec{\mathbf{R}} = \vec{\mathbf{A}} + \vec{\mathbf{B}}$ is

$$\vec{\mathbf{R}} = (A_x\hat{\mathbf{i}} + A_y\hat{\mathbf{j}}) + (B_x\hat{\mathbf{i}} + B_y\hat{\mathbf{j}})$$

or

$$\vec{\mathbf{R}} = (A_x + B_x)\hat{\mathbf{i}} + (A_y + B_y)\hat{\mathbf{j}} \quad (3.14)$$

Because $\vec{\mathbf{R}} = R_x\hat{\mathbf{i}} + R_y\hat{\mathbf{j}}$, we see that the components of the resultant vector are

$$R_x = A_x + B_x$$
$$R_y = A_y + B_y \quad (3.15)$$

The magnitude of $\vec{\mathbf{R}}$ and the angle it makes with the x axis from its components are obtained using the relationships

$$R = \sqrt{R_x^2 + R_y^2} = \sqrt{(A_x + B_x)^2 + (A_y + B_y)^2} \quad (3.16)$$

$$\tan\theta = \frac{R_y}{R_x} = \frac{A_y + B_y}{A_x + B_x} \quad (3.17)$$

We can check this addition by components with a geometric construction as shown in Figure 3.16. Remember to note the signs of the components when using either the algebraic or the graphical method.

At times, we need to consider situations involving motion in three component directions. The extension of our methods to three-dimensional vectors is straightforward. If $\vec{\mathbf{A}}$ and $\vec{\mathbf{B}}$ both have x, y, and z components, they can be expressed in the form

$$\vec{\mathbf{A}} = A_x\hat{\mathbf{i}} + A_y\hat{\mathbf{j}} + A_z\hat{\mathbf{k}} \quad (3.18)$$

$$\vec{\mathbf{B}} = B_x\hat{\mathbf{i}} + B_y\hat{\mathbf{j}} + B_z\hat{\mathbf{k}} \quad (3.19)$$

The sum of $\vec{\mathbf{A}}$ and $\vec{\mathbf{B}}$ is

$$\vec{\mathbf{R}} = (A_x + B_x)\hat{\mathbf{i}} + (A_y + B_y)\hat{\mathbf{j}} + (A_z + B_z)\hat{\mathbf{k}} \quad (3.20)$$

Notice that Equation 3.20 differs from Equation 3.14: in Equation 3.20, the resultant vector also has a z component $R_z = A_z + B_z$. If a vector $\vec{\mathbf{R}}$ has x, y, and z components, the magnitude of the vector is $R = \sqrt{R_x^2 + R_y^2 + R_z^2}$. The angle θ_x that $\vec{\mathbf{R}}$ makes with the x axis is found from the expression $\cos\theta_x = R_x/R$, with similar expressions for the angles with respect to the y and z axes.

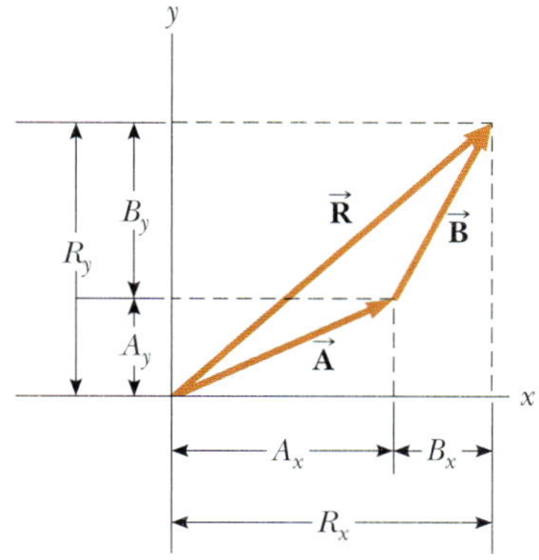

Figure 3.16 This geometric construction for the sum of two vectors shows the relationship between the components of the resultant $\vec{\mathbf{R}}$ and the components of the individual vectors.

PITFALL PREVENTION 3.4

Tangents on Calculators

Equation 3.17 involves the calculation of an angle by means of a tangent function. Generally, the inverse tangent function on calculators provides an angle between $-90°$ and $+90°$. As a consequence, if the vector you are studying lies in the second or third quadrant, the angle measured from the positive x axis will be the angle your calculator returns plus $180°$.

Quick Quiz 3.5 For which of the following vectors is the magnitude of the vector equal to one of the components of the vector? (a) $\vec{\mathbf{A}} = 2\hat{\mathbf{i}} + 5\hat{\mathbf{j}}$ (b) $\vec{\mathbf{B}} = -3\hat{\mathbf{j}}$ (c) $\vec{\mathbf{C}} = +5\hat{\mathbf{k}}$

EXAMPLE 3.3 The Sum of Two Vectors

Find the sum of two vectors $\vec{\mathbf{A}}$ and $\vec{\mathbf{B}}$ lying in the xy plane and given by

$$\vec{\mathbf{A}} = (2.0\hat{\mathbf{i}} + 2.0\hat{\mathbf{j}})\text{ m} \qquad \text{and} \qquad \vec{\mathbf{B}} = (2.0\hat{\mathbf{i}} - 4.0\hat{\mathbf{j}})\text{ m}$$

SOLUTION

Conceptualize You can conceptualize the situation by drawing the vectors on graph paper.

Categorize We categorize this example as a simple substitution problem. Comparing this expression for $\vec{\mathbf{A}}$ with the general expression $\vec{\mathbf{A}} = A_x\hat{\mathbf{i}} + A_y\hat{\mathbf{j}} + A_z\hat{\mathbf{k}}$, we see that $A_x = 2.0$ m and $A_y = 2.0$ m. Likewise, $B_x = 2.0$ m and $B_y = -4.0$ m.

Use Equation 3.14 to obtain the resultant vector $\vec{\mathbf{R}}$:

$$\vec{\mathbf{R}} = \vec{\mathbf{A}} + \vec{\mathbf{B}} = (2.0 + 2.0)\hat{\mathbf{i}}\text{ m} + (2.0 - 4.0)\hat{\mathbf{j}}\text{ m}$$

Evaluate the components of $\vec{\mathbf{R}}$:

$$R_x = 4.0\text{ m} \qquad R_y = -2.0\text{ m}$$

Use Equation 3.16 to find the magnitude of $\vec{\mathbf{R}}$: $R = \sqrt{R_x^2 + R_y^2} = \sqrt{(4.0 \text{ m})^2 + (-2.0 \text{ m})^2} = \sqrt{20} \text{ m} = \boxed{4.5 \text{ m}}$

Find the direction of $\vec{\mathbf{R}}$ from Equation 3.17: $\tan\theta = \frac{R_y}{R_x} = \frac{-2.0 \text{ m}}{4.0 \text{ m}} = -0.50$

Your calculator likely gives the answer $-27°$ for $\theta = \tan^{-1}(-0.50)$. This answer is correct if we interpret it to mean 27° clockwise from the x axis. Our standard form has been to quote the angles measured counterclockwise from the $+x$ axis, and that angle for this vector is $\theta = \boxed{333°}$

EXAMPLE 3.4 The Resultant Displacement

A particle undergoes three consecutive displacements: $\Delta\vec{\mathbf{r}}_1 = (15\hat{\mathbf{i}} + 30\hat{\mathbf{j}} + 12\hat{\mathbf{k}})$ cm, $\Delta\vec{\mathbf{r}}_2 = (23\hat{\mathbf{i}} - 14\hat{\mathbf{j}} - 5.0\hat{\mathbf{k}})$ cm, and $\Delta\vec{\mathbf{r}}_3 = (-13\hat{\mathbf{i}} + 15\hat{\mathbf{j}})$ cm. Find the components of the resultant displacement and its magnitude.

SOLUTION

Conceptualize Although x is sufficient to locate a point in one dimension, we need a vector $\vec{\mathbf{r}}$ to locate a point in two or three dimensions. The notation $\Delta\vec{\mathbf{r}}$ is a generalization of the one-dimensional displacement Δx in Equation 2.1. Three-dimensional displacements are more difficult to conceptualize than those in two dimensions because the latter can be drawn on paper.

For this problem, let us imagine that you start with your pencil at the origin of a piece of graph paper on which you have drawn x and y axes. Move your pencil 15 cm to the right along the x axis, then 30 cm upward along the y axis, and then 12 cm *perpendicularly toward you away* from the graph paper. This procedure provides the displacement described by $\Delta\vec{\mathbf{r}}_1$. From this point, move your pencil 23 cm to the right parallel to the x axis, then 14 cm parallel to the graph paper in the $-y$ direction, and then 5.0 cm perpendicularly away from you toward the graph paper. You are now at the displacement from the origin described by $\Delta\vec{\mathbf{r}}_1 + \Delta\vec{\mathbf{r}}_2$. From this point, move your pencil 13 cm to the left in the $-x$ direction, and (finally!) 15 cm parallel to the graph paper along the y axis. Your final position is at a displacement $\Delta\vec{\mathbf{r}}_1 + \Delta\vec{\mathbf{r}}_2 + \Delta\vec{\mathbf{r}}_3$ from the origin.

Categorize Despite the difficulty in conceptualizing in three dimensions, we can categorize this problem as a substitution problem because of the careful bookkeeping methods that we have developed for vectors. The mathematical manipulation keeps track of this motion along the three perpendicular axes in an organized, compact way, as we see below.

To find the resultant displacement, add the three vectors:

$$\begin{aligned}\Delta\vec{\mathbf{r}} &= \Delta\vec{\mathbf{r}}_1 + \Delta\vec{\mathbf{r}}_2 + \Delta\vec{\mathbf{r}}_3 \\ &= (15 + 23 - 13)\hat{\mathbf{i}} \text{ cm} + (30 - 14 + 15)\hat{\mathbf{j}} \text{ cm} + (12 - 5.0 + 0)\hat{\mathbf{k}} \text{ cm} \\ &= (25\hat{\mathbf{i}} + 31\hat{\mathbf{j}} + 7.0\hat{\mathbf{k}}) \text{ cm}\end{aligned}$$

Find the magnitude of the resultant vector:

$$\begin{aligned}R &= \sqrt{R_x^2 + R_y^2 + R_z^2} \\ &= \sqrt{(25 \text{ cm})^2 + (31 \text{ cm})^2 + (7.0 \text{ cm})^2} = \boxed{40 \text{ cm}}\end{aligned}$$

EXAMPLE 3.5 Taking a Hike

A hiker begins a trip by first walking 25.0 km southeast from her car. She stops and sets up her tent for the night. On the second day, she walks 40.0 km in a direction 60.0° north of east, at which point she discovers a forest ranger's tower.

(A) Determine the components of the hiker's displacement for each day.

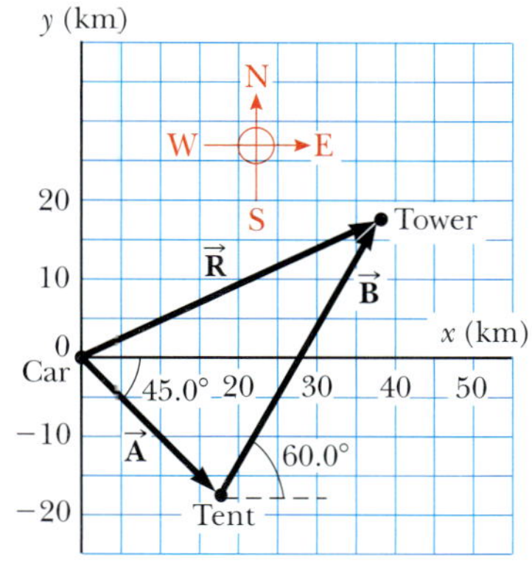

Figure 3.17 (Example 3.5) The total displacement of the hiker is the vector $\vec{R} = \vec{A} + \vec{B}$.

SOLUTION

Conceptualize We conceptualize the problem by drawing a sketch as in Figure 3.17. If we denote the displacement vectors on the first and second days by $\vec{A}$ and $\vec{B}$, respectively, and use the car as the origin of coordinates, we obtain the vectors shown in Figure 3.17.

Categorize Drawing the resultant $\vec{R}$, we can now categorize this problem as one we've solved before: an addition of two vectors. You should now have a hint of the power of categorization in that many new problems are very similar to problems we have already solved if we are careful to conceptualize them. Once we have drawn the displacement vectors and categorized the problem, this problem is no longer about a hiker, a walk, a car, a tent, or a tower. It is a problem about vector addition, one that we have already solved.

Analyze Displacement $\vec{A}$ has a magnitude of 25.0 km and is directed 45.0° below the positive x axis.

Find the components of $\vec{A}$ using Equations 3.8 and 3.9:

$$A_x = A\cos(-45.0°) = (25.0 \text{ km})(0.707) = 17.7 \text{ km}$$

$$A_y = A\sin(-45.0°) = (25.0 \text{ km})(-0.707) = -17.7 \text{ km}$$

The negative value of A_y indicates the hiker walks in the negative y direction on the first day. The signs of A_x and A_y also are evident from Figure 3.17.

Find the components of $\vec{B}$ using Equations 3.8 and 3.9:

$$B_x = B\cos 60.0° = (40.0 \text{ km})(0.500) = 20.0 \text{ km}$$

$$B_y = B\sin 60.0° = (40.0 \text{ km})(0.866) = 34.6 \text{ km}$$

(B) Determine the components of the hiker's resultant displacement $\vec{R}$ for the trip. Find an expression for $\vec{R}$ in terms of unit vectors.

SOLUTION

Use Equation 3.15 to find the components of the resultant displacement $\vec{R} = \vec{A} + \vec{B}$:

$$R_x = A_x + B_x = 17.7 \text{ km} + 20.0 \text{ km} = 37.7 \text{ km}$$

$$R_y = A_y + B_y = -17.7 \text{ km} + 34.6 \text{ km} = 16.9 \text{ km}$$

Write the total displacement in unit–vector form:

$$\vec{R} = (37.7\hat{\mathbf{i}} + 16.9\hat{\mathbf{j}}) \text{ km}$$

Finalize Looking at the graphical representation in Figure 3.17, we estimate the position of the tower to be about (38 km, 17 km), which is consistent with the components of $\vec{R}$ in our result for the final position of the hiker. Also, both components of $\vec{R}$ are positive, putting the final position in the first quadrant of the coordinate system, which is also consistent with Figure 3.17.

What If? After reaching the tower, the hiker wishes to return to her car along a single straight line. What are the components of the vector representing this hike? What should the direction of the hike be?

Answer The desired vector $\vec{R}_{car}$ is the negative of vector $\vec{R}$:

$$\vec{R}_{car} = -\vec{R} = (-37.7\hat{\mathbf{i}} - 16.9\hat{\mathbf{j}}) \text{ km}$$

The heading is found by calculating the angle that the vector makes with the x axis:

$$\tan\theta = \frac{R_{car,y}}{R_{car,x}} = \frac{-16.9 \text{ km}}{-37.7 \text{ km}} = 0.448$$

which gives an angle of $\theta = 204.1°$, or 24.1° south of west.

Summary

ThomsonNOW™ Sign in at **www.thomsonedu.com** and go to ThomsonNOW to take a practice test for this chapter.

DEFINITIONS

Scalar quantities are those that have only a numerical value and no associated direction. **Vector quantities** have both magnitude and direction and obey the laws of vector addition. The magnitude of a vector is *always* a positive number.

CONCEPTS AND PRINCIPLES

When two or more vectors are added together, they must all have the same units and all of them must be the same type of quantity. We can add two vectors $\vec{\mathbf{A}}$ and $\vec{\mathbf{B}}$ graphically. In this method (Active Fig. 3.6), the resultant vector $\vec{\mathbf{R}} = \vec{\mathbf{A}} + \vec{\mathbf{B}}$ runs from the tail of $\vec{\mathbf{A}}$ to the tip of $\vec{\mathbf{B}}$.

A second method of adding vectors involves **components** of the vectors. The x component A_x of the vector $\vec{\mathbf{A}}$ is equal to the projection of $\vec{\mathbf{A}}$ along the x axis of a coordinate system, where $A_x = A\cos\theta$. The y component A_y of $\vec{\mathbf{A}}$ is the projection of $\vec{\mathbf{A}}$ along the y axis, where $A_y = A\sin\theta$.

If a vector $\vec{\mathbf{A}}$ has an x component A_x and a y component A_y, the vector can be expressed in unit–vector form as $\vec{\mathbf{A}} = A_x\hat{\mathbf{i}} + A_y\hat{\mathbf{j}}$. In this notation, $\hat{\mathbf{i}}$ is a unit vector pointing in the positive x direction and $\hat{\mathbf{j}}$ is a unit vector pointing in the positive y direction. Because $\hat{\mathbf{i}}$ and $\hat{\mathbf{j}}$ are unit vectors, $|\hat{\mathbf{i}}| = |\hat{\mathbf{j}}| = 1$.

We can find the resultant of two or more vectors by resolving all vectors into their x and y components, adding their resultant x and y components, and then using the Pythagorean theorem to find the magnitude of the resultant vector. We can find the angle that the resultant vector makes with respect to the x axis by using a suitable trigonometric function.

Questions

☐ denotes answer available in *Student Solutions Manual/Study Guide;* **O** denotes objective question

1. **O** Yes or no: Is each of the following quantities a vector? (a) force (b) temperature (c) the volume of water in a can (d) the ratings of a TV show (e) the height of a building (f) the velocity of a sports car (g) the age of the Universe

2. A book is moved once around the perimeter of a tabletop with dimensions 1.0 m × 2.0 m. If the book ends up at its initial position, what is its displacement? What is the distance traveled?

3. **O** Figure Q3.3 shows two vectors, $\vec{\mathbf{D}}_1$ and $\vec{\mathbf{D}}_2$. Which of the possibilities (a) through (d) is the vector $\vec{\mathbf{D}}_2 - 2\vec{\mathbf{D}}_1$, or (e) is it none of them?

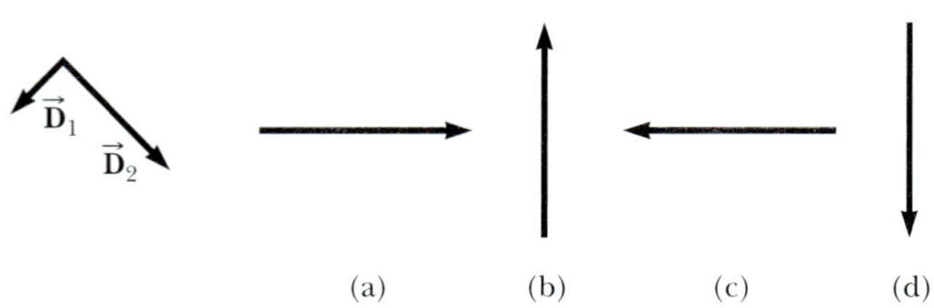

Figure Q3.3

4. **O** The cutting tool on a lathe is given two displacements, one of magnitude 4 cm and one of magnitude 3 cm, in each one of five situations (a) through (e) diagrammed in Figure Q3.4. Rank these situations according to the magnitude of the total displacement of the tool, putting the situation with the greatest resultant magnitude first. If the total displacement is the same size in two situations, give those letters equal ranks.

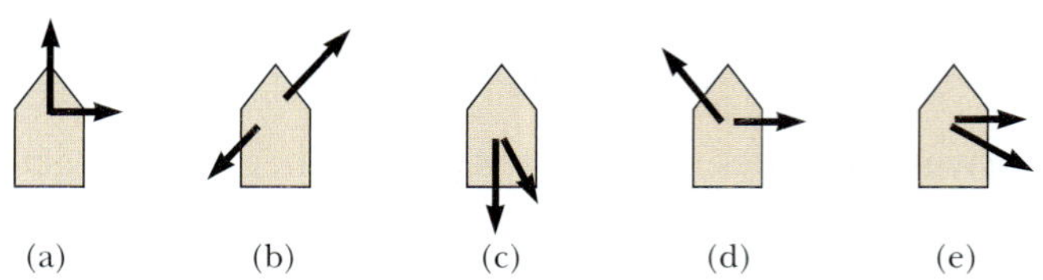

Figure Q3.4

5. **O** Let $\vec{\mathbf{A}}$ represent a velocity vector pointing from the origin into the second quadrant. (a) Is its x component positive, negative, or zero? (b) Is its y component positive, negative, or zero? Let $\vec{\mathbf{B}}$ represent a velocity vector pointing from the origin into the fourth quadrant. (c) Is its x component positive, negative, or zero? (d) Is its y component positive, negative, or zero? (e) Consider the vector $\vec{\mathbf{A}} + \vec{\mathbf{B}}$. What, if anything, can you conclude about quadrants it must be in or cannot be in? (f) Now consider the vector $\vec{\mathbf{B}} - \vec{\mathbf{A}}$. What, if anything, can you conclude about quadrants it must be in or cannot be in?

6. **O** **(i)** What is the magnitude of the vector $(10\hat{\mathbf{i}} - 10\hat{\mathbf{k}})$ m/s? (a) 0 (b) 10 m/s (c) −10 m/s (d) 10 (e) −10 (f) 14.1 m/s (g) undefined **(ii)** What is the y component of this vector? (Choose from among the same answers.)

7. **O** A submarine dives from the water surface at an angle of 30° below the horizontal, following a straight path 50 m long. How far is the submarine then below the water surface? (a) 50 m (b) sin 30° (c) cos 30° (d) tan 30° (e) (50 m)/sin 30° (f) (50 m)/cos 30° (g) (50 m)/tan 30° (h) (50 m)sin 30° (i) (50 m)cos 30° (j) (50 m)tan 30° (k) (sin 30°)/50 m (l) (cos 30°)/50 m (m) (tan 30°)/50 m (n) 30 m (o) 0 (p) none of these answers

8. **O** **(i)** What is the x component of the vector shown in Figure Q3.8? (a) 1 cm (b) 2 cm (c) 3 cm (d) 4 cm (e) 6 cm (f) −1 cm (g) −2 cm (h) −3 cm (i) −4 cm (j) −6 cm (k) none of these answers **(ii)** What is the y component of this vector? (Choose from among the same answers.)

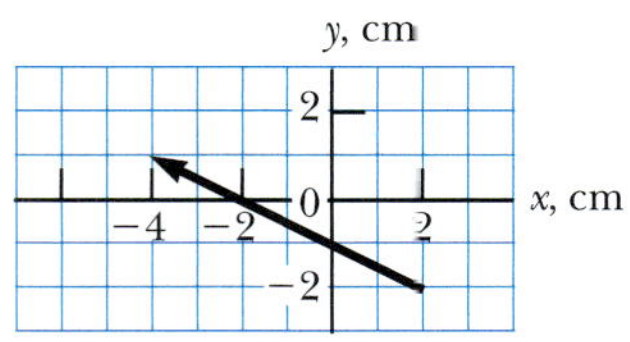

Figure Q3.8

9. **O** Vector $\vec{\mathbf{A}}$ lies in the xy plane. **(i)** Both of its components will be negative if it lies in which quadrant(s)? Choose all that apply. (a) the first quadrant (b) the second quadrant (c) the third quadrant (d) the fourth quadrant **(ii)** For what orientation(s) will its components have opposite signs? Choose from among the same possibilities.

10. If the component of vector $\vec{\mathbf{A}}$ along the direction of vector $\vec{\mathbf{B}}$ is zero, what can you conclude about the two vectors?

11. Can the magnitude of a vector have a negative value? Explain.

12. Is it possible to add a vector quantity to a scalar quantity? Explain.

Problems

WebAssign The Problems from this chapter may be assigned online in WebAssign.

ThomsonNOW™ Sign in at **www.thomsonedu.com** and go to ThomsonNOW to assess your understanding of this chapter's topics with additional quizzing and conceptual questions.

1, 2, 3 denotes straightforward, intermediate, challenging; □ denotes full solution available in *Student Solutions Manual/Study Guide*; ▲ denotes coached solution with hints available at **www.thomsonedu.com;** ▭ denotes developing symbolic reasoning; ● denotes asking for qualitative reasoning; ▪ denotes computer useful in solving problem

Section 3.1 Coordinate Systems

1. ▲ The polar coordinates of a point are $r = 5.50$ m and $\theta = 240°$. What are the Cartesian coordinates of this point?

2. Two points in a plane have polar coordinates (2.50 m, 30.0°) and (3.80 m, 120.0°). Determine (a) the Cartesian coordinates of these points and (b) the distance between them.

3. A fly lands on one wall of a room. The lower left-hand corner of the wall is selected as the origin of a two-dimensional Cartesian coordinate system. If the fly is located at the point having coordinates (2.00, 1.00) m, (a) how far is it from the corner of the room? (b) What is its location in polar coordinates?

4. The rectangular coordinates of a point are given by $(2, y)$, and its polar coordinates are $(r, 30°)$. Determine y and r.

5. Let the polar coordinates of the point (x, y) be (r, θ). Determine the polar coordinates for the points (a) $(-x, y)$, (b) $(-2x, -2y)$, and (c) $(3x, -3y)$.

Section 3.2 Vector and Scalar Quantities

Section 3.3 Some Properties of Vectors

6. A plane flies from base camp to lake A, 280 km away in the direction 20.0° north of east. After dropping off supplies it flies to lake B, which is 190 km at 30.0° west of north from lake A. Graphically determine the distance and direction from lake B to the base camp.

7. A surveyor measures the distance across a straight river by the following method: starting directly across from a tree on the opposite bank, she walks 100 m along the riverbank to establish a baseline. Then she sights across to the tree. The angle from her baseline to the tree is 35.0°. How wide is the river?

8. A force $\vec{\mathbf{F}}_1$ of magnitude 6.00 units acts on an object at the origin in a direction 30.0° above the positive x axis. A second force $\vec{\mathbf{F}}_2$ of magnitude 5.00 units acts on the object in the direction of the positive y axis. Graphically find the magnitude and direction of the resultant force $\vec{\mathbf{F}}_1 + \vec{\mathbf{F}}_2$.

9. ▲ A skater glides along a circular path of radius 5.00 m. If he coasts around one half of the circle, find (a) the magnitude of the displacement vector and (b) how far he skated. (c) What is the magnitude of the displacement if he skates all the way around the circle?

10. Arbitrarily define the "instantaneous vector height" of a person as the displacement vector from the point halfway between his or her feet to the top of the head. Make an order-of-magnitude estimate of the total vector height of all the people in a city of population 100 000 (a) at 10 o'clock on a Tuesday morning and (b) at 5 o'clock on a Saturday morning. Explain your reasoning.

11. ▲ Each of the displacement vectors $\vec{\mathbf{A}}$ and $\vec{\mathbf{B}}$ shown in Figure P3.11 has a magnitude of 3.00 m. Graphically find (a) $\vec{\mathbf{A}} + \vec{\mathbf{B}}$, (b) $\vec{\mathbf{A}} - \vec{\mathbf{B}}$, (c) $\vec{\mathbf{B}} - \vec{\mathbf{A}}$, and (d) $\vec{\mathbf{A}} - 2\vec{\mathbf{B}}$. Report all angles counterclockwise from the positive x axis.

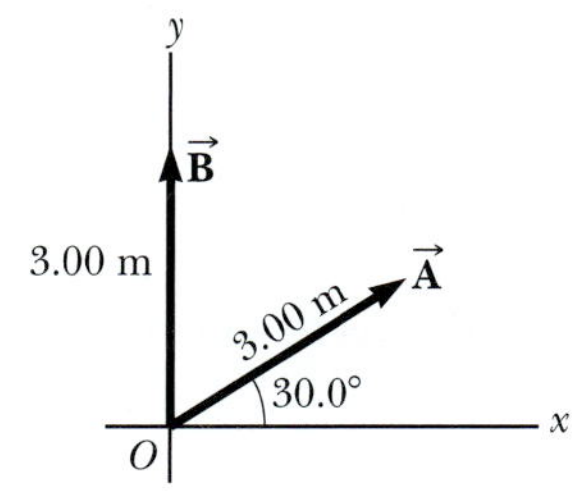

Figure P3.11 Problems 11 and 32.

12. ● Three displacements are $\vec{\mathbf{A}}$ = 200 m due south, $\vec{\mathbf{B}}$ = 250 m due west, and $\vec{\mathbf{C}}$ = 150 m at 30.0° east of north. Construct a separate diagram for each of the following possible ways of adding these vectors: $\vec{\mathbf{R}}_1 = \vec{\mathbf{A}} + \vec{\mathbf{B}} + \vec{\mathbf{C}}$; $\vec{\mathbf{R}}_2 = \vec{\mathbf{B}} + \vec{\mathbf{C}} + \vec{\mathbf{A}}$; $\vec{\mathbf{R}}_3 = \vec{\mathbf{C}} + \vec{\mathbf{B}} + \vec{\mathbf{A}}$. Explain what you can conclude from comparing the diagrams.

13. A roller-coaster car moves 200 ft horizontally and then rises 135 ft at an angle of 30.0° above the horizontal. It next travels 135 ft at an angle of 40.0° downward. What is its displacement from its starting point? Use graphical techniques.

14. ● A shopper pushing a cart through a store moves 40.0 m down one aisle, then makes a 90.0° turn and moves 15.0 m. He then makes another 90.0° turn and moves 20.0 m. (a) How far is the shopper away from his original position? (b) What angle does his total displacement make with his original direction? Notice that we have not specified whether the shopper turned right or left. Explain how many answers are possible for parts (a) and (b) and give the possible answers.

Section 3.4 Components of a Vector and Unit Vectors

15. ▲ A vector has an x component of −25.0 units and a y component of 40.0 units. Find the magnitude and direction of this vector.

16. A person walks 25.0° north of east for 3.10 km. How far would she have to walk due north and due east to arrive at the same location?

17. ● A minivan travels straight north in the right lane of a divided highway at 28.0 m/s. A camper passes the minivan and then changes from the left into the right lane. As it does so, the camper's path on the road is a straight displacement at 8.50° east of north. To avoid cutting off the minivan, the north–south distance between the camper's rear bumper and the minivan's front bumper should not decrease. Can the camper be driven to satisfy this requirement? Explain your answer.

18. A girl delivering newspapers covers her route by traveling 3.00 blocks west, 4.00 blocks north, and then 6.00 blocks east. (a) What is her resultant displacement? (b) What is the total distance she travels?

19. Obtain expressions in component form for the position vectors having the following polar coordinates: (a) 12.8 m, 150° (b) 3.30 cm, 60.0° (c) 22.0 in., 215°

20. A displacement vector lying in the xy plane has a magnitude of 50.0 m and is directed at an angle of 120° to the positive x axis. What are the rectangular components of this vector?

21. While exploring a cave, a spelunker starts at the entrance and moves the following distances. She goes 75.0 m north, 250 m east, 125 m at an angle 30.0° north of east, and 150 m south. Find her resultant displacement from the cave entrance.

22. A map suggests that Atlanta is 730 miles in a direction of 5.00° north of east from Dallas. The same map shows that Chicago is 560 miles in a direction of 21.0° west of north from Atlanta. Modeling the Earth as flat, use this information to find the displacement from Dallas to Chicago.

23. A man pushing a mop across a floor causes it to undergo two displacements. The first has a magnitude of 150 cm and makes an angle of 120° with the positive x axis. The resultant displacement has a magnitude of 140 cm and is directed at an angle of 35.0° to the positive x axis. Find the magnitude and direction of the second displacement.

24. Given the vectors $\vec{\mathbf{A}} = 2.00\hat{\mathbf{i}} + 6.00\hat{\mathbf{j}}$ and $\vec{\mathbf{B}} = 3.00\hat{\mathbf{i}} - 2.00\hat{\mathbf{j}}$, (a) draw the vector sum $\vec{\mathbf{C}} = \vec{\mathbf{A}} + \vec{\mathbf{B}}$ and the vector difference $\vec{\mathbf{D}} = \vec{\mathbf{A}} - \vec{\mathbf{B}}$. (b) Calculate $\vec{\mathbf{C}}$ and $\vec{\mathbf{D}}$, first in terms of unit vectors and then in terms of polar coordinates, with angles measured with respect to the $+x$ axis.

25. Consider the two vectors $\vec{\mathbf{A}} = 3\hat{\mathbf{i}} - 2\hat{\mathbf{j}}$ and $\vec{\mathbf{B}} = -\hat{\mathbf{i}} - 4\hat{\mathbf{j}}$. Calculate (a) $\vec{\mathbf{A}} + \vec{\mathbf{B}}$, (b) $\vec{\mathbf{A}} - \vec{\mathbf{B}}$, (c) $|\vec{\mathbf{A}} + \vec{\mathbf{B}}|$, (d) $|\vec{\mathbf{A}} - \vec{\mathbf{B}}|$, and (e) the directions of $\vec{\mathbf{A}} + \vec{\mathbf{B}}$ and $\vec{\mathbf{A}} - \vec{\mathbf{B}}$.

26. A snow-covered ski slope makes an angle of 35.0° with the horizontal. When a ski jumper plummets onto the hill, a parcel of splashed snow projects to a maximum position of 5.00 m at 20.0° from the vertical in the uphill direction as shown in Figure P3.26. Find the components of its maximum position (a) parallel to the surface and (b) perpendicular to the surface.

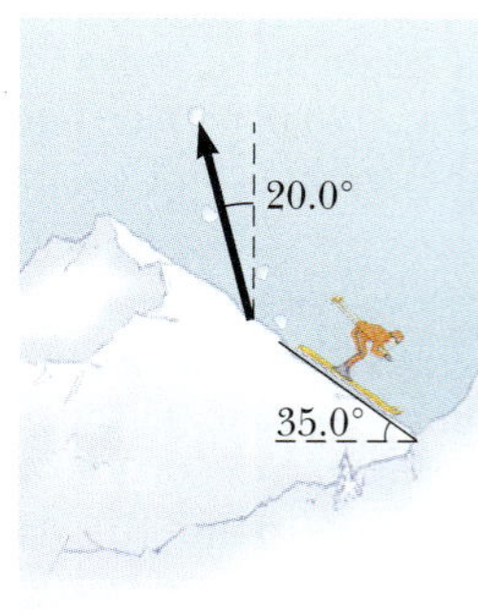

Figure P3.26

27. A particle undergoes the following consecutive displacements: 3.50 m south, 8.20 m northeast, and 15.0 m west. What is the resultant displacement?

28. In a game of American football, a quarterback takes the ball from the line of scrimmage, runs backward a distance of 10.0 yards, and then runs sideways parallel to the line of scrimmage for 15.0 yards. At this point, he throws a forward pass 50.0 yards straight downfield perpendicular to the line of scrimmage. What is the magnitude of the football's resultant displacement?

29. A novice golfer on the green takes three strokes to sink the ball. The successive displacements of the ball are 4.00 m to the north, 2.00 m northeast, and 1.00 m at 30.0° west of south. Starting at the same initial point, an expert golfer could make the hole in what single displacement?

30. Vector $\vec{\mathbf{A}}$ has x and y components of -8.70 cm and 15.0 cm, respectively; vector $\vec{\mathbf{B}}$ has x and y components of 13.2 cm and -6.60 cm, respectively. If $\vec{\mathbf{A}} - \vec{\mathbf{B}} + 3\vec{\mathbf{C}} = 0$, what are the components of $\vec{\mathbf{C}}$?

31. The helicopter view in Figure P3.31 shows two people pulling on a stubborn mule. Find (a) the single force that is equivalent to the two forces shown and (b) the force that a third person would have to exert on the mule to make the resultant force equal to zero. The forces are measured in units of newtons (symbolized N).

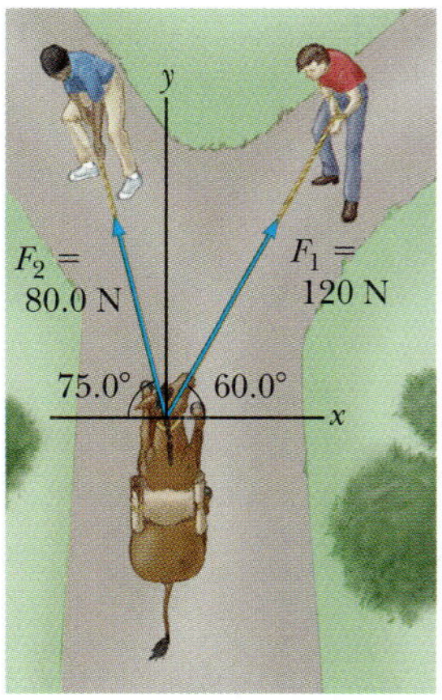

Figure P3.31

32. Use the component method to add the vectors $\vec{\mathbf{A}}$ and $\vec{\mathbf{B}}$ shown in Figure P3.11. Express the resultant $\vec{\mathbf{A}} + \vec{\mathbf{B}}$ in unit–vector notation.

33. Vector $\vec{\mathbf{B}}$ has x, y, and z components of 4.00, 6.00, and 3.00 units, respectively. Calculate the magnitude of $\vec{\mathbf{B}}$ and the angles $\vec{\mathbf{B}}$ makes with the coordinate axes.

34. Consider the three displacement vectors $\vec{\mathbf{A}} = (3\hat{\mathbf{i}} - 3\hat{\mathbf{j}})$ m, $\vec{\mathbf{B}} = (\hat{\mathbf{i}} - 4\hat{\mathbf{j}})$ m, and $\vec{\mathbf{C}} = (-2\hat{\mathbf{i}} + 5\hat{\mathbf{j}})$ m. Use the component method to determine (a) the magnitude and direction of the vector $\vec{\mathbf{D}} = \vec{\mathbf{A}} + \vec{\mathbf{B}} + \vec{\mathbf{C}}$ and (b) the magnitude and direction of $\vec{\mathbf{E}} = -\vec{\mathbf{A}} - \vec{\mathbf{B}} + \vec{\mathbf{C}}$.

35. Given the displacement vectors $\vec{\mathbf{A}} = (3\hat{\mathbf{i}} - 4\hat{\mathbf{j}} + 4\hat{\mathbf{k}})$ m and $\vec{\mathbf{B}} = (2\hat{\mathbf{i}} + 3\hat{\mathbf{j}} - 7\hat{\mathbf{k}})$ m, find the magnitudes of the vectors (a) $\vec{\mathbf{C}} = \vec{\mathbf{A}} + \vec{\mathbf{B}}$ and (b) $\vec{\mathbf{D}} = 2\vec{\mathbf{A}} - \vec{\mathbf{B}}$, also expressing each in terms of its rectangular components.

36. In an assembly operation illustrated in Figure P3.36, a robot moves an object first straight upward and then also to the east, around an arc forming one quarter of a circle of radius 4.80 cm that lies in an east–west vertical plane. The robot then moves the object upward and to the north, through one-quarter of a circle of radius 3.70 cm that lies in a north–south vertical plane. Find (a) the magnitude of the total displacement of the object and (b) the angle the total displacement makes with the vertical.

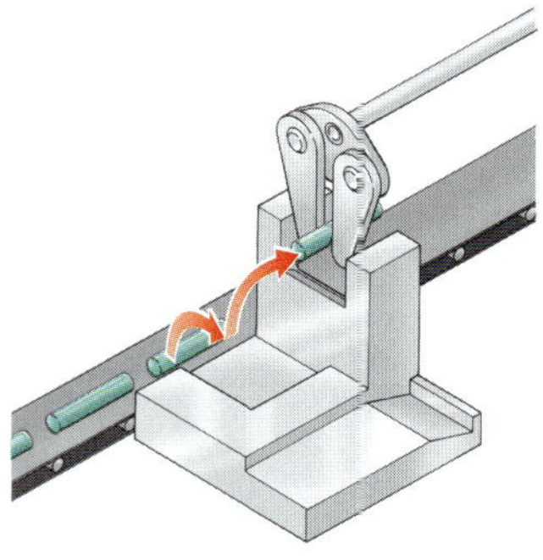

Figure P3.36

37. The vector $\vec{\mathbf{A}}$ has x, y, and z components of 8.00, 12.0, and -4.00 units, respectively. (a) Write a vector expression for $\vec{\mathbf{A}}$ in unit–vector notation. (b) Obtain a unit–vector expression for a vector $\vec{\mathbf{B}}$ one-fourth the length of $\vec{\mathbf{A}}$ pointing in the same direction as $\vec{\mathbf{A}}$. (c) Obtain a unit–vector expression for a vector $\vec{\mathbf{C}}$ three times the length of $\vec{\mathbf{A}}$ pointing in the direction opposite the direction of $\vec{\mathbf{A}}$.

38. You are standing on the ground at the origin of a coordinate system. An airplane flies over you with constant velocity parallel to the x axis and at a fixed height of 7.60×10^3 m. At time $t = 0$ the airplane is directly above you so that the vector leading from you to it is $\vec{\mathbf{P}}_0 = (7.60 \times 10^3 \text{ m})\,\hat{\mathbf{j}}$. At $t = 30.0$ s the position vector leading from you to the airplane is $\vec{\mathbf{P}}_{30} = (8.04 \times 10^3 \text{ m})\,\hat{\mathbf{i}} + (7.60 \times 10^3 \text{ m})\,\hat{\mathbf{j}}$. Determine the magnitude and orientation of the airplane's position vector at $t = 45.0$ s.

39. A radar station locates a sinking ship at range 17.3 km and bearing 136° clockwise from north. From the same station, a rescue plane is at horizontal range 19.6 km, 153° clockwise from north, with elevation 2.20 km. (a) Write the position vector for the ship relative to the plane, letting $\hat{\mathbf{i}}$ represent east, $\hat{\mathbf{j}}$ north, and $\hat{\mathbf{k}}$ up. (b) How far apart are the plane and ship?

40. (a) Vector $\vec{\mathbf{E}}$ has magnitude 17.0 cm and is directed 27.0° counterclockwise from the $+x$ axis. Express it in unit–vector notation. (b) Vector $\vec{\mathbf{F}}$ has magnitude 17.0 cm and is directed 27.0° counterclockwise from the $+y$ axis. Express it in unit–vector notation. (c) Vector $\vec{\mathbf{G}}$ has magnitude 17.0 cm and is directed 27.0° clockwise from the $-y$ axis. Express it in unit–vector notation.

41. Vector $\vec{\mathbf{A}}$ has a negative x component 3.00 units in length and a positive y component 2.00 units in length. (a) Determine an expression for $\vec{\mathbf{A}}$ in unit–vector notation. (b) Determine the magnitude and direction of $\vec{\mathbf{A}}$. (c) What vector $\vec{\mathbf{B}}$ when added to $\vec{\mathbf{A}}$ gives a resultant vector with no x component and a negative y component 4.00 units in length?

42. As it passes over Grand Bahama Island, the eye of a hurricane is moving in a direction 60.0° north of west with a speed of 41.0 km/h. Three hours later the course of the hurricane suddenly shifts due north, and its speed slows to 25.0 km/h. How far from Grand Bahama is the eye 4.50 h after it passes over the island?

43. ▲ Three displacement vectors of a croquet ball are shown in Figure P3.43, where $|\vec{\mathbf{A}}| = 20.0$ units, $|\vec{\mathbf{B}}| = 40.0$ units, and $|\vec{\mathbf{C}}| = 30.0$ units. Find (a) the resultant in unit–vector notation and (b) the magnitude and direction of the resultant displacement.

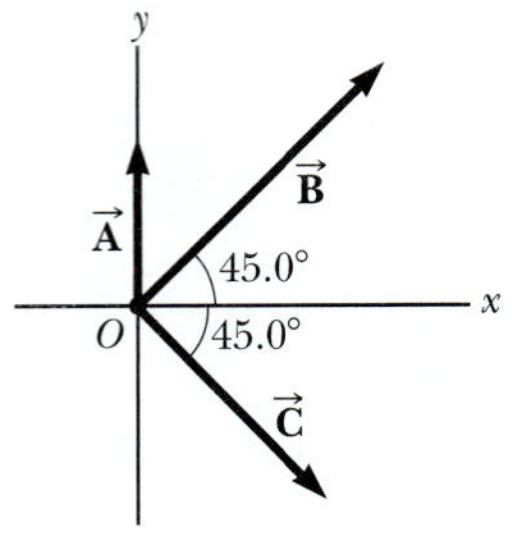

Figure P3.43

44. ● (a) Taking $\vec{A} = (6.00\hat{i} - 8.00\hat{j})$ units, $\vec{B} = (-8.00\hat{i} + 3.00\hat{j})$ units, and $\vec{C} = (26.0\hat{i} + 19.0\hat{j})$ units, determine a and b such that $a\vec{A} + b\vec{B} + \vec{C} = 0$. (b) A student has learned that a single equation cannot be solved to determine values for more than one unknown in it. How would you explain to him that both a and b can be determined from the single equation used in part (a)?

45. ● *Are we there yet?* In Figure P3.45, the line segment represents a path from the point with position vector $(5\hat{i} + 3\hat{j})$ m to the point with location $(16\hat{i} + 12\hat{j})$ m. Point A is along this path, a fraction f of the way to the destination. (a) Find the position vector of point A in terms of f. (b) Evaluate the expression from part (a) in the case $f = 0$. Explain whether the result is reasonable. (c) Evaluate the expression for $f = 1$. Explain whether the result is reasonable.

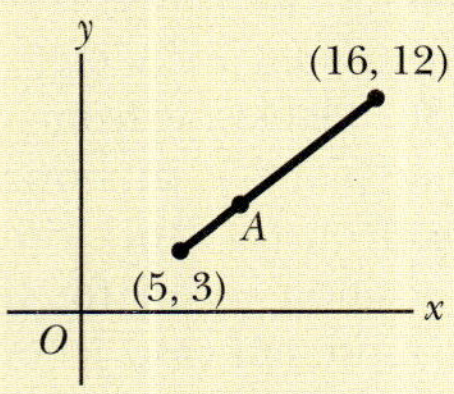

Figure P3.45 Point A is a fraction f of the distance from the initial point (5, 3) to the final point (16, 12).

Additional Problems

46. On December 1, 1955, Rosa Parks (1913–2005), an icon of the early civil rights movement, stayed seated in her bus seat when a white man demanded it. Police in Montgomery, Alabama, arrested her. On December 5, blacks began refusing to use all city buses. Under the leadership of the Montgomery Improvement Association, an efficient system of alternative transportation sprang up immediately, providing blacks with approximately 35 000 essential trips per day through volunteers, private taxis, carpooling, and ride sharing. The buses remained empty until they were integrated under court order on December 21, 1956. In picking up her riders, suppose a driver in downtown Montgomery traverses four successive displacements represented by the expression

$$(-6.30\text{b})\hat{i} - (4.00\text{b}\cos 40°)\hat{i} - (4.00\text{b}\sin 40°)\hat{j}$$
$$+ (3.00\text{b}\cos 50°)\hat{i} - (3.00\text{b}\sin 50°)\hat{j} - (5.00\text{b})\hat{j}$$

Here b represents one city block, a convenient unit of distance of uniform size; $\hat{i}$ is east; and $\hat{j}$ is north. (a) Draw a map of the successive displacements. (b) What total distance did she travel? (c) Compute the magnitude and direction of her total displacement. The logical structure of this problem and of several problems in later chapters was suggested by Alan Van Heuvelen and David Maloney, *American Journal of Physics* **67**(3) 252–256, March 1999.

47. Two vectors $\vec{A}$ and $\vec{B}$ have precisely equal magnitudes. For the magnitude of $\vec{A} + \vec{B}$ to be 100 times larger than the magnitude of $\vec{A} - \vec{B}$, what must be the angle between them?

48. Two vectors $\vec{A}$ and $\vec{B}$ have precisely equal magnitudes. For the magnitude of $\vec{A} + \vec{B}$ to be larger than the magnitude of $\vec{A} - \vec{B}$ by the factor n, what must be the angle between them?

49. An air-traffic controller observes two aircraft on his radar screen. The first is at altitude 800 m, horizontal distance 19.2 km, and 25.0° south of west. The second is at altitude 1 100 m, horizontal distance 17.6 km, and 20.0° south of west. What is the distance between the two aircraft? (Place the x axis west, the y axis south, and the z axis vertical.)

50. The biggest stuffed animal in the world is a snake 420 m long, constructed by Norwegian children. Suppose the snake is laid out in a park as shown in Figure P3.50, forming two straight sides of a 105° angle, with one side 240 m long. Olaf and Inge run a race they invent. Inge runs directly from the tail of the snake to its head, and Olaf starts from the same place at the same moment but runs along the snake. If both children run steadily at 12.0 km/h, Inge reaches the head of the snake how much earlier than Olaf?

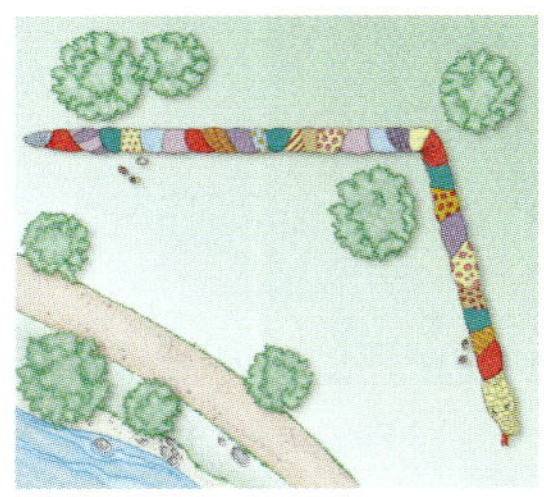

Figure P3.50

51. A ferryboat transports tourists among three islands. It sails from the first island to the second island, 4.76 km away, in a direction 37.0° north of east. It then sails from the second island to the third island in a direction 69.0° west of north. Finally, it returns to the first island, sailing in a direction 28.0° east of south. Calculate the distance between (a) the second and third islands and (b) the first and third islands.

52. A vector is given by $\vec{R} = 2\hat{i} + \hat{j} + 3\hat{k}$. Find (a) the magnitudes of the x, y, and z components, (b) the magnitude of $\vec{R}$, and (c) the angles between $\vec{R}$ and the x, y, and z axes.

53. A jet airliner, moving initially at 300 mi/h to the east, suddenly enters a region where the wind is blowing at 100 mi/h toward the direction 30.0° north of east. What are the new speed and direction of the aircraft relative to the ground?

54. ● Let $\vec{A}$ = 60.0 cm at 270° measured from the horizontal. Let $\vec{B}$ = 80.0 cm at some angle θ. (a) Find the magnitude of $\vec{A} + \vec{B}$ as a function of θ. (b) From the answer to part (a), for what value of θ does $|\vec{A} + \vec{B}|$ take on its maximum value? What is this maximum value? (c) From the answer to part (a), for what value of θ does $|\vec{\mathbf{A}} + \vec{\mathbf{B}}|$ take on its minimum value? What is this minimum value? (d) Without reference to the answer to part (a), argue that the answers to each of parts (b) and (c) do or do not make sense.

55. After a ball rolls off the edge of a horizontal table at time $t = 0$, its velocity as a function of time is given by

$$\vec{\mathbf{v}} = 1.2\hat{\mathbf{i}} \text{ m/s} - 9.8t\hat{\mathbf{j}} \text{ m/s}^2$$

The ball's displacement away from the edge of the table, during the time interval of 0.380 s during which it is in flight, is given by

$$\Delta\vec{\mathbf{r}} = \int_0^{0.380 \text{ s}} \vec{\mathbf{v}}\, dt$$

To perform the integral, you can use the calculus theorem

$$\int (A + Bf(x))\, dx = \int A\, dx + B \int f(x)\, dx$$

You can think of the units and unit vectors as constants, represented by A and B. Do the integration to calculate the displacement of the ball.

56. ● Find the sum of these four vector forces: 12.0 N to the right at 35.0° above the horizontal, 31.0 N to the left at 55.0° above the horizontal, 8.40 N to the left at 35.0° below the horizontal, and 24.0 N to the right at 55.0° below the horizontal. Follow these steps. Guided by a sketch of this situation, explain how you can simplify the calculations by making a particular choice for the directions of the x and y axes. What is your choice? Then add the vectors by the component method.

57. A person going for a walk follows the path shown in Figure P3.57. The total trip consists of four straight-line paths. At the end of the walk, what is the person's resultant displacement measured from the starting point?

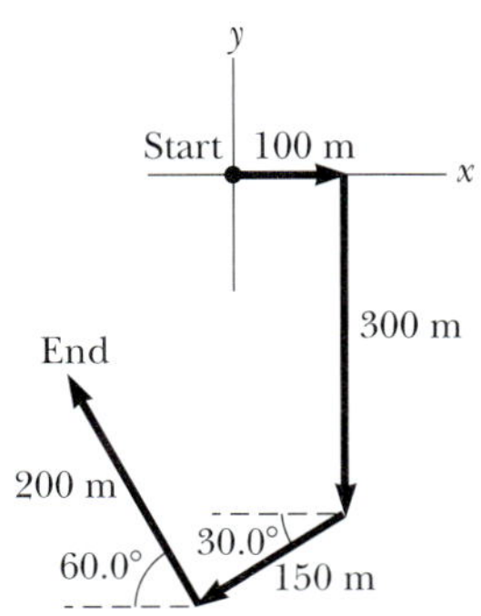

Figure P3.57

58. ● The instantaneous position of an object is specified by its position vector $\vec{\mathbf{r}}$ leading from a fixed origin to the location of the object, modeled as a particle. Suppose for a certain object the position vector is a function of time, given by $\vec{\mathbf{r}} = 4\hat{\mathbf{i}} + 3\hat{\mathbf{j}} - 2t\hat{\mathbf{k}}$, where r is in meters and t is in seconds. Evaluate $d\vec{\mathbf{r}}/dt$. What does it represent about the object?

59. ● Long John Silver, a pirate, has buried his treasure on an island with five trees, located at the points (30.0 m, −20.0 m), (60.0 m, 80.0 m), (−10.0 m, −10.0 m), (40.0 m, −30.0 m), and (−70.0 m, 60.0 m), all measured relative to some origin as shown in Figure P3.59. His ship's log instructs you to start at tree A and move toward tree B, but to cover only one-half the distance between A and B. Then move toward tree C, covering one-third the distance between your current location and C. Next move toward tree D, covering one-fourth the distance between where you are and D. Finally, move toward tree E, covering one-fifth the distance between you and E, stop, and dig. (a) Assume you have correctly determined the order in which the pirate labeled the trees as A, B, C, D, and E as shown in the figure. What are the coordinates of the point where his treasure is buried? (b) **What If?** What if you do not really know the way the pirate labeled the trees? What would happen to the answer if you rearranged the order of the trees, for instance to E(30 m, −20 m), A(60 m, 80 m), E(−10 m, −10 m), C(40 m, −30 m), and D(−70 m, 60 m)? State reasoning to show the answer does not depend on the order in which the trees are labeled.

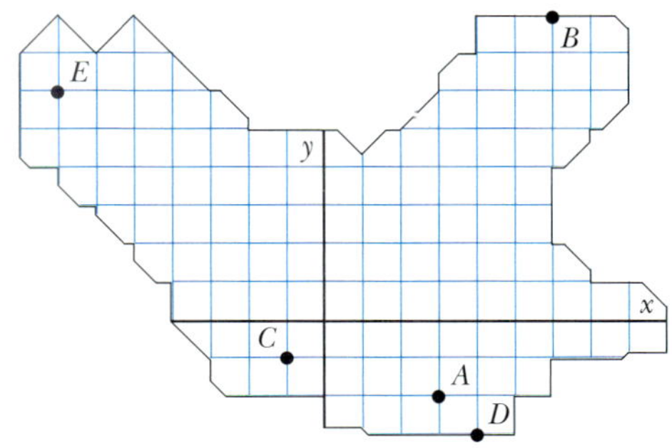

Figure P3.59

60. ● Consider a game in which N children position themselves at equal distances around the circumference of a circle. At the center of the circle is a rubber tire. Each child holds a rope attached to the tire and, at a signal, pulls on his or her rope. All children exert forces of the same magnitude F. In the case $N = 2$, it is easy to see that the net force on the tire will be zero because the two oppositely directed force vectors add to zero. Similarly, if $N = 4$, 6, or any even integer, the resultant force on the tire must be zero because the forces exerted by each pair of oppositely positioned children will cancel. When an odd number of children are around the circle, it is not as obvious whether the total force on the central tire will be zero. (a) Calculate the net force on the tire in the case $N = 3$ by adding the components of the three force vectors. Choose the x axis to lie along one of the ropes. (b) **What If?** State reasoning that will determine the net force for the general case where N is any integer, odd or even, greater than one. Proceed as follows. Assume the total force is not zero. Then it must point in some particular direction. Let every child move one position clockwise. Give a reason that the total force must then have a direction turned clockwise by $360°/N$. Argue that the total force must nevertheless be the same as before. Explain what the contradiction proves about the magnitude of the force. This problem illustrates a widely useful technique of proving a result "by symmetry," by using a bit of the mathematics of *group theory*. The particular situation is actually encountered in physics and chemistry

when an array of electric charges (ions) exerts electric forces on an atom at a central position in a molecule or in a crystal.

61. Vectors $\vec{\mathbf{A}}$ and $\vec{\mathbf{B}}$ have equal magnitudes of 5.00. The sum of $\vec{\mathbf{A}}$ and $\vec{\mathbf{B}}$ is the vector $6.00\hat{\mathbf{j}}$. Determine the angle between $\vec{\mathbf{A}}$ and $\vec{\mathbf{B}}$.

62. A rectangular parallelepiped has dimensions a, b, and c as shown in Figure P3.62. (a) Obtain a vector expression for the face diagonal vector $\vec{\mathbf{R}}_1$. What is the magnitude of this vector? (b) Obtain a vector expression for the body diagonal vector $\vec{\mathbf{R}}_2$. Notice that $\vec{\mathbf{R}}_1$, $c\hat{\mathbf{k}}$, and $\vec{\mathbf{R}}_2$ make a right triangle. Prove that the magnitude of $\vec{\mathbf{R}}_2$ is $\sqrt{a^2 + b^2 + c^2}$.

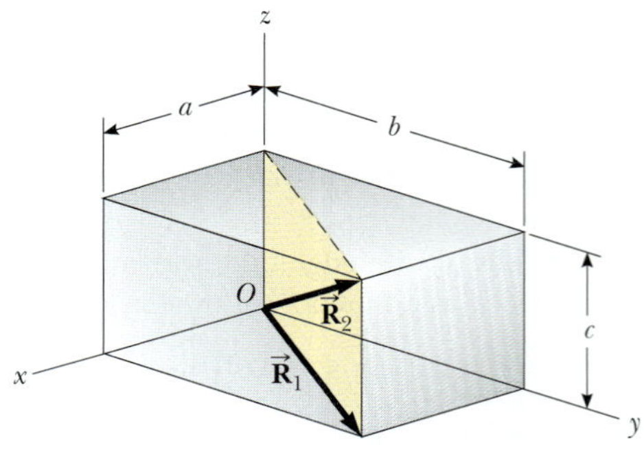

Figure P3.62

Answers to Quick Quizzes

3.1 Scalars: (a), (d), (e). None of these quantities has a direction. Vectors: (b), (c). For these quantities, the direction is necessary to specify the quantity completely.

3.2 (c). The resultant has its maximum magnitude $A + B = 12 + 8 = 20$ units when vector $\vec{\mathbf{A}}$ is oriented in the same direction as vector $\vec{\mathbf{B}}$. The resultant vector has its minimum magnitude $A - B = 12 - 8 = 4$ units when vector $\vec{\mathbf{A}}$ is oriented in the direction opposite vector $\vec{\mathbf{B}}$.

3.3 (b) and (c). To add to zero, the vectors must point in opposite directions and have the same magnitude.

3.4 (b). From the Pythagorean theorem, the magnitude of a vector is always larger than the absolute value of each component, unless there is only one nonzero component, in which case the magnitude of the vector is equal to the absolute value of that component.

3.5 (c). The magnitude of $\vec{\mathbf{C}}$ is 5 units, the same as the z component. Answer (b) is not correct because the magnitude of any vector is always a positive number, whereas the y component of $\vec{\mathbf{B}}$ is negative.

Lava spews from a volcanic eruption. Notice the parabolic paths of embers projected into the air. All projectiles follow a parabolic path in the absence of air resistance. (© Arndt/Premium Stock/PictureQuest)

4 Motion in Two Dimensions

In this chapter, we explore the kinematics of a particle moving in two dimensions. Knowing the basics of two-dimensional motion will allow us—in future chapters—to examine a variety of motions ranging from the motion of satellites in orbit to the motion of electrons in a uniform electric field. We begin by studying in greater detail the vector nature of position, velocity, and acceleration. We then treat projectile motion and uniform circular motion as special cases of motion in two dimensions. We also discuss the concept of relative motion, which shows why observers in different frames of reference may measure different positions and velocities for a given particle.

4.1 The Position, Velocity, and Acceleration Vectors

In Chapter 2, we found that the motion of a particle along a straight line is completely known if its position is known as a function of time. Let us now extend this idea to two-dimensional motion of a particle in the xy plane. We begin by describing the position of the particle by its **position vector** $\vec{\mathbf{r}}$, drawn from the origin of some coordinate system to the location of the particle in the xy plane, as in Figure 4.1 (page 72). At time t_i, the particle is at point Ⓐ, described by position vector $\vec{\mathbf{r}}_i$. At some later time t_f, it is at point Ⓑ, described by position vector $\vec{\mathbf{r}}_f$. The path from

Ⓐ to Ⓑ is not necessarily a straight line. As the particle moves from Ⓐ to Ⓑ in the time interval $\Delta t = t_f - t_i$, its position vector changes from $\vec{\mathbf{r}}_i$ to $\vec{\mathbf{r}}_f$. As we learned in Chapter 2, displacement is a vector, and the displacement of the particle is the difference between its final position and its initial position. We now define the **displacement vector** $\Delta\vec{\mathbf{r}}$ for a particle such as the one in Figure 4.1 as being the difference between its final position vector and its initial position vector:

Displacement vector ▶

$$\Delta\vec{\mathbf{r}} \equiv \vec{\mathbf{r}}_f - \vec{\mathbf{r}}_i \quad (4.1)$$

The direction of $\Delta\vec{\mathbf{r}}$ is indicated in Figure 4.1. As we see from the figure, the magnitude of $\Delta\vec{\mathbf{r}}$ is *less* than the distance traveled along the curved path followed by the particle.

As we saw in Chapter 2, it is often useful to quantify motion by looking at the ratio of a displacement divided by the time interval during which that displacement occurs, which gives the rate of change of position. Two-dimensional (or three-dimensional) kinematics is similar to one-dimensional kinematics, but we must now use full vector notation rather than positive and negative signs to indicate the direction of motion.

We define the **average velocity** $\vec{\mathbf{v}}_{\text{avg}}$ of a particle during the time interval Δt as the displacement of the particle divided by the time interval:

Average velocity ▶

$$\vec{\mathbf{v}}_{\text{avg}} \equiv \frac{\Delta\vec{\mathbf{r}}}{\Delta t} \quad (4.2)$$

Multiplying or dividing a vector quantity by a positive scalar quantity such as Δt changes only the magnitude of the vector, not its direction. Because displacement is a vector quantity and the time interval is a positive scalar quantity, we conclude that the average velocity is a vector quantity directed along $\Delta\vec{\mathbf{r}}$.

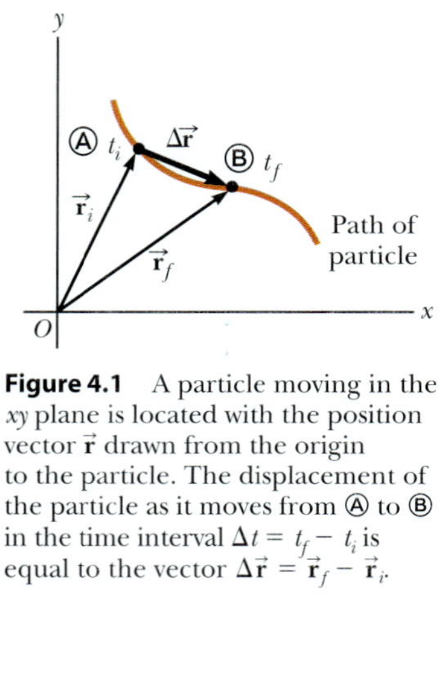

Figure 4.1 A particle moving in the *xy* plane is located with the position vector $\vec{\mathbf{r}}$ drawn from the origin to the particle. The displacement of the particle as it moves from Ⓐ to Ⓑ in the time interval $\Delta t = t_f - t_i$ is equal to the vector $\Delta\vec{\mathbf{r}} = \vec{\mathbf{r}}_f - \vec{\mathbf{r}}_i$.

The average velocity between points is *independent of the path* taken. That is because average velocity is proportional to displacement, which depends only on the initial and final position vectors and not on the path taken. As with one-dimensional motion, we conclude that if a particle starts its motion at some point and returns to this point via any path, its average velocity is zero for this trip because its displacement is zero. Consider again our basketball players on the court in Figure 2.2 (page 21). We previously considered only their one-dimensional motion back and forth between the baskets. In reality, however, they move over a two-dimensional surface, running back and forth between the baskets as well as left and right across the width of the court. Starting from one basket, a given player may follow a very complicated two-dimensional path. Upon returning to the original basket, however, a player's average velocity is zero because the player's displacement for the whole trip is zero.

Consider again the motion of a particle between two points in the *xy* plane as shown in Figure 4.2. As the time interval over which we observe the motion

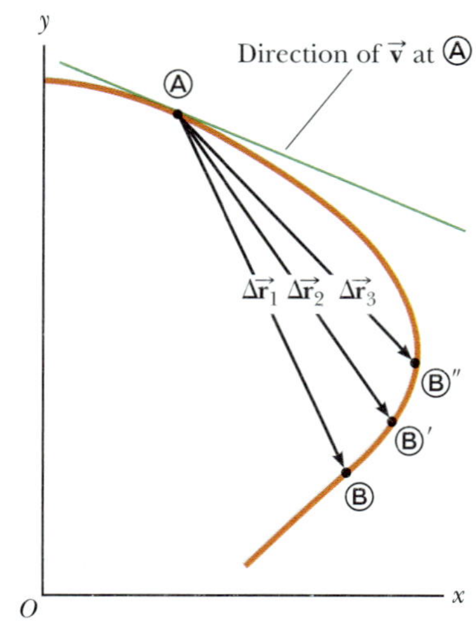

Figure 4.2 As a particle moves between two points, its average velocity is in the direction of the displacement vector $\Delta\vec{\mathbf{r}}$. As the end point of the path is moved from Ⓑ to Ⓑ′ to Ⓑ″, the respective displacements and corresponding time intervals become smaller and smaller. In the limit that the end point approaches Ⓐ, Δt approaches zero and the direction of $\Delta\vec{\mathbf{r}}$ approaches that of the line tangent to the curve at Ⓐ. By definition, the instantaneous velocity at Ⓐ is directed along this tangent line.

becomes smaller and smaller—that is, as Ⓑ is moved to Ⓑ′ and then to Ⓑ″, and so on—the direction of the displacement approaches that of the line tangent to the path at Ⓐ. The **instantaneous velocity** $\vec{\mathbf{v}}$ is defined as the limit of the average velocity $\Delta\vec{\mathbf{r}}/\Delta t$ as Δt approaches zero:

$$\vec{\mathbf{v}} \equiv \lim_{\Delta t \to 0} \frac{\Delta \vec{\mathbf{r}}}{\Delta t} = \frac{d\vec{\mathbf{r}}}{dt} \qquad (4.3)$$

◀ Instantaneous velocity

That is, the instantaneous velocity equals the derivative of the position vector with respect to time. The direction of the instantaneous velocity vector at any point in a particle's path is along a line tangent to the path at that point and in the direction of motion.

The magnitude of the instantaneous velocity vector $v = |\vec{\mathbf{v}}|$ of a particle is called the *speed* of the particle, which is a scalar quantity.

As a particle moves from one point to another along some path, its instantaneous velocity vector changes from $\vec{\mathbf{v}}_i$ at time t_i to $\vec{\mathbf{v}}_f$ at time t_f. Knowing the velocity at these points allows us to determine the average acceleration of the particle. The **average acceleration** $\vec{\mathbf{a}}_{\text{avg}}$ of a particle is defined as the change in its instantaneous velocity vector $\Delta\vec{\mathbf{v}}$ divided by the time interval Δt during which that change occurs:

$$\vec{\mathbf{a}}_{\text{avg}} \equiv \frac{\vec{\mathbf{v}}_f - \vec{\mathbf{v}}_i}{t_f - t_i} = \frac{\Delta \vec{\mathbf{v}}}{\Delta t} \qquad (4.4)$$

◀ Average acceleration

Because $\vec{\mathbf{a}}_{\text{avg}}$ is the ratio of a vector quantity $\Delta\vec{\mathbf{v}}$ and a positive scalar quantity Δt, we conclude that average acceleration is a vector quantity directed along $\Delta\vec{\mathbf{v}}$. As indicated in Figure 4.3, the direction of $\Delta\vec{\mathbf{v}}$ is found by adding the vector $-\vec{\mathbf{v}}_i$ (the negative of $\vec{\mathbf{v}}_i$) to the vector $\vec{\mathbf{v}}_f$ because, by definition, $\Delta\vec{\mathbf{v}} = \vec{\mathbf{v}}_f - \vec{\mathbf{v}}_i$.

When the average acceleration of a particle changes during different time intervals, it is useful to define its instantaneous acceleration. The **instantaneous acceleration** $\vec{\mathbf{a}}$ is defined as the limiting value of the ratio $\Delta\vec{\mathbf{v}}/\Delta t$ as Δt approaches zero:

$$\vec{\mathbf{a}} \equiv \lim_{\Delta t \to 0} \frac{\Delta \vec{\mathbf{v}}}{\Delta t} = \frac{d\vec{\mathbf{v}}}{dt} \qquad (4.5)$$

◀ Instantaneous acceleration

In other words, the instantaneous acceleration equals the derivative of the velocity vector with respect to time.

Various changes can occur when a particle accelerates. First, the magnitude of the velocity vector (the speed) may change with time as in straight-line (one-dimensional) motion. Second, the direction of the velocity vector may change with time even if its magnitude (speed) remains constant as in two-dimensional motion along a curved path. Finally, both the magnitude and the direction of the velocity vector may change simultaneously.

PITFALL PREVENTION 4.1

Vector Addition

Although the vector addition discussed in Chapter 3 involves *displacement* vectors, vector addition can be applied to *any* type of vector quantity. Figure 4.3, for example, shows the addition of *velocity* vectors using the graphical approach.

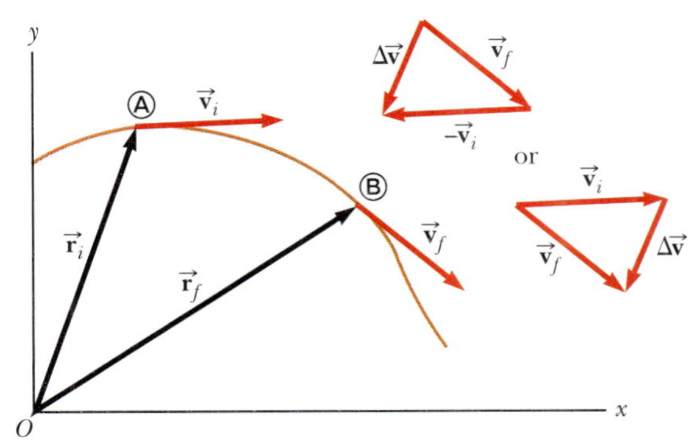

Figure 4.3 A particle moves from position Ⓐ to position Ⓑ. Its velocity vector changes from $\vec{\mathbf{v}}_i$ to $\vec{\mathbf{v}}_f$. The vector diagrams at the upper right show two ways of determining the vector $\Delta\vec{\mathbf{v}}$ from the initial and final velocities.

Quick Quiz 4.1 Consider the following controls in an automobile: gas pedal, brake, steering wheel. What are the controls in this list that cause an acceleration of the car? (a) all three controls (b) the gas pedal and the brake (c) only the brake (d) only the gas pedal

4.2 Two-Dimensional Motion with Constant Acceleration

In Section 2.5, we investigated one-dimensional motion of a particle under constant acceleration. Let us now consider two-dimensional motion during which the acceleration of a particle remains constant in both magnitude and direction. As we shall see, this approach is useful for analyzing some common types of motion.

Before embarking on this investigation, we need to emphasize an important point regarding two-dimensional motion. Imagine an air hockey puck moving in a straight line along a perfectly level, friction-free surface of an air hockey table. Figure 4.4a shows a motion diagram from an overhead point of view of this puck. Recall that in Section 2.4 we related the acceleration of an object to a force on the object. Because there are no forces on the puck in the horizontal plane, it moves with constant velocity in the x direction. Now suppose you blow a puff of air on the puck as it passes your position, with the force from your puff of air *exactly* in the y direction. Because the force from this puff of air has no component in the x direction, it causes no acceleration in the x direction. It only causes a momentary acceleration in the y direction, causing the puck to have a constant y component of velocity once the force from the puff of air is removed. After your puff of air on the puck, its velocity component in the x direction is unchanged, as shown in Figure 4.4b. The generalization of this simple experiment is that **motion in two dimensions can be modeled as two *independent* motions in each of the two perpendicular directions associated with the x and y axes. That is, any influence in the y direction does not affect the motion in the x direction and vice versa.**

The position vector for a particle moving in the xy plane can be written

$$\vec{\mathbf{r}} = x\hat{\mathbf{i}} + y\hat{\mathbf{j}} \quad (4.6)$$

where x, y, and $\vec{\mathbf{r}}$ change with time as the particle moves while the unit vectors $\hat{\mathbf{i}}$ and $\hat{\mathbf{j}}$ remain constant. If the position vector is known, the velocity of the particle can be obtained from Equations 4.3 and 4.6, which give

$$\vec{\mathbf{v}} = \frac{d\vec{\mathbf{r}}}{dt} = \frac{dx}{dt}\hat{\mathbf{i}} + \frac{dy}{dt}\hat{\mathbf{j}} = v_x\hat{\mathbf{i}} + v_y\hat{\mathbf{j}} \quad (4.7)$$

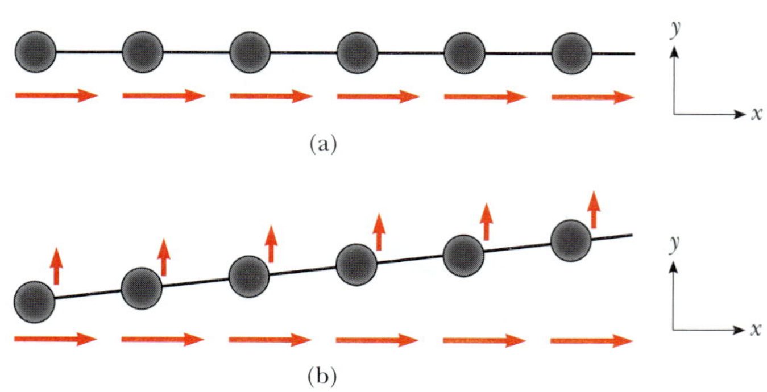

Figure 4.4 (a) A puck moves across a horizontal air hockey table at constant velocity in the x direction. (b) After a puff of air in the y direction is applied to the puck, the puck has gained a y component of velocity, but the x component is unaffected by the force in the perpendicular direction. Notice that the horizontal red vectors, representing the x component of the velocity, are the same length in both parts of the figure, which demonstrates that motion in two dimensions can be modeled as two independent motions in perpendicular directions.

Because the acceleration $\vec{\mathbf{a}}$ of the particle is assumed constant in this discussion, its components a_x and a_y also are constants. Therefore, we can model the particle as a particle under constant acceleration independently in each of the two directions and apply the equations of kinematics separately to the x and y components of the velocity vector. Substituting, from Equation 2.13, $v_{xf} = v_{xi} + a_x t$ and $v_{yf} = v_{yi} + a_y t$ into Equation 4.7 to determine the final velocity at any time t, we obtain

$$\vec{\mathbf{v}}_f = (v_{xi} + a_x t)\hat{\mathbf{i}} + (v_{yi} + a_y t)\hat{\mathbf{j}} = (v_{xi}\hat{\mathbf{i}} + v_{yi}\hat{\mathbf{j}}) + (a_x\hat{\mathbf{i}} + a_y\hat{\mathbf{j}})t$$

$$\vec{\mathbf{v}}_f = \vec{\mathbf{v}}_i + \vec{\mathbf{a}}t \tag{4.8}$$

◀ Velocity vector as a function of time

This result states that the velocity of a particle at some time t equals the vector sum of its initial velocity $\vec{\mathbf{v}}_i$ at time $t = 0$ and the additional velocity $\vec{\mathbf{a}}t$ acquired at time t as a result of constant acceleration. Equation 4.8 is the vector version of Equation 2.13.

Similarly, from Equation 2.16 we know that the x and y coordinates of a particle moving with constant acceleration are

$$x_f = x_i + v_{xi}t + \tfrac{1}{2}a_x t^2 \qquad y_f = y_i + v_{yi}t + \tfrac{1}{2}a_y t^2$$

Substituting these expressions into Equation 4.6 (and labeling the final position vector $\vec{\mathbf{r}}_f$) gives

$$\begin{aligned}\vec{\mathbf{r}}_f &= (x_i + v_{xi}t + \tfrac{1}{2}a_x t^2)\hat{\mathbf{i}} + (y_i + v_{yi}t + \tfrac{1}{2}a_y t^2)\hat{\mathbf{j}} \\ &= (x_i\hat{\mathbf{i}} + y_i\hat{\mathbf{j}}) + (v_{xi}\hat{\mathbf{i}} + v_{yi}\hat{\mathbf{j}})t + \tfrac{1}{2}(a_x\hat{\mathbf{i}} + a_y\hat{\mathbf{j}})t^2\end{aligned}$$

$$\vec{\mathbf{r}}_f = \vec{\mathbf{r}}_i + \vec{\mathbf{v}}_i t + \tfrac{1}{2}\vec{\mathbf{a}}t^2 \tag{4.9}$$

◀ Position vector as a function of time

which is the vector version of Equation 2.16. Equation 4.9 tells us that the position vector $\vec{\mathbf{r}}_f$ of a particle is the vector sum of the original position $\vec{\mathbf{r}}_i$, a displacement $\vec{\mathbf{v}}_i t$ arising from the initial velocity of the particle and a displacement $\frac{1}{2}\vec{\mathbf{a}}t^2$ resulting from the constant acceleration of the particle.

Graphical representations of Equations 4.8 and 4.9 are shown in Active Figure 4.5. The components of the position and velocity vectors are also illustrated in the figure. Notice from Active Figure 4.5a that $\vec{\mathbf{v}}_f$ is generally not along the direction of either $\vec{\mathbf{v}}_i$ or $\vec{\mathbf{a}}$ because the relationship between these quantities is a vector expression. For the same reason, from Active Figure 4.5b we see that $\vec{\mathbf{r}}_f$ is generally not along the direction of $\vec{\mathbf{v}}_i$ or $\vec{\mathbf{a}}$. Finally, notice that $\vec{\mathbf{v}}_f$ and $\vec{\mathbf{r}}_f$ are generally not in the same direction.

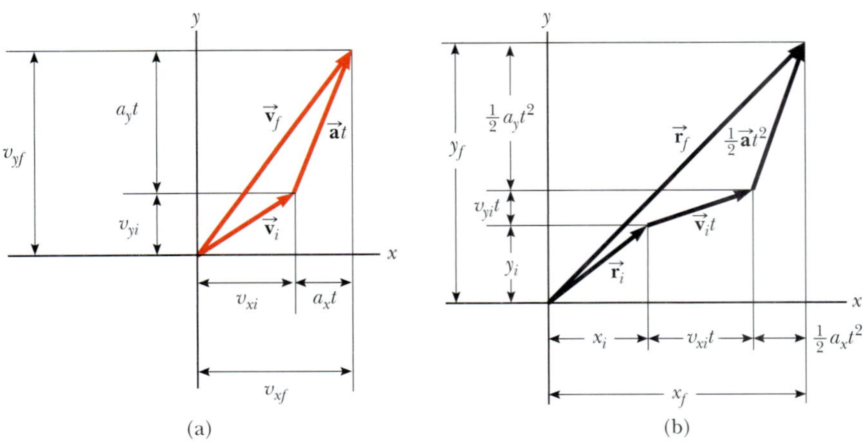

ACTIVE FIGURE 4.5

Vector representations and components of (a) the velocity and (b) the position of a particle moving with a constant acceleration $\vec{\mathbf{a}}$.

Sign in at www.thomsonedu.com and go to ThomsonNOW to investigate the effect of different initial positions and velocities on the final position and velocity (for constant acceleration).

EXAMPLE 4.1 Motion in a Plane

A particle starts from the origin at $t = 0$ with an initial velocity having an x component of 20 m/s and a y component of −15 m/s. The particle moves in the xy plane with an x component of acceleration only, given by $a_x = 4.0\ \text{m/s}^2$.

(A) Determine the total velocity vector at any time.

SOLUTION

Conceptualize The components of the initial velocity tell us that the particle starts by moving toward the right and downward. The x component of velocity starts at 20 m/s and increases by 4.0 m/s every second. The y component of velocity never changes from its initial value of −15 m/s. We sketch a motion diagram of the situation in Figure 4.6. Because the particle is accelerating in the $+x$ direction, its velocity component in this direction increases and the path curves as shown in the diagram. Notice that the spacing between successive images increases as time goes on because the speed is increasing. The placement of the acceleration and velocity vectors in Figure 4.6 helps us further conceptualize the situation.

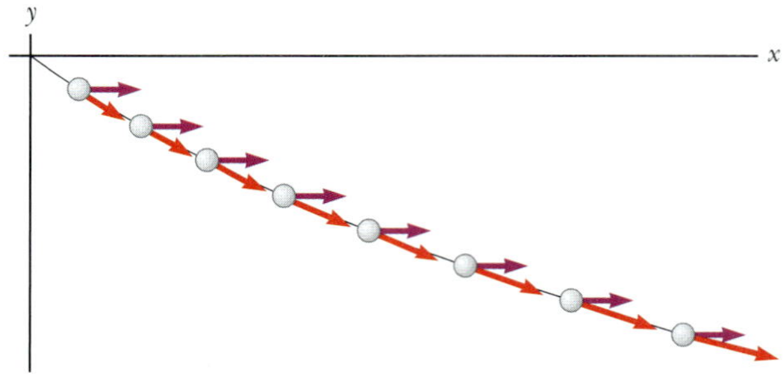

Figure 4.6 (Example 4.1) Motion diagram for the particle.

Categorize Because the initial velocity has components in both the x and y directions, we categorize this problem as one involving a particle moving in two dimensions. Because the particle only has an x component of acceleration, we model it as a particle under constant acceleration in the x direction and a particle under constant velocity in the y direction.

Analyze To begin the mathematical analysis, we set $v_{xi} = 20$ m/s, $v_{yi} = -15$ m/s, $a_x = 4.0\ \text{m/s}^2$, and $a_y = 0$.

Use Equation 4.8 for the velocity vector:

$$\vec{\mathbf{v}}_f = \vec{\mathbf{v}}_i + \vec{\mathbf{a}}t = (v_{xi} + a_xt)\,\hat{\mathbf{i}} + (v_{yi} + a_yt)\,\hat{\mathbf{j}}$$

Substitute numerical values:

$$\vec{\mathbf{v}}_f = [20\ \text{m/s} + (4.0\ \text{m/s}^2)t]\,\hat{\mathbf{i}} + [-15\ \text{m/s} + (0)t]\,\hat{\mathbf{j}}$$

$$(1)\quad \vec{\mathbf{v}}_f = [(20 + 4.0t)\,\hat{\mathbf{i}} - 15\,\hat{\mathbf{j}}]\ \text{m/s}$$

Finalize Notice that the x component of velocity increases in time while the y component remains constant; this result is consistent with what we predicted.

(B) Calculate the velocity and speed of the particle at $t = 5.0$ s.

SOLUTION

Analyze

Evaluate the result from Equation (1) at $t = 5.0$ s:

$$\vec{\mathbf{v}}_f = [(20 + 4.0(5.0))\,\hat{\mathbf{i}} - 15\,\hat{\mathbf{j}}]\ \text{m/s} = (40\,\hat{\mathbf{i}} - 15\,\hat{\mathbf{j}})\ \text{m/s}$$

Determine the angle θ that $\vec{\mathbf{v}}_f$ makes with the x axis at $t = 5.0$ s:

$$\theta = \tan^{-1}\left(\frac{v_{yf}}{v_{xf}}\right) = \tan^{-1}\left(\frac{-15\ \text{m/s}}{40\ \text{m/s}}\right) = -21°$$

Evaluate the speed of the particle as the magnitude of $\vec{\mathbf{v}}_f$:

$$v_f = |\vec{\mathbf{v}}_f| = \sqrt{v_{xf}^2 + v_{yf}^2} = \sqrt{(40)^2 + (-15)^2}\ \text{m/s} = 43\ \text{m/s}$$

Finalize The negative sign for the angle θ indicates that the velocity vector is directed at an angle of 21° below the positive x axis. Notice that if we calculate v_i from the x and y components of $\vec{\mathbf{v}}_i$, we find that $v_f > v_i$. Is that consistent with our prediction?

(C) Determine the x and y coordinates of the particle at any time t and its position vector at this time.

SOLUTION

Analyze

Use the components of Equation 4.9 with $x_i = y_i = 0$ at $t = 0$:

$$x_f = v_{xi}t + \tfrac{1}{2}a_xt^2 = (20t + 2.0t^2)\ \text{m}$$

$$y_f = v_{yi}t = (-15t)\ \text{m}$$

Express the position vector of the particle at any time t: $\vec{\mathbf{r}}_f = x_f\hat{\mathbf{i}} + y_f\hat{\mathbf{j}} = [(20t + 2.0t^2)\hat{\mathbf{i}} - 15t\hat{\mathbf{j}}]$ m

Finalize Let us now consider a limiting case for very large values of t.

What If? What if we wait a very long time and then observe the motion of the particle? How would we describe the motion of the particle for large values of the time?

Answer Looking at Figure 4.6, we see the path of the particle curving toward the x axis. There is no reason to assume that this tendency will change, which suggests that the path will become more and more parallel to the x axis as time grows large. Mathematically, Equation (1) shows that the y component of the velocity remains constant while the x component grows linearly with t. Therefore, when t is very large, the x component of the velocity will be much larger than the y component, suggesting that the velocity vector becomes more and more parallel to the x axis. Both x_f and y_f continue to grow with time, although x_f grows much faster.

4.3 Projectile Motion

Anyone who has observed a baseball in motion has observed projectile motion. The ball moves in a curved path and returns to the ground. **Projectile motion** of an object is simple to analyze if we make two assumptions: (1) the free-fall acceleration is constant over the range of motion and is directed downward,[1] and (2) the effect of air resistance is negligible.[2] With these assumptions, we find that the path of a projectile, which we call its *trajectory*, is *always* a parabola as shown in Active Figure 4.7. **We use these assumptions throughout this chapter.**

The expression for the position vector of the projectile as a function of time follows directly from Equation 4.9, with $\vec{\mathbf{a}} = \vec{\mathbf{g}}$:

$$\vec{\mathbf{r}}_f = \vec{\mathbf{r}}_i + \vec{\mathbf{v}}_i t + \tfrac{1}{2}\vec{\mathbf{g}}t^2 \qquad \textbf{(4.10)}$$

PITFALL PREVENTION 4.2

Acceleration at the Highest Point

As discussed in Pitfall Prevention 2.8, many people claim that the acceleration of a projectile at the topmost point of its trajectory is zero. This mistake arises from confusion between zero vertical velocity and zero acceleration. If the projectile were to experience zero acceleration at the highest point, its velocity at that point would not change; rather, the projectile would move horizontally at constant speed from then on! That does not happen, however, because the acceleration is *not* zero anywhere along the trajectory.

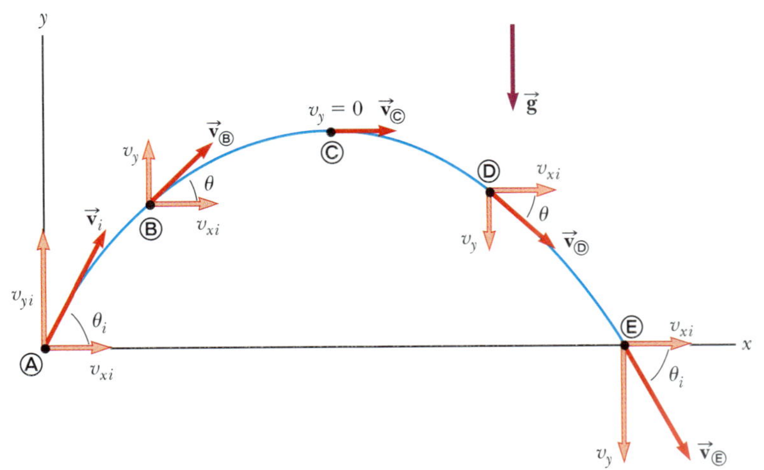

ACTIVE FIGURE 4.7

The parabolic path of a projectile that leaves the origin with a velocity $\vec{\mathbf{v}}_i$. The velocity vector $\vec{\mathbf{v}}$ changes with time in both magnitude and direction. This change is the result of acceleration in the negative y direction. The x component of velocity remains constant in time because there is no acceleration along the horizontal direction. The y component of velocity is zero at the peak of the path.

Sign in at www.thomsonedu.com and go to ThomsonNOW to change launch angle and initial speed. You can also observe the changing components of velocity along the trajectory of the projectile.

[1] This assumption is reasonable as long as the range of motion is small compared with the radius of the Earth (6.4×10^6 m). In effect, this assumption is equivalent to assuming that the Earth is flat over the range of motion considered.

[2] This assumption is generally *not* justified, especially at high velocities. In addition, any spin imparted to a projectile, such as that applied when a pitcher throws a curve ball, can give rise to some very interesting effects associated with aerodynamic forces, which will be discussed in Chapter 14.

The Telegraph Colour Library/Getty Images

A welder cuts holes through a heavy metal construction beam with a hot torch. The sparks generated in the process follow parabolic paths.

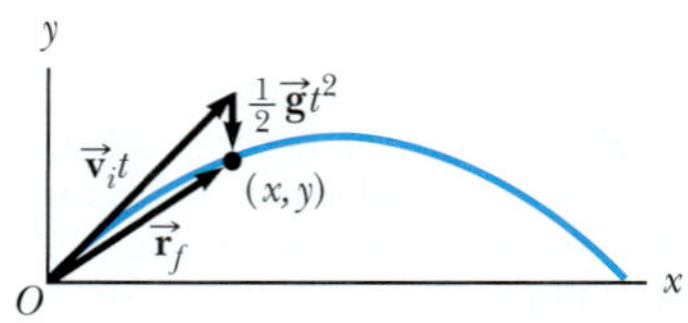

Figure 4.8 The position vector $\vec{\mathbf{r}}_f$ of a projectile launched from the origin whose initial velocity at the origin is $\vec{\mathbf{v}}_i$. The vector $\vec{\mathbf{v}}_i t$ would be the displacement of the projectile if gravity were absent, and the vector $\frac{1}{2}\vec{\mathbf{g}}t^2$ is its vertical displacement from a straight-line path due to its downward gravitational acceleration.

where the initial x and y components of the velocity of the projectile are

$$v_{xi} = v_i \cos\theta_i \qquad v_{yi} = v_i \sin\theta_i \tag{4.11}$$

The expression in Equation 4.10 is plotted in Figure 4.8, for a projectile launched from the origin, so that $\vec{\mathbf{r}}_i = 0$. The final position of a particle can be considered to be the superposition of its initial position $\vec{\mathbf{r}}_i$; the term $\vec{\mathbf{v}}_i t$, which is its displacement if no acceleration were present; and the term $\frac{1}{2}\vec{\mathbf{g}}t^2$ that arises from its acceleration due to gravity. In other words, if there were no gravitational acceleration, the particle would continue to move along a straight path in the direction of $\vec{\mathbf{v}}_i$. Therefore, the vertical distance $\frac{1}{2}\vec{\mathbf{g}}t^2$ through which the particle "falls" off the straight-line path is the same distance that an object dropped from rest would fall during the same time interval.

In Section 4.2, we stated that two-dimensional motion with constant acceleration can be analyzed as a combination of two independent motions in the x and y directions, with accelerations a_x and a_y. Projectile motion can also be handled in this way, with zero acceleration in the x direction and a constant acceleration in the y direction, $a_y = -g$. Therefore, **when analyzing projectile motion, model it to be the superposition of two motions: (1) motion of a particle under constant velocity in the horizontal direction and (2) motion of a particle under constant acceleration (free fall) in the vertical direction**. The horizontal and vertical components of a projectile's motion are completely independent of each other and can be handled separately, with time t as the common variable for both components.

Quick Quiz 4.2 **(i)** As a projectile thrown upward moves in its parabolic path (such as in Fig. 4.8), at what point along its path are the velocity and acceleration vectors for the projectile perpendicular to each other? (a) nowhere (b) the highest point (c) the launch point **(ii)** From the same choices, at what point are the velocity and acceleration vectors for the projectile parallel to each other?

Horizontal Range and Maximum Height of a Projectile

Let us assume a projectile is launched from the origin at $t_i = 0$ with a positive v_{yi} component as shown in Figure 4.9 and returns to the same horizontal level. Two points are especially interesting to analyze: the peak point Ⓐ, which has Cartesian coordinates $(R/2, h)$, and the point Ⓑ, which has coordinates $(R, 0)$. The distance R is called the *horizontal range* of the projectile, and the distance h is its *maximum height*. Let us find h and R mathematically in terms of v_i, θ_i, and g.

We can determine h by noting that at the peak $v_{y\text{Ⓐ}} = 0$. Therefore, we can use the y component of Equation 4.8 to determine the time $t_{\text{Ⓐ}}$ at which the projectile reaches the peak:

$$v_{yf} = v_{yi} + a_y t$$

$$0 = v_i \sin\theta_i - g t_{\text{Ⓐ}}$$

$$t_{\text{Ⓐ}} = \frac{v_i \sin\theta_i}{g}$$

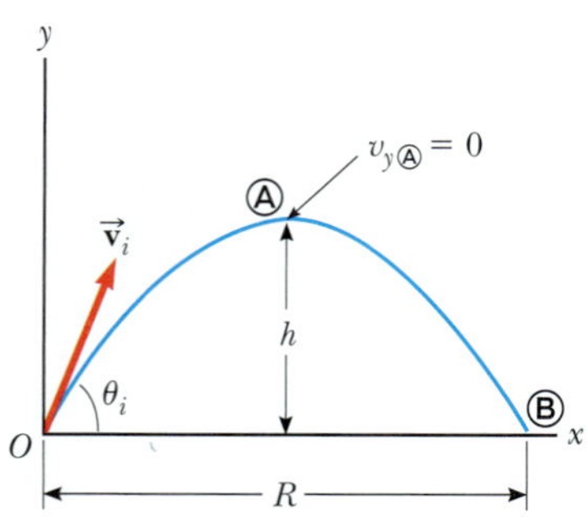

Figure 4.9 A projectile launched over a flat surface from the origin at $t_i = 0$ with an initial velocity $\vec{\mathbf{v}}_i$. The maximum height of the projectile is h, and the horizontal range is R. At Ⓐ, the peak of the trajectory, the particle has coordinates $(R/2, h)$.

Substituting this expression for $t_{\text{Ⓐ}}$ into the y component of Equation 4.9 and replacing $y = y_{\text{Ⓐ}}$ with h, we obtain an expression for h in terms of the magnitude and direction of the initial velocity vector:

$$h = (v_i \sin\theta_i)\frac{v_i \sin\theta_i}{g} - \frac{1}{2}g\left(\frac{v_i \sin\theta_i}{g}\right)^2$$

$$h = \frac{v_i^2 \sin^2\theta_i}{2g} \tag{4.12}$$

The range R is the horizontal position of the projectile at a time that is twice the time at which it reaches its peak, that is, at time $t_{\text{Ⓑ}} = 2t_{\text{Ⓐ}}$. Using the x compo-

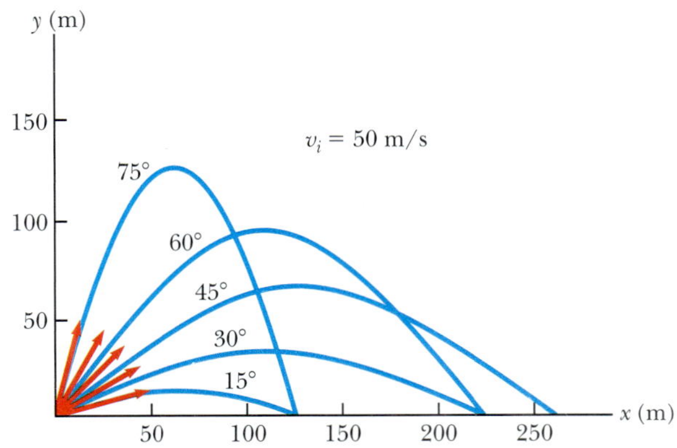

ACTIVE FIGURE 4.10

A projectile launched over a flat surface from the origin with an initial speed of 50 m/s at various angles of projection. Notice that complementary values of θ_i result in the same value of R (range of the projectile).

Sign in at www.thomsonedu.com and go to ThomsonNOW to vary the projection angle, observe the effect on the trajectory, and measure the flight time.

nent of Equation 4.9, noting that $v_{xi} = v_{x\text{Ⓑ}} = v_i \cos \theta_i$, and setting $x_{\text{Ⓑ}} = R$ at $t = 2t_{\text{Ⓐ}}$, we find that

$$R = v_{xi}t_{\text{Ⓑ}} = (v_i \cos \theta_i)2t_{\text{Ⓐ}}$$

$$= (v_i \cos \theta_i)\frac{2v_i \sin \theta_i}{g} = \frac{2v_i^2 \sin \theta_i \cos \theta_i}{g}$$

Using the identity $\sin 2\theta = 2 \sin \theta \cos \theta$ (see Appendix B.4), we can write R in the more compact form

$$R = \frac{v_i^2 \sin 2\theta_i}{g} \quad \text{(4.13)}$$

The maximum value of R from Equation 4.13 is $R_{\max} = v_i^2/g$. This result makes sense because the maximum value of $\sin 2\theta_i$ is 1, which occurs when $2\theta_i = 90°$. Therefore, R is a maximum when $\theta_i = 45°$.

Active Figure 4.10 illustrates various trajectories for a projectile having a given initial speed but launched at different angles. As you can see, the range is a maximum for $\theta_i = 45°$. In addition, for any θ_i other than 45°, a point having Cartesian coordinates $(R, 0)$ can be reached by using either one of two complementary values of θ_i, such as 75° and 15°. Of course, the maximum height and time of flight for one of these values of θ_i are different from the maximum height and time of flight for the complementary value.

PITFALL PREVENTION 4.3

The Height and Range Equations

Equation 4.13 is useful for calculating R only for a symmetric path as shown in Active Figure 4.10. If the path is not symmetric, *do not use this equation*. The general expressions given by Equations 4.8 and 4.9 are the *more important* results because they give the position and velocity components of *any* particle moving in two dimensions at *any* time t.

Quick Quiz 4.3 Rank the launch angles for the five paths in Active Figure 4.10 with respect to time of flight, from the shortest time of flight to the longest.

PROBLEM-SOLVING STRATEGY **Projectile Motion**

We suggest you use the following approach when solving projectile motion problems:

1. *Conceptualize.* Think about what is going on physically in the problem. Establish the mental representation by imagining the projectile moving along its trajectory.

2. *Categorize.* Confirm that the problem involves a particle in free fall and that air resistance is neglected. Select a coordinate system with x in the horizontal direction and y in the vertical direction.

3. *Analyze.* If the initial velocity vector is given, resolve it into x and y components. Treat the horizontal motion and the vertical motion independently. Analyze the

horizontal motion of the projectile as a particle under constant velocity. Analyze the vertical motion of the projectile as a particle under constant acceleration.

4. *Finalize.* Once you have determined your result, check to see if your answers are consistent with the mental and pictorial representations and that your results are realistic.

EXAMPLE 4.2 The Long Jump

A long jumper (Fig. 4.11) leaves the ground at an angle of 20.0° above the horizontal and at a speed of 11.0 m/s.

(A) How far does he jump in the horizontal direction?

Mike Powell/Allsport/Getty Images

Figure 4.11 (Example 4.2) Mike Powell, current holder of the world long-jump record of 8.95 m.

SOLUTION

Conceptualize The arms and legs of a long jumper move in a complicated way, but we will ignore this motion. We conceptualize the motion of the long jumper as equivalent to that of a simple projectile.

Categorize We categorize this example as a projectile motion problem. Because the initial speed and launch angle are given and because the final height is the same as the initial height, we further categorize this problem as satisfying the conditions for which Equations 4.12 and 4.13 can be used. This approach is the most direct way to analyze this problem, although the general methods that have been described will always give the correct answer.

Analyze

Use Equation 4.13 to find the range of the jumper:

$$R = \frac{v_i^2 \sin 2\theta_i}{g} = \frac{(11.0 \text{ m/s})^2 \sin 2(20.0°)}{9.80 \text{ m/s}^2} = 7.94 \text{ m}$$

(B) What is the maximum height reached?

SOLUTION

Analyze

Find the maximum height reached by using Equation 4.12:

$$h = \frac{v_i^2 \sin^2 \theta_i}{2g} = \frac{(11.0 \text{ m/s})^2 (\sin 20.0°)^2}{2(9.80 \text{ m/s}^2)} = 0.722 \text{ m}$$

Finalize Find the answers to parts (A) and (B) using the general method. The results should agree. Treating the long jumper as a particle is an oversimplification. Nevertheless, the values obtained are consistent with experience in sports. We can model a complicated system such as a long jumper as a particle and still obtain results that are reasonable.

EXAMPLE 4.3 A Bull's-Eye Every Time

In a popular lecture demonstration, a projectile is fired at a target in such a way that the projectile leaves the gun at the same time the target is dropped from rest. Show that if the gun is initially aimed at the stationary target, the projectile hits the falling target as shown in Figure 4.12a.

SOLUTION

Conceptualize We conceptualize the problem by studying Figure 4.12a. Notice that the problem does not ask for numerical values. The expected result must involve an algebraic argument.

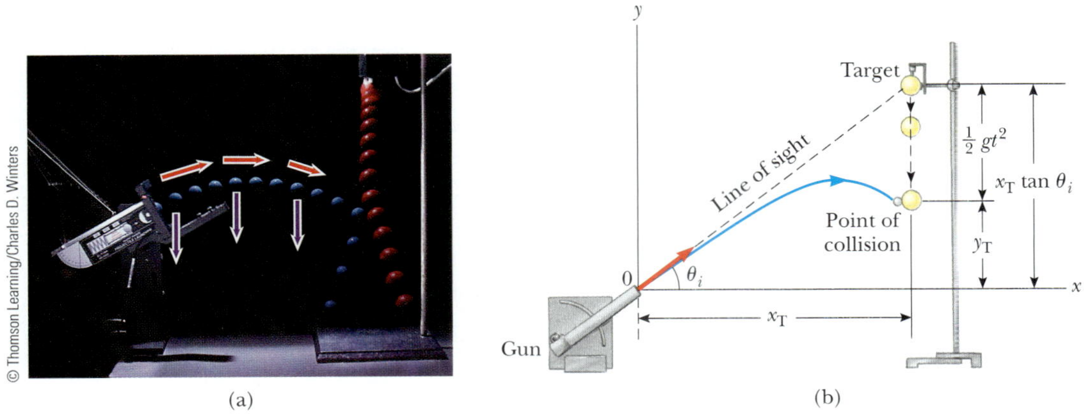

Figure 4.12 (Example 4.3) (a) Multiflash photograph of the projectile–target demonstration. If the gun is aimed directly at the target and is fired at the same instant the target begins to fall, the projectile will hit the target. Notice that the velocity of the projectile (red arrows) changes in direction and magnitude, whereas its downward acceleration (violet arrows) remains constant. (b) Schematic diagram of the projectile–target demonstration.

Categorize Because both objects are subject only to gravity, we categorize this problem as one involving two objects in free fall, the target moving in one dimension and the projectile moving in two.

Analyze The target T is modeled as a particle under constant acceleration in one dimension. Figure 4.12b shows that the initial y coordinate y_{iT} of the target is $x_T \tan \theta_i$ and its initial velocity is zero. It falls with acceleration $a_y = -g$. The projectile P is modeled as a particle under constant acceleration in the y direction and a particle under constant velocity in the x direction.

Write an expression for the y coordinate of the target at any moment after release, noting that its initial velocity is zero:

$$(1) \quad y_T = y_{iT} + (0)t - \tfrac{1}{2}gt^2 = x_T \tan \theta_i - \tfrac{1}{2}gt^2$$

Write an expression for the y coordinate of the projectile at any moment:

$$(2) \quad y_P = y_{iP} + v_{yiP}t - \tfrac{1}{2}gt^2 = 0 + (v_{iP} \sin\theta_i)t - \tfrac{1}{2}gt^2 = (v_{iP} \sin\theta_i)t - \tfrac{1}{2}gt^2$$

Write an expression for the x coordinate of the projectile at any moment:

$$x_P = x_{iP} + v_{xiP}t = 0 + (v_{iP} \cos \theta_i)t = (v_{iP} \cos \theta_i)t$$

Solve this expression for time as a function of the horizontal position of the projectile:

$$t = \frac{x_P}{v_{iP} \cos \theta_i}$$

Substitute this expression into Equation (2):

$$(3) \quad y_P = (v_{iP} \sin \theta_i)\left(\frac{x_P}{v_{iP} \cos \theta_i}\right) - \tfrac{1}{2}gt^2 = x_P \tan \theta_i - \tfrac{1}{2}gt^2$$

Compare Equations (1) and (3). We see that when the x coordinates of the projectile and target are the same—that is, when $x_T = x_P$—their y coordinates given by Equations (1) and (3) are the same and a collision results.

Finalize Note that a collision can result only when $v_{iP} \sin \theta_i \geq \sqrt{gd/2}$, where d is the initial elevation of the target above the floor. If $v_{iP} \sin \theta_i$ is less than this value, the projectile strikes the floor before reaching the target.

EXAMPLE 4.4 That's Quite an Arm!

A stone is thrown from the top of a building upward at an angle of 30.0° to the horizontal with an initial speed of 20.0 m/s as shown in Figure 4.13. The height of the building is 45.0 m.

(A) How long does it take the stone to reach the ground?

SOLUTION

Conceptualize Study Figure 4.13, in which we have indicated the trajectory and various parameters of the motion of the stone.

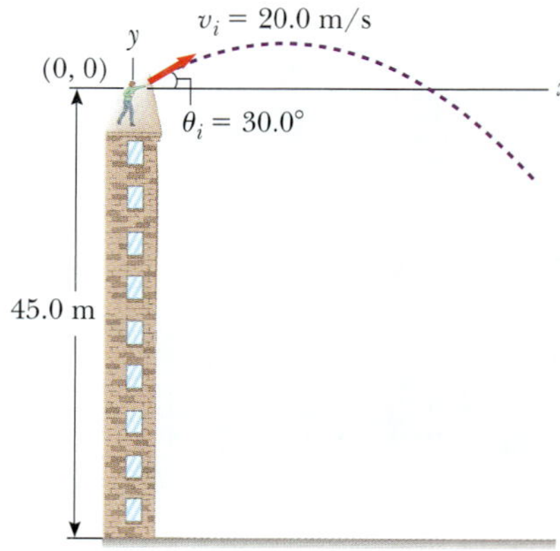

Figure 4.13 (Example 4.4) A stone is thrown from the top of a building.

Categorize We categorize this problem as a projectile motion problem. The stone is modeled as a particle under constant acceleration in the y direction and a particle under constant velocity in the x direction.

Analyze We have the information $x_i = y_i = 0$, $y_f = -45.0$ m, $a_y = -g$, and $v_i = 20.0$ m/s (the numerical value of y_f is negative because we have chosen the top of the building as the origin).

Find the initial x and y components of the stone's velocity:

$$v_{xi} = v_i \cos\theta_i = (20.0 \text{ m/s})\cos 30.0° = 17.3 \text{ m/s}$$

$$v_{yi} = v_i \sin\theta_i = (20.0 \text{ m/s})\sin 30.0° = 10.0 \text{ m/s}$$

Express the vertical position of the stone from the vertical component of Equation 4.9:

$$y_f = y_i + v_{yi}t + \tfrac{1}{2}a_y t^2$$

Substitute numerical values:

$$-45.0 \text{ m} = 0 + (10.0 \text{ m/s})t + \tfrac{1}{2}(-9.80 \text{ m/s}^2)t^2$$

Solve the quadratic equation for t:

$$t = \boxed{4.22 \text{ s}}$$

(B) What is the speed of the stone just before it strikes the ground?

SOLUTION

Use the y component of Equation 4.8 with $t = 4.22$ s to obtain the y component of the velocity of the stone just before it strikes the ground:

$$v_{yf} = v_{yi} + a_y t$$

Substitute numerical values:

$$v_{yf} = 10.0 \text{ m/s} + (-9.80 \text{ m/s}^2)(4.22 \text{ s}) = -31.3 \text{ m/s}$$

Use this component with the horizontal component $v_{xf} = v_{xi} = 17.3$ m/s to find the speed of the stone at $t = 4.22$ s:

$$v_f = \sqrt{v_{xf}^2 + v_{yf}^2} = \sqrt{(17.3 \text{ m/s})^2 + (-31.3 \text{ m/s})^2} = \boxed{35.8 \text{ m/s}}$$

Finalize Is it reasonable that the y component of the final velocity is negative? Is it reasonable that the final speed is larger than the initial speed of 20.0 m/s?

What If? What if a horizontal wind is blowing in the same direction as the stone is thrown and it causes the stone to have a horizontal acceleration component $a_x = 0.500$ m/s^2? Which part of this example, (A) or (B), will have a different answer?

Answer Recall that the motions in the x and y directions are independent. Therefore, the horizontal wind cannot affect the vertical motion. The vertical motion determines the time of the projectile in the air, so the answer to part (A) does not change. The wind causes the horizontal velocity component to increase with time, so the final speed will be larger in part (B). Taking $a_x = 0.500$ m/s^2, we find $v_{xf} = 19.4$ m/s and $v_f = 36.9$ m/s.

EXAMPLE 4.5 The End of the Ski Jump

A ski jumper leaves the ski track moving in the horizontal direction with a speed of 25.0 m/s as shown in Figure 4.14. The landing incline below her falls off with a slope of 35.0°. Where does she land on the incline?

SOLUTION

Conceptualize We can conceptualize this problem based on memories of observing winter Olympic ski competitions. We estimate the skier to be airborne for perhaps 4 s and to travel a distance of about 100 m horizontally. We should expect the value of d, the distance traveled along the incline, to be of the same order of magnitude.

Categorize We categorize the problem as one of a particle in projectile motion.

Analyze It is convenient to select the beginning of the jump as the origin. The initial velocity components are $v_{xi} = 25.0$ m/s and $v_{yi} = 0$. From the right triangle in Figure 4.14, we see that the jumper's x and y coordinates at the landing point are given by $x_f = d\cos 35.0°$ and $y_f = -d\sin 35.0°$.

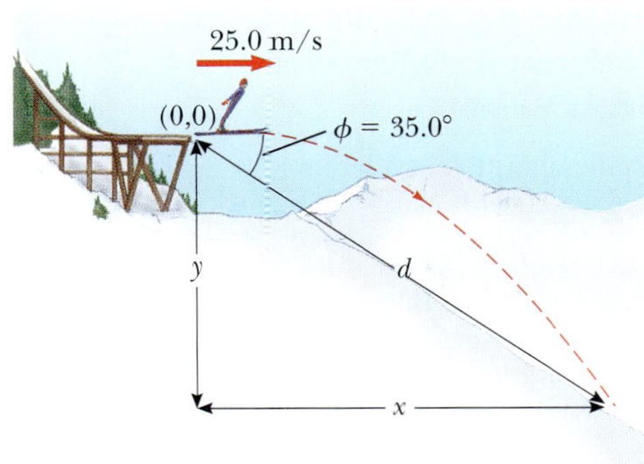

Figure 4.14 (Example 4.5) A ski jumper leaves the track moving in a horizontal direction.

Express the coordinates of the jumper as a function of time:

$$(1)\quad x_f = v_{xi}t = (25.0\text{ m/s})t$$

$$(2)\quad y_f = v_{yi}t + \tfrac{1}{2}a_y t^2 = -\tfrac{1}{2}(9.80\text{ m/s}^2)t^2$$

Substitute the values of x_f and y_f at the landing point:

$$(3)\quad d\cos 35.0° = (25.0\text{ m/s})t$$

$$(4)\quad -d\sin 35.0° = -\tfrac{1}{2}(9.80\text{ m/s}^2)t^2$$

Solve Equation (3) for t and substitute the result into Equation (4):

$$-d\sin 35.0° = -\tfrac{1}{2}(9.80\text{ m/s}^2)\left(\frac{d\cos 35.0°}{25.0\text{ m/s}}\right)^2$$

Solve for d:

$$d = 109\text{ m}$$

Evaluate the x and y coordinates of the point at which the skier lands:

$$x_f = d\cos 35.0° = (109\text{ m})\cos 35.0° = 89.3\text{ m}$$

$$y_f = -d\sin 35.0° = -(109\text{ m})\sin 35.0° = -62.5\text{ m}$$

Finalize Let us compare these results with our expectations. We expected the horizontal distance to be on the order of 100 m, and our result of 89.3 m is indeed on this order of magnitude. It might be useful to calculate the time interval that the jumper is in the air and compare it with our estimate of about 4 s.

What If? Suppose everything in this example is the same except the ski jump is curved so that the jumper is projected upward at an angle from the end of the track. Is this design better in terms of maximizing the length of the jump?

Answer If the initial velocity has an upward component, the skier will be in the air longer and should therefore travel further. Tilting the initial velocity vector upward, however, will reduce the horizontal component of the initial velocity. Therefore, angling the end of the ski track upward at a *large* angle may actually *reduce* the distance. Consider the extreme case: the skier is projected at 90° to the horizontal and simply goes up and comes back down at the end of the ski track! This argument suggests that there must be an optimal angle between 0° and 90° that represents a balance between making the flight time longer and the horizontal velocity component smaller.

Let us find this optimal angle mathematically. We modify equations (1) through (4) in the following way, assuming that the skier is projected at an angle θ with respect to the horizontal over a landing incline sloped with an arbitrary angle ϕ:

$$(1)\text{ and }(3)\ \rightarrow\ x_f = (v_i\cos\theta)t = d\cos\phi$$

$$(2)\text{ and }(4)\ \rightarrow\ y_f = (v_i\sin\theta)t - \tfrac{1}{2}gt^2 = -d\sin\phi$$

By eliminating the time t between these equations and using differentiation to maximize d in terms of θ, we arrive (after several steps; see Problem 62) at the following equation for the angle θ that gives the maximum value of d:

$$\theta = 45° - \frac{\phi}{2}$$

For the slope angle in Figure 4.14, $\phi = 35.0°$; this equation results in an optimal launch angle of $\phi = 27.5°$. For a slope angle of $\phi = 0°$, which represents a horizontal plane, this equation gives an optimal launch angle of $\theta = 45°$, as we would expect (see Active Figure 4.10).

PITFALL PREVENTION 4.4

Acceleration of a Particle in Uniform Circular Motion

Remember that acceleration in physics is defined as a change in the *velocity*, not a change in the *speed* (contrary to the everyday interpretation). In circular motion, the velocity vector is changing in direction, so there is indeed an acceleration.

4.4 The Particle in Uniform Circular Motion

Figure 4.15a shows a car moving in a circular path with *constant speed* v. Such motion, called **uniform circular motion**, occurs in many situations. Because it occurs so often, this type of motion is recognized as an analysis model called the **particle in uniform circular motion**. We discuss this model in this section.

It is often surprising to students to find that **even though an object moves at a constant speed in a circular path, it still has an acceleration.** To see why, consider the defining equation for acceleration, $\vec{\mathbf{a}} = d\vec{\mathbf{v}}/dt$ (Eq. 4.5). Notice that the acceleration depends on the change in the *velocity*. Because velocity is a vector quantity, an acceleration can occur in two ways, as mentioned in Section 4.1: by a change in the *magnitude* of the velocity and by a change in the *direction* of the velocity. The latter situation occurs for an object moving with constant speed in a circular path. The velocity vector is always tangent to the path of the object and perpendicular to the radius of the circular path.

We now show that the acceleration vector in uniform circular motion is always perpendicular to the path and always points toward the center of the circle. If that were not true, there would be a component of the acceleration parallel to the path and therefore parallel to the velocity vector. Such an acceleration component would lead to a change in the speed of the particle along the path. This situation, however, is inconsistent with our setup of the situation: the particle moves with constant speed along the path. Therefore, for *uniform* circular motion, the acceleration vector can only have a component perpendicular to the path, which is toward the center of the circle.

Let us now find the magnitude of the acceleration of the particle. Consider the diagram of the position and velocity vectors in Figure 4.15b. The figure also shows the vector representing the change in position $\Delta\vec{\mathbf{r}}$ for an arbitrary time interval. The particle follows a circular path of radius r, part of which is shown by the dashed curve. The particle is at Ⓐ at time t_i, and its velocity at that time is $\vec{\mathbf{v}}_i$; it is at Ⓑ at some later time t_f, and its velocity at that time is $\vec{\mathbf{v}}_f$. Let us also assume $\vec{\mathbf{v}}_i$ and $\vec{\mathbf{v}}_f$ differ only in direction; their magnitudes are the same (that is, $v_i = v_f = v$ because it is *uniform* circular motion).

In Figure 4.15c, the velocity vectors in Figure 4.15b have been redrawn tail to tail. The vector $\Delta\vec{\mathbf{v}}$ connects the tips of the vectors, representing the vector addition $\vec{\mathbf{v}}_f = \vec{\mathbf{v}}_i + \Delta\vec{\mathbf{v}}$. In both Figures 4.15b and 4.15c, we can identify triangles that help us analyze the motion. The angle $\Delta\theta$ between the two position vectors in Figure 4.15b is the same as the angle between the velocity vectors in Figure 4.15c because the velocity vector $\vec{\mathbf{v}}$ is always perpendicular to the position vector $\vec{\mathbf{r}}$. Therefore, the two triangles are *similar*. (Two triangles are similar if the angle between any two sides is the same for both triangles and if the ratio of the lengths of these sides is the same.) We can now write a relationship between the lengths of the sides for the two triangles in Figures 4.15b and 4.15c:

$$\frac{|\Delta\vec{\mathbf{v}}|}{v} = \frac{|\Delta\vec{\mathbf{r}}|}{r}$$

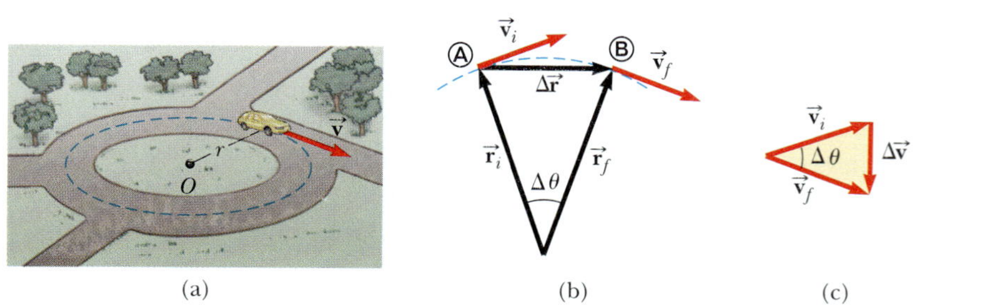

Figure 4.15 (a) A car moving along a circular path at constant speed experiences uniform circular motion. (b) As a particle moves from Ⓐ to Ⓑ, its velocity vector changes from $\vec{\mathbf{v}}_i$ to $\vec{\mathbf{v}}_f$. (c) The construction for determining the direction of the change in velocity $\Delta\vec{\mathbf{v}}$, which is toward the center of the circle for small $\Delta\vec{\mathbf{r}}$.

where $v = v_i = v_f$ and $r = r_i = r_f$. This equation can be solved for $|\Delta\vec{\mathbf{v}}|$, and the expression obtained can be substituted into Equation 4.4, $\vec{\mathbf{a}}_{avg} = \Delta\vec{\mathbf{v}}/\Delta t$, to give the magnitude of the average acceleration over the time interval for the particle to move from Ⓐ to Ⓑ:

$$|\vec{\mathbf{a}}_{avg}| = \frac{|\Delta\vec{\mathbf{v}}|}{|\Delta t|} = \frac{v}{r}\frac{|\Delta\vec{\mathbf{r}}|}{\Delta t}$$

Now imagine that points Ⓐ and Ⓑ in Figure 4.15b become extremely close together. As Ⓐ and Ⓑ approach each other, Δt approaches zero, $|\Delta\vec{\mathbf{r}}|$ approaches the distance traveled by the particle along the circular path, and the ratio $|\Delta\vec{\mathbf{r}}|/\Delta t$ approaches the speed v. In addition, the average acceleration becomes the instantaneous acceleration at point Ⓐ. Hence, in the limit $\Delta t \to 0$, the magnitude of the acceleration is

$$a_c = \frac{v^2}{r} \qquad (4.14)$$

◀ Centripetal acceleration

PITFALL PREVENTION 4.5

Centripetal Acceleration Is Not Constant

We derived the magnitude of the centripetal acceleration vector and found it to be constant for uniform circular motion, but *the centripetal acceleration vector is not constant.* It always points toward the center of the circle, but it continuously changes direction as the object moves around the circular path.

An acceleration of this nature is called a **centripetal acceleration** (*centripetal* means *center-seeking*). The subscript on the acceleration symbol reminds us that the acceleration is centripetal.

In many situations, it is convenient to describe the motion of a particle moving with constant speed in a circle of radius r in terms of the **period** T, which is defined as the time interval required for one complete revolution of the particle. In the time interval T, the particle moves a distance of $2\pi r$, which is equal to the circumference of the particle's circular path. Therefore, because its speed is equal to the circumference of the circular path divided by the period, or $v = 2\pi r/T$, it follows that

$$T = \frac{2\pi r}{v} \qquad (4.15)$$

◀ Period of circular motion

Quick Quiz 4.4 A particle moves in a circular path of radius r with speed v. It then increases its speed to $2v$ while traveling along the same circular path. **(i)** The centripetal acceleration of the particle has changed by what factor (choose one)? (a) 0.25 (b) 0.5 (c) 2 (d) 4 (e) impossible to determine **(ii)** From the same choices, by what factor has the period of the particle changed?

EXAMPLE 4.6 The Centripetal Acceleration of the Earth

What is the centripetal acceleration of the Earth as it moves in its orbit around the Sun?

SOLUTION

Conceptualize Think about a mental image of the Earth in a circular orbit around the Sun. We will model the Earth as a particle and approximate the Earth's orbit as circular (it's actually slightly elliptical, as we discuss in Chapter 13).

Categorize The Conceptualize step allows us to categorize this problem as one of a particle in uniform circular motion.

Analyze We do not know the orbital speed of the Earth to substitute into Equation 4.14. With the help of Equation 4.15, however, we can recast Equation 4.14 in terms of the period of the Earth's orbit, which we know is one year, and the radius of the Earth's orbit around the Sun, which is 1.496×10^{11} m.

Combine Equations 4.14 and 4.15:

$$a_c = \frac{v^2}{r} = \frac{\left(\frac{2\pi r}{T}\right)^2}{r} = \frac{4\pi^2 r}{T^2}$$

Substitute numerical values:

$$a_c = \frac{4\pi^2(1.496 \times 10^{11}\ \text{m})}{(1\ \text{yr})^2}\left(\frac{1\ \text{yr}}{3.156 \times 10^7\ \text{s}}\right)^2 = 5.93 \times 10^{-3}\ \text{m/s}^2$$

Finalize This acceleration is much smaller than the free-fall acceleration on the surface of the Earth. An important thing we learned here is the technique of replacing the speed v in Equation 4.14 in terms of the period T of the motion.

4.5 Tangential and Radial Acceleration

Let us consider the motion of a particle along a smooth, curved path where the velocity changes both in direction and in magnitude as described in Active Figure 4.16. In this situation, the velocity vector is always tangent to the path; the acceleration vector $\vec{\mathbf{a}}$, however, is at some angle to the path. At each of three points Ⓐ, Ⓑ, and Ⓒ in Active Figure 4.16, we draw dashed circles that represent the curvature of the actual path at each point. The radius of the circles is equal to the path's radius of curvature at each point.

As the particle moves along the curved path in Active Figure 4.16, the direction of the total acceleration vector $\vec{\mathbf{a}}$ changes from point to point. At any instant, this vector can be resolved into two components based on an origin at the center of the dashed circle corresponding to that instant: a radial component a_r along the radius of the circle and a tangential component a_t perpendicular to this radius. The *total* acceleration vector $\vec{\mathbf{a}}$ can be written as the vector sum of the component vectors:

Total acceleration ▶

$$\vec{\mathbf{a}} = \vec{\mathbf{a}}_r + \vec{\mathbf{a}}_t \tag{4.16}$$

The tangential acceleration component causes a change in the speed v of the particle. This component is parallel to the instantaneous velocity, and its magnitude is given by

Tangential acceleration ▶

$$a_t = \left|\frac{dv}{dt}\right| \tag{4.17}$$

The radial acceleration component arises from a change in direction of the velocity vector and is given by

Radial acceleration ▶

$$a_r = -a_c = -\frac{v^2}{r} \tag{4.18}$$

where r is the radius of curvature of the path at the point in question. We recognize the radial component of the acceleration as the centripetal acceleration discussed in Section 4.4. The negative sign in Equation 4.18 indicates that the direction of the centripetal acceleration is toward the center of the circle representing the radius of curvature. The direction is opposite that of the radial unit vector $\hat{\mathbf{r}}$, which always points away from the origin at the center of the circle.

Because $\vec{\mathbf{a}}_r$ and $\vec{\mathbf{a}}_t$ are perpendicular component vectors of $\vec{\mathbf{a}}$, it follows that the magnitude of $\vec{\mathbf{a}}$ is $a = \sqrt{a_r^2 + a_t^2}$. At a given speed, a_r is large when the radius of curvature is small (as at points Ⓐ and Ⓑ in Fig. 4.16) and small when r is large (as at point Ⓒ). The direction of $\vec{\mathbf{a}}_t$ is either in the same direction as $\vec{\mathbf{v}}$ (if v is increasing) or opposite $\vec{\mathbf{v}}$ (if v is decreasing).

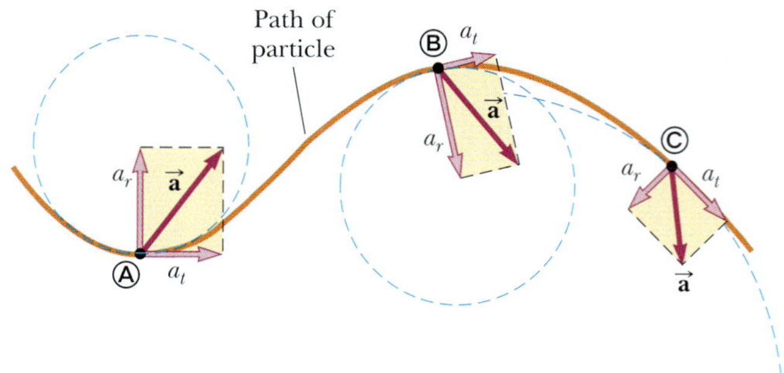

ACTIVE FIGURE 4.16

The motion of a particle along an arbitrary curved path lying in the xy plane. If the velocity vector $\vec{\mathbf{v}}$ (always tangent to the path) changes in direction and magnitude, the components of the acceleration $\vec{\mathbf{a}}$ are a tangential component a_t and a radial component a_r.

Sign in at www.thomsonedu.com and go to ThomsonNOW to study the acceleration components of a roller-coaster car.

In uniform circular motion, where v is constant, $a_t = 0$ and the acceleration is always completely radial as described in Section 4.4. In other words, uniform circular motion is a special case of motion along a general curved path. Furthermore, if the direction of $\vec{\mathbf{v}}$ does not change, there is no radial acceleration and the motion is one dimensional (in this case, $a_r = 0$, but a_t may not be zero).

Quick Quiz 4.5 A particle moves along a path and its speed increases with time. **(i)** In which of the following cases are its acceleration and velocity vectors parallel? (a) when the path is circular (b) when the path is straight (c) when the path is a parabola (d) never **(ii)** From the same choices, in which case are its acceleration and velocity vectors perpendicular everywhere along the path?

EXAMPLE 4.7 Over the Rise

A car exhibits a constant acceleration of 0.300 m/s^2 parallel to the roadway. The car passes over a rise in the roadway such that the top of the rise is shaped like a circle of radius 500 m. At the moment the car is at the top of the rise, its velocity vector is horizontal and has a magnitude of 6.00 m/s. What are the magnitude and direction of the total acceleration vector for the car at this instant?

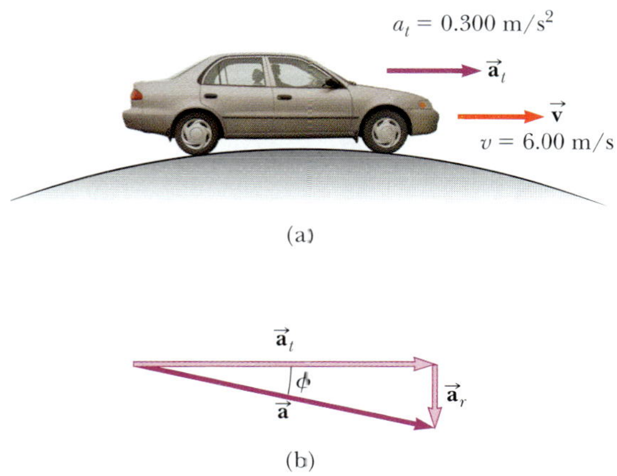

Figure 4.17 (Example 4.7) (a) A car passes over a rise that is shaped like a circle. (b) The total acceleration vector $\vec{\mathbf{a}}$ is the sum of the tangential and radial acceleration vectors $\vec{\mathbf{a}}_t$ and $\vec{\mathbf{a}}_r$.

SOLUTION

Conceptualize Conceptualize the situation using Figure 4.17a and any experiences you have had in driving over rises on a roadway.

Categorize Because the accelerating car is moving along a curved path, we categorize this problem as one involving a particle experiencing both tangential and radial acceleration. We recognize that it is a relatively simple substitution problem.

The radial acceleration is given by Equation 4.18, with $v = 6.00$ m/s and $r = 500$ m. The radial acceleration vector is directed straight downward, and the tangential acceleration vector has magnitude 0.300 m/s^2 and is horizontal.

Evaluate the radial acceleration:

$$a_r = -\frac{v^2}{r} = -\frac{(6.00 \text{ m/s})^2}{500 \text{ m}} = -0.072\,0 \text{ m/s}^2$$

Find the magnitude of $\vec{\mathbf{a}}$:

$$\sqrt{a_r^2 + a_t^2} = \sqrt{(-0.072\,0 \text{ m/s}^2)^2 + (0.300 \text{ m/s}^2)^2}$$
$$= 0.309 \text{ m/s}^2$$

Find the angle ϕ (see Fig. 4.17b) between $\vec{\mathbf{a}}$ and the horizontal:

$$\phi = \tan^{-1}\frac{a_r}{a_t} = \tan^{-1}\left(\frac{-0.072\,0 \text{ m/s}^2}{0.300 \text{ m/s}^2}\right) = -13.5°$$

4.6 Relative Velocity and Relative Acceleration

In this section, we describe how observations made by different observers in different frames of reference are related to one another. A frame of reference can be described by a Cartesian coordinate system for which an observer is at rest with respect to the origin.

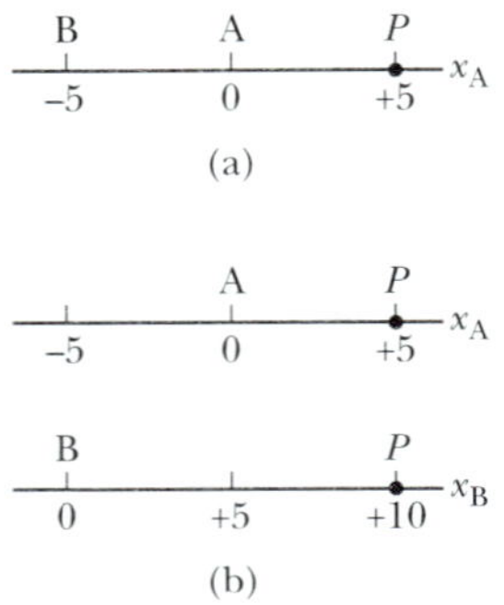

Figure 4.18 Different observers make different measurements. (a) Observer A is located at the origin, and Observer B is at a position of -5. Both observers measure the position of a particle at P. (b) If both observers see themselves at the origin of their own coordinate system, they disagree on the value of the position of the particle at P.

Figure 4.19 Two observers measure the speed of a man walking on a moving beltway. The woman standing on the beltway sees the man moving with a slower speed than does the woman observing from the stationary floor.

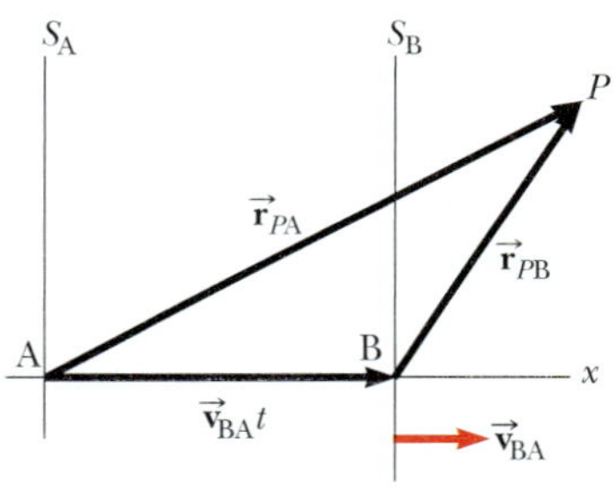

Figure 4.20 A particle located at P is described by two observers, one in the fixed frame of reference S_A and the other in the frame S_B, which moves to the right with a constant velocity $\vec{\mathbf{v}}_{BA}$. The vector $\vec{\mathbf{r}}_{PA}$ is the particle's position vector relative to S_A, and $\vec{\mathbf{r}}_{PB}$ is its position vector relative to S_B.

Let us conceptualize a sample situation in which there will be different observations for different observers. Consider the two observers A and B along the number line in Figure 4.18a. Observer A is located at the origin of a one-dimensional x_A axis, while observer B is at the position $x_A = -5$. We denote the position variable as x_A because observer A is at the origin of this axis. Both observers measure the position of point P, which is located at $x_A = +5$. Suppose observer B decides that he is located at the origin of an x_B axis as in Figure 4.18b. Notice that the two observers disagree on the value of the position of point P. Observer A claims point P is located at a position with a value of $+5$, whereas observer B claims it is located at a position with a value of $+10$. Both observers are correct, even though they make different measurements. Their measurements differ because they are making the measurement from different frames of reference.

Imagine now that observer B in Figure 4.18b is moving to the right along the x_B axis. Now the two measurements are even more different. Observer A claims point P remains at rest at a position with a value of $+5$, whereas observer B claims the position of P continuously changes with time, even passing him and moving behind him! Again, both observers are correct, with the difference in their measurements arising from their different frames of reference.

We explore this phenomenon further by considering two observers watching a man walking on a moving beltway at an airport in Figure 4.19. The woman standing on the moving beltway sees the man moving at a normal walking speed. The woman observing from the stationary floor sees the man moving with a higher speed because the beltway speed combines with his walking speed. Both observers look at the same man and arrive at different values for his speed. Both are correct; the difference in their measurements results from the relative velocity of their frames of reference.

In a more general situation, consider a particle located at point P in Figure 4.20. Imagine that the motion of this particle is being described by two observers, observer A in a reference frame S_A fixed relative to Earth and a second observer B in a reference frame S_B moving to the right relative to S_A (and therefore relative to Earth) with a constant velocity $\vec{\mathbf{v}}_{BA}$. In this discussion of relative velocity, we use a double-subscript notation; the first subscript represents what is being observed, and the second represents who is doing the observing. Therefore, the notation $\vec{\mathbf{v}}_{BA}$ means the velocity of observer B (and the attached frame S_B) as measured by observer A. With this notation, observer B measures A to be moving to the left with a velocity $\vec{\mathbf{v}}_{AB} = -\vec{\mathbf{v}}_{BA}$. For purposes of this discussion, let us place each observer at her or his respective origin.

We define the time $t = 0$ as the instant at which the origins of the two reference frames coincide in space. Therefore, at time t, the origins of the reference frames will be separated by a distance $v_{BA}t$. We label the position P of the particle relative to observer A with the position vector $\vec{\mathbf{r}}_{PA}$ and that relative to observer B with the position vector $\vec{\mathbf{r}}_{PB}$, both at time t. From Figure 4.20, we see that the vectors $\vec{\mathbf{r}}_{PA}$ and $\vec{\mathbf{r}}_{PB}$ are related to each other through the expression

$$\vec{\mathbf{r}}_{PA} = \vec{\mathbf{r}}_{PB} + \vec{\mathbf{v}}_{BA}t \tag{4.19}$$

By differentiating Equation 4.19 with respect to time, noting that $\vec{\mathbf{v}}_{BA}$ is constant, we obtain

$$\frac{d\vec{\mathbf{r}}_{PA}}{dt} = \frac{d\vec{\mathbf{r}}_{PB}}{dt} + \vec{\mathbf{v}}_{BA}$$

Galilean velocity transformation ▶

$$\vec{\mathbf{u}}_{PA} = \vec{\mathbf{u}}_{PB} + \vec{\mathbf{v}}_{BA} \tag{4.20}$$

where $\vec{\mathbf{u}}_{PA}$ is the velocity of the particle at P measured by observer A and $\vec{\mathbf{u}}_{PB}$ is its velocity measured by B. (We use the symbol $\vec{\mathbf{u}}$ for particle velocity rather than $\vec{\mathbf{v}}$, which is used for the relative velocity of two reference frames.) Equations 4.19 and 4.20 are known as **Galilean transformation equations**. They relate the position and velocity of a particle as measured by observers in relative motion. Notice the pattern of the subscripts in Equation 4.20. When relative velocities are added, the inner subscripts (B) are the same and the outer ones (P, A) match the subscripts on the velocity on the left of the equation.

Although observers in two frames measure different velocities for the particle, they measure the *same acceleration* when $\vec{\mathbf{v}}_{\mathrm{BA}}$ is constant. We can verify that by taking the time derivative of Equation 4.20:

$$\frac{d\vec{\mathbf{u}}_{P\mathrm{A}}}{dt} = \frac{d\vec{\mathbf{u}}_{P\mathrm{B}}}{dt} + \frac{d\vec{\mathbf{v}}_{\mathrm{BA}}}{dt}$$

Because $\vec{\mathbf{v}}_{\mathrm{BA}}$ is constant, $d\vec{\mathbf{v}}_{\mathrm{BA}}/dt = 0$. Therefore, we conclude that $\vec{\mathbf{a}}_{P\mathrm{A}} = \vec{\mathbf{a}}_{P\mathrm{B}}$ because $\vec{\mathbf{a}}_{P\mathrm{A}} = d\vec{\mathbf{u}}_{P\mathrm{A}}/dt$ and $\vec{\mathbf{a}}_{P\mathrm{B}} = d\vec{\mathbf{u}}_{P\mathrm{B}}/dt$. That is, **the acceleration of the particle measured by an observer in one frame of reference is the same as that measured by any other observer moving with constant velocity relative to the first frame.**

EXAMPLE 4.8 A Boat Crossing a River

A boat crossing a wide river moves with a speed of 10.0 km/h relative to the water. The water in the river has a uniform speed of 5.00 km/h due east relative to the Earth.

(A) If the boat heads due north, determine the velocity of the boat relative to an observer standing on either bank.

SOLUTION

Conceptualize Imagine moving across a river while the current pushes you down the river. You will not be able to move directly across the river, but will end up downstream as suggested in Figure 4.21a.

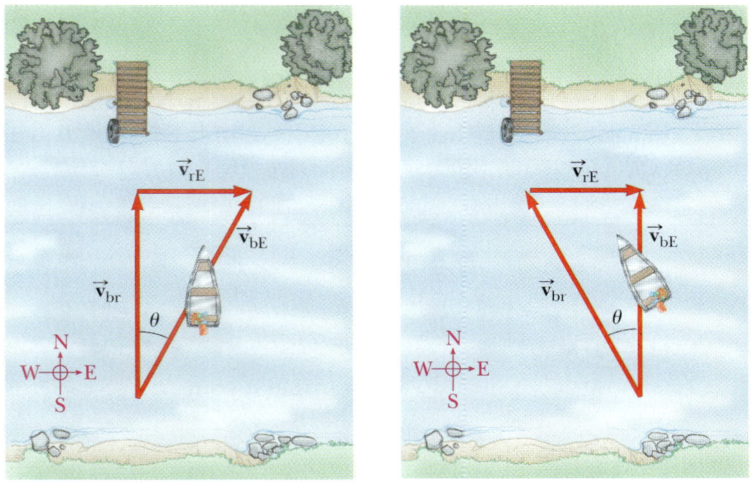

(a) (b)

Figure 4.21 (Example 4.8) (a) A boat aims directly across a river and ends up downstream. (b) To move directly across the river, the boat must aim upstream.

Categorize Because of the separate velocities of you and the river, we can categorize this problem as one involving relative velocities.

Analyze We know $\vec{\mathbf{v}}_{\mathrm{br}}$, the velocity of the *boat* relative to the *river*, and $\vec{\mathbf{v}}_{\mathrm{rE}}$, the velocity of the *river* relative to the *Earth*. What we must find is $\vec{\mathbf{v}}_{\mathrm{bE}}$, the velocity of the *boat* relative to the *Earth*. The relationship between these three quantities is $\vec{\mathbf{v}}_{\mathrm{bE}} = \vec{\mathbf{v}}_{\mathrm{br}} + \vec{\mathbf{v}}_{\mathrm{rE}}$. The terms in the equation must be manipulated as vector quantities; the vectors are shown in Figure 4.21a. The quantity $\vec{\mathbf{v}}_{\mathrm{br}}$ is due north; $\vec{\mathbf{v}}_{\mathrm{rE}}$ is due east; and the vector sum of the two, $\vec{\mathbf{v}}_{\mathrm{bE}}$, is at an angle θ as defined in Figure 4.21a.

Find the speed v_{bE} of the boat relative to the Earth using the Pythagorean theorem:

$$v_{\mathrm{bE}} = \sqrt{v_{\mathrm{br}}^2 + v_{\mathrm{rE}}^2} = \sqrt{(10.0\ \mathrm{km/h})^2 + (5.00\ \mathrm{km/h})^2}$$

$$= 11.2\ \mathrm{km/h}$$

Find the direction of $\vec{\mathbf{v}}_{\mathrm{bE}}$:

$$\theta = \tan^{-1}\left(\frac{v_{\mathrm{rE}}}{v_{\mathrm{br}}}\right) = \tan^{-1}\left(\frac{5.00}{10.0}\right) = 26.6°$$

Finalize The boat is moving at a speed of 11.2 km/h in the direction 26.6° east of north relative to the Earth. Notice that the speed of 11.2 km/h is faster than your boat speed of 10.0 km/h. The current velocity adds to yours to give you a larger speed. Notice in Figure 4.21a that your resultant velocity is at an angle to the direction straight across the river, so you will end up downstream, as we predicted.

(B) If the boat travels with the same speed of 10.0 km/h relative to the river and is to travel due north as shown in Figure 4.21b, what should its heading be?

SOLUTION

Conceptualize/Categorize This question is an extension of part (A), so we have already conceptualized and categorized the problem. One new feature of the conceptualization is that we must now aim the boat upstream so as to go straight across the river.

Analyze The analysis now involves the new triangle shown in Figure 4.21b. As in part (A), we know $\vec{\mathbf{v}}_{\text{rE}}$ and the magnitude of the vector $\vec{\mathbf{v}}_{\text{br}}$, and we want $\vec{\mathbf{v}}_{\text{bE}}$ to be directed across the river. Notice the difference between the triangle in Figure 4.21a and the one in Figure 4.21b: the hypotenuse in Figure 4.21b is no longer $\vec{\mathbf{v}}_{\text{bE}}$.

Use the Pythagorean theorem to find $\vec{\mathbf{v}}_{\text{bE}}$:

$$v_{\text{bE}} = \sqrt{v_{\text{br}}^2 - v_{\text{rE}}^2} = \sqrt{(10.0\text{ km/h})^2 - (5.00\text{ km/h})^2} = 8.66\text{ km/h}$$

Find the direction in which the boat is heading:

$$\theta = \tan^{-1}\left(\frac{v_{\text{rE}}}{v_{\text{bE}}}\right) = \tan^{-1}\left(\frac{5.00}{8.66}\right) = 30.0°$$

Finalize The boat must head upstream so as to travel directly northward across the river. For the given situation, the boat must steer a course 30.0° west of north. For faster currents, the boat must be aimed upstream at larger angles.

What If? Imagine that the two boats in parts (A) and (B) are racing across the river. Which boat arrives at the opposite bank first?

Answer In part (A), the velocity of 10 km/h is aimed directly across the river. In part (B), the velocity that is directed across the river has a magnitude of only 8.66 km/h. Therefore, the boat in part (A) has a larger velocity component directly across the river and arrives first.

Summary

ThomsonNOW™ Sign in at **www.thomsonedu.com** and go to ThomsonNOW to take a practice test for this chapter.

DEFINITIONS

The **displacement vector** $\Delta\vec{\mathbf{r}}$ for a particle is the difference between its final position vector and its initial position vector:

$$\Delta\vec{\mathbf{r}} \equiv \vec{\mathbf{r}}_f - \vec{\mathbf{r}}_i \qquad \textbf{(4.1)}$$

The **average velocity** of a particle during the time interval Δt is defined as the displacement of the particle divided by the time interval:

$$\vec{\mathbf{v}}_{\text{avg}} \equiv \frac{\Delta\vec{\mathbf{r}}}{\Delta t} \qquad \textbf{(4.2)}$$

The **instantaneous velocity** of a particle is defined as the limit of the average velocity as Δt approaches zero:

$$\vec{\mathbf{v}} \equiv \lim_{\Delta t\to 0}\frac{\Delta\vec{\mathbf{r}}}{\Delta t} = \frac{d\vec{\mathbf{r}}}{dt} \qquad \textbf{(4.3)}$$

The **average acceleration** of a particle is defined as the change in its instantaneous velocity vector divided by the time interval Δt during which that change occurs:

$$\vec{\mathbf{a}}_{\text{avg}} \equiv \frac{\vec{\mathbf{v}}_f - \vec{\mathbf{v}}_i}{t_f - t_i} = \frac{\Delta\vec{\mathbf{v}}}{\Delta t} \qquad \textbf{(4.4)}$$

The **instantaneous acceleration** of a particle is defined as the limiting value of the average acceleration as Δt approaches zero:

$$\vec{\mathbf{a}} \equiv \lim_{\Delta t\to 0}\frac{\Delta\vec{\mathbf{v}}}{\Delta t} = \frac{d\vec{\mathbf{v}}}{dt} \qquad \textbf{(4.5)}$$

Projectile motion is one type of two-dimensional motion under constant acceleration, where $a_x = 0$ and $a_y = -g$.

A particle moving in a circle of radius r with constant speed v is in **uniform circular motion**. For such a particle, the **period** of its motion is

$$T = \frac{2\pi r}{v} \qquad \textbf{(4.15)}$$

CONCEPTS AND PRINCIPLES

If a particle moves with *constant* acceleration $\vec{\mathbf{a}}$ and has velocity $\vec{\mathbf{v}}_i$ and position $\vec{\mathbf{r}}_i$ at $t = 0$, its velocity and position vectors at some later time t are

$$\vec{\mathbf{v}}_f = \vec{\mathbf{v}}_i + \vec{\mathbf{a}}t \quad (4.8)$$

$$\vec{\mathbf{r}}_f = \vec{\mathbf{r}}_i + \vec{\mathbf{v}}_i t + \tfrac{1}{2}\vec{\mathbf{a}}t^2 \quad (4.9)$$

For two-dimensional motion in the xy plane under constant acceleration, each of these vector expressions is equivalent to two component expressions: one for the motion in the x direction and one for the motion in the y direction.

It is useful to think of projectile motion in terms of a combination of two analysis models: (1) the particle under constant velocity model in the x direction and (2) the particle under constant acceleration model in the vertical direction with a constant downward acceleration of magnitude $g = 9.80\ \mathrm{m/s^2}$.

A particle in uniform circular motion undergoes a radial acceleration $\vec{\mathbf{a}}_r$ because the direction of $\vec{\mathbf{v}}$ changes in time. This acceleration is called **centripetal acceleration,** and its direction is always toward the center of the circle.

If a particle moves along a curved path in such a way that both the magnitude and the direction of $\vec{\mathbf{v}}$ change in time, the particle has an acceleration vector that can be described by two component vectors: (1) a radial component vector $\vec{\mathbf{a}}_r$ that causes the change in direction of $\vec{\mathbf{v}}$ and (2) a tangential component vector $\vec{\mathbf{a}}_t$ that causes the change in magnitude of $\vec{\mathbf{v}}$. The magnitude of $\vec{\mathbf{a}}_r$ is v^2/r, and the magnitude of $\vec{\mathbf{a}}_t$ is $|dv/dt|$.

The velocity $\vec{\mathbf{u}}_{PA}$ of a particle measured in a fixed frame of reference S_A can be related to the velocity $\vec{\mathbf{u}}_{PB}$ of the same particle measured in a moving frame of reference S_B by

$$\vec{\mathbf{u}}_{PA} = \vec{\mathbf{u}}_{PB} + \vec{\mathbf{v}}_{BA} \quad (4.20)$$

where $\vec{\mathbf{v}}_{BA}$ is the velocity of S_B relative to S_A.

ANALYSIS MODEL FOR PROBLEM SOLVING

Particle in Uniform Circular Motion If a particle moves in a circular path of radius r with a constant speed v, the magnitude of its centripetal acceleration is given by

$$a_c = \frac{v^2}{r} \quad (4.14)$$

and the period of the particle's motion is given by Equation 4.15.

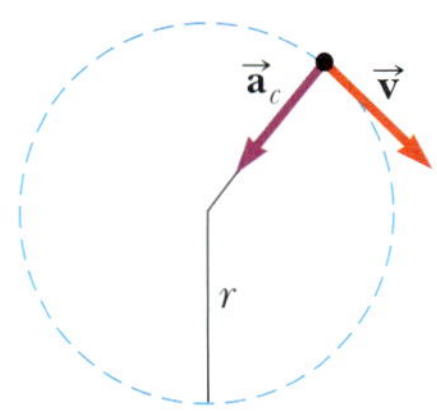

Questions

☐ denotes answer available in *Student Solutions Manual/Study Guide;* **O** denotes objective question

1. **O** Figure Q4.1 shows a bird's-eye view of a car going around a highway curve. As the car moves from point 1 to point 2, its speed doubles. Which vector (a) through (g) shows the direction of the car's average acceleration between these two points?

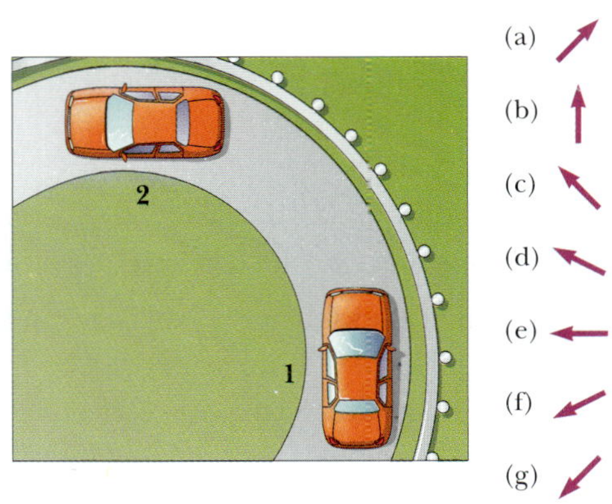

Figure Q4.1

2. If you know the position vectors of a particle at two points along its path and also know the time interval during which it moved from one point to the other, can you determine the particle's instantaneous velocity? Its average velocity? Explain.

3. Construct motion diagrams showing the velocity and acceleration of a projectile at several points along its path, assuming (a) the projectile is launched horizontally and (b) the projectile is launched at an angle θ with the horizontal.

4. **O** Entering his dorm room, a student tosses his book bag to the right and upward at an angle of 45° with the horizontal. Air resistance does not affect the bag. It moves through point Ⓐ immediately after it leaves his hand, through point Ⓑ at the top of its flight, and through point Ⓒ immediately before it lands on his top bunk bed. **(i)** Rank the following horizontal and vertical velocity components from the largest to the smallest. Note that zero is larger than a negative number. If two quantities are equal, show them as equal in your list. If any quantity is equal to zero, show that fact in your list. (a) $v_{Ⓐx}$ (b) $v_{Ⓐy}$ (c) $v_{Ⓑx}$ (d) $v_{Ⓑy}$ (e) $v_{Ⓒx}$ (f) $v_{Ⓒy}$ **(ii)** Similarly, rank the following acceleration components. (a) $a_{Ⓐx}$ (b) $a_{Ⓐy}$ (c) $a_{Ⓑx}$ (d) $a_{Ⓑy}$ (e) $a_{Ⓒx}$ (f) $a_{Ⓒy}$

5. A spacecraft drifts through space at a constant velocity. Suddenly a gas leak in the side of the spacecraft gives it a constant acceleration in a direction perpendicular to the initial velocity. The orientation of the spacecraft does not change, so the acceleration remains perpendicular to the original direction of the velocity. What is the shape of the path followed by the spacecraft in this situation?

6. **O** In which of the following situations is the moving object appropriately modeled as a projectile? Choose all correct answers. (a) A shoe is tossed in an arbitrary direction. (b) A jet airplane crosses the sky with its engines thrusting the plane forward. (c) A rocket leaves the launch pad. (d) A rocket moves through the sky, at much less than the speed of sound, after its fuel has been used up. (e) A diver throws a stone under water.

7. A projectile is launched at some angle to the horizontal with some initial speed v_i, and air resistance is negligible. Is the projectile a freely falling body? What is its acceleration in the vertical direction? What is its acceleration in the horizontal direction?

8. **O** State which of the following quantities, if any, remain constant as a projectile moves through its parabolic trajectory: (a) speed (b) acceleration (c) horizontal component of velocity (d) vertical component of velocity

9. **O** A projectile is launched on the Earth with a certain initial velocity and moves without air resistance. Another projectile is launched with the same initial velocity on the Moon, where the acceleration due to gravity is 1/6 as large. **(i)** How does the range of the projectile on the Moon compare with that of the projectile on the Earth? (a) 1/6 as large (b) the same (c) $\sqrt{6}$ times larger (d) 6 times larger (e) 36 times larger **(ii)** How does the maximum altitude of the projectile on the Moon compare with that of the projectile on the Earth? Choose from the same possibilities (a) through (e).

10. Explain whether or not the following particles have an acceleration: (a) a particle moving in a straight line with constant speed and (b) a particle moving around a curve with constant speed.

11. Describe how a driver can steer a car traveling at constant speed so that (a) the acceleration is zero or (b) the magnitude of the acceleration remains constant.

12. **O** A rubber stopper on the end of a string is swung steadily in a horizontal circle. In one trial, it moves at speed v in a circle of radius r. In a second trial, it moves at a higher speed $3v$ in a circle of radius $3r$. **(i)** In this second trial, its acceleration is (choose one) (a) the same as in the first trial (b) three times larger (c) one-third as large (d) nine times larger (e) one-ninth as large **(ii)** In the second trial, how does its period compare with its period in the first trial? Choose your answers from the same possibilities (a) through (e).

13. An ice skater is executing a figure eight, consisting of two equal, tangent circular paths. Throughout the first loop she increases her speed uniformly, and during the second loop she moves at a constant speed. Draw a motion diagram showing her velocity and acceleration vectors at several points along the path of motion.

14. **O** A certain light truck can go around a curve having a radius of 150 m with a maximum speed of 32.0 m/s. To have the same acceleration, at what maximum speed can it go around a curve having a radius of 75.0 m? (a) 64 m/s (b) 45 m/s (c) 32 m/s (d) 23 m/s (e) 16 m/s (f) 8 m/s

15. **O** Galileo suggested the idea for this question: A sailor drops a wrench from the top of a sailboat's vertical mast while the boat is moving rapidly and steadily straight forward. Where will the wrench hit the deck? (a) ahead of the base of the mast (b) at the base of the mast (c) behind the base of the mast (d) on the windward side of the base of the mast

16. **O** A girl, moving at 8 m/s on rollerblades, is overtaking a boy moving at 5 m/s as they both skate on a straight path. The boy tosses a ball backward toward the girl, giving it speed 12 m/s relative to him. What is the speed of the ball relative to the girl, who catches it? (a) (8 + 5 + 12) m/s (b) (8 − 5 − 12) m/s (c) (8 + 5 − 12) m/s (d) (8 − 5 + 12) m/s (e) (−8 + 5 + 12) m/s

Problems

WebAssign The Problems from this chapter may be assigned online in WebAssign.

ThomsonNOW Sign in at **www.thomsonedu.com** and go to ThomsonNOW to assess your understanding of this chapter's topics with additional quizzing and conceptual questions.

1, 2, 3 denotes straightforward, intermediate, challenging; □ denotes full solution available in *Student Solutions Manual/Study Guide*; ▲ denotes coached solution with hints available at **www.thomsonedu.com;** denotes developing symbolic reasoning; ● denotes asking for qualitative reasoning; denotes computer useful in solving problem

Section 4.1 The Position, Velocity, and Acceleration Vectors

1. ▲ A motorist drives south at 20.0 m/s for 3.00 min, then turns west and travels at 25.0 m/s for 2.00 min, and finally travels northwest at 30.0 m/s for 1.00 min. For this 6.00-min trip, find (a) the total vector displacement, (b) the average speed, and (c) the average velocity. Let the positive x axis point east.

2. A golf ball is hit off a tee at the edge of a cliff. Its x and y coordinates as functions of time are given by the following expressions:

$$x = (18.0\ \text{m/s})t$$

$$y = (4.00\ \text{m/s})t - (4.90\ \text{m/s}^2)t^2$$

(a) Write a vector expression for the ball's position as a function of time, using the unit vectors $\hat{\mathbf{i}}$ and $\hat{\mathbf{j}}$. By taking derivatives, obtain expressions for (b) the velocity vector $\vec{\mathbf{v}}$ as a function of time and (c) the acceleration vector $\vec{\mathbf{a}}$ as a function of time. Next use unit–vector notation to write expressions for (d) the position, (e) the velocity, and (f) the acceleration of the golf ball, all at $t = 3.00$ s.

3. When the Sun is directly overhead, a hawk dives toward the ground with a constant velocity of 5.00 m/s at 60.0° below the horizontal. Calculate the speed of its shadow on the level ground.

4. ● The coordinates of an object moving in the xy plane vary with time according to $x = -(5.00\ \text{m})\sin(\omega t)$ and $y = (4.00\ \text{m}) - (5.00\ \text{m})\cos(\omega t)$, where ω is a constant and t is in seconds. (a) Determine the components of velocity and components of acceleration of the object at $t = 0$. (b) Write expressions for the position vector, the velocity vector, and the acceleration vector of the object at any time $t > 0$. (c) Describe the path of the object in an xy plot.

Section 4.2 Two-Dimensional Motion with Constant Acceleration

5. A fish swimming in a horizontal plane has velocity $\vec{\mathbf{v}}_i = (4.00\hat{\mathbf{i}} + 1.00\hat{\mathbf{j}})$ m/s at a point in the ocean where the position relative to a certain rock is $\vec{\mathbf{r}}_i = (10.0\hat{\mathbf{i}} - 4.00\hat{\mathbf{j}})$ m. After the fish swims with constant acceleration for 20.0 s, its velocity is $\vec{\mathbf{v}} = (20.0\hat{\mathbf{i}} - 5.00\hat{\mathbf{j}})$ m/s. (a) What are the components of the acceleration? (b) What is the direction of the acceleration with respect to unit vector $\hat{\mathbf{i}}$? (c) If the fish maintains constant acceleration, where is it at $t = 25.0$ s, and in what direction is it moving?

6. The vector position of a particle varies in time according to the expression $\vec{\mathbf{r}} = (3.00\hat{\mathbf{i}} - 6.00t^2\hat{\mathbf{j}})$ m. (a) Find expressions for the velocity and acceleration of the particle as functions of time. (b) Determine the particle's position and velocity at $t = 1.00$ s.

7. *What if the acceleration is not constant?* A particle starts from the origin with velocity $5\hat{\mathbf{i}}$ m/s at $t = 0$ and moves in the xy plane with a varying acceleration given by $\vec{\mathbf{a}} = (6\sqrt{t}\,\hat{\mathbf{j}})$ m/s², where t is in s. (a) Determine the vector velocity of the particle as a function of time. (b) Determine the position of the particle as a function of time.

8. A particle initially located at the origin has an acceleration of $\vec{\mathbf{a}} = 3.00\hat{\mathbf{j}}$ m/s² and an initial velocity of $\vec{\mathbf{v}}_i = 5.00\hat{\mathbf{i}}$ m/s. Find (a) the vector position and velocity of the particle at any time t and (b) the coordinates and speed of the particle at $t = 2.00$ s.

Section 4.3 Projectile Motion

Note: Ignore air resistance in all problems. Take $g = 9.80$ m/s² at the Earth's surface.

9. ▲ In a local bar, a customer slides an empty beer mug down the counter for a refill. The bartender is momentarily distracted and does not see the mug, which slides off the counter and strikes the floor 1.40 m from the base of the counter. If the height of the counter is 0.860 m, (a) with what velocity did the mug leave the counter? (b) What was the direction of the mug's velocity just before it hit the floor?

10. In a local bar, a customer slides an empty beer mug down the counter for a refill. The bartender is just deciding to go home and rethink his life, so he does not see the mug. It slides off the counter and strikes the floor at distance d from the base of the counter. The height of the counter is h. (a) With what velocity did the mug leave the counter? (b) What was the direction of the mug's velocity just before it hit the floor?

11. To start an avalanche on a mountain slope, an artillery shell is fired with an initial velocity of 300 m/s at 55.0° above the horizontal. It explodes on the mountainside 42.0 s after firing. What are the x and y coordinates of the shell where it explodes, relative to its firing point?

12. ● A rock is thrown upward from the level ground in such a way that the maximum height of its flight is equal to its horizontal range d. (a) At what angle θ is the rock thrown? (b) **What If?** Would your answer to part (a) be different on a different planet? Explain. (c) What is the range d_{max} the rock can attain if it is launched at the same speed but at the optimal angle for maximum range?

13. A projectile is fired in such a way that its horizontal range is equal to three times its maximum height. What is the angle of projection?

14. A firefighter, a distance d from a burning building, directs a stream of water from a fire hose at angle θ_i above the horizontal as shown in Figure P4.14. If the initial speed of the stream is v_i, at what height h does the water strike the building?

Figure P4.14

15. A ball is tossed from an upper-story window of a building. The ball is given an initial velocity of 8.00 m/s at an angle of 20.0° below the horizontal. It strikes the ground 3.00 s later. (a) How far horizontally from the base of the

building does the ball strike the ground? (b) Find the height from which the ball was thrown. (c) How long does it take the ball to reach a point 10.0 m below the level of launching?

16. A landscape architect is planning an artificial waterfall in a city park. Water flowing at 1.70 m/s will leave the end of a horizontal channel at the top of a vertical wall 2.35 m high, and from there the water falls into a pool. (a) Will the space behind the waterfall be wide enough for a pedestrian walkway? (b) To sell her plan to the city council, the architect wants to build a model to standard scale, one-twelfth actual size. How fast should the water flow in the channel in the model?

17. ▲ A placekicker must kick a football from a point 36.0 m (about 40 yards) from the goal, and half the crowd hopes the ball will clear the crossbar, which is 3.05 m high. When kicked, the ball leaves the ground with a speed of 20.0 m/s at an angle of 53.0° to the horizontal. (a) By how much does the ball clear or fall short of clearing the crossbar? (b) Does the ball approach the crossbar while still rising or while falling?

18. A dive-bomber has a velocity of 280 m/s at an angle θ below the horizontal. When the altitude of the aircraft is 2.15 km, it releases a bomb, which subsequently hits a target on the ground. The magnitude of the displacement from the point of release of the bomb to the target is 3.25 km. Find the angle θ.

19. A playground is on the flat roof of a city school, 6.00 m above the street below. The vertical wall of the building is 7.00 m high, forming a 1 m-high railing around the playground. A ball has fallen to the street below, and a passerby returns it by launching it at an angle of 53.0° above the horizontal at a point 24.0 m from the base of the building wall. The ball takes 2.20 s to reach a point vertically above the wall. (a) Find the speed at which the ball was launched. (b) Find the vertical distance by which the ball clears the wall. (c) Find the distance from the wall to the point on the roof where the ball lands.

20. A basketball star covers 2.80 m horizontally in a jump to dunk the ball (Fig. P4.20a). His motion through space can be modeled precisely as that of a particle at his *center of mass*, which we will define in Chapter 9. His center of mass is at elevation 1.02 m when he leaves the floor. It reaches a maximum height of 1.85 m above the floor and is at elevation 0.900 m when he touches down again. Determine (a) his time of flight (his "hang time"), (b) his horizontal and (c) vertical velocity components at the instant of takeoff, and (d) his takeoff angle. (e) For comparison, determine the hang time of a whitetail deer making a jump (Fig. P4.20b) with center-of-mass elevations y_i = 1.20 m, y_{max} = 2.50 m, and y_f = 0.700 m.

© Ray Stubblebine/Reuters/Corbis

Bill Lee/Dembinsky Photo Associates

(a) (b)

Figure P4.20

21. A soccer player kicks a rock horizontally off a 40.0-m-high cliff into a pool of water. If the player hears the sound of the splash 3.00 s later, what was the initial speed given to the rock? Assume the speed of sound in air is 343 m/s.

22. ● The motion of a human body through space can be modeled as the motion of a particle at the body's center of mass, as we will study in Chapter 9. The components of the position of an athlete's center of mass from the beginning to the end of a certain jump are described by the two equations

$$x_f = 0 + (11.2 \text{ m/s})(\cos 18.5°)t$$

$$0.360 \text{ m} = 0.840 \text{ m} + (11.2 \text{ m/s})(\sin 18.5°)t - \tfrac{1}{2}(9.80 \text{ m/s}^2)t^2$$

where t is the time at which the athlete lands after taking off at t = 0. Identify (a) his vector position and (b) his vector velocity at the takeoff point. (c) The world long-jump record is 8.95 m. How far did the athlete jump in this problem? (d) Describe the shape of the trajectory of his center of mass.

23. A fireworks rocket explodes at height h, the peak of its vertical trajectory. It throws out burning fragments in all directions, but all at the same speed v. Pellets of solidified metal fall to the ground without air resistance. Find the smallest angle that the final velocity of an impacting fragment makes with the horizontal.

Section 4.4 The Particle in Uniform Circular Motion

Note: Problems 10 and 12 in Chapter 6 can also be assigned with this section and the next.

24. From information on the endpapers of this book, compute the radial acceleration of a point on the surface of the Earth at the equator, owing to the rotation of the Earth about its axis.

25. ▲ The athlete shown in Figure P4.25 rotates a 1.00-kg discus along a circular path of radius 1.06 m. The maximum speed of the discus is 20.0 m/s. Determine the magnitude of the maximum radial acceleration of the discus.

Sam Sargent/Liaison International

Figure P4.25

26. As their booster rockets separate, space shuttle astronauts typically feel accelerations up to $3g$, where g = 9.80 m/s². In their training, astronauts ride in a device in which they experience such an acceleration as a centripetal acceleration. Specifically, the astronaut is fastened securely at the end of a mechanical arm that then turns at constant

2 = intermediate; 3 = challenging; □ = SSM/SG; ▲ = ThomsonNOW; ▭ = symbolic reasoning; ● = qualitative reasoning

speed in a horizontal circle. Determine the rotation rate, in revolutions per second, required to give an astronaut a centripetal acceleration of $3.00g$ while in circular motion with radius 9.45 m.

27. Young David who slew Goliath experimented with slings before tackling the giant. He found he could revolve a sling of length 0.600 m at the rate of 8.00 rev/s. If he increased the length to 0.900 m, he could revolve the sling only 6.00 times per second. (a) Which rate of rotation gives the greater speed for the stone at the end of the sling? (b) What is the centripetal acceleration of the stone at 8.00 rev/s? (c) What is the centripetal acceleration at 6.00 rev/s?

Section 4.5 Tangential and Radial Acceleration

28. ● (a) Could a particle moving with instantaneous speed 3.00 m/s on a path with radius of curvature 2.00 m have an acceleration of magnitude 6.00 m/s^2? (b) Could it have $|\vec{\mathbf{a}}| = 4.00$ m/s^2? In each case, if the answer is yes, explain how it can happen; if the answer is no, explain why not.

29. A train slows down as it rounds a sharp horizontal turn, slowing from 90.0 km/h to 50.0 km/h in the 15.0 s that it takes to round the bend. The radius of the curve is 150 m. Compute the acceleration at the moment the train speed reaches 50.0 km/h. Assume it continues to slow down at this time at the same rate.

30. A ball swings in a vertical circle at the end of a rope 1.50 m long. When the ball is 36.9° past the lowest point on its way up, its total acceleration is $(-22.5\hat{\mathbf{i}} + 20.2\hat{\mathbf{j}})$ m/s^2. At that instant, (a) sketch a vector diagram showing the components of its acceleration, (b) determine the magnitude of its radial acceleration, and (c) determine the speed and velocity of the ball.

31. Figure P4.31 represents the total acceleration of a particle moving clockwise in a circle of radius 2.50 m at a certain instant of time. At this instant, find (a) the radial acceleration, (b) the speed of the particle, and (c) its tangential acceleration.

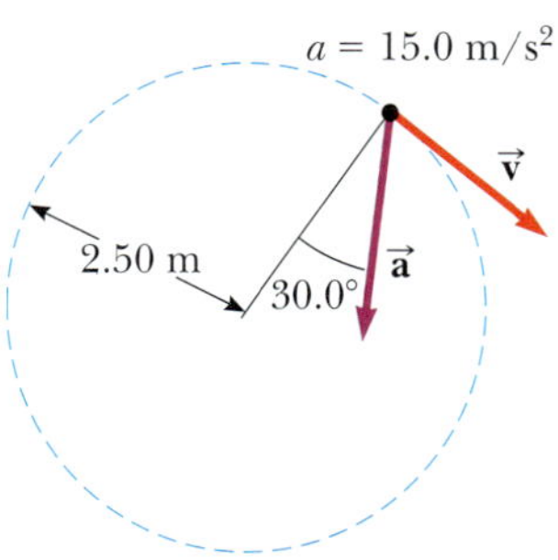

Figure P4.31

32. A race car starts from rest on a circular track. The car increases its speed at a constant rate a_t as it goes once around the track. Find the angle that the total acceleration of the car makes—with the radius connecting the center of the track and the car—at the moment the car completes the circle.

Section 4.6 Relative Velocity and Relative Acceleration

33. A car travels due east with a speed of 50.0 km/h. Raindrops are falling at a constant speed vertically with respect to the Earth. The traces of the rain on the side windows of the car make an angle of 60.0° with the vertical. Find the velocity of the rain with respect to (a) the car and (b) the Earth.

34. Heather in her Corvette accelerates at the rate of $(3.00\hat{\mathbf{i}} - 2.00\hat{\mathbf{j}})$ m/s^2, while Jill in her Jaguar accelerates at $(1.00\hat{\mathbf{i}} + 3.00\hat{\mathbf{j}})$ m/s^2. They both start from rest at the origin of an xy coordinate system. After 5.00 s, (a) what is Heather's speed with respect to Jill, (b) how far apart are they, and (c) what is Heather's acceleration relative to Jill?

35. A river has a steady speed of 0.500 m/s. A student swims upstream a distance of 1.00 km and swims back to the starting point. If the student can swim at a speed of 1.20 m/s in still water, how long does the trip take? Compare this answer with the time interval required for the trip if the water were still.

36. How long does it take an automobile traveling in the left lane at 60.0 km/h to pull alongside a car traveling in the same direction in the right lane at 40.0 km/h if the cars' front bumpers are initially 100 m apart?

37. Two swimmers, Alan and Beth, start together at the same point on the bank of a wide stream that flows with a speed v. Both move at the same speed c (where $c > v$), relative to the water. Alan swims downstream a distance L and then upstream the same distance. Beth swims so that her motion relative to the Earth is perpendicular to the banks of the stream. She swims the distance L and then back the same distance so that both swimmers return to the starting point. Which swimmer returns first? *Note:* First guess the answer.

38. ● A farm truck moves due north with a constant velocity of 9.50 m/s on a limitless horizontal stretch of road. A boy riding on the back of the truck throws a can of soda upward and catches the projectile at the same location on the truck bed, but 16.0 m farther down the road. (a) In the frame of reference of the truck, at what angle to the vertical does the boy throw the can? (b) What is the initial speed of the can relative to the truck? (c) What is the shape of the can's trajectory as seen by the boy? (d) An observer on the ground watches the boy throw the can and catch it. In this observer's ground frame of reference, describe the shape of the can's path and determine the initial velocity of the can.

39. A science student is riding on a flatcar of a train traveling along a straight horizontal track at a constant speed of 10.0 m/s. The student throws a ball into the air along a path that he judges to make an initial angle of 60.0° with the horizontal and to be in line with the track. The student's professor, who is standing on the ground nearby, observes the ball to rise vertically. How high does she see the ball rise?

40. ● A bolt drops from the ceiling of a moving train car that is accelerating northward at a rate of 2.50 m/s^2. (a) What is the acceleration of the bolt relative to the train car? (b) What is the acceleration of the bolt relative to the Earth? (c) Describe the trajectory of the bolt as seen by an observer inside the train car. (d) Describe the trajectory of the bolt as seen by an observer fixed on the Earth.

41. A Coast Guard cutter detects an unidentified ship at a distance of 20.0 km in the direction 15.0° east of north. The ship is traveling at 26.0 km/h on a course at 40.0° east of north. The Coast Guard wishes to send a speedboat to intercept the vessel and investigate it. If the speedboat travels 50.0 km/h, in what direction should it head? Express the direction as a compass bearing with respect to due north.

Additional Problems

42. *The "Vomit Comet."* In zero-gravity astronaut training and equipment testing, NASA flies a KC135A aircraft along a parabolic flight path. As shown in Figure P4.42, the aircraft climbs from 24 000 ft to 31 000 ft, where it enters the zero-g parabola with a velocity of 143 m/s nose high at 45.0° and exits with velocity 143 m/s at 45.0° nose low. During this portion of the flight, the aircraft and objects inside its padded cabin are in free fall; they have gone ballistic. The aircraft then pulls out of the dive with an upward acceleration of $0.800g$, moving in a vertical circle with radius 4.13 km. (During this portion of the flight, occupants of the aircraft perceive an acceleration of $1.8g$.) What are the aircraft's (a) speed and (b) altitude at the top of the maneuver? (c) What is the time interval spent in zero gravity? (d) What is the speed of the aircraft at the bottom of the flight path?

43. An athlete throws a basketball upward from the ground, giving it speed 10.6 m/s at an angle of 55.0° above the horizontal. (a) What is the acceleration of the basketball at the highest point in its trajectory? (b) On its way down, the basketball hits the rim of the basket, 3.05 m above the floor. It bounces straight up with one-half the speed with which it hit the rim. What height above the floor does the basketball reach on this bounce?

44. ● (a) An athlete throws a basketball toward the east, with initial speed 10.6 m/s at an angle of 55.0° above the horizontal. Just as the basketball reaches the highest point of its trajectory, it hits an eagle (the mascot of the opposing team) flying horizontally west. The ball bounces back horizontally west with 1.50 times the speed it had just before their collision. How far behind the player who threw it does the ball land? (b) This situation is not covered in the rule book, so the officials turn the clock back to repeat this part of the game. The player throws the ball in the same way. The eagle is thoroughly annoyed and this time intercepts the ball so that, at the same point in its trajectory, the ball again bounces from the bird's beak with 1.50 times its impact speed, moving west at some nonzero angle with the horizontal. Now the ball hits the player's head, at the same location where her hands had released it. Is the angle necessarily positive (that is, above the horizontal), necessarily negative (below the horizontal), or could it be either? Give a convincing argument, either mathematical or conceptual, for your answer.

45. Manny Ramírez hits a home run so that the baseball just clears the top row of bleachers, 21.0 m high, located 130 m from home plate. The ball is hit at an angle of 35.0° to the horizontal, and air resistance is negligible. Find (a) the initial speed of the ball, (b) the time interval required for the ball to reach the bleachers, and (c) the velocity components and the speed of the ball when it passes over the top row. Assume the ball is hit at a height of 1.00 m above the ground.

46. As some molten metal splashes, one droplet flies off to the east with initial velocity v_i at angle θ_i above the horizontal and another droplet flies off to the west with the same speed at the same angle above the horizontal as shown in Figure P4.46. In terms of v_i and θ_i, find the distance between the droplets as a function of time.

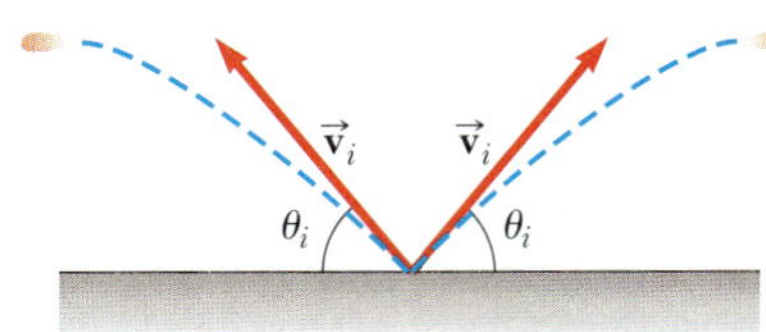

Figure P4.46

47. A pendulum with a cord of length $r = 1.00$ m swings in a vertical plane (Fig. P4.47). When the pendulum is in the two horizontal positions $\theta = 90.0°$ and $\theta = 270°$, its speed is 5.00 m/s. (a) Find the magnitude of the radial acceleration and tangential acceleration for these positions. (b) Draw vector diagrams to determine the direction of the total acceleration for these two positions. (c) Calculate the magnitude and direction of the total acceleration.

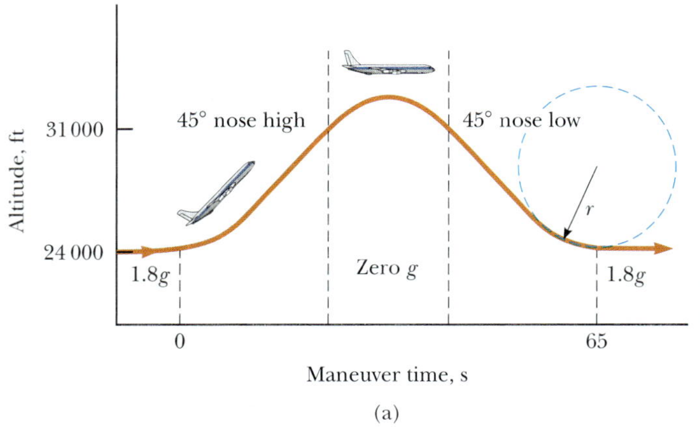

(a)

(b)

Figure P4.42

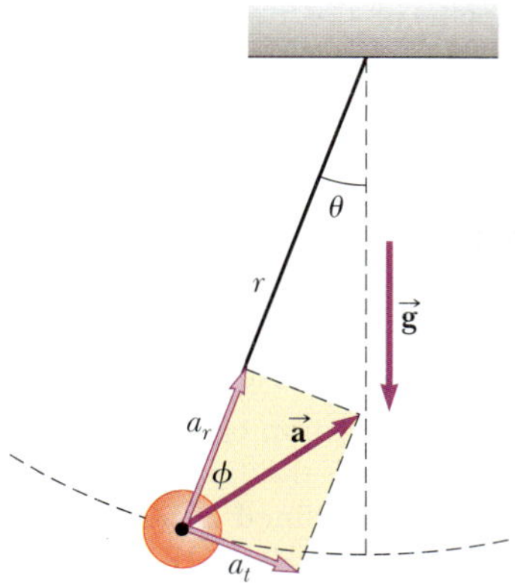

Figure P4.47

48. An astronaut on the surface of the Moon fires a cannon to launch an experiment package, which leaves the barrel moving horizontally. (a) What must be the muzzle speed of the package so that it travels completely around the Moon and returns to its original location? (b) How long does this trip around the Moon take? Assume the free-fall acceleration on the Moon is one-sixth of that on the Earth.

49. ● A projectile is launched from the point ($x = 0$, $y = 0$) with velocity $(12.0\hat{\mathbf{i}} + 49.0\hat{\mathbf{j}})$ m/s, at $t = 0$. (a) Make a table listing the projectile's distance $|\vec{\mathbf{r}}|$ from the origin at the end of each second thereafter, for $0 \le t \le 10$ s. Tabulating the x and y coordinates and the components of velocity v_x and v_y may also be useful. (b) Observe that the projectile's distance from its starting point increases with time, goes through a maximum, and starts to decrease. Prove that the distance is a maximum when the position vector is perpendicular to the velocity. *Suggestion:* Argue that if $\vec{\mathbf{v}}$ is not perpendicular to $\vec{\mathbf{r}}$, then $|\vec{\mathbf{r}}|$ must be increasing or decreasing. (c) Determine the magnitude of the maximum distance. Explain your method.

50. ● A spring cannon is located at the edge of a table that is 1.20 m above the floor. A steel ball is launched from the cannon with speed v_0 at 35.0° above the horizontal. (a) Find the horizontal displacement component of the ball to the point where it lands on the floor as a function of v_0. We write this function as $x(v_0)$. Evaluate x for (b) $v_0 = 0.100$ m/s and for (c) $v_0 = 100$ m/s. (d) Assume v_0 is close to zero but not equal to zero. Show that one term in the answer to part (a) dominates so that the function $x(v_0)$ reduces to a simpler form. (e) If v_0 is very large, what is the approximate form of $x(v_0)$? (f) Describe the overall shape of the graph of the function $x(v_0)$. *Suggestion:* As practice, you could do part (b) before doing part (a).

51. When baseball players throw the ball in from the outfield, they usually allow it to take one bounce before it reaches the infield on the theory that the ball arrives sooner that way. Suppose the angle at which a bounced ball leaves the ground is the same as the angle at which the outfielder threw it as shown in Figure P4.51, but the ball's speed after the bounce is one-half of what it was before the bounce. (a) Assume the ball is always thrown with the same initial speed. At what angle θ should the fielder throw the ball to make it go the same distance D with one bounce (blue path) as a ball thrown upward at 45.0° with no bounce (green path)? (b) Determine the ratio of the time interval for the one-bounce throw to the flight time for the no-bounce throw.

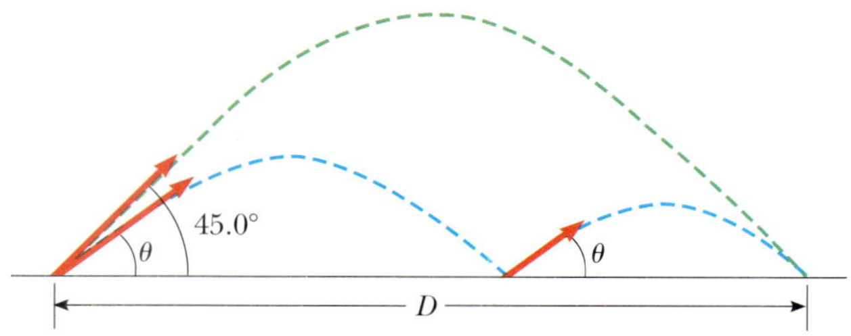

Figure P4.51

52. A truck loaded with cannonball watermelons stops suddenly to avoid running over the edge of a washed-out bridge (Fig. P4.52). The quick stop causes a number of melons to fly off the truck. One melon rolls over the edge with an initial speed $v_i = 10.0$ m/s in the horizontal direction. A cross section of the bank has the shape of the bottom half of a parabola with its vertex at the edge of the road and with the equation $y^2 = 16x$, where x and y are measured in meters. What are the x and y coordinates of the melon when it splatters on the bank?

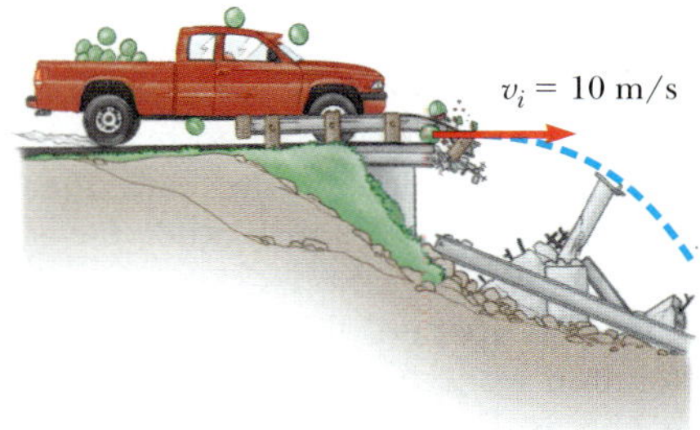

Figure P4.52

53. Your grandfather is copilot of a bomber, flying horizontally over level terrain, with a speed of 275 m/s relative to the ground, at an altitude of 3 000 m. (a) The bombardier releases one bomb. How far will the bomb travel horizontally between its release and its impact on the ground? Ignore the effects of air resistance. (b) Firing from the people on the ground suddenly incapacitates the bombardier before he can call, "Bombs away!" Consequently, the pilot maintains the plane's original course, altitude, and speed through a storm of flak. Where will the plane be when the bomb hits the ground? (c) The plane has a telescopic bombsight set so that the bomb hits the target seen in the sight at the moment of release. At what angle from the vertical was the bombsight set?

54. A person standing at the top of a hemispherical rock of radius R kicks a ball (initially at rest on the top of the rock) to give it horizontal velocity $\vec{\mathbf{v}}_i$ as shown in Figure P4.54. (a) What must be its minimum initial speed if the ball is never to hit the rock after it is kicked? (b) With this initial speed, how far from the base of the rock does the ball hit the ground?

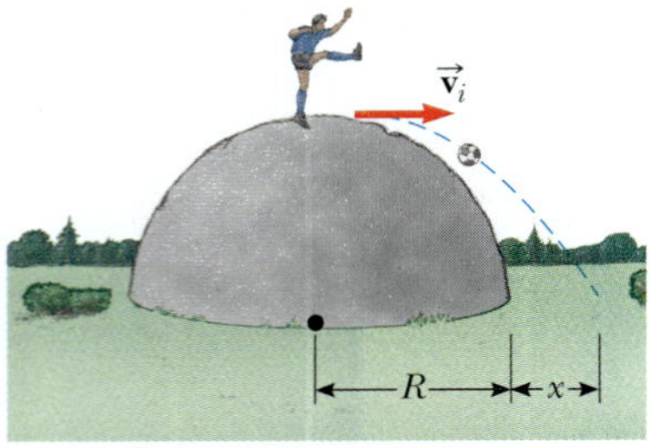

Figure P4.54

55. A hawk is flying horizontally at 10.0 m/s in a straight line, 200 m above the ground. A mouse it has been carrying struggles free from its talons. The hawk continues on its path at the same speed for 2.00 s before attempting to retrieve its prey. To accomplish the retrieval, it dives in a straight line at constant speed and recaptures the mouse 3.00 m above the ground. (a) Assuming no air resistance acts on the mouse, find the diving speed of the hawk. (b) What angle did the hawk make with the horizontal during its descent? (c) For how long did the mouse "enjoy" free fall?

56. The determined coyote is out once more in pursuit of the elusive roadrunner. The coyote wears a pair of Acme jet-powered roller skates, which provide a constant horizontal acceleration of 15.0 m/s^2 (Fig. P4.56). The coyote starts at rest 70.0 m from the brink of a cliff at the instant the roadrunner zips past in the direction of the cliff. (a) Assuming the roadrunner moves with constant speed, determine the minimum speed it must have to reach the cliff before the coyote. At the edge of the cliff, the roadrunner escapes by making a sudden turn, while the coyote continues straight ahead. The coyote's skates remain horizontal and continue to operate while the coyote is in flight, so its acceleration while in the air is $(15.0\hat{\mathbf{i}} - 9.80\hat{\mathbf{j}})$ m/s^2. (b) The cliff is 100 m above the flat floor of a canyon. Determine where the coyote lands in the canyon. (c) Determine the components of the coyote's impact velocity.

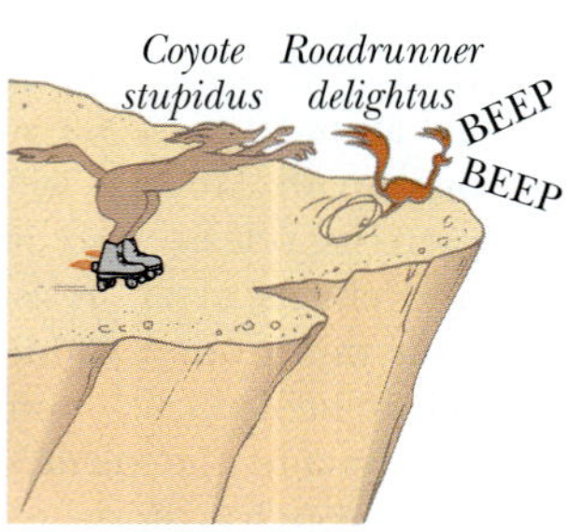

Figure P4.56

57. ▲ A car is parked on a steep incline overlooking the ocean, where the incline makes an angle of 37.0° below the horizontal. The negligent driver leaves the car in neutral, and the parking brakes are defective. Starting from rest at $t = 0$, the car rolls down the incline with a constant acceleration of 4.00 m/s^2, traveling 50.0 m to the edge of a vertical cliff. The cliff is 30.0 m above the ocean. Find (a) the speed of the car when it reaches the edge of the cliff and the time interval elapsed when it arrives there, (b) the velocity of the car when it lands in the ocean, (c) the total time interval that the car is in motion, and (d) the position of the car when it lands in the ocean, relative to the base of the cliff.

58. ● Do not hurt yourself; do not strike your hand against anything. Within these limitations, describe what you do to give your hand a large acceleration. Compute an order-of-magnitude estimate of this acceleration, stating the quantities you measure or estimate and their values.

59. ● A skier leaves the ramp of a ski jump with a velocity of 10.0 m/s, 15.0° above the horizontal, as shown in Figure P4.59. The slope is inclined at 50.0°, and air resistance is negligible. Find (a) the distance from the ramp to where the jumper lands and (b) the velocity components just before the landing. (How do you think the results might be affected if air resistance were included? Note that jumpers lean forward in the shape of an airfoil, with their hands at their sides, to increase their distance. Why does this method work?)

Figure P4.59

60. An angler sets out upstream from Metaline Falls on the Pend Oreille River in northwestern Washington State. His small boat, powered by an outboard motor, travels at a constant speed v in still water. The water flows at a lower constant speed v_w. He has traveled upstream for 2.00 km when his ice chest falls out of the boat. He notices that the chest is missing only after he has gone upstream for another 15.0 min. At that point, he turns around and heads back downstream, all the time traveling at the same speed relative to the water. He catches up with the floating ice chest just as it is about to go over the falls at his starting point. How fast is the river flowing? Solve this problem in two ways. (a) First, use the Earth as a reference frame. With respect to the Earth, the boat travels upstream at speed $v - v_w$ and downstream at $v + v_w$. (b) A second much simpler and more elegant solution is obtained by using the water as the reference frame. This approach has important applications in many more complicated problems; examples are calculating the motion of rockets and satellites and analyzing the scattering of subatomic particles from massive targets.

61. An enemy ship is on the east side of a mountainous island as shown in Figure P4.61. The enemy ship has maneuvered to within 2 500 m of the 1 800-m-high mountain peak and can shoot projectiles with an initial speed of 250 m/s. If the western shoreline is horizontally 300 m from the peak, what are the distances from the western shore at which a ship can be safe from the bombardment of the enemy ship?

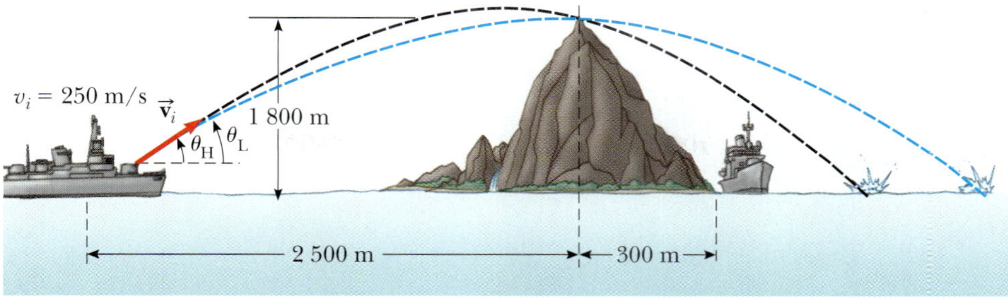

Figure P4.61

62. In the **What If?** section of Example 4.5, it was claimed that the maximum range of a ski jumper occurs for a launch angle θ given by

$$\theta = 45° - \frac{\phi}{2}$$

where ϕ is the angle that the hill makes with the horizontal in Figure 4.14. Prove this claim by deriving this equation.

Answers to Quick Quizzes

4.1 (a). Because acceleration occurs whenever the velocity changes in any way—with an increase or decrease in speed, a change in direction, or both—all three controls are accelerators. The gas pedal causes the car to speed up; the brake pedal causes the car to slow down. The steering wheel changes the direction of the velocity vector.

4.2 **(i)**, (b). At only one point—the peak of the trajectory—are the velocity and acceleration vectors perpendicular to each other. The velocity vector is horizontal at that point, and the acceleration vector is downward. **(ii)**, (a). The acceleration vector is always directed downward. The velocity vector is never vertical and parallel to the acceleration vector if the object follows a path such as that in Figure 4.8.

4.3 15°, 30°, 45°, 60°, 75°. The greater the maximum height, the longer it takes the projectile to reach that altitude and then fall back down from it. So, as the launch angle increases, the time of flight increases.

4.4 **(i)**, (d). Because the centripetal acceleration is proportional to the square of the speed of the particle, doubling the speed increases the acceleration by a factor of 4. **(ii)**, (b). The period is inversely proportional to the speed of the particle.

4.5 **(i)**, (b). The velocity vector is tangent to the path. If the acceleration vector is to be parallel to the velocity vector, it must also be tangent to the path, which requires that the acceleration vector have no component perpendicular to the path. If the path were to change direction, the acceleration vector would have a radial component, perpendicular to the path. Therefore, the path must remain straight. **(ii)**, (d). If the acceleration vector is to be perpendicular to the velocity vector, it must have no component tangent to the path. On the other hand, if the speed is changing, there *must* be a component of the acceleration tangent to the path. Therefore, the velocity and acceleration vectors are never perpendicular in this situation. They can only be perpendicular if there is no change in the speed.

A small tugboat exerts a force on a large ship, causing it to move. How can such a small boat move such a large object? (Steve Raymer/CORBIS)

5 The Laws of Motion

In Chapters 2 and 4, we described the motion of an object in terms of its position, velocity, and acceleration without considering what might influence that motion. Now we consider the external influence: What might cause one object to remain at rest and another object to accelerate? The two main factors we need to consider are the forces acting on an object and the mass of the object. In this chapter, we begin our study of *dynamics* by discussing the three basic laws of motion, which deal with forces and masses and were formulated more than three centuries ago by Isaac Newton.

5.1 The Concept of Force

Everyone has a basic understanding of the concept of force from everyday experience. When you push your empty dinner plate away, you exert a force on it. Similarly, you exert a force on a ball when you throw or kick it. In these examples, the word *force* refers to an interaction with an object by means of muscular activity and some change in the object's velocity. Forces do not always cause motion, however. For example, when you are sitting, a gravitational force acts on your body and yet you remain stationary. As a second example, you can push (in other words, exert a force) on a large boulder and not be able to move it.

What force (if any) causes the Moon to orbit the Earth? Newton answered this and related questions by stating that forces are what cause any change in the velocity of an object. The Moon's velocity is not constant because it moves in a nearly circular orbit around the Earth. This change in velocity is caused by the gravitational force exerted by the Earth on the Moon.

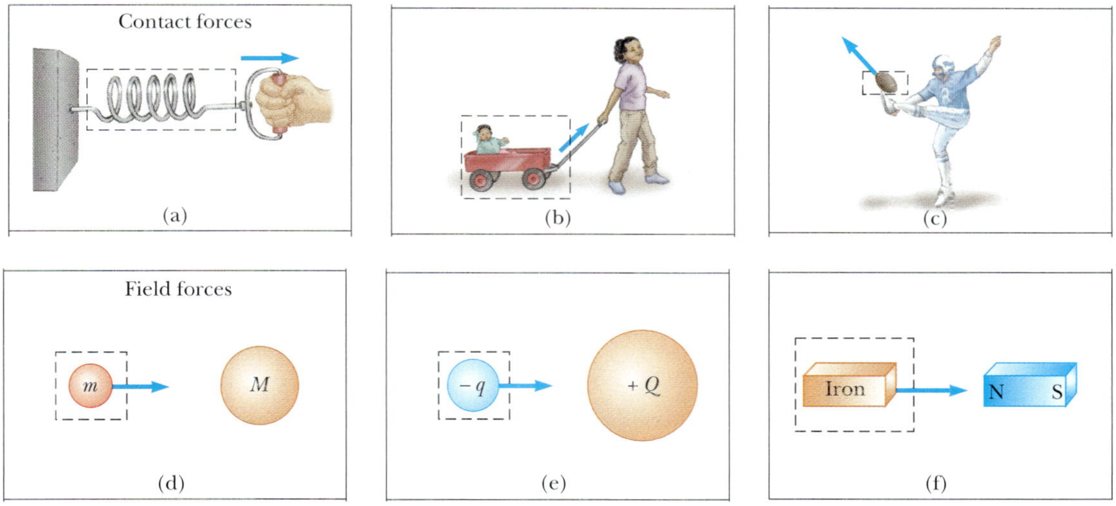

Figure 5.1 Some examples of applied forces. In each case, a force is exerted on the object within the boxed area. Some agent in the environment external to the boxed area exerts a force on the object.

When a coiled spring is pulled, as in Figure 5.1a, the spring stretches. When a stationary cart is pulled, as in Figure 5.1b, the cart moves. When a football is kicked, as in Figure 5.1c, it is both deformed and set in motion. These situations are all examples of a class of forces called *contact forces.* That is, they involve physical contact between two objects. Other examples of contact forces are the force exerted by gas molecules on the walls of a container and the force exerted by your feet on the floor.

Another class of forces, known as *field forces,* does not involve physical contact between two objects. These forces act through empty space. The gravitational force of attraction between two objects with mass, illustrated in Figure 5.1d, is an example of this class of force. The gravitational force keeps objects bound to the Earth and the planets in orbit around the Sun. Another common field force is the electric force that one electric charge exerts on another (Fig. 5.1e). As an example, these charges might be those of the electron and proton that form a hydrogen atom. A third example of a field force is the force a bar magnet exerts on a piece of iron (Fig. 5.1f).

The distinction between contact forces and field forces is not as sharp as you may have been led to believe by the previous discussion. When examined at the atomic level, all the forces we classify as contact forces turn out to be caused by electric (field) forces of the type illustrated in Figure 5.1e. Nevertheless, in developing models for macroscopic phenomena, it is convenient to use both classifications of forces. The only known *fundamental* forces in nature are all field forces: (1) *gravitational forces* between objects, (2) *electromagnetic forces* between electric charges, (3) *strong forces* between subatomic particles, and (4) *weak forces* that arise in certain radioactive decay processes. In classical physics, we are concerned only with gravitational and electromagnetic forces. We will discuss strong and weak forces in Chapter 46.

Giraudon/Art Resource

ISAAC NEWTON

English physicist and mathematician (1642–1727)

Isaac Newton was one of the most brilliant scientists in history. Before the age of 30, he formulated the basic concepts and laws of mechanics, discovered the law of universal gravitation, and invented the mathematical methods of calculus. As a consequence of his theories, Newton was able to explain the motions of the planets, the ebb and flow of the tides, and many special features of the motions of the Moon and the Earth. He also interpreted many fundamental observations concerning the nature of light. His contributions to physical theories dominated scientific thought for two centuries and remain important today.

The Vector Nature of Force

It is possible to use the deformation of a spring to measure force. Suppose a vertical force is applied to a spring scale that has a fixed upper end as shown in Figure 5.2a (page 102). The spring elongates when the force is applied, and a pointer on the scale reads the value of the applied force. We can calibrate the spring by defining a reference force $\vec{\mathbf{F}}_1$ as the force that produces a pointer reading of 1.00 cm. If we now apply a different downward force $\vec{\mathbf{F}}_2$ whose magnitude is twice that of the reference force $\vec{\mathbf{F}}_1$ as seen in Figure 5.2b, the pointer moves to 2.00 cm. Figure 5.2c shows that the combined effect of the two collinear forces is the sum of the effects of the individual forces.

Now suppose the two forces are applied simultaneously with $\vec{\mathbf{F}}_1$ downward and $\vec{\mathbf{F}}_2$ horizontal as illustrated in Figure 5.2d. In this case, the pointer reads 2.24 cm.

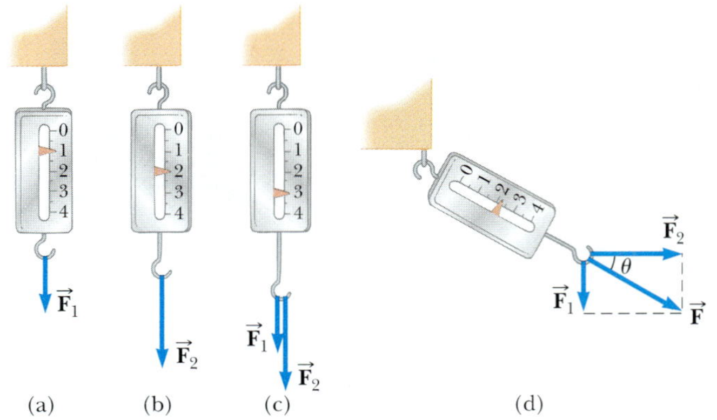

Figure 5.2 The vector nature of a force is tested with a spring scale. (a) A downward force $\vec{\mathbf{F}}_1$ elongates the spring 1.00 cm. (b) A downward force $\vec{\mathbf{F}}_2$ elongates the spring 2.00 cm. (c) When $\vec{\mathbf{F}}_1$ and $\vec{\mathbf{F}}_2$ are applied simultaneously, the spring elongates by 3.00 cm. (d) When $\vec{\mathbf{F}}_1$ is downward and $\vec{\mathbf{F}}_2$ is horizontal, the combination of the two forces elongates the spring 2.24 cm.

The single force $\vec{\mathbf{F}}$ that would produce this same reading is the sum of the two vectors $\vec{\mathbf{F}}_1$ and $\vec{\mathbf{F}}_2$ as described in Figure 5.2d. That is, $|\vec{\mathbf{F}}| = \sqrt{F_1^2 + F_2^2} = 2.24$ units, and its direction is $\theta = \tan^{-1}(-0.500) = -26.6°$. **Because forces have been experimentally verified to behave as vectors, you must use the rules of vector addition to obtain the net force on an object.**

5.2 Newton's First Law and Inertial Frames

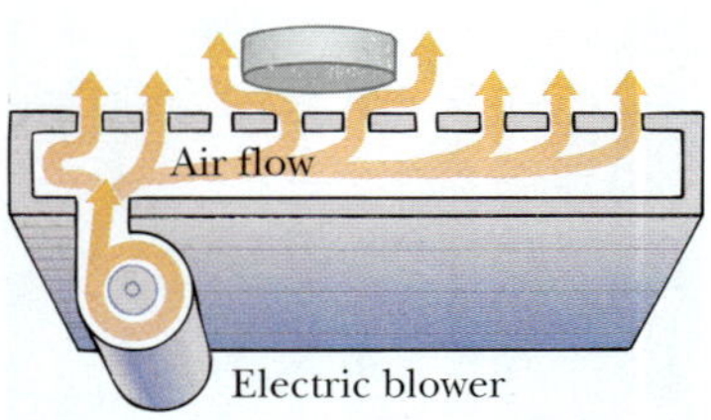

Figure 5.3 On an air hockey table, air blown through holes in the surface allows the puck to move almost without friction. If the table is not accelerating, a puck placed on the table will remain at rest.

We begin our study of forces by imagining some physical situations involving a puck on a perfectly level air hockey table (Fig. 5.3). You expect that the puck will remain where it is placed. Now imagine your air hockey table is located on a train moving with constant velocity along a perfectly smooth track. If the puck is placed on the table, the puck again remains where it is placed. If the train were to accelerate, however, the puck would start moving along the table opposite the direction of the train's acceleration, just as a set of papers on your dashboard falls onto the front seat of your car when you step on the accelerator.

As we saw in Section 4.6, a moving object can be observed from any number of reference frames. **Newton's first law of motion**, sometimes called the *law of inertia*, defines a special set of reference frames called *inertial frames*. This law can be stated as follows:

Newton's first law ▶

> If an object does not interact with other objects, it is possible to identify a reference frame in which the object has zero acceleration.

Inertial frame of reference ▶

Such a reference frame is called an **inertial frame of reference**. When the puck is on the air hockey table located on the ground, you are observing it from an inertial reference frame; there are no horizontal interactions of the puck with any other objects, and you observe it to have zero acceleration in that direction. When you are on the train moving at constant velocity, you are also observing the puck from an inertial reference frame. **Any reference frame that moves with constant velocity relative to an inertial frame is itself an inertial frame.** When you and the train accelerate, however, you are observing the puck from a **noninertial reference frame** because the train is accelerating relative to the inertial reference frame of the Earth's surface. While the puck appears to be accelerating according to your observations, a reference frame can be identified in which the puck has zero acceleration. For example, an observer standing outside the train on the ground sees the puck moving with the same velocity as the train had before it started to accelerate (because there is almost no friction to "tie" the puck and the train together). Therefore, Newton's first law is still satisfied even though your observations as a rider on the train show an apparent acceleration relative to you.

A reference frame that moves with constant velocity relative to the distant stars is the best approximation of an inertial frame, and for our purposes we can consider the Earth as being such a frame. The Earth is not really an inertial frame because of its orbital motion around the Sun and its rotational motion about its own axis, both of which involve centripetal accelerations. These accelerations are small compared with g, however, and can often be neglected. For this reason, we model the Earth as an inertial frame, along with any other frame attached to it.

Let us assume we are observing an object from an inertial reference frame. (We will return to observations made in noninertial reference frames in Section 6.3.) Before about 1600, scientists believed that the natural state of matter was the state of rest. Observations showed that moving objects eventually stopped moving. Galileo was the first to take a different approach to motion and the natural state of matter. He devised thought experiments and concluded that it is not the nature of an object to stop once set in motion: rather, it is its nature to *resist changes in its motion.* In his words, "Any velocity once imparted to a moving body will be rigidly maintained as long as the external causes of retardation are removed." For example, a spacecraft drifting through empty space with its engine turned off will keep moving forever. It would *not* seek a "natural state" of rest.

PITFALL PREVENTION 5.1
Newton's First Law

Newton's first law does *not* say what happens for an object with *zero net force,* that is, multiple forces that cancel; it says what happens *in the absence of external forces.* This subtle but important difference allows us to define force as that which causes a change in the motion. The description of an object under the effect of forces that balance is covered by Newton's second law.

Given our discussion of observations made from inertial reference frames, we can pose a more practical statement of Newton's first law of motion:

◀ Another statement of Newton's first law

In the absence of external forces and when viewed from an inertial reference frame, an object at rest remains at rest and an object in motion continues in motion with a constant velocity (that is, with a constant speed in a straight line).

In other words, **when no force acts on an object, the acceleration of the object is zero**. From the first law, we conclude that any *isolated object* (one that does not interact with its environment) is either at rest or moving with constant velocity. The tendency of an object to resist any attempt to change its velocity is called **inertia**. Given the statement of the first law above, we can conclude that an object that is accelerating must be experiencing a force. In turn, from the first law, we can define **force** as **that which causes a change in motion of an object**.

Quick Quiz 5.1 Which of the following statements is correct? (a) It is possible for an object to have motion in the absence of forces on the object. (b) It is possible to have forces on an object in the absence of motion of the object. (c) Neither (a) nor (b) is correct. (d) Both (a) and (b) are correct.

5.3 Mass

Imagine playing catch with either a basketball or a bowling ball. Which ball is more likely to keep moving when you try to catch it? Which ball requires more effort to throw it? The bowling ball requires more effort. In the language of physics, we say that the bowling ball is more resistant to changes in its velocity than the basketball. How can we quantify this concept?

◀ Definition of mass

Mass is that property of an object that specifies how much resistance an object exhibits to changes in its velocity, and as we learned in Section 1.1 the SI unit of mass is the kilogram. Experiments show that the greater the mass of an object, the less that object accelerates under the action of a given applied force.

To describe mass quantitatively, we conduct experiments in which we compare the accelerations a given force produces on different objects. Suppose a force acting on an object of mass m_1 produces an acceleration $\vec{\mathbf{a}}_1$, and the *same force* acting on an object of mass m_2 produces an acceleration $\vec{\mathbf{a}}_2$. The ratio of the two masses is defined as the *inverse* ratio of the magnitudes of the accelerations produced by the force:

$$\frac{m_1}{m_2} \equiv \frac{a_2}{a_1} \quad (5.1)$$

For example, if a given force acting on a 3-kg object produces an acceleration of 4 m/s^2, the same force applied to a 6-kg object produces an acceleration of 2 m/s^2. According to a huge number of similar observations, we conclude that **the magnitude of the acceleration of an object is inversely proportional to its mass when acted on by a given force**. If one object has a known mass, the mass of the other object can be obtained from acceleration measurements.

Mass is an inherent property of an object and is independent of the object's surroundings and of the method used to measure it. Also, **mass is a scalar quantity** and thus obeys the rules of ordinary arithmetic. For example, if you combine a 3-kg mass with a 5-kg mass, the total mass is 8 kg. This result can be verified experimentally by comparing the acceleration that a known force gives to several objects separately with the acceleration that the same force gives to the same objects combined as one unit.

Mass and weight are different quantities ▶

Mass should not be confused with weight. **Mass and weight are two different quantities.** The weight of an object is equal to the magnitude of the gravitational force exerted on the object and varies with location (see Section 5.5). For example, a person weighing 180 lb on the Earth weighs only about 30 lb on the Moon. On the other hand, the mass of an object is the same everywhere: an object having a mass of 2 kg on the Earth also has a mass of 2 kg on the Moon.

5.4 Newton's Second Law

Newton's first law explains what happens to an object when no forces act on it: it either remains at rest or moves in a straight line with constant speed. Newton's second law answers the question of what happens to an object that has one or more forces acting on it.

Imagine performing an experiment in which you push a block of fixed mass across a frictionless horizontal surface. When you exert some horizontal force $\vec{\mathbf{F}}$ on the block, it moves with some acceleration $\vec{\mathbf{a}}$. If you apply a force twice as great on the same block, the acceleration of the block doubles. If you increase the applied force to $3\vec{\mathbf{F}}$, the acceleration triples, and so on. From such observations, we conclude that **the acceleration of an object is directly proportional to the force acting on it:** $\vec{\mathbf{F}} \propto \vec{\mathbf{a}}$. This idea was first introduced in Section 2.4 when we discussed the direction of the acceleration of an object. The magnitude of the acceleration of an object is inversely proportional to its mass, as stated in the preceding section: $|\vec{\mathbf{a}}| \propto 1/m$.

PITFALL PREVENTION 5.2

Force Is the Cause of Changes in Motion

Force *does not* cause motion. We can have motion in the absence of forces as described in Newton's first law. Force is the cause of *changes* in motion as measured by acceleration.

These experimental observations are summarized in **Newton's second law:**

> When viewed from an inertial reference frame, the acceleration of an object is directly proportional to the net force acting on it and inversely proportional to its mass:
>
> $$\vec{\mathbf{a}} \propto \frac{\sum \vec{\mathbf{F}}}{m}$$

If we choose a proportionality constant of 1, we can relate mass, acceleration, and force through the following mathematical statement of Newton's second law:[1]

Newton's second law ▶

$$\sum \vec{\mathbf{F}} = m\vec{\mathbf{a}} \quad (5.2)$$

[1] Equation 5.2 is valid only when the speed of the object is much less than the speed of light. We treat the relativistic situation in Chapter 39.

In both the textual and mathematical statements of Newton's second law, we have indicated that the acceleration is due to the *net force* $\sum \vec{\mathbf{F}}$ acting on an object. The **net force** on an object is the vector sum of all forces acting on the object. (We sometimes refer to the net force as the *total force,* the *resultant force,* or the *unbalanced force.*) In solving a problem using Newton's second law, it is imperative to determine the correct net force on an object. Many forces may be acting on an object, but there is only one acceleration.

Equation 5.2 is a vector expression and hence is equivalent to three component equations:

$$\sum F_x = ma_x \qquad \sum F_y = ma_y \qquad \sum F_z = ma_z \qquad \textbf{(5.3)}$$

◀ Newton's second law: component form

Quick Quiz 5.2 An object experiences no acceleration. Which of the following *cannot* be true for the object? (a) A single force acts on the object. (b) No forces act on the object. (c) Forces act on the object, but the forces cancel.

Quick Quiz 5.3 You push an object, initially at rest, across a frictionless floor with a constant force for a time interval Δt, resulting in a final speed of v for the object. You then repeat the experiment, but with a force that is twice as large. What time interval is now required to reach the same final speed v? (a) $4\Delta t$ (b) $2\Delta t$ (c) Δt (d) $\Delta t/2$ (e) $\Delta t/4$

PITFALL PREVENTION 5.3

$m\vec{\mathbf{a}}$ Is Not a Force

Equation 5.2 does *not* say that the product $m\vec{\mathbf{a}}$ is a force. All forces on an object are added vectorially to generate the net force on the left side of the equation. This net force is then equated to the product of the mass of the object and the acceleration that results from the net force. Do *not* include an "$m\vec{\mathbf{a}}$ force" in your analysis of the forces on an object.

The SI unit of force is the **newton** (N). A force of 1 N is the force that, when acting on an object of mass 1 kg, produces an acceleration of 1 m/s². From this definition and Newton's second law, we see that the newton can be expressed in terms of the following fundamental units of mass, length, and time:

$$1 \text{ N} \equiv 1 \text{ kg} \cdot \text{m/s}^2 \qquad \textbf{(5.4)}$$

◀ Definition of the newton

In the U.S. customary system, the unit of force is the **pound** (lb). A force of 1 lb is the force that, when acting on a 1-slug mass,[2] produces an acceleration of 1 ft/s²:

$$1 \text{ lb} \equiv 1 \text{ slug} \cdot \text{ft/s}^2 \qquad \textbf{(5.5)}$$

A convenient approximation is $1 \text{ N} \approx \frac{1}{4}$ lb.

EXAMPLE 5.1 An Accelerating Hockey Puck

A hockey puck having a mass of 0.30 kg slides on the horizontal, frictionless surface of an ice rink. Two hockey sticks strike the puck simultaneously, exerting the forces on the puck shown in Figure 5.4. The force $\vec{\mathbf{F}}_1$ has a magnitude of 5.0 N, and the force $\vec{\mathbf{F}}_2$ has a magnitude of 8.0 N. Determine both the magnitude and the direction of the puck's acceleration.

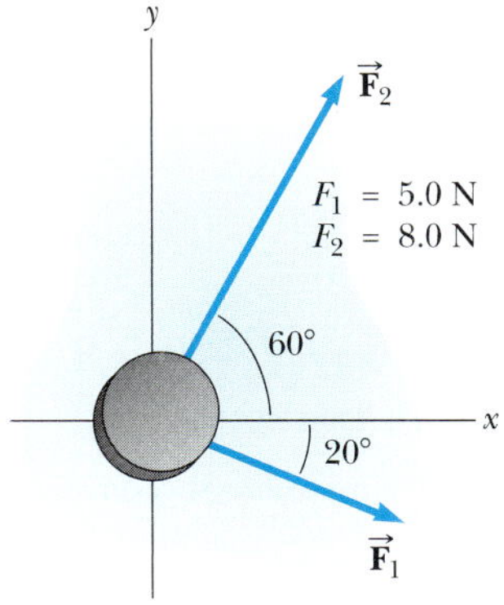

Figure 5.4 (Example 5.1) A hockey puck moving on a frictionless surface is subject to two forces $\vec{\mathbf{F}}_1$ and $\vec{\mathbf{F}}_2$.

SOLUTION

Conceptualize Study Figure 5.4. Using your expertise in vector addition from Chapter 3, predict the approximate direction of the net force vector on the puck. The acceleration of the puck will be in the same direction.

Categorize Because we can determine a net force and we want an acceleration, this problem is categorized as one that may be solved using Newton's second law.

[2] The *slug* is the unit of mass in the U.S. customary system and is that system's counterpart of the SI unit the *kilogram.* Because most of the calculations in our study of classical mechanics are in SI units, the slug is seldom used in this text.

Analyze Find the component of the net force acting on the puck in the x direction:

$$\sum F_x = F_{1x} + F_{2x} = F_1 \cos(-20°) + F_2 \cos 60°$$
$$= (5.0\text{ N})(0.940) + (8.0\text{ N})(0.500) = 8.7\text{ N}$$

Find the component of the net force acting on the puck in the y direction:

$$\sum F_y = F_{1y} + F_{2y} = F_1 \sin(-20°) + F_2 \sin 60°$$
$$= (5.0\text{ N})(-0.342) + (8.0\text{ N})(0.866) = 5.2\text{ N}$$

Use Newton's second law in component form (Eq. 5.3) to find the x and y components of the puck's acceleration:

$$a_x = \frac{\sum F_x}{m} = \frac{8.7\text{ N}}{0.30\text{ kg}} = 29\text{ m/s}^2$$

$$a_y = \frac{\sum F_y}{m} = \frac{5.2\text{ N}}{0.30\text{ kg}} = 17\text{ m/s}^2$$

Find the magnitude of the acceleration:

$$a = \sqrt{(29\text{ m/s}^2)^2 + (17\text{ m/s}^2)^2} = 34\text{ m/s}^2$$

Find the direction of the acceleration relative to the positive x axis:

$$\theta = \tan^{-1}\left(\frac{a_y}{a_x}\right) = \tan^{-1}\left(\frac{17}{29}\right) = 30°$$

Finalize The vectors in Figure 5.4 can be added graphically to check the reasonableness of our answer. Because the acceleration vector is along the direction of the resultant force, a drawing showing the resultant force vector helps us check the validity of the answer. (Try it!)

What If? Suppose three hockey sticks strike the puck simultaneously, with two of them exerting the forces shown in Figure 5.4. The result of the three forces is that the hockey puck shows *no* acceleration. What must be the components of the third force?

Answer If there is zero acceleration, the net force acting on the puck must be zero. Therefore, the three forces must cancel. We have found the components of the combination of the first two forces. The components of the third force must be of equal magnitude and opposite sign so that all the components add to zero. Therefore, $F_{3x} = -8.7$ N, $F_{3y} = -5.2$ N.

PITFALL PREVENTION 5.4
"Weight of an Object"

We are familiar with the everyday phrase, the "weight of an object." Weight, however, is not an inherent property of an object; rather, it is a measure of the gravitational force between the object and the Earth (or other planet). Therefore, weight is a property of a *system* of items: the object and the Earth.

PITFALL PREVENTION 5.5
Kilogram Is Not a Unit of Weight

You may have seen the "conversion" 1 kg = 2.2 lb. Despite popular statements of weights expressed in kilograms, the kilogram is not a unit of *weight*, it is a unit of *mass*. The conversion statement is not an equality; it is an *equivalence* that is valid only on the Earth's surface.

5.5 The Gravitational Force and Weight

All objects are attracted to the Earth. The attractive force exerted by the Earth on an object is called the **gravitational force** $\vec{\mathbf{F}}_g$. This force is directed toward the center of the Earth,[3] and its magnitude is called the **weight** of the object.

We saw in Section 2.6 that a freely falling object experiences an acceleration $\vec{\mathbf{g}}$ acting toward the center of the Earth. Applying Newton's second law $\sum \vec{\mathbf{F}} = m\vec{\mathbf{a}}$ to a freely falling object of mass m, with $\vec{\mathbf{a}} = \vec{\mathbf{g}}$ and $\sum \vec{\mathbf{F}} = \vec{\mathbf{F}}_g$, gives

$$\vec{\mathbf{F}}_g = m\vec{\mathbf{g}}$$

Therefore, the weight of an object, being defined as the magnitude of $\vec{\mathbf{F}}_g$, is equal to mg:

$$F_g = mg \tag{5.6}$$

Because it depends on g, weight varies with geographic location. Because g decreases with increasing distance from the center of the Earth, objects weigh less at higher altitudes than at sea level. For example, a 1 000-kg palette of bricks used in the construction of the Empire State Building in New York City weighed 9 800 N at street level, but weighed about 1 N less by the time it was lifted from sidewalk level to the top of the building. As another example, suppose a student has a mass of 70.0 kg. The student's weight in a location where $g = 9.80$ m/s^2 is 686 N (about 150 lb). At the top of a mountain, however, where $g = 9.77$ m/s^2, the student's

[3] This statement ignores that the mass distribution of the Earth is not perfectly spherical.

weight is only 684 N. Therefore, if you want to lose weight without going on a diet, climb a mountain or weigh yourself at 30 000 ft during an airplane flight!

Equation 5.6 quantifies the gravitational force on the object, but notice that this equation does not require the object to be moving. Even for a stationary object or for an object on which several forces act, Equation 5.6 can be used to calculate the magnitude of the gravitational force. The result is a subtle shift in the interpretation of m in the equation. The mass m in Equation 5.6 determines the strength of the gravitational attraction between the object and the Earth. This role is completely different from that previously described for mass, that of measuring the resistance to changes in motion in response to an external force. Therefore, we call m in Equation 5.6 the **gravitational mass**. Even though this quantity is different in behavior from inertial mass, it is one of the experimental conclusions in Newtonian dynamics that gravitational mass and inertial mass have the same value.

Although this discussion has focused on the gravitational force on an object due to the Earth, the concept is generally valid on any planet. The value of g will vary from one planet to the next, but the magnitude of the gravitational force will always be given by the value of mg.

NASA

The life-support unit strapped to the back of astronaut Edwin Aldrin weighed 300 lb on the Earth. During his training, a 50-lb mock-up was used. Although this strategy effectively simulated the reduced weight the unit would have on the Moon, it did not correctly mimic the unchanging mass. It was just as difficult to accelerate the unit (perhaps by jumping or twisting suddenly) on the Moon as on the Earth.

Quick Quiz 5.4 Suppose you are talking by interplanetary telephone to a friend, who lives on the Moon. He tells you that he has just won a newton of gold in a contest. Excitedly, you tell him that you entered the Earth version of the same contest and also won a newton of gold! Who is richer? (a) You are. (b) Your friend is. (c) You are equally rich.

CONCEPTUAL EXAMPLE 5.2 How Much Do You Weigh in an Elevator?

You have most likely been in an elevator that accelerates upward as it moves toward a higher floor. In this case, you feel heavier. In fact, if you are standing on a bathroom scale at the time, the scale measures a force having a magnitude that is greater than your weight. Therefore, you have tactile and measured evidence that leads you to believe you are heavier in this situation. *Are* you heavier?

SOLUTION

No; your weight is unchanged. Your experiences are due to the fact that you are in a noninertial reference frame. To provide the acceleration upward, the floor or scale must exert on your feet an upward force that is greater in magnitude than your weight. It is this greater force you feel, which you interpret as feeling heavier. The scale reads this upward force, not your weight, and so its reading increases.

5.6 Newton's Third Law

If you press against a corner of this textbook with your fingertip, the book pushes back and makes a small dent in your skin. If you push harder, the book does the same and the dent in your skin is a little larger. This simple activity illustrates that forces are *interactions* between two objects: when your finger pushes on the book, the book pushes back on your finger. This important principle is known as **Newton's third law:**

◀ Newton's third law

If two objects interact, the force $\vec{\mathbf{F}}_{12}$ exerted by object 1 on object 2 is equal in magnitude and opposite in direction to the force $\vec{\mathbf{F}}_{21}$ exerted by object 2 on object 1:

$$\vec{\mathbf{F}}_{12} = -\vec{\mathbf{F}}_{21} \quad (5.7)$$

When it is important to designate forces as interactions between two objects, we will use this subscript notation, where $\vec{\mathbf{F}}_{ab}$ means "the force exerted *by* a *on* b." The third law is illustrated in Figure 5.5a. The force that object 1 exerts on object 2 is popularly called the *action force*, and the force of object 2 on object 1 is called the

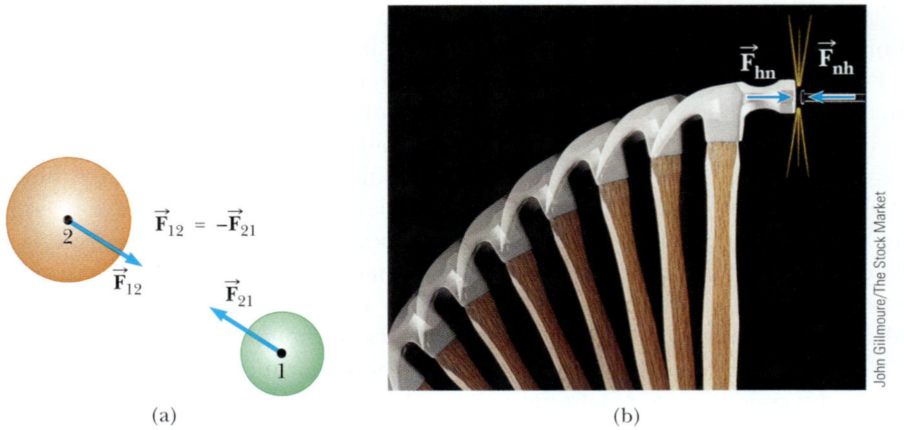

Figure 5.5 Newton's third law. (a) The force $\vec{F}_{12}$ exerted by object 1 on object 2 is equal in magnitude and opposite in direction to the force $\vec{F}_{21}$ exerted by object 2 on object 1. (b) The force $\vec{F}_{hn}$ exerted by the hammer on the nail is equal in magnitude and opposite to the force $\vec{F}_{nh}$ exerted by the nail on the hammer.

PITFALL PREVENTION 5.6
n Does Not Always Equal *mg*

In the situation shown in Figure 5.6 and in many others, we find that $n = mg$ (the normal force has the same magnitude as the gravitational force). This result, however, is *not* generally true. If an object is on an incline, if there are applied forces with vertical components, or if there is a vertical acceleration of the system, then $n \neq mg$. *Always* apply Newton's second law to find the relationship between n and mg.

reaction force. These italicized terms are not scientific terms; furthermore, either force can be labeled the action or reaction force. We will use these terms for convenience. **In all cases, the action and reaction forces act on *different* objects and must be of the same type (gravitational, electrical, etc.).** For example, the force acting on a freely falling projectile is the gravitational force exerted by the Earth on the projectile $\vec{F}_g = \vec{F}_{Ep}$ (E = Earth, p = projectile), and the magnitude of this force is mg. The reaction to this force is the gravitational force exerted by the projectile on the Earth $\vec{F}_{pE} = -\vec{F}_{Ep}$. The reaction force $\vec{F}_{pE}$ must accelerate the Earth toward the projectile just as the action force $\vec{F}_{Ep}$ accelerates the projectile toward the Earth. Because the Earth has such a large mass, however, its acceleration due to this reaction force is negligibly small.

Another example of Newton's third law is shown in Figure 5.5b. The force $\vec{F}_{hn}$ exerted by the hammer on the nail is equal in magnitude and opposite the force $\vec{F}_{nh}$ exerted by the nail on the hammer. This latter force stops the forward motion of the hammer when it strikes the nail.

Normal force ▶

Consider a computer monitor at rest on a table as in Figure 5.6a. The reaction force to the gravitational force $\vec{F}_g = \vec{F}_{Em}$ on the monitor is the force $\vec{F}_{mE} = -\vec{F}_{Em}$ exerted by the monitor on the Earth. The monitor does not accelerate because it is held up by the table. The table exerts on the monitor an upward force $\vec{n} = \vec{F}_{tm}$, called the **normal force.**[4] This force, which prevents the monitor from falling through the table, can have any value needed, up to the point of breaking the table. Because the monitor has zero acceleration, Newton's second law applied to the monitor gives us $\sum \vec{F} = \vec{n} + m\vec{g} = 0$, so $n\hat{j} - mg\hat{j} = 0$, or $n = mg$. The normal force balances the gravitational force on the monitor, so the net force on the monitor is zero. The reaction force to $\vec{n}$ is the force exerted by the monitor downward on the table, $\vec{F}_{mt} = -\vec{F}_{tm} = -\vec{n}$.

Notice that the forces acting on the monitor are $\vec{F}_g$ and $\vec{n}$ as shown in Figure 5.6b. The two forces $\vec{F}_{mE}$ and $\vec{F}_{mt}$ are exerted on objects other than the monitor.

PITFALL PREVENTION 5.7
Newton's Third Law

Remember that Newton's third law action and reaction forces act on *different* objects. For example, in Figure 5.6, $\vec{n} = \vec{F}_{tm} = -m\vec{g} = -\vec{F}_{Em}$. The forces $\vec{n}$ and $m\vec{g}$ are equal in magnitude and opposite in direction, but they do not represent an action-reaction pair because both forces act on the *same* object, the monitor.

Figure 5.6 illustrates an extremely important step in solving problems involving forces. Figure 5.6a shows many of the forces in the situation: those acting on the monitor, one acting on the table, and one acting on the Earth. Figure 5.6b, by contrast, shows only the forces acting on *one object*, the monitor. This important pictorial representation in Figure 5.6b is called a **free-body diagram.** When analyzing an object subject to forces, we are interested in the net force acting on one object, which we will model as a particle. Therefore, a free-body diagram helps us isolate only those forces on the object and eliminate the other forces from our

[4] *Normal* in this context means *perpendicular.*

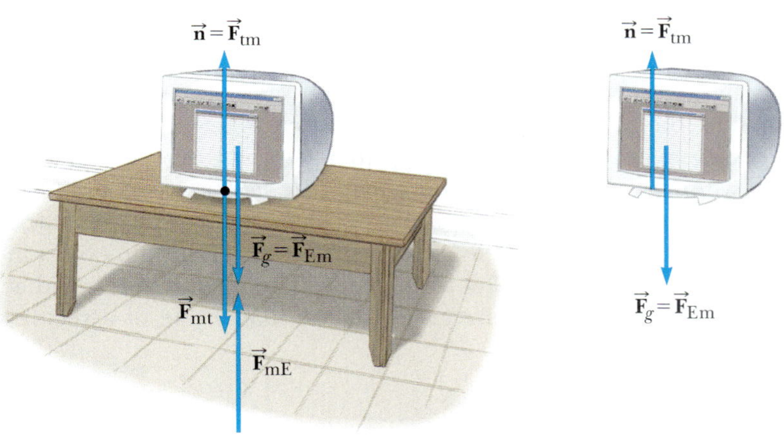

(a) (b)

Figure 5.6 (a) When a computer monitor is at rest on a table, the forces acting on the monitor are the normal force $\vec{\mathbf{n}}$ and the gravitational force $\vec{\mathbf{F}}_g$. The reaction to $\vec{\mathbf{n}}$ is the force $\vec{\mathbf{F}}_{mt}$ exerted by the monitor on the table. The reaction to $\vec{\mathbf{F}}_g$ is the force $\vec{\mathbf{F}}_{mE}$ exerted by the monitor on the Earth. (b) The free-body diagram for the monitor.

analysis. This diagram can be simplified further by representing the object (such as the monitor) as a particle simply by drawing a dot.

PITFALL PREVENTION 5.8

Free-Body Diagrams

The *most important* step in solving a problem using Newton's laws is to draw a proper sketch, the free-body diagram. Be sure to draw only those forces that act on the object you are isolating. Be sure to draw *all* forces acting on the object, including any field forces, such as the gravitational force.

Quick Quiz 5.5 **(i)** If a fly collides with the windshield of a fast-moving bus, which experiences an impact force with a larger magnitude? (a) The fly. (b) The bus. (c) The same force is experienced by both. **(ii)** Which experiences the greater acceleration? (a) The fly. (b) The bus. (c) The same acceleration is experienced by both.

CONCEPTUAL EXAMPLE 5.3 You Push Me and I'll Push You

A large man and a small boy stand facing each other on frictionless ice. They put their hands together and push against each other so that they move apart.

(A) Who moves away with the higher speed?

SOLUTION

This situation is similar to what we saw in Quick Quiz 5.5. According to Newton's third law, the force exerted by the man on the boy and the force exerted by the boy on the man are a third-law pair of forces, so they must be equal in magnitude. (A bathroom scale placed between their hands would read the same, regardless of which way it faced.) Therefore, the boy, having the smaller mass, experiences the greater acceleration. Both individuals accelerate for the same amount of time, but the greater acceleration of the boy over this time interval results in his moving away from the interaction with the higher speed.

(B) Who moves farther while their hands are in contact?

SOLUTION

Because the boy has the greater acceleration and therefore the greater average velocity, he moves farther than the man during the time interval during which their hands are in contact.

5.7 Some Applications of Newton's Laws

In this section, we discuss two analysis models for solving problems in which objects are either in equilibrium ($\vec{\mathbf{a}} = 0$) or accelerating along a straight line under the action of constant external forces. Remember that **when Newton's laws are applied to an object, we are interested only in external forces that act on the object.** If the objects are modeled as particles, we need not worry about rotational motion. For now, we also neglect the effects of friction in those problems involving

Rock climbers at rest are in equilibrium and depend on the tension forces in ropes for their safety.

motion, which is equivalent to stating that the surfaces are *frictionless.* (The friction force is discussed in Section 5.8.)

We usually neglect the mass of any ropes, strings, or cables involved. In this approximation, the magnitude of the force exerted by any element of the rope on the adjacent element is the same for all elements along the rope. In problem statements, the synonymous terms *light* and *of negligible mass* are used to indicate that a mass is to be ignored when you work the problems. When a rope attached to an object is pulling on the object, the rope exerts a force $\vec{\mathbf{T}}$ on the object in a direction away from the object, parallel to the rope. The magnitude T of that force is called the **tension** in the rope. Because it is the magnitude of a vector quantity, tension is a scalar quantity.

The Particle in Equilibrium

If the acceleration of an object modeled as a particle is zero, the object is treated with the **particle in equilibrium** model. In this model, the net force on the object is zero:

$$\sum \vec{\mathbf{F}} = 0 \qquad (5.8)$$

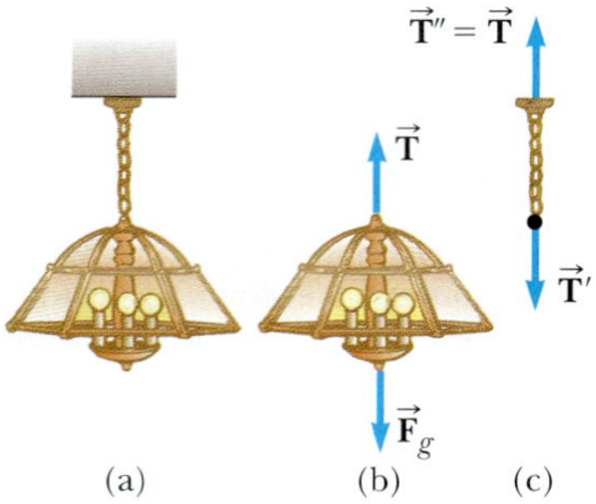

Figure 5.7 (a) A lamp suspended from a ceiling by a chain of negligible mass. (b) The forces acting on the lamp are the gravitational force $\vec{\mathbf{F}}_g$ and the force $\vec{\mathbf{T}}$ exerted by the chain. (c) The forces acting on the chain are the force $\vec{\mathbf{T}}'$ exerted by the lamp and the force $\vec{\mathbf{T}}''$ exerted by the ceiling.

Consider a lamp suspended from a light chain fastened to the ceiling as in Figure 5.7a. The free-body diagram for the lamp (Fig. 5.7b) shows that the forces acting on the lamp are the downward gravitational force $\vec{\mathbf{F}}_g$ and the upward force $\vec{\mathbf{T}}$ exerted by the chain. Because there are no forces in the x direction, $\Sigma F_x = 0$ provides no helpful information. The condition $\Sigma F_y = 0$ gives

$$\sum F_y = T - F_g = 0 \quad \text{or} \quad T = F_g$$

Again, notice that $\vec{\mathbf{T}}$ and $\vec{\mathbf{F}}_g$ are *not* an action-reaction pair because they act on the same object, the lamp. The reaction force to $\vec{\mathbf{T}}$ is $\vec{\mathbf{T}}'$, the downward force exerted by the lamp on the chain as shown in Figure 5.7c. Because the chain is a particle in equilibrium, the ceiling must exert on the chain a force $\vec{\mathbf{T}}''$ that is equal in magnitude to the magnitude of $\vec{\mathbf{T}}'$ and points in the opposite direction.

The Particle Under a Net Force

If an object experiences an acceleration, its motion can be analyzed with the **particle under a net force** model. The appropriate equation for this model is Newton's second law, Equation 5.2. Consider a crate being pulled to the right on a frictionless, horizontal surface as in Figure 5.8a. Suppose you wish to find the acceleration of the crate and the force the floor exerts on it. The forces acting on the crate are illustrated in the free-body diagram in Figure 5.8b. Notice that the horizontal force $\vec{\mathbf{T}}$ being applied to the crate acts through the rope. The magnitude of $\vec{\mathbf{T}}$ is equal to the tension in the rope. In addition to the force $\vec{\mathbf{T}}$, the free-body diagram for the crate includes the gravitational force $\vec{\mathbf{F}}_g$ and the normal force $\vec{\mathbf{n}}$ exerted by the floor on the crate.

We can now apply Newton's second law in component form to the crate. The only force acting in the x direction is $\vec{\mathbf{T}}$. Applying $\Sigma F_x = ma_x$ to the horizontal motion gives

$$\sum F_x = T = ma_x \quad \text{or} \quad a_x = \frac{T}{m}$$

No acceleration occurs in the y direction because the crate moves only horizontally. Therefore, we use the particle in equilibrium model in the y direction. Applying the y component of Equation 5.8 yields

$$\sum F_y = n + (-F_g) = 0 \quad \text{or} \quad n = F_g$$

That is, the normal force has the same magnitude as the gravitational force but acts in the opposite direction.

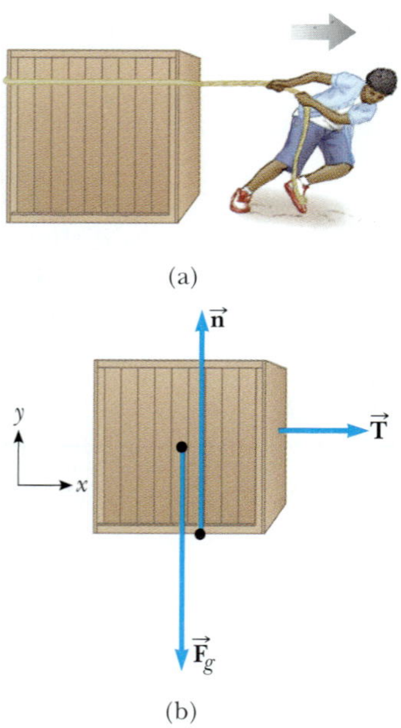

Figure 5.8 (a) A crate being pulled to the right on a frictionless surface. (b) The free-body diagram representing the external forces acting on the crate.

If $\vec{\mathbf{T}}$ is a constant force, the acceleration $a_x = T/m$ also is constant. Hence, the crate is also modeled as a particle under constant acceleration in the x direction, and the equations of kinematics from Chapter 2 can be used to obtain the crate's position x and velocity v_x as functions of time.

In the situation just described, the magnitude of the normal force $\vec{\mathbf{n}}$ is equal to the magnitude of $\vec{\mathbf{F}}_g$, but that is not always the case. For example, suppose a book is lying on a table and you push down on the book with a force $\vec{\mathbf{F}}$ as in Figure 5.9. Because the book is at rest and therefore not accelerating, $\Sigma F_y = 0$, which gives $n - F_g - F = 0$, or $n = F_g + F$. In this situation, the normal force is *greater* than the gravitational force. Other examples in which $n \neq F_g$ are presented later.

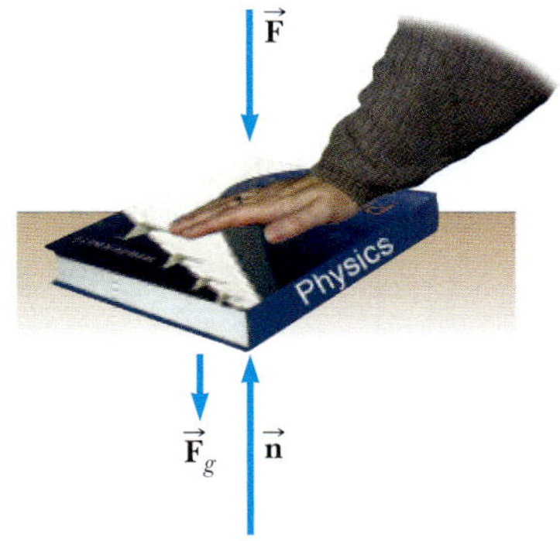

Figure 5.9 When a force $\vec{\mathbf{F}}$ pushes vertically downward on another object, the normal force $\vec{\mathbf{n}}$ on the object is greater than the gravitational force: $n = F_g + F$.

PROBLEM-SOLVING STRATEGY **Applying Newton's Laws**

The following procedure is recommended when dealing with problems involving Newton's laws:

1. *Conceptualize.* Draw a simple, neat diagram of the system. The diagram helps establish the mental representation. Establish convenient coordinate axes for each object in the system.

2. *Categorize.* If an acceleration component for an object is zero, the object is modeled as a particle in equilibrium in this direction and $\Sigma F = 0$. If not, the object is modeled as a particle under a net force in this direction and $\Sigma F = ma$.

3. *Analyze.* Isolate the object whose motion is being analyzed. Draw a free-body diagram for this object. For systems containing more than one object, draw *separate* free-body diagrams for each object. *Do not* include in the free-body diagram forces exerted by the object on its surroundings.

   Find the components of the forces along the coordinate axes. Apply the appropriate model from the Categorize step for each direction. Check your dimensions to make sure that all terms have units of force.

   Solve the component equations for the unknowns. Remember that you must have as many independent equations as you have unknowns to obtain a complete solution.

4. *Finalize.* Make sure your results are consistent with the free-body diagram. Also check the predictions of your solutions for extreme values of the variables. By doing so, you can often detect errors in your results.

EXAMPLE 5.4 **A Traffic Light at Rest**

A traffic light weighing 122 N hangs from a cable tied to two other cables fastened to a support as in Figure 5.10a. The upper cables make angles of 37.0° and 53.0° with the horizontal. These upper cables are not as strong as the vertical cable and will break if the tension in them exceeds 100 N. Does the traffic light remain hanging in this situation, or will one of the cables break?

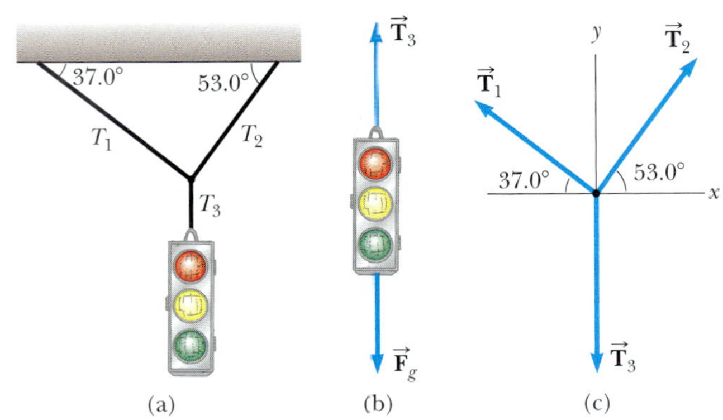

Figure 5.10 (Example 5.4) (a) A traffic light suspended by cables. (b) The free-body diagram for the traffic light. (c) The free-body diagram for the knot where the three cables are joined.

SOLUTION

Conceptualize Inspect the drawing in Figure 5.10a. Let us assume the cables do not break and that nothing is moving.

Categorize If nothing is moving, no part of the system is accelerating. We can now model the light as a particle in equilibrium on which the net force is zero. Similarly, the net force on the knot (Fig. 5.10c) is zero.

Analyze We construct two free-body diagrams: one for the traffic light, shown in Figure 5.10b, and one for the knot that holds the three cables together, shown in Figure 5.10c. This knot is a convenient object to choose because all the forces of interest act along lines passing through the knot.

Apply Equation 5.8 for the traffic light in the y direction:

$$\sum F_y = 0 \rightarrow T_3 - F_g = 0$$

$$T_3 = F_g = 122 \text{ N}$$

Choose the coordinate axes as shown in Figure 5.10c and resolve the forces acting on the knot into their components:

Force	x Component	y Component
$\vec{\mathbf{T}}_1$	$-T_1 \cos 37.0°$	$T_1 \sin 37.0°$
$\vec{\mathbf{T}}_2$	$T_2 \cos 53.0°$	$T_2 \sin 53.0°$
$\vec{\mathbf{T}}_3$	0	-122 N

Apply the particle in equilibrium model to the knot:

$$(1) \quad \sum F_x = -T_1 \cos 37.0° + T_2 \cos 53.0° = 0$$

$$(2) \quad \sum F_y = T_1 \sin 37.0° + T_2 \sin 53.0° + (-122 \text{ N}) = 0$$

Equation (1) shows that the horizontal components of $\vec{\mathbf{T}}_1$ and $\vec{\mathbf{T}}_2$ must be equal in magnitude, and Equation (2) shows that the sum of the vertical components of $\vec{\mathbf{T}}_1$ and $\vec{\mathbf{T}}_2$ must balance the downward force $\vec{\mathbf{T}}_3$, which is equal in magnitude to the weight of the light.

Solve Equation (1) for T_2 in terms of T_1:

$$(3) \quad T_2 = T_1\left(\frac{\cos 37.0°}{\cos 53.0°}\right) = 1.33T_1$$

Substitute this value for T_2 into Equation (2):

$$T_1 \sin 37.0° + (1.33T_1)(\sin 53.0°) - 122 \text{ N} = 0$$

$$T_1 = 73.4 \text{ N}$$

$$T_2 = 1.33T_1 = 97.4 \text{ N}$$

Both values are less than 100 N (just barely for T_2), so the cables will not break.

Finalize Let us finalize this problem by imagining a change in the system, as in the following **What If?**

What If? Suppose the two angles in Figure 5.10a are equal. What would be the relationship between T_1 and T_2?

Answer We can argue from the symmetry of the problem that the two tensions T_1 and T_2 would be equal to each other. Mathematically, if the equal angles are called θ, Equation (3) becomes

$$T_2 = T_1\left(\frac{\cos\theta}{\cos\theta}\right) = T_1$$

which also tells us that the tensions are equal. Without knowing the specific value of θ, we cannot find the values of T_1 and T_2. The tensions will be equal to each other, however, regardless of the value of θ.

CONCEPTUAL EXAMPLE 5.5 Forces Between Cars in a Train

Train cars are connected by *couplers*, which are under tension as the locomotive pulls the train. Imagine you are on a train speeding up with a constant acceleration. As you move through the train from the locomotive to the last car, measuring the tension in each set of couplers, does the tension increase, decrease, or stay the same? When the engineer applies the brakes, the couplers are under compression. How does this compression force vary from the locomotive to the last car? (Assume only the brakes on the wheels of the engine are applied.)

SOLUTION

As the train speeds up, tension decreases from the front of the train to the back. The coupler between the locomotive and the first car must apply enough force to accelerate the rest of the cars. As you move back along the train, each coupler is accelerating less mass behind it. The last coupler has to accelerate only the last car, and so it is under the least tension.

When the brakes are applied, the force again decreases from front to back. The coupler connecting the locomotive to the first car must apply a large force to slow down the rest of the cars, but the final coupler must apply a force large enough to slow down only the last car.

EXAMPLE 5.6 The Runaway Car

A car of mass m is on an icy driveway inclined at an angle θ as in Figure 5.11a.

(A) Find the acceleration of the car, assuming that the driveway is frictionless.

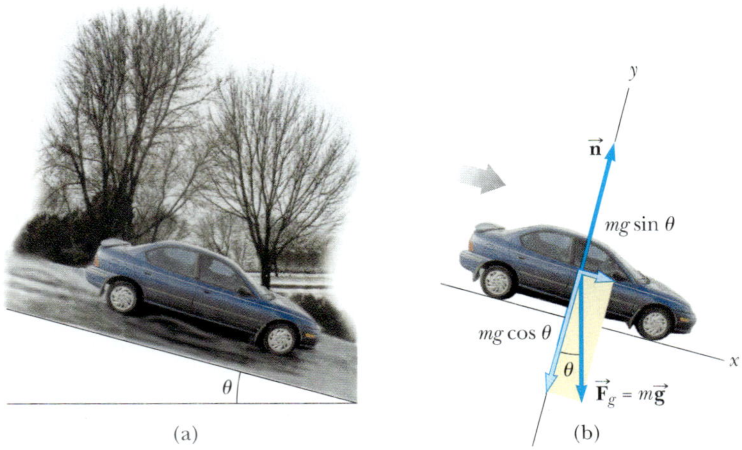

Figure 5.11 (Example 5.6) (a) A car of mass m on a frictionless incline. (b) The free-body diagram for the car.

SOLUTION

Conceptualize Use Figure 5.11a to conceptualize the situation. From everyday experience, we know that a car on an icy incline will accelerate down the incline. (The same thing happens to a car on a hill with its brakes not set.)

Categorize We categorize the car as a particle under a net force. Furthermore, this problem belongs to a very common category of problems in which an object moves under the influence of gravity on an inclined plane.

Analyze Figure 5.11b shows the free-body diagram for the car. The only forces acting on the car are the normal force $\vec{\mathbf{n}}$ exerted by the inclined plane, which acts perpendicular to the plane, and the gravitational force $\vec{\mathbf{F}}_g = m\vec{\mathbf{g}}$, which acts vertically downward. For problems involving inclined planes, it is convenient to choose the coordinate axes with x along the incline and y perpendicular to it as in Figure 5.11b. (It is possible, although inconvenient, to solve the problem with "standard" horizontal and vertical axes. You may want to try it, just for practice.) With these axes, we represent the gravitational force by a component of magnitude $mg \sin \theta$ along the positive x axis and one of magnitude $mg \cos \theta$ along the negative y axis.

Apply Newton's second law to the car in component form, noting that $a_y = 0$:

$$(1) \quad \sum F_x = mg \sin \theta = ma_x$$

$$(2) \quad \sum F_y = n - mg \cos \theta = 0$$

Solve Equation (1) for a_x:

$$(3) \quad a_x = g \sin \theta$$

Finalize Our choice of axes results in the car being modeled as a particle under a net force in the x direction and a particle in equilibrium in the y direction. Furthermore, the acceleration component a_x is independent of the mass of the car! It depends only on the angle of inclination and on g.

From Equation (2) we conclude that the component of $\vec{\mathbf{F}}_g$ perpendicular to the incline is balanced by the normal force; that is, $n = mg \cos \theta$. This situation is another case in which the normal force is *not* equal in magnitude to the weight of the object.

(B) Suppose the car is released from rest at the top of the incline and the distance from the car's front bumper to the bottom of the incline is d. How long does it take the front bumper to reach the bottom of the hill, and what is the car's speed as it arrives there?

SOLUTION

Concepualize Imagine that the car is sliding down the hill and you use a stopwatch to measure the entire time interval until it reaches the bottom.

Categorize This part of the problem belongs to kinematics rather than to dynamics, and Equation (3) shows that the acceleration a_x is constant. Therefore, you should categorize the car in this part of the problem as a particle under constant acceleration.

Analyze Defining the initial position of the front bumper as $x_i = 0$ and its final position as $x_f = d$, and recognizing that $v_{xi} = 0$, apply Equation 2.16, $x_f = x_i + v_{xi}t + \frac{1}{2}a_x t^2$:

$$d = \tfrac{1}{2}a_x t^2$$

Solve for t:

$$(4)\quad t = \sqrt{\frac{2d}{a_x}} = \sqrt{\frac{2d}{g\sin\theta}}$$

Use Equation 2.17, with $v_{xi} = 0$, to find the final velocity of the car:

$$v_{xf}^{\,2} = 2a_x d$$

$$(5)\quad v_{xf} = \sqrt{2a_x d} = \sqrt{2gd\sin\theta}$$

Finalize We see from Equations (4) and (5) that the time t at which the car reaches the bottom and its final speed v_{xf} are independent of the car's mass, as was its acceleration. Notice that we have combined techniques from Chapter 2 with new techniques from this chapter in this example. As we learn more techniques in later chapters, this process of combining information from several parts of the book will occur more often. In these cases, use the General Problem-Solving Strategy to help you identify what analysis models you will need.

What If? What previously solved problem does this situation become if $\theta = 90°$?

Answer Imagine θ going to 90° in Figure 5.11. The inclined plane becomes vertical, and the car is an object in free-fall! Equation (3) becomes

$$a_x = g\sin\theta = g\sin 90° = g$$

which is indeed the free-fall acceleration. (We find $a_x = g$ rather than $a_x = -g$ because we have chosen positive x to be downward in Fig. 5.11.) Notice also that the condition $n = mg\cos\theta$ gives us $n = mg\cos 90° = 0$. That is consistent with the car falling downward *next to* the vertical plane, in which case there is no contact force between the car and the plane.

EXAMPLE 5.7 One Block Pushes Another

Two blocks of masses m_1 and m_2, with $m_1 > m_2$, are placed in contact with each other on a frictionless, horizontal surface as in Active Figure 5.12a. A constant horizontal force $\vec{\mathbf{F}}$ is applied to m_1 as shown.

(A) Find the magnitude of the acceleration of the system.

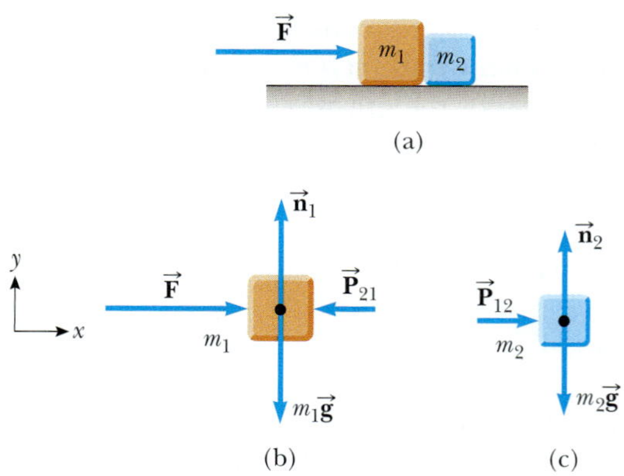

ACTIVE FIGURE 5.12

(Example 5.7) A force is applied to a block of mass m_1, which pushes on a second block of mass m_2. (b) The free-body diagram for m_1. (c) The free-body diagram for m_2.

Sign in at www.thomsonedu.com and go to ThomsonNOW to study the forces involved in this two-block system.

SOLUTION

Conceptualize Conceptualize the situation by using Active Figure 5.12a and realize that both blocks must experience the *same* acceleration because they are in contact with each other and remain in contact throughout the motion.

Categorize We categorize this problem as one involving a particle under a net force because a force is applied to a system of blocks and we are looking for the acceleration of the system.

Analyze First model the combination of two blocks as a single particle. Apply Newton's second law to the combination:

$$\sum F_x = F = (m_1 + m_2)a_x$$

$$(1)\quad a_x = \frac{F}{m_1 + m_2}$$

Finalize The acceleration given by Equation (1) is the same as that of a single object of mass $m_1 + m_2$ and subject to the same force.

(B) Determine the magnitude of the contact force between the two blocks.

SOLUTION

Conceptualize The contact force is internal to the system of two blocks. Therefore, we cannot find this force by modeling the whole system (the two blocks) as a single particle.

Categorize Now consider each of the two blocks individually by categorizing each as a particle under a net force.

Analyze We first construct a free-body diagram for each block as shown in Active Figures 5.12b and 5.12c, where the contact force is denoted by $\vec{\mathbf{P}}$. From Active Figure 5.12c we see that the only horizontal force acting on m_2 is the contact force $\vec{\mathbf{P}}_{12}$ (the force exerted by m_1 on m_2), which is directed to the right.

Apply Newton's second law to m_2:

$$(2)\quad \sum F_x = P_{12} = m_2 a_x$$

Substitute the value of the acceleration a_x given by Equation (1) into Equation (2):

$$(3)\quad P_{12} = m_2 a_x = \left(\frac{m_2}{m_1 + m_2}\right)F$$

Finalize This result shows that the contact force P_{12} is *less* than the applied force F. The force required to accelerate block 2 alone must be less than the force required to produce the same acceleration for the two-block system.

To finalize further, let us check this expression for P_{12} by considering the forces acting on m_1, shown in Active Figure 5.12b. The horizontal forces acting on m_1 are the applied force $\vec{\mathbf{F}}$ to the right and the contact force $\vec{\mathbf{P}}_{21}$ to the left (the force exerted by m_2 on m_1). From Newton's third law, $\vec{\mathbf{P}}_{21}$ is the reaction force to $\vec{\mathbf{P}}_{12}$, so $P_{21} = P_{12}$.

Apply Newton's second law to m_1:

$$(4)\quad \sum F_x = F - P_{21} = F - P_{12} = m_1 a_x$$

Solve for P_{12} and substitute the value of a_x from Equation (1):

$$P_{12} = F - m_1 a_x = F - m_1\left(\frac{F}{m_1 + m_2}\right) = \left(\frac{m_2}{m_1 + m_2}\right)F$$

This result agrees with Equation (3), as it must.

What If? Imagine that the force $\vec{\mathbf{F}}$ in Active Figure 5.12 is applied toward the left on the right-hand block of mass m_2. Is the magnitude of the force $\vec{\mathbf{P}}_{12}$ the same as it was when the force was applied toward the right on m_1?

Answer When the force is applied toward the left on m_2, the contact force must accelerate m_1. In the original situation, the contact force accelerates m_2. Because $m_1 > m_2$, more force is required, so the magnitude of $\vec{\mathbf{P}}_{12}$ is greater than in the original situation.

EXAMPLE 5.8 Weighing a Fish in an Elevator

A person weighs a fish of mass m on a spring scale attached to the ceiling of an elevator as illustrated in Figure 5.13.

(A) Show that if the elevator accelerates either upward or downward, the spring scale gives a reading that is different from the weight of the fish.

SOLUTION

Conceptualize The reading on the scale is related to the extension of the spring in the scale, which is related to the force on the end of the spring as in Figure 5.2. Imagine that the fish is hanging on a string attached to the end of the spring. In this case, the magnitude of the force exerted on the spring is equal to the tension T in the string.

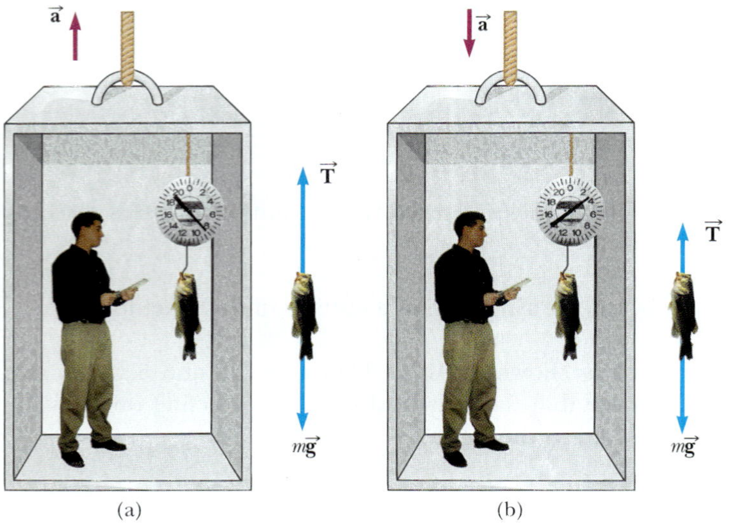

Figure 5.13 (Example 5.8) Apparent weight versus true weight. (a) When the elevator accelerates upward, the spring scale reads a value greater than the weight of the fish. (b) When the elevator accelerates downward, the spring scale reads a value less than the weight of the fish.

Therefore, we are looking for T. The force $\vec{\mathbf{T}}$ pulls down on the string and pulls up on the fish.

Categorize We can categorize this problem by identifying the fish as a particle under a net force.

Analyze Inspect the free-body diagrams for the fish in Figure 5.13 and notice that the external forces acting on the fish are the downward gravitational force $\vec{\mathbf{F}}_g = m\vec{\mathbf{g}}$ and the force $\vec{\mathbf{T}}$ exerted by the string. If the elevator is either at rest or moving at constant velocity, the fish is a particle in equilibrium, so $\sum F_y = T - F_g = 0$ or $T = F_g = mg$. (Remember that the scalar mg is the weight of the fish.)

Now suppose the elevator is moving with an acceleration $\vec{\mathbf{a}}$ relative to an observer standing outside the elevator in an inertial frame (see Fig. 5.13). The fish is now a particle under a net force.

Apply Newton's second law to the fish:

$$\sum F_y = T - mg = ma_y$$

Solve for T:

$$(1) \quad T = ma_y + mg = mg\left(\frac{a_y}{g} + 1\right) = F_g\left(\frac{a_y}{g} + 1\right)$$

where we have chosen upward as the positive y direction. We conclude from Equation (1) that the scale reading T is greater than the fish's weight mg if $\vec{\mathbf{a}}$ is upward, so a_y is positive, and that the reading is less than mg if $\vec{\mathbf{a}}$ is downward, so a_y is negative.

(B) Evaluate the scale readings for a 40.0-N fish if the elevator moves with an acceleration $a_y = \pm 2.00 \text{ m/s}^2$.

Evaluate the scale reading from Equation (1) if $\vec{\mathbf{a}}$ is upward:

$$T = (40.0 \text{ N})\left(\frac{2.00 \text{ m/s}^2}{9.80 \text{ m/s}^2} + 1\right) = 48.2 \text{ N}$$

Evaluate the scale reading from Equation (1) if $\vec{\mathbf{a}}$ is downward:

$$T = (40.0 \text{ N})\left(\frac{-2.00 \text{ m/s}^2}{9.80 \text{ m/s}^2} + 1\right) = 31.8 \text{ N}$$

Finalize Take this advice: if you buy a fish in an elevator, make sure the fish is weighed while the elevator is either at rest or accelerating downward! Furthermore, notice that from the information given here, one cannot determine the direction of motion of the elevator.

What If? Suppose the elevator cable breaks and the elevator and its contents are in free-fall. What happens to the reading on the scale?

Answer If the elevator falls freely, its acceleration is $a_y = -g$. We see from Equation (1) that the scale reading T is zero in this case; that is, the fish *appears* to be weightless.

EXAMPLE 5.9 The Atwood Machine

When two objects of unequal mass are hung vertically over a frictionless pulley of negligible mass as in Active Figure 5.14a, the arrangement is called an *Atwood machine.* The device is sometimes used in the laboratory to calculate the value of g. Determine the magnitude of the acceleration of the two objects and the tension in the lightweight cord.

SOLUTION

Conceptualize Imagine the situation pictured in Active Figure 5.14a in action: as one object moves upward, the other object moves downward. Because the objects are connected by an inextensible string, their accelerations must be of equal magnitude.

Categorize The objects in the Atwood machine are subject to the gravitational force as well as to the forces exerted by the strings connected to them. Therefore, we can categorize this problem as one involving two particles under a net force.

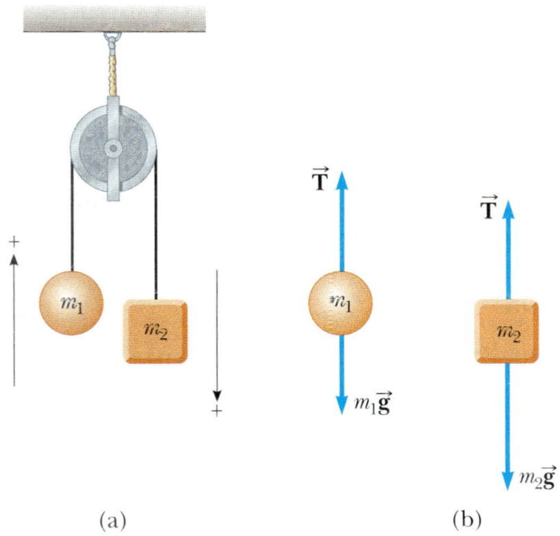

ACTIVE FIGURE 5.14

(Example 5.9) The Atwood machine. (a) Two objects connected by a massless inextensible cord over a frictionless pulley. (b) The free-body diagrams for the two objects.

Sign in at www.thomsonedu.com and go to ThomsonNOW to adjust the masses of the objects on the Atwood machine and observe the motion.

Analyze The free-body diagrams for the two objects are shown in Active Figure 5.14b. Two forces act on each object: the upward force $\vec{\mathbf{T}}$ exerted by the string and the downward gravitational force. In problems such as this one in which the pulley is modeled as massless and frictionless, the tension in the string on both sides of the pulley is the same. If the pulley has mass or is subject to friction, the tensions on either side are not the same and the situation requires techniques we will learn in Chapter 10.

We must be very careful with signs in problems such as this. In Active Figure 5.14a, notice that if object 1 accelerates upward, object 2 accelerates downward. Therefore, for consistency with signs, if we define the upward direction as positive for object 1, we must define the downward direction as positive for object 2. With this sign convention, both objects accelerate in the same direction as defined by the choice of sign. Furthermore, according to this sign convention, the y component of the net force exerted on object 1 is $T - m_1 g$, and the y component of the net force exerted on object 2 is $m_2 g - T$.

Apply Newton's second law to object 1:

$$(1) \quad \sum F_y = T - m_1 g = m_1 a_y$$

Apply Newton's second law to object 2:

$$(2) \quad \sum F_y = m_2 g - T = m_2 a_y$$

Add Equation (2) to Equation (1), noticing that T cancels:

$$-m_1 g + m_2 g = m_1 a_y + m_2 a_y$$

Solve for the acceleration:

$$(3) \quad a_y = \left(\frac{m_2 - m_1}{m_1 + m_2}\right) g$$

Substitute Equation (3) into Equation (1) to find T:

$$(4) \quad T = m_1(g + a_y) = \left(\frac{2m_1 m_2}{m_1 + m_2}\right) g$$

Finalize The acceleration given by Equation (3) can be interpreted as the ratio of the magnitude of the unbalanced force on the system $(m_2 - m_1)g$ to the total mass of the system $(m_1 + m_2)$, as expected from Newton's second law. Notice that the sign of the acceleration depends on the relative masses of the two objects.

What If? Describe the motion of the system if the objects have equal masses, that is, $m_1 = m_2$.

Answer If we have the same mass on both sides, the system is balanced and should not accelerate. Mathematically, we see that if $m_1 = m_2$, Equation (3) gives us $a_y = 0$.

What If? What if one of the masses is much larger than the other: $m_1 >> m_2$?

Answer In the case in which one mass is infinitely larger than the other, we can ignore the effect of the smaller mass. Therefore, the larger mass should simply fall as if the smaller mass were not there. We see that if $m_1 >> m_2$, Equation (3) gives us $a_y = -g$.

EXAMPLE 5.10 Acceleration of Two Objects Connected by a Cord

A ball of mass m_1 and a block of mass m_2 are attached by a lightweight cord that passes over a frictionless pulley of negligible mass as in Figure 5.15a. The block lies on a frictionless incline of angle θ. Find the magnitude of the acceleration of the two objects and the tension in the cord.

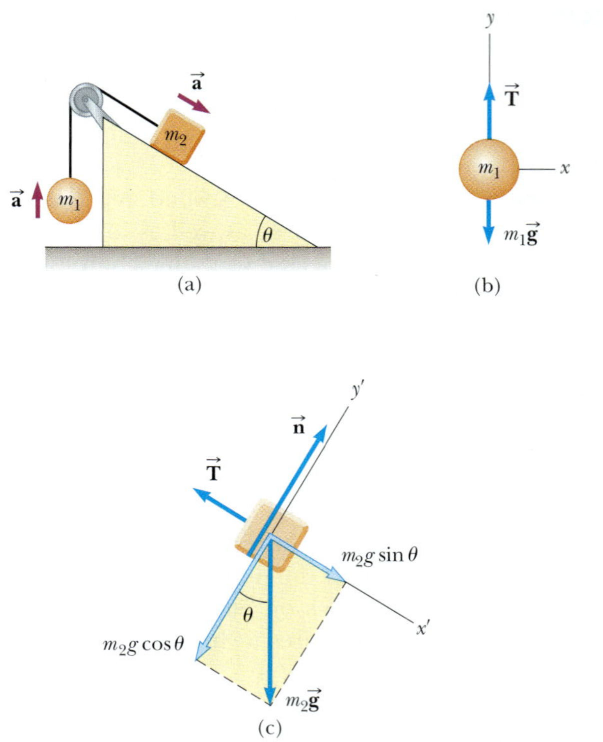

Figure 5.15 (Example 5.10) (a) Two objects connected by a lightweight cord strung over a frictionless pulley. (b) The free-body diagram for the ball. (c) The free-body diagram for the block. (The incline is frictionless.)

SOLUTION

Conceptualize Imagine the objects in Figure 5.15 in motion. If m_2 moves down the incline, m_1 moves upward. Because the objects are connected by a cord (which we assume does not stretch), their accelerations have the same magnitude.

Categorize We can identify forces on each of the two objects and we are looking for an acceleration, so we categorize the objects as particles under a net force.

Analyze Consider the free-body diagrams shown in Figures 5.15b and 5.15c.

Apply Newton's second law in component form to the ball, choosing the upward direction as positive:

$$(1)\quad \sum F_x = 0$$

$$(2)\quad \sum F_y = T - m_1 g = m_1 a_y = m_1 a$$

For the ball to accelerate upward, it is necessary that $T > m_1 g$. In Equation (2), we replaced a_y with a because the acceleration has only a y component.

For the block it is convenient to choose the positive x' axis along the incline as in Figure 5.15c. For consistency with our choice for the ball, we choose the positive direction to be down the incline.

Apply Newton's second law in component form to the block:

$$(3)\quad \sum F_{x'} = m_2 g \sin\theta - T = m_2 a_{x'} = m_2 a$$

$$(4)\quad \sum F_{y'} = n - m_2 g \cos\theta = 0$$

In Equation (3), we replaced $a_{x'}$ with a because the two objects have accelerations of equal magnitude a.

Solve Equation (2) for T:

$$(5)\quad T = m_1(g + a)$$

Substitute this expression for T into Equation (3):

$$m_2 g \sin\theta - m_1(g + a) = m_2 a$$

Solve for a:

$$(6)\quad a = \frac{m_2 g \sin\theta - m_1 g}{m_1 + m_2}$$

Substitute this expression for a into Equation (5) to find T:

$$(7)\quad T = \frac{m_1 m_2 g(\sin\theta + 1)}{m_1 + m_2}$$

Finalize The block accelerates down the incline only if $m_2 \sin\theta > m_1$. If $m_1 > m_2 \sin\theta$, the acceleration is up the incline for the block and downward for the ball. Also notice that the result for the acceleration, Equation (6), can be interpreted as the magnitude of the net external force acting on the ball–block system divided by the total mass of the system; this result is consistent with Newton's second law.

What If? What happens in this situation if $\theta = 90°$?

Answer If $\theta = 90°$, the inclined plane becomes vertical and there is no interaction between its surface and m_2. Therefore, this problem becomes the Atwood machine of Example 5.9. Letting $\theta \to 90°$ in Equations (6) and (7) causes them to reduce to Equations (3) and (4) of Example 5.9!

What If? What if $m_1 = 0$?

Answer If $m_1 = 0$, then m_2 is simply sliding down an inclined plane without interacting with m_1 through the string. Therefore, this problem becomes the sliding car problem in Example 5.6. Letting $m_1 \to 0$ in Equation (6) causes it to reduce to Equation (3) of Example 5.6!

5.8 Forces of Friction

When an object is in motion either on a surface or in a viscous medium such as air or water, there is resistance to the motion because the object interacts with its surroundings. We call such resistance a **force of friction**. Forces of friction are very important in our everyday lives. They allow us to walk or run and are necessary for the motion of wheeled vehicles.

Imagine that you are working in your garden and have filled a trash can with yard clippings. You then try to drag the trash can across the surface of your concrete patio as in Active Figure 5.16a. This surface is *real*, not an idealized, friction-

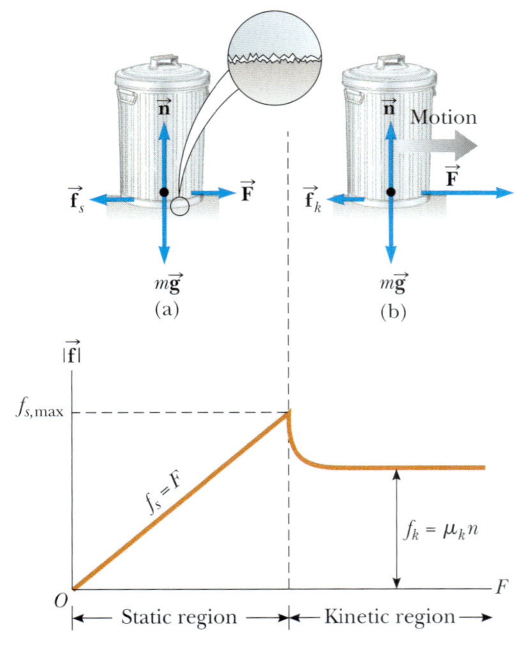

ACTIVE FIGURE 5.16

When pulling on a trash can, the direction of the force of friction $\vec{\mathbf{f}}$ between the can and a rough surface is opposite the direction of the applied force $\vec{\mathbf{F}}$. Because both surfaces are rough, contact is made only at a few points as illustrated in the "magnified" view. (a) For small applied forces, the magnitude of the force of static friction equals the magnitude of the applied force. (b) When the magnitude of the applied force exceeds the magnitude of the maximum force of static friction, the trash can breaks free. The applied force is now larger than the force of kinetic friction, and the trash can accelerates to the right. (c) A graph of friction force versus applied force. Notice that $f_{s,\,\max} > f_k$.

Sign in at www.thomsonedu.com and go to ThomsonNOW to vary the applied force on the trash can and practice sliding it on surfaces of varying roughness. Notice the effect on the trash can's motion and the corresponding behavior of the graph in (c).

less surface. If we apply an external horizontal force $\vec{\mathbf{F}}$ to the trash can, acting to the right, the trash can remains stationary when $\vec{\mathbf{F}}$ is small. The force on the trash can that counteracts $\vec{\mathbf{F}}$ and keeps it from moving acts toward the left and is called the **force of static friction** $\vec{\mathbf{f}}_s$. As long as the trash can is not moving, $f_s = F$. Therefore, if $\vec{\mathbf{F}}$ is increased, $\vec{\mathbf{f}}_s$ also increases. Likewise, if $\vec{\mathbf{F}}$ decreases, $\vec{\mathbf{f}}_s$ also decreases. Experiments show that the friction force arises from the nature of the two surfaces: because of their roughness, contact is made only at a few locations where peaks of the material touch, as shown in the magnified view of the surface in Active Figure 5.16a.

Force of static friction ▶

At these locations, the friction force arises in part because one peak physically blocks the motion of a peak from the opposing surface and in part from chemical bonding ("spot welds") of opposing peaks as they come into contact. Although the details of friction are quite complex at the atomic level, this force ultimately involves an electrical interaction between atoms or molecules.

If we increase the magnitude of $\vec{\mathbf{F}}$ as in Active Figure 5.16b, the trash can eventually slips. When the trash can is on the verge of slipping, f_s has its maximum value $f_{s,\max}$ as shown in Active Figure 5.16c. When F exceeds $f_{s,\max}$, the trash can moves and accelerates to the right. We call the friction force for an object in motion the **force of kinetic friction** $\vec{\mathbf{f}}_k$. When the trash can is in motion, the force of kinetic friction on the can is less than $f_{s,\max}$ (Active Fig. 5.16c). The net force $F - f_k$ in the x direction produces an acceleration to the right, according to Newton's second law. If $F = f_k$, the acceleration is zero and the trash can moves to the right with constant speed. If the applied force $\vec{\mathbf{F}}$ is removed from the moving can, the friction force $\vec{\mathbf{f}}_k$ acting to the left provides an acceleration of the trash can in the $-x$ direction and eventually brings it to rest, again consistent with Newton's second law.

Force of kinetic friction ▶

Experimentally, we find that, to a good approximation, both $f_{s,\max}$ and f_k are proportional to the magnitude of the normal force exerted on an object by the surface. The following descriptions of the force of friction are based on experimental observations and serve as the model we shall use for forces of friction in problem solving:

- The magnitude of the force of static friction between any two surfaces in contact can have the values

$$f_s \le \mu_s n \tag{5.9}$$

  where the dimensionless constant μ_s is called the **coefficient of static friction** and n is the magnitude of the normal force exerted by one surface on the other. The equality in Equation 5.9 holds when the surfaces are on the verge of slipping, that is, when $f_s = f_{s,\max} \equiv \mu_s n$. This situation is called *impending motion*. The inequality holds when the surfaces are not on the verge of slipping.
- The magnitude of the force of kinetic friction acting between two surfaces is

$$f_k = \mu_k n \tag{5.10}$$

  where μ_k is the **coefficient of kinetic friction**. Although the coefficient of kinetic friction can vary with speed, we shall usually neglect any such variations in this text.
- The values of μ_k and μ_s depend on the nature of the surfaces, but μ_k is generally less than μ_s. Typical values range from around 0.03 to 1.0. Table 5.1 lists some reported values.
- The direction of the friction force on an object is parallel to the surface with which the object is in contact and opposite to the actual motion (kinetic friction) or the impending motion (static friction) of the object relative to the surface.
- The coefficients of friction are nearly independent of the area of contact between the surfaces. We might expect that placing an object on the side having the most area might increase the friction force. Although this method provides more points in contact as in Active Figure 5.16a, the weight of the

PITFALL PREVENTION 5.9

The Equal Sign Is Used in Limited Situations

In Equation 5.9, the equal sign is used *only* in the case in which the surfaces are just about to break free and begin sliding. Do not fall into the common trap of using $f_s = \mu_s n$ in *any* static situation.

PITFALL PREVENTION 5.10

Friction Equations

Equations 5.9 and 5.10 are *not* vector equations. They are relationships between the *magnitudes* of the vectors representing the friction and normal forces. Because the friction and normal forces are perpendicular to each other, the vectors cannot be related by a multiplicative constant.

PITFALL PREVENTION 5.11

The Direction of the Friction Force

Sometimes, an incorrect statement about the friction force between an object and a surface is made—"the friction force on an object is opposite to its motion or impending motion"—rather than the correct phrasing, "the friction force on an object is opposite to its motion or impending motion *relative to the surface.*"

TABLE 5.1

Coefficients of Friction

	μ_s	μ_k
Rubber on concrete	1.0	0.8
Steel on steel	0.74	0.57
Aluminum on steel	0.61	0.47
Glass on glass	0.94	0.4
Copper on steel	0.53	0.36
Wood on wood	0.25–0.5	0.2
Waxed wood on wet snow	0.14	0.1
Waxed wood on dry snow	—	0.04
Metal on metal (lubricated)	0.15	0.06
Teflon on Teflon	0.04	0.04
Ice on ice	0.1	0.03
Synovial joints in humans	0.01	0.003

Note: All values are approximate. In some cases, the coefficient of friction can exceed 1.0.

object is spread out over a larger area and the individual points are not pressed together as tightly. Because these effects approximately compensate for each other, the friction force is independent of the area.

Quick Quiz 5.6 You press your physics textbook flat against a vertical wall with your hand. What is the direction of the friction force exerted by the wall on the book? (a) downward (b) upward (c) out from the wall (d) into the wall

Quick Quiz 5.7 You are playing with your daughter in the snow. She sits on a sled and asks you to slide her across a flat, horizontal field. You have a choice of (a) pushing her from behind by applying a force downward on her shoulders at 30° below the horizontal (Fig. 5.17a) or (b) attaching a rope to the front of the sled and pulling with a force at 30° above the horizontal (Fig 5.17b). Which would be easier for you and why?

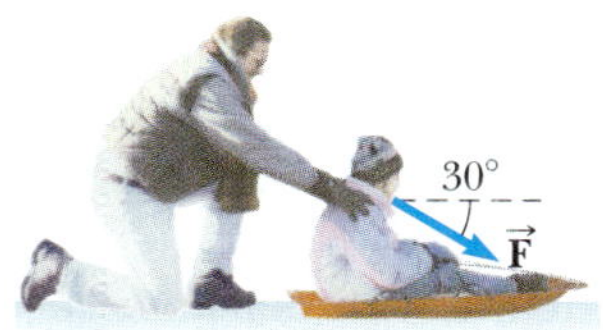

(a)

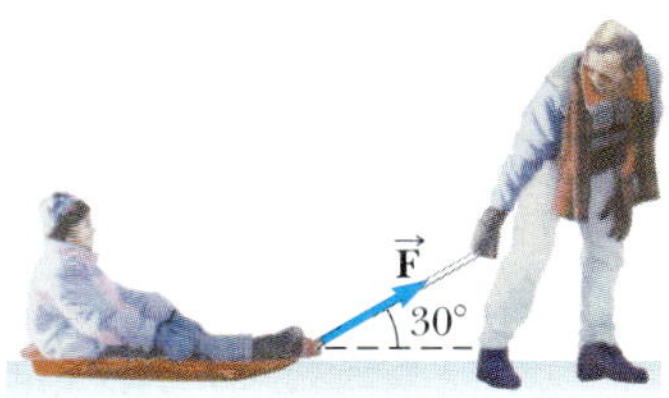

(b)

Figure 5.17 (Quick Quiz 5.7) A father slides his daughter on a sled either by (a) pushing down on her shoulders or (b) pulling up on a rope.

EXAMPLE 5.11 **Experimental Determination of μ_s and μ_k**

The following is a simple method of measuring coefficients of friction. Suppose a block is placed on a rough surface inclined relative to the horizontal as shown in Active Figure 5.18. The incline angle is increased until the block starts to move. Show that you can obtain μ_s by measuring the critical angle θ_c at which this slipping just occurs.

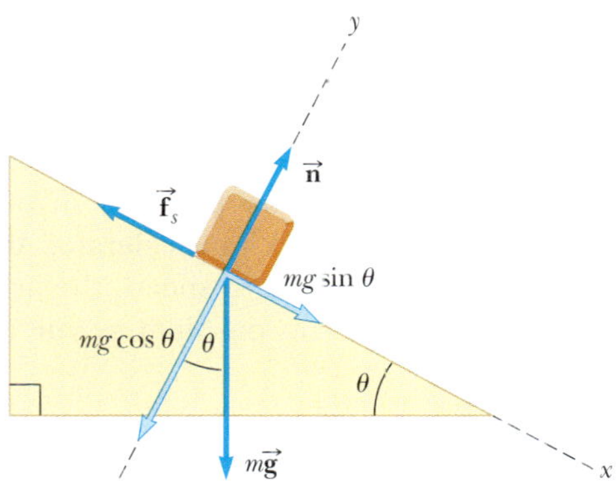

ACTIVE FIGURE 5.18

(Example 5.11) The external forces exerted on a block lying on a rough incline are the gravitational force $m\vec{g}$, the normal force $\vec{n}$, and the force of friction $\vec{f}_s$. For convenience, the gravitational force is resolved into a component $mg \sin \theta$ along the incline and a component $mg \cos \theta$ perpendicular to the incline.

Sign in at www.thomsonedu.com and go to ThomsonNOW to investigate this situation further.

SOLUTION

Conceptualize Consider the free-body diagram in Active Figure 5.18 and imagine that the block tends to slide down the incline due to the gravitational force. To simulate the situation, place a coin on this book's cover and tilt the book until the coin begins to slide.

Categorize The block is subject to various forces. Because we are raising the plane to the angle at which the block is just ready to begin to move but is not moving, we categorize the block as a particle in equilibrium.

Analyze The forces acting on the block are the gravitational force $m\vec{g}$, the normal force $\vec{n}$, and the force of static friction $\vec{f}_s$. We choose x to be parallel to the plane and y perpendicular to it.

Apply Equation 5.8 to the block:

$$(1)\quad \sum F_x = mg \sin\theta - f_s = 0$$

$$(2)\quad \sum F_y = n - mg\cos\theta = 0$$

Substitute $mg = n/\cos\theta$ from Equation (2) into Equation (1):

$$(3)\quad f_s = mg\sin\theta = \left(\frac{n}{\cos\theta}\right)\sin\theta = n\tan\theta$$

When the incline angle is increased until the block is on the verge of slipping, the force of static friction has reached its maximum value $\mu_s n$. The angle θ in this situation is the critical angle θ_c. Make these substitutions in Equation (3):

$$\mu_s n = n\tan\theta_c$$

$$\mu_s = \tan\theta_c$$

For example, if the block just slips at $\theta_c = 20.0°$, we find that $\mu_s = \tan 20.0° = 0.364$.

Finalize Once the block starts to move at $\theta \geq \theta_c$, it accelerates down the incline and the force of friction is $f_k = \mu_k n$. If θ is reduced to a value less than θ_c, however, it may be possible to find an angle θ'_c such that the block moves down the incline with constant speed as a particle in equilibrium again ($a_x = 0$). In this case, use Equations (1) and (2) with f_s replaced by f_k to find μ_k: $\mu_k = \tan\theta'_c$ where $\theta'_c < \theta_c$.

EXAMPLE 5.12 The Sliding Hockey Puck

A hockey puck on a frozen pond is given an initial speed of 20.0 m/s. If the puck always remains on the ice and slides 115 m before coming to rest, determine the coefficient of kinetic friction between the puck and ice.

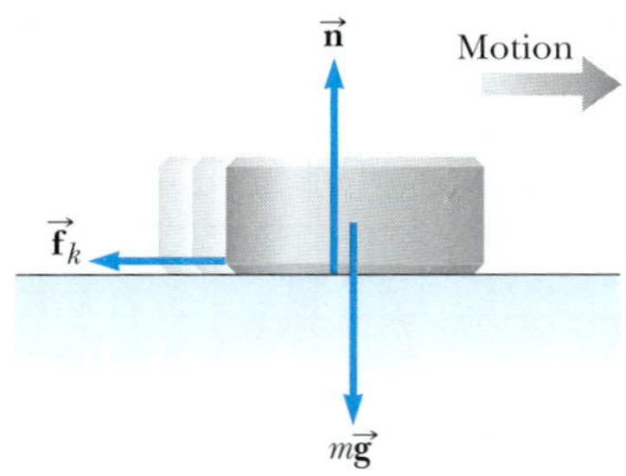

Figure 5.19 (Example 5.12) After the puck is given an initial velocity to the right, the only external forces acting on it are the gravitational force $m\vec{g}$, the normal force $\vec{n}$, and the force of kinetic friction $\vec{f}_k$.

SOLUTION

Conceptualize Imagine that the puck in Figure 5.19 slides to the right and eventually comes to rest due to the force of kinetic friction.

Categorize The forces acting on the puck are identified in Figure 5.19, but the text of the problem provides kinematic variables. Therefore, we categorize the problem in two ways. First, the problem involves a particle under a net force: kinetic friction causes the puck to accelerate. And, because we model the force of kinetic friction as independent of speed, the acceleration of the puck is constant. So, we can also categorize this problem as one involving a particle under constant acceleration.

Analyze First, we find the acceleration algebraically in terms of the coefficient of kinetic friction, using Newton's second law. Once we know the acceleration of the puck and the distance it travels, the equations of kinematics can be used to find the numerical value of the coefficient of kinetic friction.

Apply the particle under a net force model in the x direction to the puck:

$$(1)\quad \sum F_x = -f_k = ma_x$$

Apply the particle in equilibrium model in the y direction to the puck:

$$(2)\quad \sum F_y = n - mg = 0$$

Substitute $n = mg$ from Equation (2) and $f_k = \mu_k n$ into Equation (1):

$$-\mu_k n = -\mu_k mg = ma_x$$

$$a_x = -\mu_k g$$

The negative sign means the acceleration is to the left in Figure 5.19. Because the velocity of the puck is to the right, the puck is slowing down. The acceleration is independent of the mass of the puck and is constant because we assume that μ_k remains constant.

Apply the particle under constant acceleration model to the puck, using Equation 2.17, $v_{xf}^2 = v_{xi}^2 + 2a_x(x_f - x_i)$, with $x_i = 0$ and $v_f = 0$:

$$0 = v_{xi}^2 + 2a_x x_f = v_{xi}^2 - 2\mu_k g x_f$$

$$\mu_k = \frac{v_{xi}^2}{2gx_f}$$

$$\mu_k = \frac{(20.0 \text{ m/s})^2}{2(9.80 \text{ m/s}^2)(115 \text{ m})} = 0.117$$

Finalize Notice that μ_k is dimensionless, as it should be, and that it has a low value, consistent with an object sliding on ice.

EXAMPLE 5.13 Acceleration of Two Connected Objects When Friction Is Present

A block of mass m_1 on a rough, horizontal surface is connected to a ball of mass m_2 by a lightweight cord over a lightweight, frictionless pulley as shown in Figure 5.20a. A force of magnitude F at an angle θ with the horizontal is applied to the block as shown and the block slides to the right. The coefficient of kinetic friction between the block and surface is μ_k. Determine the magnitude of the acceleration of the two objects.

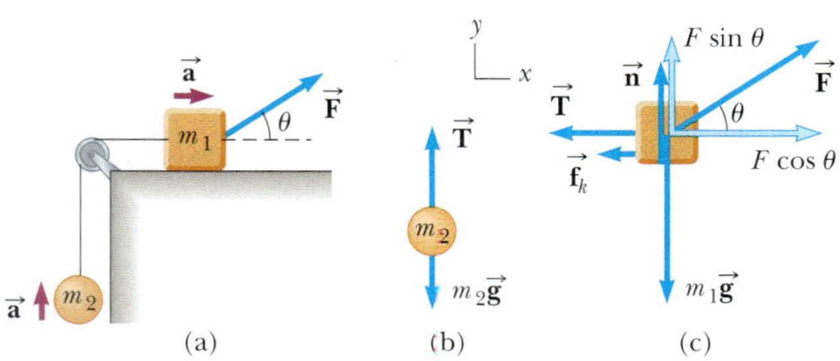

Figure 5.20 (Example 5.13) (a) The external force $\vec{\mathbf{F}}$ applied as shown can cause the block to accelerate to the right. (b, c) The free-body diagrams assuming the block accelerates to the right and the ball accelerates upward. The magnitude of the force of kinetic friction in this case is given by $f_k = \mu_k n = \mu_k(m_1 g - F \sin\theta)$.

SOLUTION

Conceptualize Imagine what happens as $\vec{\mathbf{F}}$ is applied to the block. Assuming $\vec{\mathbf{F}}$ is not large enough to lift the block, the block slides to the right and the ball rises.

Categorize We can identify forces and we want an acceleration, so we categorize this problem as one involving two particles under a net force, the ball and the block.

Analyze First draw free-body diagrams for the two objects as shown in Figures 5.20b and 5.20c. The applied force $\vec{\mathbf{F}}$ has x and y components $F\cos\theta$ and $F\sin\theta$, respectively. Because the two objects are connected, we can equate the magnitudes of the x component of the acceleration of the block and the y component of the acceleration of the ball and call them both a. Let us assume the motion of the block is to the right.

Apply the particle under a net force model to the block in the horizontal direction:

$$(1) \quad \sum F_x = F\cos\theta - f_k - T = m_1 a_x = m_1 a$$

Apply the particle in equilibrium model to the block in the vertical direction:

$$(2) \quad \sum F_y = n + F\sin\theta - m_1 g = 0$$

Apply the particle under a net force model to the ball in the vertical direction:

$$(3) \quad \sum F_y = T - m_2 g = m_2 a_y = m_2 a$$

Solve Equation (2) for n:

$$n = m_1 g - F\sin\theta$$

Substitute n into $f_k = \mu_k n$ from Equation 5.10:

$$(4) \quad f_k = \mu_k(m_1 g - F\sin\theta)$$

Substitute Equation (4) and the value of T from Equation (3) into Equation (1):

$$F\cos\theta - \mu_k(m_1 g - F\sin\theta) - m_2(a + g) = m_1 a$$

Solve for a:

$$(5) \quad a = \frac{F(\cos\theta + \mu_k \sin\theta) - (m_2 + \mu_k m_1)g}{m_1 + m_2}$$

Finalize The acceleration of the block can be either to the right or to the left depending on the sign of the numerator in Equation (5). If the motion is to the left, we must reverse the sign of f_k in Equation (1) because the force of kinetic friction must oppose the motion of the block relative to the surface. In this case, the value of a is the same as in Equation (5), with the two plus signs in the numerator changed to minus signs.

Summary

ThomsonNOW™ Sign in at **www.thomsonedu.com** and go to ThomsonNOW to take a practice test for this chapter.

DEFINITIONS

An **inertial frame of reference** is a frame in which an object that does not interact with other objects experiences zero acceleration. Any frame moving with constant velocity relative to an inertial frame is also an inertial frame.

We define **force** as **that which causes a change in motion of an object**.

CONCEPTS AND PRINCIPLES

Newton's first law states that it is possible to find an inertial frame in which an object that does not interact with other objects experiences zero acceleration, or, equivalently, in the absence of an external force, when viewed from an inertial frame, an object at rest remains at rest and an object in uniform motion in a straight line maintains that motion.
Newton's second law states that the acceleration of an object is directly proportional to the net force acting on it and inversely proportional to its mass.
Newton's third law states that if two objects interact, the force exerted by object 1 on object 2 is equal in magnitude and opposite in direction to the force exerted by object 2 on object 1.

The **gravitational force** exerted on an object is equal to the product of its mass (a scalar quantity) and the free-fall acceleration: $\vec{\mathbf{F}}_g = m\vec{\mathbf{g}}$.
The **weight** of an object is the magnitude of the gravitational force acting on the object.

The maximum **force of static friction** $\vec{\mathbf{f}}_{s,\text{max}}$ between an object and a surface is proportional to the normal force acting on the object. In general, $f_s \le \mu_s n$, where μ_s is the **coefficient of static friction** and n is the magnitude of the normal force. When an object slides over a surface, the magnitude of the **force of kinetic friction** $\vec{\mathbf{f}}_k$ is given by $f_k = \mu_k n$, where μ_k is the **coefficient of kinetic friction**. The direction of the friction force is opposite the direction of motion or impending motion of the object relative to the surface.

ANALYSIS MODELS FOR PROBLEM SOLVING

Particle Under a Net Force If a particle of mass m experiences a nonzero net force, its acceleration is related to the net force by Newton's second law:

$$\sum \vec{\mathbf{F}} = m\vec{\mathbf{a}} \quad (5.2)$$

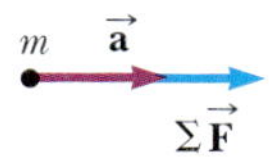

Particle in Equilibrium If a particle maintains a constant velocity (so that $\vec{\mathbf{a}} = 0$), which could include a velocity of zero, the forces on the particle balance and Newton's second law reduces to

$$\sum \vec{\mathbf{F}} = 0 \quad (5.8)$$

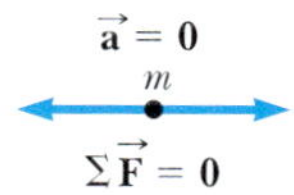

Questions

☐ denotes answer available in *Student Solutions Manual/Study Guide;* **O** denotes objective question

1. A ball is held in a person's hand. (a) Identify all the external forces acting on the ball and the reaction to each. (b) If the ball is dropped, what force is exerted on it while it is falling? Identify the reaction force in this case. (Ignore air resistance.)

2. If a car is traveling westward with a constant speed of 20 m/s, what is the resultant force acting on it?

3. **O** An experiment is performed on a puck on a level air hockey table, where friction is negligible. A constant horizontal force is applied to the puck and its acceleration is measured. Now the same puck is transported far into outer space, where both friction and gravity are negligible. The same constant force is applied to the puck (through a spring scale that stretches the same amount) and the puck's acceleration (relative to the distant stars) is measured. What is the puck's acceleration in outer space? (a) somewhat greater than its acceleration on the Earth (b) the same as its acceleration on the Earth (c) less than its acceleration on the Earth (d) infinite because neither friction nor gravity constrains it (e) very large because acceleration is inversely proportional to weight and the puck's weight is very small but not zero

4. In the motion picture *It Happened One Night* (Columbia Pictures, 1934), Clark Gable is standing inside a stationary bus in front of Claudette Colbert, who is seated. The bus suddenly starts moving forward and Clark falls into Claudette's lap. Why did that happen?

5. Your hands are wet and the restroom towel dispenser is empty. What do you do to get drops of water off your hands? How does your action exemplify one of Newton's laws? Which one?

6. A passenger sitting in the rear of a bus claims that she was injured when the driver slammed on the brakes, causing a suitcase to come flying toward her from the front of the bus. If you were the judge in this case, what disposition would you make? Why?

7. A spherical rubber balloon inflated with air is held stationary, and its opening, on the west side, is pinched shut. (a) Describe the forces exerted by the air on sections of the rubber. (b) After the balloon is released, it takes off toward the east, gaining speed rapidly. Explain this motion in terms of the forces now acting on the rubber. (c) Account for the motion of a skyrocket taking off from its launch pad.

8. If you hold a horizontal metal bar several centimeters above the ground and move it through grass, each leaf of grass bends out of the way. If you increase the speed of the bar, each leaf of grass will bend more quickly. How then does a rotary power lawn mower manage to cut grass? How can it exert enough force on a leaf of grass to shear it off?

9. A rubber ball is dropped onto the floor. What force causes the ball to bounce?

10. A child tosses a ball straight up. She says the ball is moving away from her hand because the ball feels an upward "force of the throw" as well as the gravitational force. (a) Can the "force of the throw" exceed the gravitational force? How would the ball move if it did? (b) Can the "force of the throw" be equal in magnitude to the gravitational force? Explain. (c) What strength can accurately be attributed to the force of the throw? Explain. (d) Why does the ball move away from the child's hand?

11. **O** The third graders are on one side of a schoolyard and the fourth graders on the other. The groups are throwing snowballs at each other. Between them, snowballs of various masses are moving with different velocities as shown in Figure Q5.11. Rank the snowballs (a) through (e) according to the magnitude of the total force exerted on each one. Ignore air resistance. If two snowballs rank together, make that fact clear.

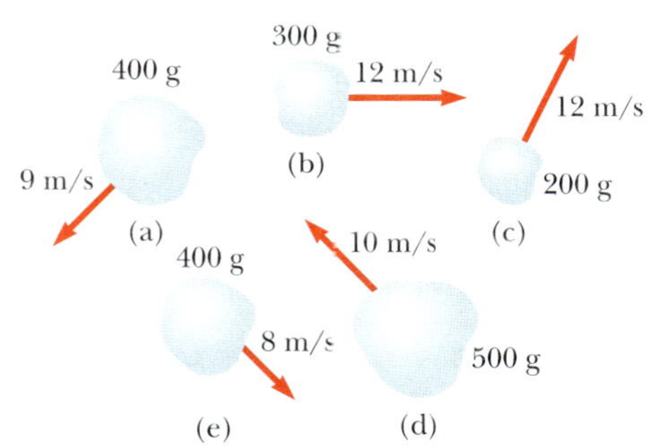

Figure Q5.11

12. The mayor of a city decides to fire some city employees because they will not remove the obvious sags from the cables that support the city traffic lights. If you were a lawyer, what defense would you give on behalf of the employees? Which side do you think would win the case in court?

13. A clip from *America's Funniest Home Videos.* Balancing carefully, three boys inch out onto a horizontal tree branch above a pond, each planning to dive in separately. The youngest and cleverest boy notices that the branch is only barely strong enough to support them. He decides to jump straight up and land back on the branch to break it, spilling all three into the pond together. When he starts to carry out his plan, at what precise moment does the branch break? Explain. *Suggestion:* Pretend to be the clever boy and imitate what he does in slow motion. If you are still unsure, stand on a bathroom scale and repeat the suggestion.

14. When you push on a box with a 200-N force instead of a 50-N force, you can feel that you are making a greater effort. When a table exerts a 200-N upward normal force instead of one of smaller magnitude, is the table really doing anything differently?

15. A weightlifter stands on a bathroom scale. He pumps a barbell up and down. What happens to the reading on the scale as he does so? **What If?** What if he is strong enough to actually *throw* the barbell upward? How does the reading on the scale vary now?

16. (a) Can a normal force be horizontal? (b) Can a normal force be directed vertically downward? (c) Consider a tennis ball in contact with a stationary floor and with nothing else. Can the normal force be different in magnitude from the gravitational force exerted on the ball? (d) Can the force exerted by the floor on the ball be different in magnitude from the force the ball exerts on the floor? Explain each of your answers.

17. Suppose a truck loaded with sand accelerates along a highway. If the driving force exerted on the truck remains constant, what happens to the truck's acceleration if its trailer leaks sand at a constant rate through a hole in its bottom?

18. **O** In Figure Q5.18, the light, taut, unstretchable cord B joins block 1 and the larger-mass block 2. Cord A exerts a force on block 1 to make it accelerate forward. (a) How does the magnitude of the force exerted by cord A on block 1 compare with the magnitude of the force exerted by cord B on block 2? Is it larger, smaller, or equal? (b) How does the acceleration of block 1 compare with the acceleration (if any) of block 2? (c) Does cord B exert a force on block 1? If so, is it forward or backward? Is it larger, smaller, or equal in magnitude to the force exerted by cord B on block 2?

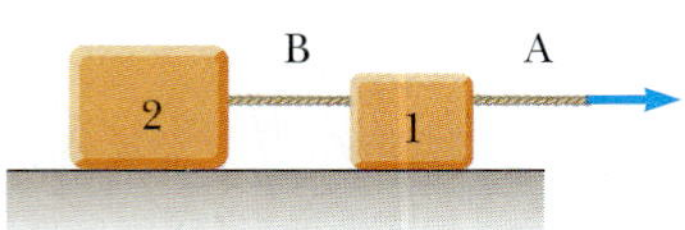

Figure Q5.18

19. Identify the action–reaction pairs in the following situations: a man takes a step, a snowball hits a girl in the back, a baseball player catches a ball, a gust of wind strikes a window.

20. **O** In an Atwood machine, illustrated in Figure 5.14, a light string that does not stretch passes over a light, frictionless pulley. On one side, block 1 hangs from the vertical string. On the other side, block 2 of larger mass hangs from the vertical string. (a) The blocks are released from rest. Is the magnitude of the acceleration of the heavier block 2 larger, smaller, or the same as the free-fall acceleration g? (b) Is the magnitude of the acceleration of block 2 larger, smaller, or the same as the acceleration of block 1? (c) Is the magnitude of the force the string exerts on block 2 larger, smaller, or the same as that of the force of the string on block 1?

21. Twenty people participate in a tug-of-war. The two teams of ten people are so evenly matched that neither team wins. After the game, the participants notice that a car is stuck in the mud. They attach the tug-of-war rope to the bumper of the car, and all the people pull on the rope. The heavy car has just moved a couple of decimeters when the rope breaks. Why did the rope break in this situation, but not when the same twenty people pulled on it in a tug-of-war?

22. **O** In Figure Q5.22, a locomotive has broken through the wall of a train station. As it did, what can be said about the force exerted by the locomotive on the wall? (a) The force exerted by the locomotive on the wall was bigger than the force the wall could exert on the locomotive. (b) The force exerted by the locomotive on the wall was the same in magnitude as the force exerted by the wall on the locomotive. (c) The force exerted by the locomotive on the wall was less than the force exerted by the wall on the locomotive. (d) The wall cannot be said to "exert" a force; after all, it broke.

Roger Viollet, Mill Valley, CA, University Science Books, 1982

Figure Q5.22

23. An athlete grips a light rope that passes over a low-friction pulley attached to the ceiling of a gym. A sack of sand precisely equal in weight to the athlete is tied to the other end of the rope. Both the sand and the athlete are initially at rest. The athlete climbs the rope, sometimes speeding up and slowing down as he does so. What happens to the sack of sand? Explain.

24. **O** A small bug is nestled between a 1-kg block and a 2-kg block on a frictionless table. A horizontal force can be applied to either of the blocks as shown in Figure Q5.24. **(i)** In which situation illustrated in the figure, (a) or (b), does the bug have a better chance of survival, or (c) does it make no difference? **(ii)** Consider the statement, "The force exerted by the larger block on the smaller one is larger in magnitude than the force exerted by the smaller block on the larger one." Is this statement true only in situation (a)? Only in situation (b)? Is it true (c) in both situations or (d) in neither? **(iii)** Consider the statement, "As the blocks move, the force exerted by the block in back on the block in front is stronger than the force exerted by the front block on the back one." Is this statement true only in situation (a), only in situation (b), in (c) both situations, or in (d) neither?

Figure Q5.24

25. Can an object exert a force on itself? Argue for your answer.

26. O The harried manager of a discount department store is pushing horizontally with a force of magnitude 200 N on a box of shirts. The box is sliding across the horizontal floor with a forward acceleration. Nothing else touches the box. What must be true about the magnitude of the force of kinetic friction acting on the box (choose one)? (a) It is greater than 200 N. (b) It is less than 200 N. (c) It is equal to 200 N. (d) None of these statements is necessarily true.

27. A car is moving forward slowly and is speeding up. A student claims "the car exerts a force on itself" or "the car's engine exerts a force on the car." Argue that this idea cannot be accurate and that friction exerted by the road is the propulsive force on the car. Make your evidence and reasoning as persuasive as possible. Is it static or kinetic friction? *Suggestions:* Consider a road covered with light gravel. Consider a sharp print of the tire tread on an asphalt road, obtained by coating the tread with dust.

28. O The driver of a speeding empty truck slams on the brakes and skids to a stop through a distance d. **(i)** If the truck now carries a load that doubles its mass, what will be the truck's "skidding distance"? (a) $4d$ (b) $2d$ (c) $\sqrt{2}d$ (d) d (e) $d/\sqrt{2}$ (f) $d/2$ (g) $d/4$ **(ii)** If the initial speed of the empty truck were halved, what would be the truck's skidding distance? Choose from the same possibilities (a) through (g).

29. O An object of mass m is sliding with speed v_0 at some instant across a level tabletop, with which its coefficient of kinetic friction is μ. It then moves through a distance d and comes to rest. Which of the following equations for the speed v_0 is reasonable (choose one)? (a) $v_0 = \sqrt{-2\mu mgd}$ (b) $v_0 = \sqrt{2\mu mgd}$ (c) $v_0 = \sqrt{-2\mu gd}$ (d) $v_0 = \sqrt{2\mu gd}$ (e) $v_0 = \sqrt{2gd/\mu}$ (f) $v_0 = \sqrt{2\mu md}$ (g) $v_0 = \sqrt{2\mu d}$

30. O A crate remains stationary after it has been placed on a ramp inclined at an angle with the horizontal. Which of the following statements is or are correct about the magnitude of the friction force that acts on the crate? Choose all that are true. (a) It is larger than the weight of the crate. (b) It is at least equal to the weight of the crate. (c) It is equal to $\mu_s n$. (d) It is greater than the component of the gravitational force acting down the ramp. (e) It is equal to the component of the gravitational force acting down the ramp. (f) It is less than the component of the gravitational force acting down the ramp.

31. Suppose you are driving a classic car. Why should you avoid slamming on your brakes when you want to stop in the shortest possible distance? (Many modern cars have antilock brakes that avoid this problem.)

32. Describe a few examples in which the force of friction exerted on an object is in the direction of motion of the object.

33. O As shown in Figure Q5.33, student A, a 55-kg girl, sits on one chair with metal runners, at rest on a classroom floor. Student B, an 80-kg boy, sits on an identical chair. Both students keep their feet off the floor. A rope runs from student A's hands around a light pulley to the hands of a teacher standing on the floor next to her. The low-friction axle of the pulley is attached to a second rope held by student B. All ropes run parallel to the chair runners. (a) If student A pulls on her end of the rope, will her chair or will B's chair slide on the floor? (b) If instead the teacher pulls on his rope end, which chair slides? (c) If student B pulls on his rope, which chair slides? (d) Now the teacher ties his rope end to student A's chair. Student A pulls on the end of the rope in her hands. Which chair slides? (Vern Rockcastle suggested the idea for this question.)

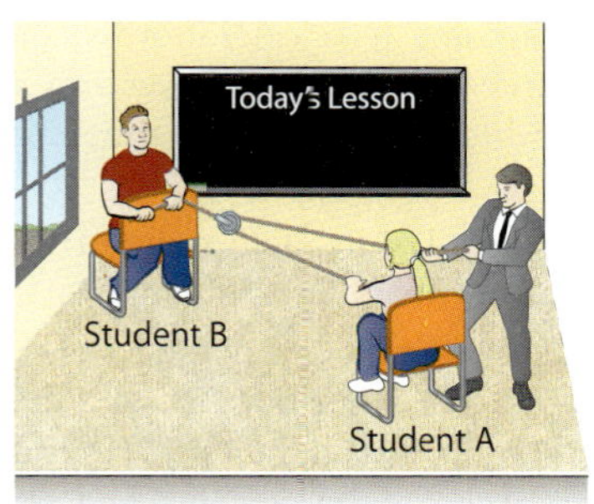

Figure Q5.33

Problems

WebAssign The Problems from this chapter may be assigned online in WebAssign.

ThomsonNOW Sign in at **www.thomsonedu.com** and go to ThomsonNOW to assess your understanding of this chapter's topics with additional quizzing and conceptual questions.

1, 2, 3 denotes straightforward, intermediate, challenging; □ denotes full solution available in *Student Solutions Manual/Study Guide*; ▲ denotes coached solution with hints available at **www.thomsonedu.com;** ▭ denotes developing symbolic reasoning; ● denotes asking for qualitative reasoning; 🖥 denotes computer useful in solving problem

Sections 5.1 through 5.6

1. A 3.00-kg object undergoes an acceleration given by $\vec{\mathbf{a}} = (2.00\hat{\mathbf{i}} + 5.00\hat{\mathbf{j}})\ \text{m/s}^2$. Find the resultant force acting on it and the magnitude of the resultant force.

2. A force $\vec{\mathbf{F}}$ applied to an object of mass m_1 produces an acceleration of 3.00 m/s^2. The same force applied to a second object of mass m_2 produces an acceleration of 1.00 m/s^2. (a) What is the value of the ratio m_1/m_2? (b) If m_1 and m_2 are combined into one object, what is its acceleration under the action of the force $\vec{\mathbf{F}}$?

3. ▲ To model a spacecraft, a toy rocket engine is securely fastened to a large puck that can glide with negligible friction over a horizontal surface, taken as the *xy* plane. The 4.00-kg puck has a velocity of $3.00\hat{\mathbf{i}}$ m/s at one instant. Eight seconds later, its velocity is to be $(8.00\hat{\mathbf{i}} + 10.0\hat{\mathbf{j}})$ m/s. Assuming the rocket engine exerts a constant horizontal force, find (a) the components of the force and (b) its magnitude.

4. The average speed of a nitrogen molecule in air is about 6.70×10^2 m/s, and its mass is 4.68×10^{-26} kg. (a) If it takes 3.00×10^{-13} s for a nitrogen molecule to hit a wall and rebound with the same speed but moving in the opposite direction, what is the average acceleration of the molecule during this time interval? (b) What average force does the molecule exert on the wall?

5. An electron of mass 9.11×10^{-31} kg has an initial speed of 3.00×10^5 m/s. It travels in a straight line, and its speed increases to 7.00×10^5 m/s in a distance of 5.00 cm. Assuming its acceleration is constant, (a) determine the force exerted on the electron and (b) compare this force with the weight of the electron, which we ignored.

6. A woman weighs 120 lb. Determine (a) her weight in newtons and (b) her mass in kilograms.

7. The distinction between mass and weight was discovered after Jean Richer transported pendulum clocks from France to French Guiana in 1671. He found that they ran slower there quite systematically. The effect was reversed when the clocks returned to France. How much weight would you personally lose when traveling from Paris, France, where $g = 9.809\ 5$ m/s^2, to Cayenne, French Guiana, where $g = 9.780\ 8$ m/s^2?

8. Besides its weight, a 2.80-kg object is subjected to one other constant force. The object starts from rest and in 1.20 s experiences a displacement of $(4.20\hat{\mathbf{i}} - 3.30\hat{\mathbf{j}})$ m, where the direction of $\hat{\mathbf{j}}$ is the upward vertical direction. Determine the other force.

9. Two forces $\vec{\mathbf{F}}_1$ and $\vec{\mathbf{F}}_2$ act on a 5.00-kg object. Taking $F_1 = 20.0$ N and $F_2 = 15.0$ N, find the accelerations in (a) and (b) of Figure P5.9.

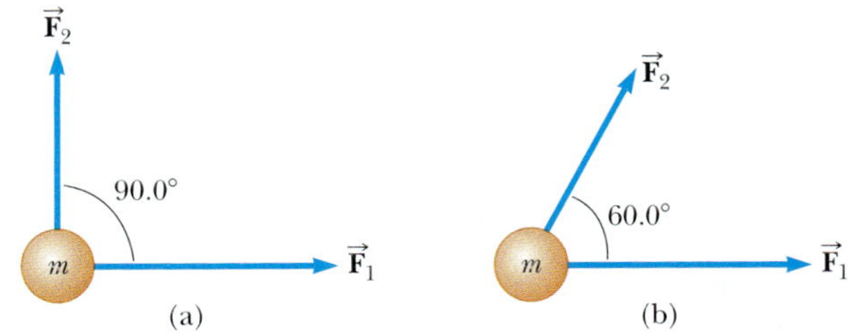

Figure P5.9

10. One or more external forces are exerted on each object enclosed in a dashed box shown in Figure 5.1. Identify the reaction to each of these forces.

11. You stand on the seat of a chair and then hop off. (a) During the time interval you are in flight down to the floor, the Earth is lurching up toward you with an acceleration of what order of magnitude? In your solution, explain your logic. Model the Earth as a perfectly solid object. (b) The Earth moves up through a distance of what order of magnitude?

12. A brick of mass M sits on a rubber pillow of mass m. Together they are sliding to the right at constant velocity on an ice-covered parking lot. (a) Draw a free-body diagram of the brick and identify each force acting on it. (b) Draw a free-body diagram of the pillow and identify each force acting on it. (c) Identify all the action–reaction pairs of forces in the brick–pillow–planet system.

13. A 15.0-lb block rests on the floor. (a) What force does the floor exert on the block? (b) A rope is tied to the block and is run vertically over a pulley. The other end of the rope is attached to a free-hanging 10.0-lb object. What is the force exerted by the floor on the 15.0-lb block? (c) If we replace the 10.0-lb object in part (b) with a 20.0-lb object, what is the force exerted by the floor on the 15.0-lb block?

14. Three forces acting on an object are given by $\vec{\mathbf{F}}_1 = (-2.00\hat{\mathbf{i}} + 2.00\hat{\mathbf{j}})$ N, $\vec{\mathbf{F}}_2 = (5.00\hat{\mathbf{i}} - 3.00\hat{\mathbf{j}})$ N, and $\vec{\mathbf{F}}_3 = (-45.0\hat{\mathbf{i}})$ N. The object experiences an acceleration of magnitude 3.75 m/s^2. (a) What is the direction of the acceleration? (b) What is the mass of the object? (c) If the object is initially at rest, what is its speed after 10.0 s? (d) What are the velocity components of the object after 10.0 s?

Section 5.7 Some Applications of Newton's Laws

15. Figure P5.15 shows a worker poling a boat—a very efficient mode of transportation—across a shallow lake. He pushes parallel to the length of the light pole, exerting on the bottom of the lake a force of 240 N. Assume the pole lies in the vertical plane containing the boat's keel.

At one moment, the pole makes an angle of 35.0° with the vertical and the water exerts a horizontal drag force of 47.5 N on the boat, opposite to its forward velocity of magnitude 0.857 m/s. The mass of the boat including its cargo and the worker is 370 kg. (a) The water exerts a buoyant force vertically upward on the boat. Find the magnitude of this force. (b) Model the forces as constant over a short interval of time to find the velocity of the boat 0.450 s after the moment described.

© Tony Arruza/CORBIS

Figure P5.15

16. A 3.00-kg object is moving in a plane, with its x and y coordinates given by $x = 5t^2 - 1$ and $y = 3t^3 + 2$, where x and y are in meters and t is in seconds. Find the magnitude of the net force acting on this object at $t = 2.00$ s.

17. The distance between two telephone poles is 50.0 m. When a 1.00-kg bird lands on the telephone wire midway between the poles, the wire sags 0.200 m. Draw a free-body diagram of the bird. How much tension does the bird produce in the wire? Ignore the weight of the wire.

18. An iron bolt of mass 65.0 g hangs from a string 35.7 cm long. The top end of the string is fixed. Without touching it, a magnet attracts the bolt so that it remains stationary, displaced horizontally 28.0 cm to the right from the previously vertical line of the string. (a) Draw a free-body diagram of the bolt. (b) Find the tension in the string. (c) Find the magnetic force on the bolt.

19. ● Figure P5.19 shows the horizontal forces acting on a sailboat moving north at constant velocity, seen from a point straight above its mast. At its particular speed, the water exerts a 220-N drag force on the sailboat's hull. (a) Choose the x direction as east and the y direction as north. Write two component equations representing Newton's second law. Solve the equations for P (the force exerted by the wind on the sail) and for n (the force exerted by the water on the keel). (b) Choose the x direction as 40.0° north of east and the y direction as 40.0° west of north. Write Newton's second law as two component equations and solve for n and P. (c) Compare your solutions. Do the results agree? Is one calculation significantly easier?

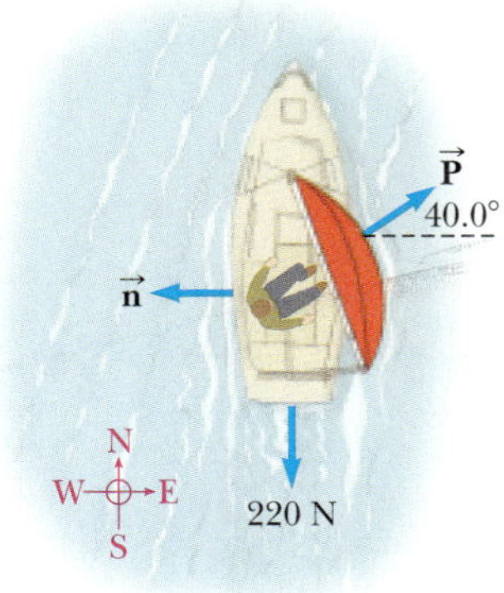

Figure P5.19

20. A bag of cement of weight 325 N hangs in equilibrium from three wires as shown in Figure P5.20. Two of the wires make angles $\theta_1 = 60.0°$ and $\theta_2 = 25.0°$ with the horizontal. Assuming the system is in equilibrium, find the tensions T_1, T_2, and T_3 in the wires.

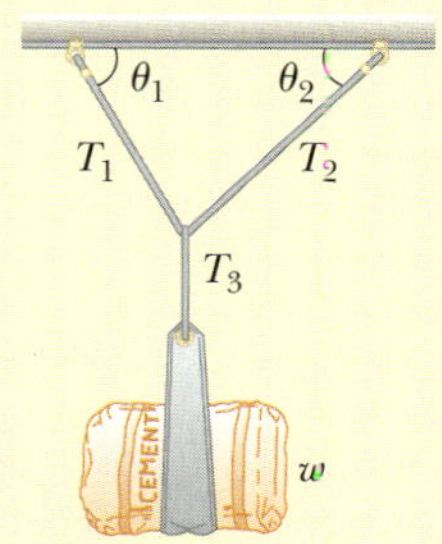

Figure P5.20 Problems 20 and 21.

21. A bag of cement of weight F_g hangs in equilibrium from three wires as shown in Figure P5.20. Two of the wires make angles θ_1 and θ_2 with the horizontal. Assuming the system is in equilibrium, show that the tension in the left-hand wire is

$$T_1 = \frac{F_g \cos \theta_2}{\sin(\theta_1 + \theta_2)}$$

22. ● You are a judge in a children's kite-flying contest, and two children will win prizes, one for the kite that pulls the most strongly on its string and one for the kite that pulls the least strongly on its string. To measure string tensions, you borrow a mass hanger, some slotted masses, and a protractor from your physics teacher, and you use the following protocol, illustrated in Figure P5.22. Wait for a child to get her kite well controlled, hook the hanger onto the kite string about 30 cm from her hand, pile on slotted masses until that section of string is horizontal, record the mass required, and record the angle between the horizontal and the string running up to the kite. (a) Explain how this method works. As you construct your explanation, imagine that the children's parents ask you about your method, that they might make false assumptions about your ability

Figure P5.22

2 = intermediate; 3 = challenging; □ = SSM/SG; ▲ = ThomsonNow; = symbolic reasoning; ● = qualitative reasoning

without concrete evidence, and that your explanation is an opportunity to give them confidence in your evaluation technique. (b) Find the string tension if the mass is 132 g and the angle of the kite string is 46.3°.

23. The systems shown in Figure P5.23 are in equilibrium. If the spring scales are calibrated in newtons, what do they read? Ignore the masses of the pulleys and strings, and assume the pulleys and the incline in part (d) are frictionless.

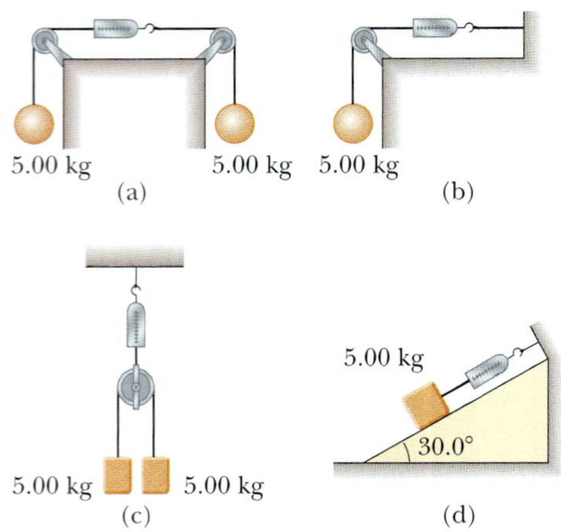

Figure P5.23

24. Draw a free-body diagram of a block that slides down a frictionless plane having an inclination of $\theta = 15.0°$. The block starts from rest at the top, and the length of the incline is 2.00 m. Find (a) the acceleration of the block and (b) its speed when it reaches the bottom of the incline.

25. ▲ A 1.00-kg object is observed to have an acceleration of 10.0 m/s² in a direction 60.0° east of north (Fig. P5.25). The force $\vec{\mathbf{F}}_2$ exerted on the object has a magnitude of 5.00 N and is directed north. Determine the magnitude and direction of the force $\vec{\mathbf{F}}_1$ acting on the object.

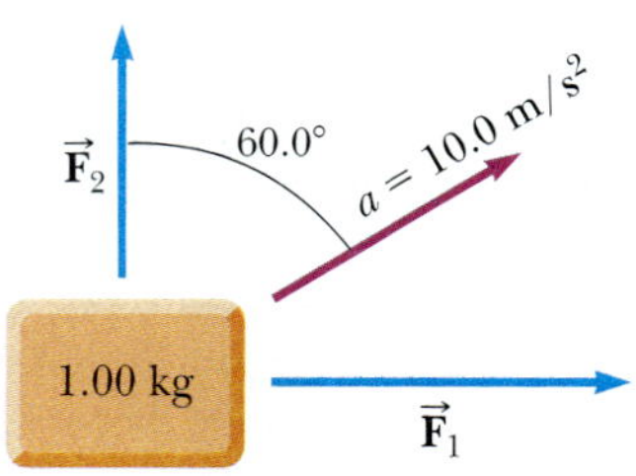

Figure P5.25

26. A 5.00-kg object placed on a frictionless, horizontal table is connected to a string that passes over a pulley and then is fastened to a hanging 9.00-kg object as shown in Figure P5.26. Draw free-body diagrams of both objects. Find the acceleration of the two objects and the tension in the string.

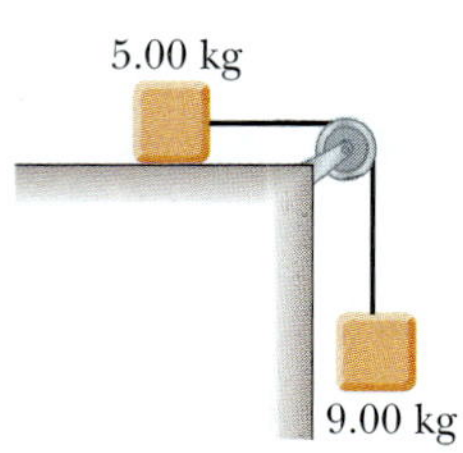

Figure P5.26 Problems 26 and 41.

27. Figure P5.27 shows the speed of a person's body as he does a chin-up. Assume the motion is vertical and the mass of the person's body is 64.0 kg. Determine the force exerted by the chin-up bar on his body at (a) time zero, (b) time 0.5 s, (c) time 1.1 s, and (d) time 1.6 s.

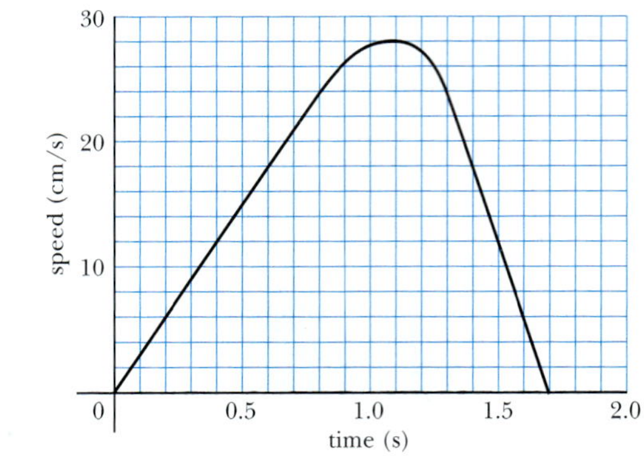

Figure P5.27

28. Two objects are connected by a light string that passes over a frictionless pulley as shown in Figure P5.28. Draw free-body diagrams of both objects. Assuming the incline is frictionless, $m_1 = 2.00$ kg, $m_2 = 6.00$ kg, and $\theta = 55.0°$, find (a) the accelerations of the objects, (b) the tension in the string, and (c) the speed of each object 2.00 s after they are released from rest.

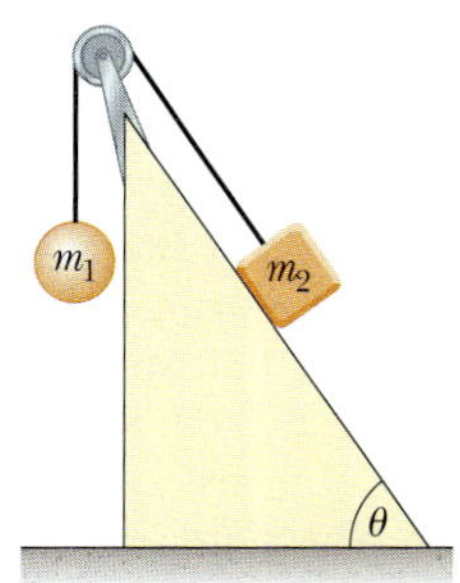

Figure P5.28

29. ▲ A block is given an initial velocity of 5.00 m/s up a frictionless 20.0° incline. How far up the incline does the block slide before coming to rest?

30. In Figure P5.30, the man and the platform together weigh 950 N. The pulley can be modeled as frictionless. Determine how hard the man has to pull on the rope to lift himself steadily upward above the ground. (Or is it impossible? If so, explain why.)

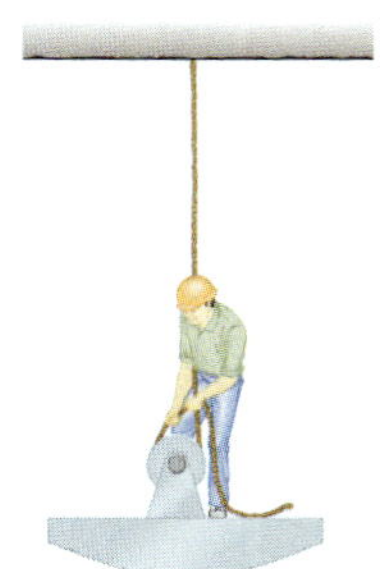

Figure P5.30

2 = intermediate; 3 = challenging; ☐ = SSM/SG; ▲ = ThomsonNow; ▭ = symbolic reasoning; ● = qualitative reasoning

31. In the system shown in Figure P5.31, a horizontal force $\vec{\mathbf{F}}_x$ acts on the 8.00-kg object. The horizontal surface is frictionless. Consider the acceleration of the sliding object as a function of F_x. (a) For what values of F_x does the 2.00-kg object accelerate upward? (b) For what values of F_x is the tension in the cord zero? (c) Plot the acceleration of the 8.00-kg object versus F_x. Include values of F_x from −100 N to +100 N.

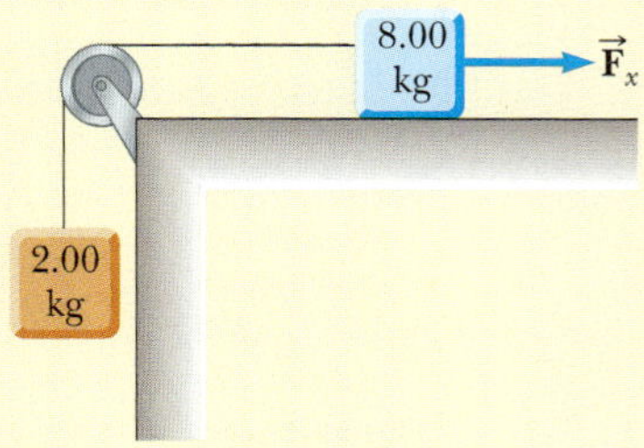

Figure P5.31

32. An object of mass m_1 on a frictionless horizontal table is connected to an object of mass m_2 through a very light pulley P_1 and a light fixed pulley P_2 as shown in Figure P5.32. (a) If a_1 and a_2 are the accelerations of m_1 and m_2, respectively, what is the relation between these accelerations? Express (b) the tensions in the strings and (c) the accelerations a_1 and a_2 in terms of g and of the masses m_1 and m_2.

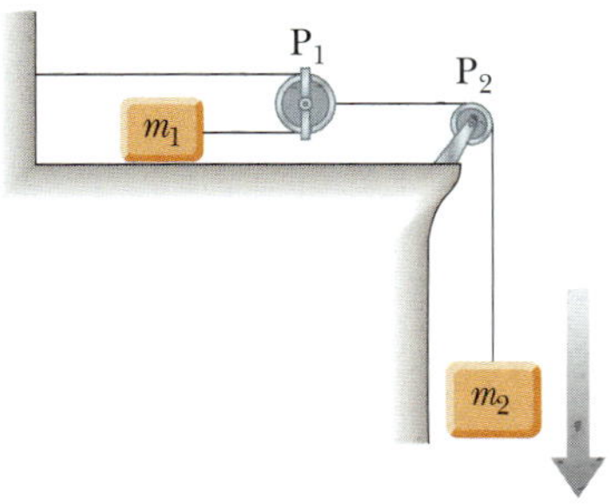

Figure P5.32

33. A 72.0-kg man stands on a spring scale in an elevator. Starting from rest, the elevator ascends, attaining its maximum speed of 1.20 m/s in 0.800 s. It travels with this constant speed for the next 5.00 s. The elevator then undergoes a uniform acceleration in the negative y direction for 1.50 s and comes to rest. What does the spring scale register (a) before the elevator starts to move, (b) during the first 0.800 s, (c) while the elevator is traveling at constant speed, and (d) during the time interval it is slowing down?

34. In the Atwood machine shown in Figure 5.14a, m_1 = 2.00 kg and m_2 = 7.00 kg. The masses of the pulley and string are negligible by comparison. The pulley turns without friction and the string does not stretch. The lighter object is released with a sharp push that sets it into motion at v_i = 2.40 m/s downward. (a) How far will m_1 descend below its initial level? (b) Find the velocity of m_1 after 1.80 seconds.

Section 5.8 Forces of Friction

35. A car is traveling at 50.0 mi/h on a horizontal highway. (a) If the coefficient of static friction between road and tires on a rainy day is 0.100, what is the minimum distance in which the car will stop? (b) What is the stopping distance when the surface is dry and μ_s = 0.600?

36. A 25.0-kg block is initially at rest on a horizontal surface. A horizontal force of 75.0 N is required to set the block in motion, after which a horizontal force of 60.0 N is required to keep the block moving with constant speed. Find the coefficients of static and kinetic friction from this information.

37. Your 3.80-kg physics book is next to you on the horizontal seat of your car. The coefficient of static friction between the book and the seat is 0.650, and the coefficient of kinetic friction is 0.550. Suppose you are traveling at 72.0 km/h = 20.0 m/s and brake to a stop over a distance of 45.0 m. (a) Will the book start to slide over the seat? (b) What force does the seat exert on the book in this process?

38. ● Before 1960, it was believed that the maximum attainable coefficient of static friction for an automobile tire was less than 1. Then, around 1962, three companies independently developed racing tires with coefficients of 1.6. Since then, tires have improved, as illustrated in this problem. According to the 1990 *Guinness Book of Records*, the fastest time interval for a piston-engine car initially at rest to cover a distance of one-quarter mile is 4.96 s. Shirley Muldowney set this record in September 1989 (a) Assume the rear wheels lifted the front wheels off the pavement as shown in Figure P5.38. What minimum value of μ_s is necessary to achieve the record time interval? (b) Suppose Muldowney were able to double her engine power, keeping other things equal. How would this change affect the time interval?

Jamie Squire/Allsport/Getty Images

Figure P5.38

39. ▲ A 3.00-kg block starts from rest at the top of a 30.0° incline and slides a distance of 2.00 m down the incline in 1.50 s. Find (a) the magnitude of the acceleration of the block, (b) the coefficient of kinetic friction between block and plane, (c) the friction force acting on the block, and (d) the speed of the block after it has slid 2.00 m.

40. A woman at an airport is towing her 20.0-kg suitcase at constant speed by pulling on a strap at an angle θ above the horizontal (Fig. P5.40). She pulls on the strap with a 35.0-N force. The friction force on the suitcase is 20.0 N. Draw a free-body diagram of the suitcase. (a) What angle does the strap make with the horizontal? (b) What normal force does the ground exert on the suitcase?

Figure P5.40

2 = intermediate; 3 = challenging; □ = SSM/SG; ▲ = ThomsonNow; = symbolic reasoning; ● = qualitative reasoning

41. A 9.00-kg hanging object is connected, by a light, inextensible cord over a light, frictionless pulley, to a 5.00-kg block that is sliding on a flat table (Fig. P5.26). Taking the coefficient of kinetic friction as 0.200, find the tension in the string.

42. Three objects are connected on a table as shown in Figure P5.42. The rough table has a coefficient of kinetic friction of 0.350. The objects have masses of 4.00 kg, 1.00 kg, and 2.00 kg, as shown, and the pulleys are frictionless. Draw a free-body diagram for each object. (a) Determine the acceleration of each object and their directions. (b) Determine the tensions in the two cords.

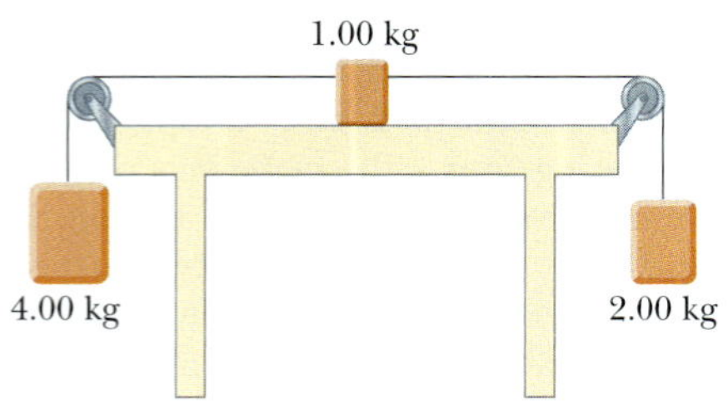

Figure P5.42

43. Two blocks connected by a rope of negligible mass are being dragged by a horizontal force (Fig. P5.43). Suppose $F = 68.0$ N, $m_1 = 12.0$ kg, $m_2 = 18.0$ kg, and the coefficient of kinetic friction between each block and the surface is 0.100. (a) Draw a free-body diagram for each block. (b) Determine the tension T and the magnitude of the acceleration of the system.

Figure P5.43

44. ● A block of mass 3.00 kg is pushed against a wall by a force $\vec{P}$ that makes a $\theta = 50.0°$ angle with the horizontal as shown in Figure P5.44. The coefficient of static friction between the block and the wall is 0.250. (a) Determine the possible values for the magnitude of $\vec{P}$ that allow the block to remain stationary. (b) Describe what happens if $|\vec{P}|$ has a larger value and what happens if it is smaller. (c) Repeat parts (a) and (b) assuming the force makes an angle of $\theta = 13.0°$ with the horizontal.

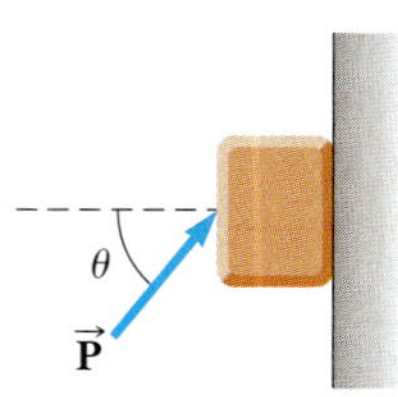

Figure P5.44

45. ● A 420-g block is at rest on a horizontal surface. The coefficient of static friction between the block and the surface is 0.720, and the coefficient of kinetic friction is 0.340. A force of magnitude P pushes the block forward and downward as shown in Figure P5.45. Assume the force is applied at an angle of 37.0° below the horizontal. (a) Find the acceleration of the block as a function of P. (b) If $P = 5.00$ N, find the acceleration and the friction force exerted on the block. (c) If $P = 10.0$ N, find the acceleration and the friction force exerted on the block. (d) Describe in words how the acceleration depends on P. Is there a definite minimum acceleration for the block? If so, what is it? Is there a definite maximum?

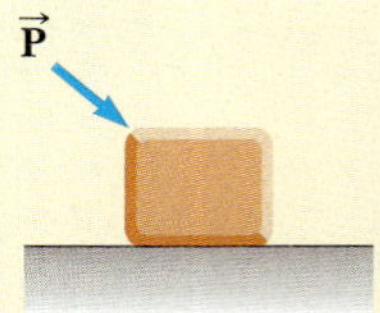

Figure P5.45

46. **Review problem.** One side of the roof of a building slopes up at 37.0°. A student throws a Frisbee onto the roof. It strikes with a speed of 15.0 m/s, does not bounce, and then slides straight up the incline. The coefficient of kinetic friction between the plastic and the roof is 0.400. The Frisbee slides 10.0 m up the roof to its peak, where it goes into free fall, following a parabolic trajectory with negligible air resistance. Determine the maximum height the Frisbee reaches above the point where it struck the roof.

47. The board sandwiched between two other boards in Figure P5.47 weighs 95.5 N. If the coefficient of friction between the boards is 0.663, what must be the magnitude of the compression forces (assumed horizontal) acting on both sides of the center board to keep it from slipping?

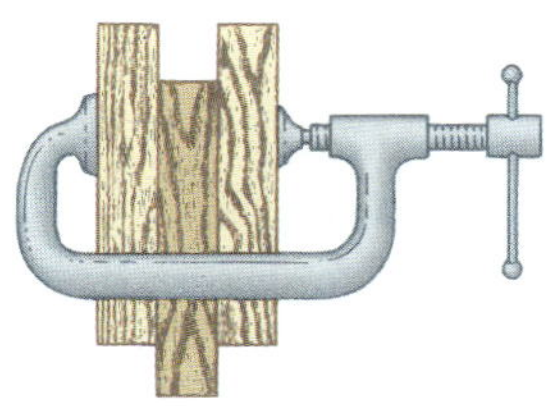
Figure P5.47

48. A magician pulls a tablecloth from under a 200-g mug located 30.0 cm from the edge of the cloth. The cloth exerts a friction force of 0.100 N on the mug, and the cloth is pulled with a constant acceleration of 3.00 m/s². How far does the mug move relative to the horizontal tabletop before the cloth is completely out from under it? Note that the cloth must move more than 30 cm relative to the tabletop during the process.

49. ● A package of dishes (mass 60.0 kg) sits on the flatbed of a pickup truck with an open tailgate. The coefficient of static friction between the package and the truck's flatbed is 0.300, and the coefficient of kinetic friction is 0.250. (a) The truck accelerates forward on level ground. What is the maximum acceleration the truck can have so that the package does not slide relative to the truck bed? (b) The truck barely exceeds this acceleration and then moves with constant acceleration, with the package sliding along its bed. What is the acceleration of the package relative to the ground? (c) The driver cleans up the frag-

ments of dishes and starts over again with an identical package at rest in the truck. The truck accelerates up a hill inclined at 10.0° with the horizontal. Now what is the maximum acceleration the truck can have such that the package does not slide relative to the flatbed? (d) When the truck exceeds this acceleration, what is the acceleration of the package relative to the ground? (e) For the truck parked at rest on a hill, what is the maximum slope the hill can have such that the package does not slide? (f) Is any piece of data unnecessary for the solution in all the parts of this problem? Explain.

Additional Problems

50. The following equations describe the motion of a system of two objects:

$$+n - (6.50 \text{ kg})(9.80 \text{ m/s}^2)\cos 13.0° = 0$$

$$f_k = 0.360n$$

$$+T + (6.50 \text{ kg})(9.80 \text{ m/s}^2)\sin 13.0° - f_k = (6.50 \text{ kg})a$$

$$-T + (3.80 \text{ kg})(9.80 \text{ m/s}^2) = (3.80 \text{ kg})a$$

(a) Solve the equations for a and T. (b) Describe a situation to which these equations apply. Draw free-body diagrams for both objects.

51. An inventive child named Pat wants to reach an apple in a tree without climbing the tree. Sitting in a chair connected to a rope that passes over a frictionless pulley (Fig. P5.51), Pat pulls on the loose end of the rope with such a force that the spring scale reads 250 N. Pat's true weight is 320 N, and the chair weighs 160 N. (a) Draw free-body diagrams for Pat and the chair considered as separate systems, and another diagram for Pat and the chair considered as one system. (b) Show that the acceleration of the system is *upward* and find its magnitude. (c) Find the force Pat exerts on the chair.

Figure P5.51 Problems 51 and 52.

52. ● In the situation described in Problem 51 and Figure P5.51, the masses of the rope, spring balance, and pulley are negligible. Pat's feet are not touching the ground. (a) Assume Pat is momentarily at rest when he stops pulling down on the rope and passes the end of the rope to another child, of weight 440 N, who is standing on the ground next to him. The rope does not break. Describe the ensuing motion. (b) Instead, assume Pat is momentarily at rest when he ties the end of the rope to a strong hook projecting from the tree trunk. Explain why this action can make the rope break.

53. A time-dependent force, $\vec{\mathbf{F}} = (8.00\hat{\mathbf{i}} - 4.00t\hat{\mathbf{j}})$ N, where t is in seconds, is exerted on a 2.00-kg object initially at rest. (a) At what time will the object be moving with a speed of 15.0 m/s? (b) How far is the object from its initial position when its speed is 15.0 m/s? (c) Through what total displacement has the object traveled at this moment?

54. ● Three blocks are in contact with one another on a frictionless, horizontal surface as shown in Figure P5.54. A horizontal force $\vec{\mathbf{F}}$ is applied to m_1. Take m_1 = 2.00 kg, m_2 = 3.00 kg, m_3 = 4.00 kg, and F = 18.0 N. Draw a separate free-body diagram for each block and find (a) the acceleration of the blocks, (b) the *resultant* force on each block, and (c) the magnitudes of the contact forces between the blocks. (d) You are working on a construction project. A coworker is nailing plasterboard on one side of a light partition, and you are on the opposite side, providing "backing" by leaning against the wall with your back pushing on it. Every hammer blow makes your back sting. The supervisor helps you to put a heavy block of wood between the wall and your back. Using the situation analyzed in parts (a), (b), and (c) as a model, explain how this change works to make your job more comfortable.

Figure P5.54

55. ● A rope with mass m_1 is attached to the bottom front edge of a block with mass 4.00 kg. Both the rope and the block rest on a horizontal frictionless surface. The rope does not stretch. The free end of the rope is pulled with a horizontal force of 12.0 N. (a) Find the acceleration of the system, as it depends on m_1. (b) Find the magnitude of the force the rope exerts on the block, as it depends on m_1. (c) Evaluate the acceleration and the force on the block for m_1 = 0.800 kg. *Suggestion:* You may find it easier to do part (c) before parts (a) and (b).

What If? (d) What happens to the force on the block as the rope's mass grows beyond all bounds? (e) What happens to the force on the block as the rope's mass approaches zero? (f) What theorem can you state about the tension in a *light* cord joining a pair of moving objects?

56. A black aluminum glider floats on a film of air above a level aluminum air track. Aluminum feels essentially no force in a magnetic field, and air resistance is negligible. A strong magnet is attached to the top of the glider, forming a total mass of 240 g. A piece of scrap iron attached to one end stop on the track attracts the magnet with a force of 0.823 N when the iron and the magnet are separated by 2.50 cm. (a) Find the acceleration of the glider at this instant. (b) The scrap iron is now attached to another green glider, forming a total mass of 120 g. Find the acceleration of each glider when they are simultaneously released at 2.50-cm separation.

57. ▲ An object of mass M is held in place by an applied force $\vec{\mathbf{F}}$ and a pulley system as shown in Figure P5.57. The pulleys are massless and frictionless. Find (a) the tension in each section of rope, T_1, T_2, T_3, T_4, and T_5 and (b) the magnitude of $\vec{\mathbf{F}}$. *Suggestion:* Draw a free-body diagram for each pulley.

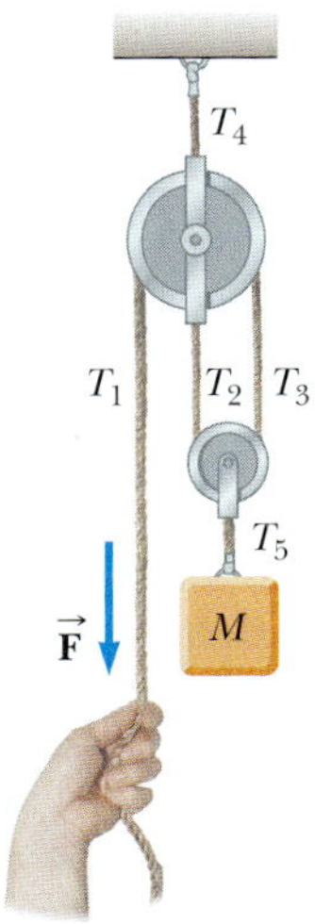

Figure P5.57

58. ● A block of mass 2.20 kg is accelerated across a rough surface by a light cord passing over a small pulley as shown in Figure P5.58. The tension T in the cord is maintained at 10.0 N, and the pulley is 0.100 m above the top of the block. The coefficient of kinetic friction is 0.400. (a) Determine the acceleration of the block when $x = 0.400$ m. (b) Describe the general behavior of the acceleration as the block slides from a location where x is large to $x = 0$. (c) Find the maximum value of the acceleration and the position x for which it occurs. (d) Find the value of x for which the acceleration is zero.

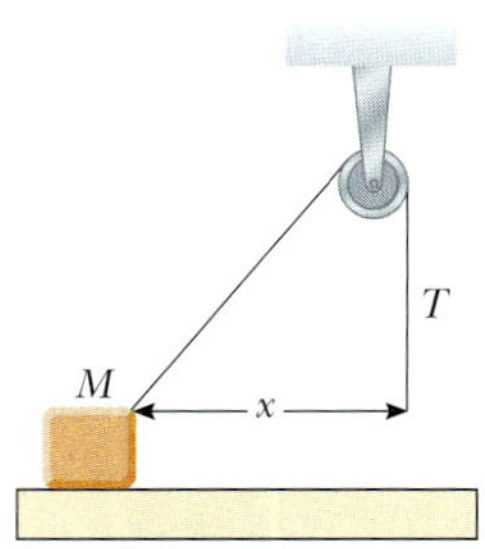

Figure P5.58

59. ● Physics students from San Diego have come in first and second in a contest and are down at the docks, watching their prizes being unloaded from a freighter. On a single light vertical cable that does not stretch, a crane is lifting a 1 207-kg Ferrari and, below it, a 1 461-kg red BMW Z8. The Ferrari is moving upward with speed 3.50 m/s and acceleration 1.25 m/s^2. (a) How do the velocity and acceleration of the BMW compare with those of the Ferrari? (b) Find the tension in the cable between the BMW and the Ferrari. (c) Find the tension in the cable above the Ferrari. (d) In our model, what is the total force exerted on the section of cable between the cars? What velocity do you predict for it 0.01 s into the future? Explain the motion of this section of cable in cause-and-effect terms.

60. A 2.00-kg aluminum block and a 6.00-kg copper block are connected by a light string over a frictionless pulley. They sit on a steel surface as shown in Figure P5.60, where $\theta = 30.0°$. When they are released from rest, will they start to move? If so, determine (a) their acceleration and (b) the tension in the string. If not, determine the sum of the magnitudes of the forces of friction acting on the blocks.

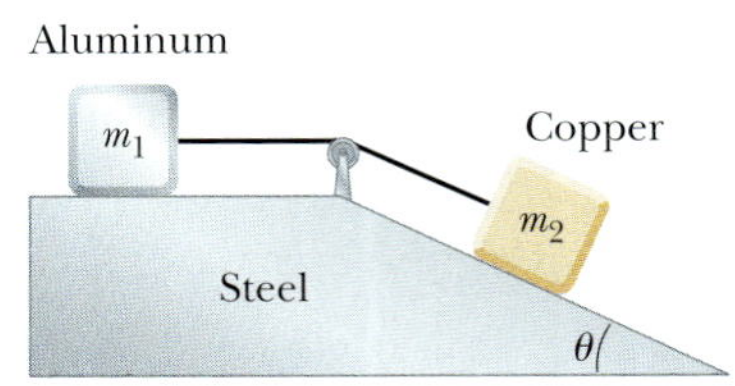

Figure P5.60

61. A crate of weight F_g is pushed by a force $\vec{\mathbf{P}}$ on a horizontal floor. (a) The coefficient of static friction is μ_s, and $\vec{\mathbf{P}}$ is directed at angle θ below the horizontal. Show that the minimum value of P that will move the crate is given by

$$P = \frac{\mu_s F_g \sec\theta}{1 - \mu_s \tan\theta}$$

(b) Find the minimum value of P that can produce motion when $\mu_s = 0.400$, $F_g = 100$ N, and $\theta = 0°$, 15.0°, 30.0°, 45.0°, and 60.0°.

62. **Review problem.** A block of mass $m = 2.00$ kg is released from rest at $h = 0.500$ m above the surface of a table, at the top of a $\theta = 30.0°$ incline as shown in Figure P5.62. The frictionless incline is fixed on a table of height $H = 2.00$ m. (a) Determine the acceleration of the block as it slides down the incline. (b) What is the velocity of the block as it leaves the incline? (c) How far from the table will the block hit the floor? (d) What time interval elapses between when the block is released and when it hits the floor? (e) Does the mass of the block affect any of the above calculations?

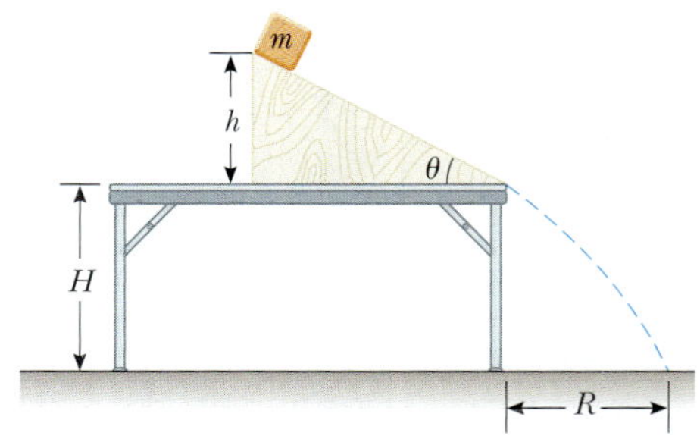

Figure P5.62 Problems 62 and 68.

63. ● A couch cushion of mass m is released from rest at the top of a building having height h. A wind blowing along the side of the building exerts a constant horizontal force of magnitude F on the cushion as it drops as shown in Figure P5.63. The air exerts no vertical force. (a) Show that the path of the cushion is a straight line. (b) Does

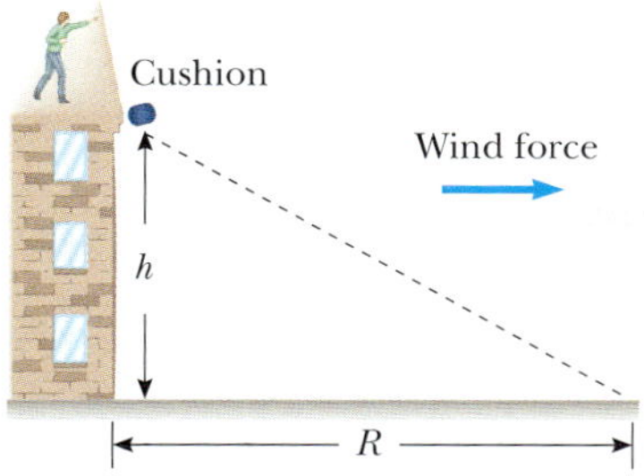

Figure P5.63

the cushion fall with constant velocity? Explain. (c) If $m = 1.20$ kg, $h = 8.00$ m, and $F = 2.40$ N, how far from the building will the cushion hit the level ground? **What If?** (d) If the cushion is thrown downward with a nonzero speed at the top of the building, what will be the shape of its trajectory? Explain.

64. A student is asked to measure the acceleration of a cart on a "frictionless" inclined plane as shown in Figure 5.11, using an air track, a stopwatch, and a meter stick. The height of the incline is measured to be 1.774 cm, and the total length of the incline is measured to be $d = 127.1$ cm. Hence, the angle of inclination θ is determined from the relation $\sin\theta = 1.774/127.1$. The cart is released from rest at the top of the incline, and its position x along the incline is measured as a function of time, where $x = 0$ refers to the cart's initial position. For x values of 10.0 cm, 20.0 cm, 35.0 cm, 50.0 cm, 75.0 cm, and 100 cm, the measured times at which these positions are reached (averaged over five runs) are 1.02 s, 1.53 s, 2.01 s, 2.64 s, 3.30 s, and 3.75 s, respectively. Construct a graph of x versus t^2, and perform a linear least-squares fit to the data. Determine the acceleration of the cart from the slope of this graph, and compare it with the value you would get using $a = g\sin\theta$, where $g = 9.80\ \text{m/s}^2$.

65. A 1.30-kg toaster is not plugged in. The coefficient of static friction between the toaster and a horizontal countertop is 0.350. To make the toaster start moving, you carelessly pull on its electric cord. (a) For the cord tension to be as small as possible, you should pull at what angle above the horizontal? (b) With this angle, how large must the tension be?

66. ● In Figure P5.66, the pulleys and the cords are light, all surfaces are frictionless, and the cords do not stretch. (a) How does the acceleration of block 1 compare with the acceleration of block 2? Explain your reasoning. (b) The mass of block 2 is 1.30 kg. Find its acceleration as it depends on the mass m_1 of block 1. (c) Evaluate your answer for $m_1 = 0.550$ kg. *Suggestion:* You may find it easier to do part (c) before part (b). **What If?** (d) What does the result of part (b) predict if m_1 is very much less than 1.30 kg? (e) What does the result of part (b) predict if m_1 approaches infinity? (f) What is the tension in the long cord in this last case? (g) Could you anticipate the answers (d), (e), and (f) without first doing part (b)? Explain.

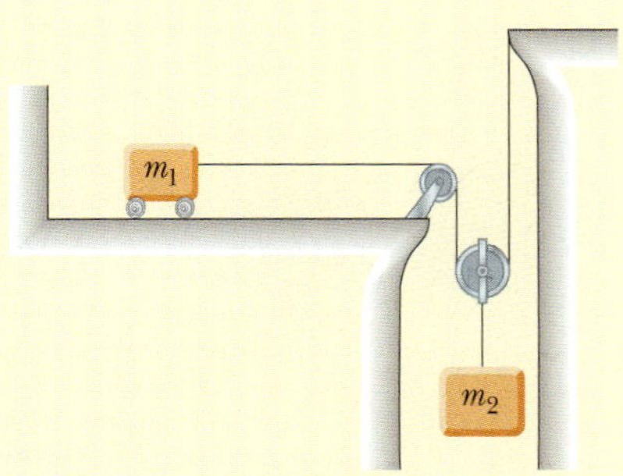

Figure P5.66

67. What horizontal force must be applied to the cart shown in Figure P5.67 so that the blocks remain stationary relative to the cart? Assume all surfaces, wheels, and pulley are frictionless. Notice that the force exerted by the string accelerates m_1.

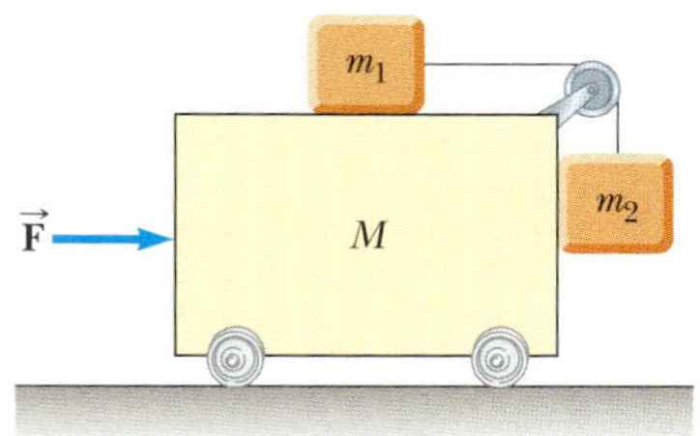

Figure P5.67

68. In Figure P5.62, the incline has mass M and is fastened to the stationary horizontal tabletop. The block of mass m is placed near the bottom of the incline and is released with a quick push that sets it sliding upward. The block stops near the top of the incline, as shown in the figure, and then slides down again, always without friction. Find the force that the tabletop exerts on the incline throughout this motion.

69. A van accelerates down a hill (Fig. P5.69), going from rest to 30.0 m/s in 6.00 s. During the acceleration, a toy ($m = 0.100$ kg) hangs by a string from the van's ceiling. The acceleration is such that the string remains perpendicular to the ceiling. Determine (a) the angle θ and (b) the tension in the string.

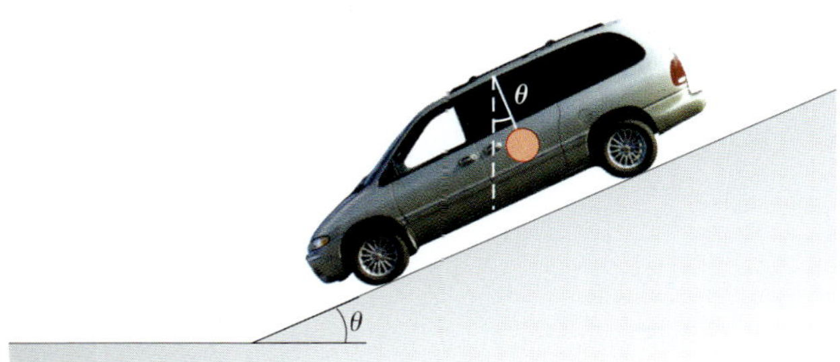

Figure P5.69

70. An 8.40-kg object slides down a fixed, frictionless inclined plane. Use a computer to determine and tabulate the normal force exerted on the object and its acceleration for a series of incline angles (measured from the horizontal) ranging from 0° to 90° in 5° increments. Plot a graph of the normal force and the acceleration as functions of the incline angle. In the limiting cases of 0° and 90°, are your results consistent with the known behavior?

71. A mobile is formed by supporting four metal butterflies of equal mass m from a string of length L. The points of support are evenly spaced a distance ℓ apart as shown in

Figure P5.71. The string forms an angle θ_1 with the ceiling at each endpoint. The center section of string is horizontal. (a) Find the tension in each section of string in terms of θ_1, m, and g. (b) Find the angle θ_2, in terms of θ_1, that the sections of string between the outside butterflies and the inside butterflies form with the horizontal. (c) Show that the distance D between the endpoints of the string is

$$D = \frac{L}{5}\left(2\cos\theta_1 + 2\cos\left[\tan^{-1}\left(\tfrac{1}{2}\tan\theta_1\right)\right] + 1\right)$$

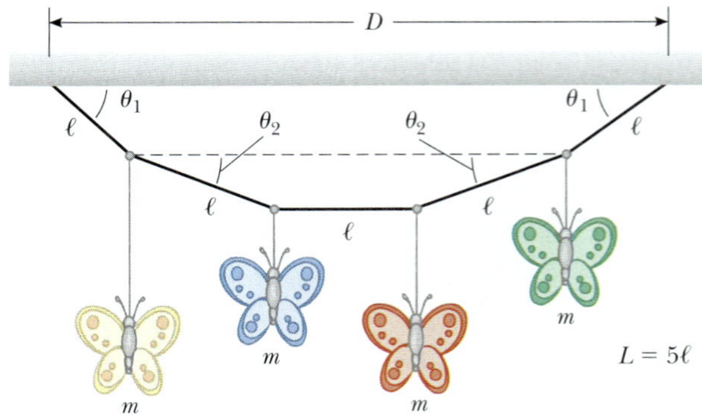

Figure P5.71

Answers to Quick Quizzes

5.1 (d). Choice (a) is true. Newton's first law tells us that motion requires no force: an object in motion continues to move at constant velocity in the absence of external forces. Choice (b) is also true. A stationary object can have several forces acting on it, but if the vector sum of all these external forces is zero, there is no net force and the object remains stationary.

5.2 (a). If a single force acts, this force constitutes the net force and there is an acceleration according to Newton's second law.

5.3 (d). With twice the force, the object will experience twice the acceleration. Because the force is constant, the acceleration is constant, and the speed of the object (starting from rest) is given by $v = at$. With twice the acceleration, the object will arrive at speed v at half the time.

5.4 (b). Because the value of g is smaller on the Moon than on the Earth, more mass of gold would be required to represent 1 newton of weight on the Moon. Therefore, your friend on the Moon is richer, by about a factor of 6!

5.5 **(i)**, (c). In accordance with Newton's third law, the fly and bus experience forces that are equal in magnitude but opposite in direction. **(ii)**, (a). Because the fly has such a small mass, Newton's second law tells us that it undergoes a very large acceleration. The large mass of the bus means that it more effectively resists any change in its motion and exhibits a small acceleration.

5.6 (b). The friction force acts opposite to the gravitational force on the book to keep the book in equilibrium. Because the gravitational force is downward, the friction force must be upward.

5.7 (b). When pulling with the rope, there is a component of your applied force that is upward, which reduces the normal force between the sled and the snow. In turn, the friction force between the sled and the snow is reduced, making the sled easier to move. If you push from behind with a force with a downward component, the normal force is larger, the friction force is larger, and the sled is harder to move.

Passengers on a "corkscrew" roller coaster experience a radial force toward the center of the circular track and a downward force due to gravity. (Robin Smith / Getty Images)

6 Circular Motion and Other Applications of Newton's Laws

In the preceding chapter, we introduced Newton's laws of motion and applied them to situations involving linear motion. Now we discuss motion that is slightly more complicated. For example, we shall apply Newton's laws to objects traveling in circular paths. We shall also discuss motion observed from an accelerating frame of reference and motion of an object through a viscous medium. For the most part, this chapter consists of a series of examples selected to illustrate the application of Newton's laws to a variety of circumstances.

6.1 Newton's Second Law for a Particle in Uniform Circular Motion

In Section 4.4, we discussed the model of a particle in uniform circular motion, in which a particle moves with constant speed v in a circular path of radius r. The particle experiences an acceleration that has a magnitude

$$a_c = \frac{v^2}{r}$$

The acceleration is called *centripetal acceleration* because $\vec{\mathbf{a}}_c$ is directed toward the center of the circle. Furthermore, $\vec{\mathbf{a}}_c$ is *always* perpendicular to $\vec{\mathbf{v}}$. (If there were a component of acceleration parallel to $\vec{\mathbf{v}}$, the particle's speed would be changing.)

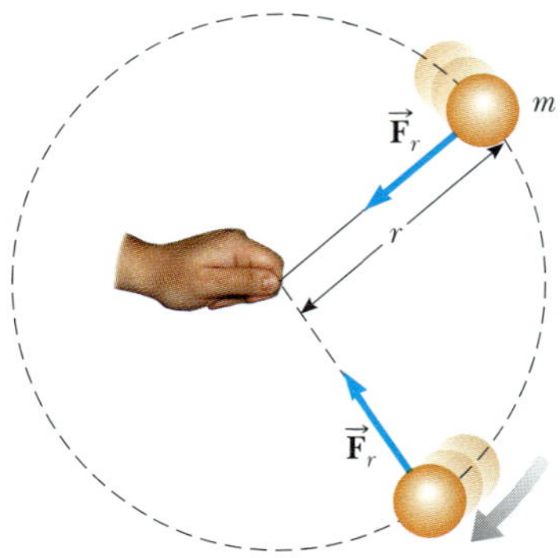

Figure 6.1 An overhead view of a ball moving in a circular path in a horizontal plane. A force $\vec{\mathbf{F}}_r$ directed toward the center of the circle keeps the ball moving in its circular path.

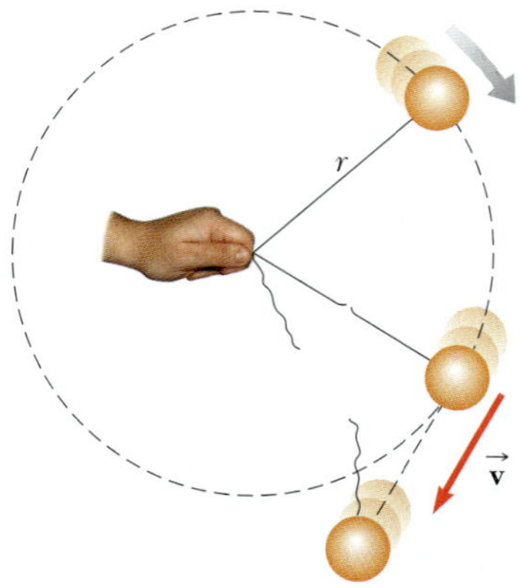

ACTIVE FIGURE 6.2

An overhead view of a ball moving in a circular path in a horizontal plane. When the string breaks, the ball moves in the direction tangent to the circle.

Sign in at www.thomsonedu.com and go to ThomsonNOW to "break" the string yourself and observe the effect on the ball's motion.

Let us now incorporate the concept of force in the particle in uniform circular motion model. Consider a ball of mass m that is tied to a string of length r and is being whirled at constant speed in a horizontal circular path as illustrated in Figure 6.1. Its weight is supported by a frictionless table. Why does the ball move in a circle? According to Newton's first law, the ball would move in a straight line if there were no force on it; the string, however, prevents motion along a straight line by exerting on the ball a radial force $\vec{\mathbf{F}}_r$ that makes it follow the circular path. This force is directed along the string toward the center of the circle as shown in Figure 6.1.

If Newton's second law is applied along the radial direction, the net force causing the centripetal acceleration can be related to the acceleration as follows:

Force causing centripetal acceleration ▶

$$\sum F = ma_c = m\frac{v^2}{r} \qquad \textbf{(6.1)}$$

A force causing a centripetal acceleration acts toward the center of the circular path and causes a change in the direction of the velocity vector. If that force should vanish, the object would no longer move in its circular path; instead, it would move along a straight-line path tangent to the circle. This idea is illustrated in Active Figure 6.2 for the ball whirling at the end of a string in a horizontal plane. If the string breaks at some instant, the ball moves along the straight-line path that is tangent to the circle at the position of the ball at this instant.

PITFALL PREVENTION 6.1

Direction of Travel When the String Is Cut

Study Active Figure 6.2 very carefully. Many students (wrongly) think that the ball will move *radially* away from the center of the circle when the string is cut. The velocity of the ball is *tangent* to the circle. By Newton's first law, the ball continues to move in the same direction in which it is moving just as the force from the string disappears.

Quick Quiz 6.1 You are riding on a Ferris wheel that is rotating with constant speed. The car in which you are riding always maintains its correct upward orientation; it does not invert. **(i)** What is the direction of the normal force on you from the seat when you are at the top of the wheel? (a) upward (b) downward (c) impossible to determine **(ii)** From the same choices, what is the direction of the net force on you when you are at the top of the wheel?

EXAMPLE 6.1 The Conical Pendulum

A small ball of mass m is suspended from a string of length L. The ball revolves with constant speed v in a horizontal circle of radius r as shown in Figure 6.3. (Because the string sweeps out the surface of a cone, the system is known as a *conical pendulum.*) Find an expression for v.

SOLUTION

Conceptualize Imagine the motion of the ball in Figure 6.3a and convince yourself that the string sweeps out a cone and that the ball moves in a circle.

Categorize The ball in Figure 6.3 does not accelerate vertically. Therefore, we model it as a particle in equilibrium in the vertical direction. It experiences a centripetal acceleration in the horizontal direction, so it is modeled as a particle in uniform circular motion in this direction.

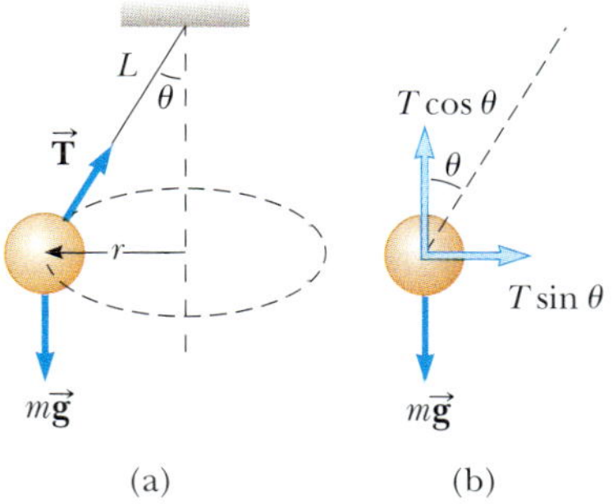

Figure 6.3 (Example 6.1) (a) A conical pendulum. The path of the object is a horizontal circle. (b) The free-body diagram for the object.

Analyze Let θ represent the angle between the string and the vertical. In the free-body diagram shown in Figure 6.3b, the force $\vec{\mathbf{T}}$ exerted by the string is resolved into a vertical component $T\cos\theta$ and a horizontal component $T\sin\theta$ acting toward the center of the circular path.

Apply the particle in equilibrium model in the vertical direction:

$$\sum F_y = T\cos\theta - mg = 0$$

$$(1) \quad T\cos\theta = mg$$

Use Equation 6.1 to express the force providing the centripetal acceleration in the horizontal direction:

$$(2) \quad \sum F_x = T\sin\theta = ma_c = \frac{mv^2}{r}$$

Divide Equation (2) by Equation (1) and use $\sin\theta/\cos\theta = \tan\theta$:

$$\tan\theta = \frac{v^2}{rg}$$

Solve for v:

$$v = \sqrt{rg\tan\theta}$$

Incorporate $r = L\sin\theta$ from the geometry in Figure 6.3a:

$$v = \sqrt{Lg\sin\theta\tan\theta}$$

Finalize Notice that the speed is independent of the mass of the ball. Consider what happens when θ goes to 90° so that the string is horizontal. Because the tangent of 90° is infinite, the speed v is infinite, which tells us the string cannot possibly be horizontal. If it were, there would be no vertical component of the force $\vec{\mathbf{T}}$ to balance the gravitational force on the ball. That is why we mentioned in regard to Figure 6.1 that the ball's weight in the figure is supported by a frictionless table.

EXAMPLE 6.2 How Fast Can It Spin?

A ball of mass 0.500 kg is attached to the end of a cord 1.50 m long. The ball is whirled in a horizontal circle as shown in Figure 6.1. If the cord can withstand a maximum tension of 50.0 N, what is the maximum speed at which the ball can be whirled before the cord breaks? Assume the string remains horizontal during the motion.

SOLUTION

Conceptualize It makes sense that the stronger the cord, the faster the ball can whirl before the cord breaks. Also, we expect a more massive ball to break the cord at a lower speed. (Imagine whirling a bowling ball on the cord!)

Categorize Because the ball moves in a circular path, we model it as a particle in uniform circular motion.

Analyze Incorporate the tension and the centripetal acceleration into Newton's second law:

$$T = m\frac{v^2}{r}$$

Solve for v:

$$(1) \quad v = \sqrt{\frac{Tr}{m}}$$

Find the maximum speed the ball can have, which corresponds to the maximum tension the string can withstand:

$$v_{max} = \sqrt{\frac{T_{max} r}{m}} = \sqrt{\frac{(50.0\text{ N})(1.50\text{ m})}{0.500\text{ kg}}} = 12.2\text{ m/s}$$

Finalize Equation (1) shows that v increases with T and decreases with larger m, as we expected from our conceptualization of the problem.

What If? Suppose the ball is whirled in a circle of larger radius at the same speed v. Is the cord more likely or less likely to break?

Answer The larger radius means that the change in the direction of the velocity vector will be smaller in a given time interval. Therefore, the acceleration is smaller and the required tension in the string is smaller. As a result, the string is less likely to break when the ball travels in a circle of larger radius.

EXAMPLE 6.3 What Is the Maximum Speed of the Car?

A 1 500-kg car moving on a flat, horizontal road negotiates a curve as shown in Figure 6.4a. If the radius of the curve is 35.0 m and the coefficient of static friction between the tires and dry pavement is 0.523, find the maximum speed the car can have and still make the turn successfully.

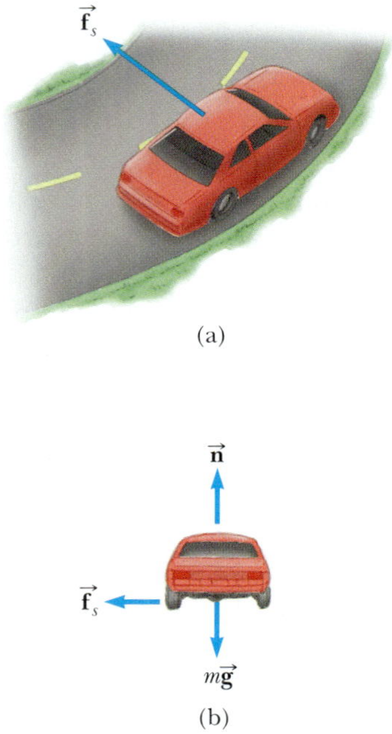

Figure 6.4 (Example 6.3) (a) The force of static friction directed toward the center of the curve keeps the car moving in a circular path. (b) The free-body diagram for the car.

SOLUTION

Conceptualize Imagine that the curved roadway is part of a large circle so that the car is moving in a circular path.

Categorize Based on the conceptualize step of the problem, we model the car as a particle in uniform circular motion in the horizontal direction. The car is not accelerating vertically, so it is modeled as a particle in equilibrium in the vertical direction.

Analyze The force that enables the car to remain in its circular path is the force of static friction. (It is *static* because no slipping occurs at the point of contact between road and tires. If this force of static friction were zero—for example, if the car were on an icy road—the car would continue in a straight line and slide off the road.) The maximum speed v_{max} the car can have around the curve is the speed at which it is on the verge of skidding outward. At this point, the friction force has its maximum value $f_{s,max} = \mu_s n$.

Apply Equation 6.1 in the radial direction for the maximum speed condition:

$$(1)\quad f_{s,max} = \mu_s n = m\frac{v_{max}^2}{r}$$

Apply the particle in equilibrium model to the car in the vertical direction:

$$\sum F_y = 0 \;\rightarrow\; n - mg = 0 \;\rightarrow\; n = mg$$

Solve Equation (1) for the maximum speed and substitute for n:

$$(2)\quad v_{max} = \sqrt{\frac{\mu_s n r}{m}} = \sqrt{\frac{\mu_s m g r}{m}} = \sqrt{\mu_s g r}$$

$$= \sqrt{(0.523)(9.80\text{ m/s}^2)(35.0\text{ m})} = 13.4\text{ m/s}$$

Finalize This speed is equivalent to 30.0 mi/h. Therefore, this roadway could benefit greatly from some banking, as in the next example! Notice that the maximum speed does not depend on the mass of the car, which is why curved highways do not need multiple speed limits to cover the various masses of vehicles using the road.

What If? Suppose a car travels this curve on a wet day and begins to skid on the curve when its speed reaches only 8.00 m/s. What can we say about the coefficient of static friction in this case?

Answer The coefficient of static friction between tires and a wet road should be smaller than that between tires and a dry road. This expectation is consistent with experience with driving because a skid is more likely on a wet road than a dry road.

To check our suspicion, we can solve Equation (2) for the coefficient of static friction:

$$\mu_s = \frac{v_{\max}^2}{gr}$$

Substituting the numerical values gives

$$\mu_s = \frac{v_{\max}^2}{gr} = \frac{(8.00 \text{ m/s})^2}{(9.80 \text{ m/s}^2)(35.0 \text{ m})} = 0.187$$

which is indeed smaller than the coefficient of 0.523 for the dry road.

EXAMPLE 6.4 The Banked Roadway

A civil engineer wishes to redesign the curved roadway in Example 6.3 in such a way that a car will not have to rely on friction to round the curve without skidding. In other words, a car moving at the designated speed can negotiate the curve even when the road is covered with ice. Such a ramp is usually *banked,* which means that the roadway is tilted toward the inside of the curve. Suppose the designated speed for the ramp is to be 13.4 m/s (30.0 mi/h) and the radius of the curve is 35.0 m. At what angle should the curve be banked?

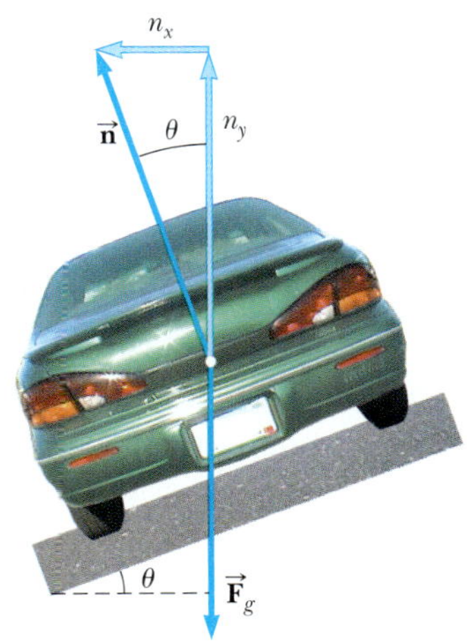

Figure 6.5 (Example 6.4) A car rounding a curve on a road banked at an angle θ to the horizontal. When friction is neglected, the force that causes the centripetal acceleration and keeps the car moving in its circular path is the horizontal component of the normal force.

SOLUTION

Conceptualize The difference between this example and Example 6.3 is that the car is no longer moving on a flat roadway. Figure 6.5 shows the banked roadway, with the center of the circular path of the car far to the left of the figure. Notice that the horizontal component of the normal force participates in causing the car's centripetal acceleration.

Categorize As in Example 6.3, the car is modeled as a particle in equilibrium in the vertical direction and a particle in uniform circular motion in the horizontal direction.

Analyze On a level (unbanked) road, the force that causes the centripetal acceleration is the force of static friction between car and road as we saw in the preceding example. If the road is banked at an angle θ as in Figure 6.5, however, the normal force $\vec{\mathbf{n}}$ has a horizontal component toward the center of the curve. Because the ramp is to be designed so that the force of static friction is zero, only the component $n_x = n \sin\theta$ causes the centripetal acceleration.

Write Newton's second law for the car in the radial direction, which is the x direction:

$$(1) \quad \sum F_r = n \sin\theta = \frac{mv^2}{r}$$

Apply the particle in equilibrium model to the car in the vertical direction:

$$\sum F_y = n \cos\theta - mg = 0$$

$$(2) \quad n \cos\theta = mg$$

Divide Equation (1) by Equation (2):

$$(3) \quad \tan\theta = \frac{v^2}{rg}$$

Solve for the angle θ:

$$\theta = \tan^{-1}\left(\frac{(13.4 \text{ m/s})^2}{(35.0 \text{ m})(9.80 \text{ m/s}^2)}\right) = 27.6°$$

Finalize Equation (3) shows that the banking angle is independent of the mass of the vehicle negotiating the curve. If a car rounds the curve at a speed less than 13.4 m/s, friction is needed to keep it from sliding down the bank (to the left in Fig. 6.5). A driver attempting to negotiate the curve at a speed greater than 13.4 m/s has to depend on friction to keep from sliding up the bank (to the right in Fig. 6.5).

What If? Imagine that this same roadway were built on Mars in the future to connect different colony centers. Could it be traveled at the same speed?

Answer The reduced gravitational force on Mars would mean that the car is not pressed as tightly to the roadway. The reduced normal force results in a smaller component of the normal force toward the center of the circle. This smaller component would not be sufficient to provide the centripetal acceleration associated with the original speed. The centripetal acceleration must be reduced, which can be done by reducing the speed v.

Mathematically, notice that Equation (3) shows that the speed v is proportional to the square root of g for a roadway of fixed radius r banked at a fixed angle θ. Therefore, if g is smaller, as it is on Mars, the speed v with which the roadway can be safely traveled is also smaller.

EXAMPLE 6.5 Let's Go Loop-the-Loop!

A pilot of mass m in a jet aircraft executes a loop-the-loop, as shown in Figure 6.6a. In this maneuver, the aircraft moves in a vertical circle of radius 2.70 km at a constant speed of 225 m/s.

(A) Determine the force exerted by the seat on the pilot at the bottom of the loop. Express your answer in terms of the weight of the pilot mg.

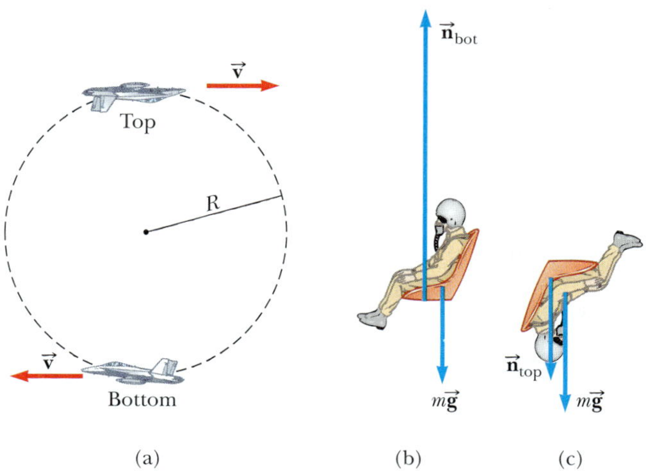

Figure 6.6 (Example 6.5) (a) An aircraft executes a loop-the-loop maneuver as it moves in a vertical circle at constant speed. (b) The free-body diagram for the pilot at the bottom of the loop. In this position the pilot experiences an apparent weight greater than his true weight. (c) The free-body diagram for the pilot at the top of the loop.

SOLUTION

Conceptualize Look carefully at Figure 6.6a. Based on experiences with driving over small hills on a road or riding at the top of a Ferris wheel, you would expect to feel lighter at the top of the path. Similarly, you would expect to feel heavier at the bottom of the path. At the bottom of the loop the normal and gravitational forces on the pilot act in *opposite* directions, whereas at the top of the loop these two forces act in the *same* direction. The vector sum of these two forces gives a force of constant magnitude that keeps the pilot moving in a circular path at a constant speed. To yield net force vectors with the same magnitude, the normal force at the bottom must be greater than that at the top.

Categorize Because the speed of the aircraft is constant (how likely is that?), we can categorize this problem as one involving a particle (the pilot) in uniform circular motion, complicated by the gravitational force acting at all times on the aircraft.

Analyze We draw a free-body diagram for the pilot at the bottom of the loop as shown in Figure 6.6b. The only forces acting on him are the downward gravitational force $\vec{F}_g = m\vec{g}$ and the upward force $\vec{n}_{bot}$ exerted by the seat. The net upward force on the pilot that provides his centripetal acceleration has a magnitude $n_{bot} - mg$.

Apply Newton's second law to the pilot in the radial direction:

$$\sum F = n_{bot} - mg = m\frac{v^2}{r}$$

Solve for the force exerted by the seat on the pilot:

$$n_{bot} = mg + m\frac{v^2}{r} = mg\left(1 + \frac{v^2}{rg}\right)$$

Substitute the values given for the speed and radius:

$$n_{bot} = mg\left(1 + \frac{(225 \text{ m/s})^2}{(2.70 \times 10^3 \text{ m})(9.80 \text{ m/s}^2)}\right)$$

$$= 2.91\,mg$$

Hence, the magnitude of the force $\vec{n}_{bot}$ exerted by the seat on the pilot is *greater* than the weight of the pilot by a factor of 2.91. So, the pilot experiences an apparent weight that is greater than his true weight by a factor of 2.91.

(B) Determine the force exerted by the seat on the pilot at the top of the loop.

SOLUTION

Analyze The free-body diagram for the pilot at the top of the loop is shown in Figure 6.6c. As noted earlier, both the gravitational force exerted by the Earth and the force $\vec{\mathbf{n}}_{\text{top}}$ exerted by the seat on the pilot act downward, so the net downward force that provides the centripetal acceleration has a magnitude $n_{\text{top}} + mg$.

Apply Newton's second law to the pilot at this position:

$$\sum F = n_{\text{top}} + mg = m\frac{v^2}{r}$$

$$n_{\text{top}} = m\frac{v^2}{r} - mg = mg\left(\frac{v^2}{rg} - 1\right)$$

$$n_{\text{top}} = mg\left(\frac{(225\text{ m/s})^2}{(2.70 \times 10^3\text{ m})(9.80\text{ m/s}^2)} - 1\right)$$

$$= 0.913mg$$

In this case, the magnitude of the force exerted by the seat on the pilot is *less* than his true weight by a factor of 0.913, and the pilot feels lighter.

Finalize The variations in the normal force are consistent with our prediction in the conceptualize step of the problem.

6.2 Nonuniform Circular Motion

In Chapter 4, we found that if a particle moves with varying speed in a circular path, there is, in addition to the radial component of acceleration, a tangential component having magnitude $|dv/dt|$. Therefore, the force acting on the particle must also have a tangential and a radial component. Because the total acceleration is $\vec{\mathbf{a}} = \vec{\mathbf{a}}_r + \vec{\mathbf{a}}_t$, the total force exerted on the particle is $\sum \vec{\mathbf{F}} = \sum \vec{\mathbf{F}}_r + \sum \vec{\mathbf{F}}_t$ as shown in Active Figure 6.7. (We express the radial and tangential forces as net forces with the summation notation because each force could consist of multiple forces that combine.) The vector $\sum \vec{\mathbf{F}}_r$ is directed toward the center of the circle and is responsible for the centripetal acceleration. The vector $\sum \vec{\mathbf{F}}_t$ tangent to the circle is responsible for the tangential acceleration, which represents a change in the particle's speed with time.

Quick Quiz 6.2 A bead slides freely along a curved wire lying on a horizontal surface at constant speed as shown in Figure 6.8. (a) Draw the vectors representing the force exerted by the wire on the bead at points Ⓐ, Ⓑ, and Ⓒ. (b) Suppose the bead in Figure 6.8 speeds up with constant tangential acceleration as it moves toward the right. Draw the vectors representing the force on the bead at points Ⓐ, Ⓑ, and Ⓒ.

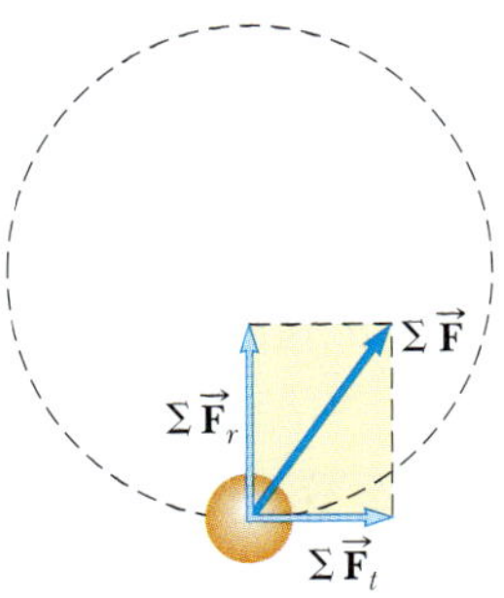

ACTIVE FIGURE 6.7

When the net force acting on a particle moving in a circular path has a tangential component $\sum F_t$, the particle's speed changes. The net force exerted on the particle in this case is the vector sum of the radial force and the tangential force. That is, $\sum \vec{\mathbf{F}} = \sum \vec{\mathbf{F}}_r + \sum \vec{\mathbf{F}}_t$.

Sign in at www.thomsonedu.com and go to ThomsonNOW to adjust the initial position of the particle and compare the component forces acting on the particle with those for a child swinging on a swing set.

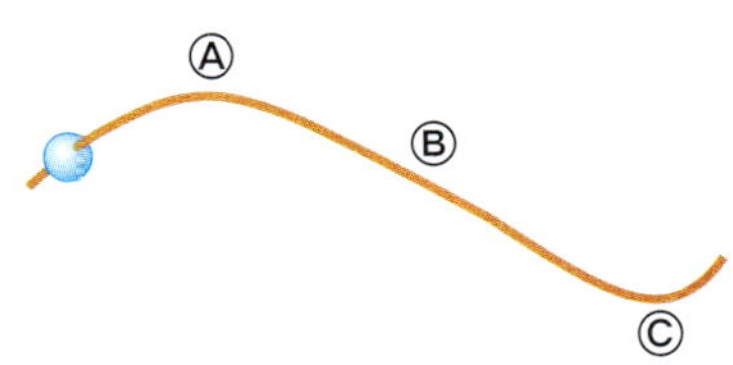

Figure 6.8 (Quick Quiz 6.2) A bead slides along a curved wire.

EXAMPLE 6.6 Keep Your Eye on the Ball

A small sphere of mass m is attached to the end of a cord of length R and set into motion in a *vertical* circle about a fixed point O as illustrated in Figure 6.9. Determine the tension in the cord at any instant when the speed of the sphere is v and the cord makes an angle θ with the vertical.

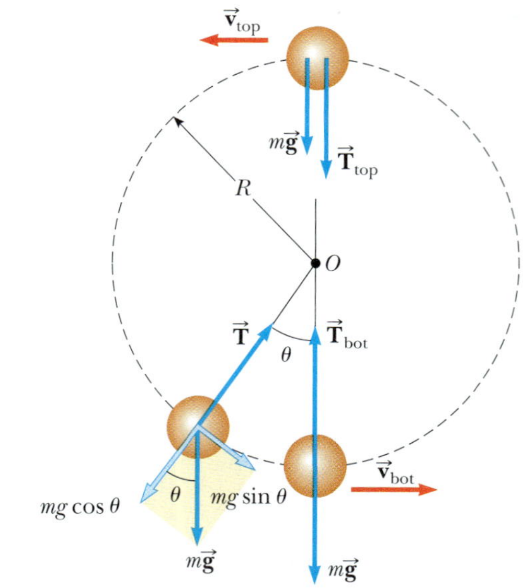

Figure 6.9 (Example 6.6) The forces acting on a sphere of mass m connected to a cord of length R and rotating in a vertical circle centered at O. Forces acting on the sphere are shown when the sphere is at the top and bottom of the circle and at an arbitrary location.

SOLUTION

Conceptualize Compare the motion of the sphere in Figure 6.9 to that of the airplane in Figure 6.6a associated with Example 6.5. Both objects travel in a circular path. Unlike the airplane in Example 6.5, however, the speed of the sphere is *not* uniform in this example because, at most points along the path, a tangential component of acceleration arises from the gravitational force exerted on the sphere.

Categorize We model the sphere as a particle under a net force and moving in a circular path, but it is not a particle in *uniform* circular motion. We need to use the techniques discussed in this section on nonuniform circular motion.

Analyze From the free-body diagram in Figure 6.9, we see that the only forces acting on the sphere are the gravitational force $\vec{\mathbf{F}}_g = m\vec{\mathbf{g}}$ exerted by the Earth and the force $\vec{\mathbf{T}}$ exerted by the cord. We resolve $\vec{\mathbf{F}}_g$ into a tangential component $mg \sin\theta$ and a radial component $mg\cos\theta$.

Apply Newton's second law to the sphere in the tangential direction:

$$\sum F_t = mg\sin\theta = ma_t$$

$$a_t = g\sin\theta$$

Apply Newton's second law to the forces acting on the sphere in the radial direction, noting that both $\vec{\mathbf{T}}$ and $\vec{\mathbf{a}}_r$ are directed toward O:

$$\sum F_r = T - mg\cos\theta = \frac{mv^2}{R}$$

$$T = mg\left(\frac{v^2}{Rg} + \cos\theta\right)$$

Finalize Let us evaluate this result at the top and bottom of the circular path (Fig. 6.9):

$$T_{top} = mg\left(\frac{v_{top}^2}{Rg} - 1\right) \qquad T_{bot} = mg\left(\frac{v_{bot}^2}{Rg} + 1\right)$$

These results have the same mathematical form as those for the normal forces n_{top} and n_{bot} on the pilot in Example 6.5, which is consistent with the normal force on the pilot playing the same physical role in Example 6.5 as the tension in the string plays in this example. Keep in mind, however, that v in the expressions above varies for different positions of the sphere, as indicated by the subscripts, whereas v in Example 6.5 is constant.

What If? What if the ball is set in motion with a slower speed? (A) What speed would the ball have as it passes over the top of the circle if the tension in the cord goes to zero instantaneously at this point?

Answer Let us set the tension equal to zero in the expression for T_{top}:

$$0 = mg\left(\frac{v_{top}^2}{Rg} - 1\right) \quad\rightarrow\quad v_{top} = \sqrt{gR}$$

(B) What if the ball is set in motion such that the speed at the top is less than this value? What happens?

Answer In this case, the ball never reaches the top of the circle. At some point on the way up, the tension in the string goes to zero and the ball becomes a projectile. It follows a segment of a parabolic path over the top of its motion, rejoining the circular path on the other side when the tension becomes nonzero again.

6.3 Motion in Accelerated Frames

Newton's laws of motion, which we introduced in Chapter 5, describe observations that are made in an inertial frame of reference. In this section, we analyze how Newton's laws are applied by an observer in a noninertial frame of reference, that is, one that is accelerating. For example, recall the discussion of the air hockey table on a train in Section 5.2. The train moving at constant velocity represents an inertial frame. An observer on the train sees the puck at rest remain at rest, and Newton's first law appears to be obeyed. The accelerating train is not an inertial frame. According to you as the observer on this train, there appears to be no force on the puck, yet it accelerates from rest toward the back of the train, appearing to violate Newton's first law. This property is a general property of observations made in noninertial frames: there appear to be unexplained accelerations of objects that are not "fastened" to the frame. Newton's first law is not violated, of course. It only *appears* to be violated because of observations made from a noninertial frame. In general, the direction of the unexplained acceleration is opposite the direction of the acceleration of the noninertial frame.

On the accelerating train, as you watch the puck accelerating toward the back of the train, you might conclude based on your belief in Newton's second law that a force has acted on the puck to cause it to accelerate. We call an apparent force such as this one a **fictitious force** because it is due to an accelerated reference frame. A fictitious force appears to act on an object in the same way as a real force. Real forces are always interactions between two objects, however, and you cannot identify a second object for a fictitious force. (What second object is interacting with the puck to cause it to accelerate?)

The train example describes a fictitious force due to a change in the train's speed. Another fictitious force is due to the change in the *direction* of the velocity vector. To understand the motion of a system that is noninertial because of a change in direction, consider a car traveling along a highway at a high speed and approaching a curved exit ramp as shown in Figure 6.10a. As the car takes the sharp left turn onto the ramp, a person sitting in the passenger seat slides to the right and hits the door. At that point the force exerted by the door on the passenger keeps her from being ejected from the car. What causes her to move toward the door? A popular but incorrect explanation is that a force acting toward the right in Figure 6.10b pushes the passenger outward from the center of the circular path. Although often called the "centrifugal force," it is a fictitious force due to the centripetal acceleration associated with the changing direction of the car's velocity vector. (The driver also experiences this effect but wisely holds on to the steering wheel to keep from sliding to the right.)

The phenomenon is correctly explained as follows. Before the car enters the ramp, the passenger is moving in a straight-line path. As the car enters the ramp and travels a curved path, the passenger tends to move along the original straight-line path, which is in accordance with Newton's first law: the natural tendency of an object is to continue moving in a straight line. If a sufficiently large force (toward the center of curvature) acts on the passenger as in Figure 6.10c, however, she moves in a curved path along with the car. This force is the force of friction between her and the car seat. If this friction force is not large enough, the seat follows a curved path while the passenger continues in the straight-line path of the car before the car began the turn. Therefore, from the point of view of an observer in the car, the passenger slides to the right relative to the seat. Eventually, she encounters the door, which provides a force large enough to enable her to follow the same curved path as the car. She slides toward the door not because of an outward force but because **the force of friction is not sufficiently great to allow her to travel along the circular path followed by the car.**

Another interesting fictitious force is the "Coriolis force." It is an apparent force caused by changing the radial position of an object in a rotating coordinate system.

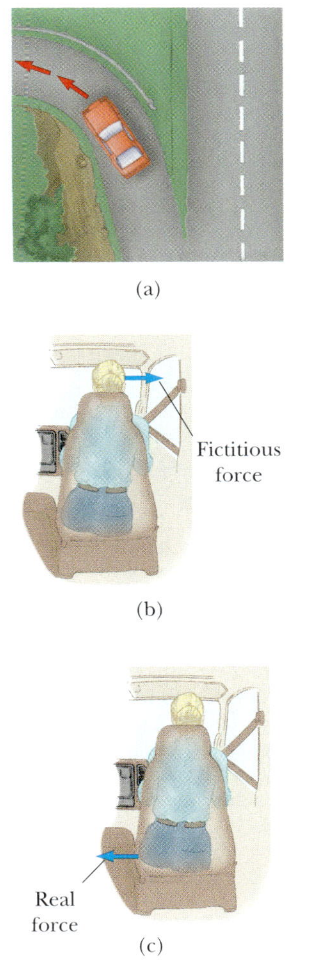

Figure 6.10 (a) A car approaching a curved exit ramp. What causes a passenger in the front seat to move toward the right-hand door? (b) From the passenger's frame of reference, a force appears to push her toward the right door, but it is a fictitious force. (c) Relative to the reference frame of the Earth, the car seat applies a real force toward the left on the passenger, causing her to change direction along with the rest of the car.

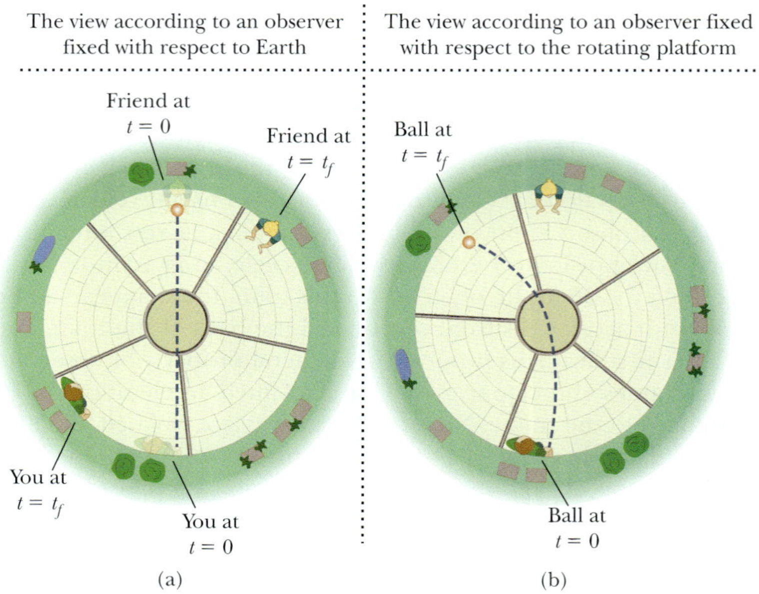

ACTIVE FIGURE 6.11

(a) You and your friend sit at the edge of a rotating turntable. In this overhead view observed by someone in an inertial reference frame attached to the Earth, you throw the ball at $t = 0$ in the direction of your friend. By the time t_f that the ball arrives at the other side of the turntable, your friend is no longer there to catch it. According to this observer, the ball follows a straight-line path, consistent with Newton's laws. (b) From your friend's point of view, the ball veers to one side during its flight. Your friend introduces a fictitious force to cause this deviation from the expected path. This fictitious force is called the "Coriolis force."

Sign in at www.thomsonedu.com and go to ThomsonNOW to observe the ball's path simultaneously from the reference frame of an inertial observer and from the reference frame of the rotating turntable.

PITFALL PREVENTION 6.2

Centrifugal Force

The commonly heard phrase "centrifugal force" is described as a force pulling *outward* on an object moving in a circular path. If you are feeling a "centrifugal force" on a rotating carnival ride, what is the other object with which you are interacting? You cannot identify another object because it is a fictitious force that occurs because you are in a noninertial reference frame.

For example, suppose you and a friend are on opposite sides of a rotating circular platform and you decide to throw a baseball to your friend. Active Figure 6.11a represents what an observer would see if the ball is viewed while the observer is hovering at rest above the rotating platform. According to this observer, who is in an inertial frame, the ball follows a straight line as it must according to Newton's first law. At $t = 0$ you throw the ball toward your friend, but by the time t_f when the ball has crossed the platform, your friend has moved to a new position. Now, however, consider the situation from your friend's viewpoint. Your friend is in a noninertial reference frame because he is undergoing a centripetal acceleration relative to the inertial frame of the Earth's surface. He starts off seeing the baseball coming toward him, but as it crosses the platform, it veers to one side as shown in Active Figure 6.11b. Therefore, your friend on the rotating platform states that the ball does not obey Newton's first law and claims that a force is causing the ball to follow a curved path. This fictitious force is called the Coriolis force.

Fictitious forces may not be real forces, but they can have real effects. An object on your dashboard *really* slides off if you press the accelerator of your car. As you ride on a merry-go-round, you feel pushed toward the outside as if due to the fictitious "centrifugal force." You are likely to fall over and injure yourself due to the Coriolis force if you walk along a radial line while a merry-go-round rotates. (One of the authors did so and suffered a separation of the ligaments from his ribs when he fell over.) The Coriolis force due to the rotation of the Earth is responsible for rotations of hurricanes and for large-scale ocean currents.

Quick Quiz 6.3 Consider the passenger in the car making a left turn in Figure 6.10. Which of the following is correct about forces in the horizontal direction if she is making contact with the right-hand door? (a) The passenger is in equilibrium between real forces acting to the right and real forces acting to the left. (b) The passenger is subject only to real forces acting to the right. (c) The passenger is subject only to real forces acting to the left. (d) None of these statements is true.

EXAMPLE 6.7 Fictitious Forces in Linear Motion

A small sphere of mass m hangs by a cord from the ceiling of a boxcar that is accelerating to the right as shown in Figure 6.12. The noninertial observer in Figure 6.12b claims that a force, which we know to be fictitious, causes the observed deviation of the cord from the vertical. How is the magnitude of this force related to the boxcar's acceleration measured by the inertial observer in Figure 6.12a?

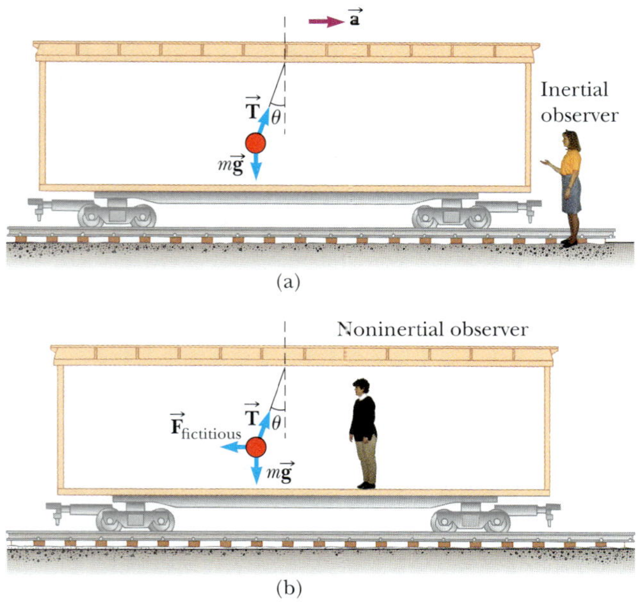

Figure 6.12 (Example 6.7) A small sphere suspended from the ceiling of a boxcar accelerating to the right is deflected as shown. (a) An inertial observer at rest outside the car claims that the acceleration of the sphere is provided by the horizontal component of $\vec{T}$. (b) A noninertial observer riding in the car says that the net force on the sphere is zero and that the deflection of the cord from the vertical is due to a fictitious force $\vec{F}_{fictitious}$ that balances the horizontal component of $\vec{T}$.

SOLUTION

Conceptualize Place yourself in the role of each of the two observers in Figure 6.12. As the inertial observer on the ground, you see the boxcar accelerating and know that the deviation of the cord is due to this acceleration. As the noninertial observer on the boxcar, imagine that you ignore any effects of the car's motion so that you are not aware of its acceleration. Because you are unaware of this acceleration, you claim that a force is pushing sideways on the sphere to cause the deviation of the cord from the vertical. To make the conceptualization more real, try running from rest while holding a hanging object on a string and notice that the string is at an angle to the vertical while you are accelerating, as if a force is pushing the object backward.

Categorize For the inertial observer, we model the sphere as a particle under a net force in the horizontal direction and a particle in equilibrium in the vertical direction. For the noninertial observer, the sphere is modeled as a particle in equilibrium for which one of the forces is fictitious.

Analyze According to the inertial observer at rest (Fig. 6.12a), the forces on the sphere are the force $\vec{T}$ exerted by the cord and the gravitational force. The inertial observer concludes that the sphere's acceleration is the same as that of the boxcar and that this acceleration is provided by the horizontal component of $\vec{T}$.

Apply Newton's second law in component form to the sphere according to the inertial observer:

$$\text{Inertial observer} \begin{cases} (1) & \sum F_x = T\sin\theta = ma \\ (2) & \sum F_y = T\cos\theta - mg = 0 \end{cases}$$

According to the noninertial observer riding in the car (Fig. 6.12b), the cord also makes an angle θ with the vertical; to that observer, however, the sphere is at rest and so its acceleration is zero. Therefore, the noninertial observer introduces a fictitious force in the horizontal direction to balance the horizontal component of $\vec{T}$ and claims that the net force on the sphere is zero.

Apply Newton's second law in component form to the sphere according to the noninertial observer:

$$\text{Noninertial observer} \begin{cases} \sum F'_x = T\sin\theta - F_{\text{fictitious}} = 0 \\ \sum F'_y = T\cos\theta - mg = 0 \end{cases}$$

These expressions are equivalent to Equations (1) and (2) if $F_{\text{fictitious}} = ma$, where a is the acceleration according to the inertial observer.

Finalize If we were to make this substitution in the equation for F'_x above, the noninertial observer obtains the same mathematical results as the inertial observer. The physical interpretation of the cord's deflection, however, differs in the two frames of reference.

What If? Suppose the inertial observer wants to measure the acceleration of the train by means of the pendulum (the sphere hanging from the cord). How could she do so?

Answer Our intuition tells us that the angle θ the cord makes with the vertical should increase as the acceleration increases. By solving Equations (1) and (2) simultaneously for a, the inertial observer can determine the magnitude of the car's acceleration by measuring the angle θ and using the relationship $a = g \tan \theta$. Because the deflection of the cord from the vertical serves as a measure of acceleration, *a simple pendulum can be used as an accelerometer.*

6.4 Motion in the Presence of Resistive Forces

In Chapter 5, we described the force of kinetic friction exerted on an object moving on some surface. We completely ignored any interaction between the object and the medium through which it moves. Now consider the effect of that medium, which can be either a liquid or a gas. The medium exerts a **resistive force** $\vec{\mathbf{R}}$ on the object moving through it. Some examples are the air resistance associated with moving vehicles (sometimes called *air drag*) and the viscous forces that act on objects moving through a liquid. The magnitude of $\vec{\mathbf{R}}$ depends on factors such as the speed of the object, and the direction of $\vec{\mathbf{R}}$ is always opposite the direction of the object's motion relative to the medium.

The magnitude of the resistive force can depend on speed in a complex way, and here we consider only two simplified models. In the first model, we assume the resistive force is proportional to the speed of the moving object; this model is valid for objects falling slowly through a liquid and for very small objects, such as dust particles, moving through air. In the second model, we assume a resistive force that is proportional to the square of the speed of the moving object; large objects, such as a skydiver moving through air in free fall, experience such a force.

Model 1: Resistive Force Proportional to Object Velocity

If we model the resistive force acting on an object moving through a liquid or gas as proportional to the object's velocity, the resistive force can be expressed as

$$\vec{\mathbf{R}} = -b\vec{\mathbf{v}} \tag{6.2}$$

where b is a constant whose value depends on the properties of the medium and on the shape and dimensions of the object and $\vec{\mathbf{v}}$ is the velocity of the object relative to the medium. The negative sign indicates that $\vec{\mathbf{R}}$ is in the opposite direction to $\vec{\mathbf{v}}$.

Consider a small sphere of mass m released from rest in a liquid as in Active Figure 6.13a. Assuming the only forces acting on the sphere are the resistive force $\vec{\mathbf{R}} = -b\vec{\mathbf{v}}$ and the gravitational force $\vec{\mathbf{F}}_g$, let us describe its motion.[1] Applying Newton's second law to the vertical motion, choosing the downward direction to be positive, and noting that $\Sigma F_y = mg - bv$, we obtain

$$mg - bv = ma = m\frac{dv}{dt} \tag{6.3}$$

where the acceleration of the sphere is downward. Solving this expression for the acceleration dv/dt gives

[1] A *buoyant force* is also acting on the submerged object. This force is constant, and its magnitude is equal to the weight of the displaced liquid. This force changes the apparent weight of the sphere by a constant factor, so we will ignore the force here. We discuss buoyant forces in Chapter 14.

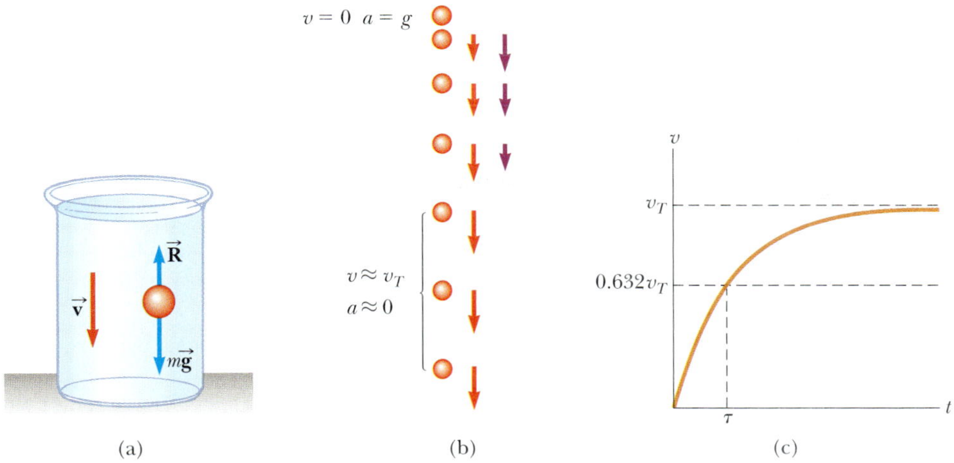

ACTIVE FIGURE 6.13

(a) A small sphere falling through a liquid. (b) A motion diagram of the sphere as it falls. Velocity vectors (red) and acceleration vectors (violet) are shown for each image after the first one. (c) A speed–time graph for the sphere. The sphere approaches a maximum (or terminal) speed v_T, and the time constant τ is the time at which it reaches a speed of $0.632v_T$.

Sign in at www.thomsonedu.com and go to ThomsonNOW to vary the size and mass of the sphere and the viscosity (resistance to flow) of the surrounding medium. Then observe the effects on the sphere's motion and its speed–time graph.

$$\frac{dv}{dt} = g - \frac{b}{m}v \tag{6.4}$$

This equation is called a *differential equation*, and the methods of solving it may not be familiar to you as yet. Notice, however, that initially when $v = 0$, the magnitude of the resistive force is also zero and the acceleration of the sphere is simply g. As t increases, the magnitude of the resistive force increases and the acceleration decreases. The acceleration approaches zero when the magnitude of the resistive force approaches the sphere's weight. In this situation, the speed of the sphere approaches its **terminal speed** v_T.

◀ Terminal speed

The terminal speed is obtained from Equation 6.3 by setting $a = dv/dt = 0$. This gives

$$mg - bv_T = 0 \quad \text{or} \quad v_T = \frac{mg}{b}$$

The expression for v that satisfies Equation 6.4 with $v = 0$ at $t = 0$ is

$$v = \frac{mg}{b}(1 - e^{-bt/m}) = v_T(1 - e^{-t/\tau}) \tag{6.5}$$

This function is plotted in Active Figure 6.13c. The symbol e represents the base of the natural logarithm and is also called *Euler's number:* $e = 2.718\ 28$. The **time constant** $\tau = m/b$ (Greek letter tau) is the time at which the sphere released from rest at $t = 0$ reaches 63.2% of its terminal speed: when $t = \tau$, Equation 6.5 yields $v = 0.632v_T$.

We can check that Equation 6.5 is a solution to Equation 6.4 by direct differentiation:

$$\frac{dv}{dt} = \frac{d}{dt}\left[\frac{mg}{b}(1 - e^{-bt/m})\right] = \frac{mg}{b}\left(0 + \frac{b}{m}e^{-bt/m}\right) = ge^{-bt/m}$$

(See Appendix Table B.4 for the derivative of e raised to some power.) Substituting into Equation 6.4 both this expression for dv/dt and the expression for v given by Equation 6.5 shows that our solution satisfies the differential equation.

EXAMPLE 6.8 **Sphere Falling in Oil**

A small sphere of mass 2.00 g is released from rest in a large vessel filled with oil, where it experiences a resistive force proportional to its speed. The sphere reaches a terminal speed of 5.00 cm/s. Determine the time constant τ and the time at which the sphere reaches 90.0% of its terminal speed.

SOLUTION

Conceptualize With the help of Active Figure 6.13, imagine dropping the sphere into the oil and watching it sink to the bottom of the vessel. If you have some thick shampoo, drop a marble in it and observe the motion of the marble.

Categorize We model the sphere as a particle under a net force, with one of the forces being a resistive force that depends on the speed of the sphere.

Analyze From $v_T = mg/b$, evaluate the coefficient b:

$$b = \frac{mg}{v_T} = \frac{(2.00\text{ g})(980\text{ cm/s}^2)}{5.00\text{ cm/s}} = 392\text{ g/s}$$

Evaluate the time constant τ:

$$\tau = \frac{m}{b} = \frac{2.00\text{ g}}{392\text{ g/s}} = 5.10 \times 10^{-3}\text{ s}$$

Find the time t at which the sphere reaches a speed of $0.900v_T$ by setting $v = 0.900v_T$ in Equation 6.5 and solving for t:

$$0.900v_T = v_T(1 - e^{-t/\tau})$$

$$1 - e^{-t/\tau} = 0.900$$

$$e^{-t/\tau} = 0.100$$

$$-\frac{t}{\tau} = \ln(0.100) = -2.30$$

$$t = 2.30\tau = 2.30(5.10 \times 10^{-3}\text{ s}) = 11.7 \times 10^{-3}\text{ s}$$

$$= 11.7\text{ ms}$$

Finalize The sphere reaches 90.0% of its terminal speed in a very short time interval. You should have also seen this behavior if you performed the activity with the marble and the shampoo.

Model 2: Resistive Force Proportional to Object Speed Squared

For objects moving at high speeds through air, such as airplanes, skydivers, cars, and baseballs, the resistive force is reasonably well modeled as proportional to the square of the speed. In these situations, the magnitude of the resistive force can be expressed as

$$R = \tfrac{1}{2}D\rho A v^2 \quad \textbf{(6.6)}$$

where D is a dimensionless empirical quantity called the *drag coefficient*, ρ is the density of air, and A is the cross-sectional area of the moving object measured in a plane perpendicular to its velocity. The drag coefficient has a value of about 0.5 for spherical objects but can have a value as great as 2 for irregularly shaped objects.

Let us analyze the motion of an object in free-fall subject to an upward air resistive force of magnitude $R = \frac{1}{2}D\rho Av^2$. Suppose an object of mass m is released from rest. As Figure 6.14 shows, the object experiences two external forces:[2] the downward gravitational force $\vec{\mathbf{F}}_g = m\vec{\mathbf{g}}$ and the upward resistive force $\vec{\mathbf{R}}$. Hence, the magnitude of the net force is

$$\sum F = mg - \tfrac{1}{2}D\rho Av^2 \quad \textbf{(6.7)}$$

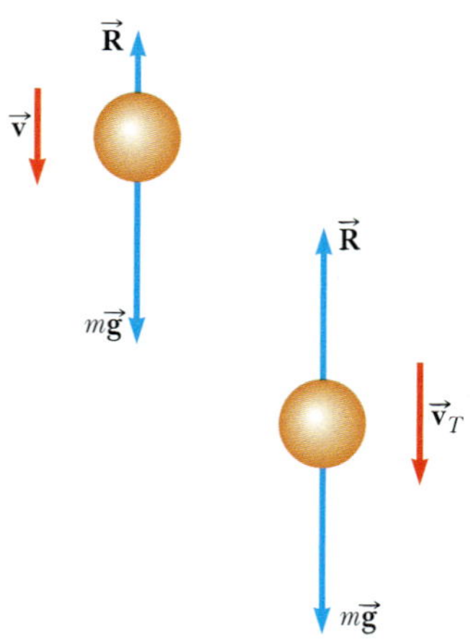

Figure 6.14 An object falling through air experiences a resistive force $\vec{\mathbf{R}}$ and a gravitational force $\vec{\mathbf{F}}_g = m\vec{\mathbf{g}}$. The object reaches terminal speed (on the right) when the net force acting on it is zero, that is, when $\vec{\mathbf{R}} = -\vec{\mathbf{F}}_g$ or $R = mg$. Before that occurs, the acceleration varies with speed according to Equation 6.8.

[2] As with Model 1, there is also an upward buoyant force that we neglect.

TABLE 6.1

Terminal Speed for Various Objects Falling Through Air

Object	Mass (kg)	Cross-Sectional Area (m^2)	v_T (m/s)
Skydiver	75	0.70	60
Baseball (radius 3.7 cm)	0.145	4.2×10^{-3}	43
Golf ball (radius 2.1 cm)	0.046	1.4×10^{-3}	44
Hailstone (radius 0.50 cm)	4.8×10^{-4}	7.9×10^{-5}	14
Raindrop (radius 0.20 cm)	3.4×10^{-5}	1.3×10^{-5}	9.0

where we have taken downward to be the positive vertical direction. Using the force in Equation 6.7 in Newton's second law, we find that the object has a downward acceleration of magnitude

$$a = g - \left(\frac{D\rho A}{2m}\right)v^2 \tag{6.8}$$

We can calculate the terminal speed v_T by noticing that when the gravitational force is balanced by the resistive force, the net force on the object is zero and therefore its acceleration is zero. Setting $a = 0$ in Equation 6.8 gives

$$g - \left(\frac{D\rho A}{2m}\right)v_T^2 = 0$$

so

$$v_T = \sqrt{\frac{2mg}{D\rho A}} \tag{6.9}$$

Table 6.1 lists the terminal speeds for several objects falling through air.

Quick Quiz 6.4 A baseball and a basketball, having the same mass, are dropped through air from rest such that their bottoms are initially at the same height above the ground, on the order of 1 m or more. Which one strikes the ground first? (a) The baseball strikes the ground first. (b) The basketball strikes the ground first. (c) Both strike the ground at the same time.

CONCEPTUAL EXAMPLE 6.9 The Skysurfer

Consider a skysurfer (Fig. 6.15) who jumps from a plane with her feet attached firmly to her surfboard, does some tricks, and then opens her parachute. Describe the forces acting on her during these maneuvers.

Jump Run Productions/Getty Images

Figure 6.15 (Conceptual Example 6.9) A skysurfer.

SOLUTION

When the surfer first steps out of the plane, she has no vertical velocity. The downward gravitational force causes her to accelerate toward the ground. As her downward speed increases, so does the upward resistive force exerted by the air on her body and the board. This upward force reduces their acceleration, and so their speed increases more slowly. Eventually, they are going so fast that the upward resistive force matches the downward gravitational force. Now the net force is zero and they no longer accelerate, but instead reach their terminal speed. At some point after reaching terminal speed, she opens her parachute, resulting in a drastic increase in the upward resistive force. The net force (and thus the acceleration) is now upward, in the direction opposite the direction of the velocity. The downward velocity therefore decreases rapidly, and the resistive force on the chute also decreases. Eventually, the upward resistive force and the downward gravitational force balance each other and a much smaller terminal speed is reached, permitting a safe landing.

(Contrary to popular belief, the velocity vector of a skydiver never points upward. You may have seen a videotape in which a skydiver appears to "rocket" upward once the chute opens. In fact, what happens is that the skydiver slows down but the person holding the camera continues falling at high speed.)

EXAMPLE 6.10 Falling Coffee Filters

The dependence of resistive force on the square of the speed is a model. Let's test the model for a specific situation. Imagine an experiment in which we drop a series of stacked coffee filters and measure their terminal speeds. Table 6.2 presents typical terminal speed data from a real experiment using these coffee filters as they fall through the air. The time constant τ is small, so a dropped filter quickly reaches terminal speed. Each filter has a mass of 1.64 g. When the filters are nested together, they stack in such a way that the front-facing surface area does not increase. Determine the relationship between the resistive force exerted by the air and the speed of the falling filters.

TABLE 6.2

Terminal Speed and Resistive Force for Stacked Coffee Filters

Number of Filters	v_T (m/s)[a]	R (N)
1	1.01	0.016 1
2	1.40	0.032 2
3	1.63	0.048 3
4	2.00	0.064 4
5	2.25	0.080 5
6	2.40	0.096 6
7	2.57	0.112 7
8	2.80	0.128 8
9	3.05	0.144 9
10	3.22	0.161 0

[a] All values of v_T are approximate.

SOLUTION

Conceptualize Imagine dropping the coffee filters through the air. (If you have some coffee filters, try dropping them.) Because of the relatively small mass of the coffee filter, you probably won't notice the time interval during which there is an acceleration. The filters will appear to fall at constant velocity immediately upon leaving your hand.

Categorize Because a filter moves at constant velocity, we model it as a particle in equilibrium.

Analyze At terminal speed, the upward resistive force on the filter balances the downward gravitational force so that $R = mg$.

Evaluate the magnitude of the resistive force:

$$R = mg = (1.64\text{ g})\left(\frac{1\text{ kg}}{1\,000\text{ g}}\right)(9.80\text{ m/s}^2) = 0.016\,1\text{ N}$$

Likewise, two filters nested together experience 0.032 2 N of resistive force, and so forth. These values of resistive force are shown in the rightmost column of Table 6.2. A graph of the resistive force on the filters as a function of terminal speed is shown in Figure 6.16a. A straight line is not a good fit, indicating that the resistive force is *not* proportional to the speed. The behavior is more clearly seen in Figure 6.16b, in which the resistive force is plotted as a function of the square of the terminal speed. This graph indicates that the resistive force is proportional to the *square* of the speed as suggested by Equation 6.6.

Finalize Here is a good opportunity for you to take some actual data at home on real coffee filters and see if you can reproduce the results shown in Figure 6.16. If you have shampoo and a marble as mentioned in Example 6.8, take data on that system too and see if the resistive force is appropriately modeled as being proportional to the speed.

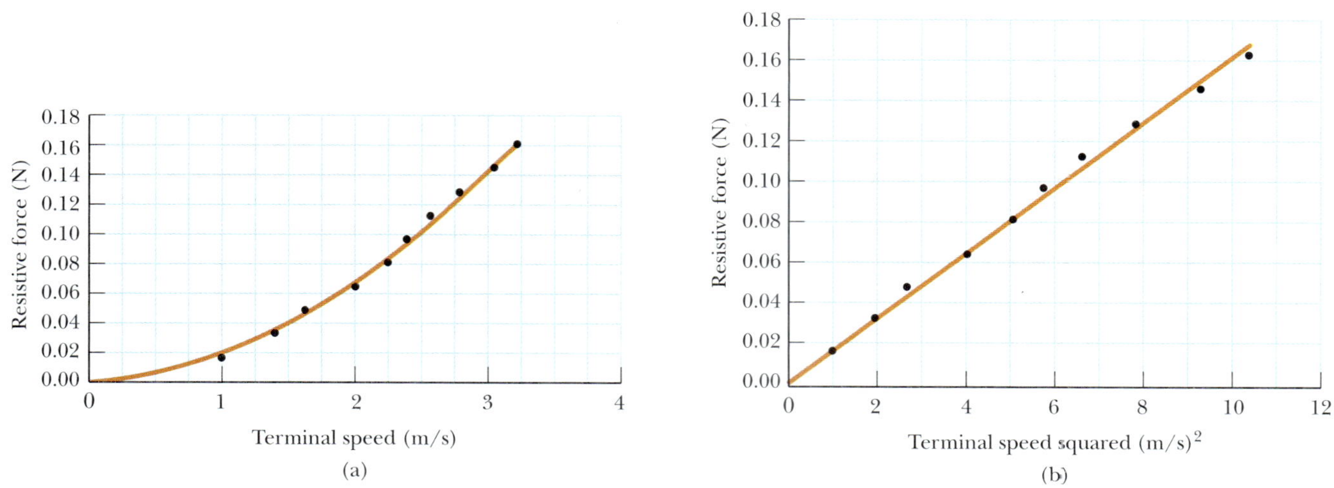

Figure 6.16 (Example 6.10) (a) Relationship between the resistive force acting on falling coffee filters and their terminal speed. The curved line is a second-order polynomial fit. (b) Graph relating the resistive force to the square of the terminal speed. The fit of the straight line to the data points indicates that the resistive force is proportional to the terminal speed squared. Can you find the proportionality constant?

EXAMPLE 6.11 Resistive Force Exerted on a Baseball

A pitcher hurls a 0.145-kg baseball past a batter at 40.2 m/s (= 90 mi/h). Find the resistive force acting on the ball at this speed.

SOLUTION

Conceptualize This example is different from the previous ones in that the object is now moving horizontally through the air instead of moving vertically under the influence of gravity and the resistive force. The resistive force causes the ball to slow down while gravity causes its trajectory to curve downward. We simplify the situation by assuming that the velocity vector is exactly horizontal at the instant it is traveling at 40.2 m/s.

Categorize In general, the ball is a particle under a net force. Because we are considering only one instant of time, however, we are not concerned about acceleration, so the problem involves only finding the value of one of the forces.

Analyze To determine the drag coefficient D, imagine we drop the baseball and allow it to reach terminal speed. Solve Equation 6.9 for D and substitute the appropriate values for m, v_T, and A from Table 6.1, taking the density of air as 1.20 kg/m³:

$$D = \frac{2mg}{v_T^2 \rho A} = \frac{2(0.145\text{ kg})(9.80\text{ m/s}^2)}{(43\text{ m/s})^2(1.20\text{ kg/m}^3)(4.2 \times 10^{-3}\text{ m}^2)}$$

$$= 0.305$$

Use this value for D in Equation 6.6 to find the magnitude of the resistive force:

$$R = \tfrac{1}{2} D\rho A v^2$$

$$= \tfrac{1}{2}(0.305)(1.20\text{ kg/m}^3)(4.2 \times 10^{-3}\text{ m}^2)(40.2\text{ m/s})^2$$

$$= 1.2\text{ N}$$

Finalize The magnitude of the resistive force is similar in magnitude to the weight of the baseball, which is about 1.4 N. Therefore, air resistance plays a major role in the motion of the ball, as evidenced by the variety of curve balls, floaters, sinkers, and the like thrown by baseball pitchers.

Summary

ThomsonNOW™ Sign in at **www.thomsonedu.com** and go to ThomsonNOW to take a practice test for this chapter.

CONCEPTS AND PRINCIPLES

A particle moving in uniform circular motion has a centripetal acceleration; this acceleration must be provided by a net force directed toward the center of the circular path.

An observer in a noninertial (accelerating) frame of reference introduces **fictitious forces** when applying Newton's second law in that frame.

An object moving through a liquid or gas experiences a speed-dependent **resistive force**. This resistive force is in a direction opposite that of the velocity of the object relative to the medium and generally increases with speed. The magnitude of the resistive force depends on the object's size and shape and on the properties of the medium through which the object is moving. In the limiting case for a falling object, when the magnitude of the resistive force equals the object's weight, the object reaches its **terminal speed**.

ANALYSIS MODEL FOR PROBLEM-SOLVING

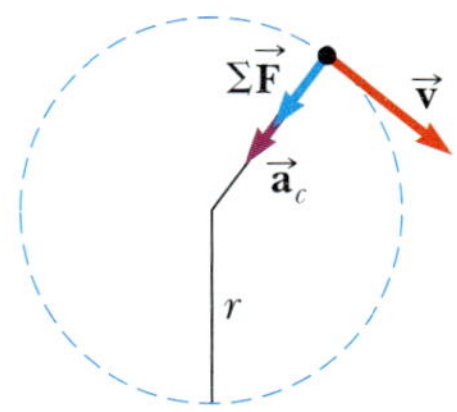

Particle in Uniform Circular Motion With our new knowledge of forces, we can add to the model of a particle in uniform circular motion, first introduced in Chapter 4. Newton's second law applied to a particle moving in uniform circular motion states that the net force causing the particle to undergo a centripetal acceleration (Eq. 4.15) is related to the acceleration according to

$$\sum F = ma_c = m\frac{v^2}{r} \tag{6.1}$$

Questions

☐ denotes answer available in *Student Solutions Manual/Study Guide;* **O** denotes objective question

1. **O** A door in a hospital has a pneumatic closer that pulls the door shut such that the doorknob moves with constant speed over most of its path. In this part of its motion, (a) does the doorknob experience a centripetal acceleration? (b) Does it experience a tangential acceleration? Hurrying to an emergency, a nurse gives a sharp push to the closed door. The door swings open against the pneumatic device, slowing down and then reversing its motion. At the moment the door is open the widest, (c) does the doorknob have a centripetal acceleration? (d) Does it have a tangential acceleration?
2. Describe the path of a moving body in the event that its acceleration is constant in magnitude at all times and (a) perpendicular to the velocity; (b) parallel to the velocity.
3. An object executes circular motion with constant speed whenever a net force of constant magnitude acts perpendicular to the velocity. What happens to the speed if the force is not perpendicular to the velocity?
4. **O** A child is practicing for a bicycle motocross race. His speed remains constant as he goes counterclockwise around a level track with two straight sections and two nearly semicircular sections as shown in the helicopter view of Figure Q6.4. (a) Rank the magnitudes of his acceleration at the points *A, B, C, D,* and *E,* from largest to smallest. If his acceleration is the same size at two points, display that fact in your ranking. If his acceleration is zero, display that fact. (b) What are the directions of his velocity at points *A, B,* and *C*? For each point choose one: north, south, east, west, or nonexistent? (c) What are the directions of his acceleration at points *A, B,* and *C*?

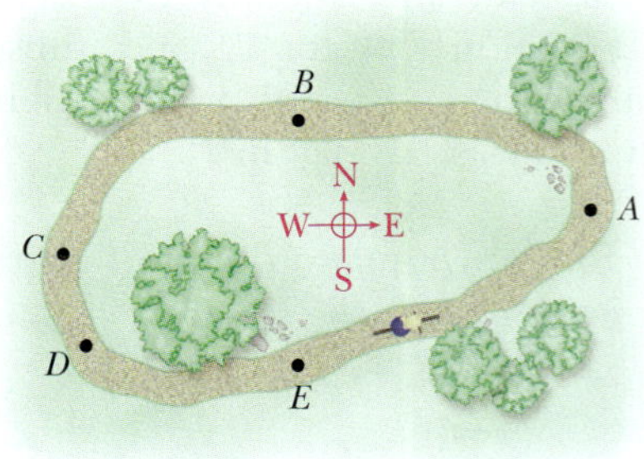

Figure Q6.4

5. **O** A pendulum consists of a small object called a bob hanging from a light cord of fixed length, with the top end of the cord fixed, as represented in Figure Q6.5. The bob moves without friction, swinging equally high on both sides. It moves from its turning point *A* through point *B* and reaches its maximum speed at point *C.* (a) Of these points, is there a point where the bob has nonzero radial acceleration and zero tangential acceleration? If so, which point? What is the direction of its total acceleration at this point? (b) Of these points, is there a point where the bob

has nonzero tangential acceleration and zero radial acceleration? If so, which point? What is the direction of its total acceleration at this point? (c) Is there a point where the bob has no acceleration? If so, which point? (d) Is there a point where the bob has both nonzero tangential and radial acceleration? If so, which point? What is the direction of its total acceleration at this point?

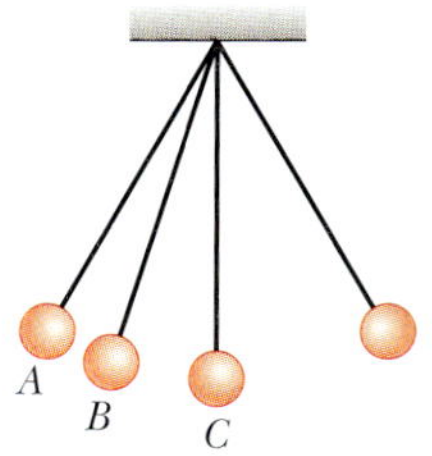

Figure Q6.5

6. If someone told you that astronauts are weightless in orbit because they are beyond the pull of gravity, would you accept the statement? Explain.

7. It has been suggested that rotating cylinders about 20 km in length and 8 km in diameter be placed in space and used as colonies. The purpose of the rotation is to simulate gravity for the inhabitants. Explain this concept for producing an effective imitation of gravity.

8. A pail of water can be whirled in a vertical path such that no water is spilled. Why does the water stay in the pail, even when the pail is above your head?

9. Why does a pilot tend to black out when pulling out of a steep dive?

10. **O** Before takeoff on an airplane, an inquisitive student on the plane takes out a ring of keys and lets it dangle on a lanyard. It hangs straight down as the plane is at rest waiting to take off. The plane then gains speed rapidly as it moves down the runway. (a) Relative to the student's hand, do the keys shift toward the front of the plane, continue to hang straight down, or shift toward the back of the plane? (b) The speed of the plane increases at a constant rate over a time interval of several seconds. During this interval, does the angle the lanyard makes with the vertical increase, stay constant, or decrease?

11. The observer in the accelerating elevator of Example 5.8 would claim that the "weight" of the fish is T, the scale reading. This answer is obviously wrong. Why does this observation differ from that of a person outside the elevator, at rest with respect to the Earth?

12. A falling skydiver reaches terminal speed with her parachute closed. After the parachute is opened, what parameters change to decrease this terminal speed?

13. What forces cause (a) an automobile, (b) a propeller-driven airplane, and (c) a rowboat to move?

14. Consider a small raindrop and a large raindrop falling through the atmosphere. Compare their terminal speeds. What are their accelerations when they reach terminal speed?

15. **O** Consider a skydiver who has stepped from a helicopter and is falling through air, before she reaches terminal speed and long before she opens her parachute. (a) Does her speed increase, decrease, or stay constant? (b) Does the magnitude of her acceleration increase, decrease, stay constant at zero, stay constant at 9.80 m/s^2, or stay constant at some other value?

16. "If the current position and velocity of every particle in the Universe were known, together with the laws describing the forces that particles exert on one another, then the whole future of the Universe could be calculated. The future is determinate and preordained. Free will is an illusion." Do you agree with this thesis? Argue for or against it.

Problems

WebAssign The Problems from this chapter may be assigned online in WebAssign.

ThomsonNOW Sign in at **www.thomsonedu.com** and go to ThomsonNOW to assess your understanding of this chapter's topics with additional quizzing and conceptual questions.

1, 2, 3 denotes straightforward, intermediate, challenging; ☐ denotes full solution available in *Student Solutions Manual/Study Guide*; ▲ denotes coached solution with hints available at **www.thomsonedu.com;** denotes developing symbolic reasoning; ● denotes asking for qualitative reasoning; denotes computer useful in solving problem

Section 6.1 Newton's Second Law for a Particle in Uniform Circular Motion

1. A light string can support a stationary hanging load of 25.0 kg before breaking. A 3.00-kg object attached to the string rotates on a horizontal, frictionless table in a circle of radius 0.800 m, and the other end of the string is held fixed. What range of speeds can the object have before the string breaks?

2. A curve in a road forms part of a horizontal circle. As a car goes around it at constant speed 14.0 m/s, the total force on the driver has magnitude 130 N. What is the total vector force on the driver if the speed is 18.0 m/s instead?

3. In the Bohr model of the hydrogen atom, the speed of the electron is approximately 2.20×10^6 m/s. Find (a) the force acting on the electron as it revolves in a circular orbit of radius 0.530×10^{-10} m and (b) the centripetal acceleration of the electron.

4. Whenever two *Apollo* astronauts were on the surface of the Moon, a third astronaut orbited the Moon. Assume the orbit to be circular and 100 km above the surface of the Moon, where the acceleration due to gravity is 1.52 m/s^2. The radius of the Moon is 1.70×10^6 m. Determine (a) the astronaut's orbital speed and (b) the period of the orbit.

5. A coin placed 30.0 cm from the center of a rotating horizontal turntable slips when its speed is 50.0 cm/s. (a) What force causes the centripetal acceleration when the coin is stationary relative to the turntable? (b) What is

the coefficient of static friction between the coin and turntable?

6. In a cyclotron (one type of particle accelerator), a deuteron (of mass 2.00 u) reaches a final speed of 10.0% of the speed of light while moving in a circular path of radius 0.480 m. The deuteron is maintained in the circular path by a magnetic force. What magnitude of force is required?

7. A space station, in the form of a wheel 120 m in diameter, rotates to provide an "artificial gravity" of 3.00 m/s^2 for persons who walk around on the inner wall of the outer rim. Find the rate of rotation of the wheel (in revolutions per minute) that will produce this effect.

8. Consider a conical pendulum (Fig. 6.3) with an 80.0-kg bob on a 10.0-m wire making an angle of $\theta = 5.00°$ with the vertical. Determine (a) the horizontal and vertical components of the force exerted by the wire on the pendulum and (b) the radial acceleration of the bob.

9. A crate of eggs is located in the middle of the flatbed of a pickup truck as the truck negotiates an unbanked curve in the road. The curve may be regarded as an arc of a circle of radius 35.0 m. If the coefficient of static friction between crate and truck is 0.600, how fast can the truck be moving without the crate sliding?

10. A car initially traveling eastward turns north by traveling in a circular path at uniform speed as shown in Figure P6.10. The length of the arc *ABC* is 235 m, and the car completes the turn in 36.0 s. (a) What is the acceleration when the car is at *B* located at an angle of 35.0°? Express your answer in terms of the unit vectors $\hat{\mathbf{i}}$ and $\hat{\mathbf{j}}$. Determine (b) the car's average speed and (c) its average acceleration during the 36.0-s interval.

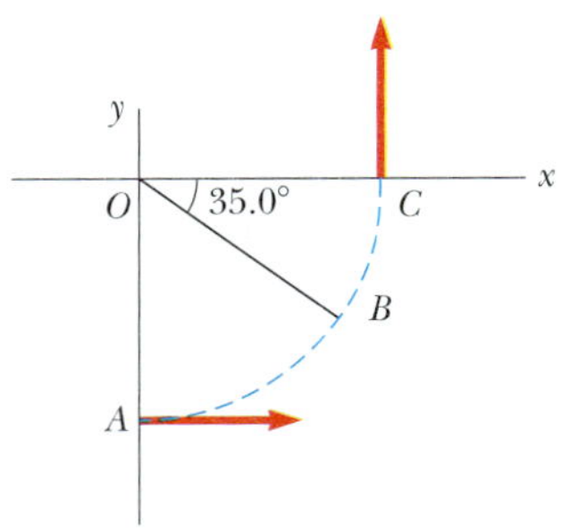

Figure P6.10

11. A 4.00-kg object is attached to a vertical rod by two strings as shown in Figure P6.11. The object rotates in a horizontal circle at constant speed 6.00 m/s. Find the tension in (a) the upper string and (b) the lower string.

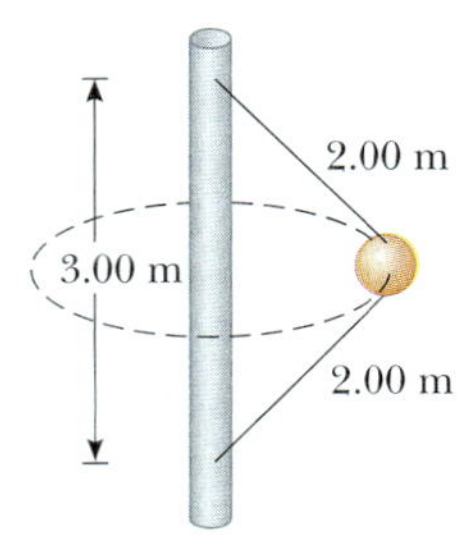

Figure P6.11

Section 6.2 Nonuniform Circular Motion

12. A hawk flies in a horizontal arc of radius 12.0 m at a constant speed of 4.00 m/s. (a) Find its centripetal acceleration. (b) It continues to fly along the same horizontal arc but increases its speed at the rate of 1.20 m/s^2. Find the acceleration (magnitude and direction) under these conditions.

13. A 40.0-kg child swings in a swing supported by two chains, each 3.00 m long. The tension in each chain at the lowest point is 350 N. Find (a) the child's speed at the lowest point and (b) the force exerted by the seat on the child at the lowest point. (Ignore the mass of the seat.)

14. A roller-coaster car (Fig. P6.14) has a mass of 500 kg when fully loaded with passengers. (a) If the vehicle has a speed of 20.0 m/s at point Ⓐ, what is the force exerted by the track on the car at this point? (b) What is the maximum speed the vehicle can have at point Ⓑ and still remain on the track?

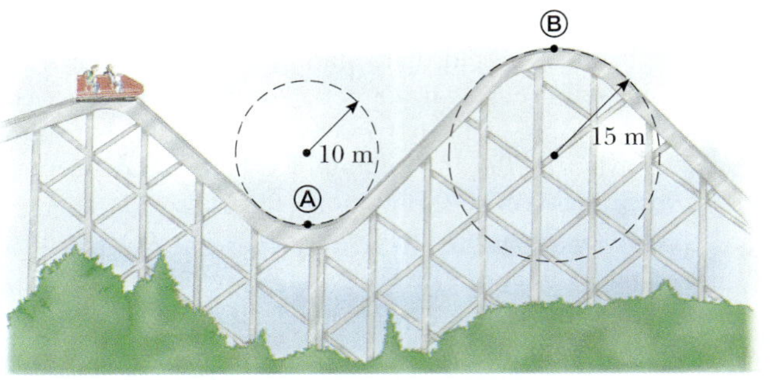

Figure P6.14

15. ▲ Tarzan (m = 85.0 kg) tries to cross a river by swinging on a vine. The vine is 10.0 m long, and his speed at the bottom of the swing (as he just clears the water) will be 8.00 m/s. Tarzan doesn't know that the vine has a breaking strength of 1 000 N. Does he make it across the river safely?

16. ● One end of a cord is fixed and a small 0.500-kg object is attached to the other end, where it swings in a section of a vertical circle of radius 2.00 m as shown in Figure 6.9. When $\theta = 20.0°$, the speed of the object is 8.00 m/s. At this instant, find (a) the tension in the string, (b) the tangential and radial components of acceleration, and (c) the total acceleration. (d) Is your answer changed if the object is swinging up instead of swinging down? Explain.

17. ▲ A pail of water is rotated in a vertical circle of radius 1.00 m. What is the pail's minimum speed at the top of the circle if no water is to spill out?

18. A roller coaster at Six Flags Great America amusement park in Gurnee, Illinois, incorporates some clever design technology and some basic physics. Each vertical loop, instead of being circular, is shaped like a teardrop (Fig. P6.18). The cars ride on the inside of the loop at the top, and the speeds are fast enough to ensure that the cars remain on the track. The biggest loop is 40.0 m high, with a maximum speed of 31.0 m/s (nearly 70 mi/h) at the bottom. Suppose the speed at the top is 13.0 m/s and the corresponding centripetal acceleration is 2g. (a) What is the radius of the arc of the teardrop at the top? (b) If the total mass of a car plus the riders is M, what force does

2 = intermediate; 3 = challenging; ☐ = SSM/SG; ▲ = ThomsonNow; = symbolic reasoning; ● = qualitative reasoning

the rail exert on the car at the top? (c) Suppose the roller coaster had a circular loop of radius 20.0 m. If the cars have the same speed, 13.0 m/s at the top, what is the centripetal acceleration at the top? Comment on the normal force at the top in this situation.

Frank Cezus/FPG International

Figure P6.18

Section 6.3 Motion in Accelerated Frames

19. ● An object of mass 5.00 kg, attached to a spring scale, rests on a frictionless, horizontal surface as shown in Figure P6.19. The spring scale, attached to the front end of a boxcar, has a constant reading of 18.0 N when the car is in motion. (a) The spring scale reads zero when the car is at rest. Determine the acceleration of the car. (b) What constant reading will the spring scale show if the car moves with constant velocity? (c) Describe the forces on the object as observed by someone in the car and by someone at rest outside the car.

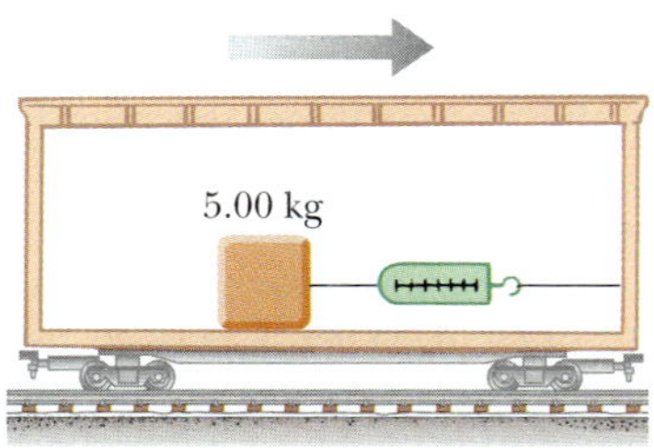

Figure P6.19

20. A small container of water is placed on a carousel inside a microwave oven at a radius of 12.0 cm from the center. The turntable rotates steadily, turning one revolution each 7.25 s. What angle does the water surface make with the horizontal?

21. A 0.500-kg object is suspended from the ceiling of an accelerating boxcar as shown in Figure 6.12. Taking $a = 3.00\ \text{m/s}^2$, find (a) the angle that the string makes with the vertical and (b) the tension in the string.

22. A student stands in an elevator that is continuously accelerating upward with acceleration a. Her backpack is sitting on the floor next to the wall. The width of the elevator car is L. The student gives her backpack a quick kick at $t = 0$, imparting to it speed v and making it slide across the elevator floor. At time t, the backpack hits the opposite wall. Find the coefficient of kinetic friction μ_k between the backpack and the elevator floor.

23. A person stands on a scale in an elevator. As the elevator starts, the scale has a constant reading of 591 N. Later, as the elevator stops, the scale reading is 391 N. Assume the magnitude of the acceleration is the same during starting and stopping. Determine (a) the weight of the person, (b) the person's mass, and (c) the acceleration of the elevator.

24. A child on vacation wakes up. She is lying on her back. The tension in the muscles on both sides of her neck is 55.0 N as she raises her head to look past her toes and out the motel window. Finally it is not raining! Ten minutes later she is screaming feet first down a water slide at terminal speed 5.70 m/s, riding high on the outside wall of a horizontal curve of radius 2.40 m (Fig. P6.24). She raises her head to look forward past her toes. Find the tension in the muscles on both sides of her neck.

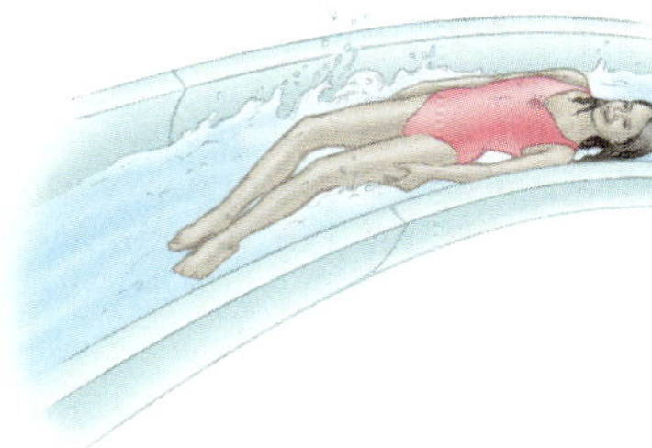

Figure P6.24

25. A plumb bob does not hang exactly along a line directed to the center of the Earth's rotation. How much does the plumb bob deviate from a radial line at 35.0° north latitude? Assume the Earth is spherical.

Section 6.4 Motion in the Presence of Resistive Forces

26. A skydiver of mass 80.0 kg jumps from a slow-moving aircraft and reaches a terminal speed of 50.0 m/s. (a) What is the acceleration of the skydiver when her speed is 30.0 m/s? What is the drag force on the skydiver when her speed is (b) 50.0 m/s? (c) When it is 30.0 m/s?

27. A small piece of Styrofoam packing material is dropped from a height of 2.00 m above the ground. Until it reaches terminal speed, the magnitude of its acceleration is given by $a = g - bv$. After falling 0.500 m, the Styrofoam effectively reaches terminal speed and then takes 5.00 s more to reach the ground. (a) What is the value of the constant b? (b) What is the acceleration at $t = 0$? (c) What is the acceleration when the speed is 0.150 m/s?

28. (a) Estimate the terminal speed of a wooden sphere (density 0.830 g/cm^3) falling through air, taking its radius as 8.00 cm and its drag coefficient as 0.500. (b) From what height would a freely falling object reach this speed in the absence of air resistance?

29. Calculate the force required to pull a copper ball of radius 2.00 cm upward through a fluid at the constant speed 9.00 cm/s. Take the drag force to be proportional to the speed, with proportionality constant 0.950 kg/s. Ignore the buoyant force.

30. The mass of a sports car is 1 200 kg. The shape of the body is such that the aerodynamic drag coefficient is 0.250 and the frontal area is 2.20 m^2. Ignoring all other sources of friction, calculate the initial acceleration the car has if it has been traveling at 100 km/h and is now shifted into neutral and allowed to coast.

31. A small, spherical bead of mass 3.00 g is released from rest at $t = 0$ in a bottle of liquid shampoo. The terminal speed is observed to be $v_T = 2.00$ cm/s. Find (a) the value of the constant b in Equation 6.2, (b) the time t at which the bead reaches $0.632v_T$, and (c) the value of the resistive force when the bead reaches terminal speed.

32. **Review problem.** An undercover police agent pulls a rubber squeegee down a very tall vertical window. The squeegee has mass 160 g and is mounted on the end of a light rod. The coefficient of kinetic friction between the squeegee and the dry glass is 0.900. The agent presses it against the window with a force having a horizontal component of 4.00 N. (a) If she pulls the squeegee down the window at constant velocity, what vertical force component must she exert? (b) The agent increases the downward force component by 25.0%, but all other forces remain the same. Find the acceleration of the squeegee in this situation. (c) The squeegee then moves into a wet portion of the window, where its motion is now resisted by a fluid drag force proportional to its velocity according to $R = -(20.0\ \text{N} \cdot \text{s/m})v$. Find the terminal velocity that the squeegee approaches, assuming the agent exerts the same force described in part (b).

33. A 9.00-kg object starting from rest falls through a viscous medium and experiences a resistive force $\vec{\mathbf{R}} = -b\vec{\mathbf{v}}$, where $\vec{\mathbf{v}}$ is the velocity of the object. The object reaches one-half of its terminal speed in 5.54 s. (a) Determine the terminal speed. (b) At what time is the speed of the object three-fourths of the terminal speed? (c) How far has the object traveled in the first 5.54 s of motion?

34. Consider an object on which the net force is a resistive force proportional to the square of its speed. For example, assume the resistive force acting on a speed skater is $f = -kmv^2$, where k is a constant and m is the skater's mass. The skater crosses the finish line of a straight-line race with speed v_0 and then slows down by coasting on his skates. Show that the skater's speed at any time t after crossing the finish line is $v(t) = v_0/(1 + ktv_0)$. This problem also provides the background for the next two problems.

35. (a) Use the result of Problem 34 to find the position x as a function of time for an object of mass m, located at $x = 0$ and moving with velocity $v_0\hat{\mathbf{i}}$ at time $t = 0$, and thereafter experiencing a net force $-kmv^2\hat{\mathbf{i}}$. (b) Find the object's velocity as a function of position.

36. At major league baseball games it is commonplace to flash on the scoreboard a speed for each pitch. This speed is determined with a radar gun aimed by an operator positioned behind home plate. The gun uses the Doppler shift of microwaves reflected from the baseball, as we will study in Chapter 39. The gun determines the speed at some particular point on the baseball's path, depending on when the operator pulls the trigger. Because the ball is subject to a drag force due to air, it slows as it travels 18.3 m toward the plate. Use the result of Problem 35(b) to find how much its speed decreases. Suppose the ball leaves the pitcher's hand at 90.0 mi/h = 40.2 m/s. Ignore its vertical motion. Use data on baseballs from Example 6.11 to determine the speed of the pitch when it crosses the plate.

37. ▲ The driver of a motorboat cuts its engine when its speed is 10.0 m/s and coasts to rest. The equation describing the motion of the motorboat during this period is $v = v_i e^{-ct}$, where v is the speed at time t, v_i is the initial speed, and c is a constant. At $t = 20.0$ s, the speed is 5.00 m/s. (a) Find the constant c. (b) What is the speed at $t = 40.0$ s? (c) Differentiate the expression for $v(t)$ and thus show that the acceleration of the boat is proportional to the speed at any time.

38. You can feel a force of air drag on your hand if you stretch your arm out of an open window of a rapidly moving car. *Note:* Do not endanger yourself. What is the order of magnitude of this force? In your solution, state the quantities you measure or estimate and their values.

Additional Problems

39. An object of mass m is projected forward along the x axis with initial speed v_0. The only force on it is a resistive force proportional to its velocity, given by $\vec{\mathbf{R}} = -b\vec{\mathbf{v}}$. For concreteness, you could visualize an airplane with pontoons landing on a lake. Newton's second law applied to the object is $-bv\hat{\mathbf{i}} = m(dv/dt)\hat{\mathbf{i}}$. From the fundamental theorem of calculus, this differential equation implies that the speed changes according to

$$\int_{\text{start}}^{\text{a later point}} \frac{dv}{v} = -\frac{b}{m}\int_0^t dt$$

Carry out the integrations to determine the speed of the object as a function of time. Sketch a graph of the speed as a function of time. Does the object come to a complete stop after a finite interval of time? Does the object travel a finite distance in stopping?

40. A 0.400-kg object is swung in a vertical circular path on a string 0.500 m long. If its speed is 4.00 m/s at the top of the circle, what is the tension in the string there?

41. (a) A luggage carousel at an airport has the form of a section of a large cone, steadily rotating about its vertical axis. Its metallic surface slopes downward toward the outside, making an angle of 20.0° with the horizontal. A piece of luggage having mass 30.0 kg is placed on the carousel, 7.46 m from the axis of rotation. The travel bag goes around once in 38.0 s. Calculate the force of static friction exerted by the carousel on the bag. (b) The drive motor is shifted to turn the carousel at a higher constant rate of rotation, and the piece of luggage is bumped to another position, 7.94 m from the axis of rotation. Now going around once in every 34.0 s, the bag is on the verge of slipping. Calculate the coefficient of static friction between the bag and the carousel.

42. In a home laundry dryer, a cylindrical tub containing wet clothes is rotated steadily about a horizontal axis as shown in Figure P6.42. So that the clothes will dry uniformly,

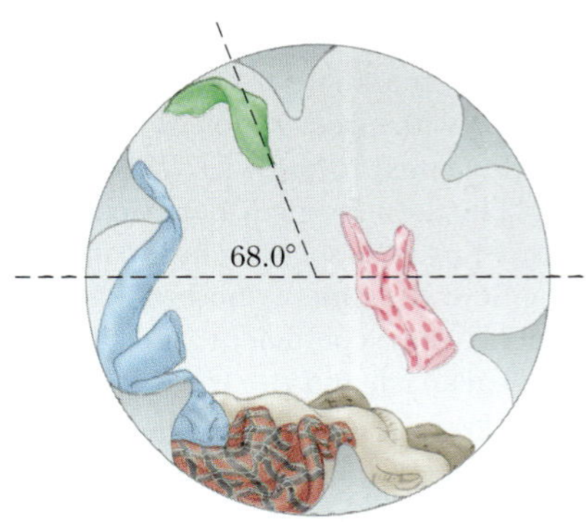

Figure P6.42

they are made to tumble. The rate of rotation of the smooth-walled tub is chosen so that a small piece of cloth will lose contact with the tub when the cloth is at an angle of 68.0° above the horizontal. If the radius of the tub is 0.330 m, what rate of revolution is needed?

43. We will study the most important work of Nobel laureate Arthur Compton in Chapter 40. Disturbed by speeding cars outside the physics building at Washington University in St. Louis, Compton designed a speed bump and had it installed. Suppose a 1 800-kg car passes over a bump in a roadway that follows the arc of a circle of radius 20.4 m as shown in Figure P6.43. (a) What force does the road exert on the car as the car passes the highest point of the bump if it travels at 30.0 km/h? (b) **What If?** What is the maximum speed the car can have as it passes this highest point without losing contact with the road?

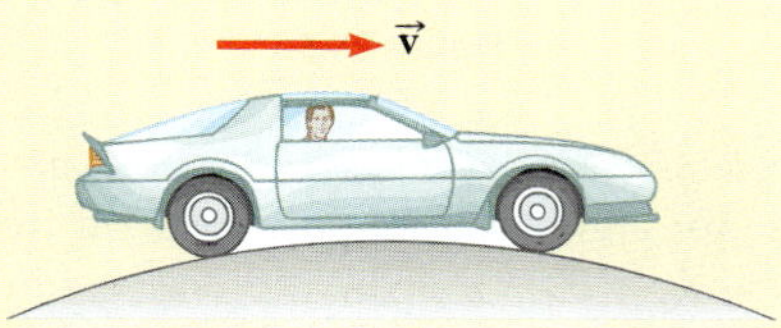

Figure P6.43 Problems 43 and 44.

44. A car of mass m passes over a bump in a road that follows the arc of a circle of radius R as shown in Figure P6.43. (a) What force does the road exert on the car as the car passes the highest point of the bump if it travels at a speed v? (b) **What If?** What is the maximum speed the car can have as it passes this highest point without losing contact with the road?

45. Interpret the graph in Figure 6.16(b). Proceed as follows. (a) Find the slope of the straight line, including its units. (b) From Equation 6.6, $R = \frac{1}{2} D\rho Av^2$, identify the theoretical slope of a graph of resistive force versus squared speed. (c) Set the experimental and theoretical slopes equal to each other and proceed to calculate the drag coefficient of the filters. Use the value for the density of air listed on the book's endpapers. Model the cross-sectional area of the filters as that of a circle of radius 10.5 cm. (d) Arbitrarily choose the eighth data point on the graph and find its vertical separation from the line of best fit. Express this scatter as a percentage. (e) In a short paragraph, state what the graph demonstrates and compare what it demonstrates to the theoretical prediction. You will need to make reference to the quantities plotted on the axes, to the shape of the graph line, to the data points, and to the results of parts (c) and (d).

46. ● A basin surrounding a drain has the shape of a circular cone opening upward, having everywhere an angle of 35.0° with the horizontal. A 25.0-g ice cube is set sliding around the cone without friction in a horizontal circle of radius R. (a) Find the speed the ice cube must have as it depends on R. (b) Is any piece of data unnecessary for the solution? Suppose R is made two times larger. (c) Will the required speed increase, decrease, or stay constant? If it changes, by what factor? (d) Will the time required for each revolution increase, decrease, or stay constant? If it changes, by what factor? (e) Do the answers to parts (c) and (d) seem contradictory? Explain how they are consistent.

47. Suppose the boxcar of Figure 6.12 is moving with constant acceleration a up a hill that makes an angle ϕ with the horizontal. If the pendulum makes a constant angle θ with the perpendicular to the ceiling, what is a?

48. The pilot of an airplane executes a constant-speed loop-the-loop maneuver in a vertical circle. The speed of the airplane is 300 mi/h; the radius of the circle is 1 200 ft. (a) What is the pilot's apparent weight at the lowest point if his true weight is 160 lb? (b) What is his apparent weight at the highest point? (c) **What If?** Describe how the pilot could experience weightlessness if both the radius and the speed can be varied. *Note:* His apparent weight is equal to the magnitude of the force exerted by the seat on his body.

49. ▲ Because the Earth rotates about its axis, a point on the equator experiences a centripetal acceleration of 0.033 7 m/s², whereas a point at the poles experiences no centripetal acceleration. (a) Show that at the equator the gravitational force on an object must exceed the normal force required to support the object. That is, show that the object's true weight exceeds its apparent weight. (b) What is the apparent weight at the equator and at the poles of a person having a mass of 75.0 kg? Assume the Earth is a uniform sphere and take $g = 9.800$ m/s².

50. ● An air puck of mass m_1 is tied to a string and allowed to revolve in a circle of radius R on a frictionless horizontal table. The other end of the string passes through a small hole in the center of the table, and a load of mass m_2 is tied to the string (Fig. P6.50). The suspended load remains in equilibrium while the puck on the tabletop revolves. What are (a) the tension in the string, (b) the radial force acting on the puck, and (c) the speed of the puck? (d) Qualitatively describe what will happen in the motion of the puck if the value of m_2 is somewhat increased by placing an additional load on it. (e) Qualitatively describe what will happen in the motion of the puck if the value of m_2 is instead decreased by removing a part of the hanging load.

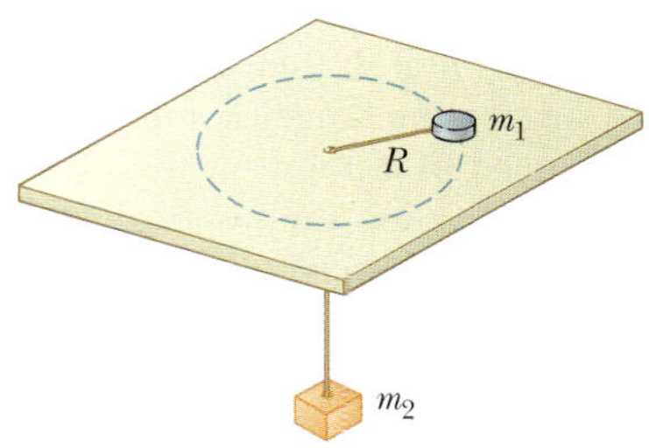

Figure P6.50

51. ● While learning to drive, you are in a 1 200-kg car moving at 20.0 m/s across a large, vacant, level parking lot. Suddenly you realize you are heading straight toward a brick sidewall of a large supermarket and are in danger of running into it. The pavement can exert a maximum horizontal force of 7 000 N on the car. (a) Explain why you should expect the force to have a well-defined maximum value. (b) Suppose you apply the brakes and do not turn the steering wheel. Find the minimum distance you must be from the wall to avoid a collision. (c) If you do not brake but instead maintain constant speed and turn the steering wheel, what is the minimum distance you must be from the wall to avoid a collision? (d) Which method,

(b) or (c), is better for avoiding a collision? Or, should you use both the brakes and the steering wheel, or neither? Explain. (e) Does the conclusion in part (d) depend on the numerical values given in this problem, or is it true in general? Explain.

52. Suppose a Ferris wheel rotates four times each minute. It carries each car around a circle of diameter 18.0 m. (a) What is the centripetal acceleration of a rider? What force does the seat exert on a 40.0-kg rider (b) at the lowest point of the ride and (c) at the highest point of the ride? (d) What force (magnitude and direction) does the seat exert on a rider when the rider is halfway between top and bottom?

53. An amusement park ride consists of a rotating circular platform 8.00 m in diameter from which 10.0-kg seats are suspended at the end of 2.50-m massless chains (Fig. P6.53). When the system rotates, the chains make an angle $\theta = 28.0°$ with the vertical. (a) What is the speed of each seat? (b) Draw a free-body diagram of a 40.0-kg child riding in a seat and find the tension in the chain.

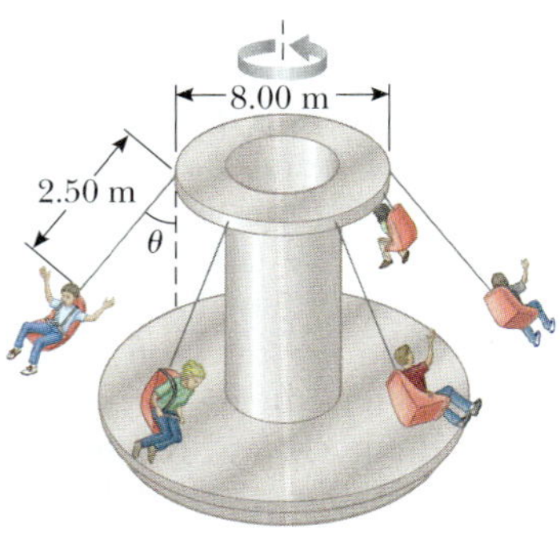

Figure P6.53

54. A piece of putty is initially located at point *A* on the rim of a grinding wheel rotating about a horizontal axis. The putty is dislodged from point *A* when the diameter through *A* is horizontal. It then rises vertically and returns to *A* at the instant the wheel completes one revolution. (a) Find the speed of a point on the rim of the wheel in terms of the acceleration due to gravity and the radius *R* of the wheel. (b) If the mass of the putty is *m*, what is the magnitude of the force that held it to the wheel?

55. ● An amusement park ride consists of a large vertical cylinder that spins about its axis fast enough that any person inside is held up against the wall when the floor drops away (Fig. P6.55). The coefficient of static friction between person and wall is μ_s, and the radius of the cylinder is *R*. (a) Show that the maximum period of revolution necessary to keep the person from falling is $T = (4\pi^2 R\mu_s/g)^{1/2}$. (b) Obtain a numerical value for *T*, taking $R = 4.00$ m and $\mu_s = 0.400$. How many revolutions per minute does the cylinder make? (c) If the rate of revolution of the cylinder is made to be somewhat larger, what happens to the magnitude of each one of the forces acting on the person? What happens in the motion of the person? (d) If instead the cylinder's rate of revolution is made to be somewhat smaller, what happens to the magnitude of each one of the forces acting on the person? What happens in the motion of the person?

Figure P6.55

56. *An example of the Coriolis effect.* Suppose air resistance is negligible for a golf ball. A golfer tees off from a location precisely at $\phi_i = 35.0°$ north latitude. He hits the ball due south, with range 285 m. The ball's initial velocity is at 48.0° above the horizontal. (a) For how long is the ball in flight? The cup is due south of the golfer's location, and he would have a hole in one if the Earth were not rotating. The Earth's rotation makes the tee move in a circle of radius $R_E \cos\phi_i = (6.37 \times 10^6 \text{ m}) \cos 35.0°$ as shown in Figure P6.56. The tee completes one revolution each day. (b) Find the eastward speed of the tee, relative to the stars. The hole is also moving east, but it is 285 m farther south and therefore at a slightly lower latitude ϕ_f. Because the hole moves in a slightly larger circle, its speed must be greater than that of the tee. (c) By how much does the hole's speed exceed that of the tee? During the time interval the ball is in flight, it moves upward and downward as well as southward with the projectile motion you studied in Chapter 4, but it also moves eastward with the speed you found in part (b). The hole moves to the east at a faster speed, however, pulling ahead of the ball with the relative speed you found in part (c). (d) How far to the west of the hole does the ball land?

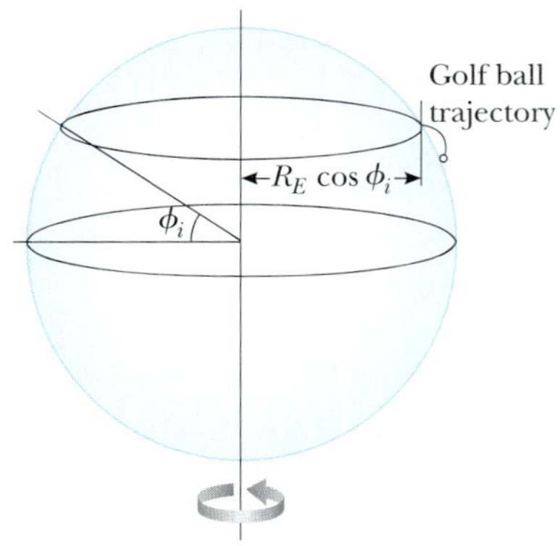

Figure P6.56

57. A car rounds a banked curve as shown in Figure 6.5. The radius of curvature of the road is R, the banking angle is θ, and the coefficient of static friction is μ_s. (a) Determine the range of speeds the car can have without slipping up or down the bank. (b) Find the minimum value for μ_s such that the minimum speed is zero. (c) What is the range of speeds possible if $R = 100$ m, $\theta = 10.0°$, and $\mu_s = 0.100$ (slippery conditions)?

58. ● A single bead can slide with negligible friction on a stiff wire that has been bent into a circular loop of radius 15.0 cm as shown in Figure P6.58. The circle is always in a vertical plane and rotates steadily about its vertical diameter with (a) a period of 0.450 s. The position of the bead is described by the angle θ that the radial line, from the center of the loop to the bead, makes with the vertical. At what angle up from the bottom of the circle can the bead stay motionless relative to the turning circle? (b) **What If?** Repeat the problem, taking the period of the circle's rotation as 0.850 s. (c) Describe how the solution to part (b) is fundamentally different from the solution to part (a). For any period or loop size, is there always an angle at which the bead can stand still relative to the loop? Are there ever more than two angles? Arnold Arons suggested the idea for this problem.

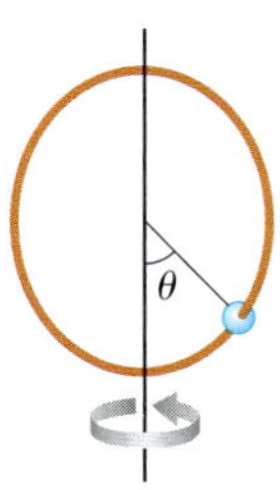

Figure P6.58

59. The expression $F = arv + br^2v^2$ gives the magnitude of the resistive force (in newtons) exerted on a sphere of radius r (in meters) by a stream of air moving at speed v (in meters per second), where a and b are constants with appropriate SI units. Their numerical values are $a = 3.10 \times 10^{-4}$ and $b = 0.870$. Using this expression, find the terminal speed for water droplets falling under their own weight in air, taking the following values for the drop radii: (a) 10.0 μm, (b) 100 μm, (c) 1.00 mm. For (a) and (c), you can obtain accurate answers without solving a quadratic equation by considering which of the two contributions to the air resistance is dominant and ignoring the lesser contribution.

60. Members of a skydiving club were given the following data to use in planning their jumps. In the table, d is the distance fallen from rest by a skydiver in a "free-fall stable spread position" versus the time of fall t. (a) Convert the distances in feet into meters. (b) Graph d (in meters) versus t. (c) Determine the value of the terminal speed v_T by finding the slope of the straight portion of the curve. Use a least-squares fit to determine this slope.

t (s)	d (ft)	t (s)	d (ft)	t (s)	d (ft)
0	0	7	652	14	1 831
1	16	8	808	15	2 005
2	62	9	971	16	2 179
3	138	10	1 138	17	2 353
4	242	11	1 309	18	2 527
5	366	12	1 483	19	2 701
6	504	13	1 657	20	2 875

61. A model airplane of mass 0.750 kg flies with a speed of 35.0 m/s in a horizontal circle at the end of a 60.0-m control wire. Compute the tension in the wire, assuming it makes a constant angle of 20.0° with the horizontal. The forces exerted on the airplane are the pull of the control wire, the gravitational force, and aerodynamic lift that acts at 20.0° inward from the vertical as shown in Figure P6.61.

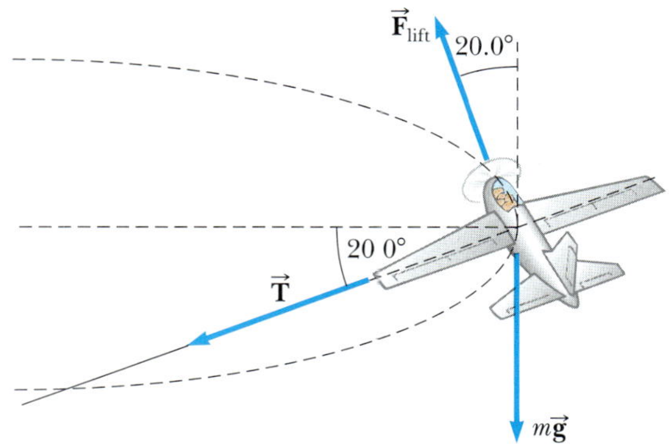

Figure P6.61

62. ● Galileo thought about whether acceleration should be defined as the rate of change of velocity over time or as the rate of change in velocity over distance. He chose the former, so let us use the name "vroomosity" for the rate of change of velocity in space. For motion of a particle on a straight line with constant acceleration, the equation $v = v_i + at$ gives its velocity v as a function of time. Similarly, for a particle's linear motion with constant vroomosity k, the equation $v = v_i + kx$ gives the velocity as a function of the position x if the particle's speed is v_i at $x = 0$. (a) Find the law describing the total force acting on this object, of mass m. Describe an example of such a motion, or explain why such a motion is unrealistic. Consider (b) the possibility of k positive and also (c) the possibility of k negative.

Answers to Quick Quizzes

6.1 **(i)**, (a). The normal force is always perpendicular to the surface that applies the force. Because your car maintains its orientation at all points on the ride, the normal force is always upward. **(ii)**, (b). Your centripetal acceleration is downward toward the center of the circle, so the net force on you must be downward.

6.2 (a) Because the speed is constant, the only direction the force can have is that of the centripetal acceleration. The force is larger at Ⓒ than at Ⓐ because the radius at Ⓒ is smaller. There is no force at Ⓑ because the wire is straight. (b) In addition to the forces in the centripetal direction in (a), there are now tangential forces to provide the tangential acceleration. The tangential force is the same at all three points because the tangential acceleration is constant.

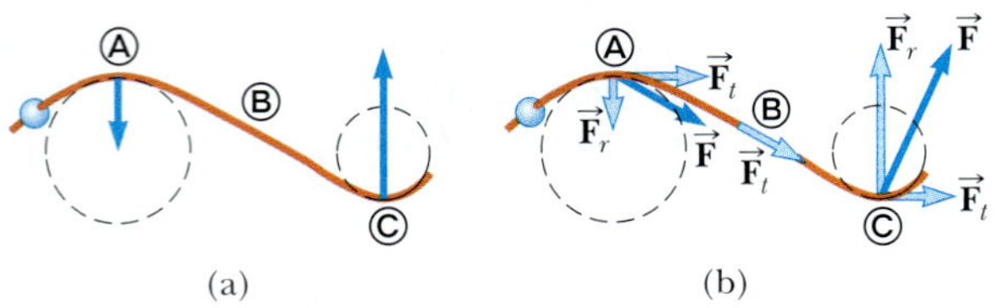

QQA 6.2

6.3 (c). The only forces acting on the passenger are the contact force with the door and the friction force from the seat. Both are real forces and both act to the left in Figure 6.10. Fictitious forces should never be drawn in a force diagram.

6.4. (a). The basketball, having a larger cross-sectional area, will have a larger force due to air resistance than the baseball, which will result in a smaller downward acceleration.

On a wind farm, the moving air does work on the blades of the windmills, causing the blades and the rotor of an electrical generator to rotate. Energy is transferred out of the system of the windmill by means of electricity. (Billy Hustace/Getty Images)

7 Energy of a System

The definitions of quantities such as position, velocity, acceleration, and force and associated principles such as Newton's second law have allowed us to solve a variety of problems. Some problems that could theoretically be solved with Newton's laws, however, are very difficult in practice, but they can be made much simpler with a different approach. Here and in the following chapters, we will investigate this new approach, which will include definitions of quantities that may not be familiar to you. Other quantities may sound familiar, but they may have more specific meanings in physics than in everyday life. We begin this discussion by exploring the notion of *energy*.

The concept of energy is one of the most important topics in science and engineering. In everyday life, we think of energy in terms of fuel for transportation and heating, electricity for lights and appliances, and foods for consumption. These ideas, however, do not truly define energy. They merely tell us that fuels are needed to do a job and that those fuels provide us with something we call energy.

Energy is present in the Universe in various forms. *Every* physical process that occurs in the Universe involves energy and energy transfers or transformations. Unfortunately, despite its extreme importance, energy cannot be easily defined. The variables in previous chapters were relatively concrete; we have everyday experience with velocities and forces, for example. Although we have *experiences* with energy, such as running out of gasoline or losing our electrical service following a violent storm, the *notion* of energy is more abstract.

The concept of energy can be applied to mechanical systems without resorting to Newton's laws. Furthermore, the energy approach allows us to understand thermal and electrical phenomena, for which Newton's laws are of no help, in later chapters of the book.

Our problem-solving techniques presented in earlier chapters were based on the motion of a particle or an object that could be modeled as a particle. These techniques used the *particle model.* We begin our new approach by focusing our attention on a *system* and developing techniques to be used in a *system model.*

7.1 Systems and Environments

PITFALL PREVENTION 7.1
Identify the System

The most important *first* step to take in solving a problem using the energy approach is to identify the appropriate system of interest.

In the system model, we focus our attention on a small portion of the Universe—the **system**—and ignore details of the rest of the Universe outside of the system. A critical skill in applying the system model to problems is *identifying the system.* A valid system

- may be a single object or particle
- may be a collection of objects or particles
- may be a region of space (such as the interior of an automobile engine combustion cylinder)
- may vary in size and shape (such as a rubber ball, which deforms upon striking a wall)

Identifying the need for a system approach to solving a problem (as opposed to a particle approach) is part of the Categorize step in the General Problem-Solving Strategy outlined in Chapter 2. Identifying the particular system is a second part of this step.

No matter what the particular system is in a given problem, we identify a **system boundary**, an imaginary surface (not necessarily coinciding with a physical surface) that divides the Universe into the system and the **environment** surrounding the system.

As an example, imagine a force applied to an object in empty space. We can define the object as the system and its surface as the system boundary. The force applied to it is an influence on the system from the environment that acts across the system boundary. We will see how to analyze this situation from a system approach in a subsequent section of this chapter.

Another example was seen in Example 5.10, where the system can be defined as the combination of the ball, the block, and the cord. The influence from the environment includes the gravitational forces on the ball and the block, the normal and friction forces on the block, and the force exerted by the pulley on the cord. The forces exerted by the cord on the ball and the block are internal to the system and therefore are not included as an influence from the environment.

There are a number of mechanisms by which a system can be influenced by its environment. The first one we shall investigate is *work.*

7.2 Work Done by a Constant Force

Almost all the terms we have used thus far—velocity, acceleration, force, and so on—convey a similar meaning in physics as they do in everyday life. Now, however, we encounter a term whose meaning in physics is distinctly different from its everyday meaning: *work.*

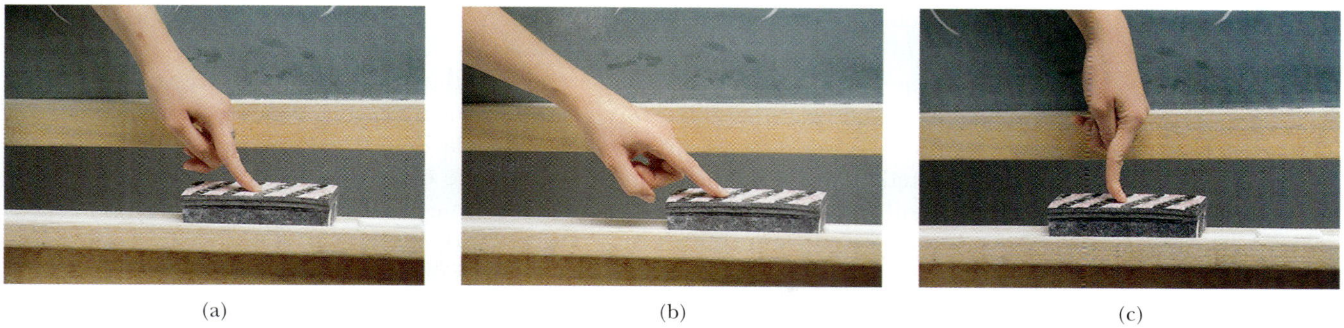

Figure 7.1 An eraser being pushed along a chalkboard tray by a force acting at different angles with respect to the horizontal direction.

To understand what work means to the physicist, consider the situation illustrated in Figure 7.1. A force $\vec{\mathbf{F}}$ is applied to a chalkboard eraser, which we identify as the system, and the eraser slides along the tray. If we want to know how effective the force is in moving the eraser, we must consider not only the magnitude of the force but also its direction. Assuming the magnitude of the applied force is the same in all three photographs, the push applied in Figure 7.1b does more to move the eraser than the push in Figure 7.1a. On the other hand, Figure 7.1c shows a situation in which the applied force does not move the eraser at all, regardless of how hard it is pushed (unless, of course, we apply a force so great that we break the chalkboard tray!). These results suggest that when analyzing forces to determine the work they do, we must consider the vector nature of forces. We must also know the displacement $\Delta\vec{\mathbf{r}}$ of the eraser as it moves along the tray if we want to determine the work done on it by the force. Moving the eraser 3 m along the tray requires more work than moving it 2 cm.

Let us examine the situation in Figure 7.2, where the object (the system) undergoes a displacement along a straight line while acted on by a constant force of magnitude F that makes an angle θ with the direction of the displacement.

PITFALL PREVENTION 7.2

What Is Being Displaced?

The displacement in Equation 7.1 is that of *the point of application of the force.* If the force is applied to a particle or a nondeformable system, this displacement is the same as the displacement of the particle or system. For deformable systems, however, these two displacements are often not the same.

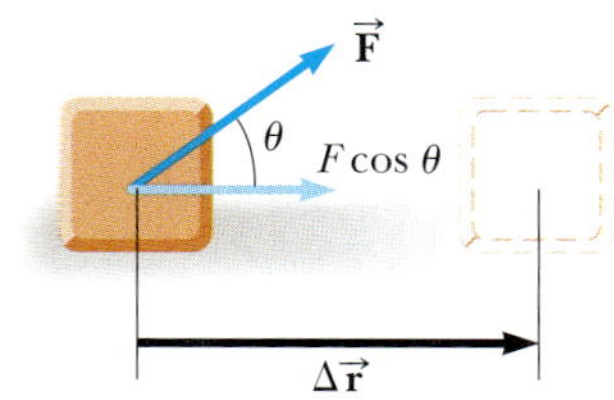

Figure 7.2 If an object undergoes a displacement $\Delta\vec{\mathbf{r}}$ under the action of a constant force $\vec{\mathbf{F}}$, the work done by the force is $F\,\Delta r\cos\theta$.

The **work** W done on a system by an agent exerting a constant force on the system is the product of the magnitude F of the force, the magnitude Δr of the displacement of the point of application of the force, and $\cos\theta$, where θ is the angle between the force and displacement vectors:

$$W \equiv F\,\Delta r\cos\theta \tag{7.1}$$

◀ Work done by a constant force

Notice in Equation 7.1 that work is a scalar, even though it is defined in terms of two vectors, a force $\vec{\mathbf{F}}$ and a displacement $\Delta\vec{\mathbf{r}}$. In Section 7.3, we explore how to combine two vectors to generate a scalar quantity.

As an example of the distinction between the definition of work and our everyday understanding of the word, consider holding a heavy chair at arm's length for 3 min. At the end of this time interval, your tired arms may lead you to think you have done a considerable amount of work on the chair. According to our definition, however, you have done no work on it whatsoever. You exert a force to support the chair, but you do not move it. A force does no work on an object if the force does not move through a displacement. If $\Delta r = 0$, Equation 7.1 gives $W = 0$, which is the situation depicted in Figure 7.1c.

Also notice from Equation 7.1 that the work done by a force on a moving object is zero when the force applied is perpendicular to the displacement of its point of application. That is, if $\theta = 90°$, then $W = 0$ because $\cos 90° = 0$. For example, in Figure 7.3, the work done by the normal force on the object and the work done by the gravitational force on the object are both zero because both forces are perpen-

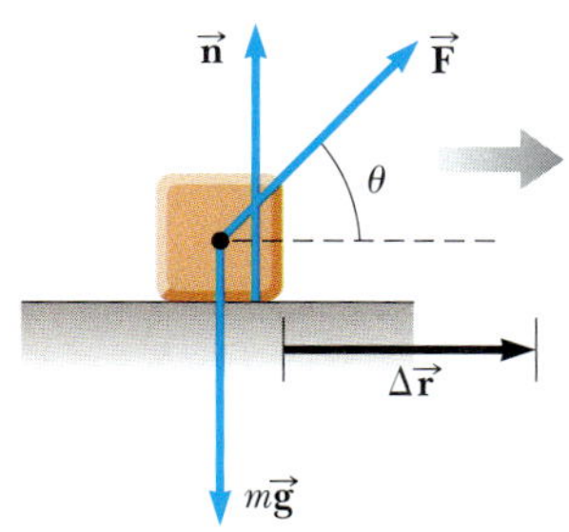

Figure 7.3 An object is displaced on a frictionless, horizontal surface. The normal force $\vec{\mathbf{n}}$ and the gravitational force $m\vec{\mathbf{g}}$ do no work on the object. In the situation shown here, $\vec{\mathbf{F}}$ is the only force doing work on the object.

PITFALL PREVENTION 7.3

Work Is Done by . . . on . . .

Not only must you identify the system, you must also identify what agent in the environment is doing work on the system. When discussing work, always use the phrase, "the work done by . . . on. . . ." After "by," insert the part of the environment that is interacting directly with the system. After "on," insert the system. For example, "the work done by the hammer on the nail" identifies the nail as the system and the force from the hammer represents the interaction with the environment.

PITFALL PREVENTION 7.4

Cause of the Displacement

We can calculate the work done by a force on an object, but that force is *not* necessarily the cause of the object's displacement. For example, if you lift an object, work is done on the object by the gravitational force, although gravity is not the cause of the object moving upward!

dicular to the displacement and have zero components along an axis in the direction of $\Delta\vec{\mathbf{r}}$.

The sign of the work also depends on the direction of $\vec{\mathbf{F}}$ relative to $\Delta\vec{\mathbf{r}}$. The work done by the applied force on a system is positive when the projection of $\vec{\mathbf{F}}$ onto $\Delta\vec{\mathbf{r}}$ is in the same direction as the displacement. For example, when an object is lifted, the work done by the applied force on the object is positive because the direction of that force is upward, in the same direction as the displacement of its point of application. When the projection of $\vec{\mathbf{F}}$ onto $\Delta\vec{\mathbf{r}}$ is in the direction opposite the displacement, W is negative. For example, as an object is lifted, the work done by the gravitational force on the object is negative. The factor $\cos\theta$ in the definition of W (Eq. 7.1) automatically takes care of the sign.

If an applied force $\vec{\mathbf{F}}$ is in the same direction as the displacement $\Delta\vec{\mathbf{r}}$, then $\theta = 0$ and $\cos 0 = 1$. In this case, Equation 7.1 gives

$$W = F\,\Delta r$$

The units of work are those of force multiplied by those of length. Therefore, the SI unit of work is the **newton·meter** ($\text{N}\cdot\text{m} = \text{kg}\cdot\text{m}^2/\text{s}^2$). This combination of units is used so frequently that it has been given a name of its own, the **joule** (J).

An important consideration for a system approach to problems is that **work is an energy transfer**. If W is the work done on a system and W is positive, energy is transferred *to* the system; if W is negative, energy is transferred *from* the system. Therefore, if a system interacts with its environment, this interaction can be described as a transfer of energy across the system boundary. The result is a change in the energy stored in the system. We will learn about the first type of energy storage in Section 7.5, after we investigate more aspects of work.

Quick Quiz 7.1 The gravitational force exerted by the Sun on the Earth holds the Earth in an orbit around the Sun. Let us assume that the orbit is perfectly circular. The work done by this gravitational force during a short time interval in which the Earth moves through a displacement in its orbital path is (a) zero (b) positive (c) negative (d) impossible to determine

Quick Quiz 7.2 Figure 7.4 shows four situations in which a force is applied to an object. In all four cases, the force has the same magnitude, and the displacement of the object is to the right and of the same magnitude. Rank the situations in order of the work done by the force on the object, from most positive to most negative.

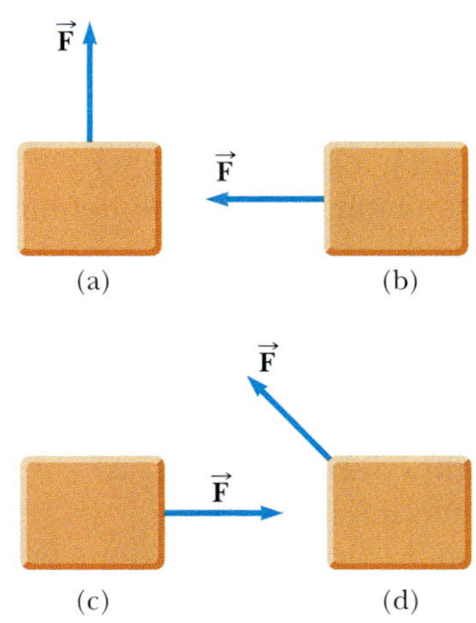

Figure 7.4 (Quick Quiz 7.2) A block is pulled by a force in four different directions. In each case, the displacement of the block is to the right and of the same magnitude.

EXAMPLE 7.1 Mr. Clean

A man cleaning a floor pulls a vacuum cleaner with a force of magnitude F = 50.0 N at an angle of 30.0° with the horizontal (Fig. 7.5). Calculate the work done by the force on the vacuum cleaner as the vacuum cleaner is displaced 3.00 m to the right.

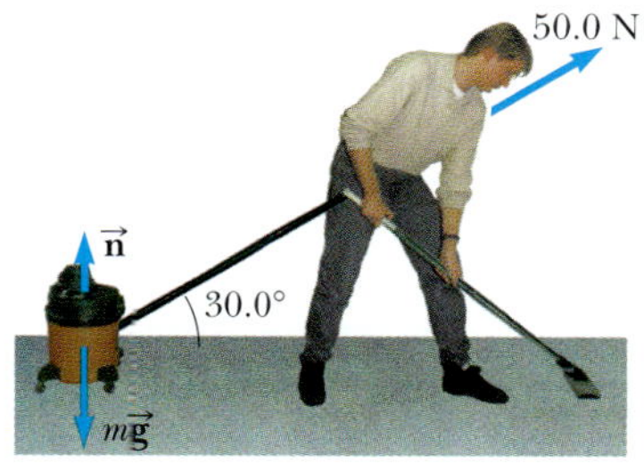

Figure 7.5 (Example 7.1) A vacuum cleaner being pulled at an angle of 30.0° from the horizontal.

SOLUTION

Conceptualize Figure 7.5 helps conceptualize the situation. Think about an experience in your life in which you pulled an object across the floor with a rope or cord.

Categorize We are given a force on an object, a displacement of the object, and the angle between the two vectors, so we categorize this example as a substitution problem. We identify the vacuum cleaner as the system.

Use the definition of work (Eq. 7.1):

$$W = F\,\Delta r\cos\theta = (50.0\text{ N})(3.00\text{ m})(\cos 30.0°)$$
$$= 130\text{ J}$$

Notice in this situation that the normal force $\vec{\mathbf{n}}$ and the gravitational $\vec{\mathbf{F}}_g = m\vec{\mathbf{g}}$ do no work on the vacuum cleaner because these forces are perpendicular to its displacement.

7.3 The Scalar Product of Two Vectors

Because of the way the force and displacement vectors are combined in Equation 7.1, it is helpful to use a convenient mathematical tool called the **scalar product** of two vectors. We write this scalar product of vectors $\vec{\mathbf{A}}$ and $\vec{\mathbf{B}}$ as $\vec{\mathbf{A}}\cdot\vec{\mathbf{B}}$. (Because of the dot symbol, the scalar product is often called the **dot product**.)

The scalar product of any two vectors $\vec{\mathbf{A}}$ and $\vec{\mathbf{B}}$ is a scalar quantity equal to the product of the magnitudes of the two vectors and the cosine of the angle θ between them:

$$\vec{\mathbf{A}}\cdot\vec{\mathbf{B}} \equiv AB\cos\theta \quad (7.2)$$

◀ Scalar product of any two vectors $\vec{\mathbf{A}}$ and $\vec{\mathbf{B}}$

As is the case with any multiplication, $\vec{\mathbf{A}}$ and $\vec{\mathbf{B}}$ need not have the same units.

By comparing this definition with Equation 7.1, we can express Equation 7.1 as a scalar product:

$$W = F\,\Delta r\cos\theta = \vec{\mathbf{F}}\cdot\Delta\vec{\mathbf{r}} \quad (7.3)$$

In other words, $\vec{\mathbf{F}}\cdot\Delta\vec{\mathbf{r}}$ is a shorthand notation for $F\,\Delta r\cos\theta$.

Before continuing with our discussion of work, let us investigate some properties of the dot product. Figure 7.6 shows two vectors $\vec{\mathbf{A}}$ and $\vec{\mathbf{B}}$ and the angle θ between them used in the definition of the dot product. In Figure 7.6, $B\cos\theta$ is the projection of $\vec{\mathbf{B}}$ onto $\vec{\mathbf{A}}$. Therefore, Equation 7.2 means that $\vec{\mathbf{A}}\cdot\vec{\mathbf{B}}$ is the product of the magnitude of $\vec{\mathbf{A}}$ and the projection of $\vec{\mathbf{B}}$ onto $\vec{\mathbf{A}}$.[1]

From the right-hand side of Equation 7.2, we also see that the scalar product is **commutative**.[2] That is,

$$\vec{\mathbf{A}}\cdot\vec{\mathbf{B}} = \vec{\mathbf{B}}\cdot\vec{\mathbf{A}}$$

Finally, the scalar product obeys the **distributive law of multiplication**, so

$$\vec{\mathbf{A}}\cdot(\vec{\mathbf{B}}+\vec{\mathbf{C}}) = \vec{\mathbf{A}}\cdot\vec{\mathbf{B}} + \vec{\mathbf{A}}\cdot\vec{\mathbf{C}}$$

PITFALL PREVENTION 7.5
Work Is a Scalar

Although Equation 7.3 defines the work in terms of two vectors, *work is a scalar*; there is no direction associated with it. *All* types of energy and energy transfer are scalars. This fact is a major advantage of the energy approach because we don't need vector calculations!

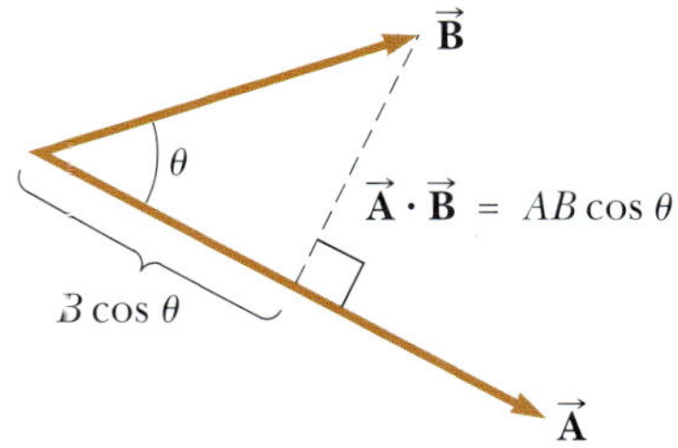

Figure 7.6 The scalar product $\vec{\mathbf{A}}\cdot\vec{\mathbf{B}}$ equals the magnitude of $\vec{\mathbf{A}}$ multiplied by $B\cos\theta$, which is the projection of $\vec{\mathbf{B}}$ onto $\vec{\mathbf{A}}$.

[1] This statement is equivalent to stating that $\vec{\mathbf{A}}\cdot\vec{\mathbf{B}}$ equals the product of the magnitude of $\vec{\mathbf{B}}$ and the projection of $\vec{\mathbf{A}}$ onto $\vec{\mathbf{B}}$.

[2] In Chapter 11, you will see another way of combining vectors that proves useful in physics and is not commutative.

The dot product is simple to evaluate from Equation 7.2 when $\vec{\mathbf{A}}$ is either perpendicular or parallel to $\vec{\mathbf{B}}$. If $\vec{\mathbf{A}}$ is perpendicular to $\vec{\mathbf{B}}$ ($\theta = 90°$), then $\vec{\mathbf{A}} \cdot \vec{\mathbf{B}} = 0$. (The equality $\vec{\mathbf{A}} \cdot \vec{\mathbf{B}} = 0$ also holds in the more trivial case in which either $\vec{\mathbf{A}}$ or $\vec{\mathbf{B}}$ is zero.) If vector $\vec{\mathbf{A}}$ is parallel to vector $\vec{\mathbf{B}}$ and the two point in the same direction ($\theta = 0$), then $\vec{\mathbf{A}} \cdot \vec{\mathbf{B}} = AB$. If vector $\vec{\mathbf{A}}$ is parallel to vector $\vec{\mathbf{B}}$ but the two point in opposite directions ($\theta = 180°$), then $\vec{\mathbf{A}} \cdot \vec{\mathbf{B}} = -AB$. The scalar product is negative when $90° < \theta \leq 180°$.

The unit vectors $\hat{\mathbf{i}}$, $\hat{\mathbf{j}}$, and $\hat{\mathbf{k}}$, which were defined in Chapter 3, lie in the positive x, y, and z directions, respectively, of a right-handed coordinate system. Therefore, it follows from the definition of $\vec{\mathbf{A}} \cdot \vec{\mathbf{B}}$ that the scalar products of these unit vectors are

Dot products of unit vectors ▶

$$\hat{\mathbf{i}} \cdot \hat{\mathbf{i}} = \hat{\mathbf{j}} \cdot \hat{\mathbf{j}} = \hat{\mathbf{k}} \cdot \hat{\mathbf{k}} = 1 \quad (7.4)$$

$$\hat{\mathbf{i}} \cdot \hat{\mathbf{j}} = \hat{\mathbf{i}} \cdot \hat{\mathbf{k}} = \hat{\mathbf{j}} \cdot \hat{\mathbf{k}} = 0 \quad (7.5)$$

Equations 3.18 and 3.19 state that two vectors $\vec{\mathbf{A}}$ and $\vec{\mathbf{B}}$ can be expressed in unit-vector form as

$$\vec{\mathbf{A}} = A_x\hat{\mathbf{i}} + A_y\hat{\mathbf{j}} + A_z\hat{\mathbf{k}}$$

$$\vec{\mathbf{B}} = B_x\hat{\mathbf{i}} + B_y\hat{\mathbf{j}} + B_z\hat{\mathbf{k}}$$

Using the information given in Equations 7.4 and 7.5 shows that the scalar product of $\vec{\mathbf{A}}$ and $\vec{\mathbf{B}}$ reduces to

$$\vec{\mathbf{A}} \cdot \vec{\mathbf{B}} = A_xB_x + A_yB_y + A_zB_z \quad (7.6)$$

(Details of the derivation are left for you in Problem 5 at the end of the chapter.) In the special case in which $\vec{\mathbf{A}} = \vec{\mathbf{B}}$, we see that

$$\vec{\mathbf{A}} \cdot \vec{\mathbf{A}} = A_x^2 + A_y^2 + A_z^2 = A^2$$

Quick Quiz 7.3 Which of the following statements is true about the relationship between the dot product of two vectors and the product of the magnitudes of the vectors? (a) $\vec{\mathbf{A}} \cdot \vec{\mathbf{B}}$ is larger than AB. (b) $\vec{\mathbf{A}} \cdot \vec{\mathbf{B}}$ is smaller than AB. (c) $\vec{\mathbf{A}} \cdot \vec{\mathbf{B}}$ could be larger or smaller than AB, depending on the angle between the vectors. (d) $\vec{\mathbf{A}} \cdot \vec{\mathbf{B}}$ could be equal to AB.

EXAMPLE 7.2 The Scalar Product

The vectors $\vec{\mathbf{A}}$ and $\vec{\mathbf{B}}$ are given by $\vec{\mathbf{A}} = 2\hat{\mathbf{i}} + 3\hat{\mathbf{j}}$ and $\vec{\mathbf{B}} = -\hat{\mathbf{i}} + 2\hat{\mathbf{j}}$.

(A) Determine the scalar product $\vec{\mathbf{A}} \cdot \vec{\mathbf{B}}$.

SOLUTION

Conceptualize There is no physical system to imagine here. Rather, it is purely a mathematical exercise involving two vectors.

Categorize Because we have a definition for the scalar product, we categorize this example as a substitution problem.

Substitute the specific vector expressions for $\vec{\mathbf{A}}$ and $\vec{\mathbf{B}}$:

$$\vec{\mathbf{A}} \cdot \vec{\mathbf{B}} = (2\hat{\mathbf{i}} + 3\hat{\mathbf{j}}) \cdot (-\hat{\mathbf{i}} + 2\hat{\mathbf{j}})$$

$$= -2\hat{\mathbf{i}} \cdot \hat{\mathbf{i}} + 2\hat{\mathbf{i}} \cdot 2\hat{\mathbf{j}} - 3\hat{\mathbf{j}} \cdot \hat{\mathbf{i}} + 3\hat{\mathbf{j}} \cdot 2\hat{\mathbf{j}}$$

$$= -2(1) + 4(0) - 3(0) + 6(1) = -2 + 6 = \boxed{4}$$

The same result is obtained when we use Equation 7.6 directly, where $A_x = 2$, $A_y = 3$, $B_x = -1$, and $B_y = 2$.

(B) Find the angle θ between $\vec{\mathbf{A}}$ and $\vec{\mathbf{B}}$.

SOLUTION

Evaluate the magnitudes of $\vec{\mathbf{A}}$ and $\vec{\mathbf{B}}$ using the Pythagorean theorem:

$$A = \sqrt{A_x^2 + A_y^2} = \sqrt{(2)^2 + (3)^2} = \sqrt{13}$$

$$B = \sqrt{B_x^2 + B_y^2} = \sqrt{(-1)^2 + (2)^2} = \sqrt{5}$$

Use Equation 7.2 and the result from part (A) to find the angle:

$$\cos\theta = \frac{\vec{\mathbf{A}}\cdot\vec{\mathbf{B}}}{AB} = \frac{4}{\sqrt{13}\sqrt{5}} = \frac{4}{\sqrt{65}}$$

$$\theta = \cos^{-1}\frac{4}{\sqrt{65}} = 60.3°$$

EXAMPLE 7.3 Work Done by a Constant Force

A particle moving in the *xy* plane undergoes a displacement given by $\Delta\vec{\mathbf{r}} = (2.0\hat{\mathbf{i}} + 3.0\hat{\mathbf{j}})$ m as a constant force $\vec{\mathbf{F}} = (5.0\hat{\mathbf{i}} + 2.0\hat{\mathbf{j}})$ N acts on the particle.

(A) Calculate the magnitudes of the force and the displacement of the particle.

SOLUTION

Conceptualize Although this example is a little more physical than the previous one in that it identifies a force and a displacement, it is similar in terms of its mathematical structure.

Categorize Because we are given two vectors and asked to find their magnitudes, we categorize this example as a substitution problem.

Use the Pythagorean theorem to find the magnitudes of the force and the displacement:

$$F = \sqrt{F_x^2 + F_y^2} = \sqrt{(5.0)^2 + (2.0)^2} = 5.4\text{ N}$$

$$\Delta r = \sqrt{(\Delta x)^2 + (\Delta y)^2} = \sqrt{(2.0)^2 + (3.0)^2} = 3.6\text{ m}$$

(B) Calculate the work done by $\vec{\mathbf{F}}$ on the particle.

SOLUTION

Substitute the expressions for $\vec{\mathbf{F}}$ and $\Delta\vec{\mathbf{r}}$ into Equation 7.3 and use Equations 7.4 and 7.5:

$$W = \vec{\mathbf{F}}\cdot\Delta\vec{\mathbf{r}} = [(5.0\hat{\mathbf{i}} + 2.0\hat{\mathbf{j}})\text{ N}]\cdot[(2.0\hat{\mathbf{i}} + 3.0\hat{\mathbf{j}})\text{ m}]$$

$$= (5.0\hat{\mathbf{i}}\cdot 2.0\hat{\mathbf{i}} + 5.0\hat{\mathbf{i}}\cdot 3.0\hat{\mathbf{j}} + 2.0\hat{\mathbf{j}}\cdot 2.0\hat{\mathbf{i}} + 2.0\hat{\mathbf{j}}\cdot 3.0\hat{\mathbf{j}})\text{ N}\cdot\text{m}$$

$$= [10 + 0 + 0 + 6]\text{ N}\cdot\text{m} = 16\text{ J}$$

7.4 Work Done by a Varying Force

Consider a particle being displaced along the x axis under the action of a force that varies with position. The particle is displaced in the direction of increasing x from $x = x_i$ to $x = x_f$. In such a situation, we cannot use $W = F\,\Delta r\cos\theta$ to calculate the work done by the force because this relationship applies only when $\vec{\mathbf{F}}$ is constant in magnitude and direction. If, however, we imagine that the particle undergoes a very small displacement Δx, shown in Figure 7.7a, the x component F_x of the force is approximately constant over this small interval; for this small displacement, we can approximate the work done on the particle by the force as

$$W \approx F_x\,\Delta x$$

which is the area of the shaded rectangle in Figure 7.7a. If we imagine the F_x versus x curve divided into a large number of such intervals, the total work done for

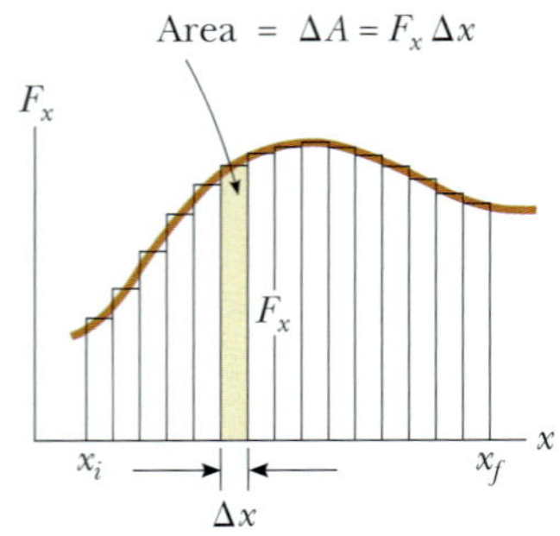

(a)

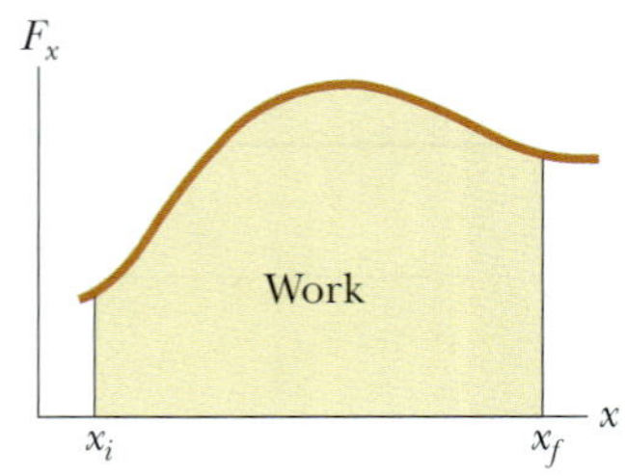

(b)

Figure 7.7 (a) The work done on a particle by the force component F_x for the small displacement Δx is $F_x \Delta x$, which equals the area of the shaded rectangle. The total work done for the displacement from x_i to x_f is approximately equal to the sum of the areas of all the rectangles. (b) The work done by the component F_x of the varying force as the particle moves from x_i to x_f is *exactly* equal to the area under this curve.

the displacement from x_i to x_f is approximately equal to the sum of a large number of such terms:

$$W \approx \sum_{x_i}^{x_f} F_x \Delta x$$

If the size of the small displacements is allowed to approach zero, the number of terms in the sum increases without limit but the value of the sum approaches a definite value equal to the area bounded by the F_x curve and the x axis:

$$\lim_{\Delta x \to 0} \sum_{x_i}^{x_f} F_x \Delta x = \int_{x_i}^{x_f} F_x \, dx$$

Therefore, we can express the work done by F_x on the particle as it moves from x_i to x_f as

$$W = \int_{x_i}^{x_f} F_x \, dx \qquad \textbf{(7.7)}$$

This equation reduces to Equation 7.1 when the component $F_x = F \cos \theta$ is constant.

If more than one force acts on a system *and the system can be modeled as a particle,* the total work done on the system is just the work done by the net force. If we express the net force in the x direction as $\sum F_x$, the total work, or *net work,* done as the particle moves from x_i to x_f is

$$\sum W = W_{\text{net}} = \int_{x_i}^{x_f} \left(\sum F_x\right) dx$$

For the general case of a net force $\Sigma \vec{\mathbf{F}}$ whose magnitude and direction may vary, we use the scalar product,

$$\sum W = W_{\text{net}} = \int \left(\sum \vec{\mathbf{F}}\right) \cdot d\vec{\mathbf{r}} \qquad \textbf{(7.8)}$$

where the integral is calculated over the path that the particle takes through space.

If the system cannot be modeled as a particle (for example, if the system consists of multiple particles that can move with respect to one another), we cannot use Equation 7.8 because different forces on the system may move through different displacements. In this case, we must evaluate the work done by each force separately and then add the works algebraically to find the net work done on the system.

EXAMPLE 7.4 Calculating Total Work Done from a Graph

A force acting on a particle varies with x as shown in Figure 7.8. Calculate the work done by the force on the particle as it moves from $x = 0$ to $x = 6.0$ m.

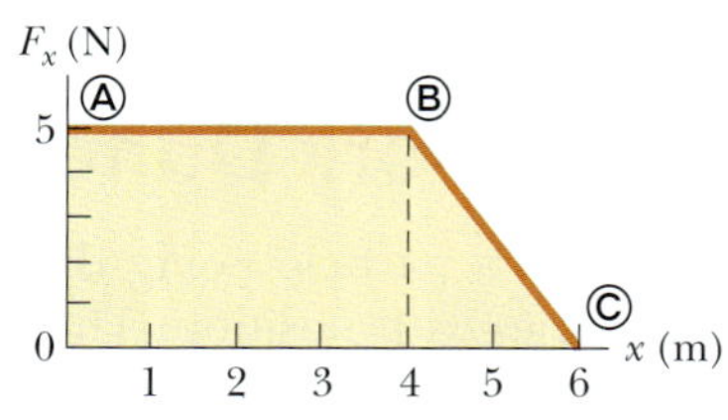

Figure 7.8 (Example 7.4) The force acting on a particle is constant for the first 4.0 m of motion and then decreases linearly with x from $x_{Ⓑ} = 4.0$ m to $x_{Ⓒ} = 6.0$ m. The net work done by this force is the area under the curve.

SOLUTION

Conceptualize Imagine a particle subject to the force in Figure 7.8. Notice that the force remains constant as the particle moves through the first 4.0 m and then decreases linearly to zero at 6.0 m.

Categorize Because the force varies during the entire motion of the particle, we must use the techniques for work done by varying forces. In this case, the graphical representation in Figure 7.8 can be used to evaluate the work done.

Analyze The work done by the force is equal to the area under the curve from $x_{Ⓐ} = 0$ to $x_{Ⓒ} = 6.0$ m. This area is equal to the area of the rectangular section from Ⓐ to Ⓑ plus the area of the triangular section from Ⓑ to Ⓒ.

Evaluate the area of the rectangle: $W_{\text{Ⓐ}\text{Ⓑ}} = (5.0\text{ N})(4.0\text{ m}) = 20\text{ J}$

Evaluate the area of the triangle: $W_{\text{Ⓑ}\text{Ⓒ}} = \frac{1}{2}(5.0\text{ N})(2.0\text{ m}) = 5.0\text{ J}$

Find the total work done by the force on the particle: $W_{\text{Ⓐ}\text{Ⓒ}} = W_{\text{Ⓐ}\text{Ⓑ}} + W_{\text{Ⓑ}\text{Ⓒ}} = 20\text{ J} + 5.0\text{ J} = 25\text{ J}$

Finalize Because the graph of the force consists of straight lines, we can use rules for finding the areas of simple geometric shapes to evaluate the total work done in this example. In a case in which the force does not vary linearly, such rules cannot be used and the force function must be integrated as in Equation 7.7 or 7.8.

Work Done by a Spring

A model of a common physical system for which the force varies with position is shown in Active Figure 7.9. A block on a horizontal, frictionless surface is connected to a spring. For many springs, if the spring is either stretched or compressed a small distance from its unstretched (equilibrium) configuration, it exerts on the block a force that can be mathematically modeled as

$$F_s = -kx \qquad \textbf{(7.9)}$$

◀ Spring force

where x is the position of the block relative to its equilibrium ($x = 0$) position and k is a positive constant called the **force constant** or the **spring constant** of the

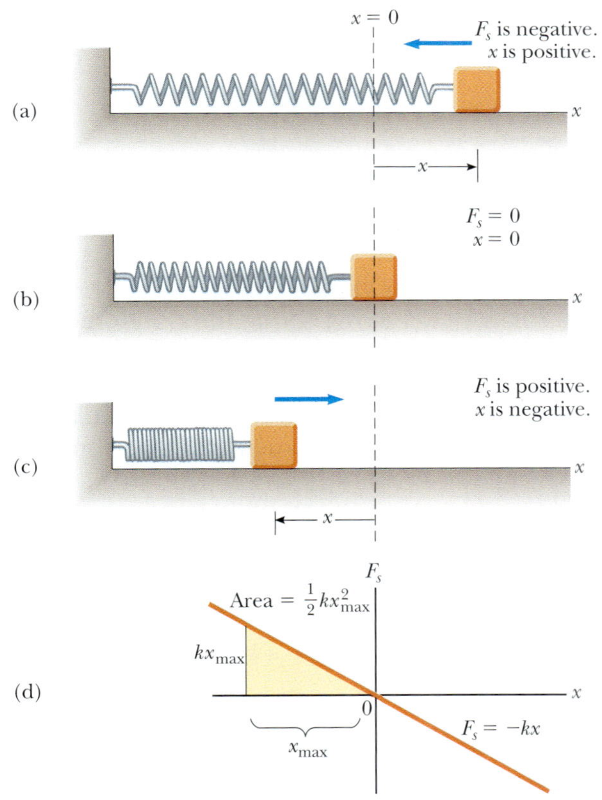

ACTIVE FIGURE 7.9

The force exerted by a spring on a block varies with the block's position x relative to the equilibrium position $x = 0$. (a) When x is positive (stretched spring), the spring force is directed to the left. (b) When x is zero (natural length of the spring), the spring force is zero. (c) When x is negative (compressed spring), the spring force is directed to the right. (d) Graph of F_s versus x for the block–spring system. The work done by the spring force on the block as it moves from $-x_{max}$ to 0 is the area of the shaded triangle, $\frac{1}{2}kx_{max}^2$.

Sign in at www.thomsonedu.com and go to ThomsonNOW to observe the block's motion for various spring constants and maximum positions of the block.

spring. In other words, the force required to stretch or compress a spring is proportional to the amount of stretch or compression x. This force law for springs is known as **Hooke's law**. The value of k is a measure of the *stiffness* of the spring. Stiff springs have large k values, and soft springs have small k values. As can be seen from Equation 7.9, the units of k are N/m.

The vector form of Equation 7.9 is

$$\vec{\mathbf{F}}_s = F_s\hat{\mathbf{i}} = -kx\hat{\mathbf{i}} \tag{7.10}$$

where we have chosen the x axis to lie along the direction the spring extends or compresses.

The negative sign in Equations 7.9 and 7.10 signifies that the force exerted by the spring is always directed *opposite* the displacement from equilibrium. When $x > 0$ as in Active Figure 7.9a so that the block is to the right of the equilibrium position, the spring force is directed to the left, in the negative x direction. When $x < 0$ as in Active Figure 7.9c, the block is to the left of equilibrium and the spring force is directed to the right, in the positive x direction. When $x = 0$ as in Active Figure 7.9b, the spring is unstretched and $F_s = 0$. Because the spring force always acts toward the equilibrium position ($x = 0$), it is sometimes called a *restoring force*.

If the spring is compressed until the block is at the point $-x_{max}$ and is then released, the block moves from $-x_{max}$ through zero to $+x_{max}$. It then reverses direction, returns to $-x_{max}$, and continues oscillating back and forth.

Suppose the block has been pushed to the left to a position $-x_{max}$ and is then released. Let us identify the block as our system and calculate the work W_s done by the spring force on the block as the block moves from $x_i = -x_{max}$ to $x_f = 0$. Applying Equation 7.8 and assuming the block may be modeled as a particle, we obtain

$$W_s = \int \vec{\mathbf{F}}_s \cdot d\vec{\mathbf{r}} = \int_{x_i}^{x_f} (-kx\hat{\mathbf{i}}) \cdot (dx\hat{\mathbf{i}}) = \int_{-x_{max}}^{0} (-kx)\,dx = \tfrac{1}{2}kx_{max}^2 \tag{7.11}$$

where we have used the integral $\int x^n dx = x^{n+1}/(n+1)$ with $n = 1$. The work done by the spring force is positive because the force is in the same direction as its displacement (both are to the right). Because the block arrives at $x = 0$ with some speed, it will continue moving until it reaches a position $+x_{max}$. The work done by the spring force on the block as it moves from $x_i = 0$ to $x_f = x_{max}$ is $W_s = -\frac{1}{2}kx_{max}^2$ because for this part of the motion the spring force is to the left and its displacement is to the right. Therefore, the *net* work done by the spring force on the block as it moves from $x_i = -x_{max}$ to $x_f = x_{max}$ is *zero*.

Active Figure 7.9d is a plot of F_s versus x. The work calculated in Equation 7.11 is the area of the shaded triangle, corresponding to the displacement from $-x_{max}$ to 0. Because the triangle has base x_{max} and height kx_{max}, its area is $\frac{1}{2}kx_{max}^2$, the work done by the spring as given by Equation 7.11.

If the block undergoes an arbitrary displacement from $x = x_i$ to $x = x_f$, the work done by the spring force on the block is

Work done by a spring ▶

$$W_s = \int_{x_i}^{x_f} (-kx)\,dx = \tfrac{1}{2}kx_i^2 - \tfrac{1}{2}kx_f^2 \tag{7.12}$$

From Equation 7.12, we see that the work done by the spring force is zero for any motion that ends where it began ($x_i = x_f$). We shall make use of this important result in Chapter 8 when we describe the motion of this system in greater detail.

Equations 7.11 and 7.12 describe the work done by the spring on the block. Now let us consider the work done on the block by an *external agent* as the agent applies a force on the block and the block moves *very slowly* from $x_i = -x_{max}$ to $x_f = 0$ as in Figure 7.10. We can calculate this work by noting that at any value of the position, the *applied force* $\vec{\mathbf{F}}_{app}$ is equal in magnitude and opposite in direction to the spring force $\vec{\mathbf{F}}_s$, so $\vec{\mathbf{F}}_{app} = F_{app}\hat{\mathbf{i}} = -\vec{\mathbf{F}}_s = -(-kx\hat{\mathbf{i}}) = kx\hat{\mathbf{i}}$. Therefore, the work done by this applied force (the external agent) on the block-spring system is

$$W_{app} = \int \vec{\mathbf{F}}_{app} \cdot d\vec{\mathbf{r}} = \int_{x_i}^{x_f} (kx\hat{\mathbf{i}}) \cdot (dx\hat{\mathbf{i}}) = \int_{-x_{max}}^{0} kx\,dx = -\tfrac{1}{2}kx_{max}^2$$

This work is equal to the negative of the work done by the spring force for this displacement (Eq. 7.11). The work is negative because the external agent must push inward on the spring to prevent it from expanding and this direction is opposite the direction of the displacement of the point of application of the force as the block moves from $-x_{max}$ to 0.

For an arbitrary displacement of the block, the work done on the system by the external agent is

$$W_{app} = \int_{x_i}^{x_f} kx\,dx = \tfrac{1}{2}kx_f^2 - \tfrac{1}{2}kx_i^2 \quad (7.13)$$

Notice that this equation is the negative of Equation 7.12.

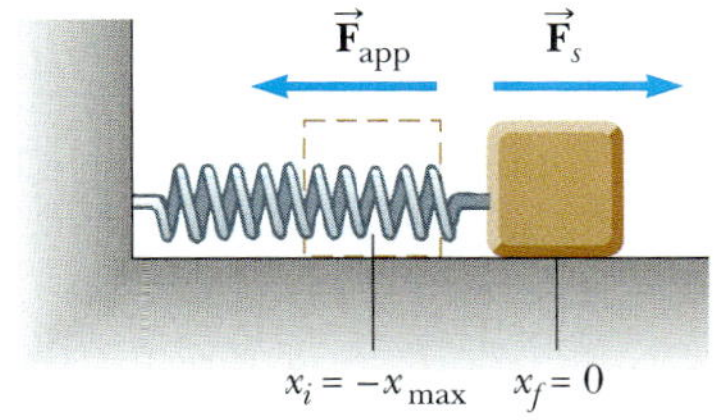

Figure 7.10 A block moves from $x_i = -x_{max}$ to $x_f = 0$ on a frictionless surface as a force $\vec{\mathbf{F}}_{app}$ is applied to the block. If the process is carried out very slowly, the applied force is equal in magnitude and opposite in direction to the spring force at all times.

Quick Quiz 7.4 A dart is loaded into a spring-loaded toy dart gun by pushing the spring in by a distance x. For the next loading, the spring is compressed a distance $2x$. How much work is required to load the second dart compared with that required to load the first? (a) four times as much (b) two times as much (c) the same (d) half as much (e) one-fourth as much

EXAMPLE 7.5 Measuring *k* for a Spring

A common technique used to measure the force constant of a spring is demonstrated by the setup in Figure 7.11. The spring is hung vertically (Fig. 7.11a), and an object of mass m is attached to its lower end. Under the action of the "load" mg, the spring stretches a distance d from its equilibrium position (Fig. 7.11b).

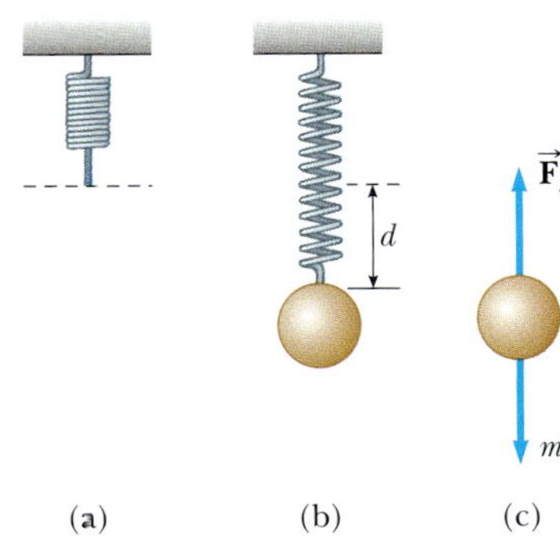

Figure 7.11 (Example 7.5) Determining the force constant k of a spring. The elongation d is caused by the attached object, which has a weight mg.

(A) If a spring is stretched 2.0 cm by a suspended object having a mass of 0.55 kg, what is the force constant of the spring?

SOLUTION

Conceptualize Consider Figure 7.11b, which shows what happens to the spring when the object is attached to it. Simulate this situation by hanging an object on a rubber band.

Categorize The object in Figure 7.11b is not accelerating, so it is modeled as a particle in equilibrium.

Analyze Because the object is in equilibrium, the net force on it is zero and the upward spring force balances the downward gravitational force $m\vec{\mathbf{g}}$ (Fig. 7.11c).

Apply Hooke's law to give $|\vec{\mathbf{F}}_s| = kd = mg$ and solve for k:

$$k = \frac{mg}{d} = \frac{(0.55\text{ kg})(9.80\text{ m/s}^2)}{2.0 \times 10^{-2}\text{ m}} = 2.7 \times 10^2\text{ N/m}$$

(B) How much work is done by the spring on the object as it stretches through this distance?

SOLUTION

Use Equation 7.12 to find the work done by the spring on the object:

$$W_s = 0 - \tfrac{1}{2}kd^2 = -\tfrac{1}{2}(2.7 \times 10^2\text{ N/m})(2.0 \times 10^{-2}\text{ m})^2$$
$$= -5.4 \times 10^{-2}\text{ J}$$

Finalize As the object moves through the 2.0-cm distance, the gravitational force also does work on it. This work is positive because the gravitational force is downward and so is the displacement of the point of application of this force. Based on Equation 7.12 and the discussion afterward, would we expect the work done by the gravitational force to be $+5.4 \times 10^{-2}$ J? Let's find out.

Evaluate the work done by the gravitational force on the object:

$$W = \vec{\mathbf{F}} \cdot \Delta\vec{\mathbf{r}} = (mg)(d)\cos 0 = mgd$$

$$= (0.55 \text{ kg})(9.80 \text{ m/s}^2)(2.0 \times 10^{-2} \text{ m}) = 1.1 \times 10^{-1} \text{ J}$$

If you expected the work done by gravity simply to be that done by the spring with a positive sign, you may be surprised by this result! To understand why that is not the case, we need to explore further, as we do in the next section.

7.5 Kinetic Energy and the Work–Kinetic Energy Theorem

We have investigated work and identified it as a mechanism for transferring energy into a system. One possible outcome of doing work on a system is that the system changes its speed. In this section, we investigate this situation and introduce our first type of energy that a system can possess, called *kinetic energy*.

Consider a system consisting of a single object. Figure 7.12 shows a block of mass m moving through a displacement directed to the right under the action of a net force $\sum \vec{\mathbf{F}}$, also directed to the right. We know from Newton's second law that the block moves with an acceleration $\vec{\mathbf{a}}$. If the block (and therefore the force) moves through a displacement $\Delta\vec{\mathbf{r}} = \Delta x\hat{\mathbf{i}} = (x_f - x_i)\hat{\mathbf{i}}$, the net work done on the block by the net force $\sum \vec{\mathbf{F}}$ is

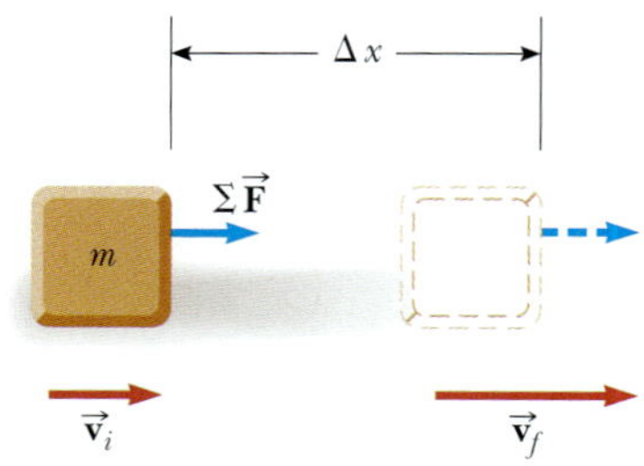

Figure 7.12 An object undergoing a displacement $\Delta\vec{\mathbf{r}} = \Delta x\hat{\mathbf{i}}$ and a change in velocity under the action of a constant net force $\Sigma\vec{\mathbf{F}}$.

$$W_{\text{net}} = \int_{x_i}^{x_f} \sum F\,dx \tag{7.14}$$

Using Newton's second law, we substitute for the magnitude of the net force $\sum F = ma$ and then perform the following chain-rule manipulations on the integrand:

$$W_{\text{net}} = \int_{x_i}^{x_f} ma\,dx = \int_{x_i}^{x_f} m\frac{dv}{dt}\,dx = \int_{x_i}^{x_f} m\frac{dv}{dx}\frac{dx}{dt}\,dx = \int_{v_i}^{v_f} mv\,dv$$

$$W_{\text{net}} = \tfrac{1}{2}mv_f^2 - \tfrac{1}{2}mv_i^2 \tag{7.15}$$

where v_i is the speed of the block when it is at $x = x_i$ and v_f is its speed at x_f.

Equation 7.15 was generated for the specific situation of one-dimensional motion, but it is a general result. It tells us that the work done by the net force on a particle of mass m is equal to the difference between the initial and final values of a quantity $\frac{1}{2}mv^2$. The quantity $\frac{1}{2}mv^2$ represents the energy associated with the motion of the particle. This quantity is so important that it has been given a special name, **kinetic energy**:

Kinetic energy ▶

$$K \equiv \tfrac{1}{2}mv^2 \tag{7.16}$$

Kinetic energy is a scalar quantity and has the same units as work. For example, a 2.0-kg object moving with a speed of 4.0 m/s has a kinetic energy of 16 J. Table 7.1 lists the kinetic energies for various objects.

Equation 7.15 states that the work done on a particle by a net force $\sum \vec{\mathbf{F}}$ acting on it equals the change in kinetic energy of the particle. It is often convenient to write Equation 7.15 in the form

$$W_{\text{net}} = K_f - K_i = \Delta K \tag{7.17}$$

Another way to write it is $K_f = K_i + W_{\text{net}}$, which tells us that the final kinetic energy of an object is equal to its initial kinetic energy plus the change due to the net work done on it.

TABLE 7.1

Kinetic Energies for Various Objects

Object	Mass (kg)	Speed (m/s)	Kinetic Energy (J)
Earth orbiting the Sun	5.98×10^{24}	2.98×10^{4}	2.66×10^{33}
Moon orbiting the Earth	7.35×10^{22}	1.02×10^{3}	3.82×10^{28}
Rocket moving at escape speed[a]	500	1.12×10^{4}	3.14×10^{10}
Automobile at 65 mi/h	2 000	29	8.4×10^{5}
Running athlete	70	10	3 500
Stone dropped from 10 m	1.0	14	98
Golf ball at terminal speed	0.046	44	45
Raindrop at terminal speed	3.5×10^{-5}	9.0	1.4×10^{-3}
Oxygen molecule in air	5.3×10^{-26}	500	6.6×10^{-21}

[a] Escape speed is the minimum speed an object must reach near the Earth's surface to move infinitely far away from the Earth.

We have generated Equation 7.17 by imagining doing work on a particle. We could also do work on a deformable system, in which parts of the system move with respect to one another. In this case, we also find that Equation 7.17 is valid as long as the net work is found by adding up the works done by each force and adding, as discussed earlier with regard to Equation 7.8.

Equation 7.17 is an important result known as the **work–kinetic energy theorem**:

When work is done on a system and the only change in the system is in its speed, the net work done on the system equals the change in kinetic energy of the system.

◄ Work–kinetic energy theorem

The work–kinetic energy theorem indicates that the speed of a system *increases* if the net work done on it is *positive* because the final kinetic energy is greater than the initial kinetic energy. The speed *decreases* if the net work is *negative* because the final kinetic energy is less than the initial kinetic energy.

Because we have so far only investigated translational motion through space, we arrived at the work–kinetic energy theorem by analyzing situations involving translational motion. Another type of motion is *rotational motion,* in which an object spins about an axis. We will study this type of motion in Chapter 10. The work–kinetic energy theorem is also valid for systems that undergo a change in the rotational speed due to work done on the system. The windmill in the photograph at the beginning of this chapter is an example of work causing rotational motion.

The work–kinetic energy theorem will clarify a result seen earlier in this chapter that may have seemed odd. In Section 7.4, we arrived at a result of zero net work done when we let a spring push a block from $x_i = -x_{\max}$ to $x_f = x_{\max}$. Notice that because the speed of the block is continually changing, it may seem complicated to analyze this process. The quantity ΔK in the work–kinetic energy theorem, however, only refers to the initial and final points for the speeds; it does not depend on details of the path followed between these points. Therefore, because the speed is zero at both the initial and final points of the motion, the net work done on the block is zero. We will often see this concept of path independence in similar approaches to problems.

Let us also return to the mystery in the Finalize step at the end of Example 7.5. Why was the work done by gravity not just the value of the work done by the spring with a positive sign? Notice that the work done by gravity is larger than the magnitude of the work done by the spring. Therefore, the total work done by all forces on the object is positive. Imagine now how to create the situation in which the *only* forces on the object are the spring force and the gravitational force. You must support the object at the highest point and then *remove* your hand and let the

PITFALL PREVENTION 7.6

Conditions for the Work–Kinetic Energy Theorem

The work–kinetic energy theorem is important but limited in its application; it is not a general principle. In many situations, other changes in the system occur besides its speed, and there are other interactions with the environment besides work. A more general principle involving energy is *conservation of energy* in Section 8.1.

PITFALL PREVENTION 7.7

The Work–Kinetic Energy Theorem: Speed, Not Velocity

The work–kinetic energy theorem relates work to a change in the *speed* of a system, not a change in its velocity. For example, if an object is in uniform circular motion, its speed is constant. Even though its velocity is changing, no work is done on the object by the force causing the circular motion.

object fall. If you do so, you know that when the object reaches a position 2.0 cm below your hand, it will be *moving*, which is consistent with Equation 7.17. Positive net work is done on the object, and the result is that it has a kinetic energy as it passes through the 2.0-cm point. The only way to prevent the object from having a kinetic energy after moving through 2.0 cm is to slowly lower it with your hand. Then, however, there is a third force doing work on the object, the normal force from your hand. If this work is calculated and added to that done by the spring force and the gravitational force, the net work done on the object is zero, which is consistent because it is not moving at the 2.0-cm point.

Earlier, we indicated that work can be considered as a mechanism for transferring energy into a system. Equation 7.17 is a mathematical statement of this concept. When work W_{net} is done on a system, the result is a transfer of energy across the boundary of the system. The result on the system, in the case of Equation 7.17, is a change ΔK in kinetic energy. In the next section, we investigate another type of energy that can be stored in a system as a result of doing work on the system.

Quick Quiz 7.5 A dart is loaded into a spring-loaded toy dart gun by pushing the spring in by a distance x. For the next loading, the spring is compressed a distance $2x$. How much faster does the second dart leave the gun compared with the first? (a) four times as fast (b) two times as fast (c) the same (d) half as fast (e) one-fourth as fast

EXAMPLE 7.6 A Block Pulled on a Frictionless Surface

A 6.0-kg block initially at rest is pulled to the right along a horizontal, frictionless surface by a constant horizontal force of 12 N. Find the block's speed after it has moved 3.0 m.

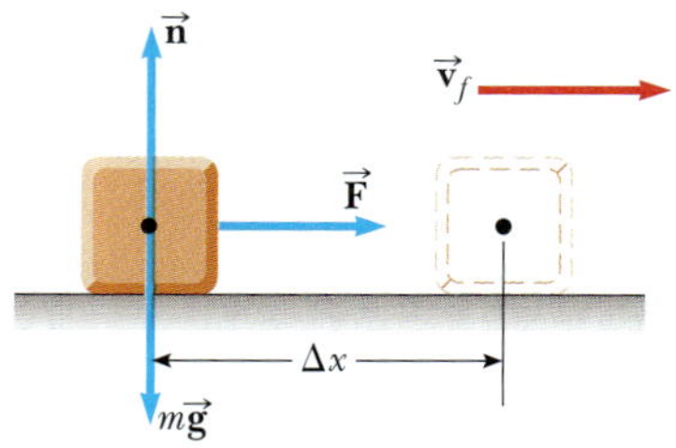

Figure 7.13 (Example 7.6) A block pulled to the right on a frictionless surface by a constant horizontal force.

SOLUTION

Conceptualize Figure 7.13 illustrates this situation. Imagine pulling a toy car across a table with a horizontal rubber band attached to the front of the car. The force is maintained constant by ensuring that the stretched rubber band always has the same length.

Categorize We could apply the equations of kinematics to determine the answer, but let us practice the energy approach. The block is the system, and three external forces act on the system. The normal force balances the gravitational force on the block, and neither of these vertically acting forces does work on the block because their points of application are horizontally displaced.

Analyze The net external force acting on the block is the horizontal 12-N force.

Find the work done by this force on the block:

$$W = F\,\Delta x = (12\text{ N})(3.0\text{ m}) = 36\text{ J}$$

Use the work–kinetic energy theorem for the block and note that its initial kinetic energy is zero:

$$W = K_f - K_i = \tfrac{1}{2}mv_f^2 - 0$$

Solve for v_f:

$$v_f = \sqrt{\frac{2W}{m}} = \sqrt{\frac{2(36\text{ J})}{6.0\text{ kg}}} = 3.5\text{ m/s}$$

Finalize It would be useful for you to solve this problem again by modeling the block as a particle under a net force to find its acceleration and then as a particle under constant acceleration to find its final velocity.

What If? Suppose the magnitude of the force in this example is doubled to $F' = 2F$. The 6.0-kg block accelerates to 3.5 m/s due to this applied force while moving through a displacement $\Delta x'$. How does the displacement $\Delta x'$ compare with the original displacement Δx?

Answer If we pull harder, the block should accelerate to a given speed in a shorter distance, so we expect that $\Delta x' < \Delta x$. In both cases, the block experiences the same change in kinetic energy ΔK. Mathematically, from the work-kinetic energy theorem, we find that

$$W = F'\Delta x' = \Delta K = F\,\Delta x$$

$$\Delta x' = \frac{F}{F'}\Delta x = \frac{F}{2F}\Delta x = \tfrac{1}{2}\Delta x$$

and the distance is shorter as suggested by our conceptual argument.

CONCEPTUAL EXAMPLE 7.7 **Does the Ramp Lessen the Work Required?**

A man wishes to load a refrigerator onto a truck using a ramp at angle θ as shown in Figure 7.14. He claims that less work would be required to load the truck if the length L of the ramp were increased. Is his claim valid?

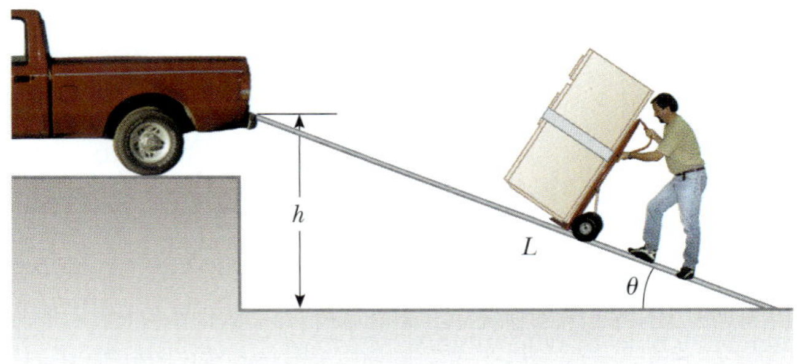

Figure 7.14 (Conceptual Example 7.7) A refrigerator attached to a frictionless, wheeled hand truck is moved up a ramp at constant speed.

SOLUTION

No. Suppose the refrigerator is wheeled on a hand truck up the ramp at constant speed. In this case, for the system of the refrigerator and the hand truck, $\Delta K = 0$. The normal force exerted by the ramp on the system is directed at 90° to the displacement of its point of application and so does no work on the system. Because $\Delta K = 0$, the work-kinetic energy theorem gives

$$W_{\text{net}} = W_{\text{by man}} + W_{\text{by gravity}} = 0$$

The work done by the gravitational force equals the product of the weight mg of the system, the distance L through which the refrigerator is displaced, and $\cos(\theta + 90°)$. Therefore,

$$W_{\text{by man}} = -W_{\text{by gravity}} = -(mg)(L)[\cos(\theta + 90°)]$$
$$= mgL\sin\theta = mgh$$

where $h = L\sin\theta$ is the height of the ramp. Therefore, the man must do the same amount of work mgh on the system *regardless* of the length of the ramp. The work depends only on the height of the ramp. Although less force is required with a longer ramp, the point of application of that force moves through a greater displacement.

7.6 Potential Energy of a System

So far in this chapter, we have defined a system in general, but have focused our attention primarily on single particles or objects under the influence of external forces. Let us now consider systems of two or more particles or objects interacting via a force that is *internal* to the system. The kinetic energy of such a system is the algebraic sum of the kinetic energies of all members of the system. There may be systems, however, in which one object is so massive that it can be modeled as stationary and its kinetic energy can be neglected. For example, if we consider a ball–Earth system as the ball falls to the Earth, the kinetic energy of the system can be considered as just the kinetic energy of the ball. The Earth moves so slowly in this process that we can ignore its kinetic energy. On the other hand, the kinetic energy of a system of two electrons must include the kinetic energies of both particles.

Let us imagine a system consisting of a book and the Earth, interacting via the gravitational force. We do some work on the system by lifting the book slowly from rest through a vertical displacement $\Delta\vec{\mathbf{r}} = (y_f - y_i)\hat{\mathbf{j}}$ as in Active Figure 7.15. According to our discussion of work as an energy transfer, this work done on the system must appear as an increase in energy of the system. The book is at rest

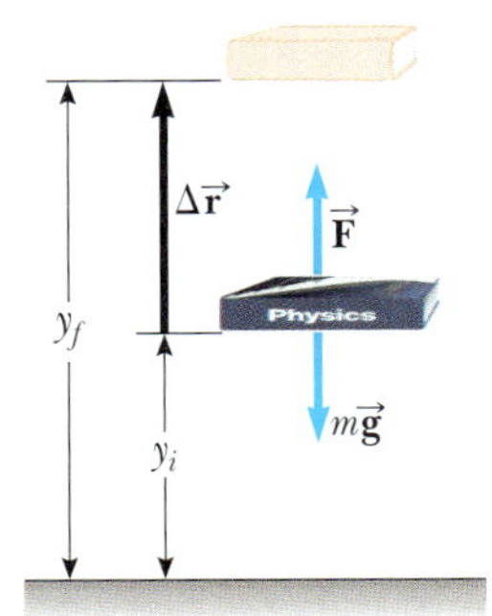

ACTIVE FIGURE 7.15

The work done by an external agent on the system of the book and the Earth as the book is lifted slowly from a height y_i to a height y_f is equal to $mgy_f - mgy_i$.

Sign in at www.thomsonedu.com and go to ThomsonNOW to move the block to various positions and determine the work done by the external agent for a general displacement.

before we perform the work and is at rest after we perform the work. Therefore, there is no change in the kinetic energy of the system.

Because the energy change of the system is not in the form of kinetic energy, it must appear as some other form of energy storage. After lifting the book, we could release it and let it fall back to the position y_i. Notice that the book (and, therefore, the system) now has kinetic energy and that its source is in the work that was done in lifting the book. While the book was at the highest point, the energy of the system had the *potential* to become kinetic energy, but it did not do so until the book was allowed to fall. Therefore, we call the energy storage mechanism before the book is released **potential energy**. We will find that the potential energy of a system can only be associated with specific types of forces acting between members of a system. The amount of potential energy in the system is determined by the *configuration* of the system. Moving members of the system to different positions or rotating them may change the configuration of the system and therefore its potential energy.

PITFALL PREVENTION 7.8
Potential Energy

The phrase *potential energy* does not refer to something that has the potential to become energy. Potential energy *is* energy.

Let us now derive an expression for the potential energy associated with an object at a given location above the surface of the Earth. Consider an external agent lifting an object of mass m from an initial height y_i above the ground to a final height y_f as in Active Figure 7.15. We assume the lifting is done slowly, with no acceleration, so the applied force from the agent can be modeled as being equal in magnitude to the gravitational force on the object: the object is modeled as a particle in equilibrium moving at constant velocity. The work done by the external agent on the system (object and the Earth) as the object undergoes this upward displacement is given by the product of the upward applied force $\vec{\mathbf{F}}_{\text{app}}$ and the upward displacement of this force, $\Delta\vec{\mathbf{r}} = \Delta y\hat{\mathbf{j}}$:

PITFALL PREVENTION 7.9
Potential Energy Belongs to a System

Potential energy is always associated with a *system* of two or more interacting objects. When a small object moves near the surface of the Earth under the influence of gravity, we may sometimes refer to the potential energy "associated with the object" rather than the more proper "associated with the system" because the Earth does not move significantly. We will not, however, refer to the potential energy "of the object" because this wording ignores the role of the Earth.

$$W_{\text{net}} = (\vec{\mathbf{F}}_{\text{app}}) \cdot \Delta\vec{\mathbf{r}} = (mg\hat{\mathbf{j}}) \cdot [(y_f - y_i)\hat{\mathbf{j}}] = mgy_f - mgy_i \quad \textbf{(7.18)}$$

where this result is the net work done on the system because the applied force is the only force on the system from the environment. Notice the similarity between Equation 7.18 and Equation 7.15. In each equation, the work done on a system equals a difference between the final and initial values of a quantity. In Equation 7.15, the work represents a transfer of energy into the system and the increase in energy of the system is kinetic in form. In Equation 7.18, the work represents a transfer of energy into the system and the system energy appears in a different form, which we have called potential energy.

Therefore, we can identify the quantity mgy as the **gravitational potential energy** U_g:

Gravitational potential energy ►

$$U_g \equiv mgy \quad \textbf{(7.19)}$$

The units of gravitational potential energy are joules, the same as the units of work and kinetic energy. Potential energy, like work and kinetic energy, is a scalar quantity. Notice that Equation 7.19 is valid only for objects near the surface of the Earth, where g is approximately constant.[3]

Using our definition of gravitational potential energy, Equation 7.18 can now be rewritten as

$$W_{\text{net}} = \Delta U_g \quad \textbf{(7.20)}$$

which mathematically describes that the net work done on the system in this situation appears as a change in the gravitational potential energy of the system.

Gravitational potential energy depends only on the vertical height of the object above the surface of the Earth. The same amount of work must be done on an object–Earth system whether the object is lifted vertically from the Earth or is pushed starting from the same point up a frictionless incline, ending up at the same height. We verified this statement for a specific situation of rolling a refrigerator up a ramp in Conceptual Example 7.7. This statement can be shown to be

[3] The assumption that g is constant is valid as long as the vertical displacement of the object is small compared with the Earth's radius.

true in general by calculating the work done on an object by an agent moving the object through a displacement having both vertical and horizontal components:

$$W_{\text{net}} = (\vec{\mathbf{F}}_{\text{app}}) \cdot \Delta\vec{\mathbf{r}} = (mg\hat{\mathbf{j}}) \cdot [(x_f - x_i)\hat{\mathbf{i}} + (y_f - y_i)\hat{\mathbf{j}}] = mgy_f - mgy_i$$

where there is no term involving x in the final result because $\hat{\mathbf{j}} \cdot \hat{\mathbf{i}} = 0$.

In solving problems, you must choose a reference configuration for which the gravitational potential energy of the system is set equal to some reference value, which is normally zero. The choice of reference configuration is completely arbitrary because the important quantity is the *difference* in potential energy, and this difference is independent of the choice of reference configuration.

It is often convenient to choose as the reference configuration for zero gravitational potential energy the configuration in which an object is at the surface of the Earth, but this choice is not essential. Often, the statement of the problem suggests a convenient configuration to use.

Quick Quiz 7.6 Choose the correct answer. The gravitational potential energy of a system (a) is always positive (b) is always negative (c) can be negative or positive

---

EXAMPLE 7.8 The Bowler and the Sore Toe

A bowling ball held by a careless bowler slips from the bowler's hands and drops on the bowler's toe. Choosing floor level as the $y = 0$ point of your coordinate system, estimate the change in gravitational potential energy of the ball–Earth system as the ball falls. Repeat the calculation, using the top of the bowler's head as the origin of coordinates.

SOLUTION

Conceptualize The bowling ball changes its vertical position with respect to the surface of the Earth. Associated with this change in position is a change in the gravitational potential energy of the system.

Categorize We evaluate a change in gravitational potential energy defined in this section, so we categorize this example as a substitution problem.

The problem statement tells us that the reference configuration of the ball–Earth system corresponding to zero potential energy is when the bottom of the ball is at the floor. To find the change in potential energy for the system, we need to estimate a few values. A bowling ball has a mass of approximately 7 kg, and the top of a person's toe is about 0.03 m above the floor. Also, we shall assume the ball falls from a height of 0.5 m.

Calculate the gravitational potential energy of the ball–Earth system just before the bowling ball is released:

$$U_i = mgy_i = (7 \text{ kg})(9.80 \text{ m/s}^2)(0.5 \text{ m}) = 34.3 \text{ J}$$

Calculate the gravitational potential energy of the ball–Earth system when the ball reaches the bowler's toe:

$$U_f = mgy_f = (7 \text{ kg})(9.80 \text{ m/s}^2)(0.03 \text{ m}) = 2.06 \text{ J}$$

Evaluate the change in gravitational potential energy of the ball–Earth system:

$$\Delta U_g = 2.06 \text{ J} - 34.3 \text{ J} = -32.24 \text{ J}$$

We should probably keep only one digit because of the roughness of our estimates; therefore, we estimate that the change in gravitational potential energy is -30 J. The system had 30 J of gravitational potential energy before the ball began its fall and approximately zero potential energy as the ball reaches the top of the toe.

The second case presented indicates that the reference configuration of the system for zero potential energy is chosen to be when the ball is at the bowler's head (even though the ball is never at this position in its motion). We estimate this position to be 1.50 m above the floor).

Calculate the gravitational potential energy of the ball–Earth system just before the bowling ball is released from its position 1 m below the bowler's head:

$$U_i = mgy_i = (7 \text{ kg})(9.80 \text{ m/s}^2)(-1 \text{ m}) = -68.6 \text{ J}$$

Calculate the gravitational potential energy of the ball–Earth system when the ball reaches the bowler's toe located 1.47 m below the bowler's head:

$$U_f = mgy_f = (7\text{ kg})(9.80\text{ m/s}^2)(-1.47\text{ m}) = -100.8\text{ J}$$

Evaluate the change in gravitational potential energy of the ball–Earth system:

$$\Delta U_g = -100.8\text{ J} - (-68.6\text{ J}) = -32.2\text{ J} \approx -30\text{ J}$$

This value is the same as before, as it must be.

Elastic Potential Energy

Now that we are familiar with gravitational potential energy of a system, let us explore a second type of potential energy that a system can possess. Consider a system consisting of a block and a spring as shown in Active Figure 7.16. The force that the spring exerts on the block is given by $F_s = -kx$ (Eq. 7.9). The work done by an external applied force F_{app} on a system consisting of a block connected to the spring is given by Equation 7.13:

$$W_{\text{app}} = \tfrac{1}{2}kx_f^2 - \tfrac{1}{2}kx_i^2 \quad \textbf{(7.21)}$$

In this situation, the initial and final x coordinates of the block are measured from its equilibrium position, $x = 0$. Again (as in the gravitational case) we see that the work done on the system is equal to the difference between the initial and final values of an expression related to the system's configuration. The **elastic potential energy** function associated with the block-spring system is defined by

Elastic potential energy ▶

$$U_s \equiv \tfrac{1}{2}kx^2 \quad \textbf{(7.22)}$$

The elastic potential energy of the system can be thought of as the energy stored in the deformed spring (one that is either compressed or stretched from its equilibrium position). The elastic potential energy stored in a spring is zero when-

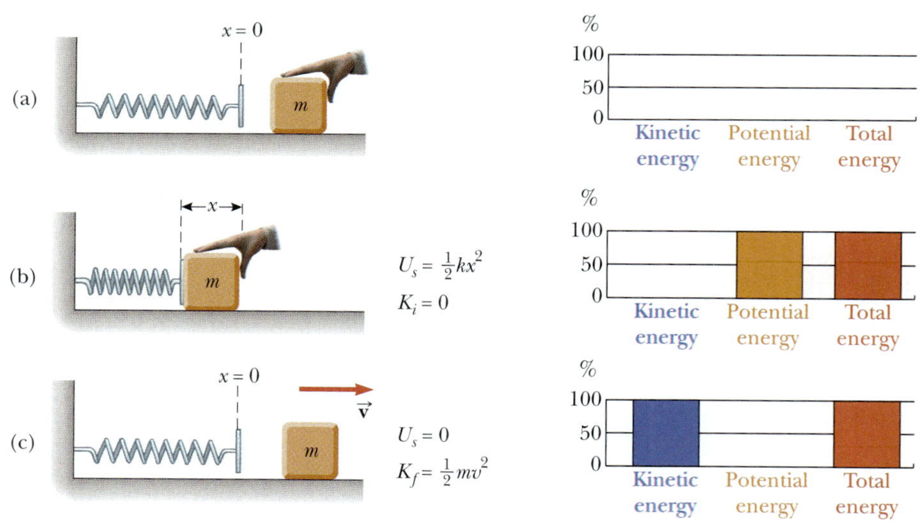

ACTIVE FIGURE 7.16

(a) An undeformed spring on a frictionless, horizontal surface. (b) A block of mass m is pushed against the spring, compressing it a distance x. Elastic potential energy is stored in the spring–block system. (c) When the block is released from rest, the elastic potential energy is transformed to kinetic energy of the block. Energy bar charts on the right of each part of the figure help keep track of the energy in the system.

Sign in at www.thomsonedu.com and go to ThomsonNOW to compress the spring by varying amounts and observe the effect on the block's speed.

ever the spring is undeformed ($x = 0$). Energy is stored in the spring only when the spring is either stretched or compressed. Because the elastic potential energy is proportional to x^2, we see that U_s is always positive in a deformed spring.

Consider Active Figure 7.16, which shows a spring on a frictionless, horizontal surface. When a block is pushed against the spring and the spring is compressed a distance x (Active Fig. 7.16b), the elastic potential energy stored in the spring is $\frac{1}{2}kx^2$. When the block is released from rest, the spring exerts a force on the block and returns to its original length. The stored elastic potential energy is transformed into kinetic energy of the block (Active Fig. 7.16c).

Active Figure 7.16 shows an important graphical representation of information related to energy of systems called an **energy bar chart**. The vertical axis represents the amount of energy of a given type in the system. The horizontal axis shows the types of energy in the system. The bar chart in Active Figure 7.16a shows that the system contains zero energy because the spring is relaxed and the block is not moving. Between Active Figure 7.16a and Active Figure 7.16b, the hand does work on the system, compressing the spring and storing elastic potential energy in the system. In Active Figure 7.16c, the spring has returned to its relaxed length and the system now contains kinetic energy associated with the moving block.

Quick Quiz 7.7 A ball is connected to a light spring suspended vertically as shown in Figure 7.17. When pulled downward from its equilibrium position and released, the ball oscillates up and down. **(i)** In the system of *the ball, the spring, and the Earth,* what forms of energy are there during the motion? (a) kinetic and elastic potential (b) kinetic and gravitational potential (c) kinetic, elastic potential, and gravitational potential (d) elastic potential and gravitational potential **(ii)** In the system of *the ball and the spring,* what forms of energy are there during the motion? Choose from the same possibilities (a) through (d).

Figure 7.17 (Quick Quiz 7.7) A ball connected to a massless spring suspended vertically. What forms of potential energy are associated with the system when the ball is displaced downward?

7.7 Conservative and Nonconservative Forces

We now introduce a third type of energy that a system can possess. Imagine that the book in Active Figure 7.18a has been accelerated by your hand and is now sliding to the right on the surface of a heavy table and slowing down due to the friction force. Suppose the *surface* is the system. Then the friction force from the sliding book does work on the surface. The force on the surface is to the right and the displacement of the point of application of the force is to the right. The work done on the surface is positive, but the surface is not moving after the book has stopped. Positive work has been done on the surface, yet there is no increase in the surface's kinetic energy or the potential energy of any system.

From your everyday experience with sliding over surfaces with friction, you can probably guess that the surface will be *warmer* after the book slides over it. (Rub your hands together briskly to find out!) The work that was done on the surface has gone into warming the surface rather than increasing its speed or changing the configuration of a system. We call the energy associated with the temperature of a system its **internal energy**, symbolized E_{int}. (We will define internal energy more generally in Chapter 20.) In this case, the work done on the surface does indeed represent energy transferred into the system, but it appears in the system as internal energy rather than kinetic or potential energy.

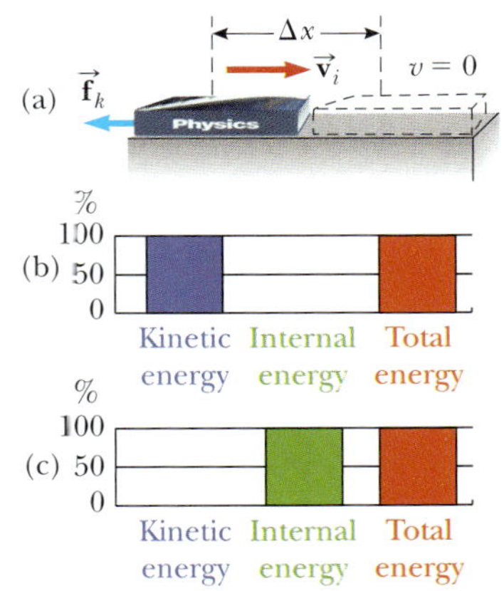

ACTIVE FIGURE 7.18

(a) A book sliding to the right on a horizontal surface slows down in the presence of a force of kinetic friction acting to the left. (b) An energy bar chart showing the energy in the system of the book and the surface at the initial instant of time. The energy of the system is all kinetic energy. (c) After the book has stopped, the energy of the system is all internal energy.

Sign in at www.thomsonedu.com and go to ThomsonNOW to slide the book with varying speeds and watch the energy transformation on an active energy bar chart.

Consider the book and the surface in Active Figure 7.18a together as a system. Initially, the system has kinetic energy because the book is moving. After the book has come to rest, the internal energy of the system has increased: the book and the surface are warmer than before. We can consider the work done by friction within the system—that is, between the book and the surface—as a *transformation mechanism* for energy. This work transforms the kinetic energy of the system into internal energy. Similarly, when a book falls straight down with no air resistance, the work done by the gravitational force within the book–Earth system transforms gravitational potential energy of the system to kinetic energy.

Active Figures 7.18b and 7.18c show energy bar charts for the situation in Active Figure 7.18a. In Active Figure 7.18b, the bar chart shows that the system contains kinetic energy at the instant the book is released by your hand. We define the reference amount of internal energy in the system as zero at this instant. In Active Figure 7.18c, after the book has stopped sliding, the kinetic energy is zero and the system now contains internal energy. Notice that the amount of internal energy in the system after the book has stopped is equal to the amount of kinetic energy in the system at the initial instant. This equality is described by an important principle called *conservation of energy*. We will explore this principle in Chapter 8.

Now consider in more detail an object moving downward near the surface of the Earth. The work done by the gravitational force on the object does not depend on whether it falls vertically or slides down a sloping incline. All that matters is the change in the object's elevation. The energy transformation to internal energy due to friction on that incline, however, depends on the distance the object slides. In other words, the path makes no difference when we consider the work done by the gravitational force, but it does make a difference when we consider the energy transformation due to friction forces. We can use this varying dependence on path to classify forces as either conservative or nonconservative. Of the two forces just mentioned, the gravitational force is conservative and the friction force is nonconservative.

Conservative Forces

Conservative forces have these two equivalent properties:

Properties of conservative forces ▶

1. The work done by a conservative force on a particle moving between any two points is independent of the path taken by the particle.
2. The work done by a conservative force on a particle moving through any closed path is zero. (A closed path is one for which the beginning point and the endpoint are identical.)

The gravitational force is one example of a conservative force; the force that an ideal spring exerts on any object attached to the spring is another. The work done by the gravitational force on an object moving between any two points near the Earth's surface is $W_g = -mg\hat{\mathbf{j}} \cdot [(y_f - y_i)\hat{\mathbf{j}}] = mgy_i - mgy_f$. From this equation, notice that W_g depends only on the initial and final y coordinates of the object and hence is independent of the path. Furthermore, W_g is zero when the object moves over any closed path (where $y_i = y_f$).

For the case of the object-spring system, the work W_s done by the spring force is given by $W_s = \frac{1}{2}kx_i^2 - \frac{1}{2}kx_f^2$ (Eq. 7.12). We see that the spring force is conservative because W_s depends only on the initial and final x coordinates of the object and is zero for any closed path.

We can associate a potential energy for a system with a force acting between members of the system, but we can do so only for conservative forces. In general, the work W_c done by a conservative force on an object that is a member of a system as the object moves from one position to another is equal to the initial value of the potential energy of the system minus the final value:

$$W_c = U_i - U_f = -\Delta U \qquad \textbf{(7.23)}$$

As an example, compare this general equation with the specific equation for the work done by the spring force (Eq. 7.12) as the extension of the spring changes.

PITFALL PREVENTION 7.10

Similar Equation Warning

Compare Equation 7.23 with Equation 7.20. These equations are similar except for the negative sign, which is a common source of confusion. Equation 7.20 tells us that positive work done *by an outside agent* on a system causes an increase in the potential energy of the system (with no change in the kinetic or internal energy). Equation 7.23 states that work done *on a component of a system by a conservative force internal to an isolated system* causes a decrease in the potential energy of the system.

Nonconservative Forces

A force is **nonconservative** if it does not satisfy properties 1 and 2 for conservative forces. We define the sum of the kinetic and potential energies of a system as the **mechanical energy** of the system:

$$E_{\text{mech}} \equiv K + U \qquad \textbf{(7.24)}$$

where K includes the kinetic energy of all moving members of the system and U includes all types of potential energy in the system. Nonconservative forces acting within a system cause a *change* in the mechanical energy of the system. For example, for a book sent sliding on a horizontal surface that is not frictionless, the mechanical energy of the book–surface system is transformed to internal energy as we discussed earlier. Only part of the book's kinetic energy is transformed to internal energy in the book. The rest appears as internal energy in the surface. (When you trip and slide across a gymnasium floor, not only does the skin on your knees warm up, so does the floor!) Because the force of kinetic friction transforms the mechanical energy of a system into internal energy, it is a nonconservative force.

As an example of the path dependence of the work for a nonconservative force, consider Figure 7.19. Suppose you displace a book between two points on a table. If the book is displaced in a straight line along the blue path between points Ⓐ and Ⓑ in Figure 7.19, you do a certain amount of work against the kinetic friction force to keep the book moving at a constant speed. Now, imagine that you push the book along the brown semicircular path in Figure 7.19. You perform more work against friction along this curved path than along the straight path because the curved path is longer. The work done on the book depends on the path, so the friction force *cannot* be conservative.

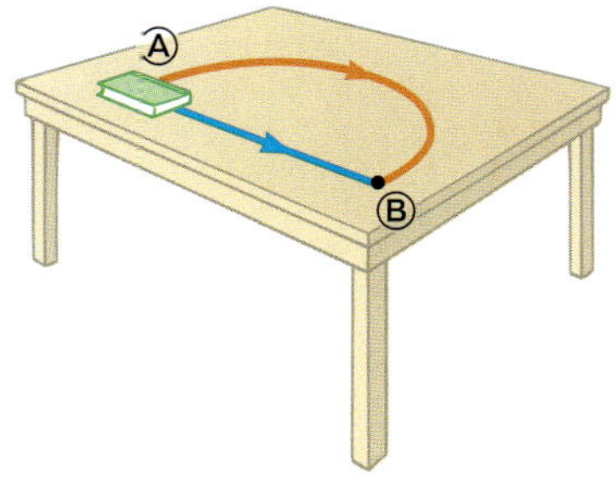

Figure 7.19 The work done against the force of kinetic friction depends on the path taken as the book is moved from Ⓐ to Ⓑ. The work is greater along the brown path than along the blue path.

7.8 Relationship Between Conservative Forces and Potential Energy

In the preceding section, we found that the work done on a member of a system by a conservative force between the members of the system does not depend on the path taken by the moving member. The work depends only on the initial and final coordinates. As a consequence, we can define a **potential energy function U** such that the work done within the system by the conservative force equals the decrease in the potential energy of the system. Let us imagine a system of particles in which the configuration changes due to the motion of one particle along the x axis. The work done by a conservative force $\vec{\mathbf{F}}$ as a particle moves along the x axis is[4]

$$W_c = \int_{x_i}^{x_f} F_x\,dx = -\Delta U \qquad \textbf{(7.25)}$$

where F_x is the component of $\vec{\mathbf{F}}$ in the direction of the displacement. That is, the work done by a conservative force acting between members of a system equals the negative of the change in the potential energy of the system associated with that force when the system's configuration changes. We can also express Equation 7.25 as

$$\Delta U = U_f - U_i = -\int_{x_i}^{x_f} F_x\,dx \qquad \textbf{(7.26)}$$

[4] For a general displacement, the work done in two or three dimensions also equals $-\Delta U$, where $U = U(x, y, z)$. We write this equation formally as $W_c = \int_i^f \vec{\mathbf{F}} \cdot d\vec{\mathbf{r}} = U_i - U_f$.

Therefore, ΔU is negative when F_x and dx are in the same direction, as when an object is lowered in a gravitational field or when a spring pushes an object toward equilibrium.

It is often convenient to establish some particular location x_i of one member of a system as representing a reference configuration and measure all potential energy differences with respect to it. We can then define the potential energy function as

$$U_f(x) = -\int_{x_i}^{x_f} F_x\,dx + U_i \tag{7.27}$$

The value of U_i is often taken to be zero for the reference configuration. It does not matter what value we assign to U_i because any nonzero value merely shifts $U_f(x)$ by a constant amount and only the *change* in potential energy is physically meaningful.

If the point of application of the force undergoes an infinitesimal displacement dx, we can express the infinitesimal change in the potential energy of the system dU as

$$dU = -F_x\,dx$$

Therefore, the conservative force is related to the potential energy function through the relationship[5]

Relation of force between members of a system to the potential energy of the system ▶

$$F_x = -\frac{dU}{dx} \tag{7.28}$$

That is, **the x component of a conservative force acting on an object within a system equals the negative derivative of the potential energy of the system with respect to x.**

We can easily check Equation 7.28 for the two examples already discussed. In the case of the deformed spring, $U_s = \frac{1}{2}kx^2$; therefore,

$$F_s = -\frac{dU_s}{dx} = -\frac{d}{dx}\left(\tfrac{1}{2}kx^2\right) = -kx$$

which corresponds to the restoring force in the spring (Hooke's law). Because the gravitational potential energy function is $U_g = mgy$, it follows from Equation 7.28 that $F_g = -mg$ when we differentiate U_g with respect to y instead of x.

We now see that U is an important function because a conservative force can be derived from it. Furthermore, Equation 7.28 should clarify that adding a constant to the potential energy is unimportant because the derivative of a constant is zero.

Quick Quiz 7.8 What does the slope of a graph of $U(x)$ versus x represent? (a) the magnitude of the force on the object (b) the negative of the magnitude of the force on the object (c) the x component of the force on the object (d) the negative of the x component of the force on the object

---

[5] In three dimensions, the expression is

$$\vec{\mathbf{F}} = -\frac{\partial U}{\partial x}\hat{\mathbf{i}} - \frac{\partial U}{\partial y}\hat{\mathbf{j}} - \frac{\partial U}{\partial z}\hat{\mathbf{k}}$$

where $(\partial U/\partial x)$ and so forth are partial derivatives. In the language of vector calculus, $\vec{\mathbf{F}}$ equals the negative of the *gradient* of the scalar quantity $U(x, y, z)$.

7.9 Energy Diagrams and Equilibrium of a System

The motion of a system can often be understood qualitatively through a graph of its potential energy versus the position of a member of the system. Consider the potential energy function for a block–spring system, given by $U_s = \frac{1}{2}kx^2$. This function is plotted versus x in Active Figure 7.20a. The force F_s exerted by the spring on the block is related to U_s through Equation 7.28:

$$F_s = -\frac{dU_s}{dx} = -kx$$

As we saw in Quick Quiz 7.8, the x component of the force is equal to the negative of the slope of the U-versus-x curve. When the block is placed at rest at the equilibrium position of the spring ($x = 0$), where $F_s = 0$, it will remain there unless some external force F_{ext} acts on it. If this external force stretches the spring from equilibrium, x is positive and the slope dU/dx is positive; therefore, the force F_s exerted by the spring is negative and the block accelerates back toward $x = 0$ when released. If the external force compresses the spring, x is negative and the slope is negative; therefore, F_s is positive and again the mass accelerates toward $x = 0$ upon release.

From this analysis, we conclude that the $x = 0$ position for a block–spring system is one of **stable equilibrium**. That is, any movement away from this position results in a force directed back toward $x = 0$. In general, **configurations of a system in stable equilibrium correspond to those for which $U(x)$ for the system is a minimum.**

If the block in Active Figure 7.20 is moved to an initial position x_{max} and then released from rest, its total energy initially is the potential energy $\frac{1}{2}kx_{max}^2$ stored in the spring. As the block starts to move, the system acquires kinetic energy and loses potential energy. The block oscillates (moves back and forth) between the two points $x = -x_{max}$ and $x = +x_{max}$, called the *turning points*. In fact, because no energy is transformed to internal energy due to friction, the block oscillates between $-x_{max}$ and $+x_{max}$ forever. (We discuss these oscillations further in Chapter 15.)

Another simple mechanical system with a configuration of stable equilibrium is a ball rolling about in the bottom of a bowl. Anytime the ball is displaced from its lowest position, it tends to return to that position when released.

Now consider a particle moving along the x axis under the influence of a conservative force F_x, where the U-versus-x curve is as shown in Figure 7.21. Once again, $F_x = 0$ at $x = 0$, and so the particle is in equilibrium at this point. This position, however, is one of **unstable equilibrium** for the following reason. Suppose the particle is displaced to the right ($x > 0$). Because the slope is negative for $x > 0$, $F_x = -dU/dx$ is positive and the particle accelerates away from $x = 0$. If instead the particle is at $x = 0$ and is displaced to the left ($x < 0$), the force is negative because the slope is positive for $x < 0$ and the particle again accelerates away from the equilibrium position. The position $x = 0$ in this situation is one of unstable equilibrium because for any displacement from this point, the force pushes the particle farther away from equilibrium and toward a position of lower potential energy. A pencil balanced on its point is in a position of unstable equilibrium. If the pencil is displaced slightly from its absolutely vertical position and is then released, it will surely fall over. In general, **configurations of a system in unstable equilibrium correspond to those for which $U(x)$ for the system is a maximum.**

Finally, a configuration called **neutral equilibrium** arises when U is constant over some region. Small displacements of an object from a position in this region produce neither restoring nor disrupting forces. A ball lying on a flat horizontal surface is an example of an object in neutral equilibrium.

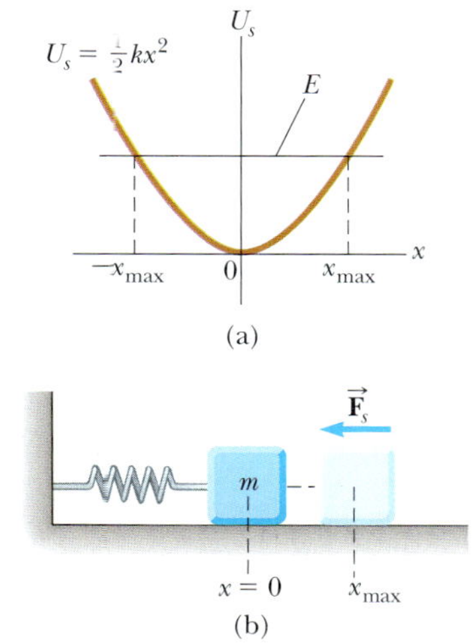

ACTIVE FIGURE 7.20

(a) Potential energy as a function of x for the frictionless block–spring system shown in (b). The block oscillates between the turning points, which have the coordinates $x = \pm x_{max}$. Notice that the restoring force exerted by the spring always acts toward $x = 0$, the position of stable equilibrium.

Sign in at www.thomsonedu.com and go to ThomsonNOW to observe the block oscillate between its turning points and trace the corresponding points on the potential energy curve for varying values of k.

PITFALL PREVENTION 7.11

Energy Diagrams

A common mistake is to think that potential energy on the graph in an energy diagram represents height. For example, that is not the case in Active Figure 7.20, where the block is only moving horizontally.

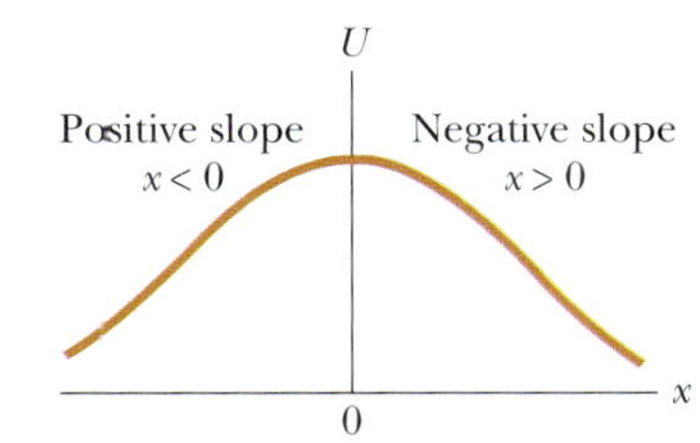

Figure 7.21 A plot of U versus x for a particle that has a position of unstable equilibrium located at $x = 0$. For any finite displacement of the particle, the force on the particle is directed away from $x = 0$.

EXAMPLE 7.9 Force and Energy on an Atomic Scale

The potential energy associated with the force between two neutral atoms in a molecule can be modeled by the Lennard–Jones potential energy function:

$$U(x) = 4\epsilon\left[\left(\frac{\sigma}{x}\right)^{12} - \left(\frac{\sigma}{x}\right)^{6}\right]$$

where x is the separation of the atoms. The function $U(x)$ contains two parameters σ and ϵ that are determined from experiments. Sample values for the interaction between two atoms in a molecule are $\sigma = 0.263$ nm and $\epsilon = 1.51 \times 10^{-22}$ J. Using a spreadsheet or similar tool, graph this function and find the most likely distance between the two atoms.

SOLUTION

Conceptualize We identify the two atoms in the molecule as a system. Based on our understanding that stable molecules exist, we expect to find stable equilibrium when the two atoms are separated by some equilibrium distance.

Categorize Because a potential energy function exists, we categorize the force between the atoms as conservative. For a conservative force, Equation 7.28 describes the relationship between the force and the potential energy function.

Analyze Stable equilibrium exists for a separation distance at which the potential energy of the system of two atoms (the molecule) is a minimum.

Take the derivative of the function $U(x)$:

$$\frac{dU(x)}{dx} = 4\epsilon\frac{d}{dx}\left[\left(\frac{\sigma}{x}\right)^{12} - \left(\frac{\sigma}{x}\right)^{6}\right] = 4\epsilon\left[\frac{-12\sigma^{12}}{x^{13}} + \frac{6\sigma^{6}}{x^{7}}\right]$$

Minimize the function $U(x)$ by setting its derivative equal to zero:

$$4\epsilon\left[\frac{-12\sigma^{12}}{x_{eq}^{13}} + \frac{6\sigma^{6}}{x_{eq}^{7}}\right] = 0 \quad\rightarrow\quad x_{eq} = (2)^{1/6}\sigma$$

Evaluate x_{eq}, the equilibrium separation of the two atoms in the molecule:

$$x_{eq} = (2)^{1/6}(0.263 \text{ nm}) = 2.95 \times 10^{-10} \text{ m}$$

We graph the Lennard–Jones function on both sides of this critical value to create our energy diagram as shown in Figure 7.22.

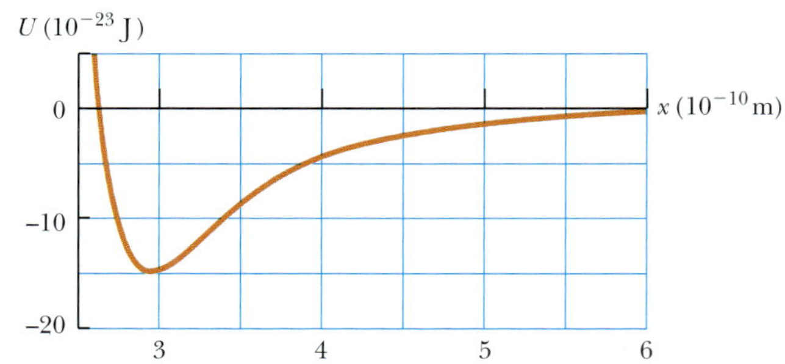

Figure 7.22 (Example 7.9) Potential energy curve associated with a molecule. The distance x is the separation between the two atoms making up the molecule.

Finalize Notice that $U(x)$ is extremely large when the atoms are very close together, is a minimum when the atoms are at their critical separation, and then increases again as the atoms move apart. When $U(x)$ is a minimum, the atoms are in stable equilibrium, indicating that the most likely separation between them occurs at this point.

Summary

ThomsonNOW™ Sign in at **www.thomsonedu.com** and go to ThomsonNOW to take a practice test for this chapter.

DEFINITIONS

A **system** is most often a single particle, a collection of particles, or a region of space, and may vary in size and shape. A **system boundary** separates the system from the **environment**.

The **work** W done on a system by an agent exerting a constant force $\vec{\mathbf{F}}$ on the system is the product of the magnitude Δr of the displacement of the point of application of the force and the component $F\cos\theta$ of the force along the direction of the displacement $\Delta\vec{\mathbf{r}}$:

$$W \equiv F\,\Delta r\cos\theta \qquad \textbf{(7.1)}$$

If a varying force does work on a particle as the particle moves along the x axis from x_i to x_f, the work done by the force on the particle is given by

$$W = \int_{x_i}^{x_f} F_x\,dx \qquad \textbf{(7.7)}$$

where F_x is the component of force in the x direction.

The **scalar product** (dot product) of two vectors $\vec{\mathbf{A}}$ and $\vec{\mathbf{B}}$ is defined by the relationship

$$\vec{\mathbf{A}}\cdot\vec{\mathbf{B}} \equiv AB\cos\theta \qquad \textbf{(7.2)}$$

where the result is a scalar quantity and θ is the angle between the two vectors. The scalar product obeys the commutative and distributive laws.

The **kinetic energy** of a particle of mass m moving with a speed v is

$$K \equiv \tfrac{1}{2}mv^2 \qquad \textbf{(7.16)}$$

If a particle of mass m is at a distance y above the Earth's surface, the **gravitational potential energy** of the particle–Earth system is

$$U_g \equiv mgy \qquad \textbf{(7.19)}$$

The **elastic potential energy** stored in a spring of force constant k is

$$U_s \equiv \tfrac{1}{2}kx^2 \qquad \textbf{(7.22)}$$

A force is **conservative** if the work it does on a particle that is a member of the system as the particle moves between two points is independent of the path the particle takes between the two points. Furthermore, a force is conservative if the work it does on a particle is zero when the particle moves through an arbitrary closed path and returns to its initial position. A force that does not meet these criteria is said to be **nonconservative.**

The **total mechanical energy of a system** is defined as the sum of the kinetic energy and the potential energy:

$$E_{\text{mech}} \equiv K + U \qquad \textbf{(7.24)}$$

CONCEPTS AND PRINCIPLES

The **work–kinetic energy theorem** states that if work is done on a system by external forces and the only change in the system is in its speed,

$$W_{\text{net}} = K_f - K_i = \Delta K = \tfrac{1}{2}mv_f^2 - \tfrac{1}{2}mv_i^2 \qquad \textbf{(7.15, 7.17)}$$

A **potential energy function** U can be associated only with a conservative force. If a conservative force $\vec{\mathbf{F}}$ acts between members of a system while one member moves along the x axis from x_i to x_f, the change in the potential energy of the system equals the negative of the work done by that force:

$$U_f - U_i = -\int_{x_i}^{x_f} F_x\,dx \qquad \textbf{(7.26)}$$

Systems can be in three types of equilibrium configurations when the net force on a member of the system is zero. Configurations of **stable equilibrium** correspond to those for which $U(x)$ is a minimum. Configurations of **unstable equilibrium** correspond to those for which $U(x)$ is a maximum. **Neutral equilibrium** arises when U is constant as a member of the system moves over some region.

Questions

☐ denotes answer available in *Student Solutions Manual/Study Guide;* **O** denotes objective question

1. Discuss whether any work is being done by each of the following agents and, if so, whether the work is positive or negative: (a) a chicken scratching the ground (b) a person studying (c) a crane lifting a bucket of concrete (d) the gravitational force on the bucket in part (c) (e) the leg muscles of a person in the act of sitting down

2. Cite two examples in which a force is exerted on an object without doing any work on the object.

3. As a simple pendulum swings back and forth, the forces acting on the suspended object are the gravitational force, the tension in the supporting cord, and air resistance. (a) Which of these forces, if any, does no work on the pendulum? (b) Which of these forces does negative work at all times during its motion? (c) Describe the work done by the gravitational force while the pendulum is swinging.

4. **O** Let $\hat{\mathbf{N}}$ represent the direction horizontally north, $\widehat{\mathbf{NE}}$ represent northeast (halfway between north and east), $\widehat{\mathbf{up}}$ represent vertically upward, and so on. Each direction specification can be thought of as a unit vector. Rank from the largest to the smallest the following dot products. Note that zero is larger than a negative number. If two quantities are equal, display that fact in your ranking. (a) $\hat{\mathbf{N}} \cdot \hat{\mathbf{N}}$ (b) $\hat{\mathbf{N}} \cdot \widehat{\mathbf{NE}}$ (c) $\hat{\mathbf{N}} \cdot \hat{\mathbf{S}}$ (d) $\hat{\mathbf{N}} \cdot \hat{\mathbf{E}}$ (e) $\hat{\mathbf{N}} \cdot \widehat{\mathbf{up}}$ (f) $\hat{\mathbf{E}} \cdot \hat{\mathbf{E}}$ (g) $\widehat{\mathbf{SE}} \cdot \hat{\mathbf{S}}$ (h) $\widehat{\mathbf{up}} \cdot \widehat{\mathbf{down}}$

5. For what values of the angle θ between two vectors is their scalar product (a) positive and (b) negative?

6. **O** Figure 7.9a shows a light extended spring exerting a force F_s to the left on a block. **(i)** Does the block exert a force on the spring? Choose every correct answer. (a) No, it does not. (b) Yes, to the left. (c) Yes, to the right. (d) Its magnitude is larger than F_s. (e) Its magnitude is equal to F_s. (f) Its magnitude is smaller than F_s. **(ii)** Does the spring exert a force on the wall? Choose every correct answer from the same list (a) through (f).

7. A certain uniform spring has spring constant k. Now the spring is cut in half. What is the relationship between k and the spring constant k' of each resulting smaller spring? Explain your reasoning.

8. Can kinetic energy be negative? Explain.

9. Discuss the work done by a pitcher throwing a baseball. What is the approximate distance through which the force acts as the ball is thrown?

10. **O** Bullet 2 has twice the mass of bullet 1. Both are fired so that they have the same speed. The kinetic energy of bullet 1 is K. The kinetic energy of bullet 2 is (a) $0.25K$ (b) $0.5K$ (c) $0.71K$ (d) K (e) $2K$ (f) $4K$

11. **O** If the speed of a particle is doubled, what happens to its kinetic energy? (a) It becomes four times larger. (b) It becomes two times larger. (c) It becomes $\sqrt{2}$ times larger. (d) It is unchanged. (e) It becomes half as large.

12. A student has the idea that the total work done on an object is equal to its final kinetic energy. Is this statement true always, sometimes, or never? If sometimes true, under what circumstances? If always or never, explain why.

13. Can a normal force do work? If not, why not? If so, give an example.

14. **O** What can be said about the speed of a particle if the net work done on it is zero? (a) It is zero. (b) It is decreased. (c) It is unchanged. (d) No conclusion can be drawn.

15. **O** A cart is set rolling across a level table, at the same speed on every trial. If it runs into a patch of sand, the cart exerts on the sand an average horizontal force of 6 N and travels a distance of 6 cm through the sand as it comes to a stop. **(i)** If instead the cart runs into a patch of gravel on which the cart exerts an average horizontal force of 9 N, how far into the gravel will the cart roll in stopping? Choose one answer. (a) 9 cm (b) 6 cm (c) 4 cm (d) 3 cm (e) none of these answers **(ii)** If instead the cart runs into a patch of flour, it rolls 18 cm before stopping. What is the average magnitude of the horizontal force that the cart exerts on the flour? (a) 2 N (b) 3 N (c) 6 N (d) 18 N (e) none of these answers **(iii)** If instead the cart runs into no obstacle at all, how far will it travel? (a) 6 cm (b) 18 cm (c) 36 cm (d) an infinite distance

16. The kinetic energy of an object depends on the frame of reference in which its motion is measured. Give an example to illustrate this point.

17. **O** Work in the amount 4 J is required to stretch a spring that is described by Hooke's law by 10 cm from its unstressed length. How much additional work is required to stretch the spring by an additional 10 cm? Choose one: (a) none (b) 2 J (c) 4 J (d) 8 J (e) 12 J (f) 16 J

18. If only one external force acts on a particle, does it necessarily change the particle's (a) kinetic energy? (b) Its velocity?

19. **O** **(i)** Rank the gravitational accelerations you would measure for (a) a 2-kg object 5 cm above the floor, (b) a 2-kg object 120 cm above the floor, (c) a 3-kg object 120 cm above the floor, and (d) a 3-kg object 80 cm above the floor. List the one with the largest-magnitude acceleration first. If two are equal, show their equality in your list. **(ii)** Rank the gravitational forces on the same four objects, largest magnitude first. **(iii)** Rank the gravitational potential energies (of the object–Earth system) for the same four objects, largest first, taking $y = 0$ at the floor.

20. You are reshelving books in a library. You lift a book from the floor to the top shelf. The kinetic energy of the book on the floor was zero and the kinetic energy of the book on the top shelf is zero, so no change occurs in the kinetic energy yet you did some work in lifting the book. Is the work–kinetic energy theorem violated?

21. Our body muscles exert forces when we lift, push, run, jump, and so forth. Are these forces conservative?

22. What shape would the graph of U versus x have if a particle were in a region of neutral equilibrium?

23. **O** An ice cube has been given a push and slides without friction on a level table. Which is correct? (a) It is in stable equilibrium. (b) It is in unstable equilibrium. (c) It is in neutral equilibrium (d) It is not in equilibrium.

24. Preparing to clean them, you pop all the removable keys off a computer keyboard. Each key has the shape of a tiny box with one side open. By accident, you spill the lot onto the floor. Explain why many more of them land letter-side down than land open-side down.

25. Who first stated the work-kinetic energy theorem? Who showed that it is useful for solving many practical problems? Do some research to answer these questions.

Problems

WebAssign The Problems from this chapter may be assigned online in WebAssign.

ThomsonNOW™ Sign in at **www.thomsonedu.com** and go to ThomsonNOW to assess your understanding of this chapter's topics with additional quizzing and conceptual questions.

1, 2, 3 denotes straightforward, intermediate, challenging; ☐ denotes full solution available in *Student Solutions Manual/Study Guide;* ▲ denotes coached solution with hints available at **www.thomsonedu.com;** ▭ denotes developing symbolic reasoning; ● denotes asking for qualitative reasoning; 🖥 denotes computer useful in solving problem

Section 7.2 Work Done by a Constant Force

1. A block of mass 2.50 kg is pushed 2.20 m along a frictionless horizontal table by a constant 16.0-N force directed 25.0° below the horizontal. Determine the work done on the block by (a) the applied force, (b) the normal force exerted by the table, and (c) the gravitational force. (d) Determine the net work done on the block.

2. A raindrop of mass 3.35×10^{-5} kg falls vertically at constant speed under the influence of gravity and air resistance. Model the drop as a particle. As it falls 100 m, what is the work done on the raindrop (a) by the gravitational force and (b) by air resistance?

3. ▲ Batman, whose mass is 80.0 kg, is dangling on the free end of a 12.0-m rope, the other end of which is fixed to a tree limb above. By repeatedly bending at the waist, he is able to get the rope in motion, eventually making it swing enough that he can reach a ledge when the rope makes a 60.0° angle with the vertical. How much work was done by the gravitational force on Batman in this maneuver?

4. ● Object 1 pushes on object 2 as the objects move together, like a bulldozer pushing a stone. Assume object 1 does 15.0 J of work on object 2. Does object 2 do work on object 1? Explain your answer. If possible, determine how much work, and explain your reasoning.

Section 7.3 The Scalar Product of Two Vectors

5. For any two vectors $\vec{\mathbf{A}}$ and $\vec{\mathbf{B}}$, show that $\vec{\mathbf{A}}\cdot\vec{\mathbf{B}} = A_xB_x + A_yB_y + A_zB_z$. *Suggestion:* Write $\vec{\mathbf{A}}$ and $\vec{\mathbf{B}}$ in unit–vector form and use Equations 7.4 and 7.5.

6. Vector $\vec{\mathbf{A}}$ has a magnitude of 5.00 units and $\vec{\mathbf{B}}$ has a magnitude of 9.00 units. The two vectors make an angle of 50.0° with each other. Find $\vec{\mathbf{A}}\cdot\vec{\mathbf{B}}$.

> *Note:* In Problems 7 through 10, calculate numerical answers to three significant figures as usual.

7. ▲ A force $\vec{\mathbf{F}} = (6\hat{\mathbf{i}} - 2\hat{\mathbf{j}})$ N acts on a particle that undergoes a displacement $\Delta\vec{\mathbf{r}} = (3\hat{\mathbf{i}} + \hat{\mathbf{j}})$ m. Find (a) the work done by the force on the particle and (b) the angle between $\vec{\mathbf{F}}$ and $\Delta\vec{\mathbf{r}}$.

8. Find the scalar product of the vectors in Figure P7.8.

9. Using the definition of the scalar product, find the angles between the following: (a) $\vec{\mathbf{A}} = 3\hat{\mathbf{i}} - 2\hat{\mathbf{j}}$ and $\vec{\mathbf{B}} = 4\hat{\mathbf{i}} - 4\hat{\mathbf{j}}$ (b) $\vec{\mathbf{A}} = -2\hat{\mathbf{i}} + 4\hat{\mathbf{j}}$ and $\vec{\mathbf{B}} = 3\hat{\mathbf{i}} - 4\hat{\mathbf{j}} + 2\hat{\mathbf{k}}$ (c) $\vec{\mathbf{A}} = \hat{\mathbf{i}} - 2\hat{\mathbf{j}} + 2\hat{\mathbf{k}}$ and $\vec{\mathbf{B}} = 3\hat{\mathbf{j}} + 4\hat{\mathbf{k}}$.

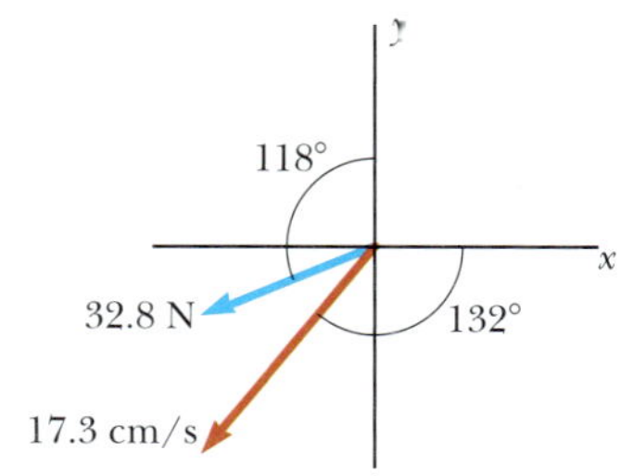

Figure P7.8

10. For the vectors $\vec{\mathbf{A}} = 3\hat{\mathbf{i}} + \hat{\mathbf{j}} - \hat{\mathbf{k}}$, $\vec{\mathbf{B}} = -\hat{\mathbf{i}} + 2\hat{\mathbf{j}} + 5\hat{\mathbf{k}}$, and $\vec{\mathbf{C}} = 2\hat{\mathbf{j}} - 3\hat{\mathbf{k}}$, find $\vec{\mathbf{C}}\cdot(\vec{\mathbf{A}} - \vec{\mathbf{B}})$.

11. Let $\vec{\mathbf{B}}$ = 5.00 m at 60.0°. Let $\vec{\mathbf{C}}$ have the same magnitude as $\vec{\mathbf{A}}$ and a direction angle greater than that of $\vec{\mathbf{A}}$ by 25.0°. Let $\vec{\mathbf{A}}\cdot\vec{\mathbf{B}} = 30.0\ \text{m}^2$ and $\vec{\mathbf{B}}\cdot\vec{\mathbf{C}} = 35.0\ \text{m}^2$. Find $\vec{\mathbf{A}}$.

Section 7.4 Work Done by a Varying Force

12. The force acting on a particle is $F_x = (8x - 16)$ N, where x is in meters. (a) Make a plot of this force versus x from $x = 0$ to $x = 3.00$ m. (b) From your graph, find the net work done by this force on the particle as it moves from $x = 0$ to $x = 3.00$ m.

13. The force acting on a particle varies as shown in Figure P7.13. Find the work done by the force on the particle as it moves (a) from $x = 0$ to $x = 8.00$ m, (b) from $x = 8.00$ m to $x = 10.0$ m, and (c) from $x = 0$ to $x = 10.0$ m.

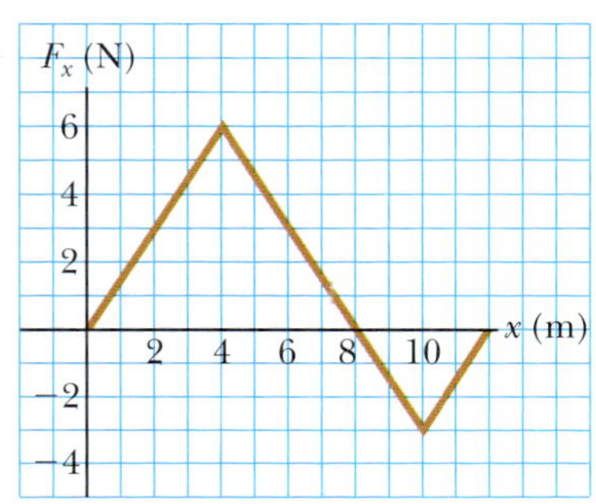

Figure P7.13

14. A force $\vec{\mathbf{F}} = (4x\hat{\mathbf{i}} + 3y\hat{\mathbf{j}})$ N acts on an object as the object moves in the x direction from the origin to $x = 5.00$ m. Find the work $W = \int\vec{\mathbf{F}}\cdot d\vec{\mathbf{r}}$ done by the force on the object.

15. ▲ A particle is subject to a force F_x that varies with position as shown in Figure P7.15. Find the work done by the force on the particle as it moves (a) from $x = 0$ to $x = 5.00$ m, (b) from $x = 5.00$ m to $x = 10.0$ m, and (c) from $x = 10.0$ m to $x = 15.0$ m. (d) What is the total work done by the force over the distance $x = 0$ to $x = 15.0$ m?

Figure P7.15 Problems 15 and 32.

16. An archer pulls her bowstring back 0.400 m by exerting a force that increases uniformly from zero to 230 N. (a) What is the equivalent spring constant of the bow? (b) How much work does the archer do in drawing the bow?

17. When a 4.00-kg object is hung vertically on a certain light spring described by Hooke's law, the spring stretches 2.50 cm. If the 4.00-kg object is removed, (a) how far will the spring stretch if a 1.50-kg block is hung on it? (b) How much work must an external agent do to stretch the same spring 4.00 cm from its unstretched position?

18. Hooke's law describes a certain light spring of unstressed length 35.0 cm. When one end is attached to the top of a door frame and a 7.50-kg object is hung from the other end, the length of the spring is 41.5 cm. (a) Find its spring constant. (b) The load and the spring are taken down. Two people pull in opposite directions on the ends of the spring, each with a force of 190 N. Find the length of the spring in this situation.

19. In a control system, an accelerometer consists of a 4.70-g object sliding on a horizontal rail. A low-mass spring attaches the object to a flange at one end of the rail. Grease on the rail makes static friction negligible, but rapidly damps out vibrations of the sliding object. When the accelerometer moves with a steady acceleration of 0.800g, the object is to assume a location 0.500 cm away from its equilibrium position. Find the force constant required for the spring.

20. A light spring with force constant 3.85 N/m is compressed by 8.00 cm as it is held between a 0.250-kg block on the left and a 0.500-kg block on the right, both resting on a horizontal surface. The spring exerts a force on each block, tending to push them apart. The blocks are simultaneously released from rest. Find the acceleration with which each block starts to move, given that the coefficient of kinetic friction between each block and the surface is (a) 0, (b) 0.100, and (c) 0.462.

21. A 6 000-kg freight car rolls along rails with negligible friction. The car is brought to rest by a combination of two coiled springs as illustrated in Figure P7.21. Both springs are described by Hooke's law with $k_1 = 1\,600$ N/m and $k_2 = 3\,400$ N/m. After the first spring compresses a distance of 30.0 cm, the second spring acts with the first to increase the force as additional compression occurs as shown in the graph. The car comes to rest 50.0 cm after first contacting the two-spring system. Find the car's initial speed.

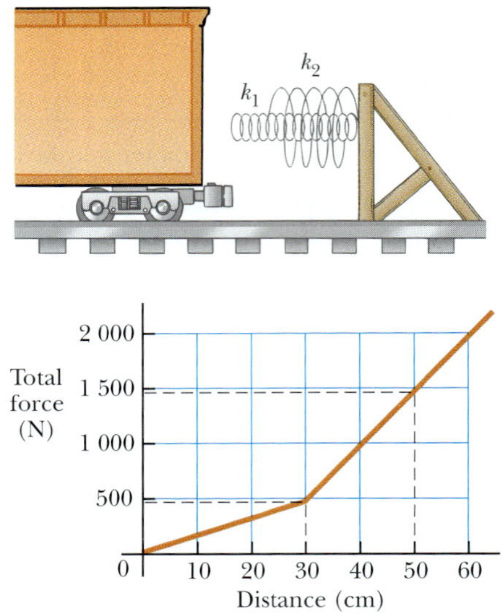

Figure P7.21

22. A 100-g bullet is fired from a rifle having a barrel 0.600 m long. Choose the origin to be at the location where the bullet begins to move. Then the force (in newtons) exerted by the expanding gas on the bullet is $15\,000 + 10\,000x - 25\,000x^2$, where x is in meters. (a) Determine the work done by the gas on the bullet as the bullet travels the length of the barrel. (b) **What If?** If the barrel is 1.00 m long, how much work is done, and how does this value compare with the work calculated in part (a)?

23. A light spring with spring constant 1 200 N/m hangs from an elevated support. From its lower end hangs a second light spring, which has spring constant 1 800 N/m. An object of mass 1.50 kg hangs at rest from the lower end of the second spring. (a) Find the total extension distance of the pair of springs. (b) Find the effective spring constant of the pair of springs as a system. We describe these springs as *in series.*

24. A light spring with spring constant k_1 hangs from an elevated support. From its lower end hangs a second light spring, which has spring constant k_2. An object of mass m hangs at rest from the lower end of the second spring. (a) Find the total extension distance of the pair of springs. (b) Find the effective spring constant of the pair of springs as a system. We describe these springs as *in series.*

25. A small particle of mass m is pulled to the top of a frictionless half-cylinder (of radius R) by a cord that passes over the top of the cylinder as illustrated in Figure P7.25. (a) Assuming the particle moves at a constant speed, show that $F = mg \cos \theta$. *Note:* If the particle moves at constant speed, the component of its acceleration tangent to the cylinder must be zero at all times. (b) By directly integrating $W = \int \vec{\mathbf{F}} \cdot d\vec{\mathbf{r}}$, find the work done in moving the particle at constant speed from the bottom to the top of the half-cylinder.

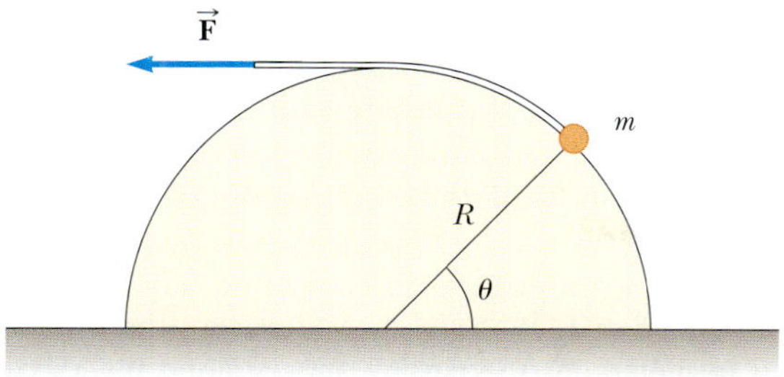

Figure P7.25

26. Express the units of the force constant of a spring in SI fundamental units.

27. **Review problem.** The graph in Figure P7.27 specifies a functional relationship between the two variables u and v.

(a) Find $\int_a^b u\,dv$. (b) Find $\int_b^a u\,dv$. (c) Find $\int_a^b v\,du$.

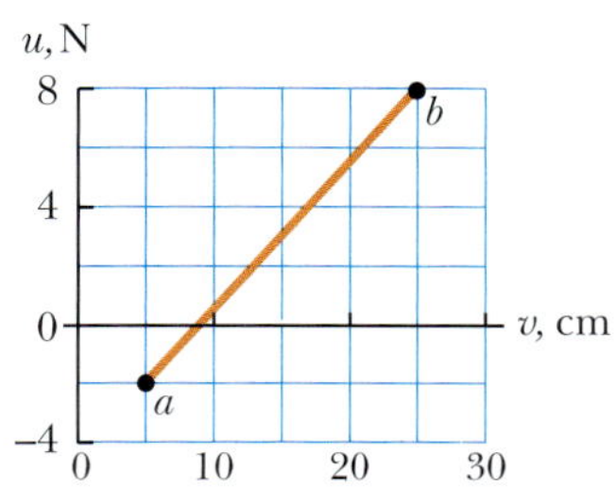

Figure P7.27

28. A cafeteria tray dispenser supports a stack of trays on a shelf that hangs from four identical spiral springs under tension, one near each corner of the shelf. Each tray is rectangular, 45.3 cm by 35.6 cm, 0.450 cm thick, and with mass 580 g. Demonstrate that the top tray in the stack can always be at the same height above the floor, however many trays are in the dispenser. Find the spring constant each spring should have for the dispenser to function in this convenient way. Is any piece of data unnecessary for this determination?

Section 7.5 Kinetic Energy and the Work–Kinetic Energy Theorem

29. A 0.600-kg particle has a speed of 2.00 m/s at point Ⓐ and kinetic energy of 7.50 J at point Ⓑ. What are (a) its kinetic energy at Ⓐ, (b) its speed at Ⓑ, and (c) the net work done on the particle as it moves from Ⓐ to Ⓑ?

30. A 0.300-kg ball has a speed of 15.0 m/s. (a) What is its kinetic energy? (b) **What If?** If its speed were doubled, what would be its kinetic energy?

31. A 3.00-kg object has a velocity of $(6.00\hat{\mathbf{i}} - 2.00\hat{\mathbf{j}})$ m/s. (a) What is its kinetic energy at this moment? (b) What is the net work done on the object if its velocity changes to $(8.00\hat{\mathbf{i}} + 4.00\hat{\mathbf{j}})$ m/s? *Note:* From the definition of the dot product, $v^2 = \vec{\mathbf{v}} \cdot \vec{\mathbf{v}}$.

32. A 4.00-kg particle is subject to a net force that varies with position as shown in Figure P7.15. The particle starts moving at $x = 0$, very nearly from rest. What is its speed at (a) $x = 5.00$ m, (b) $x = 10.0$ m, and (c) $x = 15.0$ m?

33. A 2 100-kg pile driver is used to drive a steel I-beam into the ground. The pile driver falls 5.00 m before coming into contact with the top of the beam. Then it drives the beam 12.0 cm farther into the ground as it comes to rest. Using energy considerations, calculate the average force the beam exerts on the pile driver while the pile driver is brought to rest.

34. ● A 300-g cart is rolling along a straight track with velocity $0.600\hat{\mathbf{i}}$ m/s at $x = 0$. A student holds a magnet in front of the cart to temporarily pull forward on it, and then the cart runs into a dusting of sand that turns into a small pile. These effects are represented quantitatively by the graph of the x component of the net force on the cart as a function of position in Figure P7.34. (a) Will the cart roll all the way through the pile of sand? Explain how you can tell. (b) If so, find the speed at which it exits at $x = 7.00$ cm. If not, what maximum x coordinate does it reach?

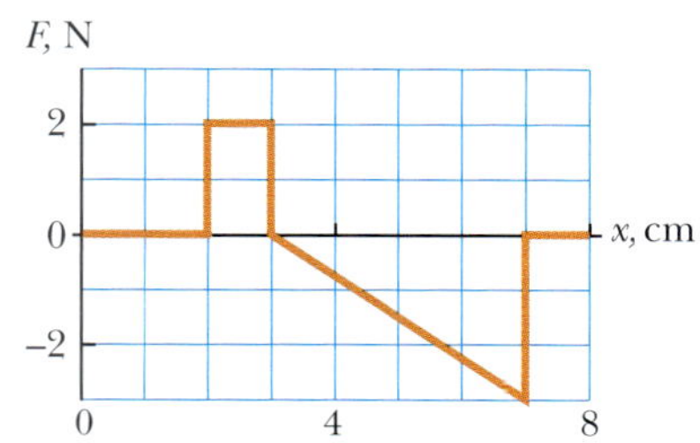

Figure P7.34

35. ● You can think of the work–kinetic energy theorem as a second theory of motion, parallel to Newton's laws in describing how outside influences affect the motion of an object. In this problem, solve parts (a) and (b) separately from parts (c) and (d) so that you can compare the predictions of the two theories. In a rifle barrel, a 15.0-g bullet is accelerated from rest to a speed of 780 m/s. (a) Find the work that is done on the bullet. (b) Assuming the rifle barrel is 72.0 cm long, find the magnitude of the average net force that acted on it, as $\Sigma F = W/(\Delta r \cos\theta)$. (c) Find the constant acceleration of a bullet that starts from rest and gains a speed of 780 m/s over a distance of 72.0 cm. (d) Assuming now the bullet has mass 15.0 g, find the net force that acted on it as $\Sigma F = ma$. (e) What conclusion can you draw from comparing your results?

36. In the neck of the picture tube of a certain black-and-white television set, an electron gun contains two charged metallic plates 2.80 cm apart. An electric force accelerates each electron in the beam from rest to 9.60% of the speed of light over this distance. (a) Determine the kinetic energy of the electron as it leaves the electron gun. Electrons carry this energy to a phosphorescent material on the inner surface of the television screen, making it glow. For an electron passing between the plates in the electron gun, determine (b) the magnitude of the constant electric force acting on the electron, (c) the acceleration, and (d) the time of flight.

Section 7.6 Potential Energy of a System

37. A 1 000-kg roller-coaster car is initially at the top of a rise, at point Ⓐ. It then moves 135 ft, at an angle of 40.0° below the horizontal, to a lower point Ⓑ. (a) Choose the car at point Ⓑ to be the zero configuration for gravitational potential energy of the roller coaster–Earth system. Find the potential energy of the system when the car is at points Ⓐ and Ⓑ and the change in potential energy as the car moves. (b) Repeat part (a), setting the zero configuration with the car at point Ⓐ.

38. A 400-N child is in a swing that is attached to ropes 2.00 m long. Find the gravitational potential energy of the child–Earth system relative to the child's lowest position when (a) the ropes are horizontal, (b) the ropes make a 30.0° angle with the vertical, and (c) the child is at the bottom of the circular arc.

Section 7.7 Conservative and Nonconservative Forces

39. ● A 4.00-kg particle moves from the origin to position *C*, having coordinates x = 5.00 m and y = 5.00 m (Fig. P7.39). One force on the particle is the gravitational force acting in the negative y direction. Using Equation 7.3, calculate the work done by the gravitational force on the particle as it goes from *O* to *C* along (a) *OAC*, (b) *OBC*, and (c) *OC*. Your results should all be identical. Why?

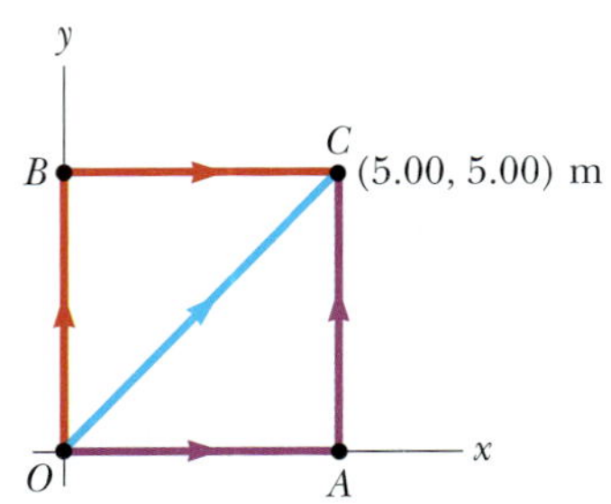

Fig. P7.39 Problems 39 through 42.

40. (a) Suppose a constant force acts on an object. The force does not vary with time or with the position or the velocity of the object. Start with the general definition for work done by a force

$$W = \int_i^f \vec{\mathbf{F}} \cdot d\vec{\mathbf{r}}$$

and show the force is conservative. (b) As a special case, suppose the force $\vec{\mathbf{F}} = (3\hat{\mathbf{i}} + 4\hat{\mathbf{j}})$ N acts on a particle that moves from *O* to *C* in Figure P7.39. Calculate the work done by $\vec{\mathbf{F}}$ on the particle as it moves along each one of the three paths *OAC*, *OBC*, and *OC*. Check that your three answers are identical.

41. ● A force acting on a particle moving in the xy plane is given by $\vec{\mathbf{F}} = (2y\hat{\mathbf{i}} + x^2\hat{\mathbf{j}})$ N, where x and y are in meters. The particle moves from the origin to a final position having coordinates x = 5.00 m and y = 5.00 m as shown in Figure P7.39. Calculate the work done by $\vec{\mathbf{F}}$ on the particle as it moves along (a) *OAC*, (b) *OBC*, and (c) *OC*. (d) Is $\vec{\mathbf{F}}$ conservative or nonconservative? Explain.

42. ● A particle moves in the xy plane in Figure P7.39 under the influence of a friction force with magnitude 3.00 N and acting in the direction opposite to the particle's displacement. Calculate the work done by the friction force on the particle as it moves along the following closed paths: (a) the path *OA* followed by the return path *AO*, (b) the path *OA* followed by *AC* and the return path *CO*, and (c) the path *OC* followed by the return path *CO*. (d) Each of your three answers should be nonzero. What is the significance of this observation?

Section 7.8 Relationship Between Conservative Forces and Potential Energy

43. ▲ A single conservative force acts on a 5.00-kg particle. The equation $F_x = (2x + 4)$ N describes the force, where x is in meters. As the particle moves along the x axis from x = 1.00 m to x = 5.00 m, calculate (a) the work done by this force on the particle, (b) the change in the potential energy of the system, and (c) the kinetic energy the particle has at x = 5.00 m if its speed is 3.00 m/s at x = 1.00 m.

44. A single conservative force acting on a particle varies as $\vec{\mathbf{F}} = (-Ax + Bx^2)\hat{\mathbf{i}}$ N, where A and B are constants and x is in meters. (a) Calculate the potential energy function $U(x)$ associated with this force, taking $U = 0$ at $x = 0$. (b) Find the change in potential energy and the change in kinetic energy of the system as the particle moves from x = 2.00 m to x = 3.00 m.

45. ▲ The potential energy of a system of two particles separated by a distance r is given by $U(r) = A/r$, where A is a constant. Find the radial force $\vec{\mathbf{F}}_r$ that each particle exerts on the other.

46. A potential energy function for a two-dimensional force is of the form $U = 3x^3y - 7x$. Find the force that acts at the point (x, y).

Section 7.9 Energy Diagrams and Equilibrium of a System

47. For the potential energy curve shown in Figure P7.47, (a) determine whether the force F_x is positive, negative, or zero at the five points indicated. (b) Indicate points of stable, unstable, and neutral equilibrium. (c) Sketch the curve for F_x versus x from $x = 0$ to x = 9.5 m.

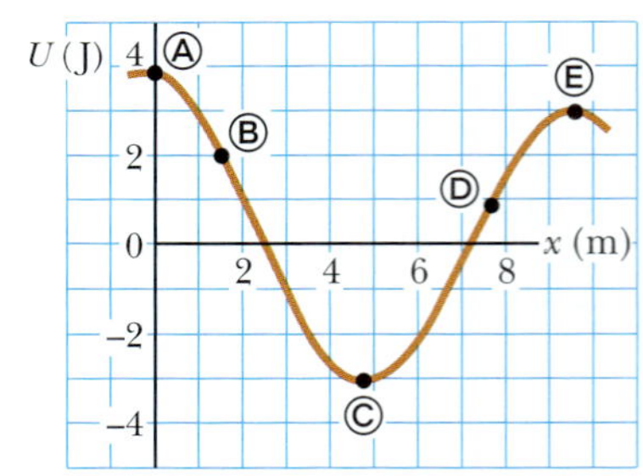

Figure P7.47

48. A right circular cone can be balanced on a horizontal surface in three different ways. Sketch these three equilibrium configurations and identify them as positions of stable, unstable, or neutral equilibrium.

49. A particle of mass 1.18 kg is attached between two identical springs on a horizontal, frictionless tabletop. Both springs have spring constant k and are initially unstressed. (a) The particle is pulled a distance x along a direction perpendicular to the initial configuration of the springs as shown in Figure P7.49. Show that the force exerted by the springs on the particle is

$$\vec{\mathbf{F}} = -2kx\left(1 - \frac{L}{\sqrt{x^2 + L^2}}\right)\hat{\mathbf{i}}$$

(b) Show that the potential energy of the system is

$$U(x) = kx^2 + 2kL\left(L - \sqrt{x^2 + L^2}\right)$$

(c) Make a plot of $U(x)$ versus x and identify all equilibrium points. Assume $L = 1.20$ m and $k = 40.0$ N/m. (d) If the particle is pulled 0.500 m to the right and then released, what is its speed when it reaches the equilibrium point $x = 0$?

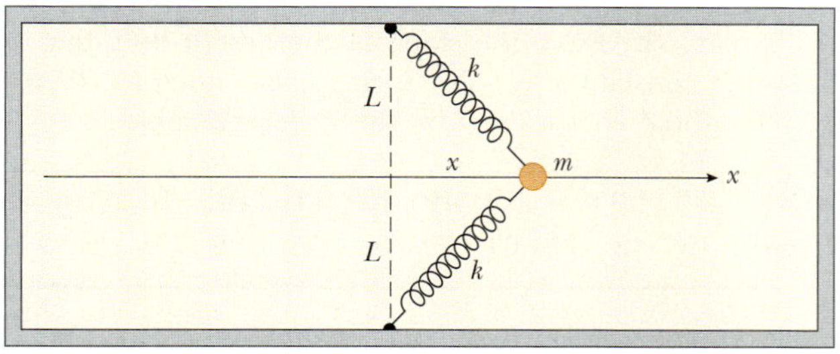

Figure P7.49

Additional Problems

50. A bead at the bottom of a bowl is one example of an object in a stable equilibrium position. When a physical system is displaced by an amount x from stable equilibrium, a restoring force acts on it, tending to return the system to its equilibrium configuration. The magnitude of the restoring force can be a complicated function of x. For example, when an ion in a crystal is displaced from its lattice site, the restoring force may not be a simple function of x. In such cases, we can generally imagine the function $F(x)$ to be expressed as a power series in x as $F(x) = -(k_1x + k_2x^2 + k_3x^3 + \ldots)$. The first term here is Hooke's law, which describes the force exerted by a simple spring for small displacements. For small excursions from equilibrium we generally ignore the higher-order terms; in some cases, however, it may be desirable to keep the second term as well. If we model the restoring force as $F = -(k_1x + k_2x^2)$, how much work is done in displacing the system from $x = 0$ to $x = x_{max}$ by an applied force $-F$?

51. A baseball outfielder throws a 0.150-kg baseball at a speed of 40.0 m/s and an initial angle of 30.0°. What is the kinetic energy of the baseball at the highest point of its trajectory?

52. The spring constant of a car's suspension spring increases with increasing load due to a spring coil that is widest at the bottom, smoothly tapering to a smaller diameter near the top. The result is a softer ride on normal road surfaces from the wider coils, but the car does not bottom out on bumps because when the lower coils collapse, the stiffer coils near the top absorb the load. For a tapered spiral spring that compresses 12.9 cm with a 1 000-N load and 31.5 cm with a 5 000-N load (a) evaluate the constants a and b in the empirical equation $F = ax^b$ and (b) find the work needed to compress the spring 25.0 cm.

53. ● A light spring has an unstressed length of 15.5 cm. It is described by Hooke's law with spring constant 4.30 N/m. One end of the horizontal spring is held on a fixed vertical axle, and the other end is attached to a puck of mass m that can move without friction over a horizontal surface. The puck is set into motion in a circle with a period of 1.30 s. (a) Find the extension of the spring x as it depends on m. Evaluate x for (b) $m = 0.070\,0$ kg, (c) $m = 0.140$ kg, (d) $m = 0.180$ kg, and (e) $m = 0.190$ kg. (f) Describe the pattern of variation of x as it depends on m.

54. Two steel balls, each of diameter 25.4 mm, moving in opposite directions at 5 m/s, run into each other head-on and bounce apart. (a) Does their interaction last only for an instant or for a nonzero time interval? State your evidence. One of the balls is squeezed in a vise while precise measurements are made of the resulting amount of compression. The results show that Hooke's law is a fair model of the ball's elastic behavior. For one datum, a force of 16 kN exerted by each jaw of the vise results in a 0.2-mm reduction in the ball's diameter. The diameter returns to its original value when the force is removed. (b) Modeling the ball as a spring, find its spring constant. (c) Compute an estimate for the kinetic energy of each of the balls before they collide. In your solution, explain your logic. (d) Compute an estimate for the maximum amount of compression each ball undergoes when they collide. (e) Compute an order-of-magnitude estimate for the time interval for which the balls are in contact. In your solution, explain your reasoning. (In Chapter 15, you will learn to calculate the contact time precisely in this model.)

55. ● Take $U = 5$ at $x = 0$ and calculate the potential energy, as a function of x, corresponding to the force $(8e^{-2x})\,\hat{\mathbf{i}}$. Explain whether the force is conservative or nonconservative and how you can tell.

56. The potential energy function for a system is given by $U(x) = -x^3 + 2x^2 + 3x$. (a) Determine the force F_x as a function of x. (b) For what values of x is the force equal to zero? (c) Plot $U(x)$ versus x and F_x versus x and indicate points of stable and unstable equilibrium.

57. The ball launcher in a pinball machine has a spring that has a force constant of 1.20 N/cm (Fig. P7.57). The surface on which the ball moves is inclined 10.0° with respect to the horizontal. The spring is initially compressed 5.00 cm. Find the launching speed of a 100-g ball when the plunger is released. Friction and the mass of the plunger are negligible.

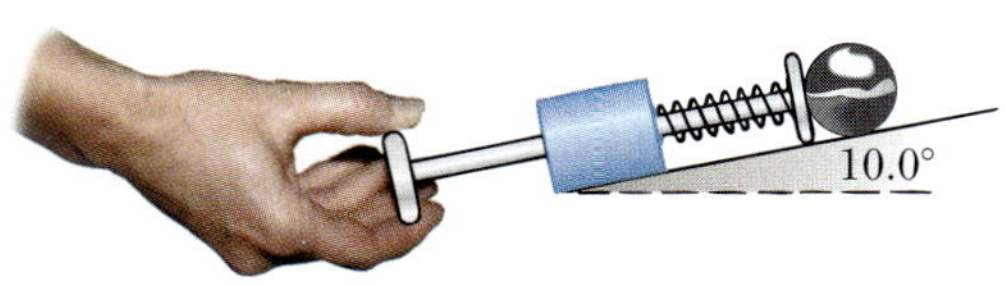

Figure P7.57

58. ● **Review problem.** Two constant forces act on a 5.00-kg object moving in the xy plane as shown in Figure P7.58. Force $\vec{\mathbf{F}}_1$ is 25.0 N at 35.0° and $\vec{\mathbf{F}}_2$ is 42.0 N at 150°. At time $t = 0$, the object is at the origin and has velocity $(4.00\hat{\mathbf{i}} + 2.50\hat{\mathbf{j}})$ m/s. (a) Express the two forces in unit-vector notation. Use unit-vector notation for your other answers. (b) Find the total force exerted on the object. (c) Find the object's acceleration. Now, considering the instant $t = 3.00$ s, (d) find the object's velocity, (e) its position, (f) its kinetic energy from $\frac{1}{2}mv_f^2$, and (g) its kinetic

energy from $\frac{1}{2}mv_i^2 + \Sigma \vec{\mathbf{F}} \cdot \Delta\vec{\mathbf{r}}$. (h) What conclusion can you draw by comparing the answers to parts (f) and (g)?

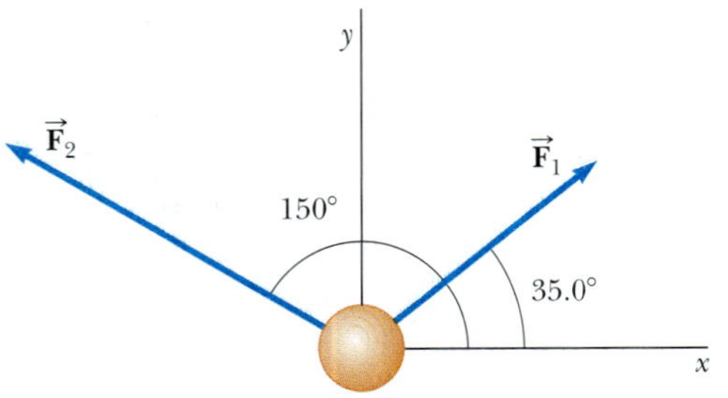

Figure P7.58

59. A particle moves along the x axis from $x = 12.8$ m to $x = 23.7$ m under the influence of a force

$$F = \frac{375}{x^3 + 3.75x}$$

where F is in newtons and x is in meters. Using numerical integration, determine the work done by this force on the particle during this displacement. Your result should be accurate to within 2%.

60. ● When different loads hang on a spring, the spring stretches to different lengths as shown in the following table. (a) Make a graph of the applied force versus the extension of the spring. By least-squares fitting, determine the straight line that best fits the data. Do you want to use all the data points, or should you ignore some of them? Explain. (b) From the slope of the best-fit line, find the spring constant k. (c) The spring is extended to 105 mm. What force does it exert on the suspended object?

***F* (N)**	2.0	4.0	6.0	8.0	10	12	14	16	18	20	22
***L* (mm)**	15	32	49	64	79	98	112	126	149	175	190

Answers to Quick Quizzes

7.1 (a). The force does no work on the Earth because the force is pointed toward the center of the circle and is therefore perpendicular to the direction of its displacement.

7.2 (c), (a), (d), (b). The work done in (c) is positive and of the largest possible value because the angle between the force and the displacement is zero. The work done in (a) is zero because the force is perpendicular to the displacement. In (d) and (b), negative work is done by the applied force because in neither case is there a component of the force in the direction of the displacement. Situation (b) is the most negative value because the angle between the force and the displacement is 180°.

7.3 (d). Because of the range of values of the cosine function, $\vec{\mathbf{A}} \cdot \vec{\mathbf{B}}$ has values that range from AB to $-AB$.

7.4 (a). Because the work done in compressing a spring is proportional to the square of the compression distance x, doubling the value of x causes the work to increase fourfold.

7.5 (b). Because the work is proportional to the square of the compression distance x and the kinetic energy is proportional to the square of the speed v, doubling the compression distance doubles the speed.

7.6 (c). The sign of the gravitational potential energy depends on your choice of zero configuration. If the two objects in the system are closer together than in the zero configuration, the potential energy is negative. If they are farther apart, the potential energy is positive.

7.7 **(i)**, (c). This system exhibits changes in kinetic energy as well as in both types of potential energy. **(ii)**, (a). Because the Earth is not included in the system, there is no gravitational potential energy associated with the system.

7.8 (d). The slope of a $U(x)$-versus-x graph is by definition $dU(x)/dx$. From Equation 7.28, we see that this expression is equal to the negative of the x component of the conservative force acting on an object that is part of the system.

As a skier coasts down a hill, the skier–snow–Earth system experiences changes in kinetic energy, related to the speed of the skier; potential energy, related to the altitude of the skier; and internal energy, related to the temperature of the skis, the snow, and the air. If the total energy of this system were to be evaluated at several instants during this process, the result would be the same at all times. One application of the *conservation of energy principle,* to be discussed in this chapter, is that the total energy of an isolated system remains constant. (Jean Y. Ruszniewski/Getty Images)

8 Conservation of Energy

In Chapter 7, we introduced three methods for storing energy in a system: kinetic energy, associated with movement of members of the system; potential energy, determined by the configuration of the system; and internal energy, which is related to the temperature of the system.

We now consider analyzing physical situations using the energy approach for two types of systems: *nonisolated* and *isolated* systems. For nonisolated systems, we shall investigate ways that energy can cross the boundary of the system, resulting in a change in the system's total energy. This analysis leads to a critically important principle called *conservation of energy.* The conservation of energy principle extends well beyond physics and can be applied to biological organisms, technological systems, and engineering situations.

In isolated systems, energy does not cross the boundary of the system. For these systems, the total energy of the system is constant. If no nonconservative forces act within the system, we can use *conservation of mechanical energy* to solve a variety of problems.

Situations involving the transformation of mechanical energy to internal energy due to nonconservative forces require special handling. We investigate the procedures for these types of problems.

Finally, we recognize that energy can cross the boundary of a system at different rates. We describe the rate of energy transfer with the quantity *power.*

8.1 The Nonisolated System: Conservation of Energy

As we have seen, an object, modeled as a particle, can be acted on by various forces, resulting in a change in its kinetic energy. This very simple situation is the first example of the model of a **nonisolated system**, for which energy crosses the boundary of the system during some time interval due to an interaction with the environment. This scenario is common in physics problems. If a system does not interact with its environment, it is an isolated system, which we will study in Section 8.2.

The work–kinetic energy theorem from Chapter 7 is our first example of an energy equation appropriate for a nonisolated system. In the case of that theorem, the interaction of the system with its environment is the work done by the external force, and the quantity in the system that changes is the kinetic energy.

So far, we have seen only one way to transfer energy into a system: work. We mention below a few other ways to transfer energy into or out of a system. The details of these processes will be studied in other sections of the book. We illustrate mechanisms to transfer energy in Figure 8.1 and summarize them as follows.

Work, as we have learned in Chapter 7, is a method of transferring energy to a system by applying a force to the system and causing a displacement of the point of application of the force (Fig. 8.1a).

Mechanical waves (Chapters 16–18) are a means of transferring energy by allowing a disturbance to propagate through air or another medium. It is the method by which energy (which you detect as sound) leaves your clock radio through the loudspeaker and enters your ears to stimulate the hearing process (Fig. 8.1b). Other examples of mechanical waves are seismic waves and ocean waves.

PITFALL PREVENTION 8.1
Heat Is Not a Form of Energy

The word *heat* is one of the most misused words in our popular language. Heat is a method of *transferring* energy, *not* a form of storing energy. Therefore, phrases such as "heat content," "the heat of the summer," and "the heat escaped" all represent uses of this word that are inconsistent with our physics definition. See Chapter 20.

Heat (Chapter 20) is a mechanism of energy transfer that is driven by a temperature difference between two regions in space. For example, the handle of a metal spoon in a cup of coffee becomes hot because fast-moving electrons and atoms in the submerged portion of the spoon bump into slower ones in the nearby part of the handle (Fig. 8.1c). These particles move faster because of the collisions and bump into the next group of slow particles. Therefore, the internal energy of the spoon handle rises from energy transfer due to this collision process.

Matter transfer (Chapter 20) involves situations in which matter physically crosses the boundary of a system, carrying energy with it. Examples include filling

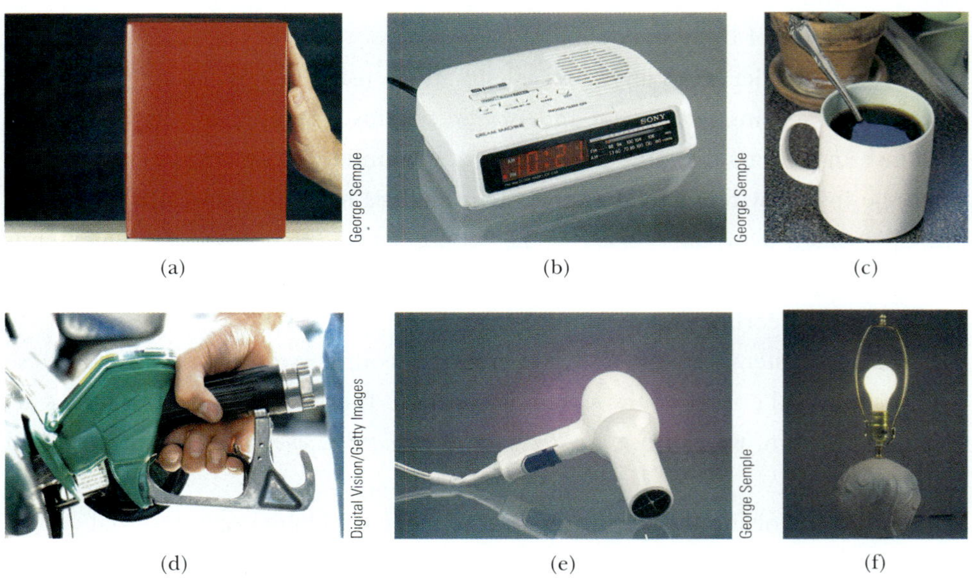

Figure 8.1 Energy transfer mechanisms. (a) Energy is transferred to the block by *work*; (b) energy leaves the radio from the speaker by *mechanical waves*; (c) energy transfers up the handle of the spoon by *heat*; (d) energy enters the automobile gas tank by *matter transfer*; (e) energy enters the hair dryer by *electrical transmission*; and (f) energy leaves the light bulb by *electromagnetic radiation*.

your automobile tank with gasoline (Fig. 8.1d) and carrying energy to the rooms of your home by circulating warm air from the furnace, a process called *convection*.

Electrical transmission (Chapters 27 and 28) involves energy transfer by means of electric currents. It is how energy transfers into your hair dryer (Fig. 8.1e), stereo system, or any other electrical device.

Electromagnetic radiation (Chapter 34) refers to electromagnetic waves such as light, microwaves, and radio waves (Fig. 8.1f). Examples of this method of transfer include cooking a baked potato in your microwave oven and light energy traveling from the Sun to the Earth through space.[1]

A central feature of the energy approach is the notion that we can neither create nor destroy energy, that energy is always *conserved*. This feature has been tested in countless experiments, and no experiment has ever shown this statement to be incorrect. Therefore, **if the total amount of energy in a system changes, it can *only* be because energy has crossed the boundary of the system by a transfer mechanism such as one of the methods listed above**. This general statement of the principle of **conservation of energy** can be described mathematically with the **conservation of energy equation** as follows:

$$\Delta E_{\text{system}} = \sum T \qquad \textbf{(8.1)}$$

◀ Conservation of energy

where E_{system} is the total energy of the system, including all methods of energy storage (kinetic, potential, and internal) and T (for *transfer*) is the amount of energy transferred across the system boundary by some mechanism. Two of our transfer mechanisms have well-established symbolic notations. For work, $T_{\text{work}} = W$ as discussed in Chapter 7, and for heat, $T_{\text{heat}} = Q$ as defined in Chapter 20. The other four members of our list do not have established symbols, so we will call them T_{MW} (mechanical waves), T_{MT} (matter transfer), T_{ET} (electrical transmission), and T_{ER} (electromagnetic radiation).

The full expansion of Equation 8.1 is

$$\Delta K + \Delta U + \Delta E_{\text{int}} = W + Q + T_{\text{MW}} + T_{\text{MT}} + T_{\text{ET}} + T_{\text{ER}} \qquad \textbf{(8.2)}$$

which is the primary mathematical representation of the energy version of the **nonisolated system model**. (We will see other versions, involving linear momentum and angular momentum, in later chapters.) In most cases, Equation 8.2 reduces to a much simpler one because some of the terms are zero. If, for a given system, all terms on the right side of the conservation of energy equation are zero, the system is an *isolated system*, which we study in the next section.

The conservation of energy equation is no more complicated in theory than the process of balancing your checking account statement. If your account is the system, the change in the account balance for a given month is the sum of all the transfers: deposits, withdrawals, fees, interest, and checks written. You may find it useful to think of energy as the *currency of nature*!

Suppose a force is applied to a nonisolated system and the point of application of the force moves through a displacement. Then suppose the only effect on the system is to change its speed. In this case, the only transfer mechanism is work (so that the right side of Equation 8.2 reduces to just W) and the only kind of energy in the system that changes is the kinetic energy (so that ΔE_{system} reduces to just ΔK). Equation 8.2 then becomes

$$\Delta K = W$$

which is the work–kinetic energy theorem. This theorem is a special case of the more general principle of conservation of energy. We shall see several more special cases in future chapters.

[1] Electromagnetic radiation and work done by field forces are the only energy transfer mechanisms that do not require molecules of the environment to be available at the system boundary. Therefore, systems surrounded by a vacuum (such as planets) can only exchange energy with the environment by means of these two possibilities.

Quick Quiz 8.1 By what transfer mechanisms does energy enter and leave (a) your television set? (b) Your gasoline-powered lawn mower? (c) Your hand-cranked pencil sharpener?

Quick Quiz 8.2 Consider a block sliding over a horizontal surface with friction. Ignore any sound the sliding might make. **(i)** If the system is the *block*, this system is (a) isolated (b) nonisolated (c) impossible to determine **(ii)** If the system is the *surface*, describe the system from the same set of choices. **(iii)** If the system is the *block and the surface*, describe the system from the same set of choices.

8.2 The Isolated System

In this section, we study another very common scenario in physics problems: an **isolated system**, for which no energy crosses the system boundary by any method. We begin by considering a gravitational situation. Think about the book–Earth system in Active Figure 7.15 in the preceding chapter. After we have lifted the book, there is gravitational potential energy stored in the system, which can be calculated from the work done by the external agent on the system, using $W = \Delta U_g$.

Let us now shift our focus to the work done on the book alone by the gravitational force (Fig. 8.2) as the book falls back to its original height. As the book falls from y_i to y_f, the work done by the gravitational force on the book is

$$W_{\text{on book}} = (m\vec{\mathbf{g}})\cdot\Delta\vec{\mathbf{r}} = (-mg\hat{\mathbf{j}})\cdot[(y_f - y_i)\hat{\mathbf{j}}] = mgy_i - mgy_f \quad \textbf{(8.3)}$$

From the work–kinetic energy theorem of Chapter 7, the work done on the book is equal to the change in the kinetic energy of the book:

$$W_{\text{on book}} = \Delta K_{\text{book}}$$

We can equate these two expressions for the work done on the book:

$$\Delta K_{\text{book}} = mgy_i - mgy_f \quad \textbf{(8.4)}$$

Let us now relate each side of this equation to the *system* of the book and the Earth. For the right-hand side,

$$mgy_i - mgy_f = -(mgy_f - mgy_i) = -\Delta U_g$$

where $U_g = mgy$ is the gravitational potential energy of the system. For the left-hand side of Equation 8.4, because the book is the only part of the system that is moving, we see that $\Delta K_{\text{book}} = \Delta K$, where K is the kinetic energy of the system. Therefore, with each side of Equation 8.4 replaced with its system equivalent, the equation becomes

$$\Delta K = -\Delta U_g \quad \textbf{(8.5)}$$

This equation can be manipulated to provide a very important general result for solving problems. First, we move the change in potential energy to the left side of the equation:

$$\Delta K + \Delta U_g = 0$$

The left side represents a sum of changes of the energy stored in the system. The right-hand side is zero because there are no transfers of energy across the boundary of the system; the book–Earth system is *isolated* from the environment. We developed this equation for a gravitational system, but it can be shown to be valid for a system with any type of potential energy. Therefore, for an isolated system,

$$\Delta K + \Delta U = 0 \quad \textbf{(8.6)}$$

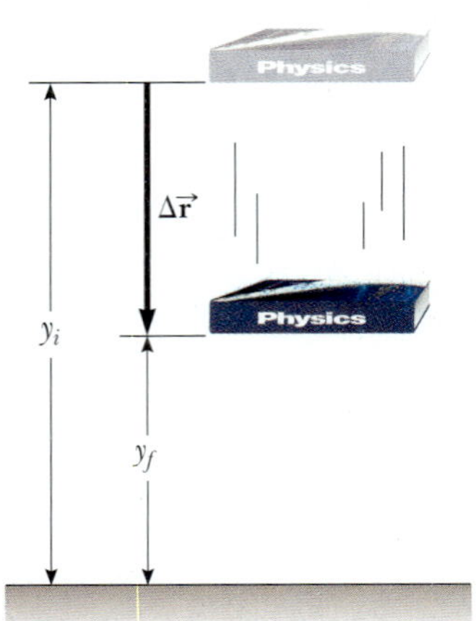

Figure 8.2 The work done by the gravitational force on the book as the book falls from y_i to a height y_f is equal to $mgy_i - mgy_f$.

We defined in Chapter 7 the sum of the kinetic and potential energies of a system as its mechanical energy:

$$E_{\text{mech}} \equiv K + U \quad \textbf{(8.7)}$$

◀ Mechanical energy of a system

where U represents the total of *all* types of potential energy. Because the system under consideration is isolated, Equations 8.6 and 8.7 tell us that the mechanical energy of the system is conserved:

$$\Delta E_{\text{mech}} = 0 \quad \textbf{(8.8)}$$

◀ The mechanical energy of an isolated system with no nonconservative forces acting is conserved.

Equation 8.8 is a statement of **conservation of mechanical energy** for an isolated system with no nonconservative forces acting. The mechanical energy in such a system is conserved: the sum of the kinetic and potential energies remains constant.

If there are nonconservative forces acting within the system, mechanical energy is transformed to internal energy as discussed in Section 7.7. If nonconservative forces act in an isolated system, the total energy of the system is conserved although the mechanical energy is not. In that case, we can express the conservation of energy of the system as

$$\Delta E_{\text{system}} = 0 \quad \textbf{(8.9)}$$

◀ The total energy of an isolated system is conserved.

where E_{system} includes all kinetic, potential, and internal energies. This equation is the most general statement of the **isolated system model**.

Let us now write the changes in energy in Equation 8.6 explicitly:

$$(K_f - K_i) + (U_f - U_i) = 0$$

$$K_f + U_f = K_i + U_i \quad \textbf{(8.10)}$$

For the gravitational situation of the falling book, Equation 8.10 can be written as

$$\tfrac{1}{2}mv_f^2 + mgy_f = \tfrac{1}{2}mv_i^2 + mgy_i$$

As the book falls to the Earth, the book–Earth system loses potential energy and gains kinetic energy such that the total of the two types of energy always remains constant.

PITFALL PREVENTION 8.2

Conditions on Equation 8.10

Equation 8.10 is only true for a system in which conservative forces act. We will see how to handle nonconservative forces in Sections 8.3 and 8.4.

Quick Quiz 8.3 A rock of mass m is dropped to the ground from a height h. A second rock, with mass $2m$, is dropped from the same height. When the second rock strikes the ground, what is its kinetic energy? (a) twice that of the first rock (b) four times that of the first rock (c) the same as that of the first rock (d) half as much as that of the first rock (e) impossible to determine

Quick Quiz 8.4 Three identical balls are thrown from the top of a building, all with the same initial speed. As shown in Active Figure 8.3, the first is thrown horizontally, the second at some angle above the horizontal, and the third at some angle below the horizontal. Neglecting air resistance, rank the speeds of the balls at the instant each hits the ground.

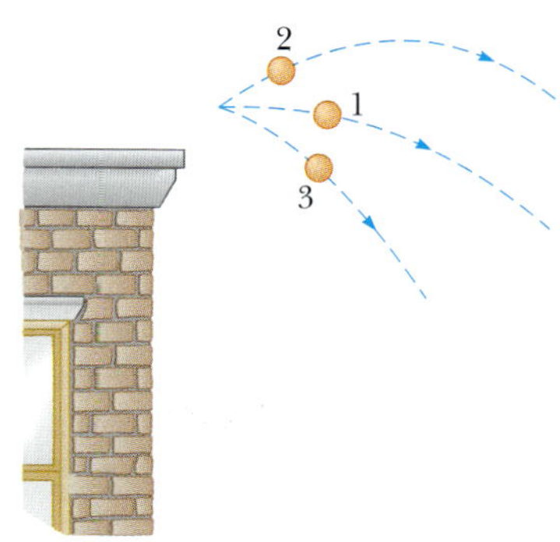

ACTIVE FIGURE 8.3

(Quick Quiz 8.4) Three identical balls are thrown with the same initial speed from the top of a building.

Sign in at www.thomsonedu.com and go to ThomsonNOW to throw balls at different angles from the top of the building and compare the trajectories and the speeds as the balls hit the ground.

PROBLEM SOLVING STRATEGY — Isolated Systems with No Nonconservative Forces: Conservation of Mechanical Energy

Many problems in physics can be solved using the principle of conservation of energy for an isolated system. The following procedure should be used when you apply this principle:

1. *Conceptualize.* Study the physical situation carefully and form a mental representation of what is happening. As you become more proficient working energy problems, you will begin to be comfortable imagining the types of energy that are changing in the system.

2. *Categorize.* Define your system, which may consist of more than one object and may or may not include springs or other possibilities for storing potential energy. Determine if any energy transfers occur across the boundary of your system. If so, use the nonisolated system model, $\Delta E_{\text{system}} = \Sigma T$, from Section 8.1. If not, use the isolated system model, $\Delta E_{\text{system}} = 0$.

   Determine whether any nonconservative forces are present within the system. If so, use the techniques of Sections 8.3 and 8.4. If not, use the principle of conservation of mechanical energy as outlined below.

3. *Analyze.* Choose configurations to represent the initial and final conditions of the system. For each object that changes elevation, select a reference position for the object that defines the zero configuration of gravitational potential energy for the system. For an object on a spring, the zero configuration for elastic potential energy is when the object is at its equilibrium position. If there is more than one conservative force, write an expression for the potential energy associated with each force.

   Write the total initial mechanical energy E_i of the system for some configuration as the sum of the kinetic and potential energies associated with the configuration. Then write a similar expression for the total mechanical energy E_f of the system for the final configuration that is of interest. Because mechanical energy is *conserved,* equate the two total energies and solve for the quantity that is unknown.

4. *Finalize.* Make sure your results are consistent with your mental representation. Also make sure the values of your results are reasonable and consistent with connections to everyday experience.

EXAMPLE 8.1 Ball in Free Fall

A ball of mass m is dropped from a height h above the ground as shown in Active Figure 8.4.

(A) Neglecting air resistance, determine the speed of the ball when it is at a height y above the ground.

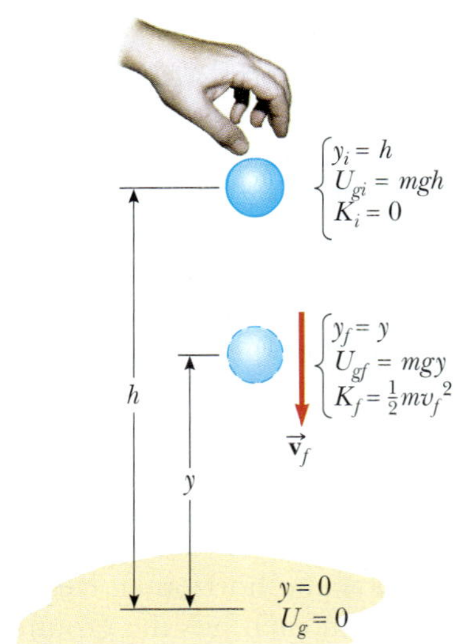

ACTIVE FIGURE 8.4

(Example 8.1) A ball is dropped from a height h above the ground. Initially, the total energy of the ball–Earth system is gravitational potential energy, equal to mgh relative to the ground. At the elevation y, the total energy is the sum of the kinetic and potential energies.

Sign in at www.thomsonedu.com and go to ThomsonNOW to drop the ball and watch energy bar charts for the ball–Earth system.

SOLUTION

Conceptualize Active Figure 8.4 and our everyday experience with falling objects allow us to conceptualize the situation. Although we can readily solve this problem with the techniques of Chapter 2, let us practice an energy approach.

Categorize We identify the system as the ball and the Earth. Because there is neither air resistance nor any other interactions between the system and the environment, the system is isolated. The only force between members of the system is the gravitational force, which is conservative.

Analyze Because the system is isolated and there are no nonconservative forces acting within the system, we apply the principle of conservation of mechanical energy to the ball–Earth system. At the instant the ball is released, its kinetic energy is $K_i = 0$ and the gravitational potential energy of the system is $U_{gi} = mgh$. When the ball is at a distance y above the ground, its kinetic energy is $K_f = \frac{1}{2}mv_f^2$ and the potential energy relative to the ground is $U_{gf} = mgy$.

Apply Equation 8.10:

$$K_f + U_{gf} = K_i + U_{gi}$$

$$\tfrac{1}{2}mv_f^2 + mgy = 0 + mgh$$

Solve for v_f: $$v_f^2 = 2g(h - y) \rightarrow v_f = \sqrt{2g(h - y)}$$

The speed is always positive. If you had been asked to find the ball's velocity, you would use the negative value of the square root as the y component to indicate the downward motion.

(B) Determine the speed of the ball at y if at the instant of release it already has an initial upward speed v_i at the initial altitude h.

SOLUTION

Analyze In this case, the initial energy includes kinetic energy equal to $\frac{1}{2}mv_i^2$.

Apply Equation 8.10: $$\tfrac{1}{2}mv_f^2 + mgy = \tfrac{1}{2}mv_i^2 + mgh$$

Solve for v_f: $$v_f^2 = v_i^2 + 2g(h - y) \rightarrow v_f = \sqrt{v_i^2 + 2g(h - y)}$$

Finalize This result for the final speed is consistent with the expression $v_{yf}^2 = v_{yi}^2 - 2g(y_f - y_i)$ from kinematics, where $y_i = h$. Furthermore, this result is valid even if the initial velocity is at an angle to the horizontal (Quick Quiz 8.4) for two reasons: (1) the kinetic energy, a scalar, depends only on the magnitude of the velocity; and (2) the change in the gravitational potential energy of the system depends only on the change in position of the ball in the vertical direction.

What If? What if the initial velocity $\vec{\mathbf{v}}_i$ in part (B) were downward? How would that affect the speed of the ball at position y?

Answer You might claim that throwing the ball downward would result in it having a higher speed at y than if you threw it upward. Conservation of mechanical energy, however, depends on kinetic and potential energies, which are scalars. Therefore, the direction of the initial velocity vector has no bearing on the final speed.

EXAMPLE 8.2 A Grand Entrance

You are designing an apparatus to support an actor of mass 65 kg who is to "fly" down to the stage during the performance of a play. You attach the actor's harness to a 130-kg sandbag by means of a lightweight steel cable running smoothly over two frictionless pulleys as in Figure 8.5a. You need 3.0 m of cable between the harness and the nearest pulley so that the pulley can be hidden behind a curtain. For the apparatus to work successfully, the sandbag must never lift above the floor as the actor swings from above the stage to the floor. Let us call the initial angle that the actor's cable makes with the vertical θ. What is the maximum value θ can have before the sandbag lifts off the floor?

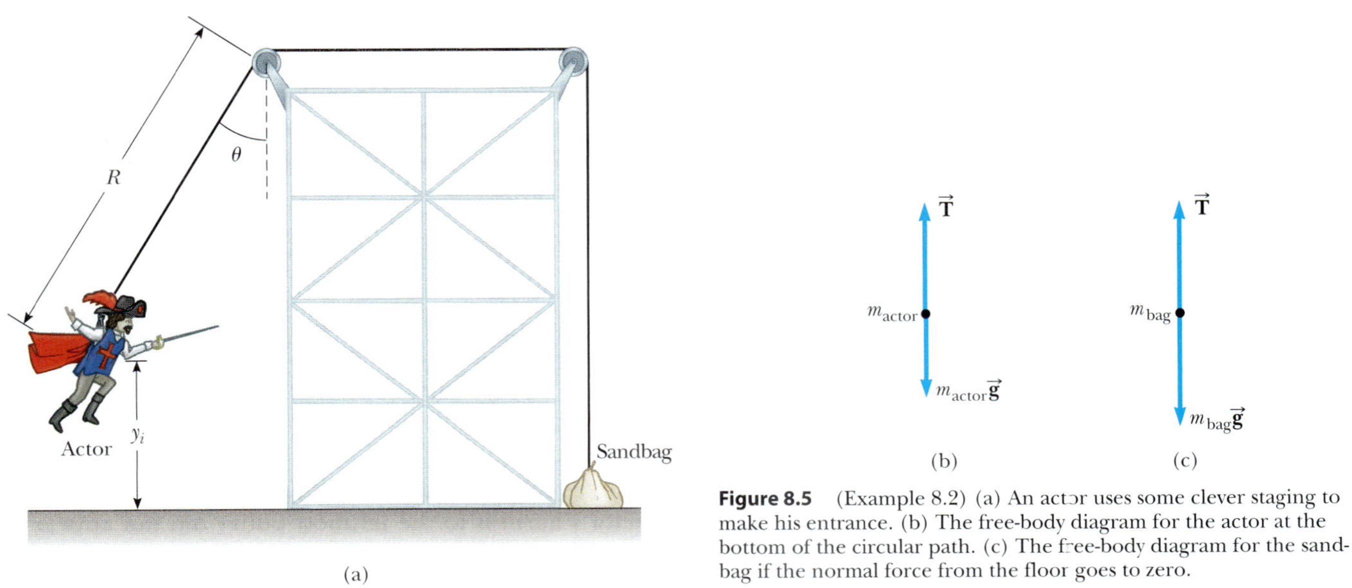

Figure 8.5 (Example 8.2) (a) An actor uses some clever staging to make his entrance. (b) The free-body diagram for the actor at the bottom of the circular path. (c) The free-body diagram for the sandbag if the normal force from the floor goes to zero.

SOLUTION

Conceptualize We must use several concepts to solve this problem. Imagine what happens as the actor approaches the bottom of the swing. At the bottom, the cable is vertical and must support his weight as well as provide centripetal acceleration of his body in the upward direction. At this point, the tension in the cable is the highest and the sandbag is most likely to lift off the floor.

Categorize Looking first at the swinging of the actor from the initial point to the lowest point, we model the actor and the Earth as an isolated system. We ignore air resistance, so there are no nonconservative forces acting. You might initially be tempted to model the system as nonisolated because of the interaction of the system with the cable, which is in the environment. The force applied to the actor by the cable, however, is always perpendicular to each element of the displacement of the actor and hence does no work. Therefore, in terms of energy transfers across the boundary, the system is isolated.

Analyze We use the principle of conservation of mechanical energy for the system to find the actor's speed as he arrives at the floor as a function of the initial angle θ and the radius R of the circular path through which he swings.

Apply conservation of mechanical energy to the actor–Earth system:

$$K_f + U_f = K_i + U_i$$

Let y_i be the initial height of the actor above the floor and v_f be his speed at the instant before he lands. (Notice that $K_i = 0$ because the actor starts from rest and that $U_f = 0$ because we define the configuration of the actor at the floor as having a gravitational potential energy of zero.)

$$(1)\quad \tfrac{1}{2}m_{\text{actor}}\, v_f^{\,2} + 0 = 0 + m_{\text{actor}}\, g y_i$$

From the geometry in Figure 8.5a, notice that $y_f = 0$, so $y_i = R - R\cos\theta = R(1 - \cos\theta)$. Use this relationship in Equation (1) and solve for $v_f^{\,2}$:

$$(2)\quad v_f^{\,2} = 2gR(1 - \cos\theta)$$

Categorize Next, focus on the instant the actor is at the lowest point. Because the tension in the cable is transferred as a force applied to the sandbag, we model the actor at this instant as a particle under a net force.

Analyze Apply Newton's second law to the actor at the bottom of his path, using the free-body diagram in Figure 8.5b as a guide:

$$\sum F_y = T - m_{\text{actor}}\, g = m_{\text{actor}}\frac{v_f^{\,2}}{R}$$

$$(3)\quad T = m_{\text{actor}}\, g + m_{\text{actor}}\frac{v_f^{\,2}}{R}$$

Categorize Finally, notice that the sandbag lifts off the floor when the upward force exerted on it by the cable exceeds the gravitational force acting on it; the normal force is zero when that happens. We do *not*, however, want the sandbag to lift off the floor. The sandbag must remain at rest, so we model it as a particle in equilibrium.

Analyze A force T of the magnitude given by Equation (3) is transmitted by the cable to the sandbag. If the sandbag remains at rest but is just ready to be lifted off the floor if any more force were applied by the cable, the normal force on it becomes zero and Newton's second law with $a = 0$ tells us that $T = m_{\text{bag}}\, g$ as in Figure 8.5c.

Use this condition together with Equations (2) and (3):

$$m_{\text{bag}}\, g = m_{\text{actor}}\, g + m_{\text{actor}}\frac{2gR(1 - \cos\theta)}{R}$$

Solve for $\cos\theta$ and substitute the given parameters:

$$\cos\theta = \frac{3m_{\text{actor}} - m_{\text{bag}}}{2m_{\text{actor}}} = \frac{3(65\text{ kg}) - 130\text{ kg}}{2(65\text{ kg})} = 0.50$$

$$\theta = 60°$$

Finalize Here we had to combine techniques from different areas of our study, energy and Newton's second law. Furthermore, notice that the length R of the cable from the actor's harness to the leftmost pulley did not appear in the final algebraic equation. Therefore, the final answer is independent of R.

EXAMPLE 8.3 The Spring-Loaded Popgun

The launching mechanism of a popgun consists of a spring of unknown spring constant (Active Fig. 8.6a). When the spring is compressed 0.120 m, the gun, when fired vertically, is able to launch a 35.0-g projectile to a maximum height of 20.0 m above the position of the projectile as it leaves the spring.

(A) Neglecting all resistive forces, determine the spring constant.

SOLUTION

Conceptualize Imagine the process illustrated in Active Figure 8.6. The projectile starts from rest, speeds up as the spring pushes upward on it, leaves the spring, and then slows down as the gravitational force pulls downward on it.

Categorize We identify the system as the projectile, the spring, and the Earth. We ignore air resistance on the projectile and friction in the gun, so we model the system as isolated with no nonconservative forces acting.

Analyze Because the projectile starts from rest, its initial kinetic energy is zero. We choose the zero configuration for the gravitational potential energy of the system to be when the projectile leaves the spring. For this configuration, the elastic potential energy is also zero.

After the gun is fired, the projectile rises to a maximum height $y_{Ⓒ}$. The final kinetic energy of the projectile is zero.

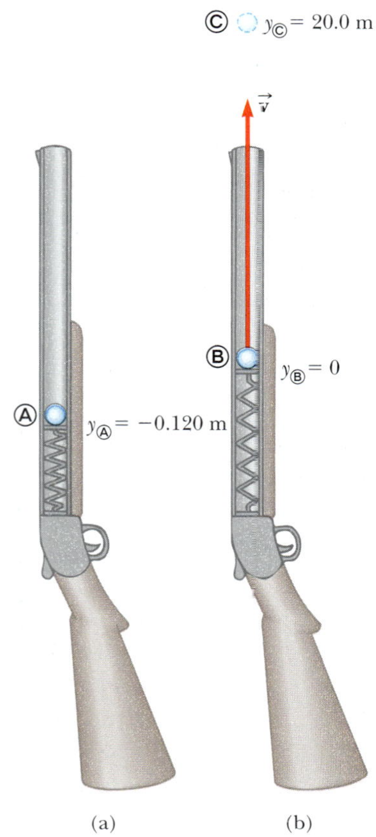

ACTIVE FIGURE 8.6

(Example 8.3) A spring-loaded popgun (a) before firing and (b) when the spring extends to its relaxed length.

Sign in at www.thomsonedu.com and go to ThomsonNOW to fire the gun and watch the energy changes in the projectile–spring–Earth system.

Write a conservation of mechanical energy equation for the system between points Ⓐ and Ⓒ:

$$K_{Ⓒ} + U_{gⒸ} + U_{sⒸ} = K_{Ⓐ} + U_{gⒶ} + U_{sⒶ}$$

Substitute for each energy:

$$0 + mgy_{Ⓒ} + 0 = 0 + mgy_{Ⓐ} + \tfrac{1}{2}kx^2$$

Solve for k:

$$k = \frac{2mg(y_{Ⓒ} - y_{Ⓐ})}{x^2}$$

Substitute numerical values:

$$k = \frac{2(0.035\,0 \text{ kg})(9.80 \text{ m/s}^2)[20.0 \text{ m} - (-0.120 \text{ m})]}{(0.120 \text{ m})^2} = 958 \text{ N/m}$$

(B) Find the speed of the projectile as it moves through the equilibrium position of the spring as shown in Active Figure 8.6b.

SOLUTION

Analyze The energy of the system as the projectile moves through the equilibrium position of the spring includes only the kinetic energy of the projectile $\frac{1}{2}mv_{Ⓑ}^2$.

Write a conservation of mechanical energy equation for the system between points Ⓐ and Ⓑ:

$$K_{Ⓑ} + U_{gⒷ} + U_{sⒷ} = K_{Ⓐ} + U_{gⒶ} + U_{sⒶ}$$

Substitute for each energy: $\frac{1}{2}mv_{Ⓑ}^2 + 0 + 0 = 0 + mgy_{Ⓐ} + \frac{1}{2}kx^2$

Solve for $v_{Ⓑ}$: $v_{Ⓑ} = \sqrt{\frac{kx^2}{m} + 2gy_{Ⓐ}}$

Substitute numerical values: $v_{Ⓑ} = \sqrt{\frac{(958\text{ N/m})(0.120\text{ m})^2}{(0.035\ 0\text{ kg})} + 2(9.80\text{ m/s}^2)(-0.120\text{ m})} = 19.8\text{ m/s}$

Finalize This example is the first one we have seen in which we must include two different types of potential energy.

8.3 Situations Involving Kinetic Friction

Consider again the book in Active Figure 7.18 sliding to the right on the surface of a heavy table and slowing down due to the friction force. Work is done by the friction force because there is a force and a displacement. Keep in mind, however, that our equations for work involve the displacement *of the point of application of the force.* A simple model of the friction force between the book and the surface is shown in Figure 8.7a. We have represented the entire friction force between the book and surface as being due to two identical teeth that have been spot-welded together.[2] One tooth projects upward from the surface, the other downward from the book, and they are welded at the points where they touch. The friction force acts at the junction of the two teeth. Imagine that the book slides a small distance d to the right as in Figure 8.7b. Because the teeth are modeled as identical, the junction of the teeth moves to the right by a distance $d/2$. Therefore, the displacement of the point of application of the friction force is $d/2$, but the displacement of the book is d!

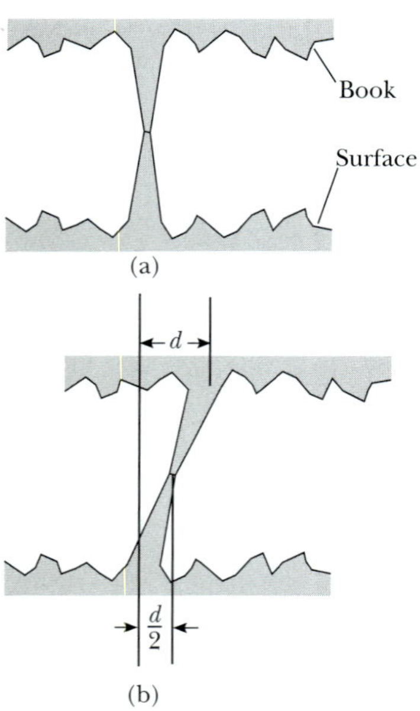

Figure 8.7 (a) A simplified model of friction between a book and a surface. The entire friction force is modeled to be applied at the interface between two identical teeth projecting from the book and the surface. (b) The book is moved to the right by a distance d. The point of application of the friction force moves through a displacement of magnitude $d/2$.

In reality, the friction force is spread out over the entire contact area of an object sliding on a surface, so the force is not localized at a point. In addition, because the magnitudes of the friction forces at various points are constantly changing as individual spot welds occur, the surface and the book deform locally, and so on, the displacement of the point of application of the friction force is not at all the same as the displacement of the book. In fact, the displacement of the point of application of the friction force is not calculable and so neither is the work done by the friction force.

The work–kinetic energy theorem is valid for a particle or an object that can be modeled as a particle. When a friction force acts, however, we cannot calculate the work done by friction. For such situations, Newton's second law is still valid for the system even though the work–kinetic energy theorem is not. The case of a nondeformable object like our book sliding on the surface[3] can be handled in a relatively straightforward way.

Starting from a situation in which forces, including friction, are applied to the book, we can follow a similar procedure to that done in developing Equation 7.17. Let us start by writing Equation 7.8 for all forces other than friction:

$$\sum W_{\text{other forces}} = \int \left(\sum \vec{\mathbf{F}}_{\text{other forces}}\right) \cdot d\vec{\mathbf{r}} \quad \textbf{(8.11)}$$

The $d\vec{\mathbf{r}}$ in this equation is the displacement of the object because for forces other than friction, under the assumption that these forces do not deform the object, this displacement is the same as the displacement of the point of application of

[2] Figure 8.7 and its discussion are inspired by a classic article on friction: B. A. Sherwood and W. H. Bernard, "Work and heat transfer in the presence of sliding friction," *American Journal of Physics,* 52:1001, 1984.

[3] The overall shape of the book remains the same, which is why we say it is nondeformable. On a microscopic level, however, there is deformation of the book's face as it slides over the surface.

the forces. To each side of Equation 8.11 let us add the integral of the scalar product of the force of kinetic friction and $d\vec{\mathbf{r}}$:

$$\sum W_{\text{other forces}} + \int \vec{\mathbf{f}}_k \cdot d\vec{\mathbf{r}} = \int \left(\sum \vec{\mathbf{F}}_{\text{other forces}}\right) \cdot d\vec{\mathbf{r}} + \int \vec{\mathbf{f}}_k \cdot d\vec{\mathbf{r}}$$

$$= \int \left(\sum \vec{\mathbf{F}}_{\text{other forces}} + \vec{\mathbf{f}}_k\right) \cdot d\vec{\mathbf{r}}$$

The integrand on the right side of this equation is the net force $\Sigma \vec{\mathbf{F}}$, so

$$\sum W_{\text{other forces}} + \int \vec{\mathbf{f}}_k \cdot d\vec{\mathbf{r}} = \int \sum \vec{\mathbf{F}} \cdot d\vec{\mathbf{r}}$$

Incorporating Newton's second law $\Sigma \vec{\mathbf{F}} = m\vec{\mathbf{a}}$ gives

$$\sum W_{\text{other forces}} + \int \vec{\mathbf{f}}_k \cdot d\vec{\mathbf{r}} = \int m\vec{\mathbf{a}} \cdot d\vec{\mathbf{r}} = \int m \frac{d\vec{\mathbf{v}}}{dt} \cdot d\vec{\mathbf{r}} = \int_{t_i}^{t_f} m \frac{d\vec{\mathbf{v}}}{dt} \cdot \vec{\mathbf{v}}\, dt \qquad \textbf{(8.12)}$$

where we have used Equation 4.3 to rewrite $d\vec{\mathbf{r}}$ as $\vec{\mathbf{v}}\, dt$. The scalar product obeys the product rule for differentiation (See Eq. B.30 in Appendix B.6), so the derivative of the scalar product of $\vec{\mathbf{v}}$ with itself can be written

$$\frac{d}{dt}(\vec{\mathbf{v}} \cdot \vec{\mathbf{v}}) = \frac{d\vec{\mathbf{v}}}{dt} \cdot \vec{\mathbf{v}} + \vec{\mathbf{v}} \cdot \frac{d\vec{\mathbf{v}}}{dt} = 2\frac{d\vec{\mathbf{v}}}{dt} \cdot \vec{\mathbf{v}}$$

where we have used the commutative property of the scalar product to justify the final expression in this equation. Consequently,

$$\frac{d\vec{\mathbf{v}}}{dt} \cdot \vec{\mathbf{v}} = \frac{1}{2}\frac{d}{dt}(\vec{\mathbf{v}} \cdot \vec{\mathbf{v}}) = \frac{1}{2}\frac{dv^2}{dt}$$

Substituting this result into Equation 8.12 gives

$$\sum W_{\text{other forces}} + \int \vec{\mathbf{f}}_k \cdot d\vec{\mathbf{r}} = \int_{t_i}^{t_f} m\left(\frac{1}{2}\frac{dv^2}{dt}\right) dt = \frac{1}{2}m \int_{v_i}^{v_f} d(v^2) = \frac{1}{2}mv_f^{\,2} - \frac{1}{2}mv_i^{\,2} = \Delta K$$

Looking at the left side of this equation, notice that in the inertial frame of the surface, $\vec{\mathbf{f}}_k$ and $d\vec{\mathbf{r}}$ will be in opposite directions for every increment $d\vec{\mathbf{r}}$ of the path followed by the object. Therefore, $\vec{\mathbf{f}}_k \cdot d\vec{\mathbf{r}} = -f_k\, dr$. The previous expression now becomes

$$\sum W_{\text{other forces}} - \int f_k\, dr = \Delta K$$

In our model for friction, the magnitude of the kinetic friction force is constant, so f_k can be brought out of the integral. The remaining integral $\int dr$ is simply the sum of increments of length along the path, which is the total path length d. Therefore,

$$\sum W_{\text{other forces}} - f_k d = \Delta K \qquad \textbf{(8.13)}$$

or

$$K_f = K_i - f_k d + \sum W_{\text{other forces}} \qquad \textbf{(8.14)}$$

Equation 8.13 is a modified form of the work–kinetic energy theorem to be used when a friction force acts on an object. The change in kinetic energy is equal to the work done by all forces other than friction minus a term $f_k d$ associated with the friction force.

Now consider the larger system of the book *and* the surface as the book slows down under the influence of a friction force alone. There is no work done across the boundary of this system because the system does not interact with the environment. There are no other types of energy transfer occurring across the boundary of the system, assuming we ignore the inevitable sound the sliding book makes! In this case, Equation 8.2 becomes

$$\Delta E_{\text{system}} = \Delta K + \Delta E_{\text{int}} = 0$$

The change in kinetic energy of this book–surface system is the same as the change in kinetic energy of the book alone because the book is the only part of the system that is moving. Therefore, incorporating Equation 8.13 gives

$$-f_k d + \Delta E_{\text{int}} = 0$$

Change in internal energy due to friction within the system ▶

$$\Delta E_{\text{int}} = f_k d \qquad \textbf{(8.15)}$$

The increase in internal energy of the system is therefore equal to the product of the friction force and the path length through which the block moves. In summary, **a friction force transforms kinetic energy in a system to internal energy, and the increase in internal energy of the system is equal to its decrease in kinetic energy.**

Quick Quiz 8.5 You are traveling along a freeway at 65 mi/h. Your car has kinetic energy. You suddenly skid to a stop because of congestion in traffic. Where is the kinetic energy your car once had? (a) It is all in internal energy in the road. (b) It is all in internal energy in the tires. (c) Some of it has transformed to internal energy and some of it transferred away by mechanical waves. (d) It is all transferred away from your car by various mechanisms.

EXAMPLE 8.4 A Block Pulled on a Rough Surface

A 6.0-kg block initially at rest is pulled to the right along a horizontal surface by a constant horizontal force of 12 N.

(A) Find the speed of the block after it has moved 3.0 m if the surfaces in contact have a coefficient of kinetic friction of 0.15.

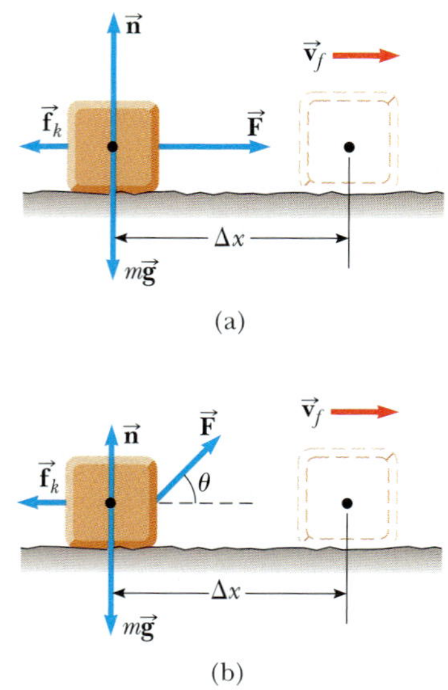

ACTIVE FIGURE 8.8

(Example 8.4) (a) A block pulled to the right on a rough surface by a constant horizontal force. (b) The applied force is at an angle θ to the horizontal.

Sign in at www.thomsonedu.com and go to ThomsonNOW to pull the block with a force oriented at different angles.

SOLUTION

Conceptualize This example is Example 7.6, modified so that the surface is no longer frictionless. The rough surface applies a friction force on the block opposite to the applied force. As a result, we expect the speed to be lower than that found in Example 7.6.

Categorize The block is pulled by a force and the surface is rough, so we model the block–surface system as nonisolated with a nonconservative force acting.

Analyze Active Figure 8.8a illustrates this situation. Neither the normal force nor the gravitational force does work on the system because their points of application are displaced horizontally.

Find the work done on the system by the applied force just as in Example 7.6:

$$W = F\,\Delta x = (12\text{ N})(3.0\text{ m}) = 36\text{ J}$$

Apply the particle in equilibrium model to the block in the vertical direction:

$$\sum F_y = 0 \;\rightarrow\; n - mg = 0 \;\rightarrow\; n = mg$$

Find the magnitude of the friction force:

$$f_k = \mu_k n = \mu_k mg = (0.15)(6.0\text{ kg})(9.80\text{ m/s}^2) = 8.82\text{ N}$$

Find the final speed of the block from Equation 8.14:

$$\tfrac{1}{2}mv_f^2 = \tfrac{1}{2}mv_i^2 - f_k d + \sum W_{\text{other forces}}$$

$$v_f = \sqrt{v_i^2 + \frac{2}{m}\left(-f_k d + \sum W_{\text{other forces}}\right)}$$

$$= \sqrt{0 + \frac{2}{6.0\text{ kg}}[-(8.82\text{ N})(3.0\text{ m}) + 36\text{ J}]} = \boxed{1.8\text{ m/s}}$$

Finalize As expected, this value is less than the 3.5 m/s found in the case of the block sliding on a frictionless surface (see Example 7.6).

(B) Suppose the force $\vec{\mathbf{F}}$ is applied at an angle θ as shown in Active Figure 8.8b. At what angle should the force be applied to achieve the largest possible speed after the block has moved 3.0 m to the right?

SOLUTION

Conceptualize You might guess that $\theta = 0$ would give the largest speed because the force would have the largest component possible in the direction parallel to the surface. Think about an arbitrary nonzero angle, however. Although the horizontal component of the force would be reduced, the vertical component of the force would reduce the normal force, in turn reducing the force of friction, which suggests that the speed could be maximized by pulling at an angle other than $\theta = 0$.

Categorize As in part (A), we model the block–surface system as nonisolated with a nonconservative force acting.

Analyze Find the work done by the applied force, noting that $\Delta x = d$ because the path followed by the block is a straight line:

$$W = F\,\Delta x \cos\theta = Fd\cos\theta$$

Apply the particle in equilibrium model to the block in the vertical direction:

$$\sum F_y = n + F\sin\theta - mg = 0$$

Solve for n:

$$n = mg - F\sin\theta$$

Use Equation 8.14 to find the final kinetic energy for this situation:

$$K_f = K_i - f_k d + \sum W_{\text{other forces}}$$

$$= 0 - \mu_k n d + Fd\cos\theta = -\mu_k(mg - F\sin\theta)d + Fd\cos\theta$$

Maximizing the speed is equivalent to maximizing the final kinetic energy. Consequently, differentiate K_f with respect to θ and set the result equal to zero:

$$\frac{d(K_f)}{d\theta} = -\mu_k(0 - F\cos\theta)d - Fd\sin\theta = 0$$

$$\mu_k\cos\theta - \sin\theta = 0$$

$$\tan\theta = \mu_k$$

Evaluate θ for $\mu_k = 0.15$:

$$\theta = \tan^{-1}(\mu_k) = \tan^{-1}(0.15) = 8.5°$$

Finalize Notice that the angle at which the speed of the block is a maximum is indeed not $\theta = 0$. When the angle exceeds 8.5°, the horizontal component of the applied force is too small to be compensated by the reduced friction force and the speed of the block begins to decrease from its maximum value.

CONCEPTUAL EXAMPLE 8.5 Useful Physics for Safer Driving

A car traveling at an initial speed v slides a distance d to a halt after its brakes lock. If the car's initial speed is instead $2v$ at the moment the brakes lock, estimate the distance it slides.

SOLUTION

Let us assume the force of kinetic friction between the car and the road surface is constant and the same for both speeds. According to Equation 8.14, the friction force multiplied by the distance d is equal to the initial kinetic energy of the car (because $K_f = 0$ and there is no work done by other forces). If the speed is doubled, as it is in this example, the kinetic energy is quadrupled. For a given friction force, the distance traveled is four times as great when the initial speed is doubled, and so the estimated distance the car slides is $4d$.

EXAMPLE 8.6 A Block–Spring System

A block of mass 1.6 kg is attached to a horizontal spring that has a force constant of 1.0×10^3 N/m as shown in Figure 8.9. The spring is compressed 2.0 cm and is then released from rest.

(A) Calculate the speed of the block as it passes through the equilibrium position $x = 0$ if the surface is frictionless.

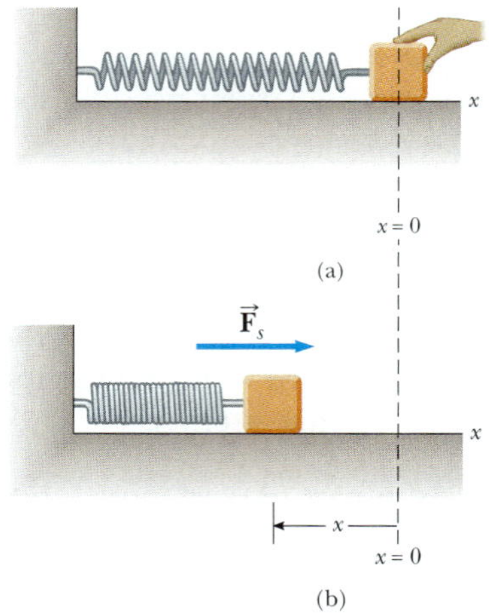

Figure 8.9 (Example 8.6) (a) A block is attached to a spring. The spring is compressed by a distance x. (b) The block is then released and the spring pushes it to the right.

SOLUTION

Conceptualize This situation has been discussed before and it is easy to visualize the block being pushed to the right by the spring and moving off with some speed.

Categorize We identify the system as the block and model the block as a nonisolated system.

Analyze In this situation, the block starts with $v_i = 0$ at $x_i = -2.0$ cm, and we want to find v_f at $x_f = 0$.

Use Equation 7.11 to find the work done by the spring with $x_{max} = x_i = -2.0 \text{ cm} = -2.0 \times 10^{-2}$ m:

$$W_s = \tfrac{1}{2}kx_{max}^2 = \tfrac{1}{2}(1.0 \times 10^3 \text{ N/m})(-2.0 \times 10^{-2}\text{ m})^2 = 0.20 \text{ J}$$

Work is done on the block and its speed changes. The conservation of energy equation, Equation 8.2, reduces to the work–kinetic energy theorem. Use that theorem to find the speed at $x = 0$:

$$W_s = \tfrac{1}{2}mv_f^2 - \tfrac{1}{2}mv_i^2$$

$$v_f = \sqrt{v_i^2 + \frac{2}{m}W_s}$$

$$= \sqrt{0 + \frac{2}{1.6 \text{ kg}}(0.20 \text{ J})} = \boxed{0.50 \text{ m/s}}$$

Finalize Although this problem could have been solved in Chapter 7, it is presented here to provide contrast with the following part (B), which requires the techniques of this chapter.

(B) Calculate the speed of the block as it passes through the equilibrium position if a constant friction force of 4.0 N retards its motion from the moment it is released.

SOLUTION

Conceptualize The correct answer must be less than that found in part (A) because the friction force retards the motion.

Categorize We identify the system as the block and the surface. The system is nonisolated because of the work done by the spring and there is a nonconservative force acting: the friction between the block and the surface.

Analyze Write Equation 8.14:

$$(1) \quad K_f = K_i - f_k d + \sum W_{\text{other forces}}$$

Evaluate $f_k d$:

$$f_k d = (4.0 \text{ N})(2.0 \times 10^{-2} \text{ m}) = 0.080 \text{ J}$$

Evaluate $\Sigma W_{\text{other forces}}$, the work done by the spring, by recalling that it was found in part (A) to be 0.20 J. Use $K_i = 0$ in Equation (1) and solve for the final speed:

$$K_f = 0 - 0.080 \text{ J} + 0.20 \text{ J} = 0.12 \text{ J} = \tfrac{1}{2}mv_f^2$$

$$v_f = \sqrt{\frac{2K_f}{m}} = \sqrt{\frac{2(0.12 \text{ J})}{1.6 \text{ kg}}} = \boxed{0.39 \text{ m/s}}$$

Finalize As expected, this value is less than the 0.50 m/s found in part (A).

What If? What if the friction force were increased to 10.0 N? What is the block's speed at $x = 0$?

Answer In this case, the value of $f_k d$ as the block moves to $x = 0$ is

$$f_k d = (10.0\text{ N})(2.0 \times 10^{-2}\text{ m}) = 0.20\text{ J}$$

which is equal in magnitude to the kinetic energy at $x = 0$ without the loss due to friction. Therefore, all the kinetic energy has been transformed by friction when the block arrives at $x = 0$, and its speed at this point is $v = 0$.

In this situation as well as that in part (B), the speed of the block reaches a maximum at some position other than $x = 0$. Problem 47 asks you to locate these positions.

8.4 Changes in Mechanical Energy for Nonconservative Forces

Consider the book sliding across the surface in the preceding section. As the book moves through a distance d, the only force that does work on it is the force of kinetic friction. This force causes a change $-f_k d$ in the kinetic energy of the book as described by Equation 8.13.

Now, however, suppose the book is part of a system that also exhibits a change in potential energy. In this case, $-f_k d$ is the amount by which the *mechanical* energy of the system changes because of the force of kinetic friction. For example, if the book moves on an incline that is not frictionless, there is a change in both the kinetic energy and the gravitational potential energy of the book–Earth system. Consequently,

$$\Delta E_{\text{mech}} = \Delta K + \Delta U_g = -f_k d$$

In general, if a friction force acts within an isolated system,

$$\Delta E_{\text{mech}} = \Delta K + \Delta U = -f_k d \qquad (8.16)$$

◀ Change in mechanical energy of a system due to friction within the system

where ΔU is the change in *all* forms of potential energy. Notice that Equation 8.16 reduces to Equation 8.10 if the friction force is zero.

If the system in which nonconservative forces act is nonisolated, the generalization of Equation 8.13 is

$$\Delta E_{\text{mech}} = -f_k d + \sum W_{\text{other forces}} \qquad (8.17)$$

PROBLEM SOLVING STRATEGY Systems with Nonconservative Forces

The following procedure should be used when you face a problem involving a system in which nonconservative forces act:

1. *Conceptualize.* Study the physical situation carefully and form a mental representation of what is happening.

2. *Categorize.* Define your system, which may consist of more than one object. The system could include springs or other possibilities for storage of potential energy. Determine whether any nonconservative forces are present. If not, use the principle of conservation of mechanical energy as outlined in Section 8.2. If so, use the procedure discussed below.

   Determine if any work is done across the boundary of your system by forces other than friction. If so, use Equation 8.17 to analyze the problem. If not, use Equation 8.16.

3. *Analyze.* Choose configurations to represent the initial and final conditions of the system. For each object that changes elevation, select a reference position for the object that defines the zero configuration of gravitational potential energy for the system. For an object on a spring, the zero configuration for elastic potential energy is when the object is at its equilibrium position. If there is more than one

conservative force, write an expression for the potential energy associated with each force.

Use either Equation 8.16 or Equation 8.17 to establish a mathematical representation of the problem. Solve for the unknown.

4. *Finalize.* Make sure your results are consistent with your mental representation. Also make sure the values of your results are reasonable and consistent with connections to everyday experience.

EXAMPLE 8.7 Crate Sliding Down a Ramp

A 3.00-kg crate slides down a ramp. The ramp is 1.00 m in length and inclined at an angle of 30.0° as shown in Figure 8.10. The crate starts from rest at the top, experiences a constant friction force of magnitude 5.00 N, and continues to move a short distance on the horizontal floor after it leaves the ramp.

(A) Use energy methods to determine the speed of the crate at the bottom of the ramp.

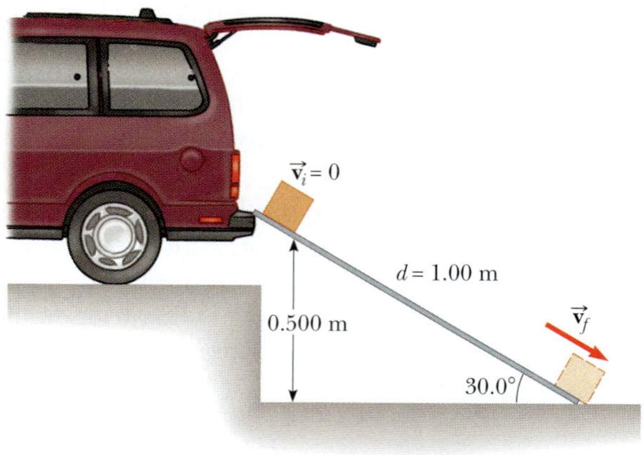

Figure 8.10 (Example 8.7) A crate slides down a ramp under the influence of gravity. The potential energy of the system decreases, whereas the kinetic energy increases.

SOLUTION

Conceptualize Imagine the crate sliding down the ramp in Figure 8.10. The larger the friction force, the more slowly the crate will slide.

Categorize We identify the crate, the surface, and the Earth as the system. The system is categorized as isolated with a nonconservative force acting.

Analyze Because $v_i = 0$, the initial kinetic energy of the system when the crate is at the top of the ramp is zero. If the y coordinate is measured from the bottom of the ramp (the final position of the crate, for which we choose the gravitational potential energy of the system to be zero) with the upward direction being positive, then $y_i = 0.500$ m.

Evaluate the total mechanical energy of the system when the crate is at the top:

$$E_i = K_i + U_i = 0 + U_i = mgy_i$$
$$= (3.00\text{ kg})(9.80\text{ m/s}^2)(0.500\text{ m}) = 14.7\text{ J}$$

Write an expression for the final mechanical energy:

$$E_f = K_f + U_f = \tfrac{1}{2}mv_f^{\,2} + 0$$

Apply Equation 8.16:

$$\Delta E_{\text{mech}} = E_f - E_i = \tfrac{1}{2}mv_f^{\,2} - mgy_i = -f_k d$$

Solve for v_f^2:

$$(1)\quad v_f^{\,2} = \frac{2}{m}(mgy_i - f_k d)$$

Substitute numerical values and solve for v_f:

$$v_f^{\,2} = \frac{2}{3.00\text{ kg}}[14.7\text{ J} - (5.00\text{ N})(1.00\text{ m})] = 6.47\text{ m}^2/\text{s}^2$$

$$v_f = \boxed{2.54\text{ m/s}}$$

(B) How far does the crate slide on the horizontal floor if it continues to experience a friction force of magnitude 5.00 N?

SOLUTION

Analyze This part of the problem is handled in exactly the same way as part (A), but in this case we can consider the mechanical energy of the system to consist only of kinetic energy because the potential energy of the system remains fixed.

Evaluate the mechanical energy of the system when the crate leaves the bottom of the ramp: $E_i = K_i = \frac{1}{2}mv_i^2 = \frac{1}{2}(3.00 \text{ kg})(2.54 \text{ m/s})^2 = 9.68 \text{ J}$

Apply Equation 8.16 with $E_f = 0$: $E_f - E_i = 0 - 9.68 \text{ J} = -f_k d$

Solve for the distance d: $d = \dfrac{9.68 \text{ J}}{f_k} = \dfrac{9.68 \text{ J}}{5.00 \text{ N}} = 1.94 \text{ m}$

Finalize For comparison, you may want to calculate the speed of the crate at the bottom of the ramp in the case in which the ramp is frictionless. Also notice that the increase in internal energy of the system as the crate slides down the ramp is 5.00 J. This energy is shared between the crate and the surface, each of which is a bit warmer than before.

Also notice that the distance d the object slides on the horizontal surface is infinite if the surface is frictionless. Is that consistent with your conceptualization of the situation?

What If? A cautious worker decides that the speed of the crate when it arrives at the bottom of the ramp may be so large that its contents may be damaged. Therefore, he replaces the ramp with a longer one such that the new ramp makes an angle of 25.0° with the ground. Does this new ramp reduce the speed of the crate as it reaches the ground?

Answer Because the ramp is longer, the friction force acts over a longer distance and transforms more of the mechanical energy into internal energy. The result is a reduction in the kinetic energy of the crate, and we expect a lower speed as it reaches the ground.

Find the length d of the new ramp: $\sin 25.0° = \dfrac{0.500 \text{ m}}{d} \rightarrow d = \dfrac{0.500 \text{ m}}{\sin 25.0°} = 1.18 \text{ m}$

Find v_f^2 from Equation (1) in part (A): $v_f^2 = \dfrac{2}{3.00 \text{ kg}}[14.7 \text{ J} - (5.00 \text{ N})(1.18 \text{ m})] = 5.87 \text{ m}^2/\text{s}^2$

$v_f = 2.42 \text{ m/s}$

The final speed is indeed lower than in the higher-angle case.

EXAMPLE 8.8 Block–Spring Collision

A block having a mass of 0.80 kg is given an initial velocity $v_{Ⓐ} = 1.2$ m/s to the right and collides with a spring whose mass is negligible and whose force constant is $k = 50$ N/m as shown in Figure 8.11.

(A) Assuming the surface to be frictionless, calculate the maximum compression of the spring after the collision.

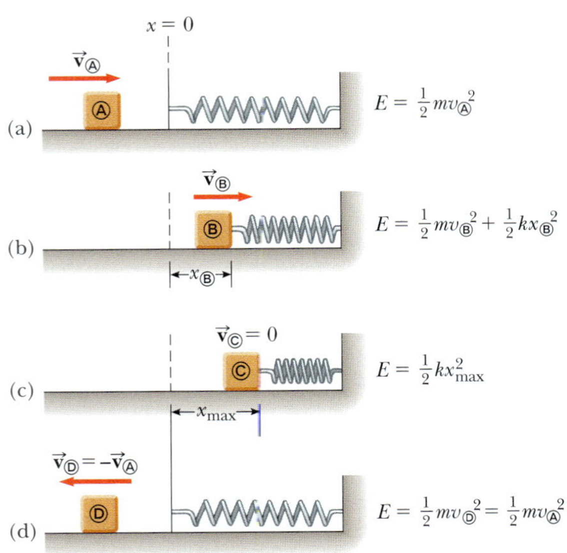

Figure 8.11 (Example 8.8) A block sliding on a smooth, horizontal surface collides with a light spring. (a) Initially, the mechanical energy is all kinetic energy. (b) The mechanical energy is the sum of the kinetic energy of the block and the elastic potential energy in the spring. (c) The energy is entirely potential energy. (d) The energy is transformed back to the kinetic energy of the block. The total energy of the system remains constant throughout the motion.

SOLUTION

Conceptualize The various parts of Figure 8.11 help us imagine what the block will do in this situation. All motion takes place in a horizontal plane, so we do not need to consider changes in gravitational potential energy.

Categorize We identify the system to be the block and the spring. The block–spring system is isolated with no nonconservative forces acting.

Analyze Before the collision, when the block is at Ⓐ, it has kinetic energy and the spring is uncompressed, so the elastic potential energy stored in the system is zero. Therefore, the total mechanical energy of the system before the collision is just $\frac{1}{2}mv_{Ⓐ}^2$. After the collision, when the block is at Ⓒ, the spring is fully compressed; now the block is at rest and so has zero kinetic energy. The elastic potential energy stored in the system, however, has its maximum value $\frac{1}{2}kx^2 = \frac{1}{2}kx_{max}^2$ where the origin of coordinates $x = 0$ is chosen to be the equilibrium position of the spring and x_{max} is the maximum compression of the spring, which in this case happens to be $x_{Ⓒ}$. The total mechanical energy of the system is conserved because no nonconservative forces act on objects within the isolated system.

Write a conservation of mechanical energy equation:

$$K_{Ⓒ} + U_{sⒸ} = K_{Ⓐ} + U_{sⒶ}$$

$$0 + \frac{1}{2}kx_{max}^2 = \frac{1}{2}mv_{Ⓐ}^2 + 0$$

Solve for x_{max} and evaluate:

$$x_{max} = \sqrt{\frac{m}{k}}\,v_{Ⓐ} = \sqrt{\frac{0.80\text{ kg}}{50\text{ N/m}}}(1.2\text{ m/s}) = 0.15\text{ m}$$

(B) Suppose a constant force of kinetic friction acts between the block and the surface, with $\mu_k = 0.50$. If the speed of the block at the moment it collides with the spring is $v_{Ⓐ} = 1.2$ m/s, what is the maximum compression $x_{Ⓒ}$ in the spring?

SOLUTION

Conceptualize Because of the friction force, we expect the compression of the spring to be smaller than in part (A) because some of the block's kinetic energy is transformed to internal energy in the block and the surface.

Categorize We identify the system as the block, the surface, and the spring. This system is isolated but now involves a nonconservative force.

Analyze In this case, the mechanical energy $E_{mech} = K + U_s$ of the system is *not* conserved because a friction force acts on the block. From the particle in equilibrium model in the vertical direction, we see that $n = mg$.

Evaluate the magnitude of the friction force:

$$f_k = \mu_k n = \mu_k mg = 0.50(0.80\text{ kg})(9.80\text{ m/s}^2) = 3.9\text{ N}$$

Write the change in the mechanical energy of the system due to friction as the block is displaced from $x = 0$ to $x_{Ⓒ}$:

$$\Delta E_{mech} = -f_k x_{Ⓒ}$$

Substitute the initial and final energies:

$$\Delta E_{mech} = E_f - E_i = (0 + \tfrac{1}{2}kx_{Ⓒ}^2) - (\tfrac{1}{2}mv_{Ⓐ}^2 + 0) = -f_k x_{Ⓒ}$$

Substitute numerical values:

$$\tfrac{1}{2}(50)x_{Ⓒ}^2 - \tfrac{1}{2}(0.80)(1.2)^2 = -3.9x_{Ⓒ}$$

$$25x_{Ⓒ}^2 + 3.9x_{Ⓒ} - 0.58 = 0$$

Solving the quadratic equation for $x_{Ⓒ}$ gives $x_{Ⓒ} = 0.093$ m and $x_{Ⓒ} = -0.25$ m. The physically meaningful root is $x_{Ⓒ} = 0.093$ m.

Finalize The negative root does not apply to this situation because the block must be to the right of the origin (positive value of x) when it comes to rest. Notice that the value of 0.093 m is less than the distance obtained in the frictionless case of part (A) as we expected.

EXAMPLE 8.9 Connected Blocks in Motion

Two blocks are connected by a light string that passes over a frictionless pulley as shown in Figure 8.12. The block of mass m_1 lies on a horizontal surface and is connected to a spring of force constant k. The system is released from rest when the spring is unstretched. If the hanging block of mass m_2 falls a distance h before coming to rest, calculate the coefficient of kinetic friction between the block of mass m_1 and the surface.

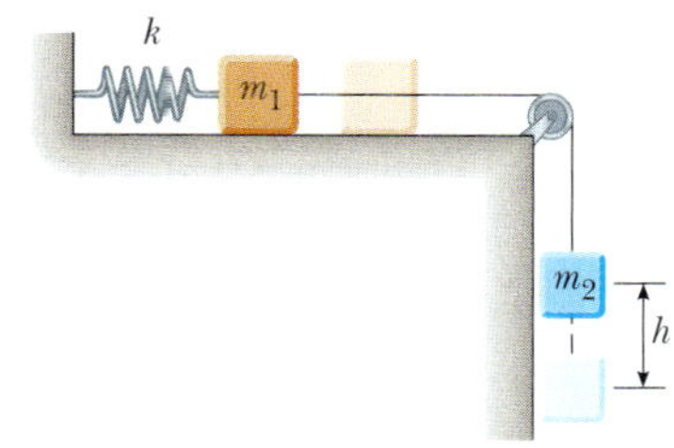

Figure 8.12 (Example 8.9) As the hanging block moves from its highest elevation to its lowest, the system loses gravitational potential energy but gains elastic potential energy in the spring. Some mechanical energy is transformed to internal energy because of friction between the sliding block and the surface.

SOLUTION

Conceptualize The key word *rest* appears twice in the problem statement. This word suggests that the configurations of the system associated with rest are good candidates for the initial and final configurations because the kinetic energy of the system is zero for these configurations.

Categorize In this situation, the system consists of the two blocks, the spring, and the Earth. The system is isolated with a nonconservative force acting. We also model the sliding block as a particle in equilibrium in the vertical direction, leading to $n = m_1 g$.

Analyze We need to consider two forms of potential energy for the system, gravitational and elastic: $\Delta U_g = U_{gf} - U_{gi}$ is the change in the system's gravitational potential energy, and $\Delta U_s = U_{sf} - U_{si}$ is the change in the system's elastic potential energy. The change in the gravitational potential energy of the system is associated with only the falling block because the vertical coordinate of the horizontally sliding block does not change. The initial and final kinetic energies of the system are zero, so $\Delta K = 0$.

Write the change in mechanical energy for the system:

$$(1) \quad \Delta E_{\text{mech}} = \Delta U_g + \Delta U_s$$

Use Equation 8.16 to find the change in mechanical energy in the system due to friction between the horizontally sliding block and the surface, noticing that as the hanging block falls a distance h, the horizontally moving block moves the same distance h to the right:

$$(2) \quad \Delta E_{\text{mech}} = -f_k h = -(\mu_k n)h = -\mu_k m_1 g h$$

Evaluate the change in gravitational potential energy of the system, choosing the configuration with the hanging block at the lowest position to represent zero potential energy:

$$(3) \quad \Delta U_g = U_{gf} - U_{gi} = 0 - m_2 g h$$

Evaluate the change in the elastic potential energy of the system:

$$(4) \quad \Delta U_s = U_{sf} - U_{si} = \tfrac{1}{2}kh^2 - 0$$

Substitute Equations (2), (3), and (4) into Equation (1):

$$-\mu_k m_1 g h = -m_2 g h + \tfrac{1}{2}kh^2$$

Solve for μ_k:

$$\mu_k = \frac{m_2 g - \frac{1}{2}kh}{m_1 g}$$

Finalize This setup represents a method of measuring the coefficient of kinetic friction between an object and some surface.

8.5 Power

Consider Conceptual Example 7.7 again, which involved rolling a refrigerator up a ramp into a truck. Suppose the man is not convinced that the work is the same regardless of the ramp's length and sets up a long ramp with a gentle rise. Although he does the same amount of work as someone using a shorter ramp, he takes longer to do the work because he has to move the refrigerator over a greater distance. Although the work done on both ramps is the same, there is *something* different about the tasks: the *time interval* during which the work is done.

The time rate of energy transfer is called the **instantaneous power** $\mathcal{P}$ and is defined as follows:

$$\mathcal{P} \equiv \frac{dE}{dt} \quad \textbf{(8.18)}$$

◀ Definition of power

We will focus on work as the energy transfer method in this discussion, but keep in mind that the notion of power is valid for *any* means of energy transfer discussed in Section 8.1. If an external force is applied to an object (which we model as a particle) and if the work done by this force on the object in the time interval Δt is W, the **average power** during this interval is

$$\mathcal{P}_{\text{avg}} = \frac{W}{\Delta t}$$

Therefore, in Example 7.7, although the same work is done in rolling the refrigerator up both ramps, less power is required for the longer ramp.

In a manner similar to the way we approached the definition of velocity and acceleration, the instantaneous power is the limiting value of the average power as Δt approaches zero:

$$\mathcal{P} = \lim_{\Delta t \to 0} \frac{W}{\Delta t} = \frac{dW}{dt}$$

where we have represented the infinitesimal value of the work done by dW. We find from Equation 7.3 that $dW = \vec{\mathbf{F}} \cdot d\vec{\mathbf{r}}$. Therefore, the instantaneous power can be written

$$\mathcal{P} = \frac{dW}{dt} = \vec{\mathbf{F}} \cdot \frac{d\vec{\mathbf{r}}}{dt} = \vec{\mathbf{F}} \cdot \vec{\mathbf{v}} \quad \textbf{(8.19)}$$

where $\vec{\mathbf{v}} = d\vec{\mathbf{r}}/dt$.

The SI unit of power is joules per second (J/s), also called the **watt** (W) after James Watt:

The watt ▶

$$1\text{ W} = 1\text{ J/s} = 1\text{ kg}\cdot\text{m}^2/\text{s}^3$$

A unit of power in the U.S. customary system is the **horsepower** (hp):

$$1\text{ hp} = 746\text{ W}$$

PITFALL PREVENTION 8.3

W, *W*, and watts

Do not confuse the symbol W for the watt with the italic symbol *W* for work. Also, remember that the watt already represents a rate of energy transfer, so "watts per second" does not make sense. The watt is *the same as* a joule per second.

A unit of energy (or work) can now be defined in terms of the unit of power. One **kilowatt-hour** (kWh) is the energy transferred in 1 h at the constant rate of 1 kW = 1 000 J/s. The amount of energy represented by 1 kWh is

$$1\text{ kWh} = (10^3\text{ W})(3\,600\text{ s}) = 3.60 \times 10^6\text{ J}$$

A kilowatt-hour is a unit of energy, not power. When you pay your electric bill, you are buying energy, and the amount of energy transferred by electrical transmission into a home during the period represented by the electric bill is usually expressed in kilowatt-hours. For example, your bill may state that you used 900 kWh of energy during a month and that you are being charged at the rate of 10¢ per kilowatt-hour. Your obligation is then $90 for this amount of energy. As another example, suppose an electric bulb is rated at 100 W. In 1.00 hour of operation, it would have energy transferred to it by electrical transmission in the amount of $(0.100\text{ kW})(1.00\text{ h}) = 0.100\text{ kWh} = 3.60 \times 10^5\text{ J}$.

EXAMPLE 8.10 Power Delivered by an Elevator Motor

An elevator car (Fig. 8.13a) has a mass of 1 600 kg and is carrying passengers having a combined mass of 200 kg. A constant friction force of 4 000 N retards its motion.

(A) How much power must a motor deliver to lift the elevator car and its passengers at a constant speed of 3.00 m/s?

SOLUTION

Conceptualize The motor must supply the force of magnitude T that pulls the elevator car upward.

Categorize The friction force increases the power necessary to lift the elevator. The problem states that the speed of the elevator is constant, which tells us that $a = 0$. We model the elevator as a particle in equilibrium.

Analyze The free-body diagram in Figure 8.13b specifies the upward direction as positive. The *total* mass M of the system (car plus passengers) is equal to 1 800 kg.

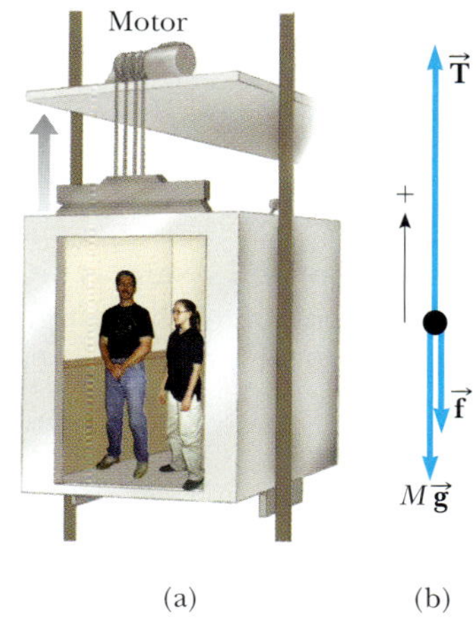

Figure 8.13 (Example 8.10) (a) The motor exerts an upward force $\vec{\mathbf{T}}$ on the elevator car. The magnitude of this force is the tension T in the cable connecting the car and motor. The downward forces acting on the car are a friction force $\vec{\mathbf{f}}$ and the gravitational force $\vec{\mathbf{F}}_g = M\vec{\mathbf{g}}$. (b) The free-body diagram for the elevator car.

Apply Newton's second law to the car:

$$\sum F_y = T - f - Mg = 0$$

Solve for T:

$$T = f + Mg$$
$$= 4.00 \times 10^3 \text{ N} + (1.80 \times 10^3 \text{ kg})(9.80 \text{ m/s}^2) = 2.16 \times 10^4 \text{ N}$$

Use Equation 8.19 and that $\vec{\mathbf{T}}$ is in the same direction as $\vec{\mathbf{v}}$ to find the power:

$$\mathcal{P} = \vec{\mathbf{T}} \cdot \vec{\mathbf{v}} = Tv$$
$$= (2.16 \times 10^4 \text{ N})(3.00 \text{ m/s}) = \boxed{6.48 \times 10^4 \text{ W}}$$

(B) What power must the motor deliver at the instant the speed of the elevator is v if the motor is designed to provide the elevator car with an upward acceleration of 1.00 m/s^2?

SOLUTION

Conceptualize In this case, the motor must supply the force of magnitude T that pulls the elevator car upward with an increasing speed. We expect that more power will be required to do that than in part (A) because the motor must now perform the additional task of accelerating the car.

Categorize In this case, we model the elevator car as a particle under a net force because it is accelerating.

Analyze Apply Newton's second law to the car:

$$\sum F_y = T - f - Mg = Ma$$

Solve for T:

$$T = M(a + g) + f$$
$$= (1.80 \times 10^3 \text{ kg})(1.00 \text{ m/s}^2 + 9.80 \text{ m/s}^2) + 4.00 \times 10^3 \text{ N}$$
$$= 2.34 \times 10^4 \text{ N}$$

Use Equation 8.19 to obtain the required power:

$$\mathcal{P} = Tv = \boxed{(2.34 \times 10^4 \text{ N})v}$$

where v is the instantaneous speed of the car in meters per second.

Finalize To compare with part (A), let $v = 3.00$ m/s, giving a power of

$$\mathcal{P} = (2.34 \times 10^4 \text{ N})(3.00 \text{ m/s}) = 7.02 \times 10^4 \text{ W}$$

which is larger than the power found in part (A), as expected.

Summary

ThomsonNOW™ Sign in at **www.thomsonedu.com** and go to ThomsonNOW to take a practice test for this chapter.

DEFINITIONS

A **nonisolated system** is one for which energy crosses the boundary of the system. An **isolated system** is one for which no energy crosses the boundary of the system.

The **instantaneous power** $\mathcal{P}$ is defined as the time rate of energy transfer:

$$\mathcal{P} \equiv \frac{dE}{dt} \quad \textbf{(8.18)}$$

CONCEPTS AND PRINCIPLES

For a nonisolated system, we can equate the change in the total energy stored in the system to the sum of all the transfers of energy across the system boundary, which is a statement of **conservation of energy.** For an isolated system, the total energy is constant.

If a system is isolated and if no nonconservative forces are acting on objects inside the system, the total mechanical energy of the system is constant:

$$K_f + U_f = K_i + U_i \quad \textbf{(8.10)}$$

If nonconservative forces (such as friction) act between objects inside a system, mechanical energy is not conserved. In these situations, the difference between the total final mechanical energy and the total initial mechanical energy of the system equals the energy transformed to internal energy by the nonconservative forces.

If a friction force acts within an isolated system, the mechanical energy of the system is reduced and the appropriate equation to be applied is

$$\Delta E_{\text{mech}} = \Delta K + \Delta U = -f_k d \quad \textbf{(8.16)}$$

If a friction force acts within a nonisolated system, the appropriate equation to be applied is

$$\Delta E_{\text{mech}} = -f_k d + \sum W_{\text{other forces}} \quad \textbf{(8.17)}$$

ANALYSIS MODELS FOR PROBLEM-SOLVING

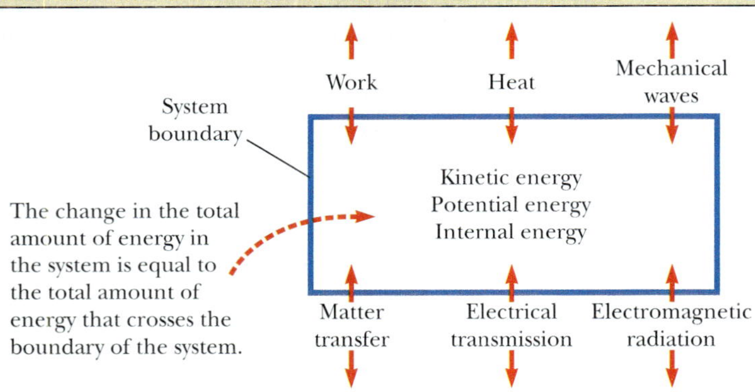

Nonisolated System (Energy). The most general statement describing the behavior of a nonisolated system is the **conservation of energy equation:**

$$\Delta E_{\text{system}} = \sum T \quad \textbf{(8.1)}$$

Including the types of energy storage and energy transfer that we have discussed gives

$$\Delta K + \Delta U + \Delta E_{\text{int}} = W + Q + T_{\text{MW}} + T_{\text{MT}} + T_{\text{ET}} + T_{\text{ER}} \quad \textbf{(8.2)}$$

For a specific problem, this equation is generally reduced to a smaller number of terms by eliminating the terms that are not appropriate to the situation.

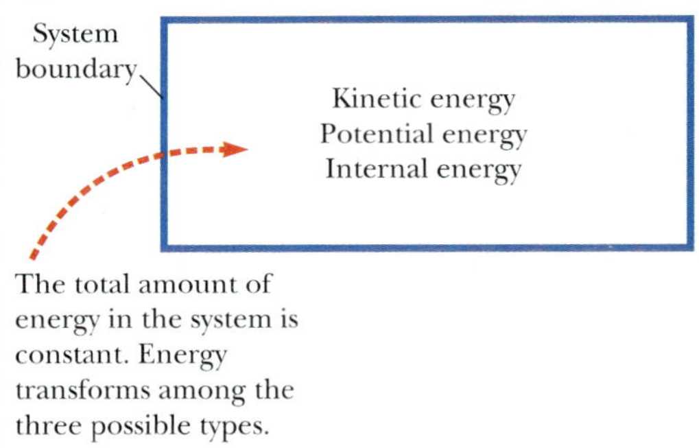

Isolated System (Energy). The total energy of an isolated system is conserved, so

$$\Delta E_{\text{system}} = 0 \quad \textbf{(8.9)}$$

If no nonconservative forces act within the isolated system, the mechanical energy of the system is conserved, so

$$\Delta E_{\text{mech}} = 0 \quad \textbf{(8.8)}$$

Questions

☐ denotes answer available in *Student Solutions Manual/Study Guide;* **O** denotes objective question

1. Does everything have energy? Give reasons for your answer.

2. **O** A pile driver is a device used to drive posts into the Earth by repeatedly dropping a heavy object on them. Assume the object is dropped from the same height each time. By what factor does the energy of the pile driver–Earth system change when the mass of the object being dropped is doubled? (a) $\frac{1}{2}$ (b) 1: the energy is the same (c) 2 (d) 4

3. **O** A curving children's slide is installed next to a backyard swimming pool. Two children climb to a platform at the top of the slide. The smaller child hops off to jump straight down into the pool and the larger child releases herself at the top of the frictionless slide. **(i)** Upon reaching the water, compared with the larger child, is the kinetic energy of the smaller child (a) greater, (b) less, or (c) equal? **(ii)** Upon reaching the water, compared with the larger child, is the speed of the smaller child (a) greater, (b) less, or (c) equal? **(iii)** During the motions from the platform to the water, compared with the larger child, is the average acceleration of the smaller child (a) greater, (b) less, or (c) equal?

4. **O** (a) Can an object–Earth system have kinetic energy and not gravitational potential energy? (b) Can it have gravitational potential energy and not kinetic energy? (c) Can it have both types of energy at the same moment? (d) Can it have neither?

5. **O** A ball of clay falls freely to the hard floor. It does not bounce noticeably, but very quickly comes to rest. What then has happened to the energy the ball had while it was falling? (a) It has been used up in producing the downward motion. (b) It has been transformed back into potential energy. (c) It has been transferred into the ball by heat. (d) It is in the ball and floor (and walls) as energy of invisible molecular motion. (e) Most of it went into sound.

6. **O** You hold a slingshot at arm's length, pull the light elastic band back to your chin, and release it to launch a pebble horizontally with speed 200 cm/s. With the same procedure, you fire a bean with speed 600 cm/s. What is the ratio of the mass of the bean to the mass of the pebble? (a) $\frac{1}{9}$ (b) $\frac{1}{3}$ (c) $1/\sqrt{3}$ (d) 1 (e) $\sqrt{3}$ (f) 3 (g) 9

[7.] One person drops a ball from the top of a building while another person at the bottom observes its motion. Will these two people agree on the value of the gravitational potential energy of the ball–Earth system? On the change in potential energy? On the kinetic energy?

8. In Chapter 7, the work–kinetic energy theorem, $W_{\text{net}} = \Delta K$, was introduced. This equation states that work done on a system appears as a change in kinetic energy. It is a special-case equation, valid if there are no changes in any other type of energy such as potential or internal. Give some examples in which work is done on a system but the change in energy of the system is not a change in kinetic energy.

[9.] You ride a bicycle. In what sense is your bicycle solar-powered?

[10.] A bowling ball is suspended from the ceiling of a lecture hall by a strong cord. The ball is drawn away from its equilibrium position and released from rest at the tip of the demonstrator's nose as shown in Figure Q8.10. The demonstrator remains stationary. Explain why the ball does not strike her on its return swing. Would this demonstrator be safe if the ball were given a push from its starting position at her nose?

Figure Q8.10

11. A block is connected to a spring that is suspended from the ceiling. Assuming the block is set into vertical motion and air resistance is ignored, describe the energy transformations that occur within the system consisting of the block, Earth, and spring.

12. **O** In a laboratory model of cars skidding to a stop, data are measured for six trials. Each of three blocks is launched at two different initial speeds v_i and slides across a level table as it comes to rest. The blocks have equal masses but differ in roughness and so have different coefficients of kinetic friction μ_k with the table. Rank the following cases (a) through (f) according to the stopping distance, from largest to smallest. If the stopping distance is the same in two cases, give them equal rank. (a) $v_i = 1$ m/s, $\mu_k = 0.2$ (b) $v_i = 1$ m/s, $\mu_k = 0.4$ (c) $v_i = 1$ m/s, $\mu_k = 0.8$ (d) $v_i = 2$ m/s, $\mu_k = 0.2$ (e) $v_i = 2$ m/s, $\mu_k = 0.4$ (f) $v_i = 2$ m/s, $\mu_k = 0.8$

13. Can a force of static friction do work? If not, why not? If so, give an example.

14. Describe human-made devices designed to produce each of the following energy transfers or transformations. Whenever you can, describe also a natural process in which the energy process occurs. Give details to defend your choices, such as identifying the system and identifying other output energy if the process has limited efficiency. (a) Chemical potential energy transforms into internal energy. (b) Energy transferred by electrical transmission becomes gravitational potential energy. (c) Elastic potential energy transfers out of a system by heat. (d) Energy transferred by mechanical waves does work on a system. (e) Energy carried by electromagnetic waves becomes kinetic energy in a system.

15. In the general conservation of energy equation, state which terms predominate in describing each of the following devices and processes. For a process going on continuously, you may consider what happens in a 10-s time interval. State which terms in the equation represent original and final forms of energy, which would be inputs, and which would be outputs. (a) a slingshot firing a pebble (b) a fire burning (c) a portable radio operating (d) a car braking to a stop (e) the surface of the sun shining visibly (f) a person jumping up onto a chair

16. **O** At the bottom of an air track tilted at angle θ, a glider of mass m is given a push to make it coast a distance d up the slope as it slows down and stops. Then the glider comes back down the track to its starting point. Now the experiment is repeated with the same original speed but with a second identical glider set on top of the first. The airflow is strong enough to support the stacked pair of gliders so that they move freely over the track. Static friction holds the second glider stationary relative to the first glider throughout the motion. The coefficient of static friction between the two gliders is μ_s. What is the change in mechanical energy of the two-glider–Earth system in the up- and downslope motion after the pair of gliders is released? Choose one. (a) $-2md$ (b) $-2\mu_s gd$ (c) $-2\mu_s md$ (d) $-2\mu_s mg$ (e) $-2mg \cos\theta$ (f) $-2mgd \cos\theta$ (g) $-2\mu_s mgd \cos\theta$ (h) $-4\mu_s mgd \cos\theta$ (i) $-\mu_s mgd \cos\theta$ (j) $-2\mu_s mgd \sin\theta$ (k) 0 (l) $+2\mu_s mgd \cos\theta$

17. A car salesperson claims that a souped-up 300-hp engine is a necessary option in a compact car in place of the conventional 130-hp engine. Suppose you intend to drive the car within speed limits ($\leq$ 65 mi/h) on flat terrain. How would you counter this sales pitch?

Problems

WebAssign The Problems from this chapter may be assigned online in WebAssign.

ThomsonNOW Sign in at **www.thomsonedu.com** and go to ThomsonNOW to assess your understanding of this chapter's topics with additional quizzing and conceptual questions.

1, 2, 3 denotes straightforward, intermediate, challenging; ☐ denotes full solution available in *Student Solutions Manual/Study Guide;* ▲ denotes coached solution with hints available at **www.thomsonedu.com;** ▭ denotes developing symbolic reasoning; ● denotes asking for qualitative reasoning; ▪ denotes computer useful in solving problem

Section 8.1 The Nonisolated System: Conservation of Energy

1. For each of the following systems and time intervals, write the appropriate reduced version of Equation 8.2, the conservation of energy equation. (a) the heating coils in your toaster during the first five seconds after you turn the toaster on (b) your automobile, from just before you fill it with gas until you pull away from the gas station at 10 mi/h (c) your body while you sit quietly and eat a peanut butter and jelly sandwich for lunch (d) your home during five minutes of a sunny afternoon while the temperature in the home remains fixed.

Section 8.2 The Isolated System

2. At 11:00 a.m. on September 7, 2001, more than one million British schoolchildren jumped up and down for 1 min. The curriculum focus of the "giant jump" was on earthquakes, but it was integrated with many other topics, such as exercise, geography, cooperation, testing hypotheses, and setting world records. Students built their own seismographs that registered local effects. (a) Find the energy converted into mechanical energy in the experiment. Assume 1 050 000 children of average mass 36.0 kg jump 12 times each, raising their centers of mass by 25.0 cm each time and briefly resting between one jump and the next. The free-fall acceleration in Britain is 9.81 m/s^2. (b) Most of the mechanical energy is converted very rapidly into internal energy within the bodies of the students and the floors of the school buildings. Of the energy that propagates into the ground, most produces high-frequency "microtremor" vibrations that are rapidly damped and cannot travel far. Assume 0.01% of the energy is carried away by a long-range seismic wave. The magnitude of an earthquake on the Richter scale is given by

$$M = \frac{\log E - 4.8}{1.5}$$

where E is the seismic wave energy in joules. According to this model, what is the magnitude of the demonstration quake? It did not register above background noise overseas or on the seismograph of the Wolverton Seismic Vault, Hampshire.

3. A bead slides without friction around a loop-the-loop (Fig. P8.3). The bead is released from a height $h = 3.50R$. (a) What is the bead's speed at point Ⓐ? (b) How large is the normal force on the bead if its mass is 5.00 g?

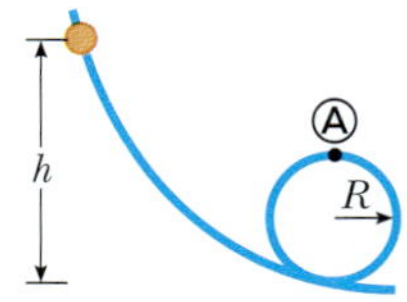

Figure P8.3

4. A particle of mass $m = 5.00$ kg is released from point Ⓐ and slides on the frictionless track shown in Figure P8.4. Determine (a) the particle's speed at points Ⓑ and Ⓒ and (b) the net work done by the gravitational force as the particle moves from Ⓐ to Ⓒ.

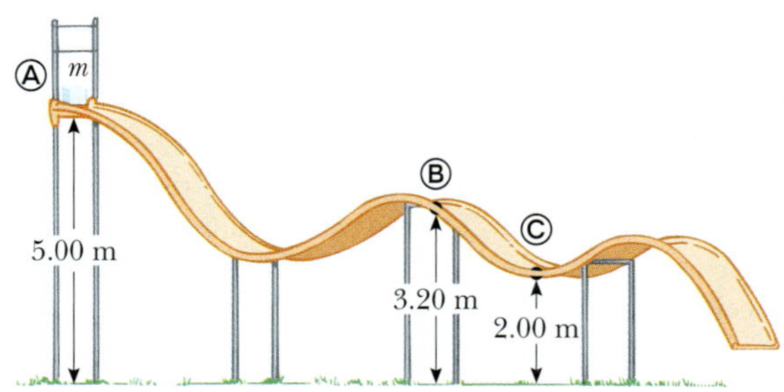

Figure P8.4

5. A block of mass 0.250 kg is placed on top of a light vertical spring of force constant 5 000 N/m and pushed downward so that the spring is compressed by 0.100 m. After the block is released from rest, it travels upward and then leaves the spring. To what maximum height above the point of release does it rise?

6. A circus trapeze consists of a bar suspended by two parallel ropes, each of length ℓ, allowing performers to swing in a vertical circular arc (Figure P8.6). Suppose a performer with mass m holds the bar and steps off an elevated platform, starting from rest with the ropes at an angle θ_i with respect to the vertical. Assume the size of the performer's body is small compared to the length ℓ, she does not pump the trapeze to swing higher, and air resistance is negligible. (a) Show that when the ropes make an angle θ with the vertical, the performer must exert a force

$$mg(3\cos\theta - 2\cos\theta_i)$$

so as to hang on. (b) Determine the angle θ_i for which the force needed to hang on at the bottom of the swing is twice as large as the gravitational force exerted on the performer.

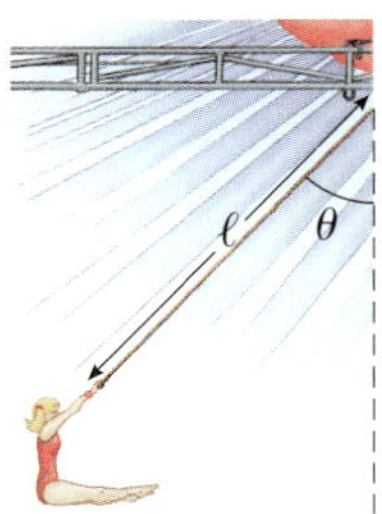

Figure P8.6

7. Two objects are connected by a light string passing over a light, frictionless pulley as shown in Figure P8.7. The object of mass 5.00 kg is released from rest. Using the isolated system model, (a) determine the speed of the 3.00-kg object just as the 5.00-kg object hits the ground. (b) Find the maximum height to which the 3.00-kg object rises.

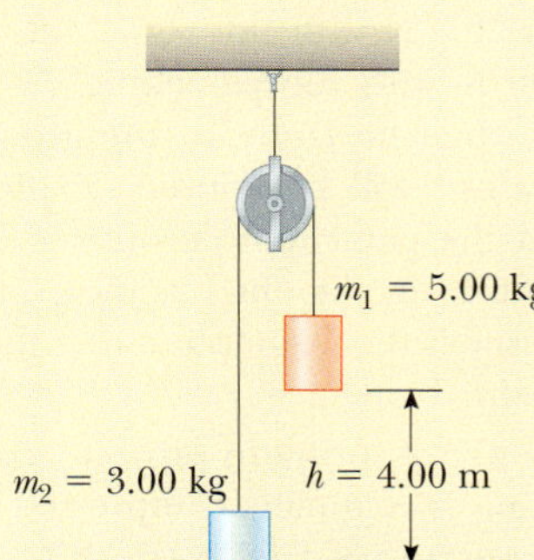

Figure P8.7 Problems 7 and 8.

8. Two objects are connected by a light string passing over a light, frictionless pulley as shown in Figure P8.7. The object of mass m_1 is released from rest at height h. Using the isolated system model, (a) determine the speed of m_2 just as m_1 hits the ground. (b) Find the maximum height to which m_2 rises.

9. A light, rigid rod is 77.0 cm long. Its top end is pivoted on a low-friction horizontal axle. The rod hangs straight down at rest with a small massive ball attached to its bottom end. You strike the ball, suddenly giving it a horizontal velocity so that it swings around in a full circle. What minimum speed at the bottom is required to make the ball go over the top of the circle?

10. A 20.0-kg cannonball is fired from a cannon with muzzle speed of 1 000 m/s at an angle of 37.0° with the horizontal. A second cannonball is fired at an angle of 90.0°. Use the isolated system model to find (a) the maximum height reached by each ball and (b) the total mechanical energy of the ball–Earth system at the maximum height for each ball. Let $y = 0$ at the cannon.

11. A daredevil plans to bungee-jump from a hot-air balloon 65.0 m above a carnival midway (Fig. P8.11). He will use a uniform elastic cord, tied to a harness around his body, to stop his fall at a point 10.0 m above the ground. Model his body as a particle and the cord as having negligible mass and obeying Hooke's law. In a preliminary test, hanging at rest from a 5.00-m length of the cord, the daredevil finds his body weight stretches the cord by 1.50 m. He intends to drop from rest at the point where the top end of a longer section of the cord is attached to the stationary hot-air balloon. (a) What length of cord should he use? (b) What maximum acceleration will he experience?

Figure P8.11 Problems 11 and 46.

12. **Review problem.** The system shown in Figure P8.12 consists of a light, inextensible cord; light, frictionless pulleys; and blocks of equal mass. It is initially held at rest so that the blocks are at the same height above the ground. The blocks are then released. Find the speed of block A at the moment when the vertical separation of the blocks is h.

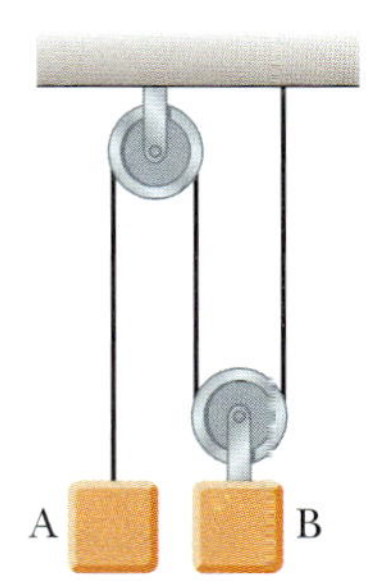

Figure P8.12

Section 8.3 Situations Involving Kinetic Friction

13. A 40.0-kg box initially at rest is pushed 5.00 m along a rough, horizontal floor with a constant applied horizontal force of 130 N. The coefficient of friction between box and floor is 0.300. Find (a) the work done by the applied force, (b) the increase in internal energy in the box–floor system as a result of friction, (c) the work done by the

normal force, (d) the work done by the gravitational force, (e) the change in kinetic energy of the box, and (f) the final speed of the box.

14. A 2.00-kg block is attached to a spring of force constant 500 N/m as shown in Active Figure 7.9. The block is pulled 5.00 cm to the right of equilibrium and released from rest. Find the speed the block has as it passes through equilibrium if (a) the horizontal surface is frictionless and (b) the coefficient of friction between block and surface is 0.350.

15. A crate of mass 10.0 kg is pulled up a rough incline with an initial speed of 1.50 m/s. The pulling force is 100 N parallel to the incline, which makes an angle of 20.0° with the horizontal. The coefficient of kinetic friction is 0.400, and the crate is pulled 5.00 m. (a) How much work is done by the gravitational force on the crate? (b) Determine the increase in internal energy of the crate–incline system owing to friction. (c) How much work is done by the 100-N force on the crate? (d) What is the change in kinetic energy of the crate? (e) What is the speed of the crate after being pulled 5.00 m?

16. ● A block of mass m is on a horizontal surface with which its coefficient of kinetic friction is μ_k. The block is pushed against the free end of a light spring with force constant k, compressing the spring by distance d. Then the block is released from rest so that the spring fires the block across the surface. Of the possible expressions (a) through (k) listed below for the speed of the block after it has slid over distance d, **(i)** which cannot be true because they are dimensionally incorrect? **(ii)** Of those remaining, which give(s) an incorrect result in the limit as k becomes very large? **(iii)** Of those remaining, which give(s) an incorrect result in the limit as μ_k goes to zero? **(iv)** Of those remaining, which can you rule out for other reasons you specify? **(v)** Which expression is correct? **(vi)** Evaluate the speed in the case $m = 250$ g, $\mu_k = 0.600$, $k = 18.0$ N/m, and $d = 12.0$ cm. You will need to explain your answer. (a) $(kd^2 - \mu_k mgd)^{1/2}$ (b) $(kd^2/m - \mu_k g)^{1/2}$ (c) $(kd/m - 2\mu_k gd)^{1/2}$ (d) $(kd^2/m - gd)^{1/2}$ (e) $(kd^2/m - \mu_k^2 gd)^{1/2}$ (f) $kd^2/m - \mu_k gd$ (g) $(\mu_k kd^2/m - gd)^{1/2}$ (h) $(kd^2/m - 2\mu_k gd)^{1/2}$ (i) $(\mu_k gd - kd^2/m)^{1/2}$ (j) $(gd - \mu_k gd)^{1/2}$ (k) $(kd^2/m + \mu_k gd)^{1/2}$

17. ▲ A sled of mass m is given a kick on a frozen pond. The kick imparts to it an initial speed of 2.00 m/s. The coefficient of kinetic friction between sled and ice is 0.100. Use energy considerations to find the distance the sled moves before it stops.

Section 8.4 Changes in Mechanical Energy for Nonconservative Forces

18. ● At time t_i, the kinetic energy of a particle is 30.0 J and the potential energy of the system to which it belongs is 10.0 J. At some later time t_f, the kinetic energy of the particle is 18.0 J. (a) If only conservative forces act on the particle, what are the potential energy and the total energy at time t_f? (b) If the potential energy of the system at time t_f is 5.00 J, are any nonconservative forces acting on the particle? Explain.

19. ▲ The coefficient of friction between the 3.00-kg block and the surface in Figure P8.19 is 0.400. The system starts from rest. What is the speed of the 5.00-kg ball when it has fallen 1.50 m?

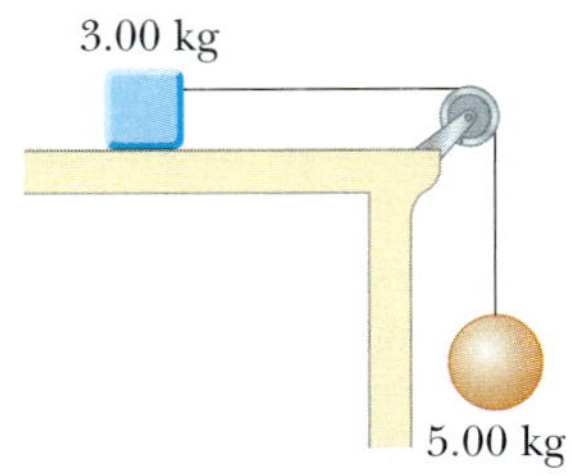

Figure P8.19

20. In her hand, a softball pitcher swings a ball of mass 0.250 kg around a vertical circular path of radius 60.0 cm before releasing it from her hand. The pitcher maintains a component of force on the ball of constant magnitude 30.0 N in the direction of motion around the complete path. The speed of the ball at the top of the circle is 15.0 m/s. If the pitcher releases the ball at the bottom of the circle, what is its speed upon release?

21. A 5.00-kg block is set into motion up an inclined plane with an initial speed of 8.00 m/s (Fig. P8.21). The block comes to rest after traveling 3.00 m along the plane, which is inclined at an angle of 30.0° to the horizontal. For this motion, determine (a) the change in the block's kinetic energy, (b) the change in the potential energy of the block–Earth system, and (c) the friction force exerted on the block (assumed to be constant). (d) What is the coefficient of kinetic friction?

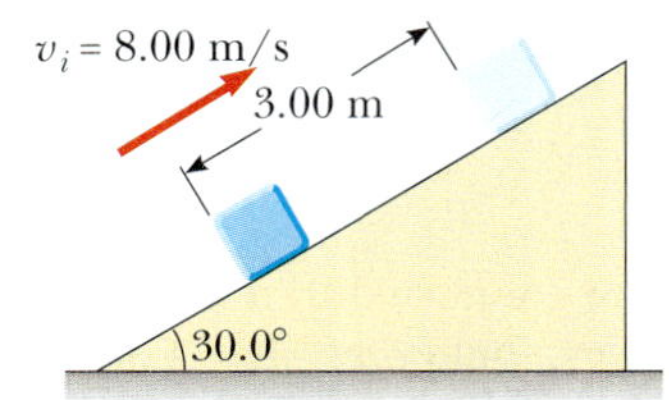

Figure P8.21

22. ● An 80.0-kg skydiver jumps out of a balloon at an altitude of 1 000 m and opens the parachute at an altitude of 200 m. (a) Assuming the total retarding force on the diver is constant at 50.0 N with the parachute closed and constant at 3 600 N with the parachute open, find the skydiver's speed when he lands on the ground. (b) Do you think the skydiver will be injured? Explain. (c) At what height should the parachute be opened so that the final speed of the skydiver when he hits the ground is 5.00 m/s? (d) How realistic is the assumption that the total retarding force is constant? Explain.

23. A toy cannon uses a spring to project a 5.30-g soft rubber ball. The spring is originally compressed by 5.00 cm and has a force constant of 8.00 N/m. When the cannon is fired, the ball moves 15.0 cm through the horizontal barrel of the cannon and the barrel exerts a constant friction force of 0.032 0 N on the ball. (a) With what speed does the projectile leave the barrel of the cannon? (b) At what point does the ball have maximum speed? (c) What is this maximum speed?

24. A particle moves along a line where the potential energy of its system depends on its position r as graphed in Figure P8.24. In the limit as r increases without bound, $U(r)$ approaches +1 J. (a) Identify each equilibrium position for this particle. Indicate whether each is a point of stable, unstable, or neutral equilibrium. (b) The particle will

be bound if the total energy of the system is in what range? Now suppose the system has energy −3 J. Determine (c) the range of positions where the particle can be found, (d) its maximum kinetic energy, (e) the location where it has maximum kinetic energy, and (f) the *binding energy* of the system, that is, the additional energy it would have to be given for the particle to move out to $r \to \infty$.

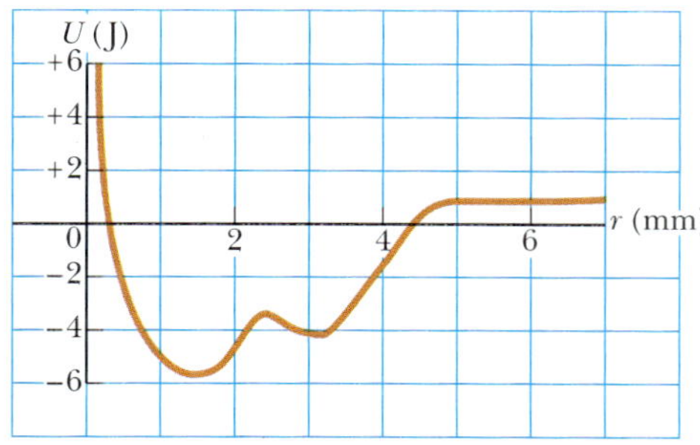

Figure P8.24

25. A 1.50-kg object is held 1.20 m above a relaxed, massless vertical spring with a force constant of 320 N/m. The object is dropped onto the spring. (a) How far does the object compress the spring? (b) **What If?** How far does the object compress the spring if the same experiment is performed on the Moon, where g = 1.63 m/s²? (c) **What If?** Repeat part (a), but this time assume a constant air-resistance force of 0.700 N acts on the object during its motion.

26. A boy in a wheelchair (total mass 47.0 kg) wins a race with a skateboarder. The boy has speed 1.40 m/s at the crest of a slope 2.60 m high and 12.4 m long. At the bottom of the slope his speed is 6.20 m/s. Assume air resistance and rolling resistance can be modeled as a constant friction force of 41.0 N. Find the work he did in pushing forward on his wheels during the downhill ride.

27. A uniform board of length L is sliding along a smooth (frictionless) horizontal plane as shown in Figure P8.27a. The board then slides across the boundary with a rough horizontal surface. The coefficient of kinetic friction between the board and the second surface is μ_k. (a) Find the acceleration of the board at the moment its front end has traveled a distance x beyond the boundary. (b) The board stops at the moment its back end reaches the boundary as shown in Figure P8.27b. Find the initial speed v of the board.

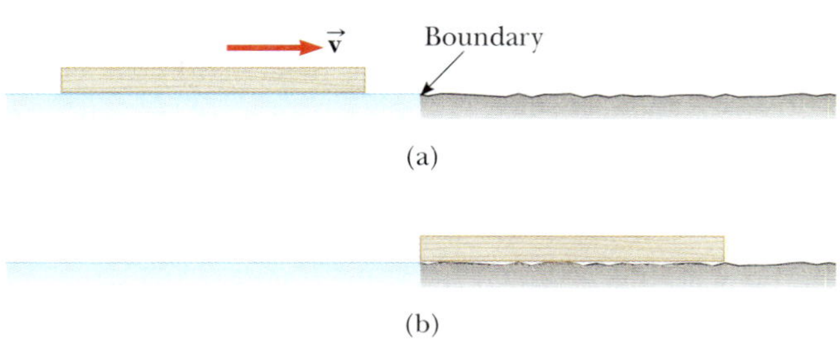

Figure P8.27

Section 8.5 Power

28. The electric motor of a model train accelerates the train from rest to 0.620 m/s in 21.0 ms. The total mass of the train is 875 g. Find the average power delivered to the train during the acceleration.

29. ▲ A 700-N Marine in basic training climbs a 10.0-m vertical rope at a constant speed in 8.00 s. What is his power output?

30. Columnist Dave Barry poked fun at the name "The Grand Cities" adopted by Grand Forks, North Dakota, and East Grand Forks, Minnesota. Residents of the prairie towns then named their next municipal building for him. At the Dave Barry Lift Station No. 16, untreated sewage is raised vertically by 5.49 m, at the rate of 1 890 000 liters each day. The waste, of density 1 050 kg/m³, enters and leaves the pump at atmospheric pressure, through pipes of equal diameter. (a) Find the output mechanical power of the lift station. (b) Assume an electric motor continuously operating with average power 5.90 kW runs the pump. Find its efficiency.

31. Make an order-of-magnitude estimate of the power a car engine contributes to speeding the car up to highway speed. For concreteness, consider your own car if you use one. In your solution, state the physical quantities you take as data and the values you measure or estimate for them. The mass of the vehicle is given in the owner's manual. If you do not wish to estimate for a car, consider a bus or truck that you specify

32. A 650-kg elevator starts from rest. It moves upward for 3.00 s with constant acceleration until it reaches its cruising speed of 1.75 m/s. (a) What is the average power of the elevator motor during this time interval? (b) How does this power compare with the motor power when the elevator moves at its cruising speed?

33. An energy-efficient lightbulb, taking in 28.0 W of power, can produce the same level of brightness as a conventional lightbulb operating at power 100 W. The lifetime of the energy-efficient lightbulb is 10 000 h and its purchase price is $17.0, whereas the conventional lightbulb has lifetime 750 h and costs $0.420 per bulb. Determine the total savings obtained by using one energy-efficient lightbulb over its lifetime as opposed to using conventional lightbulbs over the same time interval. Assume an energy cost of $0.080 0 per kilowatt-hour.

34. An electric scooter has a battery capable of supplying 120 Wh of energy. If friction forces and other losses account for 60.0% of the energy usage, what altitude change can a rider achieve when driving in hilly terrain if the rider and scooter have a combined weight of 890 N?

35. A loaded ore car has a mass of 950 kg and rolls on rails with negligible friction. It starts from rest and is pulled up a mine shaft by a cable connected to a winch. The shaft is inclined at 30.0° above the horizontal. The car accelerates uniformly to a speed of 2.20 m/s in 12.0 s and then continues at constant speed. (a) What power must the winch motor provide when the car is moving at constant speed? (b) What maximum power must the winch motor provide? (c) What total energy has transferred out of the motor by work by the time the car moves off the end of the track, which is of length 1 250 m?

36. Energy is conventionally measured in Calories as well as in joules. One Calorie in nutrition is one kilocalorie, defined as 1 kcal = 4 186 J. Metabolizing 1 g of fat can release 9.00 kcal. A student decides to try to lose weight by exercising. She plans to run up and down the stairs in a football stadium as fast as she can and as many times as

necessary. Is this activity in itself a practical way to lose weight? To evaluate the program, suppose she runs up a flight of 80 steps, each 0.150 m high, in 65.0 s. For simplicity, ignore the energy she uses in coming down (which is small). Assume a typical efficiency for human muscles is 20.0%. This statement means that when your body converts 100 J from metabolizing fat, 20 J goes into doing mechanical work (here, climbing stairs). The remainder goes into extra internal energy. Assume the student's mass is 50.0 kg. (a) How many times must she run the flight of stairs to lose 1 lb of fat? (b) What is her average power output, in watts and in horsepower, as she is running up the stairs?

Additional Problems

37. A skateboarder with his board can be modeled as a particle of mass 76.0 kg, located at his center of mass (which we will study in Chapter 9). As shown in Figure P8.37, the skateboarder starts from rest in a crouching position at one lip of a half-pipe (point Ⓐ). The half-pipe is a dry water channel, forming one half of a cylinder of radius 6.80 m with its axis horizontal. On his descent, the skateboarder moves without friction so that his center of mass moves through one quarter of a circle of radius 6.30 m. (a) Find his speed at the bottom of the half-pipe (point Ⓑ). (b) Find his centripetal acceleration. (c) Find the normal force $n_Ⓑ$ acting on the skateboarder at point Ⓑ. Immediately after passing point Ⓑ, he stands up and raises his arms, lifting his center of mass from 0.500 m to 0.950 m above the concrete (point Ⓒ). To account for the conversion of chemical into mechanical energy, model his legs as doing work by pushing him vertically up, with a constant force equal to the normal force $n_Ⓑ$, over a distance of 0.450 m. (You will be able to solve this problem with a more accurate model in Chapter 11.) (d) What is the work done on the skateboarder's body in this process? Next, the skateboarder glides upward with his center of mass moving in a quarter circle of radius 5.85 m. His body is horizontal when he passes point Ⓓ, the far lip of the half-pipe. (e) Find his speed at this location. At last he goes ballistic, twisting around while his center of mass moves vertically. (f) How high above point Ⓓ does he rise? (g) Over what time interval is he airborne before he touches down, 2.34 m below the level of point Ⓓ? *Caution:* Do not try this stunt yourself without the required skill and protective equipment or in a drainage channel to which you do not have legal access.

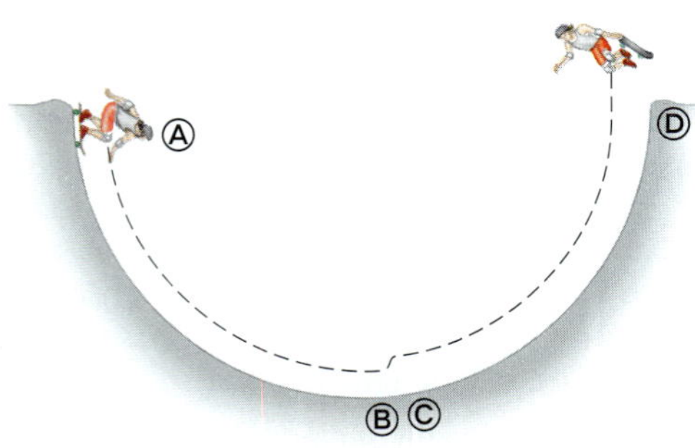

Figure P8.37

38. ● **Review problem.** As shown in Figure P8.38, a light string that does not stretch changes from horizontal to vertical as it passes over the edge of a table. The string connects a 3.50-kg block, originally at rest on the horizontal table, 1.20 m above the floor, to a hanging 1.90-kg block, originally 0.900 m above the floor. Neither the surface of the table nor its edge exerts a force of kinetic friction. The blocks start to move with negligible speed. Consider the two blocks plus the Earth as the system. (a) Does the mechanical energy of the system remain constant between the instant of release and the instant before the hanging block hits the floor? (b) Find the speed at which the sliding block leaves the edge of the table. (c) Now suppose the hanging block stops permanently as soon as it reaches the sticky floor. Does the mechanical energy of the system remain constant between the instant of release and the instant before the sliding block hits the floor? (d) Find the impact speed of the sliding block. (e) How long must the string be if it does not go taut while the sliding block is in flight? (f) Would it invalidate your speed calculation if the string does go taut? (g) Even with negligible kinetic friction, the coefficient of static friction between the heavier block and the table is 0.560. Evaluate the force of friction acting on this block before the motion begins. (h) Will the motion begin by itself, or must the experimenter give a little tap to the sliding block to get it started? Are the speed calculations still valid?

Figure P8.38

39. A 4.00-kg particle moves along the x axis. Its position varies with time according to $x = t + 2.0t^3$, where x is in meters and t is in seconds. Find (a) the kinetic energy at any time t, (b) the acceleration of the particle and the force acting on it at time t, (c) the power being delivered to the particle at time t, and (d) the work done on the particle in the interval $t = 0$ to $t = 2.00$ s.

40. ● Heedless of danger, a child leaps onto a pile of old mattresses to use them as a trampoline. His motion between two particular points is described by the energy conservation equation

$$\tfrac{1}{2}(46.0\text{ kg})(2.40\text{ m/s})^2 + (46.0\text{ kg})(9.80\text{ m/s}^2)(2.80\text{ m} + x) = \tfrac{1}{2}(1.94 \times 10^4\text{ N/m})x^2$$

(a) Solve the equation for x. (b) Compose the statement of a problem, including data, for which this equation gives the solution. Identify the physical meaning of the value of x.

41. As the driver steps on the gas pedal, a car of mass 1 160 kg accelerates from rest. During the first few seconds of motion, the car's acceleration increases with time according to the expression

$$a = (1.16\text{ m/s}^3)t - (0.210\text{ m/s}^4)t^2 + (0.240\text{ m/s}^5)t^3$$

(a) What work is done on the car by the wheels during the interval from $t = 0$ to $t = 2.50$ s? (b) What is the wheels' output power at the instant $t = 2.50$ s?

42. A 0.400-kg particle slides around a horizontal track. The track has a smooth vertical outer wall forming a circle with a radius of 1.50 m. The particle is given an initial speed of 8.00 m/s. After one revolution, its speed has dropped to 6.00 m/s because of friction with the rough floor of the track. (a) Find the energy transformed from mechanical to internal in the system as a result of friction in one revolution. (b) Calculate the coefficient of kinetic friction. (c) What is the total number of revolutions the particle makes before stopping?

43. A 200-g block is pressed against a spring of force constant 1.40 kN/m until the block compresses the spring 10.0 cm. The spring rests at the bottom of a ramp inclined at 60.0° to the horizontal. Using energy considerations, determine how far up the incline the block moves before it stops (a) if the ramp exerts no friction force on the block and (b) if the coefficient of kinetic friction is 0.400.

44. ● As it plows a parking lot, a snowplow pushes an ever-growing pile of snow in front of it. Suppose a car moving through the air is similarly modeled as a cylinder pushing a growing plug of air in front of it. The originally stationary air is set into motion at the constant speed v of the cylinder as shown in Figure P8.44. In a time interval Δt, a new disk of air of mass Δm must be moved a distance $v\,\Delta t$ and hence must be given a kinetic energy $\frac{1}{2}(\Delta m)v^2$. Using this model, show that the car's power loss owing to air resistance is $\frac{1}{2}\rho A v^3$ and that the resistive force acting on the car is $\frac{1}{2}\rho A v^2$, where ρ is the density of air. Compare this result with the empirical expression $\frac{1}{2}D\rho A v^2$ for the resistive force.

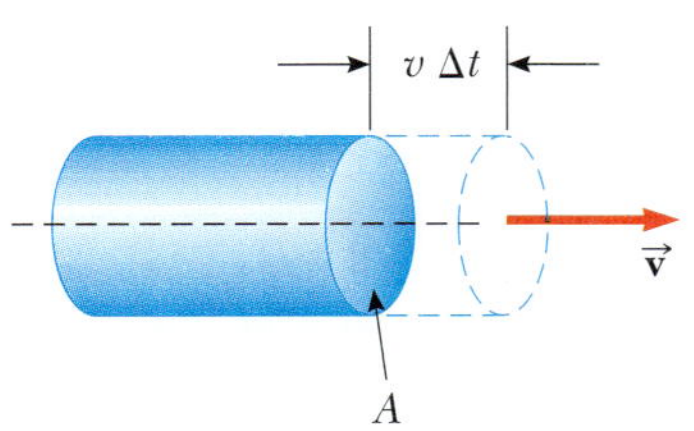

Figure P8.44

45. A windmill such as that shown in the opening photograph for Chapter 7 turns in response to a force of high-speed air resistance, $R = \frac{1}{2}D\rho A v^2$. The power available is $\mathcal{P} = Rv = \frac{1}{2}D\rho\pi r^2 v^3$, where v is the wind speed and we have assumed a circular face for the windmill, of radius r. Take the drag coefficient as $D = 1.00$ and the density of air from the front endpaper of this book. For a home windmill having $r = 1.50$ m, calculate the power available with (a) $v = 8.00$ m/s and (b) $v = 24.0$ m/s. The power delivered to the generator is limited by the efficiency of the system, about 25%. For comparison, a typical U.S. home uses about 3 kW of electric power.

46. ● Starting from rest, a 64.0-kg person bungee jumps from a tethered balloon 65.0 m above the ground (Fig. P8.11). The bungee cord has negligible mass and unstretched length 25.8 m. One end is tied to the basket of the hot-air balloon and the other end to a harness around the person's body. The cord is modeled as a spring that obeys Hooke's law with a spring constant of 81.0 N/m, and the person's body is modeled as a particle. The hot-air balloon does not move. (a) Express the gravitational potential energy of the person–Earth system as a function of the person's variable height y above the ground. (b) Express the elastic potential energy of the cord as a function of y. (c) Express the total potential energy of the person–cord–Earth system as a function of y. (d) Plot a graph of the gravitational, elastic, and total potential energies as functions of y. (e) Assume air resistance is negligible. Determine the minimum height of the person above the ground during his plunge. (f) Does the potential energy graph show any equilibrium position or positions? If so, at what elevations? Are they stable or unstable? (g) Determine the jumper's maximum speed.

47. Consider the block–spring–surface system in part (B) of Example 8.6. (a) At what position x of the block is its speed a maximum? (b) In the **What If?** section of that example, we explored the effects of an increased friction force of 10.0 N. At what position of the block does its maximum speed occur in this situation?

48. ● More than 2 300 years ago the Greek teacher Aristotle wrote the first book called *Physics*. Put into more precise terminology, this passage is from the end of its Section Eta:

> Let $\mathcal{P}$ be the power of an agent causing motion; w, the load moved; d, the distance covered; and Δt, the time interval required. Then (1) a power equal to $\mathcal{P}$ will in an interval of time equal to Δt move $w/2$ a distance $2d$, or (2) it will move $w/2$ the given distance d in the time interval $\Delta t/2$. Also, if (3) the given power $\mathcal{P}$ moves the given load w a distance $d/2$ in time interval $\Delta t/2$, then (4) $\mathcal{P}/2$ will move $w/2$ the given distance d in the given time interval Δt.

(a) Show that Aristotle's proportions are included in the equation $\mathcal{P}\Delta t = bwd$, where b is a proportionality constant. (b) Show that our theory of motion includes this part of Aristotle's theory as one special case. In particular, describe a situation in which it is true, derive the equation representing Aristotle's proportions, and determine the proportionality constant.

49. **Review problem.** The mass of a car is 1 500 kg. The shape of the car's body is such that its aerodynamic drag coefficient is $D = 0.330$ and the frontal area is 2.50 m^2. Assuming the drag force is proportional to v^2 and ignoring other sources of friction, calculate the power required to maintain a speed of 100 km/h as the car climbs a long hill sloping at 3.20°.

50. A 200-g particle is released from rest at point Ⓐ along the horizontal diameter on the inside of a frictionless, hemispherical bowl of radius $R = 30.0$ cm (Fig. P8.50). Calculate (a) the gravitational potential energy of the particle–Earth system when the particle is at point Ⓐ relative to

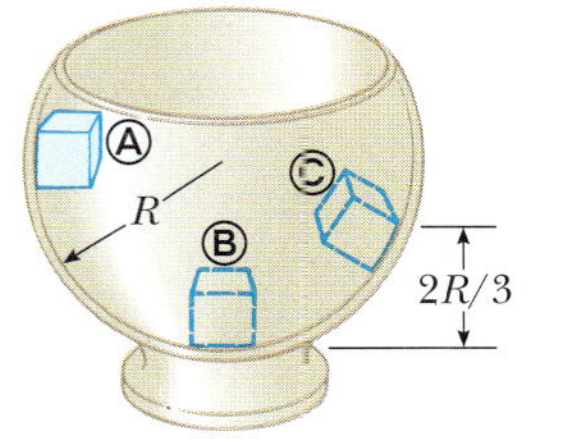

Figure P8.50 Problems 50 and 51.

point Ⓑ, (b) the kinetic energy of the particle at point Ⓑ, (c) its speed at point Ⓑ, and (d) its kinetic energy and the potential energy when the particle is at point Ⓒ.

51. ● ▲ **What If?** The particle described in Problem 50 (Fig. P8.50) is released from rest at Ⓐ, and the surface of the bowl is rough. The speed of the particle at Ⓑ is 1.50 m/s. (a) What is its kinetic energy at Ⓑ? (b) How much mechanical energy is transformed into internal energy as the particle moves from Ⓐ to Ⓑ? (c) Is it possible to determine the coefficient of friction from these results in any simple manner? Explain.

52. Assume you attend a state university that was founded as an agricultural college. Close to the center of the campus is a tall silo topped with a hemispherical cap. The cap is frictionless when wet. Someone has balanced a pumpkin at the silo's highest point. The line from the center of curvature of the cap to the pumpkin makes an angle $\theta_i = 0°$ with the vertical. While you happen to be standing nearby in the middle of a rainy night, a breath of wind makes the pumpkin start sliding downward from rest. The pumpkin loses contact with the cap when the line from the center of the hemisphere to the pumpkin makes a certain angle with the vertical. What is this angle?

53. A child's pogo stick (Fig. P8.53) stores energy in a spring with a force constant of 2.50×10^4 N/m. At position Ⓐ ($x_Ⓐ = -0.100$ m), the spring compression is a maximum and the child is momentarily at rest. At position Ⓑ ($x_Ⓑ = 0$), the spring is relaxed and the child is moving upward. At position Ⓒ, the child is again momentarily at rest at the top of the jump. The combined mass of child and pogo stick is 25.0 kg. (a) Calculate the total energy of the child–stick–Earth system, taking both gravitational and elastic potential energies as zero for $x = 0$. (b) Determine $x_Ⓒ$. (c) Calculate the speed of the child at $x = 0$. (d) Determine the value of x for which the kinetic energy of the system is a maximum. (e) Calculate the child's maximum upward speed.

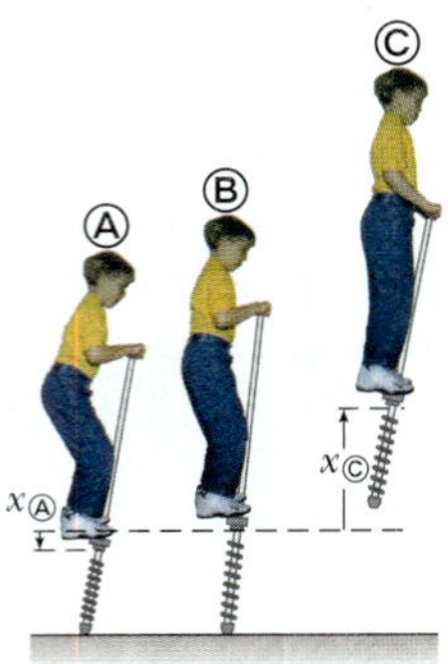

Figure P8.53

54. A 1.00-kg object slides to the right on a surface having a coefficient of kinetic friction 0.250 (Fig. P8.54). The object has a speed of $v_i = 3.00$ m/s when it makes contact with a light spring that has a force constant of 50.0 N/m. The object comes to rest after the spring has been compressed a distance d. The object is then forced toward the left by the spring and continues to move in that direction beyond the spring's unstretched position. Finally, the object comes to rest a distance D to the left of the unstretched spring. Find (a) the distance of compression d, (b) the speed v at the unstretched position when the object is moving to the left, and (c) the distance D where the object comes to rest.

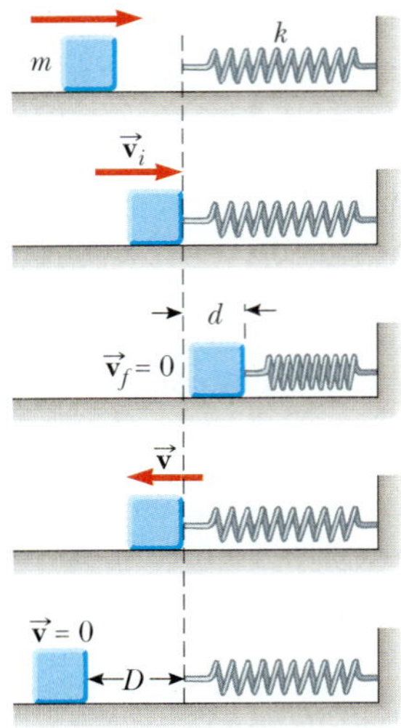

Figure P8.54

55. A 10.0-kg block is released from point Ⓐ in Figure P8.55. The track is frictionless except for the portion between points Ⓑ and Ⓒ, which has a length of 6.00 m. The block travels down the track, hits a spring of force constant 2 250 N/m, and compresses the spring 0.300 m from its equilibrium position before coming to rest momentarily. Determine the coefficient of kinetic friction between the block and the rough surface between Ⓑ and Ⓒ.

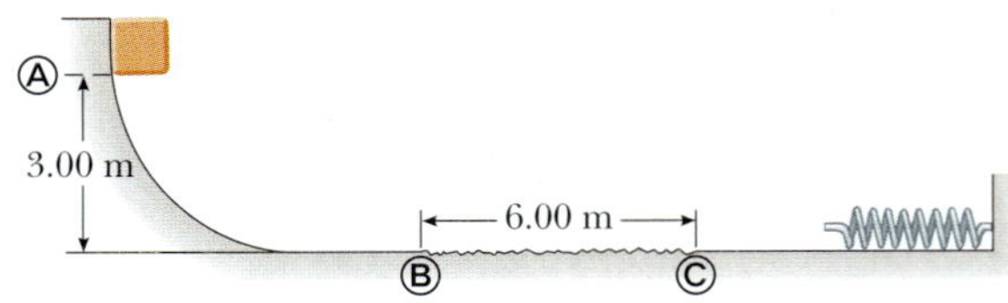

Figure P8.55

56. A uniform chain of length 8.00 m initially lies stretched out on a horizontal table. (a) Assuming the coefficient of static friction between chain and table is 0.600, show that the chain will begin to slide off the table if at least 3.00 m of it hangs over the edge of the table. (b) Determine the speed of the chain as its last link leaves the table, given that the coefficient of kinetic friction between the chain and the table is 0.400.

57. A 20.0-kg block is connected to a 30.0-kg block by a string that passes over a light, frictionless pulley. The 30.0-kg block is connected to a spring that has negligible mass and a force constant of 250 N/m as shown in Figure P8.57. The spring is unstretched when the system is as shown in the figure, and the incline is frictionless. The 20.0-kg block is pulled 20.0 cm down the incline (so that the 30.0-kg block is 40.0 cm above the floor) and released from rest. Find the speed of each block when the 30.0-kg block is 20.0 cm above the floor (that is, when the spring is unstretched).

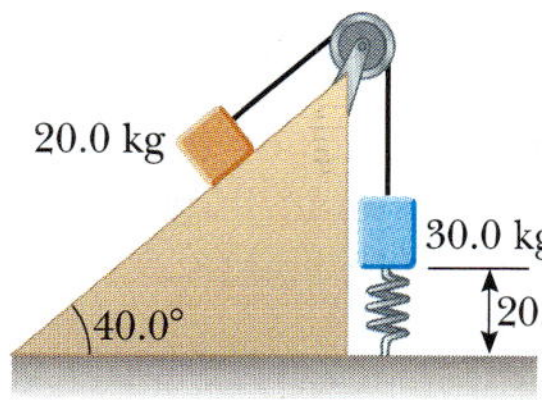

Figure P8.57

58. Jane, whose mass is 50.0 kg, needs to swing across a river (having width D) filled with person-eating crocodiles to save Tarzan from danger. She must swing into a wind exerting constant horizontal force $\vec{\mathbf{F}}$, on a vine having length L and initially making an angle θ with the vertical (Fig. P8.58). Take $D = 50.0$ m, $F = 110$ N, $L = 40.0$ m, and $\theta = 50.0°$. (a) With what minimum speed must Jane begin her swing to just make it to the other side? (b) Once the rescue is complete, Tarzan and Jane must swing back across the river. With what minimum speed must they begin their swing? Assume Tarzan has a mass of 80.0 kg.

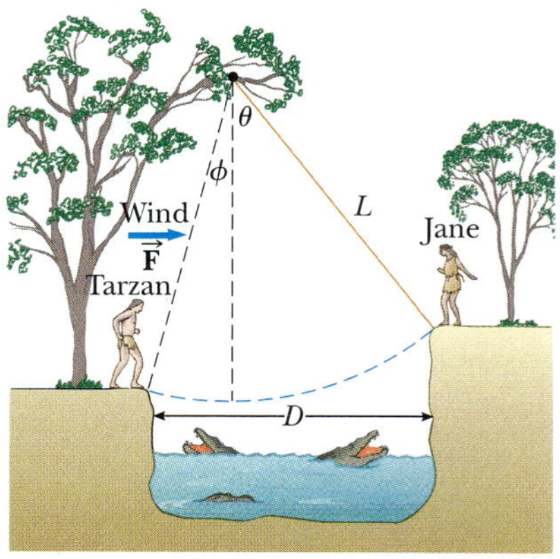

Figure P8.58

59. ● ▲ A block of mass 0.500 kg is pushed against a horizontal spring of negligible mass until the spring is compressed a distance x (Fig. P8.59). The force constant of the spring is 450 N/m. When it is released, the block travels along a frictionless, horizontal surface to point B, the bottom of a vertical circular track of radius $R = 1.00$ m, and continues to move along the track. The speed of the block at the bottom of the track is $v_B = 12.0$ m/s, and the block experiences an average friction force of 7.00 N while sliding up the track. (a) What is x? (b) What speed do you predict for the block at the top of the track? (c) Does the block actually reach the top of the track, or does it fall off before reaching the top?

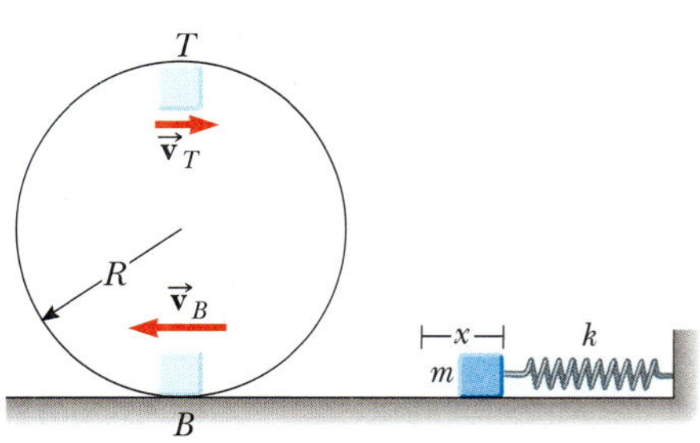

Figure P8.59

60. A ball of mass $m = 300$ g is connected by a strong string of length $L = 80.0$ cm to a pivot and held in place with the string vertical. A wind exerts constant force F to the right on the ball as shown in Figure P8.60. The ball is released from rest. The wind makes it swing up to attain maximum height H above its starting point before it swings down again. (a) Find H as a function of F. Evaluate H (b) for $F = 1.00$ N and (c) for $F = 10.0$ N. How does H behave (d) as F approaches zero (e) and as F approaches infinity? (f) Now consider the equilibrium height of the ball with the wind blowing. Determine it as a function of F. Evaluate the equilibrium height (g) for $F = 10$ N and (h) for F going to infinity.

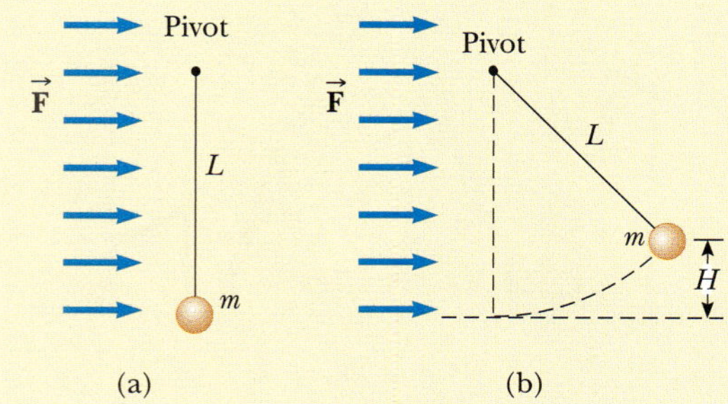

Figure P8.60

61. A block of mass M rests on a table. It is fastened to the lower end of a light, vertical spring. The upper end of the spring is fastened to a block of mass m. The upper block is pushed down by an additional force $3mg$, so the spring compression is $4mg/k$. In this configuration, the upper block is released from rest. The spring lifts the lower block off the table. In terms of m, what is the greatest possible value for M?

62. A pendulum, comprising a light string of length L and a small sphere, swings in the vertical plane. The string hits a peg located a distance d below the point of suspension (Fig. P8.62). (a) Show that if the sphere is released from a height below that of the peg, it will return to this height after the string strikes the peg. (b) Show that if the pendulum is released from the horizontal position ($\theta = 90°$) and is to swing in a complete circle centered on the peg, the minimum value of d must be $3L/5$.

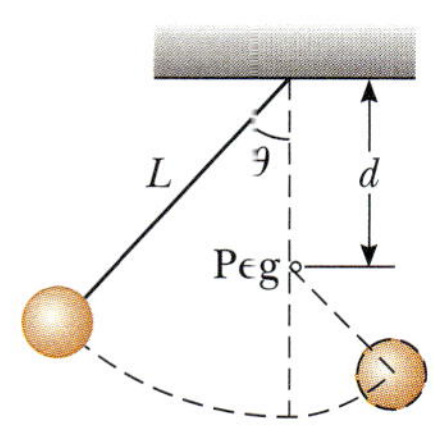

Figure P8.62

63. A ball whirls around in a vertical circle at the end of a string. The other end of the string is fixed at the center of the circle. Assuming the total energy of the ball–Earth system remains constant, show that the tension in the string at the bottom is greater than the tension at the top by six times the weight of the ball.

64. A roller-coaster car is released from rest at the top of the first rise and then moves freely with negligible friction. The roller coaster shown in Figure P8.64 has a circular loop of radius R in a vertical plane. (a) First suppose the

car barely makes it around the loop; at the top of the loop, the riders are upside down and feel weightless. Find the required height of the release point above the bottom of the loop in terms of R. (b) Now assume the release point is at or above the minimum required height. Show that the normal force on the car at the bottom of the loop exceeds the normal force at the top of the loop by six times the weight of the car. The normal force on each rider follows the same rule. Because such a large normal force is dangerous and very uncomfortable for the riders, roller coasters are not built with circular loops in vertical planes. Figure P6.18 and the photograph on page 137 show two actual designs.

Figure P8.64

65. **Review problem**. In 1887 in Bridgeport, Connecticut, C. J. Belknap built the water slide shown in Figure P8.65. A rider on a small sled, of total mass 80.0 kg, pushed off to start at the top of the slide (point Ⓐ) with a speed of 2.50 m/s. The chute was 9.76 m high at the top, 54.3 m long, and 0.51 m wide. Along its length, 725 small wheels made friction negligible. Upon leaving the chute horizontally at its bottom end (point Ⓒ), the rider skimmed across the water of Long Island Sound for as much as 50 m, "skipping along like a flat pebble," before at last coming to rest and swimming ashore, pulling his sled after him. According to *Scientific American*, "The facial expression of novices taking their first adventurous slide is quite remarkable, and the sensations felt are correspondingly novel and peculiar." (a) Find the speed of the sled and rider at point Ⓒ. (b) Model the force of water friction as a constant retarding force acting on a particle. Find the work done by water friction in stopping the sled and rider. (c) Find the magnitude of the force the water exerts on the sled. (d) Find the magnitude of the force the chute exerts on the sled at point Ⓑ. (e) At point Ⓒ, the chute is horizontal but curving in the vertical plane. Assume its radius of curvature is 20.0 m. Find the force the chute exerts on the sled at point Ⓒ.

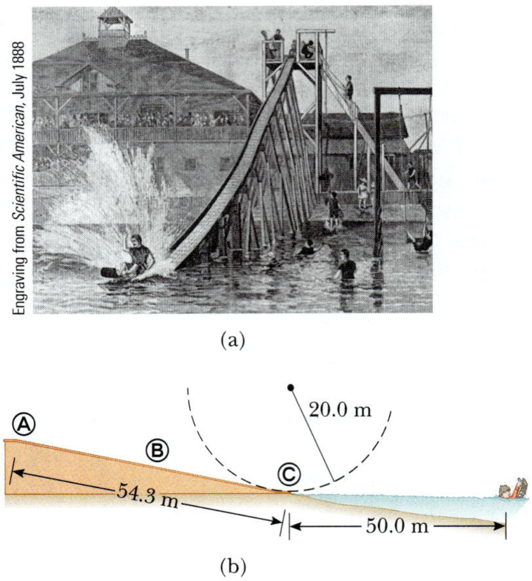

Figure P8.65

66. Consider the block–spring collision discussed in Example 8.8. (a) For the situation in part (B), in which the surface exerts a friction force on the block, show that the block never arrives back at $x = 0$. (b) What is the maximum value of the coefficient of friction that would allow the block to return to $x = 0$?

Answers to Quick Quizzes

8.1 (a) For the television set, energy enters by electrical transmission (through the power cord). Energy leaves by heat (from hot surfaces into the air), mechanical waves (sound from the speaker), and electromagnetic radiation (from the screen). (b) For the gasoline-powered lawn mower, energy enters by matter transfer (gasoline). Energy leaves by work (on the blades of grass), mechanical waves (sound), and heat (from hot surfaces into the air). (c) For the hand-cranked pencil sharpener, energy enters by work (from your hand turning the crank). Energy leaves by work (done on the pencil), mechanical waves (sound), and heat due to the temperature increase from friction.

8.2 **(i)**, (b). For the block, the friction force from the surface represents an interaction with the environment. **(ii)**, (b). For the surface, the friction force from the block represents an interaction with the environment. **(iii)**, (a). For the block and the surface, the friction force is internal to the system, so there are no interactions with the environment.

8.3 (a). The more massive rock has twice as much gravitational potential energy associated with it compared with that of the lighter rock. Because mechanical energy of an isolated system is conserved, the more massive rock will arrive at the ground with twice as much kinetic energy as the lighter rock.

8.4 $v_1 = v_2 = v_3$. The first and third balls speed up after they are thrown, whereas the second ball initially slows down but then speeds up after reaching its peak. The paths of all three balls are parabolas, and the balls take different time intervals to reach the ground because they have different initial velocities. All three balls, however, have the same speed at the moment they hit the ground because all start with the same kinetic energy and because the ball–Earth system undergoes the same change in gravitational potential energy in all three cases.

8.5 (c). The brakes and the roadway are warmer, so their internal energy has increased. In addition, the sound of the skid represents transfer of energy away by mechanical waves.

A moving bowling ball carries momentum, the topic of this chapter. In the collision between the ball and the pins, momentum is transferred to the pins. (Mark Cooper/Corbis Stock Market)

9 Linear Momentum and Collisions

Consider what happens when a bowling ball strikes a pin, as in the photograph above. The pin is given a large velocity as a result of the collision; consequently, it flies away and hits other pins or is projected toward the backstop. Because the average force exerted on the pin during the collision is large (resulting in a large acceleration), the pin achieves its large velocity very rapidly and experiences the force for a very short time interval.

Although the force and acceleration are large for the pin, they vary in time, making for a complicated situation! One of the main objectives of this chapter is to enable you to understand and analyze such events in a simple way. First, we introduce the concept of *momentum*, which is useful for describing objects in motion. The momentum of an object is related to both its mass and its velocity. The concept of momentum leads us to a second conservation law for an isolated system, that of conservation of momentum. This law is especially useful for treating problems that involve collisions between objects and for analyzing rocket propulsion. This chapter also introduces the concept of the center of mass of a system of particles. We find that the motion of a system of particles can be described by the motion of one representative particle located at the center of mass.

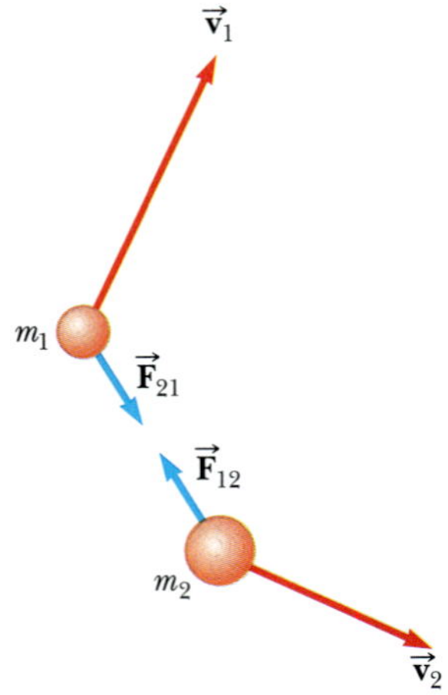

Figure 9.1 Two particles interact with each other. According to Newton's third law, we must have $\vec{F}_{12} = -\vec{F}_{21}$.

9.1 Linear Momentum and Its Conservation

In Chapter 8, we studied situations that are difficult to analyze with Newton's laws. We were able to solve problems involving these situations by identifying a system and applying a conservation principle, conservation of energy. Consider another situation in which a 60-kg archer stands on frictionless ice and fires a 0.50-kg arrow horizontally at 50 m/s. From Newton's third law, we know that the force that the bow exerts on the arrow is matched by a force in the opposite direction on the bow (and the archer). This force causes the archer to slide backward on the ice, but with what speed? We cannot answer this question directly using *either* Newton's second law or an energy approach because we do not have enough information.

Despite our inability to solve the archer problem using techniques learned so far, this problem is very simple to solve if we introduce a new quantity that describes motion, *linear momentum.* Let us apply the General Problem-Solving Strategy and *conceptualize* an isolated system of two particles (Fig. 9.1) with masses m_1 and m_2 moving with velocities $\vec{v}_1$ and $\vec{v}_2$ at an instant of time. Because the system is isolated, the only force on one particle is that from the other particle, and we can *categorize* this situation as one in which Newton's laws are useful. If a force from particle 1 (for example, a gravitational force) acts on particle 2, there must be a second force—equal in magnitude but opposite in direction—that particle 2 exerts on particle 1. That is, the forces on the particles form a Newton's third law action–reaction pair, and $\vec{F}_{12} = -\vec{F}_{21}$. We can express this condition as

$$\vec{F}_{21} + \vec{F}_{12} = 0$$

Let us further *analyze* this situation by incorporating Newton's second law. Over some time interval, the interacting particles in the system accelerate in response to the force. Therefore, replacing the force on each particle with $m\vec{a}$ for the particle gives

$$m_1\vec{a}_1 + m_2\vec{a}_2 = 0$$

Now we replace each acceleration with its definition from Equation 4.5:

$$m_1\frac{d\vec{v}_1}{dt} + m_2\frac{d\vec{v}_2}{dt} = 0$$

If the masses m_1 and m_2 are constant, we can bring them inside the derivative operation, which gives

$$\frac{d(m_1\vec{v}_1)}{dt} + \frac{d(m_2\vec{v}_2)}{dt} = 0$$

$$\frac{d}{dt}(m_1\vec{v}_1 + m_2\vec{v}_2) = 0 \quad \textbf{(9.1)}$$

To *finalize* this discussion, notice that the derivative of the sum $m_1\vec{v}_1 + m_2\vec{v}_2$ with respect to time is zero. Consequently, this sum must be constant. We learn from this discussion that the quantity $m\vec{v}$ for a particle is important in that the sum of these quantities for an isolated system of particles is conserved. We call this quantity *linear momentum:*

Definition of linear momentum of a particle ▶

The **linear momentum** of a particle or an object that can be modeled as a particle of mass m moving with a velocity $\vec{v}$ is defined to be the product of the mass and velocity of the particle:

$$\vec{p} \equiv m\vec{v} \quad \textbf{(9.2)}$$

Linear momentum is a vector quantity because it equals the product of a scalar quantity m and a vector quantity $\vec{v}$. Its direction is along $\vec{v}$, it has dimensions ML/T, and its SI unit is kg · m/s.

If a particle is moving in an arbitrary direction, $\vec{\mathbf{p}}$ has three components, and Equation 9.2 is equivalent to the component equations

$$p_x = mv_x \qquad p_y = mv_y \qquad p_z = mv_z$$

As you can see from its definition, the concept of momentum[1] provides a quantitative distinction between heavy and light particles moving at the same velocity. For example, the momentum of a bowling ball is much greater than that of a tennis ball moving at the same speed. Newton called the product $m\vec{\mathbf{v}}$ *quantity of motion;* this term is perhaps a more graphic description than our present-day word *momentum,* which comes from the Latin word for movement.

Using Newton's second law of motion, we can relate the linear momentum of a particle to the resultant force acting on the particle. We start with Newton's second law and substitute the definition of acceleration:

$$\sum \vec{\mathbf{F}} = m\vec{\mathbf{a}} = m\frac{d\vec{\mathbf{v}}}{dt}$$

In Newton's second law, the mass m is assumed to be constant. Therefore, we can bring m inside the derivative operation to give us

$$\sum \vec{\mathbf{F}} = \frac{d(m\vec{\mathbf{v}})}{dt} = \frac{d\vec{\mathbf{p}}}{dt} \qquad (9.3)$$

◀ Newton's second law for a particle

This equation shows that **the time rate of change of the linear momentum of a particle is equal to the net force acting on the particle.**

This alternative form of Newton's second law is the form in which Newton presented the law, and it is actually more general than the form introduced in Chapter 5. In addition to situations in which the velocity vector varies with time, we can use Equation 9.3 to study phenomena in which the mass changes. For example, the mass of a rocket changes as fuel is burned and ejected from the rocket. We cannot use $\Sigma\,\vec{\mathbf{F}} = m\vec{\mathbf{a}}$ to analyze rocket propulsion; we must use Equation 9.3, as we will show in Section 9.8.

Quick Quiz 9.1 Two objects have equal kinetic energies. How do the magnitudes of their momenta compare? (a) $p_1 < p_2$ (b) $p_1 = p_2$ (c) $p_1 > p_2$ (d) not enough information to tell

Quick Quiz 9.2 Your physical education teacher throws a baseball to you at a certain speed and you catch it. The teacher is next going to throw you a medicine ball whose mass is ten times the mass of the baseball. You are given the following choices: You can have the medicine ball thrown with (a) the same speed as the baseball, (b) the same momentum, or (c) the same kinetic energy. Rank these choices from easiest to hardest to catch.

Using the definition of momentum, Equation 9.1 can be written

$$\frac{d}{dt}(\vec{\mathbf{p}}_1 + \vec{\mathbf{p}}_2) = 0$$

Because the time derivative of the total momentum $\vec{\mathbf{p}}_{\text{tot}} = \vec{\mathbf{p}}_1 + \vec{\mathbf{p}}_2$ is *zero,* we conclude that the *total* momentum of the isolated system of the two particles in Figure 9.1 must remain constant:

$$\vec{\mathbf{p}}_{\text{tot}} = \text{constant} \qquad (9.4)$$

or, equivalently,

$$\vec{\mathbf{p}}_{1i} + \vec{\mathbf{p}}_{2i} = \vec{\mathbf{p}}_{1f} + \vec{\mathbf{p}}_{2f} \qquad (9.5)$$

PITFALL PREVENTION 9.1

Momentum of an Isolated *System* Is Conserved

Although the momentum of an isolated *system* is conserved, the momentum of one particle within an isolated system is not necessarily conserved because other particles in the system may be interacting with it. Always apply conservation of momentum to an isolated *system.*

[1] In this chapter, the terms *momentum* and *linear momentum* have the same meaning. Later, in Chapter 11, we shall use the term *angular momentum* for a different quantity when dealing with rotational motion.

where $\vec{\mathbf{p}}_{1i}$ and $\vec{\mathbf{p}}_{2i}$ are the initial values and $\vec{\mathbf{p}}_{1f}$ and $\vec{\mathbf{p}}_{2f}$ are the final values of the momenta for the two particles for the time interval during which the particles interact. Equation 9.5 in component form demonstrates that the total momenta in the x, y, and z directions are all independently conserved:

$$p_{1ix} + p_{2ix} = p_{1fx} + p_{2fx} \qquad p_{1iy} + p_{2iy} = p_{1fy} + p_{2fy} \qquad p_{1iz} + p_{2iz} = p_{1fz} + p_{2fz} \qquad \textbf{(9.6)}$$

This result, known as the **law of conservation of linear momentum**, can be extended to any number of particles in an isolated system. It is considered one of the most important laws of mechanics. We can state it as follows:

Conservation of momentum ▶

Whenever two or more particles in an isolated system interact, the total momentum of the system remains constant.

This law tells us that **the total momentum of an isolated system at all times equals its initial momentum.** The law is the mathematical representation of the momentum version of the **isolated system model**. We studied the energy version of the isolated system model in Chapter 8.

Notice that we have made no statement concerning the type of forces acting on the particles of the system. Furthermore, we have not specified whether the forces are conservative or nonconservative. The only requirement is that the forces must be *internal* to the system.

EXAMPLE 9.1 The Archer

Let us consider the situation proposed at the beginning of this section. A 60-kg archer stands at rest on frictionless ice and fires a 0.50-kg arrow horizontally at 50 m/s (Fig. 9.2). With what velocity does the archer move across the ice after firing the arrow?

Figure 9.2 (Example 9.1) An archer fires an arrow horizontally to the right. Because he is standing on frictionless ice, he will begin to slide to the left across the ice.

SOLUTION

Conceptualize You may have conceptualized this problem already when it was introduced at the beginning of the section. Imagine the arrow being fired one way and the archer recoiling in the opposite direction.

Categorize We *cannot* solve this problem by modeling the arrow as a particle under a net force because we have no information about the force on the arrow or its acceleration. We *cannot* solve this problem by using a system model and applying an energy approach because we do not know how much work is done in pulling the bow back or how much potential energy is stored in the bow. Nonetheless, we *can* solve this problem very easily with an approach involving momentum.

Let us take the system to consist of the archer (including the bow) and the arrow. The system is not isolated because the gravitational force and the normal force from the ice act on the system. These forces, however, are vertical and perpendicular to the motion of the system. Therefore, there are no external forces in the horizontal direction, and we can consider the system to be isolated in terms of momentum components in this direction.

Analyze The total horizontal momentum of the system before the arrow is fired is zero because nothing in the system is moving. Therefore, the total horizontal momentum of the system after the arrow is fired must also be zero. We choose the direction of firing of the arrow as the positive x direction. Identifying the archer as particle 1 and the arrow as particle 2, we have $m_1 = 60$ kg, $m_2 = 0.50$ kg, and $\vec{\mathbf{v}}_{2f} = 50\hat{\mathbf{i}}$ m/s.

Set the final momentum of the system equal to zero: $\qquad m_1\vec{\mathbf{v}}_{1f} + m_2\vec{\mathbf{v}}_{2f} = 0$

Solve this equation for $\vec{\mathbf{v}}_{1f}$ and substitute numerical values:

$$\vec{\mathbf{v}}_{1f} = -\frac{m_2}{m_1}\vec{\mathbf{v}}_{2f} = -\left(\frac{0.50 \text{ kg}}{60 \text{ kg}}\right)(50\hat{\mathbf{i}} \text{ m/s}) = -0.42\hat{\mathbf{i}} \text{ m/s}$$

Finalize The negative sign for $\vec{\mathbf{v}}_{1f}$ indicates that the archer is moving to the left in Figure 9.2 after the arrow is fired, in the direction opposite the direction of motion of the arrow, in accordance with Newton's third law. Because the archer is much more massive than the arrow, his acceleration and consequent velocity are much smaller than the acceleration and velocity of the arrow.

What If? What if the arrow were fired in a direction that makes an angle θ with the horizontal? How will that change the recoil velocity of the archer?

Answer The recoil velocity should decrease in magnitude because only a component of the velocity of the arrow is in the x direction. Conservation of momentum in the x direction gives

$$m_1 v_{1f} + m_2 v_{2f} \cos\theta = 0$$

leading to

$$v_{1f} = -\frac{m_2}{m_1} v_{2f} \cos\theta$$

For $\theta = 0$, $\cos\theta = 1$, and the final velocity of the archer reduces to the value when the arrow is fired horizontally. For nonzero values of θ, the cosine function is less than 1 and the recoil velocity is less than the value calculated for $\theta = 0$. If $\theta = 90°$, then $\cos\theta = 0$ and $v_{1f} = 0$, so there is no recoil velocity.

EXAMPLE 9.2 Can We Really Ignore the Kinetic Energy of the Earth?

In Section 7.6, we claimed that we can ignore the kinetic energy of the Earth when considering the energy of a system consisting of the Earth and a dropped ball. Verify this claim.

SOLUTION

Conceptualize Imagine dropping a ball at the surface of the Earth. From your point of view, the ball falls while the Earth remains stationary. By Newton's third law, however, the Earth experiences an upward force and therefore an upward acceleration while the ball falls. In the calculation below, we will show that this motion can be ignored.

Categorize We identify the system as the ball and the Earth. Let us ignore air resistance and any other forces on the system, so the system is isolated in terms of momentum.

Analyze We will verify this claim by setting up a ratio of the kinetic energy of the Earth to that of the ball. We identify v_E and v_b as the speeds of the Earth and the ball, respectively, after the ball has fallen through some distance.

Use the definition of kinetic energy to set up a ratio:

$$(1) \quad \frac{K_E}{K_b} = \frac{\frac{1}{2}m_E v_E^2}{\frac{1}{2}m_b v_b^2} = \left(\frac{m_E}{m_b}\right)\left(\frac{v_E}{v_b}\right)^2$$

The initial momentum of the system is zero, so set the final momentum equal to zero:

$$p_i = p_f \rightarrow 0 = m_b v_b + m_E v_E$$

Solve the equation for the ratio of speeds:

$$\frac{v_E}{v_b} = -\frac{m_b}{m_E}$$

Substitute this expression for v_E/v_b in Equation (1):

$$\frac{K_E}{K_b} = \left(\frac{m_E}{m_b}\right)\left(-\frac{m_b}{m_E}\right)^2 = \frac{m_b}{m_E}$$

Substitute order-of-magnitude numbers for the masses:
$$\frac{K_E}{K_b} = \frac{m_b}{m_E} \sim \frac{1\text{ kg}}{10^{24}\text{ kg}} \sim 10^{-24}$$

Finalize The kinetic energy of the Earth is a very small fraction of the kinetic energy of the ball, so we are justified in ignoring it in the kinetic energy of the system.

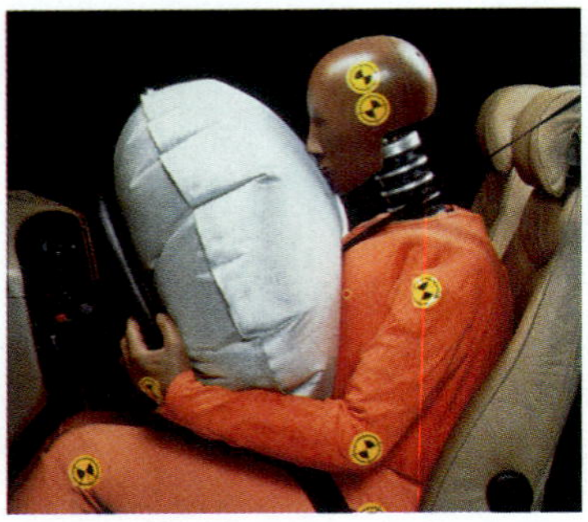
Courtesy of Saab

Air bags in automobiles have saved countless lives in accidents. The air bag increases the time interval during which the passenger is brought to rest, thereby decreasing the force on (and resultant injury to) the passenger.

9.2 Impulse and Momentum

According to Equation 9.3, the momentum of a particle changes if a net force acts on the particle. Knowing the change in momentum caused by a force is useful in solving some types of problems. To build a better understanding of this important concept, let us assume that a net force $\sum \vec{\mathbf{F}}$ acts on a particle and that this force may vary with time. According to Newton's second law, $\sum \vec{\mathbf{F}} = d\vec{\mathbf{p}}/dt$, or

$$d\vec{\mathbf{p}} = \sum \vec{\mathbf{F}}\, dt \qquad \textbf{(9.7)}$$

We can integrate[2] this expression to find the change in the momentum of a particle when the force acts over some time interval. If the momentum of the particle changes from $\vec{\mathbf{p}}_i$ at time t_i to $\vec{\mathbf{p}}_f$ at time t_f, integrating Equation 9.7 gives

$$\Delta\vec{\mathbf{p}} = \vec{\mathbf{p}}_f - \vec{\mathbf{p}}_i = \int_{t_i}^{t_f} \sum \vec{\mathbf{F}}\, dt \qquad \textbf{(9.8)}$$

To evaluate the integral, we need to know how the net force varies with time. The quantity on the right side of this equation is a vector called the **impulse** of the net force $\sum \vec{\mathbf{F}}$ acting on a particle over the time interval $\Delta t = t_f - t_i$:

Impulse of a force ▶

$$\vec{\mathbf{I}} \equiv \int_{t_i}^{t_f} \sum \vec{\mathbf{F}}\, dt \qquad \textbf{(9.9)}$$

From its definition, we see that impulse $\vec{\mathbf{I}}$ is a vector quantity having a magnitude equal to the area under the force–time curve as described in Figure 9.3a. It is assumed the force varies in time in the general manner shown in the figure and is nonzero in the time interval $\Delta t = t_f - t_i$. The direction of the impulse vector is the same as the direction of the change in momentum. Impulse has the dimensions of momentum, that is, ML/T. Impulse is *not* a property of a particle; rather, it is a measure of the degree to which an external force changes the particle's momentum.

Equation 9.8 is an important statement known as the **impulse–momentum theorem:**

Impulse–momentum theorem ▶

The change in the momentum of a particle is equal to the impulse of the net force acting on the particle:

$$\Delta\vec{\mathbf{p}} = \vec{\mathbf{I}} \qquad \textbf{(9.10)}$$

This statement is equivalent to Newton's second law. When we say that an impulse is given to a particle, we mean that momentum is transferred from an external agent to that particle. Equation 9.10 is identical in form to the conservation of energy equation, Equation 8.1. The left side of Equation 9.10 represents the change in the momentum of the system, which in this case is a single particle. The right side is a measure of how much momentum crosses the boundary of the system due to the net force being applied to the system.

Because the net force imparting an impulse to a particle can generally vary in time, it is convenient to define a time-averaged net force:

[2] Here we are integrating force with respect to time. Compare this strategy with our efforts in Chapter 7, where we integrated force with respect to position to find the work done by the force.

$$\left(\sum \vec{\mathbf{F}}\right)_{\text{avg}} \equiv \frac{1}{\Delta t}\int_{t_i}^{t_f} \sum \vec{\mathbf{F}}\, dt \qquad \textbf{(9.11)}$$

where $\Delta t = t_f - t_i$. (This equation is an application of the mean value theorem of calculus.) Therefore, we can express Equation 9.9 as

$$\vec{\mathbf{I}} = \left(\sum \vec{\mathbf{F}}\right)_{\text{avg}} \Delta t \qquad \textbf{(9.12)}$$

This time-averaged force, shown in Figure 9.3b, can be interpreted as the constant force that would give to the particle in the time interval Δt the same impulse that the time-varying force gives over this same interval.

In principle, if $\sum \vec{\mathbf{F}}$ is known as a function of time, the impulse can be calculated from Equation 9.9. The calculation becomes especially simple if the force acting on the particle is constant. In this case, $(\sum \vec{\mathbf{F}})_{\text{avg}} = \sum \vec{\mathbf{F}}$, where $\sum \vec{\mathbf{F}}$ is the constant net force, and Equation 9.12 becomes

$$\vec{\mathbf{I}} = \sum \vec{\mathbf{F}}\, \Delta t \qquad \textbf{(9.13)}$$

In many physical situations, we shall use what is called the **impulse approximation, in which we assume one of the forces exerted on a particle acts for a short time but is much greater than any other force present.** In this case, the net force $\sum \vec{\mathbf{F}}$ in Equation 9.9 is replaced with a single force $\vec{\mathbf{F}}$ to find the impulse on the particle. This approximation is especially useful in treating collisions in which the duration of the collision is very short. When this approximation is made, the single force is referred to as an *impulsive force*. For example, when a baseball is struck with a bat, the time of the collision is about 0.01 s and the average force that the bat exerts on the ball in this time is typically several thousand newtons. Because this contact force is much greater than the magnitude of the gravitational force, the impulse approximation justifies our ignoring the gravitational forces exerted on the ball and bat. When we use this approximation, it is important to remember that $\vec{\mathbf{p}}_i$ and $\vec{\mathbf{p}}_f$ represent the momenta *immediately* before and after the collision, respectively. Therefore, in any situation in which it is proper to use the impulse approximation, the particle moves very little during the collision.

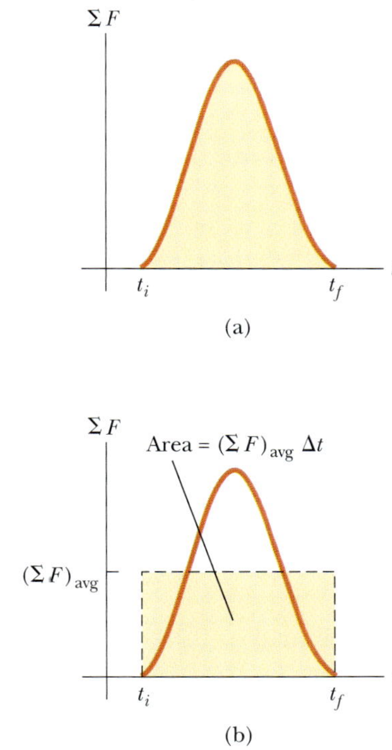

Figure 9.3 (a) A net force acting on a particle may vary in time. The impulse imparted to the particle by the force is the area under the force-versus-time curve. (b) In the time interval Δt, the time-averaged net force (horizontal dashed line) gives the same impulse to a particle as does the time-varying force described in (a).

Quick Quiz 9.3 Two objects are at rest on a frictionless surface. Object 1 has a greater mass than object 2. **(i)** When a constant force is applied to object 1, it accelerates through a distance d in a straight line. The force is removed from object 1 and is applied to object 2. At the moment when object 2 has accelerated through the same distance d, which statements are true? (a) $p_1 < p_2$ (b) $p_1 = p_2$ (c) $p_1 > p_2$ (d) $K_1 < K_2$ (e) $K_1 = K_2$ (f) $K_1 > K_2$ **(ii)** When a force is applied to object 1, it accelerates for a time interval Δt. The force is removed from object 1 and is applied to object 2. From the same list of choices, which statements are true after object 2 has accelerated for the same time interval Δt?

Quick Quiz 9.4 Rank an automobile dashboard, seat belt, and air bag in terms of (a) the impulse and (b) the average force each delivers to a front-seat passenger during a collision, from greatest to least.

EXAMPLE 9.3 How Good Are the Bumpers?

In a particular crash test, a car of mass 1 500 kg collides with a wall as shown in Figure 9.4. The initial and final velocities of the car are $\vec{\mathbf{v}}_i = -15.0\hat{\mathbf{i}}$ m/s and $\vec{\mathbf{v}}_f = 2.60\hat{\mathbf{i}}$ m/s, respectively. If the collision lasts 0.150 s, find the impulse caused by the collision and the average force exerted on the car.

SOLUTION

Conceptualize The collision time is short, so we can imagine the car being brought to rest very rapidly and then moving back in the opposite direction with a reduced speed.

Categorize Let us assume that the force exerted by the wall on the car is large compared with other forces on the car (such as friction and air resistance). Furthermore, the gravitational force and the normal force exerted by the road on the car are perpendicular to the motion and therefore do not affect the horizontal momentum. Therefore, we categorize the problem as one in which we can apply the impulse approximation in the horizontal direction.

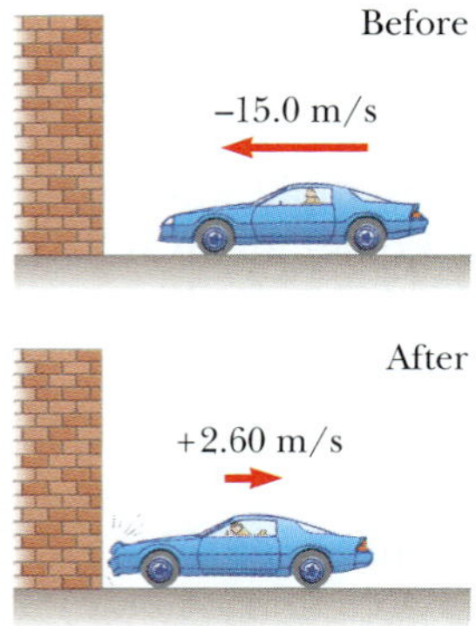

(a)

Tim Wright/CORBIS

(b)

Figure 9.4 (Example 9.3) (a) This car's momentum changes as a result of its collision with the wall. (b) In a crash test, much of the car's initial kinetic energy is transformed into energy associated with the damage to the car.

Analyze

Evaluate the initial and final momenta of the car:

$$\vec{\mathbf{p}}_i = m\vec{\mathbf{v}}_i = (1\,500 \text{ kg})(-15.0\hat{\mathbf{i}} \text{ m/s}) = -2.25 \times 10^4\hat{\mathbf{i}} \text{ kg}\cdot\text{m/s}$$

$$\vec{\mathbf{p}}_f = m\vec{\mathbf{v}}_f = (1\,500 \text{ kg})(2.60\hat{\mathbf{i}} \text{ m/s}) = 0.39 \times 10^4\hat{\mathbf{i}} \text{ kg}\cdot\text{m/s}$$

Use Equation 9.10 to find the impulse on the car:

$$\vec{\mathbf{I}} = \Delta\vec{\mathbf{p}} = \vec{\mathbf{p}}_f - \vec{\mathbf{p}}_i = 0.39 \times 10^4\hat{\mathbf{i}} \text{ kg}\cdot\text{m/s} - (-2.25 \times 10^4\hat{\mathbf{i}} \text{ kg}\cdot\text{m/s})$$

$$= 2.64 \times 10^4\hat{\mathbf{i}} \text{ kg}\cdot\text{m/s}$$

Use Equation 9.3 to evaluate the average force exerted by the wall on the car:

$$\vec{\mathbf{F}}_{\text{avg}} = \frac{\Delta\vec{\mathbf{p}}}{\Delta t} = \frac{2.64 \times 10^4\hat{\mathbf{i}} \text{ kg}\cdot\text{m/s}}{0.150 \text{ s}} = 1.76 \times 10^5\hat{\mathbf{i}} \text{ N}$$

Finalize Notice that the signs of the velocities in this example indicate the reversal of directions. What would the mathematics be describing if both the initial and final velocities had the same sign?

What If? What if the car did not rebound from the wall? Suppose the final velocity of the car is zero and the time interval of the collision remains at 0.150 s. Would that represent a larger or a smaller force by the wall on the car?

Answer In the original situation in which the car rebounds, the force by the wall on the car does two things during the time interval: (1) it stops the car, and (2) it causes the car to move away from the wall at 2.60 m/s after the collision. If the car does not rebound, the force is only doing the first of these steps—stopping the car—which requires a *smaller* force.

Mathematically, in the case of the car that does not rebound, the impulse is

$$\vec{\mathbf{I}} = \Delta\vec{\mathbf{p}} = \vec{\mathbf{p}}_f - \vec{\mathbf{p}}_i = 0 - (-2.25 \times 10^4\hat{\mathbf{i}} \text{ kg}\cdot\text{m/s}) = 2.25 \times 10^4\hat{\mathbf{i}} \text{ kg}\cdot\text{m/s}$$

The average force exerted by the wall on the car is

$$\vec{\mathbf{F}}_{\text{avg}} = \frac{\Delta\vec{\mathbf{p}}}{\Delta t} = \frac{2.25 \times 10^4\hat{\mathbf{i}} \text{ kg}\cdot\text{m/s}}{0.150 \text{ s}} = 1.50 \times 10^5\hat{\mathbf{i}} \text{ N}$$

which is indeed smaller than the previously calculated value, as was argued conceptually.

9.3 Collisions in One Dimension

In this section, we use the law of conservation of linear momentum to describe what happens when two particles collide. The term **collision** represents an event during which two particles come close to each other and interact by means of forces. The interaction forces are assumed to be much greater than any external forces present, so we can use the impulse approximation.

A collision may involve physical contact between two macroscopic objects as described in Active Figure 9.5a, but the notion of what is meant by a collision must be generalized because "physical contact" on a submicroscopic scale is ill-defined and hence meaningless. To understand this concept, consider a collision on an atomic scale (Active Fig. 9.5b) such as the collision of a proton with an alpha particle (the nucleus of a helium atom). Because the particles are both positively charged, they repel each other due to the strong electrostatic force between them at close separations and never come into "physical contact."

When two particles of masses m_1 and m_2 collide as shown in Active Figure 9.5, the impulsive forces may vary in time in complicated ways, such as that shown in Figure 9.3. Regardless of the complexity of the time behavior of the impulsive force, however, this force is internal to the system of two particles. Therefore, the two particles form an isolated system and the momentum of the system must be conserved.

In contrast, the total kinetic energy of the system of particles may or may not be conserved, depending on the type of collision. In fact, collisions are categorized as being either *elastic* or *inelastic* depending on whether or not kinetic energy is conserved.

An **elastic collision** between two objects is one in which **the total kinetic energy (as well as total momentum) of the system is the same before and after the collision.** Collisions between certain objects in the macroscopic world, such as billiard balls, are only *approximately* elastic because some deformation and loss of kinetic energy take place. For example, you can hear a billiard ball collision, so you know that some of the energy is being transferred away from the system by sound. An elastic collision must be perfectly silent! *Truly* elastic collisions occur between atomic and subatomic particles.

An **inelastic collision** is one in which **the total kinetic energy of the system is not the same before and after the collision (even though the momentum of the system is conserved).** Inelastic collisions are of two types. When the objects stick together after they collide, as happens when a meteorite collides with the Earth, the collision is called **perfectly inelastic.** When the colliding objects do not stick together but some kinetic energy is lost, as in the case of a rubber ball colliding with a hard surface, the collision is called **inelastic** (with no modifying adverb). When the rubber ball collides with the hard surface, some of the ball's kinetic energy is lost when the ball is deformed while it is in contact with the surface.

In the remainder of this section, we treat collisions in one dimension and consider the two extreme cases, perfectly inelastic and elastic collisions.

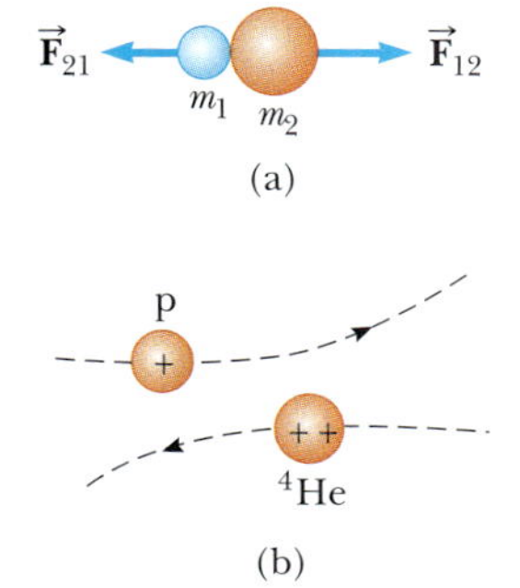

ACTIVE FIGURE 9.5

(a) The collision between two objects as the result of direct contact. (b) The "collision" between two charged particles.

Sign in at www.thomsonedu.com and go to ThomsonNOW to observe these collisions and watch the time variation of the forces on each particle.

PITFALL PREVENTION 9.2

Inelastic Collisions

Generally, inelastic collisions are hard to analyze without additional information. Lack of this information appears in the mathematical representation as having more unknowns than equations.

Perfectly Inelastic Collisions

Consider two particles of masses m_1 and m_2 moving with initial velocities $\vec{\mathbf{v}}_{1i}$ and $\vec{\mathbf{v}}_{2i}$ along the same straight line as shown in Active Figure 9.6. The two particles collide head-on, stick together, and then move with some common velocity $\vec{\mathbf{v}}_f$ after the collision. Because the momentum of an isolated system is conserved in *any* collision, we can say that the total momentum before the collision equals the total momentum of the composite system after the collision:

$$m_1\vec{\mathbf{v}}_{1i} + m_2\vec{\mathbf{v}}_{2i} = (m_1 + m_2)\vec{\mathbf{v}}_f \quad \textbf{(9.14)}$$

Solving for the final velocity gives

$$\vec{\mathbf{v}}_f = \frac{m_1\vec{\mathbf{v}}_{1i} + m_2\vec{\mathbf{v}}_{2i}}{m_1 + m_2} \quad \textbf{(9.15)}$$

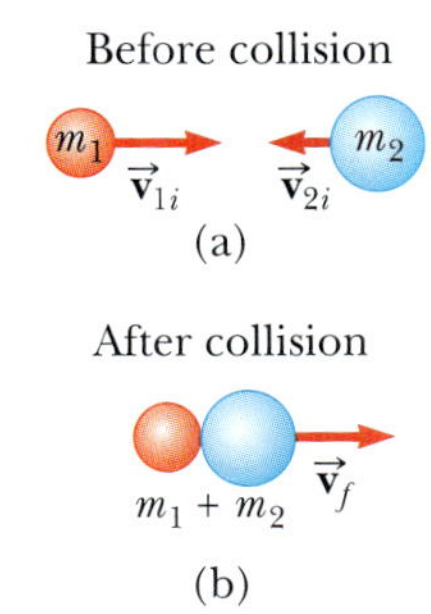

ACTIVE FIGURE 9.6

Schematic representation of a perfectly inelastic head-on collision between two particles: (a) before and (b) after collision.

Sign in at www.thomsonedu.com and go to ThomsonNOW to adjust the masses and velocities of the colliding objects and see the effect on the final velocity.

Elastic Collisions

Consider two particles of masses m_1 and m_2 moving with initial velocities $\vec{\mathbf{v}}_{1i}$ and $\vec{\mathbf{v}}_{2i}$ along the same straight line as shown in Active Figure 9.7. The two particles collide head-on and then leave the collision site with different velocities, $\vec{\mathbf{v}}_{1f}$ and $\vec{\mathbf{v}}_{2f}$. In an elastic collision, both the momentum and kinetic energy of the system are

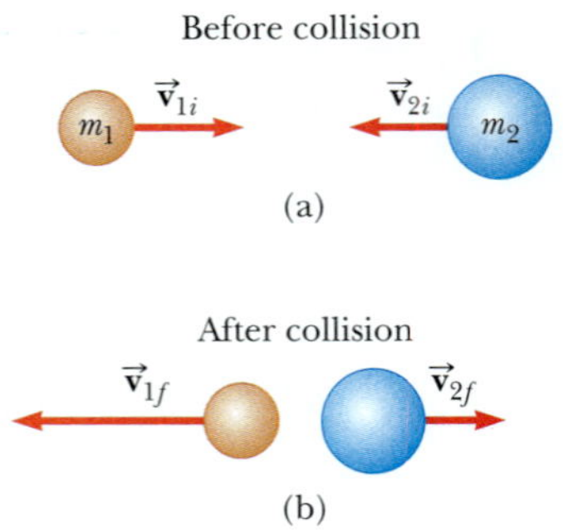

ACTIVE FIGURE 9.7

Schematic representation of an elastic head-on collision between two particles: (a) before and (b) after collision.

Sign in at www.thomsonedu.com and go to ThomsonNOW to adjust the masses and velocities of the colliding objects and see the effect on the final velocities.

conserved. Therefore, considering velocities along the horizontal direction in Active Figure 9.7, we have

$$m_1v_{1i} + m_2v_{2i} = m_1v_{1f} + m_2v_{2f} \quad \textbf{(9.16)}$$

$$\tfrac{1}{2}m_1v_{1i}^2 + \tfrac{1}{2}m_2v_{2i}^2 = \tfrac{1}{2}m_1v_{1f}^2 + \tfrac{1}{2}m_2v_{2f}^2 \quad \textbf{(9.17)}$$

Because all velocities in Active Figure 9.7 are either to the left or the right, they can be represented by the corresponding speeds along with algebraic signs indicating direction. We shall indicate v as positive if a particle moves to the right and negative if it moves to the left.

In a typical problem involving elastic collisions, there are two unknown quantities, and Equations 9.16 and 9.17 can be solved simultaneously to find them. An alternative approach, however—one that involves a little mathematical manipulation of Equation 9.17—often simplifies this process. To see how, let us cancel the factor $\frac{1}{2}$ in Equation 9.17 and rewrite it as

$$m_1(v_{1i}^2 - v_{1f}^2) = m_2(v_{2f}^2 - v_{2i}^2)$$

Factoring both sides of this equation gives

$$m_1(v_{1i} - v_{1f})(v_{1i} + v_{1f}) = m_2(v_{2f} - v_{2i})(v_{2f} + v_{2i}) \quad \textbf{(9.18)}$$

Next, let us separate the terms containing m_1 and m_2 in Equation 9.16 to obtain

$$m_1(v_{1i} - v_{1f}) = m_2(v_{2f} - v_{2i}) \quad \textbf{(9.19)}$$

To obtain our final result, we divide Equation 9.18 by Equation 9.19 and obtain

$$v_{1i} + v_{1f} = v_{2f} + v_{2i}$$

$$v_{1i} - v_{2i} = -(v_{1f} - v_{2f}) \quad \textbf{(9.20)}$$

PITFALL PREVENTION 9.3

Not a General Equation

Equation 9.20 can only be used in a very *specific* situation, a one-dimensional, elastic collision between two objects. The *general* concept is conservation of momentum (and conservation of kinetic energy if the collision is elastic) for an isolated system.

This equation, in combination with Equation 9.16, can be used to solve problems dealing with elastic collisions. According to Equation 9.20, the *relative* velocity of the two particles before the collision, $v_{1i} - v_{2i}$, equals the negative of their relative velocity after the collision, $-(v_{1f} - v_{2f})$.

Suppose the masses and initial velocities of both particles are known. Equations 9.16 and 9.20 can be solved for the final velocities in terms of the initial velocities because there are two equations and two unknowns:

$$v_{1f} = \left(\frac{m_1 - m_2}{m_1 + m_2}\right)v_{1i} + \left(\frac{2m_2}{m_1 + m_2}\right)v_{2i} \quad \textbf{(9.21)}$$

$$v_{2f} = \left(\frac{2m_1}{m_1 + m_2}\right)v_{1i} + \left(\frac{m_2 - m_1}{m_1 + m_2}\right)v_{2i} \quad \textbf{(9.22)}$$

It is important to use the appropriate signs for v_{1i} and v_{2i} in Equations 9.21 and 9.22.

Let us consider some special cases. If $m_1 = m_2$, Equations 9.21 and 9.22 show that $v_{1f} = v_{2i}$ and $v_{2f} = v_{1i}$, which means that the particles exchange velocities if they have equal masses. That is approximately what one observes in head-on billiard ball collisions: the cue ball stops and the struck ball moves away from the collision with the same velocity the cue ball had.

If particle 2 is initially at rest, then $v_{2i} = 0$, and Equations 9.21 and 9.22 become

Elastic collision: particle 2 initially at rest ▶

$$v_{1f} = \left(\frac{m_1 - m_2}{m_1 + m_2}\right)v_{1i} \quad \textbf{(9.23)}$$

$$v_{2f} = \left(\frac{2m_1}{m_1 + m_2}\right)v_{1i} \quad \textbf{(9.24)}$$

If m_1 is much greater than m_2 and $v_{2i} = 0$, we see from Equations 9.23 and 9.24 that $v_{1f} \approx v_{1i}$ and $v_{2f} \approx 2v_{1i}$. That is, when a very heavy particle collides head-on with a very light one that is initially at rest, the heavy particle continues its motion unaltered after the collision and the light particle rebounds with a speed equal to about twice the initial speed of the heavy particle. An example of such a collision is that of a moving heavy atom, such as uranium, striking a light atom, such as hydrogen.

If m_2 is much greater than m_1 and particle 2 is initially at rest, then $v_{1f} \approx -v_{1i}$ and $v_{2f} \approx 0$. That is, when a very light particle collides head-on with a very heavy particle that is initially at rest, the light particle has its velocity reversed and the heavy one remains approximately at rest.

Quick Quiz 9.5 In a perfectly inelastic one-dimensional collision between two moving objects, what condition alone is necessary so that the final kinetic energy of the system is zero after the collision? (a) The objects must have momenta with the same magnitude but opposite directions. (b) The objects must have the same mass. (c) The objects must have the same velocity. (d) The objects must have the same speed, with velocity vectors in opposite directions.

Quick Quiz 9.6 A table-tennis ball is thrown at a stationary bowling ball. The table-tennis ball makes a one-dimensional elastic collision and bounces back along the same line. Compared with the bowling ball after the collision, does the table-tennis ball have (a) a larger magnitude of momentum and more kinetic energy, (b) a smaller magnitude of momentum and more kinetic energy, (c) a larger magnitude of momentum and less kinetic energy, (d) a smaller magnitude of momentum and less kinetic energy, or (e) the same magnitude of momentum and the same kinetic energy?

PROBLEM-SOLVING STRATEGY One-Dimensional Collisions

You should use the following approach when solving collision problems in one dimension:

1. *Conceptualize.* Imagine the collision occurring in your mind. Draw simple diagrams of the particles before and after the collision and include appropriate velocity vectors. At first, you may have to guess at the directions of the final velocity vectors.

2. *Categorize.* Is the system of particles isolated? If so, categorize the collision as elastic, inelastic, or perfectly inelastic.

3. *Analyze.* Set up the appropriate mathematical representation for the problem. If the collision is perfectly inelastic, use Equation 9.15. If the collision is elastic, use Equations 9.16 and 9.20. If the collision is inelastic, use Equation 9.16. To find the final velocities in this case, you will need some additional information.

4. *Finalize.* Once you have determined your result, check to see if your answers are consistent with the mental and pictorial representations and that your results are realistic.

EXAMPLE 9.4 The Executive Stress Reliever

An ingenious device that illustrates conservation of momentum and kinetic energy is shown in Figure 9.8 (page 238). It consists of five identical hard balls supported by strings of equal lengths. When ball 1 is pulled out and released, after the almost-elastic collision between it and ball 2, ball 1 stops and ball 5 moves out as shown in Figure 9.8b. If balls 1 and 2 are pulled out and released, they stop after the collision and balls 4 and 5 swing out, and so forth. Is it ever possible that when ball 1 is released, it stops after the collision and balls 4 and 5 will swing out on the opposite side and travel with half the speed of ball 1 as in Figure 9.8c?

SOLUTION

Conceptualize With the help of Figure 9.8c, imagine one ball coming in from the left and two balls exiting the collision on the right. That is the phenomenon we want to test to see if it could ever happen.

Categorize Because of the very short time interval between the arrival of the ball from the left and the departure of the ball(s) from the right, we can use the impulse approximation to ignore the gravitational forces on the balls and categorize the system of five balls as isolated in terms of momentum and energy. Because the balls are hard, we can categorize the collisions between them as elastic for purposes of calculation.

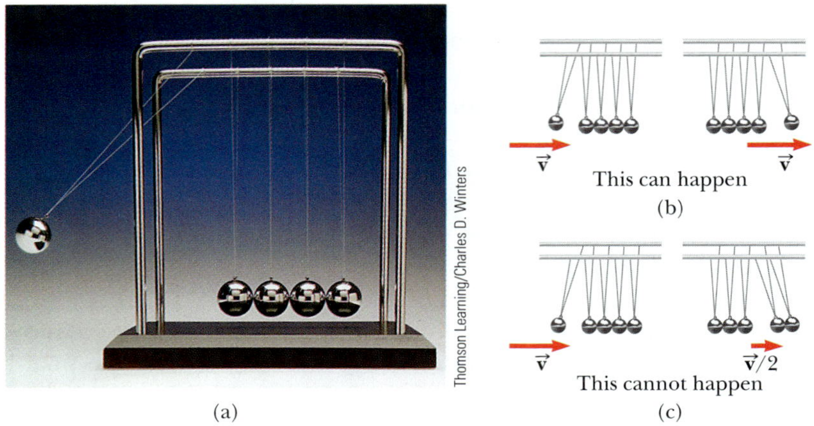

Figure 9.8 (Example 9.4) (a) An executive stress reliever. (b) If one ball swings down, we see one ball swing out at the other end. (c) Is it possible for one ball to swing down and two balls to leave the other end with half the speed of the first ball? In (b) and (c), the velocity vectors shown represent those of the balls immediately before and immediately after the collision.

Analyze The momentum of the system before the collision is mv, where m is the mass of ball 1 and v is its speed immediately before the collision. After the collision, we imagine that ball 1 stops and balls 4 and 5 swing out, each moving with speed $v/2$. The total momentum of the system after the collision would be $m(v/2) + m(v/2) = mv$. Therefore, the momentum of the system is conserved.

The kinetic energy of the system immediately before the collision is $K_i = \frac{1}{2}mv^2$ and that after the collision is $K_f = \frac{1}{2}m(v/2)^2 + \frac{1}{2}m(v/2)^2 = \frac{1}{4}mv^2$. That shows that the kinetic energy of the system is *not* conserved, which is inconsistent with our assumption that the collisions are elastic.

Finalize Our analysis shows that it is *not* possible for balls 4 and 5 to swing out when only ball 1 is released. The only way to conserve both momentum and kinetic energy of the system is for one ball to move out when one ball is released, two balls to move out when two are released, and so on.

What If? Consider what would happen if balls 4 and 5 are glued together. Now what happens when ball 1 is pulled out and released?

Answer In this situation, balls 4 and 5 must move together as a single object after the collision. We have argued that both momentum and energy of the system cannot be conserved in this case. We assumed, however, ball 1 stopped after striking ball 2. What if we do not make this assumption? Consider the conservation equations with the assumption that ball 1 moves after the collision. For conservation of momentum,

$$p_i = p_f$$

$$mv_{1i} = mv_{1f} + 2mv_{4,5}$$

where $v_{4,5}$ refers to the final speed of the ball 4–ball 5 combination. Conservation of kinetic energy gives us

$$K_i = K_f$$

$$\tfrac{1}{2}mv_{1i}^{\ 2} = \tfrac{1}{2}mv_{1f}^{\ 2} + \tfrac{1}{2}(2m)v_{4,5}^2$$

Combining these equations gives

$$v_{4,5} = \tfrac{2}{3}v_{1i} \qquad v_{1f} = -\tfrac{1}{3}v_{1i}$$

Therefore, balls 4 and 5 move together as one object after the collision while ball 1 bounces back from the collision with one third of its original speed.

EXAMPLE 9.5 **Carry Collision Insurance!**

An 1 800-kg car stopped at a traffic light is struck from the rear by a 900-kg car. The two cars become entangled, moving along the same path as that of the originally moving car. If the smaller car were moving at 20.0 m/s before the collision, what is the velocity of the entangled cars after the collision?

SOLUTION

Conceptualize This kind of collision is easily visualized, and one can predict that after the collision both cars will be moving in the same direction as that of the initially moving car. Because the initially moving car has only half the mass of the stationary car, we expect the final velocity of the cars to be relatively small.

Categorize We identify the system of two cars as isolated and apply the impulse approximation during the short time interval of the collision. The phrase "become entangled" tells us to categorize the collision as perfectly inelastic.

Analyze The magnitude of the total momentum of the system before the collision is equal to that of the smaller car because the larger car is initially at rest.

Evaluate the initial momentum of the system: $p_i = m_1 v_i = (900 \text{ kg})(20.0 \text{ m/s}) = 1.80 \times 10^4 \text{ kg} \cdot \text{m/s}$

Evaluate the final momentum of the system: $p_f = (m_1 + m_2)v_f = (2\ 700 \text{ kg})v_f$

Equate the initial and final momenta and solve for v_f: $v_f = \dfrac{p_i}{m_1 + m_2} = \dfrac{1.80 \times 10^4 \text{ kg} \cdot \text{m/s}}{2\ 700 \text{ kg}} = \boxed{6.67 \text{ m/s}}$

Finalize Because the final velocity is positive, the direction of the final velocity of the combination is the same as the velocity of the initially moving car as predicted. The speed of the combination is also much lower than the initial speed of the moving car.

What If? Suppose we reverse the masses of the cars. What if a stationary 900-kg car is struck by a moving 1 800-kg car? Is the final speed the same as before?

Answer Intuitively, we can guess that the final speed of the combination is higher than 6.67 m/s if the initially moving car is the more massive car. Mathematically, that should be the case because the system has a larger momentum if the initially moving car is the more massive one. Solving for the new final velocity, we find

$$v_f = \frac{p_i}{m_1 + m_2} = \frac{(1\ 800 \text{ kg})(20.0 \text{ m/s})}{2\ 700 \text{ kg}} = 13.3 \text{ m/s}$$

which is two times greater than the previous final velocity.

EXAMPLE 9.6 The Ballistic Pendulum

The ballistic pendulum (Fig. 9.9) is an apparatus used to measure the speed of a fast-moving projectile such as a bullet. A projectile of mass m_1 is fired into a large block of wood of mass m_2 suspended from some light wires. The projectile embeds in the block, and the entire system swings through a height h. How can we determine the speed of the projectile from a measurement of h?

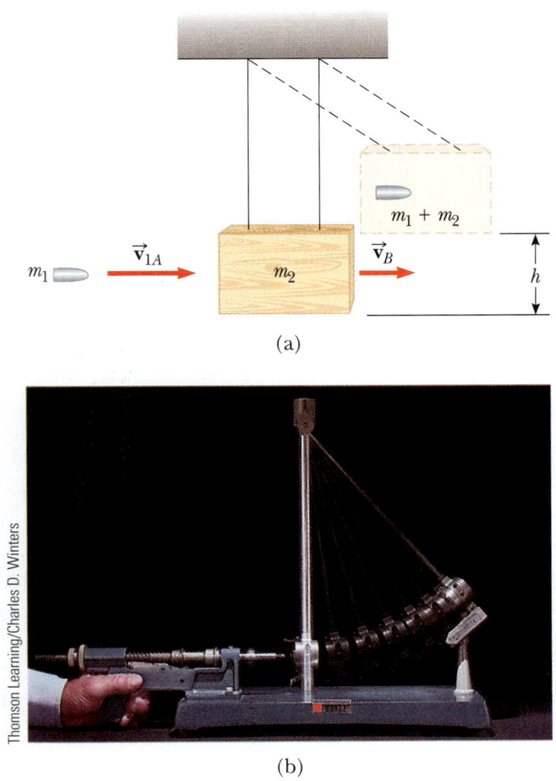

Figure 9.9 (Example 9.6) (a) Diagram of a ballistic pendulum. Notice that $\vec{v}_{1A}$ is the velocity of the projectile immediately before the collision and $\vec{v}_B$ is the velocity of the projectile–block system immediately after the perfectly inelastic collision. (b) Multiflash photograph of a ballistic pendulum used in the laboratory.

SOLUTION

Conceptualize Figure 9.9a helps conceptualize the situation. Run the animation in your mind: the projectile enters the pendulum, which swings up to some height at which it comes to rest.

Categorize The projectile and the block form an isolated system. Identify configuration A as immediately before the collision and configuration B as immediately after the collision. Because the projectile imbeds in the block, we can categorize the collision between them as perfectly inelastic.

Analyze To analyze the collision, we use Equation 9.15, which gives the speed of the system immediately after the collision when we assume the impulse approximation.

Noting that $v_{2A} = 0$, solve Equation 9.15 for v_B:

$$(1) \quad v_B = \frac{m_1 v_{1A}}{m_1 + m_2}$$

Categorize For the process during which the projectile–block combination swings upward to height h (ending at configuration C), we focus on a *different* system, that of the projectile, the block, and the Earth. We categorize this part of the problem as one involving an isolated system for energy with no nonconservative forces acting.

Analyze Write an expression for the total kinetic energy of the system immediately after the collision:

$$(2) \quad K_B = \tfrac{1}{2}(m_1 + m_2)v_B^2$$

Substitute the value of v_B from Equation (1) into Equation (2):

$$K_B = \frac{m_1^2 v_{1A}^2}{2(m_1 + m_2)}$$

This kinetic energy of the system immediately after the collision is *less* than the initial kinetic energy of the projectile as is expected in an inelastic collision.

We define the gravitational potential energy of the system for configuration B to be zero. Therefore, $U_B = 0$, whereas $U_C = (m_1 + m_2)gh$.

Apply the conservation of mechanical energy principle to the system:

$$K_B + U_B = K_C + U_C$$

$$\frac{m_1^2 v_{1A}^2}{2(m_1 + m_2)} + 0 = 0 + (m_1 + m_2)gh$$

Solve for v_{1A}:

$$v_{1A} = \left(\frac{m_1 + m_2}{m_1}\right)\sqrt{2gh}$$

Finalize We had to solve this problem in two steps. Each step involved a different system and a different conservation principle. Because the collision was assumed to be perfectly inelastic, some mechanical energy was transformed to internal energy. It would have been *incorrect* to equate the initial kinetic energy of the incoming projectile with the final gravitational potential energy of the projectile–block–Earth combination.

EXAMPLE 9.7 A Two-Body Collision with a Spring

A block of mass $m_1 = 1.60$ kg initially moving to the right with a speed of 4.00 m/s on a frictionless, horizontal track collides with a spring attached to a second block of mass $m_2 = 2.10$ kg initially moving to the left with a speed of 2.50 m/s as shown in Figure 9.10a. The spring constant is 600 N/m.

(A) Find the velocities of the two blocks after the collision.

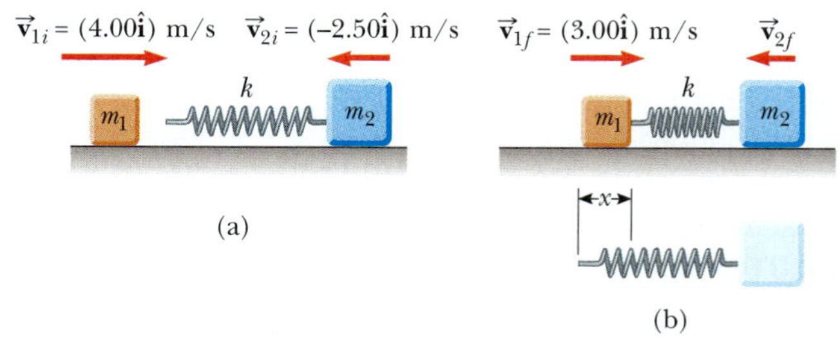

Figure 9.10 (Example 9.7) A moving block approaches a second moving block that is attached to a spring.

SOLUTION

Conceptualize With the help of Figure 9.10a, run an animation of the collision in your mind. Figure 9.10b shows an instant during the collision when the spring is compressed. Eventually, block 1 and the spring will again separate, so the system will look like Figure 9.10a again but with different velocity vectors for the two blocks.

Categorize Because the spring force is conservative, kinetic energy in the system is not transformed to internal energy during the compression of the spring. Ignoring any sound made when the block hits the spring, we can categorize the collision as being elastic.

Analyze Because momentum of the system is conserved, apply Equation 9.16:

$$m_1 v_{1i} + m_2 v_{2i} = m_1 v_{1f} + m_2 v_{2f}$$

Substitute the known values:

$$(1.60 \text{ kg})(4.00 \text{ m/s}) + (2.10 \text{ kg})(-2.50 \text{ m/s}) = (1.60 \text{ kg})v_{1f} + (2.10 \text{ kg})v_{2f}$$

$$(1) \quad 1.15 \text{ kg} \cdot \text{m/s} = (1.60 \text{ kg})v_{1f} + (2.10 \text{ kg})v_{2f}$$

Because the collision is elastic, apply Equation 9.20:

$$v_{1i} - v_{2i} = -(v_{1f} - v_{2f})$$

Substitute the known values:

$$(2) \quad 4.00 \text{ m/s} - (-2.50 \text{ m/s}) = 6.50 \text{ m/s} = -v_{1f} + v_{2f}$$

Multiply Equation (2) by 1.60 kg:

$$(3) \quad 10.4 \text{ kg} \cdot \text{m/s} = -(1.60 \text{ kg})v_{1f} + (1.60 \text{ kg})v_{2f}$$

Add Equations (1) and (3):

$$11.55 \text{ kg} \cdot \text{m/s} = (3.70 \text{ kg})v_{2f}$$

Solve for v_{2f}:

$$v_{2f} = \frac{11.55 \text{ kg} \cdot \text{m/s}}{3.70 \text{ kg}} = 3.12 \text{ m/s}$$

Use Equation (2) to find v_{1f}:

$$6.50 \text{ m/s} = -v_{1f} + 3.12 \text{ m/s}$$

$$v_{1f} = -3.38 \text{ m/s}$$

(B) During the collision, at the instant block 1 is moving to the right with a velocity of +3.00 m/s as in Figure 9.10b, determine the velocity of block 2.

SOLUTION

Conceptualize Focus your attention now on Figure 9.10b, which represents the final configuration of the system for the time interval of interest.

Categorize Because the momentum and mechanical energy of the system of two blocks are conserved *throughout* the collision for the system of two blocks, the collision can be categorized as elastic for *any* final instant of time. Let us now choose the final instant to be when block 1 is moving with a velocity of +3.00 m/s.

Analyze Apply Equation 9.16:

$$m_1v_{1i} + m_2v_{2i} = m_1v_{1f} + m_2v_{2f}$$

Substitute the known values:

$$(1.60 \text{ kg})(4.00 \text{ m/s}) + (2.10 \text{ kg})(-2.50 \text{ m/s}) = (1.60 \text{ kg})(3.00 \text{ m/s}) + (2.10 \text{ kg})v_{2f}$$

Solve for v_{2f}:

$$v_{2f} = -1.74 \text{ m/s}$$

Finalize The negative value for v_{2f} means that block 2 is still moving to the left at the instant we are considering.

(C) Determine the distance the spring is compressed at that instant.

SOLUTION

Conceptualize Once again, focus on the configuration of the system shown in Figure 9.10b.

Categorize For the system of the spring and two blocks, no friction or other nonconservative forces act within the system. Therefore, we categorize the system as isolated with no nonconservative forces acting.

Analyze We choose the initial configuration of the system to be that existing immediately before block 1 strikes the spring and the final configuration to be that when block 1 is moving to the right at 3.00 m/s.

Write a conservation of mechanical energy equation for the system:

$$K_i + U_i = K_f + U_f$$

$$\tfrac{1}{2}m_1v_{1i}^2 + \tfrac{1}{2}m_2v_{2i}^2 + 0 = \tfrac{1}{2}m_1v_{1f}^2 + \tfrac{1}{2}m_2v_{2f}^2 + \tfrac{1}{2}kx^2$$

Substitute the known values and the result of part (B):

$$\tfrac{1}{2}(1.60\text{ kg})(4.00\text{ m/s})^2 + \tfrac{1}{2}(2.10\text{ kg})(2.50\text{ m/s})^2 + 0$$
$$= \tfrac{1}{2}(1.60\text{ kg})(3.00\text{ m/s})^2 + \tfrac{1}{2}(2.10\text{ kg})(1.74\text{ m/s})^2 + \tfrac{1}{2}(600\text{ N/m})x^2$$

Solve for x:

$$x = 0.173\text{ m}$$

Finalize This answer is not the maximum compression of the spring because the two blocks are still moving toward each other at the instant shown in Figure 9.10b. Can you determine the maximum compression of the spring?

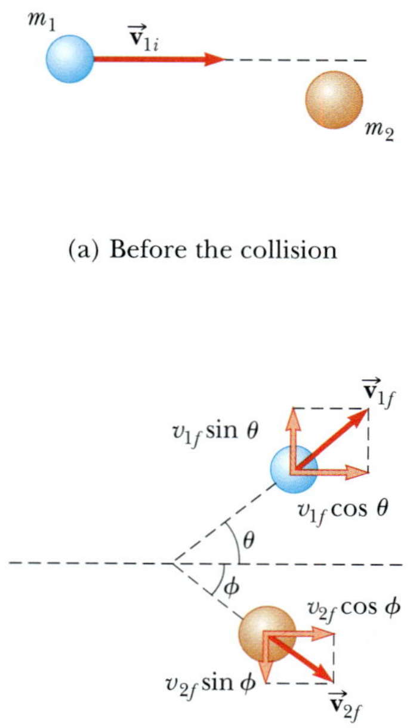

ACTIVE FIGURE 9.11

An elastic, glancing collision between two particles.

Sign in at www.thomsonedu.com and go to ThomsonNOW to adjust the speed and position of the blue particle and the masses of both particles and see the effects.

PITFALL PREVENTION 9.4

Don't Use Equation 9.20

Equation 9.20, relating the initial and final relative velocities of two colliding objects, is only valid for one-dimensional elastic collisions. Do not use this equation when analyzing two-dimensional collisions.

9.4 Collisions in Two Dimensions

In Section 9.1, we showed that the momentum of a system of two particles is conserved when the system is isolated. For any collision of two particles, this result implies that the momentum in each of the directions x, y, and z is conserved. An important subset of collisions takes place in a plane. The game of billiards is a familiar example involving multiple collisions of objects moving on a two-dimensional surface. For such two-dimensional collisions, we obtain two component equations for conservation of momentum:

$$m_1 v_{1ix} + m_2 v_{2ix} = m_1 v_{1fx} + m_2 v_{2fx}$$

$$m_1 v_{1iy} + m_2 v_{2iy} = m_1 v_{1fy} + m_2 v_{2fy}$$

where three subscripts on the velocity components in these equations represent, respectively, the identification of the object (1, 2), initial and final values (i, f), and the velocity component (x, y).

Let us consider a specific two-dimensional problem in which particle 1 of mass m_1 collides with particle 2 of mass m_2 initially at rest as in Active Figure 9.11. After the collision (Active Fig. 9.11b), particle 1 moves at an angle θ with respect to the horizontal and particle 2 moves at an angle ϕ with respect to the horizontal. This event is called a *glancing* collision. Applying the law of conservation of momentum in component form and noting that the initial y component of the momentum of the two-particle system is zero gives

$$m_1 v_{1i} = m_1 v_{1f}\cos\theta + m_2 v_{2f}\cos\phi \tag{9.25}$$

$$0 = m_1 v_{1f}\sin\theta - m_2 v_{2f}\sin\phi \tag{9.26}$$

where the minus sign in Equation 9.26 is included because after the collision particle 2 has a y component of velocity that is downward. (The symbols v in these particular equations are speeds, not velocity components. The direction of the component vector is indicated explicitly with plus or minus signs.) We now have two independent equations. As long as no more than two of the seven quantities in Equations 9.25 and 9.26 are unknown, we can solve the problem.

If the collision is elastic, we can also use Equation 9.17 (conservation of kinetic energy) with $v_{2i} = 0$:

$$\tfrac{1}{2}m_1 v_{1i}^2 = \tfrac{1}{2}m_1 v_{1f}^2 + \tfrac{1}{2}m_2 v_{2f}^2 \tag{9.27}$$

Knowing the initial speed of particle 1 and both masses, we are left with four unknowns (v_{1f}, v_{2f}, θ, and ϕ). Because we have only three equations, one of the four remaining quantities must be given to determine the motion after the elastic collision from conservation principles alone.

If the collision is inelastic, kinetic energy is *not* conserved and Equation 9.27 does *not* apply.

PROBLEM-SOLVING STRATEGY Two-Dimensional Collisions

The following procedure is recommended when dealing with problems involving collisions between two particles in two dimensions.

1. *Conceptualize.* Imagine the collisions occurring and predict the approximate directions in which the particles will move after the collision. Set up a coordinate system and define your velocities in terms of that system. It is convenient to have the x axis coincide with one of the initial velocities. Sketch the coordinate system, draw and label all velocity vectors, and include all the given information.

2. *Categorize.* Is the system of particles truly isolated? If so, categorize the collision as elastic, inelastic, or perfectly inelastic.

3. *Analyze.* Write expressions for the x and y components of the momentum of each object before and after the collision. Remember to include the appropriate signs for the components of the velocity vectors and pay careful attention to signs.

   Write expressions for the *total* momentum in the x direction *before* and *after* the collision, and equate the two. Repeat this procedure for the total momentum in the y direction.

   Proceed to solve the momentum equations for the unknown quantities. If the collision is inelastic, kinetic energy is *not* conserved and additional information is probably required. If the collision is perfectly inelastic, the final velocities of the two objects are equal.

   If the collision is elastic, kinetic energy is conserved and you can equate the total kinetic energy of the system before the collision to the total kinetic energy after the collision, providing an additional relationship between the velocity magnitudes.

4. *Finalize.* Once you have determined your result, check to see if your answers are consistent with the mental and pictorial representations and that your results are realistic.

EXAMPLE 9.8 Collision at an Intersection

A 1 500-kg car traveling east with a speed of 25.0 m/s collides at an intersection with a 2 500-kg van traveling north at a speed of 20.0 m/s as shown in Figure 9.12. Find the direction and magnitude of the velocity of the wreckage after the collision, assuming the vehicles stick together after the collision.

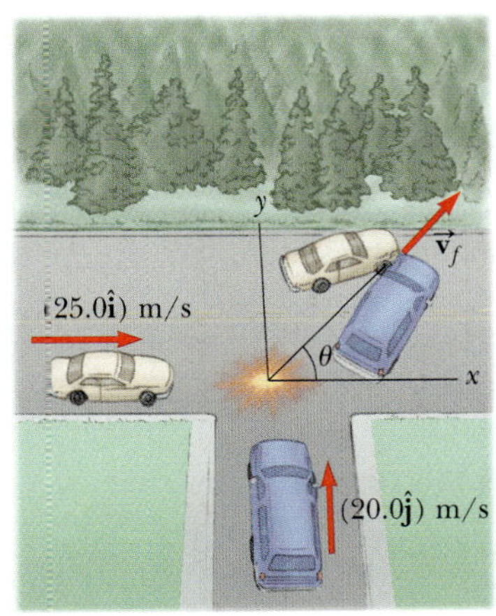

Figure 9.12 (Example 9.8) An eastbound car colliding with a northbound van.

SOLUTION

Conceptualize Figure 9.12 should help you conceptualize the situation before and after the collision. Let us choose east to be along the positive x direction and north to be along the positive y direction.

Categorize Because we consider moments immediately before and immediately after the collision as defining our time interval, we ignore the small effect that friction would have on the wheels of the car and model the system of two cars as isolated. We also ignore the cars' sizes and model them as particles. The collision is perfectly inelastic because the two cars stick together after the collision.

Analyze Before the collision, the only object having momentum in the x direction is the car. Therefore, the magnitude of the total initial momentum of the system (car plus van) in the x direction is that of only the car. Similarly, the total initial momentum of the system in the y direction is that of the van. After the collision, let us assume that the wreckage moves at an angle θ and speed v_f.

Evaluate the initial momentum of the system in the x direction:

$$\sum p_{xi} = (1\,500 \text{ kg})(25.0 \text{ m/s}) = 3.75 \times 10^4 \text{ kg}\cdot\text{m/s}$$

Write an expression for the final momentum in the x direction:

$$\sum p_{xf} = (4\,000 \text{ kg})v_f \cos\theta$$

Equate the initial and final momenta in the x direction:

$$(1)\quad 3.75 \times 10^4 \text{ kg}\cdot\text{m/s} = (4\,000 \text{ kg})v_f \cos\theta$$

Evaluate the initial momentum of the system in the y direction:

$$\sum p_{yi} = (2\,500 \text{ kg})(20.0 \text{ m/s}) = 5.00 \times 10^4 \text{ kg}\cdot\text{m/s}$$

Write an expression for the final momentum in the y direction:

$$\sum p_{yf} = (4\,000 \text{ kg})v_f \sin\theta$$

Equate the initial and final momenta in the y direction:

$$(2)\quad 5.00 \times 10^4 \text{ kg}\cdot\text{m/s} = (4\,000 \text{ kg})v_f \sin\theta$$

Divide Equation (2) by Equation (1) and solve for θ:

$$\frac{(4\,000 \text{ kg})v_f \sin\theta}{(4\,000 \text{ kg})v_f \cos\theta} = \tan\theta = \frac{5.00 \times 10^4}{3.75 \times 10^4} = 1.33$$

$$\theta = 53.1^\circ$$

Use Equation (2) to find the value of v_f:

$$v_f = \frac{5.00 \times 10^4 \text{ kg}\cdot\text{m/s}}{(4\,000 \text{ kg}) \sin 53.1^\circ} = 15.6 \text{ m/s}$$

Finalize Notice that the angle θ is qualitatively in agreement with Figure 9.12. Also notice that the final speed of the combination is less than the initial speeds of the two cars. This result is consistent with the kinetic energy of the system being reduced in an inelastic collision. It might help if you draw the momentum vectors of each vehicle before the collision and the two vehicles together after the collision.

EXAMPLE 9.9 Proton–Proton Collision

A proton collides elastically with another proton that is initially at rest. The incoming proton has an initial speed of 3.50×10^5 m/s and makes a glancing collision with the second proton as in Active Figure 9.11. (At close separations, the protons exert a repulsive electrostatic force on each other.) After the collision, one proton moves off at an angle of 37.0° to the original direction of motion and the second deflects at an angle of ϕ to the same axis. Find the final speeds of the two protons and the angle ϕ.

SOLUTION

Conceptualize This collision is like that shown in Active Figure 9.11, which will help you conceptualize the behavior of the system. We define the x axis to be along the direction of the velocity vector of the initially moving proton.

Categorize The pair of protons form an isolated system. Both momentum and kinetic energy of the system are conserved in this glancing elastic collision.

Analyze We know that $m_1 = m_2$ and $\theta = 37.0°$, and we are given that $v_{1i} = 3.50 \times 10^5$ m/s.

Enter the known values into Equations 9.25, 9.26, and 9.27:

$$(1)\quad v_{1f}\cos 37° + v_{2f}\cos\phi = 3.50 \times 10^5 \text{ m/s}$$

$$(2)\quad v_{1f}\sin 37.0° - v_{2f}\sin\phi = 0$$

$$(3)\quad v_{1f}^2 + v_{2f}^2 = (3.50 \times 10^5 \text{ m/s})^2 = 1.23 \times 10^{11} \text{ m}^2/\text{s}^2$$

Rearrange Equations (1) and (2):

$$v_{2f}\cos\phi = 3.50 \times 10^5 \text{ m/s} - v_{1f}\cos 37.0°$$

$$v_{2f}\sin\phi = v_{1f}\sin 37.0°$$

Square these two equations and add them:

$$v_{2f}^2 \cos^2 \phi + v_{2f}^2 \sin^2 \phi$$
$$= 1.23 \times 10^{11}\ \text{m}^2/\text{s}^2 - (7.00 \times 10^5\ \text{m/s})v_{1f} \cos 37.0° + v_{1f}^2 \cos^2 37.0°$$
$$+ v_{1f}^2 \sin^2 37.0°$$

$$(4) \quad v_{2f}^2 = 1.23 \times 10^{11} - (5.59 \times 10^5)v_{1f} + v_{1f}^2$$

Substitute Equation (4) into Equation (3):

$$v_{1f}^2 + [1.23 \times 10^{11} - (5.59 \times 10^5)v_{1f} + v_{1f}^2] = 1.23 \times 10^{11}$$
$$2v_{1f}^2 - (5.59 \times 10^5)v_{1f} = (2v_{1f} - 5.59 \times 10^5)v_{1f} = 0$$

One possible solution of this equation is $v_{1f} = 0$, which corresponds to a head-on collision in which the first proton stops and the second continues with the same speed in the same direction. That is not the solution we want.

Set the other factor equal to zero:

$$2v_{1f} - 5.59 \times 10^5 = 0 \quad \rightarrow \quad v_{1f} = 2.80 \times 10^5\ \text{m/s}$$

Use Equation (3) to find v_{2f}:

$$v_{2f} = \sqrt{1.23 \times 10^{11} - v_{1f}^2} = \sqrt{1.23 \times 10^{11} - (2.80 \times 10^5)^2}$$
$$= 2.11 \times 10^5\ \text{m/s}$$

Use Equation (2) to find ϕ:

$$\phi = \sin^{-1}\left(\frac{v_{1f} \sin 37.0°}{v_{2f}}\right) = \sin^{-1}\left(\frac{(2.80 \times 10^5) \sin 37.0°}{2.11 \times 10^5}\right) = 53.0°$$

Finalize It is interesting that $\theta + \phi = 90°$. This result is *not* accidental. Whenever two objects of equal mass collide elastically in a glancing collision and one of them is initially at rest, their final velocities are perpendicular to each other.

9.5 The Center of Mass

In this section, we describe the overall motion of a system in terms of a special point called the **center of mass** of the system. The system can be either a group of particles, such as a collection of atoms in a container, or an extended object, such as a gymnast leaping through the air. We shall see that the translational motion of the center of mass of the system is the same as if all the mass of the system were concentrated at that point. That is, the system moves as if the net external force were applied to a single particle located at the center of mass. This behavior is independent of other motion, such as rotation or vibration of the system. This model, the *particle model*, was introduced in Chapter 2.

Consider a system consisting of a pair of particles that have different masses and are connected by a light, rigid rod (Active Fig. 9.13). The position of the center of mass of a system can be described as being the *average position* of the system's mass. The center of mass of the system is located somewhere on the line joining the two particles and is closer to the particle having the larger mass. If a single force is applied at a point on the rod above the center of mass, the system rotates clockwise (see Active Fig. 9.13a). If the force is applied at a point on the rod below the center of mass, the system rotates counterclockwise (see Active Fig. 9.13b). If the force is applied at the center of mass, the system moves in the direction of the force without rotating (see Active Fig. 9.13c). The center of mass of an object can be located with this procedure.

The center of mass of the pair of particles described in Active Figure 9.14 (page 246) is located on the x axis and lies somewhere between the particles. Its x coordinate is given by

$$x_{\text{CM}} \equiv \frac{m_1 x_1 + m_2 x_2}{m_1 + m_2} \tag{9.28}$$

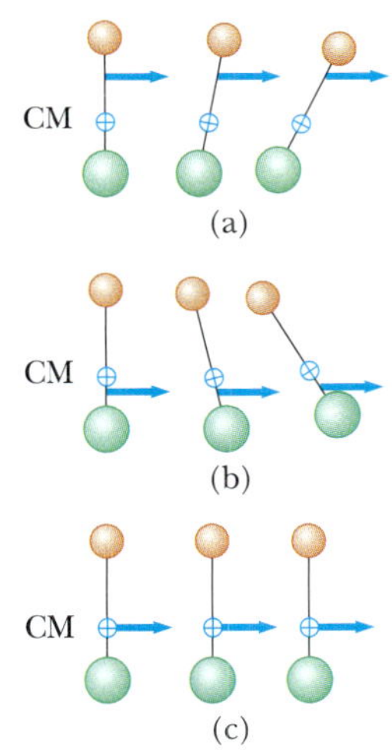

ACTIVE FIGURE 9.13

Two particles of unequal mass are connected by a light, rigid rod. (a) The system rotates clockwise when a force is applied above the center of mass. (b) The system rotates counterclockwise when a force is applied below the center of mass. (c) The system moves in the direction of the force without rotating when a force is applied at the center of mass.

Sign in at www.thomsonedu.com and go to ThomsonNOW to choose the point at which to apply the force.

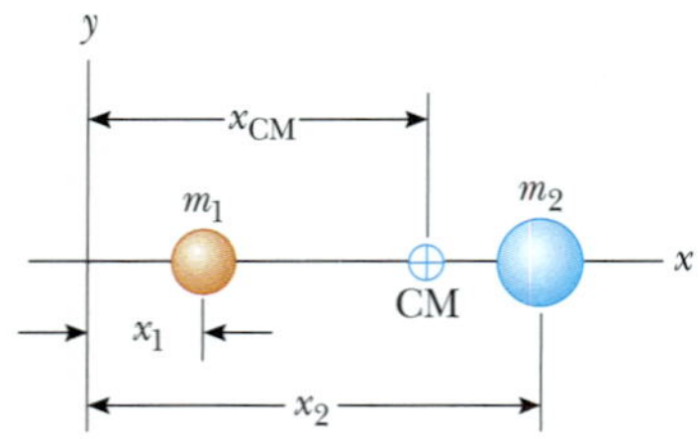

ACTIVE FIGURE 9.14

The center of mass of two particles of unequal mass on the x axis is located at x_{CM}, a point between the particles, closer to the one having the larger mass.

Sign in at www.thomsonedu.com and go to ThomsonNOW to adjust the masses and positions of the particles and see the effect on the location of the center of mass.

For example, if $x_1 = 0$, $x_2 = d$, and $m_2 = 2m_1$, we find that $x_{CM} = \frac{2}{3}d$. That is, the center of mass lies closer to the more massive particle. If the two masses are equal, the center of mass lies midway between the particles.

We can extend this concept to a system of many particles with masses m_i in three dimensions. The x coordinate of the center of mass of n particles is defined to be

$$x_{CM} \equiv \frac{m_1x_1 + m_2x_2 + m_3x_3 + \cdots + m_nx_n}{m_1 + m_2 + m_3 + \cdots + m_n} = \frac{\sum_i m_ix_i}{\sum_i m_i} = \frac{\sum_i m_ix_i}{M} = \frac{1}{M}\sum_i m_ix_i \quad \textbf{(9.29)}$$

where x_i is the x coordinate of the ith particle and the total mass is $M \equiv \sum_i m_i$ where the sum runs over all n particles. The y and z coordinates of the center of mass are similarly defined by the equations

$$y_{CM} \equiv \frac{1}{M}\sum_i m_iy_i \quad \text{and} \quad z_{CM} \equiv \frac{1}{M}\sum_i m_iz_i \quad \textbf{(9.30)}$$

The center of mass can be located in three dimensions by its position vector $\vec{\mathbf{r}}_{CM}$. The components of this vector are x_{CM}, y_{CM}, and z_{CM}, defined in Equations 9.29 and 9.30. Therefore,

$$\vec{\mathbf{r}}_{CM} = x_{CM}\hat{\mathbf{i}} + y_{CM}\hat{\mathbf{j}} + z_{CM}\hat{\mathbf{k}} = \frac{1}{M}\sum_i m_ix_i\hat{\mathbf{i}} + \frac{1}{M}\sum_i m_iy_i\hat{\mathbf{j}} + \frac{1}{M}\sum_i m_iz_i\hat{\mathbf{k}}$$

$$\vec{\mathbf{r}}_{CM} \equiv \frac{1}{M}\sum_i m_i\vec{\mathbf{r}}_i \quad \textbf{(9.31)}$$

where $\vec{\mathbf{r}}_i$ is the position vector of the ith particle, defined by

$$\vec{\mathbf{r}}_i \equiv x_i\hat{\mathbf{i}} + y_i\hat{\mathbf{j}} + z_i\hat{\mathbf{k}}$$

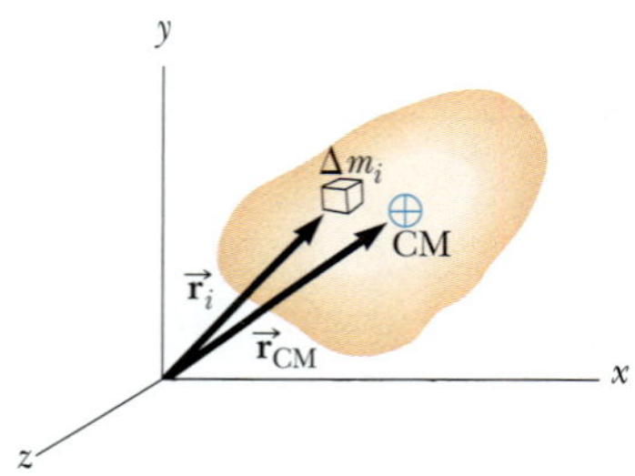

Figure 9.15 An extended object can be considered to be a distribution of small elements of mass Δm_i. The center of mass is located at the vector position $\vec{\mathbf{r}}_{CM}$, which has coordinates x_{CM}, y_{CM}, and z_{CM}.

Although locating the center of mass for an extended object is somewhat more cumbersome than locating the center of mass of a system of particles, the basic ideas we have discussed still apply. Think of an extended object as a system containing a large number of particles (Fig. 9.15). Because the particle separation is very small, the object can be considered to have a continuous mass distribution. By dividing the object into elements of mass Δm_i with coordinates x_i, y_i, z_i, we see that the x coordinate of the center of mass is approximately

$$x_{CM} \approx \frac{1}{M}\sum_i x_i\,\Delta m_i$$

with similar expressions for y_{CM} and z_{CM}. If we let the number of elements n approach infinity, the size of each element approaches zero and x_{CM} is given precisely. In this limit, we replace the sum by an integral and Δm_i by the differential element dm:

$$x_{CM} = \lim_{\Delta m_i \to 0} \frac{1}{M}\sum_i x_i\,\Delta m_i = \frac{1}{M}\int x\,dm \quad \textbf{(9.32)}$$

Likewise, for y_{CM} and z_{CM} we obtain

$$y_{CM} = \frac{1}{M}\int y\,dm \quad \text{and} \quad z_{CM} = \frac{1}{M}\int z\,dm \quad \textbf{(9.33)}$$

We can express the vector position of the center of mass of an extended object in the form

$$\vec{\mathbf{r}}_{CM} = \frac{1}{M}\int \vec{\mathbf{r}}\,dm \quad \textbf{(9.34)}$$

which is equivalent to the three expressions given by Equations 9.32 and 9.33.

The center of mass of any symmetric object lies on an axis of symmetry and on any plane of symmetry.[3] For example, the center of mass of a uniform rod lies in the rod, midway between its ends. The center of mass of a sphere or a cube lies at its geometric center.

The center of mass of an irregularly shaped object such as a wrench can be determined by suspending the object first from one point and then from another. In Figure 9.16, a wrench is hung from point A and a vertical line AB (which can be established with a plumb bob) is drawn when the wrench has stopped swinging. The wrench is then hung from point C, and a second vertical line CD is drawn. The center of mass is halfway through the thickness of the wrench, under the intersection of these two lines. In general, if the wrench is hung freely from any point, the vertical line through this point must pass through the center of mass.

Because an extended object is a continuous distribution of mass, each small mass element is acted upon by the gravitational force. The net effect of all these forces is equivalent to the effect of a single force $M\vec{\mathbf{g}}$ acting through a special point, called the **center of gravity.** If $\vec{\mathbf{g}}$ is constant over the mass distribution, the center of gravity coincides with the center of mass. If an extended object is pivoted at its center of gravity, it balances in any orientation.

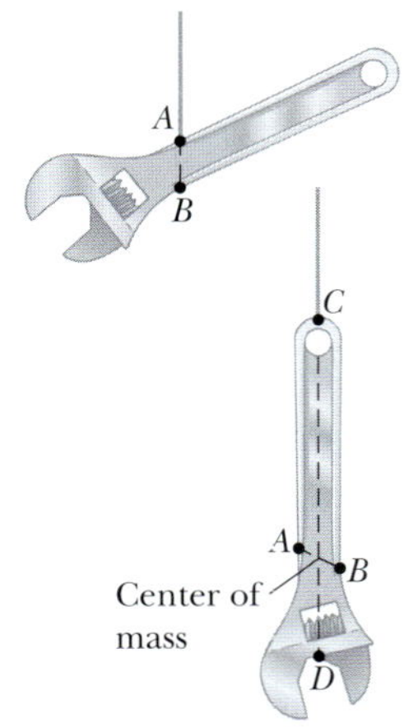

Figure 9.16 An experimental technique for determining the center of mass of a wrench. The wrench is hung freely first from point A and then from point C. The intersection of the two lines AB and CD locates the center of mass.

Quick Quiz 9.7 A baseball bat of uniform density is cut at the location of its center of mass as shown in Figure 9.17. Which piece has the smaller mass? (a) the piece on the right (b) the piece on the left (c) both pieces have the same mass (d) impossible to determine

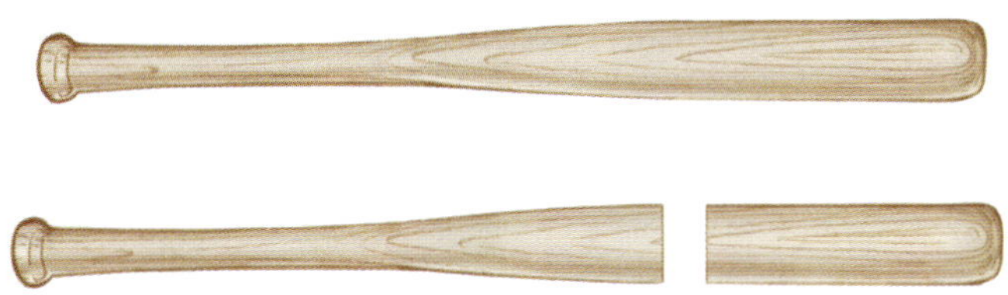

Figure 9.17 (Quick Quiz 9.7) A baseball bat cut at the location of its center of mass.

EXAMPLE 9.10 The Center of Mass of Three Particles

A system consists of three particles located as shown in Figure 9.18. Find the center of mass of the system.

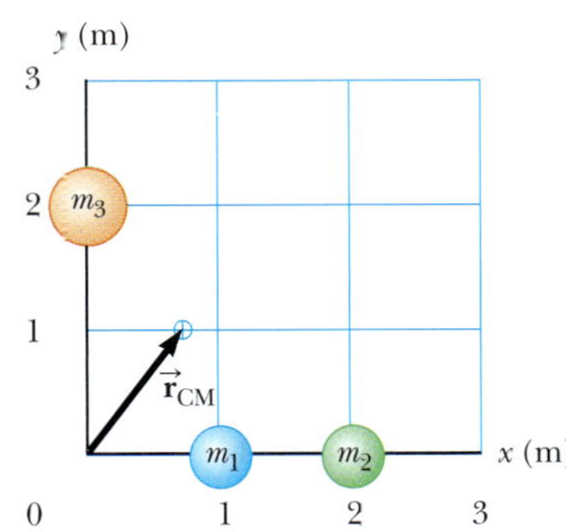

Figure 9.18 (Example 9.10) Two 1.0-kg particles are located on the x axis, and a single 2.0-kg particle is located on the y axis as shown. The vector indicates the location of the system's center of mass.

SOLUTION

Conceptualize Figure 9.18 shows the three masses. Your intuition should tell you that the center of mass is located somewhere in the region between the orange particle and the pair of particles colored blue and green as shown in the figure.

Categorize We categorize this example as a substitution problem because we will be using the equations for the center of mass developed in this section.

We set up the problem by labeling the masses of the particles as shown in the figure, with $m_1 = m_2 = 1.0$ kg and $m_3 = 2.0$ kg.

[3] This statement is valid only for objects that have a uniform density.

Use the defining equations for the coordinates of the center of mass and notice that $z_{CM} = 0$:

$$x_{CM} = \frac{1}{M}\sum_i m_i x_i = \frac{m_1x_1 + m_2x_2 + m_3x_3}{m_1 + m_2 + m_3}$$

$$= \frac{(1.0 \text{ kg})(1.0 \text{ m}) + (1.0 \text{ kg})(2.0 \text{ m}) + (2.0 \text{ kg})(0)}{1.0 \text{ kg} + 1.0 \text{ kg} + 2.0 \text{ kg}}$$

$$= \frac{3.0 \text{ kg}\cdot\text{m}}{4.0 \text{ kg}} = 0.75 \text{ m}$$

$$y_{CM} = \frac{1}{M}\sum_i m_i y_i = \frac{m_1y_1 + m_2y_2 + m_3y_3}{m_1 + m_2 + m_3}$$

$$= \frac{(1.0 \text{ kg})(0) + (1.0 \text{ kg})(0) + (2.0 \text{ kg})(2.0 \text{ m})}{4.0 \text{ kg}} = \frac{4.0 \text{ kg}\cdot\text{m}}{4.0 \text{ kg}} = 1.0 \text{ m}$$

Write the position vector of the center of mass:

$$\vec{\mathbf{r}}_{CM} \equiv x_{CM}\hat{\mathbf{i}} + y_{CM}\hat{\mathbf{j}} = (0.75\hat{\mathbf{i}} + 1.0\hat{\mathbf{j}}) \text{ m}$$

EXAMPLE 9.11 The Center of Mass of a Rod

(A) Show that the center of mass of a rod of mass M and length L lies midway between its ends, assuming the rod has a uniform mass per unit length.

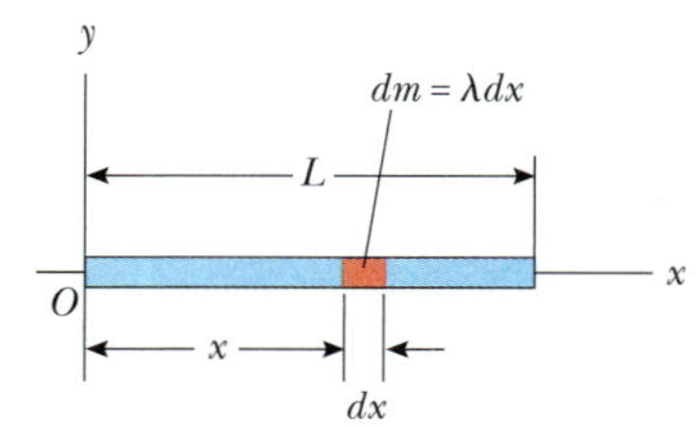

Figure 9.19 (Example 9.11) The geometry used to find the center of mass of a uniform rod.

SOLUTION

Conceptualize The rod is shown aligned along the x axis in Figure 9.19, so $y_{CM} = z_{CM} = 0$.

Categorize We categorize this example as an analysis problem because we need to divide the rod up into elements to perform the integration in Equation 9.32.

Analyze The mass per unit length (this quantity is called the *linear mass density*) can be written as $\lambda = M/L$ for the uniform rod. If the rod is divided into elements of length dx, the mass of each element is $dm = \lambda\, dx$.

Use Equation 9.32 to find an expression for x_{CM}:

$$x_{CM} = \frac{1}{M}\int x\, dm = \frac{1}{M}\int_0^L x\lambda\, dx = \frac{\lambda}{M}\frac{x^2}{2}\bigg|_0^L = \frac{\lambda L^2}{2M}$$

Substitute $\lambda = M/L$:

$$x_{CM} = \frac{L^2}{2M}\left(\frac{M}{L}\right) = \frac{L}{2}$$

One can also use symmetry arguments to obtain the same result.

(B) Suppose a rod is *nonuniform* such that its mass per unit length varies linearly with x according to the expression $\lambda = \alpha x$, where α is a constant. Find the x coordinate of the center of mass as a fraction of L.

SOLUTION

Conceptualize Because the mass per unit length is not constant but is proportional to x, elements of the rod to the right are more massive than elements near the left end of the rod.

Categorize This problem is categorized similarly to part (A), with the added twist that the linear mass density is not constant.

Analyze In this case, we replace dm in Equation 9.32 by $\lambda\, dx$, where $\lambda = \alpha x$.

Use Equation 9.32 to find an expression for x_{CM}:

$$x_{CM} = \frac{1}{M}\int x\,dm = \frac{1}{M}\int_0^L x\lambda\,dx = \frac{1}{M}\int_0^L x\alpha x\,dx$$

$$= \frac{\alpha}{M}\int_0^L x^2\,dx = \frac{\alpha L^3}{3M}$$

Find the total mass of the rod:

$$M = \int dm = \int_0^L \lambda\,dx = \int_0^L \alpha x\,dx = \frac{\alpha L^2}{2}$$

Substitute M into the expression for x_{CM}:

$$x_{CM} = \frac{\alpha L^3}{3\alpha L^2/2} = \tfrac{2}{3}L$$

Finalize Notice that the center of mass in part (B) is farther to the right than that in part (A). That result is reasonable because the elements of the rod become more massive as one moves to the right along the rod in part (B).

EXAMPLE 9.12 The Center of Mass of a Right Triangle

You have been asked to hang a metal sign from a single vertical wire. The sign has the triangular shape shown in Figure 9.20a. The bottom of the sign is to be parallel to the ground. At what distance from the left end of the sign should you attach the support wire?

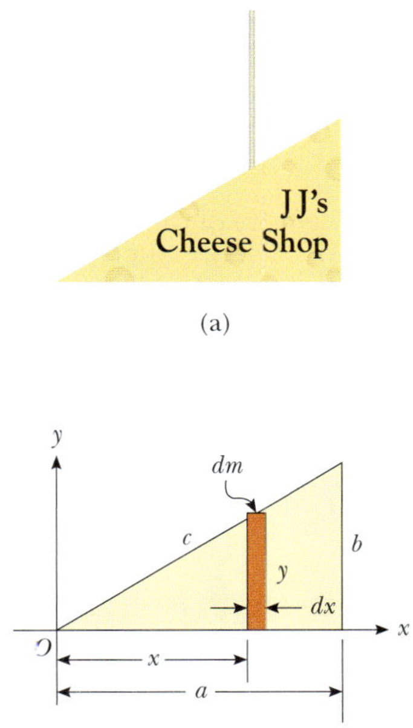

Figure 9.20 (Example 9.12) (a) A triangular sign to be hung from a single wire. (b) Geometric construction for locating the center of mass.

SOLUTION

Conceptualize Figure 9.20a shows the sign hanging from the wire. The wire must be attached at a point directly above the center of gravity of the sign, which is the same as its center of mass because it is in a uniform gravitational field.

Categorize As in the case of Example 9.11, we categorize this example as an analysis problem because it is necessary to identify infinitesimal elements of the sign to perform the integration in Equation 9.32.

Analyze We assume the triangular sign has a uniform density and total mass M. Because the sign is a continuous distribution of mass, we must use the integral expression in Equation 9.32 to find the x coordinate of the center of mass.

We divide the triangle into narrow strips of width dx and height y as shown in Figure 9.20b, where y is the height of the hypotenuse of the triangle above the x axis for a given value of x. The mass of each strip is the product of the volume of the strip and the density ρ of the material from which the sign is made: $dm = \rho yt\,dx$, where t is the thickness of the metal sign. The density of the material is the total mass of the sign divided by its total volume (area of the triangle times thickness).

Evaluate dm:

$$dm = \rho yt\,dx = \left(\frac{M}{\frac{1}{2}abt}\right)yt\,dx = \frac{2My}{ab}\,dx$$

Use Equation 9.32 to find the x coordinate of the center of mass:

$$(1)\quad x_{CM} = \frac{1}{M}\int x\,dm = \frac{1}{M}\int_0^a x\frac{2My}{ab}\,dx = \frac{2}{ab}\int_0^a xy\,dx$$

To proceed further and evaluate the integral, we must express y in terms of x. The line representing the hypotenuse of the triangle in Figure 9.20b has a slope of b/a and passes through the origin, so the equation of this line is $y = (b/a)x$.

Substitute for y in Equation (1):

$$x_{\text{CM}} = \frac{2}{ab}\int_0^a x\left(\frac{b}{a}x\right)dx = \frac{2}{a^2}\int_0^a x^2 dx = \frac{2}{a^2}\left[\frac{x^3}{3}\right]_0^a$$

$$= \tfrac{2}{3}a$$

Therefore, the wire must be attached to the sign at a distance two-thirds of the length of the bottom edge from the left end.

Finalize This answer is identical to that in part (B) of Example 9.11. For the triangular sign, the linear increase in height y with position x means that elements in the sign increase in mass linearly, reflecting the linear increase in mass density in Example 9.11. We could also find the y coordinate of the center of mass of the sign, but that is not needed to determine where the wire should be attached. You might try cutting a right triangle out of cardboard and hanging it from a string so that the long base is horizontal. Does the string need to be attached at $\frac{2}{3}a$?

9.6 Motion of a System of Particles

We can begin to understand the physical significance and utility of the center of mass concept by taking the time derivative of the position vector for the center of mass given by Equation 9.31. From Section 4.1 we know that the time derivative of a position vector is by definition the velocity vector. Assuming M remains constant for a system of particles—that is, no particles enter or leave the system—we obtain the following expression for the **velocity of the center of mass** of the system:

Velocity of the center of mass ▶

$$\vec{\mathbf{v}}_{\text{CM}} = \frac{d\vec{\mathbf{r}}_{\text{CM}}}{dt} = \frac{1}{M}\sum_i m_i \frac{d\vec{\mathbf{r}}_i}{dt} = \frac{1}{M}\sum_i m_i \vec{\mathbf{v}}_i \quad \textbf{(9.35)}$$

where $\vec{\mathbf{v}}_i$ is the velocity of the ith particle. Rearranging Equation 9.35 gives

Total momentum of a system of particles ▶

$$M\vec{\mathbf{v}}_{\text{CM}} = \sum_i m_i\vec{\mathbf{v}}_i = \sum_i \vec{\mathbf{p}}_i = \vec{\mathbf{p}}_{\text{tot}} \quad \textbf{(9.36)}$$

Therefore, the **total linear momentum of the system equals the total mass multiplied by the velocity of the center of mass.** In other words, the total linear momentum of the system is equal to that of a single particle of mass M moving with a velocity $\vec{\mathbf{v}}_{\text{CM}}$.

Differentiating Equation 9.35 with respect to time, we obtain the **acceleration of the center of mass** of the system:

Acceleration of the center of mass ▶

$$\vec{\mathbf{a}}_{\text{CM}} = \frac{d\vec{\mathbf{v}}_{\text{CM}}}{dt} = \frac{1}{M}\sum_i m_i \frac{d\vec{\mathbf{v}}_i}{dt} = \frac{1}{M}\sum_i m_i \vec{\mathbf{a}}_i \quad \textbf{(9.37)}$$

Rearranging this expression and using Newton's second law gives

$$M\vec{\mathbf{a}}_{\text{CM}} = \sum_i m_i\vec{\mathbf{a}}_i = \sum_i \vec{\mathbf{F}}_i \quad \textbf{(9.38)}$$

where $\vec{\mathbf{F}}_i$ is the net force on particle i.

The forces on any particle in the system may include both external forces (from outside the system) and internal forces (from within the system). By Newton's third law, however, the internal force exerted by particle 1 on particle 2, for example, is equal in magnitude and opposite in direction to the internal force exerted by particle 2 on particle 1. Therefore, when we sum over all internal forces in Equation 9.38, they cancel in pairs and we find that the net force on the system is caused *only* by external forces. We can then write Equation 9.38 in the form

Newton's second law for a system of particles ▶

$$\sum \vec{\mathbf{F}}_{\text{ext}} = M\vec{\mathbf{a}}_{\text{CM}} \quad \textbf{(9.39)}$$

That is, **the net external force on a system of particles equals the total mass of the system multiplied by the acceleration of the center of mass.** Comparing Equation 9.39 with Newton's second law for a single particle, we see that the particle model we have used in several chapters can be described in terms of the center of mass:

> The center of mass of a system of particles having combined mass M moves like an equivalent particle of mass M would move under the influence of the net external force on the system.

Let us integrate Equation 9.39 over a finite time interval:

$$\int \sum \vec{\mathbf{F}}_{\text{ext}}\, dt = \int M\vec{\mathbf{a}}_{\text{CM}}\, dt = \int M \frac{d\vec{\mathbf{v}}_{\text{CM}}}{dt}\, dt = M \int d\vec{\mathbf{v}}_{\text{CM}} = M\,\Delta\vec{\mathbf{v}}_{\text{CM}}$$

Notice that this equation can be written as

$$\vec{\mathbf{I}} = \Delta\vec{\mathbf{p}}_{\text{tot}} \qquad \textbf{(9.40)}$$

where $\vec{\mathbf{I}}$ is the impulse imparted to the system by external forces and $\vec{\mathbf{p}}_{\text{tot}}$ is the momentum of the system. Equation 9.40 is the generalization of the impulse-momentum theorem for a particle (Eq. 9.10) to a system of particles.

Finally, if the net external force on a system is zero, it follows from Equation 9.39 that

$$M\vec{\mathbf{a}}_{\text{CM}} = M\frac{d\vec{\mathbf{v}}_{\text{CM}}}{dt} = 0$$

so

$$M\vec{\mathbf{v}}_{\text{CM}} = \vec{\mathbf{p}}_{\text{tot}} = \text{constant} \quad \left(\text{when } \sum \vec{\mathbf{F}}_{\text{ext}} = 0\right) \qquad \textbf{(9.41)}$$

That is, the total linear momentum of a system of particles is conserved if no net external force is acting on the system. It follows that for an isolated system of particles, both the total momentum and the velocity of the center of mass are constant in time. This statement is a generalization of the law of conservation of momentum for a many-particle system.

Suppose an isolated system consisting of two or more members is at rest. The center of mass of such a system remains at rest unless acted upon by an external force. For example, consider a system of a swimmer standing on a raft, with the system initially at rest. When the swimmer dives horizontally off the raft, the raft moves in the direction opposite that of the swimmer and the center of mass of the system remains at rest (if we neglect friction between raft and water). Furthermore, the linear momentum of the diver is equal in magnitude to that of the raft, but opposite in direction.

Quick Quiz 9.8 A cruise ship is moving at constant speed through the water. The vacationers on the ship are eager to arrive at their next destination. They decide to try to speed up the cruise ship by gathering at the bow (the front) and running together toward the stern (the back) of the ship. **(i)** While they are running toward the stern, is the speed of the ship (a) higher than it was before, (b) unchanged, (c) lower than it was before, or (d) impossible to determine? **(ii)** The vacationers stop running when they reach the stern of the ship. After they have all stopped running, is the speed of the ship (a) higher than it was before they started running, (b) unchanged from what it was before they started running, (c) lower than it was before they started running, or (d) impossible to determine?

CONCEPTUAL EXAMPLE 9.13 **Exploding Projectile**

A projectile fired into the air suddenly explodes into several fragments (Active Fig. 9.21).

(A) What can be said about the motion of the center of mass of the system made up of all the fragments after the explosion?

SOLUTION

Neglecting air resistance, the only external force on the projectile is the gravitational force. Therefore, if the projectile did not explode, it would continue to move along the parabolic path indicated by the dashed line in Active Figure 9.21. Because the forces caused by the explosion are internal, they do not affect the motion of the center of mass of the system (the fragments). Therefore, after the explosion, the center of mass of the fragments follows the same parabolic path the projectile would have followed if no explosion had occurred.

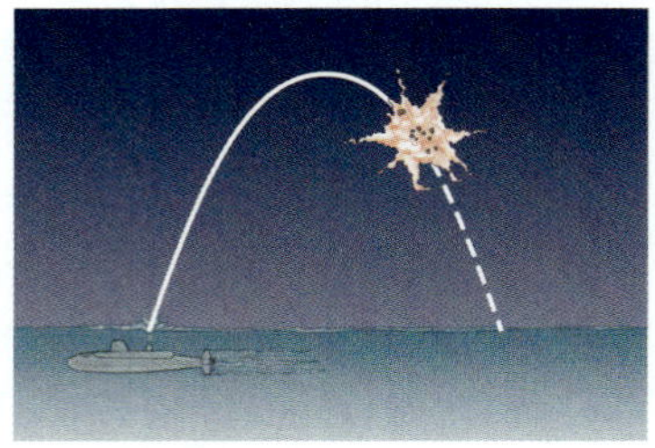

ACTIVE FIGURE 9.21

(Conceptual Example 9.13) When a projectile explodes into several fragments, the center of mass of the system made up of all the fragments follows the same parabolic path the projectile would have taken had there been no explosion.

Sign in at www.thomsonedu.com and go to ThomsonNOW to observe a variety of explosions and follow the trajectory of the center of mass.

(B) If the projectile did not explode, it would land at a distance R from its launch point. Suppose the projectile explodes and splits into two pieces of equal mass. One piece lands at a distance $2R$ from the launch point. Where does the other piece land?

SOLUTION

As discussed in part (A), the center of mass of the two-piece system lands at a distance R from the launch point. One of the pieces lands at a further distance R from the landing point (or a distance $2R$ from the launch point), to the right in Active Figure 9.21. Because the two pieces have the same mass, the other piece must land a distance R to the left of the landing point in Active Figure 9.21, which places this piece right back at the launch point!

EXAMPLE 9.14 **The Exploding Rocket**

A rocket is fired vertically upward. At the instant it reaches an altitude of 1 000 m and a speed of 300 m/s, it explodes into three fragments having equal mass. One fragment moves upward with a speed of 450 m/s following the explosion. The second fragment has a speed of 240 m/s and is moving east right after the explosion. What is the velocity of the third fragment immediately after the explosion?

SOLUTION

Conceptualize Picture the explosion in your mind, with one piece going upward and a second piece moving horizontally toward the east. Do you have an intuitive feeling about the direction in which the third piece moves?

Categorize This example is a two-dimensional problem because we have two fragments moving in perpendicular directions after the explosion as well as a third fragment moving in an unknown direction in the plane defined by the velocity vectors of the other two fragments. We assume the time interval of the explosion is very short, so we use the impulse approximation in which we ignore the gravitational force and air resistance. Because the forces of the explosion are internal to the system (the rocket), the system is modeled as isolated and the total momentum $\vec{\mathbf{p}}_i$ of the rocket immediately before the explosion must equal the total momentum $\vec{\mathbf{p}}_f$ of the fragments immediately after the explosion.

Analyze Because the three fragments have equal mass, the mass of each fragment is $M/3$, where M is the total mass of the rocket. We will let $\vec{\mathbf{v}}_f$ represent the unknown velocity of the third fragment.

Write an expression for the momentum of the system before the explosion:

$$\vec{\mathbf{p}}_i = M\vec{\mathbf{v}}_i = M(300\,\hat{\mathbf{j}}\text{ m/s})$$

Write an expression for the momentum of the system after the explosion:

$$\vec{\mathbf{p}}_f = \frac{M}{3}(240\,\hat{\mathbf{i}}\text{ m/s}) + \frac{M}{3}(450\,\hat{\mathbf{j}}\text{ m/s}) + \frac{M}{3}\vec{\mathbf{v}}_f$$

Equate these two expressions: $\frac{M}{3}\vec{\mathbf{v}}_f + \frac{M}{3}(240\,\hat{\mathbf{i}}\,\text{m/s}) + \frac{M}{3}(450\,\hat{\mathbf{j}}\,\text{m/s}) = M(300\,\hat{\mathbf{j}}\,\text{m/s})$

Solve for $\vec{\mathbf{v}}_f$: $\vec{\mathbf{v}}_f = (-240\,\hat{\mathbf{i}} + 450\,\hat{\mathbf{j}})\,\text{m/s}$

Finalize Notice that this event is the reverse of a perfectly inelastic collision. There is one object before the collision and three objects afterward. Imagine running a movie of the event backward: the three objects would come together and become a single object. In a perfectly inelastic collision, the kinetic energy of the system decreases. If you were to calculate the kinetic energy before and after the event in this example, you would find that the kinetic energy of the system increases. (Try it!) This increase in kinetic energy comes from the potential energy stored in whatever fuel exploded to cause the breakup of the rocket.

9.7 Deformable Systems

So far in our discussion of mechanics, we have analyzed the motion of particles or nondeformable systems that can be modeled as particles. The discussion in Section 9.6 can be applied to an analysis of the motion of deformable systems. For example, suppose you stand on a skateboard and push off a wall, setting yourself in motion away from the wall. How would we describe this event?

The force from the wall on your hands moves through no displacement; the force is always located at the interface between the wall and your hands. Therefore, the force does no work on the system, which is you and your skateboard. Pushing off the wall, however, does indeed result in a change in the kinetic energy of the system. If you try to use the work–kinetic energy theorem, $W = \Delta K$, to describe this event, notice that the left side of the equation is zero but the right side is not zero. The work–kinetic energy theorem is not valid for this event and is often not valid for systems that are deformable. Your body has deformed during this event: your arms were bent before the event, and they straightened out while you pushed off the wall.

To analyze the motion of deformable systems, we appeal to Equation 8.2, the conservation of energy equation, and Equation 9.40, the impulse–momentum theorem for a system. For the example of you pushing off the wall on your skateboard, Equation 8.2 gives

$$\Delta E_{\text{system}} = \sum T$$

$$\Delta K + \Delta U = 0$$

where ΔK is the change in kinetic energy due to the increased speed of the system and ΔU is the decrease in potential energy stored in the body from previous meals. This equation tells us that the system transformed potential energy into kinetic energy by virtue of the muscular exertion necessary to push off the wall.

Applying Equation 9.40 to this situation gives us

$$\vec{\mathbf{I}} = \Delta\vec{\mathbf{p}}_{\text{tot}}$$

$$\int \vec{\mathbf{F}}_{\text{wall}}\,dt = m\,\Delta\vec{\mathbf{v}}$$

where $\vec{\mathbf{F}}_{\text{wall}}$ is the force exerted by the wall on your hands, m is the mass of you and the skateboard, and $\Delta\vec{\mathbf{v}}$ is the change in the velocity of the system during the event. To evaluate the left side of this equation, we would need to know how the force from the wall varies in time. In general, this process might be complicated. In the case of constant forces, or well-behaved forces, however, the integral on the left side of the equation can be evaluated.

EXAMPLE 9.15 Pushing on a Spring[4]

As shown in Figure 9.22a, two blocks are at rest on a level, frictionless table. Both blocks have the same mass m, and they are connected by a spring of negligible mass. The separation distance of the blocks when the spring is relaxed is L. During a time interval Δt, a constant force F is applied horizontally to the left block, moving it through a distance x_1 as shown in Figure 9.22b. During this time interval, the right block moves through a distance x_2. At the end of this time interval, the force F is removed.

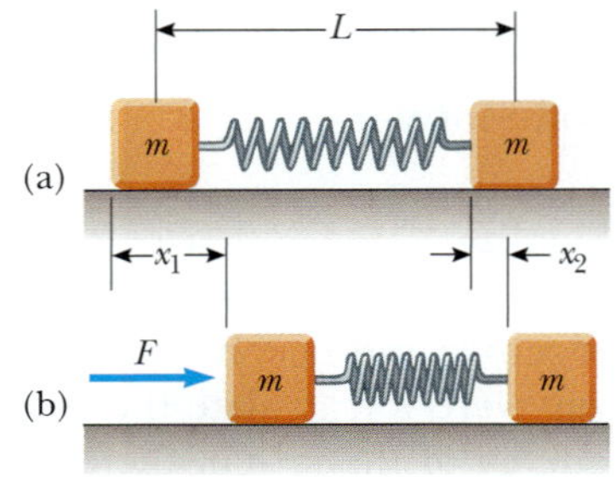

Figure 9.22 (Example 9.15) (a) Two blocks of equal mass are connected by a spring. (b) The left block is pushed with a constant force of magnitude F and moves a distance x_1 during some time interval. During this same time interval, the right block moves through a distance x_2.

(A) Find the resulting speed $\vec{\mathbf{v}}_{CM}$ of the center of mass of the system.

SOLUTION

Conceptualize Imagine what happens as you push on the left block. It begins to move to the right in Figure 9.22, and the spring begins to compress. As a result, the spring pushes to the right on the right block, which begins to move to the right. At any given time, the blocks are generally moving with different velocities. As the center of mass of the system moves to the right, the two blocks oscillate back and forth with respect to the center of mass.

Categorize The system of two blocks and a spring is a nonisolated system because work is being done on it by the applied force. It is also a deformable system. During the time interval Δt, the center of mass of the system moves a distance $\frac{1}{2}(x_1 + x_2)$. Because the applied force on the system is constant, the acceleration of its center of mass is constant and the center of mass is modeled as a particle under constant acceleration.

Analyze We apply the impulse–momentum theorem to the system of two blocks, recognizing that the force F is constant during the time interval Δt while the force is applied.

Write Equation 9.40 for the system:

$$(1) \quad F\,\Delta t = (2m)(v_{CM} - 0) = 2mv_{CM}$$

Because the center of mass is modeled as a particle under constant acceleration, the average velocity of the center of mass is the average of the initial velocity, which is zero, and the final velocity v_{CM}.

Express the time interval in terms of v_{CM}:

$$\Delta t = \frac{\frac{1}{2}(x_1 + x_2)}{v_{CM,\,avg}} = \frac{\frac{1}{2}(x_1 + x_2)}{\frac{1}{2}(0 + v_{CM})} = \frac{(x_1 + x_2)}{v_{CM}}$$

Substitute this expression into Equation (1):

$$F\frac{(x_1 + x_2)}{v_{CM}} = 2mv_{CM}$$

Solve for v_{CM}:

$$v_{CM} = \sqrt{F\frac{(x_1 + x_2)}{2m}}$$

(B) Find the total energy of the system associated with vibration relative to its center of mass after the force F is removed.

SOLUTION

Analyze The vibrational energy is all the energy of the system other than the kinetic energy associated with translational motion of the center of mass. To find the vibrational energy, we apply the conservation of energy equation. The kinetic energy of the system can be expressed as $K = K_{CM} + K_{vib}$, where K_{vib} is the kinetic energy of the blocks relative to the center of mass due to their vibration. The potential energy of the system is U_{vib}, which is the potential energy stored in the spring when the separation of the blocks is some value other than L.

Express Equation 8.2 for this system:

$$(2) \quad \Delta K_{CM} + \Delta K_{vib} + \Delta U_{vib} = W$$

[4] Example 9.15 was inspired in part by C. E. Mungan, "A primer on work–energy relationships for introductory physics," *The Physics Teacher*, 43:10, 2005.

Express Equation (2) in an alternate form, noting that $K_{\text{vib}} + U_{\text{vib}} = E_{\text{vib}}$:

$$\Delta K_{\text{CM}} + \Delta E_{\text{vib}} = W$$

The initial values of the kinetic energy of the center of mass and the vibrational energy of the system are zero:

$$K_{\text{CM}} + E_{\text{vib}} = W = Fx_1$$

Solve for the vibrational energy and use the result to part (A):

$$E_{\text{vib}} = Fx_1 - K_{\text{CM}} = Fx_1 - \tfrac{1}{2}(2m)v_{\text{CM}}^2 = F\frac{(x_1 - x_2)}{2}$$

Finalize Neither of the two answers in this example depends on the spring length, the spring constant, or the time interval. Notice also that the magnitude x_1 of the displacement of the point of application of the applied force is different from the magnitude $\frac{1}{2}(x_1 + x_2)$ of the displacement of the center of mass of the system. This difference reminds us that the displacement in the definition of work is that of the point of application of the force.

9.8 Rocket Propulsion

When ordinary vehicles such as cars are propelled, the driving force for the motion is friction. In the case of the car, the driving force is the force exerted by the road on the car. A rocket moving in space, however, has no road to push against. Therefore, the source of the propulsion of a rocket must be something other than friction. **The operation of a rocket depends on the law of conservation of linear momentum as applied to a system of particles, where the system is the rocket plus its ejected fuel.**

Rocket propulsion can be understood by first considering our archer standing on frictionless ice in Example 9.1. Imagine that the archer fires several arrows horizontally. For each arrow fired, the archer receives a compensating momentum in the opposite direction. As more arrows are fired, the archer moves faster and faster across the ice.

In a similar manner, as a rocket moves in free space, its linear momentum changes when some of its mass is ejected in the form of exhaust gases. **Because the gases are given momentum when they are ejected out of the engine, the rocket receives a compensating momentum in the opposite direction.** Therefore, the rocket is accelerated as a result of the "push," or thrust, from the exhaust gases. In free space, the center of mass of the system (rocket plus expelled gases) moves uniformly, independent of the propulsion process.[5]

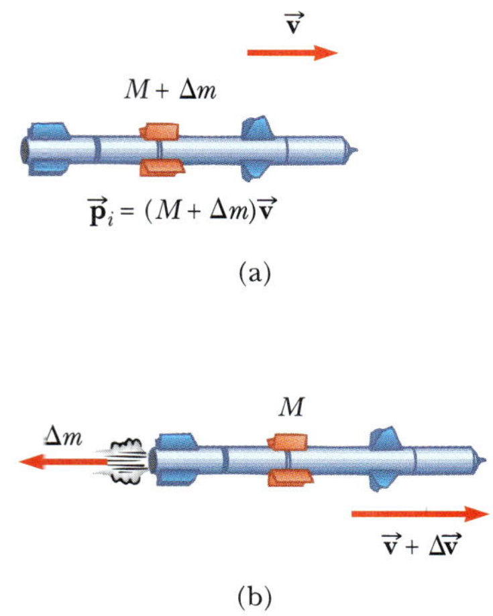

Figure 9.23 Rocket propulsion. (a) The initial mass of the rocket plus all its fuel is $M + \Delta m$ at a time t, and its speed is v. (b) At a time $t + \Delta t$, the rocket's mass has been reduced to M and an amount of fuel Δm has been ejected. The rocket's speed increases by an amount Δv.

Suppose at some time t that the magnitude of the momentum of a rocket plus its fuel is $(M + \Delta m)v$, where v is the speed of the rocket relative to the Earth (Fig. 9.23a). Over a short time interval Δt, the rocket ejects fuel of mass Δm. At the end of this interval, the rocket's mass is M and its speed is $v + \Delta v$, where Δv is the change in speed of the rocket (Fig. 9.23b). If the fuel is ejected with a speed v_e relative to the rocket (the subscript e stands for *exhaust*, and v_e is usually called the *exhaust speed*), the velocity of the fuel relative to the Earth is $v - v_e$. If we equate the total initial momentum of the system to the total final momentum, we obtain

$$(M + \Delta m)v = M(v + \Delta v) + \Delta m(v - v_e)$$

Simplifying this expression gives

$$M\Delta v = v_e \Delta m$$

If we now take the limit as Δt goes to zero, we let $\Delta v \rightarrow dv$ and $\Delta m \rightarrow dm$. Furthermore, the increase in the exhaust mass dm corresponds to an equal decrease

The force from a nitrogen-propelled hand-controlled device allows an astronaut to move about freely in space without restrictive tethers, using the thrust force from the expelled nitrogen.

[5] The rocket and the archer represent cases of the reverse of a perfectly inelastic collision: momentum is conserved, but the kinetic energy of the rocket-exhaust gas system increases (at the expense of chemical potential energy in the fuel), as does the kinetic energy of the archer-arrow system (at the expense of potential energy from the archer's previous meals).

in the rocket mass, so $dm = -dM$. Notice that dM is negative because it represents a decrease in mass, so $-dM$ is a positive number. Using this fact gives

$$M\,dv = v_e\,dm = -v_e\,dM \tag{9.42}$$

Now divide the equation by M and integrate, taking the initial mass of the rocket plus fuel to be M_i and the final mass of the rocket plus its remaining fuel to be M_f. The result is

$$\int_{v_i}^{v_f} dv = -v_e \int_{M_i}^{M_f} \frac{dM}{M}$$

Expression for rocket propulsion ▶

$$v_f - v_i = v_e \ln\left(\frac{M_i}{M_f}\right) \tag{9.43}$$

which is the basic expression for rocket propulsion. First, Equation 9.43 tells us that the increase in rocket speed is proportional to the exhaust speed v_e of the ejected gases. Therefore, the exhaust speed should be very high. Second, the increase in rocket speed is proportional to the natural logarithm of the ratio M_i/M_f. Therefore, this ratio should be as large as possible, that is, the mass of the rocket without its fuel should be as small as possible and the rocket should carry as much fuel as possible.

The **thrust** on the rocket is the force exerted on it by the ejected exhaust gases. We obtain the following expression for the thrust from Newton's second law and Equation 9.42:

$$\text{Thrust} = M\frac{dv}{dt} = \left|v_e \frac{dM}{dt}\right| \tag{9.44}$$

This expression shows that the thrust increases as the exhaust speed increases and as the rate of change of mass (called the *burn rate*) increases.

EXAMPLE 9.16 Fighting a Fire

Two firefighters must apply a total force of 600 N to steady a hose that is discharging water at the rate of 3 600 L/min. Estimate the speed of the water as it exits the nozzle.

SOLUTION

Conceptualize As the water leaves the hose, it acts in a way similar to the gases being ejected from a rocket engine. As a result, a force (thrust) acts on the firefighters in a direction opposite the direction of motion of the water. In this case, we want the end of the hose to be a particle in equilibrium rather than to accelerate as in the case of the rocket. Consequently, the firefighters must apply a force of magnitude equal to the thrust in the opposite direction to keep the end of the hose stationary.

Categorize This example is a substitution problem in which we use given values in an equation derived in this section. The water exits at 3 600 L/min, which is 60 L/s. Knowing that 1 L of water has a mass of 1 kg, we estimate that about 60 kg of water leaves the nozzle each second.

Use Equation 9.44 for the thrust: $\text{Thrust} = \left|v_e \frac{dM}{dt}\right|$

Substitute the known values: $600\text{ N} = |v_e(60\text{ kg/s})|$

Solve for the exhaust speed: $v_e = 10\text{ m/s}$

EXAMPLE 9.17 A Rocket in Space

A rocket moving in space, far from all other objects, has a speed of 3.0×10^3 m/s relative to the Earth. Its engines are turned on, and fuel is ejected in a direction opposite the rocket's motion at a speed of 5.0×10^3 m/s relative to the rocket.

(A) What is the speed of the rocket relative to the Earth once the rocket's mass is reduced to half its mass before ignition?

SOLUTION

Conceptualize From the discussion in this section and scenes from science fiction movies, we can easily imagine the rocket accelerating to a higher speed as the engine operates.

Categorize This problem is a substitution problem in which we use given values in the equations derived in this section.

Solve Equation 9.43 for the final velocity and substitute the known values:

$$v_f = v_i + v_e \ln\left(\frac{M_i}{M_f}\right)$$

$$= 3.0 \times 10^3 \text{ m/s} + (5.0 \times 10^3 \text{ m/s})\ln\left(\frac{M_i}{0.5M_i}\right)$$

$$= 6.5 \times 10^3 \text{ m/s}$$

(B) What is the thrust on the rocket if it burns fuel at the rate of 50 kg/s?

SOLUTION

Use Equation 9.44 and the result to part (A), noting that $dM/dt = 50$ kg/s:

$$\text{Thrust} = \left|v_e \frac{dM}{dt}\right| = (5.0 \times 10^3 \text{ m/s})(50 \text{ kg/s}) = 2.5 \times 10^5 \text{ N}$$

Summary

ThomsonNOW™ Sign in at **www.thomsonedu.com** and go to ThomsonNOW to take a practice test for this chapter.

DEFINITIONS

The **linear momentum** $\vec{\mathbf{p}}$ of a particle of mass m moving with a velocity $\vec{\mathbf{v}}$ is

$$\vec{\mathbf{p}} \equiv m\vec{\mathbf{v}} \quad (9.2)$$

The **impulse** imparted to a particle by a net force $\Sigma\vec{\mathbf{F}}$ is equal to the time integral of the force:

$$\vec{\mathbf{I}} \equiv \int_{t_i}^{t_f} \sum \vec{\mathbf{F}}dt \quad (9.9)$$

An **inelastic collision** is one for which the total kinetic energy of the system of colliding particles is not conserved. A **perfectly inelastic collision** is one in which the colliding particles stick together after the collision. An **elastic collision** is one in which the kinetic energy of the system is conserved.

The position vector of the **center of mass** of a system of particles is defined as

$$\vec{\mathbf{r}}_{\text{CM}} \equiv \frac{1}{M}\sum_i m_i\vec{\mathbf{r}}_i \quad (9.31)$$

where $M = \sum_i m_i$ is the total mass of the system and $\vec{\mathbf{r}}_i$ is the position vector of the ith particle.

CONCEPTS AND PRINCIPLES

The position vector of the center of mass of an extended object can be obtained from the integral expression

$$\vec{\mathbf{r}}_{\text{CM}} = \frac{1}{M}\int \vec{\mathbf{r}}\,dm \quad (9.34)$$

The velocity of the center of mass for a system of particles is

$$\vec{\mathbf{v}}_{\text{CM}} = \frac{\sum_i m_i\vec{\mathbf{v}}_i}{M} \quad (9.35)$$

The total momentum of a system of particles equals the total mass multiplied by the velocity of the center of mass.

Newton's second law applied to a system of particles is

$$\sum \vec{\mathbf{F}}_{\text{ext}} = M\vec{\mathbf{a}}_{\text{CM}} \quad (9.39)$$

where $\vec{\mathbf{a}}_{\text{CM}}$ is the acceleration of the center of mass and the sum is over all external forces. The center of mass moves like an imaginary particle of mass M under the influence of the resultant external force on the system.

ANALYSIS MODELS FOR PROBLEM SOLVING

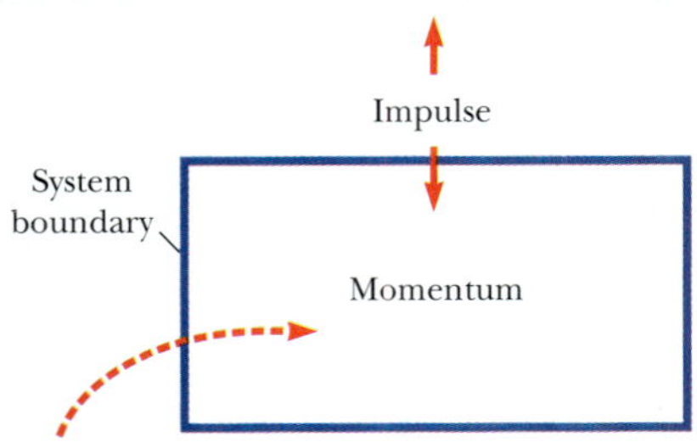

The change in the total momentum of the system is equal to the total impulse on the system.

Nonisolated System (Momentum). If a system interacts with its environment in the sense that there is an external force on the system, the behavior of the system is described by the **impulse-momentum theorem:**

$$\vec{\mathbf{I}} = \Delta\vec{\mathbf{p}}_{\text{tot}} \quad (9.40)$$

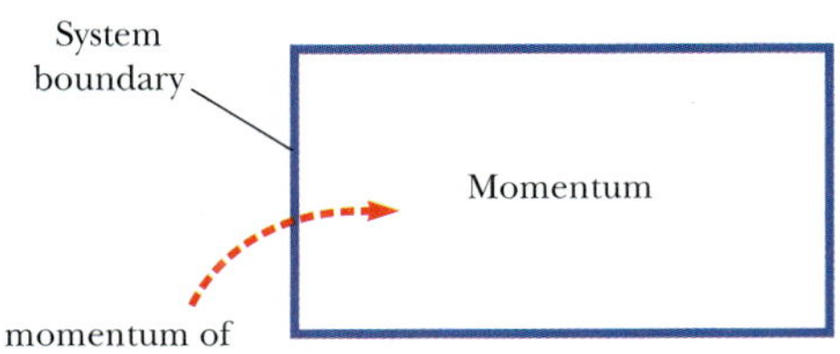

The total momentum of the system is constant.

Isolated System (Momentum). The principle of **conservation of linear momentum** indicates that the total momentum of an isolated system (no external forces) is conserved regardless of the nature of the forces between the members of the system:

$$M\vec{\mathbf{v}}_{\text{CM}} = \vec{\mathbf{p}}_{\text{tot}} = \text{constant} \left(\text{when } \sum \vec{\mathbf{F}}_{\text{ext}} = 0\right) \quad (9.41)$$

In the case of a two-particle system, this principle can be expressed as

$$\vec{\mathbf{p}}_{1i} + \vec{\mathbf{p}}_{2i} = \vec{\mathbf{p}}_{1f} + \vec{\mathbf{p}}_{2f} \quad (9.5)$$

The system may be isolated in terms of momentum but nonisolated in terms of energy, as in the case of inelastic collisions.

Questions

☐ denotes answer available in *Student Solutions Manual/Study Guide;* **O** denotes objective question

1. (a) Does a larger net force exerted on an object always produce a larger change in the momentum of the object compared with a smaller net force? Explain. (b) Does a larger net force always produce a larger change in kinetic energy than a smaller net force? Explain.

2. O (i) The momentum of a certain object is made four times larger in magnitude. By what factor is its kinetic energy changed? (a) 16 (b) 8 (c) 4 (d) 2 (e) 1 **(ii)** The kinetic energy of an object is made four times larger. By what factor is the magnitude of its momentum changed? (a) 16 (b) 8 (c) 4 (d) 2 (e) 1

3. O (i) If two particles have equal momenta, are their kinetic energies equal? (a) yes (b) no (c) if and only if their masses are equal **(ii)** If two particles have equal kinetic energies, are their momenta equal? (a) yes (b) no (c) if and only if their masses are equal (d) if and only if their masses and directions of motion are the same

4. O Two particles of different mass start from rest. The same net force acts on both of them as they move over equal distances. **(i)** How do their final kinetic energies compare? (a) The particle of larger mass has more kinetic energy. (b) The particle of smaller mass has more kinetic energy. (c) The particles have equal kinetic energies. (d) Either particle might have more kinetic energy. **(ii)** How do the magnitudes of their momentum compare? (a) The particle of larger mass has more momentum. (b) The particle of smaller mass has more momentum. (c) The particles have equal momenta. (d) Either particle might have more momentum.

5. While in motion, a pitched baseball carries kinetic energy and momentum. (a) Can we say that it carries a force that it can exert on any object it strikes? (b) Can the baseball deliver more kinetic energy to the object it strikes than the ball carries initially? (c) Can the baseball deliver to the object it strikes more momentum than the ball carries initially? Explain your answers.

6. O A basketball is tossed up into the air, falls freely, and bounces from the wooden floor. From the moment after the player releases it until the ball reaches the top of its bounce, what is the smallest system for which momentum is conserved? (a) the ball (b) the ball plus player (c) the ball plus floor (d) the ball plus Earth (e) momentum is not conserved

7. A bomb, initially at rest, explodes into several pieces. (a) Is linear momentum of the system conserved? (b) Is kinetic energy of the system conserved? Explain.

8. You are standing perfectly still and then you take a step forward. Before the step your momentum was zero, but afterward you have some momentum. Is the principle of conservation of momentum violated in this case?

9. O A massive manure spreader rolls down a country road. In a perfectly inelastic collision, a small sports car runs into the machine from behind. **(i)** Which vehicle experiences a change in momentum of larger magnitude? (a) The car does. (b) The manure spreader does. (c) Their momentum changes are the same size. (d) It could be either. **(ii)** Which vehicle experiences a larger change in kinetic energy? (a) The car does. (b) The manure spreader does. (c) Their kinetic energy changes are the same size. (d) It could be either vehicle.

10. A sharpshooter fires a rifle while standing with the butt of the gun against her shoulder. If the forward momentum of a bullet is the same as the backward momentum of the gun, why isn't it as dangerous to be hit by the gun as by the bullet?

11. O A ball is suspended by a string that is tied to a fixed point above a wooden block standing on end. The ball is pulled back as shown in Figure Q9.11 and released. In trial (a), the ball rebounds elastically from the block. In trial (b), two-sided tape causes the ball to stick to the block. **(i)** In which case, (a) or (b), is the ball more likely to knock the block over? Or (c) does it make no difference? Or (d) could it be either case depending on other factors? **(ii)** In which case, (a) or (b), is there a larger momentary temperature increase in the ball and the adjacent bit of wood? Or (c) is it the same for both? Or (d) is there no temperature increase anyway?

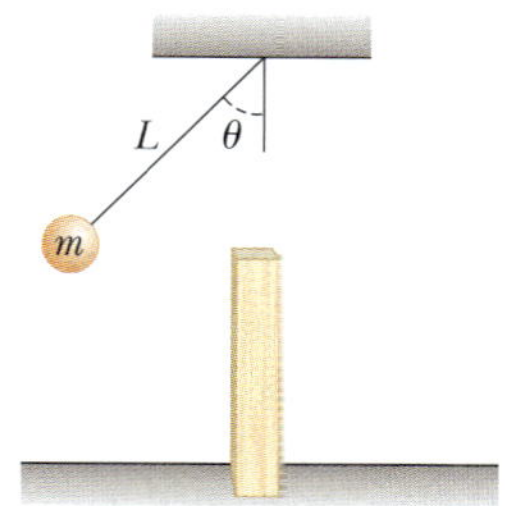

Figure Q9.11

12. A pole-vaulter falls from a height of 6.0 m onto a foam rubber pad. Can you calculate his speed immediately before he reaches the pad? Can you calculate the force exerted on him by the pad? Explain.

13. Two students hold a large bedsheet vertically between them. A third student, who happens to be the star pitcher on the school baseball team, throws a raw egg at the sheet. Explain why the egg does not break when it hits the sheet, regardless of its initial speed. (If you try this demonstration, make sure the pitcher hits the sheet near its center, and do not allow the egg to fall on the floor after it is caught.)

14. O You are standing on a saucer-shaped sled at rest in the middle of a frictionless ice rink. Your lab partner throws you a heavy Frisbee. You take different actions in successive experimental trials. Rank the following situations in order according to your final speed, from largest to smallest. If your final speed is the same in two cases, give them equal rank. (a) You catch the Frisbee and hold onto it. (b) You catch the Frisbee and throw it back to your partner. (c) You catch the Frisbee and throw it to a third person off to the side at a right angle. (d) You bobble the catch, just touching the Frisbee so that it continues in its original direction more slowly. (e) You catch the Frisbee and throw it so that it moves vertically upward above your head. (f) You catch the Frisbee as it comes from the south, spin around, and throw it north several times faster. (g) You catch the Frisbee and set it down at rest on the ice.

15. A person balances a meterstick in a horizontal position on the extended index fingers of her right and left hands. She slowly brings the two fingers together. The stick remains balanced and the two fingers always meet at the 50-cm mark regardless of their original positions. (Try it!) Explain.

16. O As a railroad train is assembled, a yard engine releases one boxcar in motion at the top of a hump. The car rolls down quietly and without friction. Switches are set to shunt it onto a straight, level track where it couples with a flatcar of smaller mass, originally at rest, so that the two cars then roll together without friction. Consider the two cars as a system from the moment of release of the boxcar until both are rolling together. (a) Is mechanical energy of the system conserved? (b) Is momentum conserved? Next, consider the process of the boxcar gaining speed as it rolls down the hump. For the boxcar and the Earth as a system, (c) is mechanical energy conserved? (d) Is momentum conserved? Finally, consider the two cars as a system as the boxcar is slowing down in the coupling process. (e) Is mechanical energy of this system conserved? (f) Is momentum conserved?

17. A juggler juggles three balls in a continuous cycle. Any one ball is in contact with his hands for one-fifth of the time. Describe the motion of the center of mass of the three balls. What average force does the juggler exert on one ball while he is touching it?

18. Does the center of mass of a rocket in free space accelerate? Explain. Can the speed of a rocket exceed the exhaust speed of the fuel? Explain.

19. On the subject of the following positions, state your own view and argue to support it. (a) The best theory of motion is that force causes acceleration. (b) The true measure of a force's effectiveness is the work it does, and the best theory of motion is that work done on an object changes its energy. (c) The true measure of a force's effect is impulse, and the best theory of motion is that impulse imparted to an object changes its momentum.

Problems

WebAssign The Problems from this chapter may be assigned online in WebAssign.

ThomsonNOW Sign in at **www.thomsonedu.com** and go to ThomsonNOW to assess your understanding of this chapter's topics with additional quizzing and conceptual questions.

1, 2, 3 denotes straightforward, intermediate, challenging; □ denotes full solution available in *Student Solutions Manual/Study Guide;* ▲ denotes coached solution with hints available at **www.thomsonedu.com;** ▭ denotes developing symbolic reasoning; ● denotes asking for qualitative reasoning; ▪ denotes computer useful in solving problem

Section 9.1 Linear Momentum and Its Conservation

1. A 3.00-kg particle has a velocity of $(3.00\hat{\mathbf{i}} - 4.00\hat{\mathbf{j}})$ m/s. (a) Find its x and y components of momentum. (b) Find the magnitude and direction of its momentum.

2. ● A 65.0-kg boy and his 40.0-kg sister, both wearing roller blades, face each other at rest. The girl pushes the boy hard, sending him backward with velocity 2.90 m/s toward the west. Ignore friction. (a) Describe the subsequent motion of the girl. (b) How much chemical energy is converted into mechanical energy in the girl's muscles? (c) Is the momentum of the boy-girl system conserved in the pushing-apart process? How can it be, with large forces acting? How can it be, with no motion beforehand and plenty of motion afterward?

3. How fast can you set the Earth moving? In particular, when you jump straight up as high as you can, what is the order of magnitude of the maximum recoil speed that you give to the Earth? Model the Earth as a perfectly solid object. In your solution, state the physical quantities you take as data and the values you measure or estimate for them.

4. ● Two blocks of masses M and $3M$ are placed on a horizontal, frictionless surface. A light spring is attached to one of them, and the blocks are pushed together with the spring between them (Fig. P9.4). A cord initially holding the blocks together is burned; after that happens, the block of mass $3M$ moves to the right with a speed of 2.00 m/s. (a) What is the velocity of the block of mass M? (b) Find the system's original elastic potential energy, taking M = 0.350 kg. (c) Is the original energy in the spring or in the cord? Explain your answer. (d) Is momentum of the system conserved in the bursting-apart process? How can it be, with large forces acting? How can it be, with no motion beforehand and plenty of motion afterward?

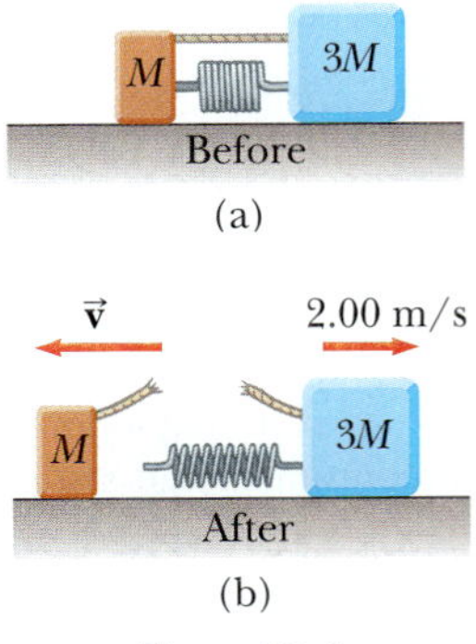

Figure P9.4

5. (a) A particle of mass m moves with momentum of magnitude p. Show that the kinetic energy of the particle is given by $K = p^2/2m$. (b) Express the magnitude of the particle's momentum in terms of its kinetic energy and mass.

Section 9.2 Impulse and Momentum

6. ● A friend claims that as long as he has his seat belt on, he can hold on to a 12.0-kg child in a 60.0 mi/h head-on

collision with a brick wall in which the car passenger compartment comes to a stop in 0.050 0 s. Is his claim true? Explain why he will experience a violent force during the collision, tearing the child from his arms. Evaluate the size of this force. (A child should always be in a toddler seat secured with a seat belt in the back seat of a car.)

7. An estimated force–time curve for a baseball struck by a bat is shown in Figure P9.7. From this curve, determine (a) the impulse delivered to the ball, (b) the average force exerted on the ball, and (c) the peak force exerted on the ball.

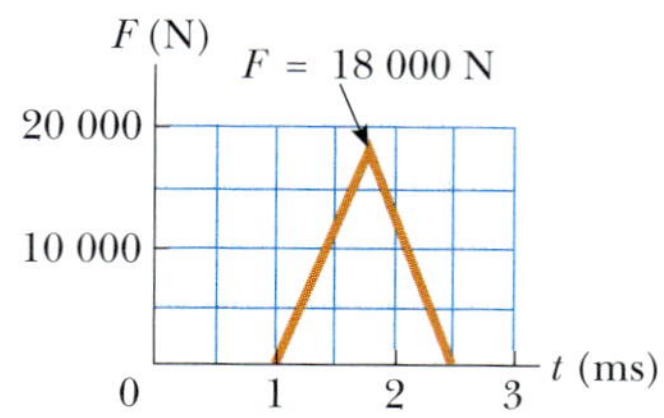

Figure P9.7

8. A ball of mass 0.150 kg is dropped from rest from a height of 1.25 m. It rebounds from the floor to reach a height of 0.960 m. What impulse was given to the ball by the floor?

9. ▲ A 3.00-kg steel ball strikes a wall with a speed of 10.0 m/s at an angle of 60.0° with the surface. It bounces off with the same speed and angle (Fig. P9.9). If the ball is in contact with the wall for 0.200 s, what is the average force exerted by the wall on the ball?

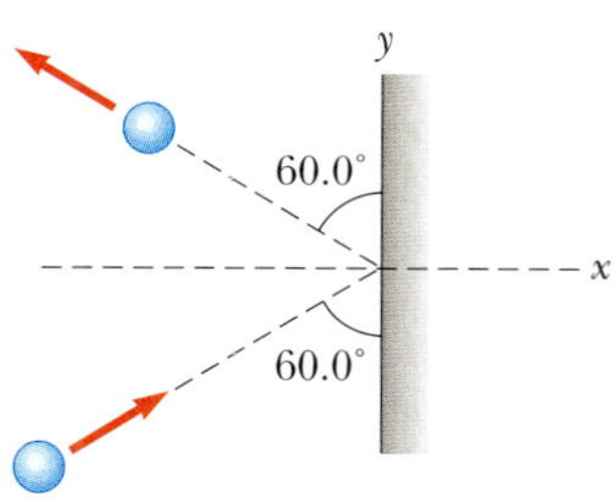

Figure P9.9

10. A tennis player receives a shot with the ball (0.060 0 kg) traveling horizontally at 50.0 m/s and returns the shot with the ball traveling horizontally at 40.0 m/s in the opposite direction. (a) What is the impulse delivered to the ball by the tennis racquet? (b) What work does the racquet do on the ball?

11. The magnitude of the net force exerted in the x direction on a 2.50-kg particle varies in time as shown in Figure P9.11. Find (a) the impulse of the force, (b) the final velocity the particle attains if it is originally at rest, (c) its final velocity if its original velocity is -2.00 m/s, and (d) the average force exerted on the particle for the time interval between 0 and 5.00 s.

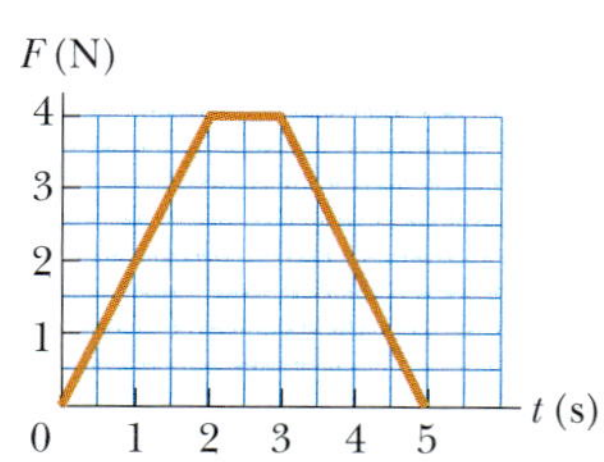

Figure P9.11

12. A *force platform* is a tool used to analyze the performance of athletes by measuring the vertical force that the athlete exerts on the ground as a function of time. Starting from rest, a 65.0-kg athlete jumps down onto the platform from a height of 0.600 m. While she is in contact with the platform during the time interval $0 < t < 0.800$ s, the force she exerts on it is described by the function

$$F = (9\,200 \text{ N/s})t - (11\,500 \text{ N/s}^2)t^2$$

(a) What impulse did the athlete receive from the platform? (b) With what speed did she reach the platform? (c) With what speed did she leave it? (d) To what height did she jump upon leaving the platform?

13. ● A glider of mass m is free to slide along a horizontal air track. It is pushed against a launcher at one end of the track. Model the launcher as a light spring of force constant k compressed by a distance x. The glider is released from rest. (a) Show that the glider attains a speed of $v = x(k/m)^{1/2}$. (b) Does a glider of large or of small mass attain a greater speed? (c) Show that the impulse imparted to the glider is given by the expression $x(km)^{1/2}$. (d) Is a greater impulse imparted to a large or a small mass? (e) Is more work done on a large or a small mass?

14. Water falls without splashing at a rate of 0.250 L/s from a height of 2.60 m into a 0.750-kg bucket on a scale. If the bucket is originally empty, what does the scale read 3.00 s after water starts to accumulate in it?

Section 9.3 Collisions in One Dimension

15. A 10.0-g bullet is fired into a stationary block of wood (m = 5.00 kg). The bullet imbeds into the block. The speed of the bullet-plus-wood combination immediately after the collision is 0.600 m/s. What was the original speed of the bullet?

16. A railroad car of mass 2.50×10^4 kg is moving with a speed of 4.00 m/s. It collides and couples with three other coupled railroad cars, each of the same mass as the single car and moving in the same direction with an initial speed of 2.00 m/s. (a) What is the speed of the four cars immediately after the collision? (b) How much energy is transformed into internal energy in the collision?

17. ● Four railroad cars, each of mass 2.50×10^4 kg, are coupled together and coasting along horizontal tracks at speed v_i toward the south. A very strong movie actor, riding on the second car, uncouples the front car and gives it a big push, increasing its speed to 4.00 m/s southward. The remaining three cars continue moving south, now at 2.00 m/s. (a) Find the initial speed of the cars. (b) How much work did the actor do? (c) State the relationship between the process described here and the process in Problem 16.

18. As shown in Figure P9.18 (page 262), a bullet of mass m and speed v passes completely through a pendulum bob of mass M. The bullet emerges with a speed of $v/2$. The pendulum bob is suspended by a stiff rod of length ℓ and negligible mass. What is the minimum value of v such that the pendulum bob will barely swing through a complete vertical circle?

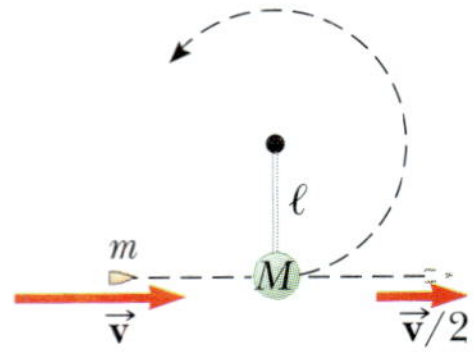

Figure P9.18

19. Two blocks are free to slide along the frictionless wooden track *ABC* shown in Figure P9.19. The block of mass $m_1 =$ 5.00 kg is released from *A*. Protruding from its front end is the north pole of a strong magnet, which is repelling the north pole of an identical magnet embedded in the back end of the block of mass $m_2 =$ 10.0 kg, initially at rest. The two blocks never touch. Calculate the maximum height to which m_1 rises after the elastic collision.

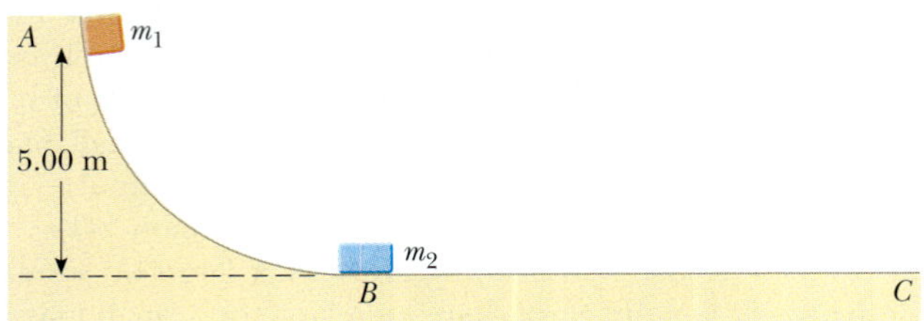

Figure P9.19

20. A tennis ball of mass 57.0 g is held just above a basketball of mass 590 g. With their centers vertically aligned, both are released from rest at the same moment, to fall through a distance of 1.20 m, as shown in Figure P9.20. (a) Find the magnitude of the downward velocity with which the basketball reaches the ground. Assume an elastic collision with the ground instantaneously reverses the velocity of the basketball while the tennis ball is still moving down. Next, the two balls meet in an elastic collision. (b) To what height does the tennis ball rebound?

Figure P9.20

21. A 45.0-kg girl is standing on a plank that has a mass of 150 kg. The plank, originally at rest, is free to slide on a frozen lake that constitutes a flat, frictionless supporting surface. The girl begins to walk along the plank at a constant velocity of $1.50\hat{\mathbf{i}}$ m/s relative to the plank. (a) What is her velocity relative to the ice surface? (b) What is the velocity of the plank relative to the ice surface?

22. A 7.00-g bullet, when fired from a gun into a 1.00-kg block of wood held in a vise, penetrates the block to a depth of 8.00 cm. This block of wood is placed on a frictionless horizontal surface, and a second 7.00-g bullet is fired from the gun into the block. To what depth does the bullet penetrate the block in this case?

23. ▲ A neutron in a nuclear reactor makes an elastic head-on collision with the nucleus of a carbon atom initially at rest. (a) What fraction of the neutron's kinetic energy is transferred to the carbon nucleus? (b) The initial kinetic energy of the neutron is 1.60×10^{-13} J. Find its final kinetic energy and the kinetic energy of the carbon nucleus after the collision. (The mass of the carbon nucleus is nearly 12.0 times the mass of the neutron.)

24. ● (a) Three carts of masses 4.00 kg, 10.0 kg, and 3.00 kg move on a frictionless, horizontal track with speeds of 5.00 m/s, 3.00 m/s, and 4.00 m/s as shown in Figure P9.24. Velcro couplers make the carts stick together after colliding. Find the final velocity of the train of three carts. (b) **What If?** Does your answer require that all the carts collide and stick together at the same moment? What if they collide in a different order?

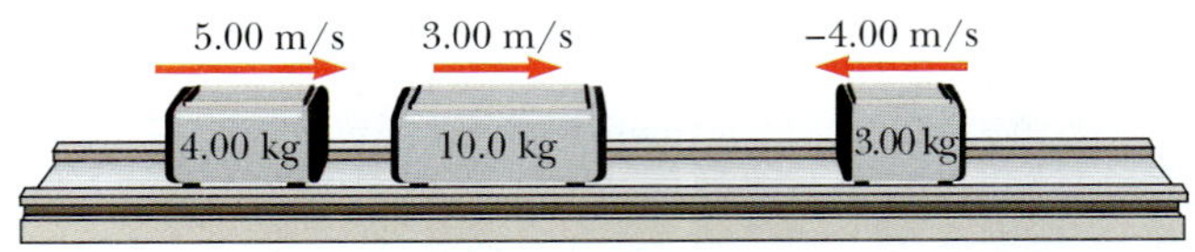

Figure P9.24

25. ▲ A 12.0-g wad of sticky clay is hurled horizontally at a 100-g wooden block initially at rest on a horizontal surface. The clay sticks to the block. After impact, the block slides 7.50 m before coming to rest. If the coefficient of friction between the block and the surface is 0.650, what was the speed of the clay immediately before impact?

Section 9.4 Collisions in Two Dimensions

26. ● In an American football game, a 90.0-kg fullback running east with a speed of 5.00 m/s is tackled by a 95.0-kg opponent running north with a speed of 3.00 m/s. (a) Explain why the successful tackle constitutes a perfectly inelastic collision. (b) Calculate the velocity of the players immediately after the tackle. (c) Determine the mechanical energy that disappears as a result of the collision. Account for the missing energy.

27. A billiard ball moving at 5.00 m/s strikes a stationary ball of the same mass. After the collision, the first ball moves, at 4.33 m/s, at an angle of 30.0° with respect to the original line of motion. Assuming an elastic collision (and ignoring friction and rotational motion), find the struck ball's velocity after the collision.

28. ● Two automobiles of equal mass approach an intersection. One vehicle is traveling with velocity 13.0 m/s toward the east, and the other is traveling north with speed v_{2i}. Neither driver sees the other. The vehicles collide in the intersection and stick together, leaving parallel skid marks at an angle of 55.0° north of east. The speed limit for both roads is 35 mi/h, and the driver of the northward-moving vehicle claims he was within the speed limit when the collision occurred. Is he telling the truth? Explain your reasoning.

29. Two shuffleboard disks of equal mass, one orange and the other yellow, are involved in an elastic, glancing collision. The yellow disk is initially at rest and is struck by the orange disk moving with a speed of 5.00 m/s. After the collision, the orange disk moves along a direction that

makes an angle of 37.0° with its initial direction of motion. The velocities of the two disks are perpendicular after the collision. Determine the final speed of each disk.

30. Two shuffleboard disks of equal mass, one orange and the other yellow, are involved in an elastic, glancing collision. The yellow disk is initially at rest and is struck by the orange disk moving with a speed v_i. After the collision, the orange disk moves along a direction that makes an angle θ with its initial direction of motion. The velocities of the two disks are perpendicular after the collision. Determine the final speed of each disk.

31. An object of mass 3.00 kg, moving with an initial velocity of $5.00\hat{\mathbf{i}}$ m/s, collides with and sticks to an object of mass 2.00 kg with an initial velocity of $-3.00\hat{\mathbf{j}}$ m/s. Find the final velocity of the composite object.

32. Two particles with masses m and $3m$ are moving toward each other along the x axis with the same initial speeds v_i. Particle m is traveling to the left, and particle $3m$ is traveling to the right. They undergo an elastic, glancing collision such that particle m is moving downward after the collision at a right angle from its initial direction. (a) Find the final speeds of the two particles. (b) What is the angle θ at which the particle $3m$ is scattered?

33. ▲ An unstable atomic nucleus of mass 17.0×10^{-27} kg initially at rest disintegrates into three particles. One of the particles, of mass 5.00×10^{-27} kg, moves in the y direction with a speed of 6.00×10^6 m/s. Another particle, of mass 8.40×10^{-27} kg, moves in the x direction with a speed of 4.00×10^6 m/s. Find (a) the velocity of the third particle and (b) the total kinetic energy increase in the process.

34. The mass of the blue puck in Figure P9.34 is 20.0% greater than the mass of the green puck. Before colliding, the pucks approach each other with momenta of equal magnitudes and opposite directions, and the green puck has an initial speed of 10.0 m/s. Find the speed each puck has after the collision if half the kinetic energy of the system becomes internal energy during the collision.

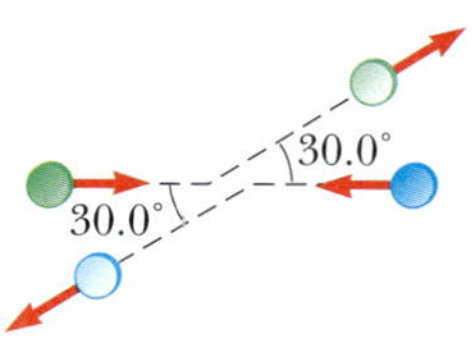

Figure P9.34

Section 9.5 The Center of Mass

35. Four objects are situated along the y axis as follows: a 2.00-kg object is located at +3.00 m, a 3.00-kg object is at +2.50 m, a 2.50-kg object is at the origin, and a 4.00-kg object is at −0.500 m. Where is the center of mass of these objects?

36. The mass of the Earth is 5.98×10^{24} kg, and the mass of the Moon is 7.36×10^{22} kg. The distance of separation, measured between their centers, is 3.84×10^8 m. Locate the center of mass of the Earth–Moon system as measured from the center of the Earth.

37. A uniform piece of sheet steel is shaped as shown in Figure P9.37. Compute the x and y coordinates of the center of mass of the piece.

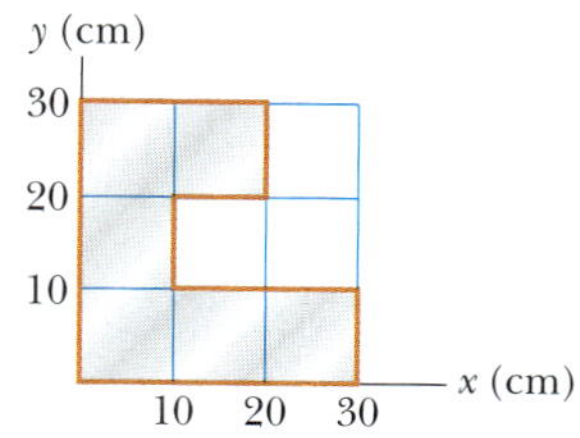

Figure P9.37

38. (a) Consider an extended object whose different portions have different elevations. Assume the free-fall acceleration is uniform over the object. Prove that the gravitational potential energy of the object–Earth system is given by $U_g = Mgy_{CM}$, where M is the total mass of the object and y_{CM} is the elevation of its center of mass above the chosen reference level. (b) Calculate the gravitational potential energy associated with a ramp constructed on level ground with stone with density 3 800 kg/m^3 and everywhere 3.60 m wide. In a side view, the ramp appears as a right triangle with height 15.7 m at the top end and base 64.8 m (Fig. P9.38).

Figure P9.38

39. A rod of length 30.0 cm has linear density (mass-per-length) given by

$$\lambda = 50.0\text{ g/m} + 20.0x\text{ g/m}^2$$

where x is the distance from one end, measured in meters. (a) What is the mass of the rod? (b) How far from the $x = 0$ end is its center of mass?

40. In the 1968 Summer Olympic Games, University of Oregon high jumper Dick Fosbury introduced a new technique of high jumping called the "Fosbury flop." It contributed to raising the world record by about 30 cm and is presently used by nearly every world-class jumper. In this technique, the jumper goes over the bar face up while arching his back as much as possible as shown in Figure P9.40a. This action places his center of mass outside his body, below his back. As his body goes over the bar, his center of mass passes below the bar. Because a given energy input implies a certain elevation for his center of mass, the action of arching his back means his body is

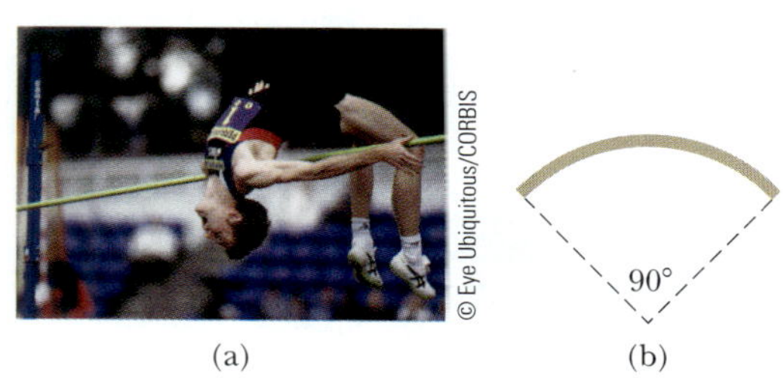

Figure P9.40

higher than if his back were straight. As a model, consider the jumper as a thin, uniform rod of length L. When the rod is straight, its center of mass is at its center. Now bend the rod in a circular arc so that it subtends an angle of 90.0° at the center of the arc as shown in Figure P9.40b. In this configuration, how far outside the rod is the center of mass?

Section 9.6 Motion of a System of Particles

41. A 2.00-kg particle has a velocity $(2.00\hat{\mathbf{i}} - 3.00\hat{\mathbf{j}})$ m/s, and a 3.00-kg particle has a velocity $(1.00\hat{\mathbf{i}} + 6.00\hat{\mathbf{j}})$ m/s. Find (a) the velocity of the center of mass and (b) the total momentum of the system.

42. The vector position of a 3.50-g particle moving in the xy plane varies in time according to $\vec{\mathbf{r}}_1 = (3\hat{\mathbf{i}} + 3\hat{\mathbf{j}})t + 2\hat{\mathbf{j}}t^2$. At the same time, the vector position of a 5.50-g particle varies as $\vec{\mathbf{r}}_2 = 3\hat{\mathbf{i}} - 2\hat{\mathbf{i}}t^2 - 6\hat{\mathbf{j}}t$, where t is in s and r is in cm. At $t = 2.50$ s, determine (a) the vector position of the center of mass, (b) the linear momentum of the system, (c) the velocity of the center of mass, (d) the acceleration of the center of mass, and (e) the net force exerted on the two-particle system.

43. Romeo (77.0 kg) entertains Juliet (55.0 kg) by playing his guitar from the rear of their boat at rest in still water, 2.70 m away from Juliet, who is in the front of the boat. After the serenade, Juliet carefully moves to the rear of the boat (away from shore) to plant a kiss on Romeo's cheek. How far does the 80.0-kg boat move toward the shore it is facing?

44. A ball of mass 0.200 kg has a velocity of $1.50\hat{\mathbf{i}}$ m/s; a ball of mass 0.300 kg has a velocity of $-0.400\hat{\mathbf{i}}$ m/s. They meet in a head-on elastic collision. (a) Find their velocities after the collision. (b) Find the velocity of their center of mass before and after the collision.

Section 9.7 Deformable Systems

45. ● For a technology project, a student has built a vehicle, of total mass 6.00 kg, that moves itself. As shown in Figure P9.45, it runs on two light caterpillar tracks that pass around four light wheels. A reel is attached to one of the axles, and a cord originally wound on the reel passes over a pulley attached to the vehicle to support an elevated load. After the vehicle is released from rest, the load descends slowly, unwinding the cord to turn the axle and make the vehicle move forward. Friction is negligible in the pulley and axle bearings. The caterpillar tread does not slip on the wheels or the floor. The reel has a conical shape so that the load descends at a constant low speed while the vehicle moves horizontally across the floor with constant acceleration, reaching final velocity $3.00\hat{\mathbf{i}}$ m/s. (a) Does the floor impart impulse to the vehicle? If so, how much? (b) Does the floor do work on the vehicle? If so, how much? (c) Does it make sense to say that the final momentum of the vehicle came from the floor? If not, from where? (d) Does it make sense to say that the final kinetic energy of the vehicle came from the floor? If not, from where? (e) Can we say that one particular force causes the forward acceleration of the vehicle? What does cause it?

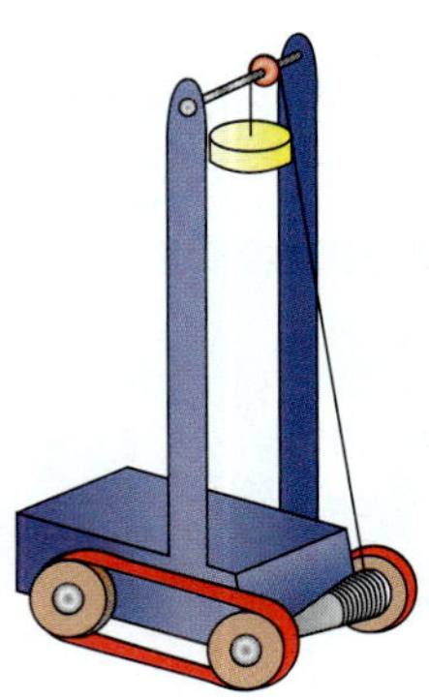

Figure P9.45

46. ● A 60.0-kg person bends his knees and then jumps straight up. After his feet leave the floor his motion is unaffected by air resistance and his center of mass rises by a maximum of 15.0 cm. Model the floor as completely solid and motionless. (a) Does the floor impart impulse to the person? (b) Does the floor do work on the person? (c) With what momentum does the person leave the floor? (d) Does it make sense to say that this momentum came from the floor? Explain. (e) With what kinetic energy does the person leave the floor? (f) Does it make sense to say that this energy came from the floor? Explain.

47. ● A particle is suspended from a post on top of a cart by a light string of length L as shown in Figure P9.47a. The cart and particle are initially moving to the right at constant speed v_i, with the string vertical. The cart suddenly comes to rest when it runs into and sticks to a bumper as shown in Figure P9.47b. The suspended particle swings through an angle θ. (a) Show that the original speed of the cart can be computed from $v_i = \sqrt{2gL(1 - \cos\theta)}$. (b) Find the initial speed implied by $L = 1.20$ m and $\theta = 35.0°$. (c) Is the bumper still exerting a horizontal force on the cart when the hanging particle is at its maximum angle from the vertical? At what moment in the observable motion does the bumper stop exerting a horizontal force on the cart?

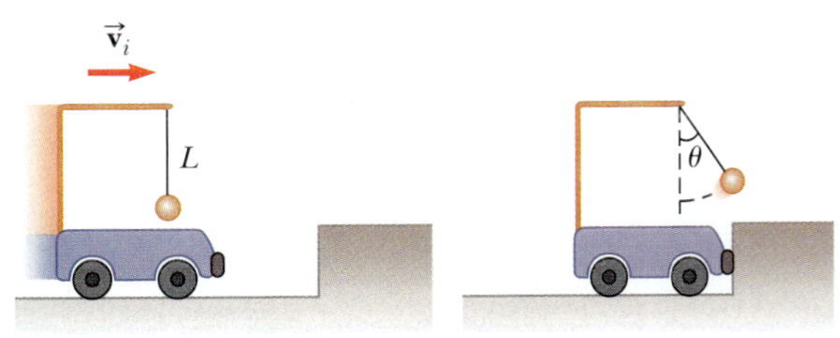

(a) (b)

Figure P9.47

48. On a horizontal air track, a glider of mass m carries a Γ-shaped post. The post supports a small dense sphere, also of mass m, hanging just above the top of the glider on a cord of length L. The glider and sphere are initially at rest with the cord vertical. (Fig. P9.47a shows a cart and a sphere similarly connected.) A constant horizontal force of magnitude F is applied to the glider, moving it through displacement x_1; then the force is removed. During the time interval when the force is applied, the sphere moves through a displacement with horizontal component x_2. (a) Find the horizontal component of the velocity of the center of mass of the glider-sphere system when the force is removed. (b) After the force is removed, the glider con-

tinues to move on the track and the sphere swings back and forth, both without friction. Find an expression for the largest angle the cord makes with the vertical.

49. ● Sand from a stationary hopper falls onto a moving conveyor belt at the rate of 5.00 kg/s as shown in Figure P9.49. The conveyor belt is supported by frictionless rollers. It moves at a constant speed of 0.750 m/s under the action of a constant horizontal external force $\vec{\mathbf{F}}_{ext}$ supplied by the motor that drives the belt. Find (a) the sand's rate of change of momentum in the horizontal direction, (b) the force of friction exerted by the belt on the sand, (c) the external force $\vec{\mathbf{F}}_{ext}$, (d) the work done by $\vec{\mathbf{F}}_{ext}$ in 1 s, and (e) the kinetic energy acquired by the falling sand each second due to the change in its horizontal motion. (f) Why are the answers to (d) and (e) different?

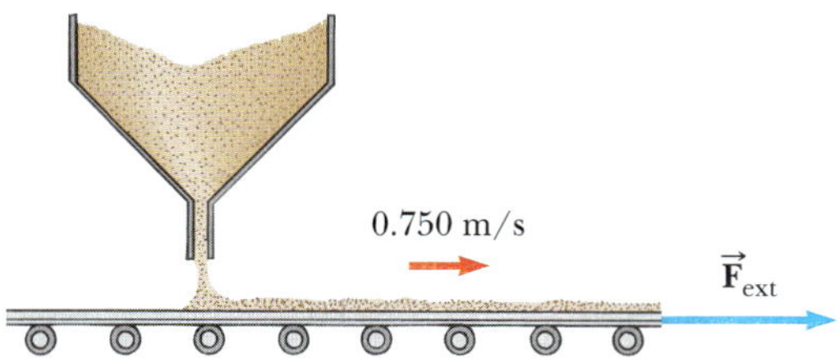

Figure P9.49

Section 9.8 Rocket Propulsion

50. Model rocket engines are sized by thrust, thrust duration, and total impulse, among other characteristics. A size C5 model rocket engine has an average thrust of 5.26 N, a fuel mass of 12.7 g, and an initial mass of 25.5 g. The duration of its burn is 1.90 s. (a) What is the average exhaust speed of the engine? (b) This engine is placed in a rocket body of mass 53.5 g. What is the final velocity of the rocket if it is fired in outer space? Assume the fuel burns at a constant rate.

51. ▲ The first stage of a Saturn V space vehicle consumed fuel and oxidizer at the rate of 1.50×10^4 kg/s, with an exhaust speed of 2.60×10^3 m/s. (a) Calculate the thrust produced by this engine. (b) Find the acceleration the vehicle had just as it lifted off the launch pad on the Earth, taking the vehicle's initial mass as 3.00×10^6 kg. *Note:* You must include the gravitational force to solve part (b).

52. *Rocket science.* A rocket has total mass $M_i = 360$ kg, including 330 kg of fuel and oxidizer. In interstellar space, it starts from rest at the position $x = 0$, turns on its engine at time $t = 0$, and puts out exhaust with relative speed $v_e =$ 1 500 m/s at the constant rate $k = 2.50$ kg/s. The fuel will last for an actual burn time of 330 kg/(2.5 kg/s) = 132 s, but define a "projected depletion time" as $T_p = M_i/k =$ 360 kg/(2.5 kg/s) = 144 s (which would be the burn time if the rocket could use its payload and fuel tanks, and even the walls of the combustion chamber as fuel). (a) Show that during the burn the velocity of the rocket as a function of time is given by

$$v(t) = -v_e \ln\left(1 - \frac{t}{T_p}\right)$$

(b) Make a graph of the velocity of the rocket as a function of time for times running from 0 to 132 s. (c) Show that the acceleration of the rocket is

$$a(t) = \frac{v_e}{T_p - t}$$

(d) Graph the acceleration as a function of time. (e) Show that the position of the rocket is

$$x(t) = v_e(T_p - t)\ln\left(1 - \frac{t}{T_p}\right) + v_e t$$

(f) Graph the position during the burn as a function of time.

53. ● A rocket for use in deep space is to be capable of boosting a total load (payload plus rocket frame and engine) of 3.00 metric tons to a speed of 10 000 m/s. (a) It has an engine and fuel designed to produce an exhaust speed of 2 000 m/s. How much fuel plus oxidizer is required? (b) If a different fuel and engine design could give an exhaust speed of 5 000 m/s, what amount of fuel and oxidizer would be required for the same task? This exhaust speed is 2.50 times higher than that in part (a). Explain why the required fuel mass is 2.50 times smaller, or larger than that, or still smaller.

Additional Problems

54. Two gliders are set in motion on an air track. A spring of force constant k is attached to the back end of the second glider. The first glider, of mass m_1, has velocity $\vec{\mathbf{v}}_1$, and the second glider, of mass m_2, moves more slowly, with velocity $\vec{\mathbf{v}}_2$, as shown in Figure P9.54. When m_1 collides with the spring attached to m_2 and compresses the spring to its maximum compression x_{max}, the velocity of the gliders is $\vec{\mathbf{v}}$. In terms of $\vec{\mathbf{v}}_1$, $\vec{\mathbf{v}}_2$, m_1, m_2, and k, find (a) the velocity $\vec{\mathbf{v}}$ at maximum compression, (b) the maximum compression x_{max}, and (c) the velocity of each glider after m_1 has lost contact with the spring.

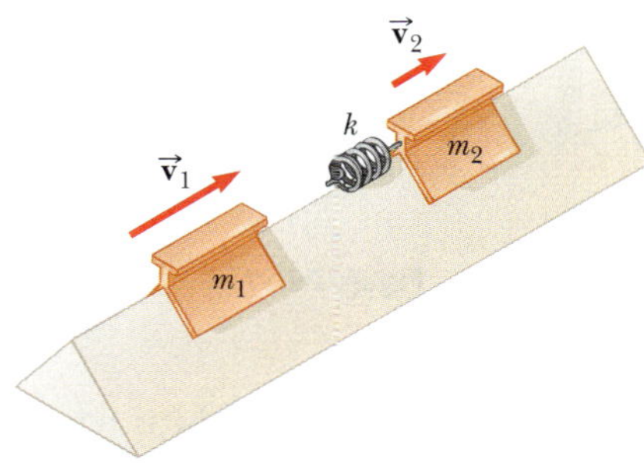

Figure P9.54

55. An 80.0-kg astronaut is taking a space walk to work on the engines of his ship, which is drifting through space with a constant velocity. The astronaut, wishing to get a better view of the Universe, pushes against the ship and much later finds himself 30.0 m behind the ship. Without a thruster or tether, the only way to return to the ship is to throw his 0.500-kg wrench directly away from the ship. If he throws the wrench with a speed of 20.0 m/s relative to the ship, after what time interval does the astronaut reach the ship?

56. ● An aging Hollywood actor (mass 80.0 kg) has been cloned, but the genetic replica is far from perfect. The clone has a different mass m, his stage presence is poor,

and he uses foul language. The clone, serving as the actor's stunt double, stands on the brink of a cliff 36.0 m high, next to a sturdy tree. The actor stands on top of a Humvee, 1.80 m above the level ground, holding a taut rope tied to a tree branch directly above the clone. When the director calls "action," the actor starts from rest and swings down on the rope without friction. The actor is momentarily hidden from the camera at the bottom of the arc, where he undergoes an elastic head-on collision with the clone, sending him over the cliff. Cursing vilely, the clone falls freely into the ocean below. The actor is prosecuted for making an obscene clone fall, and you are called as an expert witness at the sensational trial. (a) Find the horizontal component R of the clone's displacement as it depends on m. Evaluate R (b) for m = 79.0 kg and (c) for m = 81.0 kg. (d) What value of m gives a range of 30.0 m? (e) What is the maximum possible value for R, and (f) to what value of m does it correspond? What are (g) the minimum values of R and (h) the corresponding value of m? (i) For the actor–clone–Earth system, is mechanical energy conserved throughout the action sequence? Is this principle sufficient to solve the problem? Explain. (j) For the same system, is momentum conserved? Explain how this principle is used. (k) **What If?** Show that R does not depend on the value of the gravitational acceleration. Is this result remarkable? State how one might make sense of it.

57. A bullet of mass m is fired into a block of mass M initially at rest at the edge of a frictionless table of height h (Fig. P9.57). The bullet remains in the block, and after impact the block lands a distance d from the bottom of the table. Determine the initial speed of the bullet.

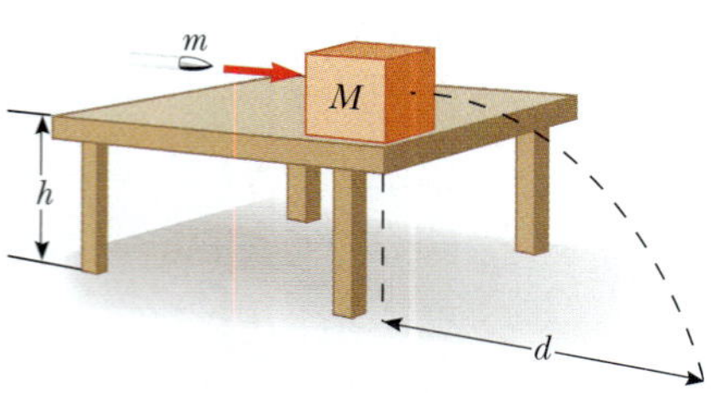

Figure P9.57

58. A small block of mass m_1 = 0.500 kg is released from rest at the top of a curve-shaped, frictionless wedge of mass m_2 = 3.00 kg, which sits on a frictionless horizontal surface as shown in Figure P9.58a. When the block leaves the wedge, its velocity is measured to be 4.00 m/s to the right as shown in the figure. (a) What is the velocity of the wedge after the block reaches the horizontal surface? (b) What is the height h of the wedge?

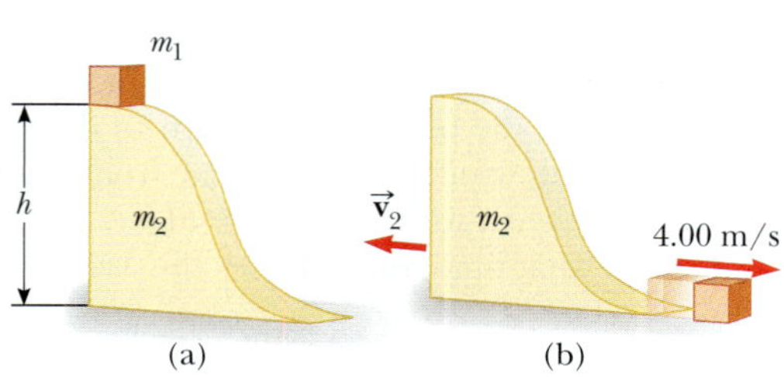

Figure P9.58

59. ● A 0.500-kg sphere moving with a velocity given by $(2.00\hat{\mathbf{i}} - 3.00\hat{\mathbf{j}} + 1.00\hat{\mathbf{k}})$ m/s strikes another sphere of mass 1.50 kg that is moving with an initial velocity of $(-1.00\hat{\mathbf{i}} + 2.00\hat{\mathbf{j}} - 3.00\hat{\mathbf{k}})$ m/s. (a) The velocity of the 0.500-kg sphere after the collision is given by $(-1.00\hat{\mathbf{i}} + 3.00\hat{\mathbf{j}} - 8.00\hat{\mathbf{k}})$ m/s. Find the final velocity of the 1.50-kg sphere and identify the kind of collision (elastic, inelastic, or perfectly inelastic). (b) Now assume the velocity of the 0.500-kg sphere after the collision is $(-0.250\hat{\mathbf{i}} + 0.750\hat{\mathbf{j}} - 2.00\hat{\mathbf{k}})$ m/s. Find the final velocity of the 1.50-kg sphere and identify the kind of collision. (c) **What If?** Take the velocity of the 0.500-kg sphere after the collision as $(-1.00\hat{\mathbf{i}} + 3.00\hat{\mathbf{j}} + a\hat{\mathbf{k}})$ m/s. Find the value of a and the velocity of the 1.50-kg sphere after an elastic collision.

60. A 75.0-kg firefighter slides down a pole while a constant friction force of 300 N retards her motion. A horizontal 20.0-kg platform is supported by a spring at the bottom of the pole to cushion the fall. The firefighter starts from rest 4.00 m above the platform, and the spring constant is 4 000 N/m. Find (a) the firefighter's speed immediately before she collides with the platform and (b) the maximum distance the spring is compressed. Assume the friction force acts during the entire motion.

61. ● George of the Jungle, with mass m, swings on a light vine hanging from a stationary tree branch. A second vine of equal length hangs from the same point, and a gorilla of larger mass M swings in the opposite direction on it. Both vines are horizontal when the primates start from rest at the same moment. George and the gorilla meet at the lowest point of their swings. Each is afraid that one vine will break, so they grab each other and hang on. They swing upward together, reaching a point where the vines make an angle of 35.0° with the vertical. (a) Find the value of the ratio m/M. (b) **What If?** Try the following experiment at home. Tie a small magnet and a steel screw to opposite ends of a string. Hold the center of the string fixed to represent the tree branch, and reproduce a model of the motions of George and the gorilla. What changes in your analysis will make it apply to this situation? **What If?** Next assume the magnet is strong so that it noticeably attracts the screw over a distance of a few centimeters. Then the screw will be moving faster immediately before it sticks to the magnet. Does this extra magnet strength make a difference?

62. ● A student performs a ballistic pendulum experiment using an apparatus similar to that shown in Figure 9.9b. She obtains the following average data: h = 8.68 cm, m_1 = 68.8 g, and m_2 = 263 g. The symbols refer to the quantities in Figure 9.9a. (a) Determine the initial speed v_{1A} of the projectile. (b) The second part of her experiment is to obtain v_{1A} by firing the same projectile horizontally (with the pendulum removed from the path) and measuring its final horizontal position x and distance of fall y (Fig. P9.62). Show that the initial speed of the projectile is related to x and y by the equation

$$v_{1A} = \frac{x}{\sqrt{2y/g}}$$

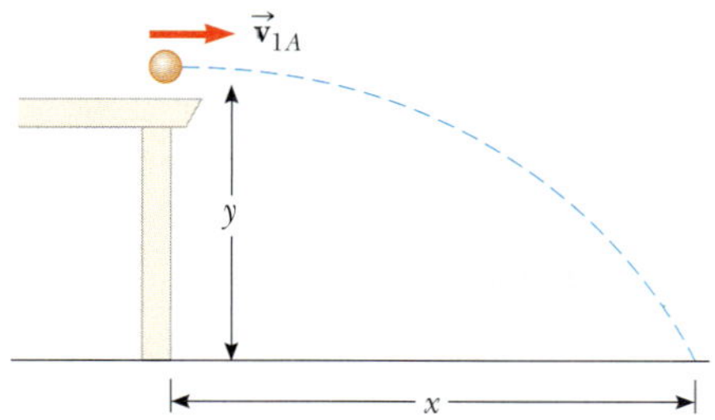

Figure P9.62

What numerical value does she obtain for v_{1A} based on her measured values of $x = 257$ cm and $y = 85.3$ cm? What factors might account for the difference in this value compared with that obtained in part (a)?

63. ● Lazarus Carnot, an artillery general, managed the military draft for Napoleon. Carnot used a ballistic pendulum to measure the firing speeds of cannonballs. In the symbols defined in Example 9.6, he proved that the ratio of the kinetic energy immediately after the collision to the kinetic energy immediately before is $m_1/(m_1+m_2)$. (a) Carry out the proof yourself. (b) If the cannonball has mass 9.60 kg and the block (a tree trunk) has mass 214 kg, what fraction of the original energy remains mechanical after the collision? (c) What is the ratio of the momentum immediately after the collision to the momentum immediately before? (d) A student believes that such a large loss of mechanical energy must be accompanied by at least a small loss of momentum. How would you convince this student of the truth? General Carnot's son Sadi was the second most important engineer in the history of ideas; we will study his work in Chapter 22.

64. ● Pursued by ferocious wolves, you are in a sleigh with no horses, gliding without friction across an ice-covered lake. You take an action described by these equations:

$$(270\text{ kg})(7.50\text{ m/s})\hat{\mathbf{i}} = (15.0\text{ kg})(-v_{1f}\hat{\mathbf{i}}) + (255\text{ kg})(v_{2f}\hat{\mathbf{i}})$$

$$v_{1f} + v_{2f} = 8.00\text{ m/s}$$

(a) Complete the statement of the problem, giving the data and identifying the unknowns. (b) Find the values of v_{1f} and v_{2f}. (c) Find the work you do.

65. **Review problem.** A light spring of force constant 3.85 N/m is compressed by 8.00 cm and held between a 0.250-kg block on the left and a 0.500-kg block on the right. Both blocks are at rest on a horizontal surface. The blocks are released simultaneously so that the spring tends to push them apart. Find the maximum velocity each block attains if the coefficient of kinetic friction between each block and the surface is (a) 0, (b) 0.100, and (c) 0.462. Assume the coefficient of static friction is greater than the coefficient of kinetic friction in every case.

66. Consider as a system the Sun with the Earth in a circular orbit around it. Find the magnitude of the change in the velocity of the Sun relative to the center of mass of the system over a 6-month period. Ignore the influence of other celestial objects. You may obtain the necessary astronomical data from the endpapers of the book.

67. A 5.00-g bullet moving with an initial speed of 400 m/s is fired into and passes through a 1.00-kg block as shown in Figure P9.67. The block, initially at rest on a frictionless, horizontal surface, is connected to a spring with force constant 900 N/m. The block moves 5.00 cm to the right after impact. Find (a) the speed at which the bullet emerges from the block and (b) the mechanical energy converted into internal energy in the collision.

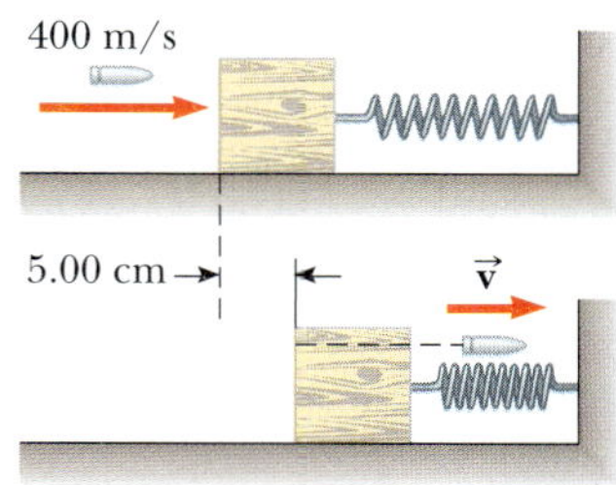

Figure P9.67

68. ● **Review problem.** There are (one can say) three coequal theories of motion: Newton's second law, stating that the total force on a particle causes its acceleration; the work–kinetic energy theorem, stating that the total work on a particle causes its change in kinetic energy; and the impulse–momentum theorem, stating that the total impulse on a particle causes its change in momentum. In this problem, you compare predictions of the three theories in one particular case. A 3.00-kg object has velocity $7.00\hat{\mathbf{j}}$ m/s. Then, a total force $12.0\hat{\mathbf{i}}$ N acts on the object for 5.00 s. (a) Calculate the object's final velocity, using the impulse–momentum theorem. (b) Calculate its acceleration from $\vec{\mathbf{a}} = (\vec{\mathbf{v}}_f - \vec{\mathbf{v}}_i)/\Delta t$. (c) Calculate its acceleration from $\vec{\mathbf{a}} = \sum \vec{\mathbf{F}}/m$. (d) Find the object's vector displacement from $\Delta\vec{\mathbf{r}} = \vec{\mathbf{v}}_i t + \frac{1}{2}\vec{\mathbf{a}}t^2$. (e) Find the work done on the object from $W = \vec{\mathbf{F}}\cdot\Delta\vec{\mathbf{r}}$. (f) Find the final kinetic energy from $\frac{1}{2}mv_f^2 = \frac{1}{2}m\vec{\mathbf{v}}_f\cdot\vec{\mathbf{v}}_f$. (g) Find the final kinetic energy from $\frac{1}{2}mv_i^2 + W$. (h) State the result of comparing the answers to parts b and c, and the answers to parts f and g.

69. A chain of length L and total mass M is released from rest with its lower end just touching the top of a table as shown in Figure P9.69a. Find the force exerted by the table on the chain after the chain has fallen through a distance x as shown in Figure P9.69b. (Assume each link comes to rest the instant it reaches the table.)

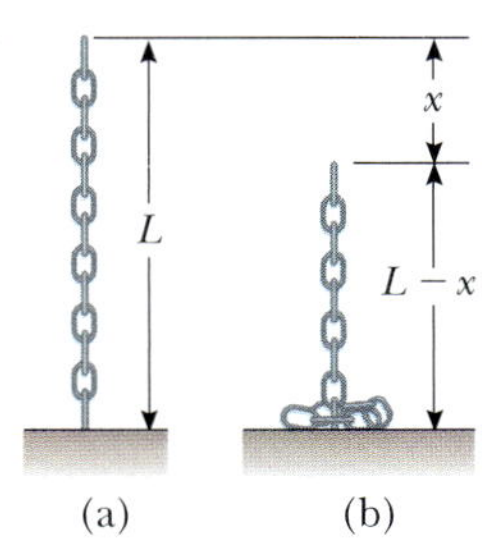

Figure P9.69

Answers to Quick Quizzes

9.1 (d). Two identical objects ($m_1 = m_2$) traveling at the same speed ($v_1 = v_2$) have the same kinetic energies and the same magnitudes of momentum. It also is possible, however, for particular combinations of masses and velocities to satisfy $K_1 = K_2$ but not $p_1 = p_2$. For example, a 1-kg object moving at 2 m/s has the same kinetic energy as a 4-kg object moving at 1 m/s, but the two clearly do not have the same momenta. Because we have no information about masses and speeds, we cannot choose among (a), (b), or (c).

9.2 (b), (c), (a). The slower the ball, the easier it is to catch. If the momentum of the medicine ball is the same as the momentum of the baseball, the speed of the medicine ball must be 1/10 the speed of the baseball because the medicine ball has 10 times the mass. If the kinetic energies are the same, the speed of the medicine ball must be $1/\sqrt{10}$ the speed of the baseball because of the squared speed term in the equation for K. The medicine ball is hardest to catch when it has the same speed as the baseball.

9.3 **(i),** (c), (e). Object 2 has a greater acceleration because of its smaller mass. Therefore, it travels the distance d in a shorter time interval. Even though the force applied to objects 1 and 2 is the same, the change in momentum is less for object 2 because Δt is smaller. The work $W = Fd$ done on both objects is the same because both F and d are the same in the two cases. Therefore, $K_1 = K_2$. **(ii),** (b), (d). The same impulse is applied to both objects, so they experience the same change in momentum. Object 2 has a larger acceleration due to its smaller mass. Therefore, the distance that object 2 covers in the time interval is larger than that for object 1. As a result, more work is done on object 2 and $K_2 > K_1$.

9.4 (a) All three are the same. Because the passenger is brought from the car's initial speed to a full stop, the change in momentum (equal to the impulse) is the same regardless of what stops the passenger. (b) Dashboard, seat belt, air bag. The dashboard stops the passenger very quickly in a front-end collision, resulting in a very large force. The seat belt takes somewhat more time, so the force is smaller. Used along with the seat belt, the air bag can extend the passenger's stopping time further, notably for his head, which would otherwise snap forward.

9.5 (a). If all the initial kinetic energy is transformed or transferred away from the system, nothing is moving after the collision. Consequently, the final momentum of the system is necessarily zero and the initial momentum of the system must therefore be zero. Although (b) and (d) *together* would satisfy the conditions, neither one *alone* does.

9.6 (b). Because momentum of the two-ball system is conserved, $\vec{\mathbf{p}}_{Ti} + 0 = \vec{\mathbf{p}}_{Tf} + \vec{\mathbf{p}}_{B}$. Because the table-tennis ball bounces back from the much more massive bowling ball with approximately the same speed, $\vec{\mathbf{p}}_{Tf} = -\vec{\mathbf{p}}_{Ti}$. As a consequence, $\vec{\mathbf{p}}_{B} = 2\vec{\mathbf{p}}_{Ti}$. Kinetic energy can be expressed as $K = p^2/2m$. Because of the much larger mass of the bowling ball, its kinetic energy is much smaller than that of the table-tennis ball.

9.7 (b). The piece with the handle will have less mass than the piece made up of the end of the bat. To see why, take the origin of coordinates as the center of mass before the bat was cut. Replace each cut piece by a small sphere located at each piece's center of mass. The sphere representing the handle piece is farther from the origin, but the product of less mass and greater distance balances the product of greater mass and less distance for the end piece as shown.

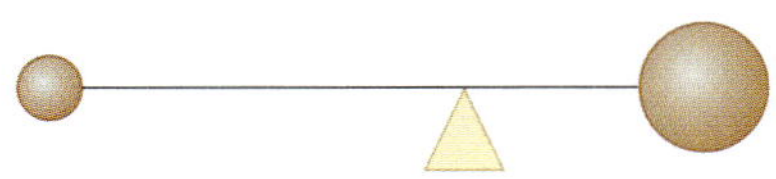

9.8 **(i),** (a). This effect is the same one as the swimmer diving off the raft that we just discussed. The vessel–passengers system is isolated. If the passengers all start running one way, the speed of the vessel increases (a small amount!) the other way. **(ii),** (b). Once they stop running, the momentum of the system is the same as it was before they started running; you cannot change the momentum of an isolated system by means of internal forces. In case you are thinking that the passengers could run to the stern repeatedly to take advantage of the speed increase *while* they are running, remember that they will slow the ship down every time they return to the bow!

The Malaysian pastime of *gasing* involves the spinning of tops that can have masses up to 5 kg. Professional spinners can spin their tops so that they might rotate for more than an hour before stopping. We will study the rotational motion of objects such as these tops in this chapter. (Courtesy Tourism Malaysia)

10 Rotation of a Rigid Object About a Fixed Axis

When an extended object such as a wheel rotates about its axis, the motion cannot be analyzed by modeling the object as a particle because at any given time different parts of the object have different linear velocities and linear accelerations. We can, however, analyze the motion of an extended object by modeling it as a collection of particles, each of which has its own linear velocity and linear acceleration.

In dealing with a rotating object, analysis is greatly simplified by assuming the object is rigid. A **rigid object** is one that is nondeformable; that is, the relative locations of all particles of which the object is composed remain constant. All real objects are deformable to some extent; our rigid-object model, however, is useful in many situations in which deformation is negligible.

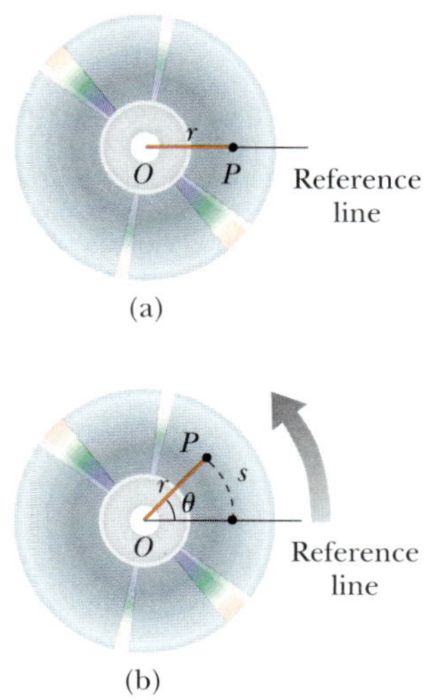

Figure 10.1 A compact disc rotating about a fixed axis through O perpendicular to the plane of the figure. (a) To define angular position for the disc, a fixed reference line is chosen. A particle at P is located at a distance r from the rotation axis at O. (b) As the disc rotates, a particle at P moves through an arc length s on a circular path of radius r.

10.1 Angular Position, Velocity, and Acceleration

Figure 10.1 illustrates an overhead view of a rotating compact disc, or CD. The disc rotates about a fixed axis perpendicular to the plane of the figure and passing through the center of the disc at O. A small element of the disc modeled as a particle at P is at a fixed distance r from the origin and rotates about it in a circle of radius r. (In fact, *every* particle on the disc undergoes circular motion about O.) It is convenient to represent the position of P with its polar coordinates (r, θ), where

r is the distance from the origin to P and θ is measured *counterclockwise* from some reference line fixed in space as shown in Figure 10.1a. In this representation, the angle θ changes in time while r remains constant. As the particle moves along the circle from the reference line, which is at angle $\theta = 0$, it moves through an arc of length s as in Figure 10.1b. The arc length s is related to the angle θ through the relationship

$$s = r\theta \quad \textbf{(10.1a)}$$

$$\theta = \frac{s}{r} \quad \textbf{(10.1b)}$$

PITFALL PREVENTION 10.1
Remember the Radian

In rotational equations, you *must* use angles expressed in radians. Don't fall into the trap of using angles measured in degrees in rotational equations.

Because θ is the ratio of an arc length and the radius of the circle, it is a pure number. Usually, however, we give θ the artificial unit **radian** (rad), where **one radian is the angle subtended by an arc length equal to the radius of the arc.** Because the circumference of a circle is $2\pi r$, it follows from Equation 10.1b that $360°$ corresponds to an angle of $(2\pi r/r)$ rad $= 2\pi$ rad. Hence, 1 rad $= 360°/2\pi \approx 57.3°$. To convert an angle in degrees to an angle in radians, we use that π rad $= 180°$, so

$$\theta(\text{rad}) = \frac{\pi}{180°}\theta(\text{deg})$$

For example, $60°$ equals $\pi/3$ rad and $45°$ equals $\pi/4$ rad.

Because the disc in Figure 10.1 is a rigid object, as the particle moves through an angle θ from the reference line, every other particle on the object rotates through the same angle θ. Therefore, **we can associate the angle θ with the entire rigid object as well as with an individual particle,** which allows us to define the *angular position* of a rigid object in its rotational motion. We choose a reference line on the object, such as a line connecting O and a chosen particle on the object. The **angular position** of the rigid object is the angle θ between this reference line on the object and the fixed reference line in space, which is often chosen as the x axis. Such identification is similar to the way we define the position of an object in translational motion as the distance x between the object and the reference position, which is the origin, $x = 0$.

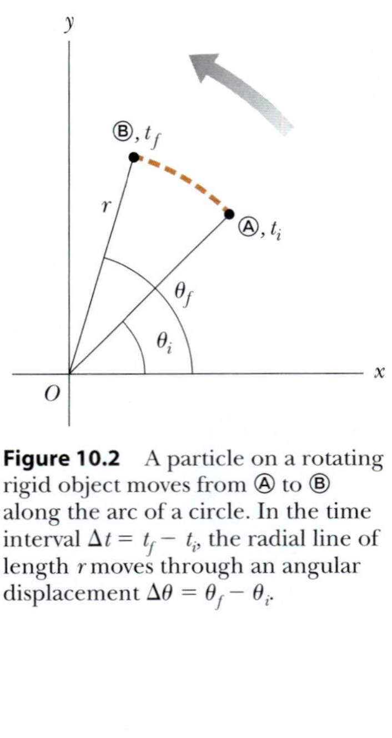

Figure 10.2 A particle on a rotating rigid object moves from Ⓐ to Ⓑ along the arc of a circle. In the time interval $\Delta t = t_f - t_i$, the radial line of length r moves through an angular displacement $\Delta\theta = \theta_f - \theta_i$.

As the particle in question on our rigid object travels from position Ⓐ to position Ⓑ in a time interval Δt as in Figure 10.2, the reference line fixed to the object sweeps out an angle $\Delta\theta = \theta_f - \theta_i$. This quantity $\Delta\theta$ is defined as the **angular displacement** of the rigid object:

$$\Delta\theta \equiv \theta_f - \theta_i$$

The rate at which this angular displacement occurs can vary. If the rigid object spins rapidly, this displacement can occur in a short time interval. If it rotates slowly, this displacement occurs in a longer time interval. These different rotation rates can be quantified by defining the **average angular speed** ω_{avg} (Greek letter omega) as the ratio of the angular displacement of a rigid object to the time interval Δt during which the displacement occurs:

Average angular speed ▶

$$\omega_{\text{avg}} \equiv \frac{\theta_f - \theta_i}{t_f - t_i} = \frac{\Delta\theta}{\Delta t} \quad \textbf{(10.2)}$$

In analogy to linear speed, the **instantaneous angular speed** ω is defined as the limit of the average angular speed as Δt approaches zero:

Instantaneous angular speed ▶

$$\omega \equiv \lim_{\Delta t \to 0} \frac{\Delta\theta}{\Delta t} = \frac{d\theta}{dt} \quad \textbf{(10.3)}$$

Angular speed has units of radians per second (rad/s), which can be written as s^{-1} because radians are not dimensional. We take ω to be positive when θ is increasing (counterclockwise motion in Figure 10.2) and negative when θ is decreasing (clockwise motion in Figure 10.2).

Quick Quiz 10.1 A rigid object rotates in a counterclockwise sense around a fixed axis. Each of the following pairs of quantities represents an initial angular position and a final angular position of the rigid object. **(i)** Which of the sets can *only* occur if the rigid object rotates through more than 180°? (a) 3 rad, 6 rad (b) −1 rad, 1 rad (c) 1 rad, 5 rad **(ii)** Suppose the change in angular position for each of these pairs of values occurs in 1 s. Which choice represents the lowest average angular speed?

If the instantaneous angular speed of an object changes from ω_i to ω_f in the time interval Δt, the object has an angular acceleration. The **average angular acceleration** α_{avg} (Greek letter alpha) of a rotating rigid object is defined as the ratio of the change in the angular speed to the time interval Δt during which the change in the angular speed occurs:

$$\alpha_{avg} \equiv \frac{\omega_f - \omega_i}{t_f - t_i} = \frac{\Delta\omega}{\Delta t} \quad (10.4)$$

◀ Average angular acceleration

In analogy to linear acceleration, the **instantaneous angular acceleration** is defined as the limit of the average angular acceleration as Δt approaches zero:

$$\alpha \equiv \lim_{\Delta t \to 0} \frac{\Delta\omega}{\Delta t} = \frac{d\omega}{dt} \quad (10.5)$$

◀ Instantaneous angular acceleration

Angular acceleration has units of radians per second squared (rad/s^2), or simply s^{-2}. Notice that α is positive when a rigid object rotating counterclockwise is speeding up or when a rigid object rotating clockwise is slowing down during some time interval.

When a rigid object is rotating about a *fixed* axis, **every particle on the object rotates through the same angle in a given time interval and has the same angular speed and the same angular acceleration.** That is, the quantities θ, ω, and α characterize the rotational motion of the entire rigid object as well as individual particles in the object.

Angular position (θ), angular speed (ω), and angular acceleration (α) are analogous to translational position (x), translational speed (v), and translational acceleration (a). The variables θ, ω, and α differ dimensionally from the variables x, v, and a only by a factor having the unit of length. (See Section 10.3.)

We have not specified any direction for angular speed and angular acceleration. Strictly speaking, ω and α are the magnitudes of the angular velocity and the angular acceleration vectors[1] $\vec{\boldsymbol{\omega}}$ and $\vec{\boldsymbol{\alpha}}$, respectively, and they should always be positive. Because we are considering rotation about a fixed axis, however, we can use nonvector notation and indicate the vectors' directions by assigning a positive or negative sign to ω and α as discussed earlier with regard to Equations 10.3 and 10.5. For rotation about a fixed axis, the only direction that uniquely specifies the rotational motion is the direction along the axis of rotation. Therefore, the directions of $\vec{\boldsymbol{\omega}}$ and $\vec{\boldsymbol{\alpha}}$ are along this axis. If a particle rotates in the xy plane as in Figure 10.2, the direction of $\vec{\boldsymbol{\omega}}$ for the particle is out of the plane of the diagram when the rotation is counterclockwise and into the plane of the diagram when the rotation is clockwise. To illustrate this convention, it is convenient to use the *right-hand rule* demonstrated in Figure 10.3. When the four fingers of the right hand are wrapped in the direction of rotation, the extended right thumb points in the direction of $\vec{\boldsymbol{\omega}}$. The direction of $\vec{\boldsymbol{\alpha}}$ follows from its definition $\vec{\boldsymbol{\alpha}} \equiv d\vec{\boldsymbol{\omega}}/dt$. It is in the same direction as $\vec{\boldsymbol{\omega}}$ if the angular speed is increasing in time, and it is antiparallel to $\vec{\boldsymbol{\omega}}$ if the angular speed is decreasing in time.

PITFALL PREVENTION 10.2

Specify Your Axis

In solving rotation problems, you must specify an axis of rotation. This new feature does not exist in our study of translational motion. The choice is arbitrary, but once you make it, you must maintain that choice consistently throughout the problem. In some problems, the physical situation suggests a natural axis, such as the center of an automobile wheel. In other problems, there may not be an obvious choice, and you must exercise judgment.

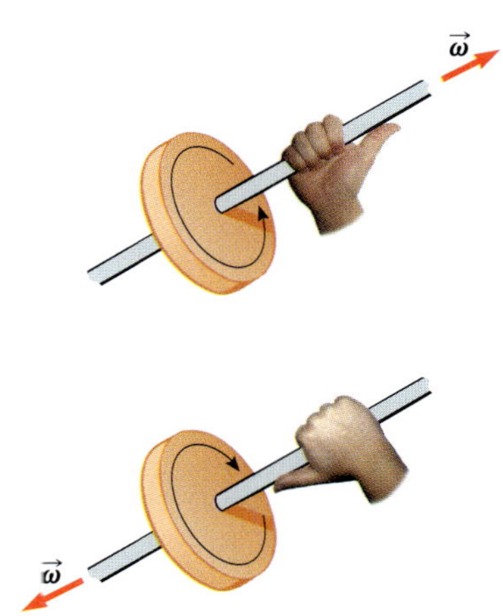

Figure 10.3 The right-hand rule for determining the direction of the angular velocity vector.

[1] Although we do not verify it here, the instantaneous angular velocity and instantaneous angular acceleration are vector quantities, but the corresponding average values are not because angular displacements do not add as vector quantities for finite rotations.

10.2 Rotational Kinematics: The Rigid Object Under Constant Angular Acceleration

When a rigid object rotates about a fixed axis, it often undergoes a constant angular acceleration. Therefore, we generate a new analysis model for rotational motion called the **rigid object under constant angular acceleration**. This model is the rotational analog to the particle under constant acceleration model. We develop kinematic relationships for this model in this section. Writing Equation 10.5 in the form $d\omega = \alpha\, dt$ and integrating from $t_i = 0$ to $t_f = t$ gives

Rotational kinematic equations ▶

$$\omega_f = \omega_i + \alpha t \quad \text{(for constant } \alpha\text{)} \tag{10.6}$$

where ω_i is the angular speed of the rigid object at time $t = 0$. Equation 10.6 allows us to find the angular speed ω_f of the object at any later time t. Substituting Equation 10.6 into Equation 10.3 and integrating once more, we obtain

$$\theta_f = \theta_i + \omega_i t + \tfrac{1}{2}\alpha t^2 \quad \text{(for constant } \alpha\text{)} \tag{10.7}$$

where θ_i is the angular position of the rigid object at time $t = 0$. Equation 10.7 allows us to find the angular position θ_f of the object at any later time t. Eliminating t from Equations 10.6 and 10.7 gives

$$\omega_f^2 = \omega_i^2 + 2\alpha(\theta_f - \theta_i) \quad \text{(for constant } \alpha\text{)} \tag{10.8}$$

This equation allows us to find the angular speed ω_f of the rigid object for any value of its angular position θ_f. If we eliminate α between Equations 10.6 and 10.7, we obtain

$$\theta_f = \theta_i + \tfrac{1}{2}(\omega_i + \omega_f)t \quad \text{(for constant } \alpha\text{)} \tag{10.9}$$

Notice that these kinematic expressions for the rigid object under constant angular acceleration are of the same mathematical form as those for a particle under constant acceleration (Chapter 2). They can be generated from the equations for translational motion by making the substitutions $x \rightarrow \theta$, $v \rightarrow \omega$, and $a \rightarrow \alpha$. Table 10.1 compares the kinematic equations for rotational and translational motion.

PITFALL PREVENTION 10.3
Just Like Translation?

Equations 10.6 to 10.9 and Table 10.1 suggest that rotational kinematics is just like translational kinematics. That is almost true, with two key differences. (1) In rotational kinematics, you must specify a rotation axis (per Pitfall Prevention 10.2). (2) In rotational motion, the object keeps returning to its original orientation; therefore, you may be asked for the number of revolutions made by a rigid object. This concept has no meaning in translational motion.

Quick Quiz 10.2 Consider again the pairs of angular positions for the rigid object in Quick Quiz 10.1. If the object starts from rest at the initial angular position, moves counterclockwise with constant angular acceleration, and arrives at the final angular position with the same angular speed in all three cases, for which choice is the angular acceleration the highest?

TABLE 10.1

Kinematic Equations for Rotational and Translational Motion Under Constant Acceleration

Rotational Motion About a Fixed Axis	Translational Motion
$\omega_f = \omega_i + \alpha t$	$v_f = v_i + at$
$\theta_f = \theta_i + \omega_i t + \frac{1}{2}\alpha t^2$	$x_f = x_i + v_i t + \frac{1}{2}at^2$
$\omega_f^2 = \omega_i^2 + 2\alpha(\theta_f - \theta_i)$	$v_f^2 = v_i^2 + 2a(x_f - x_i)$
$\theta_f = \theta_i + \frac{1}{2}(\omega_i + \omega_f)t$	$x_f = x_i + \frac{1}{2}(v_i + v_f)t$

EXAMPLE 10.1 Rotating Wheel

A wheel rotates with a constant angular acceleration of 3.50 rad/s^2.

(A) If the angular speed of the wheel is 2.00 rad/s at $t_i = 0$, through what angular displacement does the wheel rotate in 2.00 s?

SOLUTION

Conceptualize Look again at Figure 10.1. Imagine that the compact disc rotates with its angular speed increasing at a constant rate. You start your stopwatch when the disc is rotating at 2.00 rad/s. This mental image is a model for the motion of the wheel in this example.

Categorize The phrase "with a constant angular acceleration" tells us to use the rigid object under constant angular acceleration model.

Analyze Arrange Equation 10.7 so that it expresses the angular displacement of the object:

$$\Delta\theta = \theta_f - \theta_i = \omega_i t + \tfrac{1}{2}\alpha t^2$$

Substitute the known values to find the angular displacement at $t = 2.00$ s:

$$\Delta\theta = (2.00 \text{ rad/s})(2.00 \text{ s}) + \tfrac{1}{2}(3.50 \text{ rad/s}^2)(2.00 \text{ s})^2$$

$$= \boxed{11.0 \text{ rad}} = (11.0 \text{ rad})(57.3°/\text{rad}) = \boxed{630°}$$

(B) Through how many revolutions has the wheel turned during this time interval?

SOLUTION

Multiply the angular displacement found in part (A) by a conversion factor to find the number of revolutions:

$$\Delta\theta = 630°\left(\frac{1 \text{ rev}}{360°}\right) = \boxed{1.75 \text{ rev}}$$

(C) What is the angular speed of the wheel at $t = 2.00$ s?

SOLUTION

Use Equation 10.6 to find the angular speed at $t = 2.00$ s:

$$\omega_f = \omega_i + \alpha t = 2.00 \text{ rad/s} + (3.50 \text{ rad/s}^2)(2.00 \text{ s})$$

$$= \boxed{9.00 \text{ rad/s}}$$

Finalize We could also obtain this result using Equation 10.8 and the results of part (A). (Try it!)

What If? Suppose a particle moves along a straight line with a constant acceleration of 3.50 m/s^2. If the velocity of the particle is 2.00 m/s at $t_i = 0$, through what displacement does the particle move in 2.00 s? What is the velocity of the particle at $t = 2.00$ s?

Answer Notice that these questions are translational analogs to parts (A) and (C) of the original problem. The mathematical solution follows exactly the same form. For the displacement,

$$\Delta x = x_f - x_i = v_i t + \tfrac{1}{2}at^2$$

$$= (2.00 \text{ m/s})(2.00 \text{ s}) + \tfrac{1}{2}(3.50 \text{ m/s}^2)(2.00 \text{ s})^2 = 11.0 \text{ m}$$

and for the velocity,

$$v_f = v_i + at = 2.00 \text{ m/s} + (3.50 \text{ m/s}^2)(2.00 \text{ s}) = 9.00 \text{ m/s}$$

There is no translational analog to part (B) because translational motion under constant acceleration is not repetitive.

10.3 Angular and Translational Quantities

In this section, we derive some useful relationships between the angular speed and acceleration of a rotating rigid object and the translational speed and acceleration of a point in the object. To do so, we must keep in mind that when a rigid object

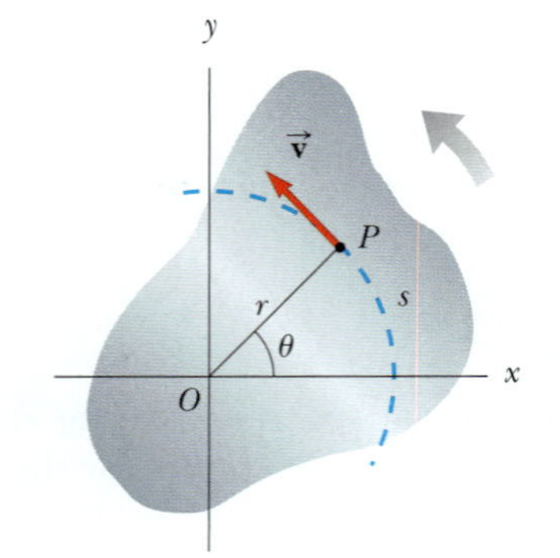

ACTIVE FIGURE 10.4

As a rigid object rotates about the fixed axis through O, the point P has a tangential velocity $\vec{\mathbf{v}}$ that is always tangent to the circular path of radius r.

Sign in at www.thomsonedu.com and go to ThomsonNOW to move point P and observe the tangential velocity as the object rotates.

rotates about a fixed axis as in Active Figure 10.4, **every particle of the object moves in a circle whose center is on the axis of rotation.**

Because point P in Active Figure 10.4 moves in a circle, the translational velocity vector $\vec{\mathbf{v}}$ is always tangent to the circular path and hence is called *tangential velocity.* The magnitude of the tangential velocity of the point P is by definition the tangential speed $v = ds/dt$, where s is the distance traveled by this point measured along the circular path. Recalling that $s = r\theta$ (Eq. 10.1a) and noting that r is constant, we obtain

$$v = \frac{ds}{dt} = r\frac{d\theta}{dt}$$

Because $d\theta/dt = \omega$ (see Eq. 10.3), it follows that

$$v = r\omega \tag{10.10}$$

That is, the tangential speed of a point on a rotating rigid object equals the perpendicular distance of that point from the axis of rotation multiplied by the angular speed. Therefore, although every point on the rigid object has the same *angular* speed, not every point has the same *tangential* speed because r is not the same for all points on the object. Equation 10.10 shows that the tangential speed of a point on the rotating object increases as one moves outward from the center of rotation, as we would intuitively expect. For example, the outer end of a swinging golf club moves much faster than the handle.

We can relate the angular acceleration of the rotating rigid object to the tangential acceleration of the point P by taking the time derivative of v:

$$a_t = \frac{dv}{dt} = r\frac{d\omega}{dt}$$

Relation between tangential and angular acceleration ▶

$$a_t = r\alpha \tag{10.11}$$

That is, the tangential component of the translational acceleration of a point on a rotating rigid object equals the point's perpendicular distance from the axis of rotation multiplied by the angular acceleration.

In Section 4.4, we found that a point moving in a circular path undergoes a radial acceleration a_r directed toward the center of rotation and whose magnitude is that of the centripetal acceleration v^2/r (Fig. 10.5). Because $v = r\omega$ for a point P on a rotating object, we can express the centripetal acceleration at that point in terms of angular speed as

$$a_c = \frac{v^2}{r} = r\omega^2 \tag{10.12}$$

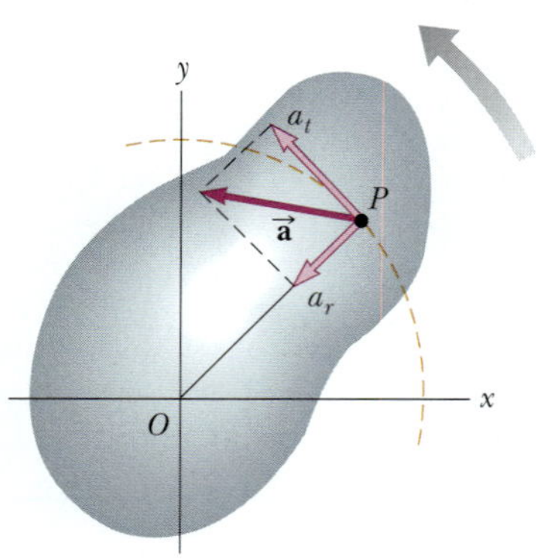

Figure 10.5 As a rigid object rotates about a fixed axis through O, the point P experiences a tangential component of translational acceleration a_t and a radial component of translational acceleration a_r. The total acceleration of this point is $\vec{\mathbf{a}} = \vec{\mathbf{a}}_t + \vec{\mathbf{a}}_r$.

The total acceleration vector at the point is $\vec{\mathbf{a}} = \vec{\mathbf{a}}_t + \vec{\mathbf{a}}_r$, where the magnitude of $\vec{\mathbf{a}}_r$ is the centripetal acceleration a_c. Because $\vec{\mathbf{a}}$ is a vector having a radial and a tangential component, the magnitude of $\vec{\mathbf{a}}$ at the point P on the rotating rigid object is

$$a = \sqrt{a_t^2 + a_r^2} = \sqrt{r^2\alpha^2 + r^2\omega^4} = r\sqrt{\alpha^2 + \omega^4} \tag{10.13}$$

Quick Quiz 10.3 Alex and Brian are riding on a merry-go-round. Alex rides on a horse at the outer rim of the circular platform, twice as far from the center of the circular platform as Brian, who rides on an inner horse. **(i)** When the merry-go-round is rotating at a constant angular speed, what is Alex's angular speed? (a) twice Brian's (b) the same as Brian's (c) half of Brian's (d) impossible to determine **(ii)** When the merry-go-round is rotating at a constant angular speed, describe Alex's tangential speed from the same list of choices.

EXAMPLE 10.2 CD Player

On a compact disc (Fig. 10.6), audio information is stored digitally in a series of pits and flat areas on the surface of the disc. The alternations between pits and flat areas on the surface represent binary ones and zeroes to be read by the CD player and converted back to sound waves. The pits and flat areas are detected by a system consisting of a laser and lenses. The length of a string of ones and zeroes representing one piece of information is the same everywhere on the disc, whether the information is near the center of the disc or near its outer edge. So that this length of ones and zeroes always passes by the laser–lens system in the same time interval, the tangential speed of the disc surface at the location of the lens must be constant. According to Equation 10.10, the angular speed must therefore vary as the laser–lens system moves radially along the disc. In a typical CD player, the constant speed of the surface at the point of the laser–lens system is 1.3 m/s.

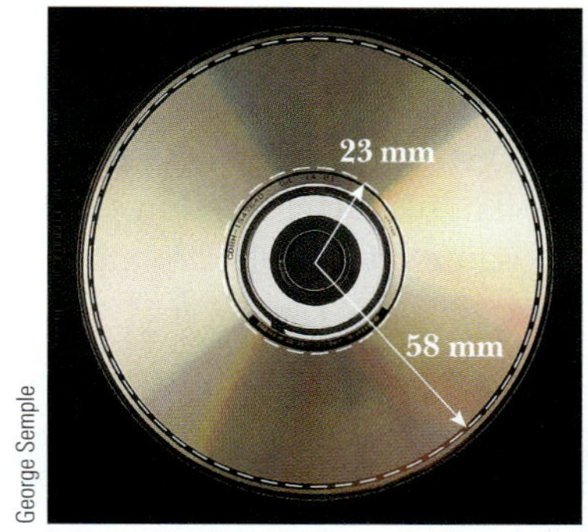

Figure 10.6 (Example 10.2) A compact disc.

(A) Find the angular speed of the disc in revolutions per minute when information is being read from the innermost first track (r = 23 mm) and the outermost final track (r = 58 mm).

SOLUTION

Conceptualize Figure 10.6 shows a photograph of a compact disc. Trace your finger around the circle marked "23 mm" in a time interval of about 3 s. Now trace your finger around the circle marked "58 mm" in the same time interval. Notice how much faster your finger is moving relative to the page around the larger circle. If your finger represents the laser reading the disc, it is moving over the surface of the disc much faster for the outer circle than for the inner circle.

Categorize This part of the example is categorized as a simple substitution problem. In later parts, we will need to identify analysis models.

Use Equation 10.10 to find the angular speed that gives the required tangential speed at the position of the inner track:

$$\omega_i = \frac{v}{r_i} = \frac{1.3 \text{ m/s}}{2.3 \times 10^{-2} \text{ m}} = 57 \text{ rad/s}$$

$$= (57 \text{ rad/s})\left(\frac{1 \text{ rev}}{2\pi \text{ rad}}\right)\left(\frac{60 \text{ s}}{1 \text{ min}}\right) = 5.4 \times 10^2 \text{ rev/min}$$

Do the same for the outer track:

$$\omega_f = \frac{v}{r_f} = \frac{1.3 \text{ m/s}}{5.8 \times 10^{-2} \text{ m}} = 22 \text{ rad/s} = 2.1 \times 10^2 \text{ rev/min}$$

The CD player adjusts the angular speed ω of the disc within this range so that information moves past the objective lens at a constant rate.

(B) The maximum playing time of a standard music disc is 74 min and 33 s. How many revolutions does the disc make during that time?

SOLUTION

Categorize From part (A), the angular speed decreases as the disc plays. Let us assume it decreases steadily, with α constant. We can then use the rigid object under constant angular acceleration model.

Analyze If $t = 0$ is the instant the disc begins rotating, with angular speed of 57 rad/s, the final value of the time t is (74 min)(60 s/min) + 33 s = 4 473 s. We are looking for the angular displacement $\Delta\theta$ during this time interval.

Use Equation 10.9 to find the angular displacement of the disc at t = 4 473 s:

$$\Delta\theta = \theta_f - \theta_i = \tfrac{1}{2}(\omega_i + \omega_f)t$$

$$= \tfrac{1}{2}(57 \text{ rad/s} + 22 \text{ rad/s})(4\,473 \text{ s}) = 1.8 \times 10^5 \text{ rad}$$

Convert this angular displacement to revolutions:

$$\Delta\theta = (1.8 \times 10^5 \text{ rad})\left(\frac{1 \text{ rev}}{2\pi \text{ rad}}\right) = 2.8 \times 10^4 \text{ rev}$$

(C) What is the angular acceleration of the compact disc over the 4 473-s time interval?

SOLUTION

Categorize We again model the disc as a rigid object under constant angular acceleration. In this case, Equation 10.6 gives the value of the constant angular acceleration. Another approach is to use Equation 10.4 to find the average angular acceleration. In this case, we are not assuming that the angular acceleration is constant. The answer is the same from both equations; only the interpretation of the result is different.

Analyze Use Equation 10.6 to find the angular acceleration:

$$\alpha = \frac{\omega_f - \omega_i}{t} = \frac{22 \text{ rad/s} - 57 \text{ rad/s}}{4\,473 \text{ s}} = -7.8 \times 10^{-3} \text{ rad/s}^2$$

Finalize The disc experiences a very gradual decrease in its rotation rate, as expected from the long time interval required for the angular speed to change from the initial value to the final value. In reality, the angular acceleration of the disc is not constant. Problem 20 allows you to explore the actual time behavior of the angular acceleration.

10.4 Rotational Kinetic Energy

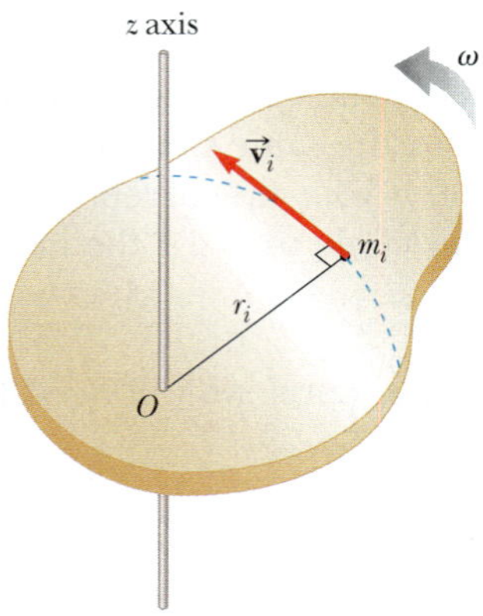

Figure 10.7 A rigid object rotating about the z axis with angular speed ω. The kinetic energy of the particle of mass m_i is $\frac{1}{2}m_i v_i^2$. The total kinetic energy of the object is called its rotational kinetic energy.

In Chapter 7, we defined the kinetic energy of an object as the energy associated with its motion through space. An object rotating about a fixed axis remains stationary in space, so there is no kinetic energy associated with translational motion. The individual particles making up the rotating object, however, are moving through space; they follow circular paths. Consequently, there is kinetic energy associated with rotational motion.

Let us consider an object as a collection of particles and assume it rotates about a fixed z axis with an angular speed ω. Figure 10.7 shows the rotating object and identifies one particle on the object located at a distance r_i from the rotation axis. If the mass of the ith particle is m_i and its tangential speed is v_i, its kinetic energy is

$$K_i = \tfrac{1}{2}m_i v_i^2$$

To proceed further, recall that although every particle in the rigid object has the same angular speed ω, the individual tangential speeds depend on the distance r_i from the axis of rotation according to Equation 10.10. The *total* kinetic energy of the rotating rigid object is the sum of the kinetic energies of the individual particles:

$$K_R = \sum_i K_i = \sum_i \tfrac{1}{2}m_i v_i^2 = \tfrac{1}{2}\sum_i m_i r_i^2 \omega^2$$

We can write this expression in the form

$$K_R = \tfrac{1}{2}\left(\sum_i m_i r_i^2\right)\omega^2 \qquad \textbf{(10.14)}$$

where we have factored ω^2 from the sum because it is common to every particle. We simplify this expression by defining the quantity in parentheses as the **moment of inertia *I*:**

Moment of inertia ▸

$$I \equiv \sum_i m_i r_i^2 \qquad \textbf{(10.15)}$$

From the definition of moment of inertia,[2] we see that it has dimensions of ML^2 (kg·m^2 in SI units). With this notation, Equation 10.14 becomes

Rotational kinetic energy ▸

$$K_R = \tfrac{1}{2}I\omega^2 \qquad \textbf{(10.16)}$$

Although we commonly refer to the quantity $\frac{1}{2}I\omega^2$ as **rotational kinetic energy,** it is not a new form of energy. It is ordinary kinetic energy because it is derived from a

[2] Civil engineers use moment of inertia to characterize the elastic properties (rigidity) of such structures as loaded beams. Hence, it is often useful even in a nonrotational context.

sum over individual kinetic energies of the particles contained in the rigid object. The mathematical form of the kinetic energy given by Equation 10.16 is convenient when we are dealing with rotational motion, provided we know how to calculate I.

It is important to recognize the analogy between kinetic energy $\frac{1}{2}mv^2$ associated with translational motion and rotational kinetic energy $\frac{1}{2}I\omega^2$. The quantities I and ω in rotational motion are analogous to m and v in translational motion, respectively. (In fact, I takes the place of m and ω takes the place of v every time we compare a translational motion equation with its rotational counterpart.) Moment of inertia is a measure of the resistance of an object to changes in its rotational motion, just as mass is a measure of the tendency of an object to resist changes in its translational motion.

PITFALL PREVENTION 10.4
No Single Moment of Inertia

There is one major difference between mass and moment of inertia. Mass is an inherent property of an object. The moment of inertia of an object depends on your choice of rotation axis. Therefore, there is no single value of the moment of inertia for an object. There is a *minimum* value of the moment of inertia, which is that calculated about an axis passing through the center of mass of the object.

EXAMPLE 10.3 Four Rotating Objects

Four tiny spheres are fastened to the ends of two rods of negligible mass lying in the xy plane (Fig. 10.8). We shall assume the radii of the spheres are small compared with the dimensions of the rods.

(A) If the system rotates about the y axis (Fig. 10.8a) with an angular speed ω, find the moment of inertia and the rotational kinetic energy of the system about this axis.

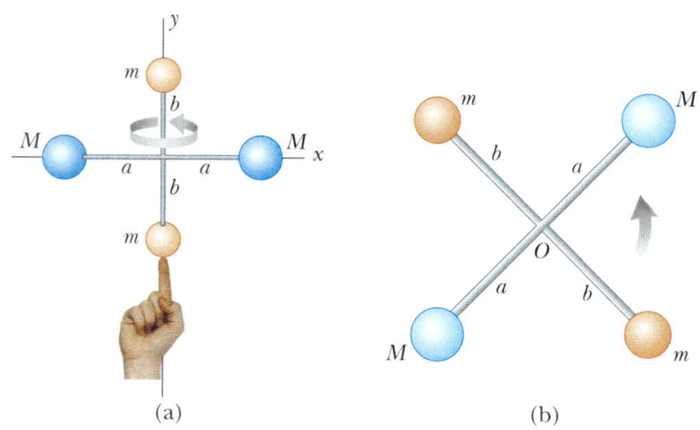

Figure 10.8 (Example 10.3) Four spheres form an unusual baton. (a) The baton is rotated about the y axis. (b) The baton is rotated about the z axis.

SOLUTION

Conceptualize Figure 10.8 is a pictorial representation that helps conceptualize the system of spheres and how it spins.

Categorize This example is a substitution problem because it is a straightforward application of the definitions discussed in this section.

Apply Equation 10.15 to the system:

$$I_y = \sum_i m_i r_i^2 = Ma^2 + Ma^2 = 2Ma^2$$

Evaluate the rotational kinetic energy using Equation 10.16:

$$K_R = \tfrac{1}{2}I_y\omega^2 = \tfrac{1}{2}(2Ma^2)\omega^2 = Ma^2\omega^2$$

That the two spheres of mass m do not enter into this result makes sense because they have no motion about the axis of rotation; hence, they have no rotational kinetic energy. By similar logic, we expect the moment of inertia about the x axis to be $I_x = 2mb^2$ with a rotational kinetic energy about that axis of $K_R = mb^2\omega^2$.

(B) Suppose the system rotates in the xy plane about an axis (the z axis) through O (Fig. 10.8b). Calculate the moment of inertia and rotational kinetic energy about this axis.

SOLUTION

Apply Equation 10.15 for this new rotation axis:

$$I_z = \sum_i m_i r_i^2 = Ma^2 + Ma^2 + mb^2 + mb^2 = 2Ma^2 + 2mb^2$$

Evaluate the rotational kinetic energy using Equation 10.16:

$$K_R = \tfrac{1}{2}I_z\omega^2 = \tfrac{1}{2}(2Ma^2 + 2mb^2)\omega^2 = (Ma^2 + mb^2)\omega^2$$

Comparing the results for parts (A) and (B), we conclude that the moment of inertia and therefore the rotational kinetic energy associated with a given angular speed depend on the axis of rotation. In part (B), we expect the result to include all four spheres and distances because all four spheres are rotating in the xy plane. Based on the work–kinetic energy theorem, that the rotational kinetic energy in part (A) is smaller than that in part (B) indicates it would require less work to set the system into rotation about the y axis than about the z axis.

What If? What if the mass M is much larger than m? How do the answers to parts (A) and (B) compare?

Answer If $M >> m$, then m can be neglected and the moment of inertia and the rotational kinetic energy in part (B) become

$$I_z = 2Ma^2 \quad \text{and} \quad K_R = Ma^2\omega^2$$

which are the same as the answers in part (A). If the masses m of the two orange spheres in Figure 10.8 are negligible, these spheres can be removed from the figure and rotations about the y and z axes are equivalent.

10.5 Calculation of Moments of Inertia

We can evaluate the moment of inertia of an extended rigid object by imagining the object to be divided into many small elements, each of which has mass Δm_i. We use the definition $I = \sum_i r_i^2 \Delta m_i$ and take the limit of this sum as $\Delta m_i \to 0$. In this limit, the sum becomes an integral over the volume of the object:

Moment of inertia of a rigid object ▶

$$I = \lim_{\Delta m_i \to 0} \sum_i r_i^2 \Delta m_i = \int r^2 dm \qquad \textbf{(10.17)}$$

It is usually easier to calculate moments of inertia in terms of the volume of the elements rather than their mass, and we can easily make that change by using

TABLE 10.2

Moments of Inertia of Homogeneous Rigid Objects with Different Geometries

Hoop or thin cylindrical shell $I_{CM} = MR^2$ 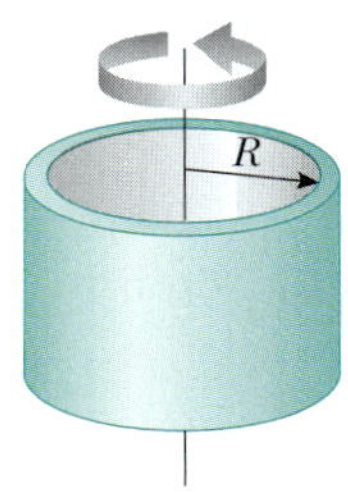	Hollow cylinder $I_{CM} = \frac{1}{2}M(R_1^2 + R_2^2)$ 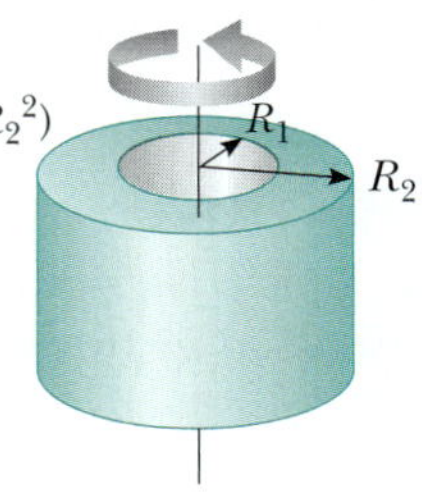
Solid cylinder or disk $I_{CM} = \frac{1}{2}MR^2$ 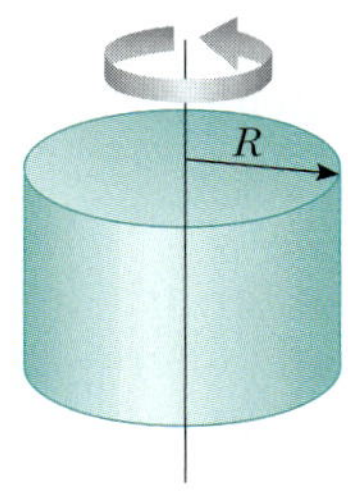	Rectangular plate $I_{CM} = \frac{1}{12}M(a^2 + b^2)$ 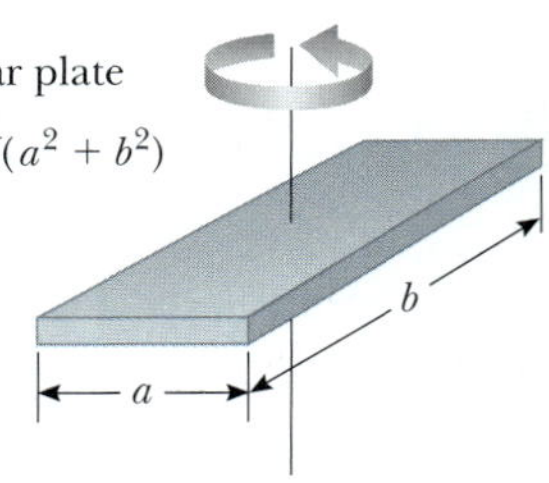
Long, thin rod with rotation axis through center $I_{CM} = \frac{1}{12}ML^2$ 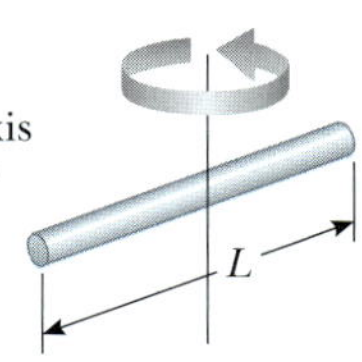	Long, thin rod with rotation axis through end $I = \frac{1}{3}ML^2$ 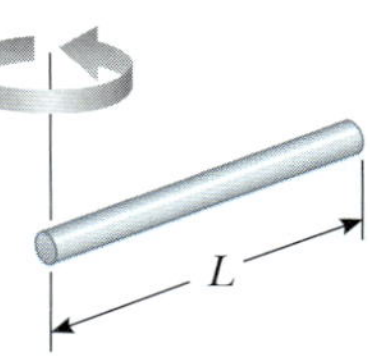
Solid sphere $I_{CM} = \frac{2}{5}MR^2$ 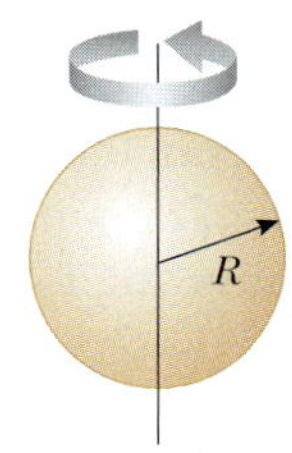	Thin spherical shell $I_{CM} = \frac{2}{3}MR^2$ 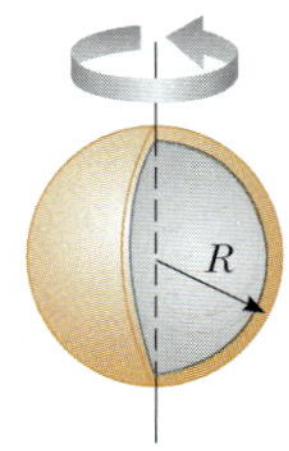

Equation 1.1, $\rho \equiv m/V$, where ρ is the density of the object and V is its volume. From this equation, the mass of a small element is $dm = \rho\, dV$. Substituting this result into Equation 10.17 gives

$$I = \int \rho r^2\, dV$$

If the object is homogeneous, ρ is constant and the integral can be evaluated for a known geometry. If ρ is not constant, its variation with position must be known to complete the integration.

The density given by $\rho = m/V$ sometimes is referred to as *volumetric mass density* because it represents mass per unit volume. Often we use other ways of expressing density. For instance, when dealing with a sheet of uniform thickness t, we can define a *surface mass density* $\sigma = \rho t$, which represents *mass per unit area*. Finally, when mass is distributed along a rod of uniform cross-sectional area A, we sometimes use *linear mass density* $\lambda = M/L = \rho A$, which is the *mass per unit length*.

Table 10.2 gives the moments of inertia for a number of objects about specific axes. The moments of inertia of rigid objects with simple geometry (high symmetry) are relatively easy to calculate provided the rotation axis coincides with an axis of symmetry, as in the examples below.

Quick Quiz 10.4 A section of hollow pipe and a solid cylinder have the same radius, mass, and length. They both rotate about their long central axes with the same angular speed. Which object has the higher rotational kinetic energy? (a) The hollow pipe does. (b) The solid cylinder does. (c) They have the same rotational kinetic energy. (d) It is impossible to determine.

EXAMPLE 10.4 Uniform Rigid Rod

Calculate the moment of inertia of a uniform rigid rod of length L and mass M (Fig. 10.9) about an axis perpendicular to the rod (the y axis) and passing through its center of mass.

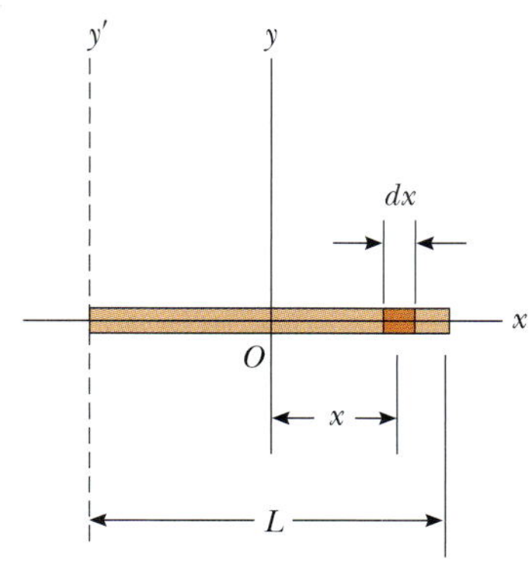

Figure 10.9 (Example 10.4) A uniform rigid rod of length L. The moment of inertia about the y axis is less than that about the y' axis. The latter axis is examined in Example 10.6.

SOLUTION

Conceptualize Imagine twirling the rod in Figure 10.9 with your fingers around its midpoint. If you have a meterstick handy, use it to simulate the spinning of a thin rod.

Categorize This example is a substitution problem, using the definition of moment of inertia in Equation 10.17. As with any calculus problem, the solution involves reducing the integrand to a single variable.

The shaded length element dx in Figure 10.9 has a mass dm equal to the mass per unit length λ multiplied by dx.

Express dm in terms of dx:

$$dm = \lambda dx = \frac{M}{L}\, dx$$

Substitute this expression into Equation 10.17, with $r^2 = x^2$:

$$I_y = \int r^2\, dm = \int_{-L/2}^{L/2} x^2 \frac{M}{L}\, dx = \frac{M}{L}\int_{-L/2}^{L/2} x^2\, dx$$

$$= \frac{M}{L}\left[\frac{x^3}{3}\right]_{-L/2}^{L/2} = \frac{1}{12}ML^2$$

Check this result in Table 10.2.

EXAMPLE 10.5 Uniform Solid Cylinder

A uniform solid cylinder has a radius R, mass M, and length L. Calculate its moment of inertia about its central axis (the z axis in Fig. 10.10).

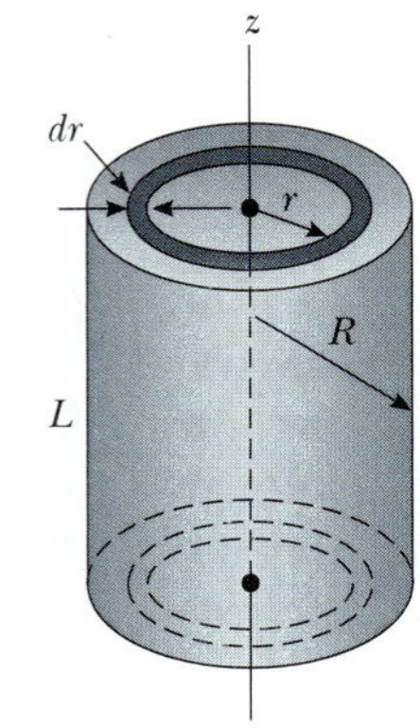

Figure 10.10 (Example 10.5) Calculating I about the z axis for a uniform solid cylinder.

SOLUTION

Conceptualize To simulate this situation, imagine twirling a can of frozen juice around its central axis.

Categorize This example is a substitution problem, using the definition of moment of inertia. As with Example 10.4, we must reduce the integrand to a single variable.

It is convenient to divide the cylinder into many cylindrical shells, each having radius r, thickness dr, and length L as shown in Figure 10.10. The density of the cylinder is ρ. The volume dV of each shell is its cross-sectional area multiplied by its length: $dV = L\,dA = L(2\pi r)\,dr$.

Express dm in terms of dr:

$$dm = \rho dV = 2\pi\rho L r\,dr$$

Substitute this expression into Equation 10.17:

$$I_z = \int r^2\,dm = \int r^2(2\pi\rho L r\,dr) = 2\pi\rho L\int_0^R r^3\,dr = \tfrac{1}{2}\pi\rho L R^4$$

Use the total volume $\pi R^2 L$ of the cylinder to express its density:

$$\rho = \frac{M}{V} = \frac{M}{\pi R^2 L}$$

Substitute this value into the expression for I_z:

$$I_z = \tfrac{1}{2}\pi\left(\frac{M}{\pi R^2 L}\right)LR^4 = \boxed{\tfrac{1}{2}MR^2}$$

Check this result in Table 10.2.

What If? What if the length of the cylinder in Figure 10.10 is increased to $2L$, while the mass M and radius R are held fixed? How does that change the moment of inertia of the cylinder?

Answer Notice that the result for the moment of inertia of a cylinder does not depend on L, the length of the cylinder. It applies equally well to a long cylinder and a flat disk having the same mass M and radius R. Therefore, the moment of inertia of the cylinder would not be affected by changing its length.

The calculation of moments of inertia of an object about an arbitrary axis can be cumbersome, even for a highly symmetric object. Fortunately, use of an important theorem, called the **parallel-axis theorem,** often simplifies the calculation.

To generate the parallel-axis theorem, suppose an object rotates about the z axis as shown in Figure 10.11. The moment of inertia does not depend on how the mass is distributed along the z axis; as we found in Example 10.5, the moment of inertia of a cylinder is independent of its length. Imagine collapsing the three-dimensional object into a planar object as in Figure 10.11b. In this imaginary process, all mass moves parallel to the z axis until it lies in the xy plane. The coordinates of the object's center of mass are now x_{CM}, y_{CM}, and $z_{CM} = 0$. Let the mass element dm have coordinates $(x, y, 0)$. Because this element is a distance $r = \sqrt{x^2 + y^2}$ from the z axis, the moment of inertia about the z axis is

$$I = \int r^2\,dm = \int (x^2 + y^2)\,dm$$

We can relate the coordinates x, y of the mass element dm to the coordinates of this same element located in a coordinate system having the object's center of mass as its origin. If the coordinates of the center of mass are x_{CM}, y_{CM}, and $z_{CM} = 0$

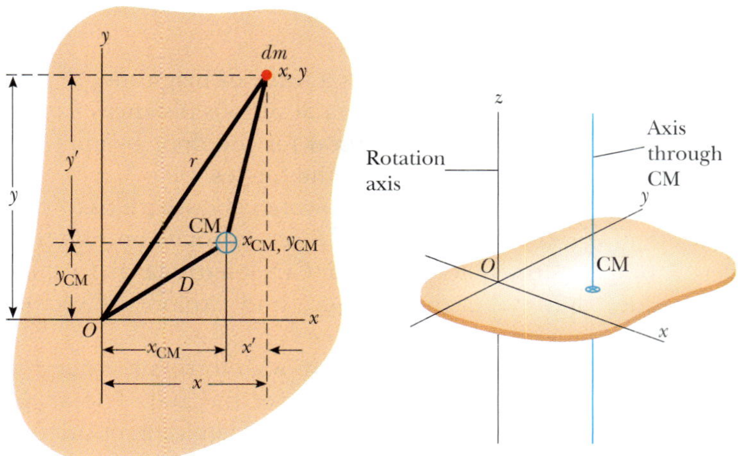

(a) (b)

Figure 10.11 (a) The parallel-axis theorem. If the moment of inertia about an axis perpendicular to the figure through the center of mass is I_{CM}, the moment of inertia about the z axis is $I_z = I_{CM} + MD^2$. (b) Perspective drawing showing the z axis (the axis of rotation) and the parallel axis through the center of mass.

in the original coordinate system centered on O, we see from Figure 10.11a that the relationships between the unprimed and primed coordinates are $x = x' + x_{CM}$, $y = y' + y_{CM}$, and $z = z' = 0$. Therefore,

$$I = \int [(x' + x_{CM})^2 - (y' + y_{CM})^2] dm$$

$$= \int [(x')^2 + (y')^2] dm + 2x_{CM} \int x' \, dm + 2y_{CM} \int y' \, dm + (x_{CM}^2 + y_{CM}^2) \int dm$$

The first integral is, by definition, the moment of inertia I_{CM} about an axis that is parallel to the z axis and passes through the center of mass. The second two integrals are zero because, by definition of the center of mass, $\int x' \, dm = \int y' \, dm = 0$. The last integral is simply MD^2 because $\int dm = M$ and $D^2 = x_{CM}^2 + y_{CM}^2$. Therefore, we conclude that

$$I = I_{CM} + MD^2 \quad \textbf{(10.18)}$$

◀ Parallel-axis theorem

EXAMPLE 10.6 Applying the Parallel-Axis Theorem

Consider once again the uniform rigid rod of mass M and length L shown in Figure 10.9. Find the moment of inertia of the rod about an axis perpendicular to the rod through one end (the y' axis in Fig. 10.9).

SOLUTION

Conceptualize Imagine twirling the rod around an endpoint rather than the midpoint. If you have a meterstick handy, try it and notice the degree of difficulty in rotating it around the end compared with rotating it around the center.

Categorize This example is a substitution problem, involving the parallel-axis theorem.

Intuitively, we expect the moment of inertia to be greater than the result $I_{CM} = \frac{1}{12}ML^2$ from Example 10.4 because there is mass up to a distance of L away from the rotation axis, whereas the farthest distance in Example 10.4 was only $L/2$. The distance between the center-of-mass axis and the y' axis is $D = L/2$.

Use the parallel-axis theorem: $I = I_{CM} + MD^2 = \frac{1}{12}ML^2 + M\left(\frac{L}{2}\right)^2 = \frac{1}{3}ML^2$

Check this result in Table 10.2.

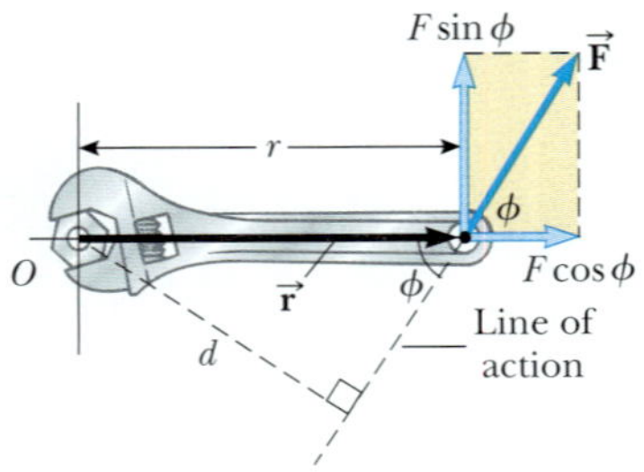

Figure 10.12 The force $\vec{F}$ has a greater rotating tendency about an axis through O as F increases and as the moment arm d increases. The component $F \sin \phi$ tends to rotate the wrench about O.

10.6 Torque

Imagine trying to rotate a door by applying a force of magnitude F perpendicular to the door surface near the hinges and then at various distances from the hinges. You will achieve a more rapid rate of rotation for the door by applying the force near the doorknob than by applying it near the hinges.

When a force is exerted on a rigid object pivoted about an axis, the object tends to rotate about that axis. The tendency of a force to rotate an object about some axis is measured by a quantity called **torque** $\vec{\tau}$ (Greek letter tau). Torque is a vector, but we will consider only its magnitude here and explore its vector nature in Chapter 11.

Consider the wrench in Figure 10.12 that we wish to rotate around an axis perpendicular to the page and through the center of the bolt. The applied force $\vec{F}$ acts at an angle ϕ to the horizontal. We define the magnitude of the torque associated with the force $\vec{F}$ by the expression

$$\tau \equiv rF \sin \phi = Fd \quad \text{(10.19)}$$

PITFALL PREVENTION 10.5

Torque Depends on Your Choice of Axis

Like moment of inertia, there is no unique value of the torque on an object. Its value depends on your choice of rotation axis.

where r is the distance between the rotation axis and the point of application of $\vec{F}$, and d is the perpendicular distance from the rotation axis to the line of action of $\vec{F}$. (The *line of action* of a force is an imaginary line extending out both ends of the vector representing the force. The dashed line extending from the tail of $\vec{F}$ in Figure 10.12 is part of the line of action of $\vec{F}$.) From the right triangle in Figure 10.12 that has the wrench as its hypotenuse, we see that $d = r \sin \phi$. The quantity d is called the **moment arm** (or *lever arm*) of $\vec{F}$.

Moment arm ▶

In Figure 10.12, the only component of $\vec{F}$ that tends to cause rotation of the wrench around an axis through O is $F \sin \phi$, the component perpendicular to a line drawn from the rotation axis to the point of application of the force. The horizontal component $F \cos \phi$, because its line of action passes through O, has no tendency to produce rotation about an axis passing through O. From the definition of torque, the rotating tendency increases as F increases and as d increases, which explains why it is easier to rotate a door if we push at the doorknob rather than at a point close to the hinges. We also want to apply our push as closely perpendicular to the door as we can so that ϕ is close to 90°. Pushing sideways on the doorknob ($\phi = 0$) will not cause the door to rotate.

If two or more forces act on a rigid object as in Active Figure 10.13, each tends to produce rotation about the axis at O. In this example, $\vec{F}_2$ tends to rotate the object clockwise and $\vec{F}_1$ tends to rotate it counterclockwise. We use the convention that the sign of the torque resulting from a force is positive if the turning tendency of the force is counterclockwise and negative if the turning tendency is clockwise. For example, in Active Figure 10.13, the torque resulting from $\vec{F}_1$, which has a moment arm d_1, is positive and equal to $+F_1 d_1$; the torque from $\vec{F}_2$ is negative and equal to $-F_2 d_2$. Hence, the *net* torque about an axis through O is

$$\sum \tau = \tau_1 + \tau_2 = F_1 d_1 - F_2 d_2$$

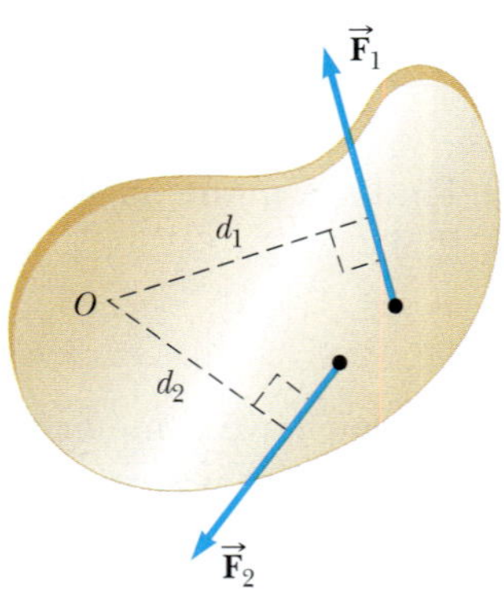

ACTIVE FIGURE 10.13

The force $\vec{F}_1$ tends to rotate the object counterclockwise about an axis through O, and $\vec{F}_2$ tends to rotate it clockwise.

Sign in at www.thomsonedu.com and go to ThomsonNOW to change the magnitudes, directions, and points of application of forces $\vec{F}_1$ and $\vec{F}_2$ and see how the object accelerates under the action of the two forces.

Torque should not be confused with force. Forces can cause a change in translational motion as described by Newton's second law. Forces can also cause a change in rotational motion, but the effectiveness of the forces in causing this change depends on both the magnitudes of the forces and the moment arms of the forces, in the combination we call *torque.* Torque has units of force times length—newton meters in SI units—and should be reported in these units. Do not confuse torque and work, which have the same units but are very different concepts.

Quick Quiz 10.5 **(i)** If you are trying to loosen a stubborn screw from a piece of wood with a screwdriver and fail, should you find a screwdriver for which the handle is (a) longer or (b) fatter? **(ii)** If you are trying to loosen a stubborn bolt from a piece of metal with a wrench and fail, should you find a wrench for which the handle is (a) longer or (b) fatter?

EXAMPLE 10.7 The Net Torque on a Cylinder

A one-piece cylinder is shaped as shown in Figure 10.14, with a core section protruding from the larger drum. The cylinder is free to rotate about the central axis shown in the drawing. A rope wrapped around the drum, which has radius R_1, exerts a force $\vec{\mathbf{T}}_1$ to the right on the cylinder. A rope wrapped around the core, which has radius R_2, exerts a force $\vec{\mathbf{T}}_2$ downward on the cylinder.

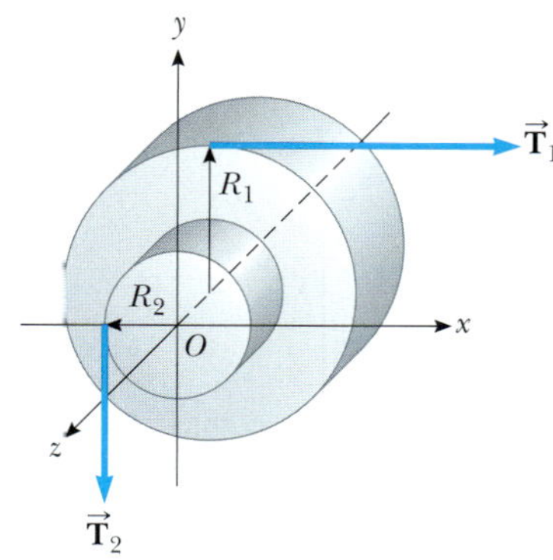

Figure 10.14 (Example 10.7) A solid cylinder pivoted about the z axis through O. The moment arm of $\vec{\mathbf{T}}_1$ is R_1, and the moment arm of $\vec{\mathbf{T}}_2$ is R_2.

(A) What is the net torque acting on the cylinder about the rotation axis (which is the z axis in Fig. 10.14)?

SOLUTION

Conceptualize Imagine that the cylinder in Figure 10.14 is a shaft in a machine. The force $\vec{\mathbf{T}}_2$ could be applied by a drive belt wrapped around the drum. The force $\vec{\mathbf{T}}_1$ could be applied by a friction brake at the surface of the core.

Categorize This example is a substitution problem in which we evaluate the net torque using Equation 10.19.

The torque due to $\vec{\mathbf{T}}_1$ about the rotation axis is $-R_1T_1$. (The sign is negative because the torque tends to produce clockwise rotation.) The torque due to $\vec{\mathbf{T}}_2$ is $+R_2T_2$. (The sign is positive because the torque tends to produce counterclockwise rotation of the cylinder.)

Evaluate the net torque about the rotation axis:

$$\sum \tau = \tau_1 + \tau_2 = R_2T_2 - R_1T_1$$

As a quick check, notice that if the two forces are of equal magnitude, the net torque is negative because $R_1 > R_2$. Starting from rest with both forces of equal magnitude acting on it, the cylinder would rotate clockwise because $\vec{\mathbf{T}}_1$ would be more effective at turning it than would $\vec{\mathbf{T}}_2$.

(B) Suppose $T_1 = 5.0$ N, $R_1 = 1.0$ m, $T_2 = 15.0$ N, and $R_2 = 0.50$ m. What is the net torque about the rotation axis, and which way does the cylinder rotate starting from rest?

SOLUTION

Substitute the given values:

$$\sum \tau = (0.50 \text{ m})(15 \text{ N}) - (1.0 \text{ m})(5.0 \text{ N}) = 2.5 \text{ N} \cdot \text{m}$$

Because this net torque is positive, the cylinder begins to rotate in the counterclockwise direction.

10.7 The Rigid Object Under a Net Torque

In Chapter 5, we learned that a net force on an object causes an acceleration of the object and that the acceleration is proportional to the net force. These facts are the basis of the particle under a net force model whose mathematical representation is Newton's second law. In this section, we show the rotational analog of Newton's second law: the angular acceleration of a rigid object rotating about a fixed axis is proportional to the net torque acting about that axis. Before discussing the more complex case of rigid-object rotation, however, it is instructive first to discuss the case of a particle moving in a circular path about some fixed point under the influence of an external force.

Consider a particle of mass m rotating in a circle of radius r under the influence of a tangential net force $\sum \vec{\mathbf{F}}_t$ and a radial net force $\sum \vec{\mathbf{F}}_r$ as shown in Figure 10.15. The radial net force causes the particle to move in the circular path with a centripetal acceleration. The tangential force provides a tangential acceleration $\vec{\mathbf{a}}_t$, and

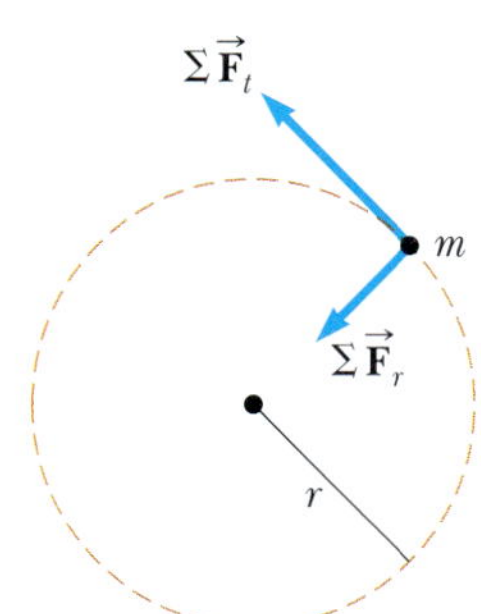

Figure 10.15 A particle rotating in a circle under the influence of a net tangential force $\sum \vec{\mathbf{F}}_t$. A net force $\sum \vec{\mathbf{F}}_r$ in the radial direction also must be present to maintain the circular motion.

$$\sum F_t = ma_t$$

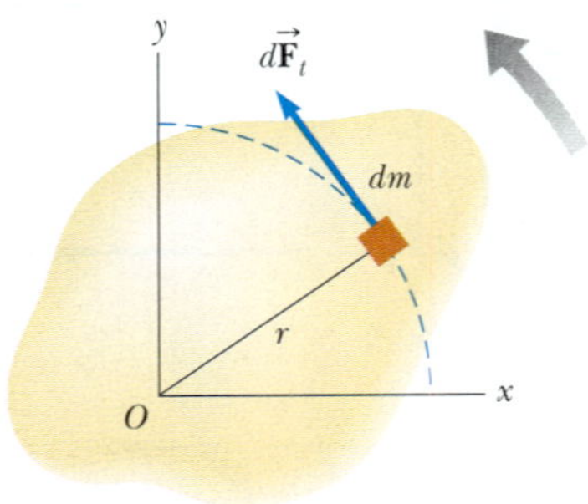

Figure 10.16 A rigid object rotating about an axis through *O*. Each mass element *dm* rotates about the axis with the same angular acceleration α.

The magnitude of the net torque due to $\sum \vec{\mathbf{F}}_t$ on the particle about an axis through the center of the circle is

$$\sum \tau = \sum F_t r = (ma_t)r$$

Because the tangential acceleration is related to the angular acceleration through the relationship $a_t = r\alpha$ (Eq. 10.11), the net torque can be expressed as

$$\sum \tau = (mr\alpha)r = (mr^2)\alpha$$

Recall from Equation 10.15 that mr^2 is the moment of inertia of the particle about the z axis passing through the origin, so that

$$\sum \tau = I\alpha \qquad \textbf{(10.20)}$$

That is, **the net torque acting on the particle is proportional to its angular acceleration,** and the proportionality constant is the moment of inertia. Notice that $\sum \tau = I\alpha$ has the same mathematical form as Newton's second law of motion, $\sum F = ma$.

Now let us extend this discussion to a rigid object of arbitrary shape rotating about a fixed axis as in Figure 10.16. The object can be regarded as an infinite number of mass elements dm of infinitesimal size. If we impose a Cartesian coordinate system on the object, each mass element rotates in a circle about the origin and each has a tangential acceleration $\vec{\mathbf{a}}_t$ produced by an external tangential force $d\vec{\mathbf{F}}_t$. For any given element, we know from Newton's second law that

$$dF_t = (dm)a_t$$

The torque $d\tau$ associated with the force $d\vec{\mathbf{F}}_t$ acts about the origin and is given by

$$d\tau = r\,dF_t = a_t r\,dm$$

Because $a_t = r\alpha$, the expression for $d\tau$ becomes

$$d\tau = \alpha r^2\,dm$$

Although each mass element of the rigid object may have a different translational acceleration $\vec{\mathbf{a}}_t$, they all have the *same* angular acceleration α. With this in mind, we can integrate the above expression to obtain the net torque $\sum\tau$ about an axis through O due to the external forces:

$$\sum \tau = \int \alpha r^2\,dm = \alpha \int r^2\,dm$$

where α can be taken outside the integral because it is common to all mass elements. From Equation 10.17, we know that $\int r^2\,dm$ is the moment of inertia of the object about the rotation axis through O, and so the expression for $\sum \tau$ becomes

Torque is proportional to angular acceleration ►

$$\sum \tau = I\alpha \qquad \textbf{(10.21)}$$

This equation for a rigid object is the same as that found for a particle moving in a circular path (Eq. 10.20). The net torque about the rotation axis is proportional to the angular acceleration of the object, with the proportionality factor being I, a quantity that depends on the axis of rotation and on the size and shape of the object. Equation 10.21 is the mathematical representation of the analysis model of a **rigid object under a net torque**, the rotational analog to the particle under a net force.

Finally, notice that the result $\sum \tau = I\alpha$ also applies when the forces acting on the mass elements have radial components as well as tangential components. That is because the line of action of all radial components must pass through the axis of rotation; hence, all radial components produce zero torque about that axis.

Quick Quiz 10.6 You turn off your electric drill and find that the time interval for the rotating bit to come to rest due to frictional torque in the drill is Δt. You replace the bit with a larger one that results in a doubling of the moment of inertia of the drill's entire rotating mechanism. When this larger bit is rotated at the same angular speed as the first and the drill is turned off, the frictional torque remains the same as that for the previous situation. What is the time interval for this second bit to come to rest? (a) $4\,\Delta t$ (b) $2\,\Delta t$ (c) Δt (d) $0.5\,\Delta t$ (e) $0.25\,\Delta t$ (f) impossible to determine.

EXAMPLE 10.8 Rotating Rod

A uniform rod of length L and mass M is attached at one end to a frictionless pivot and is free to rotate about the pivot in the vertical plane as in Figure 10.17. The rod is released from rest in the horizontal position. What are the initial angular acceleration of the rod and the initial translational acceleration of its right end?

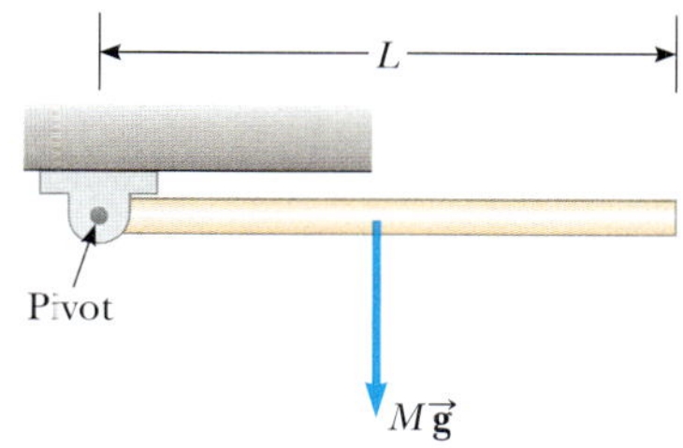

Figure 10.17 (Example 10.8) A rod is free to rotate around a pivot at the left end. The gravitational force on the rod acts at its center of mass.

SOLUTION

Conceptualize Imagine what happens to the rod in Figure 10.17 when it is released. It rotates clockwise around the pivot at the left end.

Categorize The rod is categorized as a rigid object under a net torque. The torque is due only to the gravitational force on the rod if the rotation axis is chosen to pass through the pivot in Figure 10.17. We *cannot* categorize the rod as a rigid object under constant angular acceleration because the torque exerted on the rod and therefore the angular acceleration of the rod vary with its angular position.

Analyze The only force contributing to the torque about an axis through the pivot is the gravitational force $M\vec{g}$ exerted on the rod. (The force exerted by the pivot on the rod has zero torque about the pivot because its moment arm is zero.) To compute the torque on the rod, we assume the gravitational force acts at the center of mass of the rod as shown in Figure 10.17.

Write an expression for the magnitude of the torque due to the gravitational force about an axis through the pivot:

$$\tau = Mg\left(\frac{L}{2}\right)$$

Use Equation 10.21 to obtain the angular acceleration of the rod:

$$(1)\quad \alpha = \frac{\tau}{I} = \frac{Mg(L/2)}{\frac{1}{3}ML^2} = \frac{3g}{2L}$$

Use Equation 10.11 with $r = L$ to find the initial translational acceleration of the right end of the rod:

$$a_t = L\alpha = \tfrac{3}{2}g$$

Finalize These values are the *initial* values of the angular and translational accelerations. Once the rod begins to rotate, the gravitational force is no longer perpendicular to the rod and the values of the two accelerations decrease, going to zero at the moment the rod passes through the vertical orientation.

What If? What if we were to place a penny on the end of the rod and then release the rod? Would the penny stay in contact with the rod?

Answer The result for the initial acceleration of a point on the end of the rod shows that $a_t > g$. An unsupported penny falls at acceleration g. So, if we place a penny at the end of the rod and then release the rod, the end of the rod falls faster than the penny does! The penny does not stay in contact with the rod. (Try this with a penny and a meterstick!)

The question now is to find the location on the rod at which we can place a penny that *will* stay in contact as both begin to fall. To find the translational acceleration of an arbitrary point on the rod at a distance $r < L$ from the pivot point, we combine Equation (1) with Equation 10.11:

$$a_t = r\alpha = \frac{3g}{2L}r$$

For the penny to stay in contact with the rod, the limiting case is that the translational acceleration must be equal to that due to gravity:

$$a_t = g = \frac{3g}{2L} r$$

$$r = \frac{2}{3}L$$

Therefore, a penny placed closer to the pivot than two-thirds of the length of the rod stays in contact with the falling rod, but a penny farther out than this point loses contact.

CONCEPTUAL EXAMPLE 10.9 Falling Smokestacks and Tumbling Blocks

When a tall smokestack falls over, it often breaks somewhere along its length before it hits the ground as shown in Figure 10.18. Why?

Figure 10.18 (Conceptual Example 10.9) A falling smokestack breaks at some point along its length.

SOLUTION

As the smokestack rotates around its base, each higher portion of the smokestack falls with a larger tangential acceleration than the portion below it according to Equation 10.11. The angular acceleration increases as the smokestack tips farther. Eventually, higher portions of the smokestack experience an acceleration greater than the acceleration that could result from gravity alone; this situation is similar to that described in Example 10.8. It can happen only if these portions are being pulled downward by a force in addition to the gravitational force. The force that causes that to occur is the shear force from lower portions of the smokestack. Eventually, the shear force that provides this acceleration is greater than the smokestack can withstand, and the smokestack breaks. The same thing happens with a tall tower of children's toy blocks. Borrow some blocks from a child and build such a tower. Push it over and watch it come apart at some point before it strikes the floor.

EXAMPLE 10.10 Angular Acceleration of a Wheel

A wheel of radius R, mass M, and moment of inertia I is mounted on a frictionless, horizontal axle as in Figure 10.19. A light cord wrapped around the wheel supports an object of mass m. Calculate the angular acceleration of the wheel, the linear acceleration of the object, and the tension in the cord.

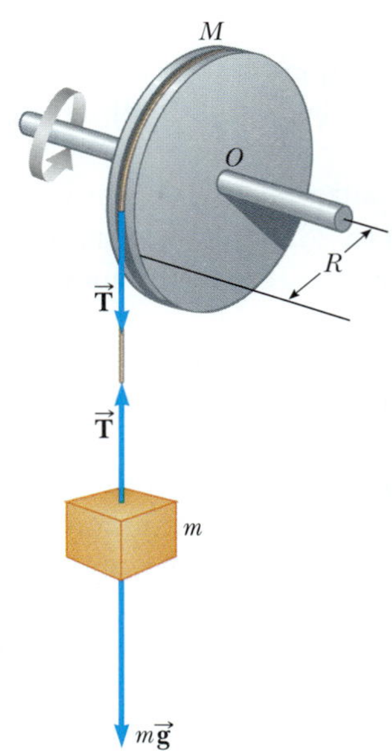

Figure 10.19 (Example 10.10) An object hangs from a cord wrapped around a wheel.

SOLUTION

Conceptualize Imagine that the object is a bucket in an old-fashioned wishing well. It is tied to a cord that passes around a cylinder equipped with a crank for raising the bucket. After the bucket has been raised, the system is released and the bucket accelerates downward while the cord unwinds off the cylinder.

Categorize The object is modeled as a particle under a net force. The wheel is modeled as a rigid object under a net torque.

Analyze The magnitude of the torque acting on the wheel about its axis of rotation is $\tau = TR$, where T is the force exerted by the cord on the rim of the wheel. (The gravitational force exerted by the Earth on the wheel and the normal force exerted by the axle on the wheel both pass through the axis of rotation and therefore produce no torque.)

Write Equation 10.21: $\sum \tau = I\alpha$

Solve for α and substitute the net torque: (1) $\alpha = \frac{\sum \tau}{I} = \frac{TR}{I}$

Apply Newton's second law to the motion of the object, taking the downward direction to be positive: $\sum F_y = mg - T = ma$

Solve for the acceleration a: (2) $a = \frac{mg - T}{m}$

Equations (1) and (2) have three unknowns: α, a, and T. Because the object and wheel are connected by a cord that does not slip, the translational acceleration of the suspended object is equal to the tangential acceleration of a point on the wheel's rim. Therefore, the angular acceleration α of the wheel and the translational acceleration of the object are related by $a = R\alpha$.

Use this fact together with Equations (1) and (2): (3) $a = R\alpha = \frac{TR^2}{I} = \frac{mg - T}{m}$

Solve for the tension T: (4) $T = \frac{mg}{1 + (mR^2/I)}$

Substitute Equation (4) into Equation (2) and solve for a: (5) $a = \frac{g}{1 + (I/mR^2)}$

Use $a = R\alpha$ and Equation (5) to solve for α: $\alpha = \frac{a}{R} = \frac{g}{R + (I/mR)}$

Finalize We finalize this problem by imagining the behavior of the system in some extreme limits.

What If? What if the wheel were to become very massive so that I becomes very large? What happens to the acceleration a of the object and the tension T?

Answer If the wheel becomes infinitely massive, we can imagine that the object of mass m will simply hang from the cord without causing the wheel to rotate.

We can show that mathematically by taking the limit $I \rightarrow \infty$. Equation (5) then becomes

$$a = \frac{g}{1 + (I/mR^2)} \rightarrow 0$$

which agrees with our conceptual conclusion that the object will hang at rest. Also, Equation (4) becomes

$$T = \frac{mg}{1 + (mR^2/I)} \rightarrow \frac{mg}{1 + 0} = mg$$

which is consistent because the object simply hangs at rest in equilibrium between the gravitational force and the tension in the string.

10.8 Energy Considerations in Rotational Motion

Up to this point in our discussion of rotational motion in this chapter, we focused primarily on an approach involving force, leading to a description of torque on a rigid object. In Section 10.4, we discussed the rotational kinetic energy of a rigid

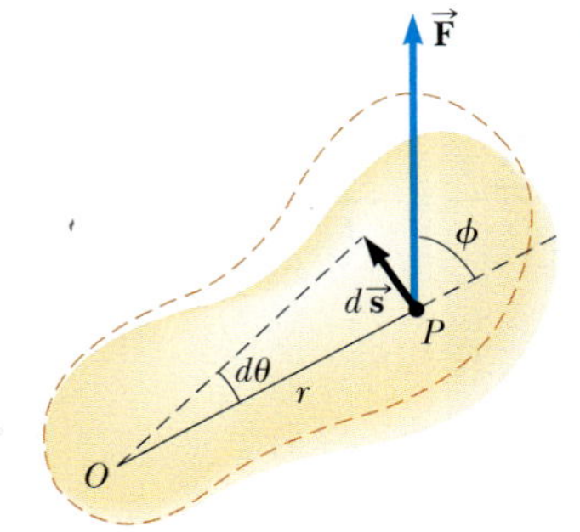

Figure 10.20 A rigid object rotates about an axis through O under the action of an external force $\vec{\mathbf{F}}$ applied at P.

object. Let us now extend that initial energy discussion and see how an energy approach can be useful in solving rotational problems.

We begin by considering the relationship between the torque acting on a rigid object and its resulting rotational motion so as to generate expressions for power and a rotational analog to the work–kinetic energy theorem. Consider the rigid object pivoted at O in Figure 10.20. Suppose a single external force $\vec{\mathbf{F}}$ is applied at P, where $\vec{\mathbf{F}}$ lies in the plane of the page. The work done on the object by $\vec{\mathbf{F}}$ as its point of application rotates through an infinitesimal distance $ds = r\,d\theta$ is

$$dW = \vec{\mathbf{F}} \cdot d\vec{\mathbf{s}} = (F \sin \phi) r\,d\theta$$

where $F \sin \phi$ is the tangential component of $\vec{\mathbf{F}}$, or, in other words, the component of the force along the displacement. Notice that the radial component vector of $\vec{\mathbf{F}}$ does no work on the object because it is perpendicular to the displacement of the point of application of $\vec{\mathbf{F}}$.

Because the magnitude of the torque due to $\vec{\mathbf{F}}$ about an axis through O is defined as $rF \sin \phi$ by Equation 10.19, we can write the work done for the infinitesimal rotation as

$$dW = \tau\,d\theta \qquad \textbf{(10.22)}$$

The rate at which work is being done by $\vec{\mathbf{F}}$ as the object rotates about the fixed axis through the angle $d\theta$ in a time interval dt is

$$\frac{dW}{dt} = \tau \frac{d\theta}{dt}$$

Because dW/dt is the instantaneous power $\mathcal{P}$ (see Section 8.5) delivered by the force and $d\theta/dt = \omega$, this expression reduces to

Power delivered to a rotating rigid object ▶

$$\mathcal{P} = \frac{dW}{dt} = \tau\omega \qquad \textbf{(10.23)}$$

This equation is analogous to $\mathcal{P} = Fv$ in the case of translational motion, and Equation 10.22 is analogous to $dW = F_x\,dx$.

In studying translational motion, models based on an energy approach can be extremely useful in describing a system's behavior. From what we learned of translational motion, we expect that when a symmetric object rotates about a fixed axis, the work done by external forces equals the change in the rotational energy of the object.

To prove that fact, let us begin with $\sum \tau = I\alpha$. Using the chain rule from calculus, we can express the net torque as

$$\sum \tau = I\alpha = I\frac{d\omega}{dt} = I\frac{d\omega}{d\theta}\frac{d\theta}{dt} = I\frac{d\omega}{d\theta}\,\omega$$

Rearranging this expression and noting that $\sum \tau\,d\theta = dW$ gives

$$\sum \tau d\theta = dW = I\omega\,d\omega$$

Integrating this expression, we obtain for the total work done by the net external force acting on a rotating system

Work–kinetic energy theorem for rotational motion ▶

$$\sum W = \int_{\omega_i}^{\omega_f} I\omega\,d\omega = \tfrac{1}{2}I\omega_f^{\,2} - \tfrac{1}{2}I\omega_i^{\,2} \qquad \textbf{(10.24)}$$

where the angular speed changes from ω_i to ω_f. Equation 10.24 is the **work–kinetic energy theorem for rotational motion.** Similar to the work–kinetic energy theorem in translational motion (Section 7.5), this theorem states that the net work done by external forces in rotating a symmetric rigid object about a fixed axis equals the change in the object's rotational energy.

This theorem is a form of the nonisolated system model discussed in Chapter 8. Work is done on the system of the rigid object, which represents a transfer of energy across the boundary of the system that appears as an increase in the object's rotational kinetic energy.

TABLE 10.3

Useful Equations in Rotational and Translational Motion

Rotational Motion About a Fixed Axis	Translational Motion
Angular speed $\omega = d\theta/dt$	Translational speed $v = dx/dt$
Angular acceleration $\alpha = d\omega/dt$	Translational acceleration $a = dv/dt$
Net torque $\Sigma\tau = I\alpha$	Net force $\Sigma F = ma$
If $\alpha =$ constant $\begin{cases} \omega_f = \omega_i + \alpha t \\ \theta_f = \theta_i + \omega_i t + \frac{1}{2}\alpha t^2 \\ \omega_f^2 = \omega_i^2 + 2\alpha(\theta_f - \theta_i) \end{cases}$	If $a =$ constant $\begin{cases} v_f = v_i + at \\ x_f = x_i + v_i t + \frac{1}{2}at^2 \\ v_f^2 = v_i^2 + 2a(x_f - x_i) \end{cases}$
Work $W = \int_{\theta_i}^{\theta_f} \tau \, d\theta$	Work $W = \int_{x_i}^{x_f} F_x \, dx$
Rotational kinetic energy $K_R = \frac{1}{2}I\omega^2$	Kinetic energy $K = \frac{1}{2}mv^2$
Power $\mathcal{P} = \tau\omega$	Power $\mathcal{P} = Fv$
Angular momentum $L = I\omega$	Linear momentum $p = mv$
Net torque $\Sigma\tau = dL/dt$	Net force $\Sigma F = dp/dt$

In general, we can combine this theorem with the translational form of the work–kinetic energy theorem from Chapter 7. Therefore, the net work done by external forces on an object is the change in its *total* kinetic energy, which is the sum of the translational and rotational kinetic energies. For example, when a pitcher throws a baseball, the work done by the pitcher's hands appears as kinetic energy associated with the ball moving through space as well as rotational kinetic energy associated with the spinning of the ball.

In addition to the work–kinetic energy theorem, other energy principles can also be applied to rotational situations. For example, if a system involving rotating objects is isolated and no nonconservative forces act within the system, the isolated system model and the principle of conservation of mechanical energy can be used to analyze the system as in Example 10.11 below.

Finally, in some situations an energy approach does not provide enough information to solve the problem and it must be combined with a momentum approach. Such a case is illustrated in Example 10.14 in Section 10.9.

Table 10.3 lists the various equations we have discussed pertaining to rotational motion together with the analogous expressions for translational motion. The last two equations in Table 10.3, involving angular momentum L, are discussed in Chapter 11 and are included here only for the sake of completeness.

EXAMPLE 10.11 Rotating Rod Revisited

A uniform rod of length L and mass M is free to rotate on a frictionless pin passing through one end (Fig 10.21). The rod is released from rest in the horizontal position.

(A) What is its angular speed when the rod reaches its lowest position?

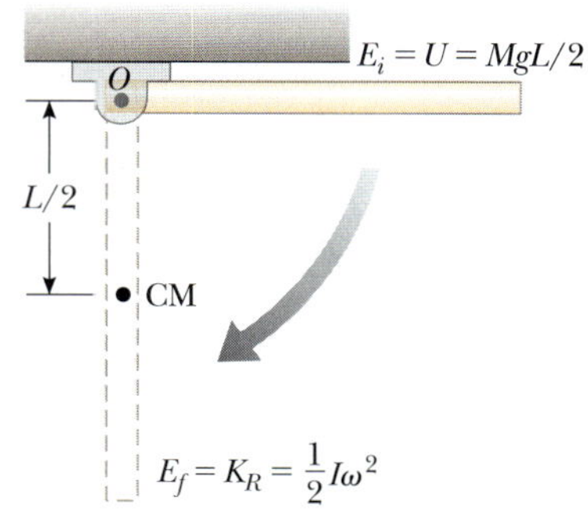

Figure 10.21 (Example 10.11) A uniform rigid rod pivoted at O rotates in a vertical plane under the action of the gravitational force.

SOLUTION

Conceptualize Consider Figure 10.21 and imagine the rod rotating downward through a quarter turn about the pivot at the left end. Also look back at Example 10.8. This physical situation is the same.

Categorize As mentioned in Example 10.8, the angular acceleration of the rod is not constant. Therefore, the kinematic equations for rotation (Section 10.2) cannot be used to solve this example. We categorize the system of the rod and the Earth as an isolated system with no nonconservative forces acting and use the principle of conservation of mechanical energy.

Analyze We choose the configuration in which the rod is hanging straight down as the reference configuration for gravitational potential energy and assign a value of zero for this configuration. When the rod is in the horizontal position, it has no rotational kinetic energy. The potential energy of the system in this configuration relative to the reference configuration is $MgL/2$ because the center of mass of the rod is at a height $L/2$ higher than its position in the reference configuration. When the rod reaches its lowest position, the energy of the system is entirely rotational energy $\frac{1}{2}I\omega^2$, where I is the moment of inertia of the rod about an axis passing through the pivot.

Write a conservation of mechanical energy equation for the system:

$$K_f + U_f = K_i + U_i$$

Substitute for each of the energies:

$$\frac{1}{2}I\omega^2 + 0 = 0 + \frac{1}{2}MgL$$

Solve for ω and use $I = \frac{1}{3}ML^2$ (see Table 10.2) for the rod:

$$\omega = \sqrt{\frac{MgL}{I}} = \sqrt{\frac{MgL}{\frac{1}{3}ML^2}} = \boxed{\sqrt{\frac{3g}{L}}}$$

(B) Determine the tangential speed of the center of mass and the tangential speed of the lowest point on the rod when it is in the vertical position.

SOLUTION

Use Equation 10.10 and the result from part (A):

$$v_{\text{CM}} = r\omega = \frac{L}{2}\omega = \boxed{\frac{1}{2}\sqrt{3gL}}$$

Because r for the lowest point on the rod is twice what it is for the center of mass, the lowest point has a tangential speed twice that of the center of mass:

$$v = 2v_{\text{CM}} = \boxed{\sqrt{3gL}}$$

Finalize The initial configuration in this example is the same as that in Example 10.8. In Example 10.8, however, we could only find the initial angular acceleration of the rod. Applying an energy approach in the current example allows us to find additional information, the angular speed of the rod at another instant of time.

EXAMPLE 10.12 Energy and the Atwood Machine

Two cylinders having different masses m_1 and m_2 are connected by a string passing over a pulley, as shown in Active Figure 10.22. The pulley has a radius R and moment of inertia I about its axis of rotation. The string does not slip on the pulley, and the system is released from rest. Find the translational speeds of the cylinders after cylinder 2 descends through a distance h, and find the angular speed of the pulley at this time.

SOLUTION

Conceptualize We have already seen examples involving the Atwood machine, so the motion of the objects in Active Figure 10.22 should be easy to visualize.

Categorize Because the string does not slip, the pulley rotates about the axle. We can neglect friction in the axle because the axle's radius is small relative to that of the pulley. Hence, the frictional torque is much smaller than the net torque applied by the two cylinders provided that their masses are significantly different. Consequently, the system consisting of the two cylinders, the pulley, and the Earth is an isolated system with no nonconservative forces acting; therefore, the mechanical energy of the system is conserved.

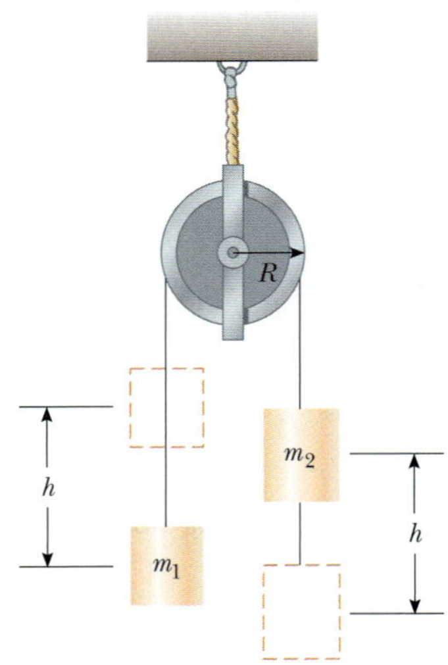

ACTIVE FIGURE 10.22

(Example 10.12) An Atwood machine with a massive pulley.

Sign in at www.thomsonedu.com and go to ThomsonNOW to change the masses of the hanging cylinders and the mass and radius of the pulley and see how the cylinders move.

Analyze We define the zero configuration for gravitational potential energy as that which exists when the system is released. From Active Figure 10.22, we see that the descent of cylinder 2 is associated with a decrease in system potential energy and that the rise of cylinder 1 represents an increase in potential energy.

Write a conservation of energy equation for the system:

$$K_f + U_f = K_i + U_i$$

Substitute for each of the energies:

$$(\tfrac{1}{2}m_1 v_f^2 + \tfrac{1}{2}m_2 v_f^2 + \tfrac{1}{2}I\omega_f^2) + (m_1 gh - m_2 gh) = 0 + 0$$

Use $v_f = R\omega_f$ to substitute for ω_f:

$$\tfrac{1}{2}m_1 v_f^2 + \tfrac{1}{2}m_2 v_f^2 + \tfrac{1}{2}\frac{I}{R^2}v_f^2 = m_2 gh - m_1 gh$$

$$\tfrac{1}{2}\left(m_1 + m_2 + \frac{I}{R^2}\right)v_f^2 = m_2 gh - m_1 gh$$

Solve for v_f:

$$(1)\quad v_f = \left[\frac{2(m_2 - m_1)gh}{m_1 + m_2 + I/R^2}\right]^{1/2}$$

Use $v_f = R\omega_f$ to solve for ω_f:

$$\omega_f = \frac{v_f}{R} = \frac{1}{R}\left[\frac{2(m_2 - m_1)gh}{m_1 - m_2 + I/R^2}\right]^{1/2}$$

Finalize Each cylinder can be modeled as a particle under constant acceleration because it experiences a constant net force. Think about what you would need to do to use Equation (1) to find the acceleration of one of the cylinders and reduce the result so that it matches the result of Example 5.9. Then do it and see if it works!

10.9 Rolling Motion of a Rigid Object

In this section, we treat the motion of a rigid object rolling along a flat surface. In general, such motion is complex. For example, suppose a cylinder is rolling on a straight path such that the axis of rotation remains parallel to its initial orientation in space. As Figure 10.23 shows, a point on the rim of the cylinder moves in a complex path called a *cycloid.* We can simplify matters, however, by focusing on the center of mass rather than on a point on the rim of the rolling object. As shown in Figure 10.23, the center of mass moves in a straight line. If an object such as a cylinder rolls without slipping on the surface (called *pure rolling motion*), a simple relationship exists between its rotational and translational motions.

Consider a uniform cylinder of radius R rolling without slipping on a horizontal surface (Fig. 10.24). As the cylinder rotates through an angle θ, its center of mass

Henry Leap and Jim Lehman

Figure 10.23 One light source at the center of a rolling cylinder and another at one point on the rim illustrate the different paths these two points take. The center moves in a straight line (green line), whereas the point on the rim moves in the path called a cycloid (red curve).

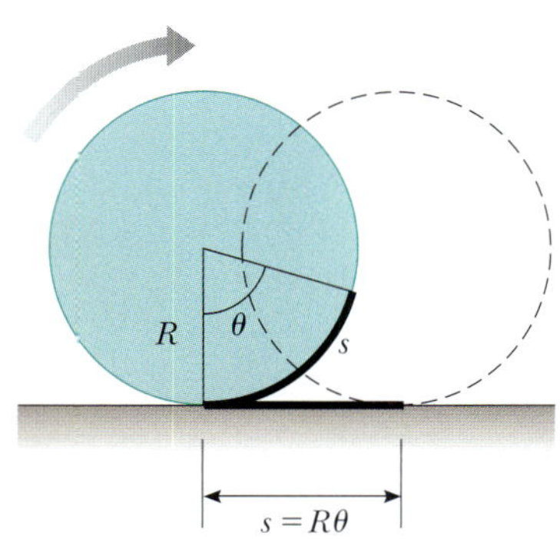

Figure 10.24 For pure rolling motion, as the cylinder rotates through an angle θ its center moves a linear distance $s = R\theta$.

PITFALL PREVENTION 10.6

Equation 10.25 Looks Familiar

Equation 10.25 looks very similar to Equation 10.10, so be sure to be clear on the difference. Equation 10.10 gives the *tangential* speed of a point on a *rotating* object located a distance r from a fixed rotation axis if the object is rotating with angular speed ω. Equation 10.25 gives the *translational* speed of the center of mass of a *rolling* object of radius R rotating with angular speed ω.

moves a linear distance $s = R\theta$ (see Eq. 10.1a). Therefore, the translational speed of the center of mass for pure rolling motion is given by

$$v_{CM} = \frac{ds}{dt} = R\frac{d\theta}{dt} = R\omega \tag{10.25}$$

where ω is the angular speed of the cylinder. Equation 10.25 holds whenever a cylinder or sphere rolls without slipping and is the **condition for pure rolling motion.** The magnitude of the linear acceleration of the center of mass for pure rolling motion is

$$a_{CM} = \frac{dv_{CM}}{dt} = R\frac{d\omega}{dt} = R\alpha \tag{10.26}$$

where α is the angular acceleration of the cylinder.

Imagine that you are moving along with a rolling object at speed v_{CM}, staying in a frame of reference at rest with respect to the center of mass of the object. As you observe the object, you will see the object in pure rotation around its center of mass. Figure 10.25a shows the velocities of points at the top, center, and bottom of the object as observed by you. In addition to these velocities, every point on the object moves in the same direction with speed v_{CM} relative to the surface on which it rolls. Figure 10.25b shows these velocities for a nonrotating object. In the reference frame at rest with respect to the surface, the velocity of a given point on the object is the sum of the velocities shown in Figures 10.25a and 10.25b. Figure 10.25c shows the results of adding these velocities.

Notice that the contact point between the surface and cylinder in Figure 10.25c has a translational speed of zero. At this instant, the rolling object is moving in exactly the same way as if the surface were removed and the object were pivoted at point P and spun about an axis passing through P. We can express the total kinetic energy of this imagined spinning object as

$$K = \tfrac{1}{2}I_P\omega^2 \tag{10.27}$$

where I_P is the moment of inertia about a rotation axis through P.

Because the motion of the imagined spinning object is the same at this instant as our actual rolling object, Equation 10.27 also gives the kinetic energy of the rolling object. Applying the parallel-axis theorem, we can substitute $I_P = I_{CM} + MR^2$ into Equation 10.27 to obtain

$$K = \tfrac{1}{2}I_{CM}\omega^2 + \tfrac{1}{2}MR^2\omega^2$$

Using $v_{CM} = R\omega$, this equation can be expressed as

Total kinetic energy of a rolling object ▶

$$K = \tfrac{1}{2}I_{CM}\omega^2 + \tfrac{1}{2}Mv_{CM}^2 \tag{10.28}$$

The term $\frac{1}{2}I_{CM}\omega^2$ represents the rotational kinetic energy of the cylinder about its center of mass, and the term $\frac{1}{2}Mv_{CM}^2$ represents the kinetic energy the cylinder

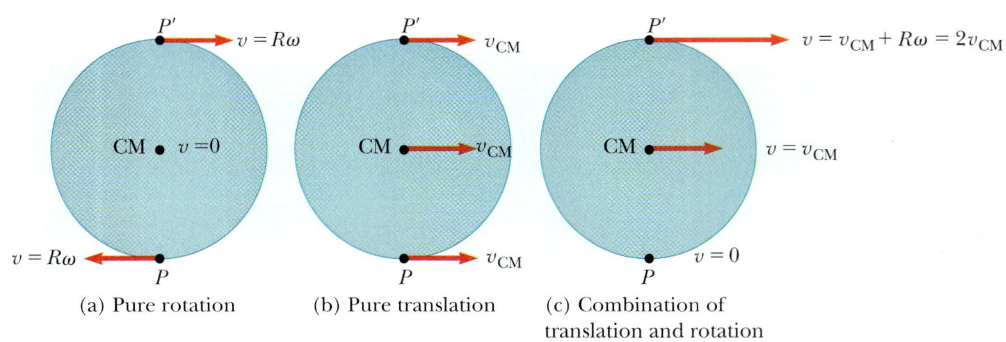

Figure 10.25 The motion of a rolling object can be modeled as a combination of pure translation and pure rotation.

would have if it were just translating through space without rotating. Therefore, the **total kinetic energy of a rolling object is the sum of the rotational kinetic energy about the center of mass and the translational kinetic energy of the center of mass.** This statement is consistent with the situation illustrated in Figure 10.25, which shows that the velocity of a point on the object is the sum of the velocity of the center of mass and the tangential velocity around the center of mass.

Energy methods can be used to treat a class of problems concerning the rolling motion of an object on a rough incline. For example, consider Active Figure 10.26, which shows a sphere rolling without slipping after being released from rest at the top of the incline. Accelerated rolling motion is possible only if a friction force is present between the sphere and the incline to produce a net torque about the center of mass. Despite the presence of friction, no loss of mechanical energy occurs because the contact point is at rest relative to the surface at any instant. (On the other hand, if the sphere were to slip, mechanical energy of the sphere–incline–Earth system would be lost due to the nonconservative force of kinetic friction.)

In reality, *rolling friction* causes mechanical energy to transform to internal energy. Rolling friction is due to deformations of the surface and the rolling object. For example, automobile tires flex as they roll on a roadway, representing a transformation of mechanical energy to internal energy. The roadway also deforms a small amount, representing additional rolling friction. In our problem-solving models, we ignore rolling friction unless stated otherwise.

Using $v_{CM} = R\omega$ for pure rolling motion, we can express Equation 10.28 as

$$K = \frac{1}{2}I_{CM}\left(\frac{v_{CM}}{R}\right)^2 + \frac{1}{2}Mv_{CM}^2$$

$$K = \frac{1}{2}\left(\frac{I_{CM}}{R^2} + M\right)v_{CM}^2 \qquad \textbf{(10.29)}$$

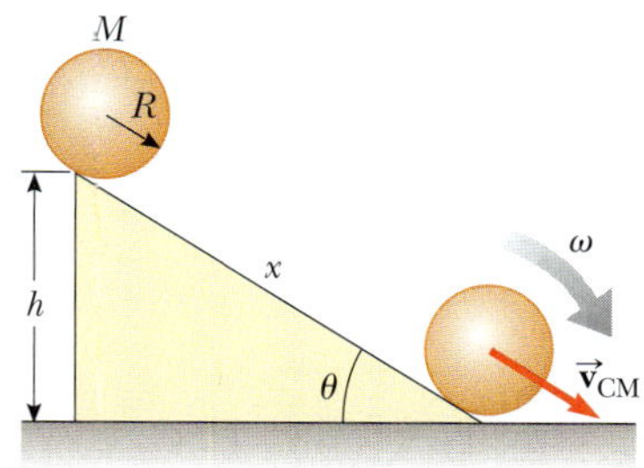

ACTIVE FIGURE 10.26

A sphere rolling down an incline. Mechanical energy of the sphere–Earth system is conserved if no slipping occurs.

Sign in at www.thomsonedu.com and go to ThomsonNOW to roll several objects down the hill and see how the final speed depends on the type of object.

For the sphere–Earth system, we define the zero configuration of gravitational potential energy to be when the sphere is at the bottom of the incline. Therefore, conservation of mechanical energy gives

$$K_f + U_f = K_i + U_i$$

$$\frac{1}{2}\left(\frac{I_{CM}}{R^2} + M\right)v_{CM}^2 + 0 = 0 + Mgh$$

$$v_{CM} = \left[\frac{2gh}{1 + (I_{CM}/MR^2)}\right]^{1/2} \qquad \textbf{(10.30)}$$

Quick Quiz 10.7 A ball rolls without slipping down incline A, starting from rest. At the same time, a box starts from rest and slides down incline B, which is identical to incline A except that it is frictionless. Which arrives at the bottom first? (a) The ball arrives first. (b) The box arrives first. (c) Both arrive at the same time. (d) It is impossible to determine.

EXAMPLE 10.13 Sphere Rolling Down an Incline

For the solid sphere shown in Active Figure 10.26, calculate the translational speed of the center of mass at the bottom of the incline and the magnitude of the translational acceleration of the center of mass.

SOLUTION

Conceptualize Imagine rolling the sphere down the incline. Compare it in your mind to a book sliding down a frictionless incline. You probably have experience with objects rolling down inclines and may be tempted to think that the sphere would move down the incline faster than the book. You do *not*, however, have experience with objects sliding down *frictionless* inclines! So, which object will reach the bottom first? (See Quick Quiz 10.7.)

Categorize We model the sphere and the Earth as an isolated system with no nonconservative forces acting. This model is the one that led to Equation 10.30, so we can use that result.

Analyze Evaluate the speed of the center of mass of the sphere from Equation 10.30:

$$(1) \quad v_{CM} = \left[\frac{2gh}{1 + \left(\frac{2}{5}MR^2/MR^2\right)}\right]^{1/2} = \left(\frac{10}{7}gh\right)^{1/2}$$

This result is less than $\sqrt{2gh}$, which is the speed an object would have if it simply slid down the incline without rotating. (Eliminate the rotation by setting $I_{CM} = 0$ in Equation 10.30.)

To calculate the translational acceleration of the center of mass, notice that the vertical displacement of the sphere is related to the distance x it moves along the incline through the relationship $h = x \sin \theta$.

Use this relationship to rewrite Equation (1):

$$v_{CM}^2 = \frac{10}{7}gx \sin \theta$$

Write Equation 2.17 for an object starting from rest and moving through a distance x:

$$v_{CM}^2 = 2a_{CM}x$$

Equate the preceding two expressions to find a_{CM}:

$$a_{CM} = \frac{5}{7}g \sin \theta$$

Finalize Both the speed and the acceleration of the center of mass are *independent* of the mass and the radius of the sphere. That is, **all homogeneous solid spheres experience the same speed and acceleration on a given incline.** Try to verify this statement experimentally with balls of different sizes, such as a marble and a croquet ball.

If we were to repeat the acceleration calculation for a hollow sphere, a solid cylinder, or a hoop, we would obtain similar results in which only the factor in front of $g \sin \theta$ would differ. The constant factors that appear in the expressions for v_{CM} and a_{CM} depend only on the moment of inertia about the center of mass for the specific object. In all cases, the acceleration of the center of mass is *less* than $g \sin \theta$, the value the acceleration would have if the incline were frictionless and no rolling occurred.

EXAMPLE 10.14 Pulling on a Spool[3]

A cylindrically symmetric spool of mass m and radius R sits at rest on a horizontal table with friction (Fig. 10.27). With your hand on a massless string wrapped around the axle of radius r, you pull on the spool with a constant horizontal force of magnitude T to the right. As a result, the spool rolls without slipping a distance L along the table with no rolling friction.

(A) Find the final translational speed of the center of mass of the spool.

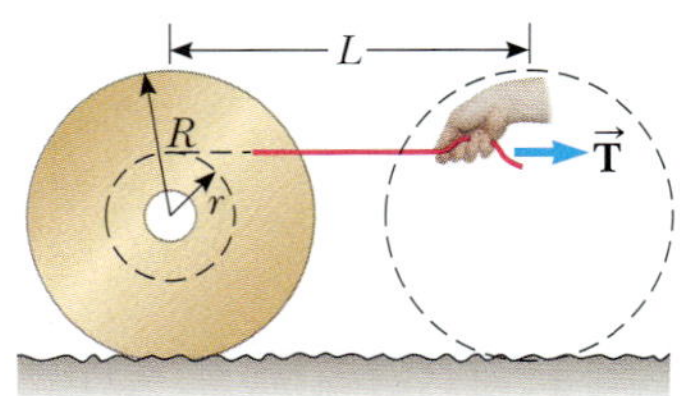

Figure 10.27 (Example 10.14) A spool rests on a horizontal table. A string is wrapped around the axle and is pulled to the right by a hand.

SOLUTION

Conceptualize Use Figure 10.27 to visualize the motion of the spool when you pull the string. For the spool to roll through a distance L, notice that your hand on the string must pull through a distance *different* from L.

Categorize The spool is a rigid object under a net torque, but the net torque includes that due to the friction force, about which we know nothing. Therefore, an approach based on the rigid object under a net torque model will not be successful. Work is done by your hand on the spool and string, which form a nonisolated system. Let's see if an approach based on the nonisolated system model is fruitful.

[3] Example 10.14 was inspired in part by C. E. Mungan, "A primer on work–energy relationships for introductory physics," *The Physics Teacher,* 43:10, 2005.

Analyze The only type of energy that changes in the system is the kinetic energy of the spool. There is no rolling friction, so there is no change in internal energy. The only way that energy crosses the system's boundary is by the work done by your hand on the string. No work is done by the static force of friction on the bottom of the spool because the point of application of the force moves through no displacement.

Write the appropriate reduction of the conservation of energy equation, Equation 8.2:

$$(1)\quad W = \Delta K = \Delta K_{\text{trans}} + \Delta K_{\text{rot}}$$

where W is the work done on the string by your hand. To find this work, we need to find the displacement of your hand during the process.

We first find the length of string that has unwound off the spool. If the spool rolls through a distance L, the total angle through which it rotates is $\theta = L/R$. The axle also rotates through this angle.

Use Equation 10.1a to find the total arc length through which the axle turns:

$$\ell = r\theta = \frac{r}{R}L$$

This result also gives the length of string pulled off the axle. Your hand will move through this distance *plus* the distance L through which the spool moves. Therefore, the magnitude of the displacement of the point of application of the force applied by your hand is $\ell + L = L(1 + r/R)$.

Evaluate the work done by your hand on the string:

$$(2)\quad W = TL\left(1 + \frac{r}{R}\right)$$

Substitute Equation (2) into Equation (1):

$$TL\left(1 + \frac{r}{R}\right) = \tfrac{1}{2}mv_{\text{CM}}^2 + \tfrac{1}{2}I\omega^2$$

where I is the moment of inertia of the spool about its center of mass and v_{CM} and ω are the final values after the wheel rolls through the distance L.

Apply the nonslip rolling condition $\omega = v_{\text{CM}}/R$:

$$TL\left(1 + \frac{r}{R}\right) = \tfrac{1}{2}mv_{\text{CM}}^2 + \tfrac{1}{2}I\frac{v_{\text{CM}}^2}{R^2}$$

Solve for v_{CM}:

$$(3)\quad v_{\text{CM}} = \sqrt{\frac{2TL(1 + r/R)}{m(1 + I/mR^2)}}$$

(B) Find the value of the friction force f.

SOLUTION

Categorize Because the friction force does no work, we cannot evaluate it from an energy approach. We model the spool as a nonisolated system, but this time in terms of momentum. The string applies a force across the boundary of the system, resulting in an impulse on the system. Because the forces on the spool are constant, we can model the spool's center of mass as a particle under constant acceleration.

Analyze Write the impulse–momentum theorem (Eq. 9.40) for the spool:

$$(4)\quad (T - f)\Delta t = m(v_{\text{CM}} - 0) = mv_{\text{CM}}$$

For a particle under constant acceleration starting from rest, Equation 2.14 tells us that the average velocity of the center of mass is half the final velocity.

Use Equation 2.2 to find the time interval for the center of mass of the spool to move a distance L from rest to a final speed v_{CM}:

$$(5)\quad \Delta t = \frac{L}{v_{CM,\,avg}} = \frac{2L}{v_{CM}}$$

Substitute Equation (5) into Equation (4):

$$(T - f)\frac{2L}{v_{CM}} = mv_{CM}$$

Solve for the friction force f:

$$f = T - \frac{mv_{CM}^2}{2L}$$

Substitute v_{CM} from Equation (3):

$$f = T - \frac{m}{2L}\left[\frac{2TL(1 + r/R)}{m(1 + I/mR^2)}\right]$$

$$= T - T\frac{(1 + r/R)}{(1 + I/mR^2)} = T\left[1 - \frac{(1 + r/R)}{(1 + I/mR^2)}\right]$$

Finalize Notice that we could use the impulse-momentum theorem for the translational motion of the spool while ignoring that the spool is rotating! This fact demonstrates the power of our growing list of approaches to solving problems.

Summary

ThomsonNOW™ Sign in at **www.thomsonedu.com** and go to ThomsonNOW to take a practice test for this chapter.

DEFINITIONS

The **angular position** of a rigid object is defined as the angle θ between a reference line attached to the object and a reference line fixed in space. The **angular displacement** of a particle moving in a circular path or a rigid object rotating about a fixed axis is $\Delta\theta \equiv \theta_f - \theta_i$.

The **instantaneous angular speed** of a particle moving in a circular path or of a rigid object rotating about a fixed axis is

$$\omega \equiv \frac{d\theta}{dt} \quad \textbf{(10.3)}$$

The **instantaneous angular acceleration** of a particle moving in a circular path or of a rigid object rotating about a fixed axis is

$$\alpha \equiv \frac{d\omega}{dt} \quad \textbf{(10.5)}$$

When a rigid object rotates about a fixed axis, every part of the object has the same angular speed and the same angular acceleration.

The **moment of inertia of a system of particles** is defined as

$$I \equiv \sum_i m_i r_i^2 \quad \textbf{(10.15)}$$

where m_i is the mass of the ith particle and r_i is its distance from the rotation axis.

The magnitude of the **torque** associated with a force $\vec{\mathbf{F}}$ acting on an object at a distance r from the rotation axis is

$$\tau \equiv rF\sin\phi = Fd \quad \textbf{(10.19)}$$

where ϕ is the angle between the position vector of the point of application of the force and the force vector, and d is the moment arm of the force, which is the perpendicular distance from the rotation axis to the line of action of the force.

CONCEPTS AND PRINCIPLES

When a rigid object rotates about a fixed axis, the angular position, angular speed, and angular acceleration are related to the translational position, translational speed, and translational acceleration through the relationships

$$s = r\theta \quad \textbf{(10.1a)}$$

$$v = r\omega \quad \textbf{(10.10)}$$

$$a_t = r\alpha \quad \textbf{(10.11)}$$

If a rigid object rotates about a fixed axis with angular speed ω, its **rotational kinetic energy** can be written

$$K_R = \tfrac{1}{2}I\omega^2 \quad \textbf{(10.16)}$$

where I is the moment of inertia about the axis of rotation.

The **moment of inertia of a rigid object** is

$$I = \int r^2\, dm \quad \textbf{(10.17)}$$

where r is the distance from the mass element dm to the axis of rotation.

The rate at which work is done by an external force in rotating a rigid object about a fixed axis, or the **power** delivered, is

$$\mathcal{P} = \tau\omega \quad \textbf{(10.23)}$$

If work is done on a rigid object and the only result of the work is rotation about a fixed axis, the net work done by external forces in rotating the object equals the change in the rotational kinetic energy of the object:

$$\sum W = \tfrac{1}{2}I\omega_f^{\,2} - \tfrac{1}{2}I\omega_i^{\,2} \quad \textbf{(10.24)}$$

The **total kinetic energy** of a rigid object rolling on a rough surface without slipping equals the rotational kinetic energy about its center of mass plus the translational kinetic energy of the center of mass:

$$K = \tfrac{1}{2}I_{\text{CM}}\omega^2 + \tfrac{1}{2}Mv_{\text{CM}}^{\,2} \quad \textbf{(10.28)}$$

ANALYSIS MODELS FOR PROBLEM SOLVING

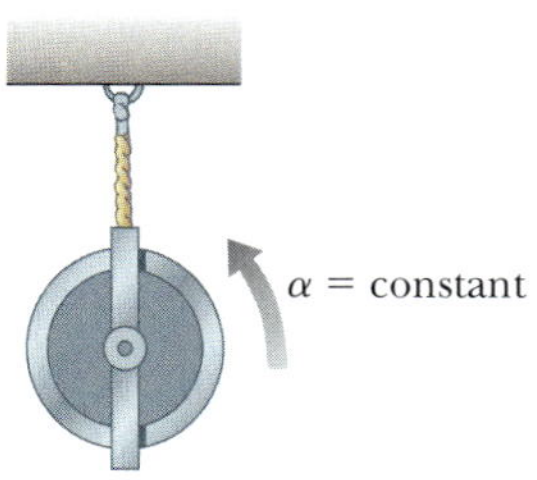

Rigid Object Under Constant Angular Acceleration. If a rigid object rotates about a fixed axis under constant angular acceleration, one can apply equations of kinematics that are analogous to those for translational motion of a particle under constant acceleration:

$$\omega_f = \omega_i + \alpha t \quad \textbf{(10.6)}$$

$$\theta_f = \theta_i + \omega_i t + \tfrac{1}{2}\alpha t^2 \quad \textbf{(10.7)}$$

$$\omega_f^{\,2} = \omega_i^{\,2} + 2\alpha(\theta_f - \theta_i) \quad \textbf{(10.8)}$$

$$\theta_f = \theta_i + \tfrac{1}{2}(\omega_i + \omega_f)t \quad \textbf{(10.9)}$$

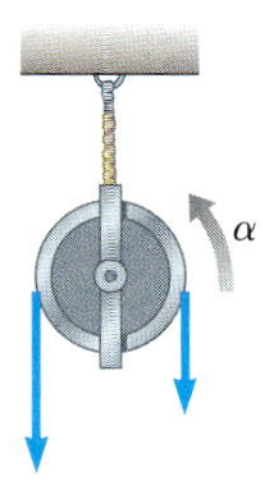

Rigid Object Under a Net Torque. If a rigid object free to rotate about a fixed axis has a net external torque acting on it, the object undergoes an angular acceleration α, where

$$\sum \tau = I\alpha \quad \textbf{(10.21)}$$

This equation is the rotational analog to Newton's second law in the particle under a net force model.

Questions

☐ denotes answer available in *Student Solutions Manual/Study Guide;* **O** denotes objective question

1. What is the angular speed of the second hand of a clock? What is the direction of $\vec{\omega}$ as you view a clock hanging on a vertical wall? What is the magnitude of the angular acceleration vector $\vec{\alpha}$ of the second hand?

2. One blade of a pair of scissors rotates counterclockwise in the xy plane. What is the direction of $\vec{\omega}$? What is the direction of $\vec{\alpha}$ if the magnitude of the angular velocity is decreasing in time?

3. **O** A wheel is moving with constant angular acceleration 3 rad/s^2. At different moments its angular speed is -2 rad/s, 0, and $+$ 2 rad/s. At these moments, analyze the magnitude of the tangential component of acceleration and the magnitude of the radial component of acceleration for a point on the rim of the wheel. Rank the following six items from largest to smallest: (a) $|a_t|$ when $\omega = -2$ rad/s (b)$|a_r|$ when $\omega = -2$ rad/s (c) $|a_t|$ when $\omega = 0$ (d) $|a_r|$ when $\omega = 0$ (e)$|a_t|$ when $\omega = 2$ rad/s (f) $|a_r|$ when $\omega = 2$ rad/s If two items are equal, show them as equal in your ranking. If a quantity is equal to zero, show that in your ranking.

4. **O (i)** Suppose a car's standard tires are replaced with tires 1.30 times larger in diameter. Then what will the speedometer reading be? (a) 1.69 times too high (b) 1.30 times too high (c) accurate (d) 1.30 times too low (e) 1.69 times too low (e) inaccurate by an unpredictable factor **(ii)** What will be the car's fuel economy in miles per gallon or km/L? (a) 1.69 times better (b) 1.30 times better (c) essentially the same (d) 1.30 times worse (e) 1.69 times worse

5. **O** Figure 10.8 shows a system of four particles joined by light, rigid rods. Assume $a = b$ and M is somewhat larger than m. **(i)** About which of the coordinate axes does the system have the smallest moment of inertia? (a) the x axis (b) the y axis (c) the z axis (d) The moment of inertia has the same small value for two axes. (e) The moment of inertia is the same for all axes. **(ii)** About which axis does the system have the largest moment of inertia? (a) the x axis (b) the y axis (c) the z axis (d) The moment of inertia has the same large value for two axes. (e) The moment of inertia is the same for all axes.

6. Suppose just two external forces act on a stationary rigid object and the two forces are equal in magnitude and opposite in direction. Under what condition does the object start to rotate?

7. **O** As shown in Figure 10.19, a cord is wrapped onto a cylindrical reel mounted on a fixed, frictionless, horizontal axle. Two experiments are conducted. (a) The cord is pulled down with a constant force of 50 N. (b) An object of weight 50 N is hung from the cord and released. Are the angular accelerations equal in the two experiments? If not, in which experiment is the angular acceleration greater in magnitude?

8. Explain how you might use the apparatus described in Example 10.10 to determine the moment of inertia of the wheel. (If the wheel does not have a uniform mass density, the moment of inertia is not necessarily equal to $\frac{1}{2}MR^2$.)

9. **O** A constant nonzero net torque is exerted on an object. Which of the following can *not* be constant? Choose all that apply. (a) angular position (b) angular velocity (c) angular acceleration (d) moment of inertia (e) kinetic energy (f) location of center of mass

10. Using the results from Example 10.10, how would you calculate the angular speed of the wheel and the linear speed of the suspended counterweight at $t = 2$ s, assuming the system is released from rest at $t = 0$? Is the expression $v = R\omega$ valid in this situation?

11. If a small sphere of mass M were placed at the end of the rod in Figure 10.21, would the result for ω be greater than, less than, or equal to the value obtained in Example 10.11?

12. **O** A solid aluminum sphere of radius R has moment of inertia I about an axis through its center. What is the moment of inertia about a central axis of a solid aluminum sphere of radius $2R$? (a) I (b) $2I$ (c) $4I$ (d) $8I$ (e) $16I$ (f) $32I$

13. Explain why changing the axis of rotation of an object changes its moment of inertia.

14. Suppose you remove two eggs from the refrigerator, one hard-boiled and the other uncooked. You wish to determine which is the hard-boiled egg without breaking the eggs. This determination can be made by spinning the two eggs on the floor and comparing the rotational motions. Which egg spins faster? Which egg rotates more uniformly? Explain.

15. Which of the entries in Table 10.2 applies to finding the moment of inertia of a long, straight sewer pipe rotating about its axis of symmetry? Of an embroidery hoop rotating about an axis through its center and perpendicular to its plane? Of a uniform door turning on its hinges? Of a coin turning about an axis through its center and perpendicular to its faces?

16. Is it possible to change the translational kinetic energy of an object without changing its rotational energy?

17. Must an object be rotating to have a nonzero moment of inertia?

18. If you see an object rotating, is there necessarily a net torque acting on it?

19. **O** A decoration hangs from the ceiling of your room at the bottom end of a string. Your bored roommate turns the decoration clockwise several times to wind up the string. When your roommate releases it, the decoration starts to spin counterclockwise, slowly at first and then faster and faster. Take counterclockwise as the positive sense and assume friction is negligible. When the string is entirely unwound, the ornament has its maximum rate of rotation. **(i)** At this moment, is its angular acceleration (a) positive, (b) negative, or (c) zero? **(ii)** The decoration continues to spin, winding the string counterclockwise as it slows down. At the moment it finally stops, is its angular acceleration (a) positive, (b) negative, or (c) zero?

20. The polar diameter of the Earth is slightly less than the equatorial diameter. How would the moment of inertia of

the Earth about its axis of rotation change if some material from near the equator were removed and transferred to the polar regions to make the Earth a perfect sphere?

21. **O** A basketball rolls across a floor without slipping, with its center of mass moving at a certain velocity. A block of ice of the same mass is set sliding across the floor with the same speed along a parallel line. **(i)** How do their energies compare? (a) The basketball has more kinetic energy. (b) The ice has more kinetic energy. (c) They have equal kinetic energies. **(ii)** How do their momenta compare? (a) The basketball has more momentum. (b) The ice has more momentum. (c) They have equal momenta. (d) Their momenta have equal magnitudes but are different vectors. **(iii)** The two objects encounter a ramp sloping upward. (a) The basketball will travel farther up the ramp. (b) The ice will travel farther up the ramp. (c) They will travel equally far up the ramp.

22. Suppose you set your textbook sliding across a gymnasium floor with a certain initial speed. It quickly stops moving because of a friction force exerted on it by the floor. Next, you start a basketball rolling with the same initial speed. It keeps rolling from one end of the gym to the other. Why does the basketball roll so far? Does friction significantly affect its motion?

23. Three objects of uniform density—a solid sphere, a solid cylinder, and a hollow cylinder—are placed at the top of an incline (Fig. Q10.23). They are all released from rest at the same elevation and roll without slipping. Which object reaches the bottom first? Which reaches it last? Try this experiment at home and notice that the result is independent of the masses and the radii of the objects.

Figure Q10.23

24. Figure Q10.24 shows a side view of a child's tricycle with rubber tires on a horizontal concrete sidewalk. If a string is attached to the upper pedal on the far side and pulled forward horizontally, the tricycle rolls forward. Instead, assume a string is attached to the lower pedal on the near side and pulled forward horizontally as shown by A. Does the tricycle start to roll? If so, which way? Answer the same questions if (b) the string is pulled forward and upward as shown by B, (c) the string is pulled straight down as shown by C, and (d) the string is pulled forward and downward as shown by D. (e) **What if** the string is instead attached to the rim of the front wheel and pulled upward and backward as shown by E? (f) Explain a pattern of reasoning, based on the diagram, that makes it easy to answer questions such as all of these. What physical quantity must you evaluate?

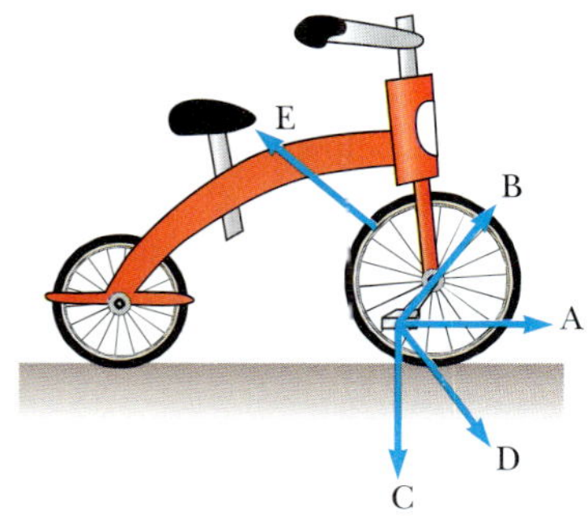

Figure Q10.24

Problems

WebAssign The Problems from this chapter may be assigned online in WebAssign.

ThomsonNOW™ Sign in at **www.thomsonedu.com** and go to ThomsonNOW to assess your understanding of this chapter's topics with additional quizzing and conceptual questions.

1, 2, 3 denotes straightforward, intermediate, challenging; □ denotes full solution available in *Student Solutions Manual/Study Guide;* ▲ denotes coached solution with hints available at **www.thomsonedu.com;** ▭ denotes developing symbolic reasoning; ● denotes asking for qualitative reasoning; ▪ denotes computer useful in solving problem

Section 10.1 Angular Position, Velocity, and Acceleration

1. During a certain period of time, the angular position of a swinging door is described by $\theta = 5.00 + 10.0t + 2.00t^2$, where θ is in radians and t is in seconds. Determine the angular position, angular speed, and angular acceleration of the door (a) at $t = 0$ and (b) at $t = 3.00$ s.

2. A bar on a hinge starts from rest and rotates with an angular acceleration $\alpha = (10 + 6t)$ rad/s^2, where t is in seconds. Determine the angle in radians through which the bar turns in the first 4.00 s.

Section 10.2 Rotational Kinematics: The Rigid Object Under Constant Angular Acceleration

3. A wheel starts from rest and rotates with constant angular acceleration to reach an angular speed of 12.0 rad/s in 3.00 s. Find (a) the magnitude of the angular acceleration of the wheel and (b) the angle in radians through which it rotates in this time interval.

4. A centrifuge in a medical laboratory rotates at an angular speed of 3 600 rev/min. When switched off, it rotates

through 50.0 revolutions before coming to rest. Find the constant angular acceleration of the centrifuge.

5. ▲ An electric motor rotating a grinding wheel at 100 rev/min is switched off. The wheel then moves with constant negative angular acceleration of magnitude 2.00 rad/s^2. (a) During what time interval does the wheel come to rest? (b) Through how many radians does it turn while it is slowing down?

6. A rotating wheel requires 3.00 s to rotate through 37.0 revolutions. Its angular speed at the end of the 3.00-s interval is 98.0 rad/s. What is the constant angular acceleration of the wheel?

7. (a) Find the angular speed of the Earth's rotation on its axis. As the Earth turns toward the east, we see the sky turning toward the west at this same rate.

(b) *The rainy Pleiads wester*
And seek beyond the sea
The head that I shall dream of
That shall not dream of me.

—A. E. Housman (© Robert E. Symons)

Cambridge, England is at longitude 0°, and Saskatoon, Saskatchewan, Canada is at longitude 107° west. How much time elapses after the Pleiades set in Cambridge until these stars fall below the western horizon in Saskatoon?

8. A merry-go-round is stationary. A dog is running on the ground just outside the merry-go-round's circumference, moving with a constant angular speed of 0.750 rad/s. The dog does not change his pace when he sees what he has been looking for: a bone resting on the edge of the merry-go-round one third of a revolution in front of him. At the instant the dog sees the bone ($t = 0$), the merry-go-round begins to move in the direction the dog is running, with a constant angular acceleration equal to 0.015 0 rad/s^2. (a) At what time will the dog reach the bone? (b) The confused dog keeps running and passes the bone. How long after the merry-go-round starts to turn do the dog and the bone draw even with each other for the second time?

9. The tub of a washing machine goes into its spin cycle, starting from rest and gaining angular speed steadily for 8.00 s, at which time it is turning at 5.00 rev/s. At this point, the person doing the laundry opens the lid and a safety switch turns off the machine. The tub smoothly slows to rest in 12.0 s. Through how many revolutions does the tub turn while it is in motion?

Section 10.3 Angular and Translational Quantities

10. A racing car travels on a circular track of radius 250 m. Assuming the car moves with a constant speed of 45.0 m/s, find (a) its angular speed and (b) the magnitude and direction of its acceleration.

11. Make an order-of-magnitude estimate of the number of revolutions through which a typical automobile tire turns in 1 yr. State the quantities you measure or estimate and their values.

12. ● Figure P10.12 shows the drive train of a bicycle that has wheels 67.3 cm in diameter and pedal cranks 17.5 cm long. The cyclist pedals at a steady cadence of 76.0 rev/min. The chain engages with a front sprocket 15.2 cm in diameter and a rear sprocket 7.00 cm in diameter. (a) Calculate the speed of a link of the chain relative to the bicycle frame. (b) Calculate the angular speed of the bicycle wheels. (c) Calculate the speed of the bicycle relative to the road. (d) What pieces of data, if any, are not necessary for the calculations?

Figure P10.12

13. A wheel 2.00 m in diameter lies in a vertical plane and rotates with a constant angular acceleration of 4.00 rad/s^2. The wheel starts at rest at $t = 0$, and the radius vector of a certain point P on the rim makes an angle of 57.3° with the horizontal at this time. At $t =$ 2.00 s, find (a) the angular speed of the wheel, (b) the tangential speed and the total acceleration of the point P, and (c) the angular position of the point P.

14. A discus thrower (Fig. P10.14) accelerates a discus from rest to a speed of 25.0 m/s by whirling it through 1.25 rev. Assume the discus moves on the arc of a circle 1.00 m in radius. (a) Calculate the final angular speed of the discus. (b) Determine the magnitude of the angular acceleration of the discus, assuming it to be constant. (c) Calculate the time interval required for the discus to accelerate from rest to 25.0 m/s.

Bruce Ayers/Stone/Getty

Figure P10.14

15. A small object with mass 4.00 kg moves counterclockwise with constant speed 4.50 m/s in a circle of radius 3.00 m centered at the origin. It starts at the point with position vector $(3.00\hat{\mathbf{i}} + 0\hat{\mathbf{j}})$ m. Then it undergoes an angular displacement of 9.00 rad. (a) What it its position vector? Use unit–vector notation for all vector answers. (b) In what quadrant is the particle located, and what angle does its position vector make with the positive x axis? (c) What is its velocity? (d) In what direction is it moving? Make a sketch of its position, velocity, and acceleration vectors. (e) What is its acceleration? (f) What total force is exerted on the object?

16. A car accelerates uniformly from rest and reaches a speed of 22.0 m/s in 9.00 s. The tires have diameter 58.0 cm and do not slip on the pavement. (a) Find the number of revolutions each tire makes during this motion. (b) What is the final angular speed of a tire in revolutions per second?

17. ▲ A disk 8.00 cm in radius rotates at a constant rate of 1 200 rev/min about its central axis. Determine (a) its angular speed, (b) the tangential speed at a point 3.00 cm from its center, (c) the radial acceleration of a point on the rim, and (d) the total distance a point on the rim moves in 2.00 s.

18. ● A straight ladder is leaning against the wall of a house. The ladder has rails 4.90 m long, joined by rungs 0.410 m long. Its bottom end is on solid but sloping ground so that the top of the ladder is 0.690 m to the left of where it should be, and the ladder is unsafe to climb. You want to put a rock under one foot of the ladder to compensate for the slope of the ground. (a) What should be the thickness of the flat rock? (b) Does using ideas from this chapter make it easier to explain the solution to part (a)? Explain your answer.

19. A car traveling on a flat (unbanked) circular track accelerates uniformly from rest with a tangential acceleration of 1.70 m/s^2. The car makes it one-quarter of the way around the circle before it skids off the track. Determine the coefficient of static friction between the car and track from these data.

20. In part (B) of Example 10.2, the compact disc was modeled as a rigid object under constant angular acceleration to find the total angular displacement during the playing time of the disc. In reality, the angular acceleration of a disc is not constant. In this problem, let us explore the actual time dependence of the angular acceleration. (a) Assume the track on the disc is a spiral such that adjacent loops of the track are separated by a small distance h. Show that the radius r of a given portion of the track is given by

$$r = r_i + \frac{h}{2\pi}\theta$$

where r_i is the radius of the innermost portion of the track and θ is the angle through which the disc turns to arrive at the location of the track of radius r. (b) Show that the rate of change of the angle θ is given by

$$\frac{d\theta}{dt} = \frac{v}{r_i + (h/2\pi)\theta}$$

where v is the constant speed with which the disc surface passes the laser. (c) From the result in part (b), use integration to find an expression for the angle θ as a function of time. (d) From the result in part (c), use differentiation to find the angular acceleration of the disc as a function of time.

Section 10.4 Rotational Kinetic Energy

21. ▲ The four particles in Figure P10.21 are connected by rigid rods of negligible mass. The origin is at the center of the rectangle. The system rotates in the xy plane about the z axis with an angular speed of 6.00 rad/s. Calculate (a) the moment of inertia of the system about the z axis and (b) the rotational kinetic energy of the system.

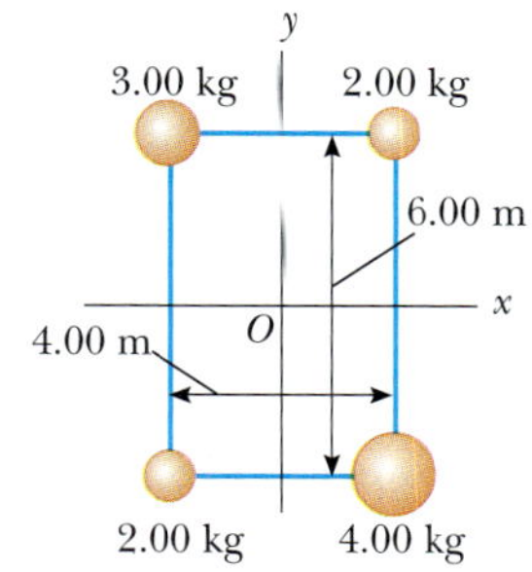

Figure P10.21

22. ● Rigid rods of negligible mass lying along the y axis connect three particles (Fig. P10.22). The system rotates about the x axis with an angular speed of 2.00 rad/s. Find (a) the moment of inertia about the x axis and the total rotational kinetic energy evaluated from $\frac{1}{2}I\omega^2$ and (b) the tangential speed of each particle and the total kinetic energy evaluated from $\Sigma\,\frac{1}{2}m_iv_i^{\,2}$. (c) Compare the answers for kinetic energy in parts (a) and (b).

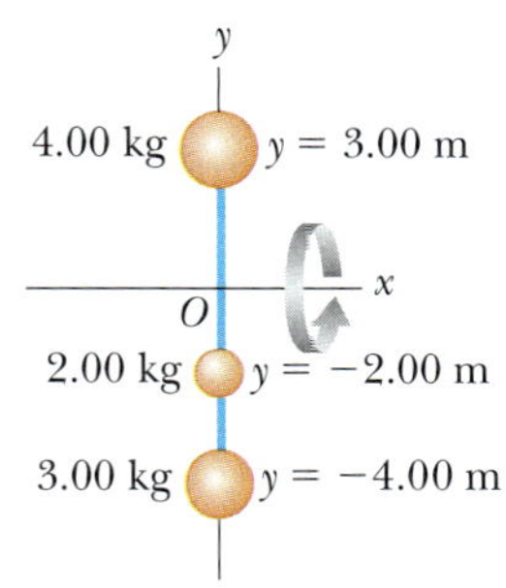

Figure P10.22

23. Two balls with masses M and m are connected by a rigid rod of length L and negligible mass as shown in Figure P10.23. For an axis perpendicular to the rod, show that the system has the minimum moment of inertia when the axis passes through the center of mass. Show that this moment of inertia is $I = \mu L^2$, where $\mu = mM/(m + M)$.

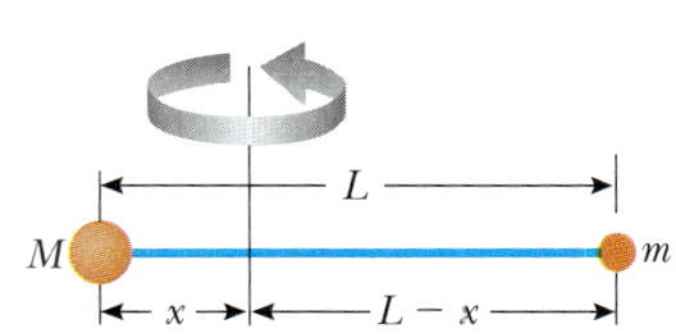

Figure P10.23

24. As a gasoline engine operates, a flywheel turning with the crankshaft stores energy after each fuel explosion to provide the energy required to compress the next charge of fuel and air. In the engine of a certain lawn tractor, suppose a flywheel must be no more than 18.0 cm in diameter. Its thickness, measured along its axis of rotation, must be no larger than 8.00 cm. The flywheel must release 60.0 J of energy when its angular speed drops from 800 rev/min to 600 rev/min. Design a sturdy steel flywheel to meet these requirements with the smallest mass you can reasonably attain. Assume the material has the density listed for iron in Table 14 1. Specify the shape and mass of the flywheel.

25. ● A *war-wolf* or *trebuchet* is a device used during the Middle Ages to throw rocks at castles and now sometimes used to fling large vegetables and pianos as a sport. A simple trebuchet is shown in Figure P10.25. Model it as a stiff rod of negligible mass, 3.00 m long, joining particles of mass 60.0 kg and 0.120 kg at its ends. It can turn on a frictionless, horizontal axle perpendicular to the rod and 14.0 cm from the large-mass particle. The rod is released from rest in a horizontal orientation. (a) Find the maximum speed that the 0.120-kg object attains. (b) While the 0.120-kg object is gaining speed, does it move with constant acceleration? Does it move with constant tangential acceleration? Does the trebuchet move with constant angular acceleration? Does it have constant momentum? Does the trebuchet-Earth system have constant mechanical energy?

Figure P10.25

Section 10.5 Calculation of Moments of Inertia

26. Three identical thin rods, each of length L and mass m, are welded perpendicular to one another as shown in Figure P10.26. The assembly is rotated about an axis that passes through the end of one rod and is parallel to another. Determine the moment of inertia of this structure.

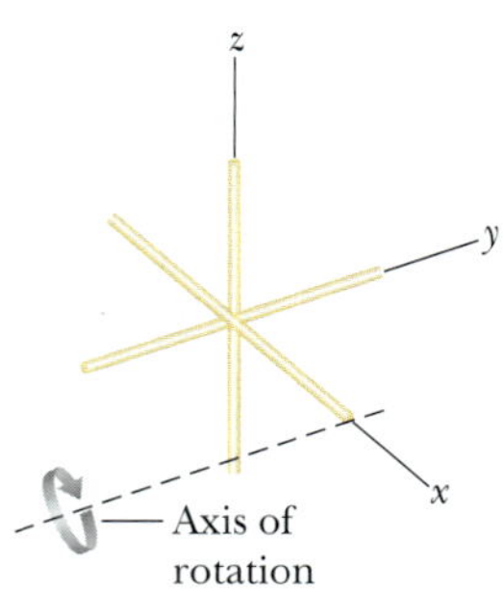

Figure P10.26

27. Figure P10.27 shows a side view of a car tire. Model it as having two sidewalls of uniform thickness 0.635 cm and a tread wall of uniform thickness 2.50 cm and width 20.0 cm. Assume the rubber has uniform density equal to 1.10×10^3 kg/m^3. Find its moment of inertia about an axis through its center.

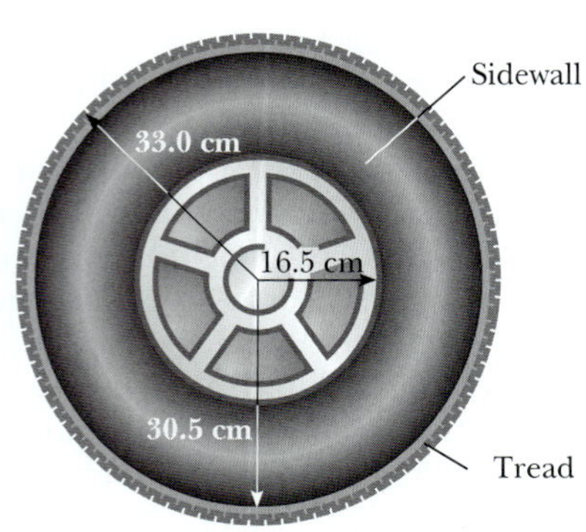

Figure P10.27

28. ● A uniform, thin solid door has height 2.20 m, width 0.870 m, and mass 23.0 kg. Find its moment of inertia for rotation on its hinges. Is any piece of data unnecessary?

29. *Attention! About face!* Compute an order-of-magnitude estimate for the moment of inertia of your body as you stand tall and turn about a vertical axis through the top of your head and the point halfway between your ankles. In your solution, state the quantities you measure or estimate and their values.

30. Many machines employ cams for various purposes such as opening and closing valves. In Figure P10.30, the cam is a circular disk rotating on a shaft that does not pass through the center of the disk. In the manufacture of the cam, a uniform solid cylinder of radius R is first machined. Then an off-center hole of radius $R/2$ is drilled, parallel to the axis of the cylinder, and centered at a point a distance $R/2$ from the cylinder's center. The cam, of mass M, is then slipped onto the circular shaft and welded into place. What is the kinetic energy of the cam when it is rotating with angular speed ω about the axis of the shaft?

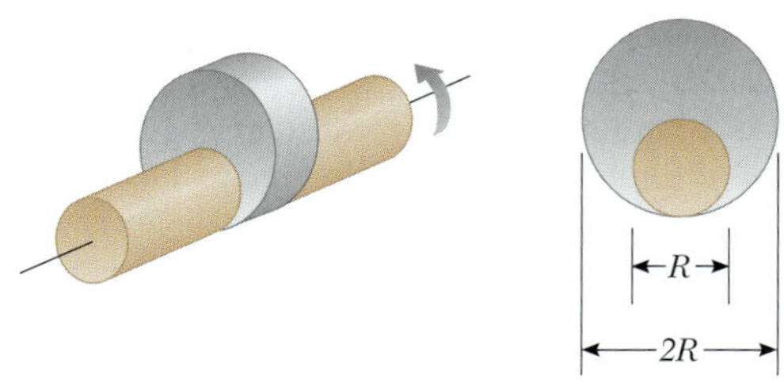

Figure P10.30

31. Following the procedure used in Example 10.4, prove that the moment of inertia about the y' axis of the rigid rod in Figure 10.9 is $\frac{1}{3}ML^2$.

Section 10.6 Torque

32. The fishing pole in Figure P10.32 makes an angle of 20.0° with the horizontal. What is the torque exerted by the fish about an axis perpendicular to the page and passing through the angler's hand?

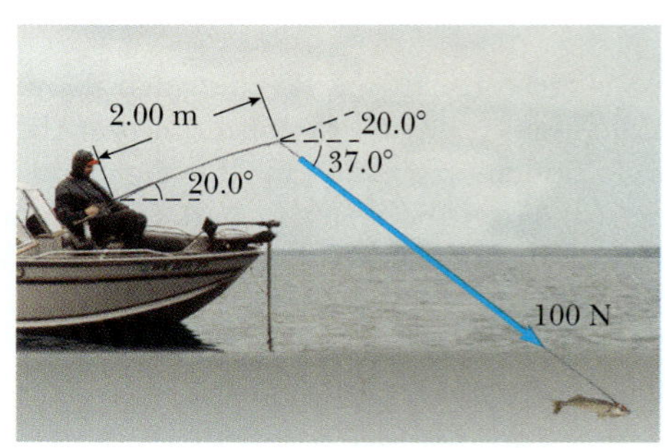

Figure P10.32

33. ▲ Find the net torque on the wheel in Figure P10.33 about the axle through O, taking a = 10.0 cm and b = 25.0 cm.

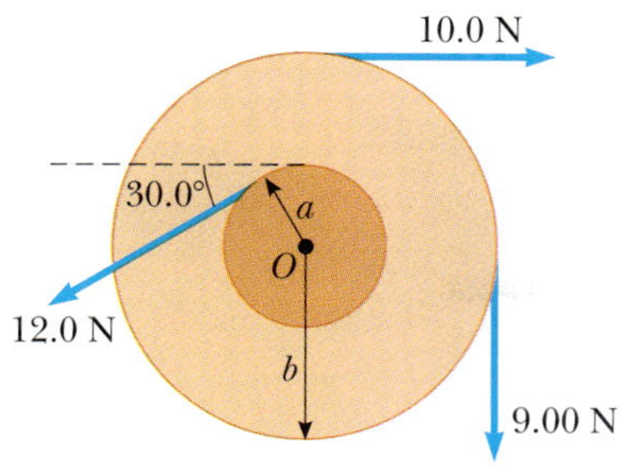

Figure P10.33

Section 10.7 The Rigid Object Under a Net Torque

34. A grinding wheel is in the form of a uniform solid disk of radius 7.00 cm and mass 2.00 kg. It starts from rest and accelerates uniformly under the action of the constant torque of 0.600 N·m that the motor exerts on the wheel. (a) How long does the wheel take to reach its final operating speed of 1 200 rev/min? (b) Through how many revolutions does it turn while accelerating?

35. ▲ A model airplane with mass 0.750 kg is tethered by a wire so that it flies in a circle 30.0 m in radius. The airplane engine provides a net thrust of 0.800 N perpendicular to the tethering wire. (a) Find the torque the net thrust produces about the center of the circle. (b) Find the angular acceleration of the airplane when it is in level flight. (c) Find the translational acceleration of the airplane tangent to its flight path.

36. The combination of an applied force and a friction force produces a constant total torque of 36.0 N·m on a wheel rotating about a fixed axis. The applied force acts for 6.00 s. During this time, the angular speed of the wheel increases from 0 to 10.0 rad/s. The applied force is then removed, and the wheel comes to rest in 60.0 s. Find (a) the moment of inertia of the wheel, (b) the magnitude of the frictional torque, and (c) the total number of revolutions of the wheel.

37. A block of mass $m_1 = 2.00$ kg and a block of mass $m_2 = 6.00$ kg are connected by a massless string over a pulley in the shape of a solid disk having radius $R = 0.250$ m and mass $M = 10.0$ kg. These blocks are allowed to move on a fixed wedge of angle $\theta = 30.0°$ as shown in Figure P10.37. The coefficient of kinetic friction is 0.360 for both blocks. Draw free-body diagrams of both blocks and of the pulley. Determine (a) the acceleration of the two blocks and (b) the tensions in the string on both sides of the pulley.

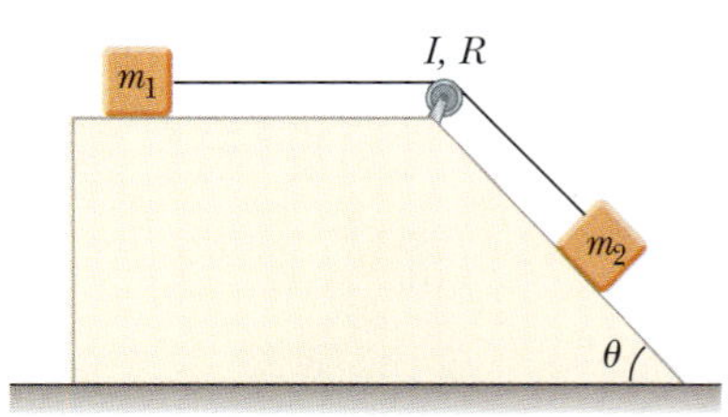

Figure P10.37

38. A potter's wheel—a thick stone disk of radius 0.500 m and mass 100 kg—is freely rotating at 50.0 rev/min. The potter can stop the wheel in 6.00 s by pressing a wet rag against the rim and exerting a radially inward force of 70.0 N. Find the effective coefficient of kinetic friction between wheel and rag.

39. An electric motor turns a flywheel through a drive belt that joins a pulley on the motor and a pulley that is rigidly attached to the flywheel as shown in Figure P10.39. The flywheel is a solid disk with a mass of 80.0 kg and a diameter of 1.25 m. It turns on a frictionless axle. Its pulley has much smaller mass and a radius of 0.230 m. The tension in the upper (taut) segment of the belt is 135 N, and the flywheel has a clockwise angular acceleration of 1.67 rad/s^2. Find the tension in the lower (slack) segment of the belt.

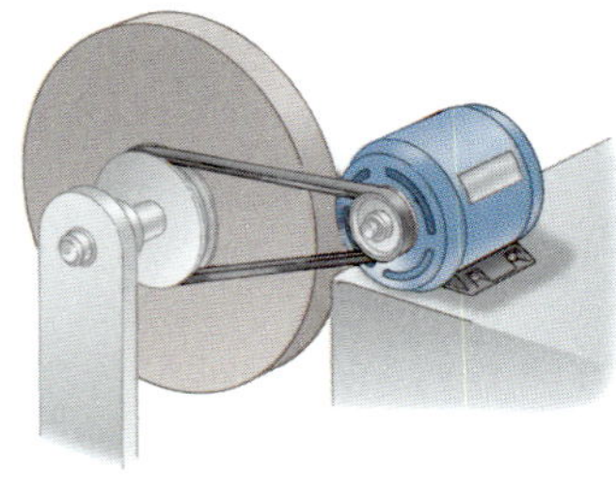

Figure P10.39

40. ● A disk having moment of inertia 100 kg·m^2 is free to rotate without friction, starting from rest, about a fixed axis through its center as shown at the top of Figure 10.19. A tangential force whose magnitude can range from $T = 0$ to $T = 50.0$ N can be applied at any distance ranging from $R = 0$ to $R = 3.00$ m from the axis of rotation. Find a pair of values of T and R that cause the disk to complete 2.00 revolutions in 10.0 s. Does one answer exist, or no answer, or two answers, or more than two, or many, or an infinite number?

Section 10.8 Energy Considerations in Rotational Motion

41. In a city with an air-pollution problem, a bus has no combustion engine. It runs on energy drawn from a large, rapidly rotating flywheel under the floor of the bus. At the bus terminal, the flywheel is spun up to its maximum rotation rate of 4 000 rev/min by an electric motor. Every time the bus speeds up, the flywheel slows down slightly. The bus is equipped with regenerative braking so that the flywheel can speed up when the bus slows down. The flywheel is a uniform solid cylinder with mass 1 600 kg and radius 0.650 m. The bus body does work against air resistance and rolling resistance at the average rate of 18.0 hp as it travels with an average speed of 40.0 km/h. How far can the bus travel before the flywheel has to be spun up to speed again?

42. Big Ben, the Parliament tower clock in London, has an hour hand 2.70 m long with a mass of 60.0 kg and a minute hand 4.50 m long with a mass of 100 kg (Fig.

Figure P10.42 Problem 42 and 76.

2 = intermediate; 3 = challenging; □ = SSM/SG; ▲ = ThomsonNOW; = symbolic reasoning; ● = qualitative reasoning

P10.42). Calculate the total rotational kinetic energy of the two hands about the axis of rotation. (You may model the hands as long, thin rods.)

43. The top in Figure P10.43 has a moment of inertia equal to 4.00×10^{-4} kg · m^2 and is initially at rest. It is free to rotate about the stationary axis AA'. A string, wrapped around a peg along the axis of the top, is pulled in such a manner as to maintain a constant tension of 5.57 N. If the string does not slip while it is unwound from the peg, what is the angular speed of the top after 80.0 cm of string has been pulled off the peg?

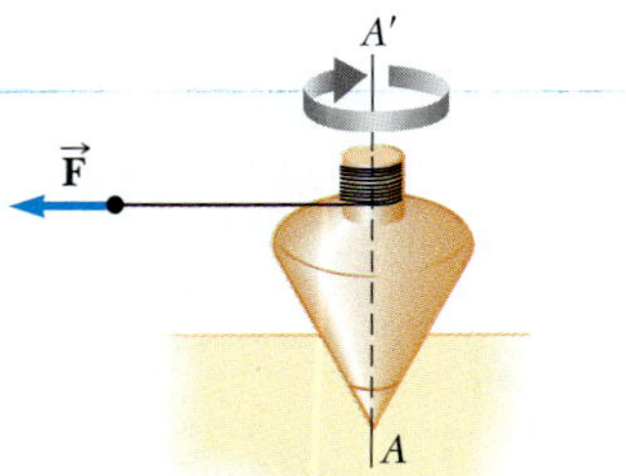

Figure P10.43

44. ● Consider the system shown in Figure P10.44 with $m_1 = 20.0$ kg, $m_2 = 12.5$ kg, $R = 0.200$ m, and the mass of the uniform pulley $M = 5.00$ kg. Object m_2 is resting on the floor, and object m_1 is 4.00 m above the floor when it is released from rest. The pulley axis is frictionless. The cord is light, does not stretch, and does not slip on the pulley. Calculate the time interval required for m_1 to hit the floor. How would your answer change if the pulley were massless?

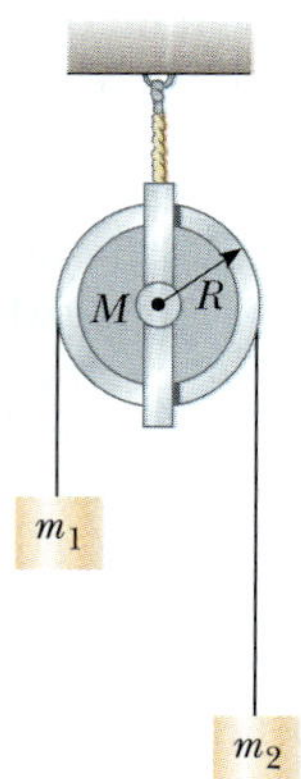

Figure P10.44

45. In Figure P10.45, the sliding block has a mass of 0.850 kg, the counterweight has a mass of 0.420 kg, and the pulley is a hollow cylinder with a mass of 0.350 kg, an inner radius of 0.020 0 m, and an outer radius of 0.030 0 m. The coefficient of kinetic friction between the block and the horizontal surface is 0.250. The pulley turns without friction on its axle. The light cord does not stretch and does not slip on the pulley. The block has a velocity of 0.820 m/s toward the pulley when it passes through a photogate. (a) Use energy methods to predict its speed after it has moved to a second photogate, 0.700 m away. (b) Find the angular speed of the pulley at the same moment.

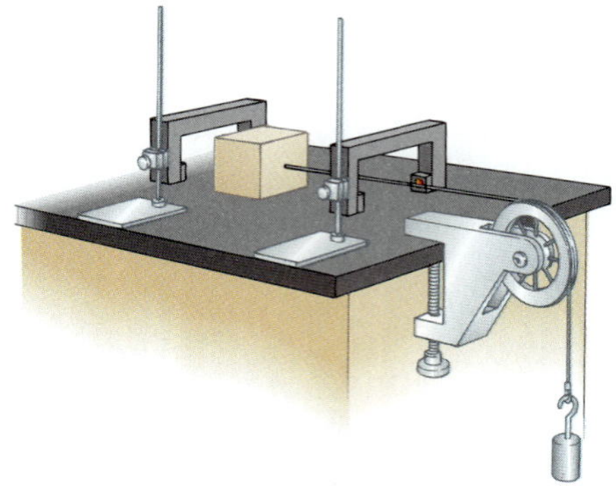

Figure P10.45

46. A cylindrical rod 24.0 cm long with mass 1.20 kg and radius 1.50 cm has a ball of diameter 8.00 cm and mass 2.00 kg attached to one end. The arrangement is originally vertical and stationary, with the ball at the top. The system is free to pivot about the bottom end of the rod after being given a slight nudge. (a) After the rod rotates through 90°, what is its rotational kinetic energy? (b) What is the angular speed of the rod and ball? (c) What is the linear speed of the ball? (d) How does this speed compare with the speed if the ball had fallen freely through the same distance of 28 cm?

47. An object with a weight of 50.0 N is attached to the free end of a light string wrapped around a reel of radius 0.250 m and mass 3.00 kg. The reel is a solid disk, free to rotate in a vertical plane about the horizontal axis passing through its center. The suspended object is released 6.00 m above the floor. (a) Determine the tension in the string, the acceleration of the object, and the speed with which the object hits the floor. (b) Verify your last answer by using the principle of conservation of energy to find the speed with which the object hits the floor.

48. A horizontal 800-N merry-go-round is a solid disk of radius 1.50 m, started from rest by a constant horizontal force of 50.0 N applied tangentially to the edge of the disk. Find the kinetic energy of the disk after 3.00 s.

49. This problem describes one experimental method for determining the moment of inertia of an irregularly shaped object such as the payload for a satellite. Figure P10.49 shows a counterweight of mass m suspended by a cord wound around a spool of radius r, forming part of a turntable supporting the object. The turntable can rotate without friction. When the counterweight is released from rest, it descends through a distance h, acquiring a speed v. Show that the moment of inertia I of the rotating apparatus (including the turntable) is $mr^2(2gh/v^2 - 1)$.

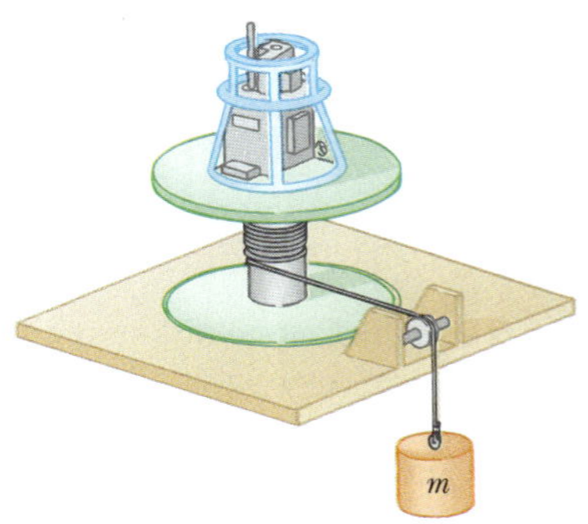

Figure P10.49

50. The head of a grass string trimmer has 100 g of cord wound in a light cylindrical spool with inside diameter 3.00 cm and outside diameter 18.0 cm, as shown in Figure P10.50. The cord has a linear density of 10.0 g/m. A single strand of the cord extends 16.0 cm from the outer edge of the spool. (a) When switched on, the trimmer speeds up from 0 to 2 500 rev/min in 0.215 s. (a) What average power is delivered to the head by the trimmer motor while it is accelerating? (b) When the trimmer is cutting grass, it spins at 2 000 rev/min and the grass exerts an average tangential force of 7.65 N on the outer end of the cord, which is still at a radial distance of 16.0 cm from the outer edge of the spool. What is the power delivered to the head under load?

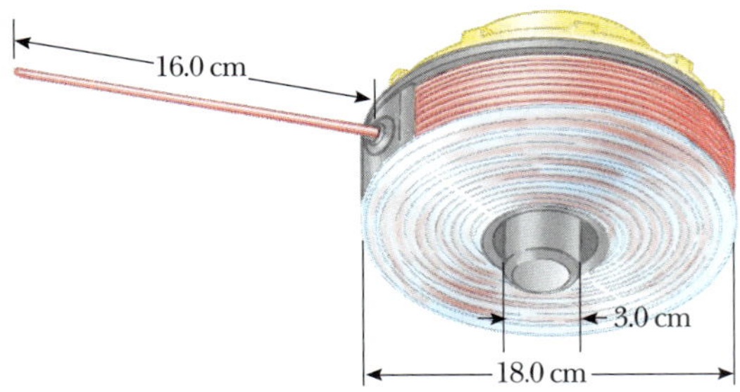

Figure P10.50

[51.] (a) A uniform solid disk of radius R and mass M is free to rotate on a frictionless pivot through a point on its rim (Fig. P10.51). If the disk is released from rest in the position shown by the blue circle, what is the speed of its center of mass when the disk reaches the position indicated by the dashed circle? (b) What is the speed of the lowest point on the disk in the dashed position? (c) **What If?** Repeat part (a) using a uniform hoop.

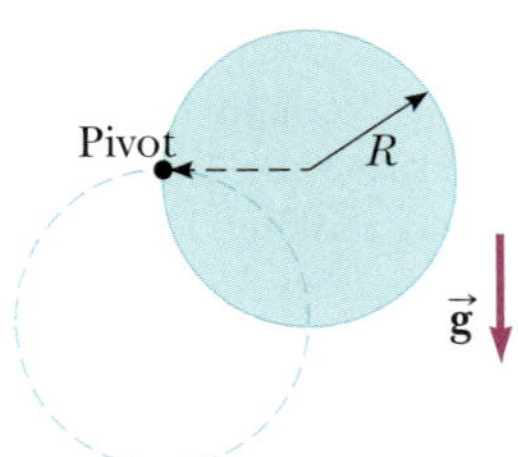

Figure P10.51

Section 10.9 Rolling Motion of a Rigid Object

52. ● A solid sphere is released from height h from the top of an incline making an angle θ with the horizontal. Calculate the speed of the sphere when it reaches the bottom of the incline (a) in the case that it rolls without slipping and (b) in the case that it slides frictionlessly without rolling. (c) Compare the time intervals required to reach the bottom in cases (a) and (b).

[53.] ▲ A cylinder of mass 10.0 kg rolls without slipping on a horizontal surface. At a certain instant its center of mass has a speed of 10.0 m/s. Determine (a) the translational kinetic energy of its center of mass, (b) the rotational kinetic energy about its center of mass, and (c) its total energy.

54. ● A smooth cube of mass m and edge length r slides with speed v on a horizontal surface with negligible friction. The cube then moves up a smooth incline that makes an angle θ with the horizontal. A cylinder of mass m and radius r rolls without slipping with its center of mass moving with speed v and encounters an incline of the same angle of inclination but with sufficient friction that the cylinder continues to roll without slipping. (a) Which object will go the greater distance up the incline? (b) Find the difference between the maximum distances the objects travel up the incline. (c) Explain what accounts for this difference in distances traveled.

[55.] (a) Determine the acceleration of the center of mass of a uniform solid disk rolling down an incline making angle θ with the horizontal. Compare this acceleration with that of a uniform hoop. (b) What is the minimum coefficient of friction required to maintain pure rolling motion for the disk?

56. A uniform solid disk and a uniform hoop are placed side by side at the top of an incline of height h. If they are released from rest at the same time and roll without slipping, which object reaches the bottom first? Verify your answer by calculating their speeds when they reach the bottom in terms of h.

57. ● A metal can containing condensed mushroom soup has mass 215 g, height 10.8 cm, and diameter 6.38 cm. It is placed at rest on its side at the top of a 3.00-m-long incline that is at 25.0° to the horizontal and is then released to roll straight down. It reaches the bottom of the incline after 1.50 s. Assuming mechanical energy conservation, calculate the moment of inertia of the can. Which pieces of data, if any, are unnecessary for calculating the solution?

58. ● A tennis ball is a hollow sphere with a thin wall. It is set rolling without slipping at 4.03 m/s on a horizontal section of a track as shown in Figure P10.58. It rolls around the inside of a vertical circular loop 90.0 cm in diameter and finally leaves the track at a point 20.0 cm below the horizontal section. (a) Find the speed of the ball at the top of the loop. Demonstrate that it will not fall from the track. (b) Find its speed as it leaves the track. **What If?** (c) Suppose static friction between ball and track were negligible so that the ball slid instead of rolling. Would its speed then be higher, lower, or the same at the top of the loop? Explain.

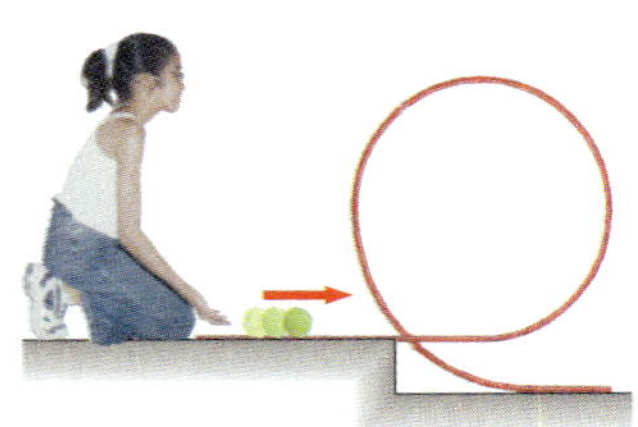

Figure P10.58

Additional Problems

59. As shown in Figure P10.59, toppling chimneys often break apart in midfall because the mortar between the bricks cannot withstand much shear stress. As the chimney begins to fall, shear forces must act on the topmost sections to accelerate them tangentially so that they can keep up with the rotation of the lower part of the stack. For simplicity, let us model the chimney as a uniform rod of

length ℓ pivoted at the lower end. The rod starts at rest in a vertical position (with the frictionless pivot at the bottom) and falls over under the influence of gravity. What fraction of the length of the rod has a tangential acceleration greater than $g \sin \theta$, where θ is the angle the chimney makes with the vertical axis?

Jerry Wachter/Photo Researchers, Inc.

Figure P10.59 A building demolition site in Baltimore, Maryland. At the left is a chimney, mostly concealed by the building, that has broken apart on its way down. Compare with Figure 10.18.

60. **Review problem.** A mixing beater consists of three thin rods, each 10.0 cm long. The rods diverge from a central hub, separated from one another by 120°, and all turn in the same plane. A ball is attached to the end of each rod. Each ball has cross-sectional area 4.00 cm^2 and is so shaped that it has a drag coefficient of 0.600. Calculate the power input required to spin the beater at 1 000 rev/min (a) in air and (b) in water.

61. A 4.00-m length of light nylon cord is wound around a uniform cylindrical spool of radius 0.500 m and mass 1.00 kg. The spool is mounted on a frictionless axle and is initially at rest. The cord is pulled from the spool with a constant acceleration of magnitude 2.50 m/s^2. (a) How much work has been done on the spool when it reaches an angular speed of 8.00 rad/s? (b) Assuming there is enough cord on the spool, how long does it take the spool to reach this angular speed? (c) Is there enough cord on the spool?

62. ● An elevator system in a tall building consists of an 800-kg car and a 950-kg counterweight, joined by a cable that passes over a pulley of mass 280 kg. The pulley, called a sheave, is a solid cylinder of radius 0.700 m turning on a horizontal axle. The cable has comparatively small mass and constant length. It does not slip on the sheave. The car and the counterweight move vertically, next to each other inside the same shaft. A number n of people, each of mass 80.0 kg, are riding in the elevator car, moving upward at 3.00 m/s and approaching the floor where the car should stop. As an energy-conservation measure, a computer disconnects the electric elevator motor at just the right moment so that the sheave-car-counterweight system then coasts freely without friction and comes to rest at the floor desired. There it is caught by a simple latch rather than by a massive brake. (a) Determine the distance d the car coasts upward as a function of n. Evaluate the distance for (b) $n = 2$, (c) $n = 12$, and (d) $n = 0$. (e) Does the expression in part (a) apply for all integer values of n or only for what values? Explain. (f) Describe the shape of a graph of d versus n. (g) Is any piece of data unnecessary for the solution? Explain. (h) Contrast the meaning of energy conservation as it is used in the statement of this problem and as it is used in Chapter 8. (i) Find the magnitude of the acceleration of the coasting elevator car, as it depends on n.

63. ● Figure P10.63 is a photograph of a lawn sprinkler. Its rotor consists of three metal tubes that fill with water when a hose is connected to the base. As water sprays out of the holes at the ends of the arms and the hole near the center of each arm, the assembly with the three arms rotates. To analyze this situation, let us make the following assumptions: (1) The arms can be modeled as thin, straight rods, each of length L. (2) The water coming from the hole at distance ℓ from the center sprays out horizontally, parallel to the ground and perpendicular to the arm. (3) The water emitted from the holes at the ends of the arms sprays out radially away from the center of the rotor. When filled with water, each arm has mass m. The center of the assembly is massless. The water ejected from a hole at distance ℓ from the center causes a thrust force F on the arm containing the hole. The mounting for the three-arm rotor assembly exerts a frictional torque that is described by $\tau = -b\omega$, where ω is the angular speed of the assembly. (a) Imagine that the sprinkler is in operation. Find an expression for the constant angular speed with which the assembly rotates *after* it completes an initial period of angular acceleration. Your expression should be in terms of F, ℓ, and b. (b) Imagine that the sprinkler has been at rest and is just turned on. Find an expression for the *initial* angular acceleration of the rotor, that is, the angular acceleration when the arms are filled with water and the assembly just begins to move from rest. Your expression should be in terms of F, ℓ, m, and L. (c) Now, take a step toward reality from the simplified model. The arms are actually bent as shown in the photograph. Therefore, the water from the ends of the arms is not actually sprayed radially. How will this fact affect the constant angular speed with which the assembly rotates in part (a)? In reality, will it be larger, smaller, or unchanged? Provide a convincing argument for your response. (d) How will the bend in the arms, described in part (c), affect the angular acceleration in part (b)? In reality, will it be larger, smaller, or unchanged? Provide a convincing argument for your response.

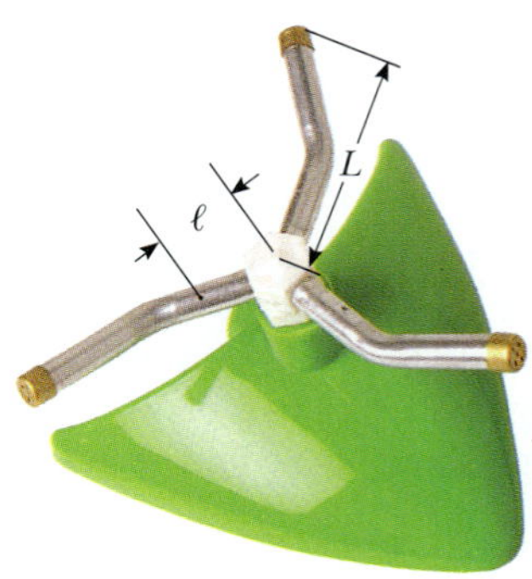

Figure P10.63

64. A shaft is turning at 65.0 rad/s at time $t = 0$. Thereafter, its angular acceleration is given by

$$\alpha = -10.0 \text{ rad/s}^2 - 5.00t \text{ rad/s}^3$$

where t is the elapsed time. (a) Find its angular speed at $t = 3.00$ s. (b) How far does it turn in these 3 s?

65. A long, uniform rod of length L and mass M is pivoted about a horizontal, frictionless pin through one end. The rod is released, almost from rest in a vertical position as shown in Figure P10.65. At the instant the rod is horizontal, find (a) its angular speed, (b) the magnitude of its angular acceleration, (c) the x and y components of the acceleration of its center of mass, and (d) the components of the reaction force at the pivot.

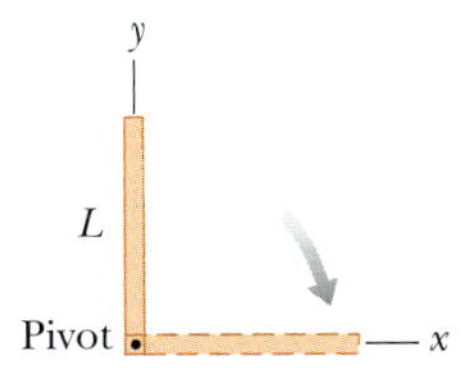

Figure P10.65

66. A cord is wrapped around a pulley of mass m and radius r. The free end of the cord is connected to a block of mass M. The block starts from rest and then slides down an incline that makes an angle θ with the horizontal. The coefficient of kinetic friction between block and incline is μ. (a) Use energy methods to show that the block's speed as a function of position d down the incline is

$$v = \sqrt{\frac{4gdM(\sin\theta - \mu\cos\theta)}{m + 2M}}$$

(b) Find the magnitude of the acceleration of the block in terms of μ, m, M, g, and θ.

67. A bicycle is turned upside down while its owner repairs a flat tire. A friend spins the other wheel, of radius 0.381 m, and observes that drops of water fly off tangentially. She measures the height reached by drops moving vertically (Fig. P10.67). A drop that breaks loose from the tire on one turn rises $h = 54.0$ cm above the tangent point. A drop that breaks loose on the next turn rises 51.0 cm above the tangent point. The height to which the drops rise decreases because the angular speed of the wheel decreases. From this information, determine the magnitude of the average angular acceleration of the wheel.

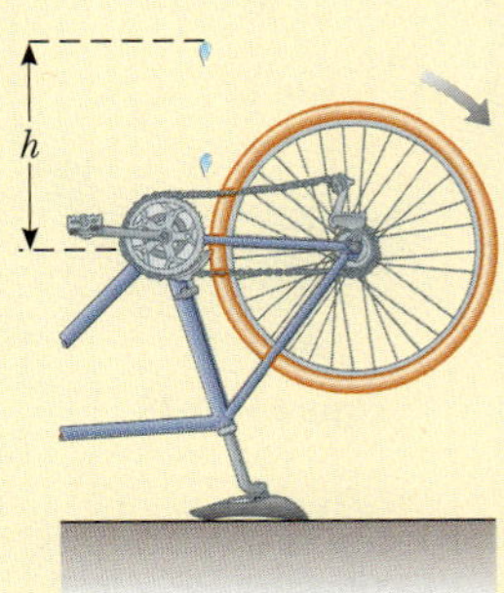

Figure P10.67 Problems 67 and 68.

68. A bicycle is turned upside down while its owner repairs a flat tire. A friend spins the other wheel, of radius R, and observes that drops of water fly off tangentially. She measures the height reached by drops moving vertically (Fig. P10.67). A drop that breaks loose from the tire on one turn rises a distance h_1 above the tangent point. A drop that breaks loose on the next turn rises a distance $h_2 < h_1$ above the tangent point. The height to which the drops rise decreases because the angular speed of the wheel decreases. From this information, determine the magnitude of the average angular acceleration of the wheel.

69. A uniform, hollow, cylindrical spool has inside radius $R/2$, outside radius R, and mass M (Fig. P10.69). It is mounted so that it rotates on a fixed, horizontal axle. A counterweight of mass m is connected to the end of a string wound around the spool. The counterweight falls from rest at $t = 0$ to a position y at time t. Show that the torque due to the friction forces between spool and axle is

$$\tau_f = R\left[m\left(g - \frac{2y}{t^2}\right) - M\frac{5y}{4t^2}\right]$$

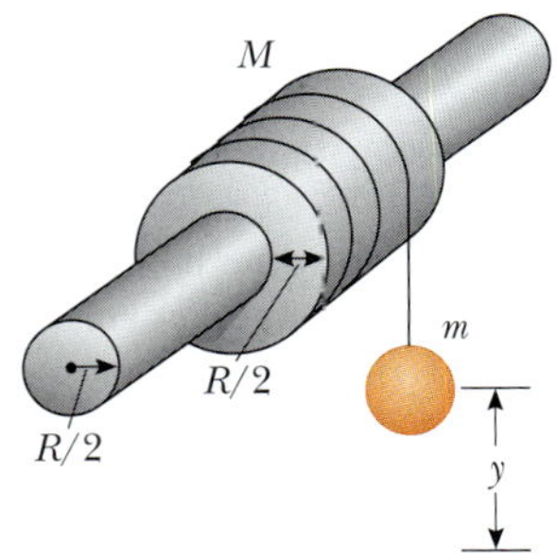

Figure P10.69

70. (a) What is the rotational kinetic energy of the Earth about its spin axis? Model the Earth as a uniform sphere and use data from the endpapers. (b) The rotational kinetic energy of the Earth is decreasing steadily because of tidal friction. Find the change in one day, assuming the rotational period increases by 10.0 μs each year.

71. Two blocks as shown in Figure P10.71 are connected by a string of negligible mass passing over a pulley of radius 0.250 m and moment of inertia I. The block on the frictionless incline is moving up with a constant acceleration of 2.00 m/s^2. (a) Determine T_1 and T_2, the tensions in the two parts of the string. (b) Find the moment of inertia of the pulley.

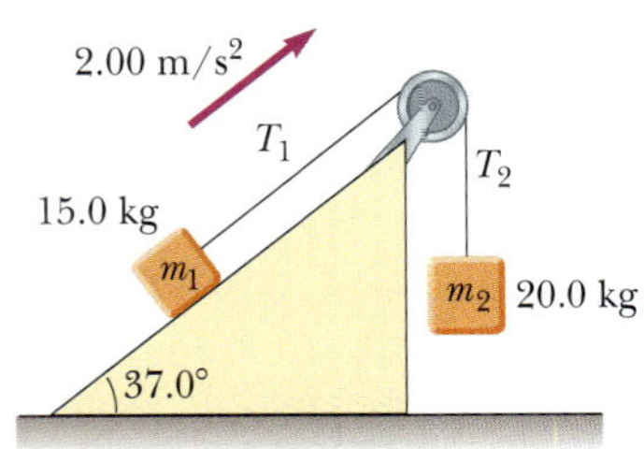

Figure P10.71

72. The reel shown in Figure P10.72 has radius R and moment of inertia I. One end of the block of mass m is connected to a spring of force constant k, and the other end is fastened to a cord wrapped around the reel. The reel axle and the incline are frictionless. The reel is

wound counterclockwise so that the spring stretches a distance d from its unstretched position and the reel is then released from rest. (a) Find the angular speed of the reel when the spring is again unstretched. (b) Evaluate the angular speed numerically at this point, taking $I = 1.00\ \text{kg}\cdot\text{m}^2$, $R = 0.300$ m, $k = 50.0$ N/m, $m = 0.500$ kg, $d = 0.200$ m, and $\theta = 37.0°$.

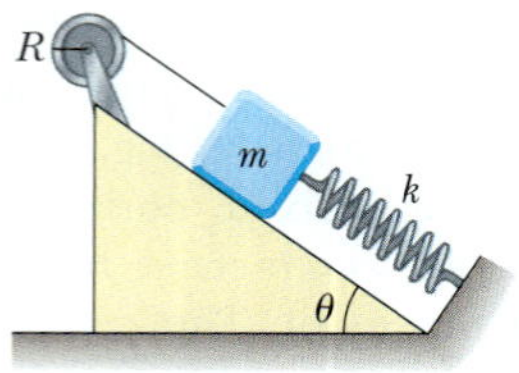

Figure P10.72

73. As a result of friction, the angular speed of a wheel changes with time according to

$$\frac{d\theta}{dt} = \omega_0 e^{-\sigma t}$$

where ω_0 and σ are constants. The angular speed changes from 3.50 rad/s at $t = 0$ to 2.00 rad/s at $t = 9.30$ s. Use this information to determine σ and ω_0. Then determine (a) the magnitude of the angular acceleration at $t = 3.00$ s, (b) the number of revolutions the wheel makes in the first 2.50 s, and (c) the number of revolutions it makes before coming to rest.

74. A common demonstration, illustrated in Figure P10.74, consists of a ball resting at one end of a uniform board of length ℓ, hinged at the other end, and elevated at an angle θ. A light cup is attached to the board at r_c so that it will catch the ball when the support stick is suddenly removed. (a) Show that the ball will lag behind the falling board when θ is less than 35.3°. (b) Assuming the board is 1.00 m long and is supported at this limiting angle, show that the cup must be 18.4 cm from the moving end.

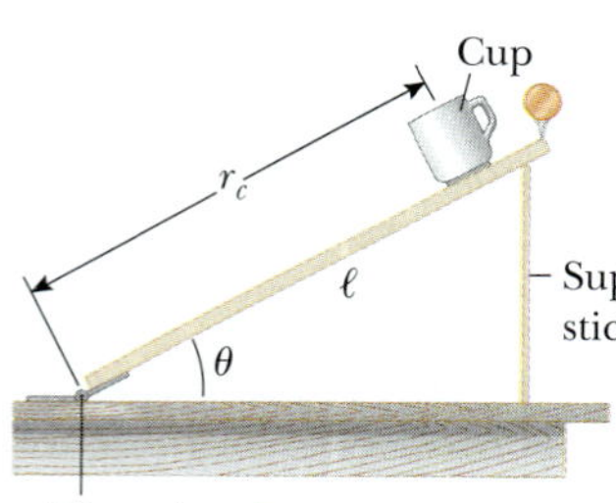

Figure P10.74

75. A tall building is located on the Earth's equator. As the Earth rotates, a person on the top floor of the building moves faster than someone on the ground with respect to an inertial reference frame because the latter person is closer to the Earth's axis. Consequently, if an object is dropped from the top floor to the ground a distance h below, it lands east of the point vertically below where it was dropped. (a) How far to the east will the object land? Express your answer in terms of h, g, and the angular speed ω of the Earth. Ignore air resistance, and assume the free-fall acceleration is constant over this range of heights. (b) Evaluate the eastward displacement for $h = 50.0$ m. (c) In your judgment, were we justified in ignoring this aspect of the *Coriolis effect* in our previous study of free fall?

76. The hour hand and the minute hand of Big Ben, the Parliament tower clock in London, are 2.70 m and 4.50 m long and have masses of 60.0 kg and 100 kg, respectively (see Figure P10.42). **(i)** Determine the total torque due to the weight of these hands about the axis of rotation when the time reads (a) 3:00, (b) 5:15, (c) 6:00, (d) 8:20, and (e) 9:45. (You may model the hands as long, thin uniform rods.) **(ii)** Determine all times when the total torque about the axis of rotation is zero. Determine the times to the nearest second, solving a transcendental equation numerically.

77. A string is wound around a uniform disk of radius R and mass M. The disk is released from rest with the string vertical and its top end tied to a fixed bar (Fig. P10.77). Show that (a) the tension in the string is one-third of the weight of the disk, (b) the magnitude of the acceleration of the center of mass is $2g/3$, and (c) the speed of the center of mass is $(4gh/3)^{1/2}$ after the disk has descended through distance h. Verify your answer to part (c) using the energy approach.

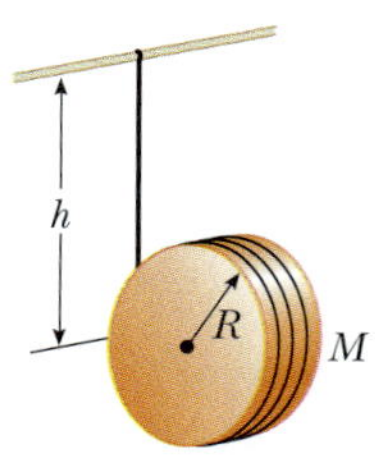

Figure P10.77

78. A uniform solid sphere of radius r is placed on the inside surface of a hemispherical bowl with much larger radius R. The sphere is released from rest at an angle θ to the vertical and rolls without slipping (Fig. P10.78). Determine the angular speed of the sphere when it reaches the bottom of the bowl.

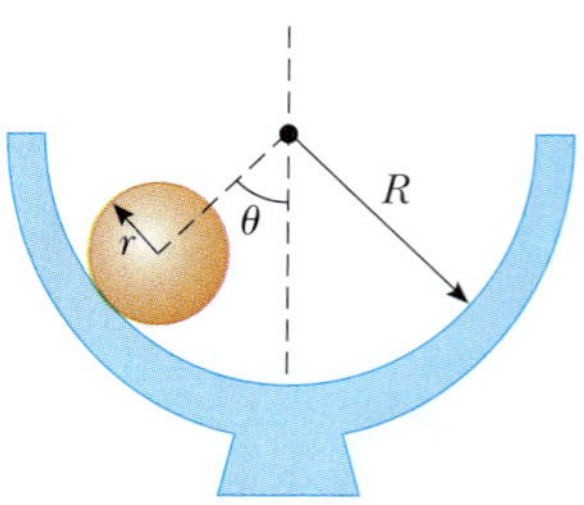

Figure P10.78

79. A solid sphere of mass m and radius r rolls without slipping along the track shown in Figure P10.79. It starts from rest with the lowest point of the sphere at height h

above the bottom of the loop of radius R, much larger than r. (a) What is the minimum value of h (in terms of R) such that the sphere completes the loop? (b) What are the components of the net force on the sphere at the point P if $h = 3R$?

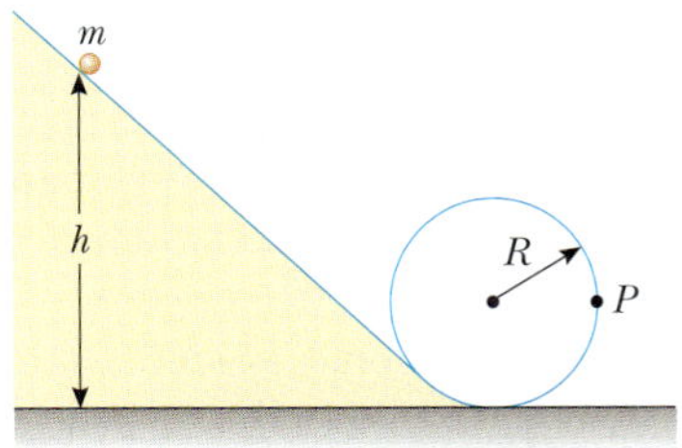

Figure P10.79

80. A thin rod of mass 0.630 kg and length 1.24 m is at rest, hanging vertically from a strong fixed hinge at its top end. Suddenly a horizontal impulsive force $(14.7\hat{\mathbf{i}})$ N is applied to it. (a) Suppose the force acts at the bottom end of the rod. Find the acceleration of its center of mass and the horizontal force the hinge exerts. (b) Suppose the force acts at the midpoint of the rod. Find the acceleration of this point and the horizontal hinge reaction. (c) Where can the impulse be applied so that the hinge will exert no horizontal force? This point is called the *center of percussion.*

81. (a) A thin rod of length h and mass M is held vertically with its lower end resting on a frictionless horizontal surface. The rod is then released to fall freely. Determine the speed of its center of mass just before it hits the horizontal surface. (b) **What If?** Now suppose the rod has a fixed pivot at its lower end. Determine the speed of the rod's center of mass just before it hits the surface.

82. Following Thanksgiving dinner your uncle falls into a deep sleep, sitting straight up facing the television set. A naughty grandchild balances a small spherical grape at the top of his bald head, which itself has the shape of a sphere. After all the children have had time to giggle, the grape starts from rest and rolls down without slipping. The grape loses contact with your uncle's scalp when the radial line joining it to the center of curvature makes what angle with the vertical?

83. A spool of wire of mass M and radius R is unwound under a constant force $\vec{\mathbf{F}}$ (Fig. P10.83). Assuming the spool is a uniform solid cylinder that doesn't slip, show that (a) the acceleration of the center of mass is $4\vec{\mathbf{F}}/3M$ and (b) the

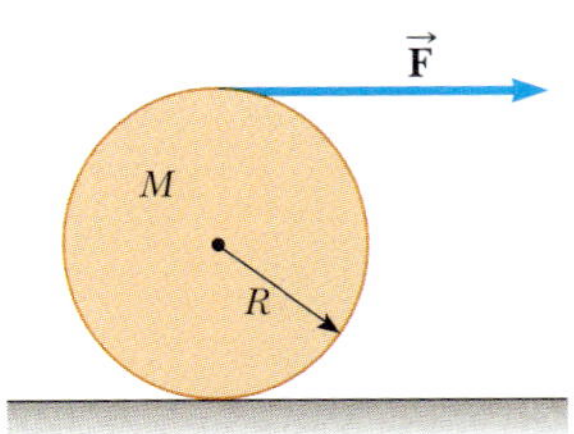

Figure P10.83

force of friction is to the *right* and equal in magnitude to $F/3$. (c) If the cylinder starts from rest and rolls without slipping, what is the speed of its center of mass after it has rolled through a distance d?

84. A plank with a mass $M = 6.00$ kg rides on top of two identical solid cylindrical rollers that have $R = 5.00$ cm and $m = 2.00$ kg (Fig. P10.84). The plank is pulled by a constant horizontal force $\vec{\mathbf{F}}$ of magnitude 6.00 N applied to the end of the plank and perpendicular to the axes of the cylinders (which are parallel). The cylinders roll without slipping on a flat surface. There is also no slipping between the cylinders and the plank. (a) Find the acceleration of the plank and of the rollers. (b) What friction forces are acting?

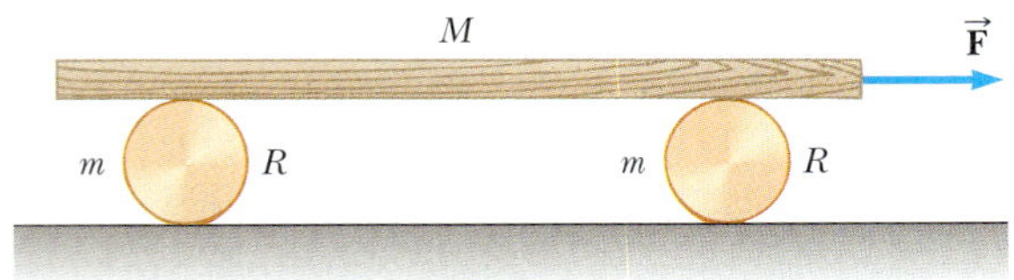

Figure P10.84

85. A spool of thread consists of a cylinder of radius R_1 with end caps of radius R_2 as shown in the end view illustrated in Figure P10.85. The mass of the spool, including the thread, is m, and its moment of inertia about an axis through its center is I. The spool is placed on a rough horizontal surface so that it rolls without slipping when a force $\vec{\mathbf{T}}$ acting to the right is applied to the free end of the thread. Show that the magnitude of the friction force exerted by the surface on the spool is given by

$$f = \left(\frac{I + mR_1R_2}{I + mR_2^2}\right)T$$

Determine the direction of the force of friction.

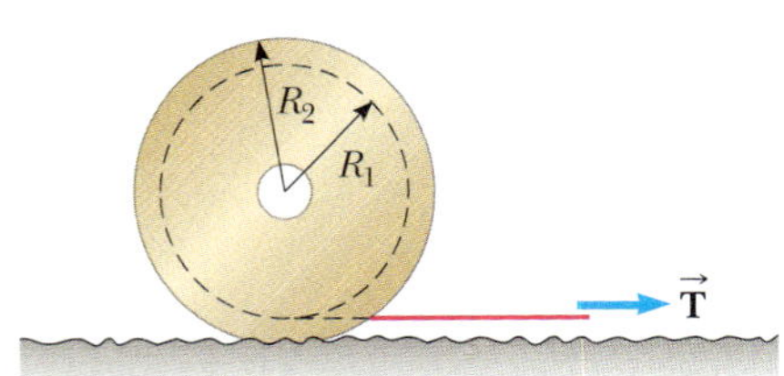

Figure P10.85

86. ● A large, cylindrical roll of tissue paper of initial radius R lies on a long, horizontal surface with the outside end of the paper nailed to the surface. The roll is given a slight shove ($v_i \approx 0$) and commences to unroll. Assume the roll has a uniform density and that mechanical energy is conserved in the process. (a) Determine the speed of the center of mass of the roll when its radius has diminished to r. (b) Calculate a numerical value for this speed at $r = 1.00$ mm, assuming $R = 6.00$ m. (c) **What If?** What happens to the energy of the system when the paper is completely unrolled?

Answers to Quick Quizzes

10.1 **(i),** (c). For a rotation of more than 180°, the angular displacement must be larger than $\pi = 3.14$ rad. The angular displacements in the three choices are (a) 6 rad − 3 rad = 3 rad, (b) 1 rad − (−1) rad = 2 rad, and (c) 5 rad − 1 rad = 4 rad. **(ii),** (b). Because all angular displacements occur in the same time interval, the displacement with the lowest value will be associated with the lowest average angular speed.

10.2 (b). In Equation 10.8, both the initial and final angular speeds are the same in all three cases. As a result, the angular acceleration is inversely proportional to the angular displacement. Therefore, the highest angular acceleration is associated with the lowest angular displacement.

10.3 **(i),** (b). The system of the platform, Alex, and Brian is a rigid object, so all points on the rigid object have the same angular speed. **(ii),** (a). The tangential speed is proportional to the radial distance from the rotation axis.

10.4 (a). Almost all the mass of the pipe is at the same distance from the rotation axis, so it has a larger moment of inertia than the solid cylinder.

10.5 **(i),** (b). The fatter handle of the screwdriver gives you a larger moment arm and increases the torque you can apply with a given force from your hand. **(ii),** (a). The longer handle of the wrench gives you a larger moment arm and increases the torque you can apply with a given force from your hand.

10.6 (b). With twice the moment of inertia and the same frictional torque, there is half the angular acceleration. With half the angular acceleration, it will require twice as long to change the speed to zero.

10.7 (b). All the gravitational potential energy of the box–Earth system is transformed to kinetic energy of translation. For the ball, some of the gravitational potential energy of the ball–Earth system is transformed to rotational kinetic energy, leaving less for translational kinetic energy, so the ball moves downhill more slowly than the box does.

A competitive diver undergoes a rotation during a dive. She spins at a higher rate when she folds her body into a smaller package due to the principle of conservation of angular momentum, as discussed in this chapter. (The Image Bank/Getty Images)

11 Angular Momentum

The central topic of this chapter is angular momentum, a quantity that plays a key role in rotational dynamics. In analogy to the principle of conservation of linear momentum for an isolated system, the angular momentum of a system is conserved if no external torques act on the system. Like the law of conservation of linear momentum, the law of conservation of angular momentum is a fundamental law of physics, equally valid for relativistic and quantum systems.

11.1 The Vector Product and Torque

An important consideration in defining angular momentum is the process of multiplying two vectors by means of the operation called the *vector product.* We will introduce the vector product by considering the vector nature of torque.

Consider a force $\vec{\mathbf{F}}$ acting on a rigid object at the vector position $\vec{\mathbf{r}}$ (Active Fig. 11.1). As we saw in Section 10.6, the *magnitude* of the torque due to this force about an axis through the origin is $rF \sin \phi$, where ϕ is the angle between $\vec{\mathbf{r}}$ and $\vec{\mathbf{F}}$. The axis about which $\vec{\mathbf{F}}$ tends to produce rotation is perpendicular to the plane formed by $\vec{\mathbf{r}}$ and $\vec{\mathbf{F}}$.

The torque vector $\vec{\boldsymbol{\tau}}$ is related to the two vectors $\vec{\mathbf{r}}$ and $\vec{\mathbf{F}}$. We can establish a mathematical relationship between $\vec{\boldsymbol{\tau}}$, $\vec{\mathbf{r}}$, and $\vec{\mathbf{F}}$ using a mathematical operation called the **vector product**, or **cross product:**

$$\vec{\boldsymbol{\tau}} \equiv \vec{\mathbf{r}} \times \vec{\mathbf{F}} \tag{11.1}$$

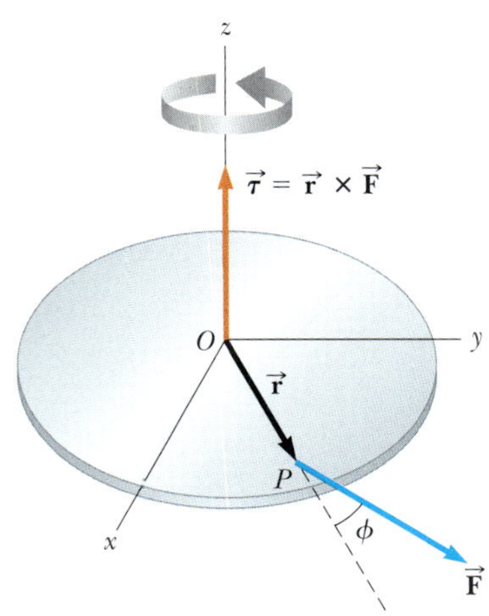

ACTIVE FIGURE 11.1

The torque vector $\vec{\boldsymbol{\tau}}$ lies in a direction perpendicular to the plane formed by the position vector $\vec{\mathbf{r}}$ and the applied force vector $\vec{\mathbf{F}}$.

Sign in at www.thomsonedu.com and go to ThomsonNOW to move point P and change the force vector $\vec{\mathbf{F}}$ to see the effect on the torque vector.

PITFALL PREVENTION 11.1
The Cross Product Is a Vector

Remember that the result of taking a cross product between two vectors is *a third vector*. Equation 11.3 gives only the magnitude of this vector.

We now give a formal definition of the vector product. Given any two vectors $\vec{\mathbf{A}}$ and $\vec{\mathbf{B}}$, the **vector product** $\vec{\mathbf{A}} \times \vec{\mathbf{B}}$ is defined as a third vector $\vec{\mathbf{C}}$, which has a magnitude of $AB \sin \theta$, where θ is the angle between $\vec{\mathbf{A}}$ and $\vec{\mathbf{B}}$. That is, if $\vec{\mathbf{C}}$ is given by

$$\vec{\mathbf{C}} = \vec{\mathbf{A}} \times \vec{\mathbf{B}} \tag{11.2}$$

its magnitude is

$$C = AB \sin \theta \tag{11.3}$$

The quantity $AB \sin \theta$ is equal to the area of the parallelogram formed by $\vec{\mathbf{A}}$ and $\vec{\mathbf{B}}$ as shown in Figure 11.2. The *direction* of $\vec{\mathbf{C}}$ is perpendicular to the plane formed by $\vec{\mathbf{A}}$ and $\vec{\mathbf{B}}$, and the best way to determine this direction is to use the right-hand rule illustrated in Figure 11.2. The four fingers of the right hand are pointed along $\vec{\mathbf{A}}$ and then "wrapped" into $\vec{\mathbf{B}}$ through the angle θ. The direction of the upright thumb is the direction of $\vec{\mathbf{A}} \times \vec{\mathbf{B}} = \vec{\mathbf{C}}$. Because of the notation, $\vec{\mathbf{A}} \times \vec{\mathbf{B}}$ is often read "$\vec{\mathbf{A}}$ cross $\vec{\mathbf{B}}$," hence the term *cross product.*

Some properties of the vector product that follow from its definition are as follows:

Properties of the vector product ▶

1. Unlike the scalar product, the vector product is *not* commutative. Instead, the order in which the two vectors are multiplied in a cross product is important:

$$\vec{\mathbf{A}} \times \vec{\mathbf{B}} = -\vec{\mathbf{B}} \times \vec{\mathbf{A}} \tag{11.4}$$

Therefore, if you change the order of the vectors in a cross product, you must change the sign. You can easily verify this relationship with the right-hand rule.

2. If $\vec{\mathbf{A}}$ is parallel to $\vec{\mathbf{B}}$ ($\theta = 0$ or $180°$), then $\vec{\mathbf{A}} \times \vec{\mathbf{B}} = 0$; therefore, it follows that $\vec{\mathbf{A}} \times \vec{\mathbf{A}} = 0$.
3. If $\vec{\mathbf{A}}$ is perpendicular to $\vec{\mathbf{B}}$, then $|\vec{\mathbf{A}} \times \vec{\mathbf{B}}| = AB$.
4. The vector product obeys the distributive law:

$$\vec{\mathbf{A}} \times (\vec{\mathbf{B}} + \vec{\mathbf{C}}) = \vec{\mathbf{A}} \times \vec{\mathbf{B}} + \vec{\mathbf{A}} \times \vec{\mathbf{C}} \tag{11.5}$$

5. The derivative of the cross product with respect to some variable such as t is

$$\frac{d}{dt}(\vec{\mathbf{A}} \times \vec{\mathbf{B}}) = \frac{d\vec{\mathbf{A}}}{dt} \times \vec{\mathbf{B}} + \vec{\mathbf{A}} \times \frac{d\vec{\mathbf{B}}}{dt} \tag{11.6}$$

where it is important to preserve the multiplicative order of the terms on the right side in view of Equation 11.4.

It is left as an exercise (Problem 10) to show from Equations 11.3 and 11.4 and from the definition of unit vectors that the cross products of the unit vectors $\hat{\mathbf{i}}$, $\hat{\mathbf{j}}$, and $\hat{\mathbf{k}}$ obey the following rules:

Cross products of unit vectors ▶

$$\hat{\mathbf{i}} \times \hat{\mathbf{i}} = \hat{\mathbf{j}} \times \hat{\mathbf{j}} = \hat{\mathbf{k}} \times \hat{\mathbf{k}} = 0 \tag{11.7a}$$

$$\hat{\mathbf{i}} \times \hat{\mathbf{j}} = -\hat{\mathbf{j}} \times \hat{\mathbf{i}} = \hat{\mathbf{k}} \tag{11.7b}$$

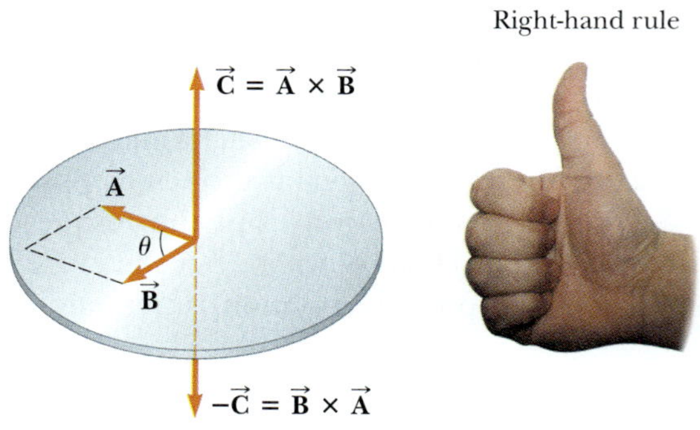

Figure 11.2 The vector product $\vec{\mathbf{A}} \times \vec{\mathbf{B}}$ is a third vector $\vec{\mathbf{C}}$ having a magnitude $AB \sin \theta$ equal to the area of the parallelogram shown. The direction of $\vec{\mathbf{C}}$ is perpendicular to the plane formed by $\vec{\mathbf{A}}$ and $\vec{\mathbf{B}}$, and this direction is determined by the right-hand rule.

$$\hat{\mathbf{j}} \times \hat{\mathbf{k}} = -\hat{\mathbf{k}} \times \hat{\mathbf{j}} = \hat{\mathbf{i}} \qquad \textbf{(11.7c)}$$

$$\hat{\mathbf{k}} \times \hat{\mathbf{i}} = -\hat{\mathbf{i}} \times \hat{\mathbf{k}} = \hat{\mathbf{j}} \qquad \textbf{(11.7d)}$$

Signs are interchangeable in cross products. For example, $\vec{\mathbf{A}} \times (-\vec{\mathbf{B}}) = -\vec{\mathbf{A}} \times \vec{\mathbf{B}}$ and $\hat{\mathbf{i}} \times (-\hat{\mathbf{j}}) = -\hat{\mathbf{i}} \times \hat{\mathbf{j}}$.

The cross product of any two vectors $\vec{\mathbf{A}}$ and $\vec{\mathbf{B}}$ can be expressed in the following determinant form:

$$\vec{\mathbf{A}} \times \vec{\mathbf{B}} = \begin{vmatrix} \hat{\mathbf{i}} & \hat{\mathbf{j}} & \hat{\mathbf{k}} \\ A_x & A_y & A_z \\ B_x & B_y & B_z \end{vmatrix} = \begin{vmatrix} A_y & A_z \\ B_y & B_z \end{vmatrix} \hat{\mathbf{i}} + \begin{vmatrix} A_z & A_x \\ B_z & B_x \end{vmatrix} \hat{\mathbf{j}} + \begin{vmatrix} A_x & A_y \\ B_x & B_y \end{vmatrix} \hat{\mathbf{k}}$$

Expanding these determinants gives the result

$$\vec{\mathbf{A}} \times \vec{\mathbf{B}} = (A_yB_z - A_zB_y)\hat{\mathbf{i}} + (A_zB_x - A_xB_z)\hat{\mathbf{j}} + (A_xB_y - A_yB_x)\hat{\mathbf{k}} \qquad \textbf{(11.8)}$$

Given the definition of the cross product, we can now assign a direction to the torque vector. If the force lies in the xy plane, as in Active Figure 11.1, the torque $\vec{\boldsymbol{\tau}}$ is represented by a vector parallel to the z axis. The force in Active Figure 11.1 creates a torque that tends to rotate the object counterclockwise about the z axis; the direction of $\vec{\boldsymbol{\tau}}$ is toward increasing z, and $\vec{\boldsymbol{\tau}}$ is therefore in the positive z direction. If we reversed the direction of $\vec{\mathbf{F}}$ in Active Figure 11.1, $\vec{\boldsymbol{\tau}}$ would be in the negative z direction.

Quick Quiz 11.1 Which of the following statements about the relationship between the magnitude of the cross product of two vectors and the product of the magnitudes of the vectors is true? (a) $|\vec{\mathbf{A}} \times \vec{\mathbf{B}}|$ is larger than AB. (b) $|\vec{\mathbf{A}} \times \vec{\mathbf{B}}|$ is smaller than AB. (c) $|\vec{\mathbf{A}} \times \vec{\mathbf{B}}|$ could be larger or smaller than AB, depending on the angle between the vectors. (d) $|\vec{\mathbf{A}} \times \vec{\mathbf{B}}|$ could be equal to AB.

EXAMPLE 11.1 The Vector Product

Two vectors lying in the xy plane are given by the equations $\vec{\mathbf{A}} = 2\hat{\mathbf{i}} + 3\hat{\mathbf{j}}$ and $\vec{\mathbf{B}} = -\hat{\mathbf{i}} + 2\hat{\mathbf{j}}$. Find $\vec{\mathbf{A}} \times \vec{\mathbf{B}}$ and verify that $\vec{\mathbf{A}} \times \vec{\mathbf{B}} = -\vec{\mathbf{B}} \times \vec{\mathbf{A}}$.

SOLUTION

Conceptualize Given the unit–vector notations of the vectors, think about the directions the vectors point in space. Imagine the parallelogram shown in Figure 11.2 for these vectors.

Categorize Because we use the definition of the cross product discussed in this section, we categorize this example as a substitution problem.

Write the cross product of the two vectors:

$$\vec{\mathbf{A}} \times \vec{\mathbf{B}} = (2\hat{\mathbf{i}} + 3\hat{\mathbf{j}}) \times (-\hat{\mathbf{i}} + 2\hat{\mathbf{j}})$$

Perform the multiplication:

$$\vec{\mathbf{A}} \times \vec{\mathbf{B}} = 2\hat{\mathbf{i}} \times (-\hat{\mathbf{i}}) + 2\hat{\mathbf{i}} \times 2\hat{\mathbf{j}} + 3\hat{\mathbf{j}} \times (-\hat{\mathbf{i}}) + 3\hat{\mathbf{j}} \times 2\hat{\mathbf{j}}$$

Use Equations 11.7a through 11.7d to evaluate the various terms:

$$\vec{\mathbf{A}} \times \vec{\mathbf{B}} = 0 + 4\hat{\mathbf{k}} + 3\hat{\mathbf{k}} + 0 = 7\hat{\mathbf{k}}$$

To verify that $\vec{\mathbf{A}} \times \vec{\mathbf{B}} = -\vec{\mathbf{B}} \times \vec{\mathbf{A}}$, evaluate $\vec{\mathbf{B}} \times \vec{\mathbf{A}}$:

$$\vec{\mathbf{B}} \times \vec{\mathbf{A}} = (-\hat{\mathbf{i}} + 2\hat{\mathbf{j}}) \times (2\hat{\mathbf{i}} + 3\hat{\mathbf{j}})$$

Perform the multiplication:

$$\vec{\mathbf{B}} \times \vec{\mathbf{A}} = (-\hat{\mathbf{i}}) \times 2\hat{\mathbf{i}} + (-\hat{\mathbf{i}}) \times 3\hat{\mathbf{j}} - 2\hat{\mathbf{j}} \times 2\hat{\mathbf{i}} + 2\hat{\mathbf{j}} \times 3\hat{\mathbf{j}}$$

Use Equations 11.7a through 11.7d to evaluate the various terms:

$$\vec{\mathbf{B}} \times \vec{\mathbf{A}} = 0 - 3\hat{\mathbf{k}} - 4\hat{\mathbf{k}} + 0 = -7\hat{\mathbf{k}}$$

Therefore, $\vec{\mathbf{A}} \times \vec{\mathbf{B}} = -\vec{\mathbf{B}} \times \vec{\mathbf{A}}$. As an alternative method for finding $\vec{\mathbf{A}} \times \vec{\mathbf{B}}$, you could use Equation 11.8. Try it!

EXAMPLE 11.2 The Torque Vector

A force of $\vec{\mathbf{F}} = (2.00\hat{\mathbf{i}} + 3.00\hat{\mathbf{j}})$ N is applied to an object that is pivoted about a fixed axis aligned along the z coordinate axis. The force is applied at a point located at $\vec{\mathbf{r}} = (4.00\hat{\mathbf{i}} + 5.00\hat{\mathbf{j}})$ m. Find the torque vector $\vec{\boldsymbol{\tau}}$.

SOLUTION

Conceptualize Given the unit-vector notations, think about the directions of the force and position vectors. If this force were applied at this position, in what direction would an object pivoted at the origin turn?

Categorize Because we use the definition of the cross product discussed in this section, we categorize this example as a substitution problem.

Set up the torque vector using Equation 11.1:

$$\vec{\boldsymbol{\tau}} = \vec{\mathbf{r}} \times \vec{\mathbf{F}} = [(4.00\hat{\mathbf{i}} + 5.00\hat{\mathbf{j}})\text{ m}] \times [(2.00\hat{\mathbf{i}} + 3.00\hat{\mathbf{j}})\text{ N}]$$

Perform the multiplication:

$$\vec{\boldsymbol{\tau}} = [(4.00)(2.00)\hat{\mathbf{i}} \times \hat{\mathbf{i}} + (4.00)(3.00)\hat{\mathbf{i}} \times \hat{\mathbf{j}} + (5.00)(2.00)\hat{\mathbf{j}} \times \hat{\mathbf{i}} + (5.00)(3.00)\hat{\mathbf{j}} \times \hat{\mathbf{j}}]\text{ N}\cdot\text{m}$$

Use Equations 11.7a through 11.7d to evaluate the various terms:

$$\vec{\boldsymbol{\tau}} = [0 + 12.0\hat{\mathbf{k}} - 10.0\hat{\mathbf{k}} + 0]\text{ N}\cdot\text{m} = 2.0\hat{\mathbf{k}}\text{ N}\cdot\text{m}$$

Notice that both $\vec{\mathbf{r}}$ and $\vec{\mathbf{F}}$ are in the xy plane. As expected, the torque vector is perpendicular to this plane, having only a z component. We have followed the rules for significant figures discussed in Section 1.6, which lead to an answer with two significant figures. We have lost some precision because we ended up subtracting two numbers that are close.

11.2 Angular Momentum: The Nonisolated System

ACTIVE FIGURE 11.3

As the skater passes the pole, she grabs hold of it, which causes her to swing around the pole rapidly in a circular path.

Sign in at www.thomsonedu.com and go to ThomsonNOW to change the speed of the skater and her distance to the pole and watch her spin when she grabs the pole.

Imagine a rigid pole sticking up through the ice on a frozen pond (Active Fig. 11.3). A skater glides rapidly toward the pole, aiming a little to the side so that she does not hit it. As she passes the pole, she reaches out to her side and grabs it, an action that causes her to move in a circular path around the pole. Just as the idea of linear momentum helps us analyze translational motion, a rotational analog—*angular momentum*—helps us analyze the motion of this skater and other objects undergoing rotational motion.

In Chapter 9, we developed the mathematical form of linear momentum and then proceeded to show how this new quantity was valuable in problem solving. We will follow a similar procedure for angular momentum.

Consider a particle of mass m located at the vector position $\vec{\mathbf{r}}$ and moving with linear momentum $\vec{\mathbf{p}}$ as in Active Figure 11.4. In describing translational motion, we found that the net force on the particle equals the time rate of change of its linear momentum, $\sum \vec{\mathbf{F}} = d\vec{\mathbf{p}}/dt$ (see Eq. 9.3). Let us take the cross product of each side of Equation 9.3 with $\vec{\mathbf{r}}$, which gives the net torque on the particle on the left side of the equation:

$$\vec{\mathbf{r}} \times \sum \vec{\mathbf{F}} = \sum \vec{\boldsymbol{\tau}} = \vec{\mathbf{r}} \times \frac{d\vec{\mathbf{p}}}{dt}$$

Now let's add to the right side the term $(d\vec{\mathbf{r}}/dt) \times \vec{\mathbf{p}}$, which is zero because $d\vec{\mathbf{r}}/dt = \vec{\mathbf{v}}$ and $\vec{\mathbf{v}}$ and $\vec{\mathbf{p}}$ are parallel. Therefore,

$$\sum \vec{\boldsymbol{\tau}} = \vec{\mathbf{r}} \times \frac{d\vec{\mathbf{p}}}{dt} + \frac{d\vec{\mathbf{r}}}{dt} \times \vec{\mathbf{p}}$$

We recognize the right side of this equation as the derivative of $\vec{\mathbf{r}} \times \vec{\mathbf{p}}$ (see Eq. 11.6). Therefore,

$$\sum \vec{\boldsymbol{\tau}} = \frac{d(\vec{\mathbf{r}} \times \vec{\mathbf{p}})}{dt} \tag{11.9}$$

which looks very similar in form to Equation 9.3, $\sum \vec{\mathbf{F}} = d\vec{\mathbf{p}}/dt$. This result suggests that the combination $\vec{\mathbf{r}} \times \vec{\mathbf{p}}$ should play the same role in rotational motion that $\vec{\mathbf{p}}$ plays in translational motion. We call this combination the *angular momentum* of the particle:

The instantaneous **angular momentum** $\vec{\mathbf{L}}$ of a particle relative to an axis through the origin O is defined by the cross product of the particle's instantaneous position vector $\vec{\mathbf{r}}$ and its instantaneous linear momentum $\vec{\mathbf{p}}$:

$$\vec{\mathbf{L}} \equiv \vec{\mathbf{r}} \times \vec{\mathbf{p}} \tag{11.10}$$

◀ Angular momentum of a particle

We can now write Equation 11.9 as

$$\sum \vec{\boldsymbol{\tau}} = \frac{d\vec{\mathbf{L}}}{dt} \tag{11.11}$$

which is the rotational analog of Newton's second law, $\sum \vec{\mathbf{F}} = d\vec{\mathbf{p}}/dt$. Torque causes the angular momentum $\vec{\mathbf{L}}$ to change just as force causes linear momentum $\vec{\mathbf{p}}$ to change. Equation 11.11 states that **the torque acting on a particle is equal to the time rate of change of the particle's angular momentum.**

Notice that Equation 11.11 is valid only if $\sum \vec{\boldsymbol{\tau}}$ and $\vec{\mathbf{L}}$ are measured about the same axis. Furthermore, **the expression is valid for any axis fixed in an inertial frame.**

The SI unit of angular momentum is $\text{kg} \cdot \text{m}^2/\text{s}$. Notice also that both the magnitude and the direction of $\vec{\mathbf{L}}$ depend on the choice of axis. Following the right-hand rule, we see that the direction of $\vec{\mathbf{L}}$ is perpendicular to the plane formed by $\vec{\mathbf{r}}$ and $\vec{\mathbf{p}}$. In Active Figure 11.4, $\vec{\mathbf{r}}$ and $\vec{\mathbf{p}}$ are in the xy plane, so $\vec{\mathbf{L}}$ points in the z direction. Because $\vec{\mathbf{p}} = m\vec{\mathbf{v}}$, the magnitude of $\vec{\mathbf{L}}$ is

$$L = mvr \sin \phi \tag{11.12}$$

where ϕ is the angle between $\vec{\mathbf{r}}$ and $\vec{\mathbf{p}}$. It follows that L is zero when $\vec{\mathbf{r}}$ is parallel to $\vec{\mathbf{p}}$ ($\phi = 0$ or $180°$). In other words, when the translational velocity of the particle is along a line that passes through the axis, the particle has zero angular momentum with respect to the axis. On the other hand, if $\vec{\mathbf{r}}$ is perpendicular to $\vec{\mathbf{p}}$ ($\phi = 90°$), then $L = mvr$. At that instant, the particle moves exactly as if it were on the rim of a wheel rotating about the axis in a plane defined by $\vec{\mathbf{r}}$ and $\vec{\mathbf{p}}$.

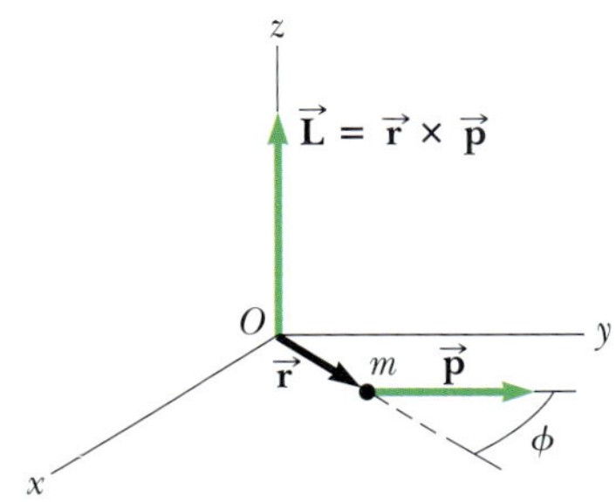

ACTIVE FIGURE 11.4

The angular momentum $\vec{\mathbf{L}}$ of a particle with linear momentum $\vec{\mathbf{p}}$ located at the vector position $\vec{\mathbf{r}}$ is a vector given by $\vec{\mathbf{L}} = \vec{\mathbf{r}} \times \vec{\mathbf{p}}$. The value of $\vec{\mathbf{L}}$ depends on the axis about which it is measured and is a vector perpendicular to both $\vec{\mathbf{r}}$ and $\vec{\mathbf{p}}$.

Sign in at www.thomsonedu.com and go to ThomsonNOW to change the position vector $\vec{\mathbf{r}}$ and the momentum vector $\vec{\mathbf{p}}$ and see the effect on the angular momentum vector.

PITFALL PREVENTION 11.2

Is Rotation Necessary for Angular Momentum?

We can define angular momentum even if the particle is not moving in a circular path. Even a particle moving in a straight line has angular momentum about any axis displaced from the path of the particle.

Quick Quiz 11.2 Recall the skater described at the beginning of this section. Let her mass be m. **(i)** What would be her angular momentum relative to the pole at the instant she is a distance d from the pole if she were skating directly toward it at speed v? (a) zero (b) mvd (c) impossible to determine **(ii)** What would be her angular momentum relative to the pole at the instant she is a distance d from the pole if she were skating at speed v along a straight path that is a perpendicular distance a from the pole? (a) zero (b) mvd (c) mva (d) impossible to determine

EXAMPLE 11.3 Angular Momentum of a Particle in Circular Motion

A particle moves in the xy plane in a circular path of radius r as shown in Figure 11.5. Find the magnitude and direction of its angular momentum relative to an axis through O when its velocity is $\vec{\mathbf{v}}$.

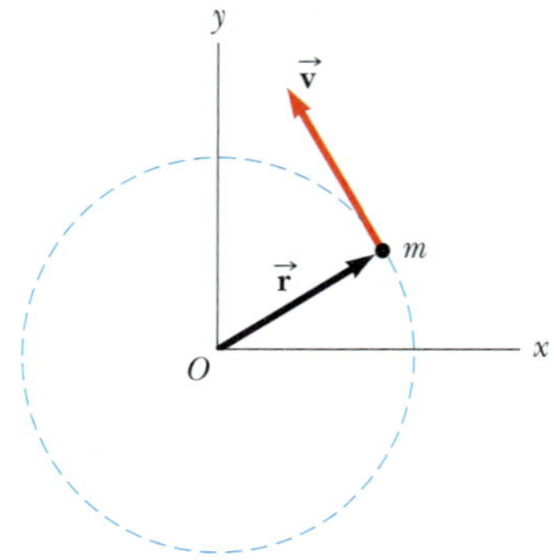

Figure 11.5 (Example 11.3) A particle moving in a circle of radius r has an angular momentum about an axis through O that has magnitude mvr. The vector $\vec{\mathbf{L}} = \vec{\mathbf{r}} \times \vec{\mathbf{p}}$ points *out* of the page.

SOLUTION

Conceptualize The linear momentum of the particle is changing in direction (but not in magnitude). You might therefore be tempted to conclude that the angular momentum of the particle is always changing. In this situation, however, that is not the case. Let's see why.

Categorize We use the definition of the angular momentum of a particle discussed in this section, so we categorize this example as a substitution problem.

Use Equation 11.12 to evaluate the magnitude of $\vec{\mathbf{L}}$: $L = mvr \sin 90° = mvr$

This value of L is constant because all three factors on the right are constant. The direction of $\vec{\mathbf{L}}$ also is constant, even though the direction of $\vec{\mathbf{p}} = m\vec{\mathbf{v}}$ keeps changing. To verify this statement, apply the right-hand rule to find the direction of $\vec{\mathbf{L}} = \vec{\mathbf{r}} \times \vec{\mathbf{p}} = m\vec{\mathbf{r}} \times \vec{\mathbf{v}}$ in Figure 11.5. Your thumb points upward and away from the page, so that is the direction of $\vec{\mathbf{L}}$. Hence, we can write the vector expression $\vec{\mathbf{L}} = (mvr)\hat{\mathbf{k}}$. If the particle were to move clockwise, $\vec{\mathbf{L}}$ would point downward and into the page and $\vec{\mathbf{L}} = -(mvr)\hat{\mathbf{k}}$. **A particle in uniform circular motion has a constant angular momentum about an axis through the center of its path.**

Angular Momentum of a System of Particles

In Section 9.6, we showed that Newton's second law for a particle could be extended to a system of particles, resulting in

$$\sum \vec{\mathbf{F}}_{\text{ext}} = \frac{d\vec{\mathbf{p}}_{\text{tot}}}{dt}$$

This equation states that the net external force on a system of particles is equal to the time rate of change of the total linear momentum of the system. Let's see if a similar statement can be made for rotational motion. The total angular momentum of a system of particles about some axis is defined as the vector sum of the angular momenta of the individual particles:

$$\vec{\mathbf{L}}_{\text{tot}} = \vec{\mathbf{L}}_1 + \vec{\mathbf{L}}_2 + \cdots + \vec{\mathbf{L}}_n = \sum_i \vec{\mathbf{L}}_i$$

where the vector sum is over all n particles in the system.

Differentiating this equation with respect to time gives

$$\frac{d\vec{\mathbf{L}}_{\text{tot}}}{dt} = \sum_i \frac{d\vec{\mathbf{L}}_i}{dt} = \sum_i \vec{\boldsymbol{\tau}}_i$$

where we have used Equation 11.11 to replace the time rate of change of the angular momentum of each particle with the net torque on the particle.

The torques acting on the particles of the system are those associated with internal forces between particles and those associated with external forces. The net torque associated with all internal forces, however, is zero. Recall that Newton's third law tells us that internal forces between particles of the system are equal in magnitude and opposite in direction. If we assume these forces lie along the line of separation of each pair of particles, the total torque around some axis passing through an origin O due to each action–reaction force pair is zero (that is, the moment arm d from O to the line of action of the forces is equal for both particles and the forces are in opposite directions). In the summation, therefore, the net internal torque is zero. We conclude that the total angular momentum of a system can vary with time only if a net external torque is acting on the system:

$$\sum \vec{\tau}_{\text{ext}} = \frac{d\vec{\mathbf{L}}_{\text{tot}}}{dt} \quad (11.13)$$

◀ The net external torque on a system equals the time rate of change of angular momentum of the system

This equation is indeed the rotational analog of $\sum \vec{\mathbf{F}}_{\text{ext}} = d\vec{\mathbf{p}}_{\text{tot}}/dt$ for a system of particles. Equation 11.13 is the mathematical representation of the **angular momentum version of the nonisolated system model. If a system is nonisolated in the sense that there is a net torque on it, the torque is equal to the time rate of change of angular momentum.**

Although we do not prove it here, this statement is true regardless of the motion of the center of mass. It applies even if the center of mass is accelerating, provided that the torque and angular momentum are evaluated relative to an axis through the center of mass.

EXAMPLE 11.4 A System of Objects

A sphere of mass m_1 and a block of mass m_2 are connected by a light cord that passes over a pulley as shown in Figure 11.6. The radius of the pulley is R, and the mass of the thin rim is M. The spokes of the pulley have negligible mass. The block slides on a frictionless, horizontal surface. Find an expression for the linear acceleration of the two objects, using the concepts of angular momentum and torque.

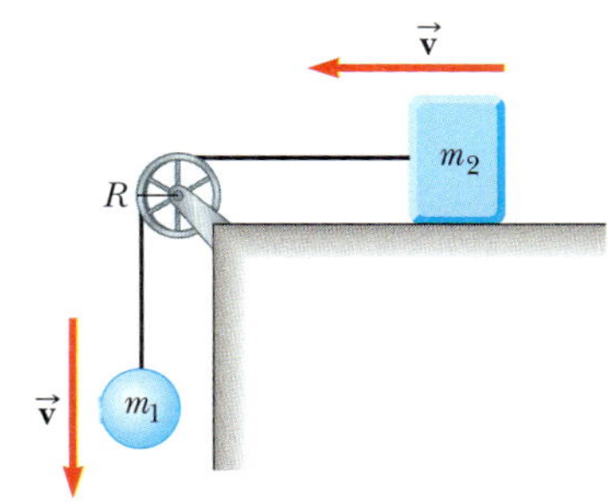

Figure 11.6 (Example 11.4) When the system is released, the sphere moves downward and the block moves to the left.

SOLUTION

Conceptualize When the system is released, the block slides to the left, the sphere drops downward, and the pulley rotates counterclockwise. This situation is similar to problems we have solved earlier except that now we want to use an angular momentum approach.

Categorize We identify the block, pulley, and sphere as a nonisolated system, subject to the external torque due to the gravitational force on the sphere. We shall calculate the angular momentum about an axis that coincides with the axle of the pulley. The angular momentum of the system includes that of two objects moving translationally (the sphere and the block) and one object undergoing pure rotation (the pulley).

Analyze At any instant of time, the sphere and the block have a common speed v, so the angular momentum of the sphere is m_1vR and that of the block is m_2vR. At the same instant, all points on the rim of the pulley also move with speed v, so the angular momentum of the pulley is MvR.

Now let's address the total external torque acting on the system about the pulley axle. Because it has a moment arm of zero, the force exerted by the axle on the pulley does not contribute to the torque. Furthermore, the normal force acting on the block is balanced by the gravitational force $m_2\vec{\mathbf{g}}$, so these forces do not contribute to the torque. The gravitational force $m_1\vec{\mathbf{g}}$ acting on the sphere produces a torque about the axle equal in magnitude to m_1gR, where R is the moment arm of the force about the axle. This result is the total external torque about the pulley axle; that is, $\sum \tau_{\text{ext}} = m_1gR$.

Write an expression for the total angular momentum of the system:

$$(1) \quad L = m_1vR + m_2vR + MvR = (m_1 + m_2 + M)vR$$

Substitute this expression and the total external torque into Equation 11.13:

$$\sum \tau_{\text{ext}} = \frac{dL}{dt}$$

$$m_1gR = \frac{d}{dt}[(m_1 + m_2 + M)vR]$$

$$(2) \quad m_1gR = (m_1 + m_2 + M)R\frac{dv}{dt}$$

Recognizing that $dv/dt = a$, solve Equation (2) for a:

$$a = \frac{m_1 g}{m_1 + m_2 + M}$$

Finalize Evaluating the net torque about the axle, we did not include the forces that the cord exerts on the objects because these forces are internal to the system under consideration. Instead we analyzed the system as a whole. Only *external* torques contribute to the change in the system's angular momentum.

11.3 Angular Momentum of a Rotating Rigid Object

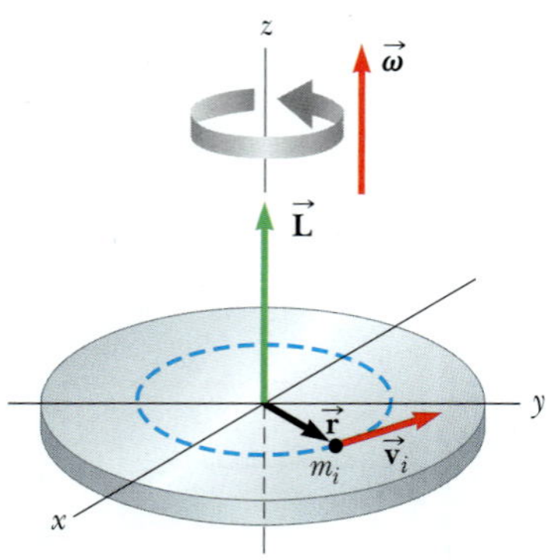

Figure 11.7 When a rigid object rotates about an axis, the angular momentum $\vec{L}$ is in the same direction as the angular velocity $\vec{\omega}$ according to the expression $\vec{L} = I\vec{\omega}$.

In Example 11.4, we considered the angular momentum of a deformable system. Let us now restrict our attention to a nondeformable system, a rigid object. Consider a rigid object rotating about a fixed axis that coincides with the z axis of a coordinate system as shown in Figure 11.7. Let's determine the angular momentum of this object. Each *particle* of the object rotates in the xy plane about the z axis with an angular speed ω. The magnitude of the angular momentum of a particle of mass m_i about the z axis is $m_i v_i r_i$. Because $v_i = r_i\omega$ (Eq. 10.10), we can express the magnitude of the angular momentum of this particle as

$$L_i = m_i r_i^2 \omega$$

The vector $\vec{L}_i$ is directed along the z axis, as is the vector $\vec{\omega}$.

We can now find the angular momentum (which in this situation has only a z component) of the whole object by taking the sum of L_i over all particles:

$$L_z = \sum_i L_i = \sum_i m_i r_i^2 \omega = \left(\sum_i m_i r_i^2\right)\omega$$

$$L_z = I\omega \tag{11.14}$$

where $\sum_i m_i r_i^2$ is the moment of inertia I of the object about the z axis (Eq. 10.15).

Now let's differentiate Equation 11.14 with respect to time, noting that I is constant for a rigid object:

$$\frac{dL_z}{dt} = I\frac{d\omega}{dt} = I\alpha \tag{11.15}$$

where α is the angular acceleration relative to the axis of rotation. Because dL_z/dt is equal to the net external torque (see Eq. 11.13), we can express Equation 11.15 as

Rotational form of Newton's second law ►

$$\sum \tau_{\text{ext}} = I\alpha \tag{11.16}$$

That is, the net external torque acting on a rigid object rotating about a fixed axis equals the moment of inertia about the rotation axis multiplied by the object's angular acceleration relative to that axis. This result is the same as Equation 10.21, which was derived using a force approach, but we derived Equation 11.16 using the concept of angular momentum. This equation is also valid for a rigid object rotating about a moving axis provided the moving axis (1) passes through the center of mass and (2) is a symmetry axis.

If a symmetrical object rotates about a fixed axis passing through its center of mass, you can write Equation 11.14 in vector form as $\vec{L} = I\vec{\omega}$, where $\vec{L}$ is the total angular momentum of the object measured with respect to the axis of rotation. Furthermore, the expression is valid for any object, regardless of its symmetry, if $\vec{L}$ stands for the component of angular momentum along the axis of rotation.[1]

[1] In general, the expression $\vec{L} = I\vec{\omega}$ is not always valid. If a rigid object rotates about an *arbitrary* axis, then $\vec{L}$ and $\vec{\omega}$ may point in different directions. In this case, the moment of inertia cannot be treated as a scalar. Strictly speaking, $\vec{L} = I\vec{\omega}$ applies only to rigid objects of any shape that rotate about one of three mutually perpendicular axes (called *principal axes*) through the center of mass. This concept is discussed in more advanced texts on mechanics.

Quick Quiz 11.3 A solid sphere and a hollow sphere have the same mass and radius. They are rotating with the same angular speed. Which one has the higher angular momentum? (a) the solid sphere (b) the hollow sphere (c) both have the same angular momentum (d) impossible to determine

EXAMPLE 11.5 Bowling Ball

Estimate the magnitude of the angular momentum of a bowling ball spinning at 10 rev/s as shown in Figure 11.8.

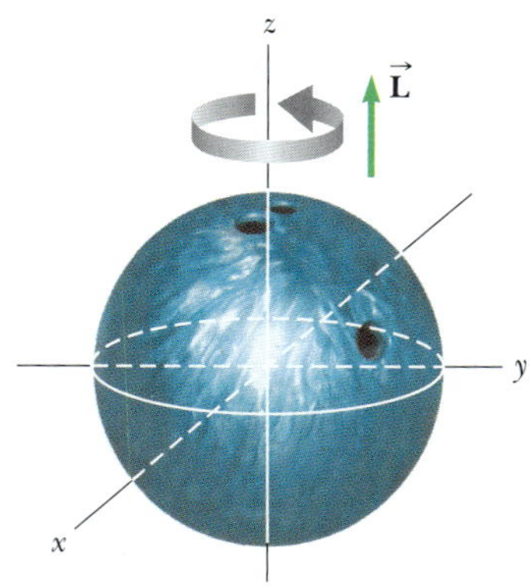

Figure 11.8 (Example 11.5) A bowling ball that rotates about the z axis in the direction shown has an angular momentum $\vec{L}$ in the positive z direction. If the direction of rotation is reversed, $\vec{L}$ points in the negative z direction.

SOLUTION

Conceptualize Imagine spinning a bowling ball on the smooth floor of a bowling alley. Because a bowling ball is relatively heavy, the angular momentum should be relatively large.

Categorize We evaluate the angular momentum using Equation 11.14, so we categorize this example as a substitution problem.

We start by making some estimates of the relevant physical parameters and model the ball as a uniform solid sphere. A typical bowling ball might have a mass of 7.0 kg and a radius of 12 cm.

Evaluate the moment of inertia of the ball about an axis through its center from Table 10.2:

$$I = \frac{2}{5}MR^2 = \frac{2}{5}(7.0\text{ kg})(0.12\text{ m})^2 = 0.040\text{ kg}\cdot\text{m}^2$$

Evaluate the magnitude of the angular momentum from Equation 11.14:

$$L_z = I\omega = (0.040\text{ kg}\cdot\text{m}^2)(10\text{ rev/s})(2\pi\text{ rad/rev}) = 2.53\text{ kg}\cdot\text{m}^2/\text{s}$$

Because of the roughness of our estimates, we should keep only one significant figure, so $L_z = 3\text{ kg}\cdot\text{m}^2/\text{s}$

EXAMPLE 11.6 The Seesaw

A father of mass m_f and his daughter of mass m_d sit on opposite ends of a seesaw at equal distances from the pivot at the center (Fig. 11.9). The seesaw is modeled as a rigid rod of mass M and length ℓ and is pivoted without friction. At a given moment, the combination rotates in a vertical plane with an angular speed ω.

(A) Find an expression for the magnitude of the system's angular momentum.

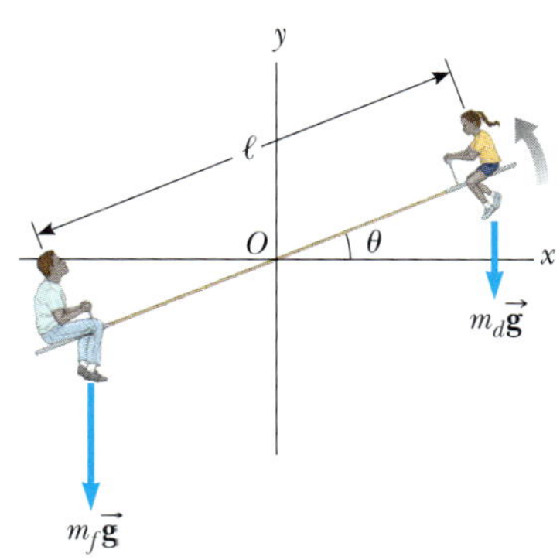

Figure 11.9 (Example 11.6) A father and daughter demonstrate angular momentum on a seesaw.

SOLUTION

Conceptualize Imagine an axis of rotation passing through the pivot at O in Figure 11.9. The rotating system has angular momentum about that axis.

Categorize Ignore any movement of arms or legs of the father and daughter and model them both as particles. The system is therefore modeled as a rigid object. This first part of the example is categorized as a substitution problem.

The moment of inertia of the system equals the sum of the moments of inertia of the three components: the seesaw and the two individuals. We can refer to Table 10.2 to obtain the expression for the moment of inertia of the rod and use the particle expression $I = mr^2$ for each person.

Find the total moment of inertia of the system about the z axis through O:

$$I = \tfrac{1}{12}M\ell^2 + m_f\left(\frac{\ell}{2}\right)^2 + m_d\left(\frac{\ell}{2}\right)^2 = \frac{\ell^2}{4}\left(\frac{M}{3} + m_f + m_d\right)$$

Find the magnitude of the angular momentum of the system:

$$L = I\omega = \frac{\ell^2}{4}\left(\frac{M}{3} + m_f + m_d\right)\omega$$

(B) Find an expression for the magnitude of the angular acceleration of the system when the seesaw makes an angle θ with the horizontal.

SOLUTION

Conceptualize Generally, fathers are more massive than daughters, so the system is not in equilibrium and has an angular acceleration. We expect the angular acceleration to be positive in Figure 11.9.

Categorize We identify the system as nonisolated because of the external torque associated with the gravitational force. We again identify an axis of rotation passing through the pivot at O in Figure 11.9.

Analyze To find the angular acceleration of the system at any angle θ, we first calculate the net torque on the system and then use $\sum \tau_{\text{ext}} = I\alpha$ to obtain an expression for α.

Evaluate the torque due to the gravitational force on the father:

$$\tau_f = m_f g\frac{\ell}{2}\cos\theta \quad (\vec{\boldsymbol{\tau}}_f \text{ out of page})$$

Evaluate the torque due to the gravitational force on the daughter:

$$\tau_d = -m_d g\frac{\ell}{2}\cos\theta \quad (\vec{\boldsymbol{\tau}}_d \text{ into page})$$

Evaluate the net torque exerted on the system:

$$\sum \tau_{\text{ext}} = \tau_f + \tau_d = \tfrac{1}{2}(m_f - m_d)g\ell\cos\theta$$

Use Equation 11.16 and I from part (A) to find α:

$$\alpha = \frac{\sum \tau_{\text{ext}}}{I} = \frac{2(m_f - m_d)g\cos\theta}{\ell\left[(M/3) + m_f + m_d\right]}$$

Finalize For a father more massive than his daughter, the angular acceleration is positive as expected. If the seesaw begins in a horizontal orientation ($\theta = 0$) and is released, the rotation is counterclockwise in Figure 11.9 and the father's end of the seesaw drops, which is consistent with everyday experience.

What If? Imagine the father moves inward on the seesaw to a distance d from the pivot to try to balance the two sides. What is the angular acceleration of the system in this case when it is released from an arbitrary angle θ?

Answer The angular acceleration of the system should decrease if the system is more balanced.

Find the total moment of inertia about the z axis through O for the modified system:

$$I = \tfrac{1}{12}M\ell^2 + m_f d^2 + m_d\left(\frac{\ell}{2}\right)^2 = \frac{\ell^2}{4}\left(\frac{M}{3} + m_d\right) + m_f d^2$$

Find the net torque exerted on the system about an axis through O:

$$\sum \tau_{\text{ext}} = \tau_f + \tau_d = m_f g d\cos\theta - \tfrac{1}{2}m_d g\ell\cos\theta$$

Find the new angular acceleration of the system:

$$\alpha = \frac{\sum \tau_{\text{ext}}}{I} = \frac{m_f g d\cos\theta - \tfrac{1}{2}m_d g\ell\cos\theta}{(\ell^2/4)\left[(M/3) + m_d\right] + m_f d^2}$$

The seesaw is balanced when the angular acceleration is zero. In this situation, both father and daughter can push off the ground and rise to the highest possible point.

Find the required position of the father by setting $\alpha = 0$:

$$\alpha = \frac{m_f g d \cos\theta - \frac{1}{2} m_d g \ell \cos\theta}{(\ell^2/4)[(M/3) + m_d] + m_f d^2} = 0$$

$$m_f g d \cos\theta - \tfrac{1}{2} m_d g \ell \cos\theta = 0 \quad \rightarrow \quad d = \left(\frac{m_d}{m_f}\right)\tfrac{1}{2}\ell$$

In the rare case that the father and daughter have the same mass, the father is located at the end of the seesaw, $d = \ell/2$.

11.4 The Isolated System: Conservation of Angular Momentum

In Chapter 9, we found that the total linear momentum of a system of particles remains constant if the system is isolated, that is, if the net external force acting on the system is zero. We have an analogous conservation law in rotational motion:

> The total angular momentum of a system is constant in both magnitude and direction if the net external torque acting on the system is zero, that is, if the system is isolated.

◀ Conservation of angular momentum

This statement is the principle of **conservation of angular momentum** and is the basis of the **angular momentum version of the isolated system model.** This principle follows directly from Equation 11.13, which indicates that if

$$\sum \vec{\boldsymbol{\tau}}_{\text{ext}} = \frac{d\vec{\mathbf{L}}_{\text{tot}}}{dt} = 0 \qquad \textbf{(11.17)}$$

then

$$\vec{\mathbf{L}}_{\text{tot}} = \text{constant} \quad \text{or} \quad \vec{\mathbf{L}}_i = \vec{\mathbf{L}}_f \qquad \textbf{(11.18)}$$

For an isolated system consisting of a number of particles, we write this conservation law as $\vec{\mathbf{L}}_{\text{tot}} = \Sigma \vec{\mathbf{L}}_n = \text{constant}$, where the index n denotes the nth particle in the system.

If an isolated rotating system is deformable so that its mass undergoes redistribution in some way, the system's moment of inertia changes. Because the magnitude of the angular momentum of the system is $L = I\omega$ (Eq. 11.14), conservation of angular momentum requires that the product of I and ω must remain constant. Therefore, a change in I for an isolated system requires a change in ω. In this case, we can express the principle of conservation of angular momentum as

$$I_i\omega_i = I_f\omega_f = \text{constant} \qquad \textbf{(11.19)}$$

This expression is valid both for rotation about a fixed axis and for rotation about an axis through the center of mass of a moving system as long as that axis remains fixed in direction. We require only that the net external torque be zero.

Many examples demonstrate conservation of angular momentum for a deformable system. You may have observed a figure skater spinning in the finale of a program (Fig. 11.10). The angular speed of the skater is large when his hands and feet are close to the trunk of his body. Ignoring friction between skater and ice, there are no external torques on the skater. The moment of inertia of his body increases as his hands and feet are moved away from his body at the finish of the spin. According to the principle of conservation of angular momentum, his angular speed must decrease. In a similar way, when divers or acrobats wish to make

© Stuart Franklin/Getty Images

© Stuart Franklin/Getty Images

Figure 11.10 Angular momentum is conserved as Russian figure skater Evgeni Plushenko performs during the 2004 World Figure Skating Championships. When his arms and legs are close to his body, his moment of inertia is small and his angular speed is large. To slow down for the finish of his spin, he moves his arms and legs outward, increasing his moment of inertia.

several somersaults, they pull their hands and feet close to their bodies to rotate at a higher rate, as in the opening photograph of this chapter. In these cases, the external force due to gravity acts through the center of mass and hence exerts no torque about an axis through this point. Therefore, the angular momentum about the center of mass must be conserved; that is, $I_i\omega_i = I_f\omega_f$. For example, when divers wish to double their angular speed, they must reduce their moment of inertia to half its initial value.

In Equation 11.18, we have a third version of the isolated system model. We can now state that the energy, linear momentum, and angular momentum of an isolated system are all conserved:

$$\text{For an isolated system}\begin{cases} E_i = E_f & \text{(if there are no energy transfers)} \\ \vec{\mathbf{p}}_i = \vec{\mathbf{p}}_f & \text{(if the net external force is zero)} \\ \vec{\mathbf{L}}_i = \vec{\mathbf{L}}_f & \text{(if the net external torque is zero)} \end{cases}$$

Quick Quiz 11.4 A competitive diver leaves the diving board and falls toward the water with her body straight and rotating slowly. She pulls her arms and legs into a tight tuck position. **(i)** What happens to her angular speed? (a) It increases. (b) It decreases. (c) It stays the same. (d) It is impossible to determine. **(ii)** From the same list of choices, what happens to the rotational kinetic energy of her body?

EXAMPLE 11.7 Formation of a Neutron Star

A star rotates with a period of 30 days about an axis through its center. After the star undergoes a supernova explosion, the stellar core, which had a radius of 1.0×10^4 km, collapses into a neutron star of radius 3.0 km. Determine the period of rotation of the neutron star.

SOLUTION

Conceptualize The change in the neutron star's motion is similar to that of the skater described above, but in the reverse direction. As the mass of the star moves closer to the rotation axis, we expect the star to spin faster.

Categorize Let us assume that during the collapse of the stellar core, (1) no external torque acts on it, (2) it remains spherical with the same relative mass distribution, and (3) its mass remains constant. We categorize the star as an isolated system. We do not know the mass distribution of the star, but we have assumed the distribution is symmetric, so the moment of inertia can be expressed as kMR^2, where k is some numerical constant. (From Table 10.2, for example, we see that $k = \frac{2}{5}$ for a solid sphere and $k = \frac{2}{3}$ for a spherical shell.)

Analyze Let's use the symbol T for the period, with T_i being the initial period of the star and T_f being the period of the neutron star. The period is the time interval required for a point on the star's equator to make one complete revolution around the axis of rotation. The star's angular speed is given by $\omega = 2\pi/T$.

Write Equation 11.19 for the star:

$$I_i\omega_i = I_f\omega_f$$

Use $\omega = 2\pi/T$ to rewrite this equation in terms of the initial and final periods:

$$I_i\left(\frac{2\pi}{T_i}\right) = I_f\left(\frac{2\pi}{T_f}\right)$$

Substitute the moments of inertia in the preceding equation:

$$kMR_i^2\left(\frac{2\pi}{T_i}\right) = kMR_f^2\left(\frac{2\pi}{T_f}\right)$$

Solve for the final period of the star:

$$T_f = \left(\frac{R_f}{R_i}\right)^2 T_i$$

Substitute numerical values: $$T_f = \left(\frac{3.0 \text{ km}}{1.0 \times 10^4 \text{ km}}\right)^2 (30 \text{ days}) = 2.7 \times 10^{-6} \text{ days} = 0.23 \text{ s}$$

Finalize The neutron star does indeed rotate faster after it collapses, as predicted. It moves very fast, in fact, rotating about four times each second.

EXAMPLE 11.8 The Merry-Go-Round

A horizontal platform in the shape of a circular disk rotates freely in a horizontal plane about a frictionless vertical axle (Fig. 11.11). The platform has a mass $M = 100$ kg and a radius $R = 2.0$ m. A student whose mass is $m = 60$ kg walks slowly from the rim of the disk toward its center. If the angular speed of the system is 2.0 rad/s when the student is at the rim, what is the angular speed when she reaches a point $r = 0.50$ m from the center?

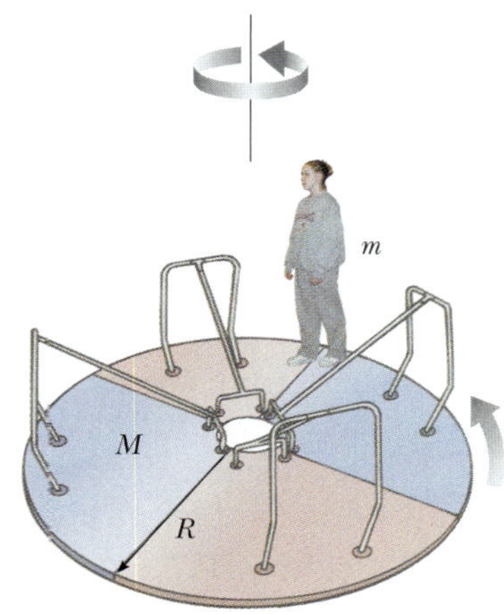

Figure 11.11 (Example 11.8) As the student walks toward the center of the rotating platform, the angular speed of the system increases because the angular momentum of the system remains constant.

SOLUTION

Conceptualize The speed change here is similar to those of the spinning skater and the neutron star in preceding discussions. This problem is different because part of the moment of inertia of the system changes (that of the student) while part remains fixed (that of the platform).

Categorize Because the platform rotates on a frictionless axle, we identify the system of the student and the platform as an isolated system.

Analyze Let us denote the moment of inertia of the platform as I_p and that of the student as I_s. We model the student as a particle.

Find the initial moment of inertia I_i of the system (student plus platform) about the axis of rotation: $$I_i = I_{pi} + I_{si} = \tfrac{1}{2}MR^2 + mR^2$$

Find the moment of inertia of the system when the student walks to the position $r < R$: $$I_f = I_{pf} + I_{sf} = \tfrac{1}{2}MR^2 + mr^2$$

Apply the law of conservation of angular momentum to the system: $$I_i\omega_i = I_f\omega_f$$

Substitute the moments of inertia: $$(\tfrac{1}{2}MR^2 + mR^2)\omega_i = (\tfrac{1}{2}MR^2 + mr^2)\omega_f$$

Solve for the final angular speed: $$\omega_f = \left(\frac{\frac{1}{2}MR^2 + mR^2}{\frac{1}{2}MR^2 + mr^2}\right)\omega_i$$

Substitute numerical values: $$\omega_f = \left[\frac{\frac{1}{2}(100 \text{ kg})(2.0 \text{ m})^2 + (60 \text{ kg})(2.0 \text{ m})^2}{\frac{1}{2}(100 \text{ kg})(2.0 \text{ m})^2 + (60 \text{ kg})(0.50 \text{ m})^2}\right](2.0 \text{ rad/s})$$

$$\omega_f = \left(\frac{440 \text{ kg}\cdot\text{m}^2}{215 \text{ kg}\cdot\text{m}^2}\right)(2.0 \text{ rad/s}) = 4.1 \text{ rad/s}$$

Finalize As expected, the angular speed increases. The fastest that this system could spin would be when the student moves to the center of the platform. Do this calculation to show that this maximum angular speed is 4.4 rad/s. Notice that the activity described in this problem is dangerous as discussed with regard to the Coriolis force in Section 6.3.

What If? What if you measured the kinetic energy of the system before and after the student walks inward? Are the initial kinetic energy and the final kinetic energy the same?

Answer You may be tempted to say yes because the system is isolated. Remember, however, that energy can be transformed among several forms, so we have to handle an energy question carefully.

Find the initial kinetic energy: $K_i = \frac{1}{2}I_i\omega_i^2 = \frac{1}{2}(440 \text{ kg}\cdot\text{m}^2)(2.0 \text{ rad/s})^2 = 880 \text{ J}$

Find the final kinetic energy: $K_f = \frac{1}{2}I_f\omega_f^2 = \frac{1}{2}(215 \text{ kg}\cdot\text{m}^2)(4.1 \text{ rad/s})^2 = 1.81 \times 10^3 \text{ J}$

Therefore, the kinetic energy of the system *increases*. The student must do work to move herself closer to the center of rotation, so this extra kinetic energy comes from chemical potential energy in the student's body.

EXAMPLE 11.9 Disk and Stick Collision

A 2.0-kg disk traveling at 3.0 m/s strikes a 1.0-kg stick of length 4.0 m that is lying flat on nearly frictionless ice as shown in the overhead view of Figure 11.12a. Assume the collision is elastic and the disk does not deviate from its original line of motion. Find the translational speed of the disk, the translational speed of the stick, and the angular speed of the stick after the collision. The moment of inertia of the stick about its center of mass is 1.33 kg · m^2.

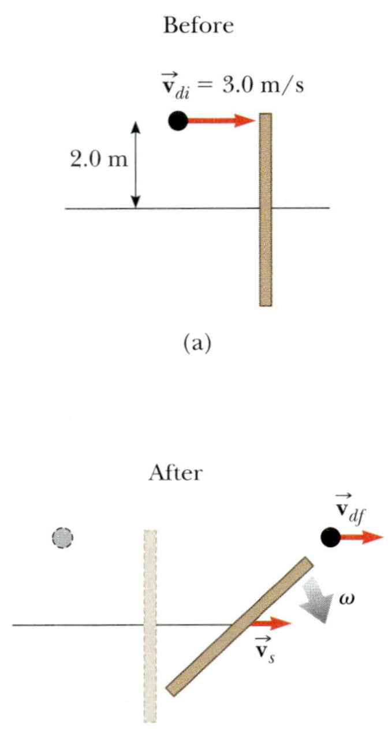

Figure 11.12 (Example 11.9) Overhead view of a disk striking a stick in an elastic collision. (a) Before the collision, the disk moves toward the stick. (b) The collision causes the stick to rotate and move to the right.

SOLUTION

Conceptualize Examine Figure 11.12a and imagine what happens after the disk hits the stick. Figure 11.12b shows what you might expect: the disk continues to move at a slower speed and the stick is in both translational and rotational motion. We assume the disk does not deviate from its original line of motion because the force exerted by the stick on the disk is parallel to the original path of the disk.

Categorize Because the ice is frictionless, the disk and stick form an isolated system. Also, because the collision is assumed to be elastic, the energy, linear momentum, and angular momentum of the system are all conserved.

Analyze First notice that we have three unknowns, so we need three equations to solve simultaneously.

Apply the law of the conservation of linear momentum to the system: $m_d v_{di} = m_d v_{df} + m_s v_s$

Substitute the known values: $(2.0 \text{ kg})(3.0 \text{ m/s}) = (2.0 \text{ kg})v_{df} + (1.0 \text{ kg})v_s$

Rearrange the equation: (1) $6.0 \text{ kg}\cdot\text{m/s} - (2.0 \text{ kg})v_{df} = (1.0 \text{ kg})v_s$

Now we apply the law of conservation of angular momentum for the system, using an axis passing through the center of the stick as our rotation axis. The component of angular momentum of the disk along the axis perpendicular to the plane of the ice is negative. (The right-hand rule shows that $\vec{\mathbf{L}}_d$ points into the ice.)

Apply conservation of angular momentum to the system: $-rm_d v_{di} = -rm_d v_{df} + I\omega$

Substitute the known values:

$$-(2.0\text{ m})(2.0\text{ kg})(3.0\text{ m/s}) = -(2.0\text{ m})(2.0\text{ kg})v_{df} + (1.33\text{ kg}\cdot\text{m}^2)\omega$$

$$-12\text{ kg}\cdot\text{m}^2/\text{s} = -(4.0\text{ kg}\cdot\text{m})v_{df} + (1.33\text{ kg}\cdot\text{m}^2)\omega$$

Divide the equation by $1.33\text{ kg}\cdot\text{m}^2$ and rearrange:

$$(2)\quad -9.0\text{ rad/s} + (3.0\text{ rad/m})v_{df} = \omega$$

Finally, the elastic nature of the collision tells us that the kinetic energy of the system is conserved; in this case, the kinetic energy consists of translational and rotational forms.

Apply conservation of kinetic energy to the system:

$$\tfrac{1}{2}m_d v_{di}^2 = \tfrac{1}{2}m_d v_{df}^2 + \tfrac{1}{2}m_s v_s^2 + \tfrac{1}{2}I\omega^2$$

Substitute the known values:

$$\tfrac{1}{2}(2.0\text{ kg})(3.0\text{ m/s})^2 = \tfrac{1}{2}(2.0\text{ kg})v_{df}^2 + \tfrac{1}{2}(1.0\text{ kg})v_s^2 + \tfrac{1}{2}(1.33\text{ kg}\cdot\text{m}^2)\omega^2$$

$$(3)\quad 18\text{ m}^2/\text{s}^2 = 2.0v_{df}^2 + v_s^2 + (1.33\text{ m}^2)\omega^2$$

Solving Equations (1), (2), and (3) simultaneously, we find that $v_d = 2.3\text{ m/s}$, $v_s = 1.3\text{ m/s}$, and $\omega = -2.0\text{ rad/s}$.

Finalize These values seem reasonable. The disk is moving more slowly after the collision than it was before the collision, and the stick has a small translational speed. Table 11.1 summarizes the initial and final values of variables for the disk and the stick, and verifies the conservation of linear momentum, angular momentum, and kinetic energy for the system.

TABLE 11.1

Comparison of Values in Example 11.9 Before and After the Collision

	v (m/s)	ω (rad/s)	p (kg·m/s)	L (kg·m²/s)	K_{trans} (J)	K_{rot} (J)
Before						
Disk	3.0	—	6.0	−12	9.0	—
Stick	0	0	0	0	0	0
Total for system	—	—	6.0	−12	9.0	0
After						
Disk	2.3	—	4.7	−9.3	5.4	—
Stick	1.3	−2.0	1.3	−2.7	0.9	2.7
Total for system	—	—	6.0	−12	6.3	2.7

Note: Linear momentum, angular momentum, and total kinetic energy of the system are all conserved.

What If? What if the collision between the disk and the stick is perfectly inelastic? How does that change the analysis?

Answer In this case, the disk adheres to the end of the stick upon collision.

Alter the conservation of linear momentum principle leading to Equation (1):

$$m_d v_{di} = (m_d + m_s)v_{CM}$$

$$(2.0\text{ kg})(3.0\text{ m/s}) = (2.0\text{ kg} + 1.0\text{ kg})v_{CM}$$

$$v_{CM} = 2.0\text{ m/s}$$

Choose the center of the stick as the origin and find the y position of the center of mass along the vertical stick:

$$y_{CM} = \frac{(2.0\text{ kg})(2.0\text{ m}) + (1.0\text{ kg})(0)}{(2.0\text{ kg} + 1.0\text{ kg})} = 1.33\text{ m}$$

Therefore, the center of mass of the system is 2.0 m − 1.33 m = 0.67 m from the upper end of the stick.

Alter the conservation of angular momentum principle leading to Equation (2), evaluating angular momenta around an axis passing through the center of mass of the system:

$$-rm_d v_{di} = I_d\omega + I_s\omega$$

$$(4)\quad -(0.67\text{ m})m_d v_{di} = [m_d(0.67\text{ m})^2]\omega + I_s\omega$$

Find the moment of inertia of the stick around the center of mass *of the system* from the parallel-axis theorem:

$$I_s = I_{\text{CM}} + MD^2$$

$$= 1.33\text{ kg}\cdot\text{m}^2 + (1.0\text{ kg})(1.33\text{ m})^2 = 3.1\text{ kg}\cdot\text{m}^2$$

Use these results in Equation (4):

$$-(0.67\text{ m})(2.0\text{ kg})(3.0\text{ m/s}) = [(2.0\text{ kg})(0.67\text{ m})^2]\omega + (3.1\text{ kg}\cdot\text{m}^2)\omega$$

$$-4.0\text{ kg}\cdot\text{m}^2/\text{s} = (4.0\text{ kg}\cdot\text{m}^2)\omega$$

$$\omega = \frac{-4.0\text{ kg}\cdot\text{m}^2/\text{s}}{4.0\text{ kg}\cdot\text{m}^2} = -1.0\text{ rad/s}$$

Evaluating the total kinetic energy of the system after the collision shows that it is less than that before the collision because kinetic energy is not conserved in an inelastic collision.

11.5 The Motion of Gyroscopes and Tops

Precessional motion ▶

An unusual and fascinating type of motion you have probably observed is that of a top spinning about its axis of symmetry as shown in Figure 11.13a. If the top spins rapidly, the symmetry axis rotates about the z axis, sweeping out a cone (see Fig. 11.13b). The motion of the symmetry axis about the vertical—known as **precessional motion**—is usually slow relative to the spinning motion of the top.

It is quite natural to wonder why the top does not fall over. Because the center of mass is not directly above the pivot point O, a net torque is acting on the top about an axis passing through O, a torque resulting from the gravitational force $M\vec{\mathbf{g}}$. The top would certainly fall over if it were not spinning. Because it is spinning, however, it has an angular momentum $\vec{\mathbf{L}}$ directed along its symmetry axis. We shall show that this symmetry axis moves about the z axis (precessional motion occurs) because the torque produces a change in the *direction* of the symmetry axis. This illustration is an excellent example of the importance of the directional nature of angular momentum.

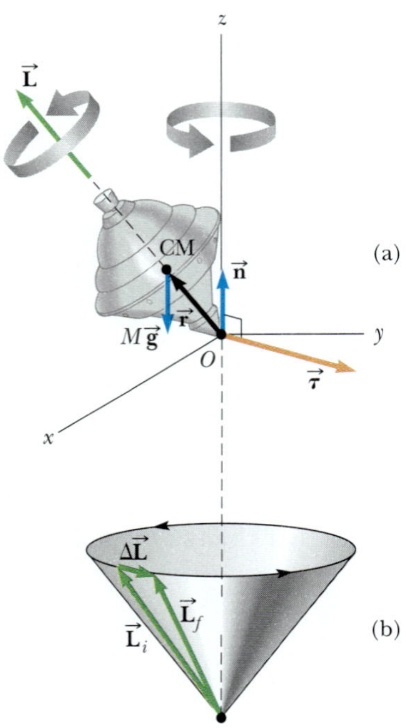

Figure 11.13 Precessional motion of a top spinning about its symmetry axis. (a) The only external forces acting on the top are the normal force $\vec{\mathbf{n}}$ and the gravitational force $M\vec{\mathbf{g}}$. The direction of the angular momentum $\vec{\mathbf{L}}$ is along the axis of symmetry. The right-hand rule indicates that $\vec{\boldsymbol{\tau}} = \vec{\mathbf{r}} \times \vec{\mathbf{F}} = \vec{\mathbf{r}} \times M\vec{\mathbf{g}}$ is in the xy plane. (b). The direction of $\Delta\vec{\mathbf{L}}$ is parallel to that of $\vec{\boldsymbol{\tau}}$ in (a). Because $\vec{\mathbf{L}}_f = \Delta\vec{\mathbf{L}} + \vec{\mathbf{L}}_i$, the top precesses about the z axis.

The essential features of precessional motion can be illustrated by considering the simple gyroscope shown in Figure 11.14a. The two forces acting on the gyroscope are the downward gravitational force $M\vec{\mathbf{g}}$ and the normal force $\vec{\mathbf{n}}$ acting upward at the pivot point O. The normal force produces no torque about an axis passing through the pivot because its moment arm through that point is zero. The gravitational force, however, produces a torque $\vec{\boldsymbol{\tau}} = \vec{\mathbf{r}} \times M\vec{\mathbf{g}}$ about an axis passing through O, where the direction of $\vec{\boldsymbol{\tau}}$ is perpendicular to the plane formed by $\vec{\mathbf{r}}$ and $M\vec{\mathbf{g}}$. By necessity, the vector $\vec{\boldsymbol{\tau}}$ lies in a horizontal xy plane perpendicular to the angular momentum vector. The net torque and angular momentum of the gyroscope are related through Equation 11.13:

$$\vec{\boldsymbol{\tau}} = \frac{d\vec{\mathbf{L}}}{dt}$$

This expression shows that in the infinitesimal time interval dt, the nonzero torque produces a change in angular momentum $d\vec{\mathbf{L}}$, a change that is in the same direction as $\vec{\boldsymbol{\tau}}$. Therefore, like the torque vector, $d\vec{\mathbf{L}}$ must also be perpendicular to $\vec{\mathbf{L}}$. Figure 11.14b illustrates the resulting precessional motion of the symmetry axis of the gyroscope. In a time interval dt, the change in angular momentum is $d\vec{\mathbf{L}} = \vec{\mathbf{L}}_f - \vec{\mathbf{L}}_i = \vec{\boldsymbol{\tau}}dt$. Because $d\vec{\mathbf{L}}$ is perpendicular to $\vec{\mathbf{L}}$, the magnitude of $\vec{\mathbf{L}}$ does not change ($|\vec{\mathbf{L}}_i| = |\vec{\mathbf{L}}_f|$). Rather, what is changing is the *direction* of $\vec{\mathbf{L}}$. Because the

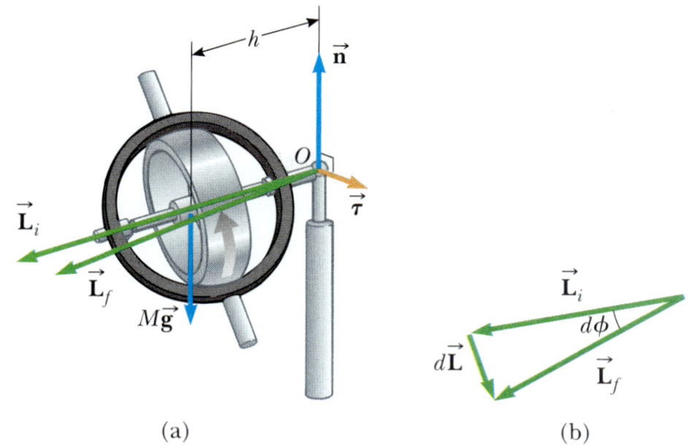

Figure 11.14 (a) The motion of a simple gyroscope pivoted a distance h from its center of mass. The gravitational force $M\vec{g}$ produces a torque about the pivot, and this torque is perpendicular to the axle. (b) Overhead view of the initial and final angular momentum vectors. The torque results in a change in angular momentum $d\vec{L}$ in a direction perpendicular to the axle. The axle sweeps out an angle $d\phi$ in a time interval dt.

change in angular momentum $d\vec{L}$ is in the direction of $\vec{\tau}$, which lies in the xy plane, the gyroscope undergoes precessional motion.

To simplify the description of the system, we assume that the total angular momentum of the precessing wheel is the sum of the angular momentum $I\vec{\omega}$ due to the spinning and the angular momentum due to the motion of the center of mass about the pivot. In our treatment, we shall neglect the contribution from the center-of-mass motion and take the total angular momentum to be simply $I\vec{\omega}$. In practice, this approximation is good if $\vec{\omega}$ is made very large.

The vector diagram in Figure 11.14b shows that in the time interval dt, the angular momentum vector rotates through an angle $d\phi$, which is also the angle through which the axle rotates. From the vector triangle formed by the vectors $\vec{L}_i$, $\vec{L}_f$, and $d\vec{L}$, we see that

$$\sin(d\phi) \approx d\phi = \frac{dL}{L} = \frac{\tau\,dt}{L} = \frac{(Mgh)\,dt}{L}$$

where we have used that, for small values of any angle θ, $\sin\theta \approx \theta$. Dividing through by dt and using the relationship $L = I\omega$, we find that the rate at which the axle rotates about the vertical axis is

$$\omega_p = \frac{d\phi}{dt} = \frac{Mgh}{I\omega} \qquad \textbf{(11.20)}$$

The angular speed ω_p is called the **precessional frequency.** This result is valid only when $\omega_p << \omega$. Otherwise, a much more complicated motion is involved. As you can see from Equation 11.20, the condition $\omega_p << \omega$ is met when ω is large, that is, when the wheel spins rapidly. Furthermore, notice that the precessional frequency decreases as ω increases, that is, as the wheel spins faster about its axis of symmetry.

As an example of the usefulness of gyroscopes, suppose you are in a spacecraft in deep space and you need to alter your trajectory. To fire the engines in the correct direction, you need to turn the spacecraft around. How, though, do you turn a spacecraft around in empty space? One way is to have small rocket engines that fire perpendicularly out the side of the spacecraft, providing a torque around its center of mass. Such a setup is desirable, and many spacecraft have such rockets.

Let us consider another method, however, that is related to angular momentum and does not require the consumption of rocket fuel. Suppose the spacecraft carries a gyroscope that is not rotating as in Figure 11.15a. In this case, the angular momentum of the spacecraft about its center of mass is zero. Suppose the gyroscope is set into rotation, giving the gyroscope a nonzero angular momentum. There is no external torque on the isolated system (spacecraft and gyroscope), so the angular momentum of this system must remain zero according to the principle of conservation of angular momentum. This principle can be satisfied if the

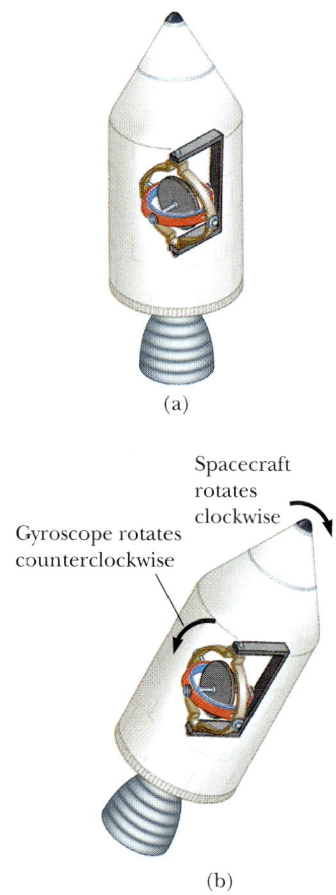

Figure 11.15 (a) A spacecraft carries a gyroscope that is not spinning. (b) When the gyroscope is set into rotation, the spacecraft turns the other way so that the angular momentum of the system is conserved.

spacecraft rotates in the direction opposite that of the gyroscope so that the angular momentum vectors of the gyroscope and the spacecraft cancel, resulting in no angular momentum of the system. The result of rotating the gyroscope, as in Figure 11.15b, is that the spacecraft turns around! By including three gyroscopes with mutually perpendicular axles, any desired rotation in space can be achieved.

This effect created an undesirable situation with the *Voyager 2* spacecraft during its flight. The spacecraft carried a tape recorder whose reels rotated at high speeds. Each time the tape recorder was turned on, the reels acted as gyroscopes and the spacecraft started an undesirable rotation in the opposite direction. This rotation had to be counteracted by Mission Control by using the sideward-firing jets to stop the rotation!

Summary

ThomsonNOW™ Sign in at **www.thomsonedu.com** and go to ThomsonNOW to take a practice test for this chapter.

DEFINITIONS

Given two vectors $\vec{\mathbf{A}}$ and $\vec{\mathbf{B}}$, the cross product $\vec{\mathbf{A}} \times \vec{\mathbf{B}}$ is a vector $\vec{\mathbf{C}}$ having a magnitude

$$C = AB\sin\theta \tag{11.3}$$

where θ is the angle between $\vec{\mathbf{A}}$ and $\vec{\mathbf{B}}$. The direction of the vector $\vec{\mathbf{C}} = \vec{\mathbf{A}} \times \vec{\mathbf{B}}$ is perpendicular to the plane formed by $\vec{\mathbf{A}}$ and $\vec{\mathbf{B}}$, and this direction is determined by the right-hand rule.

The **torque** $\vec{\boldsymbol{\tau}}$ due to a force $\vec{\mathbf{F}}$ about an axis through the origin in an inertial frame is defined to be

$$\vec{\boldsymbol{\tau}} \equiv \vec{\mathbf{r}} \times \vec{\mathbf{F}} \tag{11.1}$$

The **angular momentum** $\vec{\mathbf{L}}$ of a particle having linear momentum $\vec{\mathbf{p}} = m\vec{\mathbf{v}}$ about an axis through the origin is

$$\vec{\mathbf{L}} \equiv \vec{\mathbf{r}} \times \vec{\mathbf{p}} \tag{11.10}$$

where $\vec{\mathbf{r}}$ is the vector position of the particle relative to the origin.

CONCEPTS AND PRINCIPLES

The z component of angular momentum of a rigid object rotating about a fixed z axis is

$$L_z = I\omega \tag{11.14}$$

where I is the moment of inertia of the object about the axis of rotation and ω is its angular speed.

ANALYSIS MODELS FOR PROBLEM SOLVING

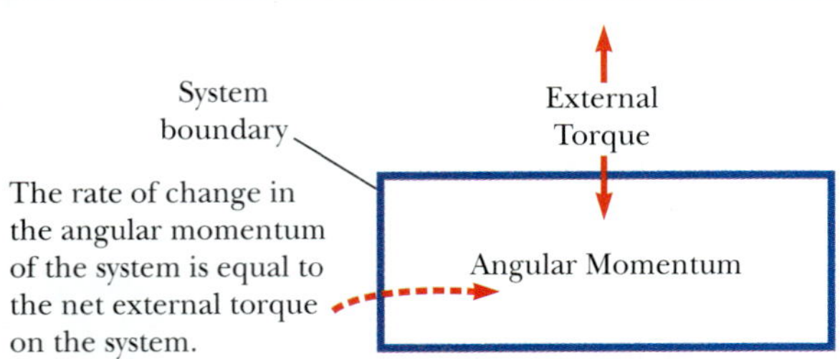

Nonisolated System (Angular Momentum). If a system interacts with its environment in the sense that there is an external torque on the system, the net external torque acting on a system is equal to the time rate of change of its angular momentum:

$$\sum \vec{\boldsymbol{\tau}}_{\text{ext}} = \frac{d\vec{\mathbf{L}}_{\text{tot}}}{dt} \tag{11.13}$$

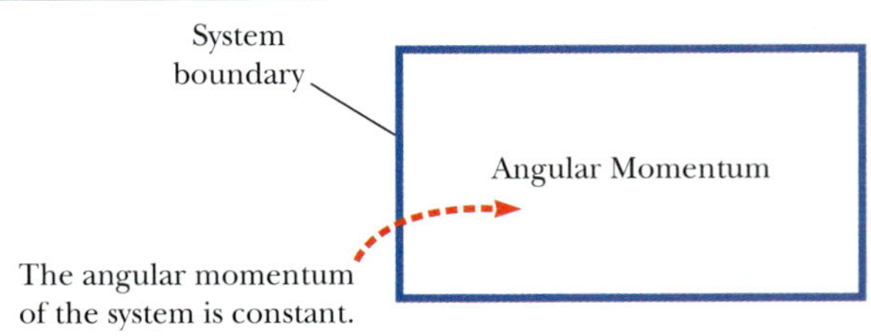

Isolated System (Angular Momentum). If a system experiences no external torque from the environment, the total angular momentum of the system is conserved:

$$\vec{\mathbf{L}}_i = \vec{\mathbf{L}}_f \tag{11.18}$$

Applying this law of conservation of angular momentum to a system whose moment of inertia changes gives

$$I_i\omega_i = I_f\omega_f = \text{constant} \tag{11.19}$$

Questions

☐ denotes answer available in *Student Solutions Manual/Study Guide;* **O** denotes objective question

1. **O** Is it possible to calculate the torque acting on a rigid object without specifying an axis of rotation? Is the torque independent of the location of the axis of rotation?

2. **O** Vector $\vec{\mathbf{A}}$ is in the negative y direction and vector $\vec{\mathbf{B}}$ is in the negative x direction. **(i)** What is the direction of $\vec{\mathbf{A}} \times \vec{\mathbf{B}}$? (a) y (b) $-y$ (c) x (d) $-x$ (e) z (f) $-z$ (g) no direction because it is zero (h) no direction because it is a scalar **(ii)** What is the direction of $\vec{\mathbf{B}} \times \vec{\mathbf{A}}$? Choose from the same possibilities (a) through (h).

3. **O** Let us name three perpendicular directions as right, up, and toward you, as you might name them when you are facing a television screen that lies in a vertical plane. Unit vectors for these directions are $\hat{\mathbf{r}}$, $\hat{\mathbf{u}}$, and $\hat{\mathbf{t}}$, respectively. For the quantity $(-3\hat{\mathbf{u}}\text{ m} \times 2\hat{\mathbf{t}}\text{ N})$, identify the magnitude, unit, and direction, if any. **(i)** The magnitude is (a) 6 (b) 3 (c) 2 (d) 0 **(ii)** The unit is (a) newton meters (b) newtons (c) meters (d) no unit **(iii)** The direction is (a) down (b) toward you (c) no direction (d) up (e) away from you (f) left (g) right

4. **O** Let the four horizontal compass directions north, east, south, and west be represented by units vectors $\hat{\mathbf{n}}$, $\hat{\mathbf{e}}$, $\hat{\mathbf{s}}$, and $\hat{\mathbf{w}}$, respectively. Vertically up and down are represented as $\hat{\mathbf{u}}$ and $\hat{\mathbf{d}}$. Let us also identify unit vectors that are halfway between these directions, such as $\widehat{\mathbf{ne}}$ for northeast. Rank the magnitudes of the following cross products from the largest to the smallest. If any are equal in magnitude, or equal to zero, show that in your ranking. (a) $\hat{\mathbf{n}} \times \hat{\mathbf{n}}$ (b) $\hat{\mathbf{w}} \times \widehat{\mathbf{ne}}$ (c) $\hat{\mathbf{u}} \times \widehat{\mathbf{ne}}$ (d) $\hat{\mathbf{n}} \times \widehat{\mathbf{nw}}$ (e) $\hat{\mathbf{n}} \times \hat{\mathbf{e}}$

5. If the torque acting on a particle about a certain origin is zero, what can you say about its angular momentum about that origin?

6. A ball is thrown in such a way that it does not spin about its own axis. Does this statement imply that the angular momentum is zero about an arbitrary origin? Explain.

7. **O** Compound terms can sometimes be confusing. For example, an ant lion is not a kind of lion but rather a different kind of insect. (a) Is rotational kinetic energy a kind of kinetic energy? (b) Is torque a kind of force? (c) Is angular momentum a kind of momentum?

8. ☐ Why does a long pole help a tightrope walker stay balanced?

9. **O** An ice skater starts a spin with her arms stretched out to the sides. She balances on the tip of one skate to turn without friction. She then pulls her arms in so that her moment of inertia decreases by a factor of two. In the process of her doing so, what happens to her kinetic energy? (a) It increases by a factor of four. (b) It increases by a factor of two. (c) It remains constant. (d) It decreases by a factor of two. (e) It decreases by a factor of four. (f) It is zero because her center of mass is stationary. (g) It undergoes a change by an amount that obviously depends on how fast the skater pulls her arms in.

10. ☐ In a tape recorder, the tape is pulled past the read-and-write heads at a constant speed by the drive mechanism. Consider the reel from which the tape is pulled. As the tape is pulled from it, the radius of the roll of remaining tape decreases. How does the torque on the reel change with time? How does the angular speed of the reel change in time? If the drive mechanism is switched on so that the tape is suddenly jerked with a large force, is the tape more likely to break when it is being pulled from a nearly full reel or from a nearly empty reel?

11. **O** A pet mouse sleeps near the eastern edge of a stationary, horizontal turntable that is supported by a frictionless, vertical axle through its center. The mouse wakes up and starts to walk north on the turntable. **(i)** As it takes its first steps, what is the mouse's displacement relative to the stationary ground below? (A) north (B) south (C) none **(ii)** In this process, the spot on the turntable where the mouse had been snoozing undergoes what displacement relative to the ground below? (A) north (B) south (C) none **(iii)** In this process for the mouse-turntable system, is mechanical energy conserved? **(iv)** Is momentum conserved? **(v)** Is angular momentum conserved?

12. **O** An employee party for a very successful company features a merry-go-round with real animals. The horizontal turntable has no motor, but is turning freely on a vertical, frictionless axle through its center. Two ponies of equal mass are tethered at diametrically opposite points on the rim. Children untie them, and the placid beasts simultaneously start plodding toward each other across the turntable. **(i)** As they walk, what happens to the angular speed of the carousel? (a) It increases. (b) It stays constant. (c) It decreases. Consider the ponies–turntable system in this process. **(ii)** Is its mechanical energy conserved? **(iii)** Is its momentum conserved? **(iv)** Is its angular momentum conserved?

13. **O** A horizontal disk with moment of inertia I_1 rotates with angular velocity ω_0 on a vertical, frictionless axle. A second horizontal disk, having moment of inertia I_2 and initially not rotating, drops onto the first. Because of friction between the surfaces of the disks, the two reach the same angular velocity. What is it? (a) $I_1\omega_0/I_2$ (b) $I_2\omega_0/I_1$ (c) $I_1\omega_0/(I_1 + I_2)$ (d) $I_2\omega_0/(I_1 + I_2)$ (e) $(I_1 + I_2)\omega_0/I_1$ (f) $(I_1 + I_2)\omega_0/I_2$

14. In some motorcycle races, the riders drive over small hills and the motorcycle becomes airborne for a short time interval. If the motorcycle racer keeps the throttle open while leaving the hill and going into the air, the motorcycle tends to nose upward. Why?

15. ☐ If global warming continues over the next one hundred years, it is likely that some polar ice will melt and the water will be distributed closer to the Equator. How would that change the moment of inertia of the Earth? Would the duration of the day (one revolution) increase or decrease?

16. A scientist arriving at a hotel asks a bellhop to carry a heavy suitcase. When the bellhop rounds a corner, the suitcase suddenly swings away from him for some unknown reason. The alarmed bellhop drops the suitcase and runs away. What might be in the suitcase?

Problems

WebAssign The Problems from this chapter may be assigned online in WebAssign.

ThomsonNOW Sign in at **www.thomsonedu.com** and go to ThomsonNOW to assess your understanding of this chapter's topics with additional quizzing and conceptual questions.

1, 2, 3 denotes straightforward, intermediate, challenging; □ denotes full solution available in *Student Solutions Manual/Study Guide;* ▲ denotes coached solution with hints available at **www.thomsonedu.com;** ▭ denotes developing symbolic reasoning; ● denotes asking for qualitative reasoning; ▪ denotes computer useful in solving problem

Section 11.1 The Vector Product and Torque

1. Given $\vec{\mathbf{M}} = 6\hat{\mathbf{i}} + 2\hat{\mathbf{j}} - \hat{\mathbf{k}}$ and $\vec{\mathbf{N}} = 2\hat{\mathbf{i}} - \hat{\mathbf{j}} - 3\hat{\mathbf{k}}$, calculate the vector product $\vec{\mathbf{M}} \times \vec{\mathbf{N}}$.

2. The vectors 42.0 cm at 15.0° and 23.0 cm at 65.0° both start from the origin. Both angles are measured counterclockwise from the x axis. The vectors form two sides of a parallelogram. (a) Find the area of the parallelogram. (b) Find the length of its longer diagonal.

3. ▲ Two vectors are given by $\vec{\mathbf{A}} = -3\hat{\mathbf{i}} + 4\hat{\mathbf{j}}$ and $\vec{\mathbf{B}} = 2\hat{\mathbf{i}} + 3\hat{\mathbf{j}}$. Find (a) $\vec{\mathbf{A}} \times \vec{\mathbf{B}}$ and (b) the angle between $\vec{\mathbf{A}}$ and $\vec{\mathbf{B}}$.

4. Two vectors are given by $\vec{\mathbf{A}} = -3\hat{\mathbf{i}} + 7\hat{\mathbf{j}} - 4\hat{\mathbf{k}}$ and $\vec{\mathbf{B}} = 6\hat{\mathbf{i}} - 10\hat{\mathbf{j}} + 9\hat{\mathbf{k}}$. Evaluate the quantities (a) $\cos^{-1}[\vec{\mathbf{A}}\cdot\vec{\mathbf{B}}/AB]$ and (b) $\sin^{-1}[|\vec{\mathbf{A}} \times \vec{\mathbf{B}}|/AB]$. (c) Which give(s) the angle between the vectors?

5. The wind exerts on a flower the force 0.785 N horizontally to the east. The stem of the flower is 0.450 m long and tilts toward the east, making an angle of 14.0° with the vertical. Find the vector torque of the wind force about the base of the stem.

6. ● A student claims that he has found a vector $\vec{\mathbf{A}}$ such that $(2\hat{\mathbf{i}} - 3\hat{\mathbf{j}} + 4\hat{\mathbf{k}}) \times \vec{\mathbf{A}} = (4\hat{\mathbf{i}} + 3\hat{\mathbf{j}} - \hat{\mathbf{k}})$. Do you believe this claim? Explain.

7. Assume $|\vec{\mathbf{A}} \times \vec{\mathbf{B}}| = \vec{\mathbf{A}}\cdot\vec{\mathbf{B}}$. What is the angle between $\vec{\mathbf{A}}$ and $\vec{\mathbf{B}}$?

8. ● A particle is located at the vector position $\vec{\mathbf{r}} = (4.00\hat{\mathbf{i}} + 6.00\hat{\mathbf{j}})$ m and a force exerted on it is given by $\vec{\mathbf{F}} = (3.00\hat{\mathbf{i}} + 2.00\hat{\mathbf{j}})$ N. (a) What is the torque acting on the particle about the origin? (b) Can there be another point about which the torque caused by this force on this particle will be in the opposite direction and half as large in magnitude? Can there be more than one such point? Can such a point lie on the y axis? Can more than one such point lie on the y axis? Determine the position vector of such a point.

9. Two forces $\vec{\mathbf{F}}_1$ and $\vec{\mathbf{F}}_2$ act along the two sides of an equilateral triangle as shown in Figure P11.9. Point O is the intersection of the altitudes of the triangle. Find a third force $\vec{\mathbf{F}}_3$ to be applied at B and along BC that will make the total torque zero about the point O. **What If?** Will the total torque change if $\vec{\mathbf{F}}_3$ is applied not at B but at any other point along BC?

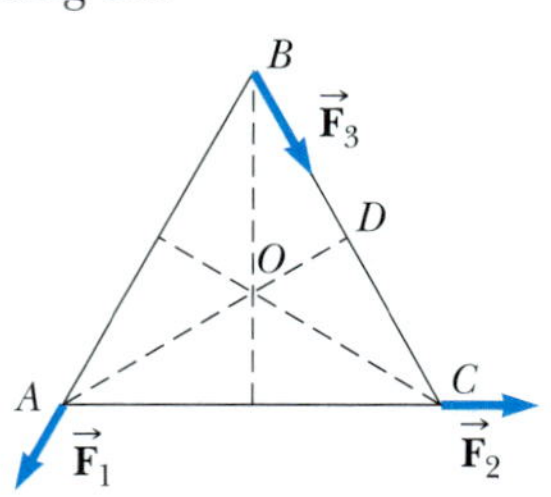

Figure P11.9

10. Use the definition of the vector product and the definitions of the unit vectors $\hat{\mathbf{i}}$, $\hat{\mathbf{j}}$, and $\hat{\mathbf{k}}$ to prove Equations 11.7. You may assume the x axis points to the right, the y axis up, and the z axis horizontally toward you (not away from you). This choice is said to make the coordinate system a *right-handed* system.

Section 11.2 Angular Momentum: The Nonisolated System

11. A light, rigid rod 1.00 m in length joins two particles, with masses 4.00 kg and 3.00 kg, at its ends. The combination rotates in the xy plane about a pivot through the center of the rod (Fig. P11.11). Determine the angular momentum of the system about the origin when the speed of each particle is 5.00 m/s.

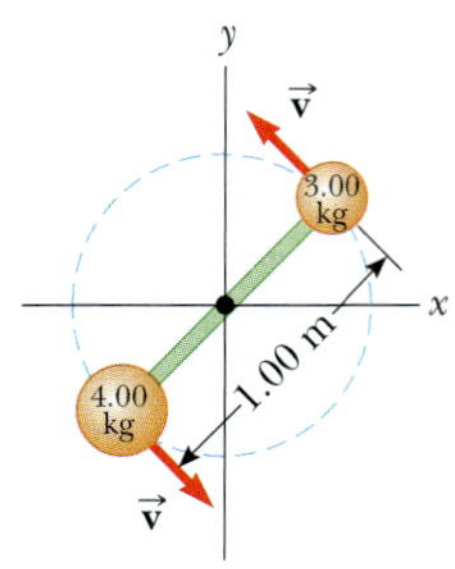

Figure P11.11

12. A 1.50-kg particle moves in the xy plane with a velocity of $\vec{\mathbf{v}} = (4.20\hat{\mathbf{i}} - 3.60\hat{\mathbf{j}})$ m/s. Determine the angular momentum of the particle about the origin when its position vector is $\vec{\mathbf{r}} = (1.50\hat{\mathbf{i}} + 2.20\hat{\mathbf{j}})$ m.

13. ▲ The position vector of a particle of mass 2.00 kg as a function of time is given by $\vec{\mathbf{r}} = (6.00\hat{\mathbf{i}} + 5.00t\hat{\mathbf{j}})$ m. Determine the angular momentum of the particle about the origin as a function of time.

14. A conical pendulum consists of a bob of mass m in motion in a circular path in a horizontal plane as shown in Figure P11.14. During the motion, the supporting wire of length ℓ maintains the constant angle θ with the vertical. Show that the magnitude of the angular momentum of the bob about the circle's center is

$$L = \left(\frac{m^2 g \ell^3 \sin^4 \theta}{\cos \theta}\right)^{1/2}$$

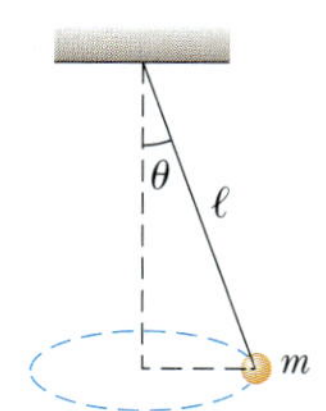

Figure P11.14

15. A particle of mass m moves in a circle of radius R at a constant speed v as shown in Figure P11.15. The motion begins at point Q at time $t = 0$. Determine the angular momentum of the particle about point P as a function of time.

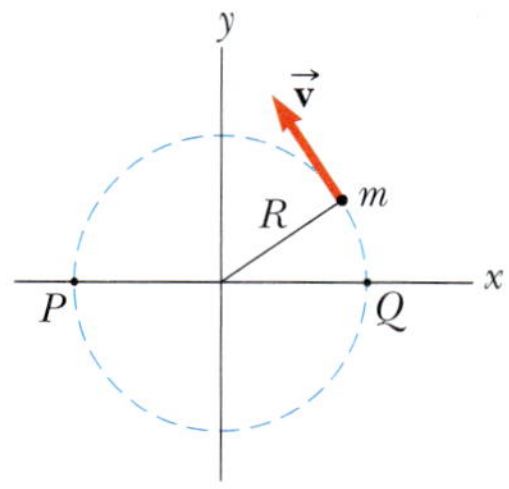

Figure P11.15 Problems 15 and 30.

16. A 4.00-kg counterweight is attached to a light cord that is wound around a reel (refer to Fig. 10.19). The reel is a uniform solid cylinder of radius 8.00 cm and mass 2.00 kg. (a) What is the net torque on the system about the point O? (b) When the counterweight has a speed v, the reel has an angular speed $\omega = v/R$. Determine the total angular momentum of the system about O. (c) Using that $\vec{\boldsymbol{\tau}} = d\vec{\mathbf{L}}/dt$ and your result from part (b), calculate the acceleration of the counterweight.

17. ● A particle of mass m is shot with an initial velocity $\vec{\mathbf{v}}_i$ making an angle θ with the horizontal as shown in Figure P11.17. The particle moves in the gravitational field of the Earth. Find the angular momentum of the particle about the origin when the particle is (a) at the origin, (b) at the highest point of its trajectory, and (c) just before it hits the ground. (d) What torque causes its angular momentum to change?

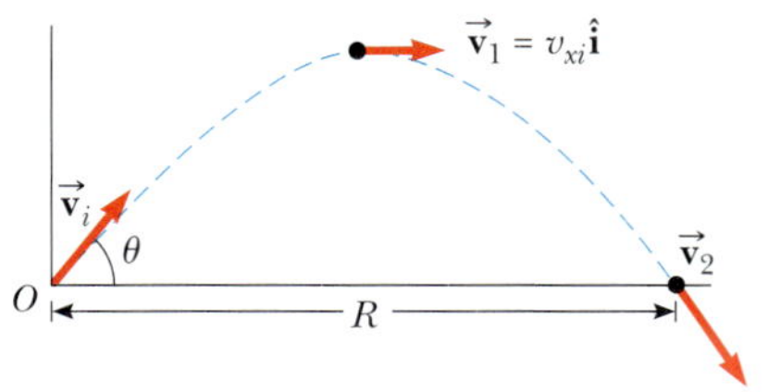

Figure P11.17

18. Heading straight toward the summit of Pikes Peak, an airplane of mass 12 000 kg flies over the plains of Kansas at nearly constant altitude 4.30 km with constant velocity 175 m/s west. (a) What is the airplane's vector angular momentum relative to a wheat farmer on the ground directly below the airplane? (b) Does this value change as the airplane continues its motion along a straight line? (c) **What If?** What is its angular momentum relative to the summit of Pikes Peak?

19. ● A ball having mass m is fastened at the end of a flagpole that is connected to the side of a tall building at point P shown in Figure P11.19. The length of the flagpole is ℓ, and it makes an angle θ with the x axis. The ball becomes loose and starts to fall with acceleration $-g\hat{\mathbf{j}}$. (a) Determine the angular momentum of the ball about point P as a function of time. (b) For what physical reason does the angular momentum change? (c) What is its rate of change?

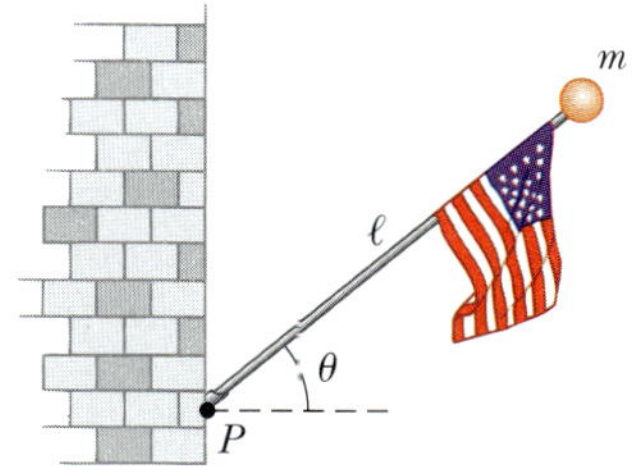

Figure P11.19

20. ● A 5.00-kg particle starts from the origin at time zero. Its velocity as a function of time is given by

$$\vec{\mathbf{v}} = (6\text{ m/s}^3)t^2\hat{\mathbf{i}} + (2\text{ m/s}^2)t\hat{\mathbf{j}}$$

(a) Find its position as a function of time. (b) Describe its motion qualitatively. (c) Find its acceleration as a function of time. (d) Find the net force exerted on the particle as a function of time. (e) Find the net torque about the origin exerted on the particle as a function of time. (f) Find the angular momentum of the particle as a function of time. (g) Find the kinetic energy of the particle as a function of time. (h) Find the power injected into the particle as a function of time.

Section 11.3 Angular Momentum of a Rotating Rigid Object

21. Show that the kinetic energy of an object rotating about a fixed axis with angular momentum $L = I\omega$ can be written as $K = L^2/2I$.

22. A uniform solid sphere of radius 0.500 m and mass 15.0 kg turns counterclockwise about a vertical axis through its center. Find its vector angular momentum when its angular speed is 3.00 rad/s.

23. A uniform solid disk of mass 3.00 kg and radius 0.200 m rotates about a fixed axis perpendicular to its face with angular frequency 6.00 rad/s. Calculate the angular momentum of the disk when the axis of rotation (a) passes through its center of mass and (b) passes through a point midway between the center and the rim.

24. ● (a) Model the Earth as a uniform sphere. Calculate the angular momentum of the Earth due to its spinning motion about its axis. (b) Calculate the angular momentum of the Earth due to its orbital motion about the Sun. (c) Are the two quantities of angular momentum nearly equal or quite different? Which is larger in magnitude? By what factor?

25. A particle of mass 0.400 kg is attached to the 100-cm mark of a meterstick of mass 0.100 kg. The meterstick rotates on a horizontal, frictionless table with an angular speed of 4.00 rad/s. Calculate the angular momentum of the system when the meterstick is pivoted about an axis (a) perpendicular to the table through the 50.0-cm mark and (b) perpendicular to the table through the 0-cm mark.

26. Big Ben (Fig. P10.42), the Parliament tower clock in London, has hour and minute hands with lengths of 2.70 m and 4.50 m and masses of 60.0 kg and 100 kg, respectively. Calculate the total angular momentum of these hands about the center point. Treat the hands as long, thin, uniform rods.

27. A space station is constructed in the shape of a hollow ring of mass 5.00×10^4 kg. Members of the crew walk on a deck formed by the inner surface of the outer cylindrical wall of the ring with radius 100 m. At rest when constructed, the ring is set rotating about its axis so that the people inside experience an effective free-fall acceleration equal to g. (Fig. P11.27 shows the ring together with some other parts that make a negligible contribution to the total moment of inertia.) The rotation is achieved by firing two small rockets attached tangentially to opposite points on the outside of the ring. (a) What angular momentum does the space station acquire? (b) For what time interval must the rockets be fired if each exerts a thrust of 125 N? (c) Prove that the total torque on the ring, multiplied by the time interval found in part (b), is equal to the change in angular momentum, found in part (a). This equality represents the *angular impulse–angular momentum theorem.*

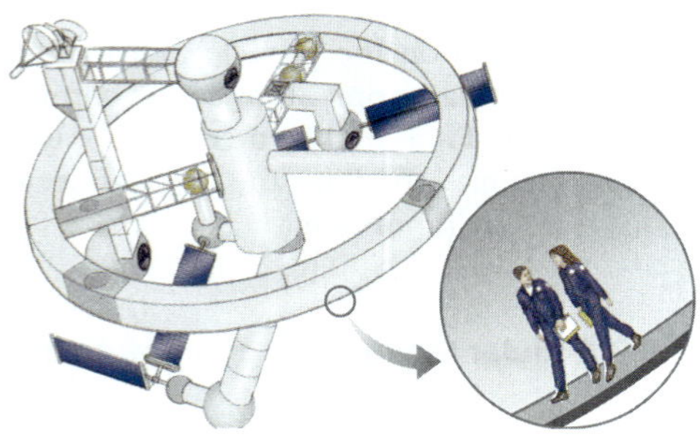

Figure P11.27 Problems 27 and 38.

28. The distance between the centers of the wheels of a motorcycle is 155 cm. The center of mass of the motorcycle, including the biker, is 88.0 cm above the ground and halfway between the wheels. Assume the mass of each wheel is small compared with the body of the motorcycle. The engine drives the rear wheel only. What horizontal acceleration of the motorcycle will make the front wheel rise off the ground?

Section 11.4 The Isolated System: Conservation of Angular Momentum

29. A cylinder with moment of inertia I_1 rotates about a vertical, frictionless axle with angular speed ω_i. A second cylinder, this one having moment of inertia I_2 and initially not rotating, drops onto the first cylinder (Fig. P11.29). Because of friction between the surfaces, the two eventually reach the same angular speed ω_f. (a) Calculate ω_f. (b) Show that the kinetic energy of the system decreases in this interaction, and calculate the ratio of the final to the initial rotational energy.

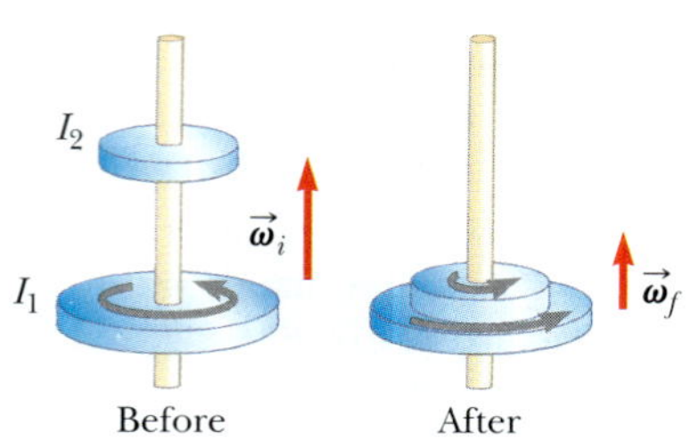

Figure P11.29

30. ● Figure P11.15 represents a small, flat puck with mass $m = 2.40$ kg sliding on a frictionless, horizontal surface. It is held in a circular orbit about a fixed axis by a rod with negligible mass and length $R = 1.50$ m, pivoted at one end. Initially, the puck has a speed of $v = 5.00$ m/s. A 1.30-kg ball of putty is dropped vertically onto the puck from a small distance above it and immediately sticks to the puck. (a) What is the new period of rotation? (b) Is angular momentum of the puck-putty system about the axis of rotation conserved in this process? (c) Is momentum of the system conserved in the process of the putty sticking to the puck? (d) Is mechanical energy of the system conserved in the process?

31. ● A uniform cylindrical turntable of radius 1.90 m and mass 30.0 kg rotates counterclockwise in a horizontal plane with an initial angular speed of 4π rad/s. The fixed turntable bearing is frictionless. A lump of clay of mass 2.25 kg and negligible size is dropped onto the turntable from a small distance above it and immediately sticks to the turntable at a point 1.80 m to the east of the axis. (a) Find the final angular speed of the clay and turntable. (b) Is mechanical energy of the turntable-clay system conserved in this process? Explain and use numerical results to verify your answer. (c) Is momentum of the system conserved in this process? Explain your answer.

32. A student sits on a freely rotating stool holding two dumbbells, each of mass 3.00 kg (Fig. P11.32). When the student's arms are extended horizontally (Fig. P11.32a), the dumbbells are 1.00 m from the axis of rotation and the student rotates with an angular speed of 0.750 rad/s. The moment of inertia of the student plus stool is $3.00 \text{ kg} \cdot \text{m}^2$ and is assumed to be constant. The student pulls the dumbbells inward horizontally to a position 0.300 m from the rotation axis (Fig. P11.32b). (a) Find the new angular speed of the student. (b) Find the kinetic energy of the rotating system before and after he pulls the dumbbells inward.

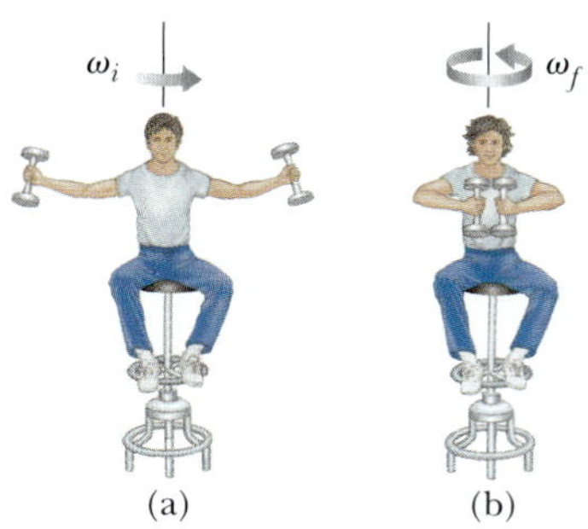

Figure P11.32

33. A playground merry-go-round of radius $R = 2.00$ m has a moment of inertia $I = 250 \text{ kg} \cdot \text{m}^2$ and is rotating at 10.0 rev/min about a frictionless vertical axle. Facing the axle, a 25.0-kg child hops onto the merry-go-round and manages to sit down on the edge. What is the new angular speed of the merry-go-round?

34. ● A uniform rod of mass 300 g and length 50.0 cm rotates in a horizontal plane about a fixed, vertical, frictionless pin through its center. Two small, dense beads, each of mass m, are mounted on the rod so that they can slide without friction along its length. Initially, the beads are held by catches at positions 10.0 cm on each side of the center, and the system is rotating at an angular speed of 36.0 rad/s. The catches are released simultaneously, and the beads slide outward along the rod. (a) Find the angular speed ω_f of the system at the instant the beads slide off the ends of the rod as it depends on m. (b) What are the maximum and minimum possible values for ω_f and the values of m to which they correspond? Describe the shape of a graph of ω_f versus m.

35. ● ▲ A 60.0-kg woman stands at the western rim of a horizontal turntable having a moment of inertia of 500 kg · m^2 and a radius of 2.00 m. The turntable is initially at rest and is free to rotate about a frictionless, vertical axle through its center. The woman then starts walking around the rim clockwise (as viewed from above the system) at a constant speed of 1.50 m/s relative to the Earth. (a) Consider the woman–turntable system as motion begins. Is mechanical energy of the system conserved? Is momentum of the system conserved? Is angular momentum of the system conserved? (b) In what direction and with what angular speed does the turntable rotate? (c) How much chemical energy does the woman's body convert into mechanical energy as the woman sets herself and the turntable into motion?

36. A puck of mass 80.0 g and radius 4.00 cm glides across an air table at a speed of 1.50 m/s as shown in Figure P11.36a. It makes a glancing collision with a second puck of radius 6.00 cm and mass 120 g (initially at rest) such that their rims just touch. Because their rims are coated with instant-acting glue, the pucks stick together and spin after the collision (Fig. P11.36b). (a) What is the angular momentum of the system relative to the center of mass? (b) What is the angular speed about the center of mass?

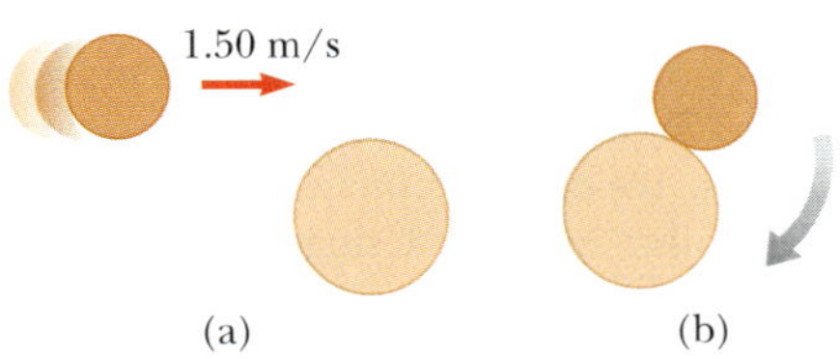

Figure P11.36

37. A wooden block of mass M resting on a frictionless, horizontal surface is attached to a rigid rod of length ℓ and of negligible mass (Fig. P11.37). The rod is pivoted at the other end. A bullet of mass m traveling parallel to the horizontal surface and perpendicular to the rod with speed v hits the block and becomes embedded in it. (a) What is the angular momentum of the bullet–block system? (b) What fraction of the original kinetic energy is converted into internal energy in the collision?

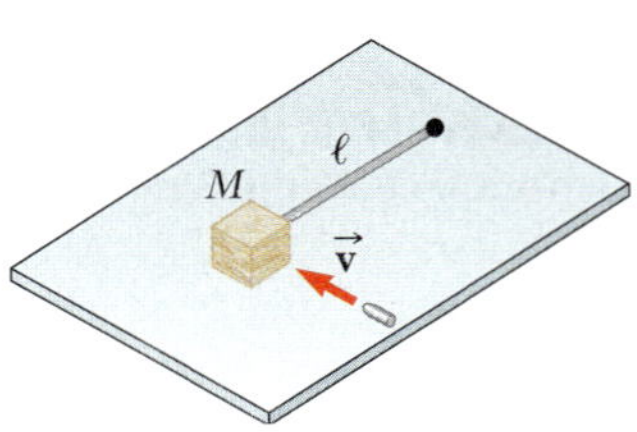

Figure P11.37

38. A space station shaped like a giant wheel has a radius of 100 m and a moment of inertia of 5.00×10^8 kg · m^2. A crew of 150 is living on the rim, and the station's rotation causes the crew to experience an apparent free-fall acceleration of g (Fig. P11.27). When 100 people move to the center of the station for a union meeting, the angular speed changes. What apparent free-fall acceleration is experienced by the managers remaining at the rim? Assume the average mass of each inhabitant is 65.0 kg.

39. ● A wad of sticky clay with mass m and velocity $\vec{\mathbf{v}}_i$ is fired at a solid cylinder of mass M and radius R (Figure P11.39). The cylinder is initially at rest and is mounted on a fixed horizontal axle that runs through its center of mass. The line of motion of the projectile is perpendicular to the axle and at a distance $d < R$ from the center. (a) Find the angular speed of the system just after the clay strikes and sticks to the surface of the cylinder. (b) Is mechanical energy of the clay–cylinder system conserved in this process? Explain your answer. (c) Is momentum of the clay–cylinder system conserved in this process? Explain your answer.

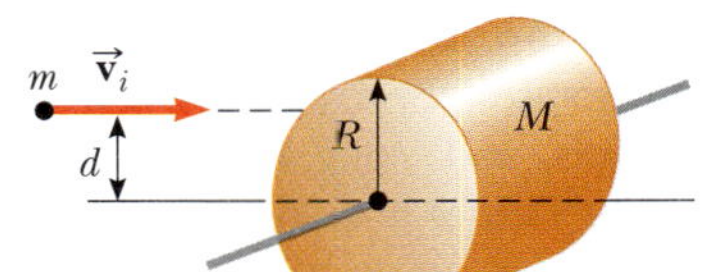

Figure P11.39

40. A thin uniform rectangular signboard hangs vertically above the door of a shop. The sign is hinged to a stationary horizontal rod along its top edge. The mass of the sign is 2.40 kg, and its vertical dimension is 50.0 cm. The sign is swinging without friction, so it is a tempting target for children armed with snowballs. The maximum angular displacement of the sign is 25.0° on both sides of the vertical. At a moment when the sign is vertical and moving to the left, a snowball of mass 400 g, traveling horizontally with a velocity of 160 cm/s to the right, strikes perpendicularly at the lower edge of the sign and sticks there. (a) Calculate the angular speed of the sign immediately before the impact. (b) Calculate its angular speed immediately after the impact. (c) The spattered sign will swing up through what maximum angle?

41. Suppose a meteor of mass 3.00×10^{13} kg, moving at 30.0 km/s relative to the center of the Earth, strikes the Earth. What is the order of magnitude of the maximum possible decrease in the angular speed of the Earth due to this collision? Explain your answer.

Section 11.5 The Motion of Gyroscopes and Tops

42. A spacecraft is in empty space. It carries on board a gyroscope with a moment of inertia of $I_g = 20.0$ kg · m^2 about the axis of the gyroscope. The moment of inertia of the spacecraft around the same axis is $I_s = 5.00 \times 10^5$ kg · m^2. Neither the spacecraft nor the gyroscope is originally rotating. The gyroscope can be powered up in a negligible period of time to an angular speed of 100 s^{-1}. If the orientation of the spacecraft is to be changed by 30.0°, for what time interval should the gyroscope be operated?

43. The angular momentum vector of a precessing gyroscope sweeps out a cone as shown in Figure 11.13b. Its angular

speed, called its precessional frequency, is given by $\omega_p = \tau/L$, where τ is the magnitude of the torque on the gyroscope and L is the magnitude of its angular momentum. In the motion called *precession of the equinoxes*, the Earth's axis of rotation precesses about the perpendicular to its orbital plane with a period of 2.58×10^4 yr. Model the Earth as a uniform sphere and calculate the torque on the Earth that is causing this precession.

Additional Problems

44. We have all complained that there aren't enough hours in a day. In an attempt to fix that, suppose all the people in the world line up at the equator, and all start running east at 2.50 m/s relative to the surface of the Earth. By how much does the length of a day increase? Assume the world population to be 7.00×10^9 people with an average mass of 55.0 kg each and the Earth to be a solid homogeneous sphere. In addition, you may use the approximation $1/(1 - x) \approx 1 + x$ for small x.

45. An American football game has been canceled because of bad weather in Cleveland, and two retired players are sliding like children on a frictionless ice-covered parking lot. William "Refrigerator" Perry, mass 162 kg, is gliding to the right at 8.00 m/s, and Doug Flutie, mass 81.0 kg, is gliding to the left at 11.0 m/s along the same line. When they meet, they grab each other and hang on. (a) What is their velocity immediately thereafter? (b) What fraction of their original kinetic energy is still mechanical energy after their collision? The athletes had so much fun that they repeat the collision with the same original velocities, this time moving along parallel lines 1.20 m apart. At closest approach they lock arms and start rotating about their common center of mass. Model the men as particles and their arms as a cord that does not stretch. (c) Find the velocity of their center of mass. (d) Find their angular speed. (e) What fraction of their original kinetic energy is still mechanical energy after they link arms?

46. ● A skateboarder with his board can be modeled as a particle of mass 76.0 kg, located at his center of mass. As shown in Figure P8.37 of Chapter 8, the skateboarder starts from rest in a crouching position at one lip of a half-pipe (point Ⓐ). The half-pipe forms one half of a cylinder of radius 6.80 m with its axis horizontal. On his descent, the skateboarder moves without friction and maintains his crouch so that his center of mass moves through one quarter of a circle of radius 6.30 m. (a) Find his speed at the bottom of the half-pipe (point Ⓑ). (b) Find his angular momentum about the center of curvature. (c) Immediately after passing point Ⓑ, he stands up and raises his arms, lifting his center of gravity from 0.500 m to 0.950 m above the concrete (point Ⓒ). Explain why his angular momentum is constant in this maneuver, whereas his linear momentum and his mechanical energy are not constant. (d) Find his speed immediately after he stands up, when his center of mass is moving in a quarter circle of radius 5.85 m. (e) How much chemical energy in the skateboarder's legs was converted into mechanical energy as he stood up? Next, the skateboarder glides upward with his center of mass moving in a quarter circle of radius 5.85 m. His body is horizontal when he passes point Ⓓ, the far lip of the half-pipe. (f) Find his speed at this location. At last he goes ballistic, twisting around while his center of mass moves vertically. (g) How high above point Ⓓ does he rise? (h) Over what time interval is he airborne before he touches down, facing downward and again in a crouch, 2.34 m below the level of point Ⓓ? (i) Compare the solution to this problem with the solution to Problem 8.37. Which is more accurate? Why? *Caution:* Do not try this stunt yourself without the required skill and protective equipment, or in a drainage channel to which you do not have legal access.

47. A rigid, massless rod has three particles with equal masses attached to it as shown in Figure P11.47. The rod is free to rotate in a vertical plane about a frictionless axle perpendicular to the rod through the point P and is released from rest in the horizontal position at $t = 0$. Assuming m and d are known, find (a) the moment of inertia of the system (rod plus particles) about the pivot, (b) the torque acting on the system at $t = 0$, (c) the angular acceleration of the system at $t = 0$, (d) the linear acceleration of the particle labeled 3 at $t = 0$, (e) the maximum kinetic energy of the system, (f) the maximum angular speed reached by the rod, (g) the maximum angular momentum of the system, and (h) the maximum speed reached by the particle labeled 2.

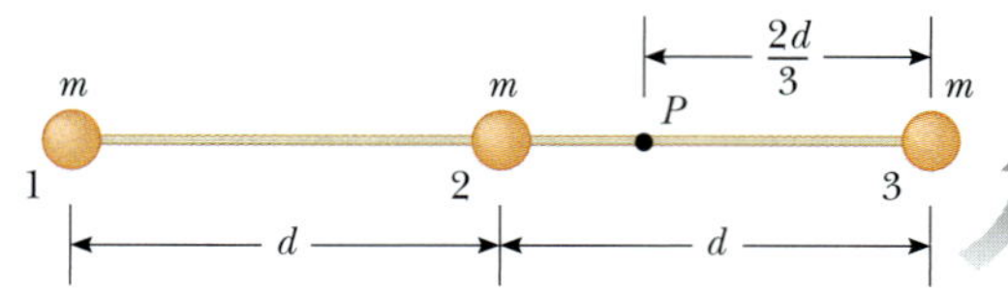

Figure P11.47

48. ● A light rope passes over a light, frictionless pulley. One end is fastened to a bunch of bananas of mass M, and a monkey of mass M clings to the other end (Fig. P11.48). The monkey climbs the rope in an attempt to reach the bananas. (a) Treating the system as consisting of the monkey, bananas, rope, and pulley, evaluate the net torque about the pulley axis. (b) Using the results of (a), determine the total angular momentum about the pulley axis and describe the motion of the system. Will the monkey reach the bananas?

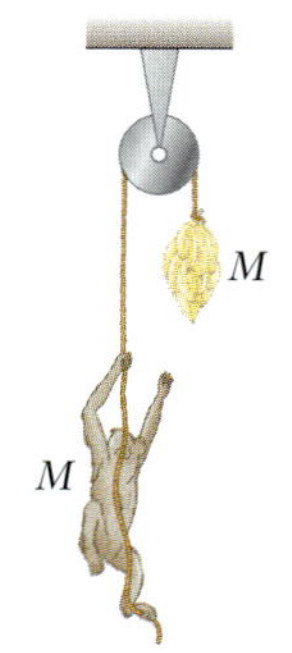

Figure P11.48

49. Comet Halley moves about the Sun in an elliptical orbit, with its closest approach to the Sun being about 0.590 AU and its greatest distance 35.0 AU (1 AU = the Earth–Sun distance). The comet's speed at closest approach is 54.0 km/s. What is its speed when it is farthest from the Sun? The angular momentum of the comet about the Sun is conserved because no torque acts on the comet.

The gravitational force exerted by the Sun has zero moment arm.

50. A projectile of mass m moves to the right with a speed v_i (Fig. P11.50a). The projectile strikes and sticks to the end of a stationary rod of mass M, length d, pivoted about a frictionless axle through its center (Fig. P11.50b). (a) Find the angular speed of the system right after the collision. (b) Determine the fractional loss in mechanical energy due to the collision.

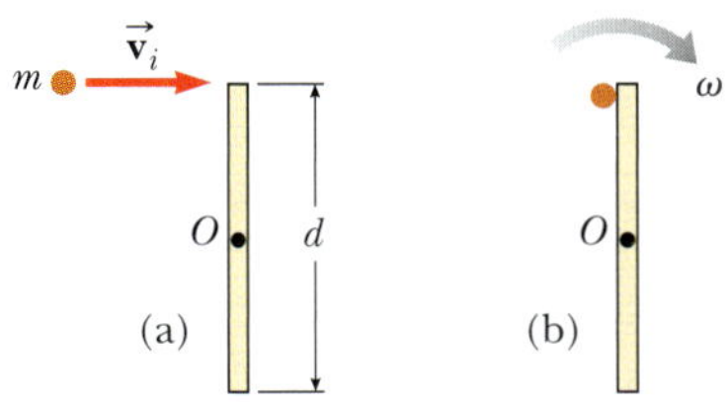

Figure P11.50

51. A puck of mass m is attached to a cord passing through a small hole in a frictionless, horizontal surface (Fig. P11.51). The puck is initially orbiting with speed v_i in a circle of radius r_i. The cord is then slowly pulled from below, decreasing the radius of the circle to r. (a) What is the speed of the puck when the radius is r? (b) Find the tension in the cord as a function of r. (c) How much work W is done in moving m from r_i to r? *Note:* The tension depends on r. (d) Obtain numerical values for v, T, and W when $r = 0.100$ m, $m = 50.0$ g, $r_i = 0.300$ m, and $v_i = 1.50$ m/s.

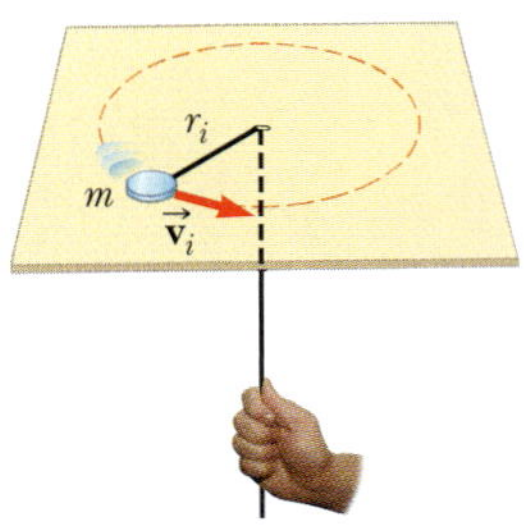

Figure P11.51

52. ● Two children are playing on stools at a restaurant counter. Their feet do not reach the footrests, and the tops of the stools are free to rotate without friction on pedestals fixed to the floor. One of the children catches a tossed ball in a process described by the equation

$$(0.730 \text{ kg}\cdot\text{m}^2)(2.40\hat{\mathbf{j}} \text{ rad/s}) + (0.120 \text{ kg})(0.350\hat{\mathbf{i}} \text{ m}) \times (4.30\hat{\mathbf{k}} \text{ m/s}) = [0.730 \text{ kg}\cdot\text{m}^2 + (0.120 \text{ kg})(0.350 \text{ m})^2]\vec{\boldsymbol{\omega}}$$

(a) Solve the equation for the unknown $\vec{\boldsymbol{\omega}}$. (b) Complete the statement of the problem to which this equation applies. Your statement must include the given numerical information and specification of the unknown to be determined. (c) Could the equation equally well describe the other child throwing the ball? Explain your answer.

53. ▲ Two astronauts (Fig. P11.53), each having a mass of 75.0 kg, are connected by a 10.0-m rope of negligible mass. They are isolated in space, orbiting their center of mass at speeds of 5.00 m/s. Treating the astronauts as particles, calculate (a) the magnitude of the angular momentum of the system and (b) the rotational energy of the system. By pulling on the rope, one astronaut shortens the distance between them to 5.00 m. (c) What is the new angular momentum of the system? (d) What are the astronauts' new speeds? (e) What is the new rotational energy of the system? (f) How much work does the astronaut do in shortening the rope?

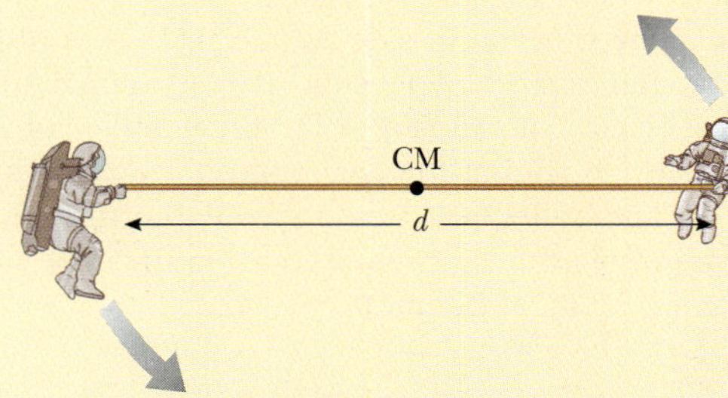

Figure P11.53 Problems 53 and 54.

54. Two astronauts (Fig. P11.53), each having a mass M, are connected by a rope of length d having negligible mass. They are isolated in space, orbiting their center of mass at speeds v. Treating the astronauts as particles, calculate (a) the magnitude of the angular momentum of the system and (b) the rotational energy of the system. By pulling on the rope, one astronaut shortens the distance between them to $d/2$. (c) What is the new angular momentum of the system? (d) What are the astronauts' new speeds? (e) What is the new rotational energy of the system? (f) How much work does the astronaut do in shortening the rope?

55. ● Native people throughout North and South America used a *bola* to hunt for birds and animals. A bola can consist of three stones, each with mass m, at the ends of three light cords, each with length ℓ. The other ends of the cords are tied together to form a Y. The hunter holds one stone and swings the other two stones above her head (Fig. P11.55a). Both stones move together in a horizontal circle of radius 2ℓ with speed v_0. At a moment when the horizontal component of their velocity is directed toward the quarry, the hunter releases the stone in her hand. As the bola flies through the air, the cords quickly take a stable arrangement with constant 120-degree angles between them (Fig. P11.55b). In the vertical direction, the bola is in free fall. Gravitational forces exerted by the Earth make the junction of the cords move with the downward acceleration $\vec{\mathbf{g}}$. You may ignore the vertical motion as you proceed to describe the horizontal motion of the bola. In

(a) (b)

Figure P11.55

terms of m, ℓ, and v_0, calculate (a) the magnitude of the momentum, (b) the horizontal speed of the center of mass, (c) the angular momentum about the center of mass, and (d) the angular speed of the bola about its center of mass. Calculate the kinetic energy of the bola (e) at the instant of release and (f) in its stable Y shape. (g) Explain how the conservation laws apply to the bola as its configuration changes. Robert Beichner suggested the idea for this problem.

56. A solid cube of wood of side $2a$ and mass M is resting on a horizontal surface. The cube is constrained to rotate about a fixed axis AB (Fig. P11.56). A bullet of mass m and speed v is shot at the face opposite $ABCD$ at a height of $4a/3$. The bullet becomes embedded in the cube. Find the minimum value of v required to tip the cube so that it falls on face $ABCD$. Assume $m << M$.

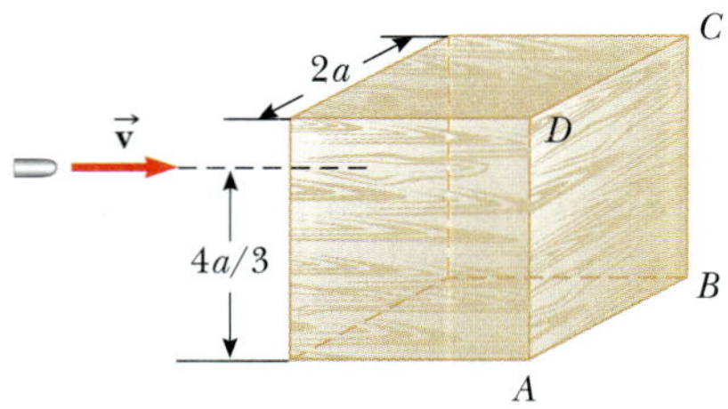

Figure P11.56

57. ● Global warming is a cause for concern because even small changes in the Earth's temperature can have significant consequences. For example, if the Earth's polar ice caps were to melt entirely, the resulting additional water in the oceans would flood many coastal areas. Calculate the resulting change in the duration of one day. Model the polar ice as having mass 2.30×10^{19} kg and forming two flat disks of radius 6.00×10^5 m. Assume the water spreads into an unbroken thin spherical shell after it melts. Is the change in the duration of a day appreciable?

58. A uniform solid disk is set into rotation with an angular speed ω_i about an axis through its center. While still rotating at this speed, the disk is placed into contact with a horizontal surface and released as shown in Figure P11.58. (a) What is the angular speed of the disk once pure rolling takes place? (b) Find the fractional loss in kinetic energy from the moment the disk is released until pure rolling occurs. *Suggestion:* Consider torques about the center of mass.

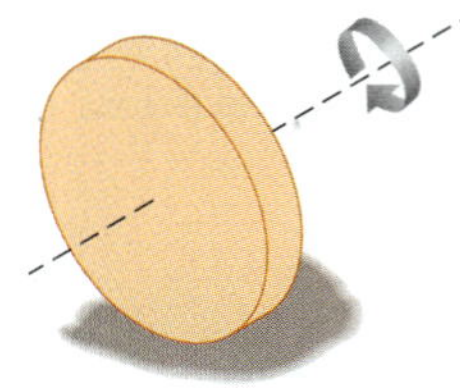

Figure P11.58 Problems 58 and 59.

59. Suppose a solid disk of radius R is given an angular speed ω_i about an axis through its center and then lowered to a horizontal surface and released as shown in Figure P11.58. Furthermore, assume the coefficient of friction between disk and surface is μ. (a) Show that the time interval before pure rolling motion occurs is $R\omega_i/3\mu g$. (b) Show that the distance the disk travels before pure rolling occurs is $R^2\omega_i^2/18\mu g$.

60. A solid cube of side $2a$ and mass M is sliding on a frictionless surface with uniform velocity $\vec{v}$ as shown in Figure P11.60a. It hits a small obstacle at the end of the table that causes the cube to tilt as shown in Figure P11.60b. Find the minimum value of $\vec{v}$ such that the cube falls off the table. The moment of inertia of the cube about an axis along one of its edges is $8Ma^2/3$. *Note:* The cube undergoes an inelastic collision at the edge.

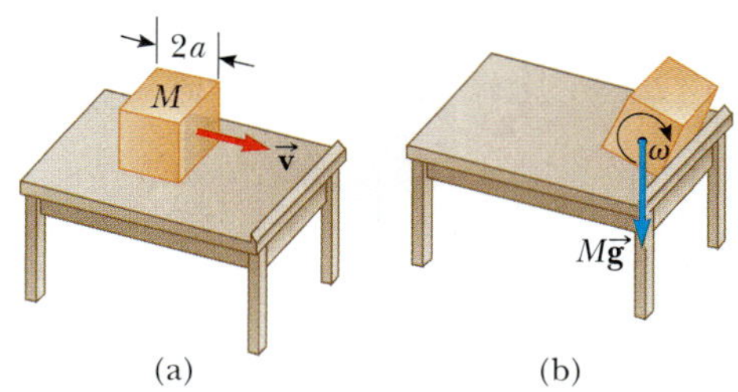

Figure P11.60

Answers to Quick Quizzes

11.1 (d). Because of the $\sin\theta$ function, $|\vec{A} \times \vec{B}|$ is either equal to or smaller than AB, depending on the angle θ.

11.2 **(i)**, (a). If $\vec{p}$ and $\vec{r}$ are parallel or antiparallel, the angular momentum is zero. For a nonzero angular momentum, the linear momentum vector must be offset from the rotation axis. **(ii)**, (c). The angular momentum is the product of the linear momentum and the perpendicular distance from the rotation axis to the line along which the linear momentum vector lies.

11.3 (b). The hollow sphere has a larger moment of inertia than the solid sphere.

11.4 **(i)**, (a). The diver is an isolated system, so the product $I\omega$ remains constant. Because her moment of inertia decreases, her angular speed increases. **(ii)**, (a). As the moment of inertia of the diver decreases, the angular speed increases by the same factor. For example, if I goes down by a factor of 2, then ω goes up by a factor of 2. The rotational kinetic energy varies as the square of ω. If I is halved, then ω^2 increases by a factor of 4 and the energy increases by a factor of 2.

Balanced Rock in Arches National Park, Utah, is a 3 000 000-kg boulder that has been in stable equilibrium for several millennia. It had a smaller companion nearby, called "Chip Off the Old Block," that fell during the winter of 1975. Balanced Rock appeared in an early scene of the movie *Indiana Jones and the Last Crusade.* We will study the conditions under which an object is in equilibrium in this chapter. (John W. Jewett Jr.)

12 Static Equilibrium and Elasticity

In Chapters 10 and 11, we studied the dynamics of rigid objects. Part of this chapter addresses the conditions under which a rigid object is in equilibrium. The term *equilibrium* implies either that the object is at rest or that its center of mass moves with constant velocity relative to an observer in an inertial reference frame. We deal here only with the former case, in which the object is in *static equilibrium.* Static equilibrium represents a common situation in engineering practice, and the principles it involves are of special interest to civil engineers, architects, and mechanical engineers. If you are an engineering student, you will undoubtedly take an advanced course in statics in the near future.

The last section of this chapter deals with how objects deform under load conditions. An *elastic* object returns to its original shape when the deforming forces are removed. Several elastic constants are defined, each corresponding to a different type of deformation.

12.1 The Rigid Object in Equilibrium

In Chapter 5, we discussed the particle in equilibrium model, in which a particle moves with constant velocity because the net force acting on it is zero. The situation with real (extended) objects is more complex because these objects often cannot be modeled as particles. For an extended object to be in equilibrium, a second condition must be satisfied. This second condition involves the net torque acting on the extended object.

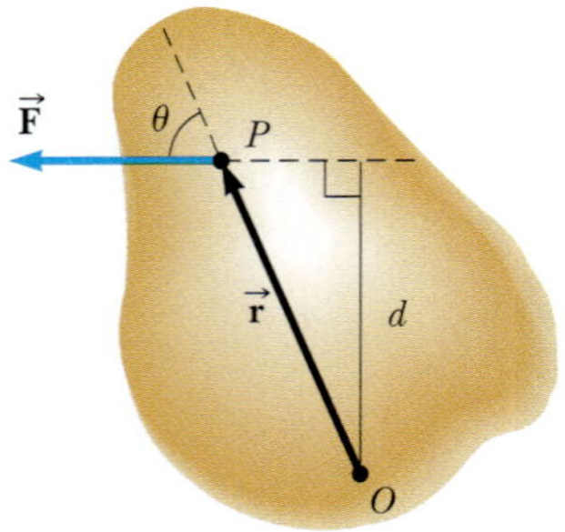

Figure 12.1 A single force $\vec{\mathbf{F}}$ acts on a rigid object at the point *P*.

Consider a single force $\vec{\mathbf{F}}$ acting on a rigid object as shown in Figure 12.1. The effect of the force depends on the location of its point of application *P*. If $\vec{\mathbf{r}}$ is the position vector of this point relative to *O*, the torque associated with the force $\vec{\mathbf{F}}$ about an axis through *O* is given by Equation 11.1:

$$\vec{\boldsymbol{\tau}} = \vec{\mathbf{r}} \times \vec{\mathbf{F}}$$

Recall from the discussion of the vector product in Section 11.1 that the vector $\vec{\boldsymbol{\tau}}$ is perpendicular to the plane formed by $\vec{\mathbf{r}}$ and $\vec{\mathbf{F}}$. You can use the right-hand rule to determine the direction of $\vec{\boldsymbol{\tau}}$ as shown in Figure 11.2. Hence, in Figure 12.1, $\vec{\boldsymbol{\tau}}$ is directed toward you out of the page.

As you can see from Figure 12.1, the tendency of $\vec{\mathbf{F}}$ to rotate the object about an axis through *O* depends on the moment arm *d* as well as the magnitude of $\vec{\mathbf{F}}$. Recall that the magnitude of $\vec{\boldsymbol{\tau}}$ is *Fd* (see Eq. 10.19). According to Equation 10.21, the net torque on a rigid object causes it to undergo an angular acceleration.

In this discussion, we investigate those rotational situations in which the angular acceleration of a rigid object is zero. Such an object is in **rotational equilibrium.** Because $\sum \tau = I\alpha$ for rotation about a fixed axis, the necessary condition for rotational equilibrium is that **the net torque about any axis must be zero.** We now have two necessary conditions for equilibrium of an object:

1. The net external force on the object must equal zero:

$$\sum \vec{\mathbf{F}} = 0 \qquad (12.1)$$

2. The net external torque on the object about *any* axis must be zero:

$$\sum \vec{\boldsymbol{\tau}} = 0 \qquad (12.2)$$

PITFALL PREVENTION 12.1
Zero Torque

Zero net torque does not mean an absence of rotational motion. An object that is rotating at a constant angular speed can be under the influence of a net torque of zero. This possibility is analogous to the translational situation: zero net force does not mean an absence of translational motion.

These conditions describe the **rigid object in equilibrium** analysis model. The first condition is a statement of translational equilibrium; it states that the translational acceleration of the object's center of mass must be zero when viewed from an inertial reference frame. The second condition is a statement of rotational equilibrium; it states that the angular acceleration about any axis must be zero. In the special case of **static equilibrium,** which is the main subject of this chapter, the object in equilibrium is at rest relative to the observer and so has no translational or angular speed (that is, $v_{CM} = 0$ and $\omega = 0$).

Quick Quiz 12.1 Consider the object subject to the two forces in Figure 12.2. Choose the correct statement with regard to this situation. (a) The object is in force equilibrium but not torque equilibrium. (b) The object is in torque equilibrium but not force equilibrium. (c) The object is in both force equilibrium and torque equilibrium. (d) The object is in neither force equilibrium nor torque equilibrium.

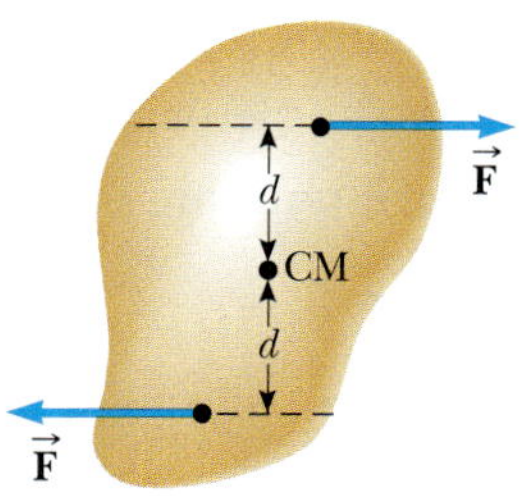

Figure 12.2 (Quick Quiz 12.1) Two forces of equal magnitude are applied at equal distances from the center of mass of a rigid object.

Quick Quiz 12.2 Consider the object subject to the three forces in Figure 12.3. Choose the correct statement with regard to this situation. (a) The object is in force equilibrium but not torque equilibrium. (b) The object is in torque equilibrium but not force equilibrium. (c) The object is in both force equilibrium and torque equilibrium. (d) The object is in neither force equilibrium nor torque equilibrium.

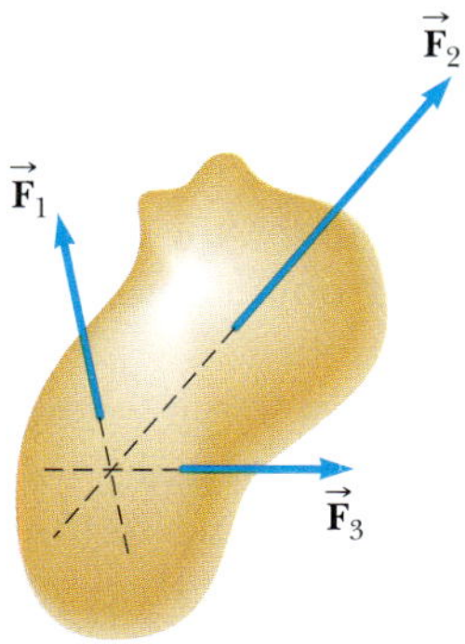

Figure 12.3 (Quick Quiz 12.2) Three forces act on an object. Notice that the lines of action of all three forces pass through a common point.

The two vector expressions given by Equations 12.1 and 12.2 are equivalent, in general, to six scalar equations: three from the first condition for equilibrium and three from the second (corresponding to *x*, *y*, and *z* components). Hence, in a complex system involving several forces acting in various directions, you could be faced with solving a set of equations with many unknowns. Here, we restrict our discussion to situations in which all the forces lie in the *xy* plane. (Forces whose vector representations are in the same plane are said to be *coplanar.*) With this restriction, we must deal with only three scalar equations. Two come from balancing the forces in the *x* and *y* directions. The third comes from the torque equation, namely that the net torque about a perpendicular axis through *any* point in the *xy* plane must be zero. Hence, the two conditions of the rigid object in equilibrium model provide the equations

$$\sum F_x = 0 \qquad \sum F_y = 0 \qquad \sum \tau_z = 0 \qquad \textbf{(12.3)}$$

where the location of the axis of the torque equation is arbitrary, as we now show.

Regardless of the number of forces that are acting, if an object is in translational equilibrium and the net torque is zero about one axis, the net torque must also be zero about any other axis. The axis can pass through a point that is inside or outside the object's boundaries. Consider an object being acted on by several forces such that the resultant force $\sum \vec{\mathbf{F}} = \vec{\mathbf{F}}_1 + \vec{\mathbf{F}}_2 + \vec{\mathbf{F}}_3 + \cdots = 0$. Figure 12.4 describes this situation (for clarity, only four forces are shown). The point of application of $\vec{\mathbf{F}}_1$ relative to O is specified by the position vector $\vec{\mathbf{r}}_1$. Similarly, the points of application of $\vec{\mathbf{F}}_2, \vec{\mathbf{F}}_3, \ldots$ are specified by $\vec{\mathbf{r}}_2, \vec{\mathbf{r}}_3, \ldots$ (not shown). The net torque about an axis through O is

$$\sum \vec{\boldsymbol{\tau}}_O = \vec{\mathbf{r}}_1 \times \vec{\mathbf{F}}_1 + \vec{\mathbf{r}}_2 \times \vec{\mathbf{F}}_2 + \vec{\mathbf{r}}_3 \times \vec{\mathbf{F}}_3 + \cdots$$

Now consider another arbitrary point O' having a position vector $\vec{\mathbf{r}}\,'$ relative to O. The point of application of $\vec{\mathbf{F}}_1$ relative to O' is identified by the vector $\vec{\mathbf{r}}_1 - \vec{\mathbf{r}}\,'$. Likewise, the point of application of $\vec{\mathbf{F}}_2$ relative to O' is $\vec{\mathbf{r}}_2 - \vec{\mathbf{r}}\,'$ and so forth. Therefore, the torque about an axis through O' is

$$\begin{aligned}\sum \vec{\boldsymbol{\tau}}_{O'} &= (\vec{\mathbf{r}}_1 - \vec{\mathbf{r}}\,') \times \vec{\mathbf{F}}_1 + (\vec{\mathbf{r}}_2 - \vec{\mathbf{r}}\,') \times \vec{\mathbf{F}}_2 + (\vec{\mathbf{r}}_3 - \vec{\mathbf{r}}\,') \times \vec{\mathbf{F}}_3 + \cdots \\ &= \vec{\mathbf{r}}_1 \times \vec{\mathbf{F}}_1 + \vec{\mathbf{r}}_2 \times \vec{\mathbf{F}}_2 + \vec{\mathbf{r}}_3 \times \vec{\mathbf{F}}_3 + \cdots - \vec{\mathbf{r}}\,' \times (\vec{\mathbf{F}}_1 + \vec{\mathbf{F}}_2 + \vec{\mathbf{F}}_3 + \cdots)\end{aligned}$$

Because the net force is assumed to be zero (given that the object is in translational equilibrium), the last term vanishes, and we see that the torque about an axis through O' is equal to the torque about an axis through O. Hence, **if an object is in translational equilibrium and the net torque is zero about one axis, the net torque must be zero about any other axis.**

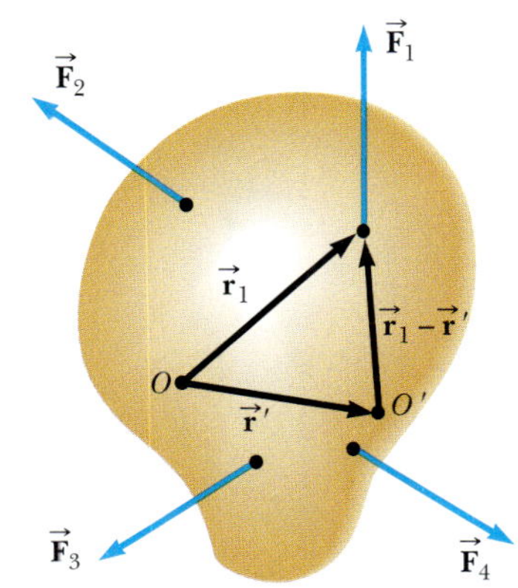

Figure 12.4 Construction showing that if the net torque is zero about an axis through the origin O, it is also zero about an axis through any other origin, such as O'.

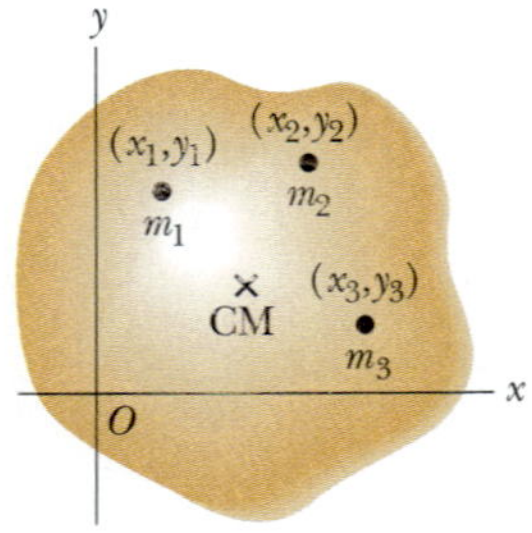

Figure 12.5 An object can be divided into many small particles, each having a specific mass and specific coordinates. These particles can be used to locate the center of mass.

12.2 More on the Center of Gravity

Whenever we deal with a rigid object, one of the forces we must consider is the gravitational force acting on it, and we must know the point of application of this force. As we learned in Section 9.5, associated with every object is a special point called its center of gravity. The combination of the various gravitational forces acting on all the various mass elements of the object is equivalent to a single gravitational force acting through this point. Therefore, to compute the torque due to the gravitational force on an object of mass M, we need only consider the force $M\vec{\mathbf{g}}$ acting at the object's center of gravity.

How do we find this special point? As mentioned in Section 9.5, if we assume that $\vec{\mathbf{g}}$ is uniform over the object, the center of gravity of the object coincides with its center of mass. To see why, consider an object of arbitrary shape lying in the xy plane as illustrated in Figure 12.5. Suppose the object is divided into a large number of particles of masses $m_1, m_2, m_3, \ldots$ having coordinates (x_1, y_1), (x_2, y_2), (x_3, y_3), ... In Equation 9.28, we defined the x coordinate of the center of mass of such an object to be

$$x_{\text{CM}} = \frac{m_1x_1 + m_2x_2 + m_3x_3 + \cdots}{m_1 + m_2 + m_3 + \cdots} = \frac{\sum_i m_ix_i}{\sum_i m_i}$$

We use a similar equation to define the y coordinate of the center of mass, replacing each x with its y counterpart.

Let us now examine the situation from another point of view by considering the gravitational force exerted on each particle as shown in Figure 12.6. Each particle contributes a torque about an axis through the origin equal in magnitude to the particle's weight mg multiplied by its moment arm. For example, the magnitude of the torque due to the force $m_1\vec{\mathbf{g}}_1$ is $m_1g_1x_1$, where g_1 is the value of the gravitational acceleration at the position of the particle of mass m_1. We wish to locate the center of gravity, the point at which application of the single gravitational force $M\vec{\mathbf{g}}_{\text{CG}}$ (where $M = m_1 + m_2 + m_3 + \cdots$ is the total mass of the object and $\vec{\mathbf{g}}_{\text{CG}}$ is the acceleration due to gravity at the location of the center of gravity) has the same effect on rotation as does the combined effect of all the individual gravitational forces $m_i\vec{\mathbf{g}}_i$. Equating the torque resulting from $M\vec{\mathbf{g}}_{\text{CG}}$ acting at the center of gravity to the sum of the torques acting on the individual particles gives

$$(m_1 + m_2 + m_3 + \cdots)g_{\text{CG}}x_{\text{CG}} = m_1g_1x_1 + m_2g_2x_2 + m_3g_3x_3 + \cdots$$

This expression accounts for the possibility that the value of g can in general vary over the object. If we assume uniform g over the object (as is usually the case), the g terms cancel and we obtain

$$x_{\text{CG}} = \frac{m_1x_1 + m_2x_2 + m_3x_3 + \cdots}{m_1 + m_2 + m_3 + \cdots} \qquad \textbf{(12.4)}$$

Comparing this result with Equation 9.29 shows that **the center of gravity is located at the center of mass as long as $\vec{\mathbf{g}}$ is uniform over the entire object.** Several examples in the next section deal with homogeneous, symmetric objects. The center of gravity for any such object coincides with its geometric center.

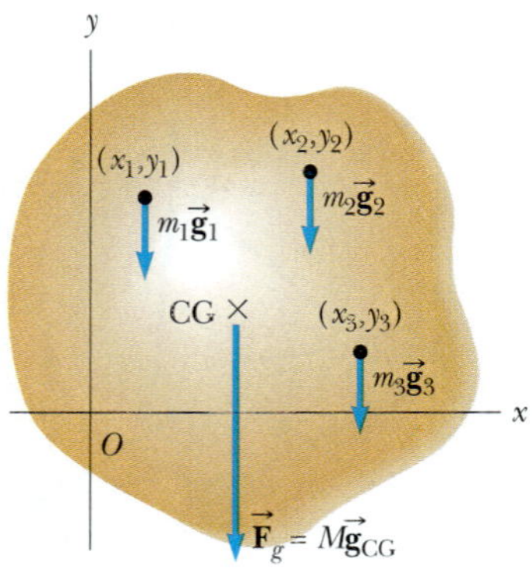

Figure 12.6 The gravitational force on an object is located at the center of gravity, which is the average position of the gravitational forces on all particles from which the object is made.

Quick Quiz 12.3 A meterstick is hung from a string tied at the 25-cm mark. A 0.50-kg object is hung from the zero end of the meterstick, and the meterstick is balanced horizontally. What is the mass of the meterstick? (a) 0.25 kg (b) 0.50 kg (c) 0.75 kg (d) 1.0 kg (e) 2.0 kg (f) impossible to determine

12.3 Examples of Rigid Objects in Static Equilibrium

The photograph of the one-bottle wine holder in Figure 12.7 shows one example of a balanced mechanical system that seems to defy gravity. For the system (wine holder plus bottle) to be in equilibrium, the net external force must be zero (see Eq. 12.1) and the net external torque must be zero (see Eq. 12.2). The second condition can be satisfied only when the center of gravity of the system is directly over the support point.

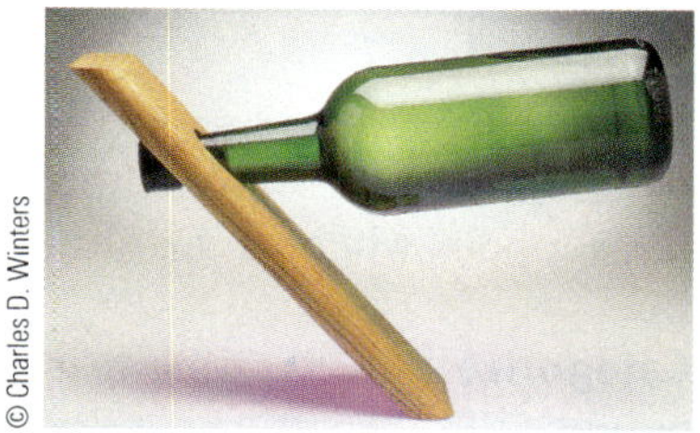

Figure 12.7 This one-bottle wine holder is a surprising display of static equilibrium. The center of gravity of the system (bottle plus holder) is directly over the support point.

PROBLEM-SOLVING STRATEGY Rigid Object in Equilibrium

When analyzing a rigid object in equilibrium under the action of several external forces, use the following procedure.

1. *Conceptualize.* Think about the object that is in equilibrium and identify all the forces on it. Imagine what effect each force would have on the rotation of the object if it were the only force acting.

2. *Categorize.* Confirm that the object under consideration is indeed a rigid object in equilibrium.

3. *Analyze.* Draw a free-body diagram and label all external forces acting on the object. Try to guess the correct direction for each force.

   Resolve all forces into rectangular components, choosing a convenient coordinate system. Then apply the first condition for equilibrium, Equation 12.1. Remember to keep track of the signs of the various force components.

   Choose a convenient axis for calculating the net torque on the rigid object. Remember that the choice of the axis for the torque equation is arbitrary; therefore, choose an axis that simplifies your calculation as much as possible. Usually, the most convenient axis for calculating torques is one through a point at which several forces act, so their torques around this axis are zero. If you don't know a force or don't need to know a force, it is often beneficial to choose an axis through the point at which this force acts. Apply the second condition for equilibrium, Equation 12.2.

   Solve the simultaneous equations for the unknowns in terms of the known quantities.

4. *Finalize.* Make sure your results are consistent with the free-body diagram. If you selected a direction that leads to a negative sign in your solution for a force, do not be alarmed; it merely means that the direction of the force is the opposite of what you guessed. Add up the vertical and horizontal forces on the object and confirm that each set of components adds to zero. Add up the torques on the object and confirm that the sum equals zero.

EXAMPLE 12.1 The Seesaw Revisited

A seesaw consisting of a uniform board of mass M and length ℓ supports at rest a father and daughter with masses m_f and m_d, respectively, as shown in Figure 12.8. The support (called the *fulcrum*) is under the center of gravity of the board, the father is a distance d from the center, and the daughter is a distance $\ell/2$ from the center.

(A) Determine the magnitude of the upward force $\vec{\mathbf{n}}$ exerted by the support on the board.

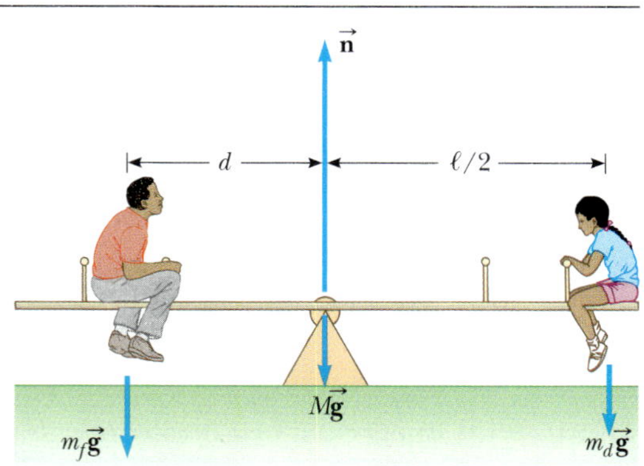

Figure 12.8 (Example 12.1) A balanced system.

SOLUTION

Conceptualize Let us focus our attention on the board and consider the gravitational forces on the father and daughter as forces applied directly to the board. The daughter would cause a clockwise rotation of the board around the support, whereas the father would cause a counterclockwise rotation.

Categorize Because the text of the problem states that the system is at rest, we model the board as a rigid object in equilibrium. Because we will only need the first condition of equilibrium to solve this part of the problem, however, we model the board as a particle in equilibrium.

Analyze Define upward as the positive y direction and substitute the forces on the board into Equation 12.1:

$$n - m_f g - m_d g - Mg = 0$$

Solve for the magnitude of the force $\vec{\mathbf{n}}$:

$$n = m_f g + m_d g + Mg = (m_f + m_d + M)g$$

(B) Determine where the father should sit to balance the system at rest.

SOLUTION

Categorize This part of the problem requires the introduction of torque to find the position of the father, so we model the board as a rigid object in equilibrium.

Analyze The board's center of gravity is at its geometric center because we are told that the board is uniform. If we choose a rotation axis perpendicular to the page through the center of gravity of the board, the torques produced by $\vec{\mathbf{n}}$ and the gravitational force about this axis are zero.

Substitute expressions for the torques on the board due to the father and daughter into Equation 12.2:

$$(m_f g)(d) - (m_d g)\frac{\ell}{2} = 0$$

Solve for d:

$$d = \left(\frac{m_d}{m_f}\right)\tfrac{1}{2}\ell$$

Finalize This result is the same one we obtained in Example 11.6 by evaluating the angular acceleration of the system and setting the angular acceleration equal to zero.

What If? Suppose we had chosen another point through which the rotation axis were to pass. For example, suppose the axis is perpendicular to the page and passes through the location of the father. Does that change the results to parts (A) and (B)?

Answer Part (A) is unaffected because the calculation of the net force does not involve a rotation axis. In part (B), we would conceptually expect there to be no change if a different rotation axis is chosen because the second condition of equilibrium claims that the torque is zero about any rotation axis.

Let's verify this answer mathematically. Recall that the sign of the torque associated with a force is positive if that force tends to rotate the system counterclockwise, whereas the sign of the torque is negative if the force tends to rotate the system clockwise. Let's choose a rotation axis passing through the location of the father.

Apply the condition for rotational equilibrium, $\sum \tau = 0$:

$$n(d) - (Mg)(d) - (m_d g)\left(d + \frac{\ell}{2}\right) = 0$$

Substitute $n = (m_f + m_d + M)g$ from part (A) and solve for d:

$$(m_f + m_d + M)g(d) - (Mg)(d) - (m_d g)\left(d + \frac{\ell}{2}\right) = 0$$

$$(m_f g)(d) - (m_d g)\left(\frac{\ell}{2}\right) = 0 \quad \rightarrow \quad d = \left(\frac{m_d}{m_f}\right)\tfrac{1}{2}\ell$$

This result is in agreement with the one obtained in part (B).

EXAMPLE 12.2 Standing on a Horizontal Beam

A uniform horizontal beam with a length of 8.00 m and a weight of 200 N is attached to a wall by a pin connection. Its far end is supported by a cable that makes an angle of 53.0° with the beam (Fig. 12.9a). A 600-N person stands 2.00 m from the wall. Find the tension in the cable as well as the magnitude and direction of the force exerted by the wall on the beam.

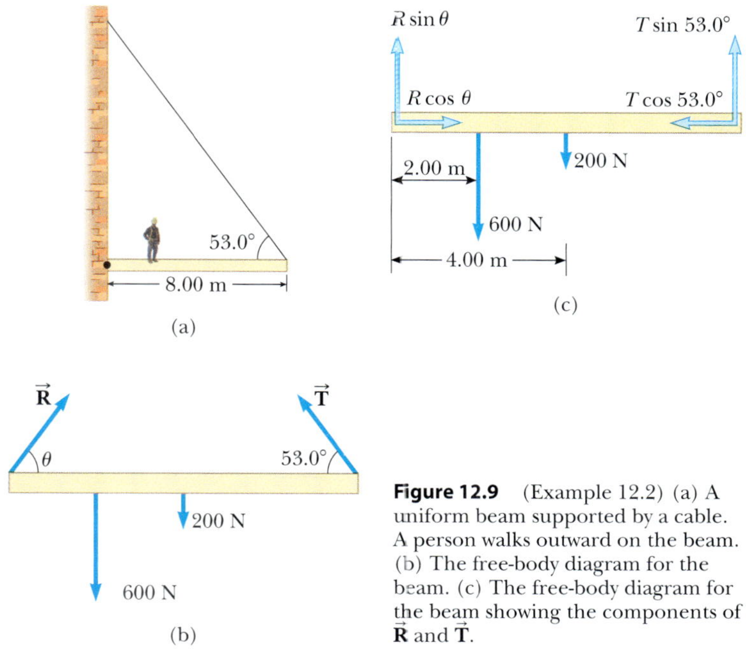

Figure 12.9 (Example 12.2) (a) A uniform beam supported by a cable. A person walks outward on the beam. (b) The free-body diagram for the beam. (c) The free-body diagram for the beam showing the components of $\vec{\mathbf{R}}$ and $\vec{\mathbf{T}}$.

SOLUTION

Conceptualize Imagine that the person in Figure 12.9a moves outward on the beam. It seems reasonable that the farther he moves outward, the larger the torque he applies about the pivot and the larger the tension in the cable must be to balance this torque.

Categorize Because the system is at rest, we categorize the beam as a rigid object in equilibrium.

Analyze We identify all the external forces acting on the beam: the 200-N gravitational force, the force $\vec{\mathbf{T}}$ exerted by the cable, the force $\vec{\mathbf{R}}$ exerted by the wall at the pivot, and the 600-N force that the person exerts on the beam. These forces are all indicated in the free-body diagram for the beam shown in Figure 12.9b. When we assign directions for forces, it is sometimes helpful to imagine what would happen if a force were suddenly removed. For example, if the wall were to vanish suddenly, the left end of the beam would move to the left as it begins to fall. This scenario tells us that the wall is not only holding the beam up but is also pressing outward against it. Therefore, we draw the vector $\vec{\mathbf{R}}$ as shown in Figure 12.9b. Figure 12.9c shows the horizontal and vertical components of $\vec{\mathbf{T}}$ and $\vec{\mathbf{R}}$.

Substitute expressions for the forces on the beam into Equation 12.1:

$$(1)\quad \sum F_x = R\cos\theta - T\cos 53.0° = 0$$

$$(2)\quad \sum F_y = R\sin\theta + T\sin 53.0° - 600\text{ N} - 200\text{ N} = 0$$

where we have chosen rightward and upward as our positive directions. Because R, T, and θ are all unknown, we cannot obtain a solution from these expressions alone. (To solve for the unknowns, the number of simultaneous equations must equal the number of unknowns.)

Now let's invoke the condition for rotational equilibrium. A convenient axis to choose for our torque equation is the one that passes through the pin connection. The feature that makes this axis so convenient is that the force $\vec{\mathbf{R}}$ and the horizontal component of $\vec{\mathbf{T}}$ both have a moment arm of zero; hence, these forces produce no torque about this axis. The moment arms of the 600-N, 200-N, and $T\sin 53.0°$ forces about this axis are 2.00 m, 4.00 m, and 8.00 m, respectively.

Substitute expressions for the torques on the beam into Equation 12.2:

$$(3)\quad \sum \tau = (T\sin 53.0°)(8.00\text{ m}) - (600\text{ N})(2.00\text{ m}) - (200\text{ N})(4.00\text{ m}) = 0$$

Solve for T:

$$T = 313\text{ N}$$

Substitute this value into Equations (1) and (2):

$$R\cos\theta - (313\text{ N})\cos 53.0° = 0$$

$$R\sin\theta + (313\text{ N})\sin 53.0° - 600\text{ N} - 200\text{ N} = 0$$

Solve for $R\cos\theta$ and $R\sin\theta$:

$$(4)\quad R\cos\theta = (313\text{ N})\cos 53.0° = 188\text{ N}$$

$$(5)\quad R\sin\theta = 600\text{ N} + 200\text{ N} - (313\text{ N})\sin 53.0° = 550\text{ N}$$

Divide Equation (5) by Equation (4):

$$\frac{R\sin\theta}{R\cos\theta} = \tan\theta = \frac{550\text{ N}}{188\text{ N}} = 2.93$$

Determine the angle θ: $\theta = 71.1°$

Solve Equation (4) for R: $R = \frac{188 \text{ N}}{\cos\theta} = \frac{188 \text{ N}}{\cos 71.1°} = 580 \text{ N}$

Finalize The positive value for the angle θ indicates that our estimate of the direction of $\vec{\mathbf{R}}$ was accurate.

Had we selected some other axis for the torque equation, the solution might differ in the details but the answers would be the same. For example, had we chosen an axis through the center of gravity of the beam, the torque equation would involve both T and R. This equation, coupled with Equations (1) and (2), however, could still be solved for the unknowns. Try it!

What If? What if the person walks farther out on the beam? Does T change? Does R change? Does θ change?

Answer T must increase because the weight of the person exerts a larger torque about the pin connection, which must be countered by a larger torque in the opposite direction due to an increased value of T. If T increases, the vertical component of $\vec{\mathbf{R}}$ decreases to maintain force equilibrium in the vertical direction. Force equilibrium in the horizontal direction, however, requires an increased horizontal component of $\vec{\mathbf{R}}$ to balance the horizontal component of the increased $\vec{\mathbf{T}}$. This fact suggests that θ becomes smaller, but it is hard to predict what happens to R. Problem 20 asks you to explore the behavior of R.

EXAMPLE 12.3 The Leaning Ladder

A uniform ladder of length ℓ rests against a smooth, vertical wall (Fig. 12.10a). The mass of the ladder is m, and the coefficient of static friction between the ladder and the ground is $\mu_s = 0.40$. Find the minimum angle θ_{min} at which the ladder does not slip.

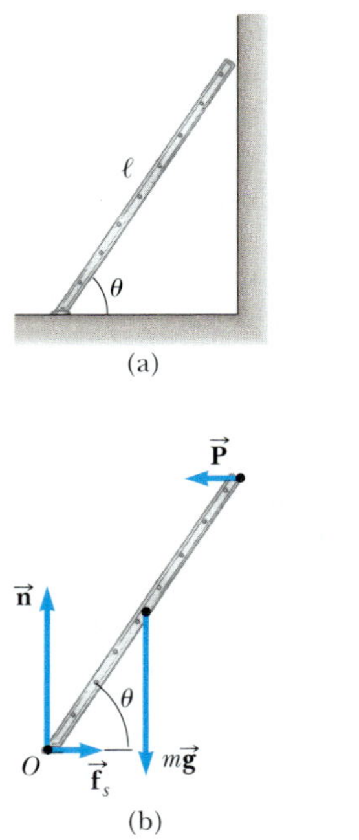

Figure 12.10 (Example 12.3) (a) A uniform ladder at rest, leaning against a smooth wall. The ground is rough. (b) The free-body diagram for the ladder.

SOLUTION

Conceptualize Think about any ladders you have climbed. Do you want a large friction force between the bottom of the ladder and the surface or a small one? If the friction force is zero, will the ladder stay up? Simulate a ladder with a ruler leaning against a vertical surface. Does the ruler slip at some angles and stay up at others?

Categorize We do not wish the ladder to slip, so we model it as a rigid object in equilibrium.

Analyze The free-body diagram showing all the external forces acting on the ladder is illustrated in Figure 12.10b. The force exerted by the ground on the ladder is the vector sum of a normal force $\vec{\mathbf{n}}$ and the force of static friction $\vec{\mathbf{f}}_s$. The force $\vec{\mathbf{P}}$ exerted by the wall on the ladder is horizontal because the wall is frictionless.

Apply the first condition for equilibrium to the ladder:

$$(1) \quad \sum F_x = f_s - P = 0$$

$$(2) \quad \sum F_y = n - mg = 0$$

Solve Equation (1) for P: $(3) \quad P = f_s$

Solve Equation (2) for n: $(4) \quad n = mg$

When the ladder is on the verge of slipping, the force of static friction must have its maximum value, which is given by $f_{s,\max} = \mu_s n$. Combine this equation with Equations (3) and (4):

$$P = f_{s,\max} = \mu_s n = \mu_s mg$$

Apply the second condition for equilibrium to the ladder, taking torques about an axis through O:

$$(5)\quad \sum \tau_O = P\ell \sin\theta_{\min} - mg\frac{\ell}{2}\cos\theta_{\min} = 0$$

Solve for $\tan\theta_{\min}$:

$$\frac{\sin\theta_{\min}}{\cos\theta_{\min}} = \tan\theta_{\min} = \frac{mg}{2P} = \frac{mg}{2\mu_s mg} = \frac{1}{2\mu_s} = 1.25$$

Solve for the angle $\theta_{\min}$:

$$\theta_{\min} = \tan^{-1}(1.25) = 51^\circ$$

Finalize Notice that the angle depends only on the coefficient of friction, not on the mass or length of the ladder.

EXAMPLE 12.4 Negotiating a Curb

(A) Estimate the magnitude of the force $\vec{\mathbf{F}}$ a person must apply to a wheelchair's main wheel to roll up over a sidewalk curb (Fig. 12.11a). This main wheel that comes in contact with the curb has a radius r, and the height of the curb is h.

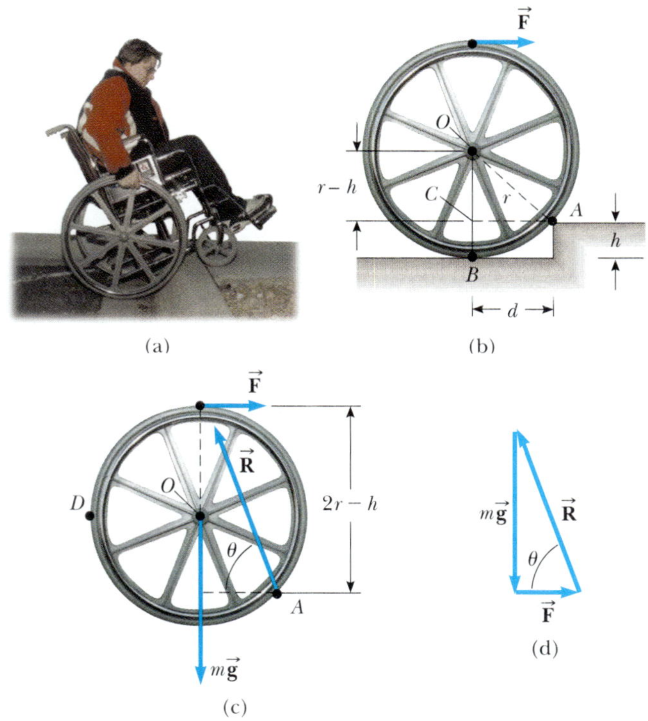

Figure 12.11 (Example 12.4) (a) A person in a wheelchair attempts to roll up over a curb. (b) Details of the wheel and curb. The person applies a force $\vec{\mathbf{F}}$ to the top of the wheel. (c) The free-body diagram for the wheel when it is just about to be raised. Three forces act on the wheel at this instant: $\vec{\mathbf{F}}$, which is exerted by the hand; $\vec{\mathbf{R}}$, which is exerted by the curb; and the gravitational force $m\vec{\mathbf{g}}$. (d) The vector sum of the three external forces acting on the wheel is zero.

SOLUTION

Conceptualize Think about wheelchair access to buildings. Generally, there are ramps built for individuals in wheelchairs. Steplike structures such as curbs are serious barriers to a wheelchair.

Categorize Imagine that the person exerts enough force so that the bottom of the wheel just loses contact with the lower surface and hovers at rest. We model the wheel in this situation as a rigid object in equilibrium.

Analyze Usually, the person's hands supply the required force to a slightly smaller wheel that is concentric with the main wheel. For simplicity, let's assume the radius of this second wheel is the same as the radius of the main wheel. Let's estimate a combined weight of $mg = 1\,400$ N for the person and the wheelchair and choose a wheel radius of $r = 30$ cm. We also pick a curb height of $h = 10$ cm. Let's also assume the wheelchair and occupant are symmetric and each wheel supports a weight of 700 N. We then proceed to analyze only one of the wheels. Figure 12.11b shows the geometry for a single wheel.

When the wheel is just about to be raised from the street, the normal force exerted by the ground on the wheel at point B goes to zero. Hence, at this time only three forces act on the wheel as shown in the free-body diagram in Figure 12.11c. The force $\vec{\mathbf{R}}$, which is the force exerted by the curb on the wheel, acts at point A, so if we choose to have our axis of rotation pass through point A, we do not need to include $\vec{\mathbf{R}}$ in our torque equation. The moment arm of $\vec{\mathbf{F}}$ relative to an axis through A is $2r - h$ (see Fig. 12.11c).

Use the triangle OAC in Figure 12.11b to find the moment arm d of the gravitational force $m\vec{\mathbf{g}}$ acting on the wheel relative to an axis through point A:

$$(1)\quad d = \sqrt{r^2 - (r - h)^2} = \sqrt{2rh - h^2}$$

Apply the second condition for equilibrium to the wheel, taking torques about an axis through A:

$$(2)\quad \sum \tau_A = mgd - F(2r - h) = 0$$

Substitute for d from Equation (1):

$$mg\sqrt{2rh - h^2} - F(2r - h) = 0$$

Solve for F:

$$F = \frac{mg\sqrt{2rh - h^2}}{2r - h}$$

Substitute the known values:

$$F = \frac{(700\text{ N})\sqrt{2(0.3\text{ m})(0.1\text{ m}) - (0.1\text{ m})^2}}{2(0.3\text{ m}) - 0.1\text{ m}}$$

$$= 3 \times 10^2\text{ N}$$

(B) Determine the magnitude and direction of $\vec{\mathbf{R}}$.

SOLUTION

Apply the first condition for equilibrium to the wheel:

$$\sum F_x = F - R\cos\theta = 0$$

$$\sum F_y = R\sin\theta - mg = 0$$

Divide the second equation by the first:

$$\frac{R\sin\theta}{R\cos\theta} = \tan\theta = \frac{mg}{F} = \frac{700\text{ N}}{300\text{ N}} = 2.33$$

Solve for the angle θ:

$$\theta = \tan^{-1}(2.33) = 70°$$

Use the right triangle shown in Figure 12.11d to obtain R:

$$R = \sqrt{(mg)^2 + F^2} = \sqrt{(700\text{ N})^2 + (300\text{ N})^2} = 8 \times 10^2\text{ N}$$

Finalize Notice that we have kept only one digit as significant. (We have written the angle as 70° because 7×10^{1}° is awkward!) The results indicate that the force that must be applied to each wheel is substantial. You may want to estimate the force required to roll a wheelchair up a typical sidewalk accessibility ramp for comparison.

What If? Would it be easier to negotiate the curb if the person grabbed the wheel at point D in Figure 12.11c and pulled *upward?*

Answer If the force $\vec{\mathbf{F}}$ in Figure 12.11c is rotated counterclockwise by 90° and applied at D, its moment arm is $d + r$. Let's call the magnitude of this new force F'.

Modify Equation (2) for this situation:

$$\sum \tau_A = mgd - F'(d + r) = 0$$

Solve this equation for F' and substitute for d:

$$F' = \frac{mgd}{d + r} = \frac{mg\sqrt{2rh - h^2}}{\sqrt{2rh - h^2} + r}$$

Take the ratio of this force to the original force that we calculated and express the result in terms of h/r, the ratio of the curb height to the wheel radius:

$$\frac{F'}{F} = \frac{\dfrac{mg\sqrt{2rh - h^2}}{\sqrt{2rh - h^2} + r}}{\dfrac{mg\sqrt{2rh - h^2}}{2r - h}} = \frac{2r - h}{\sqrt{2rh - h^2} + r} = \frac{2 - \left(\dfrac{h}{r}\right)}{\sqrt{2\left(\dfrac{h}{r}\right) - \left(\dfrac{h}{r}\right)^2} + 1}$$

Substitute the ratio $h/r = 0.33$ from the given values:

$$\frac{F'}{F} = \frac{2 - 0.33}{\sqrt{2(0.33) - (0.33)^2} + 1} = 0.96$$

This result tells us that, *for these values*, it is slightly easier to pull upward at D than horizontally at the top of the wheel. For very high curbs, so that h/r is close to 1, the ratio F'/F drops to about 0.5 because point A is located near the right edge of the wheel in Figure 12.11b. The force at D is applied at distance of about $2r$ from A, whereas the force at the top of the wheel has a moment arm of only about r. For high curbs, then, it is best to pull upward at D, although a large value of the force is required. For small curbs, it is best to apply the force at the top of the wheel. The ratio F'/F becomes larger than 1 at about $h/r = 0.3$ because point A is now close to the bottom of the wheel and the force applied at the top of the wheel has a larger moment arm than when applied at D.

Finally, let's comment on the validity of these mathematical results. Consider Figure 12.11d and imagine that the vector $\vec{\mathbf{F}}$ is upward instead of to the right. There is no way the three vectors can add to equal zero as required by the first equilibrium condition. Therefore, our results above may be qualitatively valid, but not exact quantitatively. To cancel the horizontal component of $\vec{\mathbf{R}}$, the force at D must be applied at an angle to the vertical rather than straight upward. This feature makes the calculation more complicated and requires both conditions of equilibrium.

12.4 Elastic Properties of Solids

Except for our discussion about springs in earlier chapters, we have assumed that objects remain rigid when external forces act on them. In Section 9.7, we explored deformable systems. In reality, all objects are deformable to some extent. That is, it is possible to change the shape or the size (or both) of an object by applying external forces. As these changes take place, however, internal forces in the object resist the deformation.

We shall discuss the deformation of solids in terms of the concepts of *stress* and *strain.* **Stress** is a quantity that is proportional to the force causing a deformation; more specifically, stress is the external force acting on an object per unit cross-sectional area. The result of a stress is **strain,** which is a measure of the degree of deformation. It is found that, for sufficiently small stresses, **stress is proportional to strain;** the constant of proportionality depends on the material being deformed and on the nature of the deformation. We call this proportionality constant the **elastic modulus.** The elastic modulus is therefore defined as the ratio of the stress to the resulting strain:

$$\text{Elastic modulus} \equiv \frac{\text{stress}}{\text{strain}} \quad \textbf{(12.5)}$$

The elastic modulus in general relates what is done to a solid object (a force is applied) to how that object responds (it deforms to some extent). It is similar to the spring constant k in Hooke's law (Eq. 7.9) that relates a force applied to a spring and the resultant deformation of the spring, measured by its extension or compression.

We consider three types of deformation and define an elastic modulus for each:

1. **Young's modulus** measures the resistance of a solid to a change in its length.
2. **Shear modulus** measures the resistance to motion of the planes within a solid parallel to each other.
3. **Bulk modulus** measures the resistance of solids or liquids to changes in their volume.

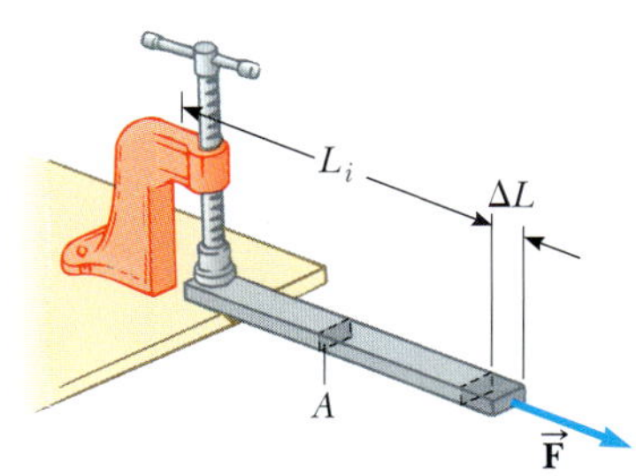

ACTIVE FIGURE 12.12

A long bar clamped at one end is stretched by an amount ΔL under the action of a force $\vec{\mathbf{F}}$.

Sign in at www.thomsonedu.com and go to ThomsonNOW to adjust the values of the applied force and Young's modulus and observe the change in length of the bar.

Young's Modulus: Elasticity in Length

Consider a long bar of cross-sectional area A and initial length L_i that is clamped at one end as in Active Figure 12.12. When an external force is applied perpendicular to the cross section, internal forces in the bar resist distortion ("stretching"), but the bar reaches an equilibrium situation in which its final length L_f is greater

TABLE 12.1

Typical Values for Elastic Moduli

Substance	Young's Modulus (N/m²)	Shear Modulus (N/m²)	Bulk Modulus (N/m²)
Tungsten	35×10^{10}	14×10^{10}	20×10^{10}
Steel	20×10^{10}	8.4×10^{10}	6×10^{10}
Copper	11×10^{10}	4.2×10^{10}	14×10^{10}
Brass	9.1×10^{10}	3.5×10^{10}	6.1×10^{10}
Aluminum	7.0×10^{10}	2.5×10^{10}	7.0×10^{10}
Glass	$6.5–7.8 \times 10^{10}$	$2.6–3.2 \times 10^{10}$	$5.0–5.5 \times 10^{10}$
Quartz	5.6×10^{10}	2.6×10^{10}	2.7×10^{10}
Water	—	—	0.21×10^{10}
Mercury	—	—	2.8×10^{10}

than L_i and in which the external force is exactly balanced by internal forces. In such a situation, the bar is said to be stressed. We define the **tensile stress** as the ratio of the magnitude of the external force F to the cross-sectional area A. The **tensile strain** in this case is defined as the ratio of the change in length ΔL to the original length L_i. We define **Young's modulus** by a combination of these two ratios:

Young's modulus ▶

$$Y \equiv \frac{\text{tensile stress}}{\text{tensile strain}} = \frac{F/A}{\Delta L/L_i} \tag{12.6}$$

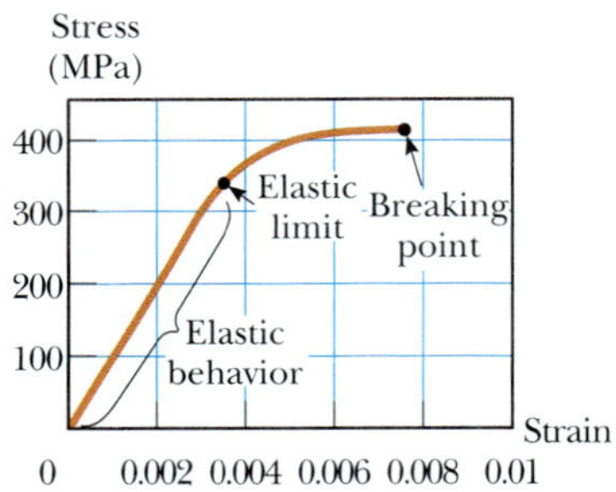

Figure 12.13 Stress-versus-strain curve for an elastic solid.

Young's modulus is typically used to characterize a rod or wire stressed under either tension or compression. Because strain is a dimensionless quantity, Y has units of force per unit area. Typical values are given in Table 12.1.

For relatively small stresses, the bar returns to its initial length when the force is removed. The **elastic limit** of a substance is defined as the maximum stress that can be applied to the substance before it becomes permanently deformed and does not return to its initial length. It is possible to exceed the elastic limit of a substance by applying a sufficiently large stress as seen in Figure 12.13. Initially, a stress-versus-strain curve is a straight line. As the stress increases, however, the curve is no longer a straight line. When the stress exceeds the elastic limit, the object is permanently distorted and does not return to its original shape after the stress is removed. As the stress is increased even further, the material ultimately breaks.

Shear Modulus: Elasticity of Shape

Another type of deformation occurs when an object is subjected to a force parallel to one of its faces while the opposite face is held fixed by another force (Active Fig. 12.14a). The stress in this case is called a shear stress. If the object is originally a rectangular block, a shear stress results in a shape whose cross section is a parallelogram. A book pushed sideways as shown in Active Figure 12.14b is an example of an object subjected to a shear stress. To a first approximation (for small distortions), no change in volume occurs with this deformation.

We define the **shear stress** as F/A, the ratio of the tangential force to the area A of the face being sheared. The **shear strain** is defined as the ratio $\Delta x/h$, where Δx is the horizontal distance that the sheared face moves and h is the height of the object. In terms of these quantities, the **shear modulus** is

Shear modulus ▶

$$S \equiv \frac{\text{shear stress}}{\text{shear strain}} = \frac{F/A}{\Delta x/h} \tag{12.7}$$

Values of the shear modulus for some representative materials are given in Table 12.1. Like Young's modulus, the unit of shear modulus is the ratio of that for force to that for area.

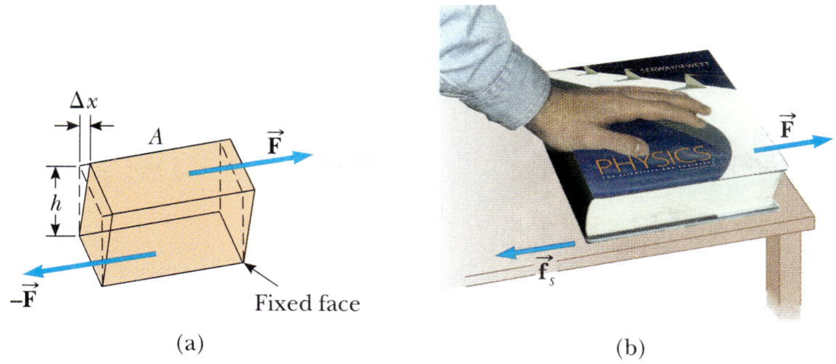

ACTIVE FIGURE 12.14

(a) A shear deformation in which a rectangular block is distorted by two forces of equal magnitude but opposite directions applied to two parallel faces. (b) A book is under shear stress when a hand placed on the cover applies a horizontal force away from the spine.

Sign in at www.thomsonedu.com and go to ThomsonNOW to adjust the values of the applied force and the shear modulus and observe the change in shape of the block in (a).

Bulk Modulus: Volume Elasticity

Bulk modulus characterizes the response of an object to changes in a force of uniform magnitude applied perpendicularly over the entire surface of the object as shown in Active Figure 12.15. (We assume here that the object is made of a single substance.) As we shall see in Chapter 14, such a uniform distribution of forces occurs when an object is immersed in a fluid. An object subject to this type of deformation undergoes a change in volume but no change in shape. The **volume stress** is defined as the ratio of the magnitude of the total force F exerted on a surface to the area A of the surface. The quantity $P = F/A$ is called **pressure,** which we shall study in more detail in Chapter 14. If the pressure on an object changes by an amount $\Delta P = \Delta F/A$, the object experiences a volume change ΔV. The **volume strain** is equal to the change in volume ΔV divided by the initial volume V_i. Therefore, from Equation 12.5, we can characterize a volume ("bulk") compression in terms of the **bulk modulus,** which is defined as

$$B \equiv \frac{\text{volume stress}}{\text{volume strain}} = -\frac{\Delta F/A}{\Delta V/V_i} = -\frac{\Delta P}{\Delta V/V_i} \qquad \textbf{(12.8)}$$

◀ Bulk modulus

A negative sign is inserted in this defining equation so that B is a positive number. This maneuver is necessary because an increase in pressure (positive ΔP) causes a decrease in volume (negative ΔV) and vice versa.

Table 12.1 lists bulk moduli for some materials. If you look up such values in a different source, you may find the reciprocal of the bulk modulus listed. The reciprocal of the bulk modulus is called the **compressibility** of the material.

Notice from Table 12.1 that both solids and liquids have a bulk modulus. No shear modulus and no Young's modulus are given for liquids, however, because a liquid does not sustain a shearing stress or a tensile stress. If a shearing force or a tensile force is applied to a liquid, the liquid simply flows in response.

Quick Quiz 12.4 For the three parts of this Quick Quiz, choose from the following choices the correct answer for the elastic modulus that describes the relationship between stress and strain for the system of interest, which is in italics: (a) Young's modulus (b) shear modulus (c) bulk modulus (d) none of these choices **(i)** A *block of iron* is sliding across a horizontal floor. The friction force between the block and the floor causes the block to deform. **(ii)** A trapeze artist swings through a circular arc. At the bottom of the swing, the *wires* supporting the trapeze are longer than when the trapeze artist simply hangs from the trapeze due to the increased tension in them. **(iii)** A spacecraft carries a *steel sphere* to a planet on which atmospheric pressure is much higher than on the Earth. The higher pressure causes the radius of the sphere to decrease.

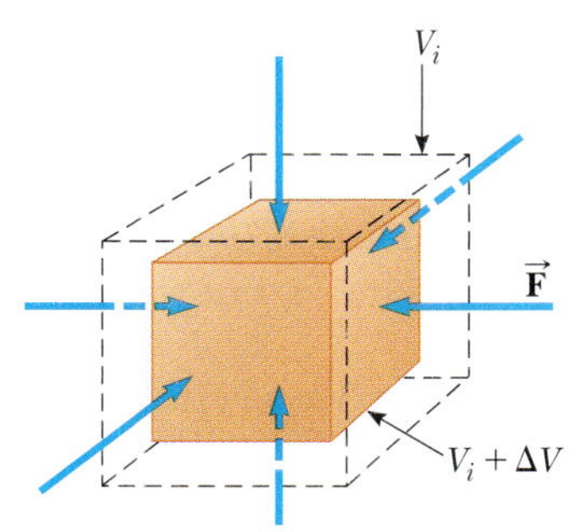

ACTIVE FIGURE 12.15

When a solid is under uniform pressure, it undergoes a change in volume but no change in shape. This cube is compressed on all sides by forces normal to its six faces.

Sign in at www.thomsonedu.com and go to ThomsonNOW to adjust the values of the applied force and the bulk modulus and observe the change in volume of the cube.

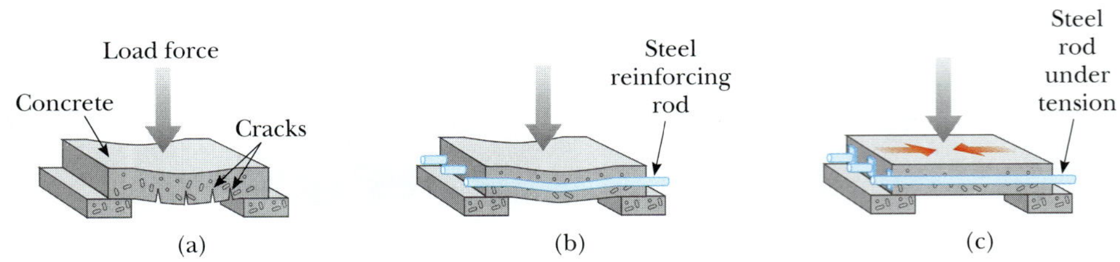

Figure 12.16 (a) A concrete slab with no reinforcement tends to crack under a heavy load. (b) The strength of the concrete is increased by using steel reinforcement rods. (c) The concrete is further strengthened by prestressing it with steel rods under tension.

Prestressed Concrete

If the stress on a solid object exceeds a certain value, the object fractures. The maximum stress that can be applied before fracture occurs—called the *tensile strength, compressive strength,* or *shear strength*—depends on the nature of the material and on the type of applied stress. For example, concrete has a tensile strength of about 2×10^6 N/m^2, a compressive strength of 20×10^6 N/m^2, and a shear strength of 2×10^6 N/m^2. If the applied stress exceeds these values, the concrete fractures. It is common practice to use large safety factors to prevent failure in concrete structures.

Concrete is normally very brittle when it is cast in thin sections. Therefore, concrete slabs tend to sag and crack at unsupported areas as shown in Figure 12.16a. The slab can be strengthened by the use of steel rods to reinforce the concrete as illustrated in Figure 12.16b. Because concrete is much stronger under compression (squeezing) than under tension (stretching) or shear, vertical columns of concrete can support very heavy loads, whereas horizontal beams of concrete tend to sag and crack. A significant increase in shear strength is achieved, however, if the reinforced concrete is prestressed as shown in Figure 12.16c. As the concrete is being poured, the steel rods are held under tension by external forces. The external forces are released after the concrete cures; the result is a permanent tension in the steel and hence a compressive stress on the concrete. The concrete slab can now support a much heavier load.

EXAMPLE 12.5 Stage Design

In Example 8.2, we analyzed a cable used to support an actor as he swung onto the stage. Now suppose the tension in the cable is 940 N as the actor reaches the lowest point. What diameter should a 10-m-long steel cable have if we do not want it to stretch more than 0.50 cm under these conditions?

SOLUTION

Conceptualize Look back at Example 8.2 to recall what is happening in this situation. We ignored any stretching of the cable there, but we wish to address this phenomenon in this example.

Categorize We perform a simple calculation involving Equation 12.6, so we categorize this example as a substitution problem.

Solve Equation 12.6 for the cross-sectional area of the cable:

$$A = \frac{FL_i}{Y\Delta L}$$

Substitute the known values:

$$A = \frac{(940\text{ N})(10\text{ m})}{(20 \times 10^{10}\text{ N/m}^2)(0.005\,0\text{ m})} = 9.4 \times 10^{-6}\text{ m}^2$$

Assuming that the cross section is circular, find the radius of the cable from $A = \pi r^2$:

$$r = \sqrt{\frac{A}{\pi}} = \sqrt{\frac{9.4 \times 10^{-6}\ \text{m}^2}{\pi}} = 1.7 \times 10^{-3}\ \text{m} = 1.7\ \text{mm}$$

Find the diameter of the cable:

$$d = 2r = 2(1.7\ \text{mm}) = 3.4\ \text{mm}$$

To provide a large margin of safety, you would probably use a flexible cable made up of many smaller wires having a total cross-sectional area substantially greater than our calculated value.

EXAMPLE 12.6 Squeezing a Brass Sphere

A solid brass sphere is initially surrounded by air, and the air pressure exerted on it is $1.0 \times 10^5\ \text{N/m}^2$ (normal atmospheric pressure). The sphere is lowered into the ocean to a depth where the pressure is $2.0 \times 10^7\ \text{N/m}^2$. The volume of the sphere in air is $0.50\ \text{m}^3$. By how much does this volume change once the sphere is submerged?

SOLUTION

Conceptualize Think about movies or television shows you have seen in which divers go to great depths in the water in submersible vessels. These vessels must be very strong to withstand the large pressure under water. This pressure squeezes the vessel and reduces its volume.

Categorize We perform a simple calculation involving Equation 12.8, so we categorize this example as a substitution problem.

Solve Equation 12.8 for the volume change of the sphere:

$$\Delta V = -\frac{V_i \Delta P}{B}$$

Substitute the numerical values:

$$\Delta V = -\frac{(0.50\ \text{m}^3)(2.0 \times 10^7\ \text{N/m}^2 - 1.0 \times 10^5\ \text{N/m}^2)}{6.1 \times 10^{10}\ \text{N/m}^2}$$

$$= -1.6 \times 10^{-4}\ \text{m}^3$$

The negative sign indicates that the volume of the sphere decreases.

Summary

ThomsonNOW™ Sign in at **www.thomsonedu.com** and go to ThomsonNOW to take a practice test for this chapter.

DEFINITIONS

The gravitational force exerted on an object can be considered as acting at a single point called the **center of gravity.** An object's center of gravity coincides with its center of mass if the object is in a uniform gravitational field.

We can describe the elastic properties of a substance using the concepts of stress and strain. **Stress** is a quantity proportional to the force producing a deformation; **strain** is a measure of the degree of deformation. Stress is proportional to strain, and the constant of proportionality is the **elastic modulus:**

$$\text{Elastic modulus} \equiv \frac{\text{stress}}{\text{strain}} \quad \textbf{(12.5)}$$

CONCEPTS AND PRINCIPLES

Three common types of deformation are represented by (1) the resistance of a solid to elongation under a load, characterized by **Young's modulus** Y; (2) the resistance of a solid to the motion of internal planes sliding past each other, characterized by the **shear modulus** S; and (3) the resistance of a solid or fluid to a volume change, characterized by the **bulk modulus** B.

ANALYSIS MODEL FOR PROBLEM SOLVING

Rigid Object in Equilibrium A rigid object in equilibrium exhibits no translational or angular acceleration. The net external force acting on it is zero, and the net external torque on it is zero about any axis:

$$\sum \vec{\mathbf{F}} = 0 \quad \textbf{(12.1)}$$

$$\sum \vec{\boldsymbol{\tau}} = 0 \quad \textbf{(12.2)}$$

The first condition is the condition for translational equilibrium, and the second is the condition for rotational equilibrium.

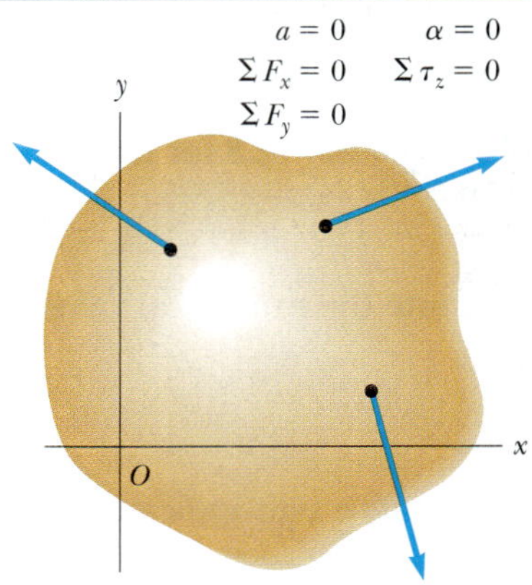

Questions

☐ denotes answer available in *Student Solutions Manual/Study Guide;* **O** denotes objective question

1. **O** Assume a single 300-N force is exerted on a bicycle frame as shown in Figure Q12.1. Consider the torque produced by this force about axes perpendicular to the plane of the paper and through each of the points A through F, where F is the center of mass of the frame. Rank the torques τ_A, τ_B, τ_C, τ_D, τ_E, and τ_F from largest to smallest, noting that zero is greater than a negative quantity. If two torques are equal, note their equality in your ranking.

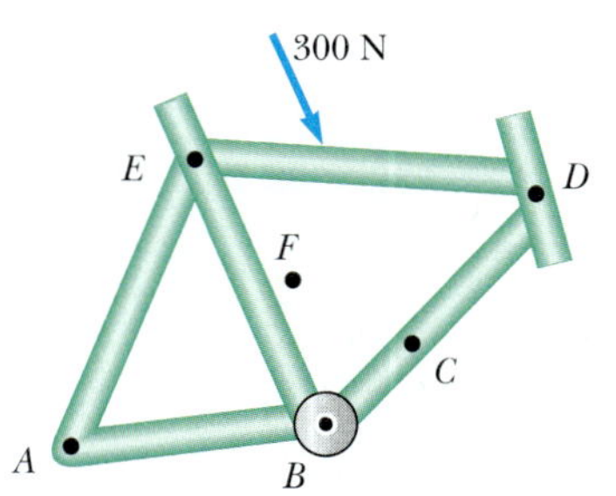

Figure Q12.1

2. Stand with your back against a wall. Why can't you put your heels firmly against the wall and then bend forward without falling?
3. Can an object be in equilibrium if it is in motion? Explain.
4. (a) Give an example in which the net force acting on an object is zero and yet the net torque is nonzero. (b) Give an example in which the net torque acting on an object is zero and yet the net force is nonzero.
5. **O** Consider the object in Figure 12.1. A single force is exerted on the object. The line of action of the force does

not pass through the object's center of mass. The acceleration of the center of mass of the object due to this force (a) is the same as if the force were applied at the center of mass, (b) is larger than the acceleration would be if the force were applied at the center of mass, (c) is smaller than the acceleration would be if the force were applied at the center of mass, or (d) is zero because the force causes only angular acceleration about the center of mass.

6. The center of gravity of an object may be located outside the object. Give a few examples for which this case is true.

7. Assume you are given an arbitrarily shaped piece of plywood, together with a hammer, nail, and plumb bob. How could you use these items to locate the center of gravity of the plywood? *Suggestion:* Use the nail to suspend the plywood.

8. O In the cabin of a ship, a soda can rests in a saucer-shaped indentation in a built-in counter. The can tilts as the ship slowly rolls. In which case is the can most stable against tipping over? (a) It is most stable when it is full. (b) It is most stable when it is half full. (c) It is most stable when it is empty. (d) It is most stable in two of these cases. (e) It is equally stable in all cases.

9. O The acceleration due to gravity becomes weaker by about three parts in ten million for each meter of increased elevation above the Earth's surface. Suppose a skyscraper is 100 stories tall, with the same floor plan for each story and with uniform average density. Compare the location of the building's center of mass and the location of its center of gravity. Choose one. (a) Its center of mass is higher by a distance of several meters. (b) Its center of mass is higher by a distance of several millimeters. (c) Its center of mass is higher by an infinitesimally small amount. (d) Its center of mass and its center of gravity are in the same location. (e) Its center of gravity is higher by a distance of several millimeters. (f) Its center of gravity is higher by a distance of several meters.

10. A girl has a large, docile dog she wishes to weigh on a small bathroom scale. She reasons that she can determine her dog's weight with the following method. First she puts the dog's two front feet on the scale and records the scale reading. Then she places the dog's two back feet on the scale and records the reading. She thinks that the sum of the readings will be the dog's weight. Is she correct? Explain your answer.

11. O The center of gravity of an ax is on the centerline of the handle, close to the head. Assume you saw across the handle through the center of gravity and weigh the two parts. What will you discover? (a) The handle side is heavier than the head side. (b) The head side is heavier than the handle side. (c) The two parts are equally heavy. (d) Their comparative weights cannot be predicted.

[12.] A ladder stands on the ground, leaning against a wall. Would you feel safer climbing up the ladder if you were told that the ground is frictionless but the wall is rough or if you were told that the wall is frictionless but the ground is rough? Justify your answer.

13. O In analyzing the equilibrium of a flat, rigid object, consider the step of choosing an axis about which to calculate torques. (a) No choice needs to be made. (b) The axis should pass through the object's center of mass. (c) The axis should pass through one end of the object. (d) The axis should be either the x axis or the y axis. (e) The axis needs to be an axle, hinge pin, pivot point, or fulcrum. (f) The axis should pass through any point within the object. (g) Any axis within or outside the object can be chosen.

14. O A certain wire, 3 m long, stretches by 1.2 mm when under tension 200 N. **(i)** An equally thick wire 6 m long, made of the same material and under the same tension, stretches by (a) 4.8 mm (b) 2.4 mm (c) 1.2 mm (d) 0.6 mm (e) 0.3 mm (f) 0 **(ii)** A wire with twice the diameter, 3 m long, made of the same material and under the same tension, stretches by what amount? Choose from the same possibilities (a) through (f).

15. What kind of deformation does a cube of Jell-O exhibit when it jiggles?

Problems

WebAssign The Problems from this chapter may be assigned online in WebAssign.

ThomsonNOW Sign in at **www.thomsonedu.com** and go to ThomsonNOW to assess your understanding of this chapter's topics with additional quizzing and conceptual questions.

1, 2, 3 denotes straightforward, intermediate, challenging; ☐ denotes full solution available in *Student Solutions Manual/Study Guide;* ▲ denotes coached solution with hints available at **www.thomsonedu.com;** denotes developing symbolic reasoning; ● denotes asking for qualitative reasoning; denotes computer useful in solving problem

Section 12.1 The Rigid Object in Equilibrium

[1.] ▲ A uniform beam of mass m_b and length ℓ supports blocks with masses m_1 and m_2 at two positions as shown in Figure P12.1. The beam rests on two knife edges. For what value of x will the beam be balanced at P such that the normal force at O is zero?

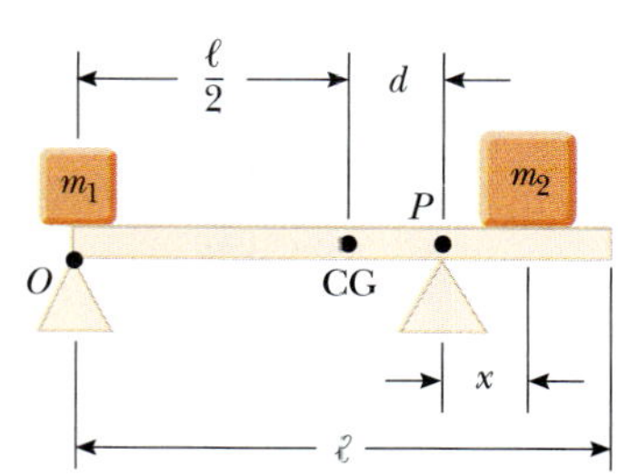

Figure P12.1

2. Write the necessary conditions for equilibrium of the object shown in Figure P12.2. Calculate torques about an axis through point O.

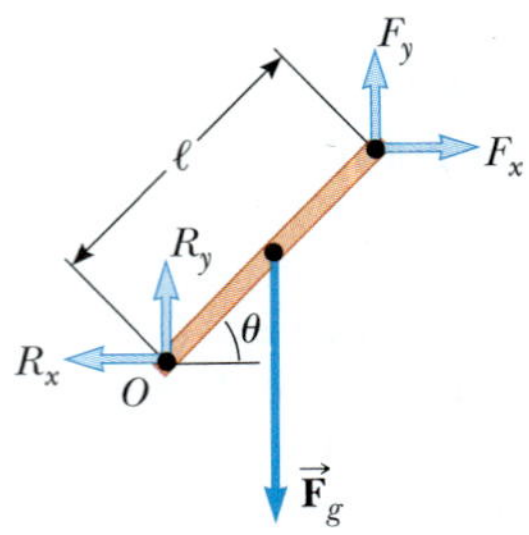

Figure P12.2

Section 12.2 More on the Center of Gravity

> Problems 35, 37, 39, and 40 in Chapter 9 can also be assigned with this section.

3. A carpenter's square has the shape of an L as shown in Figure P12.3. Locate its center of gravity.

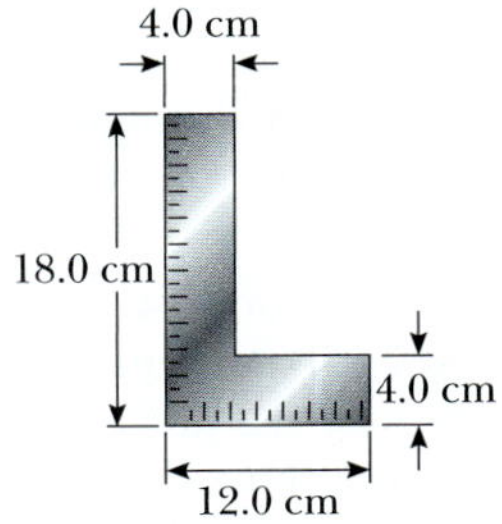

Figure P12.3

4. A circular pizza of radius R has a circular piece of radius $R/2$ removed from one side as shown in Figure P12.4. The center of gravity has moved from C to C' along the x axis. Show that the distance from C to C' is $R/6$. Assume the thickness and density of the pizza are uniform throughout.

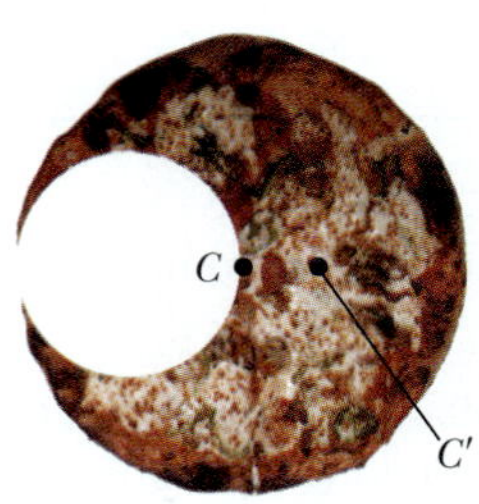

Figure P12.4

5. ▲ Consider the following distribution of objects: a 5.00-kg object with its center of gravity at (0, 0) m, a 3.00-kg object at (0, 4.00) m, and a 4.00-kg object at (3.00, 0) m. Where should a fourth object of mass 8.00 kg be placed so that the center of gravity of the four-object arrangement will be at (0, 0)?

6. Pat builds a track for his model car out of solid wood as shown in Figure P12.6. The track is 5.00 cm wide, 1.00 m high, and 3.00 m long. The runway is cut so that it forms a parabola with the equation $y = (x - 3)^2/9$. Locate the horizontal coordinate of the center of gravity of this track.

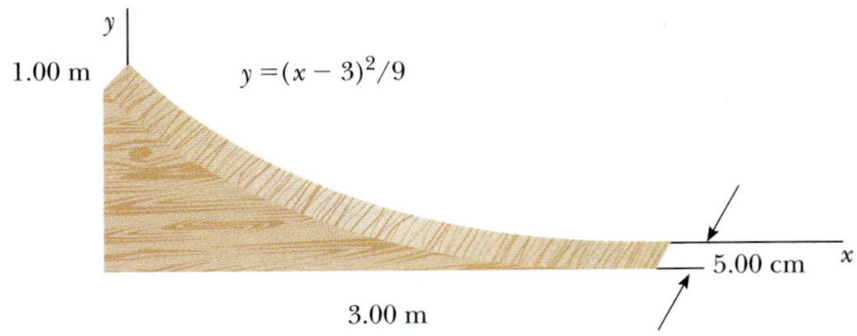

Figure P12.6

7. Figure P12.7 shows three uniform objects: a rod, a right triangle, and a square. Their masses and their coordinates in meters are given. Determine the center of gravity for the three-object system.

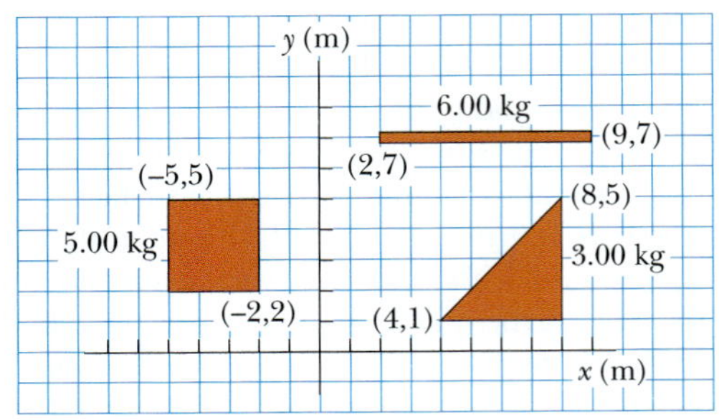

Figure P12.7

Section 12.3 Examples of Rigid Objects in Static Equilibrium

> Problems 12, 17, 18, 19, 20, 21, 22, 23, 30, 40, 44, 47, 57, 61, 65, and 71 in Chapter 5 can also be assigned with this section.

8. A mobile is constructed of light rods, light strings, and beach souvenirs as shown in Figure P12.8. Determine the masses of the objects (a) m_1, (b) m_2, and (c) m_3.

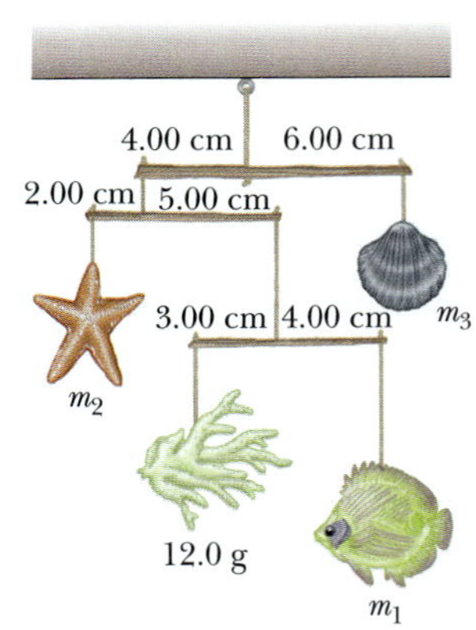

Figure P12.8

9. Find the mass m of the counterweight needed to balance the 1 500-kg truck on the incline shown in Figure P12.9. Assume all pulleys are frictionless and massless.

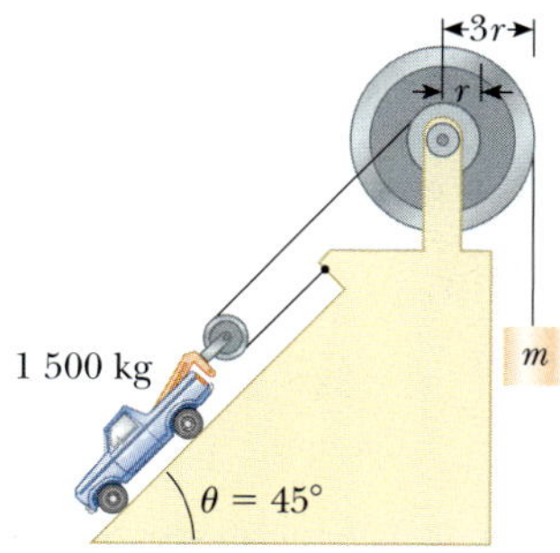

Figure P12.9

10. Figure P12.10 shows a claw hammer being used to pull a nail out of a horizontal board. A force of 150 N is exerted horizontally as shown. Find (a) the force exerted by the hammer claws on the nail and (b) the force exerted by the surface on the point of contact with the hammer head. Assume the force the hammer exerts on the nail is parallel to the nail.

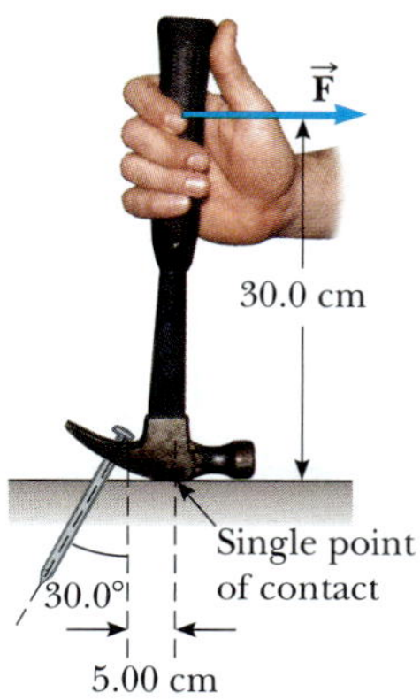

Figure P12.10

11. A 15.0-m uniform ladder weighing 500 N rests against a frictionless wall. The ladder makes a 60.0° angle with the horizontal. (a) Find the horizontal and vertical forces the ground exerts on the base of the ladder when an 800-N firefighter is 4.00 m from the bottom. (b) If the ladder is just on the verge of slipping when the firefighter is 9.00 m up, what is the coefficient of static friction between ladder and ground?

12. A uniform ladder of length L and mass m_1 rests against a frictionless wall. The ladder makes an angle θ with the horizontal. (a) Find the horizontal and vertical forces the ground exerts on the base of the ladder when a firefighter of mass m_2 is a distance x from the bottom. (b) If the ladder is just on the verge of slipping when the firefighter is a distance d from the bottom, what is the coefficient of static friction between ladder and ground?

13. A 1 500-kg automobile has a wheel base (the distance between the axles) of 3.00 m. The automobile's center of mass is on the centerline at a point 1.20 m behind the front axle. Find the force exerted by the ground on each wheel.

14. ● A 20.0-kg floodlight in a park is supported at the end of a horizontal beam of negligible mass that is hinged to a pole as shown in Figure P12.14. A cable at an angle of 30.0° with the beam helps support the light. Consider the equilibrium of the beam, drawing a free-body diagram of that object. Compute torques about an axis at the hinge at its left-hand end. Find (a) the tension in the cable, (b) the horizontal component of the force exerted by the pole on the beam, and (c) the vertical component of this force. Now solve the same problem from the same free-body diagram by computing torques around the junction between the cable and the beam at the right-hand end of the beam. Find (d) the vertical component of the force exerted by the pole on the beam, (e) the tension in the cable, and (f) the horizontal component of the force exerted by the pole on the beam. (g) Compare the solution to parts (a) through (c) with the solution to parts (d) through (f). Is either solution more accurate? Simpler? Taking together the whole set of equations read from the free-body diagram in both solutions, how many equations do you have? How many unknown quantities can be determined?

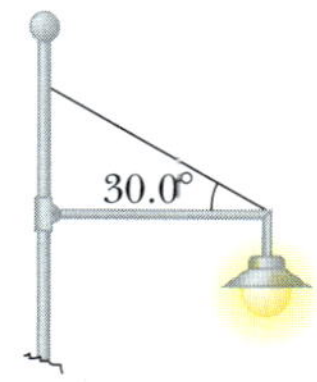

Figure P12.14

15. A flexible chain weighing 40.0 N hangs between two hooks located at the same height (Fig. P12.15). At each hook, the tangent to the chain makes an angle $\theta = 42.0°$ with the horizontal. Find (a) the magnitude of the force each hook exerts on the chain and (b) the tension in the chain at its midpoint. *Suggestion:* for part (b), make a free-body diagram for half of the chain.

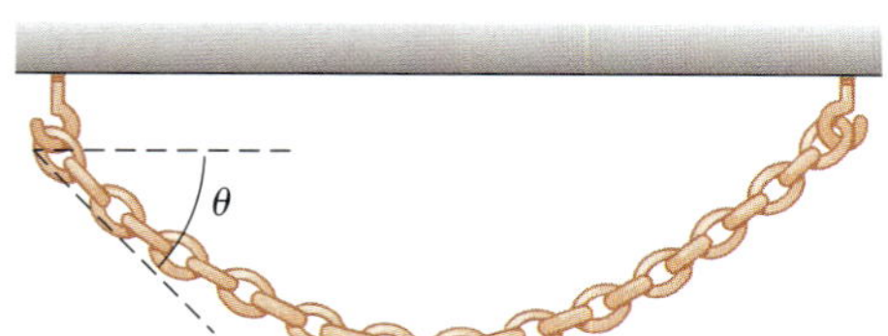

Figure P12.15

16. Sir Lost-a-Lot dons his armor and sets out from the castle on his trusty steed in his quest to improve communication between damsels and dragons (Fig. P12.16). Unfortunately, his squire lowered the drawbridge too far and finally stopped it 20.0° below the horizontal. Lost-a-Lot

Figure P12.16 Problems 16 and 17.

and his horse stop when their combined center of mass is 1.00 m from the end of the bridge. The uniform bridge is 8.00 m long and has mass 2 000 kg. The lift cable is attached to the bridge 5.00 m from the hinge at the castle end and to a point on the castle wall 12.0 m above the bridge. Lost-a-Lot's mass combined with his armor and steed is 1 000 kg. Determine (a) the tension in the cable and the (b) horizontal and (c) vertical force components acting on the bridge at the hinge.

17. **Review problem.** In the situation described in Problem 16 and illustrated in Figure P12.16, the lift cable suddenly breaks! The hinge between the castle wall and the bridge is frictionless, and the bridge swings freely until it is vertical. (a) Find the angular acceleration of the bridge once it starts to move. (b) Find the angular speed of the bridge when it strikes the vertical castle wall below the hinge. (c) Find the force exerted by the hinge on the bridge immediately after the cable breaks. (d) Find the force exerted by the hinge on the bridge immediately before it strikes the castle wall.

18. Stephen is pushing his sister Joyce in a wheelbarrow when it is stopped by a brick 8.00 cm high (Fig. P12.18). The wheelbarrow handles make an angle of 15.0° below the horizontal. A downward force of 400 N is exerted on the wheel, which has a radius of 20.0 cm. (a) What force must Stephen apply along the handles to just start the wheel over the brick? (b) What is the force (magnitude and direction) that the brick exerts on the wheel just as the wheel begins to lift over the brick? In both parts (a) and (b), assume the brick remains fixed and does not slide along the ground.

Figure P12.18

19. One end of a uniform 4.00-m-long rod of weight F_g is supported by a cable. The other end rests against the wall, where it is held by friction, as shown in Figure P12.19. The coefficient of static friction between the wall and the rod is $\mu_s = 0.500$. Determine the minimum distance x from point A at which an additional object, also with the same weight F_g, can be hung without causing the rod to slip at point A.

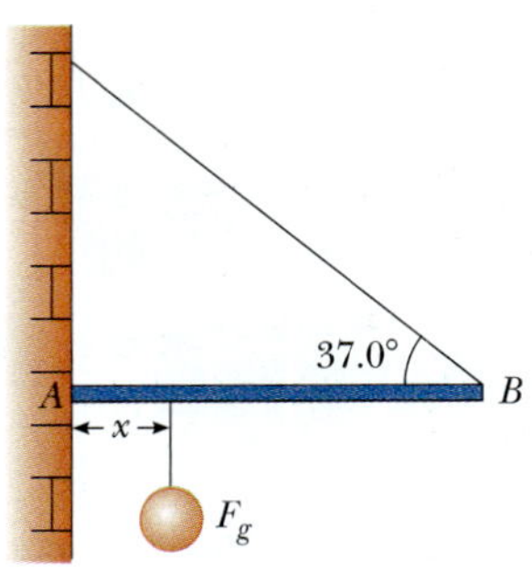

Figure P12.19

20. In the **What If?** section of Example 12.2, let x represent the distance in meters between the person and the hinge at the left end of the beam. (a) Show that the cable tension in newtons is given by $T = 93.9x + 125$. Argue that T increases as x increases. (b) Show that the direction angle θ of the hinge force is described by

$$\tan\theta = \left(\frac{32}{3x+4} - 1\right)\tan 53.0°$$

How does θ change as x increases? (c) Show that the magnitude of the hinge force is given by

$$R = \sqrt{8.82 \times 10^3 x^2 - 9.65 \times 10^4 x + 4.96 \times 10^5}$$

How does R change as x increases?

21. A vaulter holds a 29.4-N pole in equilibrium by exerting an upward force $\vec{\mathbf{U}}$ with her leading hand and a downward force $\vec{\mathbf{D}}$ with her trailing hand as shown in Figure P12.21. Point C is the center of gravity of the pole. What are the magnitudes of $\vec{\mathbf{U}}$ and $\vec{\mathbf{D}}$?

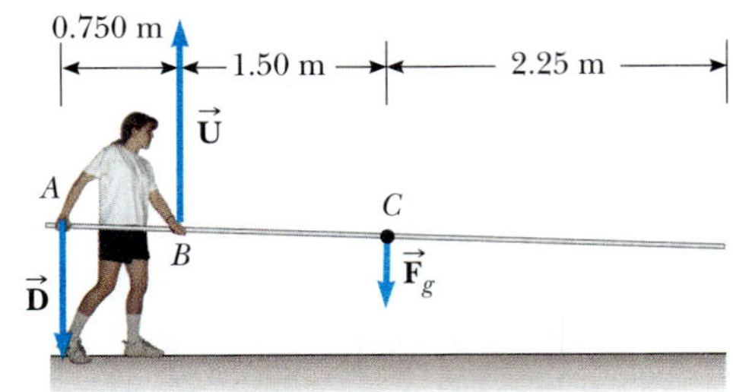

Figure P12.21

Section 12.4 Elastic Properties of Solids

22. Evaluate Young's modulus for the material whose stress-strain curve is shown in Figure 12.13.

23. A 200-kg load is hung on a wire of length 4.00 m, cross-sectional area 0.200×10^{-4} m^2, and Young's modulus 8.00×10^{10} N/m^2. What is its increase in length?

24. Assume Young's modulus for bone is 1.50×10^{10} N/m^2. The bone breaks if stress greater than 1.50×10^8 N/m^2 is imposed on it. (a) What is the maximum force that can be exerted on the femur bone in the leg if it has a minimum effective diameter of 2.50 cm? (b) If this much force is applied compressively, by how much does the 25.0-cm-long bone shorten?

25. A child slides across a floor in a pair of rubber-soled shoes. The friction force acting on each foot is 20.0 N. The footprint area of each shoe sole is 14.0 cm^2, and the thickness of each sole is 5.00 mm. Find the horizontal distance by which the upper and lower surfaces of each sole are offset. The shear modulus of the rubber is 3.00 MN/m^2.

26. A steel wire of diameter 1 mm can support a tension of 0.2 kN. A cable to support a tension of 20 kN should have diameter of what order of magnitude?

27. ▲ Assume that if the shear stress in steel exceeds about 4.00×10^8 N/m^2, the steel ruptures. Determine the shearing force necessary to (a) shear a steel bolt 1.00 cm in diameter and (b) punch a 1.00-cm-diameter hole in a steel plate 0.500 cm thick.

28. **Review problem.** A 30.0-kg hammer, moving with speed 20.0 m/s, strikes a steel spike 2.30 cm in diameter. The hammer rebounds with speed 10.0 m/s after 0.110 s. What is the average strain in the spike during the impact?

29. When water freezes, it expands by about 9.00%. What pressure increase would occur inside your automobile engine block if the water in it froze? (The bulk modulus of ice is 2.00×10^9 N/m^2.)

30. **Review problem.** A 2.00-m-long cylindrical steel wire with a cross-sectional diameter of 4.00 mm is placed over a light, frictionless pulley, with one end of the wire connected to a 5.00-kg object and the other end connected to a 3.00-kg object. By how much does the wire stretch while the objects are in motion?

31. A walkway suspended across a hotel lobby is supported at numerous points along its edges by a vertical cable above each point and a vertical column underneath. The steel cable is 1.27 cm in diameter and is 5.75 m long before loading. The aluminum column is a hollow cylinder with an inside diameter of 16.14 cm, an outside diameter of 16.24 cm, and unloaded length of 3.25 m. When the walkway exerts a load force of 8 500 N on one of the support points, how much does the point move down?

32. ● The deepest point in any ocean is in the Mariana Trench, which is about 11 km deep, in the Pacific. The pressure at this depth is huge, about 1.13×10^8 N/m^2. (a) Calculate the change in volume of 1.00 m^3 of seawater carried from the surface to this deepest point. (b) The density of seawater at the surface is 1.03×10^3 kg/m^3. Find its density at the bottom. (c) Explain whether or when it is a good approximation to think of water as incompressible.

Additional Problems

33. A bridge of length 50.0 m and mass 8.00×10^4 kg is supported on a smooth pier at each end as shown in Figure P12.33. A truck of mass 3.00×10^4 kg is located 15.0 m from one end. What are the forces on the bridge at the points of support?

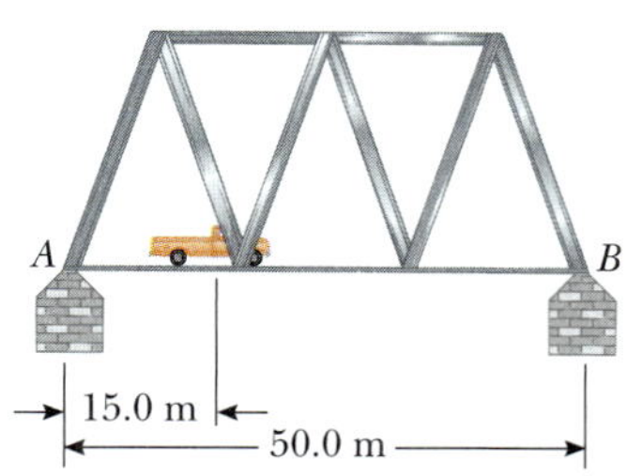

Figure P12.33

34. ● A new General Electric kitchen stove has a mass of 68.0 kg and the dimensions shown in Figure P12.34. The stove comes with a warning that it can tip forward if a person stands or sits on the oven door when it is open. What can you conclude about the weight of such a person? Could it be a child? List the assumptions you make in solving this problem. The stove is supplied with a wall bracket to prevent the accident.

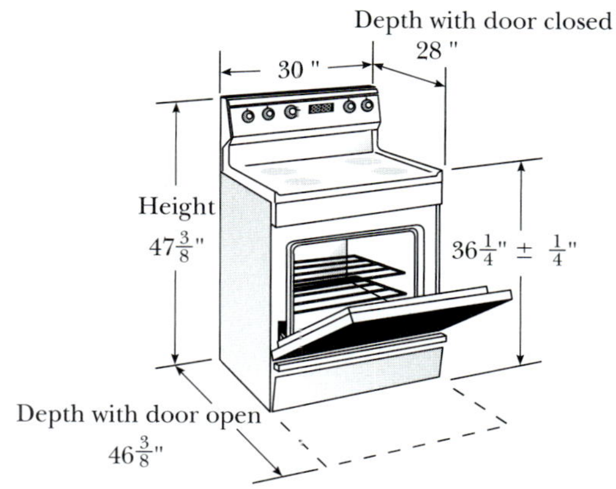

Figure P12.34

35. A uniform pole is propped between the floor and the ceiling of a room. The height of the room is 7.80 ft, and the coefficient of static friction between the pole and the ceiling is 0.576. The coefficient of static friction between the pole and the floor is greater than that. What is the length of the longest pole that can be propped between the floor and the ceiling?

36. Refer to Figure 12.16c. A lintel of prestressed reinforced concrete is 1.50 m long. The cross-sectional area of the concrete is 50.0 cm^2. The concrete encloses one steel reinforcing rod with cross-sectional area 1.50 cm^2. The rod joins two strong end plates. Young's modulus for the concrete is 30.0×10^9 N/m^2. After the concrete cures and the original tension T_1 in the rod is released, the concrete is to be under compressive stress 8.00×10^6 N/m^2. (a) By what distance will the rod compress the concrete when the original tension in the rod is released? (b) What is the new tension T_2 in the rod? (c) The rod will then be how much longer than its unstressed length? (d) When the concrete was poured, the rod should have been stretched by what extension distance from its unstressed length? (e) Find the required original tension T_1 in the rod.

37. A hungry bear weighing 700 N walks out on a beam in an attempt to retrieve a basket of food hanging at the end of the beam (Fig. P12.37). The beam is uniform, weighs 200 N, and is 6.00 m long; the basket weighs 80.0 N. (a) Draw a free-body diagram for the beam. (b) When the bear is at $x = 1.00$ m, find the tension in the wire and the components of the force exerted by the wall on the left end of the beam. (c) **What If?** If the wire can withstand a maximum tension of 900 N, what is the maximum distance the bear can walk before the wire breaks?

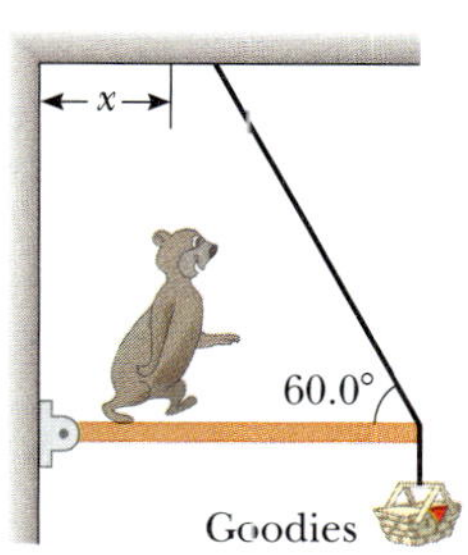

Figure P12.37

2 = intermediate; 3 = challenging; □ = SSM/SG; ▲ = ThomsonNOW; = symbolic reasoning; ● = qualitative reasoning

38. The following equations are obtained from a free-body diagram of a rectangular farm gate, supported by two hinges on the left-hand side. A bucket of grain is hanging from the latch.

$$-A + C = 0$$

$$+B - 392\text{ N} - 50.0\text{ N} = 0$$

$$A(0) + B(0) + C(1.80\text{ m}) - 392\text{ N}(1.50\text{ m}) - 50.0\text{ N}(3.00\text{ m}) = 0$$

(a) Draw the free-body diagram and complete the statement of the problem, specifying the unknowns. (b) Determine the values of the unknowns and state the physical meaning of each.

39. ▲ A uniform sign of weight F_g and width $2L$ hangs from a light, horizontal beam hinged at the wall and supported by a cable (Fig. P12.39). Determine (a) the tension in the cable and (b) the components of the reaction force exerted by the wall on the beam, in terms of F_g, d, L, and θ.

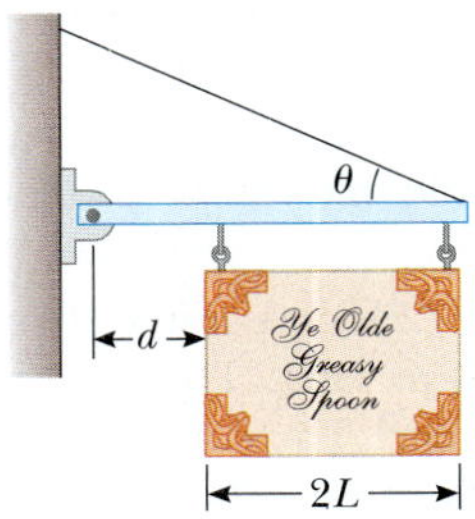

Figure P12.39

40. A 1 200-N uniform boom is supported by a cable as shown in Figure P12.40. The boom is pivoted at the bottom, and a 2 000-N object hangs from its top. Find the tension in the cable and the components of the reaction force exerted by the floor on the boom.

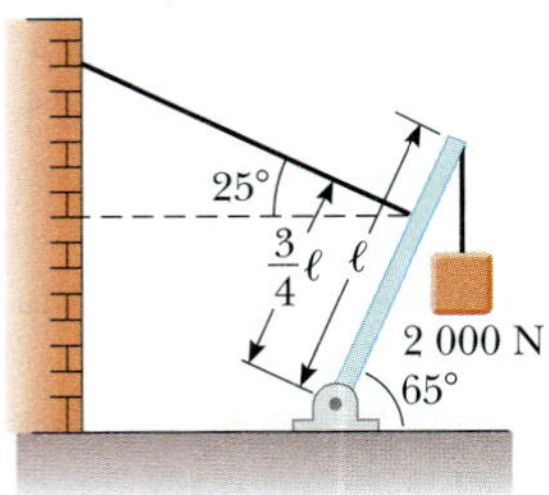

Figure P12.40

41. A crane of mass 3 000 kg supports a load of 10 000 kg as shown in Figure P12.41. The crane is pivoted with a frictionless pin at A and rests against a smooth support at B. Find the reaction forces at A and B.

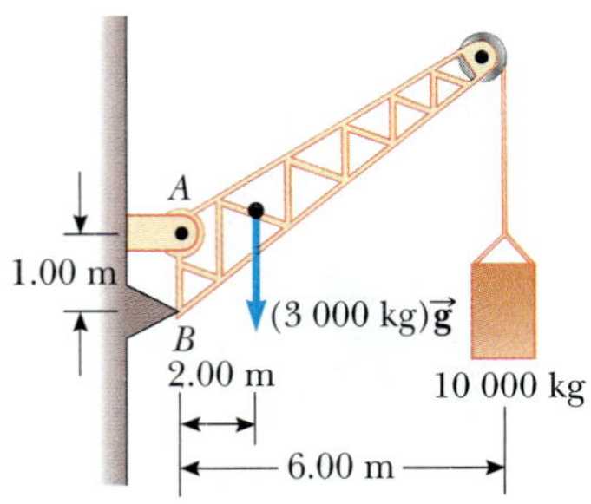

Figure P12.41

42. ● Assume a person bends forward to lift a load "with his back" as shown in Figure P12.42a. The person's spine pivots mainly at the fifth lumbar vertebra, with the principal supporting force provided by the erector spinalis muscle in the back. To estimate the magnitude of the forces involved, consider the model shown in Figure P12.42b for a person bending forward to lift a 200-N object. The person's spine and upper body are represented as a uniform horizontal rod of weight 350 N, pivoted at the base of the spine. The erector spinalis muscle, attached at a point two thirds of the way up the spine, maintains the position of the back. The angle between the spine and this muscle is 12.0°. Find (a) the tension in the back muscle and (b) the compressional force in the spine. (c) Is this method a good way to lift a load? Explain your answer, using the results of parts (a) and (b). It can be instructive to compare a human to other animals. Can you suggest a better method to lift a load?

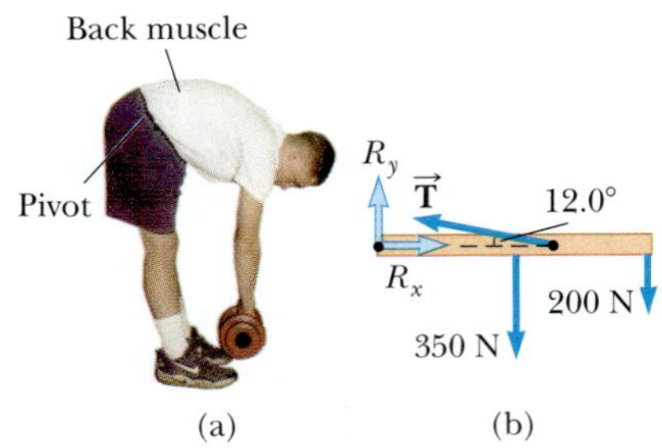

Figure P12.42

43. A 10 000-N shark is supported by a cable attached to a 4.00 m rod that can pivot at the base. Calculate the tension in the tie rope between the rod and the wall, assuming the tie rope is holding the system in the position shown in Figure P12.43. Find the horizontal and vertical forces exerted on the base of the rod. Ignore the weight of the rod.

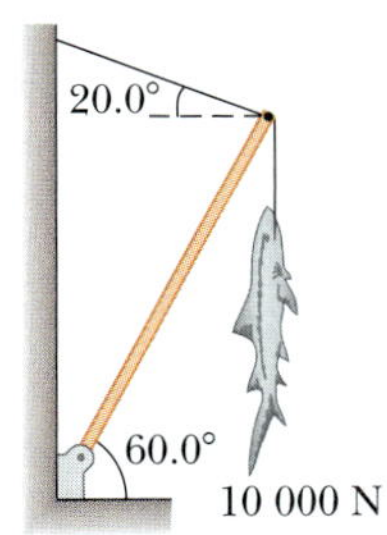

Figure P12.43

44. A uniform rod of weight F_g and length L is supported at its ends by a frictionless trough as shown in Figure P12.44. (a) Show that the center of gravity of the rod must be vertically over point O when the rod is in equilibrium. (b) Determine the equilibrium value of the angle θ.

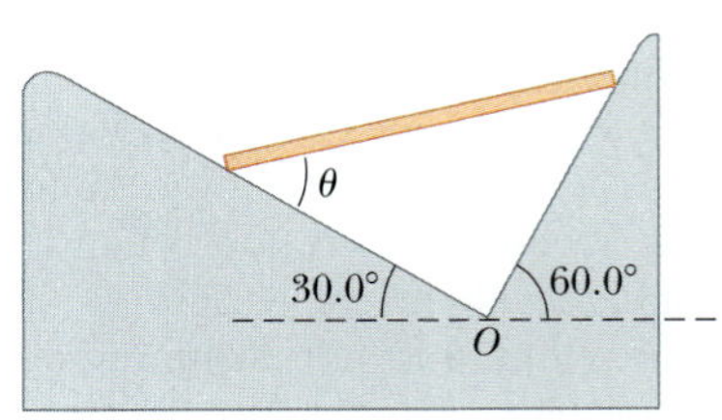

Figure P12.44

45. A force is exerted on a uniform rectangular cabinet weighing 400 N as shown in Figure P12.45. (a) The cabinet slides with constant speed when $F = 200$ N and $h = 0.400$ m. Find the coefficient of kinetic friction and the position of the resultant normal force. (b) Taking $F = 300$ N, find the value of h for which the cabinet just begins to tip.

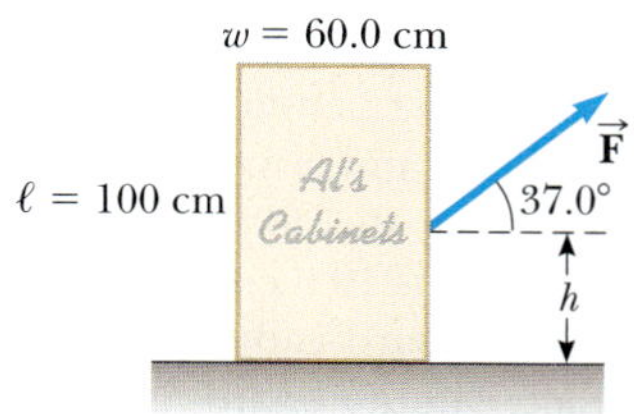

Figure P12.45

46. Consider the rectangular cabinet of Problem 45, but with a force $\vec{\mathbf{F}}$ applied horizontally at the upper edge. (a) What is the minimum force required to start to tip the cabinet? (b) What is the minimum coefficient of static friction required for the cabinet not to slide with the application of a force of this magnitude? (c) Find the magnitude and direction of the minimum force required to tip the cabinet if the point of application can be chosen anywhere on the cabinet.

47. A uniform beam of mass m is inclined at an angle θ to the horizontal. Its upper end produces a 90° bend in a very rough rope tied to a wall, and its lower end rests on a rough floor (Fig. P12.47). (a) Let μ_s represent the coefficient of static friction between beam and floor. Assume μ_s is less than the cotangent of θ. Determine an expression for the maximum mass M that can be suspended from the top before the beam slips. (b) Determine the magnitude of the reaction force at the floor and the magnitude of the force exerted by the beam on the rope at P in terms of m, M, and μ_s.

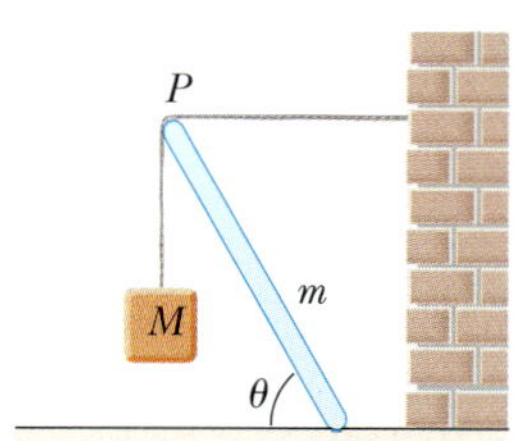

Figure P12.47

48. ● Consider a light truss, with weight negligible compared with the load it supports. Suppose it is formed from struts lying in a plane and joined by smooth hinge pins at their ends. External forces act on the truss only at the joints. Figure P12.48 shows one example of the simplest truss, with three struts and three pins. State reasoning to prove that the force any strut exerts on a pin must be directed along the length of the strut, as a force of tension or compression.

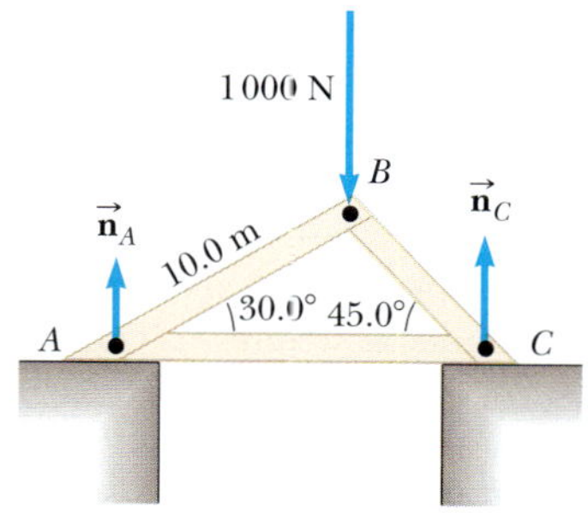

Figure P12.48

49. Figure P12.48 shows a truss that supports a downward force of 1 000 N applied at the point B. The truss has negligible weight. The piers at A and C are smooth. (a) Apply the conditions of equilibrium to prove that $n_A = 366$ N and $n_C = 634$ N. (b) Use the result proved in Problem 48 to identify the directions of the forces that the bars exert on the pins joining them. Find the force of tension or of compression in each of the three bars.

50. One side of a plant shelf is supported by a bracket mounted on a vertical wall by a single screw as shown in Figure P12.50. Ignore the weight of the bracket. (a) Find the horizontal component of the force that the screw exerts on the bracket when an 80.0 N vertical force is applied as shown. (b) As your grandfather waters his geraniums, the 80.0-N load force is increasing at the rate 0.150 N/s. At what rate is the force exerted by the screw changing? *Suggestions:* Imagine that the bracket is slightly loose. You can do parts (a) and (b) most efficiently if you call the load force W and solve symbolically for the screw force F.

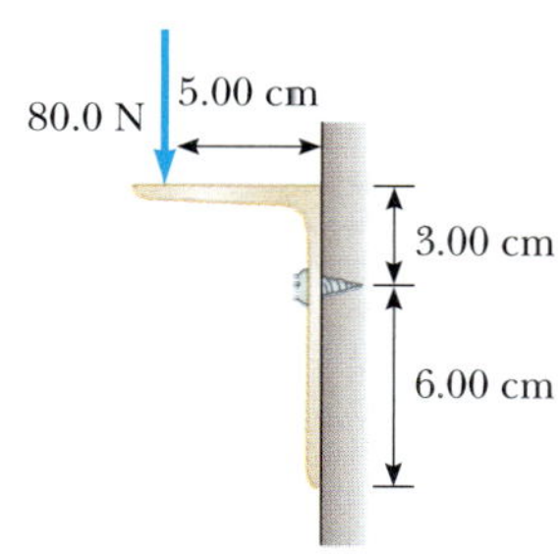

Figure P12.50

51. A stepladder of negligible weight is constructed as shown in Figure P12.51. A painter of mass 70.0 kg stands on the ladder 3.00 m from the bottom. Assume the floor is frictionless. Find (a) the tension in the horizontal bar

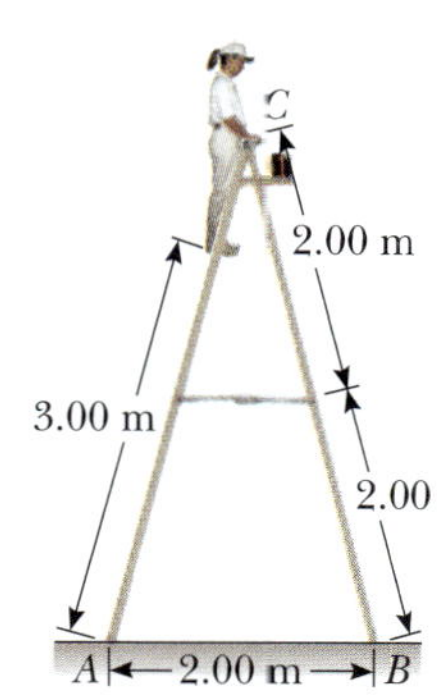

Figure P12.51

connecting the two halves of the ladder, (b) the normal forces at A and B, and (c) the components of the reaction force at the single hinge C that the left half of the ladder exerts on the right half. *Suggestion:* Treat the ladder as a single object, but also each half of the ladder separately.

52. ● Figure P12.52 shows a vertical force applied tangentially to a uniform cylinder of weight F_g. The coefficient of static friction between the cylinder and all surfaces is 0.500. In terms of F_g, find the maximum force P that can be applied without causing the cylinder to rotate. As a first step, explain why both friction forces will be at their maximum values when the cylinder is on the verge of slipping.

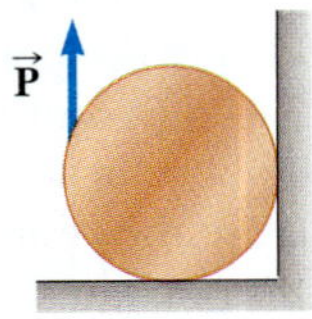

Figure P12.52

53. ▲ **Review problem.** A wire of length L, Young's modulus Y, and cross-sectional area A is stretched elastically by an amount ΔL. By Hooke's law, the restoring force is $-k\,\Delta L$. (a) Show that $k = YA/L$. (b) Show that the work done in stretching the wire by an amount ΔL is

$$W = \tfrac{1}{2}YA\frac{(\Delta L)^2}{L}$$

54. Two racquetballs each having a mass of 170 g are placed in a glass jar as shown in Figure P12.54. Their centers and the point A lie on a straight line. Assume the walls are frictionless. (a) Determine P_1, P_2, and P_3. (b) Determine the magnitude of the force exerted by the left ball on the right ball.

Figure P12.54

55. In exercise physiology studies, it is sometimes important to determine the location of a person's center of mass. This determination can be done with the arrangement shown in Figure P12.55. A light plank rests on two scales, which read $F_{g1} = 380$ N and $F_{g2} = 320$ N. A distance of 2.00 m separates the scales. How far from the woman's feet is her center of mass?

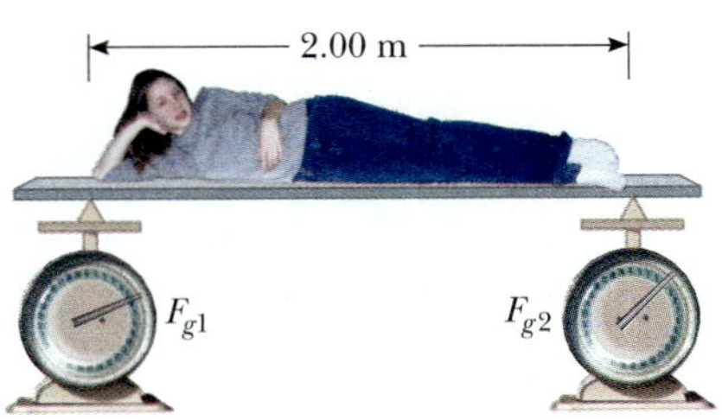

Figure P12.55

56. A steel cable 3.00 cm^2 in cross-sectional area has a mass of 2.40 kg per meter of length. If 500 m of the cable is hung over a vertical cliff, how much does the cable stretch under its own weight? Take $Y_{\text{steel}} = 2.00 \times 10^{11}$ N/m^2.

57. (a) Estimate the force with which a karate master strikes a board, assuming the hand's speed at the moment of impact is 10.0 m/s, decreasing to 1.00 m/s during a 0.002 00-s time interval of contact between the hand and the board. The mass of his hand and arm is 1.00 kg. (b) Estimate the shear stress, assuming this force is exerted on a 1.00-cm-thick pine board that is 10.0 cm wide. (c) If the maximum shear stress a pine board can support before breaking is 3.60×10^6 N/m^2, will the board break?

58. **Review problem.** An aluminum wire is 0.850 m long and has a circular cross section of diameter 0.780 mm. Fixed at the top end, the wire supports a 1.20-kg object that swings in a horizontal circle. Determine the angular velocity required to produce a strain of 1.00×10^{-3}.

59. **Review problem.** A trailer with loaded weight $\vec{\mathbf{F}}_g$ is being pulled by a vehicle with a force $\vec{\mathbf{P}}$ as shown in Figure P12.59. The trailer is loaded such that its center of mass is located as shown. Ignore the force of rolling friction and let a represent the x component of the acceleration of the trailer. (a) Find the vertical component of $\vec{\mathbf{P}}$ in terms of the given parameters. (b) Assume $a = 2.00$ m/s^2 and $h = 1.50$ m. What must be the value of d so that $P_y = 0$ (no vertical load on the vehicle)? (c) Find the values of P_x and P_y given that $F_g = 1\,500$ N, $d = 0.800$ m, $L = 3.00$ m, $h = 1.50$ m, and $a = -2.00$ m/s^2.

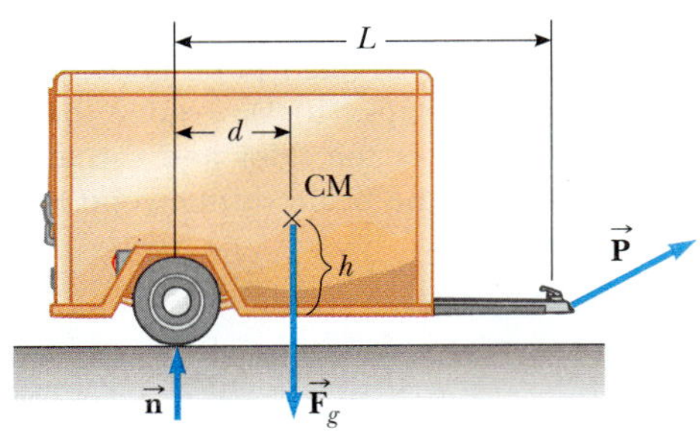

Figure P12.59

60. **Review problem.** A car moves with speed v on a horizontal circular track of radius R. A head-on view of the car is shown in Figure P12.60. The height of the car's center of mass above the ground is h, and the separation between its inner and outer wheels is d. The road is dry, and the

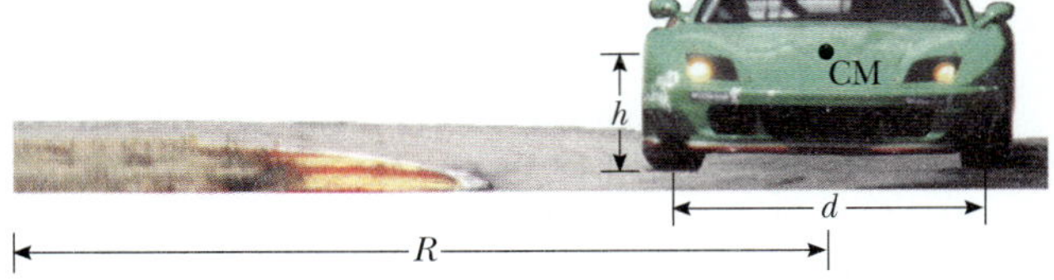

Figure P12.60

2 = intermediate; 3 = challenging; □ = SSM/SG; ▲ = ThomsonNOW; ▭ = symbolic reasoning; ● = qualitative reasoning

car does not skid. Show that the maximum speed the car can have without overturning is given by

$$v_{\max} = \sqrt{\frac{gRd}{2h}}$$

To reduce the risk of rollover, should one increase or decrease h? Should one increase or decrease the width d of the wheel base?

Answers to Quick Quizzes

12.1 (a). The unbalanced torques due to the forces in Figure 12.2 cause an angular acceleration even though the translational acceleration is zero.

12.2 (b). The lines of action of all the forces in Figure 12.3 intersect at a common point. Therefore, the net torque about this point is zero. This zero value of the net torque is independent of the values of the forces. Because no force has a downward component, there is a net force and the object is not in force equilibrium.

12.3 (b). Both the object and the center of gravity of the meterstick are 25 cm from the pivot point. Therefore, the meterstick and the object must have the same mass for the system to be balanced.

12.4 **(i),** (b). The friction force on the block as it slides along the surface is parallel to the lower surface and will cause the block to undergo a shear deformation. **(ii),** (a). The stretching of the wire due to the increased tension is described by Young's modulus. **(iii),** (c). The pressure of the atmosphere results in a force of uniform magnitude perpendicular at all points on the surface of the sphere.

The red supergiant V838 Monocerotis is 20 000 lightyears from Earth. In 2002, the star exhibited a major outburst of energy typical of a nova event. Following the outburst, however, the variable behavior of infrared radiation from the star did not follow the typical nova pattern. Models of gravitational interaction leading to the merging of the star with a binary companion or its own planets have been proposed to explain the unusual behavior. (©STScI/NASA/Corbis)

13 Universal Gravitation

Before 1687, a large amount of data had been collected on the motions of the Moon and the planets, but a clear understanding of the forces related to these motions was not available. In that year, Isaac Newton provided the key that unlocked the secrets of the heavens. He knew, from his first law, that a net force had to be acting on the Moon because without such a force the Moon would move in a straight-line path rather than in its almost circular orbit. Newton reasoned that this force was the gravitational attraction exerted by the Earth on the Moon. He realized that the forces involved in the Earth–Moon attraction and in the Sun–planet attraction were not something special to those systems, but rather were particular cases of a general and universal attraction between objects. In other words, Newton saw that the same force of attraction that causes the Moon to follow its path around the Earth also causes an apple to fall from a tree. It was the first time that "earthly" and "heavenly" motions were unified.

In this chapter, we study the law of universal gravitation. We emphasize a description of planetary motion because astronomical data provide an important test of this law's validity. We then show that the laws of planetary motion developed by Johannes Kepler follow from the law of universal gravitation and the principle of conservation of angular momentum. We conclude by deriving a general expression for gravitational potential energy and examining the energetics of planetary and satellite motion.

13.1 Newton's Law of Universal Gravitation

You may have heard the legend that, while napping under a tree, Newton was struck on the head by a falling apple. This alleged accident supposedly prompted him to imagine that perhaps all objects in the Universe were attracted to each other in the same way the apple was attracted to the Earth. Newton analyzed astronomical data on the motion of the Moon around the Earth. From that analysis, he made the bold assertion that the force law governing the motion of planets was the *same* as the force law that attracted a falling apple to the Earth.

In 1687, Newton published his work on the law of gravity in his treatise *Mathematical Principles of Natural Philosophy.* **Newton's law of universal gravitation** states that

> every particle in the Universe attracts every other particle with a force that is directly proportional to the product of their masses and inversely proportional to the square of the distance between them.

◀ The law of universal gravitation

If the particles have masses m_1 and m_2 and are separated by a distance r, the magnitude of this gravitational force is

$$F_g = G\frac{m_1 m_2}{r^2} \tag{13.1}$$

where G is a constant, called the *universal gravitational constant.* Its value in SI units is

$$G = 6.673 \times 10^{-11}\ \mathrm{N \cdot m^2/kg^2} \tag{13.2}$$

Henry Cavendish (1731–1810) measured the universal gravitational constant in an important 1798 experiment. Cavendish's apparatus consists of two small spheres, each of mass m, fixed to the ends of a light, horizontal rod suspended by a fine fiber or thin metal wire as illustrated in Figure 13.1. When two large spheres, each of mass M, are placed near the smaller ones, the attractive force between smaller and larger spheres causes the rod to rotate and twist the wire suspension to a new equilibrium orientation. The angle of rotation is measured by the deflection of a light beam reflected from a mirror attached to the vertical suspension.

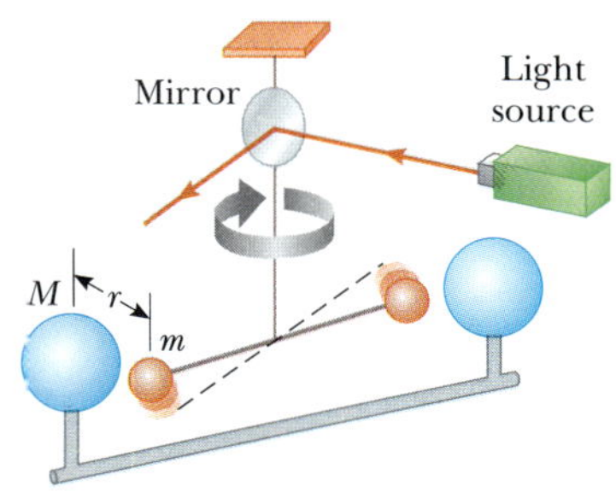

Figure 13.1 Cavendish apparatus for measuring G. The dashed line represents the original position of the rod.

The form of the force law given by Equation 13.1 is often referred to as an **inverse-square law** because the magnitude of the force varies as the inverse square of the separation of the particles.[1] We shall see other examples of this type of force law in subsequent chapters. We can express this force in vector form by defining a unit vector $\hat{\mathbf{r}}_{12}$ (Active Fig. 13.2). Because this unit vector is directed from particle 1 toward particle 2, the force exerted by particle 1 on particle 2 is

$$\vec{\mathbf{F}}_{12} = -G\frac{m_1 m_2}{r^2}\,\hat{\mathbf{r}}_{12} \tag{13.3}$$

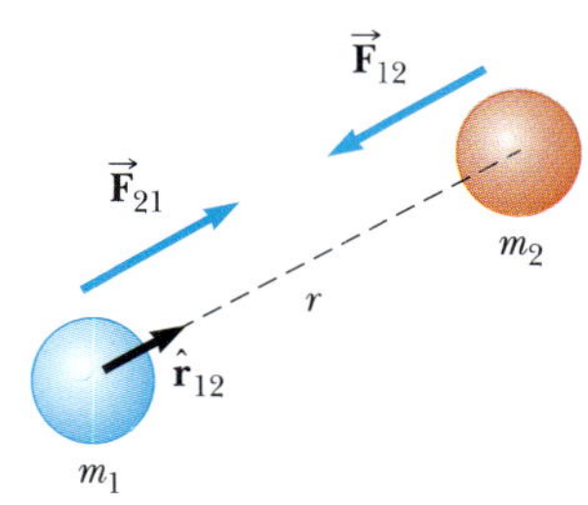

ACTIVE FIGURE 13.2

The gravitational force between two particles is attractive. The unit vector $\hat{\mathbf{r}}_{12}$ is directed from particle 1 toward particle 2. Notice that $\vec{\mathbf{F}}_{21} = -\vec{\mathbf{F}}_{12}$.

Sign in at www.thomsonedu.com and go to ThomsonNOW to change the masses of the particles and the separation distance between the particles and see the effect on the gravitational force.

where the negative sign indicates that particle 2 is attracted to particle 1; hence, the force on particle 2 must be directed toward particle 1. By Newton's third law, the force exerted by particle 2 on particle 1, designated $\vec{\mathbf{F}}_{21}$, is equal in magnitude to $\vec{\mathbf{F}}_{12}$ and in the opposite direction. That is, these forces form an action–reaction pair, and $\vec{\mathbf{F}}_{21} = -\vec{\mathbf{F}}_{12}$.

Two features of Equation 13.3 deserve mention. First, the gravitational force is a field force that always exists between two particles, regardless of the medium that separates them. Because the force varies as the inverse square of the distance between the particles, it decreases rapidly with increasing separation.

Equation 13.3 can also be used to show that **the gravitational force exerted by a finite-size, spherically symmetric mass distribution on a particle outside the**

[1] An *inverse* proportionality between two quantities x and y is one in which $y = k/x$, where k is a constant. A *direct* proportion between x and y exists when $y = kx$.

PITFALL PREVENTION 13.1
Be Clear on *g* and *G*

The symbol *g* represents the magnitude of the free-fall acceleration near a planet. At the surface of the Earth, *g* has an average value of $9.80\ \text{m/s}^2$. On the other hand, *G* is a universal constant that has the same value everywhere in the Universe.

distribution is the same as if the entire mass of the distribution were concentrated at the center. For example, the magnitude of the force exerted by the Earth on a particle of mass m near the Earth's surface is

$$F_g = G\frac{M_E m}{R_E^2} \tag{13.4}$$

where M_E is the Earth's mass and R_E its radius. This force is directed toward the center of the Earth.

Quick Quiz 13.1 A planet has two moons of equal mass. Moon 1 is in a circular orbit of radius r. Moon 2 is in a circular orbit of radius $2r$. What is the magnitude of the gravitational force exerted by the planet on Moon 2? (a) four times as large as that on Moon 1 (b) twice as large as that on Moon 1 (c) equal to that on Moon 1 (d) half as large as that on Moon 1 (e) one-fourth as large as that on Moon 1

EXAMPLE 13.1 Billiards, Anyone?

Three 0.300-kg billiard balls are placed on a table at the corners of a right triangle as shown in Figure 13.3. The sides of the triangle are of lengths $a = 0.400$ m, $b = 0.300$ m, and $c = 0.500$ m. Calculate the gravitational force vector on the cue ball (designated m_1) resulting from the other two balls as well as the magnitude and direction of this force.

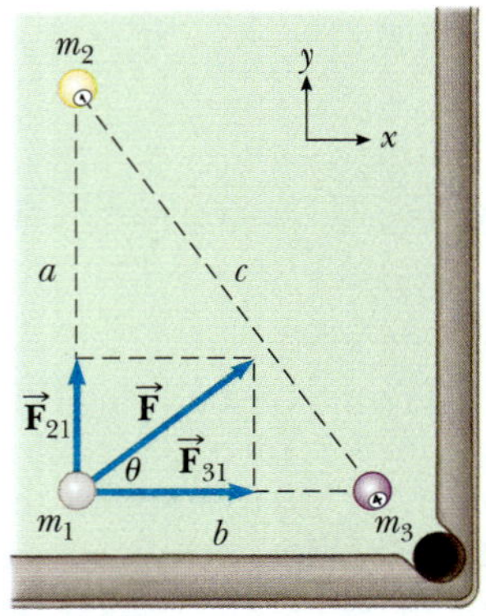

Figure 13.3 (Example 13.1) The resultant gravitational force acting on the cue ball is the vector sum $\vec{\mathbf{F}}_{21} + \vec{\mathbf{F}}_{31}$.

SOLUTION

Conceptualize Notice in Figure 13.3 that the cue ball is attracted to both other balls by the gravitational force. We can see graphically that the net force should point upward and toward the right. We locate our coordinate axes as shown in Figure 13.3, placing our origin at the position of the cue ball.

Categorize This problem involves evaluating the gravitational forces on the cue ball using Equation 13.3. Once these forces are evaluated, it becomes a vector addition problem to find the net force.

Analyze Find the force exerted by m_2 on the cue ball:

$$\vec{\mathbf{F}}_{21} = G\frac{m_2 m_1}{r_{21}^{\,2}}\hat{\mathbf{j}}$$

$$= (6.67 \times 10^{-11}\ \text{N}\cdot\text{m}^2/\text{kg}^2)\frac{(0.300\ \text{kg})(0.300\ \text{kg})}{(0.400\ \text{m})^2}\hat{\mathbf{j}}$$

$$= 3.75 \times 10^{-11}\hat{\mathbf{j}}\ \text{N}$$

Find the force exerted by m_3 on the cue ball:

$$\vec{\mathbf{F}}_{31} = G\frac{m_3 m_1}{r_{31}^{\,2}}\hat{\mathbf{i}}$$

$$= (6.67 \times 10^{-11}\ \text{N}\cdot\text{m}^2/\text{kg}^2)\frac{(0.300\ \text{kg})(0.300\ \text{kg})}{(0.300\ \text{m})^2}\hat{\mathbf{i}}$$

$$= 6.67 \times 10^{-11}\hat{\mathbf{i}}\ \text{N}$$

Find the net gravitational force on the cue ball by adding these force vectors:

$$\vec{\mathbf{F}} = \vec{\mathbf{F}}_{31} + \vec{\mathbf{F}}_{21} = (6.67\hat{\mathbf{i}} + 3.75\hat{\mathbf{j}}) \times 10^{-11}\ \text{N}$$

Find the magnitude of this force:

$$F = \sqrt{F_{31}^{\,2} + F_{21}^{\,2}} = \sqrt{(6.67)^2 + (3.75)^2} \times 10^{-11}\ \text{N}$$

$$= 7.65 \times 10^{-11}\ \text{N}$$

Find the tangent of the angle θ for the net force vector:

$$\tan\theta = \frac{F_y}{F_x} = \frac{F_{21}}{F_{31}} = \frac{3.75 \times 10^{-11}\text{ N}}{6.67 \times 10^{-11}\text{ N}} = 0.562$$

Evaluate the angle θ:

$$\theta = \tan^{-1}(0.562) = 29.3°$$

Finalize The result for F shows that the gravitational forces between everyday objects have extremely small magnitudes.

13.2 Free-Fall Acceleration and the Gravitational Force

Because the magnitude of the force acting on a freely falling object of mass m near the Earth's surface is given by Equation 13.4, we can equate this force to that given by Equation 5.6, $F_g = mg$, to obtain

$$mg = G\frac{M_E m}{R_E^2}$$

$$g = G\frac{M_E}{R_E^2} \tag{13.5}$$

Now consider an object of mass m located a distance h above the Earth's surface or a distance r from the Earth's center, where $r = R_E + h$. The magnitude of the gravitational force acting on this object is

$$F_g = G\frac{M_E m}{r^2} = G\frac{M_E m}{(R_E + h)^2}$$

The magnitude of the gravitational force acting on the object at this position is also $F_g = mg$, where g is the value of the free-fall acceleration at the altitude h. Substituting this expression for F_g into the last equation shows that g is given by

$$g = \frac{GM_E}{r^2} = \frac{GM_E}{(R_E + h)^2} \tag{13.6}$$

◀ Variation of g with altitude

Therefore, it follows that g *decreases* with *increasing altitude*. Values of g at various altitudes are listed in Table 13.1. Because an object's weight is mg, we see that as $r \rightarrow \infty$, the weight approaches zero.

TABLE 13.1

Free-Fall Acceleration g at Various Altitudes Above the Earth's Surface

Altitude h (km)	g (m/s²)
1 000	7.33
2 000	5.68
3 000	4.53
4 000	3.70
5 000	3.08
6 000	2.60
7 000	2.23
8 000	1.93
9 000	1.69
10 000	1.49
50 000	0.13
∞	0

Quick Quiz 13.2 Superman stands on top of a very tall mountain and throws a baseball horizontally with a speed such that the baseball goes into a circular orbit around the Earth. While the baseball is in orbit, what is the magnitude of the acceleration of the ball? (a) It depends on how fast the baseball is thrown. (b) It is zero because the ball does not fall to the ground. (c) It is slightly less than 9.80 m/s². (d) It is equal to 9.80 m/s².

EXAMPLE 13.2 Variation of g with Altitude h

The International Space Station operates at an altitude of 350 km. Plans for the final construction show that 4.22×10^6 N of material, measured at the Earth's surface, will have been lifted off the surface by various spacecraft. What is the weight of the space station when in orbit?

SOLUTION

Conceptualize The mass of the space station is fixed; it is independent of its location. Based on the discussion in this section, we realize that the value of g will be reduced at the height of the space station's orbit. Therefore, its weight will be smaller than that at the surface of the Earth.

Categorize This example is a relatively simple substitution problem.

Find the mass of the space station from its weight at the surface of the Earth:

$$m = \frac{F_g}{g} = \frac{4.22 \times 10^6 \text{ N}}{9.80 \text{ m/s}^2} = 4.31 \times 10^5 \text{ kg}$$

Use Equation 13.6 with h = 350 km to find g at the orbital location:

$$g = \frac{GM_E}{(R_E + h)^2}$$

$$= \frac{(6.67 \times 10^{-11} \text{ N}\cdot\text{m}^2/\text{kg}^2)(5.98 \times 10^{24} \text{ kg})}{(6.37 \times 10^6 \text{ m} + 0.350 \times 10^6 \text{ m})^2} = 8.83 \text{ m/s}^2$$

Use this value of g to find the space station's weight in orbit:

$$mg = (4.31 \times 10^5 \text{ kg})(8.83 \text{ m/s}^2) = 3.80 \times 10^6 \text{ N}$$

EXAMPLE 13.3 The Density of the Earth

Using the known radius of the Earth and that $g = 9.80 \text{ m/s}^2$ at the Earth's surface, find the average density of the Earth.

SOLUTION

Conceptualize Assume the Earth is a perfect sphere. The density of material in the Earth varies, but let's adopt a simplified model in which we assume the density to be uniform throughout the Earth. The resulting density is the average density of the Earth.

Categorize This example is a relatively simple substitution problem.

Solve Equation 13.5 for the mass of the Earth:

$$M_E = \frac{gR_E^{\,2}}{G}$$

Substitute this mass into the definition of density (Eq. 1.1):

$$\rho_E = \frac{M_E}{V_E} = \frac{(gR_E^{\,2}/G)}{\frac{4}{3}\pi R_E^{\,3}} = \frac{3}{4}\frac{g}{\pi G R_E}$$

$$= \frac{3}{4}\frac{9.80 \text{ m/s}^2}{\pi(6.67 \times 10^{-11} \text{ N}\cdot\text{m}^2/\text{kg}^2)(6.37 \times 10^6 \text{ m})} = 5.51 \times 10^3 \text{ kg/m}^3$$

What If? What if you were told that a typical density of granite at the Earth's surface were $2.75 \times 10^3 \text{ kg/m}^3$. What would you conclude about the density of the material in the Earth's interior?

Answer Because this value is about half the density we calculated as an average for the entire Earth, we would conclude that the inner core of the Earth has a density much higher than the average value. It is most amazing that the Cavendish experiment—which determines G and can be done on a tabletop—combined with simple free-fall measurements of g, provides information about the core of the Earth!

13.3 Kepler's Laws and the Motion of Planets

Humans have observed the movements of the planets, stars, and other celestial objects for thousands of years. In early history, these observations led scientists to regard the Earth as the center of the Universe. This *geocentric model* was elaborated and formalized by the Greek astronomer Claudius Ptolemy (c. 100–c. 170) in the second century and was accepted for the next 1 400 years. In 1543, Polish astronomer Nicolaus Copernicus (1473–1543) suggested that the Earth and the other planets revolved in circular orbits around the Sun (the *heliocentric model*).

Danish astronomer Tycho Brahe (1546–1601) wanted to determine how the heavens were constructed and pursued a project to determine the positions of both stars and planets. Those observations of the planets and 777 stars visible to the naked eye were carried out with only a large sextant and a compass. (The telescope had not yet been invented.)

German astronomer Johannes Kepler was Brahe's assistant for a short while before Brahe's death, whereupon he acquired his mentor's astronomical data and spent 16 years trying to deduce a mathematical model for the motion of the planets. Such data are difficult to sort out because the moving planets are observed from a moving Earth. After many laborious calculations, Kepler found that Brahe's data on the revolution of Mars around the Sun led to a successful model.

Kepler's complete analysis of planetary motion is summarized in three statements known as **Kepler's laws:**

Art Resource

JOHANNES KEPLER
German astronomer (1571–1630)
Kepler is best known for developing the laws of planetary motion based on the careful observations of Tycho Brahe.

1. All planets move in elliptical orbits with the Sun at one focus.
2. The radius vector drawn from the Sun to a planet sweeps out equal areas in equal time intervals.
3. The square of the orbital period of any planet is proportional to the cube of the semimajor axis of the elliptical orbit.

◄ Kepler's laws

Kepler's First Law

We are familiar with circular orbits of objects around gravitational force centers from our discussions in this chapter. Kepler's first law indicates that the circular orbit is a very special case and elliptical orbits are the general situation. This notion was difficult for scientists of the time to accept because they believed that perfect circular orbits of the planets reflected the perfection of heaven.

Active Figure 13.4 shows the geometry of an ellipse, which serves as our model for the elliptical orbit of a planet. An ellipse is mathematically defined by choosing two points F_1 and F_2, each of which is a called a **focus,** and then drawing a curve through points for which the sum of the distances r_1 and r_2 from F_1 and F_2, respectively, is a constant. The longest distance through the center between points on the ellipse (and passing through each focus) is called the **major axis,** and this distance is $2a$. In Active Figure 13.4, the major axis is drawn along the x direction. The distance a is called the **semimajor axis.** Similarly, the shortest distance through the center between points on the ellipse is called the **minor axis** of length $2b$, where the distance b is the **semiminor axis.** Either focus of the ellipse is located at a distance c from the center of the ellipse, where $a^2 = b^2 + c^2$. In the elliptical orbit of a planet around the Sun, the Sun is at one focus of the ellipse. There is nothing at the other focus.

The **eccentricity** of an ellipse is defined as $e = c/a$, and it describes the general shape of the ellipse. For a circle, $c = 0$, and the eccentricity is therefore zero. The smaller b is compared to a, the shorter the ellipse is along the y direction compared with its extent in the x direction in Active Figure 13.4. As b decreases, c increases and the eccentricity e increases. Therefore, higher values of eccentricity correspond to longer and thinner ellipses. The range of values of the eccentricity for an ellipse is $0 < e < 1$.

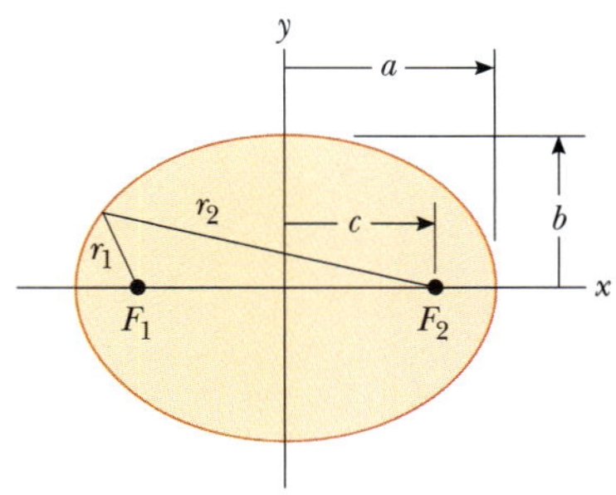

ACTIVE FIGURE 13.4

Plot of an ellipse. The semimajor axis has length a, and the semiminor axis has length b. Each focus is located at a distance c from the center on each side of the center.

Sign in at www.thomsonedu.com and go to ThomsonNOW to move the focal points or enter values for a, b, c, and the eccentricity $e = c/a$ and see the resulting elliptical shape.

PITFALL PREVENTION 13.2

Where Is the Sun?

The Sun is located at one focus of the elliptical orbit of a planet. It is *not* located at the center of the ellipse.

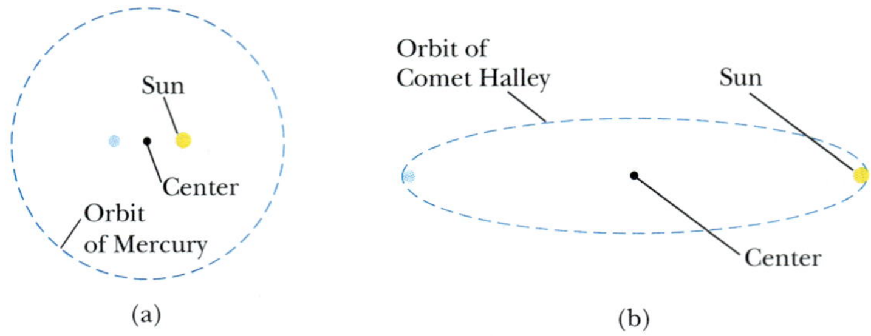

Figure 13.5 (a) The shape of the orbit of Mercury, which has the highest eccentricity ($e = 0.21$) among the eight planets in the solar system. The Sun is located at the large yellow dot, which is a focus of the ellipse. There is nothing physical located at the center (the small dot) or the other focus (the blue dot). (b) The shape of the orbit of Comet Halley.

Eccentricities for planetary orbits vary widely in the solar system. The eccentricity of the Earth's orbit is 0.017, which makes it nearly circular. On the other hand, the eccentricity of Mercury's orbit is 0.21, the highest of the eight planets. Figure 13.5a shows an ellipse with an eccentricity equal to that of Mercury's orbit. Notice that even this highest-eccentricity orbit is difficult to distinguish from a circle, which is one reason Kepler's first law is an admirable accomplishment. The eccentricity of the orbit of Comet Halley is 0.97, describing an orbit whose major axis is much longer than its minor axis, as shown in Figure 13.5b. As a result, Comet Halley spends much of its 76-year period far from the Sun and invisible from the Earth. It is only visible to the naked eye during a small part of its orbit when it is near the Sun.

Now imagine a planet in an elliptical orbit such as that shown in Active Figure 13.4, with the Sun at focus F_2. When the planet is at the far left in the diagram, the distance between the planet and the Sun is $a + c$. At this point, called the *aphelion*, the planet is at its maximum distance from the Sun. (For an object in orbit around the Earth, this point is called the *apogee*.) Conversely, when the planet is at the right end of the ellipse, the distance between the planet and the Sun is $a - c$. At this point, called the *perihelion* (for an Earth orbit, the *perigee*), the planet is at its minimum distance from the Sun.

Kepler's first law is a direct result of the inverse square nature of the gravitational force. We have already discussed circular and elliptical orbits, the allowed shapes of orbits for objects that are *bound* to the gravitational force center. These objects include planets, asteroids, and comets that move repeatedly around the Sun, as well as moons orbiting a planet. There are also *unbound* objects, such as a meteoroid from deep space that might pass by the Sun once and then never return. The gravitational force between the Sun and these objects also varies as the inverse square of the separation distance, and the allowed paths for these objects include parabolas ($e = 1$) and hyperbolas ($e > 1$).

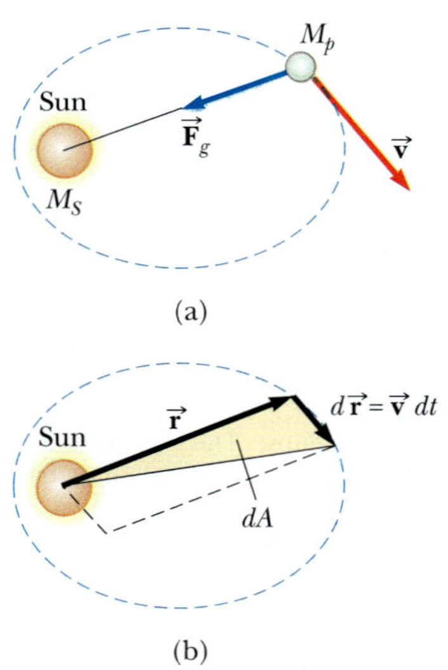

ACTIVE FIGURE 13.6

(a) The gravitational force acting on a planet is directed toward the Sun. (b) As a planet orbits the Sun, the area swept out by the radius vector in a time interval dt is equal to half the area of the parallelogram formed by the vectors $\vec{r}$ and $d\vec{r} = \vec{v}\,dt$.

Sign in at www.thomsonedu.com and go to ThomsonNOW to assign a value of the eccentricity and see the resulting motion of the planet around the Sun.

Kepler's Second Law

Kepler's second law can be shown to be a consequence of angular momentum conservation as follows. Consider a planet of mass M_p moving about the Sun in an elliptical orbit (Active Fig. 13.6a). Let us consider the planet as a system. We model the Sun to be so much more massive than the planet that the Sun does not move. The gravitational force exerted by the Sun on the planet is a central force, always along the radius vector, directed toward the Sun (Active Fig. 13.6a). The torque on the planet due to this central force is clearly zero because $\vec{F}_g$ is parallel to $\vec{r}$.

Recall that the external net torque on a system equals the time rate of change of angular momentum of the system; that is, $\sum \vec{\tau} = d\vec{L}/dt$ (Eq. 11.13). Therefore, because the external torque on the planet is zero, it is modeled as an isolated system for angular momentum and **the angular momentum $\vec{L}$ of the planet is a constant of the motion:**

$$\vec{L} = \vec{r} \times \vec{p} = M_p\vec{r} \times \vec{v} = \text{constant}$$

We can relate this result to the following geometric consideration. In a time interval dt, the radius vector $\vec{\mathbf{r}}$ in Active Figure 13.6b sweeps out the area dA, which equals half the area $|\vec{\mathbf{r}} \times d\vec{\mathbf{r}}|$ of the parallelogram formed by the vectors $\vec{\mathbf{r}}$ and $d\vec{\mathbf{r}}$. Because the displacement of the planet in the time interval dt is given by $d\vec{\mathbf{r}} = \vec{\mathbf{v}}dt$,

$$dA = \tfrac{1}{2}|\vec{\mathbf{r}} \times d\vec{\mathbf{r}}| = \tfrac{1}{2}|\vec{\mathbf{r}} \times \vec{\mathbf{v}}\,dt| = \frac{L}{2M_p}\,dt$$

$$\frac{dA}{dt} = \frac{L}{2M_p} \tag{13.7}$$

where L and M_p are both constants. This result shows that that **the radius vector from the Sun to any planet sweeps out equal areas in equal times.**

This conclusion is a result of the gravitational force being a central force, which in turn implies that angular momentum of the planet is constant. Therefore, the law applies to *any* situation that involves a central force, whether inverse square or not.

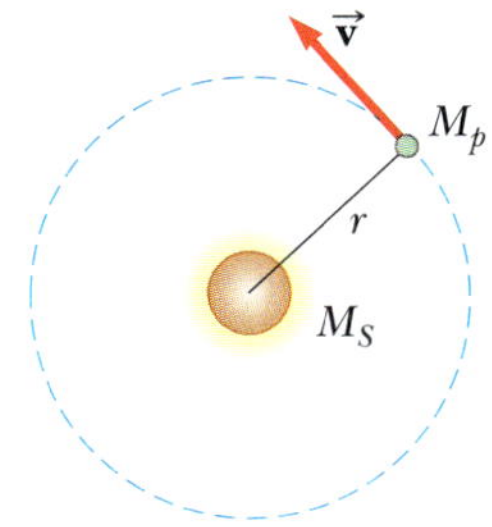

Figure 13.7 A planet of mass M_p moving in a circular orbit around the Sun. The orbits of all planets except Mercury are nearly circular.

Kepler's Third Law

Kepler's third law can be predicted from the inverse-square law for circular orbits. Consider a planet of mass M_p that is assumed to be moving about the Sun (mass M_S) in a circular orbit as in Figure 13.7. Because the gravitational force provides the centripetal acceleration of the planet as it moves in a circle, we use Newton's second law for a particle in uniform circular motion,

$$F_g = \frac{GM_SM_p}{r^2} = M_pa = \frac{M_pv^2}{r}$$

The orbital speed of the planet is $2\pi r/T$, where T is the period; therefore, the preceding expression becomes

$$\frac{GM_S}{r^2} = \frac{(2\pi r/T)^2}{r}$$

$$T^2 = \left(\frac{4\pi^2}{GM_S}\right)r^3 = K_Sr^3$$

where K_S is a constant given by

$$K_S = \frac{4\pi^2}{GM_S} = 2.97 \times 10^{-19}\ \mathrm{s^2/m^3}$$

This equation is also valid for elliptical orbits if we replace r with the length a of the semimajor axis (Active Fig. 13.4):

$$T^2 = \left(\frac{4\pi^2}{GM_S}\right)a^3 = K_Sa^3 \tag{13.8}$$

◀ Kepler's third law

Equation 13.8 is Kepler's third law. Because the semimajor axis of a circular orbit is its radius, this equation is valid for both circular and elliptical orbits. Notice that the constant of proportionality K_S is independent of the mass of the planet. Equation 13.8 is therefore valid for *any* planet.[2] If we were to consider the orbit of a satellite such as the Moon about the Earth, the constant would have a different value, with the Sun's mass replaced by the Earth's mass, that is, $K_E = 4\pi^2/GM_E$.

Table 13.2 is a collection of useful data for planets and other objects in the solar system. The far-right column verifies that the ratio T^2/r^3 is constant for all objects orbiting the Sun. The small variations in the values in this column are the result of uncertainties in the data measured for the periods and semimajor axes of the objects.

Recent astronomical work has revealed the existence of a large number of solar system objects beyond the orbit of Neptune. In general, these objects lie in the

[2] Equation 13.8 is indeed a proportion because the ratio of the two quantities T^2 and a^3 is a constant. The variables in a proportion are not required to be limited to the first power only.

TABLE 13.2

Useful Planetary Data

Body	Mass (kg)	Mean Radius (m)	Period of Revolution (s)	Mean Distance from the Sun (m)	$\frac{T^2}{r^3}$ (s^2/m^3)
Mercury	3.18×10^{23}	2.43×10^{6}	7.60×10^{6}	5.79×10^{10}	2.97×10^{-19}
Venus	4.88×10^{24}	6.06×10^{6}	1.94×10^{7}	1.08×10^{11}	2.99×10^{-19}
Earth	5.98×10^{24}	6.37×10^{6}	3.156×10^{7}	1.496×10^{11}	2.97×10^{-19}
Mars	6.42×10^{23}	3.37×10^{6}	5.94×10^{7}	2.28×10^{11}	2.98×10^{-19}
Jupiter	1.90×10^{27}	6.99×10^{7}	3.74×10^{8}	7.78×10^{11}	2.97×10^{-19}
Saturn	5.68×10^{26}	5.85×10^{7}	9.35×10^{8}	1.43×10^{12}	2.99×10^{-19}
Uranus	8.68×10^{25}	2.33×10^{7}	2.64×10^{9}	2.87×10^{12}	2.95×10^{-19}
Neptune	1.03×10^{26}	2.21×10^{7}	5.22×10^{9}	4.50×10^{12}	2.99×10^{-19}
Pluto[a]	$\approx 1.4 \times 10^{22}$	$\approx 1.5 \times 10^{6}$	7.82×10^{9}	5.91×10^{12}	2.96×10^{-19}
Moon	7.36×10^{22}	1.74×10^{6}	—	—	—
Sun	1.991×10^{30}	6.96×10^{8}	—	—	—

[a] In August, 2006, the International Astronomical Union adopted a definition of a planet that separates Pluto from the other eight planets. Pluto is now defined as a "dwarf planet" like the asteroid Ceres.

Kuiper belt, a region that extends from about 30 AU (the orbital radius of Neptune) to 50 AU. (An AU is an *astronomical unit,* equal to the radius of the Earth's orbit.) Current estimates identify at least 70 000 objects in this region with diameters larger than 100 km. The first Kuiper belt object (KBO) is Pluto, discovered in 1930, and formerly classified as a planet. Starting in 1992, many more have been detected, such as Varuna (diameter about 900–1 000 km, discovered in 2000), Ixion (diameter about 900–1 000 km, discovered in 2001), and Quaoar (diameter about 800 km, discovered in 2002). Others do not yet have names, but are currently indicated by their date of discovery, such as 2003 EL61, 2004 DW, and 2005 FY9. One KBO, 2003 UP313, is thought to be larger than Pluto.

A subset of about 1 400 KBOs are called "Plutinos" because, like Pluto, they exhibit a resonance phenomenon, orbiting the Sun two times in the same time interval as Neptune revolves three times. The contemporary application of Kepler's laws and such exotic proposals as planetary angular momentum exchange and migrating planets[3] suggest the excitement of this active area of current research.

Quick Quiz 13.3 An asteroid is in a highly eccentric elliptical orbit around the Sun. The period of the asteroid's orbit is 90 days. Which of the following statements is true about the possibility of a collision between this asteroid and the Earth? (a) There is no possible danger of a collision. (b) There is a possibility of a collision. (c) There is not enough information to determine whether there is danger of a collision.

EXAMPLE 13.4 The Mass of the Sun

Calculate the mass of the Sun noting that the period of the Earth's orbit around the Sun is 3.156×10^7 s and its distance from the Sun is 1.496×10^{11} m.

SOLUTION

Conceptualize Based on Kepler's third law, we realize that the mass of the Sun is related to the orbital size and period of a planet.

Categorize This example is a relatively simple substitution problem.

[3] R. Malhotra, "Migrating Planets," *Scientific American,* **281**(3): 56–63, September 1999.

Solve Equation 13.8 for the mass of the Sun:

$$M_S = \frac{4\pi^2 r^3}{GT^2}$$

Substitute the known values:

$$M_S = \frac{4\pi^2(1.496 \times 10^{11}\ \text{m})^3}{(6.67 \times 10^{-11}\ \text{N}\cdot\text{m}^2/\text{kg}^2)(3.156 \times 10^7\ \text{s})^2} = 1.99 \times 10^{30}\ \text{kg}$$

In Example 13.3, an understanding of gravitational forces enabled us to find out something about the density of the Earth's core, and now we have used this understanding to determine the mass of the Sun!

EXAMPLE 13.5 A Geosynchronous Satellite

Consider a satellite of mass m moving in a circular orbit around the Earth at a constant speed v and at an altitude h above the Earth's surface as illustrated in Figure 13.8.

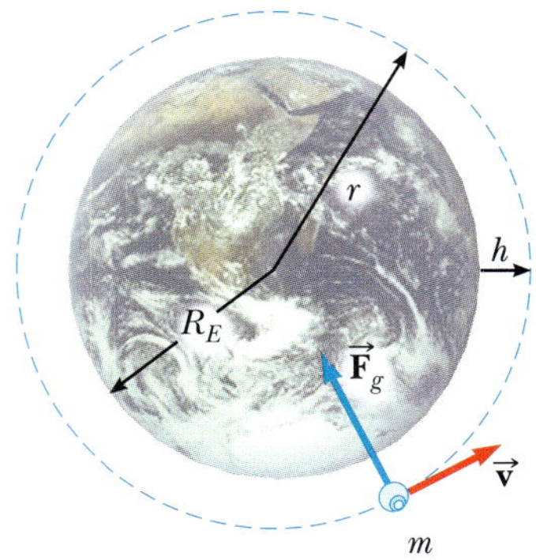

Figure 13.8 (Example 13.5) A satellite of mass m moving around the Earth in a circular orbit of radius r with constant speed v. The only force acting on the satellite is the gravitational force $\vec{F}_g$. (Not drawn to scale.)

(A) Determine the speed of the satellite in terms of G, h, R_E (the radius of the Earth), and M_E (the mass of the Earth).

SOLUTION

Conceptualize Imagine the satellite moving around the Earth in a circular orbit under the influence of the gravitational force.

Categorize The satellite must have a centripetal acceleration. Therefore, we categorize the satellite as a particle under a net force and a particle in uniform circular motion.

Analyze The only external force acting on the satellite is the gravitational force, which acts toward the center of the Earth and keeps the satellite in its circular orbit.

Apply Newton's second law to the satellite:

$$F_g = G\frac{M_E m}{r^2} = ma = m\frac{v^2}{r}$$

Solve for v, noting that the distance r from the center of the Earth to the satellite is $r = R_E + h$:

$$(1)\quad v = \sqrt{\frac{GM_E}{r}} = \sqrt{\frac{GM_E}{R_E + h}}$$

(B) If the satellite is to be *geosynchronous* (that is, appearing to remain over a fixed position on the Earth), how fast is it moving through space?

SOLUTION

To appear to remain over a fixed position on the Earth, the period of the satellite must be 24 h = 86 400 s and the satellite must be in orbit directly over the equator.

Solve Kepler's third law (with $a = r$ and $M_S \rightarrow M_E$) for r:

$$r = \left(\frac{GM_E T^2}{4\pi^2}\right)^{1/3}$$

Substitute numerical values:

$$r = \left[\frac{(6.67 \times 10^{-11}\ \text{N}\cdot\text{m}^2/\text{kg}^2)(5.98 \times 10^{24}\ \text{kg})(86\,400\ \text{s})^2}{4\pi^2}\right]^{1/3}$$

$$= 4.23 \times 10^7\ \text{m}$$

Use Equation (1) to find the speed of the satellite:

$$v = \sqrt{\frac{(6.67 \times 10^{-11}\ \text{N}\cdot\text{m}^2/\text{kg}^2)(5.98 \times 10^{24}\ \text{kg})}{4.23 \times 10^7\ \text{m}}}$$

$$= 3.07 \times 10^3\ \text{m/s}$$

Finalize The value of r calculated here translates to a height of the satellite above the surface of the Earth of almost 36 000 km. Therefore, geosynchronous satellites have the advantage of allowing an earthbound antenna to be aimed in a fixed direction, but there is a disadvantage in that the signals between Earth and the satellite must travel a long distance. It is difficult to use geosynchronous satellites for optical observation of the Earth's surface because of their high altitude.

What If? What if the satellite motion in part (A) were taking place at height h above the surface of another planet more massive than the Earth but of the same radius? Would the satellite be moving at a higher speed or a lower speed than it does around the Earth?

Answer If the planet exerts a larger gravitational force on the satellite due to its larger mass, the satellite must move with a higher speed to avoid moving toward the surface. This conclusion is consistent with the predictions of Equation (1), which shows that because the speed v is proportional to the square root of the mass of the planet, the speed increases as the mass of the planet increases.

13.4 The Gravitational Field

When Newton published his theory of universal gravitation, it was considered a success because it satisfactorily explained the motion of the planets. Since 1687, the same theory has been used to account for the motions of comets, the deflection of a Cavendish balance, the orbits of binary stars, and the rotation of galaxies. Nevertheless, both Newton's contemporaries and his successors found it difficult to accept the concept of a force that acts at a distance. They asked how it was possible for two objects to interact when they were not in contact with each other. Newton himself could not answer that question.

An approach to describing interactions between objects that are not in contact came well after Newton's death. This approach enables us to look at the gravitational interaction in a different way, using the concept of a **gravitational field** that exists at every point in space. When a particle of mass m is placed at a point where the gravitational field is $\vec{\mathbf{g}}$, the particle experiences a force $\vec{\mathbf{F}}_g = m\vec{\mathbf{g}}$. In other words, we imagine that the field exerts a force on the particle rather than consider a direct interaction between two particles. The gravitational field $\vec{\mathbf{g}}$ is defined as

Gravitational field ▶

$$\vec{\mathbf{g}} \equiv \frac{\vec{\mathbf{F}}_g}{m} \qquad (13.9)$$

That is, the gravitational field at a point in space equals the gravitational force experienced by a *test particle* placed at that point divided by the mass of the test particle. We call the object creating the field the *source particle.* (Although the Earth is not a particle, it is possible to show that we can model the Earth as a particle for the purpose of finding the gravitational field that it creates.) Notice that the presence of the test particle is not necessary for the field to exist: the source particle creates the gravitational field. We can detect the presence of the field and measure its strength by placing a test particle in the field and noting the force exerted on it. In essence, we are describing the "effect" that any object (in this case, the Earth) has on the empty space around itself in terms of the force that *would* be present *if* a second object were somewhere in that space.[4]

[4] We shall return to this idea of mass affecting the space around it when we discuss Einstein's theory of gravitation in Chapter 39.

As an example of how the field concept works, consider an object of mass m near the Earth's surface. Because the gravitational force acting on the object has a magnitude $GM_E m/r^2$ (see Eq. 13.4), the field $\vec{\mathbf{g}}$ at a distance r from the center of the Earth is

$$\vec{\mathbf{g}} = \frac{\vec{\mathbf{F}}_g}{m} = -\frac{GM_E}{r^2}\hat{\mathbf{r}} \qquad (13.10)$$

where $\hat{\mathbf{r}}$ is a unit vector pointing radially outward from the Earth and the negative sign indicates that the field points toward the center of the Earth as illustrated in Figure 13.9a. The field vectors at different points surrounding the Earth vary in both direction and magnitude. In a small region near the Earth's surface, the downward field $\vec{\mathbf{g}}$ is approximately constant and uniform as indicated in Figure 13.9b. Equation 13.10 is valid at all points *outside* the Earth's surface, assuming the Earth is spherical. At the Earth's surface, where $r = R_E$, $\vec{\mathbf{g}}$ has a magnitude of 9.80 N/kg. (The unit N/kg is the same as m/s^2.)

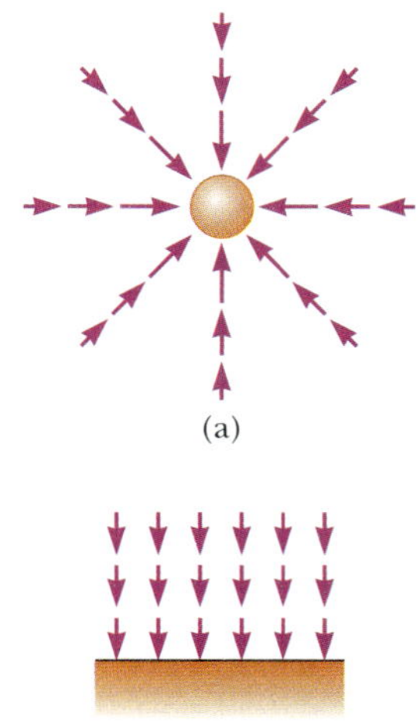

Figure 13.9 (a) The gravitational field vectors in the vicinity of a uniform spherical mass such as the Earth vary in both direction and magnitude. The vectors point in the direction of the acceleration a particle would experience if it were placed in the field. The magnitude of the field vector at any location is the magnitude of the free-fall acceleration at that location. (b) The gravitational field vectors in a small region near the Earth's surface are uniform in both direction and magnitude.

13.5 Gravitational Potential Energy

In Chapter 8, we introduced the concept of gravitational potential energy, which is the energy associated with the configuration of a system of objects interacting via the gravitational force. We emphasized that the gravitational potential energy function mgy for a particle–Earth system is valid only when the particle is near the Earth's surface, where the gravitational force is constant. Because the gravitational force between two particles varies as $1/r^2$, we expect that a more general potential energy function—one that is valid without the restriction of having to be near the Earth's surface—will be different from $U = mgy$.

Recall from Equation 7.26 that the change in the gravitational potential energy of a system associated with a given displacement of a member of the system is defined as the negative of the work done by the gravitational force on that member during the displacement:

$$\Delta U = U_f - U_i = -\int_{r_i}^{r_f} F(r)\,dr \qquad (13.11)$$

We can use this result to evaluate the gravitational potential energy function. Consider a particle of mass m moving between two points Ⓐ and Ⓑ above the Earth's surface (Fig. 13.10). The particle is subject to the gravitational force given by Equation 13.1. We can express this force as

$$F(r) = -\frac{GM_E m}{r^2}$$

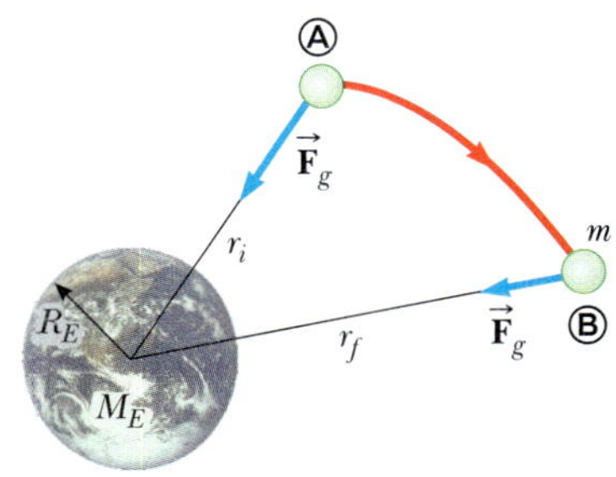

Figure 13.10 As a particle of mass m moves from Ⓐ to Ⓑ above the Earth's surface, the gravitational potential energy of the particle–Earth system changes according to Equation 13.12.

where the negative sign indicates that the force is attractive. Substituting this expression for $F(r)$ into Equation 13.11, we can compute the change in the gravitational potential energy function for the particle–Earth system:

$$U_f - U_i = GM_E m \int_{r_i}^{r_f} \frac{dr}{r^2} = GM_E m\left[-\frac{1}{r}\right]_{r_i}^{r_f}$$

$$U_f - U_i = -GM_E m\left(\frac{1}{r_f} - \frac{1}{r_i}\right) \qquad (13.12)$$

As always, the choice of a reference configuration for the potential energy is completely arbitrary. It is customary to choose the reference configuration for zero potential energy to be the same as that for which the force is zero. Taking $U_i = 0$ at $r_i = \infty$, we obtain the important result

$$U(r) = -\frac{GM_E m}{r} \qquad (13.13)$$

◀ Gravitational potential energy of the Earth–particle system

Figure 13.11 Graph of the gravitational potential energy U versus r for the system of an object above the Earth's surface. The potential energy goes to zero as r approaches infinity.

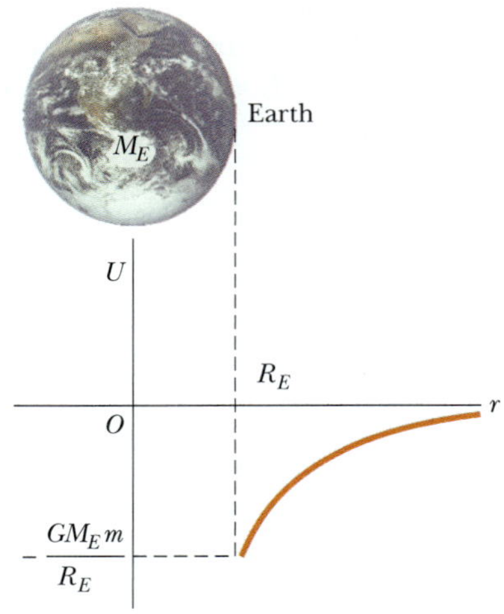

This expression applies when the particle is separated from the center of the Earth by a distance r, provided that $r \geq R_E$. The result is not valid for particles inside the Earth, where $r < R_E$. Because of our choice of U_i, the function U is always negative (Fig. 13.11).

Although Equation 13.13 was derived for the particle–Earth system, it can be applied to any two particles. That is, the gravitational potential energy associated with any pair of particles of masses m_1 and m_2 separated by a distance r is

$$U = -\frac{Gm_1 m_2}{r} \tag{13.14}$$

This expression shows that the gravitational potential energy for any pair of particles varies as $1/r$, whereas the force between them varies as $1/r^2$. Furthermore, the potential energy is negative because the force is attractive and we have chosen the potential energy as zero when the particle separation is infinite. Because the force between the particles is attractive, an external agent must do positive work to increase the separation between them. The work done by the external agent produces an increase in the potential energy as the two particles are separated. That is, U becomes less negative as r increases.

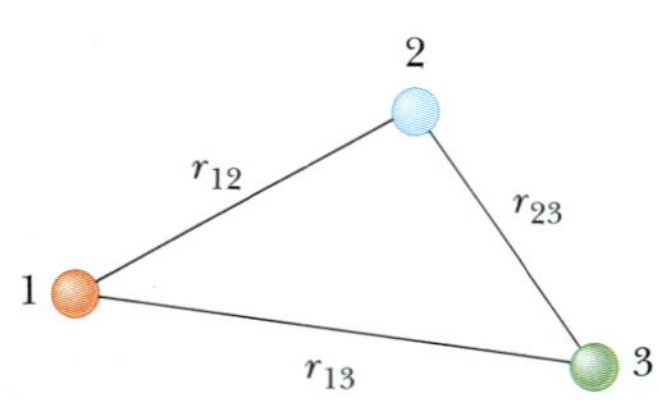

Figure 13.12 Three interacting particles.

When two particles are at rest and separated by a distance r, an external agent has to supply an energy at least equal to $+Gm_1 m_2/r$ to separate the particles to an infinite distance. It is therefore convenient to think of the absolute value of the potential energy as the *binding energy* of the system. If the external agent supplies an energy greater than the binding energy, the excess energy of the system is in the form of kinetic energy of the particles when the particles are at an infinite separation.

We can extend this concept to three or more particles. In this case, the total potential energy of the system is the sum over all pairs of particles. Each pair contributes a term of the form given by Equation 13.14. For example, if the system contains three particles as in Figure 13.12,

$$U_{\text{total}} = U_{12} + U_{13} + U_{23} = -G\left(\frac{m_1 m_2}{r_{12}} + \frac{m_1 m_3}{r_{13}} + \frac{m_2 m_3}{r_{23}}\right) \tag{13.15}$$

The absolute value of U_{total} represents the work needed to separate the particles by an infinite distance.

EXAMPLE 13.6 The Change in Potential Energy

A particle of mass m is displaced through a small vertical distance Δy near the Earth's surface. Show that in this situation the general expression for the change in gravitational potential energy given by Equation 13.12 reduces to the familiar relationship $\Delta U = mg\,\Delta y$.

SOLUTION

Conceptualize Compare the two different situations for which we have developed expressions for gravitational potential energy: (1) a planet and an object that are far apart for which the energy expression is Equation 13.12 and (2) a small object at the surface of a planet for which the energy expression is Equation 7.19. We wish to show that these two expressions are equivalent.

Categorize This example is a substitution problem.

Combine the fractions in Equation 13.12:

$$(1)\quad \Delta U = -GM_E m\left(\frac{1}{r_f} - \frac{1}{r_i}\right) = GM_E m\left(\frac{r_f - r_i}{r_i r_f}\right)$$

Evaluate $r_f - r_i$ and $r_i r_f$ if both the initial and final positions of the particle are close to the Earth's surface:

$$r_f - r_i = \Delta y \qquad r_i r_f \approx R_E^2$$

Substitute these expressions into Equation (1):

$$\Delta U \approx \frac{GM_E m}{R_E^2}\,\Delta y = mg\,\Delta y$$

where $g = GM_E/R_E^2$ (Eq. 13.5).

What If? Suppose you are performing upper-atmosphere studies and are asked by your supervisor to find the height in the Earth's atmosphere at which the "surface equation" $\Delta U = mg\,\Delta y$ gives a 1.0% error in the change in the potential energy. What is this height?

Answer Because the surface equation assumes a constant value for g, it will give a ΔU value that is larger than the value given by the general equation, Equation 13.12.

Set up a ratio reflecting a 1.0% error:

$$\frac{\Delta U_{\text{surface}}}{\Delta U_{\text{general}}} = 1.010$$

Substitute the expressions for each of these changes ΔU:

$$\frac{mg\,\Delta y}{GM_E m(\Delta y/r_i r_f)} = \frac{g r_i r_f}{GM_E} = 1.010$$

Substitute for r_i, r_f, and g from Equation 13.5:

$$\frac{(GM_E/R_E^2)R_E(R_E + \Delta y)}{GM_E} = \frac{R_E + \Delta y}{R_E} = 1 + \frac{\Delta y}{R_E} = 1.010$$

Solve for Δy:

$$\Delta y = 0.010R_E = 0.010(6.37 \times 10^6 \text{ m}) = 6.37 \times 10^4 \text{ m} = 63.7 \text{ km}$$

13.6 Energy Considerations in Planetary and Satellite Motion

Consider an object of mass m moving with a speed v in the vicinity of a massive object of mass M, where $M >> m$. The system might be a planet moving around the Sun, a satellite in orbit around the Earth, or a comet making a one-time flyby of the Sun. If we assume the object of mass M is at rest in an inertial reference frame, the total mechanical energy E of the two-object system when the objects are separated by a distance r is the sum of the kinetic energy of the object of mass m and the potential energy of the system, given by Equation 13.14:

$$E = K + U$$

$$E = \tfrac{1}{2}mv^2 - \frac{GMm}{r} \qquad \textbf{(13.16)}$$

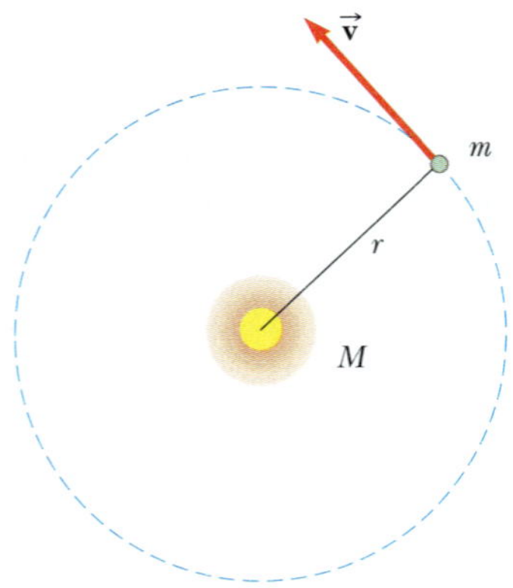

Figure 13.13 An object of mass m moving in a circular orbit about a much larger object of mass M.

Equation 13.16 shows that E may be positive, negative, or zero, depending on the value of v. For a bound system such as the Earth–Sun system, however, E is necessarily *less than zero* because we have chosen the convention that $U \rightarrow 0$ as $r \rightarrow \infty$.

We can easily establish that $E < 0$ for the system consisting of an object of mass m moving in a circular orbit about an object of mass $M >> m$ (Fig. 13.13). Newton's second law applied to the object of mass m gives

$$F_g = \frac{GMm}{r^2} = ma = \frac{mv^2}{r}$$

Multiplying both sides by r and dividing by 2 gives

$$\tfrac{1}{2}mv^2 = \frac{GMm}{2r} \tag{13.17}$$

Substituting this equation into Equation 13.16, we obtain

$$E = \frac{GMm}{2r} - \frac{GMm}{r}$$

Total energy for circular orbits ▶

$$E = -\frac{GMm}{2r} \quad \text{(circular orbits)} \tag{13.18}$$

This result shows that **the total mechanical energy is negative in the case of circular orbits.** Notice that **the kinetic energy is positive and equal to half the absolute value of the potential energy.** The absolute value of E is also equal to the binding energy of the system because this amount of energy must be provided to the system to move the two objects infinitely far apart.

The total mechanical energy is also negative in the case of elliptical orbits. The expression for E for elliptical orbits is the same as Equation 13.18 with r replaced by the semimajor axis length a:

Total energy for elliptical orbits ▶

$$E = -\frac{GMm}{2a} \quad \text{(elliptical orbits)} \tag{13.19}$$

Furthermore, the total energy is constant if we assume the system is isolated. Therefore, as the object of mass m moves from Ⓐ to Ⓑ in Figure 13.10, the total energy remains constant and Equation 13.16 gives

$$E = \tfrac{1}{2}mv_i^2 - \frac{GMm}{r_i} = \tfrac{1}{2}mv_f^2 - \frac{GMm}{r_f} \tag{13.20}$$

Combining this statement of energy conservation with our earlier discussion of conservation of angular momentum, we see that **both the total energy and the total angular momentum of a gravitationally bound, two-object system are constants of the motion.**

Quick Quiz 13.4 A comet moves in an elliptical orbit around the Sun. Which point in its orbit (perihelion or aphelion) represents the highest value of (a) the speed of the comet, (b) the potential energy of the comet–Sun system, (c) the kinetic energy of the comet, and (d) the total energy of the comet–Sun system?

EXAMPLE 13.7 Changing the Orbit of a Satellite

A space transportation vehicle releases a 470-kg communications satellite while in an orbit 280 km above the surface of the Earth. A rocket engine on the satellite boosts it into a geosynchronous orbit. How much energy does the engine have to provide?

SOLUTION

Conceptualize Notice that the height of 280 km is much lower than that for a geosynchronous satellite, 36 000 km, as mentioned in Example 13.5. Therefore, energy must be expended to raise the satellite to this much higher position.

Categorize This example is a substitution problem.

Find the initial radius of the satellite's orbit when it is still in the shuttle's cargo bay:

$$r_i = R_E + 280 \text{ km} = 6.65 \times 10^6 \text{ m}$$

Use Equation 13.18 to find the difference in energies for the satellite–Earth system with the satellite at the initial and final radii:

$$\Delta E = E_f - E_i = -\frac{GM_E m}{2r_f} - \left(-\frac{GM_E m}{2r_i}\right) = -\frac{GM_E m}{2}\left(\frac{1}{r_f} - \frac{1}{r_i}\right)$$

Substitute numerical values, using $r_f = 4.23 \times 10^7$ m from Example 13.5:

$$\Delta E = -\frac{(6.67 \times 10^{-11} \text{ N}\cdot\text{m}^2/\text{kg}^2)(5.98 \times 10^{24} \text{ kg})(470 \text{ kg})}{2}$$

$$\times \left(\frac{1}{4.23 \times 10^7 \text{ m}} - \frac{1}{6.65 \times 10^6 \text{ m}}\right)$$

$$= 1.19 \times 10^{10} \text{ J}$$

which is the energy equivalent of 89 gal of gasoline. NASA engineers must account for the changing mass of the spacecraft as it ejects burned fuel, something we have not done here. Would you expect the calculation that includes the effect of this changing mass to yield a greater or a lesser amount of energy required from the engine?

Escape Speed

Suppose an object of mass m is projected vertically upward from the Earth's surface with an initial speed v_i as illustrated in Figure 13.14. We can use energy considerations to find the minimum value of the initial speed needed to allow the object to move infinitely far away from the Earth. Equation 13.16 gives the total energy of the system at any point. At the surface of the Earth, $v = v_i$ and $r = r_i = R_E$. When the object reaches its maximum altitude, $v = v_f = 0$ and $r = r_f = r_{max}$. Because the total energy of the object–Earth system is conserved, substituting these conditions into Equation 13.20 gives

$$\frac{1}{2}mv_i^2 - \frac{GM_E m}{R_E} = -\frac{GM_E m}{r_{max}}$$

Solving for v_i^2 gives

$$v_i^2 = 2GM_E\left(\frac{1}{R_E} - \frac{1}{r_{max}}\right) \qquad \textbf{(13.21)}$$

For a given maximum altitude $h = r_{max} - R_E$, we can use this equation to find the required initial speed.

We are now in a position to calculate **escape speed,** which is the minimum speed the object must have at the Earth's surface to approach an infinite separation distance from the Earth. Traveling at this minimum speed, the object continues to move farther and farther away from the Earth as its speed asymptotically approaches zero. Letting $r_{max} \to \infty$ in Equation 13.21 and taking $v_i = v_{esc}$ gives

$$v_{esc} = \sqrt{\frac{2GM_E}{R_E}} \qquad \textbf{(13.22)}$$

This expression for v_{esc} is independent of the mass of the object. In other words, a spacecraft has the same escape speed as a molecule. Furthermore, the result is independent of the direction of the velocity and ignores air resistance.

If the object is given an initial speed equal to v_{esc}, the total energy of the system is equal to zero. Notice that when $r \to \infty$, the object's kinetic energy and the potential energy of the system are both zero. If v_i is greater than v_{esc}, the total energy of the system is greater than zero and the object has some residual kinetic energy as $r \to \infty$.

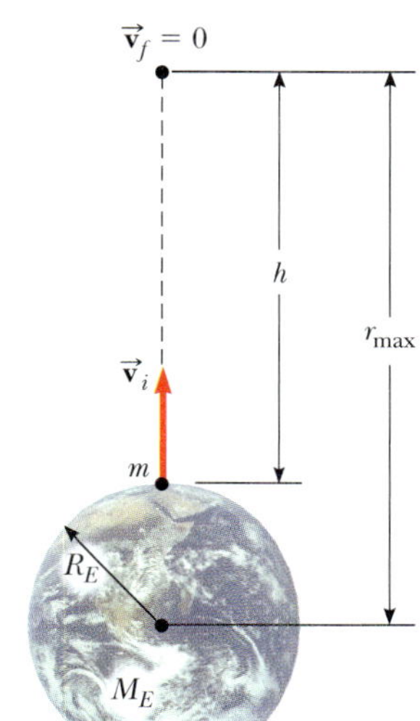

Figure 13.14 An object of mass m projected upward from the Earth's surface with an initial speed v_i reaches a maximum altitude h.

PITFALL PREVENTION 13.3

You Can't Really Escape

Although Equation 13.22 provides the "escape speed" from the Earth, *complete* escape from the Earth's gravitational influence is impossible because the gravitational force is of infinite range. No matter how far away you are, you will always feel some gravitational force due to the Earth.

EXAMPLE 13.8 Escape Speed of a Rocket

Calculate the escape speed from the Earth for a 5 000-kg spacecraft and determine the kinetic energy it must have at the Earth's surface to move infinitely far away from the Earth.

SOLUTION

Conceptualize Imagine projecting the spacecraft from the Earth's surface so that it moves farther and farther away, traveling more and more slowly, with its speed approaching zero. Its speed will never reach zero, however, so the object will never turn around and come back.

Categorize This example is a substitution problem.

Use Equation 13.22 to find the escape speed:

$$v_{\text{esc}} = \sqrt{\frac{2GM_E}{R_E}} = \sqrt{\frac{2(6.67 \times 10^{-11}\ \text{N}\cdot\text{m}^2/\text{kg}^2)(5.98 \times 10^{24}\ \text{kg})}{6.37 \times 10^6\ \text{m}}}$$

$$= 1.12 \times 10^4\ \text{m/s}$$

Evaluate the kinetic energy of the spacecraft from Equation 7.16:

$$K = \tfrac{1}{2}mv_{\text{esc}}^2 = \tfrac{1}{2}(5.00 \times 10^3\ \text{kg})(1.12 \times 10^4\ \text{m/s})^2$$

$$= 3.14 \times 10^{11}\ \text{J}$$

The calculated escape speed corresponds to about 25 000 mi/h. The kinetic energy of the spacecraft is equivalent to the energy released by the combustion of about 2 300 gal of gasoline.

What If? What if you want to launch a 1 000-kg spacecraft at the escape speed? How much energy would that require?

Answer In Equation 13.22, the mass of the object moving with the escape speed does not appear. Therefore, the escape speed for the 1 000-kg spacecraft is the same as that for the 5 000-kg spacecraft. The only change in the kinetic energy is due to the mass, so the 1 000-kg spacecraft requires one-fifth of the energy of the 5 000-kg spacecraft:

$$K = \tfrac{1}{5}(3.14 \times 10^{11}\ \text{J}) = 6.28 \times 10^{10}\ \text{J}$$

Equations 13.21 and 13.22 can be applied to objects projected from any planet. That is, in general, the escape speed from the surface of any planet of mass M and radius R is

$$v_{\text{esc}} = \sqrt{\frac{2GM}{R}} \tag{13.23}$$

Escape speeds for the planets, the Moon, and the Sun are provided in Table 13.3. The values vary from 2.3 km/s for the Moon to about 618 km/s for the Sun. These results, together with some ideas from the kinetic theory of gases (see Chapter 21), explain why some planets have atmospheres and others do not. As we shall see later, at a given temperature the average kinetic energy of a gas molecule depends only on the mass of the molecule. Lighter molecules, such as hydrogen and helium, have a higher average speed than heavier molecules at the same temperature. When the average speed of the lighter molecules is not much less than the escape speed of a planet, a significant fraction of them have a chance to escape.

This mechanism also explains why the Earth does not retain hydrogen molecules and helium atoms in its atmosphere but does retain heavier molecules, such as oxygen and nitrogen. On the other hand, the very large escape speed for Jupiter enables that planet to retain hydrogen, the primary constituent of its atmosphere.

TABLE 13.3

Escape Speeds from the Surfaces of the Planets, Moon, and Sun

Planet	v_{esc} (km/s)
Mercury	4.3
Venus	10.3
Earth	11.2
Mars	5.0
Jupiter	60
Saturn	36
Uranus	22
Neptune	24
Moon	2.3
Sun	618

Black Holes

In Example 11.7, we briefly described a rare event called a supernova, the catastrophic explosion of a very massive star. The material that remains in the central core of such an object continues to collapse, and the core's ultimate fate depends on its mass. If the core has a mass less than 1.4 times the mass of our Sun, it gradually cools down and ends its life as a white dwarf star. If the core's mass is greater than this value, however, it may collapse further due to gravitational forces. What remains is a neutron star, discussed in Example 11.7, in which the mass of a star is compressed to a radius of about 10 km. (On the Earth, a teaspoon of this material would weigh about 5 billion tons!)

An even more unusual star death may occur when the core has a mass greater than about three solar masses. The collapse may continue until the star becomes a very small object in space, commonly referred to as a **black hole.** In effect, black holes are remains of stars that have collapsed under their own gravitational force. If an object such as a spacecraft comes close to a black hole, the object experiences an extremely strong gravitational force and is trapped forever.

The escape speed for a black hole is very high because of the concentration of the star's mass into a sphere of very small radius (see Eq. 13.23). If the escape speed exceeds the speed of light c, radiation from the object (such as visible light) cannot escape and the object appears to be black (hence the origin of the terminology "black hole"). The critical radius R_S at which the escape speed is c is called the **Schwarzschild radius** (Fig. 13.15). The imaginary surface of a sphere of this radius surrounding the black hole is called the **event horizon,** which is the limit of how close you can approach the black hole and hope to escape.

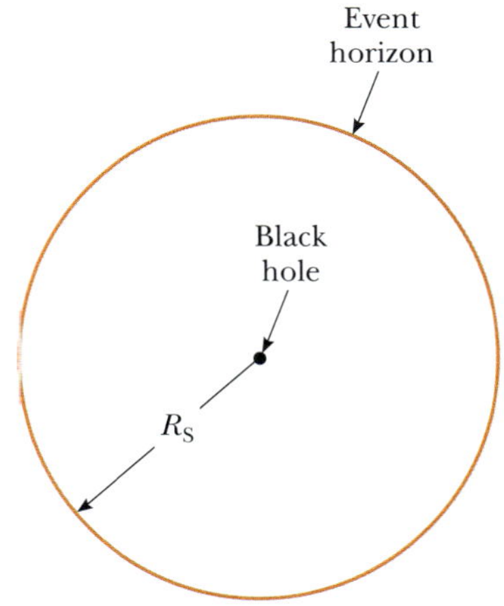

Figure 13.15 A black hole. The distance R_S equals the Schwarzschild radius. Any event occurring within the boundary of radius R_S, called the event horizon, is invisible to an outside observer.

Although light from a black hole cannot escape, light from events taking place near the black hole should be visible. For example, it is possible for a binary star system to consist of one normal star and one black hole. Material surrounding the ordinary star can be pulled into the black hole, forming an **accretion disk** around the black hole as suggested in Figure 13.16. Friction among particles in the accretion disk results in transformation of mechanical energy into internal energy. As a result, the temperature of the material above the event horizon rises. This high-temperature material emits a large amount of radiation, extending well into the x-ray region of the electromagnetic spectrum. These x-rays are characteristic of a black hole. Several possible candidates for black holes have been identified by observation of these x-rays.

There is also evidence that supermassive black holes exist at the centers of galaxies, with masses very much larger than the Sun. (There is strong evidence of a supermassive black hole of mass 2–3 million solar masses at the center of our galaxy.) Theoretical models for these bizarre objects predict that jets of material should be evident along the rotation axis of the black hole. Figure 13.17 (page 380) shows a Hubble Space Telescope photograph of galaxy M87. The jet of material coming from this galaxy is believed to be evidence for a supermassive black hole at the center of the galaxy.

Figure 13.16 A binary star system consisting of an ordinary star on the left and a black hole on the right. Matter pulled from the ordinary star forms an accretion disk around the black hole, in which matter is raised to very high temperatures, resulting in the emission of x-rays.

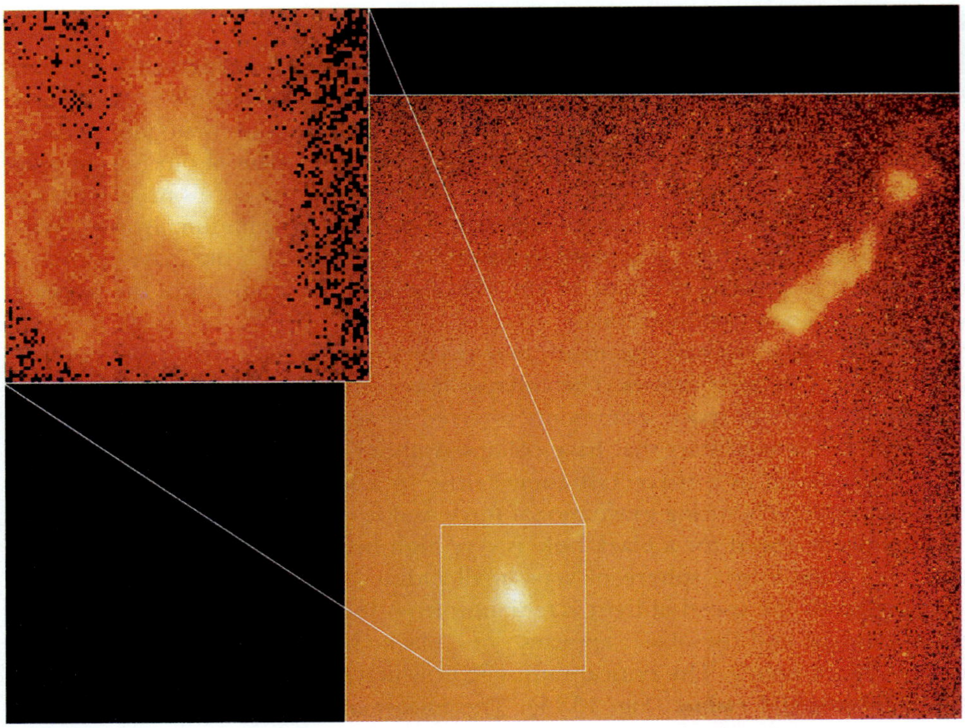

Figure 13.17 Hubble Space Telescope images of the galaxy M87. The inset shows the center of the galaxy. The wider view shows a jet of material moving away from the center of the galaxy toward the upper right of the figure at about one tenth of the speed of light. Such jets are believed to be evidence of a supermassive black hole at the galaxy center.

Summary

ThomsonNOW™ Sign in at **www.thomsonedu.com** and go to ThomsonNOW to take a practice test for this chapter.

DEFINITIONS

The **gravitational field** at a point in space is defined as the gravitational force experienced by any test particle located at that point divided by the mass of the test particle:

$$\vec{\mathbf{g}} \equiv \frac{\vec{\mathbf{F}}_g}{m} \tag{13.9}$$

CONCEPTS AND PRINCIPLES

Newton's law of universal gravitation states that the gravitational force of attraction between any two particles of masses m_1 and m_2 separated by a distance r has the magnitude

$$F_g = G\frac{m_1 m_2}{r^2} \tag{13.1}$$

where $G = 6.673 \times 10^{-11}\ \text{N}\cdot\text{m}^2/\text{kg}^2$ is the **universal gravitational constant.** This equation enables us to calculate the force of attraction between masses under many circumstances.

An object at a distance h above the Earth's surface experiences a gravitational force of magnitude mg, where g is the free-fall acceleration at that elevation:

$$g = \frac{GM_E}{r^2} = \frac{GM_E}{(R_E + h)^2} \tag{13.6}$$

In this expression, M_E is the mass of the Earth and R_E is its radius. Therefore, the weight of an object decreases as the object moves away from the Earth's surface.

(*continued*)

Kepler's laws of planetary motion state:

1. All planets move in elliptical orbits with the Sun at one focus.
2. The radius vector drawn from the Sun to a planet sweeps out equal areas in equal time intervals.
3. The square of the orbital period of any planet is proportional to the cube of the semimajor axis of the elliptical orbit.

Kepler's third law can be expressed as

$$T^2 = \left(\frac{4\pi^2}{GM_S}\right)a^3 \qquad \textbf{(13.8)}$$

where M_S is the mass of the Sun and a is the semimajor axis. For a circular orbit, a can be replaced in Equation 13.8 by the radius r. Most planets have nearly circular orbits around the Sun.

The **gravitational potential energy** associated with two particles separated by a distance r is

$$U = -\frac{Gm_1m_2}{r} \qquad \textbf{(13.14)}$$

where U is taken to be zero as $r \to \infty$.

If an isolated system consists of an object of mass m moving with a speed v in the vicinity of a massive object of mass M, the total energy E of the system is the sum of the kinetic and potential energies:

$$E = \tfrac{1}{2}mv^2 - \frac{GMm}{r} \qquad \textbf{(13.16)}$$

The total energy of the system is a constant of the motion. If the object moves in an elliptical orbit of semimajor axis a around the massive object and $M >> m$, the total energy of the system is

$$E = -\frac{GMm}{2a} \qquad \textbf{(13.19)}$$

For a circular orbit, this same equation applies with $a = r$.

The **escape speed** for an object projected from the surface of a planet of mass M and radius R is

$$v_{esc} = \sqrt{\frac{2GM}{R}} \qquad \textbf{(13.23)}$$

Questions

☐ denotes answer available in *Student Solutions Manual/Study Guide;* **O** denotes objective question

1. **O** Rank the magnitudes of the following gravitational forces from largest to smallest. If two forces are equal, show their equality in your list. (a) The force exerted by a 2-kg object on a 3-kg object 1 m away. (b) The force exerted by a 2-kg object on a 9-kg object 1 m away. (c) The force exerted by a 2-kg object on a 9-kg object 2 m away. (d) The force exerted by a 9-kg object on a 2-kg object 2 m away. (e) The force exerted by a 4-kg object on another 4-kg object 2 m away.

2. **O** The gravitational force exerted on an astronaut on the Earth's surface is 650 N directed downward. When she is in the International Space Station, what is the gravitational force on her? (a) several times larger (b) slightly larger (c) precisely the same (d) slightly smaller (e) several times smaller (f) nearly but not exactly zero (g) precisely zero (h) up instead of down

3. **O** Imagine that nitrogen and other atmospheric gases were more soluble in water so that the atmosphere of the Earth were entirely absorbed by the oceans. Atmospheric pressure would then be zero, and outer space would start at the planet's surface. Would the Earth then have a gravitational field? (a) yes; at the surface it would be larger in magnitude than 9.8 N/kg (b) yes, essentially the same as the current value (c) yes, somewhat less than 9.8 N/kg (d) yes, much less than 9.8 N/kg (e) no

4. The gravitational force exerted by the Sun on you is downward into the Earth at night and upward into the sky during the day. If you had a sensitive enough bathroom scale, would you expect to weigh more at night than during the day? Note also that you are farther away from the Sun at night than during the day. Would you expect to weigh less?

5. **O** Suppose the gravitational acceleration at the surface of a certain satellite A of Jupiter is 2 m/s^2. Satellite B has twice the mass and twice the radius of satellite A. What is the gravitational acceleration at its surface? (a) 16 m/s^2 (b) 8 m/s^2 (c) 4 m/s^2 (d) 2 m/s^2 (e) 1 m/s^2 (f) 0.5 m/s^2 (g) 0.25 m/s^2

6. **O** A satellite originally moves in a circular orbit of radius R around the Earth. Suppose it is moved into a circular orbit of radius $4R$. **(i)** What does the force exerted on the satellite then become? (a) 16 times larger (b) 8 times larger (c) 4 times larger (d) 2 times larger (e) unchanged (f) 1/2 as large (g) 1/4 as large (h) 1/8 as large (i) 1/16 as large **(ii)** What happens to the speed of the satellite? Choose from the same possibilities (a) through (i). **(iii)** What happens to its period? Choose from the same possibilities (a) through (i).

7. **O** The vernal equinox and the autumnal equinox are associated with two points 180° apart in the Earth's orbit.

That is, the Earth is on precisely opposite sides of the Sun when it passes through these two points. From the vernal equinox, 185.4 days elapse before the autumnal equinox. Only 179.8 days elapse from the autumnal equinox until the next vernal equinox. In the year 2007, for example, the vernal equinox is 8 minutes after midnight Greenwich Mean Time on March 21, 2007, and the autumnal equinox is 9:51 p.m. September 23. Why is the interval from the March to the September equinox (which contains the summer solstice) longer than the interval from the September to the March equinox, rather than being equal to that interval? (a) They are really the same, but the Earth spins faster during the "summer" interval, so the days are shorter. (b) Over the "summer" interval the Earth moves slower because it is farther from the Sun. (c) Over the March-to-September interval the Earth moves slower because it is closer to the Sun. (d) The Earth has less kinetic energy when it is warmer. (e) The Earth has less orbital angular momentum when it is warmer. (f) Other objects do work to speed up and slow down the Earth's orbital motion.

8. A satellite in orbit around the Earth is not truly traveling through a vacuum. Rather, it moves through very thin air. Does the resulting air friction cause the satellite to slow down?

9. **O** A system consists of five particles. How many terms appear in the expression for the total gravitational potential energy? (a) 4 (b) 5 (c) 10 (d) 20 (e) 25 (f) 120

10. Explain why it takes more fuel for a spacecraft to travel from the Earth to the Moon than for the return trip. Estimate the difference.

11. **O** Rank the following quantities of energy from the largest to the smallest. State if any are equal. (a) the absolute value of the average potential energy of the Sun–Earth system (b) the average kinetic energy of the Earth in its orbital motion relative to the Sun (c) the absolute value of the total energy of the Sun–Earth system

12. Why don't we put a geosynchronous weather satellite in orbit around the 45th parallel? Wouldn't such a satellite be more useful in the United States than one in orbit around the equator?

13. Explain why the force exerted on a particle by a uniform sphere must be directed toward the center of the sphere. Would this statement be true if the mass distribution of the sphere were not spherically symmetric?

14. At what position in its elliptical orbit is the speed of a planet a maximum? At what position is the speed a minimum?

15. You are given the mass and radius of planet X. How would you calculate the free-fall acceleration on the surface of this planet?

16. If a hole could be dug to the center of the Earth, would the force on an object of mass m still obey Equation 13.1 there? What do you think the force on m would be at the center of the Earth?

17. In his 1798 experiment, Cavendish was said to have "weighed the Earth." Explain this statement.

18. Is the gravitational force a conservative or a nonconservative force? Each *Voyager* spacecraft was accelerated toward escape speed from the Sun by the gravitational force exerted by Jupiter on the spacecraft. Does the interaction of the spacecraft with Jupiter meet the definition of an elastic collision? How could the spacecraft be moving faster after the collision?

Problems

WebAssign The Problems from this chapter may be assigned online in WebAssign.

ThomsonNOW™ Sign in at **www.thomsonedu.com** and go to ThomsonNOW to assess your understanding of this chapter's topics with additional quizzing and conceptual questions.

1, 2, 3 denotes straightforward, intermediate, challenging; □ denotes full solution available in *Student Solutions Manual/Study Guide;* ▲ denotes coached solution with hints available at **www.thomsonedu.com;** ▭ denotes developing symbolic reasoning; ● denotes asking for qualitative reasoning; ▪ denotes computer useful in solving problem

Section 13.1 Newton's Law of Universal Gravitation

1. Determine the order of magnitude of the gravitational force that you exert on another person 2 m away. In your solution, state the quantities you measure or estimate and their values.

2. Two ocean liners, each with a mass of 40 000 metric tons, are moving on parallel courses 100 m apart. What is the magnitude of the acceleration of one of the liners toward the other due to their mutual gravitational attraction? Model the ships as particles.

3. A 200-kg object and a 500-kg object are separated by 0.400 m. (a) Find the net gravitational force exerted by these objects on a 50.0-kg object placed midway between them. (b) At what position (other than an infinitely remote one) can the 50.0-kg object be placed so as to experience a net force of zero?

4. Two objects attract each other with a gravitational force of magnitude 1.00×10^{-8} N when separated by 20.0 cm. If the total mass of the two objects is 5.00 kg, what is the mass of each?

5. Three uniform spheres of mass 2.00 kg, 4.00 kg, and 6.00 kg are placed at the corners of a right triangle as shown in Figure P13.5. Calculate the resultant gravitational force on the 4.00-kg object, assuming the spheres are isolated from the rest of the Universe.

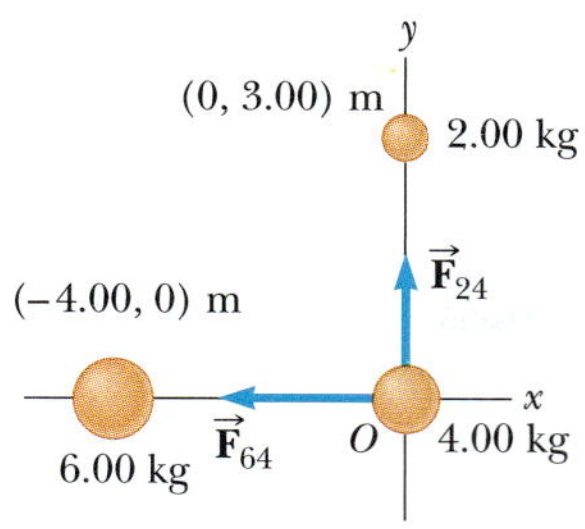

Figure P13.5

6. ● During a solar eclipse, the Moon, Earth, and Sun all lie on the same line, with the Moon between the Earth and the Sun. (a) What force is exerted by the Sun on the Moon? (b) What force is exerted by the Earth on the Moon? (c) What force is exerted by the Sun on the Earth? (d) Compare the answers to parts (a) and (b). Why doesn't the Sun capture the Moon away from the Earth?

7. ▲ In introductory physics laboratories, a typical Cavendish balance for measuring the gravitational constant G uses lead spheres with masses of 1.50 kg and 15.0 g whose centers are separated by about 4.50 cm. Calculate the gravitational force between these spheres, treating each as a particle located at the center of the sphere.

8. ● A student proposes to measure the gravitational constant G by suspending two spherical objects from the ceiling of a tall cathedral and measuring the deflection of the cables from the vertical. Draw a free-body diagram of one of the objects. Assume two 100.0-kg objects are suspended at the lower ends of cables 45.00 m long and the cables are attached to the ceiling 1.000 m apart. What is the separation of the objects? Is there more than one equilibrium separation distance? Explain.

Section 13.2 Free-Fall Acceleration and the Gravitational Force

9. ▲ When a falling meteoroid is at a distance above the Earth's surface of 3.00 times the Earth's radius, what is its acceleration due to the Earth's gravitation?

10. **Review problem.** Miranda, a satellite of Uranus, is shown in Figure P13.10a. It can be modeled as a sphere of radius 242 km and mass 6.68×10^{19} kg. (a) Find the free-fall acceleration on its surface. (b) A cliff on Miranda is 5.00 km high. It appears on the limb at the 11 o'clock position in Figure P13.10a and is magnified in Figure P13.10b. A devotee of extreme sports runs horizontally off the top of the cliff at 8.50 m/s. For what time interval is he in flight? (Or is he in orbit?) (c) How far from the base of the vertical cliff does he strike the icy surface of Miranda? (d) What is his vector impact velocity?

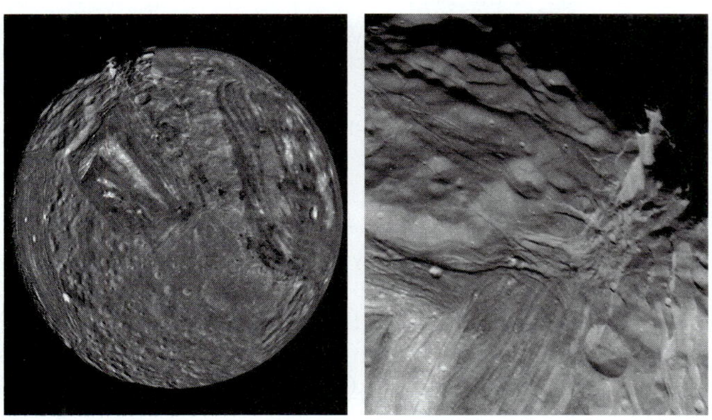

Courtesy of NASA/JPL

(a) (b)

Figure P13.10

11. The free-fall acceleration on the surface of the Moon is about one-sixth that on the surface of the Earth. The radius of the Moon is about $0.250R_E$. Find the ratio of their average densities, ρ_{Moon}/ρ_{Earth}.

Section 13.3 Kepler's Laws and the Motion of Planets

12. ● A particle of mass m moves along a straight line with constant speed in the x direction, a distance b from the x axis (Fig. P13.12). Does the particle possess any angular momentum about the origin? Explain why the amount of its angular momentum should change or should stay constant. Show that Kepler's second law is satisfied by showing that the two shaded triangles in the figure have the same area when $t_4 - t_3 = t_2 - t_1$.

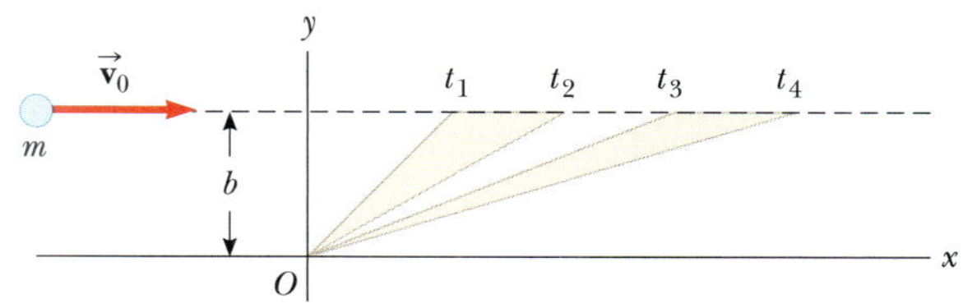

Figure P13.12

13. Plaskett's binary system consists of two stars that revolve in a circular orbit about a center of mass midway between them. This statement implies that the masses of the two stars are equal (Fig. P13.13). Assume the orbital speed of each star is 220 km/s and the orbital period of each is 14.4 days. Find the mass M of each star. (For comparison, the mass of our Sun is 1.99×10^{30} kg.)

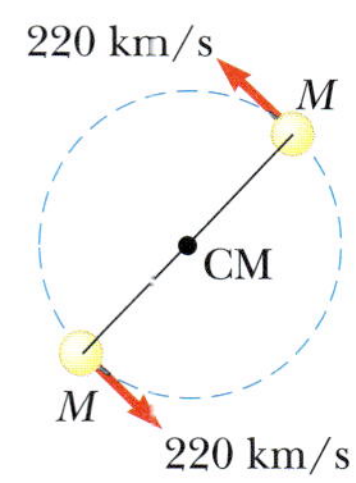

Figure P13.13

14. Comet Halley (Fig. P13.14) approaches the Sun to within 0.570 AU, and its orbital period is 75.6 years. (AU is the symbol for astronomical unit, where 1 AU = 1.50×10^{11} m is the mean Earth–Sun distance.) How far from the

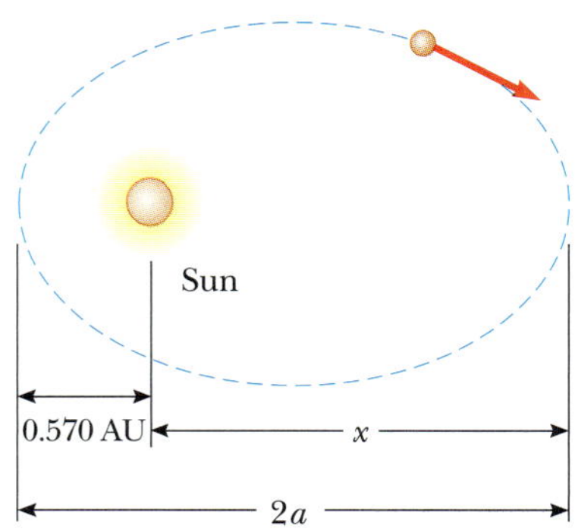

Figure P13.14

2 = intermediate; 3 = challenging; □ = SSM/SG; ▲ = ThomsonNOW; = symbolic reasoning; ● = qualitative reasoning

Sun will Halley's comet travel before it starts its return journey?

15. ▲ Io, a satellite of Jupiter, has an orbital period of 1.77 days and an orbital radius of 4.22×10^5 km. From these data, determine the mass of Jupiter.

16. Two planets X and Y travel counterclockwise in circular orbits about a star as shown in Figure P13.16. The radii of their orbits are in the ratio 3:1. At one moment, they are aligned as shown in Figure P13.16a, making a straight line with the star. During the next five years the angular displacement of planet X is 90.0° as shown in Figure P13.16b. Where is planet Y at this moment?

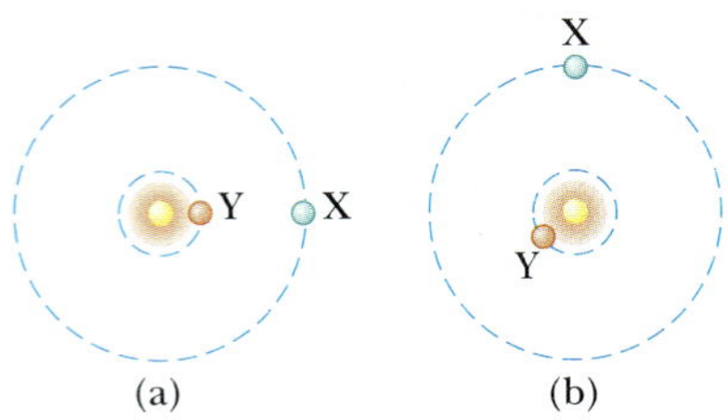

Figure P13.16

17. A synchronous satellite, which always remains above the same point on a planet's equator, is put in orbit around Jupiter to study the famous red spot. Jupiter rotates once every 9.84 h. Use the data of Table 13.2 to find the altitude of the satellite.

18. Neutron stars are extremely dense objects formed from the remnants of supernova explosions. Many rotate very rapidly. Suppose the mass of a certain spherical neutron star is twice the mass of the Sun and its radius is 10.0 km. Determine the greatest possible angular speed it can have so that the matter at the surface of the star on its equator is just held in orbit by the gravitational force.

19. Suppose the Sun's gravity were switched off. Objects in the solar system would leave their orbits and fly away in straight lines as described by Newton's first law. Would Mercury ever be farther from the Sun than Pluto? If so, find how long it would take for Mercury to achieve this passage. If not, give a convincing argument that Pluto is always farther from the Sun.

20. ● Given that the period of the Moon's orbit about the Earth is 27.32 d and the nearly constant distance between the center of the Earth and the center of the Moon is 3.84×10^8 m, use Equation 13.8 to calculate the mass of the Earth. Why is the value you calculate a bit too large?

Section 13.4 The Gravitational Field

21. Three objects of equal mass are located at three corners of a square of edge length ℓ as shown in Figure P13.21. Find the gravitational field at the fourth corner due to these objects.

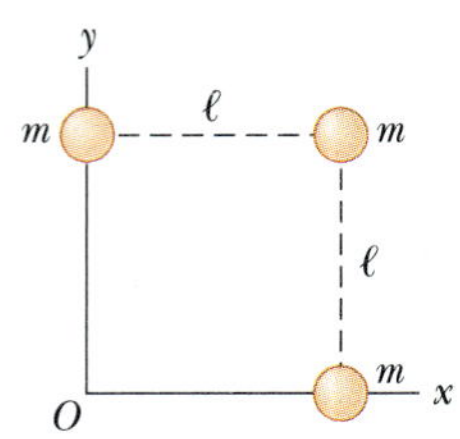

Figure P13.21

22. A spacecraft in the shape of a long cylinder has a length of 100 m, and its mass with occupants is 1 000 kg. It has strayed too close to a black hole having a mass 100 times that of the Sun (Fig. P13.22). The nose of the spacecraft points toward the black hole, and the distance between the nose and the center of the black hole is 10.0 km. (a) Determine the total force on the spacecraft. (b) What is the difference in the gravitational fields acting on the occupants in the nose of the ship and on those in the rear of the ship, farthest from the black hole? This difference in accelerations grows rapidly as the ship approaches the black hole. It puts the body of the ship under extreme tension and eventually tears it apart.

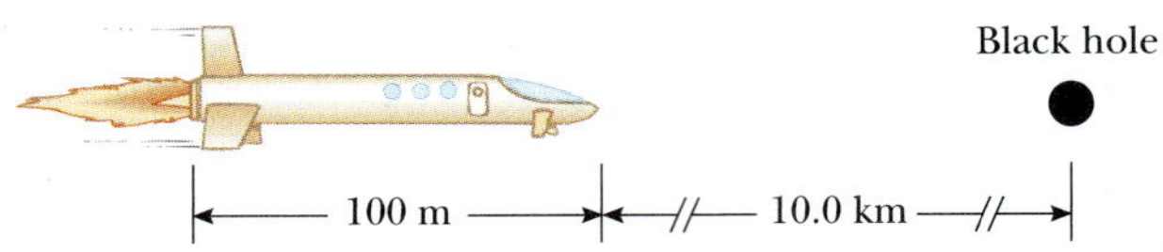

Figure P13.22

23. ● (a) Compute the vector gravitational field at a point P on the perpendicular bisector of the line joining two objects of equal mass separated by a distance $2a$ as shown in Figure P13.23. (b) Explain physically why the field should approach zero as $r \to 0$. (c) Prove mathematically that the answer to part (a) behaves in this way. (d) Explain physically why the magnitude of the field should approach $2GM/r^2$ as $r \to \infty$. (e) Prove mathematically that the answer to part (a) behaves correctly in this limit.

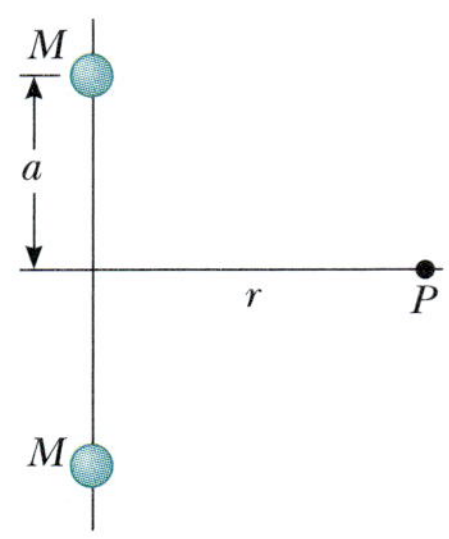

Figure P13.23

Section 13.5 Gravitational Potential Energy

In problems 24–39, assume $U = 0$ at $r = \infty$.

24. A satellite of the Earth has a mass of 100 kg and is at an altitude of 2.00×10^6 m. (a) What is the potential energy of the satellite–Earth system? (b) What is the magnitude of the gravitational force exerted by the Earth on the satellite? (c) **What If?** What force, if any, does the satellite exert on the Earth?

25. After our Sun exhausts its nuclear fuel, its ultimate fate may be to collapse to a *white dwarf* state. In this state, it would have approximately the same mass as it has now but a radius equal to the radius of the Earth. Calculate (a) the average density of the white dwarf, (b) the surface free-fall acceleration, and (c) the gravitational potential energy associated with a 1.00-kg object at its surface.

26. At the Earth's surface a projectile is launched straight up at a speed of 10.0 km/s. To what height will it rise? Ignore air resistance and the rotation of the Earth.

27. ● A system consists of three particles, each of mass 5.00 g, located at the corners of an equilateral triangle with sides of 30.0 cm. (a) Calculate the potential energy of the system. (b) Assume the particles are released simultaneously. Describe the subsequent motion of each. Will any collisions take place? Explain.

28. How much work is done by the Moon's gravitational field on a 1 000-kg meteor as it comes in from outer space and impacts on the Moon's surface?

29. An object is released from rest at an altitude h above the surface of the Earth. (a) Show that its speed at a distance r from the Earth's center, where $R_E \le r \le R_E + h$, is

$$v = \sqrt{2GM_E\left(\frac{1}{r} - \frac{1}{R_E + h}\right)}$$

(b) Assume the release altitude is 500 km. Perform the integral

$$\Delta t = \int_i^f dt = -\int_i^f \frac{dr}{v}$$

to find the time of fall as the object moves from the release point to the Earth's surface. The negative sign appears because the object is moving opposite to the radial direction, so its speed is $v = -dr/dt$. Perform the integral numerically.

Section 13.6 Energy Considerations in Planetary and Satellite Motion

30. (a) What is the minimum speed, relative to the Sun, necessary for a spacecraft to escape the solar system, if it starts at the Earth's orbit? (b) *Voyager 1* achieved a maximum speed of 125 000 km/h on its way to photograph Jupiter. Beyond what distance from the Sun is this speed sufficient to escape the solar system?

31. ▲ A space probe is fired as a projectile from the Earth's surface with an initial speed of 2.00×10^4 m/s. What will its speed be when it is very far from the Earth? Ignore friction and the rotation of the Earth.

32. ● A 1 000-kg satellite orbits the Earth at a constant altitude of 100 km. How much energy must be added to the system to move the satellite into a circular orbit with altitude 200 km? Discuss the changes in kinetic energy, potential energy, and total energy.

33. A "treetop satellite" moves in a circular orbit just above the surface of a planet, assumed to offer no air resistance. Show that its orbital speed v and the escape speed from the planet are related by the expression $v_{esc} = \sqrt{2}v$.

34. ● Ganymede is the largest of Jupiter's moons. Consider a rocket on the surface of Ganymede, at the point farthest from the planet (Fig. P13.34). Does the presence of Ganymede make Jupiter exert a larger, smaller, or same-size force on the rocket compared with the force it would exert if Ganymede were not interposed? Determine the escape speed for the rocket from the planet–satellite system. The radius of Ganymede is 2.64×10^6 m, and its mass is 1.495×10^{23} kg. The distance between Jupiter and Ganymede is 1.071×10^9 m, and the mass of Jupiter is 1.90×10^{27} kg. Ignore the motion of Jupiter and Ganymede as they revolve about their center of mass.

Figure P13.34

35. A satellite of mass 200 kg is placed in Earth orbit at a height of 200 km above the surface. (a) Assuming a circular orbit, how long does the satellite take to complete one orbit? (b) What is the satellite's speed? (c) Starting from the satellite on the Earth's surface, what is the minimum energy input necessary to place this satellite in orbit? Ignore air resistance, but include the effect of the planet's daily rotation.

36. ● A satellite of mass m, originally on the surface of the Earth, is placed into Earth orbit at an altitude h. (a) Assuming a circular orbit, how long does the satellite take to complete one orbit? (b) What is the satellite's speed? (c) What is the minimum energy input necessary to place this satellite in orbit? Ignore air resistance, but include the effect of the planet's daily rotation. At what location on the Earth's surface and in what direction should the satellite be launched to minimize the required energy investment? Represent the mass and radius of the Earth as M_E and R_E.

37. An object is fired vertically upward from the surface of the Earth (of radius R_E) with an initial speed v_i that is comparable to but less than the escape speed v_{esc}. (a) Show that the object attains a maximum height h given by

$$h = \frac{R_E v_i^2}{v_{esc}^2 - v_i^2}$$

(b) A space vehicle is launched vertically upward from the Earth's surface with an initial speed of 8.76 km/s, which is less than the escape speed of 11.2 km/s. What maximum height does it attain? (c) A meteorite falls toward the Earth. It is essentially at rest with respect to the Earth when it is at a height of 2.51×10^7 m. With what speed does the meteorite strike the Earth? (d) **What If?** Assume a baseball is tossed up with an initial speed that is very small compared with the escape speed. Show that the equation from part (a) is consistent with Equation 4.12.

38. A satellite moves around the Earth in a circular orbit of radius r. (a) What is the speed v_0 of the satellite? Suddenly, an explosion breaks the satellite into two pieces, with masses m and $4m$. Immediately after the explosion the smaller piece of mass m is stationary with respect to the Earth and falls directly toward the Earth. (b) What is the speed v_i of the larger piece immediately after the explosion? (c) Because of the increase in its speed, this larger piece now moves in a new elliptical orbit. Find its distance away from the center of the Earth when it reaches the other end of the ellipse.

39. A comet of mass 1.20×10^{10} kg moves in an elliptical orbit around the Sun. Its distance from the Sun ranges between 0.500 AU and 50.0 AU. (a) What is the eccentricity of its orbit? (b) What is its period? (c) At aphelion what is the

potential energy of the comet–Sun system? *Note:* 1 AU = one astronomical unit = the average distance from Sun to Earth = 1.496×10^{11} m.

Additional Problems

40. ● Assume you are agile enough to run across a horizontal surface at 8.50 m/s, independently of the value of the gravitational field. What would be (a) the radius and (b) the mass of an airless spherical asteroid of uniform density 1.10×10^3 kg/m^3 on which you could launch yourself into orbit by running? (c) What would be your period? (d) Would your running significantly affect the rotation of the asteroid? Explain.

41. The Solar and Heliospheric Observatory (SOHO) spacecraft has a special orbit, chosen so that its view of the Sun is never eclipsed and it is always close enough to the Earth to transmit data easily. It moves in a near-circle around the Sun that is smaller than the Earth's circular orbit. Its period, however, is not less than 1 yr but rather is just equal to 1 yr. It is always located between the Earth and the Sun along the line joining them. Both objects exert gravitational forces on the observatory. Show that its distance from the Earth must be between 1.47×10^9 m and 1.48×10^9 m. In 1772, Joseph Louis Lagrange determined theoretically the special location allowing this orbit. The SOHO spacecraft took this position on February 14, 1996. *Suggestion:* Use data that are precise to four digits. The mass of the Earth is 5.983×10^{24} kg.

42. Let Δg_M represent the difference in the gravitational fields produced by the Moon at the points on the Earth's surface nearest to and farthest from the Moon. Find the fraction $\Delta g_M/g$, where g is the Earth's gravitational field. (This difference is responsible for the occurrence of the *lunar tides* on the Earth.)

43. Review problem. Two identical hard spheres, each of mass m and radius r, are released from rest in otherwise empty space with their centers separated by the distance R. They are allowed to collide under the influence of their gravitational attraction. (a) Show that the magnitude of the impulse received by each sphere before they make contact is given by $[Gm^3(1/2r - 1/R)]^{1/2}$. (b) **What If?** Find the magnitude of the impulse each receives during their contact if they collide elastically.

44. Two spheres having masses M and $2M$ and radii R and $3R$, respectively, are released from rest when the distance between their centers is $12R$. How fast will each sphere be moving when they collide? Assume the two spheres interact only with each other.

45. A ring of matter is a familiar structure in planetary and stellar astronomy. Examples include Saturn's rings and a ring nebula. Consider a large uniform ring having a mass of 2.36×10^{20} kg and radius 1.00×10^8 m. An object of mass 1 000 kg is placed at a point A on the axis of the ring, 2.00×10^8 m from the center of the ring (Fig. P13.45). When the object is released, the attraction of the ring makes the object move along the axis toward the center of the ring (point B). (a) Calculate the gravitational potential energy of the object–ring system when the object is at A. (b) Calculate the gravitational potential energy of the system when the object is at B. (c) Calculate the object's speed as it passes through B.

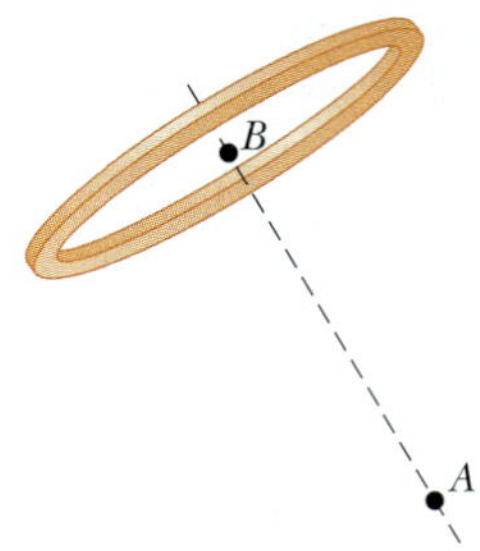

Figure P13.45

46. (a) Show that the rate of change of the free-fall acceleration with distance above the Earth's surface is

$$\frac{dg}{dr} = -\frac{2GM_E}{R_E^3}$$

This rate of change over distance is called a *gradient.* (b) Assuming that h is small in comparison to the radius of the Earth, show that the difference in free-fall acceleration between two points separated by vertical distance h is

$$|\Delta g| = \frac{2GM_E h}{R_E^3}$$

(c) Evaluate this difference for $h = 6.00$ m, a typical height for a two-story building.

47. As an astronaut, you observe a small planet to be spherical. After landing on the planet, you set off, walking always straight ahead, and find yourself returning to your spacecraft from the opposite side after completing a lap of 25.0 km. You hold a hammer and a falcon feather at a height of 1.40 m, release them, and observe that they fall together to the surface in 29.2 s. Determine the mass of the planet.

48. A certain quaternary star system consists of three stars, each of mass m, moving in the same circular orbit of radius r about a central star of mass M. The stars orbit in the same sense and are positioned one-third of a revolution apart from one another. Show that the period of each of the three stars is

$$T = 2\pi\sqrt{\frac{r^3}{G(M + m/\sqrt{3})}}$$

49. Review problem. A cylindrical habitat in space 6.00 km in diameter and 30 km long has been proposed (by G. K. O'Neill, 1974). Such a habitat would have cities, land, and lakes on the inside surface and air and clouds in the center. They would all be held in place by rotation of the cylinder about its long axis. How fast would the cylinder have to rotate to imitate the Earth's gravitational field at the walls of the cylinder?

50. ● Many people assume air resistance acting on a moving object will always make the object slow down. It can, however, actually be responsible for making the object speed up. Consider a 100-kg Earth satellite in a circular orbit at an altitude of 200 km. A small force of air resistance makes the satellite drop into a circular orbit with an altitude of 100 km. (a) Calculate its initial speed. (b) Calculate its final speed in this process. (c) Calculate the initial energy of the satellite–Earth system. (d) Calculate the final energy of the system. (e) Show that the system has lost mechanical energy and find the amount of the loss due to friction. (f) What force makes the satellite's speed increase? You will find a free-body diagram to be useful in explaining your answer.

51. ▲ Two hypothetical planets of masses m_1 and m_2 and radii r_1 and r_2, respectively, are nearly at rest when they are an infinite distance apart. Because of their gravitational attraction, they head toward each other on a collision course. (a) When their center-to-center separation is d, find expressions for the speed of each planet and for their relative speed. (b) Find the kinetic energy of each planet just before they collide, taking $m_1 = 2.00 \times 10^{24}$ kg, $m_2 = 8.00 \times 10^{24}$ kg, $r_1 = 3.00 \times 10^6$ m, and $r_2 = 5.00 \times 10^6$ m. *Note:* Both energy and momentum of the system are conserved.

52. The maximum distance from the Earth to the Sun (at aphelion) is 1.521×10^{11} m, and the distance of closest approach (at perihelion) is 1.471×10^{11} m. The Earth's orbital speed at perihelion is 3.027×10^4 m/s. Determine (a) the Earth's orbital speed at aphelion, (b) the kinetic and potential energies of the Earth–Sun system at perihelion, and (c) the kinetic and potential energies at aphelion. Is the total energy of the system constant? (Ignore the effect of the Moon and other planets.)

53. Studies of the relationship of the Sun to its galaxy—the Milky Way—have revealed that the Sun is located near the outer edge of the galactic disk, about 30 000 ly from the center. The Sun has an orbital speed of approximately 250 km/s around the galactic center. (a) What is the period of the Sun's galactic motion? (b) What is the order of magnitude of the mass of the Milky Way galaxy? Suppose the galaxy is made mostly of stars of which the Sun is typical. What is the order of magnitude of the number of stars in the Milky Way?

54. X-ray pulses from Cygnus X-1, a celestial x-ray source, have been recorded during high-altitude rocket flights. The signals can be interpreted as originating when a blob of ionized matter orbits a black hole with a period of 5.0 ms. If the blob is in a circular orbit about a black hole whose mass is $20M_{\text{Sun}}$, what is the orbit radius?

55. Astronomers detect a distant meteoroid moving along a straight line that, if extended, would pass at a distance $3R_E$ from the center of the Earth, where R_E is the radius of the Earth. What minimum speed must the meteoroid have if the Earth's gravitation is not to deflect the meteoroid to make it strike the Earth?

56. The oldest artificial satellite in orbit is *Vanguard I*, launched March 3, 1958. Its mass is 1.60 kg. In its initial orbit, its minimum distance from the center of the Earth was 7.02 Mm and its speed at this perigee point was 8.23 km/s. (a) Find the total energy of the satellite–Earth system. (b) Find the magnitude of the angular momentum of the satellite. (c) At apogee, find its speed and its distance from the center of the Earth. (d) Find the semimajor axis of its orbit. (e) Determine its period.

57. Two stars of masses M and m, separated by a distance d, revolve in circular orbits about their center of mass (Fig. P13.57). Show that each star has a period given by

$$T^2 = \frac{4\pi^2 d^3}{G(M + m)}$$

Proceed by applying Newton's second law to each star. Note that the center-of-mass condition requires that $Mr_2 = mr_1$, where $r_1 + r_2 = d$.

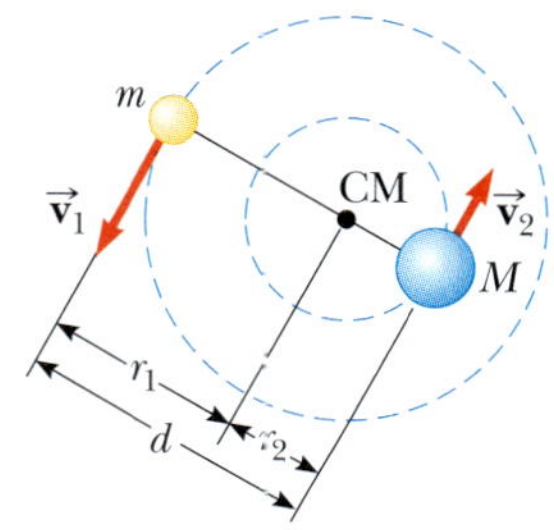

Figure P13.57

58. Show that the minimum period for a satellite in orbit around a spherical planet of uniform density ρ is

$$T_{\min} = \sqrt{\frac{3\pi}{G\rho}}$$

independent of the radius of the planet.

59. Two identical particles, each of mass 1 000 kg, are coasting in free space along the same path. At one instant their separation is 20.0 m and each has precisely the same velocity of $800\hat{\mathbf{i}}$ m/s. What are their velocities when they are 2.00 m apart?

60. (a) Consider an object of mass m, not necessarily small compared with the mass of the Earth, released at a distance of 1.20×10^7 m from the center of the Earth. Assume the objects behave as a pair of particles, isolated from the rest of the Universe. Find the magnitude of the acceleration a_{rel} with which each starts to move relative to the other. Evaluate the acceleration (b) for $m = 5.00$ kg, (c) for $m = 2\,000$ kg, and (d) for $m = 2.00 \times 10^{24}$ kg. (e) Describe the pattern of variation of a_{rel} with m.

61. As thermonuclear fusion proceeds in its core, the Sun loses mass at a rate of 3.64×10^9 kg/s. During the 5 000-yr period of recorded history, by how much has the length of the year changed due to the loss of mass from the Sun? *Suggestions*: Assume the Earth's orbit is circular. No external torque acts on the Earth–Sun system, so angular momentum is conserved. If x is small compared with 1, then $(1 + x)^n$ is nearly equal to $1 + nx$.

Answers to Quick Quizzes

13.1 (e). The gravitational force follows an inverse-square behavior, so doubling the distance causes the force to be one-fourth as large.

13.2 (c). An object in orbit is simply falling while it moves around the Earth. The acceleration of the object is that due to gravity. Because the object was launched from a very tall mountain, the value for g is slightly less than that at the surface.

13.3 (a). From Kepler's third law and the given period, the major axis of the asteroid can be calculated. It is found to be 1.2×10^{11} m. Because this value is smaller than the Earth–Sun distance, the asteroid cannot possibly collide with the Earth.

13.4 (a) Perihelion. Because of conservation of angular momentum, the speed of the comet is highest at its closest position to the Sun. (b) Aphelion. The potential energy of the comet–Sun system is highest when the comet is at its farthest distance from the Sun. (c) Perihelion. The kinetic energy is highest at the point at which the speed of the comet is highest. (d) All points. The total energy of the system is the same regardless of where the comet is in its orbit.

In the Dead Sea, a lake between Jordan and Israel, the high percentage of salt dissolved in the water raises the fluid's density, dramatically increasing the buoyant force on objects in the water. Bathers can kick back and enjoy a good read, dispensing with floating lounge chairs. (© Alison Wright/ Corbis)

14 Fluid Mechanics

Matter is normally classified as being in one of three states: solid, liquid, or gas. From everyday experience we know that a solid has a definite volume and shape, a liquid has a definite volume but no definite shape, and an unconfined gas has neither a definite volume nor a definite shape. These descriptions help us picture the states of matter, but they are somewhat artificial. For example, asphalt and plastics are normally considered solids, but over long time intervals they tend to flow like liquids. Likewise, most substances can be a solid, a liquid, or a gas (or a combination of any of these three), depending on the temperature and pressure. In general, the time interval required for a particular substance to change its shape in response to an external force determines whether we treat the substance as a solid, a liquid, or a gas.

A **fluid** is a collection of molecules that are randomly arranged and held together by weak cohesive forces and by forces exerted by the walls of a container. Both liquids and gases are fluids.

In our treatment of the mechanics of fluids, we'll be applying principles we have already discussed. First, we consider the mechanics of a fluid at rest, that is, *fluid statics*, and then study fluids in motion, that is, *fluid dynamics*.

14.1 Pressure

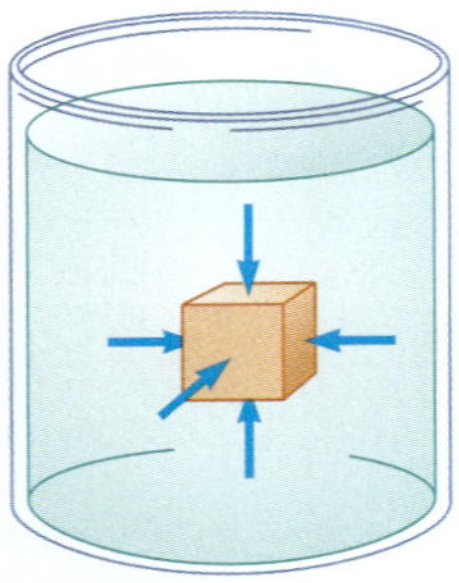

Figure 14.1 At any point on the surface of a submerged object, the force exerted by the fluid is perpendicular to the surface of the object. The force exerted by the fluid on the walls of the container is perpendicular to the walls at all points.

Fluids do not sustain shearing stresses or tensile stresses; therefore, the only stress that can be exerted on an object submerged in a static fluid is one that tends to compress the object from all sides. In other words, the force exerted by a static fluid on an object is always perpendicular to the surfaces of the object as shown in Figure 14.1.

The pressure in a fluid can be measured with the device pictured in Figure 14.2. The device consists of an evacuated cylinder that encloses a light piston connected to a spring. As the device is submerged in a fluid, the fluid presses on the top of the piston and compresses the spring until the inward force exerted by the fluid is balanced by the outward force exerted by the spring. The fluid pressure can be measured directly if the spring is calibrated in advance. If F is the magnitude of the force exerted on the piston and A is the surface area of the piston, the **pressure** P of the fluid at the level to which the device has been submerged is defined as the ratio of the force to the area:

Definition of pressure ▶

$$P \equiv \frac{F}{A} \tag{14.1}$$

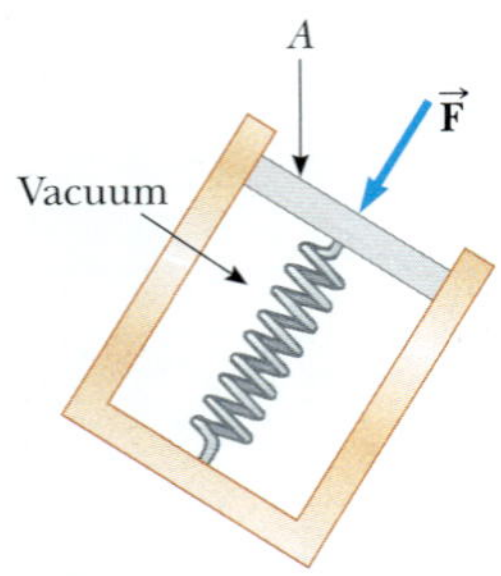

Figure 14.2 A simple device for measuring the pressure exerted by a fluid.

Pressure is a scalar quantity because it is proportional to the magnitude of the force on the piston.

If the pressure varies over an area, the infinitesimal force dF on an infinitesimal surface element of area dA is

$$dF = P\,dA \tag{14.2}$$

where P is the pressure at the location of the area dA. To calculate the total force exerted on a surface of a container, we must integrate Equation 14.2 over the surface.

The units of pressure are newtons per square meter (N/m^2) in the SI system. Another name for the SI unit of pressure is the **pascal** (Pa):

$$1 \text{ Pa} \equiv 1 \text{ N/m}^2 \tag{14.3}$$

For a tactile demonstration of the definition of pressure, hold a tack between your thumb and forefinger, with the point of the tack on your thumb and the head of the tack on your forefinger. Now *gently* press your thumb and forefinger together. Your thumb will begin to feel pain immediately while your forefinger will not. The tack is exerting the same force on both your thumb and forefinger, but the pressure on your thumb is much larger because of the small area over which the force is applied.

PITFALL PREVENTION 14.1

Force and Pressure

Equations 14.1 and 14.2 make a clear distinction between force and pressure. Another important distinction is that *force is a vector* and *pressure is a scalar.* There is no direction associated with pressure, but the direction of the force associated with the pressure is perpendicular to the surface on which the pressure acts.

Quick Quiz 14.1 Suppose you are standing directly behind someone who steps back and accidentally stomps on your foot with the heel of one shoe. Would you be better off if that person were (a) a large, male professional basketball player wearing sneakers or (b) a petite woman wearing spike-heeled shoes?

EXAMPLE 14.1 The Water Bed

The mattress of a water bed is 2.00 m long by 2.00 m wide and 30.0 cm deep.

(A) Find the weight of the water in the mattress.

SOLUTION

Conceptualize Think about carrying a jug of water and how heavy it is. Now imagine a sample of water the size of a water bed. We expect the weight to be relatively large.

Categorize This example is a substitution problem.

Find the volume of the water filling the mattress:

$$V = (2.00\text{ m})(2.00\text{ m})(0.300\text{ m}) = 1.20\text{ m}^3$$

Use Equation 1.1 and the density of fresh water (see Table 14.1) to find the mass of the water bed:

$$M = \rho V = (1\,000\text{ kg/m}^3)(1.20\text{ m}^3) = 1.20 \times 10^3\text{ kg}$$

Find the weight of the bed:

$$Mg = (1.20 \times 10^3\text{ kg})(9.80\text{ m/s}^2) = \boxed{1.18 \times 10^4\text{ N}}$$

which is approximately 2 650 lb. (A regular bed, including mattress, box springs, and metal frame, weighs approximately 300 lb.) Because this load is so great, it is best to place a water bed in the basement or on a sturdy, well-supported floor.

(B) Find the pressure exerted by the water on the floor when the water bed rests in its normal position. Assume the entire lower surface of the bed makes contact with the floor.

SOLUTION

When the water bed is in its normal position, the area in contact with the floor is 4.00 m². Use Equation 14.1 to find the pressure:

$$P = \frac{1.18 \times 10^4\text{ N}}{4.00\text{ m}^2} = \boxed{2.94 \times 10^3\text{ Pa}}$$

What If? What if the water bed is replaced by a 300-lb regular bed that is supported by four legs? Each leg has a circular cross section of radius 2.00 cm. What pressure does this bed exert on the floor?

Answer The weight of the regular bed is distributed over four circular cross sections at the bottom of the legs. Therefore, the pressure is

$$P = \frac{F}{A} = \frac{mg}{4(\pi r^2)} = \frac{300\text{ lb}}{4\pi(0.020\,0\text{ m})^2}\left(\frac{1\text{ N}}{0.225\text{ lb}}\right)$$
$$= 2.65 \times 10^5\text{ Pa}$$

This result is almost 100 times larger than the pressure due to the water bed! The weight of the regular bed, even though it is much less than the weight of the water bed, is applied over the very small area of the four legs. The high pressure on the floor at the feet of a regular bed could cause dents in wood floors or permanently crush carpet pile.

14.2 Variation of Pressure with Depth

As divers well know, water pressure increases with depth. Likewise, atmospheric pressure decreases with increasing altitude; for this reason, aircraft flying at high altitudes must have pressurized cabins for the comfort of the passengers.

TABLE 14.1

Densities of Some Common Substances at Standard Temperature (0°C) and Pressure (Atmospheric)

Substance	ρ (kg/m³)	Substance	ρ (kg/m³)
Air	1.29	Ice	0.917×10^3
Aluminum	2.70×10^3	Iron	7.86×10^3
Benzene	0.879×10^3	Lead	11.3×10^3
Copper	8.92×10^3	Mercury	13.6×10^3
Ethyl alcohol	0.806×10^3	Oak	0.710×10^3
Fresh water	1.00×10^3	Oxygen gas	1.43
Glycerin	1.26×10^3	Pine	0.373×10^3
Gold	19.3×10^3	Platinum	21.4×10^3
Helium gas	1.79×10^{-1}	Seawater	1.03×10^3
Hydrogen gas	8.99×10^{-2}	Silver	10.5×10^3

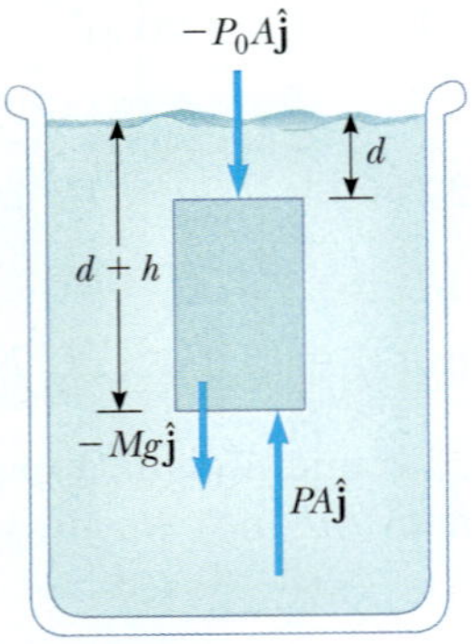

Figure 14.3 A parcel of fluid (darker region) in a larger volume of fluid is singled out. The net force exerted on the parcel of fluid must be zero because it is in equilibrium.

We now show how the pressure in a liquid increases with depth. As Equation 1.1 describes, the *density* of a substance is defined as its mass per unit volume; Table 14.1 lists the densities of various substances. These values vary slightly with temperature because the volume of a substance is dependent on temperature (as shown in Chapter 19). Under standard conditions (at 0°C and at atmospheric pressure), the densities of gases are about $\frac{1}{1\,000}$ the densities of solids and liquids. This difference in densities implies that the average molecular spacing in a gas under these conditions is about ten times greater than that in a solid or liquid.

Now consider a liquid of density ρ at rest as shown in Figure 14.3. We assume ρ is uniform throughout the liquid, which means the liquid is incompressible. Let us select a sample of the liquid contained within an imaginary cylinder of cross-sectional area A extending from depth d to depth $d + h$. The liquid external to our sample exerts forces at all points on the surface of the sample, perpendicular to the surface. The pressure exerted by the liquid on the bottom face of the sample is P, and the pressure on the top face is P_0. Therefore, the upward force exerted by the outside fluid on the bottom of the cylinder has a magnitude PA, and the downward force exerted on the top has a magnitude P_0A. The mass of liquid in the cylinder is $M = \rho V = \rho Ah$; therefore, the weight of the liquid in the cylinder is $Mg = \rho Ahg$. Because the cylinder is in equilibrium, the net force acting on it must be zero. Choosing upward to be the positive y direction, we see that

$$\sum \vec{\mathbf{F}} = PA\hat{\mathbf{j}} - P_0A\hat{\mathbf{j}} - Mg\hat{\mathbf{j}} = 0$$

or

$$PA - P_0A - \rho Ahg = 0$$

Variation of pressure with depth ▶

$$P = P_0 + \rho gh \qquad (14.4)$$

That is, **the pressure P at a depth h below a point in the liquid at which the pressure is P_0 is greater by an amount ρgh.** If the liquid is open to the atmosphere and P_0 is the pressure at the surface of the liquid, then P_0 is atmospheric pressure. In our calculations and working of end-of-chapter problems, we usually take atmospheric pressure to be

$$P_0 = 1.00 \text{ atm} = 1.013 \times 10^5 \text{ Pa}$$

Equation 14.4 implies that the pressure is the same at all points having the same depth, independent of the shape of the container.

Because the pressure in a fluid depends on depth and on the value of P_0, any increase in pressure at the surface must be transmitted to every other point in the fluid. This concept was first recognized by French scientist Blaise Pascal (1623–1662) and is called **Pascal's law: a change in the pressure applied to a fluid is transmitted undiminished to every point of the fluid and to the walls of the container.**

Pascal's law ▶

An important application of Pascal's law is the hydraulic press illustrated in Figure 14.4a. A force of magnitude F_1 is applied to a small piston of surface area A_1. The pressure is transmitted through an incompressible liquid to a larger piston of surface area A_2. Because the pressure must be the same on both sides, $P = F_1/A_1 = F_2/A_2$. Therefore, the force F_2 is greater than the force F_1 by a factor A_2/A_1. By designing a hydraulic press with appropriate areas A_1 and A_2, a large output force can be applied by means of a small input force. Hydraulic brakes, car lifts, hydraulic jacks, and forklifts all make use of this principle (Fig. 14.4b).

Because liquid is neither added to nor removed from the system, the volume of liquid pushed down on the left in Figure 14.4a as the piston moves downward through a displacement Δx_1 equals the volume of liquid pushed up on the right as the right piston moves upward through a displacement Δx_2. That is, $A_1\,\Delta x_1 = A_2\,\Delta x_2$; therefore, $A_2/A_1 = \Delta x_1/\Delta x_2$. We have already shown that $A_2/A_1 = F_2/F_1$. Therefore, $F_2/F_1 = \Delta x_1/\Delta x_2$, so $F_1\,\Delta x_1 = F_2\,\Delta x_2$. Each side of this equation is the work done by the force on its respective piston. Therefore, the work done by $\vec{\mathbf{F}}_1$ on the input piston equals the work done by $\vec{\mathbf{F}}_2$ on the output piston, as it must to conserve energy.

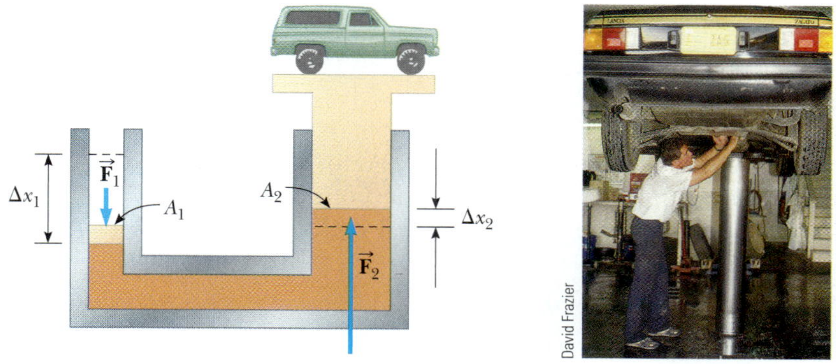

Figure 14.4 (a) Diagram of a hydraulic press. Because the increase in pressure is the same on the two sides, a small force $\vec{\mathbf{F}}_1$ at the left produces a much greater force $\vec{\mathbf{F}}_2$ at the right. (b) A vehicle undergoing repair is supported by a hydraulic lift in a garage.

Quick Quiz 14.2 The pressure at the bottom of a filled glass of water (ρ = 1 000 kg/m^3) is P. The water is poured out, and the glass is filled with ethyl alcohol (ρ = 806 kg/m^3). What is the pressure at the bottom of the glass? (a) smaller than P (b) equal to P (c) larger than P (d) indeterminate

---

EXAMPLE 14.2 The Car Lift

In a car lift used in a service station, compressed air exerts a force on a small piston that has a circular cross section and a radius of 5.00 cm. This pressure is transmitted by a liquid to a piston that has a radius of 15.0 cm. What force must the compressed air exert to lift a car weighing 13 300 N? What air pressure produces this force?

SOLUTION

Conceptualize Review the material just discussed about Pascal's law to understand the operation of a car lift.

Categorize This example is a substitution problem.

Solve $F_1/A_1 = F_2/A_2$ for F_1:

$$F_1 = \left(\frac{A_1}{A_2}\right)F_2 = \frac{\pi(5.00 \times 10^{-2}\text{ m})^2}{\pi(15.0 \times 10^{-2}\text{ m})^2}(1.33 \times 10^4\text{ N})$$

$$= \boxed{1.48 \times 10^3\text{ N}}$$

Use Equation 14.1 to find the air pressure that produces this force:

$$P = \frac{F_1}{A_1} = \frac{1.48 \times 10^3\text{ N}}{\pi(5.00 \times 10^{-2}\text{ m})^2}$$

$$= \boxed{1.88 \times 10^5\text{ Pa}}$$

This pressure is approximately twice atmospheric pressure.

EXAMPLE 14.3 A Pain in Your Ear

Estimate the force exerted on your eardrum due to the water when you are swimming at the bottom of a pool that is 5.0 m deep.

SOLUTION

Conceptualize As you descend in the water, the pressure increases. You may have noticed this increased pressure in your ears while diving in a swimming pool, a pond, or the ocean. We can find the pressure difference exerted on the eardrum from the depth given in the problem; then, after estimating the eardrum's surface area, we can determine the net force the water exerts on it.

Categorize This example is a substitution problem.

The air inside the middle ear is normally at atmospheric pressure P_0. Therefore, to find the net force on the eardrum, we must consider the difference between the total pressure at the bottom of the pool and atmospheric pressure. Let's estimate the surface area of the eardrum to be approximately $1\ \text{cm}^2 = 1 \times 10^{-4}\ \text{m}^2$.

Use Equation 14.4 to find this pressure difference:

$$P_{\text{bot}} - P_0 = \rho g h$$
$$= (1.00 \times 10^3\ \text{kg/m}^3)(9.80\ \text{m/s}^2)(5.0\ \text{m}) = 4.9 \times 10^4\ \text{Pa}$$

Use Equation 14.1 to find the net force on the ear:

$$F = (P_{\text{bot}} - P_0)A = (4.9 \times 10^4\ \text{Pa})(1 \times 10^{-4}\ \text{m}^2) \approx \boxed{5\ \text{N}}$$

Because a force of this magnitude on the eardrum is extremely uncomfortable, swimmers often "pop their ears" while under water, an action that pushes air from the lungs into the middle ear. Using this technique equalizes the pressure on the two sides of the eardrum and relieves the discomfort.

EXAMPLE 14.4 The Force on a Dam

Water is filled to a height H behind a dam of width w (Fig. 14.5). Determine the resultant force exerted by the water on the dam.

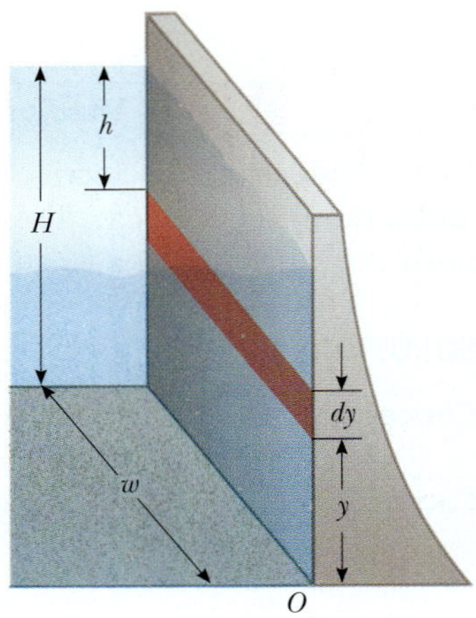

Figure 14.5 (Example 14.4) Water exerts a force on a dam.

SOLUTION

Conceptualize Because pressure varies with depth, we cannot calculate the force simply by multiplying the area by the pressure.

Categorize Because of the variation of pressure with depth, we must use integration to solve this example, so we categorize it as an analysis problem.

Analyze Let's imagine a vertical y axis, with $y = 0$ at the bottom of the dam. We divide the face of the dam into narrow horizontal strips at a distance y above the bottom, such as the red strip in Figure 14.5. The pressure on each such strip is due only to the water; atmospheric pressure acts on both sides of the dam.

Use Equation 14.4 to calculate the pressure due to the water at the depth h:

$$P = \rho g h = \rho g(H - y)$$

Use Equation 14.2 to find the force exerted on the shaded strip of area $dA = w\,dy$:

$$dF = P\,dA = \rho g(H - y)w\,dy$$

Integrate to find the total force on the dam:

$$F = \int P\,dA = \int_0^H \rho g(H - y)w\,dy = \boxed{\tfrac{1}{2}\rho g w H^2}$$

Finalize Notice that the thickness of the dam shown in Figure 14.5 increases with depth. This design accounts for the greater pressure the water exerts on the dam at greater depths.

What If? What if you were asked to find this force without using calculus? How could you determine its value?

Answer We know from Equation 14.4 that the pressure varies linearly with depth. Therefore, the average pressure due to the water over the face of the dam is the average of the pressure at the top and the pressure at the bottom:

$$P_{\text{avg}} = \frac{P_{\text{top}} + P_{\text{bottom}}}{2} = \frac{0 + \rho g H}{2} = \tfrac{1}{2}\rho g H$$

The total force on the dam is equal to the product of the average pressure and the area of the face of the dam:

$$F = P_{\text{avg}}A = \left(\tfrac{1}{2}\rho gH\right)(Hw) = \tfrac{1}{2}\rho gwH^2$$

which is the same result we obtained using calculus.

14.3 Pressure Measurements

During the weather report on a television news program, the *barometric pressure* is often provided. This reading is the current pressure of the atmosphere, which varies over a small range from the standard value provided earlier. How is this pressure measured?

One instrument used to measure atmospheric pressure is the common barometer, invented by Evangelista Torricelli (1608–1647). A long tube closed at one end is filled with mercury and then inverted into a dish of mercury (Fig. 14.6a). The closed end of the tube is nearly a vacuum, so the pressure at the top of the mercury column can be taken as zero. In Figure 14.6a, the pressure at point A, due to the column of mercury, must equal the pressure at point B, due to the atmosphere. If that were not the case, there would be a net force that would move mercury from one point to the other until equilibrium is established. Therefore, $P_0 = \rho_{\text{Hg}}gh$, where ρ_{Hg} is the density of the mercury and h is the height of the mercury column. As atmospheric pressure varies, the height of the mercury column varies, so the height can be calibrated to measure atmospheric pressure. Let us determine the height of a mercury column for one atmosphere of pressure, $P_0 = 1 \text{ atm} = 1.013 \times 10^5 \text{ Pa}$:

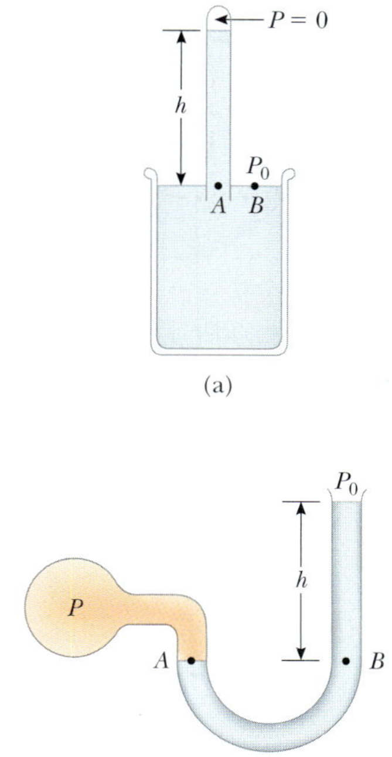

Figure 14.6 Two devices for measuring pressure: (a) a mercury barometer and (b) an open-tube manometer.

$$P_0 = \rho_{\text{Hg}}gh \quad \rightarrow \quad h = \frac{P_0}{\rho_{\text{Hg}}g} = \frac{1.013 \times 10^5 \text{ Pa}}{(13.6 \times 10^3 \text{ kg/m}^3)(9.80 \text{ m/s}^2)} = 0.760 \text{ m}$$

Based on such a calculation, one atmosphere of pressure is defined to be the pressure equivalent of a column of mercury that is exactly 0.760 0 m in height at 0°C.

A device for measuring the pressure of a gas contained in a vessel is the open-tube manometer illustrated in Figure 14.6b. One end of a U-shaped tube containing a liquid is open to the atmosphere, and the other end is connected to a system of unknown pressure P. In an equilibrium situation, the pressures at points A and B must be the same (otherwise, the curved portion of the liquid would experience a net force and would accelerate), and the pressure at A is the unknown pressure of the gas. Therefore, equating the unknown pressure P to the pressure at point B, we see that $P = P_0 + \rho gh$. The difference in pressure $P - P_0$ is equal to ρgh. The pressure P is called the **absolute pressure,** and the difference $P - P_0$ is called the **gauge pressure.** For example, the pressure you measure in your bicycle tire is gauge pressure.

Quick Quiz 14.3 Several common barometers are built, with a variety of fluids. For which of the following fluids will the column of fluid in the barometer be the highest? (a) mercury (b) water (c) ethyl alcohol (d) benzene

14.4 Buoyant Forces and Archimedes's Principle

Have you ever tried to push a beach ball down under water (Fig. 14.7a, page 396)? It is extremely difficult to do because of the large upward force exerted by the water on the ball. The upward force exerted by a fluid on any immersed object is called a **buoyant force.** We can determine the magnitude of a buoyant force by

ARCHIMEDES
Greek Mathematician, Physicist, and Engineer (c. 287–212 BC)
Archimedes was perhaps the greatest scientist of antiquity. He was the first to compute accurately the ratio of a circle's circumference to its diameter, and he also showed how to calculate the volume and surface area of spheres, cylinders, and other geometric shapes. He is well known for discovering the nature of the buoyant force and was also a gifted inventor. One of his practical inventions, still in use today, is Archimedes's screw, an inclined, rotating, coiled tube used originally to lift water from the holds of ships. He also invented the catapult and devised systems of levers, pulleys, and weights for raising heavy loads. Such inventions were successfully used to defend his native city, Syracuse, during a two-year siege by Romans.

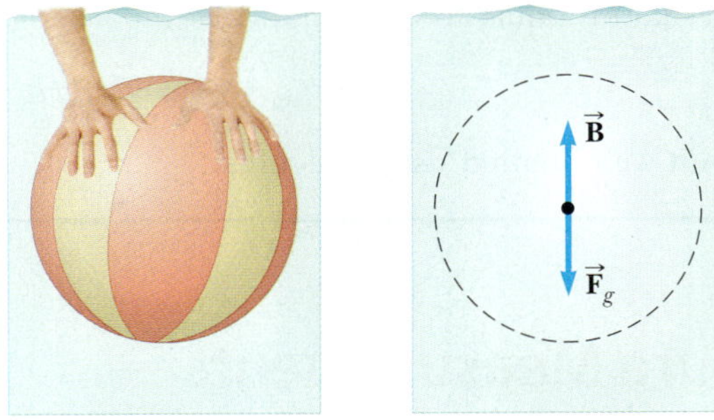

(a) (b)

Figure 14.7 (a) A swimmer pushes a beach ball underwater. (b) The forces on a beach ball–sized parcel of water. The buoyant force $\vec{\mathbf{B}}$ on a beach ball that replaces this parcel is exactly the same as the buoyant force on the parcel.

applying some logic. Imagine a beach ball–sized parcel of water beneath the water surface as in Figure 14.7b. Because this parcel is in equilibrium, there must be an upward force that balances the downward gravitational force on the parcel. This upward force is the buoyant force, and its magnitude is equal to the weight of the water in the parcel. The buoyant force is the resultant force on the parcel due to all forces applied by the fluid surrounding the parcel.

Now imagine replacing the beach ball–sized parcel of water with a beach ball of the same size. The net force applied by the fluid surrounding the beach ball is the same, regardless of whether it is applied to a beach ball or to a parcel of water. Consequently, **the magnitude of the buoyant force on an object always equals the weight of the fluid displaced by the object.** This statement is known as **Archimedes's principle.**

With the beach ball under water, the buoyant force, equal to the weight of a beach ball–sized parcel of water, is much larger than the weight of the beach ball. Therefore, there is a large net upward force, which explains why it is so hard to hold the beach ball under the water. Note that Archimedes's principle does not refer to the makeup of the object experiencing the buoyant force. The object's composition is not a factor in the buoyant force because the buoyant force is exerted by the fluid.

To better understand the origin of the buoyant force, consider a cube immersed in a liquid as in Figure 14.8. According to Equation 14.4, the pressure P_{bot} at the bottom of the cube is greater than the pressure P_{top} at the top by an amount $\rho_{\text{fluid}}gh$, where h is the height of the cube and ρ_{fluid} is the density of the fluid. The pressure at the bottom of the cube causes an *upward* force equal to $P_{\text{bot}}A$, where A is the area of the bottom face. The pressure at the top of the cube causes a *downward* force equal to $P_{\text{top}}A$. The resultant of these two forces is the buoyant force $\vec{\mathbf{B}}$ with magnitude

$$B = (P_{\text{bot}} - P_{\text{top}})A = (\rho_{\text{fluid}}gh)A$$

Buoyant force ▶

$$B = \rho_{\text{fluid}}gV \qquad \textbf{(14.5)}$$

where $V = Ah$ is the volume of the fluid displaced by the cube. Because the product $\rho_{\text{fluid}}V$ is equal to the mass of fluid displaced by the object,

$$B = Mg$$

where Mg is the weight of the fluid displaced by the cube. This result is consistent with our initial statement about Archimedes's principle above, based on the discussion of the beach ball.

Under normal conditions, the weight of a fish is slightly greater than the buoyant force on the fish. Hence, the fish would sink if it did not have some mechanism for adjusting the buoyant force. The fish accomplishes that by internally regulating the size of its air-filled swim bladder to increase its volume and the magnitude of the buoyant force acting on it, according to Equation 14.5. In this manner, fish are able to swim to various depths.

Before we proceed with a few examples, it is instructive to discuss two common situations: a totally submerged object and a floating (partly submerged) object.

Case 1: Totally Submerged Object When an object is totally submerged in a fluid of density ρ_{fluid}, the magnitude of the upward buoyant force is $B = \rho_{\text{fluid}} g V = \rho_{\text{fluid}} g V_{\text{obj}}$, where V_{obj} is the volume of the object. If the object has a mass M and density ρ_{obj}, its weight is equal to $F_g = Mg = \rho_{\text{obj}} g V_{\text{obj}}$, and the net force on the object is $B - F_g = (\rho_{\text{fluid}} - \rho_{\text{obj}}) g V_{\text{obj}}$. Hence, if the density of the object is less than the density of the fluid, the downward gravitational force is less than the buoyant force and the unsupported object accelerates upward (Active Fig. 14.9a). If the density of the object is greater than the density of the fluid, the upward buoyant force is less than the downward gravitational force and the unsupported object sinks (Active Fig. 14.9b). If the density of the submerged object equals the density of the fluid, the net force on the object is zero and the object remains in equilibrium. Therefore, **the direction of motion of an object submerged in a fluid is determined only by the densities of the object and the fluid.**

Case 2: Floating Object Now consider an object of volume V_{obj} and density $\rho_{\text{obj}} < \rho_{\text{fluid}}$ in static equilibrium floating on the surface of a fluid, that is, an object that is only *partially* submerged (Active Fig. 14.10). In this case, the upward buoyant force is balanced by the downward gravitational force acting on the object. If V_{fluid} is the volume of the fluid displaced by the object (this volume is the same as the volume of that part of the object beneath the surface of the fluid), the buoyant force has a magnitude $B = \rho_{\text{fluid}} g V_{\text{fluid}}$. Because the weight of the object is $F_g = Mg = \rho_{\text{obj}} g V_{\text{obj}}$ and because $F_g = B$, we see that $\rho_{\text{fluid}} g V_{\text{fluid}} = \rho_{\text{obj}} g V_{\text{obj}}$, or

$$\frac{V_{\text{fluid}}}{V_{\text{obj}}} = \frac{\rho_{\text{obj}}}{\rho_{\text{fluid}}} \tag{14.6}$$

This equation shows that **the fraction of the volume of a floating object that is below the fluid surface is equal to the ratio of the density of the object to that of the fluid.**

Quick Quiz 14.4 You are shipwrecked and floating in the middle of the ocean on a raft. Your cargo on the raft includes a treasure chest full of gold that you found before your ship sank, and the raft is just barely afloat. To keep you floating as high as possible in the water, should you (a) leave the treasure chest on top of the raft, (b) secure the treasure chest to the underside of the raft, or (c) hang the treasure chest in the water with a rope attached to the raft? (Assume throwing the treasure chest overboard is not an option you wish to consider.)

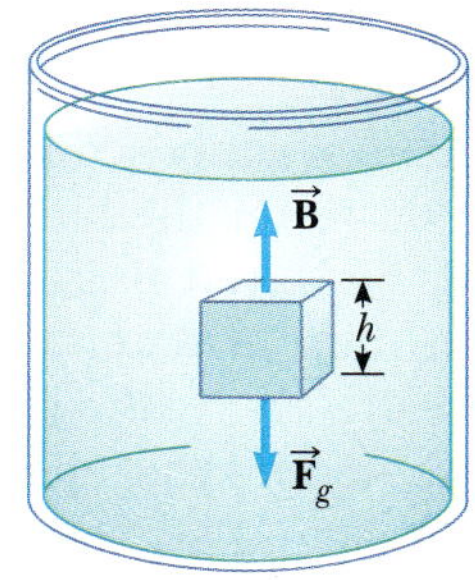

Figure 14.8 The external forces acting on the cube of liquid are the gravitational force $\vec{F}_g$ and the buoyant force $\vec{B}$. Under equilibrium conditions, $B = F_g$.

PITFALL PREVENTION 14.2

Buoyant Force Is Exerted by the Fluid

Remember that **the buoyant force is exerted by the fluid.** It is not determined by properties of the object except for the amount of fluid displaced by the object. Therefore, if several objects of different densities but the same volume are immersed in a fluid, they will all experience the same buoyant force. Whether they sink or float is determined by the relationship between the buoyant force and the gravitational force.

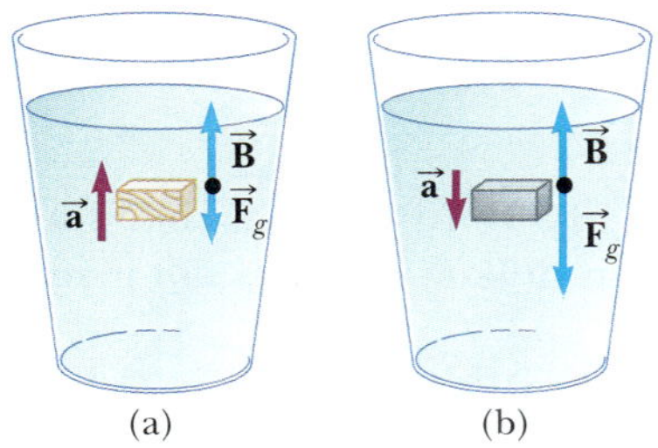

ACTIVE FIGURE 14.9

(a) A totally submerged object that is less dense than the fluid in which it is submerged experiences a net upward force. (b) A totally submerged object that is denser than the fluid experiences a net downward force and sinks.

Sign in at www.thomsonedu.com and go to ThomsonNOW to move the object to new positions as well as change the density of the object and see the results.

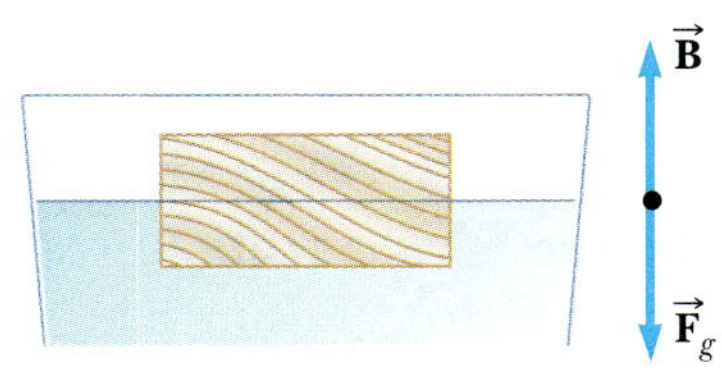

ACTIVE FIGURE 14.10

An object floating on the surface of a fluid experiences two forces, the gravitational force $\vec{F}_g$ and the buoyant force $\vec{B}$. Because the object floats in equilibrium, $B = F_g$.

Sign in at www.thomsonedu.com and go to ThomsonNOW to change the densities of the object and the fluid.

EXAMPLE 14.5 Eureka!

Archimedes supposedly was asked to determine whether a crown made for the king consisted of pure gold. According to legend, he solved this problem by weighing the crown first in air and then in water as shown in Figure 14.11. Suppose the scale read 7.84 N when the crown was in air and 6.84 N when it was in water. What should Archimedes have told the king?

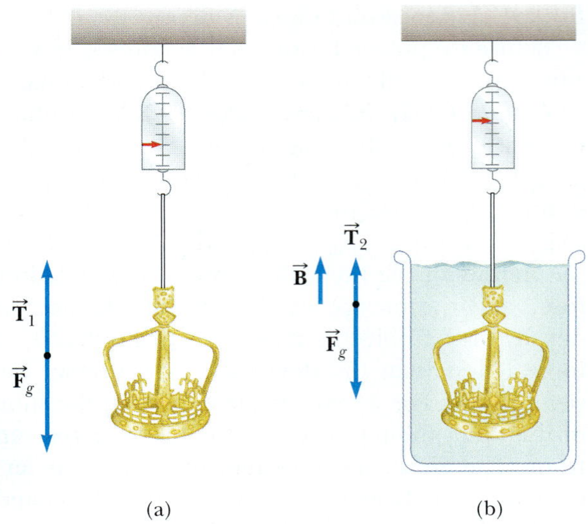

Figure 14.11 (Example 14.5) (a) When the crown is suspended in air, the scale reads its true weight because $T_1 = F_g$ (the buoyancy of air is negligible). (b) When the crown is immersed in water, the buoyant force $\vec{\mathbf{B}}$ changes the scale reading to a lower value $T_2 = F_g - B$.

SOLUTION

Conceptualize Figure 14.11 helps us imagine what is happening in this example. Because of the buoyant force, the scale reading is smaller in Figure 14.11b than in Figure 14.11a.

Categorize This problem is an example of Case 1 discussed earlier because the crown is completely submerged. The scale reading is a measure of one of the forces on the crown, and the crown is stationary. Therefore, we can categorize the crown as a particle in equilibrium.

Analyze When the crown is suspended in air, the scale reads the true weight $T_1 = F_g$ (neglecting the small buoyant force due to the surrounding air). When the crown is immersed in water, the buoyant force $\vec{\mathbf{B}}$ reduces the scale reading to an *apparent* weight of $T_2 = F_g - B$.

Apply the force equilibrium condition to the crown in water:

$$\sum F = B + T_2 - F_g = 0$$

Solve for B and substitute the known values:

$$B = F_g - T_2 = 7.84\ \text{N} - 6.84\ \text{N} = 1.00\ \text{N}$$

Because this buoyant force is equal in magnitude to the weight of the displaced water, $\rho_w g V_w = 1.00$ N, where V_w is the volume of the displaced water and ρ_w is its density. Also, the volume of the crown V_c is equal to the volume of the displaced water because the crown is completely submerged.

Find the volume of the crown:

$$V_c = V_w = \frac{1.00\ \text{N}}{\rho_w g} = \frac{1.00\ \text{N}}{(1\,000\ \text{kg/m}^3)(9.80\ \text{m/s}^2)} = 1.02 \times 10^{-4}\ \text{m}^3$$

Find the density of the crown from Equation 1.1:

$$\rho_c = \frac{m_c}{V_c} = \frac{m_c g}{V_c g} = \frac{7.84\ \text{N}}{(1.02 \times 10^{-4}\ \text{m}^3)(9.80\ \text{m/s}^2)}$$

$$= 7.84 \times 10^3\ \text{kg/m}^3$$

Finalize From Table 14.1 we see that the density of gold is 19.3×10^3 kg/m^3. Therefore, Archimedes should have reported that the king had been cheated. Either the crown was hollow, or it was not made of pure gold.

What If? Suppose the crown has the same weight but is indeed pure gold and not hollow. What would the scale reading be when the crown is immersed in water?

Answer Find the volume of the solid gold crown:

$$V_c = \frac{m_c}{\rho_c} = \frac{m_c g}{\rho_c g} = \frac{7.84\ \text{N}}{(19.3 \times 10^3\ \text{kg/m}^3)(9.80\ \text{m/s}^2)}$$

$$= 4.15 \times 10^{-5}\ \text{m}^3$$

Find the buoyant force on the crown:

$$B = \rho_w g V_w = \rho_w g V_c$$
$$= (1.00 \times 10^3 \text{ kg/m}^3)(9.80 \text{ m/s}^2)(4.15 \times 10^{-5} \text{ m}^3) = 0.406 \text{ N}$$

Find the tension in the string hanging from the scale:

$$T_2 = F_g - B = 7.84 \text{ N} - 0.406 \text{ N} = 7.43 \text{ N}$$

EXAMPLE 14.6 **A Titanic Surprise**

An iceberg floating in seawater as shown in Figure 14.12a is extremely dangerous because most of the ice is below the surface. This hidden ice can damage a ship that is still a considerable distance from the visible ice. What fraction of the iceberg lies below the water level?

Geraldine Prentice/Getty

(a) (b)

Figure 14.12 (Example 14.6) (a) Much of the volume of this iceberg is beneath the water. (b) A ship can be damaged even when it is not near the visible ice.

SOLUTION

Conceptualize You are likely familiar with the phrase, "That's only the tip of the iceberg." The origin of this popular saying is that most of the volume of a floating iceberg is beneath the surface of the water (Fig. 14.12b).

Categorize This example corresponds to Case 2. It is also a simple substitution problem involving Equation 14.6.

Evaluate Equation 14.6 using the densities of ice and seawater (Table 14.1):

$$f = \frac{V_{\text{seawater}}}{V_{\text{ice}}} = \frac{\rho_{\text{ice}}}{\rho_{\text{seawater}}} = \frac{917 \text{ kg/m}^3}{1\,030 \text{ kg/m}^3} = \boxed{0.890 \text{ or } 89.0\%}$$

Therefore, the visible fraction of ice above the water's surface is about 11%. It is the unseen 89% below the water that represents the danger to a passing ship.

14.5 Fluid Dynamics

Thus far, our study of fluids has been restricted to fluids at rest. We now turn our attention to fluids in motion. When fluid is in motion, its flow can be characterized as being one of two main types. The flow is said to be **steady,** or **laminar,** if each particle of the fluid follows a smooth path such that the paths of different particles never cross each other as shown in Figure 14.13. In steady flow, every fluid particle arriving at a given point has the same velocity.

Above a certain critical speed, fluid flow becomes **turbulent.** Turbulent flow is irregular flow characterized by small whirlpool-like regions as shown in Figure 14.14 (page 400).

The term **viscosity** is commonly used in the description of fluid flow to characterize the degree of internal friction in the fluid. This internal friction, or *viscous force,* is associated with the resistance that two adjacent layers of fluid have to moving relative to each other. Viscosity causes part of the fluid's kinetic energy to be converted to internal energy. This mechanism is similar to the one by which an object sliding on a rough horizontal surface loses kinetic energy.

Andy Sacks/Getty

Figure 14.13 Laminar flow around an automobile in a test wind tunnel.

Figure 14.14 Hot gases from a cigarette made visible by smoke particles. The smoke first moves in laminar flow at the bottom and then in turbulent flow above.

Because the motion of real fluids is very complex and not fully understood, we make some simplifying assumptions in our approach. In our model of **ideal fluid flow,** we make the following four assumptions:

1. **The fluid is nonviscous.** In a nonviscous fluid, internal friction is neglected. An object moving through the fluid experiences no viscous force.
2. **The flow is steady.** In steady (laminar) flow, all particles passing through a point have the same velocity.
3. **The fluid is incompressible.** The density of an incompressible fluid is constant.
4. **The flow is irrotational.** In irrotational flow, the fluid has no angular momentum about any point. If a small paddle wheel placed anywhere in the fluid does not rotate about the wheel's center of mass, the flow is irrotational.

The path taken by a fluid particle under steady flow is called a **streamline.** The velocity of the particle is always tangent to the streamline as shown in Figure 14.15. A set of streamlines like the ones shown in Figure 14.15 form a *tube of flow.* Fluid particles cannot flow into or out of the sides of this tube; if they could, the streamlines would cross one another.

Consider ideal fluid flow through a pipe of nonuniform size as illustrated in Figure 14.16. The particles in the fluid move along streamlines in steady flow. In a time interval Δt, a short element of the fluid at the bottom end of the pipe moves a distance $\Delta x_1 = v_1 \Delta t$. If A_1 is the cross-sectional area in this region, the mass of fluid contained in the left shaded region in Figure 14.16 is $m_1 = \rho A_1 \Delta x_1 = \rho A_1 v_1 \Delta t$, where ρ is the (unchanging) density of the ideal fluid. Similarly, the fluid that moves through the upper end of the pipe in the time interval Δt has a mass $m_2 = \rho A_2 v_2 \Delta t$. Because the fluid is incompressible and the flow is steady, however, the mass of fluid that crosses A_1 in a time interval Δt must equal the mass that crosses A_2 in the same time interval. That is, $m_1 = m_2$ or $\rho A_1 v_1 = \rho A_2 v_2$, which means that

$$A_1 v_1 = A_2 v_2 = \text{constant} \tag{14.7}$$

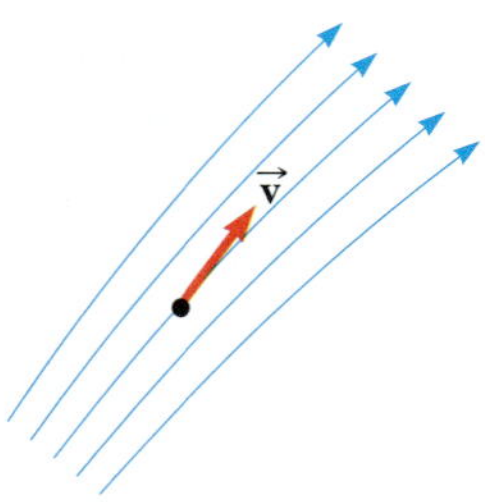

Figure 14.15 A particle in laminar flow follows a streamline, and at each point along its path the particle's velocity is tangent to the streamline.

This expression is called the **equation of continuity for fluids.** It states that the product of the area and the fluid speed at all points along a pipe is constant for an incompressible fluid. Equation 14.7 shows that the speed is high where the tube is constricted (small A) and low where the tube is wide (large A). The product Av, which has the dimensions of volume per unit time, is called either the *volume flux* or the *flow rate.* The condition $Av =$ constant is equivalent to the statement that the volume of fluid that enters one end of a tube in a given time interval equals the volume leaving the other end of the tube in the same time interval if no leaks are present.

You demonstrate the equation of continuity each time you water your garden with your thumb over the end of a garden hose as in Figure 14.17. By partially blocking the opening with your thumb, you reduce the cross-sectional area through which the water passes. As a result, the speed of the water increases as it exits the hose, and it can be sprayed over a long distance.

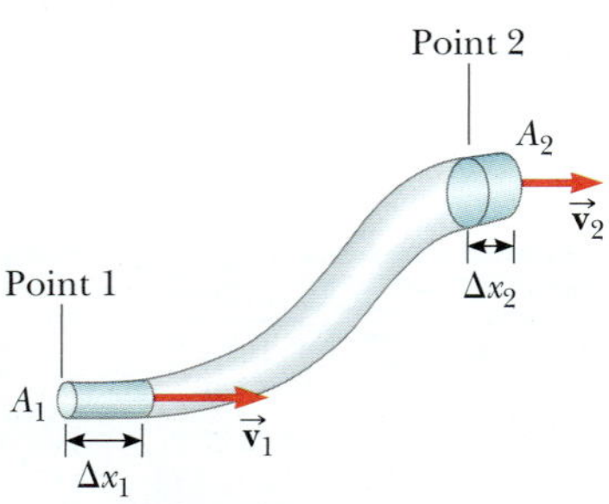

Figure 14.16 A fluid moving with steady flow through a pipe of varying cross-sectional area. The volume of fluid flowing through area A_1 in a time interval Δt must equal the volume flowing through area A_2 in the same time interval. Therefore, $A_1 v_1 = A_2 v_2$.

Figure 14.17 The speed of water spraying from the end of a garden hose increases as the size of the opening is decreased with the thumb.

EXAMPLE 14.7 Watering a Garden

A gardener uses a water hose 2.50 cm in diameter to fill a 30.0-L bucket. The gardener notes that it takes 1.00 min to fill the bucket. A nozzle with an opening of cross-sectional area 0.500 cm^2 is then attached to the hose. The nozzle is held so that water is projected horizontally from a point 1.00 m above the ground. Over what horizontal distance can the water be projected?

SOLUTION

Conceptualize Imagine any past experience you have with projecting water from a hose or a pipe. The faster the water is traveling as it leaves the hose, the farther it will land on the ground from the end of the hose.

Categorize Once the water leaves the hose, it is in free fall. Therefore, we categorize a given element of the water as a projectile. The element is modeled as a particle under constant acceleration (due to gravity) in the vertical direction and a particle under constant velocity in the horizontal direction. The horizontal distance over which the element is projected depends on the speed with which it is projected. This example involves a change in area for the pipe, so we also categorize it as one in which we use the equation of continuity for fluids.

Analyze We first find the speed of the water in the hose from the bucket-filling information.

Find the cross-sectional area of the hose:

$$A = \pi r^2 = \pi \frac{d^2}{4} = \pi \left[\frac{(2.50 \text{ cm})^2}{4}\right] = 4.91 \text{ cm}^2$$

Evaluate the volume flow rate in cm^3/s:

$$Av = 30.0 \text{ L/min} = \frac{30.0 \times 10^3 \text{ cm}^3}{60.0 \text{ s}} = 500 \text{ cm}^3\text{/s}$$

Solve for the speed of the water in the hose:

$$v_1 = \frac{500 \text{ cm}^3\text{/s}}{A} = \frac{500 \text{ cm}^3\text{/s}}{4.91 \text{ cm}^2} = 102 \text{ cm/s} = 1.02 \text{ m/s}$$

We have labeled this speed v_1 because we identify point 1 within the hose. We identify point 2 in the air just outside the nozzle. We must find the speed $v_2 = v_{xi}$ with which the water exits the nozzle. The subscript i anticipates that it will be the *initial* velocity component of the water projected from the hose, and the subscript x indicates that the initial velocity vector of the projected water is horizontal.

Solve the continuity equation for fluids for v_2:

$$v_2 = v_{xi} = \frac{A_1}{A_2} v_1$$

Substitute numerical values:

$$v_{xi} = \frac{4.91 \text{ cm}^2}{0.500 \text{ cm}^2}(1.02 \text{ m/s}) = 10.0 \text{ m/s}$$

We now shift our thinking away from fluids and to projectile motion. In the vertical direction, an element of the water starts from rest and falls through a vertical distance of 1.00 m.

Write Equation 2.16 for the vertical position of an element of water, modeled as a particle under constant acceleration:

$$y_f = y_i + v_{yi}t - \tfrac{1}{2}gt^2$$

Substitute numerical values:

$$-1.00 \text{ m} = 0 + 0 - \tfrac{1}{2}(9.80 \text{ m/s}^2)t^2$$

Solve for the time at which the element of water lands on the ground:

$$t = \sqrt{\frac{2(1.00 \text{ m})}{9.80 \text{ m/s}^2}} = 0.452 \text{ s}$$

Use Equation 2.7 to find the horizontal position of the element at this time, modeled as a particle under constant velocity:

$$x_f = x_i + v_{xi}t = 0 + (10.0 \text{ m/s})(0.452 \text{ s}) = 4.52 \text{ m}$$

Finalize The time interval for the element of water to fall to the ground is unchanged if the projection speed is changed because the projection is horizontal. Increasing the projection speed results in the water hitting the ground farther from the end of the hose, but requires the same time interval to strike the ground.

DANIEL BERNOULLI
Swiss physicist (1700–1782)

Bernoulli made important discoveries in fluid dynamics. Born into a family of mathematicians, he was the only member of his family to make a mark in physics.

Bernoulli's most famous work, *Hydrodynamica,* was published in 1738; it is both a theoretical and a practical study of equilibrium, pressure, and speed in fluids. He showed that as the speed of a fluid increases, its pressure decreases. Referred to as "Bernoulli's principle," Bernoulli's work is used to produce a partial vacuum in chemical laboratories by connecting a vessel to a tube through which water is running rapidly.

In *Hydrodynamica,* Bernoulli also attempted the first explanation of the behavior of gases with changing pressure and temperature; this step was the beginning of the kinetic theory of gases, a topic we study in Chapter 21.

14.6 Bernoulli's Equation

You have probably experienced driving on a highway and having a large truck pass you at high speed. In this situation, you may have had the frightening feeling that your car was being pulled in toward the truck as it passed. We will investigate the origin of this effect in this section.

As a fluid moves through a region where its speed or elevation above the Earth's surface changes, the pressure in the fluid varies with these changes. The relationship between fluid speed, pressure, and elevation was first derived in 1738 by Swiss physicist Daniel Bernoulli. Consider the flow of a segment of an ideal fluid through a nonuniform pipe in a time interval Δt as illustrated in Figure 14.18. At the beginning of the time interval, the segment of fluid consists of the blue shaded portion (portion 1) at the left and the unshaded portion. During the time interval, the left end of the segment moves to the right by a distance Δx_1, which is the length of the blue shaded portion at the left. Meanwhile, the right end of the segment moves to the right through a distance Δx_2, which is the length of the blue shaded portion (portion 2) at the upper right of Figure 14.18. Therefore, at the end of the time interval, the segment of fluid consists of the unshaded portion and the blue shaded portion at the upper right.

Now consider forces exerted on this segment by fluid to the left and the right of the segment. The force exerted by the fluid on the left end has a magnitude P_1A_1. The work done by this force on the segment in a time interval Δt is $W_1 = F_1\,\Delta x_1 = P_1A_1\,\Delta x_1 = P_1V$, where V is the volume of portion 1. In a similar manner, the work done by the fluid to the right of the segment in the same time interval Δt is $W_2 = -P_2A_2\,\Delta x_2 = -P_2V$. (The volume of portion 1 equals the volume of portion 2 because the fluid is incompressible.) This work is negative because the force on the segment of fluid is to the left and the displacement is to the right. Therefore, the net work done on the segment by these forces in the time interval Δt is

$$W = (P_1 - P_2)V$$

Part of this work goes into changing the kinetic energy of the segment of fluid, and part goes into changing the gravitational potential energy of the segment–Earth system. Because we are assuming streamline flow, the kinetic energy K_{uns} of the unshaded portion of the segment in Figure 14.18 is unchanged during the time interval. Therefore, the change in the kinetic energy of the segment of fluid is

$$\Delta K = \left(\tfrac{1}{2}mv_2^{\,2} + K_{\text{uns}}\right) - \left(\tfrac{1}{2}mv_1^{\,2} + K_{\text{uns}}\right) = \tfrac{1}{2}mv_2^{\,2} - \tfrac{1}{2}mv_1^{\,2}$$

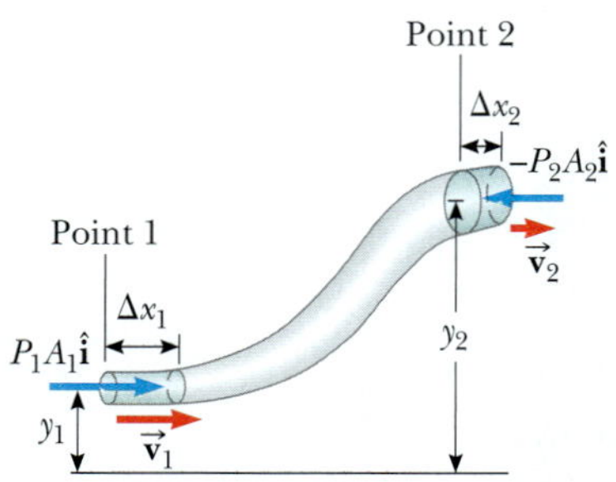

Figure 14.18 A fluid in laminar flow through a constricted pipe. The volume of the shaded portion on the left is equal to the volume of the shaded portion on the right.

where m is the mass of both portion 1 and portion 2. (Because the volumes of both portions are the same, they also have the same mass.)

Considering the gravitational potential energy of the segment–Earth system, once again there is no change during the time interval for the gravitational potential energy U_{uns} associated with the unshaded portion of the fluid. Consequently, the change in gravitational potential energy is

$$\Delta U = (mgy_2 + U_{\text{uns}}) - (mgy_1 + U_{\text{uns}}) = mgy_2 - mgy_1$$

From Equation 8.2, the total work done on the system by the fluid outside the segment is equal to the change in mechanical energy of the system: $W = \Delta K + \Delta U$. Substituting for each of these terms gives

$$(P_1 - P_2)V = \tfrac{1}{2}mv_2^{\,2} - \tfrac{1}{2}mv_1^{\,2} + mgy_2 - mgy_1$$

If we divide each term by the portion volume V and recall that $\rho = m/V$, this expression reduces to

$$P_1 - P_2 = \tfrac{1}{2}\rho v_2^2 - \tfrac{1}{2}\rho v_1^2 + \rho g y_2 - \rho g y_1$$

Rearranging terms gives

$$P_1 + \tfrac{1}{2}\rho v_1^2 + \rho g y_1 = P_2 + \tfrac{1}{2}\rho v_2^2 + \rho g y_2 \qquad \textbf{(14.8)}$$

which is **Bernoulli's equation** as applied to an ideal fluid. This equation is often expressed as

$$P + \tfrac{1}{2}\rho v^2 + \rho g y = \text{constant} \qquad \textbf{(14.9)}$$

◀ Bernoulli's equation

Bernoulli's equation shows that the pressure of a fluid decreases as the speed of the fluid increases. In addition, the pressure decreases as the elevation increases. This latter point explains why water pressure from faucets on the upper floors of a tall building is weak unless measures are taken to provide higher pressure for these upper floors.

When the fluid is at rest, $v_1 = v_2 = 0$ and Equation 14.8 becomes

$$P_1 - P_2 = \rho g(y_2 - y_1) = \rho g h$$

This result is in agreement with Equation 14.4.

Although Equation 14.9 was derived for an incompressible fluid, the general behavior of pressure with speed is true even for gases: as the speed increases, the pressure decreases. This *Bernoulli effect* explains the experience with the truck on the highway at the opening of this section. As air passes between you and the truck, it must pass through a relatively narrow channel. According to the continuity equation, the speed of the air is higher. According to the Bernoulli effect, this higher-speed air exerts less pressure on your car than the slower-moving air on the other side of your car. Therefore, there is a net force pushing you toward the truck!

Quick Quiz 14.5 You observe two helium balloons floating next to each other at the ends of strings secured to a table. The facing surfaces of the balloons are separated by 1–2 cm. You blow through the small space between the balloons. What happens to the balloons? (a) They move toward each other. (b) They move away from each other. (c) They are unaffected.

EXAMPLE 14.8 The Venturi Tube

The horizontal constricted pipe illustrated in Figure 14.19, known as a *Venturi tube*, can be used to measure the flow speed of an incompressible fluid. Determine the flow speed at point 2 of Figure 14.19a if the pressure difference $P_1 - P_2$ is known.

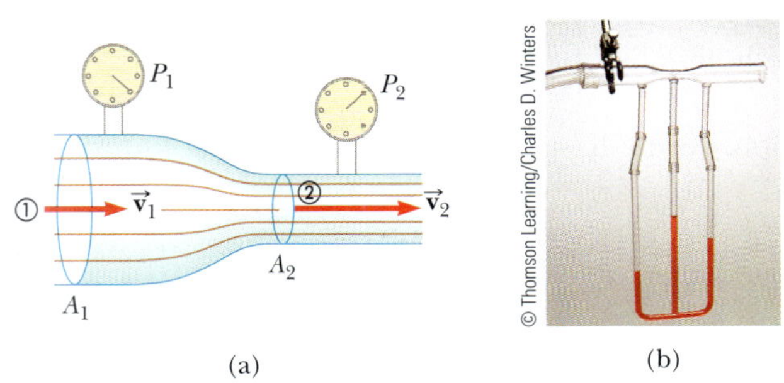

Figure 14.19 (Example 14.8) (a) Pressure P_1 is greater than pressure P_2 because $v_1 < v_2$. This device can be used to measure the speed of fluid flow. (b) A Venturi tube, located at the top of the photograph. The higher level of fluid in the middle column shows that the pressure at the top of the column, which is in the constricted region of the Venturi tube, is lower.

SOLUTION

Conceptualize Bernoulli's equation shows how the pressure of a fluid decreases as its speed increases. Therefore, we should be able to calibrate a device to give us the fluid speed if we can measure pressure.

Categorize Because the problem states that the fluid is incompressible, we can categorize it as one in which we can use the equation of continuity for fluids and Bernoulli's equation.

Analyze Apply Equation 14.8 to points 1 and 2, noting that $y_1 = y_2$ because the pipe is horizontal:

$$(1) \quad P_1 + \tfrac{1}{2}\rho v_1^2 = P_2 + \tfrac{1}{2}\rho v_2^2$$

Solve the equation of continuity for v_1:

$$v_1 = \frac{A_2}{A_1} v_2$$

Substitute this expression into Equation (1):

$$P_1 + \tfrac{1}{2}\rho\left(\frac{A_2}{A_1}\right)^2 v_2^2 = P_2 + \tfrac{1}{2}\rho v_2^2$$

Solve for v_2:

$$v_2 = A_1\sqrt{\frac{2(P_1 - P_2)}{\rho(A_1^2 - A_2^2)}}$$

Finalize From the design of the tube (areas A_1 and A_2) and measurements of the pressure difference $P_1 - P_2$, we can calculate the speed of the fluid with this equation. To see the relationship between fluid speed and pressure difference, place two empty soda cans on their sides about 2 cm apart on a table. Gently blow a stream of air horizontally between the cans and watch them roll together slowly due to a modest pressure difference between the stagnant air on their outside edges and the moving air between them. Now blow more strongly and watch the increased pressure difference move the cans together more rapidly.

EXAMPLE 14.9 Torricelli's Law

An enclosed tank containing a liquid of density ρ has a hole in its side at a distance y_1 from the tank's bottom (Fig. 14.20). The hole is open to the atmosphere, and its diameter is much smaller than the diameter of the tank. The air above the liquid is maintained at a pressure P. Determine the speed of the liquid as it leaves the hole when the liquid's level is a distance h above the hole.

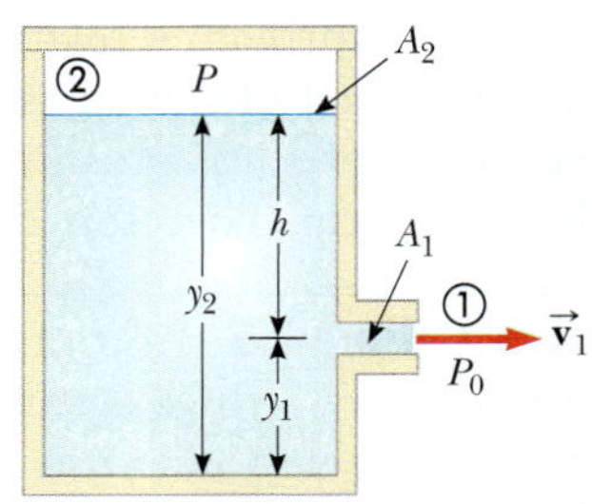

Figure 14.20 (Example 14.9) A liquid leaves a hole in a tank at speed v_1.

SOLUTION

Conceptualize Imagine that the tank is a fire extinguisher. When the hole is opened, liquid leaves the hole with a certain speed. If the pressure P at the top of the liquid is increased, the liquid leaves with a higher speed. If the pressure P falls too low, the liquid leaves with a low speed and the extinguisher must be replaced.

Categorize Looking at Figure 14.20, we know the pressure at two points and the velocity at one of those points. We wish to find the velocity at the second point. Therefore, we can categorize this example as one in which we can apply Bernoulli's equation.

Analyze Because $A_2 >> A_1$, the liquid is approximately at rest at the top of the tank, where the pressure is P. At the hole P_1 is equal to atmospheric pressure P_0.

Apply Bernoulli's equation between points 1 and 2:

$$P_0 + \tfrac{1}{2}\rho v_1^2 + \rho g y_1 = P + \rho g y_2$$

Solve for v_1, noting that $y_2 - y_1 = h$:

$$v_1 = \sqrt{\frac{2(P - P_0)}{\rho} + 2gh}$$

Finalize When P is much greater than P_0 (so that the term $2gh$ can be neglected), the exit speed of the water is mainly a function of P. If the tank is open to the atmosphere, then $P = P_0$ and $v_1 = \sqrt{2gh}$. In other words, for an open tank, the speed of liquid leaving a hole a distance h below the surface is equal to that acquired by an object falling freely through a vertical distance h. This phenomenon is known as **Torricelli's law.**

What If? What if the position of the hole in Figure 14.20 could be adjusted vertically? If the tank is open to the atmosphere and sitting on a table, what position of the hole would cause the water to land on the table at the farthest distance from the tank?

Answer Model a parcel of water exiting the hole as a projectile. Find the time at which the parcel strikes the table from a hole at an arbitrary position:

$$y_f = y_i + v_{yi}t - \tfrac{1}{2}gt^2$$

$$0 = y_1 + 0 - \tfrac{1}{2}gt^2$$

$$t = \sqrt{\frac{2y_1}{g}}$$

Find the horizontal position of the parcel at the time it strikes the table:

$$x_f = x_i + v_{xi}t = 0 + \sqrt{2g(y_2 - y_1)}\sqrt{\frac{2y_1}{g}}$$

$$= 2\sqrt{(y_2y_1 - y_1^2)}$$

Maximize the horizontal position by taking the derivative of x_f with respect to y_1 (because y_1, the height of the hole, is the variable that can be adjusted) and setting it equal to zero:

$$\frac{dx_f}{dy_1} = \tfrac{1}{2}(2)(y_2y_1 - y_1^2)^{-1/2}(y_2 - 2y_1) = 0$$

Solve for y_1:

$$y_1 = \tfrac{1}{2}y_2$$

Therefore, to maximize the horizontal distance, the hole should be halfway between the bottom of the tank and the upper surface of the water. Below this location, the water is projected at a higher speed but falls for a short time interval, reducing the horizontal range. Above this point, the water is in the air for a longer time interval but is projected with a smaller horizontal speed.

14.7 Other Applications of Fluid Dynamics

Consider the streamlines that flow around an airplane wing as shown in Figure 14.21. Let's assume the airstream approaches the wing horizontally from the right with a velocity $\vec{\mathbf{v}}_1$. The tilt of the wing causes the airstream to be deflected downward with a velocity $\vec{\mathbf{v}}_2$. Because the airstream is deflected by the wing, the wing must exert a force on the airstream. According to Newton's third law, the airstream exerts a force $\vec{\mathbf{F}}$ on the wing that is equal in magnitude and opposite in direction. This force has a vertical component called **lift** (or aerodynamic lift) and a horizontal component called **drag.** The lift depends on several factors, such as the speed of the airplane, the area of the wing, the wing's curvature, and the angle between the wing and the horizontal. The curvature of the wing surfaces causes the pressure above the wing to be lower than that below the wing due to the Bernoulli effect. This pressure difference assists with the lift on the wing. As the angle between the wing and the horizontal increases, turbulent flow can set in above the wing to reduce the lift.

In general, an object moving through a fluid experiences lift as the result of any effect that causes the fluid to change its direction as it flows past the object. Some factors that influence lift are the shape of the object, its orientation with respect to the fluid flow, any spinning motion it might have, and the texture of its surface. For example, a golf ball struck with a club is given a rapid backspin due to the slant of the club. The dimples on the ball increase the friction force between the ball and the air so that air adheres to the ball's surface. Figure 14.22 (page 406) shows air adhering to the ball and being deflected downward as a result. Because the ball pushes the air down, the air must push up on the ball. Without the dimples, the friction force is lower and the golf ball does not travel as far. It may seem counterintuitive to increase the range by increasing the friction force, but the lift gained by spinning the ball more than compensates for the loss of range due to

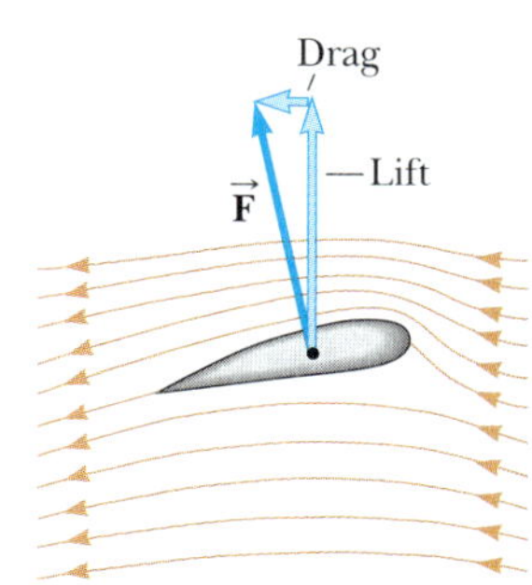

Figure 14.21 Streamline flow around a moving airplane wing. The air approaching from the right is deflected downward by the wing. By Newton's third law, this deflection must coincide with an upward force on the wing from the air: *lift.* Because of air resistance, there is also a force opposite the velocity of the wing: *drag.*

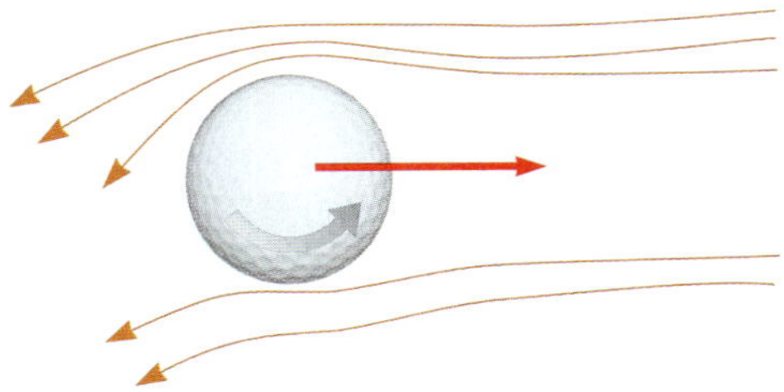

Figure 14.22 Because of the deflection of air, a spinning golf ball experiences a lifting force that allows it to travel much farther than it would if it were not spinning.

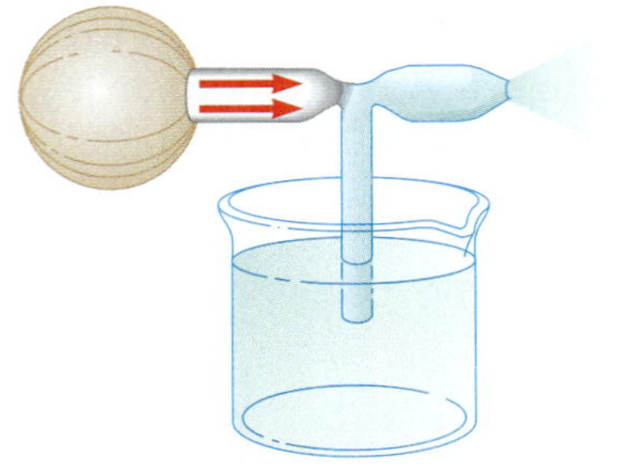

Figure 14.23 A stream of air passing over a tube dipped into a liquid causes the liquid to rise in the tube.

the effect of friction on the translational motion of the ball. For the same reason, a baseball's cover helps the spinning ball "grab" the air rushing by and helps deflect it when a "curve ball" is thrown.

A number of devices operate by means of the pressure differentials that result from differences in a fluid's speed. For example, a stream of air passing over one end of an open tube, the other end of which is immersed in a liquid, reduces the pressure above the tube as illustrated in Figure 14.23. This reduction in pressure causes the liquid to rise into the air stream. The liquid is then dispersed into a fine spray of droplets. You might recognize that this so-called atomizer is used in perfume bottles and paint sprayers.

Summary

ThomsonNOW™ Sign in at **www.thomsonedu.com** and go to ThomsonNOW to take a practice test for this chapter.

DEFINITIONS

The **pressure** P in a fluid is the force per unit area exerted by the fluid on a surface:

$$P \equiv \frac{F}{A} \qquad \textbf{(14.1)}$$

In the SI system, pressure has units of newtons per square meter (N/m^2), and 1 N/m^2 = 1 **pascal** (Pa).

CONCEPTS AND PRINCIPLES

The pressure in a fluid at rest varies with depth h in the fluid according to the expression

$$P = P_0 + \rho g h \qquad \textbf{(14.4)}$$

where P_0 is the pressure at $h = 0$ and ρ is the density of the fluid, assumed uniform.

Pascal's law states that when pressure is applied to an enclosed fluid, the pressure is transmitted undiminished to every point in the fluid and to every point on the walls of the container.

When an object is partially or fully submerged in a fluid, the fluid exerts on the object an upward force called the **buoyant force.** According to **Archimedes's principle,** the magnitude of the buoyant force is equal to the weight of the fluid displaced by the object:

$$B = \rho_{\text{fluid}} g V \qquad \textbf{(14.5)}$$

The flow rate (volume flux) through a pipe that varies in cross-sectional area is constant; that is equivalent to stating that the product of the cross-sectional area A and the speed v at any point is a constant. This result is expressed in the **equation of continuity for fluids:**

$$A_1 v_1 = A_2 v_2 = \text{constant} \qquad \textbf{(14.7)}$$

The sum of the pressure, kinetic energy per unit volume, and gravitational potential energy per unit volume has the same value at all points along a streamline for an ideal fluid. This result is summarized in **Bernoulli's equation:**

$$P + \tfrac{1}{2}\rho v^2 + \rho g y = \text{constant} \qquad \textbf{(14.9)}$$

Questions

☐ denotes answer available in *Student Solutions Manual/Study Guide;* **O** denotes objective question

1. **O** Figure Q14.1 shows aerial views from directly above two dams. Both dams are equally wide (the vertical dimension in the diagram) and equally high (into the page in the diagram). The dam on the left holds back a very large lake, and the dam on the right holds back a narrow river. Which dam has to be built more strongly? (a) the dam on the left (b) the dam on the right (c) both the same (d) cannot be predicted

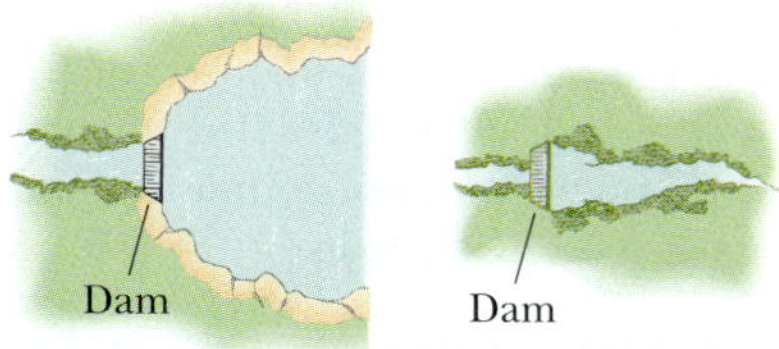

Figure Q14.1

2. Two thin-walled drinking glasses having equal base areas but different shapes, with very different cross-sectional areas above the base, are filled to the same level with water. According to the expression $P = P_0 + \rho gh$, the pressure is the same at the bottom of both glasses. In view of this equality, why does one glass weigh more than the other?

3. Because atmospheric pressure is about 10^5 N/m^2 and the area of a person's chest is about 0.13 m^2, the force of the atmosphere on one's chest is around 13 000 N. In view of this enormous force, why don't our bodies collapse?

4. A fish rests on the bottom of a bucket of water while the bucket is being weighed on a scale. When the fish begins to swim around, does the scale reading change?

5. You are a passenger on a spacecraft. For your survival and comfort, the interior contains air just like that at the surface of the Earth. The spacecraft is coasting through a very empty region of space. That is, a nearly perfect vacuum exists just outside the wall. Suddenly, a meteoroid pokes a hole, about the size of a large coin, right through the wall next to your seat. What happens? Is there anything you can or should do about it?

6. Does a ship float higher in the water of an inland lake or in the ocean? Why?

7. **O** An apple is held completely submerged just below the surface of water in a container. The apple is then moved to a deeper point in the water. Compared with the force needed to hold the apple just below the surface, what is the force needed to hold it at the deeper point? (a) larger (b) the same (c) smaller (d) impossible to determine

8. When an object is immersed in a liquid at rest, why is the net force on the object in the horizontal direction equal to zero?

9. A barge is carrying a load of gravel along a river. The barge approaches a low bridge, and the captain realizes that the top of the pile of gravel is not going to make it under the bridge. The captain orders the crew to shovel gravel from the pile into the water. Is that a good decision?

10. An empty metal soap dish barely floats in water. A bar of Ivory soap floats in water. When the soap is stuck in the soap dish, the combination sinks. Explain why.

11. **O** A beach ball is made of thin plastic. It has been inflated with air, but the plastic is not stretched. By swimming with fins on, you manage to take the ball from the surface of a pool to the bottom. Once the ball is completely submerged, what happens to the buoyant force exerted on the beach ball as you take it deeper? (a) increases (b) remains constant (c) decreases (d) is impossible to determine

12. If you release a ball while inside a freely falling elevator, the ball remains in front of you rather than falling to the floor because the ball, the elevator, and you all experience the same downward gravitational acceleration. What happens if you repeat this experiment with a helium-filled balloon? (This question is tricky.)

13. **O** A small piece of steel is tied to a block of wood. When the wood is placed in a tub of water with the steel on top, half of the block is submerged. Now the block is inverted so that the steel is under water. **(i)** Does the amount of the block submerged (a) increase, (b) decrease, or (c) remain the same? **(ii)** What happens to the water level in the tub when the block is inverted? (a) It rises. (b) It falls. (c) It remains the same.

14. How would you determine the density of an irregularly shaped rock?

15. **O** Rank the buoyant forces exerted on the following seven objects, from the largest to the smallest. Assume the objects have been dropped into a swimming pool and allowed to come to mechanical equilibrium. If any buoyant forces are equal, state that in your ranking. (a) a block of solid oak (b) an aluminum block of equal volume to the wood (c) a beach ball made of thin plastic and inflated with air, of equal volume (d) an iron block of equal volume (e) a thin-walled, sealed bottle of water equal in volume to the wood (f) an aluminum block having the same mass as the wood (g) an iron block of equal mass

16. **O** A person in a boat floating in a small pond throws an anchor overboard. What happens to the level of the pond? (a) It rises. (b) It falls. (c) It remains the same.

17. Is the buoyant force a conservative force? Is a potential energy associated with it? Explain your answers.

18. An unopened can of diet cola floats when placed in a tank of water, whereas a can of regular cola of the same brand sinks in the tank. What do you suppose could explain this behavior?

19. **O** A piece of unpainted porous wood floats in a container partly filled with water. The container is sealed and pressurized above atmospheric pressure. What happens to the wood? (a) It rises. (b) It falls. (c) It remains at the same level.

20. The water supply for a city is often provided from reservoirs built on high ground. Water flows from the reservoir, through pipes, and into your home when you turn the tap on your faucet. Why is the water flow more rapid out of a faucet on the first floor of a building than in an apartment on a higher floor?

21. If the airstream from a hair dryer is directed over a table-tennis ball, the ball can be levitated. Explain.

22. When ski jumpers are airborne (Fig. Q14.22), they bend their bodies forward and keep their hands at their sides. Why?

© TempSport/CORBIS

Figure Q14.22

23. Why do airplane pilots prefer to take off with the airplane facing into the wind?

24. **O** A water supply maintains a constant rate of flow for water in a hose. You want to change the opening of the nozzle so that water leaving the nozzle reaches a height that is four times the current maximum height the water reaches with the nozzle vertical. To do so, what should you do? (a) decrease the area of the opening by a factor of 16 (b) decrease the area by a factor of 8 (c) decrease the area by a factor of 4 (d) decrease the area by a factor of 2 (e) give up because it cannot be done

25. Prairie dogs (Fig. Q14.25) ventilate their burrows by building a mound around one entrance, which is open to a stream of air when wind blows from any direction. A second entrance at ground level is open to almost stagnant air. How does this construction create an airflow through the burrow?

Pamela Zilly

Figure Q14.25

26. In Figure Q14.26, an airstream moves from right to left through a tube that is constricted at the middle. Three table-tennis balls are levitated in equilibrium above the vertical columns through which the air escapes. (a) Why is the ball at the right higher than the one in the middle? (b) Why is the ball at the left lower than the ball at the right even though the horizontal tube has the same dimensions at these two points?

Henry Leap and Jim Lehman

Figure Q14.26

27. **O** **(i)** A glass of water contains floating ice cubes. When the ice melts, does the water level in the glass (a) go up, (b) go down, or (c) remain the same? **(ii)** One of the predicted problems due to global warming is that ice in the polar ice caps will melt and raise sea level everywhere in the world. Is that more of a worry for ice (a) at the north pole, where most of the ice floats on water; (b) at the south pole, where most of the ice sits on land; (c) both at the north and the south poles equally; or (d) at neither pole?

Problems

WebAssign The Problems from this chapter may be assigned online in WebAssign.

ThomsonNOW Sign in at **www.thomsonedu.com** and go to ThomsonNOW to assess your understanding of this chapter's topics with additional quizzing and conceptual questions.

1, 2, 3 denotes straightforward, intermediate, challenging; ☐ denotes full solution available in *Student Solutions Manual/Study Guide;* ▲ denotes coached solution with hints available at **www.thomsonedu.com;** ▭ denotes developing symbolic reasoning; ● denotes asking for qualitative reasoning; ▪ denotes computer useful in solving problem

Section 14.1 Pressure

☐**1.** Calculate the mass of a solid iron sphere that has a diameter of 3.00 cm.

2. Find the order of magnitude of the density of the nucleus of an atom. What does this result suggest concerning the structure of matter? Model a nucleus as consisting of protons and neutrons closely packed together. Each has mass 1.67×10^{-27} kg and radius on the order of 10^{-15} m.

☐**3.** A 50.0-kg woman balances on one heel of a pair of high-heeled shoes. If the heel is circular and has a radius of 0.500 cm, what pressure does she exert on the floor?

4. What is the total mass of the Earth's atmosphere? (The radius of the Earth is 6.37×10^6 m, and atmospheric pressure at the surface is 1.013×10^5 N/m^2.)

Section 14.2 Variation of Pressure with Depth

5. The spring of the pressure gauge shown in Figure 14.2 has a force constant of 1 000 N/m, and the piston has a diameter of 2.00 cm. As the gauge is lowered into water, what change in depth causes the piston to move in by 0.500 cm?

6. (a) Calculate the absolute pressure at an ocean depth of 1 000 m. Assume the density of seawater is 1 024 kg/m^3 and the air above exerts a pressure of 101.3 kPa. (b) At this depth, what force must the frame around a circular submarine porthole having a diameter of 30.0 cm exert to counterbalance the force exerted by the water?

7. ▲ What must be the contact area between a suction cup (completely exhausted) and a ceiling if the cup is to support the weight of an 80.0-kg student?

8. The small piston of a hydraulic lift has a cross-sectional area of 3.00 cm^2 and its large piston has a cross-sectional area of 200 cm^2 (Fig. 14.4a). What force must be applied to the small piston for the lift to raise a load of 15.0 kN? (In service stations, this force is usually exerted by compressed air.)

9. For the basement of a new house, a hole is dug in the ground, with vertical sides going down 2.40 m. A concrete foundation wall is built across the 9.60-m width of the excavation. This foundation wall is 0.183 m from the front of the basement hole. During a rainstorm, drainage from the street fills up the space in front of the concrete wall, but not the basement behind the wall. The water does not soak into the clay soil. Find the force the water causes on the foundation wall. For comparison, the gravitational force exerted on the water is $(2.40 \text{ m})(9.60 \text{ m})(0.183 \text{ m})(1\,000 \text{ kg/m}^3)(9.80 \text{ m/s}^2) = 41.3 \text{ kN}$.

10. (a) A powerful vacuum cleaner has a hose 2.86 cm in diameter. With no nozzle on the hose, what is the weight of the heaviest brick that the cleaner can lift (Fig. P14.10a)? (b) **What If?** An octopus uses one sucker of diameter 2.86 cm on each of the two shells of a clam in an attempt to pull the shells apart (Fig. P14.10b). Find the greatest force the octopus can exert in seawater 32.3 m deep. *Caution*: Experimental verification can be interesting, but do not drop a brick on your foot. Do not overheat the motor of a vacuum cleaner. Do not get an octopus mad at you.

(a) (b)

Figure P14.10

11. A swimming pool has dimensions 30.0 m × 10.0 m and a flat bottom. When the pool is filled to a depth of 2.00 m with fresh water, what is the force caused by the water on the bottom? On each end? On each side?

12. The tank in Figure P14.12 is filled with water 2.00 m deep. At the bottom of one sidewall is a rectangular hatch 1.00 m high and 2.00 m wide that is hinged at the top of the hatch. (a) Determine the force the water causes on the hatch. (b) Find the torque caused by the water about the hinges.

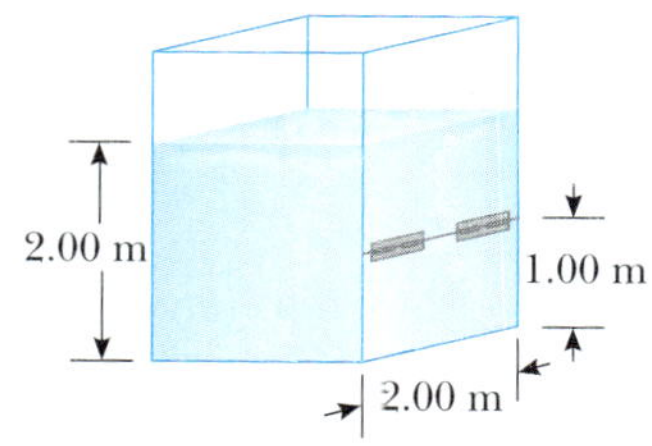

Figure P14.12

13. **Review problem.** The Abbott of Aberbrothock paid for a bell moored to the Inchcape Rock to warn sailors away. Assume the bell was 3.00 m in diameter and cast from brass with a bulk modulus of 14.0×10^{10} N/m^2. The pirate Ralph the Rover cut loose the bell and threw it into the ocean. By how much did the diameter of the bell decrease as it sank to a depth of 10.0 km? Years later, the klutz drowned when his ship collided with the rock. *Note:* The brass is compressed uniformly, so you may model the bell as a sphere of diameter 3.00 m.

Section 14.3 Pressure Measurements

14. Figure P14.14 shows Superman attempting to drink water through a very long straw. With his great strength he achieves maximum possible suction. The walls of the tubular straw do not collapse. (a) Find the maximum height through which he can lift the water. (b) **What If?** Still thirsty, the Man of Steel repeats his attempt on the Moon, which has no atmosphere. Find the difference between the water levels inside and outside the straw.

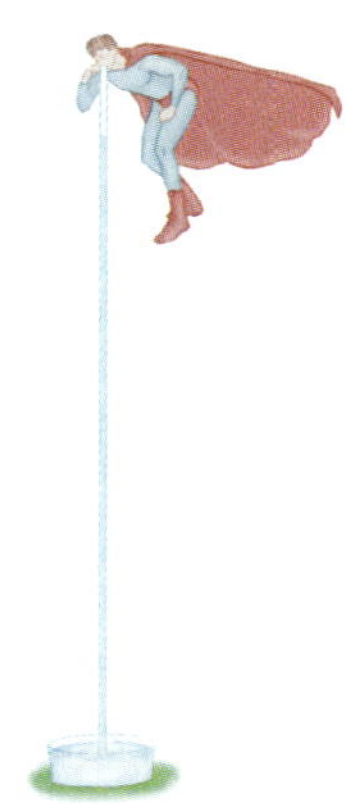

Figure P14.14

15. ▲ ● Blaise Pascal duplicated Torricelli's barometer using a red Bordeaux wine, of density 984 kg/m^3, as the working liquid (Fig. P14.15). What was the height h of the wine column for normal atmospheric pressure? Would

you expect the vacuum above the column to be as good as for mercury?

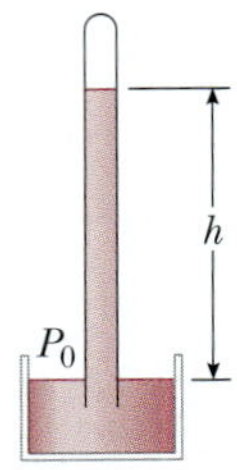

Figure P14.15

16. Mercury is poured into a U-tube as shown in Figure P14.16a. The left arm of the tube has cross-sectional area A_1 of 10.0 cm^2, and the right arm has a cross-sectional area A_2 of 5.00 cm^2. One hundred grams of water are then poured into the right arm as shown in Figure P14.16b. (a) Determine the length of the water column in the right arm of the U-tube. (b) Given that the density of mercury is 13.6 g/cm^3, what distance h does the mercury rise in the left arm?

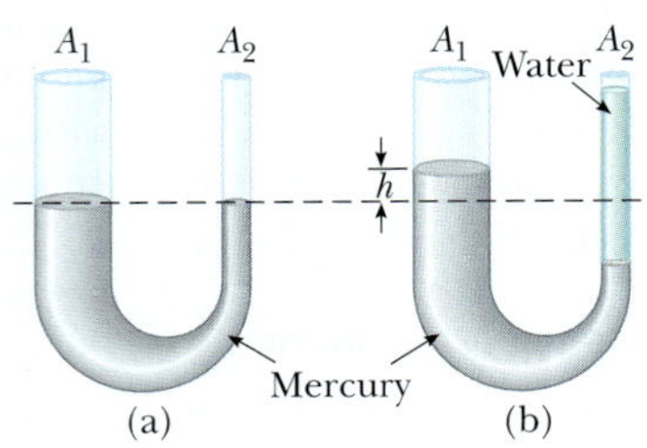

Figure P14.16

17. Normal atmospheric pressure is 1.013×10^5 Pa. The approach of a storm causes the height of a mercury barometer to drop by 20.0 mm from the normal height. What is the atmospheric pressure? (The density of mercury is 13.59 g/cm^3.)

18. A tank with a flat bottom of area A and vertical sides is filled to a depth h with water. The pressure is 1 atm at the top surface. (a) What is the absolute pressure at the bottom of the tank? (b) Suppose an object of mass M and density less than the density of water is placed in the tank and floats. No water overflows. What is the resulting increase in pressure at the bottom of the tank? (c) Evaluate your results for a backyard swimming pool with depth 1.50 m and a circular base with diameter 6.00 m. Two persons with combined mass 150 kg enter the pool and float quietly there. Find the original absolute pressure and the pressure increase at the bottom of the pool.

19. ● The human brain and spinal cord are immersed in the cerebrospinal fluid. The fluid is normally continuous between the cranial and spinal cavities and exerts a pressure of 100 to 200 mm of H_2O above the prevailing atmospheric pressure. In medical work, pressures are often measured in units of millimeters of H_2O because body fluids, including the cerebrospinal fluid, typically have the same density as water. The pressure of the cerebrospinal fluid can be measured by means of a *spinal tap* as illustrated in Figure P14.19. A hollow tube is inserted into the spinal column, and the height to which the fluid rises is observed. If the fluid rises to a height of 160 mm, we write its gauge pressure as 160 mm H_2O. (a) Express this pressure in pascals, in atmospheres, and in millimeters of mercury. (b) Sometimes it is necessary to determine whether an accident victim has suffered a crushed vertebra that is blocking flow of the cerebrospinal fluid in the spinal column. In other cases, a physician may suspect that a tumor or other growth is blocking the spinal column and inhibiting flow of cerebrospinal fluid. Such conditions can be investigated by means of *Queckenstedt's test.* In this procedure, the veins in the patient's neck are compressed to make the blood pressure rise in the brain. The increase in pressure in the blood vessels is transmitted to the cerebrospinal fluid. What should be the normal effect on the height of the fluid in the spinal tap? (c) Suppose compressing the veins had no effect on the fluid level. What might account for this result?

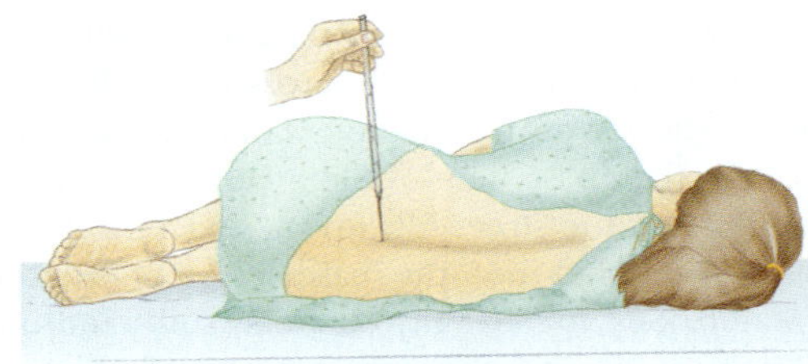

Figure P14.19

Section 14.4 Buoyant Forces and Archimedes's Principle

20. (a) A light balloon is filled with 400 m^3 of helium. At 0°C, the balloon can lift a payload of what mass? (b) **What If?** In Table 14.1, observe that the density of hydrogen is nearly one-half the density of helium. What load can the balloon lift if filled with hydrogen?

21. A table-tennis ball has a diameter of 3.80 cm and average density of 0.084 0 g/cm^3. What force is required to hold it completely submerged under water?

22. The gravitational force exerted on a solid object is 5.00 N. When the object is suspended from a spring scale and submerged in water, the scale reads 3.50 N (Fig. P14.22). Find the density of the object.

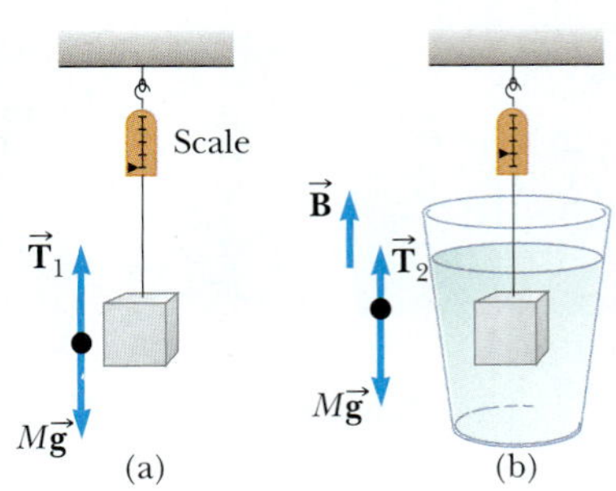

Figure P14.22 Problems 22 and 23.

23. A 10.0-kg block of metal measuring 12.0 cm × 10.0 cm × 10.0 cm is suspended from a scale and immersed in water as shown in Figure P14.22b. The 12.0-cm dimension is vertical, and the top of the block is 5.00 cm below the surface of the water. (a) What are the forces acting on the top and on the bottom of the block? (Take P_0 = 101.30 kPa.) (b) What is the reading of the spring scale? (c) Show that the buoyant force equals the difference between the forces at the top and bottom of the block.

24. ● The weight of a rectangular block of low-density material is 15.0 N. With a thin string, the center of the horizontal bottom face of the block is tied to the bottom of a beaker partly filled with water. When 25.0% of the block's volume is submerged, the tension in the string is 10.0 N. (a) Sketch a free-body diagram for the block, showing all forces acting on it. (b) Find the buoyant force on the block. (c) Oil of density 800 kg/m^3 is now steadily added to the beaker, forming a layer above the water and surrounding the block. The oil exerts forces on each of the four sidewalls of the block that the oil touches. What are the directions of these forces? (d) What happens to the string tension as the oil is added? Explain how the oil has this effect on the string tension. (e) The string breaks when its tension reaches 60.0 N. At this moment, 25.0% of the block's volume is still below the waterline. What additional fraction of the block's volume is below the top surface of the oil? (f) After the string breaks, the block comes to a new equilibrium position in the beaker. It is now in contact only with the oil. What fraction of the block's volume is submerged?

25. Preparing to anchor a buoy at the edge of a swimming area, a worker uses a rope to lower a cubical concrete block, 0.250 m on each edge, into ocean water. The block moves down at a constant speed of 1.90 m/s. You can accurately model the concrete and the water as incompressible. (a) At what rate is the force the water exerts on one face of the block increasing? (b) At what rate is the buoyant force on the block increasing?

26. To an order of magnitude, how many helium-filled toy balloons would be required to lift you? Because helium is an irreplaceable resource, develop a theoretical answer rather than an experimental answer. In your solution, state what physical quantities you take as data and the values you measure or estimate for them.

27. ▲ A cube of wood having an edge dimension of 20.0 cm and a density of 650 kg/m^3 floats on water. (a) What is the distance from the horizontal top surface of the cube to the water level? (b) What mass of lead should be placed on the cube so that the top of the cube will be just level with the water?

28. A spherical aluminum ball of mass 1.26 kg contains an empty spherical cavity that is concentric with the ball. The ball barely floats in water. Calculate (a) the outer radius of the ball and (b) the radius of the cavity.

29. Determination of the density of a fluid has many important applications. A car battery contains sulfuric acid, for which density is a measure of concentration. For the battery to function properly, the density must be within a range specified by the manufacturer. Similarly, the effectiveness of antifreeze in your car's engine coolant depends on the density of the mixture (usually ethylene glycol and water). When you donate blood to a blood bank, its screening includes determination of the density of the blood because higher density correlates with higher hemoglobin content. A *hydrometer* is an instrument used to determine liquid density. A simple one is sketched in Figure P14.29. The bulb of a syringe is squeezed and released to let the atmosphere lift a sample of the liquid of interest into a tube containing a calibrated rod of known density. The rod, of length L and average density ρ_0, floats partially immersed in the liquid of density ρ. A length h of the rod protrudes above the surface of the liquid. Show that the density of the liquid is

$$\rho = \frac{\rho_0 L}{L - h}$$

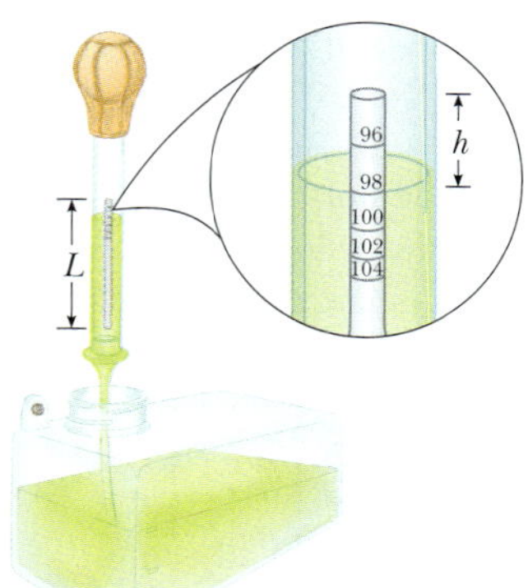

Figure P14.29 Problems 29 and 30.

30. ● Refer to Problem 29 and Figure P14.29. A hydrometer is to be constructed with a cylindrical floating rod. Nine fiduciary marks are to be placed along the rod to indicate densities having values of 0.98 g/cm^3, 1.00 g/cm^3, 1.02 g/cm^3, 1.04 g/cm^3, . . ., 1.14 g/cm^3. The row of marks is to start 0.200 cm from the top end of the rod and end 1.80 cm from the top end. (a) What is the required length of the rod? (b) What must be its average density? (c) Should the marks be equally spaced? Explain your answer.

31. How many cubic meters of helium are required to lift a balloon with a 400-kg payload to a height of 8 000 m? (Take ρ_{He} = 0.180 kg/m^3.) Assume the balloon maintains a constant volume and the density of air decreases with the altitude z according to the expression $\rho_{air} = \rho_0 e^{-z/8\,000}$, where z is in meters and ρ_0 = 1.25 kg/m^3 is the density of air at sea level.

32. A bathysphere used for deep-sea exploration has a radius of 1.50 m and a mass of 1.20 × 10^4 kg. To dive, this submarine takes on mass in the form of seawater. Determine the amount of mass the submarine must take on if it is to descend at a constant speed of 1.20 m/s, when the resistive force on it is 1 100 N in the upward direction. The density of seawater is 1.03 × 10^3 kg/m^3.

33. A plastic sphere floats in water with 50.0% of its volume submerged. This same sphere floats in glycerin with 40.0% of its volume submerged. Determine the densities of the glycerin and the sphere.

34. The United States possesses the eight largest warships in the world—aircraft carriers of the *Nimitz* class—and is building two more. Suppose one of the ships bobs up to float 11.0 cm higher in the water when 50 fighter planes take off from it in 25 minutes, at a location where the free-fall acceleration is 9.78 m/s^2. Bristling with bombs and missiles, the planes have an average mass of 29 000 kg. Find the horizontal area enclosed by the waterline of the $4-billion ship. By comparison, its flight deck has area 18 000 m^2. Below decks are passageways hundreds of meters long, so narrow that two large men cannot pass each other.

Section 14.5 Fluid Dynamics

Section 14.6 Bernoulli's Equation

35. A large storage tank, open at the top and filled with water, develops a small hole in its side at a point 16.0 m below the water level. The rate of flow from the leak is 2.50×10^{-3} m^3/min. Determine (a) the speed at which the water leaves the hole and (b) the diameter of the hole.

36. A village maintains a large tank with an open top, containing water for emergencies. The water can drain from the tank through a hose of diameter 6.60 cm. The hose ends with a nozzle of diameter 2.20 cm. A rubber stopper is inserted into the nozzle. The water level in the tank is kept 7.50 m above the nozzle. (a) Calculate the friction force exerted on the stopper by the nozzle. (b) The stopper is removed. What mass of water flows from the nozzle in 2.00 h? (c) Calculate the gauge pressure of the flowing water in the hose just behind the nozzle.

37. Water flows through a fire hose of diameter 6.35 cm at a rate of 0.012 0 m^3/s. The fire hose ends in a nozzle of inner diameter 2.20 cm. What is the speed with which the water exits the nozzle?

38. Water moves through a constricted pipe in steady, ideal flow. At one point as shown in Figure 14.16 where the pressure is 2.50×10^4 Pa, the diameter is 8.00 cm. At another point 0.500 m higher, the pressure is equal to 1.50×10^4 Pa and the diameter is 4.00 cm. Find the speed of flow (a) in the lower section and (b) in the upper section. (c) Find the volume flow rate through the pipe.

39. Figure P14.39 shows a stream of water in steady flow from a kitchen faucet. At the faucet, the diameter of the stream is 0.960 cm. The stream fills a 125-cm^3 container in 16.3 s. Find the diameter of the stream 13.0 cm below the opening of the faucet.

40. Water falls over a dam of height h with a mass flow rate of R, in units of kilograms per second. (a) Show that the power available from the water is

$$\mathcal{P} = Rgh$$

where g is the free-fall acceleration. (b) Each hydroelectric unit at the Grand Coulee Dam takes in water at a rate of 8.50×10^5 kg/s from a height of 87.0 m. The power developed by the falling water is converted to electric power with an efficiency of 85.0%. How much electric power does each hydroelectric unit produce?

41. A legendary Dutch boy saved Holland by plugging a 1.20-cm diameter hole in a dike with his finger. If the hole was 2.00 m below the surface of the North Sea (density 1 030 kg/m^3), (a) what was the force on his finger? (b) If he pulled his finger out of the hole, during what time interval would the released water fill 1 acre of land to a depth of 1 ft? Assume the hole remained constant in size. (A typical U.S. family of four uses 1 acre-foot of water, 1 234 m^3, in 1 year.)

42. ● In ideal flow, a liquid of density 850 kg/m^3 moves from a horizontal tube of radius 1.00 cm into a second horizontal tube of radius 0.500 cm. A pressure difference ΔP exists between the tubes. (a) Find the volume flow rate as a function of ΔP. Evaluate the volume flow rate (b) for $\Delta P = 6.00$ kPa and (c) for $\Delta P = 12.0$ kPa. (d) State how the volume flow rate depends on ΔP.

43. Water is pumped up from the Colorado River to supply Grand Canyon Village, located on the rim of the canyon. The river is at an elevation of 564 m, and the village is at an elevation of 2 096 m. Imagine that the water is pumped through a single long pipe 15.0 cm in diameter, driven by a single pump at the bottom end. (a) What is the minimum pressure at which the water must be pumped if it is to arrive at the village? (b) If 4 500 m^3 of water is pumped per day, what is the speed of the water in the pipe? (c) What additional pressure is necessary to deliver this flow? *Note:* Assume the free-fall acceleration and the density of air are constant over this range of elevations. The pressures you calculate are too high for an ordinary pipe. The water is actually lifted in stages by several pumps through shorter pipes.

44. ● Old Faithful Geyser in Yellowstone National Park erupts at approximately 1-h intervals, and the height of the water column reaches 40.0 m (Fig. P14.44). (a) Model the rising stream as a series of separate drops. Analyze the free-fall motion of one of the drops to determine the speed at which the water leaves the ground. (b) **What If?** Model the rising stream as an ideal fluid in streamline

George Semple

Figure P14.39

Stan Osolinski/Dembinsky Photo Associates

Figure P14.44

2 = intermediate; 3 = challenging; □ = SSM/SG; ▲ = ThomsonNOW; = symbolic reasoning; ● = qualitative reasoning

flow. Use Bernoulli's equation to determine the speed of the water as it leaves ground level. (c) How does the answer from part (a) compare with the answer from part (b)? (d) What is the pressure (above atmospheric) in the heated underground chamber if its depth is 175 m? Assume the chamber is large compared with the geyser's vent.

45. A Venturi tube may be used as a fluid flowmeter (see Fig. 14.19). Taking the difference in pressure as $P_1 - P_2 =$ 21.0 kPa, find the fluid flow rate in cubic meters per second given that the radius of the outlet tube is 1.00 cm, the radius of the inlet tube is 2.00 cm, and the fluid is gasoline ($\rho = 700$ kg/m^3).

Section 14.7 Other Applications of Fluid Dynamics

46. An airplane has a mass of 1.60×10^4 kg, and each wing has an area of 40.0 m^2. During level flight, the pressure on the lower wing surface is 7.00×10^4 Pa. Determine the pressure on the upper wing surface.

47. A siphon of uniform diameter is used to drain water from a tank as illustrated in Figure P14.47. Assume steady flow without friction. (a) If h = 1.00 m, find the speed of outflow at the end of the siphon. (b) **What If?** What is the limitation on the height of the top of the siphon above the water surface? (For the flow of the liquid to be continuous, the pressure must not drop below the vapor pressure of the liquid.)

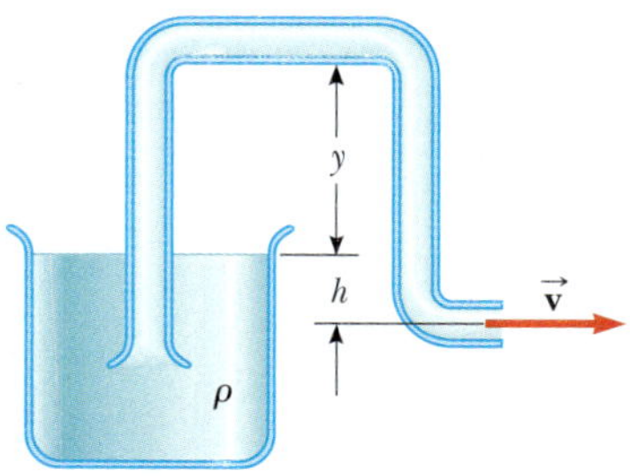

Figure P14.47

48. An airplane is cruising at altitude 10 km. The pressure outside the craft is 0.287 atm; within the passenger compartment, the pressure is 1.00 atm and the temperature is 20°C. A small leak occurs in one of the window seals in the passenger compartment. Model the air as an ideal fluid to find the speed of the stream of air flowing through the leak.

49. A hypodermic syringe contains a medicine having the density of water (Fig. P14.49). The barrel of the syringe has a cross-sectional area $A = 2.50 \times 10^{-5}$ m^2, and the needle has a cross-sectional area $a = 1.00 \times 10^{-8}$ m^2. In the absence of a force on the plunger, the pressure everywhere is 1 atm. A force $\vec{F}$ of magnitude 2.00 N acts on the plunger, making medicine squirt horizontally from the needle. Determine the speed of the medicine as it leaves the needle's tip.

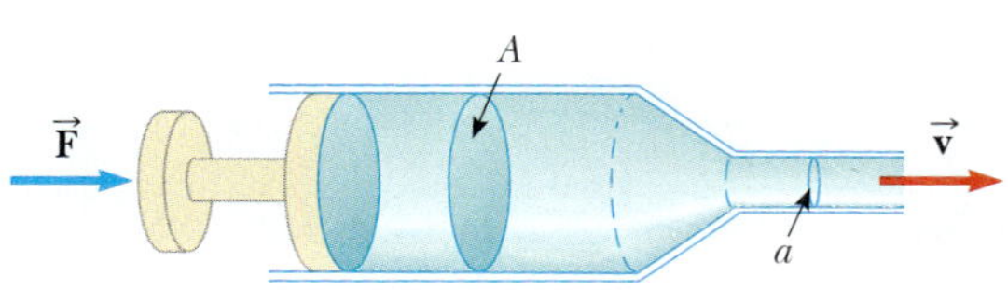

Figure P14.49

50. The Bernoulli effect can have important consequences for the design of buildings. For example, wind can blow around a skyscraper at remarkably high speed, creating low pressure. The higher atmospheric pressure in the still air inside the buildings can cause windows to pop out. As originally constructed, the John Hancock Building in Boston popped windowpanes that fell many stories to the sidewalk below. (a) Suppose a horizontal wind blows with a speed of 11.2 m/s outside a large pane of plate glass with dimensions 4.00 m × 1.50 m. Assume the density of the air to be 1.30 kg/m^3. The air inside the building is at atmospheric pressure. What is the total force exerted by air on the windowpane? (b) **What If?** If a second skyscraper is built nearby, the airspeed can be especially high where wind passes through the narrow separation between the buildings. Solve part (a) again with a wind speed of 22.4 m/s, twice as high.

Additional Problems

51. A helium-filled balloon is tied to a 2.00-m-long, 0.050 0-kg uniform string. The balloon is spherical with a radius of 0.400 m. When released, it lifts a length h of string and then remains in equilibrium as shown in Figure P14.51. Determine the value of h. The envelope of the balloon has a mass of 0.250 kg.

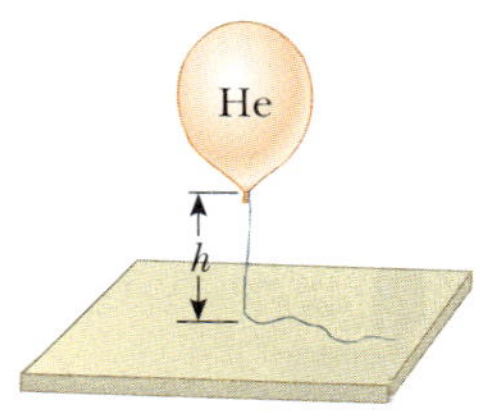

Figure P14.51

52. Figure P14.52 shows a water tank with a valve at the bottom. If this valve is opened, what is the maximum height attained by the water stream coming out of the right side of the tank? Assume h = 10.0 m, L = 2.00 m, and θ = 30.0° and assume the cross-sectional area at A is very large compared with that at B.

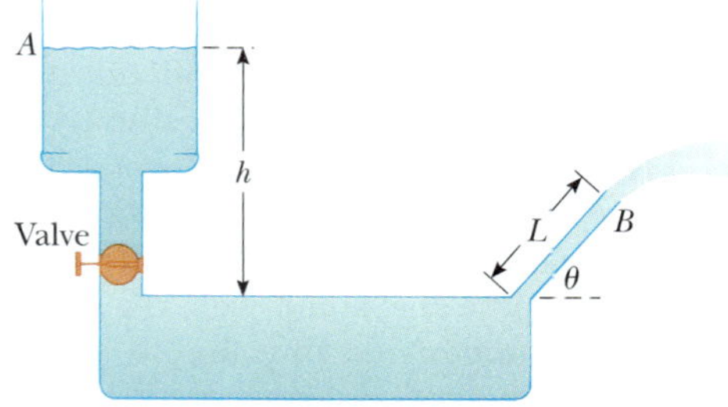

Figure P14.52

53. The true weight of an object can be measured in a vacuum, where buoyant forces are absent. An object of volume V is weighed in air on an equal-arm balance with the use of counterweights of density ρ. Representing the density of air as ρ_{air} and the balance reading as F'_g, show that the true weight F_g is

$$F_g = F'_g + \left(V - \frac{F'_g}{\rho g}\right)\rho_{air} g$$

54. Water is forced out of a fire extinguisher by air pressure as shown in Figure P14.54. How much gauge air pressure in the tank (above atmospheric) is required for the water jet to have a speed of 30.0 m/s when the water level is 0.500 m below the nozzle?

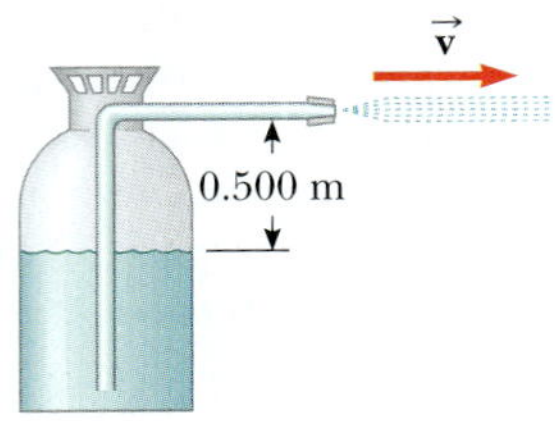

Figure P14.54

55. A light spring of constant $k = 90.0$ N/m is attached vertically to a table (Fig. P14.55a). A 2.00-g balloon is filled with helium (density $= 0.180$ kg/m^3) to a volume of 5.00 m^3 and is then connected to the spring, causing the spring to stretch as shown in Figure P14.55b. Determine the extension distance L when the balloon is in equilibrium.

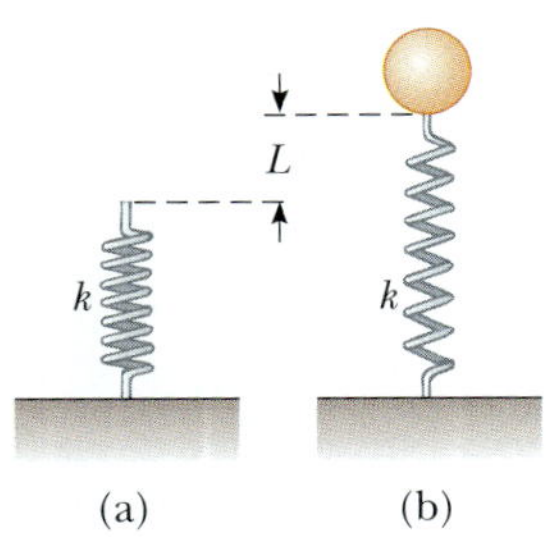

Figure P14.55

56. ● *We can't call it Flubber.* Assume a certain liquid, with density 1 230 kg/m^3, exerts no friction force on spherical objects. A ball of mass 2.10 kg and radius 9.00 cm is dropped from rest into a deep tank of this liquid from a height of 3.30 m above the surface. (a) Find the speed at which the ball enters the liquid. (b) What two forces are exerted on the ball as it moves through the liquid? (c) Explain why the ball moves down only a limited distance into the liquid and calculate this distance. (d) With what speed does the ball pop up out of the liquid? (e) How does the time interval Δt_{down}, during which the ball moves from the surface down to its lowest point, compare with the time interval Δt_{up} for the return trip between the same two points? (f) **What If?** Now modify the model to suppose the liquid exerts a small friction force on the ball, opposite in direction to its motion. In this case, how do the time intervals Δt_{down} and Δt_{up} compare? Explain your answer with a conceptual argument rather than a numerical calculation.

57. ● As a 950-kg helicopter hovers, its horizontal rotor pushes a column of air downward at 40.0 m/s. What can you say about the quantity of this air? Explain your answer. You may model the air motion as ideal flow.

58. Evangelista Torricelli was the first person to realize that we live at the bottom of an ocean of air. He correctly surmised that the pressure of our atmosphere is attributable to the weight of the air. The density of air at 0°C at the Earth's surface is 1.29 kg/m^3. The density decreases with increasing altitude (as the atmosphere thins). On the other hand, if we assume the density is constant at 1.29 kg/m^3 up to some altitude h and is zero above that altitude, then h would represent the depth of the ocean of air. Use this model to determine the value of h that gives a pressure of 1.00 atm at the surface of the Earth. Would the peak of Mount Everest rise above the surface of such an atmosphere?

59. ▲ **Review problem.** With reference to Figure 14.5, show that the total torque exerted by the water behind the dam about a horizontal axis through O is $\frac{1}{6}\rho g w H^3$. Show that the effective line of action of the total force exerted by the water is at a distance $\frac{1}{3}H$ above O.

60. In about 1657, Otto von Guericke, inventor of the air pump, evacuated a sphere made of two brass hemispheres. Two teams of eight horses each could pull the hemispheres apart only on some trials and then "with greatest difficulty," with the resulting sound likened to a cannon firing (Fig. P14.60). (a) Show that the force F required to pull the thin-walled evacuated hemispheres apart is $\pi R^2(P_0 - P)$, where R is the radius of the hemispheres and P is the pressure inside the hemispheres, which is much less than P_0. (b) Determine the force for $P = 0.100P_0$ and $R = 0.300$ m.

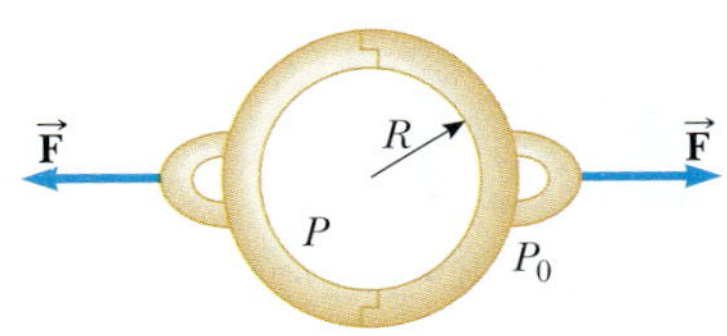

The Granger Collection

Figure P14.60 The colored engraving, dated 1672, illustrates Otto von Guericke's demonstration of the force due to air pressure as it might have been performed before Emperor Ferdinand III.

61. A 1.00-kg beaker containing 2.00 kg of oil (density = 916.0 kg/m^3) rests on a scale. A 2.00-kg block of iron suspended from a spring scale is completely submerged in the oil as shown in Figure P14.61. Determine the equilibrium readings of both scales.

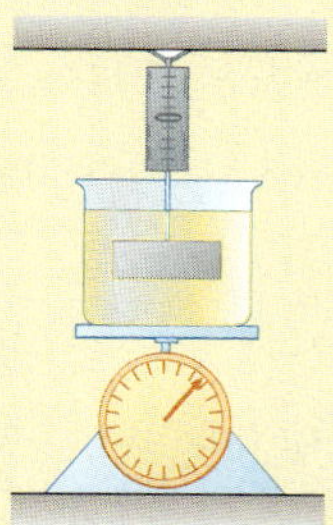

Figure P14.61 Problems 61 and 62.

62. A beaker of mass m_b containing oil of mass m_o and density ρ_o rests on a scale. A block of iron of mass m_{Fe} suspended from a spring scale is completely submerged in the oil as shown in Figure P14.61. Determine the equilibrium readings of both scales.

63. In 1983, the United States began coining the cent piece out of copper-clad zinc rather than pure copper. The mass of the old copper penny is 3.083 g and that of the new cent is 2.517 g. Calculate the percent of zinc (by volume) in the new cent. The density of copper is 8.960 g/cm^3 and that of zinc is 7.133 g/cm^3. The new and old coins have the same volume.

64. Show that the variation of atmospheric pressure with altitude is given by $P = P_0 e^{-\alpha y}$, where $\alpha = \rho_0 g/P_0$, P_0 is atmospheric pressure at some reference level $y = 0$, and ρ_0 is the atmospheric density at this level. Assume the decrease in atmospheric pressure over an infinitesimal change in altitude (so that the density is approximately uniform) is given by $dP = -\rho g\, dy$ and that the density of air is proportional to the pressure.

65. **Review problem.** A uniform disk of mass 10.0 kg and radius 0.250 m spins at 300 rev/min on a low-friction axle. It must be brought to a stop in 1.00 min by a brake pad that makes contact with the disk at an average distance of 0.220 m from the axis. The coefficient of friction between the pad and the disk is 0.500. A piston in a cylinder of diameter 5.00 cm presses the brake pad against the disk. Find the pressure required for the brake fluid in the cylinder.

66. A cube of ice whose edges measure 20.0 mm is floating in a glass of ice-cold water, and one of the ice cube's faces is parallel to the water's surface. (a) How far below the water surface is the bottom face of the ice cube? (b) Ice-cold ethyl alcohol is gently poured onto the water surface to form a layer 5.00 mm thick above the water. The alcohol does not mix with the water. When the ice cube again attains hydrostatic equilibrium, what is the distance from the top of the water to the bottom face of the block? (c) Additional cold ethyl alcohol is poured onto the water's surface until the top surface of the alcohol coincides with the top surface of the ice cube (in hydrostatic equilibrium). How thick is the required layer of ethyl alcohol?

67. An incompressible, nonviscous fluid is initially at rest in the vertical portion of the pipe shown in Figure P14.67a, where L = 2.00 m. When the valve is opened, the fluid flows into the horizontal section of the pipe. What is the speed of the fluid when it is all in the horizontal section as shown in Figure P14.67b? Assume the cross-sectional area of the entire pipe is constant.

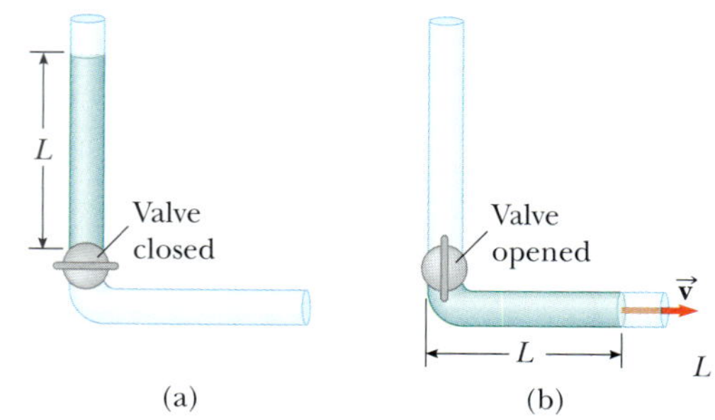

Figure P14.67

68. The water supply of a building is fed through a main pipe 6.00 cm in diameter. A 2.00-cm-diameter faucet tap, located 2.00 m above the main pipe, is observed to fill a 25.0-L container in 30.0 s. (a) What is the speed at which the water leaves the faucet? (b) What is the gauge pressure in the 6-cm main pipe? (Assume the faucet is the only "leak" in the building.)

69. A U-tube open at both ends is partially filled with water (Fig. P14.69a). Oil having a density 750 kg/m^3 is then poured into the right arm and forms a column L = 5.00 cm high (Fig. P14.69b). (a) Determine the difference h in the heights of the two liquid surfaces. (b) The right arm is then shielded from any air motion while air is blown across the top of the left arm until the surfaces of the two liquids are at the same height (Fig. P14.69c). Determine the speed of the air being blown across the left arm. Take the density of air as 1.29 kg/m^3.

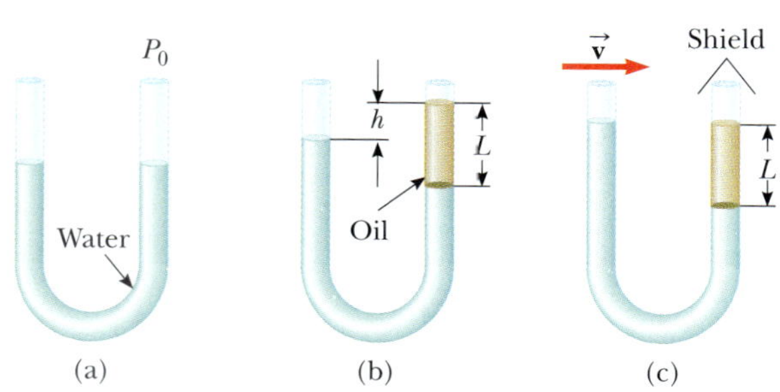

Figure P14.69

70. A woman is draining her fish tank by siphoning the water into an outdoor drain as shown in Figure P14.70 (page 416). The rectangular tank has footprint area A and depth h. The drain is located a distance d below the surface of the water in the tank, where $d >> h$. The cross-sectional area of the siphon tube is A'. Model the water as

flowing without friction. (a) Show that the time interval required to empty the tank is

$$\Delta t = \frac{Ah}{A'\sqrt{2gd}}$$

(b) Evaluate the time interval required to empty the tank if it is a cube 0.500 m on each edge, taking $A' = 2.00$ cm^2 and $d = 10.0$ m.

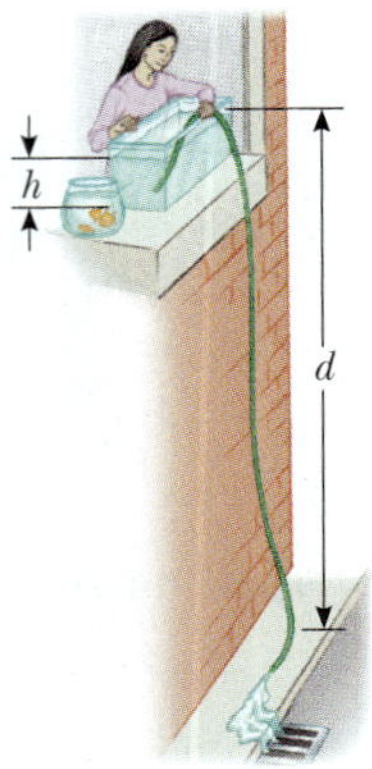

Figure P14.70

71. The hull of an experimental boat is to be lifted above the water by a hydrofoil mounted below its keel as shown in Figure P14.71. The hydrofoil is shaped like an airplane wing. Its area projected onto a horizontal surface is A. When the boat is towed at sufficiently high speed, water of density ρ moves in streamline flow so that its average speed at the top of the hydrofoil is n times larger than its speed v_b below the hydrofoil. (a) Ignoring the buoyant force, show that the upward lift force exerted by the water on the hydrofoil has a magnitude

$$F \approx \tfrac{1}{2}(n^2 - 1)\rho v_b^2 A$$

(b) The boat has mass M. Show that the liftoff speed is

$$v \approx \sqrt{\frac{2Mg}{(n^2 - 1)A\rho}}$$

(c) Assume an 800-kg boat is to lift off at 9.50 m/s. Evaluate the area A required for the hydrofoil if its design yields $n = 1.05$.

Figure P14.71

Answers to Quick Quizzes

14.1 (a). Because the basketball player's weight is distributed over the larger surface area of the shoe, the pressure (F/A) he applies is relatively small. The woman's lesser weight is distributed over the very small cross-sectional area of the spiked heel, so the pressure is high.

14.2 (a). Because both fluids have the same depth, the one with the smaller density (alcohol) will exert the smaller pressure.

14.3 (c). All barometers will have the same pressure at the bottom of the column of fluid: atmospheric pressure. Therefore, the barometer with the highest column will be the one with the fluid of lowest density.

14.4 (b) or (c). In all three cases, the weight of the treasure chest causes a downward force on the raft that makes the raft sink into the water. In (b) and (c), however, the treasure chest also displaces water, which provides a buoyant force in the upward direction, reducing the effect of the chest's weight.

14.5 (a). The high-speed air between the balloons results in low pressure in this region. The higher pressure on the outer surfaces of the balloons pushes them toward each other.

PART 2

Oscillations and Mechanical Waves

We begin this new part of the text by studying a special type of motion called *periodic* motion, the repeating motion of an object in which it continues to return to a given position after a fixed time interval. The repetitive movements of such an object are called *oscillations*. We will focus our attention on a special case of periodic motion called *simple harmonic motion*. All periodic motions can be modeled as combinations of simple harmonic motions.

Simple harmonic motion also forms the basis for our understanding of *mechanical waves*. Sound waves, seismic waves, waves on stretched strings, and water waves are all produced by some source of oscillation. As a sound wave travels through the air, elements of the air oscillate back and forth; as a water wave travels across a pond, elements of the water oscillate up and down and backward and forward. The motion of the elements of the medium bears a strong resemblance to the periodic motion of an oscillating pendulum or an object attached to a spring.

To explain many other phenomena in nature, we must understand the concepts of oscillations and waves. For instance, although skyscrapers and bridges appear to be rigid, they actually oscillate, something the architects and engineers who design and build them must take into account. To understand how radio and television work, we must understand the origin and nature of electromagnetic waves and how they propagate through space. Finally, much of what scientists have learned about atomic structure has come from information carried by waves. Therefore, we must first study oscillations and waves if we are to understand the concepts and theories of atomic physics.

Drops of water fall from a leaf into a pond. The disturbance caused by the falling water causes the water surface to oscillate. These oscillations are associated with waves moving away from the point at which the water fell. In Part 2 of the text, we will explore the principles related to oscillations and waves. (Don Bonsey/Getty Images)

To reduce swaying in tall buildings because of the wind, tuned dampers are placed near the top of the building. These mechanisms include an object of large mass that oscillates under computer control at the same frequency as the building, reducing the swaying. The large sphere in the photograph on the left is part of the tuned damper system of the building in the photograph on the right, called Taipei 101, in Taiwan. The building, also called the Taipei Financial Center, was completed in 2004, at which time it held the record as the world's tallest building. (left, Courtesy of Motioneering, Inc.; right, © Simon Kwang/Reuters/CORBIS)

15 Oscillatory Motion

Periodic motion **is motion of an object that regularly returns to a given position** after a fixed time interval. With a little thought, we can identify several types of periodic motion in everyday life. Your car returns to the driveway each afternoon. You return to the dinner table each night to eat. A bumped chandelier swings back and forth, returning to the same position at a regular rate. The Earth returns to the same position in its orbit around the Sun each year, resulting in the variation among the four seasons.

In addition to these everyday examples, numerous other systems exhibit periodic motion. The molecules in a solid oscillate about their equilibrium positions; electromagnetic waves, such as light waves, radar, and radio waves, are characterized by oscillating electric and magnetic field vectors; and in alternating-current electrical circuits, voltage, current, and electric charge vary periodically with time.

A special kind of periodic motion occurs in mechanical systems when the force acting on an object is proportional to the position of the object relative to some equilibrium position. If this force is always directed toward the equilibrium position, the motion is called *simple harmonic motion*, which is the primary focus of this chapter.

15.1 Motion of an Object Attached to a Spring

As a model for simple harmonic motion, consider a block of mass m attached to the end of a spring, with the block free to move on a horizontal, frictionless surface (Active Fig. 15.1). When the spring is neither stretched nor compressed, the block is at rest at the position called the **equilibrium position** of the system, which we identify as $x = 0$. We know from experience that such a system oscillates back and forth if disturbed from its equilibrium position.

We can understand the oscillating motion of the block in Active Figure 15.1 qualitatively by first recalling that when the block is displaced to a position x, the spring exerts on the block a force that is proportional to the position and given by **Hooke's law** (see Section 7.4):

$$F_s = -kx \qquad (15.1)$$

◀ Hooke's law

We call F_s a **restoring force** because it is always directed toward the equilibrium position and therefore *opposite* the displacement of the block from equilibrium. That is, when the block is displaced to the right of $x = 0$ in Active Figure 15.1a, the position is positive and the restoring force is directed to the left. Figure 15.1b shows the block at $x = 0$, where the force on the block is zero. When the block is displaced to the left of $x = 0$ as in Figure 15.1c, the position is negative and the restoring force is directed to the right.

Applying Newton's second law to the motion of the block, with Equation 15.1 providing the net force in the x direction, we obtain

$$-kx = ma_x$$

$$a_x = -\frac{k}{m}x \qquad (15.2)$$

That is, the acceleration of the block is proportional to its position, and the direction of the acceleration is opposite the direction of the displacement of the block from equilibrium. Systems that behave in this way are said to exhibit **simple harmonic motion.** An object moves with simple harmonic motion whenever its acceleration is proportional to its position and is oppositely directed to the displacement from equilibrium.

If the block in Active Figure 15.1 is displaced to a position $x = A$ and released from rest, its *initial* acceleration is $-kA/m$. When the block passes through the equilibrium position $x = 0$, its acceleration is zero. At this instant, its speed is a maximum because the acceleration changes sign. The block then continues to travel to the left of equilibrium with a positive acceleration and finally reaches $x = -A$, at which time its acceleration is $+kA/m$ and its speed is again zero as discussed in Sections 7.4 and 7.9. The block completes a full cycle of its motion by returning to the original position, again passing through $x = 0$ with maximum speed. Therefore,

PITFALL PREVENTION 15.1

The Orientation of the Spring

Active Figure 15.1 shows a *horizontal* spring, with an attached block sliding on a frictionless surface. Another possibility is a block hanging from a *vertical* spring. All the results we discuss for the horizontal spring are the same for the vertical spring with one exception: when the block is placed on the vertical spring, its weight causes the spring to extend. If the resting position of the block is defined as $x = 0$, the results of this chapter also apply to this vertical system.

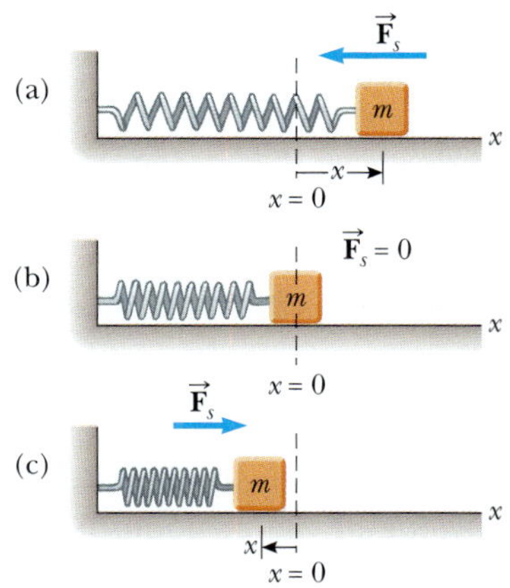

ACTIVE FIGURE 15.1

A block attached to a spring moving on a frictionless surface. (a) When the block is displaced to the right of equilibrium ($x > 0$), the force exerted by the spring acts to the left. (b) When the block is at its equilibrium position ($x = 0$), the force exerted by the spring is zero. (c) When the block is displaced to the left of equilibrium ($x < 0$), the force exerted by the spring acts to the right.

Sign in at www.thomsonedu.com and go to ThomsonNOW to choose the spring constant and the initial position and velocity of the block and see the resulting simple harmonic motion.

the block oscillates between the turning points $x = \pm A$. In the absence of friction, this idealized motion will continue forever because the force exerted by the spring is conservative. Real systems are generally subject to friction, so they do not oscillate forever. We shall explore the details of the situation with friction in Section 15.6.

Quick Quiz 15.1 A block on the end of a spring is pulled to position $x = A$ and released from rest. In one full cycle of its motion, through what total distance does it travel? (a) $A/2$ (b) A (c) $2A$ (d) $4A$

15.2 The Particle in Simple Harmonic Motion

The motion described in the preceding section occurs so often that we identify the **particle in simple harmonic motion** model to represent such situations. To develop a mathematical representation for this model, first recognize that the block is a particle under a net force as described in Equation 15.1. We will generally choose x as the axis along which the oscillation occurs; hence, we will drop the subscript-x notation in this discussion. Recall that, by definition, $a = dv/dt = d^2x/dt^2$, and so we can express Equation 15.2 as

$$\frac{d^2x}{dt^2} = -\frac{k}{m}x \tag{15.3}$$

PITFALL PREVENTION 15.2
A Nonconstant Acceleration

The acceleration of a particle in simple harmonic motion is not constant. Equation 15.3 shows that its acceleration varies with position x. Therefore, we *cannot* apply the kinematic equations of Chapter 2 in this situation.

If we denote the ratio k/m with the symbol ω^2 (we choose ω^2 rather than ω so as to make the solution we develop below simpler in form), then

$$\omega^2 = \frac{k}{m} \tag{15.4}$$

and Equation 15.3 can be written in the form

$$\frac{d^2x}{dt^2} = -\omega^2 x \tag{15.5}$$

Let's now find a mathematical solution to Equation 15.5, that is, a function $x(t)$ that satisfies this second-order differential equation and is a mathematical representation of the position of the particle as a function of time. We seek a function whose second derivative is the same as the original function with a negative sign and multiplied by ω^2. The trigonometric functions sine and cosine exhibit this behavior, so we can build a solution around one or both of them. The following cosine function is a solution to the differential equation:

Position versus time for an object in simple harmonic motion ▶

$$x(t) = A\cos(\omega t + \phi) \tag{15.6}$$

where A, ω, and ϕ are constants. To show explicitly that this solution satisfies Equation 15.5, notice that

$$\frac{dx}{dt} = A\frac{d}{dt}\cos(\omega t + \phi) = -\omega A\sin(\omega t + \phi) \tag{15.7}$$

$$\frac{d^2x}{dt^2} = -\omega A\frac{d}{dt}\sin(\omega t + \phi) = -\omega^2 A\cos(\omega t + \phi) \tag{15.8}$$

PITFALL PREVENTION 15.3
Where's the Triangle?

Equation 15.6 includes a trigonometric function, a *mathematical function* that can be used whether it refers to a triangle or not. In this case, the cosine function happens to have the correct behavior for representing the position of a particle in simple harmonic motion.

Comparing Equations 15.6 and 15.8, we see that $d^2x/dt^2 = -\omega^2 x$ and Equation 15.5 is satisfied.

The parameters A, ω, and ϕ are constants of the motion. To give physical significance to these constants, it is convenient to form a graphical representation of the motion by plotting x as a function of t as in Active Figure 15.2a. First, A, called the **amplitude** of the motion, is simply **the maximum value of the position of the particle in either the positive or negative x direction.** The constant ω is called the

angular frequency, and it has units[1] of rad/s. It is a measure of how rapidly the oscillations are occurring; the more oscillations per unit time, the higher the value of ω. From Equation 15.4, the angular frequency is

$$\omega = \sqrt{\frac{k}{m}} \tag{15.9}$$

The constant angle ϕ is called the **phase constant** (or initial phase angle) and, along with the amplitude A, is determined uniquely by the position and velocity of the particle at $t = 0$. If the particle is at its maximum position $x = A$ at $t = 0$, the phase constant is $\phi = 0$ and the graphical representation of the motion is as shown in Active Figure 15.2b. The quantity $(\omega t + \phi)$ is called the **phase** of the motion. Notice that the function $x(t)$ is periodic and its value is the same each time ωt increases by 2π radians.

Equations 15.1, 15.5, and 15.6 form the basis of the mathematical representation of the particle in simple harmonic motion model. If you are analyzing a situation and find that the force on a particle is of the mathematical form of Equation 15.1, you know the motion is that of a simple harmonic oscillator and the position of the particle is described by Equation 15.6. If you analyze a system and find that it is described by a differential equation of the form of Equation 15.5, the motion is that of a simple harmonic oscillator. If you analyze a situation and find that the position of a particle is described by Equation 15.6, you know the particle undergoes simple harmonic motion.

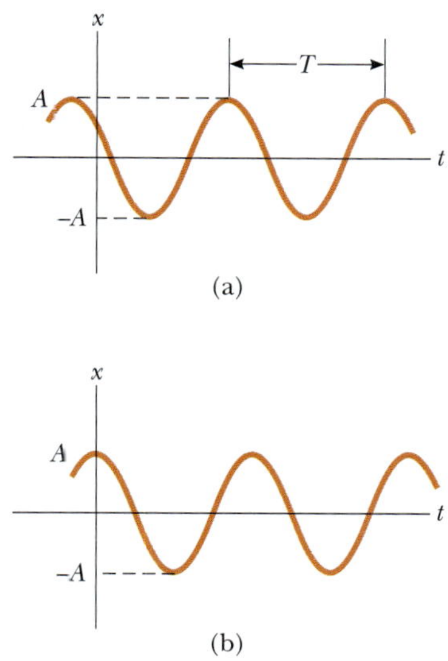

ACTIVE FIGURE 15.2

(a) An x–t graph for an object undergoing simple harmonic motion. The amplitude of the motion is A, the period (defined in Eq. 15.10) is T. (b) The x–t graph in the special case in which $x = A$ at $t = 0$ and hence $\phi = 0$.

Sign in at www.thomsonedu.com and go to ThomsonNOW to adjust the graphical representation and see the resulting simple harmonic motion of the block in Active Figure 15.1.

Quick Quiz 15.2 Consider a graphical representation (Fig. 15.3) of simple harmonic motion as described mathematically in Equation 15.6. When the object is at point Ⓐ on the graph, what can you say about its position and velocity? (a) The position and velocity are both positive. (b) The position and velocity are both negative. (c) The position is positive, and its velocity is zero. (d) The position is negative, and its velocity is zero. (e) The position is positive, and its velocity is negative. (f) The position is negative, and its velocity is positive.

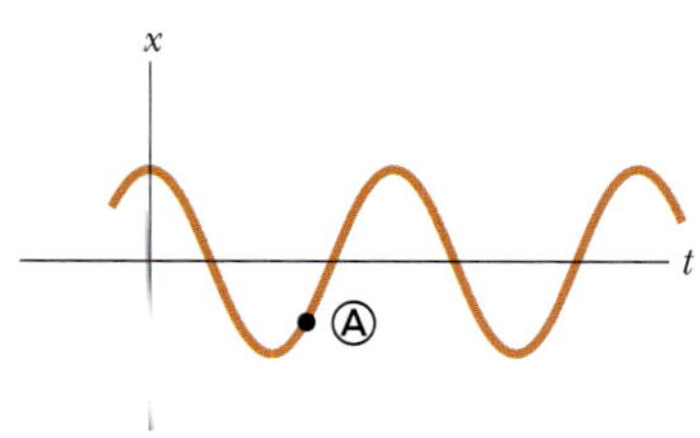

Figure 15.3 (Quick Quiz 15.2) An x–t graph for an object undergoing simple harmonic motion. At a particular time, the object's position is indicated by Ⓐ in the graph.

Quick Quiz 15.3 Figure 15.4 shows two curves representing objects undergoing simple harmonic motion. The correct description of these two motions is that the simple harmonic motion of object B is (a) of larger angular frequency and larger amplitude than that of object A, (b) of larger angular frequency and smaller amplitude than that of object A, (c) of smaller angular frequency and larger amplitude than that of object A, or (d) of smaller angular frequency and smaller amplitude than that of object A.

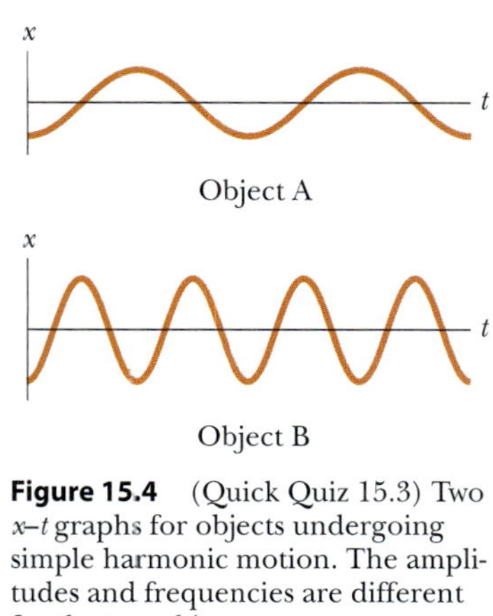

Figure 15.4 (Quick Quiz 15.3) Two x–t graphs for objects undergoing simple harmonic motion. The amplitudes and frequencies are different for the two objects.

Let us investigate further the mathematical description of simple harmonic motion. The **period** T of the motion is the time interval required for the particle to go through one full cycle of its motion (Active Fig. 15.2a). That is, the values of x and v for the particle at time t equal the values of x and v at time $t + T$. Because the phase increases by 2π radians in a time interval of T,

$$[\omega(t + T) + \phi] - (\omega t + \phi) = 2\pi$$

Simplifying this expression gives $\omega T = 2\pi$, or

$$T = \frac{2\pi}{\omega} \tag{15.10}$$

[1] We have seen many examples in earlier chapters in which we evaluate a trigonometric function of an angle. The argument of a trigonometric function, such as sine or cosine, *must* be a pure number. The radian is a pure number because it is a ratio of lengths. Angles in degrees are pure numbers because the degree is an artificial "unit"; it is not related to measurements of lengths. The argument of the trigonometric function in Equation 15.6 must be a pure number. Therefore, ω *must* be expressed in rad/s (and not, for example, in revolutions per second) if t is expressed in seconds. Furthermore, other types of functions such as logarithms and exponential functions require arguments that are pure numbers.

PITFALL PREVENTION 15.4
Two Kinds of Frequency

We identify two kinds of frequency for a simple harmonic oscillator: f, called simply the *frequency*, is measured in hertz, and ω, the *angular frequency*, is measured in radians per second. Be sure you are clear about which frequency is being discussed or requested in a given problem. Equations 15.11 and 15.12 show the relationship between the two frequencies.

The inverse of the period is called the **frequency** f of the motion. Whereas the period is the time interval per oscillation, the frequency represents the **number of oscillations the particle undergoes per unit time interval:**

$$f = \frac{1}{T} = \frac{\omega}{2\pi} \qquad (15.11)$$

The units of f are cycles per second, or **hertz** (Hz). Rearranging Equation 15.11 gives

$$\omega = 2\pi f = \frac{2\pi}{T} \qquad (15.12)$$

Equations 15.9, 15.10, and 15.11 can be used to express the period and frequency of the motion for the particle in simple harmonic motion in terms of the characteristics m and k of the system as

Period ▶

$$T = \frac{2\pi}{\omega} = 2\pi\sqrt{\frac{m}{k}} \qquad (15.13)$$

Frequency ▶

$$f = \frac{1}{T} = \frac{1}{2\pi}\sqrt{\frac{k}{m}} \qquad (15.14)$$

That is, the period and frequency depend *only* on the mass of the particle and the force constant of the spring and *not* on the parameters of the motion, such as A or ϕ. As we might expect, the frequency is larger for a stiffer spring (larger value of k) and decreases with increasing mass of the particle.

We can obtain the velocity and acceleration[2] of a particle undergoing simple harmonic motion from Equations 15.7 and 15.8:

Velocity of an object in simple harmonic motion ▶

$$v = \frac{dx}{dt} = -\omega A \sin(\omega t + \phi) \qquad (15.15)$$

Acceleration of an object in simple harmonic motion ▶

$$a = \frac{d^2x}{dt^2} = -\omega^2 A \cos(\omega t + \phi) \qquad (15.16)$$

From Equation 15.15 we see that, because the sine and cosine functions oscillate between ± 1, the extreme values of the velocity v are $\pm\omega A$. Likewise, Equation 15.16 shows that the extreme values of the acceleration a are $\pm\omega^2 A$. Therefore, the *maximum* values of the magnitudes of the velocity and acceleration are

Maximum magnitudes of velocity and acceleration in simple harmonic motion ▶

$$v_{\text{max}} = \omega A = \sqrt{\frac{k}{m}}A \qquad (15.17)$$

$$a_{\text{max}} = \omega^2 A = \frac{k}{m}A \qquad (15.18)$$

Figure 15.5a plots position versus time for an arbitrary value of the phase constant. The associated velocity–time and acceleration–time curves are illustrated in Figures 15.5b and 15.5c. They show that the phase of the velocity differs from the phase of the position by $\pi/2$ rad, or 90°. That is, when x is a maximum or a minimum, the velocity is zero. Likewise, when x is zero, the speed is a maximum. Furthermore, notice that the phase of the acceleration differs from the phase of the position by π radians, or 180°. For example, when x is a maximum, a has a maximum magnitude in the opposite direction.

Quick Quiz 15.4 An object of mass m is hung from a spring and set into oscillation. The period of the oscillation is measured and recorded as T. The object of

[2] Because the motion of a simple harmonic oscillator takes place in one dimension, we denote velocity as v and acceleration as a, with the direction indicated by a positive or negative sign as in Chapter 2.

mass m is removed and replaced with an object of mass $2m$. When this object is set into oscillation, what is the period of the motion? (a) $2T$ (b) $\sqrt{2}T$ (c) T (d) $T/\sqrt{2}$ (e) $T/2$

Equation 15.6 describes simple harmonic motion of a particle in general. Let's now see how to evaluate the constants of the motion. The angular frequency ω is evaluated using Equation 15.9. The constants A and ϕ are evaluated from the initial conditions, that is, the state of the oscillator at $t = 0$.

Suppose the particle is set into motion by pulling it from equilibrium by a distance A and releasing it from rest at $t = 0$ as in Active Figure 15.6. We must then require our solutions for $x(t)$ and $v(t)$ (Eqs. 15.6 and 15.15) to obey the initial conditions that $x(0) = A$ and $v(0) = 0$:

$$x(0) = A\cos\phi = A$$

$$v(0) = -\omega A\sin\phi = 0$$

These conditions are met if $\phi = 0$, giving $x = A\cos\omega t$ as our solution. To check this solution, notice that it satisfies the condition that $x(0) = A$ because $\cos 0 = 1$.

The position, velocity, and acceleration versus time are plotted in Figure 15.7a for this special case. The acceleration reaches extreme values of $\mp\omega^2 A$ when the position has extreme values of $\pm A$. Furthermore, the velocity has extreme values of $\pm\omega A$, which both occur at $x = 0$. Hence, the quantitative solution agrees with our qualitative description of this system.

Let's consider another possibility. Suppose the system is oscillating and we define $t = 0$ as the instant the particle passes through the unstretched position of the spring while moving to the right (Active Fig. 15.8). In this case, our solutions for $x(t)$ and $v(t)$ must obey the initial conditions that $x(0) = 0$ and $v(0) = v_i$:

$$x(0) = A\cos\phi = 0$$

$$v(0) = -\omega A\sin\phi = v_i$$

The first of these conditions tells us that $\phi = \pm\pi/2$. With these choices for ϕ, the second condition tells us that $A = \mp v_i/\omega$. Because the initial velocity is positive and the amplitude must be positive, we must have $\phi = -\pi/2$. Hence, the solution is

$$x = \frac{v_i}{\omega}\cos\left(\omega t - \frac{\pi}{2}\right)$$

The graphs of position, velocity, and acceleration versus time for this choice of $t = 0$ are shown in Figure 15.7b. Notice that these curves are the same as those in Figure 15.7a, but shifted to the right by one fourth of a cycle. This shift is described mathematically by the phase constant $\phi = -\pi/2$, which is one fourth of a full cycle of 2π.

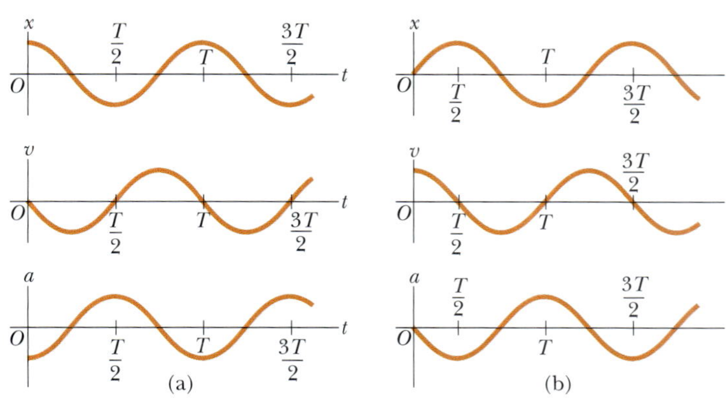

Figure 15.7 (a) Position, velocity, and acceleration versus time for a block undergoing simple harmonic motion under the initial conditions that at $t = 0$, $x(0) = A$ and $v(0) = 0$. (b) Position, velocity, and acceleration versus time for a block undergoing simple harmonic motion under the initial conditions that at $t = 0$, $x(0) = 0$ and $v(0) = v_i$.

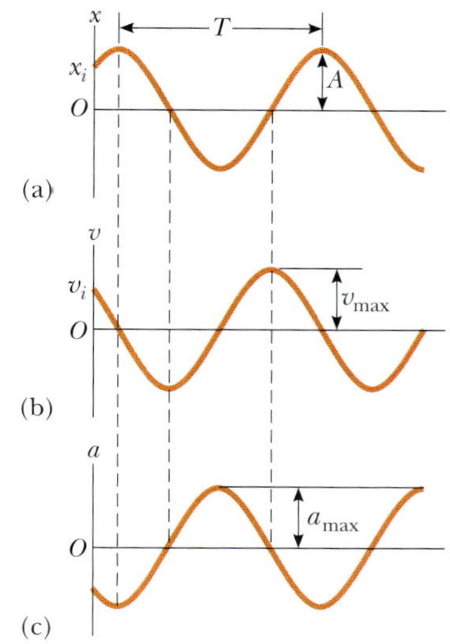

Figure 15.5 Graphical representation of simple harmonic motion. (a) Position versus time. (b) Velocity versus time. (c) Acceleration versus time. Notice that at any specified time the velocity is 90° out of phase with the position and the acceleration is 180° out of phase with the position.

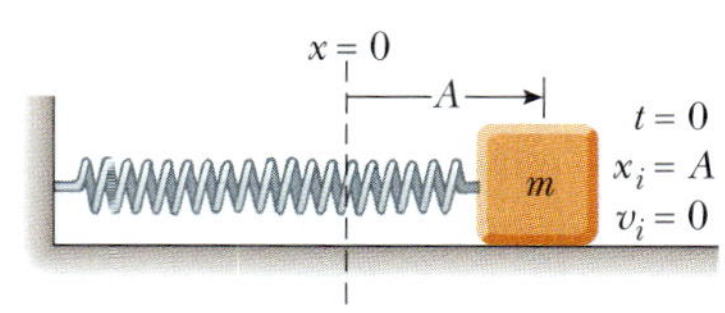

ACTIVE FIGURE 15.6

A block–spring system that begins its motion from rest with the block at $x = A$ at $t = 0$. In this case, $\phi = 0$; therefore, $x = A\cos\omega t$.

Sign in at www.thomsonedu.com and go to ThomsonNOW to compare the oscillations of two blocks starting from different initial positions and see that the frequency is independent of the amplitude.

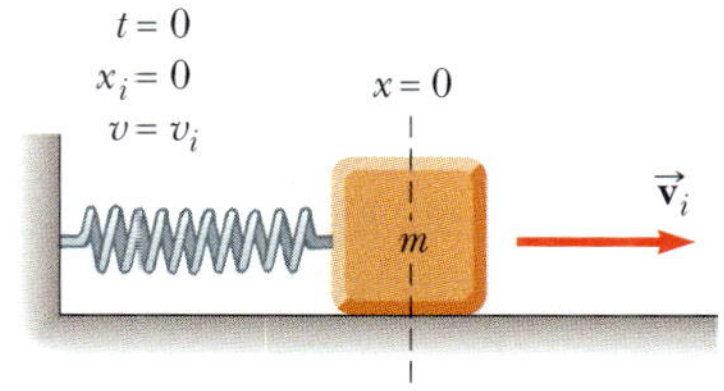

ACTIVE FIGURE 15.8

The block–spring system is undergoing oscillation, and $t = 0$ is defined at an instant when the block passes through the equilibrium position $x = 0$ and is moving to the right with speed v_i.

Sign in at www.thomsonedu.com and go to ThomsonNOW to compare the oscillations of two blocks with different velocities at $t = 0$ and see that the frequency is independent of the amplitude.

EXAMPLE 15.1 A Block-Spring System

A 200-g block connected to a light spring for which the force constant is 5.00 N/m is free to oscillate on a horizontal, frictionless surface. The block is displaced 5.00 cm from equilibrium and released from rest as in Active Figure 15.6.

(A) Find the period of its motion.

SOLUTION

Conceptualize Study Active Figure 15.6 and imagine the block moving back and forth in simple harmonic motion once it is released. Set up an experimental model in the vertical direction by hanging a heavy object such as a stapler from a strong rubber band.

Categorize The block is modeled as a particle in simple harmonic motion. We find values from equations developed in this section for the particle in simple harmonic motion model, so we categorize this example as a substitution problem.

Use Equation 15.9 to find the angular frequency of the block-spring system:

$$\omega = \sqrt{\frac{k}{m}} = \sqrt{\frac{5.00\ \text{N/m}}{200 \times 10^{-3}\ \text{kg}}} = 5.00\ \text{rad/s}$$

Use Equation 15.13 to find the period of the system:

$$T = \frac{2\pi}{\omega} = \frac{2\pi}{5.00\ \text{rad/s}} = \boxed{1.26\ \text{s}}$$

(B) Determine the maximum speed of the block.

SOLUTION

Use Equation 15.17 to find v_{max}:

$$v_{\text{max}} = \omega A = (5.00\ \text{rad/s})(5.00 \times 10^{-2}\ \text{m}) = \boxed{0.250\ \text{m/s}}$$

(C) What is the maximum acceleration of the block?

SOLUTION

Use Equation 15.18 to find a_{max}:

$$a_{\text{max}} = \omega^2 A = (5.00\ \text{rad/s})^2(5.00 \times 10^{-2}\ \text{m}) = \boxed{1.25\ \text{m/s}^2}$$

(D) Express the position, velocity, and acceleration as functions of time.

SOLUTION

Find the phase constant from the initial condition that $x = A$ at $t = 0$:

$$x(0) = A\cos\phi = A \quad \rightarrow \quad \phi = 0$$

Use Equation 15.6 to write an expression for $x(t)$:

$$x = A\cos(\omega t + \phi) = \boxed{(0.050\,0\ \text{m})\cos 5.00t}$$

Use Equation 15.15 to write an expression for $v(t)$:

$$v = -\omega A\sin(\omega t + \phi) = \boxed{-(0.250\ \text{m/s})\sin 5.00t}$$

Use Equation 15.16 to write an expression for $a(t)$:

$$a = -\omega^2 A\cos(\omega t + \phi) = \boxed{-(1.25\ \text{m/s}^2)\cos 5.00t}$$

What If? What if the block were released from the same initial position, $x_i = 5.00$ cm, but with an initial velocity of $v_i = -0.100$ m/s? Which parts of the solution change and what are the new answers for those that do change?

Answers Part (A) does not change because the period is independent of how the oscillator is set into motion. Parts (B), (C), and (D) will change.

Write position and velocity expressions for the initial conditions:

$$(1) \quad x(0) = A\cos\phi = x_i$$

$$(2) \quad v(0) = -\omega A\sin\phi = v_i$$

Divide Equation (2) by Equation (1) to find the phase constant:

$$\frac{-\omega A \sin \phi}{A \cos \phi} = \frac{v_i}{x_i}$$

$$\tan \phi = -\frac{v_i}{\omega x_i} = -\frac{-0.100 \text{ m/s}}{(5.00 \text{ rad/s})(0.050\,0 \text{ m})} = 0.400$$

$$\phi = 0.127\pi$$

Use Equation (1) to find A:

$$A = \frac{x_i}{\cos \phi} = \frac{0.050\,0 \text{ m}}{\cos (0.127\pi)} = 0.054\,3 \text{ m}$$

Find the new maximum speed:

$$v_{\max} = \omega A = (5.00 \text{ rad/s})(5.43 \times 10^{-2} \text{ m}) = 0.271 \text{ m/s}$$

Find the new magnitude of the maximum acceleration:

$$a_{\max} = \omega^2 A = (5.00 \text{ rad/s})^2(5.43 \times 10^{-2} \text{ m}) = 1.36 \text{ m/s}^2$$

Find new expressions for position, velocity, and acceleration:

$$x = (0.054\,3 \text{ m}) \cos (5.00t + 0.127\pi)$$

$$v = -(0.271 \text{ m/s}) \sin (5.00t + 0.127\pi)$$

$$a = -(1.36 \text{ m/s}^2) \cos (5.00t + 0.127\pi)$$

As we saw in Chapters 7 and 8, many problems are easier to solve using an energy approach rather than one based on variables of motion. This particular **What If?** is easier to solve from an energy approach. Therefore, we shall investigate the energy of the simple harmonic oscillator in the next section.

EXAMPLE 15.2 Watch Out for Potholes!

A car with a mass of 1 300 kg is constructed so that its frame is supported by four springs. Each spring has a force constant of 20 000 N/m. Two people riding in the car have a combined mass of 160 kg. Find the frequency of vibration of the car after it is driven over a pothole in the road.

SOLUTION

Conceptualize Think about your experiences with automobiles. When you sit in a car, it moves downward a small distance because your weight is compressing the springs further. If you push down on the front bumper and release it, the front of the car oscillates a few times.

Categorize We imagine the car as being supported by a single spring and model the car as a particle in simple harmonic motion.

Analyze First, let's determine the effective spring constant of the four springs combined. For a given extension x of the springs, the combined force on the car is the sum of the forces from the individual springs.

Find an expression for the total force on the car:

$$F_{\text{total}} = \sum (-kx) = -\left(\sum k\right)x$$

In this expression, x has been factored from the sum because it is the same for all four springs. The effective spring constant for the combined springs is the sum of the individual spring constants.

Evaluate the effective spring constant:

$$k_{\text{eff}} = \sum k = 4 \times 20\,000 \text{ N/m} = 80\,000 \text{ N/m}$$

Use Equation 15.14 to find the frequency of vibration:

$$f = \frac{1}{2\pi}\sqrt{\frac{k_{\text{eff}}}{m}} = \frac{1}{2\pi}\sqrt{\frac{80\,000 \text{ N/m}}{1\,460 \text{ kg}}} = 1.18 \text{ Hz}$$

Finalize The mass we used here is that of the car plus the people because that is the total mass that is oscillating. Also notice that we have explored only up-and-down motion of the car. If an oscillation is established in which the car rocks back and forth such that the front end goes up when the back end goes down, the frequency will be different.

What If? Suppose the car stops on the side of the road and the two people exit the car. One of them pushes downward on the car and releases it so that it oscillates vertically. Is the frequency of the oscillation the same as the value we just calculated?

Answer The suspension system of the car is the same, but the mass that is oscillating is smaller: it no longer includes the mass of the two people. Therefore, the frequency should be higher. Let's calculate the new frequency taking the mass to be 1 300 kg:

$$f = \frac{1}{2\pi}\sqrt{\frac{k_{\text{eff}}}{m}} = \frac{1}{2\pi}\sqrt{\frac{80\,000\text{ N/m}}{1\,300\text{ kg}}} = 1.25\text{ Hz}$$

As predicted, the new frequency is a bit higher.

15.3 Energy of the Simple Harmonic Oscillator

Let us examine the mechanical energy of the block-spring system illustrated in Active Figure 15.1. Because the surface is frictionless, the system is isolated and we expect the total mechanical energy of the system to be constant. We assume a massless spring, so the kinetic energy of the system corresponds only to that of the block. We can use Equation 15.15 to express the kinetic energy of the block as

Kinetic energy of a simple harmonic oscillator ▶

$$K = \tfrac{1}{2}mv^2 = \tfrac{1}{2}m\omega^2A^2\sin^2(\omega t + \phi) \qquad \textbf{(15.19)}$$

The elastic potential energy stored in the spring for any elongation x is given by $\frac{1}{2}kx^2$ (see Eq. 7.22). Using Equation 15.6 gives

Potential energy of a simple harmonic oscillator ▶

$$U = \tfrac{1}{2}kx^2 = \tfrac{1}{2}kA^2\cos^2(\omega t + \phi) \qquad \textbf{(15.20)}$$

We see that K and U are *always* positive quantities or zero. Because $\omega^2 = k/m$, we can express the total mechanical energy of the simple harmonic oscillator as

$$E = K + U = \tfrac{1}{2}kA^2[\sin^2(\omega t + \phi) + \cos^2(\omega t + \phi)]$$

From the identity $\sin^2\theta + \cos^2\theta = 1$, we see that the quantity in square brackets is unity. Therefore, this equation reduces to

Total energy of a simple harmonic oscillator ▶

$$E = \tfrac{1}{2}kA^2 \qquad \textbf{(15.21)}$$

That is, **the total mechanical energy of a simple harmonic oscillator is a constant of the motion and is proportional to the square of the amplitude.** The total mechanical energy is equal to the maximum potential energy stored in the spring when $x = \pm A$ because $v = 0$ at these points and there is no kinetic energy. At the equilibrium position, where $U = 0$ because $x = 0$, the total energy, all in the form of kinetic energy, is again $\frac{1}{2}kA^2$.

Plots of the kinetic and potential energies versus time appear in Active Figure 15.9a, where we have taken $\phi = 0$. At all times, the sum of the kinetic and potential energies is a constant equal to $\frac{1}{2}kA^2$, the total energy of the system.

The variations of K and U with the position x of the block are plotted in Active Figure 15.9b. Energy is continuously being transformed between potential energy stored in the spring and kinetic energy of the block.

Active Figure 15.10 illustrates the position, velocity, acceleration, kinetic energy, and potential energy of the block-spring system for one full period of the motion. Most of the ideas discussed so far are incorporated in this important figure. Study it carefully.

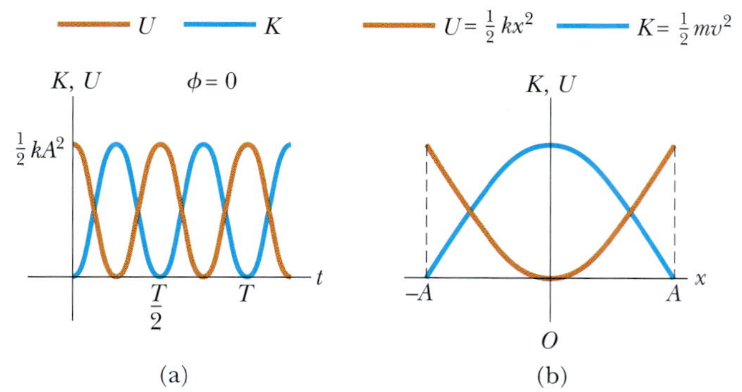

ACTIVE FIGURE 15.9

(a) Kinetic energy and potential energy versus time for a simple harmonic oscillator with $\phi = 0$. (b) Kinetic energy and potential energy versus position for a simple harmonic oscillator. In either plot, notice that $K + U =$ constant.

Sign in at www.thomsonedu.com and go to ThomsonNOW to compare the physical oscillation of a block with energy graphs in this figure as well as with energy bar graphs.

Finally, we can obtain the velocity of the block at an arbitrary position by expressing the total energy of the system at some arbitrary position x as

$$E = K + U = \tfrac{1}{2}mv^2 + \tfrac{1}{2}kx^2 = \tfrac{1}{2}kA^2$$

$$v = \pm\sqrt{\frac{k}{m}(A^2 - x^2)} = \pm\,\omega\sqrt{A^2 - x^2} \qquad \textbf{(15.22)}$$

◀ Velocity as a function of position for a simple harmonic oscillator

When you check Equation 15.22 to see whether it agrees with known cases, you find that it verifies that the speed is a maximum at $x = 0$ and is zero at the turning points $x = \pm A$.

You may wonder why we are spending so much time studying simple harmonic oscillators. We do so because they are good models of a wide variety of physical phenomena. For example, recall the Lennard–Jones potential discussed in Example 7.9. This complicated function describes the forces holding atoms together. Figure 15.11a (page 428) shows that for small displacements from the equilibrium position, the potential energy curve for this function approximates a parabola, which represents the potential energy function for a simple harmonic oscillator. Therefore, we can model the complex atomic binding forces as being due to tiny springs as depicted in Figure 15.11b.

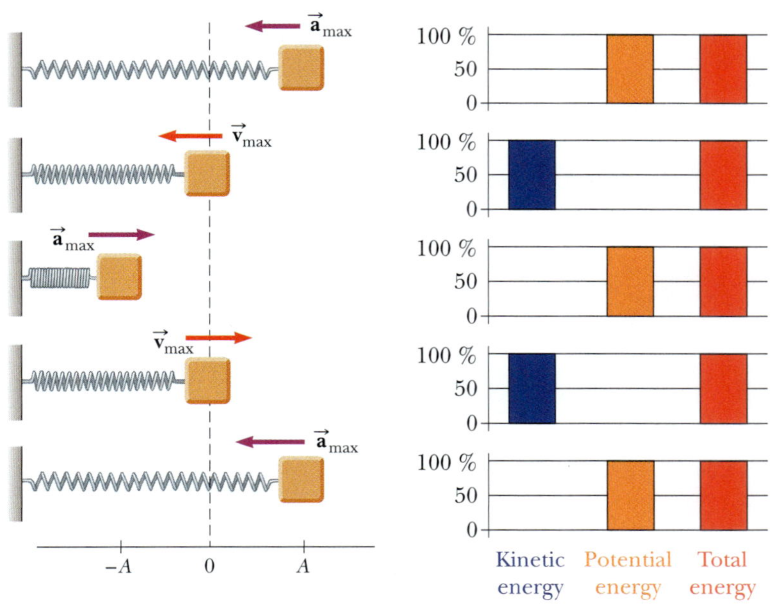

t	x	v	a	K	U
0	A	0	$-\omega^2 A$	0	$\frac{1}{2}kA^2$
$\frac{T}{4}$	0	$-\omega A$	0	$\frac{1}{2}kA^2$	0
$\frac{T}{2}$	$-A$	0	$\omega^2 A$	0	$\frac{1}{2}kA^2$
$\frac{3T}{4}$	0	ωA	0	$\frac{1}{2}kA^2$	0
T	A	0	$-\omega^2 A$	0	$\frac{1}{2}kA^2$

ACTIVE FIGURE 15.10

Several instants in the simple harmonic motion for a block–spring system. Energy bar graphs show the distribution of the energy of the system at each instant. The parameters in the table at the right refer to the block–spring system, assuming that at $t = 0$, $x = A$; hence, $x = A\cos\omega t$.

Sign in at www.thomsonedu.com and go to ThomsonNOW to set the initial position of the block and see the block–spring system and the analogous energy bar graphs.

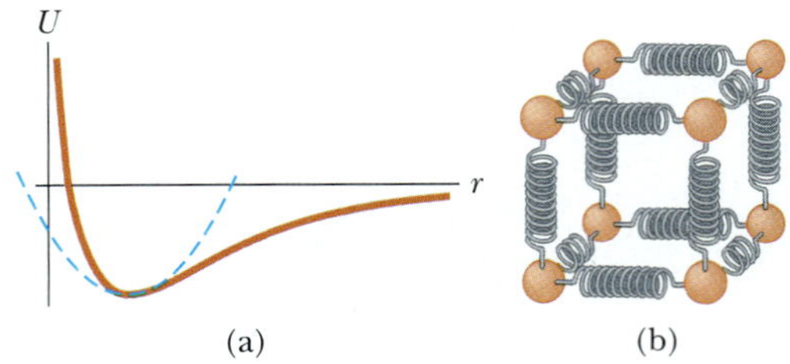

Figure 15.11 (a) If the atoms in a molecule do not move too far from their equilibrium positions, a graph of potential energy versus separation distance between atoms is similar to the graph of potential energy versus position for a simple harmonic oscillator (dashed blue curve). (b) The forces between atoms in a solid can be modeled by imagining springs between neighboring atoms.

The ideas presented in this chapter apply not only to block-spring systems and atoms, but also to a wide range of situations that include bungee jumping, tuning in a television station, and viewing the light emitted by a laser. You will see more examples of simple harmonic oscillators as you work through this book.

EXAMPLE 15.3 Oscillations on a Horizontal Surface

A 0.500-kg cart connected to a light spring for which the force constant is 20.0 N/m oscillates on a horizontal, frictionless air track.

(A) Calculate the total energy of the system and the maximum speed of the cart if the amplitude of the motion is 3.00 cm.

SOLUTION

Conceptualize The system oscillates in exactly the same way as the block in Active Figure 15.10.

Categorize The cart is modeled as a particle in simple harmonic motion.

Analyze Use Equation 15.21 to find the energy of the oscillator:

$$E = \tfrac{1}{2}kA^2 = \tfrac{1}{2}(20.0\text{ N/m})(3.00 \times 10^{-2}\text{ m})^2$$
$$= \boxed{9.00 \times 10^{-3}\text{ J}}$$

When the cart is at $x = 0$, the energy of the oscillator is entirely kinetic, so set $E = \frac{1}{2}mv_{\max}^2$:

$$\tfrac{1}{2}mv_{\max}^2 = 9.00 \times 10^{-3}\text{ J}$$

Solve for the maximum speed:

$$v_{\max} = \sqrt{\frac{2(9.00 \times 10^{-3}\text{ J})}{0.500\text{ kg}}} = \boxed{0.190\text{ m/s}}$$

(B) What is the velocity of the cart when the position is 2.00 cm?

SOLUTION

Use Equation 15.22 to evaluate the velocity:

$$v = \pm\sqrt{\frac{k}{m}(A^2 - x^2)}$$
$$= \pm\sqrt{\frac{20.0\text{ N/m}}{0.500\text{ kg}}[(0.030\,0\text{ m})^2 - (0.020\,0\text{ m})^2]}$$
$$= \boxed{\pm 0.141\text{ m/s}}$$

The positive and negative signs indicate that the cart could be moving to either the right or the left at this instant.

(C) Compute the kinetic and potential energies of the system when the position is 2.00 cm.

SOLUTION

Use the result of part (B) to evaluate the kinetic energy at $x = 0.020\ 0$ m:

$$K = \frac{1}{2}mv^2 = \frac{1}{2}(0.500\ \text{kg})(0.141\ \text{m/s})^2 = 5.00 \times 10^{-3}\ \text{J}$$

Evaluate the elastic potential energy at $x = 0.020\ 0$ m:

$$U = \frac{1}{2}kx^2 = \frac{1}{2}(20.0\ \text{N/m})(0.0200\ \text{m})^2 = 4.00 \times 10^{-3}\ \text{J}$$

Finalize Notice that the sum of the kinetic and potential energies in part (C) is equal to the total energy found in part (A). That must be true for *any* position of the cart.

What If? The cart in this example could have been set into motion by releasing the cart from rest at $x = 3.00$ cm. What if the cart were released from the same position, but with an initial velocity of $v = -0.100$ m/s? What are the new amplitude and maximum speed of the cart?

Answer This question is of the same type we asked at the end of Example 15.1, but here we apply an energy approach.

First calculate the total energy of the system at $t = 0$:

$$E = \frac{1}{2}mv^2 + \frac{1}{2}kx^2$$
$$= \frac{1}{2}(0.500\ \text{kg})(-0.100\ \text{m/s})^2 + \frac{1}{2}(20.0\ \text{N/m})(0.030\ 0\ \text{m})^2$$
$$= 1.15 \times 10^{-2}\ \text{J}$$

Equate this total energy to the potential energy when the cart is at the end point of the motion:

$$E = \frac{1}{2}kA^2$$

Solve for the amplitude A:

$$A = \sqrt{\frac{2E}{k}} = \sqrt{\frac{2(1.15 \times 10^{-2}\ \text{J})}{20.0\ \text{N/m}}} = 0.033\ 9\ \text{m}$$

Find the new maximum speed by equating the total energy to the kinetic energy when the cart is at the equilibrium position:

$$E = \frac{1}{2}mv_{\text{max}}^2$$

Solve for the maximum speed:

$$v_{\text{max}} = \sqrt{\frac{2E}{m}} = \sqrt{\frac{2(1.15 \times 10^{-2}\ \text{J})}{0.500\ \text{kg}}} = 0.214\ \text{m/s}$$

The amplitude and maximum velocity are larger than the previous values because the cart was given an initial velocity at $t = 0$.

15.4 Comparing Simple Harmonic Motion with Uniform Circular Motion

Some common devices in our everyday life exhibit a relationship between oscillatory motion and circular motion. For example, the pistons in an automobile engine (Fig. 15.12a, page 430) go up and down—oscillatory motion—yet the net result of this motion is circular motion of the wheels. In an old-fashioned locomotive (Fig. 15.12b), the drive shaft goes back and forth in oscillatory motion, causing a circular motion of the wheels. In this section, we explore this interesting relationship between these two types of motion.

Active Figure 15.13 (page 430) is a view of an experimental arrangement that shows this relationship. A ball is attached to the rim of a turntable of radius A, which is illuminated from the side by a lamp. The ball casts a shadow on a screen. **As the turntable rotates with constant angular speed, the shadow of the ball moves back and forth in simple harmonic motion.**

Figure 15.12 (*Left*) The pistons of an automobile engine move in periodic motion along a single dimension as shown in this cutaway view of two of these pistons. This motion is converted to circular motion of the crankshaft, at the lower right, and ultimately of the wheels of the automobile. (*Right*) The back-and-forth motion of pistons (in the curved housing at the left) in an old-fashioned locomotive is converted to circular motion of the wheels.

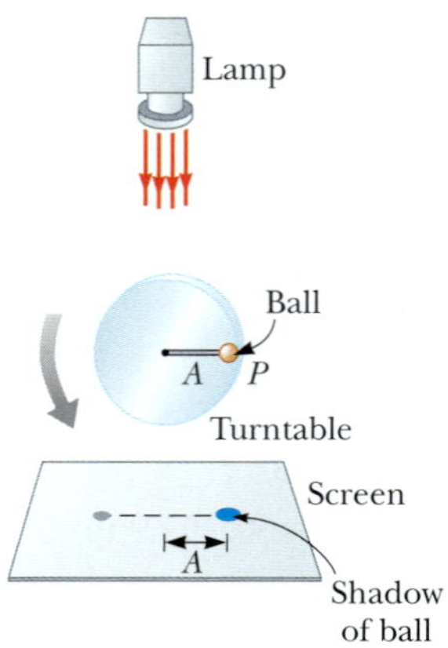

ACTIVE FIGURE 15.13

An experimental setup for demonstrating the connection between simple harmonic motion and uniform circular motion. As the ball rotates on the turntable with constant angular speed, its shadow on the screen moves back and forth in simple harmonic motion.

Sign in at www.thomsonedu.com and go to ThomsonNOW to adjust the frequency and radial position of the ball and see the resulting simple harmonic motion of the shadow.

Consider a particle located at point P on the circumference of a circle of radius A as in Figure 15.14a, with the line OP making an angle ϕ with the x axis at $t = 0$. We call this circle a *reference circle* for comparing simple harmonic motion with uniform circular motion, and we choose the position of P at $t = 0$ as our reference position. If the particle moves along the circle with constant angular speed ω until OP makes an angle θ with the x axis as in Figure 15.14b, at some time $t > 0$ the angle between OP and the x axis is $\theta = \omega t + \phi$. As the particle moves along the circle, the projection of P on the x axis, labeled point Q, moves back and forth along the x axis between the limits $x = \pm A$.

Notice that points P and Q always have the same x coordinate. From the right triangle OPQ, we see that this x coordinate is

$$x(t) = A\cos(\omega t + \phi) \tag{15.23}$$

This expression is the same as Equation 15.6 and shows that the point Q moves with simple harmonic motion along the x axis. Therefore, **simple harmonic motion along a straight line can be represented by the projection of uniform circular motion along a diameter of a reference circle.**

This geometric interpretation shows that the time interval for one complete revolution of the point P on the reference circle is equal to the period of motion T for simple harmonic motion between $x = \pm A$. That is, the angular speed ω of P is the same as the angular frequency ω of simple harmonic motion along the x axis

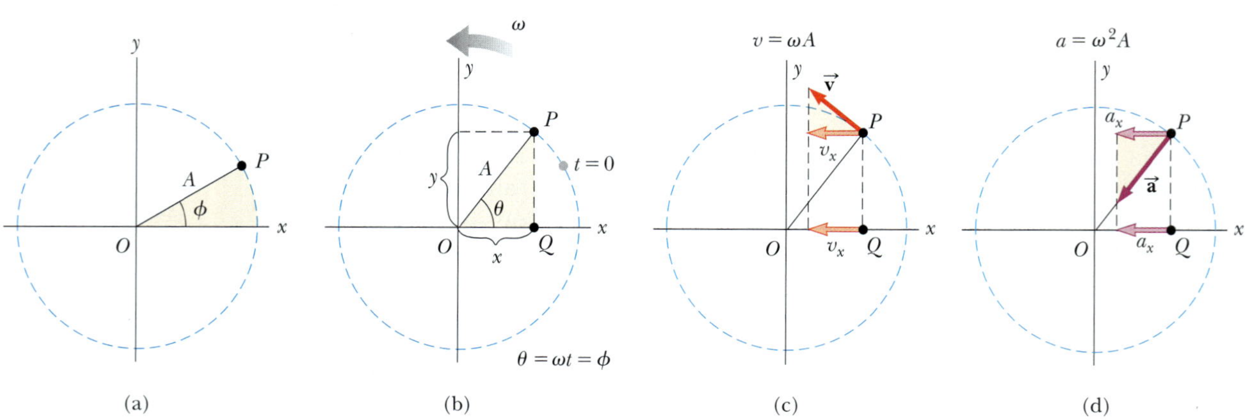

Figure 15.14 Relationship between the uniform circular motion of a point P and the simple harmonic motion of a point Q. A particle at P moves in a circle of radius A with constant angular speed ω. (a) A reference circle showing the position of P at $t = 0$. (b) The x coordinates of points P and Q are equal and vary in time according to the expression $x = A\cos(\omega t + \phi)$. (c) The x component of the velocity of P equals the velocity of Q. (d) The x component of the acceleration of P equals the acceleration of Q.

(which is why we use the same symbol). The phase constant ϕ for simple harmonic motion corresponds to the initial angle OP makes with the x axis. The radius A of the reference circle equals the amplitude of the simple harmonic motion.

Because the relationship between linear and angular speed for circular motion is $v = r\omega$ (see Eq. 10.10), the particle moving on the reference circle of radius A has a velocity of magnitude ωA. From the geometry in Figure 15.14c, we see that the x component of this velocity is $-\omega A \sin(\omega t + \phi)$. By definition, point Q has a velocity given by dx/dt. Differentiating Equation 15.23 with respect to time, we find that the velocity of Q is the same as the x component of the velocity of P.

The acceleration of P on the reference circle is directed radially inward toward O and has a magnitude $v^2/A = \omega^2 A$. From the geometry in Figure 15.14d, we see that the x component of this acceleration is $-\omega^2 A \cos(\omega t + \phi)$. This value is also the acceleration of the projected point Q along the x axis, as you can verify by taking the second derivative of Equation 15.23.

Quick Quiz 15.5 Figure 15.15 shows the position of an object in uniform circular motion at $t = 0$. A light shines from above and projects a shadow of the object on a screen below the circular motion. What are the correct values for the *amplitude* and *phase constant* (relative to an x axis to the right) of the simple harmonic motion of the shadow? (a) 0.50 m and 0 (b) 1.00 m and 0 (c) 0.50 m and π (d) 1.00 m and π

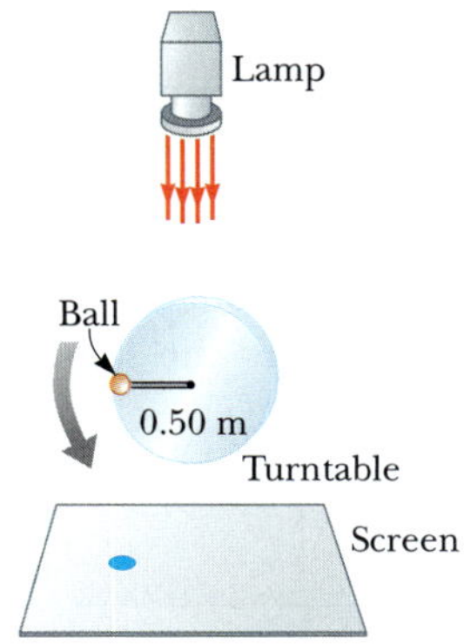

Figure 15.15 (Quick Quiz 15.5) An object moves in circular motion, casting a shadow on the screen below. Its position at an instant of time is shown.

EXAMPLE 15.4 Circular Motion with Constant Angular Speed

A particle rotates counterclockwise in a circle of radius 3.00 m with a constant angular speed of 8.00 rad/s. At $t = 0$, the particle has an x coordinate of 2.00 m and is moving to the right.

(A) Determine the x coordinate of the particle as a function of time.

SOLUTION

Conceptualize Be sure you understand the relationship between circular motion of a particle and simple harmonic motion of its shadow as described in Active Figure 15.13.

Categorize The particle on the circle is a particle under constant angular speed. The shadow is a particle in simple harmonic motion.

Analyze Use Equation 15.23 to write an expression for the x coordinate of the rotating particle with $\omega = 8.00$ rad/s:

$$x = A\cos(\omega t + \phi) = (3.00\text{ m})\cos(8.00t + \phi)$$

Evaluate ϕ by using the initial condition $x = 2.00$ m at $t = 0$:

$$2.00\text{ m} = (3.00\text{ m})\cos(0 + \phi)$$

Solve for ϕ:

$$\phi = \cos^{-1}\left(\frac{2.00\text{ m}}{3.00\text{ m}}\right) = \cos^{-1}(0.667) = \pm 48.2° = \pm 0.841\text{ rad}$$

If we were to take $\phi = +0.841$ rad as our answer, the particle would be moving to the left at $t = 0$. Because the particle is moving to the right at $t = 0$, we must choose $\phi = -0.841$ rad.

Write the x coordinate as a function of time:

$$x = (3.00\text{ m})\cos(8.00t - 0.841)$$

(B) Find the x components of the particle's velocity and acceleration at any time t.

SOLUTION

Differentiate the x coordinate with respect to time to find the velocity at any time:

$$v_x = \frac{dx}{dt} = (-3.00\text{ m})(8.00\text{ rad/s})\sin(8.00t - 0.841)$$
$$= -(24.0\text{ m/s})\sin(8.00t - 0.841)$$

Differentiate the velocity with respect to time to find the acceleration at any time:

$$a_x = \frac{dv_x}{dt} = (-24.0\text{ m/s})(8.00\text{ rad/s})\cos(8.00t - 0.841)$$
$$= -(192\text{ m/s}^2)\cos(8.00t - 0.841)$$

Finalize Although we have evaluated these results for the particle moving in the circle, remember that these same results apply to the shadow, which is moving in simple harmonic motion.

15.5 The Pendulum

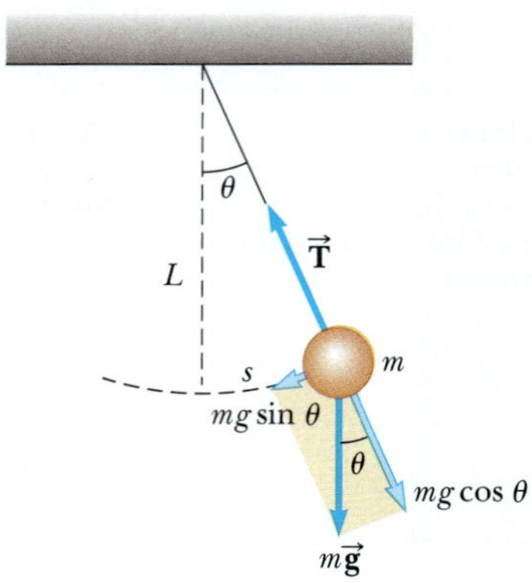

ACTIVE FIGURE 15.16

The restoring force is $-mg\sin\theta$, the component of the gravitational force tangent to the arc. When θ is small, a simple pendulum oscillates in simple harmonic motion about the equilibrium position $\theta = 0$.

Sign in at www.thomsonedu.com and go to ThomsonNOW to adjust the mass of the bob, the length of the string, and the initial angle and see the resulting oscillation of the pendulum.

The **simple pendulum** is another mechanical system that exhibits periodic motion. It consists of a particle-like bob of mass m suspended by a light string of length L that is fixed at the upper end as shown in Active Figure 15.16. The motion occurs in the vertical plane and is driven by the gravitational force. We shall show that, provided the angle θ is small (less than about 10°), the motion is very close to that of a simple harmonic oscillator.

The forces acting on the bob are the force $\vec{\mathbf{T}}$ exerted by the string and the gravitational force $m\vec{\mathbf{g}}$. The tangential component $mg\sin\theta$ of the gravitational force always acts toward $\theta = 0$, opposite the displacement of the bob from the lowest position. Therefore, the tangential component is a restoring force, and we can apply Newton's second law for motion in the tangential direction:

$$F_t = -mg\sin\theta = m\frac{d^2s}{dt^2}$$

where s is the bob's position measured along the arc and the negative sign indicates that the tangential force acts toward the equilibrium (vertical) position. Because $s = L\theta$ (Eq. 10.1a) and L is constant, this equation reduces to

$$\frac{d^2\theta}{dt^2} = -\frac{g}{L}\sin\theta$$

Considering θ as the position, let us compare this equation to Equation 15.3. Does it have the same mathematical form? The right side is proportional to $\sin\theta$ rather than to θ; hence, we would not expect simple harmonic motion because this expression is not of the form of Equation 15.3. If we assume θ is *small* (less than about 10° or 0.2 rad), however, we can use the **small angle approximation**, in which $\sin\theta \approx \theta$, where θ is measured in radians. Table 15.1 shows angles in degrees and radians and the sines of these angles. As long as θ is less than approximately 10°, the angle in radians and its sine are the same to within an accuracy of less than 1.0%.

PITFALL PREVENTION 15.5

Not True Simple Harmonic Motion

The pendulum *does not* exhibit true simple harmonic motion for *any* angle. If the angle is less than about 10°, the motion is close to and can be *modeled* as simple harmonic.

Therefore, for small angles, the equation of motion becomes

$$\frac{d^2\theta}{dt^2} = -\frac{g}{L}\theta \quad \text{(for small values of } \theta\text{)} \qquad \textbf{(15.24)}$$

Equation 15.24 has the same form as Equation 15.3, so we conclude that the motion for small amplitudes of oscillation can be modeled as simple harmonic motion. Therefore, the solution of Equation 15.24 is $\theta = \theta_{\max}\cos(\omega t + \phi)$, where $\theta_{\max}$ is the *maximum angular position* and the angular frequency ω is

Angular frequency for a simple pendulum ▶

$$\omega = \sqrt{\frac{g}{L}} \qquad \textbf{(15.25)}$$

TABLE 15.1

Angles and Sines of Angles

Angle in Degrees	Angle in Radians	Sine of Angle	Percent Difference
0°	0.000 0	0.000 0	0.0%
1°	0.017 5	0.017 5	0.0%
2°	0.034 9	0.034 9	0.0%
3°	0.052 4	0.052 3	0.0%
5°	0.087 3	0.087 2	0.1%
10°	0.174 5	0.173 6	0.5%
15°	0.261 8	0.258 8	1.2%
20°	0.349 1	0.342 0	2.1%
30°	0.523 6	0.500 0	4.7%

The period of the motion is

$$T = \frac{2\pi}{\omega} = 2\pi\sqrt{\frac{L}{g}} \tag{15.26}$$

◀ Period of a simple pendulum

In other words, **the period and frequency of a simple pendulum depend only on the length of the string and the acceleration due to gravity.** Because the period is independent of the mass, we conclude that all simple pendula that are of equal length and are at the same location (so that g is constant) oscillate with the same period.

The simple pendulum can be used as a timekeeper because its period depends only on its length and the local value of g. It is also a convenient device for making precise measurements of the free-fall acceleration. Such measurements are important because variations in local values of g can provide information on the location of oil and other valuable underground resources.

Quick Quiz 15.6 A grandfather clock depends on the period of a pendulum to keep correct time. **(i)** Suppose a grandfather clock is calibrated correctly and then a mischievous child slides the bob of the pendulum downward on the oscillating rod. Does the grandfather clock run (a) slow, (b) fast, or (c) correctly? **(ii)** Suppose a grandfather clock is calibrated correctly at sea level and is then taken to the top of a very tall mountain. Does the grandfather clock now run (a) slow, (b) fast, or (c) correctly?

EXAMPLE 15.5 A Connection Between Length and Time

Christian Huygens (1629–1695), the greatest clockmaker in history, suggested that an international unit of length could be defined as the length of a simple pendulum having a period of exactly 1 s. How much shorter would our length unit be if his suggestion had been followed?

SOLUTION

Conceptualize Imagine a pendulum that swings back and forth in exactly 1 second. Based on your experience in observing swinging objects, can you make an estimate of the required length? Hang a small object from a string and simulate the 1-s pendulum.

Categorize This example involves a simple pendulum, so we categorize it as an application of the concepts introduced in this section.

Analyze Solve Equation 15.26 for the length and substitute the known values:

$$L = \frac{T^2 g}{4\pi^2} = \frac{(1.00\text{ s})^2(9.80\text{ m/s}^2)}{4\pi^2} = 0.248\text{ m}$$

Finalize The meter's length would be slightly less than one-fourth of its current length. Also, the number of significant digits depends only on how precisely we know g because the time has been defined to be exactly 1 s.

What If? What if Huygens had been born on another planet? What would the value for g have to be on that planet such that the meter based on Huygens's pendulum would have the same value as our meter?

Answer Solve Equation 15.26 for g:

$$g = \frac{4\pi^2 L}{T^2} = \frac{4\pi^2(1.00\text{ m})}{(1.00\text{ s})^2} = 4\pi^2\text{ m/s}^2 = 39.5\text{ m/s}^2$$

No planet in our solar system has an acceleration due to gravity that large.

Physical Pendulum

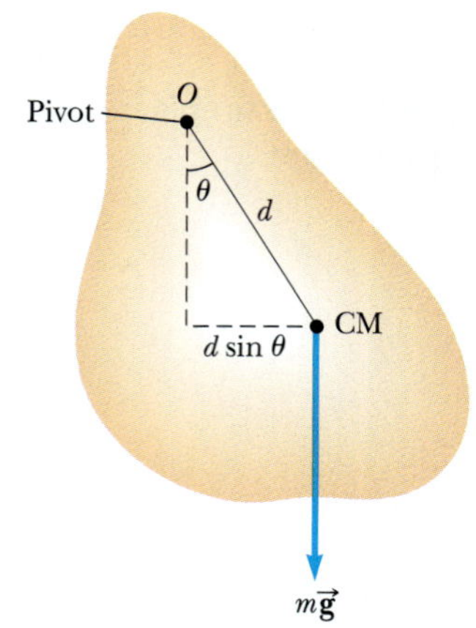

Figure 15.17 A physical pendulum pivoted at O.

Suppose you balance a wire coat hanger so that the hook is supported by your extended index finger. When you give the hanger a small angular displacement (with your other hand) and then release it, it oscillates. If a hanging object oscillates about a fixed axis that does not pass through its center of mass and the object cannot be approximated as a point mass, we cannot treat the system as a simple pendulum. In this case, the system is called a **physical pendulum**.

Consider a rigid object pivoted at a point O that is a distance d from the center of mass (Fig. 15.17). The gravitational force provides a torque about an axis through O, and the magnitude of that torque is $mgd \sin\theta$, where θ is as shown in Figure 15.17. We model the object as a rigid object under a net torque and use the rotational form of Newton's second law, $\sum \tau = I\alpha$, where I is the moment of inertia of the object about the axis through O. The result is

$$-mgd \sin\theta = I\frac{d^2\theta}{dt^2}$$

The negative sign indicates that the torque about O tends to decrease θ. That is, the gravitational force produces a restoring torque. If we again assume θ is small, the approximation $\sin\theta \approx \theta$ is valid and the equation of motion reduces to

$$\frac{d^2\theta}{dt^2} = -\left(\frac{mgd}{I}\right)\theta = -\omega^2\theta \qquad \textbf{(15.27)}$$

Because this equation is of the same form as Equation 15.3, its solution is that of the simple harmonic oscillator. That is, the solution of Equation 15.27 is given by $\theta = \theta_{\text{max}} \cos(\omega t + \phi)$, where θ_{max} is the maximum angular position and

$$\omega = \sqrt{\frac{mgd}{I}}$$

The period is

Period of a physical pendulum ▶

$$T = \frac{2\pi}{\omega} = 2\pi\sqrt{\frac{I}{mgd}} \qquad \textbf{(15.28)}$$

This result can be used to measure the moment of inertia of a flat rigid object. If the location of the center of mass—and hence the value of d—is known, the moment of inertia can be obtained by measuring the period. Finally, notice that Equation 15.28 reduces to the period of a simple pendulum (Eq. 15.26) when $I = md^2$, that is, when all the mass is concentrated at the center of mass.

EXAMPLE 15.6 A Swinging Rod

A uniform rod of mass M and length L is pivoted about one end and oscillates in a vertical plane (Fig. 15.18). Find the period of oscillation if the amplitude of the motion is small.

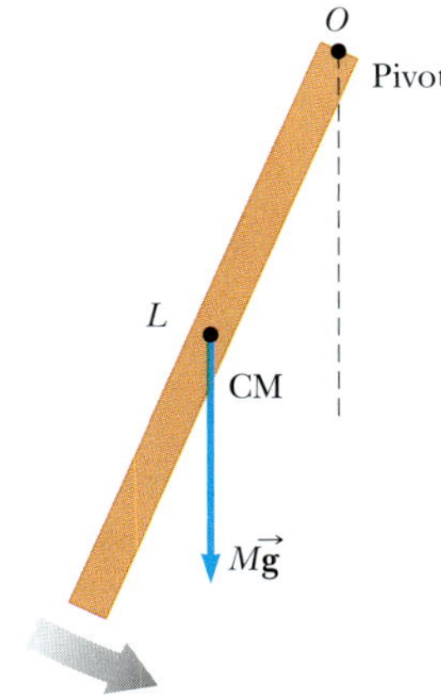

Figure 15.18 (Example 15.6) A rigid rod oscillating about a pivot through one end is a physical pendulum with $d = L/2$ and, from Table 10.2, $I = \frac{1}{3}ML^2$.

SOLUTION

Conceptualize Imagine a rod swinging back and forth when pivoted at one end. Try it with a meterstick or a scrap piece of wood.

Categorize Because the rod is not a point particle, we categorize it as a physical pendulum.

Analyze In Chapter 10, we found that the moment of inertia of a uniform rod about an axis through one end is $\frac{1}{3}ML^2$. The distance d from the pivot to the center of mass of the rod is $L/2$.

Substitute these quantities into Equation 15.28:

$$T = 2\pi\sqrt{\frac{\frac{1}{3}ML^2}{Mg(L/2)}} = 2\pi\sqrt{\frac{2L}{3g}}$$

Finalize In one of the Moon landings, an astronaut walking on the Moon's surface had a belt hanging from his space suit, and the belt oscillated as a physical pendulum. A scientist on the Earth observed this motion on television and used it to estimate the free-fall acceleration on the Moon. How did the scientist make this calculation?

Torsional Pendulum

Figure 15.19 shows a rigid object suspended by a wire attached at the top to a fixed support. When the object is twisted through some angle θ, the twisted wire exerts on the object a restoring torque that is proportional to the angular position. That is,

$$\tau = -\kappa\theta$$

where κ (Greek letter kappa) is called the *torsion constant* of the support wire. The value of κ can be obtained by applying a known torque to twist the wire through a measurable angle θ. Applying Newton's second law for rotational motion, we find that

$$\tau = -\kappa\theta = I\frac{d^2\theta}{dt^2}$$

$$\frac{d^2\theta}{dt^2} = -\frac{\kappa}{I}\theta \qquad (15.29)$$

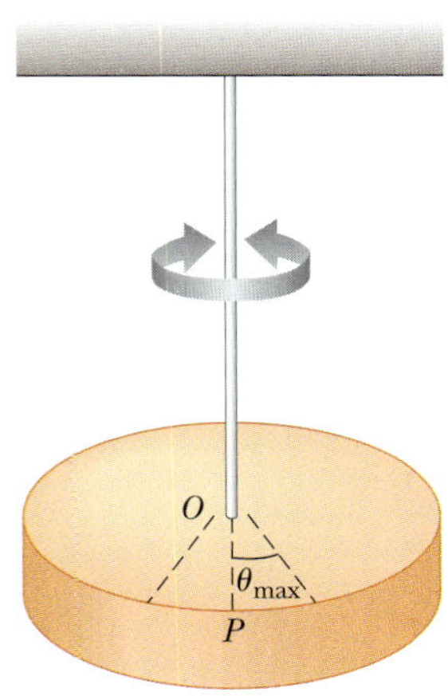

Figure 15.19 A torsional pendulum consists of a rigid object suspended by a wire attached to a rigid support. The object oscillates about the line OP with an amplitude θ_{max}.

Again, this result is the equation of motion for a simple harmonic oscillator, with $\omega = \sqrt{\kappa/I}$ and a period

$$T = 2\pi\sqrt{\frac{I}{\kappa}} \qquad (15.30)$$

◀ Period of a torsional pendulum

This system is called a *torsional pendulum.* There is no small-angle restriction in this situation as long as the elastic limit of the wire is not exceeded.

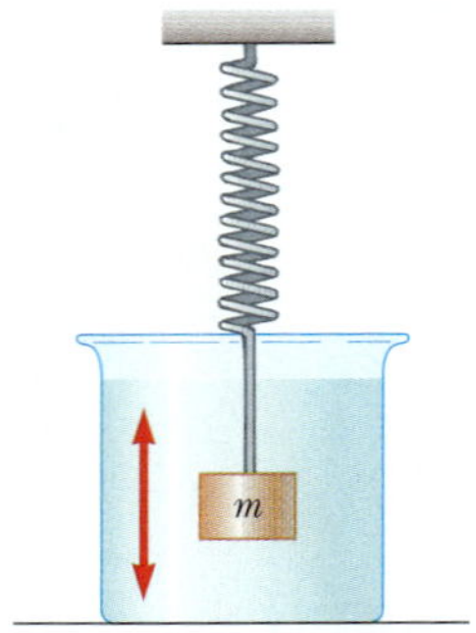

Figure 15.20 One example of a damped oscillator is an object attached to a spring and submersed in a viscous liquid.

15.6 Damped Oscillations

The oscillatory motions we have considered so far have been for ideal systems, that is, systems that oscillate indefinitely under the action of only one force, a linear restoring force. In many real systems, nonconservative forces such as friction retard the motion. Consequently, the mechanical energy of the system diminishes in time, and the motion is said to be *damped.* The lost mechanical energy is transformed into internal energy in the object and the retarding medium. Figure 15.20 depicts one such system: an object attached to a spring and submersed in a viscous liquid.

One common type of retarding force is that discussed in Section 6.4, where the force is proportional to the speed of the moving object and acts in the direction opposite the velocity of the object with respect to the medium. This retarding force is often observed when an object moves through air, for instance. Because the retarding force can be expressed as $\vec{\mathbf{R}} = -b\vec{\mathbf{v}}$ (where b is a constant called the *damping coefficient*) and the restoring force of the system is $-kx$, we can write Newton's second law as

$$\sum F_x = -kx - bv_x = ma_x$$

$$-kx - b\frac{dx}{dt} = m\frac{d^2x}{dt^2} \qquad \textbf{(15.31)}$$

The solution to this equation requires mathematics that may be unfamiliar to you; we simply state it here without proof. When the retarding force is small compared with the maximum restoring force—that is, when b is small—the solution to Equation 15.31 is

$$x = Ae^{-(b/2m)t}\cos(\omega t + \phi) \qquad \textbf{(15.32)}$$

where the angular frequency of oscillation is

$$\omega = \sqrt{\frac{k}{m} - \left(\frac{b}{2m}\right)^2} \qquad \textbf{(15.33)}$$

This result can be verified by substituting Equation 15.32 into Equation 15.31. It is convenient to express the angular frequency of a damped oscillator in the form

$$\omega = \sqrt{\omega_0^{\,2} - \left(\frac{b}{2m}\right)^2}$$

where $\omega_0 = \sqrt{k/m}$ represents the angular frequency in the absence of a retarding force (the undamped oscillator) and is called the **natural frequency** of the system.

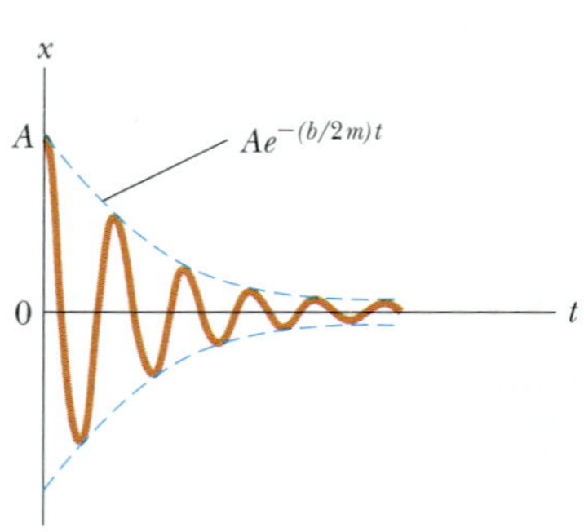

ACTIVE FIGURE 15.21

Graph of position versus time for a damped oscillator. Notice the decrease in amplitude with time.

Sign in at www.thomsonedu.com and go to ThomsonNOW to adjust the spring constant, the mass of the object, and the damping constant and see the resulting damped oscillation of the object.

Active Figure 15.21 shows the position as a function of time for an object oscillating in the presence of a retarding force. **When the retarding force is small, the oscillatory character of the motion is preserved but the amplitude decreases in time, with the result that the motion ultimately ceases.** Any system that behaves in this way is known as a **damped oscillator.** The dashed blue lines in Active Figure 15.21, which define the *envelope* of the oscillatory curve, represent the exponential factor in Equation 15.32. This envelope shows that **the amplitude decays exponentially with time.** For motion with a given spring constant and object mass, the oscillations dampen more rapidly for larger values of the retarding force.

When the magnitude of the retarding force is small such that $b/2m < \omega_0$, the system is said to be **underdamped.** The resulting motion is represented by the blue curve in Figure 15.22. As the value of b increases, the amplitude of the oscillations decreases more and more rapidly. When b reaches a critical value b_c such that $b_c/2m = \omega_0$, the system does not oscillate and is said to be **critically damped.** In this case, the system, once released from rest at some nonequilibrium position, approaches but does not pass through the equilibrium position. The graph of position versus time for this case is the red curve in Figure 15.22.

If the medium is so viscous that the retarding force is large compared to the restoring force—that is, if $b/2m > \omega_0$—the system is **overdamped.** Again, the displaced system, when free to move, does not oscillate but rather simply returns to its equilibrium position. As the damping increases, the time interval required for the system to approach equilibrium also increases as indicated by the black curve in Figure 15.22. For critically damped and overdamped systems, there is no angular frequency ω and the solution in Equation 15.32 is not valid.

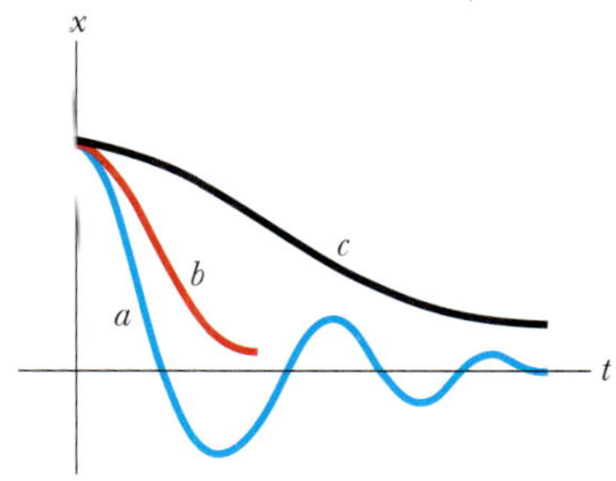

Figure 15.22 Graphs of position versus time for an underdamped oscillator (blue, curve *a*), a critically damped oscillator (red, curve *b*), and an overdamped oscillator (black, curve *c*).

15.7 Forced Oscillations

We have seen that the mechanical energy of a damped oscillator decreases in time as a result of the resistive force. It is possible to compensate for this energy decrease by applying an external force that does positive work on the system. At any instant, energy can be transferred into the system by an applied force that acts in the direction of motion of the oscillator. For example, a child on a swing can be kept in motion by appropriately timed "pushes." The amplitude of motion remains constant if the energy input per cycle of motion exactly equals the decrease in mechanical energy in each cycle that results from resistive forces.

A common example of a forced oscillator is a damped oscillator driven by an external force that varies periodically, such as $F(t) = F_0 \sin \omega t$, where F_0 is a constant and ω is the angular frequency of the driving force. In general, the frequency ω of the driving force is variable, whereas the natural frequency ω_0 of the oscillator is fixed by the values of k and m. Newton's second law in this situation gives

$$\sum F = ma \rightarrow F_0 \sin \omega t - b\frac{dx}{dt} - kx = m\frac{d^2x}{dt^2} \tag{15.34}$$

Again, the solution of this equation is rather lengthy and will not be presented. After the driving force on an initially stationary object begins to act, the amplitude of the oscillation will increase. After a sufficiently long period of time, when the energy input per cycle from the driving force equals the amount of mechanical energy transformed to internal energy for each cycle, a steady-state condition is reached in which the oscillations proceed with constant amplitude. In this situation, the solution of Equation 15.34 is

$$x = A \cos (\omega t + \phi) \tag{15.35}$$

where

$$A = \frac{F_0/m}{\sqrt{(\omega^2 - \omega_0^{\,2})^2 + \left(\dfrac{b\omega}{m}\right)^2}} \tag{15.36}$$

◀ Amplitude of a driven oscillator

and where $\omega_0 = \sqrt{k/m}$ is the natural frequency of the undamped oscillator ($b = 0$).

Equations 15.35 and 15.36 show that the forced oscillator vibrates at the frequency of the driving force and that the amplitude of the oscillator is constant for a given driving force because it is being driven in steady-state by an external force. For small damping, the amplitude is large when the frequency of the driving force is near the natural frequency of oscillation, or when $\omega \approx \omega_0$. The dramatic increase in amplitude near the natural frequency is called **resonance,** and the natural frequency ω_0 is also called the **resonance frequency** of the system.

The reason for large-amplitude oscillations at the resonance frequency is that energy is being transferred to the system under the most favorable conditions. We can better understand this concept by taking the first time derivative of x in Equation 15.35, which gives an expression for the velocity of the oscillator. We find that v is proportional to $\sin (\omega t + \phi)$, which is the same trigonometric function as that describing the driving force. Therefore, the applied force $\vec{\mathbf{F}}$ is in phase with the velocity. The rate at which work is done on the oscillator by $\vec{\mathbf{F}}$ equals the dot product

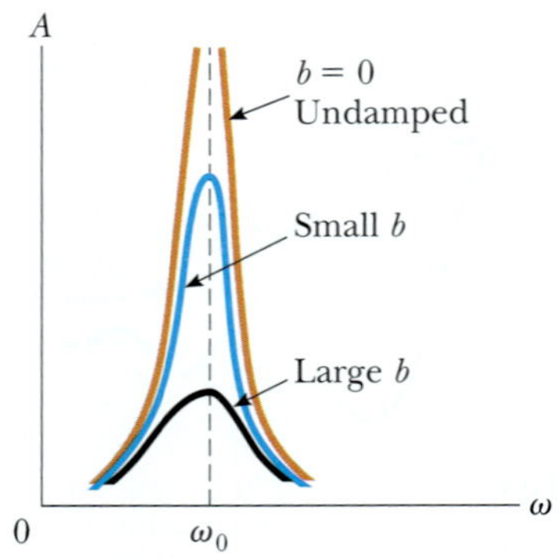

Figure 15.23 Graph of amplitude versus frequency for a damped oscillator when a periodic driving force is present. When the frequency ω of the driving force equals the natural frequency ω_0 of the oscillator, resonance occurs. Notice that the shape of the resonance curve depends on the size of the damping coefficient b.

$\vec{\mathbf{F}} \cdot \vec{\mathbf{v}}$; this rate is the power delivered to the oscillator. Because the product $\vec{\mathbf{F}} \cdot \vec{\mathbf{v}}$ is a maximum when $\vec{\mathbf{F}}$ and $\vec{\mathbf{v}}$ are in phase, we conclude that **at resonance, the applied force is in phase with the velocity and the power transferred to the oscillator is a maximum.**

Figure 15.23 is a graph of amplitude as a function of frequency for a forced oscillator with and without damping. Notice that the amplitude increases with decreasing damping ($b \rightarrow 0$) and that the resonance curve broadens as the damping increases. In the absence of a damping force ($b = 0$), we see from Equation 15.36 that the steady-state amplitude approaches infinity as ω approaches ω_0. In other words, if there are no losses in the system and we continue to drive an initially motionless oscillator with a periodic force that is in phase with the velocity, the amplitude of motion builds without limit (see the brown curve in Fig. 15.23). This limitless building does not occur in practice because some damping is always present in reality.

Later in this book we shall see that resonance appears in other areas of physics. For example, certain electric circuits have natural frequencies. A bridge has natural frequencies that can be set into resonance by an appropriate driving force. A dramatic example of such resonance occurred in 1940 when the Tacoma Narrows Bridge in the state of Washington was destroyed by resonant vibrations. Although the winds were not particularly strong on that occasion, the "flapping" of the wind across the roadway (think of the "flapping" of a flag in a strong wind) provided a periodic driving force whose frequency matched that of the bridge. The resulting oscillations of the bridge caused it to ultimately collapse (Fig. 15.24) because the bridge design had inadequate built-in safety features.

Many other examples of resonant vibrations can be cited. A resonant vibration you may have experienced is the "singing" of telephone wires in the wind. Machines often break if one vibrating part is in resonance with some other moving part. Soldiers marching in cadence across a bridge have been known to set up resonant vibrations in the structure and thereby cause it to collapse. Whenever any real physical system is driven near its resonance frequency, you can expect oscillations of very large amplitudes.

(a) (b)

Figure 15.24 (a) In 1940, turbulent winds set up torsional vibrations in the Tacoma Narrows Bridge, causing it to oscillate at a frequency near one of the natural frequencies of the bridge structure. (b) Once established, this resonance condition led to the bridge's collapse. *(UPI/Bettmann Newsphotos)*

Summary

ThomsonNOW™ Sign in at **www.thomsonedu.com** and go to ThomsonNOW to take a practice test for this chapter.

CONCEPTS AND PRINCIPLES

The kinetic energy and potential energy for an object of mass m oscillating at the end of a spring of force constant k vary with time and are given by

$$K = \tfrac{1}{2}mv^2 = \tfrac{1}{2}m\omega^2A^2 \sin^2(\omega t + \phi) \quad \textbf{(15.19)}$$

$$U = \tfrac{1}{2}kx^2 = \tfrac{1}{2}kA^2 \cos^2(\omega t + \phi) \quad \textbf{(15.20)}$$

The total energy of a simple harmonic oscillator is a constant of the motion and is given by

$$E = \tfrac{1}{2}kA^2 \quad \textbf{(15.21)}$$

A **simple pendulum** of length L moves in simple harmonic motion for small angular displacements from the vertical. Its period is

$$T = 2\pi\sqrt{\frac{L}{g}} \quad \textbf{(15.26)}$$

For small angular displacements from the vertical, a **physical pendulum** moves in simple harmonic motion about a pivot that does not go through the center of mass. The period of this motion is

$$T = 2\pi\sqrt{\frac{I}{mgd}} \quad \textbf{(15.28)}$$

where I is the moment of inertia about an axis through the pivot and d is the distance from the pivot to the center of mass.

If an oscillator experiences a damping force $\vec{\mathbf{R}} = -b\vec{\mathbf{v}}$, its position for small damping is described by

$$x = Ae^{-(b/2m)t}\cos(\omega t + \phi) \quad \textbf{(15.32)}$$

where

$$\omega = \sqrt{\frac{k}{m} - \left(\frac{b}{2m}\right)^2} \quad \textbf{(15.33)}$$

If an oscillator is subject to a sinusoidal driving force $F(t) = F_0 \sin \omega t$, it exhibits **resonance,** in which the amplitude is largest when the driving frequency ω matches the natural frequency $\omega_0 = \sqrt{k/m}$ of the oscillator.

ANALYSIS MODEL FOR PROBLEM SOLVING

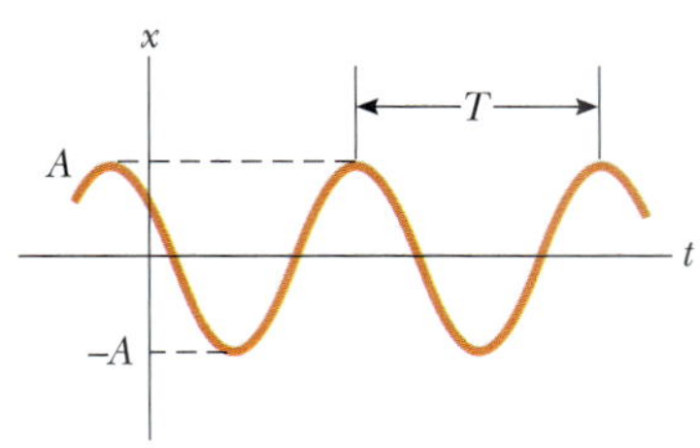

Particle in Simple Harmonic Motion If a particle is subject to a force of the form of Hooke's law $F = -kx$, the particle exhibits **simple harmonic motion**. Its position is described by

$$x(t) = A\cos(\omega t + \phi) \quad \textbf{(15.6)}$$

where A is the **amplitude** of the motion, ω is the **angular frequency,** and ϕ is the **phase constant.** The value of ϕ depends on the initial position and initial velocity of the oscillator.

The **period** of the oscillation is

$$T = \frac{2\pi}{\omega} = 2\pi\sqrt{\frac{m}{k}} \quad \textbf{(15.13)}$$

and the inverse of the period is the **frequency.**

Questions

☐ denotes answer available in *Student Solutions Manual/Study Guide;* **O** denotes objective question

1. Is a bouncing ball an example of simple harmonic motion? Is the daily movement of a student from home to school and back simple harmonic motion? Why or why not?

2. O A particle on a spring moves in simple harmonic motion along the x axis between turning points at $x_1 = 100$ cm and $x_2 = 140$ cm. **(i)** At which of the following positions does the particle have maximum speed? (a) 100 cm (b) 110 cm (c) 120 cm (d) some other position (e) The same greatest value occurs at multiple points. **(ii)** At which position does it have maximum acceleration? Choose from the same possibilities. **(iii)** At which position is the greatest net force exerted on the particle? **(iv)** At which position does the particle have the greatest magnitude of momentum? **(v)** At which position does the particle have greatest kinetic energy? **(vi)** At which position does the particle-spring system have the greatest total energy?

☐3. If the coordinate of a particle varies as $x = -A \cos \omega t$, what is the phase constant in Equation 15.6? At what position is the particle at $t = 0$?

4. O Rank the periods of the following oscillating systems from the greatest to the smallest. If any periods are equal, show their equality in your ranking. Each system differs in only one way from system (a), which is a 0.1-kg glider on a horizontal, frictionless surface, oscillating with amplitude 0.1 m on a spring with force constant 10 N/m. In situation (b), the amplitude is 0.2 m. In situation (c), the mass is 0.2 kg. In situation (d), the spring has stiffness constant 20 N/m. Situation (e) is just like situation (a) except for being in a gravitational field of 4.9 m/s^2 instead of 9.8 m/s^2. Situation (f) is just like situation (a) except that the object bounces in simple harmonic motion on the bottom end of the spring hanging vertically. Situation (g) is just like situation (a) except that a small resistive force makes the motion underdamped.

☐5. O For a simple harmonic oscillator, the position is measured as the displacement from equilibrium. (a) Can the quantities position and velocity be in the same direction? (b) Can velocity and acceleration be in the same direction? (c) Can position and acceleration be in the same direction?

6. O The top end of a spring is held fixed. A block is hung on the bottom end and the frequency f of the oscillation of the system is measured. The block, a second identical block, and the spring are carried up in a space shuttle to Earth orbit. The two blocks are attached to the ends of the spring. The spring is compressed, without making adjacent coils touch, and the system is released to oscillate while floating within the shuttle cabin. What is the frequency of oscillation for this system in terms of f? (a) $f/4$ (b) $f/2$ (c) $f/\sqrt{2}$ (d) f (e) $\sqrt{2}f$ (f) $2f$ (g) $4f$

7. O You attach a block to the bottom end of a spring hanging vertically. You slowly let the block move down and find that it hangs at rest with the spring stretched by 15.0 cm. Next, you lift the block back up and release it from rest with the spring unstretched. What maximum distance does it move down? (a) 7.5 cm (b) 15.0 cm (c) 30.0 cm (d) 60.0 cm (e) The distance cannot be determined without knowing the mass and spring constant.

8. The equations listed in Table 2.2 give position as a function of time, velocity as a function of time, and velocity as function of position for an object moving in a straight line with constant acceleration. The quantity v_{xi} appears in every equation. Do any of these equations apply to an object moving in a straight line with simple harmonic motion? Using a similar format, make a table of equations describing simple harmonic motion. Include equations giving acceleration as a function of time and acceleration as a function of position. State the equations in such a form that they apply equally to a block-spring system, to a pendulum, and to other vibrating systems. What quantity appears in every equation?

9. O A simple pendulum has a period of 2.5 s. **(i)** What is its period if its length is made four times larger? (a) 0.625 s (b) 1.25 s (c) 2.5 s (d) 3.54 s (e) 5 s (f) 10 s **(ii)** What is its period if, instead of changing its length, the mass of the suspended bob is made four times larger? Choose from the same possibilities.

10. O A simple pendulum is suspended from the ceiling of a stationary elevator, and the period is determined. **(i)** When the elevator accelerates upward, is the period (a) greater, (b) smaller, or (c) unchanged? **(ii)** When the elevator has a downward acceleration, is the period (a) greater, (b) smaller, or (c) unchanged? **(iii)** When the elevator moves with constant upward velocity, is the period of the pendulum (a) greater, (b) smaller, or (c) unchanged?

11. Figure Q15.11 shows graphs of the potential energy of four different systems versus the position of a particle in each system. Each particle is set into motion with a push at an arbitrarily chosen location. Describe its subsequent motion in each case (a), (b), (c), and (d).

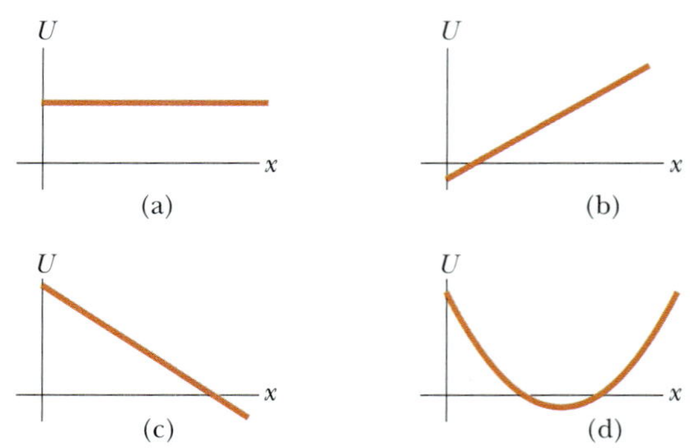

Figure Q15.11

12. A simple pendulum can be modeled as exhibiting simple harmonic motion when θ is small. Is the motion periodic when θ is large? How does the period of motion change as θ increases?

13. The mechanical energy of an undamped block-spring system is constant as kinetic energy transforms to elastic potential energy and vice versa. For comparison, explain in the same terms what happens to the energy of a damped oscillator.

14. A student thinks that any real vibration must be damped. Is the student correct? If so, give convincing reasoning. If not, give an example of a real vibration that keeps constant amplitude forever if the system is isolated.

15. Will damped oscillations occur for any values of b and k? Explain.

16. Is it possible to have damped oscillations when a system is at resonance? Explain.

17. You stand on the end of a diving board and bounce to set it into oscillation. You find a maximum response, in terms of the amplitude of oscillation of the end of the board, when you bounce at frequency f. You now move to the middle of the board and repeat the experiment. Is the resonance frequency for forced oscillations at this point higher, lower, or the same as f? Why?

18. You are looking at a small, leafy tree. You do not notice any breeze, and most of the leaves on the tree are motionless. One leaf, however, is fluttering back and forth wildly. After a while, that leaf stops moving and you notice a different leaf moving much more than all the others. Explain what could cause the large motion of one particular leaf.

19. The bob of a certain pendulum is a sphere filled with water. What would happen to the frequency of vibration of this pendulum if there were a hole in the sphere that allowed the water to leak out slowly?

Problems

WebAssign The Problems from this chapter may be assigned online in WebAssign.

ThomsonNOW Sign in at **www.thomsonedu.com** and go to ThomsonNOW to assess your understanding of this chapter's topics with additional quizzing and conceptual questions.

1, 2, 3 denotes straightforward, intermediate, challenging; ☐ denotes full solution available in *Student Solutions Manual/Study Guide;* ▲ denotes coached solution with hints available at **www.thomsonedu.com;** ▭ denotes developing symbolic reasoning; ● denotes asking for qualitative reasoning; 🖥 denotes computer useful in solving problem

Note: Ignore the mass of every spring, except in Problems 62 and 64.

Section 15.1 Motion of an Object Attached to a Spring

Problems 16, 17, 18, 26, and 60 in Chapter 7 can also be assigned with this section.

1. ● A ball dropped from a height of 4.00 m makes an elastic collision with the ground. Assuming no mechanical energy is lost due to air resistance, (a) show that the ensuing motion is periodic and (b) determine the period of the motion. (c) Is the motion simple harmonic? Explain.

Section 15.2 The Particle in Simple Harmonic Motion

2. In an engine, a piston oscillates with simple harmonic motion so that its position varies according to the expression

$$x = (5.00\text{ cm})\cos\left(2t + \frac{\pi}{6}\right)$$

where x is in centimeters and t is in seconds. At $t = 0$, find (a) the position of the particle, (b) its velocity, and (c) its acceleration. (d) Find the period and amplitude of the motion.

3. The position of a particle is given by the expression $x = (4.00\text{ m})\cos(3.00\pi t + \pi)$, where x is in meters and t is in seconds. Determine (a) the frequency and period of the motion, (b) the amplitude of the motion, (c) the phase constant, and (d) the position of the particle at $t = 0.250$ s.

4. ● (a) A hanging spring stretches by 35.0 cm when an object of mass 450 g is hung on it at rest. In this situation, we define its position as $x = 0$. The object is pulled down an additional 18.0 cm and released from rest to oscillate without friction. What is its position x at a moment 84.4 s later? (b) **What If?** Another hanging spring stretches by 35.5 cm when an object of mass 440 g is hung on it at rest. We define this new position as $x = 0$. This object is also pulled down an additional 18.0 cm and released from rest to oscillate without friction. Find its position 84.4 s later. (c) Why are the answers to parts (a) and (b) different by such a large percentage when the data are so similar? Does this circumstance reveal a fundamental difficulty in calculating the future? (d) Find the distance traveled by the vibrating object in part (a). (e) Find the distance traveled by the object in part (b).

5. ▲ A particle moving along the x axis in simple harmonic motion starts from its equilibrium position, the origin, at $t = 0$ and moves to the right. The amplitude of its motion is 2.00 cm, and the frequency is 1.50 Hz. (a) Show that the position of the particle is given by

$$x = (2.00\text{ cm})\sin(3.00\pi t)$$

Determine (b) the maximum speed and the earliest time ($t > 0$) at which the particle has this speed, (c) the maximum acceleration and the earliest time ($t > 0$) at which the particle has this acceleration, and (d) the total distance traveled between $t = 0$ and $t = 1.00$ s.

6. A simple harmonic oscillator takes 12.0 s to undergo five complete vibrations. Find (a) the period of its motion, (b) the frequency in hertz, and (c) the angular frequency in radians per second.

7. A 7.00-kg object is hung from the bottom end of a vertical spring fastened to an overhead beam. The object is set into vertical oscillations having a period of 2.60 s. Find the force constant of the spring.

8. **Review problem.** A particle moves along the x axis. It is initially at the position 0.270 m, moving with velocity

0.140 m/s and acceleration -0.320 m/s^2. Suppose it moves with constant acceleration for 4.50 s. Find (a) its position and (b) its velocity at the end of this time interval. Next, assume it moves with simple harmonic motion for 4.50 s and $x = 0$ is its equilibrium position. Find (c) its position and (d) its velocity at the end of this time interval.

9. A piston in a gasoline engine is in simple harmonic motion. Taking the extremes of its position relative to its center point as ± 5.00 cm, find the maximum velocity and acceleration of the piston when the engine is running at the rate of 3 600 rev/min.

10. A 1.00-kg glider attached to a spring with a force constant of 25.0 N/m oscillates on a horizontal, frictionless air track. At $t = 0$, the glider is released from rest at $x = -3.00$ cm. (That is, the spring is compressed by 3.00 cm.) Find (a) the period of its motion, (b) the maximum values of its speed and acceleration, and (c) the position, velocity, and acceleration as functions of time.

11. A 0.500-kg object attached to a spring with a force constant of 8.00 N/m vibrates in simple harmonic motion with an amplitude of 10.0 cm. Calculate (a) the maximum value of its speed and acceleration, (b) the speed and acceleration when the object is 6.00 cm from the equilibrium position, and (c) the time interval required for the object to move from $x = 0$ to $x = 8.00$ cm.

12. ● You attach an object to the bottom end of a hanging vertical spring. It hangs at rest after extending the spring 18.3 cm. You then set the object vibrating. Do you have enough information to find its period? Explain your answer and state whatever you can about its period.

13. A 1.00-kg object is attached to a horizontal spring. The spring is initially stretched by 0.100 m, and the object is released from rest there. It proceeds to move without friction. The next time the speed of the object is zero is 0.500 s later. What is the maximum speed of the object?

Section 15.3 Energy of the Simple Harmonic Oscillator

14. A 200-g block is attached to a horizontal spring and executes simple harmonic motion with a period of 0.250 s. The total energy of the system is 2.00 J. Find (a) the force constant of the spring and (b) the amplitude of the motion.

15. ▲ An automobile having a mass of 1 000 kg is driven into a brick wall in a safety test. The car's bumper behaves like a spring of constant 5.00×10^6 N/m and compresses 3.16 cm as the car is brought to rest. What was the speed of the car before impact, assuming that no mechanical energy is lost during impact with the wall?

16. A block-spring system oscillates with an amplitude of 3.50 cm. The spring constant is 250 N/m, and the mass of the block is 0.500 kg. Determine (a) the mechanical energy of the system, (b) the maximum speed of the block, and (c) the maximum acceleration.

17. A 50.0-g object connected to a spring with a force constant of 35.0 N/m oscillates on a horizontal, frictionless surface with an amplitude of 4.00 cm. Find (a) the total energy of the system and (b) the speed of the object when the position is 1.00 cm. Find (c) the kinetic energy and (d) the potential energy when the position is 3.00 cm.

18. A 2.00-kg object is attached to a spring and placed on a horizontal, smooth surface. A horizontal force of 20.0 N is required to hold the object at rest when it is pulled 0.200 m from its equilibrium position (the origin of the x axis). The object is now released from rest with an initial position of $x_i = 0.200$ m, and it subsequently undergoes simple harmonic oscillations. Find (a) the force constant of the spring, (b) the frequency of the oscillations, and (c) the maximum speed of the object. Where does this maximum speed occur? (d) Find the maximum acceleration of the object. Where does it occur? (e) Find the total energy of the oscillating system. Find (f) the speed and (g) the acceleration of the object when its position is equal to one-third of the maximum value.

19. A particle executes simple harmonic motion with an amplitude of 3.00 cm. At what position does its speed equal one half of its maximum speed?

20. A 65.0-kg bungee jumper steps off a bridge with a light bungee cord tied to her and to the bridge (Fig. P15.20). The unstretched length of the cord is 11.0 m. The jumper reaches the bottom of her motion 36.0 m below the bridge before bouncing back. Her motion can be separated into an 11.0-m free fall and a 25.0-m section of simple harmonic oscillation. (a) For what time interval is she in free fall? (b) Use the principle of conservation of energy to find the spring constant of the bungee cord. (c) What is the location of the equilibrium point where the spring force balances the gravitational force exerted on the jumper? This point is taken as the origin in our mathematical description of simple harmonic oscillation. (d) What is the angular frequency of the oscillation? (e) What time interval is required for the cord to stretch by 25.0 m? (f) What is the total time interval for the entire 36.0 m drop?

Telegraph Colour Library/FPG International

Figure P15.20 Problems 20 and 54.

21. A cart attached to a spring with constant 3.24 N/m vibrates such that its position is given by the function $x = (5.00 \text{ cm}) \cos (3.60t \text{ rad/s})$. (a) During the first cycle, for $0 < t < 1.75$ s, at what value of t is the system's potential energy changing most rapidly into kinetic energy? (b) What is the maximum rate of energy transformation?

Section 15.4 Comparing Simple Harmonic Motion with Uniform Circular Motion

22. ● Consider the simplified single-piston engine in Figure P15.22. Assuming the wheel rotates with constant angular speed, explain why the piston rod oscillates in simple harmonic motion.

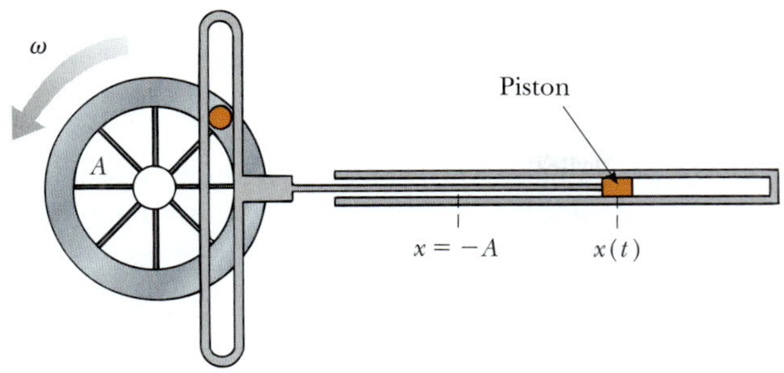

Figure P15.22

23. ● While riding behind a car traveling at 3.00 m/s, you notice that one of the car's tires has a small hemispherical bump on its rim as shown in Figure P15.23. (a) Explain why the bump, from your viewpoint behind the car, executes simple harmonic motion. (b) If the radii of the car's tires are 0.300 m, what is the bump's period of oscillation?

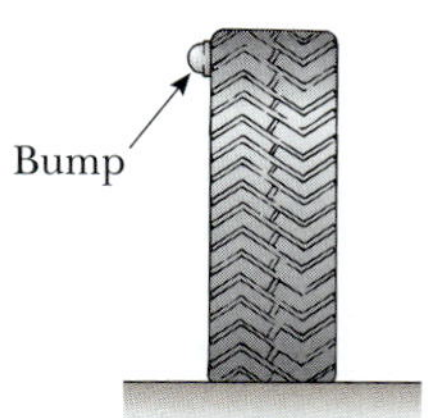

Figure P15.23

Section 15.5 The Pendulum

Problem 52 in Chapter 1 can also be assigned with this section.

24. A "seconds pendulum" is one that moves through its equilibrium position once each second. (The period of the pendulum is precisely 2 s.) The length of a seconds pendulum is 0.992 7 m at Tokyo, Japan, and 0.994 2 m at Cambridge, England. What is the ratio of the free-fall accelerations at these two locations?

25. ▲ ● A simple pendulum has a mass of 0.250 kg and a length of 1.00 m. It is displaced through an angle of 15.0° and then released. What are (a) the maximum speed, (b) the maximum angular acceleration, and (c) the maximum restoring force? **What If?** Solve this problem by using the simple harmonic motion model for the motion of the pendulum and then solve the problem by using more general principles. Compare the answers.

26. The angular position of a pendulum is represented by the equation $\theta = (0.032\ 0 \text{ rad}) \cos \omega t$, where θ is in radians and $\omega = 4.43$ rad/s. Determine the period and length of the pendulum.

27. A particle of mass m slides without friction inside a hemispherical bowl of radius R. Show that if the particle starts from rest with a small displacement from equilibrium, it moves in simple harmonic motion with an angular frequency equal to that of a simple pendulum of length R. That is, $\omega = \sqrt{g/R}$.

28. **Review problem.** A simple pendulum is 5.00 m long. (a) What is the period of small oscillations for this pendulum if it is located in an elevator accelerating upward at 5.00 m/s^2? (b) What is its period if the elevator is accelerating downward at 5.00 m/s^2? (c) What is the period of this pendulum if it is placed in a truck that is accelerating horizontally at 5.00 m/s^2?

29. A physical pendulum in the form of a planar object moves in simple harmonic motion with a frequency of 0.450 Hz. The pendulum has a mass of 2.20 kg, and the pivot is located 0.350 m from the center of mass. Determine the moment of inertia of the pendulum about the pivot point.

30. A small object is attached to the end of a string to form a simple pendulum. The period of its harmonic motion is measured for small angular displacements and three lengths. For each length, the time interval for 50 oscillations is measured with a stopwatch. For lengths of 1.000 m, 0.750 m, and 0.500 m, total time intervals of 99.8 s, 86.6 s, and 71.1 s are measured for 50 oscillations. (a) Determine the period of motion for each length. (b) Determine the mean value of g obtained from these three independent measurements and compare it with the accepted value. (c) Plot T^2 versus L and obtain a value for g from the slope of your best-fit straight-line graph. Compare this value with that obtained in part (b).

31. Consider the physical pendulum of Figure 15.17. (a) Represent its moment of inertia about an axis passing through its center of mass and parallel to the axis passing through its pivot point as I_{CM}. Show that its period is

$$T = 2\pi\sqrt{\frac{I_{CM} + md^2}{mgd}}$$

where d is the distance between the pivot point and center of mass. (b) Show that the period has a minimum value when d satisfies $md^2 = I_{CM}$.

32. A very light rigid rod with a length of 0.500 m extends straight out from one end of a meterstick. The meterstick is suspended from a pivot at the far end of the rod and is set into oscillation. (a) Determine the period of oscillation. *Suggestion:* Use the parallel-axis theorem from Section 10.5. (b) By what percentage does the period differ from the period of a simple pendulum 1.00 m long?

33. A clock balance wheel (Fig. P15.33) has a period of oscillation of 0.250 s. The wheel is constructed so that its mass of 20.0 g is concentrated around a rim of radius 0.500 cm. What are (a) the wheel's moment of inertia and (b) the torsion constant of the attached spring?

Figure P15.33

Section 15.6 Damped Oscillations

34. Show that the time rate of change of mechanical energy for a damped, undriven oscillator is given by $dE/dt = -bv^2$ and hence is always negative. To do so, differentiate the expression for the mechanical energy of an oscillator, $E = \frac{1}{2}mv^2 + \frac{1}{2}kx^2$, and use Equation 15.31.

35. A pendulum with a length of 1.00 m is released from an initial angle of 15.0°. After 1 000 s, its amplitude has been reduced by friction to 5.50°. What is the value of $b/2m$?

36. Show that Equation 15.32 is a solution of Equation 15.31 provided $b^2 < 4mk$.

37. A 10.6-kg object oscillates at the end of a vertical spring that has a spring constant of 2.05×10^4 N/m. The effect of air resistance is represented by the damping coefficient $b = 3.00$ N·s/m. (a) Calculate the frequency of the damped oscillation. (b) By what percentage does the amplitude of the oscillation decrease in each cycle? (c) Find the time interval that elapses while the energy of the system drops to 5.00% of its initial value.

Section 15.7 Forced Oscillations

38. The front of her sleeper wet from teething, a baby rejoices in the day by crowing and bouncing up and down in her crib. Her mass is 12.5 kg, and the crib mattress can be modeled as a light spring with force constant 4.30 kN/m. (a) The baby soon learns to bounce with maximum amplitude and minimum effort by bending her knees at what frequency? (b) She learns to use the mattress as a trampoline—losing contact with it for part of each cycle—when her amplitude exceeds what value?

39. A 2.00-kg object attached to a spring moves without friction and is driven by an external force given by $F = (3.00 \text{ N}) \sin (2\pi t)$. The force constant of the spring is 20.0 N/m. Determine (a) the period and (b) the amplitude of the motion.

40. Considering an undamped, forced oscillator ($b = 0$), show that Equation 15.35 is a solution of Equation 15.34, with an amplitude given by Equation 15.36.

41. A block weighing 40.0 N is suspended from a spring that has a force constant of 200 N/m. The system is undamped and is subjected to a harmonic driving force of frequency 10.0 Hz, resulting in a forced-motion amplitude of 2.00 cm. Determine the maximum value of the driving force.

42. Damping is negligible for a 0.150-kg object hanging from a light 6.30-N/m spring. A sinusoidal force with an amplitude of 1.70 N drives the system. At what frequency will the force make the object vibrate with an amplitude of 0.440 m?

43. You are a research biologist. Even though your emergency pager's batteries are getting low, you take the pager along to a fine restaurant. You switch the small pager to vibrate instead of beep, and you put it into a side pocket of your suit coat. The arm of your chair presses the light cloth against your body at one spot. Fabric with a length of 8.21 cm hangs freely below that spot, with the pager at the bottom. A coworker urgently needs instructions and pages you from the laboratory. The motion of the pager makes the hanging part of your coat swing back and forth with remarkably large amplitude. The waiter, maître d', wine steward, and nearby diners notice immediately and fall silent. Your daughter pipes up and says, accurately enough, "Daddy, look! Your cockroaches must have gotten out again!" Find the frequency at which your pager vibrates.

Additional Problems

44. ● **Review problem.** The problem extends the reasoning of Problem 54 in Chapter 9. Two gliders are set in motion on an air track. Glider one has mass $m_1 = 0.240$ kg and velocity $0.740\hat{\mathbf{i}}$ m/s. It will have a rear-end collision with glider number two, of mass $m_2 = 0.360$ kg, which has original velocity $0.120\hat{\mathbf{i}}$ m/s. A light spring of force constant 45.0 N/m is attached to the back end of glider two as shown in Figure P9.54. When glider one touches the spring, superglue instantly and permanently makes it stick to its end of the spring. (a) Find the common velocity the two gliders have when the spring compression is a maximum. (b) Find the maximum spring compression distance. (c) Argue that the motion after the gliders become attached consists of the center of mass of the two-glider system moving with the constant velocity found in part (a) while both gliders oscillate in simple harmonic motion relative to the center of mass. (d) Find the energy of the center-of-mass motion. (e) Find the energy of the oscillation.

45. ● An object of mass m moves in simple harmonic motion with amplitude 12.0 cm on a light spring. Its maximum acceleration is 108 cm/s². Regard m as a variable. (a) Find the period T of the object. (b) Find its frequency f. (c) Find the maximum speed v_{max} of the object. (d) Find the energy E of the vibration. (e) Find the force constant k of the spring. (f) Describe the pattern of dependence of each of the quantities T, f, v_{max}, E, and k on m.

46. ● **Review problem.** A rock rests on a concrete sidewalk. An earthquake strikes, making the ground move vertically in harmonic motion with a constant frequency of 2.40 Hz and with gradually increasing amplitude. (a) With what amplitude does the ground vibrate when the rock begins to lose contact with the sidewalk? Another rock is sitting on the concrete bottom of a swimming pool full of water. The earthquake produces only vertical motion, so the water does not slosh from side to side. (b) Present a convincing argument that when the ground vibrates with the amplitude found in part (a), the submerged rock also barely loses contact with the floor of the swimming pool.

47. A small ball of mass M is attached to the end of a uniform rod of equal mass M and length L that is pivoted at the top (Fig. P15.47). (a) Determine the tensions in the rod at the pivot and at the point P when the system is stationary. (b) Calculate the period of oscillation for small displacements from equilibrium and determine this period for $L = 2.00$ m. *Suggestions:* Model the object at the end of the rod as a particle and use Eq. 15.28.

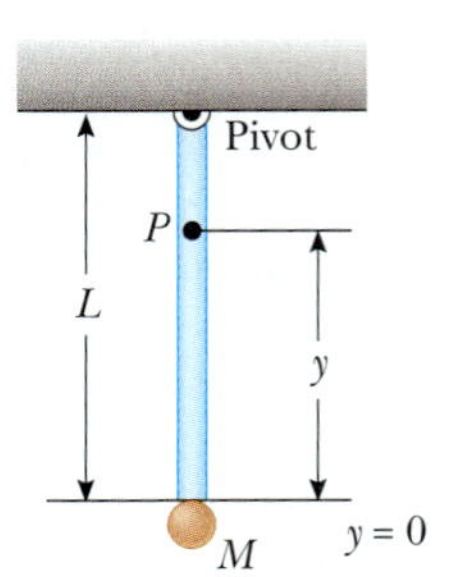

Figure P15.47

48. An object of mass $m_1 = 9.00$ kg is in equilibrium, connected to a light spring of constant $k = 100$ N/m that is fastened to a wall as shown in Figure P15.48a. A second object, $m_2 = 7.00$ kg, is slowly pushed up against m_1, compressing the spring by the amount $A = 0.200$ m (see Fig. P15.48b). The system is then released, and both objects start moving to the right on the frictionless surface. (a) When m_1 reaches the equilibrium point, m_2 loses contact with m_1 (see Fig. P15.48c) and moves to the right with speed v. Determine the value of v. (b) How far apart are the objects when the spring is fully stretched for the first time (D in Fig. P15.48d)? *Suggestion:* First determine the period of oscillation and the amplitude of the m_1–spring system after m_2 loses contact with m_1.

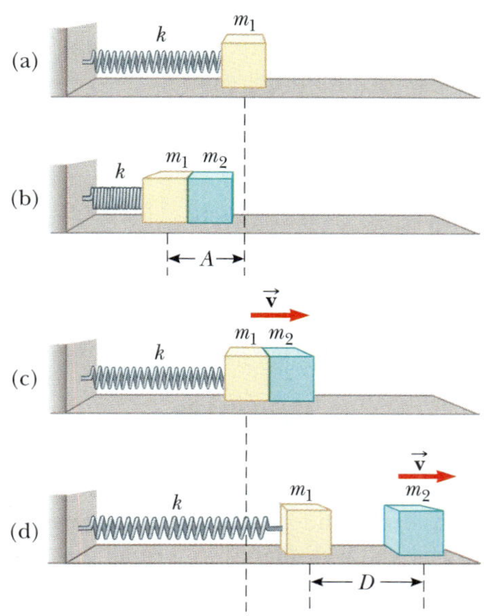

Figure P15.48

49. ▲ A large block P executes horizontal simple harmonic motion as it slides across a frictionless surface with a frequency $f = 1.50$ Hz. Block B rests on it as shown in Figure P15.49, and the coefficient of static friction between the two is $\mu_s = 0.600$. What maximum amplitude of oscillation can the system have if block B is not to slip?

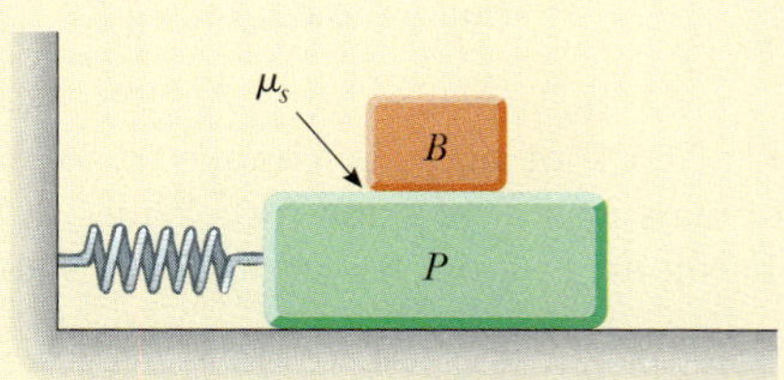

Figure P15.49 Problems 49 and 50.

50. A large block P executes horizontal simple harmonic motion as it slides across a frictionless surface with a frequency f. Block B rests on it as shown in Figure P15.49, and the coefficient of static friction between the two is μ_s. What maximum amplitude of oscillation can the system have if the upper block is not to slip?

51. The mass of the deuterium molecule (D_2) is twice that of the hydrogen molecule (H_2). If the vibrational frequency of H_2 is 1.30×10^{14} Hz, what is the vibrational frequency of D_2? Assume the "spring constant" of attracting forces is the same for the two molecules.

52. ● *You can now more completely analyze the situation in Problem 54 of Chapter 7.* Two steel balls, each of diameter 25.4 mm, move in opposite directions at 5.00 m/s. They collide head-on and bounce apart elastically. (a) Does their interaction last only for an instant or for a nonzero time interval? State your evidence. (b) One of the balls is squeezed in a vise while precise measurements are made of the resulting amount of compression. Assume Hooke's law is a good model of the ball's elastic behavior. For one datum, a force of 16.0 kN exerted by each jaw of the vise reduces the diameter by 0.200 mm. Modeling the ball as a spring, find its spring constant. (c) Assume the balls have the density of iron. Compute the kinetic energy of each ball before the balls collide. (d) Model each ball as a particle with a massless spring as its front bumper. Let the particle have the initial kinetic energy found in part (c) and the bumper have the spring constant found in part (b). Compute the maximum amount of compression each ball undergoes when the balls collide. (e) Model the motion of each ball, while the balls are in contact, as one half of a cycle of simple harmonic motion. Compute the time interval for which the balls are in contact.

53. A light, cubical container of volume a^3 is initially filled with a liquid of mass density ρ. The cube is initially supported by a light string to form a simple pendulum of length L_i, measured from the center of mass of the filled container, where $L_i >> a$. The liquid is allowed to flow from the bottom of the container at a constant rate (dM/dt). At any time t, the level of the fluid in the container is h and the length of the pendulum is L (measured relative to the instantaneous center of mass). (a) Sketch the apparatus and label the dimensions a, h, L_i, and L. (b) Find the time rate of change of the period as a function of time t. (c) Find the period as a function of time.

54. After a thrilling plunge, bungee jumpers bounce freely on the bungee cord through many cycles (Fig. P15.20). After the first few cycles, the cord does not go slack. Your younger brother can make a pest of himself by figuring out the mass of each person, using a proportion that you set up by solving this problem: An object of mass m is oscillating freely on a vertical spring with a period T. Another object of unknown mass m' on the same spring oscillates with a period T'. Determine (a) the spring constant and (b) the unknown mass.

55. A pendulum of length L and mass M has a spring of force constant k connected to it at a distance h below its point of suspension (Fig. P15.55). Find the frequency of vibration

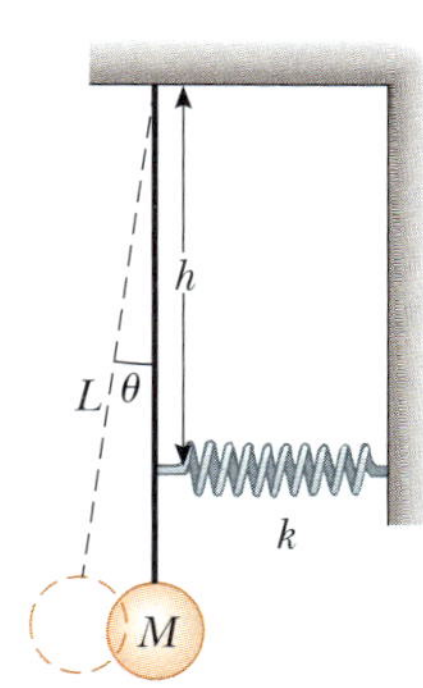

Figure P15.55

of the system for small values of the amplitude (small θ). Assume the vertical suspension rod of length L is rigid, but ignore its mass.

56. A particle with a mass of 0.500 kg is attached to a spring with a force constant of 50.0 N/m. At the moment $t = 0$, the particle has its maximum speed of 20.0 m/s and is moving to the left. (a) Determine the particle's equation of motion, specifying its position as a function of time. (b) Where in the motion is the potential energy three times the kinetic energy? (c) Find the length of a simple pendulum with the same period. (d) Find the minimum time interval required for the particle to move from $x = 0$ to $x = 1.00$ m.

57. A horizontal plank of mass m and length L is pivoted at one end. The plank's other end is supported by a spring of force constant k (Fig. P15.57). The moment of inertia of the plank about the pivot is $\frac{1}{3}mL^2$. The plank is displaced by a small angle θ from its horizontal equilibrium position and released. (a) Show that the plank moves with simple harmonic motion with an angular frequency $\omega = \sqrt{3k/m}$. (b) Evaluate the frequency, taking the mass as 5.00 kg and the spring force constant as 100 N/m.

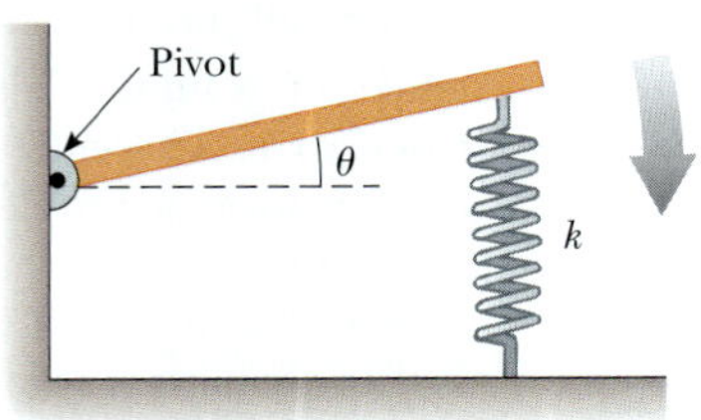

Figure P15.57

58. ● **Review problem.** A particle of mass 4.00 kg is attached to a spring with a force constant of 100 N/m. It is oscillating on a horizontal, frictionless surface with an amplitude of 2.00 m. A 6.00-kg object is dropped vertically on top of the 4.00-kg object as it passes through its equilibrium point. The two objects stick together. (a) By how much does the amplitude of the vibrating system change as a result of the collision? (b) By how much does the period change? (c) By how much does the energy change? (d) Account for the change in energy.

59. A simple pendulum with a length of 2.23 m and a mass of 6.74 kg is given an initial speed of 2.06 m/s at its equilibrium position. Assume it undergoes simple harmonic motion. Determine its (a) period, (b) total energy, and (c) maximum angular displacement.

60. **Review problem.** One end of a light spring with force constant 100 N/m is attached to a vertical wall. A light string is tied to the other end of the horizontal spring. The string changes from horizontal to vertical as it passes over a solid pulley of diameter 4.00 cm. The pulley is free to turn on a fixed, smooth axle. The vertical section of the string supports a 200-g object. The string does not slip at its contact with the pulley. Find the frequency of oscillation of the object, assuming the mass of the pulley is (a) negligible, (b) 250 g, and (c) 750 g.

61. ● People who ride motorcycles and bicycles learn to look out for bumps in the road and especially for *washboarding*, a condition in which many equally spaced ridges are worn into the road. What is so bad about washboarding? A motorcycle has several springs and shock absorbers in its suspension, but you can model it as a single spring supporting a block. You can estimate the force constant by thinking about how far the spring compresses when a heavy rider sits on the seat. A motorcyclist traveling at highway speed must be particularly careful of washboard bumps that are a certain distance apart. What is the order of magnitude of their separation distance? State the quantities you take as data and the values you measure or estimate for them.

62. A block of mass M is connected to a spring of mass m and oscillates in simple harmonic motion on a horizontal, frictionless track (Fig. P15.62). The force constant of the spring is k, and the equilibrium length is ℓ. Assume all portions of the spring oscillate in phase and the velocity of a segment dx is proportional to the distance x from the fixed end; that is, $v_x = (x/\ell)v$. Also, notice that the mass of a segment of the spring is $dm = (m/\ell)\,dx$. Find (a) the kinetic energy of the system when the block has a speed v and (b) the period of oscillation.

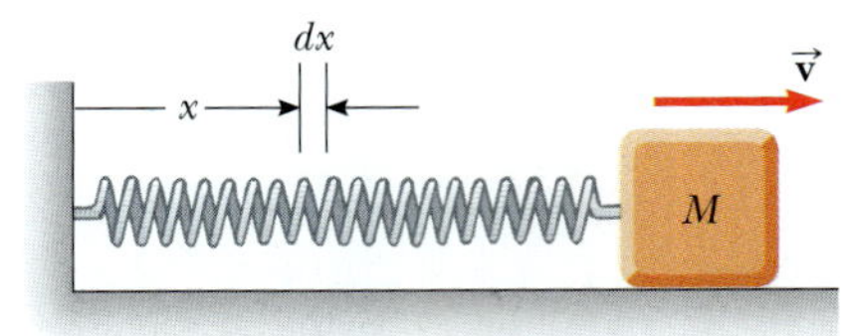

Figure P15.62

63. ▲ A ball of mass m is connected to two rubber bands of length L, each under tension T as shown in Figure P15.63. The ball is displaced by a small distance y perpendicular to the length of the rubber bands. Assuming the tension does not change, show that (a) the restoring force is $-(2T/L)y$ and (b) the system exhibits simple harmonic motion with an angular frequency $\omega = \sqrt{2T/mL}$.

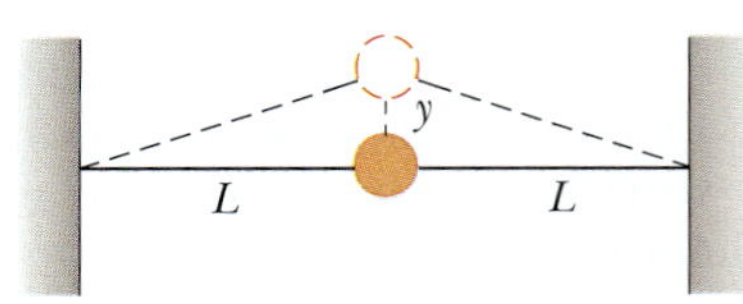

Figure P15.63

64. When a block of mass M, connected to the end of a spring of mass $m_s = 7.40$ g and force constant k, is set into simple harmonic motion, the period of its motion is

$$T = 2\pi\sqrt{\frac{M + (m_s/3)}{k}}$$

A two-part experiment is conducted with the use of blocks of various masses suspended vertically from the spring as shown in Figure P15.64. (a) Static extensions of 17.0, 29.3, 35.3, 41.3, 47.1, and 49.3 cm are measured for M values of 20.0, 40.0, 50.0, 60.0, 70.0, and 80.0 g, respectively. Construct a graph of Mg versus x and perform a linear least-squares fit to the data. From the slope of your graph, determine a value for k for this spring. (b) The system is now set into simple harmonic motion, and periods are measured with a stopwatch. With $M = 80.0$ g, the total time interval required for ten oscillations is measured to be 13.41 s. The experiment is repeated with M values of 70.0, 60.0, 50.0, 40.0, and 20.0 g, with corresponding time intervals for ten oscillations of 12.52, 11.67, 10.67, 9.62, and 7.03 s. Compute the experimental value for T from each of these measurements. Plot a graph of T^2 versus M and determine a value for k from the slope of the linear least-squares fit through the data points. Compare this value of k with that obtained in part (a). (c) Obtain a value for m_s from your graph and compare it with the given value of 7.40 g.

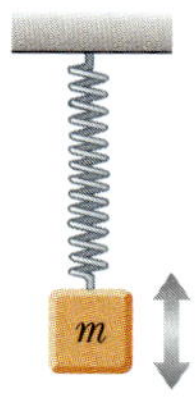

Figure P15.64

65. A smaller disk of radius r and mass m is attached rigidly to the face of a second larger disk of radius R and mass M as shown in Figure P15.65. The center of the small disk is located at the edge of the large disk. The large disk is mounted at its center on a frictionless axle. The assembly is rotated through a small angle θ from its equilibrium position and released. (a) Show that the speed of the center of the small disk as it passes through the equilibrium position is

$$v = 2\left[\frac{Rg(1-\cos\theta)}{(M/m)+(r/R)^2+2}\right]^{1/2}$$

(b) Show that the period of the motion is

$$T = 2\pi\left[\frac{(M+2m)R^2+mr^2}{2mgR}\right]^{1/2}$$

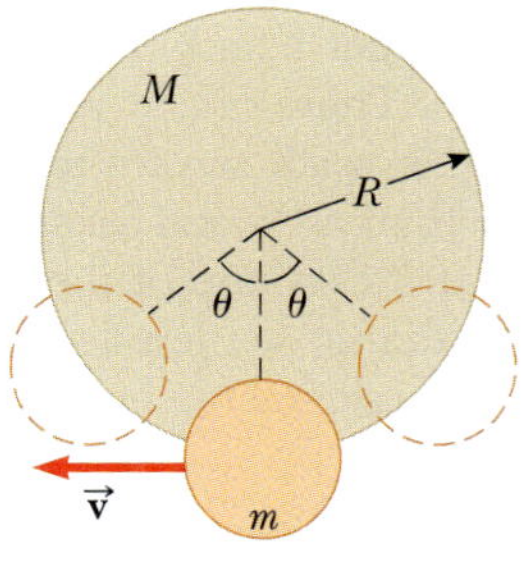

Figure P15.65

66. Consider a damped oscillator illustrated in Figures 15.20 and 15.21. The mass of the object is 375 g, the spring constant is 100 N/m, and b = 0.100 N · s/m. (a) Over what time interval does the amplitude drop to half its initial value? (b) **What If?** Over what time interval does the mechanical energy drop to half its initial value? (c) Show that, in general, the fractional rate at which the amplitude decreases in a damped harmonic oscillator is one-half the fractional rate at which the mechanical energy decreases.

67. A block of mass m is connected to two springs of force constants k_1 and k_2 in two ways as shown in Figures P15.67a and P15.67b. In both cases, the block moves on a frictionless table after it is displaced from equilibrium and released. Show that in the two cases the block exhibits simple harmonic motion with periods

(a) $T = 2\pi\sqrt{\dfrac{m(k_1+k_2)}{k_1k_2}}$ and (b) $T = 2\pi\sqrt{\dfrac{m}{k_1+k_2}}$

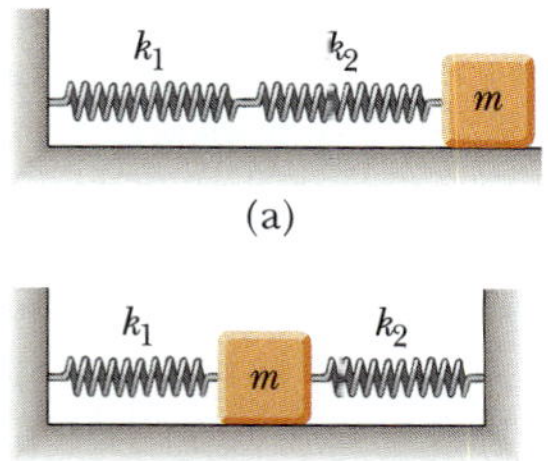

Figure P15.67

68. A lobsterman's buoy is a solid wooden cylinder of radius r and mass M. It is weighted at one end so that it floats upright in calm seawater, having density ρ. A passing shark tugs on the slack rope mooring the buoy to a lobster trap, pulling the buoy down a distance x from its equilibrium position and releasing it. Show that the buoy will execute simple harmonic motion if the resistive effects of the water are ignored and determine the period of the oscillations.

69. **Review problem.** Imagine that a hole is drilled through the center of the Earth to the other side. An object of mass m at a distance r from the center of the Earth is pulled toward the center of the Earth only by the mass within the sphere of radius r (the reddish region in Fig. P15.69). (a) Write Newton's law of gravitation for an object at the distance r from the center of the Earth and show that the force on it is of Hooke's law form, $F = -kr$, where the effective force constant is $k = \frac{4}{3}\pi\rho Gm$. Here ρ is the density of the Earth, assumed uniform, and G is the constant. (b) Show that a sack of mail dropped into the hole will execute simple harmonic motion if it moves without friction. When will it arrive at the other side of the Earth?

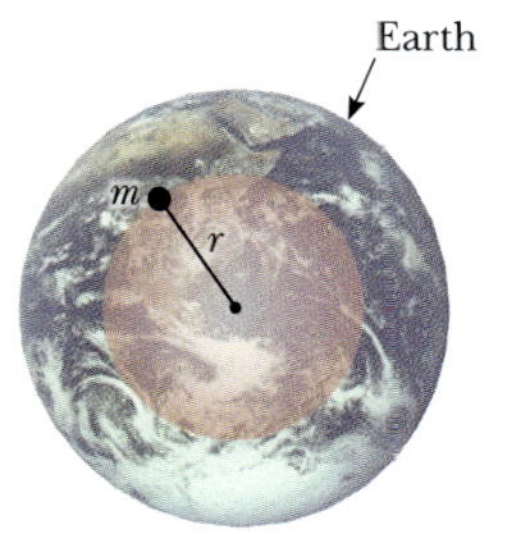

Figure P15.69

70. Your thumb squeaks on a plate you have just washed. Your sneakers squeak on the gym floor. Car tires squeal when you start or stop abruptly. Mortise joints groan in an old barn. The concertmaster's violin sings out over a full orchestra. You can make a goblet sing by wiping your moistened finger around its rim. As you slide it across the table, a Styrofoam cup may not make much sound, but it

makes the surface of some water inside it dance in a complicated resonance vibration. When chalk squeaks on a blackboard, you can see that it makes a row of regularly spaced dashes. As these examples suggest, vibration commonly results when friction acts on a moving elastic object. The oscillation is not simple harmonic motion, but is called *stick and slip*. This problem models stick-and-slip motion.

A block of mass m is attached to a fixed support by a horizontal spring with force constant k and negligible mass (Fig. P15.70). Hooke's law describes the spring both in extension and in compression. The block sits on a long horizontal board, with which it has coefficient of static friction μ_s and a smaller coefficient of kinetic friction μ_k. The board moves to the right at constant speed v. Assume the block spends most of its time sticking to the board and moving to the right, so the speed v is small in comparison to the average speed the block has as it slips back toward the left. (a) Show that the maximum extension of the spring from its unstressed position is very nearly given by $\mu_s mg/k$. (b) Show that the block oscillates around an equilibrium position at which the spring is stretched by $\mu_k mg/k$. (c) Graph the block's position versus time. (d) Show that the amplitude of the block's motion is

$$A = \frac{(\mu_s - \mu_k)mg}{k}$$

(e) Show that the period of the block's motion is

$$T = \frac{2(\mu_s - \mu_k)mg}{vk} + \pi\sqrt{\frac{m}{k}}$$

(f) Evaluate the frequency of the motion, taking $\mu_s = 0.400$, $\mu_k = 0.250$, $m = 0.300$ kg, $k = 12.0$ N/m, and $v = 2.40$ cm/s. (g) **What If?** What happens to the frequency if the mass increases? (h) If the spring constant increases? (i) If the speed of the board increases? (j) If the coefficient of static friction increases relative to the coefficient of kinetic friction? It is the excess of static over kinetic friction that is important for the vibration. "The squeaky wheel gets the grease" because even a viscous fluid cannot exert a force of static friction.

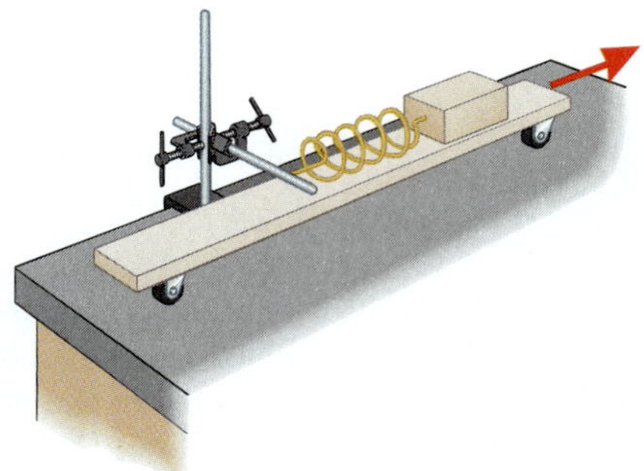

Figure P15.70

Answers to Quick Quizzes

15.1 (d). From its maximum positive position to the equilibrium position, the block travels a distance A. Next, it goes an equal distance past the equilibrium position to its maximum negative position. It then repeats these two motions in the reverse direction to return to its original position and complete one cycle.

15.2 (f). The object is in the region $x < 0$, so the position is negative. Because the object is moving back toward the origin in this region, the velocity is positive.

15.3 (a). The amplitude is larger because the curve for object B shows that the displacement from the origin (the vertical axis on the graph) is larger. The frequency is larger for object B because there are more oscillations per unit time interval.

15.4 (b). According to Equation 15.13, the period is proportional to the square root of the mass.

15.5 (c). The amplitude of the simple harmonic motion is the same as the radius of the circular motion. The initial position of the object in its circular motion is π radians from the positive x axis.

15.6 **(i)**, (a). With a longer length, the period of the pendulum will increase. Therefore, it will take longer to execute each swing, so each second according to the clock will take longer than an actual second and the clock will run slow. **(ii)**, (a). At the top of the mountain, the value of g is less than that at sea level. As a result, the period of the pendulum will increase and the clock will run slow.

Ocean waves combine properties of both transverse and longitudinal waves. With proper balance and timing, a surfer can capture a wave and take it for a ride. (© Rick Doyle/Corbis)

16 Wave Motion

Most of us experienced waves as children when we dropped a pebble into a pond. At the point the pebble hits the water's surface, waves are created. These waves move outward from the creation point in expanding circles until they reach the shore. If you were to examine carefully the motion of a small object floating on the disturbed water, you would see that the object moves vertically and horizontally about its original position but does not undergo any net displacement away from or toward the point the pebble hit the water. The small elements of water in contact with the object, as well as all the other water elements on the pond's surface, behave in the same way. That is, the water *wave* moves from the point of origin to the shore, but the water is not carried with it.

The world is full of waves, the two main types being *mechanical* waves and *electromagnetic* waves. In the case of mechanical waves, some physical medium is being disturbed; in our pebble example, elements of water are disturbed. Electromagnetic waves do not require a medium to propagate; some examples of electromagnetic waves are visible light, radio waves, television signals, and x-rays. Here, in this part of the book, we study only mechanical waves.

Consider again the small object floating on the water. We have caused the object to move at one point in the water by dropping a pebble at another location. The object has gained kinetic energy from our action, so energy must have transferred from the point at which the pebble is dropped to the position of the object. This feature is central to wave motion: *energy* is transferred over a distance, but *matter* is not.

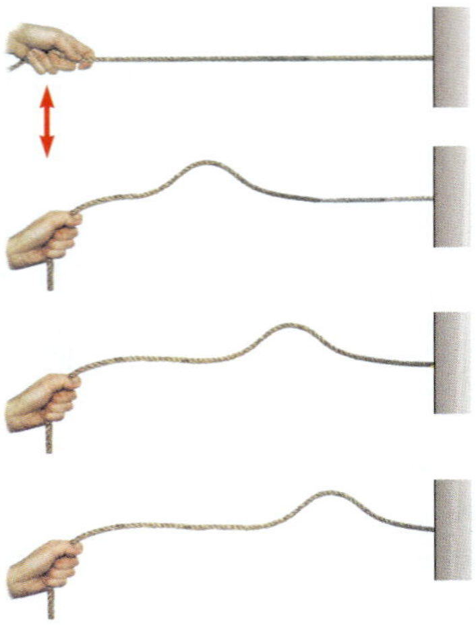

Figure 16.1 A pulse traveling down a stretched string. The shape of the pulse is approximately unchanged as it travels along the string.

16.1 Propagation of a Disturbance

The introduction to this chapter alluded to the essence of wave motion: the transfer of energy through space without the accompanying transfer of matter. In the list of energy transfer mechanisms in Chapter 8, two mechanisms—mechanical waves and electromagnetic radiation—depend on waves. By contrast, in another mechanism, matter transfer, the energy transfer is accompanied by a movement of matter through space.

All mechanical waves require (1) some source of disturbance, (2) a medium containing elements that can be disturbed, and (3) some physical mechanism through which elements of the medium can influence each other. One way to demonstrate wave motion is to flick one end of a long string that is under tension and has its opposite end fixed as shown in Figure 16.1. In this manner, a single bump (called a *pulse*) is formed and travels along the string with a definite speed. Figure 16.1 represents four consecutive "snapshots" of the creation and propagation of the traveling pulse. The string is the medium through which the pulse travels. The pulse has a definite height and a definite speed of propagation along the medium (the string). The shape of the pulse changes very little as it travels along the string.[1]

We shall first focus on a pulse traveling through a medium. Once we have explored the behavior of a pulse, we will then turn our attention to a *wave*, which is a *periodic* disturbance traveling through a medium. We create a pulse on our string by flicking the end of the string once as in Figure 16.1. If we were to move the end of the string up and down repeatedly, we would create a traveling wave, which has characteristics a pulse does not have. We shall explore these characteristics in Section 16.2.

As the pulse in Figure 16.1 travels, each disturbed element of the string moves in a direction *perpendicular* to the direction of propagation. Figure 16.2 illustrates this point for one particular element, labeled *P*. Notice that no part of the string ever moves in the direction of the propagation. A traveling wave or pulse that

Figure 16.2 A transverse pulse traveling on a stretched string. The direction of motion of any element *P* of the string (blue arrows) is perpendicular to the direction of propagation (red arrows).

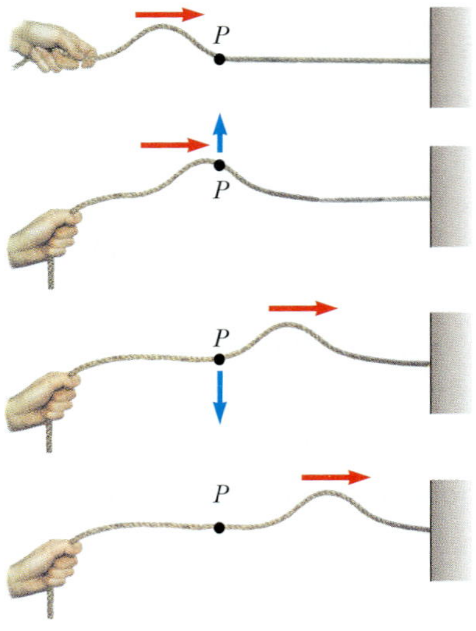

[1] In reality, the pulse changes shape and gradually spreads out during the motion. This effect, called *dispersion*, is common to many mechanical waves as well as to electromagnetic waves. We do not consider dispersion in this chapter.

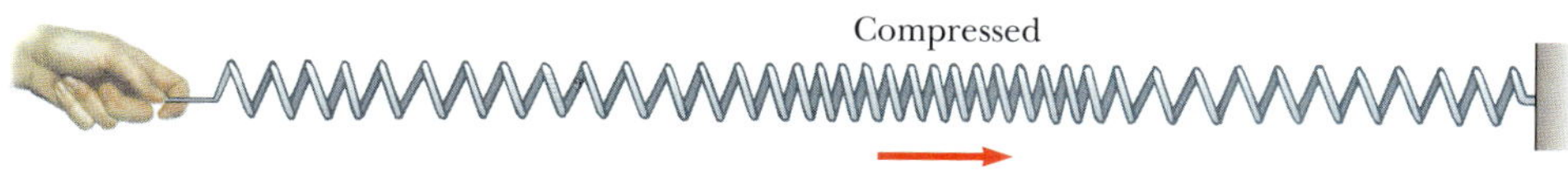

Figure 16.3 A longitudinal pulse along a stretched spring. The displacement of the coils is parallel to the direction of the propagation.

causes the elements of the disturbed medium to move perpendicular to the direction of propagation is called a **transverse wave.**

Compare this wave with another type of pulse, one moving down a long, stretched spring as shown in Figure 16.3. The left end of the spring is pushed briefly to the right and then pulled briefly to the left. This movement creates a sudden compression of a region of the coils. The compressed region travels along the spring (to the right in Fig. 16.3). Notice that the direction of the displacement of the coils is *parallel* to the direction of propagation of the compressed region. A traveling wave or pulse that causes the elements of the medium to move parallel to the direction of propagation is called a **longitudinal wave.**

Sound waves, which we shall discuss in Chapter 17, are another example of longitudinal waves. The disturbance in a sound wave is a series of high-pressure and low-pressure regions that travel through air.

Some waves in nature exhibit a combination of transverse and longitudinal displacements. Surface-water waves are a good example. When a water wave travels on the surface of deep water, elements of water at the surface move in nearly circular paths as shown in Active Figure 16.4. The disturbance has both transverse and longitudinal components. The transverse displacements seen in Active Figure 16.4 represent the variations in vertical position of the water elements. The longitudinal displacements represent elements of water moving back and forth in a horizontal direction.

The three-dimensional waves that travel out from a point under the Earth's surface at which an earthquake occurs are of both types, transverse and longitudinal. The longitudinal waves are the faster of the two, traveling at speeds in the range of 7 to 8 km/s near the surface. They are called **P waves**, with "P" standing for *primary*, because they travel faster than the transverse waves and arrive first at a seismograph (a device used to detect waves due to earthquakes). The slower transverse waves, called **S waves,** with "S" standing for *secondary*, travel through the Earth at 4 to 5 km/s near the surface. By recording the time interval between the arrivals of these two types of waves at a seismograph, the distance from the seismograph to the point of origin of the waves can be determined. A single measurement establishes an imaginary sphere centered on the seismograph, with the sphere's radius determined by the difference in arrival times of the P and S waves. The origin of the waves is located somewhere on that sphere. The imaginary spheres from three or more monitoring stations located far apart from one another intersect at one region of the Earth, and this region is where the earthquake occurred.

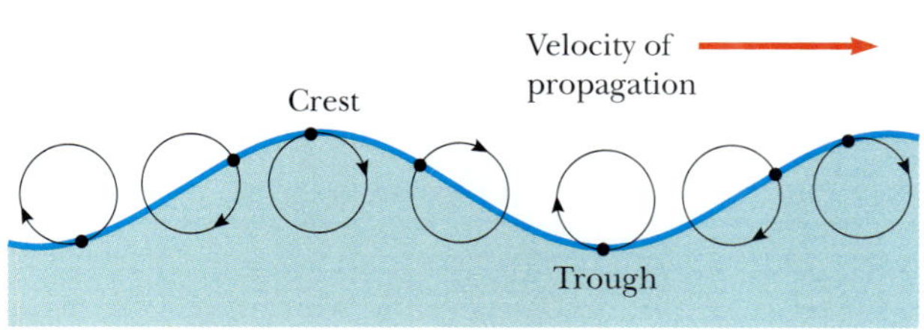

ACTIVE FIGURE 16.4

The motion of water elements on the surface of deep water in which a wave is propagating is a combination of transverse and longitudinal displacements. The result is that elements at the surface move in nearly circular paths. Each element is displaced both horizontally and vertically from its equilibrium position.

Sign in at www.thomsonedu.com and go to ThomsonNOW to observe the displacement of water elements at the surface of the moving waves.

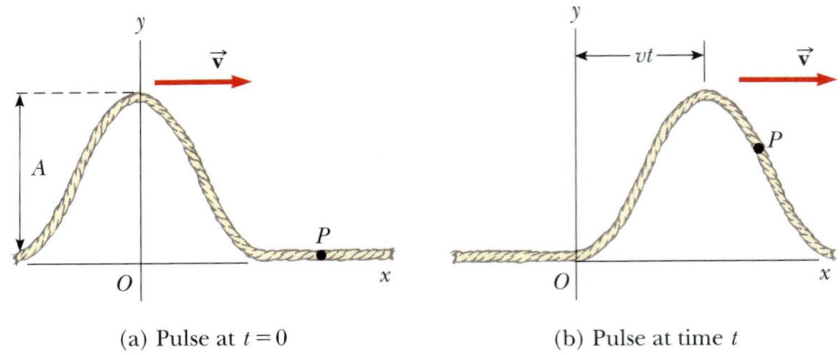

Figure 16.5 A one-dimensional pulse traveling to the right with a speed v. (a) At $t = 0$, the shape of the pulse is given by $y = f(x)$. (b) At some later time t, the shape remains unchanged and the vertical position of an element of the medium at any point P is given by $y = f(x - vt)$.

Consider a pulse traveling to the right on a long string as shown in Figure 16.5. Figure 16.5a represents the shape and position of the pulse at time $t = 0$. At this time, the shape of the pulse, whatever it may be, can be represented by some mathematical function that we will write as $y(x, 0) = f(x)$. This function describes the transverse position y of the element of the string located at each value of x at time $t = 0$. Because the speed of the pulse is v, the pulse has traveled to the right a distance vt at the time t (Fig. 16.5b). We assume the shape of the pulse does not change with time. Therefore, at time t, the shape of the pulse is the same as it was at time $t = 0$ as in Figure 16.5a. Consequently, an element of the string at x at this time has the same y position as an element located at $x - vt$ had at time $t = 0$:

$$y(x, t) = y(x - vt, 0)$$

In general, then, we can represent the transverse position y for all positions and times, measured in a stationary frame with the origin at O, as

Pulse traveling to the right ▶

$$y(x, t) = f(x - vt) \qquad \textbf{(16.1)}$$

Similarly, if the pulse travels to the left, the transverse positions of elements of the string are described by

Pulse traveling to the left ▶

$$y(x, t) = f(x + vt) \qquad \textbf{(16.2)}$$

The function y, sometimes called the **wave function,** depends on the two variables x and t. For this reason, it is often written $y(x, t)$, which is read "y as a function of x and t."

It is important to understand the meaning of y. Consider an element of the string at point P, identified by a particular value of its x coordinate. As the pulse passes through P, the y coordinate of this element increases, reaches a maximum, and then decreases to zero. **The wave function $y(x, t)$ represents the y coordinate—the transverse position—of any element located at position x at any time t.** Furthermore, if t is fixed (as, for example, in the case of taking a snapshot of the pulse), the wave function $y(x)$, sometimes called the **waveform,** defines a curve representing the geometric shape of the pulse at that time.

Quick Quiz 16.1 **(i)** In a long line of people waiting to buy tickets, the first person leaves and a pulse of motion occurs as people step forward to fill the gap. As each person steps forward, the gap moves through the line. Is the propagation of this gap (a) transverse or (b) longitudinal? **(ii)** Consider the "wave" at a baseball game: people stand up and raise their arms as the wave arrives at their location, and the resultant pulse moves around the stadium. Is this wave (a) transverse or (b) longitudinal?

EXAMPLE 16.1 A Pulse Moving to the Right

A pulse moving to the right along the x axis is represented by the wave function

$$y(x, t) = \frac{2}{(x - 3.0t)^2 + 1}$$

where x and y are measured in centimeters and t is measured in seconds. Find expressions for the wave function at $t = 0$, $t = 1.0$ s, and $t = 2.0$ s.

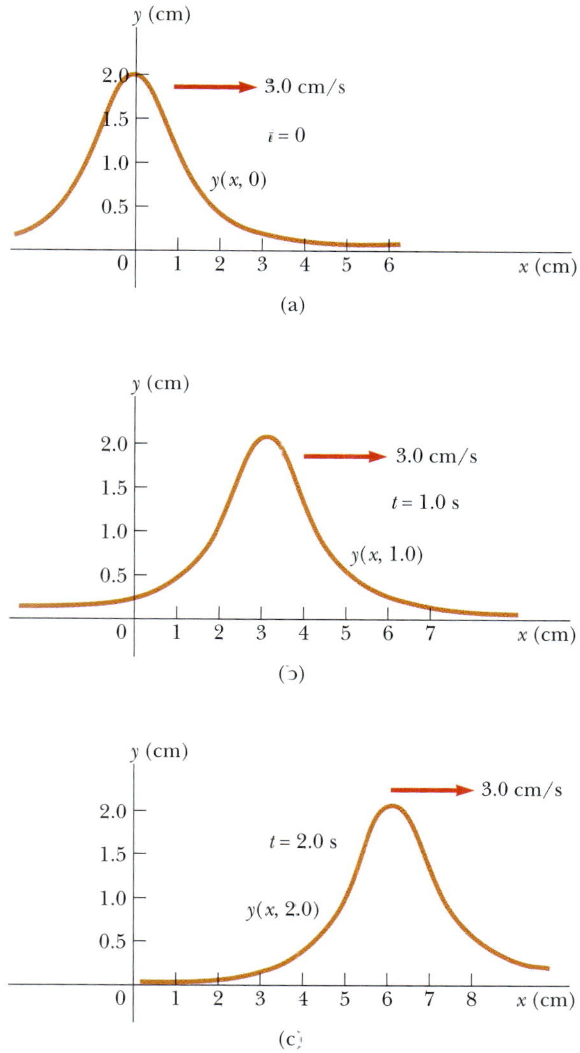

Figure 16.6 (Example 16.1) Graphs of the function $y(x, t) = 2/[(x - 3.0t)^2 + 1]$ at (a) $t = 0$, (b) $t = 1.0$ s, and (c) $t = 2.0$ s.

SOLUTION

Conceptualize Figure 16.6a shows the pulse represented by this wave function at $t = 0$. Imagine this pulse moving to the right and maintaining its shape as suggested by Figures 16.6b and 16.6c.

Categorize We categorize this example as a relatively simple analysis problem in which we interpret the mathematical representation of a pulse.

Analyze The wave function is of the form $y = f(x - vt)$. Inspection of the expression for $y(x, t)$ reveals that the wave speed is $v = 3.0$ cm/s. Furthermore, by letting $x - 3.0t = 0$, we find that the maximum value of y is given by $A = 2.0$ cm.

Write the wave function expression at $t = 0$:

$$y(x, 0) = \frac{2}{x^2 + 1}$$

Write the wave function expression at $t = 1.0$ s:

$$y(x, 1.0) = \frac{2}{(x - 3.0)^2 + 1}$$

Write the wave function expression at $t = 2.0$ s:

$$y(x, 2.0) = \frac{2}{(x - 6.0)^2 + 1}$$

For each of these expressions, we can substitute various values of x and plot the wave function. This procedure yields the wave functions shown in the three parts of Figure 16.6.

Finalize These snapshots show that the pulse moves to the right without changing its shape and that it has a constant speed of 3.0 cm/s.

What If? What if the wave function were

$$y(x, t) = \frac{4}{(x + 3.0t)^2 + 1}$$

How would that change the situation?

Answer One new feature in this expression is the plus sign in the denominator rather than the minus sign. The new expression represents a pulse with the same shape as that in Figure 16.6, but moving to the left as time progresses. Another new feature here is the numerator of 4 rather than 2. Therefore, the new expression represents a pulse with twice the height of that in Figure 16.6.

16.2 The Traveling Wave Model

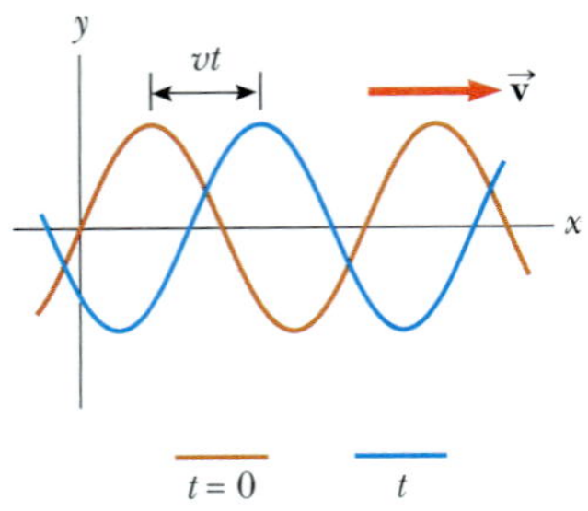

ACTIVE FIGURE 16.7

A one-dimensional sinusoidal wave traveling to the right with a speed v. The brown curve represents a snapshot of the wave at $t = 0$, and the blue curve represents a snapshot at some later time t.

Sign in at www.thomsonedu.com and go to ThomsonNOW to watch the wave move and take snapshots of it at various times.

In this section, we introduce an important wave function whose shape is shown in Active Figure 16.7. The wave represented by this curve is called a **sinusoidal wave** because the curve is the same as that of the function sin θ plotted against θ. A sinusoidal wave could be established on a rope by shaking the end of the rope up and down in simple harmonic motion.

The sinusoidal wave is the simplest example of a periodic continuous wave and can be used to build more complex waves (see Section 18.8). The brown curve in Active Figure 16.7 represents a snapshot of a traveling sinusoidal wave at $t = 0$, and the blue curve represents a snapshot of the wave at some later time t. Imagine two types of motion that can occur. First, the entire waveform in Active Figure 16.7 moves to the right so that the brown curve moves toward the right and eventually reaches the position of the blue curve. This movement is the motion of the *wave*. If we focus on one element of the medium, such as the element at $x = 0$, we see that each element moves up and down along the y axis in simple harmonic motion. This movement is the motion of the *elements of the medium*. It is important to differentiate between the motion of the wave and the motion of the elements of the medium.

In the early chapters of this book, we developed several analysis models based on the particle model. With our introduction to waves, we can develop a new simplification model, the **wave model,** that will allow us to explore more analysis models for solving problems. An ideal particle has zero size. We can build physical objects with nonzero size as combinations of particles. Therefore, the particle can be considered a basic building block. An ideal wave has a single frequency and is infinitely long; that is, the wave exists throughout the Universe. (An unbounded wave of finite length must necessarily have a mixture of frequencies.) When this concept is explored in Section 18.8, we will find that ideal waves can be combined, just as we combined particles.

In what follows, we will develop the principal features and mathematical representations of the analysis model of a **traveling wave.** This model is used in situations in which a wave moves through space without interacting with other waves or particles.

PITFALL PREVENTION 16.1

What's the Difference Between Active Figures 16.8a and 16.8b?

Notice the visual similarity between Active Figures 16.8a and 16.8b. The shapes are the same, but (a) is a graph of vertical position versus horizontal position, whereas (b) is vertical position versus time. Active Figure 16.8a is a pictorial representation of the wave *for a series of particles of the medium*; it is what you would see at an instant of time. Active Figure 16.8b is a graphical representation of the position of *one element of the medium* as a function of time. That both figures have the identical shape represents Equation 16.1: a wave is the *same* function of both x and t.

Active Figure 16.8a shows a snapshot of a wave moving through a medium. Active Figure 16.8b shows a graph of the position of one element of the medium as a function of time. A point in Active Figure 16.8a at which the displacement of the element from its normal position is highest is called the **crest** of the wave. The lowest point is called the **trough.** The distance from one crest to the next is called the **wavelength** λ (Greek letter lambda). More generally, **the wavelength is the minimum distance between any two identical points on adjacent waves** as shown in Active Figure 16.8a.

If you count the number of seconds between the arrivals of two adjacent crests at a given point in space, you measure the **period** T of the waves. In general, **the period is the time interval required for two identical points of adjacent waves to pass by a point** as shown in Active Figure 16.8b. The period of the wave is the same as the period of the simple harmonic oscillation of one element of the medium.

The same information is more often given by the inverse of the period, which is called the **frequency** f. In general, **the frequency of a periodic wave is the number of crests (or troughs, or any other point on the wave) that pass a given point in a unit time interval.** The frequency of a sinusoidal wave is related to the period by the expression

$$f = \frac{1}{T} \tag{16.3}$$

The frequency of the wave is the same as the frequency of the simple harmonic oscillation of one element of the medium. The most common unit for frequency, as we learned in Chapter 15, is s^{-1}, or **hertz** (Hz). The corresponding unit for T is seconds.

The maximum position of an element of the medium relative to its equilibrium position is called the **amplitude** A of the wave.

Waves travel with a specific speed, and this speed depends on the properties of the medium being disturbed. For instance, sound waves travel through room-temperature air with a speed of about 343 m/s (781 mi/h), whereas they travel through most solids with a speed greater than 343 m/s.

Consider the sinusoidal wave in Active Figure 16.8a, which shows the position of the wave at $t = 0$. Because the wave is sinusoidal, we expect the wave function at this instant to be expressed as $y(x, 0) = A \sin ax$, where A is the amplitude and a is a constant to be determined. At $x = 0$, we see that $y(0, 0) = A \sin a(0) = 0$, consistent with Active Figure 16.8a. The next value of x for which y is zero is $x = \lambda/2$. Therefore,

$$y\left(\frac{\lambda}{2}, 0\right) = A \sin\left(a\frac{\lambda}{2}\right) = 0$$

For this equation to be true, we must have $a\lambda/2 = \pi$, or $a = 2\pi/\lambda$. Therefore, the function describing the positions of the elements of the medium through which the sinusoidal wave is traveling can be written

$$y(x, 0) = A \sin\left(\frac{2\pi}{\lambda}x\right) \tag{16.4}$$

where the constant A represents the wave amplitude and the constant λ is the wavelength. Notice that the vertical position of an element of the medium is the same whenever x is increased by an integral multiple of λ. If the wave moves to the right with a speed v, the wave function at some later time t is

$$y(x, t) = A \sin\left[\frac{2\pi}{\lambda}(x - vt)\right] \tag{16.5}$$

The wave function has the form $f(x - vt)$ (Eq. 16.1). If the wave were traveling to the left, the quantity $x - vt$ would be replaced by $x + vt$ as we learned when we developed Equations 16.1 and 16.2.

By definition, the wave travels through a displacement Δx equal to one wavelength λ in a time interval Δt of one period T. Therefore, the wave speed, wavelength, and period are related by the expression

$$v = \frac{\Delta x}{\Delta t} = \frac{\lambda}{T} \tag{16.6}$$

Substituting this expression for v into Equation 16.5 gives

$$y = A \sin\left[2\pi\left(\frac{x}{\lambda} - \frac{t}{T}\right)\right] \tag{16.7}$$

This form of the wave function shows the *periodic* nature of y. Note that we will often use y rather than $y(x, t)$ as a shorthand notation. At any given time t, y has the *same* value at the positions x, $x + \lambda$, $x + 2\lambda$, and so on. Furthermore, at any given position x, the value of y is the same at times t, $t + T$, $t + 2T$, and so on.

We can express the wave function in a convenient form by defining two other quantities, the **angular wave number** k (usually called simply the **wave number**) and the **angular frequency** ω:

$$k \equiv \frac{2\pi}{\lambda} \tag{16.8}$$

◀ Angular wave number

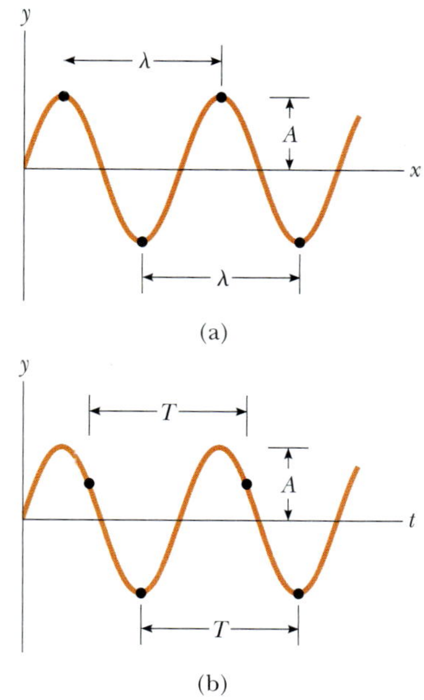

ACTIVE FIGURE 16.8

(a) A snapshot of a sinusoidal wave. The wavelength λ of a wave is the distance between adjacent crests or adjacent troughs. (b) The position of one element of the medium as a function of time. The period T of a wave is the time interval required for the element to complete one cycle of its oscillation and for the wave to travel one wavelength.

Sign in at www.thomsonedu.com and go to ThomsonNOW to change the parameters to see the effect on the wave function.

Angular frequency ▶

$$\omega \equiv \frac{2\pi}{T} = 2\pi f \tag{16.9}$$

Using these definitions, Equation 16.7 can be written in the more compact form

Wave function for a sinusoidal wave ▶

$$y = A \sin (kx - \omega t) \tag{16.10}$$

Using Equations 16.3, 16.8, and 16.9, the wave speed v originally given in Equation 16.6 can be expressed in the following alternative forms:

$$v = \frac{\omega}{k} \tag{16.11}$$

Speed of a sinusoidal wave ▶

$$v = \lambda f \tag{16.12}$$

The wave function given by Equation 16.10 assumes the vertical position y of an element of the medium is zero at $x = 0$ and $t = 0$. That need not be the case. If it is not, we generally express the wave function in the form

General expression for a sinusoidal wave ▶

$$y = A \sin (kx - \omega t + \phi) \tag{16.13}$$

where ϕ is the **phase constant,** just as we learned in our study of periodic motion in Chapter 15. This constant can be determined from the initial conditions.

Quick Quiz 16.2 A sinusoidal wave of frequency f is traveling along a stretched string. The string is brought to rest, and a second traveling wave of frequency $2f$ is established on the string. **(i)** What is the wave speed of the second wave? (a) twice that of the first wave (b) half that of the first wave (c) the same as that of the first wave (d) impossible to determine **(ii)** From the same choices, describe the wavelength of the second wave. **(iii)** From the same choices, describe the amplitude of the second wave.

EXAMPLE 16.2 A Traveling Sinusoidal Wave

A sinusoidal wave traveling in the positive x direction has an amplitude of 15.0 cm, a wavelength of 40.0 cm, and a frequency of 8.00 Hz. The vertical position of an element of the medium at $t = 0$ and $x = 0$ is also 15.0 cm as shown in Figure 16.9.

(A) Find the wave number k, period T, angular frequency ω, and speed v of the wave.

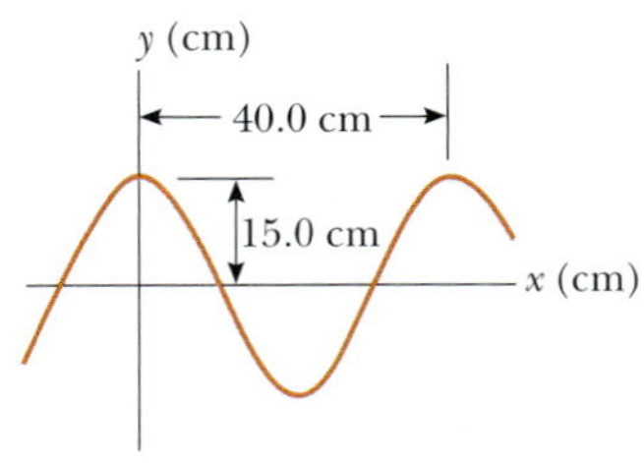

Figure 16.9 (Example 16.2) A sinusoidal wave of wavelength λ = 40.0 cm and amplitude A = 15.0 cm. The wave function can be written in the form $y = A \cos (kx - \omega t)$.

SOLUTION

Conceptualize Figure 16.9 shows the wave at $t = 0$. Imagine this wave moving to the right and maintaining its shape.

Categorize We will evaluate parameters of the wave using equations generated in the preceding discussion, so we categorize this example as a substitution problem.

Evaluate the wave number from Equation 16.8:

$$k = \frac{2\pi}{\lambda} = \frac{2\pi \text{ rad}}{40.0 \text{ cm}} = 0.157 \text{ rad/cm}$$

Evaluate the period of the wave from Equation 16.3:

$$T = \frac{1}{f} = \frac{1}{8.00 \text{ s}^{-1}} = 0.125 \text{ s}$$

Evaluate the angular frequency of the wave from Equation 16.9:

$$\omega = 2\pi f = 2\pi(8.00 \text{ s}^{-1}) = 50.3 \text{ rad/s}$$

Evaluate the wave speed from Equation 16.12:

$$v = \lambda f = (40.0 \text{ cm})(8.00 \text{ s}^{-1}) = 320 \text{ cm/s}$$

(B) Determine the phase constant ϕ and write a general expression for the wave function.

SOLUTION

Substitute $A = 15.0$ cm, $y = 15.0$ cm, $x = 0$, and $t = 0$ into Equation 16.13:

$$15.0 = (15.0)\sin\phi \rightarrow \sin\phi = 1 \rightarrow \phi = \frac{\pi}{2}\text{ rad}$$

Write the wave function:

$$y = A\sin\left(kx - \omega t + \frac{\pi}{2}\right) = A\cos(kx - \omega t)$$

Substitute the values for A, k, and ω into this expression:

$$y = (15.0\text{ cm})\cos(0.157x - 50.3t)$$

Sinusoidal Waves on Strings

In Figure 16.1, we demonstrated how to create a pulse by jerking a taut string up and down once. To create a series of such pulses—a wave—let's replace the hand with an oscillating blade vibrating in simple harmonic motion. Active Figure 16.10 represents snapshots of the wave created in this way at intervals of $T/4$. Because the end of the blade oscillates in simple harmonic motion, **each element of the string, such as that at *P*, also oscillates vertically with simple harmonic motion.** That must be the case because each element follows the simple harmonic motion of the blade. Therefore, every element of the string can be treated as a simple harmonic oscillator vibrating with a frequency equal to the frequency of oscillation of the blade.[2] Notice that although each element oscillates in the y direction, the wave travels in the x direction with a speed v. Of course, that is the definition of a transverse wave.

If the wave at $t = 0$ is as described in Active Figure 16.10b, the wave function can be written as

$$y = A\sin(kx - \omega t)$$

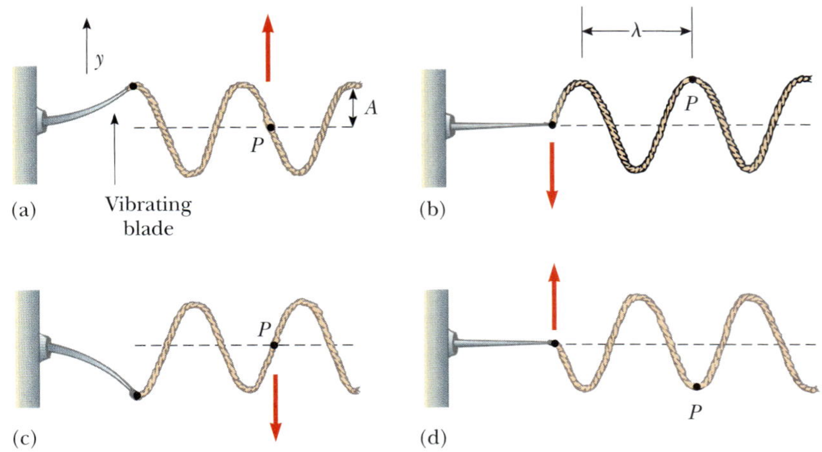

ACTIVE FIGURE 16.10

One method for producing a sinusoidal wave on a string. The left end of the string is connected to a blade that is set into oscillation. Every element of the string, such as that at point *P*, oscillates with simple harmonic motion in the vertical direction.

Sign in at www.thomsonedu.com and go to ThomsonNOW to adjust the frequency of the blade.

[2] In this arrangement, we are assuming that a string element always oscillates in a vertical line. The tension in the string would vary if an element were allowed to move sideways. Such motion would make the analysis very complex.

We can use this expression to describe the motion of any element of the string. An element at point P (or any other element of the string) moves only vertically, and so its x coordinate remains constant. Therefore, the **transverse speed** v_y (not to be confused with the wave speed v) and the **transverse acceleration** a_y of elements of the string are

$$v_y = \left.\frac{dy}{dt}\right]_{x=\text{constant}} = \frac{\partial y}{\partial t} = -\omega A \cos(kx - \omega t) \qquad \textbf{(16.14)}$$

$$a_y = \left.\frac{dv_y}{dt}\right]_{x=\text{constant}} = \frac{\partial v_y}{\partial t} = -\omega^2 A \sin(kx - \omega t) \qquad \textbf{(16.15)}$$

PITFALL PREVENTION 16.2
Two Kinds of Speed/Velocity

Do not confuse v, the speed of the wave as it propagates along the string, with v_y, the transverse velocity of a point on the string. The speed v is constant for a uniform medium, whereas v_y varies sinusoidally.

These expressions incorporate partial derivatives (see Section 7.8) because y depends on both x and t. In the operation $\partial y/\partial t$, for example, we take a derivative with respect to t while holding x constant. The maximum values of the transverse speed and transverse acceleration are simply the absolute values of the coefficients of the cosine and sine functions:

$$v_{y,\,\max} = \omega A \qquad \textbf{(16.16)}$$

$$a_{y,\,\max} = \omega^2 A \qquad \textbf{(16.17)}$$

The transverse speed and transverse acceleration of elements of the string do not reach their maximum values simultaneously. The transverse speed reaches its maximum value (ωA) when $y = 0$, whereas the magnitude of the transverse acceleration reaches its maximum value ($\omega^2 A$) when $y = \pm A$. Finally, Equations 16.16 and 16.17 are identical in mathematical form to the corresponding equations for simple harmonic motion, Equations 15.17 and 15.18.

Quick Quiz 16.3 The amplitude of a wave is doubled, with no other changes made to the wave. As a result of this doubling, which of the following statements is correct? (a) The speed of the wave changes. (b) The frequency of the wave changes. (c) The maximum transverse speed of an element of the medium changes. (d) Statements (a) through (c) are all true. (e) None of statements (a) through (c) is true.

16.3 The Speed of Waves on Strings

In this section, we determine the speed of a transverse pulse traveling on a taut string. Let's first conceptually predict the parameters that determine the speed. If a string under tension is pulled sideways and then released, the force of tension is responsible for accelerating a particular element of the string back toward its equilibrium position. According to Newton's second law, the acceleration of the element increases with increasing tension. If the element returns to equilibrium more rapidly due to this increased acceleration, we would intuitively argue that the wave speed is greater. Therefore, we expect the wave speed to increase with increasing tension.

Likewise, because it is more difficult to accelerate a massive element of the string than a light element, the wave speed should decrease as the mass per unit length of the string increases. If the tension in the string is T and its mass per unit length is μ (Greek letter mu), the wave speed, as we shall show, is

Speed of a wave on a stretched string ▶

$$v = \sqrt{\frac{T}{\mu}} \qquad \textbf{(16.18)}$$

Let us use a mechanical analysis to derive Equation 16.18. Consider a pulse moving on a taut string to the right with a uniform speed v measured relative to a stationary frame of reference. Instead of staying in this reference frame, it is more convenient to choose a different inertial reference frame that moves along with the pulse with the same speed as the pulse so that the pulse is at rest within the frame. This change of reference frame is permitted because Newton's laws are valid in either a stationary frame or one that moves with constant velocity. In our new reference frame, all elements of the string move to the left: a given element of the string initially to the right of the pulse moves to the left, rises up and follows the shape of the pulse, and then continues to move to the left. Figure 16.11a shows such an element at the instant it is located at the top of the pulse.

The small element of the string of length Δs shown in Figure 16.11a, and magnified in Figure 16.11b, forms an approximate arc of a circle of radius R. In the moving frame of reference (which moves to the right at a speed v along with the pulse), the shaded element moves to the left with a speed v. This element has a centripetal acceleration equal to v^2/R, which is supplied by components of the force $\vec{\mathbf{T}}$ whose magnitude is the tension in the string. The force $\vec{\mathbf{T}}$ acts on both sides of the element and is tangent to the arc as shown in Figure 16.11b. The horizontal components of $\vec{\mathbf{T}}$ cancel, and each vertical component $T \sin \theta$ acts radially toward the arc's center. Hence, the total radial force on the element is $2T \sin \theta$. Because the element is small, θ is small, and we can therefore use the small-angle approximation $\sin \theta \approx \theta$. So, the total radial force is

$$F_r = 2T \sin \theta \approx 2T\theta$$

The element has a mass $m = \mu \Delta s$. Because the element forms part of a circle and subtends an angle 2θ at the center, $\Delta s = R(2\theta)$, and

$$m = \mu \Delta s = 2\mu R\theta$$

Applying Newton's second law to this element in the radial direction gives

$$F_r = ma = \frac{mv^2}{R}$$

$$2T\theta = \frac{2\mu R\theta v^2}{R} \quad \rightarrow \quad v = \sqrt{\frac{T}{\mu}}$$

This expression for v is Equation 16.18.

Notice that this derivation is based on the assumption that the pulse height is small relative to the length of the string. Using this assumption, we were able to use the approximation $\sin \theta \approx \theta$. Furthermore, the model assumes the tension T is not affected by the presence of the pulse; therefore, T is the same at all points on the string. Finally, this proof does *not* assume any particular shape for the pulse. Therefore, a pulse of *any shape* travels along the string with speed $v = \sqrt{T/\mu}$ without any change in pulse shape.

Quick Quiz 16.4 Suppose you create a pulse by moving the free end of a taut string up and down once with your hand beginning at $t = 0$. The string is attached at its other end to a distant wall. The pulse reaches the wall at time t. Which of the following actions, taken by itself, decreases the time interval required for the pulse to reach the wall? More than one choice may be correct. (a) moving your hand more quickly, but still only up and down once by the same amount (b) moving your hand more slowly, but still only up and down once by the same amount (c) moving your hand a greater distance up and down in the same amount of time (d) moving your hand a lesser distance up and down in the same amount of time (e) using a heavier string of the same length and under the same tension (f) using a lighter string of the same length and under the same tension (g) using a string of the same linear mass density but under decreased tension (h) using a string of the same linear mass density but under increased tension

PITFALL PREVENTION 16.3
Multiple *T*s

Do not confuse the T in Equation 16.18 for the tension with the symbol T used in this chapter for the period of a wave. The context of the equation should help you identify which quantity is meant. There simply aren't enough letters in the alphabet to assign a unique letter to each variable!

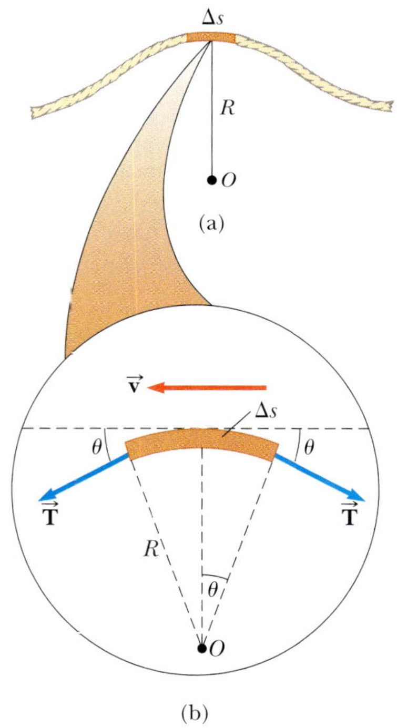

Figure 16.11 (a) To obtain the speed v of a wave on a stretched string, it is convenient to describe the motion of a small element of the string in a moving frame of reference. (b) In the moving frame of reference, the small element of length Δs moves to the left with speed v. The net force on the element is in the radial direction because the horizontal components of the tension force cancel.

EXAMPLE 16.3 The Speed of a Pulse on a Cord

A uniform string has a mass of 0.300 kg and a length of 6.00 m (Fig. 16.12). The string passes over a pulley and supports a 2.00-kg object. Find the speed of a pulse traveling along this string.

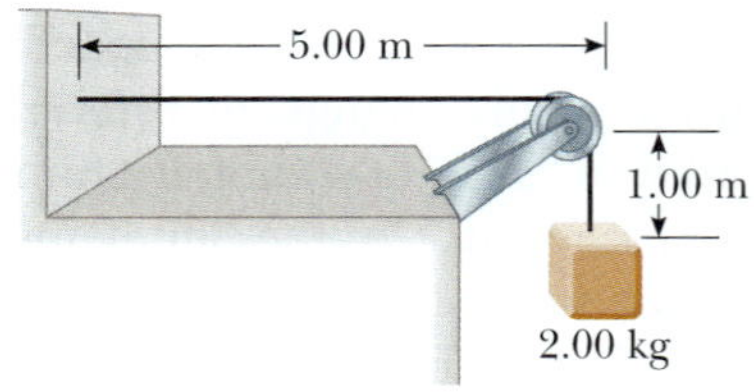

Figure 16.12 (Example 16.3) The tension T in the cord is maintained by the suspended object. The speed of any wave traveling along the cord is given by $v = \sqrt{T/\mu}$.

SOLUTION

Conceptualize In Figure 16.12, the hanging block establishes a tension in the horizontal string. This tension determines the speed with which waves move on the string.

Categorize To find the tension in the string, we model the hanging block as a particle in equilibrium. Then we use the tension to evaluate the wave speed on the string using Equation 16.18.

Analyze Apply the particle in equilibrium model to the block:

$$\sum F_y = T - m_{\text{block}}g = 0$$

Solve for the tension in the string:

$$T = m_{\text{block}}g$$

Use Equation 16.18 to find the wave speed, using $\mu = m_{\text{string}}/\ell$ for the linear mass density of the string:

$$v = \sqrt{\frac{T}{\mu}} = \sqrt{\frac{m_{\text{block}}g\ell}{m_{\text{string}}}}$$

Evaluate the wave speed:

$$v = \sqrt{\frac{(2.00 \text{ kg})(9.80 \text{ m/s}^2)(6.00 \text{ m})}{0.300 \text{ kg}}} = 19.8 \text{ m/s}$$

Finalize The calculation of the tension neglects the small mass of the string. Strictly speaking, the string can never be exactly straight; therefore, the tension is not uniform.

What If? What if the block were swinging back and forth with respect to the vertical? How would that affect the wave speed on the string?

Answer The swinging block is categorized as a particle under a net force. The magnitude of one of the forces on the block is the tension in the string, which determines the wave speed. As the block swings, the tension changes, so the wave speed changes.

When the block is at the bottom of the swing, the string is vertical and the tension is larger than the weight of the block because the net force must be upward to provide the centripetal acceleration of the block. Therefore, the wave speed must be greater than 19.8 m/s.

When the block is at its highest point at the end of a swing, it is momentarily at rest, so there is no centripetal acceleration at that instant. The block is a particle in equilibrium in the radial direction. The tension is balanced by a component of the gravitational force on the block. Therefore, the tension is smaller than the weight and the wave speed is less than 19.8 m/s.

EXAMPLE 16.4 Rescuing the Hiker

An 80.0-kg hiker is trapped on a mountain ledge following a storm. A helicopter rescues the hiker by hovering above him and lowering a cable to him. The mass of the cable is 8.00 kg, and its length is 15.0 m. A sling of mass 70.0 kg is attached to the end of the cable. The hiker attaches himself to the sling, and the helicopter then accelerates upward. Terrified by hanging from the cable in midair, the hiker tries to signal the pilot by sending transverse pulses up the cable. A pulse takes 0.250 s to travel the length of the cable. What is the acceleration of the helicopter?

SOLUTION

Conceptualize Imagine the effect of the acceleration of the helicopter on the cable. The greater the upward acceleration, the larger the tension in the cable. In turn, the larger the tension, the higher the speed of pulses on the cable.

Categorize This problem is a combination of one involving the speed of pulses on a string and one in which the hiker and sling are modeled as a particle under a net force.

Analyze Use the time interval for the pulse to travel from the hiker to the helicopter to find the speed of the pulses on the cable:

$$v = \frac{\Delta x}{\Delta t} = \frac{15.0 \text{ m}}{0.250 \text{ s}} = 60.0 \text{ m/s}$$

Solve Equation 16.18 for the tension in the cable:

$$v = \sqrt{\frac{T}{\mu}} \rightarrow T = \mu v^2$$

Model the hiker and sling as a particle under a net force, noting that the acceleration of this particle of mass m is the same as the acceleration of the helicopter:

$$\sum F = ma \rightarrow T - mg = ma$$

Solve for the acceleration:

$$a = \frac{T}{m} - g = \frac{\mu v^2}{m} - g = \frac{m_{\text{cable}} v^2}{\ell_{\text{cable}} m} - g$$

Substitute numerical values:

$$a = \frac{(8.00 \text{ kg})(60.0 \text{ m/s})^2}{(15.0 \text{ m})(150.0 \text{ kg})} - 9.80 \text{ m/s}^2 = 3.00 \text{ m/s}^2$$

Finalize A real cable has stiffness in addition to tension. Stiffness tends to return a wire to its original straight-line shape even when it is not under tension. For example, a piano wire straightens if released from a curved shape; package-wrapping string does not.

Stiffness represents a restoring force in addition to tension and increases the wave speed. Consequently, for a real cable, the speed of 60.0 m/s that we determined is most likely associated with a smaller acceleration of the helicopter.

16.4 Reflection and Transmission

The traveling wave model describes waves traveling through a uniform medium without interacting with anything along the way. We now consider how a traveling wave is affected when it encounters a change in the medium. For example, consider a pulse traveling on a string that is rigidly attached to a support at one end as in Active Figure 16.13. When the pulse reaches the support, a severe change in the medium occurs: the string ends. As a result, the pulse undergoes **reflection;** that is, the pulse moves back along the string in the opposite direction.

Notice that the reflected pulse is *inverted.* This inversion can be explained as follows. When the pulse reaches the fixed end of the string, the string produces an upward force on the support. By Newton's third law, the support must exert an equal-magnitude and oppositely directed (downward) reaction force on the string. This downward force causes the pulse to invert upon reflection.

Now consider another case. This time, the pulse arrives at the end of a string that is free to move vertically as in Active Figure 16.14 (page 462). The tension at the free end is maintained because the string is tied to a ring of negligible mass that is free to slide vertically on a smooth post without friction. Again, the pulse is reflected, but this time it is not inverted. When it reaches the post, the pulse exerts a force on the free end of the string, causing the ring to accelerate upward. The ring rises as high as the incoming pulse, and then the downward component of the tension force pulls the ring back down. This movement of the ring produces a reflected pulse that is not inverted and that has the same amplitude as the incoming pulse.

Finally, consider a situation in which the boundary is intermediate between these two extremes. In this case, part of the energy in the incident pulse is reflected and part undergoes **transmission;** that is, some of the energy passes through the boundary. For instance, suppose a light string is attached to a heavier

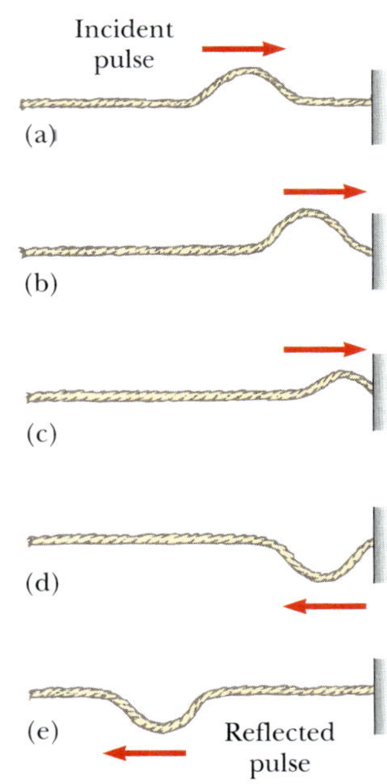

ACTIVE FIGURE 16.13

The reflection of a traveling pulse at the fixed end of a stretched string. The reflected pulse is inverted, but its shape is otherwise unchanged.

Sign in at www.thomsonedu.com and go to ThomsonNOW to adjust the linear mass density of the string and the transverse direction of the initial pulse.

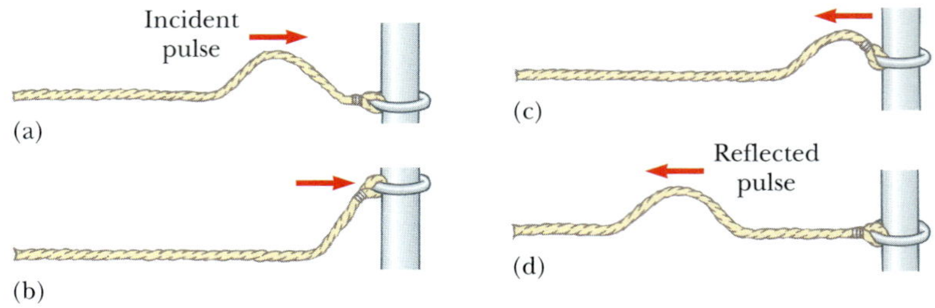

ACTIVE FIGURE 16.14

The reflection of a traveling pulse at the free end of a stretched string. The reflected pulse is not inverted.

Sign in at www.thomsonedu.com and go to ThomsonNOW to adjust the linear mass density of the string and the transverse direction of the initial pulse.

string as in Active Figure 16.15. When a pulse traveling on the light string reaches the boundary between the two strings, part of the pulse is reflected and inverted and part is transmitted to the heavier string. The reflected pulse is inverted for the same reasons described earlier in the case of the string rigidly attached to a support.

The reflected pulse has a smaller amplitude than the incident pulse. In Section 16.5, we show that the energy carried by a wave is related to its amplitude. According to the principle of the conservation of energy, when the pulse breaks up into a reflected pulse and a transmitted pulse at the boundary, the sum of the energies of these two pulses must equal the energy of the incident pulse. Because the reflected pulse contains only part of the energy of the incident pulse, its amplitude must be smaller.

When a pulse traveling on a heavy string strikes the boundary between the heavy string and a lighter one as in Active Figure 16.16, again part is reflected and part is transmitted. In this case, the reflected pulse is not inverted.

In either case, the relative heights of the reflected and transmitted pulses depend on the relative densities of the two strings. If the strings are identical, there is no discontinuity at the boundary and no reflection takes place.

According to Equation 16.18, the speed of a wave on a string increases as the mass per unit length of the string decreases. In other words, a wave travels more slowly on a heavy string than on a light string if both are under the same tension. The following general rules apply to reflected waves: **when a wave or pulse travels from medium A to medium B and $v_A > v_B$ (that is, when B is denser than A), it is inverted upon reflection. When a wave or pulse travels from medium A to medium B and $v_A < v_B$ (that is, when A is denser than B), it is not inverted upon reflection.**

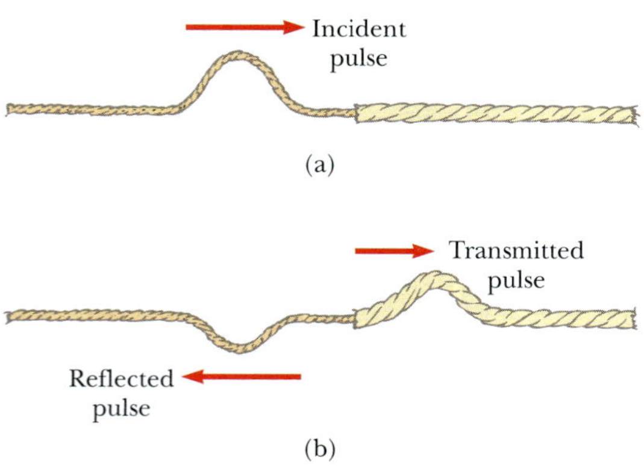

ACTIVE FIGURE 16.15

(a) A pulse traveling to the right on a light string attached to a heavier string. (b) Part of the incident pulse is reflected (and inverted), and part is transmitted to the heavier string.

Sign in at www.thomsonedu.com and go to ThomsonNOW to adjust the linear mass densities of the strings and the transverse direction of the initial pulse.

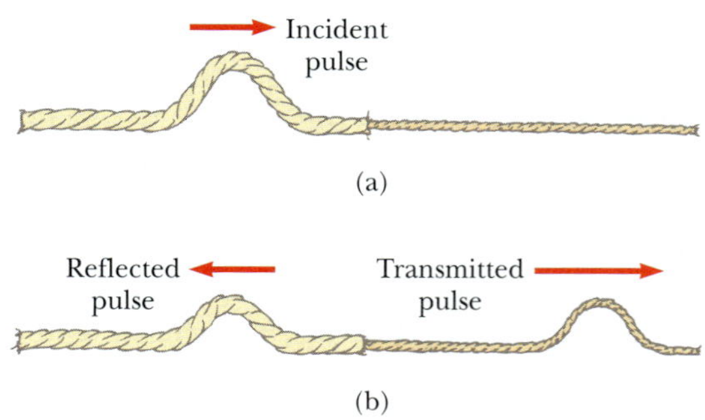

ACTIVE FIGURE 16.16

(a) A pulse traveling to the right on a heavy string attached to a lighter string. (b) The incident pulse is partially reflected and partially transmitted, and the reflected pulse is not inverted.

Sign in at www.thomsonedu.com and go to ThomsonNOW to adjust the linear mass densities of the strings and the transverse direction of the initial pulse.

16.5 Rate of Energy Transfer by Sinusoidal Waves on Strings

Waves transport energy through a medium as they propagate. For example, suppose an object is hanging on a stretched string and a pulse is sent down the string as in Figure 16.17a. When the pulse meets the suspended object, the object is momentarily displaced upward as in Figure 16.17b. In the process, energy is transferred to the object and appears as an increase in the gravitational potential energy of the object–Earth system. This section examines the rate at which energy is transported along a string. We shall assume a one-dimensional sinusoidal wave in the calculation of the energy transferred.

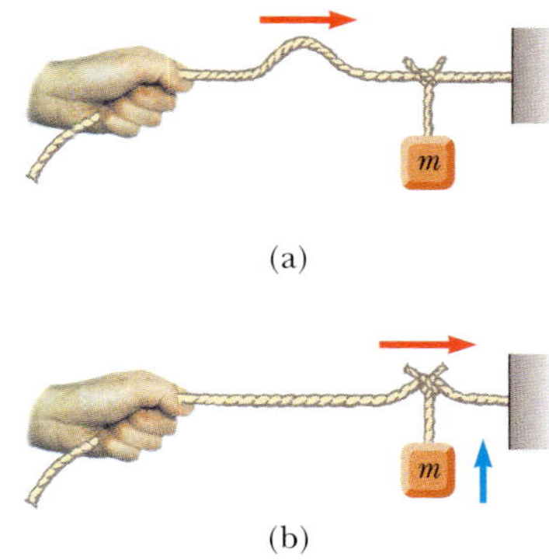

Figure 16.17 (a) A pulse traveling to the right on a stretched string that has an object suspended from it. (b) Energy is transmitted to the suspended object when the pulse arrives.

Consider a sinusoidal wave traveling on a string (Fig. 16.18). The source of the energy is some external agent at the left end of the string, which does work in producing the oscillations. We can consider the string to be a nonisolated system. As the external agent performs work on the end of the string, moving it up and down, energy enters the system of the string and propagates along its length. Let's focus our attention on an infinitesimal element of the string of length dx and mass dm. Each such element moves vertically with simple harmonic motion. Therefore, we can model each element of the string as a simple harmonic oscillator, with the oscillation in the y direction. All elements have the same angular frequency ω and the same amplitude A. The kinetic energy K associated with a moving particle is $K = \frac{1}{2}mv^2$. If we apply this equation to the infinitesimal element, the kinetic energy dK of this element is

$$dK = \tfrac{1}{2}(dm)v_y^{\,2}$$

where v_y is the transverse speed of the element. If μ is the mass per unit length of the string, the mass dm of the element of length dx is equal to $\mu\ dx$. Hence, we can express the kinetic energy of an element of the string as

$$dK = \tfrac{1}{2}(\mu\, dx)v_y^{\,2} \qquad \textbf{(16.19)}$$

Substituting for the general transverse speed of a simple harmonic oscillator using Equation 16.14 gives

$$dK = \tfrac{1}{2}\mu[-\omega A \cos(kx - \omega t)]^2\, dx = \tfrac{1}{2}\mu\omega^2 A^2 \cos^2(kx - \omega t)\, dx$$

If we take a snapshot of the wave at time $t = 0$, the kinetic energy of a given element is

$$dK = \tfrac{1}{2}\mu\omega^2 A^2 \cos^2(kx)\, dx$$

Integrating this expression over all the string elements in a wavelength of the wave gives the total kinetic energy K_λ in one wavelength:

$$K_\lambda = \int dK = \int_0^\lambda \tfrac{1}{2}\mu\omega^2 A^2 \cos^2(kx)\, dx = \tfrac{1}{2}\mu\omega^2 A^2 \int_0^\lambda \cos^2(kx)\, dx$$

$$= \tfrac{1}{2}\mu\omega^2 A^2 \left[\tfrac{1}{2}x + \frac{1}{4k}\sin 2kx\right]_0^\lambda = \tfrac{1}{2}\mu\omega^2 A^2\left[\tfrac{1}{2}\lambda\right] = \tfrac{1}{4}\mu\omega^2 A^2 \lambda$$

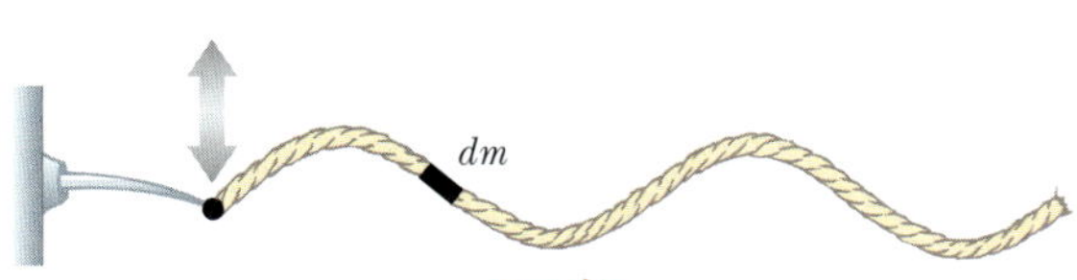

Figure 16.18 A sinusoidal wave traveling along the x axis on a stretched string. Every element of the string moves vertically, and every element has the same total energy.

In addition to kinetic energy, there is potential energy associated with each element of the string due to its displacement from the equilibrium position and the restoring forces from neighboring elements. A similar analysis to that above for the total potential energy U_λ in one wavelength gives exactly the same result:

$$U_\lambda = \tfrac{1}{4}\mu\omega^2 A^2 \lambda$$

The total energy in one wavelength of the wave is the sum of the potential and kinetic energies:

$$E_\lambda = U_\lambda + K_\lambda = \tfrac{1}{2}\mu\omega^2 A^2 \lambda \qquad \textbf{(16.20)}$$

As the wave moves along the string, this amount of energy passes by a given point on the string during a time interval of one period of the oscillation. Therefore, the power $\mathcal{P}$, or rate of energy transfer T_{MW} associated with the mechanical wave, is

$$\mathcal{P} = \frac{T_{\text{MW}}}{\Delta t} = \frac{E_\lambda}{T} = \frac{\frac{1}{2}\mu\omega^2 A^2 \lambda}{T} = \tfrac{1}{2}\mu\omega^2 A^2 \left(\frac{\lambda}{T}\right)$$

Power of a wave ▶

$$\mathcal{P} = \tfrac{1}{2}\mu\omega^2 A^2 v \qquad \textbf{(16.21)}$$

Equation 16.21 shows that the rate of energy transfer by a sinusoidal wave on a string is proportional to (a) the square of the frequency, (b) the square of the amplitude, and (c) the wave speed. In fact, **the rate of energy transfer in any sinusoidal wave is proportional to the square of the angular frequency and to the square of the amplitude.**

Quick Quiz 16.5 Which of the following, taken by itself, would be most effective in increasing the rate at which energy is transferred by a wave traveling along a string? (a) reducing the linear mass density of the string by one half (b) doubling the wavelength of the wave (c) doubling the tension in the string (d) doubling the amplitude of the wave

---

EXAMPLE 16.5 Power Supplied to a Vibrating String

A taut string for which $\mu = 5.00 \times 10^{-2}$ kg/m is under a tension of 80.0 N. How much power must be supplied to the string to generate sinusoidal waves at a frequency of 60.0 Hz and an amplitude of 6.00 cm?

SOLUTION

Conceptualize Consider Active Figure 16.10 again and notice that the vibrating blade supplies energy to the string at a certain rate. This energy then propagates to the right along the string.

Categorize We evaluate quantities from equations developed in the chapter, so we categorize this example as a substitution problem.

Evaluate the wave speed on the string from Equation 16.18:

$$v = \sqrt{\frac{T}{\mu}} = \sqrt{\frac{80.0\text{ N}}{5.00 \times 10^{-2}\text{ kg/m}}} = 40.0\text{ m/s}$$

Evaluate the angular frequency ω of the sinusoidal waves on the string from Equation 16.9:

$$\omega = 2\pi f = 2\pi(60.0\text{ Hz}) = 377\text{ s}^{-1}$$

Use these values and $A = 6.00 \times 10^{-2}$ m in Equation 16.21 to evaluate the power:

$$\mathcal{P} = \tfrac{1}{2}\mu\omega^2 A^2 v$$

$$= \tfrac{1}{2}(5.00 \times 10^{-2}\text{ kg/m})(377\text{ s}^{-1})^2(6.00 \times 10^{-2}\text{ m})^2(40.0\text{ m/s})$$

$$= 512\text{ W}$$

What If? What if the string is to transfer energy at a rate of 1 000 W? What must be the required amplitude if all other parameters remain the same?

Answer Let us set up a ratio of the new and old power, reflecting only a change in the amplitude:

$$\frac{\mathcal{P}_{\text{new}}}{\mathcal{P}_{\text{old}}} = \frac{\frac{1}{2}\mu\omega^2 A_{\text{new}}^2 v}{\frac{1}{2}\mu\omega^2 A_{\text{old}}^2 v} = \frac{A_{\text{new}}^2}{A_{\text{old}}^2}$$

Solving for the new amplitude gives

$$A_{\text{new}} = A_{\text{old}}\sqrt{\frac{\mathcal{P}_{\text{new}}}{\mathcal{P}_{\text{old}}}} = (6.00\text{ cm})\sqrt{\frac{1\,000\text{ W}}{512\text{ W}}} = 8.39\text{ cm}$$

16.6 The Linear Wave Equation

In Section 16.1, we introduced the concept of the wave function to represent waves traveling on a string. All wave functions $y(x, t)$ represent solutions of an equation called the *linear wave equation.* This equation gives a complete description of the wave motion, and from it one can derive an expression for the wave speed. Furthermore, the linear wave equation is basic to many forms of wave motion. In this section, we derive this equation as applied to waves on strings.

Suppose a traveling wave is propagating along a string that is under a tension T. Let's consider one small string element of length Δx (Fig. 16.19). The ends of the element make small angles θ_A and θ_B with the x axis. The net force acting on the element in the vertical direction is

$$\sum F_y = T\sin\theta_B - T\sin\theta_A = T(\sin\theta_B - \sin\theta_A)$$

Because the angles are small, we can use the small-angle approximation $\sin\theta \approx \tan\theta$ to express the net force as

$$\sum F_y \approx T(\tan\theta_B - \tan\theta_A) \qquad \textbf{(16.22)}$$

Imagine undergoing an infinitesimal displacement outward from the end of the rope element in Figure 16.19 along the blue line representing the force $\vec{\mathbf{T}}$. This displacement has infinitesimal x and y components and can be represented by the vector $dx\hat{\mathbf{i}} + dy\hat{\mathbf{j}}$. The tangent of the angle with respect to the x axis for this displacement is dy/dx. Because we evaluate this tangent at a particular instant of time, we must express it in partial form as $\partial y/\partial x$. Substituting for the tangents in Equation 16.22 gives

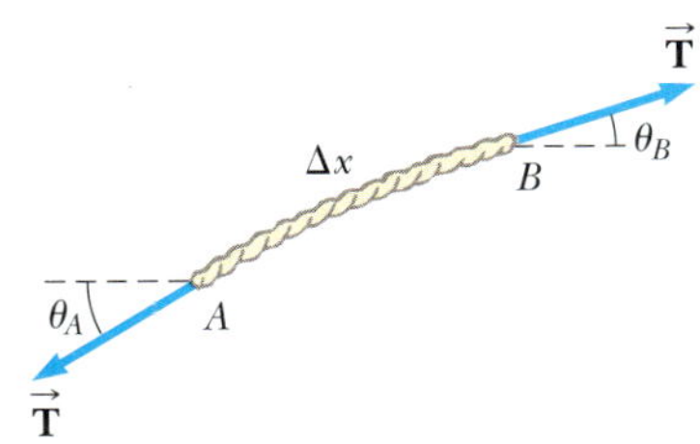

Figure 16.19 An element of a string under tension T.

$$\sum F_y \approx T\left[\left(\frac{\partial y}{\partial x}\right)_B - \left(\frac{\partial y}{\partial x}\right)_A\right] \qquad \textbf{(16.23)}$$

Now let's apply Newton's second law to the element, with the mass of the element given by $m = \mu\,\Delta x$:

$$\sum F_y = ma_y = \mu\Delta x\left(\frac{\partial^2 y}{\partial t^2}\right) \qquad \textbf{(16.24)}$$

Combining Equation 16.23 with Equation 16.24 gives

$$\mu\Delta x\left(\frac{\partial^2 y}{\partial t^2}\right) = T\left[\left(\frac{\partial y}{\partial x}\right)_B - \left(\frac{\partial y}{\partial x}\right)_A\right]$$

$$\frac{\mu}{T}\frac{\partial^2 y}{\partial t^2} = \frac{(\partial y/\partial x)_B - (\partial y/dx)_A}{\Delta x} \qquad \textbf{(16.25)}$$

The right side of Equation 16.25 can be expressed in a different form if we note that the partial derivative of any function is defined as

$$\frac{\partial f}{\partial x} \equiv \lim_{\Delta x \to 0} \frac{f(x + \Delta x) - f(x)}{\Delta x}$$

Associating $f(x + \Delta x)$ with $(\partial y/\partial x)_B$ and $f(x)$ with $(\partial y/\partial x)_A$, we see that, in the limit $\Delta x \to 0$, Equation 16.25 becomes

Linear wave equation for a string ▶

$$\frac{\mu}{T}\frac{\partial^2 y}{\partial t^2} = \frac{\partial^2 y}{\partial x^2} \qquad \textbf{(16.26)}$$

This expression is the linear wave equation as it applies to waves on a string.

The linear wave equation (Eq. 16.26) is often written in the form

Linear wave equation in general ▶

$$\frac{\partial^2 y}{\partial x^2} = \frac{1}{v^2}\frac{\partial^2 y}{\partial t^2} \qquad \textbf{(16.27)}$$

Equation 16.27 applies in general to various types of traveling waves. For waves on strings, y represents the vertical position of elements of the string. For sound waves, y corresponds to longitudinal position of elements of air from equilibrium or variations in either the pressure or the density of the gas through which the sound waves are propagating. In the case of electromagnetic waves, y corresponds to electric or magnetic field components.

We have shown that the sinusoidal wave function (Eq. 16.10) is one solution of the linear wave equation (Eq. 16.27). Although we do not prove it here, the linear wave equation is satisfied by *any* wave function having the form $y = f(x \pm vt)$. Furthermore, we have seen that the linear wave equation is a direct consequence of Newton's second law applied to any element of a string carrying a traveling wave.

Summary

ThomsonNOW™ Sign in at **www.thomsonedu.com** and go to ThomsonNOW to take a practice test for this chapter.

DEFINITIONS

A one-dimensional **sinusoidal wave** is one for which the positions of the elements of the medium vary sinusoidally. A sinusoidal wave traveling to the right can be expressed with a **wave function**

$$y(x, t) = A \sin\left[\frac{2\pi}{\lambda}(x - vt)\right] \qquad \textbf{(16.5)}$$

where A is the **amplitude,** λ is the **wavelength,** and v is the **wave speed.**

The **angular wave number** k and **angular frequency** ω of a wave are defined as follows:

$$k \equiv \frac{2\pi}{\lambda} \qquad \textbf{(16.8)}$$

$$\omega \equiv \frac{2\pi}{T} = 2\pi f \qquad \textbf{(16.9)}$$

where T is the **period** of the wave and f is its **frequency.**

A **transverse wave** is one in which the elements of the medium move in a direction *perpendicular* to the direction of propagation. A **longitudinal wave** is one in which the elements of the medium move in a direction *parallel* to the direction of propagation.

(*continued*)

CONCEPTS AND PRINCIPLES

Any one-dimensional wave traveling with a speed v in the x direction can be represented by a wave function of the form

$$y(x, t) = f(x \pm vt) \qquad \textbf{(16.1, 16.2)}$$

where the positive sign applies to a wave traveling in the negative x direction and the negative sign applies to a wave traveling in the positive x direction. The shape of the wave at any instant in time (a snapshot of the wave) is obtained by holding t constant.

The speed of a wave traveling on a taut string of mass per unit length μ and tension T is

$$v = \sqrt{\frac{T}{\mu}} \qquad \textbf{(16.18)}$$

A wave is totally or partially reflected when it reaches the end of the medium in which it propagates or when it reaches a boundary where its speed changes discontinuously. If a wave traveling on a string meets a fixed end, the wave is reflected and inverted. If the wave reaches a free end, it is reflected but not inverted.

The **power** transmitted by a sinusoidal wave on a stretched string is

$$\mathcal{P} = \tfrac{1}{2}\mu\omega^2A^2v \qquad \textbf{(16.21)}$$

Wave functions are solutions to a differential equation called the **linear wave equation:**

$$\frac{\partial^2 y}{\partial x^2} = \frac{1}{v^2}\frac{\partial^2 y}{\partial t^2} \qquad \textbf{(16.27)}$$

ANALYSIS MODEL FOR PROBLEM SOLVING

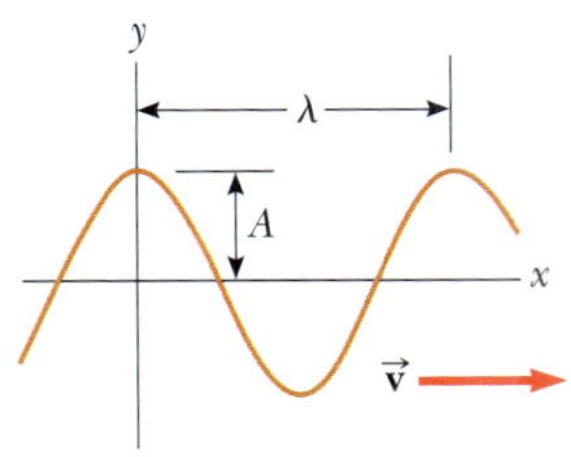

Traveling Wave. The wave speed of a sinusoidal wave is

$$v = \frac{\lambda}{T} = \lambda f \qquad \textbf{(16.6, 16.12)}$$

A sinusoidal wave can be expressed as

$$y = A\sin(kx - \omega t) \qquad \textbf{(16.10)}$$

Questions

☐ denotes answer available in *Student Solutions Manual/Study Guide;* **O** denotes objective question

1. Why is a pulse on a string considered to be transverse?
2. How would you create a longitudinal wave in a stretched spring? Would it be possible to create a transverse wave in a spring?
3. **O (i)** Rank the waves represented by the following functions according to their amplitudes from the largest to the smallest. If two waves have the same amplitude, show them as having equal rank.
   (a) $y = 2\sin(3x - 15t + 2)$ (b) $y = 4\sin(3x - 15t)$
   (c) $y = 6\cos(3x + 15t - 2)$ (d) $y = 8\sin(2x + 15t)$
   (e) $y = 8\cos(4x + 20t)$ (f) $y = 7\sin(6x - 24t)$
   (ii) Rank the same waves according to their wavelengths from largest to smallest. **(iii)** Rank the same waves according to their frequencies from largest to smallest. **(iv)** Rank the same waves according to their periods from largest to smallest. **(v)** Rank the same waves according to their speeds from largest to smallest.
4. **O** If the string does not stretch, by what factor would you have to multiply the tension in a taut string so as to double the wave speed? (a) 8 (b) 4 (c) 2 (d) 0.5 (e) You could not change the speed by a predictable factor by changing the tension.

5. **O** When all the strings on a guitar are stretched to the same tension, will the speed of a wave along the most massive bass string be (a) faster, (b) slower, or (c) the same as the speed of a wave on the lighter strings? Alternatively, (d) is the speed on the bass string not necessarily any of these answers?

6. **O** If you stretch a rubber hose and pluck it, you can observe a pulse traveling up and down the hose. **(i)** What happens to the speed of the pulse if you stretch the hose more tightly? (a) It increases. (b) It decreases. (c) It is constant. (d) It changes unpredictably. **(ii)** What happens to the speed if you fill the hose with water? Choose from the same possibilities.

7. When a pulse travels on a taut string, does it always invert upon reflection? Explain.

8. Does the vertical speed of a segment of a horizontal taut string, through which a wave is traveling, depend on the wave speed?

9. **O** (a) Can a wave on a string move with a wave speed that is greater than the maximum transverse speed $v_{y,\,\text{max}}$ of an element of the string? (b) Can the wave speed be much greater than the maximum element speed? (c) Can the wave speed be equal to the maximum element speed? (d) Can the wave speed be less than $v_{y,\,\text{max}}$?

10. If you shake one end of a taut rope steadily three times each second, what would be the period of the sinusoidal wave set up in the rope?

11. If a long rope is hung from a ceiling and waves are sent up the rope from its lower end, they do not ascend with constant speed. Explain.

12. **O** A source vibrating at constant frequency generates a sinusoidal wave on a string under constant tension. If the power delivered to the string is doubled, by what factor does the amplitude change? (a) 4 (b) 2 (c) $\sqrt{2}$ (d) 1 (e) 0.707 (f) cannot be predicted

13. **O** If one end of a heavy rope is attached to one end of a light rope, a wave can move from the heavy rope into the lighter one. **(i)** What happens to the speed of the wave? (a) It increases. (b) It decreases. (c) It is constant. (d) It changes unpredictably. **(ii)** What happens to the frequency? Choose from the same possibilities. **(iii)** What happens to the wavelength? Choose from the same possibilities.

14. A solid can transport both longitudinal waves and transverse waves, but a homogeneous fluid can transport only longitudinal waves. Why?

15. In an earthquake both S (transverse) and P (longitudinal) waves propagate from the focus of the earthquake. The focus is in the ground below the epicenter on the surface. Assume the waves move in straight lines through uniform material. The S waves travel through the Earth more slowly than the P waves (at about 5 km/s versus 8 km/s). By detecting the time of arrival of the waves, how can one determine the distance to the focus of the earthquake? How many detection stations are necessary to locate the focus unambiguously?

16. In mechanics, massless strings are often assumed. Why is that not a good assumption when discussing waves on strings?

Problems

WebAssign The Problems from this chapter may be assigned online in WebAssign.

ThomsonNOW Sign in at **www.thomsonedu.com** and go to ThomsonNOW to assess your understanding of this chapter's topics with additional quizzing and conceptual questions.

1, 2, 3 denotes straightforward, intermediate, challenging; ☐ denotes full solution available in *Student Solutions Manual/Study Guide;* ▲ denotes coached solution with hints available at **www.thomsonedu.com;** ▭ denotes developing symbolic reasoning; ● denotes asking for qualitative reasoning; ▪ denotes computer useful in solving problem

Section 16.1 Propagation of a Disturbance

1. At $t = 0$, a transverse pulse in a wire is described by the function

$$y = \frac{6}{x^2 + 3}$$

where x and y are in meters. Write the function $y(x, t)$ that describes this pulse if it is traveling in the positive x direction with a speed of 4.50 m/s.

2. ● Ocean waves with a crest-to-crest distance of 10.0 m can be described by the wave function

$$y(x, t) = (0.800\text{ m})\sin\left[0.628(x - vt)\right]$$

where $v = 1.20$ m/s. (a) Sketch $y(x, t)$ at $t = 0$. (b) Sketch $y(x, t)$ at $t = 2.00$ s. Compare this graph with that for part (a) and explain similarities and differences. What has the wave done between picture (a) and picture (b)?

3. Two points A and B on the surface of the Earth are at the same longitude and 60.0° apart in latitude. Suppose an earthquake at point A creates a P wave that reaches point B by traveling straight through the body of the Earth at a constant speed of 7.80 km/s. The earthquake also radiates a Rayleigh wave that travels along the surface of the Earth at 4.50 km/s. (a) Which of these two seismic waves arrives at B first? (b) What is the time difference between the arrivals of these two waves at B? Take the radius of the Earth to be 6 370 km.

4. A seismographic station receives S and P waves from an earthquake, 17.3 s apart. Assume the waves have traveled over the same path at speeds of 4.50 km/s and 7.80 km/s. Find the distance from the seismograph to the hypocenter of the earthquake.

Section 16.2 The Traveling Wave Model

5. ▲ The wave function for a traveling wave on a taut string is (in SI units)

$$y(x, t) = (0.350\text{ m})\sin\left(10\pi t - 3\pi x + \frac{\pi}{4}\right)$$

(a) What are the speed and direction of travel of the wave? (b) What is the vertical position of an element of the string at $t = 0$, $x = 0.100$ m? (c) What are the wavelength and frequency of the wave? (d) What is the maximum transverse speed of an element of the string?

6. ● A certain uniform string is held under constant tension. (a) Draw a side-view snapshot of a sinusoidal wave on a string as shown in diagrams in the text. (b) Immediately below diagram (a), draw the same wave at a moment later by one quarter of the period of the wave. (c) Then, draw a wave with an amplitude 1.5 times larger than the wave in diagram (a). (d) Next, draw a wave differing from the one in your diagram (a) just by having a wavelength 1.5 times larger. (e) Finally, draw a wave differing from that in diagram (a) just by having a frequency 1.5 times larger.

7. A sinusoidal wave is traveling along a rope. The oscillator that generates the wave completes 40.0 vibrations in 30.0 s. Also, a given maximum travels 425 cm along the rope in 10.0 s. What is the wavelength of the wave?

8. For a certain transverse wave, the distance between two successive crests is 1.20 m, and eight crests pass a given point along the direction of travel every 12.0 s. Calculate the wave speed.

9. A wave is described by $y = (2.00 \text{ cm}) \sin (kx - \omega t)$, where $k = 2.11$ rad/m, $\omega = 3.62$ rad/s, x is in meters, and t is in seconds. Determine the amplitude, wavelength, frequency, and speed of the wave.

10. When a particular wire is vibrating with a frequency of 4.00 Hz, a transverse wave of wavelength 60.0 cm is produced. Determine the speed of waves along the wire.

11. The string shown in Active Figure 16.10 is driven at a frequency of 5.00 Hz. The amplitude of the motion is 12.0 cm, and the wave speed is 20.0 m/s. Furthermore, the wave is such that $y = 0$ at $x = 0$ and $t = 0$. Determine (a) the angular frequency and (b) wave number for this wave. (c) Write an expression for the wave function. Calculate (d) the maximum transverse speed and (e) the maximum transverse acceleration of a point on the string.

12. Consider the sinusoidal wave of Example 16.2 with the wave function

$$y = (15.0 \text{ cm}) \cos (0.157x - 50.3t)$$

At a certain instant, let point A be at the origin and point B be the first point along the x axis where the wave is 60.0° out of phase with A. What is the coordinate of B?

13. A sinusoidal wave is described by the wave function

$$y = (0.25 \text{ m}) \sin (0.30x - 40t)$$

where x and y are in meters and t is in seconds. Determine for this wave the (a) amplitude, (b) angular frequency, (c) angular wave number, (d) wavelength, (e) wave speed, and (f) direction of motion.

14. ● (a) Plot y versus t at $x = 0$ for a sinusoidal wave of the form $y = (15.0 \text{ cm}) \cos (0.157x - 50.3t)$, where x and y are in centimeters and t is in seconds. (b) Determine the period of vibration from this plot. State how your result compares with the value found in Example 16.2.

15. ▲ (a) Write the expression for y as a function of x and t for a sinusoidal wave traveling along a rope in the *negative* x direction with the following characteristics: $A = 8.00$ cm, $\lambda = 80.0$ cm, $f = 3.00$ Hz, and $y(0, t) = 0$ at $t = 0$. (b) **What If?** Write the expression for y as a function of x and t for the wave in part (a) assuming that $y(x, 0) = 0$ at the point $x = 10.0$ cm.

16. A sinusoidal wave traveling in the $-x$ direction (to the left) has an amplitude of 20.0 cm, a wavelength of 35.0 cm, and a frequency of 12.0 Hz. The transverse position of an element of the medium at $t = 0$, $x = 0$ is $y = -3.00$ cm, and the element has a positive velocity here. (a) Sketch the wave at $t = 0$. (b) Find the angular wave number, period, angular frequency, and wave speed of the wave. (c) Write an expression for the wave function $y(x, t)$.

17. A transverse wave on a string is described by the wave function

$$y = (0.120 \text{ m}) \sin \left(\frac{\pi}{8} x + 4\pi t\right)$$

(a) Determine the transverse speed and acceleration of the string at $t = 0.200$ s for the point on the string located at $x = 1.60$ m. (b) What are the wavelength, period, and speed of propagation of this wave?

18. A transverse sinusoidal wave on a string has a period $T = 25.0$ ms and travels in the negative x direction with a speed of 30.0 m/s. At $t = 0$, an element of the string at $x = 0$ has a transverse position of 2.00 cm and is traveling downward with a speed of 2.00 m/s. (a) What is the amplitude of the wave? (b) What is the initial phase angle? (c) What is the maximum transverse speed of an element of the string? (d) Write the wave function for the wave.

19. A sinusoidal wave of wavelength 2.00 m and amplitude 0.100 m travels on a string with a speed of 1.00 m/s to the right. Initially, the left end of the string is at the origin. Find (a) the frequency and angular frequency, (b) the angular wave number, and (c) the wave function for this wave. Determine the equation of motion for (d) the left end of the string and (e) the point on the string at $x = 1.50$ m to the right of the left end. (f) What is the maximum speed of any point on the string?

20. A wave on a string is described by the wave function $y = (0.100 \text{ m}) \sin (0.50x - 20t)$. (a) Show that an element of the string at $x = 2.00$ m executes harmonic motion. (b) Determine the frequency of oscillation of this particular point.

Section 16.3 The Speed of Waves on Strings

21. A telephone cord is 4.00 m long. The cord has a mass of 0.200 kg. A transverse pulse is produced by plucking one end of the taut cord. The pulse makes four trips down and back along the cord in 0.800 s. What is the tension in the cord?

22. A transverse traveling wave on a taut wire has an amplitude of 0.200 mm and a frequency of 500 Hz. It travels with a speed of 196 m/s. (a) Write an equation in SI units of the form $y = A \sin (kx - \omega t)$ for this wave. (b) The mass per unit length of this wire is 4.10 g/m. Find the tension in the wire.

23. A piano string having a mass per unit length equal to 5.00×10^{-3} kg/m is under a tension of 1 350 N. Find the speed with which a wave travels on this string.

24. Transverse pulses travel with a speed of 200 m/s along a taut copper wire whose diameter is 1.50 mm. What is the tension in the wire? (The density of copper is 8.92 g/cm^3.)

25. An astronaut on the Moon wishes to measure the local value of the free-fall acceleration by timing pulses traveling down a wire that has an object of large mass suspended from it. Assume a wire has a mass of 4.00 g and a length of 1.60 m and assume a 3.00-kg object is suspended from it. A pulse requires 36.1 ms to traverse the length of the wire. Calculate g_{Moon} from these data. (You may ignore the mass of the wire when calculating the tension in it.)

26. A simple pendulum consists of a ball of mass M hanging from a uniform string of mass m and length L, with $m << M$. Let T represent the period of oscillations for the pendulum. Determine the speed of a transverse wave in the string when the pendulum hangs at rest.

27. Transverse waves travel with a speed of 20.0 m/s in a string under a tension of 6.00 N. What tension is required for a wave speed of 30.0 m/s in the same string?

28. **Review problem.** A light string with a mass per unit length of 8.00 g/m has its ends tied to two walls separated by a distance equal to three-fourths the length of the string (Fig. P16.28). An object of mass m is suspended from the center of the string, putting a tension in the string. (a) Find an expression for the transverse wave speed in the string as a function of the mass of the hanging object. (b) What should be the mass of the object suspended from the string if the wave speed is to be 60.0 m/s?

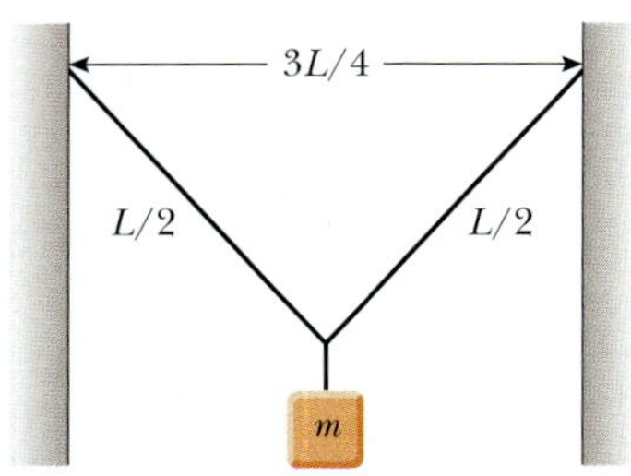

Figure P16.28

29. The elastic limit of a piece of steel wire is 2.70×10^8 Pa. What is the maximum speed at which transverse wave pulses can propagate along this wire without exceeding this stress? (The density of steel is 7.86×10^3 kg/m^3.)

30. ● A student taking a quiz finds on a reference sheet the two equations

$$f = \frac{1}{T} \quad \text{and} \quad v = \sqrt{\frac{T}{\mu}}$$

She has forgotten what T represents in each equation. (a) Use dimensional analysis to determine the units required for T in each equation. (b) Explain how you can identify the physical quantity each T represents from the units.

31. ▲ A steel wire of length 30.0 m and a copper wire of length 20.0 m, both with 1.00-mm diameters, are connected end to end and stretched to a tension of 150 N. During what time interval will a transverse wave travel the entire length of the two wires?

Section 16.5 Rate of Energy Transfer by Sinusoidal Waves on Strings

32. A taut rope has a mass of 0.180 kg and a length of 3.60 m. What power must be supplied to the rope so as to generate sinusoidal waves having an amplitude of 0.100 m and a wavelength of 0.500 m and traveling with a speed of 30.0 m/s?

33. A two-dimensional water wave spreads in circular ripples. Show that the amplitude A at a distance r from the initial disturbance is proportional to $1/\sqrt{r}$. *Suggestion:* Consider the energy carried by one outward-moving ripple.

34. Transverse waves are being generated on a rope under constant tension. By what factor is the required power increased or decreased if (a) the length of the rope is doubled and the angular frequency remains constant, (b) the amplitude is doubled and the angular frequency is halved, (c) both the wavelength and the amplitude are doubled, and (d) both the length of the rope and the wavelength are halved?

35. ▲ Sinusoidal waves 5.00 cm in amplitude are to be transmitted along a string that has a linear mass density of 4.00×10^{-2} kg/m. The source can deliver a maximum power of 300 W and the string is under a tension of 100 N. What is the highest frequency at which the source can operate?

36. A 6.00-m segment of a long string contains four complete waves and has a mass of 180 g. The string vibrates sinusoidally with a frequency of 50.0 Hz and a peak-to-valley displacement of 15.0 cm. (The "peak-to-valley" distance is the vertical distance from the farthest positive position to the farthest negative position.) (a) Write the function that describes this wave traveling in the positive x direction. (b) Determine the power being supplied to the string.

37. A sinusoidal wave on a string is described by the wave function

$$y = (0.15\ \text{m}) \sin (0.80x - 50t)$$

where x and y are in meters and t is in seconds. The mass per unit length of this string is 12.0 g/m. Determine (a) the speed of the wave, (b) the wavelength, (c) the frequency, and (d) the power transmitted to the wave.

38. The wave function for a wave on a taut string is

$$y(x, t) = (0.350\ \text{m}) \sin \left(10\pi t - 3\pi x + \frac{\pi}{4}\right)$$

where x is in meters and t is in seconds. (a) What is the average rate at which energy is transmitted along the string if the linear mass density is 75.0 g/m? (b) What is the energy contained in each cycle of the wave?

39. A horizontal string can transmit a maximum power $\mathcal{P}_0$ (without breaking) if a wave with amplitude A and angular frequency ω is traveling along it. To increase this maximum power, a student folds the string and uses this "double string" as a medium. Determine the maximum power that can be transmitted along the "double string," assuming that the tension in the two strands together is the same as the original tension in the single string.

40. ● In a region far from the epicenter of an earthquake, a seismic wave can be modeled as transporting energy in a single direction without absorption, just as a string wave

does. Suppose the seismic wave moves from granite into mudfill with similar density but with a much smaller bulk modulus. Assume the speed of the wave gradually drops by a factor of 25.0, with negligible reflection of the wave. Explain whether the amplitude of the ground shaking will increase or decrease. Does it change by a predictable factor? This phenomenon led to the collapse of part of the Nimitz Freeway in Oakland, California, during the Loma Prieta earthquake of 1989.

Section 16.6 The Linear Wave Equation

41. ● (a) Evaluate A in the scalar equality $(7 + 3)4 = A$. (b) Evaluate A, B, and C in the vector equality $7.00\hat{\mathbf{i}} + 3.00\hat{\mathbf{k}} = A\hat{\mathbf{i}} + B\hat{\mathbf{j}} + C\hat{\mathbf{k}}$. Explain how you arrive at the answers to convince a student who thinks that you cannot solve a single equation for three different unknowns. (c) **What If?** The functional equality or identity

$$A + B\cos(Cx + Dt + E) = (7.00 \text{ mm})\cos(3x + 4t + 2)$$

is true for all values of the variables x and t, measured in meters and in seconds, respectively. Evaluate the constants A, B, C, D, and E. Explain how you arrive at the answers.

42. Show that the wave function $y = e^{b(x-vt)}$ is a solution of the linear wave equation (Eq. 16.27), where b is a constant.

43. Show that the wave function $y = \ln[b(x - vt)]$ is a solution to Equation 16.27, where b is a constant.

44. (a) Show that the function $y(x, t) = x^2 + v^2t^2$ is a solution to the wave equation. (b) Show that the function in part (a) can be written as $f(x + vt) + g(x - vt)$ and determine the functional forms for f and g. (c) **What If?** Repeat parts (a) and (b) for the function $y(x, t) = \sin(x)\cos(vt)$.

Additional Problems

45. The "wave" is a particular type of pulse that can propagate through a large crowd gathered at a sports arena (Fig. P16.45). The elements of the medium are the spectators, with zero position corresponding to their being seated and maximum position corresponding to their standing and raising their arms. When a large fraction of the spectators participate in the wave motion, a somewhat stable pulse shape can develop. The wave speed depends on people's reaction time, which is typically on the order of 0.1 s. Estimate the order of magnitude, in minutes, of the time interval required for such a pulse to make one circuit around a large sports stadium. State the quantities you measure or estimate and their values.

Figure P16.45

46. ● A sinusoidal wave in a string is described by the wave function

$$y = (0.150 \text{ m})\sin(0.800x - 50.0t)$$

where x is in meters and t is in seconds. The mass per length of the string is 12.0 g/m. (a) Find the maximum transverse acceleration of an element on this string. (b) Determine the maximum transverse force on a 1.00-cm segment of the string. State how this force compares with the tension in the string.

47. Motion picture film is projected at 24.0 frames per second. Each frame is a photograph 19.0 mm high. At what constant speed does the film pass into the projector?

48. A transverse wave on a string is described by the wave function

$$y(x, t) = (0.350 \text{ m})\sin[(1.25 \text{ rad/m})x + (99.6 \text{ rad/s})t]$$

Consider the element of the string at $x = 0$. (a) What is the time interval between the first two instants when this element has a position of $y = 0.175$ m? (b) What distance does the wave travel during this time interval?

49. **Review problem.** A 2.00-kg block hangs from a rubber cord, being supported so that the cord is not stretched. The unstretched length of the cord is 0.500 m, and its mass is 5.00 g. The "spring constant" for the cord is 100 N/m. The block is released and stops at the lowest point. (a) Determine the tension in the cord when the block is at this lowest point. (b) What is the length of the cord in this "stretched" position? (c) Find the speed of a transverse wave in the cord if the block is held in this lowest position.

50. **Review problem.** A block of mass M hangs from a rubber cord. The block is supported so that the cord is not stretched. The unstretched length of the cord is L_0, and its mass is m, much less than M. The "spring constant" for the cord is k. The block is released and stops at the lowest point. (a) Determine the tension in the string when the block is at this lowest point. (b) What is the length of the cord in this "stretched" position? (c) Find the speed of a transverse wave in the cord if the block is held in this lowest position.

51. ● An earthquake or a landslide can produce an ocean wave of short duration carrying great energy, called a tsunami. When its wavelength is large compared to the ocean depth d, the speed of a water wave is given approximately by $v = \sqrt{gd}$. (a) Explain why the amplitude of the wave increases as the wave approaches shore. What can you consider to be constant in the motion of any one wave crest? (b) Assume an earthquake occurs all along a tectonic plate boundary running north to south and produces a straight tsunami wave crest moving everywhere to the west. If the wave has amplitude 1.80 m when its speed is 200 m/s, what will be its amplitude where the water is 9.00 m deep? (c) Explain why the amplitude at the shore should be expected to be still greater, but cannot be meaningfully predicted by your model.

52. **Review problem.** A block of mass M, supported by a string, rests on a frictionless incline making an angle θ with the horizontal (Fig. P16.52). The length of the string is L, and its mass is $m << M$. Derive an expression for the time interval required for a transverse wave to travel from one end of the string to the other.

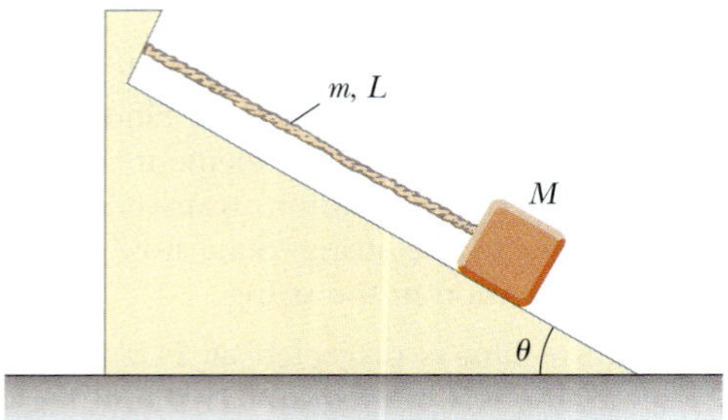

Figure P16.52

53. ● A string with linear density 0.500 g/m is held under tension 20.0 N. As a transverse sinusoidal wave propagates on the string, elements of the string move with maximum speed $v_{y,\,max}$. (a) Determine the power transmitted by the wave as a function of $v_{y,\,max}$. (b) State how the power depends on $v_{y,\,max}$. (c) Find the energy contained in a section of string 3.00 m long. Express it as a function of $v_{y,\,max}$ and the mass m_3 of this section. (d) Find the energy that the wave carries past a point in 6.00 s.

54. A sinusoidal wave in a rope is described by the wave function

$$y = (0.20 \text{ m}) \sin (0.75\pi x + 18\pi t)$$

where x and y are in meters and t is in seconds. The rope has a linear mass density of 0.250 kg/m. The tension in the rope is provided by an arrangement like the one illustrated in Figure 16.12. What is the mass of the suspended object?

55. A block of mass 0.450 kg is attached to one end of a cord of mass 0.003 20 kg; the other end of the cord is attached to a fixed point. The block rotates with constant angular speed in a circle on a horizontal, frictionless table. Through what angle does the block rotate in the time interval during which a transverse wave travels along the string from the center of the circle to the block?

56. A wire of density ρ is tapered so that its cross-sectional area varies with x according to

$$A = (1.0 \times 10^{-3}x + 0.010) \text{ cm}^2$$

(a) The tension in the wire is T. Derive a relationship for the speed of a wave as a function of position. (b) **What If?** Assume the wire is aluminum and is under a tension of 24.0 N. Determine the wave speed at the origin and at $x =$ 10.0 m.

57. A rope of total mass m and length L is suspended vertically. Show that a transverse pulse travels the length of the rope in a time interval $\Delta t = 2\sqrt{L/g}$. *Suggestion:* First find an expression for the wave speed at any point a distance x from the lower end by considering the rope's tension as resulting from the weight of the segment below that point.

58. Assume an object of mass M is suspended from the bottom of the rope in Problem 57. (a) Show that the time interval for a transverse pulse to travel the length of the rope is

$$\Delta t = 2\sqrt{\frac{L}{mg}}\left(\sqrt{M + m} - \sqrt{M}\right)$$

What If? (b) Show that the expression in part (a) reduces to the result of Problem 57 when $M = 0$. (c) Show that for $m << M$, the expression in part (a) reduces to

$$\Delta t = \sqrt{\frac{mL}{Mg}}$$

59. It is stated in Problem 57 that a pulse travels from the bottom to the top of a hanging rope of length L in a time interval $\Delta t = 2\sqrt{L/g}$. Use this result to answer the following questions. (It is not necessary to set up any new integrations.) (a) Over what time interval does a pulse travel halfway up the rope? Give your answer as a fraction of the quantity $2\sqrt{L/g}$. (b) A pulse starts traveling up the rope. How far has it traveled after a time interval $\sqrt{L/g}$?

60. If a loop of chain is spun at high speed, it can roll along the ground like a circular hoop without collapsing. Consider a chain of uniform linear mass density μ whose center of mass travels to the right at a high speed v_0. (a) Determine the tension in the chain in terms of μ and v_0. (b) If the loop rolls over a bump, the resulting deformation of the chain causes two transverse pulses to propagate along the chain, one moving clockwise and one moving counterclockwise. What is the speed of the pulses traveling along the chain? (c) Through what angle does each pulse travel during the time interval over which the loop makes one revolution?

61. **Review problem.** An aluminum wire is clamped at each end under zero tension at room temperature. Reducing the temperature, which results in a decrease in the wire's equilibrium length, increases the tension in the wire. What strain ($\Delta L/L$) results in a transverse wave speed of 100 m/s? Take the cross-sectional area of the wire to be equal to 5.00×10^{-6} m^2, the density to be 2.70×10^3 kg/m^3, and Young's modulus to be 7.00×10^{10} N/m^2.

62. (a) Show that the speed of longitudinal waves along a spring of force constant k is $v = \sqrt{kL/\mu}$, where L is the unstretched length of the spring and μ is the mass per unit length. (b) A spring with a mass of 0.400 kg has an unstretched length of 2.00 m and a force constant of 100 N/m. Using the result you obtained in part (a), determine the speed of longitudinal waves along this spring.

63. A pulse traveling along a string of linear mass density μ is described by the wave function

$$y = [A_0 e^{-bx}] \sin (kx - \omega t)$$

where the factor in brackets is said to be the amplitude. (a) What is the power $\mathcal{P}(x)$ carried by this wave at a point x? (b) What is the power carried by this wave at the origin? (c) Compute the ratio $\mathcal{P}(x)/\mathcal{P}(0)$.

64. An earthquake on the ocean floor in the Gulf of Alaska produces a tsunami that reaches Hilo, Hawaii, 4 450 km away, in a time interval of 9 h 30 min. Tsunamis have enormous wavelengths (100 to 200 km), and the propagation speed for these waves is $v \approx \sqrt{g\bar{d}}$, where $\bar{d}$ is the average depth of the water. From the information given, find the average wave speed and the average ocean depth between Alaska and Hawaii. (This method was used in 1856 to estimate the average depth of the Pacific Ocean

long before soundings were made to give a direct determination.)

65. A string on a musical instrument is held under tension T and extends from the point $x = 0$ to the point $x = L$. The string is overwound with wire in such a way that its mass per unit length $\mu(x)$ increases uniformly from μ_0 at $x = 0$ to μ_L at $x = L$. (a) Find an expression for $\mu(x)$ as a function of x over the range $0 \le x \le L$. (b) Show that the time interval required for a transverse pulse to travel the length of the string is given by

$$\Delta t = \frac{2L\left(\mu_L + \mu_0 + \sqrt{\mu_L \mu_0}\right)}{3\sqrt{T}\left(\sqrt{\mu_L} + \sqrt{\mu_0}\right)}$$

Answers to Quick Quizzes

16.1 **(i),** (b). It is longitudinal because the disturbance (the shift of position of the people) is parallel to the direction in which the wave travels. **(ii),** (a). It is transverse because the people stand up and sit down (vertical motion), whereas the wave moves either to the left or to the right.

16.2 **(i),** (c). The wave speed is determined by the medium, so it is unaffected by changing the frequency. **(ii),** (b). Because the wave speed remains the same, the result of doubling the frequency is that the wavelength is half as large. **(iii),** (d). The amplitude of a wave is unrelated to the wave speed, so we cannot determine the new amplitude without further information.

16.3 (c). With a larger amplitude, an element of the string has more energy associated with its simple harmonic motion, so the element passes through the equilibrium position with a higher maximum transverse speed.

16.4 Only answers (f) and (h) are correct. Choices (a) and (b) affect the transverse speed of a particle of the string, but not the wave speed along the string. Choices (c) and (d) change the amplitude. Choices (e) and (g) increase the time interval by decreasing the wave speed.

16.5 (d). Doubling the amplitude of the wave causes the power to be larger by a factor of 4. In choice (a), halving the linear mass density of the string causes the power to change by a factor of 0.71, and the rate decreases. In choice (b), doubling the wavelength of the wave halves the frequency and causes the power to change by a factor of 0.25, and the rate decreases. In choice (c), doubling the tension in the string changes the wave speed and causes the power to change by a factor of 1.4, which is not as large as in choice (d).

Human ears have evolved to detect sound waves and interpret them as music or speech. Some animals, such as this young bat-eared fox, have ears adapted for the detection of very weak sounds. (Getty Images)

17 Sound Waves

Sound waves travel through any material medium with a speed that depends on the properties of the medium. As sound waves travel through air, the elements of air vibrate to produce changes in density and pressure along the direction of motion of the wave. If the source of the sound waves vibrates sinusoidally, the pressure variations are also sinusoidal. The mathematical description of sinusoidal sound waves is very similar to that of sinusoidal waves on strings, which were discussed in Chapter 16.

Sound waves are divided into three categories that cover different frequency ranges. (1) *Audible waves* lie within the range of sensitivity of the human ear. They can be generated in a variety of ways, such as by musical instruments, human voices, or loudspeakers. (2) *Infrasonic waves* have frequencies below the audible range. Elephants can use infrasonic waves to communicate with one another, even when separated by many kilometers. (3) *Ultrasonic waves* have frequencies above the audible range. You may have used a "silent" whistle to retrieve your dog. Dogs easily hear the ultrasonic sound this whistle emits, although humans cannot detect it at all. Ultrasonic waves are also used in medical imaging.

This chapter begins with a discussion of the speed of sound waves and then wave intensity, which is a function of wave amplitude. We then provide an alternative description of the intensity of sound waves that compresses the wide range of intensities to which the ear is sensitive into a smaller range for convenience. The effects of the motion of sources and listeners on the frequency of a sound are also investigated. Finally, we explore digital reproduction of sound, focusing in particular on sound systems used in modern motion pictures.

17.1 Speed of Sound Waves

Let us describe pictorially the motion of a one-dimensional longitudinal pulse moving through a long tube containing a compressible gas as shown in Figure 17.1. A piston at the left end can be moved to the right to compress the gas and create the pulse. Before the piston is moved, the gas is undisturbed and of uniform density as represented by the uniformly shaded region in Figure 17.1a. When the piston is suddenly pushed to the right (Fig. 17.1b), the gas just in front of it is compressed (as represented by the more heavily shaded region); the pressure and density in this region are now higher than they were before the piston moved. When the piston comes to rest (Fig. 17.1c), the compressed region of the gas continues to move to the right, corresponding to a longitudinal pulse traveling through the tube with speed v.

The speed of sound waves in a medium depends on the compressibility and density of the medium. If the medium is a liquid or a gas and has a bulk modulus B (see Section 12.4) and density ρ, the speed of sound waves in that medium is

$$v = \sqrt{\frac{B}{\rho}} \quad \textbf{(17.1)}$$

It is interesting to compare this expression with Equation 16.18 for the speed of transverse waves on a string, $v = \sqrt{T/\mu}$. In both cases, the wave speed depends on an elastic property of the medium (bulk modulus B or string tension T) and on an inertial property of the medium (ρ or μ). In fact, the speed of all mechanical waves follows an expression of the general form

$$v = \sqrt{\frac{\text{elastic property}}{\text{inertial property}}}$$

For longitudinal sound waves in a solid rod of material, for example, the speed of sound depends on Young's modulus Y and the density ρ. Table 17.1 (page 476) provides the speed of sound in several different materials.

The speed of sound also depends on the temperature of the medium. For sound traveling through air, the relationship between wave speed and air temperature is

$$v = (331\ \text{m/s})\sqrt{1 + \frac{T_C}{273^\circ\text{C}}}$$

where 331 m/s is the speed of sound in air at 0°C and T_C is the air temperature in degrees Celsius. Using this equation, one finds that at 20°C, the speed of sound in air is approximately 343 m/s.

This information provides a convenient way to estimate the distance to a thunderstorm. First count the number of seconds between seeing the flash of lightning and hearing the thunder. Dividing this time by 3 gives the approximate distance to the lightning in kilometers because 343 m/s is approximately $\frac{1}{3}$ km/s. Dividing the time in seconds by 5 gives the approximate distance to the lightning in miles because the speed of sound is approximately $\frac{1}{5}$ mi/s.

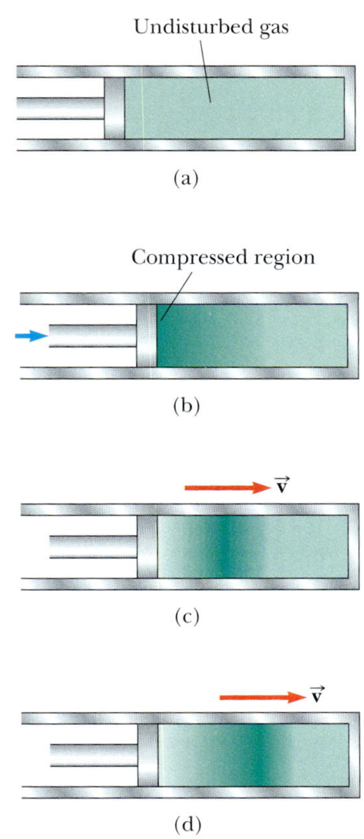

Figure 17.1 Motion of a longitudinal pulse through a compressible gas. The compression (darker region) is produced by the moving piston.

TABLE 17.1

Speed of Sound in Various Media

Medium	v (m/s)	Medium	v (m/s)	Medium	v (m/s)
Gases		**Liquids at 25°C**		**Solids**[a]	
Hydrogen (0°C)	1 286	Glycerol	1 904	Pyrex glass	5 640
Helium (0°C)	972	Seawater	1 533	Iron	5 950
Air (20°C)	343	Water	1 493	Aluminum	6 420
Air (0°C)	331	Mercury	1 450	Brass	4 700
Oxygen (0°C)	317	Kerosene	1 324	Copper	5 010
		Methyl alcohol	1 143	Gold	3 240
		Carbon tetrachloride	926	Lucite	2 680
				Lead	1 960
				Rubber	1 600

[a] Values given are for propagation of longitudinal waves in bulk media. Speeds for longitudinal waves in thin rods are smaller, and speeds of transverse waves in bulk are smaller yet.

17.2 Periodic Sound Waves

One can produce a one-dimensional periodic sound wave in a long, narrow tube containing a gas by means of an oscillating piston at one end as shown in Active Figure 17.2. The darker parts of the colored areas in this figure represent regions in which the gas is compressed and the density and pressure are above their equilibrium values. A compressed region is formed whenever the piston is pushed into the tube. This compressed region, called a **compression**, moves through the tube, continuously compressing the region just in front of itself. When the piston is pulled back, the gas in front of it expands and the pressure and density in this region fall below their equilibrium values (represented by the lighter parts of the colored areas in Active Fig. 17.2). These low-pressure regions, called **rarefactions**, also propagate along the tube, following the compressions. Both regions move at the speed of sound in the medium.

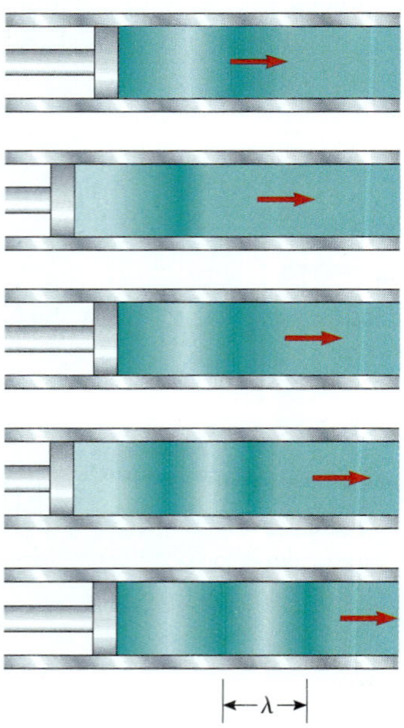

ACTIVE FIGURE 17.2

A longitudinal wave propagating through a gas-filled tube. The source of the wave is an oscillating piston at the left.

Sign in at www.thomsonedu.com and go to ThomsonNOW to adjust the frequency of the piston.

As the piston oscillates sinusoidally, regions of compression and rarefaction are continuously set up. The distance between two successive compressions (or two successive rarefactions) equals the wavelength λ of the sound wave. As these regions travel through the tube, any small element of the medium moves with simple harmonic motion parallel to the direction of the wave. If $s(x, t)$ is the position of a small element relative to its equilibrium position,[1] we can express this harmonic position function as

$$s(x, t) = s_{\max} \cos (kx - \omega t) \tag{17.2}$$

where $s_{\max}$ is the maximum position of the element relative to equilibrium. This parameter is often called the **displacement amplitude** of the wave. The parameter k is the wave number, and ω is the angular frequency of the wave. Notice that the displacement of the element is along x, in the direction of propagation of the sound wave, which means we are describing a longitudinal wave.

The variation in the gas pressure ΔP measured from the equilibrium value is also periodic. For the position function in Equation 17.2, ΔP is given by

$$\Delta P = \Delta P_{\max} \sin (kx - \omega t) \tag{17.3}$$

[1] We use $s(x, t)$ here instead of $y(x, t)$ because the displacement of elements of the medium is not perpendicular to the x direction.

where **the pressure amplitude ΔP_{max}**—which is the maximum change in pressure from the equilibrium value—is given by

$$\Delta P_{max} = \rho v \omega s_{max} \quad \textbf{(17.4)}$$

Equation 17.3 is derived in Example 17.1.

A sound wave may be considered to be either a displacement wave or a pressure wave. A comparison of Equations 17.2 and 17.3 shows that **the pressure wave is 90° out of phase with the displacement wave**. Graphs of these functions are shown in Figure 17.3. The pressure variation is a maximum when the displacement from equilibrium is zero, and the displacement from equilibrium is a maximum when the pressure variation is zero.

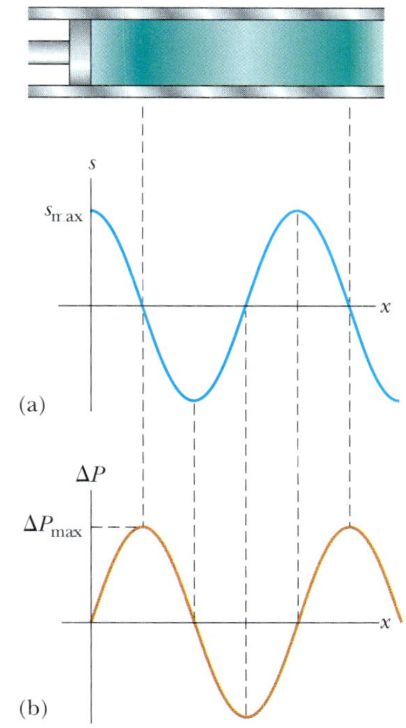

Figure 17.3 (a) Displacement amplitude and (b) pressure amplitude versus position for a sinusoidal longitudinal wave.

Quick Quiz 17.1 If you blow across the top of an empty soft-drink bottle, a pulse of sound travels down through the air in the bottle. At the moment the pulse reaches the bottom of the bottle, what is the correct description of the displacement of elements of air from their equilibrium positions and the pressure of the air at this point? (a) The displacement and pressure are both at a maximum. (b) The displacement and pressure are both at a minimum. (c) The displacement is zero, and the pressure is a maximum. (d) The displacement is zero, and the pressure is a minimum.

EXAMPLE 17.1 Derivation of Equation 17.3

Derive the expression for the pressure variation in a sound wave given by Equation 17.3.

SOLUTION

Conceptualize Consider a thin, disk-shaped element of gas whose flat faces are parallel to the piston in Active Figure 17.2. This element will undergo changes in position, pressure, and density as a sound wave propagates through the gas.

Categorize This derivation combines elastic properties of a gas (Chapter 12) with the wave phenomena discussed in this chapter.

Analyze The element of gas has a thickness Δx in the horizontal direction and a cross-sectional area A, so its volume is $V_i = A\,\Delta x$. When a sound wave displaces the element, the disk's two flat faces move through different distances s. The change in volume ΔV of the element when a sound wave displaces the element is equal to $A\,\Delta s$, where Δs is the difference between the values of s between the two flat faces of the disk.

From the definition of bulk modulus (see Eq. 12.8), express the pressure variation in the element of gas as a function of its change in volume:

$$\Delta P = -B\frac{\Delta V}{V_i}$$

Substitute for the initial volume and the change in volume of the element:

$$\Delta P = -B\frac{A\,\Delta s}{A\,\Delta x} = -B\frac{\Delta s}{\Delta x}$$

Let the thickness Δx of the disk approach zero so that the ratio $\Delta s/\Delta x$ becomes a partial derivative:

$$\Delta P = -B\frac{\partial s}{\partial x}$$

Substitute the position function given by Equation 17.2:

$$\Delta P = -B\frac{\partial}{\partial x}[s_{max}\cos(kx - \omega t)] = Bs_{max}k\sin(kx - \omega t)$$

Use Equation 17.1 to express the bulk modulus as $B = \rho v^2$ and substitute:

$$\Delta P = \rho v^2 s_{\max} k \sin (kx - \omega t)$$

Use Equation 16.11 in the form $k = \omega / v$ and substitute:

$$\Delta P = \rho v \omega s_{\max} \sin (kx - \omega t)$$

Because the sine function has a maximum value of 1, identify the maximum value of the pressure variation as $\Delta P_{\max} = \rho v \omega s_{\max}$ (see Eq. 17.4) and substitute for this combination in the previous expression:

$$\Delta P = \Delta P_{\max} \sin (kx - \omega t)$$

Finalize This final expression for the pressure variation of the air in a sound wave matches Equation 17.3.

17.3 Intensity of Periodic Sound Waves

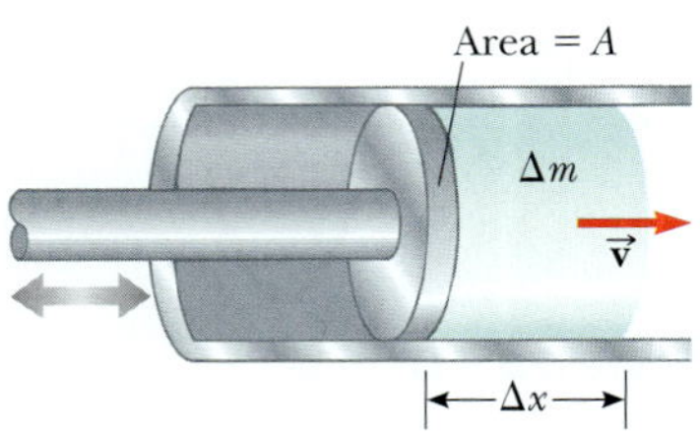

Figure 17.4 An oscillating piston transfers energy to the air in the tube, causing the element of air of length Δx and mass Δm to oscillate with an amplitude $s_{\max}$.

In Chapter 16, we showed that a wave traveling on a taut string transports energy. The same concept applies to sound waves. Consider an element of air of mass Δm and length Δx in front of a piston of area A oscillating with a frequency ω as shown in Figure 17.4. The piston transmits energy to this element of air in the tube, and the energy is propagated away from the piston by the sound wave. To evaluate the rate of energy transfer for the sound wave, let's evaluate the kinetic energy of this element of air, which is undergoing simple harmonic motion. A procedure similar to that in Section 16.5 in which we evaluated the rate of energy transfer for a wave on a string shows that the kinetic energy in one wavelength of the sound wave is

$$K_\lambda = \tfrac{1}{4}(\rho A)\omega^2 s_{\max}^2 \lambda$$

As in the case of the string wave in Section 16.5, the total potential energy for one wavelength has the same value as the total kinetic energy; therefore, the total mechanical energy for one wavelength is

$$E_\lambda = K_\lambda + U_\lambda = \tfrac{1}{2}(\rho A)\omega^2 s_{\max}^2 \lambda$$

As the sound wave moves through the air, this amount of energy passes by a given point during one period of oscillation. Hence, the rate of energy transfer is

$$\mathcal{P} = \frac{E_\lambda}{T} = \frac{\frac{1}{2}(\rho A)\omega^2 s_{\max}^2 \lambda}{T} = \tfrac{1}{2}(\rho A)\omega^2 s_{\max}^2 \left(\frac{\lambda}{T}\right) = \tfrac{1}{2}\rho A v \omega^2 s_{\max}^2$$

where v is the speed of sound in air. Compare this expression with Equation 16.21 for a wave on a string.

We define the **intensity** I of a wave, or the power per unit area, as the rate at which the energy transported by the wave transfers through a unit area A perpendicular to the direction of travel of the wave:

$$I \equiv \frac{\mathcal{P}}{A} \qquad \textbf{(17.5)}$$

In this case, the intensity is therefore

Intensity of a sound wave ▶

$$I = \tfrac{1}{2}\rho v(\omega s_{\max})^2$$

Hence, the intensity of a periodic sound wave is proportional to the square of the displacement amplitude and to the square of the angular frequency. This expression can also be written in terms of the pressure amplitude $\Delta P_{\max}$; in this case, we use Equation 17.4 to obtain

$$I = \frac{(\Delta P_{\max})^2}{2\rho v} \qquad \textbf{(17.6)}$$

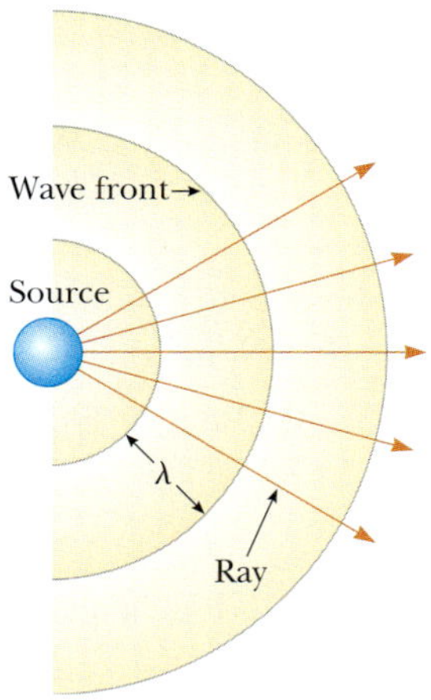

Figure 17.5 Spherical waves emitted by a point source. The circular arcs represent the spherical wave fronts that are concentric with the source. The rays are radial lines pointing outward from the source, perpendicular to the wave fronts.

Now consider a point source emitting sound waves equally in all directions. From everyday experience, we know that the intensity of sound decreases as we move farther from the source. When a source emits sound equally in all directions, the result is a **spherical wave**. Figure 17.5 shows these spherical waves as a series of circular arcs concentric with the source. Each arc represents a surface over which the phase of the wave is constant. We call such a surface of constant phase a **wave front**. The distance between adjacent wave fronts that have the same phase is the wavelength λ of the wave. The radial lines pointing outward from the source are called **rays**.

The average power $\mathcal{P}_{avg}$ emitted by the source must be distributed uniformly over each spherical wave front of area $4\pi r^2$. Hence, the wave intensity at a distance r from the source is

$$I = \frac{\mathcal{P}_{avg}}{A} = \frac{\mathcal{P}_{avg}}{4\pi r^2} \quad \textbf{(17.7)}$$

◀ Inverse-square behavior of intensity for a point source

This inverse-square law, which is reminiscent of the behavior of gravity in Chapter 13, states that the intensity decreases in proportion to the square of the distance from the source.

Quick Quiz 17.2 A vibrating guitar string makes very little sound if it is not mounted on the guitar body. Why does the sound have greater intensity if the string is attached to the guitar body? (a) The string vibrates with more energy. (b) The energy leaves the guitar at a greater rate. (c) The sound power is spread over a larger area at the listener's position. (d) The sound power is concentrated over a smaller area at the listener's position. (e) The speed of sound is higher in the material of the guitar body. (f) None of these answers is correct.

EXAMPLE 17.2 Hearing Limits

The faintest sounds the human ear can detect at a frequency of 1 000 Hz correspond to an intensity of about 1.00×10^{-12} W/m^2, which is called *threshold of hearing*. The loudest sounds the ear can tolerate at this frequency correspond to an intensity of about 1.00 W/m^2, the *threshold of pain*. Determine the pressure amplitude and displacement amplitude associated with these two limits.

SOLUTION

Conceptualize Think about the quietest environment you have ever experienced. It is likely that the intensity of sound in even this quietest environment is higher than the threshold of hearing.

Categorize Because we are given intensities and asked to calculate pressure and displacement amplitudes, this problem requires the concepts discussed in this section.

Analyze To find the pressure amplitude at the threshold of hearing, use Equation 17.6, taking the speed of sound waves in air to be $v = 343$ m/s and the density of air to be $\rho = 1.20\ \text{kg/m}^3$:

$$\Delta P_{\max} = \sqrt{2\rho v I}$$

$$= \sqrt{2(1.20\ \text{kg/m}^3)(343\ \text{m/s})(1.00 \times 10^{-12}\ \text{W/m}^2)}$$

$$= 2.87 \times 10^{-5}\ \text{N/m}^2$$

Calculate the corresponding displacement amplitude using Equation 17.4, recalling that $\omega = 2\pi f$ (Eq. 16.9):

$$s_{\max} = \frac{\Delta P_{\max}}{\rho v \omega} = \frac{2.87 \times 10^{-5}\ \text{N/m}^2}{(1.20\ \text{kg/m}^3)(343\ \text{m/s})(2\pi \times 1\,000\ \text{Hz})}$$

$$= 1.11 \times 10^{-11}\ \text{m}$$

In a similar manner, one finds that the loudest sounds the human ear can tolerate correspond to a pressure amplitude of $28.7\ \text{N/m}^2$ and a displacement amplitude equal to 1.11×10^{-5} m.

Finalize Because atmospheric pressure is about $10^5\ \text{N/m}^2$, the result for the pressure amplitude tells us that the ear is sensitive to pressure fluctuations as small as 3 parts in 10^{10}! The displacement amplitude is also a remarkably small number! If we compare this result for $s_{\max}$ to the size of an atom (about 10^{-10} m), we see that the ear is an extremely sensitive detector of sound waves.

EXAMPLE 17.3 **Intensity Variations of a Point Source**

A point source emits sound waves with an average power output of 80.0 W.

(A) Find the intensity 3.00 m from the source.

SOLUTION

Conceptualize Imagine a small loudspeaker sending sound out at an average rate of 80.0 W uniformly in all directions. You are standing 3.00 m away from the speakers. As the sound propagates, the energy of the sound waves is spread out over an ever-expanding sphere.

Categorize We evaluate the intensity from a given equation, so we categorize this example as a substitution problem.

Because a point source emits energy in the form of spherical waves, use Equation 17.7 to find the intensity:

$$I = \frac{\mathcal{P}_{\text{avg}}}{4\pi r^2} = \frac{80.0\ \text{W}}{4\pi(3.00\ \text{m})^2} = 0.707\ \text{W/m}^2$$

This intensity is close to the threshold of pain.

(B) Find the distance at which the intensity of the sound is $1.00 \times 10^{-8}\ \text{W/m}^2$.

SOLUTION

Solve for r in Equation 17.7 and use the given value for I:

$$r = \sqrt{\frac{\mathcal{P}_{\text{avg}}}{4\pi I}} = \sqrt{\frac{80.0\ \text{W}}{4\pi(1.00 \times 10^{-8}\ \text{W/m}^2)}}$$

$$= 2.52 \times 10^4\ \text{m}$$

Sound Level in Decibels

Example 17.2 illustrates the wide range of intensities the human ear can detect. Because this range is so wide, it is convenient to use a logarithmic scale, where the **sound level** β (Greek letter beta) is defined by the equation

Sound level in decibels ▶

$$\beta \equiv 10 \log \left(\frac{I}{I_0}\right) \quad \textbf{(17.8)}$$

The constant I_0 is the *reference intensity,* taken to be at the threshold of hearing ($I_0 = 1.00 \times 10^{-12}$ W/m^2), and I is the intensity in watts per square meter to which the sound level β corresponds, where β is measured[2] in **decibels** (dB). On this scale, the threshold of pain ($I = 1.00$ W/m^2) corresponds to a sound level of $\beta = 10 \log [(1 \text{ W/m}^2)/(10^{-12} \text{ W/m}^2)] = 10 \log (10^{12}) = 120$ dB, and the threshold of hearing corresponds to $\beta = 10 \log [(10^{-12} \text{ W/m}^2)/(10^{-12} \text{ W/m}^2)] = 0$ dB.

Prolonged exposure to high sound levels may seriously damage the human ear. Ear plugs are recommended whenever sound levels exceed 90 dB. Recent evidence suggests that "noise pollution" may be a contributing factor to high blood pressure, anxiety, and nervousness. Table 17.2 gives some typical sound levels.

Quick Quiz 17.3 Increasing the intensity of a sound by a factor of 100 causes the sound level to increase by what amount? (a) 100 dB (b) 20 dB (c) 10 dB (d) 2 dB

TABLE 17.2
Sound Levels

Source of Sound	β (dB)
Nearby jet airplane	150
Jackhammer; machine gun	130
Siren; rock concert	120
Subway; power lawn mower	100
Busy traffic	80
Vacuum cleaner	70
Normal conversation	50
Mosquito buzzing	40
Whisper	30
Rustling leaves	10
Threshold of hearing	0

EXAMPLE 17.4 Sound Levels

Two identical machines are positioned the same distance from a worker. The intensity of sound delivered by each operating machine at the worker's location is 2.0×10^{-7} W/m^2.

(A) Find the sound level heard by the worker when one machine is operating.

SOLUTION

Conceptualize Imagine a situation in which one source of sound is active and is then joined by a second identical source, such as one person speaking and then a second person speaking at the same time or one musical instrument playing and then being joined by a second instrument.

Categorize Because we are asked for a sound level, we will perform calculations with Equation 17.8.

Analyze Use Equation 17.8 to calculate the sound level at the worker's location with one machine operating:

$$\beta_1 = 10 \log \left(\frac{2.0 \times 10^{-7} \text{ W/m}^2}{1.00 \times 10^{-12} \text{ W/m}^2} \right) = 10 \log (2.0 \times 10^5) = 53 \text{ dB}$$

(B) Find the sound level heard by the worker when two machines are operating.

SOLUTION

Use Equation 17.8 to calculate the sound level at the worker's location with double the intensity:

$$\beta_2 = 10 \log \left(\frac{4.0 \times 10^{-7} \text{ W/m}^2}{1.00 \times 10^{-12} \text{ W/m}^2} \right) = 10 \log (4.0 \times 10^5) = 56 \text{ dB}$$

Finalize These results show that when the intensity is doubled, the sound level increases by only 3 dB.

What If? *Loudness* is a psychological response to a sound. It depends on both the intensity and the frequency of the sound. As a rule of thumb, a doubling in loudness is approximately associated with an increase in sound level of 10 dB. (This rule of thumb is relatively inaccurate at very low or very high frequencies.) If the loudness of the machines in this example is to be doubled, how many machines at the same distance from the worker must be running?

[2] The unit *bel* is named after the inventor of the telephone, Alexander Graham Bell (1847–1922). The prefix *deci-* is the SI prefix that stands for 10^{-1}.

Answer Using the rule of thumb, a doubling of loudness corresponds to a sound level increase of 10 dB. Therefore,

$$\beta_2 - \beta_1 = 10 \text{ dB} = 10 \log\left(\frac{I_2}{I_0}\right) - 10 \log\left(\frac{I_1}{I_0}\right) = 10 \log\left(\frac{I_2}{I_1}\right)$$

$$\log\left(\frac{I_2}{I_1}\right) = 1 \quad \rightarrow \quad I_2 = 10 I_1$$

Therefore, ten machines must be operating to double the loudness.

Loudness and Frequency

The discussion of sound level in decibels relates to a *physical* measurement of the strength of a sound. Let us now extend our discussion from Example 17.4 concerning the *psychological* "measurement" of the strength of a sound.

Of course, we don't have instruments in our bodies that can display numerical values of our reactions to stimuli. We have to "calibrate" our reactions somehow by comparing different sounds to a reference sound, but that is not easy to accomplish. For example, earlier we mentioned that the threshold intensity is 10^{-12} W/m^2, corresponding to an intensity level of 0 dB. In reality, this value is the threshold only for a sound of frequency 1 000 Hz, which is a standard reference frequency in acoustics. If we perform an experiment to measure the threshold intensity at other frequencies, we find a distinct variation of this threshold as a function of frequency. For example, at 100 Hz, a barely audible sound must have an intensity level of about 30 dB! Unfortunately, there is no simple relationship between physical measurements and psychological "measurements." The 100-Hz, 30-dB sound is psychologically "equal" to the 1 000-Hz, 0-dB sound (both are just barely audible), but they are not physically equal (30 dB $\neq$ 0 dB).

By using test subjects, the human response to sound has been studied, and the results are shown in the white area of Figure 17.6 along with the approximate frequency and sound-level ranges of other sound sources. The lower curve of the white area corresponds to the threshold of hearing. Its variation with frequency is clear from this diagram. Notice that humans are sensitive to frequencies ranging from about 20 Hz to about 20 000 Hz. The upper bound of the white area is the

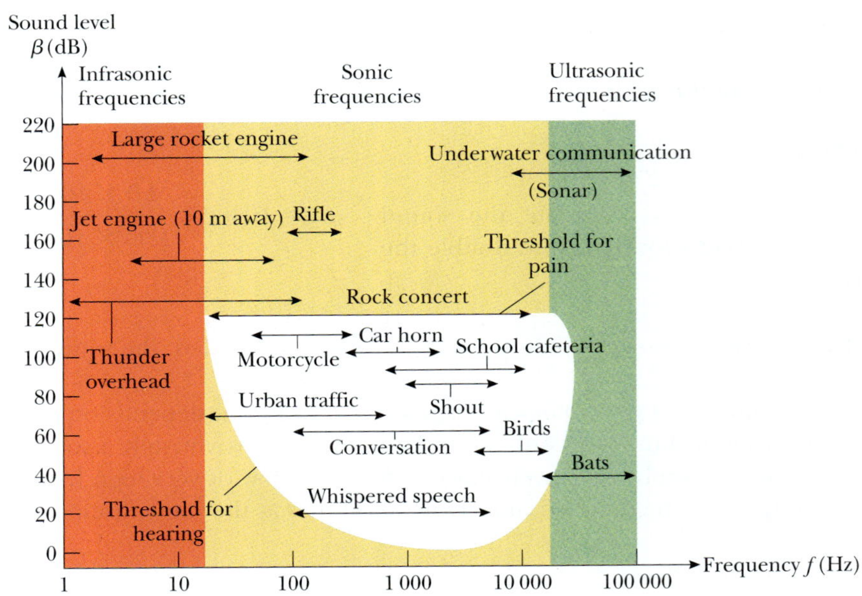

Figure 17.6 Approximate ranges of frequency and sound level of various sources and that of normal human hearing, shown by the white area. (From R. L. Reese, *University Physics*, Pacific Grove, Brooks/Cole, 2000.)

threshold of pain. Here the boundary of the white area is straight because the psychological response is relatively independent of frequency at this high sound level.

The most dramatic change with frequency is in the lower left region of the white area, for low frequencies and low intensity levels. Our ears are particularly insensitive in this region. If you are listening to your stereo and the bass (low frequencies) and treble (high frequencies) sound balanced at a high volume, try turning the volume down and listening again. You will probably notice that the bass seems weak, which is due to the insensitivity of the ear to low frequencies at low sound levels as shown in Figure 17.6.

17.4 The Doppler Effect

Perhaps you have noticed how the sound of a vehicle's horn changes as the vehicle moves past you. The frequency of the sound you hear as the vehicle approaches you is higher than the frequency you hear as it moves away from you. This experience is one example of the **Doppler effect.**[3]

To see what causes this apparent frequency change, imagine you are in a boat that is lying at anchor on a gentle sea where the waves have a period of $T = 3.0$ s. Hence, every 3.0 s a crest hits your boat. Figure 17.7a shows this situation, with the water waves moving toward the left. If you set your watch to $t = 0$ just as one crest hits, the watch reads 3.0 s when the next crest hits, 6.0 s when the third crest hits, and so on. From these observations, you conclude that the wave frequency is $f = 1/T = 1/(3.0 \text{ s}) = 0.33$ Hz. Now suppose you start your motor and head directly into the oncoming waves as in Figure 17.7b. Again you set your watch to $t = 0$ as a crest hits the front (the bow) of your boat. Now, however, because you are moving toward the next wave crest as it moves toward you, it hits you less than 3.0 s after the first hit. In other words, the period you observe is shorter than the 3.0-s period you observed when you were stationary. Because $f = 1/T$, you observe a higher wave frequency than when you were at rest.

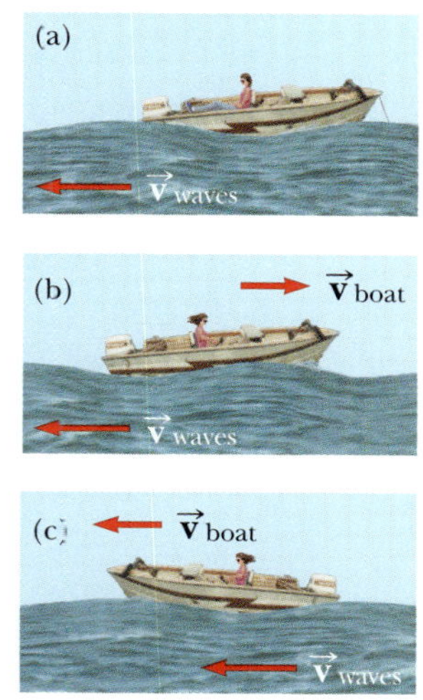

Figure 17.7 (a) Waves moving toward a stationary boat. The waves travel to the left, and their source is far to the right of the boat, out of the frame of the photograph. (b) The boat moving toward the wave source. (c) The boat moving away from the wave source.

If you turn around and move in the same direction as the waves (Fig. 17.7c), you observe the opposite effect. You set your watch to $t = 0$ as a crest hits the back (the stern) of the boat. Because you are now moving away from the next crest, more than 3.0 s has elapsed on your watch by the time that crest catches you. Therefore, you observe a lower frequency than when you were at rest.

These effects occur because the *relative* speed between your boat and the waves depends on the direction of travel and on the speed of your boat. When you are moving toward the right in Figure 17.7b, this relative speed is higher than that of the wave speed, which leads to the observation of an increased frequency. When you turn around and move to the left, the relative speed is lower, as is the observed frequency of the water waves.

Let's now examine an analogous situation with sound waves in which the water waves become sound waves, the water becomes the air, and the person on the boat becomes an observer listening to the sound. In this case, an observer O is moving and a sound source S is stationary. For simplicity, we assume the air is also stationary and the observer moves directly toward the source (Active Fig. 17.8). The observer moves with a speed v_O toward a stationary point source ($v_S = 0$), where *stationary* means at rest with respect to the medium, air.

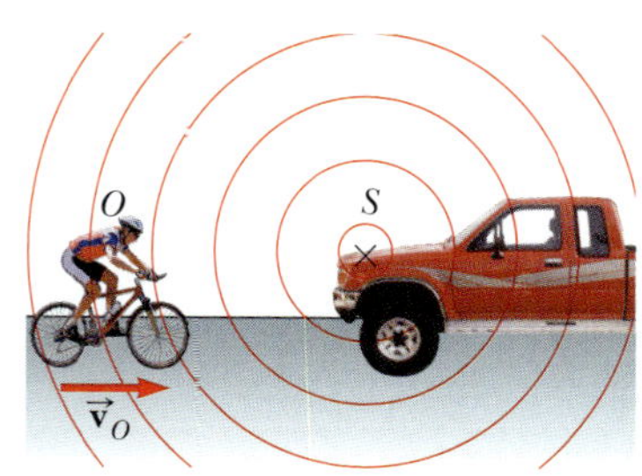

ACTIVE FIGURE 17.8

An observer O (the cyclist) moves with a speed v_O toward a stationary point source S, the horn of a parked truck. The observer hears a frequency f' that is greater than the source frequency.

Sign in at www.thomsonedu.com and go to ThomsonNOW to adjust the speed of the observer.

If a point source emits sound waves and the medium is uniform, the waves move at the same speed in all directions radially away from the source; the result is a spherical wave as mentioned in Section 17.3. The distance between adjacent wave fronts equals the wavelength λ. In Active Figure 17.8, the circles are the intersections of these three-dimensional wave fronts with the two-dimensional paper.

We take the frequency of the source in Active Figure 17.8 to be f, the wavelength to be λ, and the speed of sound to be v. If the observer were also stationary,

[3] Named after Austrian physicist Christian Johann Doppler (1803–1853), who in 1842 predicted the effect for both sound waves and light waves.

he would detect wave fronts at a frequency f. (That is, when $v_O = 0$ and $v_S = 0$, the observed frequency equals the source frequency.) When the observer moves toward the source, the speed of the waves relative to the observer is $v' = v + v_O$, as in the case of the boat in Figure 17.7, but the wavelength λ is unchanged. Hence, using Equation 16.12, $v = \lambda f$, we can say that the frequency f' heard by the observer is *increased* and is given by

$$f' = \frac{v'}{\lambda} = \frac{v + v_O}{\lambda}$$

Because $\lambda = v/f$, we can express f' as

$$f' = \left(\frac{v + v_O}{v}\right)f \quad \text{(observer moving toward source)} \tag{17.9}$$

If the observer is moving away from the source, the speed of the wave relative to the observer is $v' = v - v_O$. The frequency heard by the observer in this case is *decreased* and is given by

$$f' = \left(\frac{v - v_O}{v}\right)f \quad \text{(observer moving away from source)} \tag{17.10}$$

In general, whenever an observer moves with a speed v_O relative to a stationary source, the frequency heard by the observer is given by Equation 17.9, with a sign convention: a positive value is substituted for v_O when the observer moves toward the source, and a negative value is substituted when the observer moves away from the source.

Now suppose the source is in motion and the observer is at rest. If the source moves directly toward observer A in Active Figure 17.9a, the wave fronts heard by the observer are closer together than they would be if the source were not moving. As a result, the wavelength λ' measured by observer A is shorter than the wavelength λ of the source. During each vibration, which lasts for a time interval T (the period), the source moves a distance $v_S T = v_S/f$ and the wavelength is *shortened* by this amount. Therefore, the observed wavelength λ' is

$$\lambda' = \lambda - \Delta\lambda = \lambda - \frac{v_S}{f}$$

Because $\lambda = v/f$, the frequency f' heard by observer A is

$$f' = \frac{v}{\lambda'} = \frac{v}{\lambda - (v_S/f)} = \frac{v}{(v/f) - (v_S/f)}$$

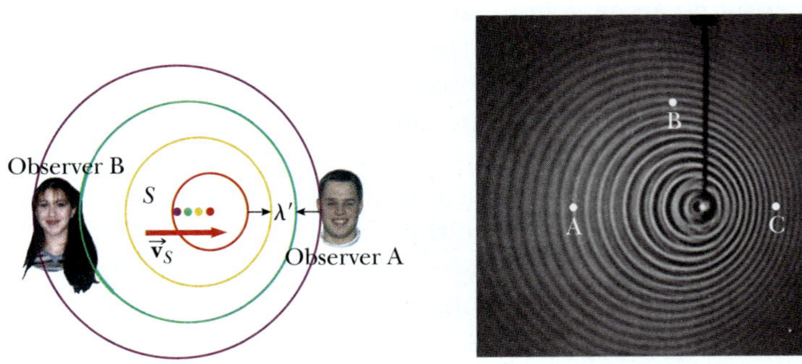

(a) (b)

Courtesy of the Educational Development Center, Newton, MA

ACTIVE FIGURE 17.9

(a) A source S moving with a speed v_S toward a stationary observer A and away from a stationary observer B. Observer A hears an increased frequency, and observer B hears a decreased frequency. (b) The Doppler effect in water, observed in a ripple tank. A point source is moving to the right with speed v_S. Letters shown in the photo refer to Quick Quiz 17.4.

Sign in at www.thomsonedu.com and go to ThomsonNOW to adjust the speed of the source.

$$f' = \left(\frac{v}{v - v_S}\right)f \quad \text{(source moving toward observer)} \tag{17.11}$$

That is, the observed frequency is *increased* whenever the source is moving toward the observer.

When the source moves away from a stationary observer, as is the case for observer B in Active Figure 17.9a, the observer measures a wavelength λ' that is *greater* than λ and hears a *decreased* frequency:

$$f' = \left(\frac{v}{v + v_S}\right)f \quad \text{(source moving away from observer)} \tag{17.12}$$

We can express the general relationship for the observed frequency when a source is moving and an observer is at rest as Equation 17.11, with the same sign convention applied to v_S as was applied to v_O: a positive value is substituted for v_S when the source moves toward the observer, and a negative value is substituted when the source moves away from the observer.

Finally, combining Equations 17.9 and 17.11 gives the following general relationship for the observed frequency:

$$f' = \left(\frac{v + v_O}{v - v_S}\right)f \tag{17.13}$$

◄ General Doppler-shift expression

In this expression, the signs for the values substituted for v_O and v_S depend on the direction of the velocity. A positive value is used for motion of the observer or the source *toward* the other (associated with an *increase* in observed frequency), and a negative value is used for motion of one *away from* the other (associated with a *decrease* in observed frequency).

Although the Doppler effect is most typically experienced with sound waves, it is a phenomenon common to all waves. For example, the relative motion of source and observer produces a frequency shift in light waves. The Doppler effect is used in police radar systems to measure the speeds of motor vehicles. Likewise, astronomers use the effect to determine the speeds of stars, galaxies, and other celestial objects relative to the Earth.

PITFALL PREVENTION 17.1

Doppler Effect Does Not Depend on Distance

Some people think that the Doppler effect depends on the distance between the source and the observer. Although the intensity of a sound varies as the distance changes, the apparent frequency depends only on the relative speed of source and observer. As you listen to an approaching source, you will detect increasing intensity but constant frequency. As the source passes, you will hear the frequency suddenly drop to a new constant value and the intensity begin to decrease.

Quick Quiz 17.4 Consider detectors of water waves at three locations A, B, and C in Active Figure 17.9b. Which of the following statements is true? (a) The wave speed is highest at location A. (b) The wave speed is highest at location C. (c) The detected wavelength is largest at location B. (d) The detected wavelength is largest at location C. (e) The detected frequency is highest at location C. (f) The detected frequency is highest at location A.

Quick Quiz 17.5 You stand on a platform at a train station and listen to a train approaching the station at a constant velocity. While the train approaches, but before it arrives, what do you hear? (a) the intensity and the frequency of the sound both increasing (b) the intensity and the frequency of the sound both decreasing (c) the intensity increasing and the frequency decreasing (d) the intensity decreasing and the frequency increasing (e) the intensity increasing and the frequency remaining the same (f) the intensity decreasing and the frequency remaining the same

EXAMPLE 17.5 The Broken Clock Radio

Your clock radio awakens you with a steady and irritating sound of frequency 600 Hz. One morning, it malfunctions and cannot be turned off. In frustration, you drop the clock radio out of your fourth-story dorm window, 15.0 m from the ground. Assume the speed of sound is 343 m/s. As you listen to the falling clock radio, what frequency do you hear just before you hear it striking the ground?

SOLUTION

Conceptualize The speed of the clock radio increases as it falls. Therefore, it is a source of sound moving away from you with an increasing speed so the frequency you hear should be less than 600 Hz.

Categorize We categorize this problem as one in which we must combine our understanding of falling objects with that of the frequency shift due to the Doppler effect.

Analyze Because the clock radio is modeled as a particle under constant acceleration due to gravity, use Equation 2.13 to express the speed of the source of sound:

$$v_S = v_{yi} + a_y t = 0 - gt = -gt$$

Use Equation 17.13 to determine the Doppler-shifted frequency heard from the falling clock radio:

$$(1)\quad f' = \left[\frac{v + 0}{v - (-gt)}\right] f = \left(\frac{v}{v + gt}\right) f$$

From Equation 2.16, find the time at which the clock radio strikes the ground:

$$y_f = y_i + v_{yi}t - \tfrac{1}{2}gt^2$$

$$-15.0 \text{ m} = 0 + 0 - \tfrac{1}{2}(9.80 \text{ m/s}^2)t^2$$

$$t = 1.75 \text{ s}$$

From Equation (1), evaluate the Doppler-shifted frequency just as the clock radio strikes the ground:

$$f' = \left[\frac{343 \text{ m/s}}{343 \text{ m/s} + (9.80 \text{ m/s}^2)(1.75 \text{ s})}\right](600 \text{ Hz})$$

$$= 571 \text{ Hz}$$

Finalize The frequency is lower than the actual frequency of 600 Hz because the clock radio is moving away from you. If it were to fall from a higher floor so that it passes below $y = -15.0$ m, the clock radio would continue to accelerate and the frequency would continue to drop.

EXAMPLE 17.6 Doppler Submarines

A submarine (sub A) travels through water at a speed of 8.00 m/s, emitting a sonar wave at a frequency of 1 400 Hz. The speed of sound in the water is 1 533 m/s. A second submarine (sub B) is located such that both submarines are traveling directly toward each other. The second submarine is moving at 9.00 m/s.

(A) What frequency is detected by an observer riding on sub B as the subs approach each other?

SOLUTION

Conceptualize Even though the problem involves subs moving in water, there is a Doppler effect just like there is when you are in a moving car and listening to a sound moving through the air from another car.

Categorize Because both subs are moving, we categorize this problem as one involving the Doppler effect for both a moving source and a moving observer.

Analyze Use Equation 17.13 to find the Doppler-shifted frequency heard by the observer in sub B, being careful with the signs of the source and observer velocities:

$$f' = \left(\frac{v + v_O}{v - v_S}\right) f$$

$$f' = \left[\frac{1\,533 \text{ m/s} + (+9.00 \text{ m/s})}{1\,533 \text{ m/s} - (+8.00 \text{ m/s})}\right](1\,400 \text{ Hz}) = 1\,416 \text{ Hz}$$

(B) The subs barely miss each other and pass. What frequency is detected by an observer riding on sub B as the subs recede from each other?

SOLUTION

Use Equation 17.13 to find the Doppler-shifted frequency heard by the observer in sub B, again being careful with the signs of the source and observer velocities:

$$f' = \left(\frac{v + v_O}{v - v_S}\right) f$$

$$f' = \left[\frac{1\,533 \text{ m/s} + (-9.00 \text{ m/s})}{1\,533 \text{ m/s} - (-8.00 \text{ m/s})}\right](1\,400 \text{ Hz}) = 1\,385 \text{ Hz}$$

Finalize Notice that the frequency drops from 1 416 Hz to 1 385 Hz as the subs pass. This effect is similar to the drop in frequency you hear when a car passes by you while blowing its horn.

What If? While the subs are approaching each other, some of the sound from sub A reflects from sub B and returns to sub A. If this sound were to be detected by an observer on sub A, what is its frequency?

Answer The sound of apparent frequency 1 416 Hz found in part (A) is reflected from a moving source (sub B) and then detected by a moving observer (sub A). Therefore, the frequency detected by sub A is

$$f'' = \left(\frac{v + v_O}{v - v_S}\right) f'$$

$$= \left[\frac{1\,533 \text{ m/s} + (+8.00 \text{ m/s})}{1\,533 \text{ m/s} - (+9.00 \text{ m/s})}\right](1\,416 \text{ Hz}) = 1\,432 \text{ Hz}$$

This technique is used by police officers to measure the speed of a moving car. Microwaves are emitted from the police car and reflected by the moving car. By detecting the Doppler-shifted frequency of the reflected microwaves, the police officer can determine the speed of the moving car.

Shock Waves

Now consider what happens when the speed v_S of a source *exceeds* the wave speed v. This situation is depicted graphically in Figure 17.10a. The circles represent spherical wave fronts emitted by the source at various times during its motion. At $t = 0$, the source is at S_0, and at a later time t, the source is at S_n. At the time t, the wave front centered at S_0 reaches a radius of vt. In this same time interval, the

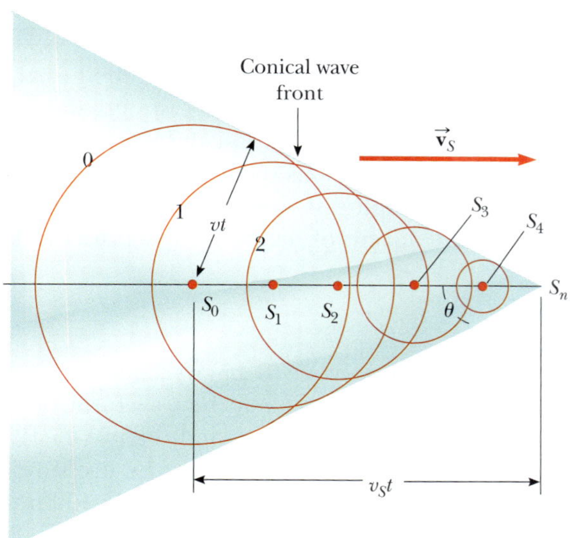

(a)

(b)

Figure 17.10 (a) A representation of a shock wave produced when a source moves from S_0 to S_n with a speed v_S, which is greater than the wave speed v in the medium. The envelope of the wave fronts forms a cone whose apex half-angle is given by $\sin\theta = v/v_S$. (b) A stroboscopic photograph of a bullet moving at supersonic speed through the hot air above a candle. Notice the shock wave in the vicinity of the bullet.

Figure 17.11 The V-shaped bow wave of a boat is formed because the boat speed is greater than the speed of the water waves it generates. A bow wave is analogous to a shock wave formed by an airplane traveling faster than sound.

source travels a distance $v_S t$ to S_n. At the instant the source is at S_n, waves are just beginning to be generated at this location; hence, the wave front has zero radius at this point. The tangent line drawn from S_n to the wave front centered on S_0 is tangent to all other wave fronts generated at intermediate times. Therefore, the envelope of these wave fronts is a cone whose apex half-angle θ (the "Mach angle") is given by

$$\sin\theta = \frac{vt}{v_S t} = \frac{v}{v_S}$$

The ratio v_S/v is referred to as the *Mach number,* and the conical wave front produced when $v_S > v$ (supersonic speeds) is known as a *shock wave.* An interesting analogy to shock waves is the V-shaped wave fronts produced by a boat (the bow wave) when the boat's speed exceeds the speed of the surface-water waves (Fig. 17.11).

Jet airplanes traveling at supersonic speeds produce shock waves, which are responsible for the loud "sonic boom" one hears. The shock wave carries a great deal of energy concentrated on the surface of the cone, with correspondingly great pressure variations. Such shock waves are unpleasant to hear and can cause damage to buildings when aircraft fly supersonically at low altitudes. In fact, an airplane flying at supersonic speeds produces a double boom because two shock waves are formed, one from the nose of the plane and one from the tail. People near the path of a space shuttle as it glides toward its landing point often report hearing what sounds like two very closely spaced cracks of thunder.

Quick Quiz 17.6 An airplane flying with a constant velocity moves from a cold air mass into a warm air mass. Does the Mach number (a) increase, (b) decrease, or (c) stay the same?

17.5 Digital Sound Recording

The first sound recording device, the phonograph, was invented by Thomas Edison in the 19th century. Sound waves were recorded in early phonographs by encoding the sound waveforms as variations in the depth of a continuous groove cut into tin foil wrapped around a cylinder. During playback, as a needle follows along the groove of the rotating cylinder, the needle is pushed back and forth according to the sound waves encoded on the record. The needle is attached to a diaphragm and a horn, making the sound intense enough to be heard.

As the development of the phonograph continued, sound was recorded on cardboard cylinders coated with wax. During the last decade of the 19th century and the first half of the 20th century, sound was recorded on disks made of shellac and clay. In 1948, the plastic phonograph disk was introduced and dominated the recording industry market until the advent of digital compact discs in the 1980s.

Digital Recording

In digital recording, information is converted to binary code (ones and zeros), similar to the dots and dashes of Morse code. First, the waveform of the sound is *sampled,* typically at the rate of 44 100 times per second. Figure 17.12a illustrates this process. Between each pair of blue lines in the figure, the pressure of the wave is measured and converted to a voltage. Therefore, there are 44 100 numbers associated with each second of the sound being sampled. The sampling frequency is much higher than the upper range of human hearing, about 20 000 Hz, so all frequencies of audible sound are adequately sampled at this rate.

These measurements are then converted to *binary numbers,* which are numbers expressed using base 2 rather than base 10. Table 17.3 shows some sample binary numbers. Generally, voltage measurements are recorded in 16-bit "words," where

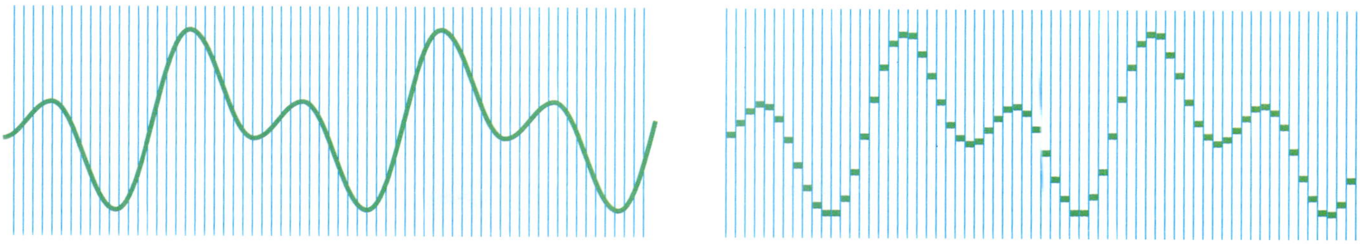

Figure 17.12 (a) Sound is digitized by electronically sampling the sound waveform at periodic intervals. During each time interval between the blue lines, a number is recorded for the average voltage during the interval. The sampling rate shown here is much slower than the actual sampling rate of 44 100 samples per second. (b) The reconstruction of the sound wave sampled in (a). Notice the stepwise reconstruction rather than the continuous waveform in (a).

TABLE 17.3

Sample Binary Numbers

Number in Base 10	Number in Binary	Sum
1	0000000000000001	1
2	0000000000000010	2+0
3	0000000000000011	2+1
10	0000000000001010	8+0+2+0
37	0000000000100101	32+0+0+4+0+1
275	0000000100010011	256+0+0+0+16+0+0+2+1

each bit is a one or a zero. Therefore, the number of different voltage levels that can be assigned codes is $2^{16} = 65\,536$. The number of bits in one second of sound is $16 \times 44\,100 = 705\,600$. It is these strings of ones and zeros, in 16-bit words, that are recorded on the surface of a compact disc.

Figure 17.13 shows a magnification of the surface of a compact disc. Two types of areas—*lands* and *pits*—are detected by the laser playback system. The lands are untouched regions of the disc surface that are highly reflective. The pits, which are areas burned into the surface, scatter light rather than reflecting it back to the detection system. The playback system samples the reflected light 705 600 times per second. When the laser moves from a pit to a flat or from a flat to a pit, the reflected light changes during the sampling and the bit is recorded as a one. If there is no change during the sampling, the bit is recorded as a zero.

The binary numbers read from the compact disc are converted back to voltages, and the waveform is reconstructed as shown in Figure 17.12b. Because the sampling rate is so high, it is not evident in the sound that the waveform is constructed from step-wise discrete voltages.

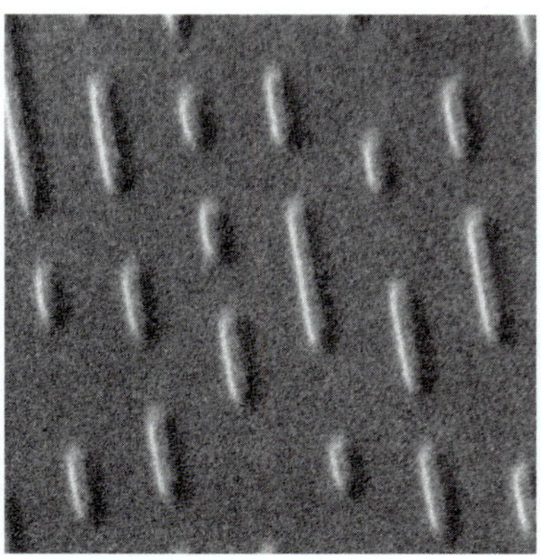

Courtesy of University of Miami, Music Engineering

Figure 17.13 The surface of a compact disc, showing the pits. Transitions between pits and lands correspond to binary ones. Regions without transitions correspond to binary zeros.

The advantage of digital recording is in the high fidelity of the sound. With analog recording, any small imperfection in the record surface or the recording equipment can cause a distortion of the waveform. For example, clipping all peaks of a waveform by 10% has a major effect on the spectrum of the sound in an analog recording. With digital recording, however, it takes a major imperfection to turn a one into a zero. If an imperfection causes the magnitude of a one to be 90% of the original value, it still registers as a one and there is no distortion. Another advantage of digital recording is that the information is extracted optically, so there is no mechanical wear on the disc.

EXAMPLE 17.7 How Big Are the Pits?

In Example 10.2, we mentioned that the speed with which the surface of a compact disc passes the laser is 1.3 m/s. What is the average length of the audio track on a compact disc associated with each bit of the audio information?

SOLUTION

Conceptualize Imagine the surface of the disc passing by the laser at 1.3 m/s. In one second, a 1.3-m length of audio track passes by the laser. This length includes 705 600 bits of audio information.

Categorize This example is a simple substitution problem.

From knowing the number of bits in a length of 1.3 m, find the average length per bit:

$$\text{Length per bit} = \frac{1.3 \text{ m}}{705\,600 \text{ bits}} = 1.8 \times 10^{-6} \text{ m/bit}$$

$$= 1.8 \mu\text{m/bit}$$

The average length per bit of *total* information on the compact disc is smaller than this value because there is additional information on the disc besides the audio information. This information includes error correction codes, song numbers, and timing codes. As a result, the shortest length per bit is actually about 0.8 μm.

EXAMPLE 17.8 What's the Number?

Audio data on a compact disc undergoes complicated processing so as to reduce a variety of errors in reading the data. Therefore, an audio "word" is not laid out linearly on the disc. Suppose data has been read from the disc, the error encoding has been removed, and the resulting audio word is

1 0 1 1 1 0 1 1 1 0 1 1 1 0 1 1

What is the decimal number represented by this 16-bit word?

SOLUTION

Conceptualize When looking at the binary number above, it is most likely that, based on your lack of experience with binary representations, you will not be able to immediately identify the number. Remember, however, that it is just a string of multipliers of powers of 2, just like the numbers with which you are familiar are strings of multipliers of powers of 10.

Categorize This example is a straightforward problem in which we change a representation from binary code to decimal code.

We convert each of these bits to a power of 2 and add the results:

$1 \times 2^{15} = 32\,768$	$1 \times 2^{9} = 512$	$1 \times 2^{3} = 8$
$0 \times 2^{14} = 0$	$1 \times 2^{8} = 256$	$0 \times 2^{2} = 0$
$1 \times 2^{13} = 8\,192$	$1 \times 2^{7} = 128$	$1 \times 2^{1} = 2$
$1 \times 2^{12} = 4\,096$	$0 \times 2^{6} = 0$	$1 \times 2^{0} = 1$
$1 \times 2^{11} = 2\,048$	$1 \times 2^{5} = 32$	
$0 \times 2^{10} = 0$	$1 \times 2^{4} = 16$	sum = 48 059

This number is converted by the compact disc player into a voltage, representing one of the 44 100 values that is used to build one second of the electronic waveform representing the recorded sound.

17.6 Motion Picture Sound

Another interesting application of digital sound is the soundtrack of a motion picture. Early 20th-century movies recorded sound on phonograph records, which were synchronized with the action on the screen. Beginning with early newsreel films, the *variable-area optical soundtrack* process was introduced in which sound was recorded on an optical track on the film. The width of the transparent portion of the track varied according to the sound wave that was recorded. A photocell detecting light passing through the track converted the varying light intensity to a sound wave. As with phonograph recording, there are a number of difficulties with this recording system. For example, dirt or fingerprints on the film cause fluctuations in intensity and loss of fidelity.

Digital recording on film first appeared with *Dick Tracy* (1990), using the Cinema Digital Sound, or CDS, system. This system suffered from lack of an analog backup system in case of equipment failure and is no longer used in the film industry. It did, however, introduce the use of 5.1 channels of sound: left, center, right, right surround, left surround, and low frequency effects (LFE). The LFE channel, which is the "0.1 channel" of 5.1, carries very low frequencies for dramatic sound from explosions, earthquakes, and the like.

Current motion pictures are produced with three systems of digital sound recording:

Dolby digital. In Dolby digital format, 5.1 channels of digital sound are optically stored between the sprocket holes of the film. There is an analog optical backup in case the digital system fails. The first film to use this technique was *Batman Returns* (1992).

Digital theater sound (DTS). In DTS, 5.1 channels of sound are stored on a separate CD-ROM, which is synchronized to the film print by time codes on the film. There is an analog optical backup in case the digital system fails. The first film to use this technique was *Jurassic Park* (1993).

Sony dynamic digital sound (SDDS). In SDDS, eight full channels of digital sound are optically stored outside the sprocket holes on both sides of film. There is an analog optical backup in case the digital system fails. The first film to use this technique was *Last Action Hero* (1993). The existence of information on both sides of the film is a system of redundancy; in case one side is damaged, the system still operates. SDDS employs a full-spectrum LFE channel and two additional channels (left center and right center behind the screen).

Summary

ThomsonNOW™ Sign in at **www.thomsonedu.com** and go to ThomsonNOW to take a practice test for this chapter.

DEFINITIONS

The **intensity** of a periodic sound wave, which is the power per unit area, is

$$I \equiv \frac{\mathcal{P}}{A} = \frac{(\Delta P_{\max})^2}{2\rho v} \quad \textbf{(17.5, 17.6)}$$

The **sound level** of a sound wave in decibels is

$$\beta \equiv 10 \log \left(\frac{I}{I_0}\right) \quad \textbf{(17.8)}$$

The constant I_0 is a reference intensity, usually taken to be at the threshold of hearing (1.00×10^{-12} W/m^2), and I is the intensity of the sound wave in watts per square meter.

CONCEPTS AND PRINCIPLES

Sound waves are longitudinal and travel through a compressible medium with a speed that depends on the elastic and inertial properties of that medium. The speed of sound in a liquid or gas having a bulk modulus B and density ρ is

$$v = \sqrt{\frac{B}{\rho}} \quad \textbf{(17.1)}$$

For sinusoidal sound waves, the variation in the position of an element of the medium is

$$s(x, t) = s_{\max} \cos (kx - \omega t) \quad \textbf{(17.2)}$$

and the variation in pressure from the equilibrium value is

$$\Delta P = \Delta P_{\max} \sin (kx - \omega t) \quad \textbf{(17.3)}$$

where $\Delta P_{\max}$ is the **pressure amplitude**. The pressure wave is 90° out of phase with the displacement wave. The relationship between $s_{\max}$ and $\Delta P_{\max}$ is

$$\Delta P_{\max} = \rho v \omega s_{\max} \quad \textbf{(17.4)}$$

The change in frequency heard by an observer whenever there is relative motion between a source of sound waves and the observer is called the **Doppler effect**. The observed frequency is

$$f' = \left(\frac{v + v_O}{v - v_S}\right) f \quad \textbf{(17.13)}$$

In this expression, the signs for the values substituted for v_O and v_S depend on the direction of the velocity. A positive value for the velocity of the observer or source is substituted if the velocity of one is toward the other, whereas a negative value represents a velocity of one away from the other.

In digital recording of sound, the sound waveform is sampled 44 100 times per second. The pressure of the wave for each sampling is measured and converted to a binary number. In playback, these binary numbers are read and used to build the original waveform.

Questions

☐ denotes answer available in *Student Solutions Manual/Study Guide;* **O** denotes objective question

1. **O** Table 17.1 shows that the speed of sound is typically an order of magnitude larger in solids than in gases. To what can this higher value be most directly attributed? (a) the difference in density between solids and gases (b) the difference in compressibility between solids and gases (c) the limited size of a solid object compared to a free gas (d) the impossibility of holding a gas under significant tension

2. ☐ If an alarm clock is placed in a good vacuum and then activated, no sound is heard. Explain.

3. A sonic ranger is a device that determines the distance to an object by sending out an ultrasonic sound pulse and measuring the time interval required for the wave to return by reflection from the object. Typically these devices cannot reliably detect an object that is less than half a meter from the sensor. Why is that?

4. A friend sitting in her car far down the road waves to you and beeps her horn at the same moment. How far away must she be for you to calculate the speed of sound to two significant figures by measuring the time interval required for the sound to reach you?

5. **O** Assume a change at the source of sound reduces the wavelength of a sound wave in air by a factor of 2. **(i)** What happens to its frequency? (a) It increases by a factor of 4. (b) It increases by a factor of 2. (c) It is unchanged. (d) It decreases by a factor of 2. (e) It changes by an unpredictable factor. **(ii)** What happens to its speed? Choose from the same possibilities.

6. **O** A sound wave travels in air with a frequency of 500 Hz. If the wave travels from the air into water, **(i)** what happens to its frequency? (a) It increases. (b) It decreases. (c) It is unchanged. **(ii)** What happens to its wavelength? Choose from the same possibilities.

7. By listening to a band or orchestra, how can you determine that the speed of sound is the same for all frequencies?

8. **O** A point source broadcasts sound into a uniform medium. If the distance from the source is tripled, how does the intensity change? (a) It becomes one-ninth as large. (b) It becomes one-third as large. (c) It is unchanged. (d) It becomes three times larger. (e) It becomes nine times larger.

9. **O** A church bell in a steeple rings once. At 300 m in front of the church, the maximum sound intensity is 2 $\mu W/m^2$. At 950 m behind the church, the maximum intensity is 0.2 $\mu W/m^2$. What is the main reason for the difference in the intensity? (a) Most of the sound is absorbed by the air before it gets far away from the source. (b) Most of the sound is absorbed by the ground as it travels away from the source. (c) The bell broadcasts the sound mostly toward the front. (d) At a larger distance, the power is spread over a larger area. (e) At a larger distance, the power is spread throughout a larger spherical volume.

10. **O** Of the following sounds, which is most likely to have a sound level of 60 dB? (a) a rock concert (b) the turning of a page in this textbook (c) dinner-table conversation (d) a cheering crowd at a football game

11. **O** With a sensitive sound level meter you measure the sound of a running spider as −10 dB. What does the negative sign imply? (a) The spider is moving away from you. (b) The frequency of the sound is too low to be audible to humans. (c) The intensity of the sound is too faint to be audible to humans. (d) You have made a mistake; negative signs do not fit with logarithms.

12. *The Tunguska event.* On June 30, 1908, a meteor burned up and exploded in the atmosphere above the Tunguska River valley in Siberia. It knocked down trees over thousands of square kilometers and started a forest fire, but produced no crater and apparently caused no human casualties. A witness sitting on his doorstep outside the zone of falling trees recalled events in the following sequence. He saw a moving light in the sky, brighter than the sun and descending at a low angle to the horizon. He felt his face become warm. He felt the ground shake. An invisible agent picked him up and immediately dropped him about a meter farther away from where the light had been. He heard a very loud protracted rumbling. Suggest an explanation for these observations and for the order in which they happened.

13. Explain what happens to the frequency of the echo of your car horn as you drive toward the wall of a canyon. What happens to the frequency as you move away from the wall?

14. **O** A source of sound vibrates with constant frequency. Rank the frequency of sound observed in the following cases from the highest to the lowest. If two frequencies are equal, show their equality in your ranking. Only one thing is moving at a time, and all the motions mentioned have the same speed, 25 m/s. (a) Source and observer are stationary in stationary air. (b) The source is moving toward the observer in still air. (c) The source is moving away from the observer in still air. (d) The observer is moving toward the source in still air. (e) The observer is moving away from the source in still air. (f) Source and observer are stationary, with a steady wind blowing from the source toward the observer. (g) Source and observer are stationary, with a steady wind blowing from the observer toward the source.

15. **O** Suppose an observer and a source of sound are both at rest and a strong wind is blowing away from the source toward the observer. **(i)** What effect does the wind have on the observed frequency? (a) It causes an increase. (b) It causes a decrease. (c) It causes no change. **(ii)** What effect does the wind have on the observed wavelength? Choose from the same possibilities. **(iii)** What effect does the wind have on the observed speed of the wave? Choose from the same possibilities.

16. How can an object move with respect to an observer so that the sound from it is not shifted in frequency?

Problems

WebAssign The Problems from this chapter may be assigned online in WebAssign.

ThomsonNOW Sign in at **www.thomsonedu.com** and go to ThomsonNOW to assess your understanding of this chapter's topics with additional quizzing and conceptual questions.

1, 2, 3 denotes straightforward, intermediate, challenging; □ denotes full solution available in *Student Solutions Manual/Study Guide;* ▲ denotes coached solution with hints available at **www.thomsonedu.com;** denotes developing symbolic reasoning; ● denotes asking for qualitative reasoning; denotes computer useful in solving problem

Section 17.1 Speed of Sound Waves

Problem 60 in Chapter 2 can also be assigned with this section.

1. ● Suppose you hear a clap of thunder 16.2 s after seeing the associated lightning stroke. The speed of sound in air is 343 m/s, and the speed of light in air is 3.00×10^8 m/s. How far are you from the lightning stroke? Do you need to know the value of the speed of light to answer? Explain.

2. Find the speed of sound in mercury, which has a bulk modulus of approximately 2.80×10^{10} N/m^2 and a density of 13 600 kg/m^3.

3. A dolphin in seawater at a temperature of 25°C makes a chirp. How much time passes before it hears an echo from the bottom of the ocean, 150 m below?

4. The speed of sound in air (in meters per second) depends on temperature according to the approximate expression

$$v = 331.5 + 0.607T_C$$

where T_C is the Celsius temperature. In dry air, the temperature decreases about 1°C for every 150 m rise in altitude. (a) Assume this change is constant up to an altitude of 9 000 m. What time interval is required for the sound from an airplane flying at 9 000 m to reach the ground on a day when the ground temperature is 30°C? (b) **What If?** Compare your answer with the time interval required if the air were uniformly at 30°C. Which time interval is longer?

5. A flowerpot is knocked off a balcony 20.0 m above the sidewalk and falls toward an unsuspecting 1.75-m-tall man who is standing below. How close to the sidewalk can the flowerpot fall before it is too late for a warning shouted from the balcony to reach the man in time? Assume the man below requires 0.300 s to respond to the warning. The ambient temperature is 20°C.

6. A rescue plane flies horizontally at a constant speed searching for a disabled boat. When the plane is directly above the boat, the boat's crew blows a loud horn. By the time the plane's sound detector receives the horn's sound, the plane has traveled a distance equal to half its altitude above the ocean. Assuming it takes the sound 2.00 s to reach the plane, determine (a) the speed of the plane and (b) its altitude. Take the speed of sound to be 343 m/s.

7. A cowboy stands on horizontal ground between two parallel vertical cliffs. He is not midway between the cliffs. He fires a shot and hears its echoes. The second echo arrives 1.92 s after the first and 1.47 s before the third. Consider only the sound traveling parallel to the ground and reflecting from the cliffs. Take the speed of sound as 340 m/s. (a) What is the distance between the cliffs? (b) **What If?** If he can hear a fourth echo, how long after the third echo does it arrive?

Section 17.2 Periodic Sound Waves

Note: Use the following values as needed unless otherwise specified. The equilibrium density of air at 20°C is $\rho = 1.20$ kg/m^3 and the speed of sound is $v = 343$ m/s. Pressure variations ΔP are measured relative to atmospheric pressure, 1.013×10^5 N/m^2.

8. ● A sound wave propagates in air at 27°C with frequency 4.00 kHz. It passes through a region where the temperature gradually changes, and then it moves through air at 0°C. (a) What happens to the speed of the wave? (b) What happens to its frequency? (c) What happens to its wavelength? Give numerical answers to these questions to the extent possible and state your reasoning about what happens to the wave physically.

9. Ultrasound is used in medicine both for diagnostic imaging and for therapy. For diagnosis, short pulses of ultrasound are passed through the patient's body. An echo reflected from a structure of interest is recorded, and the distance to the structure can be determined from the time delay for the echo's return. A single transducer emits and detects the ultrasound. An image of the structure is obtained by reducing the data with a computer. With sound of low intensity, this technique is noninvasive and harmless. It is used to examine fetuses, tumors, aneurysms, gallstones, and many other structures. To reveal detail, the wavelength of the reflected ultrasound must be small compared to the size of the object reflecting the wave. (a) What is the wavelength of ultrasound with a frequency of 2.40 MHz, used in echocardiography to map the beating human heart? (b) In the whole set of imaging techniques, frequencies in the range 1.00 to 20.0 MHz are used. What is the range of wavelengths corresponding to this range of frequencies? The speed of ultrasound in human tissue is about 1 500 m/s (nearly the same as the speed of sound in water).

10. A sound wave in air has a pressure amplitude equal to 4.00×10^{-3} N/m^2. Calculate the displacement amplitude of the wave at a frequency of 10.0 kHz.

11. A sinusoidal sound wave is described by the displacement wave function

$$s(x, t) = (2.00\ \mu\text{m})\cos\left[(15.7\ \text{m}^{-1})x - (858\ \text{s}^{-1})t\right]$$

(a) Find the amplitude, wavelength, and speed of this wave. (b) Determine the instantaneous displacement from equilibrium of the elements of air at the position $x = 0.050\ 0$ m at $t = 3.00$ ms. (c) Determine the maximum speed of the element's oscillatory motion.

12. As a certain sound wave travels through the air, it produces pressure variations (above and below atmospheric pressure) given by $\Delta P = 1.27 \sin(\pi x - 340\pi t)$ in SI units. Find (a) the amplitude of the pressure variations, (b) the frequency, (c) the wavelength in air, and (d) the speed of the sound wave.

13. Write an expression that describes the pressure variation as a function of position and time for a sinusoidal sound wave in air. Assume $\lambda = 0.100$ m and $\Delta P_{max} = 0.200$ N/m^2.

14. The tensile stress in a thick copper bar is 99.5% of its elastic breaking point of 13.0×10^{10} N/m^2. If a 500-Hz sound wave is transmitted through the material, (a) what displacement amplitude will cause the bar to break? (b) What is the maximum speed of the elements of copper at this moment? (c) What is the sound intensity in the bar?

15. ▲ An experimenter wishes to generate in air a sound wave that has a displacement amplitude of 5.50×10^{-6} m. The pressure amplitude is to be limited to 0.840 N/m^2. What is the minimum wavelength the sound wave can have?

Section 17.3 Intensity of Periodic Sound Waves

16. The area of a typical eardrum is about 5.00×10^{-5} m^2. Calculate the sound power incident on an eardrum at (a) the threshold of hearing and (b) the threshold of pain.

17. Calculate the sound level (in decibels) of a sound wave that has an intensity of 4.00 μW/m^2.

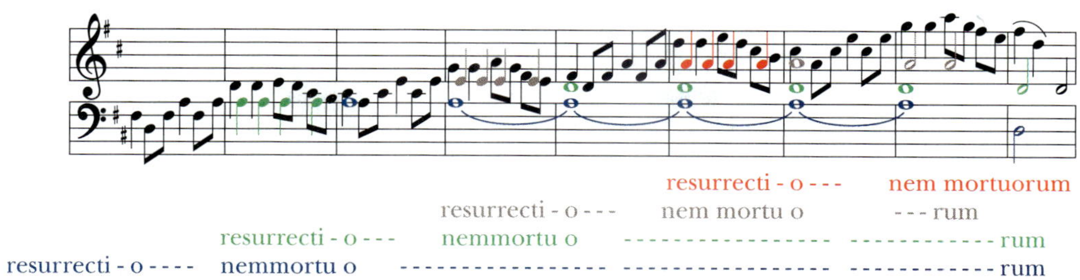

Figure P17.21 Bass (blue), tenor (green), alto (brown), and first soprano (red) parts for a portion of Bach's Mass in B Minor. The basses sing the foreground melody for two measures, then the tenors for two measures, then the altos, and then the first sopranos. For emphasis, this line is printed in black throughout. Parts for the second sopranos, violins, viola, flutes, oboes, and continuo are omitted. The tenor part is written as it is sung.

18. The tube depicted in Active Figure 17.2 is filled with air at 20°C and equilibrium pressure 1 atm. The diameter of the tube is 8.00 cm. The piston is driven at a frequency of 600 Hz with an amplitude of 0.120 cm. What power must be supplied to maintain the oscillation of the piston?

19. The intensity of a sound wave at a fixed distance from a speaker vibrating at 1.00 kHz is 0.600 W/m². (a) Determine the intensity that results if the frequency is increased to 2.50 kHz while a constant displacement amplitude is maintained. (b) Calculate the intensity if the frequency is reduced to 0.500 kHz and the displacement amplitude is doubled.

20. The intensity of a sound wave at a fixed distance from a speaker vibrating at a frequency f is I. (a) Determine the intensity that results if the frequency is increased to f' while a constant displacement amplitude is maintained. (b) Calculate the intensity if the frequency is reduced to $f/2$ and the displacement amplitude is doubled.

21. The most soaring vocal melody is in Johann Sebastian Bach's Mass in B Minor. A portion of the score for the Credo section, number 9, bars 25 to 33, appears in Figure P17.21. The repeating syllable O in the phrase "resurrectionem mortuorum" (the resurrection of the dead) is seamlessly passed from basses to tenors to altos to first sopranos, like a baton in a relay. Each voice carries the foreground melody up through a rising passage encompassing an octave or more. Together the voices carry it from D below middle C to A above a tenor's high C. In concert pitch, these notes are now assigned frequencies of 146.8 Hz and 880.0 Hz. (a) Find the wavelengths of the initial and final notes. (b) Assume the chorus sings the melody with a uniform sound level of 75.0 dB. Find the pressure amplitudes of the initial and final notes. (c) Find the displacement amplitudes of the initial and final notes. (d) **What If?** In Bach's time, before the invention of the tuning fork, frequencies were assigned to notes as a matter of immediate local convenience. Assume the rising melody was sung starting from 134.3 Hz and ending at 804.9 Hz. How would the answers to parts (a) through (c) change?

22. Show that the difference between decibel levels β_1 and β_2 of a sound is related to the ratio of the distances r_1 and r_2 from the sound source by

$$\beta_2 - \beta_1 = 20 \log\left(\frac{r_1}{r_2}\right)$$

23. ▲ A family ice show is held at an enclosed arena. The skaters perform to music with level 80.0 dB. This level is too loud for your baby, who yells at 75.0 dB. (a) What total sound intensity engulfs you? (b) What is the combined sound level?

24. A jackhammer, operated continuously at a construction site, behaves as a point source of spherical sound waves. A construction supervisor stands 50.0 m due north of this sound source and begins to walk due west. How far does she have to walk for the amplitude of the wave function to drop by a factor of 2.00?

25. The power output of a certain public address speaker is 6.00 W. Suppose it broadcasts equally in all directions. (a) Within what distance from the speaker would the sound be painful to the ear? (b) At what distance from the speaker would the sound be barely audible?

26. Two small speakers emit sound waves of different frequencies equally in all directions. Speaker A has an output of 1.00 mW, and speaker B has an output of 1.50 mW. Determine the sound level (in decibels) at point C in Figure P17.26 assuming (a) only speaker A emits sound, (b) only speaker B emits sound, and (c) both speakers emit sound.

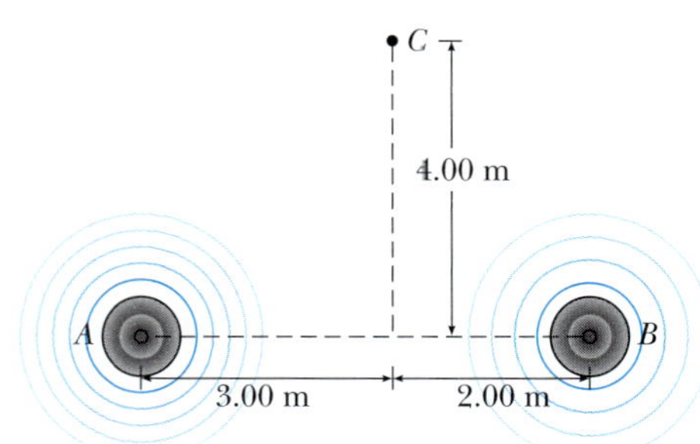

Figure P17.26

27. A firework charge is detonated many meters above the ground. At a distance of 400 m from the explosion, the acoustic pressure reaches a maximum of 10.0 N/m². Assume the speed of sound is constant at 343 m/s throughout the atmosphere over the region considered, the ground absorbs all the sound falling on it, and the air absorbs sound energy as described by the rate 7.00 dB/km. What is the sound level (in decibels) at 4.00 km from the explosion?

28. A fireworks rocket explodes at a height of 100 m above the ground. An observer on the ground directly under the explosion experiences an average sound intensity of 7.00×10^{-2} W/m² for 0.200 s. (a) What is the total sound energy of the explosion? (b) What is the sound level (in decibels) heard by the observer?

29. The sound level at a distance of 3.00 m from a source is 120 dB. At what distance is the sound level (a) 100 dB and (b) 10.0 dB?

30. The smallest change in sound level that a person can distinguish is approximately 1 dB. When you are standing next to your power lawn mower as it is running, can you hear the steady roar of your neighbor's lawn mower? Perform an order-of-magnitude calculation to substantiate your answer, stating the data you measure or estimate.

31. As the people sing in church, the sound level everywhere inside is 101 dB. No sound is transmitted through the massive walls, but all the windows and doors are open on a summer morning. Their total area is 22.0 m^2. (a) How much sound energy is radiated in 20.0 min? (b) Suppose the ground is a good reflector and sound radiates uniformly in all horizontal and upward directions. Find the sound level 1.00 km away.

Section 17.4 The Doppler Effect

32. Expectant parents are thrilled to hear their unborn baby's heartbeat, revealed by an ultrasonic motion detector. Suppose the fetus's ventricular wall moves in simple harmonic motion with an amplitude of 1.80 mm and a frequency of 115 per minute. (a) Find the maximum linear speed of the heart wall. Suppose the motion detector in contact with the mother's abdomen produces sound at 2 000 000.0 Hz, which travels through tissue at 1.50 km/s. (b) Find the maximum frequency at which sound arrives at the wall of the baby's heart. (c) Find the maximum frequency at which reflected sound is received by the motion detector. By electronically "listening" for echoes at a frequency different from the broadcast frequency, the motion detector can produce beeps of audible sound in synchronization with the fetal heartbeat.

33. A driver travels northbound on a highway at a speed of 25.0 m/s. A police car, traveling southbound at a speed of 40.0 m/s, approaches with its siren producing sound at a frequency of 2 500 Hz. (a) What frequency does the driver observe as the police car approaches? (b) What frequency does the driver detect after the police car passes him? (c) Repeat parts (a) and (b) for the case when the police car is traveling northbound.

34. A block with a speaker bolted to it is connected to a spring having spring constant $k = 20.0$ N/m as shown in Figure P17.34. The total mass of the block and speaker is 5.00 kg, and the amplitude of this unit's motion is 0.500 m. (a) The speaker emits sound waves of frequency 440 Hz. Determine the highest and lowest frequencies heard by the person to the right of the speaker. (b) If the maximum sound level heard by the person is 60.0 dB when he is closest to the speaker, 1.00 m away, what is the minimum sound level heard by the observer? Assume the speed of sound is 343 m/s.

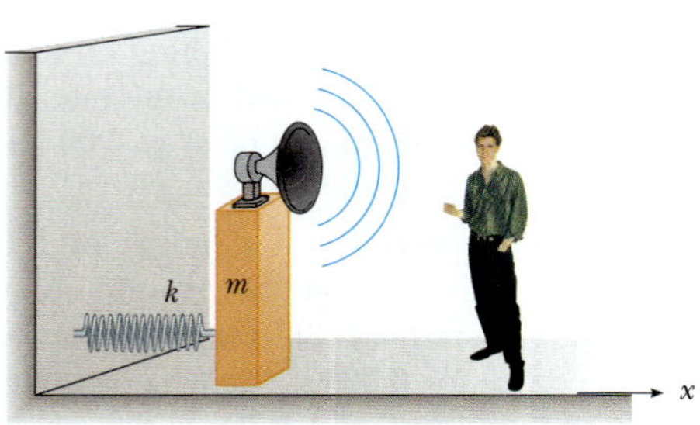

Figure P17.34

35. ▲ Standing at a crosswalk, you hear a frequency of 560 Hz from the siren of an approaching ambulance. After the ambulance passes, the observed frequency of the siren is 480 Hz. Determine the ambulance's speed from these observations.

36. At the Winter Olympics, an athlete rides her luge down the track while a bell just above the wall of the chute rings continuously. When her sled passes the bell, she hears the frequency of the bell fall by the musical interval called a minor third. That is, the frequency she hears drops to five-sixths its original value. (a) Find the speed of sound in air at the ambient temperature −10.0°C. (b) Find the speed of the athlete.

37. A tuning fork vibrating at 512 Hz falls from rest and accelerates at 9.80 m/s^2. How far below the point of release is the tuning fork when waves of frequency 485 Hz reach the release point? Take the speed of sound in air to be 340 m/s.

38. A siren mounted on the roof of a firehouse emits sound at a frequency of 900 Hz. A steady wind is blowing with a speed of 15.0 m/s. Taking the speed of sound in calm air to be 343 m/s, find the wavelength of the sound (a) upwind of the siren and (b) downwind of the siren. Firefighters are approaching the siren from various directions at 15.0 m/s. What frequency does a firefighter hear (c) if she is approaching from an upwind position so that she is moving in the direction in which the wind is blowing and (d) if she is approaching from a downwind position and moving against the wind?

39. ▲ A supersonic jet traveling at Mach 3.00 at an altitude of 20 000 m is directly over a person at time $t = 0$ as shown in Figure P17.39. (a) At what time will the person encounter the shock wave? (b) Where will the plane be when the "boom" is finally heard? Assume the speed of sound in air is 335 m/s.

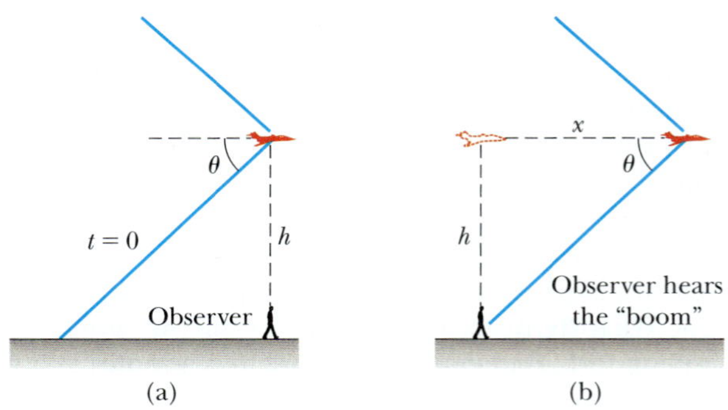

Figure P17.39

40. The loop of a circus ringmaster's whip travels at Mach 1.38 (that is, $v_S/v = 1.38$). What angle does the shock front make with the direction of the whip's motion?

41. When high-energy charged particles move through a transparent medium with a speed greater than the speed of light in that medium, a shock wave, or bow wave, of light is produced. This phenomenon is called the *Cerenkov effect*. When a nuclear reactor is shielded by a large pool of water, Cerenkov radiation can be seen as a

blue glow in the vicinity of the reactor core due to high-speed electrons moving through the water. In a particular case, the Cerenkov radiation produces a wave front with an apex half-angle of 53.0°. Calculate the speed of the electrons in the water. The speed of light in water is 2.25×10^8 m/s.

Section 17.5 Digital Sound Recording

Section 17.6 Motion Picture Sound

42. ● This problem represents a possible (but not recommended) way to code instantaneous pressures in a sound wave into 16-bit digital words. Example 17.2 mentions that the pressure amplitude of a 120-dB sound is 28.7 N/m^2. Let this pressure variation be represented by the digital code 65 536. Let the digital word 0 on the recording represent zero pressure variation. Let other intermediate pressures be represented by digital words of intermediate size, in direct proportion to the pressure. (a) What digital word would represent the maximum pressure in a 40-dB sound? (b) Explain why this scheme works poorly for soft sounds. (c) Explain how this coding scheme would clip off half of the waveform of any sound, ignoring the actual shape of the wave and turning it into a string of zeros. By introducing sharp corners into every recorded waveform, this coding scheme would make everything sound like a buzzer or a kazoo.

Additional Problems

43. ● A 150-g glider moving at 2.30 m/s on an air track undergoes a completely inelastic collision with an originally stationary 200-g glider, and the two gliders latch together over a time interval of 7.00 ms. A student suggests that roughly half the missing mechanical energy goes into sound. Is this suggestion reasonable? To evaluate the idea, find the implied level of the sound 0.800 m from the gliders. If the student's idea is unreasonable, suggest a better idea.

44. ● Explain how the wave function

$$\Delta P(r, t) = \left(\frac{25.0 \text{ Pa} \cdot \text{m}}{r}\right) \sin (1.36r \text{ rad/m} - 2\,030t \text{ rad/s})$$

can apply to a wave radiating from a small source, with r being the radial distance from the center of the source to any point outside the source. Give the most detailed description of the wave that you can. Include answers to such questions as the following. Does the wave move more toward the right or the left? As it moves away from the source, what happens to its amplitude? Its speed? Its frequency? Its wavelength? Its power? Its intensity? What are representative values for each of these quantities? What can you say about the source of the wave? About the medium through which it travels?

45. ● A large set of unoccupied football bleachers has solid seats and risers. You stand on the field in front of the bleachers and sharply clap two wooden boards together once. The sound pulse you produce has no definite frequency and no wavelength. The sound you hear reflected from the bleachers has an identifiable frequency and may remind you of a short toot on a trumpet or of a buzzer or kazoo. Account for this sound. (a) Compute order-of-magnitude estimates for the frequency, wavelength, and duration of the sound, on the basis of data you specify. (b) Each face of a great Mayan pyramid is like a steep stairway with very narrow steps. Can it produce an echo of a handclap that sounds like the call of a bird? Explain your answer.

46. ● Spherical waves of wavelength 45.0 cm propagate outward from a point source. (a) Explain how the intensity at a distance of 240 cm compares with the intensity at a distance of 60.0 cm. (b) Explain how the amplitude at a distance of 240 cm compares with the amplitude at a distance of 60.0 cm. (c) Explain how the phase of the wave at a distance of 240 cm compares with the phase at 60.0 cm at the same moment.

47. A sound wave in a cylinder is described by Equations 17.2 through 17.4. Show that $\Delta P = \pm \rho v \omega \sqrt{s_{\text{max}}^2 - s^2}$.

48. Many artists sing very high notes in ad-lib ornaments and cadenzas. The highest note written for a singer in a published score was F-sharp above high C, 1.480 kHz, for Zerbinetta in the original version of Richard Strauss's opera *Ariadne auf Naxos.* (a) Find the wavelength of this sound in air. (b) Suppose people in the fourth row of seats hear this note with level 81.0 dB. Find the displacement amplitude of the sound. (c) **What If?** In response to complaints, Strauss later transposed the note down to F above high C, 1.397 kHz. By what increment did the wavelength change? (The Queen of the Night in Mozart's *Magic Flute* also sings F above high C.)

49. On a Saturday morning, pickup trucks and sport utility vehicles carrying garbage to the town dump form a nearly steady procession on a country road, all traveling at 19.7 m/s. From one direction, two trucks arrive at the dump every 3 min. A bicyclist is also traveling toward the dump, at 4.47 m/s. (a) With what frequency do the trucks pass the cyclist? (b) **What If?** A hill does not slow down the trucks, but makes the out-of-shape cyclist's speed drop to 1.56 m/s. How often do noisy, smelly, inefficient, garbage-dripping, road-hogging trucks whiz past the cyclist now?

50. **Review problem.** For a certain type of steel, stress is always proportional to strain with Young's modulus as shown in Table 12.1. The steel has the density listed for iron in Table 14.1. It will fail by bending permanently if subjected to compressive stress greater than its yield strength σ_y = 400 MPa. A rod 80.0 cm long, made of this steel, is fired at 12.0 m/s straight at a very hard wall or at another identical rod moving in the opposite direction. (a) The speed of a one-dimensional compressional wave moving along the rod is given by $v = \sqrt{Y/\rho}$, where Y is Young's modulus for the rod and ρ is the density. Calculate this speed. (b) After the front end of the rod hits the wall and stops, the back end of the rod keeps moving as described by Newton's first law until it is stopped by excess pressure in a sound wave moving back through the rod. What time interval elapses before the back end of the rod receives the message that it should stop? (c) How far has the back end of the rod moved in this time interval? Find (d) the strain and (e) the stress in the rod. (f) If it is not to fail, show that the maximum impact speed a rod can have is given by the expression $v = \sigma_y/\sqrt{\rho Y}$.

51. To permit measurement of her speed, a skydiver carries a buzzer emitting a steady tone at 1 800 Hz. A friend on the

ground at the landing site directly below listens to the amplified sound he receives. Assume the air is calm and the sound speed is 343 m/s, independent of altitude. While the skydiver is falling at terminal speed, her friend on the ground receives waves of frequency 2 150 Hz. (a) What is the skydiver's speed of descent? (b) **What If?** Suppose the skydiver can hear the sound of the buzzer reflected from the ground. What frequency does she receive?

52. Prove that sound waves propagate with a speed given by Equation 17.1. Proceed as follows. In Active Figure 17.2, consider a thin, cylindrical layer of air in the cylinder, with face area A and thickness Δx. Draw a free-body diagram of this thin layer. Show that $\Sigma F_x = ma_x$ implies that

$$-\frac{\partial(\Delta P)}{\partial x} A\,\Delta x = \rho A\,\Delta x \frac{\partial^2 s}{\partial t^2}$$

By substituting $\Delta P = -B(\partial s/\partial x)$, derive the following wave equation for sound:

$$\frac{B}{\rho}\frac{\partial^2 s}{\partial x^2} = \frac{\partial^2 s}{\partial t^2}$$

To a mathematical physicist, this equation demonstrates the existence of sound waves and determines their speed. As a physics student, you must take another step or two. Substitute into the wave equation the trial solution $s(x, t) = s_{\max}\cos(kx - \omega t)$. Show that this function satisfies the wave equation provided that $\omega/k = \sqrt{B/\rho}$. This result reveals that sound waves exist provided they move with the speed $v = f\lambda = (2\pi f)(\lambda/2\pi) = \omega/k = \sqrt{B/\rho}$.

53. Two ships are moving along a line due east. The trailing vessel has a speed relative to a land-based observation point of 64.0 km/h, and the leading ship has a speed of 45.0 km/h relative to that point. The two ships are in a region of the ocean where the current is moving uniformly due west at 10.0 km/h. The trailing ship transmits a sonar signal at a frequency of 1 200.0 Hz. What frequency is monitored by the leading ship? Use 1 520 m/s as the speed of sound in ocean water.

54. A bat, moving at 5.00 m/s, is chasing a flying insect. If the bat emits a 40.0-kHz chirp and receives back an echo at 40.4 kHz, at what speed is the insect moving toward or away from the bat? (Take the speed of sound in air to be $v = 340$ m/s.)

55. Assume a loudspeaker broadcasts sound equally in all directions and produces sound with a level of 103 dB at a distance of 1.60 m from its center. (a) Find its sound power output. (b) If a salesperson claims to be giving you 150 W per channel, he is referring to the electrical power input to the speaker. Find the efficiency of the speaker, that is, the fraction of input power that is converted into useful output power.

56. A police car is traveling east at 40.0 m/s along a straight road, overtaking a car ahead of it moving east at 30.0 m/s. The police car has a malfunctioning siren that is stuck at 1 000 Hz. (a) Sketch the appearance of the wave fronts of the sound produced by the siren. Show the wave fronts both to the east and the west of the police car. (b) What would be the wavelength in air of the siren sound if the police car were at rest? (c) What is the wavelength in front of the police car? (d) What is it behind the police car? (e) What is the frequency heard by the driver being chased?

57. ● The speed of a one-dimensional compressional wave traveling along a thin copper rod is 3.56 km/s. A copper bar is given a sharp hammer blow at one end. A listener at the far end of the bar hears the sound twice, transmitted through the metal and through air at 0°C, with a time interval Δt between the two pulses. (a) Which sound arrives first? (b) Find the length of the bar as a function of Δt. (c) Evaluate the length of the bar if $\Delta t = 127$ ms. (d) Imagine that the copper were replaced by a much stiffer material through which sound would travel much faster. How would the answer to part (b) change? Would it go to a well-defined limit as the signal speed in the rod goes to infinity? Explain your answer.

58. An interstate highway has been built though a poor neighborhood in a city. In the afternoon, the sound level in a rented room is 80.0 dB as 100 cars pass outside the window every minute. Late at night, when the room's tenant is at work in a factory, the traffic flow is only five cars per minute. What is the average late-night sound level?

59. A meteoroid the size of a truck enters the earth's atmosphere at a speed of 20.0 km/s and is not significantly slowed before entering the ocean. (a) What is the Mach angle of the shock wave from the meteoroid in the atmosphere? (Use 331 m/s as the sound speed.) (b) Assuming the meteoroid survives the impact with the ocean surface, what is the (initial) Mach angle of the shock wave the meteoroid produces in the water? (Use the wave speed for seawater given in Table 17.1.)

60. Equation 17.7 states that at distance r away from a point source with power $\mathcal{P}_{\text{avg}}$, the wave intensity is

$$I = \frac{\mathcal{P}_{\text{avg}}}{4\pi r^2}$$

Study Active Figure 17.9 and prove that at distance r straight in front of a point source with power $\mathcal{P}_{\text{avg}}$ moving with constant speed v_S the wave intensity is

$$I = \frac{\mathcal{P}_{\text{avg}}}{4\pi r^2}\left(\frac{v - v_S}{v}\right)$$

61. ▲ With particular experimental methods, it is possible to produce and observe in a long, thin rod both a longitudinal wave and a transverse wave whose speed depends primarily on tension in the rod. The speed of the longitudinal wave is determined by Young's modulus and the density of the material according to the expression $v = \sqrt{Y/\rho}$. The transverse wave can be modeled as a wave in a stretched string. A particular metal rod is 150 cm long and has a radius of 0.200 cm and a mass of 50.9 g. Young's modulus for the material is 6.80×10^{10} N/m^2. What must the tension in the rod be if the ratio of the speed of longitudinal waves to the speed of transverse waves is 8.00?

62. The Doppler equation presented in the text is valid when the motion between the observer and the source occurs on a straight line so that the source and observer are moving either directly toward or directly away from each other. If this restriction is relaxed, one must use the more general Doppler equation

$$f' = \left(\frac{v + v_O \cos \theta_O}{v - v_S \cos \theta_S}\right) f$$

where θ_O and θ_S are defined in Figure P17.62a. (a) Show that if the observer and source are moving directly away from each other, the preceding equation reduces to Equation 17.13 with negative values for both v_O and v_S. (b) Use the preceding equation to solve the following problem. A train moves at a constant speed of 25.0 m/s toward the intersection shown in Figure P17.62b. A car is stopped near the crossing, 30.0 m from the tracks. If the train's horn emits a frequency of 500 Hz, what is the frequency heard by the passengers in the car when the train is 40.0 m from the intersection? Take the speed of sound to be 343 m/s.

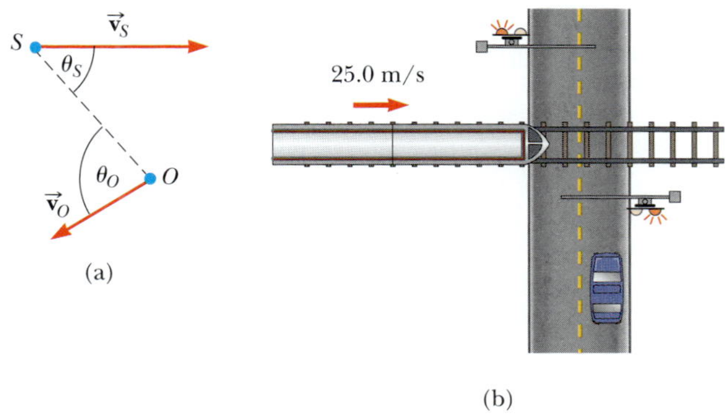

Figure P17.62

63. Three metal rods are located relative to each other as shown in Figure P17.63, where $L_1 + L_2 = L_3$. The speed of sound in a rod is given by $v = \sqrt{Y/\rho}$, where Y is Young's modulus for the rod and ρ is the density. Values of density and Young's modulus for the three materials are $\rho_1 = 2.70 \times 10^3$ kg/m³, $Y_1 = 7.00 \times 10^{10}$ N/m², $\rho_2 = 11.3 \times 10^3$ kg/m³, $Y_2 = 1.60 \times 10^{10}$ N/m², $\rho_3 = 8.80 \times 10^3$ kg/m³, and $Y_3 = 11.0 \times 10^{10}$ N/m². (a) If $L_3 = 1.50$ m, what must the ratio L_1/L_2 be if a sound wave is to travel the length of rods 1 and 2 in the same time interval required for the wave to travel the length of rod 3? (b) The frequency of the source is 4.00 kHz. Determine the phase difference between the wave traveling along rods 1 and 2 and the one traveling along rod 3.

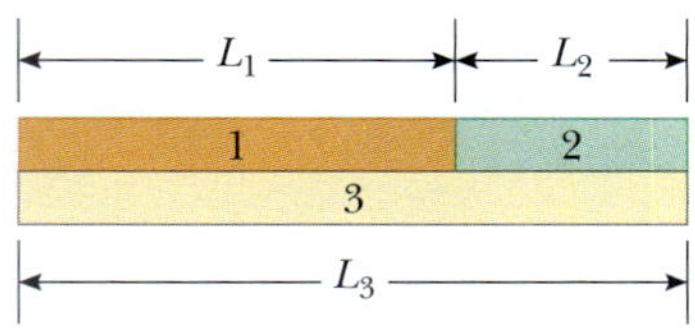

Figure P17.63

Answers to Quick Quizzes

17.1 (c). Because the bottom of the bottle is a rigid barrier, the displacement of elements of air at the bottom is zero. Because the pressure variation is a minimum or a maximum when the displacement is zero and because the pulse is moving downward, the pressure variation at the bottom is a maximum.

17.2 (b). The large area of the guitar body sets many elements of air into oscillation and allows the energy to leave the system by mechanical waves at a much larger rate than from the thin vibrating string.

17.3 (b). The factor of 100 is two powers of 10. The logarithm of 100 is 2, which multiplied by 10 gives 20 dB.

17.4 (e). The wave speed cannot be changed by moving the source, so choices (a) and (b) are incorrect. The detected wavelength is largest at A, so choices (c) and (d) are incorrect. Choice (f) is incorrect because the detected frequency is lowest at A.

17.5 (e). The intensity of the sound increases because the train is moving closer to you. Because the train moves at a constant velocity, the Doppler-shifted frequency remains fixed.

17.6 (b). The Mach number is the ratio of the plane's speed (which does not change) to the speed of sound, which is greater in the warm air than in the cold. The denominator of this ratio increases, whereas the numerator stays constant. Therefore, the ratio as a whole—the Mach number—decreases.

Guitarist Carlos Santana takes advantage of standing waves on strings. He changes to higher notes on the guitar by pushing the strings against the frets on the fingerboard, shortening the lengths of the portions of the strings that vibrate. (Bettmann/Corbis)

18 Superposition and Standing Waves

The wave model was introduced in the previous two chapters. We have seen that waves are very different from particles. A particle is of zero size, whereas a wave has a characteristic size, its wavelength. Another important difference between waves and particles is that we can explore the possibility of two or more waves combining at one point in the same medium. Particles can be combined to form extended objects, but the particles must be at *different* locations. In contrast, two waves can both be present at the same location. The ramifications of this possibility are explored in this chapter.

When waves are combined in systems with boundary conditions, only certain allowed frequencies can exist and we say the frequencies are *quantized*. Quantization is a notion that is at the heart of quantum mechanics, a subject introduced formally in Chapter 40. There we show that waves under boundary conditions explain many of the quantum phenomena. In this chapter, we use quantization to understand the behavior of the wide array of musical instruments that are based on strings and air columns.

We also consider the combination of waves having different frequencies. When two sound waves having nearly the same frequency interfere, we hear variations in the loudness called *beats*. Finally, we discuss how any nonsinusoidal periodic wave can be described as a sum of sine and cosine functions.

18.1 Superposition and Interference

Many interesting wave phenomena in nature cannot be described by a single traveling wave. Instead, one must analyze these phenomena in terms of a combination of traveling waves. To analyze such wave combinations, we make use of the **superposition principle**:

> If two or more traveling waves are moving through a medium, the resultant value of the wave function at any point is the algebraic sum of the values of the wave functions of the individual waves.

◀ Superposition principle

Waves that obey this principle are called *linear waves.* In the case of mechanical waves, linear waves are generally characterized by having amplitudes much smaller than their wavelengths. Waves that violate the superposition principle are called *nonlinear waves* and are often characterized by large amplitudes. In this book, we deal only with linear waves.

One consequence of the superposition principle is that **two traveling waves can pass through each other without being destroyed or even altered**. For instance, when two pebbles are thrown into a pond and hit the surface at different locations, the expanding circular surface waves from the two locations do not destroy each other but rather pass through each other. The resulting complex pattern can be viewed as two independent sets of expanding circles.

Active Figure 18.1 (page 502) is a pictorial representation of the superposition of two pulses. The wave function for the pulse moving to the right is y_1, and the wave function for the pulse moving to the left is y_2. The pulses have the same speed but different shapes, and the displacement of the elements of the medium is in the positive y direction for both pulses. When the waves begin to overlap (Active Fig. 18.1b), the wave function for the resulting complex wave is given by $y_1 + y_2$. When the crests of the pulses coincide (Active Fig. 18.1c), the resulting wave given by $y_1 + y_2$ has a larger amplitude than that of the individual pulses. The two pulses finally separate and continue moving in their original directions (Active Fig. 18.1d). Notice that the pulse shapes remain unchanged after the interaction, as if the two pulses had never met!

PITFALL PREVENTION 18.1

Do Waves Actually *Interfere*?

In popular usage, the term *interfere* implies that an agent affects a situation in some way so as to preclude something from happening. For example, in American football, *pass interference* means that a defending player has affected the receiver so that the receiver is unable to catch the ball. This usage is very different from its use in physics, where waves pass through each other and interfere, but do not affect each other in any way. In physics, interference is similar to the notion of *combination* as described in this chapter.

The combination of separate waves in the same region of space to produce a resultant wave is called **interference**. For the two pulses shown in Active Figure 18.1, the displacement of the elements of the medium is in the positive y direction for both pulses, and the resultant pulse (created when the individual pulses overlap) exhibits an amplitude greater than that of either individual pulse. Because the displacements caused by the two pulses are in the same direction, we refer to their superposition as **constructive interference**.

◀ Constructive interference

Now consider two pulses traveling in opposite directions on a taut string where one pulse is inverted relative to the other as illustrated in Active Figure 18.2 (page 502). When these pulses begin to overlap, the resultant pulse is given by $y_1 + y_2$, but the values of the function y_2 are negative. Again, the two pulses pass through each other; because the displacements caused by the two pulses are in opposite directions, however, we refer to their superposition as **destructive interference**.

◀ Destructive interference

The superposition principle is the centerpiece of the **waves in interference model**. In many situations, both in acoustics and optics, waves combine according to this principle and exhibit interesting phenomena with practical applications.

Quick Quiz 18.1 Two pulses move in opposite directions on a string and are identical in shape except that one has positive displacements of the elements of the string and the other has negative displacements. At the moment the two pulses completely overlap on the string, what happens? (a) The energy associated with the pulses has disappeared. (b) The string is not moving. (c) The string forms a straight line. (d) The pulses have vanished and will not reappear.

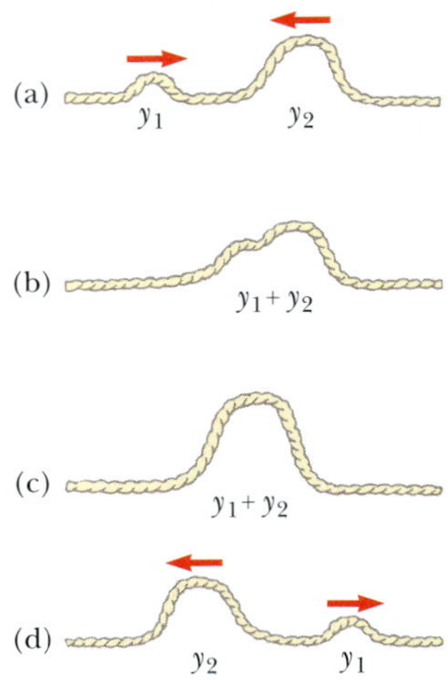

ACTIVE FIGURE 18.1

(a–d) Two pulses traveling on a stretched string in opposite directions pass through each other. When the pulses overlap, as shown in (b) and (c), the net displacement of the string equals the sum of the displacements produced by each pulse. Because each pulse produces positive displacements of the string, we refer to their superposition as *constructive interference.*

Sign in at www.thomsonedu.com and go to ThomsonNOW to choose the amplitude and orientation of each of the pulses and study the interference between them as they pass each other.

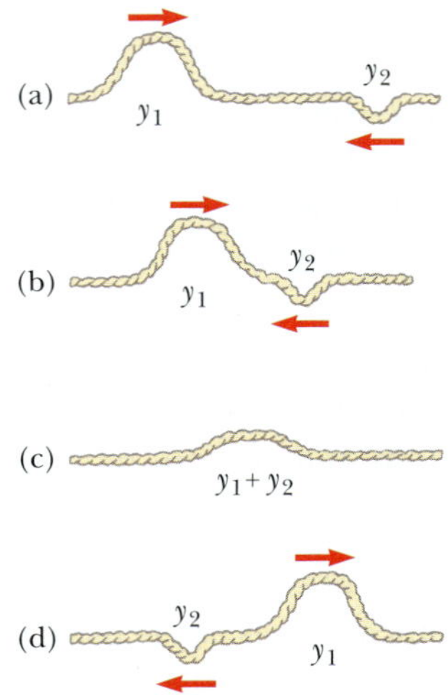

ACTIVE FIGURE 18.2

(a–d) Two pulses traveling in opposite directions and having displacements that are inverted relative to each other. When the two overlap in (c), their displacements partially cancel each other.

Sign in at www.thomsonedu.com and go to ThomsonNOW to choose the amplitude and orientation of each of the pulses and watch the interference as they pass each other.

Superposition of Sinusoidal Waves

Let us now apply the principle of superposition to two sinusoidal waves traveling in the same direction in a linear medium. If the two waves are traveling to the right and have the same frequency, wavelength, and amplitude but differ in phase, we can express their individual wave functions as

$$y_1 = A \sin (kx - \omega t) \qquad y_2 = A \sin (kx - \omega t + \phi)$$

where, as usual, $k = 2\pi/\lambda$, $\omega = 2\pi f$, and ϕ is the phase constant as discussed in Section 16.2. Hence, the resultant wave function y is

$$y = y_1 + y_2 = A[\sin (kx - \omega t) + \sin (kx - \omega t + \phi)]$$

To simplify this expression, we use the trigonometric identity

$$\sin a + \sin b = 2 \cos \left(\frac{a - b}{2}\right) \sin \left(\frac{a + b}{2}\right)$$

Letting $a = kx - \omega t$ and $b = kx - \omega t + \phi$, we find that the resultant wave function y reduces to

Resultant of two traveling sinusoidal waves ▶

$$y = 2A \cos \left(\frac{\phi}{2}\right) \sin \left(kx - \omega t + \frac{\phi}{2}\right)$$

This result has several important features. The resultant wave function y also is sinusoidal and has the same frequency and wavelength as the individual waves because the sine function incorporates the same values of k and ω that appear in the original wave functions. The amplitude of the resultant wave is $2A \cos (\phi/2)$, and its phase is $\phi/2$. If the phase constant ϕ equals 0, then $\cos (\phi/2) = \cos 0 = 1$ and the amplitude of the resultant wave is $2A$, twice the amplitude of either individual wave. In this case, the waves are said to be everywhere *in phase* and therefore interfere constructively. That is, the crests and troughs of the individual waves y_1

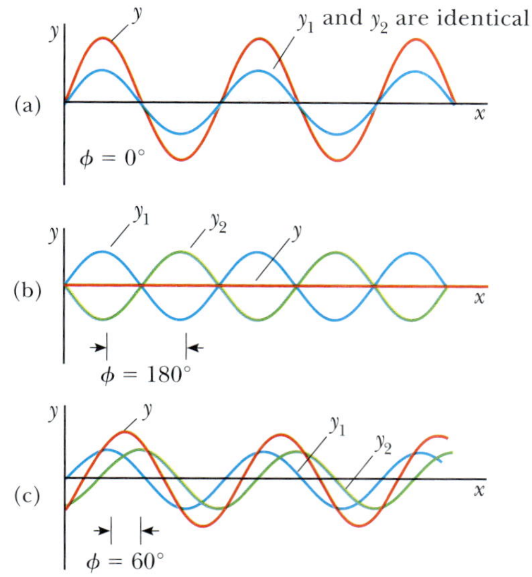

ACTIVE FIGURE 18.3

The superposition of two identical waves y_1 and y_2 (blue and green, respectively) to yield a resultant wave (red). (a) When y_1 and y_2 are in phase, the result is constructive interference. (b) When y_1 and y_2 are π rad out of phase, the result is destructive interference. (c) When the phase angle has a value other than 0 or π rad, the resultant wave y falls somewhere between the extremes shown in (a) and (b).

Sign in at www.thomsonedu.com and go to ThomsonNOW to change the phase relationship between the waves and observe the wave representing the superposition.

and y_2 occur at the same positions and combine to form the red curve y of amplitude $2A$ shown in Active Figure 18.3a. Because the individual waves are in phase, they are indistinguishable in Active Figure 18.3a, in which they appear as a single blue curve. In general, constructive interference occurs when $\cos(\phi/2) = \pm 1$. That is true, for example, when $\phi = 0, 2\pi, 4\pi, \ldots$ rad, that is, when ϕ is an *even* multiple of π.

When ϕ is equal to π rad or to any *odd* multiple of π, then $\cos(\phi/2) = \cos(\pi/2) = 0$ and the crests of one wave occur at the same positions as the troughs of the second wave (Active Fig. 18.3b). Therefore, as a consequence of destructive interference, the resultant wave has *zero* amplitude everywhere. Finally, when the phase constant has an arbitrary value other than 0 or an integer multiple of π rad (Active Fig. 18.3c), the resultant wave has an amplitude whose value is somewhere between 0 and $2A$.

In the more general case in which the waves have the same wavelength but different amplitudes, the results are similar with the following exceptions. In the in-phase case, the amplitude of the resultant wave is not twice that of a single wave, but rather is the sum of the amplitudes of the two waves. When the waves are π rad out of phase, they do not completely cancel as in Active Figure 18.3b. The result is a wave whose amplitude is the difference in the amplitudes of the individual waves.

Interference of Sound Waves

One simple device for demonstrating interference of sound waves is illustrated in Figure 18.4. Sound from a loudspeaker S is sent into a tube at point P, where there is a T-shaped junction. Half the sound energy travels in one direction, and half travels in the opposite direction. Therefore, the sound waves that reach the receiver R can travel along either of the two paths. The distance along any path from speaker to receiver is called the **path length** r. The lower path length r_1 is fixed, but the upper path length r_2 can be varied by sliding the U-shaped tube, which is similar to that on a slide trombone. When the difference in the path lengths $\Delta r = |r_2 - r_1|$ is either zero or some integer multiple of the wavelength λ (that is, $\Delta r = n\lambda$, where $n = 0, 1, 2, 3, \ldots$), the two waves reaching the receiver at

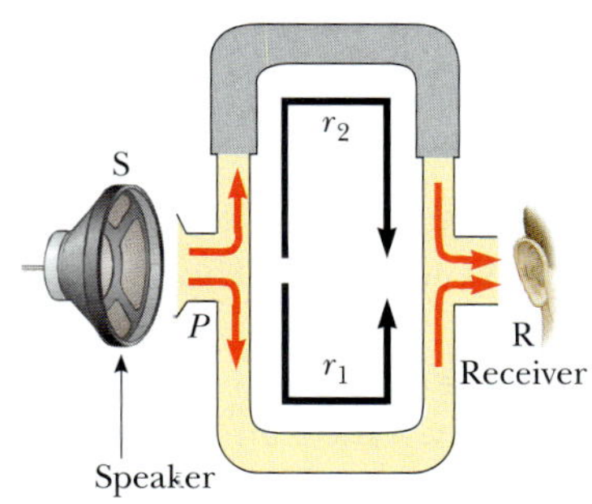

Figure 18.4 An acoustical system for demonstrating interference of sound waves. A sound wave from the speaker (S) propagates into the tube and splits into two parts at point P. The two waves, which combine at the opposite side, are detected at the receiver (R). The upper path length r_2 can be varied by sliding the upper section.

any instant are in phase and interfere constructively as shown in Active Figure 18.3a. For this case, a maximum in the sound intensity is detected at the receiver. If the path length r_2 is adjusted such that the path difference $\Delta r = \lambda/2, 3\lambda/2, \ldots, n\lambda/2$ (for n odd), the two waves are exactly π rad, or 180°, out of phase at the receiver and hence cancel each other. In this case of destructive interference, no sound is detected at the receiver. This simple experiment demonstrates that a phase difference may arise between two waves generated by the same source when they travel along paths of unequal lengths. This important phenomenon will be indispensable in our investigation of the interference of light waves in Chapter 37.

EXAMPLE 18.1 Two Speakers Driven by the Same Source

Two identical loudspeakers placed 3.00 m apart are driven by the same oscillator (Fig. 18.5). A listener is originally at point O, located 8.00 m from the center of the line connecting the two speakers. The listener then moves to point P, which is a perpendicular distance 0.350 m from O, and she experiences the *first minimum* in sound intensity. What is the frequency of the oscillator?

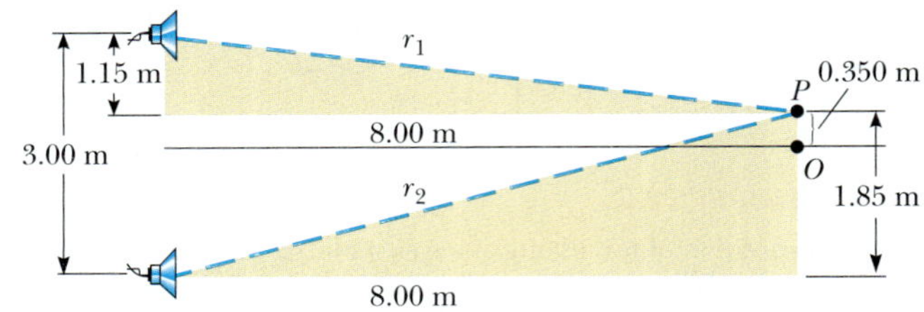

Figure 18.5 (Example 18.1) Two identical loudspeakers emit sound waves to a listener at P.

SOLUTION

Conceptualize In Figure 18.4, a sound wave enters a tube and is then *acoustically* split into two different paths before recombining at the other end. In this example, a signal representing the sound is *electrically* split and sent to two different loudspeakers. After leaving the speakers, the sound waves recombine at the position of the listener. Despite the difference in how the splitting occurs, the path difference discussion related to Figure 18.4 can be applied here.

Categorize Because the sound waves from two separate sources combine, we apply the waves in interference analysis model.

Analyze Figure 18.5 shows the physical arrangement of the speakers, along with two shaded right triangles that can be drawn on the basis of the lengths described in the problem. The first minimum occurs when the two waves reaching the listener at point P are 180° out of phase, in other words, when their path difference Δr equals $\lambda/2$.

From the shaded triangles, find the path lengths from the speakers to the listener:

$$r_1 = \sqrt{(8.00\text{ m})^2 + (1.15\text{ m})^2} = 8.08\text{ m}$$

$$r_2 = \sqrt{(8.00\text{ m})^2 + (1.85\text{ m})^2} = 8.21\text{ m}$$

Hence, the path difference is $r_2 - r_1 = 0.13$ m. Because this path difference must equal $\lambda/2$ for the first minimum, $\lambda = 0.26$ m.

To obtain the oscillator frequency, use Equation 16.12, $v = \lambda f$, where v is the speed of sound in air, 343 m/s:

$$f = \frac{v}{\lambda} = \frac{343\text{ m/s}}{0.26\text{ m}} = \boxed{1.3\text{ kHz}}$$

Finalize This example enables us to understand why the speaker wires in a stereo system should be connected properly. When connected the wrong way—that is, when the positive (or red) wire is connected to the negative (or black) terminal on one of the speakers and the other is correctly wired—the speakers are said to be "out of phase," with one speaker moving outward while the other moves inward. As a consequence, the sound wave coming from one speaker destructively interferes with the wave coming from the other at point O in Figure 18.5. A rarefaction region due to one speaker is superposed on a compression region from the other speaker. Although the two sounds probably do not completely cancel each other (because the left and right stereo signals are usually not identical), a substantial loss of sound quality occurs at point O.

What If? What if the speakers were connected out of phase? What happens at point P in Figure 18.5?

Answer In this situation, the path difference of $\lambda/2$ combines with a phase difference of $\lambda/2$ due to the incorrect wiring to give a full phase difference of λ. As a result, the waves are in phase and there is a *maximum* intensity at point *P*.

18.2 Standing Waves

The sound waves from the pair of loudspeakers in Example 18.1 leave the speakers in the forward direction, and we considered interference at a point in front of the speakers. Suppose we turn the speakers so that they face each other and then have them emit sound of the same frequency and amplitude. In this situation, two identical waves travel in opposite directions in the same medium as in Figure 18.6. These waves combine in accordance with the waves in interference model.

Figure 18.6 Two identical loudspeakers emit sound waves toward each other. When they overlap, identical waves traveling in opposite directions will combine to form standing waves.

We can analyze such a situation by considering wave functions for two transverse sinusoidal waves having the same amplitude, frequency, and wavelength but traveling in opposite directions in the same medium:

$$y_1 = A \sin (kx - \omega t) \qquad y_2 = A \sin (kx + \omega t)$$

where y_1 represents a wave traveling in the $+x$ direction and y_2 represents one traveling in the $-x$ direction. Adding these two functions gives the resultant wave function y:

$$y = y_1 + y_2 = A \sin (kx - \omega t) + A \sin (kx + \omega t)$$

When we use the trigonometric identity $\sin (a \pm b) = \sin (a) \cos (b) \pm \cos (a) \sin (b)$, this expression reduces to

$$y = (2A \sin kx) \cos \omega t \tag{18.1}$$

Equation 18.1 represents the wave function of a **standing wave**. A standing wave such as the one on a string shown in Figure 18.7 is an oscillation pattern *with a stationary outline* that results from the superposition of two identical waves traveling in opposite directions.

Notice that Equation 18.1 does not contain a function of $kx - \omega t$. Therefore, it is not an expression for a single traveling wave. When you observe a standing wave, there is no sense of motion in the direction of propagation of either original wave. Comparing Equation 18.1 with Equation 15.6, we see that it describes a special kind of simple harmonic motion. Every element of the medium oscillates in simple harmonic motion with the same angular frequency ω (according to the $\cos \omega t$ factor in the equation). The amplitude of the simple harmonic motion of a given

PITFALL PREVENTION 18.2

Three Types of Amplitude

We need to distinguish carefully here between the **amplitude of the individual waves**, which is A, and the **amplitude of the simple harmonic motion of the elements of the medium**, which is $2A \sin kx$. A given element in a standing wave vibrates within the constraints of the *envelope* function $2A \sin kx$, where x is that element's position in the medium. Such vibration is in contrast to traveling sinusoidal waves, in which all elements oscillate with the same amplitude and the same frequency and the amplitude A of the wave is the same as the amplitude A of the simple harmonic motion of the elements. Furthermore, we can identify the **amplitude of the standing wave** as $2A$.

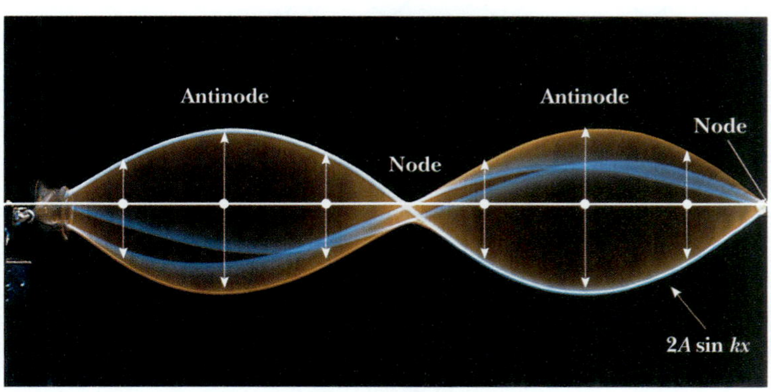

Figure 18.7 Multiflash photograph of a standing wave on a string. The time behavior of the vertical displacement from equilibrium of an individual element of the string is given by $\cos \omega t$. That is, each element vibrates at an angular frequency ω. The amplitude of the vertical oscillation of any element of the string depends on the horizontal position of the element. Each element vibrates within the confines of the envelope function $2A \sin kx$.

element (given by the factor $2A \sin kx$, the coefficient of the cosine function) depends on the location x of the element in the medium, however.

The amplitude of the simple harmonic motion of an element of the medium has a minimum value of zero when x satisfies the condition $\sin kx = 0$, that is, when

$$kx = 0, \pi, 2\pi, 3\pi, \ldots$$

Because $k = 2\pi/\lambda$, these values for kx give

Positions of nodes ►

$$x = 0, \frac{\lambda}{2}, \lambda, \frac{3\lambda}{2}, \ldots = \frac{n\lambda}{2} \quad n = 0, 1, 2, 3, \ldots \tag{18.2}$$

These points of zero amplitude are called **nodes**.

The element of the medium with the *greatest* possible displacement from equilibrium has an amplitude of $2A$, which we define as the amplitude of the standing wave. The positions in the medium at which this maximum displacement occurs are called **antinodes**. The antinodes are located at positions for which the coordinate x satisfies the condition $\sin kx = \pm 1$, that is, when

$$kx = \frac{\pi}{2}, \frac{3\pi}{2}, \frac{5\pi}{2}, \ldots$$

Therefore, the positions of the antinodes are given by

Positions of antinodes ►

$$x = \frac{\lambda}{4}, \frac{3\lambda}{4}, \frac{5\lambda}{4}, \ldots = \frac{n\lambda}{4} \quad n = 1, 3, 5, \ldots \tag{18.3}$$

Two nodes and two antinodes are labeled in the standing wave in Figure 18.7. The light blue curve labeled $2A \sin kx$ in Figure 18.7 represents one wavelength of the traveling waves that combine to form the standing wave. Figure 18.7 and Equations 18.2 and 18.3 provide the following important features of the locations of nodes and antinodes:

The distance between adjacent antinodes is equal to $\lambda/2$.
The distance between adjacent nodes is equal to $\lambda/2$.
The distance between a node and an adjacent antinode is $\lambda/4$.

Wave patterns of the elements of the medium produced at various times by two transverse traveling waves moving in opposite directions are shown in Active Figure 18.8. The blue and green curves are the wave patterns for the individual travel-

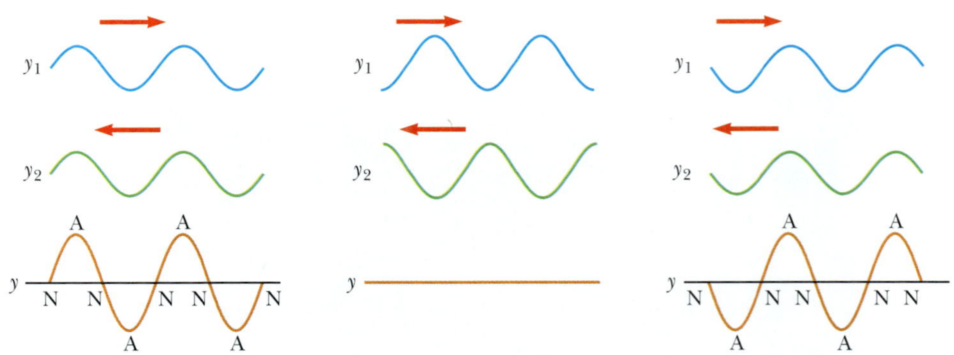

(a) $t = 0$ (b) $t = T/4$ (c) $t = T/2$

ACTIVE FIGURE 18.8

Standing-wave patterns produced at various times by two waves of equal amplitude traveling in opposite directions. For the resultant wave y, the nodes (N) are points of zero displacement and the antinodes (A) are points of maximum displacement.

Sign in at www.thomsonedu.com and go to ThomsonNOW to choose the wavelength of the waves and see the standing wave that results.

ing waves, and the brown curves are the wave patterns for the resultant standing wave. At $t = 0$ (Active Fig. 18.8a), the two traveling waves are in phase, giving a wave pattern in which each element of the medium is at rest and experiencing its maximum displacement from equilibrium. One quarter of a period later, at $t = T/4$ (Active Fig. 18.8b), the traveling waves have moved one quarter of a wavelength (one to the right and the other to the left). At this time, the traveling waves are out of phase, and each element of the medium is passing through the equilibrium position in its simple harmonic motion. The result is zero displacement for elements at all values of x; that is, the wave pattern is a straight line. At $t = T/2$ (Active Fig. 18.8c), the traveling waves are again in phase, producing a wave pattern that is inverted relative to the $t = 0$ pattern. In the standing wave, the elements of the medium alternate in time between the extremes shown in Active Figure 18.8a and c.

Quick Quiz 18.2 Consider a standing wave on a string as shown in Active Figure 18.8. Define the velocity of elements of the string as positive if they are moving upward in the figure. **(i)** At the moment the string has the shape shown by the brown curve in Active Figure 18.8a, what is the instantaneous velocity of elements along the string? (a) zero for all elements (b) positive for all elements (c) negative for all elements (d) varies with the position of the element **(ii)** From the same choices, at the moment the string has the shape shown by the brown curve in Active Figure 18.8b, what is the instantaneous velocity of elements along the string?

EXAMPLE 18.2 Formation of a Standing Wave

Two waves traveling in opposite directions produce a standing wave. The individual wave functions are

$$y_1 = (4.0 \text{ cm}) \sin (3.0x - 2.0t)$$

$$y_2 = (4.0 \text{ cm}) \sin (3.0x + 2.0t)$$

where x and y are measured in centimeters.

(A) Find the amplitude of the simple harmonic motion of the element of the medium located at $x = 2.3$ cm.

SOLUTION

Conceptualize The waves described by the given equations are identical except for their directions of travel, so they indeed combine to form a standing wave as discussed in this section.

Categorize We will substitute values into equations developed in this section, so we categorize this example as a substitution problem.

From the equations for the waves, we see that $A = 4.0$ cm, $k = 3.0$ rad/cm, and $\omega = 2.0$ rad/s. Use Equation 18.1 to write an expression for the standing wave:

$$y = (2A \sin kx) \cos \omega t = [(8.0 \text{ cm}) \sin 3.0x] \cos 2.0t$$

Find the amplitude of the simple harmonic motion of the element at the position $x = 2.3$ cm by evaluating the coefficient of the cosine function at this position:

$$y_{\text{max}} = (8.0 \text{ cm}) \sin 3.0x \big|_{x=2.3}$$

$$= (8.0 \text{ cm}) \sin (6.9 \text{ rad}) = \boxed{4.6 \text{ cm}}$$

(B) Find the positions of the nodes and antinodes if one end of the string is at $x = 0$.

SOLUTION

Find the wavelength of the traveling waves:

$$k = \frac{2\pi}{\lambda} = 3.0 \text{ rad/cm} \quad \rightarrow \quad \lambda = \frac{2\pi}{3.0} \text{ cm}$$

Use Equation 18.2 to find the locations of the nodes: $x = n\frac{\lambda}{2} = n\left(\frac{\pi}{3}\right)$ cm $\quad n = 0, 1, 2, 3, \ldots$

Use Equation 18.3 to find the locations of the antinodes: $x = n\frac{\lambda}{4} = n\left(\frac{\pi}{6}\right)$ cm $\quad n = 1, 3, 5, 7, \ldots$

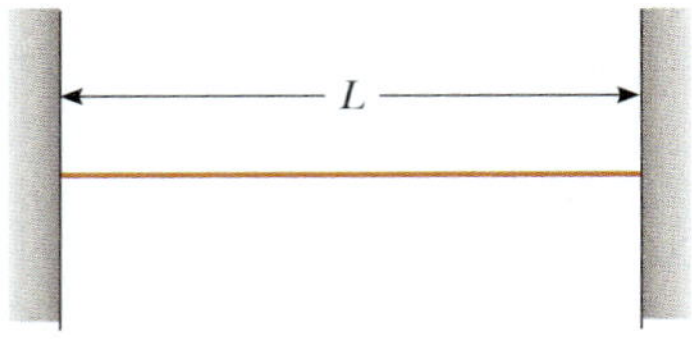

Figure 18.9 A string of length L fixed at both ends.

18.3 Standing Waves in a String Fixed at Both Ends

Consider a string of length L fixed at both ends as shown in Figure 18.9. We will use this system as a model for a guitar string or piano string. Standing waves can be set up in the string by a continuous superposition of waves incident on and reflected from the ends. Notice that there is a boundary condition for the waves on the string. Because the ends of the string are fixed, they must necessarily have zero displacement and are therefore nodes by definition. This boundary condition results in the string having a number of discrete natural patterns of oscillation, called **normal modes**, each of which has a characteristic frequency that is easily calculated. This situation in which only certain frequencies of oscillation are allowed is called **quantization**. Quantization is a common occurrence when waves are subject to boundary conditions and is a central feature in our discussions of quantum physics in the extended version of this text. Notice in Active Figure 18.8 that there are no boundary conditions, so standing waves of *any* frequency can be established; there is no quantization without boundary conditions. Because boundary conditions occur so often for waves, we identify an analysis model called the **waves under boundary conditions model** for the discussion that follows.

The normal modes of oscillation for the string in Figure 18.9 can be described by imposing the boundary conditions that the ends be nodes and that the nodes and antinodes be separated by one-fourth of a wavelength. The first normal mode that is consistent with these requirements, shown in Active Figure 18.10a, has nodes at its ends and one antinode in the middle. This normal mode is the longest-wavelength mode that is consistent with our boundary conditions. The first normal mode occurs when the wavelength λ_1 is equal to twice the length of the string, or $\lambda_1 = 2L$. The section of a standing wave from one node to the next node is called a *loop*. In the first normal mode, the string is vibrating in one loop. In the second normal mode (see Active Fig. 18.10b), the string vibrates in two loops. In this case, the wavelength λ_2 is equal to the length of the string, as expressed by $\lambda_2 = L$. The third normal mode (see Active Fig. 18.10c) corresponds to the case in which $\lambda_3 = 2L/3$, and our string vibrates in three loops. In general, the wavelengths of the various normal modes for a string of length L fixed at both ends are

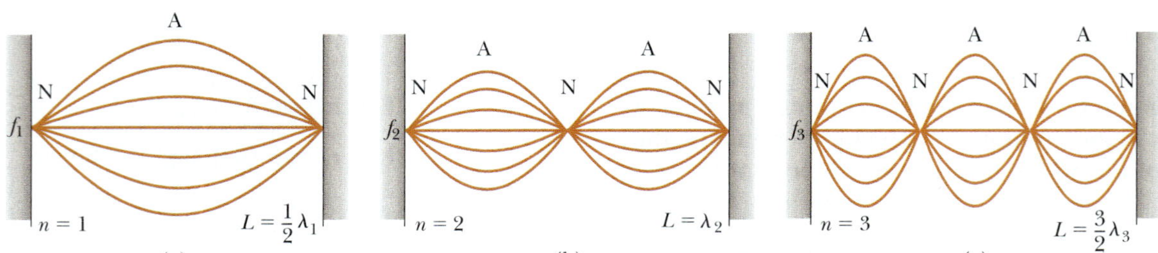

ACTIVE FIGURE 18.10

The normal modes of vibration of the string in Figure 18.9 form a harmonic series: (a) the fundamental, or first harmonic; (b) the second harmonic; (c) the third harmonic.

Sign in at www.thomsonedu.com and go to ThomsonNOW to choose the mode number and see the corresponding standing wave.

$$\lambda_n = \frac{2L}{n} \qquad n = 1, 2, 3, \ldots \qquad (18.4)$$

◀ Wavelengths of normal modes

where the index n refers to the nth normal mode of oscillation. These nodes are the *possible* modes of oscillation for the string. The *actual* modes that are excited on a string are discussed shortly.

The natural frequencies associated with the modes of oscillation are obtained from the relationship $f = v/\lambda$, where the wave speed v is the same for all frequencies. Using Equation 18.4, we find that the natural frequencies f_n of the normal modes are

$$f_n = \frac{v}{\lambda_n} = n\frac{v}{2L} \qquad n = 1, 2, 3, \ldots \qquad (18.5)$$

◀ Frequencies of normal modes as functions of wave speed and length of string

These natural frequencies are also called the *quantized frequencies* associated with the vibrating string fixed at both ends.

Because $v = \sqrt{T/\mu}$ (see Eq. 16.18) for waves on a string, where T is the tension in the string and μ is its linear mass density, we can also express the natural frequencies of a taut string as

$$f_n = \frac{n}{2L}\sqrt{\frac{T}{\mu}} \qquad n = 1, 2, 3, \ldots \qquad (18.6)$$

◀ Frequencies of normal modes as functions of string tension and linear mass density

The lowest frequency f_1, which corresponds to $n = 1$, is called either the **fundamental** or the **fundamental frequency** and is given by

$$f_1 = \frac{1}{2L}\sqrt{\frac{T}{\mu}} \qquad (18.7)$$

◀ Fundamental frequency of a taut string

The frequencies of the remaining normal modes are integer multiples of the fundamental frequency. Frequencies of normal modes that exhibit an integer-multiple relationship such as this form a **harmonic series**, and the normal modes are called **harmonics**. The fundamental frequency f_1 is the frequency of the first harmonic, the frequency $f_2 = 2f_1$ is the frequency of the second harmonic, and the frequency $f_n = nf_1$ is the frequency of the nth harmonic. Other oscillating systems, such as a drumhead, exhibit normal modes, but the frequencies are not related as integer multiples of a fundamental (see Section 18.6). Therefore, we do not use the term *harmonic* in association with these types of systems.

Let us examine further how the various harmonics are created in a string. To excite only a single harmonic, the string must be distorted into a shape that corresponds to that of the desired harmonic. After being released, the string vibrates at the frequency of that harmonic. This maneuver is difficult to perform, however, and is not how a string of a musical instrument is excited. If the string is distorted such that its shape is not that of just one harmonic, the resulting vibration includes a combination of various harmonics. Such a distortion occurs in musical instruments when the string is plucked (as in a guitar), bowed (as in a cello), or struck (as in a piano). When the string is distorted into a nonsinusoidal shape, only waves that satisfy the boundary conditions can persist on the string. These waves are the harmonics.

The frequency of a string that defines the musical note that it plays is that of the fundamental. The string's frequency can be varied by changing either the string's tension or its length. For example, the tension in guitar and violin strings is varied by a screw adjustment mechanism or by tuning pegs located on the neck of the instrument. As the tension is increased, the frequency of the normal modes increases in accordance with Equation 18.6. Once the instrument is "tuned," players vary the frequency by moving their fingers along the neck, thereby changing the length of the oscillating portion of the string. As the length is shortened, the frequency increases because, as Equation 18.6 specifies, the normal-mode frequencies are inversely proportional to string length.

Quick Quiz 18.3 When a standing wave is set up on a string fixed at both ends, which of the following statements is true? (a) The number of nodes is equal to the number of antinodes. (b) The wavelength is equal to the length of the string divided by an integer. (c) The frequency is equal to the number of nodes times the fundamental frequency. (d) The shape of the string at any instant shows a symmetry about the midpoint of the string.

EXAMPLE 18.3 Give Me a C Note!

Middle C on a piano has a fundamental frequency of 262 Hz, and the first A above middle C has a fundamental frequency of 440 Hz.

(A) Calculate the frequencies of the next two harmonics of the C string.

SOLUTION

Conceptualize Remember that the harmonics of a vibrating string have frequencies that are related by integer multiples of the fundamental.

Categorize This first part of the example is a simple substitution problem.

Knowing that the fundamental frequency is $f_1 = 262$ Hz, find the frequencies of the next harmonics by multiplying by integers:

$$f_2 = 2f_1 = \boxed{524 \text{ Hz}}$$

$$f_3 = 3f_1 = \boxed{786 \text{ Hz}}$$

(B) If the A and C strings have the same linear mass density μ and length L, determine the ratio of tensions in the two strings.

SOLUTION

Categorize This part of the example is more of an analysis problem than is part (A).

Analyze Use Equation 18.7 to write expressions for the fundamental frequencies of the two strings:

$$f_{1\text{A}} = \frac{1}{2L}\sqrt{\frac{T_\text{A}}{\mu}} \quad \text{and} \quad f_{1\text{C}} = \frac{1}{2L}\sqrt{\frac{T_\text{C}}{\mu}}$$

Divide the first equation by the second and solve for the ratio of tensions:

$$\frac{f_{1\text{A}}}{f_{1\text{C}}} = \sqrt{\frac{T_\text{A}}{T_\text{C}}} \rightarrow \frac{T_\text{A}}{T_\text{C}} = \left(\frac{f_{1\text{A}}}{f_{1\text{C}}}\right)^2 = \left(\frac{440}{262}\right)^2 = \boxed{2.82}$$

Finalize If the frequencies of piano strings were determined solely by tension, this result suggests that the ratio of tensions from the lowest string to the highest string on the piano would be enormous. Such large tensions would make it difficult to design a frame to support the strings. In reality, the frequencies of piano strings vary due to additional parameters, including the mass per unit length and the length of the string. The **What If?** below explores a variation in length.

What If? If you look inside a real piano, you'll see that the assumption made in part (B) is only partially true. The strings are not likely to have the same length. The string densities are equal, but suppose the length of the A string is only 64% of the length of the C string. What is the ratio of their tensions?

Answer Using Equation 18.7 again, we set up the ratio of frequencies:

$$\frac{f_{1\text{A}}}{f_{1\text{C}}} = \frac{L_\text{C}}{L_\text{A}}\sqrt{\frac{T_\text{A}}{T_\text{C}}} \rightarrow \frac{T_\text{A}}{T_\text{C}} = \left(\frac{L_\text{A}}{L_\text{C}}\right)^2\left(\frac{f_{1\text{A}}}{f_{1\text{C}}}\right)^2$$

$$\frac{T_\text{A}}{T_\text{C}} = (0.64)^2\left(\frac{440}{262}\right)^2 = 1.16$$

Notice that this result represents only a 16% increase in tension, compared with the 182% increase in part (B).

EXAMPLE 18.4 Changing String Vibration with Water

One end of a horizontal string is attached to a vibrating blade, and the other end passes over a pulley as in Figure 18.11a. A sphere of mass 2.00 kg hangs on the end of the string. The string is vibrating in its second harmonic. A container of water is raised under the sphere so that the sphere is completely submerged. In this configuration, the string vibrates in its fifth harmonic as shown in Figure 18.11b. What is the radius of the sphere?

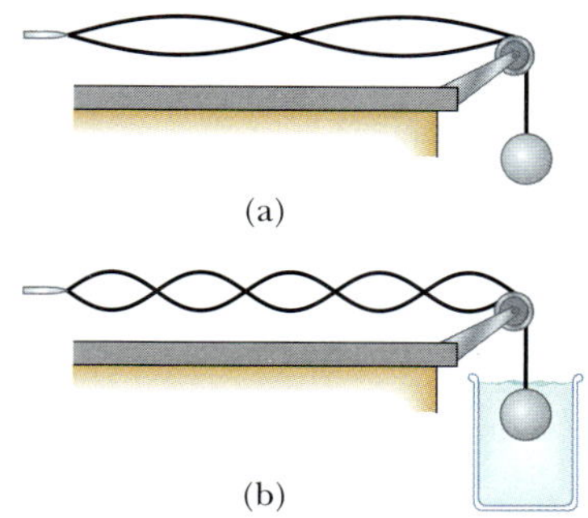

Figure 18.11 (Example 18.4) (a) When the sphere hangs in air, the string vibrates in its second harmonic. (b) When the sphere is immersed in water, the string vibrates in its fifth harmonic.

SOLUTION

Conceptualize Imagine what happens when the sphere is immersed in the water. The buoyant force acts upward on the sphere, reducing the tension in the string. The change in tension causes a change in the speed of waves on the string, which in turn causes a change in the wavelength. This altered wavelength results in the string vibrating in its fifth normal mode rather than the second.

Categorize The hanging sphere is modeled as a particle in equilibrium. One of the forces acting on it is the buoyant force from the water. We also apply the waves under boundary conditions model to the string.

Analyze Apply the particle in equilibrium model to the sphere in Figure 18.11a, identifying T_1 as the tension in the string as the sphere hangs in air:

$$\sum F = T_1 - mg = 0$$

$$T_1 = mg = (2.00\text{ kg})(9.80\text{ m/s}^2) = 19.6\text{ N}$$

Apply the particle in equilibrium model to the sphere in Figure 18.11b, where T_2 is the tension in the string as the sphere is immersed in water:

$$T_2 + B - mg = 0$$

$$(1)\quad B = mg - T_2$$

The desired quantity, the radius of the sphere, will appear in the expression for the buoyant force B. Before proceeding in this direction, however, we must evaluate T_2 from the information about the standing wave.

Write the equation for the frequency of a standing wave on a string (Eq. 18.6) twice, once before the sphere is immersed and once after. Notice that the frequency f is the same in both cases because it is determined by the vibrating blade. In addition, the linear mass density μ and the length L of the vibrating portion of the string are the same in both cases. Divide the equations:

$$\begin{matrix} f = \dfrac{n_1}{2L}\sqrt{\dfrac{T_1}{\mu}} \\ f = \dfrac{n_2}{2L}\sqrt{\dfrac{T_2}{\mu}} \end{matrix} \quad \rightarrow \quad 1 = \frac{n_1}{n_2}\sqrt{\frac{T_1}{T_2}}$$

Solve for T_2:

$$T_2 = \left(\frac{n_1}{n_2}\right)^2 T_1 = \left(\frac{2}{5}\right)^2 (19.6\text{ N}) = 3.14\text{ N}$$

Substitute this result into Equation (1):

$$B = mg - T_2 = 19.6\text{ N} - 3.14\text{ N} = 16.5\text{ N}$$

Using Equation 14.5, express the buoyant force in terms of the radius of the sphere:

$$B = \rho_{\text{water}} g V_{\text{sphere}} = \rho_{\text{water}} g \left(\tfrac{4}{3}\pi r^3\right)$$

Solve for the radius of the sphere:

$$r = \left(\frac{3B}{4\pi\rho_{\text{water}} g}\right)^{1/3} = \left(\frac{3(16.5\text{ N})}{4\pi(1\,000\text{ kg/m}^3)(9.80\text{ m/s}^2)}\right)^{1/3}$$

$$= 7.38 \times 10^{-2}\text{ m} = 7.38\text{ cm}$$

Finalize Notice that only certain radii of the sphere will result in the string vibrating in a normal mode; the speed of waves on the string must be changed to a value such that the length of the string is an integer multiple of half wavelengths. This limitation is a feature of the *quantization* that was introduced earlier in this chapter: the sphere radii that cause the string to vibrate in a normal mode are *quantized*.

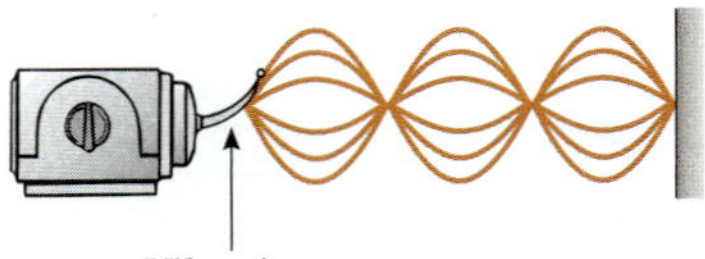

Figure 18.12 Standing waves are set up in a string when one end is connected to a vibrating blade. When the blade vibrates at one of the natural frequencies of the string, large-amplitude standing waves are created.

18.4 Resonance

We have seen that a system such as a taut string is capable of oscillating in one or more normal modes of oscillation. **If a periodic force is applied to such a system, the amplitude of the resulting motion is greatest when the frequency of the applied force is equal to one of the natural frequencies of the system.** This phenomenon, known as *resonance,* was discussed in Section 15.7. Although a block–spring system or a simple pendulum has only one natural frequency, standing-wave systems have a whole set of natural frequencies, such as that given by Equation 18.6 for a string. Because an oscillating system exhibits a large amplitude when driven at any of its natural frequencies, these frequencies are often referred to as **resonance frequencies.**

Consider a taut string fixed at one end and connected at the opposite end to an oscillating blade as illustrated in Figure 18.12. The fixed end is a node, and the end connected to the blade is very nearly a node because the amplitude of the blade's motion is small compared with that of the elements of the string. As the blade oscillates, transverse waves sent down the string are reflected from the fixed end. As we learned in Section 18.3, the string has natural frequencies that are determined by its length, tension, and linear mass density (see Eq. 18.6). When the frequency of the blade equals one of the natural frequencies of the string, standing waves are produced and the string oscillates with a large amplitude. In this resonance case, the wave generated by the oscillating blade is in phase with the reflected wave and the string absorbs energy from the blade. If the string is driven at a frequency that is not one of its natural frequencies, the oscillations are of low amplitude and exhibit no stable pattern.

Resonance is very important in the excitation of musical instruments based on air columns. We shall discuss this application of resonance in Section 18.5.

18.5 Standing Waves in Air Columns

The waves under boundary conditions model can also be applied to sound waves in a column of air such as that inside an organ pipe. Standing waves are the result of interference between longitudinal sound waves traveling in opposite directions.

In a pipe closed at one end, **the closed end is a displacement node because the rigid barrier at this end does not allow longitudinal motion of the air.** Because the pressure wave is 90° out of phase with the displacement wave (see Section 17.2), **the closed end of an air column corresponds to a pressure antinode** (that is, a point of maximum pressure variation).

The open end of an air column is approximately a displacement antinode[1] and a pressure node. We can understand why no pressure variation occurs at an open end by noting that the end of the air column is open to the atmosphere; therefore, the pressure at this end must remain constant at atmospheric pressure.

You may wonder how a sound wave can reflect from an open end because there may not appear to be a change in the medium at this point: the medium through which the sound wave moves is air both inside and outside the pipe. Sound is a pressure wave, however, and a compression region of the sound wave is constrained by the sides of the pipe as long as the region is inside the pipe. As the compression region exits at the open end of the pipe, the constraint of the pipe is removed and the compressed air is free to expand into the atmosphere. Therefore, there is a change in the *character* of the medium between the inside of the pipe and the outside even though there is no change in the *material* of the medium. This change in character is sufficient to allow some reflection.

[1] Strictly speaking, the open end of an air column is not exactly a displacement antinode. A compression reaching an open end does not reflect until it passes beyond the end. For a tube of circular cross section, an end correction equal to approximately $0.6R$, where R is the tube's radius, must be added to the length of the air column. Hence, the effective length of the air column is longer than the true length L. We ignore this end correction in this discussion.

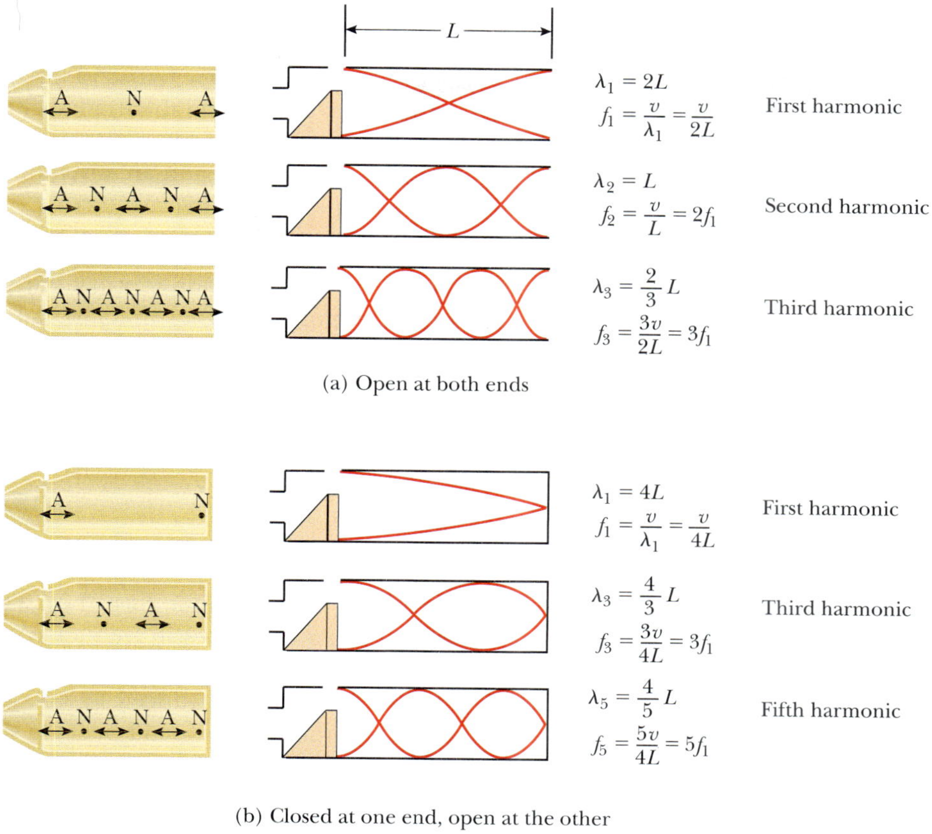

Figure 18.13 Motion of elements of air in standing longitudinal waves in a pipe, along with schematic representations of the waves. In the schematic representations, the structure at the left end has the purpose of exciting the air column into a normal mode. The hole in the upper edge of the column ensures that the left end acts as an open end. The graphs represent the displacement amplitudes, not the pressure amplitudes. (a) In a pipe open at both ends, the harmonic series created consists of all integer multiples of the fundamental frequency: $f_1, 2f_1, 3f_1, \ldots$. (b) In a pipe closed at one end and open at the other, the harmonic series created consists of only odd-integer multiples of the fundamental frequency: $f_1, 3f_1, 5f_1, \ldots$.

With the boundary conditions of nodes or antinodes at the ends of the air column, we have a set of normal modes of oscillation as is the case for the string fixed at both ends. Therefore, the air column has quantized frequencies.

The first three normal modes of oscillation of a pipe open at both ends are shown in Figure 18.13a. Notice that both ends are displacement antinodes (approximately). In the first normal mode, the standing wave extends between two adjacent antinodes, which is a distance of half a wavelength. Therefore, the wavelength is twice the length of the pipe, and the fundamental frequency is $f_1 = v/2L$. As Figure 18.13a shows, the frequencies of the higher harmonics are $2f_1, 3f_1, \ldots$.

In a pipe open at both ends, the natural frequencies of oscillation form a harmonic series that includes all integral multiples of the fundamental frequency.

Because all harmonics are present and because the fundamental frequency is given by the same expression as that for a string (see Eq. 18.5), we can express the natural frequencies of oscillation as

$$f_n = n\frac{v}{2L} \qquad n = 1, 2, 3, \ldots \qquad \textbf{(18.8)}$$

◀ Natural frequencies of a pipe open at both ends

Despite the similarity between Equations 18.5 and 18.8, you must remember that v in Equation 18.5 is the speed of waves on the string, whereas v in Equation 18.8 is the speed of sound in air.

PITFALL PREVENTION 18.3

Sound Waves in Air Are Longitudinal, Not Transverse

The standing longitudinal waves are drawn as transverse waves in Figure 18.13. Because they are in the same direction as the propagation, it is difficult to draw longitudinal displacements. Therefore, it is best to interpret the red curves in Figure 18.13 as a graphical representation of the waves (our diagrams of string waves are pictorial representations), with the vertical axis representing horizontal displacement of the elements of the medium.

If a pipe is closed at one end and open at the other, the closed end is a displacement node (see Fig. 18.13b). In this case, the standing wave for the fundamental mode extends from an antinode to the adjacent node, which is one-fourth of a wavelength. Hence, the wavelength for the first normal mode is $4L$, and the fundamental frequency is $f_1 = v/4L$. As Figure 18.13b shows, the higher-frequency waves that satisfy our conditions are those that have a node at the closed end and an antinode at the open end; hence, the higher harmonics have frequencies $3f_1, 5f_1, \ldots$.

In a pipe closed at one end, the natural frequencies of oscillation form a harmonic series that includes only odd integral multiples of the fundamental frequency.

We express this result mathematically as

Natural frequencies of a pipe closed at one end and open at the other ▶

$$f_n = n\frac{v}{4L} \qquad n = 1, 3, 5, \ldots \tag{18.9}$$

It is interesting to investigate what happens to the frequencies of instruments based on air columns and strings during a concert as the temperature rises. The sound emitted by a flute, for example, becomes sharp (increases in frequency) as the flute warms up because the speed of sound increases in the increasingly warmer air inside the flute (consider Eq. 18.8). The sound produced by a violin becomes flat (decreases in frequency) as the strings thermally expand because the expansion causes their tension to decrease (see Eq. 18.6).

Musical instruments based on air columns are generally excited by resonance. The air column is presented with a sound wave that is rich in many frequencies. The air column then responds with a large-amplitude oscillation to the frequencies that match the quantized frequencies in its set of harmonics. In many woodwind instruments, the initial rich sound is provided by a vibrating reed. In brass instruments, this excitation is provided by the sound coming from the vibration of the player's lips. In a flute, the initial excitation comes from blowing over an edge at the mouthpiece of the instrument in a manner similar to blowing across the opening of a bottle with a narrow neck. The sound of the air rushing across the edge has many frequencies, including one that sets the air cavity in the bottle into resonance.

Quick Quiz 18.4 A pipe open at both ends resonates at a fundamental frequency f_{open}. When one end is covered and the pipe is again made to resonate, the fundamental frequency is f_{closed}. Which of the following expressions describes how these two resonant frequencies compare? (a) $f_{\text{closed}} = f_{\text{open}}$ (b) $f_{\text{closed}} = \frac{1}{2}f_{\text{open}}$ (c) $f_{\text{closed}} = 2f_{\text{open}}$ (d) $f_{\text{closed}} = \frac{3}{2}f_{\text{open}}$

Quick Quiz 18.5 Balboa Park in San Diego has an outdoor organ. When the air temperature increases, the fundamental frequency of one of the organ pipes (a) stays the same, (b) goes down, (c) goes up, or (d) is impossible to determine.

EXAMPLE 18.5 Wind in a Culvert

A section of drainage culvert 1.23 m in length makes a howling noise when the wind blows across its open ends.

(A) Determine the frequencies of the first three harmonics of the culvert if it is cylindrical in shape and open at both ends. Take $v = 343$ m/s as the speed of sound in air.

SOLUTION

Conceptualize The sound of the wind blowing across the end of the pipe contains many frequencies, and the culvert responds to the sound by vibrating at the natural frequencies of the air column.

Categorize This example is a relatively simple substitution problem.

Find the frequency of the first harmonic of the culvert, modeling it as an air column open at both ends:

$$f_1 = \frac{v}{2L} = \frac{343 \text{ m/s}}{2(1.23 \text{ m})} = \boxed{139 \text{ Hz}}$$

Find the next harmonics by multiplying by integers:

$$f_2 = 2f_1 = \boxed{278 \text{ Hz}}$$

$$f_3 = 3f_1 = \boxed{417 \text{ Hz}}$$

(B) What are the three lowest natural frequencies of the culvert if it is blocked at one end?

SOLUTION

Find the frequency of the first harmonic of the culvert, modeling it as an air column closed at one end:

$$f_1 = \frac{v}{4L} = \frac{343 \text{ m/s}}{4(1.23 \text{ m})} = \boxed{69.7 \text{ Hz}}$$

Find the next two harmonics by multiplying by odd integers:

$$f_3 = 3f_1 = \boxed{209 \text{ Hz}}$$

$$f_5 = 5f_1 = \boxed{349 \text{ Hz}}$$

EXAMPLE 18.6 Measuring the Frequency of a Tuning Fork

A simple apparatus for demonstrating resonance in an air column is depicted in Figure 18.14. A vertical pipe open at both ends is partially submerged in water, and a tuning fork vibrating at an unknown frequency is placed near the top of the pipe. The length L of the air column can be adjusted by moving the pipe vertically. The sound waves generated by the fork are reinforced when L corresponds to one of the resonance frequencies of the pipe. For a certain pipe, the smallest value of L for which a peak occurs in the sound intensity is 9.00 cm.

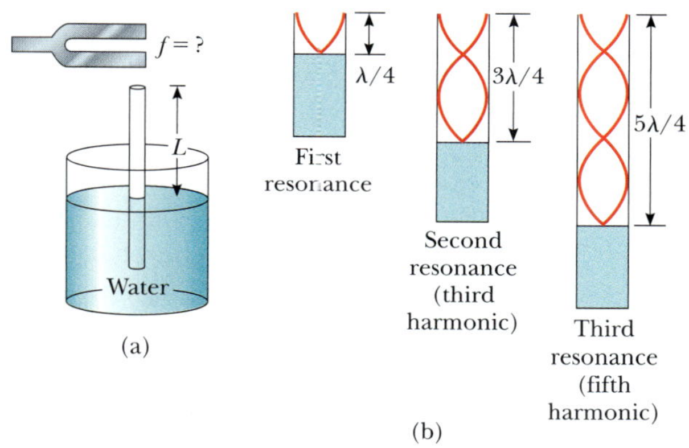

Figure 18.14 (Example 18.6) (a) Apparatus for demonstrating the resonance of sound waves in a pipe closed at one end. The length L of the air column is varied by moving the pipe vertically while it is partially submerged in water. (b) The first three normal modes of the system shown in (a).

(A) What is the frequency of the tuning fork?

SOLUTION

Conceptualize Consider how this problem differs from the preceding example. In the culvert, the length was fixed and the air column was presented with a mixture of very many frequencies. The pipe in this example is presented with one single frequency from the tuning fork, and the length of the pipe is varied until resonance is achieved.

Categorize This example is a simple substitution problem. Although the pipe is open at its lower end to allow the water to enter, the water's surface acts like a barrier. Therefore, this setup can be modeled as an air column closed at one end.

Use Equation 18.9 to find the fundamental frequency for $L = 0.090\,0$ m:

$$f_1 = \frac{v}{4L} = \frac{343 \text{ m/s}}{4(0.090\,0 \text{ m})} = \boxed{953 \text{ Hz}}$$

Because the tuning fork causes the air column to resonate at this frequency, this frequency must also be that of the tuning fork.

(B) What are the values of L for the next two resonance conditions?

SOLUTION

Use Equation 16.12 to find the wavelength of the sound wave from the tuning fork:

$$\lambda = \frac{v}{f} = \frac{343 \text{ m/s}}{953 \text{ Hz}} = 0.360 \text{ m}$$

Notice from Figure 18.14b that the length of the air column for the second resonance is $3\lambda/4$:

$$L = 3\lambda/4 = 0.270 \text{ m}$$

Notice from Figure 18.14b that the length of the air column for the third resonance is $5\lambda/4$:

$$L = 5\lambda/4 = 0.450 \text{ m}$$

18.6 Standing Waves in Rods and Membranes

Standing waves can also be set up in rods and membranes. A rod clamped in the middle and stroked parallel to the rod at one end oscillates as depicted in Figure 18.15a. The oscillations of the elements of the rod are longitudinal, and so the red curves in Figure 18.15 represent *longitudinal* displacements of various parts of the rod. For clarity, the displacements have been drawn in the transverse direction as they were for air columns. The midpoint is a displacement node because it is fixed by the clamp, whereas the ends are displacement antinodes because they are free to oscillate. The oscillations in this setup are analogous to those in a pipe open at both ends. The red lines in Figure 18.15a represent the first normal mode, for which the wavelength is $2L$ and the frequency is $f = v/2L$, where v is the speed of longitudinal waves in the rod. Other normal modes may be excited by clamping the rod at different points. For example, the second normal mode (Fig. 18.15b) is excited by clamping the rod a distance $L/4$ away from one end.

It is also possible to set up transverse standing waves in rods. Musical instruments that depend on transverse standing waves in rods include triangles, marimbas, xylophones, glockenspiels, chimes, and vibraphones. Other devices that make sounds from vibrating bars include music boxes and wind chimes.

Two-dimensional oscillations can be set up in a flexible membrane stretched over a circular hoop such as that in a drumhead. As the membrane is struck at some point, waves that arrive at the fixed boundary are reflected many times. The resulting sound is not harmonic because the standing waves have frequencies that are *not* related by integer multiples. Without this relationship, the sound may be more correctly described as *noise* rather than as music. The production of noise is in contrast to the situation in wind and stringed instruments, which produce sounds that we describe as musical.

Some possible normal modes of oscillation for a two-dimensional circular membrane are shown in Figure 18.16. Whereas nodes are *points* in one-dimensional standing waves on strings and in air columns, a two-dimensional oscillator has *curves* along which there is no displacement of the elements of the medium. The lowest normal mode, which has a frequency f_1, contains only one nodal curve; this curve runs around the outer edge of the membrane. The other possible normal modes show additional nodal curves that are circles and straight lines across the diameter of the membrane.

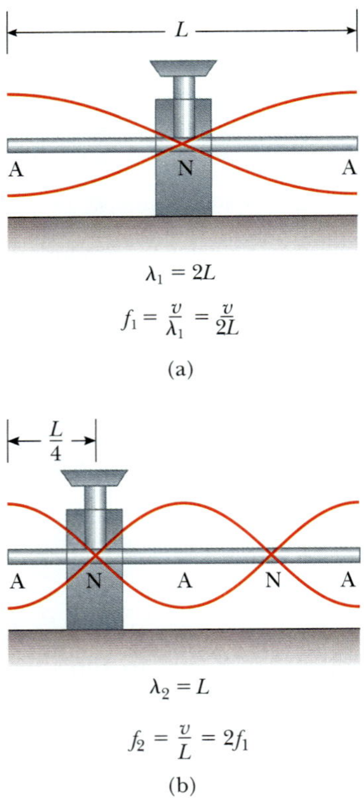

Figure 18.15 Normal-mode longitudinal vibrations of a rod of length L (a) clamped at the middle to produce the first normal mode and (b) clamped at a distance $L/4$ from one end to produce the second normal mode. Notice that the red curves represent oscillations parallel to the rod (longitudinal waves).

18.7 Beats: Interference in Time

The interference phenomena we have studied so far involve the superposition of two or more waves having the same frequency. Because the amplitude of the oscillation of elements of the medium varies with the position in space of the element

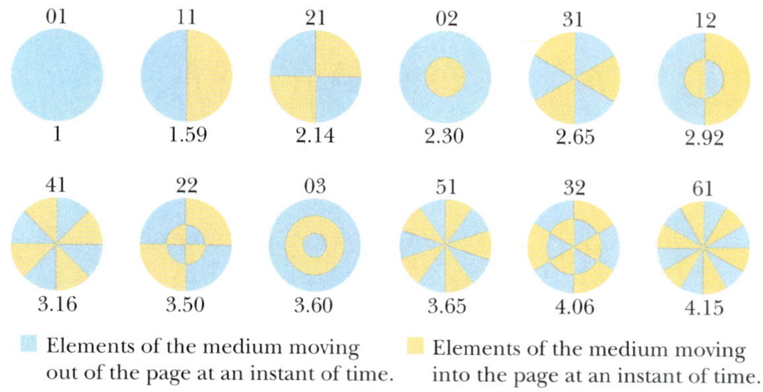

Figure 18.16 Representation of some of the normal modes possible in a circular membrane fixed at its perimeter. The pair of numbers above each pattern corresponds to the number of radial nodes and the number of circular nodes, respectively. Below each pattern is a factor by which the frequency of the mode is larger than that of the 01 mode. The frequencies of oscillation do not form a harmonic series because these factors are not integers. In each diagram, elements of the membrane on either side of a nodal line move in opposite directions, as indicated by the colors. (*Adapted from T. D. Rossing,* The Science of Sound, 2nd ed., *Reading, Massachusetts, Addison-Wesley Publishing Co., 1990*)

in such a wave, we refer to the phenomenon as *spatial interference.* Standing waves in strings and pipes are common examples of spatial interference.

Now let's consider another type of interference, one that results from the superposition of two waves having slightly *different* frequencies. In this case, when the two waves are observed at a point in space, they are periodically in and out of phase. That is, there is a *temporal* (time) alternation between constructive and destructive interference. As a consequence, we refer to this phenomenon as *interference in time* or *temporal interference.* For example, if two tuning forks of slightly different frequencies are struck, one hears a sound of periodically varying amplitude. This phenomenon is called **beating**.

Beating is the periodic variation in amplitude at a given point due to the superposition of two waves having slightly different frequencies.

◀ Definition of beating

The number of amplitude maxima one hears per second, or the *beat frequency,* equals the difference in frequency between the two sources as we shall show below. The maximum beat frequency that the human ear can detect is about 20 beats/s. When the beat frequency exceeds this value, the beats blend indistinguishably with the sounds producing them.

Consider two sound waves of equal amplitude traveling through a medium with slightly different frequencies f_1 and f_2. We use equations similar to Equation 16.10 to represent the wave functions for these two waves at a point that we choose so that $kx = \pi/2$:

$$y_1 = A \sin\left(\frac{\pi}{2} - \omega_1 t\right) = A \cos\left(2\pi f_1 t\right)$$

$$y_2 = A \sin\left(\frac{\pi}{2} - \omega_2 t\right) = A \cos\left(2\pi f_2 t\right)$$

Using the superposition principle, we find that the resultant wave function at this point is

$$y = y_1 + y_2 = A\left(\cos 2\pi f_1 t + \cos 2\pi f_2 t\right)$$

The trigonometric identity

$$\cos a + \cos b = 2 \cos\left(\frac{a-b}{2}\right) \cos\left(\frac{a+b}{2}\right)$$

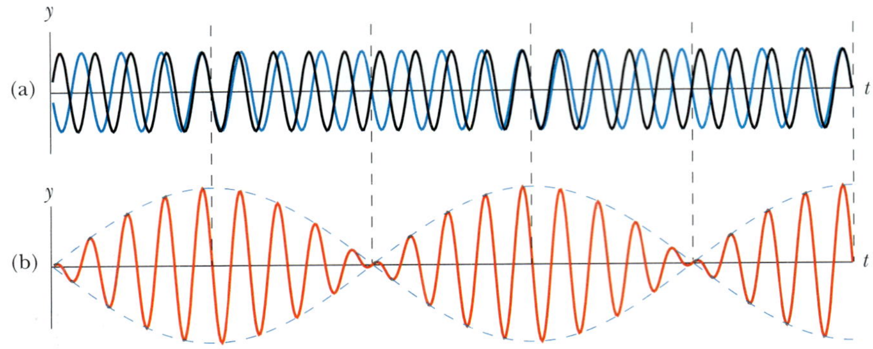

ACTIVE FIGURE 18.17

Beats are formed by the combination of two waves of slightly different frequencies. (a) The individual waves. (b) The combined wave. The envelope wave (dashed line) represents the beating of the combined sounds.

Sign in at www.thomsonedu.com and go to ThomsonNOW to choose the two frequencies and see the corresponding beats.

allows us to write the expression for y as

Resultant of two waves of different frequencies but equal amplitude ▶

$$y = \left[2A\cos 2\pi\left(\frac{f_1 - f_2}{2}\right)t\right]\cos 2\pi\left(\frac{f_1 + f_2}{2}\right)t \qquad \textbf{(18.10)}$$

Graphs of the individual waves and the resultant wave are shown in Active Figure 18.17. From the factors in Equation 18.10, we see that the resultant wave has an effective frequency equal to the average frequency $(f_1 + f_2)/2$. This wave is multiplied by an envelope wave given by the expression in the square brackets:

$$y_{\text{envelope}} = 2A\cos 2\pi\left(\frac{f_1 - f_2}{2}\right)t \qquad \textbf{(18.11)}$$

That is, the **amplitude and therefore the intensity of the resultant sound vary in time**. The dashed blue line in Active Figure 18.17b is a graphical representation of the envelope wave in Equation 18.11 and is a sine wave varying with frequency $(f_1 - f_2)/2$.

A maximum in the amplitude of the resultant sound wave is detected whenever

$$\cos 2\pi\left(\frac{f_1 - f_2}{2}\right)t = \pm 1$$

Hence, there are *two* maxima in each period of the envelope wave. Because the amplitude varies with frequency as $(f_1 - f_2)/2$, the number of beats per second, or the beat frequency f_{beat}, is twice this value. That is,

Beat frequency ▶

$$f_{\text{beat}} = |f_1 - f_2| \qquad \textbf{(18.12)}$$

For instance, if one tuning fork vibrates at 438 Hz and a second one vibrates at 442 Hz, the resultant sound wave of the combination has a frequency of 440 Hz (the musical note A) and a beat frequency of 4 Hz. A listener would hear a 440-Hz sound wave go through an intensity maximum four times every second.

EXAMPLE 18.7 The Mistuned Piano Strings

Two identical piano strings of length 0.750 m are each tuned exactly to 440 Hz. The tension in one of the strings is then increased by 1.0%. If they are now struck, what is the beat frequency between the fundamentals of the two strings?

SOLUTION

Conceptualize As the tension in one of the strings is changed, its fundamental frequency changes. Therefore, when both strings are played, they will have different frequencies and beats will be heard.

Categorize We must combine our understanding of the waves under boundary conditions model for strings with our new knowledge of beats.

Analyze Set up a ratio of the fundamental frequencies of the two strings using Equation 18.5:

$$\frac{f_2}{f_1} = \frac{(v_2/2L)}{(v_1/2L)} = \frac{v_2}{v_1}$$

Use Equation 16.18 to substitute for the wave speeds on the strings:

$$\frac{f_2}{f_1} = \frac{\sqrt{T_2/\mu}}{\sqrt{T_1/\mu}} = \sqrt{\frac{T_2}{T_1}}$$

Incorporate that the tension in one string is 1.0% larger than the other; that is, $T_2 = 1.010T_1$:

$$\frac{f_2}{f_1} = \sqrt{\frac{1.010T_1}{T_1}} = 1.005$$

Solve for the frequency of the tightened string:

$$f_2 = 1.005f_1 = 1.005(440 \text{ Hz}) = 442 \text{ Hz}$$

Find the beat frequency using Equation 18.12:

$$f_{\text{beat}} = 442 \text{ Hz} - 440 \text{ Hz} = 2 \text{ Hz}$$

Finalize Notice that a 1.0% mistuning in tension leads to an easily audible beat frequency of 2 Hz. A piano tuner can use beats to tune a stringed instrument by "beating" a note against a reference tone of known frequency. The tuner can then adjust the string tension until the frequency of the sound it emits equals the frequency of the reference tone. The tuner does so by tightening or loosening the string until the beats produced by it and the reference source become too infrequent to notice.

18.8 Nonsinusoidal Wave Patterns

It is relatively easy to distinguish the sounds coming from a violin and a saxophone even when they are both playing the same note. On the other hand, a person untrained in music may have difficulty distinguishing a note played on a clarinet from the same note played on an oboe. We can use the pattern of the sound waves from various sources to explain these effects.

When frequencies that are integer multiples of a fundamental frequency are combined to make a sound, the result is a *musical* sound. A listener can assign a pitch to the sound based on the fundamental frequency. Pitch is a psychological reaction to a sound that allows the listener to place the sound on a scale of low to high (bass to treble). Combinations of frequencies that are not integer multiples of a fundamental result in a *noise* rather than a musical sound. It is much harder for a listener to assign a pitch to a noise than to a musical sound.

The wave patterns produced by a musical instrument are the result of the superposition of frequencies that are integer multiples of a fundamental. This superposition results in the corresponding richness of musical tones. The human perceptive response associated with various mixtures of harmonics is the *quality* or *timbre* of the sound. For instance, the sound of the trumpet is perceived to have a "brassy" quality (that is, we have learned to associate the adjective *brassy* with that sound); this quality enables us to distinguish the sound of the trumpet from that of the saxophone, whose quality is perceived as "reedy." The clarinet and oboe, however, both contain air columns excited by reeds; because of this similarity, they have similar mixtures of frequencies and it is more difficult for the human ear to distinguish them on the basis of their sound quality.

The sound wave patterns produced by the majority of musical instruments are nonsinusoidal. Characteristic patterns produced by a tuning fork, a flute, and a clarinet, each playing the same note, are shown in Figure 18.18 (page 520). Each instrument has its own characteristic pattern. Notice, however, that despite the

PITFALL PREVENTION 18.4
Pitch Versus Frequency

Do not confuse the term *pitch* with *frequency*. Frequency is the physical measurement of the number of oscillations per second. Pitch is a psychological reaction to sound that enables a person to place the sound on a scale from high to low or from treble to bass. Therefore, frequency is the stimulus and pitch is the response. Although pitch is related mostly (but not completely) to frequency, they are not the same. A phrase such as "the pitch of the sound" is incorrect because pitch is not a physical property of the sound.

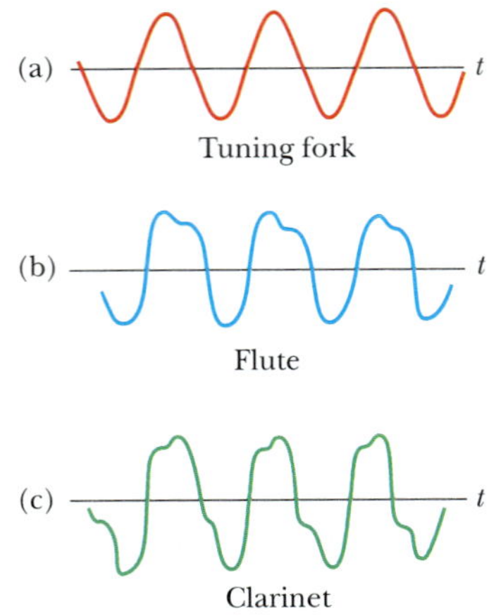

Figure 18.18 Sound wave patterns produced by (a) a tuning fork, (b) a flute, and (c) a clarinet, each at approximately the same frequency.

differences in the patterns, each pattern is periodic. This point is important for our analysis of these waves.

The problem of analyzing nonsinusoidal wave patterns appears at first sight to be a formidable task. If the wave pattern is periodic, however, **it can be represented as closely as desired by the combination of a sufficiently large number of sinusoidal waves that form a harmonic series**. In fact, we can represent any periodic function as a series of sine and cosine terms by using a mathematical technique based on **Fourier's theorem.**[2] The corresponding sum of terms that represents the periodic wave pattern is called a **Fourier series**. Let $y(t)$ be any function that is periodic in time with period T such that $y(t + T) = y(t)$. Fourier's theorem states that this function can be written as

Fourier's theorem ►

$$y(t) = \sum (A_n \sin 2\pi f_n t + B_n \cos 2\pi f_n t) \qquad \textbf{(18.13)}$$

where the lowest frequency is $f_1 = 1/T$. The higher frequencies are integer multiples of the fundamental, $f_n = nf_1$, and the coefficients A_n and B_n represent the amplitudes of the various waves. Figure 18.19 represents a harmonic analysis of the wave patterns shown in Figure 18.18. Each bar in the graph represents one of the terms in the series in Equation 18.13. Notice that a struck tuning fork pro-

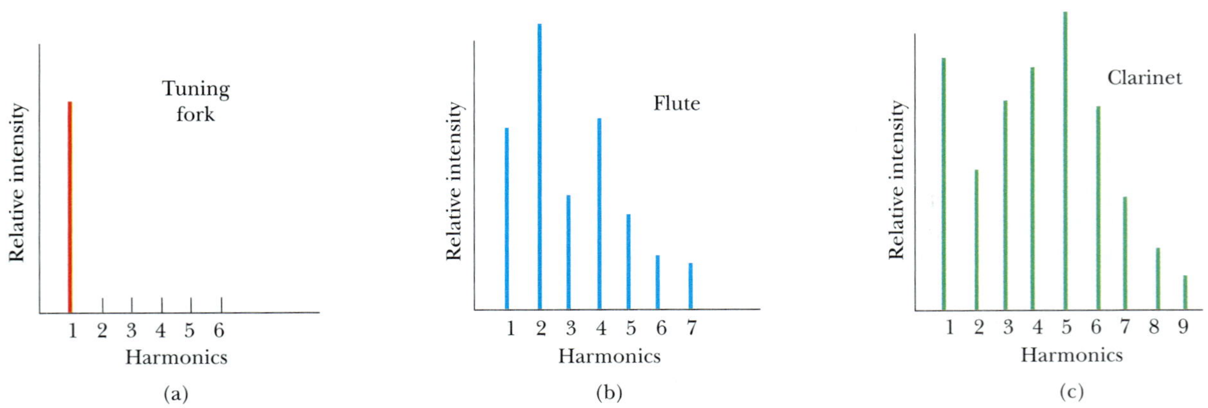

Figure 18.19 Harmonics of the wave patterns shown in Figure 18.18. Notice the variations in intensity of the various harmonics. Parts (a), (b), and (c) correspond to those in Figure 18.18.

[2] Developed by Jean Baptiste Joseph Fourier (1786–1830).

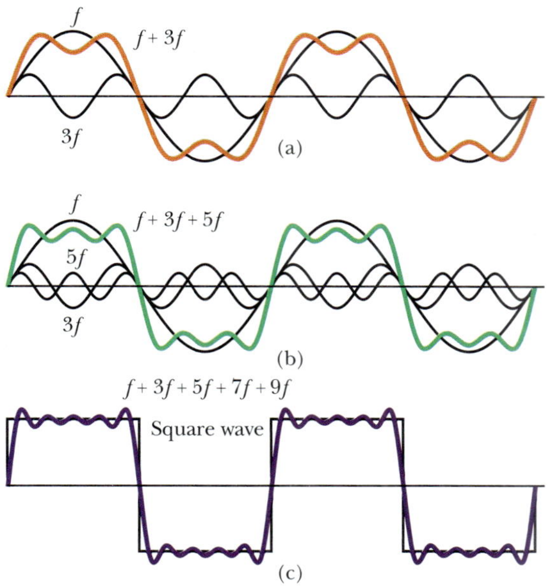

ACTIVE FIGURE 18.20

Fourier synthesis of a square wave, which is represented by the sum of odd multiples of the first harmonic, which has frequency f. (a) Waves of frequency f and $3f$ are added. (b) One more odd harmonic of frequency $5f$ is added. (c) The synthesis curve approaches closer to the square wave when odd frequencies up to $9f$ are added.

Sign in at www.thomsonedu.com and go to ThomsonNOW to add in harmonics with frequencies higher than $9f$ to try to synthesize a square wave.

duces only one harmonic (the first), whereas the flute and clarinet produce the first harmonic and many higher ones.

Notice the variation in relative intensity of the various harmonics for the flute and the clarinet. In general, any musical sound consists of a fundamental frequency f plus other frequencies that are integer multiples of f, all having different intensities.

We have discussed the *analysis* of a wave pattern using Fourier's theorem. The analysis involves determining the coefficients of the harmonics in Equation 18.13 from a knowledge of the wave pattern. The reverse process, called *Fourier synthesis,* can also be performed. In this process, the various harmonics are added together to form a resultant wave pattern. As an example of Fourier synthesis, consider the building of a square wave as shown in Active Figure 18.20. The symmetry of the square wave results in only odd multiples of the fundamental frequency combining in its synthesis. In Active Figure 18.20a, the orange curve shows the combination of f and $3f$. In Active Figure 18.20b, we have added $5f$ to the combination and obtained the green curve. Notice how the general shape of the square wave is approximated, even though the upper and lower portions are not flat as they should be.

Active Figure 18.20c shows the result of adding odd frequencies up to $9f$. This approximation (purple curve) to the square wave is better than the approximations in Active Figures 18.20a and 18.20b. To approximate the square wave as closely as possible, we must add all odd multiples of the fundamental frequency, up to infinite frequency.

Using modern technology, musical sounds can be generated electronically by mixing different amplitudes of any number of harmonics. These widely used electronic music synthesizers are capable of producing an infinite variety of musical tones.

Summary

ThomsonNOW™ Sign in at **www.thomsonedu.com** and go to ThomsonNOW to take a practice test for this chapter.

CONCEPTS AND PRINCIPLES

The **superposition principle** specifies that when two or more waves move through a medium, the value of the resultant wave function equals the algebraic sum of the values of the individual wave functions.

The phenomenon of **beating** is the periodic variation in intensity at a given point due to the superposition of two waves having slightly different frequencies.

Standing waves are formed from the combination of two sinusoidal waves having the same frequency, amplitude, and wavelength but traveling in opposite directions. The resultant standing wave is described by the wave function

$$y = (2A \sin kx) \cos \omega t \qquad \textbf{(18.1)}$$

Hence, the amplitude of the standing wave is $2A$, and the amplitude of the simple harmonic motion of any particle of the medium varies according to its position as $2A \sin kx$. The points of zero amplitude (called **nodes**) occur at $x = n\lambda/2$ ($n = 0, 1, 2, 3, \ldots$). The maximum amplitude points (called **antinodes**) occur at $x = n\lambda/4$ ($n = 1, 3, 5, \ldots$). Adjacent antinodes are separated by a distance $\lambda/2$. Adjacent nodes also are separated by a distance $\lambda/2$.

ANALYSIS MODELS FOR PROBLEM SOLVING

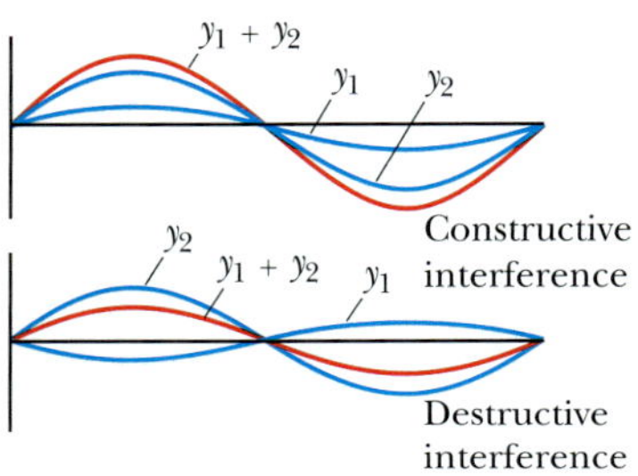

Waves in Interference. When two traveling waves having equal frequencies superimpose, the resultant wave has an amplitude that depends on the phase angle ϕ between the two waves. **Constructive interference** occurs when the two waves are in phase, corresponding to $\phi = 0, 2\pi, 4\pi, \ldots$ rad. **Destructive interference** occurs when the two waves are 180° out of phase, corresponding to $\phi = \pi, 3\pi, 5\pi, \ldots$ rad.

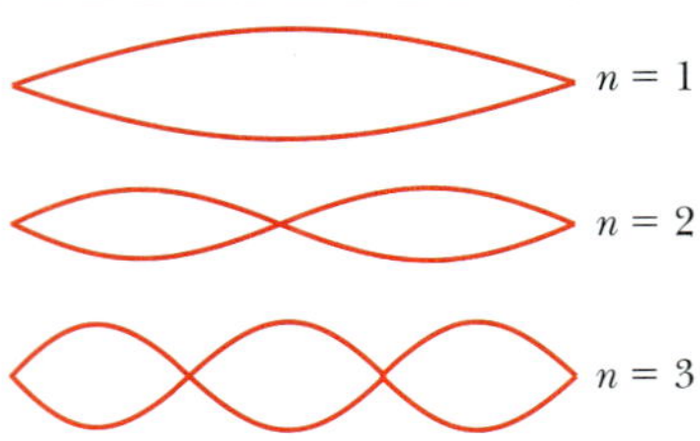

Waves Under Boundary Conditions. When a wave is subject to boundary conditions, only certain natural frequencies are allowed; we say that the frequencies are quantized.

For waves on a string fixed at both ends, the natural frequencies are

$$f_n = \frac{n}{2L}\sqrt{\frac{T}{\mu}} \qquad n = 1, 2, 3, \ldots \qquad \textbf{(18.6)}$$

where T is the tension in the string and μ is its linear mass density.

For sound waves in an air column open at both ends, the natural frequencies are

$$f_n = n\frac{v}{2L} \qquad n = 1, 2, 3, \ldots \qquad \textbf{(18.8)}$$

If an air column is open at one end and closed at the other, only odd harmonics are present and the natural frequencies are

$$f_n = n\frac{v}{4L} \qquad n = 1, 3, 5, \ldots \qquad \textbf{(18.9)}$$

Questions

□ denotes answer available in *Student Solutions Manual/Study Guide;* **O** denotes objective question

1. Does the phenomenon of wave interference apply only to sinusoidal waves?

2. O A series of pulses, each of amplitude 0.1 m, is sent down a string that is attached to a post at one end. The pulses are reflected at the post and travel back along the string without loss of amplitude. What is the net displacement at a point on the string where two pulses are crossing? **(i)** First answer the question making the assumption the string is rigidly attached to the post. (a) 0.4 m (b) 0.2 m (c) 0.1 m (d) 0 **(ii)** Next assume the end at which reflection occurs is free to slide up and down. Choose your answer from the same possibilities.

3. O In Figure 18.4, a sound wave of wavelength 0.8 m divides into two equal parts that recombine to interfere constructively, with the original difference between their path lengths being $|r_2 - r_1| = 0.8$ m. Rank the following situations according to the intensity of sound at the receiver from the highest to the lowest. Assume the tube walls absorb no sound energy. Give equal ranks to situations in which the intensity is equal. (a) From its original position, the sliding section is moved out by 0.1 m. (b) Next it slides out an additional 0.1 m. (c) It slides out still another 0.1 m. (d) It slides out 0.1 m more.

4. When two waves interfere constructively or destructively, is there any gain or loss in energy? Explain.

5. O In Example 18.1, we investigated an oscillator at 1.3 kHz driving two identical side-by-side speakers. We found that a listener at the point *O* hears sound with maximum intensity, whereas a listener at point *P* hears a minimum. What is the intensity at *P*? (a) less than but close to the intensity at *O* (b) half the intensity at *O* (c) very low but not zero (d) zero

6. What limits the amplitude of motion of a real vibrating system that is driven at one of its resonant frequencies?

7. O Suppose all six strings of an acoustic guitar are played without fingering, that is, without being pressed down at any frets. What quantities are the same for all six strings? Choose each and every correct answer. (a) the fundamental frequency (b) the fundamental wavelength of the string wave (c) the fundamental wavelength of the sound emitted (d) the speed of the string wave (e) the speed of the sound emitted

8. O A string of length *L*, mass per unit length μ, and tension *T* is vibrating at its fundamental frequency. **(i)** If the length of the string is doubled, with all other factors held constant, what is the effect on the fundamental frequency? (a) It becomes four times larger. (b) It becomes two times larger. (c) It becomes $\sqrt{2}$ times larger. (d) It is unchanged. (e) It becomes $1/\sqrt{2}$ times as large. (f) It becomes one-half as large. (g) It becomes one-fourth as large. **(ii)** If the mass per unit length is doubled, with all other factors held constant, what is the effect on the fundamental frequency? Choose from the same possibilities. **(iii)** If the tension is doubled, with all other factors held constant, what is the effect on the fundamental frequency? Choose from the same possibilities.

9. O As oppositely moving pulses of the same shape (one upward, one downward) on a string pass through each other, at one particular instant the string shows no displacement from the equilibrium position at any point. What has happened to the energy carried by the pulses at this instant of time? (a) It was used up in producing the previous motion. (b) It is all potential energy. (c) It is all internal energy. (d) It is all kinetic energy. (e) It is momentum. (f) The positive energy of one pulse adds to zero with the negative energy of the other pulse. (g) Each pulse separately has zero total energy.

10. O Assume two identical sinusoidal waves are moving through the same medium in the same direction. Under what condition will the amplitude of the resultant wave be greater than either of the two original waves? (a) in all cases (b) only if the waves have no difference in phase (c) only if the phase difference is less than 90° (d) only if the phase difference is less than 120° (e) only if the phase difference is less than 180°

11. Explain how a musical instrument such as a piano may be tuned by using the phenomenon of beats.

12. O An archer shoots an arrow horizontally from the center of the string of a bow held vertically. After the arrow leaves it, the string of the bow will vibrate as a superposition of what standing-wave harmonics? (a) It vibrates only in harmonic number 1, the fundamental. (b) It vibrates only in the second harmonic. (c) It vibrates only in the odd-numbered harmonics 1, 3, 5, 7, ... (d) It vibrates only in the even-numbered harmonics 2, 4, 6, 8, ... (e) It vibrates in all harmonics. (f) None; it vibrates as a traveling wave rather than a standing wave. (g) None; it does not vibrate if the arrow leaves it with perfect symmetry as described.

13. A tuning fork by itself produces a faint sound. Try each one of the following four methods for obtaining a louder sound from it. Explain how each method works. Explain also any effect on the time interval for which the fork vibrates audibly. (a) Hold the edge of a sheet of paper against one vibrating tine. (b) Press the handle of the tuning fork against a chalkboard or a tabletop. (c) Hold the tuning fork above a column of air of properly chosen length as in Example 18.6. (d) Hold the tuning fork close to an open slot cut in a sheet of foam plastic or cardboard as shown in Figure Q18.13. The slot should be similar in

Alexandra Heder

Figure Q18.13

size and shape to one tine of the fork. The motion of the tines should be perpendicular to the sheet.

14. Despite a reasonably steady hand, a person often spills his coffee when carrying it to his seat. Discuss resonance as a possible cause of this difficulty and devise a means for preventing the spills.

15. **O** A tuning fork is known to vibrate with frequency 262 Hz. When it is sounded along with a mandolin string, four beats are heard every second. Next, a bit of tape is put onto each tine of the tuning fork, and the tuning fork now produces five beats per second with the same mandolin string. What is the frequency of the string? (a) 257 Hz (b) 258 Hz (c) 262 Hz (d) 266 Hz (e) 267 Hz (f) This sequence of events could not happen.

16. An airplane mechanic notices that the sound from a twin-engine aircraft rapidly varies in loudness when both engines are running. What could be causing this variation from loud to soft?

Problems

WebAssign The Problems from this chapter may be assigned online in WebAssign.

ThomsonNOW Sign in at **www.thomsonedu.com** and go to ThomsonNOW to assess your understanding of this chapter's topics with additional quizzing and conceptual questions.

1, 2, 3 denotes straightforward, intermediate, challenging; □ denotes full solution available in *Student Solutions Manual/Study Guide;* ▲ denotes coached solution with hints available at **www.thomsonedu.com;** denotes developing symbolic reasoning; ● denotes asking for qualitative reasoning; denotes computer useful in solving problem

Section 18.1 Superposition and Interference

1. Two waves in one string are described by the wave functions

$$y_1 = (3.0\text{ cm})\cos(4.0x - 1.6t)$$

$$y_2 = (4.0\text{ cm})\sin(5.0x - 2.0t)$$

where y and x are in centimeters and t is in seconds. Find the superposition of the waves $y_1 + y_2$ at the points (a) $x = 1.00$, $t = 1.00$; (b) $x = 1.00$, $t = 0.500$; and (c) $x = 0.500$, $t = 0$. (Remember that the arguments of the trigonometric functions are in radians.)

2. Two pulses A and B are moving in opposite directions along a taut string with a speed of 2.00 cm/s. The amplitude of A is twice the amplitude of B. The pulses are shown in Figure P18.2 at $t = 0$. Sketch the shape of the string at $t = 1$, 1.5, 2, 2.5, and 3 s.

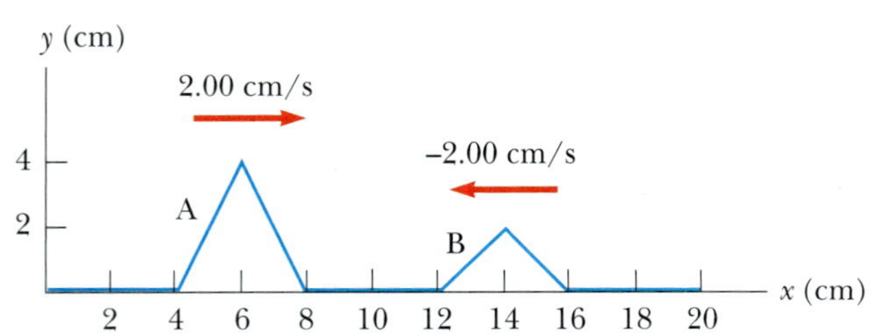

Figure P18.2

3. Two pulses traveling on the same string are described by

$$y_1 = \frac{5}{(3x - 4t)^2 + 2} \qquad y_2 = \frac{-5}{(3x + 4t - 6)^2 + 2}$$

(a) In which direction does each pulse travel? (b) At what instant do the two cancel everywhere? (c) At what point do the two pulses always cancel?

4. Two waves are traveling in the same direction along a stretched string. The waves are 90.0° out of phase. Each wave has an amplitude of 4.00 cm. Find the amplitude of the resultant wave.

5. ▲ Two traveling sinusoidal waves are described by the wave functions

$$y_1 = (5.00\text{ m})\sin[\pi(4.00x - 1\,200t)]$$

$$y_2 = (5.00\text{ m})\sin[\pi(4.00x - 1\,200t - 0.250)]$$

where x, y_1, and y_2 are in meters and t is in seconds. (a) What is the amplitude of the resultant wave? (b) What is the frequency of the resultant wave?

6. Two identical loudspeakers are placed on a wall 2.00 m apart. A listener stands 3.00 m from the wall directly in front of one of the speakers. A single oscillator is driving the speakers at a frequency of 300 Hz. (a) What is the phase difference between the two waves when they reach the observer? (b) **What If?** What is the frequency closest to 300 Hz to which the oscillator may be adjusted such that the observer hears minimal sound?

7. Two identical loudspeakers are driven by the same oscillator of frequency 200 Hz. The speakers are located on a vertical pole a distance of 4.00 m from each other. A man walks straight toward the lower speaker in a direction perpendicular to the pole as shown in Figure P18.7. (a) How many times will he hear a minimum in sound intensity? (b) How far is he from the pole at these moments? Take the speed of sound to be 330 m/s and ignore any sound reflection from the ground.

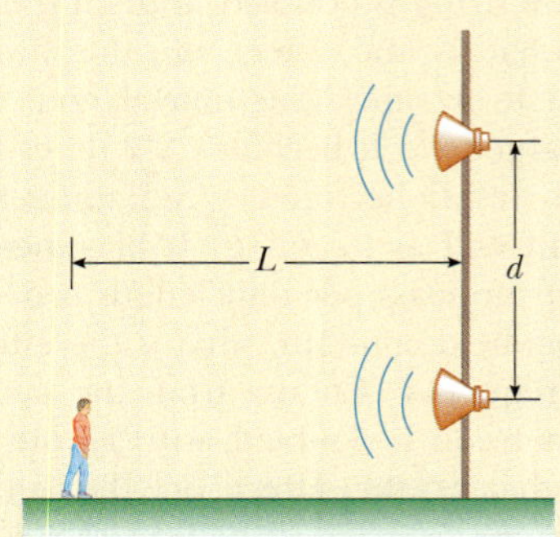

Figure P18.7 Problems 7 and 8.

8. Two identical loudspeakers are driven by the same oscillator of frequency f. The speakers are located a distance d from each other on a vertical pole. A man walks straight toward the lower speaker in a direction perpendicular to the pole as shown in Figure P18.7. (a) How many times will he hear a minimum in sound intensity? (b) How far is he from the pole at these moments? Let v represent the speed of sound and assume the ground does not reflect sound.

9. ▲ Two sinusoidal waves in a string are defined by the functions

$$y_1 = (2.00 \text{ cm}) \sin (20.0x - 32.0t)$$

$$y_2 = (2.00 \text{ cm}) \sin (25.0x - 40.0t)$$

where y_1, y_2, and x are in centimeters and t is in seconds. (a) What is the phase difference between these two waves at the point $x = 5.00$ cm at $t = 2.00$ s? (b) What is the positive x value closest to the origin for which the two phases differ by $\pm\pi$ at $t = 2.00$ s? (That is a location where the two waves add to zero.)

10. ● In air where the speed of sound is 344 m/s, two identical loudspeakers 10.0 m apart are driven by the same oscillator with a frequency of $f = 21.5$ Hz (Fig. P18.10). (a) Explain why a receiver at point A records a minimum in sound intensity from the two speakers. (b) If the receiver is moved in the plane of the speakers, what path should it take so that the intensity remains at a minimum? That is, determine the relationship between x and y (the coordinates of the receiver) that causes the receiver to record a minimum in sound intensity. (c) Can the receiver remain at a minimum and move far away from the two sources? If so, determine the limiting form of the path it must take. If not, explain how far it can go.

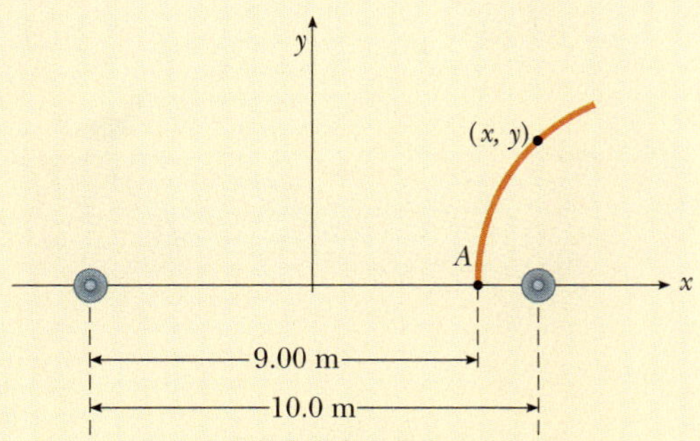

Figure P18.10

Section 18.2 Standing Waves

11. Two sinusoidal waves traveling in opposite directions interfere to produce a standing wave with the wave function

$$y = (1.50 \text{ m}) \sin (0.400x) \cos (200t)$$

where x is in meters and t is in seconds. Determine the wavelength, frequency, and speed of the interfering waves.

12. Verify by direct substitution that the wave function for a standing wave given in Equation 18.1,

$$y = (2A \sin kx) \cos \omega t$$

is a solution of the general linear wave equation, Equation 16.27:

$$\frac{\partial^2 y}{\partial x^2} = \frac{1}{v^2}\frac{\partial^2 y}{\partial t^2}$$

13. ▲ Two identical loudspeakers are driven in phase by a common oscillator at 800 Hz and face each other at a distance of 1.25 m. Locate the points along the line joining the two speakers where relative minima of sound pressure amplitude would be expected. (Use $v = 343$ m/s.)

14. ● A standing wave is described by the function

$$y = 6 \sin \left(\frac{\pi}{2}x\right) \cos (100\pi t)$$

where x and y are in meters and t is in seconds. (a) Prepare a graph showing y as a function of x for $t = 0$, for $t = 5$ ms, for $t = 10$ ms, for $t = 15$ ms, and for $t = 20$ ms. (b) From the graph, identify the wavelength of the wave and explain how you do it. (c) From the graph, identify the frequency of the wave and explain how you do it. (d) From the equation, directly identify the wavelength of the wave and explain how you do it. (e) From the equation, directly identify the frequency and explain how you do it.

15. Two sinusoidal waves combining in a medium are described by the wave functions

$$y_1 = (3.0 \text{ cm}) \sin \pi(x + 0.60t)$$

$$y_2 = (3.0 \text{ cm}) \sin \pi(x - 0.60t)$$

where x is in centimeters and t is in seconds. Determine the maximum transverse position of an element of the medium at (a) $x = 0.250$ cm, (b) $x = 0.500$ cm, and (c) $x = 1.50$ cm. (d) Find the three smallest values of x corresponding to antinodes.

16. ● Two waves simultaneously present in a long string are given by the wave functions

$$y_1 = A \sin (kx - \omega t + \phi) \qquad y_2 = A \sin (kx + \omega t)$$

(a) Do the two traveling waves add to give a standing wave? Explain. (b) Is it still true that the nodes are one-half wavelength apart? Argue for your answer. (c) Are the nodes different in any way from the way they would be if ϕ were zero? Explain.

Section 18.3 Standing Waves in a String Fixed at Both Ends

17. Find the fundamental frequency and the next three frequencies that could cause standing wave patterns on a string that is 30.0 m long, has a mass per unit length of 9.00×10^{-3} kg/m, and is stretched to a tension of 20.0 N.

18. A string with a mass of 8.00 g and a length of 5.00 m has one end attached to a wall; the other end is draped over a small fixed pulley and attached to a hanging object with a mass of 4.00 kg. If the string is plucked, what is the fundamental frequency of its vibration?

19. In the arrangement shown in Figure P18.19, an object can be hung from a string (with linear mass density $\mu = 0.002\ 00$ kg/m) that passes over a light pulley. The string is connected to a vibrator (of constant frequency f), and the length of the string between point P and the pulley is $L = 2.00$ m. When the mass m of the object is either 16.0 kg or 25.0 kg, standing waves are observed; no standing waves are observed with any mass between these values, however. (a) What is the frequency of the vibrator? *Note:* The greater the tension in the string, the smaller the number of nodes in the standing wave. (b) What is the

largest object mass for which standing waves could be observed?

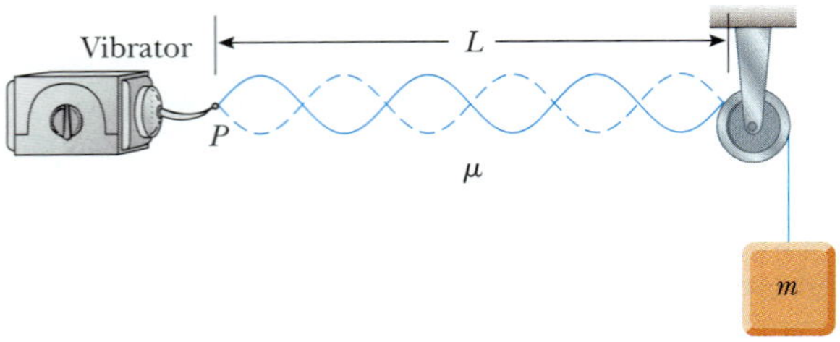

Figure P18.19

20. The top string of a guitar has a fundamental frequency of 330 Hz when it is allowed to vibrate as a whole, along all its 64.0-cm length from the neck to the bridge. A fret is provided for limiting vibration to just the lower two-thirds of the string. If the string is pressed down at this fret and plucked, what is the new fundamental frequency? (b) **What If?** The guitarist can play a "natural harmonic" by gently touching the string at the location of this fret and plucking the string at about one-sixth of the way along its length from the bridge. What frequency will be heard then?

21. The A string on a cello vibrates in its first normal mode with a frequency of 220 Hz. The vibrating segment is 70.0 cm long and has a mass of 1.20 g. (a) Find the tension in the string. (b) Determine the frequency of vibration when the string vibrates in three segments.

22. A violin string has a length of 0.350 m and is tuned to concert G, with f_G = 392 Hz. Where must the violinist place her finger to play concert A, with f_A = 440 Hz? If this position is to remain correct to one-half the width of a finger (that is, to within 0.600 cm), what is the maximum allowable percentage change in the string tension?

23. **Review problem.** A sphere of mass M is supported by a string that passes over a light horizontal rod of length L (Fig. P18.23). Given that the angle is θ and that f represents the fundamental frequency of standing waves in the portion of the string above the rod, determine the mass of this portion of the string.

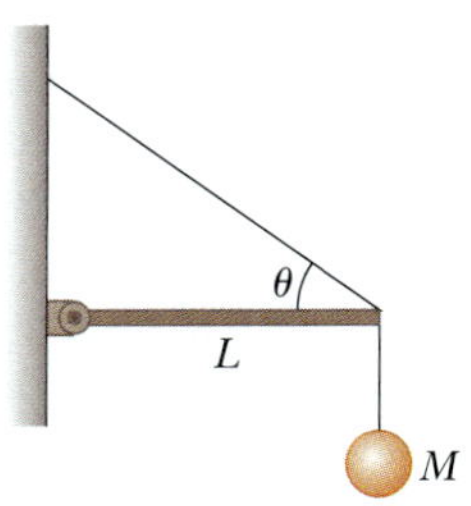

Figure P18.23

24. **Review problem.** A copper cylinder hangs at the bottom of a steel wire of negligible mass. The top end of the wire is fixed. When the wire is struck, it emits sound with a fundamental frequency of 300 Hz. The copper cylinder is then submerged in water so that half its volume is below the water line. Determine the new fundamental frequency.

25. A standing wave pattern is observed in a thin wire with a length of 3.00 m. The wave function is

$$y = (0.002\text{ m})\sin(\pi x)\cos(100\pi t)$$

where x is in meters and t is in seconds. (a) How many loops does this pattern exhibit? (b) What is the fundamental frequency of vibration of the wire? (c) **What If?** If the original frequency is held constant and the tension in the wire is increased by a factor of 9, how many loops are present in the new pattern?

Section 18.4 Resonance

26. ● The Bay of Fundy, Nova Scotia, has the highest tides in the world. Assume in midocean and at the mouth of the bay the Moon's gravity gradient and the Earth's rotation make the water surface oscillate with an amplitude of a few centimeters and a period of 12 h 24 min. At the head of the bay, the amplitude is several meters. Argue for or against the proposition that the tide is magnified by standing-wave resonance. Assume the bay has a length of 210 km and a uniform depth of 36.1 m. The speed of long-wavelength water waves is given by $\sqrt{gd}$, where d is the water's depth.

27. An earthquake can produce a *seiche* in a lake in which the water sloshes back and forth from end to end with remarkably large amplitude and long period. Consider a seiche produced in a rectangular farm pond as shown in the cross-sectional view of Figure P18.27. (The figure is not drawn to scale.) Suppose the pond is 9.15 m long and of uniform width and depth. You measure that a pulse produced at one end reaches the other end in 2.50 s. (a) What is the wave speed? (b) To produce the seiche, several people stand on the bank at one end and paddle together with snow shovels, moving them in simple harmonic motion. What should be the frequency of this motion?

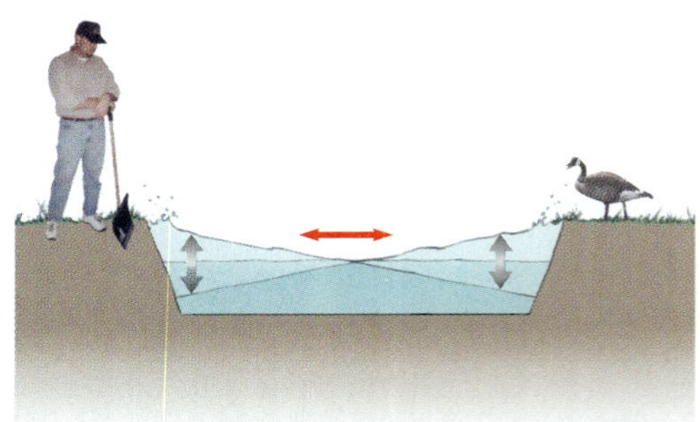

Figure P18.27

28. Figure P18.28a is a photograph of a vibrating wine glass. A special technique makes black and white stripes appear where the glass is moving, with closer spacing where the

Courtesy Professor T. D. Rossing, Northern Illinois University

Steve Bronstein/Image Bank/Getty Images

(a) (b)

Figure P18.28

amplitude is larger. Six nodes and six antinodes alternate around the rim of the glass in the vibration photographed, but consider instead the case of a standing-wave vibration with four nodes and four antinodes equally spaced around the 20.0-cm circumference of the rim of a goblet. If transverse waves move around the glass at 900 m/s, an opera singer would have to produce a high harmonic with what frequency to shatter the glass with a resonant vibration as shown in Figure P18.28b?

Section 18.5 Standing Waves in Air Columns

Note: Unless otherwise specified, assume the speed of sound in air is 343 m/s at 20°C and is described by

$$v = (331 \text{ m/s})\sqrt{1 + \frac{T_C}{273°}}$$

at any Celsius temperature T_C.

29. Calculate the length of a pipe that has a fundamental frequency of 240 Hz assuming the pipe is (a) closed at one end and (b) open at both ends.

30. The overall length of a piccolo is 32.0 cm. The resonating air column vibrates as in a pipe open at both ends. (a) Find the frequency of the lowest note a piccolo can sound, assuming the speed of sound in air is 340 m/s. (b) Opening holes in the side effectively shortens the length of the resonant column. Assume the highest note a piccolo can sound is 4 000 Hz. Find the distance between adjacent antinodes for this mode of vibration.

31. The fundamental frequency of an open organ pipe corresponds to middle C (261.6 Hz on the chromatic musical scale). The third resonance of a closed organ pipe has the same frequency. What is the length of each pipe?

32. ● Do not stick anything into your ear! Estimate the length of your ear canal from its opening at the external ear to the eardrum. If you regard the canal as a narrow tube that is open at one end and closed at the other, at approximately what fundamental frequency would you expect your hearing to be most sensitive? Explain why you can hear especially soft sounds just around this frequency.

33. ▲ A shower stall has dimensions 86.0 cm × 86.0 cm × 210 cm. If you were singing in this shower, which frequencies would sound the richest (because of resonance)? Assume the stall acts as a pipe closed at both ends, with nodes at opposite sides. Assume the voices of various singers range from 130 Hz to 2 000 Hz. Let the speed of sound in the hot air be 355 m/s.

34. As shown in Figure P18.34, water is pumped into a tall vertical cylinder at a volume flow rate R. The radius of the cylinder is r, and at the open top of the cylinder a tuning fork is vibrating with a frequency f. As the water rises, what time interval elapses between successive resonances?

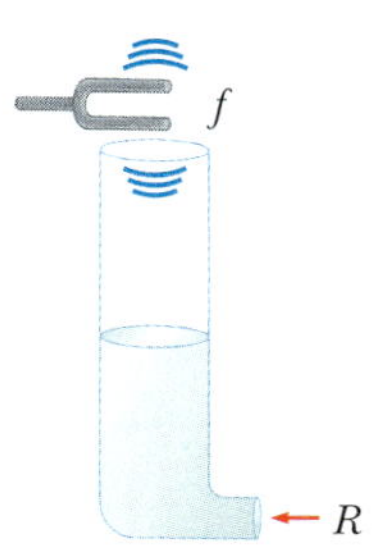

Figure P18.34

35. ▲ Two adjacent natural frequencies of an organ pipe are determined to be 550 Hz and 650 Hz. Calculate the fundamental frequency and length of this pipe. (Use v = 340 m/s.)

36. ● A tunnel under a river is 2.00 km long. (a) At what frequencies can the air in the tunnel resonate? (b) Explain whether it would be good to make a rule against blowing your car horn when you are in the tunnel.

37. An air column in a glass tube is open at one end and closed at the other by a movable piston. The air in the tube is warmed above room temperature, and a 384-Hz tuning fork is held at the open end. Resonance is heard when the piston is 22.8 cm from the open end and again when it is 68.3 cm from the open end. (a) What speed of sound is implied by these data? (b) How far from the open end will the piston be when the next resonance is heard?

38. A tuning fork with a frequency of 512 Hz is placed near the top of the tube shown in Figure 18.14a. The water level is lowered so that the length L slowly increases from an initial value of 20.0 cm. Determine the next two values of L that correspond to resonant modes.

39. ● A student uses an audio oscillator of adjustable frequency to measure the depth of a water well. The student reports hearing two successive resonances at 51.5 Hz and 60.0 Hz. How deep is the well? Explain the precision you can ascribe to your answer.

40. With a particular fingering, a flute sounds a note with frequency 880 Hz at 20.0°C. The flute is open at both ends. (a) Find the air column length. (b) Find the frequency the flute produces at the beginning of the halftime performance at a late-season American football game, when the ambient temperature is −5.00°C and the musician has not had a chance to warm up the flute.

Section 18.6 Standing Waves in Rods and Membranes

41. An aluminum rod 1.60 m long is held at its center. It is stroked with a rosin-coated cloth to set up a longitudinal vibration. The speed of sound in a thin rod of aluminum is 5 100 m/s. (a) What is the fundamental frequency of the waves established in the rod? (b) What harmonics are set up in the rod held in this manner? (c) **What If?** What would be the fundamental frequency if the rod were copper, in which the speed of sound is 3 560 m/s?

42. An aluminum rod is clamped one-quarter of the way along its length and set into longitudinal vibration by a variable-frequency driving source. The lowest frequency that produces resonance is 4 400 Hz. The speed of sound in an aluminum rod is 5 100 m/s. Determine the length of the rod.

Section 18.7 Beats: Interference in Time

43. ▲ In certain ranges of a piano keyboard, more than one string is tuned to the same note to provide extra loudness. For example, the note at 110 Hz has two strings at this frequency. If one string slips from its normal tension of 600 N to 540 N, what beat frequency is heard when the hammer strikes the two strings simultaneously?

44. While attempting to tune the note C at 523 Hz, a piano tuner hears 2.00 beats/s between a reference oscillator and the string. (a) What are the possible frequencies of the string? (b) When she tightens the string slightly, she hears 3.00 beats/s. What is the frequency of the string now? (c) By what percentage should the piano tuner now change the tension in the string to bring it into tune?

45. A student holds a tuning fork oscillating at 256 Hz. He walks toward a wall at a constant speed of 1.33 m/s. (a) What beat frequency does he observe between the tuning fork and its echo? (b) How fast must he walk away from the wall to observe a beat frequency of 5.00 Hz?

46. When beats occur at a rate higher than about 20 per second, they are not heard individually but rather as a steady hum, called a *combination tone.* The player of a typical pipe organ can press a single key and make the organ produce sound with different fundamental frequencies. She can select and pull out different stops to make the same key for the note C produce sound at the following frequencies: 65.4 Hz from a so-called 8-foot pipe; $2 \times 65.4 = 131$ Hz from a 4-foot pipe; $3 \times 65.4 = 196$ Hz from a $2\frac{2}{3}$-foot pipe; $4 \times 65.4 = 262$ Hz from a 2-foot pipe; or any combination of these sounds. With notes at low frequencies, she obtains sound with the richest quality by pulling out all the stops. When an air leak develops in one of the pipes, that pipe cannot be used. If a leak occurs in an 8-foot pipe, playing a combination of other pipes can create the sensation of sound at the frequency that the 8-foot pipe would produce. Which sets of stops, among those listed, could be pulled out to do that?

Section 18.8 Nonsinusoidal Wave Patterns

47. An A-major chord consists of the notes called A, C#, and E. It can be played on a piano by simultaneously striking strings with fundamental frequencies of 440.00 Hz, 554.37 Hz, and 659.26 Hz. The rich consonance of the chord is associated with near equality of the frequencies of some of the higher harmonics of the three tones. Consider the first five harmonics of each string and determine which harmonics show near equality.

48. Suppose a flutist plays a 523-Hz C note with first harmonic displacement amplitude $A_1 = 100$ nm. From Figure 18.19b, read, by proportion, the displacement amplitudes of harmonics 2 through 7. Take them as the values A_2 through A_7 in the Fourier analysis of the sound and assume $B_1 = B_2 = \ldots = B_7 = 0$. Construct a graph of the waveform of the sound. Your waveform will not look exactly like the flute waveform in Figure 18.18b because you simplify by ignoring cosine terms; nevertheless, it produces the same sensation to human hearing.

Additional Problems

49. ● **Review problem.** The top end of a yo-yo string is held stationary. The yo-yo itself is much more massive than the string. It starts from rest and moves down with constant acceleration 0.800 m/s^2 as it unwinds from the string. The rubbing of the string against the edge of the yo-yo excites transverse standing-wave vibrations in the string. Both ends of the string are nodes even as the length of the string increases. Consider the instant 1.20 s after the motion begins. (a) Show that the rate of change with time of the wavelength of the fundamental mode of oscillation is 1.92 m/s. (b) **What If?** Is the rate of change of the wavelength of the second harmonic also 1.92 m/s at this moment? Explain your answer. (c) **What If?** The experiment is repeated after more mass has been added to the yo-yo body. The mass distribution is kept the same so that the yo-yo still moves with downward acceleration 0.800 m/s^2. At the 1.20-s point in this case, is the rate of change of the fundamental wavelength of the string vibration still equal to 1.92 m/s? Explain. Is the rate of change of second harmonic wavelength the same as in part (b)? Explain.

50. ● A loudspeaker at the front of a room and an identical loudspeaker at the rear of the room are being driven by the same oscillator at 456 Hz. A student walks at a uniform rate of 1.50 m/s along the length of the room. She hears a single tone, repeatedly becoming louder and softer. (a) Model these variations as beats between the Doppler-shifted sounds the student receives. Calculate the number of beats the student hears each second. (b) **What If?** Model the two speakers as producing a standing wave in the room and the student as walking between antinodes. Calculate the number of intensity maxima the student hears each second. (c) Explain how the answers to parts (a) and (b) compare with each other.

51. On a marimba (Fig. P18.51), the wooden bar that sounds a tone when struck vibrates in a transverse standing wave having three antinodes and two nodes. The lowest frequency note is 87.0 Hz, produced by a bar 40.0 cm long. (a) Find the speed of transverse waves on the bar. (b) A resonant pipe suspended vertically below the center of the bar enhances the loudness of the emitted sound. If the pipe is open at the top end only and the speed of sound in air is 340 m/s, what is the length of the pipe required to resonate with the bar in part (a)?

© SuperStock, Inc.

Figure P18.51 Marimba players in Mexico City.

52. A nylon string has mass 5.50 g and length 86.0 cm. One end is tied to the floor and the other end to a small magnet, with a mass negligible compared to that of the string. A magnetic field (which we will study in Chapter 29) exerts an upward force of 1.30 N on the magnet wherever the magnet is located. At equilibrium, the string is vertical and motionless, with the magnet at the top. When it is carrying a small-amplitude wave, you may assume the string is always under uniform tension 1.30 N. (a) Find the speed of transverse waves on the string. (b) The string's vibration possibilities are a set of standing-wave states, each with a node at the fixed bottom end and an antinode at the free top end. Find the node–antinode distances for each one of the three simplest states. (c) Find the frequency of each of these states.

53. Two train whistles have identical frequencies of 180 Hz. When one train is at rest in the station and the other is moving nearby, a commuter standing on the station platform hears beats with a frequency of 2.00 beats/s when the whistles operate together. What are the two possible speeds and directions the moving train can have?

54. A string fixed at both ends and having a mass of 4.80 g, a length of 2.00 m, and a tension of 48.0 N vibrates in its second ($n = 2$) normal mode. What is the wavelength in air of the sound emitted by this vibrating string?

55. Two wires are welded together end to end. The wires are made of the same material, but the diameter of one is twice that of the other. They are subjected to a tension of 4.60 N. The thin wire has a length of 40.0 cm and a linear mass density of 2.00 g/m. The combination is fixed at both ends and vibrated in such a way that two antinodes are present, with the node between them being precisely at the weld. (a) What is the frequency of vibration? (b) What is the length of the thick wire?

56. A string of linear density 1.60 g/m is stretched between clamps 48.0 cm apart. The string does not stretch appreciably as the tension in it is steadily raised from 15.0 N at $t = 0$ to 25.0 N at $t = 3.50$ s. Therefore, the tension as a function of time is given by the expression $T = 15.0\text{ N} + (10.0\text{ N})t/3.50\text{ s}$. The string is vibrating in its fundamental mode throughout this process. Find the number of oscillations it completes during the 3.50-s interval.

57. A standing wave is set up in a string of variable length and tension by a vibrator of variable frequency. Both ends of the string are fixed. When the vibrator has a frequency f, in a string of length L and under tension T, n antinodes are set up in the string. (a) If the length of the string is doubled, by what factor should the frequency be changed so that the same number of antinodes is produced? (b) If the frequency and length are held constant, what tension will produce $n + 1$ antinodes? (c) If the frequency is tripled and the length of the string is halved, by what factor should the tension be changed so that twice as many antinodes are produced?

58. **Review problem**. For the arrangement shown in Figure P18.58, $\theta = 30.0°$, the inclined plane and the small pulley are frictionless, the string supports the object of mass M at the bottom of the plane, and the string has mass m that is small compared with M. The system is in equilibrium, and the vertical part of the string has a length h. Standing waves are set up in the vertical section of the string. (a) Find the tension in the string. (b) Model the shape of the string as one leg and the hypotenuse of a right triangle. Find the whole length of the string. (c) Find the mass per unit length of the string. (d) Find the speed of waves on the string. (e) Find the lowest frequency for a standing wave. (f) Find the period of the standing wave having three nodes. (g) Find the wavelength of the standing wave having three nodes. (h) Find the frequency of the beats resulting from the interference of the sound wave of lowest frequency generated by the string with another sound wave having a frequency that is 2.00% greater.

59. Two waves are described by the wave functions

$$y_1(x, t) = (5.0\text{ m})\sin(2.0x - 10t)$$

$$y_2(x, t) = (10\text{ m})\cos(2.0x - 10t)$$

where y_1, y_2, and x are in meters and t is in seconds. Show that the wave resulting from their superposition is also sinusoidal. Determine the amplitude and phase of this sinusoidal wave.

60. A quartz watch contains a crystal oscillator in the form of a block of quartz that vibrates by contracting and expanding. Two opposite faces of the block, 7.05 mm apart, are antinodes, moving alternately toward each other and away from each other. The plane halfway between these two faces is a node of the vibration. The speed of sound in quartz is 3.70 km/s. Find the frequency of the vibration. An oscillating electric voltage accompanies the mechanical oscillation; the quartz is described as *piezoelectric*. An electric circuit feeds in energy to maintain the oscillation and also counts the voltage pulses to keep time.

61. **Review problem**. A 12.0-kg object hangs in equilibrium from a string with a total length of $L = 5.00$ m and a linear mass density of $\mu = 0.001\,00$ kg/m. The string is wrapped around two light, frictionless pulleys that are separated by a distance of $d = 2.00$ m (Fig. P18.61a). (a) Determine the tension in the string. (b) At what frequency must the string between the pulleys vibrate to form the standing wave pattern shown in Figure P18.61b?

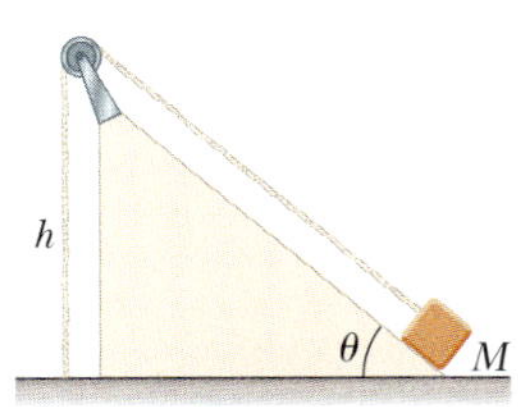

Figure P18.58

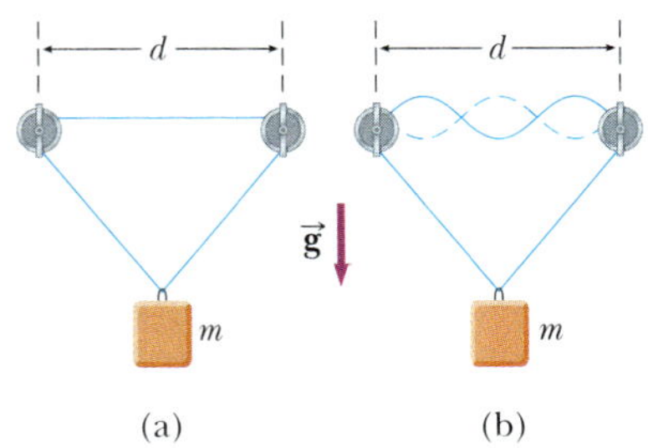

Figure P18.61

Answers to Quick Quizzes

18.1 (c). The pulses completely cancel each other in terms of displacement of elements of the string from equilibrium, but the string is still moving. A short time later, the string will be displaced again and the pulses will have passed each other.

18.2 **(i)**, (a). The pattern shown at the bottom of Active Figure 18.8a corresponds to the extreme position of the string. All elements of the string have momentarily come to rest. **(ii)**, (d). Near a nodal point, elements on one side of the point are moving upward at this instant and elements on the other side are moving downward.

18.3 (d). Choice (a) is incorrect because the number of nodes is one greater than the number of antinodes. Choice (b) is only true for half of the modes; it is not true for any odd-numbered mode. Choice (c) would be correct if we replace the word *nodes* with *antinodes.*

18.4 (b). With both ends open, the pipe has a fundamental frequency given by Equation 18.8: $f_{\text{open}} = v/2L$. With one end closed, the pipe has a fundamental frequency given by Equation 18.9:

$$f_{\text{closed}} = \frac{v}{4L} = \frac{1}{2}\frac{v}{2L} = \frac{1}{2}f_{\text{open}}$$

18.5 (c). The increase in temperature causes the speed of sound to go up. According to Equation 18.8, the result is an increase in the fundamental frequency of a given organ pipe.

PART 3

Thermodynamics

We now direct our attention to the study of thermodynamics, which involves situations in which the temperature or state (solid, liquid, gas) of a system changes due to energy transfers. As we shall see, thermodynamics is very successful in explaining the bulk properties of matter and the correlation between these properties and the mechanics of atoms and molecules.

Historically, the development of thermodynamics paralleled the development of the atomic theory of matter. By the 1820s, chemical experiments had provided solid evidence for the existence of atoms. At that time, scientists recognized that a connection between thermodynamics and the structure of matter must exist. In 1827, botanist Robert Brown reported that grains of pollen suspended in a liquid move erratically from one place to another as if under constant agitation. In 1905, Albert Einstein used kinetic theory to explain the cause of this erratic motion, known today as *Brownian motion*. Einstein explained this phenomenon by assuming the grains are under constant bombardment by "invisible" molecules in the liquid, which themselves move erratically. This explanation gave scientists insight into the concept of molecular motion and gave credence to the idea that matter is made up of atoms. A connection was thus forged between the everyday world and the tiny, invisible building blocks that make up this world.

Thermodynamics also addresses more practical questions. Have you ever wondered how a refrigerator is able to cool its contents, or what types of transformations occur in a power plant or in the engine of your automobile, or what happens to the kinetic energy of a moving object when the object comes to rest? The laws of thermodynamics can be used to provide explanations for these and other phenomena.

The Alyeska oil pipeline near the Tazlina River in Alaska. The oil in the pipeline is warm, and energy transferring from the pipeline could melt environmentally sensitive permafrost in the ground. The finned structures on top of the support posts are thermal radiators that allow the energy to be transferred into the air to protect the permafrost. (Topham Picturepoint/The Image Works)

Why would someone designing a pipeline include these strange loops? Pipelines carrying liquids often contain such loops to allow for expansion and contraction as the temperature changes. We will study thermal expansion in this chapter. (Lowell Georgia/CORBIS)

19 Temperature

In our study of mechanics, we carefully defined such concepts as *mass, force,* and *kinetic energy* to facilitate our quantitative approach. Likewise, a quantitative description of thermal phenomena requires careful definitions of such important terms as *temperature, heat,* and *internal energy.* This chapter begins with a discussion of temperature.

Next, we consider the importance when studying thermal phenomena of the particular substance we are investigating. For example, gases expand appreciably when heated, whereas liquids and solids expand only slightly.

This chapter concludes with a study of ideal gases on the macroscopic scale. Here, we are concerned with the relationships among such quantities as pressure, volume, and temperature of a gas. In Chapter 21, we shall examine gases on a microscopic scale, using a model that represents the components of a gas as small particles.

19.1 Temperature and the Zeroth Law of Thermodynamics

We often associate the concept of temperature with how hot or cold an object feels when we touch it. In this way, our senses provide us with a qualitative indication of temperature. Our senses, however, are unreliable and often mislead us. For

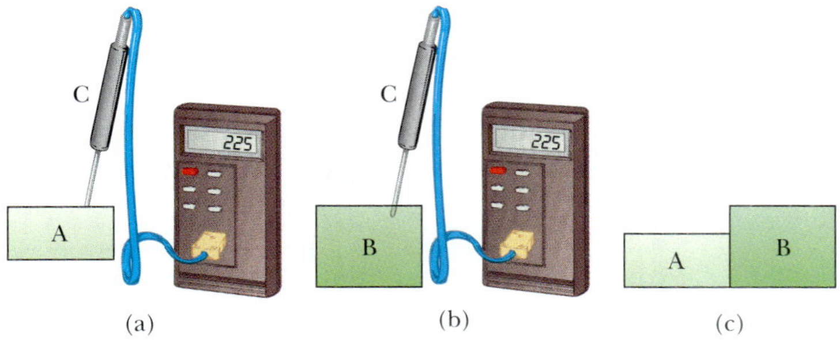

Figure 19.1 The zeroth law of thermodynamics. (a, b) If the temperatures of A and B are measured to be the same by placing them in thermal contact with a thermometer (object C), no energy will be exchanged between them when they are placed in thermal contact with each other (c).

example, if you remove a metal ice tray and a cardboard box of frozen vegetables from the freezer, the ice tray feels colder than the box *even though both are at the same temperature.* The two objects feel different because metal transfers energy by heat at a higher rate than cardboard does. What we need is a reliable and reproducible method for measuring the relative hotness or coldness of objects rather than the rate of energy transfer. Scientists have developed a variety of thermometers for making such quantitative measurements.

Two objects at different initial temperatures eventually reach some intermediate temperature when placed in contact with each other. For example, when hot water and cold water are mixed in a bathtub, the final temperature of the mixture is somewhere between the initial hot and cold temperatures. Likewise, when an ice cube is dropped into a cup of hot coffee, the ice cube melts and the coffee's temperature decreases.

Imagine that two objects are placed in an insulated container such that they interact with each other but not with the environment. If the objects are at different temperatures, energy is transferred between them, even if they are initially not in physical contact with each other. The energy transfer mechanisms from Chapter 8 that we will focus on are heat and electromagnetic radiation. For purposes of this discussion, let's assume two objects are in **thermal contact** with each other if energy can be exchanged between them by these processes due to a temperature difference. **Thermal equilibrium** is a situation in which two objects would not exchange energy by heat or electromagnetic radiation if they were placed in thermal contact.

Let's consider two objects A and B, which are not in thermal contact, and a third object C, which is our thermometer. We wish to determine whether A and B are in thermal equilibrium with each other. The thermometer (object C) is first placed in thermal contact with object A until thermal equilibrium is reached[1] as shown in Figure 19.1a. From that moment on, the thermometer's reading remains constant and we record this reading. The thermometer is then removed from object A and placed in thermal contact with object B as shown in Figure 19.1b. The reading is again recorded after thermal equilibrium is reached. If the two readings are the same, object A and object B are in thermal equilibrium with each other. If they are placed in contact with each other as in Figure 19.1c, there is no exchange of energy between them.

We can summarize these results in a statement known as the **zeroth law of thermodynamics** (the law of equilibrium):

◀ Zeroth law of thermodynamics

> If objects A and B are separately in thermal equilibrium with a third object C, then A and B are in thermal equilibrium with each other.

[1] We assume a negligible amount of energy transfers between the thermometer and object A during the equilibrium process. Without this assumption, which is also made for the thermometer and object B, the measurement of the temperature of an object disturbs the system so that the measured temperature is different from the initial temperature of the object. In practice, whenever you measure a temperature with a thermometer, you measure the disturbed system, not the original system.

This statement can easily be proved experimentally and is very important because it enables us to define temperature. We can think of **temperature** as the property that determines whether an object is in thermal equilibrium with other objects. **Two objects in thermal equilibrium with each other are at the same temperature.** Conversely, if two objects have different temperatures, they are not in thermal equilibrium with each other.

Quick Quiz 19.1 Two objects, with different sizes, masses, and temperatures, are placed in thermal contact. In which direction does the energy travel? (a) Energy travels from the larger object to the smaller object. (b) Energy travels from the object with more mass to the one with less mass. (c) Energy travels from the object at higher temperature to the object at lower temperature.

19.2 Thermometers and the Celsius Temperature Scale

Thermometers are devices used to measure the temperature of a system. All thermometers are based on the principle that some physical property of a system changes as the system's temperature changes. Some physical properties that change with temperature are (1) the volume of a liquid, (2) the dimensions of a solid, (3) the pressure of a gas at constant volume, (4) the volume of a gas at constant pressure, (5) the electric resistance of a conductor, and (6) the color of an object.

A common thermometer in everyday use consists of a mass of liquid—usually mercury or alcohol—that expands into a glass capillary tube when heated (Fig. 19.2). In this case, the physical property that changes is the volume of a liquid. Any temperature change in the range of the thermometer can be defined as being proportional to the change in length of the liquid column. The thermometer can be calibrated by placing it in thermal contact with a natural system that remains at constant temperature. One such system is a mixture of water and ice in thermal equilibrium at atmospheric pressure. On the **Celsius temperature scale,** this mixture is defined to have a temperature of zero degrees Celsius, which is written as 0°C; this temperature is called the *ice point* of water. Another commonly used system is a mixture of water and steam in thermal equilibrium at atmospheric pressure; its temperature is defined as 100°C, which is the *steam point* of water. Once the liquid levels in the thermometer have been established at these two points, the

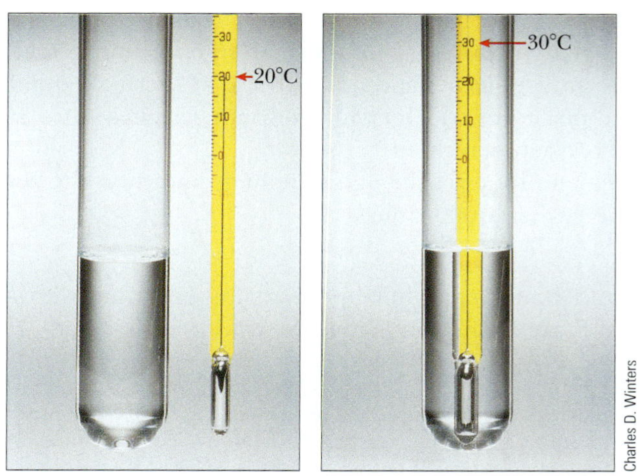

Figure 19.2 As a result of thermal expansion, the level of the mercury in the thermometer rises as the mercury is heated by water in the test tube.

length of the liquid column between the two points is divided into 100 equal segments to create the Celsius scale. Therefore, each segment denotes a change in temperature of one Celsius degree.

Thermometers calibrated in this way present problems when extremely accurate readings are needed. For instance, the readings given by an alcohol thermometer calibrated at the ice and steam points of water might agree with those given by a mercury thermometer only at the calibration points. Because mercury and alcohol have different thermal expansion properties, when one thermometer reads a temperature of, for example, 50°C, the other may indicate a slightly different value. The discrepancies between thermometers are especially large when the temperatures to be measured are far from the calibration points.[2]

An additional practical problem of any thermometer is the limited range of temperatures over which it can be used. A mercury thermometer, for example, cannot be used below the freezing point of mercury, which is −39°C, and an alcohol thermometer is not useful for measuring temperatures above 85°C, the boiling point of alcohol. To surmount this problem, we need a universal thermometer whose readings are independent of the substance used in it. The gas thermometer, discussed in the next section, approaches this requirement.

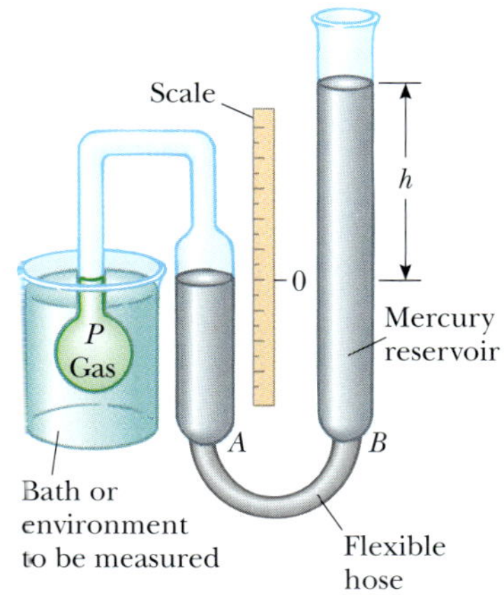

Figure 19.3 A constant-volume gas thermometer measures the pressure of the gas contained in the flask immersed in the bath. The volume of gas in the flask is kept constant by raising or lowering reservoir B to keep the mercury level in column A constant.

19.3 The Constant-Volume Gas Thermometer and the Absolute Temperature Scale

One version of a gas thermometer is the constant-volume apparatus shown in Figure 19.3. The physical change exploited in this device is the variation of pressure of a fixed volume of gas with temperature. The flask is immersed in an ice-water bath, and mercury reservoir B is raised or lowered until the top of the mercury in column A is at the zero point on the scale. The height h, the difference between the mercury levels in reservoir B and column A, indicates the pressure in the flask at 0°C.

The flask is then immersed in water at the steam point. Reservoir B is readjusted until the top of the mercury in column A is again at zero on the scale, which ensures that the gas's volume is the same as it was when the flask was in the ice bath (hence the designation "constant volume"). This adjustment of reservoir B gives a value for the gas pressure at 100°C. These two pressure and temperature values are then plotted as shown in Figure 19.4. The line connecting the two points serves as a calibration curve for unknown temperatures. (Other experiments show that a linear relationship between pressure and temperature is a very good assumption.) To measure the temperature of a substance, the gas flask of Figure 19.3 is placed in thermal contact with the substance and the height of reservoir B is adjusted until the top of the mercury column in A is at zero on the scale. The height of the mercury column in B indicates the pressure of the gas; knowing the pressure, the temperature of the substance is found using the graph in Figure 19.4.

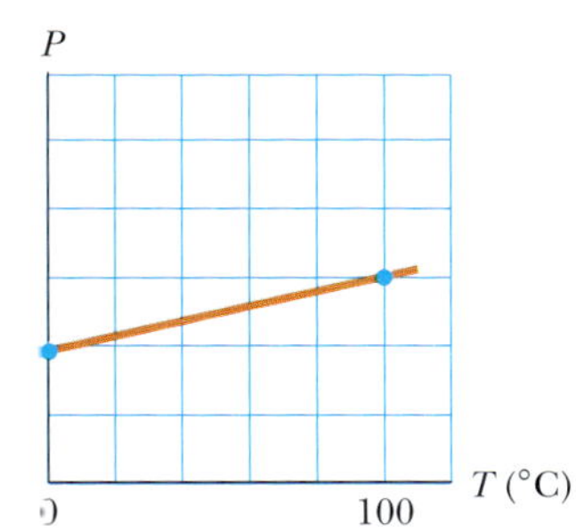

Figure 19.4 A typical graph of pressure versus temperature taken with a constant-volume gas thermometer. The two dots represent known reference temperatures (the ice and steam points of water).

Now suppose temperatures of different gases at different initial pressures are measured with gas thermometers. Experiments show that the thermometer readings are nearly independent of the type of gas used as long as the gas pressure is low and the temperature is well above the point at which the gas liquefies (Fig. 19.5). The agreement among thermometers using various gases improves as the pressure is reduced.

If we extend the straight lines in Figure 19.5 toward negative temperatures, we find a remarkable result: **in every case, the pressure is zero when the temperature is −273.15°C!** This finding suggests some special role that this particular temperature must play. It is used as the basis for the **absolute temperature scale,** which sets

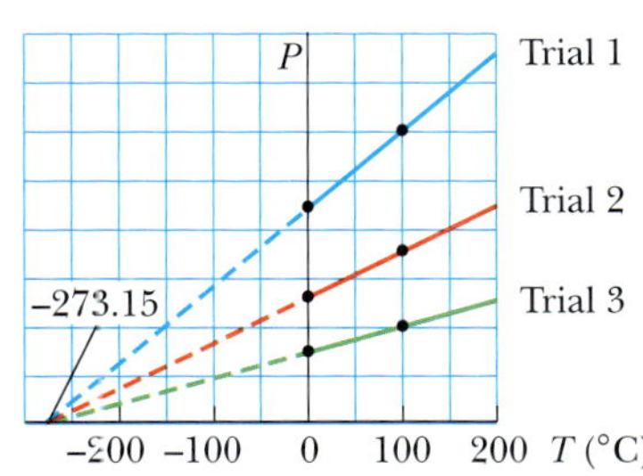

Figure 19.5 Pressure versus temperature for experimental trials in which gases have different pressures in a constant-volume gas thermometer. Notice that, for all three trials, the pressure extrapolates to zero at the temperature −273.15°C.

[2] Two thermometers that use the same liquid may also give different readings, due in part to difficulties in constructing uniform-bore glass capillary tubes.

PITFALL PREVENTION 19.1

A Matter of Degree

Notations for temperatures in the Kelvin scale do not use the degree sign. The unit for a Kelvin temperature is simply "kelvins" and not "degrees Kelvin."

$-273.15°C$ as its zero point. This temperature is often referred to as **absolute zero.** It is indicated as a zero because at a lower temperature, the pressure of the gas would become negative, which is meaningless. The size of one degree on the absolute temperature scale is chosen to be identical to the size of one degree on the Celsius scale. Therefore, the conversion between these temperatures is

$$T_C = T - 273.15 \tag{19.1}$$

where T_C is the Celsius temperature and T is the absolute temperature.

Because the ice and steam points are experimentally difficult to duplicate and depend on atmospheric pressure, an absolute temperature scale based on two new fixed points was adopted in 1954 by the International Committee on Weights and Measures. The first point is absolute zero. The second reference temperature for this new scale was chosen as the **triple point of water,** which is the single combination of temperature and pressure at which liquid water, gaseous water, and ice (solid water) coexist in equilibrium. This triple point occurs at a temperature of 0.01°C and a pressure of 4.58 mm of mercury. On the new scale, which uses the unit *kelvin,* the temperature of water at the triple point was set at 273.16 kelvins, abbreviated 273.16 K. This choice was made so that the old absolute temperature scale based on the ice and steam points would agree closely with the new scale based on the triple point. This new absolute temperature scale (also called the **Kelvin scale**) employs the SI unit of absolute temperature, the **kelvin,** which is defined to be **1/273.16 of the difference between absolute zero and the temperature of the triple point of water.**

Figure 19.6 gives the absolute temperature for various physical processes and structures. The temperature of absolute zero (0 K) cannot be achieved, although laboratory experiments have come very close, reaching temperatures of less than one nanokelvin.

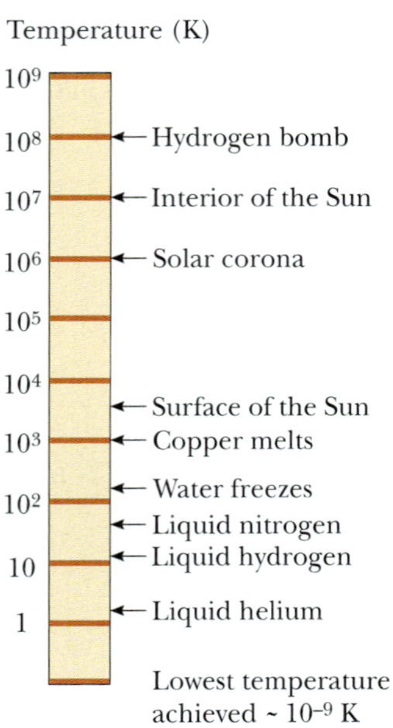

Figure 19.6 Absolute temperatures at which various physical processes occur. Notice that the scale is logarithmic.

The Celsius, Fahrenheit, and Kelvin Temperature Scales[3]

Equation 19.1 shows that the Celsius temperature T_C is shifted from the absolute (Kelvin) temperature T by 273.15°. Because the size of one degree is the same on the two scales, a temperature difference of 5°C is equal to a temperature difference of 5 K. The two scales differ only in the choice of the zero point. Therefore, the ice-point temperature on the Kelvin scale, 273.15 K, corresponds to 0.00°C, and the Kelvin-scale steam point, 373.15 K, is equivalent to 100.00°C.

A common temperature scale in everyday use in the United States is the **Fahrenheit scale**. This scale sets the temperature of the ice point at 32°F and the temperature of the steam point at 212°F. The relationship between the Celsius and Fahrenheit temperature scales is

$$T_F = \tfrac{9}{5}T_C + 32°F \tag{19.2}$$

We can use Equations 19.1 and 19.2 to find a relationship between changes in temperature on the Celsius, Kelvin, and Fahrenheit scales:

$$\Delta T_C = \Delta T = \tfrac{5}{9}\,\Delta T_F \tag{19.3}$$

Of these three temperature scales, only the Kelvin scale is based on a true zero value of temperature. The Celsius and Fahrenheit scales are based on an arbitrary zero associated with one particular substance, water, on one particular planet, Earth. Therefore, if you encounter an equation that calls for a temperature T or that involves a ratio of temperatures, you *must* convert all temperatures to kelvins. If the equation contains a change in temperature ΔT, using Celsius temperatures will give you the correct answer, in light of Equation 19.3, but it is always *safest* to convert temperatures to the Kelvin scale.

[3] Named after Anders Celsius (1701–1744), Daniel Gabriel Fahrenheit (1686–1736), and William Thomson, Lord Kelvin (1824–1907), respectively.

Quick Quiz 19.2 Consider the following pairs of materials. Which pair represents two materials, one of which is twice as hot as the other? (a) boiling water at 100°C, a glass of water at 50°C (b) boiling water at 100°C, frozen methane at −50°C (c) an ice cube at −20°C, flames from a circus fire-eater at 233°C (d) none of these pairs

EXAMPLE 19.1 **Converting Temperatures**

On a day when the temperature reaches 50°F, what is the temperature in degrees Celsius and in kelvins?

SOLUTION

Conceptualize In the United States, a temperature of 50°F is well understood. In many other parts of the world, however, this temperature might be meaningless because people are familiar with the Celsius temperature scale.

Categorize This example is a simple substitution problem.

Substitute the given temperature into Equation 19.2:

$$T_C = \tfrac{5}{9}(T_F - 32) = \tfrac{5}{9}(50 - 32) = 10°\text{C}$$

Use Equation 19.1 to find the Kelvin temperature:

$$T = T_C + 273.15 = 10°\text{C} + 273.15 = 283\ \text{K}$$

A convenient set of weather-related temperature equivalents to keep in mind is that 0°C is (literally) freezing at 32°F, 10°C is cool at 50°F, 20°C is room temperature, 30°C is warm at 86°F, and 40°C is a hot day at 104°F.

19.4 Thermal Expansion of Solids and Liquids

Our discussion of the liquid thermometer makes use of one of the best-known changes in a substance: as its temperature increases, its volume increases. This phenomenon, known as **thermal expansion,** plays an important role in numerous engineering applications. For example, thermal-expansion joints such as those shown in Figure 19.7 must be included in buildings, concrete highways, railroad tracks, brick walls, and bridges to compensate for dimensional changes that occur as the temperature changes.

Thermal expansion is a consequence of the change in the *average* separation between the atoms in an object. To understand this concept, let's model the atoms as being connected by stiff springs as discussed in Section 15.3 and shown

George Semple

George Semple

(a) (b)

Figure 19.7 (a) Thermal-expansion joints are used to separate sections of roadways on bridges. Without these joints, the surfaces would buckle due to thermal expansion on very hot days or crack due to contraction on very cold days. (b) The long, vertical joint is filled with a soft material that allows the wall to expand and contract as the temperature of the bricks changes.

in Figure 15.11b. At ordinary temperatures, the atoms in a solid oscillate about their equilibrium positions with an amplitude of approximately 10^{-11} m and a frequency of approximately 10^{13} Hz. The average spacing between the atoms is about 10^{-10} m. As the temperature of the solid increases, the atoms oscillate with greater amplitudes; as a result, the average separation between them increases.[4] Consequently, the object expands.

If thermal expansion is sufficiently small relative to an object's initial dimensions, the change in any dimension is, to a good approximation, proportional to the first power of the temperature change. Suppose an object has an initial length L_i along some direction at some temperature and the length increases by an amount ΔL for a change in temperature ΔT. Because it is convenient to consider the fractional change in length per degree of temperature change, we define the **average coefficient of linear expansion** as

$$\alpha \equiv \frac{\Delta L/L_i}{\Delta T}$$

PITFALL PREVENTION 19.2

Do Holes Become Larger or Smaller?

When an object's temperature is raised, every linear dimension increases in size. That includes any holes in the material, which expand in the same way as if the hole were filled with the material as shown in Active Figure 19.8. Keep in mind the notion of thermal expansion as being similar to a photographic enlargement.

Experiments show that α is constant for small changes in temperature. For purposes of calculation, this equation is usually rewritten as

$$\Delta L = \alpha L_i \, \Delta T \tag{19.4}$$

or as

$$L_f - L_i = \alpha L_i (T_f - T_i) \tag{19.5}$$

where L_f is the final length, T_i and T_f are the initial and final temperatures, respectively, and the proportionality constant α is the average coefficient of linear expansion for a given material and has units of $(^\circ\text{C})^{-1}$.

It may be helpful to think of thermal expansion as an effective magnification or as a photographic enlargement of an object. For example, as a metal washer is heated (Active Fig. 19.8), all dimensions, including the radius of the hole, increase according to Equation 19.4. **A cavity in a piece of material expands in the same way as if the cavity were filled with the material.**

Table 19.1 lists the average coefficients of linear expansion for various materials. For these materials, α is positive, indicating an increase in length with increasing temperature. That is not always the case, however. Some substances—calcite ($CaCO_3$) is one example—expand along one dimension (positive α) and contract along another (negative α) as their temperatures are increased.

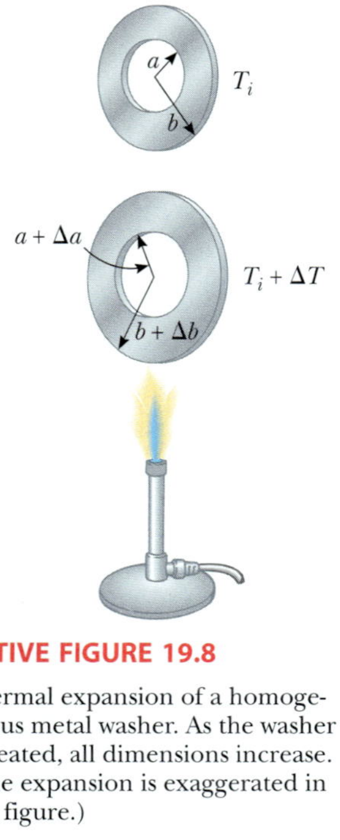

ACTIVE FIGURE 19.8

Thermal expansion of a homogeneous metal washer. As the washer is heated, all dimensions increase. (The expansion is exaggerated in this figure.)

Sign in at www.thomsonedu.com and go to ThomsonNOW to compare expansions for various temperatures of the burner and materials from which the washer is made.

Because the linear dimensions of an object change with temperature, it follows that surface area and volume change as well. The change in volume is proportional to the initial volume V_i and to the change in temperature according to the relationship

$$\Delta V = \beta V_i \, \Delta T \tag{19.6}$$

where β is the **average coefficient of volume expansion**. To find the relationship between β and α, assume the average coefficient of linear expansion of the solid is the same in all directions; that is, assume the material is *isotropic*. Consider a solid box of dimensions ℓ, w, and h. Its volume at some temperature T_i is $V_i = \ell w h$. If the temperature changes to $T_i + \Delta T$, its volume changes to $V_i + \Delta V$, where each dimension changes according to Equation 19.4. Therefore,

$$\begin{aligned} V_i + \Delta V &= (\ell + \Delta \ell)(w + \Delta w)(h + \Delta h) \\ &= (\ell + \alpha \ell \, \Delta T)(w + \alpha w \, \Delta T)(h + \alpha h \, \Delta T) \\ &= \ell w h (1 + \alpha \, \Delta T)^3 \\ &= V_i [1 + 3\alpha \, \Delta T + 3(\alpha \, \Delta T)^2 + (\alpha \, \Delta T)^3] \end{aligned}$$

[4] More precisely, thermal expansion arises from the *asymmetrical* nature of the potential energy curve for the atoms in a solid as shown in Figure 15.11a. If the oscillators were truly harmonic, the average atomic separations would not change regardless of the amplitude of vibration.

TABLE 19.1

Average Expansion Coefficients for Some Materials Near Room Temperature

Material	Average Linear Expansion Coefficient (α) $(°C)^{-1}$	Material	Average Volume Expansion Coefficient (β) $(°C)^{-1}$
Aluminum	24×10^{-6}	Alcohol, ethyl	1.12×10^{-4}
Brass and bronze	19×10^{-6}	Benzene	1.24×10^{-4}
Copper	17×10^{-6}	Acetone	1.5×10^{-4}
Glass (ordinary)	9×10^{-6}	Glycerin	4.85×10^{-4}
Glass (Pyrex)	3.2×10^{-6}	Mercury	1.82×10^{-4}
Lead	29×10^{-6}	Turpentine	9.0×10^{-4}
Steel	11×10^{-6}	Gasoline	9.6×10^{-4}
Invar (Ni–Fe alloy)	0.9×10^{-6}	Air[a] at 0°C	3.67×10^{-3}
Concrete	12×10^{-6}	Helium[a]	3.665×10^{-3}

[a] Gases do not have a specific value for the volume expansion coefficient because the amount of expansion depends on the type of process through which the gas is taken. The values given here assume the gas undergoes an expansion at constant pressure.

Dividing both sides by V_i and isolating the term $\Delta V/V_i$, we obtain the fractional change in volume:

$$\frac{\Delta V}{V_i} = 3\alpha\,\Delta T + 3(\alpha\,\Delta T)^2 + (\alpha\,\Delta T)^3$$

Because $\alpha\,\Delta T << 1$ for typical values of ΔT ($< \sim 100°C$), we can neglect the terms $3(\alpha\,\Delta T)^2$ and $(\alpha\,\Delta T)^3$. Upon making this approximation, we see that

$$\frac{\Delta V}{V_i} = 3\alpha\,\Delta T \quad \rightarrow \quad \Delta V = (3\alpha)V_i\,\Delta T$$

Comparing this expression to Equation 19.6 shows that

$$\beta = 3\alpha$$

In a similar way, you can show that the change in area of a rectangular plate is given by $\Delta A = 2\alpha A_i\,\Delta T$ (see Problem 41).

As Table 19.1 indicates, each substance has its own characteristic average coefficient of expansion. A simple mechanism called a *bimetallic strip,* found in practical devices such as thermostats, uses the difference in coefficients of expansion for different materials. It consists of two thin strips of dissimilar metals bonded together. As the temperature of the strip increases, the two metals expand by different amounts and the strip bends as shown in Figure 19.9.

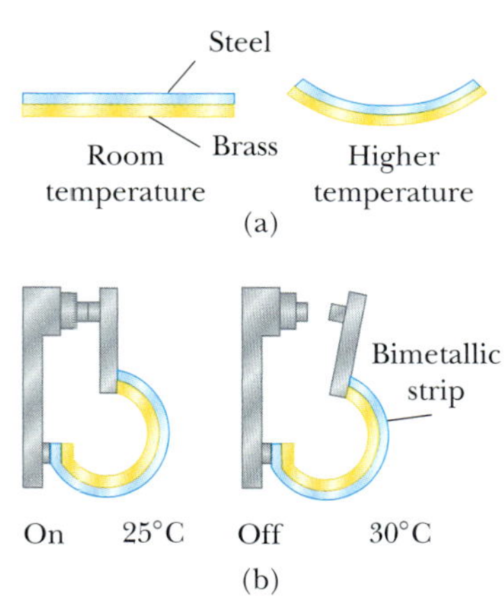

Figure 19.9 (a) A bimetallic strip bends as the temperature changes because the two metals have different expansion coefficients. (b) A bimetallic strip used in a thermostat to break or make electrical contact.

Quick Quiz 19.3 If you are asked to make a very sensitive glass thermometer, which of the following working liquids would you choose? (a) mercury (b) alcohol (c) gasoline (d) glycerin

Quick Quiz 19.4 Two spheres are made of the same metal and have the same radius, but one is hollow and the other is solid. The spheres are taken through the same temperature increase. Which sphere expands more? (a) The solid sphere expands more. (b) The hollow sphere expands more. (c) They expand by the same amount. (d) There is not enough information to say.

EXAMPLE 19.2 Expansion of a Railroad Track

A segment of steel railroad track has a length of 30.000 m when the temperature is 0.0°C.

(A) What is its length when the temperature is 40.0°C?

SOLUTION

Conceptualize Because the rail is relatively long, we expect to obtain a measurable increase in length for a 40°C temperature increase.

Categorize We will evaluate a length increase using the discussion of this section, so this example is a substitution problem.

Use Equation 19.4 and the value of the coefficient of linear expansion from Table 19.1:

$$\Delta L = \alpha L_i \, \Delta T = [11 \times 10^{-6} (°\text{C})^{-1}](30.000 \text{ m})(40.0°\text{C}) = 0.013 \text{ m}$$

Find the new length of the track:

$$L_f = 30.000 \text{ m} + 0.013 \text{ m} = \boxed{30.013 \text{ m}}$$

(B) Suppose the ends of the rail are rigidly clamped at 0.0°C so that expansion is prevented. What is the thermal stress set up in the rail if its temperature is raised to 40.0°C?

SOLUTION

Categorize This part of the example is an analysis problem because we need to use concepts from another chapter.

Analyze The thermal stress is the same as the tensile stress in the situation in which the rail expands freely and is then compressed with a mechanical force F back to its original length.

Find the tensile stress from Equation 12.6 using Young's modulus for steel from Table 12.1:

$$\text{Tensile stress} = \frac{F}{A} = Y \frac{\Delta L}{L_i}$$

$$\frac{F}{A} = (20 \times 10^{10} \text{ N/m}^2)\left(\frac{0.013 \text{ m}}{30.000 \text{ m}}\right) = \boxed{8.7 \times 10^7 \text{ N/m}^2}$$

Finalize The expansion in part (A) is 1.3 cm. This expansion is indeed measurable as predicted in the Conceptualize step. The thermal stress in part (B) can be avoided by leaving small expansion gaps between the rails.

What If? What if the temperature drops to −40.0° C? What is the length of the unclamped segment?

Answer The expression for the change in length in Equation 19.4 is the same whether the temperature increases or decreases. Therefore, if there is an increase in length of 0.013 m when the temperature increases by 40°C, there is a decrease in length of 0.013 m when the temperature decreases by 40°C. (We assume α is constant over the entire range of temperatures.) The new length at the colder temperature is 30.000 m − 0.013 m = 29.987 m.

EXAMPLE 19.3 The Thermal Electrical Short

A poorly designed electronic device has two bolts attached to different parts of the device that almost touch each other in its interior as in Figure 19.10. The steel and brass bolts are at different electric potentials, and if they touch, a short circuit will develop, damaging the device. (We will study electric potential in Chapter 25.) The initial gap between the ends of the bolts is 5.0 μm at 27°C. At what temperature will the bolts touch? Assume that the distance between the walls of the device is not affected by the temperature change.

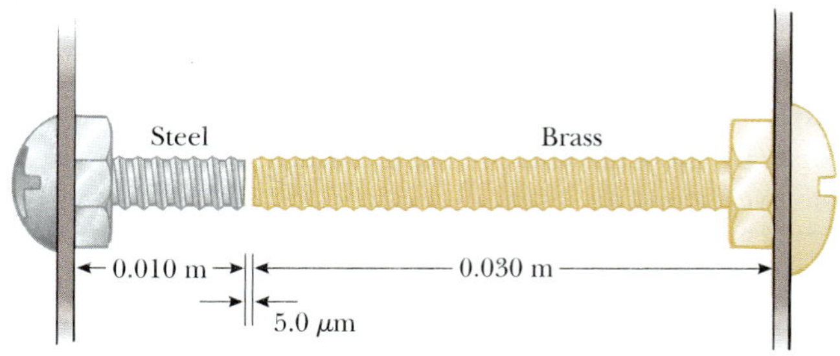

Figure 19.10 (Example 19.3) Two bolts attached to different parts of an electrical device are almost touching when the temperature is 27°C. As the temperature increases, the ends of the bolts move toward each other.

SOLUTION

Conceptualize Imagine the ends of both bolts expanding into the gap between them as the temperature rises.

Categorize We categorize this example as a thermal expansion problem in which the *sum* of the changes in length of the two bolts must equal the length of the initial gap between the ends.

Analyze Set the sum of the length changes equal to the width of the gap:

$$\Delta L_{\text{br}} + \Delta L_{\text{st}} = \alpha_{\text{br}} L_{i,\text{br}}\ \Delta T + \alpha_{\text{st}} L_{i,\text{st}}\ \Delta T = 5.0 \times 10^{-6}\ \text{m}$$

Solve for ΔT:

$$\Delta T = \frac{5.0 \times 10^{-6}\ \text{m}}{\alpha_{\text{br}} L_{i,\text{br}} + \alpha_{\text{st}} L_{i,\text{st}}}$$

$$= \frac{5.0 \times 10^{-6}\ \text{m}}{[19 \times 10^{-6}(^\circ\text{C})^{-1}](0.030\ \text{m}) + [11 \times 10^{-6}(^\circ\text{C})^{-1}](0.010\ \text{m})} = 7.4^\circ\text{C}$$

Find the temperature at which the bolts touch:

$$T = 27^\circ\text{C} + 7.4^\circ\text{C} = \boxed{34^\circ\text{C}}$$

Finalize This temperature is possible if the air conditioning in the building housing the device fails for a long period on a very hot summer day.

The Unusual Behavior of Water

Liquids generally increase in volume with increasing temperature and have average coefficients of volume expansion about ten times greater than those of solids. Cold water is an exception to this rule as you can see from its density-versus-temperature curve shown in Figure 19.11. As the temperature increases from 0°C to 4°C, water contracts and its density therefore increases. Above 4°C, water expands with increasing temperature and so its density decreases. Therefore, the density of water reaches a maximum value of 1.000 g/cm^3 at 4°C.

We can use this unusual thermal-expansion behavior of water to explain why a pond begins freezing at the surface rather than at the bottom. When the air temperature drops from, for example, 7°C to 6°C, the surface water also cools and consequently decreases in volume. The surface water is denser than the water below it, which has not cooled and decreased in volume. As a result, the surface water sinks, and warmer water from below is forced to the surface to be cooled. When the air temperature is between 4°C and 0°C, however, the surface water

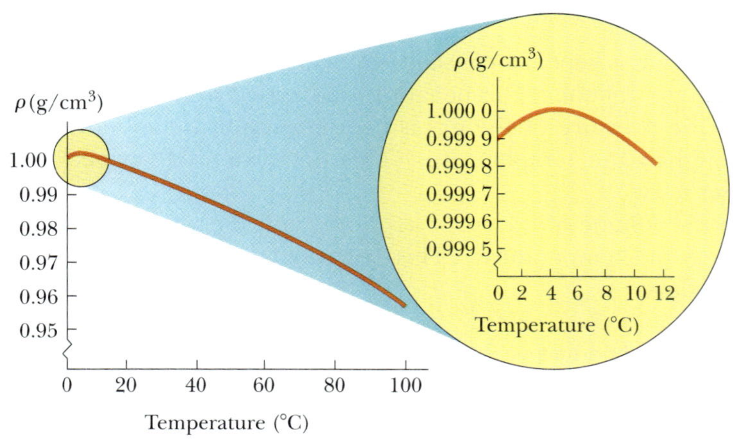

Figure 19.11 The variation in the density of water at atmospheric pressure with temperature. The inset at the right shows that the maximum density of water occurs at 4°C.

expands as it cools, becoming less dense than the water below it. The mixing process stops, and eventually the surface water freezes. As the water freezes, the ice remains on the surface because ice is less dense than water. The ice continues to build up at the surface, while water near the bottom remains at 4°C. If that were not the case, fish and other forms of marine life would not survive.

19.5 Macroscopic Description of an Ideal Gas

The volume expansion equation $\Delta V = \beta V_i \Delta T$ is based on the assumption that the material has an initial volume V_i before the temperature change occurs. Such is the case for solids and liquids because they have a fixed volume at a given temperature.

The case for gases is completely different. The interatomic forces within gases are very weak, and, in many cases, we can imagine these forces to be nonexistent and still make very good approximations. Therefore, *there is no equilibrium separation* for the atoms and no "standard" volume at a given temperature; the volume depends on the size of the container. As a result, we cannot express changes in volume ΔV in a process on a gas with Equation 19.6 because we have no defined volume V_i at the beginning of the process. Equations involving gases contain the volume V, rather than a *change* in the volume from an initial value, as a variable.

For a gas, it is useful to know how the quantities volume V, pressure P, and temperature T are related for a sample of gas of mass m. In general, the equation that interrelates these quantities, called the *equation of state*, is very complicated. If the gas is maintained at a very low pressure (or low density), however, the equation of state is quite simple and can be found experimentally. Such a low-density gas is commonly referred to as an **ideal gas**.[5] We can use the **ideal gas model** to make predictions that are adequate to describe the behavior of real gases at low pressures.

It is convenient to express the amount of gas in a given volume in terms of the number of moles n. One **mole** of any substance is that amount of the substance that contains **Avogadro's number** $N_A = 6.022 \times 10^{23}$ of constituent particles (atoms or molecules). The number of moles n of a substance is related to its mass m through the expression

$$n = \frac{m}{M} \tag{19.7}$$

where M is the molar mass of the substance. The molar mass of each chemical element is the atomic mass (from the periodic table; see Appendix C) expressed in grams per mole. For example, the mass of one He atom is 4.00 u (atomic mass units), so the molar mass of He is 4.00 g/mol.

Now suppose an ideal gas is confined to a cylindrical container whose volume can be varied by means of a movable piston as in Active Figure 19.12. If we assume the cylinder does not leak, the mass (or the number of moles) of the gas remains constant. For such a system, experiments provide the following information:

- When the gas is kept at a constant temperature, its pressure is inversely proportional to the volume. (This behavior is described historically as Boyle's law.)
- When the pressure of the gas is kept constant, the volume is directly proportional to the temperature. (This behavior is described historically as Charles's law.)
- When the volume of the gas is kept constant, the pressure is directly proportional to the temperature. (This behavior is described historically as Gay–Lussac's law.)

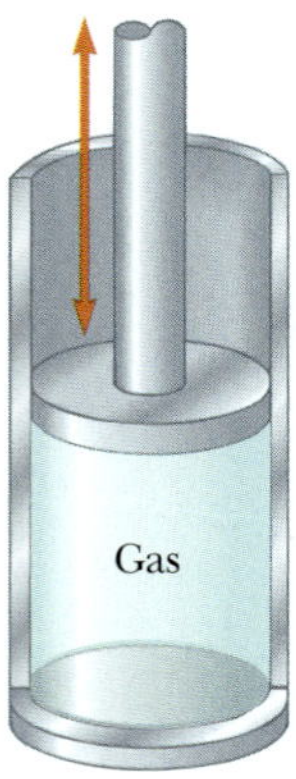

ACTIVE FIGURE 19.12

An ideal gas confined to a cylinder whose volume can be varied by means of a movable piston.

Sign in at www.thomsonedu.com and go to ThomsonNOW to choose to keep either the temperature or the pressure constant and verify Boyle's law and Charles's law.

[5] To be more specific, the assumptions here are that the temperature of the gas must not be too low (the gas must not condense into a liquid) or too high and that the pressure must be low. The concept of an ideal gas implies that the gas molecules do not interact except upon collision and that the molecular volume is negligible compared with the volume of the container. In reality, an ideal gas does not exist. Nonetheless, the concept of an ideal gas is very useful because real gases at low pressures behave as ideal gases do.

These observations are summarized by the **equation of state for an ideal gas:**

$$PV = nRT \tag{19.8}$$

◀ Equation of state for an ideal gas

In this expression, also known as the **ideal gas law,** n is the number of moles of gas in the sample and R is a constant. Experiments on numerous gases show that as the pressure approaches zero, the quantity PV/nT approaches the same value R for all gases. For this reason, R is called the **universal gas constant**. In SI units, in which pressure is expressed in pascals ($1\ \text{Pa} = 1\ \text{N/m}^2$) and volume in cubic meters, the product PV has units of newton·meters, or joules, and R has the value

$$R = 8.314\ \text{J/mol}\cdot\text{K} \tag{19.9}$$

If the pressure is expressed in atmospheres and the volume in liters ($1\ \text{L} = 10^3\ \text{cm}^3 = 10^{-3}\ \text{m}^3$), then R has the value

$$R = 0.082\ 06\ \text{L}\cdot\text{atm/mol}\cdot\text{K}$$

Using this value of R and Equation 19.8 shows that the volume occupied by 1 mol of any gas at atmospheric pressure and at 0°C (273 K) is 22.4 L.

The ideal gas law states that if the volume and temperature of a fixed amount of gas do not change, the pressure also remains constant. Consider a bottle of champagne that is shaken and then spews liquid when opened as shown in Figure 19.13. A common misconception is that the pressure inside the bottle is increased when the bottle is shaken. On the contrary, because the temperature of the bottle and its contents remains constant as long as the bottle is sealed, so does the pressure, as can be shown by replacing the cork with a pressure gauge. The correct explanation is as follows. Carbon dioxide gas resides in the volume between the liquid surface and the cork. The pressure of the gas in this volume is set higher than atmospheric pressure in the bottling process. Shaking the bottle displaces some of the carbon dioxide gas into the liquid, where it forms bubbles, and these bubbles become attached to the inside of the bottle. (No new gas is generated by shaking.) When the bottle is opened, the pressure is reduced to atmospheric pressure, which causes the volume of the bubbles to increase suddenly. If the bubbles are attached to the bottle (beneath the liquid surface), their rapid expansion expels liquid from the bottle. If the sides and bottom of the bottle are first tapped until no bubbles remain beneath the surface, however, the drop in pressure does not force liquid from the bottle when the champagne is opened.

Steve Niedorf/Getty Images

Figure 19.13 A bottle of champagne is shaken and opened. Liquid spews out of the opening. A common misconception is that the pressure inside the bottle is increased by the shaking.

The ideal gas law is often expressed in terms of the total number of molecules N. Because the number of moles n equals the ratio of the total number of molecules and Avogadro's number N_A, we can write Equation 19.8 as

$$PV = nRT = \frac{N}{N_A}RT$$

$$PV = Nk_BT \tag{19.10}$$

where k_B is **Boltzmann's constant,** which has the value

$$k_B = \frac{R}{N_A} = 1.38 \times 10^{-23}\ \text{J/K} \tag{19.11}$$

◀ Boltzmann's constant

It is common to call quantities such as P, V, and T the **thermodynamic variables** of an ideal gas. If the equation of state is known, one of the variables can always be expressed as some function of the other two.

PITFALL PREVENTION 19.3

So Many *k*s

There are a variety of physical quantities for which the letter k is used. Two we have seen previously are the force constant for a spring (Chapter 15) and the wave number for a mechanical wave (Chapter 16). Boltzmann's constant is another k, and we will see k used for thermal conductivity in Chapter 20 and for an electrical constant in Chapter 23. To make some sense of this confusing state of affairs, we use a subscript B for Boltzmann's constant to help us recognize it. In this book, you will see Boltzmann's constant as k_B, but you may see Boltzmann's constant in other resources as simply k.

Quick Quiz 19.5 A common material for cushioning objects in packages is made by trapping bubbles of air between sheets of plastic. This material is more effective at keeping the contents of the package from moving around inside the package on (a) a hot day (b) a cold day (c) either hot or cold days.

Quick Quiz 19.6 On a winter day, you turn on your furnace and the temperature of the air inside your home increases. Assume your home has the normal amount of leakage between inside air and outside air. Is the number of moles of air in your room at the higher temperature (a) larger than before, (b) smaller than before, or (c) the same as before?

EXAMPLE 19.4 Heating a Spray Can

A spray can containing a propellant gas at twice atmospheric pressure (202 kPa) and having a volume of 125.00 cm^3 is at 22°C. It is then tossed into an open fire. When the temperature of the gas in the can reaches 195°C, what is the pressure inside the can? Assume any change in the volume of the can is negligible.

SOLUTION

Conceptualize Intuitively, you should expect that the pressure of the gas in the container increases because of the increasing temperature.

Categorize We model the gas in the can as ideal and use the ideal gas law to calculate the new pressure.

Analyze Rearrange Equation 19.8:

$$(1) \quad \frac{PV}{T} = nR$$

No air escapes during the compression, so that n, and therefore nR, remains constant. Hence, set the initial value of the left side of Equation (1) equal to the final value:

$$(2) \quad \frac{P_iV_i}{T_i} = \frac{P_fV_f}{T_f}$$

Because the initial and final volumes of the gas are assumed to be equal, cancel the volumes:

$$(3) \quad \frac{P_i}{T_i} = \frac{P_f}{T_f}$$

Solve for P_f:

$$P_f = \left(\frac{T_f}{T_i}\right)P_i = \left(\frac{468\text{ K}}{295\text{ K}}\right)(202\text{ kPa}) = \boxed{320\text{ kPa}}$$

Finalize The higher the temperature, the higher the pressure exerted by the trapped gas as expected. If the pressure increases sufficiently, the can may explode. Because of this possibility, you should never dispose of spray cans in a fire.

What If? Suppose we include a volume change due to thermal expansion of the steel can as the temperature increases. Does that alter our answer for the final pressure significantly?

Answer Because the thermal expansion coefficient of steel is very small, we do not expect much of an effect on our final answer.

Find the change in the volume of the can using Equation 19.6 and the value for α for steel from Table 19.1:

$$\Delta V = \beta V_i\,\Delta T = 3\alpha V_i\,\Delta T$$
$$= 3[11 \times 10^{-6}(°\text{C})^{-1}](125.00\text{ cm}^3)(173°\text{C}) = 0.71\text{ cm}^3$$

Start from Equation (2) again and find an equation for the final pressure:

$$P_f = \left(\frac{T_f}{T_i}\right)\left(\frac{V_i}{V_f}\right)P_i$$

This result differs from Equation (3) only in the factor V_i/V_f. Evaluate this factor:

$$\frac{V_i}{V_f} = \frac{125.00\text{ cm}^3}{(125.00\text{ cm}^3 + 0.71\text{ cm}^3)} = 0.994 = 99.4\%$$

Therefore, the final pressure will differ by only 0.6% from the value calculated without considering the thermal expansion of the can. Taking 99.4% of the previous final pressure, the final pressure including thermal expansion is 318 kPa.

Summary

ThomsonNOW™ Sign in at **www.thomsonedu.com** and go to ThomsonNOW to take a practice test for this chapter.

DEFINITIONS

Two objects are in **thermal equilibrium** with each other if they do not exchange energy when in thermal contact.

Temperature is the property that determines whether an object is in thermal equilibrium with other objects. **Two objects in thermal equilibrium with each other are at the same temperature.** The SI unit of absolute temperature is the **kelvin,** which is defined to be 1/273.16 of the difference between absolute zero and the temperature of the triple point of water.

CONCEPTS AND PRINCIPLES

The **zeroth law of thermodynamics** states that if objects A and B are separately in thermal equilibrium with a third object C, then objects A and B are in thermal equilibrium with each other.

When the temperature of an object is changed by an amount ΔT, its length changes by an amount ΔL that is proportional to ΔT and to its initial length L_i:

$$\Delta L = \alpha L_i \, \Delta T \qquad \textbf{(19.4)}$$

where the constant α is the **average coefficient of linear expansion.** The **average coefficient of volume expansion** β for a solid is approximately equal to 3α.

An **ideal gas** is one for which PV/nT is constant. An ideal gas is described by the **equation of state,**

$$PV = nRT \qquad \textbf{(19.8)}$$

where n equals the number of moles of the gas, P is its pressure, V is its volume, R is the universal gas constant (8.314 J/mol · K), and T is the absolute temperature of the gas. A real gas behaves approximately as an ideal gas if it has a low density.

Questions

☐ denotes answer available in *Student Solutions Manual/Study Guide;* **O** denotes objective question

1. Is it possible for two objects to be in thermal equilibrium if they are not in contact with each other? Explain.

2. ☐ A piece of copper is dropped into a beaker of water. If the water's temperature rises, what happens to the temperature of the copper? Under what conditions are the water and copper in thermal equilibrium?

3. In describing his upcoming trip to the Moon and as portrayed in the movie *Apollo 13* (Universal, 1995), astronaut Jim Lovell said, "I'll be walking in a place where there's a 400-degree difference between sunlight and shadow." What is it that is hot in sunlight and cold in shadow? Suppose an astronaut standing on the Moon holds a thermometer in his gloved hand. Is the thermometer reading the temperature of the vacuum at the Moon's surface? Does it read any temperature? If so, what object or substance has that temperature?

4. **O** What would happen if the glass of a thermometer expanded more on warming than did the liquid in the tube? (a) The thermometer would break. (b) It could not be used for measuring temperature. (c) It could be used for temperatures only below room temperature. (d) You would have to hold it with the bulb on top. (e) Larger numbers would be found closer to the bulb. (f) The numbers would not be evenly spaced.

5. **O** Suppose you empty a tray of ice cubes into a bowl partly full of water and cover the bowl. After one-half hour, the contents of the bowl come to thermal equilibrium, with more liquid water and less ice than you started with. Which of the following is true? (a) The temperature of the liquid water is higher than the temperature of the remaining ice. (b) The temperature of the liquid water is the same as that of the ice. (c) The temperature of the liquid water is less than that of the ice. (d) The comparative temperatures of the liquid water and ice depend on the amounts present.

6. **O** The coefficient of linear expansion of copper is 17×10^{-6} (°C)$^{-1}$. The Statue of Liberty is 93 m tall on a summer morning when the temperature is 25°C. Assume the copper plates covering the statue are mounted edge to edge without expansion joints and do not buckle or bind on the framework supporting them as the day grows hot. What is the order of magnitude of the statue's increase in height? (a) 0.1 mm (b) 1 mm (c) 1 cm (d) 10 cm (e) 1 m (f) 10 m (g) none of these answers

7. Markings to indicate length are placed on a steel tape in a room that has a temperature of 22°C. Are measurements made with the tape on a day when the temperature is 27°C too long, too short, or accurate? Defend your answer.

8. Use a periodic table of the elements (see Appendix C) to determine the number of grams in one mole of (a) hydrogen, which has diatomic molecules; (b) helium; and (c) carbon monoxide.

9. What does the ideal gas law predict about the volume of a sample of gas at absolute zero? Why is this prediction incorrect?

10. **O** A rubber balloon is filled with 1 L of air at 1 atm and 300 K and is then put into a cryogenic refrigerator at 100 K. The rubber remains flexible as it cools. **(i)** What happens to the volume of the balloon? (a) It decreases to $\frac{1}{6}$ L. (b) It decreases to $\frac{1}{3}$ L. (c) It decreases to $1/\sqrt{3}$ L. (d) It is constant. (e) It increases. **(ii)** What happens to the pressure of the air in the balloon? (a) It decreases to $\frac{1}{6}$ atm. (b) It decreases to $\frac{1}{3}$ atm. (c) It decreases to $1/\sqrt{3}$ atm. (d) It is constant. (e) It increases.

11. **O** Two cylinders at the same temperature contain the same quantity of the same kind of gas. Is it possible that cylinder A has three times the volume of cylinder B? If so, what can you conclude about the pressures the gases exert? (a) The situation is not possible. (b) It is possible, but we can conclude nothing about the pressure. (c) It is possible only if the pressure in A is three times the pressure in B. (d) The pressures must be equal. (e) The pressure in A must be one-third the pressure in B.

12. **O** Choose every correct answer. The graph of pressure versus temperature in Figure 19.5 shows what for each sample of gas? (a) The pressure is proportional to the Celsius temperature. (b) The pressure is a linear function of the temperature. (c) The pressure increases at the same rate as the temperature. (d) The pressure increases with temperature at a constant rate.

13. **O** A cylinder with a piston contains a sample of a thin gas. The kind of gas and the sample size can be changed. The cylinder can be placed in different constant-temperature baths, and the piston can be held in different positions. Rank the following cases according to the pressure of the gas from the highest to the lowest, displaying any cases of equality. (a) A 2-mmol sample of oxygen is held at 300 K in a 100-cm^3 container. (b) A 2-mmol sample of oxygen is held at 600 K in a 200-cm^3 container. (c) A 2-mmol sample of oxygen is held at 600 K in a 300-cm^3 container. (d) A 4-mmol sample of helium is held at 300 K in a 200-cm^3 container. (e) A 4-mmol sample of helium is held at 250 K in a 200-cm^3 container.

14. The pendulum of a certain pendulum clock is made of brass. When the temperature increases, does the period of the clock increase, decrease, or remain the same? Explain.

15. An automobile radiator is filled to the brim with water when the engine is cool. What happens to the water when the engine is running and the water has been raised to a high temperature? What do modern automobiles have in their cooling systems to prevent the loss of coolants?

16. Metal lids on glass jars can often be loosened by running hot water over them. Why does that work?

17. When the metal ring and metal sphere in Figure Q19.17 are both at room temperature, the sphere can barely be passed through the ring. After the sphere is warmed in a flame, it cannot be passed through the ring. Explain. **What If?** What if the ring is warmed and the sphere is left at room temperature? Does the sphere pass through the ring?

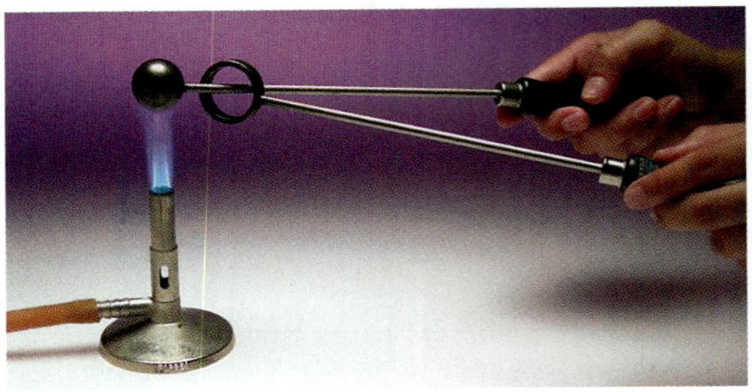

Figure Q19.17

Problems

WebAssign The Problems from this chapter may be assigned online in WebAssign.

ThomsonNOW Sign in at **www.thomsonedu.com** and go to ThomsonNOW to assess your understanding of this chapter's topics with additional quizzing and conceptual questions.

1, 2, 3 denotes straightforward, intermediate, challenging; ☐ denotes full solution available in *Student Solutions Manual/Study Guide;* ▲ denotes coached solution with hints available at **www.thomsonedu.com;** ▭ denotes developing symbolic reasoning; ● denotes asking for qualitative reasoning; 🖥 denotes computer useful in solving problem

Section 19.2 Thermometers and the Celsius Temperature Scale

Section 19.3 The Constant-Volume Gas Thermometer and the Absolute Temperature Scale

1. ▲ A constant-volume gas thermometer is calibrated in dry ice (that is, evaporating carbon dioxide in the solid state, with a temperature of −80.0°C) and in boiling ethyl alcohol (78.0°C). The two pressures are 0.900 atm and 1.635 atm. (a) What Celsius value of absolute zero does the calibration yield? What is the pressure at (b) the freezing point of water and (c) the boiling point of water?

2. The temperature difference between the inside and the outside of an automobile engine is 450°C. Express this temperature difference on (a) the Fahrenheit scale and (b) the Kelvin scale.

3. Liquid nitrogen has a boiling point of −195.81°C at atmospheric pressure. Express this temperature (a) in degrees Fahrenheit and (b) in kelvins.

4. The melting point of gold is 1 064°C, and its boiling point is 2 660°C. (a) Express these temperatures in kelvins. (b) Compute the difference between these temperatures in Celsius degrees and kelvins.

Section 19.4 Thermal Expansion of Solids and Liquids

Note: Table 19.1 is available for use in solving problems in this section.

5. A copper telephone wire has essentially no sag between poles 35.0 m apart on a winter day when the temperature is −20.0°C. How much longer is the wire on a summer day when $T_C = 35.0$°C?

6. The concrete sections of a certain superhighway are designed to have a length of 25.0 m. The sections are poured and cured at 10.0°C. What minimum spacing should the engineer leave between the sections to eliminate buckling if the concrete is to reach a temperature of 50.0°C?

7. ▲ The active element of a certain laser is made of a glass rod 30.0 cm long and 1.50 cm in diameter. If the temperature of the rod increases by 65.0°C, what is the increase in (a) its length, (b) its diameter, and (c) its volume? Assume the average coefficient of linear expansion of the glass is 9.00×10^{-6} (°C)$^{-1}$.

8. **Review problem.** Inside the wall of a house, an L-shaped section of hot water pipe consists of a straight, horizontal piece 28.0 cm long, an elbow, and a straight vertical piece 134 cm long (Fig. P19.8). A stud and a second-story floorboard hold stationary the ends of this section of copper pipe. Find the magnitude and direction of the displacement of the pipe elbow when the water flow is turned on, raising the temperature of the pipe from 18.0°C to 46.5°C.

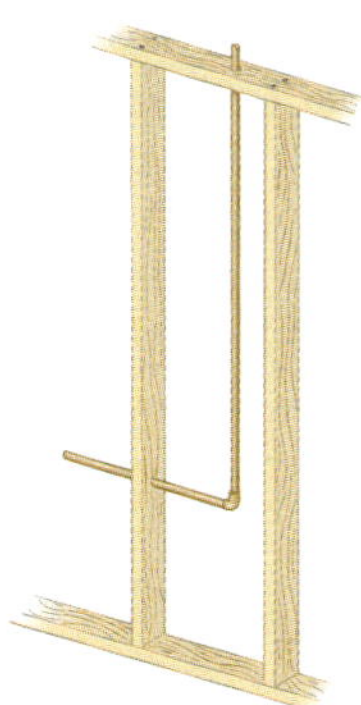

Figure P19.8

9. ● A thin brass ring of inner diameter 10.00 cm at 20.0°C is warmed and slipped over an aluminum rod of diameter 10.01 cm at 20.0°C. Assuming the average coefficients of linear expansion are constant, (a) to what temperature must this combination be cooled to separate the parts? Explain whether this separation is attainable. (b) **What If?** What if the aluminum rod were 10.02 cm in diameter?

10. ● At 20.0°C, an aluminum ring has an inner diameter of 5.000 0 cm and a brass rod has a diameter of 5.050 0 cm. (a) If only the ring is warmed, what temperature must it reach so that it will just slip over the rod? (b) **What If?** If both the ring and the rod are warmed together, what temperature must they both reach so that the ring barely slips over the rod? Would this latter process work? Explain.

11. ● A volumetric flask made of Pyrex is calibrated at 20.0°C. It is filled to the 100-mL mark with 35.0°C acetone. (a) What is the volume of the acetone when it cools to 20.0°C? (b) How significant is the change in volume of the flask?

12. On a day that the temperature is 20.0°C, a concrete walk is poured in such a way that the ends of the walk are unable to move. (a) What is the stress in the cement on a hot day of 50.0°C? (b) Does the concrete fracture? Take Young's modulus for concrete to be 7.00×10^9 N/m^2 and the compressive strength to be 2.00×10^9 N/m^2.

13. A hollow aluminum cylinder 20.0 cm deep has an internal capacity of 2.000 L at 20.0°C. It is completely filled with turpentine and then slowly warmed to 80.0°C. (a) How much turpentine overflows? (b) If the cylinder is then cooled back to 20.0°C, how far below the cylinder's rim does the turpentine's surface recede?

14. ● The Golden Gate Bridge in San Francisco has a main span of length 1.28 km, one of the longest in the world. Imagine that a taut steel wire with this length and a cross-sectional area of 4.00×10^{-6} m^2 is laid on the bridge deck with its ends attached to the towers of the bridge and that on this summer day the temperature of the wire is 35.0°C. (a) When winter arrives, the towers stay the same distance apart and the bridge deck keeps the same shape as its expansion joints open. When the temperature drops to −10.0°C, what is the tension in the wire? Take Young's modulus for steel to be 20.0×10^{10} N/m^2. (b) Permanent deformation occurs if the stress in the steel exceeds its elastic limit of 3.00×10^8 N/m^2. At what temperature would the wire reach its elastic limit? (c) **What If?** Explain how your answers to parts (a) and (b) would change if the Golden Gate Bridge were twice as long.

15. A certain telescope forms an image of part of a cluster of stars on a square silicon charge-coupled detector chip 2.00 cm on each side. A star field is focused on the chip when it is first turned on, and its temperature is 20.0°C. The star field contains 5 342 stars scattered uniformly. To make the detector more sensitive, it is cooled to −100°C. How many star images then fit onto the chip? The average coefficient of linear expansion of silicon is 4.68×10^{-6} (°C)$^{-1}$.

Section 19.5 Macroscopic Description of an Ideal Gas

16. On your wedding day your lover gives you a gold ring of mass 3.80 g. Fifty years later its mass is 3.35 g. On the average, how many atoms were abraded from the ring during each second of your marriage? The molar mass of gold is 197 g/mol.

17. An automobile tire is inflated with air originally at 10.0°C and normal atmospheric pressure. During the process, the air is compressed to 28.0% of its original volume and the temperature is increased to 40.0°C. (a) What is the tire pressure? (b) After the car is driven at high speed, the tire's air temperature rises to 85.0°C and the tire's

interior volume increases by 2.00%. What is the new tire pressure (absolute) in pascals?

18. Gas is contained in an 8.00-L vessel at a temperature of 20.0°C and a pressure of 9.00 atm. (a) Determine the number of moles of gas in the vessel. (b) How many molecules are in the vessel?

19. An auditorium has dimensions 10.0 m × 20.0 m × 30.0 m. How many molecules of air fill the auditorium at 20.0°C and a pressure of 101 kPa?

20. A cook puts 9.00 g of water in a 2.00-L pressure cooker and warms it to 500°C. What is the pressure inside the container?

21. ▲ The mass of a hot-air balloon and its cargo (not including the air inside) is 200 kg. The air outside is at 10.0°C and 101 kPa. The volume of the balloon is 400 m^3. To what temperature must the air in the balloon be warmed before the balloon will lift off? (Air density at 10.0°C is 1.25 kg/m^3.)

22. ● *Male-pattern dumbness.* Your father and your younger brother are confronted with the same puzzle. Your father's garden sprayer and your brother's water cannon both have tanks with a capacity of 5.00 L (Fig. P19.22). Your father puts a negligible amount of concentrated fertilizer into his tank. They both pour in 4.00 L of water and seal up their tanks, so the tanks also contain air at atmospheric pressure. Next, each uses a hand-operated piston pump to inject more air until the absolute pressure in the tank reaches 2.40 atm and it becomes too difficult to move the pump handle. Now each uses his device to spray out water—not air—until the stream becomes feeble as it does when the pressure in the tank reaches 1.20 atm. Then he must pump it up again, spray again, and so on. To accomplish spraying out all the water, each finds he must pump up the tank three times. Here is the puzzle: most of the water sprays out as a result of the second pumping. The first and the third pumping-up processes seem just as difficult as the second but result in a disappointingly small amount of water coming out. Account for this phenomenon.

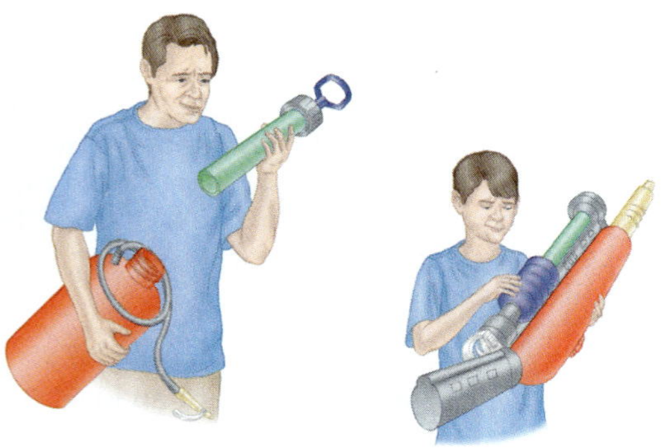

Figure P19.22

23. ● (a) Find the number of moles in one cubic meter of an ideal gas at 20.0°C and atmospheric pressure. (b) For air, Avogadro's number of molecules has mass 28.9 g. Calculate the mass of one cubic meter of air. State how the result compares with the tabulated density of air.

24. At 25.0 m below the surface of the sea (density = 1 025 kg/m^3), where the temperature is 5.00°C, a diver exhales an air bubble having a volume of 1.00 cm^3. If the surface temperature of the sea is 20.0°C, what is the volume of the bubble just before it breaks the surface?

25. ● A cube 10.0 cm on each edge contains air (with equivalent molar mass 28.9 g/mol) at atmospheric pressure and temperature 300 K. Find (a) the mass of the gas, (b) the gravitational force exerted on it, and (c) the force it exerts on each face of the cube. (d) Comment on the physical reason such a small sample can exert such a great force.

26. Estimate the mass of the air in your bedroom. State the quantities you take as data and the value you measure or estimate for each.

27. The pressure gauge on a tank registers the gauge pressure, which is the difference between the interior and exterior pressure. When the tank is full of oxygen (O_2), it contains 12.0 kg of the gas at a gauge pressure of 40.0 atm. Determine the mass of oxygen that has been withdrawn from the tank when the pressure reading is 25.0 atm. Assume the temperature of the tank remains constant.

28. In state-of-the-art vacuum systems, pressures as low as 10^{-9} Pa are being attained. Calculate the number of molecules in a 1.00-m^3 vessel at this pressure and a temperature of 27.0°C.

29. ● *How much water will a shearwater shear?* To measure how far below the ocean's surface a bird dives to catch a fish, Will Mackin used a method originated by Lord Kelvin for soundings by the British Navy. Mackin dusted the interiors of thin plastic tubes with powdered sugar and then sealed one end of each tube. Charging around on a rocky beach at night with a miner's headlamp, he would grab an Audubon's shearwater in its nest and attach a tube to its back. He would then catch the same bird the next night and remove the tube. After hundreds of captures, the birds thoroughly disliked him but were not permanently frightened away from the rookery. Assume in one trial, with a tube 6.50 cm long, he found that water had entered the tube to wash away the sugar over a distance of 2.70 cm from the open end. (a) Find the greatest depth to which the shearwater dove, assuming the air in the tube stayed at constant temperature. (b) Must the tube be attached to the bird in any particular orientation for this method to work? (Audubon's shearwater can dive to more than twice the depth you calculate, and larger species can dive nearly ten times deeper.)

30. A room of volume V contains air having equivalent molar mass M (in grams per mole). If the temperature of the room is raised from T_1 to T_2, what mass of air will leave the room? Assume the air pressure in the room is maintained at P_0.

Additional Problems

31. A student measures the length of a brass rod with a steel tape at 20.0°C. The reading is 95.00 cm. What will the tape indicate for the length of the rod when the rod and the tape are at (a) −15.0°C and (b) 55.0°C?

32. The density of gasoline is 730 kg/m^3 at 0°C. Its average coefficient of volume expansion is 9.60×10^{-4} $(°C)^{-1}$. Assume 1.00 gal of gasoline occupies 0.003 80 m^3. How many extra kilograms of gasoline would you get if you bought 10.0 gal of gasoline at 0°C rather than at 20.0°C from a pump that is not temperature compensated?

33. A mercury thermometer is constructed as shown in Figure P19.33. The capillary tube has a diameter of 0.004 00 cm, and the bulb has a diameter of 0.250 cm. Ignoring the expansion of the glass, find the change in height of the mercury column that occurs with a temperature change of 30.0°C.

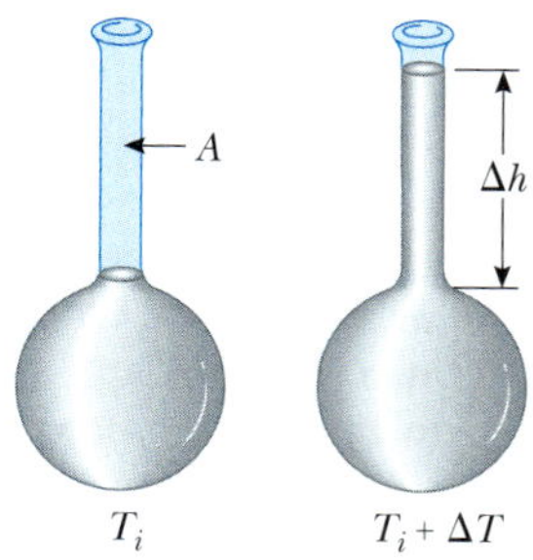

Figure P19.33 Problems 33 and 34.

34. ● A liquid with a coefficient of volume expansion β just fills a spherical shell of volume V_i at a temperature of T_i (Fig. P19.33). The shell is made of a material with an average coefficient of linear expansion α. The liquid is free to expand into an open capillary of area A projecting from the top of the sphere. (a) Assuming the temperature increases by ΔT, show that the liquid rises in the capillary by the amount Δh given by the equation $\Delta h = (V_i/A)(\beta - 3\alpha)\,\Delta T$. (b) For a typical system such as a mercury thermometer, why is it a good approximation to ignore the expansion of the shell?

35. **Review problem.** An aluminum pipe, 0.655 m long at 20.0°C and open at both ends, is used as a flute. The pipe is cooled to a low temperature, but then filled with air at 20.0°C as soon as you start to play it. After that, by how much does its fundamental frequency change as the metal rises in temperature from 5.00°C to 20.0°C?

36. Two metal bars are made of invar and one is made of aluminum. At 0°C, each of the three bars is drilled with two holes 40.0 cm apart. Pins are put through the holes to assemble the bars into an equilateral triangle. (a) First ignore the expansion of the invar. Find the angle between the invar bars as a function of Celsius temperature. (b) Is your answer accurate for negative as well as positive temperatures? Is it accurate for 0°C? (c) Solve the problem again, including the expansion of the invar. (d) Aluminum melts at 660°C and invar at 1 427°C. Assume the tabulated expansion coefficients are constant. What are the greatest and smallest attainable angles between the invar bars?

37. ● ▲ A liquid has a density ρ. (a) Show that the fractional change in density for a change in temperature ΔT is $\Delta\rho/\rho = -\beta\,\Delta T$. What does the negative sign signify? (b) Fresh water has a maximum density of 1.000 0 g/cm^3 at 4.0°C. At 10.0°C, its density is 0.999 7 g/cm^3. What is β for water over this temperature interval?

38. A cylinder is closed by a piston connected to a spring of constant 2.00×10^3 N/m (Fig. P19.38). With the spring relaxed, the cylinder is filled with 5.00 L of gas at a pressure of 1.00 atm and a temperature of 20.0°C. (a) If the piston has a cross-sectional area of 0.010 0 m^2 and negligible mass, how high will it rise when the temperature is raised to 250°C? (b) What is the pressure of the gas at 250°C?

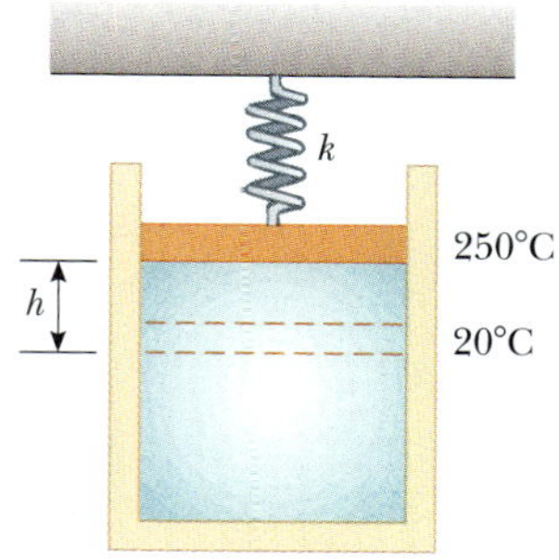

Figure P19.38

39. ▲ A vertical cylinder of cross-sectional area A is fitted with a tight-fitting, frictionless piston of mass m (Fig. P19.39). (a) If n moles of an ideal gas are in the cylinder at a temperature of T, what is the height h at which the piston is in equilibrium under its own weight? (b) What is the value for h if $n = 0.200$ mol, $T = 400$ K, $A = 0.008\,00$ m^2, and $m = 20.0$ kg?

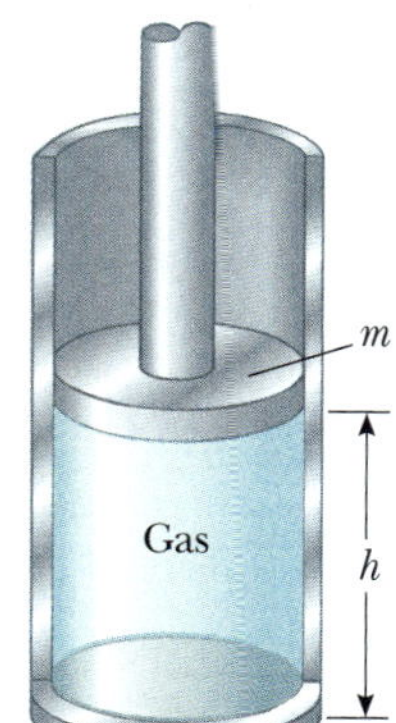

Figure P19.39

40. A bimetallic strip is made of two ribbons of different metals bonded together. (a) First assume the strip is originally straight. As the strip is warmed, the metal with the greater average coefficient of expansion expands more than the other, forcing the strip into an arc with the outer radius having a greater circumference (Fig. P19.40a, page 550). Derive an expression for the angle of bending θ as a function of the initial length of the strips, their average coefficients of linear expansion, the change in temperature, and the separation of the centers of the strips ($\Delta r = r_2 - r_1$). (b) Show that the angle of bending decreases to zero when ΔT decreases to zero and also when the two average coefficients of expansion become equal. (c) **What If?** What happens if the strip is cooled? (d) Figure P19.40b shows a compact spiral bimetallic strip in a home thermostat. If θ is interpreted as the angle of additional bending caused by a change in temperature, the equation from part (a) applies to it as well. The inner end of the spiral strip is fixed, and the outer end is free to move. Assume the metals are bronze and invar, the thickness of the strip is $2\,\Delta r = 0.500$ mm, and the overall length of the spiral strip is 20.0 cm. Find the angle through which the free end of the strip turns when the temperature changes by

1°C. The free end of the strip supports a capsule partly filled with mercury, visible above the strip in Figure P19.40b. When the capsule tilts, the mercury shifts from one end to the other to make or break an electrical contact switching the furnace on or off.

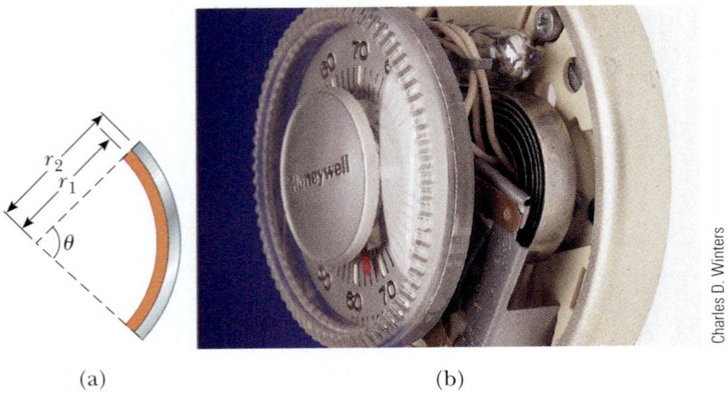

Figure P19.40

41. ● The rectangular plate shown in Figure P19.41 has an area A_i equal to ℓw. If the temperature increases by ΔT, each dimension increases according to the equation $\Delta L = \alpha L_i \Delta T$, where α is the average coefficient of linear expansion. Show that the increase in area is $\Delta A = 2\alpha A_i \Delta T$. What approximation does this expression assume?

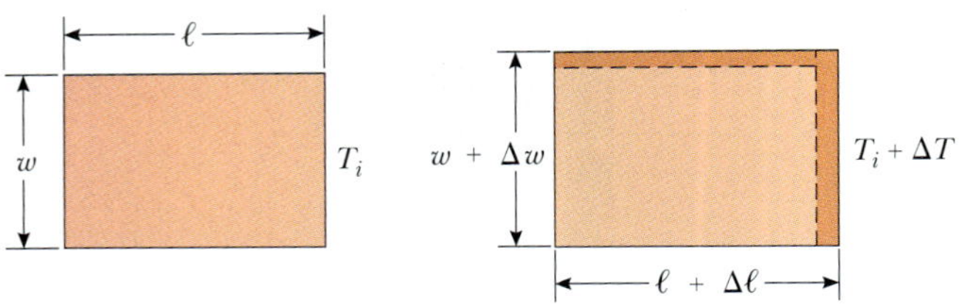

Figure P19.41

42. ● The measurement of the average coefficient of volume expansion for a liquid is complicated because the container also changes size with temperature. Figure P19.42 shows a simple means for overcoming this difficulty. With this apparatus, one arm of a U-tube is maintained at 0°C in an ice-water bath, and the other arm is maintained at a different temperature T_C in a constant-temperature bath. The connecting tube is horizontal. (a) Explain how use of this equipment permits determination of β for the liquid from measurements of the column heights h_0 and h_t of the liquid columns in the U-tube, without having to correct for expansion of the apparatus. (b) Derive the expression for β in terms of h_0, h_t, and T_C.

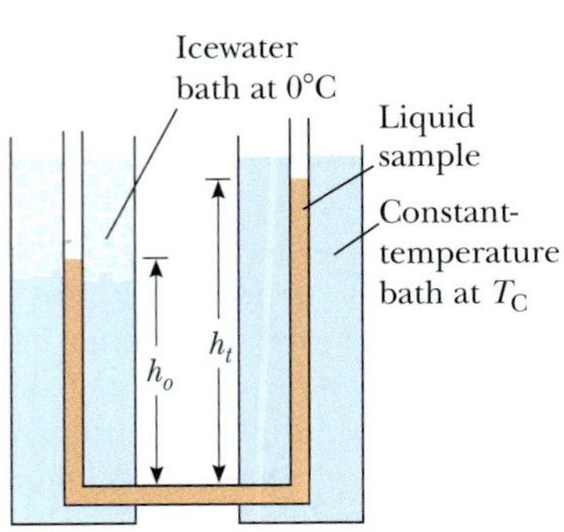

Figure P19.42

43. ● A copper rod and a steel rod are different in length by 5.00 cm at 0°C. The rods are warmed and cooled together. Is it possible that the length difference remains constant at all temperatures? Explain. Describe the lengths at 0°C as precisely as you can. Can you tell which rod is longer? Can you tell the lengths of the rods?

44. **Review problem.** A clock with a brass pendulum has a period of 1.000 s at 20.0°C. If the temperature increases to 30.0°C, (a) by how much does the period change and (b) how much time does the clock gain or lose in one week?

45. **Review problem.** Consider an object with any one of the shapes displayed in Table 10.2. What is the percentage increase in the moment of inertia of the object when it is warmed from 0°C to 100°C if it is composed of (a) copper or (b) aluminum? Assume the average linear expansion coefficients shown in Table 19.1 do not vary between 0°C and 100°C.

46. **Review problem.** (a) Derive an expression for the buoyant force on a spherical balloon, submerged in water, as a function of the depth below the surface, the volume of the balloon at the surface, the pressure at the surface, and the density of the water. (Assume the water temperature does not change with depth.) (b) Does the buoyant force increase or decrease as the balloon is submerged? (c) At what depth is the buoyant force one-half the surface value?

47. Two concrete spans of a 250-m-long bridge are placed end to end so that no room is allowed for expansion (Fig. P19.47a). If a temperature increase of 20.0°C occurs, what is the height y to which the spans rise when they buckle (Fig. P19.47b)?

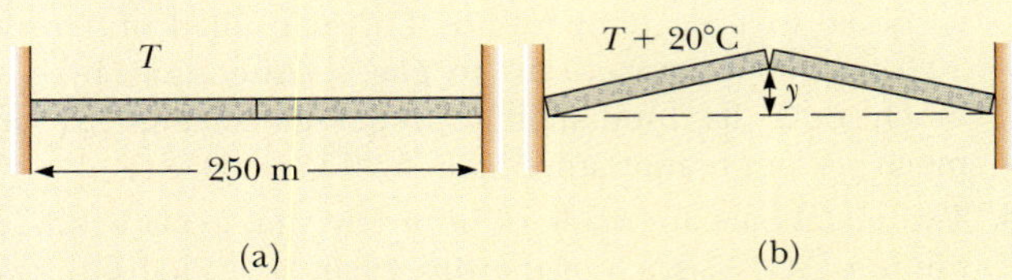

Figure P19.47 Problems 47 and 48.

48. Two concrete spans that form a bridge of length L are placed end to end so that no room is allowed for expansion (Fig. P19.47a). If a temperature increase of ΔT occurs, what is the height y to which the spans rise when they buckle (Fig. P19.47b)?

49. (a) Show that the density of an ideal gas occupying a volume V is given by $\rho = PM/RT$, where M is the molar mass. (b) Determine the density of oxygen gas at atmospheric pressure and 20.0°C.

50. ● (a) Take the definition of the coefficient of volume expansion to be

$$\beta = \frac{1}{V}\frac{dV}{dT}\bigg|_{P=\text{constant}} = \frac{1}{V}\frac{\partial V}{\partial T}$$

Use the equation of state for an ideal gas to show that the coefficient of volume expansion for an ideal gas at constant pressure is given by $\beta = 1/T$, where T is the absolute temperature. (b) What value does this expression predict for β at 0°C? State how this result compares with the experimental values for helium and air in Table

19.1. Notice that these values are much larger than the coefficients of volume expansion for most liquids and solids.

51. Starting with Equation 19.10, show that the total pressure P in a container filled with a mixture of several ideal gases is $P = P_1 + P_2 + P_3 + \ldots$, where $P_1, P_2, \ldots$, are the pressures that each gas would exert if it alone filled the container. (These individual pressures are called the *partial pressures* of the respective gases.) This result is known as *Dalton's law of partial pressures.*

52. ● **Review problem.** Following a collision in outer space, a copper disk at 850°C is rotating about its axis with an angular speed of 25.0 rad/s. As the disk radiates infrared light, its temperature falls to 20.0°C. No external torque acts on the disk. (a) Does the angular speed change as the disk cools? Explain how it changes or why it does not. (b) What is its angular speed at the lower temperature?

53. ● Helium gas is sold in steel tanks. If the helium is used to inflate a balloon, could the balloon lift the spherical tank the helium came in? Justify your answer. Steel will rupture if subjected to tensile stress greater than its yield strength of 5×10^8 N/m^2. *Suggestion:* You may consider a steel shell of radius r and thickness t having the density of iron and containing helium at high pressure and on the verge of breaking apart into two hemispheres.

54. A cylinder that has a 40.0-cm radius and is 50.0 cm deep is filled with air at 20.0°C and 1.00 atm (Fig. P19.54a). A 20.0-kg piston is now lowered into the cylinder, compressing the air trapped inside as it takes equilibrium height h_i (Fig. P19.54b). Finally, a 75.0-kg dog stands on the piston, further compressing the air, which remains at 20°C (Fig. P19.54c). (a) How far down (Δh) does the piston move when the dog steps onto it? (b) To what temperature should the gas be warmed to raise the piston and dog back to h_i?

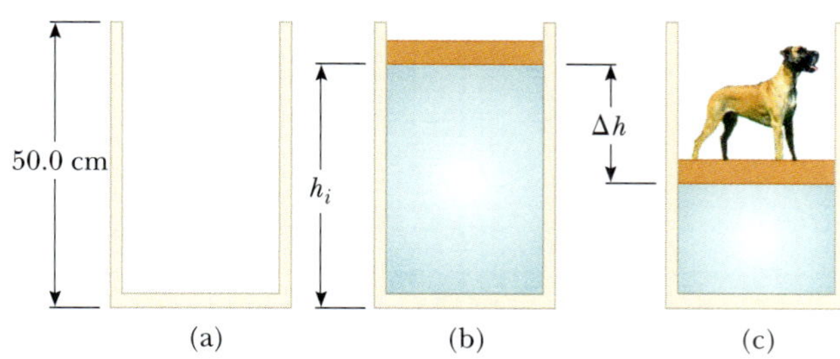

Figure P19.54

55. The relationship $L_f = L_i(1 + \alpha\,\Delta T)$ is a valid approximation when the average coefficient of expansion is small. If α is large, one must integrate the relationship $dL/dT = \alpha L$ to determine the final length. (a) Assuming that the coefficient of linear expansion is constant as L varies, determine a general expression for the final length. (b) Given a rod of length 1.00 m and a temperature change of 100.0°C, determine the error caused by the approximation when $\alpha = 2.00 \times 10^{-5}\ (°\text{C})^{-1}$ (a typical value for a metal) and when $\alpha = 0.020\,0\ (°\text{C})^{-1}$ (an unrealistically large value for comparison).

56. A steel wire and a copper wire, each of diameter 2.000 mm, are joined end to end. At 40.0°C, each has an unstretched length of 2.000 m. The wires are connected between two fixed supports 4.000 m apart on a tabletop. The steel wire extends from $x = -2.000$ m to $x = 0$, the copper wire extends from $x = 0$ to $x = 2.000$ m, and the tension is negligible. The temperature is then lowered to 20.0°C. At this lower temperature, find the tension in the wire and the x coordinate of the junction between the wires. (Refer to Tables 12.1 and 19.1.)

57. **Review problem.** A guitar string made of steel with a diameter of 1.00 mm is stretched between supports 80.0 cm apart. The temperature is 0.0°C. (a) Find the mass per unit length of this string. (Use the value 7.86×10^3 kg/m^3 for the density.) (b) The fundamental frequency of transverse oscillations of the string is 200 Hz. What is the tension in the string? (c) The temperature is raised to 30.0°C. Find the resulting values of the tension and the fundamental frequency. Assume both the Young's modulus (Table 12.1) and the average coefficient of expansion (Table 19.1) have constant values between 0.0°C and 30.0°C.

58. In a chemical processing plant, a reaction chamber of fixed volume V_0 is connected to a reservoir chamber of fixed volume $4V_0$ by a passage containing a thermally insulating porous plug. The plug permits the chambers to be at different temperatures. It allows gas to pass from either chamber to the other, ensuring that the pressure is the same in both. At one point in the processing, both chambers contain gas at a pressure of 1.00 atm and a temperature of 27.0°C. Intake and exhaust valves to the pair of chambers are closed. The reservoir is maintained at 27.0°C while the reaction chamber is warmed to 400°C. What is the pressure in both chambers after these temperatures are achieved?

59. ▲ A 1.00-km steel railroad rail is fastened securely at both ends when the temperature is 20.0°C. As the temperature increases, the rail buckles, taking the shape of an arc of a vertical circle. Find the height h of the center of the rail when the temperature is 25.0°C. You will need to solve a transcendental equation.

60. ● **Review problem.** A perfectly plane house roof makes an angle θ with the horizontal. When its temperature changes, between T_c before dawn each day and T_h in the middle of each afternoon, the roof expands and contracts uniformly with a coefficient of thermal expansion α_1. Resting on the roof is a flat, rectangular metal plate with expansion coefficient α_2, greater than α_1. The length of the plate is L, measured along the slope of the roof. The component of the plate's weight perpendicular to the roof is supported by a normal force uniformly distributed over the area of the plate. The coefficient of kinetic friction between the plate and the roof is μ_k. The plate is always at the same temperature as the roof, so we assume its temperature is continuously changing. Because of the difference in expansion coefficients, each bit of the plate is moving relative to the roof below it, except for points along a certain horizontal line running across the plate called the stationary line. If the temperature is rising, parts of the plate below the stationary line are moving down relative to the roof and feel a force of kinetic friction acting up the roof. Elements of area above the stationary line are sliding up the roof, and on them kinetic friction acts downward parallel to the roof. The stationary line occupies no area, so we assume no force of static fric-

tion acts on the plate while the temperature is changing. The plate as a whole is very nearly in equilibrium, so the net frictional force on it must be equal to the component of its weight acting down the incline. (a) Prove that the stationary line is at a distance of

$$\frac{L}{2}\left(1 - \frac{\tan\theta}{\mu_k}\right)$$

below the top edge of the plate. (b) Analyze the forces that act on the plate when the temperature is falling and prove that the stationary line is at that same distance above the bottom edge of the plate. (c) Show that the plate steps down the roof like an inchworm, moving each day by the distance

$$\frac{L}{\mu_k}(\alpha_2 - \alpha_1)(T_h - T_c)\tan\theta$$

(d) Evaluate the distance an aluminum plate moves each day if its length is 1.20 m, the temperature cycles between 4.00°C and 36.0°C, and the roof has slope 18.5°, coefficient of linear expansion 1.50×10^{-5} $(°C)^{-1}$, and coefficient of friction 0.420 with the plate. (e) **What If?** What if the expansion coefficient of the plate is less than that of the roof? Will the plate creep up the roof?

Answers to Quick Quizzes

19.1 (c). The direction of the transfer of energy depends only on temperature and not on the size of the object or on which object has more mass.

19.2 (c). The phrase "twice as hot" refers to a ratio of temperatures. When the given temperatures are converted to kelvins, only those in part (c) are in the correct ratio.

19.3 (c). Gasoline has the largest average coefficient of volume expansion.

19.4 (c). A cavity in a material expands in the same way as if it were filled with that material.

19.5 (a). On a cold day, the trapped air in the bubbles is reduced in pressure according to the ideal gas law. Therefore, the volume of the bubbles may be smaller than on a hot day and the package contents can shift more.

19.6 (b). Because of the increased temperature, the air expands. Consequently, some of the air leaks to the outside, leaving less air in the house.

In this photograph of Bow Lake in Banff National Park, Alberta, we see evidence of water in all three phases. In the lake is liquid water, and solid water in the form of snow appears on the ground. The clouds in the sky consist of liquid water droplets that have condensed from the gaseous water vapor in the air. Changes of a substance from one phase to another are a result of energy transfer. (Jacob Taposchaner/Getty Images)

20 The First Law of Thermodynamics

Until about 1850, the fields of thermodynamics and mechanics were considered to be two distinct branches of science. The law of conservation of energy seemed to describe only certain kinds of mechanical systems. Mid-19th-century experiments performed by Englishman James Joule and others, however, showed a strong connection between the transfer of energy by heat in thermal processes and the transfer of energy by work in mechanical processes. Today we know that mechanical energy can be transformed to internal energy, which is formally defined in this chapter. Once the concept of energy was generalized from mechanics to include internal energy, the law of conservation of energy emerged as a universal law of nature.

This chapter focuses on the concept of internal energy, the first law of thermodynamics, and some important applications of the first law. The first law of thermodynamics describes systems in which the only energy change is that of internal energy and the transfers of energy are by heat and work. A major difference in our discussion of work in this chapter from that in most of the chapters on mechanics is that we will consider work done on *deformable* systems.

20.1 Heat and Internal Energy

PITFALL PREVENTION 20.1

Internal Energy, Thermal Energy, and Bond Energy

When reading other physics books, you may see terms such as *thermal energy* and *bond energy*. Thermal energy can be interpreted as that part of the internal energy associated with random motion of molecules and, therefore, related to temperature. Bond energy is the intermolecular potential energy. Therefore,

Internal energy = thermal energy + bond energy

Although this breakdown is presented here for clarification with regard to other books, we will not use these terms because there is no need for them.

PITFALL PREVENTION 20.2

Heat, Temperature, and Internal Energy Are Different

As you read the newspaper or listen to the radio, be alert for incorrectly used phrases including the word *heat* and think about the proper word to be used in place of *heat*. Incorrect examples include "As the truck braked to a stop, a large amount of heat was generated by friction" and "The heat of a hot summer day . . ."

At the outset, it is important to make a major distinction between internal energy and heat, terms that are often incorrectly used interchangeably in popular language. **Internal energy is all the energy of a system that is associated with its microscopic components—atoms and molecules—when viewed from a reference frame at rest with respect to the center of mass of the system.** The last part of this sentence ensures that any bulk kinetic energy of the system due to its motion through space is not included in internal energy. Internal energy includes kinetic energy of random translational, rotational, and vibrational motion of molecules; vibrational potential energy associated with forces between atoms in molecules; and electric potential energy associated with forces between molecules. It is useful to relate internal energy to the temperature of an object, but this relationship is limited. We show in Section 20.3 that internal energy changes can also occur in the absence of temperature changes.

Heat is defined as the transfer of energy across the boundary of a system due to a temperature difference between the system and its surroundings. When you *heat* a substance, you are transferring energy into it by placing it in contact with surroundings that have a higher temperature. Such is the case, for example, when you place a pan of cold water on a stove burner. The burner is at a higher temperature than the water, and so the water gains energy. We shall also use the term *heat* to represent the amount of energy transferred by this method.

As an analogy to the distinction between heat and internal energy, consider the distinction between work and mechanical energy discussed in Chapter 7. The work done on a system is a measure of the amount of energy transferred to the system from its surroundings, whereas the mechanical energy (kinetic energy plus potential energy) of a system is a consequence of the motion and configuration of the system. Therefore, when a person does work on a system, energy is transferred from the person to the system. It makes no sense to talk about the work *of* a system; one can refer only to the work done *on* or *by* a system when some process has occurred in which energy has been transferred to or from the system. Likewise, it makes no sense to talk about the heat *of* a system; one can refer to heat only when energy has been transferred as a result of a temperature difference. Both heat and work are ways of changing the energy of a system.

Units of Heat

Early studies of heat focused on the resultant increase in temperature of a substance, which was often water. Initial notions of heat were based on a fluid called *caloric* that flowed from one substance to another and caused changes in temperature. From the name of this mythical fluid came an energy unit related to thermal processes, the **calorie (cal),** which is defined as **the amount of energy transfer necessary to raise the temperature of 1 g of water from 14.5°C to 15.5°C.**[1] (The "Calorie," written with a capital "C" and used in describing the energy content of foods, is actually a kilocalorie.) The unit of energy in the U.S. customary system is the **British thermal unit (Btu),** which is defined as **the amount of energy transfer required to raise the temperature of 1 lb of water from 63°F to 64°F.**

Once the relationship between energy in thermal and mechanical processes became clear, there was no need for a separate unit related to thermal processes. The *joule* has already been defined as an energy unit based on mechanical processes. Scientists are increasingly turning away from the calorie and the Btu and are using the joule when describing thermal processes. In this textbook, heat, work, and internal energy are usually measured in joules.

[1] Originally, the calorie was defined as the energy transfer necessary to raise the temperature of 1 g of water by 1°C. Careful measurements, however, showed that the amount of energy required to produce a 1°C change depends somewhat on the initial temperature; hence, a more precise definition evolved.

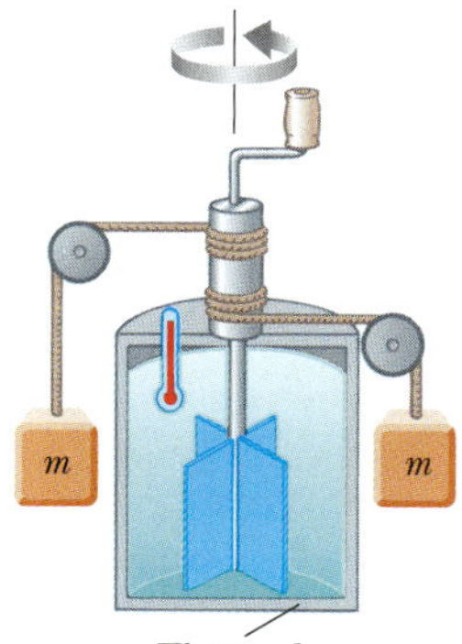

Figure 20.1 Joule's experiment for determining the mechanical equivalent of heat. The falling blocks rotate the paddles, causing the temperature of the water to increase.

The Mechanical Equivalent of Heat

In Chapters 7 and 8, we found that whenever friction is present in a mechanical system, the mechanical energy in the system decreases; in other words, mechanical energy is not conserved in the presence of nonconservative forces. Various experiments show that this mechanical energy does not simply disappear but is transformed into internal energy. You can perform such an experiment at home by hammering a nail into a scrap piece of wood. What happens to all the kinetic energy of the hammer once you have finished? Some of it is now in the nail as internal energy, as demonstrated by the nail being measurably warmer. Although this connection between mechanical and internal energy was first suggested by Benjamin Thompson, it was James Prescott Joule who established the equivalence of the decrease in mechanical energy and the increase in internal energy.

A schematic diagram of Joule's most famous experiment is shown in Figure 20.1. The system of interest is the water in a thermally insulated container. Work is done on the water by a rotating paddle wheel, which is driven by heavy blocks falling at a constant speed. If the energy lost in the bearings and through the walls is neglected, the loss in potential energy of the blocks–Earth system as the blocks fall equals the work done by the paddle wheel on the water. If the two blocks fall through a distance h, the loss in potential energy is $2mgh$, where m is the mass of one block; this energy causes the temperature of the water to increase due to friction between the paddles and the water. By varying the conditions of the experiment, Joule found that the loss in mechanical energy is proportional to the product of the mass of the water and the increase in water temperature. The proportionality constant was found to be approximately 4.18 J/g · °C. Hence, 4.18 J of mechanical energy raises the temperature of 1 g of water by 1°C. More precise measurements taken later demonstrated the proportionality to be 4.186 J/g · °C when the temperature of the water was raised from 14.5°C to 15.5°C. We adopt this "15-degree calorie" value:

$$1 \text{ cal} = 4.186 \text{ J} \tag{20.1}$$

This equality is known, for purely historical reasons, as the **mechanical equivalent of heat**.

By kind permission of the President and Council of the Royal Society

JAMES PRESCOTT JOULE
British physicist (1818–1889)
Joule received some formal education in mathematics, philosophy, and chemistry from John Dalton but was in large part self-educated. Joule's research led to the establishment of the principle of conservation of energy. His study of the quantitative relationship among electrical, mechanical, and chemical effects of heat culminated in his announcement in 1843 of the amount of work required to produce a unit of energy, called the mechanical equivalent of heat.

EXAMPLE 20.1 Losing Weight the Hard Way

A student eats a dinner rated at 2 000 Calories. He wishes to do an equivalent amount of work in the gymnasium by lifting a 50.0-kg barbell. How many times must he raise the barbell to expend this much energy? Assume he raises the barbell 2.00 m each time he lifts it and he regains no energy when he lowers the barbell.

SOLUTION

Conceptualize Imagine the student raising the barbell. He is doing work on the system of the barbell and the Earth, so energy is leaving his body. The total amount of work that the student must do is 2 000 Calories.

Categorize We model the system of the barbell and the Earth as a nonisolated system.

Analyze Reduce the conservation of energy equation, Equation 8.2, to the appropriate expression for the system of the barbell and the Earth:

$$(1)\quad \Delta U_{\text{total}} = W_{\text{total}}$$

Express the change in gravitational potential energy of the system after the barbell is raised once:

$$\Delta U = mgh$$

Express the total amount of energy that must be transferred into the system by work for lifting the barbell n times, assuming energy is not regained when the barbell is lowered:

$$(2)\quad \Delta U_{\text{total}} = nmgh$$

Substitute Equation (2) into Equation (1):

$$nmgh = W_{\text{total}}$$

Solve for n:

$$n = \frac{W_{\text{total}}}{mgh}$$

$$= \frac{(2\,000\text{ Cal})}{(50.0\text{ kg})(9.80\text{ m/s}^2)(2.00\text{ m})}\left(\frac{1.00\times10^3\text{ cal}}{\text{Calorie}}\right)\left(\frac{4.186\text{ J}}{1\text{ cal}}\right)$$

$$= 8.54\times10^3\text{ times}$$

Finalize If the student is in good shape and lifts the barbell once every 5 s, it will take him about 12 h to perform this feat. Clearly, it is much easier for this student to lose weight by dieting.

In reality, the human body is not 100% efficient. Therefore, not all the energy transformed within the body from the dinner transfers out of the body by work done on the barbell. Some of this energy is used to pump blood and perform other functions within the body. Therefore, the 2 000 Calories can be worked off in less time than 12 h when these other energy requirements are included.

20.2 Specific Heat and Calorimetry

When energy is added to a system and there is no change in the kinetic or potential energy of the system, the temperature of the system usually rises. (An exception to this statement is the case in which a system undergoes a change of state—also called a *phase transition*—as discussed in the next section.) If the system consists of a sample of a substance, we find that the quantity of energy required to raise the temperature of a given mass of the substance by some amount varies from one substance to another. For example, the quantity of energy required to raise the temperature of 1 kg of water by 1°C is 4 186 J, but the quantity of energy required to raise the temperature of 1 kg of copper by 1°C is only 387 J. In the discussion that follows, we shall use heat as our example of energy transfer, but keep in mind that the temperature of the system could be changed by means of any method of energy transfer.

The **heat capacity** C of a particular sample is defined as the amount of energy needed to raise the temperature of that sample by 1°C. From this definition, we see that if energy Q produces a change ΔT in the temperature of a sample, then

$$Q = C\,\Delta T \tag{20.2}$$

The **specific heat** c of a substance is the heat capacity per unit mass. Therefore, if energy Q transfers to a sample of a substance with mass m and the temperature of the sample changes by ΔT, the specific heat of the substance is

TABLE 20.1

Specific Heats of Some Substances at 25°C and Atmospheric Pressure

Substance	Specific Heat c J/kg · °C	Specific Heat c cal/g · °C	Substance	Specific Heat c J/kg · °C	Specific Heat c cal/g · °C
Elemental solids			*Other solids*		
Aluminum	900	0.215	Brass	380	0.092
Beryllium	1 830	0.436	Glass	837	0.200
Cadmium	230	0.055	Ice (−5°C)	2 090	0.50
Copper	387	0.092 4	Marble	860	0.21
Germanium	322	0.077	Wood	1 700	0.41
Gold	129	0.030 8	*Liquids*		
Iron	448	0.107	Alcohol (ethyl)	2 400	0.58
Lead	128	0.030 5	Mercury	140	0.033
Silicon	703	0.168	Water (15°C)	4 186	1.00
Silver	234	0.056	*Gas*		
			Steam (100°C)	2 010	0.48

$$c \equiv \frac{Q}{m\,\Delta T} \qquad (20.3)$$

◀ Specific heat

Specific heat is essentially a measure of how thermally insensitive a substance is to the addition of energy. The greater a material's specific heat, the more energy must be added to a given mass of the material to cause a particular temperature change. Table 20.1 lists representative specific heats.

From this definition, we can relate the energy Q transferred between a sample of mass m of a material and its surroundings to a temperature change ΔT as

$$Q = mc\,\Delta T \qquad (20.4)$$

For example, the energy required to raise the temperature of 0.500 kg of water by 3.00°C is $Q = (0.500\text{ kg})(4\ 186\text{ J/kg}\cdot{}^\circ\text{C})(3.00^\circ\text{C}) = 6.28 \times 10^3$ J. Notice that when the temperature increases, Q and ΔT are taken to be positive and energy transfers into the system. When the temperature decreases, Q and ΔT are negative and energy transfers out of the system.

Specific heat varies with temperature. If, however, temperature intervals are not too great, the temperature variation can be ignored and c can be treated as a constant.[2] For example, the specific heat of water varies by only about 1% from 0°C to 100°C at atmospheric pressure. Unless stated otherwise, we shall neglect such variations.

Quick Quiz 20.1 Imagine you have 1 kg each of iron, glass, and water, and all three samples are at 10°C. (a) Rank the samples from lowest to highest temperature after 100 J of energy is added to each sample. (b) Rank the samples from least to greatest amount of energy transferred by heat if each sample increases in temperature by 20°C.

Notice from Table 20.1 that water has the highest specific heat of common materials. This high specific heat is in part responsible for the moderate temperatures found near large bodies of water. As the temperature of a body of water decreases during the winter, energy is transferred from the cooling water to the air by heat, increasing the internal energy of the air. Because of the high specific heat

PITFALL PREVENTION 20.3
An Unfortunate Choice of Terminology

The name *specific heat* is an unfortunate holdover from the days when thermodynamics and mechanics developed separately. A better name would be *specific energy transfer*, but the existing term is too entrenched to be replaced.

PITFALL PREVENTION 20.4
Energy Can Be Transferred by Any Method

The symbol Q represents the amount of energy transferred, but keep in mind that the energy transfer in Equation 20.4 could be by *any* of the methods introduced in Chapter 8; it does not have to be heat. For example, repeatedly bending a wire coat hanger raises the temperature at the bending point by *work*.

[2] The definition given by Equation 20.4 assumes the specific heat does not vary with temperature over the interval $\Delta T = T_f - T_i$. In general, if c varies with temperature over the interval, the correct expression for Q is $Q = m\int_{T_i}^{T_f} c\,dT$.

of water, a relatively large amount of energy is transferred to the air for even modest temperature changes of the water. The prevailing winds on the West Coast of the United States are toward the land (eastward). Hence, the energy liberated by the Pacific Ocean as it cools keeps coastal areas much warmer than they would otherwise be. As a result, West Coast states generally have more favorable winter weather than East Coast states, where the prevailing winds do not tend to carry the energy toward land.

Calorimetry

One technique for measuring specific heat involves heating a sample to some known temperature T_x, placing it in a vessel containing water of known mass and temperature $T_w < T_x$, and measuring the temperature of the water after equilibrium has been reached. This technique is called **calorimetry,** and devices in which this energy transfer occurs are called **calorimeters**. If the system of the sample and the water is isolated, the principle of conservation of energy requires that the amount of energy that leaves the sample (of unknown specific heat) equal the amount of energy that enters the water.[3] Conservation of energy allows us to write the mathematical representation of this energy statement as

PITFALL PREVENTION 20.5
Remember the Negative Sign

It is *critical* to include the negative sign in Equation 20.5. The negative sign in the equation is necessary for consistency with our sign convention for energy transfer. The energy transfer Q_{hot} has a negative value because energy is leaving the hot substance. The negative sign in the equation ensures that the right side is a positive number, consistent with the left side, which is positive because energy is entering the cold water.

$$Q_{\text{cold}} = -Q_{\text{hot}} \qquad \textbf{(20.5)}$$

Suppose m_x is the mass of a sample of some substance whose specific heat we wish to determine. Let's call its specific heat c_x and its initial temperature T_x. Likewise, let m_w, c_w, and T_w represent corresponding values for the water. If T_f is the final equilibrium temperature after everything is mixed, Equation 20.4 shows that the energy transfer for the water is $m_w c_w(T_f - T_w)$, which is positive because $T_f > T_w$, and that the energy transfer for the sample of unknown specific heat is $m_x c_x(T_f - T_x)$, which is negative. Substituting these expressions into Equation 20.5 gives

$$m_w c_w(T_f - T_w) = -m_x c_x(T_f - T_x)$$

Solving for c_x gives

$$c_x = \frac{m_w c_w(T_f - T_w)}{m_x(T_x - T_f)}$$

EXAMPLE 20.2 Cooling a Hot Ingot

A 0.050 0-kg ingot of metal is heated to 200.0°C and then dropped into a calorimeter containing 0.400 kg of water initially at 20.0°C. The final equilibrium temperature of the mixed system is 22.4°C. Find the specific heat of the metal.

SOLUTION

Conceptualize Imagine the process occurring in the system. Energy is leaving the hot ingot and going into the cold water, so the ingot cools off and the water warms up. Once both are at the same temperature, the energy transfer stops.

Categorize We use an equation developed in this section, so we categorize this example as a substitution problem.

Use Equation 20.4 to evaluate each side of Equation 20.5:

$$m_w c_w(T_f - T_w) = -m_x c_x(T_f - T_x)$$

$$(0.400\text{ kg})(4\,186\text{ J/kg}\cdot{}^\circ\text{C})(22.4^\circ\text{C} - 20.0^\circ\text{C})$$
$$= -(0.050\,0\text{ kg})(c_x)(22.4^\circ\text{C} - 200.0^\circ\text{C})$$

[3] For precise measurements, the water container should be included in our calculations because it also exchanges energy with the sample. Doing so would require that we know the container's mass and composition, however. If the mass of the water is much greater than that of the container, we can neglect the effects of the container.

Solve for the specific heat of the metal:

$$c_x = \boxed{453 \text{ J/kg} \cdot {}^\circ\text{C}}$$

The ingot is most likely iron as you can see by comparing this result with the data given in Table 20.1. The temperature of the ingot is initially above the steam point. Therefore, some of the water may vaporize when the ingot is dropped into the water. We assume the system is sealed and this steam cannot escape. Because the final equilibrium temperature is lower than the steam point, any steam that does result recondenses back into water.

What If? Suppose you are performing an experiment in the laboratory that uses this technique to determine the specific heat of a sample and you wish to decrease the overall uncertainty in your final result for c_x. Of the data given in this example, changing which value would be most effective in decreasing the uncertainty?

Answer The largest experimental uncertainty is associated with the small difference in temperature of 2.4°C for the water. For example, using the rules for propagation of uncertainty in Appendix Section B.8, an uncertainty of 0.1°C in each of T_f and T_w leads to an 8% uncertainty in their difference. For this temperature difference to be larger experimentally, the most effective change is to *decrease the amount of water.*

EXAMPLE 20.3 Fun Time for a Cowboy

A cowboy fires a silver bullet with a muzzle speed of 200 m/s into the pine wall of a saloon. Assume all the internal energy generated by the impact remains with the bullet. What is the temperature change of the bullet?

SOLUTION

Conceptualize Imagine similar experiences you may have had in which mechanical energy is transformed to internal energy when a moving object is stopped. For example, as mentioned in Section 20.1, a nail becomes warm after it is hit a few times with a hammer.

Categorize The bullet is modeled as an isolated system. No work is done on the system because the force from the wall moves through no displacement. This example is similar to the skateboarder pushing off a wall in Section 9.7. There, no work is done on the skateboarder by the wall, and potential energy stored in the body from previous meals is transformed to kinetic energy. Here, no work is done by the wall on the bullet, and kinetic energy is transformed to internal energy.

Analyze Reduce the conservation of energy equation, Equation 8.2, to the appropriate expression for the system of the bullet:

$$(1)\quad \Delta K + \Delta E_{\text{int}} = 0$$

The change in the bullet's internal energy is the same as that which would take place if energy were transferred by heat from a stove to the bullet. Using this concept, evaluate the change in internal energy of the bullet:

$$(2)\quad \Delta E_{\text{int}} = Q = mc\,\Delta T$$

Substitute Equation (2) into Equation (1):

$$(0 - \tfrac{1}{2}mv^2) + mc\,\Delta T = 0$$

Solve for ΔT, using 234 J/kg · °C as the specific heat of silver (see Table 20.1):

$$(3)\quad \Delta T = \frac{\frac{1}{2}mv^2}{mc} = \frac{v^2}{2c} = \frac{(200 \text{ m/s})^2}{2(234 \text{ J/kg} \cdot {}^\circ\text{C})} = \boxed{85.5^\circ\text{C}}$$

Finalize Notice that the result does not depend on the mass of the bullet.

What If? Suppose the cowboy runs out of silver bullets and fires a lead bullet at the same speed into the wall. Will the temperature change of the bullet be larger or smaller?

Answer Table 20.1 shows that the specific heat of lead is 128 J/kg · °C, which is smaller than that for silver. Therefore, a given amount of energy input or transformation raises lead to a higher temperature than silver and the final temperature of the lead bullet will be larger. In Equation (3), let's substitute the new value for the specific heat:

$$\Delta T = \frac{v^2}{2c} = \frac{(200 \text{ m/s})^2}{2(128 \text{ J/kg} \cdot {}^\circ\text{C})} = 156^\circ\text{C}$$

There is no requirement that the silver and lead bullets have the same mass to determine this change in temperature. The only requirement is that they have the same speed.

20.3 Latent Heat

As we have seen in the preceding section, a substance can undergo a change in temperature when energy is transferred between it and its surroundings. In some situations, however, the transfer of energy does not result in a change in temperature. That is the case whenever the physical characteristics of the substance change from one form to another; such a change is commonly referred to as a **phase change**. Two common phase changes are from solid to liquid (melting) and from liquid to gas (boiling); another is a change in the crystalline structure of a solid. All such phase changes involve a change in the system's internal energy but no change in its temperature. The increase in internal energy in boiling, for example, is represented by the breaking of bonds between molecules in the liquid state; this bond breaking allows the molecules to move farther apart in the gaseous state, with a corresponding increase in intermolecular potential energy.

PITFALL PREVENTION 20.6
Signs Are Critical

Sign errors occur very often when students apply calorimetry equations. For phase changes, use the correct explicit sign in Equation 20.7, depending on whether you are adding or removing energy from the substance. In Equation 20.4, there is no explicit sign to consider, but be sure your ΔT is *always* the final temperature minus the initial temperature. In addition, you must *always* include the negative sign on the right side of Equation 20.5.

As you might expect, different substances respond differently to the addition or removal of energy as they change phase because their internal molecular arrangements vary. Also, the amount of energy transferred during a phase change depends on the amount of substance involved. (It takes less energy to melt an ice cube than it does to thaw a frozen lake.) If a quantity Q of energy transfer is required to change the phase of a mass m of a substance, the **latent heat** of the substance is defined as

$$L \equiv \frac{Q}{m} \quad \textbf{(20.6)}$$

This parameter is called latent heat (literally, the "hidden" heat) because this added or removed energy does not result in a temperature change. The value of L for a substance depends on the nature of the phase change as well as on the properties of the substance.

From the definition of latent heat, and again choosing heat as our energy transfer mechanism, the energy required to change the phase of a given mass m of a pure substance is

Latent heat ▶

$$Q = \pm mL \quad \textbf{(20.7)}$$

Latent heat of fusion L_f is the term used when the phase change is from solid to liquid (*to fuse* means "to combine by melting"), and **latent heat of vaporization** L_v is the term used when the phase change is from liquid to gas (the liquid "vaporizes").[4] The latent heats of various substances vary considerably as data in Table 20.2 show. The positive sign in Equation 20.7 is used when energy enters a system, causing melting or vaporization. The negative sign corresponds to energy leaving a system such that the system freezes or condenses.

[4] When a gas cools, it eventually *condenses*; that is, it returns to the liquid phase. The energy given up per unit mass is called the *latent heat of condensation* and is numerically equal to the latent heat of vaporization. Likewise, when a liquid cools, it eventually solidifies, and the *latent heat of solidification* is numerically equal to the latent heat of fusion.

TABLE 20.2

Latent Heats of Fusion and Vaporization

Substance	Melting Point (°C)	Latent Heat of Fusion (J/kg)	Boiling Point (°C)	Latent Heat of Vaporization (J/kg)
Helium	−269.65	5.23×10^3	−268.93	2.09×10^4
Nitrogen	−209.97	2.55×10^4	−195.81	2.01×10^5
Oxygen	−218.79	1.38×10^4	−182.97	2.13×10^5
Ethyl alcohol	−114	1.04×10^5	78	8.54×10^5
Water	0.00	3.33×10^5	100.00	2.26×10^6
Sulfur	119	3.81×10^4	444.60	3.26×10^5
Lead	327.3	2.45×10^4	1 750	8.70×10^5
Aluminum	660	3.97×10^5	2 450	1.14×10^7
Silver	960.80	8.82×10^4	2 193	2.33×10^6
Gold	1 063.00	6.44×10^4	2 660	1.58×10^6
Copper	1 083	1.34×10^5	1 187	5.06×10^6

To understand the role of latent heat in phase changes, consider the energy required to convert a 1.00-g cube of ice at −30.0°C to steam at 120.0°C. Figure 20.2 indicates the experimental results obtained when energy is gradually added to the ice. The results are presented as a graph of temperature of the system of the ice cube versus energy added to the system. Let's examine each portion of the red curve.

Part A. On this portion of the curve, the temperature of the ice changes from −30.0°C to 0.0°C. Equation 20.4 indicates that the temperature varies linearly with the energy added, so the experimental result is a straight line on the graph. Because the specific heat of ice is 2 090 J/kg · °C, we can calculate the amount of energy added by using Equation 20.4:

$$Q = m_i c_i \, \Delta T = (1.00 \times 10^{-3}\ \text{kg})(2\,090\ \text{J/kg}\cdot{}^\circ\text{C})(30.0^\circ\text{C}) = 62.7\ \text{J}$$

Part B. When the temperature of the ice reaches 0.0°C, the ice-water mixture remains at this temperature—even though energy is being added—until all the ice melts. The energy required to melt 1.00 g of ice at 0.0°C is, from Equation 20.7,

$$Q = m_i L_f = (1.00 \times 10^{-3}\ \text{kg})(3.33 \times 10^5\ \text{J/kg}) = 333\ \text{J}$$

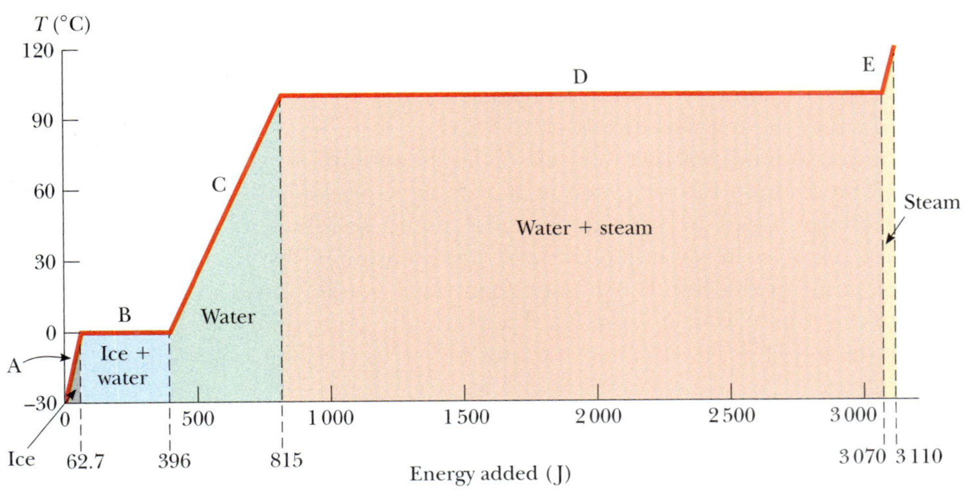

Figure 20.2 A plot of temperature versus energy added when 1.00 g of ice initially at −30.0°C is converted to steam at 120.0°C.

At this point, we have moved to the 396 J (= 62.7 J + 333 J) mark on the energy axis in Figure 20.2.

Part C. Between 0.0°C and 100.0°C, nothing surprising happens. No phase change occurs, and so all energy added to the water is used to increase its temperature. The amount of energy necessary to increase the temperature from 0.0°C to 100.0°C is

$$Q = m_w c_w \,\Delta T = (1.00 \times 10^{-3}\text{ kg})(4.19 \times 10^{3}\text{ J/kg}\cdot{}^\circ\text{C})(100.0^\circ\text{C}) = 419\text{ J}$$

Part D. At 100.0°C, another phase change occurs as the water changes from water at 100.0°C to steam at 100.0°C. Similar to the ice-water mixture in part B, the water-steam mixture remains at 100.0°C—even though energy is being added—until all the liquid has been converted to steam. The energy required to convert 1.00 g of water to steam at 100.0°C is

$$Q = m_w L_v = (1.00 \times 10^{-3}\text{ kg})(2.26 \times 10^{6}\text{ J/kg}) = 2.26 \times 10^{3}\text{ J}$$

Part E. On this portion of the curve, as in parts A and C, no phase change occurs; therefore, all energy added is used to increase the temperature of the steam. The energy that must be added to raise the temperature of the steam from 100.0°C to 120.0°C is

$$Q = m_s c_s \,\Delta T = (1.00 \times 10^{-3}\text{ kg})(2.01 \times 10^{3}\text{ J/kg}\cdot{}^\circ\text{C})(20.0^\circ\text{C}) = 40.2\text{ J}$$

The total amount of energy that must be added to change 1 g of ice at −30.0°C to steam at 120.0°C is the sum of the results from all five parts of the curve, which is 3.11×10^3 J. Conversely, to cool 1 g of steam at 120.0°C to ice at −30.0°C, we must remove 3.11×10^3 J of energy.

Notice in Figure 20.2 the relatively large amount of energy that is transferred into the water to vaporize it to steam. Imagine reversing this process, with a large amount of energy transferred out of steam to condense it into water. That is why a burn to your skin from steam at 100°C is much more damaging than exposure of your skin to water at 100°C. A very large amount of energy enters your skin from the steam, and the steam remains at 100°C for a long time while it condenses. Conversely, when your skin makes contact with water at 100°C, the water immediately begins to drop in temperature as energy transfers from the water to your skin.

If liquid water is held perfectly still in a very clean container, it is possible for the water to drop below 0°C without freezing into ice. This phenomenon, called **supercooling,** arises because the water requires a disturbance of some sort for the molecules to move apart and start forming the large open ice structure that makes the density of ice lower than that of water as discussed in Section 19.4. If supercooled water is disturbed, it suddenly freezes. The system drops into the lower-energy configuration of bound molecules of the ice structure, and the energy released raises the temperature back to 0°C.

Commercial hand warmers consist of liquid sodium acetate in a sealed plastic pouch. The solution in the pouch is in a stable supercooled state. When a disk in the pouch is clicked by your fingers, the liquid solidifies and the temperature increases, just like the supercooled water just mentioned. In this case, however, the freezing point of the liquid is higher than body temperature, so the pouch feels warm to the touch. To reuse the hand warmer, the pouch must be boiled until the solid liquefies. Then, as it cools, it passes below its freezing point into the supercooled state.

It is also possible to create **superheating**. For example, clean water in a very clean cup placed in a microwave oven can sometimes rise in temperature beyond 100°C without boiling because the formation of a bubble of steam in the water requires scratches in the cup or some type of impurity in the water to serve as a nucleation site. When the cup is removed from the microwave oven, the superheated water can become explosive as bubbles form immediately and the hot water is forced upward out of the cup.

Quick Quiz 20.2 Suppose the same process of adding energy to the ice cube is performed as discussed above, but instead we graph the internal energy of the system as a function of energy input. What would this graph look like?

EXAMPLE 20.4 Cooling the Steam

What mass of steam initially at 130°C is needed to warm 200 g of water in a 100-g glass container from 20.0°C to 50.0°C?

SOLUTION

Conceptualize Imagine placing water and steam together in a closed insulated container. The system eventually reaches a uniform state of water with a final temperature of 50.0°C.

Categorize Based on our conceptualization of this situation, we categorize this example as one involving calorimetry in which a phase change occurs.

Analyze Write Equation 20.5 to describe the calorimetry process:

$$(1)\quad Q_{\text{cold}} = -Q_{\text{hot}}$$

The steam undergoes three processes: a decrease in temperature to 100°C, condensation into liquid water, and finally a decrease in temperature of the water to 50.0°C. Find the energy transfer in the first process using the unknown mass m_s of the steam:

$$Q_1 = m_s c_s\,\Delta T = m_s(2.01 \times 10^3\ \text{J/kg}\cdot{}^\circ\text{C})(-30.0^\circ\text{C})$$
$$= -m_s(6.03 \times 10^4\ \text{J/kg})$$

Find the energy transfer in the second process, being sure to use a negative sign in Equation 20.7 because energy is leaving the steam:

$$Q_2 = -m_s(2.26 \times 10^6\ \text{J/kg})$$

Find the energy transfer in the third process:

$$Q_3 = m_s c_w\,\Delta T = m_s(4.19 \times 10^3\ \text{J/kg}\cdot{}^\circ\text{C})\ (-50.0^\circ\text{C})$$
$$= -m_s(2.09 \times 10^5\ \text{J/kg})$$

Add the energy transfers in these three stages:

$$Q_{\text{hot}} = Q_1 + Q_2 + Q_3$$
$$Q_{\text{hot}} = -m_s[6.03 \times 10^4\ \text{J/kg} + 2.26 \times 10^6\ \text{J/kg} + 2.09 \times 10^5\ \text{J/kg}]$$
$$(2)\quad Q_{\text{hot}} = -m_s(2.53 \times 10^6\ \text{J/kg})$$

The 20.0°C water and the glass undergo only one process, an increase in temperature to 50.0°C. Find the energy transfer in this process:

$$Q_{\text{cold}} = (0.200\ \text{kg})(4.19 \times 10^3\ \text{J/kg}\cdot{}^\circ\text{C})(30.0^\circ\text{C})$$
$$+ (0.100\ \text{kg})(837\ \text{J/kg}\cdot{}^\circ\text{C})(30.0^\circ\text{C})$$
$$(3)\quad Q_{\text{cold}} = 2.77 \times 10^4\ \text{J}$$

Substitute Equations (2) and (3) into Equation (1) and solve for m_s:

$$2.77 \times 10^4\ \text{J} = -[-m_s(2.53 \times 10^6\ \text{J/kg})]$$
$$m_s = 1.09 \times 10^{-2}\ \text{kg} = 10.9\ \text{g}$$

What If? What if the final state of the system is water at 100°C? Would we need more steam or less steam? How would the analysis above change?

Answer More steam would be needed to raise the temperature of the water and glass to 100°C instead of 50.0°C. There would be two major changes in the analysis. First, we would not have a term Q_3 for the steam because the water that condenses from the steam does not cool below 100°C. Second, in Q_{cold}, the temperature change would be 80.0°C instead of 30.0°C. For practice, show that the result is a required mass of steam of 31.8 g.

20.4 Work and Heat in Thermodynamic Processes

In thermodynamics, we describe the *state* of a system using such variables as pressure, volume, temperature, and internal energy. As a result, these quantities belong to a category called **state variables**. For any given configuration of the system, we can identify values of the state variables. (For mechanical systems, the state variables include kinetic energy K and potential energy U.) A state of a system can be specified only if the system is in thermal equilibrium internally. In the case of a gas in a container, internal thermal equilibrium requires that every part of the gas be at the same pressure and temperature.

A second category of variables in situations involving energy is **transfer variables**. These variables are those that appear on the right side of the conservation of energy equation, Equation 8.2. Such a variable has a nonzero value if a process occurs in which energy is transferred across the system's boundary. The transfer variable is positive or negative, depending on whether energy is entering or leaving the system. Because a transfer of energy across the boundary represents a change in the system, transfer variables are not associated with a given state of the system but, rather, with a *change* in the state of the system.

In the previous sections, we discussed heat as a transfer variable. In this section, we study another important transfer variable for thermodynamic systems, work. Work performed on particles was studied extensively in Chapter 7, and here we investigate the work done on a deformable system, a gas. Consider a gas contained in a cylinder fitted with a movable piston (Fig. 20.3). At equilibrium, the gas occupies a volume V and exerts a uniform pressure P on the cylinder's walls and on the piston. If the piston has a cross-sectional area A, the force exerted by the gas on the piston is $F = PA$. Now let's assume that we push the piston inward and compress the gas **quasi-statically,** that is, slowly enough to allow the system to remain essentially in internal thermal equilibrium at all times. As the piston is pushed downward by an external force $\vec{\mathbf{F}} = -F\hat{\mathbf{j}}$ through a displacement of $d\vec{\mathbf{r}} = dy\hat{\mathbf{j}}$ (Fig. 20.3b), the work done on the gas is, according to our definition of work in Chapter 7,

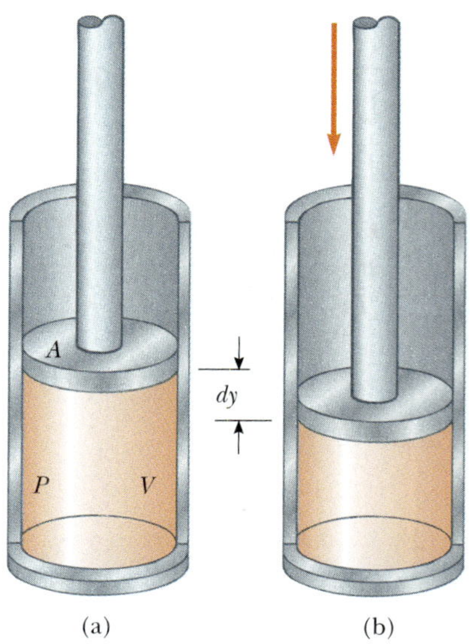

Figure 20.3 Work is done on a gas contained in a cylinder at a pressure P as the piston is pushed downward so that the gas is compressed.

$$dW = \vec{\mathbf{F}} \cdot d\vec{\mathbf{r}} = -F\hat{\mathbf{j}} \cdot dy\hat{\mathbf{j}} = -F\,dy = -PA\,dy$$

where the magnitude F of the external force is equal to PA because the piston is always in equilibrium between the external force and the force from the gas. The mass of the piston is assumed to be negligible in this discussion. Because $A\,dy$ is the change in volume of the gas dV, we can express the work done on the gas as

$$dW = -P\,dV \qquad \textbf{(20.8)}$$

If the gas is compressed, dV is negative and the work done on the gas is positive. If the gas expands, dV is positive and the work done on the gas is negative. If the volume remains constant, the work done on the gas is zero. The total work done on the gas as its volume changes from V_i to V_f is given by the integral of Equation 20.8:

$$W = -\int_{V_i}^{V_f} P\,dV \qquad \textbf{(20.9)}$$

◀ Work done on a gas

To evaluate this integral, you must know how the pressure varies with volume during the process.

In general, the pressure is not constant during a process followed by a gas, but depends on the volume and temperature. If the pressure and volume are known at each step of the process, the state of the gas at each step can be plotted on a graphical representation called a ***PV* diagram** as in Active Figure 20.4. This type of diagram allows us to visualize a process through which a gas is progressing. The curve on a *PV* diagram is called the *path* taken between the initial and final states.

Notice that the integral in Equation 20.9 is equal to the area under a curve on a *PV* diagram. Therefore, we can identify an important use for *PV* diagrams:

The work done on a gas in a quasi-static process that takes the gas from an initial state to a final state is the negative of the area under the curve on a *PV* diagram, evaluated between the initial and final states.

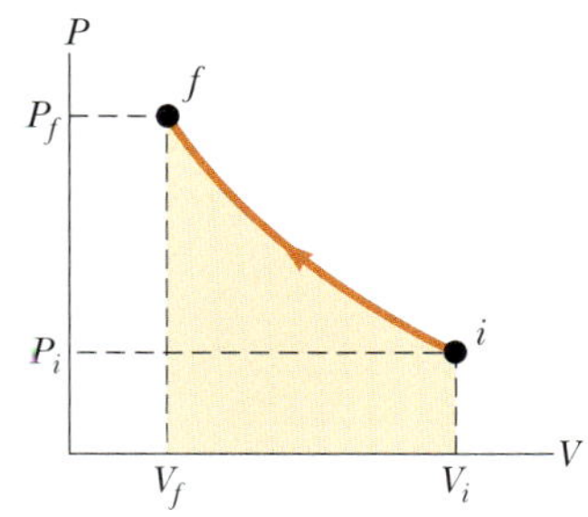

ACTIVE FIGURE 20.4

A gas is compressed quasi-statically (slowly) from state i to state f. The work done on the gas equals the negative of the area under the *PV* curve. The volume is decreasing, so this area is negative. Then the work done on the gas is positive. An outside agent must do positive work on the gas to compress it.

Sign in at www.thomsonedu.com and go to ThomsonNOW to compress the piston in Figure 20.3 and see the result on the *PV* diagram in this figure.

For the process of compressing a gas in a cylinder, the work done depends on the particular path taken between the initial and final states as Active Figure 20.4 suggests. To illustrate this important point, consider several different paths connecting i and f (Active Fig. 20.5). In the process depicted in Active Figure 20.5a, the volume of the gas is first reduced from V_i to V_f at constant pressure P_i and the pressure of the gas then increases from P_i to P_f by heating at constant volume V_f. The work done on the gas along this path is $-P_i(V_f - V_i)$. In Active Figure 20.5b, the pressure of the gas is increased from P_i to P_f at constant volume V_i and then the volume of the gas is reduced from V_i to V_f at constant pressure P_f. The work done on the gas is $-P_f(V_f - V_i)$. This value is greater than that for the process

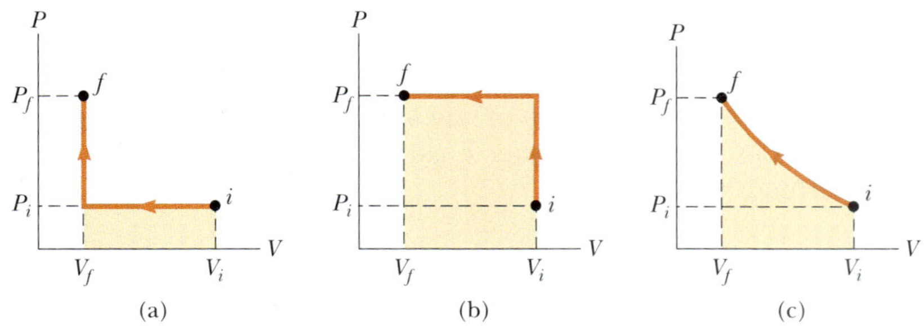

ACTIVE FIGURE 20.5

The work done on a gas as it is taken from an initial state to a final state depends on the path between these states.

Sign in at www.thomsonedu.com and go to ThomsonNOW to choose one of the three paths and see the movement of the piston in Figure 20.3 and of a point on the *PV* diagram in this figure.

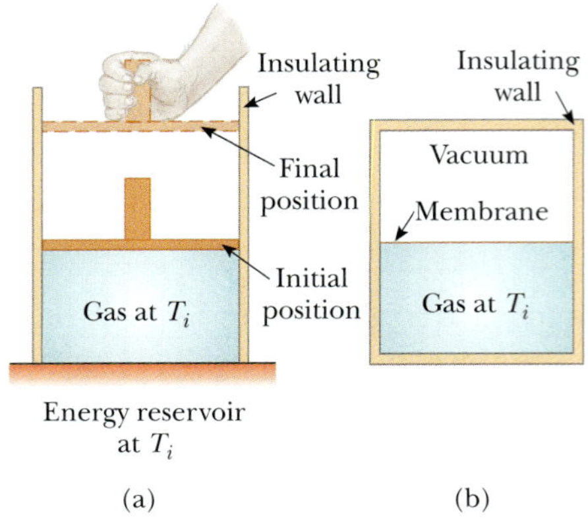

Figure 20.6 (a) A gas at temperature T_i expands slowly while absorbing energy from a reservoir to maintain a constant temperature. (b) A gas expands rapidly into an evacuated region after a membrane is broken.

described in Active Figure 20.5a because the piston is moved through the same displacement by a larger force. Finally, for the process described in Active Figure 20.5c, where both P and V change continuously, the work done on the gas has some value between the values obtained in the first two processes. To evaluate the work in this case, the function $P(V)$ must be known so that we can evaluate the integral in Equation 20.9.

The energy transfer Q into or out of a system by heat also depends on the process. Consider the situations depicted in Figure 20.6. In each case, the gas has the same initial volume, temperature, and pressure, and is assumed to be ideal. In Figure 20.6a, the gas is thermally insulated from its surroundings except at the bottom of the gas-filled region, where it is in thermal contact with an energy reservoir. An *energy reservoir* is a source of energy that is considered to be so great that a finite transfer of energy to or from the reservoir does not change its temperature. The piston is held at its initial position by an external agent such as a hand. When the force holding the piston is reduced slightly, the piston rises very slowly to its final position. Because the piston is moving upward, the gas is doing work on the piston. During this expansion to the final volume V_f, just enough energy is transferred by heat from the reservoir to the gas to maintain a constant temperature T_i.

Now consider the completely thermally insulated system shown in Figure 20.6b. When the membrane is broken, the gas expands rapidly into the vacuum until it occupies a volume V_f and is at a pressure P_f. In this case, the gas does no work because it does not apply a force; no force is required to expand into a vacuum. Furthermore, no energy is transferred by heat through the insulating wall.

The initial and final states of the ideal gas in Figure 20.6a are identical to the initial and final states in Figure 20.6b, but the paths are different. In the first case, the gas does work on the piston and energy is transferred slowly to the gas by heat. In the second case, no energy is transferred by heat and the value of the work done is zero. Therefore, **energy transfer by heat, like work done, depends on the initial, final, and intermediate states of the system**. In other words, because heat and work depend on the path, neither quantity is determined solely by the endpoints of a thermodynamic process.

20.5 The First Law of Thermodynamics

When we introduced the law of conservation of energy in Chapter 8, we stated that the change in the energy of a system is equal to the sum of all transfers of energy across the system's boundary. The **first law of thermodynamics** is a special case of the law of conservation of energy that describes processes in which only the internal energy[5] changes and the only energy transfers are by heat and work:

$$\Delta E_{int} = Q + W \quad (20.10)$$

◀ First law of thermodynamics

An important consequence of the first law of thermodynamics is that there exists a quantity known as internal energy whose value is determined by the state of the system. The internal energy is therefore a state variable like pressure, volume, and temperature.

When a system undergoes an infinitesimal change in state in which a small amount of energy dQ is transferred by heat and a small amount of work dW is done, the internal energy changes by a small amount dE_{int}. Therefore, for infinitesimal processes we can express the first law as[6]

$$dE_{int} = dQ + dW$$

Let us investigate some special cases in which the first law can be applied. First, consider an *isolated system*, that is, one that does not interact with its surroundings. In this case, no energy transfer by heat takes place and the work done on the system is zero; hence, the internal energy remains constant. That is, because $Q = W = 0$, it follows that $\Delta E_{int} = 0$; therefore, $E_{int,i} = E_{int,f}$. We conclude that **the internal energy E_{int} of an isolated system remains constant.**

Next, consider the case of a system that can exchange energy with its surroundings and is taken through a **cyclic process**, that is, a process that starts and ends at the same state. In this case, the change in the internal energy must again be zero because E_{int} is a state variable; therefore, the energy Q added to the system must equal the negative of the work W done on the system during the cycle. That is, in a cyclic process,

$$\Delta E_{int} = 0 \quad \text{and} \quad Q = -W \quad \text{(cyclic process)}$$

On a PV diagram, a cyclic process appears as a closed curve. (The processes described in Active Figure 20.5 are represented by open curves because the initial and final states differ.) It can be shown that **in a cyclic process, the net work done on the system per cycle equals the area enclosed by the path representing the process on a PV diagram.**

PITFALL PREVENTION 20.7
Dual Sign Conventions

Some physics and engineering books present the first law as $\Delta E_{int} = Q - W$, with a minus sign between the heat and work. The reason is that work is defined in these treatments as the work done *by* the gas rather than *on* the gas, as in our treatment. The equivalent equation to Equation 20.9 in these treatments defines work as $W = \int_{V_i}^{V_f} P\,dV$. Therefore, if positive work is done by the gas, energy is leaving the system, leading to the negative sign in the first law.

In your studies in other chemistry or engineering courses, or in your reading of other physics books, be sure to note which sign convention is being used for the first law.

PITFALL PREVENTION 20.8
The First Law

With our approach to energy in this book, the first law of thermodynamics is a special case of Equation 8.2. Some physicists argue that the first law is the general equation for energy conservation, equivalent to Equation 8.2. In this approach, the first law is applied to a closed system (so that there is no matter transfer), heat is interpreted so as to include electromagnetic radiation, and work is interpreted so as to include electrical transmission ("electrical work") and mechanical waves ("molecular work"). Keep that in mind if you run across the first law in your reading of other physics books.

20.6 Some Applications of the First Law of Thermodynamics

In this section, we consider applications of the first law to processes through which a gas is taken. As a model, let's consider the sample of gas contained in the piston-cylinder apparatus in Active Figure 20.7 (page 568). This figure shows work being done on the gas and energy transferring in by heat, so the internal energy of the gas is rising. In the following discussion of various processes, refer back to this figure and mentally alter the directions of the transfer of energy to reflect what is happening in the process.

Before we apply the first law of thermodynamics to specific systems, it is useful to first define some idealized thermodynamic processes. An **adiabatic process** is one during which no energy enters or leaves the system by heat; that is, $Q = 0$. An

[5] It is an unfortunate accident of history that the traditional symbol for internal energy is U, which is also the traditional symbol for potential energy as introduced in Chapter 7. To avoid confusion between potential energy and internal energy, we use the symbol E_{int} for internal energy in this book. If you take an advanced course in thermodynamics, however, be prepared to see U used as the symbol for internal energy in the first law.

[6] Notice that dQ and dW are not true differential quantities because Q and W are not state variables, but dE_{int} is. Because dQ and dW are *inexact differentials*, they are often represented by the symbols $đQ$ and $đW$. For further details on this point, see an advanced text on thermodynamics.

adiabatic process can be achieved either by thermally insulating the walls of the system or by performing the process rapidly so that there is negligible time for energy to transfer by heat. Applying the first law of thermodynamics to an adiabatic process gives

$$\Delta E_{\text{int}} = W \quad \text{(adiabatic process)} \tag{20.11}$$

This result shows that if a gas is compressed adiabatically such that W is positive, then ΔE_{int} is positive and the temperature of the gas increases. Conversely, the temperature of a gas decreases when the gas expands adiabatically.

Adiabatic processes are very important in engineering practice. Some common examples are the expansion of hot gases in an internal combustion engine, the liquefaction of gases in a cooling system, and the compression stroke in a diesel engine.

The process described in Figure 20.6b, called an **adiabatic free expansion,** is unique. The process is adiabatic because it takes place in an insulated container. Because the gas expands into a vacuum, it does not apply a force on a piston as was depicted in Figure 20.6a, so no work is done on or by the gas. Therefore, in this adiabatic process, both $Q = 0$ and $W = 0$. As a result, $\Delta E_{\text{int}} = 0$ for this process as can be seen from the first law. That is, **the initial and final internal energies of a gas are equal in an adiabatic free expansion**. As we shall see in Chapter 21, the internal energy of an ideal gas depends only on its temperature. Therefore, we expect no change in temperature during an adiabatic free expansion. This prediction is in accord with the results of experiments performed at low pressures. (Experiments performed at high pressures for real gases show a slight change in temperature after the expansion due to intermolecular interactions, which represent a deviation from the model of an ideal gas.)

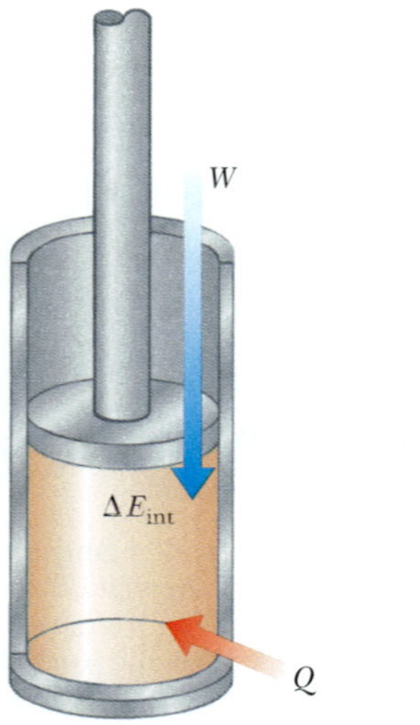

ACTIVE FIGURE 20.7

The first law of thermodynamics equates the change in internal energy E_{int} in a system to the net energy transfer to the system by heat Q and work W. In the situation shown here, the internal energy of the gas increases.

Sign in at www.thomsonedu.com and go to ThomsonNOW to choose one of the four processes for the gas discussed in this section and see the movement of the piston and of a point on a PV diagram.

A process that occurs at constant pressure is called an **isobaric process**. In Active Figure 20.7, an isobaric process could be established by allowing the piston to move freely so that it is always in equilibrium between the net force from the gas pushing upward and the weight of the piston plus the force due to atmospheric pressure pushing downward. The first process in Active Figure 20.5a and the second process in Active Figure 20.5b are both isobaric.

In such a process, the values of the heat and the work are both usually nonzero. The work done on the gas in an isobaric process is simply

Isobaric process ▶

$$W = -P(V_f - V_i) \quad \text{(isobaric process)} \tag{20.12}$$

where P is the constant pressure of the gas during the process.

A process that takes place at constant volume is called an **isovolumetric process**. In Active Figure 20.7, clamping the piston at a fixed position would ensure an isovolumetric process. The second process in Active Figure 20.5a and the first process in Active Figure 20.5b are both isovolumetric.

Because the volume of the gas does not change in such a process, the work given by Equation 20.9 is zero. Hence, from the first law we see that in an isovolumetric process, because $W = 0$,

Isovolumetric process ▶

$$\Delta E_{\text{int}} = Q \quad \text{(isovolumetric process)} \tag{20.13}$$

This expression specifies that **if energy is added by heat to a system kept at constant volume, all the transferred energy remains in the system as an increase in its internal energy**. For example, when a can of spray paint is thrown into a fire, energy enters the system (the gas in the can) by heat through the metal walls of the can. Consequently, the temperature, and therefore the pressure, in the can increases until the can possibly explodes.

Isothermal process ▶

A process that occurs at constant temperature is called an **isothermal process**. This process can be established by immersing the cylinder in Active Figure 20.7 in an ice-water bath or by putting the cylinder in contact with some other constant-temperature reservoir. A plot of P versus V at constant temperature for an ideal gas yields a hyperbolic curve called an *isotherm*. The internal energy of an ideal gas is a function of temperature only. Hence, in an isothermal process involving an ideal gas, $\Delta E_{\text{int}} = 0$. For an isothermal process, we conclude from the first law that the energy transfer Q must be equal to the negative of the work done on the gas; that is, $Q = -W$. Any energy that enters the system by heat is transferred out of the system by work; as a result, no change in the internal energy of the system occurs in an isothermal process.

PITFALL PREVENTION 20.9

$Q \neq 0$ in an Isothermal Process

Do not fall into the common trap of thinking there must be no transfer of energy by heat if the temperature does not change as is the case in an isothermal process. Because the cause of temperature change can be either heat *or* work, the temperature can remain constant even if energy enters the gas by heat, which can only happen if the energy entering the gas by heat leaves by work.

Quick Quiz 20.3 In the last three columns of the following table, fill in the boxes with the correct signs ($-$, $+$, or 0) for Q, W, and ΔE_{int}. For each situation, the system to be considered is identified.

Situation	System	Q	W	ΔE_{int}
(a) Rapidly pumping up a bicycle tire	Air in the pump			
(b) Pan of room-temperature water sitting on a hot stove	Water in the pan			
(c) Air quickly leaking out of a balloon	Air originally in the balloon			

Isothermal Expansion of an Ideal Gas

Suppose an ideal gas is allowed to expand quasi-statically at constant temperature. This process is described by the PV diagram shown in Figure 20.8. The curve is a hyperbola (see Appendix B, Eq. B.23), and the ideal gas law with T constant indicates that the equation of this curve is $PV =$ constant.

Let's calculate the work done on the gas in the expansion from state i to state f. The work done on the gas is given by Equation 20.9. Because the gas is ideal and the process is quasi-static, the ideal gas law is valid for each point on the path. Therefore,

$$W = -\int_{V_i}^{V_f} P\,dV = -\int_{V_i}^{V_f} \frac{nRT}{V}\,dV$$

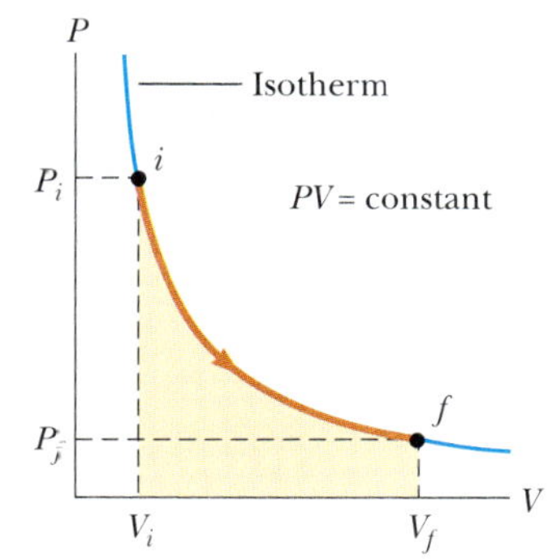

Figure 20.8 The PV diagram for an isothermal expansion of an ideal gas from an initial state to a final state. The curve is a hyperbola.

Because T is constant in this case, it can be removed from the integral along with n and R:

$$W = -nRT\int_{V_i}^{V_f} \frac{dV}{V} = -nRT \ln V \Big|_{V_i}^{V_f}$$

To evaluate the integral, we used $\int (dx/x) = \ln x$. (See Appendix B.) Evaluating the result at the initial and final volumes gives

$$W = nRT \ln\left(\frac{V_i}{V_f}\right) \qquad \textbf{(20.14)}$$

Numerically, this work W equals the negative of the shaded area under the PV curve shown in Figure 20.8. Because the gas expands, $V_f > V_i$ and the value for the work done on the gas is negative as we expect. If the gas is compressed, then $V_f < V_i$ and the work done on the gas is positive.

Quick Quiz 20.4 Characterize the paths in Figure 20.9 as isobaric, isovolumetric, isothermal, or adiabatic. For path B, $Q = 0$.

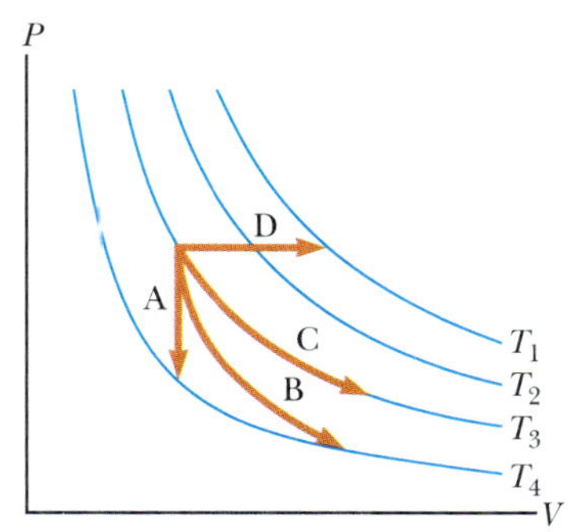

Figure 20.9 (Quick Quiz 20.4) Identify the nature of paths A, B, C, and D.

EXAMPLE 20.5 An Isothermal Expansion

A 1.0-mol sample of an ideal gas is kept at 0.0°C during an expansion from 3.0 L to 10.0 L.

(A) How much work is done on the gas during the expansion?

SOLUTION

Conceptualize Run the process in your mind: the cylinder in Active Figure 20.7 is immersed in an ice-water bath, and the piston moves outward so that the volume of the gas increases.

Categorize We will evaluate parameters using equations developed in the preceding sections, so we categorize this example as a substitution problem. Because the temperature of the gas is fixed, the process is isothermal.

Substitute the given values into Equation 20.14:

$$W = nRT\ln\left(\frac{V_i}{V_f}\right)$$

$$= (1.0\text{ mol})(8.31\text{ J/mol}\cdot\text{K})(273\text{ K})\ln\left(\frac{3.0\text{ L}}{10.0\text{ L}}\right)$$

$$= \boxed{-2.7\times10^3\text{ J}}$$

(B) How much energy transfer by heat occurs between the gas and its surroundings in this process?

SOLUTION

Find the heat from the first law:

$$\Delta E_{\text{int}} = Q + W$$

$$0 = Q + W$$

$$Q = -W = \boxed{2.7\times10^3\text{ J}}$$

(C) If the gas is returned to the original volume by means of an isobaric process, how much work is done on the gas?

SOLUTION

Use Equation 20.12. The pressure is not given, so incorporate the ideal gas law:

$$W = -P(V_f - V_i) = -\frac{nRT_i}{V_i}(V_f - V_i)$$

$$= -\frac{(1.0\text{ mol})(8.31\text{ J/mol}\cdot\text{K})(273\text{ K})}{10.0\times10^{-3}\text{ m}^3}(3.0\times10^{-3}\text{ m}^3 - 10.0\times10^{-3}\text{ m}^3)$$

$$= \boxed{1.6\times10^3\text{ J}}$$

We used the initial temperature and volume to calculate the work done because the final temperature was unknown. The work done on the gas is positive because the gas is being compressed.

EXAMPLE 20.6 Boiling Water

Suppose 1.00 g of water vaporizes isobarically at atmospheric pressure (1.013×10^5 Pa). Its volume in the liquid state is $V_i = V_{\text{liquid}} = 1.00\text{ cm}^3$, and its volume in the vapor state is $V_f = V_{\text{vapor}} = 1\,671\text{ cm}^3$. Find the work done in the expansion and the change in internal energy of the system. Ignore any mixing of the steam and the surrounding air; imagine that the steam simply pushes the surrounding air out of the way.

SOLUTION

Conceptualize Notice that the temperature of the system does not change. There is a phase change occurring as the water evaporates to steam.

Categorize Because the expansion takes place at constant pressure, we categorize the process as isobaric. We will use equations developed in the preceding sections, so we categorize this example as a substitution problem.

Use Equation 20.12 to find the work done on the system as the air is pushed out of the way:

$$W = -P(V_f - V_i)$$

$$= -(1.013\times10^5\text{ Pa})(1\,671\times10^{-6}\text{ m}^3 - 1.00\times10^{-6}\text{ m}^3)$$

$$= \boxed{-169\text{ J}}$$

Use Equation 20.7 and the latent heat of vaporization for water to find the energy transferred into the system by heat:

$$Q = mL_v = (1.00 \times 10^{-3}\text{ kg})(2.26 \times 10^6\text{ J/kg}) = 2\,260\text{ J}$$

Use the first law to find the change in internal energy of the system:

$$\Delta E_{\text{int}} = Q + W = 2\,260\text{ J} + (-169\text{ J}) = \boxed{2.09\text{ kJ}}$$

The positive value for ΔE_{int} indicates that the internal energy of the system increases. The largest fraction of the energy (2 090 J/ 2 260 J = 93%) transferred to the liquid goes into increasing the internal energy of the system. The remaining 7% of the energy transferred leaves the system by work done by the steam on the surrounding atmosphere.

EXAMPLE 20.7 Heating a Solid

A 1.0-kg bar of copper is heated at atmospheric pressure so that its temperature increases from 20°C to 50°C.

(A) What is the work done on the copper bar by the surrounding atmosphere?

SOLUTION

Conceptualize This example involves a solid, whereas the preceding two examples involved liquids and gases. For a solid, the change in volume due to thermal expansion is very small.

Categorize Because the expansion takes place at constant atmospheric pressure, we categorize the process as isobaric.

Analyze Calculate the change in volume of the copper bar using Equation 19.6, the average linear expansion coefficient for copper given in Table 19.1, and that $\beta = 3\alpha$:

$$\Delta V = \beta V_i\,\Delta T = 3\alpha V_i\,\Delta T$$
$$= 3[1.7 \times 10^{-5}(°\text{C})^{-1}]V_i(50°\text{C} - 20°\text{C}) = 1.5 \times 10^{-3}\,V_i$$

Use Equation 1.1 to express the initial volume of the bar in terms of the mass of the bar and the density of copper from Table 14.1:

$$\Delta V = (1.5 \times 10^{-3})\left(\frac{m}{\rho}\right) = (1.5 \times 10^{-3})\left(\frac{1.0\text{ kg}}{8.92 \times 10^3\text{ kg/m}^3}\right)$$
$$= 1.7 \times 10^{-7}\text{ m}^3$$

Find the work done on the copper bar using Equation 20.12:

$$W = -P\,\Delta V = -(1.013 \times 10^5\text{ N/m}^2)(1.7 \times 10^{-7}\text{ m}^3)$$
$$= \boxed{-1.7 \times 10^{-2}\text{ J}}$$

Because this work is negative, work is done *by* the copper bar on the atmosphere.

(B) How much energy is transferred to the copper bar by heat?

SOLUTION

Use Equation 20.4 and the specific heat of copper from Table 20.1:

$$Q = mc\,\Delta T = (1.0\text{ kg})(387\text{ J/kg}\cdot°\text{C})(50°\text{C} - 20°\text{C})$$
$$= \boxed{1.2 \times 10^4\text{ J}}$$

(C) What is the increase in internal energy of the copper bar?

SOLUTION

Use the first law of thermodynamics:

$$\Delta E_{\text{int}} = Q + W = 1.2 \times 10^4\text{ J} + (-1.7 \times 10^{-2}\text{ J})$$
$$= \boxed{1.2 \times 10^4\text{ J}}$$

Finalize Most of the energy transferred into the system by heat goes into increasing the internal energy of the copper bar. The fraction of energy used to do work on the surrounding atmosphere is only about 10^{-6}. Hence, when the thermal expansion of a solid or a liquid is analyzed, the small amount of work done on or by the system is usually ignored.

TABLE 20.3

Thermal Conductivities

Substance	Thermal Conductivity (W/m · °C)
Metals (at 25°C)	
Aluminum	238
Copper	397
Gold	314
Iron	79.5
Lead	34.7
Silver	427
Nonmetals (approximate values)	
Asbestos	0.08
Concrete	0.8
Diamond	2 300
Glass	0.8
Ice	2
Rubber	0.2
Water	0.6
Wood	0.08
Gases (at 20°C)	
Air	0.023 4
Helium	0.138
Hydrogen	0.172
Nitrogen	0.023 4
Oxygen	0.023 8

20.7 Energy Transfer Mechanisms

In Chapter 8, we introduced a global approach to the energy analysis of physical processes through Equation 8.1, $\Delta E_{\text{system}} = \Sigma T$, where T represents energy transfer, which can occur by several mechanisms. Earlier in this chapter, we discussed two of the terms on the right side of this equation, work W and heat Q. In this section, we explore more details about heat as a means of energy transfer and two other energy transfer methods often related to temperature changes: convection (a form of matter transfer T_{MT}) and electromagnetic radiation T_{ER}.

Thermal Conduction

The process of energy transfer by heat can also be called **conduction** or **thermal conduction**. In this process, the transfer can be represented on an atomic scale as an exchange of kinetic energy between microscopic particles—molecules, atoms, and free electrons—in which less-energetic particles gain energy in collisions with more-energetic particles. For example, if you hold one end of a long metal bar and insert the other end into a flame, you will find that the temperature of the metal in your hand soon increases. The energy reaches your hand by means of conduction. Initially, before the rod is inserted into the flame, the microscopic particles in the metal are vibrating about their equilibrium positions. As the flame raises the temperature of the rod, the particles near the flame begin to vibrate with greater and greater amplitudes. These particles, in turn, collide with their neighbors and transfer some of their energy in the collisions. Slowly, the amplitudes of vibration of metal atoms and electrons farther and farther from the flame increase until eventually those in the metal near your hand are affected. This increased vibration is detected by an increase in the temperature of the metal and of your potentially burned hand.

The rate of thermal conduction depends on the properties of the substance being heated. For example, it is possible to hold a piece of asbestos in a flame indefinitely, which implies that very little energy is conducted through the asbestos. In general, metals are good thermal conductors and materials such as asbestos, cork, paper, and fiberglass are poor conductors. Gases also are poor conductors because the separation distance between the particles is so great. Metals are good thermal conductors because they contain large numbers of electrons that are relatively free to move through the metal and so can transport energy over large distances. Therefore, in a good conductor such as copper, conduction takes place by means of both the vibration of atoms and the motion of free electrons.

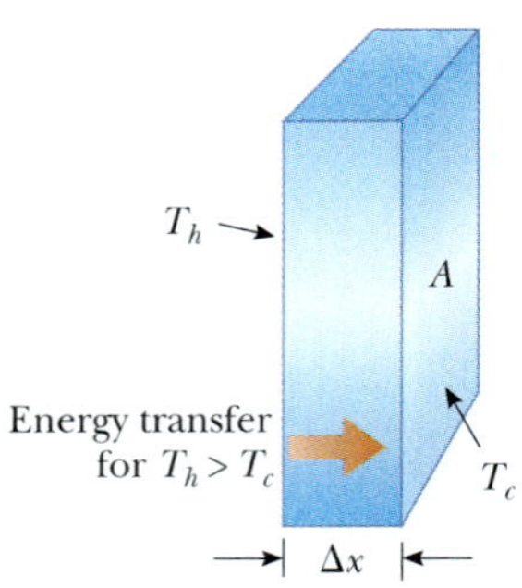

Figure 20.10 Energy transfer through a conducting slab with a cross-sectional area A and a thickness Δx. The opposite faces are at different temperatures T_c and T_h.

Conduction occurs only if there is a difference in temperature between two parts of the conducting medium. Consider a slab of material of thickness Δx and cross-sectional area A. One face of the slab is at a temperature T_c, and the other face is at a temperature $T_h > T_c$ (Fig. 20.10). Experimentally, it is found that energy Q transfers in a time interval Δt from the hotter face to the colder one. The rate $\mathcal{P} = Q/\Delta t$ at which this energy transfer occurs is found to be proportional to the cross-sectional area and the temperature difference $\Delta T = T_h - T_c$ and inversely proportional to the thickness:

$$\mathcal{P} = \frac{Q}{\Delta t} \propto A\frac{\Delta T}{\Delta x}$$

Notice that $\mathcal{P}$ has units of watts when Q is in joules and Δt is in seconds. That is not surprising because $\mathcal{P}$ is *power*, the rate of energy transfer by heat. For a slab of infinitesimal thickness dx and temperature difference dT, we can write the **law of thermal conduction** as

Law of thermal conduction ▶

$$\mathcal{P} = kA\left|\frac{dT}{dx}\right| \qquad \textbf{(20.15)}$$

where the proportionality constant k is the **thermal conductivity** of the material and $|dT/dx|$ is the **temperature gradient** (the rate at which temperature varies with position).

Suppose a long, uniform rod of length L is thermally insulated so that energy cannot escape by heat from its surface except at the ends as shown in Figure 20.11. One end is in thermal contact with an energy reservoir at temperature T_c, and the other end is in thermal contact with a reservoir at temperature $T_h > T_c$. When a steady state has been reached, the temperature at each point along the rod is constant in time. In this case, if we assume k is not a function of temperature, the temperature gradient is the same everywhere along the rod and is

$$\left|\frac{dT}{dx}\right| = \frac{T_h - T_c}{L}$$

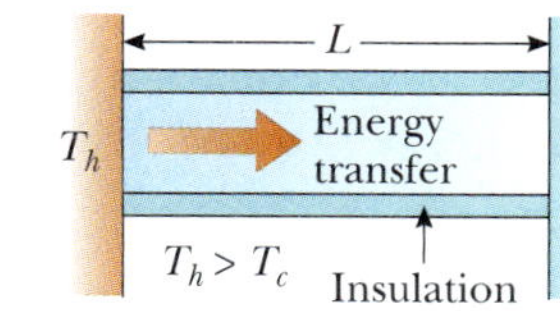

Figure 20.11 Conduction of energy through a uniform, insulated rod of length L. The opposite ends are in thermal contact with energy reservoirs at different temperatures.

Therefore, the rate of energy transfer by conduction through the rod is

$$\mathcal{P} = kA\left(\frac{T_h - T_c}{L}\right) \qquad \textbf{(20.16)}$$

Substances that are good thermal conductors have large thermal conductivity values, whereas good thermal insulators have low thermal conductivity values. Table 20.3 lists thermal conductivities for various substances. Notice that metals are generally better thermal conductors than nonmetals.

For a compound slab containing several materials of thicknesses $L_1, L_2, \ldots$ and thermal conductivities $k_1, k_2, \ldots$, the rate of energy transfer through the slab at steady state is

$$\mathcal{P} = \frac{A(T_h - T_c)}{\sum_i (L_i/k_i)} \qquad \textbf{(20.17)}$$

where T_c and T_h are the temperatures of the outer surfaces (which are held constant) and the summation is over all slabs. Example 20.8 shows how Equation 20.17 results from a consideration of two thicknesses of materials.

Quick Quiz 20.5 You have two rods of the same length and diameter, but they are formed from different materials. The rods are used to connect two regions at different temperatures so that energy transfers through the rods by heat. They can be connected in series as in Figure 20.12a or in parallel as in Figure 20.12b. In which case is the rate of energy transfer by heat larger? (a) The rate is larger when the rods are in series. (b) The rate is larger when the rods are in parallel. (c) The rate is the same in both cases.

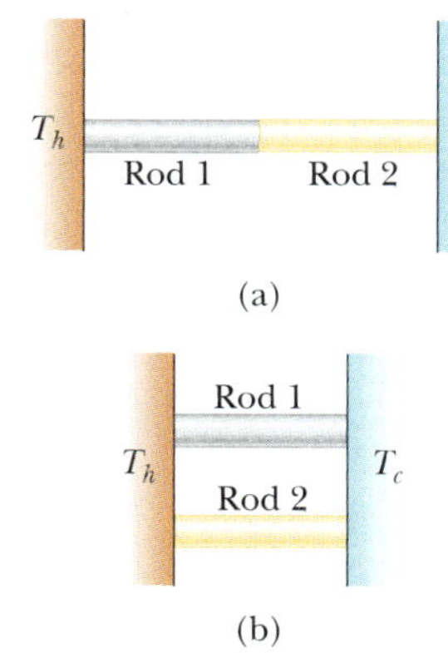

Figure 20.12 (Quick Quiz 20.5) In which case is the rate of energy transfer larger?

EXAMPLE 20.8 Energy Transfer Through Two Slabs

Two slabs of thickness L_1 and L_2 and thermal conductivities k_1 and k_2 are in thermal contact with each other as shown in Figure 20.13. The temperatures of their outer surfaces are T_c and T_h, respectively, and $T_h > T_c$. Determine the temperature at the interface and the rate of energy transfer by conduction through the slabs in the steady-state condition.

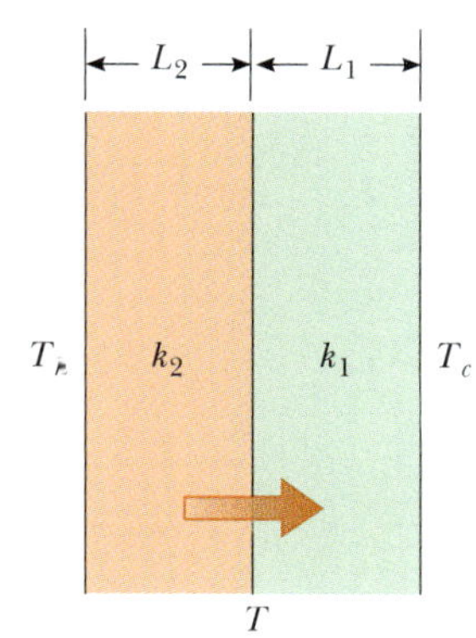

Figure 20.13 (Example 20.8) Energy transfer by conduction through two slabs in thermal contact with each other. At steady state, the rate of energy transfer through slab 1 equals the rate of energy transfer through slab 2.

SOLUTION

Conceptualize Notice the phrase "in the steady-state condition." We interpret this phrase to mean that energy transfers through the compound slab at the same rate at all points. Otherwise, energy would be building up or disappearing at some point. Furthermore, the temperature varies with position in the two slabs, most likely at different rates in each part of the compound slab. When the system is in steady state, the interface is at some fixed temperature T.

Categorize We categorize this example as an equilibrium thermal conduction problem and impose the condition that the power is the same in both slabs of material.

Analyze Use Equation 20.16 to express the rate at which energy is transferred through slab 1:

$$(1) \quad \mathcal{P}_1 = k_1 A\left(\frac{T - T_c}{L_1}\right)$$

Express the rate at which energy is transferred through slab 2:

$$(2) \quad \mathcal{P}_2 = k_2 A\left(\frac{T_h - T}{L_2}\right)$$

Set these two rates equal to represent the equilibrium situation:

$$k_1 A\left(\frac{T - T_c}{L_1}\right) = k_2 A\left(\frac{T_h - T}{L_2}\right)$$

Solve for T:

$$(3) \quad T = \frac{k_1 L_2 T_c + k_2 L_1 T_h}{k_1 L_2 + k_2 L_1}$$

Substitute Equation (3) into either Equation (1) or Equation (2):

$$(4) \quad \mathcal{P} = \frac{A(T_h - T_c)}{(L_1/k_1) + (L_2/k_2)}$$

Finalize Extension of this procedure to several slabs of materials leads to Equation 20.17.

What If? Suppose you are building an insulated container with two layers of insulation and the rate of energy transfer determined by Equation (4) turns out to be too high. You have enough room to increase the thickness of one of the two layers by 20%. How would you decide which layer to choose?

Answer To decrease the power as much as possible, you must increase the denominator in Equation (4) as much as possible. Whichever thickness you choose to increase, L_1 or L_2, you increase the corresponding term L/k in the denominator by 20%. For this percentage change to represent the largest absolute change, you want to take 20% of the larger term. Therefore, you should increase the thickness of the layer that has the larger value of L/k.

Home Insulation

In engineering practice, the term L/k for a particular substance is referred to as the ***R*-value** of the material. Therefore, Equation 20.17 reduces to

$$\mathcal{P} = \frac{A(T_h - T_c)}{\sum_i R_i} \qquad \textbf{(20.18)}$$

where $R_i = L_i/k_i$. The R-values for a few common building materials are given in Table 20.4. In the United States, the insulating properties of materials used in

TABLE 20.4

R-Values for Some Common Building Materials

Material	R-value ($ft^2 \cdot °F \cdot h/Btu$)
Hardwood siding (1 in. thick)	0.91
Wood shingles (lapped)	0.87
Brick (4 in. thick)	4.00
Concrete block (filled cores)	1.93
Fiberglass insulation (3.5 in. thick)	10.90
Fiberglass insulation (6 in. thick)	18.80
Fiberglass board (1 in. thick)	4.35
Cellulose fiber (1 in. thick)	3.70
Flat glass (0.125 in. thick)	0.89
Insulating glass (0.25-in. space)	1.54
Air space (3.5 in. thick)	1.01
Stagnant air layer	0.17
Drywall (0.5 in. thick)	0.45
Sheathing (0.5 in. thick)	1.32

buildings are usually expressed in U.S. customary units, not SI units. Therefore, in Table 20.4, R-values are given as a combination of British thermal units, feet, hours, and degrees Fahrenheit.

At any vertical surface open to the air, a very thin stagnant layer of air adheres to the surface. One must consider this layer when determining the R-value for a wall. The thickness of this stagnant layer on an outside wall depends on the speed of the wind. Energy transfer through the walls of a house on a windy day is greater than that on a day when the air is calm. A representative R-value for this stagnant layer of air is given in Table 20.4.

EXAMPLE 20.9 The *R*-Value of a Typical Wall

Calculate the total R-value for a wall constructed as shown in Figure 20.14a. Starting outside the house (toward the front in the figure) and moving inward, the wall consists of 4 in. of brick, 0.5 in. of sheathing, an air space 3.5 in. thick, and 0.5 in. of drywall.

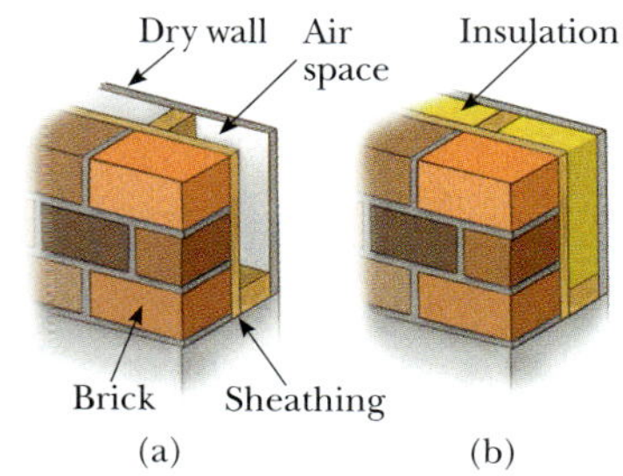

Figure 20.14 (Example 20.9) An exterior house wall containing (a) an air space and (b) insulation.

SOLUTION

Conceptualize Use Figure 20.14 to help conceptualize the structure of the wall. Do not forget the stagnant air layers inside and outside the house.

Categorize We will use specific equations developed in this section on home insulation, so we categorize this example as a substitution problem.

Use Table 20.4 to find the R-value of each layer:

$$R_1\,(\text{outside stagnant air layer}) = 0.17\ \text{ft}^2\cdot{}^\circ\text{F}\cdot\text{h/Btu}$$
$$R_2\,(\text{brick}) = 4.00\ \text{ft}^2\cdot{}^\circ\text{F}\cdot\text{h/Btu}$$
$$R_3\,(\text{sheathing}) = 1.32\ \text{ft}^2\cdot{}^\circ\text{F}\cdot\text{h/Btu}$$
$$R_4\,(\text{air space}) = 1.01\ \text{ft}^2\cdot{}^\circ\text{F}\cdot\text{h/Btu}$$
$$R_5\,(\text{drywall}) = 0.45\ \text{ft}^2\cdot{}^\circ\text{F}\cdot\text{h/Btu}$$
$$R_6\,(\text{inside stagnant air layer}) = 0.17\ \text{ft}^2\cdot{}^\circ\text{F}\cdot\text{h/Btu}$$

Add the R-values to obtain the total R-value for the wall:

$$R_{\text{total}} = R_1 + R_2 + R_3 + R_4 + R_5 + R_6 = 7.12\ \text{ft}^2\cdot{}^\circ\text{F}\cdot\text{h/Btu}$$

What If? Suppose you are not happy with this total R-value for the wall. You cannot change the overall structure, but you can fill the air space as in Figure 20.14b. To *maximize* the total R-value, what material should you choose to fill the air space?

Answer Looking at Table 20.4, we see that 3.5 in. of fiberglass insulation is more than ten times as effective as 3.5 in. of air. Therefore, we should fill the air space with fiberglass insulation. The result is that we add 10.90 $\text{ft}^2\cdot{}^\circ\text{F}\cdot\text{h/Btu}$ of R-value, and we lose 1.01 $\text{ft}^2\cdot{}^\circ\text{F}\cdot\text{h/Btu}$ due to the air space we have replaced. The new total R-value is equal to $7.12\ \text{ft}^2\cdot{}^\circ\text{F}\cdot\text{h/Btu} + 9.89\ \text{ft}^2\cdot{}^\circ\text{F}\cdot\text{h/Btu} = 17.01\ \text{ft}^2\cdot{}^\circ\text{F}\cdot\text{h/Btu}$.

Convection

At one time or another, you probably have warmed your hands by holding them over an open flame. In this situation, the air directly above the flame is heated and expands. As a result, the density of this air decreases and the air rises. This hot air warms your hands as it flows by. **Energy transferred by the movement of a warm substance is said to have been transferred by convection.** When resulting from differences in density, as with air around a fire, the process is referred to as *natural convection.* Airflow at a beach is an example of natural convection, as is the mixing

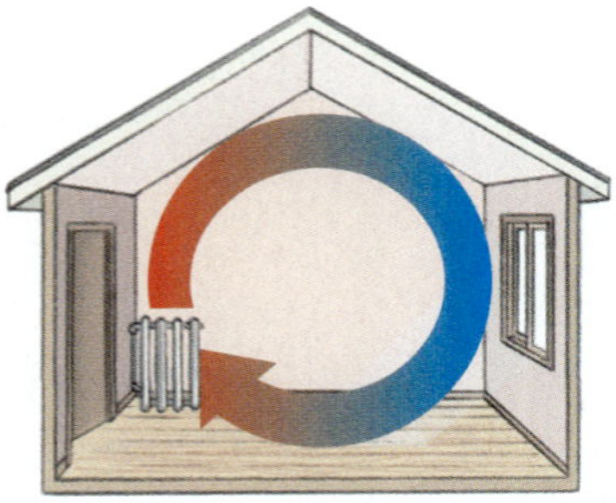

Figure 20.15 Convection currents are set up in a room warmed by a radiator.

that occurs as surface water in a lake cools and sinks (see Section 19.4). When the heated substance is forced to move by a fan or pump, as in some hot-air and hot-water heating systems, the process is called *forced convection.*

If it were not for convection currents, it would be very difficult to boil water. As water is heated in a teakettle, the lower layers are warmed first. This water expands and rises to the top because its density is lowered. At the same time, the denser, cool water at the surface sinks to the bottom of the kettle and is heated.

The same process occurs when a room is heated by a radiator. The hot radiator warms the air in the lower regions of the room. The warm air expands and rises to the ceiling because of its lower density. The denser, cooler air from above sinks, and the continuous air current pattern shown in Figure 20.15 is established.

Radiation

The third means of energy transfer we shall discuss is **thermal radiation**. All objects radiate energy continuously in the form of electromagnetic waves (see Chapter 34) produced by thermal vibrations of the molecules. You are likely familiar with electromagnetic radiation in the form of the orange glow from an electric stove burner, an electric space heater, or the coils of a toaster.

The rate at which an object radiates energy is proportional to the fourth power of its absolute temperature. Known as **Stefan's law**, this behavior is expressed in equation form as

Stefan's law ▶

$$\mathcal{P} = \sigma A e T^4 \tag{20.19}$$

where $\mathcal{P}$ is the power in watts of electromagnetic waves radiated from the surface of the object, σ is a constant equal to $5.669\,6 \times 10^{-8}\ \mathrm{W/m^2 \cdot K^4}$, A is the surface area of the object in square meters, e is the **emissivity**, and T is the surface temperature in kelvins. The value of e can vary between zero and unity depending on the properties of the surface of the object. The emissivity is equal to the **absorptivity**, which is the fraction of the incoming radiation that the surface absorbs. A mirror has very low absorptivity because it reflects almost all incident light. Therefore, a mirror surface also has a very low emissivity. At the other extreme, a black surface has high absorptivity and high emissivity. An **ideal absorber** is defined as an object that absorbs all the energy incident on it, and for such an object, $e = 1$. An object for which $e = 1$ is often referred to as a **black body**. We shall investigate experimental and theoretical approaches to radiation from a black body in Chapter 40.

Every second, approximately 1 370 J of electromagnetic radiation from the Sun passes perpendicularly through each 1 $\mathrm{m^2}$ at the top of the Earth's atmosphere. This radiation is primarily visible and infrared light accompanied by a significant amount of ultraviolet radiation. We shall study these types of radiation in detail in Chapter 34. Enough energy arrives at the surface of the Earth each day to supply all our energy needs on this planet hundreds of times over, if only it could be captured and used efficiently. The growth in the number of solar energy–powered houses built in the United States reflects the increasing efforts being made to use this abundant energy.

What happens to the atmospheric temperature at night is another example of the effects of energy transfer by radiation. If there is a cloud cover above the Earth, the water vapor in the clouds absorbs part of the infrared radiation emitted by the Earth and re-emits it back to the surface. Consequently, temperature levels at the surface remain moderate. In the absence of this cloud cover, there is less in the way to prevent this radiation from escaping into space; therefore, the temperature decreases more on a clear night than on a cloudy one.

As an object radiates energy at a rate given by Equation 20.19, it also absorbs electromagnetic radiation from the surroundings, which consist of other objects that radiate energy. If the latter process did not occur, an object would eventually radiate all its energy and its temperature would reach absolute zero. If an object is at a temperature T and its surroundings are at an average temperature T_0, the net rate of energy gained or lost by the object as a result of radiation is

$$\mathcal{P}_{\text{net}} = \sigma A e(T^4 - T_0^{\,4}) \qquad (20.20)$$

When an object is in equilibrium with its surroundings, it radiates and absorbs energy at the same rate and its temperature remains constant. When an object is hotter than its surroundings, it radiates more energy than it absorbs and its temperature decreases.

The Dewar Flask

The *Dewar flask*[7] is a container designed to minimize energy transfers by conduction, convection, and radiation. Such a container is used to store cold or hot liquids for long periods of time. (An insulated bottle, such as a Thermos, is a common household equivalent of a Dewar flask.) The standard construction (Fig. 20.16) consists of a double-walled Pyrex glass vessel with silvered walls. The space between the walls is evacuated to minimize energy transfer by conduction and convection. The silvered surfaces minimize energy transfer by radiation because silver is a very good reflector and has very low emissivity. A further reduction in energy loss is obtained by reducing the size of the neck. Dewar flasks are commonly used to store liquid nitrogen (boiling point 77 K) and liquid oxygen (boiling point 90 K).

To confine liquid helium (boiling point 4.2 K), which has a very low heat of vaporization, it is often necessary to use a double Dewar system in which the Dewar flask containing the liquid is surrounded by a second Dewar flask. The space between the two flasks is filled with liquid nitrogen.

Newer designs of storage containers use "super insulation" that consists of many layers of reflecting material separated by fiberglass. All this material is in a vacuum, and no liquid nitrogen is needed with this design.

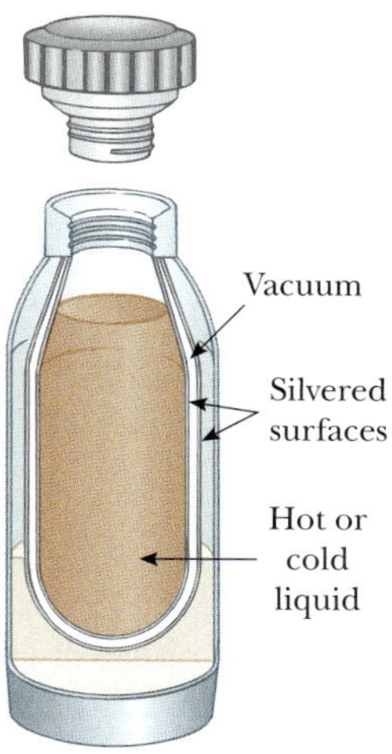

Figure 20.16 A cross-sectional view of a Dewar flask, which is used to store hot or cold substances.

[7] Invented by Sir James Dewar (1842–1923).

Summary

ThomsonNOW™ Sign in at **www.thomsonedu.com** and go to ThomsonNOW to take a practice test for this chapter.

DEFINITIONS

Internal energy is all a system's energy that is associated with the system's microscopic components. Internal energy includes kinetic energy of random translation, rotation, and vibration of molecules, vibrational potential energy within molecules, and potential energy between molecules.

Heat is the transfer of energy across the boundary of a system resulting from a temperature difference between the system and its surroundings. The symbol Q represents the amount of energy transferred by this process.

A **calorie** is the amount of energy necessary to raise the temperature of 1 g of water from 14.5°C to 15.5°C.

The **heat capacity** C of any sample is the amount of energy needed to raise the temperature of the sample by 1°C.

The **specific heat** c of a substance is the heat capacity per unit mass:

$$c \equiv \frac{Q}{m\,\Delta T} \qquad (20.3)$$

The **latent heat** of a substance is defined as the ratio of the energy necessary to cause a phase change to the mass of the substance:

$$L \equiv \frac{Q}{m} \qquad (20.6)$$

(*continued*)

CONCEPTS AND PRINCIPLES

The energy Q required to change the temperature of a mass m of a substance by an amount ΔT is

$$Q = mc\,\Delta T \tag{20.4}$$

where c is the specific heat of the substance.

The energy required to change the phase of a pure substance of mass m is

$$Q = \pm mL \tag{20.7}$$

where L is the latent heat of the substance and depends on the nature of the phase change and the substance. The positive sign is used if energy is entering the system, and the negative sign is used if energy is leaving the system.

The **work** done on a gas as its volume changes from some initial value V_i to some final value V_f is

$$W = -\int_{V_i}^{V_f} P\,dV \tag{20.9}$$

where P is the pressure of the gas, which may vary during the process. To evaluate W, the process must be fully specified; that is, P and V must be known during each step. The work done depends on the path taken between the initial and final states.

The **first law of thermodynamics** states that when a system undergoes a change from one state to another, the change in its internal energy is

$$\Delta E_{\text{int}} = Q + W \tag{20.10}$$

where Q is the energy transferred into the system by heat and W is the work done on the system. Although Q and W both depend on the path taken from the initial state to the final state, the quantity ΔE_{int} does not depend on the path.

In a **cyclic process** (one that originates and terminates at the same state), $\Delta E_{\text{int}} = 0$ and therefore $Q = -W$. That is, the energy transferred into the system by heat equals the negative of the work done on the system during the process.

In an **adiabatic process**, no energy is transferred by heat between the system and its surroundings ($Q = 0$). In this case, the first law gives $\Delta E_{\text{int}} = W$. In the **adiabatic free expansion** of a gas, $Q = 0$ and $W = 0$, so $\Delta E_{\text{int}} = 0$. That is, the internal energy of the gas does not change in such a process.

An **isobaric process** is one that occurs at constant pressure. The work done on a gas in such a process is $W = -P(V_f - V_i)$.

An **isovolumetric process** is one that occurs at constant volume. No work is done in such a process, so $\Delta E_{\text{int}} = Q$.

An **isothermal process** is one that occurs at constant temperature. The work done on an ideal gas during an isothermal process is

$$W = nRT\ln\left(\frac{V_i}{V_f}\right) \tag{20.14}$$

Conduction can be viewed as an exchange of kinetic energy between colliding molecules or electrons. The rate of energy transfer by conduction through a slab of area A is

$$\mathcal{P} = kA\left|\frac{dT}{dx}\right| \tag{20.15}$$

where k is the **thermal conductivity** of the material from which the slab is made and $|dT/dx|$ is the **temperature gradient**.

In **convection**, a warm substance transfers energy from one location to another.

All objects emit **thermal radiation** in the form of electromagnetic waves at the rate

$$\mathcal{P} = \sigma AeT^4 \tag{20.19}$$

Questions

☐ denotes answer available in *Student Solutions Manual/Study Guide;* **O** denotes objective question

1. Clearly distinguish among temperature, heat, and internal energy.

2. **O** Ethyl alcohol has about half the specific heat of water. Assume equal amounts of energy are transferred by heat into equal-mass liquid samples of alcohol and water in separate insulated containers. The water rises in temperature by 25°C. How much will the alcohol rise in temperature? (a) 12°C (b) 25°C (c) 50°C (d) It depends on the rate of energy transfer. (e) It will not rise in temperature.

3. What is wrong with the following statement: "Given any two bodies, the one with the higher temperature contains more heat."

4. **O** Beryllium has roughly one-half the specific heat of liquid water (H_2O). Rank the quantities of energy input required to produce the following changes from the largest to the smallest. In your ranking, note any cases of equality. (a) raising the temperature of 1 kg of H_2O from 20°C to 26°C (b) raising the temperature of 2 kg of H_2O from 20°C to 23°C (c) raising the temperature of 2 kg of H_2O from 1°C to 4°C (d) raising the temperature of 2 kg of beryllium from −1°C to 2°C (e) raising the temperature of 2 kg of H_2O from −1°C to 2°C

5. Why is a person able to remove a piece of dry aluminum foil from a hot oven with bare fingers, whereas a burn results if there is moisture on the foil?

6. The air temperature above coastal areas is profoundly influenced by the large specific heat of water. One reason is that the energy released when 1 m^3 of water cools by 1°C will raise the temperature of a much larger volume of air by 1°C. Find this volume of air. The specific heat of air is approximately 1 kJ/kg · °C. Take the density of air to be 1.3 kg/m^3.

7. **O** Assume you are measuring the specific heat of a sample of originally hot metal by the method of mixtures as described in Example 20.2. Because your calorimeter is not perfectly insulating, energy can transfer by heat between the contents of the calorimeter and the room. To obtain the most accurate result for the specific heat of the metal, you should use water with which initial temperature? (a) slightly lower than room temperature (b) the same as room temperature (c) slightly above room temperature (d) whatever you like because the initial temperature makes no difference.

8. Using the first law of thermodynamics, explain why the *total* energy of an isolated system is always constant.

9. **O** A person shakes a sealed insulated bottle containing hot coffee for a few minutes. **(i)** What is the change in the temperature of the coffee? (a) a large decrease (b) a slight decrease (c) no change (d) a slight increase (e) a large increase **(ii)** What is the change in the internal energy of the coffee? Choose from the same possibilities.

10. Is it possible to convert internal energy to mechanical energy? Explain with examples.

11. A tile floor in a bathroom may feel uncomfortably cold to your bare feet, but a carpeted floor in an adjoining room at the same temperature will feel warm. Why?

12. It is the morning of a day that will become hot. You just purchased drinks for a picnic and are loading them, with ice, into a chest in the back of your car. You have a wool blanket. Should you wrap it around the chest? Would doing so help to keep the beverages cool, or should you expect the wool blanket to warm them up? Your little sister tells you emphatically that she would not like to be wrapped up in a wool blanket on a hot day. Explain your answers and your response to her.

13. When camping in a canyon on a still night, a camper notices that as soon as the sun strikes the surrounding peaks, a breeze begins to stir. What causes the breeze?

14. **O** A poker is a stiff, nonflammable rod used to push burning logs around in a fireplace. For ease of use and safety, the poker should be made from a material (a) with high specific heat and high thermal conductivity, (b) with low specific heat and low thermal conductivity, (c) with low specific heat and high thermal conductivity, or (d) with high specific heat and low thermal conductivity.

15. **O** Star *A* has twice the radius and twice the absolute temperature of star *B*. What is the ratio of the power output of star *A* to that of star *B*? The emissivity of both stars is essentially 1. (a) 4 (b) 8 (c) 16 (d) 32 (e) 64

16. If water is a poor thermal conductor, why can the temperature throughout a pot of water be raised quickly when it is placed over a flame?

17. You need to pick up a very hot cooking pot in your kitchen. You have a pair of hot pads. To be able to pick up the pot most comfortably, should you soak the pads in cold water or keep them dry?

18. Suppose you pour hot coffee for your guests, and one of them wants to drink it with cream, several minutes later, and then as warm as possible. To have the warmest coffee, should the person add the cream just after the coffee is poured or just before drinking? Explain.

19. **O** Warning signs seen on highways just before a bridge are "Caution—Bridge freezes before road surface," or "Bridge may be icy." Which of the three energy transfer processes discussed in Section 20.7 is most important in causing ice to form on a bridge surface before it does on the rest of the road surface on very cold days? (a) conduction (b) convection (c) radiation (d) none of these choices because the ice freezes without a change in temperature

20. A physics teacher drops one marshmallow into a flask of liquid nitrogen, waits for the most energetic boiling to stop, fishes it out with tongs, shakes it off, pops it into his mouth, chews it up, and swallows it. Clouds of ice crystals issue from his mouth as he crunches noisily and comments on the sweet taste. How can he do that without injury? *Caution:* Liquid nitrogen can be a dangerous substance. You should *not* try this demonstration yourself. The teacher might be badly injured if he did not shake the marshmallow off, if he touched the tongs to a tooth, or if he did not start with a mouthful of saliva.

21. In 1801, Humphry Davy rubbed together pieces of ice inside an icehouse. He made sure that nothing in the environment was at a higher temperature than the rubbed pieces. He observed the production of drops of liquid water. Make a table listing this and other experiments or processes to illustrate each of the following situations. (a) A system can absorb energy by heat, increase in internal energy, and increase in temperature. (b) A system can absorb energy by heat and increase in internal energy without an increase in temperature. (c) A system can absorb energy by heat without increasing in temperature or in internal energy. (d) A system can increase in internal energy and in temperature without absorbing energy by heat. (e) A system can increase in internal energy without absorbing energy by heat or increasing in temperature. (f) **What If?** If a system's temperature increases, is it necessarily true that its internal energy increases?

22. Consider the opening photograph for Part 3 (page 531). Discuss the roles of conduction, convection, and radiation in the operation of the cooling fins on the support posts of the Alaskan oil pipeline.

Problems

WebAssign The Problems from this chapter may be assigned online in WebAssign.

ThomsonNOW™ Sign in at **www.thomsonedu.com** and go to ThomsonNOW to assess your understanding of this chapter's topics with additional quizzing and conceptual questions.

1, 2, 3 denotes straightforward, intermediate, challenging; ☐ denotes full solution available in *Student Solutions Manual/Study Guide;* ▲ denotes coached solution with hints available at **www.thomsonedu.com;** ▭ denotes developing symbolic reasoning; ● denotes asking for qualitative reasoning; 🖥 denotes computer useful in solving problem

Section 20.1 Heat and Internal Energy

1. ☐ On his honeymoon, James Joule tested the conversion of mechanical energy into internal energy by measuring temperatures of falling water. If water at the top of a Swiss waterfall has a temperature of 10.0°C and then falls 50.0 m, what maximum temperature at the bottom could Joule expect? He did not succeed in measuring the temperature change, partly because evaporation cooled the falling water and also because his thermometer was not sufficiently sensitive.

2. Consider Joule's apparatus described in Figure 20.1. The mass of each of the two blocks is 1.50 kg, and the insulated tank is filled with 200 g of water. What is the increase in the temperature of the water after the blocks fall through a distance of 3.00 m?

Section 20.2 Specific Heat and Calorimetry

3. ☐ The temperature of a silver bar rises by 10.0°C when it absorbs 1.23 kJ of energy by heat. The mass of the bar is 525 g. Determine the specific heat of silver.

4. ● The *Nova* laser at Lawrence Livermore National Laboratory in California was used in early studies of initiating controlled nuclear fusion (Section 45.4). It delivered a power of 1.60×10^{13} W over a time interval of 2.50 ns. Explain how its energy output in one such time interval compares with the energy required to make a pot of tea by warming 0.800 kg of water from 20.0°C to 100°C.

5. Systematic use of solar energy can yield a large saving in the cost of winter space heating for a typical house in the northern United States. If the house has good insulation, you may model it as losing energy by heat steadily at the rate 6 000 W on a day in April when the average exterior temperature is 4°C and when the conventional heating system is not used at all. The passive solar energy collector can consist simply of very large windows in a room facing south. Sunlight shining in during the daytime is absorbed by the floor, interior walls, and objects in the room, raising their temperature to 38.0°C. As the sun goes down, insulating draperies or shutters are closed over the windows. During the period between 5:00 p.m. and 7:00 a.m., the temperature of the house will drop and a sufficiently large "thermal mass" is required to keep it from dropping too far. The thermal mass can be a large quantity of stone (with specific heat 850 J/kg · °C) in the floor and the interior walls exposed to sunlight. What mass of stone is required if the temperature is not to drop below 18.0°C overnight?

6. An aluminum cup of mass 200 g contains 800 g of water in thermal equilibrium at 80.0°C. The combination of cup and water is cooled uniformly so that the temperature decreases by 1.50°C per minute. At what rate is energy being removed by heat? Express your answer in watts.

7. ☐ ▲ A 1.50-kg iron horseshoe initially at 600°C is dropped into a bucket containing 20.0 kg of water at 25.0°C. What is the final temperature? (Ignore the heat capacity of the container and assume a negligible amount of water boils away.)

8. ● An electric drill with a steel drill bit of mass 27.0 g and diameter 0.635 cm is used to drill into a cubical steel block of mass 240 g. Assume steel has the same properties as iron. The cutting process can be modeled as happening at one point on the circumference of the bit. This point moves in a spiral at constant speed 40.0 m/s and exerts a force of constant magnitude 3.20 N on the block. As shown in Figure P20.8, a groove in the bit carries the chips up to the top of the block, where they form a pile around the hole. The block is held in a clamp made of material of low thermal conductivity, and the drill bit is held in a chuck also made of this material. We consider turning the drill on for a time interval of 15.0 s. This time interval is sufficiently short that the steel objects lose only a negligible amount of energy by conduction, convection, and radiation into their environment. Nevertheless, 15 s is long enough for conduction within the steel to bring it all to a uniform temperature. The temperature is promptly measured with a thermometer probe, shown in the side of the block in the figure. (a) Suppose the drill bit is sharp and cuts three-quarters of the way through the block during 15 s. Find the temperature change of the whole quantity of steel. (b) **What If?** Now suppose the drill bit is dull and cuts only one-eighth of the way through the block. Identify the temperature change of the whole quantity of steel in this case. (c) What pieces of data, if any, are unnecessary for the solution? Explain.

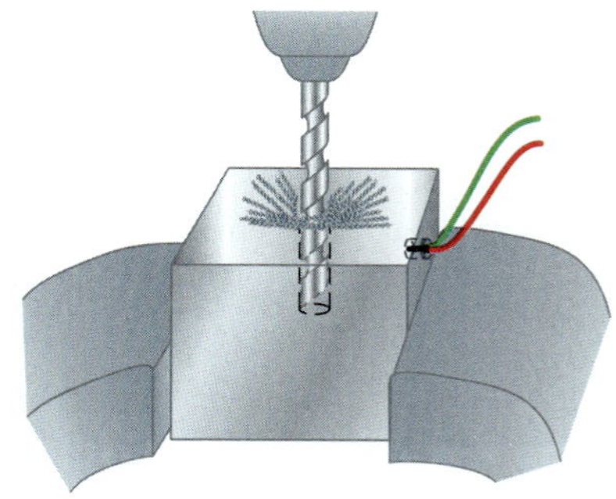

Figure P20.8

9. ● An aluminum calorimeter with a mass of 100 g contains 250 g of water. The calorimeter and water are in thermal equilibrium at 10.0°C. Two metallic blocks are placed into the water. One is a 50.0-g piece of copper at 80.0°C. The other has a mass of 70.0 g and is originally at a temperature of 100°C. The entire system stabilizes at a final temperature of 20.0°C. (a) Determine the specific heat of the unknown sample. (b) Using the data in Table 20.1, can you make a positive identification of the unknown material? Can you identify a possible material? Explain your answers.

10. ● A 3.00-g copper penny at 25.0°C drops 50.0 m to the ground. (a) Assuming 60.0% of the change in potential energy of the penny–Earth system goes into increasing the internal energy of the penny, determine the penny's final temperature. (b) **What If?** Does the result depend on the mass of the penny? Explain.

11. A combination of 0.250 kg of water at 20.0°C, 0.400 kg of aluminum at 26.0°C, and 0.100 kg of copper at 100°C is mixed in an insulated container and allowed to come to thermal equilibrium. Ignore any energy transfer to or from the container and determine the final temperature of the mixture.

12. Two thermally insulated vessels are connected by a narrow tube fitted with a valve that is initially closed. One vessel of volume 16.8 L contains oxygen at a temperature of 300 K and a pressure of 1.75 atm. The other vessel of volume 22.4 L contains oxygen at a temperature of 450 K and a pressure of 2.25 atm. When the valve is opened, the gases in the two vessels mix and the temperature and pressure become uniform throughout. (a) What is the final temperature? (b) What is the final pressure?

Section 20.3 Latent Heat

13. How much energy is required to change a 40.0-g ice cube from ice at −10.0°C to steam at 110°C?

14. A 50.0-g copper calorimeter contains 250 g of water at 20.0°C. How much steam must be condensed into the water if the final temperature of the system is to reach 50.0°C?

15. A 3.00-g lead bullet at 30.0°C is fired at a speed of 240 m/s into a large block of ice at 0°C, in which it becomes embedded. What quantity of ice melts?

16. Steam at 100°C is added to ice at 0°C. (a) Find the amount of ice melted and the final temperature when the mass of steam is 10.0 g and the mass of ice is 50.0 g. (b) **What If?** Repeat when the mass of steam is 1.00 g and the mass of ice is 50.0 g.

17. A 1.00-kg block of copper at 20.0°C is dropped into a large vessel of liquid nitrogen at 77.3 K. How many kilograms of nitrogen boil away by the time the copper reaches 77.3 K? (The specific heat of copper is 0.092 0 cal/g · °C. The latent heat of vaporization of nitrogen is 48.0 cal/g.)

18. ● An automobile has a mass of 1 500 kg, and its aluminum brakes have an overall mass of 6.00 kg. (a) Assume all the mechanical energy that disappears when the car stops is deposited in the brakes and no energy is transferred out of the brakes by heat. The brakes are originally at 20.0°C. How many times can the car be stopped from 25.0 m/s before the brakes start to melt? (b) Identify some effects ignored in part (a) that are important in a more realistic assessment of the warming of the brakes.

19. ▲ In an insulated vessel, 250 g of ice at 0°C is added to 600 g of water at 18.0°C. (a) What is the final temperature of the system? (b) How much ice remains when the system reaches equilibrium?

20. **Review problem.** The following equation describes a process that occurs so rapidly that negligible energy is transferred between the system and the environment by conduction, convection, or radiation:

$$\tfrac{1}{2}(0.012\,0\text{ kg})(300\text{ m/s})^2 + \tfrac{1}{2}(0.008\,00\text{ kg})(400\text{ m/s})^2$$
$$= \tfrac{1}{2}(0.020\,0\text{ kg})(20\text{ m/s})^2$$
$$+ (0.020\,0\text{ kg})(128\text{ J/kg}\cdot{}^\circ\text{C})(327.3^\circ\text{C} - 30.0^\circ\text{C})$$
$$+ m_\ell(2.45 \times 10^4\text{ J/kg})$$

(a) Write a problem for which the equation will appear in the solution. Give the data, describe the system, and describe the process going on. Let the problem end with the statement, "Describe the state of the system immediately thereafter." (b) Solve the problem, including calculating the unknown in the equation and identifying its physical meaning.

Section 20.4 Work and Heat in Thermodynamic Processes

Problems 4 and 27 in Chapter 7 can also be assigned with this section.

21. ▲ A sample of ideal gas is expanded to twice its original volume of 1.00 m³ in a quasi-static process for which $P = \alpha V^2$, with $\alpha = 5.00\text{ atm/m}^6$, as shown in Figure P20.21. How much work is done on the expanding gas?

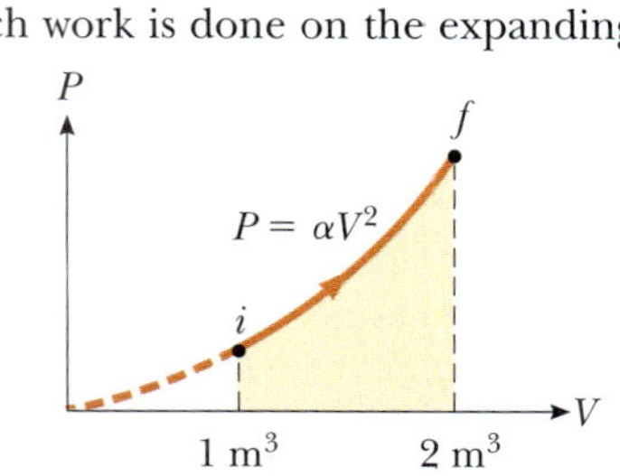

Figure P20.21

2 = intermediate; 3 = challenging; □ = SSM/SG; ▲ = ThomsonNOW; = symbolic reasoning; ● = qualitative reasoning

22. (a) Determine the work done on a fluid that expands from i to f as indicated in Figure P20.22. (b) **What If?** How much work is performed on the fluid if it is compressed from f to i along the same path?

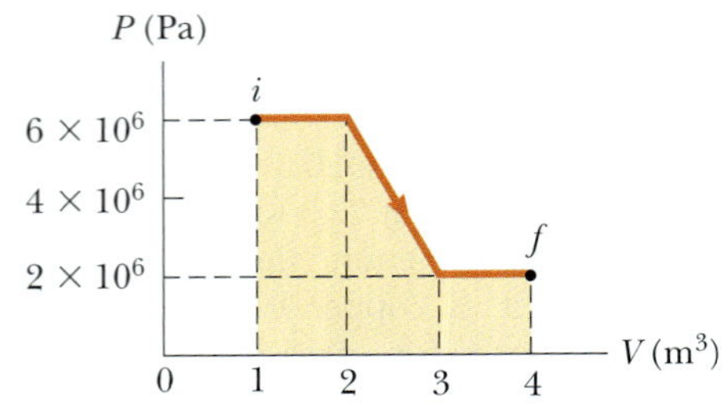

Figure P20.22

23. ▲ An ideal gas is enclosed in a cylinder with a movable piston on top of it. The piston has a mass of 8 000 g and an area of 5.00 cm^2 and is free to slide up and down, keeping the pressure of the gas constant. How much work is done on the gas as the temperature of 0.200 mol of the gas is raised from 20.0°C to 300°C?

24. An ideal gas is enclosed in a cylinder that has a movable piston on top. The piston has a mass m and an area A and is free to slide up and down, keeping the pressure of the gas constant. How much work is done on the gas as the temperature of n mol of the gas is raised from T_1 to T_2?

25. ● One mole of an ideal gas is warmed slowly so that it goes from the PV state (P_i, V_i), to $(3P_i, 3V_i)$, in such a way that the pressure of the gas is directly proportional to the volume. (a) How much work is done on the gas in the process? (b) How is the temperature of the gas related to its volume during this process?

Section 20.5 The First Law of Thermodynamics

26. A gas is taken through the cyclic process described in Figure P20.26. (a) Find the net energy transferred to the system by heat during one complete cycle. (b) **What If?** If the cycle is reversed—that is, the process follows the path $ACBA$—what is the net energy input per cycle by heat?

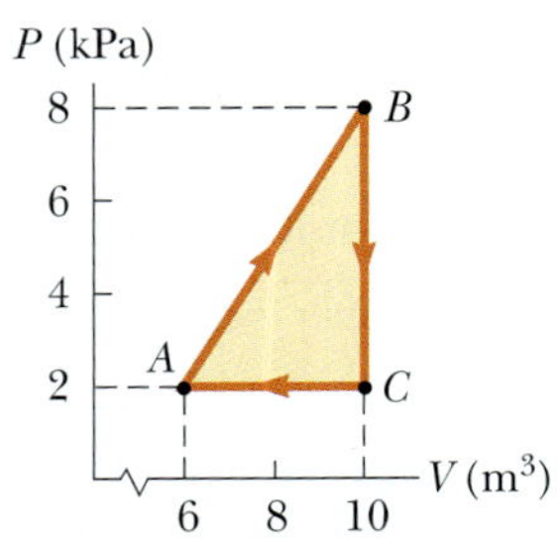

Figure P20.26 Problems 26 and 29.

27. A thermodynamic system undergoes a process in which its internal energy decreases by 500 J. Over the same time interval, 220 J of work is done on the system. Find the energy transferred to or from it by heat.

28. A sample of an ideal gas goes through the process shown in Figure P20.28. From A to B, the process is adiabatic; from B to C, it is isobaric with 100 kJ of energy entering the system by heat. From C to D, the process is isothermal; from D to A, it is isobaric with 150 kJ of energy leaving the system by heat. Determine the difference in internal energy $E_{\text{int},B} - E_{\text{int},A}$.

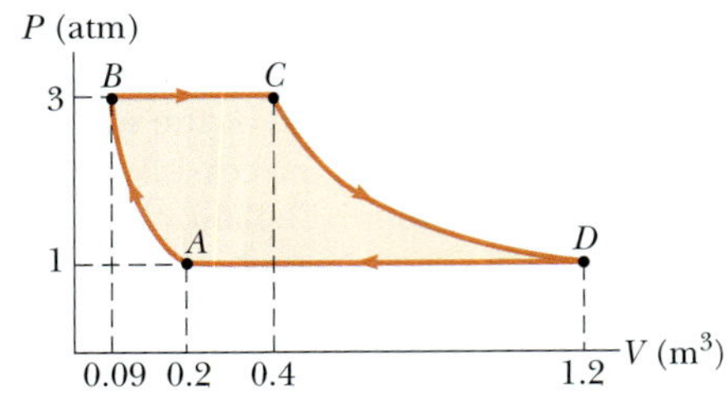

Figure P20.28

29. Consider the cyclic process depicted in Figure P20.26. If Q is negative for the process BC and ΔE_{int} is negative for the process CA, what are the signs of Q, W, and ΔE_{int} that are associated with each process?

Section 20.6 Some Applications of the First Law of Thermodynamics

30. One mole of an ideal gas does 3 000 J of work on its surroundings as it expands isothermally to a final pressure of 1.00 atm and volume of 25.0 L. Determine (a) the initial volume and (b) the temperature of the gas.

31. An ideal gas initially at 300 K undergoes an isobaric expansion at 2.50 kPa. If the volume increases from 1.00 m^3 to 3.00 m^3 and 12.5 kJ is transferred to the gas by heat, what are (a) the change in its internal energy and (b) its final temperature?

32. A 1.00-kg block of aluminum is warmed at atmospheric pressure so that its temperature increases from 22.0°C to 40.0°C. Find (a) the work done on the aluminum, (b) the energy added to it by heat, and (c) the change in its internal energy.

33. How much work is done on the steam when 1.00 mol of water at 100°C boils and becomes 1.00 mol of steam at 100°C at 1.00 atm pressure? Assume the steam to behave as an ideal gas. Determine the change in internal energy of the material as it vaporizes.

34. An ideal gas initially at P_i, V_i, and T_i is taken through a cycle as shown in Figure P20.34. (a) Find the net work done on the gas per cycle. (b) What is the net energy added by heat to the system per cycle? (c) Obtain a numerical value for the net work done per cycle for 1.00 mol of gas initially at 0°C.

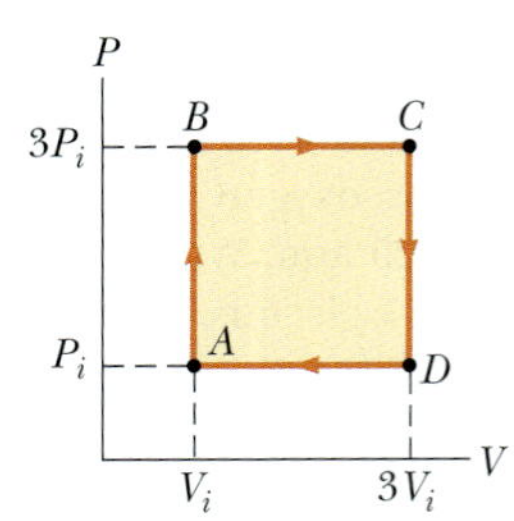

Figure P20.34

35. A 2.00-mol sample of helium gas initially at 300 K and 0.400 atm is compressed isothermally to 1.20 atm. Noting that the helium behaves as an ideal gas, find (a) the final volume of the gas, (b) the work done on the gas, and (c) the energy transferred by heat.

36. In Figure P20.36, the change in internal energy of a gas that is taken from A to C is $+800$ J. The work done on the gas along path ABC is -500 J. (a) How much energy must be added to the system by heat as it goes from A through B to C? (b) If the pressure at point A is five times that of point C, what is the work done on the system in going from C to D? (c) What is the energy exchanged with the surroundings by heat as the cycle goes from C to A along the green path? (d) If the change in internal energy in going from point D to point A is $+500$ J, how much energy must be added to the system by heat as it goes from point C to point D?

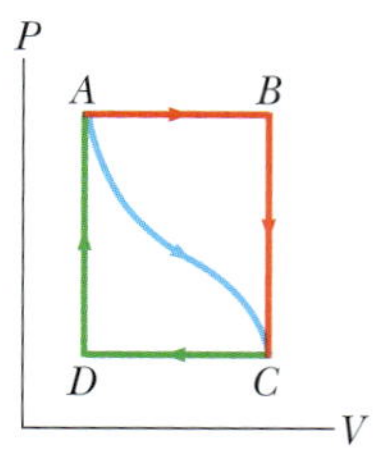

Figure P20.36

Section 20.7 Energy Transfer Mechanisms

37. A glass windowpane has an area of 3.00 m^2 and a thickness of 0.600 cm. If the temperature difference between its faces is 25.0°C, what is the rate of energy transfer by conduction through the window?

38. A thermal window with an area of 6.00 m^2 is constructed of two layers of glass, each 4.00 mm thick, separated from each other by an air space of 5.00 mm. If the inside surface is at 20.0°C and the outside is at -30.0°C, what is the rate of energy transfer by conduction through the window?

39. A bar of gold (Au) is in thermal contact with a bar of silver (Ag) of the same length and area (Fig. P20.39). One end of the compound bar is maintained at 80.0°C, and the opposite end is at 30.0°C. When the energy transfer reaches steady state, what is the temperature at the junction?

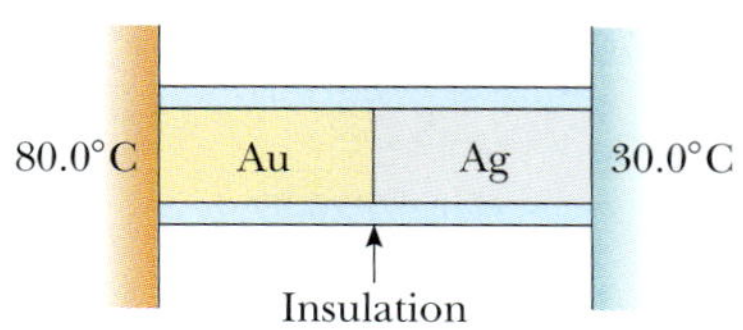

Figure P20.39

40. Calculate the R-value of (a) a window made of a single pane of flat glass $\frac{1}{8}$ in. thick and (b) a thermal window made of two single panes each $\frac{1}{8}$ in. thick and separated by a $\frac{1}{4}$-in. air space. (c) By what factor is the transfer of energy by heat through the window reduced by using the thermal window instead of the single-pane window?

41. A student is trying to decide what to wear. His bedroom is at 20.0°C. His skin temperature is 35.0°C. The area of his exposed skin is 1.50 m^2. People all over the world have skin that is dark in the infrared, with emissivity about 0.900. Find the net energy loss from his body by radiation in 10.0 min.

42. The surface of the Sun has a temperature of about 5 800 K. The radius of the Sun is 6.96×10^8 m. Calculate the total energy radiated by the Sun each second. Assume the emissivity is 0.986.

43. ● For bacteriological testing of water supplies and in medical clinics, samples must routinely be incubated for 24 h at 37°C. A standard constant-temperature bath with electric heating and thermostatic control is not practical in war-torn places and developing countries without continuously operating electric power lines. Peace Corps volunteer and MIT engineer Amy Smith invented a low-cost, low-maintenance incubator to fill the need. It consists of a foam-insulated box containing several packets of a waxy material that melts at 37.0°C, interspersed among tubes, dishes, or bottles containing the test samples and growth medium (bacteria food). Outside the box, the waxy material is first melted by a stove or solar energy collector. Then the waxy material is put into the box to keep the test samples warm as it solidifies. The heat of fusion of the phase-change material is 205 kJ/kg. Model the insulation as a panel with surface area 0.490 m^2, thickness 4.50 cm, and conductivity 0.012 0 W/m · °C. Assume the exterior temperature is 23.0°C for 12.0 h and 16.0°C for 12.0 h. (a) What mass of the waxy material is required to conduct the bacteriological test? (b) Explain why your calculation can be done without knowing the mass of the test samples or of the insulation.

44. A large, hot pizza floats in outer space after being jettisoned as refuse from a Vogon spacecraft. What is the order of magnitude (a) of its rate of energy loss and (b) of its rate of temperature change? List the quantities you estimate and the value you estimate for each.

45. The tungsten filament of a certain 100-W lightbulb radiates 2.00 W of light. (The other 98 W is carried away by convection and conduction.) The filament has a surface area of 0.250 mm^2 and an emissivity of 0.950. Find the filament's temperature. (The melting point of tungsten is 3 683 K.)

46. At high noon, the Sun delivers 1 000 W to each square meter of a blacktop road. If the hot asphalt loses energy only by radiation, what is its steady-state temperature?

47. ● At our distance from the Sun, the intensity of solar radiation is 1 370 W/m^2. The temperature of the Earth is affected by the so-called greenhouse effect of the atmosphere, which makes our planet's emissivity for visible light higher than its emissivity for infrared light. For comparison, consider a spherical object of radius r with no atmosphere at the same distance from the Sun as the Earth. Assume its emissivity is the same for all kinds of electromagnetic waves and its temperature is uniform over its surface. Explain why the projected area over which it absorbs sunlight is πr^2 and the surface area over which it radiates is $4\pi r^2$. Compute its steady-state temperature. Is it chilly? Your calculation applies to (1) the average temperature of the Moon, (2) astronauts in mortal danger aboard the crippled *Apollo 13* spacecraft, and (3) global catastrophe on the Earth if widespread fires caused a layer of soot to accumulate throughout the upper atmosphere so that most of the radiation from the Sun were absorbed there rather than at the surface below the atmosphere.

48. Two lightbulbs have cylindrical filaments much greater in length than in diameter. The evacuated lightbulbs are identical except that one operates at a filament temperature of

2 100°C and the other operates at 2 000°C. (a) Find the ratio of the power emitted by the hotter lightbulb to that emitted by the cooler lightbulb. (b) With the lightbulbs operating at the same respective temperatures, the cooler one is to be altered so that it emits the same power as the hotter one, by making the filament of the cooler lightbulb thicker. By what factor should the radius of this filament be increased?

Additional Problems

49. A 75.0-kg cross-country skier moves horizontally across snow at 0°C. The coefficient of friction between the skis and the snow is 0.200. Assume all the internal energy generated by friction is added to the snow, which sticks to her skis until it melts. How far does she have to ski to melt 1.00 kg of snow?

50. On a cold winter day you buy roasted chestnuts from a street vendor. You put the change he gives you—coins constituting 9.00 g of copper at −12.0°C—into the pocket of your down parka. Your pocket already contains 14.0 g of silver coins at 30.0°C. After a short time interval, the temperature of the copper coins is 4.00°C and is increasing at a rate of 0.500°C/s. At this moment, (a) what is the temperature of the silver coins and (b) at what rate is it changing?

51. An aluminum rod 0.500 m in length and with a cross-sectional area of 2.50 cm^2 is inserted into a thermally insulated vessel containing liquid helium at 4.20 K. The rod is initially at 300 K. (a) If one half of the rod is inserted into the helium, how many liters of helium boil off by the time the inserted half cools to 4.20 K? (Assume the upper half does not yet cool.) (b) If the upper end of the rod is maintained at 300 K, what is the approximate boil-off rate of liquid helium after the lower half has reached 4.20 K? (Aluminum has thermal conductivity of 31.0 J/s · cm · K at 4.2 K; ignore its temperature variation. Aluminum has a specific heat of 0.210 cal/g · °C and density of 2.70 g/cm^3. The density of liquid helium is 0.125 g/cm^3.)

52. ● One mole of an ideal gas is contained in a cylinder with a movable piston. The initial pressure, volume, and temperature are P_i, V_i, and T_i, respectively. Find the work done on the gas in the following processes. In operational terms, describe how to carry out each process. Show each process on a *PV* diagram: (a) an isobaric compression in which the final volume is one-half the initial volume (b) an isothermal compression in which the final pressure is four times the initial pressure (c) an isovolumetric process in which the final pressure is three times the initial pressure

53. A *flow calorimeter* is an apparatus used to measure the specific heat of a liquid. The technique of flow calorimetry involves measuring the temperature difference between the input and output points of a flowing stream of the liquid while energy is added by heat at a known rate. A liquid of density ρ flows through the calorimeter with volume flow rate R. At steady state, a temperature difference ΔT is established between the input and output points when energy is supplied at the rate $\mathcal{P}$. What is the specific heat of the liquid?

54. Review problem. Continue the analysis of Problem 52 in Chapter 19. Following a collision between a large spacecraft and an asteroid, a copper disk of radius 28.0 m and thickness 1.20 m, at a temperature of 850°C, is floating in space, rotating about its axis with an angular speed of 25.0 rad/s. As the disk radiates infrared light, its temperature falls to 20.0°C. No external torque acts on the disk. (a) Find the change in kinetic energy of the disk. (b) Find the change in internal energy of the disk. (c) Find the amount of energy it radiates.

55. Review problem. A 670-kg meteorite happens to be composed of aluminum. When it is far from the Earth, its temperature is −15°C and it moves with a speed of 14.0 km/s relative to the planet. As it crashes into the Earth, assume the resulting additional internal energy is shared equally between the meteor and the planet and all the material of the meteor rises momentarily to the same final temperature. Find this temperature. Assume the specific heat of liquid and of gaseous aluminum is 1 170 J/kg · °C.

56. Water in an electric teakettle is boiling. The power absorbed by the water is 1.00 kW. Assuming the pressure of vapor in the kettle equals atmospheric pressure, determine the speed of effusion of vapor from the kettle's spout if the spout has a cross-sectional area of 2.00 cm^2.

57. ▲ A solar cooker consists of a curved reflecting surface that concentrates sunlight onto the object to be warmed (Fig. P20.57). The solar power per unit area reaching the Earth's surface at the location is 600 W/m^2. The cooker faces the Sun and has a face diameter of 0.600 m. Assume 40.0% of the incident energy is transferred to 0.500 L of water in an open container, initially at 20.0°C. Over what time interval does the water completely boil away? (Ignore the heat capacity of the container.)

Figure P20.57

58. (a) In air at 0°C, a 1.60-kg copper block at 0°C is set sliding at 2.50 m/s over a sheet of ice at 0°C. Friction brings the block to rest. Find the mass of the ice that melts. To describe the process of slowing down, identify the energy input Q, the work input W, the change in internal energy ΔE_{int}, and the change in mechanical energy ΔK for the block and also for the ice. (b) A 1.60-kg block of ice at 0°C is set sliding at 2.50 m/s over a sheet of copper at 0°C. Friction brings the block to rest. Find the mass of the ice that melts. Identify Q, W, ΔE_{int}, and ΔK for the block and for the metal sheet during the process. (c) A thin 1.60-kg slab of copper at 20°C is set sliding at 2.50 m/s over an identical stationary slab at the same temperature. Friction quickly stops the motion. Assuming no energy is lost to the environment by heat, find the change in temperature of both objects. Identify Q, W, ΔE_{int}, and ΔK for each object during the process.

59. A cooking vessel on a slow burner contains 10.0 kg of water and an unknown mass of ice in equilibrium at 0°C at time $t = 0$. The temperature of the mixture is measured at various times, and the result is plotted in Figure P20.59. During the first 50.0 minutes, the mixture remains at 0°C. From 50.0 min to 60.0 min, the temperature increases to 2.00°C. Ignoring the heat capacity of the vessel, determine the initial mass of the ice.

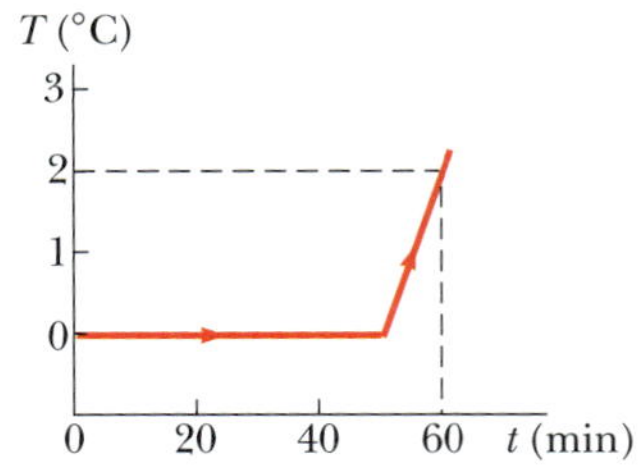

Figure P20.59

60. A pond of water at 0°C is covered with a layer of ice 4.00 cm thick. If the air temperature stays constant at −10.0°C, what time interval is required for the ice thickness to increase to 8.00 cm? *Suggestion:* Use Equation 20.16 in the form

$$\frac{dQ}{dt} = kA\frac{\Delta T}{x}$$

and note that the incremental energy dQ extracted from the water through the thickness x of ice is the amount required to freeze a thickness dx of ice. That is, $dQ = L\rho A\,dx$, where ρ is the density of the ice, A is the area, and L is the latent heat of fusion.

61. The average thermal conductivity of the walls (including the windows) and roof of the house depicted in Figure P20.61 is 0.480 W/m · °C, and their average thickness is 21.0 cm. The house is kept warm with natural gas having a heat of combustion (that is, the energy provided per cubic meter of gas burned) of 9 300 kcal/m^3. How many cubic meters of gas must be burned each day to maintain an inside temperature of 25.0°C if the outside temperature is 0.0°C? Disregard radiation and the energy lost by heat through the ground.

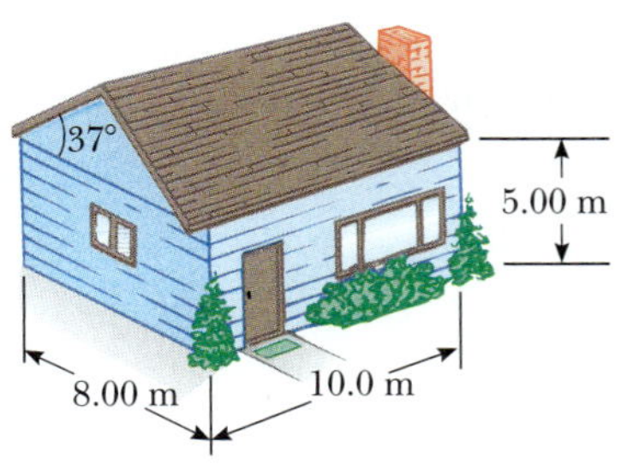

Figure P20.61

62. The inside of a hollow cylinder is maintained at a temperature T_a while the outside is at a lower temperature, T_b (Fig. P20.62). The wall of the cylinder has a thermal conductivity k. Ignoring end effects, show that the rate of energy conduction from the inner to the outer surface in the radial direction is

$$\frac{dQ}{dt} = 2\pi Lk\left[\frac{T_a - T_b}{\ln(b/a)}\right]$$

Suggestions: The temperature gradient is dT/dr. Notice that a radial energy current passes through a concentric cylinder of area $2\pi rL$.

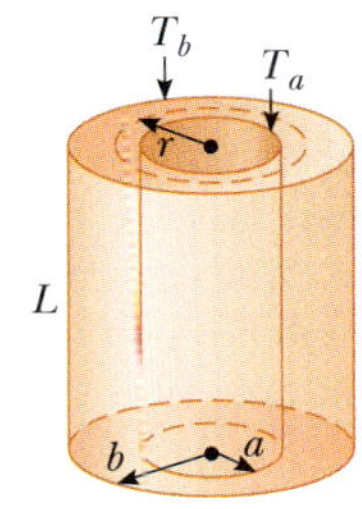

Figure P20.62

63. The passenger section of a jet airliner is in the shape of a cylindrical tube with a length of 35.0 m and an inner radius of 2.50 m. Its walls are lined with an insulating material 6.00 cm in thickness and having a thermal conductivity of 4.00×10^{-5} cal/s · cm · °C. A heater must maintain the interior temperature at 25.0°C while the outside temperature is −35.0°C. What power must be supplied to the heater? (You may use the result of Problem 62.)

64. ● A student measures the following data in a calorimetry experiment designed to determine the specific heat of aluminum:

Initial temperature of water and calorimeter:	70°C
Mass of water:	0.400 kg
Mass of calorimeter:	0.040 kg
Specific heat of calorimeter:	0.63 kJ/kg · °C
Initial temperature of aluminum:	27°C
Mass of aluminum:	0.200 kg
Final temperature of mixture:	66.3°C

Use these data to determine the specific heat of aluminum. Explain whether your result is within 15% of the value listed in Table 20.1.

65. ● A spherical shell has inner radius 3.00 cm and outer radius 7.00 cm. It is made of material with thermal conductivity $k = 0.800$ W/m · °C. The interior is maintained at temperature 5°C and the exterior at 40°C. After an interval of time, the shell reaches a steady state with the temperature at each point within it remaining constant in time. (a) Explain why the rate of energy transfer $\mathcal{P}$ must be the same through each spherical surface, of radius r, within the shell and must satisfy

$$\frac{dT}{dr} = \frac{\mathcal{P}}{4\pi k r^2}$$

(b) Next, prove that

$$\int_{5°C}^{40°C} dT = \frac{\mathcal{P}}{4\pi k}\int_{3\text{ cm}}^{7\text{ cm}} r^{-2}\,dr$$

(c) Find the rate of energy transfer through the shell. (d) Prove that

$$\int_{5°C}^{T} dT = (1.84\text{ m}\cdot°C)\int_{3\text{ cm}}^{r} r^{-2}\,dr$$

2 = intermediate; 3 = challenging; □ = SSM/SG; ▲ = ThomsonNOW; ■ = symbolic reasoning; ● = qualitative reasoning

(e) Find the temperature within the shell as a function of radius. (f) Find the temperature at $r = 5.00$ cm, halfway through the shell.

66. ● During periods of high activity, the Sun has more sunspots than usual. Sunspots are cooler than the rest of the luminous layer of the Sun's atmosphere (the photosphere). Paradoxically, the total power output of the active Sun is not lower than average but is the same or slightly higher than average. Work out the details of the following crude model of this phenomenon. Consider a patch of the photosphere with an area of 5.10×10^{14} m^2. Its emissivity is 0.965. (a) Find the power it radiates if its temperature is uniformly 5 800 K, corresponding to the quiet Sun. (b) To represent a sunspot, assume 10.0% of the patch area is at 4 800 K and the other 90.0% is at 5 890 K. That is, a section with the surface area of the Earth is 1 000 K cooler than before and a section nine times larger is 90 K warmer. Find the average temperature of the patch. State how it compares with 5 800 K. (c) Find the power output of the patch. State how it compares with the answer to part (a). (The next sunspot maximum is expected around the year 2012.)

Answers to Quick Quizzes

20.1 **(i)** Water, glass, iron. Because water has the highest specific heat (4 186 J/kg · °C), it has the smallest change in temperature. Glass is next (837 J/kg · °C), and iron is last (448 J/kg · °C). **(i)** Iron, glass, water. For a given temperature increase, the energy transfer by heat is proportional to the specific heat.

20.2 The figure shows a graphical representation of the internal energy of the ice as a function of energy added. Notice that this graph looks quite different from Figure 20.2 in that it doesn't have the flat portions during the phase changes. Regardless of how the temperature is varying in Figure 20.2, the internal energy of the system simply increases linearly with energy input.

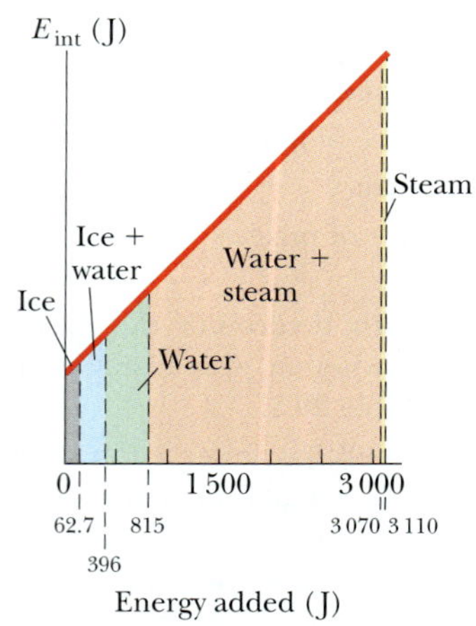

20.3

Situation	System	Q	W	ΔE_{int}
(a) Rapidly pumping up a bicycle tire	Air in the pump	0	+	+
(b) Pan of room-temperature water sitting on a hot stove	Water in the pan	+	0	+
(c) Air quickly leaking out of a balloon	Air originally in the balloon	0	−	−

(a) Because the pumping is rapid, no energy enters or leaves the system by heat. Because $W > 0$ when work is done *on* the system, it is positive here. Therefore, $\Delta E_{int} = Q + W$ must be positive. The air in the pump is warmer. (b) There is no work done either on or by the system, but energy transfers into the water by heat from the hot burner, making both Q and ΔE_{int} positive. (c) Again no energy transfers into or out of the system by heat, but the air molecules escaping from the balloon do work on the surrounding air molecules as they push them out of the way. Therefore, W is negative and ΔE_{int} is negative. The decrease in internal energy is evident because the escaping air becomes cooler.

20.4 Path A is isovolumetric, path B is adiabatic, path C is isothermal, and path D is isobaric.

20.5 (b). In parallel, the rods present a larger area through which energy can transfer and a smaller length.

Dogs do not have sweat glands like humans do. In hot weather, a dog pants to promote evaporation from the tongue. In this chapter, we show that evaporation is a cooling process based on the removal of molecules with high kinetic energy from a liquid. (Frank Oberle/Getty Images)

21 The Kinetic Theory of Gases

In Chapter 19, we discussed the properties of an ideal gas by using such macro-scopic variables as pressure, volume, and temperature. Such large-scale properties can be related to a description on a microscopic scale, where matter is treated as a collection of molecules. Applying Newton's laws of motion in a statistical manner to a collection of particles provides a reasonable description of thermodynamic processes. To keep the mathematics relatively simple, we shall consider primarily the behavior of gases because in gases the interactions between molecules are much weaker than they are in liquids or solids.

21.1 Molecular Model of an Ideal Gas

We begin this chapter by developing a microscopic model of an ideal gas, called **kinetic theory**. In developing this model, we make the following assumptions:

◄ Assumptions of the molecular model of an ideal gas

1. **The number of molecules in the gas is large, and the average separation between them is large compared with their dimensions**. In other words, the molecules occupy a negligible volume in the container. That is consistent with the ideal gas model, in which we model the molecules as particles.
2. **The molecules obey Newton's laws of motion, but as a whole they move randomly**. By "randomly" we mean that any molecule can move in any direction with any speed.

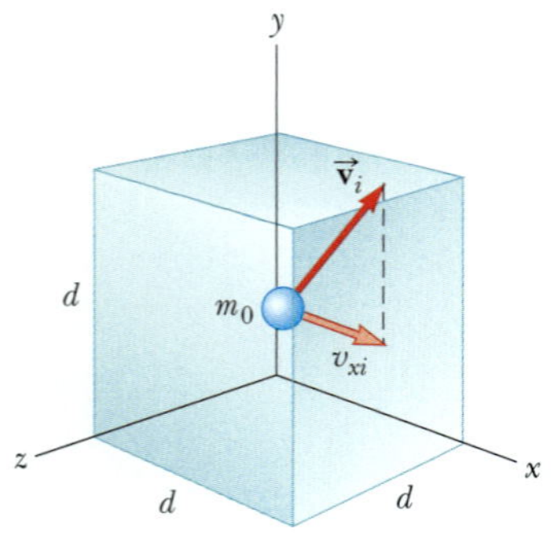

Figure 21.1 A cubical box with sides of length d containing an ideal gas. The molecule shown moves with velocity $\vec{\mathbf{v}}_i$.

3. **The molecules interact only by short-range forces during elastic collisions.** That is consistent with the ideal gas model, in which the molecules exert no long-range forces on each other.
4. **The molecules make elastic collisions with the walls.** These collisions lead to the macroscopic pressure on the walls of the container.
5. **The gas under consideration is a pure substance; that is, all molecules are identical.**

Although we often picture an ideal gas as consisting of single atoms, the behavior of molecular gases approximates that of ideal gases rather well at low pressures. Usually, molecular rotations or vibrations have no effect on the motions considered here.

For our first application of kinetic theory, let us derive an expression for the pressure of N molecules of an ideal gas in a container of volume V in terms of microscopic quantities. The container is a cube with edges of length d (Fig. 21.1). We shall first focus our attention on one of these molecules of mass m_0 and assume it is moving so that its component of velocity in the x direction is v_{xi} as in Active Figure 21.2. (The subscript i here refers to the ith molecule, not to an initial value. We will combine the effects of all the molecules shortly.) As the molecule collides elastically with any wall (assumption 4), its velocity component perpendicular to the wall is reversed because the mass of the wall is far greater than the mass of the molecule. Because the momentum component p_{xi} of the molecule is $m_0 v_{xi}$ before the collision and $-m_0 v_{xi}$ after the collision, the change in the x component of the momentum of the molecule is

$$\Delta p_{xi} = -m_0 v_{xi} - (m_0 v_{xi}) = -2m_0 v_{xi}$$

Because the molecules obey Newton's laws (assumption 2), we can apply the impulse-momentum theorem (Eq. 9.8) to the molecule to give

$$\bar{F}_{i,\text{ on molecule}}\, \Delta t_{\text{collision}} = \Delta p_{xi} = -2m_0 v_{xi}$$

where $\bar{F}_{i,\text{ on molecule}}$ is the x component of the average force[1] the wall exerts on the molecule during the collision and $\Delta t_{\text{collision}}$ is the duration of the collision. For the molecule to make another collision with the same wall after this first collision, it must travel a distance of $2d$ in the x direction (across the container and back). Therefore, the time interval between two collisions with the same wall is

$$\Delta t = \frac{2d}{v_{xi}}$$

The force that causes the change in momentum of the molecule in the collision with the wall occurs only during the collision. We can, however, average the force over the time interval for the molecule to move across the cube and back. Sometime during this time interval the collision occurs, so the change in momentum for this time interval is the same as that for the short duration of the collision. Therefore, we can rewrite the impulse-momentum theorem as

$$\bar{F}_i\, \Delta t = -2m_0 v_{xi}$$

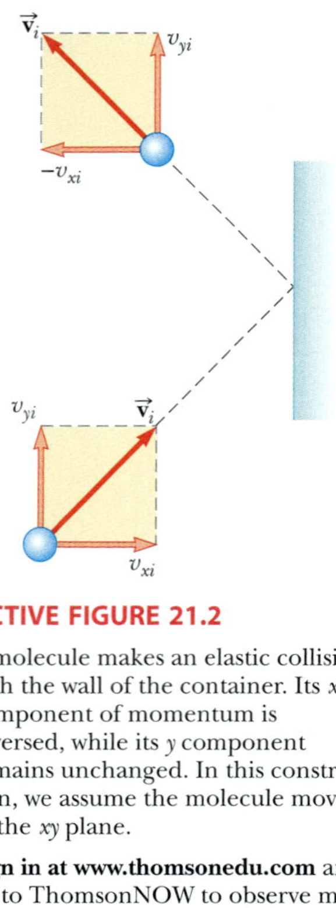

ACTIVE FIGURE 21.2

A molecule makes an elastic collision with the wall of the container. Its x component of momentum is reversed, while its y component remains unchanged. In this construction, we assume the molecule moves in the xy plane.

Sign in at www.thomsonedu.com and go to ThomsonNOW to observe molecules within a container making collisions with the walls of the container and with each other.

where $\bar{F}_i$ is the average force component over the time interval for the molecule to move across the cube and back. Because exactly one collision occurs for each such time interval, this result is also the long-term average force on the molecule over long time intervals containing any number of multiples of Δt.

This equation and the preceding one enable us to express the x component of the long-term average force exerted by the wall on the molecule as

$$\bar{F}_i = -\frac{2m_0 v_{xi}}{\Delta t} = -\frac{2m_0 v_{xi}^2}{2d} = -\frac{m_0 v_{xi}^2}{d}$$

[1] For this discussion, we use a bar over a variable to represent the average value of the variable, such as $\bar{F}$ for the average force, rather than the subscript "avg" that we have used before. This notation is to save confusion because we already have a number of subscripts on variables.

Now, by Newton's third law, the x component of the long-term average force exerted by the *molecule* on the *wall* is equal in magnitude and opposite in direction:

$$\overline{F}_{i,\text{ on wall}} = -\overline{F}_i = -\left(-\frac{m_0 v_{xi}^{\,2}}{d}\right) = \frac{m_0 v_{xi}^{\,2}}{d}$$

The total average force $\overline{F}$ exerted by the gas on the wall is found by adding the average forces exerted by the individual molecules. Adding terms such as that above for all molecules gives

$$\overline{F} = \sum_{i=1}^{N} \frac{m_0 v_{xi}^{\,2}}{d} = \frac{m_0}{d} \sum_{i=1}^{N} v_{xi}^{\,2}$$

where we have factored out the length of the box and the mass m_0 because assumption 5 tells us that all the molecules are the same. We now impose assumption 1, that the number of molecules is large. For a small number of molecules, the actual force on the wall would vary with time. It would be nonzero during the short interval of a collision of a molecule with the wall and zero when no molecule happens to be hitting the wall. For a very large number of molecules such as Avogadro's number, however, these variations in force are smoothed out so that the average force given above is the same over *any* time interval. Therefore, the *constant* force F on the wall due to the molecular collisions is

$$F = \frac{m_0}{d} \sum_{i=1}^{N} v_{xi}^{\,2}$$

To proceed further, let's consider how to express the average value of the square of the x component of the velocity for N molecules. The traditional average of a set of values is the sum of the values over the number of values:

$$\overline{v_x^{\,2}} = \frac{\sum_{i=1}^{N} v_{xi}^{\,2}}{N}$$

The numerator of this expression is contained in the right side of the preceding equation. Therefore, by combining the two expressions the total force on the wall can be written

$$F = \frac{m_0}{d} N\overline{v_x^{\,2}} \qquad \textbf{(21.1)}$$

Now let's focus again on one molecule with velocity components v_{xi}, v_{yi}, and v_{zi}. The Pythagorean theorem relates the square of the speed of the molecule to the squares of the velocity components:

$$v_i^{\,2} = v_{xi}^{\,2} + v_{yi}^{\,2} + v_{zi}^{\,2}$$

Hence, the average value of v^2 for all the molecules in the container is related to the average values of $v_x^{\,2}$, $v_y^{\,2}$, and $v_z^{\,2}$ according to the expression

$$\overline{v^2} = \overline{v_x^{\,2}} + \overline{v_y^{\,2}} + \overline{v_z^{\,2}}$$

Because the motion is completely random (assumption 2), the average values $\overline{v_x^{\,2}}$, $\overline{v_y^{\,2}}$, and $\overline{v_z^{\,2}}$ are equal to each other. Using this fact and the preceding equation, we find that

$$\overline{v^2} = 3\overline{v_x^{\,2}}$$

Therefore, from Equation 21.1, the total force exerted on the wall is

$$F = \tfrac{1}{3} N \frac{m_0 \overline{v^2}}{d}$$

Using this expression, we can find the total pressure exerted on the wall:

$$P = \frac{F}{A} = \frac{F}{d^2} = \tfrac{1}{3}N\frac{m_0\overline{v^2}}{d^3} = \tfrac{1}{3}\left(\frac{N}{V}\right)m_0\overline{v^2}$$

Relationship between pressure and molecular kinetic energy ▶

$$P = \tfrac{2}{3}\left(\frac{N}{V}\right)\left(\tfrac{1}{2}m_0\overline{v^2}\right) \qquad \textbf{(21.2)}$$

This result indicates that **the pressure of a gas is proportional to the number of molecules per unit volume and to the average translational kinetic energy of the molecules,** $\frac{1}{2}m_0\overline{v^2}$. In analyzing this simplified model of an ideal gas, we obtain an important result that relates the macroscopic quantity of pressure to a microscopic quantity, the average value of the square of the molecular speed. Therefore, a key link between the molecular world and the large-scale world has been established.

Notice that Equation 21.2 verifies some features of pressure with which you are probably familiar. One way to increase the pressure inside a container is to increase the number of molecules per unit volume N/V in the container. That is what you do when you add air to a tire. The pressure in the tire can also be increased by increasing the average translational kinetic energy of the air molecules in the tire. That can be accomplished by increasing the temperature of that air, which is why the pressure inside a tire increases as the tire warms up during long road trips. The continuous flexing of the tire as it moves along the road surface results in work done on the rubber as parts of the tire distort, causing an increase in internal energy of the rubber. The increased temperature of the rubber results in the transfer of energy by heat into the air inside the tire. This transfer increases the air's temperature, and this increase in temperature in turn produces an increase in pressure.

Molecular Interpretation of Temperature

We can gain some insight into the meaning of temperature by first writing Equation 21.2 in the form

$$PV = \tfrac{2}{3}N\left(\tfrac{1}{2}m_0\overline{v^2}\right)$$

Let's now compare this expression with the equation of state for an ideal gas (Eq. 19.10):

$$PV = Nk_BT$$

Recall that the equation of state is based on experimental facts concerning the macroscopic behavior of gases. Equating the right sides of these expressions gives

Temperature is proportional to average kinetic energy ▶

$$T = \frac{2}{3k_B}\left(\tfrac{1}{2}m_0\overline{v^2}\right) \qquad \textbf{(21.3)}$$

This result tells us that temperature is a direct measure of average molecular kinetic energy. By rearranging Equation 21.3, we can relate the translational molecular kinetic energy to the temperature:

Average kinetic energy per molecule ▶

$$\tfrac{1}{2}m_0\overline{v^2} = \tfrac{3}{2}k_BT \qquad \textbf{(21.4)}$$

That is, the average translational kinetic energy per molecule is $\frac{3}{2}k_BT$. Because $\overline{v_x^2} = \frac{1}{3}\overline{v^2}$, it follows that

$$\tfrac{1}{2}m_0\overline{v_x^2} = \tfrac{1}{2}k_BT \qquad \textbf{(21.5)}$$

In a similar manner, for the y and z directions,

$$\tfrac{1}{2}m_0\overline{v_y^2} = \tfrac{1}{2}k_BT \quad \text{and} \quad \tfrac{1}{2}m_0\overline{v_z^2} = \tfrac{1}{2}k_BT$$

Therefore, each translational degree of freedom contributes an equal amount of energy, $\frac{1}{2}k_BT$, to the gas. (In general, a "degree of freedom" refers to an indepen-

TABLE 21.1

Some Root-Mean-Square (rms) Speeds

Gas	Molar Mass (g/mol)	v_{rms} at 20°C (m/s)	Gas	Molar Mass (g/mol)	v_{rms} at 20°C (m/s)
H_2	2.02	1902	NO	30.0	494
He	4.00	1352	O_2	32.0	478
H_2O	18.0	637	CO_2	44.0	408
Ne	20.2	602	SO_2	64.1	338
N_2 or CO	28.0	511			

dent means by which a molecule can possess energy.) A generalization of this result, known as the **theorem of equipartition of energy**, is as follows:

Each degree of freedom contributes $\frac{1}{2}k_BT$ to the energy of a system, where possible degrees of freedom are those associated with translation, rotation, and vibration of molecules.

◀ Theorem of equipartition of energy

The total translational kinetic energy of N molecules of gas is simply N times the average energy per molecule, which is given by Equation 21.4:

$$K_{\text{tot trans}} = N\left(\tfrac{1}{2}m_0\overline{v^2}\right) = \tfrac{3}{2}Nk_BT = \tfrac{3}{2}nRT \qquad \textbf{(21.6)}$$

◀ Total translational kinetic energy of N molecules

where we have used $k_B = R/N_A$ for Boltzmann's constant and $n = N/N_A$ for the number of moles of gas. If the gas molecules possess only translational kinetic energy, Equation 21.6 represents the internal energy of the gas. This result implies that **the internal energy of an ideal gas depends only on the temperature**. We will follow up on this point in Section 21.2.

The square root of $\overline{v^2}$ is called the *root-mean-square* (rms) *speed* of the molecules. From Equation 21.4, we find that the rms speed is

$$v_{rms} = \sqrt{\overline{v^2}} = \sqrt{\frac{3k_BT}{m_0}} = \sqrt{\frac{3RT}{M}} \qquad \textbf{(21.7)}$$

◀ Root-mean-square speed

where M is the molar mass in kilograms per mole and is equal to m_0N_A. This expression shows that, at a given temperature, lighter molecules move faster, on the average, than do heavier molecules. For example, at a given temperature, hydrogen molecules, whose molar mass is 2.02×10^{-3} kg/mol, have an average speed approximately four times that of oxygen molecules, whose molar mass is 32.0×10^{-3} kg/mol. Table 21.1 lists the rms speeds for various molecules at 20°C.

PITFALL PREVENTION 21.1

The Square Root of the Square?

Taking the square root of $\overline{v^2}$ does not "undo" the square because we have taken an average *between* squaring and taking the square root. Although the square root of $(\overline{v})^2$ is $\overline{v} = v_{avg}$ because the squaring is done after the averaging, the square root of $\overline{v^2}$ is *not* v_{avg}, but rather v_{rms}.

Quick Quiz 21.1 Two containers hold an ideal gas at the same temperature and pressure. Both containers hold the same type of gas, but container B has twice the volume of container A. **(i)** What is the average translational kinetic energy per molecule in container B? (a) twice that of container A (b) the same as that of container A (c) half that of container A (d) impossible to determine **(ii)** From the same choices, describe the internal energy of the gas in container B.

EXAMPLE 21.1 **A Tank of Helium**

A tank used for filling helium balloons has a volume of 0.300 m^3 and contains 2.00 mol of helium gas at 20.0°C. Assume the helium behaves like an ideal gas.

(A) What is the total translational kinetic energy of the gas molecules?

SOLUTION

Conceptualize Imagine a microscopic model of a gas in which you can watch the molecules move about the container more rapidly as the temperature increases.

Categorize We evaluate parameters with equations developed in the preceding discussion, so this example is a substitution problem.

Use Equation 21.6 with $n = 2.00$ mol and $T = 293$ K:

$$K_{\text{tot trans}} = \tfrac{3}{2}nRT = \tfrac{3}{2}(2.00\text{ mol})(8.31\text{ J/mol}\cdot\text{K})(293\text{ K})$$
$$= 7.30 \times 10^3\text{ J}$$

(B) What is the average kinetic energy per molecule?

SOLUTION

Use Equation 21.4:

$$\tfrac{1}{2}m_0\overline{v^2} = \tfrac{3}{2}k_BT = \tfrac{3}{2}(1.38 \times 10^{-23}\text{ J/K})(293\text{ K})$$
$$= 6.07 \times 10^{-21}\text{ J}$$

What If? What if the temperature is raised from 20.0°C to 40.0°C? Because 40.0 is twice as large as 20.0, is the total translational energy of the molecules of the gas twice as large at the higher temperature?

Answer The expression for the total translational energy depends on the temperature, and the value for the temperature must be expressed in kelvins, not in degrees Celsius. Therefore, the ratio of 40.0 to 20.0 is *not* the appropriate ratio. Converting the Celsius temperatures to kelvins, 20.0°C is 293 K and 40.0°C is 313 K. Therefore, the total translational energy increases by a factor of only 313 K/293 K = 1.07.

21.2 Molar Specific Heat of an Ideal Gas

Consider an ideal gas undergoing several processes such that the change in temperature is $\Delta T = T_f - T_i$ for all processes. The temperature change can be achieved by taking a variety of paths from one isotherm to another as shown in Figure 21.3. Because ΔT is the same for each path, the change in internal energy ΔE_{int} is the same for all paths. The work W done on the gas (the negative of the area under the curves) is different for each path. Therefore, from the first law of thermodynamics, the heat associated with a given change in temperature does *not* have a unique value as discussed in Section 20.4.

We can address this difficulty by defining specific heats for two special processes: isovolumetric and isobaric. Because the number of moles is a convenient measure of the amount of gas, we define the **molar specific heats** associated with these processes as follows:

$$Q = nC_V\,\Delta T \quad \text{(constant volume)} \tag{21.8}$$

$$Q = nC_P\,\Delta T \quad \text{(constant pressure)} \tag{21.9}$$

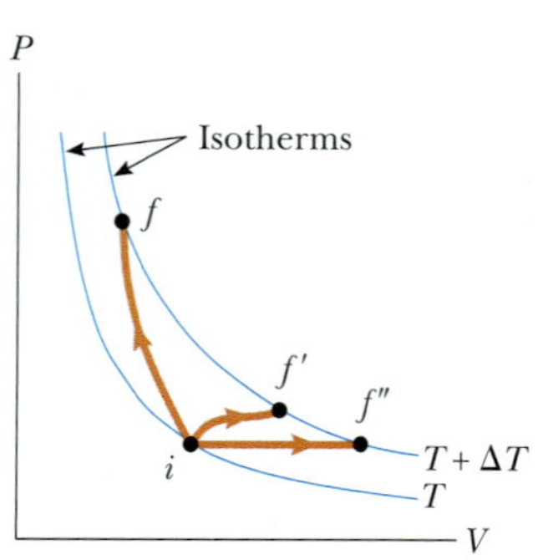

Figure 21.3 An ideal gas is taken from one isotherm at temperature T to another at temperature $T + \Delta T$ along three different paths.

where C_V is the **molar specific heat at constant volume** and C_P is the **molar specific heat at constant pressure**. When energy is added to a gas by heat at constant pressure, not only does the internal energy of the gas increase, but (negative) work is done on the gas because of the change in volume. Therefore, the heat Q in Equation 21.9 must account for both the increase in internal energy and the transfer of energy out of the system by work. For this reason, Q is greater in Equation 21.9 than in Equation 21.8 for given values of n and ΔT. Therefore, C_P is greater than C_V.

In the previous section, we found that the temperature of a gas is a measure of the average translational kinetic energy of the gas molecules. This kinetic energy is associated with the motion of the center of mass of each molecule. It does not include the energy associated with the internal motion of the molecule, namely, vibrations and rotations about the center of mass. That should not be surprising because the simple kinetic theory model assumes a structureless molecule.

So, let's first consider the simplest case of an ideal monatomic gas, that is, a gas containing one atom per molecule such as helium, neon, or argon. When energy is added to a monatomic gas in a container of fixed volume, all the added energy goes into increasing the translational kinetic energy of the atoms. There is no other way to store the energy in a monatomic gas. Therefore, from Equation 21.6, we see that the internal energy E_{int} of N molecules (or n mol) of an ideal monatomic gas is

$$E_{\text{int}} = K_{\text{tot trans}} = \tfrac{3}{2}Nk_BT = \tfrac{3}{2}nRT \qquad (21.10)$$

◀ Internal energy of an ideal monatomic gas

For a monatomic ideal gas, E_{int} is a function of T only and the functional relationship is given by Equation 21.10. In general, the internal energy of any ideal gas is a function of T only and the exact relationship depends on the type of gas.

If energy is transferred by heat to a system at constant volume, no work is done on the system. That is, $W = -\int P\,dV = 0$ for a constant-volume process. Hence, from the first law of thermodynamics,

$$Q = \Delta E_{\text{int}} \qquad (21.11)$$

In other words, all the energy transferred by heat goes into increasing the internal energy of the system. A constant-volume process from i to f for an ideal gas is described in Active Figure 21.4, where ΔT is the temperature difference between the two isotherms. Substituting the expression for Q given by Equation 21.8 into Equation 21.11, we obtain

$$\Delta E_{\text{int}} = nC_V\,\Delta T \qquad (21.12)$$

If the molar specific heat is constant, we can express the internal energy of a gas as

$$E_{\text{int}} = nC_VT$$

This equation applies to all ideal gases, those gases having more than one atom per molecule as well as monatomic ideal gases. In the limit of infinitesimal changes, we can use Equation 21.12 to express the molar specific heat at constant volume as

$$C_V = \frac{1}{n}\frac{dE_{\text{int}}}{dT} \qquad (21.13)$$

Let's now apply the results of this discussion to a monatomic gas. Substituting the internal energy from Equation 21.10 into Equation 21.13 gives

$$C_V = \tfrac{3}{2}R \qquad (21.14)$$

This expression predicts a value of $C_V = \tfrac{3}{2}R = 12.5\ \text{J/mol}\cdot\text{K}$ for *all* monatomic gases. This prediction is in excellent agreement with measured values of molar specific heats for such gases as helium, neon, argon, and xenon over a wide range of temperatures (Table 21.2, page 594). Small variations in Table 21.2 from the predicted values are because real gases are not ideal gases. In real gases, weak intermolecular interactions occur, which are not addressed in our ideal gas model.

Now suppose the gas is taken along the constant-pressure path $i \rightarrow f'$ shown in Active Figure 21.4. Along this path, the temperature again increases by ΔT. The energy that must be transferred by heat to the gas in this process is $Q = nC_P\,\Delta T$. Because the volume changes in this process, the work done on the gas is $W = -P\,\Delta V$, where P is the constant pressure at which the process occurs. Applying the first law of thermodynamics to this process, we have

$$\Delta E_{\text{int}} = Q + W = nC_P\,\Delta T + (-P\,\Delta V) \qquad (21.15)$$

In this case, the energy added to the gas by heat is channeled as follows. Part of it leaves the system by work (that is, the gas moves a piston through a displacement), and the remainder appears as an increase in the internal energy of the gas. The

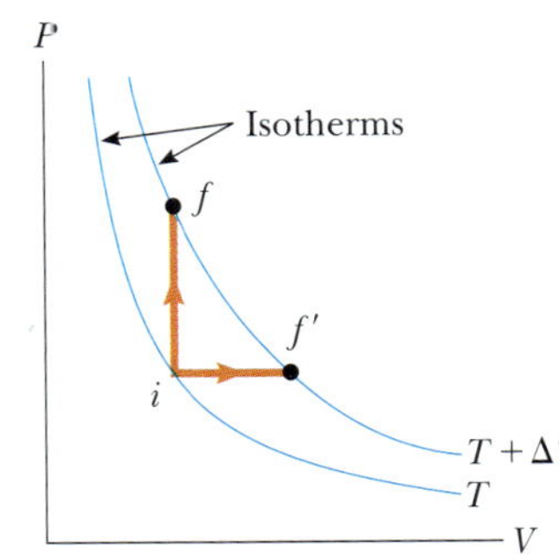

ACTIVE FIGURE 21.4

Energy is transferred by heat to an ideal gas in two ways. For the constant-volume path $i \rightarrow f$, all the energy goes into increasing the internal energy of the gas because no work is done. Along the constant-pressure path $i \rightarrow f'$, part of the energy transferred in by heat is transferred out by work.

Sign in at www.thomsonedu.com and go to ThomsonNOW to choose initial and final temperatures for one mole of an ideal gas undergoing constant-volume and constant-pressure processes and measure Q, W, ΔE_{int}, C_V, and C_P.

TABLE 21.2

Molar Specific Heats of Various Gases

	Molar Specific Heat (J/mol · K)[a]			
Gas	C_P	C_V	$C_P - C_V$	$\gamma = C_P/C_V$
Monatomic gases				
He	20.8	12.5	8.33	1.67
Ar	20.8	12.5	8.33	1.67
Ne	20.8	12.7	8.12	1.64
Kr	20.8	12.3	8.49	1.69
Diatomic gases				
H_2	28.8	20.4	8.33	1.41
N_2	29.1	20.8	8.33	1.40
O_2	29.4	21.1	8.33	1.40
CO	29.3	21.0	8.33	1.40
Cl_2	34.7	25.7	8.96	1.35
Polyatomic gases				
CO_2	37.0	28.5	8.50	1.30
SO_2	40.4	31.4	9.00	1.29
H_2O	35.4	27.0	8.37	1.30
CH_4	35.5	27.1	8.41	1.31

[a] All values except that for water were obtained at 300 K.

change in internal energy for the process $i \to f'$, however, is equal to that for the process $i \to f$ because E_{int} depends only on temperature for an ideal gas and ΔT is the same for both processes. In addition, because $PV = nRT$, note that for a constant-pressure process, $P\Delta V = nR\,\Delta T$. Substituting this value for $P\Delta V$ into Equation 21.15 with $\Delta E_{\text{int}} = nC_V\,\Delta T$ (Eq. 21.12) gives

$$nC_V\,\Delta T = nC_P\,\Delta T - nR\,\Delta T$$

$$C_P - C_V = R \qquad \textbf{(21.16)}$$

This expression applies to *any* ideal gas. It predicts that the molar specific heat of an ideal gas at constant pressure is greater than the molar specific heat at constant volume by an amount R, the universal gas constant (which has the value 8.31 J/mol · K). This expression is applicable to real gases as the data in Table 21.2 show.

Because $C_V = \frac{3}{2}R$ for a monatomic ideal gas, Equation 21.16 predicts a value $C_P = \frac{5}{2}R = 20.8$ J/mol·K for the molar specific heat of a monatomic gas at constant pressure. The ratio of these molar specific heats is a dimensionless quantity γ (Greek letter gamma):

Ratio of molar specific heats for a monatomic ideal gas ▶

$$\gamma = \frac{C_P}{C_V} = \frac{5R/2}{3R/2} = \frac{5}{3} = 1.67 \qquad \textbf{(21.17)}$$

Theoretical values of C_V, C_P, and γ are in excellent agreement with experimental values obtained for monatomic gases, but they are in serious disagreement with the values for the more complex gases (see Table 21.2). That is not surprising; the value $C_V = \frac{3}{2}R$ was derived for a monatomic ideal gas, and we expect some additional contribution to the molar specific heat from the internal structure of the more complex molecules. In Section 21.4, we describe the effect of molecular structure on the molar specific heat of a gas. The internal energy—and hence the molar specific heat—of a complex gas must include contributions from the rotational and the vibrational motions of the molecule.

In the case of solids and liquids heated at constant pressure, very little work is done because the thermal expansion is small. Consequently, C_P and C_V are approximately equal for solids and liquids.

Quick Quiz 21.2 **(i)** How does the internal energy of an ideal gas change as it follows path $i \rightarrow f$ in Active Figure 21.4? (a) E_{int} increases. (b) E_{int} decreases. (c) E_{int} stays the same. (d) There is not enough information to determine how E_{int} changes. **(ii)** From the same choices, how does the internal energy of an ideal gas change as it follows path $f \rightarrow f'$ along the isotherm labeled $T + \Delta T$ in Active Figure 21.4?

EXAMPLE 21.2 Heating a Cylinder of Helium

A cylinder contains 3.00 mol of helium gas at a temperature of 300 K.

(A) If the gas is heated at constant volume, how much energy must be transferred by heat to the gas for its temperature to increase to 500 K?

SOLUTION

Conceptualize Run the process in your mind with the help of the piston–cylinder arrangement in Figure 19.12.

Categorize Because the gas maintains a constant volume, the piston in Figure 19.12 is locked in place. We evaluate parameters with equations developed in the preceding discussion, so this example is a substitution problem.

Use Equation 21.8 to find the energy transfer:

$$Q_1 = nC_V\,\Delta T$$

Substitute the given values:

$$Q_1 = (3.00\text{ mol})(12.5\text{ J/mol}\cdot\text{K})(500\text{ K} - 300\text{ K})$$
$$= 7.50 \times 10^3\text{ J}$$

(B) How much energy must be transferred by heat to the gas at constant pressure to raise the temperature to 500 K?

SOLUTION

Categorize Because the gas maintains a constant pressure, the piston in Figure 19.12 is free to move, so the piston is modeled as a particle in equilibrium.

Use Equation 21.9 to find the energy transfer:

$$Q_2 = nC_P\,\Delta T$$

Substitute the given values:

$$Q_2 = (3.00\text{ mol})(20.8\text{ J/mol}\cdot\text{K})(500\text{ K} - 300\text{ K})$$
$$= 12.5 \times 10^3\text{ J}$$

This value is larger than Q_1 because of the transfer of energy out of the gas by work in the constant pressure process.

21.3 Adiabatic Processes for an Ideal Gas

As noted in Section 20.6, an **adiabatic process** is one in which no energy is transferred by heat between a system and its surroundings. For example, if a gas is compressed (or expanded) rapidly, very little energy is transferred out of (or into) the system by heat, so the process is nearly adiabatic. Such processes occur in the cycle of a gasoline engine, which is discussed in detail in Chapter 22. Another example of an adiabatic process is the slow expansion of a gas that is thermally insulated from its surroundings. All three variables in the ideal gas law—P, V, and T—change during an adiabatic process.

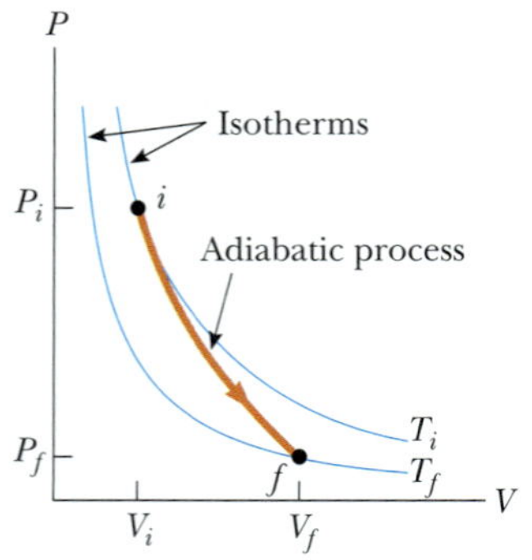

Figure 21.5 The *PV* diagram for an adiabatic expansion of an ideal gas. Notice that $T_f < T_i$ in this process, so the temperature of the gas decreases.

Let's imagine an adiabatic gas process involving an infinitesimal change in volume dV and an accompanying infinitesimal change in temperature dT. The work done on the gas is $-P\,dV$. Because the internal energy of an ideal gas depends only on temperature, the change in the internal energy in an adiabatic process is the same as that for an isovolumetric process between the same temperatures, $dE_{\text{int}} = nC_V\,dT$ (Eq. 21.12). Hence, the first law of thermodynamics, $\Delta E_{\text{int}} = Q + W$, with $Q = 0$ becomes

$$dE_{\text{int}} = nC_V\,dT = -P\,dV$$

Taking the total differential of the equation of state of an ideal gas, $PV = nRT$, gives

$$P\,dV + V\,dP = nR\,dT$$

Eliminating dT from these two equations, we find that

$$P\,dV + V\,dP = -\frac{R}{C_V}P\,dV$$

Substituting $R = C_P - C_V$ and dividing by PV gives

$$\frac{dV}{V} + \frac{dP}{P} = -\left(\frac{C_P - C_V}{C_V}\right)\frac{dV}{V} = (1-\gamma)\frac{dV}{V}$$

$$\frac{dP}{P} + \gamma\frac{dV}{V} = 0$$

Integrating this expression, we have

$$\ln P + \gamma \ln V = \text{constant}$$

which is equivalent to

Relationship between *P* and *V* for an adiabatic process involving an ideal gas ▶

$$PV^\gamma = \text{constant} \tag{21.18}$$

The *PV* diagram for an adiabatic expansion is shown in Figure 21.5. Because $\gamma > 1$, the *PV* curve is steeper than it would be for an isothermal expansion. By the definition of an adiabatic process, no energy is transferred by heat into or out of the system. Hence, from the first law, we see that ΔE_{int} is negative (work is done *by* the gas, so its internal energy decreases) and so ΔT also is negative. Therefore, the temperature of the gas decreases ($T_f < T_i$) during an adiabatic expansion.[2] Conversely, the temperature increases if the gas is compressed adiabatically. Applying Equation 21.18 to the initial and final states, we see that

$$P_iV_i^\gamma = P_fV_f^\gamma \tag{21.19}$$

Using the ideal gas law, we can express Equation 21.19 as

Relationship between *T* and *V* for an adiabatic process involving an ideal gas ▶

$$T_iV_i^{\gamma-1} = T_fV_f^{\gamma-1} \tag{21.20}$$

EXAMPLE 21.3 A Diesel Engine Cylinder

Air at 20.0°C in the cylinder of a diesel engine is compressed from an initial pressure of 1.00 atm and volume of 800.0 cm^3 to a volume of 60.0 cm^3. Assume air behaves as an ideal gas with $\gamma = 1.40$ and the compression is adiabatic. Find the final pressure and temperature of the air.

SOLUTION

Conceptualize Imagine what happens if a gas is compressed into a smaller volume. Our discussion above and Figure 21.5 tell us that the pressure and temperature both increase.

Categorize We categorize this example as a problem involving an adiabatic process.

[2] In the adiabatic free expansion discussed in Section 20.6, the temperature remains constant. In this unique process, no work is done because the gas expands into a vacuum. In general, the temperature decreases in an adiabatic expansion in which work is done.

Analyze Use Equation 21.19 to find the final pressure:

$$P_f = P_i\left(\frac{V_i}{V_f}\right)^{\gamma} = (1.00\text{ atm})\left(\frac{800.0\text{ cm}^3}{60.0\text{ cm}^3}\right)^{1.40}$$

$$= 37.6\text{ atm}$$

Use the ideal gas law to find the final temperature:

$$\frac{P_iV_i}{T_i} = \frac{P_fV_f}{T_f}$$

$$T_f = \frac{P_fV_f}{P_iV_i}T_i = \frac{(37.6\text{ atm})(60.0\text{ cm}^3)}{(1.00\text{ atm})(800.0\text{ cm}^3)}(293\text{ K})$$

$$= 826\text{ K} = 553°\text{C}$$

Finalize The temperature of the gas increases by a factor of 826 K/293 K = 2.82. The high compression in a diesel engine raises the temperature of the fuel enough to cause its combustion without the use of spark plugs.

21.4 The Equipartition of Energy

Predictions based on our model for molar specific heat agree quite well with the behavior of monatomic gases, but not with the behavior of complex gases (see Table 21.2). The value predicted by the model for the quantity $C_P - C_V = R$, however, is the same for all gases. This similarity is not surprising because this difference is the result of the work done on the gas, which is independent of its molecular structure.

To clarify the variations in C_V and C_P in gases more complex than monatomic gases, let's explore further the origin of molar specific heat. So far, we have assumed the sole contribution to the internal energy of a gas is the translational kinetic energy of the molecules. The internal energy of a gas, however, includes contributions from the translational, vibrational, and rotational motion of the molecules. The rotational and vibrational motions of molecules can be activated by collisions and therefore are "coupled" to the translational motion of the molecules. The branch of physics known as *statistical mechanics* has shown that, for a large number of particles obeying the laws of Newtonian mechanics, the available energy is, on average, shared equally by each independent degree of freedom. Recall from Section 21.1 that the equipartition theorem states that, at equilibrium, each degree of freedom contributes $\frac{1}{2}k_BT$ of energy per molecule.

Let's consider a diatomic gas whose molecules have the shape of a dumbbell (Fig. 21.6). In this model, the center of mass of the molecule can translate in the x, y, and z directions (Fig. 21.6a). In addition, the molecule can rotate about three mutually perpendicular axes (Fig. 21.6b). The rotation about the y axis can be neglected because the molecule's moment of inertia I_y and its rotational energy $\frac{1}{2}I_y\omega^2$ about this axis are negligible compared with those associated with the x and z axes. (If the two atoms are modeled as particles, then I_y is identically zero.) Therefore, there are five degrees of freedom for translation and rotation: three associated with the translational motion and two associated with the rotational motion. Because each degree of freedom contributes, on average, $\frac{1}{2}k_BT$ of energy per molecule, the internal energy for a system of N molecules, ignoring vibration for now, is

$$E_{\text{int}} = 3N(\tfrac{1}{2}k_BT) + 2N(\tfrac{1}{2}k_BT) = \tfrac{5}{2}Nk_BT = \tfrac{5}{2}nRT$$

We can use this result and Equation 21.13 to find the molar specific heat at constant volume:

$$C_V = \frac{1}{n}\frac{dE_{\text{int}}}{dT} = \frac{1}{n}\frac{d}{dT}(\tfrac{5}{2}nRT) = \tfrac{5}{2}R \qquad \textbf{(21.21)}$$

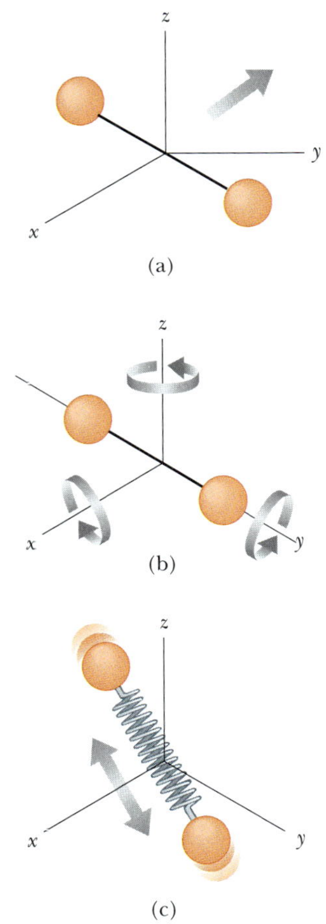

Figure 21.6 Possible motions of a diatomic molecule: (a) translational motion of the center of mass, (b) rotational motion about the various axes, and (c) vibrational motion along the molecular axis.

From Equations 21.16 and 21.17, we find that

$$C_P = C_V + R = \tfrac{7}{2}R$$

$$\gamma = \frac{C_P}{C_V} = \frac{\frac{7}{2}R}{\frac{5}{2}R} = \frac{7}{5} = 1.40$$

These results agree quite well with most of the data for diatomic molecules given in Table 21.2. That is rather surprising because we have not yet accounted for the possible vibrations of the molecule.

In the model for vibration, the two atoms are joined by an imaginary spring (see Fig. 21.6c). The vibrational motion adds two more degrees of freedom, which correspond to the kinetic energy and the potential energy associated with vibrations along the length of the molecule. Hence, a model that includes all three types of motion predicts a total internal energy of

$$E_{\text{int}} = 3N(\tfrac{1}{2}k_BT) + 2N(\tfrac{1}{2}k_BT) + 2N(\tfrac{1}{2}k_BT) = \tfrac{7}{2}Nk_BT = \tfrac{7}{2}nRT$$

and a molar specific heat at constant volume of

$$C_V = \frac{1}{n}\frac{dE_{\text{int}}}{dT} = \frac{1}{n}\frac{d}{dT}(\tfrac{7}{2}nRT) = \tfrac{7}{2}R \quad \textbf{(21.22)}$$

This value is inconsistent with experimental data for molecules such as H_2 and N_2 (see Table 21.2) and suggests a breakdown of our model based on classical physics.

It might seem that our model is a failure for predicting molar specific heats for diatomic gases. We can claim some success for our model, however, if measurements of molar specific heat are made over a wide temperature range rather than at the single temperature that gives us the values in Table 21.2. Figure 21.7 shows the molar specific heat of hydrogen as a function of temperature. The remarkable feature about the three plateaus in the graph's curve is that they are at the values of the molar specific heat predicted by Equations 21.14, 21.21, and 21.22! For low temperatures, the diatomic hydrogen gas behaves like a monatomic gas. As the temperature rises to room temperature, its molar specific heat rises to a value for a diatomic gas, consistent with the inclusion of rotation but not vibration. For high temperatures, the molar specific heat is consistent with a model including all types of motion.

Before addressing the reason for this mysterious behavior, let's make some brief remarks about polyatomic gases. For molecules with more than two atoms, the vibrations are more complex than for diatomic molecules and the number of degrees of freedom is even larger. The result is an even higher predicted molar specific heat, which is in qualitative agreement with experiment. The molar specific heats for the polyatomic gases in Table 21.2 are higher than those for

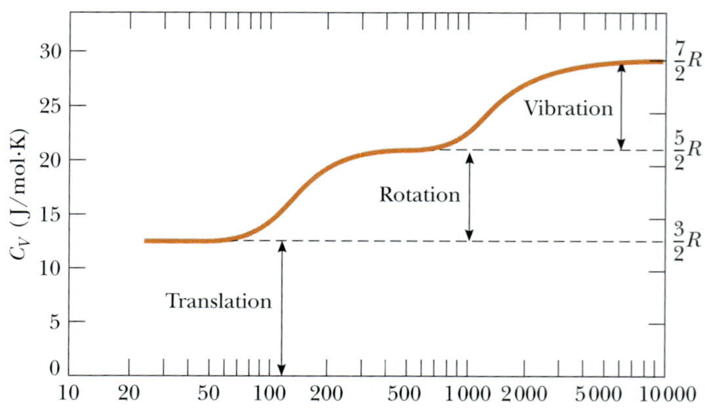

Figure 21.7 The molar specific heat of hydrogen as a function of temperature. The horizontal scale is logarithmic. Notice that hydrogen liquefies at 20 K.

diatomic gases. The more degrees of freedom available to a molecule, the more "ways" there are to store energy, resulting in a higher molar specific heat.

A Hint of Energy Quantization

Our model for molar specific heats has been based so far on purely classical notions. It predicts a value of the specific heat for a diatomic gas that, according to Figure 21.7, only agrees with experimental measurements made at high temperatures. To explain why this value is only true at high temperatures and why the plateaus in Figure 21.7 exist, we must go beyond classical physics and introduce some quantum physics into the model. In Chapter 18, we discussed quantization of frequency for vibrating strings and air columns; only certain frequencies of standing waves can exist. That is a natural result whenever waves are subject to boundary conditions.

Quantum physics (Chapters 40 through 43) shows that atoms and molecules can be described by the physics of waves under boundary conditions. Consequently, these waves have quantized frequencies. Furthermore, in quantum physics, the energy of a system is proportional to the frequency of the wave representing the system. Hence, **the energies of atoms and molecules are quantized**.

For a molecule, quantum physics tells us that the rotational and vibrational energies are quantized. Figure 21.8 shows an **energy-level diagram** for the rotational and vibrational quantum states of a diatomic molecule. The lowest allowed state is called the **ground state**. Notice that vibrational states are separated by larger energy gaps than are rotational states.

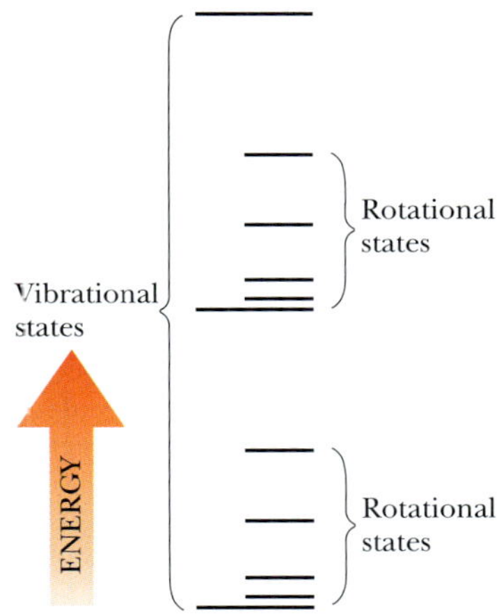

Figure 21.8 An energy-level diagram for vibrational and rotational states of a diatomic molecule. Notice that the rotational states lie closer together in energy than do the vibrational states.

At low temperatures, the energy a molecule gains in collisions with its neighbors is generally not large enough to raise it to the first excited state of either rotation or vibration. Therefore, even though rotation and vibration are allowed according to classical physics, they do not occur in reality at low temperatures. All molecules are in the ground state for rotation and vibration. The only contribution to the molecules' average energy is from translation, and the specific heat is that predicted by Equation 21.14.

As the temperature is raised, the average energy of the molecules increases. In some collisions, a molecule may have enough energy transferred to it from another molecule to excite the first rotational state. As the temperature is raised further, more molecules can be excited to this state. The result is that rotation begins to contribute to the internal energy and the molar specific heat rises. At about room temperature in Figure 21.7, the second plateau has been reached and rotation contributes fully to the molar specific heat. The molar specific heat is now equal to the value predicted by Equation 21.21.

There is no contribution at room temperature from vibration because the molecules are still in the ground vibrational state. The temperature must be raised even further to excite the first vibrational state, which happens in Figure 21.7 between 1 000 K and 10 000 K. At 10 000 K on the right side of the figure, vibration is contributing fully to the internal energy and the molar specific heat has the value predicted by Equation 21.22.

The predictions of this model are supportive of the theorem of equipartition of energy. In addition, the inclusion in the model of energy quantization from quantum physics allows a full understanding of Figure 21.7.

Quick Quiz 21.3 The molar specific heat of a diatomic gas is measured at constant volume and found to be 29.1 J/mol · K. What are the types of energy that are contributing to the molar specific heat? (a) translation only (b) translation and rotation only (c) translation and vibration only (d) translation, rotation, and vibration

---

Quick Quiz 21.4 The molar specific heat of a gas is measured at constant volume and found to be $11R/2$. Is the gas most likely to be (a) monatomic, (b) diatomic, or (c) polyatomic?

PITFALL PREVENTION 21.2
The Distribution Function

The distribution function $n_V(E)$ is defined in terms of the number of molecules with energy in the range E to $E + dE$ rather than in terms of the number of molecules with energy E. Because the number of molecules is finite and the number of possible values of the energy is infinite, the number of molecules with an *exact* energy E may be zero.

21.5 Distribution of Molecular Speeds

Thus far, we have considered only average values of the energies of molecules in a gas and have not addressed the distribution of energies among molecules. In reality, the motion of the molecules is extremely chaotic. Any individual molecule collides with others at an enormous rate, typically a billion times per second. Each collision results in a change in the speed and direction of motion of each of the participant molecules. Equation 21.7 shows that rms molecular speeds increase with increasing temperature. What is the relative number of molecules that possess some characteristic such as energy within a certain range?

We shall address this question by considering the **number density** $n_V(E)$. This quantity, called a *distribution function*, is defined so that $n_V(E)\,dE$ is the number of molecules per unit volume with energy between E and $E + dE$. (The ratio of the number of molecules that have the desired characteristic to the total number of molecules is the probability that a particular molecule has that characteristic.) In general, the number density is found from statistical mechanics to be

Boltzmann distribution law ▶

$$n_V(E) = n_0 e^{-E/k_B T} \qquad \textbf{(21.23)}$$

where n_0 is defined such that $n_0\,dE$ is the number of molecules per unit volume having energy between $E = 0$ and $E = dE$. This equation, known as the **Boltzmann distribution law**, is important in describing the statistical mechanics of a large number of molecules. It states that **the probability of finding the molecules in a particular energy state varies exponentially as the negative of the energy divided by $k_B T$.** All the molecules would fall into the lowest energy level if the thermal agitation at a temperature T did not excite the molecules to higher energy levels.

EXAMPLE 21.4 Thermal Excitation of Atomic Energy Levels

As discussed in Section 21.4, atoms can occupy only certain discrete energy levels. Consider a gas at a temperature of 2 500 K whose atoms can occupy only two energy levels separated by 1.50 eV, where 1 eV (electron volt) is an energy unit equal to 1.60×10^{-19} J (Fig. 21.9). Determine the ratio of the number of atoms in the higher energy level to the number in the lower energy level.

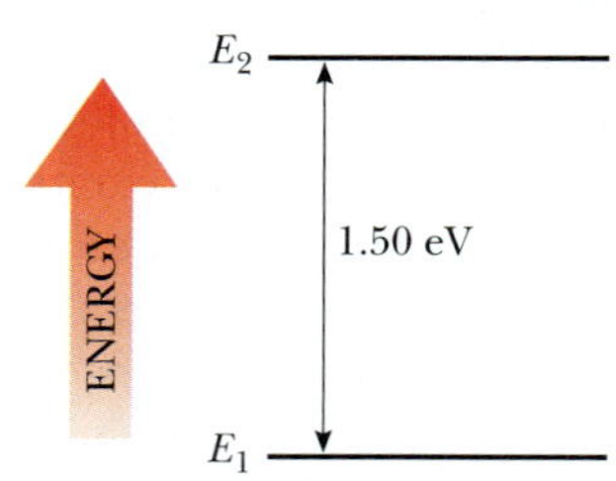

Figure 21.9 (Example 21.4) Energy-level diagram for a gas whose atoms can occupy two energy states.

SOLUTION

Conceptualize In your mental representation of this example, remember that only two possible states are allowed for the system of the atom. Figure 21.9 helps you visualize the two states on an energy-level diagram. In this case, the atom has two possible energies, E_1 and E_2, where $E_1 < E_2$.

Categorize We categorize this example as one in which we apply the Boltzmann distribution law to a quantized system.

Analyze Set up the ratio of the number of atoms in the higher energy level to the number in the lower energy level and use Equation 21.23 to express each number:

$$(1) \quad \frac{n_V(E_2)}{n_V(E_1)} = \frac{n_0 e^{-E_2/k_B T}}{n_0 e^{-E_1/k_B T}} = e^{-(E_2 - E_1)/k_B T}$$

Evaluate $k_B T$ in the exponent:

$$k_B T = (1.38 \times 10^{-23}\ \text{J/K})(2\,500\ \text{K})\left(\frac{1\ \text{eV}}{1.60 \times 10^{-19}\ \text{J}}\right) = 0.216\ \text{eV}$$

Substitute this value into Equation (1):

$$\frac{n_V(E_2)}{n_V(E_1)} = e^{-1.50\ \text{eV}/0.216\ \text{eV}} = e^{-6.96} = 9.52 \times 10^{-4}$$

Finalize This result indicates that at $T = 2\ 500$ K, only a small fraction of the atoms are in the higher energy level. In fact, for every atom in the higher energy level, there are about 1 000 atoms in the lower level. The number of atoms in the higher level increases at even higher temperatures, but the distribution law specifies that at equilibrium there are always more atoms in the lower level than in the higher level.

What If? What if the energy levels in Figure 21.9 were closer together in energy? Would that increase or decrease the fraction of the atoms in the upper energy level?

Answer If the excited level is lower in energy than that in Figure 21.9, it would be easier for thermal agitation to excite atoms to this level and the fraction of atoms in this energy level would be larger. Let us see this mathematically by expressing Equation (1) as

$$r_2 = e^{-(E_2-E_1)/k_BT}$$

where r_2 is the ratio of atoms having energy E_2 to those with energy E_1. Differentiating with respect to E_2, we find

$$\frac{dr_2}{dE_2} = \frac{d}{dE_2}\left(e^{-(E_2-E_1)/k_BT}\right) = -\frac{E_2}{k_BT}e^{-(E_2-E_1)/k_BT} < 0$$

Because the derivative has a negative value, as E_2 decreases, r_2 increases.

Now that we have discussed the distribution of energies, let's think about the distribution of molecular speeds. In 1860, James Clerk Maxwell (1831–1879) derived an expression that describes the distribution of molecular speeds in a very definite manner. His work and subsequent developments by other scientists were highly controversial because direct detection of molecules could not be achieved experimentally at that time. About 60 years later, however, experiments were devised that confirmed Maxwell's predictions.

Let's consider a container of gas whose molecules have some distribution of speeds. Suppose we want to determine how many gas molecules have a speed in the range from, for example, 400 to 401 m/s. Intuitively, we expect the speed distribution to depend on temperature. Furthermore, we expect the distribution to peak in the vicinity of v_{rms}. That is, few molecules are expected to have speeds much less than or much greater than v_{rms} because these extreme speeds result only from an unlikely chain of collisions.

The observed speed distribution of gas molecules in thermal equilibrium is shown in Active Figure 21.10 (page 602). The quantity N_v, called the **Maxwell–Boltzmann speed distribution function**, is defined as follows. If N is the total number of molecules, the number of molecules with speeds between v and $v + dv$ is $dN = N_v\,dv$. This number is also equal to the area of the shaded rectangle in Active Figure 21.10. Furthermore, the fraction of molecules with speeds between v and $v + dv$ is $(N_v\,dv)/N$. This fraction is also equal to the probability that a molecule has a speed in the range v to $v + dv$.

The fundamental expression that describes the distribution of speeds of N gas molecules is

$$N_v = 4\pi N\left(\frac{m_0}{2\pi k_BT}\right)^{3/2} v^2 e^{-m_0v^2/2k_BT} \tag{21.24}$$

where m_0 is the mass of a gas molecule, k_B is Boltzmann's constant, and T is the absolute temperature.[3] Observe the appearance of the Boltzmann factor e^{-E/k_BT} with $E = \frac{1}{2}m_0v^2$.

[3] For the derivation of this expression, see an advanced textbook on thermodynamics.

Courtesy of AIP Niels Bohr Library, Lande Collection

LUDWIG BOLTZMANN
Austrian physicist (1844–1906)
Boltzmann made many important contributions to the development of the kinetic theory of gases, electromagnetism, and thermodynamics. His pioneering work in the field of kinetic theory led to the branch of physics known as statistical mechanics.

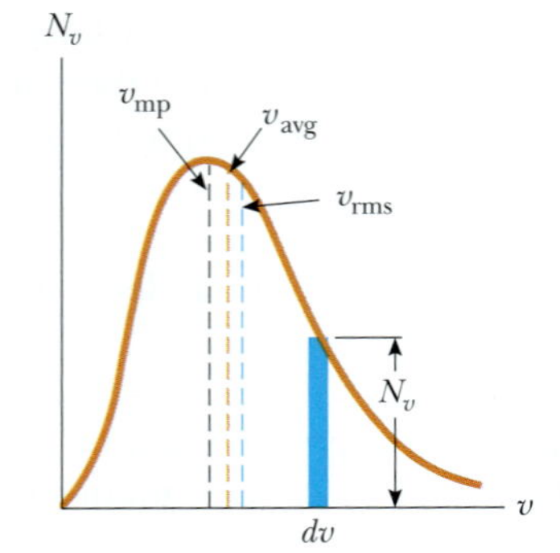

ACTIVE FIGURE 21.10

The speed distribution of gas molecules at some temperature. The number of molecules having speeds in the range v to $v + dv$ is equal to the area of the shaded rectangle, $N_v\,dv$. The function N_v approaches zero as v approaches infinity.

Sign in at www.thomsonedu.com and go to ThomsonNOW to move the blue rectangle and measure the number of molecules with speeds within a small range.

As indicated in Active Figure 21.10, the average speed is somewhat lower than the rms speed. The *most probable speed* v_{mp} is the speed at which the distribution curve reaches a peak. Using Equation 21.24, we find that

$$v_{rms} = \sqrt{\overline{v^2}} = \sqrt{\frac{3k_BT}{m_0}} = 1.73\sqrt{\frac{k_BT}{m_0}} \qquad \textbf{(21.25)}$$

$$v_{avg} = \sqrt{\frac{8k_BT}{\pi m_0}} = 1.60\sqrt{\frac{k_BT}{m_0}} \qquad \textbf{(21.26)}$$

$$v_{mp} = \sqrt{\frac{2k_BT}{m_0}} = 1.41\sqrt{\frac{k_BT}{m_0}} \qquad \textbf{(21.27)}$$

Equation 21.25 has previously appeared as Equation 21.7. The details of the derivations of these equations from Equation 21.24 are left for the student (see Problems 33 and 57). From these equations, we see that

$$v_{rms} > v_{avg} > v_{mp}$$

Active Figure 21.11 represents speed distribution curves for nitrogen, N_2. The curves were obtained by using Equation 21.24 to evaluate the distribution function at various speeds and at two temperatures. Notice that the peak in each curve shifts to the right as T increases, indicating that the average speed increases with increasing temperature, as expected. Because the lowest speed possible is zero and the upper classical limit of the speed is infinity, the curves are asymmetrical. (In Chapter 39, we show that the actual upper limit is the speed of light.)

Equation 21.24 shows that the distribution of molecular speeds in a gas depends both on mass and on temperature. At a given temperature, the fraction of molecules with speeds exceeding a fixed value increases as the mass decreases. Hence, lighter molecules such as H_2 and He escape more readily from the Earth's atmosphere than do heavier molecules such as N_2 and O_2. (See the discussion of escape speed in Chapter 13. Gas molecules escape even more readily from the Moon's surface than from the Earth's because the escape speed on the Moon is lower than that on the Earth.)

The speed distribution curves for molecules in a liquid are similar to those shown in Active Figure 21.11. We can understand the phenomenon of evaporation of a liquid from this distribution in speeds, given that some molecules in the liquid are more energetic than others. Some of the faster-moving molecules in the liquid penetrate the surface and even leave the liquid at temperatures well below

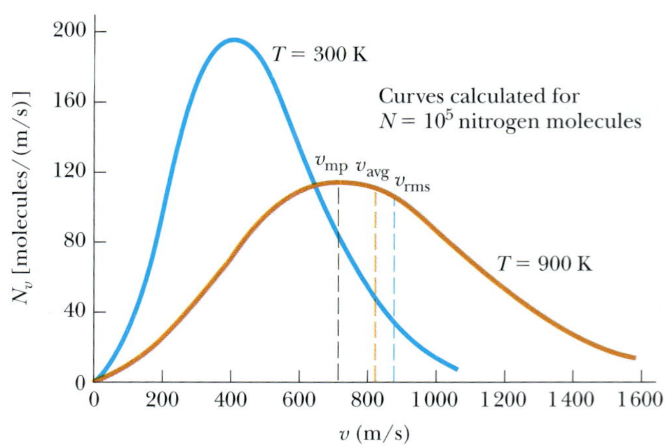

ACTIVE FIGURE 21.11

The speed distribution function for 10^5 nitrogen molecules at 300 K and 900 K. The total area under either curve is equal to the total number of molecules, which in this case equals 10^5. Notice that $v_{rms} > v_{avg} > v_{mp}$.

Sign in at www.thomsonedu.com and go to ThomsonNOW to set the desired temperature and see the effect on the distribution curve.

the boiling point. The molecules that escape the liquid by evaporation are those that have sufficient energy to overcome the attractive forces of the molecules in the liquid phase. Consequently, the molecules left behind in the liquid phase have a lower average kinetic energy; as a result, the temperature of the liquid decreases. Hence, evaporation is a cooling process. For example, an alcohol-soaked cloth can be placed on a feverish head to cool and comfort a patient.

EXAMPLE 21.5 A System of Nine Particles

Nine particles have speeds of 5.00, 8.00, 12.0, 12.0, 12.0, 14.0, 14.0, 17.0, and 20.0 m/s.

(A) Find the particles' average speed.

SOLUTION

Conceptualize Imagine a small number of particles moving in random directions with the few speeds listed.

Categorize Because we are dealing with a small number of particles, we can calculate the average speed directly.

Analyze Find the average speed of the particles by dividing the sum of the speeds by the total number of particles:

$$v_{\text{avg}} = \frac{(5.00 + 8.00 + 12.0 + 12.0 + 12.0 + 14.0 + 14.0 + 17.0 + 20.0)\ \text{m/s}}{9}$$

$$= \boxed{12.7\ \text{m/s}}$$

(B) What is the rms speed of the particles?

SOLUTION

Find the average speed squared of the particles by dividing the sum of the speeds squared by the total number of particles:

$$\overline{v^2} = \frac{(5.00^2 + 8.00^2 + 12.0^2 + 12.0^2 + 12.0^2 + 14.0^2 + 14\ 0^2 + 17.0^2 + 20.0^2)\ \text{m}^2/\text{s}^2}{9}$$

$$= 178\ \text{m}^2/\text{s}^2$$

Find the rms speed of the particles by taking the square root:

$$v_{\text{rms}} = \sqrt{\overline{v^2}} = \sqrt{178\ \text{m}^2/\text{s}^2} = \boxed{13.3\ \text{m/s}}$$

(C) What is the most probable speed of the particles?

SOLUTION

Three of the particles have a speed of 12.0 m/s, two have a speed of 14.0 m/s, and the remaining four have different speeds. Hence, the most probable speed v_{mp} is 12.0 m/s.

Finalize Compare this example, in which the number of particles is small and we know the individual particle speeds, with the next example.

EXAMPLE 21.6 Molecular Speeds in a Hydrogen Gas

A 0.500-mol sample of hydrogen gas is at 300 K.

(A) Find the average speed, the rms speed, and the most probable speed of the hydrogen molecules.

SOLUTION

Conceptualize Imagine a huge number of particles in a real gas, all moving in random directions with different speeds.

Categorize We cannot calculate the averages as was done in Example 21.5 because the individual speeds of the particles are not known. We are dealing with a very large number of particles, however, so we can use the Maxwell-Boltzmann speed distribution function.

Analyze Use Equation 21.26 to find the average speed:

$$v_{\text{avg}} = 1.60\sqrt{\frac{k_B T}{m_0}} = 1.60\sqrt{\frac{(1.38 \times 10^{-23}\text{ J/K})(300\text{ K})}{2(1.67 \times 10^{-27}\text{ kg})}}$$

$$= \boxed{1.78 \times 10^3\text{ m/s}}$$

Use Equation 21.25 to find the rms speed:

$$v_{\text{rms}} = 1.73\sqrt{\frac{k_B T}{m_0}} = 1.73\sqrt{\frac{(1.38 \times 10^{-23}\text{ J/K})(300\text{ K})}{2(1.67 \times 10^{-27}\text{ kg})}}$$

$$= \boxed{1.93 \times 10^3\text{ m/s}}$$

Use Equation 21.27 to find the most probable speed:

$$v_{\text{mp}} = 1.41\sqrt{\frac{k_B T}{m_0}} = 1.41\sqrt{\frac{(1.38 \times 10^{-23}\text{ J/K})(300\text{ K})}{2(1.67 \times 10^{-27}\text{ kg})}}$$

$$= \boxed{1.57 \times 10^3\text{ m/s}}$$

(B) Find the number of molecules with speeds between 400 m/s and 401 m/s.

SOLUTION

Use Equation 21.24 to evaluate the number of molecules in a narrow speed range between v and $v + dv$:

$$(1)\quad N_v\,dv = 4\pi N\left(\frac{m_0}{2\pi k_B T}\right)^{3/2} v^2 e^{-m_0 v^2/2k_B T}\,dv$$

Evaluate the constant in front of v^2:

$$4\pi N\left(\frac{m_0}{2\pi k_B T}\right)^{3/2} = 4\pi n N_A\left(\frac{m_0}{2\pi k_B T}\right)^{3/2}$$

$$= 4\pi(0.500\text{ mol})(6.02 \times 10^{23}\text{ mol}^{-1})\left[\frac{2(1.67 \times 10^{-27}\text{ kg})}{2\pi(1.38 \times 10^{-23}\text{ J/K})(300\text{ K})}\right]^{3/2}$$

$$= 1.74 \times 10^{14}\text{ s}^3/\text{m}^3$$

Evaluate the exponent of e:

$$-\frac{m_0 v^2}{2k_B T} = -\frac{2(1.67 \times 10^{-27}\text{ kg})(400\text{ m/s})^2}{2(1.38 \times 10^{-23}\text{ J/K})(300\text{ K})} = -0.064\,5$$

Evaluate $N_v\,dv$ using Equation (1):

$$N_v\,dv = (1.74 \times 10^{14}\text{ s}^3/\text{m}^3)(400\text{ m/s})^2 e^{-0.064\,5}(1\text{ m/s})$$

$$= \boxed{2.61 \times 10^{19}\text{ molecules}}$$

Finalize In this evaluation, we could calculate the result without integration because $dv = 1$ m/s is much smaller than $v = 400$ m/s. Had we sought the number of particles between, say, 400 m/s and 500 m/s, we would need to integrate Equation (1) between these speed limits.

Summary

ThomsonNOW™ Sign in at **www.thomsonedu.com** and go to ThomsonNOW to take a practice test for this chapter.

CONCEPTS AND PRINCIPLES

The pressure of N molecules of an ideal gas contained in a volume V is

$$P = \frac{2}{3}\left(\frac{N}{V}\right)\left(\frac{1}{2}m_0\overline{v^2}\right) \qquad \textbf{(21.2)}$$

The average translational kinetic energy per molecule of a gas, $\frac{1}{2}m_0\overline{v^2}$, is related to the temperature T of the gas through the expression

$$\frac{1}{2}m_0\overline{v^2} = \frac{3}{2}k_BT \qquad \textbf{(21.4)}$$

where k_B is Boltzmann's constant. Each translational degree of freedom (x, y, or z) has $\frac{1}{2}k_BT$ of energy associated with it.

The internal energy of N molecules (or n mol) of an ideal monatomic gas is

$$E_{int} = \frac{3}{2}Nk_BT = \frac{3}{2}nRT \qquad \textbf{(21.10)}$$

The change in internal energy for n mol of any ideal gas that undergoes a change in temperature ΔT is

$$\Delta E_{int} = nC_V\,\Delta T \qquad \textbf{(21.12)}$$

where C_V is the **molar specific heat at constant volume.**

The molar specific heat of an ideal monatomic gas at constant volume is $C_V = \frac{3}{2}R$; the molar specific heat at constant pressure is $C_P = \frac{5}{2}R$. The ratio of specific heats is given by $\gamma = C_P/C_V = \frac{5}{3}$.

If an ideal gas undergoes an adiabatic expansion or compression, the first law of thermodynamics, together with the equation of state, shows that

$$PV^{\gamma} = \text{constant} \qquad \textbf{(21.18)}$$

The **Boltzmann distribution law** describes the distribution of particles among available energy states. The relative number of particles having energy between E and $E + dE$ is $n_V(E)\,dE$, where

$$n_V(E) = n_0e^{-E/k_BT} \qquad \textbf{(21.23)}$$

The **Maxwell–Boltzmann speed distribution function** describes the distribution of speeds of molecules in a gas:

$$N_v = 4\pi N\left(\frac{m_0}{2\pi k_BT}\right)^{3/2} v^2e^{-m_0v^2/2k_BT} \qquad \textbf{(21.24)}$$

Equation 21.24 enables us to calculate the **root-mean-square speed**, the **average speed**, and the **most probable speed** of molecules in a gas:

$$v_{rms} = \sqrt{\overline{v^2}} = \sqrt{\frac{3k_BT}{m_0}} = 1.73\sqrt{\frac{k_BT}{m_0}} \qquad \textbf{(21.25)}$$

$$v_{avg} = \sqrt{\frac{8k_BT}{\pi m_0}} = 1.60\sqrt{\frac{k_BT}{m_0}} \qquad \textbf{(21.26)}$$

$$v_{mp} = \sqrt{\frac{2k_BT}{m_0}} = 1.41\sqrt{\frac{k_BT}{m_0}} \qquad \textbf{(21.27)}$$

Questions

☐ denotes answer available in *Student Solutions Manual/Study Guide;* **O** denotes objective question

1. Dalton's law of partial pressures states that the total pressure of a mixture of gases is equal to the sum of the pressures that each gas in the mixture would exert if it were alone in the container. Give a convincing argument for this law based on the kinetic theory of gases.

2. One container is filled with helium gas and another with argon gas. Both containers are at the same temperature. Which molecules have the higher rms speed? Explain.

3. ☐ When alcohol is rubbed on your body, it lowers your skin temperature. Explain this effect.

4. **O** A student is asked to give a step-by-step account of what makes the temperature of a sample of gas increase. His response: When a sample of a gas is held over a hotplate, (a) the molecules speed up. (b) Then the molecules collide with one another more often. (c) Internal friction makes the collisions inelastic. (d) Heat is produced in the collisions. (e) As soon as we put in a thermometer, we see that the temperature has gone up. (f) The same process can take place without the use of a hotplate if you quickly push in the piston in an insulated cylinder containing the gas. **(i)** Which of the parts (a) through (f) of this account are correct statements necessary for a clear and complete explanation? **(ii)** Which are correct statements that are not necessary to account for the higher thermometer reading? **(iii)** Which are incorrect statements?

5. **O** A helium-filled balloon initially at room temperature is placed in a freezer. The rubber remains flexible. **(i)** Does its volume (a) increase, (b) decrease, or (c) remain the same? **(ii)** Does its pressure (a) increase, (b) decrease, or (c) remain the same?

6. **O** A gas is at 200 K. If we wish to double the rms speed of the molecules of the gas, to what must we raise its temperature? (a) 283 K (b) 400 K (c) 566 K (d) 800 K (e) 1 130 K

7. **O** Rank the following from the largest to the smallest, noting any cases of equality. (a) The average speed of molecules in a particular sample of ideal gas. (b) The most probable speed. (c) The root-mean-square speed. (d) The average vector velocity of the molecules. (e) The speed of sound in the gas.

8. **O** Two samples of the same ideal gas have the same pressure and density. Sample B has twice the volume of sample A. What is the rms speed of the molecules in sample B? (a) twice that in sample A (b) equal to that in sample A (c) half that in sample A (d) impossible to determine

9. Which is denser, dry air or air saturated with water vapor? Explain.

10. What happens to a helium-filled balloon released into the air? Does it expand or contract? Does it stop rising at some height?

11. Why does a diatomic gas have a greater energy content per mole than a monatomic gas at the same temperature?

12. **O** An ideal gas is contained in a vessel at 300 K. If the temperature is increased to 900 K, what is the factor of change in **(i)** the average kinetic energy of the molecules? (a) 9 (b) 3 (c) $\sqrt{3}$ (d) 1 (e) $\frac{1}{3}$ **(ii)** What is the factor of change in the rms molecular speed? Choose from the same possibilities. **(iii)** What is the factor of change in the average momentum change that one molecule undergoes in a collision with one particular wall? **(iv)** What is the factor of change in the rate of collisions of molecules with walls? **(v)** What is the factor of change in the pressure of the gas? Choose from the same possibilities (a) through (e).

13. Hot air rises, so why does it generally become cooler as you climb a mountain? *Note:* Air has low thermal conductivity.

14. **O** The brown curve in Active Figure 21.11 shows the speed distribution for 100 000 nitrogen molecules at 900 K. Krypton has very nearly three times the molecular mass of nitrogen. The blue curve, disregarding its label, shows the speed distribution for which of the following? (a) 100 000 krypton molecules at 900 K (b) 100 000 krypton molecules at 520 K (c) 100 000 krypton molecules at 300 K (d) 100 000 krypton molecules at 100 K (e) 33 000 krypton molecules at 900 K (f) 33 000 krypton molecules at 300 K (g) This distribution cannot be attributed to any sample of krypton, which does not exist on Earth.

Problems

WebAssign The Problems from this chapter may be assigned online in WebAssign.

ThomsonNOW Sign in at **www.thomsonedu.com** and go to ThomsonNOW to assess your understanding of this chapter's topics with additional quizzing and conceptual questions.

1, 2, 3 denotes straightforward, intermediate, challenging; ☐ denotes full solution available in *Student Solutions Manual/Study Guide;* ▲ denotes coached solution with hints available at **www.thomsonedu.com;** ▭ denotes developing symbolic reasoning; ● denotes asking for qualitative reasoning; ▪ denotes computer useful in solving problem

Section 21.1 Molecular Model of an Ideal Gas

Note: Problem 25 in Chapter 19 can be assigned with this section.

1. ▲ Calculate the mass of an atom of (a) helium, (b) iron, and (c) lead. Give your answers in grams. The atomic masses of these atoms are 4.00 u, 55.9 u, and 207 u, respectively.

2. ● We use m to represent the mass of a sample, m_0 to represent the mass of one molecule, M for the molar mass, n for the number of moles in a sample, N for the number of molecules, and N_A to represent Avogadro's number. Explain why each of the following equations is true:

$$N = nN_A \qquad m = nM = Nm_0 \qquad M = m_0N_A$$

Are the equations true only for ideal gases? Only for gases? Only for pure chemical elements? Only for pure chemical compounds? For mixtures?

3. In a 30.0-s interval, 500 hailstones strike a glass window of area 0.600 m^2 at an angle of 45.0° to the window surface. Each hailstone has a mass of 5.00 g and a speed of 8.00 m/s. Assuming the collisions are elastic, find the average force and pressure on the window.

4. In a period of 1.00 s, 5.00×10^{23} nitrogen molecules strike a wall with an area of 8.00 cm^2. Assume the molecules move with a speed of 300 m/s and strike the wall head-on in elastic collisions. What is the pressure exerted on the wall? (The mass of one N_2 molecule is 4.68×10^{-26} kg.)

5. In an ultrahigh vacuum system, the pressure is measured to be 1.00×10^{-10} torr (where 1 torr = 133 Pa). Assum-

ing the temperature is 300 K, find the number of molecules in a volume of $1.00\ m^3$.

6. A 2.00-mol sample of oxygen gas is confined to a 5.00-L vessel at a pressure of 8.00 atm. Find the average translational kinetic energy of an oxygen molecule under these conditions.

7. A spherical balloon of volume $4\,000\ cm^3$ contains helium at an (inside) pressure of 1.20×10^5 Pa. How many moles of helium are in the balloon if the average kinetic energy of each helium atom is 3.60×10^{-22} J?

8. A 5.00-L vessel contains nitrogen gas at 27.0°C and 3.00 atm. (a) Find the total translational kinetic energy of the gas molecules and (b) the average kinetic energy per molecule.

9. (a) How many atoms of helium gas fill a balloon of diameter 30.0 cm at 20.0°C and 1.00 atm? (b) What is the average kinetic energy of the helium atoms? (c) What is the root-mean-square speed of the helium atoms?

10. (a) Show that 1 Pa = 1 J/m^3. (b) Show that the density in space of the translational kinetic energy of an ideal gas (the energy per volume) is $3P/2$.

11. ▲ A cylinder contains a mixture of helium and argon gas in equilibrium at 150°C. (a) What is the average kinetic energy for each type of gas molecule? (b) What is the root-mean-square speed of each type of molecule?

Section 21.2 Molar Specific Heat of an Ideal Gas

> *Note:* You may use data in Table 21.2 about particular gases. Here we define a "monatomic ideal gas" to have molar specific heats $C_V = 3R/2$ and $C_P = 5R/2$, and a "diatomic ideal gas" to have $C_V = 5R/2$ and $C_P = 7R/2$.

12. In a constant-volume process, 209 J of energy is transferred by heat to 1.00 mol of an ideal monatomic gas initially at 300 K. Find (a) the increase in internal energy of the gas, (b) the work done on it, and (c) its final temperature.

13. ▲ A 1.00-mol sample of hydrogen gas is heated at constant pressure from 300 K to 420 K. Calculate (a) the energy transferred to the gas by heat, (b) the increase in its internal energy, and (c) the work done on the gas.

14. A house has well-insulated walls. It contains a volume of $100\ m^3$ of air at 300 K. (a) Calculate the energy required to increase the temperature of this diatomic ideal gas by 1.00°C. (b) **What If?** If this energy could be used to lift an object of mass m through a height of 2.00 m, what is the value of m?

15. A 1-L insulated bottle is full of tea at 90°C. You pour out one cup and immediately screw the stopper back on the bottle. Make an order-of-magnitude estimate of the change in temperature of the tea remaining in the bottle that results from the admission of air at room temperature. State the quantities you take as data and the values you measure or estimate for them.

16. A vertical cylinder with a heavy piston contains air at 300 K. The initial pressure is 200 kPa, and the initial volume is $0.350\ m^3$. Take the molar mass of air as 28.9 g/mol and assume $C_V = 5R/2$. (a) Find the specific heat of air at constant volume in units of J/kg · °C. (b) Calculate the mass of the air in the cylinder. (c) Suppose the piston is held fixed. Find the energy input required to raise the temperature of the air to 700 K. (d) **What If?** Assume again the conditions of the initial state and assume the heavy piston is free to move. Find the energy input required to raise the temperature to 700 K.

17. A 1.00-mol sample of a diatomic ideal gas has pressure P and volume V. When the gas is warmed, its pressure triples and its volume doubles. This warming process includes two steps, the first at constant pressure and the second at constant volume. Determine the amount of energy transferred to the gas by heat.

Section 21.3 Adiabatic Processes for an Ideal Gas

18. During the compression stroke of a certain gasoline engine, the pressure increases from 1.00 atm to 20.0 atm. If the process is adiabatic and the fuel–air mixture behaves as a diatomic ideal gas, (a) by what factor does the volume change and (b) by what factor does the temperature change? (c) Assuming the compression starts with 0.016 0 mol of gas at 27.0°C, find the values of Q, W, and ΔE_{int} that characterize the process.

19. A 2.00-mol sample of a diatomic ideal gas expands slowly and adiabatically from a pressure of 5.00 atm and a volume of 12.0 L to a final volume of 30.0 L. (a) What is the final pressure of the gas? (b) What are the initial and final temperatures? (c) Find Q, W, and ΔE_{int}.

20. Air (a diatomic ideal gas) at 27.0°C and atmospheric pressure is drawn into a bicycle pump that has a cylinder with an inner diameter of 2.50 cm and length 50.0 cm. The downstroke adiabatically compresses the air, which reaches a gauge pressure of 800 kPa before entering the tire (Fig. P21.20). Determine (a) the volume of the compressed air and (b) the temperature of the compressed air. (c) **What If?** The pump is made of steel and has an inner wall that is 2.00 mm thick. Assume 4.00 cm of the

George Semple

Figure P21.20

cylinder's length is allowed to come to thermal equilibrium with the air. What will be the increase in wall temperature?

21. Air in a thundercloud expands as it rises. If its initial temperature is 300 K and no energy is lost by thermal conduction on expansion, what is its temperature when the initial volume has doubled?

22. During the power stroke in a four-stroke automobile engine, the piston is forced down as the mixture of combustion products and air undergoes an adiabatic expansion. Assume (1) the engine is running at 2 500 cycles/min; (2) the gauge pressure right before the expansion is 20.0 atm; (3) the volumes of the mixture right before and after the expansion are 50.0 and 400 cm^3, respectively (Fig. P21.22); (4) the time interval for the expansion is one-fourth that of the total cycle; and (5) the mixture behaves like an ideal gas with specific heat ratio 1.40. Find the average power generated during the expansion stroke.

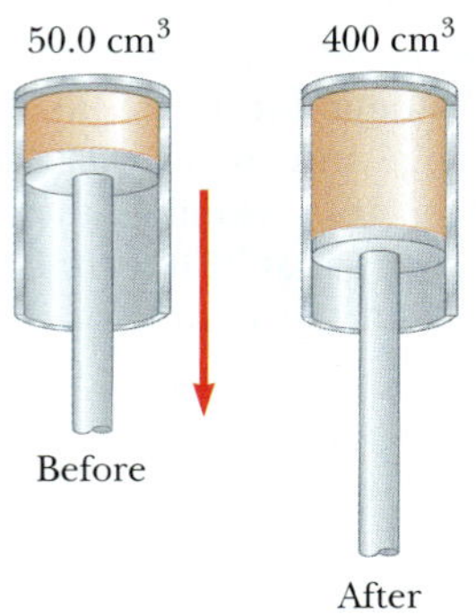

Figure P21.22

23. A 4.00-L sample of a diatomic ideal gas with specific heat ratio 1.40, confined to a cylinder, is carried through a closed cycle. The gas is initially at 1.00 atm and at 300 K. First, its pressure is tripled under constant volume. Then, it expands adiabatically to its original pressure. Finally, the gas is compressed isobarically to its original volume. (a) Draw a *PV* diagram of this cycle. (b) Determine the volume of the gas at the end of the adiabatic expansion. (c) Find the temperature of the gas at the start of the adiabatic expansion. (d) Find the temperature at the end of the cycle. (e) What was the net work done on the gas for this cycle?

24. A diatomic ideal gas (γ = 1.40) confined to a cylinder is put through a closed cycle. Initially, the gas is at P_i, V_i, and T_i. First, its pressure is tripled under constant volume. It then expands adiabatically to its original pressure and finally is compressed isobarically to its original volume. (a) Draw a *PV* diagram of this cycle. (b) Determine the volume at the end of the adiabatic expansion. Find (c) the temperature of the gas at the start of the adiabatic expansion and (d) the temperature at the end of the cycle. (e) What was the net work done on the gas for this cycle?

25. How much work is required to compress 5.00 mol of air at 20.0°C and 1.00 atm to one-tenth of the original volume (a) by an isothermal process? (b) **What If?** How much work is required to produce the same compression in an adiabatic process? (c) What is the final pressure in each of these two cases?

Section 21.4 The Equipartition of Energy

26. A certain molecule has *f* degrees of freedom. Show that an ideal gas consisting of such molecules has the following properties: (1) its total internal energy is $fnRT/2$, (2) its molar specific heat at constant volume is $fR/2$, (3) its molar specific heat at constant pressure is $(f + 2)R/2$, and (4) its specific heat ratio is $\gamma = C_P/C_V = (f + 2)/f$.

27. ▲ Consider 2.00 mol of an ideal diatomic gas. (a) Find the total heat capacity as defined by Equation 20.2 at constant volume and the total heat capacity at constant pressure, assuming the molecules rotate but do not vibrate. (b) **What If?** Repeat part (a), assuming the molecules both rotate and vibrate.

28. In a crude model (Fig. P21.28) of a rotating diatomic molecule of chlorine (Cl_2), the two Cl atoms are 2.00×10^{-10} m apart and rotate about their center of mass with angular speed $\omega = 2.00 \times 10^{12}$ rad/s. What is the rotational kinetic energy of one molecule of Cl_2, which has a molar mass of 70.0 g/mol?

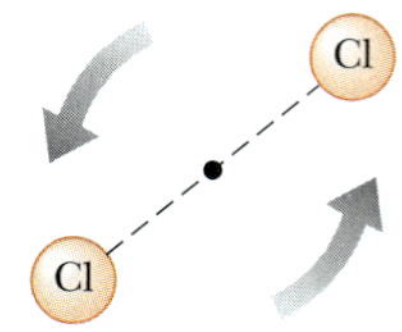

Figure P21.28

29. ● Examine the data for polyatomic gases in Table 21.2 and give a reason why sulfur dioxide has a higher specific heat at constant volume than the other polyatomic gases at 300 K.

30. ● A triatomic molecule can have the three atoms lying along one line, as does CO_2, or it can be nonlinear, like H_2O. Suppose the temperature of a gas of triatomic molecules is sufficiently low that vibrational motion is negligible. What is the molar heat capacity at constant volume, expressed as a multiple of the universal gas constant, (a) if the molecules are linear and (b) if the molecules are nonlinear? At high temperatures, a triatomic molecule has two modes of vibration, and each contributes $\frac{1}{2}R$ to the molar heat capacity for its kinetic energy and another $\frac{1}{2}R$ for its potential energy. Identify the high-temperature molar heat capacity at constant volume for a triatomic ideal gas of (c) linear molecules and of (d) nonlinear molecules. (e) Explain how specific heat data can be used to determine whether a triatomic molecule is linear or nonlinear. Are the data in Table 21.2 sufficient to make this determination?

Section 21.5 Distribution of Molecular Speeds

31. Fifteen identical particles have various speeds: one has a speed of 2.00 m/s, two have speeds of 3.00 m/s, three have speeds of 5.00 m/s, four have speeds of 7.00 m/s, three have speeds of 9.00 m/s, and two have speeds of 12.0 m/s. Find (a) the average speed, (b) the rms speed, and (c) the most probable speed of these particles.

32. One cubic meter of atomic hydrogen at 0°C at atmospheric pressure contains approximately 2.70×10^{25} atoms. The first excited state of the hydrogen atom has an energy of 10.2 eV above the lowest energy level, called the ground state. Use the Boltzmann factor to find the number of atoms in the first excited state at 0°C and at 10 000°C.

33. From the Maxwell–Boltzmann speed distribution, show that the most probable speed of a gas molecule is given by Equation 21.27. Notice that the most probable speed corresponds to the point at which the slope of the speed distribution curve dN_v/dv is zero.

34. Two gases in a mixture diffuse through a filter at rates proportional to the gases' rms speeds. (a) Find the ratio of speeds for the two isotopes of chlorine, ^{35}Cl and ^{37}Cl, as they diffuse through the air. (b) Which isotope moves faster?

35. **Review problem.** At what temperature would the average speed of helium atoms equal (a) the escape speed from the Earth, 1.12×10^4 m/s and (b) the escape speed from the Moon, 2.37×10^3 m/s? (See Chapter 13 for a discussion of escape speed.) *Note:* The mass of a helium atom is 6.64×10^{-27} kg.

36. ● Calculate (a) the most probable speed, (b) the average speed, and (c) the rms speed for nitrogen gas molecules at 900 K. (d) State how your results compare with the values displayed in Active Figure 21.11.

37. Assume the Earth's atmosphere has a uniform temperature of 20°C and uniform composition, with an effective molar mass of 28.9 g/mol. (a) Show that the number density of molecules depends on height y above sea level according to

$$n_V(y) = n_0 e^{-m_0 g y / k_B T}$$

where n_0 is the number density at sea level (where $y = 0$). This result is called the *law of atmospheres.* (b) Commercial jetliners typically cruise at an altitude of 11.0 km. Find the ratio of the atmospheric density there to the density at sea level.

38. *If you can't walk to outer space, can you at least walk halfway?* Use the law of atmospheres from Problem 37. The average height of a molecule in the Earth's atmosphere is given by

$$\bar{y} = \frac{\int_0^\infty y n_V(y)\, dy}{\int_0^\infty n_V(y)\, dy} = \frac{\int_0^\infty y e^{-m_0 g y / k_B T}\, dy}{\int_0^\infty e^{-m_0 g y / k_B T}\, dy}$$

(a) Prove that this average height is equal to $k_B T / m_0 g$. (b) Evaluate the average height, assuming the temperature is 10°C and the molecular mass is 28.9 u, both uniform throughout the atmosphere.

Additional Problems

39. The function $E_{int} = 3.50nRT$ describes the internal energy of a certain ideal gas. A 2.00-mol sample of the gas always starts at pressure 100 kPa and temperature 300 K. For each of the following processes, determine the final pressure, volume, and temperature; the change in internal energy of the gas; the energy added to the gas by heat; and the work done on the gas. (a) The gas is heated at constant pressure to 400 K. (b) The gas is heated at constant volume to 400 K. (c) The gas is compressed at constant temperature to 120 kPa. (d) The gas is compressed adiabatically to 120 kPa.

40. ● The dimensions of a classroom are 4.20 m × 3.00 m × 2.50 m. (a) Find the number of molecules of air in it at atmospheric pressure and 20.0°C. (b) Find the mass of this air, assuming the air consists of diatomic molecules with molar mass 28.9 g/mol. (c) Find the average kinetic energy of one molecule. (d) Find the root-mean-square molecular speed. (e) On the assumption that the molar specific heat is a constant independent of temperature, $E_{int} = 5nRT/2$. Find the internal energy in the air. (f) **What If?** Find the internal energy of the air in the room at 25.0°C. Explain how it compares with the result at 20.0°C and how it happens that way.

41. ● In a sample of a solid metal, each atom is free to vibrate about some equilibrium position. The atom's energy consists of kinetic energy for motion in the x, y, and z directions plus elastic potential energy associated with the Hooke's law forces exerted by neighboring atoms on it in the x, y, and z directions. According to the theorem of equipartition of energy, assume the average energy of each atom is $\frac{1}{2}k_B T$ for each degree of freedom. (a) Prove that the molar specific heat of the solid is $3R$. The *Dulong–Petit law* states that this result generally describes pure solids at sufficiently high temperatures. (You may ignore the difference between the specific heat at constant pressure and the specific heat at constant volume.) (b) Evaluate the specific heat c of iron. Explain how it compares with the value listed in Table 20.1. (c) Repeat the evaluation and comparison for gold.

42. Twenty particles, each of mass m_0 and confined to a volume V, have various speeds: two have speed v, three have speed $2v$, five have speed $3v$, four have speed $4v$, three have speed $5v$, two have speed $6v$, and one has speed $7v$. Find (a) the average speed, (b) the rms speed, (c) the most probable speed, (d) the pressure they exert on the walls of the vessel, and (e) the average kinetic energy per particle.

43. ● ▲ A cylinder containing n mol of an ideal gas undergoes an adiabatic process. (a) Starting with the expression $W = -\int P\, dV$ and using the condition PV^γ = constant, show that the work done on the gas is

$$W = \left(\frac{1}{\gamma - 1}\right)(P_f V_f - P_i V_i)$$

(b) Starting with the first law of thermodynamics in differential form, show that the work done on the gas is equal to $nC_V(T_f - T_i)$. Explain whether these two results are consistent with each other.

44. As a 1.00-mol sample of a monatomic ideal gas expands adiabatically, the work done on it is −2 500 J. The initial temperature and pressure of the gas are 500 K and 3.60 atm. Calculate (a) the final temperature and (b) the final pressure. You may use the result of Problem 43.

45. A cylinder is closed at both ends and has insulating walls. It is divided into two compartments by an adiabatic partition that is perpendicular to the axis of the cylinder. Each compartment contains 1.00 mol of oxygen that behaves as an ideal gas with $\gamma = \frac{7}{5}$. Initially, the two compartments have equal volumes and their temperatures are 550 K and 250 K. The partition is then allowed to move slowly until the pressures on its two sides are equal. Find the final temperatures in the two compartments. You may use the result of Problem 43.

46. An air rifle shoots a lead pellet by allowing high-pressure air to expand, propelling the pellet down the rifle barrel. Because this process happens very quickly, no appreciable

thermal conduction occurs and the expansion is essentially adiabatic. Suppose the rifle starts with 12.0 cm^3 of compressed air, which behaves as an ideal gas with $\gamma = 1.40$. The expanding air pushes a 1.10-g pellet as a piston with cross-sectional area 0.030 0 cm^2 along the gun barrel, 50.0 cm long. The pellet emerges with a muzzle speed of 120 m/s. Use the result of Problem 43 to find the initial pressure required.

47. **Review problem.** Oxygen at pressures much greater than 1 atm is toxic to lung cells. Assume a deep-sea diver breathes a mixture of oxygen (O_2) and helium (He). By weight, what ratio of helium to oxygen must be used if the diver is at an ocean depth of 50.0 m?

48. A vessel contains 1.00×10^4 oxygen molecules at 500 K. (a) Make an accurate graph of the Maxwell speed distribution function versus speed with points at speed intervals of 100 m/s. (b) Determine the most probable speed from this graph. (c) Calculate the average and rms speeds for the molecules and label these points on your graph. (d) From the graph, estimate the fraction of molecules with speeds in the range 300 m/s to 600 m/s.

49. ▲ The compressibility κ of a substance is defined as the fractional change in volume of that substance for a given change in pressure:

$$\kappa = -\frac{1}{V}\frac{dV}{dP}$$

(a) Explain why the negative sign in this expression ensures κ is always positive. (b) Show that if an ideal gas is compressed isothermally, its compressibility is given by $\kappa_1 = 1/P$. (c) **What If?** Show that if an ideal gas is compressed adiabatically, its compressibility is given by $\kappa_2 = 1/\gamma P$. (d) Determine values for κ_1 and κ_2 for a monatomic ideal gas at a pressure of 2.00 atm.

50. ● **Review problem.** (a) Show that the speed of sound in an ideal gas is

$$v = \sqrt{\frac{\gamma RT}{M}}$$

where M is the molar mass. Use the general expression for the speed of sound in a fluid from Section 17.1; the definition of the bulk modulus from Section 12.4; and the result of Problem 49 above. As a sound wave passes through a gas, the compressions are either so rapid or so far apart that thermal conduction is prevented by a negligible time interval or by effective thickness of insulation. The compressions and rarefactions are adiabatic. (b) Compute the theoretical speed of sound in air at 20°C and state how it compares with the value in Table 17.1. Take $M = 28.9$ g/mol. (c) Show that the speed of sound in an ideal gas is

$$v = \sqrt{\frac{\gamma k_B T}{m_0}}$$

where m_0 is the mass of one molecule. State how it compares with the most probable, average, and rms molecular speeds.

51. ● The latent heat of vaporization for water at room temperature is 2 430 J/g. Consider one particular molecule at the surface of a glass of liquid water, moving upward with sufficiently high speed that it will be the next molecule to join the vapor. (a) Find its translational kinetic energy. (b) Find its speed. (c) Now consider a thin gas made just of molecules like that one. What is its temperature? Why are you not burned by evaporating water?

52. ● *Brownian motion.* Molecular motion is invisible in itself. When a small particle is suspended in a fluid, bombardment by molecules makes the particle jitter about at random. Robert Brown discovered this motion in 1827 while studying plant fertilization. Albert Einstein analyzed it in 1905, and Jean Perrin used it for an early measurement of Avogadro's number. The visible particle's average kinetic energy can be taken as $\frac{3}{2}k_B T$, the same as that of a molecule in an ideal gas. Consider a spherical particle of density 1 000 kg/m^3 in water at 20°C. (a) For a particle of diameter d, evaluate the rms speed. (b) The particle's actual motion is a random walk, but imagine that it moves with constant velocity equal in magnitude to its rms speed. In what time interval would it move by a distance equal to its own diameter? (c) Evaluate the rms speed and the time interval for a particle of diameter 3.00 μm. (d) Evaluate the rms speed and the time interval for a sphere of mass 70.0 kg, modeling your own body. (e) Find the diameter of a particle whose rms speed is equal to its own diameter divided by 1 s. (f) Explain whether your results suggest that there is an optimum particle size for observation of Brownian motion.

53. Model air as a diatomic ideal gas with $M = 28.9$ g/mol. A cylinder with a piston contains 1.20 kg of air at 25.0°C and 200 kPa. Energy is transferred by heat into the system as it is permitted to expand, with the pressure rising to 400 kPa. Throughout the expansion, the relationship between pressure and volume is given by

$$P = CV^{1/2}$$

where C is a constant. (a) Find the initial volume. (b) Find the final volume. (c) Find the final temperature. (d) Find the work done on the air. (e) Find the energy transferred by heat.

54. *Smokin'!* A pitcher throws a 0.142-kg baseball at 47.2 m/s. As it travels 19.4 m to home plate, the ball slows down to 42.5 m/s because of air resistance. Find the change in temperature of the air through which it passes. To find the greatest possible temperature change, you may make the following assumptions. Air has a molar specific heat of $C_P = 7R/2$ and an equivalent molar mass of 28.9 g/mol. The process is so rapid that the cover of the baseball acts as thermal insulation, and the temperature of the ball itself does not change. A change in temperature happens initially only for the air in a cylinder 19.4 m in length and 3.70 cm in radius. This air is initially at 20.0°C.

55. For a Maxwellian gas, use a computer or programmable calculator to find the numerical value of the ratio $N_v(v)/N_v(v_{mp})$ for the following values of v: $v = (v_{mp}/50)$, $(v_{mp}/10)$, $(v_{mp}/2)$, v_{mp}, $2v_{mp}$, $10v_{mp}$, $50v_{mp}$. Give your results to three significant figures.

56. Consider the particles in a gas centrifuge, a device used to separate particles of different mass by whirling them in a circular path of radius r at angular speed ω. The centripetal force acting on a particle is $m_0\omega^2 r$. (a) Discuss how a gas centrifuge can be used to separate particles of different mass. (b) Show that the density of the particles as a function of r is

$$n(r) = n_0 e^{m_0 r^2 \omega^2 / 2k_B T}$$

57. Verify Equations 21.25 and 21.26 for the rms and average speed of the molecules of a gas at a temperature T. The average value of v^n is

$$\overline{v^n} = \frac{1}{N}\int_0^\infty v^n N_v \, dv$$

Use the table of integrals B.6 in Appendix B.

58. On the PV diagram for an ideal gas, one isothermal curve and one adiabatic curve pass through each point. Prove that the slope of the adiabatic curve is steeper than the slope of the isotherm by the factor γ.

59. ● A sample of monatomic ideal gas occupies 5.00 L at atmospheric pressure and 300 K (point A in Fig. P21.59). It is warmed at constant volume to 3.00 atm (point B). Then it is allowed to expand isothermally to 1.00 atm (point C) and at last compressed isobarically to its original state. (a) Find the number of moles in the sample. (b) Find the temperature at points B and C and the volume at point C. (c) Assume the molar specific heat does not depend on temperature so that $E_{int} = 3nRT/2$. Find the internal energy at points A, B, and C. (d) Tabulate P, V, T, and E_{int} at the states at points A, B, and C. (e) Now consider the processes $A \to B$, $B \to C$, and $C \to A$. Describe how to carry out each process experimentally. (f) Find Q, W, and ΔE_{int} for each of the processes. (g) For the whole cycle $A \to B \to C \to A$, find Q, W, and ΔE_{int}.

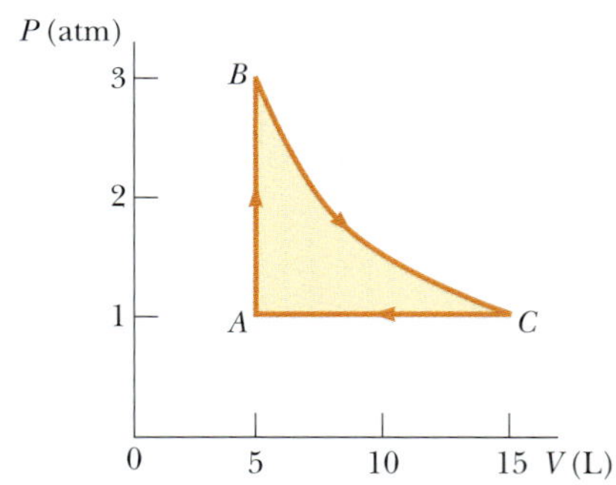

Figure P21.59

60. This problem can help you to think about the size of molecules. In Beijing, a restaurant keeps a pot of chicken broth simmering continuously. Every morning it is topped up to contain 10.0 L of water, along with a fresh chicken, vegetables, and spices. The soup is thoroughly stirred. The molar mass of water is 18.0 g/mol. (a) Find the number of molecules of water in the pot. (b) During a certain month, 90.0% of the broth was served each day to people who then emigrated immediately. Of the water molecules in the pot on the first day of the month, when was the last one likely to have been ladled out of the pot? (c) The broth has been simmering for centuries, through wars, earthquakes, and stove repairs. Suppose the water that was in the pot long ago has thoroughly mixed into the Earth's hydrosphere, of mass 1.32×10^{21} kg. How many of the water molecules originally in the pot are likely to be present in it again today?

61. **Review problem.** (a) If it has enough kinetic energy, a molecule at the surface of the Earth can "escape the Earth's gravitation" in the sense that it can continue to move away from the Earth forever as discussed in Section 13.6. Using the principle of conservation of energy, show that the minimum kinetic energy needed for "escape" is $m_0 g R_E$, where m_0 is the mass of the molecule, g is the free-fall acceleration at the surface, and R_E is the radius of the Earth. (b) Calculate the temperature for which the minimum escape kinetic energy is ten times the average kinetic energy of an oxygen molecule.

62. Using multiple laser beams, physicists have been able to cool and trap sodium atoms in a small region. In one experiment, the temperature of the atoms was reduced to 0.240 mK. (a) Determine the rms speed of the sodium atoms at this temperature. The atoms can be trapped for about 1.00 s. The trap has a linear dimension of roughly 1.00 cm. (b) Over what approximate time interval would an atom wander out of the trap region if there were no trapping action?

Answers to Quick Quizzes

21.1 **(i)**, (b). The average translational kinetic energy per molecule is a function only of temperature. **(ii)**, (a). Because there are twice as many molecules and the temperature of both containers is the same, the total energy in B is twice that in A.

21.2 **(i)**, (a). According to Equation 21.10, E_{int} is a function of temperature only. Because the temperature increases, the internal energy increases. **(ii)**, (c). Along an isotherm, T is constant by definition. Therefore, the internal energy of the gas does not change.

21.3 (d). The value of 29.1 J/mol · K is $7R/2$. According to Figure 21.7, this result suggests that all three types of motion are occurring.

21.4 (c). The highest possible value of C_V for a diatomic gas is $7R/2$, so the gas must be polyatomic.

This computer artwork shows the interior of an automobile engine cylinder at the moment the spark plug (upper left) fires and ignites the air-fuel mixture. The expanding gases push downward on the piston (lower right), ultimately resulting in energy provided to the drive wheels of the automobile. An automobile engine is one example of a heat engine, which we study in this chapter. (Roger Harris/Science Photo Library)

22 Heat Engines, Entropy, and the Second Law of Thermodynamics

J.-L. Charmet/SPL/Photo Researchers, Inc.

LORD KELVIN
British physicist and mathematician (1824–1907)
Born William Thomson in Belfast, Kelvin was the first to propose the use of an absolute scale of temperature. The Kelvin temperature scale is named in his honor. Kelvin's work in thermodynamics led to the idea that energy cannot pass spontaneously from a colder object to a hotter object.

The first law of thermodynamics, which we studied in Chapter 20, is a statement of conservation of energy. This law states that a change in internal energy in a system can occur as a result of energy transfer by heat, by work, or by both. Although the first law of thermodynamics is very important, it makes no distinction between processes that occur spontaneously and those that do not. Only certain types of energy-conversion and energy-transfer processes actually take place in nature, however. The *second law of thermodynamics,* the major topic in this chapter, establishes which processes do and do not occur. The following are examples of processes that do not violate the first law of thermodynamics if they proceed in either direction, but are observed in reality to proceed in only one direction:

- When two objects at different temperatures are placed in thermal contact with each other, the net transfer of energy by heat is always from the warmer object to the cooler object, never from the cooler to the warmer.
- A rubber ball dropped to the ground bounces several times and eventually comes to rest, but a ball lying on the ground never gathers internal energy from the ground and begins bouncing on its own.
- An oscillating pendulum eventually comes to rest because of collisions with air molecules and friction at the point of suspension. The mechanical energy of the system is converted to internal energy in the air, the pendulum, and the suspension; the reverse conversion of energy never occurs.

All these processes are *irreversible;* that is, they are processes that occur naturally in one direction only. No irreversible process has ever been observed to run backward. If it were to do so, it would violate the second law of thermodynamics.[1]

22.1 Heat Engines and the Second Law of Thermodynamics

A **heat engine** is a device that takes in energy by heat[2] and, operating in a cyclic process, expels a fraction of that energy by means of work. For instance, in a typical process by which a power plant produces electricity, a fuel such as coal is burned and the high-temperature gases produced are used to convert liquid water to steam. This steam is directed at the blades of a turbine, setting it into rotation. The mechanical energy associated with this rotation is used to drive an electric generator. Another device that can be modeled as a heat engine is the internal combustion engine in an automobile. This device uses energy from a burning fuel to perform work on pistons that results in the motion of the automobile.

A heat engine carries some working substance through a cyclic process during which (1) the working substance absorbs energy by heat from a high-temperature energy reservoir, (2) work is done by the engine, and (3) energy is expelled by heat to a lower-temperature reservoir. As an example, consider the operation of a steam engine (Fig. 22.1), which uses water as the working substance. The water in a boiler absorbs energy from burning fuel and evaporates to steam, which then does work by expanding against a piston. After the steam cools and condenses, the liquid water produced returns to the boiler and the cycle repeats.

It is useful to represent a heat engine schematically as in Active Figure 22.2. The engine absorbs a quantity of energy $|Q_h|$ from the hot reservoir. For the mathematical discussion of heat engines, we use absolute values to make all energy transfers positive and the direction of transfer is indicated with an explicit positive or negative sign. The engine does work W_{eng} (so that *negative* work $W = -W_{\text{eng}}$ is done *on* the engine) and then gives up a quantity of energy $|Q_c|$ to the cold reservoir. Because the working substance goes through a cycle, its initial and final internal energies are equal: $\Delta E_{\text{int}} = 0$. Hence, from the first law of thermodynamics, $\Delta E_{\text{int}} = Q + W = Q - W_{\text{eng}} = 0$, and **the net work W_{eng} done by a heat engine is**

Figure 22.1 This steam-driven locomotive runs from Durango to Silverton, Colorado. It obtains its energy by burning wood or coal. The generated energy vaporizes water into steam, which powers the locomotive. (This locomotive must take on water from tanks located along the route to replace steam lost through the funnel.) Modern locomotives use diesel fuel instead of wood or coal. Whether old-fashioned or modern, such locomotives can be modeled as heat engines, which extract energy from a burning fuel and convert a fraction of it to mechanical energy.

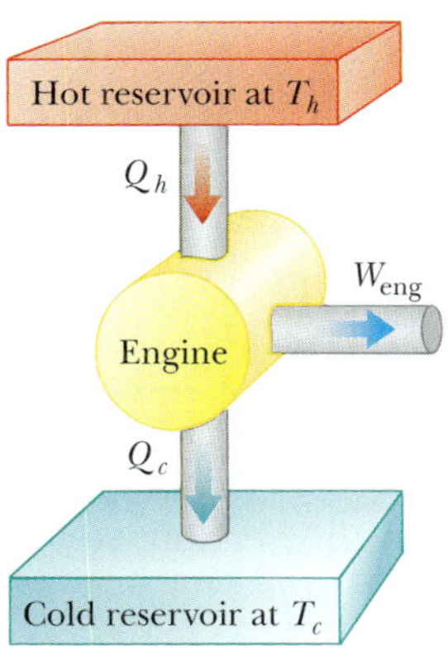

ACTIVE FIGURE 22.2

Schematic representation of a heat engine. The engine does work W_{eng}. The arrow at the top represents energy $Q_h > 0$ entering the engine. At the bottom, $Q_c < 0$ represents energy leaving the engine.

Sign in at www.thomsonedu.com and go to ThomsonNOW to select the efficiency of the engine and observe the transfer of energy.

PITFALL PREVENTION 22.1

The First and Second Laws

Notice the distinction between the first and second laws of thermodynamics. If a gas undergoes a *one-time isothermal process,* then $\Delta E_{\text{int}} = Q + W = 0$ and $W = -Q$. Therefore, the first law allows *all* energy input by heat to be expelled by work. In a heat engine, however, in which a substance undergoes a *cyclic* process, only a *portion* of the energy input by heat can be expelled by work according to the second law.

[1] Although a process occurring in the time-reversed sense has never been *observed,* it is *possible* for it to occur. As we shall see later in this chapter, however, the probability of such a process occurring is infinitesimally small. From this viewpoint, processes occur with a vastly greater probability in one direction than in the opposite direction.

[2] We use heat as our model for energy transfer into a heat engine. Other methods of energy transfer are possible in the model of a heat engine, however. For example, the Earth's atmosphere can be modeled as a heat engine in which the input energy transfer is by means of electromagnetic radiation from the Sun. The output of the atmospheric heat engine causes the wind structure in the atmosphere.

equal to the net energy Q_{net} transferred to it. As you can see from Active Figure 22.2, $Q_{net} = |Q_h| - |Q_c|$; therefore,

$$W_{eng} = |Q_h| - |Q_c| \tag{22.1}$$

The **thermal efficiency** e of a heat engine is defined as the ratio of the net work done by the engine during one cycle to the energy input at the higher temperature during the cycle:

Thermal efficiency of a heat engine ▶

$$e \equiv \frac{W_{eng}}{|Q_h|} = \frac{|Q_h| - |Q_c|}{|Q_h|} = 1 - \frac{|Q_c|}{|Q_h|} \tag{22.2}$$

You can think of the efficiency as the ratio of what you gain (work) to what you give (energy transfer at the higher temperature). In practice, all heat engines expel only a fraction of the input energy Q_h by mechanical work; consequently, their efficiency is always less than 100%. For example, a good automobile engine has an efficiency of about 20%, and diesel engines have efficiencies ranging from 35% to 40%.

Equation 22.2 shows that a heat engine has 100% efficiency ($e = 1$) only if $|Q_c| = 0$, that is, if no energy is expelled to the cold reservoir. In other words, a heat engine with perfect efficiency would have to expel all the input energy by work. Because efficiencies of real engines are well below 100%, the **Kelvin–Planck form of the second law of thermodynamics** states the following:

> It is impossible to construct a heat engine that, operating in a cycle, produces no effect other than the input of energy by heat from a reservoir and the performance of an equal amount of work.

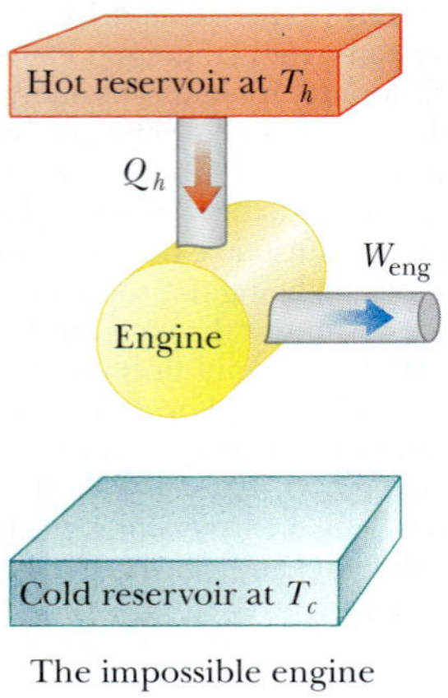

Figure 22.3 Schematic diagram of a heat engine that takes in energy from a hot reservoir and does an equivalent amount of work. It is impossible to construct such a perfect engine.

This statement of the second law means that during the operation of a heat engine, W_{eng} can never be equal to $|Q_h|$ or, alternatively, that some energy $|Q_c|$ must be rejected to the environment. Figure 22.3 is a schematic diagram of the impossible "perfect" heat engine.

Quick Quiz 22.1 The energy input to an engine is 3.00 times greater than the work it performs. **(i)** What is its thermal efficiency? (a) 3.00 (b) 1.00 (c) 0.333 (d) impossible to determine **(ii)** What fraction of the energy input is expelled to the cold reservoir? (a) 0.333 (b) 0.667 (c) 1.00 (d) impossible to determine

---

EXAMPLE 22.1 The Efficiency of an Engine

An engine transfers 2.00×10^3 J of energy from a hot reservoir during a cycle and transfers 1.50×10^3 J as exhaust to a cold reservoir.

(A) Find the efficiency of the engine.

SOLUTION

Conceptualize Review Active Figure 22.2; think about energy going into the engine from the hot reservoir and splitting, with part coming out by work and part by heat into the cold reservoir.

Categorize This example involves evaluation of quantities from the equations introduced in this section, so we categorize it as a substitution problem.

Find the efficiency of the engine from Equation 22.2:

$$e = 1 - \frac{|Q_c|}{|Q_h|} = 1 - \frac{1.50 \times 10^3\ \text{J}}{2.00 \times 10^3\ \text{J}} = 0.250, \text{ or } 25.0\%$$

(B) How much work does this engine do in one cycle?

SOLUTION

Find the work done by the engine by taking the difference between the input and output energies:

$$W_{eng} = |Q_h| - |Q_c| = 2.00 \times 10^3\ \text{J} - 1.50 \times 10^3\ \text{J}$$
$$= 5.0 \times 10^2\ \text{J}$$

What If? Suppose you were asked for the power output of this engine. Do you have sufficient information to answer this question?

Answer No, you do not have enough information. The power of an engine is the *rate* at which work is done by the engine. You know how much work is done per cycle, but you have no information about the time interval associated with one cycle. If you were told that the engine operates at 2 000 rpm (revolutions per minute), however, you could relate this rate to the period of rotation T of the mechanism of the engine. Assuming there is one thermodynamic cycle per revolution, the power is

$$\mathcal{P} = \frac{W_{\text{eng}}}{T} = \frac{5.0 \times 10^2\ \text{J}}{\left(\frac{1}{2\,000}\ \text{min}\right)}\left(\frac{1\ \text{min}}{60\ \text{s}}\right) = 1.7 \times 10^4\ \text{W}$$

22.2 Heat Pumps and Refrigerators

In a heat engine, the direction of energy transfer is from the hot reservoir to the cold reservoir, which is the natural direction. The role of the heat engine is to process the energy from the hot reservoir so as to do useful work. What if we wanted to transfer energy from the cold reservoir to the hot reservoir? Because that is not the natural direction of energy transfer, we must put some energy into a device to be successful. Devices that perform this task are called **heat pumps** and **refrigerators**. For example, homes in summer are cooled using heat pumps called *air conditioners*. The air conditioner transfers energy from the cool room in the home to the warm air outside.

In a refrigerator or a heat pump, the engine takes in energy $|Q_c|$ from a cold reservoir and expels energy $|Q_h|$ to a hot reservoir (Active Fig. 22.4), which can be accomplished only if work is done *on* the engine. From the first law, we know that the energy given up to the hot reservoir must equal the sum of the work done and the energy taken in from the cold reservoir. Therefore, the refrigerator or heat pump transfers energy from a colder body (for example, the contents of a kitchen refrigerator or the winter air outside a building) to a hotter body (the air in the kitchen or a room in the building). In practice, it is desirable to carry out this process with a minimum of work. If the process could be accomplished without doing any work, the refrigerator or heat pump would be "perfect" (Fig. 22.5). Again, the existence of such a device would be in violation of the second law of thermodynamics, which in the form of the **Clausius statement**[3] states:

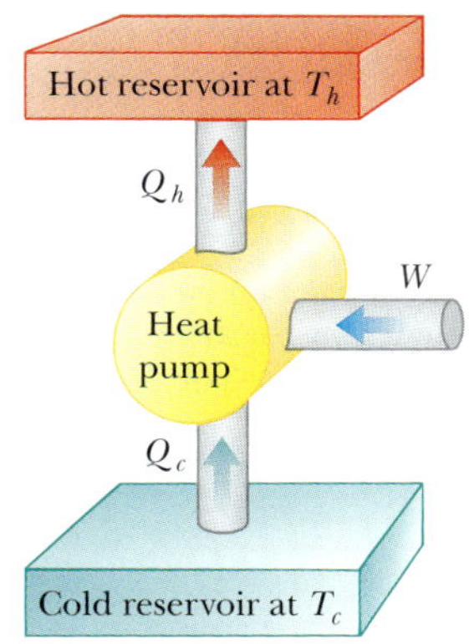

ACTIVE FIGURE 22.4

Schematic diagram of a heat pump, which takes in energy $Q_c > 0$ from a cold reservoir and expels energy $Q_h < 0$ to a hot reservoir. Work W is done *on* the heat pump. A refrigerator works the same way.

Sign in at www.thomsonedu.com and go to ThomsonNOW to select the COP of the heat pump and observe the transfer of energy.

> It is impossible to construct a cyclical machine whose sole effect is to transfer energy continuously by heat from one object to another object at a higher temperature without the input of energy by work.

In simpler terms, **energy does not transfer spontaneously by heat from a cold object to a hot object.**

The Clausius and Kelvin–Planck statements of the second law of thermodynamics appear at first sight to be unrelated, but in fact they are equivalent in all respects. Although we do not prove so here, if either statement is false, so is the other.[4]

In practice, a heat pump includes a circulating fluid that passes through two sets of metal coils that can exchange energy with the surroundings. The fluid is cold and at low pressure when it is in the coils located in a cool environment, where it absorbs energy by heat. The resulting warm fluid is then compressed and enters the other coils as a hot, high-pressure fluid. There it releases its stored energy to the warm surroundings. In an air conditioner, energy is absorbed into the fluid in coils located in a building's interior; after the fluid is compressed, energy leaves the fluid through coils located outdoors. In a refrigerator, the external coils are behind or

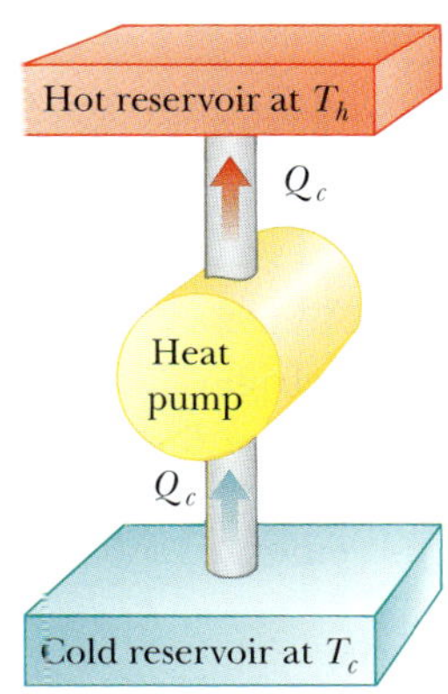

Impossible heat pump

Figure 22.5 Schematic diagram of an impossible heat pump or refrigerator, that is, one that takes in energy from a cold reservoir and expels an equivalent amount of energy to a hot reservoir without the input of energy by work.

[3] First expressed by Rudolf Clausius (1822–1888).

[4] See an advanced textbook on thermodynamics for this proof.

Charles D. Winters

Figure 22.6 The coils on the back of a refrigerator transfer energy by heat to the air. Due to the input of energy by work, this amount of energy must be greater than the amount of energy removed from the contents of the refrigerator.

underneath the unit (Fig. 22.6). The internal coils are in the walls of the refrigerator and absorb energy from the food.

The effectiveness of a heat pump is described in terms of a number called the **coefficient of performance** (COP). The COP is similar to the thermal efficiency for a heat engine in that it is a ratio of what you gain (energy transferred to or from a reservoir) to what you give (work input). For a heat pump operating in the cooling mode, "what you gain" is energy removed from the cold reservoir. The most effective refrigerator or air conditioner is one that removes the greatest amount of energy from the cold reservoir in exchange for the least amount of work. Therefore, for these devices operating in the cooling mode, we define the COP in terms of $|Q_c|$:

$$\text{COP (cooling mode)} = \frac{|Q_c|}{W} \tag{22.3}$$

A good refrigerator should have a high COP, typically 5 or 6.

In addition to cooling applications, heat pumps are becoming increasingly popular for heating purposes. The energy-absorbing coils for a heat pump are located outside a building, in contact with the air or buried in the ground. The other set of coils are in the building's interior. The circulating fluid flowing through the coils absorbs energy from the outside and releases it to the interior of the building from the interior coils.

In the heating mode, the COP of a heat pump is defined as the ratio of the energy transferred to the hot reservoir to the work required to transfer that energy:

$$\text{COP (heating mode)} = \frac{\text{energy transferred at high temperature}}{\text{work done on heat pump}} = \frac{|Q_h|}{W} \tag{22.4}$$

If the outside temperature is 25°F (−4°C) or higher, a typical value of the COP for a heat pump is about 4. That is, the amount of energy transferred to the building is about four times greater than the work done by the motor in the heat pump. As the outside temperature decreases, however, it becomes more difficult for the heat pump to extract sufficient energy from the air and so the COP decreases. Therefore, the use of heat pumps that extract energy from the air, although satisfactory in moderate climates, is not appropriate in areas where winter temperatures are very low. It is possible to use heat pumps in colder areas by burying the external coils deep in the ground. In that case, the energy is extracted from the ground, which tends to be warmer than the air in the winter.

Quick Quiz 22.2 The energy entering an electric heater by electrical transmission can be converted to internal energy with an efficiency of 100%. By what factor does the cost of heating your home change when you replace your electric heating system with an electric heat pump that has a COP of 4.00? Assume the motor running the heat pump is 100% efficient. (a) 4.00 (b) 2.00 (c) 0.500 (d) 0.250

EXAMPLE 22.2 Freezing Water

A certain refrigerator has a COP of 5.00. When the refrigerator is running, its power input is 500 W. A sample of water of mass 500 g and temperature 20.0°C is placed in the freezer compartment. How long does it take to freeze the water to ice at 0°C? Assume all other parts of the refrigerator stay at the same temperature and there is no leakage of energy from the exterior, so the operation of the refrigerator results only in energy being extracted from the water.

SOLUTION

Conceptualize Energy leaves the water, reducing its temperature and then freezing it into ice. The time interval required for this entire process is related to the rate at which energy is withdrawn from the water, which, in turn, is related to the power input of the refrigerator.

Categorize We categorize this example as one that combines our understanding of temperature changes and phase changes from Chapter 20 and our understanding of heat pumps from this chapter.

Analyze Using Equations 20.4 and 20.7, find the amount of energy that must be extracted from 500 g of water at 20°C to turn it into ice at 0°C:

$$|Q_c| = |mc\,\Delta T - mL_f| = m|c\,\Delta T - L_f|$$
$$= |(0.500\text{ kg})[(4\,186\text{ J/kg}\cdot{}^\circ\text{C})(-20.0^\circ\text{C}) - 3.33\times10^5\text{ J/kg}]|$$
$$= 2.08\times10^5\text{ J}$$

Use Equation 22.3 to find how much energy must be provided to the refrigerator to extract this much energy from the water:

$$\text{COP} = \frac{|Q_c|}{W} \rightarrow W = \frac{|Q_c|}{\text{COP}} = \frac{2.08\times10^5\text{ J}}{5.00}$$
$$W = 4.17\times10^4\text{ J}$$

Use the power rating of the refrigerator to find the time interval required for the freezing process to occur:

$$\mathcal{P} = \frac{W}{\Delta t} \rightarrow \Delta t = \frac{W}{\mathcal{P}} = \frac{4.17\times10^4\text{ J}}{500\text{ W}} = 83.3\text{ s}$$

Finalize In reality, the time interval for the water to freeze in a refrigerator is much longer than 83.3 s, which suggests that the assumptions of our model are not valid. Only a small part of the energy extracted from the refrigerator interior in a given time interval comes from the water. Energy must also be extracted from the container in which the water is placed, and energy that continuously leaks into the interior from the exterior must be extracted.

22.3 Reversible and Irreversible Processes

In the next section, we will discuss a theoretical heat engine that is the most efficient possible. To understand its nature, we must first examine the meaning of reversible and irreversible processes. In a **reversible** process, the system undergoing the process can be returned to its initial conditions along the same path on a *PV* diagram, and every point along this path is an equilibrium state. A process that does not satisfy these requirements is **irreversible.**

PITFALL PREVENTION 22.2
All Real Processes Are Irreversible
The reversible process is an idealization; all real processes on the Earth are irreversible.

All natural processes are known to be irreversible. Let's examine the adiabatic free expansion of a gas, which was already discussed in Section 20.6, and show that it cannot be reversible. Consider a gas in a thermally insulated container as shown in Figure 22.7. A membrane separates the gas from a vacuum. When the membrane is punctured, the gas expands freely into the vacuum. As a result of the puncture, the system has changed because it occupies a greater volume after the expansion. Because the gas does not exert a force through a displacement, it does no work on the surroundings as it expands. In addition, no energy is transferred to or from the gas by heat because the container is insulated from its surroundings. Therefore, in this adiabatic process, the system has changed but the surroundings have not.

For this process to be reversible, we must return the gas to its original volume and temperature without changing the surroundings. Imagine trying to reverse the process by compressing the gas to its original volume. To do so, we fit the container with a piston and use an engine to force the piston inward. During this process, the surroundings change because work is being done by an outside agent on the system. In addition, the system changes because the compression increases the temperature of the gas. The temperature of the gas can be lowered by allowing it to come into contact with an external energy reservoir. Although this step returns the gas to its original conditions, the surroundings are again affected because energy is being added to the surroundings from the gas. If this energy could be used to drive the engine that compressed the gas, the net energy transfer to the surroundings would be zero. In this way, the system and its surroundings could be returned to their initial conditions and we could identify the process as reversible. The Kelvin–Planck statement of the second law, however, specifies that the energy removed from the gas to return the temperature to its original value cannot be completely converted to mechanical energy in the form of the work done by the engine in compressing the gas. Therefore, we must conclude that the process is irreversible.

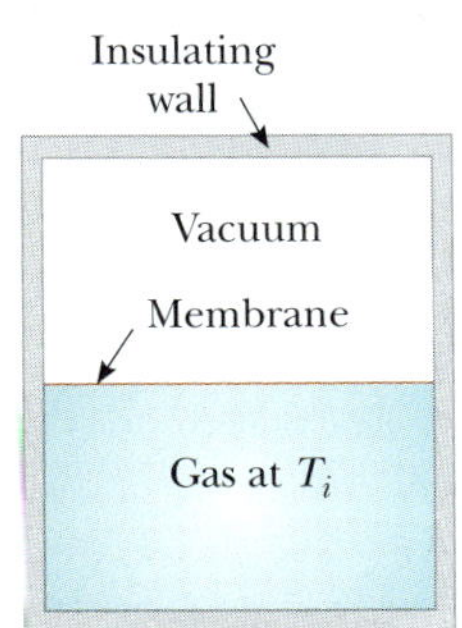

Figure 22.7 Adiabatic free expansion of a gas.

We could also argue that the adiabatic free expansion is irreversible by relying on the portion of the definition of a reversible process that refers to equilibrium

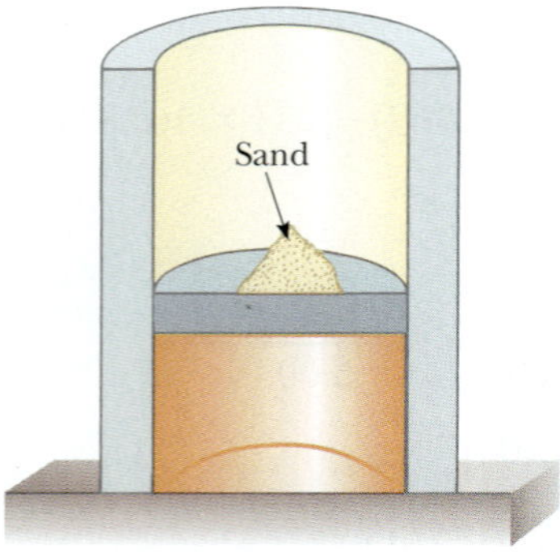

Figure 22.8 A gas in thermal contact with an energy reservoir is compressed slowly as individual grains of sand drop onto the piston. The compression is isothermal and reversible.

states. For example, during the sudden expansion, significant variations in pressure occur throughout the gas. Therefore, there is no well-defined value of the pressure for the entire system at any time between the initial and final states. In fact, the process cannot even be represented as a path on a *PV* diagram. The *PV* diagram for an adiabatic free expansion would show the initial and final conditions as points, but these points would not be connected by a path. Therefore, because the intermediate conditions between the initial and final states are not equilibrium states, the process is irreversible.

Although all real processes are irreversible, some are almost reversible. If a real process occurs very slowly such that the system is always very nearly in an equilibrium state, the process can be approximated as being reversible. Suppose a gas is compressed isothermally in a piston-cylinder arrangement in which the gas is in thermal contact with an energy reservoir and we continuously transfer just enough energy from the gas to the reservoir to keep the temperature constant. For example, imagine that the gas is compressed very slowly by dropping grains of sand onto a frictionless piston as shown in Figure 22.8. As each grain lands on the piston and compresses the gas a small amount, the system deviates from an equilibrium state, but it is so close to one that it achieves a new equilibrium state in a relatively short time interval. Each grain added represents a change to a new equilibrium state, but the differences between states are so small that the entire process can be approximated as occurring through continuous equilibrium states. The process can be reversed by slowly removing grains from the piston.

A general characteristic of a reversible process is that no dissipative effects (such as turbulence or friction) that convert mechanical energy to internal energy can be present. Such effects can be impossible to eliminate completely. Hence, it is not surprising that real processes in nature are irreversible.

22.4 The Carnot Engine

In 1824, a French engineer named Sadi Carnot described a theoretical engine, now called a **Carnot engine**, that is of great importance from both practical and theoretical viewpoints. He showed that a heat engine operating in an ideal, reversible cycle—called a **Carnot cycle**—between two energy reservoirs is the most efficient engine possible. Such an ideal engine establishes an upper limit on the efficiencies of all other engines. That is, the net work done by a working substance taken through the Carnot cycle is the greatest amount of work possible for a given amount of energy supplied to the substance at the higher temperature. **Carnot's theorem** can be stated as follows:

> No real heat engine operating between two energy reservoirs can be more efficient than a Carnot engine operating between the same two reservoirs.

SADI CARNOT
French engineer (1796–1832)
Carnot was the first to show the quantitative relationship between work and heat. In 1824, he published his only work, *Reflections on the Motive Power of Heat,* which reviewed the industrial, political, and economic importance of the steam engine. In it, he defined work as "weight lifted through a height."

To prove the validity of this theorem, imagine two heat engines operating between the *same* energy reservoirs. One is a Carnot engine with efficiency e_C, and the other is an engine with efficiency e, where we assume $e > e_C$. Because the cycle in the Carnot engine is reversible, the engine can operate in reverse as a refrigerator. The more efficient engine is used to drive the Carnot engine as a Carnot refrigerator. The output by work of the more efficient engine is matched to the input by work of the Carnot refrigerator. For the *combination* of the engine and refrigerator, no exchange by work with the surroundings occurs. Because we have assumed the engine is more efficient than the refrigerator, the net result of the combination is a transfer of energy from the cold to the hot reservoir without work being done on the combination. According to the Clausius statement of the second law, this process is impossible. Hence, the assumption that $e > e_C$ must be false. **All real engines are less efficient than the Carnot engine because they do not operate through a reversible cycle**. The efficiency of a real engine is further reduced by such practical difficulties as friction and energy losses by conduction.

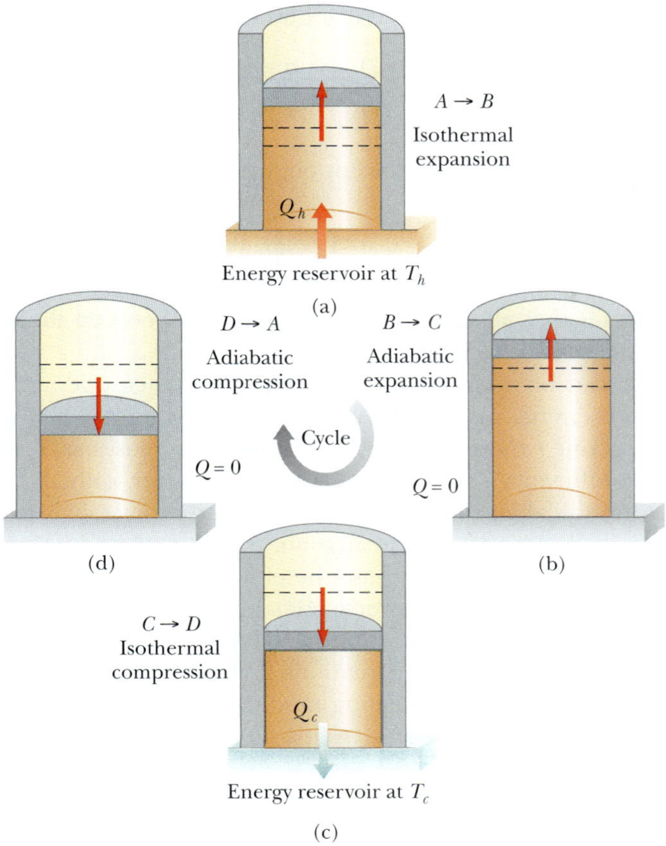

ACTIVE FIGURE 22.9

The Carnot cycle. (a) In process $A \rightarrow B$, the gas expands isothermally while in contact with a reservoir at T_h. (b) In process $B \rightarrow C$, the gas expands adiabatically ($Q = 0$). (c) In process $C \rightarrow D$, the gas is compressed isothermally while in contact with a reservoir at $T_c < T_h$. (d) In process $D \rightarrow A$, the gas is compressed adiabatically. The arrows on the piston indicate the direction of its motion during each process.

Sign in at www.thomsonedu.com and go to ThomsonNOW to observe the motion of the piston in the Carnot cycle while you also observe the cycle on the *PV* diagram of Active Figure 22.10.

PITFALL PREVENTION 22.3

Don't Shop for a Carnot Engine

The Carnot engine is an idealization; do not expect a Carnot engine to be developed for commercial use. We explore the Carnot engine only for theoretical considerations.

To describe the Carnot cycle taking place between temperatures T_c and T_h, let's assume the working substance is an ideal gas contained in a cylinder fitted with a movable piston at one end. The cylinder's walls and the piston are thermally nonconducting. Four stages of the Carnot cycle are shown in Active Figure 22.9, and the *PV* diagram for the cycle is shown in Active Figure 22.10. The Carnot cycle consists of two adiabatic processes and two isothermal processes, all reversible:

1. Process $A \rightarrow B$ (Active Fig. 22.9a) is an isothermal expansion at temperature T_h. The gas is placed in thermal contact with an energy reservoir at temperature T_h. During the expansion, the gas absorbs energy $|Q_h|$ from the reservoir through the base of the cylinder and does work W_{AB} in raising the piston.
2. In process $B \rightarrow C$ (Active Fig. 22.9b), the base of the cylinder is replaced by a thermally nonconducting wall and the gas expands adiabatically; that is, no energy enters or leaves the system by heat. During the expansion, the temperature of the gas decreases from T_h to T_c and the gas does work W_{BC} in raising the piston.
3. In process $C \rightarrow D$ (Active Fig. 22.9c), the gas is placed in thermal contact with an energy reservoir at temperature T_c and is compressed isothermally at temperature T_c. During this time, the gas expels energy $|Q_c|$ to the reservoir and the work done by the piston on the gas is W_{CD}.
4. In the final process $D \rightarrow A$ (Active Fig. 22.9d), the base of the cylinder is replaced by a nonconducting wall and the gas is compressed adiabatically. The temperature of the gas increases to T_h, and the work done by the piston on the gas is W_{DA}.

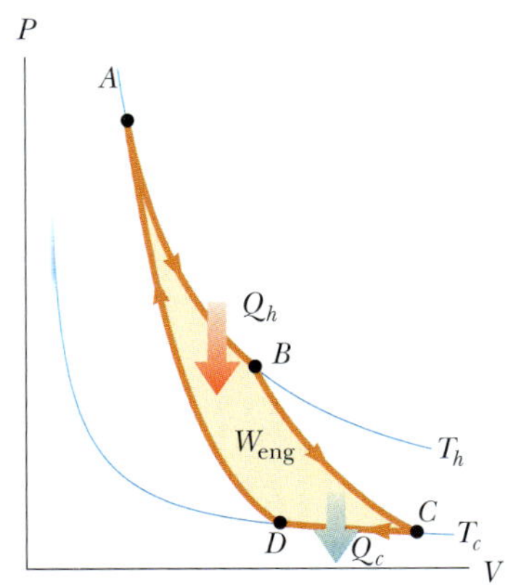

ACTIVE FIGURE 22.10

PV diagram for the Carnot cycle. The net work done W_{eng} equals the net energy transferred into the Carnot engine in one cycle, $|Q_h| - |Q_c|$. As with any cycle, the work done during the cycle is the area enclosed by the path on the *PV* diagram. Notice that $\Delta E_{int} = 0$ for the cycle.

Sign in at www.thomsonedu.com and go to ThomsonNOW to observe the Carnot cycle on the *PV* diagram while you also observe the motion of the piston in Active Figure 22.9.

The thermal efficiency of the engine is given by Equation 22.2:

$$e = \frac{W_{eng}}{|Q_h|} = \frac{|Q_h| - |Q_c|}{|Q_h|} = 1 - \frac{|Q_c|}{|Q_h|}$$

In Example 22.3, we show that for a Carnot cycle,

$$\frac{|Q_c|}{|Q_h|} = \frac{T_c}{T_h} \tag{22.5}$$

Hence, the thermal efficiency of a Carnot engine is

Efficiency of a Carnot engine ▶

$$e_C = 1 - \frac{T_c}{T_h} \tag{22.6}$$

This result indicates that **all Carnot engines operating between the same two temperatures have the same efficiency.**[5]

Equation 22.6 can be applied to any working substance operating in a Carnot cycle between two energy reservoirs. According to this equation, the efficiency is zero if $T_c = T_h$, as one would expect. The efficiency increases as T_c is lowered and T_h is raised. The efficiency can be unity (100%), however, only if $T_c = 0$ K. Such reservoirs are not available; therefore, the maximum efficiency is always less than 100%. In most practical cases, T_c is near room temperature, which is about 300 K. Therefore, one usually strives to increase the efficiency by raising T_h. Theoretically, a Carnot-cycle heat engine run in reverse constitutes the most effective heat pump possible, and it determines the maximum COP for a given combination of hot and cold reservoir temperatures. Using Equations 22.1 and 22.4, we see that the maximum COP for a heat pump in its heating mode is

$$\text{COP}_\text{C}\ (\text{heating mode}) = \frac{|Q_h|}{W}$$

$$= \frac{|Q_h|}{|Q_h| - |Q_c|} = \frac{1}{1 - \dfrac{|Q_c|}{|Q_h|}} = \frac{1}{1 - \dfrac{T_c}{T_h}} = \frac{T_h}{T_h - T_c}$$

The Carnot COP for a heat pump in the cooling mode is

$$\text{COP}_\text{C}\ (\text{cooling mode}) = \frac{T_c}{T_h - T_c}$$

As the difference between the temperatures of the two reservoirs approaches zero in this expression, the theoretical COP approaches infinity. In practice, the low temperature of the cooling coils and the high temperature at the compressor limit the COP to values below 10.

Quick Quiz 22.3 Three engines operate between reservoirs separated in temperature by 300 K. The reservoir temperatures are as follows: Engine A: T_h = 1 000 K, T_c = 700 K; Engine B: T_h = 800 K, T_c = 500 K; Engine C: T_h = 600 K, T_c = 300 K. Rank the engines in order of theoretically possible efficiency from highest to lowest.

---

[5] For the processes in the Carnot cycle to be reversible, they must be carried out infinitesimally slowly. Therefore, although the Carnot engine is the most efficient engine possible, it has zero power output because it takes an infinite time interval to complete one cycle! For a real engine, the short time interval for each cycle results in the working substance reaching a high temperature lower than that of the hot reservoir and a low temperature higher than that of the cold reservoir. An engine undergoing a Carnot cycle between this narrower temperature range was analyzed by F. L. Curzon and B. Ahlborn (*Am. J. Phys.* **43**(1), 22, 1975), who found that the efficiency at maximum power output depends only on the reservoir temperatures T_c and T_h and is given by $e_{\text{C-A}} = 1 - (T_c/T_h)^{1/2}$. The Curzon–Ahlborn efficiency $e_{\text{C-A}}$ provides a closer approximation to the efficiencies of real engines than does the Carnot efficiency.

EXAMPLE 22.3 Efficiency of the Carnot Engine

Show that the ratio of energy transfers by heat in a Carnot engine is equal to the ratio of reservoir temperatures, as given by Equation 22.5.

SOLUTION

Conceptualize Make use of Active Figures 22.9 and 22.10 to help you visualize the processes in the Carnot cycle.

Categorize Because of our understanding of the Carnot cycle, we can categorize the processes in the cycle as isothermal and adiabatic.

Analyze For the isothermal expansion (process $A \rightarrow B$ in Active Fig. 22.9), find the energy transfer by heat from the hot reservoir using Equation 20.14 and the first law of thermodynamics:

$$|Q_h| = |\Delta E_{\text{int}} - W_{AB}| = |0 - W_{AB}| = W_{AB} = nRT_h \ln \frac{V_B}{V_A}$$

In a similar manner, find the energy transfer to the cold reservoir during the isothermal compression $C \rightarrow D$:

$$|Q_c| = |\Delta E_{\text{int}} - W_{CD}| = |0 - W_{CD}| = W_{CD} = nRT_c \ln \frac{V_C}{V_D}$$

Divide the second expression by the first:

$$(1) \quad \frac{|Q_c|}{|Q_h|} = \frac{T_c}{T_h} \frac{\ln (V_C/V_D)}{\ln (V_B/V_A)}$$

Apply Equation 21.20 to the adiabatic processes $B \rightarrow C$ and $D \rightarrow A$:

$$T_h V_B^{\gamma-1} = T_c V_C^{\gamma-1}$$

$$T_h V_A^{\gamma-1} = T_c V_D^{\gamma-1}$$

Divide the first equation by the second:

$$\left(\frac{V_B}{V_A}\right)^{\gamma-1} = \left(\frac{V_C}{V_D}\right)^{\gamma-1}$$

$$(2) \quad \frac{V_B}{V_A} = \frac{V_C}{V_D}$$

Substitute Equation (2) into Equation (1):

$$\frac{|Q_c|}{|Q_h|} = \frac{T_c}{T_h} \frac{\ln (V_C/V_D)}{\ln (V_B/V_A)} = \frac{T_c}{T_h} \frac{\ln (V_C/V_D)}{\ln (V_C/V_D)} = \frac{T_c}{T_h}$$

Finalize This last equation is Equation 22.5, the one we set out to prove.

EXAMPLE 22.4 The Steam Engine

A steam engine has a boiler that operates at 500 K. The energy from the burning fuel changes water to steam, and this steam then drives a piston. The cold reservoir's temperature is that of the outside air, approximately 300 K. What is the maximum thermal efficiency of this steam engine?

SOLUTION

Conceptualize In a steam engine, the gas pushing on the piston in Active Figure 22.9 is steam. A real steam engine does not operate in a Carnot cycle, but, to find the maximum possible efficiency, imagine a Carnot steam engine.

Categorize We calculate an efficiency using Equation 22.6, so we categorize this example as a substitution problem.

Substitute the reservoir temperatures into Equation 22.6:

$$e_C = 1 - \frac{T_c}{T_h} = 1 - \frac{300 \text{ K}}{500 \text{ K}} = 0.400 \quad \text{or} \quad 40.0\%$$

This result is the highest *theoretical* efficiency of the engine. In practice, the efficiency is considerably lower.

What If? Suppose we wished to increase the theoretical efficiency of this engine. This increase can be achieved by raising T_h by ΔT or by decreasing T_c by the same ΔT. Which would be more effective?

Answer A given ΔT would have a larger fractional effect on a smaller temperature, so you would expect a larger change in efficiency if you alter T_c by ΔT. Let's test that numerically. Raising T_h by 50 K, corresponding to $T_h = 550$ K, would give a maximum efficiency of

$$e_C = 1 - \frac{T_c}{T_h} = 1 - \frac{300\text{ K}}{550\text{ K}} = 0.455$$

Decreasing T_c by 50 K, corresponding to $T_c = 250$ K, would give a maximum efficiency of

$$e_C = 1 - \frac{T_c}{T_h} = 1 - \frac{250\text{ K}}{500\text{ K}} = 0.500$$

Although changing T_c is *mathematically* more effective, often changing T_h is *practically* more feasible.

22.5 Gasoline and Diesel Engines

In a gasoline engine, six processes occur in each cycle; five of them are illustrated in Active Figure 22.11. In this discussion, let's consider the interior of the cylinder above the piston to be the system that is taken through repeated cycles in the engine's operation. For a given cycle, the piston moves up and down twice, which represents a four-stroke cycle consisting of two upstrokes and two downstrokes. The processes in the cycle can be approximated by the **Otto cycle** shown in the *PV* diagram in Active Figure 22.12. In the following discussion, refer to Active Figure 22.11 for the pictorial representation of the strokes and Active Figure 22.12 for the significance on the *PV* diagram of the letter designations below:

1. During the *intake stroke* $O \rightarrow A$ (Active Fig. 22.11a), the piston moves downward and a gaseous mixture of air and fuel is drawn into the cylinder at atmospheric pressure. In this process, the volume increases from V_2 to V_1. That is the energy input part of the cycle: energy enters the system (the interior of the cylinder) by matter transfer as potential energy stored in the fuel.
2. During the *compression stroke* $A \rightarrow B$ (Active Fig. 22.11b), the piston moves upward, the air-fuel mixture is compressed adiabatically from volume V_1 to volume V_2, and the temperature increases from T_A to T_B. The work done on the gas is positive, and its value is equal to the negative of the area under the curve *AB* in Active Figure 22.12.

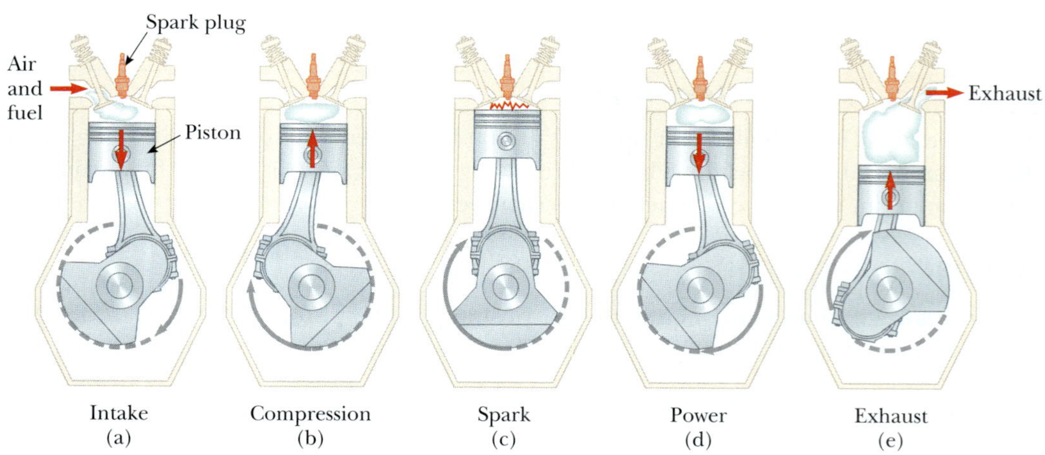

ACTIVE FIGURE 22.11

The four-stroke cycle of a conventional gasoline engine. The arrows on the piston indicate the direction of its motion during each process. (a) In the intake stroke, air and fuel enter the cylinder. (b) The intake valve is then closed, and the air–fuel mixture is compressed by the piston. (c) The mixture is ignited by the spark plug, with the result that the temperature of the mixture increases at essentially constant volume. (d) In the power stroke, the gas expands against the piston. (e) Finally, the residual gases are expelled and the cycle repeats.

Sign in at www.thomsonedu.com and go to ThomsonNOW to observe the motion of the piston and crankshaft while you also observe the cycle on the *PV* diagram of Active Figure 22.12.

3. In process $B \rightarrow C$, combustion occurs when the spark plug fires (Active Fig. 22.11c). That is not one of the strokes of the cycle because it occurs in a very short time interval while the piston is at its highest position. The combustion represents a rapid transformation from potential energy stored in chemical bonds in the fuel to internal energy associated with molecular motion, which is related to temperature. During this time interval, the mixture's pressure and temperature increase rapidly, with the temperature rising from T_B to T_C. The volume, however, remains approximately constant because of the short time interval. As a result, approximately no work is done on or by the gas. We can model this process in the *PV* diagram (Active Fig. 22.12) as that process in which the energy $|Q_h|$ enters the system. (In reality, however, this process is a *conversion* of energy already in the cylinder from process $O \rightarrow A$.)
4. In the *power stroke* $C \rightarrow D$ (Active Fig. 22.11d), the gas expands adiabatically from V_2 to V_1. This expansion causes the temperature to drop from T_C to T_D. Work is done by the gas in pushing the piston downward, and the value of this work is equal to the area under the curve *CD*.
5. In the process $D \rightarrow A$ (not shown in Active Fig. 22.11), an exhaust valve is opened as the piston reaches the bottom of its travel and the pressure suddenly drops for a short time interval. During this time interval, the piston is almost stationary and the volume is approximately constant. Energy is expelled from the interior of the cylinder and continues to be expelled during the next process.
6. In the final process, the *exhaust stroke* $A \rightarrow O$ (Active Fig. 22.11e), the piston moves upward while the exhaust valve remains open. Residual gases are exhausted at atmospheric pressure, and the volume decreases from V_1 to V_2. The cycle then repeats.

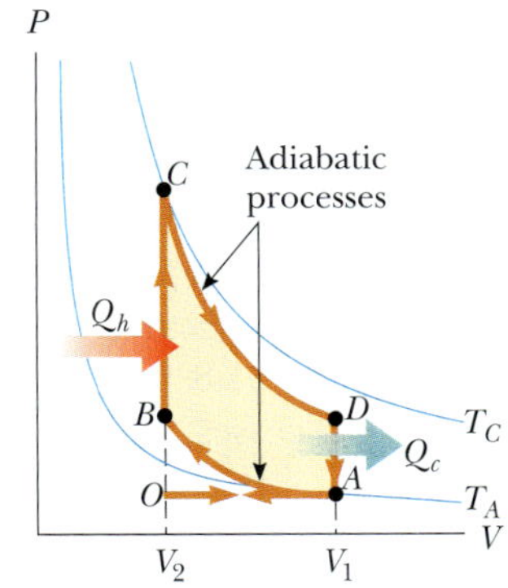

ACTIVE FIGURE 22.12

PV diagram for the Otto cycle, which approximately represents the processes occurring in an internal combustion engine.

Sign in at www.thomsonedu.com and go to ThomsonNOW to observe the Otto cycle on the *PV* diagram while you observe the motion of the piston and crankshaft in Active Figure 22.11.

If the air–fuel mixture is assumed to be an ideal gas, the efficiency of the Otto cycle is

$$e = 1 - \frac{1}{(V_1/V_2)^{\gamma-1}} \quad \text{(Otto cycle)} \qquad \textbf{(22.7)}$$

where V_1/V_2 is the **compression ratio** and γ is the ratio of the molar specific heats C_P/C_V for the fuel–air mixture. Equation 22.7, which is derived in Example 22.5, shows that the efficiency increases as the compression ratio increases. For a typical compression ratio of 8 and with $\gamma = 1.4$, Equation 22.7 predicts a theoretical efficiency of 56% for an engine operating in the idealized Otto cycle. This value is much greater than that achieved in real engines (15% to 20%) because of such effects as friction, energy transfer by conduction through the cylinder walls, and incomplete combustion of the air–fuel mixture.

Diesel engines operate on a cycle similar to the Otto cycle, but they do not employ a spark plug. The compression ratio for a diesel engine is much greater than that for a gasoline engine. Air in the cylinder is compressed to a very small volume, and, as a consequence, the cylinder temperature at the end of the compression stroke is very high. At this point, fuel is injected into the cylinder. The temperature is high enough for the fuel–air mixture to ignite without the assistance of a spark plug. Diesel engines are more efficient than gasoline engines because of their greater compression ratios and resulting higher combustion temperatures.

EXAMPLE 22.5 Efficiency of the Otto Cycle

Show that the thermal efficiency of an engine operating in an idealized Otto cycle (see Active Figs. 22.11 and 22.12) is given by Equation 22.7. Treat the working substance as an ideal gas.

SOLUTION

Conceptualize Study Active Figures 22.11 and 22.12 to make sure you understand the working of the Otto cycle.

Categorize As seen in Active Figure 22.12, we categorize the processes in the Otto cycle as isovolumetric and adiabatic.

Analyze Model the energy input and output as occurring by heat in processes $B \rightarrow C$ and $D \rightarrow A$. (In reality, most of the energy enters and leaves by matter transfer as the fuel–air mixture enters and leaves the cylinder.) Use Equation 21.8 to find the energy transfers by heat for these processes, which take place at constant volume:

$$B \rightarrow C \quad |Q_h| = nC_V(T_C - T_B)$$

$$D \rightarrow A \quad |Q_c| = nC_V(T_D - T_A)$$

Substitute these expressions into Equation 22.2:

$$(1) \quad e = \frac{W_{\text{eng}}}{|Q_h|} = 1 - \frac{|Q_c|}{|Q_h|} = 1 - \frac{T_D - T_A}{T_C - T_B}$$

Apply Equation 21.20 to the adiabatic processes $A \rightarrow B$ and $C \rightarrow D$:

$$A \rightarrow B \quad T_A V_A^{\gamma-1} = T_B V_B^{\gamma-1}$$

$$C \rightarrow D \quad T_C V_C^{\gamma-1} = T_D V_D^{\gamma-1}$$

Solve these equations for the temperatures T_A and T_D, noting that $V_A = V_D = V_1$ and $V_B = V_C = V_2$:

$$(2) \quad T_A = T_B\left(\frac{V_B}{V_A}\right)^{\gamma-1} = T_B\left(\frac{V_2}{V_1}\right)^{\gamma-1}$$

$$(3) \quad T_D = T_C\left(\frac{V_C}{V_D}\right)^{\gamma-1} = T_C\left(\frac{V_2}{V_1}\right)^{\gamma-1}$$

Subtract Equation (2) from Equation (3) and rearrange:

$$(4) \quad \frac{T_D - T_A}{T_C - T_B} = \left(\frac{V_2}{V_1}\right)^{\gamma-1}$$

Substitute Equation (4) into Equation (1):

$$e = 1 - \frac{1}{(V_1/V_2)^{\gamma-1}}$$

Finalize This final expression is Equation 22.7.

PITFALL PREVENTION 22.4

Entropy Is Abstract

Entropy is one of the most abstract notions in physics, so follow the discussion in this and the subsequent sections very carefully. Do not confuse energy with entropy. Even though the names sound similar, they are very different concepts.

22.6 Entropy

The zeroth law of thermodynamics involves the concept of temperature, and the first law involves the concept of internal energy. Temperature and internal energy are both state variables; that is, the value of each depends only on the thermodynamic state of a system, not on the process that brought it to that state. Another state variable—this one related to the second law of thermodynamics—is **entropy** S. In this section, we define entropy on a macroscopic scale as it was first expressed by Clausius in 1865.

Entropy was originally formulated as a useful concept in thermodynamics. Its importance grew, however, as the field of statistical mechanics developed because the analytical techniques of statistical mechanics provide an alternative means of interpreting entropy and a more global significance to the concept. In statistical mechanics, the behavior of a substance is described in terms of the statistical behavior of its atoms and molecules. An important finding in these studies is that **isolated systems tend toward disorder and entropy is a measure of this disorder.** For example, consider the molecules of a gas in the air in your room. If half the gas molecules had velocity vectors of equal magnitude directed toward the left and the other half had velocity vectors of the same magnitude directed toward the right, the situation would be very ordered. Such a situation is extremely unlikely, however. If you could view the molecules, you would see that they move haphazardly in all directions, bumping into one another, changing speed upon collision, some going fast and others going slowly. This situation is highly disordered.

The cause of the tendency of an isolated system toward disorder is easily explained. To do so, let's distinguish between *microstates* and *macrostates* of a system. A **microstate** is a particular configuration of the individual constituents of the system. For example, the description of the ordered velocity vectors of the air molecules in your room refers to a particular microstate, and the more likely haphazard motion is another microstate, one that represents disorder. A **macrostate** is a description of the system's conditions from a macroscopic point of view and makes use of macroscopic variables such as pressure, density, and temperature for gases.

For any given macrostate of the system, a number of microstates are possible. For example, the macrostate of a 4 on a pair of dice can be formed from the possible microstates 1–3, 2–2, and 3–1. It is assumed that all microstates are equally probable. When all possible macrostates are examined, however, it is found that macrostates associated with disorder have far more possible microstates than those associated with order. For example, there is only one microstate associated with the macrostate of a royal flush in a poker hand of five spades, laid out in order from ten to ace (Fig. 22.13a). This combination of cards is a highly ordered hand. Many microstates (the set of five individual cards in a poker hand), however, are associated with a worthless hand in poker (Fig. 22.13b).

The probability of being dealt the royal flush in spades is exactly the same as the probability of being dealt any *particular* worthless hand. Because there are so many worthless hands, however, the probability of a macrostate of a worthless hand is far larger than the probability of a macrostate of the royal flush in spades.

George Semple

(a)

George Semple

(b)

Figure 22.13 (a) A royal flush is a highly ordered poker hand with low probability of occurring. (b) A disordered and worthless poker hand. The probability of this *particular* hand occurring is the same as that of the royal flush in spades. There are so many worthless hands, however, that the probability of being dealt a worthless hand is much higher than that of being dealt a royal flush in spades.

Quick Quiz 22.4 (a) Suppose you select four cards at random from a standard deck of playing cards and end up with a macrostate of four deuces. How many microstates are associated with this macrostate? (b) Suppose you pick up two cards and end up with a macrostate of two aces. How many microstates are associated with this macrostate?

We can also imagine ordered macrostates and disordered macrostates in physical processes, not just in games of dice and poker. The probability of a system moving in time from an ordered macrostate to a disordered macrostate is far greater than the probability of the reverse because there are more microstates in a disordered macrostate.

If we consider a system and its surroundings to include the entire Universe, the Universe is always moving toward a macrostate corresponding to greater disorder. Because entropy is a measure of disorder, an alternative way of stating this law is that **the entropy of the Universe increases in all real processes**. This statement is yet another wording of the second law of thermodynamics that can be shown to be equivalent to the Kelvin–Planck and Clausius statements.

◀ Entropy statement of the second law of thermodynamics

The original formulation of entropy in thermodynamics involves the transfer of energy by heat during a reversible process. Consider any infinitesimal process in which a system changes from one equilibrium state to another. If dQ_r is the amount of energy transferred by heat when the system follows a reversible path between the states, the change in entropy dS is equal to this amount of energy for the reversible process divided by the absolute temperature of the system:

$$dS = \frac{dQ_r}{T} \tag{22.8}$$

We have assumed the temperature is constant because the process is infinitesimal. Because entropy is a state variable, **the change in entropy during a process depends only on the endpoints and therefore is independent of the actual path followed. Consequently, the entropy change for an irreversible process can be determined by calculating the entropy change for a reversible process that connects the same initial and final states.**

The subscript r on the quantity dQ_r is a reminder that the transferred energy is to be measured along a reversible path even though the system may actually have followed some irreversible path. When energy is absorbed by the system, dQ_r is positive and the entropy of the system increases. When energy is expelled by the system, dQ_r is negative and the entropy of the system decreases. Notice that Equation 22.8 does not define entropy but, rather, the *change* in entropy. Hence, the meaningful quantity in describing a process is the *change* in entropy.

To calculate the change in entropy for a *finite* process, first recognize that T is generally not constant. Therefore, we must integrate Equation 22.8:

Change in entropy for a finite process ▶

$$\Delta S = \int_i^f dS = \int_i^f \frac{dQ_r}{T} \tag{22.9}$$

As with an infinitesimal process, the change in entropy ΔS of a system going from one state to another has the same value for *all* paths connecting the two states. That is, the finite change in entropy ΔS of a system depends only on the properties of the initial and final equilibrium states. Therefore, we are free to choose a particular reversible path over which to evaluate the entropy in place of the actual path as long as the initial and final states are the same for both paths. This point is explored further in Section 22.7.

Quick Quiz 22.5 An ideal gas is taken from an initial temperature T_i to a higher final temperature T_f along two different reversible paths. Path A is at constant pressure and path B is at constant volume. What is the relation between the entropy changes of the gas for these paths? (a) $\Delta S_A > \Delta S_B$ (b) $\Delta S_A = \Delta S_B$ (c) $\Delta S_A < \Delta S_B$

EXAMPLE 22.6 Change in Entropy: Melting

A solid that has a latent heat of fusion L_f melts at a temperature T_m. Calculate the change in entropy of this substance when a mass m of the substance melts.

SOLUTION

Conceptualize Imagine placing the substance in a warm environment so that energy enters the substance by heat. The process can be reversed by placing the substance in a cool environment so that energy leaves the substance by heat.

Categorize Because the melting takes place at a fixed temperature, we categorize the process as isothermal.

Analyze Use Equation 20.7 in Equation 22.9, noting that the temperature remains fixed:

$$\Delta S = \int \frac{dQ_r}{T} = \frac{1}{T_m}\int dQ_r = \frac{Q_r}{T_m} = \frac{mL_f}{T_m}$$

Finalize Notice that ΔS is positive, representing that energy is added to the ice cube.

What If? Suppose you did not have Equation 22.9 available to calculate an entropy change. How could you argue from the statistical description of entropy that the changes in entropy should be positive?

Answer When a solid melts, its entropy increases because the molecules are much more disordered in the liquid state than they are in the solid state. The positive value for ΔS also means that the substance in its liquid state does not spontaneously transfer energy from itself to the warm surroundings and freeze because to do so would involve a spontaneous increase in order and a decrease in entropy.

Let's consider the changes in entropy that occur in a Carnot heat engine that operates between the temperatures T_c and T_h. In one cycle, the engine takes in energy $|Q_h|$ from the hot reservoir and expels energy $|Q_c|$ to the cold reservoir. These energy transfers occur only during the isothermal portions of the Carnot cycle; therefore, the constant temperature can be brought out in front of the integral sign in Equation 22.9. The integral then simply has the value of the total amount of energy transferred by heat. Therefore, the total change in entropy for one cycle is

$$\Delta S = \frac{|Q_h|}{T_h} - \frac{|Q_c|}{T_c}$$

where the minus sign represents that energy is leaving the engine. In Example 22.3, we showed that for a Carnot engine,

$$\frac{|Q_c|}{|Q_h|} = \frac{T_c}{T_h}$$

Using this result in the previous expression for ΔS, we find that the total change in entropy for a Carnot engine operating in a cycle is *zero*:

$$\Delta S = 0$$

Now consider a system taken through an arbitrary (non-Carnot) reversible cycle. Because entropy is a state variable—and hence depends only on the properties of a given equilibrium state—we conclude that $\Delta S = 0$ for *any* reversible cycle. In general, we can write this condition as

$$\oint \frac{dQ_r}{T} = 0 \quad \text{(reversible cycle)} \qquad \textbf{(22.10)}$$

where the symbol $\oint$ indicates that the integration is over a closed path.

22.7 Entropy Changes in Irreversible Processes

By definition, a calculation of the change in entropy for a system requires information about a reversible path connecting the initial and final equilibrium states. To calculate changes in entropy for real (irreversible) processes, remember that entropy (like internal energy) depends only on the *state* of the system. That is, entropy is a state variable, and the change in entropy depends only on the initial and final states.

You can calculate the entropy change in some irreversible process between two equilibrium states by devising a reversible process (or series of reversible processes) between the same two states and computing $\Delta S = \int dQ_r/T$ for the reversible process. In irreversible processes, it is important to distinguish between Q, the actual energy transfer in the process, and Q_r, the energy that would have been transferred by heat along a reversible path. Only Q_r is the correct value to be used in calculating the entropy change.

The change in entropy for a system and its surroundings is always positive for an irreversible process. In general, the total entropy—and therefore the disorder—always increases in an irreversible process. Keeping these considerations in mind, we can state the second law of thermodynamics as follows:

The total entropy of an isolated system that undergoes a change cannot decrease.

Furthermore, **if the process is irreversible, the total entropy of an isolated system always increases. In a reversible process, the total entropy of an isolated system remains constant.**

When dealing with a system that is not isolated from its surroundings, remember that the increase in entropy described in the second law is that of the system *and* its surroundings. When a system and its surroundings interact in an irreversible process, the increase in entropy of one is greater than the decrease in entropy of the other. Hence, **the change in entropy of the Universe must be greater than zero for an irreversible process and equal to zero for a reversible process.** Ultimately, the entropy of the Universe should reach a maximum value. At this value, the Universe will be in a state of uniform temperature and density. All physical, chemical, and biological processes will cease because a state of perfect disorder implies that no energy is available for doing work. This gloomy state of affairs is sometimes referred to as the heat death of the Universe.

Quick Quiz 22.6 True or False: The entropy change in an adiabatic process must be zero because $Q = 0$.

Entropy Change in Thermal Conduction

Let's now consider a system consisting of a hot reservoir and a cold reservoir that are in thermal contact with each other and isolated from the rest of the Universe. A process occurs during which energy Q is transferred by heat from the hot reservoir at temperature T_h to the cold reservoir at temperature T_c. The process as described is irreversible, so we must find an equivalent reversible process. Because the temperature of a reservoir does not change during the process, we can replace the real process for each reservoir with a reversible, isothermal process in which the same amount of energy is transferred by heat. Consequently, for a reservoir, the entropy change does not depend on whether the process is reversible or irreversible.

Because the cold reservoir absorbs energy Q, its entropy increases by Q/T_c. At the same time, the hot reservoir loses energy Q, so its entropy change is $-Q/T_h$. Because $T_h > T_c$, the increase in entropy of the cold reservoir is greater than the decrease in entropy of the hot reservoir. Therefore, the change in entropy of the system (and of the Universe) is greater than zero:

$$\Delta S_U = \frac{Q}{T_c} + \frac{-Q}{T_h} > 0$$

Suppose energy were to transfer spontaneously from a cold object to a hot object, in violation of the second law. This impossible energy transfer can be described in terms of disorder. Before the transfer, a certain degree of order is associated with the different temperatures of the objects. The hot object's molecules have a higher average energy than the cold object's molecules. If energy spontaneously transfers from the cold object to the hot object, the cold object becomes colder over a time interval and the hot object becomes hotter. The difference in average molecular energy becomes even greater, which would represent an increase in order for the system and a violation of the second law.

In comparison, the process that does occur naturally is the transfer of energy from the hot object to the cold object. In this process, the difference in average molecular energy decreases, which represents a more random distribution of energy and an increase in disorder.

Entropy Change in a Free Expansion

Let's again consider the adiabatic free expansion of a gas occupying an initial volume V_i (Fig. 22.14). In this situation, a membrane separating the gas from an evacuated region is broken and the gas expands (irreversibly) to a volume V_f. What are the changes in entropy of the gas and of the Universe during this process? The process is neither reversible nor quasi-static. As shown in Section 20.6, the initial and final temperatures of the gas are the same.

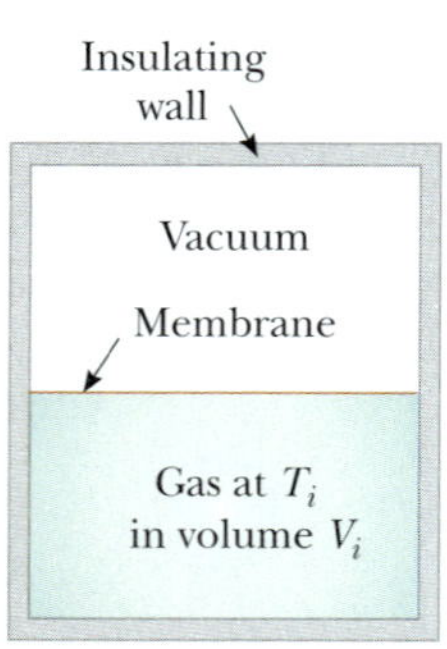

Figure 22.14 Adiabatic free expansion of a gas. When the membrane separating the gas from the evacuated region is ruptured, the gas expands freely and irreversibly. As a result, it occupies a greater final volume. The container is thermally insulated from its surroundings; therefore, $Q = 0$.

To apply Equation 22.9, we cannot take $Q = 0$, the value for the irreversible process, but must instead find Q_r; that is, we must find an equivalent reversible path that shares the same initial and final states. A simple choice is an isothermal, reversible expansion in which the gas pushes slowly against a piston while energy enters the gas by heat from a reservoir to hold the temperature constant. Because T is constant in this process, Equation 22.9 gives

$$\Delta S = \int_i^f \frac{dQ_r}{T} = \frac{1}{T}\int_i^f dQ_r$$

For an isothermal process, the first law of thermodynamics specifies that $\int_i^f dQ_r$ is equal to the negative of the work done on the gas during the expansion from V_i to

V_f, which is given by Equation 20.14. Using this result, we find that the entropy change for the gas is

$$\Delta S = nR\ln\left(\frac{V_f}{V_i}\right) \tag{22.11}$$

Because $V_f > V_i$, we conclude that ΔS is positive. This positive result indicates that both the entropy and the disorder of the gas *increase* as a result of the irreversible, adiabatic expansion.

It is easy to see that the gas is more disordered after the expansion. Instead of being concentrated in a relatively small space, the molecules are scattered over a larger region.

Because the free expansion takes place in an insulated container, no energy is transferred by heat from the surroundings. (Remember that the isothermal, reversible expansion is only a *replacement* process used to calculate the entropy change for the gas; it is not the *actual* process.) Therefore, the free expansion has no effect on the surroundings, and the entropy change of the surroundings is zero.

22.8 Entropy on a Microscopic Scale

As we have seen, entropy can be approached by relying on macroscopic concepts. Entropy can also be treated from a microscopic viewpoint through statistical analysis of molecular motions. Let's use a microscopic model to investigate once again the free expansion of an ideal gas, which was discussed from a macroscopic point of view in Section 22.7.

In the kinetic theory of gases, gas molecules are represented as particles moving randomly. Suppose the gas is initially confined to the volume V_i shown in Figure 22.14. When the membrane is removed, the molecules eventually are distributed throughout the greater volume V_f of the entire container. For a given uniform distribution of gas in the volume, there are a large number of equivalent microstates, and the entropy of the gas can be related to the number of microstates corresponding to a given macrostate.

Let's count the number of microstates by considering the variety of molecular locations available to the molecules. At the instant after the partition is removed (and before the molecules have had a chance to rush into the other half of the container), all the molecules are in the initial volume. Let's assume each molecule occupies some microscopic volume V_m. The total number of possible locations of a single molecule in a macroscopic initial volume V_i is the ratio $w_i = V_i/V_m$, which is a huge number. We use w_i here to represent either the number of *ways* the molecule can be placed in the initial volume or the number of microstates, which is equivalent to the number of available locations. We assume the probabilities of a molecule occupying any of these locations are equal.

As more molecules are added to the system, the number of possible ways the molecules can be positioned in the volume multiplies. For example, if you consider two molecules, for every possible placement of the first, all possible placements of the second are available. Therefore, there are w_i ways of locating the first molecule, and for each way, there are w_i ways of locating the second molecule. The total number of ways of locating the two molecules is $w_iw_i = w_i^2$.

Neglecting the very small probability of having two molecules occupy the same location, each molecule may go into any of the V_i/V_m locations, and so the number of ways of locating N molecules in the volume becomes $W_i = w_i^N = (V_i/V_m)^N$. ($W_i$ is not to be confused with work.) Similarly, when the volume is increased to V_f, the number of ways of locating N molecules increases to $W_f = w_f^N = (V_f/V_m)^N$. The ratio of the number of ways of placing the molecules in the volume for the initial and final configurations is

$$\frac{W_f}{W_i} = \frac{(V_f/V_m)^N}{(V_i/V_m)^N} = \left(\frac{V_f}{V_i}\right)^N$$

Taking the natural logarithm of this equation and multiplying by Boltzmann's constant gives

$$k_B \ln\left(\frac{W_f}{W_i}\right) = k_B \ln\left(\frac{V_f}{V_i}\right)^N = nN_A k_B \ln\left(\frac{V_f}{V_i}\right)$$

where we have used the equality $N = nN_A$. We know from Equation 19.11 that $N_A k_B$ is the universal gas constant R; therefore, we can write this equation as

$$k_B \ln W_f - k_B \ln W_i = nR \ln\left(\frac{V_f}{V_i}\right) \quad \textbf{(22.12)}$$

From Equation 22.11, we know that when a gas undergoes a free expansion from V_i to V_f, the change in entropy is

$$S_f - S_i = nR \ln\left(\frac{V_f}{V_i}\right) \quad \textbf{(22.13)}$$

Notice that the right sides of Equations 22.12 and 22.13 are identical. Therefore, from the left sides, we make the following important connection between entropy and the number of microstates for a given macrostate:

Entropy (microscopic definition) ▶

$$S \equiv k_B \ln W \quad \textbf{(22.14)}$$

The more microstates there are that correspond to a given macrostate, the greater the entropy of that macrostate. As discussed previously, there are many more microstates associated with disordered macrostates than with ordered macrostates. Therefore, Equation 22.14 indicates mathematically that **entropy is a measure of disorder**. Although our discussion used the specific example of the free expansion of an ideal gas, a more rigorous development of the statistical interpretation of entropy would lead us to the same conclusion.

We have stated that individual microstates are equally probable. Because there are far more microstates associated with a disordered macrostate than with an ordered macrostate, however, a disordered macrostate is much more probable than an ordered one.

Let's explore this concept by considering 100 molecules in a container. At any given moment, the probability of one molecule being in the left part of the container shown in Active Figure 22.15a as a result of random motion is $\frac{1}{2}$. If there are two molecules as shown in Active Figure 22.15b, the probability of both being in the left part is $(\frac{1}{2})^2$, or 1 in 4. If there are three molecules (Active Fig. 22.15c), the probability of them all being in the left portion at the same moment is $(\frac{1}{2})^3$, or 1 in 8. For 100 independently moving molecules, the probability that the 50 fastest ones will be found in the left part at any moment is $(\frac{1}{2})^{50}$. Likewise, the probability that the remaining 50 slower molecules will be found in the right part at any moment is $(\frac{1}{2})^{50}$. Therefore, the probability of finding this fast-slow separation as a result of random motion is the product $(\frac{1}{2})^{50}(\frac{1}{2})^{50} = (\frac{1}{2})^{100}$, which corresponds to about 1 in 10^{30}. When this calculation is extrapolated from 100 molecules to the number in 1 mol of gas (6.02×10^{23}), the ordered arrangement is found to be *extremely* improbable!

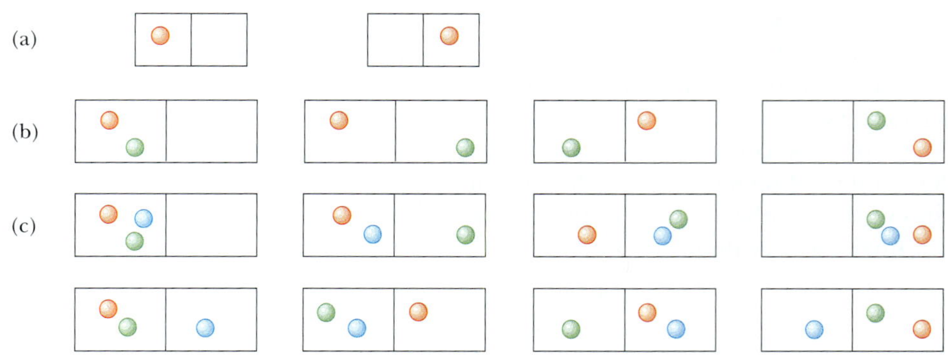

ACTIVE FIGURE 22.15

(a) One molecule in a container has a 1-in-2 chance of being on the left side. (b) Two molecules have a 1-in-4 chance of being on the left side at the same time. (c) Three molecules have a 1-in-8 chance of being on the left side at the same time.

Sign in at www.thomsonedu.com and go to ThomsonNOW to choose the number of molecules to put in the container and measure the probability of them all being on the left side.

CONCEPTUAL EXAMPLE 22.7 Let's Play Marbles!

Suppose you have a bag of 100 marbles of which 50 are red and 50 are green. You are allowed to draw four marbles from the bag according to the following rules. Draw one marble, record its color, and return it to the bag. Shake the bag and then draw another marble. Continue this process until you have drawn and returned four marbles. What are the possible macrostates for this set of events? What is the most likely macrostate? What is the least likely macrostate?

TABLE 22.1
Possible Results of Drawing Four Marbles from a Bag

Macrostate	Possible Microstates	Total Number of Microstates
All R	RRRR	1
1G, 3R	RRRG, RRGR, RGRR, GRRR	4
2G, 2R	RRGG, RGRG, GRRG, RGGR, GRGR, GGRR	6
3G, 1R	GGGR, GGRG, GRGG, RGGG	4
All G	GGGG	1

SOLUTION

Because each marble is returned to the bag before the next one is drawn and the bag is then shaken, the probability of drawing a red marble is always the same as the probability of drawing a green one. All the possible microstates and macrostates are shown in Table 22.1. As this table indicates, there is only one way to draw a macrostate of four red marbles, so there is only one microstate for that macrostate. There are, however, four possible microstates that correspond to the macrostate of one green marble and three red marbles, six microstates that correspond to two green marbles and two red marbles, four microstates that correspond to three green marbles and one red marble, and one microstate that corresponds to four green marbles. The most likely, and most disordered, macrostate—two red marbles and two green marbles—corresponds to the largest number of microstates. The least likely, most ordered macrostates—four red marbles or four green marbles—correspond to the smallest number of microstates.

EXAMPLE 22.8 Adiabatic Free Expansion: One Last Time

Let's verify that the macroscopic and microscopic approaches to the calculation of entropy lead to the same conclusion for the adiabatic free expansion of an ideal gas. Suppose an ideal gas expands to four times its initial volume. As we have seen for this process, the initial and final temperatures are the same.

(A) Using a macroscopic approach, calculate the entropy change for the gas.

SOLUTION

Conceptualize Look back at Figure 22.14, which is a diagram of the system before the adiabatic free expansion. Imagine breaking the membrane so that the gas moves into the evacuated area. The expansion is irreversible.

Categorize We can replace the irreversible process with a reversible isothermal process between the same initial and final states. This approach is macroscopic, so we use thermodynamics state variables such as P, V, and T.

Analyze Use Equation 22.11 to evaluate the entropy change:

$$\Delta S = nR \ln\left(\frac{V_f}{V_i}\right) = nR \ln\left(\frac{4V_i}{V_i}\right) = nR \ln 4$$

(B) Using statistical considerations, calculate the change in entropy for the gas and show that it agrees with the answer you obtained in part (A).

SOLUTION

Categorize This approach is microscopic, so we use variables related to the individual molecules.

Analyze The number of microstates available to a single molecule in the initial volume V_i is $w_i = V_i/V_m$. Use this number to find the number of available microstates for N molecules:

$$W_i = w_i^N = \left(\frac{V_i}{V_m}\right)^N$$

Find the number of available microstates for N molecules in the final volume $V_f = 4V_i$:

$$W_f = \left(\frac{V_f}{V_m}\right)^N = \left(\frac{4V_i}{V_m}\right)^N$$

Use Equation 22.14 to find the entropy change:

$$\Delta S = k_B \ln W_f - k_B \ln W_i = k_B \ln\left(\frac{W_f}{W_i}\right)$$

$$= k_B \ln\left(\frac{4V_i}{V_i}\right)^N = k_B \ln(4^N) = Nk_B \ln 4 = nR \ln 4$$

Finalize The answer is the same as that for part (A), which dealt with macroscopic parameters.

What If? In part (A), we used Equation 22.11, which was based on a reversible isothermal process connecting the initial and final states. Would you arrive at the same result if you chose a different reversible process?

Answer You *must* arrive at the same result because entropy is a state variable. For example, consider the two-step process in Figure 22.16: a reversible adiabatic expansion from V_i to $4V_i$, ($A \rightarrow B$) during which the temperature drops from T_1 to T_2 and a reversible isovolumetric process ($B \rightarrow C$) that takes the gas back to the initial temperature T_1. During the reversible adiabatic process, $\Delta S = 0$ because $Q_r = 0$.

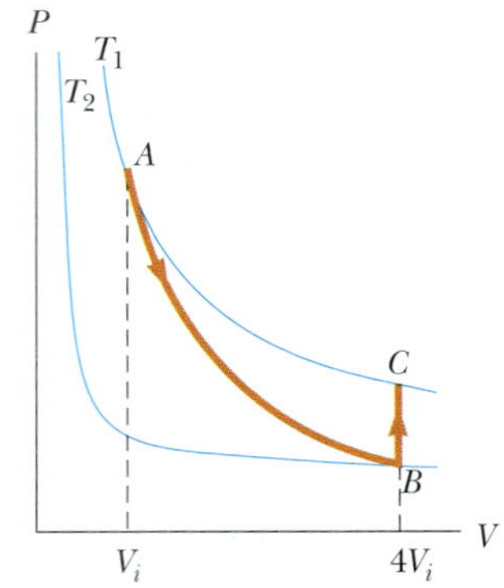

Figure 22.16 (Example 22.8) A gas expands to four times its initial volume and back to the initial temperature by means of a two-step process.

For the reversible isovolumetric process ($B \rightarrow C$), use Equation 22.9:

$$\Delta S = \int_i^f \frac{dQ_r}{T} = \int_{T_2}^{T_1} \frac{nC_V dT}{T} = nC_V \ln\left(\frac{T_1}{T_2}\right)$$

Find the ratio of temperature T_2 to T_1 from Equation 21.20 for the adiabatic process:

$$\frac{T_1}{T_2} = \left(\frac{4V_i}{V_i}\right)^{\gamma-1} = (4)^{\gamma-1}$$

Substitute to find ΔS:

$$\Delta S = nC_V \ln(4)^{\gamma-1} = nC_V(\gamma - 1)\ln 4$$

$$= nC_V\left(\frac{C_P}{C_V} - 1\right)\ln 4 = n(C_P - C_V)\ln 4 = nR \ln 4$$

and you do indeed obtain the exact same result for the entropy change.

Summary

ThomsonNOW™ Sign in at **www.thomsonedu.com** and go to ThomsonNOW to take a practice test for this chapter.

DEFINITIONS

The **thermal efficiency** e of a heat engine is

$$e \equiv \frac{W_{eng}}{|Q_h|} = \frac{|Q_h| - |Q_c|}{|Q_h|} = 1 - \frac{|Q_c|}{|Q_h|} \quad \textbf{(22.2)}$$

From a microscopic viewpoint, the entropy of a given macrostate is defined as

$$S \equiv k_B \ln W \quad \textbf{(22.14)}$$

where k_B is Boltzmann's constant and W is the number of microstates of the system corresponding to the macrostate.

In a **reversible** process, the system can be returned to its initial conditions along the same path on a PV diagram, and every point along this path is an equilibrium state. A process that does not satisfy these requirements is **irreversible**.

(*continued*)

CONCEPTS AND PRINCIPLES

A **heat engine** is a device that takes in energy by heat and, operating in a cyclic process, expels a fraction of that energy by means of work. The net work done by a heat engine in carrying a working substance through a cyclic process ($\Delta E_{\text{int}} = 0$) is

$$W_{\text{eng}} = |Q_h| - |Q_c| \quad \textbf{(22.1)}$$

where $|Q_h|$ is the energy taken in from a hot reservoir and $|Q_c|$ is the energy expelled to a cold reservoir.

Two ways the **second law of thermodynamics** can be stated are as follows:

- It is impossible to construct a heat engine that, operating in a cycle, produces no effect other than the input of energy by heat from a reservoir and the performance of an equal amount of work (the Kelvin–Planck statement).
- It is impossible to construct a cyclical machine whose sole effect is to transfer energy continuously by heat from one object to another object at a higher temperature without the input of energy by work (the Clausius statement).

Carnot's theorem states that no real heat engine operating (irreversibly) between the temperatures T_c and T_h can be more efficient than an engine operating reversibly in a Carnot cycle between the same two temperatures.

The thermal efficiency of a heat engine operating in the Carnot cycle is

$$e_{\text{C}} = 1 - \frac{T_c}{T_h} \quad \textbf{(22.6)}$$

The second law of thermodynamics states that when real (irreversible) processes occur, the degree of disorder in the system plus the surroundings increases. When a process occurs in an isolated system, the state of the system becomes more disordered. The measure of disorder in a system is called **entropy** S. Therefore, yet another way the second law can be stated is as follows:

- The entropy of the Universe increases in all real processes.

The **change in entropy** dS of a system during a process between two infinitesimally separated equilibrium states is

$$dS = \frac{dQ_r}{T} \quad \textbf{(22.8)}$$

where dQ_r is the energy transfer by heat for the system for a reversible process that connects the initial and final states.

The change in entropy of a system during an arbitrary process between an initial state and a final state is

$$\Delta S = \int_i^f \frac{dQ_r}{T} \quad \textbf{(22.9)}$$

The value of ΔS for the system is the same for all paths connecting the initial and final states. The change in entropy for a system undergoing any reversible, cyclic process is zero, and when such a process occurs, the entropy of the Universe remains constant.

Questions

☐ denotes answer available in *Student Solutions Manual/Study Guide;* **O** denotes objective question

1. ☐ What are some factors that affect the efficiency of automobile engines?

2. **O** Consider cyclic processes completely characterized by each of the following net energy inputs and outputs. In each case, the energy transfers listed are the *only* ones occurring. Classify each process as (a) possible, (b) impossible according to the first law of thermodynamics, (c) impossible according to the second law of thermodynamics, or (d) impossible according to both the first and second laws. **(i)** Input is 5 J of work and output is 4 J of work. **(ii)** Input is 5 J of work and output is 5 J of energy transferred by heat. **(iii)** Input is 5 J of energy transferred by electrical transmission and output is 6 J of work. **(iv)** Input is 5 J of energy transferred by heat and output is 5 J of energy transferred by heat. **(v)** Input is 5 J of energy transferred by heat and output is 5 J of work. **(vi)** Input is 5 J of energy transferred by heat and output is 3 J of work plus 2 J of energy transferred by heat. **(vii)** Input is 5 J of energy transferred by heat and output is 3 J of work plus 2 J of energy transferred by mechanical waves. **(viii)** Input is 5 J of energy transferred by heat and output is 3 J of work plus 1 J of energy transferred by electromagnetic radiation.

3. A steam-driven turbine is one major component of an electric power plant. Why is it advantageous to have the temperature of the steam as high as possible?

4. Does the second law of thermodynamics contradict or correct the first law? Argue for your answer.

5. "The first law of thermodynamics says you can't really win, and the second law says you can't even break even." Explain how this statement applies to a particular device or process; alternatively, argue against the statement.

6. **O** The arrow *OA* in the *PV* diagram shown in Figure Q22.6 represents a reversible adiabatic expansion of an ideal gas. The same sample of gas, starting from the same state *O*, now undergoes an adiabatic free expansion to the same final volume. What point on the diagram could represent the final state of the gas? (a) the same point *A* as for the reversible expansion (b) point *B* (c) point *C* (d) any of these choices (e) none of these choices

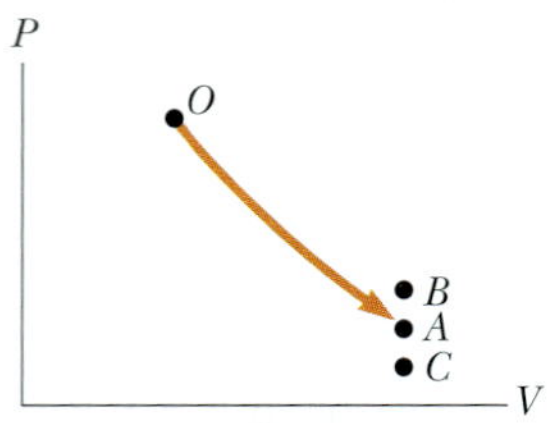

Figure Q22.6

7. Give various examples of irreversible processes that occur in nature. Give an example of a process in nature that is nearly reversible.

8. The device shown in Figure Q22.8, called a thermoelectric converter, uses a series of semiconductor cells to transform internal energy to electric potential energy, which we will study in Chapter 25. In the photograph at the left, both legs of the device are at the same temperature and no electric potential energy is produced. When one leg is at a higher temperature than the other as shown in the photograph on the right, however, electric potential energy is produced as the device extracts energy from the hot reservoir and drives a small electric motor. (a) Why is the difference in temperature necessary to produce electric potential energy in this demonstration? (b) In what sense does this intriguing experiment demonstrate the second law of thermodynamics?

Figure Q22.8

9. **O** Suppose you come home to a small, hot apartment in a well-insulated building on a summer afternoon. The appliance store has just delivered your new air conditioner, but you are too tired to install it properly. Until the sun sets, it will be hotter outside than inside, so you do not open a window. **(i)** You take the air conditioner out of its box, set it on the dining room table, plug it in, and turn it on. What happens to the temperature of the apartment? (a) It increases. (b) It decreases. (c) It remains constant. **(ii)** Suppose instead you quickly take all the ice cubes and frozen vegetables from the refrigerator's freezing compartment, put them into a bowl on the table, and close the refrigerator. What happens to the apartment temperature now? Choose from the same possibilities.

10. **O** **(i)** The second law of thermodynamics implies that the coefficient of performance of a refrigerator must be (a) less than 1, (b) less than or equal to 1, (c) greater than or equal to 1, (d) less than infinity, or (e) greater than 0. **(ii)** What does the second law of thermodynamics imply that the coefficient of performance of a heat pump must be? Choose from the same possibilities.

11. Discuss three different common examples of natural processes that involve an increase in entropy. Be sure to account for all parts of each system under consideration.

12. Discuss the change in entropy of a gas that expands (a) at constant temperature and (b) adiabatically.

13. **O** A thermodynamic process occurs in which the entropy of a system changes by -8 J/K. According to the second law of thermodynamics, what can you conclude about the entropy change of the environment? (a) It must be -8 J/K or less. (b) It must be equal to -8 J/K. (c) It must be between -8 J/K and 0. (d) It must be 0. (e) It must be between 0 and $+8$ J/K. (e) It must be equal to $+8$ J/K (f) It must be $+8$ J/K or more. (g) We would need to know the nature of the process to reach a conclusion. (h) It is impossible for the system to have a negative entropy change.

14. **O** A sample of a monatomic ideal gas is contained in a cylinder with a piston. Its state is represented by the dot in the *PV* diagram shown in Figure Q22.14. Arrows *A* through *H* represent isothermal, isobaric, isovolumetric, and adiabatic processes that the sample can undergo. In each compression and expansion, the volume changes by a factor of 2. (a) Rank these processes according to the work $W_{eng} = +\int P\,dV$ done by the gas from the largest positive value to the largest-magnitude negative value. In your ranking, display any cases of equality. (b) Rank the same processes according to the change in internal energy of the gas from the largest positive value to the largest-magnitude negative value. (c) Rank the same processes according to the energy transferred to the sample by heat.

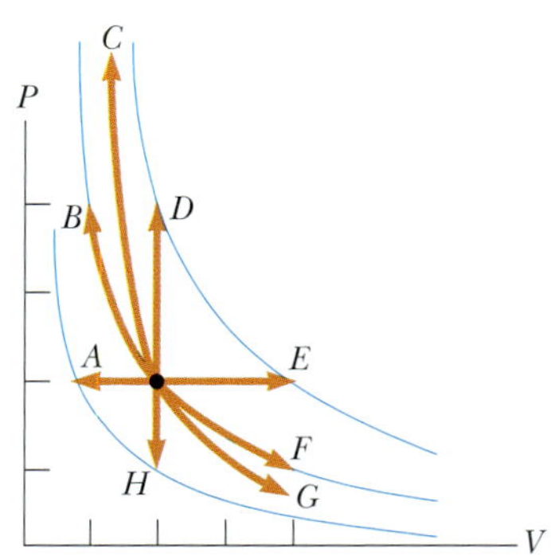

Figure Q22.14

15. **O** Consider the processes shown in Figure Q22.14 and described in Question 14. Rank the processes *A* through *H* according to the change in entropy of the monatomic ideal gas sample from the largest positive value to the largest-magnitude negative value.

16. The energy exhaust from a certain coal-fired electric generating station is carried by "cooling water" into Lake Ontario. The water is warm from the viewpoint of living things in the lake. Some of them congregate around the outlet port and can impede the water flow. (a) Use the theory of heat engines to explain why this action can reduce the electric power output of the station. (b) An engineer says that the electric output is reduced because of "higher back pressure on the turbine blades." Comment on the accuracy of this statement.

17. **O** Assume a sample of an ideal gas is at room temperature. What action will necessarily make the entropy of the sample increase? (a) transfer energy into it by heat (b) transfer energy into it irreversibly by heat (c) do work on it (d) increase either its temperature or its volume, without letting the other variable decrease (e) none of these choices

18. Suppose your roommate is "Mr. Clean" and tidies up your messy room after a big party. Because your roommate is creating more order, does this process represent a violation of the second law of thermodynamics?

19. "Energy is the mistress of the Universe and entropy is her shadow." Writing for an audience of general readers, argue for this statement with examples. Alternatively, argue for the view that entropy is like a decisive hands-on executive instantly determining what will happen, whereas energy is like a wretched back-office bookkeeper telling us how little we can afford. (Arnold Sommerfeld suggested the idea for this question.)

20. If you shake a jar full of jelly beans of different sizes, the larger jelly beans tend to appear near the top and the smaller ones tend to fall to the bottom. Why? Does this process violate the second law of thermodynamics?

Problems

WebAssign The Problems from this chapter may be assigned online in WebAssign.

ThomsonNOW™ Sign in at **www.thomsonedu.com** and go to ThomsonNOW to assess your understanding of this chapter's topics with additional quizzing and conceptual questions.

1, 2, 3 denotes straightforward, intermediate, challenging; ☐ denotes full solution available in *Student Solutions Manual/Study Guide;* ▲ denotes coached solution with hints available at **www.thomsonedu.com;** denotes developing symbolic reasoning; ● denotes asking for qualitative reasoning; denotes computer useful in solving problem

Section 22.1 Heat Engines and the Second Law of Thermodynamics

1. A heat engine takes in 360 J of energy from a hot reservoir and performs 25.0 J of work in each cycle. Find (a) the efficiency of the engine and (b) the energy expelled to the cold reservoir in each cycle.

2. A gun is a heat engine. In particular, it is an internal combustion piston engine that does not operate in a cycle, but comes apart during its adiabatic expansion process. A certain gun consists of 1.80 kg of iron. It fires one 2.40-g bullet at 320 m/s with an energy efficiency of 1.10%. Assume the body of the gun absorbs all the energy exhaust—the other 98.9%—and increases uniformly in temperature for a short time interval before it loses any energy by heat into the environment. Find its temperature increase.

3. A particular heat engine has a mechanical power output of 5.00 kW and an efficiency of 25.0%. The engine expels 8 000 J of exhaust energy in each cycle. Find (a) the energy taken in during each cycle and (b) the time interval for each cycle.

4. A multicylinder gasoline engine in an airplane, operating at 2 500 rev/min, takes in energy 7.89×10^3 J and exhausts 4.58×10^3 J for each revolution of the crankshaft. (a) How many liters of fuel does it consume in 1.00 h of operation if the heat of combustion is 4.03×10^7 J/L? (b) What is the mechanical power output of the engine? Ignore friction and express the answer in horsepower. (c) What is the torque exerted by the crankshaft on the load? (d) What power must the exhaust and cooling system transfer out of the engine?

5. Suppose a heat engine is connected to two energy reservoirs, one a pool of molten aluminum (660°C) and the other a block of solid mercury (−38.9°C). The engine runs by freezing 1.00 g of aluminum and melting 15.0 g of mercury during each cycle. The heat of fusion of aluminum is 3.97×10^5 J/kg; the heat of fusion of mercury is 1.18×10^4 J/kg. What is the efficiency of this engine?

Section 22.2 Heat Pumps and Refrigerators

6. A refrigerator has a coefficient of performance equal to 5.00. The refrigerator takes in 120 J of energy from a cold reservoir in each cycle. Find (a) the work required in each cycle and (b) the energy expelled to the hot reservoir.

7. A refrigerator has a coefficient of performance of 3.00. The ice tray compartment is at −20.0°C, and the room temperature is 22.0°C. The refrigerator can convert 30.0 g of water at 22.0°C to 30.0 g of ice at −20.0°C each minute. What input power is required? Give your answer in watts.

8. ● In 1993, the U.S. government instituted a requirement that all room air conditioners sold in the United States must have an energy efficiency ratio (EER) of 10 or higher. The EER is defined as the ratio of the cooling capacity of the air conditioner, measured in British thermal units per hour, or Btu/h, to its electrical power requirement in watts. (a) Convert the EER of 10.0 to dimensionless form, using the conversion 1 Btu = 1 055 J. (b) What is the appropriate name for this dimensionless quantity? (c) In the 1970s, it was common to find room air conditioners with EERs of 5 or lower. State how the operating costs compare for 10 000-Btu/h air conditioners with EERs of 5.00 and 10.0. Assume each air conditioner operates for 1 500 h during the summer in a city where electricity costs 10.0¢ per kWh.

Section 22.3 Reversible and Irreversible Processes

Section 22.4 The Carnot Engine

9. One of the most efficient heat engines ever built is a steam turbine in the Ohio River valley, operating between

430°C and 1 870°C on energy from West Virginia coal to produce electricity for the Midwest. (a) What is its maximum theoretical efficiency? (b) The actual efficiency of the engine is 42.0%. How much mechanical power does the engine deliver if it takes in 1.40×10^5 J of energy each second from its hot reservoir?

10. A Carnot engine has a power output of 150 kW. The engine operates between two reservoirs at 20.0°C and 500°C. (a) How much energy does it take in per hour? (b) How much energy is lost per hour in its exhaust?

11. ● An engine operates in a cycle, taking in energy by heat at 180°C and putting out exhaust at 100°C. In each cycle, the exhaust energy is 2.00×10^4 J and the engine does 1.50×10^3 J of work. Explain how the actual efficiency of the engine compares with the efficiency of a reversible engine operating between the same temperatures.

12. A Carnot heat engine operates between temperatures T_h and T_c. (a) If $T_h = 500$ K and $T_c = 350$ K, what is the efficiency of the engine? (b) What is the change in its efficiency for each degree of increase in T_h above 500 K? (c) What is the change in its efficiency for each degree of decrease in T_c below 350 K?

13. ▲ An ideal gas is taken through a Carnot cycle. The isothermal expansion occurs at 250°C, and the isothermal compression takes place at 50.0°C. The gas takes in 1 200 J of energy from the hot reservoir during the isothermal expansion. Find (a) the energy expelled to the cold reservoir in each cycle and (b) the net work done by the gas in each cycle.

14. A power plant operates at a 32.0% efficiency during the summer when the seawater used for cooling is at 20.0°C. The plant uses 350°C steam to drive turbines. If the plant's efficiency changes in the same proportion as the ideal efficiency, what is the plant's efficiency in the winter, when the seawater is 10.0°C?

15. Argon enters a turbine at a rate of 80.0 kg/min, a temperature of 800°C, and a pressure of 1.50 MPa. It expands adiabatically as it pushes on the turbine blades and exits at pressure 300 kPa. (a) Calculate its temperature at exit. (b) Calculate the (maximum) power output of the turning turbine. (c) The turbine is one component of a model closed-cycle gas turbine engine. Calculate the maximum efficiency of the engine.

16. ● An electric power plant that would make use of the temperature gradient in the ocean has been proposed. The system is to operate between 20.0°C (surface-water temperature) and 5.00°C (water temperature at a depth of about 1 km). (a) What is the maximum efficiency of such a system? (b) If the electric power output of the plant is 75.0 MW, how much energy is taken in from the warm reservoir per hour? (c) In view of your answer to part (a), explain whether you think such a system is worthwhile. Note that the "fuel" is free.

17. ● Suppose you build a two-engine device with the exhaust energy output from one heat engine supplying the input energy for a second heat engine. We say that the two engines are running *in series*. Let e_1 and e_2 represent the efficiencies of the two engines. (a) The overall efficiency of the two-engine device is defined as the total work output divided by the energy put into the first engine by heat. Show that the overall efficiency is given by

$$e = e_1 + e_2 - e_1 e_2$$

(b) **What If?** Assume the two engines are Carnot engines. Engine 1 operates between temperatures T_h and T_i. The gas in engine 2 varies in temperature between T_i and T_c. In terms of the temperatures, what is the efficiency of the combination engine? Does an improvement in net efficiency result from the use of two engines instead of one? (c) What value of the intermediate temperature T_i results in equal work being done by each of the two engines in series? (d) What value of T_i results in each of the two engines in series having the same efficiency?

18. ● An electric generating station is designed to have an electric output power 1.40 MW using a turbine with two-thirds the efficiency of a Carnot engine. The exhaust energy is transferred by heat into a cooling tower at 110°C. (a) Find the rate at which the station exhausts energy by heat, as a function of the fuel combustion temperature T_h. If the firebox is modified to run hotter by using more advanced combustion technology, how does the amount of energy exhaust change? (b) Find the exhaust power for $T_h = 800$°C. (c) Find the value of T_h for which the exhaust power would be only half as large as in part (b). (d) Find the value of T_h for which the exhaust power would be one quarter as large as in part (b).

19. What is the coefficient of performance of a refrigerator that operates with Carnot efficiency between temperatures −3.00°C and +27.0°C?

20. At point *A* in a Carnot cycle, 2.34 mol of a monatomic ideal gas has a pressure of 1 400 kPa, a volume of 10.0 L, and a temperature of 720 K. The gas expands isothermally to point *B* and then expands adiabatically to point *C*, where its volume is 24.0 L. An isothermal compression brings it to point *D*, where its volume is 15.0 L. An adiabatic process returns the gas to point *A*. (a) Determine all the unknown pressures, volumes, and temperatures as you fill in the following table:

	P	*V*	*T*
A	1 400 kPa	10.0 L	720 K
B			
C		24.0 L	
D		15.0 L	

(b) Find the energy added by heat, the work done by the engine, and the change in internal energy for each of the steps $A \to B$, $B \to C$, $C \to D$, and $D \to A$. (c) Calculate the efficiency $W_{\text{net}}/|Q_h|$. Show that it is equal to $1 - T_C/T_A$, the Carnot efficiency.

21. An ideal refrigerator or ideal heat pump is equivalent to a Carnot engine running in reverse. That is, energy $|Q_c|$ is taken in from a cold reservoir and energy $|Q_h|$ is rejected to a hot reservoir. (a) Show that the work that must be supplied to run the refrigerator or heat pump is

$$W = \frac{T_h - T_c}{T_c}|Q_c|$$

(b) Show that the coefficient of performance of the ideal refrigerator is

$$\text{COP} = \frac{T_c}{T_h - T_c}$$

22. What is the maximum possible coefficient of performance of a heat pump that brings energy from outdoors at

$-3.00°C$ into a $22.0°C$ house? *Note:* The work done to run the heat pump is also available to warm up the house.

23. ▲ How much work does an ideal Carnot refrigerator require to remove 1.00 J of energy from liquid helium at 4.00 K and reject this energy to a room-temperature (293 K) environment?

24. A heat pump used for heating shown in Figure P22.24 is essentially an air conditioner installed backward. It extracts energy from colder air outside and deposits it in a warmer room. Suppose the ratio of the actual energy entering the room to the work done by the device's motor is 10.0% of the theoretical maximum ratio. Determine the energy entering the room per joule of work done by the motor, given that the inside temperature is 20.0°C and the outside temperature is $-5.00°C$.

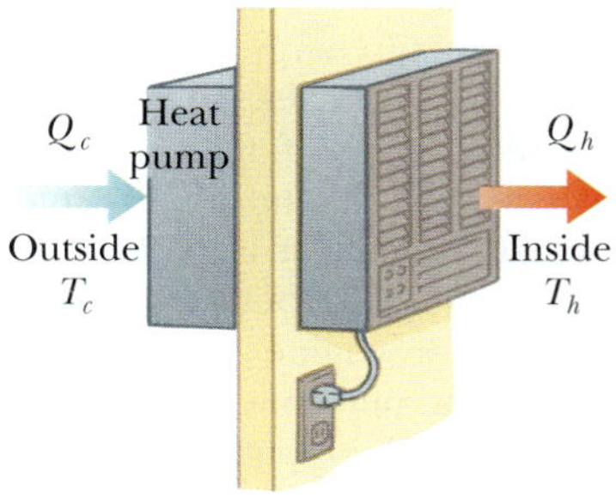

Figure P22.24

25. An ideal (Carnot) freezer in a kitchen has a constant temperature of 260 K, whereas the air in the kitchen has a constant temperature of 300 K. Suppose the insulation for the freezer is not perfect and conducts energy into the freezer at a rate of 0.150 W. Determine the average power required for the freezer's motor to maintain the constant temperature in the freezer.

26. If a 35.0%-efficient Carnot heat engine (Active Fig. 22.2) is run in reverse so as to form a refrigerator (Active Fig. 22.4), what would be this refrigerator's coefficient of performance?

Section 22.5 Gasoline and Diesel Engines

27. In a cylinder of an automobile engine, immediately after combustion, the gas is confined to a volume of 50.0 cm^3 and has an initial pressure of 3.00×10^6 Pa. The piston moves outward to a final volume of 300 cm^3, and the gas expands without energy loss by heat. (a) If $\gamma = 1.40$ for the gas, what is the final pressure? (b) How much work is done by the gas in expanding?

28. The compression ratio of an Otto cycle as shown in Active Figure 22.12 is $V_A/V_B = 8.00$. At the beginning A of the compression process, 500 cm^3 of gas is at 100 kPa and 20.0°C. At the beginning of the adiabatic expansion, the temperature is $T_C = 750°C$. Model the working fluid as an ideal gas with $E_{int} = nC_VT = 2.50\,nRT$ and $\gamma = 1.40$. (a) Fill in this table to follow the states of the gas:

	T (K)	P (kPa)	V (cm^3)	E_{int} (J)
A	293	100	500	
B				
C	1 023			
D				
A				

(b) Fill in this table to follow the processes:

	Q (input)	W (output)	ΔE_{int}
$A \to B$			
$B \to C$			
$C \to D$			
$D \to A$			
$ABCDA$			

(c) Identify the energy input $|Q_h|$, the energy exhaust $|Q_c|$, and the net output work W_{eng}. (d) Calculate the thermal efficiency. (e) Find the number of crankshaft revolutions per minute required for a one-cylinder engine to have an output power of 1.00 kW = 1.34 hp. *Note:* The thermodynamic cycle involves four piston strokes.

29. A gasoline engine has a compression ratio of 6.00 and uses a gas for which $\gamma = 1.40$. (a) What is the efficiency of the engine if it operates in an idealized Otto cycle? (b) **What If?** If the actual efficiency is 15.0%, what fraction of the fuel is wasted as a result of friction and energy losses by heat that could by avoided in a reversible engine? Assume complete combustion of the air-fuel mixture.

Section 22.6 Entropy

30. An ice tray contains 500 g of liquid water at 0°C. Calculate the change in entropy of the water as it freezes slowly and completely at 0°C.

31. A sample consisting of a mass m of a substance with specific heat c is warmed from temperature T_i to temperature T_f. Imagine that it absorbs energy by heat successively from reservoirs at incrementally higher temperatures $T_i + \delta$, $T_i + 2\delta$, $T_i + 3\delta, \ldots, T_f$. Prove that the change in entropy of the sample is given by $mc \ln (T_f/T_i)$.

32. ● In making raspberry jelly, 900 g of raspberry juice is combined with 930 g of sugar. The mixture starts at room temperature, 23.0°C, and is slowly warmed on a stove until it reaches 220°F. It is then poured into hot jars and allowed to cool. Assume the juice has the same specific heat as water. The specific heat of sucrose is 0.299 cal/g · °C. Consider the warming process. (a) Which of the following terms describe(s) this process: adiabatic, isobaric, isothermal, isovolumetric, cyclic, reversible, isentropic? Explain your answer. (b) How much energy does the mixture absorb? (c) What is the minimum change in entropy of the jelly while it is warmed? You may use the result of Problem 31.

33. Calculate the change in entropy of 250 g of water warmed slowly from 20.0°C to 80.0°C. You may use the result of Problem 31.

Section 22.7 Entropy Changes in Irreversible Processes

34. The temperature at the surface of the Sun is approximately 5 700 K, and the temperature at the surface of the Earth is approximately 290 K. What entropy change occurs when 1 000 J of energy is transferred by radiation from the Sun to the Earth?

35. ▲ A 1 500-kg car is moving at 20.0 m/s. The driver brakes to a stop. The brakes cool to the temperature of the surrounding air, which is nearly constant at 20.0°C. What is the total entropy change? You may use the result of Problem 31.

36. Calculate the increase in entropy of the Universe when you add 20.0 g of 5.00°C cream to 200 g of 60.0°C coffee.

Assume the specific heats of cream and coffee are both $4.20\ \text{J/g} \cdot {}^\circ\text{C}$. You may use the result of Problem 31.

37. How fast are you personally making the entropy of the Universe increase right now? Compute an order-of-magnitude estimate, stating what quantities you take as data and the values you measure or estimate for them.

38. A 1.00-kg iron horseshoe is taken from a forge at 900°C and dropped into 4.00 kg of water at 10.0°C. Assuming no energy is lost by heat to the surroundings, determine the total entropy change of the horseshoe-plus-water system. You may use the result of Problem 31.

39. A 1.00-mol sample of H_2 gas is contained in the left side of the container shown in Figure P22.39, which has equal volumes left and right. The right side is evacuated. When the valve is opened, the gas streams into the right side. What is the final entropy change of the gas? Does the temperature of the gas change? Assume the container is so large that the hydrogen behaves as an ideal gas.

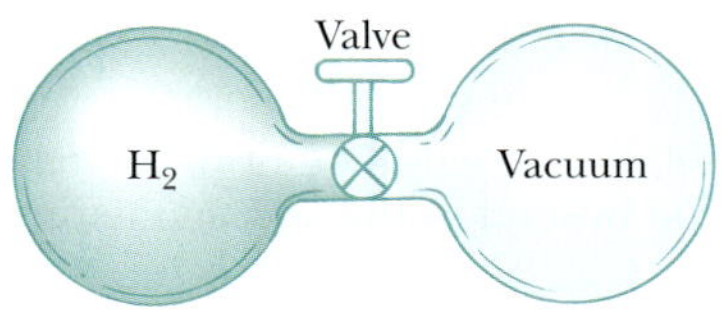

Figure P22.39

40. A 2.00-L container has a center partition that divides it into two equal parts as shown in Figure P22.40. The left side contains H_2 gas, and the right side contains O_2 gas. Both gases are at room temperature and at atmospheric pressure. The partition is removed and the gases are allowed to mix. What is the entropy increase of the system?

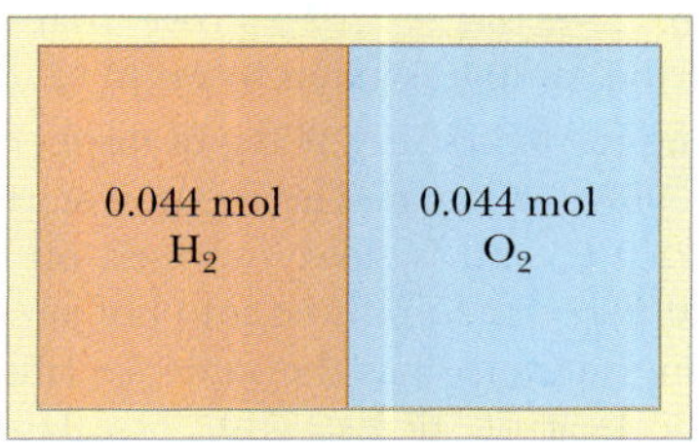

Figure P22.40

Section 22.8 Entropy on a Microscopic Scale

41. If you roll two dice, what is the total number of ways in which you can obtain (a) a 12? (b) a 7?

42. Prepare a table like Table 22.1 for the following occurrence. You toss four coins into the air simultaneously and then record the results of your tosses in terms of the numbers of heads and tails that result. For example, HHTH and HTHH are two possible ways in which three heads and one tail can be achieved. (a) On the basis of your table, what is the most probable result recorded for a toss? In terms of entropy, (b) what is the most ordered macrostate, and (c) what is the most disordered?

43. Repeat the procedure used to construct Table 22.1 (a) for the case in which you draw three marbles from your bag rather than four and (b) for the case in which you draw five rather than four.

Additional Problems

44. Every second at Niagara Falls, some 5 000 m^3 of water falls a distance of 50.0 m. What is the increase in entropy per second due to the falling water? Assume the mass of the surroundings is so great that its temperature and that of the water stay nearly constant at 20.0°C. Also assume a negligible amount of water evaporates.

45. ● A firebox is at 750 K, and the ambient temperature is 300 K. The efficiency of a Carnot engine doing 150 J of work as it transports energy between these constant-temperature baths is 60.0%. The Carnot engine must take in energy 150 J/0.600 = 250 J from the hot reservoir and must put out 100 J of energy by heat into the environment. To follow Carnot's reasoning, suppose some other heat engine S could have an efficiency of 70.0%. (a) Find the energy input and wasted energy output of engine S as it does 150 J of work. (b) Let engine S operate as in part (a) and run the Carnot engine in reverse. Find the total energy the firebox puts out as both engines operate together and the total energy transferred to the environment. Explain how the results show that the Clausius statement of the second law of thermodynamics is violated. (c) Find the energy input and work output of engine S as it puts out exhaust energy of 100 J. (d) Let engine S operate as in part (c) and contribute 150 J of its work output to running the Carnot engine in reverse. Find the total energy the firebox puts out as both engines operate together, the total work output, and the total energy transferred to the environment. Explain how the results show that the Kelvin–Planck statement of the second law is violated. Therefore, our assumption about the efficiency of engine S must be false. (e) Let the engines operate together through one cycle as in part (d). Find the change in entropy of the Universe. Explain how the result shows that the entropy statement of the second law is violated.

46. **Review problem.** This problem complements Problem 24 in Chapter 10. In the operation of a single-cylinder internal combustion piston engine, one charge of fuel explodes to drive the piston outward in the so-called power stroke. Part of its energy output is stored in a turning flywheel. This energy is then used to push the piston inward to compress the next charge of fuel and air. In this compression process, assume an original volume of 0.120 L of a diatomic ideal gas at atmospheric pressure is compressed adiabatically to one-eighth of its original volume. (a) Find the work input required to compress the gas. (b) Assume the flywheel is a solid disk of mass 5.10 kg and radius 8.50 cm, turning freely without friction between the power stroke and the compression stroke. How fast must the flywheel turn immediately after the power stroke? This situation represents the minimum angular speed at which the engine can operate; it is on the point of stalling. (c) When the engine's operation is well above the point of stalling, assume the flywheel puts 5.00% of its maximum energy into compressing the next charge of fuel and air. Find its maximum angular speed in this case.

47. ▲ A house loses energy through the exterior walls and roof at a rate of 5 000 J/s = 5.00 kW when the interior temperature is 22.0°C and the outside temperature is −5.00°C. (a) Calculate the electric power required to maintain the interior temperature at 22.0°C if the electric power is used in electric resistance heaters that convert all the energy transferred in by electrical transmission into internal energy. (b) **What If?** Calculate the electric power required to maintain the interior temperature at 22.0°C if the electric power is used to drive an electric motor that operates the compressor of a heat pump that has a coefficient of performance equal to 60.0% of the Carnot-cycle value.

48. A heat engine operates between two reservoirs at T_2 = 600 K and T_1 = 350 K. It takes in 1 000 J of energy from the higher-temperature reservoir and performs 250 J of work. Find (a) the entropy change of the Universe ΔS_U for this process and (b) the work W that could have been done by an ideal Carnot engine operating between these two reservoirs. (c) Show that the difference between the amounts of work done in parts (a) and (b) is $T_1 \Delta S_U$.

49. ▲ In 1816, Robert Stirling, a Scottish clergyman, patented the *Stirling engine*, which has found a wide variety of applications ever since. Fuel is burned externally to warm one of the engine's two cylinders. A fixed quantity of inert gas moves cyclically between the cylinders, expanding in the hot one and contracting in the cold one. Figure P22.49 represents a model for its thermodynamic cycle. Consider n mol of an ideal monatomic gas being taken once through the cycle, consisting of two isothermal processes at temperatures $3T_i$ and T_i and two constant-volume processes. Determine in terms of n, R, and T_i (a) the net energy transferred by heat to the gas and (b) the efficiency of the engine. A Stirling engine is easier to manufacture than an internal combustion engine or a turbine. It can run on burning garbage. It can run on the energy of sunlight and produce no material exhaust.

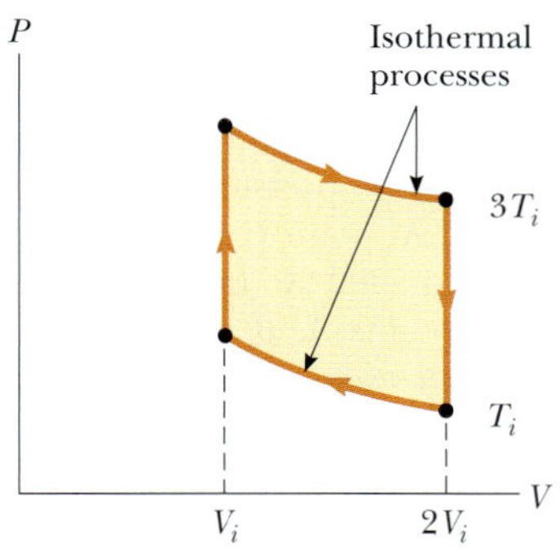

Figure P22.49

50. ● An athlete whose mass is 70.0 kg drinks 16 ounces (453.6 g) of refrigerated water. The water is at a temperature of 35.0°F. (a) Ignoring the temperature change of the body that results from the water intake (so that the body is regarded as a reservoir always at 98.6°F), find the entropy increase of the entire system. (b) **What If?** Assume the entire body is cooled by the drink and the average specific heat of a person is equal to the specific heat of liquid water. Ignoring any other energy transfers by heat and any metabolic energy release, find the athlete's temperature after she drinks the cold water, given an initial body temperature of 98.6°F. Under these assumptions, what is the entropy increase of the entire system? State how this result compares with the one you obtained in part (a).

51. A power plant, having a Carnot efficiency, produces 1 000 MW of electrical power from turbines that take in steam at 500 K and reject water at 300 K into a flowing river. The water downstream is 6.00 K warmer due to the output of the power plant. Determine the flow rate of the river.

52. A power plant, having a Carnot efficiency, produces electric power $\mathcal{P}$ from turbines that take in energy from steam at temperature T_h and discharge energy at temperature T_c through a heat exchanger into a flowing river. The water downstream is warmer by ΔT due to the output of the power plant. Determine the flow rate of the river.

53. A biology laboratory is maintained at a constant temperature of 7.00°C by an air conditioner, which is vented to the air outside. On a typical hot summer day, the outside temperature is 27.0°C and the air-conditioning unit emits energy to the outside at a rate of 10.0 kW. Model the unit as having a coefficient of performance equal to 40.0% of the coefficient of performance of an ideal Carnot device. (a) At what rate does the air conditioner remove energy from the laboratory? (b) Calculate the power required for the work input. (c) Find the change in entropy produced by the air conditioner in 1.00 h. (d) **What If?** The outside temperature increases to 32.0°C. Find the fractional change in the coefficient of performance of the air conditioner.

54. ● A 1.00-mol sample of an ideal monatomic gas is taken through the cycle shown in Figure P22.54. The process $A \rightarrow B$ is a reversible isothermal expansion. Calculate (a) the net work done by the gas, (b) the energy added to the gas by heat, (c) the energy exhausted from the gas by heat, and (d) the efficiency of the cycle. (e) Explain how the efficiency compares with that of a Carnot engine operating between the same temperature extremes.

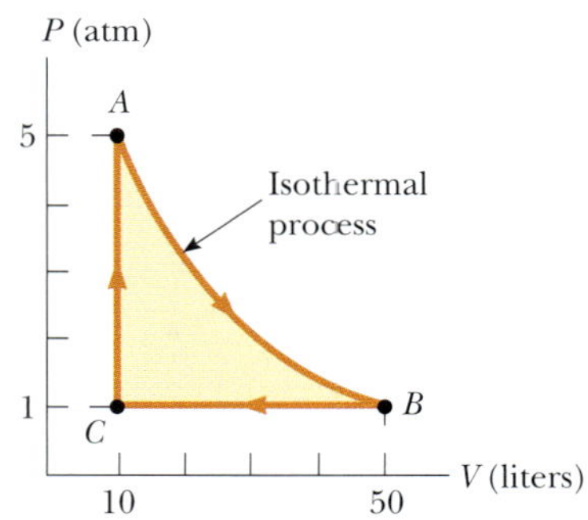

Figure P22.54

55. ● A 1.00-mol sample of a monatomic ideal gas is taken through the cycle shown in Figure P22.55. At point A, the pressure, volume, and temperature are P_i, V_i, and T_i, respectively. In terms of R and T_i, find (a) the total energy entering the system by heat per cycle, (b) the total energy leaving the system by heat per cycle, and (c) the efficiency of an engine operating in this cycle. (d) Explain how the efficiency compares with that of an engine operating in a Carnot cycle between the same temperature extremes.

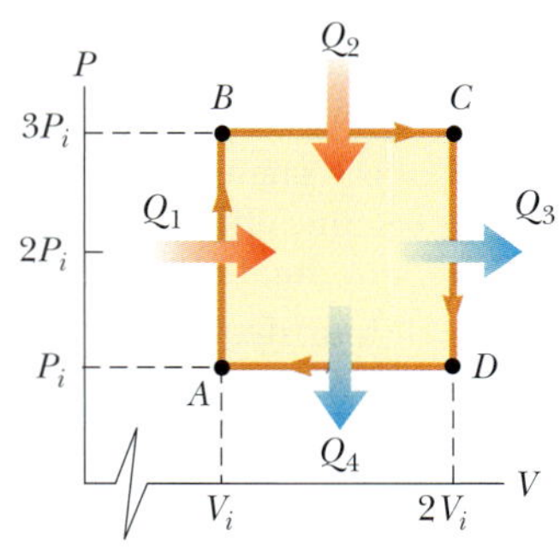

Figure P22.55

56. A sample consisting of n mol of an ideal gas undergoes a reversible isobaric expansion from volume V_i to volume $3V_i$. Find the change in entropy of the gas by calculating $\int_i^f dQ/T$, where $dQ = nC_P\,dT$.

57. ● A system consisting of n mol of an ideal gas undergoes two reversible processes. It starts with pressure P_i and volume V_i, expands isothermally, and then contracts adiabatically to reach a final state with pressure P_i and volume $3V_i$. (a) Find its change in entropy in the isothermal process. (The entropy does not change in the adiabatic process.) (b) **What If?** Explain why the answer to part (a) must be the same as the answer to Problem 56.

58. ● A 1.00-mol sample of an ideal gas expands isothermally, doubling in volume. (a) Show that the work it does in expanding is $W_{\text{eng}} = RT \ln 2$. (b) Because the internal energy E_{int} of an ideal gas depends solely on its temperature, the change in internal energy is zero during the expansion. It follows from the first law that the energy input to the gas by heat during the expansion is equal to the energy output by work. Does this process have 100% efficiency in converting energy input by heat into work output? Does this conversion violate the second law? Explain your answers.

59. An idealized diesel engine operates in a cycle known as the *air-standard diesel cycle* shown in Figure P22.59. Fuel is sprayed into the cylinder at the point of maximum compression, B. Combustion occurs during the expansion $B \to C$, which is modeled as an isobaric process. Show that the efficiency of an engine operating in this idealized diesel cycle is

$$e = 1 - \frac{1}{\gamma}\left(\frac{T_D - T_A}{T_C - T_B}\right)$$

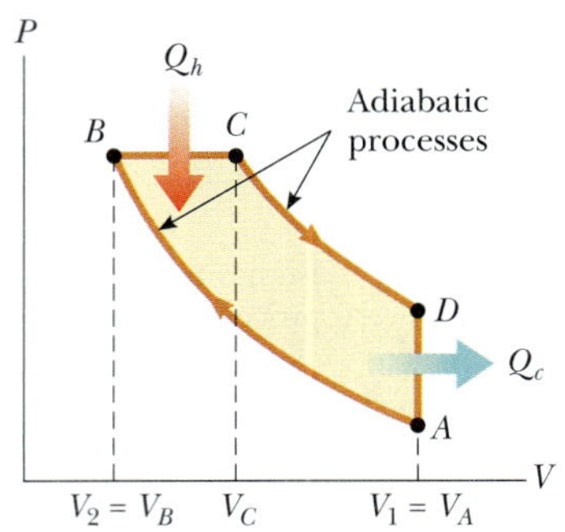

Figure P22.59

60. ● Suppose you are working in a patent office and an inventor comes to you with the claim that her heat engine, which employs water as a working substance, has a thermodynamic efficiency of 0.61. She explains that it operates between energy reservoirs at 4°C and 0°C. It is a very complicated device, with many pistons, gears, and pulleys, and the cycle involves freezing and melting. Does her claim that $e = 0.61$ warrant serious consideration? Explain your answer.

61. ● Suppose 1.00 kg of water at 10.0°C is mixed with 1.00 kg of water at 30.0°C at constant pressure. When the mixture has reached equilibrium, (a) what is the final temperature? (b) Take $c_P = 4.19$ kJ/kg · K for water and show that the entropy of the system increases by

$$\Delta S = 4.19 \ln\left[\left(\frac{293}{283}\right)\left(\frac{293}{303}\right)\right] \text{kJ/K}$$

You may use the result of Problem 31. (c) Verify numerically that $\Delta S > 0$. (d) Is the mixing an irreversible process? Explain how you know.

62. A 1.00-mol sample of an ideal gas ($\gamma = 1.40$) is carried through the Carnot cycle described in Active Figure 22.10. At point A, the pressure is 25.0 atm and the temperature is 600 K. At point C, the pressure is 1.00 atm and the temperature is 400 K. (a) Determine the pressures and volumes at points A, B, C, and D. (b) Calculate the net work done per cycle. (c) Determine the efficiency of an engine operating in this cycle.

Answers to Quick Quizzes

22.1 **(i)**, (c). Equation 22.2 gives this result directly. **(ii)**, (b). The work represents one third of the input energy. The remaining two thirds must be expelled to the cold reservoir.

22.2 (d). The COP of 4.00 for the heat pump means that you are receiving four times as much energy as that entering by electrical transmission. With four times as much energy per unit of energy from electricity, you need only one-fourth as much electricity.

22.3 C, B, A. Although all three engines operate over a 300-K temperature difference, the efficiency depends on the ratio of temperatures, not the difference.

22.4 (a) One microstate: all four deuces. (b) Six microstates: club–diamond, club–heart, club–spade, diamond–heart, diamond–spade, heart–spade. The macrostate of two aces is more probable than that of four deuces in part (a) because there are six times as many microstates for this particular macrostate compared with the macrostate of four deuces. Therefore, in a hand of poker, two of a kind is less valuable than four of a kind.

22.5 (a). From the first law of thermodynamics, for these two reversible processes, $Q_r = \Delta E_{\text{int}} - W$. During the constant-volume process, $W = 0$, while the work W is nonzero and negative during the constant-pressure expansion. Therefore, Q_r is larger for the constant-pressure process, leading to a larger value for the change in entropy. In terms of entropy as disorder, during the constant-pressure process the gas must expand. The increase in volume results in more ways of locating the molecules of the gas in a container, resulting in a larger increase in entropy.

22.6 False. The determining factor for the entropy change is Q_r, not Q. If the adiabatic process is not reversible, the entropy change is not necessarily zero because a reversible path between the same initial and final states may involve energy transfer by heat.

PART 4

Electricity and Magnetism

We now study the branch of physics concerned with electric and magnetic phenomena. The laws of electricity and magnetism play a central role in the operation of such devices as MP3 players, televisions, electric motors, computers, high-energy accelerators, and other electronic devices. More fundamentally, the interatomic and intermolecular forces responsible for the formation of solids and liquids are electric in origin.

Evidence in Chinese documents suggests magnetism was observed as early as 2000 BC. The ancient Greeks observed electric and magnetic phenomena possibly as early as 700 BC. The Greeks knew about magnetic forces from observations that the naturally occurring stone *magnetite* (Fe_3O_4) is attracted to iron. (The word *electric* comes from *elecktron*, the Greek word for "amber." The word *magnetic* comes from *Magnesia*, the name of the district of Greece where magnetite was first found.)

Not until the early part of the nineteenth century did scientists establish that electricity and magnetism are related phenomena. In 1819, Hans Oersted discovered that a compass needle is deflected when placed near a circuit carrying an electric current. In 1831, Michael Faraday and, almost simultaneously, Joseph Henry showed that when a wire is moved near a magnet (or, equivalently, when a magnet is moved near a wire), an electric current is established in the wire. In 1873, James Clerk Maxwell used these observations and other experimental facts as a basis for formulating the laws of electromagnetism as we know them today. (*Electromagnetism* is a name given to the combined study of electricity and magnetism.)

Maxwell's contributions to the field of electromagnetism were especially significant because the laws he formulated are basic to *all* forms of electromagnetic phenomena. His work is as important as Newton's work on the laws of motion and the theory of gravitation.

Lightning is a dramatic example of electrical phenomena occurring in nature. Although we are most familiar with lightning originating from thunderclouds, it can occur in other situations such as in a volcanic eruption (here, the Sakurajima volcano, Japan). (M. Zhilin/M. Newman/Photo Researchers, Inc.)

Mother and daughter are both enjoying the effects of electrically charging their bodies. Each individual hair on their heads becomes charged and exerts a repulsive force on the other hairs, resulting in the "stand-up" hair-dos seen here. (Courtesy of Resonance Research Corporation)

23 Electric Fields

The electromagnetic force between charged particles is one of the fundamental forces of nature. We begin this chapter by describing some basic properties of one manifestation of the electromagnetic force, the electric force. We then discuss Coulomb's law, which is the fundamental law governing the electric force between any two charged particles. Next, we introduce the concept of an electric field associated with a charge distribution and describe its effect on other charged particles. We then show how to use Coulomb's law to calculate the electric field for a given charge distribution. The chapter concludes with a discussion of the motion of a charged particle in a uniform electric field.

23.1 Properties of Electric Charges

A number of simple experiments demonstrate the existence of electric forces. For example, after rubbing a balloon on your hair on a dry day, you will find that the balloon attracts bits of paper. The attractive force is often strong enough to suspend the paper from the balloon.

When materials behave in this way, they are said to be *electrified* or to have become **electrically charged**. You can easily electrify your body by vigorously rubbing your shoes on a wool rug. Evidence of the electric charge on your body can be detected by lightly touching (and startling) a friend. Under the right conditions, you will see a spark when you touch and both of you will feel a slight tingle.

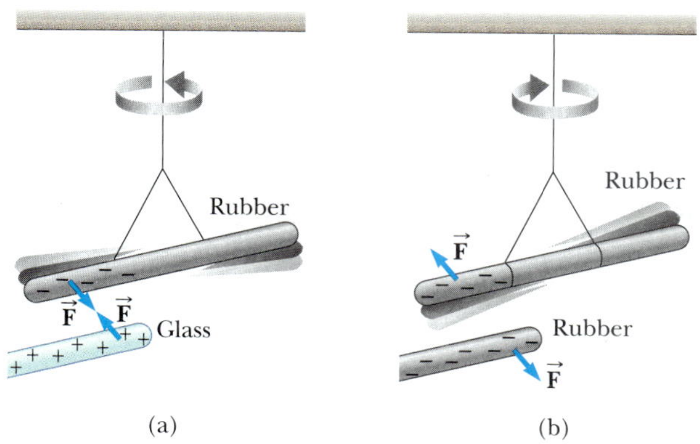

Figure 23.1 (a) A negatively charged rubber rod suspended by a thread is attracted to a positively charged glass rod. (b) A negatively charged rubber rod is repelled by another negatively charged rubber rod.

(Experiments such as these work best on a dry day because an excessive amount of moisture in the air can cause any charge you build up to "leak" from your body to the Earth.)

In a series of simple experiments, it was found that there are two kinds of electric charges, which were given the names **positive** and **negative** by Benjamin Franklin (1706–1790). Electrons are identified as having negative charge, while protons are positively charged. To verify that there are two types of charge, suppose a hard rubber rod that has been rubbed on fur is suspended by a sewing thread as shown in Figure 23.1. When a glass rod that has been rubbed on silk is brought near the rubber rod, the two attract each other (Fig. 23.1a). On the other hand, if two charged rubber rods (or two charged glass rods) are brought near each other as shown in Figure 23.1b, the two repel each other. This observation shows that the rubber and glass have two different types of charge on them. On the basis of these observations, we conclude that **charges of the same sign repel one another and charges with opposite signs attract one another**.

Using the convention suggested by Franklin, the electric charge on the glass rod is called positive and that on the rubber rod is called negative. Therefore, any charged object attracted to a charged rubber rod (or repelled by a charged glass rod) must have a positive charge, and any charged object repelled by a charged rubber rod (or attracted to a charged glass rod) must have a negative charge.

◀ Electric charge is conserved

Another important aspect of electricity that arises from experimental observations is that **electric charge is always conserved** in an isolated system. That is, when one object is rubbed against another, charge is not created in the process. The electrified state is due to a *transfer* of charge from one object to the other. One object gains some amount of negative charge while the other gains an equal amount of positive charge. For example, when a glass rod is rubbed on silk as in Figure 23.2, the silk obtains a negative charge equal in magnitude to the positive charge on the glass rod. We now know from our understanding of atomic structure that electrons are transferred in the rubbing process from the glass to the silk. Similarly, when rubber is rubbed on fur, electrons are transferred from the fur to the rubber, giving the rubber a net negative charge and the fur a net positive charge. This process is consistent with the fact that neutral, uncharged matter contains as many positive charges (protons within atomic nuclei) as negative charges (electrons).

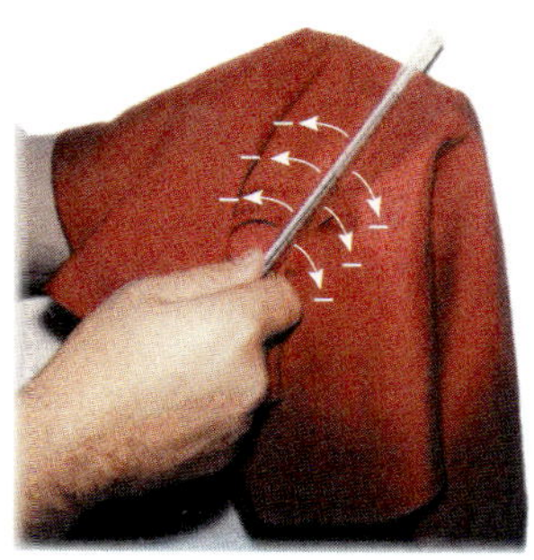

Figure 23.2 When a glass rod is rubbed with silk, electrons are transferred from the glass to the silk. Because of conservation of charge, each electron adds negative charge to the silk and an equal positive charge is left behind on the rod. Also, because the charges are transferred in discrete bundles, the charges on the two objects are $\pm e$, or $\pm 2e$, or $\pm 3e$, and so on.

In 1909, Robert Millikan (1868–1953) discovered that electric charge always occurs as integral multiples of a fundamental amount of charge e (see Section 25.7). In modern terms, the electric charge q is said to be **quantized,** where q is the standard symbol used for charge as a variable. That is, electric charge exists as discrete "packets," and we can write $q = \pm Ne$, where N is some integer. Other experiments in the same period showed that the electron has a charge $-e$ and the proton has a charge of equal magnitude but opposite sign $+e$. Some particles, such as the neutron, have no charge.

Quick Quiz 23.1 Three objects are brought close to each other, two at a time. When objects A and B are brought together, they repel. When objects B and C are brought together, they also repel. Which of the following are true? (a) Objects A and C possess charges of the same sign. (b) Objects A and C possess charges of opposite sign. (c) All three objects possess charges of the same sign. (d) One object is neutral. (e) Additional experiments must be performed to determine the signs of the charges.

23.2 Charging Objects by Induction

It is convenient to classify materials in terms of the ability of electrons to move through the material:

> Electrical **conductors** are materials in which some of the electrons are free electrons[1] that are not bound to atoms and can move relatively freely through the material; electrical **insulators** are materials in which all electrons are bound to atoms and cannot move freely through the material.

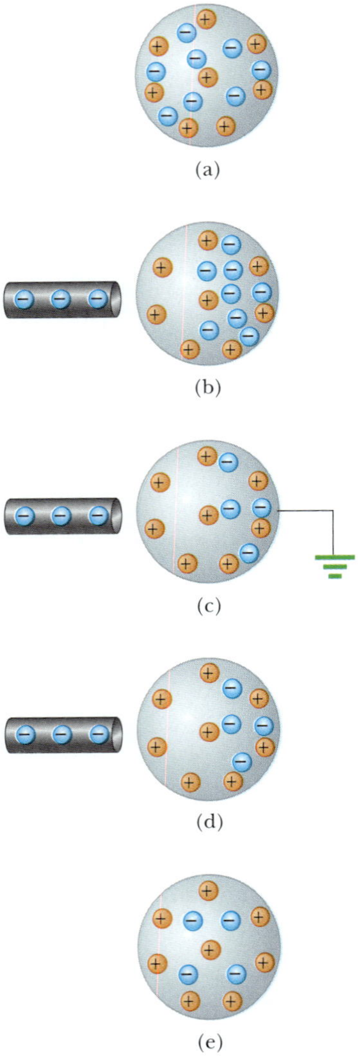

Figure 23.3 Charging a metallic object by *induction* (that is, the two objects never touch each other). (a) A neutral metallic sphere, with equal numbers of positive and negative charges. (b) The electrons on the neutral sphere are redistributed when a charged rubber rod is placed near the sphere. (c) When the sphere is grounded, some of its electrons leave through the ground wire. (d) When the ground connection is removed, the sphere has excess positive charge that is nonuniformly distributed. (e) When the rod is removed, the remaining electrons redistribute uniformly and there is a net uniform distribution of positive charge on the sphere.

Materials such as glass, rubber, and dry wood fall into the category of electrical insulators. When such materials are charged by rubbing, only the area rubbed becomes charged and the charged particles are unable to move to other regions of the material.

In contrast, materials such as copper, aluminum, and silver are good electrical conductors. When such materials are charged in some small region, the charge readily distributes itself over the entire surface of the material.

Semiconductors are a third class of materials, and their electrical properties are somewhere between those of insulators and those of conductors. Silicon and germanium are well-known examples of semiconductors commonly used in the fabrication of a variety of electronic chips used in computers, cellular telephones, and stereo systems. The electrical properties of semiconductors can be changed over many orders of magnitude by the addition of controlled amounts of certain atoms to the materials.

To understand how to charge a conductor by a process known as **induction,** consider a neutral (uncharged) conducting sphere insulated from the ground as shown in Figure 23.3a. There are an equal number of electrons and protons in the sphere if the charge on the sphere is exactly zero. When a negatively charged rubber rod is brought near the sphere, electrons in the region nearest the rod experience a repulsive force and migrate to the opposite side of the sphere. This migration leaves the side of the sphere near the rod with an effective positive charge because of the diminished number of electrons as in Figure 23.3b. (The left side of the sphere in Figure 23.3b is positively charged *as if* positive charges moved into this region, but remember that it is only electrons that are free to move.) This process occurs even if the rod never actually touches the sphere. If the same experiment is performed with a conducting wire connected from the sphere to the Earth (Fig. 23.3c), some of the electrons in the conductor are so strongly repelled by the presence of the negative charge in the rod that they move out of the sphere through the wire and into the Earth. The symbol ⏚ at the end of the wire in Figure 23.3c indicates that the wire is connected to **ground,** which means a reservoir, such as the Earth, that can accept or provide electrons freely with negligible effect on its electrical characteristics. If the wire to ground is then removed

[1] A metal atom contains one or more outer electrons, which are weakly bound to the nucleus. When many atoms combine to form a metal, the *free electrons* are these outer electrons, which are not bound to any one atom. These electrons move about the metal in a manner similar to that of gas molecules moving in a container.

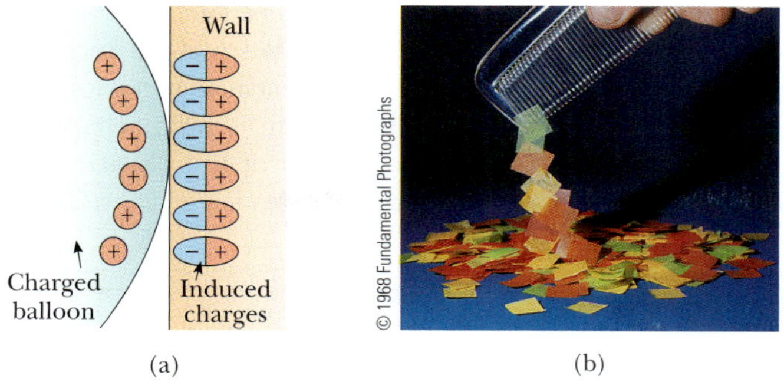

Figure 23.4 (a) The charged object on the left induces a charge distribution on the surface of an insulator due to realignment of charges in the molecules. (b) A charged comb attracts bits of paper because charges in molecules in the paper are realigned.

(Fig. 23.3d), the conducting sphere contains an excess of *induced* positive charge because it has fewer electrons than it needs to cancel out the positive charge of the protons. When the rubber rod is removed from the vicinity of the sphere (Fig. 23.3e), this induced positive charge remains on the ungrounded sphere. Notice that the rubber rod loses none of its negative charge during this process.

Charging an object by induction requires no contact with the object inducing the charge. That is in contrast to charging an object by rubbing (that is, by *conduction*), which does require contact between the two objects.

A process similar to induction in conductors takes place in insulators. In most neutral molecules, the center of positive charge coincides with the center of negative charge. In the presence of a charged object, however, these centers inside each molecule in an insulator may shift slightly, resulting in more positive charge on one side of the molecule than on the other. This realignment of charge within individual molecules produces a layer of charge on the surface of the insulator as shown in Figure 23.4a. Your knowledge of induction in insulators should help you explain why a comb that has been drawn through your hair attracts bits of electrically neutral paper as shown in Figure 23.4b.

Quick Quiz 23.2 Three objects are brought close to one another, two at a time. When objects A and B are brought together, they attract. When objects B and C are brought together, they repel. Which of the following are necessarily true? (a) Objects A and C possess charges of the same sign. (b) Objects A and C possess charges of opposite sign. (c) All three objects possess charges of the same sign. (d) One object is neutral. (e) Additional experiments must be performed to determine information about the charges on the objects.

23.3 Coulomb's Law

Charles Coulomb measured the magnitudes of the electric forces between charged objects using the torsion balance, which he invented (Fig. 23.5). The operating principle of the torsion balance is the same as that of the apparatus used by Cavendish to measure the gravitational constant (see Section 13.1), with the electrically neutral spheres replaced by charged ones. The electric force between charged spheres A and B in Figure 23.5 causes the spheres to either attract or repel each other, and the resulting motion causes the suspended fiber to twist. Because the restoring torque of the twisted fiber is proportional to the angle through which the fiber rotates, a measurement of this angle provides a quantitative measure of the electric force of attraction or repulsion. Once the spheres are charged by rubbing, the electric force between them is very large compared with the gravitational attraction, and so the gravitational force can be neglected.

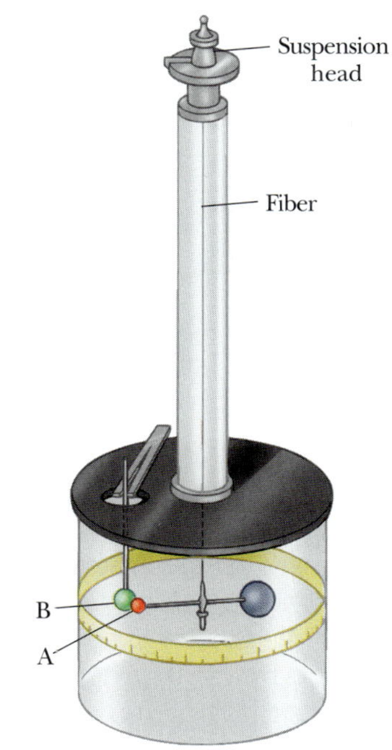

Figure 23.5 Coulomb's torsion balance, used to establish the inverse-square law for the electric force between two charges.

From Coulomb's experiments, we can generalize the properties of the **electric force** between two stationary charged particles. We use the term **point charge** to refer to a charged particle of zero size. The electrical behavior of electrons and protons is very well described by modeling them as point charges. From experimental observations, we find that the magnitude of the electric force (sometimes called the *Coulomb force*) between two point charges is given by **Coulomb's law:**

Coulomb's law ▶

$$F_e = k_e \frac{|q_1||q_2|}{r^2} \quad (23.1)$$

where k_e is a constant called the **Coulomb constant**. In his experiments, Coulomb was able to show that the value of the exponent of r was 2 to within an uncertainty of a few percent. Modern experiments have shown that the exponent is 2 to within an uncertainty of a few parts in 10^{16}. Experiments also show that the electric force, like the gravitational force, is conservative.

The value of the Coulomb constant depends on the choice of units. The SI unit of charge is the **coulomb** (C). The Coulomb constant k_e in SI units has the value

Coulomb constant ▶

$$k_e = 8.987\,6 \times 10^9 \text{ N}\cdot\text{m}^2/\text{C}^2 \quad (23.2)$$

This constant is also written in the form

$$k_e = \frac{1}{4\pi\epsilon_0} \quad (23.3)$$

where the constant ϵ_0 (Greek letter epsilon) is known as the **permittivity of free space** and has the value

$$\epsilon_0 = 8.854\,2 \times 10^{-12} \text{ C}^2/\text{N}\cdot\text{m}^2 \quad (23.4)$$

The smallest unit of free charge e known in nature,[2] the charge on an electron $(-e)$ or a proton $(+e)$, has a magnitude

$$e = 1.602\,18 \times 10^{-19} \text{ C} \quad (23.5)$$

Therefore, 1 C of charge is approximately equal to the charge of 6.24×10^{18} electrons or protons. This number is very small when compared with the number of free electrons in 1 cm^3 of copper, which is on the order of 10^{23}. Nevertheless, 1 C is a substantial amount of charge. In typical experiments in which a rubber or glass rod is charged by friction, a net charge on the order of 10^{-6} C is obtained. In other words, only a very small fraction of the total available charge is transferred between the rod and the rubbing material.

The charges and masses of the electron, proton, and neutron are given in Table 23.1.

Photo courtesy of AIP Niels Bohr Library/E. Scott Barr Collection

CHARLES COULOMB
French physicist (1736–1806)
Coulomb's major contributions to science were in the areas of electrostatics and magnetism. During his lifetime, he also investigated the strengths of materials and determined the forces that affect objects on beams, thereby contributing to the field of structural mechanics. In the field of ergonomics, his research provided a fundamental understanding of the ways in which people and animals can best do work.

TABLE 23.1

Charge and Mass of the Electron, Proton, and Neutron

Particle	Charge (C)	Mass (kg)
Electron (e)	$-1.602\,176\,5 \times 10^{-19}$	$9.109\,4 \times 10^{-31}$
Proton (p)	$+1.602\,176\,5 \times 10^{-19}$	$1.672\,62 \times 10^{-27}$
Neutron (n)	0	$1.674\,93 \times 10^{-27}$

[2] No unit of charge smaller than e has been detected on a free particle; current theories, however, propose the existence of particles called *quarks* having charges $-e/3$ and $2e/3$. Although there is considerable experimental evidence for such particles inside nuclear matter, *free* quarks have never been detected. We discuss other properties of quarks in Chapter 46.

EXAMPLE 23.1 The Hydrogen Atom

The electron and proton of a hydrogen atom are separated (on the average) by a distance of approximately 5.3×10^{-11} m. Find the magnitudes of the electric force and the gravitational force between the two particles.

SOLUTION

Conceptualize Think about the two particles separated by the very small distance given in the problem statement. In Chapter 13, we found the gravitational force between small objects to be weak, so we expect the gravitational force between the electron and proton to be significantly smaller than the electric force.

Categorize The electric and gravitational forces will be evaluated from universal force laws, so we categorize this example as a substitution problem.

Use Coulomb's law to find the magnitude of the electric force:

$$F_e = k_e \frac{|e||-e|}{r^2} = (8.99 \times 10^9 \text{ N}\cdot\text{m}^2/\text{C}^2)\frac{(1.60 \times 10^{-19}\text{ C})^2}{(5.3 \times 10^{-11}\text{ m})^2}$$

$$= 8.2 \times 10^{-8}\text{ N}$$

Use Newton's law of universal gravitation and Table 23.1 (for the particle masses) to find the magnitude of the gravitational force:

$$F_g = G\frac{m_e m_p}{r^2}$$

$$= (6.67 \times 10^{-11}\text{ N}\cdot\text{m}^2/\text{kg}^2)\frac{(9.11 \times 10^{-31}\text{ kg})(1.67 \times 10^{-27}\text{ kg})}{(5.3 \times 10^{-11}\text{ m})^2}$$

$$= 3.6 \times 10^{-47}\text{ N}$$

The ratio $F_e/F_g \approx 2 \times 10^{39}$. Therefore, the gravitational force between charged atomic particles is negligible when compared with the electric force. Notice the similar forms of Newton's law of universal gravitation and Coulomb's law of electric forces. Other than magnitude, what is a fundamental difference between the two forces?

When dealing with Coulomb's law, remember that force is a vector quantity and must be treated accordingly. Coulomb's law expressed in vector form for the electric force exerted by a charge q_1 on a second charge q_2, written $\vec{\mathbf{F}}_{12}$, is

$$\vec{\mathbf{F}}_{12} = k_e \frac{q_1 q_2}{r^2}\hat{\mathbf{r}}_{12} \quad \textbf{(23.6)}$$

◀ Vector form of Coulomb's law

where $\hat{\mathbf{r}}_{12}$ is a unit vector directed from q_1 toward q_2 as shown in Active Figure 23.6a. Because the electric force obeys Newton's third law, the electric force exerted by q_2 on q_1 is equal in magnitude to the force exerted by q_1 on q_2 and in

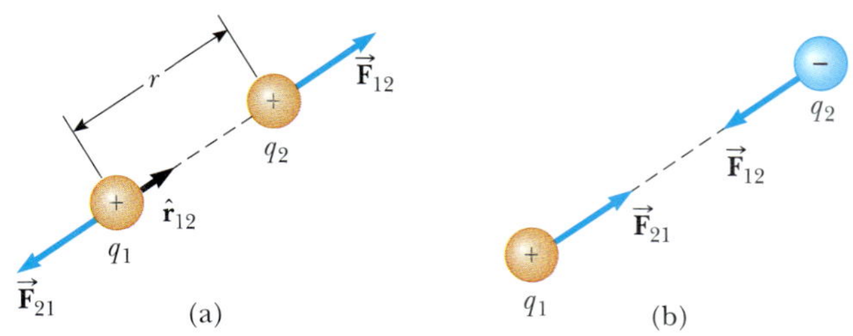

ACTIVE FIGURE 23.6

Two point charges separated by a distance r exert a force on each other that is given by Coulomb's law. The force $\vec{\mathbf{F}}_{21}$ exerted by q_2 on q_1 is equal in magnitude and opposite in direction to the force $\vec{\mathbf{F}}_{12}$ exerted by q_1 on q_2. (a) When the charges are of the same sign, the force is repulsive. (b) When the charges are of opposite signs, the force is attractive.

Sign in at www.thomsonedu.com and go to ThomsonNOW to move the charges to any position in two-dimensional space and observe the electric forces on them.

the opposite direction; that is, $\vec{\mathbf{F}}_{21} = -\vec{\mathbf{F}}_{12}$. Finally, Equation 23.6 shows that if q_1 and q_2 have the same sign as in Active Figure 23.6a, the product q_1q_2 is positive. If q_1 and q_2 are of opposite sign as shown in Active Figure 23.6b, the product q_1q_2 is negative. These signs describe the *relative* direction of the force but not the *absolute* direction. A negative product indicates an attractive force, and each charge experiences a force toward the other. A positive product indicates a repulsive force such that each charge experiences a force away from the other. The *absolute* direction of the force on a charge depends on the location of the other charge. For example, if an x axis lies along the two charges in Active Figure 23.6a, the product q_1q_2 is positive, but $\vec{\mathbf{F}}_{12}$ points in the $+x$ direction and $\vec{\mathbf{F}}_{21}$ points in the $-x$ direction.

When more than two charges are present, the force between any pair of them is given by Equation 23.6. Therefore, the resultant force on any one of them equals the vector sum of the forces exerted by the other individual charges. For example, if four charges are present, the resultant force exerted by particles 2, 3, and 4 on particle 1 is

$$\vec{\mathbf{F}}_1 = \vec{\mathbf{F}}_{21} + \vec{\mathbf{F}}_{31} + \vec{\mathbf{F}}_{41}$$

Quick Quiz 23.3 Object A has a charge of $+2\ \mu$C, and object B has a charge of $+6\ \mu$C. Which statement is true about the electric forces on the objects? (a) $\vec{\mathbf{F}}_{AB} = -3\vec{\mathbf{F}}_{BA}$ (b) $\vec{\mathbf{F}}_{AB} = -\vec{\mathbf{F}}_{BA}$ (c) $3\vec{\mathbf{F}}_{AB} = -\vec{\mathbf{F}}_{BA}$ (d) $\vec{\mathbf{F}}_{AB} = 3\vec{\mathbf{F}}_{BA}$ (e) $\vec{\mathbf{F}}_{AB} = \vec{\mathbf{F}}_{BA}$ (f) $3\vec{\mathbf{F}}_{AB} = \vec{\mathbf{F}}_{BA}$

EXAMPLE 23.2 Find the Resultant Force

Consider three point charges located at the corners of a right triangle as shown in Figure 23.7, where $q_1 = q_3 = 5.0\ \mu$C, $q_2 = -2.0\ \mu$C, and $a = 0.10$ m. Find the resultant force exerted on q_3.

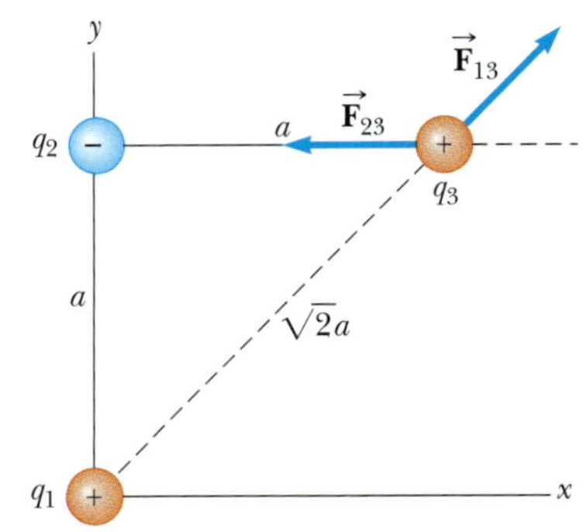

Figure 23.7 (Example 23.2) The force exerted by q_1 on q_3 is $\vec{\mathbf{F}}_{13}$. The force exerted by q_2 on q_3 is $\vec{\mathbf{F}}_{23}$. The resultant force $\vec{\mathbf{F}}_3$ exerted on q_3 is the vector sum $\vec{\mathbf{F}}_{13} + \vec{\mathbf{F}}_{23}$.

SOLUTION

Conceptualize Think about the net force on q_3. Because charge q_3 is near two other charges, it will experience two electric forces.

Categorize Because two forces are exerted on charge q_3, we categorize this example as a vector addition problem.

Analyze The directions of the individual forces exerted by q_1 and q_2 on q_3 are shown in Figure 23.7. The force $\vec{\mathbf{F}}_{23}$ exerted by q_2 on q_3 is attractive because q_2 and q_3 have opposite signs. In the coordinate system shown in Figure 23.7, the attractive force $\vec{\mathbf{F}}_{23}$ is to the left (in the negative x direction).

The force $\vec{\mathbf{F}}_{13}$ exerted by q_1 on q_3 is repulsive because both charges are positive. The repulsive force $\vec{\mathbf{F}}_{13}$ makes an angle of 45° with the x axis.

Use Equation 23.1 to find the magnitude of $\vec{\mathbf{F}}_{23}$:

$$F_{23} = k_e \frac{|q_2||q_3|}{a^2}$$

$$= (8.99 \times 10^9\ \mathrm{N \cdot m^2/C^2}) \frac{(2.0 \times 10^{-6}\ \mathrm{C})(5.0 \times 10^{-6}\ \mathrm{C})}{(0.10\ \mathrm{m})^2} = 9.0\ \mathrm{N}$$

Find the magnitude of the force $\vec{\mathbf{F}}_{13}$:

$$F_{13} = k_e \frac{|q_1||q_3|}{(\sqrt{2}a)^2}$$

$$= (8.99 \times 10^9\ \mathrm{N \cdot m^2/C^2}) \frac{(5.0 \times 10^{-6}\ \mathrm{C})(5.0 \times 10^{-6}\ \mathrm{C})}{2(0.10\ \mathrm{m})^2} = 11\ \mathrm{N}$$

Find the x and y components of the force $\vec{\mathbf{F}}_{13}$:

$$F_{13x} = F_{13} \cos 45° = 7.9 \text{ N}$$

$$F_{13y} = F_{13} \sin 45° = 7.9 \text{ N}$$

Find the components of the resultant force acting on q_3:

$$F_{3x} = F_{13x} + F_{23x} = 7.9 \text{ N} + (-9.0 \text{ N}) = -1.1 \text{ N}$$

$$F_{3y} = F_{13y} + F_{23y} = 7.9 \text{ N} + 0 = 7.9 \text{ N}$$

Express the resultant force acting on q_3 in unit–vector form:

$$\vec{\mathbf{F}}_3 = (-1.1\hat{\mathbf{i}} + 7.9\hat{\mathbf{j}}) \text{ N}$$

Finalize The net force on q_3 is upward and toward the left in Figure 23.7. If q_3 moves in response to the net force, the distances between q_3 and the other charges change, so the net force changes. Therefore, q_3 can be modeled as a particle under a net force as long as it is recognized that the force exerted on q_3 is *not* constant.

What If? What if the signs of all three charges were changed to the opposite signs? How would that affect the result for $\vec{\mathbf{F}}_3$?

Answer The charge q_3 would still be attracted toward q_2 and repelled from q_1 with forces of the same magnitude. Therefore, the final result for $\vec{\mathbf{F}}_3$ would be the same.

EXAMPLE 23.3 Where Is the Net Force Zero?

Three point charges lie along the x axis as shown in Figure 23.8. The positive charge $q_1 = 15.0\ \mu\text{C}$ is at $x = 2.00$ m, the positive charge $q_2 = 6.00\ \mu\text{C}$ is at the origin, and the net force acting on q_3 is zero. What is the x coordinate of q_3?

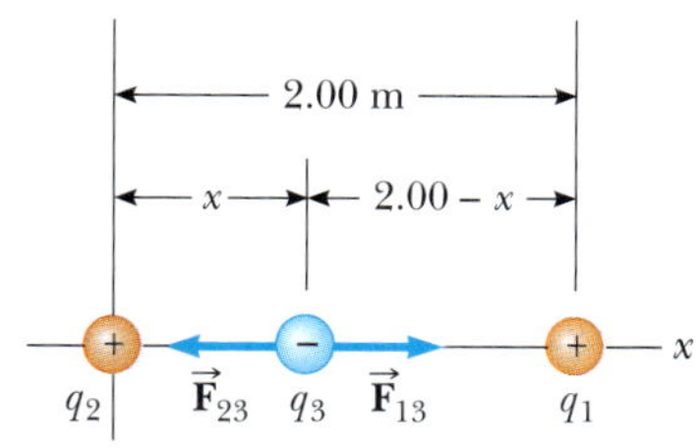

Figure 23.8 (Example 23.3) Three point charges are placed along the x axis. If the resultant force acting on q_3 is zero, the force $\vec{\mathbf{F}}_{13}$ exerted by q_1 on q_3 must be equal in magnitude and opposite in direction to the force $\vec{\mathbf{F}}_{23}$ exerted by q_2 on q_3.

SOLUTION

Conceptualize Because q_3 is near two other charges, it experiences two electric forces. Unlike the preceding example, however, the forces lie along the same line in this problem as indicated in Figure 23.8. Because q_3 is negative while q_1 and q_2 are positive, the forces $\vec{\mathbf{F}}_{13}$ and $\vec{\mathbf{F}}_{23}$ are both attractive.

Categorize Because the net force on q_3 is zero, we model the point charge as a particle in equilibrium.

Analyze Write an expression for the net force on charge q_3 when it is in equilibrium:

$$\vec{\mathbf{F}}_3 = \vec{\mathbf{F}}_{23} + \vec{\mathbf{F}}_{13} = -k_e \frac{|q_2||q_3|}{x^2}\hat{\mathbf{i}} + k_e \frac{|q_1||q_3|}{(2.00 - x)^2}\hat{\mathbf{i}} = 0$$

Move the second term to the right side of the equation and set the coefficients of the unit vector $\hat{\mathbf{i}}$ equal:

$$k_e \frac{|q_2||q_3|}{x^2} = k_e \frac{|q_1||q_3|}{(2.00 - x)^2}$$

Eliminate k_e and $|q_3|$ and rearrange the equation:

$$(2.00 - x)^2|q_2| = x^2|q_1|$$

$$(4.00 - 4.00x + x^2)(6.00 \times 10^{-6} \text{ C}) = x^2(15.0 \times 10^{-6} \text{ C})$$

Reduce the quadratic equation to a simpler form:

$$3.00x^2 + 8.00x - 8.00 = 0$$

Solve the quadratic equation for the positive root:

$$x = 0.775 \text{ m}$$

Finalize The second root to the quadratic equation is $x = -3.44$ m. That is another location where the *magnitudes* of the forces on q_3 are equal, but both forces are in the same direction.

What If? Suppose q_3 is constrained to move only along the x axis. From its initial position at $x = 0.775$ m, it is pulled a small distance along the x axis. When released, does it return to equilibrium, or is it pulled further from equilibrium? That is, is the equilibrium stable or unstable?

Answer If q_3 is moved to the right, $\vec{\mathbf{F}}_{13}$ becomes larger and $\vec{\mathbf{F}}_{23}$ becomes smaller. The result is a net force to the right, in the same direction as the displacement. Therefore, the charge q_3 would continue to move to the right and the equilibrium is *unstable.* (See Section 7.9 for a review of stable and unstable equilibrium.)

If q_3 is constrained to stay at a *fixed* x coordinate but allowed to move up and down in Figure 23.8, the equilibrium is stable. In this case, if the charge is pulled upward (or downward) and released, it moves back toward the equilibrium position and oscillates about this point.

EXAMPLE 23.4 Find the Charge on the Spheres

Two identical small charged spheres, each having a mass of 3.0×10^{-2} kg, hang in equilibrium as shown in Figure 23.9a. The length of each string is 0.15 m, and the angle θ is 5.0°. Find the magnitude of the charge on each sphere.

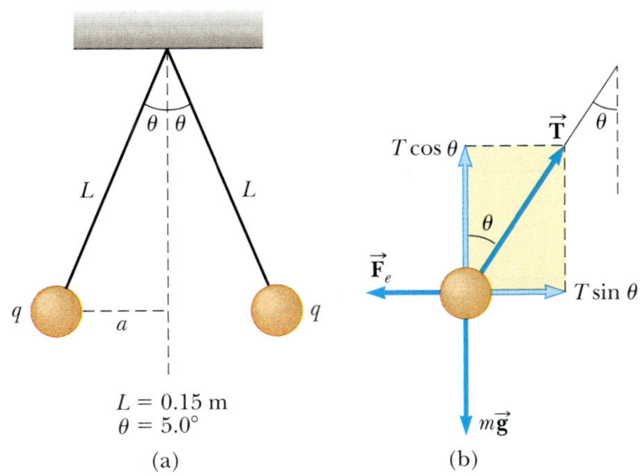

Figure 23.9 (Example 23.4) (a) Two identical spheres, each carrying the same charge q, suspended in equilibrium. (b) The free-body diagram for the sphere on the left of part (a).

SOLUTION

Conceptualize Figure 23.9a helps us conceptualize this example. The two spheres exert repulsive forces on each other. If they are held close to each other and released, they move outward from the center and settle into the configuration in Figure 23.9a after the oscillations have vanished due to air resistance.

Categorize The key phrase "in equilibrium" helps us model each sphere as a particle in equilibrium. This example is similar to the particle in equilibrium problems in Chapter 5 with the added feature that one of the forces on a sphere is an electric force.

Analyze The free-body diagram for the left-hand sphere is shown in Figure 23.9b. The sphere is in equilibrium under the application of the forces $\vec{\mathbf{T}}$ from the string, the electric force $\vec{\mathbf{F}}_e$ from the other sphere, and the gravitational force $m\vec{\mathbf{g}}$.

Write Newton's second law for the left-hand sphere in component form:

$$(1)\quad \sum F_x = T\sin\theta - F_e = 0 \rightarrow T\sin\theta = F_e$$

$$(2)\quad \sum F_y = T\cos\theta - mg = 0 \rightarrow T\cos\theta = mg$$

Divide Equation (1) by Equation (2) to find F_e:

$$\tan\theta = \frac{F_e}{mg} \rightarrow F_e = mg\tan\theta$$

Evaluate the electric force numerically:

$$F_e = (3.0 \times 10^{-2}\text{ kg})(9.80\text{ m/s}^2)\tan(5.0°) = 2.6 \times 10^{-2}\text{ N}$$

Use the geometry of the right triangle in Figure 23.9a to find a relationship between a, L, and θ:

$$\sin\theta = \frac{a}{L} \rightarrow a = L\sin\theta$$

Evaluate a:

$$a = (0.15\text{ m})\sin(5.0°) = 0.013\text{ m}$$

Solve Coulomb's law (Eq. 23.1) for the charge $|q|$ on each sphere:

$$F_e = k_e\frac{|q|^2}{r^2} \rightarrow |q| = \sqrt{\frac{F_e r^2}{k_e}} = \sqrt{\frac{F_e(2a)^2}{k_e}}$$

Substitute numerical values:

$$|q| = \sqrt{\frac{(2.6 \times 10^{-2}\ \text{N})[2(0.013\ \text{m})]^2}{8.99 \times 10^9\ \text{N}\cdot\text{m}^2/\text{C}^2}} = 4.4 \times 10^{-8}\ \text{C}$$

Finalize We cannot determine the sign of the charge from the information given. In fact, the sign of the charge is not important. The situation is the same whether both spheres are positively charged or negatively charged.

What If? Suppose your roommate proposes solving this problem without the assumption that the charges are of equal magnitude. She claims the symmetry of the problem is destroyed if the charges are not equal, so the strings would make two different angles with the vertical and the problem would be much more complicated. How would you respond?

Answer The symmetry is not destroyed and the angles are not different. Newton's third law requires the magnitudes of the electric forces on the two charges to be the same, regardless of the equality or nonequality of the charges. The solution to the example remains the same with one change: the value of $|q|^2$ in the solution is replaced by $|q_1 q_2|$ in the new situation, where q_1 and q_2 are the values of the charges on the two spheres. The symmetry of the problem would be destroyed if the *masses* of the spheres were not the same. In this case, the strings would make different angles with the vertical and the problem would be more complicated.

23.4 The Electric Field

© Johnny Autery

This dramatic photograph captures a lightning bolt striking a tree near some rural homes. Lightning is associated with very strong electric fields in the atmosphere.

Two field forces—the gravitational force in Chapter 13 and the electric force here—have been introduced into our discussions so far. As pointed out earlier, field forces can act through space, producing an effect even when no physical contact occurs between interacting objects. The gravitational field $\vec{\mathbf{g}}$ at a point in space due to a source particle was defined in Section 13.4 to be equal to the gravitational force $\vec{\mathbf{F}}_g$ acting on a test particle of mass m divided by that mass: $\vec{\mathbf{g}} \equiv \vec{\mathbf{F}}_g/m$. The concept of a field was developed by Michael Faraday (1791–1867) in the context of electric forces and is of such practical value that we shall devote much attention to it in the next several chapters. In this approach, an **electric field** is said to exist in the region of space around a charged object, the **source charge**. When another charged object—the **test charge**—enters this electric field, an electric force acts on it. As an example, consider Figure 23.10, which shows a small positive test charge q_0 placed near a second object carrying a much greater positive charge Q. We define the electric field due to the source charge at the location of the test charge to be the electric force on the test charge *per unit charge*, or, to be more specific, **the electric field vector $\vec{\mathbf{E}}$** at a point in space is defined as the electric force $\vec{\mathbf{F}}_e$ acting on a positive test charge q_0 placed at that point divided by the test charge:[3]

$$\vec{\mathbf{E}} \equiv \frac{\vec{\mathbf{F}}_e}{q_0} \qquad (23.7)$$

◀ Definition of electric field

The vector $\vec{\mathbf{E}}$ has the SI units of newtons per coulomb (N/C). Note that $\vec{\mathbf{E}}$ is the field produced by some charge or charge distribution *separate from* the test charge; it is not the field produced by the test charge itself. Also note that the existence of an electric field is a property of its source; the presence of the test charge is not necessary for the field to exist. The test charge serves as a *detector* of the electric field.

The direction of $\vec{\mathbf{E}}$ as shown in Figure 23.10 is the direction of the force a positive test charge experiences when placed in the field. **An electric field exists at a point if a test charge at that point experiences an electric force.**

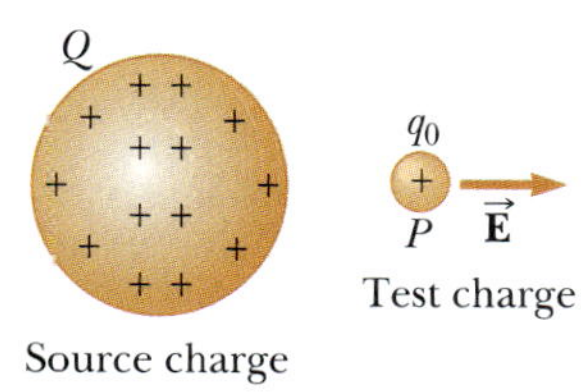

Figure 23.10 A small positive test charge q_0 placed at point P near an object carrying a much larger positive charge Q experiences an electric field $\vec{\mathbf{E}}$ at point P established by the source charge Q.

[3] When using Equation 23.7, we must assume the test charge q_0 is small enough that it does not disturb the charge distribution responsible for the electric field. If the test charge is great enough, the charge on the metallic sphere is redistributed and the electric field it sets up is different from the field it sets up in the presence of the much smaller test charge.

PITFALL PREVENTION 23.1
Particles Only

Equation 23.8 is only valid for a *particle* of charge q, that is, an object of zero size. For a charged *object* of finite size in an electric field, the field may vary in magnitude and direction over the size of the object, so the corresponding force equation may be more complicated.

Equation 23.7 can be rearranged as

$$\vec{\mathbf{F}}_e = q\vec{\mathbf{E}} \qquad (23.8)$$

This equation gives us the force on a charged particle q placed in an electric field. If q is positive, the force is in the same direction as the field. If q is negative, the force and the field are in opposite directions. Notice the similarity between Equation 23.8 and the corresponding equation for a particle with mass placed in a gravitational field, $\vec{\mathbf{F}}_g = m\vec{\mathbf{g}}$ (Section 5.5). Once the magnitude and direction of the electric field are known at some point, the electric force exerted on *any* charged particle placed at that point can be calculated from Equation 23.8.

To determine the direction of an electric field, consider a point charge q as a source charge. This charge creates an electric field at all points in space surrounding it. A test charge q_0 is placed at point P, a distance r from the source charge, as in Active Figure 23.11a. We imagine using the test charge to determine the direction of the electric force and therefore that of the electric field. According to Coulomb's law, the force exerted by q on the test charge is

$$\vec{\mathbf{F}}_e = k_e \frac{qq_0}{r^2}\hat{\mathbf{r}}$$

where $\hat{\mathbf{r}}$ is a unit vector directed from q toward q_0. This force in Active Figure 23.11a is directed away from the source charge q. Because the electric field at P, the position of the test charge, is defined by $\vec{\mathbf{E}} = \vec{\mathbf{F}}_e/q_0$, the electric field at P created by q is

$$\vec{\mathbf{E}} = k_e \frac{q}{r^2}\hat{\mathbf{r}} \qquad (23.9)$$

If the source charge q is positive, Active Figure 23.11b shows the situation with the test charge removed: the source charge sets up an electric field at P, directed away from q. If q is negative, as in Active Figure 23.11c, the force on the test charge is toward the source charge, so the electric field at P is directed toward the source charge as in Active Figure 23.11d.

To calculate the electric field at a point P due to a group of point charges, we first calculate the electric field vectors at P individually using Equation 23.9 and then add them vectorially. In other words, at any point P, the total electric field due to a group of source charges equals the vector sum of the electric fields of all the charges. This superposition principle applied to fields follows directly from the vector addition of electric forces. Therefore, the electric field at point P due to a group of source charges can be expressed as the vector sum

Electric field due to a finite number of point charges ►

$$\vec{\mathbf{E}} = k_e \sum_i \frac{q_i}{r_i^2}\hat{\mathbf{r}}_i \qquad (23.10)$$

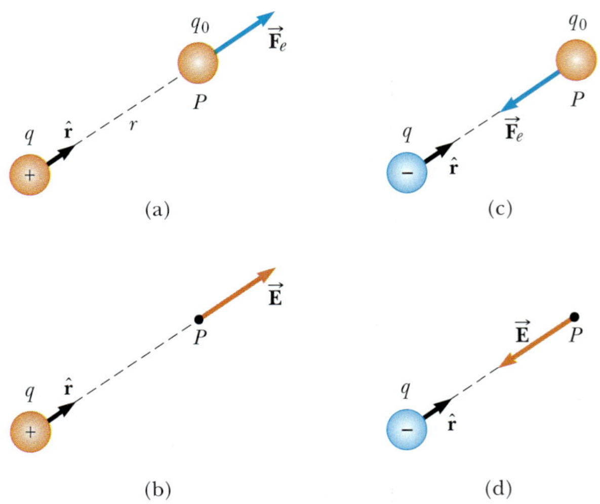

ACTIVE FIGURE 23.11

A test charge q_0 at point P is a distance r from a point charge q. (a) If q is positive, the force on the test charge is directed away from q. (b) For the positive source charge, the electric field at P points radially outward from q. (c) If q is negative, the force on the test charge is directed toward q. (d) For the negative source charge, the electric field at P points radially inward toward q.

Sign in at www.thomsonedu.com and go to ThomsonNOW to move point P to any position in two-dimensional space and observe the electric field due to q.

where r_i is the distance from the ith source charge q_i to the point P and $\hat{\mathbf{r}}_i$ is a unit vector directed from q_i toward P.

In Example 23.5, we explore the electric field due to two charges using the superposition principle. Part (B) of the example focuses on an **electric dipole**, which is defined as a positive charge q and a negative charge $-q$ separated by a distance $2a$. The electric dipole is a good model of many molecules, such as hydrochloric acid (HCl). Neutral atoms and molecules behave as dipoles when placed in an external electric field. Furthermore, many molecules, such as HCl, are permanent dipoles. The effect of such dipoles on the behavior of materials subjected to electric fields is discussed in Chapter 26.

Quick Quiz 23.4 A test charge of $+3\ \mu C$ is at a point P where an external electric field is directed to the right and has a magnitude of 4×10^6 N/C. If the test charge is replaced with another test charge of $-3\ \mu C$, what happens to the external electric field at P? (a) It is unaffected. (b) It reverses direction. (c) It changes in a way that cannot be determined.

EXAMPLE 23.5 Electric Field Due to Two Charges

Charges q_1 and q_2 are located on the x axis, at distances a and b, respectively, from the origin as shown in Figure 23.12.

(A) Find the components of the net electric field at the point P, which is on the y axis.

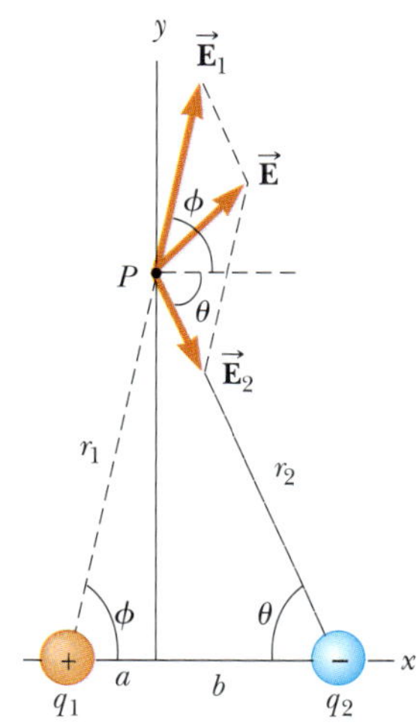

Figure 23.12 (Example 23.5) The total electric field $\vec{\mathbf{E}}$ at P equals the vector sum $\vec{\mathbf{E}}_1 + \vec{\mathbf{E}}_2$, where $\vec{\mathbf{E}}_1$ is the field due to the positive charge q_1 and $\vec{\mathbf{E}}_2$ is the field due to the negative charge q_2.

SOLUTION

Conceptualize Compare this example to Example 23.2. There, we add vector forces to find the net force on a charged particle. Here, we add electric field vectors to find the net electric field at a point in space.

Categorize We have two source charges and wish to find the resultant electric field, so we categorize this example as one in which we can use the superposition principle represented by Equation 23.10.

Analyze Find the magnitude of the electric field at P due to charge q_1:

$$E_1 = k_e \frac{|q_1|}{r_1^2} = k_e \frac{|q_1|}{(a^2 + y^2)}$$

Find the magnitude of the electric field at P due to charge q_2:

$$E_2 = k_e \frac{|q_2|}{r_2^2} = k_e \frac{|q_2|}{(b^2 + y^2)}$$

Write the electric field vectors for each charge in unit–vector form:

$$\vec{\mathbf{E}}_1 = k_e \frac{|q_1|}{(a^2 + y^2)} \cos\phi\,\hat{\mathbf{i}} + k_e \frac{|q_1|}{(a^2 + y^2)} \sin\phi\,\hat{\mathbf{j}}$$

$$\vec{\mathbf{E}}_2 = k_e \frac{|q_2|}{(b^2 + y^2)} \cos\theta\,\hat{\mathbf{i}} - k_e \frac{|q_2|}{(b^2 + y^2)} \sin\theta\,\hat{\mathbf{j}}$$

Write the components of the net electric field vector:

$$(1)\quad E_x = E_{1x} + E_{2x} = k_e \frac{|q_1|}{(a^2 + y^2)} \cos\phi + k_e \frac{|q_2|}{(b^2 + y^2)} \cos\theta$$

$$(2)\quad E_y = E_{1y} + E_{2y} = k_e \frac{|q_1|}{(a^2 + y^2)} \sin\phi - k_e \frac{|q_2|}{(b^2 + y^2)} \sin\theta$$

(B) Evaluate the electric field at point P in the special case that $|q_1| = |q_2|$ and $a = b$.

SOLUTION

Conceptualize Figure 23.13 shows the situation in this special case. Notice the symmetry in the situation and that the charge distribution is now an electric dipole.

Categorize Because Figure 23.13 is a special case of the general case shown in Figure 23.12, we can categorize this example as one in which we can take the result of part (A) and substitute the appropriate values of the variables.

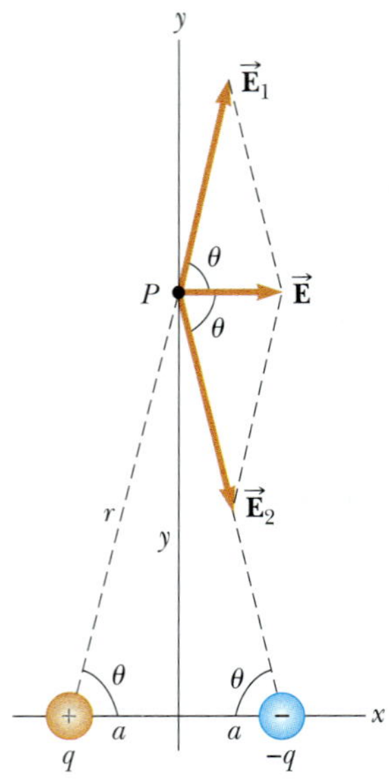

Figure 23.13 (Example 23.5) When the charges in Figure 23.12 are of equal magnitude and equidistant from the origin, the situation becomes symmetric as shown here.

Analyze Based on the symmetry in Figure 23.13, evaluate Equations (1) and (2) from part (A) with $a = b$, $|q_1| = |q_2| = q$, and $\phi = \theta$:

$$(3)\quad E_x = k_e \frac{q}{(a^2 + y^2)} \cos\theta + k_e \frac{q}{(a^2 + y^2)} \cos\theta = 2k_e \frac{q}{(a^2 + y^2)} \cos\theta$$

$$E_y = k_e \frac{q}{(a^2 + y^2)} \sin\theta - k_e \frac{q}{(a^2 + y^2)} \sin\theta = 0$$

From the geometry in Figure 23.13, evaluate $\cos\theta$:

$$(4)\quad \cos\theta = \frac{a}{r} = \frac{a}{(a^2 + y^2)^{1/2}}$$

Substitute Equation (4) into Equation (3):

$$E_x = 2k_e \frac{q}{(a^2 + y^2)} \frac{a}{(a^2 + y^2)^{1/2}} = k_e \frac{2qa}{(a^2 + y^2)^{3/2}}$$

(C) Find the electric field due to the electric dipole when point P is a distance $y \gg a$ from the origin.

SOLUTION

In the solution to part (B), because $y \gg a$, neglect a^2 compared with y^2 and write the expression for E in this case:

$$(5)\quad E \approx k_e \frac{2qa}{y^3}$$

Finalize From Equation (5), we see that at points far from a dipole but along the perpendicular bisector of the line joining the two charges, the magnitude of the electric field created by the dipole varies as $1/r^3$, whereas the more slowly varying field of a point charge varies as $1/r^2$ (see Eq. 23.9). That is because at distant points, the fields of the two charges of equal magnitude and opposite sign almost cancel each other. The $1/r^3$ variation in E for the dipole also is obtained for a distant point along the x axis (see Problem 18) and for any general distant point.

23.5 Electric Field of a Continuous Charge Distribution

Very often, the distances between charges in a group of charges are much smaller than the distance from the group to a point where the electric field is to be calculated. In such situations, the system of charges can be modeled as continuous. That is, the system of closely spaced charges is equivalent to a total charge that is continuously distributed along some line, over some surface, or throughout some volume.

To set up the process for evaluating the electric field created by a continuous charge distribution, let's use the following procedure. First, divide the charge distribution into small elements, each of which contains a small charge Δq as shown in Figure 23.14. Next, use Equation 23.9 to calculate the electric field due to one of these elements at a point *P*. Finally, evaluate the total electric field at *P* due to the charge distribution by summing the contributions of all the charge elements (that is, by applying the superposition principle).

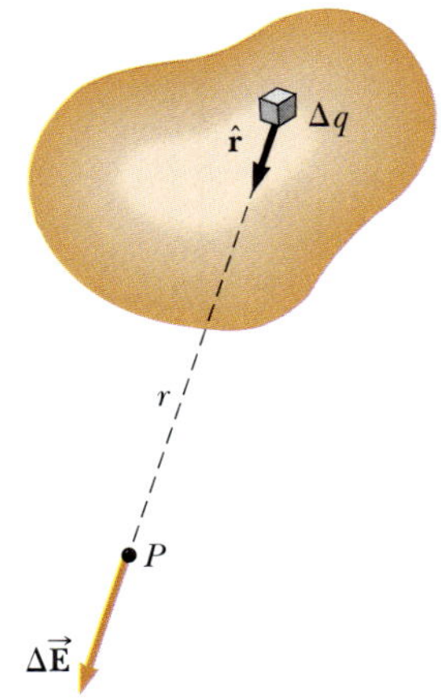

Figure 23.14 The electric field at *P* due to a continuous charge distribution is the vector sum of the fields $\Delta\vec{\mathbf{E}}$ due to all the elements Δq of the charge distribution.

The electric field at *P* due to one charge element carrying charge Δq is

$$\Delta\vec{\mathbf{E}} = k_e \frac{\Delta q}{r^2}\hat{\mathbf{r}}$$

where r is the distance from the charge element to point *P* and $\hat{\mathbf{r}}$ is a unit vector directed from the element toward *P*. The total electric field at *P* due to all elements in the charge distribution is approximately

$$\vec{\mathbf{E}} \approx k_e \sum_i \frac{\Delta q_i}{r_i^2}\hat{\mathbf{r}}_i$$

where the index i refers to the ith element in the distribution. Because the charge distribution is modeled as continuous, the total field at *P* in the limit $\Delta q_i \to 0$ is

$$\vec{\mathbf{E}} = k_e \lim_{\Delta q_i \to 0} \sum_i \frac{\Delta q_i}{r_i^2}\hat{\mathbf{r}}_i = k_e \int \frac{dq}{r^2}\hat{\mathbf{r}} \qquad \textbf{(23.11)}$$

◀ Electric field due to a continuous charge distribution

where the integration is over the entire charge distribution. The integration in Equation 23.11 is a vector operation and must be treated appropriately.

Let's illustrate this type of calculation with several examples in which the charge is distributed on a line, on a surface, or throughout a volume. When performing such calculations, it is convenient to use the concept of a *charge density* along with the following notations:

- If a charge Q is uniformly distributed throughout a volume V, the **volume charge density** ρ is defined by

  $$\rho \equiv \frac{Q}{V}$$

  ◀ Volume charge density

  where ρ has units of coulombs per cubic meter (C/m^3).

- If a charge Q is uniformly distributed on a surface of area A, the **surface charge density** σ (Greek letter sigma) is defined by

  $$\sigma \equiv \frac{Q}{A}$$

  ◀ Surface charge density

  where σ has units of coulombs per square meter (C/m^2).

- If a charge Q is uniformly distributed along a line of length ℓ, the **linear charge density** λ is defined by

  $$\lambda \equiv \frac{Q}{\ell}$$

  ◀ Linear charge density

  where λ has units of coulombs per meter (C/m).

- If the charge is nonuniformly distributed over a volume, surface, or line, the amounts of charge dq in a small volume, surface, or length element are

  $$dq = \rho\, dV \qquad dq = \sigma\, dA \qquad dq = \lambda\, d\ell$$

PROBLEM-SOLVING STRATEGY Calculating the Electric Field

The following procedure is recommended for solving problems that involve the determination of an electric field due to individual charges or a charge distribution:

1. *Conceptualize.* Establish a mental representation of the problem: think carefully about the individual charges or the charge distribution and imagine what type of electric field they would create. Appeal to any symmetry in the arrangement of charges to help you visualize the electric field.

2. *Categorize.* Are you analyzing a group of individual charges or a continuous charge distribution? The answer to this question tells you how to proceed in the Analyze step.

3. *Analyze.*

   (a) *If you are analyzing a group of individual charge,* use the superposition principle: When several point charges are present, the resultant field at a point in space is the *vector sum* of the individual fields due to the individual charges (Eq. 23.10). Be very careful in the manipulation of vector quantities. It may be useful to review the material on vector addition in Chapter 3. Example 23.5 demonstrated this procedure.

   (b) *If you are analyzing a continuous charge distribution,* replace the vector sums for evaluating the total electric field from individual charges by vector integrals. The charge distribution is divided into infinitesimal pieces, and the vector sum is carried out by integrating over the entire charge distribution (Eq. 23.11). Examples 23.6 through 23.8 demonstrate such procedures.

   Consider symmetry when dealing with either a distribution of point charges or a continuous charge distribution. Take advantage of any symmetry in the system you observed in the Conceptualize step to simplify your calculations. The cancellation of field components perpendicular to the axis in Example 23.7 is an example of the application of symmetry.

4. *Finalize.* Check to see if your electric field expression is consistent with the mental representation and if it reflects any symmetry that you noted previously. Imagine varying parameters such as the distance of the observation point from the charges or the radius of any circular objects to see if the mathematical result changes in a reasonable way.

EXAMPLE 23.6 The Electric Field Due to a Charged Rod

A rod of length ℓ has a uniform positive charge per unit length λ and a total charge Q. Calculate the electric field at a point P that is located along the long axis of the rod and a distance a from one end (Fig. 23.15).

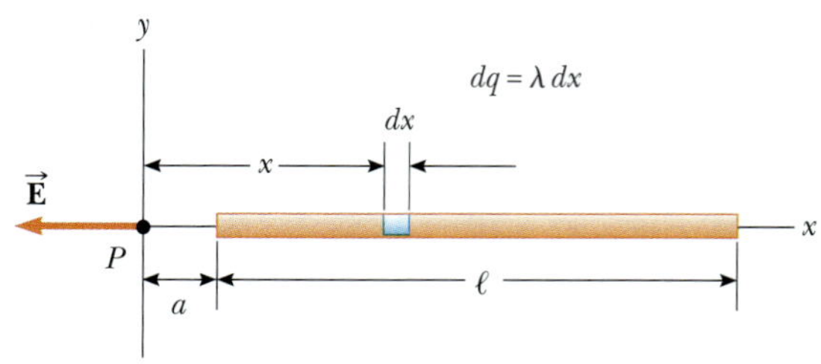

Figure 23.15 (Example 23.6) The electric field at P due to a uniformly charged rod lying along the x axis. The magnitude of the field at P due to the segment of charge dq is $k_e\,dq/x^2$. The total field at P is the vector sum over all segments of the rod.

SOLUTION

Conceptualize The field $d\vec{\mathbf{E}}$ at P due to each segment of charge on the rod is in the negative x direction because every segment carries a positive charge.

Categorize Because the rod is continuous, we are evaluating the field due to a continuous charge distribution rather than a group of individual charges. Because every segment of the rod produces an electric field in the negative x direction, the sum of their contributions can be handled without the need to add vectors.

Analyze Let's assume the rod is lying along the x axis, dx is the length of one small segment, and dq is the charge on that segment. Because the rod has a charge per unit length λ, the charge dq on the small segment is $dq = \lambda\,dx$.

Find the magnitude of the electric field at P due to one segment of the rod having a charge dq:

$$dE = k_e \frac{dq}{x^2} = k_e \frac{\lambda\, dx}{x^2}$$

Find the total field at P using[4] Equation 23.11:

$$E = \int_a^{\ell + a} k_e \lambda \frac{dx}{x^2}$$

Noting that k_e and $\lambda = Q/\ell$ are constants and can be removed from the integral, evaluate the integral:

$$E = k_e \lambda \int_a^{\ell + a} \frac{dx}{x^2} = k_e \lambda \left[-\frac{1}{x} \right]_a^{\ell + a}$$

$$(1) \quad E = k_e \frac{Q}{\ell} \left(\frac{1}{a} - \frac{1}{\ell + a} \right) = \frac{k_e Q}{a(\ell + a)}$$

Finalize If ℓ goes to zero, Equation (1) reduces to the electric field due to a point charge as given by Equation 23.9, which is what we expect because the rod has shrunk to zero size.

What If? Suppose point P is very far away from the rod. What is the nature of the electric field at such a point?

Answer If P is far from the rod ($a \gg \ell$), then ℓ in the denominator of Equation (1) can be neglected and $E \approx k_e Q/a^2$. That is exactly the form you would expect for a point charge. Therefore, at large values of a/ℓ, the charge distribution appears to be a point charge of magnitude Q; the point P is so far away from the rod we cannot distinguish that it has a size. The use of the limiting technique ($a/\ell \to \infty$) is often a good method for checking a mathematical expression.

[4]To carry out integrations such as this one, first express the charge element dq in terms of the other variables in the integral. (In this example, there is one variable, x, so we made the change $dq = \lambda\, dx$.) The integral must be over scalar quantities; therefore, express the electric field in terms of components, if necessary. (In this example, the field has only an x component, so this detail is of no concern.) Then, reduce your expression to an integral over a single variable (or to multiple integrals, each over a single variable). In examples that have spherical or cylindrical symmetry, the single variable is a radial coordinate.

EXAMPLE 23.7 The Electric Field of a Uniform Ring of Charge

A ring of radius a carries a uniformly distributed positive total charge Q. Calculate the electric field due to the ring at a point P lying a distance x from its center along the central axis perpendicular to the plane of the ring (Fig. 23.16a).

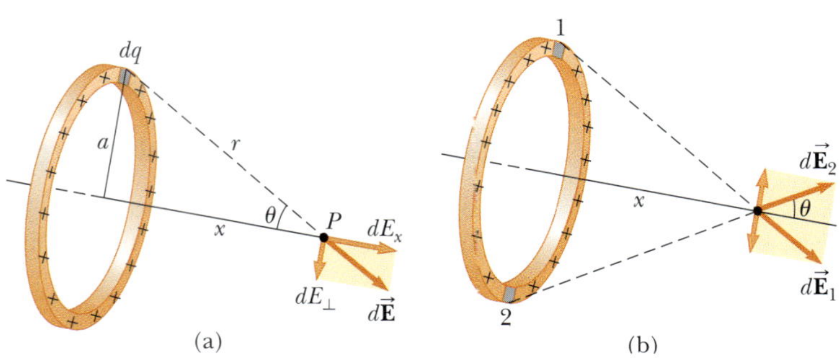

Figure 23.16 (Example 23.7) A uniformly charged ring of radius a. (a) The field at P on the x axis due to an element of charge dq. (b) The total electric field at P is along the x axis. The perpendicular component of the field at P due to segment 1 is canceled by the perpendicular component due to segment 2.

SOLUTION

Conceptualize Figure 23.16a shows the electric field contribution $d\vec{\mathbf{E}}$ at P due to a single segment of charge at the top of the ring. This field vector can be resolved into components dE_x parallel to the axis of the ring and $dE_\perp$ perpendicular to the axis. Figure 23.16b shows the electric field contributions from two segments on opposite sides of the ring. Because of the symmetry of the situation, the perpendicular components of the field cancel. That is true for all pairs of segments around the ring, so we can ignore the perpendicular component of the field and focus solely on the parallel components, which simply add.

Categorize Because the ring is continuous, we are evaluating the field due to a continuous charge distribution rather than a group of individual charges.

Analyze Evaluate the parallel component of an electric field contribution from a segment of charge dq on the ring:

$$(1)\quad dE_x = k_e \frac{dq}{r^2}\cos\theta = k_e \frac{dq}{(a^2 + x^2)}\cos\theta$$

From the geometry in Figure 23.16a, evaluate $\cos\theta$:

$$(2)\quad \cos\theta = \frac{x}{r} = \frac{x}{(a^2 + x^2)^{1/2}}$$

Substitute Equation (2) into Equation (1):

$$dE_x = k_e \frac{dq}{(a^2 + x^2)} \frac{x}{(a^2 + x^2)^{1/2}} = \frac{k_e x}{(a^2 + x^2)^{3/2}}\, dq$$

All segments of the ring make the same contribution to the field at P because they are all equidistant from this point. Integrate to obtain the total field at P:

$$E_x = \int \frac{k_e x}{(a^2 + x^2)^{3/2}}\, dq = \frac{k_e x}{(a^2 + x^2)^{3/2}} \int dq$$

$$(3)\quad E = \frac{k_e x}{(a^2 + x^2)^{3/2}}\, Q$$

Finalize This result shows that the field is zero at $x = 0$. Is that consistent with the symmetry in the problem? Furthermore, notice that Equation (3) reduces to $k_e Q/x^2$ if $x \gg a$, so the ring acts like a point charge for locations far away from the ring.

What If? Suppose a negative charge is placed at the center of the ring in Figure 23.16 and displaced slightly by a distance $x \ll a$ along the x axis. When the charge is released, what type of motion does it exhibit?

Answer In the expression for the field due to a ring of charge, let $x \ll a$, which results in

$$E_x = \frac{k_e Q}{a^3}\, x$$

Therefore, from Equation 23.8, the force on a charge $-q$ placed near the center of the ring is

$$F_x = -\frac{k_e q Q}{a^3}\, x$$

Because this force has the form of Hooke's law (Eq. 15.1), the motion of the negative charge is *simple harmonic*!

EXAMPLE 23.8 The Electric Field of a Uniformly Charged Disk

A disk of radius R has a uniform surface charge density σ. Calculate the electric field at a point P that lies along the central perpendicular axis of the disk and a distance x from the center of the disk (Fig. 23.17).

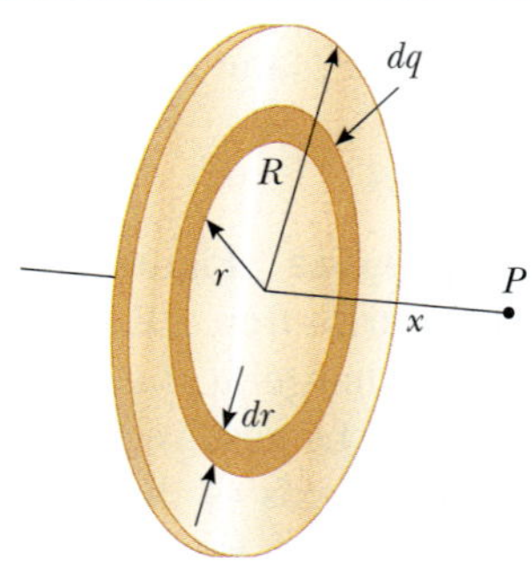

Figure 23.17 (Example 23.8) A uniformly charged disk of radius R. The electric field at an axial point P is directed along the central axis, perpendicular to the plane of the disk.

SOLUTION

Conceptualize If the disk is considered to be a set of concentric rings, we can use our result from Example 23.7—which gives the field created by a ring of radius a—and sum the contributions of all rings making up the disk. By symmetry, the field at an axial point must be along the central axis.

Categorize Because the disk is continuous, we are evaluating the field due to a continuous charge distribution rather than a group of individual charges.

Analyze Find the amount of charge dq on a ring of radius r and width dr as shown in Figure 23.17:

$$dq = \sigma\, dA = \sigma(2\pi r\, dr) = 2\pi\sigma r\, dr$$

Use this result in the equation given for E_x in Example 23.7 (with a replaced by r and Q replaced by dq) to find the field due to the ring:

$$dE_x = \frac{k_e x}{(r^2 + x^2)^{3/2}}(2\pi\sigma r\, dr)$$

To obtain the total field at P, integrate this expression over the limits $r = 0$ to $r = R$, noting that x is a constant in this situation:

$$E_x = k_e x \pi\sigma \int_0^R \frac{2r\, dr}{(r^2 + x^2)^{3/2}}$$

$$= k_e x \pi\sigma \int_0^R (r^2 + x^2)^{-3/2}\, d(r^2)$$

$$= k_e x \pi\sigma \left[\frac{(r^2 + x^2)^{-1/2}}{-1/2}\right]_0^R = 2\pi k_e \sigma \left[1 - \frac{x}{(R^2 + x^2)^{1/2}}\right]$$

Finalize This result is valid for all values of $x > 0$. We can calculate the field close to the disk along the axis by assuming that $R \gg x$; therefore, the expression in brackets reduces to unity to give us the near-field approximation

$$E_x = 2\pi k_e \sigma = \frac{\sigma}{2\epsilon_0}$$

where ϵ_0 is the permittivity of free space. In Chapter 24, we obtain the same result for the field created by an infinite plane of charge with uniform surface charge density.

23.6 Electric Field Lines

We have defined the electric field mathematically through Equation 23.7. Let's now explore a means of visualizing the electric field in a pictorial representation. A convenient way of visualizing electric field patterns is to draw lines, called **electric field lines** and first introduced by Faraday, that are related to the electric field in a region of space in the following manner:

- The electric field vector $\vec{\mathbf{E}}$ is tangent to the electric field line at each point. The line has a direction, indicated by an arrowhead, that is the same as that of the electric field vector. The direction of the line is that of the force on a positive test charge placed in the field.
- The number of lines per unit area through a surface perpendicular to the lines is proportional to the magnitude of the electric field in that region. Therefore, the field lines are close together where the electric field is strong and far apart where the field is weak.

These properties are illustrated in Figure 23.18. The density of field lines through surface A is greater than the density of lines through surface B. Therefore, the magnitude of the electric field is larger on surface A than on surface B. Furthermore, because the lines at different locations point in different directions, the field is nonuniform.

Is this relationship between strength of the electric field and the density of field lines consistent with Equation 23.9, the expression we obtained for E using Coulomb's law? To answer this question, consider an imaginary spherical surface of radius r concentric with a point charge. From symmetry, we see that the magnitude of the electric field is the same everywhere on the surface of the sphere. The number of lines N that emerge from the charge is equal to the number that penetrate the spherical surface. Hence, the number of lines per unit area on the sphere is $N/4\pi r^2$ (where the surface area of the sphere is $4\pi r^2$). Because E is proportional to the number of lines per unit area, we see that E varies as $1/r^2$; this finding is consistent with Equation 23.9.

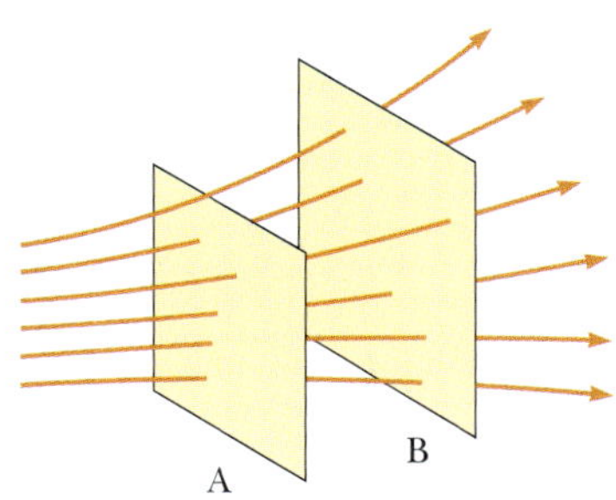

Figure 23.18 Electric field lines penetrating two surfaces. The magnitude of the field is greater on surface A than on surface B.

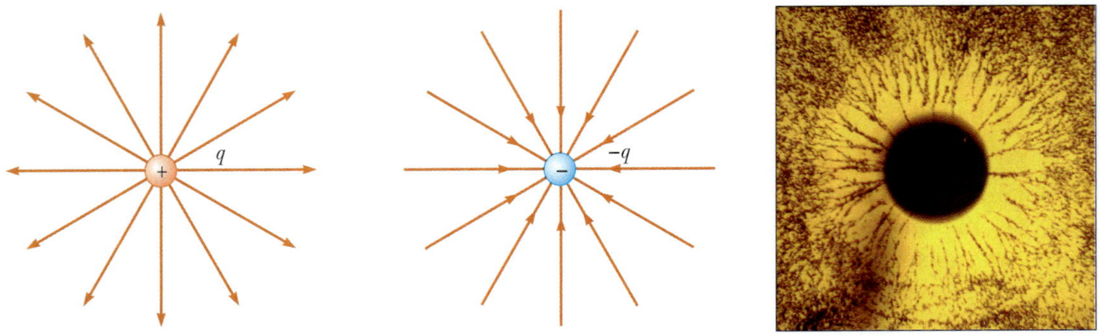

Douglas C. Johnson, California State Polytechnic University, Pomona

(a) (b) (c)

Figure 23.19 The electric field lines for a point charge. (a) For a positive point charge, the lines are directed radially outward. (b) For a negative point charge, the lines are directed radially inward. Notice that the figures show only those field lines that lie in the plane of the page. (c) The dark areas are small particles suspended in oil, which align with the electric field produced by a charged conductor at the center.

PITFALL PREVENTION 23.2
Electric Field Lines Are Not Paths of Particles!

Electric field lines represent the field at various locations. Except in very special cases, they *do not* represent the path of a charged particle moving in an electric field.

PITFALL PREVENTION 23.3
Electric Field Lines Are Not Real

Electric field lines are not material objects. They are used only as a pictorial representation to provide a qualitative description of the electric field. Only a finite number of lines from each charge can be drawn, which makes it appear as if the field were quantized and exists only in certain parts of space. The field, in fact, is continuous, existing at every point. You should avoid obtaining the wrong impression from a two-dimensional drawing of field lines used to describe a three-dimensional situation.

Representative electric field lines for the field due to a single positive point charge are shown in Figure 23.19a. This two-dimensional drawing shows only the field lines that lie in the plane containing the point charge. The lines are actually directed radially outward from the charge in all directions; therefore, instead of the flat "wheel" of lines shown, you should picture an entire spherical distribution of lines. Because a positive test charge placed in this field would be repelled by the positive source charge, the lines are directed radially away from the source charge. The electric field lines representing the field due to a single negative point charge are directed toward the charge (Fig. 23.19b). In either case, the lines are along the radial direction and extend all the way to infinity. Notice that the lines become closer together as they approach the charge, indicating that the strength of the field increases as we move toward the source charge.

The rules for drawing electric field lines are as follows:

- The lines must begin on a positive charge and terminate on a negative charge. In the case of an excess of one type of charge, some lines will begin or end infinitely far away.
- The number of lines drawn leaving a positive charge or approaching a negative charge is proportional to the magnitude of the charge.
- No two field lines can cross.

We choose the number of field lines starting from any positively charged object to be Cq and the number of lines ending on any negatively charged object to be $C|q|$, where C is an arbitrary proportionality constant. Once C is chosen, the number of lines is fixed. For example, in a two-charge system, if object 1 has charge Q_1 and object 2 has charge Q_2, the ratio of number of lines in contact with the

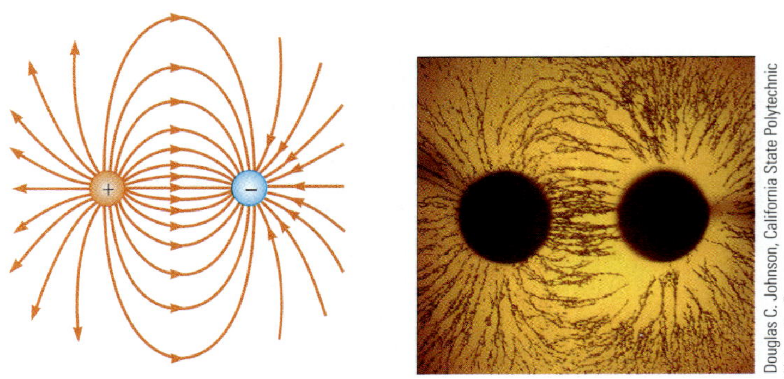

Douglas C. Johnson, California State Polytechnic University, Pomona

(a) (b)

Figure 23.20 (a) The electric field lines for two point charges of equal magnitude and opposite sign (an electric dipole). The number of lines leaving the positive charge equals the number terminating at the negative charge. (b) Small particles suspended in oil align with the electric field.

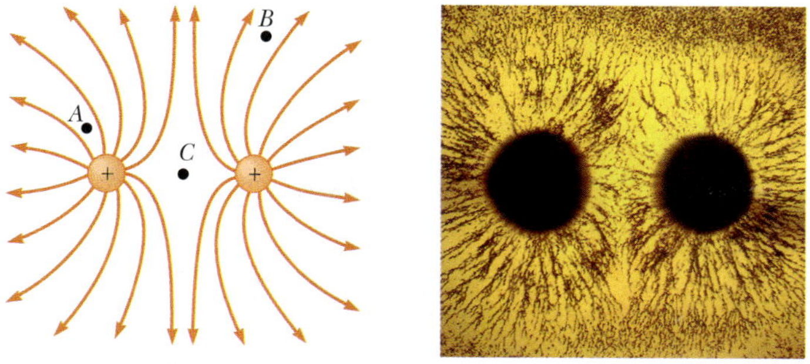

(a) (b)

Douglas C. Johnson, California State Polytechnic University, Pomona

Figure 23.21 (a) The electric field lines for two positive point charges. (The locations *A*, *B*, and *C* are discussed in Quick Quiz 23.5.) (b) Small particles suspended in oil align with the electric field.

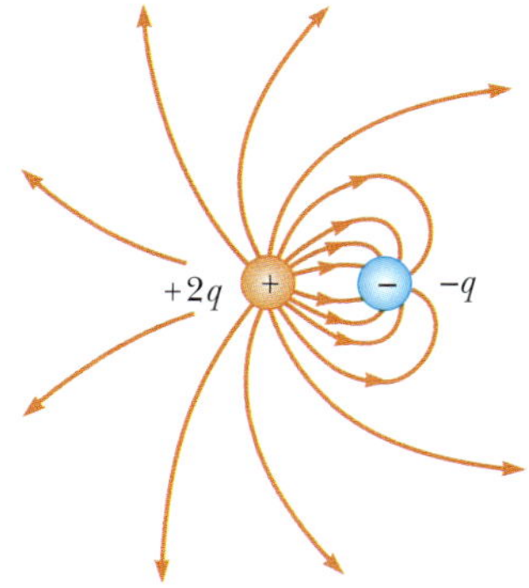

ACTIVE FIGURE 23.22

The electric field lines for a point charge $+2q$ and a second point charge $-q$. Notice that two lines leave $+2q$ for every one that terminates on $-q$.

Sign in at www.thomsonedu.com and go to ThomsonNOW to choose the values and signs for the two charges and observe the electric field lines for the configuration you have chosen.

charges is $N_2/N_1 = Q_2/Q_1$. The electric field lines for two point charges of equal magnitude but opposite signs (an electric dipole) are shown in Figure 23.20. Because the charges are of equal magnitude, the number of lines that begin at the positive charge must equal the number that terminate at the negative charge. At points very near the charges, the lines are nearly radial. The high density of lines between the charges indicates a region of strong electric field.

Figure 23.21 shows the electric field lines in the vicinity of two equal positive point charges. Again, the lines are nearly radial at points close to either charge, and the same number of lines emerge from each charge because the charges are equal in magnitude. At great distances from the charges, the field is approximately equal to that of a single point charge of magnitude $2q$.

Finally, in Active Figure 23.22, we sketch the electric field lines associated with a positive charge $+2q$ and a negative charge $-q$. In this case, the number of lines leaving $+2q$ is twice the number terminating at $-q$. Hence, only half the lines that leave the positive charge reach the negative charge. The remaining half terminate on a negative charge we assume to be at infinity. At distances much greater than the charge separation, the electric field lines are equivalent to those of a single charge $+q$.

Quick Quiz 23.5 Rank the magnitudes of the electric field at points *A*, *B*, and *C* shown in Figure 23.21a (greatest magnitude first).

23.7 Motion of a Charged Particle in a Uniform Electric Field

When a particle of charge q and mass m is placed in an electric field $\vec{\mathbf{E}}$, the electric force exerted on the charge is $q\vec{\mathbf{E}}$ according to Equation 23.8. If that is the only force exerted on the particle, it must be the net force and it causes the particle to accelerate according to the particle under a net force model. Therefore,

$$\vec{\mathbf{F}}_e = q\vec{\mathbf{E}} = m\vec{\mathbf{a}}$$

and the acceleration of the particle is

$$\vec{\mathbf{a}} = \frac{q\vec{\mathbf{E}}}{m} \tag{23.12}$$

If $\vec{\mathbf{E}}$ is uniform (that is, constant in magnitude and direction), the electric force on the particle is constant and we can apply the particle under constant acceleration model. If the particle has a positive charge, its acceleration is in the direction of the electric field. If the particle has a negative charge, its acceleration is in the direction opposite the electric field.

PITFALL PREVENTION 23.4

Just Another Force

Electric forces and fields may seem abstract to you. Once $\vec{\mathbf{F}}_e$ is evaluated, however, it causes a particle to move according to our well-established models of forces and motion from Chapters 2 through 6. Keeping this link with the past in mind should help you solve problems in this chapter.

EXAMPLE 23.9 An Accelerating Positive Charge

A uniform electric field $\vec{\mathbf{E}}$ is directed along the x axis between parallel plates of charge separated by a distance d as shown in Figure 23.23. A positive point charge q of mass m is released from rest at a point Ⓐ next to the positive plate and accelerates to a point Ⓑ next to the negative plate.

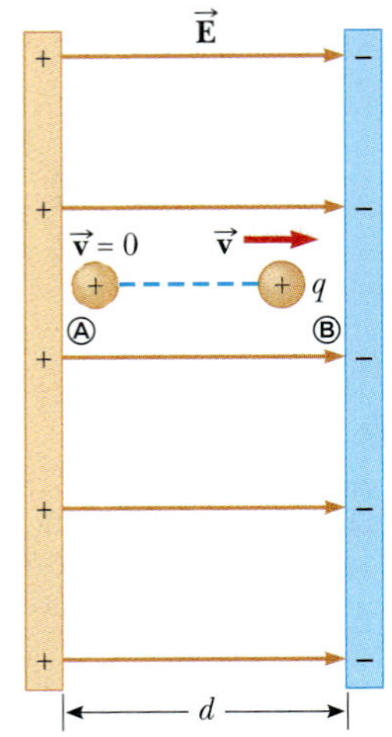

Figure 23.23 (Example 23.9) A positive point charge q in a uniform electric field $\vec{\mathbf{E}}$ undergoes constant acceleration in the direction of the field.

(A) Find the speed of the particle at Ⓑ by modeling it as a particle under constant acceleration.

SOLUTION

Conceptualize When the positive charge is placed at Ⓐ, it experiences an electric force toward the right in Figure 23.23 due to the electric field directed toward the right.

Categorize Because the electric field is uniform, a constant electric force acts on the charge. Therefore, the example involves a charged particle under constant acceleration.

Analyze Use Equation 2.17 to express the velocity of the particle as a function of position:

$$v_f^2 = v_i^2 + 2a(x_f - x_i) = 0 + 2a(d - 0) = 2ad$$

Solve for v_f and substitute for the magnitude of the acceleration from Equation 23.12:

$$v_f = \sqrt{2ad} = \sqrt{2\left(\frac{qE}{m}\right)d} = \boxed{\sqrt{\frac{2qEd}{m}}}$$

(B) Find the speed of the particle at Ⓑ by modeling it as a nonisolated system.

SOLUTION

Categorize The problem statement tells us that the charge is a nonisolated system. Energy is transferred to this charge by work done by the electric force exerted on the charge. The initial configuration of the system is when the particle is at Ⓐ, and the final configuration is when it is at Ⓑ.

Analyze Write the appropriate reduction of the conservation of energy equation, Equation 8.2, for the system of the charged particle:

$$W = \Delta K$$

Replace the work and kinetic energies with values appropriate for this situation:

$$F_e\,\Delta x = K_{Ⓑ} - K_{Ⓐ} = \tfrac{1}{2}mv_f^2 - 0 \;\rightarrow\; v_f = \sqrt{\frac{2F_e\,\Delta x}{m}}$$

Substitute for the electric force F_e and the displacement Δx:

$$v_f = \sqrt{\frac{2(qE)(d)}{m}} = \boxed{\sqrt{\frac{2qEd}{m}}}$$

Finalize The answer to part (B) is the same as that for part (A), as we expect.

EXAMPLE 23.10 An Accelerated Electron

An electron enters the region of a uniform electric field as shown in Active Figure 23.24, with $v_i = 3.00 \times 10^6$ m/s and $E = 200$ N/C. The horizontal length of the plates is $\ell = 0.100$ m.

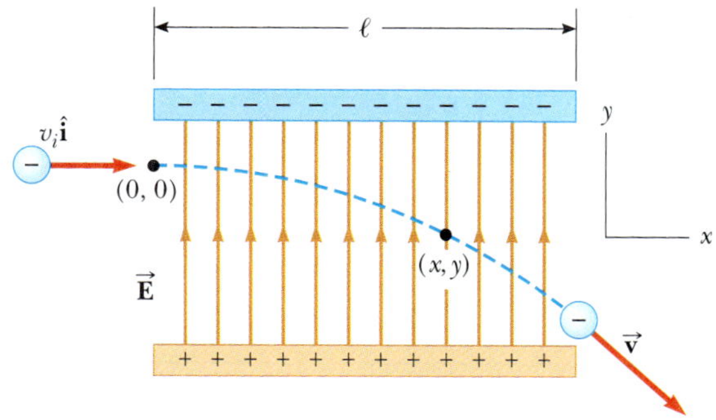

ACTIVE FIGURE 23.24

(Example 23.10) An electron is projected horizontally into a uniform electric field produced by two charged plates. The electron undergoes a downward acceleration (opposite $\vec{\mathbf{E}}$), and its motion is parabolic while it is between the plates.

Sign in at www.thomsonedu.com and go to ThomsonNOW to choose the strength of the electric field and the mass and charge of the projected particle.

(A) Find the acceleration of the electron while it is in the electric field.

SOLUTION

Conceptualize This example differs from the preceding one because the velocity of the charged particle is initially perpendicular to the electric field lines. In Example 23.9, the velocity of the charged particle is always parallel to the electric field lines. As a result, the electron in this example follows a curved path as shown in Active Figure 23.24.

Categorize Because the electric field is uniform, a constant electric force is exerted on the electron. To find the acceleration of the electron, we can model it as a particle under a net force.

Analyze The direction of the electron's acceleration is downward in Active Figure 23.24, opposite the direction of the electric field lines.

Combine Newton's second law with the magnitude of the electric force given by Equation 23.8 to find the y component of the acceleration of the electron:

$$\sum F_y = ma_y \rightarrow a_y = \frac{\sum F_y}{m} = -\frac{eE}{m_e}$$

Substitute numerical values:

$$a_y = -\frac{(1.60 \times 10^{-19}\ \text{C})(200\ \text{N/C})}{9.11 \times 10^{-31}\ \text{kg}} = \boxed{-3.51 \times 10^{13}\ \text{m/s}^2}$$

(B) Assuming the electron enters the field at time $t = 0$, find the time at which it leaves the field.

SOLUTION

Categorize Because the electric force acts only in the vertical direction in Active Figure 23.24, the motion of the particle in the horizontal direction can be analyzed by modeling it as a particle under constant velocity.

Analyze Solve Equation 2.7 for the time at which the electron arrives at the right edges of the plates:

$$x_f = x_i + v_x t \rightarrow t = \frac{x_f - x_i}{v_x}$$

Substitute numerical values:

$$t = \frac{\ell - 0}{v_x} = \frac{0.100\ \text{m}}{3.00 \times 10^6\ \text{m/s}} = \boxed{3.33 \times 10^{-8}\ \text{s}}$$

(C) Assuming the vertical position of the electron as it enters the field is $y_i = 0$, what is its vertical position when it leaves the field?

SOLUTION

Categorize Because the electric force is constant in Active Figure 23.24, the motion of the particle in the vertical direction can be analyzed by modeling it as a particle under constant acceleration.

Analyze Use Equation 2.16 to describe the position of the particle at any time t:

$$y_f = y_i + v_{yi}t + \tfrac{1}{2}a_y t^2$$

Substitute numerical values:

$$y_f = 0 + 0 + \tfrac{1}{2}(-3.51 \times 10^{13}\ \text{m/s}^2)(3.33 \times 10^{-8}\ \text{s})^2$$
$$= -0.019\,5\ \text{m} = -1.95\ \text{cm}$$

Finalize If the electron enters just below the negative plate in Active Figure 23.24 and the separation between the plates is less than the value just calculated, the electron will strike the positive plate.

We have neglected the gravitational force acting on the electron, which represents a good approximation when dealing with atomic particles. For an electric field of 200 N/C, the ratio of the magnitude of the electric force eE to the magnitude of the gravitational force mg is on the order of 10^{12} for an electron and on the order of 10^9 for a proton.

Summary

ThomsonNOW™ Sign in at **www.thomsonedu.com** and go to ThomsonNOW to take a practice test for this chapter.

DEFINITIONS

The **electric field** $\vec{\mathbf{E}}$ at some point in space is defined as the electric force $\vec{\mathbf{F}}_e$ that acts on a small positive test charge placed at that point divided by the magnitude q_0 of the test charge:

$$\vec{\mathbf{E}} \equiv \frac{\vec{\mathbf{F}}_e}{q_0} \quad \textbf{(23.7)}$$

CONCEPTS AND PRINCIPLES

Electric charges have the following important properties:

- Charges of opposite sign attract one another, and charges of the same sign repel one another.
- The total charge in an isolated system is conserved.
- Charge is quantized.

Conductors are materials in which electrons move freely. **Insulators** are materials in which electrons do not move freely.

Coulomb's law states that the electric force exerted by a point charge q_1 on a second point charge q_2 is

$$\vec{\mathbf{F}}_{12} = k_e \frac{q_1 q_2}{r^2}\hat{\mathbf{r}}_{12} \quad \textbf{(23.6)}$$

where r is the distance between the two charges and $\hat{\mathbf{r}}_{12}$ is a unit vector directed from q_1 toward q_2. The constant k_e, which is called the **Coulomb constant,** has the value $k_e = 8.99 \times 10^9\ \text{N}\cdot\text{m}^2/\text{C}^2$.

The electric force on a charge q placed in an electric field $\vec{\mathbf{E}}$ is

$$\vec{\mathbf{F}}_e = q\vec{\mathbf{E}} \quad \textbf{(23.8)}$$

At a distance r from a point charge q, the electric field due to the charge is

$$\vec{\mathbf{E}} = k_e \frac{q}{r^2}\hat{\mathbf{r}} \quad \textbf{(23.9)}$$

where $\hat{\mathbf{r}}$ is a unit vector directed from the charge toward the point in question. The electric field is directed radially outward from a positive charge and radially inward toward a negative charge.

The electric field due to a group of point charges can be obtained by using the superposition principle. That is, the total electric field at some point equals the vector sum of the electric fields of all the charges:

$$\vec{\mathbf{E}} = k_e \sum_i \frac{q_i}{r_i^2}\hat{\mathbf{r}}_i \quad \textbf{(23.10)}$$

The electric field at some point due to a continuous charge distribution is

$$\vec{\mathbf{E}} = k_e \int \frac{dq}{r^2}\hat{\mathbf{r}} \quad \textbf{(23.11)}$$

where dq is the charge on one element of the charge distribution and r is the distance from the element to the point in question.

Questions

☐ denotes answer available in *Student Solutions Manual/Study Guide;* **O** denotes objective question

1. Explain what is meant by the term "a neutral atom." Explain what "a negatively charged atom" means.

2. **O (i)** A metallic coin is given a positive electric charge. Does its mass (a) increase measurably, (b) increase by an amount too small to measure directly, (c) remain unchanged, (d) decrease by an amount too small to measure directly, or (e) decrease measurably? **(ii)** Now the coin is given a negative electric charge. What happens to its mass? Choose from the same possibilities.

3. A student who grew up in a tropical country and is studying in the United States may have no experience with static electricity sparks and shocks until their first American winter. Explain.

4. Explain the similarities and differences between Newton's law of universal gravitation and Coulomb's law.

☐5. A balloon is negatively charged by rubbing, and then it clings to a wall. Does that mean the wall is positively charged? Why does the balloon eventually fall?

6. **O** In Figure 23.8, assume the objects with charges q_2 and q_3 are fixed. Notice that there is no sightline from the location of object 2 to the location of object 1. We could say that a bug standing on q_1 is unable to see q_2 because it is behind q_3. How would you calculate the force exerted on the object with charge q_1? (a) Find only the force exerted by q_2 on charge q_1. (b) Find only the force exerted by q_3 on charge q_1. (c) Add the force that q_2 would exert by itself on charge q_1 to the force that q_3 would exert by itself on charge q_1. (d) Add the force that q_3 would exert by itself to a certain fraction of the force that q_2 would exert by itself. (e) There is no definite way to find the force on charge q_1.

7. **O** A charged particle is at the origin of coordinates. The particle produces an electric field of $4\hat{\mathbf{i}}$ kN/C at the point with position vector $36\hat{\mathbf{i}}$ cm. **(i)** At what location does the field have the value $1\hat{\mathbf{i}}$ kN/C? (a) $9\hat{\mathbf{i}}$ cm (b) $18\hat{\mathbf{i}}$ cm (c) $72\hat{\mathbf{i}}$ cm (d) $144\hat{\mathbf{i}}$ cm (e) nowhere **(ii)** At what location is the value $16\hat{\mathbf{i}}$ kN/C? Choose from the same possibilities.

8. Is it possible for an electric field to exist in empty space? Explain. Consider point *A* in Figure 23.21a. Does charge exist at this point? Does a force exist at this point? Does a field exist at this point?

9. **O (i)** Rank the magnitude of the forces charged particle A exerts on charged particle B, located at distance r away from A, from the largest to the smallest in the following cases. In your ranking, note any cases of equality. (a) q_A = 20 nC, q_B = 20 nC, r = 2 cm (b) q_A = 30 nC, q_B = 10 nC, r = 2 cm (c) q_A = 10 nC, q_B = 30 nC, r = 2 cm (d) q_A = 30 nC, q_B = 20 nC, r = 3 cm (e) q_A = 45 nC, q_B = 20 nC, r = 3 cm **(ii)** Rank the magnitudes of the electric fields charged particle A creates at the location of charged particle B, a distance r away from A, from the largest to the smallest in the same cases. In your ranking, note any cases of equality.

10. **O** Three charged particles are arranged on corners of a square as shown in Figure Q23.10, with charge $-Q$ on both the particle at the upper left corner and the particle at the lower right corner, and charge $+2Q$ on the particle at the lower left corner. **(i)** What is the direction of the electric field at the upper right corner, which is a point in empty space? (a) It is upward and to the right. (b) It is straight to the right. (c) It is straight downward. (d) It is downward and to the left. (e) It is perpendicular to the plane of the picture and outward. (f) There is no direction; no field exists at that corner because no charge is there. (g) There is no direction; the total field there is zero. **(ii)** Suppose the $+2Q$ charge at the lower left corner is removed. Then does the magnitude of the field at the upper right corner (a) become larger, (b) become smaller, (c) stay the same, or (d) change unpredictably?

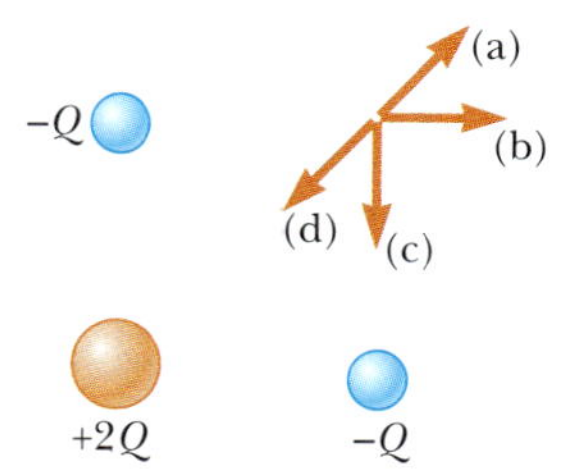

Figure Q23.10

11. **O** Two charged particles A and B are alone in the Universe, 8 cm apart. The charge of A is 40 nC. The net electric field at one certain point 4 cm from A is zero. What can you conclude about the charge of B? Choose every correct answer. (a) It can be 40 nC. (b) It can be 120 nC. (c) It can be 360 nC. (d) It can be −40 nC. (e) It can be −120 nC. (f) It can be −360 nC. (g) It can have any of an infinite number of values. (h) It can have any of several values. (i) It must have one of three values. (j) It must have one of two values. (k) It must have one certain value. (l) No possible value for q_B exists; the situation is impossible.

12. Explain why electric field lines never cross. *Suggestion:* Begin by explaining why the electric field at a particular point must have only one direction.

13. Figures 23.12 and 23.13 show three electric field vectors at the same point. With a little extrapolation, Figure 23.19 would show many electric field lines at the same point. Is it really true that "no two field lines can cross"? Are the diagrams drawn correctly? Explain your answers.

14. **O** A circular ring of charge with radius b has total charge q uniformly distributed around it. What is the magnitude of the electric field at the center of the ring? (a) 0 (b) $k_e q/b^2$ (c) $k_e q^2/b^2$ (d) $k_e q^2/b$ (c) none of these answers

15. **O** Assume a uniformly charged ring of radius R and charge Q produces an electric field E_{ring} at a point P on its axis, at distance x away from the center of the ring. Now the charge Q is spread uniformly over the circular area the ring encloses, forming a flat disk of charge with the same radius. How does the field E_{disk} produced by the disk at P compare to the field produced by the ring at the same point? (a) $E_{\text{disk}} < E_{\text{ring}}$ (b) $E_{\text{disk}} = E_{\text{ring}}$ (c) $E_{\text{disk}} > E_{\text{ring}}$ (d) impossible to determine

16. **O** A free electron and a free proton are released in identical electric fields. **(i)** How do the magnitudes of the electric force exerted on the two particles compare? (a) It is millions of times greater for the electron. (b) It is thousands of times greater for the electron. (c) They are equal. (d) It is thousands of times smaller for the electron. (e) It is millions of times smaller for the electron. (f) It is zero for the electron. (g) It is zero for the proton. **(ii)** Compare the magnitudes of their accelerations. Choose from the same possibilities.

17. **O** An object with negative charge is placed in a region of space where the electric field is directed vertically upward. What is the direction of the electric force exerted on this charge? (a) It is up. (b) It is down. (c) There is no force. (d) The force can be in any direction.

18. Explain the differences between linear, surface, and volume charge densities and give examples of when each would be used.

19. Would life be different if the electron were positively charged and the proton were negatively charged? Does the choice of signs have any bearing on physical and chemical interactions? Explain.

20. Consider two electric dipoles in empty space. Each dipole has zero net charge. Does an electric force exist between the dipoles; that is, can two objects with zero net charge exert electric forces on each other? If so, is the force one of attraction or of repulsion?

Problems

WebAssign The Problems from this chapter may be assigned online in WebAssign.

ThomsonNOW Sign in at **www.thomsonedu.com** and go to ThomsonNOW to assess your understanding of this chapter's topics with additional quizzing and conceptual questions.

1, 2, 3 denotes straightforward, intermediate, challenging; □ denotes full solution available in *Student Solutions Manual/Study Guide;* ▲ denotes coached solution with hints available at **www.thomsonedu.com;** ▢ denotes developing symbolic reasoning; ● denotes asking for qualitative reasoning; ▪ denotes computer useful in solving problem

Section 23.1 Properties of Electric Charges

1. (a) Find to three significant digits the charge and the mass of an ionized hydrogen atom, represented as H^+. *Suggestion:* Begin by looking up the mass of a neutral atom on the periodic table of the elements in Appendix C. (b) Find the charge and the mass of Na^+, a singly ionized sodium atom. (c) Find the charge and the average mass of a chloride ion Cl^- that joins with the Na^+ to make one molecule of table salt. (d) Find the charge and the mass of $Ca^{++} = Ca^{2+}$, a doubly ionized calcium atom. (e) You can model the center of an ammonia molecule as an N^{3-} ion. Find its charge and mass. (f) The plasma in a hot star contains quadruply ionized nitrogen atoms, N^{4+}. Find their charge and mass. (g) Find the charge and the mass of the nucleus of a nitrogen atom. (h) Find the charge and the mass of the molecular ion H_2O^-.

2. (a) Calculate the number of electrons in a small, electrically neutral silver pin that has a mass of 10.0 g. Silver has 47 electrons per atom, and its molar mass is 107.87 g/mol. (b) Imagine adding electrons to the pin until the negative charge has the very large value 1.00 mC. How many electrons are added for every 10^9 electrons already present?

Section 23.2 Charging Objects by Induction

Section 23.3 Coulomb's Law

3. ▲ Nobel laureate Richard Feynman (1918–1988) once said that if two persons stood at arm's length from each other and each person had 1% more electrons than protons, the force of repulsion between them would be enough to lift a "weight" equal to that of the entire Earth. Carry out an order-of-magnitude calculation to substantiate this assertion.

4. A charged particle A exerts a force of 2.62 μN to the right on charged particle B when the particles are 13.7 mm apart. Particle B moves straight away from A to make the distance between them 17.7 mm. What vector force does it then exert on A?

5. ● (a) Two protons in a molecule are 3.80×10^{-10} m apart. Find the electrical force exerted by one proton on the other. (b) State how the magnitude of this force compares with the magnitude of the gravitational force exerted by one proton on the other. (c) **What If?** What must be a particle's charge-to-mass ratio if the magnitude of the gravitational force between two of these particles is equal to the magnitude of electrical force between them?

6. Two small silver spheres, each with a mass of 10.0 g, are separated by 1.00 m. Calculate the fraction of the electrons in one sphere that must be transferred to the other to produce an attractive force of 1.00×10^4 N (about 1 ton) between the spheres. (The number of electrons per atom of silver is 47, and the number of atoms per gram is Avogadro's number divided by the molar mass of silver, 107.87 g/mol.)

7. Three charged particles are located at the corners of an equilateral triangle as shown in Figure P23.7. Calculate the total electric force on the 7.00-μC charge.

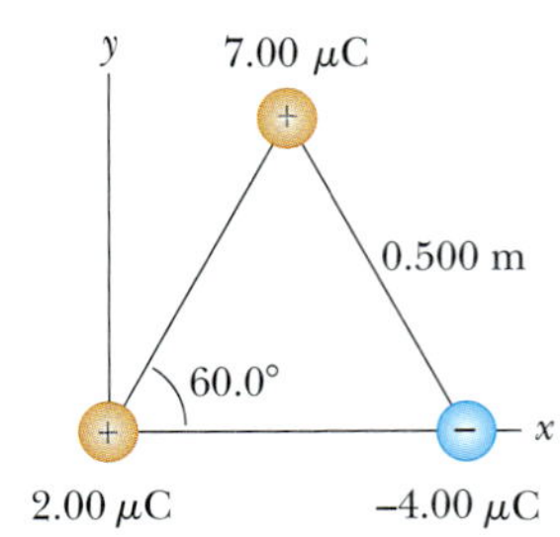

Figure P23.7 Problems 7 and 14.

8. ● Two small beads having positive charges $3q$ and q are fixed at the opposite ends of a horizontal insulating rod, extending from the origin to the point $x = d$. As shown in Figure P23.8, a third small charged bead is free to slide on the rod. At what position is the third bead in equilibrium? Explain whether it can be in stable equilibrium.

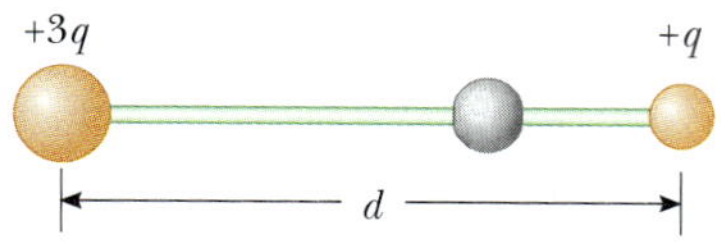

Figure P23.8

9. Two identical conducting small spheres are placed with their centers 0.300 m apart. One is given a charge of 12.0 nC and the other a charge of −18.0 nC. (a) Find the electric force exerted by one sphere on the other. (b) **What If?** The spheres are connected by a conducting wire. Find the electric force each exerts on the other after they have come to equilibrium.

10. **Review problem.** Two identical particles, each having charge $+q$, are fixed in space and separated by a distance d. A third particle with charge $-Q$ is free to move and lies initially at rest on the perpendicular bisector of the two fixed charges a distance x from the midpoint between the two fixed charges (Fig. P23.10). (a) Show that if x is small compared with d, the motion of $-Q$ is simple harmonic along the perpendicular bisector. Determine the period of that motion. (b) How fast will the charge $-Q$ be moving when it is at the midpoint between the two fixed charges if initially it is released at a distance $a \ll d$ from the midpoint?

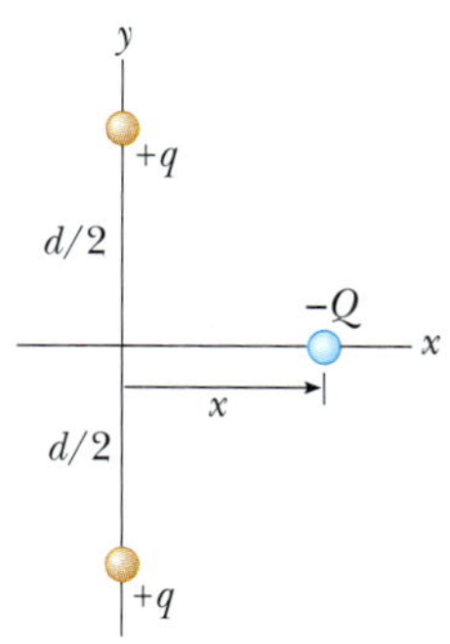

Figure P23.10

11. **Review problem.** In the Bohr theory of the hydrogen atom, an electron moves in a circular orbit about a proton, where the radius of the orbit is 0.529×10^{-10} m. (a) Find the electric force exerted on each particle. (b) If this force causes the centripetal acceleration of the electron, what is the speed of the electron?

Section 23.4 The Electric Field

12. In Figure P23.12, determine the point (other than infinity) at which the electric field is zero.

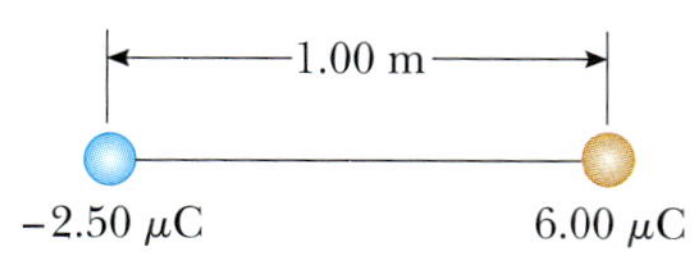

Figure P23.12

13. What are the magnitude and direction of the electric field that balances the weight of (a) an electron and (b) a proton? You may use the data in Table 23.1.

14. Three charged particles are at the corners of an equilateral triangle as shown in Figure P23.7. (a) Calculate the electric field at the position of the 2.00-μC charge due to the 7.00-μC and −4.00-μC charges. (b) Use your answer to part (a) to determine the force on the 2.00-μC charge.

15. ● Two charged particles are located on the x axis. The first is a charge $+Q$ at $x = -a$. The second is an unknown charge located at $x = -3a$. The net electric field these charges produce at the origin has a magnitude of $2k_eQ/a^2$. Explain how many values are possible for the unknown charge and find the possible values.

16. Two 2.00-μC charged particles are located on the x axis. One is at $x = 1.00$ m, and the other is at $x = -1.00$ m. (a) Determine the electric field on the y axis at $y = 0.500$ m. (b) Calculate the electric force on a −3.00-μC charge placed on the y axis at $y = 0.500$ m.

17. Four charged particles are at the corners of a square of side a as shown in Figure P23.17. (a) Determine the magnitude and direction of the electric field at the location of charge q. (b) What is the total electric force exerted on q?

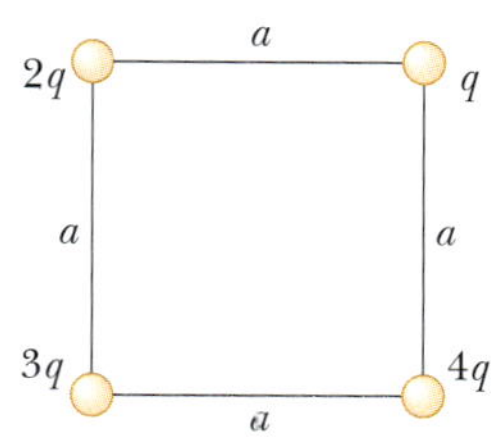

Figure P23.17

18. Consider the electric dipole shown in Figure P23.18. Show that the electric field at a *distant* point on the $+x$ axis is $E_x \approx -4k_eqa/x^3$.

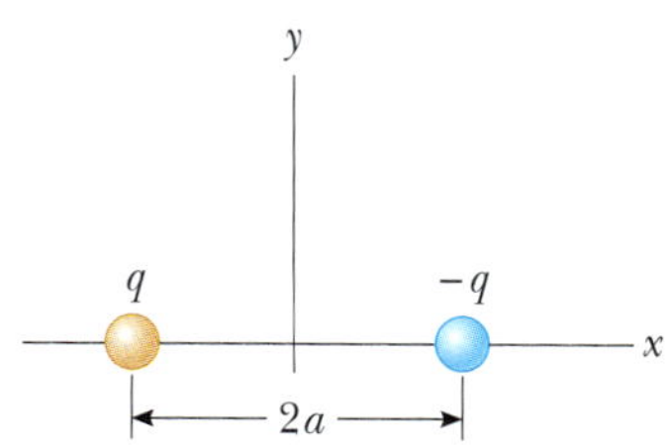

Figure P23.18

19. ● Consider n equal positive charged particles each of magnitude Q/n placed symmetrically around a circle of radius R. (a) Calculate the magnitude of the electric field at a point a distance x from the center of the circle and on the line passing through the center and perpendicular to the plane of the circle. (b) Explain why this result is identical to the result of the calculation done in Example 23.7.

Section 23.5 Electric Field of a Continuous Charge Distribution

20. A continuous line of charge lies along the x axis, extending from $x = +x_0$ to positive infinity. The line carries charge with a uniform linear charge density λ_0. What are

the magnitude and direction of the electric field at the origin?

21. A rod 14.0 cm long is uniformly charged and has a total charge of -22.0 μC. Determine the magnitude and direction of the electric field along the axis of the rod at a point 36.0 cm from its center.

22. Show that the maximum magnitude E_{max} of the electric field along the axis of a uniformly charged ring occurs at $x = a/\sqrt{2}$ (see Fig. 23.16) and has the value $Q/(6\sqrt{3}\pi\epsilon_0 a^2)$.

23. A uniformly charged ring of radius 10.0 cm has a total charge of 75.0 μC. Find the electric field on the axis of the ring at (a) 1.00 cm, (b) 5.00 cm, (c) 30.0 cm, and (d) 100 cm from the center of the ring.

24. A uniformly charged disk of radius 35.0 cm carries charge with a density of 7.90×10^{-3} C/m^2. Calculate the electric field on the axis of the disk at (a) 5.00 cm, (b) 10.0 cm, (c) 50.0 cm, and (d) 200 cm from the center of the disk.

25. ● Example 23.8 derives the exact expression for the electric field at a point on the axis of a uniformly charged disk. Consider a disk of radius $R = 3.00$ cm having a uniformly distributed charge of $+5.20$ μC. (a) Using the result of Example 23.8, compute the electric field at a point on the axis and 3.00 mm from the center. **What If?** Explain how this answer compares with the field computed from the near-field approximation $E = \sigma/2\epsilon_0$. (b) Using the result of Example 23.8, compute the electric field at a point on the axis and 30.0 cm from the center of the disk. **What If?** Explain how this answer compares with the electric field obtained by treating the disk as a $+5.20$-μC charged particle at a distance of 30.0 cm.

26. The electric field along the axis of a uniformly charged disk of radius R and total charge Q was calculated in Example 23.8. Show that the electric field at distances x that are large compared with R approaches that of a particle with charge $Q = \sigma\pi R^2$. *Suggestion:* First show that $x/(x^2 + R^2)^{1/2} = (1 + R^2/x^2)^{-1/2}$ and use the binomial expansion $(1 + \delta)^n \approx 1 + n\delta$ when $\delta \ll 1$.

27. ▲ A uniformly charged insulating rod of length 14.0 cm is bent into the shape of a semicircle as shown in Figure P23.27. The rod has a total charge of -7.50 μC. Find the magnitude and direction of the electric field at O, the center of the semicircle.

Figure P23.27

28. (a) Consider a uniformly charged thin-walled right circular cylindrical shell having total charge Q, radius R, and height h. Determine the electric field at a point a distance d from the right side of the cylinder as shown in Figure P23.28. *Suggestion:* Use the result of Example 23.7 and treat the cylinder as a collection of ring charges. (b) **What If?** Consider now a solid cylinder with the same dimensions and carrying the same charge, uniformly distributed through its volume. Use the result of Example 23.8 to find the field it creates at the same point.

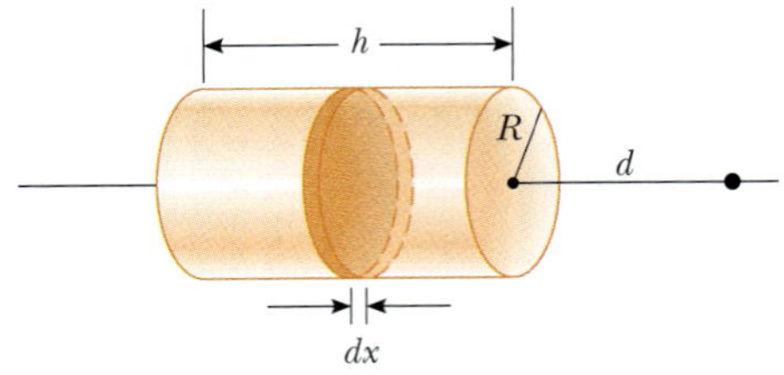

Figure P23.28

29. A thin rod of length ℓ and uniform charge per unit length λ lies along the x axis as shown in Figure P23.29. (a) Show that the electric field at P, a distance y from the rod along its perpendicular bisector, has no x component and is given by $E = 2k_e\lambda \sin\theta_0/y$. (b) **What If?** Using your result to part (a), show that the field of a rod of infinite length is $E = 2k_e\lambda/y$. *Suggestion:* First calculate the field at P due to an element of length dx, which has a charge $\lambda\, dx$. Then change variables from x to θ, using the relationships $x = y \tan\theta$ and $dx = y \sec^2\theta\, d\theta$, and integrate over θ.

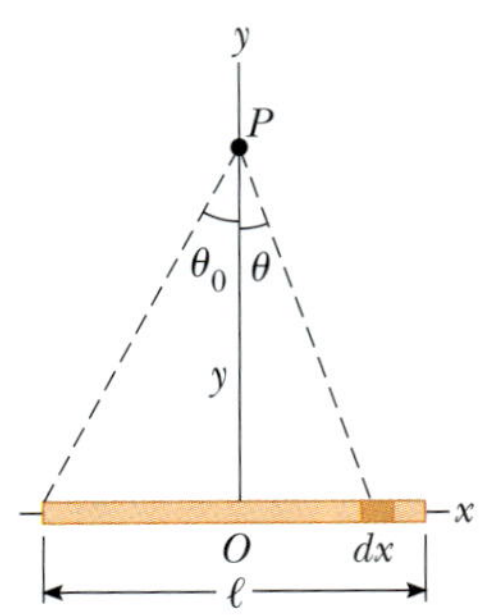

Figure P23.29

30. Three solid plastic cylinders all have radius 2.50 cm and length 6.00 cm. One (a) carries charge with uniform density 15.0 nC/m^2 everywhere on its surface. Another (b) carries charge with the same uniform density on its curved lateral surface only. The third (c) carries charge with uniform density 500 nC/m^3 throughout the plastic. Find the charge of each cylinder.

31. Eight solid plastic cubes, each 3.00 cm on each edge, are glued together to form each one of the objects (i, ii, iii, and iv) shown in Figure P23.31. (a) Assuming each object carries charge with uniform density 400 nC/m^3 throughout its volume, find the charge of each object. (b) Assuming each object carries charge with uniform density 15.0 nC/m^2 everywhere on its exposed surface, find the charge on each object. (c) Assuming charge is placed

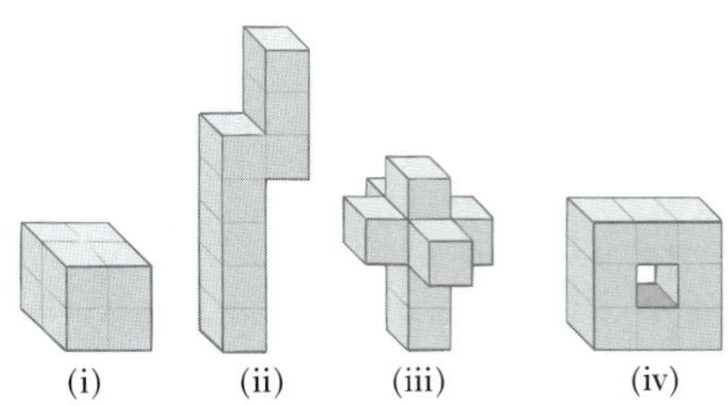

Figure P23.31

only on the edges where perpendicular surfaces meet, with uniform density 80.0 pC/m, find the charge of each object.

Section 23.6 Electric Field Lines

32. A positively charged disk has a uniform charge per unit area as described in Example 23.8. Sketch the electric field lines in a plane perpendicular to the plane of the disk passing through its center.

33. A negatively charged rod of finite length carries charge with a uniform charge per unit length. Sketch the electric field lines in a plane containing the rod.

34. Figure P23.34 shows the electric field lines for two charged particles separated by a small distance. (a) Determine the ratio q_1/q_2. (b) What are the signs of q_1 and q_2?

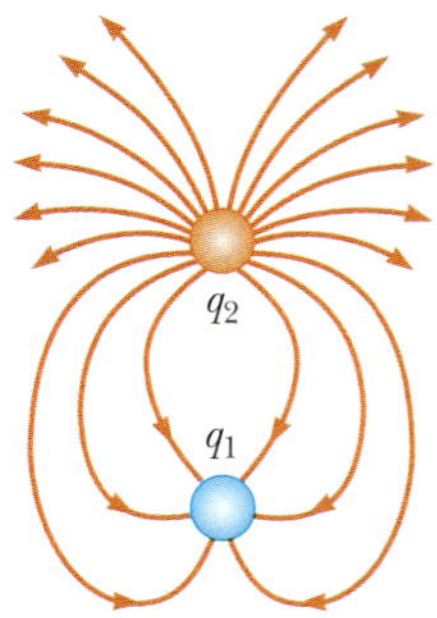

Figure P23.34

35. ▲ Three equal positive charges q are at the corners of an equilateral triangle of side a as shown in Figure P23.35. (a) Assume the three charges together create an electric field. Sketch the field lines in the plane of the charges. Find the location of one point (other than ∞) where the electric field is zero. (b) What are the magnitude and direction of the electric field at P due to the two charges at the base?

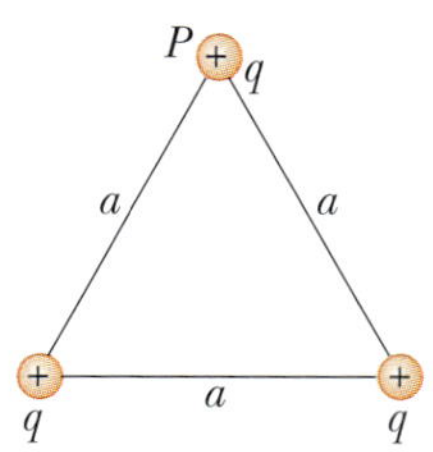

Figure P23.35 Problems 35 and 58.

Section 23.7 Motion of Charged Particles in a Uniform Electric Field

36. A proton is projected in the positive x direction into a region of a uniform electric field $\vec{\mathbf{E}} = -6.00 \times 10^5\,\hat{\mathbf{i}}$ N/C at $t = 0$. The proton travels 7.00 cm as it comes to rest. Determine (a) the acceleration of the proton, (b) its initial speed, and (c) the time interval over which the proton comes to rest.

37. A proton accelerates from rest in a uniform electric field of 640 N/C. At one later moment, its speed is 1.20 Mm/s (nonrelativistic because v is much less than the speed of light). (a) Find the acceleration of the proton. (b) Over what time interval does the proton reach this speed? (c) How far does it move in this time interval? (d) What is its kinetic energy at the end of this interval?

38. ● Two horizontal metal plates, each 100 mm square, are aligned 10.0 mm apart, with one above the other. They are given equal-magnitude charges of opposite sign so that a uniform downward electric field of 2 000 N/C exists in the region between them. A particle of mass 2.00×10^{-16} kg and with a positive charge of 1.00×10^{-6} C leaves the center of the bottom negative plate with an initial speed of 1.00×10^5 m/s at an angle of 37.0° above the horizontal. Describe the trajectory of the particle. Which plate does it strike? Where does it strike, relative to its starting point?

39. ▲ The electrons in a particle beam each have a kinetic energy K. What are the magnitude and direction of the electric field that will stop these electrons in a distance d?

40. Protons are projected with initial speed $v_i = 9.55$ km/s into a region where a uniform electric field $\vec{\mathbf{E}} = (-720\,\hat{\mathbf{j}})$ N/C is present as shown in Figure P23.40. The protons are to hit a target that lies at a horizontal distance of 1.27 mm from the point where the protons cross the plane and enter the electric field. Find (a) the two projection angles θ that will result in a hit and (b) the time of flight (the time interval during which the proton is above the plane in Fig. P23.40) for each trajectory.

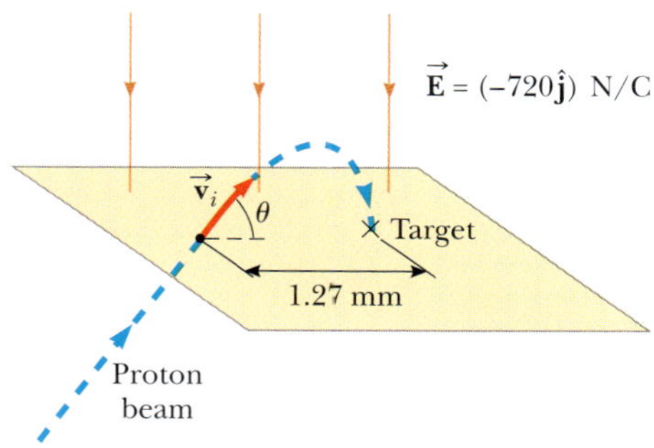

Figure P23.40

41. A proton moves at 4.50×10^5 m/s in the horizontal direction. It enters a uniform vertical electric field with a magnitude of 9.60×10^3 N/C. Ignoring any gravitational effects, find (a) the time interval required for the proton to travel 5.00 cm horizontally, (b) its vertical displacement during the time interval in which it travels 5.00 cm horizontally, and (c) the horizontal and vertical components of its velocity after it has traveled 5.00 cm horizontally.

Additional Problems

42. ● Two known charges, −12.0 μC and 45.0 μC, and a third unknown charge are located on the x axis. The charge −12.0 μC is at the origin, and the charge 45.0 μC is at $x = 15.0$ cm. The unknown charge is to be placed so that each charge is in equilibrium under the action of the electric forces exerted by the other two charges. Is this situation possible? Is it possible in more than one way? Explain. Find the required location, magnitude, and sign of the unknown charge.

43. A uniform electric field of magnitude 640 N/C exists between two parallel plates that are 4.00 cm apart. A proton is released from the positive plate at the same instant an electron is released from the negative plate. (a) Determine the distance from the positive plate at which the two pass each other. (Ignore the electrical attraction between the proton and electron.) (b) **What If?** Repeat part (a) for a sodium ion (Na^+) and a chloride ion (Cl^-).

44. Three charged particles are aligned along the x axis as shown in Figure P23.44. Find the electric field at (a) the position (2.00, 0) and (b) the position (0, 2.00).

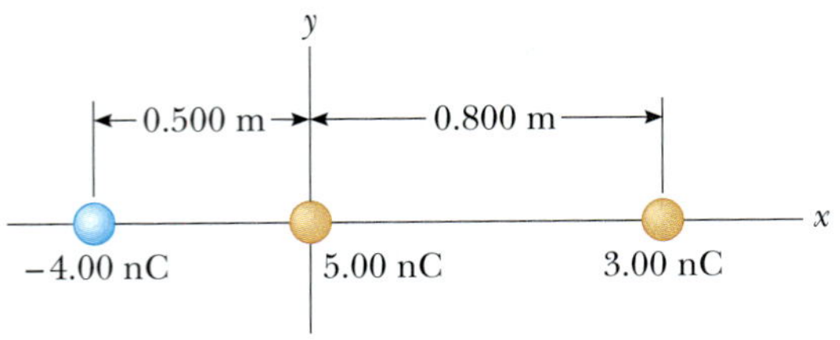

Figure P23.44

45. ▲ A charged cork ball of mass 1.00 g is suspended on a light string in the presence of a uniform electric field as shown in Figure P23.45. When $\vec{\mathbf{E}} = (3.00\hat{\mathbf{i}} + 5.00\hat{\mathbf{j}}) \times 10^5$ N/C, the ball is in equilibrium at $\theta = 37.0°$. Find (a) the charge on the ball and (b) the tension in the string.

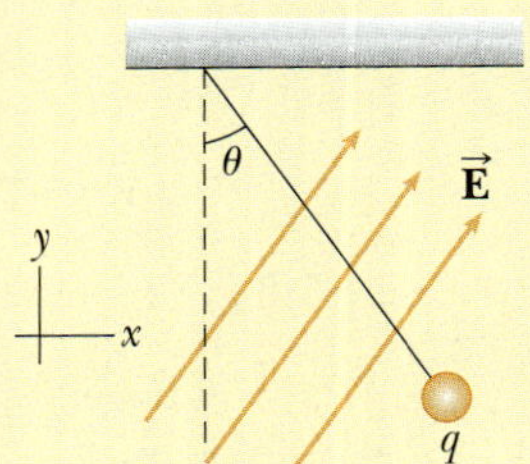

Figure P23.45 Problems 45 and 46.

46. A charged cork ball of mass m is suspended on a light string in the presence of a uniform electric field as shown in Figure P23.45. When $\vec{\mathbf{E}} = (A\hat{\mathbf{i}} + B\hat{\mathbf{j}})$ N/C, where A and B are positive numbers, the ball is in equilibrium at the angle θ. Find (a) the charge on the ball and (b) the tension in the string.

47. Four identical charged particles ($q = +10.0\ \mu$C) are located on the corners of a rectangle as shown in Figure P23.47. The dimensions of the rectangle are $L = 60.0$ cm and $W = 15.0$ cm. Calculate the magnitude and direction of the total electric force exerted on the charge at the lower left corner by the other three charges.

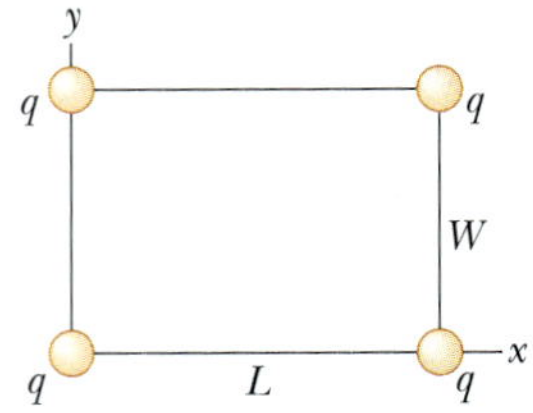

Figure P23.47

48. Inez is putting up decorations for her sister's quinceañera (fifteenth birthday party). She ties three light silk ribbons together to the top of a gateway and hangs a rubber balloon from each ribbon (Fig. P23.48). To include the effects of the gravitational and buoyant forces on it, each balloon can be modeled as a particle of mass 2.00 g, with its center 50.0 cm from the point of support. To show off the colors of the balloons, Inez rubs the whole surface of each balloon with her woolen scarf, making them hang separately with gaps between them. The centers of the hanging balloons form a horizontal equilateral triangle with sides 30.0 cm long. What is the common charge each balloon carries?

Figure P23.48

49. **Review problem.** Two identical metallic blocks resting on a frictionless horizontal surface are connected by a light metallic spring having a spring constant k and an unstretched length L_i as shown in Figure P23.49a. A total charge Q is slowly placed on the system, causing the spring to stretch to an equilibrium length L as shown in Figure P23.49b. Determine the value of Q, assuming all the charge resides on the blocks and modeling the blocks as charged particles.

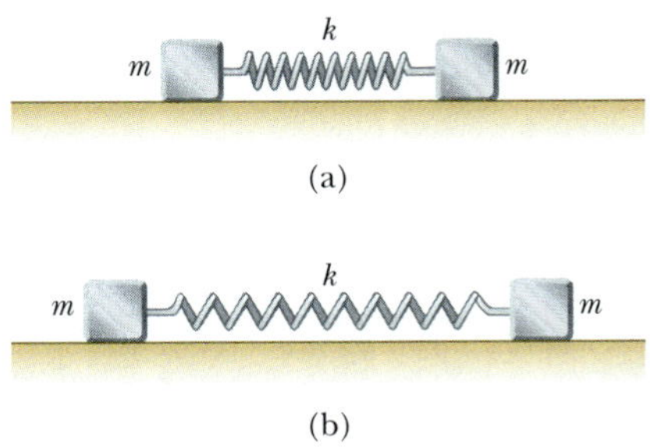

Figure P23.49

50. Consider a regular polygon with 29 sides. The distance from the center to each vertex is a. Identical charges q are placed at 28 vertices of the polygon. A single charge Q is placed at the center of the polygon. What is the magnitude and direction of the force experienced by the charge Q? *Suggestion:* You may use the result of Problem 60 in Chapter 3.

51. Identical thin rods of length $2a$ carry equal charges $+Q$ uniformly distributed along their lengths. The rods lie along the x axis with their centers separated by a distance $b > 2a$ (Fig. P23.51). Show that the magnitude of the force exerted by the left rod on the right one is

$$F = \left(\frac{k_e Q^2}{4a^2}\right) \ln\left(\frac{b^2}{b^2 - 4a^2}\right)$$

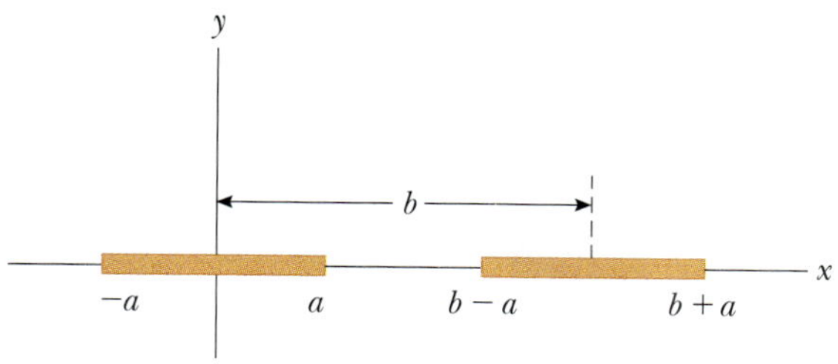

Figure P23.51

52. Two small spheres hang in equilibrium at the bottom ends of threads, 40.0 cm long, that have their top ends tied to the same fixed point. One sphere has mass 2.40 g and charge +300 nC. The other sphere has the same mass and a charge of +200 nC. Find the distance between the centers of the spheres. You will need to solve an equation numerically.

53. A line of positive charge is formed into a semicircle of radius R = 60.0 cm as shown in Figure P23.53. The charge per unit length along the semicircle is described by the expression $\lambda = \lambda_0 \cos\theta$. The total charge on the semicircle is 12.0 μC. Calculate the total force on a charge of 3.00 μC placed at the center of curvature.

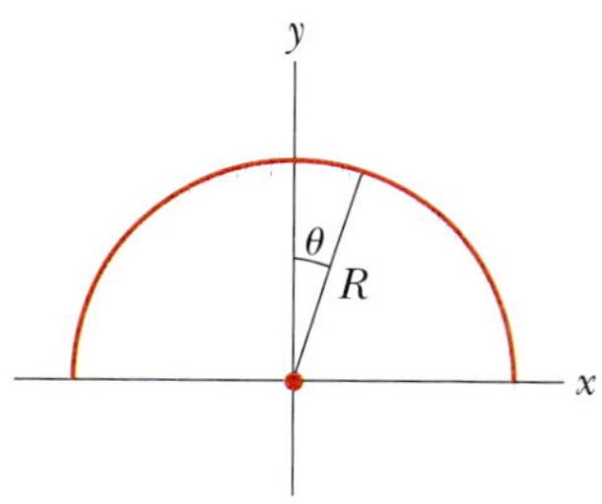

Figure P23.53

54. Two particles, each with charge 52.0 nC, are located on the y axis at y = 25.0 cm and y = −25.0 cm. (a) Find the vector electric field at a point on the x axis as a function of x. (b) Find the field at x = 36.0 cm. (c) At what location is the field $1.00\hat{\mathbf{i}}$ kN/C? You may need to solve an equation numerically. (d) At what location is the field $16.0\hat{\mathbf{i}}$ kN/C? (e) Compare this problem with Question 7. Describe the similarities and explain the differences.

55. Two small spheres of mass m are suspended from strings of length ℓ that are connected at a common point. One sphere has charge Q and the other has charge $2Q$. The strings make angles θ_1 and θ_2 with the vertical. (a) Explain how θ_1 and θ_2 are related. (b) Assume θ_1 and θ_2 are small. Show that the distance r between the spheres is approximately

$$r \approx \left(\frac{4k_e Q^2 \ell}{mg}\right)^{1/3}$$

56. Two identical beads each have a mass m and charge q. When placed in a hemispherical bowl of radius R with frictionless, nonconducting walls, the beads move, and at equilibrium they are a distance R apart (Fig. P23.56). Determine the charge on each bead.

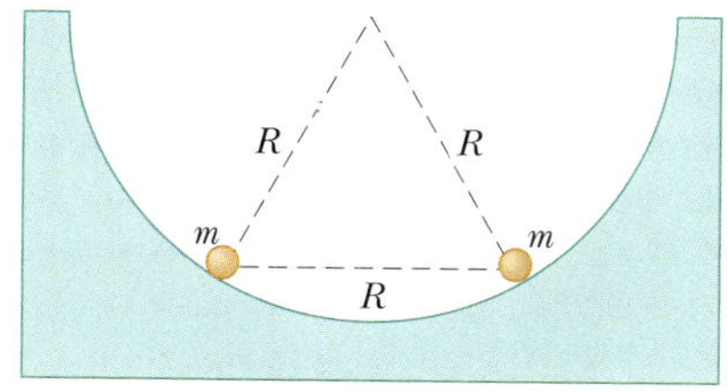

Figure P23.56

57. **Review problem.** A 1.00-g cork ball with charge 2.00 μC is suspended vertically on a 0.500-m-long light string in the presence of a uniform, downward-directed electric field of magnitude $E = 1.00 \times 10^5$ N/C. If the ball is displaced slightly from the vertical, it oscillates like a simple pendulum. (a) Determine the period of this oscillation. (b) Should the effect of gravitation be included in the calculation for part (a)? Explain.

58. Figure P23.35 shows three equal positive charges at the corners of an equilateral triangle of side a = 3.00 cm. Add a vertical line through the top charge at P, bisecting the triangle. Along this line label points A, B, C, D, E, and F, with A just below the charge at P; B at the center of the triangle; B, C, D, and E in order and close together with E at the center of the bottom side of the triangle; and F close below E. (a) Identify the direction of the total electric field at A, E, and F. Identify the electric field at B. Identify the direction of the electric field at C. (b) Argue that the answers to part (a) imply that the electric field must be zero at a point close to D. (c) Find the distance from point E on the bottom side of the triangle to the point around D where the electric field is zero. You will need to solve a transcendental equation.

59. Eight charged particles, each of magnitude q, are located on the corners of a cube of edge s as shown in Figure P23.59. (a) Determine the x, y, and z components of the total force exerted by the other charges on the charge located at point A. (b) What are the magnitude and direction of this total force?

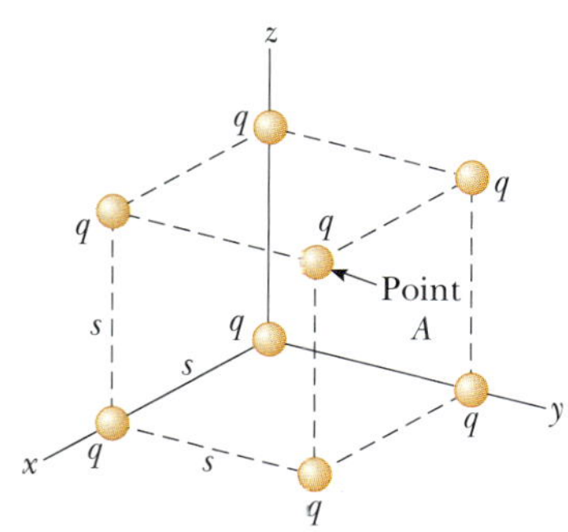

Figure P23.59 Problems 59 and 60.

60. Consider the charge distribution shown in Figure P23.59. (a) Show that the magnitude of the electric field at the center of any face of the cube has a value of $2.18k_e q/s^2$. (b) What is the direction of the electric field at the center of the top face of the cube?

61. **Review problem.** A negatively charged particle $-q$ is placed at the center of a uniformly charged ring, where the ring has a total positive charge Q as shown in Example 23.7. The particle, confined to move along the x axis,

is moved a small distance x along the axis (where $x \ll a$) and released. Show that the particle oscillates in simple harmonic motion with a frequency given by

$$f = \frac{1}{2\pi}\left(\frac{k_e qQ}{ma^3}\right)^{1/2}$$

62. A line of charge with uniform density 35.0 nC/m lies along the line $y = -15.0$ cm between the points with coordinates $x = 0$ and $x = 40.0$ cm. Find the electric field it creates at the origin.

63. **Review problem**. An electric dipole in a uniform electric field is displaced slightly from its equilibrium position as shown in Figure P23.63, where θ is small. The separation of the charges is $2a$, and the moment of inertia of the dipole is I. Assuming the dipole is released from this position, show that its angular orientation exhibits simple harmonic motion with a frequency

$$f = \frac{1}{2\pi}\sqrt{\frac{2qaE}{I}}$$

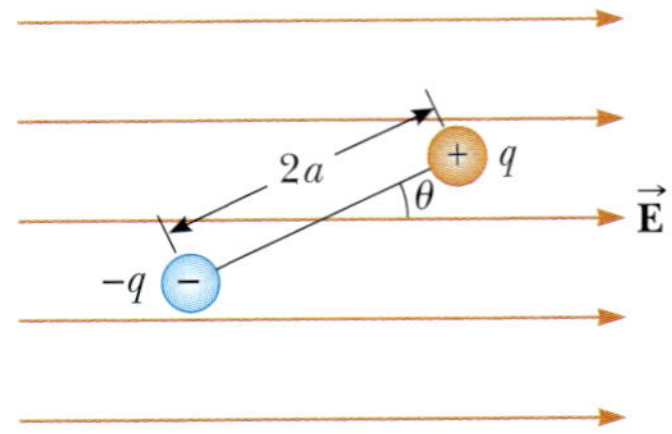

Figure P23.63

64. Consider an infinite number of identical particles, each with charge q, placed along the x axis at distances a, $2a$, $3a$, $4a$, . . . , from the origin. What is the electric field at the origin due to this distribution? *Suggestion:* Use the fact that

$$1 + \frac{1}{2^2} + \frac{1}{3^2} + \frac{1}{4^2} + \cdots = \frac{\pi^2}{6}$$

65. A line of charge starts at $x = +x_0$ and extends to positive infinity. The linear charge density is $\lambda = \lambda_0 x_0 / x$, where λ_0 is a constant. Determine the electric field at the origin.

Answers to Quick Quizzes

23.1 (a), (c), (e). The experiment shows that A and B have charges of the same sign, as do objects B and C. Therefore, all three objects have charges of the same sign. We cannot determine from this information, however, if the charges are positive or negative.

23.2 (e). In the first experiment, objects A and B may have charges with opposite signs or one of the objects may be neutral. The second experiment shows that B and C have charges with the same signs, so B must be charged. We still do not know, however, if A is charged or neutral.

23.3 (b). From Newton's third law, the electric force exerted by object B on object A is equal in magnitude to the force exerted by object A on object B and in the opposite direction.

23.4 (a). There is no effect on the electric field if we assume the source charge producing the field is not disturbed by our actions. Remember that the electric field is created by the source charge(s) (unseen in this case), not the test charge(s).

23.5 *A, B, C.* The field is greatest at point *A* because that is where the field lines are closest together. The absence of lines near point *C* indicates that the electric field there is zero.

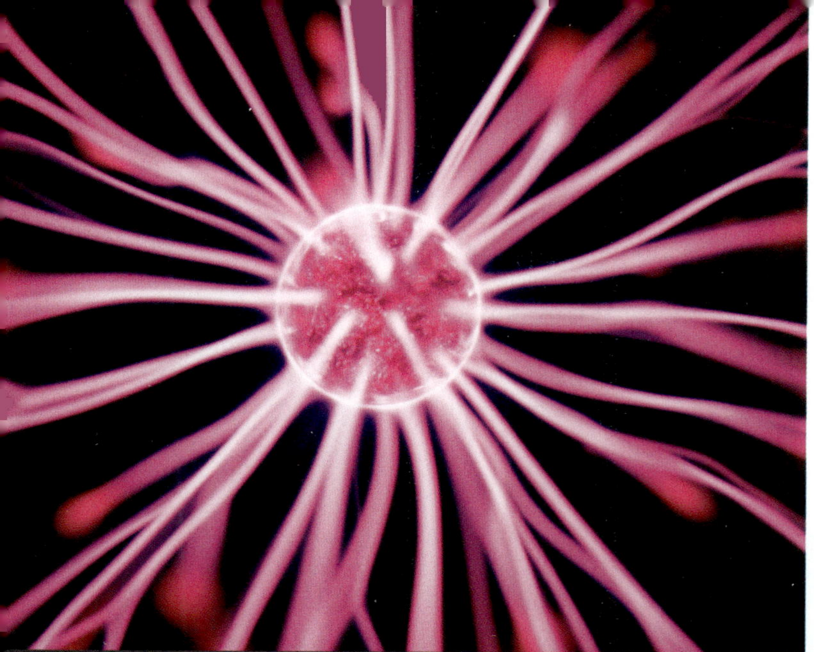

In a tabletop plasma ball, the colorful lines emanating from the sphere give evidence of strong electric fields. Using Gauss's law, we show in this chapter that the electric field surrounding a uniformly charged sphere is identical to that of a point charge. (Getty Images)

24 Gauss's Law

In Chapter 23, we showed how to calculate the electric field due to a given charge distribution. In this chapter, we describe *Gauss's law* and an alternative procedure for calculating electric fields. Gauss's law is based on the inverse-square behavior of the electric force between point charges. Although Gauss's law is a consequence of Coulomb's law, it is more convenient for calculating the electric fields of highly symmetric charge distributions and makes it possible to deal with complicated problems using qualitative reasoning.

24.1 Electric Flux

The concept of electric field lines was described qualitatively in Chapter 23. We now treat electric field lines in a more quantitative way.

Consider an electric field that is uniform in both magnitude and direction as shown in Figure 24.1. The field lines penetrate a rectangular surface of area A, whose plane is oriented perpendicular to the field. Recall from Section 23.6 that the number of lines per unit area (in other words, the *line density*) is proportional to the magnitude of the electric field. Therefore, the total number of lines penetrating the surface is proportional to the product EA. This product of the magnitude of the electric field E and surface area A perpendicular to the field is called the **electric flux** Φ_E (uppercase Greek letter phi):

$$\Phi_E = EA \tag{24.1}$$

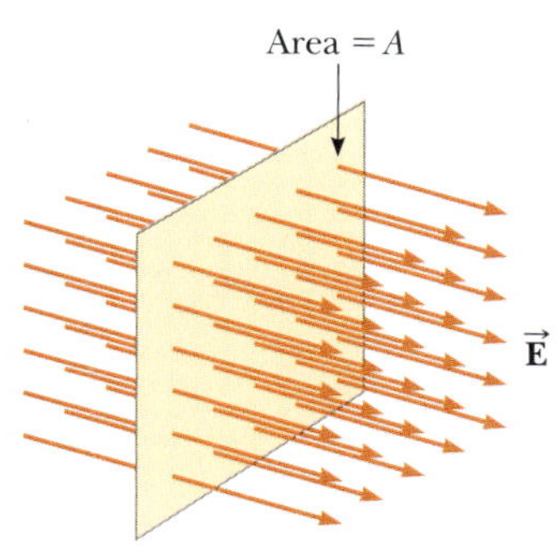

Figure 24.1 Field lines representing a uniform electric field penetrating a plane of area A perpendicular to the field. The electric flux Φ_E through this area is equal to EA.

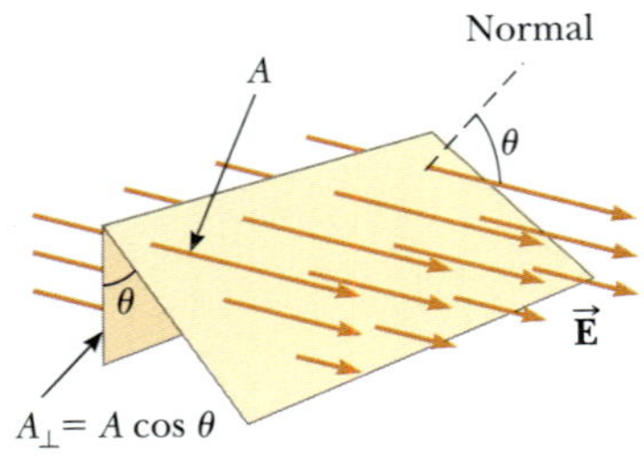

Figure 24.2 Field lines representing a uniform electric field penetrating an area A that is at an angle θ to the field. Because the number of lines that go through the area $A_\perp$ is the same as the number that go through A, the flux through $A_\perp$ is equal to the flux through A and is given by $\Phi_E = EA\cos\theta$.

From the SI units of E and A, we see that Φ_E has units of newton meters squared per coulomb ($\mathrm{N \cdot m^2/C}$). **Electric flux is proportional to the number of electric field lines penetrating some surface.**

If the surface under consideration is not perpendicular to the field, the flux through it must be less than that given by Equation 24.1. Consider Figure 24.2, where the normal to the surface of area A is at an angle θ to the uniform electric field. Notice that the number of lines that cross this area A is equal to the number of lines that cross the area $A_\perp$, which is a projection of area A onto a plane oriented perpendicular to the field. Figure 24.2 shows that the two areas are related by $A_\perp = A\cos\theta$. Because the flux through A equals the flux through $A_\perp$, the flux through A is

$$\Phi_E = EA_\perp = EA\cos\theta \qquad \textbf{(24.2)}$$

From this result, we see that the flux through a surface of fixed area A has a maximum value EA when the surface is perpendicular to the field (when the normal to the surface is parallel to the field, that is, when $\theta = 0°$ in Fig. 24.2); the flux is zero when the surface is parallel to the field (when the normal to the surface is perpendicular to the field, that is, when $\theta = 90°$).

We assumed a uniform electric field in the preceding discussion. In more general situations, the electric field may vary over a large surface. Therefore, the definition of flux given by Equation 24.2 has meaning only for a small element of area over which the field is approximately constant. Consider a general surface divided into a large number of small elements, each of area ΔA. It is convenient to define a vector $\Delta\vec{\mathbf{A}}_i$ whose magnitude represents the area of the ith element of the large surface and whose direction is defined to be *perpendicular* to the surface element as shown in Figure 24.3. The electric field $\vec{\mathbf{E}}_i$ at the location of this element makes an angle θ_i with the vector $\Delta\vec{\mathbf{A}}_i$. The electric flux $\Delta\Phi_E$ through this element is

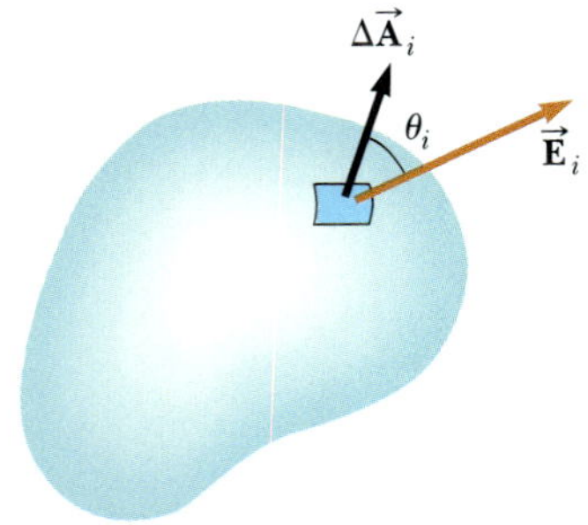

Figure 24.3 A small element of surface area ΔA_i. The electric field makes an angle θ_i with the vector $\Delta\vec{\mathbf{A}}_i$, defined as being normal to the surface element, and the flux through the element is equal to $E_i\,\Delta A_i\cos\theta_i$.

$$\Delta\Phi_E = E_i\,\Delta A_i\cos\theta_i = \vec{\mathbf{E}}_i \cdot \Delta\vec{\mathbf{A}}_i$$

where we have used the definition of the scalar product of two vectors ($\vec{\mathbf{A}} \cdot \vec{\mathbf{B}} = AB\cos\theta$; see Chapter 7). Summing the contributions of all elements gives an approximation to the total flux through the surface:

$$\Phi_E \approx \sum \vec{\mathbf{E}}_i \cdot \Delta\vec{\mathbf{A}}_i$$

If the area of each element approaches zero, the number of elements approaches infinity and the sum is replaced by an integral. Therefore, the general definition of electric flux is

Definition of electric flux ▶

$$\Phi_E \equiv \int_{\text{surface}} \vec{\mathbf{E}} \cdot d\vec{\mathbf{A}} \qquad \textbf{(24.3)}$$

Equation 24.3 is a *surface integral,* which means it must be evaluated over the surface in question. In general, the value of Φ_E depends both on the field pattern and on the surface.

We are often interested in evaluating the flux through a *closed surface,* defined as a surface that divides space into an inside and an outside region so that one cannot move from one region to the other without crossing the surface. The surface of a sphere, for example, is a closed surface.

Consider the closed surface in Active Figure 24.4. The vectors $\Delta\vec{\mathbf{A}}_i$ point in different directions for the various surface elements, but at each point they are normal to the surface and, by convention, always point outward. At the element labeled ①, the field lines are crossing the surface from the inside to the outside and $\theta < 90°$; hence, the flux $\Delta\Phi_E = \vec{\mathbf{E}} \cdot \Delta\vec{\mathbf{A}}_1$ through this element is positive. For element ②, the field lines graze the surface (perpendicular to the vector $\Delta\vec{\mathbf{A}}_2$); therefore, $\theta = 90°$ and the flux is zero. For elements such as ③, where the field lines are crossing the surface from outside to inside, $180° > \theta > 90°$ and the flux is negative because $\cos\theta$ is negative. The *net* flux through the surface is proportional to the net number of lines leaving the surface, where the net number

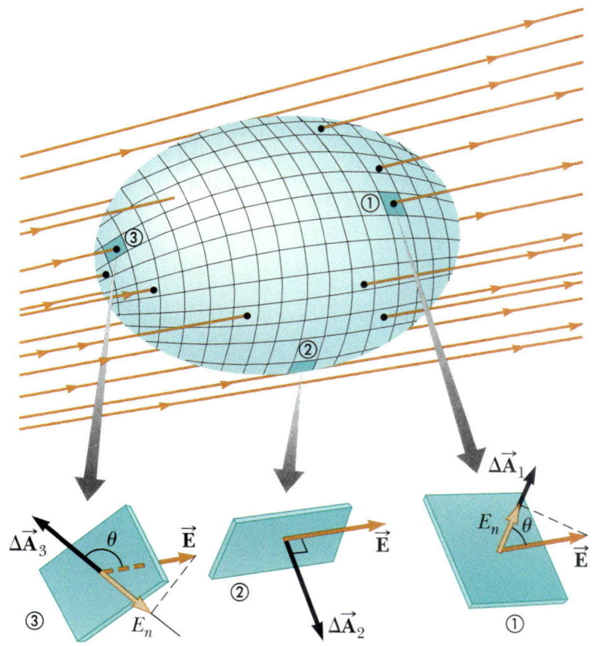

ACTIVE FIGURE 24.4

A closed surface in an electric field. The area vectors are, by convention, normal to the surface and point outward. The flux through an area element can be positive (element ①), zero (element ②), or negative (element ③).

Sign in at www.thomsonedu.com and go to ThomsonNOW to select any segment on the surface and see the relationship between the electric field vector $\vec{\mathbf{E}}$ and the area vector $\Delta\vec{\mathbf{A}}_i$.

means *the number of lines leaving the surface minus the number of lines entering the surface.* If more lines are leaving than entering, the net flux is positive. If more lines are entering than leaving, the net flux is negative. Using the symbol $\oint$ to represent an integral over a closed surface, we can write the net flux Φ_E through a closed surface as

$$\Phi_E = \oint \vec{\mathbf{E}} \cdot d\vec{\mathbf{A}} = \oint E_n \, dA \quad \textbf{(24.4)}$$

where E_n represents the component of the electric field normal to the surface.

Quick Quiz 24.1 Suppose a point charge is located at the center of a spherical surface. The electric field at the surface of the sphere and the total flux through the sphere are determined. Now the radius of the sphere is halved. What happens to the flux through the sphere and the magnitude of the electric field at the surface of the sphere? (a) The flux and field both increase. (b) The flux and field both decrease. (c) The flux increases, and the field decreases. (d) The flux decreases, and the field increases. (e) The flux remains the same, and the field increases. (f) The flux decreases, and the field remains the same.

EXAMPLE 24.1 Flux Through a Cube

Consider a uniform electric field $\vec{\mathbf{E}}$ oriented in the x direction in empty space. Find the net electric flux through the surface of a cube of edge length ℓ, oriented as shown in Figure 24.5.

SOLUTION

Conceptualize Examine Figure 24.5 carefully. Notice that the electric field lines pass through two faces perpendicularly and are parallel to four other faces of the cube.

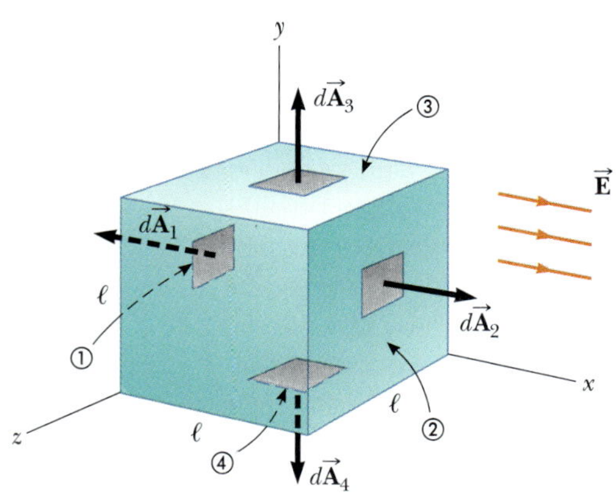

Figure 24.5 (Example 24.1) A closed surface in the shape of a cube in a uniform electric field oriented parallel to the x axis. Side ④ is the bottom of the cube, and side ① is opposite side ②.

Categorize We evaluate the flux from its definition, so we categorize this example as a substitution problem.

The flux through four of the faces (③, ④, and the unnumbered ones) is zero because $\vec{\mathbf{E}}$ is parallel to the four faces and therefore perpendicular to $d\vec{\mathbf{A}}$ on these faces.

Write the integrals for the net flux through faces ① and ②:

$$\Phi_E = \int_1 \vec{\mathbf{E}} \cdot d\vec{\mathbf{A}} + \int_2 \vec{\mathbf{E}} \cdot d\vec{\mathbf{A}}$$

For face ①, $\vec{\mathbf{E}}$ is constant and directed inward but $d\vec{\mathbf{A}}_1$ is directed outward ($\theta = 180°$). Find the flux through this face:

$$\int_1 \vec{\mathbf{E}} \cdot d\vec{\mathbf{A}} = \int_1 E(\cos 180°)\, dA = -E \int_1 dA = -EA = -E\ell^2$$

For face ②, $\vec{\mathbf{E}}$ is constant and outward and in the same direction as $d\vec{\mathbf{A}}_2$ ($\theta = 0°$). Find the flux through this face:

$$\int_2 \vec{\mathbf{E}} \cdot d\vec{\mathbf{A}} = \int_2 E(\cos 0°)\, dA = E \int_2 dA = +EA = E\ell^2$$

Find the net flux by adding the flux over all six faces:

$$\Phi_E = -E\ell^2 + E\ell^2 + 0 + 0 + 0 + 0 = 0$$

24.2 Gauss's Law

In this section, we describe a general relationship between the net electric flux through a closed surface (often called a *gaussian surface*) and the charge enclosed by the surface. This relationship, known as *Gauss's law,* is of fundamental importance in the study of electric fields.

Consider a positive point charge q located at the center of a sphere of radius r as shown in Figure 24.6. From Equation 23.9, we know that the magnitude of the electric field everywhere on the surface of the sphere is $E = k_e q/r^2$. The field lines are directed radially outward and hence are perpendicular to the surface at every point on the surface. That is, at each surface point, $\vec{\mathbf{E}}$ is parallel to the vector $\Delta\vec{\mathbf{A}}_i$ representing a local element of area ΔA_i surrounding the surface point. Therefore,

$$\vec{\mathbf{E}} \cdot \Delta\vec{\mathbf{A}}_i = E\,\Delta A_i$$

and, from Equation 24.4, we find that the net flux through the gaussian surface is

$$\Phi_E = \oint \vec{\mathbf{E}} \cdot d\vec{\mathbf{A}} = \oint E\, dA = E \oint dA$$

where we have moved E outside of the integral because, by symmetry, E is constant over the surface. The value of E is given by $E = k_e q/r^2$. Furthermore, because the surface is spherical, $\oint dA = A = 4\pi r^2$. Hence, the net flux through the gaussian surface is

$$\Phi_E = k_e \frac{q}{r^2}(4\pi r^2) = 4\pi k_e q$$

Recalling from Section 23.3 that $k_e = 1/4\pi\epsilon_0$, we can write this equation in the form

$$\Phi_E = \frac{q}{\epsilon_0} \qquad \textbf{(24.5)}$$

Equation 24.5 shows that the net flux through the spherical surface is proportional to the charge inside the surface. The flux is independent of the radius r because the area of the spherical surface is proportional to r^2, whereas the electric field is proportional to $1/r^2$. Therefore, in the product of area and electric field, the dependence on r cancels.

KARL FRIEDRICH GAUSS
German mathematician and astronomer (1777–1855)
Gauss received a doctoral degree in mathematics from the University of Helmstedt in 1799. In addition to his work in electromagnetism, he made contributions to mathematics and science in number theory, statistics, non-Euclidean geometry, and cometary orbital mechanics. He was a founder of the German Magnetic Union, which studies the Earth's magnetic field on a continual basis.

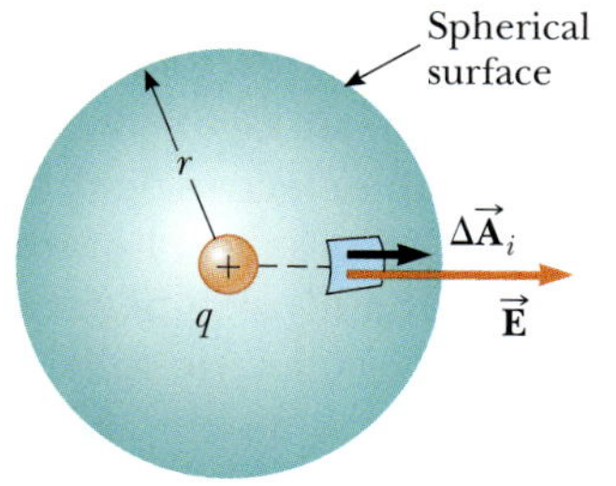

Figure 24.6 A spherical gaussian surface of radius r surrounding a point charge q. When the charge is at the center of the sphere, the electric field is everywhere normal to the surface and constant in magnitude.

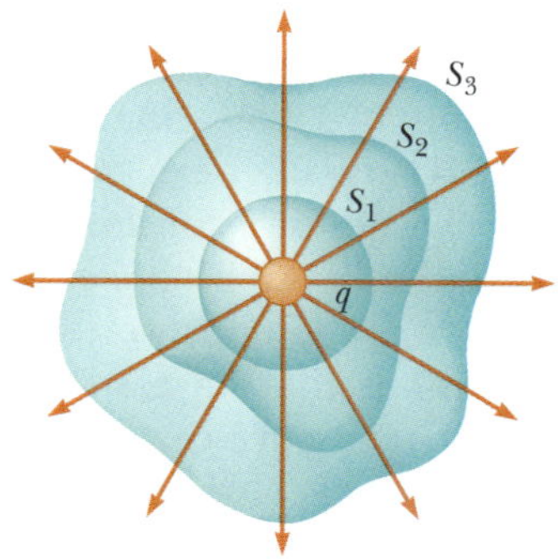

Figure 24.7 Closed surfaces of various shapes surrounding a charge q. The net electric flux is the same through all surfaces.

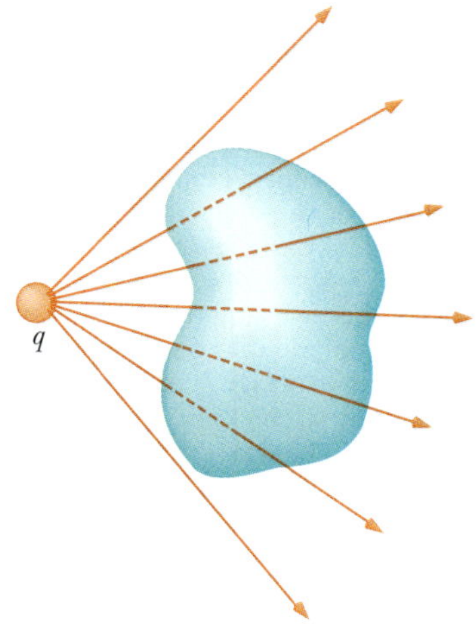

Figure 24.8 A point charge located *outside* a closed surface. The number of lines entering the surface equals the number leaving the surface.

Now consider several closed surfaces surrounding a charge q as shown in Figure 24.7. Surface S_1 is spherical, but surfaces S_2 and S_3 are not. From Equation 24.5, the flux that passes through S_1 has the value q/ϵ_0. As discussed in the preceding section, flux is proportional to the number of electric field lines passing through a surface. The construction shown in Figure 24.7 shows that the number of lines through S_1 is equal to the number of lines through the nonspherical surfaces S_2 and S_3. Therefore, **the net flux through *any* closed surface surrounding a point charge q is given by q/ϵ_0 and is independent of the shape of that surface.**

Now consider a point charge located *outside* a closed surface of arbitrary shape as shown in Figure 24.8. As can be seen from this construction, any electric field line entering the surface leaves the surface at another point. The number of electric field lines entering the surface equals the number leaving the surface. Therefore, **the net electric flux through a closed surface that surrounds no charge is zero.** Applying this result to Example 24.1, we see that the net flux through the cube is zero because there is no charge inside the cube.

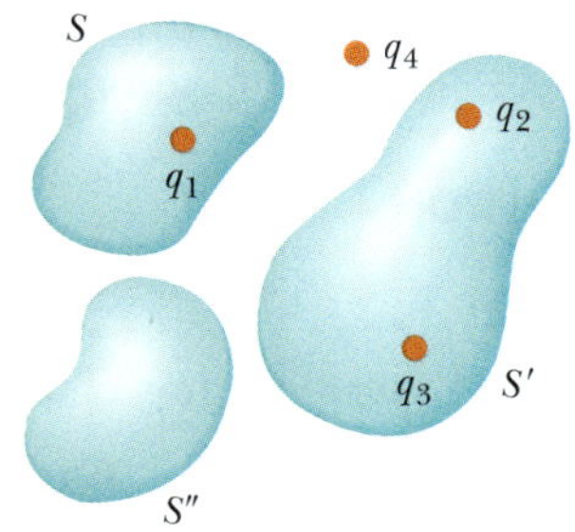

ACTIVE FIGURE 24.9

The net electric flux through any closed surface depends only on the charge *inside* that surface. The net flux through surface S is q_1/ϵ_0, the net flux through surface S' is $(q_2 + q_3)/\epsilon_0$, and the net flux through surface S'' is zero. Charge q_4 does not contribute to the flux through any surface because it is outside all surfaces.

Sign in at www.thomsonedu.com and go to ThomsonNOW to change the size and shape of a closed surface and see the effect on the electric flux of surrounding combinations of charge with that surface.

Let's extend these arguments to two generalized cases: (1) that of many point charges and (2) that of a continuous distribution of charge. We once again use the superposition principle, which states that **the electric field due to many charges is the vector sum of the electric fields produced by the individual charges.** Therefore, the flux through any closed surface can be expressed as

$$\oint \vec{\mathbf{E}} \cdot d\vec{\mathbf{A}} = \oint (\vec{\mathbf{E}}_1 + \vec{\mathbf{E}}_2 + \cdots) \cdot d\vec{\mathbf{A}}$$

where $\vec{\mathbf{E}}$ is the total electric field at any point on the surface produced by the vector addition of the electric fields at that point due to the individual charges. Consider the system of charges shown in Active Figure 24.9. The surface S surrounds only one charge, q_1; hence, the net flux through S is q_1/ϵ_0. The flux through S due to charges q_2, q_3, and q_4 outside it is zero because each electric field line from these charges that enters S at one point leaves it at another. The surface S' surrounds charges q_2 and q_3; hence, the net flux through it is $(q_2 + q_3)/\epsilon_0$. Finally, the net flux through surface S'' is zero because there is no charge inside this surface. That is, *all* the electric field lines that enter S'' at one point leave at another. Charge q_4 does not contribute to the net flux through any of the surfaces because it is outside all the surfaces.

Gauss's law is a generalization of what we have just described and states that the net flux through *any* closed surface is

$$\Phi_E = \oint \vec{\mathbf{E}} \cdot d\vec{\mathbf{A}} = \frac{q_{\text{in}}}{\epsilon_0} \qquad \textbf{(24.6)}$$

◀ Gauss's law

where $\vec{\mathbf{E}}$ represents the electric field at any point on the surface and q_{in} represents the net charge inside the surface.

PITFALL PREVENTION 24.1

Zero Flux Is Not Zero Field

In two situations, there is zero flux through a closed surface: either there are no charged particles enclosed by the surface or there are charged particles enclosed, but the net charge inside the surface is zero. For either situation, it is *incorrect* to conclude that the electric field on the surface is zero. Gauss's law states that the electric *flux* is proportional to the enclosed charge, not the electric *field*.

When using Equation 24.6, you should note that although the charge q_{in} is the net charge inside the gaussian surface, $\vec{\mathbf{E}}$ represents the *total electric field,* which includes contributions from charges both inside and outside the surface.

In principle, Gauss's law can be solved for $\vec{\mathbf{E}}$ to determine the electric field due to a system of charges or a continuous distribution of charge. In practice, however, this type of solution is applicable only in a limited number of highly symmetric situations. In the next section, we use Gauss's law to evaluate the electric field for charge distributions that have spherical, cylindrical, or planar symmetry. If one chooses the gaussian surface surrounding the charge distribution carefully, the integral in Equation 24.6 can be simplified.

Quick Quiz 24.2 If the net flux through a gaussian surface is *zero,* the following four statements *could be true.* Which of the statements *must be true?* (a) There are no charges inside the surface. (b) The net charge inside the surface is zero. (c) The electric field is zero everywhere on the surface. (d) The number of electric field lines entering the surface equals the number leaving the surface.

CONCEPTUAL EXAMPLE 24.2 **Flux Due to a Point Charge**

A spherical gaussian surface surrounds a point charge q. Describe what happens to the total flux through the surface if **(A)** the charge is tripled, **(B)** the radius of the sphere is doubled, **(C)** the surface is changed to a cube, and **(D)** the charge is moved to another location inside the surface.

SOLUTION

(A) The flux through the surface is tripled because flux is proportional to the amount of charge inside the surface.

(B) The flux does not change because all electric field lines from the charge pass through the sphere, regardless of its radius.

(C) The flux does not change when the shape of the gaussian surface changes because all electric field lines from the charge pass through the surface, regardless of its shape.

(D) The flux does not change when the charge is moved to another location inside that surface because Gauss's law refers to the total charge enclosed, regardless of where the charge is located inside the surface.

24.3 Application of Gauss's Law to Various Charge Distributions

PITFALL PREVENTION 24.2

Gaussian Surfaces Are Not Real

A gaussian surface is an imaginary surface you construct to satisfy the conditions listed here. It does not have to coincide with a physical surface in the situation.

As mentioned earlier, Gauss's law is useful for determining electric fields when the charge distribution is highly symmetric. The following examples demonstrate ways of choosing the gaussian surface over which the surface integral given by Equation 24.6 can be simplified and the electric field determined. In choosing the surface, always take advantage of the symmetry of the charge distribution so that E can be removed from the integral. The goal in this type of calculation is to determine a surface for which each portion of the surface satisfies one or more of the following conditions:

1. The value of the electric field can be argued by symmetry to be constant over the portion of the surface.
2. The dot product in Equation 24.6 can be expressed as a simple algebraic product $E\,dA$ because $\vec{\mathbf{E}}$ and $d\vec{\mathbf{A}}$ are parallel.
3. The dot product in Equation 24.6 is zero because $\vec{\mathbf{E}}$ and $d\vec{\mathbf{A}}$ are perpendicular.
4. The electric field is zero over the portion of the surface.

Different portions of the gaussian surface can satisfy different conditions as long as every portion satisfies at least one condition. All four conditions are used in examples throughout the remainder of this chapter.

EXAMPLE 24.3 A Spherically Symmetric Charge Distribution

An insulating solid sphere of radius a has a uniform volume charge density ρ and carries a total positive charge Q (Fig. 24.10).

(A) Calculate the magnitude of the electric field at a point outside the sphere.

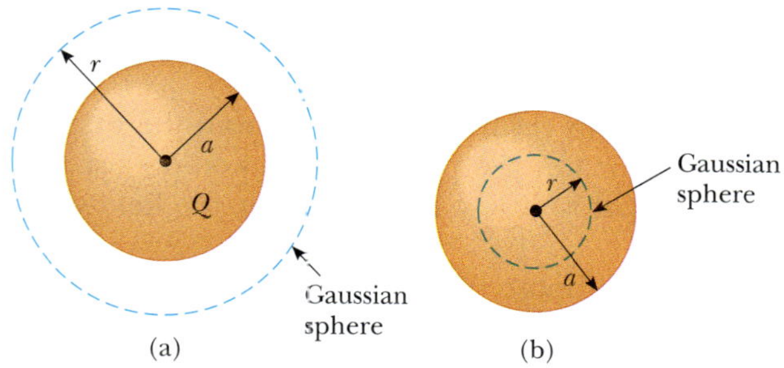

Figure 24.10 (Example 24.3) A uniformly charged insulating sphere of radius a and total charge Q. (a) For points outside the sphere, a large, spherical gaussian surface is drawn concentric with the sphere. In diagrams such as this one, the dotted line represents the intersection of the gaussian surface with the plane of the page. (b) For points inside the sphere, a spherical gaussian surface smaller than the sphere is drawn.

SOLUTION

Conceptualize Note how this problem differs from our previous discussion of Gauss's law. The electric field due to point charges was discussed in Section 24.2. Now we are considering the electric field due to a distribution of charge. We found the field for various distributions of charge in Chapter 23 by integrating over the distribution. In this chapter, we find the electric field using Gauss's law.

Categorize Because the charge is distributed uniformly throughout the sphere, the charge distribution has spherical symmetry and we can apply Gauss's law to find the electric field.

Analyze To reflect the spherical symmetry, let's choose a spherical gaussian surface of radius r, concentric with the sphere, as shown in Figure 24.10a. For this choice, condition (2) is satisfied everywhere on the surface and $\vec{\mathbf{E}} \cdot d\vec{\mathbf{A}} = E\,dA$.

Replace $\vec{\mathbf{E}} \cdot d\vec{\mathbf{A}}$ in Gauss's law with $E\,dA$:

$$\Phi_E = \oint \vec{\mathbf{E}} \cdot d\vec{\mathbf{A}} = \oint E\,dA = \frac{Q}{\epsilon_0}$$

By symmetry, E is constant everywhere on the surface, which satisfies condition (1), so we can remove E from the integral:

$$\oint E\,dA = E\oint dA = E(4\pi r^2) = \frac{Q}{\epsilon_0}$$

Solve for E:

$$(1)\quad E = \frac{Q}{4\pi\epsilon_0 r^2} = k_e\frac{Q}{r^2} \quad (\text{for } r > a)$$

Finalize This field is identical to that for a point charge. Therefore, **the electric field due to a uniformly charged sphere in the region external to the sphere is *equivalent* to that of a point charge located at the center of the sphere.**

(B) Find the magnitude of the electric field at a point inside the sphere.

SOLUTION

Analyze In this case, let's choose a spherical gaussian surface having radius $r < a$, concentric with the insulating sphere (Fig. 24.10b). Let V' be the volume of this smaller sphere. To apply Gauss's law in this situation, recognize that the charge q_{in} within the gaussian surface of volume V' is less than Q.

Calculate q_{in} by using $q_{\text{in}} = \rho V'$:

$$q_{\text{in}} = \rho V' = \rho\left(\tfrac{4}{3}\pi r^3\right)$$

Notice that conditions (1) and (2) are satisfied everywhere on the gaussian surface in Figure 24.10b. Apply Gauss's law in the region $r < a$:

$$\oint E\,dA = E\oint dA = E(4\pi r^2) = \frac{q_{\text{in}}}{\epsilon_0}$$

Solve for E and substitute for q_{in}:

$$E = \frac{q_{\text{in}}}{4\pi\epsilon_0 r^2} = \frac{\rho\left(\frac{4}{3}\pi r^3\right)}{4\pi\epsilon_0 r^2} = \frac{\rho}{3\epsilon_0}r$$

Substitute $\rho = Q/\frac{4}{3}\pi a^3$ and $\epsilon_0 = 1/4\pi k_e$:

$$(2)\quad E = \frac{(Q/\frac{4}{3}\pi a^3)}{3(1/4\pi k_e)} r = k_e \frac{Q}{a^3} r \quad (\text{for } r < a)$$

Finalize This result for E differs from the one obtained in part (A). It shows that $E \rightarrow 0$ as $r \rightarrow 0$. Therefore, the result eliminates the problem that would exist at $r = 0$ if E varied as $1/r^2$ inside the sphere as it does outside the sphere. That is, if $E \propto 1/r^2$ for $r < a$, the field would be infinite at $r = 0$, which is physically impossible.

What If? Suppose the radial position $r = a$ is approached from inside the sphere and from outside. Do we obtain the same value of the electric field from both directions?

Answer Equation (1) shows that the electric field approaches a value from the outside given by

$$E = \lim_{r \rightarrow a} \left(k_e \frac{Q}{r^2} \right) = k_e \frac{Q}{a^2}$$

From the inside, Equation (2) gives

$$E = \lim_{r \rightarrow a} \left(k_e \frac{Q}{a^3} r \right) = k_e \frac{Q}{a^3} a = k_e \frac{Q}{a^2}$$

Therefore, the value of the field is the same as the surface is approached from both directions. A plot of E versus r is shown in Figure 24.11. Notice that the magnitude of the field is continuous.

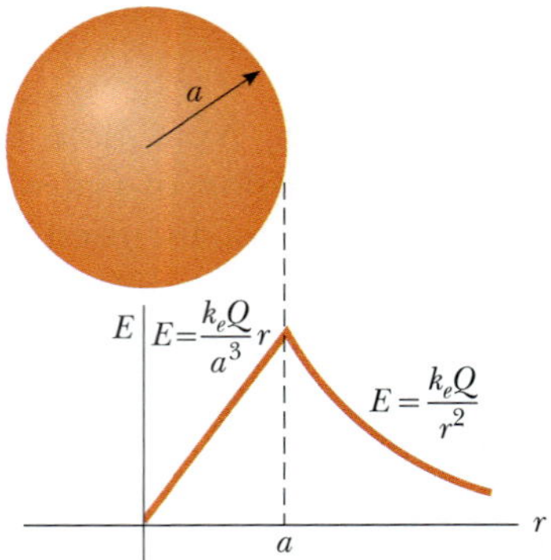

Figure 24.11 (Example 24.3) A plot of E versus r for a uniformly charged insulating sphere. The electric field inside the sphere ($r < a$) varies linearly with r. The field outside the sphere ($r > a$) is the same as that of a point charge Q located at $r = 0$.

EXAMPLE 24.4 A Cylindrically Symmetric Charge Distribution

Find the electric field a distance r from a line of positive charge of infinite length and constant charge per unit length λ (Fig. 24.12a).

SOLUTION

Conceptualize The line of charge is *infinitely* long. Therefore, the field is the same at all points equidistant from the line, regardless of the vertical position of the point in Figure 24.12a.

Categorize Because the charge is distributed uniformly along the line, the charge distribution has cylindrical symmetry and we can apply Gauss's law to find the electric field.

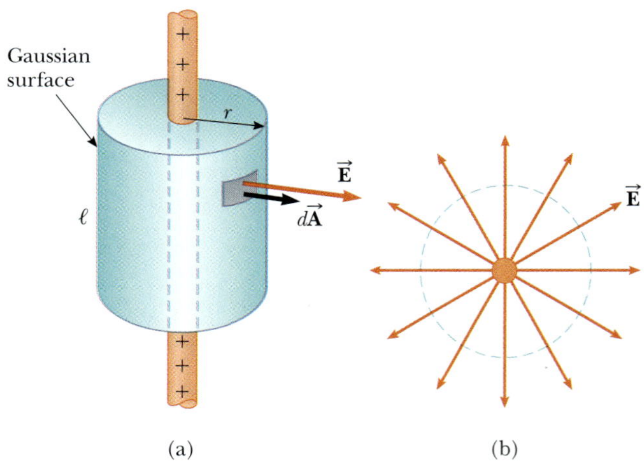

Figure 24.12 (Example 24.4) (a) An infinite line of charge surrounded by a cylindrical gaussian surface concentric with the line. (b) An end view shows that the electric field at the cylindrical surface is constant in magnitude and perpendicular to the surface.

Analyze The symmetry of the charge distribution requires that $\vec{E}$ be perpendicular to the line charge and directed outward as shown in Figures 24.12a and b. To reflect the symmetry of the charge distribution, let's choose a cylindrical gaussian surface of radius r and length ℓ that is coaxial with the line charge. For the curved part of this surface, $\vec{E}$ is constant in magnitude and perpendicular to the surface at each point, satisfying conditions (1) and (2). Furthermore, the flux through the ends of the gaussian cylinder is zero because $\vec{E}$ is parallel to these surfaces. That is the first application we have seen of condition (3).

We must take the surface integral in Gauss's law over the entire gaussian surface. Because $\vec{E} \cdot d\vec{A}$ is zero for the flat ends of the cylinder, however, we restrict our attention to only the curved surface of the cylinder.

Apply Gauss's law and conditions (1) and (2) for the curved surface, noting that the total charge inside our gaussian surface is $\lambda\ell$:

$$\Phi_E = \oint \vec{E} \cdot d\vec{A} = E \oint dA = EA = \frac{q_{\text{in}}}{\epsilon_0} = \frac{\lambda \ell}{\epsilon_0}$$

Substitute the area $A = 2\pi r\ell$ of the curved surface:

$$E(2\pi r\ell) = \frac{\lambda\ell}{\epsilon_0}$$

Solve for the magnitude of the electric field:

$$E = \frac{\lambda}{2\pi\epsilon_0 r} = 2k_e\frac{\lambda}{r} \qquad \textbf{(24.7)}$$

Finalize This result shows that the electric field due to a cylindrically symmetric charge distribution varies as $1/r$, whereas the field external to a spherically symmetric charge distribution varies as $1/r^2$. Equation 24.7 can also be derived by direct integration over the charge distribution. (See Problem 29 in Chapter 23.)

What If? What if the line segment in this example were not infinitely long?

Answer If the line charge in this example were of finite length, the electric field would not be given by Equation 24.7. A finite line charge does not possess sufficient symmetry to make use of Gauss's law because the magnitude of the electric field is no longer constant over the surface of the gaussian cylinder: the field near the ends of the line would be different from that far from the ends. Therefore, condition (1) would not be satisfied in this situation. Furthermore, $\vec{\mathbf{E}}$ is not perpendicular to the cylindrical surface at all points: the field vectors near the ends would have a component parallel to the line. Therefore, condition (2) would not be satisfied. For points close to a finite line charge and far from the ends, Equation 24.7 gives a good approximation of the value of the field.

It is left for you to show (see Problem 27) that the electric field inside a uniformly charged rod of finite radius and infinite length is proportional to r.

EXAMPLE 24.5 A Plane of Charge

Find the electric field due to an infinite plane of positive charge with uniform surface charge density σ.

SOLUTION

Conceptualize Note that the plane of charge is *infinitely* large. Therefore, the electric field should be the same at all points near the plane.

Categorize Because the charge is distributed uniformly on the plane, the charge distribution is symmetric; hence, we can use Gauss's law to find the electric field.

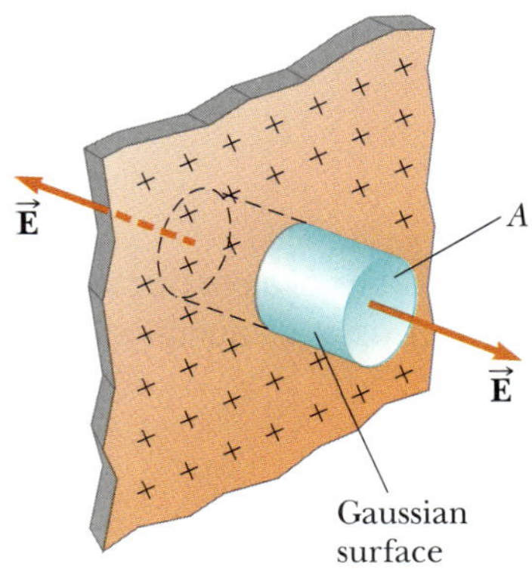

Figure 24.13 (Example 24.5) A cylindrical gaussian surface penetrating an infinite plane of charge. The flux is EA through each end of the gaussian surface and zero through its curved surface.

Analyze By symmetry, $\vec{\mathbf{E}}$ must be perpendicular to the plane at all points. The direction of $\vec{\mathbf{E}}$ is away from positive charges, indicating that the direction of $\vec{\mathbf{E}}$ on one side of the plane must be opposite its direction on the other side as shown in Figure 24.13. A gaussian surface that reflects the symmetry is a small cylinder whose axis is perpendicular to the plane and whose ends each have an area A and are equidistant from the plane. Because $\vec{\mathbf{E}}$ is parallel to the curved surface—and therefore perpendicular to $d\vec{\mathbf{A}}$ everywhere on the surface—condition (3) is satisfied and there is no contribution to the surface integral from this surface. For the flat ends of the cylinder, conditions (1) and (2) are satisfied. The flux through each end of the cylinder is EA; hence, the total flux through the entire gaussian surface is just that through the ends, $\Phi_E = 2EA$.

Write Gauss's law for this surface, noting that the enclosed charge is $q_{\text{in}} = \sigma A$:

$$\Phi_E = 2EA = \frac{q_{\text{in}}}{\epsilon_0} = \frac{\sigma A}{\epsilon_0}$$

Solve for E:

$$E = \frac{\sigma}{2\epsilon_0} \qquad \textbf{(24.8)}$$

Finalize Because the distance from each flat end of the cylinder to the plane does not appear in Equation 24.8, we conclude that $E = \sigma/2\epsilon_0$ at *any* distance from the plane. That is, the field is uniform everywhere.

What If? Suppose two infinite planes of charge are parallel to each other, one positively charged and the other negatively charged. Both planes have the same surface charge density. What does the electric field look like in this situation?

Answer The electric fields due to the two planes add in the region between the planes, resulting in a uniform field of magnitude σ/ϵ_0, and cancel elsewhere to give a field of zero. This method is a practical way to achieve uniform electric fields.

CONCEPTUAL EXAMPLE 24.6 **Don't Use Gauss's Law Here!**

Explain why Gauss's law cannot be used to calculate the electric field near an electric dipole, a charged disk, or a triangle with a point charge at each corner.

SOLUTION

The charge distributions of all these configurations do not have sufficient symmetry to make the use of Gauss's law practical. We cannot find a closed surface surrounding any of these distributions that satisfies one or more of conditions (1) through (4) listed at the beginning of this section.

24.4 Conductors in Electrostatic Equilibrium

As we learned in Section 23.2, a good electrical conductor contains charges (electrons) that are not bound to any atom and therefore are free to move about within the material. When there is no net motion of charge within a conductor, the conductor is in **electrostatic equilibrium.** A conductor in electrostatic equilibrium has the following properties:

Properties of a conductor in electrostatic equilibrium ▶

1. The electric field is zero everywhere inside the conductor, whether the conductor is solid or hollow.
2. If an isolated conductor carries a charge, the charge resides on its surface.
3. The electric field just outside a charged conductor is perpendicular to the surface of the conductor and has a magnitude σ/ϵ_0, where σ is the surface charge density at that point.
4. On an irregularly shaped conductor, the surface charge density is greatest at locations where the radius of curvature of the surface is smallest.

We verify the first three properties in the discussion that follows. The fourth property is presented here (but not verified until Chapter 25) to provide a complete list of properties for conductors in electrostatic equilibrium.

We can understand the first property by considering a conducting slab placed in an external field $\vec{\mathbf{E}}$ (Fig. 24.14). The electric field inside the conductor *must* be zero assuming electrostatic equilibrium exists. If the field were not zero, free electrons in the conductor would experience an electric force ($\vec{\mathbf{F}} = q\vec{\mathbf{E}}$) and would accelerate due to this force. This motion of electrons, however, would mean that the conductor is not in electrostatic equilibrium. Therefore, the existence of electrostatic equilibrium is consistent only with a zero field in the conductor.

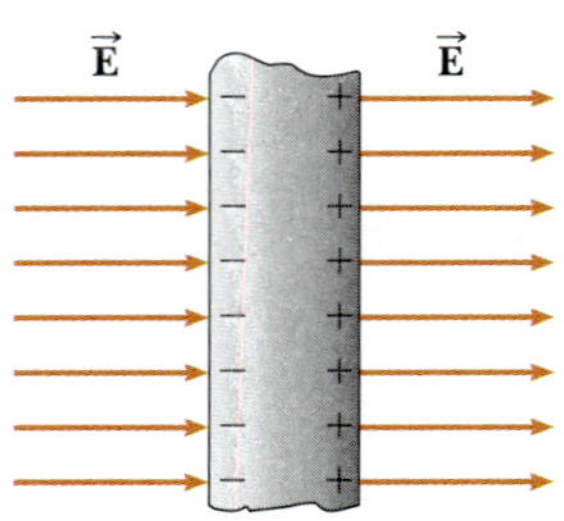

Figure 24.14 A conducting slab in an external electric field $\vec{\mathbf{E}}$. The charges induced on the two surfaces of the slab produce an electric field that opposes the external field, giving a resultant field of zero inside the slab.

Let's investigate how this zero field is accomplished. Before the external field is applied, free electrons are uniformly distributed throughout the conductor. When the external field is applied, the free electrons accelerate to the left in Figure 24.14, causing a plane of negative charge to accumulate on the left surface. The movement of electrons to the left results in a plane of positive charge on the right surface. These planes of charge create an additional electric field inside the conductor that opposes the external field. As the electrons move, the surface charge densities on the left and right surfaces increase until the magnitude of the internal field equals that of the external field, resulting in a net field of zero inside the conductor. The time it takes a good conductor to reach equilibrium is on the order of 10^{-16} s, which for most purposes can be considered instantaneous.

If the conductor is hollow, the electric field inside the conductor is also zero, whether we consider points in the conductor or in the cavity within the conductor. The zero value of the electric field in the cavity is easiest to argue with the concept of electric potential, so we will address this issue in Section 25.6.

Gauss's law can be used to verify the second property of a conductor in electrostatic equilibrium. Figure 24.15 shows an arbitrarily shaped conductor. A gaussian surface is drawn inside the conductor and can be very close to the conductor's surface. As we have just shown, the electric field everywhere inside the conductor is zero when it is in electrostatic equilibrium. Therefore, the electric field must be zero at every point on the gaussian surface, in accordance with condition (4) in Section 24.3, and the net flux through this gaussian surface is zero. From this result and Gauss's law, we conclude that the net charge inside the gaussian surface is zero. Because there can be no net charge inside the gaussian surface (which is arbitrarily close to the conductor's surface), **any net charge on the conductor must reside on its surface.** Gauss's law does not indicate how this excess charge is distributed on the conductor's surface, only that it resides exclusively on the surface.

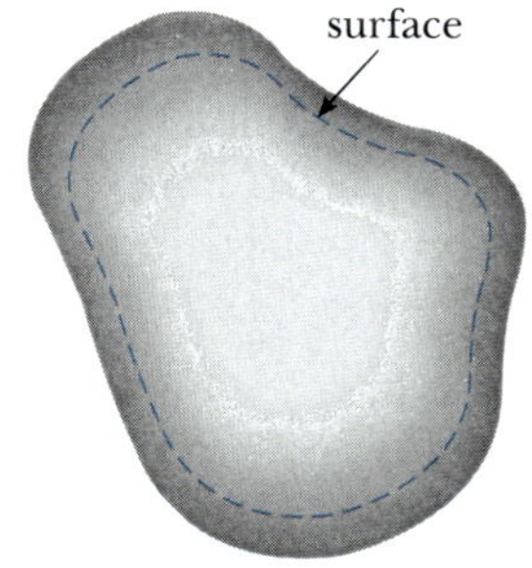

Figure 24.15 A conductor of arbitrary shape. The broken line represents a gaussian surface that can be just inside the conductor's surface.

Let's verify the third property. If the field vector $\vec{\mathbf{E}}$ had a component parallel to the conductor's surface, free electrons would experience an electric force and move along the surface; in such a case, the conductor would not be in equilibrium. Therefore, the field vector must be perpendicular to the surface. To determine the magnitude of the electric field, we use Gauss's law and draw a gaussian surface in the shape of a small cylinder whose end faces are parallel to the conductor's surface (Fig. 24.16). Part of the cylinder is just outside the conductor, and part is inside. The field is perpendicular to the conductor's surface from the condition of electrostatic equilibrium. Therefore, condition (3) in Section 24.3 is satisfied for the curved part of the cylindrical gaussian surface: there is no flux through this part of the gaussian surface because $\vec{\mathbf{E}}$ is parallel to the surface. There is no flux through the flat face of the cylinder inside the conductor because here $\vec{\mathbf{E}} = 0$, which satisfies condition (4). Hence, the net flux through the gaussian surface is equal to that through only the flat face outside the conductor, where the field is perpendicular to the gaussian surface. Using conditions (1) and (2) for this face, the flux is EA, where E is the electric field just outside the conductor and A is the area of the cylinder's face. Applying Gauss's law to this surface gives

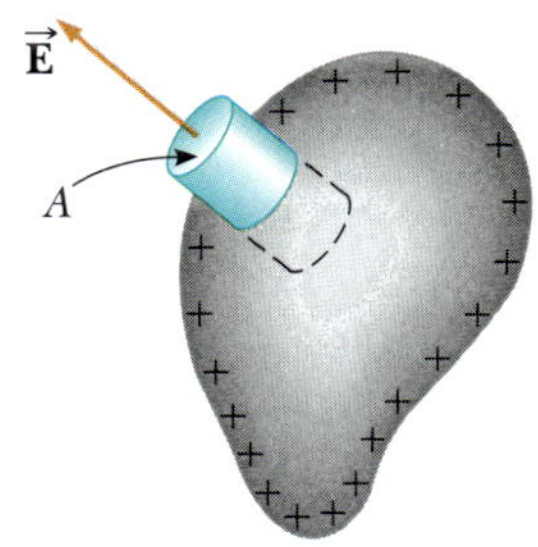

Figure 24.16 A gaussian surface in the shape of a small cylinder is used to calculate the electric field immediately outside a charged conductor. The flux through the gaussian surface is EA. Remember that $\vec{\mathbf{E}}$ is zero inside the conductor.

$$\Phi_E = \oint E\,dA = EA = \frac{q_{\text{in}}}{\epsilon_0} = \frac{\sigma A}{\epsilon_0}$$

where we have used $q_{\text{in}} = \sigma A$. Solving for E gives for the electric field immediately outside a charged conductor

$$E = \frac{\sigma}{\epsilon_0} \quad (24.9)$$

Quick Quiz 24.3 Your younger brother likes to rub his feet on the carpet and then touch you to give you a shock. While you are trying to escape the shock treatment, you discover a hollow metal cylinder in your basement, large enough to climb inside. In which of the following cases will you *not* be shocked? (a) You climb inside the cylinder, making contact with the inner surface, and your charged brother touches the outer metal surface. (b) Your charged brother is inside touching the inner metal surface and you are outside, touching the outer metal surface. (c) Both of you are outside the cylinder, touching its outer metal surface but not touching each other directly.

EXAMPLE 24.7 A Sphere Inside a Spherical Shell

A solid insulating sphere of radius a carries a net positive charge Q uniformly distributed throughout its volume. A conducting spherical shell of inner radius b and outer radius c is concentric with the solid sphere and carries a net charge $-2Q$. Using Gauss's law, find the electric field in the regions labeled ①, ②, ③, and ④ in Active Figure 24.17 and the charge distribution on the shell when the entire system is in electrostatic equilibrium.

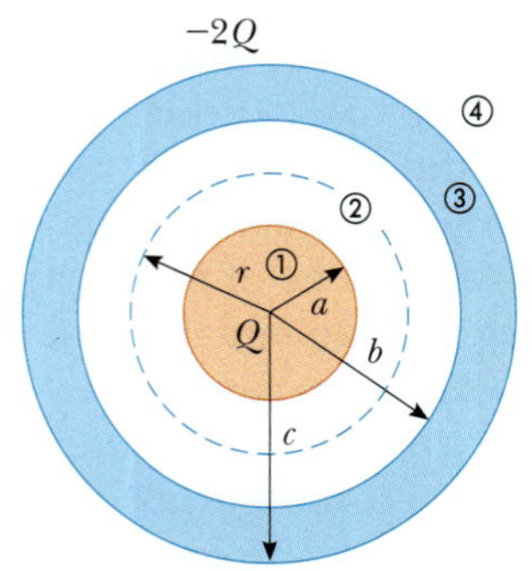

ACTIVE FIGURE 24.17

(Example 24.7) An insulating sphere of radius a and carrying a charge Q surrounded by a conducting spherical shell carrying a charge $-2Q$.

Sign in at www.thomsonedu.com and go to ThomsonNOW to vary the charges on the sphere and shell and see the effect on the electric field.

SOLUTION

Conceptualize Note how this problem differs from Example 24.3. The charged sphere in Figure 24.10 is now surrounded by a shell carrying a charge of $-2Q$.

Categorize The charge is distributed uniformly throughout the sphere, and we know that the charge on the conducting shell distributes itself uniformly on the surfaces. Therefore, the system has spherical symmetry and we can apply Gauss's law to find the electric field.

Analyze In region ②—between the surface of the solid sphere and the inner surface of the shell—we construct a spherical gaussian surface of radius r, where $a < r < b$, noting that the charge inside this surface is $+Q$ (the charge on the solid sphere). Because of the spherical symmetry, the electric field lines must be directed radially outward and be constant in magnitude on the gaussian surface.

The charge on the conducting shell creates zero electric field in the region $r < b$, so the shell has no effect on the field due to the sphere. Therefore, write an expression for the field in region ② as that due to the sphere from part (A) of Example 24.3:

$$E_2 = k_e \frac{Q}{r^2} \quad (\text{for } a < r < b)$$

Because the conducting shell creates zero field inside itself, it also has no effect on the field inside the sphere. Therefore, write an expression for the field in region ① as that due to the sphere from part (B) of Example 24.3:

$$E_1 = k_e \frac{Q}{a^3} r \quad (\text{for } r < a)$$

In region ④, where $r > c$, construct a spherical gaussian surface; this surface surrounds a total charge of $q_{\text{in}} = Q + (-2Q) = -Q$. Therefore, model the charge distribution as a sphere with charge $-Q$ and write an expression for the field in region ④ from part (A) of Example 24.3:

$$E_4 = -k_e \frac{Q}{r^2} \quad (\text{for } r > c)$$

In region ③, the electric field must be zero because the spherical shell is a conductor in equilibrium:

$$E_3 = 0 \quad (\text{for } b < r < c)$$

Construct a gaussian surface of radius r, where $b < r < c$, and note that q_{in} must be zero because $E_3 = 0$. Find the amount of charge q_{inner} on the inner surface of the shell:

$$q_{\text{in}} = q_{\text{sphere}} + q_{\text{inner}}$$

$$q_{\text{inner}} = q_{\text{in}} - q_{\text{sphere}} = 0 - Q = -Q$$

Finalize The charge on the inner surface of the spherical shell must be $-Q$ to cancel the charge $+Q$ on the solid sphere and give zero electric field in the material of the shell. Because the net charge on the shell is $-2Q$, its outer surface must carry a charge $-Q$.

What If? How would the results of this problem differ if the sphere were conducting instead of insulating?

Answer The only change would be in region ①, where $r < a$. Because there can be no charge inside a conductor in electrostatic equilibrium, $q_{in} = 0$ for a gaussian surface of radius $r < a$; therefore, on the basis of Gauss's law and symmetry, $E_1 = 0$. In regions ②, ③, and ④, there would be no way to determine whether the sphere is conducting or insulating.

Summary

ThomsonNOW™ Sign in at **www.thomsonedu.com** and go to ThomsonNOW to take a practice test for this chapter.

DEFINITION

Electric flux is proportional to the number of electric field lines that penetrate a surface. If the electric field is uniform and makes an angle θ with the normal to a surface of area A, the electric flux through the surface is

$$\Phi_E = EA\cos\theta \qquad \textbf{(24.2)}$$

In general, the electric flux through a surface is

$$\Phi_E \equiv \int_{\text{surface}} \vec{\mathbf{E}} \cdot d\vec{\mathbf{A}} \qquad \textbf{(24.3)}$$

CONCEPTS AND PRINCIPLES

Gauss's law says that the net electric flux Φ_E through any closed gaussian surface is equal to the *net* charge q_{in} inside the surface divided by ϵ_0:

$$\Phi_E = \oint \vec{\mathbf{E}} \cdot d\vec{\mathbf{A}} = \frac{q_{in}}{\epsilon_0} \qquad \textbf{(24.6)}$$

Using Gauss's law, you can calculate the electric field due to various symmetric charge distributions.

A conductor in **electrostatic equilibrium** has the following properties:

1. The electric field is zero everywhere inside the conductor, whether the conductor is solid or hollow.
2. If an isolated conductor carries a charge, the charge resides on its surface.
3. The electric field just outside a charged conductor is perpendicular to the surface of the conductor and has a magnitude σ/ϵ_0, where σ is the surface charge density at that point.
4. On an irregularly shaped conductor, the surface charge density is greatest at locations where the radius of curvature of the surface is smallest.

Questions

☐ denotes answer available in *Student Solutions Manual/Study Guide;* **O** denotes objective question

1. The Sun is lower in the sky during the winter than it is during the summer. How does this change affect the flux of sunlight hitting a given area on the surface of the Earth? How does this change affect the weather?
2. If more electric field lines leave a gaussian surface than enter it, what can you conclude about the net charge enclosed by that surface?
3. A uniform electric field exists in a region of space containing no charges. What can you conclude about the net electric flux through a gaussian surface placed in this region of space?
4. **O** A particle with charge q is located inside a cubical gaussian surface. No other charges are nearby. **(i)** If the particle is at the center of the cube, what is the flux through each one of the faces of the cube? (a) 0 (b) q/ϵ_0 (c) $q/2\epsilon_0$ (d) $q/4\epsilon_0$ (e) $q/6\epsilon_0$ (f) $q/8\epsilon_0$ (g) depends on the size of the cube **(ii)** If the particle can be moved to any point within the cube, what maximum value can the flux through one face approach? Choose from the same possibilities. **(iii)** If the particle can be moved anywhere within the cube or on its surface, what is the minimum possible flux through one face? Choose from the same possibilities.
5. **O** A cubical gaussian surface surrounds a long, straight, charged filament that passes perpendicularly through two opposite faces. No other charges are nearby. **(i)** Over how many of the cube's faces is the electric field zero? (a) 0

(b) 2 (c) 4 (d) 6 **(ii)** Through how many of the cube's faces is the electric flux zero? Choose from the same possibilities.

6. O A cubical gaussian surface is bisected by a large sheet of charge, parallel to its top and bottom faces. No other charges are nearby. **(i)** Over how many of the cube's faces is the electric field zero? (a) 0 (b) 2 (c) 4 (d) 6 **(ii)** Through how many of the cube's faces is the electric flux zero? Choose from the same possibilities.

7. O Two solid spheres, both of radius 5 cm, carry identical total charges of 2 μC. Sphere A is a good conductor. Sphere B is an insulator, and its charge is distributed uniformly throughout its volume. **(i)** How do the magnitudes of the electric fields they separately create at a radial distance of 6 cm compare? (a) $E_A > E_B = 0$ (b) $E_A > E_B > 0$ (c) $E_A = E_B > 0$ (d) $E_A = E_B = 0$ (e) $0 < E_A < E_B$ (f) $0 = E_A < E_B$ **(ii)** How do the magnitudes of the electric fields they separately create at radius 4 cm compare? Choose from the same possibilities.

8. If the total charge inside a closed surface is known but the distribution of the charge is unspecified, can you use Gauss's law to find the electric field? Explain.

9. Explain why the electric flux through a closed surface with a given enclosed charge is independent of the size or shape of the surface.

10. On the basis of the repulsive nature of the force between like charges and the freedom of motion of charge within a conductor, explain why excess charge on an isolated conductor must reside on its surface.

11. O A solid insulating sphere of radius 5 cm carries electric charge uniformly distributed throughout its volume. Concentric with the sphere is a conducting spherical shell with no net charge as shown in Figure Q24.11. The inner radius of the shell is 10 cm, and the outer radius is 15 cm. No other charges are nearby. (a) Rank the magnitude of the electric field at points A (at radius 4 cm), B (radius 8 cm), C (radius 12 cm), and D (radius 16 cm) from largest to smallest. Display any cases of equality in your ranking. (b) Similarly rank the electric flux through concentric spherical surfaces through points A, B, C, and D.

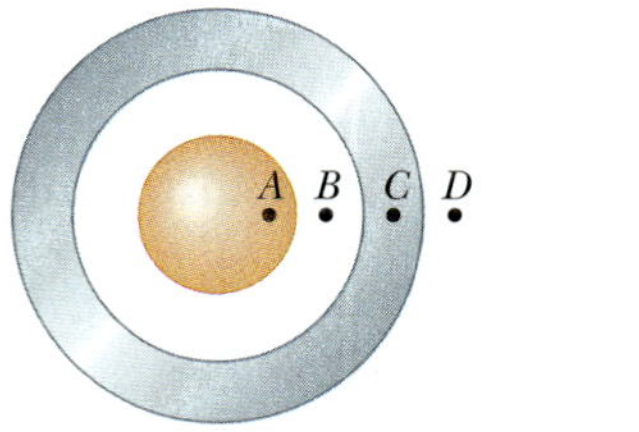

Figure Q24.11 Question 11 and Problem 44.

12. O A coaxial cable consists of a long, straight filament surrounded by a long, coaxial, cylindrical conducting shell. Assume charge Q is on the filament, zero net charge is on the shell, and the electric field is $E_1\hat{\mathbf{i}}$ at a particular point P midway between the filament and the inner surface of the shell. Next, you place the cable into a uniform external field $-E_1\hat{\mathbf{i}}$. What is the x component of the electric field at P then? (a) 0 (b) between 0 and E_1 (c) E_1 (d) greater than E_1 (e) between 0 and $-E_1$ (f) $-E_1$ (g) less than $-E_1$

13. A person is placed in a large, hollow, metallic sphere that is insulated from ground. If a large charge is placed on the sphere, will the person be harmed upon touching the inside of the sphere? Explain what will happen if the person also has an initial charge whose sign is opposite that of the charge on the sphere.

14. O A large, metallic, spherical shell has no net charge. It is supported on an insulating stand and has a small hole at the top. A small tack with charge Q is lowered on a silk thread through the hole into the interior of the shell. **(i)** What is the charge on the inner surface of the shell now? (a) Q (b) $Q/2$ (c) 0 (d) $-Q/2$ (e) $-Q$ Choose your answers to the following parts from the same possibilities. **(ii)** What is the charge on the outer surface of the shell? **(iii)** The tack is now allowed to touch the interior surface of the shell. After this contact, what is the charge on the tack? **(iv)** What is the charge on the inner surface of the shell now? **(v)** What is the charge on the outer surface of the shell now?

15. A common demonstration involves charging a rubber balloon, which is an insulator, by rubbing it on your hair and then touching the balloon to a ceiling or wall, which is also an insulator. Because of the electrical attraction between the charged balloon and the neutral wall, the balloon sticks to the wall. Imagine now that we have two infinitely large flat sheets of insulating material. One is charged, and the other is neutral. If these sheets are brought into contact, does an attractive force exist between them, as there was for the balloon and the wall?

Problems

WebAssign The Problems from this chapter may be assigned online in WebAssign.

ThomsonNOW Sign in at **www.thomsonedu.com** and go to ThomsonNOW to assess your understanding of this chapter's topics with additional quizzing and conceptual questions.

1, 2, 3 denotes straightforward, intermediate, challenging; ☐ denotes full solution available in *Student Solutions Manual/Study Guide;* ▲ denotes coached solution with hints available at **www.thomsonedu.com;** ▭ denotes developing symbolic reasoning; ● denotes asking for qualitative reasoning; ▪ denotes computer useful in solving problem

Section 24.1 Electric Flux

1. A 40.0-cm-diameter loop is rotated in a uniform electric field until the position of maximum electric flux is found. The flux in this position is 5.20×10^5 N·m^2/C. What is the magnitude of the electric field?

2. A vertical electric field of magnitude 2.00×10^4 N/C exists above the Earth's surface on a day when a thunderstorm is brewing. A car with a rectangular size of 6.00 m by 3.00 m is traveling along a dry gravel roadway sloping downward at 10.0°. Determine the electric flux through the bottom of the car.

3. A uniform electric field $a\hat{\mathbf{i}} + b\hat{\mathbf{j}}$ intersects a surface of area A. What is the flux through this area if the surface lies (a) in the yz plane, (b) in the xz plane, and (c) in the xy plane?

4. Consider a closed triangular box resting within a horizontal electric field of magnitude $E = 7.80 \times 10^4$ N/C as shown in Figure P24.4. Calculate the electric flux through (a) the vertical rectangular surface, (b) the slanted surface, and (c) the entire surface of the box.

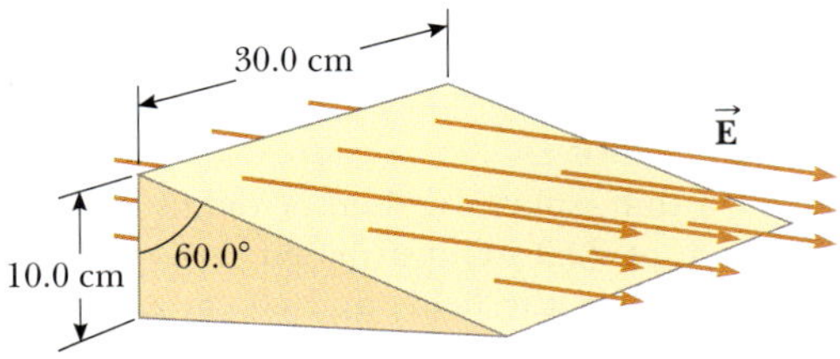

Figure P24.4

5. A pyramid with horizontal square base, 6.00 m on each side, and a height of 4.00 m is placed in a vertical electric field of 52.0 N/C. Calculate the total electric flux through the pyramid's four slanted surfaces.

Section 24.2 Gauss's Law

6. ● The electric field everywhere on the surface of a thin, spherical shell of radius 0.750 m is measured to be 890 N/C and points radially toward the center of the sphere. (a) What is the net charge within the sphere's surface? (b) What can you conclude about the nature and distribution of the charge inside the spherical shell?

7. The following charges are located inside a submarine: 5.00 μC, -9.00 μC, 27.0 μC, and -84.0 μC. (a) Calculate the net electric flux through the hull of the submarine. (b) Is the number of electric field lines leaving the submarine greater than, equal to, or less than the number entering it?

8. ● (a) A particle with charge q is located a distance d from an infinite plane. Determine the electric flux through the plane due to the charged particle. (b) **What If?** A particle with charge q is located a *very small* distance from the center of a *very large* square on the line perpendicular to the square and going through its center. Determine the approximate electric flux through the square due to the charged particle. (c) Explain why the answers to parts (a) and (b) are identical.

9. Four closed surfaces, S_1 through S_4, together with the charges $-2Q$, Q, and $-Q$ are sketched in Figure P24.9.

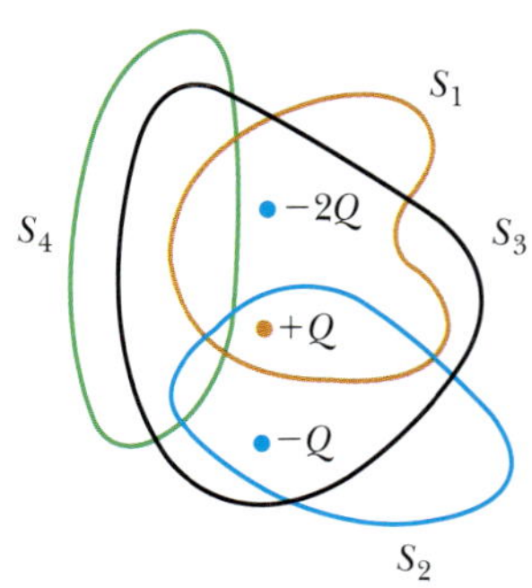

Figure P24.9

(The colored lines are the intersections of the surfaces with the page.) Find the electric flux through each surface.

10. ● A particle with charge of 12.0 μC is placed at the center of a spherical shell of radius 22.0 cm. What is the total electric flux through (a) the surface of the shell and (b) any hemispherical surface of the shell? (c) Do the results depend on the radius? Explain.

11. ▲ A particle with charge Q is located immediately above the center of the flat face of a hemisphere of radius R as shown in Figure P24.11. What is the electric flux (a) through the curved surface and (b) through the flat face?

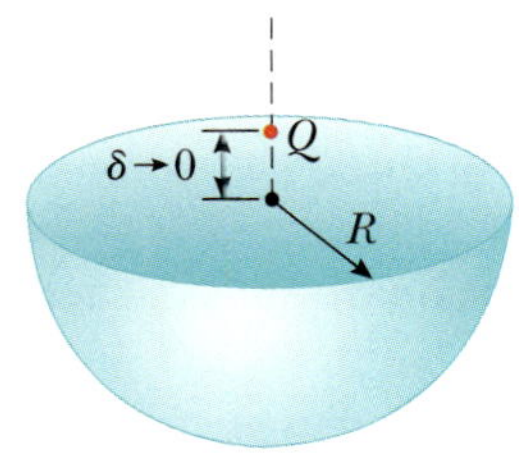

Figure P24.11

12. In the air over a particular region at an altitude of 500 m above the ground, the electric field is 120 N/C directed downward. At 600 m above the ground, the electric field is 100 N/C downward. What is the average volume charge density in the layer of air between these two elevations? Is it positive or negative?

13. A particle with charge $Q = 5.00$ μC is located at the center of a cube of edge $L = 0.100$ m. In addition, six other identical charged particles having $q = -1.00$ μC are positioned symmetrically around Q as shown in Figure P24.13. Determine the electric flux through one face of the cube.

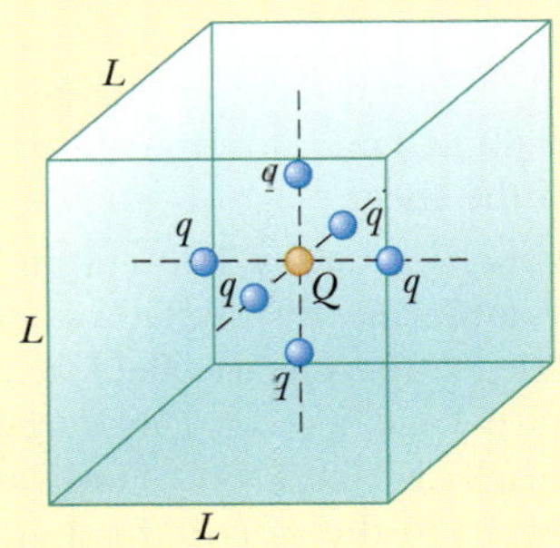

Figure P24.13 Problems 13 and 14.

14. A particle with charge Q is located at the center of a cube of edge L. In addition, six other identical negative charged particles $-q$ are positioned symmetrically around Q as shown in Figure P24.13. Determine the electric flux through one face of the cube.

15. An infinitely long line charge having a uniform charge per unit length λ lies a distance d from point O as shown in Figure P24.15. Determine the total electric flux

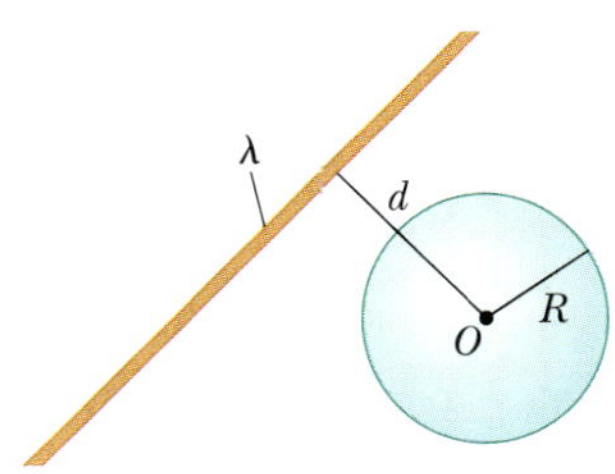

Figure P24.15

2 = intermediate; 3 = challenging; ☐ = SSM/SG; ▲ = ThomsonNOW; ☐ = symbolic reasoning; ● = qualitative reasoning

through the surface of a sphere of radius R centered at O resulting from this line charge. Consider both cases, where $R < d$ and $R > d$.

16. An uncharged nonconducting hollow sphere of radius 10.0 cm surrounds a 10.0-μC charge located at the origin of a cartesian coordinate system. A drill with a radius of 1.00 mm is aligned along the z axis, and a hole is drilled in the sphere. Calculate the electric flux through the hole.

17. ● A charge of 170 μC is at the center of a cube of edge 80.0 cm. No other charges are nearby. (a) Find the flux through each face of the cube. (b) Find the flux through the whole surface of the cube. (c) **What If?** Would your answers to parts (a) or (b) change if the charge were not at the center? Explain.

Section 24.3 Application of Gauss's Law to Various Charge Distributions

18. A solid sphere of radius 40.0 cm has a total positive charge of 26.0 μC uniformly distributed throughout its volume. Calculate the magnitude of the electric field (a) 0 cm, (b) 10.0 cm, (c) 40.0 cm, and (d) 60.0 cm from the center of the sphere.

19. Determine the magnitude of the electric field at the surface of a lead-208 nucleus, which contains 82 protons and 126 neutrons. Assume the lead nucleus has a volume 208 times that of one proton and consider a proton to be a sphere of radius 1.20×10^{-15} m.

20. The charge per unit length on a long, straight filament is $-90.0\ \mu$C/m. Find the electric field (a) 10.0 cm, (b) 20.0 cm, and (c) 100 cm from the filament, where distances are measured perpendicular to the length of the filament.

21. A large, flat, horizontal sheet of charge has a charge per unit area of 9.00 μC/m^2. Find the electric field just above the middle of the sheet.

22. A cylindrical shell of radius 7.00 cm and length 240 cm has its charge uniformly distributed on its curved surface. The magnitude of the electric field at a point 19.0 cm radially outward from its axis (measured from the midpoint of the shell) is 36.0 kN/C. Find (a) the net charge on the shell and (b) the electric field at a point 4.00 cm from the axis, measured radially outward from the midpoint of the shell.

23. A 10.0-g piece of Styrofoam carries a net charge of $-0.700\ \mu$C and floats above the center of a large, horizontal sheet of plastic that has a uniform charge density on its surface. What is the charge per unit area on the plastic sheet?

24. (a) Write a problem for which the following equation gives the solution. Include the required data in your problem statement and identify the one unknown.

$$2\pi(3\text{ cm})(8\text{ cm})E\cos 0^\circ + 0 + 0 = \frac{\pi(2\text{ cm})^2(8\text{ cm})(5\times 10^{-6}\text{ C/m}^3)}{8.85\times 10^{-12}\text{ C}^2/\text{N}\cdot\text{m}^2}$$

(b) Solve the equation for the unknown.

25. **Review problem.** A particle with a charge of -60.0 nC is placed at the center of a nonconducting spherical shell of inner radius 20.0 cm and outer radius 25.0 cm. The spherical shell carries charge with a uniform density of $-1.33\ \mu$C/m^3. A proton moves in a circular orbit just outside the spherical shell. Calculate the speed of the proton.

26. ● A nonconducting wall carries charge with a uniform density of 8.60 μC/cm^2. What is the electric field 7.00 cm in front of the wall? Explain whether your result changes as the distance from the wall is varied.

27. ▲ Consider a long, cylindrical charge distribution of radius R with a uniform charge density ρ. Find the electric field at distance r from the axis, where $r < R$.

28. In nuclear fission, a nucleus of uranium-238, containing 92 protons, can divide into two smaller spheres, each having 46 protons and a radius of 5.90×10^{-15} m. What is the magnitude of the repulsive electric force pushing the two spheres apart?

29. Consider a thin, spherical shell of radius 14.0 cm with a total charge of 32.0 μC distributed uniformly on its surface. Find the electric field (a) 10.0 cm and (b) 20.0 cm from the center of the charge distribution.

30. Fill two rubber balloons with air. Suspend both of them from the same point and let them hang down on strings of equal length. Rub each with wool or on your hair so that the balloons hang apart with a noticeable separation between them. Make order-of-magnitude estimates of (a) the force on each, (b) the charge on each, (c) the field each creates at the center of the other, and (d) the total flux of electric field created by each balloon. In your solution, state the quantities you take as data and the values you measure or estimate for them.

31. A uniformly charged, straight filament 7.00 m in length has a total positive charge of 2.00 μC. An uncharged cardboard cylinder 2.00 cm in length and 10.0 cm in radius surrounds the filament at its center, with the filament as the axis of the cylinder. Using reasonable approximations, find (a) the electric field at the surface of the cylinder and (b) the total electric flux through the cylinder.

Section 24.4 Conductors in Electrostatic Equilibrium

32. A very large, thin, flat plate of aluminum of area A has a total charge Q uniformly distributed over its surfaces. Assuming the same charge is spread uniformly over the *upper* surface of an otherwise identical glass plate, compare the electric fields just above the center of the upper surface of each plate.

33. A long, straight metal rod has a radius of 5.00 cm and a charge per unit length of 30.0 nC/m. Find the electric field (a) 3.00 cm, (b) 10.0 cm, and (c) 100 cm from the axis of the rod, where distances are measured perpendicular to the rod.

34. ● A solid copper sphere of radius 15.0 cm carries a charge of 40.0 nC. Find the electric field (a) 12.0 cm, (b) 17.0 cm, and (c) 75.0 cm from the center of the sphere. (d) **What If?** Explain how your answers would change if the sphere were hollow.

35. A square plate of copper with 50.0-cm sides has no net charge and is placed in a region of uniform electric field of 80.0 kN/C directed perpendicularly to the plate. Find (a) the charge density of each face of the plate and (b) the total charge on each face.

36. In a certain region of space, the electric field is $\vec{\mathbf{E}} = 6\,000x^2\hat{\mathbf{i}}$ N/C·m^2. Find the volume density of electric charge at $x = 0.300$ m. *Suggestion:* Apply Gauss's law to a box between $x = 0.300$ m and $x = 0.300 \text{ m} + dx$.

37. Two identical conducting spheres each having a radius of 0.500 cm are connected by a light 2.00-m-long conducting wire. A charge of 60.0 μC is placed on one of the conductors. Assume the surface distribution of charge on each sphere is uniform. Determine the tension in the wire.

38. ● A solid metallic sphere of radius a carries total charge Q. No other charges are nearby. The electric field just outside its surface is k_eQ/a^2 radially outward. Is the electric field here also given by σ/ϵ_0? By $\sigma/2\epsilon_0$? Explain whether you should expect it to be equal to either of these quantities.

39. A long, straight wire is surrounded by a hollow metal cylinder whose axis coincides with that of the wire. The wire has a charge per unit length of λ, and the cylinder has a net charge per unit length of 2λ. From this information, use Gauss's law to find (a) the charge per unit length on the inner and outer surfaces of the cylinder and (b) the electric field outside the cylinder, a distance r from the axis.

40. A positively charged particle is at a distance $R/2$ from the center of an uncharged thin, conducting, spherical shell of radius R. Sketch the electric field lines set up by this arrangement both inside and outside the shell.

41. ▲ A thin, square, conducting plate 50.0 cm on a side lies in the xy plane. A total charge of 4.00×10^{-8} C is placed on the plate. Find (a) the charge density on the plate, (b) the electric field just above the plate, and (c) the electric field just below the plate. You may assume the charge density is uniform.

Additional Problems

42. A nonuniform electric field is given by the expression

$$\vec{\mathbf{E}} = ay\hat{\mathbf{i}} + bz\hat{\mathbf{j}} + cx\hat{\mathbf{k}}$$

where a, b, and c are constants. Determine the electric flux through a rectangular surface in the xy plane, extending from $x = 0$ to $x = w$ and from $y = 0$ to $y = h$.

43. A sphere of radius R surrounds a particle with charge Q, located at its center. (a) Show that the electric flux through a circular cap of half-angle θ (Fig. P24.43) is

$$\Phi_E = \frac{Q}{2\epsilon_0}(1 - \cos\theta)$$

What is the flux for (b) $\theta = 90°$ and (c) $\theta = 180°$?

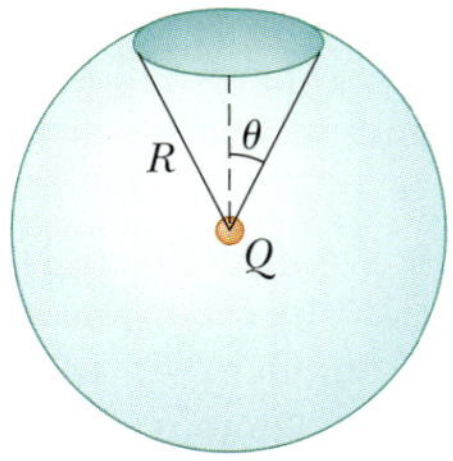

Figure P24.43

44. A solid insulating sphere of radius 5.00 cm carries a net positive charge of 3.00 μC, uniformly distributed throughout its volume. Concentric with this sphere is a conducting spherical shell with inner radius 10.0 cm and outer radius 15.0 cm, having net charge -1.00 μC, as shown in Figure Q24.11. (a) Consider a spherical gaussian surface of radius 16.0 cm and find the net charge enclosed by this surface. (b) What is the direction of the electric field at point D, to the right of the shell and at radius 16 cm? (c) Find the magnitude of the electric field at point D. (d) Find the vector electric field at point C, at radius 12.0 cm. (e) Consider a spherical gaussian surface through point C and find the net charge enclosed by this surface. (f) Consider a spherical gaussian surface of radius 8.00 cm and find the net charge enclosed by this surface. (g) Find the vector electric field at point B, at radius 8 cm. (h) Consider a spherical gaussian surface through point A, at radius 4.00 cm, and find the net charge enclosed by this surface. (i) Find the vector electric field at point A. (j) Determine the charge on the inner surface of the conducting shell. (k) Determine the charge on the outer surface of the conducting shell. (l) Sketch a graph of the magnitude of the electric field versus r.

45. ● A hollow, metallic, spherical shell has exterior radius 0.750 m, carries no net charge, and is supported on an insulating stand. The electric field everywhere just outside its surface is 890 N/C radially toward the center of the sphere. (a) Explain what you can conclude about the amount of charge on the exterior surface of the sphere and the distribution of this charge. (b) Explain what you can conclude about the amount of charge on the interior surface of the sphere and its distribution. (c) Explain what you can conclude about the amount of charge inside the shell and its distribution.

46. ● Consider two identical conducting spheres whose surfaces are separated by a small distance. One sphere is given a large net positive charge, and the other is given a small net positive charge. It is found that the force between the spheres is attractive even though they both have net charges of the same sign. Explain how this attraction is possible.

47. ▲ A solid, insulating sphere of radius a has a uniform charge density ρ and a total charge Q. Concentric with this sphere is an uncharged, conducting, hollow sphere whose inner and outer radii are b and c as shown in Figure P24.47. (a) Find the magnitude of the electric field in the regions $r < a$, $a < r < b$, $b < r < c$, and $r > c$. (b) Determine the induced charge per unit area on the inner and outer surfaces of the hollow sphere.

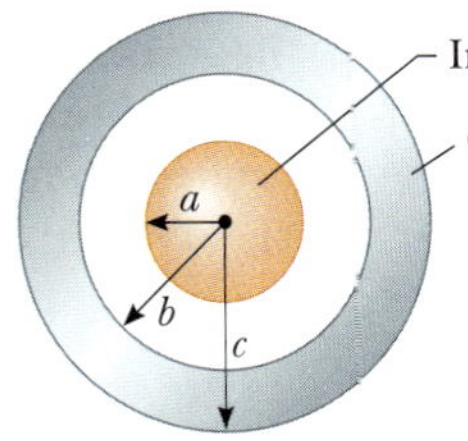

Figure P24.47 Problems 47 and 63.

48. **Review problem.** An early (incorrect) model of the hydrogen atom, suggested by J. J. Thomson, proposed that a positive cloud of charge $+e$ was uniformly distributed throughout the volume of a sphere of radius R, with the electron (an equal-magnitude negatively charged particle $-e$) at the center. (a) Using Gauss's law, show that the

electron would be in equilibrium at the center and, if displaced from the center a distance $r < R$, would experience a restoring force of the form $F = -Kr$, where K is a constant. (b) Show that $K = k_e e^2/R^3$. (c) Find an expression for the frequency f of simple harmonic oscillations that an electron of mass m_e would undergo if displaced a small distance ($< R$) from the center and released. (d) Calculate a numerical value for R that would result in a frequency of 2.47×10^{15} Hz, the frequency of the light radiated in the most intense line in the hydrogen spectrum.

49. A particle of mass m and charge q moves at high speed along the x axis. It is initially near $x = -\infty$, and it ends up near $x = +\infty$. A second charge Q is fixed at the point $x = 0$, $y = -d$. As the moving charge passes the stationary charge, its x component of velocity does not change appreciably, but it acquires a small velocity in the y direction. Determine the angle through which the moving charge is deflected. *Suggestion:* The integral you encounter in determining v_y can be evaluated by applying Gauss's law to a long cylinder of radius d, centered on the stationary charge.

50. Two infinite, nonconducting sheets of charge are parallel to each other as shown in Figure P24.50. The sheet on the left has a uniform surface charge density σ, and the one on the right has a uniform charge density $-\sigma$. Calculate the electric field at points (a) to the left of, (b) in between, and (c) to the right of the two sheets.

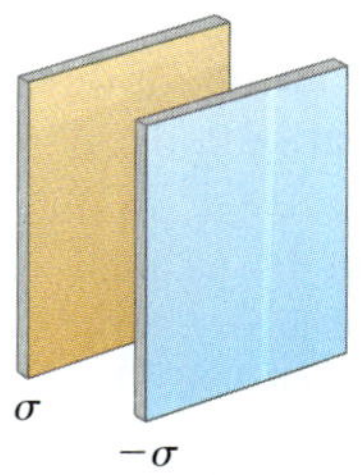

Figure P24.50

51. ▲ **What If?** Repeat the calculations for Problem 50 when both sheets have *positive* uniform surface charge densities of value σ.

52. A sphere of radius $2a$ is made of a nonconducting material that has a uniform volume charge density ρ. (Assume the material does not affect the electric field.) A spherical cavity of radius a is now removed from the sphere as shown in Figure P24.52. Show that the electric field within the cavity is uniform and is given by $E_x = 0$ and $E_y = \rho a/3\epsilon_0$. *Suggestion:* The field within the cavity is the superposition of the field due to the original uncut sphere plus the field due to a sphere the size of the cavity with a uniform negative charge density $-\rho$.

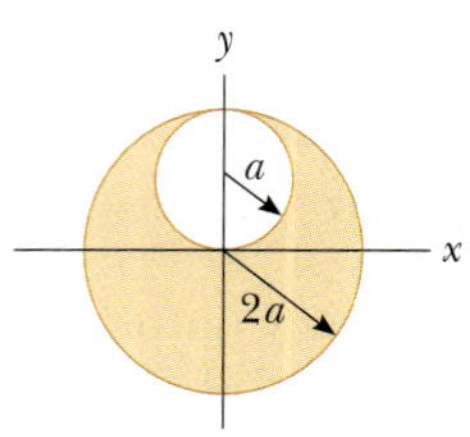

Figure P24.52

53. A uniformly charged spherical shell with surface charge density σ contains a circular hole in its surface. The radius of the hole is small compared with the radius of the sphere. What is the electric field at the center of the hole? *Suggestion:* This problem, like Problem 52, can be solved by using the idea of superposition.

54. A closed surface with dimensions $a = b = 0.400$ m and $c = 0.600$ m is located as shown in Figure P24.54. The left edge of the closed surface is located at position $x = a$. The electric field throughout the region is nonuniform and is given by $\vec{\mathbf{E}} = (3.0 + 2.0x^2)\hat{\mathbf{i}}$ N/C, where x is in meters. Calculate the net electric flux leaving the closed surface. What net charge is enclosed by the surface?

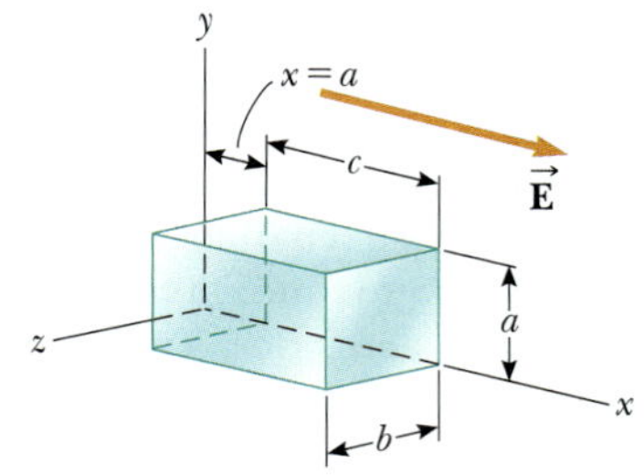

Figure P24.54

55. A solid insulating sphere of radius R has a nonuniform charge density that varies with r according to the expression $\rho = Ar^2$, where A is a constant and $r < R$ is measured from the center of the sphere. (a) Show that the magnitude of the electric field outside ($r > R$) the sphere is $E = AR^5/5\epsilon_0 r^2$. (b) Show that the magnitude of the electric field inside ($r < R$) the sphere is $E = Ar^3/5\epsilon_0$. *Suggestion:* The total charge Q on the sphere is equal to the integral of $\rho\, dV$, where r extends from 0 to R; also, the charge q within a radius $r < R$ is less than Q. To evaluate the integrals, note that the volume element dV for a spherical shell of radius r and thickness dr is equal to $4\pi r^2\, dr$.

56. A particle with charge Q is located on the axis of a disk of radius R at a distance b from the plane of the disk (Fig. P24.56). Show that if one fourth of the electric flux from the charge passes through the disk, then $R = \sqrt{3}b$.

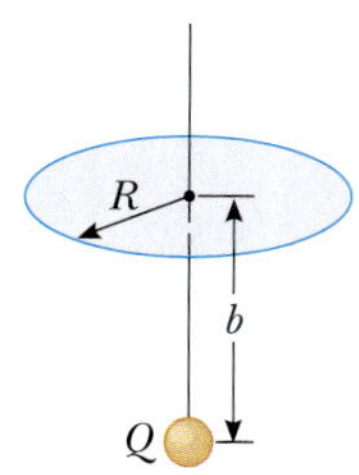

Figure P24.56

57. A spherically symmetric charge distribution has a charge density given by $\rho = a/r$, where a is constant. Find the electric field as a function of r. *Suggestion:* The charge within a sphere of radius R is equal to the integral of $\rho\, dV$, where r extends from 0 to R. To evaluate the integral, note that the volume element dV for a spherical shell of radius r and thickness dr is equal to $4\pi r^2\, dr$.

58. An infinitely long insulating cylinder of radius R has a volume charge density that varies with the radius as

$$\rho = \rho_0\left(a - \frac{r}{b}\right)$$

where ρ_0, a, and b are positive constants and r is the distance from the axis of the cylinder. Use Gauss's law to determine the magnitude of the electric field at radial distances (a) $r < R$ and (b) $r > R$.

59. **Review problem.** A slab of insulating material (infinite in two of its three dimensions) has a uniform positive charge density ρ. An edge view of the slab is shown in Figure P24.59. (a) Show that the magnitude of the electric field a distance x from its center and inside the slab is $E = \rho x/\epsilon_0$. (b) **What If?** Suppose an electron of charge $-e$ and mass m_e can move freely within the slab. It is released from rest at a distance x from the center. Show that the electron exhibits simple harmonic motion with a frequency

$$f = \frac{1}{2\pi}\sqrt{\frac{\rho e}{m_e \epsilon_0}}$$

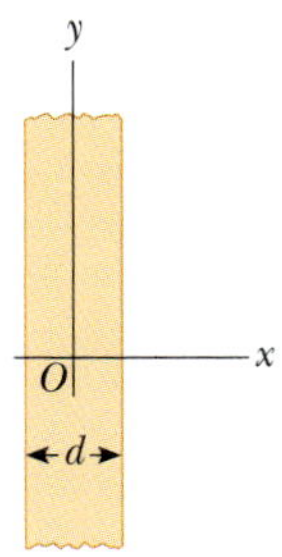

Figure P24.59 Problems 59 and 60.

60. A slab of insulating material has a nonuniform positive charge density $\rho = Cx^2$, where x is measured from the center of the slab as shown in Figure P24.59 and C is a constant. The slab is infinite in the y and z directions. Derive expressions for the electric field in (a) the exterior regions and (b) the interior region of the slab ($-d/2 < x < d/2$).

61. (a) Using the mathematical similarity between Coulomb's law and Newton's law of universal gravitation, show that Gauss's law for gravitation can be written as

$$\oint \vec{\mathbf{g}} \cdot d\vec{\mathbf{A}} = -4\pi G m_{\text{in}}$$

where m_{in} is the net mass inside the gaussian surface and $\vec{\mathbf{g}} = \vec{\mathbf{F}}_g/m$ represents the gravitational field at any point on the gaussian surface. (b) Determine the gravitational field at a distance r from the center of the Earth where $r < R_E$, assuming the Earth's mass density is uniform.

62. An insulating solid sphere of radius a has a uniform volume charge density and carries a total positive charge Q. A spherical gaussian surface of radius r, which shares a common center with the insulating sphere, is inflated starting from $r = 0$. (a) Find an expression for the electric flux passing through the surface of the gaussian sphere as a function of r for $r < a$. (b) Find an expression for the electric flux for $r > a$. (c) Plot the flux versus r.

63. For the configuration shown in Figure P24.47, suppose $a = 5.00$ cm, $b = 20.0$ cm, and $c = 25.0$ cm. Furthermore, suppose the electric field at a point 10.0 cm from the center is measured to be 3.60×10^3 N/C radially inward and the electric field at a point 50.0 cm from the center is 2.00×10^2 N/C radially outward. From this information, find (a) the charge on the insulating sphere, (b) the net charge on the hollow conducting sphere, and (c) the charges on the inner and outer surfaces of the hollow conducting sphere.

64. An infinitely long cylindrical insulating shell of inner radius a and outer radius b has a uniform volume charge density ρ. A line of uniform linear charge density λ is placed along the axis of the shell. Determine the electric field everywhere.

65. Consider an electric field that is uniform in direction throughout a certain volume. Can it be uniform in magnitude? Must it be uniform in magnitude? Answer these questions (a) assuming the volume is filled with an insulating material carrying charge described by a volume charge density and (b) assuming the volume is empty space. State reasoning to prove your answers.

Answers to Quick Quizzes

24.1 (e). The same number of field lines pass through a sphere of any size. Because points on the surface of the sphere are closer to the charge, the field is stronger.

24.2 (b) and (d). Statement (a) is not necessarily true because an equal number of positive and negative charges could be present inside the surface. Statement (c) is not necessarily true as can be seen from Figure 24.8: a nonzero electric field exists everywhere on the surface, but the charge is not enclosed within the surface and the net flux is therefore zero.

24.3 (a). Charges added to the metal cylinder by your brother will reside on the outer surface of the conducting cylinder. If you are on the inside, these charges cannot transfer to you from the inner surface. For this same reason, you are safe in a metal automobile during a lightning storm.

Processes occurring during thunderstorms cause large differences in electric potential between a thundercloud and the ground. The result of this potential difference is an electrical discharge that we call lightning, such as this display over Tucson, Arizona. (© Keith Kent/Photo Researchers, Inc.)

25 Electric Potential

The concept of potential energy was introduced in Chapter 7 in connection with such conservative forces as the gravitational force and the elastic force exerted by a spring. By using the law of conservation of energy when solving various problems in mechanics, we were able to avoid working directly with forces. The concept of potential energy is also of great value in the study of electricity. Because the electrostatic force is conservative, electrostatic phenomena can be conveniently described in terms of an electric potential energy. This idea enables us to define a quantity known as *electric potential.* Because the electric potential at any point in an electric field is a scalar quantity, we can use it to describe electrostatic phenomena more simply than if we were to rely only on the electric field and electric forces. The concept of electric potential is of great practical value in the operation of electric circuits and devices that we will study in later chapters.

25.1 Electric Potential and Potential Difference

When a test charge q_0 is placed in an electric field $\vec{\mathbf{E}}$ created by some source charge distribution, the electric force acting on the test charge is $q_0\vec{\mathbf{E}}$. The force $q_0\vec{\mathbf{E}}$ is conservative because the force between charges described by Coulomb's law is conservative. When the test charge is moved in the field by some external agent, the work done by the field on the charge is equal to the negative of the work done by the external agent causing the displacement. This situation is analogous to that of lifting an object with mass in a gravitational field: the work done by the external agent is *mgh*, and the work done by the gravitational force is $-mgh$.

When analyzing electric and magnetic fields, it is common practice to use the notation $d\vec{\mathbf{s}}$ to represent an infinitesimal displacement vector that is oriented tangent to a path through space. This path may be straight or curved, and an integral performed along this path is called either a *path integral* or a *line integral* (the two terms are synonymous).

For an infinitesimal displacement $d\vec{\mathbf{s}}$ of a point charge q_0 immersed in an electric field, the work done by the electric field on the charge is $\vec{\mathbf{F}} \cdot d\vec{\mathbf{s}} = q_0\vec{\mathbf{E}} \cdot d\vec{\mathbf{s}}$. As this amount of work is done by the field, the potential energy of the charge–field system is changed by an amount $dU = -q_0\vec{\mathbf{E}} \cdot d\vec{\mathbf{s}}$. For a finite displacement of the charge from point Ⓐ to point Ⓑ, the change in potential energy of the system $\Delta U = U_{Ⓑ} - U_{Ⓐ}$ is

$$\Delta U = -q_0 \int_{Ⓐ}^{Ⓑ} \vec{\mathbf{E}} \cdot d\vec{\mathbf{s}} \qquad \textbf{(25.1)}$$

◀ Change in electric potential energy of a system

The integration is performed along the path that q_0 follows as it moves from Ⓐ to Ⓑ. Because the force $q_0\vec{\mathbf{E}}$ is conservative, **this line integral does not depend on the path taken from Ⓐ to Ⓑ.**

For a given position of the test charge in the field, the charge–field system has a potential energy U relative to the configuration of the system that is defined as $U = 0$. Dividing the potential energy by the test charge gives a physical quantity that depends only on the source charge distribution and has a value at every point in an electric field. This quantity is called the **electric potential** (or simply the **potential**) V:

$$V = \frac{U}{q_0} \qquad \textbf{(25.2)}$$

Because potential energy is a scalar quantity, electric potential also is a scalar quantity.

PITFALL PREVENTION 25.1
Potential and Potential Energy

The *potential is characteristic of the field only*, independent of a charged test particle that may be placed in the field. *Potential energy is characteristic of the charge-field system* due to an interaction between the field and a charged particle placed in the field.

As described by Equation 25.1, if the test charge is moved between two positions Ⓐ and Ⓑ in an electric field, the charge–field system experiences a change in potential energy. The **potential difference** $\Delta V = V_{Ⓑ} - V_{Ⓐ}$ between two points Ⓐ and Ⓑ in an electric field is defined as the change in potential energy of the system when a test charge q_0 is moved between the points divided by the test charge:

$$\Delta V \equiv \frac{\Delta U}{q_0} = -\int_{Ⓐ}^{Ⓑ} \vec{\mathbf{E}} \cdot d\vec{\mathbf{s}} \qquad \textbf{(25.3)}$$

◀ Potential difference between two points

Just as with potential energy, only *differences* in electric potential are meaningful. We often take the value of the electric potential to be zero at some convenient point in an electric field.

Potential difference should not be confused with difference in potential energy. The potential difference between Ⓐ and Ⓑ depends only on the source charge distribution (consider points Ⓐ and Ⓑ *without* the presence of the test charge), whereas the difference in potential energy exists only if a test charge is moved between the points.

If an external agent moves a test charge from Ⓐ to Ⓑ without changing the kinetic energy of the test charge, the agent performs work that changes the potential energy of the system: $W = \Delta U$. Imagine an arbitrary charge q located in an electric field. From Equation 25.3, the work done by an external agent in moving a charge q through an electric field at constant velocity is

$$W = q\Delta V \qquad \textbf{(25.4)}$$

PITFALL PREVENTION 25.2

Voltage

A variety of phrases are used to describe the potential difference between two points, the most common being **voltage,** arising from the unit for potential. A voltage *applied* to a device, such as a television, or *across* a device is the same as the potential difference across the device.

PITFALL PREVENTION 25.3

The Electron Volt

The electron volt is a unit of *energy,* NOT of potential. The energy of any system may be expressed in eV, but this unit is most convenient for describing the emission and absorption of visible light from atoms. Energies of nuclear processes are often expressed in MeV.

Because electric potential is a measure of potential energy per unit charge, the SI unit of both electric potential and potential difference is joules per coulomb, which is defined as a **volt** (V):

$$1\text{ V} \equiv 1\text{ J/C}$$

That is, 1 J of work must be done to move a 1-C charge through a potential difference of 1 V.

Equation 25.3 shows that potential difference also has units of electric field times distance. It follows that the SI unit of electric field (N/C) can also be expressed in volts per meter:

$$1\text{ N/C} = 1\text{ V/m}$$

Therefore, **we can interpret the electric field as a measure of the rate of change with position of the electric potential.**

A unit of energy commonly used in atomic and nuclear physics is the **electron volt** (eV), which is defined as **the energy a charge–field system gains or loses when a charge of magnitude *e* (that is, an electron or a proton) is moved through a potential difference of 1 V.** Because 1 V = 1 J/C and the fundamental charge is 1.60×10^{-19} C, the electron volt is related to the joule as follows:

$$1\text{ eV} = 1.60 \times 10^{-19}\text{ C}\cdot\text{V} = 1.60 \times 10^{-19}\text{ J} \qquad \textbf{(25.5)}$$

For instance, an electron in the beam of a typical television picture tube may have a speed of 3.0×10^7 m/s. This speed corresponds to a kinetic energy equal to 4.1×10^{-16} J, which is equivalent to 2.6×10^3 eV. Such an electron has to be accelerated from rest through a potential difference of 2.6 kV to reach this speed.

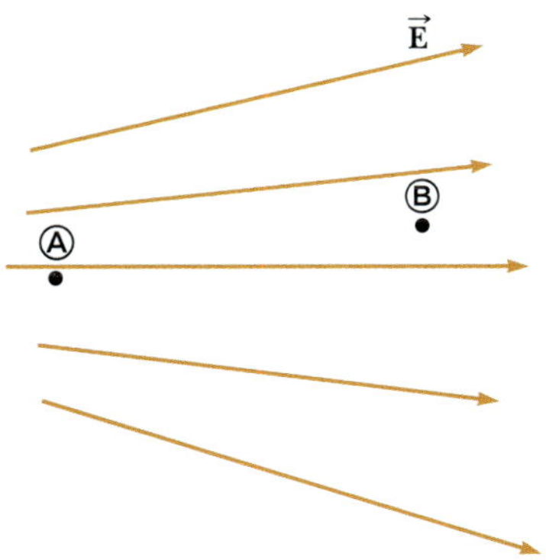

Figure 25.1 (Quick Quiz 25.1) Two points in an electric field.

Quick Quiz 25.1 In Figure 25.1, two points Ⓐ and Ⓑ are located within a region in which there is an electric field. **(i)** How would you describe the potential difference $\Delta V = V_{\text{Ⓑ}} - V_{\text{Ⓐ}}$? (a) It is positive. (b) It is negative. (c) It is zero. **(ii)** A negative charge is placed at Ⓐ and then moved to Ⓑ. How would you describe the change in potential energy of the charge-field system for this process? Choose from the same possibilities.

25.2 Potential Difference in a Uniform Electric Field

Equations 25.1 and 25.3 hold in all electric fields, whether uniform or varying, but they can be simplified for a uniform field. First, consider a uniform electric field directed along the negative *y* axis as shown in Active Figure 25.2a. Let's calculate the potential difference between two points Ⓐ and Ⓑ separated by a distance $|\vec{\mathbf{s}}| = d$, where $\vec{\mathbf{s}}$ is parallel to the field lines. Equation 25.3 gives

$$V_{\text{Ⓑ}} - V_{\text{Ⓐ}} = \Delta V = -\int_{\text{Ⓐ}}^{\text{Ⓑ}} \vec{\mathbf{E}}\cdot d\vec{\mathbf{s}} = -\int_{\text{Ⓐ}}^{\text{Ⓑ}} (E\cos 0^\circ)\,ds = -\int_{\text{Ⓐ}}^{\text{Ⓑ}} E\,ds$$

Because *E* is constant, it can be removed from the integral sign, which gives

Potential difference between two points in a uniform electric field ▶

$$\Delta V = -E\int_{\text{Ⓐ}}^{\text{Ⓑ}} ds = -Ed \qquad \textbf{(25.6)}$$

The negative sign indicates that the electric potential at point Ⓑ is lower than at point Ⓐ; that is, $V_{\text{Ⓑ}} < V_{\text{Ⓐ}}$. **Electric field lines always point in the direction of decreasing electric potential** as shown in Active Figure 25.2a.

Now suppose a test charge q_0 moves from Ⓐ to Ⓑ. We can calculate the change in the potential energy of the charge–field system from Equations 25.3 and 25.6:

$$\Delta U = q_0\,\Delta V = -q_0Ed \qquad \textbf{(25.7)}$$

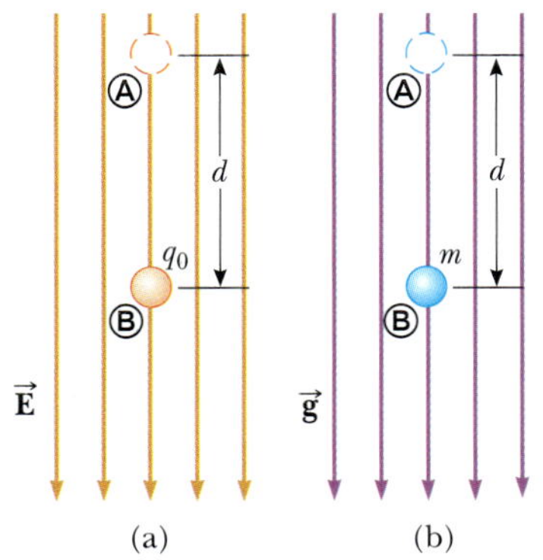

ACTIVE FIGURE 25.2

(a) When the electric field $\vec{\mathbf{E}}$ is directed downward, point Ⓑ is at a lower electric potential than point Ⓐ. When a positive test charge moves from point Ⓐ to point Ⓑ, the electric potential energy of the charge–field system decreases. (b) When an object of mass m moves downward in the direction of the gravitational field $\vec{\mathbf{g}}$, the gravitational potential energy of the object–field system decreases.

Sign in at www.thomsonedu.com and go to ThomsonNOW to observe and compare the motion of the charged object in an electric field and an object with mass in a gravitational field.

This result shows that if q_0 is positive, then U is negative. Therefore, **a system consisting of a positive charge and an electric field loses electric potential energy when the charge moves in the direction of the field.** Equivalently, an electric field does work on a positive charge when the charge moves in the direction of the electric field. (That is analogous to the work done by the gravitational field on a falling object as shown in Active Fig. 25.2b.) If a positive test charge is released from rest in this electric field, it experiences an electric force $q_0\vec{\mathbf{E}}$ in the direction of $\vec{\mathbf{E}}$ (downward in Active Fig. 25.2a). Therefore, it accelerates downward, gaining kinetic energy. **As the charged particle gains kinetic energy, the charge–field system loses an equal amount of potential energy.** This equivalence should not be surprising; it is simply conservation of mechanical energy in an isolated system as introduced in Chapter 8.

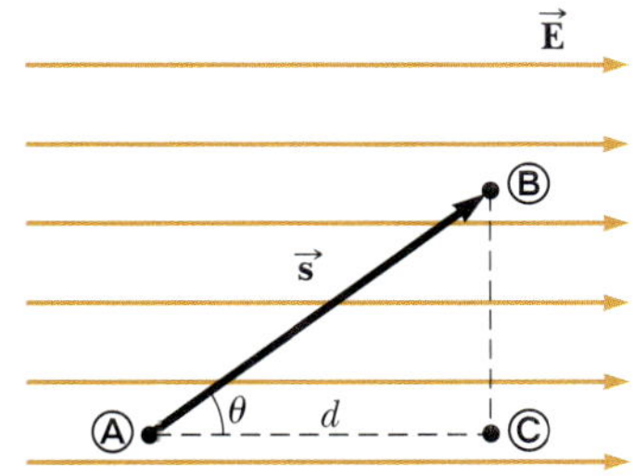

Figure 25.3 A uniform electric field directed along the positive x axis. Point Ⓑ is at a lower electric potential than point Ⓐ. Points Ⓑ and Ⓒ are at the *same* electric potential.

If q_0 is negative, then ΔU in Equation 25.7 is positive and the situation is reversed. **A system consisting of a negative charge and an electric field gains electric potential energy when the charge moves in the direction of the field.** If a negative charge is released from rest in an electric field, it accelerates in a direction opposite the direction of the field. For the negative charge to move in the direction of the field, an external agent must apply a force and do positive work on the charge.

Now consider the more general case of a charged particle that moves between Ⓐ and Ⓑ in a uniform electric field such that the vector $\vec{\mathbf{s}}$ is not parallel to the field lines as shown in Figure 25.3. In this case, Equation 25.3 gives

$$\Delta V = -\int_{Ⓐ}^{Ⓑ} \vec{\mathbf{E}} \cdot d\vec{\mathbf{s}} = -\vec{\mathbf{E}} \cdot \int_{Ⓐ}^{Ⓑ} d\vec{\mathbf{s}} = -\vec{\mathbf{E}} \cdot \vec{\mathbf{s}} \quad \textbf{(25.8)}$$

◀ Change in potential energy when a charged particle is moved in a uniform electric field

where again $\vec{\mathbf{E}}$ was removed from the integral because it is constant. The change in potential energy of the charge-field system is

$$\Delta U = q_0\, \Delta V = -q_0 \vec{\mathbf{E}} \cdot \vec{\mathbf{s}} \quad \textbf{(25.9)}$$

Finally, we conclude from Equation 25.8 that all points in a plane perpendicular to a uniform electric field are at the same electric potential. We can see that in Figure 25.3, where the potential difference $V_Ⓑ - V_Ⓐ$ is equal to the potential difference $V_Ⓒ - V_Ⓐ$. (Prove this fact to yourself by working out two dot products for $\vec{\mathbf{E}} \cdot \vec{\mathbf{s}}$: one for $\vec{\mathbf{s}}_{Ⓐ\to Ⓑ}$, where the angle θ between $\vec{\mathbf{E}}$ and $\vec{\mathbf{s}}$ is arbitrary as shown in Figure 25.3, and one for $\vec{\mathbf{s}}_{Ⓐ\to Ⓒ}$, where $\theta = 0$.) Therefore, $V_Ⓑ = V_Ⓒ$. **The name equipotential surface is given to any surface consisting of a continuous distribution of points having the same electric potential.**

The equipotential surfaces associated with a uniform electric field consist of a family of parallel planes that are all perpendicular to the field. Equipotential surfaces associated with fields having other symmetries are described in later sections.

Quick Quiz 25.2 The labeled points in Figure 25.4 are on a series of equipotential surfaces associated with an electric field. Rank (from greatest to least) the work done by the electric field on a positively charged particle that moves from Ⓐ to Ⓑ, from Ⓑ to Ⓒ, from Ⓒ to Ⓓ, and from Ⓓ to Ⓔ.

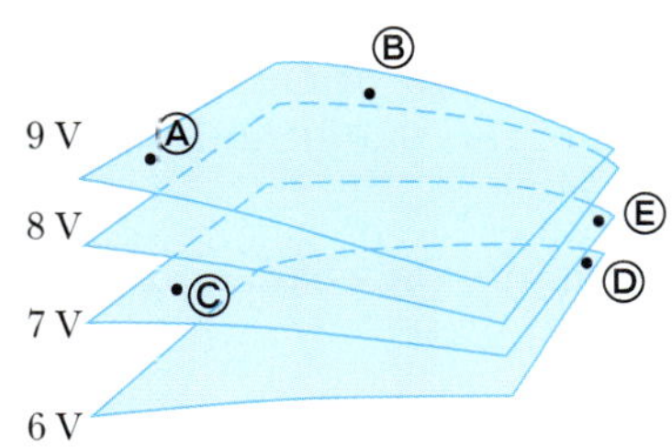

Figure 25.4 (Quick Quiz 25.2) Four equipotential surfaces.

EXAMPLE 25.1 The Electric Field Between Two Parallel Plates of Opposite Charge

A battery has a specified potential difference ΔV between its terminals and establishes that potential difference between conductors attached to the terminals. A 12-V battery is connected between two parallel plates as shown in Figure 25.5. The separation between the plates is $d = 0.30$ cm, and we assume the electric field between the plates to be uniform. (This assumption is reasonable if the plate separation is small relative to the plate dimensions and we do not consider locations near the plate edges.) Find the magnitude of the electric field between the plates.

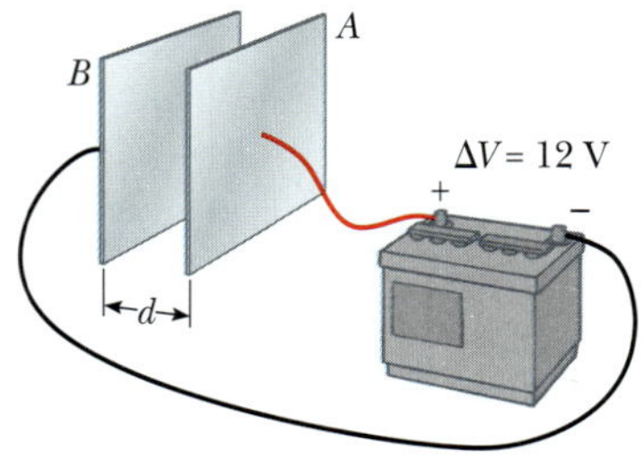

Figure 25.5 (Example 25.1) A 12-V battery connected to two parallel plates. The electric field between the plates has a magnitude given by the potential difference ΔV divided by the plate separation d.

SOLUTION

Conceptualize In earlier chapters, we investigated the uniform electric field between parallel plates. The new feature to this problem is that the electric field is related to the new concept of electric potential.

Categorize The electric field is evaluated from a relationship between field and potential given in this section, so we categorize this example as a substitution problem.

Use Equation 25.6 to evaluate the magnitude of the electric field between the plates:

$$E = \frac{|V_B - V_A|}{d} = \frac{12\text{ V}}{0.30 \times 10^{-2}\text{ m}} = 4.0 \times 10^3\text{ V/m}$$

The configuration of plates in Figure 25.5 is called a *parallel-plate capacitor* and is examined in greater detail in Chapter 26.

EXAMPLE 25.2 Motion of a Proton in a Uniform Electric Field

A proton is released from rest at point Ⓐ in a uniform electric field that has a magnitude of 8.0×10^4 V/m (Fig. 25.6). The proton undergoes a displacement of 0.50 m to point Ⓑ in the direction of $\vec{\mathbf{E}}$. Find the speed of the proton after completing the 0.50 m displacement.

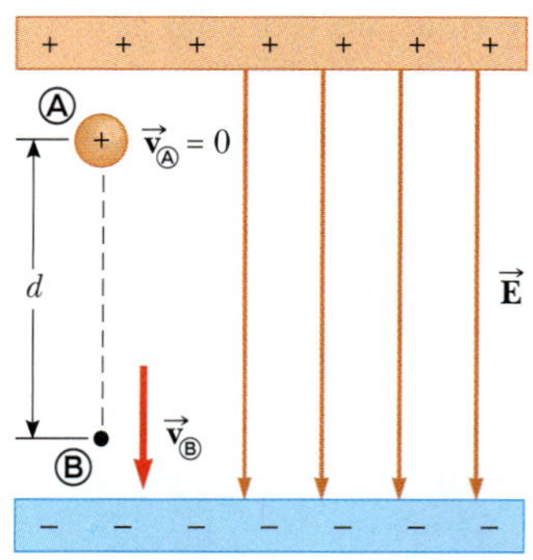

Figure 25.6 (Example 25.2) A proton accelerates from Ⓐ to Ⓑ in the direction of the electric field.

SOLUTION

Conceptualize Visualize the proton in Figure 25.6 moving downward through the potential difference. The situation is analogous to an object falling through a gravitational field.

Categorize The system of the proton and the two plates in Figure 25.6 does not interact with the environment, so we model it as an isolated system.

Analyze Use Equation 25.6 to find the potential difference between points Ⓐ and Ⓑ:

$$\Delta V = -Ed = -(8.0 \times 10^4\text{ V/m})(0.50\text{ m}) = -4.0 \times 10^4\text{ V}$$

Write the appropriate reduction of Equation 8.2, the conservation of energy equation, for the isolated system of the charge and the electric field:

$$\Delta K + \Delta U = 0$$

Substitute the changes in energy for both terms:

$$(\tfrac{1}{2}mv^2 - 0) + e\,\Delta V = 0$$

Solve for the final speed of the proton:

$$v = \sqrt{\frac{-2e\,\Delta V}{m}}$$

Substitute numerical values:

$$v = \sqrt{\frac{-2(1.6 \times 10^{-19}\ \text{C})(-4.0 \times 10^4\ \text{V})}{1.67 \times 10^{-27}\ \text{kg}}}$$

$$= 2.8 \times 10^6\ \text{m/s}$$

Finalize Because ΔV is negative, ΔU is also negative. The negative value of ΔU means the potential energy of the system decreases as the proton moves in the direction of the electric field. As the proton accelerates in the direction of the field, it gains kinetic energy and the system loses electric potential energy at the same time.

Figure 25.6 is oriented so that the proton falls downward. The proton's motion is analogous to that of an object falling in a gravitational field. Although the gravitational field is always downward at the surface of the Earth, an electric field can be in any direction, depending on the orientation of the plates creating the field. Therefore, Figure 25.6 could be rotated 90° or 180° and the proton could fall horizontally or upward in the electric field!

25.3 Electric Potential and Potential Energy Due to Point Charges

As discussed in Section 23.4, an isolated positive point charge q produces an electric field directed radially outward from the charge. To find the electric potential at a point located a distance r from the charge, let's begin with the general expression for potential difference,

$$V_{Ⓑ} - V_{Ⓐ} = -\int_{Ⓐ}^{Ⓑ} \vec{\mathbf{E}} \cdot d\vec{\mathbf{s}}$$

where Ⓐ and Ⓑ are the two arbitrary points shown in Figure 25.7. At any point in space, the electric field due to the point charge is $\vec{\mathbf{E}} = (k_e q/r^2)\hat{\mathbf{r}}$ (Eq. 23.9), where $\hat{\mathbf{r}}$ is a unit vector directed from the charge toward the point. The quantity $\vec{\mathbf{E}} \cdot d\vec{\mathbf{s}}$ can be expressed as

$$\vec{\mathbf{E}} \cdot d\vec{\mathbf{s}} = k_e \frac{q}{r^2}\,\hat{\mathbf{r}} \cdot d\vec{\mathbf{s}}$$

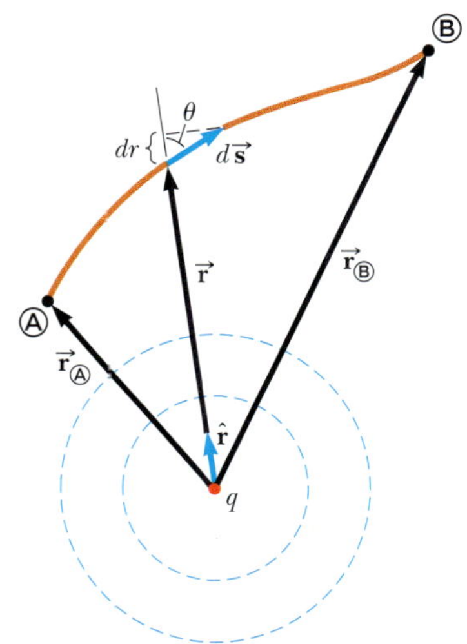

Figure 25.7 The potential difference between points Ⓐ and Ⓑ due to a point charge q depends *only* on the initial and final radial coordinates $r_{Ⓐ}$ and $r_{Ⓑ}$. The two dashed circles represent intersections of spherical equipotential surfaces with the page.

Because the magnitude of $\hat{\mathbf{r}}$ is 1, the dot product $\hat{\mathbf{r}} \cdot d\vec{\mathbf{s}} = ds \cos\theta$, where θ is the angle between $\hat{\mathbf{r}}$ and $d\vec{\mathbf{s}}$. Furthermore, $ds \cos\theta$ is the projection of $d\vec{\mathbf{s}}$ onto $\vec{\mathbf{r}}$; therefore, $ds \cos\theta = dr$. That is, any displacement $d\vec{\mathbf{s}}$ along the path from point Ⓐ to point Ⓑ produces a change dr in the magnitude of $\vec{\mathbf{r}}$, the position vector of the point relative to the charge creating the field. Making these substitutions, we find that $\vec{\mathbf{E}} \cdot d\vec{\mathbf{s}} = (k_e q/r^2)dr$; hence, the expression for the potential difference becomes

$$V_{Ⓑ} - V_{Ⓐ} = -k_e q \int_{r_Ⓐ}^{r_Ⓑ} \frac{dr}{r^2} = k_e \frac{q}{r}\Bigg|_{r_Ⓐ}^{r_Ⓑ}$$

$$V_{Ⓑ} - V_{Ⓐ} = k_e q \left[\frac{1}{r_Ⓑ} - \frac{1}{r_Ⓐ}\right] \qquad \textbf{(25.10)}$$

Equation 25.10 shows us that the integral of $\vec{\mathbf{E}} \cdot d\vec{\mathbf{s}}$ is *independent* of the path between points Ⓐ and Ⓑ. Multiplying by a charge q_0 that moves between points Ⓐ and Ⓑ, we see that the integral of $q_0\vec{\mathbf{E}} \cdot d\vec{\mathbf{s}}$ is also independent of path. This latter integral, which is the work done by the electric force, shows that the electric force is conservative (see Section 7.7). We define a field that is related to a conservative force as a **conservative field.** Therefore, Equation 25.10 tells us that the electric field of a fixed point charge is conservative. Furthermore, Equation 25.10 expresses the important result that the potential difference between any two

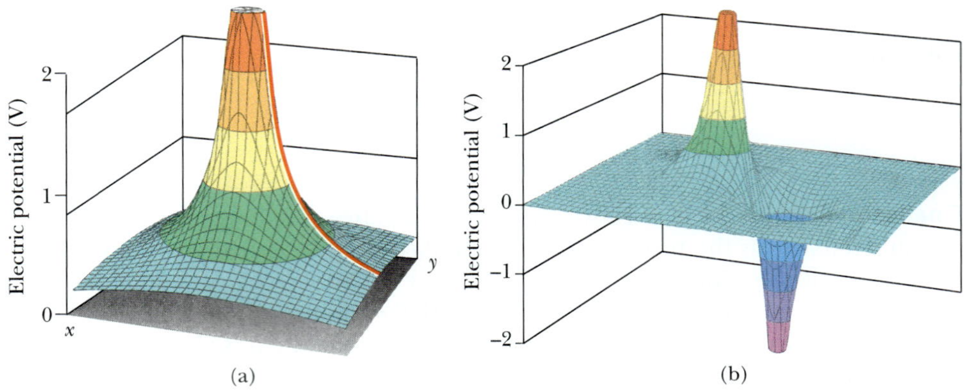

Figure 25.8 (a) The electric potential in the plane around a single positive charge is plotted on the vertical axis. (The electric potential function for a negative charge would look like a hole instead of a hill.) The red line shows the $1/r$ nature of the electric potential as given by Equation 25.11. (b) The electric potential in the plane containing a dipole.

PITFALL PREVENTION 25.4
Similar Equation Warning

Do not confuse Equation 25.11 for the electric potential of a point charge with Equation 23.9 for the electric field of a point charge. Potential is proportional to $1/r$, whereas the field is proportional to $1/r^2$. The effect of a charge on the space surrounding it can be described in two ways. The charge sets up a vector electric field $\vec{\mathbf{E}}$, which is related to the force experienced by a test charge placed in the field. It also sets up a scalar potential V, which is related to the potential energy of the two-charge system when a test charge is placed in the field.

points Ⓐ and Ⓑ in a field created by a point charge depends only on the radial coordinates $r_Ⓐ$ and $r_Ⓑ$. It is customary to choose the reference of electric potential for a point charge to be $V = 0$ at $r_Ⓐ = \infty$. With this reference choice, the electric potential created by a point charge at any distance r from the charge is

$$V = k_e \frac{q}{r} \tag{25.11}$$

Figure 25.8a shows a plot of the electric potential on the vertical axis for a positive charge located in the xy plane. Consider the following analogy to gravitational potential. Imagine trying to roll a marble toward the top of a hill shaped like the surface in Figure 25.8a. Pushing the marble up the hill is analogous to pushing one positively charged object toward another positively charged object. Similarly, the electric potential graph of the region surrounding a negative charge is analogous to a "hole" with respect to any approaching positively charged objects. A charged object must be infinitely distant from another charge before the surface in Figure 25.8a is "flat" and has an electric potential of zero.

We obtain the electric potential resulting from two or more point charges by applying the superposition principle. That is, the total electric potential at some point P due to several point charges is the sum of the potentials due to the individual charges. For a group of point charges, we can write the total electric potential at P as

Electric potential due to several point charges ▶

$$V = k_e \sum_i \frac{q_i}{r_i} \tag{25.12}$$

where the potential is again taken to be zero at infinity and r_i is the distance from the point P to the charge q_i. Notice that the sum in Equation 25.12 is an algebraic sum of scalars rather than a vector sum (which we use to calculate the electric field of a group of charges). Therefore, it is often much easier to evaluate V than $\vec{\mathbf{E}}$. The electric potential around a dipole is illustrated in Figure 25.8b. Notice the steep slope of the potential between the charges, representing a region of strong electric field.

Now consider the potential energy of a system of two charged particles. If V_2 is the electric potential at a point P due to charge q_2, the work an external agent must do to bring a second charge q_1 from infinity to P without acceleration is $q_1 V_2$. This work represents a transfer of energy into the system, and the energy appears in the system as potential energy U when the particles are separated by a distance

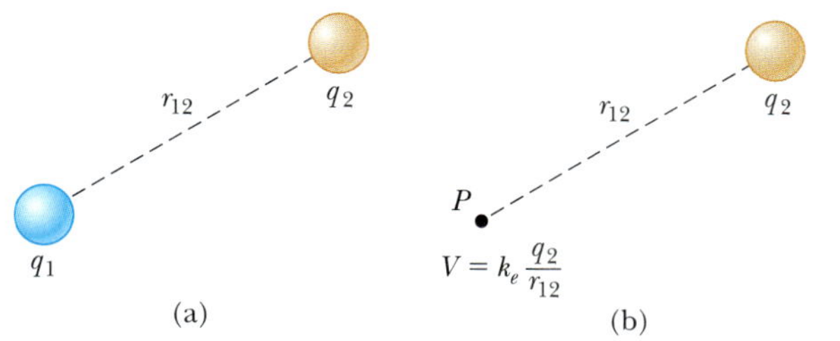

ACTIVE FIGURE 25.9

(a) If two point charges are separated by a distance r_{12}, the potential energy of the pair of charges is given by $k_eq_1q_2/r_{12}$. (b) If charge q_1 is removed, a potential k_eq_2/r_{12} exists at point P due to charge q_2.

Sign in at www.thomsonedu.com and go to ThomsonNOW to move charge q_1 or point P and see the result on the electric potential energy of the system for part (a) and the electric potential due to charge q_2 for part (b).

r_{12} (Active Fig. 25.9a). Therefore, the potential energy of the system can be expressed as[1]

$$U = k_e \frac{q_1 q_2}{r_{12}} \tag{25.13}$$

If the charges are of the same sign, U is positive. Positive work must be done by an external agent on the system to bring the two charges near each other (because charges of the same sign repel). If the charges are of opposite sign, U is negative. Negative work is done by an external agent against the attractive force between the charges of opposite sign as they are brought near each other; a force must be applied opposite the displacement to prevent q_1 from accelerating toward q_2.

In Active Figure 25.9b, we have removed the charge q_1. At the position this charge previously occupied, point P, Equations 25.2 and 25.13 can be used to define a potential due to charge q_2 as $V = U/q_1 = k_eq_2/r_{12}$. This expression is consistent with Equation 25.11.

If the system consists of more than two charged particles, we can obtain the total potential energy of the system by calculating U for every pair of charges and summing the terms algebraically. For example, the total potential energy of the system of three charges shown in Figure 25.10 is

$$U = k_e\left(\frac{q_1q_2}{r_{12}} + \frac{q_1q_3}{r_{13}} + \frac{q_2q_3}{r_{23}}\right) \tag{25.14}$$

Physically, this result can be interpreted as follows. Imagine q_1 is fixed at the position shown in Figure 25.10 but q_2 and q_3 are at infinity. The work an external agent must do to bring q_2 from infinity to its position near q_1 is $k_eq_1q_2/r_{12}$, which is the first term in Equation 25.14. The last two terms represent the work required to bring q_3 from infinity to its position near q_1 and q_2. (The result is independent of the order in which the charges are transported.)

PITFALL PREVENTION 25.5

Which Work?

There is a difference between work done *by one member of a system on another member* and work done *on a system by an external agent.* In the discussion related to Equation 25.14, we consider the group of charges to be the system; an external agent is doing work on the system to move the charges from an infinite separation to a small separation.

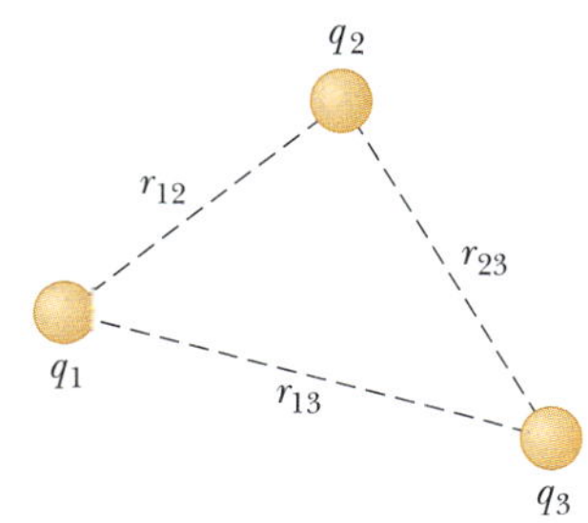

Figure 25.10 Three point charges are fixed at the positions shown. The potential energy of this system of charges is given by Equation 25.14.

Quick Quiz 25.3 In Active Figure 25.9a, take q_1 to be a negative source charge and q_2 to be the test charge. **(i)** If q_2 is initially positive and is changed to a charge of the same magnitude but negative, what happens to the potential at the position of q_2 due to q_1? (a) It increases. (b) It decreases. (c) It remains the same. **(ii)** When q_2 is changed from positive to negative, what happens to the potential energy of the two-charge system? Choose from the same possibilities.

[1] The expression for the electric potential energy of a system made up of two point charges, Equation 25.13, is of the *same* form as the equation for the gravitational potential energy of a system made up of two point masses, $-Gm_1m_2/r$ (see Chapter 13). The similarity is not surprising considering that both expressions are derived from an inverse-square force law.

EXAMPLE 25.3 The Electric Potential Due to Two Point Charges

As shown in Figure 25.11a, a charge $q_1 = 2.00\ \mu C$ is located at the origin and a charge $q_2 = -6.00\ \mu C$ is located at (0, 3.00) m.

(A) Find the total electric potential due to these charges at the point *P*, whose coordinates are (4.00, 0) m.

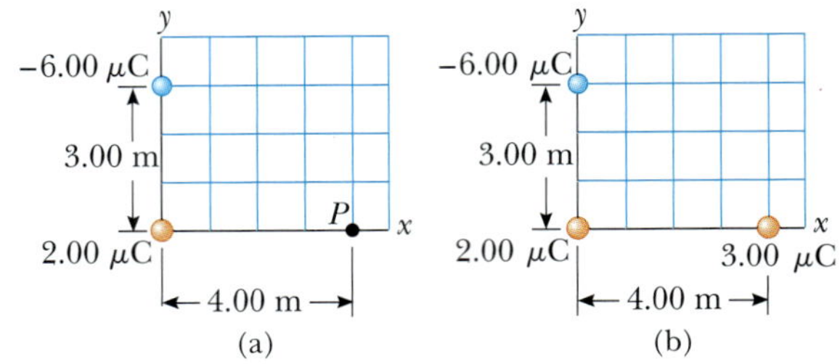

Figure 25.11 (Example 25.3) (a) The electric potential at *P* due to the two charges q_1 and q_2 is the algebraic sum of the potentials due to the individual charges. (b) A third charge $q_3 = 3.00\ \mu C$ is brought from infinity to point *P*.

SOLUTION

Conceptualize Recognize that the 2.00 μC and −6.00 μC charges are source charges and set up an electric field as well as a potential at all points in space, including point *P*.

Categorize The potential is evaluated using an equation developed in this chapter, so we categorize this example as a substitution problem.

Use Equation 25.12 for the system of two source charges:

$$V_P = k_e\left(\frac{q_1}{r_1} + \frac{q_2}{r_2}\right)$$

Substitute numerical values:

$$V_P = (8.99 \times 10^9\ \text{N}\cdot\text{m}^2/\text{C}^2)\left(\frac{2.00 \times 10^{-6}\ \text{C}}{4.00\ \text{m}} + \frac{-6.00 \times 10^{-6}\ \text{C}}{5.00\ \text{m}}\right)$$

$$= \boxed{-6.29 \times 10^3\ \text{V}}$$

(B) Find the change in potential energy of the system of two charges plus a third charge $q_3 = 3.00\ \mu C$ as the latter charge moves from infinity to point *P* (Fig. 25.11b).

SOLUTION

Assign $U_i = 0$ for the system to the configuration in which the charge q_3 is at infinity. Use Equation 25.2 to evaluate the potential energy for the configuration in which the charge is at *P*:

$$U_f = q_3 V_P$$

Substitute numerical values to evaluate ΔU:

$$\Delta U = U_f - U_i = q_3 V_P - 0 = (3.00 \times 10^{-6}\ \text{C})(-6.29 \times 10^3\ \text{V})$$

$$= \boxed{-1.89 \times 10^{-2}\ \text{J}}$$

Therefore, because the potential energy of the system has decreased, an external agent has to do positive work to remove the charge from point *P* back to infinity.

What If? You are working through this example with a classmate and she says, "Wait a minute! In part (B), we ignored the potential energy associated with the pair of charges q_1 and q_2!" How would you respond?

Answer Given the statement of the problem, it is not necessary to include this potential energy because part (B) asks for the *change* in potential energy of the system as q_3 is brought in from infinity. Because the configuration of charges q_1 and q_2 does not change in the process, there is no ΔU associated with these charges. Had part (B) asked to find the change in potential energy when *all three* charges start out infinitely far apart and are then brought to the positions in Figure 25.11b, however, you would have to calculate the change using Equation 25.14.

25.4 Obtaining the Value of the Electric Field from the Electric Potential

The electric field $\vec{\mathbf{E}}$ and the electric potential V are related as shown in Equation 25.3, which tells us how to find ΔV if the electric field $\vec{\mathbf{E}}$ is known. We now show how to calculate the value of the electric field if the electric potential is known in a certain region.

From Equation 25.3, we can express the potential difference dV between two points a distance ds apart as

$$dV = -\vec{\mathbf{E}} \cdot d\vec{\mathbf{s}} \tag{25.15}$$

If the electric field has only one component E_x, then $\vec{\mathbf{E}} \cdot d\vec{\mathbf{s}} = E_x\,dx$. Therefore, Equation 25.15 becomes $dV = -E_x\,dx$, or

$$E_x = -\frac{dV}{dx} \tag{25.16}$$

That is, the x component of the electric field is equal to the negative of the derivative of the electric potential with respect to x. Similar statements can be made about the y and z components. Equation 25.16 is the mathematical statement of the electric field being a measure of the rate of change with position of the electric potential as mentioned in Section 25.1.

Experimentally, electric potential and position can be measured easily with a voltmeter (see Section 28.5) and a meterstick. Consequently, an electric field can be determined by measuring the electric potential at several positions in the field and making a graph of the results. According to Equation 25.16, the slope of a graph of V versus x at a given point provides the magnitude of the electric field at that point.

When a test charge undergoes a displacement $d\vec{\mathbf{s}}$ along an equipotential surface, then $dV = 0$ because the potential is constant along an equipotential surface. From Equation 25.15, we see that $dV = -\vec{\mathbf{E}} \cdot d\vec{\mathbf{s}} = 0$; therefore, $\vec{\mathbf{E}}$ must be perpendicular to the displacement along the equipotential surface. This result shows that the **equipotential surfaces must always be perpendicular to the electric field lines passing through them.**

As mentioned at the end of Section 25.2, the equipotential surfaces associated with a uniform electric field consist of a family of planes perpendicular to the field lines. Figure 25.12a shows some representative equipotential surfaces for this situation.

If the charge distribution creating an electric field has spherical symmetry such that the volume charge density depends only on the radial distance r, the electric field is radial. In this case, $\vec{\mathbf{E}} \cdot d\vec{\mathbf{s}} = E_r\,dr$, and we can express dV as $dV = -E_r\,dr$. Therefore,

$$E_r = -\frac{dV}{dr} \tag{25.17}$$

For example, the electric potential of a point charge is $V = k_e q/r$. Because V is a function of r only, the potential function has spherical symmetry. Applying Equation 25.17, we find that the electric field due to the point charge is $E_r = k_e q/r^2$, a familiar result. Notice that the potential changes only in the radial direction, not in any direction perpendicular to r. Therefore, V (like E_r) is a function only of r, which is again consistent with the idea that **equipotential surfaces are perpendicular to field lines.** In this case, the equipotential surfaces are a family of spheres concentric with the spherically symmetric charge distribution (Fig. 25.12b). The equipotential surfaces for an electric dipole are sketched in Figure 25.12c.

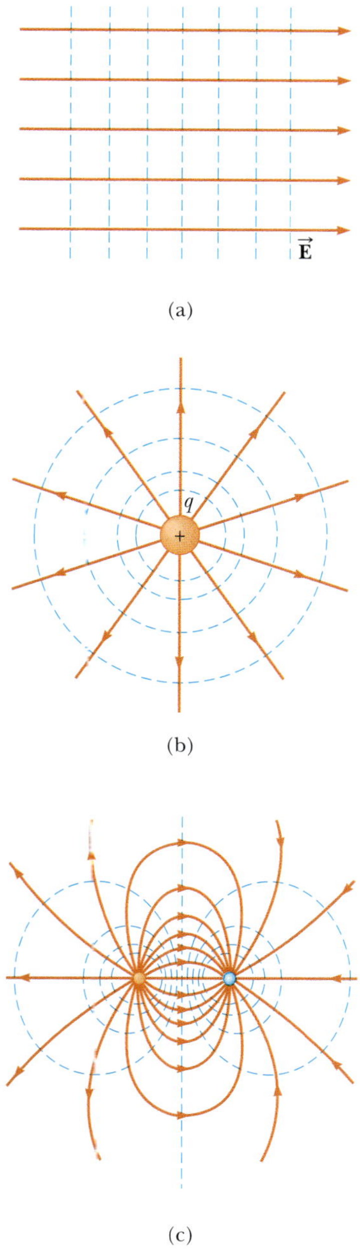

Figure 25.12 Equipotential surfaces (the dashed blue lines are intersections of these surfaces with the page) and electric field lines for (a) a uniform electric field produced by an infinite sheet of charge, (b) a point charge, and (c) an electric dipole. In all cases, the equipotential surfaces are *perpendicular* to the electric field lines at every point.

In general, the electric potential is a function of all three spatial coordinates. If $V(r)$ is given in terms of the Cartesian coordinates, the electric field components E_x, E_y, and E_z can readily be found from $V(x, y, z)$ as the partial derivatives[2]

Finding the electric field from the potential ▶

$$E_x = -\frac{\partial V}{\partial x} \qquad E_y = -\frac{\partial V}{\partial y} \qquad E_z = -\frac{\partial V}{\partial z} \tag{25.18}$$

Quick Quiz 25.4 In a certain region of space, the electric potential is zero everywhere along the x axis. From this information, you can conclude that the x component of the electric field in this region is (a) zero, (b) in the $+x$ direction, or (c) in the $-x$ direction.

EXAMPLE 25.4 The Electric Potential Due to a Dipole

An electric dipole consists of two charges of equal magnitude and opposite sign separated by a distance $2a$ as shown in Figure 25.13. The dipole is along the x axis and is centered at the origin.

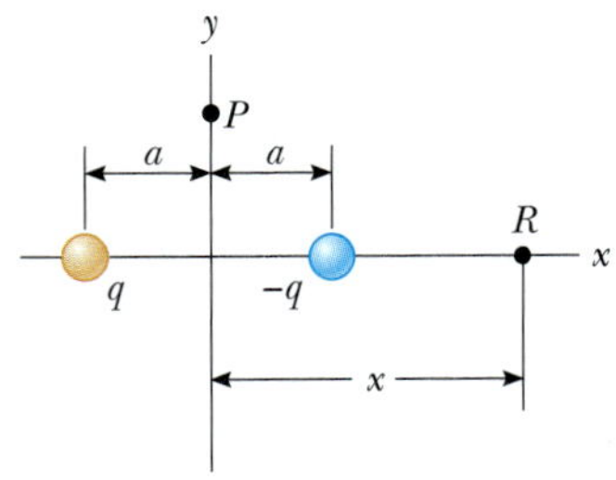

Figure 25.13 (Example 25.4) An electric dipole located on the x axis.

(A) Calculate the electric potential at point P on the y axis.

SOLUTION

Conceptualize Compare this situation to that in part (B) of Example 23.5. It is the same situation, but here we are seeking the electric potential rather than the electric field.

Categorize Because the dipole consists of only two source charges, the electric potential can be evaluated by summing the potentials due to the individual charges.

Analyze Use Equation 25.12 to find the electric potential at P due to the two charges:

$$V_P = k_e \sum_i \frac{q_i}{r_i} = k_e\left(\frac{q}{\sqrt{a^2 + y^2}} + \frac{-q}{\sqrt{a^2 + y^2}}\right) = 0$$

(B) Calculate the electric potential at point R on the $+x$ axis.

SOLUTION

Use Equation 25.12 to find the electric potential at R due to the two charges:

$$V_R = k_e \sum_i \frac{q_i}{r_i} = k_e\left(\frac{-q}{x - a} + \frac{q}{x + a}\right) = -\frac{2k_e qa}{x^2 - a^2}$$

(C) Calculate V and E_x at a point on the x axis far from the dipole.

SOLUTION

For point R far from the dipole such that $x \gg a$, neglect a^2 in the denominator of the answer to part (B) and write V in this limit:

$$V_R = \lim_{x \gg a}\left(-\frac{2k_e qa}{x^2 - a^2}\right) \approx -\frac{2k_e qa}{x^2} \quad (x \gg a)$$

[2] In vector notation, $\vec{\mathbf{E}}$ is often written in Cartesian coordinate systems as

$$\vec{\mathbf{E}} = -\nabla V = -\left(\hat{\mathbf{i}}\frac{\partial}{\partial x} + \hat{\mathbf{j}}\frac{\partial}{\partial y} + \hat{\mathbf{k}}\frac{\partial}{\partial z}\right)V$$

where ∇ is called the *gradient operator*.

Use Equation 25.16 and this result to calculate the x component of the electric field at a point on the x axis far from the dipole:

$$E_x = -\frac{dV}{dx} = -\frac{d}{dx}\left(-\frac{2k_e qa}{x^2}\right)$$

$$= 2k_e qa \frac{d}{dx}\left(\frac{1}{x^2}\right) = -\frac{4k_e qa}{x^3} \quad (x \gg a)$$

Finalize The potentials in parts (B) and (C) are negative because points on the $+x$ axis are closer to the negative charge than to the positive charge. For the same reason, the x component of the electric field is negative. Compare the result of part (C) to that of Problem 18 in Chapter 23, in which the electric field on the x axis due to a dipole was calculated directly.

What If? Suppose you want to find the electric field at a point P on the y axis. In part (A), the electric potential was found to be zero for all values of y. Is the electric field zero at all points on the y axis?

Answer No. That there is no change in the potential along the y axis tells us only that the y component of the electric field is zero. Look back at Figure 23.13 in Example 23.5. We showed there that the electric field of a dipole on the y axis has only an x component. We could not find the x component in the current example because we do not have an expression for the potential near the y axis as a function of x.

25.5 Electric Potential Due to Continuous Charge Distributions

The electric potential due to a continuous charge distribution can be calculated in two ways. If the charge distribution is known, we consider the potential due to a small charge element dq, treating this element as a point charge (Fig. 25.14). From Equation 25.11, the electric potential dV at some point P due to the charge element dq is

$$dV = k_e \frac{dq}{r} \quad \textbf{(25.19)}$$

where r is the distance from the charge element to point P. To obtain the total potential at point P, we integrate Equation 25.19 to include contributions from all elements of the charge distribution. Because each element is, in general, a different distance from point P and k_e is constant, we can express V as

$$V = k_e \int \frac{dq}{r} \quad \textbf{(25.20)}$$

◀ Electric potential due to a continuous charge distribution

In effect, we have replaced the sum in Equation 25.12 with an integral. In this expression for V, the electric potential is taken to be zero when point P is infinitely far from the charge distribution.

If the electric field is already known from other considerations such as Gauss's law, we can calculate the electric potential due to a continuous charge distribution using Equation 25.3. If the charge distribution has sufficient symmetry, we first evaluate $\vec{\mathbf{E}}$ using Gauss's law and then substitute the value obtained into Equation 25.3 to determine the potential difference ΔV between any two points. We then choose the electric potential V to be zero at some convenient point.

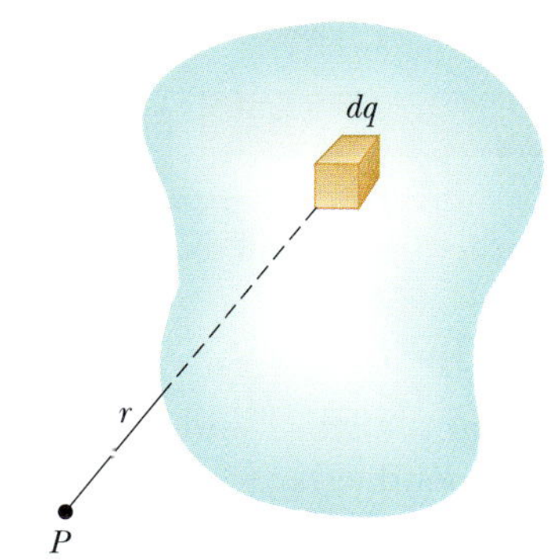

Figure 25.14 The electric potential at point P due to a continuous charge distribution can be calculated by dividing the charge distribution into elements of charge dq and summing the electric potential contributions over all elements.

PROBLEM-SOLVING STRATEGY Calculating Electric Potential

The following procedure is recommended for solving problems that involve the determination of an electric potential due to a charge distribution.

1. *Conceptualize.* Think carefully about the individual charges or the charge distribution you have in the problem and imagine what type of potential would be created. Appeal to any symmetry in the arrangement of charges to help you visualize the potential.

2. *Categorize.* Are you analyzing a group of individual charges or a continuous charge distribution? The answer to this question will tell you how to proceed in the *Analyze* step.

3. *Analyze.* When working problems involving electric potential, remember that it is a *scalar quantity,* so there are no components to consider. Therefore, when using the superposition principle to evaluate the electric potential at a point, simply take the algebraic sum of the potentials due to each charge. You must keep track of signs, however.

   As with potential energy in mechanics, only *changes* in electric potential are significant; hence, the point where the potential is set at zero is arbitrary. When dealing with point charges or a finite-sized charge distribution, we usually define $V = 0$ to be at a point infinitely far from the charges. If the charge distribution itself extends to infinity, however, some other nearby point must be selected as the reference point.

   (*a*) *If you are analyzing a group of individual charges:* Use the superposition principle, which states that when several point charges are present, the resultant potential at a point in space is the *algebraic sum* of the individual potentials due to the individual charges (Eq. 25.12). Example 25.4 demonstrated this procedure.

   (*b*) *If you are analyzing a continuous charge distribution:* Replace the sums for evaluating the total potential at some point P from individual charges by integrals (Eq. 25.20). The charge distribution is divided into infinitesimal elements of charge dq located at a distance r from the point P. An element is then treated as a point charge, so the potential at P due to the element is $dV = k_e\, dq/r$. The total potential at P is obtained by integrating over the entire charge distribution. For many problems, it is possible in performing the integration to express dq and r in terms of a single variable. To simplify the integration, give careful consideration to the geometry involved in the problem. Examples 25.5 through 25.7 demonstrate such a procedure.

   To obtain the potential from the electric field: Another method used to obtain the potential is to start with the definition of the potential difference given by Equation 25.3. If $\vec{\mathbf{E}}$ is known or can be obtained easily (such as from Gauss's law), the line integral of $\vec{\mathbf{E}} \cdot d\vec{\mathbf{s}}$ can be evaluated.

4. *Finalize.* Check to see if your expression for the potential is consistent with the mental representation and reflects any symmetry you noted previously. Imagine varying parameters such as the distance of the observation point from the charges or the radius of any circular objects to see if the mathematical result changes in a reasonable way.

EXAMPLE 25.5 Electric Potential Due to a Uniformly Charged Ring

(A) Find an expression for the electric potential at a point P located on the perpendicular central axis of a uniformly charged ring of radius a and total charge Q.

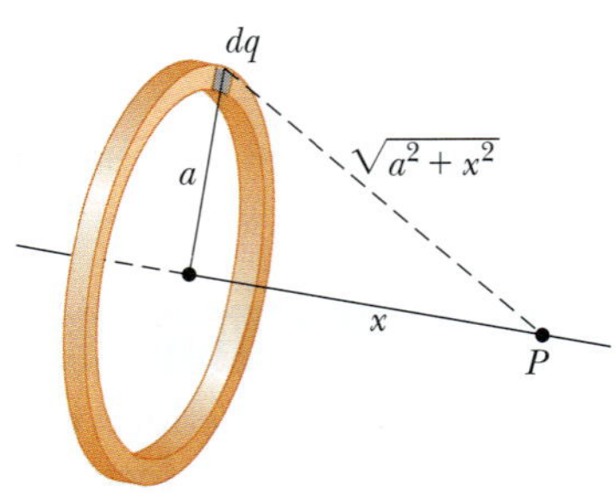

Figure 25.15 (Example 25.5) A uniformly charged ring of radius a lies in a plane perpendicular to the x axis. All elements dq of the ring are the same distance from a point P lying on the x axis.

SOLUTION

Conceptualize Study Figure 25.15, in which the ring is oriented so that its plane is perpendicular to the x axis and its center is at the origin.

Categorize Because the ring consists of a continuous distribution of charge rather than a set of discrete charges, we must use the integration technique represented by Equation 25.20 in this example.

Analyze We take point P to be at a distance x from the center of the ring as shown in Figure 25.15. Notice that all charge elements dq are at the same distance $\sqrt{a^2 + x^2}$ from point P.

Use Equation 25.20 to express V in terms of the geometry:

$$V = k_e \int \frac{dq}{r} = k_e \int \frac{dq}{\sqrt{a^2 + x^2}}$$

Noting that a and x are constants, bring $\sqrt{a^2 + x^2}$ in front of the integral sign and integrate over the ring:

$$V = \frac{k_e}{\sqrt{a^2 + x^2}} \int dq = \frac{k_e Q}{\sqrt{a^2 + x^2}} \tag{25.21}$$

(B) Find an expression for the magnitude of the electric field at point P.

SOLUTION

From symmetry, notice that along the x axis $\vec{\mathbf{E}}$ can have only an x component. Therefore, apply Equation 25.16 to Equation 25.21:

$$E_x = -\frac{dV}{dx} = -k_e Q \frac{d}{dx}(a^2 + x^2)^{-1/2}$$

$$= -k_e Q(-\tfrac{1}{2})(a^2 + x^2)^{-3/2}(2x)$$

$$E_x = \frac{k_e x}{(a^2 + x^2)^{3/2}} Q \tag{25.22}$$

Finalize The only variable in the expressions for V and E_x is x. That is not surprising because our calculation is valid only for points along the x axis, where y and z are both zero. This result for the electric field agrees with that obtained by direct integration (see Example 23.7).

EXAMPLE 25.6 Electric Potential Due to a Uniformly Charged Disk

A uniformly charged disk has radius R and surface charge density σ.

(A) Find the electric potential at a point P along the perpendicular central axis of the disk.

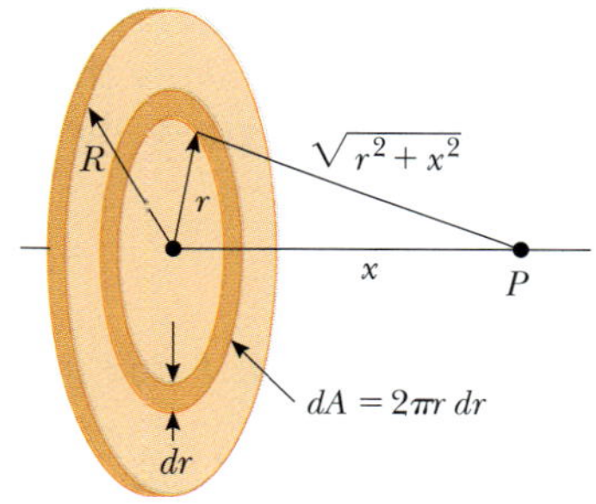

Figure 25.16 (Example 25.6) A uniformly charged disk of radius R lies in a plane perpendicular to the x axis. The calculation of the electric potential at any point P on the x axis is simplified by dividing the disk into many rings of radius r and width dr, with area $2\pi r\, dr$.

SOLUTION

Conceptualize If we consider the disk to be a set of concentric rings, we can use our result from Example 25.5—which gives the potential created by a ring of radius a—and sum the contributions of all rings making up the disk.

Categorize Because the disk is continuous, we evaluate the potential due to a continuous charge distribution rather than a group of individual charges.

Analyze Find the amount of charge dq on a ring of radius r and width dr as shown in Figure 25.16:

$$dq = \sigma\, dA = \sigma(2\pi r\, dr) = 2\pi\sigma r\, dr$$

Use this result in the equation given for V in Example 25.5 (with a replaced by r and Q replaced by dq) to find the potential due to the ring:

$$dV = \frac{k_e\, dq}{\sqrt{r^2 + x^2}} = \frac{k_e 2\pi\sigma r\, dr}{\sqrt{r^2 + x^2}}$$

To obtain the total potential at P, integrate this expression over the limits $r = 0$ to $r = R$, noting that x is a constant:

$$V = \pi k_e \sigma \int_0^R \frac{2r\, dr}{\sqrt{r^2 + x^2}} = \pi k_e \sigma \int_0^R (r^2 + x^2)^{-1/2}\, 2r\, dr$$

This integral is of the common form $\int u^n\, du$ and has the value $u^{n+1}/(n + 1)$, where $n = -\frac{1}{2}$ and $u = r^2 + x^2$. Use this result to evaluate the integral:

$$V = 2\pi k_e \sigma[(R^2 + x^2)^{1/2} - x] \tag{25.23}$$

(B) Find the x component of the electric field at a point P along the perpendicular central axis of the disk.

SOLUTION

As in Example 25.5, use Equation 25.16 to find the electric field at any axial point:

$$E_x = -\frac{dV}{dx} = 2\pi k_e \sigma \left[1 - \frac{x}{(R^2 + x^2)^{1/2}}\right] \quad \textbf{(25.24)}$$

Finalize Compare Equation 25.24 with the result of Example 23.8. The calculation of V and $\vec{\mathbf{E}}$ for an arbitrary point off the x axis is more difficult to perform, and we do not treat that situation in this book.

EXAMPLE 25.7 Electric Potential Due to a Finite Line of Charge

A rod of length ℓ located along the x axis has a total charge Q and a uniform linear charge density $\lambda = Q/\ell$. Find the electric potential at a point P located on the y axis a distance a from the origin (Fig. 25.17).

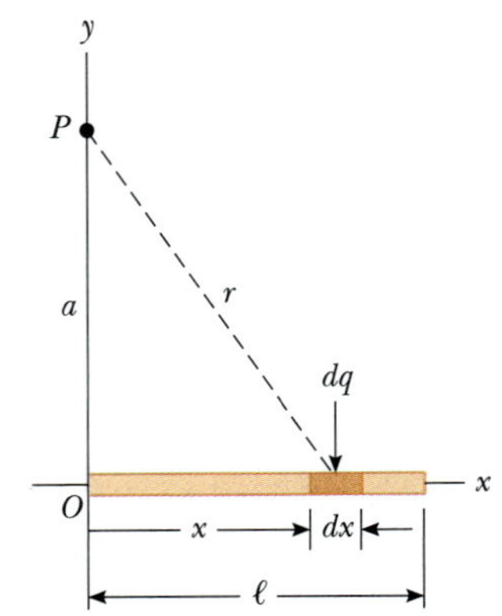

Figure 25.17 (Example 25.7) A uniform line charge of length ℓ located along the x axis. To calculate the electric potential at P, the line charge is divided into segments each of length dx and each carrying a charge $dq = \lambda\, dx$.

SOLUTION

Conceptualize The potential at P due to every segment of charge on the rod is positive because every segment carries a positive charge.

Categorize Because the rod is continuous, we evaluate the potential due to a continuous charge distribution rather than a group of individual charges.

Analyze In Figure 25.17, the rod lies along the x axis, dx is the length of one small segment, and dq is the charge on that segment. Because the rod has a charge per unit length λ, the charge dq on the small segment is $dq = \lambda\, dx$.

Find the potential at P due to one segment of the rod:

$$dV = k_e \frac{dq}{r} = k_e \frac{\lambda\, dx}{\sqrt{a^2 + x^2}}$$

Find the total potential at P by integrating this expression over the limits $x = 0$ to $x = \ell$:

$$V = \int_0^\ell k_e \frac{\lambda\, dx}{\sqrt{a^2 + x^2}}$$

Noting that k_e and $\lambda = Q/\ell$ are constants and can be removed from the integral, evaluate the integral with the help of Appendix B:

$$V = k_e \lambda \int_0^\ell \frac{dx}{\sqrt{a^2 + x^2}} = k_e \frac{Q}{\ell} \ln\left(x + \sqrt{a^2 + x^2}\right)\Big|_0^\ell$$

Evaluate the result between the limits:

$$V = k_e \frac{Q}{\ell}\left[\ln\left(\ell + \sqrt{a^2 + \ell^2}\right) - \ln a\right] = k_e \frac{Q}{\ell} \ln\left(\frac{\ell + \sqrt{a^2 + \ell^2}}{a}\right) \quad \textbf{(25.25)}$$

What If? What if you were asked to find the electric field at point P? Would that be a simple calculation?

Answer Calculating the electric field by means of Equation 23.11 would be a little messy. There is no symmetry to appeal to, and the integration over the line of charge would represent a vector addition of electric fields at point P. Using Equation 25.18, you could find E_y by replacing a with y in Equation 25.25 and performing the differentiation with respect to y. Because the charged rod in Figure 25.17 lies entirely to the right of $x = 0$, the electric field at point P would have an x component to the left if the rod is charged positively. You cannot use Equation 25.18 to find the x

component of the field, however, because the potential due to the rod was evaluated at a specific value of x ($x = 0$) rather than a general value of x. You would have to find the potential as a function of both x and y to be able to find the x and y components of the electric field using Equation 25.25.

25.6 Electric Potential Due to a Charged Conductor

In Section 24.4, we found that when a solid conductor in equilibrium carries a net charge, the charge resides on the conductor's outer surface. Furthermore, the electric field just outside the conductor is perpendicular to the surface and the field inside is zero.

We now show that **every point on the surface of a charged conductor in equilibrium is at the same electric potential.** Consider two points Ⓐ and Ⓑ on the surface of a charged conductor as shown in Figure 25.18. Along a surface path connecting these points, $\vec{\mathbf{E}}$ is always perpendicular to the displacement $d\vec{\mathbf{s}}$; therefore $\vec{\mathbf{E}} \cdot d\vec{\mathbf{s}} = 0$. Using this result and Equation 25.3, we conclude that the potential difference between Ⓐ and Ⓑ is necessarily zero:

$$V_{\text{Ⓑ}} - V_{\text{Ⓐ}} = -\int_{\text{Ⓐ}}^{\text{Ⓑ}} \vec{\mathbf{E}} \cdot d\vec{\mathbf{s}} = 0$$

This result applies to any two points on the surface. Therefore, V is constant everywhere on the surface of a charged conductor in equilibrium. That is,

> the surface of any charged conductor in electrostatic equilibrium is an equipotential surface. Furthermore, because the electric field is zero inside the conductor, the electric potential is constant everywhere inside the conductor and equal to its value at the surface.

PITFALL PREVENTION 25.6

Potential May Not Be Zero

The electric potential inside the conductor is not necessarily zero in Figure 25.18, even though the electric field is zero. Equation 25.15 shows that a zero value of the field results in no *change* in the potential from one point to another inside the conductor. Therefore, the potential everywhere inside the conductor, including the surface, has the same value, which may or may not be zero, depending on where the zero of potential is defined.

Because of the constant value of the potential, no work is required to move a test charge from the interior of a charged conductor to its surface.

Consider a solid metal conducting sphere of radius R and total positive charge Q as shown in Figure 25.19a. As determined in part (A) of Example 24.3, the electric field outside the sphere is k_eQ/r^2 and points radially outward. Because the field outside of a spherically symmetric charge distribution is identical to that of a point charge, we expect the potential to also be that of a point charge, k_eQ/r. At the surface of the conducting sphere in Figure 25.19a, the potential must be k_eQ/R. Because the entire sphere must be at the same potential, the potential at any point within the sphere must also be k_eQ/R. Figure 25.19b is a plot of the electric potential as a function of r, and Figure 25.19c shows how the electric field varies with r.

When a net charge is placed on a spherical conductor, the surface charge density is uniform as indicated in Figure 25.19a. If the conductor is nonspherical as in

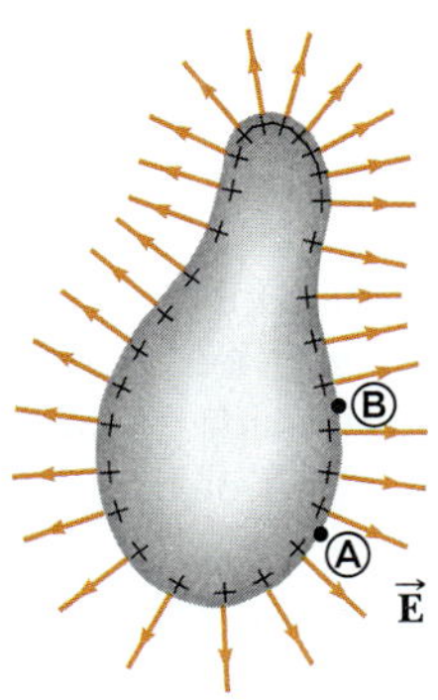

Figure 25.18 An arbitrarily shaped conductor carrying a positive charge. When the conductor is in electrostatic equilibrium, all the charge resides at the surface, $\vec{\mathbf{E}} = 0$ inside the conductor, and the direction of $\vec{\mathbf{E}}$ immediately outside the conductor is perpendicular to the surface. The electric potential is constant inside the conductor and is equal to the potential at the surface. Notice from the spacing of the positive signs that the surface charge density is nonuniform.

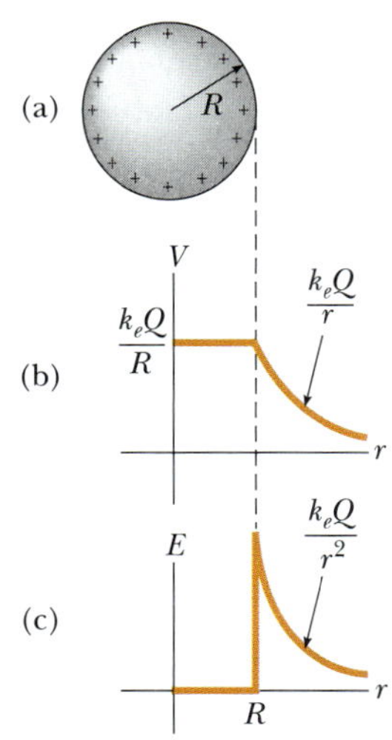

Figure 25.19 (a) The excess charge on a conducting sphere of radius R is uniformly distributed on its surface. (b) Electric potential versus distance r from the center of the charged conducting sphere. (c) Electric field magnitude versus distance r from the center of the charged conducting sphere.

Figure 25.18, however, the surface charge density is high where the radius of curvature is small (as noted in Section 24.4) and low where the radius of curvature is large. Because the electric field immediately outside the conductor is proportional to the surface charge density, **the electric field is large near convex points having small radii of curvature and reaches very high values at sharp points.** In Example 25.8, the relationship between electric field and radius of curvature is explored mathematically.

EXAMPLE 25.8 Two Connected Charged Spheres

Two spherical conductors of radii r_1 and r_2 are separated by a distance much greater than the radius of either sphere. The spheres are connected by a conducting wire as shown in Figure 25.20. The charges on the spheres in equilibrium are q_1 and q_2, respectively, and they are uniformly charged. Find the ratio of the magnitudes of the electric fields at the surfaces of the spheres.

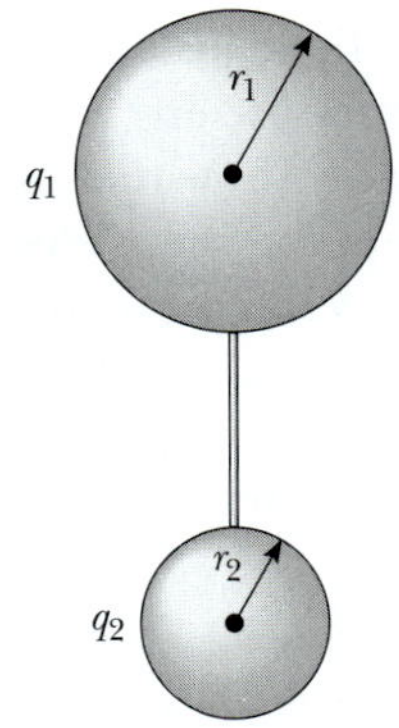

Figure 25.20 (Example 25.8) Two charged spherical conductors connected by a conducting wire. The spheres are at the *same* electric potential V.

SOLUTION

Conceptualize Imagine that the spheres are much farther apart than shown in Figure 25.20. Because they are so far apart, the field of one does not affect the charge distribution on the other. The conducting wire between them ensures that both spheres have the same electric potential.

Categorize Because the spheres are so far apart, we model the charge distribution on them as spherically symmetric, and we can model the field and potential outside the spheres to be that due to point charges.

Analyze Set the electric potentials at the surfaces of the spheres equal to each other:

$$V = k_e \frac{q_1}{r_1} = k_e \frac{q_2}{r_2}$$

Solve for the ratio of charges on the spheres:

$$(1) \quad \frac{q_1}{q_2} = \frac{r_1}{r_2}$$

Write expressions for the magnitudes of the electric fields at the surfaces of the spheres:

$$E_1 = k_e \frac{q_1}{r_1^2} \quad \text{and} \quad E_2 = k_e \frac{q_2}{r_2^2}$$

Evaluate the ratio of these two fields:

$$\frac{E_1}{E_2} = \frac{q_1}{q_2}\frac{r_2^2}{r_1^2}$$

Substitute for the ratio of charges from Equation (1):

$$(2) \quad \frac{E_1}{E_2} = \frac{r_1}{r_2}\frac{r_2^2}{r_1^2} = \frac{r_2}{r_1}$$

Finalize The field is stronger in the vicinity of the smaller sphere even though the electric potentials at the surfaces of both spheres are the same.

A Cavity Within a Conductor

Suppose a conductor of arbitrary shape contains a cavity as shown in Figure 25.21. Let's assume no charges are inside the cavity. **In this case, the electric field inside the cavity must be zero** regardless of the charge distribution on the outside surface of the conductor as we mentioned in Section 24.4. Furthermore, the field in the cavity is zero even if an electric field exists outside the conductor.

To prove this point, remember that every point on the conductor is at the same electric potential; therefore, any two points Ⓐ and Ⓑ on the cavity's surface must be at the same potential. Now imagine a field $\vec{\mathbf{E}}$ exists in the cavity and evaluate the potential difference $V_{Ⓑ} - V_{Ⓐ}$ defined by Equation 25.3:

$$V_{Ⓑ} - V_{Ⓐ} = -\int_{Ⓐ}^{Ⓑ} \vec{\mathbf{E}} \cdot d\vec{\mathbf{s}}$$

Because $V_{Ⓑ} - V_{Ⓐ} = 0$, the integral of $\vec{\mathbf{E}} \cdot d\vec{\mathbf{s}}$ must be zero for all paths between any two points Ⓐ and Ⓑ on the conductor. The only way that can be true for *all* paths is if $\vec{\mathbf{E}}$ is zero *everywhere* in the cavity. Therefore, **a cavity surrounded by conducting walls is a field-free region as long as no charges are inside the cavity.**

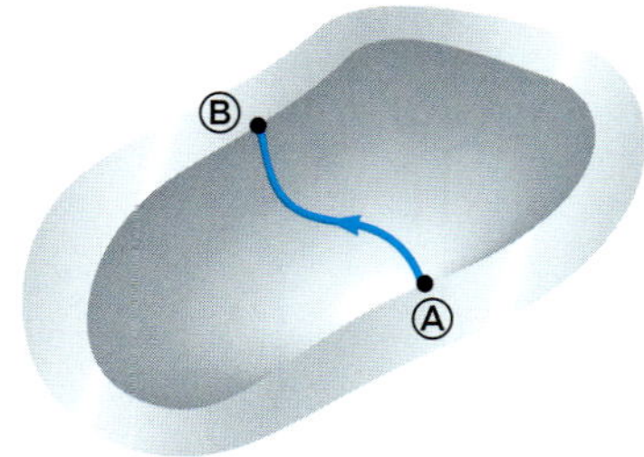

Figure 25.21 A conductor in electrostatic equilibrium containing a cavity. The electric field in the cavity is zero, regardless of the charge on the conductor.

Corona Discharge

A phenomenon known as **corona discharge** is often observed near a conductor such as a high-voltage power line. When the electric field in the vicinity of the conductor is sufficiently strong, electrons resulting from random ionizations of air molecules near the conductor accelerate away from their parent molecules. These rapidly moving electrons can ionize additional molecules near the conductor, creating more free electrons. The observed glow (or corona discharge) results from the recombination of these free electrons with the ionized air molecules. If a conductor has an irregular shape, the electric field can be very high near sharp points or edges of the conductor; consequently, the ionization process and corona discharge are most likely to occur around such points.

Corona discharge is used in the electrical transmission industry to locate broken or faulty components. For example, a broken insulator on a transmission tower has sharp edges where corona discharge is likely to occur. Similarly, corona discharge will occur at the sharp end of a broken conductor strand. Observation of these discharges is difficult because the visible radiation emitted is weak and most of the radiation is in the ultraviolet. (We will discuss ultraviolet radiation and other portions of the electromagnetic spectrum in Section 34.7.) Even use of traditional ultraviolet cameras is of little help because the radiation from the corona discharge is overwhelmed by ultraviolet radiation from the Sun. Newly developed dual-spectrum devices combine a narrow-band ultraviolet camera with a visible-light camera to show a daylight view of the corona discharge in the actual location on the transmission tower or cable. The ultraviolet part of the camera is designed to operate in a wavelength range in which radiation from the Sun is very weak.

25.7 The Millikan Oil-Drop Experiment

Robert Millikan performed a brilliant set of experiments from 1909 to 1913 in which he measured e, the magnitude of the elementary charge on an electron, and demonstrated the quantized nature of this charge. His apparatus, diagrammed in Active Figure 25.22, contains two parallel metallic plates. Oil droplets

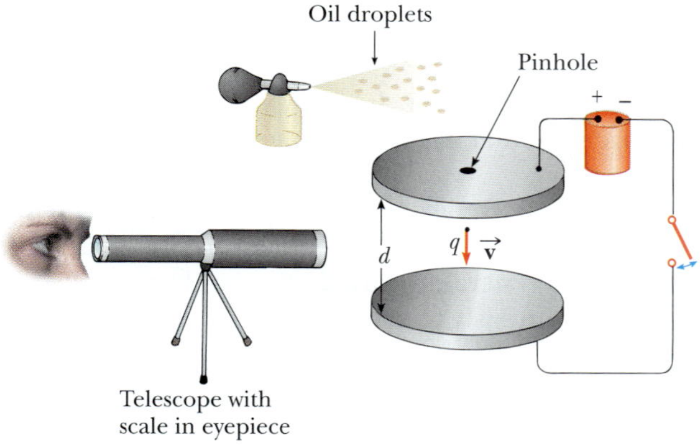

ACTIVE FIGURE 25.22

Schematic drawing of the Millikan oil-drop apparatus.

Sign in at www.thomsonedu.com and go to ThomsonNOW to perform a simplified version of the experiment yourself. You will be able to take data on a number of oil drops and determine the elementary charge from your data.

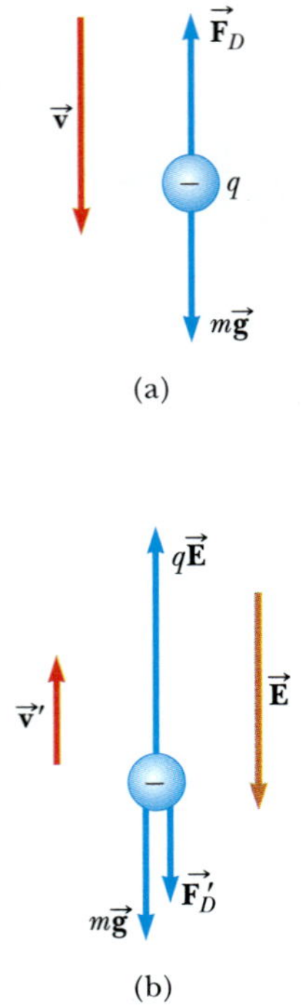

Figure 25.23 The forces acting on a negatively charged oil droplet in the Millikan experiment. (a) With the electric field off, the droplet falls at terminal velocity $\vec{v}$ under the influence of the gravitational and drag forces. (b) When the electric field is turned on, the droplet moves upward at terminal velocity $\vec{v}'$ under the influence of the electric, gravitational, and drag forces.

from an atomizer are allowed to pass through a small hole in the upper plate. Millikan used x-rays to ionize the air in the chamber so that freed electrons would adhere to the oil drops, giving them a negative charge. A horizontally directed light beam is used to illuminate the oil droplets, which are viewed through a telescope whose long axis is perpendicular to the light beam. When viewed in this manner, the droplets appear as shining stars against a dark background and the rate at which individual drops fall can be determined.

Let's assume a single drop having a mass m and carrying a charge q is being viewed and its charge is negative. If no electric field is present between the plates, the two forces acting on the charge are the gravitational force $m\vec{g}$ acting downward[3] and a viscous drag force $\vec{F}_D$ acting upward as indicated in Figure 25.23a. The drag force is proportional to the drop's speed as discussed in Section 6.4. When the drop reaches its terminal speed v, the two forces balance each other ($mg = F_D$).

Now suppose a battery connected to the plates sets up an electric field between the plates such that the upper plate is at the higher electric potential. In this case, a third force $q\vec{E}$ acts on the charged drop. Because q is negative and $\vec{E}$ is directed downward, this electric force is directed upward as shown in Figure 25.23b. If this upward force is strong enough, the drop moves upward and the drag force $\vec{F}'_D$ acts downward. When the upward electric force $q\vec{E}$ balances the sum of the gravitational force and the downward drag force $\vec{F}'_D$, the drop reaches a new terminal speed v' in the upward direction.

With the field turned on, a drop moves slowly upward, typically at rates of hundredths of a centimeter per second. The rate of fall in the absence of a field is comparable. Hence, one can follow a single droplet for hours, alternately rising and falling, by simply turning the electric field on and off.

After recording measurements on thousands of droplets, Millikan and his coworkers found that all droplets, to within about 1% precision, had a charge equal to some integer multiple of the elementary charge e:

$$q = ne \qquad n = 0, -1, -2, -3, \ldots$$

where $e = 1.60 \times 10^{-19}$ C. Millikan's experiment yields conclusive evidence that charge is quantized. For this work, he was awarded the Nobel Prize in Physics in 1923.

25.8 Applications of Electrostatics

The practical application of electrostatics is represented by such devices as lightning rods and electrostatic precipitators and by such processes as xerography and the painting of automobiles. Scientific devices based on the principles of electrostatics include electrostatic generators, the field-ion microscope, and ion-drive rocket engines.

The Van de Graaff Generator

Experimental results show that when a charged conductor is placed in contact with the inside of a hollow conductor, all the charge on the charged conductor is transferred to the hollow conductor. In principle, the charge on the hollow conductor and its electric potential can be increased without limit by repetition of the process.

In 1929, Robert J. Van de Graaff (1901–1967) used this principle to design and build an electrostatic generator, and a schematic representation of it is given in Figure 25.24. This type of generator is used extensively in nuclear physics research. Charge is delivered continuously to a high-potential electrode by means of a mov-

[3] There is also a buoyant force on the oil drop due to the surrounding air. This force can be incorporated as a correction in the gravitational force $m\vec{g}$ on the drop, so we will not consider it in our analysis.

ing belt of insulating material. The high-voltage electrode is a hollow metal dome mounted on an insulating column. The belt is charged at point Ⓐ by means of a corona discharge between comb-like metallic needles and a grounded grid. The needles are maintained at a positive electric potential of typically 10^4 V. The positive charge on the moving belt is transferred to the dome by a second comb of needles at point Ⓑ. Because the electric field inside the dome is negligible, the positive charge on the belt is easily transferred to the conductor regardless of its potential. In practice, it is possible to increase the electric potential of the dome until electrical discharge occurs through the air. Because the "breakdown" electric field in air is about 3×10^6 V/m, a sphere 1 m in radius can be raised to a maximum potential of 3×10^6 V. The potential can be increased further by increasing the dome's radius and placing the entire system in a container filled with high-pressure gas.

Van de Graaff generators can produce potential differences as large as 20 million volts. Protons accelerated through such large potential differences receive enough energy to initiate nuclear reactions between themselves and various target nuclei. Smaller generators are often seen in science classrooms and museums. If a person insulated from the ground touches the sphere of a Van de Graaff generator, his or her body can be brought to a high electric potential. The person's hair acquires a net positive charge, and each strand is repelled by all the others as in the opening photograph of Chapter 23.

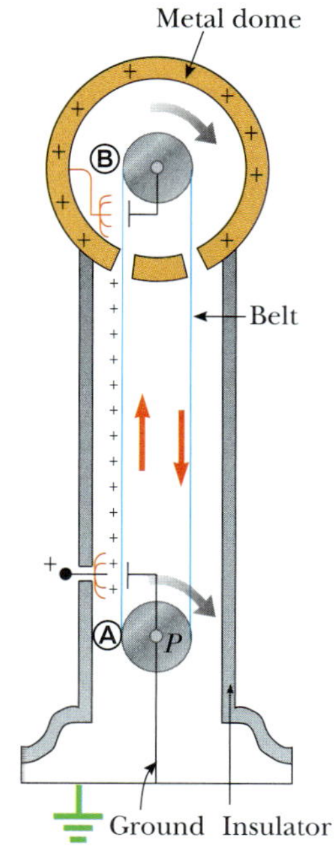

Figure 25.24 Schematic diagram of a Van de Graaff generator. Charge is transferred to the metal dome at the top by means of a moving belt. The charge is deposited on the belt at point Ⓐ and transferred to the hollow conductor at point Ⓑ.

The Electrostatic Precipitator

One important application of electrical discharge in gases is the *electrostatic precipitator.* This device removes particulate matter from combustion gases, thereby reducing air pollution. Precipitators are especially useful in coal-burning power plants and industrial operations that generate large quantities of smoke. Current systems are able to eliminate more than 99% of the ash from smoke.

Figure 25.25a shows a schematic diagram of an electrostatic precipitator. A high potential difference (typically 40 to 100 kV) is maintained between a wire running down the center of a duct and the walls of the duct, which are grounded. The wire is maintained at a negative electric potential with respect to the walls, so the electric field is directed toward the wire. The values of the field near the wire become high enough to cause a corona discharge around the wire; the air near the wire contains positive ions, electrons, and such negative ions as O_2^-. The air to be cleaned enters the duct and moves near the wire. As the electrons and negative

Figure 25.25 (a) Schematic diagram of an electrostatic precipitator. The high negative electric potential maintained on the central coiled wire creates a corona discharge in the vicinity of the wire. Compare the air pollution when the electrostatic precipitator is (b) operating and (c) turned off.

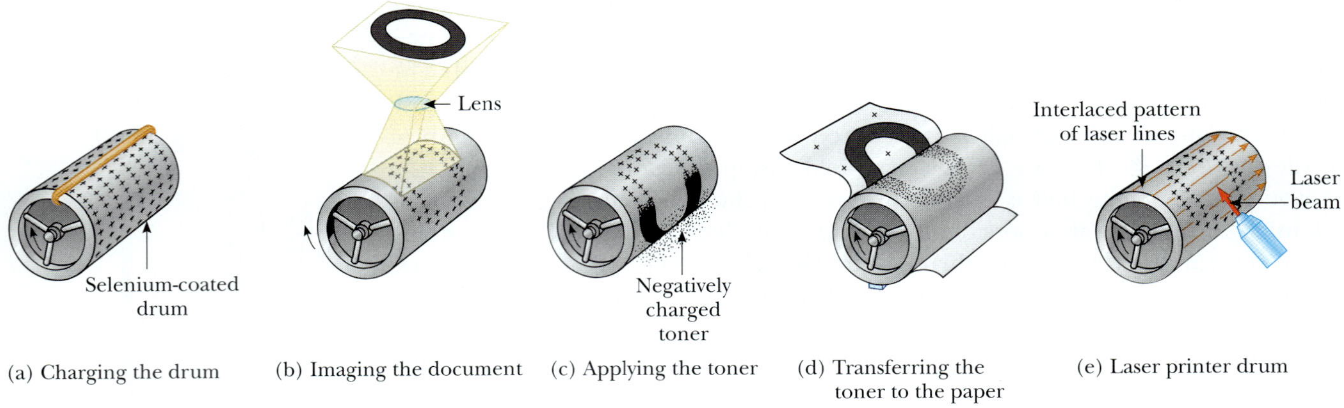

Figure 25.26 The xerographic process. (a) The photoconductive surface of the drum is positively charged. (b) Through the use of a light source and lens, an image is created on the surface in the form of positive charges. (c) The surface containing the image is covered with a negatively charged powder, which adheres only to the image area. (d) A piece of paper is placed over the surface and given a positive charge, which transfers the image to the paper as the negatively charged powder particles migrate to the paper. The paper is then thermally treated to "fix" the powder. (e) A laser printer operates similarly except that the image is produced by turning a laser beam on and off as it sweeps across the selenium-coated drum.

ions created by the discharge are accelerated toward the outer wall by the electric field, the dirt particles in the air become charged by collisions and ion capture. Because most of the charged dirt particles are negative, they too are drawn to the duct walls by the electric field. When the duct is periodically shaken, the particles break loose and are collected at the bottom.

In addition to reducing the level of particulate matter in the atmosphere (compare Figs. 25.25b and c), the electrostatic precipitator recovers valuable materials in the form of metal oxides.

Xerography and Laser Printers

The basic idea of xerography[4] was developed by Chester Carlson, who was granted a patent for the xerographic process in 1940. The unique feature of this process is the use of a photoconductive material to form an image. (A *photoconductor* is a material that is a poor electrical conductor in the dark but becomes a good electrical conductor when exposed to light.)

The xerographic process is illustrated in parts (a) through (d) of Figure 25.26. First, the surface of a plate or drum that has been coated with a thin film of photoconductive material (usually selenium or some compound of selenium) is given a positive electrostatic charge in the dark. An image of the page to be copied is then focused by a lens onto the charged surface. The photoconducting surface becomes conducting only in areas where light strikes it. In such areas, the light produces charge carriers in the photoconductor that move the positive charge off the drum. Positive charges, however, remain on those areas of the photoconductor not exposed to light, leaving a latent image of the object in the form of a positive surface charge distribution.

Next, a negatively charged powder called a *toner* is dusted onto the photoconducting surface. The charged powder adheres only to those areas of the surface that contain the positively charged image. The toner (and hence the image) is then transferred to the surface of a sheet of positively charged paper.

Finally, the toner is "fixed" to the surface of the paper as the toner melts while passing through high-temperature rollers. The result is a permanent copy of the original.

A laser printer (Fig. 25.26e) operates by the same principle, with the exception that a computer-directed laser beam is used to illuminate the photoconductor instead of a lens.

[4] The prefix *xero-* is from the Greek word meaning "dry." Note that liquid ink is not used in xerography.

Summary

ThomsonNOW™ Sign in at **www.thomsonedu.com** and go to ThomsonNOW to take a practice test for this chapter.

DEFINITIONS

The **potential difference** ΔV between points Ⓐ and Ⓑ in an electric field $\vec{\mathbf{E}}$ is defined as

$$\Delta V \equiv \frac{\Delta U}{q_0} = -\int_{Ⓐ}^{Ⓑ} \vec{\mathbf{E}} \cdot d\vec{\mathbf{s}} \qquad (25.3)$$

where ΔU is given by Equation 25.1 below. The **electric potential** $V = U/q_0$ is a scalar quantity and has the units of joules per coulomb, where $1\ \text{J/C} \equiv 1\ \text{V}$.

An **equipotential surface** is one on which all points are at the same electric potential. Equipotential surfaces are perpendicular to electric field lines.

CONCEPTS AND PRINCIPLES

When a positive test charge q_0 is moved between points Ⓐ and Ⓑ in an electric field $\vec{\mathbf{E}}$, the **change in the potential energy of the charge–field system** is

$$\Delta U = -q_0 \int_{Ⓐ}^{Ⓑ} \vec{\mathbf{E}} \cdot d\vec{\mathbf{s}} \qquad (25.1)$$

The potential difference between two points Ⓐ and Ⓑ separated by a distance d in a uniform electric field $\vec{\mathbf{E}}$, where $\vec{\mathbf{s}}$ is a vector that points from Ⓐ toward Ⓔ and is parallel to $\vec{\mathbf{E}}$, is

$$\Delta V = -E \int_{Ⓐ}^{Ⓑ} ds = -Ed \qquad (25.6)$$

If we define $V = 0$ at $r = \infty$, the electric potential due to a point charge at any distance r from the charge is

$$V = k_e \frac{q}{r} \qquad (25.11)$$

The electric potential associated with a group of point charges is obtained by summing the potentials due to the individual charges.

The **potential energy associated with a pair of point charges** separated by a distance r_{12} is

$$U = k_e \frac{q_1 q_2}{r_{12}} \qquad (25.13)$$

We obtain the potential energy of a distribution of point charges by summing terms like Equation 25.13 over all pairs of particles.

If the electric potential is known as a function of coordinates x, y, and z, we can obtain the components of the electric field by taking the negative derivative of the electric potential with respect to the coordinates. For example, the x component of the electric field is

$$E_x = -\frac{dV}{dx} \qquad (25.16)$$

The **electric potential due to a continuous charge distribution** is

$$V = k_e \int \frac{dq}{r} \qquad (25.20)$$

Every point on the surface of a charged conductor in electrostatic equilibrium is at the same electric potential. The potential is constant everywhere inside the conductor and equal to its value at the surface.

Questions

□ denotes answer available in *Student Solutions Manual/Study Guide;* **O** denotes objective question

1. Distinguish between electric potential and electric potential energy.

2. **O** In a certain region of space, a uniform electric field is in the x direction. A particle with negative charge is carried from $x = 20$ cm to $x = 60$ cm. **(i)** Does the potential energy of the charge-field system (a) increase, (b) remain constant, (c) decrease, or (d) change unpredictably? **(ii)** Does the particle move to a position where the potential is (a) higher than before, (b) unchanged, (c) lower than before, or (d) unpredictable?

3. **O** Consider the equipotential surfaces shown in Figure 25.4. In this region of space, what is the approximate direction of the electric field? (a) out of the page (b) into the page (c) toward the right (d) toward the left (e) toward the top of the page (f) toward the bottom of the page (g) the field is zero

4. **O** A particle with charge −40 nC is on the x axis at the point with coordinate $x = 0$. A second particle, with charge −20 nC, is on the x axis at $x = 500$ mm. **(i)** Is there a point at a finite distance where the electric field is zero? (a) Yes; it is to the left of $x = 0$. (b) Yes; it is between $x = 0$ and $x = 500$ mm. (c) Yes; it is to the right of $x = 500$ mm. (d) No. **(ii)** Is the electric potential zero at this point? (a) No; it is positive. (b) Yes. (c) No; it is negative. (d) No such point exists. **(iii)** Is there a point at a finite distance where the electric potential is zero? (a) Yes; it is to the left of $x = 0$. (b) Yes; it is between $x = 0$ and $x = 500$ mm. (c) Yes; it is to the right of $x = 500$ mm. (d) No. **(iv)** Is the electric field zero at this point? (a) No; it points to the right. (b) Yes. (c) No; it points to the left. (d) No such point exists.

5. The potential energy of a pair of charged particles with the same sign is positive, whereas the potential energy of a pair of charged particles with opposite signs is negative. Give a physical explanation of this statement.

6. Describe the equipotential surfaces for (a) an infinite line of charge and (b) a uniformly charged sphere.

7. **O** In a certain region of space, the electric field is zero. From this fact, what can you conclude about the electric potential in this region? (a) It is zero. (b) It is constant. (c) It is positive. (d) It is negative. (e) None of these answers is necessarily true.

8. **O** A filament running along the x axis from the origin to $x = 80$ cm carries electric charge with uniform density. At the point P with coordinates ($x = 80$ cm, $y = 80$ cm), this filament creates potential 100 V. Now we add another filament along the y axis, running from the origin to $y = 80$ cm, carrying the same amount of charge with the same uniform density. At the same point P, does the pair of filaments create potential (a) greater than 200 V, (b) 200 V, (c) between 141 V and 200 V, (d) 141 V, (e) between 100 V and 141 V, (f) 100 V, (g) between 0 and 100 V, or (h) 0?

9. **O** In different experimental trials, an electron, a proton, or a doubly charged oxygen atom (O^{--}) is fired within a vacuum tube. The particle's trajectory carries it through a point where the electric potential is 40 V and then through a point at a different potential. Rank each of the following cases according to the change in kinetic energy of the particle over this part of its flight, from the largest increase to the largest decrease in kinetic energy. (a) An electron moves from 40 V to 60 V. (b) An electron moves from 40 V to 20 V. (c) A proton moves from 40 V to 20 V. (d) A proton moves from 40 V to 10 V. (e) An O^{--} ion moves from 40 V to 50 V. (f) An O^{--} ion moves from 40 V to 60 V. For comparison, include also in your ranking (g) zero change and (h) +10 electron volts of change in kinetic energy. In your ranking, display any cases of equality.

10. What determines the maximum potential to which the dome of a Van de Graaff generator can be raised?

11. **O** **(i)** A metallic sphere A of radius 1 cm is several centimeters away from a metallic spherical shell B of radius 2 cm. Charge 450 nC is placed on A, with no charge on B or anywhere nearby. Next, the two objects are joined by a long, thin, metallic wire (as shown in Fig. 25.20), and finally the wire is removed. How is the charge shared between A and B? (a) 0 on A, 450 nC on B (b) 50 nC on A and 400 nC on B, with equal volume charge densities (c) 90 nC on A and 360 nC on B, with equal surface charge densities (d) 150 nC on A and 300 nC on B (e) 225 nC on A and 225 nC on B (f) 450 nC on A and 0 on B (g) in some other predictable way (h) in some unpredictable way **(ii)** A metallic sphere A of radius 1 cm with charge 450 nC hangs on an insulating thread inside an uncharged thin metallic spherical shell B of radius 2 cm. Next, A is made temporarily to touch the inner surface of B. How is the charge then shared between them? Choose from the same possibilities. Arnold Arons, the only physics teacher yet to have his picture on the cover of *Time* magazine, suggested the idea for this question.

12. Study Figure 23.3 and the accompanying text discussion of charging by induction. When the grounding wire is touched to the rightmost point on the sphere in Figure 23.3c, electrons are drained away from the sphere to leave the sphere positively charged. Suppose the grounding wire is touched to the leftmost point on the sphere instead. Will electrons still drain away, moving closer to the negatively charged rod as they do so? What kind of charge, if any, remains on the sphere?

Problems

WebAssign The Problems from this chapter may be assigned online in WebAssign.

ThomsonNOW™ Sign in at **www.thomsonedu.com** and go to ThomsonNOW to assess your understanding of this chapter's topics with additional quizzing and conceptual questions.

1, 2, 3 denotes straightforward, intermediate, challenging; ☐ denotes full solution available in *Student Solutions Manual/Study Guide;* ▲ denotes coached solution with hints available at **www.thomsonedu.com;** ▭ denotes developing symbolic reasoning; ● denotes asking for qualitative reasoning; ▪ denotes computer useful in solving problem

Section 25.1 Electric Potential and Potential Difference

1. (a) Calculate the speed of a proton that is accelerated from rest through a potential difference of 120 V. (b) Calculate the speed of an electron that is accelerated through the same potential difference.

2. How much work is done (by a battery, generator, or some other source of potential difference) in moving Avogadro's number of electrons from an initial point where the electric potential is 9.00 V to a point where the potential is -5.00 V? (The potential in each case is measured relative to a common reference point.)

Section 25.2 Potential Difference in a Uniform Electric Field

3. The difference in potential between the accelerating plates in the electron gun of a television picture tube is about 25 000 V. If the distance between these plates is 1.50 cm, what is the magnitude of the uniform electric field in this region?

4. A uniform electric field of magnitude 325 V/m is directed in the negative y direction in Figure P25.4. The coordinates of point A are $(-0.200, -0.300)$ m and those of point B are $(0.400, 0.500)$ m. Calculate the potential difference $V_B - V_A$, using the blue path.

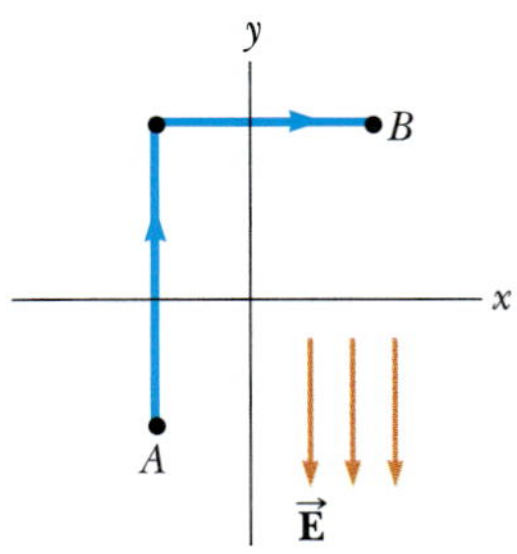

Figure P25.4

5. ▲ An electron moving parallel to the x axis has an initial speed of 3.70×10^6 m/s at the origin. Its speed is reduced to 1.40×10^5 m/s at the point $x = 2.00$ cm. Calculate the potential difference between the origin and that point. Which point is at the higher potential?

6. ● Starting with the definition of work, prove that at every point on an equipotential surface the surface must be perpendicular to the electric field there.

7. **Review problem.** A block having mass m and charge $+Q$ is connected to an insulating spring having constant k. The block lies on a frictionless, insulating horizontal track, and the system is immersed in a uniform electric field of magnitude E directed as shown in Figure P25.7. If the block is released from rest when the spring is unstretched (at $x = 0$), (a) by what maximum amount does the spring expand? (b) What is the equilibrium position of the block? (c) Show that the block's motion is simple harmonic and determine its period. (d) **What If?** Repeat part (a), assuming the coefficient of kinetic friction between block and surface is μ_k.

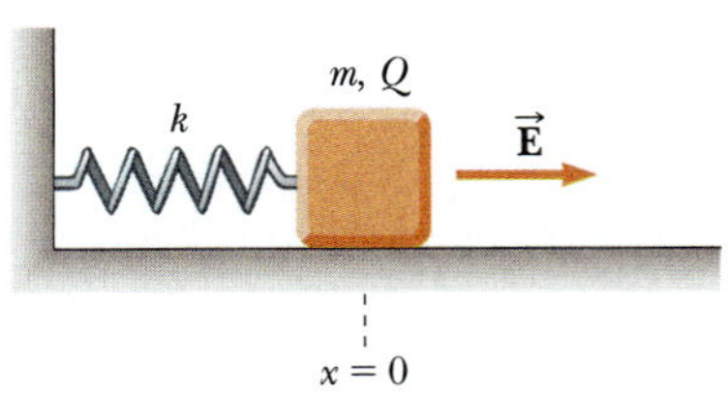

Figure P25.7

8. A particle having charge $q = +2.00\ \mu$C and mass $m = 0.010\,0$ kg is connected to a string that is $L = 1.50$ m long and tied to the pivot point P in Figure P25.8. The particle, string, and pivot point all lie on a frictionless, horizontal table. The particle is released from rest when the string makes an angle $\theta = 60.0°$ with a uniform electric field of magnitude $E = 300$ V/m. Determine the speed of the particle when the string is parallel to the electric field (point a in Fig. P25.8).

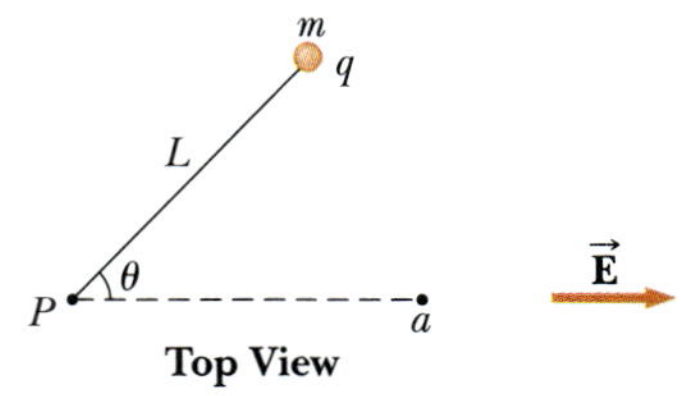

Figure P25.8

9. ● An insulating rod having linear charge density $\lambda = 40.0\ \mu$C/m and linear mass density $\mu = 0.100$ kg/m is released from rest in a uniform electric field $E = 100$ V/m directed perpendicular to the rod (Fig. P25.9). (a) Determine the speed of the rod after it has traveled 2.00 m. (b) **What If?** How does your answer to part (a) change if the electric field is not perpendicular to the rod? Explain.

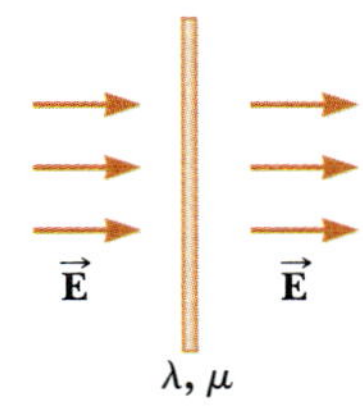

Figure P25.9

Section 25.3 Electric Potential and Potential Energy Due to Point Charges

Note: Unless stated otherwise, assume the reference level of potential is $V = 0$ at $r = \infty$.

10. Given two particles with 2.00-μC charges as shown in Figure P25.10 and a particle with charge $q = 1.28 \times 10^{-18}$ C at the origin, (a) what is the net force exerted by the two 2.00-μC charges on the test charge q? (b) What is the electric field at the origin due to the two 2.00-μC particles? (c) What is the electric potential at the origin due to the two 2.00-μC particles?

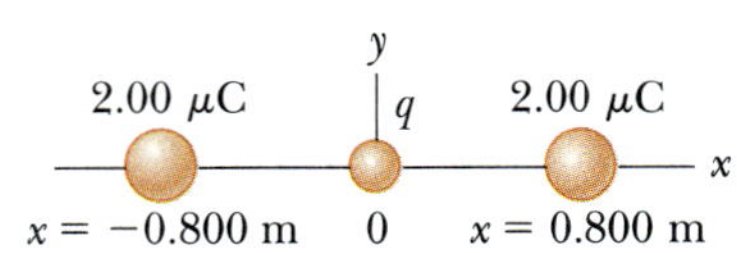

Figure P25.10

11. (a) Find the potential at a distance of 1.00 cm from a proton. (b) What is the potential difference between two points that are 1.00 cm and 2.00 cm from a proton? (c) **What If?** Repeat parts (a) and (b) for an electron.

12. A particle with charge $+q$ is at the origin. A particle with charge $-2q$ is at $x = 2.00$ m on the x axis. (a) For what finite value(s) of x is the electric field zero? (b) For what finite value(s) of x is the electric potential zero?

13. At a certain distance from a charged particle, the magnitude of the electric field is 500 V/m and the electric potential is -3.00 kV. (a) What is the distance to the particle? (b) What is the magnitude of the charge?

14. ● Two charged particles, $Q_1 = +5.00$ nC and $Q_2 = -3.00$ nC, are separated by 35.0 cm. (a) What is the potential energy of the pair? Explain the significance of the algebraic sign of your answer. (b) What is the electric potential at a point midway between the charged particles?

15. The three charged particles in Figure P25.15 are at the vertices of an isosceles triangle. Calculate the electric potential at the midpoint of the base, taking $q = 7.00\ \mu$C.

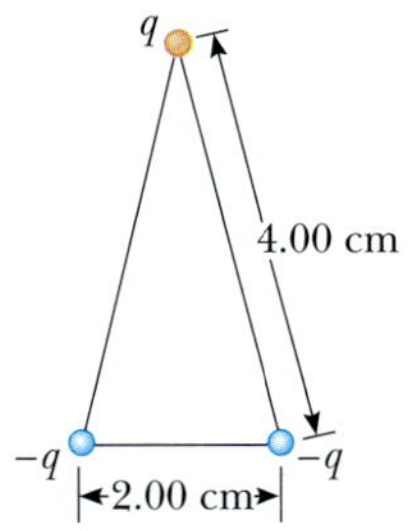

Figure P25.15

16. *Compare this problem with Problem 16 in Chapter 23.* Two charged particles each of magnitude 2.00 μC are located on the x axis. One is at $x = 1.00$ m, and the other is at $x = -1.00$ m. (a) Determine the electric potential on the y axis at $y = 0.500$ m. (b) Calculate the change in electric potential energy of the system as a third charged particle of $-3.00\ \mu$C is brought from infinitely far away to a position on the y axis at $y = 0.500$ m.

17. *Compare this problem with Problem 47 in Chapter 23.* Four identical charged particles ($q = +10.0\ \mu$C) are located on the corners of a rectangle as shown in Figure P23.47. The dimensions of the rectangle are $L = 60.0$ cm and $W = 15.0$ cm. Calculate the change in electric potential energy of the system as the particle at the lower left corner in Figure P23.47 is brought to this position from infinitely far away. Assume the other three particles in Figure P23.47 remain fixed in position.

18. Two charged particles have effects at the origin, described by the expressions

$$8.99 \times 10^9\ \text{N}\cdot\text{m}^2/\text{C}^2\left[-\frac{7 \times 10^{-9}\ \text{C}}{(0.07\ \text{m})^2}\cos 70°\,\hat{\mathbf{i}} - \frac{7 \times 10^{-9}\ \text{C}}{(0.07\ \text{m})^2}\sin 70°\,\hat{\mathbf{j}} + \frac{8 \times 10^{-9}\ \text{C}}{(0.03\ \text{m})^2}\hat{\mathbf{j}}\right]$$

and

$$8.99 \times 10^9\ \text{N}\cdot\text{m}^2/\text{C}^2\left[\frac{7 \times 10^{-9}\ \text{C}}{0.07\ \text{m}} - \frac{8 \times 10^{-9}\ \text{C}}{0.03\ \text{m}}\right]$$

(a) Identify the locations of the particles and the charges on them. (b) Find the force on a particle with charge -16.0 nC placed at the origin. (c) Find the work required to move this third charged particle to the origin from a very distant point.

19. ▲ Show that the amount of work required to assemble four identical charged particles of magnitude Q at the corners of a square of side s is $5.41k_eQ^2/s$.

20. *Compare this problem with Problem 19 in Chapter 23.* Five equal negative charged particles $-q$ are placed symmetrically around a circle of radius R. Calculate the electric potential at the center of the circle.

21. *Compare this problem with Problem 35 in Chapter 23.* Three particles with equal positive charges q are at the corners of an equilateral triangle of side a as shown in Figure P23.35. (a) At what point, if any, in the plane of the particles is the electric potential zero? (b) What is the electric potential at the point P due to the two particles at the base of the triangle?

22. Two charged particles of equal magnitude are located along the y axis equal distances above and below the x axis as shown in Figure P25.22. (a) Plot a graph of the potential at points along the x axis over the interval $-3a < x < 3a$. You should plot the potential in units of k_eQ/a. (b) Let the charge of the particle located at $y = -a$ be negative. Plot the potential along the y axis over the interval $-4a < y < 4a$.

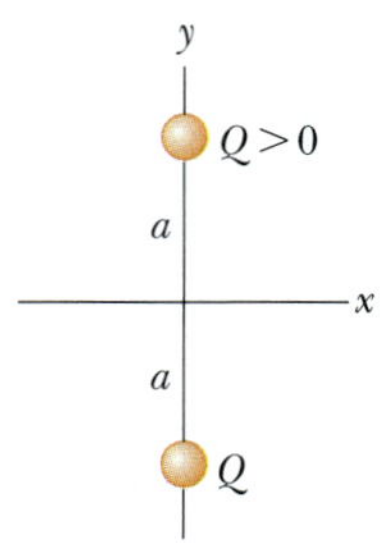

Figure P25.22

23. ● **Review problem.** Two insulating spheres have radii 0.300 cm and 0.500 cm, masses 0.100 kg and 0.700 kg, and uniformly distributed charges of $-2.00\ \mu$C and 3.00 μC. They are released from rest when their centers are separated by 1.00 m. (a) How fast will each be moving when they collide? *Suggestion:* Consider conservation of energy and of linear momentum. (b) **What If?** If the spheres were conductors, would the speeds be greater or less than those calculated in part (a)? Explain.

24. ● **Review problem.** Two insulating spheres have radii r_1 and r_2, masses m_1 and m_2, and uniformly distributed charges $-q_1$ and q_2. They are released from rest when their centers are separated by a distance d. (a) How fast is each moving when they collide? *Suggestion:* Consider conservation of energy and conservation of linear momentum. (b) **What If?** If the spheres were conductors, would their speeds be greater or less than those calculated in part (a)? Explain.

25. **Review problem.** A light, unstressed spring has length d. Two identical particles, each with charge q, are connected to the opposite ends of the spring. The particles are held stationary a distance d apart and then released at the same moment. The system then oscillates on a horizontal, frictionless table. The spring has a bit of internal kinetic friction, so the oscillation is damped. The particles eventually stop vibrating when the distance between them is $3d$. Find the increase in internal energy that appears in the spring during the oscillations. Assume the system of the spring and two charged particles is isolated.

26. In 1911, Ernest Rutherford and his assistants Geiger and Marsden conducted an experiment in which they scattered alpha particles from thin sheets of gold. An alpha particle, having charge $+2e$ and mass 6.64×10^{-27} kg, is a product of certain radioactive decays. The results of the experiment led Rutherford to the idea that most of the mass of an atom is in a very small nucleus, with electrons in orbit around it, his planetary model of the atom. Assume an alpha particle, initially very far from a gold nucleus, is fired with a velocity of 2.00×10^7 m/s directly toward the nucleus (charge $+79e$). How close does the alpha particle get to the nucleus before turning around? Assume the gold nucleus remains stationary.

27. Four identical particles each have charge q and mass m. They are released from rest at the vertices of a square of side L. How fast is each particle moving when their distance from the center of the square doubles?

28. How much work is required to assemble eight identical charged particles, each of magnitude q, at the corners of a cube of side s?

Section 25.4 Obtaining the Value of the Electric Field from the Electric Potential

29. The potential in a region between $x = 0$ and $x = 6.00$ m is $V = a + bx$, where $a = 10.0$ V and $b = -7.00$ V/m. Determine (a) the potential at $x = 0$, 3.00 m, and 6.00 m and (b) the magnitude and direction of the electric field at $x = 0$, 3.00 m, and 6.00 m.

30. The electric potential inside a charged spherical conductor of radius R is given by $V = k_eQ/R$, and the potential outside is given by $V = k_eQ/r$. Using $E_r = -dV/dr$, derive the electric field (a) inside and (b) outside this charge distribution.

31. ▲ Over a certain region of space, the electric potential is $V = 5x - 3x^2y + 2yz^2$. Find the expressions for the x, y, and z components of the electric field over this region. What is the magnitude of the field at the point P that has coordinates (1, 0, −2) m?

32. ● Figure P25.32 shows several equipotential lines, each labeled by its potential in volts. The distance between the lines of the square grid represents 1.00 cm. (a) Is the magnitude of the field larger at A or at B? Explain how you can tell. (b) Explain what you can determine about $\vec{\mathbf{E}}$ at B. (c) Represent what the field looks like by drawing at least eight field lines.

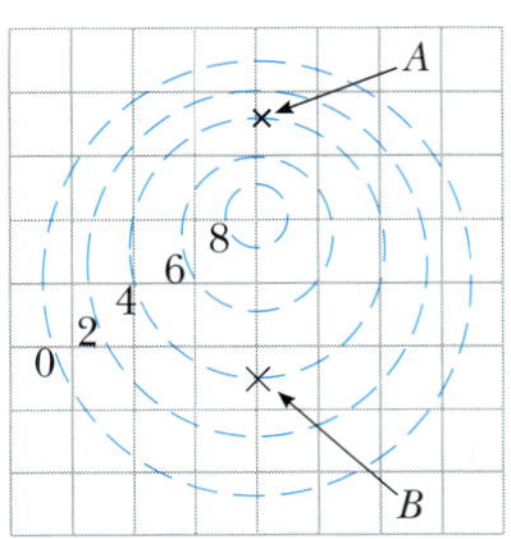

Figure P25.32

33. It is shown in Example 25.7 that the potential at a point P a distance a above one end of a uniformly charged rod of length ℓ lying along the x axis is

$$V = k_e \frac{Q}{\ell} \ln\left(\frac{\ell + \sqrt{a^2 + \ell^2}}{a}\right)$$

Use this result to derive an expression for the y component of the electric field at P. *Suggestion:* Replace a with y.

Section 25.5 Electric Potential Due to Continuous Charge Distributions

34. Consider a ring of radius R with the total charge Q spread uniformly over its perimeter. What is the potential difference between the point at the center of the ring and a point on its axis a distance $2R$ from the center?

35. A rod of length L (Fig. P25.35) lies along the x axis with its left end at the origin. It has a nonuniform charge density $\lambda = \alpha x$, where α is a positive constant. (a) What are the units of α? (b) Calculate the electric potential at A.

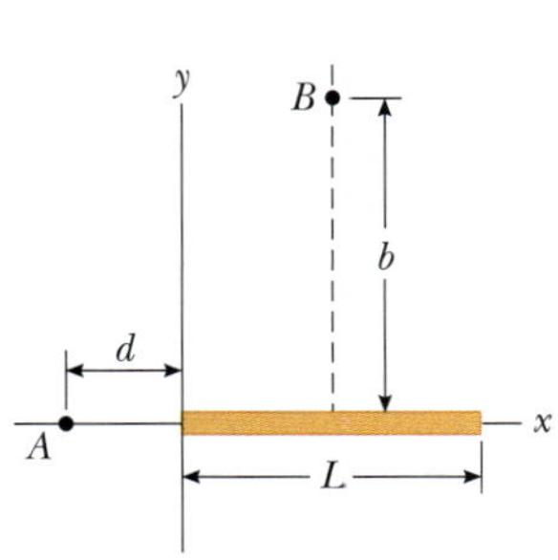

Figure P25.35 Problems 35 and 36.

36. For the arrangement described in Problem 35, calculate the electric potential at point B, which lies on the perpendicular bisector of the rod a distance b above the x axis.

37. *Compare this problem with Problem 27 in Chapter 23.* A uniformly charged insulating rod of length 14.0 cm is bent into the shape of a semicircle as shown in Figure P23.27. The rod has a total charge of −7.50 μC. Find the electric potential at O, the center of the semicircle.

38. A wire having a uniform linear charge density λ is bent into the shape shown in Figure P25.38. Find the electric potential at point O.

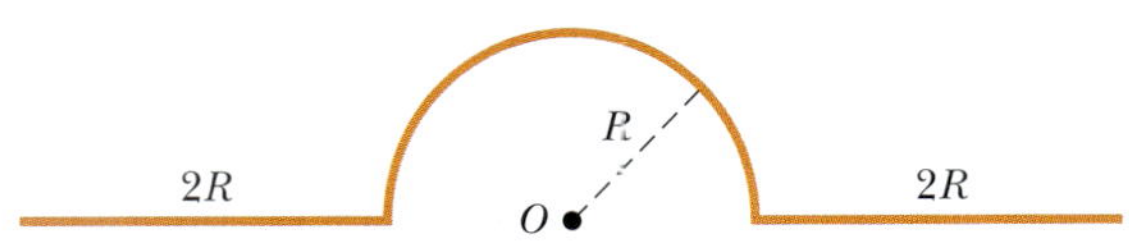

Figure P25.38

Section 25.6 Electric Potential Due to a Charged Conductor

39. ▲ A spherical conductor has a radius of 14.0 cm and charge of 26.0 μC. Calculate the electric field and the electric potential at (a) $r = 10.0$ cm, (b) $r = 20.0$ cm, and (c) $r = 14.0$ cm from the center.

40. How many electrons should be removed from an initially uncharged spherical conductor of radius 0.300 m to produce a potential of 7.50 kV at the surface?

41. The electric field on the surface of an irregularly shaped conductor varies from 56.0 kN/C to 28.0 kN/C. Calculate the local surface charge density at the point on the surface where the radius of curvature of the surface is (a) greatest and (b) smallest.

42. Electric charge can accumulate on an airplane in flight. You may have observed needle-shaped metal extensions on the wing tips and tail of an airplane. Their purpose is to allow charge to leak off before much of it accumulates. The electric field around the needle is much larger than the field around the body of the airplane and can become large enough to produce dielectric breakdown of the air, discharging the airplane. To model this process, assume two charged spherical conductors are connected by a long conducting wire and a charge of 1.20 μC is placed on the combination. One sphere, representing the body of the airplane, has a radius of 6.00 cm, and the other, representing the tip of the needle, has a radius of 2.00 cm. (a) What is the electric potential of each sphere? (b) What is the electric field at the surface of each sphere?

Section 25.8 Applications of Electrostatics

43. Lightning can be studied with a Van de Graaff generator, essentially consisting of a spherical dome on which charge is continuously deposited by a moving belt. Charge can be added until the electric field at the surface of the dome becomes equal to the dielectric strength of air. Any more charge leaks off in sparks as shown in Figure P25.43. Assume the dome has a diameter of 30.0 cm and is surrounded by dry air with dielectric strength 3.00×10^6 V/m. (a) What is the maximum potential of the dome? (b) What is the maximum charge on the dome?

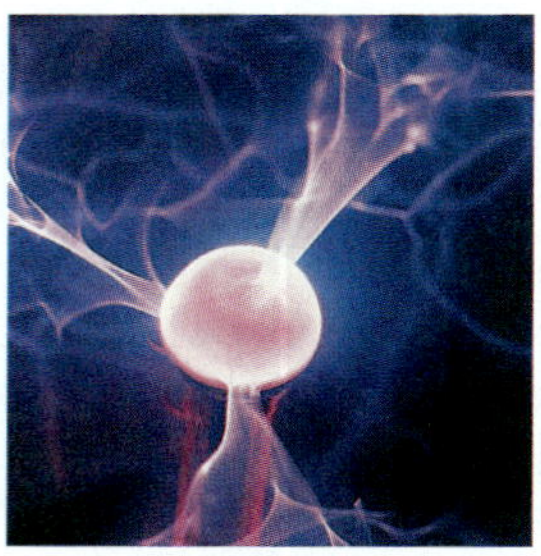

E. R. Degginer/H. Armstrong Roberts

Figure P25.43

44. A Geiger-Mueller tube is a radiation detector that consists of a closed, hollow, metal cylinder (the cathode) of inner radius r_a and a coaxial cylindrical wire (the anode) of radius r_b (Fig. P25.44). The charge per unit length on the anode is λ, and the charge per unit length on the cathode is $-\lambda$. A gas fills the space between the electrodes. When a high-energy elementary particle passes through this space, it can ionize an atom of the gas. The strong electric field makes the resulting ion and electron accelerate in opposite directions. They strike other molecules of the gas to ionize them, producing an avalanche of electrical discharge. The pulse of electric current between the wire and the cylinder is counted by an external circuit. (a) Show that the magnitude of the potential difference between the wire and the cylinder is

$$\Delta V = 2k_e\lambda \ln\left(\frac{r_a}{r_b}\right)$$

(b) Show that the magnitude of the electric field in the space between cathode and anode is

$$E = \frac{\Delta V}{\ln(r_a/r_b)}\left(\frac{1}{r}\right)$$

where r is the distance from the axis of the anode to the point where the field is to be calculated.

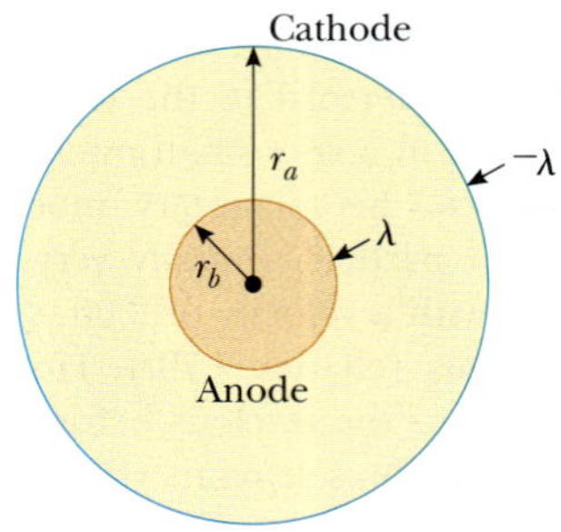

Figure P25.44 Problems 44 and 45.

45. The results of Problem 44 apply also to an electrostatic precipitator (Figs. 25.25 and P25.44). An applied potential difference $\Delta V = V_a - V_b = 50.0$ kV is to produce an electric field of magnitude 5.50 MV/m at the surface of the central wire. Assume the outer cylindrical wall has uniform radius $r_a = 0.850$ m. (a) What should be the radius r_b of the central wire? You will need to solve a transcendental equation. (b) What is the magnitude of the electric field at the outer wall?

Additional Problems

46. ● **Review problem.** From a large distance away, a particle of mass 2.00 g and charge 15.0 μC is fired at $21.0\hat{\mathbf{i}}$ m/s straight toward a second particle, originally stationary but free to move, with mass 5.00 g and charge 8.50 μC. (a) At the instant of closest approach, both particles will be moving at the same velocity. Explain why. (b) Find this velocity. (c) Find the distance of closest approach. (d) Find the velocities of both particles after they separate again.

47. The liquid-drop model of the atomic nucleus suggests high-energy oscillations of certain nuclei can split the nucleus into two unequal fragments plus a few neutrons. The fission products acquire kinetic energy from their mutual Coulomb repulsion. Calculate the electric potential energy (in electron volts) of two spherical fragments from a uranium nucleus having the following charges and radii: $38e$ and 5.50×10^{-15} m, $54e$ and 6.20×10^{-15} m. Assume the charge is distributed uniformly throughout the volume of each spherical fragment and, immediately before separating, each fragment is at rest and their surfaces are in contact. The electrons surrounding the nucleus can be ignored.

48. In fair weather, the electric field in the air at a particular location immediately above the Earth's surface is 120 N/C directed downward. (a) What is the surface charge density on the ground? Is it positive or negative? (b) Imagine the atmosphere is stripped off and the surface charge density is uniform over the planet. What then is the charge of the whole surface of the Earth? (c) What is the Earth's electric potential? (d) What is the difference in potential between the head and the feet of a person 1.75 m tall? (e) Imagine the Moon, with 27.3% of the radius of the Earth, had a charge 27.3% as large, with the same sign. Find the electric force that the Earth would then exert on the Moon. (f) State how the answer to part (e) compares with the gravitational force the Earth exerts on the Moon. (g) A

dust particle of mass 6.00 mg is in the air near the surface of the spherical Earth. What charge must the dust particle carry to be suspended in equilibrium between the electric and gravitational forces exerted on it? Ignore buoyancy. (h) The Earth is not perfectly spherical. It has an equatorial bulge due to its rotation, so the radius of curvature of the ground is slightly larger at the poles than at the equator. Would the dust particle in part (g) require more charge or less charge to be suspended at the equator compared with being suspended at one of the poles? Explain your answer with reference to variations in both the electric force and the gravitational force.

49. The Bohr model of the hydrogen atom states that the single electron can exist only in certain allowed orbits around the proton. The radius of each Bohr orbit is $r = n^2(0.052\ 9\text{ nm})$, where $n = 1, 2, 3, \ldots$. Calculate the electric potential energy of a hydrogen atom when the electron (a) is in the first allowed orbit, with $n = 1$, (b) is in the second allowed orbit, with $n = 2$, and (c) has escaped from the atom, with $r = \infty$. Express your answers in electron volts.

50. On a dry winter day, you scuff your leather-soled shoes across a carpet and get a shock when you extend the tip of one finger toward a metal doorknob. In a dark room, you see a spark perhaps 5 mm long. Make order-of-magnitude estimates of (a) your electric potential and (b) the charge on your body before you touch the doorknob. Explain your reasoning.

51. The electric potential immediately outside a charged conducting sphere is 200 V, and 10.0 cm farther from the center of the sphere the potential is 150 V. (a) Is this information sufficient to determine the charge on the sphere and its radius? Explain. (b) The electric potential immediately outside another charged conducting sphere is 210 V, and 10.0 cm farther from the center the magnitude of the electric field is 400 V/m. Is this information sufficient to determine the charge on the sphere and its radius? Explain.

52. As shown in Figure P25.52, two large, parallel, vertical conducting plates separated by distance d are charged so that their potentials are $+V_0$ and $-V_0$. A small conducting ball of mass m and radius R (where $R \ll d$) is hung midway between the plates. The thread of length L supporting the ball is a conducting wire connected to ground, so the potential of the ball is fixed at $V = 0$. The ball hangs straight down in stable equilibrium when V_0 is sufficiently small. Show that the equilibrium of the ball is unstable if V_0 exceeds the critical value $k_e d^2 mg/(4RL)$. *Suggestion:* Consider the forces on the ball when it is displaced a distance $x \ll L$.

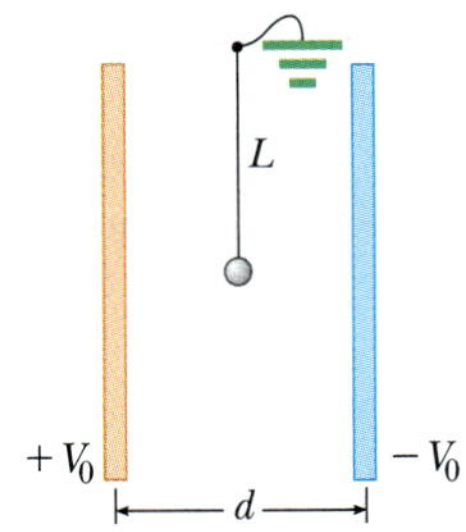

Figure P25.52

53. The electric potential everywhere on the xy plane is given by

$$V = \frac{36}{\sqrt{(x+1)^2 + y^2}} - \frac{45}{\sqrt{x^2 + (y-2)^2}}$$

where V is in volts and x and y are in meters. Determine the position and charge on each of the particles that create this potential.

54. *Compare this problem with Problem 28 in Chapter 23.* (a) A uniformly charged cylindrical shell has total charge Q, radius R, and height h. Determine the electric potential at a point a distance d from the right end of the cylinder as shown in Figure P25.54. *Suggestion:* Use the result of Example 25.5 by treating the cylinder as a collection of ring charges. (b) **What If?** Use the result of Example 25.6 to solve the same problem for a solid cylinder.

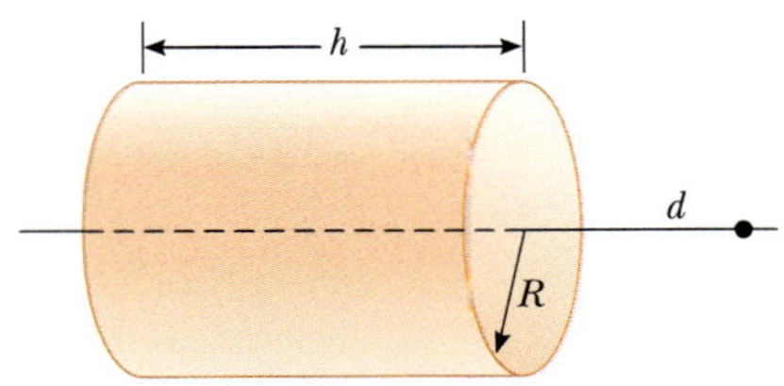

Figure P25.54

55. Calculate the work that must be done to charge a spherical shell of radius R to a total charge Q.

56. ● (a) Use the exact result from Example 25.4 to find the electric potential created by the dipole described at the point $(3a, 0)$. (b) Explain how this answer compares with the result of the approximate expression that is valid when x is much greater than a.

57. ▲ From Gauss's law, the electric field set up by a uniform line of charge is

$$\vec{\mathbf{E}} = \left(\frac{\lambda}{2\pi\epsilon_0 r}\right)\hat{\mathbf{r}}$$

where $\hat{\mathbf{r}}$ is a unit vector pointing radially away from the line and λ is the linear charge density along the line. Derive an expression for the potential difference between $r = r_1$ and $r = r_2$.

58. Four balls, each with mass m, are connected by four nonconducting strings to form a square with side a as shown in Figure P25.58. The assembly is placed on a horizontal, nonconducting, frictionless surface. Balls 1 and 2 each have charge q, and balls 3 and 4 are uncharged. Find the maximum speed of balls 3 and 4 after the string connecting balls 1 and 2 is cut.

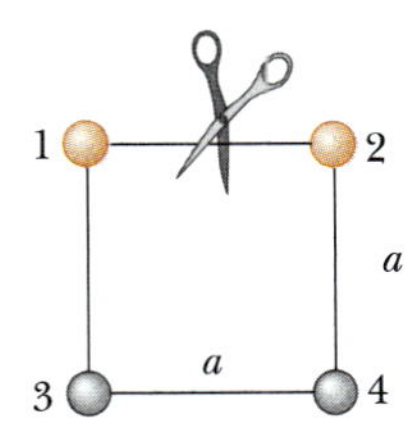

Figure P25.58

59. The x axis is the symmetry axis of a stationary, uniformly charged ring of radius R and charge Q (Fig. P25.59). A particle with charge Q and mass M is located initially at

the center of the ring. When it is displaced slightly, the particle accelerates along the x axis to infinity. Show that the ultimate speed of the particle is

$$v = \left(\frac{2k_eQ^2}{MR}\right)^{1/2}$$

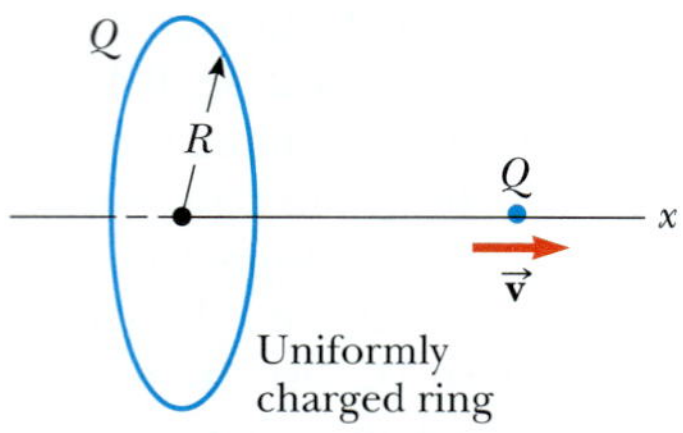

Figure P25.59

60. The thin, uniformly charged rod shown in Figure P25.60 has a linear charge density λ. Find an expression for the electric potential at P.

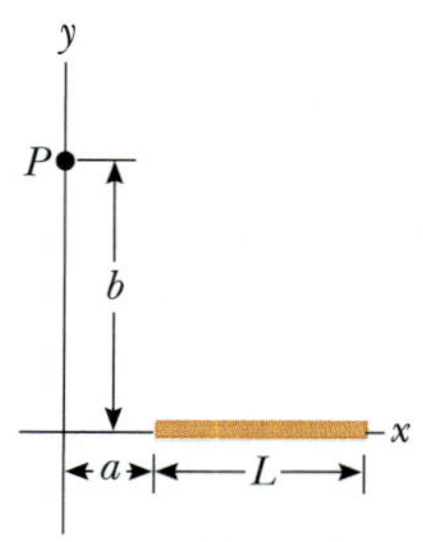

Figure P25.60

61. An electric dipole is located along the y axis as shown in Figure P25.61. The magnitude of its electric dipole moment is defined as $p = 2qa$. (a) At a point P, which is far from the dipole ($r \gg a$), show that the electric potential is

$$V = \frac{k_e p \cos\theta}{r^2}$$

(b) Calculate the radial component E_r and the perpendicular component E_θ of the associated electric field. Note that $E_\theta = -(1/r)(\partial V/\partial\theta)$. Do these results seem reasonable for $\theta = 90°$ and $0°$? For $r = 0$? (c) For the dipole arrangement shown, express V in terms of Cartesian coordinates using $r = (x^2 + y^2)^{1/2}$ and

$$\cos\theta = \frac{y}{(x^2 + y^2)^{1/2}}$$

Using these results and again taking $r \gg a$, calculate the field components E_x and E_y.

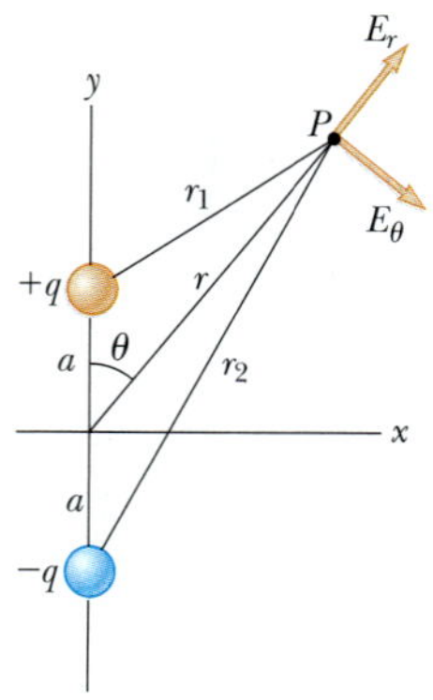

Figure P25.61

62. A solid sphere of radius R has a uniform charge density ρ and total charge Q. Derive an expression for its total electric potential energy. *Suggestion:* Imagine the sphere is constructed by adding successive layers of concentric shells of charge $dq = (4\pi r^2\, dr)\rho$ and use $dU = V\, dq$.

63. A disk of radius R (Fig. P25.63) has a nonuniform surface charge density $\sigma = Cr$, where C is a constant and r is measured from the center of the disk to a point on the surface of the disk. Find (by direct integration) the potential at P.

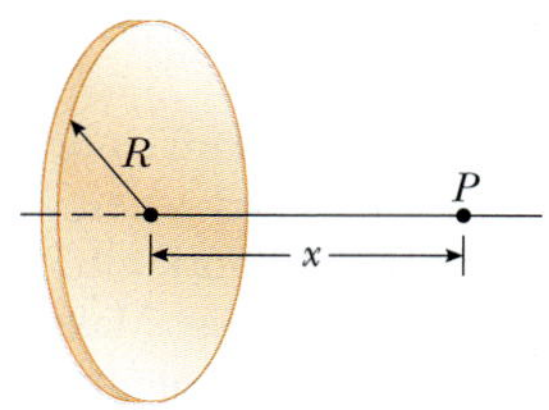

Figure P25.63

64. ● A uniformly charged filament lies along the x axis between $x = a = 1.00$ m and $x = a + \ell = 3.00$ m as shown in Figure 23.15. The total charge on the filament is 1.60 nC. Calculate successive approximations for the electric potential at the origin by modeling the filament as (a) a single charged particle at $x = 2.00$ m, (b) two 0.800-nC charged particles at $x = 1.5$ m and $x = 2.5$ m, and (c) four 0.400-nC charged particles at $x = 1.25$ m, $x = 1.75$ m, $x = 2.25$ m, and $x = 2.75$ m. Next, write and execute a computer program that will reproduce the results of parts (a), (b), and (c) and extend your calculation to (d) 32 and (e) 64 equally spaced charged particles. (f) Explain how the results compare with the potential given by the exact expression

$$V = \frac{k_eQ}{\ell}\ln\left(\frac{\ell + a}{a}\right)$$

65. Two parallel plates having charges of equal magnitude but opposite sign are separated by 12.0 cm. Each plate has a surface charge density of 36.0 nC/m^2. A proton is released from rest at the positive plate. Determine (a) the potential difference between the plates, (b) the kinetic energy of the proton when it reaches the negative plate, (c) the speed of the proton just before it strikes the negative plate, (d) the acceleration of the proton, and (e) the force on the proton. (f) From the force, find the magnitude of the electric field and show that it is equal to the electric field found from the charge densities on the plates.

66. A particle with charge q is located at $x = -R$, and a particle with charge $-2q$ is located at the origin. Prove that the equipotential surface that has zero potential is a sphere centered at $(-4R/3, 0, 0)$ and having a radius $r = 2R/3$.

67. When an uncharged conducting sphere of radius a is placed at the origin of an xyz coordinate system that lies in an initially uniform electric field $\vec{\mathbf{E}} = E_0\hat{\mathbf{k}}$, the resulting electric potential is $V(x, y, z) = V_0$ for points inside the sphere and

$$V(x, y, z) = V_0 - E_0 z + \frac{E_0 a^3 z}{(x^2 + y^2 + z^2)^{3/2}}$$

for points outside the sphere, where V_0 is the (constant) electric potential on the conductor. Use this equation to determine the x, y, and z components of the resulting electric field.

Answers to Quick Quizzes

25.1 **(i),** (b). When moving straight from Ⓐ to Ⓑ, $\vec{\mathbf{E}}$ and $d\vec{\mathbf{s}}$ both point toward the right. Therefore, the dot product $\vec{\mathbf{E}} \cdot d\vec{\mathbf{s}}$ in Equation 25.3 is positive and ΔV is negative. **(ii),** (a). From Equation 25.3, $\Delta U = q_0\,\Delta V$, so if a negative test charge is moved through a negative potential difference, the change in potential energy is positive. Work must be done to move the charge in the direction opposite to the electric force on it.

25.2 Ⓑ to Ⓒ, Ⓒ to Ⓓ, Ⓐ to Ⓑ, Ⓓ to Ⓔ. Moving from Ⓑ to Ⓒ decreases the electric potential by 2 V, so the electric field performs 2 J of work on each coulomb of positive charge that moves. Moving from Ⓒ to Ⓓ decreases the electric potential by 1 V, so 1 J of work is done by the field. It takes no work to move the charge from Ⓐ to Ⓑ because the electric potential does not change. Moving from Ⓓ to Ⓔ increases the electric potential by 1 V; therefore, the field does −1 J of work per unit of positive charge that moves.

25.3 **(i),** (c). The potential is established only by the source charge and is independent of the test charge. **(ii),** (a). The potential energy of the two-charge system is initially negative due to the product of charges of opposite sign in Equation 25.13. When the sign of q_2 is changed, both charges are negative and the potential energy of the system is positive.

25.4 (a). If the potential is constant (zero in this case), its derivative along this direction is zero.

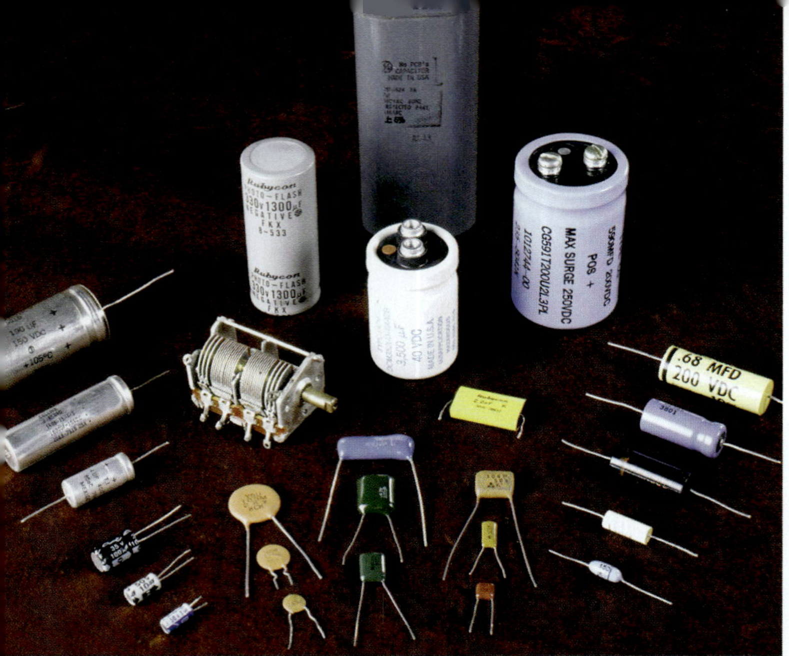

All these devices are capacitors, which store electric charge and energy. A capacitor is one type of circuit element that we can combine with others to make electric circuits. (Paul Silverman/Fundamental Photographs)

26 Capacitance and Dielectrics

In this chapter, we introduce the first of three simple *circuit elements* that can be connected with wires to form an electric circuit. Electric circuits are the basis for the vast majority of the devices used in our society. Here we shall discuss *capacitors,* devices that store electric charge. This discussion is followed by the study of *resistors* in Chapter 27 and *inductors* in Chapter 32. In later chapters, we will study more sophisticated circuit elements such as *diodes* and *transistors.*

Capacitors are commonly used in a variety of electric circuits. For instance, they are used to tune the frequency of radio receivers, as filters in power supplies, to eliminate sparking in automobile ignition systems, and as energy-storing devices in electronic flash units.

PITFALL PREVENTION 26.1
Capacitance Is a Capacity

To understand capacitance, think of similar notions that use a similar word. The *capacity* of a milk carton is the volume of milk it can store. The *heat capacity* of an object is the amount of energy an object can store per unit of temperature difference. The *capacitance* of a capacitor is the amount of charge the capacitor can store per unit of potential difference.

26.1 Definition of Capacitance

Consider two conductors as shown in Figure 26.1. Such a combination of two conductors is called a **capacitor.** The conductors are called *plates.* If the conductors carry charges of equal magnitude and opposite sign, a potential difference ΔV exists between them.

What determines how much charge is on the plates of a capacitor for a given voltage? Experiments show that the quantity of charge Q on a capacitor[1] is linearly pro-

[1] Although the total charge on the capacitor is zero (because there is as much excess positive charge on one conductor as there is excess negative charge on the other), it is common practice to refer to the magnitude of the charge on either conductor as "the charge on the capacitor."

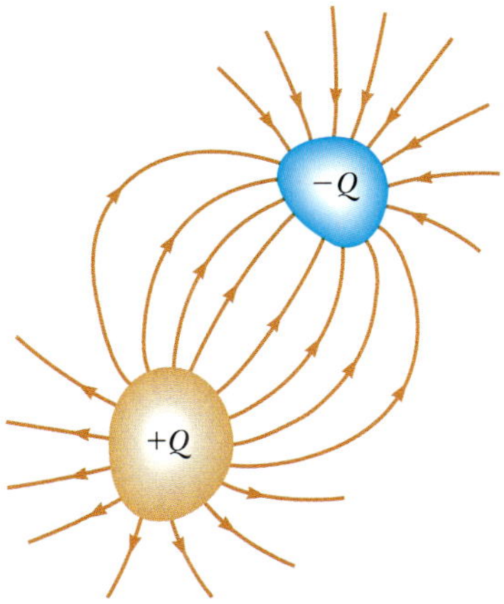

Figure 26.1 A capacitor consists of two conductors. When the capacitor is charged, the conductors carry charges of equal magnitude and opposite sign.

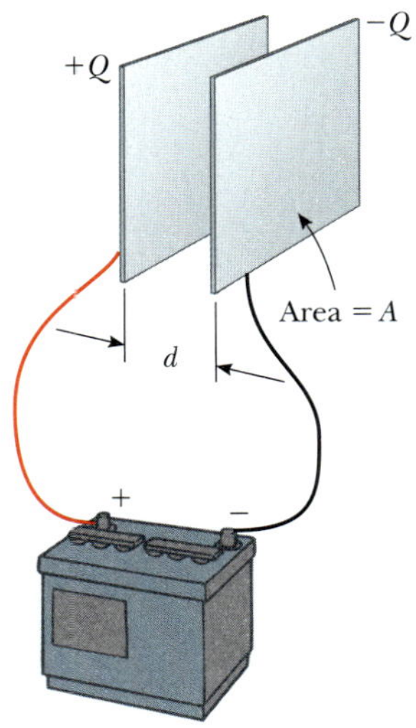

Figure 26.2 A parallel-plate capacitor consists of two parallel conducting plates, each of area A, separated by a distance d. When the capacitor is charged by connecting the plates to the terminals of a battery, the plates carry equal amounts of charge. One plate carries positive charge, and the other carries negative charge.

portional to the potential difference between the conductors; that is, $Q \propto \Delta V$. The proportionality constant depends on the shape and separation of the conductors.[2] This relationship can be written as $Q = C\Delta V$ if we define capacitance as follows:

The **capacitance** C of a capacitor is defined as the ratio of the magnitude of the charge on either conductor to the magnitude of the potential difference between the conductors:

$$C \equiv \frac{Q}{\Delta V} \tag{26.1}$$

◀ Definition of capacitance

By definition *capacitance is always a positive quantity.* Furthermore, the charge Q and the potential difference ΔV are always expressed in Equation 26.1 as positive quantities.

From Equation 26.1, we see that capacitance has SI units of coulombs per volt. Named in honor of Michael Faraday, the SI unit of capacitance is the **farad** (F):

$$1 \text{ F} = 1 \text{ C/V}$$

The farad is a very large unit of capacitance. In practice, typical devices have capacitances ranging from microfarads (10^{-6} F) to picofarads (10^{-12} F). We shall use the symbol μF to represent microfarads. In practice, to avoid the use of Greek letters, physical capacitors are often labeled "mF" for microfarads and "mmF" for micromicrofarads or, equivalently, "pF" for picofarads.

Let's consider a capacitor formed from a pair of parallel plates as shown in Figure 26.2. Each plate is connected to one terminal of a battery, which acts as a source of potential difference. If the capacitor is initially uncharged, the battery establishes an electric field in the connecting wires when the connections are made. Let's focus on the plate connected to the negative terminal of the battery. The electric field in the wire applies a force on electrons in the wire immediately outside this plate; this force causes the electrons to move onto the plate. The movement continues until the plate, the wire, and the terminal are all at the same electric potential. Once this equilibrium situation is attained, a potential difference no longer exists between the terminal and the plate; as a result no electric field is present in the wire and the electrons stop moving. The plate now carries a negative charge. A similar process occurs at the other capacitor plate, where electrons move from the plate to the wire, leaving the plate positively charged. In this final configuration, the potential difference across the capacitor plates is the same as that between the terminals of the battery.

PITFALL PREVENTION 26.2
Potential Difference Is ΔV, Not V

We use the symbol ΔV for the potential difference across a circuit element or a device because this notation is consistent with our definition of potential difference and with the meaning of the delta sign. It is a common but confusing practice to use the symbol V without the delta sign for both a potential and a potential difference! Keep that in mind if you consult other texts.

PITFALL PREVENTION 26.3
Too Many Cs

Do not confuse an italic C for capacitance with a nonitalic C for the unit coulomb.

[2] The proportionality between ΔV and Q can be proven from Coulomb's law or by experiment.

Quick Quiz 26.1 A capacitor stores charge Q at a potential difference ΔV. What happens if the voltage applied to the capacitor by a battery is doubled to $2\Delta V$? (a) The capacitance falls to half its initial value, and the charge remains the same. (b) The capacitance and the charge both fall to half their initial values. (c) The capacitance and the charge both double. (d) The capacitance remains the same, and the charge doubles.

26.2 Calculating Capacitance

We can derive an expression for the capacitance of a pair of oppositely charged conductors having a charge of magnitude Q in the following manner. First we calculate the potential difference using the techniques described in Chapter 25. We then use the expression $C = Q/\Delta V$ to evaluate the capacitance. The calculation is relatively easy if the geometry of the capacitor is simple.

Although the most common situation is that of two conductors, a single conductor also has a capacitance. For example, imagine a spherical, charged conductor. The electric field lines around this conductor are exactly the same as if there were a conducting, spherical shell of infinite radius, concentric with the sphere and carrying a charge of the same magnitude but opposite sign. Therefore, we can identify the imaginary shell as the second conductor of a two-conductor capacitor. The electric potential of the sphere of radius a is simply k_eQ/a, and setting $V = 0$ for the infinitely large shell gives

Capacitance of an isolated charged sphere ▶

$$C = \frac{Q}{\Delta V} = \frac{Q}{k_eQ/a} = \frac{a}{k_e} = 4\pi\epsilon_0 a \qquad \textbf{(26.2)}$$

This expression shows that the capacitance of an isolated, charged sphere is proportional to its radius and is independent of both the charge on the sphere and the potential difference.

The capacitance of a pair of conductors is illustrated below with three familiar geometries, namely, parallel plates, concentric cylinders, and concentric spheres. In these calculations, we assume the charged conductors are separated by a vacuum.

Parallel-Plate Capacitors

Two parallel, metallic plates of equal area A are separated by a distance d as shown in Figure 26.2. One plate carries a charge $+Q$, and the other carries a charge $-Q$. The surface charge density on each plate is $\sigma = Q/A$. If the plates are very close together (in comparison with their length and width), we can assume the electric field is uniform between the plates and zero elsewhere. According to the **What If?** feature of Example 24.5, the value of the electric field between the plates is

$$E = \frac{\sigma}{\epsilon_0} = \frac{Q}{\epsilon_0 A}$$

Because the field between the plates is uniform, the magnitude of the potential difference between the plates equals Ed (see Eq. 25.6); therefore,

$$\Delta V = Ed = \frac{Qd}{\epsilon_0 A}$$

Substituting this result into Equation 26.1, we find that the capacitance is

$$C = \frac{Q}{\Delta V} = \frac{Q}{Qd/\epsilon_0 A}$$

Capacitance of parallel plates ▶

$$C = \frac{\epsilon_0 A}{d} \qquad \textbf{(26.3)}$$

That is, **the capacitance of a parallel-plate capacitor is proportional to the area of its plates and inversely proportional to the plate separation.**

Let's consider how the geometry of these conductors influences the capacity of the pair of plates to store charge. As a capacitor is being charged by a battery, electrons flow into the negative plate and out of the positive plate. If the capacitor plates are large, the accumulated charges are able to distribute themselves over a substantial area and the amount of charge that can be stored on a plate for a given potential difference increases as the plate area is increased. Therefore, it is reasonable that the capacitance is proportional to the plate area A as in Equation 26.3.

Now consider the region that separates the plates. Imagine moving the plates closer together. Consider the situation before any charges have had a chance to move in response to this change. Because no charges have moved, the electric field between the plates has the same value but extends over a shorter distance. Therefore, the magnitude of the potential difference between the plates $\Delta V = Ed$ (Eq. 25.6) is smaller. The difference between this new capacitor voltage and the terminal voltage of the battery appears as a potential difference across the wires connecting the battery to the capacitor, resulting in an electric field in the wires that drives more charge onto the plates and increases the potential difference between the plates. When the potential difference between the plates again matches that of the battery, the flow of charge stops. Therefore, moving the plates closer together causes the charge on the capacitor to increase. If d is increased, the charge decreases. As a result, the inverse relationship between C and d in Equation 26.3 is reasonable.

Quick Quiz 26.2 Many computer keyboard buttons are constructed of capacitors as shown in Figure 26.3. When a key is pushed down, the soft insulator between the movable plate and the fixed plate is compressed. When the key is pressed, what happens to the capacitance? (a) It increases. (b) It decreases. (c) It changes in a way you cannot determine because the electric circuit connected to the keyboard button may cause a change in ΔV.

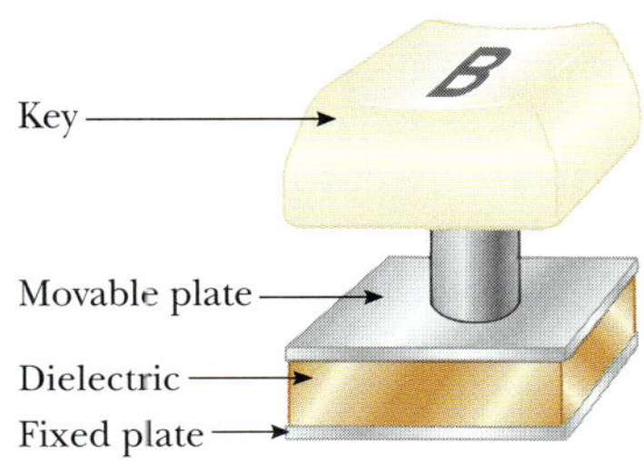

Figure 26.3 (Quick Quiz 26.2) One type of computer keyboard button.

EXAMPLE 26.1 **The Cylindrical Capacitor**

A solid, cylindrical conductor of radius a and charge Q is coaxial with a cylindrical shell of negligible thickness, radius $b > a$, and charge $-Q$ (Fig. 26.4a). Find the capacitance of this cylindrical capacitor if its length is ℓ.

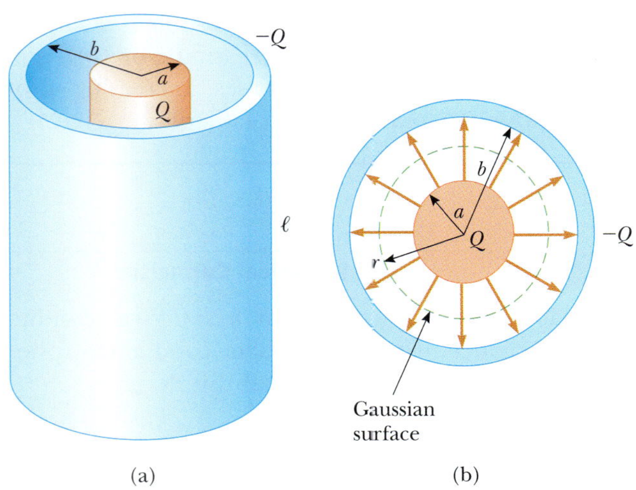

Figure 26.4 (Example 26.1) (a) A cylindrical capacitor consists of a solid cylindrical conductor of radius a and length ℓ surrounded by a coaxial cylindrical shell of radius b. (b) End view. The electric field lines are radial. The dashed line represents the end of the cylindrical gaussian surface of radius r and length ℓ.

SOLUTION

Conceptualize Recall that any pair of conductors qualifies as a capacitor, so the system described in this example therefore qualifies. Figure 26.4b helps visualize the electric field between the conductors.

Categorize Because of the cylindrical symmetry of the system, we can use results from previous studies of cylindrical systems to find the capacitance.

Analyze Assuming ℓ is much greater than a and b, we can neglect end effects. In this case, the electric field is perpendicular to the long axis of the cylinders and is confined to the region between them (Fig. 26.4b).

Write an expression for the potential difference between the two cylinders from Equation 25.3:

$$V_b - V_a = -\int_a^b \vec{\mathbf{E}} \cdot d\vec{\mathbf{s}}$$

Apply Equation 24.7 for the electric field outside a cylindrically symmetric charge distribution and notice from Figure 26.4b that $\vec{\mathbf{E}}$ is parallel to $d\vec{\mathbf{s}}$ along a radial line:

$$V_b - V_a = -\int_a^b E_r\,dr = -2k_e\lambda \int_a^b \frac{dr}{r} = -2k_e\lambda \ln\left(\frac{b}{a}\right)$$

Substitute the absolute value of ΔV into Equation 26.1 and use $\lambda = Q/\ell$:

$$C = \frac{Q}{\Delta V} = \frac{Q}{(2k_e Q/\ell)\ln(b/a)} = \frac{\ell}{2k_e \ln(b/a)} \quad \textbf{(26.4)}$$

Finalize The capacitance is proportional to the length of the cylinders. As you might expect, the capacitance also depends on the radii of the two cylindrical conductors. Equation 26.4 shows that the capacitance per unit length of a combination of concentric cylindrical conductors is

$$\frac{C}{\ell} = \frac{1}{2k_e \ln(b/a)} \quad \textbf{(26.5)}$$

An example of this type of geometric arrangement is a *coaxial cable*, which consists of two concentric cylindrical conductors separated by an insulator. You probably have a coaxial cable attached to your television set or VCR if you are a subscriber to cable television. The coaxial cable is especially useful for shielding electrical signals from any possible external influences.

What If? Suppose $b = 2.00a$ for the cylindrical capacitor. You would like to increase the capacitance, and you can do so by choosing to increase either ℓ by 10% or a by 10%. Which choice is more effective at increasing the capacitance?

Answer According to Equation 26.4, C is proportional to ℓ, so increasing ℓ by 10% results in a 10% increase in C. For the result of the change in a, let's use Equation 26.4 to set up a ratio of the capacitance C' for the enlarged cylinder radius a' to the original capacitance:

$$\frac{C'}{C} = \frac{\ell/2k_e \ln(b/a')}{\ell/2k_e \ln(b/a)} = \frac{\ln(b/a)}{\ln(b/a')}$$

We now substitute $b = 2.00a$ and $a' = 1.10a$, representing a 10% increase in a:

$$\frac{C'}{C} = \frac{\ln(2.00a/a)}{\ln(2.00a/1.10a)} = \frac{\ln 2.00}{\ln 1.82} = 1.16$$

which corresponds to a 16% increase in capacitance. Therefore, it is more effective to increase a than to increase ℓ.

Note two more extensions of this problem. First, it is advantageous to increase a only for a range of relationships between a and b. If $b > 2.85a$, increasing ℓ by 10% is more effective than increasing a (see Problem 66). Second, if b decreases, the capacitance increases. Increasing a or decreasing b has the effect of bringing the plates closer together, which increases the capacitance.

EXAMPLE 26.2 The Spherical Capacitor

A spherical capacitor consists of a spherical conducting shell of radius b and charge $-Q$ concentric with a smaller conducting sphere of radius a and charge Q (Fig. 26.5). Find the capacitance of this device.

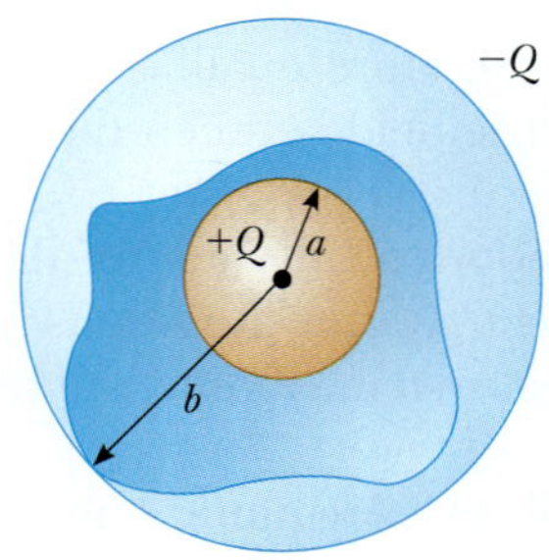

Figure 26.5 (Example 26.2) A spherical capacitor consists of an inner sphere of radius a surrounded by a concentric spherical shell of radius b. The electric field between the spheres is directed radially outward when the inner sphere is positively charged.

SOLUTION

Conceptualize As with Example 26.1, this system involves a pair of conductors and qualifies as a capacitor.

Categorize Because of the spherical symmetry of the system, we can use results from previous studies of spherical systems to find the capacitance.

Analyze As shown in Chapter 24, the magnitude of the electric field outside a spherically symmetric charge distribution is radial and given by the expression $E = k_e Q/r^2$. In this case, this result applies to the field *between* the spheres ($a < r < b$).

Write an expression for the potential difference between the two conductors from Equation 25.3:

$$V_b - V_a = -\int_a^b \vec{\mathbf{E}} \cdot d\vec{\mathbf{s}}$$

Apply the result of Example 24.3 for the electric field outside a spherically symmetric charge distribution and note that $\vec{\mathbf{E}}$ is parallel to $d\vec{\mathbf{s}}$ along a radial line:

$$V_b - V_a = -\int_a^b E_r\,dr = -k_eQ\int_a^b \frac{dr}{r^2} = k_eQ\left[\frac{1}{r}\right]_a^b$$

$$(1)\quad V_b - V_a = k_eQ\left(\frac{1}{b} - \frac{1}{a}\right) = k_eQ\frac{a-b}{ab}$$

Substitute the absolute value of ΔV into Equation 26.1:

$$C = \frac{Q}{\Delta V} = \frac{Q}{|V_b - V_a|} = \frac{ab}{k_e(b-a)} \qquad (26.6)$$

Finalize The potential difference between the spheres in Equation (1) is negative because of the choice of signs on the spheres. Therefore, in Equation 26.6, when we take the absolute value, we change $a - b$ to $b - a$. The result is a positive number because $b > a$.

What If? If the radius b of the outer sphere approaches infinity, what does the capacitance become?

Answer In Equation 26.6, we let $b \to \infty$:

$$C = \lim_{b\to\infty}\frac{ab}{k_e(b-a)} = \frac{ab}{k_e(b)} = \frac{a}{k_e} = 4\pi\epsilon_0 a$$

Notice that this expression is the same as Equation 26.2, the capacitance of an isolated spherical conductor.

26.3 Combinations of Capacitors

Two or more capacitors often are combined in electric circuits. We can calculate the equivalent capacitance of certain combinations using methods described in this section. Throughout this section, we assume the capacitors to be combined are initially uncharged.

In studying electric circuits, we use a simplified pictorial representation called a **circuit diagram.** Such a diagram uses **circuit symbols** to represent various circuit elements. The circuit symbols are connected by straight lines that represent the wires between the circuit elements. The circuit symbols for capacitors, batteries, and switches as well as the color codes used for them in this text are given in Figure 26.6. The symbol for the capacitor reflects the geometry of the most common model for a capacitor, a pair of parallel plates. The positive terminal of the battery is at the higher potential and is represented in the circuit symbol by the longer line.

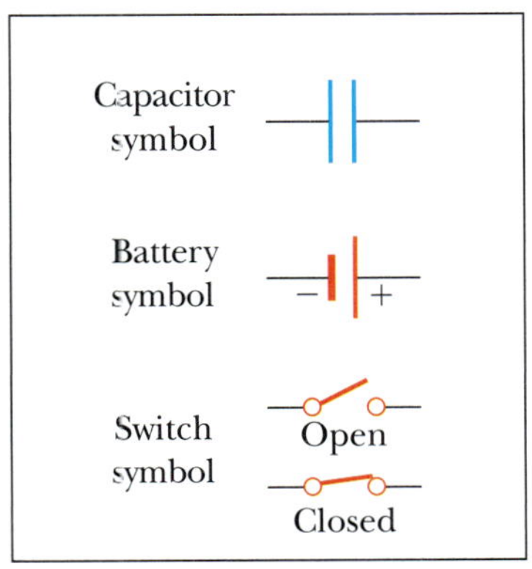

Figure 26.6 Circuit symbols for capacitors, batteries, and switches. Notice that capacitors are in blue and batteries and switches are in red. The closed switch can carry current, whereas the open one cannot.

Parallel Combination

Two capacitors connected as shown in Active Figure 26.7a (page 728) are known as a **parallel combination** of capacitors. Active Figure 26.7b shows a circuit diagram for this combination of capacitors. The left plates of the capacitors are connected to the positive terminal of the battery by a conducting wire and are therefore both at the same electric potential as the positive terminal. Likewise, the right plates are connected to the negative terminal and so are both at the same potential as the negative terminal. Therefore, **the individual potential differences across capacitors connected in parallel are the same and are equal to the potential difference applied across the combination.** That is,

$$\Delta V_1 = \Delta V_2 = \Delta V$$

where ΔV is the battery terminal voltage.

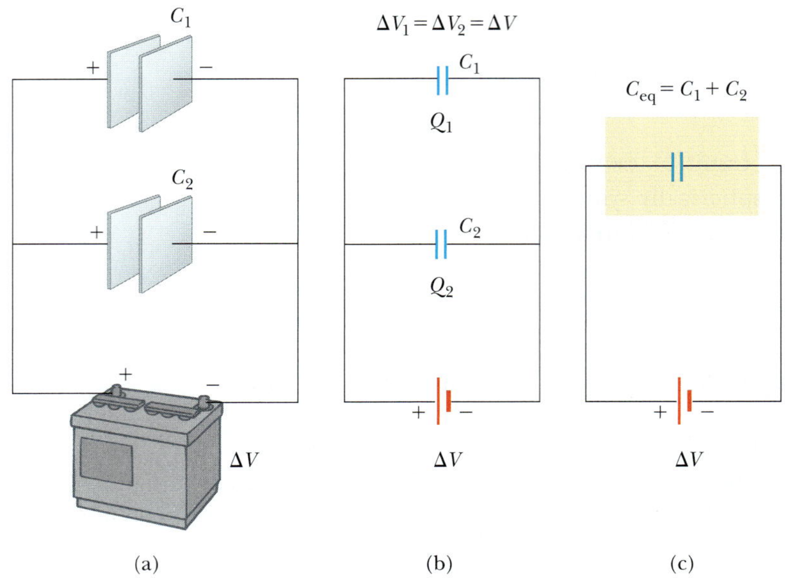

ACTIVE FIGURE 26.7

(a) A parallel combination of two capacitors in an electric circuit in which the potential difference across the battery terminals is ΔV. (b) The circuit diagram for the parallel combination. (c) The equivalent capacitance is given by Equation 26.8.

Sign in at www.thomsonedu.com and go to ThomsonNOW to adjust the battery voltage and the individual capacitances and see the resulting charges and voltages on the capacitors. You can combine up to four capacitors in parallel.

After the battery is attached to the circuit, the capacitors quickly reach their maximum charge. Let's call the maximum charges on the two capacitors Q_1 and Q_2. The *total charge* Q_{tot} stored by the two capacitors is

$$Q_{tot} = Q_1 + Q_2 \qquad \textbf{(26.7)}$$

That is, **the total charge on capacitors connected in parallel is the sum of the charges on the individual capacitors.**

Suppose you wish to replace these two capacitors by one *equivalent capacitor* having a capacitance C_{eq} as in Active Figure 26.7c. The effect this equivalent capacitor has on the circuit must be exactly the same as the effect of the combination of the two individual capacitors. That is, the equivalent capacitor must store charge Q_{tot} when connected to the battery. Active Figure 26.7c shows that the voltage across the equivalent capacitor is ΔV because the equivalent capacitor is connected directly across the battery terminals. Therefore, for the equivalent capacitor,

$$Q_{tot} = C_{eq}\,\Delta V$$

Substituting for the charges in Equation 26.7 gives

$$C_{eq}\,\Delta V = C_1\,\Delta V_1 + C_2\,\Delta V_2$$

$$C_{eq} = C_1 + C_2 \quad \text{(parallel combination)}$$

where we have canceled the voltages because they are all the same. If this treatment is extended to three or more capacitors connected in parallel, the equivalent capacitance is found to be

Capacitors in parallel ▶

$$C_{eq} = C_1 + C_2 + C_3 + \cdots \quad \text{(parallel combination)} \qquad \textbf{(26.8)}$$

Therefore, **the equivalent capacitance of a parallel combination of capacitors is (1) the algebraic sum of the individual capacitances and (2) greater than any of the individual capacitances.** Statement (2) makes sense because we are essentially combining the areas of all the capacitor plates when they are connected with conducting wire, and capacitance of parallel plates is proportional to area (Eq. 26.3).

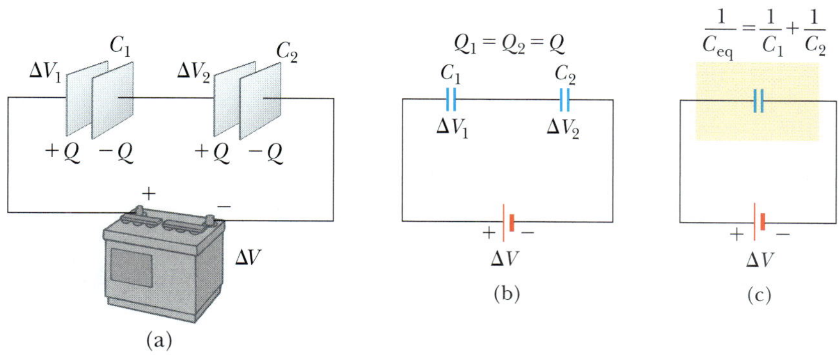

ACTIVE FIGURE 26.8

(a) A series combination of two capacitors. The charges on the two capacitors are the same. (b) The circuit diagram for the series combination. (c) The equivalent capacitance can be calculated from Equation 26.10.

Sign in at www.thomsonedu.com and go to ThomsonNOW to adjust the battery voltage and the individual capacitances and see the resulting charges and voltages on the capacitors. You can combine up to four capacitors in series.

Series Combination

Two capacitors connected as shown in Active Figure 26.8a and the equivalent circuit diagram in Active Figure 26.8b are known as a **series combination** of capacitors. The left plate of capacitor 1 and the right plate of capacitor 2 are connected to the terminals of a battery. The other two plates are connected to each other and to nothing else; hence, they form an isolated system that is initially uncharged and must continue to have zero net charge. To analyze this combination, let's first consider the uncharged capacitors and then follow what happens immediately after a battery is connected to the circuit. When the battery is connected, electrons are transferred out of the left plate of C_1 and into the right plate of C_2. As this negative charge accumulates on the right plate of C_2, an equivalent amount of negative charge is forced off the left plate of C_2, and this left plate therefore has an excess positive charge. The negative charge leaving the left plate of C_2 causes negative charges to accumulate on the right plate of C_1. As a result, all the right plates end up with a charge $-Q$ and all the left plates end up with a charge $+Q$. Therefore, **the charges on capacitors connected in series are the same:**

$$Q_1 = Q_2 = Q$$

where Q is the charge that moved between a wire and the connected outside plate of one of the capacitors.

Active Figure 26.8a shows that the total voltage ΔV_{tot} across the combination is split between the two capacitors:

$$\Delta V_{tot} = \Delta V_1 + \Delta V_2 \qquad \textbf{(26.9)}$$

where ΔV_1 and ΔV_2 are the potential differences across capacitors C_1 and C_2, respectively. In general, **the total potential difference across any number of capacitors connected in series is the sum of the potential differences across the individual capacitors.**

Suppose the equivalent single capacitor in Active Figure 26.8c has the same effect on the circuit as the series combination when it is connected to the battery. After it is fully charged, the equivalent capacitor must have a charge of $-Q$ on its right plate and a charge of $+Q$ on its left plate. Applying the definition of capacitance to the circuit in Active Figure 26.8c gives

$$\Delta V_{tot} = \frac{Q}{C_{eq}}$$

Substituting for the voltages in Equation 26.9, we have

$$\frac{Q}{C_{eq}} = \frac{Q_1}{C_1} + \frac{Q_2}{C_2}$$

Canceling the charges because they are all the same gives

$$\frac{1}{C_{eq}} = \frac{1}{C_1} + \frac{1}{C_2} \quad \text{(series combination)}$$

When this analysis is applied to three or more capacitors connected in series, the relationship for the equivalent capacitance is

Capacitors in series ▶

$$\frac{1}{C_{eq}} = \frac{1}{C_1} + \frac{1}{C_2} + \frac{1}{C_3} + \cdots \quad \text{(series combination)} \qquad \textbf{(26.10)}$$

This expression shows that **(1) the inverse of the equivalent capacitance is the algebraic sum of the inverses of the individual capacitances and (2) the equivalent capacitance of a series combination is always less than any individual capacitance in the combination.**

Quick Quiz 26.3 Two capacitors are identical. They can be connected in series or in parallel. If you want the *smallest* equivalent capacitance for the combination, how should you connect them? (a) in series (b) in parallel (c) either way because both combinations have the same capacitance

EXAMPLE 26.3 Equivalent Capacitance

Find the equivalent capacitance between a and b for the combination of capacitors shown in Figure 26.9a. All capacitances are in microfarads.

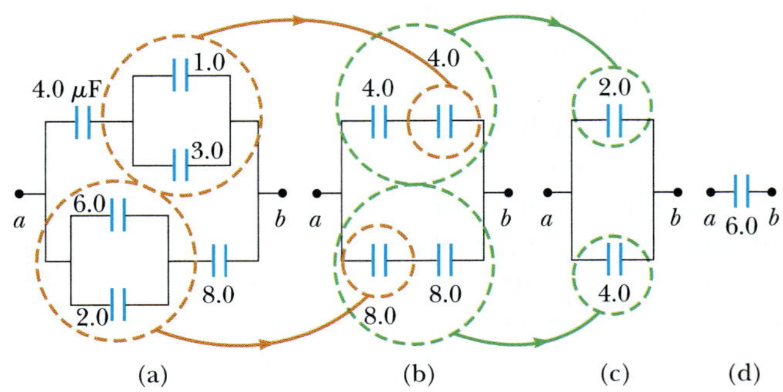

Figure 26.9 (Example 26.3) To find the equivalent capacitance of the capacitors in (a), we reduce the various combinations in steps as indicated in (b), (c), and (d), using the series and parallel rules described in the text.

SOLUTION

Conceptualize Study Figure 26.9a carefully and make sure you understand how the capacitors are connected.

Categorize Figure 26.9a shows that the circuit contains both series and parallel connections, so we use the rules for series and parallel combinations discussed in this section.

Analyze Using Equations 26.8 and 26.10, we reduce the combination step by step as indicated in the figure.

The 1.0-μF and 3.0-μF capacitors in Figure 26.9a are in parallel. Find the equivalent capacitance from Equation 26.8:

$$C_{eq} = C_1 + C_2 = 4.0\,\mu\text{F}$$

The 2.0-μF and 6.0-μF capacitors in Figure 26.9a are also in parallel:

$$C_{eq} = C_1 + C_2 = 8.0\,\mu\text{F}$$

The circuit now looks like Figure 26.9b. The two 4.0-μF capacitors in the upper branch are in series. Find the equivalent capacitance from Equation 26.10:

$$\frac{1}{C_{eq}} = \frac{1}{C_1} + \frac{1}{C_2} = \frac{1}{4.0\,\mu\text{F}} + \frac{1}{4.0\,\mu\text{F}} = \frac{1}{2.0\,\mu\text{F}}$$

$$C_{eq} = 2.0\,\mu\text{F}$$

The two 8.0-μF capacitors in the lower branch are also in series. Find the equivalent capacitance from Equation 26.10:

$$\frac{1}{C_{eq}} = \frac{1}{C_1} + \frac{1}{C_2} = \frac{1}{8.0\,\mu\text{F}} + \frac{1}{8.0\,\mu\text{F}} = \frac{1}{4.0\,\mu\text{F}}$$

$$C_{eq} = 4.0\,\mu\text{F}$$

The circuit now looks like Figure 26.9c. The 2.0-μF and 4.0-μF capacitors are in parallel:

$$C_{eq} = C_1 + C_2 = 6.0\ \mu\text{F}$$

Finalize This final value is that of the single equivalent capacitor shown in Figure 26.9d. For further practice in treating circuits with combinations of capacitors, imagine that a battery is connected between points a and b so that a potential difference ΔV is established across the combination. Can you find the voltage across and the charge on each capacitor?

26.4 Energy Stored in a Charged Capacitor

Because positive and negative charges are separated in the system of two conductors in a capacitor, electric potential energy is stored in the system. Many of those who work with electronic equipment have at some time verified that a capacitor can store energy. If the plates of a charged capacitor are connected by a conductor such as a wire, charge moves between each plate and its connecting wire until the capacitor is uncharged. The discharge can often be observed as a visible spark. If you accidentally touch the opposite plates of a charged capacitor, your fingers act as a pathway for discharge and the result is an electric shock. The degree of shock you receive depends on the capacitance and the voltage applied to the capacitor. Such a shock could be fatal if high voltages are present, as in the power supply of a television set. Because the charges can be stored in a capacitor even when the set is turned off, unplugging the television does not make it safe to open the case and touch the components inside.

Active Figure 26.10a shows a battery connected to a single parallel-plate capacitor with a switch in the circuit. Let us identify the circuit as a system. When the switch is closed (Active Fig. 26.10b), the battery establishes an electric field in the wires and charges flow between the wires and the capacitor. As that occurs, there is

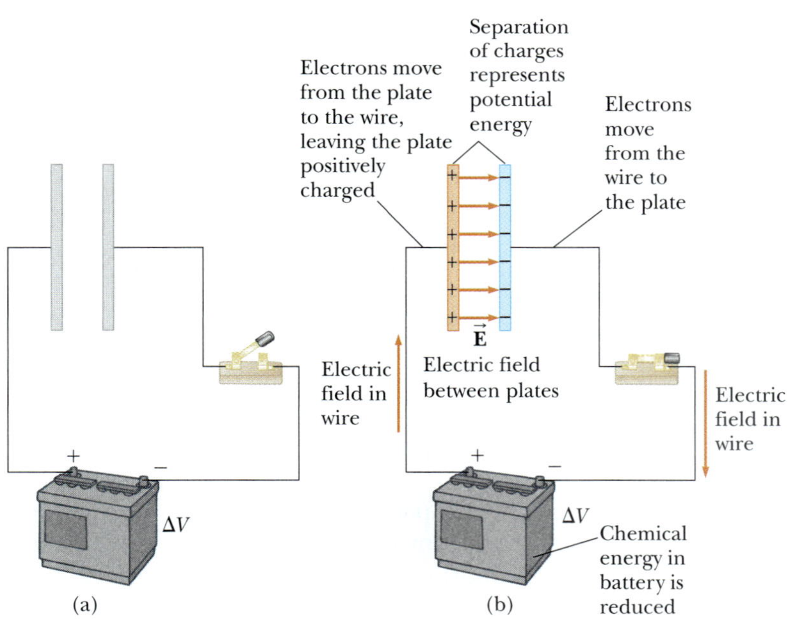

ACTIVE FIGURE 26.10

(a) A circuit consisting of a capacitor, a battery, and a switch. (b) When the switch is closed, the battery establishes an electric field in the wire that causes electrons to move from the left plate into the wire and into the right plate from the wire. As a result, a separation of charge exists on the plates, which represents an increase in electric potential energy of the system of the circuit. This energy in the system has been transformed from chemical energy in the battery.

Sign in at www.thomsonedu.com and go to ThomsonNOW to adjust the battery voltage and see the resulting charge on the plates and the electric field between the plates.

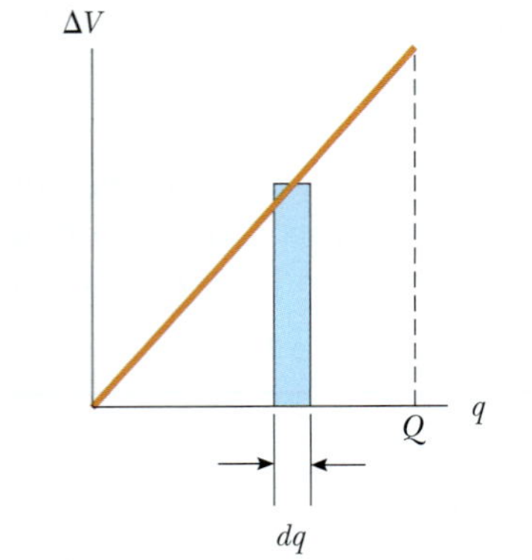

Figure 26.11 A plot of potential difference versus charge for a capacitor is a straight line having slope $1/C$. The work required to move charge dq through the potential difference ΔV existing at the time across the capacitor plates is given approximately by the area of the shaded rectangle. The total work required to charge the capacitor to a final charge Q is the triangular area under the straight line, $W = \frac{1}{2}Q\,\Delta V$. (Don't forget that 1 V = 1 J/C; hence, the unit for the triangular area is the joule.)

a transformation of energy within the system. Before the switch is closed, energy is stored as chemical energy in the battery. This energy is transformed during the chemical reaction that occurs within the battery when it is operating in an electric circuit. When the switch is closed, some of the chemical energy in the battery is converted to electric potential energy associated with the separation of positive and negative charges on the plates.

To calculate the energy stored in the capacitor, we shall assume a charging process that is different from the actual process described in Section 26.1 but that gives the same final result. This assumption is justified because the energy in the final configuration does not depend on the actual charge-transfer process.[3] Imagine that you transfer the charge mechanically through the space between the plates as follows. You grab a small amount of positive charge on the plate connected to the negative terminal and apply a force that causes this positive charge to move over to the plate connected to the positive terminal. Therefore, you do work on the charge as it is transferred from one plate to the other. At first, no work is required to transfer a small amount of charge dq from one plate to the other,[4] but once this charge has been transferred, a small potential difference exists between the plates. Therefore, work must be done to move additional charge through this potential difference. As more and more charge is transferred from one plate to the other, the potential difference increases in proportion and more work is required.

Suppose q is the charge on the capacitor at some instant during the charging process. At the same instant, the potential difference across the capacitor is $\Delta V = q/C$. From Section 25.1, we know that the work necessary to transfer an increment of charge dq from the plate carrying charge $-q$ to the plate carrying charge q (which is at the higher electric potential) is

$$dW = \Delta V\,dq = \frac{q}{C}\,dq$$

This situation is illustrated in Figure 26.11. The total work required to charge the capacitor from $q = 0$ to some final charge $q = Q$ is

$$W = \int_0^Q \frac{q}{C}\,dq = \frac{1}{C}\int_0^Q q\,dq = \frac{Q^2}{2C}$$

The work done in charging the capacitor appears as electric potential energy U stored in the capacitor. Using Equation 26.1, we can express the potential energy stored in a charged capacitor as

Energy stored in a charged capacitor ▶

$$U = \frac{Q^2}{2C} = \tfrac{1}{2}Q\,\Delta V = \tfrac{1}{2}C(\Delta V)^2 \qquad \textbf{(26.11)}$$

This result applies to any capacitor, regardless of its geometry. For a given capacitance, the stored energy increases as the charge and the potential difference increase. In practice, there is a limit to the maximum energy (or charge) that can be stored because, at a sufficiently large value of ΔV, discharge ultimately occurs between the plates. For this reason, capacitors are usually labeled with a maximum operating voltage.

We can consider the energy in a capacitor to be stored in the electric field created between the plates as the capacitor is charged. This description is reasonable because the electric field is proportional to the charge on the capacitor. For a parallel-plate capacitor, the potential difference is related to the electric field

[3] This discussion is similar to that of state variables in thermodynamics. The change in a state variable such as temperature is independent of the path followed between the initial and final states. The potential energy of a capacitor (or any system) is also a state variable, so it does not depend on the actual process followed to charge the capacitor.

[4] We shall use lowercase q for the time-varying charge on the capacitor while it is charging to distinguish it from uppercase Q, which is the total charge on the capacitor after it is completely charged.

through the relationship $\Delta V = Ed$. Furthermore, its capacitance is $C = \epsilon_0 A/d$ (Eq. 26.3). Substituting these expressions into Equation 26.11 gives

$$U = \frac{1}{2}\frac{\epsilon_0 A}{d}(E^2 d^2) = \frac{1}{2}(\epsilon_0 Ad)E^2 \tag{26.12}$$

Because the volume occupied by the electric field is Ad, the *energy per unit volume* $u_E = U/Ad$, known as the *energy density,* is

$$u_E = \frac{1}{2}\epsilon_0 E^2 \tag{26.13}$$

◀ Energy density in an electric field

Although Equation 26.13 was derived for a parallel-plate capacitor, the expression is generally valid regardless of the source of the electric field. That is, **the energy density in any electric field is proportional to the square of the magnitude of the electric field at a given point.**

PITFALL PREVENTION 26.4
Not a New Kind of Energy

The energy given by Equation 26.13 is not a new kind of energy. The equation describes familiar electric potential energy associated with a system of separated source charges. Equation 26.13 provides a new *interpretation*, or a new way of *modeling* the energy. Furthermore, the equation correctly describes the energy associated with *any* electric field, regardless of the source.

Quick Quiz 26.4 You have three capacitors and a battery. In which of the following combinations of the three capacitors is the maximum possible energy stored when the combination is attached to the battery? (a) series (b) parallel (c) no difference because both combinations store the same amount of energy

---

EXAMPLE 26.4 Rewiring Two Charged Capacitors

Two capacitors C_1 and C_2 (where $C_1 > C_2$) are charged to the same initial potential difference ΔV_i. The charged capacitors are removed from the battery, and their plates are connected with opposite polarity as in Figure 26.12a. The switches S_1 and S_2 are then closed as in Figure 26.12b.

(A) Find the final potential difference ΔV_f between a and b after the switches are closed.

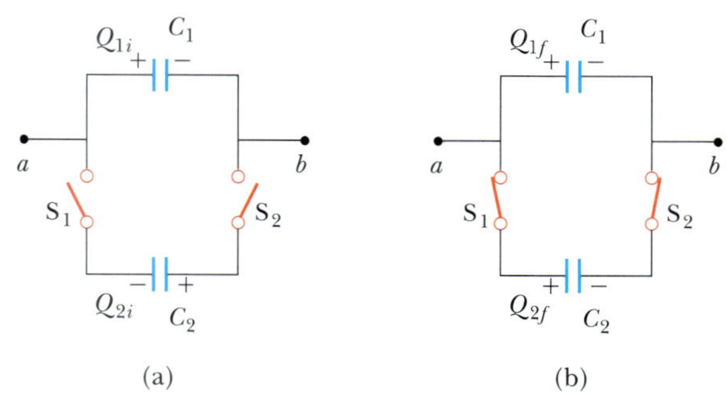

Figure 26.12 (Example 26.4) (a) Two capacitors are charged to the same initial potential difference and connected together with plates of opposite sign to be in contact when the switches are closed. (b) When the switches are closed, the charges redistribute.

SOLUTION

Conceptualize Figure 26.12 helps us understand the initial and final configurations of the system.

Categorize In Figure 26.12b, it might appear as if the capacitors are connected in parallel, but there is no battery in this circuit to apply a voltage across the combination. Therefore, we *cannot* categorize this problem as one in which capacitors are connected in parallel. We *can* categorize it as a problem involving an isolated system for electric charge. The left-hand plates of the capacitors form an isolated system because they are not connected to the right-hand plates by conductors.

Analyze Write an expression for the total charge on the left-hand plates of the system before the switches are closed, noting that a negative sign for Q_{2i} is necessary because the charge on the left plate of capacitor C_2 is negative:

$$(1) \quad Q_i = Q_{1i} + Q_{2i} = C_1\,\Delta V_i - C_2\,\Delta V_i = (C_1 - C_2)\Delta V_i$$

After the switches are closed, the charges on the individual capacitors change to new values Q_{1f} and Q_{2f} such that the potential difference is again the same across both capacitors, ΔV_f. Write an expression for the total charge on the left-hand plates of the system after the switches are closed:

$$(2) \quad Q_f = Q_{1f} + Q_{2f} = C_1\,\Delta V_f + C_2\,\Delta V_f = (C_1 + C_2)\,\Delta V_f$$

Because the system is isolated, the initial and final charges on the system must be the same. Use this condition and Equations (1) and (2) to solve for ΔV_f:

$$Q_f = Q_i \rightarrow (C_1 + C_2)\Delta V_f = (C_1 - C_2)\Delta V_i$$

$$(3) \quad \Delta V_f = \left(\frac{C_1 - C_2}{C_1 + C_2}\right)\Delta V_i$$

(B) Find the total energy stored in the capacitors before and after the switches are closed and determine the ratio of the final energy to the initial energy.

SOLUTION

Use Equation 26.11 to find an expression for the total energy stored in the capacitors before the switches are closed:

$$(4) \quad U_i = \tfrac{1}{2}C_1(\Delta V_i)^2 + \tfrac{1}{2}C_2(\Delta V_i)^2 = \tfrac{1}{2}(C_1 + C_2)(\Delta V_i)^2$$

Write an expression for the total energy stored in the capacitors after the switches are closed:

$$U_f = \tfrac{1}{2}C_1(\Delta V_f)^2 + \tfrac{1}{2}C_2(\Delta V_f)^2 = \tfrac{1}{2}(C_1 + C_2)(\Delta V_f)^2$$

Use the results of part (A) to rewrite this expression in terms of ΔV_i:

$$(5) \quad U_f = \tfrac{1}{2}(C_1 + C_2)\left[\left(\frac{C_1 - C_2}{C_1 + C_2}\right)\Delta V_i\right]^2 = \tfrac{1}{2}\frac{(C_1 - C_2)^2(\Delta V_i)^2}{(C_1 + C_2)}$$

Divide Equation (5) by Equation (4) to obtain the ratio of the energies stored in the system:

$$\frac{U_f}{U_i} = \frac{\tfrac{1}{2}(C_1 - C_2)^2(\Delta V_i)^2/(C_1 + C_2)}{\tfrac{1}{2}(C_1 + C_2)(\Delta V_i)^2}$$

$$(6) \quad \frac{U_f}{U_i} = \left(\frac{C_1 - C_2}{C_1 + C_2}\right)^2$$

Finalize The ratio of energies is *less* than unity, indicating that the final energy is *less* than the initial energy. At first, you might think the law of energy conservation has been violated, but that is not the case. The "missing" energy is transferred out of the system by the mechanism of electromagnetic waves (T_{ER} in Equation 8.2), as we shall see in Chapter 34.

What If? What if the two capacitors have the same capacitance? What would you expect to happen when the switches are closed?

Answer Because both capacitors have the same initial potential difference applied to them, the charges on the capacitors have the same magnitude. When the capacitors with opposite polarities are connected together, the equal-magnitude charges should cancel each other, leaving the capacitors uncharged.

Let's test our results to see if that is the case mathematically. In Equation (1), because the capacitances are equal, the initial charge Q_i on the system of left-hand plates is zero. Equation (3) shows that $\Delta V_f = 0$, which is consistent with uncharged capacitors. Finally, Equation (5) shows that $U_f = 0$, which is also consistent with uncharged capacitors.

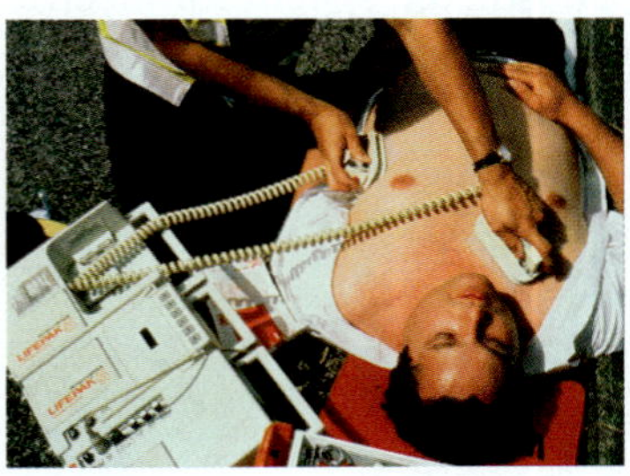

Adam Hart-Davis/SPL/Custom Medical Stock

Figure 26.13 In a hospital or at an emergency scene, you might see a patient being revived with a portable defibrillator. The defibrillator's paddles are applied to the patient's chest, and an electric shock is sent through the chest cavity. The aim of this technique is to restore the heart's normal rhythm pattern.

One device in which capacitors have an important role is the portable *defibrillator* (Fig. 26.13). When cardiac fibrillation (random contractions) occurs, the heart produces a rapid, irregular pattern of beats. A fast discharge of energy through the heart can return the organ to its normal beat pattern. Emergency medical teams use portable defibrillators that contain batteries capable of charging a capacitor to a high voltage. (The circuitry actually permits the capacitor to be charged to a much higher voltage than that of the battery.) Up to 360 J is stored in the electric field of a large capacitor in a defibrillator when it is fully charged. The stored energy is released through the heart by conducting electrodes, called paddles, which are placed on both sides of the victim's chest. The defibrillator can deliver the energy to a patient in about 2 ms (roughly equivalent to 3 000 times the power delivered to a 60-W lightbulb!). The paramedics must wait between applications of the energy because of the time necessary for the capacitors to become fully charged. In this application and others (e.g., camera flash units and lasers used for fusion experiments), capacitors serve as energy reservoirs that can be slowly charged and then quickly discharged to provide large amounts of energy in a short pulse.

26.5 Capacitors with Dielectrics

A **dielectric** is a nonconducting material such as rubber, glass, or waxed paper. We can perform the following experiment to illustrate the effect of a dielectric in a capacitor. Consider a parallel-plate capacitor that without a dielectric has a charge Q_0 and a capacitance C_0. The potential difference across the capacitor is $\Delta V_0 = Q_0/C_0$. Figure 26.14a illustrates this situation. The potential difference is measured by a *voltmeter*, a device discussed in greater detail in Chapter 28. Notice that no battery is shown in the figure; also, we must assume no charge can flow through an ideal voltmeter. Hence, there is no path by which charge can flow and alter the charge on the capacitor. If a dielectric is now inserted between the plates as in Figure 26.14b, the voltmeter indicates that the voltage between the plates decreases to a value ΔV. The voltages with and without the dielectric are related by a factor κ as follows:

$$\Delta V = \frac{\Delta V_0}{\kappa}$$

Because $\Delta V < \Delta V_0$, we see that $\kappa > 1$. The dimensionless factor κ is called the **dielectric constant** of the material. The dielectric constant varies from one material to another. In this section, we analyze this change in capacitance in terms of electrical parameters such as electric charge, electric field, and potential difference; Section 26.7 describes the microscopic origin of these changes.

Because the charge Q_0 on the capacitor does not change, the capacitance must change to the value

$$C = \frac{Q_0}{\Delta V} = \frac{Q_0}{\Delta V_0/\kappa} = \kappa\frac{Q_0}{\Delta V_0}$$

$$C = \kappa C_0 \qquad (26.14)$$

◀ Capacitance of a capacitor filled with a material of dielectric constant κ

That is, the capacitance *increases* by the factor κ when the dielectric completely fills the region between the plates.[5] Because $C_0 = \epsilon_0 A/d$ (Eq. 26.3) for a parallel-plate

PITFALL PREVENTION 26.5

Is the Capacitor Connected to a Battery?

For problems in which a capacitor is modified (by insertion of a dielectric, for example), you must note whether modifications to the capacitor are being made while the capacitor is connected to a battery or after it is disconnected. If the capacitor remains connected to the battery, the voltage across the capacitor necessarily remains the same. If you disconnect the capacitor from the battery before making any modifications to the capacitor, the capacitor is an isolated system and its charge remains the same.

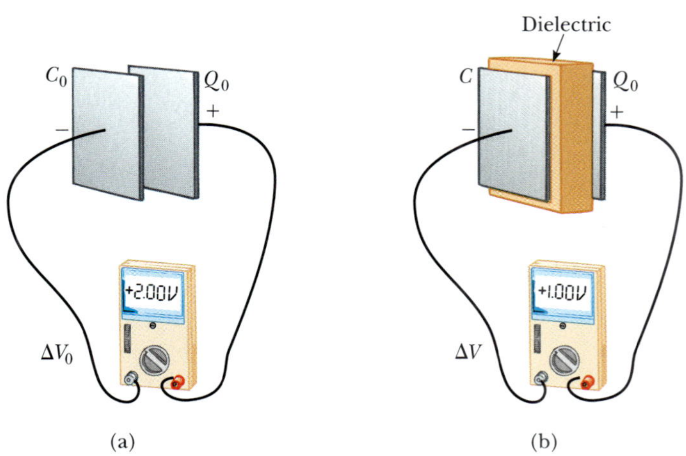

Figure 26.14 A charged capacitor (a) before and (b) after insertion of a dielectric between the plates. The charge on the plates remains unchanged, but the potential difference decreases from ΔV_0 to $\Delta V = \Delta V_0/\kappa$. Therefore, the capacitance increases from C_0 to κC_0.

[5] If the dielectric is introduced while the potential difference is held constant by a battery, the charge increases to a value $Q = \kappa Q_0$. The additional charge comes from the wires attached to the capacitor, and the capacitance again increases by the factor κ.

capacitor, we can express the capacitance of a parallel-plate capacitor filled with a dielectric as

$$C = \kappa \frac{\epsilon_0 A}{d} \tag{26.15}$$

From Equations 26.3 and 26.15, it would appear that the capacitance could be made very large by decreasing d, the distance between the plates. In practice, the lowest value of d is limited by the electric discharge that could occur through the dielectric medium separating the plates. For any given separation d, the maximum voltage that can be applied to a capacitor without causing a discharge depends on the **dielectric strength** (maximum electric field) of the dielectric. If the magnitude of the electric field in the dielectric exceeds the dielectric strength, the insulating properties break down and the dielectric begins to conduct.

Physical capacitors have a specification called by a variety of names, including *working voltage, breakdown voltage,* and *rated voltage.* This parameter represents the largest voltage that can be applied to the capacitor without exceeding the dielectric strength of the dielectric material in the capacitor. Consequently, when selecting a capacitor for a given application, you must consider its capacitance as well as the expected voltage across the capacitor in the circuit, making sure the expected voltage is smaller than the rated voltage of the capacitor. You can see the rated voltage on several of the capacitors in this chapter's opening photograph.

Insulating materials have values of κ greater than unity and dielectric strengths greater than that of air as Table 26.1 indicates. Therefore, a dielectric provides the following advantages:

- An increase in capacitance
- An increase in maximum operating voltage
- Possible mechanical support between the plates, which allows the plates to be close together without touching, thereby decreasing d and increasing C

TABLE 26.1

Approximate Dielectric Constants and Dielectric Strengths of Various Materials at Room Temperature

Material	Dielectric Constant κ	Dielectric Strength[a] (10^6 V/m)
Air (dry)	1.000 59	3
Bakelite	4.9	24
Fused quartz	3.78	8
Mylar	3.2	7
Neoprene rubber	6.7	12
Nylon	3.4	14
Paper	3.7	16
Paraffin-impregnated paper	3.5	11
Polystyrene	2.56	24
Polyvinyl chloride	3.4	40
Porcelain	6	12
Pyrex glass	5.6	14
Silicone oil	2.5	15
Strontium titanate	233	8
Teflon	2.1	60
Vacuum	1.000 00	—
Water	80	—

[a] The dielectric strength equals the maximum electric field that can exist in a dielectric without electrical breakdown. These values depend strongly on the presence of impurities and flaws in the materials.

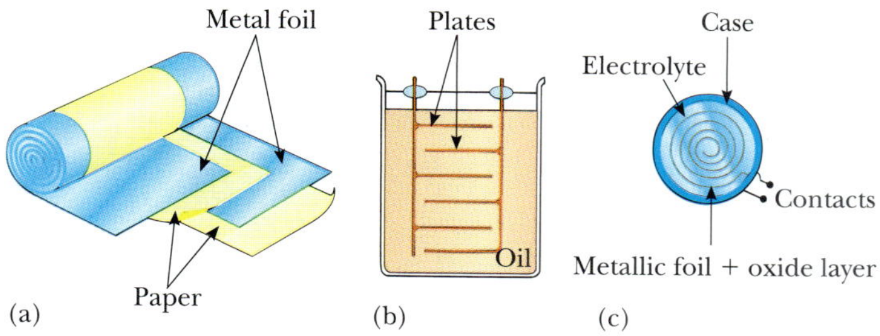

Figure 26.15 Three commercial capacitor designs. (a) A tubular capacitor, whose plates are separated by paper and then rolled into a cylinder. (b) A high-voltage capacitor consisting of many parallel plates separated by insulating oil. (c) An electrolytic capacitor.

© Thomson Learning/George Semple

Figure 26.16 A variable capacitor. When one set of metal plates is rotated so as to lie between a fixed set of plates, the capacitance of the device changes.

Types of Capacitors

Commercial capacitors are often made from metallic foil interlaced with thin sheets of either paraffin-impregnated paper or Mylar as the dielectric material. These alternate layers of metallic foil and dielectric are rolled into a cylinder to form a small package (Fig. 26.15a). High-voltage capacitors commonly consist of a number of interwoven metallic plates immersed in silicone oil (Fig. 26.15b). Small capacitors are often constructed from ceramic materials.

Often, an *electrolytic capacitor* is used to store large amounts of charge at relatively low voltages. This device, shown in Figure 26.15c, consists of a metallic foil in contact with an *electrolyte,* a solution that conducts electricity by virtue of the motion of ions contained in the solution. When a voltage is applied between the foil and the electrolyte, a thin layer of metal oxide (an insulator) is formed on the foil, and this layer serves as the dielectric. Very large values of capacitance can be obtained in an electrolytic capacitor because the dielectric layer is very thin and therefore the plate separation is very small.

Electrolytic capacitors are not reversible as are many other capacitors. They have a polarity, which is indicated by positive and negative signs marked on the device. When electrolytic capacitors are used in circuits, the polarity must be correct. If the polarity of the applied voltage is the opposite of what is intended, the oxide layer is removed and the capacitor conducts electricity instead of storing charge.

Variable capacitors (typically 10 to 500 pF) usually consist of two interwoven sets of metallic plates, one fixed and the other movable, and contain air as the dielectric (Fig. 26.16). These types of capacitors are often used in radio tuning circuits.

Quick Quiz 26.5 If you have ever tried to hang a picture or a mirror, you know it can be difficult to locate a wooden stud in which to anchor your nail or screw. A carpenter's stud finder is a capacitor with its plates arranged side by side instead of facing each other as shown in Figure 26.17. When the device is moved over a stud, does the capacitance (a) increase or (b) decrease?

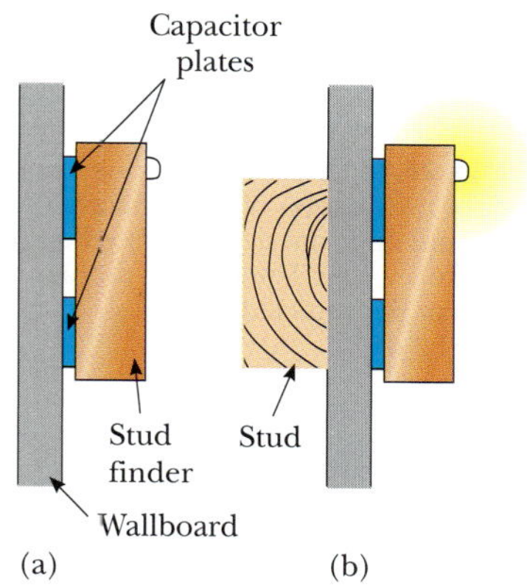

Figure 26.17 (Quick Quiz 26.5) A stud finder. (a) The materials between the plates of the capacitor are the wallboard and air. (b) When the capacitor moves across a stud in the wall, the materials between the plates are the wallboard and the wood. The change in the dielectric constant causes a signal light to illuminate.

EXAMPLE 26.5 **Energy Stored Before and After**

A parallel-plate capacitor is charged with a battery to a charge Q_0. The battery is then removed, and a slab of material that has a dielectric constant κ is inserted between the plates. Identify the system as the capacitor and the dielectric. Find the energy stored in the system before and after the dielectric is inserted.

SOLUTION

Conceptualize Think about what happens when the dielectric is inserted between the plates. Because the battery has been removed, the charge on the capacitor must remain the same. We know from our earlier discussion, however, that the capacitance must change. Therefore, we expect a change in the energy of the system.

Categorize Because we expect the energy of the system to change, we model it as a nonisolated system.

Analyze From Equation 26.11, find the energy stored in the absence of the dielectric:

$$U_0 = \frac{Q_0^2}{2C_0}$$

Find the energy stored in the capacitor after the dielectric is inserted between the plates:

$$U = \frac{Q_0^2}{2C}$$

Use Equation 26.14 to replace the capacitance C:

$$U = \frac{Q_0^2}{2\kappa C_0} = \frac{U_0}{\kappa}$$

Finalize Because $\kappa > 1$, the final energy is less than the initial energy. We can account for the "missing" energy by noting that the dielectric, when inserted, is pulled into the device. To keep the dielectric from accelerating, an external agent must do negative work (W in Eq. 8.2) on the dielectric, which is simply the difference $U - U_0$.

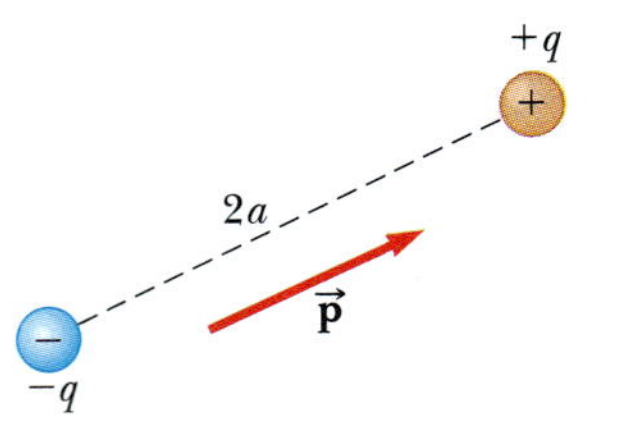

Figure 26.18 An electric dipole consists of two charges of equal magnitude and opposite sign separated by a distance of $2a$. The electric dipole moment $\vec{\mathbf{p}}$ is directed from $-q$ toward $+q$.

26.6 Electric Dipole in an Electric Field

We have discussed the effect on the capacitance of placing a dielectric between the plates of a capacitor. In Section 26.7, we shall describe the microscopic origin of this effect. Before we can do so, however, let's expand the discussion of the electric dipole introduced in Section 23.4 (see Example 23.5). The electric dipole consists of two charges of equal magnitude and opposite sign separated by a distance $2a$ as shown in Figure 26.18. The **electric dipole moment** of this configuration is defined as the vector $\vec{\mathbf{p}}$ directed from $-q$ toward $+q$ along the line joining the charges and having magnitude $2aq$:

$$p \equiv 2aq \tag{26.16}$$

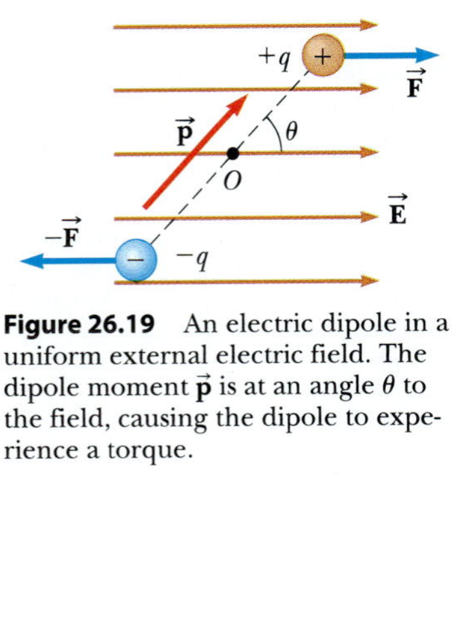

Figure 26.19 An electric dipole in a uniform external electric field. The dipole moment $\vec{\mathbf{p}}$ is at an angle θ to the field, causing the dipole to experience a torque.

Now suppose an electric dipole is placed in a uniform electric field $\vec{\mathbf{E}}$ and makes an angle θ with the field as shown in Figure 26.19. We identify $\vec{\mathbf{E}}$ as the field *external* to the dipole, established by some other charge distribution, to distinguish it from the field *due to* the dipole, which we discussed in Section 23.4.

The electric forces acting on the two charges are equal in magnitude ($F = qE$) and opposite in direction as shown in Figure 26.19. Therefore, the net force on the dipole is zero. The two forces produce a net torque on the dipole, however; as a result, the dipole rotates in the direction that brings the dipole moment vector into greater alignment with the field. The torque due to the force on the positive charge about an axis through O in Figure 26.19 has magnitude $Fa \sin \theta$, where $a \sin \theta$ is the moment arm of F about O. This force tends to produce a clockwise rotation. The torque about O on the negative charge is also of magnitude $Fa \sin \theta$; here again, the force tends to produce a clockwise rotation. Therefore, the magnitude of the net torque about O is

$$\tau = 2Fa \sin \theta$$

Because $F = qE$ and $p = 2aq$, we can express τ as

$$\tau = 2aqE \sin \theta = pE \sin \theta \tag{26.17}$$

It is convenient to express the torque in vector form as the cross product of the vectors $\vec{\mathbf{p}}$ and $\vec{\mathbf{E}}$:

Torque on an electric dipole in an external electric field ▶

$$\vec{\boldsymbol{\tau}} = \vec{\mathbf{p}} \times \vec{\mathbf{E}} \tag{26.18}$$

Let's determine the potential energy of the system of an electric dipole in an external electric field as a function of the dipole's orientation with respect to the field. To do so, recognize that work must be done by an external agent to rotate the dipole through an angle so as to cause the dipole moment vector to become less aligned with the field. The work done is then stored as potential energy in the system. The work dW required to rotate the dipole through an angle $d\theta$ is $dW = \tau d\theta$ (Eq. 10.22). Because $\tau = pE \sin\theta$ and the work results in an increase in the potential energy U, we find that for a rotation from θ_i to θ_f, the change in potential energy of the system is

$$U_f - U_i = \int_{\theta_i}^{\theta_f} \tau\, d\theta = \int_{\theta_i}^{\theta_f} pE \sin\theta\, d\theta = pE \int_{\theta_i}^{\theta_f} \sin\theta\, d\theta$$

$$= pE[-\cos\theta]_{\theta_i}^{\theta_f} = pE(\cos\theta_i - \cos\theta_f)$$

The term that contains $\cos\theta_i$ is a constant that depends on the initial orientation of the dipole. It is convenient to choose a reference angle of $\theta_i = 90°$ so that $\cos\theta_i = \cos 90° = 0$. Furthermore, let's choose $U_i = 0$ at $\theta_i = 90°$ as our reference of potential energy. Hence, we can express a general value of $U = U_f$ as

$$U = -pE\cos\theta \qquad \textbf{(26.19)}$$

We can write this expression for the potential energy of a dipole in an electric field as the dot product of the vectors $\vec{\mathbf{p}}$ and $\vec{\mathbf{E}}$:

$$U = -\vec{\mathbf{p}} \cdot \vec{\mathbf{E}} \qquad \textbf{(26.20)}$$

Potential energy of the system of an electric dipole in an external electric field

To develop a conceptual understanding of Equation 26.19, compare it with the expression for the potential energy of the system of an object in the Earth's gravitational field, $U = mgy$ (see Chapter 7). This expression includes a parameter associated with the object placed in the gravitational field, its mass m. Likewise, Equation 26.19 includes a parameter of the object in the electric field, its dipole moment p. The gravitational expression includes the magnitude of the gravitational field g. Similarly, Equation 26.19 includes the magnitude of the electric field E. So far, these two contributions to the potential energy expressions appear analogous. The final contribution, however, is somewhat different in the two cases. In the gravitational expression, the potential energy depends on the vertical position of the object, measured by y. In Equation 26.19, the potential energy depends on the angle θ through which the dipole is rotated. In both cases, the configuration of the system is being changed. In the gravitational case, the change involves moving an object in a *translational* sense, whereas in the electrical case, the change involves moving an object in a *rotational* sense. In both cases, however, once the change is made, the system tends to return to the original configuration when the object is released: the object of mass m falls toward the ground, and the dipole begins to rotate back toward the configuration in which it is aligned with the field. Therefore, apart from the type of motion, the expressions for potential energy in these two cases are similar.

Molecules are said to be *polarized* when a separation exists between the average position of the negative charges and the average position of the positive charges in the molecule. In some molecules such as water, this condition is always present; such molecules are called **polar molecules.** Molecules that do not possess a permanent polarization are called **nonpolar molecules.**

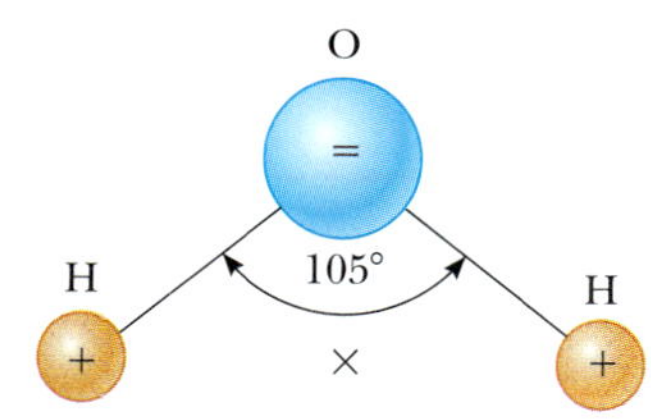

Figure 26.20 The water molecule, H_2O, has a permanent polarization resulting from its nonlinear geometry. The center of the positive charge distribution is at the point ×.

We can understand the permanent polarization of water by inspecting the geometry of the water molecule. The oxygen atom in the water molecule is bonded to the hydrogen atoms such that an angle of 105° is formed between the two bonds (Fig. 26.20). The center of the negative charge distribution is near the oxygen atom, and the center of the positive charge distribution lies at a point midway along the line joining the hydrogen atoms (the point labeled × in Fig. 26.20). We can model the water molecule and other polar molecules as dipoles because the average positions of the positive and negative charges act as point charges. As a result, we can apply our discussion of dipoles to the behavior of polar molecules.

Washing with soap and water is a household scenario in which the dipole structure of water is exploited. Grease and oil are made up of nonpolar molecules,

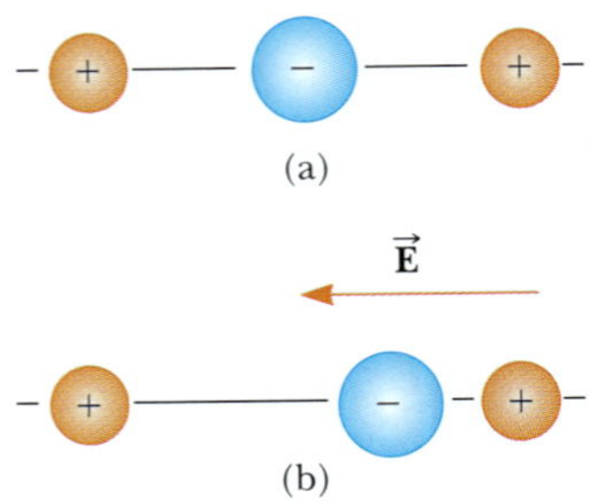

Figure 26.21 (a) A linear symmetric molecule has no permanent polarization. (b) An external electric field induces a polarization in the molecule.

which are generally not attracted to water. Plain water is not very useful for removing this type of grime. Soap contains long molecules called *surfactants.* In a long molecule, the polarity characteristics of one end of the molecule can be different from those at the other end. In a surfactant molecule, one end acts like a nonpolar molecule and the other acts like a polar molecule. The nonpolar end can attach to a grease or oil molecule, and the polar end can attach to a water molecule. Therefore, the soap serves as a chain, linking the dirt and water molecules together. When the water is rinsed away, the grease and oil go with it.

A symmetric molecule (Fig. 26.21a) has no permanent polarization, but polarization can be induced by placing the molecule in an electric field. A field directed to the left as in Figure 26.21b causes the center of the positive charge distribution to shift to the left from its initial position and the center of the negative charge distribution to shift to the right. This *induced polarization* is the effect that predominates in most materials used as dielectrics in capacitors.

EXAMPLE 26.6 The H_2O Molecule

The water (H_2O) molecule has an electric dipole moment of 6.3×10^{-30} C · m. A sample contains 10^{21} water molecules, with the dipole moments all oriented in the direction of an electric field of magnitude 2.5×10^5 N/C. How much work is required to rotate the dipoles from this orientation ($\theta = 0°$) to one in which all the moments are perpendicular to the field ($\theta = 90°$)?

SOLUTION

Conceptualize When all the dipoles are aligned with the electric field, the dipoles–electric field system has the minimum potential energy. This energy has a negative value given by the product of the right side of Equation 26.19, evaluated at 0°, and the number N of dipoles. Work must be done to rotate all the dipoles of the system by 90° because the system's potential energy is raised to a higher value of zero.

Categorize We use Equation 26.19 to evaluate the potential energy, so we categorize this example as a substitution problem.

Write the appropriate reduction of the conservation of energy equation, Equation 8.2, for this situation:

$$(1) \quad \Delta U = W$$

Use Equation 26.19 to evaluate the initial and final potential energies of the system and Equation (1) to calculate the work required to rotate the dipoles:

$$W = U_{90°} - U_{0°} = (-NpE\cos 90°) - (-NpE\cos 0°)$$
$$= NpE = (10^{21})(6.3 \times 10^{-30}\ \text{C}\cdot\text{m})(2.5 \times 10^5\ \text{N/C})$$
$$= 1.6 \times 10^{-3}\ \text{J}$$

26.7 An Atomic Description of Dielectrics

In Section 26.5, we found that the potential difference ΔV_0 between the plates of a capacitor is reduced to $\Delta V_0/\kappa$ when a dielectric is introduced. The potential difference is reduced because the magnitude of the electric field decreases between the plates. In particular, if $\vec{\mathbf{E}}_0$ is the electric field without the dielectric, the field in the presence of a dielectric is

$$\vec{\mathbf{E}} = \frac{\vec{\mathbf{E}}_0}{\kappa} \qquad \textbf{(26.21)}$$

First consider a dielectric made up of polar molecules placed in the electric field between the plates of a capacitor. The dipoles (that is, the polar molecules making up the dielectric) are randomly oriented in the absence of an electric field

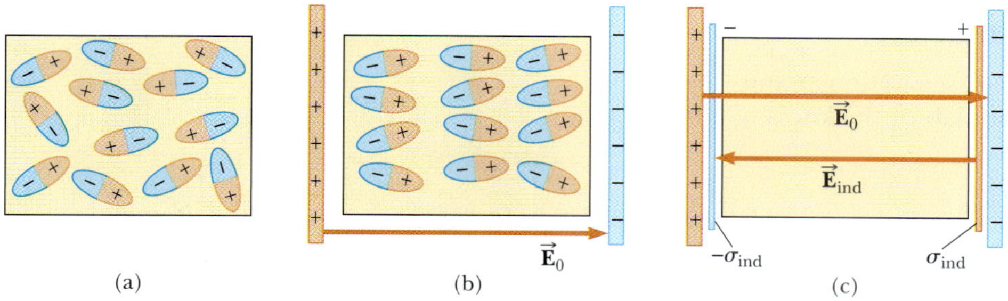

Figure 26.22 (a) Polar molecules are randomly oriented in the absence of an external electric field. (b) When an external electric field is applied, the molecules partially align with the field. (c) The charged edges of the dielectric can be modeled as an additional pair of parallel plates establishing an electric field $\vec{\mathbf{E}}_{\text{ind}}$ in the direction opposite that of $\vec{\mathbf{E}}_0$.

as shown in Figure 26.22a. When an external field $\vec{\mathbf{E}}_0$ due to charges on the capacitor plates is applied, a torque is exerted on the dipoles, causing them to partially align with the field as shown in Figure 26.22b. The dielectric is now polarized. The degree of alignment of the molecules with the electric field depends on temperature and the magnitude of the field. In general, the alignment increases with decreasing temperature and with increasing electric field.

If the molecules of the dielectric are nonpolar, the electric field due to the plates produces an induced polarization in the molecule. These induced dipole moments tend to align with the external field, and the dielectric is polarized. Therefore, a dielectric can be polarized by an external field regardless of whether the molecules in the dielectric are polar or nonpolar.

With these ideas in mind, consider a slab of dielectric material placed between the plates of a capacitor so that it is in a uniform electric field $\vec{\mathbf{E}}_0$ as shown in Figure 26.22b. The electric field due to the plates is directed to the right and polarizes the dielectric. The net effect on the dielectric is the formation of an *induced* positive surface charge density σ_{ind} on the right face and an equal-magnitude negative surface charge density $-\sigma_{\text{ind}}$ on the left face as shown in Figure 26.22c. Because we can model these surface charge distributions as being due to charged parallel plates, the induced surface charges on the dielectric give rise to an induced electric field $\vec{\mathbf{E}}_{\text{ind}}$ in the direction opposite the external field $\vec{\mathbf{E}}_0$. Therefore, the net electric field $\vec{\mathbf{E}}$ in the dielectric has a magnitude

$$E = E_0 - E_{\text{ind}} \tag{26.22}$$

In the parallel-plate capacitor shown in Figure 26.23, the external field E_0 is related to the charge density σ on the plates through the relationship $E_0 = \sigma/\epsilon_0$. The induced electric field in the dielectric is related to the induced charge density σ_{ind} through the relationship $E_{\text{ind}} = \sigma_{\text{ind}}/\epsilon_0$. Because $E = E_0/\kappa = \sigma/\kappa\epsilon_0$, substitution into Equation 26.22 gives

$$\frac{\sigma}{\kappa\epsilon_0} = \frac{\sigma}{\epsilon_0} - \frac{\sigma_{\text{ind}}}{\epsilon_0}$$

$$\sigma_{\text{ind}} = \left(\frac{\kappa - 1}{\kappa}\right)\sigma \tag{26.23}$$

Because $\kappa > 1$, this expression shows that the charge density σ_{ind} induced on the dielectric is less than the charge density σ on the plates. For instance, if $\kappa = 3$, the induced charge density is two-thirds the charge density on the plates. If no dielectric is present, then $\kappa = 1$ and $\sigma_{\text{ind}} = 0$ as expected. If the dielectric is replaced by an electrical conductor for which $E = 0$, however, Equation 26.22 indicates that $E_0 = E_{\text{ind}}$, which corresponds to $\sigma_{\text{ind}} = \sigma$. That is, the surface charge induced on the conductor is equal in magnitude but opposite in sign to that on the plates, resulting in a net electric field of zero in the conductor (see Fig. 24.14).

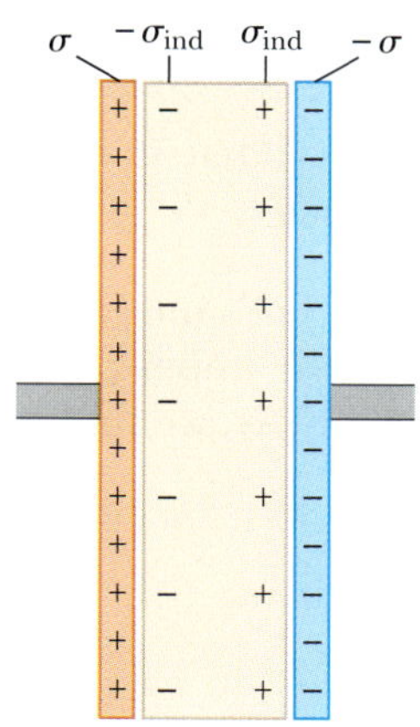

Figure 26.23 Induced charge on a dielectric placed between the plates of a charged capacitor. Notice that the induced charge density on the dielectric is *less* than the charge density on the plates.

EXAMPLE 26.7 Effect of a Metallic Slab

A parallel-plate capacitor has a plate separation d and plate area A. An uncharged metallic slab of thickness a is inserted midway between the plates.

(A) Find the capacitance of the device.

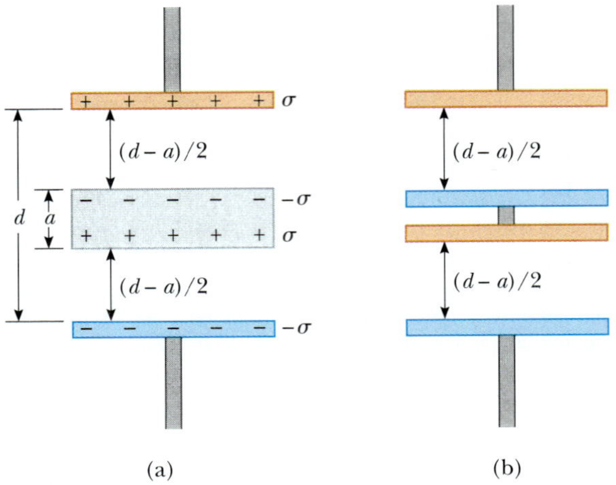

Figure 26.24 (Example 26.7) (a) A parallel-plate capacitor of plate separation d partially filled with a metallic slab of thickness a. (b) The equivalent circuit of the device in (a) consists of two capacitors in series, each having a plate separation $(d - a)/2$.

SOLUTION

Conceptualize Figure 26.24a shows the metallic slab between the plates of the capacitor. Any charge that appears on one plate of the capacitor must induce a charge of equal magnitude and opposite sign on the near side of the slab as shown in Figure 26.24a. Consequently, the net charge on the slab remains zero and the electric field inside the slab is zero.

Categorize The planes of charge on the metallic slab's upper and lower edges are identical to the distribution of charges on the plates of a capacitor. The metal between the slab's edges serves only to make an electrical connection between the edges. Therefore, we can model the edges of the slab as conducting planes and the bulk of the slab as a wire. As a result, the capacitor in Figure 26.24a is equivalent to two capacitors in series, each having a plate separation $(d - a)/2$ as shown in Figure 26.24b.

Analyze Use Equation 26.3 and the rule for adding two capacitors in series (Eq. 26.10) to find the equivalent capacitance in Figure 26.24b:

$$\frac{1}{C} = \frac{1}{C_1} + \frac{1}{C_2} = \frac{1}{\left[\dfrac{\epsilon_0 A}{(d-a)/2}\right]} + \frac{1}{\left[\dfrac{\epsilon_0 A}{(d-a)/2}\right]}$$

$$C = \frac{\epsilon_0 A}{d - a}$$

(B) Show that the capacitance of the original capacitor is unaffected by the insertion of the metallic slab if the slab is infinitesimally thin.

SOLUTION

In the result for part (A), let $a \to 0$:

$$C = \lim_{a \to 0}\left(\frac{\epsilon_0 A}{d - a}\right) = \frac{\epsilon_0 A}{d}$$

Finalize The result of part (B) is the original capacitance before the slab is inserted, which tells us that we can insert an infinitesimally thin metallic sheet between the plates of a capacitor without affecting the capacitance. We use this fact in the next example.

What If? What if the metallic slab in part (A) is not midway between the plates? How would that affect the capacitance?

Answer Let's imagine moving the slab in Figure 26.24a upward so that the distance between the upper edge of the slab and the upper plate is b. Then, the distance between the lower edge of the slab and the lower plate is $d - b - a$. As in part (A), we find the total capacitance of the series combination:

$$\frac{1}{C} = \frac{1}{C_1} + \frac{1}{C_2} = \frac{1}{\epsilon_0 A/b} + \frac{1}{\epsilon_0 A/(d - b - a)}$$

$$= \frac{b}{\epsilon_0 A} + \frac{d - b - a}{\epsilon_0 A} = \frac{d - a}{\epsilon_0 A} \quad \rightarrow \quad C = \frac{\epsilon_0 A}{d - a}$$

which is the same result as found in part (A). The capacitance is independent of the value of b, so it does not matter where the slab is located. In Figure 26.24b, when the central structure is moved up or down, the decrease in plate separation of one capacitor is compensated by the increase in plate separation for the other.

EXAMPLE 26.8 A Partially Filled Capacitor

A parallel-plate capacitor with a plate separation d has a capacitance C_0 in the absence of a dielectric. What is the capacitance when a slab of dielectric material of dielectric constant κ and thickness fd is inserted between the plates (Fig. 26.25a), where f is a fraction between 0 and 1?

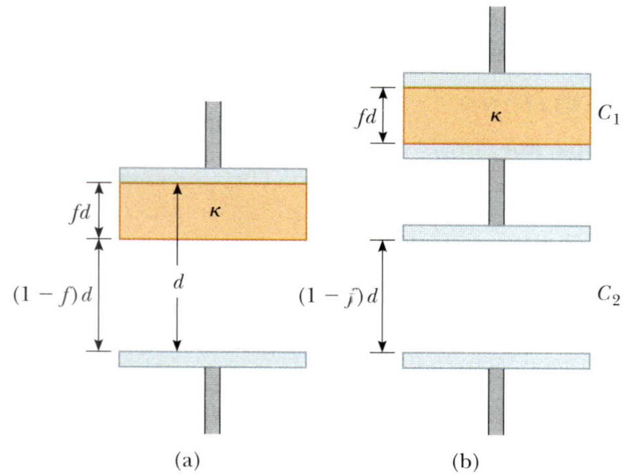

Figure 26.25 (Example 26.8) (a) A parallel-plate capacitor of plate separation d partially filled with a dielectric of thickness fd. (b) The equivalent circuit of the capacitor consists of two capacitors connected in series.

SOLUTION

Conceptualize In our previous discussions of dielectrics between the plates of a capacitor, the dielectric filled the volume between the plates. In this example, only part of the volume between the plates contains the dielectric material.

Categorize In Example 26.7, we found that an infinitesimally thin metallic sheet inserted between the plates of a capacitor does not affect the capacitance. Imagine sliding an infinitesimally thin metallic slab along the bottom face of the dielectric shown in Figure 26.25a. We can model this system as a series combination of two capacitors as shown in Figure 26.25b. One capacitor has a plate separation fd and is filled with a dielectric; the other has a plate separation $(1 - f)d$ and has air between its plates.

Analyze Evaluate the two capacitances in Figure 26.25b from Equation 26.15:

$$C_1 = \frac{\kappa\epsilon_0 A}{fd} \quad \text{and} \quad C_2 = \frac{\epsilon_0 A}{(1-f)d}$$

Find the equivalent capacitance C from Equation 26.10 for two capacitors combined in series:

$$\frac{1}{C} = \frac{1}{C_1} + \frac{1}{C_2} = \frac{fd}{\kappa\epsilon_0 A} + \frac{(1-f)d}{\epsilon_0 A}$$

$$\frac{1}{C} = \frac{fd}{\kappa\epsilon_0 A} + \frac{\kappa(1-f)d}{\kappa\epsilon_0 A} = \frac{f + \kappa(1-f)}{\kappa}\frac{d}{\epsilon_0 A}$$

Invert and substitute for the capacitance without the dielectric, $C_0 = \epsilon_0 A/d$:

$$C = \frac{\kappa}{f + \kappa(1-f)}\frac{\epsilon_0 A}{d} = \boxed{\frac{\kappa}{f + \kappa(1-f)} C_0}$$

Finalize Let's test this result for some known limits. If $f \rightarrow 0$, the dielectric should disappear. In this limit, $C \rightarrow C_0$, which is consistent with a capacitor with air between the plates. If $f \rightarrow 1$, the dielectric fills the volume between the plates. In this limit, $C \rightarrow \kappa C_0$, which is consistent with Equation 26.14.

Summary

ThomsonNOW™ Sign in at **www.thomsonedu.com** and go to ThomsonNOW to take a practice test for this chapter.

DEFINITIONS

A **capacitor** consists of two conductors carrying charges of equal magnitude and opposite sign. The **capacitance** C of any capacitor is the ratio of the charge Q on either conductor to the potential difference ΔV between them:

$$C \equiv \frac{Q}{\Delta V} \quad \textbf{(26.1)}$$

The capacitance depends only on the geometry of the conductors and not on an external source of charge or potential difference. The SI unit of capacitance is coulombs per volt, or the **farad** (F): 1 F = 1 C/V.

The **electric dipole moment** $\vec{\mathbf{p}}$ of an electric dipole has a magnitude

$$p \equiv 2aq \quad \textbf{(26.16)}$$

where $2a$ is the distance between the charges q and $-q$. The direction of the electric dipole moment vector is from the negative charge toward the positive charge.

CONCEPTS AND PRINCIPLES

If two or more capacitors are connected in parallel, the potential difference is the same across all capacitors. The equivalent capacitance of a **parallel combination** of capacitors is

$$C_{\text{eq}} = C_1 + C_2 + C_3 + \cdots \quad \textbf{(26.8)}$$

If two or more capacitors are connected in series, the charge is the same on all capacitors, and the equivalent capacitance of the **series combination** is given by

$$\frac{1}{C_{\text{eq}}} = \frac{1}{C_1} + \frac{1}{C_2} + \frac{1}{C_3} + \cdots \quad \textbf{(26.10)}$$

These two equations enable you to simplify many electric circuits by replacing multiple capacitors with a single equivalent capacitance.

Energy is stored in a capacitor because the charging process is equivalent to the transfer of charges from one conductor at a lower electric potential to another conductor at a higher potential. The energy stored in a capacitor with charge Q is

$$U = \frac{Q^2}{2C} = \tfrac{1}{2}Q\,\Delta V = \tfrac{1}{2}C(\Delta V)^2 \quad \textbf{(26.11)}$$

When a dielectric material is inserted between the plates of a capacitor, the capacitance increases by a dimensionless factor κ, called the **dielectric constant:**

$$C = \kappa C_0 \quad \textbf{(26.14)}$$

where C_0 is the capacitance in the absence of the dielectric.

The torque acting on an electric dipole in a uniform electric field $\vec{\mathbf{E}}$ is

$$\vec{\boldsymbol{\tau}} = \vec{\mathbf{p}} \times \vec{\mathbf{E}} \quad \textbf{(26.18)}$$

The potential energy of the system of an electric dipole in a uniform external electric field $\vec{\mathbf{E}}$ is

$$U = -\vec{\mathbf{p}} \cdot \vec{\mathbf{E}} \quad \textbf{(26.20)}$$

Questions

☐ denotes answer available in *Student Solutions Manual/Study Guide;* **O** denotes objective question

1. **O** True or False? (a) From the definition of capacitance $C = Q/\Delta V$, it follows that an uncharged capacitor has a capacitance of zero. (b) As described by the definition of capacitance, the potential difference across an uncharged capacitor is zero.

2. If you are given three different capacitors C_1, C_2, and C_3, how many different combinations of capacitance can you produce?

3. **O** By what factor is the capacitance of a metal sphere multiplied if its volume is tripled? (a) 9 (b) 3 (c) $3^{2/3}$ (d) $3^{1/3}$ (e) 1 (f) $3^{-1/3}$ (g) $3^{-2/3}$ (h) $\frac{1}{3}$

4. **O** A capacitor with very large capacitance is in series with another capacitor with very small capacitance. What is the equivalent capacitance of the combination? (a) slightly greater than the capacitance of the large capacitor (b) slightly less than the capacitance of the large capacitor (c) slightly greater than the capacitance of the small capacitor (d) slightly less than the capacitance of the small capacitor.

5. **O** **(i)** Rank the following six capacitors in order from greatest to smallest capacitance, noting any cases of equality. (a) a 20-μF capacitor with a 4-V potential difference between its plates (b) a 30-μF capacitor with charges of magnitude 90 μC on each plate (c) a capacitor with charges of magnitude 80 μC on its plates, differing by 2 V in potential (d) a 10-μF capacitor storing 125 μJ (e) a capacitor storing energy 250 μJ with a 10-V potential difference (f) a capacitor storing charge 120 μC and energy 360 μJ **(ii)** Rank the same capacitors from largest to smallest according to the potential difference between the plates. **(iii)** Rank the capacitors in the order of the magnitudes of the charges on their plates. **(iv)** Rank the capacitors in the order of the energy they store.

6. The sum of the charges on both plates of a capacitor is zero. What does a capacitor store?

7. **O** **(i)** What happens to the magnitude of the charge on each plate of a capacitor if the potential difference between the conductors is doubled? (a) It becomes four times larger. (b) It becomes two times larger. (c) It is unchanged. (d) It becomes one-half as large. (e) It becomes one-fourth as large. **(ii)** If the potential difference across a capacitor is doubled, what happens to the energy stored? Choose from the same possibilities.

8. **O** A parallel-plate capacitor is charged and then is disconnected from the battery. By what factor does the stored energy change when the plate separation is then doubled? (a) It becomes four times larger. (b) It becomes two times larger. (c) It stays the same. (d) It becomes one-half as large (e) It becomes one-fourth as large.

9. **O** You charge a parallel-plate capacitor, remove it from the battery, and prevent the wires connected to the plates from touching each other. When you increase the plate separation, does each of the following quantities (a) increase, (b) decrease, or (c) stay the same? **(i)** C **(ii)** Q **(iii)** E between the plates **(iv)** ΔV **(v)** the energy stored in the capacitor

10. **O** Repeat Question 9, but this time answer for the situation in which the battery remains connected to the capacitor while you increase the plate separation.

11. Because the charges on the plates of a parallel-plate capacitor are opposite in sign, they attract each other. Hence, it would take positive work to increase the plate separation. What type of energy in the system changes due to the external work done in this process?

12. Explain why the work needed to move a particle with charge Q through a potential difference ΔV is $W = Q\,\Delta V$, whereas the energy stored in a charged capacitor is $U = \frac{1}{2}Q\,\Delta V$. Where does the $\frac{1}{2}$ factor come from?

13. **O** Assume a device is designed to obtain a large potential difference by first charging a bank of capacitors connected in parallel and then activating a switch arrangement that in effect disconnects the capacitors from the charging source and from each other and reconnects them all in a series arrangement. The group of charged capacitors is then discharged in series. What is the maximum potential difference that can be obtained in this manner by using ten capacitors each of 500 μF and a charging source of 800 V? (a) 80 kV (b) 8 kV (c) 2.5 kV (d) 800 V (e) 80 V (f) 8 V (g) 0

14. An air-filled capacitor is charged, then disconnected from the power supply, and finally connected to a voltmeter. Explain how and why the potential difference changes when a dielectric is inserted between the plates of the capacitor.

15. **O** A fully charged parallel-plate capacitor remains connected to a battery while you slide a dielectric between the plates. Do the following quantities (a) increase, (b) decrease, or (c) stay the same? **(i)** C **(ii)** Q **(iii)** E between the plates **(iv)** ΔV **(v)** the energy stored in the capacitor

16. Assume you want to increase the maximum operating voltage of a parallel-plate capacitor. Describe how you can do that with a fixed plate separation.

17. If you were asked to design a capacitor in which small size and large capacitance were required, what factors would be important in your design?

Problems

WebAssign The Problems from this chapter may be assigned online in WebAssign.

ThomsonNOW™ Sign in at **www.thomsonedu.com** and go to ThomsonNOW to assess your understanding of this chapter's topics with additional quizzing and conceptual questions.

1, 2, 3 denotes straightforward, intermediate, challenging; □ denotes full solution available in *Student Solutions Manual/Study Guide;* ▲ denotes coached solution with hints available at **www.thomsonedu.com;** ▭ denotes developing symbolic reasoning; ● denotes asking for qualitative reasoning; ▪ denotes computer useful in solving problem

Section 26.1 Definition of Capacitance

1. (a) How much charge is on each plate of a 4.00-μF capacitor when it is connected to a 12.0-V battery? (b) If this same capacitor is connected to a 1.50-V battery, what charge is stored?

2. Two conductors having net charges of $+10.0$ μC and -10.0 μC have a potential difference of 10.0 V between them. (a) Determine the capacitance of the system. (b) What is the potential difference between the two conductors if the charges on each are increased to $+100$ μC and -100 μC?

Section 26.2 Calculating Capacitance

3. An isolated, charged conducting sphere of radius 12.0 cm creates an electric field of 4.90×10^4 N/C at a distance 21.0 cm from its center. (a) What is its surface charge density? (b) What is its capacitance?

4. Regarding the Earth and a cloud layer 800 m above the Earth as the "plates" of a capacitor, calculate the capacitance of the Earth-cloud layer system. Assume the cloud layer has an area of 1.00 km^2 and the air between the cloud and the ground is pure and dry. Assume charge builds up on the cloud and on the ground until a uniform electric field of 3.00×10^6 N/C throughout the space between them makes the air break down and conduct electricity as a lightning bolt. What is the maximum charge the cloud can hold?

5. ▲ An air-filled capacitor consists of two parallel plates, each with an area of 7.60 cm^2, separated by a distance of 1.80 mm. A 20.0-V potential difference is applied to these plates. Calculate (a) the electric field between the plates, (b) the surface charge density, (c) the capacitance, and (d) the charge on each plate.

6. A variable air capacitor used in a radio tuning circuit is made of N semicircular plates each of radius R and positioned a distance d from its neighbors, to which it is electrically connected. As shown in Figures 26.16 and P26.6, a second identical set of plates is enmeshed with the first set. Each plate in the second set is halfway between two plates of the first set. The second set can rotate as a unit. Determine the capacitance as a function of the angle of rotation θ, where $\theta = 0$ corresponds to the maximum capacitance.

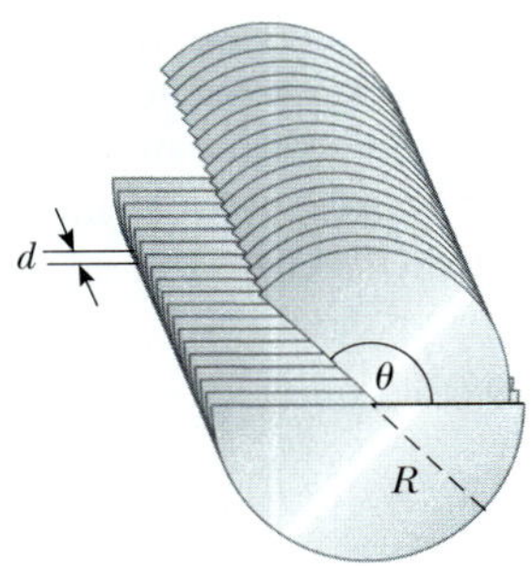

Figure P26.6

7. When a potential difference of 150 V is applied to the plates of a parallel-plate capacitor, the plates carry a surface charge density of 30.0 nC/cm^2. What is the spacing between the plates?

8. A small object of mass m carries a charge q and is suspended by a thread between the vertical plates of a parallel-plate capacitor. The plate separation is d. If the thread makes an angle θ with the vertical, what is the potential difference between the plates?

9. ▲ A 50.0-m length of coaxial cable has an inner conductor that has a diameter of 2.58 mm and carries a charge of 8.10 μC. The surrounding conductor has an inner diameter of 7.27 mm and a charge of -8.10 μC. (a) What is the capacitance of this cable? (b) What is the potential difference between the two conductors? Assume the region between the conductors is air.

10. ● A 10.0-μF capacitor has plates with vacuum between them. Each plate carries a charge of magnitude 1 000 μC. A particle with charge -3.00 μC and mass 2.00×10^{-16} kg is fired from the positive plate toward the negative plate with an initial speed of 2.00×10^6 m/s. Does the particle reach the negative plate? Explain how you can tell. If it does, what is its impact speed? If it does not, what fraction of the way across the capacitor does it travel?

11. An air-filled spherical capacitor is constructed with inner and outer shell radii of 7.00 and 14.0 cm, respectively. (a) Calculate the capacitance of the device. (b) What potential difference between the spheres results in a charge of 4.00 μC on the capacitor?

Section 26.3 Combinations of Capacitors

12. Two capacitors, $C_1 = 5.00$ μF and $C_2 = 12.0$ μF, are connected in parallel, and the resulting combination is connected to a 9.00-V battery. Find (a) the equivalent capacitance of the combination, (b) the potential difference across each capacitor, and (c) the charge stored on each capacitor.

13. **What If?** The two capacitors of Problem 12 are now connected in series and to a 9.00-V battery. Find (a) the equivalent capacitance of the combination, (b) the potential difference across each capacitor, and (c) the charge on each capacitor.

14. ● Three capacitors are connected to a battery as shown in Figure P26.14. Their capacitances are $C_1 = 3C$, $C_2 = C$, and $C_3 = 5C$. (a) What is the equivalent capacitance of this set of capacitors? (b) State the ranking of the capacitors according to the charge they store, from largest to smallest. (c) Rank the capacitors according to the potential differences across them, from largest to smallest. (d) **What If?** Assume C_3 is increased. Explain what happens to the charge stored by each of the capacitors.

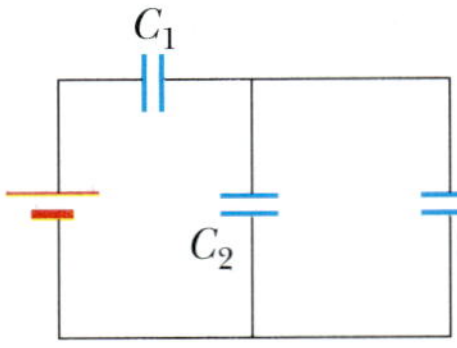

Figure P26.14

15. Two capacitors give an equivalent capacitance of 9.00 pF when connected in parallel and give an equivalent capacitance of 2.00 pF when connected in series. What is the capacitance of each capacitor?

16. Two capacitors give an equivalent capacitance of C_p when connected in parallel and an equivalent capacitance of C_s when connected in series. What is the capacitance of each capacitor?

17. ▲ Four capacitors are connected as shown in Figure P26.17. (a) Find the equivalent capacitance between

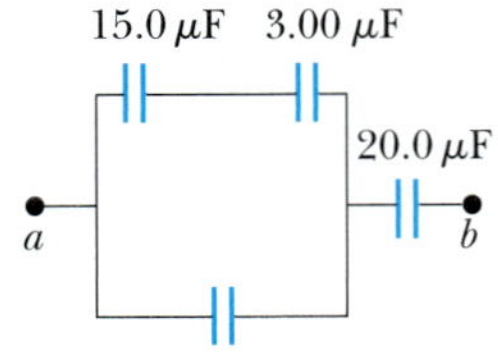

Figure P26.17

points a and b. (b) Calculate the charge on each capacitor, taking $\Delta V_{ab} = 15.0$ V.

18. According to its design specification, the timer circuit delaying the closing of an elevator door is to have a capacitance of 32.0 μF between two points A and B. (a) When one circuit is being constructed, the inexpensive but durable capacitor installed between these two points is found to have capacitance 34.8 μF. To meet the specification, one additional capacitor can be placed between the two points. Should it be in series or in parallel with the 34.8-μF capacitor? What should be its capacitance? (b) **What If?** The next circuit comes down the assembly line with capacitance 29.8 μF between A and B. To meet the specification, what additional capacitor should be installed in series or in parallel in that circuit?

19. Consider the circuit shown in Figure P26.19, where $C_1 = 6.00\ \mu$F, $C_2 = 3.00\ \mu$F, and $\Delta V = 20.0$ V. Capacitor C_1 is first charged by closing switch S_1. Switch S_1 is then opened, and the charged capacitor is connected to the uncharged capacitor by closing S_2. Calculate the initial charge acquired by C_1 and the final charge on each capacitor.

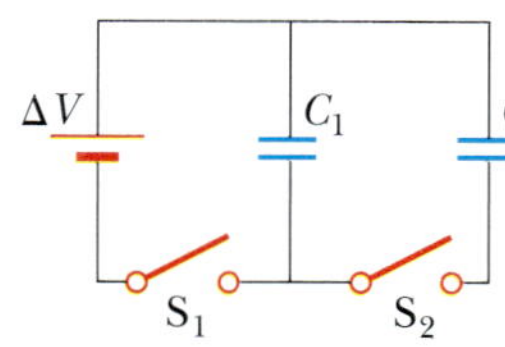

Figure P26.19

20. Consider three capacitors C_1, C_2, C_3, and a battery. If C_1 is connected to the battery, the charge on C_1 is 30.8 μC. Now C_1 is disconnected, discharged, and connected in series with C_2. When the series combination of C_2 and C_1 is connected across the battery, the charge on C_1 is 23.1 μC. The circuit is disconnected and the capacitors discharged. Capacitor C_3, capacitor C_1, and the battery are connected in series, resulting in a charge on C_1 of 25.2 μC. If, after being disconnected and discharged, C_1, C_2, and C_3 are connected in series with one another and with the battery, what is the charge on C_1?

21. A group of identical capacitors is connected first in series and then in parallel. The combined capacitance in parallel is 100 times larger than for the series connection. How many capacitors are in the group?

22. Some physical systems possessing capacitance continuously distributed over space can be modeled as an infinite array of discrete circuit elements. Examples are a microwave waveguide and the axon of a nerve cell. To practice analysis of an infinite array, determine the equivalent capacitance C between terminals X and Y of the infinite set of capacitors represented in Figure P26.22. Each capacitor has capacitance C_0. *Suggestion:* Imagine that the ladder is cut at the line AB and note that the equivalent capacitance of the infinite section to the right of AB is also C.

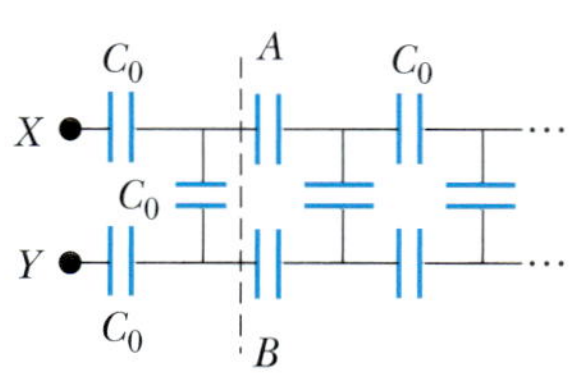

Figure P26.22

23. Find the equivalent capacitance between points a and b for the group of capacitors connected as shown in Figure P26.23. Take $C_1 = 5.00\ \mu$F, $C_2 = 10.0\ \mu$F, and $C_3 = 2.00\ \mu$F.

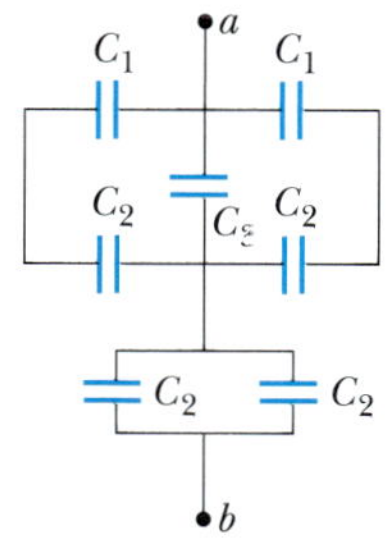

Figure P26.23 Problems 23 and 24.

24. For the network described in Problem 23, what charge is stored on C_3 if the potential difference between points a and b is 60.0 V?

25. Find the equivalent capacitance between points a and b in the combination of capacitors shown in Figure P26.25.

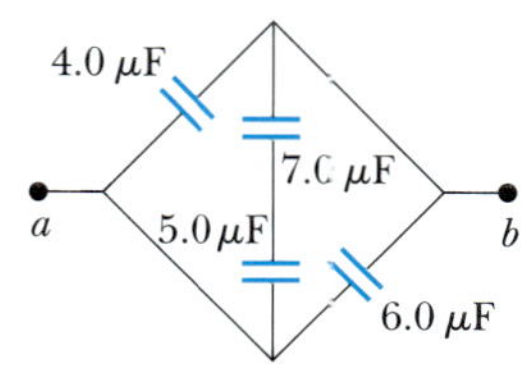

Figure P26.25

Section 26.4 Energy Stored in a Charged Capacitor

26. The immediate cause of many deaths is ventricular fibrillation, which is an uncoordinated quivering of the heart. An electric shock to the chest can cause momentary paralysis of the heart muscle, after which the heart sometimes resumes its proper beating. One type of *defibrillator* (Fig. 26.13) applies a strong electric shock to the chest over a time interval of a few milliseconds. This device contains a capacitor of several microfarads, charged to several thousand volts. Electrodes called paddles, about 8 cm across and coated with conducting paste, are held against the chest on both sides of the heart. Their handles are insulated to prevent injury to the operator, who calls "Clear!" and pushes a button on one paddle to discharge the capacitor through the patient's chest. Assume an energy of 300 J is to be delivered from a 30.0-μF capacitor. To what potential difference must it be charged?

27. (a) A 3.00-μF capacitor is connected to a 12.0-V battery. How much energy is stored in the capacitor? (b) Had the capacitor been connected to a 6.00-V battery, how much energy would have been stored?

28. Two capacitors, $C_1 = 25.0\ \mu$F and $C_2 = 5.00\ \mu$F, are connected in parallel and charged with a 100-V power supply. (a) Draw a circuit diagram and calculate the total energy stored in the two capacitors. (b) **What If?** What potential difference would be required across the same two capacitors connected in series for the combination to store the same amount of energy as in part (a)? Draw a circuit diagram of this circuit.

29. ▲ A parallel-plate capacitor has a charge Q and plates of area A. What force acts on one plate to attract it toward the other plate? Because the electric field between the plates is $E = Q/A\epsilon_0$, you might think the force is $F = QE = Q^2/A\epsilon_0$. This conclusion is wrong because the field E includes contributions from both plates and the field created by the positive plate cannot exert any force on the positive plate. Show that the force exerted on each plate is actually $F = Q^2/2\epsilon_0 A$. *Suggestion:* Let $C = \epsilon_0 A/x$ for an arbitrary plate separation x and note that the work done in separating the two charged plates is $W = \int F\,dx$.

30. The circuit in Figure P26.30 consists of two identical, parallel metal plates connected by identical metal springs to a 100-V battery. With the switch open, the plates are uncharged, are separated by a distance $d = 8.00$ mm, and have a capacitance $C = 2.00\ \mu$F. When the switch is closed, the distance between the plates decreases by a factor of 0.500. (a) How much charge collects on each plate? (b) What is the spring constant for each spring? *Suggestion:* Use the result of Problem 29.

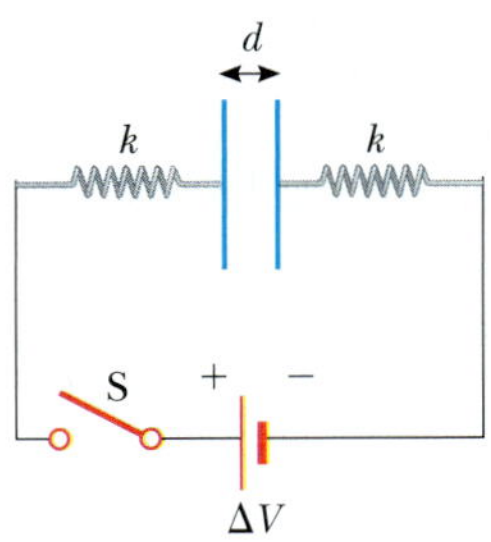

Figure P26.30

31. As a person moves about in a dry environment, electric charge accumulates on the person's body. Once it is at high voltage, either positive or negative, the body can discharge via sometimes noticeable sparks and shocks. Consider a human body isolated from ground, with the typical capacitance 150 pF. (a) What charge on the body will produce a potential of 10.0 kV? (b) Sensitive electronic devices can be destroyed by electrostatic discharge from a person. A particular device can be destroyed by a discharge releasing an energy of 250 μJ. To what voltage on the body does this situation correspond?

32. ● Two identical parallel-plate capacitors, each with capacitance C, are charged to potential difference ΔV and connected in parallel. Then the plate separation in one of the capacitors is doubled. (a) Find the total energy of the system of two capacitors *before* the plate separation is doubled. (b) Find the potential difference across each capacitor *after* the plate separation is doubled. (c) Find the total energy of the system *after* the plate separation is doubled. (d) Reconcile the difference in the answers to parts (a) and (c) with the law of conservation of energy.

33. Show that the energy associated with a conducting sphere of radius R and charge Q surrounded by a vacuum is $U = k_e Q^2/2R$.

34. Consider two conducting spheres with radii R_1 and R_2 separated by a distance much greater than either radius. A total charge Q is shared between the spheres, subject to the condition that the electric potential energy of the system has the smallest possible value. The total charge Q is equal to $q_1 + q_2$, where q_1 represents the charge on the first sphere and q_2 the charge on the second. Because the spheres are very far apart, you can assume the charge of each is uniformly distributed over its surface. You may use the result of Problem 33. (a) Determine the values of q_1 and q_2 in terms of Q, R_1, and R_2. (b) Show that the potential difference between the spheres is zero. (We saw in Chapter 25 that two conductors joined by a conducting wire are at the same potential in a static situation. This problem illustrates the general principle that charge on a conductor distributes itself so that the electric potential energy of the system is a minimum.)

35. **Review problem.** A certain storm cloud has a potential of 1.00×10^8 V relative to a tree. If, during a lightning storm, 50.0 C of charge is transferred through this potential difference and 1.00% of the energy is absorbed by the tree, how much sap in the tree can be boiled away? Model the sap as water initially at 30.0°C. Water has a specific heat of 4 186 J/kg · °C, a boiling point of 100°C, and a latent heat of vaporization of 2.26×10^6 J/kg.

Section 26.5 Capacitors with Dielectrics

36. (a) How much charge can be placed on a capacitor with air between the plates before it breaks down if the area of each of the plates is 5.00 cm^2? (b) **What If?** Find the maximum charge if polystyrene is used between the plates instead of air.

37. Determine (a) the capacitance and (b) the maximum potential difference that can be applied to a Teflon-filled parallel-plate capacitor having a plate area of 1.75 cm^2 and plate separation of 0.040 0 mm.

38. A supermarket sells rolls of aluminum foil, of plastic wrap, and of waxed paper. Describe a capacitor made from such materials. Compute order-of-magnitude estimates for its capacitance and its breakdown voltage.

39. A commercial capacitor is to be constructed as shown in Figure 26.15a. This particular capacitor is made from two strips of aluminum separated by a strip of paraffin-coated paper. Each strip of foil and paper is 7.00 cm wide. The foil is 0.004 00 mm thick, and the paper is 0.025 0 mm thick and has a dielectric constant of 3.70. What length should the strips have if a capacitance of 9.50×10^{-8} F is desired before the capacitor is rolled up? (Adding a second strip of paper and rolling the capacitor effectively doubles its capacitance by allowing charge storage on both sides of each strip of foil.)

40. A parallel-plate capacitor in air has a plate separation of 1.50 cm and a plate area of 25.0 cm^2. The plates are charged to a potential difference of 250 V and disconnected from the source. The capacitor is then immersed in distilled water. Determine (a) the charge on the plates before and after immersion, (b) the capacitance and potential difference after immersion, and (c) the change in energy of the capacitor. Assume the liquid is an insulator.

41. Each capacitor in the combination shown in Figure P26.41 has a breakdown voltage of 15.0 V. What is the breakdown voltage of the combination?

2 = intermediate; 3 = challenging; □ = SSM/SG; ▲ = ThomsonNOW; = symbolic reasoning; ● = qualitative reasoning

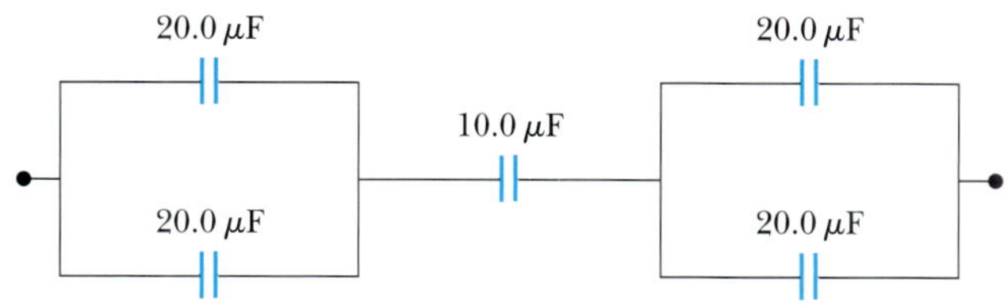

Figure P26.41

Section 26.6 Electric Dipole in an Electric Field

42. A small, rigid object carries positive and negative 3.50-nC charges. It is oriented so that the positive charge has coordinates (−1.20 mm, 1.10 mm) and the negative charge is at the point (1.40 mm, −1.30 mm). (a) Find the electric dipole moment of the object. The object is placed in an electric field $\vec{\mathbf{E}} = (7\,800\hat{\mathbf{i}} - 4\,900\hat{\mathbf{j}})$ N/C. (b) Find the torque acting on the object. (c) Find the potential energy of the object–field system when the object is in this orientation. (d) Assuming the orientation of the object can change, find the difference between the maximum and minimum potential energies of the system.

43. A small object with electric dipole moment $\vec{\mathbf{p}}$ is placed in a nonuniform electric field $\vec{\mathbf{E}} = E(x)\hat{\mathbf{i}}$. That is, the field is in the x direction and its magnitude depends on the coordinate x. Let θ represent the angle between the dipole moment and the x direction. (a) Prove that the net force on the dipole is

$$F = p\left(\frac{dE}{dx}\right)\cos\theta$$

acting in the direction of increasing field. (b) Consider a spherical balloon centered at the origin, with radius 15.0 cm and carrying charge 2.00 μC. Evaluate dE/dx at the point (16 cm, 0, 0). Assume a water droplet at this point has an induced dipole moment of $6.30\hat{\mathbf{i}}$ nC · m. Find the net force exerted on it.

Section 26.7 An Atomic Description of Dielectrics

44. The general form of Gauss's law describes how a charge creates an electric field in a material, as well as in vacuum:

$$\oint \vec{\mathbf{E}} \cdot d\vec{\mathbf{A}} = \frac{q_{\text{in}}}{\epsilon}$$

where $\epsilon = \kappa\epsilon_0$ is the permittivity of the material. (a) A sheet with charge Q uniformly distributed over its area A is surrounded by a dielectric. Show that the sheet creates a uniform electric field at nearby points, with magnitude $E = Q/2A\epsilon$. (b) Two large sheets of area A, carrying opposite charges of equal magnitude Q, are a small distance d apart. Show that they create uniform electric field in the space between them, with magnitude $E = Q/A\epsilon$. (c) Assume the negative plate is at zero potential. Show that the positive plate is at potential $Qd/A\epsilon$. (d) Show that the capacitance of the pair of plates is $A\epsilon/d = \kappa A\epsilon_0/d$.

45. The inner conductor of a coaxial cable has a radius of 0.800 mm, and the outer conductor's inside radius is 3.00 mm. The space between the conductors is filled with polyethylene, which has a dielectric constant of 2.30 and a dielectric strength of 18.0×10^6 V/m. What is the maximum potential difference that this cable can withstand?

Additional Problems

46. Two large, parallel metal plates each of area A are oriented horizontally and separated by a distance $3d$. A grounded conducting wire joins them, and initially each plate carries no charge. Now a third identical plate carrying charge Q is inserted between the two plates, parallel to them and located a distance d from the upper plate as shown in Figure P26.46. (a) What induced charge appears on each of the two original plates? (b) What potential difference appears between the middle plate and each of the other plates?

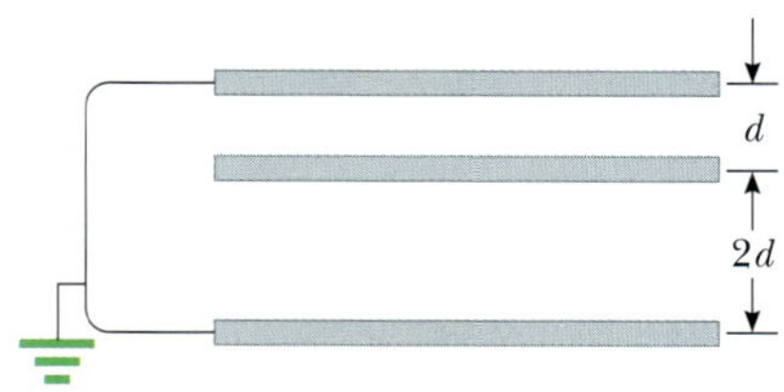

Figure P26.46

47. Four parallel metal plates P_1, P_2, P_3, and P_4, each of area 7.50 cm^2, are separated successively by a distance d = 1.19 mm as shown in Figure P26.47. P_1 is connected to the negative terminal of a battery, and P_2 is connected to the positive terminal. The battery maintains a potential difference of 12.0 V. (a) If P_3 is connected to the negative terminal, what is the capacitance of the three-plate system $P_1P_2P_3$? (b) What is the charge on P_2? (c) If P_4 is now connected to the positive terminal, what is the capacitance of the four-plate system $P_1P_2P_3P_4$? (d) What is the charge on P_4?

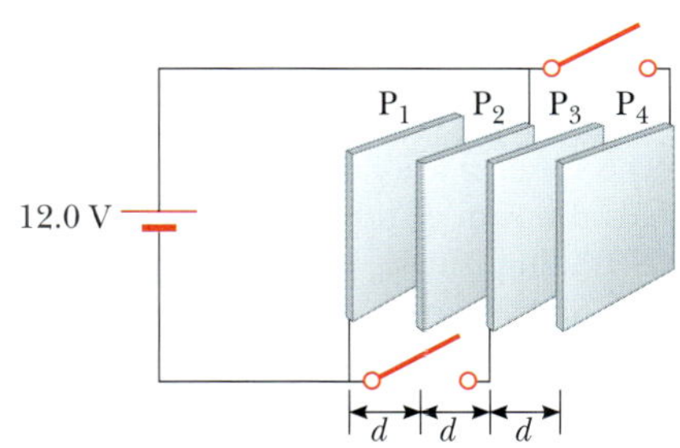

Figure P26.47

48. One conductor of an overhead electric transmission line is a long aluminum wire 2.40 cm in radius. Suppose it carries a charge per length of 1.40 μC/m at a particular moment and its potential is 345 kV. Find the potential 12.0 m below the wire. Ignore the other conductors of the transmission line and assume the electric field is radial everywhere.

49. A 2.00-nF parallel-plate capacitor is charged to an initial potential difference ΔV_i = 100 V and is then isolated. The dielectric material between the plates is mica, with a dielectric constant of 5.00. (a) How much work is required to withdraw the mica sheet? (b) What is the potential difference across the capacitor after the mica is withdrawn?

50. (a) Draw a circuit diagram showing four capacitors between two points a and b for which the following expression determines the equivalent capacitance:

$$\frac{1}{\dfrac{1}{30\ \mu\text{F}}+\dfrac{1}{20\ \mu\text{F}+C_1}}+50\ \mu\text{F}=70\ \mu\text{F}$$

(b) Find the value of C_1. (c) Assume a 6.00-V battery is connected between a and b. Find the potential difference across each of the individual capacitors and the charge on each.

51. ▲ A parallel-plate capacitor is constructed using a dielectric material whose dielectric constant is 3.00 and whose dielectric strength is 2.00×10^8 V/m. The desired capacitance is 0.250 μF, and the capacitor must withstand a maximum potential difference of 4.00 kV. Find the minimum area of the capacitor plates.

52. ● A horizontal, parallel-plate capacitor with vacuum between its plates has a capacitance of 25.0 μF. A nonconducting liquid with dielectric constant 6.50 is poured into the space between the plates, filling up a fraction f of its volume. (a) Find the new capacitance as a function of f. (b) What should you expect the capacitance to be when $f = 0$? Does your expression from part (a) agree with your answer? (c) What capacitance should you expect when $f = 1$? Does the expression from part (a) agree with your answer? (d) Charges of magnitude 300 μC are placed on the plates of the partially filled capacitor. What can you determine about the induced charge on the free upper surface of the liquid? How does this charge depend on f?

53. (a) Two spheres have radii a and b, and their centers are a distance d apart. Show that the capacitance of this system is

$$C=\frac{4\pi\epsilon_0}{\dfrac{1}{a}+\dfrac{1}{b}-\dfrac{2}{d}}$$

provided d is large compared with a and b. *Suggestion:* Because the spheres are far apart, assume the potential of each equals the sum of the potentials due to each sphere. When calculating those potentials, assume $V = k_eQ/r$ applies. (b) Show that as d approaches infinity, the above result reduces to that of two spherical capacitors in series.

54. A 10.0-μF capacitor is charged to 15.0 V. It is next connected in series with an uncharged 5.00-μF capacitor. The series combination is finally connected across a 50.0-V battery as diagrammed in Figure P26.54. Find the new potential differences across the 5.00-μF and 10.0-μF capacitors.

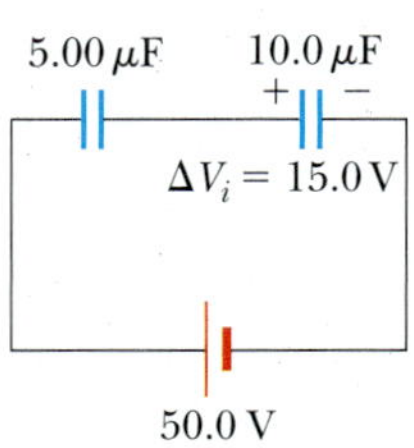

Figure P26.54

55. ● When considering the energy supply for an automobile, the energy per unit mass (in joules per kilogram) of the energy source is an important parameter. Using the following data, explain how the energy per unit mass compares among gasoline, lead-acid batteries, and capacitors. (The ampere A will be introduced in Chapter 27 as the SI unit of electric current. 1 A = 1 C/s.)

Gasoline: 126 000 Btu/gal; density = 670 kg/m^3.

Lead-acid battery: 12.0 V; 100 A·h; mass = 16.0 kg.

Capacitor: potential difference at full charge = 12.0 V; capacitance = 0.100 F; mass = 0.100 kg.

56. A capacitor is constructed from two square, metallic plates of sides ℓ and separation d. Charges $+Q$ and $-Q$ are placed on the plates, and the power supply is then removed. A material of dielectric constant κ is inserted a distance x into the capacitor as shown in Figure P26.56. Assume d is much smaller than x. (a) Find the equivalent capacitance of the device. (b) Calculate the energy stored in the capacitor. (c) Find the direction and magnitude of the force exerted by the plates on the dielectric. (d) Obtain a numerical value for the force when $x = \ell/2$, assuming $\ell = 5.00$ cm, $d = 2.00$ mm, the dielectric is glass ($\kappa = 4.50$), and the capacitor was charged to 2 000 V before the dielectric was inserted. *Suggestion:* The system can be considered as two capacitors connected in parallel.

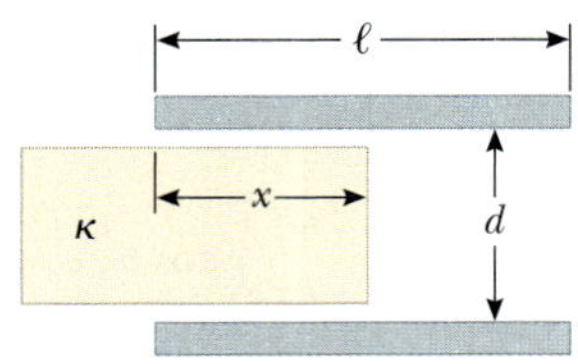

Figure P26.56 Problems 56 and 57.

57. ● Two square plates of sides ℓ are placed parallel to each other with separation d as suggested in Figure P26.56. You may assume d is much less than ℓ. The plates carry uniformly distributed static charges $+Q_0$ and $-Q_0$. A block of metal has width ℓ, length ℓ, and thickness slightly less than d. It is inserted a distance x into the space between the plates. The charges on the plates remain uniformly distributed as the block slides in. In a static situation, a metal prevents an electric field from penetrating inside it. The metal can be thought of as a perfect dielectric, with $\kappa \to \infty$. (a) Calculate the stored energy as a function of x. (b) Find the direction and magnitude of the force that acts on the metallic block. (c) The area of the advancing front face of the block is essentially equal to ℓd. Considering the force on the block as acting on this face, find the stress (force per area) on it. (d) Express the energy density in the electric field between the charged plates in terms of Q_0, ℓ, d, and ϵ_0. Explain how the answers to parts (c) and (d) compare with each other.

58. ● To repair a power supply for a stereo amplifier, an electronics technician needs a 100-μF capacitor capable of withstanding a potential difference of 90 V between the plates. The immediately available supply is a box of five 100-μF capacitors, each having a maximum voltage capability of 50 V. Can the technician use one of the capacitors from the box? Can she substitute a combination of these capacitors that has the proper electrical characteristics? Will the technician use all the capacitors in the box? Explain your answers. In a combination of capacitors, what will be the maximum voltage across each of the capacitors used?

59. An isolated capacitor of unknown capacitance has been charged to a potential difference of 100 V. When the charged capacitor is then connected in parallel to an uncharged 10.0-μF capacitor, the potential difference across the combination is 30.0 V. Calculate the unknown capacitance.

60. A parallel-plate capacitor with plates of area LW and plate separation t has the region between its plates filled with wedges of two dielectric materials as shown in Figure P26.60. Assume t is much less than both L and W. (a) Determine its capacitance. (b) Should the capacitance be the same if the labels κ_1 and κ_2 are interchanged? Demonstrate that your expression does or does not have this property. (c) Show that if κ_1 and κ_2 approach equality to a common value κ, your result becomes the same as the capacitance of a capacitor containing a single dielectric: $C = \kappa\epsilon_0 LW/t$.

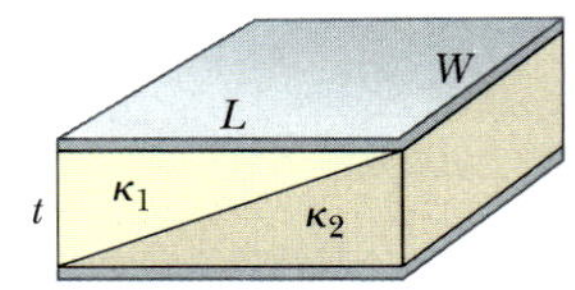

Figure P26.60

61. ● A parallel-plate capacitor of plate separation d is charged to a potential difference ΔV_0. A dielectric slab of thickness d and dielectric constant κ is introduced between the plates while the battery remains connected to the plates. (a) Show that the ratio of energy stored after the dielectric is introduced to the energy stored in the empty capacitor is $U/U_0 = \kappa$. Give a physical explanation for this increase in stored energy. (b) What happens to the charge on the capacitor? (Notice that this situation is not the same as in Example 26.5, in which the battery was removed from the circuit before the dielectric was introduced.)

62. Calculate the equivalent capacitance between points a and b in Figure P26.62. Notice that this system is not a simple series or parallel combination. *Suggestion:* Assume a potential difference ΔV between points a and b. Write expressions for ΔV_{ab} in terms of the charges and capacitances for the various possible pathways from a to b and require conservation of charge for those capacitor plates that are connected to each other.

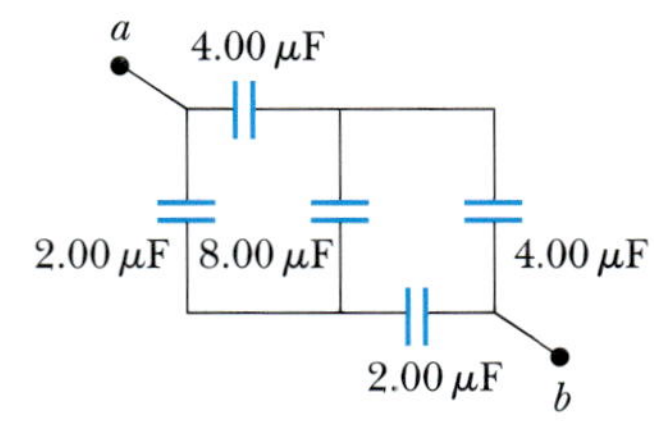

Figure P26.62

63. Capacitors $C_1 = 6.00\ \mu$F and $C_2 = 2.00\ \mu$F are charged as a parallel combination across a 250-V battery. The capacitors are disconnected from the battery and from each other. They are then connected positive plate to negative plate and negative plate to positive plate. Calculate the resulting charge on each capacitor.

64. Consider two long, parallel, and oppositely charged wires of radius r with their centers separated by a distance D that is much larger than r. Assuming the charge is distributed uniformly on the surface of each wire, show that the capacitance per unit length of this pair of wires is

$$\frac{C}{\ell} = \frac{\pi\epsilon_0}{\ln(D/r)}$$

65. Determine the equivalent capacitance of the combination shown in Figure P26.65. *Suggestion:* Consider the symmetry involved.

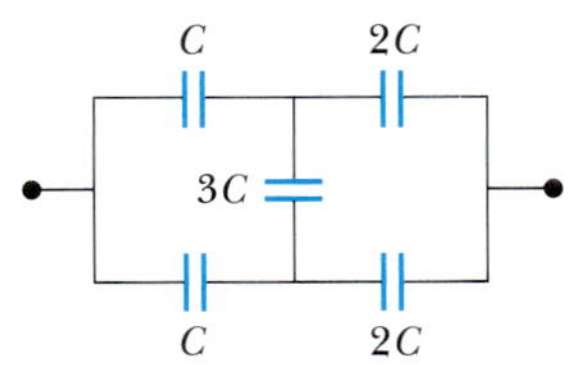

Figure P26.65

66. Example 26.1 explored a cylindrical capacitor of length ℓ with radii a and b for the two conductors. In the **What If?** section of that example, it was claimed that increasing ℓ by 10% is more effective in terms of increasing the capacitance than increasing a by 10% if $b > 2.85a$. Verify this claim mathematically.

Answers to Quick Quizzes

26.1 (d). The capacitance is a property of the physical system and does not vary with applied voltage. According to Equation 26.1, if the voltage is doubled, the charge is doubled.

26.2 (a). When the key is pressed, the plate separation is decreased and the capacitance increases. Capacitance depends only on how a capacitor is constructed and not on the external circuit.

26.3 (a). When connecting capacitors in series, the inverses of the capacitances add, resulting in a smaller overall equivalent capacitance.

26.4 (b). For a given voltage, the energy stored in a capacitor is proportional to C according to $U = C(\Delta V)^2/2$. Therefore, you want to maximize the equivalent capacitance. You do that by connecting the three capacitors in parallel so that the capacitances add.

26.5 (a). The dielectric constant of wood (and of all other insulating materials, for that matter) is greater than 1; therefore, the capacitance increases (Eq. 26.14). This increase is sensed by the stud finder's special circuitry, which causes an indicator on the device to light up.

2 = intermediate; 3 = challenging; □ = SSM/SG; ▲ = ThomsonNOW; = symbolic reasoning; ● = qualitative reasoning

These power lines transfer energy from the electric company to homes and businesses. The energy is transferred at a very high voltage, possibly hundreds of thousands of volts in some cases. Even though it makes power lines very dangerous, the high voltage results in less loss of energy due to resistance in the wires. (Telegraph Colour Library/FPG)

27 Current and Resistance

We now consider situations involving electric charges that are in motion through some region of space. We use the term *electric current,* or simply *current,* to describe the rate of flow of charge. Most practical applications of electricity deal with electric currents. For example, the battery in a flashlight produces a current in the filament of the bulb when the switch is turned on. A variety of home appliances operate on alternating current. In these common situations, current exists in a conductor such as a copper wire. Currents can also exist outside a conductor. For instance, a beam of electrons in a television picture tube constitutes a current.

This chapter begins with the definition of current. A microscopic description of current is given, and some factors that contribute to the opposition to the flow of charge in conductors are discussed. A classical model is used to describe electrical conduction in metals, and some limitations of this model are cited. We also define electrical resistance and introduce a new circuit element, the resistor. We conclude by discussing the rate at which energy is transferred to a device in an electric circuit.

27.1 Electric Current

In this section, we study the flow of electric charges through a piece of material. The amount of flow depends on both the material through which the charges are passing and the potential difference across the material. Whenever there is a net flow of charge through some region, an electric **current** is said to exist.

It is instructive to draw an analogy between water flow and current. In many localities, it is common practice to install low-flow showerheads in homes as a water-conservation measure. We quantify the flow of water from these and similar devices by specifying the amount of water that emerges during a given time interval, often measured in liters per minute. On a grander scale, we can characterize a river current by describing the rate at which the water flows past a particular location. For example, the flow over the brink at Niagara Falls is maintained at rates between 1 400 m^3/s and 2 800 m^3/s.

There is also an analogy between thermal conduction and current. In Section 20.7, we discussed the flow of energy by heat through a sample of material. The rate of energy flow is determined by the material as well as the temperature difference across the material as described by Equation 20.15.

To define current more precisely, suppose charges are moving perpendicular to a surface of area A as shown in Figure 27.1. (This area could be the cross-sectional area of a wire, for example.) **The current is the rate at which charge flows through this surface.** If ΔQ is the amount of charge that passes through this surface in a time interval Δt, the **average current** I_{avg} is equal to the charge that passes through A per unit time:

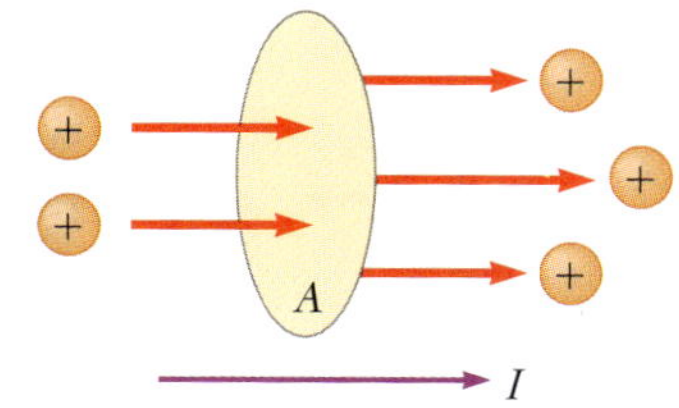

Figure 27.1 Charges in motion through an area A. The time rate at which charge flows through the area is defined as the current I. The direction of the current is the direction in which positive charges flow when free to do so.

$$I_{avg} = \frac{\Delta Q}{\Delta t} \qquad (27.1)$$

If the rate at which charge flows varies in time, the current varies in time; we define the **instantaneous current** I as the differential limit of average current:

$$I \equiv \frac{dQ}{dt} \qquad (27.2)$$

◄ Electric current

The SI unit of current is the **ampere** (A):

$$1 \text{ A} = 1 \text{ C/s} \qquad (27.3)$$

That is, 1 A of current is equivalent to 1 C of charge passing through a surface in 1 s.

The charged particles passing through the surface in Figure 27.1 can be positive, negative, or both. **It is conventional to assign to the current the same direction as the flow of positive charge.** In electrical conductors such as copper or aluminum, the current results from the motion of negatively charged electrons. Therefore, in an ordinary conductor, **the direction of the current is opposite the direction of flow of electrons.** For a beam of positively charged protons in an accelerator, however, the current is in the direction of motion of the protons. In some cases—such as those involving gases and electrolytes, for instance—the current is the result of the flow of both positive and negative charges. It is common to refer to a moving charge (positive or negative) as a mobile **charge carrier.**

PITFALL PREVENTION 27.1

"Current Flow" Is Redundant

The phrase *current flow* is commonly used, although it is technically incorrect because current *is* a flow (of charge). This wording is similar to the phrase *heat transfer*, which is also redundant because heat *is* a transfer (of energy). We will avoid this phrase and speak of *flow of charge* or *charge flow*.

If the ends of a conducting wire are connected to form a loop, all points on the loop are at the same electric potential; hence, the electric field is zero within and at the surface of the conductor. Because the electric field is zero, there is no net transport of charge through the wire; therefore, there is no current. If the ends of the conducting wire are connected to a battery, however, all points on the loop are not at the same potential. The battery sets up a potential difference between the ends of the loop, creating an electric field within the wire. The electric field exerts forces on the conduction electrons in the wire, causing them to move in the wire and therefore creating a current.

PITFALL PREVENTION 27.2

Batteries Do Not Supply Electrons

A battery does not supply electrons to the circuit. It establishes the electric field that exerts a force on electrons already in the wires and elements of the circuit.

Microscopic Model of Current

We can relate current to the motion of the charge carriers by describing a microscopic model of conduction in a metal. Consider the current in a conductor of

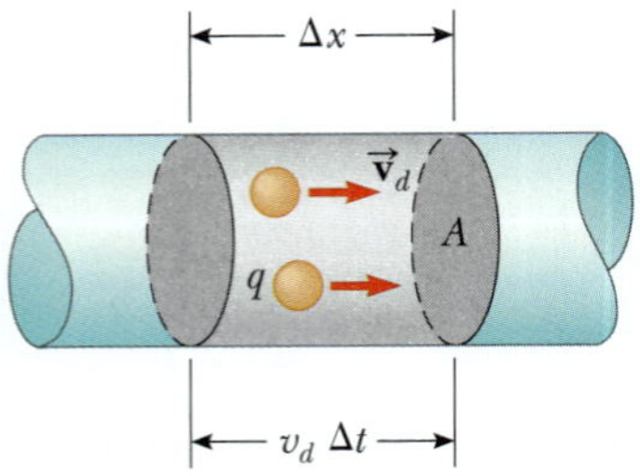

Figure 27.2 A section of a uniform conductor of cross-sectional area A. The mobile charge carriers move with a speed v_d, and the displacement they experience in the x direction in a time interval Δt is $\Delta x = v_d\,\Delta t$. If we choose Δt to be the time interval during which the charges are displaced, on average, by the length of the cylinder, the number of carriers in the section of length Δx is $nAv_d\,\Delta t$, where n is the number of carriers per unit volume.

cross-sectional area A (Fig. 27.2). The volume of a section of the conductor of length Δx (the gray region of the conductor shown in Fig. 27.2) is $A\,\Delta x$. If n represents the number of mobile charge carriers per unit volume (in other words, the charge carrier density), the number of carriers in the gray section is $nA\,\Delta x$. Therefore, the total charge ΔQ in this section is

$$\Delta Q = (nA\,\Delta x)q$$

where q is the charge on each carrier. If the carriers move with a speed v_d, the displacement they experience in the x direction in a time interval Δt is $\Delta x = v_d\,\Delta t$. Let Δt be the time interval required for the charge carriers in the cylinder to move through a displacement whose magnitude is equal to the length of the cylinder. This time interval is also the same as that required for all the charge carriers in the cylinder to pass through the circular area at one end. With this choice, we can write ΔQ as

$$\Delta Q = (nAv_d\,\Delta t)q$$

Dividing both sides of this equation by Δt, we find that the average current in the conductor is

Current in a conductor in terms of microscopic quantities ▶

$$I_{\text{avg}} = \frac{\Delta Q}{\Delta t} = nqv_dA \qquad \textbf{(27.4)}$$

The speed of the charge carriers v_d is an average speed called the **drift speed.** To understand the meaning of drift speed, consider a conductor in which the charge carriers are free electrons. If the conductor is isolated—that is, the potential difference across it is zero—these electrons undergo random motion that is analogous to the motion of gas molecules. The electrons collide repeatedly with the metal atoms, and their resultant motion is complicated and zigzagged as in Active Figure 27.3a. As discussed earlier, when a potential difference is applied across the conductor (for example, by means of a battery), an electric field is set up in the conductor; this field exerts an electric force on the electrons, producing a current. In addition to the zigzag motion due to the collisions with the metal atoms, the electrons move slowly along the conductor (in a direction opposite that of $\vec{\mathbf{E}}$) at the drift velocity $\vec{\mathbf{v}}_d$ as shown in Active Figure 27.3b.

You can think of the atom–electron collisions in a conductor as an effective internal friction (or drag force) similar to that experienced by a liquid's molecules flowing through a pipe stuffed with steel wool. The energy transferred from the electrons to the metal atoms during collisions causes an increase in the atom's vibrational energy and a corresponding increase in the conductor's temperature.

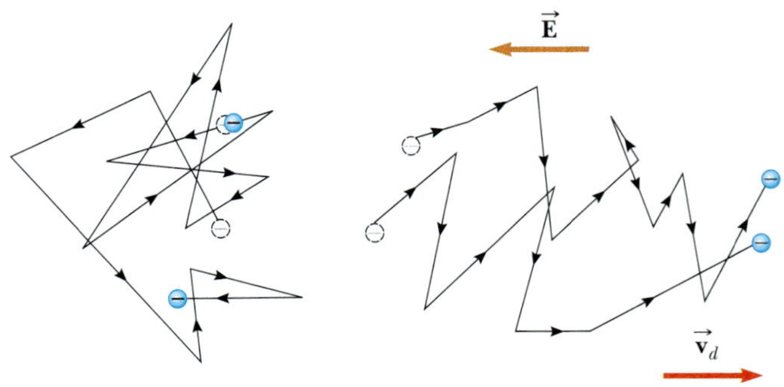

(a) (b)

ACTIVE FIGURE 27.3

(a) A schematic diagram of the random motion of two charge carriers in a conductor in the absence of an electric field. The drift velocity is zero. (b) The motion of the charge carriers in a conductor in the presence of an electric field. Notice that the random motion is modified by the field and the charge carriers have a drift velocity opposite the direction of the electric field. Because of the acceleration of the charge carriers due to the electric force, the paths are actually parabolic. The drift speed, however, is much smaller than the average speed, so the parabolic shape is not visible on this scale.

Sign in at www.thomsonedu.com and go to ThomsonNOW to adjust the electric field and see the resulting effect on the motion of an electron.

Quick Quiz 27.1 Consider positive and negative charges moving horizontally through the four regions shown in Figure 27.4. Rank the current in these four regions from lowest to highest.

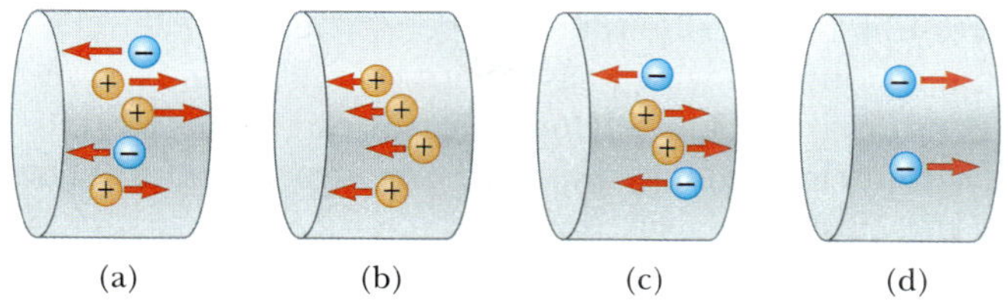

Figure 27.4 (Quick Quiz 27.1) Charges move through four regions.

EXAMPLE 27.1 Drift Speed in a Copper Wire

The 12-gauge copper wire in a typical residential building has a cross-sectional area of 3.31×10^{-6} m^2. It carries a constant current of 10.0 A. What is the drift speed of the electrons in the wire? Assume each copper atom contributes one free electron to the current. The density of copper is 8.92 g/cm^3.

SOLUTION

Conceptualize Imagine electrons following a zigzag motion such as that in Active Figure 27.3a, with a drift motion parallel to the wire superimposed on the motion as in Active Figure 27.3b. As mentioned earlier, the drift speed is small, and this example helps us quantify the speed.

Categorize We evaluate the drift speed using Equation 27.4. Because the current is constant, the average current during any time interval is the same as the constant current: $I_{avg} = I$.

Analyze The periodic table of the elements in Appendix C shows that the molar mass of copper is 63.5 g/mol. Recall that 1 mol of any substance contains Avogadro's number of atoms (6.02×10^{23}).

Use the molar mass and the density of copper to find the volume of 1 mole of copper:

$$V = \frac{m}{\rho} = \frac{63.5\text{ g}}{8.92\text{ g/cm}^3} = 7.12\text{ cm}^3$$

From the assumption that each copper atom contributes one free electron to the current, find the electron density in copper:

$$n = \frac{6.02 \times 10^{23}\text{ electrons}}{7.12\text{ cm}^3}\left(\frac{1.00 \times 10^6\text{ cm}^3}{1\text{ m}^3}\right)$$
$$= 8.46 \times 10^{28}\text{ electrons/m}^3$$

Solve Equation 27.4 for the drift speed:

$$v_d = \frac{I_{avg}}{nqA} = \frac{I}{nqA}$$

Substitute numerical values:

$$v_d = \frac{I}{neA} = \frac{10.0\text{ A}}{(8.46 \times 10^{28}\text{ m}^{-3})(1.60 \times 10^{-19}\text{ C})(3.31 \times 10^{-6}\text{ m}^2)}$$
$$= 2.23 \times 10^{-4}\text{ m/s}$$

Finalize This result shows that typical drift speeds are very small. For instance, electrons traveling with a speed of 2.23×10^{-4} m/s would take about 75 min to travel 1 m! You might therefore wonder why a light turns on almost instantaneously when its switch is thrown. In a conductor, changes in the electric field that drives the free electrons travel through the conductor with a speed close to that of light. So, when you flip on a light switch, electrons already in the filament of the lightbulb experience electric forces and begin moving after a time interval on the order of nanoseconds.

27.2 Resistance

In Chapter 24, we found that the electric field inside a conductor is zero. This statement is true, however, *only* if the conductor is in static equilibrium. The purpose of this section is to describe what happens when the charges in the conductor are not in equilibrium, in which case there is an electric field in the conductor.

Consider a conductor of cross-sectional area A carrying a current I. The **current density** J in the conductor is defined as the current per unit area. Because the current $I = nqv_dA$, the current density is

Current density ►

$$J \equiv \frac{I}{A} = nqv_d \qquad (27.5)$$

where J has SI units of amperes per meter squared. This expression is valid only if the current density is uniform and only if the surface of cross-sectional area A is perpendicular to the direction of the current.

A current density and an electric field are established in a conductor whenever a potential difference is maintained across the conductor. In some materials, the current density is proportional to the electric field:

$$J = \sigma E \qquad (27.6)$$

where the constant of proportionality σ is called the **conductivity** of the conductor.[1] Materials that obey Equation 27.6 are said to follow **Ohm's law,** named after Georg Simon Ohm. More specifically, Ohm's law states the following:

GEORG SIMON OHM
German physicist (1789–1854)
Ohm, a high school teacher and later a professor at the University of Munich, formulated the concept of resistance and discovered the proportionalities expressed in Equations 27.6 and 27.7.

For many materials (including most metals), the ratio of the current density to the electric field is a constant σ that is independent of the electric field producing the current.

Materials that obey Ohm's law and hence demonstrate this simple relationship between E and J are said to be *ohmic*. Experimentally, however, it is found that not all materials have this property. Materials and devices that do not obey Ohm's law are said to be *nonohmic*. Ohm's law is not a fundamental law of nature; rather, it is an empirical relationship valid only for certain materials.

We can obtain an equation useful in practical applications by considering a segment of straight wire of uniform cross-sectional area A and length ℓ as shown in Figure 27.5. A potential difference $\Delta V = V_b - V_a$ is maintained across the wire, creating in the wire an electric field and a current. If the field is assumed to be uniform, the potential difference is related to the field through the relationship[2]

$$\Delta V = E\ell$$

Therefore, we can express the current density in the wire as

$$J = \sigma E = \sigma \frac{\Delta V}{\ell}$$

Because $J = I/A$, the potential difference across the wire is

$$\Delta V = \frac{\ell}{\sigma} J = \left(\frac{\ell}{\sigma A}\right) I = RI$$

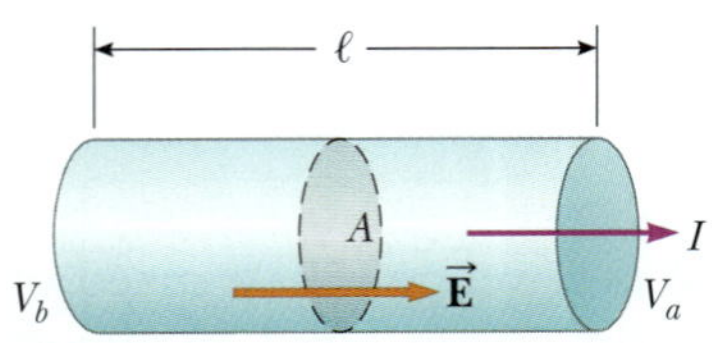

Figure 27.5 A uniform conductor of length ℓ and cross-sectional area A. A potential difference $\Delta V = V_b - V_a$ maintained across the conductor sets up an electric field $\vec{\mathbf{E}}$, and this field produces a current I that is proportional to the potential difference.

[1] Do not confuse conductivity σ with surface charge density, for which the same symbol is used.

[2] This result follows from the definition of potential difference:

$$V_b - V_a = -\int_a^b \vec{\mathbf{E}} \cdot d\vec{\mathbf{s}} = E\int_a^b dx = E\ell$$

The quantity $R = \ell/\sigma A$ is called the **resistance** of the conductor. We define the resistance as the ratio of the potential difference across a conductor to the current in the conductor:

$$R \equiv \frac{\Delta V}{I} \qquad (27.7)$$

We will use this equation again and again when studying electric circuits. This result shows that resistance has SI units of volts per ampere. One volt per ampere is defined to be one **ohm** (Ω):

$$1\ \Omega \equiv 1\ \text{V/A} \qquad (27.8)$$

This expression shows that if a potential difference of 1 V across a conductor causes a current of 1 A, the resistance of the conductor is 1 Ω. For example, if an electrical appliance connected to a 120-V source of potential difference carries a current of 6 A, its resistance is 20 Ω.

Most electric circuits use circuit elements called **resistors** to control the current in the various parts of the circuit. Two common types are the *composition resistor*, which contains carbon, and the *wire-wound resistor*, which consists of a coil of wire. Values of resistors in ohms are normally indicated by color coding as shown in Figure 27.6 and Table 27.1.

The inverse of conductivity is **resistivity**[3] ρ:

$$\rho = \frac{1}{\sigma} \qquad (27.9)$$

◀ Resistivity is the inverse of conductivity

where ρ has the units ohm meters ($\Omega \cdot \text{m}$). Because $R = \ell/\sigma A$, we can express the resistance of a uniform block of material along the length ℓ as

$$R = \rho \frac{\ell}{A} \qquad (27.10)$$

◀ Resistance of a uniform material along the length ℓ

Every ohmic material has a characteristic resistivity that depends on the properties of the material and on temperature. In addition, as you can see from Equation 27.10, the resistance of a sample depends on geometry as well as on resistivity. Table 27.2 (page 758) gives the resistivities of a variety of materials at 20°C. Notice the enormous range, from very low values for good conductors such as copper and silver to very high values for good insulators such as glass and rubber. An ideal conductor would have zero resistivity, and an ideal insulator would have infinite resistivity.

PITFALL PREVENTION 27.3

Equation 27.7 Is Not Ohm's Law

Many individuals call Equation 27.7 Ohm's law, but that is incorrect. This equation is simply the definition of resistance, and it provides an important relationship between voltage, current, and resistance. Ohm's law is related to a proportionality of J to E (Eq. 27.6) or, equivalently, of I to ΔV, which, from Equation 27.7, indicates that the resistance is constant, independent of the applied voltage.

TABLE 27.1

Color Coding for Resistors

Color	Number	Multiplier	Tolerance
Black	0	1	
Brown	1	10^1	
Red	2	10^2	
Orange	3	10^3	
Yellow	4	10^4	
Green	5	10^5	
Blue	6	10^6	
Violet	7	10^7	
Gray	8	10^8	
White	9	10^9	
Gold		10^{-1}	5%
Silver		10^{-2}	10%
Colorless			20%

SuperStock

Figure 27.6 The colored bands on a resistor represent a code for determining resistance. The first two colors give the first two digits in the resistance value. The third color represents the power of 10 for the multiplier of the resistance value. The last color is the tolerance of the resistance value. As an example, the four colors on the circled resistors are red (= 2), black (= 0), orange (= 10^3), and gold (= 5%), and so the resistance value is $20 \times 10^3\ \Omega = 20\ \text{k}\Omega$ with a tolerance value of 5% = 1 kΩ. (The values for the colors are from Table 27.1.)

[3] Do not confuse resistivity ρ with mass density or charge density, for which the same symbol is used.

TABLE 27.2

Resistivities and Temperature Coefficients of Resistivity for Various Materials

Material	Resistivity[a] ($\Omega \cdot$ m)	Temperature Coefficient[b] $\alpha[(°C)^{-1}]$
Silver	1.59×10^{-8}	3.8×10^{-3}
Copper	1.7×10^{-8}	3.9×10^{-3}
Gold	2.44×10^{-8}	3.4×10^{-3}
Aluminum	2.82×10^{-8}	3.9×10^{-3}
Tungsten	5.6×10^{-8}	4.5×10^{-3}
Iron	10×10^{-8}	5.0×10^{-3}
Platinum	11×10^{-8}	3.92×10^{-3}
Lead	22×10^{-8}	3.9×10^{-3}
Nichrome[c]	1.50×10^{-6}	0.4×10^{-3}
Carbon	3.5×10^{-5}	-0.5×10^{-3}
Germanium	0.46	-48×10^{-3}
Silicon[d]	2.3×10^{3}	-75×10^{-3}
Glass	10^{10} to 10^{14}	
Hard rubber	$\sim 10^{13}$	
Sulfur	10^{15}	
Quartz (fused)	75×10^{16}	

[a] All values at 20°C. All elements in this table are assumed to be free of impurities.

[b] See Section 27.4.

[c] A nickel–chromium alloy commonly used in heating elements.

[d] The resistivity of silicon is very sensitive to purity. The value can be changed by several orders of magnitude when it is doped with other atoms.

PITFALL PREVENTION 27.4

Resistance and Resistivity

Resistivity is a property of a *substance,* whereas resistance is a property of an *object.* We have seen similar pairs of variables before. For example, density is a property of a substance, whereas mass is a property of an object. Equation 27.10 relates resistance to resistivity and Equation 1.1 relates mass to density.

Equation 27.10 shows that the resistance of a given cylindrical conductor such as a wire is proportional to its length and inversely proportional to its cross-sectional area. If the length of a wire is doubled, its resistance doubles. If its cross-sectional area is doubled, its resistance decreases by one half. The situation is analogous to the flow of a liquid through a pipe. As the pipe's length is increased, the resistance to flow increases. As the pipe's cross-sectional area is increased, more liquid crosses a given cross section of the pipe per unit time interval. Therefore, more liquid flows for the same pressure differential applied to the pipe, and the resistance to flow decreases.

Ohmic materials and devices have a linear current–potential difference relationship over a broad range of applied potential differences (Fig. 27.7a). The slope of the I-versus-ΔV curve in the linear region yields a value for $1/R$. Nonohmic materials have a nonlinear current–potential difference relationship. One common semiconducting device with nonlinear I-versus-ΔV characteristics is the *junction diode* (Fig. 27.7b). The resistance of this device is low for currents in one direction (positive ΔV) and high for currents in the reverse direction (negative ΔV). In fact, most modern electronic devices, such as transistors, have nonlin-

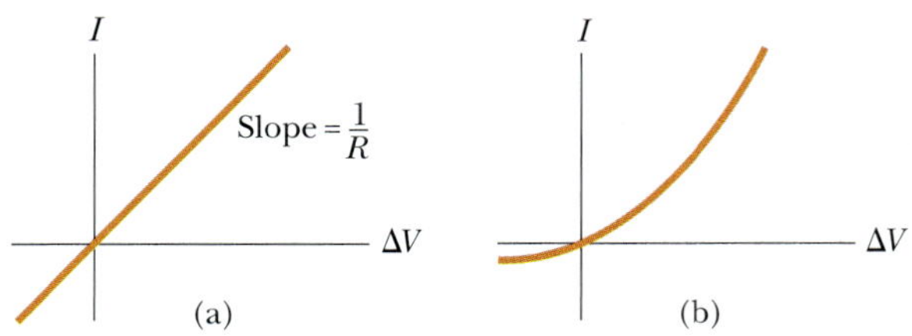

Figure 27.7 (a) The current–potential difference curve for an ohmic material. The curve is linear, and the slope is equal to the inverse of the resistance of the conductor. (b) A nonlinear current–potential difference curve for a junction diode. This device does not obey Ohm's law.

ear current–potential difference relationships; their proper operation depends on the particular way they violate Ohm's law.

Quick Quiz 27.2 A cylindrical wire has a radius r and length ℓ. If both r and ℓ are doubled, does the resistance of the wire (a) increase, (b) decrease, or (c) remain the same?

Quick Quiz 27.3 In Figure 27.7b, as the applied voltage increases, does the resistance of the diode (a) increase, (b) decrease, or (c) remain the same?

EXAMPLE 27.2 The Resistance of Nichrome Wire

The radius of 22-gauge Nichrome wire is 0.321 mm. **(A)** Calculate the resistance per unit length of this wire.

SOLUTION

Conceptualize Table 27.2 shows that Nichrome has a resistivity two orders of magnitude larger than the best conductors in the table. Therefore, we expect it to have some special practical applications that the best conductors may not have.

Categorize We model the wire as a cylinder so that a simple geometric analysis can be applied to find the resistance.

Analyze Use Equation 27.10 and the resistivity of Nichrome from Table 27.2 to find the resistance per unit length:

$$\frac{R}{\ell} = \frac{\rho}{A} = \frac{\rho}{\pi r^2} = \frac{1.5 \times 10^{-6}\ \Omega \cdot \text{m}}{\pi(0.321 \times 10^{-3}\ \text{m})^2} = \boxed{4.6\ \Omega/\text{m}}$$

(B) If a potential difference of 10 V is maintained across a 1.0-m length of the Nichrome wire, what is the current in the wire?

SOLUTION

Analyze Use Equation 27.7 to find the current:

$$I = \frac{\Delta V}{R} = \frac{\Delta V}{(4.6\ \Omega/\text{m})\ell} = \frac{10\ \text{V}}{(4.6\ \Omega/\text{m})(1.0\ \text{m})} = \boxed{2.2\ \text{A}}$$

Finalize A copper wire of the same radius would have a resistance per unit length of only 0.053 Ω/m. A 1.0-m length of copper wire of the same radius would carry the same current (2.2 A) with an applied potential difference of only 0.12 V.

Because of its high resistivity and resistance to oxidation, Nichrome is often used for heating elements in toasters, irons, and electric heaters.

EXAMPLE 27.3 The Radial Resistance of a Coaxial Cable

Coaxial cables are used extensively for cable television and other electronic applications. A coaxial cable consists of two concentric cylindrical conductors. The region between the conductors is completely filled with polyethylene plastic as shown in Figure 27.8a. Current leakage through the plastic, in the *radial* direction, is unwanted. (The cable is designed to conduct current along its length, but that is *not* the current being considered here.) The radius of the inner conductor is a = 0.500 cm, the radius of the outer conductor is b = 1.75 cm, and the length is L = 15.0 cm. The resistivity of the plastic is $1.0 \times 10^{13}\ \Omega \cdot \text{m}$. Calculate the resistance of the plastic between the two conductors.

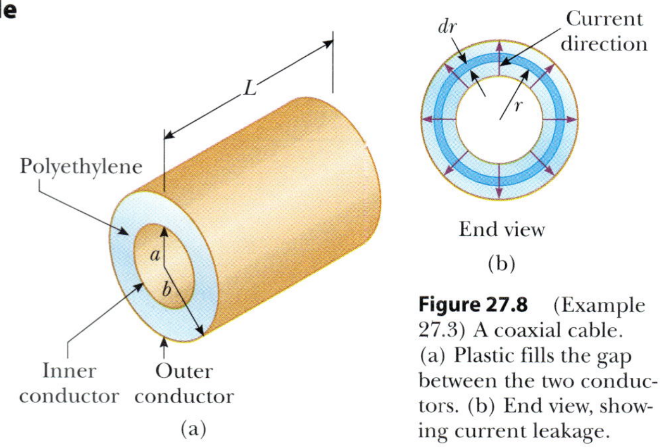

Figure 27.8 (Example 27.3) A coaxial cable. (a) Plastic fills the gap between the two conductors. (b) End view, showing current leakage.

SOLUTION

Conceptualize Imagine two currents as suggested in the text of the problem. The desired current is along the cable, carried within the conductors. The undesired current corresponds to charge leakage through the plastic, and its direction is radial.

Categorize Because the resistivity and the geometry of the plastic are known, we categorize this problem as one in which we find the resistance of the plastic from these parameters, using Equation 27.10. Because the area through which the charges pass depends on the radial position, we must use integral calculus to determine the answer.

Analyze We divide the plastic into concentric elements of infinitesimal thickness dr (Fig. 27.8b). Use the differential form of Equation 27.10, replacing ℓ with r for the distance variable: $dR = \rho\, dr/A$, where dR is the resistance of an element of plastic of thickness dr and surface area A. In this example, our representative element is a concentric hollow plastic cylinder of radius r, thickness dr, and length L as in Figure 27.8. Any charge passing from the inner to the outer conductor must move radially through this concentric element. The area through which this charge passes is $A = 2\pi rL$ (the curved surface area—circumference multiplied by length—of our hollow plastic cylinder of thickness dr).

Write an expression for the resistance of our hollow cylinder of plastic:

$$dR = \frac{\rho}{2\pi rL}\, dr$$

Integrate this expression from $r = a$ to $r = b$:

$$(1) \quad R = \int dR = \frac{\rho}{2\pi L}\int_a^b \frac{dr}{r} = \frac{\rho}{2\pi L}\ln\left(\frac{b}{a}\right)$$

Substitute the values given:

$$R = \frac{1.0 \times 10^{13}\ \Omega\cdot\text{m}}{2\pi(0.150\ \text{m})}\ln\left(\frac{1.75\ \text{cm}}{0.500\ \text{cm}}\right) = \boxed{1.33 \times 10^{13}\ \Omega}$$

Finalize Let's compare this resistance to that of the inner copper conductor of the cable along the 15.0-cm length.

Use Equation 27.10 to find the resistance of the copper cylinder:

$$R = \rho\frac{\ell}{A} = (1.7 \times 10^{-8}\ \Omega\cdot\text{m})\left[\frac{0.150\ \text{m}}{\pi(5.00 \times 10^{-3}\ \text{m})^2}\right]$$

$$= 3.2 \times 10^{-5}\ \Omega$$

This resistance is 18 orders of magnitude smaller than the radial resistance. Therefore, almost all the current corresponds to charge moving along the length of the cable, with a very small fraction leaking in the radial direction.

What If? Suppose the coaxial cable is enlarged to twice the overall diameter with two possible choices: (1) the ratio b/a is held fixed or (2) the difference $b - a$ is held fixed. For which choice does the leakage current between the inner and outer conductors increase when the voltage is applied between them?

Answer For the current to increase, the resistance must decrease. For choice (1), in which b/a is held fixed, Equation (1) shows that the resistance is unaffected. For choice (2), we do not have an equation involving the difference $b - a$ to inspect. Looking at Figure 27.8b, however, we see that increasing b and a while holding the voltage constant results in charge flowing through the same thickness of plastic but through a larger area perpendicular to the flow. This larger area results in lower resistance and a higher current.

27.3 A Model for Electrical Conduction

In this section, we describe a classical model of electrical conduction in metals that was first proposed by Paul Drude (1863–1906) in 1900. This model leads to Ohm's law and shows that resistivity can be related to the motion of electrons in metals. Although the Drude model described here has limitations, it introduces concepts that are applied in more elaborate treatments.

Consider a conductor as a regular array of atoms plus a collection of free electrons, which are sometimes called *conduction* electrons. The conduction electrons, although bound to their respective atoms when the atoms are not part of a solid, become free when the atoms condense into a solid. In the absence of an electric field, the conduction electrons move in random directions through the conductor with average speeds on the order of 10^6 m/s (Active Fig. 27.3a). The situation is similar to the motion of gas molecules confined in a vessel. In fact, some scientists refer to conduction electrons in a metal as an *electron gas.*

When an electric field is applied, the free electrons drift slowly in a direction opposite that of the electric field (Active Fig. 27.3b), with an average drift speed v_d that is much smaller (typically 10^{-4} m/s) than their average speed between collisions (typically 10^6 m/s).

In our model, we make the following assumptions:

1. The electron's motion after a collision is independent of its motion before the collision.
2. The excess energy acquired by the electrons in the electric field is lost to the atoms of the conductor when the electrons and atoms collide.

With regard to assumption (2), the energy given up to the atoms increases their vibrational energy, which causes the temperature of the conductor to increase.

We are now in a position to derive an expression for the drift velocity. When a free electron of mass m_e and charge q $(= -e)$ is subjected to an electric field $\vec{\mathbf{E}}$, it experiences a force $\vec{\mathbf{F}} = q\vec{\mathbf{E}}$. The electron is a particle under a net force, and its acceleration can be found from Newton's second law, $\sum \vec{\mathbf{F}} = m\vec{\mathbf{a}}$:

$$\vec{\mathbf{a}} = \frac{\sum \vec{\mathbf{F}}}{m} = \frac{q\vec{\mathbf{E}}}{m_e} \quad \textbf{(27.11)}$$

Because the electric field is uniform, the electron's acceleration is constant, so the electron can be modeled as a particle under constant acceleration. If $\vec{\mathbf{v}}_i$ is the electron's initial velocity the instant after a collision (which occurs at a time defined as $t = 0$), the velocity of the electron at a very short time t later (immediately before the next collision occurs) is, from Equation 4.8,

$$\vec{\mathbf{v}}_f = \vec{\mathbf{v}}_i + \vec{\mathbf{a}}t = \vec{\mathbf{v}}_i + \frac{q\vec{\mathbf{E}}}{m_e}t \quad \textbf{(27.12)}$$

Let's now take the average value of $\vec{\mathbf{v}}_f$ for all the electrons in the wire over all possible collision times t and all possible values of $\vec{\mathbf{v}}_i$. Assuming the initial velocities are randomly distributed over all possible values, the average value of $\vec{\mathbf{v}}_i$ is zero. The average value of the second term of Equation 27.12 is $(q\vec{\mathbf{E}}/m_e)\tau$, where τ is the *average time interval between successive collisions.* Because the average value of $\vec{\mathbf{v}}_f$ is equal to the drift velocity,

$$\vec{\mathbf{v}}_{f,\text{avg}} = \vec{\mathbf{v}}_d = \frac{q\vec{\mathbf{E}}}{m_e}\tau \quad \textbf{(27.13)}$$

◀ Drift velocity in terms of microscopic quantities

The value of τ depends on the size of the metal atoms and the number of electrons per unit volume. We can relate this expression for drift velocity in Equation 27.13 to the current in the conductor. Substituting the magnitude of the velocity from Equation 27.13 into Equation 27.5, the current density becomes

$$J = nqv_d = \frac{nq^2E}{m_e}\tau \quad \textbf{(27.14)}$$

◀ Current density in terms of microscopic quantities

where n is the number of electrons per unit volume. Comparing this expression with Ohm's law, $J = \sigma E$, we obtain the following relationships for conductivity and resistivity of a conductor:

$$\sigma = \frac{nq^2\tau}{m_e} \quad \textbf{(27.15)}$$

◀ Conductivity in terms of microscopic quantities

Resistivity in terms of microscopic quantities ▶

$$\rho = \frac{1}{\sigma} = \frac{m_e}{nq^2\tau} \quad (27.16)$$

According to this classical model, conductivity and resistivity do not depend on the strength of the electric field. This feature is characteristic of a conductor obeying Ohm's law.

27.4 Resistance and Temperature

Over a limited temperature range, the resistivity of a conductor varies approximately linearly with temperature according to the expression

Variation of ρ with temperature ▶

$$\rho = \rho_0[1 + \alpha(T - T_0)] \quad (27.17)$$

where ρ is the resistivity at some temperature T (in degrees Celsius), ρ_0 is the resistivity at some reference temperature T_0 (usually taken to be 20°C), and α is the **temperature coefficient of resistivity.** From Equation 27.17, the temperature coefficient of resistivity can be expressed as

Temperature coefficient of resistivity ▶

$$\alpha = \frac{1}{\rho_0}\frac{\Delta\rho}{\Delta T} \quad (27.18)$$

where $\Delta\rho = \rho - \rho_0$ is the change in resistivity in the temperature interval $\Delta T = T - T_0$.

The temperature coefficients of resistivity for various materials are given in Table 27.2. Notice that the unit for α is degrees Celsius^{-1} $[(°C)^{-1}]$. Because resistance is proportional to resistivity (Eq. 27.10), the variation of resistance of a sample is

$$R = R_0[1 + \alpha(T - T_0)] \quad (27.19)$$

where R_0 is the resistance at temperature T_0. Use of this property enables precise temperature measurements through careful monitoring of the resistance of a probe made from a particular material.

For some metals such as copper, resistivity is nearly proportional to temperature as shown in Figure 27.9. A nonlinear region always exists at very low temperatures, however, and the resistivity usually reaches some finite value as the temperature approaches absolute zero. This residual resistivity near absolute zero is caused primarily by the collision of electrons with impurities and imperfections in the metal. In contrast, high-temperature resistivity (the linear region) is predominantly characterized by collisions between electrons and metal atoms.

Notice that three of the α values in Table 27.2 are negative, indicating that the resistivity of these materials decreases with increasing temperature. This behavior is indicative of a class of materials called *semiconductors,* first introduced in Section 23.2, and is due to an increase in the density of charge carriers at higher temperatures.

Because the charge carriers in a semiconductor are often associated with impurity atoms, the resistivity of these materials is very sensitive to the type and concentration of such impurities.

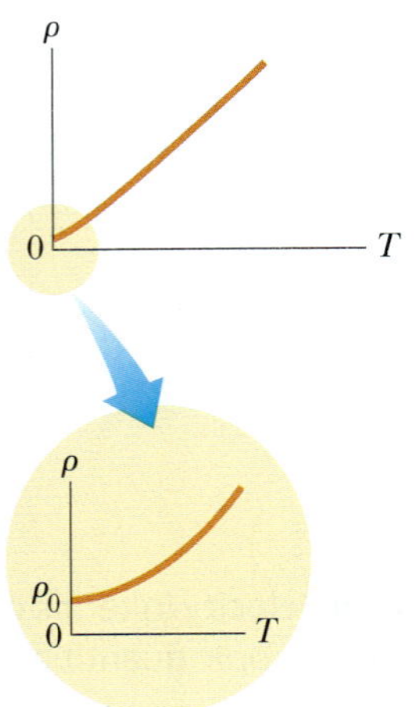

Figure 27.9 Resistivity versus temperature for a metal such as copper. The curve is linear over a wide range of temperatures, and ρ increases with increasing temperature. As T approaches absolute zero (inset), the resistivity approaches a finite value ρ_0.

Quick Quiz 27.4 When does a lightbulb carry more current, (a) immediately after it is turned on and the glow of the metal filament is increasing or (b) after it has been on for a few milliseconds and the glow is steady?

27.5 Superconductors

There is a class of metals and compounds whose resistance decreases to zero when they are below a certain temperature T_c, known as the **critical temperature.** These materials are known as **superconductors.** The resistance–temperature graph for a superconductor follows that of a normal metal at temperatures above T_c (Fig. 27.10).

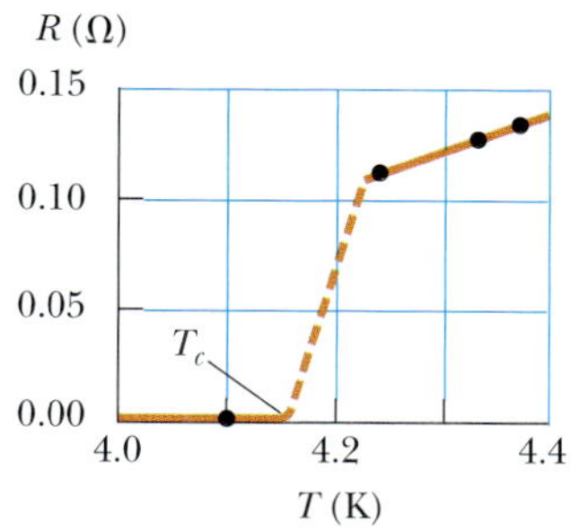

Figure 27.10 Resistance versus temperature for a sample of mercury (Hg). The graph follows that of a normal metal above the critical temperature T_c. The resistance drops to zero at T_c, which is 4.2 K for mercury.

TABLE 27.3

Critical Temperatures for Various Superconductors

Material	T_c (K)
$HgBa_2Ca_2Cu_3O_8$	134
Tl—Ba—Ca—Cu—O	125
Bi—Sr—Ca—Cu—O	105
$YBa_2Cu_3O_7$	92
Nb_3Ge	23.2
Nb_3Sn	18.05
Nb	9.46
Pb	7.18
Hg	4.15
Sn	3.72
Al	1.19
Zn	0.88

Courtesy of IBM Research Laboratory

A small permanent magnet levitated above a disk of the superconductor $YBa_2Cu_3O_7$, which is in liquid nitrogen at 77 K.

When the temperature is at or below T_c, the resistivity drops suddenly to zero. This phenomenon was discovered in 1911 by Dutch physicist Heike Kamerlingh-Onnes (1853–1926) as he worked with mercury, which is a superconductor below 4.2 K. Measurements have shown that the resistivities of superconductors below their T_c values are less than $4 \times 10^{-25}\ \Omega \cdot \text{m}$, or approximately 10^{17} times smaller than the resistivity of copper. In practice, these resistivities are considered to be zero.

Today, thousands of superconductors are known, and as Table 27.3 illustrates, the critical temperatures of recently discovered superconductors are substantially higher than initially thought possible. Two kinds of superconductors are recognized. The more recently identified ones are essentially ceramics with high critical temperatures, whereas superconducting materials such as those observed by Kamerlingh-Onnes are metals. If a room-temperature superconductor is ever identified, its effect on technology could be tremendous.

The value of T_c is sensitive to chemical composition, pressure, and molecular structure. Copper, silver, and gold, which are excellent conductors, do not exhibit superconductivity.

One truly remarkable feature of superconductors is that once a current is set up in them, it persists *without any applied potential difference* (because $R = 0$). Steady currents have been observed to persist in superconducting loops for several years with no apparent decay!

An important and useful application of superconductivity is in the development of superconducting magnets, in which the magnitudes of the magnetic field are approximately ten times greater than those produced by the best normal electromagnets. Such superconducting magnets are being considered as a means of storing energy. Superconducting magnets are currently used in medical magnetic resonance imaging, or MRI, units, which produce high-quality images of internal organs without the need for excessive exposure of patients to x-rays or other harmful radiation.

27.6 Electrical Power

In typical electric circuits, energy is transferred from a source such as a battery, to some device such as a lightbulb or a radio receiver. Let's determine an expression that will allow us to calculate the rate of this energy transfer. First, consider the simple circuit in Active Figure 27.11 (page 764), where energy is delivered to a resistor. (Resistors are designated by the circuit symbol .) Because

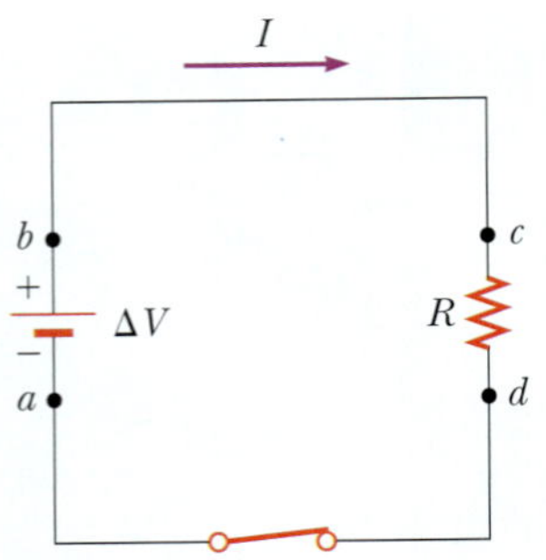

ACTIVE FIGURE 27.11

A circuit consisting of a resistor of resistance R and a battery having a potential difference ΔV across its terminals. Positive charge flows in the clockwise direction.

Sign in at www.thomsonedu.com and go to ThomsonNOW to adjust the battery voltage and the resistance and see the resulting current in the circuit and power delivered to the resistor.

PITFALL PREVENTION 27.5

Charges Do Not Move All the Way Around a Circuit in a Short Time

Because of the very small magnitude of the drift velocity, it might take *hours* for a single electron to make one complete trip around the circuit. In terms of understanding the energy transfer in a circuit, however, it is useful to *imagine* a charge moving all the way around the circuit.

PITFALL PREVENTION 27.6

Misconceptions About Current

Several common misconceptions are associated with current in a circuit like that in Active Figure 27.11. One is that current comes out of one terminal of the battery and is then "used up" as it passes through the resistor, leaving current in only one part of the circuit. The current is actually the same *everywhere* in the circuit. A related misconception has the current coming out of the resistor being smaller than that going in because some of the current is "used up." Yet another misconception has current coming out of both terminals of the battery, in opposite directions, and then "clashing" in the resistor, delivering the energy in this manner. That is not the case; charges flow in the same rotational sense at *all* points in the circuit.

the connecting wires also have resistance, some energy is delivered to the wires and some to the resistor. Unless noted otherwise, we shall assume the resistance of the wires is small compared with the resistance of the circuit element so that the energy delivered to the wires is negligible.

Imagine following a positive quantity of charge Q moving clockwise around the circuit in Active Figure 27.11 from point a through the battery and resistor back to point a. We identify the entire circuit as our system. As the charge moves from a to b through the battery, the electric potential energy of the system *increases* by an amount $Q\,\Delta V$ while the chemical potential energy in the battery *decreases* by the same amount. (Recall from Eq. 25.3 that $\Delta U = q\,\Delta V$.) As the charge moves from c to d through the resistor, however, the system *loses* this electric potential energy during collisions of electrons with atoms in the resistor. In this process, the energy is transformed to internal energy corresponding to increased vibrational motion of the atoms in the resistor. Because the resistance of the interconnecting wires is neglected, no energy transformation occurs for paths bc and da. When the charge returns to point a, the net result is that some of the chemical energy in the battery has been delivered to the resistor and resides in the resistor as internal energy associated with molecular vibration.

The resistor is normally in contact with air, so its increased temperature results in a transfer of energy by heat into the air. In addition, the resistor emits thermal radiation, representing another means of escape for the energy. After some time interval has passed, the resistor reaches a constant temperature. At this time, the input of energy from the battery is balanced by the output of energy from the resistor by heat and radiation. Some electrical devices include *heat sinks*[4] connected to parts of the circuit to prevent these parts from reaching dangerously high temperatures. Heat sinks are pieces of metal with many fins. Because the metal's high thermal conductivity provides a rapid transfer of energy by heat away from the hot component and the large number of fins provides a large surface area in contact with the air, energy can transfer by radiation and into the air by heat at a high rate.

Let's now investigate the rate at which the system loses electric potential energy as the charge Q passes through the resistor:

$$\frac{dU}{dt} = \frac{d}{dt}(Q\,\Delta V) = \frac{dQ}{dt}\Delta V = I\,\Delta V$$

where I is the current in the circuit. The system regains this potential energy when the charge passes through the battery, at the expense of chemical energy in the battery. The rate at which the system loses potential energy as the charge passes through the resistor is equal to the rate at which the system gains internal energy in the resistor. Therefore, the power $\mathcal{P}$, representing the rate at which energy is delivered to the resistor, is

$$\mathcal{P} = I\,\Delta V \tag{27.20}$$

We derived this result by considering a battery delivering energy to a resistor. Equation 27.20, however, can be used to calculate the power delivered by a voltage source to *any* device carrying a current I and having a potential difference ΔV between its terminals.

Using Equation 27.20 and $\Delta V = IR$ for a resistor, we can express the power delivered to the resistor in the alternative forms

$$\mathcal{P} = I^2R = \frac{(\Delta V)^2}{R} \tag{27.21}$$

[4] This usage is another misuse of the word *heat* that is ingrained in our common language.

When I is expressed in amperes, ΔV in volts, and R in ohms, the SI unit of power is the watt, as it was in Chapter 8 in our discussion of mechanical power. The process by which power is lost as internal energy in a conductor of resistance R is often called *joule heating*;[5] this transformation is also often referred to as an I^2R loss.

When transporting energy by electricity through power lines such as those shown in the opening photograph for this chapter, you should not assume that the lines have zero resistance. Real power lines do indeed have resistance, and power is delivered to the resistance of these wires. Utility companies seek to minimize the energy transformed to internal energy in the lines and maximize the energy delivered to the consumer. Because $\mathcal{P} = I\,\Delta V$, the same amount of energy can be transported either at high currents and low potential differences or at low currents and high potential differences. Utility companies choose to transport energy at low currents and high potential differences primarily for economic reasons. Copper wire is very expensive, so it is cheaper to use high-resistance wire (that is, wire having a small cross-sectional area; see Eq. 27.10). Therefore, in the expression for the power delivered to a resistor, $\mathcal{P} = I^2R$, the resistance of the wire is fixed at a relatively high value for economic considerations. The I^2R loss can be reduced by keeping the current I as low as possible, which means transferring the energy at a high voltage. In some instances, power is transported at potential differences as great as 765 kV. At the destination of the energy, the potential difference is usually reduced to 4 kV by a device called a *transformer.* Another transformer drops the potential difference to 240 V for use in your home. Of course, each time the potential difference decreases, the current increases by the same factor and the power remains the same. We shall discuss transformers in greater detail in Chapter 33.

PITFALL PREVENTION 27.7
Energy Is Not "Dissipated"

In some books, you may see Equation 27.21 described as the power "dissipated in" a resistor, suggesting that energy disappears. Instead, we say energy is "delivered to" a resistor. The notion of *dissipation* arises because a warm resistor expels energy by radiation and heat, so energy delivered by the battery leaves the circuit. (It does not disappear!)

Quick Quiz 27.5 For the two lightbulbs shown in Figure 27.12, rank the current values at points a through f from greatest to least.

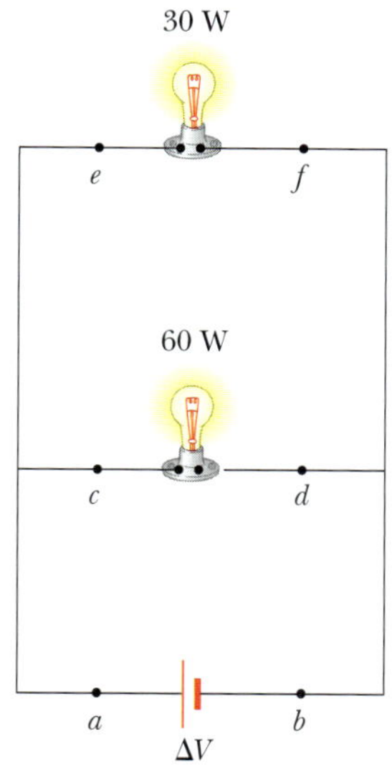

Figure 27.12 (Quick Quiz 27.5) Two lightbulbs connected across the same potential difference.

EXAMPLE 27.4 Power in an Electric Heater

An electric heater is constructed by applying a potential difference of 120 V across a Nichrome wire that has a total resistance of 8.00 Ω. Find the current carried by the wire and the power rating of the heater.

SOLUTION

Conceptualize As discussed in Example 27.2, Nichrome wire has high resistivity and is often used for heating elements in toasters, irons, and electric heaters. Therefore, we expect the power delivered to the wire to be relatively high.

Categorize We evaluate the power from Equation 27.21, so we categorize this example as a substitution problem.

Use Equation 27.7 to find the current in the wire:

$$I = \frac{\Delta V}{R} = \frac{120\text{ V}}{8.00\ \Omega} = 15.0\text{ A}$$

Find the power rating using the expression $\mathcal{P} = I^2R$ from Equation 27.21:

$$\mathcal{P} = I^2R = (15.0\text{ A})^2(8.00\ \Omega) = 1.80 \times 10^3\text{ W} = 1.80\text{ kW}$$

What If? What if the heater were accidentally connected to a 240-V supply? (That is difficult to do because the shape and orientation of the metal contacts in 240-V plugs are different from those in 120-V plugs.) How would that affect the current carried by the heater and the power rating of the heater?

[5] It is commonly called *joule heating* even though the process of heat does not occur when energy delivered to a resistor appears as internal energy. This is another example of incorrect usage of the word *heat* that has become entrenched in our language.

Answer If the applied potential difference were doubled, Equation 27.7 shows that the current would double. According to Equation 27.21, $\mathcal{P} = (\Delta V)^2/R$, the power would be four times larger.

EXAMPLE 27.5 **Linking Electricity and Thermodynamics**

An immersion heater must increase the temperature of 1.50 kg of water from 10.0°C to 50.0°C in 10.0 min while operating at 110 V.

(A) What is the required resistance of the heater?

SOLUTION

Conceptualize An immersion heater is a resistor that is inserted into a container of water. As energy is delivered to the immersion heater, raising its temperature, energy leaves the surface of the resistor by heat, going into the water. When the immersion heater reaches a constant temperature, the rate of energy delivered to the resistance by electrical transmission is equal to the rate of energy delivered by heat to the water.

Categorize This example allows us to link our new understanding of power in electricity with our experience with specific heat in thermodynamics (Chapter 20). The water is a nonisolated system. Its internal energy is rising because of energy transferred into the water by heat from the resistor: $\Delta E_{\text{int}} = Q$. In our model, we assume the energy that enters the water from the heater remains in the water.

Analyze To simplify the analysis, let's ignore the initial period during which the temperature of the resistor increases and also ignore any variation of resistance with temperature. Therefore, we imagine a constant rate of energy transfer for the entire 10.0 min.

Set the rate of energy delivered to the resistor equal to the rate of energy Q entering the water by heat:

$$\mathcal{P} = \frac{(\Delta V)^2}{R} = \frac{Q}{\Delta t}$$

Use Equation 20.4, $Q = mc\,\Delta T$, to relate the energy input by heat to the resulting temperature change of the water and solve for the resistance:

$$\frac{(\Delta V)^2}{R} = \frac{mc\,\Delta T}{\Delta t} \quad\rightarrow\quad R = \frac{(\Delta V)^2 \Delta t}{mc\,\Delta T}$$

Substitute the values given in the statement of the problem:

$$R = \frac{(110\text{ V})^2(600\text{ s})}{(1.50\text{ kg})(4\,186\text{ J/kg}\cdot{}^\circ\text{C})(50.0^\circ\text{C} - 10.0^\circ\text{C})} = 28.9\ \Omega$$

(B) Estimate the cost of heating the water.

SOLUTION

Multiply the power by the time interval to find the amount of energy transferred:

$$\mathcal{P}\,\Delta t = \frac{(\Delta V)^2}{R}\Delta t = \frac{(110\text{ V})^2}{28.9\ \Omega}(10.0\text{ min})\left(\frac{1\text{ h}}{60.0\text{ min}}\right)$$
$$= 69.8\text{ Wh} = 0.069\,8\text{ kWh}$$

Find the cost knowing that energy is purchased at an estimated price of 10¢ per kilowatt-hour:

$$\text{Cost} = (0.069\,8\text{ kWh})(\$0.1/\text{kWh}) = \$0.007 = 0.7¢$$

Finalize The cost to heat the water is very low, less than one cent. In reality, the cost is higher because some energy is transferred from the water into the surroundings by heat and electromagnetic radiation while its temperature is increasing. If you have electrical devices in your home with power ratings on them, use this power rating and an approximate time interval of use to estimate the cost for one use of the device.

Summary

ThomsonNOW™ Sign in at **www.thomsonedu.com** and go to ThomsonNOW to take a practice test for this chapter.

DEFINITIONS

The electric **current** I in a conductor is defined as

$$I \equiv \frac{dQ}{dt} \quad \textbf{(27.2)}$$

where dQ is the charge that passes through a cross section of the conductor in a time interval dt. The SI unit of current is the **ampere** (A), where 1 A = 1 C/s.

The **current density** J in a conductor is the current per unit area:

$$J \equiv \frac{I}{A} \quad \textbf{(27.5)}$$

The **resistance** R of a conductor is defined as

$$R \equiv \frac{\Delta V}{I} \quad \textbf{(27.7)}$$

where ΔV is the potential difference across it and I is the current it carries. The SI unit of resistance is volts per ampere, which is defined to be 1 **ohm** (Ω); that is, $1\ \Omega = 1\ \text{V/A}$.

CONCEPTS AND PRINCIPLES

The average current in a conductor is related to the motion of the charge carriers through the relationship

$$I_{\text{avg}} = nqv_dA \quad \textbf{(27.4)}$$

where n is the density of charge carriers, q is the charge on each carrier, v_d is the drift speed, and A is the cross-sectional area of the conductor.

The current density in an ohmic conductor is proportional to the electric field according to the expression

$$J = \sigma E \quad \textbf{(27.6)}$$

The proportionality constant σ is called the **conductivity** of the material of which the conductor is made. The inverse of σ is known as **resistivity** ρ (that is, $\rho = 1/\sigma$). Equation 27.6 is known as **Ohm's law,** and a material is said to obey this law if the ratio of its current density to its applied electric field is a constant that is independent of the applied field.

For a uniform block of material of cross-sectional area A and length ℓ, the resistance over the length ℓ is

$$R = \rho\frac{\ell}{A} \quad \textbf{(27.10)}$$

where ρ is the resistivity of the material.

In a classical model of electrical conduction in metals, the electrons are treated as molecules of a gas. In the absence of an electric field, the average velocity of the electrons is zero. When an electric field is applied, the electrons move (on average) with a **drift velocity** $\vec{\mathbf{v}}_d$ that is opposite the electric field. The drift velocity is given by

$$\vec{\mathbf{v}}_d = \frac{q\vec{\mathbf{E}}}{m_e}\tau \quad \textbf{(27.13)}$$

where q is the electron's charge, m_e is the mass of the electron, and τ is the average time interval between electron–atom collisions. According to this model, the resistivity of the metal is

$$\rho = \frac{m_e}{nq^2\tau} \quad \textbf{(27.16)}$$

where n is the number of free electrons per unit volume.

The resistivity of a conductor varies approximately linearly with temperature according to the expression

$$\rho = \rho_0[1 + \alpha(T - T_0)] \quad \textbf{(27.17)}$$

where ρ_0 is the resistivity at some reference temperature T_0 and α is the **temperature coefficient of resistivity.**

If a potential difference ΔV is maintained across a circuit element, the **power,** or rate at which energy is supplied to the element, is

$$\mathcal{P} = I\,\Delta V \quad \textbf{(27.20)}$$

Because the potential difference across a resistor is given by $\Delta V = IR$, we can express the power delivered to a resistor as

$$\mathcal{P} = I^2R = \frac{(\Delta V)^2}{R} \quad \textbf{(27.21)}$$

The energy delivered to a resistor by electrical transmission appears in the form of internal energy in the resistor.

Questions

☐ denotes answer available in *Student Solutions Manual/Study Guide;* **O** denotes objective question

1. Newspaper articles often contain a statement such as "10 000 volts of electricity surged through the victim's body." What is wrong with this statement?

2. What factors affect the resistance of a conductor?

[3.] **O** Two wires A and B with circular cross sections are made of the same metal and have equal lengths, but the resistance of wire A is three times greater than that of wire B. **(i)** What is the ratio of the cross-sectional area of A to that of B? (a) 9 (b) 3 (c) $\sqrt{3}$ (d) 1 (e) $1/\sqrt{3}$ (f) $\frac{1}{3}$ (g) $\frac{1}{9}$ (h) None of these answers is necessarily true. **(ii)** What is the ratio of the radius of A to that of B? Choose from the same possibilities.

4. **O** A metal wire of resistance R is cut into three equal pieces that are then braided together side by side to form a new cable with a length equal to one-third the original length. What is the resistance of this new wire? (a) $R/27$ (b) $R/9$ (c) $R/3$ (d) R (e) $3R$ (f) $9R$ (g) $27R$

5. When the potential difference across a certain conductor is doubled, the current is observed to increase by a factor of three. What can you conclude about the conductor?

[6.] Use the atomic theory of matter to explain why the resistance of a material should increase as its temperature increases.

7. **O** A current-carrying ohmic metal wire has a cross-sectional area that gradually becomes smaller from one end of the wire to the other. The current has the same value for each section of the wire, so charge does not accumulate at any one point. **(i)** How does the drift speed vary along the wire as the area becomes smaller? (a) It increases. (b) It decreases. (c) It remains constant. **(ii)** How does the resistance per unit length vary along the wire as the area becomes smaller? Choose from the same possibilities.

8. How does the resistance for copper and for silicon change with temperature? Why are the behaviors of these two materials different?

9. Over the time interval after a difference in potential is applied between the ends of a wire, what would happen to the drift velocity of the electrons in a wire and to the current in the wire if the electrons could move freely without resistance through the wire?

[10.] If charges flow very slowly through a metal, why does it not require several hours for a light to come on when you throw a switch?

11. **O** A cylindrical metal wire at room temperature is carrying electric current between its ends. One end is at potential $V_A = 50$ V, and the other end at potential $V_B = 0$ V. Rank the following actions in terms of the change that each one separately would produce in the current, from the greatest increase to the greatest decrease. In your ranking, note any cases of equality. (a) Make $V_A = 150$ V with $V_B = 0$ V. (b) Make $V_A = 150$ V with $V_B = 100$ V. (c) Adjust V_A to triple the power with which the wire converts electrically transmitted energy into internal energy. (d) Double the radius of the wire. (e) Double the length of the wire. (f) Double the Celsius temperature of the wire. (g) Change the material to an insulator.

12. **O** Two conductors made of the same material are connected across the same potential difference. Conductor A has twice the diameter and twice the length of conductor B. What is the ratio of the power delivered to A to the power delivered to B? (a) 32 (b) 16 (c) 8 (d) 4 (e) 2 (f) 1 (g) $\frac{1}{2}$ (h) $\frac{1}{4}$

13. **O** Two conducting wires A and B of the same length and radius are connected across the same potential difference. Conductor A has twice the resistivity of conductor B. What is the ratio of the power delivered to A to the power delivered to B? (a) 4 (b) 2 (c) $\sqrt{2}$ (d) 1 (e) $1/\sqrt{2}$ (f) $\frac{1}{2}$ (g) $\frac{1}{4}$ (h) None of these answers is necessarily correct.

14. **O** Two lightbulbs both operate from 120 V. One has a power of 25 W and the other 100 W. **(i)** Which lightbulb has higher resistance? (a) The dim 25-W lightbulb does. (b) The bright 100-W lightbulb does. (c) Both are the same. **(ii)** Which lightbulb carries more current? Choose from the same possibilities.

15. **O** Car batteries are often rated in ampere-hours. Does this information designate the amount of (a) current, (b) power, (c) energy, (d) charge, or (e) potential that the battery can supply?

16. If you were to design an electric heater using Nichrome wire as the heating element, what parameters of the wire could you vary to meet a specific power output such as 1 000 W?

Problems

WebAssign The Problems from this chapter may be assigned online in WebAssign.

ThomsonNOW™ Sign in at **www.thomsonedu.com** and go to ThomsonNOW to assess your understanding of this chapter's topics with additional quizzing and conceptual questions.

1, 2, 3 denotes straightforward, intermediate, challenging; ☐ denotes full solution available in *Student Solutions Manual/Study Guide;* ▲ denotes coached solution with hints available at **www.thomsonedu.com;** ▭ denotes developing symbolic reasoning; ● denotes asking for qualitative reasoning; ▪ denotes computer useful in solving problem

Section 27.1 Electric Current

1. In a particular cathode-ray tube, the measured beam current is 30.0 μA. How many electrons strike the tube screen every 40.0 s?

2. A teapot with a surface area of 700 cm^2 is to be plated with silver. It is attached to the negative electrode of an electrolytic cell containing silver nitrate ($Ag^+ NO_3^-$). The cell is powered by a 12.0-V battery and has a resistance of 1.80 Ω. Over what time interval does a 0.133-mm layer of silver build up on the teapot? (The density of silver is 10.5×10^3 kg/m^3.)

3. ▲ Suppose the current in a conductor decreases exponentially with time according to the equation $I(t) = I_0 e^{-t/\tau}$, where I_0 is the initial current (at $t = 0$) and τ is a constant having dimensions of time. Consider a fixed observation point within the conductor. (a) How much charge passes this point between $t = 0$ and $t = \tau$? (b) How much charge passes this point between $t = 0$ and $t = 10\tau$? (c) **What If?** How much charge passes this point between $t = 0$ and $t = \infty$?

4. A small sphere that carries a charge q is whirled in a circle at the end of an insulating string. The angular frequency of rotation is ω. What average current does this rotating charge represent?

5. The quantity of charge q (in coulombs) that has passed through a surface of area 2.00 cm^2 varies with time according to the equation $q = 4t^3 + 5t + 6$, where t is in seconds. (a) What is the instantaneous current through the surface at $t = 1.00$ s? (b) What is the value of the current density?

6. An electric current is given by the expression $I(t) = 100 \sin(120\pi t)$, where I is in amperes and t is in seconds. What is the total charge carried by the current from $t = 0$ to $t = \frac{1}{240}$ s?

7. The electron beam emerging from a certain high-energy electron accelerator has a circular cross section of radius 1.00 mm. (a) The beam current is 8.00 μA. Find the current density in the beam assuming it is uniform throughout. (b) The speed of the electrons is so close to the speed of light that their speed can be taken as 300 Mm/s with negligible error. Find the electron density in the beam. (c) Over what time interval does Avogadro's number of electrons emerge from the accelerator?

8. ● Figure P27.8 represents a section of a circular conductor of nonuniform diameter carrying a current of 5.00 A. The radius of cross-section A_1 is 0.400 cm. (a) What is the magnitude of the current density across A_1? (b) The radius at A_2 is larger than the radius at A_1. Is the current at A_2 larger, smaller, or the same? Is the current density larger, smaller, or the same? Assume one of these two quantities is different at A_2 by a factor of 4 from its value at A_1. Specify the current, current density, and radius at A_2.

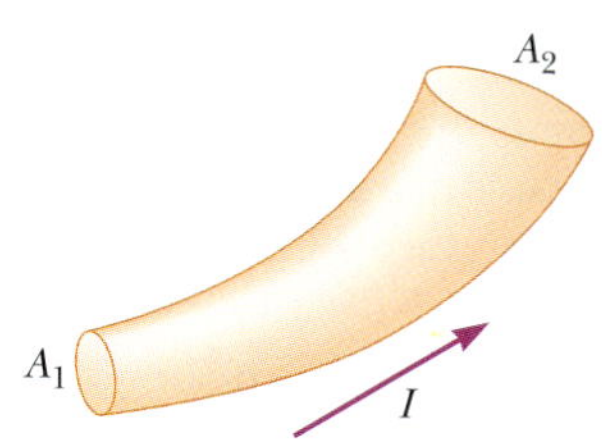

Figure P27.8

9. ● A Van de Graaff generator produces a beam of 2.00-MeV deuterons, which are heavy hydrogen nuclei containing a proton and a neutron. (a) If the beam current is 10.0 μA, how far apart are the deuterons? (b) Is the electrical force of repulsion among them a significant factor in beam stability? Explain.

10. An aluminum wire having a cross-sectional area of 4.00×10^{-6} m^2 carries a current of 5.00 A. Find the drift speed of the electrons in the wire. The density of aluminum is 2.70 g/cm^3. Assume each aluminum atom supplies one conduction electron.

Section 27.2 Resistance

11. ▲ A 0.900-V potential difference is maintained across a 1.50-m length of tungsten wire that has a cross-sectional area of 0.600 mm^2. What is the current in the wire?

12. A lightbulb has a resistance of 240 Ω when operating with a potential difference of 120 V across it. What is the current in the lightbulb?

13. Suppose you wish to fabricate a uniform wire from 1.00 g of copper. If the wire is to have a resistance of $R = 0.500$ Ω and all the copper is to be used, what must be (a) the length and (b) the diameter of this wire?

14. (a) Make an order-of-magnitude estimate of the resistance between the ends of a rubber band. (b) Make an order-of-magnitude estimate of the resistance between the "heads" and "tails" sides of a penny. In each case, state what quantities you take as data and the values you measure or estimate for them. (c) WARNING: Do not try this part at home! What is the order of magnitude of the current that each would carry if it were connected across a 120-V power supply?

15. A current density of 6.00×10^{-13} A/m^2 exists in the atmosphere at a location where the electric field is 100 V/m. Calculate the electrical conductivity of the Earth's atmosphere in this region.

Section 27.3 A Model for Electrical Conduction

16. ● If the current carried by a conductor is doubled, what happens to (a) the charge carrier density, (b) the current density, (c) the electron drift velocity, and (d) the average time interval between collisions? Explain your answers.

17. ▲ If the magnitude of the drift velocity of free electrons in a copper wire is 7.84×10^{-4} m/s, what is the electric field in the conductor?

Section 27.4 Resistance and Temperature

18. A certain lightbulb has a tungsten filament with a resistance of 19.0 Ω when cold and 140 Ω when hot. Assume the resistivity of tungsten varies linearly with temperature even over the large temperature range involved here. Find the temperature of the hot filament. Assume an initial temperature of 20.0°C.

19. An aluminum wire with a diameter of 0.100 mm has a uniform electric field of 0.200 V/m imposed along its entire length. The temperature of the wire is 50.0°C. Assume one free electron per atom. (a) Use the information in Table 27.2 and determine the resistivity. (b) What is the current density in the wire? (c) What is the total current in the wire? (d) What is the drift speed of the conduction electrons? (e) What potential difference must exist between the ends of a 2.00-m length of the wire to produce the stated electric field?

20. An engineer needs a resistor with zero overall temperature coefficient of resistance at 20°C. She designs a pair of circular cylinders, one of carbon and one of Nichrome as shown in Figure P27.20. The device must have an overall resistance of $R_1 + R_2 = 10.0\ \Omega$ independent of temperature and a uniform radius of $r = 1.50$ mm. Can she meet the design goal with this method? If so, state what you can determine about the lengths ℓ_1 and ℓ_2 of each segment. You may ignore thermal expansion of the cylinders and assume both are always at the same temperature.

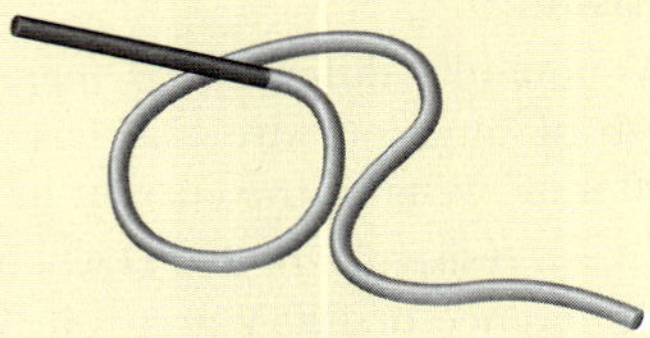

Figure P27.20

21. What is the fractional change in the resistance of an iron filament when its temperature changes from 25.0°C to 50.0°C?

22. **Review problem.** An aluminum rod has a resistance of 1.234 Ω at 20.0°C. Calculate the resistance of the rod at 120°C by accounting for the changes in both the resistivity and the dimensions of the rod.

Section 27.6 Electrical Power

23. A toaster is rated at 600 W when connected to a 120-V source. What current does the toaster carry and what is its resistance?

24. A Van de Graaff generator (see Fig. 25.24) is operating so that the potential difference between the high-potential electrode Ⓑ and the charging needles at Ⓐ is 15.0 kV. Calculate the power required to drive the belt against electrical forces at an instant when the effective current delivered to the high-potential electrode is 500 μA.

25. ▲ A well-insulated electric water heater warms 109 kg of water from 20.0°C to 49.0°C in 25.0 min. Find the resistance of its heating element, which is connected across a 220-V potential difference.

26. A 120-V motor has mechanical power output of 2.50 hp. It is 90.0% efficient in converting power that it takes in by electrical transmission into mechanical power. (a) Find the current in the motor. (b) Find the energy delivered to the motor by electrical transmission in 3.00 h of operation. (c) If the electric company charges $0.160/kWh, what does it cost to run the motor for 3.00 h?

27. Suppose a voltage surge produces 140 V for a moment. By what percentage does the power output of a 120-V, 100-W lightbulb increase? Assume its resistance does not change.

28. One rechargeable battery of mass 15.0 g delivers an average current of 18.0 mA to a compact disc player at 1.60 V for 2.40 h before the battery needs to be recharged. The recharger maintains a potential difference of 2.30 V across the battery and delivers a charging current of 13.5 mA for 4.20 h. (a) What is the efficiency of the battery as an energy storage device? (b) How much internal energy is produced in the battery during one charge–discharge cycle? (b) If the battery is surrounded by ideal thermal insulation and has an overall effective specific heat of 975 J/kg · °C, by how much will its temperature increase during the cycle?

29. A 500-W heating coil designed to operate from 110 V is made of Nichrome wire 0.500 mm in diameter. (a) Assuming the resistivity of the Nichrome remains constant at its 20.0°C value, find the length of wire used. (b) **What If?** Now consider the variation of resistivity with temperature. What power is delivered to the coil of part (a) when it is warmed to 1 200°C?

30. A coil of Nichrome wire is 25.0 m long. The wire has a diameter of 0.400 mm and is at 20.0°C. If it carries a current of 0.500 A, what are (a) the magnitude of the electric field in the wire and (b) the power delivered to it? (c) **What If?** If the temperature is increased to 340°C and the potential difference across the wire remains constant, what is the power delivered?

31. Batteries are rated in terms of ampere-hours (A · h). For example, a battery that can produce a current of 2.00 A for 3.00 h is rated at 6.00 A · h. (a) What is the total energy, in kilowatt-hours, stored in a 12.0-V battery rated at 55.0 A · h? (b) At $0.060 0 per kilowatt-hour, what is the value of the electricity produced by this battery?

32. ● Residential building codes typically require the use of 12-gauge copper wire (diameter 0.205 3 cm) for wiring receptacles. Such circuits carry currents as large as 20 A. If a wire of smaller diameter (with a higher gauge number) carried that much current, the wire could rise to a high temperature and cause a fire. (a) Calculate the rate at which internal energy is produced in 1.00 m of 12-gauge copper wire carrying 20.0 A. (b) **What If?** Repeat the calculation for an aluminum wire. Explain whether a 12-gauge aluminum wire would be as safe as a copper wire.

33. An 11.0-W energy-efficient fluorescent lamp is designed to produce the same illumination as a conventional 40.0-W incandescent lightbulb. How much money does the user of the energy-efficient lamp save during 100 h of use? Assume a cost of $0.080 0/kWh for energy from the electric company.

34. We estimate that 270 million plug-in electric clocks are in the United States, approximately one clock for each person. The clocks convert energy at the average rate 2.50 W. To supply this energy, how many metric tons of coal are burned per hour in coal-fired power plants that are, on average, 25.0% efficient? The heat of combustion for coal is 33.0 MJ/kg.

35. Compute the cost per day of operating a lamp that draws a current of 1.70 A from a 110-V line. Assume the cost of energy from the electric company is $0.060 0/kWh.

36. **Review problem.** The heating element of an electric coffee maker operates at 120 V and carries a current of 2.00 A. Assuming the water absorbs all the energy delivered to the resistor, calculate the time interval during which the temperature of 0.500 kg of water rises from room temperature (23.0°C) to the boiling point.

37. A certain toaster has a heating element made of Nichrome wire. When the toaster is first connected to a 120-V source (and the wire is at a temperature of 20.0°C),

the initial current is 1.80 A. The current decreases as the heating element warms up. When the toaster reaches its final operating temperature, the current is 1.53 A. (a) Find the power delivered to the toaster when it is at its operating temperature. (b) What is the final temperature of the heating element?

38. The cost of electricity varies widely through the United States; \$0.120/kWh is one typical value. At this unit price, calculate the cost of (a) leaving a 40.0-W porch light on for two weeks while you are on vacation, (b) making a piece of dark toast in 3.00 min with a 970-W toaster, and (c) drying a load of clothes in 40.0 min in a 5 200-W dryer.

39. Make an order-of-magnitude estimate of the cost of one person's routine use of a handheld hair dryer for 1 yr. If you do not use a hair dryer yourself, observe or interview someone who does. State the quantities you estimate and their values.

Additional Problems

40. ● One lightbulb is marked "25 W 120 V," and another is marked "100 W 120 V." These labels mean that each lightbulb has its respective power delivered to it when it is connected to a constant 120-V source. (a) Find the resistance of each lightbulb. (b) During what time interval does 1.00 C pass into the dim lightbulb? Is this charge different upon its exit versus its entry into the lightbulb? Explain. (c) In what time interval does 1.00 J pass into the dim lightbulb? By what mechanisms does this energy enter and exit the lightbulb? Explain. (d) Find the cost of running the dim lightbulb continuously for 30.0 days, assuming the electric company sells its product at \$0.070 0 per kWh. What product *does* the electric company sell? What is its price for one SI unit of this quantity?

41. An office worker uses an immersion heater to warm 250 g of water in a light, covered, insulated cup from 20°C to 100°C in 4.00 min. In electrical terms, the heater is a Nichrome resistance wire connected to a 120-V power supply. Specify a diameter and a length that the wire can have. Can it be made from less than 0.5 cm³ of Nichrome? You may assume the wire is at 100°C throughout the time interval.

42. A charge Q is placed on a capacitor of capacitance C. The capacitor is connected into the circuit shown in Figure P27.42, with an open switch, a resistor, and an initially uncharged capacitor of capacitance $3C$. The switch is then closed, and the circuit comes to equilibrium. In terms of Q and C, find (a) the final potential difference between the plates of each capacitor, (b) the charge on each capacitor, and (c) the final energy stored in each capacitor. (d) Find the internal energy appearing in the resistor.

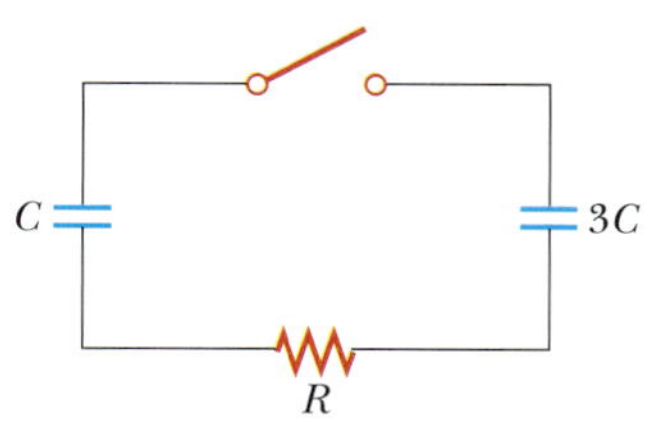

Figure P27.42

43. A more general definition of the temperature coefficient of resistivity is

$$\alpha = \frac{1}{\rho}\frac{d\rho}{dT}$$

where ρ is the resistivity at temperature T. (a) Assuming α is constant, show that

$$\rho = \rho_0 e^{\alpha(T-T_0)}$$

where ρ_0 is the resistivity at temperature T_0. (b) Using the series expansion $e^x \approx 1 + x$ for $x \ll 1$, show that the resistivity is given approximately by the expression $\rho = \rho_0[1 + \alpha(T - T_0)]$ for $\alpha(T - T_0) \ll 1$.

44. A high-voltage transmission line with a diameter of 2.00 cm and a length of 200 km carries a steady current of 1 000 A. If the conductor is copper wire with a free charge density of 8.46×10^{28} electrons/m³, over what time interval does one electron travel the full length of the line?

45. ▲ ● An experiment is conducted to measure the electrical resistivity of Nichrome in the form of wires with different lengths and cross-sectional areas. For one set of measurements, a student uses 30-gauge wire, which has a cross-sectional area of 7.30×10^{-8} m². The student measures the potential difference across the wire and the current in the wire with a voltmeter and an ammeter, respectively. For each set of measurements given in the table taken on wires of three different lengths, calculate the resistance of the wires and the corresponding values of the resistivity. What is the average value of the resistivity? Explain how this value compares with the value given in Table 27.2.

L (m)	ΔV (V)	I (A)	R (Ω)	ρ (Ω · m)
0.540	5.22	0.500		
1.028	5.82	0.276		
1.543	5.94	0.187		

46. An electric utility company supplies a customer's house from the main power lines (120 V) with two copper wires, each of which is 50.0 m long and has a resistance of 0.108 Ω per 300 m. (a) Find the potential difference at the customer's house for a load current of 110 A. For this load current, find (b) the power delivered to the customer and (c) the rate at which internal energy is produced in the copper wires.

47. A straight, cylindrical wire lying along the x axis has a length of 0.500 m and a diameter of 0.200 mm. It is made of a material described by Ohm's law with a resistivity of $\rho = 4.00 \times 10^{-8}$ Ω · m. Assume a potential of 4.00 V is maintained at $x = 0$. Also assume $V = 0$ at $x = 0.500$ m. Find (a) the electric field in the wire, (b) the resistance of the wire, (c) the electric current in the wire, and (d) the current density in the wire. State the direction of the electric field and of the current. (e) Show that $E = \rho J$.

48. A straight, cylindrical wire lying along the x axis has a length L and a diameter d. It is made of a material described by Ohm's law with a resistivity ρ. Assume potential V is maintained at $x = 0$. Also assume the potential is zero at $x = L$. In terms of L, d, V, ρ, and physical constants, derive expressions for (a) the electric field in the wire, (b) the resistance of the wire, (c) the electric current in the wire, and (d) the current density in the wire. State the direction of the field and of the current. (e) Prove that $E = \rho J$.

49. An all-electric car (not a hybrid) is designed to run from a bank of 12.0-V batteries with total energy storage of 2.00×10^7 J. (a) If the electric motor draws 8.00 kW, what is the current delivered to the motor? (b) If the electric motor draws 8.00 kW as the car moves at a steady speed of 20.0 m/s, how far can the car travel before it is "out of juice"?

50. ● **Review problem.** When a straight wire is warmed, its resistance is given by $R = R_0\,[1 + \alpha\,(T - T_0)]$ according to Equation 27.19, where α is the temperature coefficient of resistivity. (a) Show that a more precise result, one that includes that the length and area of the wire change when it is warmed, is

$$R = \frac{R_0[1 + \alpha(T - T_0)][1 + \alpha'(T - T_0)]}{[1 + 2\alpha'(T - T_0)]}$$

where α' is the coefficient of linear expansion (see Chapter 19). (b) Explain how these two results compare for a 2.00-m-long copper wire of radius 0.100 mm, first at 20.0°C and then warmed to 100.0°C.

51. The temperature coefficients of resistivity in Table 27.2 were determined at a temperature of 20°C. What would they be at 0°C? Note that the temperature coefficient of resistivity at 20°C satisfies $\rho = \rho_0[1 + \alpha(T - T_0)]$, where ρ_0 is the resistivity of the material at $T_0 = 20°\text{C}$. The temperature coefficient of resistivity α' at 0°C must satisfy the expression $\rho = \rho'_0[1 + \alpha' T]$, where ρ'_0 is the resistivity of the material at 0°C.

52. An oceanographer is studying how the ion concentration in seawater depends on depth. She makes a measurement by lowering into the water a pair of concentric metallic cylinders (Fig. P27.52) at the end of a cable and taking data to determine the resistance between these electrodes as a function of depth. The water between the two cylinders forms a cylindrical shell of inner radius r_a, outer radius r_b, and length L much larger than r_b. The scientist applies a potential difference ΔV between the inner and outer surfaces, producing an outward radial current I. Let ρ represent the resistivity of the water. (a) Find the resistance of the water between the cylinders in terms of L, ρ, r_a, and r_b. (b) Express the resistivity of the water in terms of the measured quantities L, r_a, r_b, ΔV, and I.

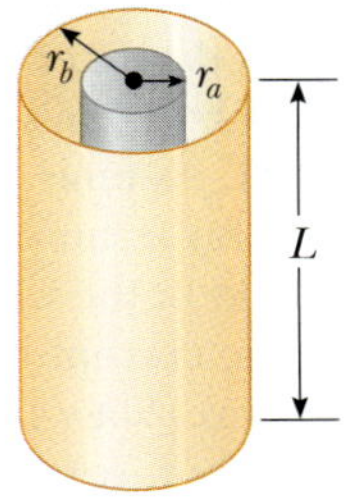

Figure P27.52

53. ● The strain in a wire can be monitored and computed by measuring the resistance of the wire. Let L_i represent the original length of the wire, A_i its original cross-sectional area, $R_i = \rho L_i/A_i$ the original resistance between its ends, and $\delta = \Delta L/L_i = (L - L_i)/L_i$ the strain resulting from the application of tension. Assume the resistivity and the volume of the wire do not change as the wire stretches. Show that the resistance between the ends of the wire under strain is given by $R = R_i(1 + 2\delta + \delta^2)$. If the assumptions are precisely true, is this result exact or approximate? Explain your answer.

54. ● In a certain stereo system, each speaker has a resistance of 4.00 Ω. The system is rated at 60.0 W in each channel, and each speaker circuit includes a fuse rated 4.00 A. Is this system adequately protected against overload? Explain your reasoning.

55. ● A close analogy exists between the flow of energy by heat because of a temperature difference (see Section 20.7) and the flow of electric charge because of a potential difference. In a metal, energy dQ and electrical charge dq are both transported by free electrons. Consequently, a good electrical conductor is usually a good thermal conductor as well. Consider a thin conducting slab of thickness dx, area A, and electrical conductivity σ, with a potential difference dV between opposite faces. (a) Show that the current $I = dq/dt$ is given by the equation on the left:

Charge conduction	Thermal conduction	
$\dfrac{dq}{dt} = \sigma A \left\lvert \dfrac{dV}{dx} \right\rvert$	$\dfrac{dQ}{dt} = kA \left\lvert \dfrac{dT}{dx} \right\rvert$	**(Eq. 20.15)**

In the analogous thermal conduction equation on the right, the rate of energy flow dQ/dt (in SI units of joules per second) is due to a temperature gradient dT/dx, in a material of thermal conductivity k. (b) State analogous rules relating the direction of the electric current to the change in potential and relating the direction of energy flow to the change in temperature.

56. A material of resistivity ρ is formed into the shape of a truncated cone of altitude h as shown in Figure P27.56. The bottom end has radius b, and the top end has radius a. Assume the current is distributed uniformly over any circular cross section of the cone so that the current density does not depend on radial position. (The current density does vary with position along the axis of the cone.) Show that the resistance between the two ends is

$$R = \frac{\rho}{\pi}\left(\frac{h}{ab}\right)$$

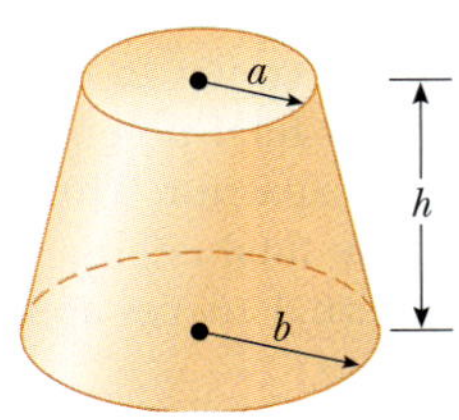

Figure P27.56

57. Material with uniform resistivity ρ is formed into a wedge as shown in Figure P27.57. Show that the resistance between face A and face B of this wedge is

$$R = \rho\,\frac{L}{w(y_2 - y_1)}\ln\left(\frac{y_2}{y_1}\right)$$

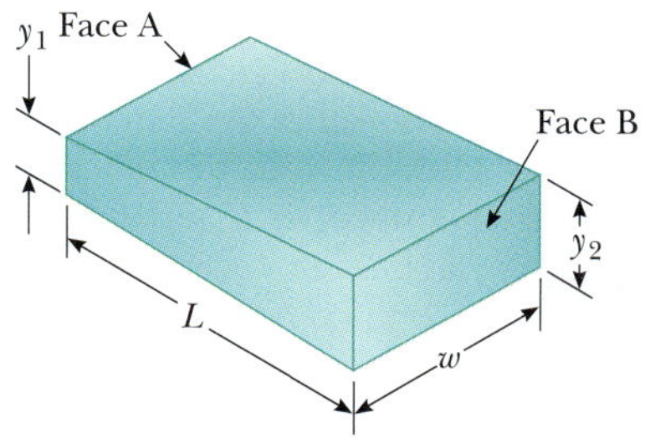

Figure P27.57

58. A spherical shell with inner radius r_a and outer radius r_b is formed from a material of resistivity ρ. It carries current radially, with uniform density in all directions. Show that its resistance is

$$R = \frac{\rho}{4\pi}\left(\frac{1}{r_a} - \frac{1}{r_b}\right)$$

59. ● Problems 56, 57, and 58 deal with calculating the resistance between specified surfaces of an oddly shaped resistor. To verify the results experimentally, a potential difference may be applied to the indicated surfaces and the resulting current measured. The resistance can then be calculated from its definition. Describe a method to ensure that the electric potential is uniform over the surface. Explain whether you can then be sure that the current is spread out over the whole surfaces where it enters and exits.

60. The dielectric material between the plates of a parallel-plate capacitor always has some nonzero conductivity σ. Let A represent the area of each plate and d the distance between them. Let κ represent the dielectric constant of the material. (a) Show that the resistance R and the capacitance C of the capacitor are related by

$$RC = \frac{\kappa \epsilon_0}{\sigma}$$

(b) Find the resistance between the plates of a 14.0-nF capacitor with a fused quartz dielectric.

61. **Review problem.** A parallel-plate capacitor consists of square plates of edge length ℓ that are separated by a distance d, where $d \ll \ell$. A potential difference ΔV is maintained between the plates. A material of dielectric constant κ fills half the space between the plates. The dielectric slab is withdrawn from the capacitor as shown in Figure P27.61. (a) Find the capacitance when the left edge of the dielectric is at a distance x from the center of the capacitor. (b) If the dielectric is removed at a constant speed v, what is the current in the circuit as the dielectric is being withdrawn?

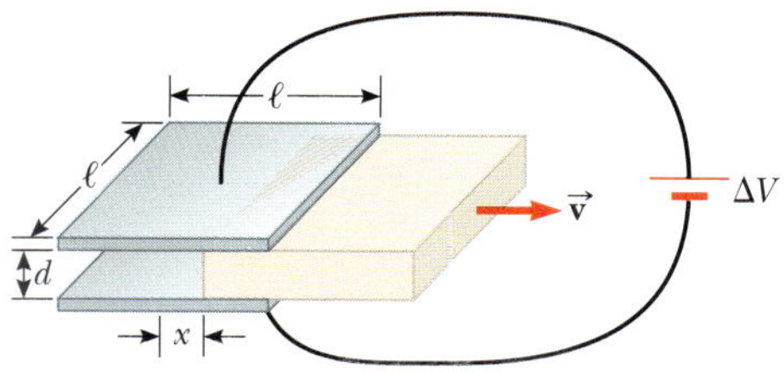

Figure P27.61

62. The current–voltage characteristic curve for a semiconductor diode as a function of temperature T is given by

$$I = I_0\left(e^{e\Delta V/k_B T} - 1\right)$$

Here the first symbol e represents Euler's number, the base of natural logarithms. The second e is the magnitude of the electron charge. The k_B stands for Boltzmann's constant, and T is the absolute temperature. Set up a spreadsheet to calculate I and $R = \Delta V/I$ for $\Delta V = 0.400$ V to 0.600 V in increments of 0.005 V. Assume $I_0 = 1.00$ nA. Plot R versus ΔV for $T = 280$ K, 300 K, and 320 K.

63. Gold is the most ductile of all metals. For example, one gram of gold can be drawn into a wire 2.40 km long. What is the resistance of such a wire at 20°C? You can find the necessary reference information in this textbook.

64. One wire in a high-voltage transmission line carries 1 000 A starting at 700 kV for a distance of 100 mi. If the resistance in the wire is 0.500 Ω/mi, what is the power loss due to the resistance of the wire?

65. The potential difference across the filament of a lamp is maintained at a constant value while equilibrium temperature is being reached. It is observed that the steady-state current in the lamp is only one tenth of the current drawn by the lamp when it is first turned on. If the temperature coefficient of resistivity for the lamp at 20.0°C is 0.004 50 $(°C)^{-1}$ and the resistance increases linearly with increasing temperature, what is the final operating temperature of the filament?

Answers to Quick Quizzes

27.1 (d), (b) = (c), (a). The current in part (d) is equivalent to two positive charges moving to the left. Parts (b) and (c) each represent four positive charges moving in the same direction because negative charges moving to the left are equivalent to positive charges moving to the right. The current in part (a) is equivalent to five positive charges moving to the right.

27.2 (b). The doubling of the radius causes the area A to be four times as large, so Equation 27.10 tells us that the resistance decreases.

27.3 (b). According to Equation 27.7, resistance is the ratio of voltage across a device to current in the device. In Figure 27.7b, a line drawn from the origin to a point on the curve will have a slope equal to $I/\Delta V$, which is the inverse of resistance. As ΔV increases, the slope of the line also increases, so the resistance decreases.

27.4 (a). When the filament is at room temperature, its resistance is low and the current is therefore relatively large. As the filament warms up, its resistance increases and the current decreases. Older lightbulbs often fail just as they are turned on because this large, initial current "spike" produces a rapid temperature increase and mechanical stress on the filament, causing it to break.

27.5 $I_a = I_b > I_c = I_d > I_e = I_f$. The current I_a leaves the positive terminal of the battery and then splits to flow through the two lightbulbs; therefore, $I_a = I_c + I_e$. Looking at Equation 27.21, we see that power rating is inversely related to resistance. Therefore, we know that the current in the 60-W lightbulb is greater than that in the 30-W lightbulb. Because charge does not build up in the lightbulbs, we know that the same amount of charge flowing into a lightbulb from the left must flow out on the right; consequently, $I_c = I_d$ and $I_e = I_f$. The two currents leaving the lightbulbs recombine to form the current back into the battery, $I_f + I_d = I_b$.

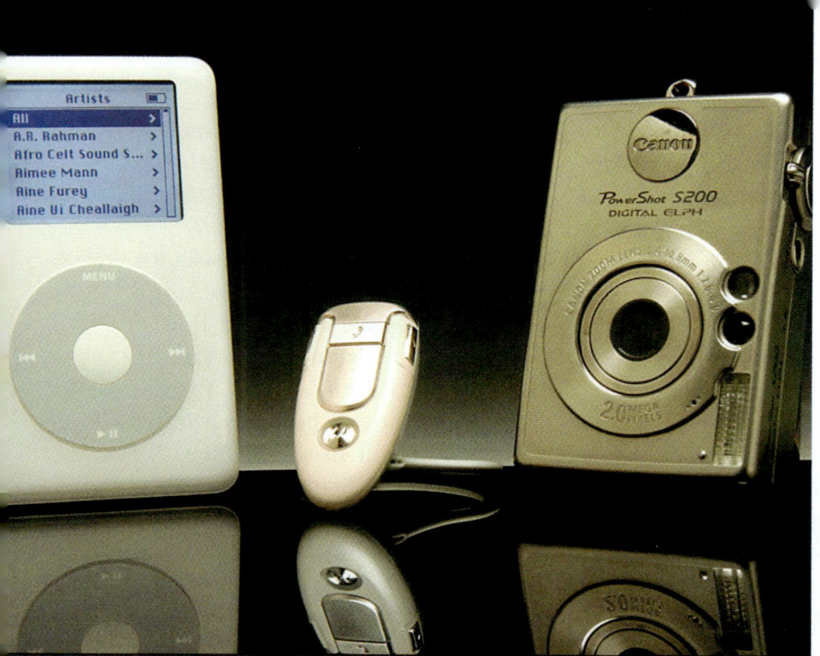

In our lives today, we use many items that are generally called "personal electronics," such as MP3 players, cell phones, and digital cameras. These items as well as dozens of others that we use contain electric circuits powered by batteries. In this chapter, we study simple types of circuits and learn how to analyze them. (© Thomson Learning/Charles D. Winters)

28 Direct Current Circuits

In this chapter, we analyze simple electric circuits that contain batteries, resistors, and capacitors in various combinations. Some circuits contain resistors that can be combined using simple rules. The analysis of more complicated circuits is simplified using *Kirchhoff's rules,* which follow from the laws of conservation of energy and conservation of electric charge for isolated systems. Most of the circuits analyzed are assumed to be in *steady state,* which means that currents in the circuit are constant in magnitude and direction. A current that is constant in direction is called a *direct current* (DC). We will study *alternating current* (AC), in which the current changes direction periodically, in Chapter 33. Finally, we describe electrical meters for measuring current and potential difference and then discuss electrical circuits in the home.

28.1 Electromotive Force

In Section 27.6, we discussed a circuit in which a battery produces a current. We will generally use a battery as a source of energy for circuits in our discussion. Because the potential difference at the battery terminals is constant in a particular circuit, the current in the circuit is constant in magnitude and direction and is called **direct current.** A battery is called either a *source of electromotive force* or, more commonly, a *source of emf.* (The phrase *electromotive force* is an unfortunate historical term, describing not a force, but rather a potential difference in volts.) **The emf $\mathcal{E}$**

of a battery is the maximum possible voltage the battery can provide between its terminals. You can think of a source of emf as a "charge pump." When an electric potential difference exists between two points, the source moves charges "uphill" from the lower potential to the higher.

We shall generally assume the connecting wires in a circuit have no resistance. The positive terminal of a battery is at a higher potential than the negative terminal. Because a real battery is made of matter, there is resistance to the flow of charge within the battery. This resistance is called **internal resistance** r. For an idealized battery with zero internal resistance, the potential difference across the battery (called its *terminal voltage*) equals its emf. For a real battery, however, the terminal voltage is *not* equal to the emf for a battery in a circuit in which there is a current. To understand why, consider the circuit diagram in Active Figure 28.1a. The battery in this diagram is represented by the dashed rectangle containing an ideal, resistance-free emf $\mathcal{E}$ in series with an internal resistance r. A resistor of resistance R is connected across the terminals of the battery. Now imagine moving through the battery from a to d and measuring the electric potential at various locations. Passing from the negative terminal to the positive terminal, the potential increases by an amount $\mathcal{E}$. As we move through the resistance r, however, the potential *decreases* by an amount Ir, where I is the current in the circuit. Therefore, the terminal voltage of the battery $\Delta V = V_d - V_a$ is

$$\Delta V = \mathcal{E} - Ir \quad \textbf{(28.1)}$$

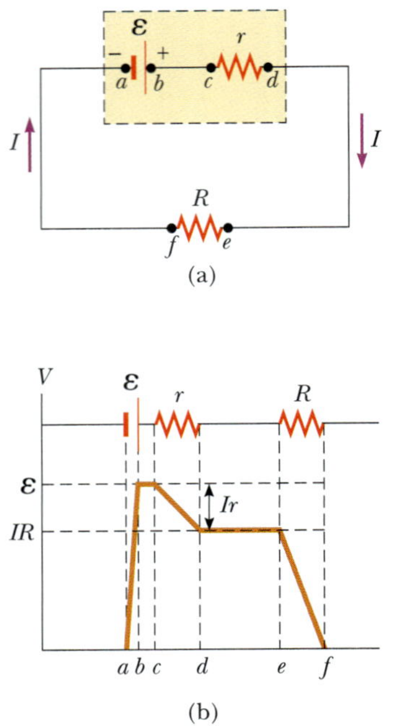

ACTIVE FIGURE 28.1

(a) Circuit diagram of a source of emf $\mathcal{E}$ (in this case, a battery), of internal resistance r, connected to an external resistor of resistance R. (b) Graphical representation showing how the electric potential changes as the circuit in (a) is traversed clockwise.

Sign in at www.thomsonedu.com and go to ThomsonNOW to adjust the emf and resistances r and R and see the effect on the current and on the graph in part (b).

From this expression, notice that $\mathcal{E}$ is equivalent to the **open-circuit voltage,** that is, the terminal voltage when the current is zero. The emf is the voltage labeled on a battery; for example, the emf of a D cell is 1.5 V. The actual potential difference between a battery's terminals depends on the current in the battery as described by Equation 28.1.

Active Figure 28.1b is a graphical representation of the changes in electric potential as the circuit is traversed in the clockwise direction. Active Figure 28.1a shows that the terminal voltage ΔV must equal the potential difference across the external resistance R, often called the **load resistance.** The load resistor might be a simple resistive circuit element as in Active Figure 28.1a, or it could be the resistance of some electrical device (such as a toaster, electric heater, or lightbulb) connected to the battery (or, in the case of household devices, to the wall outlet). The resistor represents a *load* on the battery because the battery must supply energy to operate the device containing the resistance. The potential difference across the load resistance is $\Delta V = IR$. Combining this expression with Equation 28.1, we see that

$$\mathcal{E} = IR + Ir \quad \textbf{(28.2)}$$

Solving for the current gives

$$I = \frac{\mathcal{E}}{R + r} \quad \textbf{(28.3)}$$

Equation 28.3 shows that the current in this simple circuit depends on both the load resistance R external to the battery and the internal resistance r. If R is much greater than r, as it is in many real-world circuits, we can neglect r.

Multiplying Equation 28.2 by the current I in the circuit gives

$$I\mathcal{E} = I^2R + I^2r \quad \textbf{(28.4)}$$

Equation 28.4 indicates that because power $\mathcal{P} = I\,\Delta V$ (see Eq. 27.20), the total power output $I\mathcal{E}$ of the battery is delivered to the external load resistance in the amount I^2R and to the internal resistance in the amount I^2r.

PITFALL PREVENTION 28.1

What Is Constant in a Battery?

It is a common misconception that a battery is a source of constant current. Equation 28.3 shows that is not true. The current in the circuit depends on the resistance R connected to the battery. It is also not true that a battery is a source of constant terminal voltage as shown by Equation 28.1. **A battery is a source of constant emf.**

Quick Quiz 28.1 To maximize the percentage of the power that is delivered from a battery to a device, what should the internal resistance of the battery be? (a) It should be as low as possible. (b) It should be as high as possible. (c) The percentage does not depend on the internal resistance.

EXAMPLE 28.1 Terminal Voltage of a Battery

A battery has an emf of 12.0 V and an internal resistance of 0.05 Ω. Its terminals are connected to a load resistance of 3.00 Ω.

(A) Find the current in the circuit and the terminal voltage of the battery.

SOLUTION

Conceptualize Study Active Figure 28.1a, which shows a circuit consistent with the problem statement. The battery delivers energy to the load resistor.

Categorize This example involves simple calculations from this section, so we categorize it as a substitution problem.

Use Equation 28.3 to find the current in the circuit:

$$I = \frac{\mathcal{E}}{R + r} = \frac{12.0\text{ V}}{(3.00\ \Omega + 0.05\ \Omega)} = \boxed{3.93\text{ A}}$$

Use Equation 28.1 to find the terminal voltage:

$$\Delta V = \mathcal{E} - Ir = 12.0\text{ V} - (3.93\text{ A})(0.05\ \Omega) = \boxed{11.8\text{ V}}$$

To check this result, calculate the voltage across the load resistance R:

$$\Delta V = IR = (3.93\text{ A})(3.00\ \Omega) = 11.8\text{ V}$$

(B) Calculate the power delivered to the load resistor, the power delivered to the internal resistance of the battery, and the power delivered by the battery.

SOLUTION

Use Equation 27.21 to find the power delivered to the load resistor:

$$\mathcal{P}_R = I^2R = (3.93\text{ A})^2(3.00\ \Omega) = \boxed{46.3\text{ W}}$$

Find the power delivered to the internal resistance:

$$\mathcal{P}_r = I^2r = (3.93\text{ A})^2(0.05\ \Omega) = \boxed{0.772\text{ W}}$$

Find the power delivered by the battery by adding these quantities:

$$\mathcal{P} = \mathcal{P}_R + \mathcal{P}_r = 46.3\text{ W} + 0.772\text{ W} = \boxed{47.1\text{ W}}$$

What If? As a battery ages, its internal resistance increases. Suppose the internal resistance of this battery rises to 2.00 Ω toward the end of its useful life. How does that alter the battery's ability to deliver energy?

Answer Let's connect the same 3.00-Ω load resistor to the battery.

Find the new current in the battery:

$$I = \frac{\mathcal{E}}{R + r} = \frac{12.0\text{ V}}{(3.00\ \Omega + 2.00\ \Omega)} = 2.40\text{ A}$$

Find the new terminal voltage:

$$\Delta V = \mathcal{E} - Ir = 12.0\text{ V} - (2.40\text{ A})(2.00\ \Omega) = 7.2\text{ V}$$

Find the new powers delivered to the load resistor and internal resistance:

$$\mathcal{P}_R = I^2R = (2.40\text{ A})^2(3.00\ \Omega) = 17.3\text{ W}$$

$$\mathcal{P}_r = I^2r = (2.40\text{ A})^2(2.00\ \Omega) = 11.5\text{ W}$$

The terminal voltage is only 60% of the emf. Notice that 40% of the power from the battery is delivered to the internal resistance when r is 2.00 Ω. When r is 0.05 Ω as in part (B), this percentage is only 1.6%. Consequently, even though the emf remains fixed, the increasing internal resistance of the battery significantly reduces the battery's ability to deliver energy.

EXAMPLE 28.2 Matching the Load

Find the load resistance R for which the maximum power is delivered to the load resistance in Active Figure 28.1a.

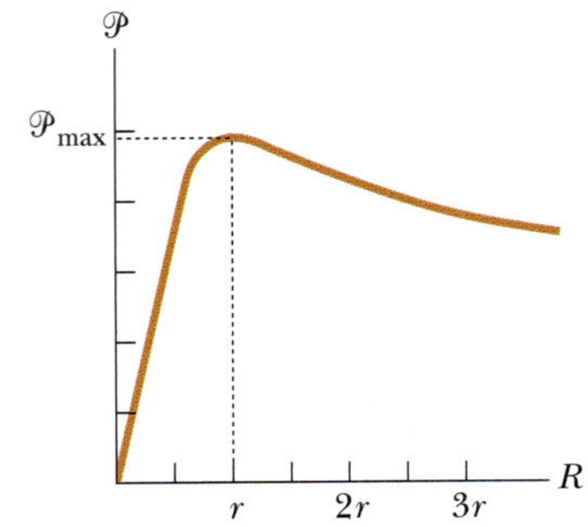

Figure 28.2 (Example 28.2) Graph of the power $\mathcal{P}$ delivered by a battery to a load resistor of resistance R as a function of R. The power delivered to the resistor is a maximum when the load resistance equals the internal resistance of the battery.

SOLUTION

Conceptualize Think about varying the load resistance in Active Figure 28.1a and the effect on the power delivered to the load resistance. When R is large, there is very little current, so the power I^2R delivered to the load resistor is small. When R is small, the current is large and there is significant loss of power I^2r as energy is delivered to the internal resistance. Therefore, the power delivered to the load resistor is small again. For some intermediate value of the resistance R, the power must maximize.

Categorize The circuit is the same as that in Example 28.1. The load resistance R in this case, however, is a variable.

Analyze Find the power delivered to the load resistance using Equation 27.21, with I given by Equation 28.3:

$$(1) \quad \mathcal{P} = I^2R = \frac{\mathcal{E}^2R}{(R+r)^2}$$

Differentiate the power with respect to the load resistance R and set the derivative equal to zero to maximize the power:

$$\frac{d\mathcal{P}}{dR} = \frac{d}{dR}\left[\frac{\mathcal{E}^2R}{(R+r)^2}\right] = \frac{d}{dR}\left[\mathcal{E}^2R(R+r)^{-2}\right] = 0$$

$$\left[\mathcal{E}^2(R+r)^{-2}\right] + \left[\mathcal{E}^2R(-2)(R+r)^{-3}\right] = 0$$

$$\frac{\mathcal{E}^2(R+r)}{(R+r)^3} - \frac{2\mathcal{E}^2R}{(R+r)^3} = \frac{\mathcal{E}^2(r-R)}{(R+r)^3} = 0$$

Solve for R:

$$R = r$$

Finalize To check this result, let's plot $\mathcal{P}$ versus R as in Figure 28.2. The graph shows that $\mathcal{P}$ reaches a maximum value at $R = r$. Equation (1) shows that this maximum value is $\mathcal{P}_{max} = \mathcal{E}^2/4r$.

28.2 Resistors in Series and Parallel

When two or more resistors are connected together as are the lightbulbs in Active Figure 28.3a, they are said to be in a **series combination.** Active Figure 28.3b is the circuit diagram for the lightbulbs, shown as resistors, and the battery. In a series connection, if an amount of charge Q exits resistor R_1, charge Q must also enter the second resistor R_2. Otherwise, charge would accumulate on the wire between the resistors. Therefore, the same amount of charge passes through both resistors in a given time interval and the currents are the same in both resistors:

$$I = I_1 = I_2$$

where I is the current leaving the battery, I_1 is the current in resistor R_1, and I_2 is the current in resistor R_2.

The potential difference applied across the series combination of resistors divides between the resistors. In Active Figure 28.3b, because the voltage drop[1]

[1] The term *voltage drop* is synonymous with a decrease in electric potential across a resistor. It is often used by individuals working with electric circuits.

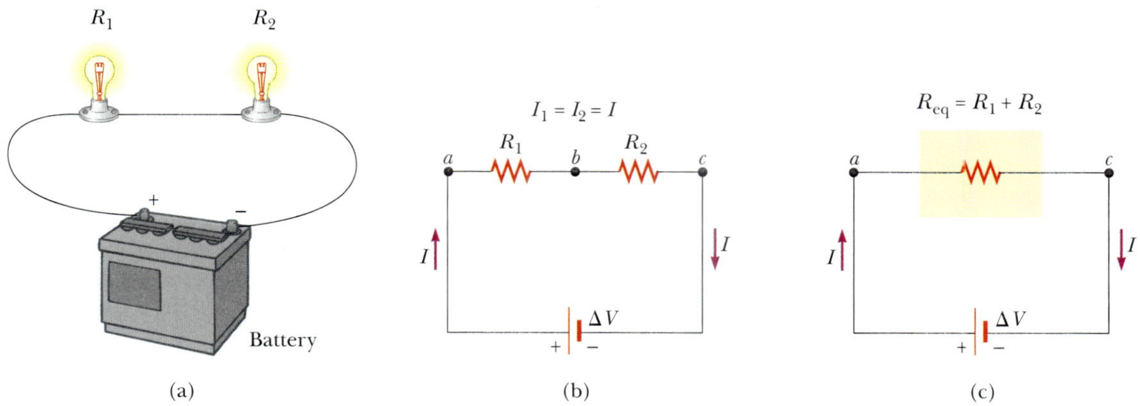

ACTIVE FIGURE 28.3

(a) A series combination of two lightbulbs with resistances R_1 and R_2. (b) Circuit diagram for the two-resistor circuit. The current in R_1 is the same as that in R_2. (c) The resistors replaced with a single resistor having an equivalent resistance $R_{eq} = R_1 + R_2$.

Sign in at www.thomsonedu.com and go to ThomsonNOW to adjust the battery voltage and resistances R_1 and R_2 and see the effect on the currents and voltages in the individual resistors.

from a to b equals I_1R_1 and the voltage drop from b to c equals I_2R_2, the voltage drop from a to c is

$$\Delta V = I_1R_1 + I_2R_2$$

The potential difference across the battery is also applied to the **equivalent resistance** R_{eq} in Active Figure 28.3c:

$$\Delta V = IR_{eq}$$

where the equivalent resistance has the same effect on the circuit as the series combination because it results in the same current I in the battery. Combining these equations for ΔV, we see that we can replace the two resistors in series with a single equivalent resistance whose value is the *sum* of the individual resistances:

$$\Delta V = IR_{eq} = I_1R_1 + I_2R_2 \quad \rightarrow \quad R_{eq} = R_1 + R_2 \tag{28.5}$$

where we have canceled the currents I, I_1, and I_2 because they are all the same.

The equivalent resistance of three or more resistors connected in series is

$$R_{eq} = R_1 + R_2 + R_3 + \cdots \tag{28.6}$$

◀ The equivalent resistance of a series combination of resistors

This relationship indicates that **the equivalent resistance of a series combination of resistors is the numerical sum of the individual resistances and is always greater than any individual resistance.**

Looking back at Equation 28.3, we see that the denominator is the simple algebraic sum of the external and internal resistances. That is consistent with the internal and external resistances being in series in Active Figure 28.1a.

If the filament of one lightbulb in Active Figure 28.3 were to fail, the circuit would no longer be complete (resulting in an open-circuit condition) and the second lightbulb would also go out. This fact is a general feature of a series circuit: if one device in the series creates an open circuit, all devices are inoperative.

PITFALL PREVENTION 28.2

Lightbulbs Don't Burn

We will describe the end of the life of a lightbulb by saying *the filament fails* rather than by saying the lightbulb "burns out." The word *burn* suggests a combustion process, which is not what occurs in a lightbulb. The failure of a lightbulb results from the slow sublimation of tungsten from the very hot filament over the life of the lightbulb. The filament eventually becomes very thin because of this process. The mechanical stress from a sudden temperature increase when the lightbulb is turned on causes the thin filament to break.

Quick Quiz 28.2 With the switch in the circuit of Figure 28.4a (page 780) closed, there is no current in R_2 because the current has an alternate zero-resistance path through the switch. There is current in R_1, and this current is measured with the ammeter (a device for measuring current) at the bottom of the circuit. If the switch is opened (Fig. 28.4b), there is current in R_2. What happens to the

PITFALL PREVENTION 28.3
Local and Global Changes

A local change in one part of a circuit may result in a global change throughout the circuit. For example, if a single resistor is changed in a circuit containing several resistors and batteries, the currents in all resistors and batteries, the terminal voltages of all batteries, and the voltages across all resistors may change as a result.

reading on the ammeter when the switch is opened? (a) The reading goes up. (b) The reading goes down. (c) The reading does not change.

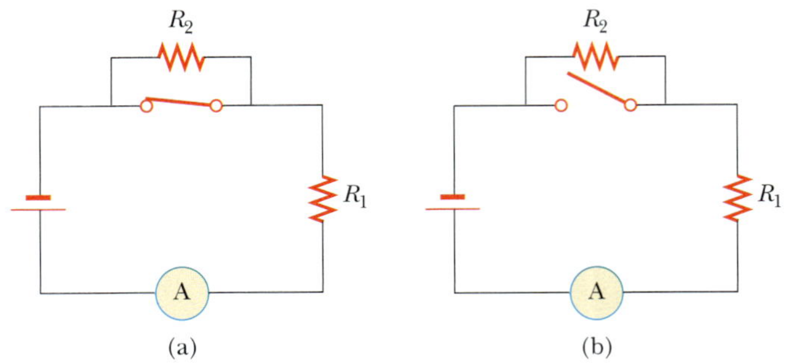

Figure 28.4 (Quick Quiz 28.2) What happens when the switch is opened?

Now consider two resistors in a **parallel combination** as shown in Active Figure 28.5. Notice that both resistors are connected directly across the terminals of the battery. Therefore, the potential differences across the resistors are the same:

$$\Delta V = \Delta V_1 = \Delta V_2$$

where ΔV is the terminal voltage of the battery.

When charges reach point a in Active Figure 28.5b, they split into two parts, with some going toward R_1 and the rest going toward R_2. A **junction** is any such point in a circuit where a current can split. This split results in less current in each individual resistor than the current leaving the battery. Because electric charge is conserved, the current I that enters point a must equal the total current leaving that point:

$$I = I_1 + I_2$$

where I_1 is the current in R_1 and I_2 is the current in R_2.

The current in the **equivalent resistance** R_{eq} in Active Figure 28.5c is

$$I = \frac{\Delta V}{R_{eq}}$$

where the equivalent resistance has the same effect on the circuit as the two resistors in parallel; that is, the equivalent resistance draws the same current I from the

PITFALL PREVENTION 28.4
Current Does Not Take the Path of Least Resistance

You may have heard the phrase "current takes the path of least resistance" (or similar wording) in reference to a parallel combination of current paths such that there are two or more paths for the current to take. Such wording is incorrect. The current takes *all* paths. Those paths with lower resistance have larger currents, but even very high resistance paths carry *some* of the current. In theory, if current has a choice between a zero-resistance path and a finite resistance path, all the current takes the path of zero-resistance; a path with zero resistance, however, is an idealization.

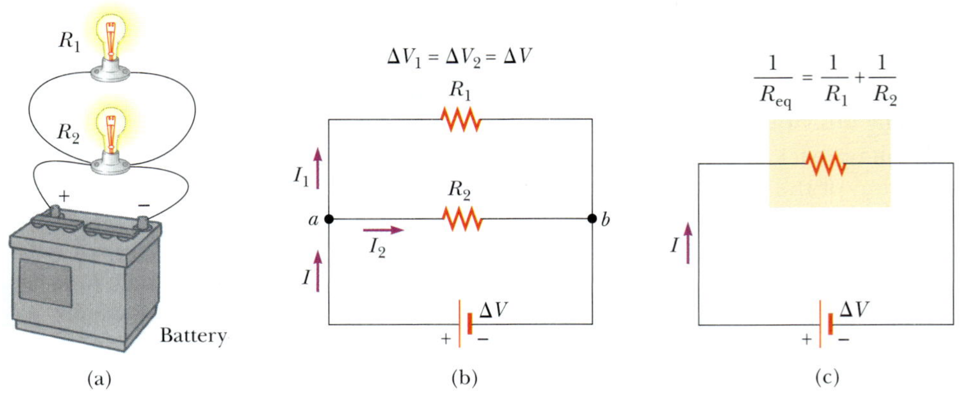

ACTIVE FIGURE 28.5

(a) A parallel combination of two lightbulbs with resistances R_1 and R_2. (b) Circuit diagram for the two-resistor circuit. The potential difference across R_1 is the same as that across R_2. (c) The resistors replaced with a single resistor having an equivalent resistance given by Equation 28.7.

Sign in at www.thomsonedu.com and go to ThomsonNOW to adjust the battery voltage and resistances R_1 and R_2 and see the effect on the currents and voltages in the individual resistors.

battery. Combining these equations for I, we see that the equivalent resistance of two resistors in parallel is given by

$$I = \frac{\Delta V}{R_{eq}} = \frac{\Delta V_1}{R_1} + \frac{\Delta V_2}{R_2} \rightarrow \frac{1}{R_{eq}} = \frac{1}{R_1} + \frac{1}{R_2} \tag{28.7}$$

where we have canceled ΔV, ΔV_1, and ΔV_2 because they are all the same.

An extension of this analysis to three or more resistors in parallel gives

$$\frac{1}{R_{eq}} = \frac{1}{R_1} + \frac{1}{R_2} + \frac{1}{R_3} + \cdots \tag{28.8}$$

◀ The equivalent resistance o⁻ a parallel combination of resistors

This expression shows that **the inverse of the equivalent resistance of two or more resistors in a parallel combination is equal to the sum of the inverses of the individual resistances. Furthermore, the equivalent resistance is always less than the smallest resistance in the group.**

Household circuits are always wired such that the appliances are connected in parallel. Each device operates independently of the others so that if one is switched off, the others remain on. In addition, in this type of connection, all the devices operate on the same voltage.

Let's consider two examples of practical applications of series and parallel circuits. Figure 28.6 illustrates how a three-way lightbulb is constructed to provide three levels of light intensity.[2] The socket of the lamp is equipped with a three-way switch for selecting different light intensities. The lightbulb contains two filaments. When the lamp is connected to a 120-V source, one filament receives 100 W of power and the other receives 75 W. The three light intensities are made possible by applying the 120 V to one filament alone, to the other filament alone, or to the two filaments in parallel. When switch S_1 is closed and switch S_2 is opened, current exists only in the 75-W filament. When switch S_1 is open and switch S_2 is closed, current exists only in the 100-W filament. When both switches are closed, current exists in both filaments and the total power is 175 W.

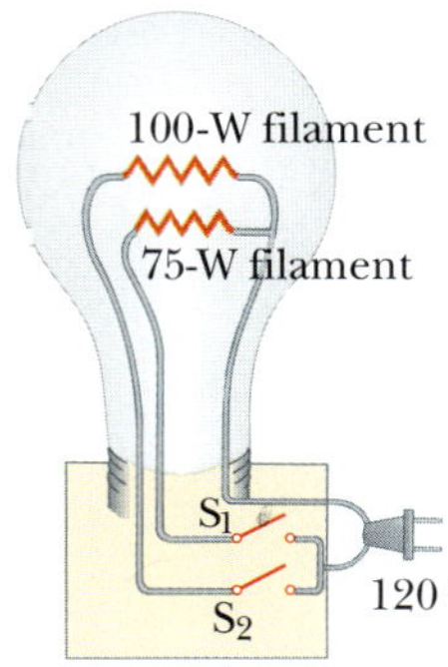

Figure 28.6 A three-way lightbulb.

If the filaments were connected in series and one of them were to break, no charges could pass through the lightbulb and it would not glow, regardless of the switch position. If, however, the filaments were connected in parallel and one of them (for example, the 75-W filament) were to break, the lightbulb would continue to glow in two of the switch positions because current exists in the other (100-W) filament.

As a second example, consider strings of lights that are used for many ornamental purposes such as decorating Christmas trees. Over the years, both parallel and series connections have been used for strings of lights. Because series-wired lightbulbs operate with less energy per bulb and at a lower temperature, they are safer than parallel-wired lightbulbs for indoor Christmas-tree use. If, however, the filament of a single lightbulb in a series-wired string were to fail (or if the lightbulb were removed from its socket), all the lights on the string would go out. The popularity of series-wired light strings diminished because troubleshooting a failed lightbulb is a tedious, time-consuming chore that involves trial-and-error substitution of a good lightbulb in each socket along the string until the defective one is found.

In a parallel-wired string, each lightbulb operates at 120 V. By design, the lightbulbs are brighter and hotter than those on a series-wired string. As a result, they are inherently more dangerous (more likely to start a fire, for instance), but if one lightbulb in a parallel-wired string fails or is removed, the rest of the lightbulbs continue to glow.

To prevent the failure of one lightbulb from causing the entire string to go out, a new design was developed for so-called miniature lights wired in series. When the filament breaks in one of these miniature lightbulbs, the break in the filament

[2] The three-way lightbulb and other household devices actually operate on alternating current (AC), to be introduced in Chapter 33.

Figure 28.7 (a) Schematic diagram of a modern "miniature" holiday lightbulb, with a jumper connection to provide a current path if the filament breaks. When the filament is intact, charges flow in the filament. (b) A holiday lightbulb with a broken filament. In this case, charges flow in the jumper connection. (c) A Christmas-tree lightbulb.

represents the largest resistance in the series, much larger than that of the intact filaments. As a result, most of the applied 120 V appears across the lightbulb with the broken filament. Inside the lightbulb, a small jumper loop covered by an insulating material is wrapped around the filament leads. When the filament fails and 120 V appears across the lightbulb, an arc burns the insulation on the jumper and connects the filament leads. This connection now completes the circuit through the lightbulb even though its filament is no longer active (Fig. 28.7).

When a lightbulb fails, the resistance across its terminals is reduced to almost zero because of the alternate jumper connection mentioned in the preceding paragraph. All the other lightbulbs not only stay on, but they glow more brightly because the total resistance of the string is reduced and consequently the current in each lightbulb increases. Each lightbulb operates at a slightly higher temperature than before. As more lightbulbs fail, the current keeps rising, the filament of each lightbulb operates at a higher temperature, and the lifetime of the lightbulb is reduced. For this reason, you should check for failed (nonglowing) lightbulbs in such a series-wired string and replace them as soon as possible, thereby maximizing the lifetimes of all the lightbulbs.

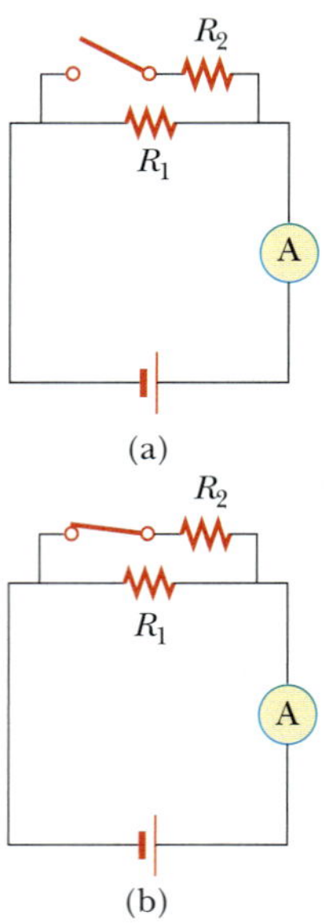

Figure 28.8 (Quick Quiz 28.3) What happens when the switch is closed?

Quick Quiz 28.3 With the switch in the circuit of Figure 28.8a open, there is no current in R_2. There is current in R_1, however, and it is measured with the ammeter at the right side of the circuit. If the switch is closed (Fig. 28.8b), there is current in R_2. What happens to the reading on the ammeter when the switch is closed? (a) The reading increases. (b) The reading decreases. (c) The reading does not change.

Quick Quiz 28.4 Consider the following choices: (a) increases, (b) decreases, (c) remains the same. From these choices, choose the best answer for the following situations. **(i)** In Active Figure 28.3, a third resistor is added in series with the first two. What happens to the current in the battery? **(ii)** What happens to the terminal voltage of the battery? **(iii)** In Active Figure 28.5, a third resistor is added in parallel with the first two. What happens to the current in the battery? **(iv)** What happens to the terminal voltage of the battery?

CONCEPTUAL EXAMPLE 28.3 Landscape Lights

A homeowner wishes to install low-voltage landscape lighting in his back yard. To save money, he purchases inexpensive 18-gauge cable, which has a relatively high resistance per unit length. This cable consists of two side-by-side wires separated by insulation, like the cord on an appliance. He runs a 200-foot length of this cable from the power supply to the farthest point at which he plans to position a light fixture. He attaches light fixtures across the two wires on the cable at 10-foot intervals so that the light fixtures are in parallel. Because of the cable's resistance, the brightness of the lightbulbs in the fixtures is not as desired. Which of the following problems does the homeowner have? (a) All the lightbulbs glow equally less brightly than they would if lower-resistance cable had been used. (b) The brightness of the lightbulbs decreases as you move farther from the power supply.

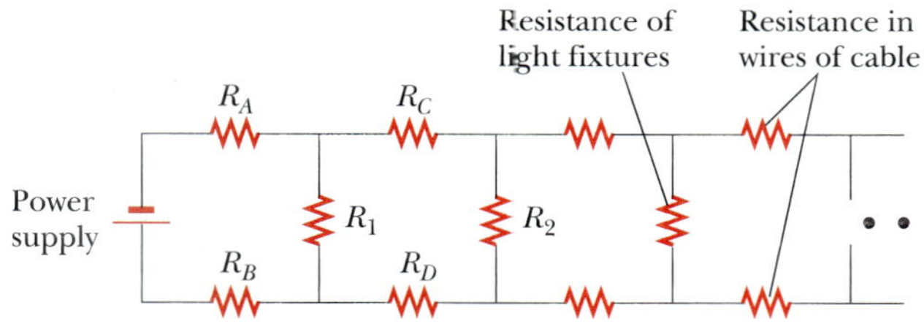

Figure 28.9 (Conceptual Example 28.3) The circuit diagram for a set of landscape light fixtures connected in parallel across the two wires of a two-wire cable. The horizontal resistors represent resistance in the wires of the cable. The vertical resistors represent the light fixtures.

SOLUTION

A circuit diagram for the system appears in Figure 28.9. The horizontal resistors with letter subscripts (such as R_A) represent the resistance of the wires in the cable between the light fixtures, and the vertical resistors with number subscripts (such as R_1) represent the resistance of the light fixtures themselves. Part of the terminal voltage of the power supply is dropped across resistors R_A and R_B. Therefore, the voltage across light fixture R_1 is less than the terminal voltage. There is a further voltage drop across resistors R_C and R_D. Consequently, the voltage across light fixture R_2 is smaller than that across R_1. This pattern continues down the line of light fixtures, so the correct choice is (b). Each successive light fixture has a smaller voltage across it and glows less brightly than the one before.

EXAMPLE 28.4 Find the Equivalent Resistance

Four resistors are connected as shown in Figure 28.10a.

(A) Find the equivalent resistance between points *a* and *c*.

SOLUTION

Conceptualize Imagine charges flowing into this combination from the left. All charges must pass through the first two resistors, but the charges split into two different paths when encountering the combination of the 6.0-Ω and the 3.0-Ω resistors.

Categorize Because of the simple nature of the combination of resistors in Figure 28.10, we categorize this example as one for which we can use the rules for series and parallel combinations of resistors.

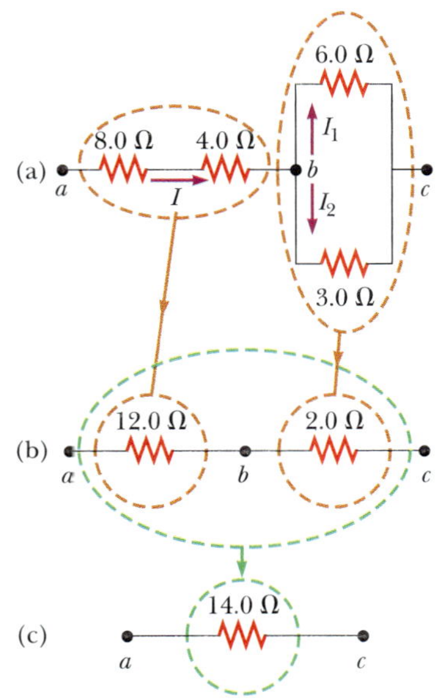

Figure 28.10 (Example 28.4) The original network of resistors is reduced to a single equivalent resistance.

Analyze The combination of resistors can be reduced in steps as shown in Figure 28.10.

Find the equivalent resistance between *a* and *b* of the 8.0-Ω and 4.0-Ω resistors, which are in series:

$$R_{eq} = 8.0\ \Omega + 4.0\ \Omega = 12.0\ \Omega$$

Find the equivalent resistance between *b* and *c* of the 6.0-Ω and 3.0-Ω resistors, which are in parallel:

$$\frac{1}{R_{eq}} = \frac{1}{6.0\ \Omega} + \frac{1}{3.0\ \Omega} = \frac{3}{6.0\ \Omega}$$

$$R_{eq} = \frac{6.0\ \Omega}{3} = 2.0\ \Omega$$

The circuit of equivalent resistances now looks like Figure 28.10b. Find the equivalent resistance from *a* to *c*:

$$R_{eq} = 12.0\ \Omega + 2.0\ \Omega = \boxed{14.0\ \Omega}$$

This resistance is that of the single equivalent resistor in Figure 28.10c.

(B) What is the current in each resistor if a potential difference of 42 V is maintained between *a* and *c*?

SOLUTION

The currents in the 8.0-Ω and 4.0-Ω resistors are the same because they are in series. In addition, they carry the same current that would exist in the 14.0-Ω equivalent resistor subject to the 42-V potential difference.

Use Equation 27.7 ($R = \Delta V/I$) and the result from part (A) to find the current in the 8.0-Ω and 4.0-Ω resistors:

$$I = \frac{\Delta V_{ac}}{R_{eq}} = \frac{42\text{ V}}{14.0\ \Omega} = 3.0\text{ A}$$

Set the voltages across the resistors in parallel in Figure 28.10a equal to find a relationship between the currents:

$$\Delta V_1 = \Delta V_2 \rightarrow (6.0\ \Omega)I_1 = (3.0\ \Omega)I_2 \rightarrow I_2 = 2I_1$$

Use $I_1 + I_2 = 3.0$ A to find I_1:

$$I_1 + I_2 = 3.0\text{ A} \rightarrow I_1 + 2I_1 = 3.0\text{ A} \rightarrow I_1 = 1.0\text{ A}$$

Find I_2:

$$I_2 = 2I_1 = 2(1.0\text{ A}) = 2.0\text{ A}$$

Finalize As a final check of our results, note that $\Delta V_{bc} = (6.0\ \Omega)I_1 = (3.0\ \Omega)I_2 = 6.0$ V and $\Delta V_{ab} = (12.0\ \Omega)I = 36$ V; therefore, $\Delta V_{ac} = \Delta V_{ab} + \Delta V_{bc} = 42$ V, as it must.

EXAMPLE 28.5 Three Resistors in Parallel

Three resistors are connected in parallel as shown in Figure 28.11a. A potential difference of 18.0 V is maintained between points *a* and *b*.

(A) Calculate the equivalent resistance of the circuit.

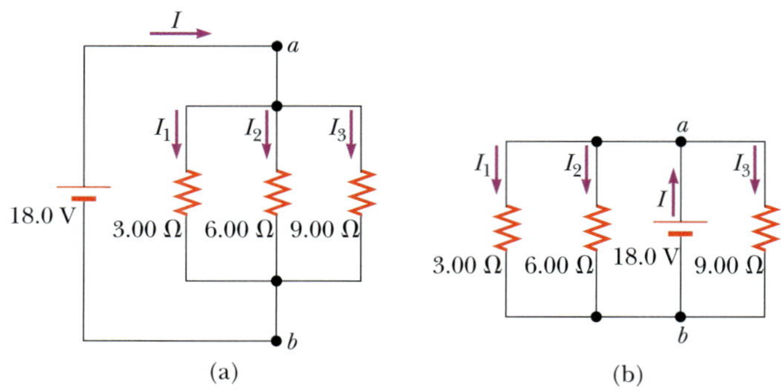

Figure 28.11 (Example 28.5) (a) Three resistors connected in parallel. The voltage across each resistor is 18.0 V. (b) Another circuit with three resistors and a battery. Is it equivalent to the circuit in (a)?

SOLUTION

Conceptualize Figure 28.11a shows that we are dealing with a simple parallel combination of three resistors.

Categorize Because the three resistors are connected in parallel, we can use Equation 28.8 to evaluate the equivalent resistance.

Analyze Use Equation 28.8 to find R_{eq}:

$$\frac{1}{R_{eq}} = \frac{1}{3.00\ \Omega} + \frac{1}{6.00\ \Omega} + \frac{1}{9.00\ \Omega} = \frac{11.0}{18.0\ \Omega}$$

$$R_{eq} = \frac{18.0\ \Omega}{11.0} = 1.64\ \Omega$$

(B) Find the current in each resistor.

SOLUTION

The potential difference across each resistor is 18.0 V. Apply the relationship $\Delta V = IR$ to find the currents:

$$I_1 = \frac{\Delta V}{R_1} = \frac{18.0\text{ V}}{3.00\ \Omega} = 6.00\text{ A}$$

$$I_2 = \frac{\Delta V}{R_2} = \frac{18.0\text{ V}}{6.00\ \Omega} = 3.00\text{ A}$$

$$I_3 = \frac{\Delta V}{R_3} = \frac{18.0\text{ V}}{9.00\ \Omega} = 2.00\text{ A}$$

(C) Calculate the power delivered to each resistor and the total power delivered to the combination of resistors.

SOLUTION

Apply the relationship $\mathcal{P} = I^2R$ to each resistor using the currents calculated in part (B):

$$3.00\text{-}\Omega: \quad \mathcal{P}_1 = I_1{}^2R_1 = (6.00\text{ A})^2(3.00\ \Omega) = \boxed{108\text{ W}}$$

$$6.00\text{-}\Omega: \quad \mathcal{P}_2 = I_2{}^2R_2 = (3.00\text{ A})^2(6.00\ \Omega) = \boxed{54.0\text{ W}}$$

$$9.00\text{-}\Omega: \quad \mathcal{P}_3 = I_3{}^2R_3 = (2.00\text{ A})^2(9.00\ \Omega) = \boxed{36.0\text{ W}}$$

Finalize Part (C) shows that the smallest resistor receives the most power. Summing the three quantities gives a total power of 198 W. We could have calculated this final result from part (A) by considering the equivalent resistance as follows: $\mathcal{P} = (\Delta V)^2/R_{eq} = (18.0\text{ V})^2/\ 1.64\ \Omega = 198\text{ W}$.

What If? What if the circuit were as shown in Figure 28.11b instead of as in Figure 28.11a? How would that affect the calculation?

Answer There would be no effect on the calculation. The physical placement of the battery is not important. In Figure 28.11b, the battery still maintains a potential difference of 18.0 V between points *a* and *b*, so the two circuits in the figure are electrically identical.

28.3 Kirchhoff's Rules

As we saw in the preceding section, combinations of resistors can be simplified and analyzed using the expression $\Delta V = IR$ and the rules for series and parallel combinations of resistors. Very often, however, it is not possible to reduce a circuit to a single loop. The procedure for analyzing more complex circuits is made possible by using the two following principles, called **Kirchhoff's rules.**

1. **Junction rule.** At any junction, the sum of the currents must equal zero:

$$\sum_{\text{junction}} I = 0 \tag{28.9}$$

2. **Loop rule.** The sum of the potential differences across all elements around any closed circuit loop must be zero:

$$\sum_{\text{closed loop}} \Delta V = 0 \tag{28.10}$$

Kirchhoff's first rule is a statement of conservation of electric charge. All charges that enter a given point in a circuit must leave that point because charge cannot build up at a point. Currents directed into the junction are entered into the junction rule as $+I$, whereas currents directed out of a junction are entered as $-I$. Applying this rule to the junction in Figure 28.12a gives

$$I_1 - I_2 - I_3 = 0$$

Figure 28.12b represents a mechanical analog of this situation, in which water flows through a branched pipe having no leaks. Because water does not build up anywhere in the pipe, the flow rate into the pipe on the left equals the total flow rate out of the two branches on the right.

Kirchhoff's second rule follows from the law of conservation of energy. Let's imagine moving a charge around a closed loop of a circuit. When the charge returns to the starting point, the charge-circuit system must have the same total energy as it had before the charge was moved. The sum of the increases in energy

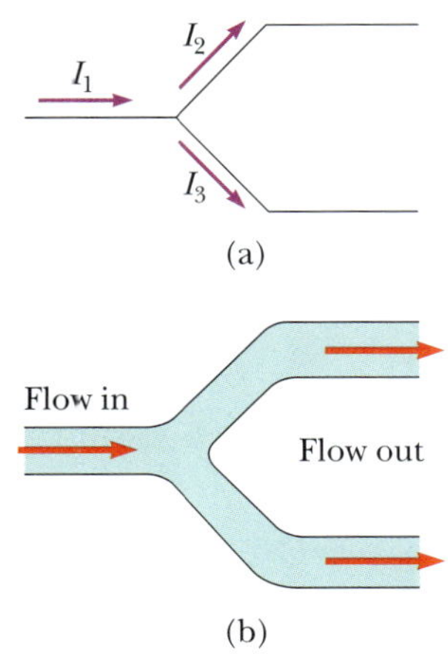

Figure 28.12 (a) Kirchhoff's junction rule. Conservation of charge requires that all charges entering a junction must leave that junction. Therefore, $I_1 - I_2 - I_3 = 0$. (b) A mechanical analog of the junction rule. The amount of water flowing out of the branches on the right must equal the amount flowing into the single branch on the left.

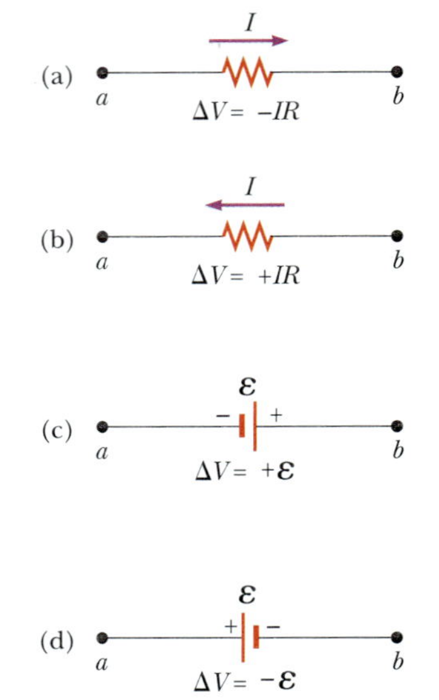

Figure 28.13 Rules for determining the potential differences across a resistor and a battery. (The battery is assumed to have no internal resistance.) Each circuit element is traversed from *a* to *b*, left to right.

as the charge passes through some circuit elements must equal the sum of the decreases in energy as it passes through other elements. The potential energy decreases whenever the charge moves through a potential drop $-IR$ across a resistor or whenever it moves in the reverse direction through a source of emf. The potential energy increases whenever the charge passes through a battery from the negative terminal to the positive terminal.

When applying Kirchhoff's second rule, imagine *traveling* around the loop and consider changes in *electric potential* rather than the changes in *potential energy* described in the preceding paragraph. Imagine traveling through the circuit elements in Figure 28.13 toward the right. The following sign conventions apply when using the second rule:

- Charges move from the high-potential end of a resistor toward the low-potential end, so if a resistor is traversed in the direction of the current, the potential difference ΔV across the resistor is $-IR$ (Fig. 28.13a).
- If a resistor is traversed in the direction *opposite* the current, the potential difference ΔV across the resistor is $+IR$ (Fig. 28.13b).
- If a source of emf (assumed to have zero internal resistance) is traversed in the direction of the emf (from negative to positive), the potential difference ΔV is $+\mathcal{E}$ (Fig. 28.13c).
- If a source of emf (assumed to have zero internal resistance) is traversed in the direction opposite the emf (from positive to negative), the potential difference ΔV is $-\mathcal{E}$ (Fig. 28.13d).

There are limits on the numbers of times you can usefully apply Kirchhoff's rules in analyzing a circuit. You can use the junction rule as often as you need as long as you include in it a current that has not been used in a preceding junction-rule equation. In general, the number of times you can use the junction rule is one fewer than the number of junction points in the circuit. You can apply the loop rule as often as needed as long as a new circuit element (resistor or battery) or a new current appears in each new equation. In general, **to solve a particular circuit problem, the number of independent equations you need to obtain from the two rules equals the number of unknown currents.**

Complex networks containing many loops and junctions generate great numbers of independent linear equations and a correspondingly great number of unknowns. Such situations can be handled formally through the use of matrix algebra. Computer software can also be used to solve for the unknowns.

The following examples illustrate how to use Kirchhoff's rules. In all cases, it is assumed the circuits have reached steady-state conditions; in other words, the currents in the various branches are constant. **Any capacitor acts as an open branch in a circuit;** that is, the current in the branch containing the capacitor is zero under steady-state conditions.

AIP ESVA/W. F. Meggers Collection

GUSTAV KIRCHHOFF
German Physicist (1824–1887)
Kirchhoff, a professor at Heidelberg, and Robert Bunsen invented the spectroscope and founded the science of spectroscopy, which we shall study in Chapter 42. They discovered the elements cesium and rubidium and invented astronomical spectroscopy.

PROBLEM-SOLVING STRATEGY Kirchhoff's Rules

The following procedure is recommended for solving problems that involve circuits that cannot be reduced by the rules for combining resistors in series or parallel.

1. *Conceptualize.* Study the circuit diagram and make sure you recognize all elements in the circuit. Identify the polarity of each battery and try to imagine the directions in which the current would exist in the batteries.
2. *Categorize.* Determine whether the circuit can be reduced by means of combining series and parallel resistors. If so, use the techniques of Section 28.2. If not, apply Kirchhoff's rules according to the *Analyze* step below.
3. *Analyze.* Assign labels to all known quantities and symbols to all unknown quantities. You must assign *directions* to the currents in each part of the circuit.

Although the assignment of current directions is arbitrary, you must adhere *rigorously* to the directions you assign when you apply Kirchhoff's rules.

Apply the junction rule (Kirchhoff's first rule) to all junctions in the circuit except one. Now apply the loop rule (Kirchhoff's second rule) to as many loops in the circuit as are needed to obtain, in combination with the equations from the junction rule, as many equations as there are unknowns. To apply this rule, you must choose a direction in which to travel around the loop (either clockwise or counterclockwise) and correctly identify the change in potential as you cross each element. Be careful with signs!

Solve the equations simultaneously for the unknown quantities.

4. *Finalize.* Check your numerical answers for consistency. Do not be alarmed if any of the resulting currents have a negative value. That only means you have guessed the direction of that current incorrectly, but *its magnitude will be correct.*

EXAMPLE 28.6 A Single-Loop Circuit

A single-loop circuit contains two resistors and two batteries as shown in Figure 28.14. (Neglect the internal resistances of the batteries.) Find the current in the circuit.

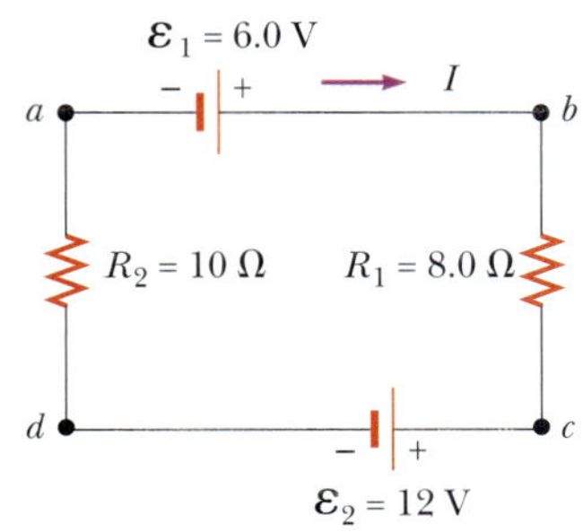

Figure 28.14 (Example 28.6) A series circuit containing two batteries and two resistors, where the polarities of the batteries are in opposition.

SOLUTION

Conceptualize Figure 28.14 shows the polarities of the batteries and a guess at the direction of the current.

Categorize We do not need Kirchhoff's rules to analyze this simple circuit, but let's use them anyway simply to see how they are applied. There are no junctions in this single-loop circuit; therefore, the current is the same in all elements.

Analyze Let's assume the current is clockwise as shown in Figure 28.14. Traversing the circuit in the clockwise direction, starting at a, we see that $a \rightarrow b$ represents a potential difference of $+\mathcal{E}_1$, $b \rightarrow c$ represents a potential difference of $-IR_1$, $c \rightarrow d$ represents a potential difference of $-\mathcal{E}_2$, and $d \rightarrow a$ represents a potential difference of $-IR_2$.

Apply Kirchhoff's loop rule to the single loop in the circuit:

$$\sum \Delta V = 0 \quad \rightarrow \quad \mathcal{E}_1 - IR_1 - \mathcal{E}_2 - IR_2 = 0$$

Solve for I and use the values given in Figure 28.14:

$$(1) \quad I = \frac{\mathcal{E}_1 - \mathcal{E}_2}{R_1 + R_2} = \frac{6.0\text{ V} - 12\text{ V}}{8.0\ \Omega + 10\ \Omega} = -0.33\text{ A}$$

Finalize The negative sign for I indicates that the direction of the current is opposite the assumed direction. The emfs in the numerator subtract because the batteries in Figure 28.14 have opposite polarities. The resistances in the denominator add because the two resistors are in series.

What If? What if the polarity of the 12.0-V battery were reversed? How would that affect the circuit?

Answer Although we could repeat the Kirchhoff's rules calculation, let's instead examine Equation (1) and modify it accordingly. Because the polarities of the two batteries are now in the same direction, the signs of $\mathcal{E}_1$ and $\mathcal{E}_2$ are the same and Equation (1) becomes

$$I = \frac{\mathcal{E}_1 + \mathcal{E}_2}{R_1 + R_2} = \frac{6.0\text{ V} + 12\text{ V}}{8.0\ \Omega + 10\ \Omega} = 1.0\text{ A}$$

EXAMPLE 28.7 A Multiloop Circuit

Find the currents I_1, I_2, and I_3 in the circuit shown in Figure 28.15.

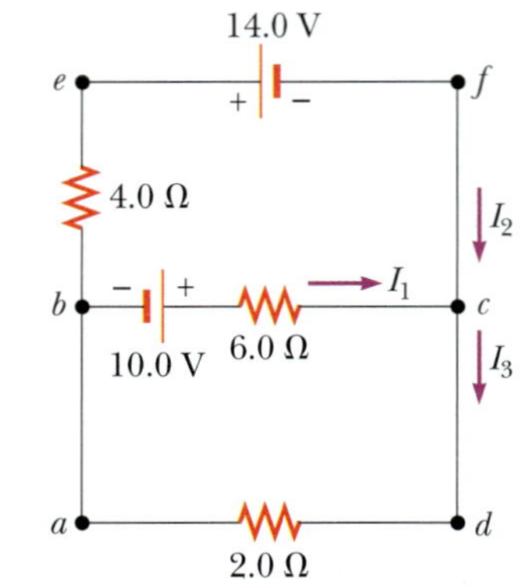

Figure 28.15 (Example 28.7) A circuit containing different branches.

SOLUTION

Conceptualize We cannot simplify the circuit by the rules associated with combining resistances in series and in parallel. (If the 10.0-V battery were not present, we could reduce the remaining circuit with series and parallel combinations.)

Categorize Because the circuit is not a simple series and parallel combination of resistances, this problem is one in which we must use Kirchhoff's rules.

Analyze We arbitrarily choose the directions of the currents as labeled in Figure 28.15.

Apply Kirchhoff's junction rule to junction c:

$$(1) \quad I_1 + I_2 - I_3 = 0$$

We now have one equation with three unknowns: I_1, I_2, and I_3. There are three loops in the circuit: *abcda, befcb,* and *aefda.* We need only two loop equations to determine the unknown currents. (The third loop equation would give no new information.) Let's choose to traverse these loops in the clockwise direction. Apply Kirchhoff's loop rule to loops *abcda* and *befcb*:

$$abcda\text{:} \quad (2) \quad 10.0\text{ V} - (6.0\ \Omega)I_1 - (2.0\ \Omega)I_3 = 0$$

$$befcb\text{:} \quad -(4.0\ \Omega)I_2 - 14.0\text{ V} + (6.0\ \Omega)I_1 - 10.0\text{ V} = 0$$

$$(3) \quad -24.0\text{ V} + (6.0\ \Omega)I_1 - (4.0\ \Omega)I_2 = 0$$

Solve Equation (1) for I_3 and substitute into Equation (2):

$$10.0\text{ V} - (6.0\ \Omega)I_1 - (2.0\ \Omega)(I_1 + I_2) = 0$$

$$(4) \quad 10.0\text{ V} - (8.0\ \Omega)I_1 - (2.0\ \Omega)I_2 = 0$$

Multiply each term in Equation (3) by 4 and each term in Equation (4) by 3:

$$(5) \quad -96.0\text{ V} + (24.0\ \Omega)I_1 - (16.0\ \Omega)I_2 = 0$$

$$(6) \quad 30.0\text{ V} - (24.0\ \Omega)I_1 - (6.0\ \Omega)I_2 = 0$$

Add Equation (6) to Equation (5) to eliminate I_1 and find I_2:

$$-66.0\text{ V} - (22.0\ \Omega)I_2 = 0$$

$$I_2 = -3.0\text{ A}$$

Use this value of I_2 in Equation (3) to find I_1:

$$-24.0\text{ V} + (6.0\ \Omega)I_1 - (4.0\ \Omega)(-3.0\text{ A}) = 0$$

$$-24.0\text{ V} + (6.0\ \Omega)I_1 + 12.0\text{ V} = 0$$

$$I_1 = 2.0\text{ A}$$

Use Equation (1) to find I_3:

$$I_3 = I_1 + I_2 = 2.0\text{ A} - 3.0\text{ A} = -1.0\text{ A}$$

Finalize Because our values for I_2 and I_3 are negative, the directions of these currents are opposite those indicated in Figure 28.15. The numerical values for the currents are correct. Despite the incorrect direction, we *must* continue to use these negative values in subsequent calculations because our equations were established with our original choice of direction. What would have happened had we left the current directions as labeled in Figure 28.15 but traversed the loops in the opposite direction?

28.4 *RC* Circuits

So far, we have analyzed direct current circuits in which the current is constant. In DC circuits containing capacitors, the current is always in the same direction but may vary in time. A circuit containing a series combination of a resistor and a capacitor is called an ***RC* circuit.**

Charging a Capacitor

Active Figure 28.16 shows a simple series *RC* circuit. Let's assume the capacitor in this circuit is initially uncharged. There is no current while the switch is open (Active Fig. 28.16a). If the switch is thrown to position a at $t = 0$ (Active Fig. 28.16b), however, charge begins to flow, setting up a current in the circuit, and the capacitor begins to charge.[3] Notice that during charging, charges do not jump across the capacitor plates because the gap between the plates represents an open circuit. Instead, charge is transferred between each plate and its connecting wires due to the electric field established in the wires by the battery until the capacitor is fully charged. As the plates are being charged, the potential difference across the capacitor increases. The value of the maximum charge on the plates depends on the voltage of the battery. Once the maximum charge is reached, the current in the circuit is zero because the potential difference across the capacitor matches that supplied by the battery.

To analyze this circuit quantitatively, let's apply Kirchhoff's loop rule to the circuit after the switch is thrown to position a. Traversing the loop in Active Figure 28.16b clockwise gives

$$\mathcal{E} - \frac{q}{C} - IR = 0 \qquad \textbf{(28.11)}$$

where q/C is the potential difference across the capacitor and IR is the potential difference across the resistor. We have used the sign conventions discussed earlier for the signs on $\mathcal{E}$ and IR. The capacitor is traversed in the direction from the positive plate to the negative plate, which represents a decrease in potential. Therefore, we use a negative sign for this potential difference in Equation 28.11. Note that q and I are *instantaneous* values that depend on time (as opposed to steady-state values) as the capacitor is being charged.

We can use Equation 28.11 to find the initial current in the circuit and the maximum charge on the capacitor. At the instant the switch is thrown to position a ($t = 0$), the charge on the capacitor is zero. Equation 28.11 shows that the initial current I_i in the circuit is a maximum and is given by

$$I_i = \frac{\mathcal{E}}{R} \quad \text{(current at } t = 0\text{)} \qquad \textbf{(28.12)}$$

At this time, the potential difference from the battery terminals appears entirely across the resistor. Later, when the capacitor is charged to its maximum value Q, charges cease to flow, the current in the circuit is zero, and the potential difference from the battery terminals appears entirely across the capacitor. Substituting $I = 0$ into Equation 28.11 gives the maximum charge on the capacitor:

$$Q = C\mathcal{E} \quad \text{(maximum charge)} \qquad \textbf{(28.13)}$$

To determine analytical expressions for the time dependence of the charge and current, we must solve Equation 28.11, a single equation containing two variables q and I. The current in all parts of the series circuit must be the same. Therefore, the current in the resistance R must be the same as the current between each capacitor plate and the wire connected to it. This current is equal to the time rate of change of the charge on the capacitor plates. Therefore, we substitute $I = dq/dt$ into Equation 28.11 and rearrange the equation:

$$\frac{dq}{dt} = \frac{\mathcal{E}}{R} - \frac{q}{RC}$$

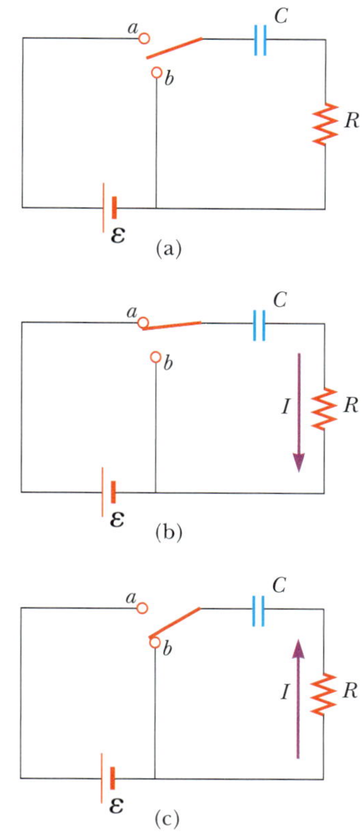

ACTIVE FIGURE 28.16

(a) A capacitor in series with a resistor, switch, and battery. (b) When the switch is thrown to position a, the capacitor begins to charge up. (c) When the switch is thrown to position b, the capacitor discharges.

Sign in at www.thomsonedu.com and go to ThomsonNOW to adjust the values of R and C and see the effect on the charging and discharging of the capacitor.

[3] In previous discussions of capacitors, we assumed a steady-state situation, in which no current was present in any branch of the circuit containing a capacitor. Now we are considering the case *before* the steady-state condition is realized; in this situation, charges are moving and a current exists in the wires connected to the capacitor.

To find an expression for q, we solve this separable differential equation as follows. First combine the terms on the right-hand side:

$$\frac{dq}{dt} = \frac{C\mathcal{E}}{RC} - \frac{q}{RC} = -\frac{q - C\mathcal{E}}{RC}$$

Multiply this equation by dt and divide by $q - C\mathcal{E}$:

$$\frac{dq}{q - C\mathcal{E}} = -\frac{1}{RC}\,dt$$

Integrate this expression, using $q = 0$ at $t = 0$:

$$\int_0^q \frac{dq}{q - C\mathcal{E}} = -\frac{1}{RC}\int_0^t dt$$

$$\ln\left(\frac{q - C\mathcal{E}}{-C\mathcal{E}}\right) = -\frac{t}{RC}$$

From the definition of the natural logarithm, we can write this expression as

Charge as a function of time for a capacitor being charged ▶

$$q(t) = C\mathcal{E}(1 - e^{-t/RC}) = Q(1 - e^{-t/RC}) \quad \textbf{(28.14)}$$

where e is the base of the natural logarithm and we have made the substitution from Equation 28.13.

We can find an expression for the charging current by differentiating Equation 28.14 with respect to time. Using $I = dq/dt$, we find that

Current as a function of time for a capacitor being charged ▶

$$I(t) = \frac{\mathcal{E}}{R}e^{-t/RC} \quad \textbf{(28.15)}$$

Plots of capacitor charge and circuit current versus time are shown in Figure 28.17. Notice that the charge is zero at $t = 0$ and approaches the maximum value $C\mathcal{E}$ as $t \to \infty$. The current has its maximum value $I_i = \mathcal{E}/R$ at $t = 0$ and decays exponentially to zero as $t \to \infty$. The quantity RC, which appears in the exponents of Equations 28.14 and 28.15, is called the **time constant** τ of the circuit:

$$\tau = RC \quad \textbf{(28.16)}$$

The time constant represents the time interval during which the current decreases to $1/e$ of its initial value; that is, after a time interval τ, the current decreases to $I = e^{-1}I_i = 0.368I_i$. After a time interval 2τ, the current decreases to $I = e^{-2}I_i = 0.135I_i$, and so forth. Likewise, in a time interval τ, the charge increases from zero to $C\mathcal{E}[1 - e^{-1}] = 0.632C\mathcal{E}$.

The following dimensional analysis shows that τ has units of time:

$$[\tau] = [RC] = \left[\left(\frac{\Delta V}{I}\right)\left(\frac{Q}{\Delta V}\right)\right] = \left[\frac{Q}{Q/\Delta t}\right] = [\Delta t] = \text{T}$$

Because $\tau = RC$ has units of time, the combination t/RC is dimensionless, as it must be to be an exponent of e in Equations 28.14 and 28.15.

The energy output of the battery as the capacitor is fully charged is $Q\mathcal{E} = C\mathcal{E}^2$. After the capacitor is fully charged, the energy stored in the capacitor is $\frac{1}{2}Q\mathcal{E} = \frac{1}{2}C\mathcal{E}^2$, which is only half the energy output of the battery. It is left as a problem (Problem 52) to show that the remaining half of the energy supplied by the battery appears as internal energy in the resistor.

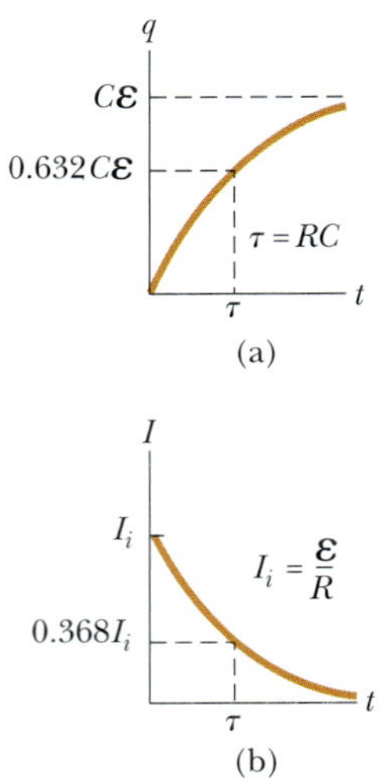

Figure 28.17 (a) Plot of capacitor charge versus time for the circuit shown in Active Figure 28.16. After a time interval equal to one time constant τ has passed, the charge is 63.2% of the maximum value $C\mathcal{E}$. The charge approaches its maximum value as t approaches infinity. (b) Plot of current versus time for the circuit shown in Active Figure 28.16. The current has its maximum value $I_i = \mathcal{E}/R$ at $t = 0$ and decays to zero exponentially as t approaches infinity. After a time interval equal to one time constant τ has passed, the current is 36.8% of its initial value.

Discharging a Capacitor

Imagine that the capacitor in Active Figure 28.16b is completely charged. A potential difference Q/C exists across the capacitor and there is zero potential difference across the resistor because $I = 0$. If the switch is now thrown to position b at $t = 0$ (Active Fig. 28.16c), the capacitor begins to discharge through the resistor.

At some time t during the discharge, the current in the circuit is I and the charge on the capacitor is q. The circuit in Active Figure 28.16c is the same as the circuit in Active Figure 28.16b except for the absence of the battery. Therefore, we eliminate the emf $\mathcal{E}$ from Equation 28.11 to obtain the appropriate loop equation for the circuit in Active Figure 28.16c:

$$-\frac{q}{C} - IR = 0 \tag{28.17}$$

When we substitute $I = dq/dt$ into this expression, it becomes

$$-R\frac{dq}{dt} = \frac{q}{C}$$

$$\frac{dq}{q} = -\frac{1}{RC}\,dt$$

Integrating this expression using $q = Q$ at $t = 0$ gives

$$\int_Q^q \frac{dq}{q} = -\frac{1}{RC}\int_0^t dt$$

$$\ln\left(\frac{q}{Q}\right) = -\frac{t}{RC}$$

$$q(t) = Qe^{-t/RC} \tag{28.18}$$

◀ Charge as a function of time for a discharging capacitor

Differentiating Equation 28.18 with respect to time gives the instantaneous current as a function of time:

$$I(t) = -\frac{Q}{RC}e^{-t/RC} \tag{28.19}$$

◀ Current as a function of time for a discharging capacitor

where $Q/RC = I_i$ is the initial current. The negative sign indicates that as the capacitor discharges, the current direction is opposite its direction when the capacitor was being charged. (Compare the current directions in Figs. 28.16b and 28.16c.) Both the charge on the capacitor and the current decay exponentially at a rate characterized by the time constant $\tau = RC$.

Quick Quiz 28.5 Consider the circuit in Figure 28.18 and assume the battery has no internal resistance. **(i)** Just after the switch is closed, what is the current in the battery? (a) 0 (b) $\mathcal{E}/2R$ (c) $2\mathcal{E}/R$ (d) $\mathcal{E}/R$ (e) impossible to determine **(ii)** After a very long time, what is the current in the battery? Choose from the same choices.

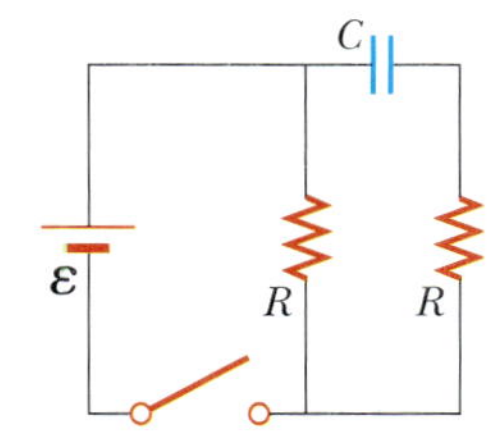

Figure 28.18 (Quick Quiz 28.5) How does the current vary after the switch is closed?

CONCEPTUAL EXAMPLE 28.8 Intermittent Windshield Wipers

Many automobiles are equipped with windshield wipers that can operate intermittently during a light rainfall. How does the operation of such wipers depend on the charging and discharging of a capacitor?

SOLUTION

The wipers are part of an RC circuit whose time constant can be varied by selecting different values of R through a multiposition switch. As the voltage across the capacitor increases, the capacitor reaches a point at which it discharges and triggers the wipers. The circuit then begins another charging cycle. The time interval between the individual sweeps of the wipers is determined by the value of the time constant.

EXAMPLE 28.9 Charging a Capacitor in an *RC* Circuit

An uncharged capacitor and a resistor are connected in series to a battery as shown in Active Figure 28.16, where $\mathcal{E} = 12.0$ V, $C = 5.00\ \mu$F, and $R = 8.00 \times 10^5\ \Omega$. The switch is thrown to position *a*. Find the time constant of the circuit, the maximum charge on the capacitor, the maximum current in the circuit, and the charge and current as functions of time.

SOLUTION

Conceptualize Study Active Figure 28.16 and imagine throwing the switch to position *a* as shown in Active Figure 28.16b. Upon doing so, the capacitor begins to charge.

Categorize We evaluate our results using equations developed in this section, so we categorize this example as a substitution problem.

Evaluate the time constant of the circuit from Equation 28.16:

$$\tau = RC = (8.00 \times 10^5\ \Omega)(5.00 \times 10^{-6}\ \text{F}) = 4.00\ \text{s}$$

Evaluate the maximum charge on the capacitor from Equation 28.13:

$$Q = C\mathcal{E} = (5.00\ \mu\text{F})(12.0\ \text{V}) = 60.0\ \mu\text{C}$$

Evaluate the maximum current in the circuit from Equation 28.12:

$$I_i = \frac{\mathcal{E}}{R} = \frac{12.0\ \text{V}}{8.00 \times 10^5\ \Omega} = 15.0\ \mu\text{A}$$

Use these values in Equations 28.14 and 28.15 to find the charge and current as functions of time:

$$q(t) = (60.0\mu\text{C})(1 - e^{-t/4.00\ \text{s}})$$

$$I(t) = (15.0\ \mu\text{A})e^{-t/4.00\ \text{s}}$$

EXAMPLE 28.10 Discharging a Capacitor in an *RC* Circuit

Consider a capacitor of capacitance *C* that is being discharged through a resistor of resistance *R* as shown in Active Figure 28.16c.

(A) After how many time constants is the charge on the capacitor one-fourth its initial value?

SOLUTION

Conceptualize Study Active Figure 28.16 and imagine throwing the switch to position *b* as shown in Active Figure 28.16c. Upon doing so, the capacitor begins to discharge.

Categorize We categorize the example as one involving a discharging capacitor and use the appropriate equations.

Analyze Substitute $q(t) = Q/4$ into Equation 28.18:

$$\frac{Q}{4} = Qe^{-t/RC}$$

$$\tfrac{1}{4} = e^{-t/RC}$$

Take the logarithm of both sides of the equation and solve for *t*:

$$-\ln 4 = -\frac{t}{RC}$$

$$t = RC \ln 4 = 1.39RC = 1.39\tau$$

(B) The energy stored in the capacitor decreases with time as the capacitor discharges. After how many time constants is this stored energy one-fourth its initial value?

SOLUTION

Use Equations 26.11 and 28.18 to express the energy stored in the capacitor at any time t:

$$(1)\quad U(t) = \frac{q^2}{2C} = \frac{Q^2}{2C}e^{-2t/RC}$$

Substitute $U(t) = \frac{1}{4}(Q^2/2C)$ into Equation (1):

$$\frac{1}{4}\frac{Q^2}{2C} = \frac{Q^2}{2C}e^{-2t/RC}$$

$$\frac{1}{4} = e^{-2t/RC}$$

Take the logarithm of both sides of the equation and solve for t:

$$-\ln 4 = -\frac{2t}{RC}$$

$$t = \frac{1}{2}RC\ln 4 = 0.693RC = 0.693\tau$$

Finalize Notice that because the energy depends on the square of the charge, the energy in the capacitor drops more rapidly than the charge on the capacitor.

What If? What if you want to describe the circuit in terms of the time interval required for the charge to fall to one-half its original value rather than by the time constant τ? That would give a parameter for the circuit called its *half-life* $t_{1/2}$. How is the half-life related to the time constant?

Answer In one half-life, the charge falls from Q to $Q/2$. Therefore, from Equation 28.18,

$$\frac{Q}{2} = Qe^{-t_{1/2}/RC} \quad\rightarrow\quad \frac{1}{2} = e^{-t_{1/2}/RC}$$

which leads to

$$t_{1/2} = 0.693\tau$$

The concept of half-life will be important to us when we study nuclear decay in Chapter 44. The radioactive decay of an unstable sample behaves in a mathematically similar manner to a discharging capacitor in an *RC* circuit.

EXAMPLE 28.11 Energy Delivered to a Resistor

A 5.00-μF capacitor is charged to a potential difference of 800 V and then discharged through a resistor. How much energy is delivered to the resistor in the time interval required to fully discharge the capacitor?

SOLUTION

Conceptualize In Example 28.10, we considered the energy decrease in a discharging capacitor to a value of one-fourth of the initial energy. In this example, the capacitor fully discharges.

Categorize We solve this example using two approaches. The first approach is to model the circuit as an isolated system. Because energy in an isolated system is conserved, the initial electric potential energy U_C stored in the capacitor is transformed into internal energy $E_{int} = E_R$ in the resistor. The second approach is to model the resistor as a nonisolated system. Energy enters the resistor by electrical transmission from the capacitor, causing an increase in the resistor's internal energy.

Analyze We begin with the isolated system approach.

Write the appropriate reduction of the conservation of energy equation, Equation 8.2:

$$\Delta U + \Delta E_{int} = 0$$

Substitute the initial and final values of the energies:

$$(0 - U_C) + (E_{int} - 0) = 0 \quad\rightarrow\quad E_R = U_C$$

Use Equation 26.11 for the electric potential energy in the capacitor:

$$E_R = \tfrac{1}{2}C\mathcal{E}^2$$

Substitute numerical values:

$$E_R = \tfrac{1}{2}(5.00 \times 10^{-6}\text{ F})(800\text{ V})^2 = \boxed{1.60\text{ J}}$$

The second approach, which is more difficult but perhaps more instructive, is to note that as the capacitor discharges through the resistor, the rate at which energy is delivered to the resistor by electrical transmission is I^2R, where I is the instantaneous current given by Equation 28.19.

Evaluate the energy delivered to the resistor by integrating the power over all time because it takes an infinite time interval for the capacitor to completely discharge:

$$\mathcal{P} = \frac{dE}{dt} \rightarrow E_R = \int_0^\infty \mathcal{P}\,dt$$

Substitute for the power delivered to the resistor:

$$E_R = \int_0^\infty I^2R\,dt$$

Substitute for the current from Equation 28.19:

$$E_R = \int_0^\infty \left(-\frac{Q}{RC}e^{-t/RC}\right)^2 R\,dt = \frac{Q^2}{RC^2}\int_0^\infty e^{-2t/RC}\,dt = \frac{\mathcal{E}^2}{R}\int_0^\infty e^{-2t/RC}\,dt$$

Substitute the value of the integral, which is $RC/2$ (see Problem 30):

$$E_R = \frac{\mathcal{E}^2}{R}\left(\frac{RC}{2}\right) = \tfrac{1}{2}C\mathcal{E}^2$$

Finalize This result agrees with that obtained using the isolated system approach, as it must. We can use this second approach to find the total energy delivered to the resistor at *any* time after the switch is closed by simply replacing the upper limit in the integral with that specific value of t.

28.5 Electrical Meters

In this section, we discuss various electrical meters that are used in the electrical and electronics industries to make electrical measurements.

The Galvanometer

The **galvanometer** is the main component in analog meters for measuring current and voltage. (Many analog meters are still in use, although digital meters, which operate on a different principle, are currently more common.) One type, called the *D'Arsonval galvanometer,* consists of a coil of wire mounted so that it is free to rotate on a pivot in a magnetic field provided by a permanent magnet. The deflection of a needle attached to the coil is proportional to the current in the galvanometer. Once the instrument is properly calibrated, it can be used in conjunction with other circuit elements to measure either currents or potential differences.

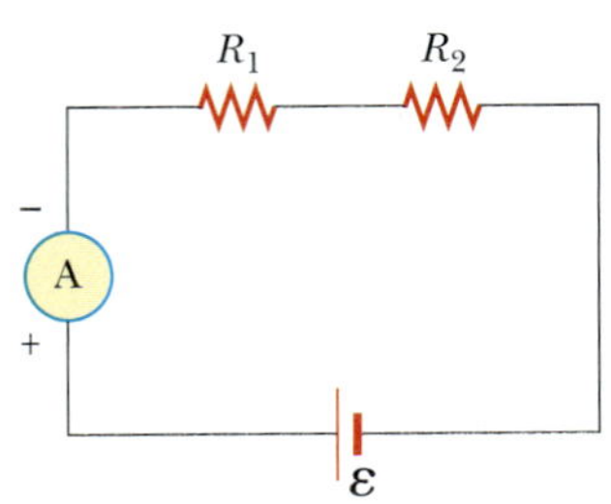

Figure 28.19 Current can be measured with an ammeter connected in series with the elements in which the measurement of a current is desired. An ideal ammeter has zero resistance.

The Ammeter

A device that measures current is called an **ammeter.** Because the charges constituting the current to be measured must pass directly through the ammeter, the ammeter must be connected in series with other elements in the circuit as shown in Figure 28.19. When using an ammeter to measure direct currents, you must connect it so that charges enter the instrument at the positive terminal and exit at the negative terminal.

Ideally, an ammeter should have zero resistance so that the current being measured is not altered. In the circuit shown in Figure 28.19, this condition requires that the resistance of the ammeter be much less than $R_1 + R_2$. Because any ammeter always has some internal resistance, the presence of the ammeter in the circuit slightly reduces the current from the value it would have in the meter's absence.

A typical off-the-shelf galvanometer is often not suitable for use as an ammeter primarily because it has a resistance of about 60 Ω. An ammeter resistance this great considerably alters the current in a circuit. Consider the following example. The current in a simple series circuit containing a 3-V battery and a 3-Ω resistor is 1 A. If you insert a 60-Ω galvanometer in this circuit to measure the current, the total resistance becomes 63 Ω and the current is reduced to 0.048 A!

A second factor that limits the use of a galvanometer as an ammeter is that a typical galvanometer gives a full-scale deflection for currents on the order of 1 mA or less. Consequently, such a galvanometer cannot be used directly to measure currents greater than this value. It can, however, be converted to a useful ammeter by placing a shunt resistor R_p in parallel with the galvanometer as shown in Active Figure 28.20. The value of R_p must be much less than the galvanometer resistance so that most of the current to be measured is directed to the shunt resistor.

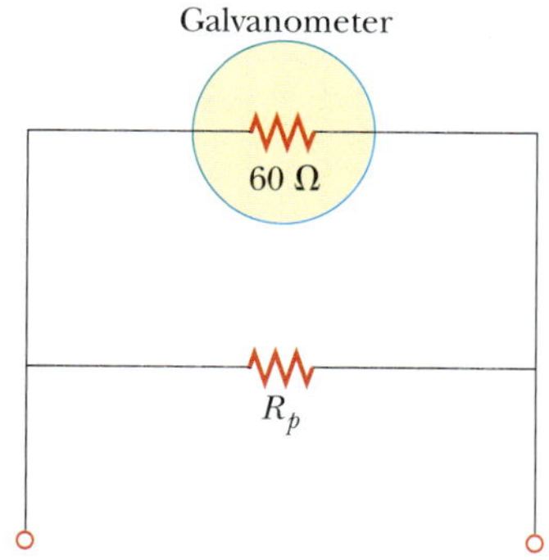

ACTIVE FIGURE 28.20

A galvanometer is represented here by its internal resistance of 60 Ω. When a galvanometer is to be used as an ammeter, a shunt resistor R_p is connected in parallel with the galvanometer.

Sign in at www.thomsonedu.com and go to ThomsonNOW to predict the value of R_p needed to cause full-scale deflection in the circuit of Figure 28.19 and test your result.

The Voltmeter

A device that measures potential difference is called a **voltmeter.** The potential difference between any two points in a circuit can be measured by attaching the terminals of the voltmeter between these points without breaking the circuit as shown in Figure 28.21. The potential difference across resistor R_2 is measured by connecting the voltmeter in parallel with R_2. Again, it is necessary to observe the instrument's polarity. The voltmeter's positive terminal must be connected to the end of the resistor that is at the higher potential, and its negative terminal must be connected to the end of the resistor at the lower potential.

An ideal voltmeter has infinite resistance so that no current exists in it. In Figure 28.21, this condition requires that the voltmeter have a resistance much greater than R_2. In practice, corrections should be made for the known resistance of the voltmeter if this condition is not met.

A galvanometer can also be used as a voltmeter by adding an external resistor R_s in series with it as shown in Active Figure 28.22. In this case, the external resistor must have a value much greater than the resistance of the galvanometer to ensure that the galvanometer does not significantly alter the voltage being measured.

A digital multimeter is used to measure a voltage across a circuit element.

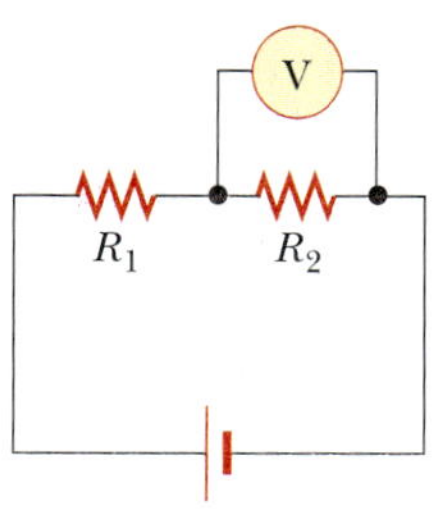

Figure 28.21 The potential difference across a resistor can be measured with a voltmeter connected in parallel with the resistor. An ideal voltmeter has infinite resistance.

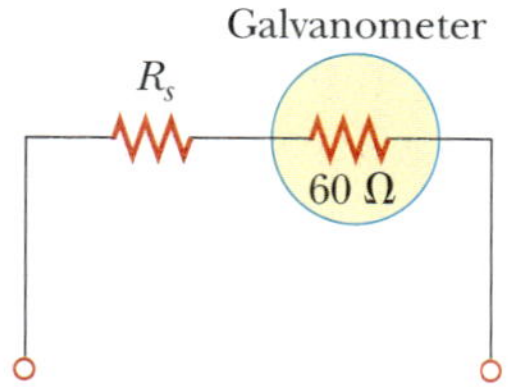

ACTIVE FIGURE 28.22

When the galvanometer is used as a voltmeter, a resistor R_s is connected in series with the galvanometer.

Sign in at www.thomsonedu.com and go to ThomsonNOW to predict the value of R_s needed to cause full-scale deflection in the circuit of Figure 28.21 and test your result.

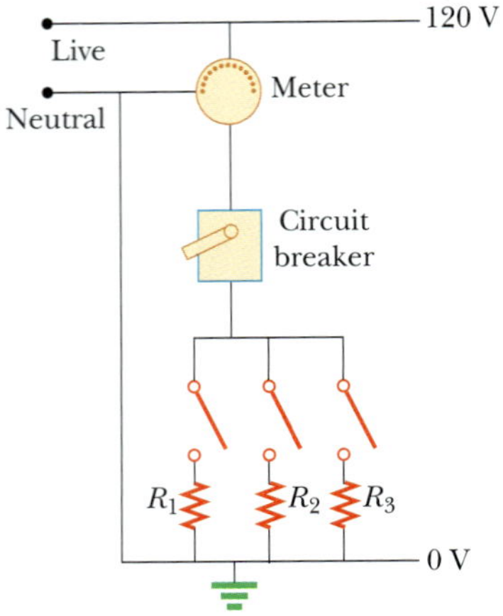

Figure 28.23 Wiring diagram for a household circuit. The resistances represent appliances or other electrical devices that operate with an applied voltage of 120 V.

28.6 Household Wiring and Electrical Safety

Many considerations are important in the design of an electrical system of a home that will provide adequate electrical service for the occupants while maximizing their safety. We discuss some aspects of a home electrical system in this section.

Household Wiring

Household circuits represent a practical application of some of the ideas presented in this chapter. In our world of electrical appliances, it is useful to understand the power requirements and limitations of conventional electrical systems and the safety measures that prevent accidents.

In a conventional installation, the utility company distributes electric power to individual homes by means of a pair of wires, with each home connected in parallel to these wires. One wire is called the *live wire*[4] as illustrated in Figure 28.23, and the other is called the *neutral wire*. The neutral wire is grounded; that is, its electric potential is taken to be zero. The potential difference between the live and neutral wires is approximately 120 V. This voltage alternates in time, and the potential of the live wire oscillates relative to ground. Much of what we have learned so far for the constant-emf situation (direct current) can also be applied to the alternating current that power companies supply to businesses and households. (Alternating voltage and current are discussed in Chapter 33.)

To record a household's energy consumption, a meter is connected in series with the live wire entering the house. After the meter, the wire splits so that there are several separate circuits in parallel distributed throughout the house. Each circuit contains a circuit breaker (or, in older installations, a fuse). The wire and circuit breaker for each circuit are carefully selected to meet the current requirements for that circuit. If a circuit is to carry currents as large as 30 A, a heavy wire and an appropriate circuit breaker must be selected to handle this current. A circuit used to power only lamps and small appliances often requires only 20 A. Each circuit has its own circuit breaker to provide protection for that part of the entire electrical system of the house.

As an example, consider a circuit in which a toaster oven, a microwave oven, and a coffee maker are connected (corresponding to R_1, R_2, and R_3 in Fig. 28.23). We can calculate the current in each appliance by using the expression $\mathcal{P} = I\,\Delta V$. The toaster oven, rated at 1 000 W, draws a current of 1 000 W/120 V = 8.33 A. The microwave oven, rated at 1 300 W, draws 10.8 A, and the coffee maker, rated at 800 W, draws 6.67 A. When the three appliances are operated simultaneously, they draw a total current of 25.8 A. Therefore, the circuit must be wired to handle at least this much current. If the rating of the circuit breaker protecting the circuit is too small—say, 20 A—the breaker will be tripped when the third appliance is turned on, preventing all three appliances from operating. To avoid this situation, the toaster oven and coffee maker can be operated on one 20-A circuit and the microwave oven on a separate 20-A circuit.

Many heavy-duty appliances such as electric ranges and clothes dryers require 240 V for their operation. The power company supplies this voltage by providing a third wire that is 120 V below ground potential (Fig. 28.24). The potential difference between this live wire and the other live wire (which is 120 V above ground potential) is 240 V. An appliance that operates from a 240-V line requires half as much current compared with operating it at 120 V; therefore, smaller wires can be used in the higher-voltage circuit without overheating.

(a)

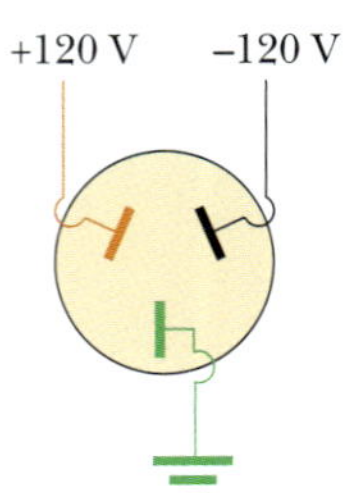

(b)

Figure 28.24 (a) An outlet for connection to a 240-V supply. (b) The connections for each of the openings in a 240-V outlet.

Electrical Safety

When the live wire of an electrical outlet is connected directly to ground, the circuit is completed and a *short-circuit condition* exists. A short circuit occurs when

[4] *Live wire* is a common expression for a conductor whose electric potential is above or below ground potential.

almost zero resistance exists between two points at different potentials, and the result is a very large current. When that happens accidentally, a properly operating circuit breaker opens the circuit and no damage is done. A person in contact with ground, however, can be electrocuted by touching the live wire of a frayed cord or other exposed conductor. An exceptionally effective (and dangerous!) ground contact is made when the person either touches a water pipe (normally at ground potential) or stands on the ground with wet feet. The latter situation represents effective ground contact because normal, nondistilled water is a conductor due to the large number of ions associated with impurities. This situation should be avoided at all cost.

Electric shock can result in fatal burns or can cause the muscles of vital organs such as the heart to malfunction. The degree of damage to the body depends on the magnitude of the current, the length of time it acts, the part of the body touched by the live wire, and the part of the body in which the current exists. Currents of 5 mA or less cause a sensation of shock, but ordinarily do little or no damage. If the current is larger than about 10 mA, the muscles contract and the person may be unable to release the live wire. If the body carries a current of about 100 mA for only a few seconds, the result can be fatal. Such a large current paralyzes the respiratory muscles and prevents breathing. In some cases, currents of approximately 1 A can produce serious (and sometimes fatal) burns. In practice, no contact with live wires is regarded as safe whenever the voltage is greater than 24 V.

Many 120-V outlets are designed to accept a three-pronged power cord. (This feature is required in all new electrical installations.) One of these prongs is the live wire at a nominal potential of 120 V. The second is the neutral wire, nominally at 0 V, which carries current to ground. Figure 28.25a shows a connection to an electric drill with only these two wires. If the live wire accidentally makes contact with the casing of the electric drill (which can occur if the wire insulation wears off), current can be carried to ground by way of the person, resulting in an electric shock. The third wire in a three-pronged power cord, the round prong, is a

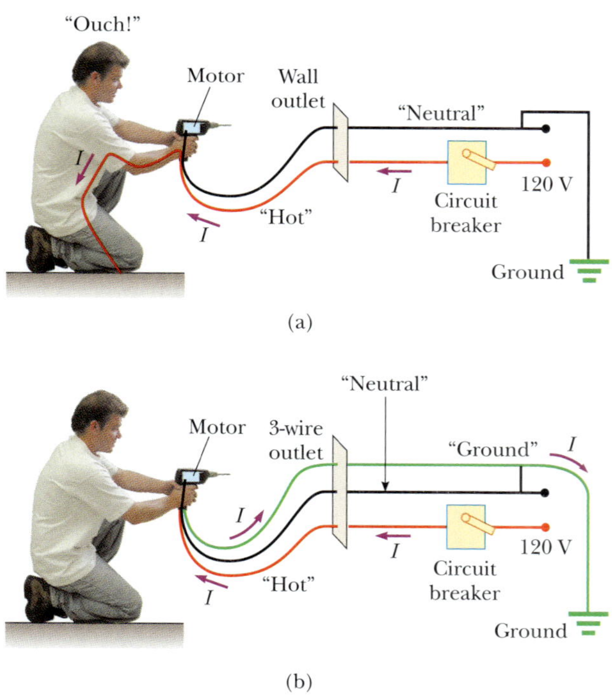

Figure 28.25 (a) A diagram of the circuit for an electric drill with only two connecting wires. The normal current path is from the live wire through the motor connections and back to ground through the neutral wire. In the situation shown, the live wire has come into contact with the drill case. As a result, the person holding the drill acts as a current path to ground and receives an electric shock. (b) This shock can be avoided by connecting the drill case to ground through a third ground wire. In this situation, the drill case remains at ground potential and no current exists in the person.

safety ground wire that normally carries no current. It is both grounded and connected directly to the casing of the appliance. If the live wire is accidentally shorted to the casing in this situation, most of the current takes the low-resistance path through the appliance to ground as shown in Figure 28.25b.

Special power outlets called *ground-fault interrupters,* or GFIs, are used in kitchens, bathrooms, basements, exterior outlets, and other hazardous areas of homes. These devices are designed to protect persons from electric shock by sensing small currents (< 5 mA) leaking to ground. (The principle of their operation is described in Chapter 31.) When an excessive leakage current is detected, the current is shut off in less than 1 ms.

Summary

ThomsonNOW™ Sign in at **www.thomsonedu.com** and go to ThomsonNOW to take a practice test for this chapter.

DEFINITIONS

The **emf** of a battery is equal to the voltage across its terminals when the current is zero. That is, the emf is equivalent to the **open-circuit voltage** of the battery.

CONCEPTS AND PRINCIPLES

The **equivalent resistance** of a set of resistors connected in a **series combination** is

$$R_{eq} = R_1 + R_2 + R_3 + \cdots \quad \textbf{(28.6)}$$

The **equivalent resistance** of a set of resistors connected in a **parallel combination** is found from the relationship

$$\frac{1}{R_{eq}} = \frac{1}{R_1} + \frac{1}{R_2} + \frac{1}{R_3} + \cdots \quad \textbf{(28.8)}$$

Circuits involving more than one loop are conveniently analyzed with the use of **Kirchhoff's rules:**

1. **Junction rule.** At any junction, the sum of the currents must equal zero:

$$\sum_{\text{junction}} I = 0 \quad \textbf{(28.9)}$$

2. **Loop rule.** The sum of the potential differences across all elements around any circuit loop must be zero:

$$\sum_{\text{closed loop}} \Delta V = 0 \quad \textbf{(28.10)}$$

When a resistor is traversed in the direction of the current, the potential difference ΔV across the resistor is $-IR$. When a resistor is traversed in the direction opposite the current, $\Delta V = +IR$. When a source of emf is traversed in the direction of the emf (negative terminal to positive terminal), the potential difference is $+\mathcal{E}$. When a source of emf is traversed opposite the emf (positive to negative), the potential difference is $-\mathcal{E}$.

If a capacitor is charged with a battery through a resistor of resistance R, the charge on the capacitor and the current in the circuit vary in time according to the expressions

$$q(t) = Q(1 - e^{-t/RC}) \quad \textbf{(28.14)}$$

$$I(t) = \frac{\mathcal{E}}{R} e^{-t/RC} \quad \textbf{(28.15)}$$

where $Q = C\mathcal{E}$ is the maximum charge on the capacitor. The product RC is called the **time constant** τ of the circuit.

If a charged capacitor is discharged through a resistor of resistance R, the charge and current decrease exponentially in time according to the expressions

$$q(t) = Qe^{-t/RC} \quad \textbf{(28.18)}$$

$$I(t) = -I_i e^{-t/RC} \quad \textbf{(28.19)}$$

where Q is the initial charge on the capacitor and $I_i = Q/RC$ is the initial current in the circuit.

Questions

☐ denotes answer available in *Student Solutions Manual/Study Guide;* **O** denotes objective question

☐ **1.** Is the direction of current in a battery always from the negative terminal to the positive terminal? Explain.

2. O A certain battery has some internal resistance. **(i)** Can the potential difference across the terminals of a battery be equal to its emf? (a) No. (b) Yes, if the battery is absorbing energy by electrical transmission. (c) Yes, if more than one wire is connected to each terminal. (d) Yes, if the current in the battery is zero. (e) Yes; no special condition is required. **(ii)** Can the terminal voltage exceed the emf? Choose your answer from the same possibilities.

3. Given three lightbulbs and a battery, sketch as many different electric circuits as you can.

4. O When resistors with different resistances are connected in series, which of the following must be the same for each resistor? Choose all correct answers. (a) potential difference (b) current (c) power delivered (d) charge entering (e) none of these answers

5. O When resistors with different resistances are connected in parallel, which of the following must be the same for each resistor? Choose all correct answers. (a) potential difference (b) current (c) power delivered (d) charge entering (e) none of these answers

☐ **6.** Why is it possible for a bird to sit on a high-voltage wire without being electrocuted?

7. O Are the two headlights of a car wired (a) in series with each other, (b) in parallel, (c) neither in series or in parallel, or (d) is it impossible to tell?

8. A student claims that the second of two lightbulbs in series is less bright than the first, because the first lightbulb uses up some of the current. How would you respond to this statement?

9. O Is a circuit breaker wired (a) in series with the device it is protecting, (b) in parallel, (c) neither in series or in parallel, or (d) is it impossible to tell?

10. O In the circuit shown in Figure Q28.10, each battery is delivering energy to the circuit by electrical transmission. All the resistors have equal resistance. **(i)** Rank the electric potentials at points *a, b, c, d, e, f, g,* and *h* from highest to lowest, noting any cases of equality in the ranking. **(ii)** Rank the magnitudes of the currents at the same points from greatest to least, noting any cases of equality.

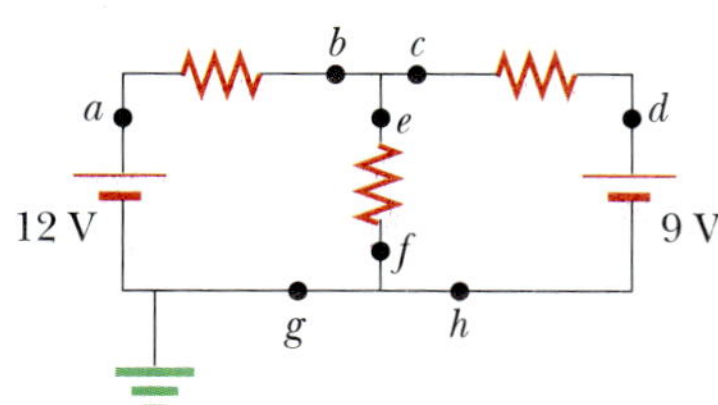

Figure Q28.10

11. O A series circuit consists of three identical lamps connected to a battery as shown in Figure Q28.11. The switch S, originally open, is closed. **(i)** What then happens to the brightness of lamp B? (a) It increases. (b) It decreases somewhat. (c) It does not change. (d) It drops to zero. **(ii)** What happens to the brightness of lamp C? Choose from the same possibilities. **(iii)** What happens to the current in the battery? Choose from the same possibilities. **(iv)** What happens to the potential difference across lamp A? **(v)** What happens to the potential difference across lamp C? **(vi)** What happens to the total power delivered to the lamps by the battery? Choose in each case from the same possibilities (a) through (d).

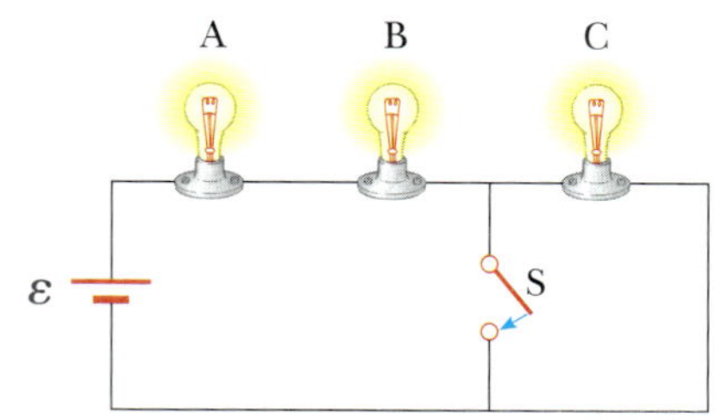

Figure Q28.11

12. O A circuit consists of three identical lamps connected to a battery having some internal resistance as in Figure Q28.12. The switch S, originally open, is closed. **(i)** What then happens to the brightness of lamp B? (a) It increases. (b) It decreases somewhat. (c) It does not change. (d) It drops to zero. **(ii)** What happens to the brightness of lamp C? Choose from the same possibilities. **(iii)** What happens to the current in the battery? Choose from the same possibilities. **(iv)** What happens to the potential difference across lamp A? **(v)** What happens to the potential difference across lamp C? **(vi)** What happens to the total power delivered to the lamps by the battery? Choose in each case from the same possibilities (a) through (d).

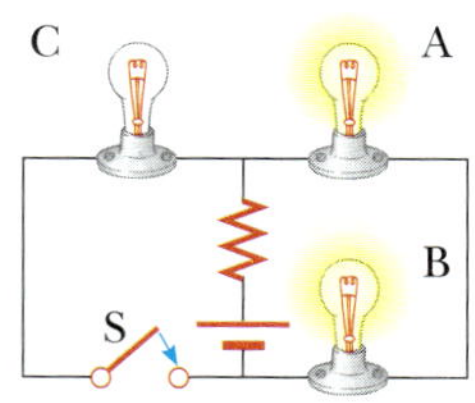

Figure Q28.12

13. A ski resort consists of a few chairlifts and several interconnected downhill runs on the side of a mountain, with a lodge at the bottom. The chairlifts are analogous to batteries, and the runs are analogous to resistors. Describe how two runs can be in series. Describe how three runs can be in parallel. Sketch a junction between one chairlift and two runs. State Kirchhoff's junction rule for ski resorts. One of the skiers happens to be carrying a skydiver's altimeter. She never takes the same set of chairlifts and runs twice, but keeps passing you at the fixed location where you are working. State Kirchhoff's loop rule for ski resorts.

14. Referring to Figure Q28.14, describe what happens to the lightbulb after the switch is closed. Assume the capacitor has a large capacitance and is initially uncharged and assume the light illuminates when connected directly across the battery terminals.

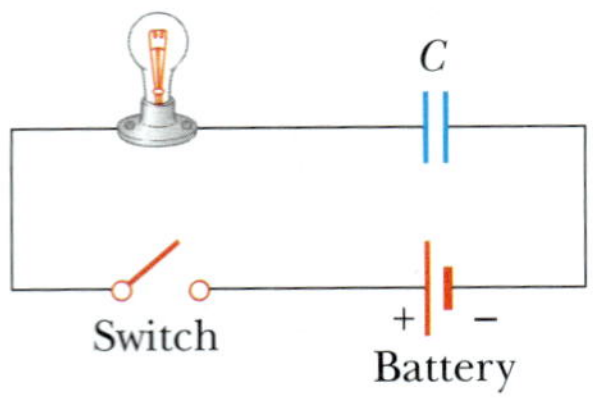

Figure Q28.14

15. So that your grandmother can listen to *A Prairie Home Companion*, you take her bedside radio to the hospital where she is staying. You are required to have a maintenance worker test the radio for electrical safety. Finding that it develops 120 V on one of its knobs, he does not let you take it up to your grandmother's room. She complains that she has had the radio for many years and nobody has ever gotten a shock from it. You end up having to buy a new plastic radio. Is that fair? Will the old radio be safe back in her bedroom?

16. What advantage does 120-V operation offer over 240 V? What disadvantages does it have?

Problems

WebAssign The Problems from this chapter may be assigned online in WebAssign.

ThomsonNOW Sign in at **www.thomsonedu.com** and go to ThomsonNOW to assess your understanding of this chapter's topics with additional quizzing and conceptual questions.

1, 2, 3 denotes straightforward, intermediate, challenging; □ denotes full solution available in *Student Solutions Manual/Study Guide;* ▲ denotes coached solution with hints available at **www.thomsonedu.com;** denotes developing symbolic reasoning; ● denotes asking for qualitative reasoning; denotes computer useful in solving problem

Section 28.1 Electromotive Force

1. ▲ A battery has an emf of 15.0 V. The terminal voltage of the battery is 11.6 V when it is delivering 20.0 W of power to an external load resistor R. (a) What is the value of R? (b) What is the internal resistance of the battery?

2. Two 1.50-V batteries—with their positive terminals in the same direction—are inserted in series into the barrel of a flashlight. One battery has an internal resistance of 0.255 Ω, the other an internal resistance of 0.153 Ω. When the switch is closed, a current of 600 mA occurs in the lamp. (a) What is the lamp's resistance? (b) What fraction of the chemical energy transformed appears as internal energy in the batteries?

3. An automobile battery has an emf of 12.6 V and an internal resistance of 0.080 0 Ω. The headlights together have an equivalent resistance of 5.00 Ω (assumed constant). What is the potential difference across the headlight bulbs (a) when they are the only load on the battery and (b) when the starter motor is operated, requiring an additional 35.0 A from the battery?

4. ● As in Example 28.2, consider a power supply with fixed emf $\mathcal{E}$ and internal resistance r causing current in a load resistance R. In this problem, R is fixed and r is a variable. The efficiency is defined as the energy delivered to the load divided by the energy delivered by the emf. (a) When the internal resistance is adjusted for maximum power transfer, what is the efficiency? (b) What should be the internal resistance for maximum possible efficiency? (c) When the electric company sells energy to a customer, does it have a goal of high efficiency or of maximum power transfer? Explain. (d) When a student connects a loudspeaker to an amplifier, does she most want high efficiency or high power transfer? Explain.

Section 28.2 Resistors in Series and Parallel

5. (a) Find the equivalent resistance between points a and b in Figure P28.5. (b) A potential difference of 34.0 V is applied between points a and b. Calculate the current in each resistor.

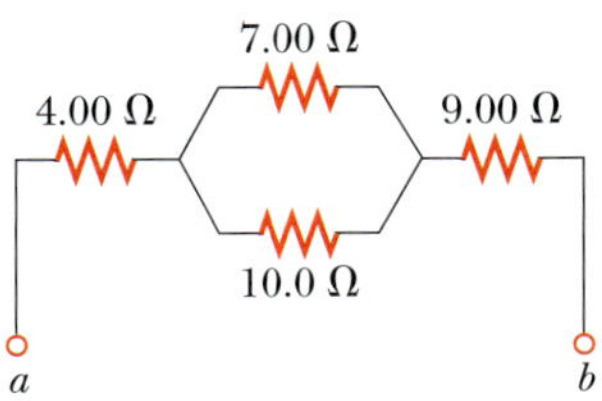

Figure P28.5

6. ● A lightbulb marked "75 W [at] 120 V" is screwed into a socket at one end of a long extension cord, in which each of the two conductors has resistance 0.800 Ω. The other end of the extension cord is plugged into a 120-V outlet. (a) Explain why the actual power delivered to the lightbulb cannot be 75 W in this situation. (b) What can you reasonably model as constant about the lightbulb? Draw a circuit diagram and find the actual power delivered to the lightbulb in this circuit.

7. ▲ Consider the circuit shown in Figure P28.7. Find (a) the current in the 20.0-Ω resistor and (b) the potential difference between points a and b.

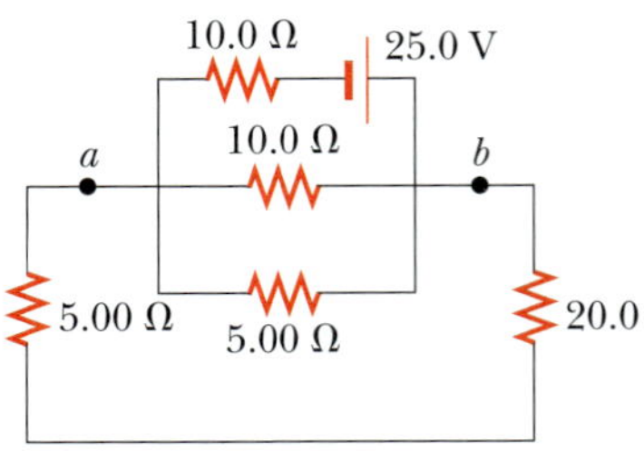

Figure P28.7

8. For the purpose of measuring the electric resistance of shoes through the body of the wearer to a metal ground

plate, the American National Standards Institute (ANSI) specifies the circuit shown in Figure P28.8. The potential difference ΔV across the 1.00-MΩ resistor is measured with a high-resistance voltmeter. (a) Show that the resistance of the footwear is

$$R_{\text{shoes}} = 1.00\ \text{M}\Omega\left(\frac{50.0\ \text{V} - \Delta V}{\Delta V}\right)$$

(b) In a medical test, a current through the human body should not exceed 150 μA. Can the current delivered by the ANSI-specified circuit exceed 150 μA? To decide, consider a person standing barefoot on the ground plate.

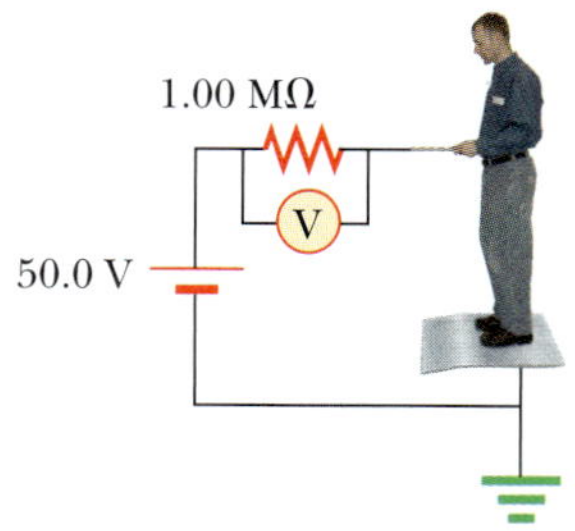

Figure P28.8

9. Three 100-Ω resistors are connected as shown in Figure P28.9. The maximum power that can safely be delivered to any one resistor is 25.0 W. (a) What is the maximum potential difference that can be applied to the terminals a and b? (b) For the voltage determined in part (a), what is the power delivered to each resistor? What is the total power delivered?

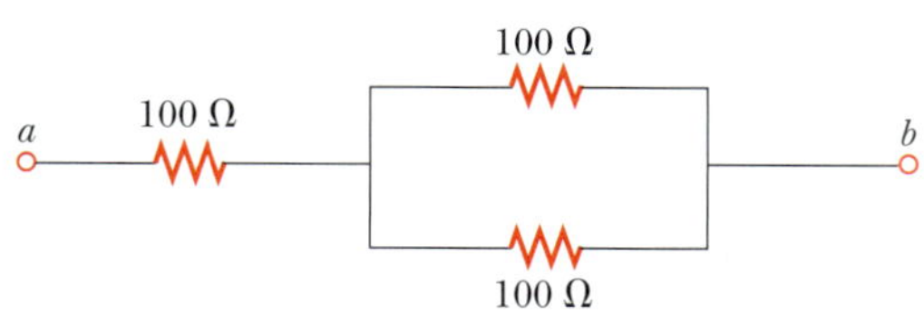

Figure P28.9

10. Using only three resistors—2.00 Ω, 3.00 Ω, and 4.00 Ω—find 17 resistance values that may be obtained by various combinations of one or more resistors. Tabulate the combinations in order of increasing resistance.

11. A 6.00-V battery supplies current to the circuit shown in Figure P28.11. When the double-throw switch S is open as shown in the figure, the current in the battery is 1.00 mA. When the switch is closed in position a, the current in the battery is 1.20 mA. When the switch is closed in position b, the current in the battery is 2.00 mA. Find the resistances R_1, R_2, and R_3.

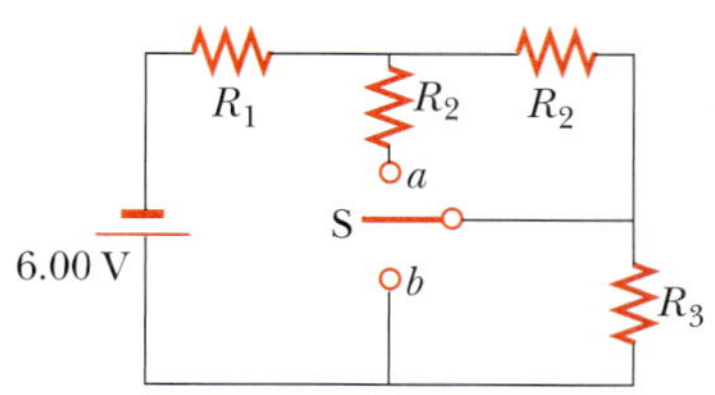

Figure P28.11

12. Two resistors connected in series have an equivalent resistance of 690 Ω. When they are connected in parallel, their equivalent resistance is 150 Ω. Find the resistance of each resistor.

13. ● When the switch S in the circuit of Figure P28.13 is closed, will the equivalent resistance between points a and b increase or decrease? State your reasoning. Assume the equivalent resistance changes by a factor of 2. Determine the value of R.

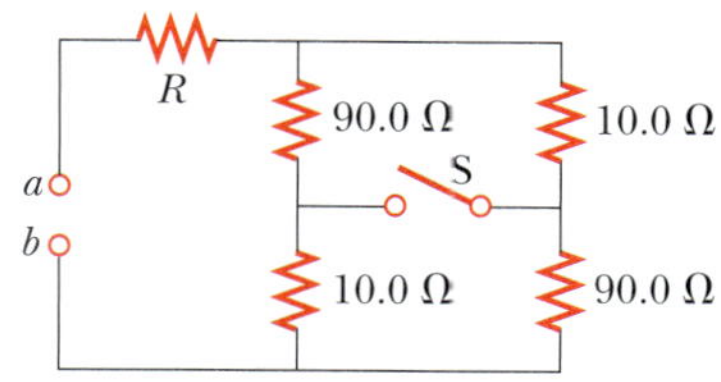

Figure P28.13

14. ● Four resistors are connected to a battery as shown in Figure P28.14. The current in the battery is I; the battery emf is $\mathcal{E}$; and the resistor values are $R_1 = R$, $R_2 = 2R$, $R_3 = 4R$, and $R_4 = 3R$. (a) Rank the resistors according to the potential difference across them from largest to smallest. Note any cases of equal potential differences. (b) Determine the potential difference across each resistor in terms of $\mathcal{E}$. (c) Rank the resistors according to the current in them from largest to smallest. Note any cases of equal currents. (d) Determine the current in each resistor in terms of I. (e) **What If?** If R_3 is increased, explain what happens to the current in each of the resistors. (f) In the limit that $R_3 \to \infty$, what are the new values of the current in each resistor in terms of I, the original current in the battery?

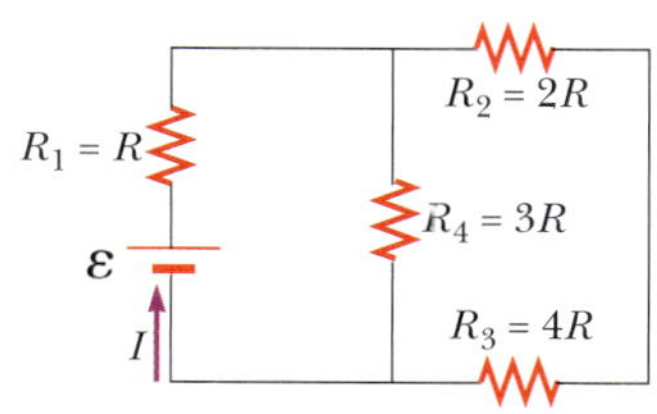

Figure P28.14

15. Calculate the power delivered to each resistor in the circuit shown in Figure P28.15.

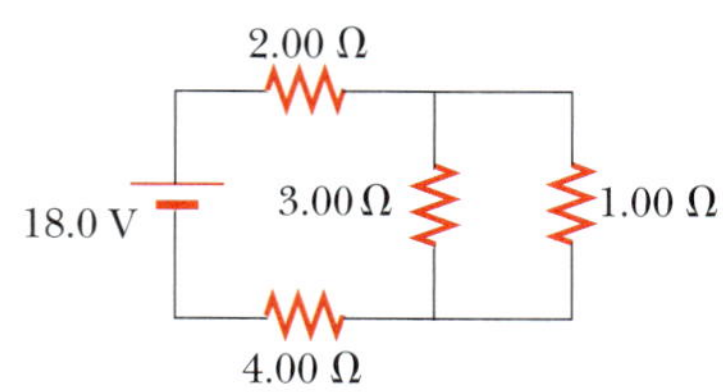

Figure P28.15

Section 28.3 Kirchhoff's Rules

16. The ammeter shown in Figure P28.16 reads 2.00 A. Find I_1, I_2, and $\mathcal{E}$.

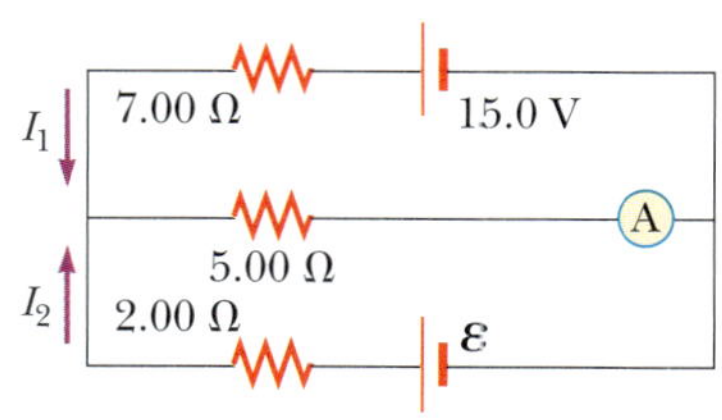

Figure P28.16

17. ▲ Determine the current in each branch of the circuit shown in Figure P28.17.

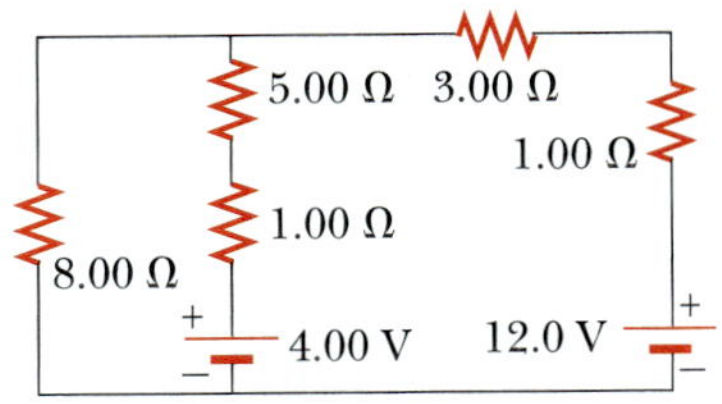

Figure P28.17 Problems 17, 18, and 19.

18. In Figure P28.17, show how to add only enough ammeters to measure every different current. Show how to add only enough voltmeters to measure the potential difference across each resistor and across each battery.

19. ● The circuit considered in Problem 17 and shown in Figure P28.17 is connected for 2.00 min. (a) Find the energy delivered by each battery. (b) Find the energy delivered to each resistor. (c) Identify the net energy transformation that occurs in the operation of the circuit. Find the total amount of energy transformed.

20. The following equations describe an electric circuit:

$$-I_1(220\ \Omega) + 5.80\text{ V} - I_2(370\ \Omega) = 0$$

$$I_2(370\ \Omega) + I_3(150\ \Omega) - 3.10\text{ V} = 0$$

$$I_1 + I_3 - I_2 = 0$$

(a) Draw a diagram of the circuit. (b) Calculate the unknowns and identify the physical meaning of each unknown.

21. Consider the circuit shown in Figure P28.21. What are the expected readings of the ideal ammeter and ideal voltmeter?

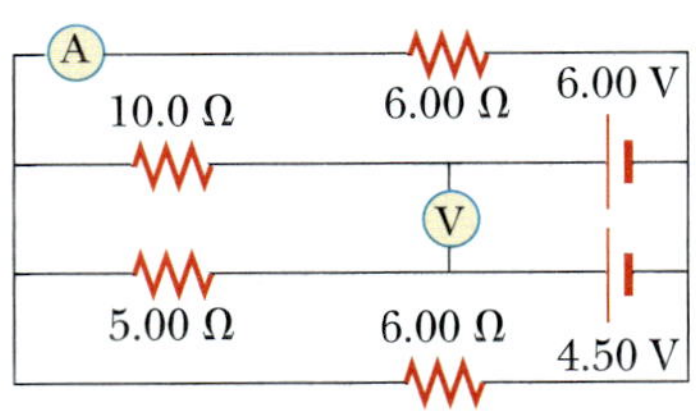

Figure P28.21

22. Taking $R = 1.00\text{ k}\Omega$ and $\mathcal{E} = 250$ V in Figure P28.22, determine the direction and magnitude of the current in the horizontal wire between *a* and *e*.

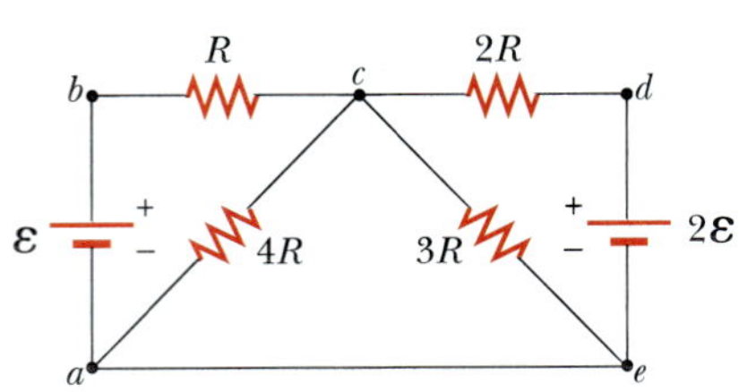

Figure P28.22

23. In the circuit of Figure P28.23, determine the current in each resistor and the potential difference across the 200-Ω resistor.

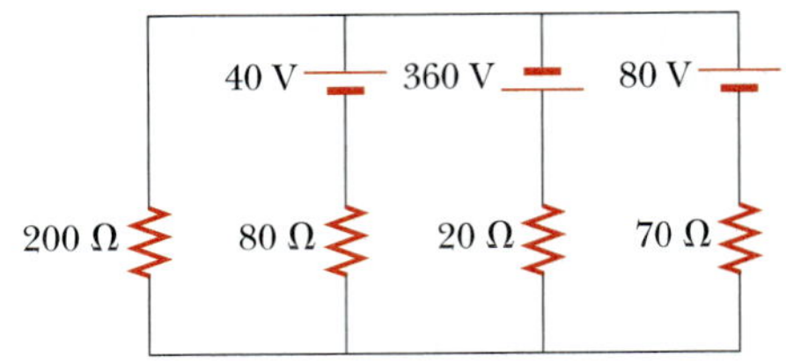

Figure P28.23

24. A dead battery is charged by connecting it to the live battery of another car with jumper cables (Fig. P28.24). Determine the current in the starter and in the dead battery.

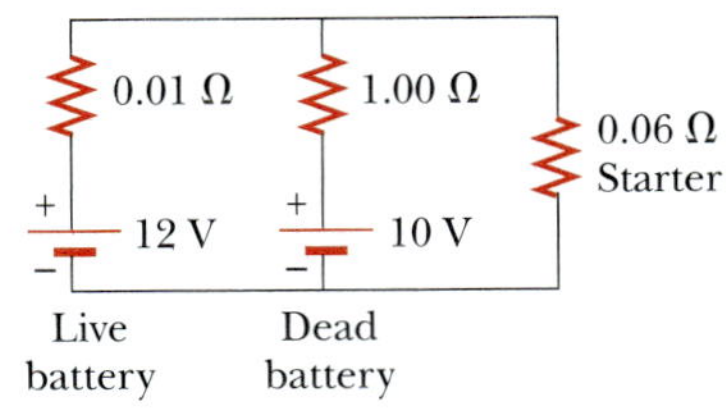

Figure P28.24

25. For the circuit shown in Figure P28.25, calculate (a) the current in the 2.00-Ω resistor and (b) the potential difference between points *a* and *b*.

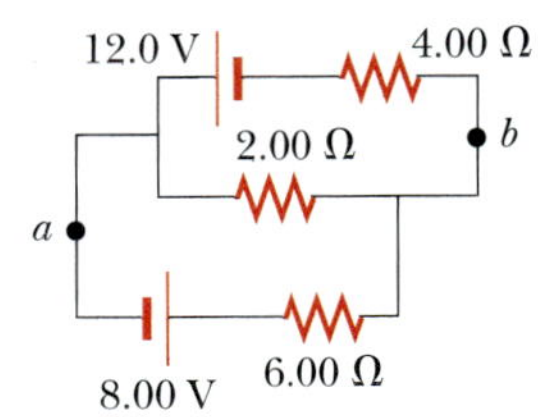

Figure P28.25

26. For the network shown in Figure P28.26, show that the resistance $R_{ab} = \frac{27}{17}\ \Omega$.

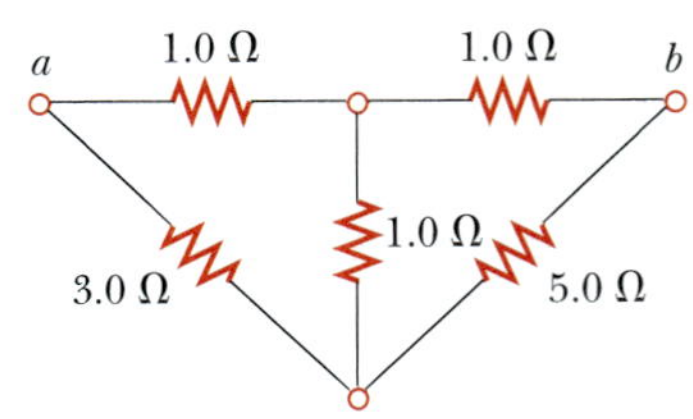

Figure P28.26

Section 28.4 *RC* Circuits

27. ▲ Consider a series *RC* circuit (see Active Fig. 28.16) for which $R = 1.00\text{ M}\Omega$, $C = 5.00\ \mu\text{F}$, and $\mathcal{E} = 30.0$ V. Find (a) the time constant of the circuit and (b) the maximum charge on the capacitor after the switch is thrown to *a*, connecting the capacitor to the battery. (c) Find the current in the resistor 10.0 s after the switch is thrown to *a*.

28. A 10.0-μF capacitor is charged by a 10.0-V battery through a resistance *R*. The capacitor reaches a potential difference of 4.00 V in a time interval of 3.00 s after charging begins. Find *R*.

29. A 2.00-nF capacitor with an initial charge of 5.10 μC is discharged through a 1.30-kΩ resistor. (a) Calculate the current in the resistor 9.00 μs after the resistor is connected across the terminals of the capacitor. (b) What charge remains on the capacitor after 8.00 μs? (c) What is the maximum current in the resistor?

30. Show that the integral $\int_0^\infty e^{-2t/RC}dt$ in Example 28.11 has the value $RC/2$.

31. The circuit in Figure P28.31 has been connected for a long time. (a) What is the potential difference across the capacitor? (b) If the battery is disconnected from the circuit, over what time interval does the capacitor discharge to one-tenth of its initial voltage?

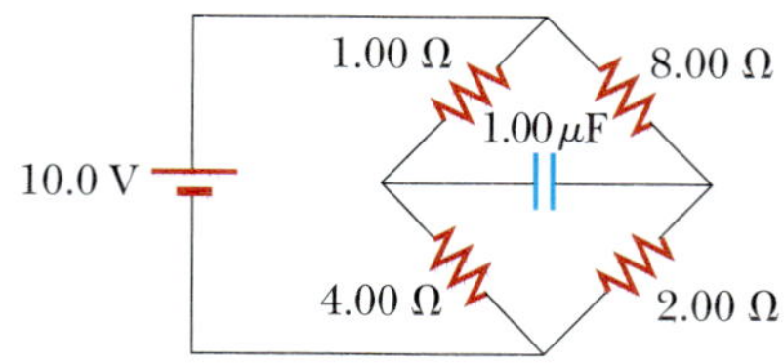

Figure P28.31

32. In the circuit of Figure P28.32, the switch S has been open for a long time. It is then suddenly closed. Determine the time constant (a) before the switch is closed and (b) after the switch is closed. (c) Let the switch be closed at $t = 0$. Determine the current in the switch as a function of time.

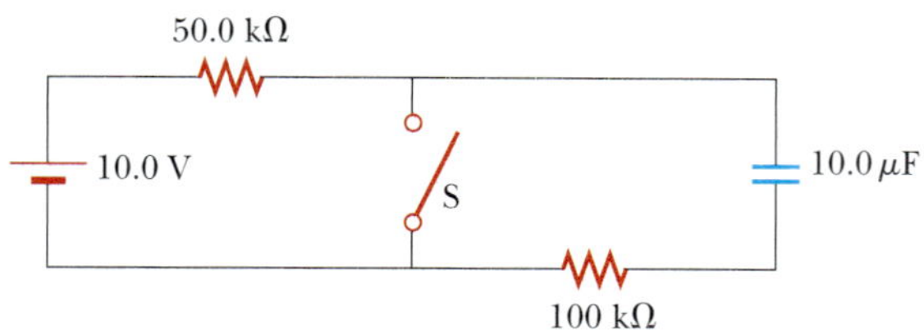

Figure P28.32

Section 28.5 Electrical Meters

33. Assume a galvanometer has an internal resistance of 60.0 Ω and requires a current of 0.500 mA to produce full-scale deflection. What resistance must be connected in parallel with the galvanometer if the combination is to serve as an ammeter that has a full-scale deflection for a current of 0.100 A?

34. A particular galvanometer serves as a 2.00-V full-scale voltmeter when a 2 500-Ω resistor is connected in series with it. It serves as a 0.500-A full-scale ammeter when a 0.220-Ω resistor is connected in parallel with it. Determine the internal resistance of the galvanometer and the current required to produce full-scale deflection.

35. A particular galvanometer requires a current of 1.50 mA for full-scale deflection and has a resistance of 75.0 Ω. It can be used to measure voltages by wiring a large resistor in series with the galvanometer as suggested in Active Figure 28.22. The effect is to limit the current in the galvanometer when large voltages are applied. Calculate the value of the resistor that allows the galvanometer to measure an applied voltage of 25.0 V at full-scale deflection.

36. ● *Meter loading.* Work this problem to five-digit precision. Refer to Figure P28.36. (a) When a 180.00-Ω resistor is connected across a battery of emf 6.000 0 V and internal resistance 20.000 Ω, what is the current in the resistor? What is the potential difference across it? (b) Suppose now an ammeter of resistance 0.500 00 Ω and a voltmeter of resistance 20 000 Ω are added to the circuit as shown in Figure P28.36b. Find the reading of each. (c) **What If?** Now one terminal of one wire is moved as shown in Figure P28.36c. Find the new meter readings. (d) Explain whether one of the voltmeter–ammeter connections is significantly better for taking data to determine the resistance of the resistor.

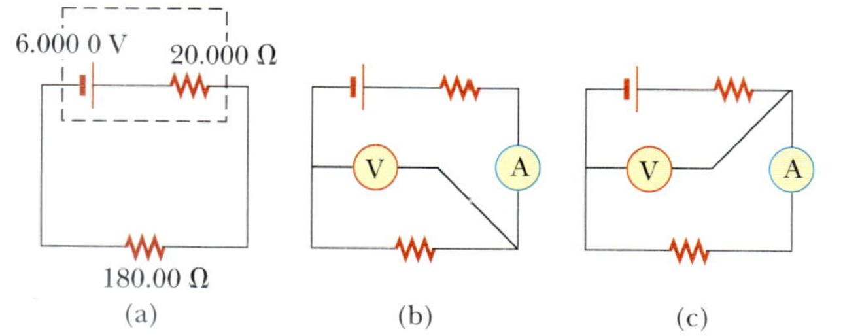

Figure P28.36

Section 28.6 Household Wiring and Electrical Safety

37. ● ▲ An electric heater is rated at 1 500 W, a toaster at 750 W, and an electric grill at 1 000 W. The three appliances are connected to a common 120-V household circuit. (a) How much current does each draw? (b) Is a circuit with a 25.0-A circuit breaker sufficient in this situation? Explain your answer.

38. Turn on your desk lamp. Pick up the cord, with your thumb and index finger spanning the width of the cord. (a) Compute an order-of-magnitude estimate for the current in your hand. Assume the conductor inside the lamp cord next to your thumb is at potential ~ 10^2 V at a typical instant and the conductor next to your index finger is at ground potential (0 V). The resistance of your hand depends strongly on the thickness and the moisture content of the outer layers of your skin. Assume the resistance of your hand between fingertip and thumb tip is ~10^4 Ω. You may model the cord as having rubber insulation. State the other quantities you measure or estimate and their values. Explain your reasoning. (b) Suppose your body is isolated from any other charges or currents. In order-of-magnitude terms, describe the potential of your thumb where it contacts the cord and the potential of your finger where it touches the cord.

Additional Problems

39. The circuit in Figure P28.39 has been connected for several seconds. Find the current (a) in the 4.00-V battery,

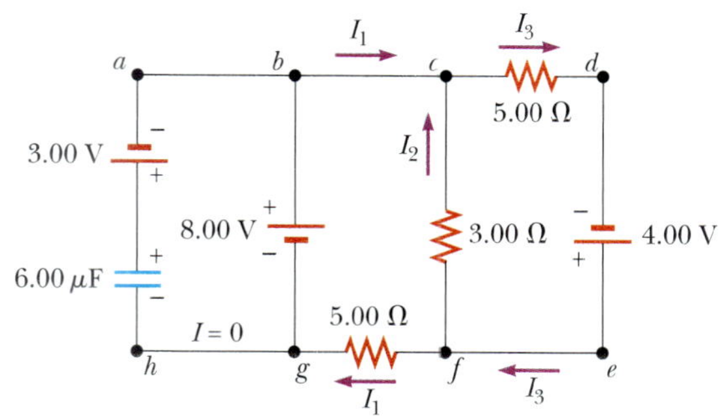

Figure P28.39

(b) in the 3.00-Ω resistor, (c) in the 8.00-V battery, and (d) in the 3.00-V battery. Find (e) the charge on the capacitor.

40. ● The circuit in Figure P28.40a consists of three resistors and one battery with no internal resistance. (a) Find the current in the 5.00-Ω resistor. (b) Find the power delivered to the 5.00-Ω resistor. (c) In each of the circuits in Figures P28.40b, P28.40c, and P28.40d, an additional 15.0-V battery has been inserted into the circuit. Which diagram or diagrams represent a circuit that requires the use of Kirchhoff's rules to find the currents? Explain why. In which of these three circuits is the smallest amount of power delivered to the 10.0-Ω resistor? You need not calculate the power in each circuit if you explain your answer.

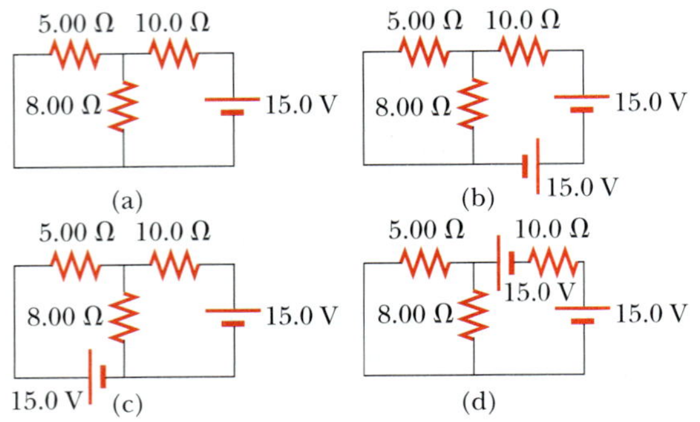

Figure P28.40

41. Four 1.50-V AA batteries in series are used to power a transistor radio. If the batteries can move a charge of 240 C, how long will they last if the radio has a resistance of 200 Ω?

42. ● A battery has an emf of 9.20 V and an internal resistance of 1.20 Ω. (a) What resistance across the battery will extract from it a power of 12.8 W? (b) A power of 21.2 W? Explain your answers.

43. Calculate the potential difference between points *a* and *b* in Figure P28.43 and identify which point is at the higher potential.

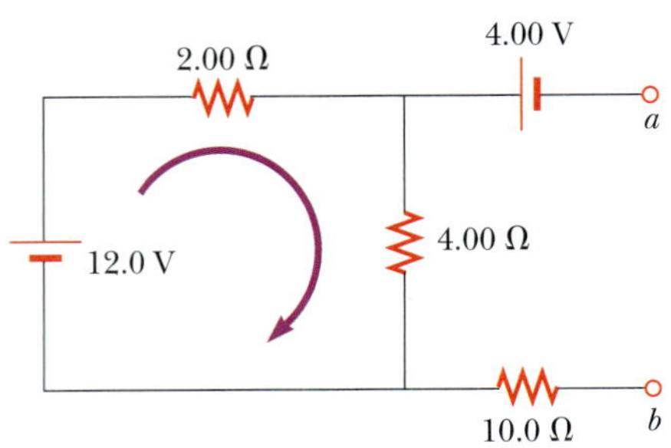

Figure P28.43

44. Assume you have a battery of emf $\mathcal{E}$ and three identical lightbulbs, each having constant resistance *R*. What is the total power delivered by the battery if the lightbulbs are connected (a) in series? (b) In parallel? (c) For which connection will the lightbulbs shine the brightest?

45. A rechargeable battery has a constant emf of 13.2 V and an internal resistance of 0.850 Ω. It is charged by a 14.7-V power supply for a time interval of 1.80 h. After charging, the battery returns to its original state as it delivers a constant current to a load resistor over 7.30 h. Find the efficiency of the battery as an energy storage device. (The efficiency here is defined as the energy delivered to the load during discharge divided by the energy delivered by the 14.7-V power supply during the charging process.)

46. A power supply has an open-circuit voltage of 40.0 V and an internal resistance of 2.00 Ω. It is used to charge two storage batteries connected in series, each having an emf of 6.00 V and internal resistance of 0.300 Ω. If the charging current is to be 4.00 A, (a) what additional resistance should be added in series? (b) At what rate does the internal energy increase in the supply, in the batteries, and in the added series resistance? (c) At what rate does the chemical energy increase in the batteries?

47. When two unknown resistors are connected in series with a battery, the battery delivers 225 W and carries a total current of 5.00 A. For the same total current, 50.0 W is delivered when the resistors are connected in parallel. Determine the value of each resistor.

48. When two unknown resistors are connected in series with a battery, the battery delivers total power $\mathcal{P}_s$ and carries a total current of *I*. For the same total current, a total power $\mathcal{P}_p$ is delivered when the resistors are connected in parallel. Determine the value of each resistor.

49. Two resistors R_1 and R_2 are in parallel with each other. Together they carry total current *I*. (a) Determine the current in each resistor. (b) Prove that this division of the total current *I* between the two resistors results in less power delivered to the combination than any other division. It is a general principle that *current in a direct current circuit distributes itself so that the total power delivered to the circuit is a minimum.*

50. ● (a) Determine the equilibrium charge on the capacitor in the circuit of Figure P28.50 as a function of *R*. (b) Evaluate the charge when $R = 10.0\ \Omega$. (c) Can the charge on the capacitor be zero? If so, for what value of *R*? (d) What is the maximum possible magnitude of the charge on the capacitor? For what value of *R* is it achieved? (e) Is it experimentally meaningful to take $R = \infty$? Explain your answer. If so, what charge magnitude does it imply? *Suggestion:* You may do part (b) before part (a), as practice for it.

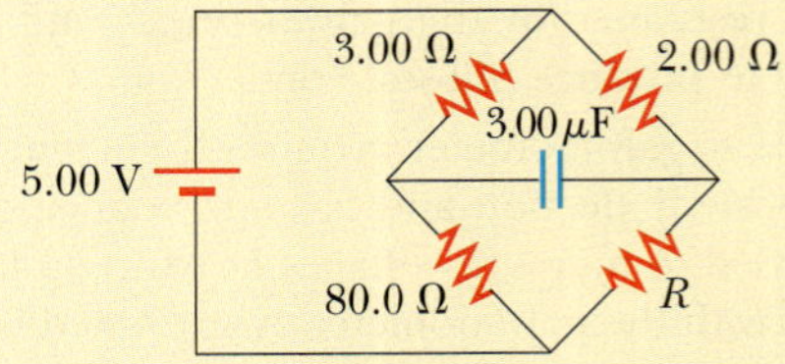

Figure P28.50

51. The value of a resistor *R* is to be determined using the ammeter–voltmeter setup shown in Figure P28.51. The ammeter has a resistance of 0.500 Ω, and the voltmeter has a resistance of 20.0 kΩ. Within what range of actual values of *R* will the measured values be correct to within

5.00% if the measurement is made using the circuit shown in (a) Figure P28.51a and (b) Figure P28.51b?

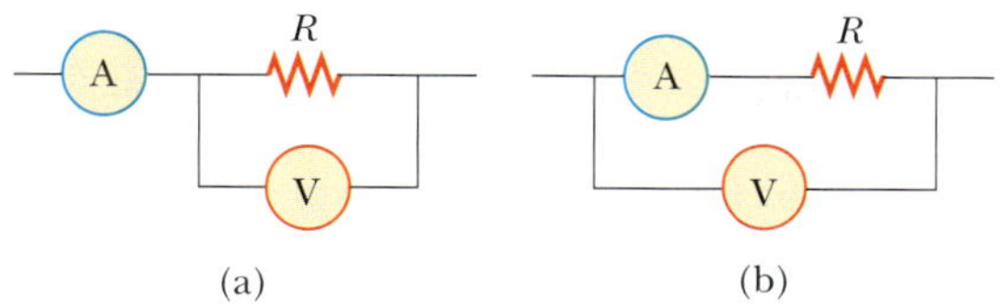

Figure P28.51

52. A battery is used to charge a capacitor through a resistor as shown in Active Figure 28.16b. Show that half the energy supplied by the battery appears as internal energy in the resistor and half is stored in the capacitor.

53. The values of the components in a simple series *RC* circuit containing a switch (Active Fig. 28.16b) are $C = 1.00\ \mu F$, $R = 2.00 \times 10^6\ \Omega$, and $\mathcal{E} = 10.0$ V. At the instant 10.0 s after the switch is thrown to *a*, calculate (a) the charge on the capacitor, (b) the current in the resistor, (c) the rate at which energy is being stored in the capacitor, and (d) the rate at which energy is being delivered by the battery.

54. A young man owns a canister vacuum cleaner marked 535 W at 120 V and a Volkswagen Beetle, which he wishes to clean. He parks the car in his apartment parking lot and uses an inexpensive extension cord 15.0 m long to plug in the vacuum cleaner. You may assume the cleaner has constant resistance. (a) If the resistance of each of the two conductors in the extension cord is 0.900 Ω, what is the actual power delivered to the cleaner? (b) If instead the power is to be at least 525 W, what must be the diameter of each of two identical copper conductors in the cord he buys? (c) Repeat part (b) assuming the power is to be at least 532 W. *Suggestion:* A symbolic solution can simplify the calculations.

55. Three 60.0-W, 120-V lightbulbs are connected across a 120-V power source as shown in Figure P28.55. Find (a) the total power delivered to the three lightbulbs and (b) the potential difference across each. Assume the resistance of each lightbulb is constant (even though in reality the resistance might increase markedly with current).

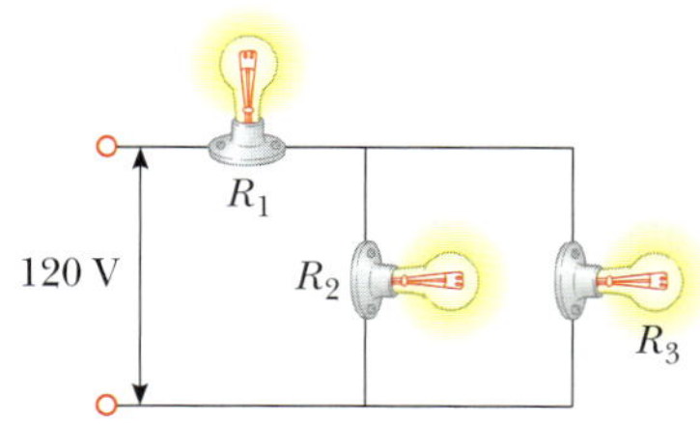

Figure P28.55

56. Switch S shown in Figure P28.56 has been closed for a long time and the electric circuit carries a constant current. Take $C_1 = 3.00\ \mu F$, $C_2 = 6.00\ \mu F$, $R_1 = 4.00$ kΩ, and $R_2 = 7.00$ kΩ. The power delivered to R_2 is 2.40 W. (a) Find the charge on C_1. (b) Now the switch is opened. After many milliseconds, by how much has the charge on C_2 changed?

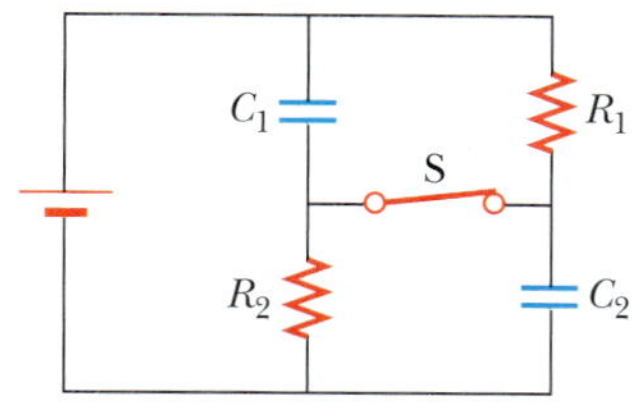

Figure P28.56

57. ● An ideal voltmeter connected across a certain fresh battery reads 9.30 V, and an ideal ammeter briefly connected across the same battery reads 3.70 A. We say that the battery has an open-circuit voltage of 9.30 V and a short-circuit current of 3.70 A. (a) Model the battery as a source of emf $\mathcal{E}$ in series with an internal resistance *r*. Determine both $\mathcal{E}$ and *r*. (b) An irresponsible experimenter connects 20 of these identical batteries together as suggested in Figure P28.57. Do not try this experiment yourself! Find the open-circuit voltage and the short-circuit current of the set of connected batteries. (c) Assume the resistance between the palms of the experimenter's two hands is 120 Ω. Find the current in his body that would result if his palms touched the two exposed terminals of the set of connected batteries. (d) Find the power that would be delivered to his body in this situation. (e) Thinking it is safe to do so, the experimenter threads a copper wire inside his shirt between his hands, like a mitten string. To reduce the current in his body to 5.00 mA when he presses the ends of the wire against the battery poles, what should the resistance of the copper wire be? (f) Find the power delivered to his body in this situation. (g) Find the power delivered to the copper wire. (h) Explain why the sum of the two powers in parts (f) and (g) is much less than the power calculated in part (d). Is it meaningful to ask where the rest of the power is going?

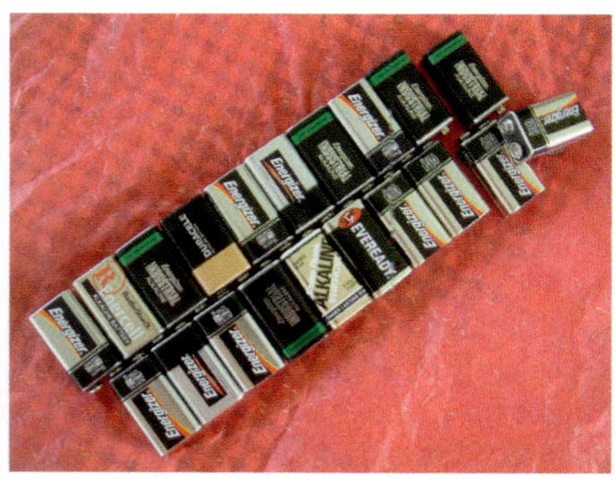

Richard McGrew

Figure P28.57

58. ● Four resistors are connected in parallel across a 9.20-V battery. They carry currents of 150 mA, 45.0 mA, 14.0 mA, and 4.00 mA. (a) If the resistor with the largest resistance is replaced with one having twice the resistance, what is the ratio of the new current in the battery to the original current? (b) **What If?** If instead the resistor with the smallest resistance is replaced with one having twice the resistance, what is the ratio of the new total current to the original current? (c) On a February night, energy leaves a house by several energy leaks, including the following: 1 500 W by conduction through the ceiling, 450 W by infiltration (airflow) around the windows, 140 W by conduction through the basement wall above the foundation sill, and 40.0 W by conduction through the plywood door to the attic. To produce the biggest saving in heating bills, which one of these energy transfers should be reduced first? Explain how you decide. Clifford Swartz suggested the idea for this problem.

59. Figure P28.59 shows a circuit model for the transmission of an electrical signal such as cable TV to a large number of subscribers. Each subscriber connects a load resistance R_L between the transmission line and the ground. The ground is assumed to be at zero potential and able to carry any current between any ground connections with negligible resistance. The resistance of the transmission line between the connection points of different subscribers is modeled as the constant resistance R_T. Show that the equivalent resistance across the signal source is

$$R_{eq} = \tfrac{1}{2}[(4R_T R_L + R_T^2)^{1/2} + R_T]$$

Suggestion: Because the number of subscribers is large, the equivalent resistance would not change noticeably if the first subscriber canceled the service. Consequently, the equivalent resistance of the section of the circuit to the right of the first load resistor is nearly equal to R_{eq}.

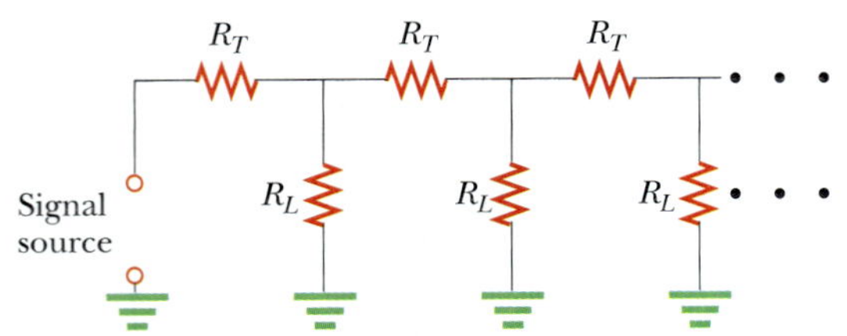

Figure P28.59

60. A regular tetrahedron is a pyramid with a triangular base. Six 10.0-Ω resistors are placed along its six edges, with junctions at its four vertices. A 12.0-V battery is connected to any two of the vertices. Find (a) the equivalent resistance of the tetrahedron between these vertices and (b) the current in the battery.

61. In Figure P28.61, suppose the switch has been closed for a time interval sufficiently long for the capacitor to become fully charged. Find (a) the steady-state current in each resistor and (b) the charge Q on the capacitor. (c) The switch is now opened at $t = 0$. Write an equation for the current I_{R_2} in R_2 as a function of time and (d) find the time interval required for the charge on the capacitor to fall to one-fifth its initial value.

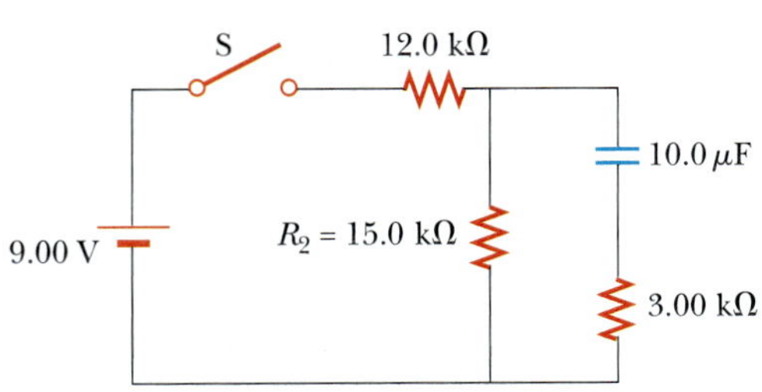

Figure P28.61

62. The circuit shown in Figure P28.62 is set up in the laboratory to measure an unknown capacitance C with the use of a voltmeter of resistance $R = 10.0$ MΩ and a battery whose emf is 6.19 V. The data given in the table are the measured voltages across the capacitor as a function of time, where $t = 0$ represents the instant at which the switch is opened. (a) Construct a graph of $\ln(\mathcal{E}/\Delta V)$ versus t and perform a linear least-squares fit to the data. (b) From the slope of your graph, obtain a value for the time constant of the circuit and a value for the capacitance.

ΔV (V)	t (s)	$\ln(\mathcal{E}/\Delta V)$
6.19	0	
5.55	4.87	
4.93	11.1	
4.34	19.4	
3.72	30.8	
3.09	46.6	
2.47	67.3	
1.83	102.2	

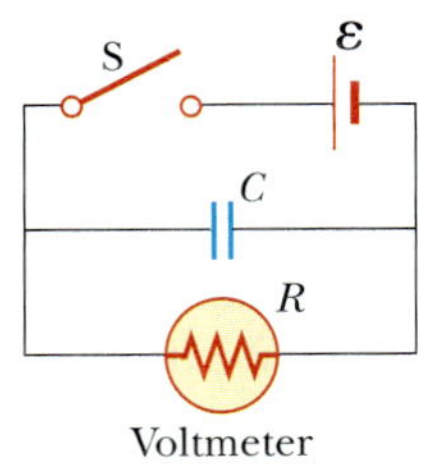

Figure P28.62

63. The student engineer of a campus radio station wishes to verify the effectiveness of the lightning rod on the antenna mast (Fig. P28.63). The unknown resistance R_x is between points C and E. Point E is a true ground, but it is inaccessible for direct measurement because this stratum is several meters below the Earth's surface. Two identical rods are driven into the ground at A and B, introducing an unknown resistance R_y. The procedure is as follows. Measure resistance R_1 between points A and B, then connect A and B with a heavy conducting wire and measure resistance R_2 between points A and C. (a) Derive an equation for R_x in terms of the observable resistances, R_1 and R_2. (b) A satisfactory ground resistance would be $R_x < 2.00$ Ω. Is the grounding of the station adequate if measurements give $R_1 = 13.0$ Ω and $R_2 = 6.00$ Ω?

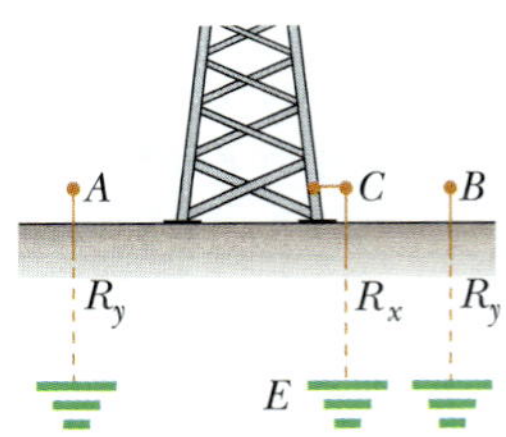

Figure P28.63

64. The switch in Figure P28.64a closes when $\Delta V_c > 2\,\Delta V/3$ and opens when $\Delta V_c < \Delta V/3$. The voltmeter reads a

potential difference as plotted in Figure P28.64b. What is the period T of the waveform in terms of R_1, R_2, and C?

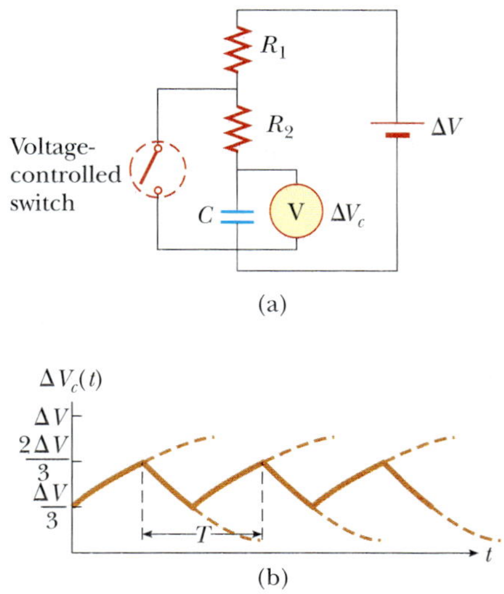

Figure P28.64

65. An electric teakettle has a multiposition switch and two heating coils. When only one coil is switched on, the well-insulated kettle brings a full pot of water to a boil over the time interval Δt. When only the other coil is switched on, it takes a time interval of $2\Delta t$ to boil the same amount of water. Find the time interval required to boil the same amount of water if both coils are switched on (a) in a parallel connection and (b) in a series connection.

66. In places such as hospital operating rooms or factories for electronic circuit boards, electric sparks must be avoided. A person standing on a grounded floor and touching nothing else can typically have a body capacitance of 150 pF, in parallel with a foot capacitance of 80.0 pF produced by the dielectric soles of his or her shoes. The person acquires static electric charge from interactions with furniture, clothing, equipment, packaging materials, and essentially everything else. The static charge flows to ground through the equivalent resistance of the two shoe soles in parallel with each other. A pair of rubber-soled street shoes can present an equivalent resistance of 5 000 MΩ. A pair of shoes with special static-dissipative soles can have an equivalent resistance of 1.00 MΩ. Consider the person's body and shoes as forming an RC circuit with the ground. (a) How long does it take the rubber-soled shoes to reduce a person's potential from 3 000 V to 100 V? (b) How long does it take the static-dissipative shoes to do the same thing?

Answers to Quick Quizzes

28.1 (a). Power is delivered to the internal resistance of a battery, so decreasing the internal resistance decreases this "lost" power and increase the percentage of the power delivered to the device.

28.2 (b). When the switch is opened, resistors R_1 and R_2 are in series, so the total circuit resistance is larger than when the switch was closed. As a result, the current decreases.

28.3 (a). When the switch is closed, resistors R_1 and R_2 are in parallel, so the total circuit resistance is smaller than when the switch was open. As a result, the current increases.

28.4 **(i),** (b). Adding another series resistor increases the total resistance of the circuit and therefore reduces the current in the circuit. **(ii),** (a). The potential difference across the battery terminals increases because the reduced current results in a smaller voltage decrease across the internal resistance. **(iii),** (a). If a third resistor were connected in parallel, the total resistance of the circuit would decrease and the current in the battery would increase. **(iv),** (b). The potential difference across the terminals would decrease because the increased current results in a greater voltage drop across the internal resistance.

28.5 **(i),** (c). Just after the switch is closed, there is no charge on the capacitor. Current exists in both branches of the circuit as the capacitor begins to charge, so the right half of the circuit is equivalent to two resistances R in parallel for an equivalent resistance of $\frac{1}{2}R$. **(ii),** (d). After a long time, the capacitor is fully charged and the current in the right-hand branch drops to zero. Now, current exists only in a resistance R across the battery.

Magnetic fingerprinting allows fingerprints to be seen on surfaces that otherwise would not allow prints to be lifted. The powder spread on the surface is coated with an organic material that adheres to the greasy residue in a fingerprint. A magnetic "brush" removes the excess powder and makes the fingerprint visible. (James King-Holmes/Photo Researchers, Inc.)

29 Magnetic Fields

Many historians of science believe that the compass, which uses a magnetic needle, was used in China as early as the 13th century BC, its invention being of Arabic or Indian origin. The early Greeks knew about magnetism as early as 800 BC. They discovered that the stone magnetite (Fe_3O_4) attracts pieces of iron. Legend ascribes the name *magnetite* to the shepherd Magnes, the nails of whose shoes and the tip of whose staff stuck fast to chunks of magnetite while he pastured his flocks.

In 1269, Pierre de Maricourt of France found that the directions of a needle near a spherical natural magnet formed lines that encircled the sphere and passed through two points diametrically opposite each other, which he called the *poles* of the magnet. Subsequent experiments showed that every magnet, regardless of its shape, has two poles, called *north* (N) and *south* (S) poles, that exert forces on other magnetic poles similar to the way electric charges exert forces on one another. That is, like poles (N–N or S–S) repel each other, and opposite poles (N–S) attract each other.

The poles received their names because of the way a magnet, such as that in a compass, behaves in the presence of the Earth's magnetic field. If a bar magnet is suspended from its midpoint and can swing freely in a horizontal plane, it will rotate until its north pole points to the Earth's geographic North Pole and its south pole points to the Earth's geographic South Pole.[1]

In 1600, William Gilbert (1540–1603) extended de Maricourt's experiments to a variety of materials. He knew that a compass needle orients in preferred directions, so he suggested that the Earth itself is a large, permanent magnet. In 1750, experimenters used a torsion balance to show that magnetic poles exert attractive or repulsive forces on each other and that these forces vary as the inverse square of the distance between interacting poles. Although the force between two magnetic poles is otherwise similar to the force between two electric charges, electric charges can be isolated (witness the electron and proton), whereas **a single magnetic pole has never been isolated.** That is, **magnetic poles are always found in pairs.** All attempts thus far to detect an isolated magnetic pole have been unsuccessful. No matter how many times a permanent magnet is cut in two, each piece always has a north and a south pole.[2]

The relationship between magnetism and electricity was discovered in 1819 when, during a lecture demonstration, Hans Christian Oersted found that an electric current in a wire deflected a nearby compass needle.[3] In the 1820s, further connections between electricity and magnetism were demonstrated independently by Faraday and Joseph Henry (1797–1878). They showed that an electric current can be produced in a circuit either by moving a magnet near the circuit or by changing the current in a nearby circuit. These observations demonstrate that a changing magnetic field creates an electric field. Years later, theoretical work by Maxwell showed that the reverse is also true: a changing electric field creates a magnetic field.

This chapter examines the forces that act on moving charges and on current-carrying wires in the presence of a magnetic field. The source of the magnetic field is described in Chapter 30.

North Wind Picture Archives

HANS CHRISTIAN OERSTED
Danish Physicist and Chemist (1777–1851)
Oersted s best known for observing that a compass needle deflects when placed near a wire carrying a current. This important discovery was the first evidence of the connection between electric and magnetic phenomena. Oersted was also the first to prepare pure aluminum.

29.1 Magnetic Fields and Forces

In our study of electricity, we described the interactions between charged objects in terms of electric fields. Recall that an electric field surrounds any electric charge. In addition to containing an electric field, the region of space surrounding any *moving* electric charge also contains a magnetic field. A magnetic field also surrounds a magnetic substance making up a permanent magnet.

Historically, the symbol $\vec{\mathbf{B}}$ has been used to represent a magnetic field, and we use this notation in this book. The direction of the magnetic field $\vec{\mathbf{B}}$ at any location is the direction in which a compass needle points at that location. As with the electric field, we can represent the magnetic field by means of drawings with *magnetic field lines.*

Active Figure 29.1 shows how the magnetic field lines of a bar magnet can be traced with the aid of a compass. Notice that the magnetic field lines outside the

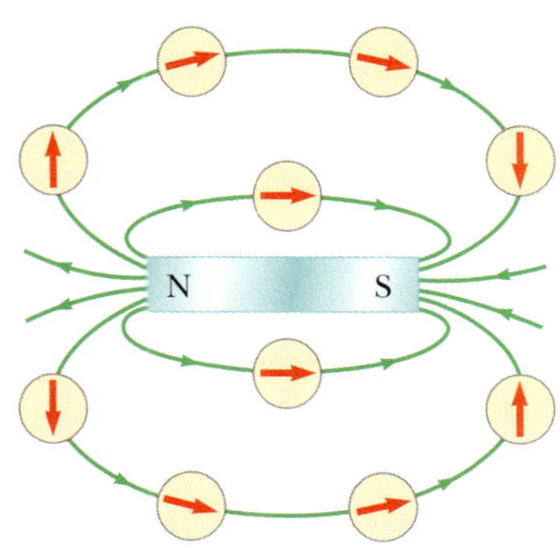

ACTIVE FIGURE 29.1

Compass needles can be used to trace the magnetic field lines in the region outside a bar magnet.

Sign in at www.thomsonedu.com and go to ThomsonNOW to move the compass around and trace the magnetic field lines for yourself.

[1] The Earth's geographic North Pole is magnetically a south pole, whereas the Earth's geographic South Pole is magnetically a north pole. Because *opposite* magnetic poles attract each other, the pole on a magnet that is attracted to the Earth's geographic North Pole is the magnet's *north* pole and the pole attracted to the Earth's geographic South Pole is the magnet's *south* pole.

[2] There is some theoretical basis for speculating that magnetic *monopoles*—isolated north or south poles—may exist in nature, and attempts to detect them are an active experimental field of investigation.

[3] The same discovery was reported in 1802 by an Italian jurist, Gian Domenico Romagnosi, but was overlooked, probably because it was published in an obscure journal.

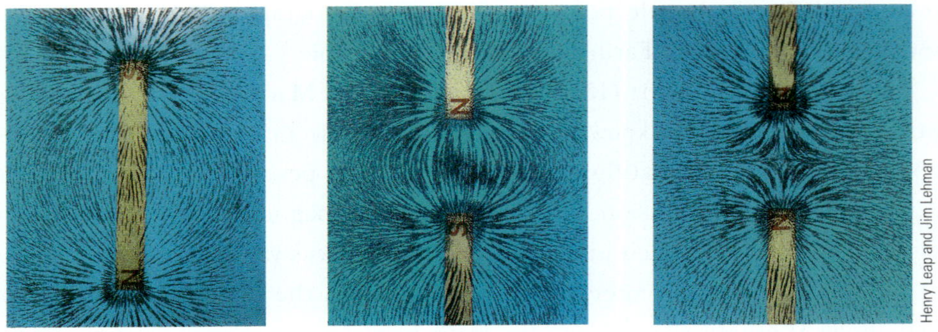

(a) (b) (c)

Figure 29.2 (a) Magnetic field pattern surrounding a bar magnet as displayed with iron filings. (b) Magnetic field pattern between *opposite* poles (N–S) of two bar magnets. (c) Magnetic field pattern between *like* poles (N–N) of two bar magnets.

magnet point away from the north pole and toward the south pole. One can display magnetic field patterns of a bar magnet using small iron filings as shown in Figure 29.2.

We can define a magnetic field $\vec{\mathbf{B}}$ at some point in space in terms of the magnetic force $\vec{\mathbf{F}}_B$ that the field exerts on a charged particle moving with a velocity $\vec{\mathbf{v}}$, which we call the test object. For the time being, let's assume no electric or gravitational fields are present at the location of the test object. Experiments on various charged particles moving in a magnetic field give the following results:

Properties of the magnetic force on a charge moving in a magnetic field ▶

- The magnitude F_B of the magnetic force exerted on the particle is proportional to the charge q and to the speed v of the particle.
- When a charged particle moves parallel to the magnetic field vector, the magnetic force acting on the particle is zero.
- When the particle's velocity vector makes any angle $\theta \neq 0$ with the magnetic field, the magnetic force acts in a direction perpendicular to both $\vec{\mathbf{v}}$ and $\vec{\mathbf{B}}$; that is, $\vec{\mathbf{F}}_B$ is perpendicular to the plane formed by $\vec{\mathbf{v}}$ and $\vec{\mathbf{B}}$ (Fig. 29.3a).
- The magnetic force exerted on a positive charge is in the direction opposite the direction of the magnetic force exerted on a negative charge moving in the same direction (Fig. 29.3b).
- The magnitude of the magnetic force exerted on the moving particle is proportional to $\sin\theta$, where θ is the angle the particle's velocity vector makes with the direction of $\vec{\mathbf{B}}$.

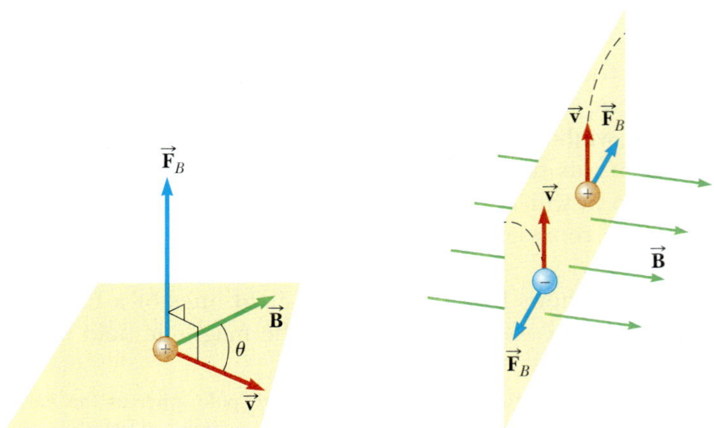

(a) (b)

Figure 29.3 The direction of the magnetic force $\vec{\mathbf{F}}_B$ acting on a charged particle moving with a velocity $\vec{\mathbf{v}}$ in the presence of a magnetic field $\vec{\mathbf{B}}$. (a) The magnetic force is perpendicular to both $\vec{\mathbf{v}}$ and $\vec{\mathbf{B}}$. (b) Oppositely directed magnetic forces $\vec{\mathbf{F}}_B$ are exerted on two oppositely charged particles moving at the same velocity in a magnetic field. The dashed lines show the paths of the particles, which are investigated in Section 29.2.

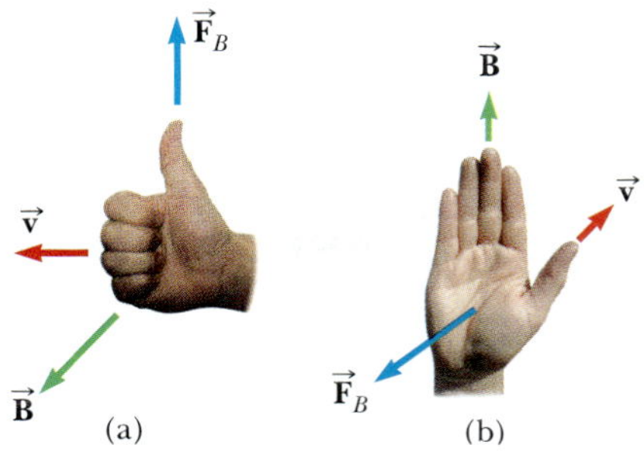

Figure 29.4 Two right-hand rules for determining the direction of the magnetic force $\vec{\mathbf{F}}_B = q\vec{\mathbf{v}} \times \vec{\mathbf{B}}$ acting on a particle with charge q moving with a velocity $\vec{\mathbf{v}}$ in a magnetic field $\vec{\mathbf{B}}$. (a) In this rule, your fingers point in the direction of $\vec{\mathbf{v}}$, with $\vec{\mathbf{B}}$ coming out of your palm, so that you can curl your fingers in the direction of $\vec{\mathbf{B}}$. The direction of $\vec{\mathbf{v}} \times \vec{\mathbf{B}}$, and the force on a positive charge, is the direction in which your thumb points. (b) In this rule, the vector $\vec{\mathbf{v}}$ is in the direction of your thumb and $\vec{\mathbf{B}}$ in the direction of your fingers. The force $\vec{\mathbf{F}}_B$ on a positive charge is in the direction of your palm, as if you are pushing the particle with your hand.

We can summarize these observations by writing the magnetic force in the form

$$\vec{\mathbf{F}}_B = q\vec{\mathbf{v}} \times \vec{\mathbf{B}} \tag{29.1}$$

◀ Vector expression for the magnetic force on a charged particle moving in a magnetic field

which by definition of the cross product (see Section 11.1) is perpendicular to both $\vec{\mathbf{v}}$ and $\vec{\mathbf{B}}$. We can regard this equation as an operational definition of the magnetic field at some point in space. That is, the magnetic field is defined in terms of the force acting on a moving charged particle.

Figure 29.4 reviews two right-hand rules for determining the direction of the cross product $\vec{\mathbf{v}} \times \vec{\mathbf{B}}$ and determining the direction of $\vec{\mathbf{F}}_B$. The rule in Figure 29.4a depends on our right-hand rule for the cross product in Figure 11.2. Point the four fingers of your right hand along the direction of $\vec{\mathbf{v}}$ with the palm facing $\vec{\mathbf{B}}$ and curl them toward $\vec{\mathbf{B}}$. Your extended thumb, which is at a right angle to your fingers, points in the direction of $\vec{\mathbf{v}} \times \vec{\mathbf{B}}$. Because $\vec{\mathbf{F}}_B = q\vec{\mathbf{v}} \times \vec{\mathbf{B}}$, $\vec{\mathbf{F}}_B$ is in the direction of your thumb if q is positive and is opposite the direction of your thumb if q is negative. (If you need more help understanding the cross product, you should review Section 11.1, including Fig. 11.2.)

An alternative rule is shown in Figure 29.4b. Here the thumb points in the direction of $\vec{\mathbf{v}}$ and the extended fingers in the direction of $\vec{\mathbf{B}}$. Now, the force $\vec{\mathbf{F}}_B$ on a positive charge extends outward from the palm. The advantage of this rule is that the force on the charge is in the direction that you would push on something with your hand: outward from your palm. The force on a negative charge is in the opposite direction. You can use either of these two right-hand rules.

The magnitude of the magnetic force on a charged particle is

$$F_B = |q|vB\sin\theta \tag{29.2}$$

◀ Magnitude of the magnetic force on a charged particle moving in a magnetic field

where θ is the smaller angle between $\vec{\mathbf{v}}$ and $\vec{\mathbf{B}}$. From this expression, we see that F_B is zero when $\vec{\mathbf{v}}$ is parallel or antiparallel to $\vec{\mathbf{B}}$ ($\theta = 0$ or $180°$) and maximum when $\vec{\mathbf{v}}$ is perpendicular to $\vec{\mathbf{B}}$ ($\theta = 90°$).

Electric and magnetic forces have several important differences:

- The electric force vector is along the direction of the electric field, whereas the magnetic force vector is perpendicular to the magnetic field.
- The electric force acts on a charged particle regardless of whether the particle is moving, whereas the magnetic force acts on a charged particle only when the particle is in motion.
- The electric force does work in displacing a charged particle, whereas the magnetic force associated with a steady magnetic field does no work when a particle is displaced because the force is perpendicular to the displacement.

From the last statement and on the basis of the work–kinetic energy theorem, we conclude that the kinetic energy of a charged particle moving through a magnetic field cannot be altered by the magnetic field alone. The field can alter the

TABLE 29.1

Some Approximate Magnetic Field Magnitudes

Source of Field	Field Magnitude (T)
Strong superconducting laboratory magnet	30
Strong conventional laboratory magnet	2
Medical MRI unit	1.5
Bar magnet	10^{-2}
Surface of the Sun	10^{-2}
Surface of the Earth	0.5×10^{-4}
Inside human brain (due to nerve impulses)	10^{-13}

direction of the velocity vector, but it cannot change the speed or kinetic energy of the particle.

From Equation 29.2, we see that the SI unit of magnetic field is the newton per coulomb-meter per second, which is called the **tesla** (T):

The tesla ▶

$$1\ \text{T} = 1\ \frac{\text{N}}{\text{C}\cdot\text{m/s}}$$

Because a coulomb per second is defined to be an ampere,

$$1\ \text{T} = 1\ \frac{\text{N}}{\text{A}\cdot\text{m}}$$

A non-SI magnetic-field unit in common use, called the *gauss* (G), is related to the tesla through the conversion $1\ \text{T} = 10^4\ \text{G}$. Table 29.1 shows some typical values of magnetic fields.

Quick Quiz 29.1 An electron moves in the plane of this paper toward the top of the page. A magnetic field is also in the plane of the page and directed toward the right. What is the direction of the magnetic force on the electron? (a) toward the top of the page (b) toward the bottom of the page (c) toward the left edge of the page (d) toward the right edge of the page (e) upward out of the page (f) downward into the page

EXAMPLE 29.1 An Electron Moving in a Magnetic Field

An electron in a television picture tube moves toward the front of the tube with a speed of 8.0×10^6 m/s along the x axis (Fig. 29.5). Surrounding the neck of the tube are coils of wire that create a magnetic field of magnitude 0.025 T, directed at an angle of 60° to the x axis and lying in the xy plane. Calculate the magnetic force on the electron.

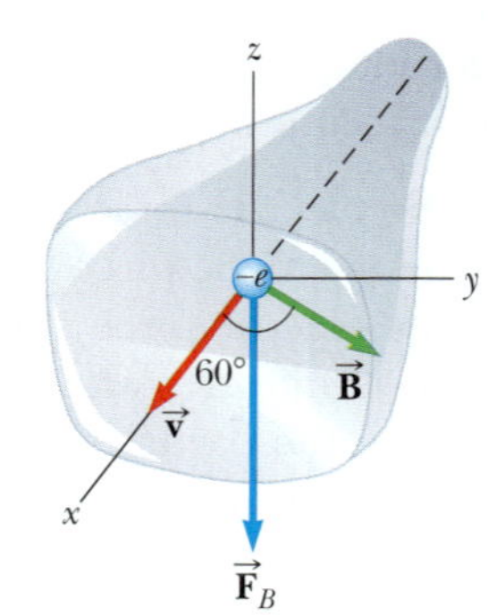

Figure 29.5 (Example 29.1) The magnetic force $\vec{\mathbf{F}}_B$ acting on the electron is in the negative z direction when $\vec{\mathbf{v}}$ and $\vec{\mathbf{B}}$ lie in the xy plane.

SOLUTION

Conceptualize Recall that the magnetic force on a charged particle is perpendicular to the plane formed by the velocity and magnetic field vectors. Use the right-hand rule in Figure 29.4 to convince yourself that the direction of the force on the electron is downward in Figure 29.5.

Categorize We evaluate the magnetic force using an equation developed in this section, so we categorize this example as a substitution problem.

Use Equation 29.2 to find the magnitude of the magnetic force:

$$F_B = |q|vB\sin\theta$$
$$= (1.6 \times 10^{-19}\ \text{C})(8.0 \times 10^{6}\ \text{m/s})(0.025\ \text{T})(\sin 60°)$$
$$= 2.8 \times 10^{-14}\ \text{N}$$

For practice using the vector product, evaluate this force in vector notation using Equation 29.1.

29.2 Motion of a Charged Particle in a Uniform Magnetic Field

Before we continue our discussion, some explanation of the notation used in this book is in order. To indicate the direction of $\vec{\mathbf{B}}$ in illustrations, we sometimes present perspective views such as those in Figure 29.5. If $\vec{\mathbf{B}}$ lies in the plane of the page or is present in a perspective drawing, we use green vectors or green field lines with arrowheads. In nonperspective illustrations, we depict a magnetic field perpendicular to and directed out of the page with a series of green dots, which represent the tips of arrows coming toward you (see Fig. 29.6a). In this case, the field is labeled $\vec{\mathbf{B}}_{\text{out}}$. If $\vec{\mathbf{B}}$ is directed perpendicularly into the page, we use green crosses, which represent the feathered tails of arrows fired away from you, as in Figure 29.6b. In this case, the field is labeled $\vec{\mathbf{B}}_{\text{in}}$, where the subscript "in" indicates "into the page." The same notation with crosses and dots is also used for other quantities that might be perpendicular to the page such as forces and current directions.

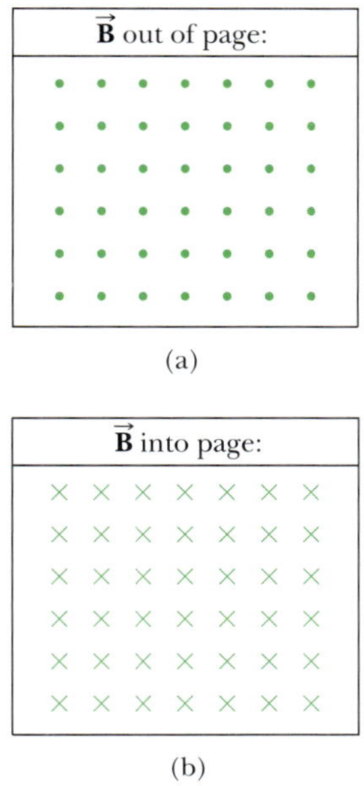

Figure 29.6 (a) Magnetic field lines coming out of the paper are indicated by dots, representing the tips of arrows coming outward. (b) Magnetic field lines going into the paper are indicated by crosses, representing the feathers of arrows going inward.

In Section 29.1, we found that the magnetic force acting on a charged particle moving in a magnetic field is perpendicular to the particle's velocity and consequently the work done by the magnetic force on the particle is zero. Now consider the special case of a positively charged particle moving in a uniform magnetic field with the initial velocity vector of the particle perpendicular to the field. Let's assume that the direction of the magnetic field is into the page as in Active Figure 29.7. As the particle changes the direction of its velocity in response to the magnetic force, the magnetic force remains perpendicular to the velocity. As we found in Section 6.1, if the force is always perpendicular to the velocity, the path of the particle is a circle! Active Figure 29.7 shows the particle moving in a circle in a plane perpendicular to the magnetic field.

The particle moves in a circle because the magnetic force $\vec{\mathbf{F}}_B$ is perpendicular to $\vec{\mathbf{v}}$ and $\vec{\mathbf{B}}$ and has a constant magnitude qvB. As Active Figure 29.7 illustrates, the rotation is counterclockwise for a positive charge in a magnetic field directed into the page. If q were negative, the rotation would be clockwise. We use the particle under a net force model to write Newton's second law for the particle:

$$\sum F = F_B = ma$$

Because the particle moves in a circle, we also model it as a particle in uniform circular motion and we replace the acceleration with centripetal acceleration:

$$F_B = qvB = \frac{mv^2}{r}$$

This expression leads to the following equation for the radius of the circular path:

$$r = \frac{mv}{qB} \tag{29.3}$$

That is, the radius of the path is proportional to the linear momentum mv of the particle and inversely proportional to the magnitude of the charge on the particle and to the magnitude of the magnetic field. The angular speed of the particle (from Eq. 10.10) is

$$\omega = \frac{v}{r} = \frac{qB}{m} \tag{29.4}$$

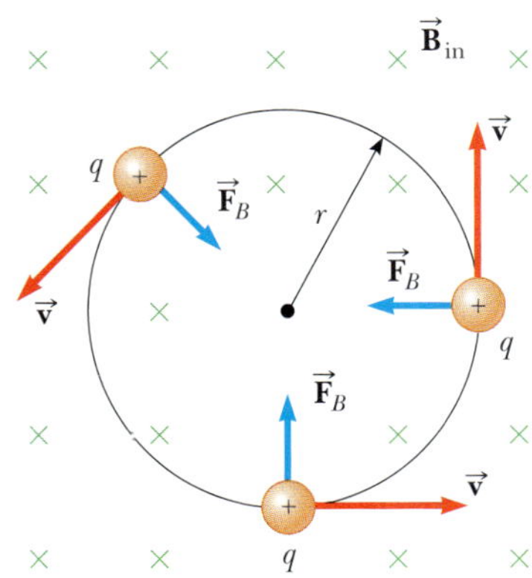

ACTIVE FIGURE 29.7

When the velocity of a charged particle is perpendicular to a uniform magnetic field, the particle moves in a circular path in a plane perpendicular to $\vec{\mathbf{B}}$. The magnetic force $\vec{\mathbf{F}}_B$ acting on the charge is always directed toward the center of the circle.

Sign in at www.thomsonedu.com and go to ThomsonNOW to adjust the mass, speed, and charge of the particle and the magnitude of the magnetic field and observe the resulting circular motion.

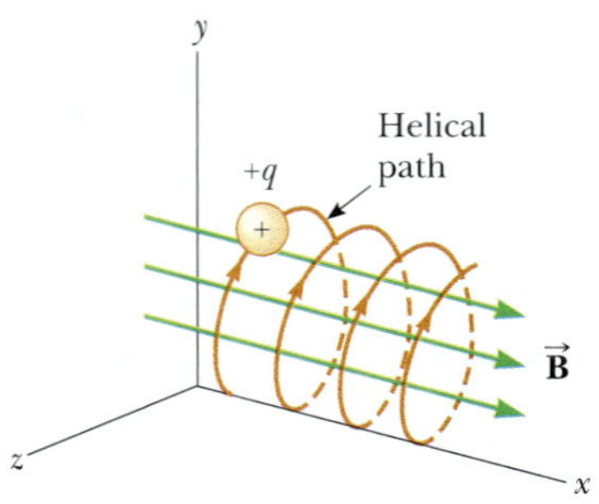

ACTIVE FIGURE 29.8

A charged particle having a velocity vector that has a component parallel to a uniform magnetic field moves in a helical path.

Sign in at www.thomsonedu.com and go to ThomsonNOW to adjust the x component of the velocity of the particle and observe the resulting helical motion.

The period of the motion (the time interval the particle requires to complete one revolution) is equal to the circumference of the circle divided by the speed of the particle:

$$T = \frac{2\pi r}{v} = \frac{2\pi}{\omega} = \frac{2\pi m}{qB} \qquad \textbf{(29.5)}$$

These results show that the angular speed of the particle and the period of the circular motion do not depend on the speed of the particle or on the radius of the orbit. The angular speed ω is often referred to as the **cyclotron frequency** because charged particles circulate at this angular frequency in the type of accelerator called a *cyclotron,* which is discussed in Section 29.3.

If a charged particle moves in a uniform magnetic field with its velocity at some arbitrary angle with respect to $\vec{\mathbf{B}}$, its path is a helix. For example, if the field is directed in the x direction as shown in Active Figure 29.8, there is no component of force in the x direction. As a result, $a_x = 0$, and the x component of velocity remains constant. The magnetic force $q\vec{\mathbf{v}} \times \vec{\mathbf{B}}$ causes the components v_y and v_z to change in time, however, and the resulting motion is a helix whose axis is parallel to the magnetic field. The projection of the path onto the yz plane (viewed along the x axis) is a circle. (The projections of the path onto the xy and xz planes are sinusoids!) Equations 29.3 to 29.5 still apply provided v is replaced by $v_\perp = \sqrt{v_y^2 + v_z^2}$.

Quick Quiz 29.2 A charged particle is moving perpendicular to a magnetic field in a circle with a radius r. **(i)** An identical particle enters the field, with $\vec{\mathbf{v}}$ perpendicular to $\vec{\mathbf{B}}$, but with a higher speed than the first particle. Compared with the radius of the circle for the first particle, is the radius of the circular path for the second particle (a) smaller, (b) larger, or (c) equal in size? **(ii)** The magnitude of the magnetic field is increased. From the same choices, compare the radius of the new circular path of the first particle with the radius of its initial path.

EXAMPLE 29.2 A Proton Moving Perpendicular to a Uniform Magnetic Field

A proton is moving in a circular orbit of radius 14 cm in a uniform 0.35-T magnetic field perpendicular to the velocity of the proton. Find the speed of the proton.

SOLUTION

Conceptualize From our discussion in this section, we know that the proton follows a circular path when moving in a uniform magnetic field.

Categorize We evaluate the speed of the proton using an equation developed in this section, so we categorize this example as a substitution problem.

Solve Equation 29.3 for the speed of the particle:

$$v = \frac{qBr}{m_p}$$

Substitute numerical values:

$$v = \frac{(1.60 \times 10^{-19}\ \text{C})(0.35\ \text{T})(0.14\ \text{m})}{1.67 \times 10^{-27}\ \text{kg}}$$

$$= 4.7 \times 10^6\ \text{m/s}$$

What If? What if an electron, rather than a proton, moves in a direction perpendicular to the same magnetic field with this same speed? Will the radius of its orbit be different?

Answer An electron has a much smaller mass than a proton, so the magnetic force should be able to change its velocity much more easily than that for the proton. Therefore, we expect the radius to be smaller. Equation 29.3 shows that r is proportional to m with q, B, and v the same for the electron as for the proton. Consequently, the radius will be smaller by the same factor as the ratio of masses m_e/m_p.

EXAMPLE 29.3 Bending an Electron Beam

In an experiment designed to measure the magnitude of a uniform magnetic field, electrons are accelerated from rest through a potential difference of 350 V and then enter a uniform magnetic field that is perpendicular to the velocity vector of the electrons. The electrons travel along a curved path because of the magnetic force exerted on them, and the radius of the path is measured to be 7.5 cm. (Such a curved beam of electrons is shown in Fig. 29.9.)

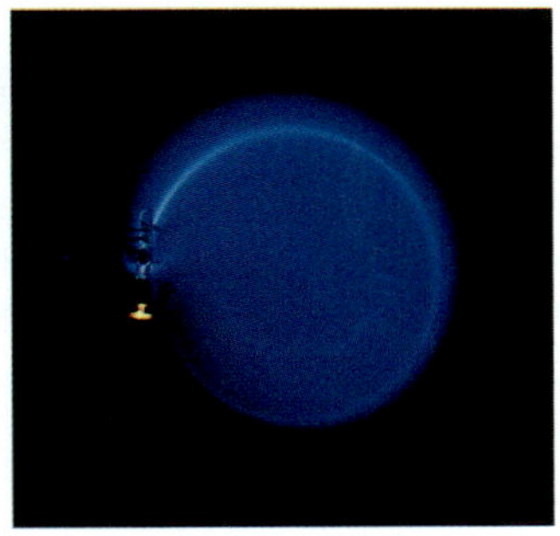

Henry Leap and Jim Lehman

Figure 29.9 (Example 29.3) The bending of an electron beam in a magnetic field.

(A) What is the magnitude of the magnetic field?

SOLUTION

Conceptualize With the help of Figures 29.7 and 29.9, visualize the circular motion of the electrons.

Categorize This example involves electrons accelerating from rest due to an electric force and then moving in a circular path due to a magnetic force. Equation 29.3 shows that we need the speed v of the electron to find the magnetic field magnitude, and v is not given. Consequently, we must find the speed of the electron based on the potential difference through which it is accelerated. To do so, we categorize the first part of the problem by modeling an electron and the electric field as an isolated system. Once the electron enters the magnetic field, we categorize the second part of the problem as one similar to those we have studied in this section.

Analyze Write the appropriate reduction of the conservation of energy equation, Equation 8.2, for the electron–electric field system:

$$\Delta K + \Delta U = 0$$

Substitute the appropriate initial and final energies:

$$(\tfrac{1}{2}m_e v^2 - 0) + (q\Delta V) = 0$$

Solve for the speed of the electron:

$$v = \sqrt{\frac{-2q\,\Delta V}{m_e}}$$

Substitute numerical values:

$$v = \sqrt{\frac{-2(-1.60 \times 10^{-19}\ \mathrm{C})(350\ \mathrm{V})}{9.11 \times 10^{-31}\ \mathrm{kg}}} = 1.11 \times 10^{7}\ \mathrm{m/s}$$

Now imagine the electron entering the magnetic field with this speed. Solve Equation 29.3 for the magnitude of the magnetic field:

$$B = \frac{m_e v}{er}$$

Substitute numerical values:

$$B = \frac{(9.11 \times 10^{-31}\ \mathrm{kg})(1.11 \times 10^{7}\ \mathrm{m/s})}{(1.60 \times 10^{-19}\ \mathrm{C})(0.075\ \mathrm{m})} = 8.4 \times 10^{-4}\ \mathrm{T}$$

(B) What is the angular speed of the electrons?

SOLUTION

Use Equation 10.10:

$$\omega = \frac{v}{r} = \frac{1.11 \times 10^{7}\ \mathrm{m/s}}{0.075\ \mathrm{m}} = 1.5 \times 10^{8}\ \mathrm{rad/s}$$

Finalize The angular speed can be represented as $\omega = (1.5 \times 10^8 \text{ rad/s})(1 \text{ rev}/2\pi \text{ rad}) = 2.4 \times 10^7 \text{ rev/s}$. The electrons travel around the circle 24 million times per second! This answer is consistent with the very high speed found in part (A).

What If? What if a sudden voltage surge causes the accelerating voltage to increase to 400 V? How does that affect the angular speed of the electrons, assuming the magnetic field remains constant?

Answer The increase in accelerating voltage ΔV causes the electrons to enter the magnetic field with a higher speed v. This higher speed causes them to travel in a circle with a larger radius r. The angular speed is the ratio of v to r. Both v and r increase by the same factor, so the effects cancel and the angular speed remains the same. Equation 29.4 is an expression for the cyclotron frequency, which is the same as the angular speed of the electrons. The cyclotron frequency depends only on the charge q, the magnetic field B, and the mass m_e, none of which have changed. Therefore, the voltage surge has no effect on the angular speed. (In reality, however, the voltage surge may also increase the magnetic field if the magnetic field is powered by the same source as the accelerating voltage. In that case, the angular speed increases according to Equation 29.4.)

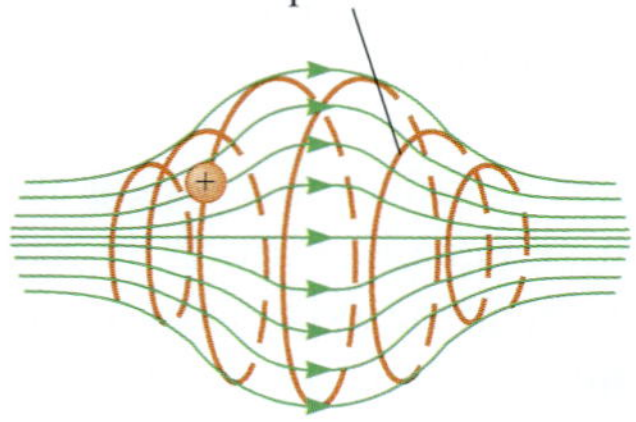

Figure 29.10 A charged particle moving in a nonuniform magnetic field (a magnetic bottle) spirals about the field and oscillates between the endpoints. The magnetic force exerted on the particle near either end of the bottle has a component that causes the particle to spiral back toward the center.

When charged particles move in a nonuniform magnetic field, the motion is complex. For example, in a magnetic field that is strong at the ends and weak in the middle such as that shown in Figure 29.10, the particles can oscillate between two positions. A charged particle starting at one end spirals along the field lines until it reaches the other end, where it reverses its path and spirals back. This configuration is known as a *magnetic bottle* because charged particles can be trapped within it. The magnetic bottle has been used to confine a *plasma,* a gas consisting of ions and electrons. Such a plasma-confinement scheme could fulfill a crucial role in the control of nuclear fusion, a process that could supply us in the future with an almost endless source of energy. Unfortunately, the magnetic bottle has its problems. If a large number of particles are trapped, collisions between them cause the particles to eventually leak from the system.

The Van Allen radiation belts consist of charged particles (mostly electrons and protons) surrounding the Earth in doughnut-shaped regions (Fig. 29.11). The particles, trapped by the Earth's nonuniform magnetic field, spiral around the field lines from pole to pole, covering the distance in only a few seconds. These particles originate mainly from the Sun, but some come from stars and other heavenly objects. For this reason, the particles are called *cosmic rays.* Most cosmic rays are deflected by the Earth's magnetic field and never reach the atmosphere. Some of the particles become trapped, however, and it is these particles that make up the Van Allen belts. When the particles are located over the poles, they sometimes collide with atoms in the atmosphere, causing the atoms to emit visible light. Such collisions are the origin of the beautiful Aurora Borealis, or Northern Lights, in the northern hemisphere and the Aurora Australis in the southern hemisphere. Auroras are usually confined to the polar regions because the Van Allen belts are nearest the Earth's surface there. Occasionally, though, solar activity causes larger numbers of charged particles to enter the belts and significantly distort the normal magnetic field lines associated with the Earth. In these situations, an aurora can sometimes be seen at lower latitudes.

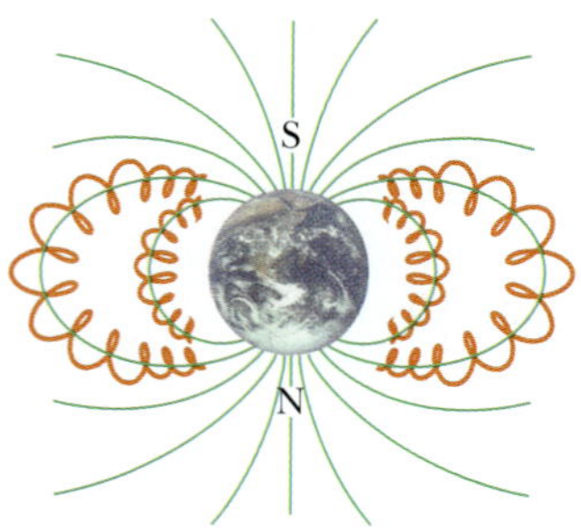

Figure 29.11 The Van Allen belts are made up of charged particles trapped by the Earth's nonuniform magnetic field. The magnetic field lines are in green, and the particle paths are in brown.

29.3 Applications Involving Charged Particles Moving in a Magnetic Field

A charge moving with a velocity $\vec{\mathbf{v}}$ in the presence of both an electric field $\vec{\mathbf{E}}$ and a magnetic field $\vec{\mathbf{B}}$ experiences both an electric force $q\vec{\mathbf{E}}$ and a magnetic force $q\vec{\mathbf{v}} \times \vec{\mathbf{B}}$. The total force (called the Lorentz force) acting on the charge is

Lorentz force ▶

$$\vec{\mathbf{F}} = q\vec{\mathbf{E}} + q\vec{\mathbf{v}} \times \vec{\mathbf{B}} \tag{29.6}$$

Velocity Selector

In many experiments involving moving charged particles, it is important that all particles move with essentially the same velocity, which can be achieved by applying a combination of an electric field and a magnetic field oriented as shown in Active Figure 29.12. A uniform electric field is directed to the right (in the plane of the page in Active Fig. 29.12), and a uniform magnetic field is applied in the direction perpendicular to the electric field (into the page in Active Fig. 29.12). If q is positive and the velocity $\vec{\mathbf{v}}$ is upward, the magnetic force $q\vec{\mathbf{v}} \times \vec{\mathbf{B}}$ is to the left and the electric force $q\vec{\mathbf{E}}$ is to the right. When the magnitudes of the two fields are chosen so that $qE = qvB$, the charged particle is modeled as a particle in equilibrium and moves in a straight vertical line through the region of the fields. From the expression $qE = qvB$, we find that

$$v = \frac{E}{B} \tag{29.7}$$

Only those particles having this speed pass undeflected through the mutually perpendicular electric and magnetic fields. The magnetic force exerted on particles moving at speeds greater than that is stronger than the electric force, and the particles are deflected to the left. Those moving at slower speeds are deflected to the right.

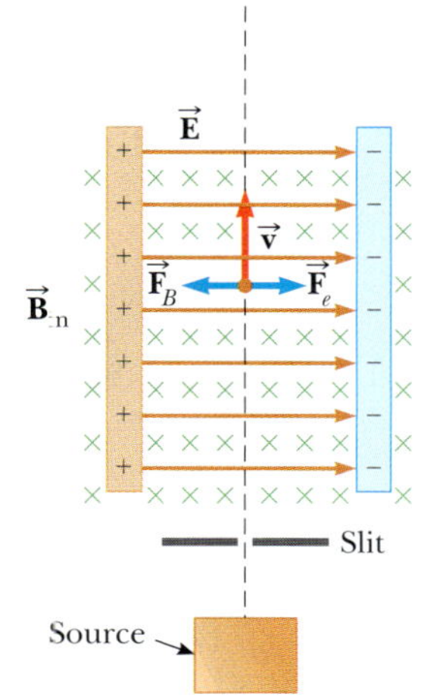

ACTIVE FIGURE 29.12

A velocity selector. When a positively charged particle is moving with velocity $\vec{\mathbf{v}}$ in the presence of a magnetic field directed into the page and an electric field directed to the right, it experiences an electric force $q\vec{\mathbf{E}}$ to the right and a magnetic force $q\vec{\mathbf{v}} \times \vec{\mathbf{B}}$ to the left.

Sign in at www.thomsonedu.com and go to ThomsonNOW to adjust the electric and magnetic fields and try to achieve straight-line motion for the charge

The Mass Spectrometer

A **mass spectrometer** separates ions according to their mass-to-charge ratio. In one version of this device, known as the *Bainbridge mass spectrometer,* a beam of ions first passes through a velocity selector and then enters a second uniform magnetic field $\vec{\mathbf{B}}_0$ that has the same direction as the magnetic field in the selector (Active Fig. 29.13). Upon entering the second magnetic field, the ions move in a semicircle of radius r before striking a detector array at P. If the ions are positively charged, the beam deflects to the left as Active Figure 29.13 shows. If the ions are negatively charged, the beam deflects to the right. From Equation 29.3, we can express the ratio m/q as

$$\frac{m}{q} = \frac{rB_0}{v}$$

Using Equation 29.7 gives

$$\frac{m}{q} = \frac{rB_0B}{E} \tag{29.8}$$

Therefore, we can determine m/q by measuring the radius of curvature and knowing the field magnitudes B, B_0, and E. In practice, one usually measures the masses of various isotopes of a given ion, with the ions all carrying the same charge q. In this way, the mass ratios can be determined even if q is unknown.

A variation of this technique was used by J. J. Thomson (1856–1940) in 1897 to measure the ratio e/m_e for electrons. Figure 29.14a (page 818) shows the basic apparatus he used. Electrons are accelerated from the cathode and pass through two slits. They then drift into a region of perpendicular electric and magnetic fields. The magnitudes of the two fields are first adjusted to produce an undeflected beam. When the magnetic field is turned off, the electric field produces a measurable beam deflection that is recorded on the fluorescent screen. From the size of the deflection and the measured values of E and B, the charge-to-mass ratio can be determined. The results of this crucial experiment represent the discovery of the electron as a fundamental particle of nature.

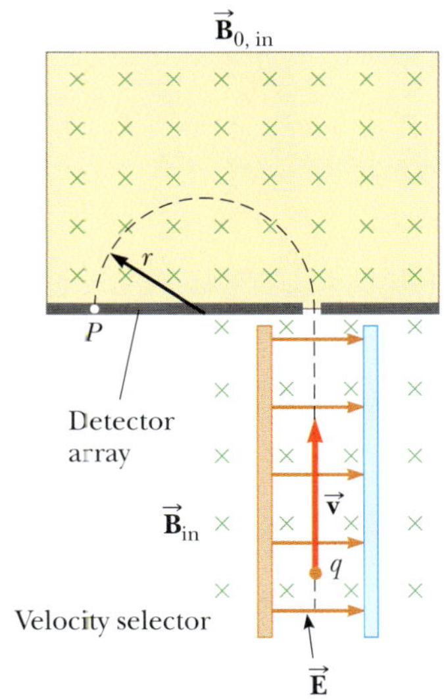

ACTIVE FIGURE 29.13

A mass spectrometer. Positively charged particles are sent first through a velocity selector and then into a region where the magnetic field $\vec{\mathbf{B}}_0$ causes the particles to move in a semicircular path and strike a detector array at P.

Sign in at www.thomsonedu.com and go to ThomsonNOW to predict where particles will strike the detector array.

The Cyclotron

A **cyclotron** is a device that can accelerate charged particles to very high speeds. The energetic particles produced are used to bombard atomic nuclei and thereby

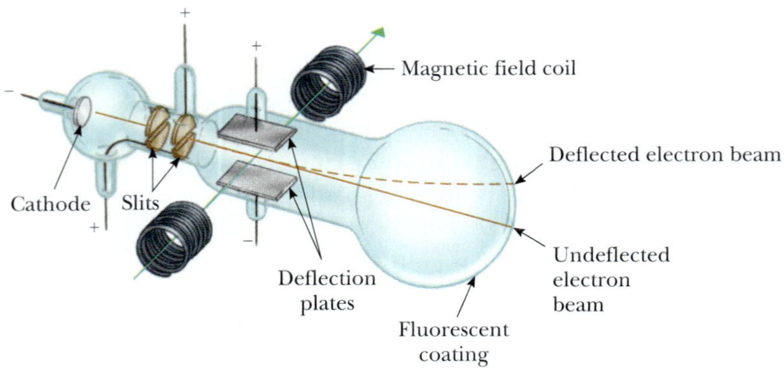

(a)

(b)

Figure 29.14 (a) Thomson's apparatus for measuring e/m_e. Electrons are accelerated from the cathode, pass through two slits, and are deflected by both an electric field and a magnetic field (directed perpendicular to the electric field). The beam of electrons then strikes a fluorescent screen. (b) J. J. Thomson *(left)* in the Cavendish Laboratory, University of Cambridge. The man on the right, Frank Baldwin Jewett, is a distant relative of John W. Jewett, Jr., coauthor of this text.

PITFALL PREVENTION 29.1

The Cyclotron Is Not State-of-the-Art Technology

The cyclotron is important historically because it was the first particle accelerator to produce particles with very high speeds. Cyclotrons are still in use in medical applications, but most accelerators currently in research use are not cyclotrons. Research accelerators work on a different principle and are generally called *synchrotrons.*

produce nuclear reactions of interest to researchers. A number of hospitals use cyclotron facilities to produce radioactive substances for diagnosis and treatment.

Both electric and magnetic forces play a key role in the operation of a cyclotron, a schematic drawing of which is shown in Figure 29.15a. The charges move inside two semicircular containers D_1 and D_2, referred to as *dees* because of their shape like the letter D. A high-frequency alternating potential difference is applied to the dees, and a uniform magnetic field is directed perpendicular to them. A positive ion released at P near the center of the magnet in one dee moves in a semicircular path (indicated by the dashed brown line in the drawing) and arrives back at the gap in a time interval $T/2$, where T is the time interval needed to make one complete trip around the two dees, given by Equation 29.5. The frequency of the applied potential difference is adjusted so that the polarity of the dees is reversed in the same time interval during which the ion travels around one dee. If the applied potential difference is adjusted such that D_2 is at a lower electric potential than D_1 by an amount ΔV, the ion accelerates across the gap to D_2 and its kinetic energy increases by an amount $q\,\Delta V$. It then moves around D_2 in a semicircular path of greater radius (because its speed has increased). After a time interval $T/2$, it again arrives at the gap between the dees. By this time, the polarity across the dees has again been reversed and the ion is given another "kick" across

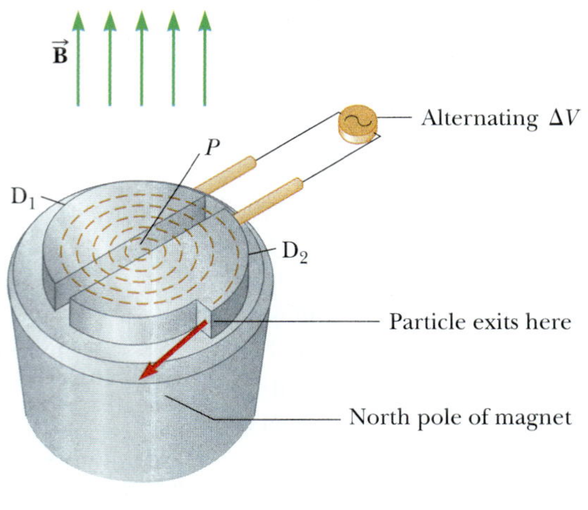

(a)

(b)

Figure 29.15 (a) A cyclotron consists of an ion source at P, two dees D_1 and D_2 across which an alternating potential difference is applied, and a uniform magnetic field. (The south pole of the magnet is not shown.) The brown, dashed, curved lines represent the path of the particles. (b) The first cyclotron, invented by E. O. Lawrence and M. S. Livingston in 1934.

the gap. The motion continues so that for each half-circle trip around one dee, the ion gains additional kinetic energy equal to $q\,\Delta V$. When the radius of its path is nearly that of the dees, the energetic ion leaves the system through the exit slit. The cyclotron's operation depends on T being independent of the speed of the ion and of the radius of the circular path (Eq. 29.5).

We can obtain an expression for the kinetic energy of the ion when it exits the cyclotron in terms of the radius R of the dees. From Equation 29.3 we know that $v = qBR/m$. Hence, the kinetic energy is

$$K = \tfrac{1}{2}mv^2 = \frac{q^2B^2R^2}{2m} \qquad \textbf{(29.9)}$$

When the energy of the ions in a cyclotron exceeds about 20 MeV, relativistic effects come into play. (Such effects are discussed in Chapter 39.) Observations show that T increases and the moving ions do not remain in phase with the applied potential difference. Some accelerators overcome this problem by modifying the period of the applied potential difference so that it remains in phase with the moving ions.

29.4 Magnetic Force Acting on a Current-Carrying Conductor

If a magnetic force is exerted on a single charged particle when the particle moves through a magnetic field, it should not surprise you that a current-carrying wire also experiences a force when placed in a magnetic field. The current is a collection of many charged particles in motion; hence, the resultant force exerted by the field on the wire is the vector sum of the individual forces exerted on all the charged particles making up the current. The force exerted on the particles is transmitted to the wire when the particles collide with the atoms making up the wire.

One can demonstrate the magnetic force acting on a current-carrying conductor by hanging a wire between the poles of a magnet as shown in Figure 29.16a. For ease in visualization, part of the horseshoe magnet in part (a) is removed to show the end face of the south pole in parts (b), (c), and (d) of Figure 29.16. The magnetic field is directed into the page and covers the region within the shaded squares. When the current in the wire is zero, the wire remains vertical as in Figure 29.16b. When the wire carries a current directed upward as in Figure 29.16c, however, the wire deflects to the left. If the current is reversed as in Figure 29.16d, the wire deflects to the right.

Let's quantify this discussion by considering a straight segment of wire of length L and cross-sectional area A carrying a current I in a uniform magnetic field $\vec{\mathbf{B}}$ as in

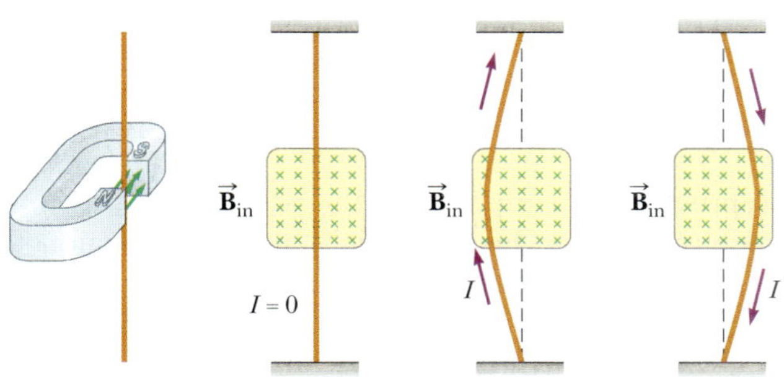

(a) (b) (c) (d)

Figure 29.16 (a) A wire suspended vertically between the poles of a magnet. (b) The setup shown in (a) as seen looking at the south pole of the magnet so that the magnetic field (green crosses) is directed into the page. When there is no current in the wire, the wire remains vertical. (c) When the current is upward, the wire deflects to the left. (d) When the current is downward, the wire deflects to the right.

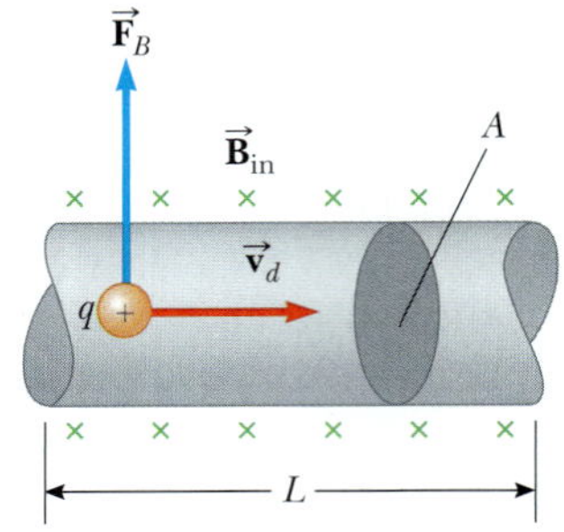

Figure 29.17 A segment of a current-carrying wire in a magnetic field $\vec{B}$. The magnetic force exerted on each charge making up the current is $q\vec{v}_d \times \vec{B}$, and the net force on the segment of length L is $I\vec{L} \times \vec{B}$.

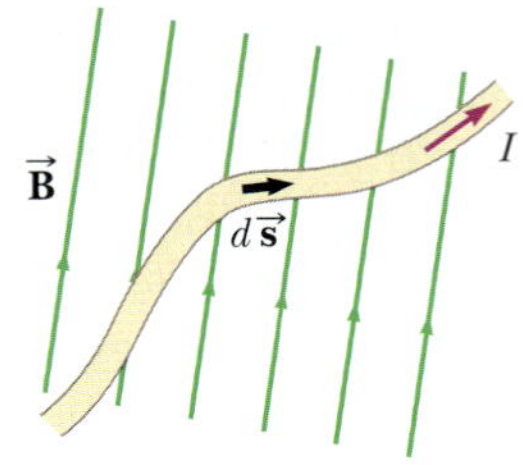

Figure 29.18 A wire segment of arbitrary shape carrying a current I in a magnetic field $\vec{B}$ experiences a magnetic force. The magnetic force on any segment $d\vec{s}$ is $I\,d\vec{s} \times \vec{B}$ and is directed out of the page. You should use the right-hand rule to confirm this force direction.

Figure 29.17. The magnetic force exerted on a charge q moving with a drift velocity $\vec{v}_d$ is $q\vec{v}_d \times \vec{B}$. To find the total force acting on the wire, we multiply the force $q\vec{v}_d \times \vec{B}$ exerted on one charge by the number of charges in the segment. Because the volume of the segment is AL, the number of charges in the segment is nAL, where n is the number of charges per unit volume. Hence, the total magnetic force on the wire of length L is

$$\vec{F}_B = (q\vec{v}_d \times \vec{B})nAL$$

We can write this expression in a more convenient form by noting that, from Equation 27.4, the current in the wire is $I = nqv_dA$. Therefore,

Force on a segment of current-carrying wire in a uniform magnetic field ▶

$$\vec{F}_B = I\vec{L} \times \vec{B} \quad (29.10)$$

where $\vec{L}$ is a vector that points in the direction of the current I and has a magnitude equal to the length L of the segment. This expression applies only to a straight segment of wire in a uniform magnetic field.

Now consider an arbitrarily shaped wire segment of uniform cross section in a magnetic field as shown in Figure 29.18. It follows from Equation 29.10 that the magnetic force exerted on a small segment of vector length $d\vec{s}$ in the presence of a field $\vec{B}$ is

$$d\vec{F}_B = I\,d\vec{s} \times \vec{B} \quad (29.11)$$

where $d\vec{F}_B$ is directed out of the page for the directions of $\vec{B}$ and $d\vec{s}$ in Figure 29.18. Equation 29.11 can be considered as an alternative definition of $\vec{B}$. That is, we can define the magnetic field $\vec{B}$ in terms of a measurable force exerted on a current element, where the force is a maximum when $\vec{B}$ is perpendicular to the element and zero when $\vec{B}$ is parallel to the element.

To calculate the total force $\vec{F}_B$ acting on the wire shown in Figure 29.18, we integrate Equation 29.11 over the length of the wire:

$$\vec{F}_B = I\int_a^b d\vec{s} \times \vec{B} \quad (29.12)$$

where a and b represent the endpoints of the wire. When this integration is carried out, the magnitude of the magnetic field and the direction the field makes with the vector $d\vec{s}$ may differ at different points.

Quick Quiz 29.3 A wire carries current in the plane of this paper toward the top of the page. The wire experiences a magnetic force toward the right edge of the page. Is the direction of the magnetic field causing this force (a) in the plane of the page and toward the left edge, (b) in the plane of the page and toward the bottom edge, (c) upward out of the page, or (d) downward into the page?

EXAMPLE 29.4 Force on a Semicircular Conductor

A wire bent into a semicircle of radius R forms a closed circuit and carries a current I. The wire lies in the xy plane, and a uniform magnetic field is directed along the positive y axis as in Figure 29.19. Find the magnitude and direction of the magnetic force acting on the straight portion of the wire and on the curved portion.

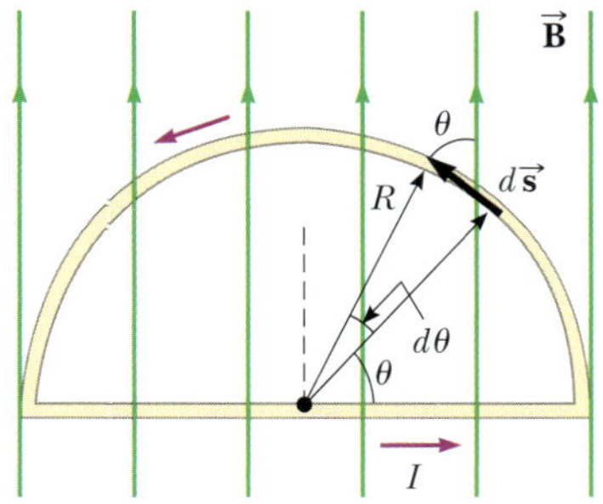

Figure 29.19 (Example 29.4) The magnetic force on the straight portion of the loop is directed out of the page, and the magnetic force on the curved portion is directed into the page.

SOLUTION

Conceptualize Using the right-hand rule for cross products, we see that the force $\vec{\mathbf{F}}_1$ on the straight portion of the wire is out of the page and the force $\vec{\mathbf{F}}_2$ on the curved portion is into the page. Is $\vec{\mathbf{F}}_2$ larger in magnitude than $\vec{\mathbf{F}}_1$ because the length of the curved portion is longer than that of the straight portion?

Categorize Because we are dealing with a current-carrying wire in a magnetic field rather than a single charged particle, we must use Equation 29.12 to find the total force on each portion of the wire.

Analyze Note that $d\vec{\mathbf{s}}$ is perpendicular to $\vec{\mathbf{B}}$ everywhere on the straight portion of the wire. Use Equation 29.12 to find the force on this portion:

$$\vec{\mathbf{F}}_1 = I\int_a^b d\vec{\mathbf{s}} \times \vec{\mathbf{B}} = I\int_0^{2R} B\,ds\,\hat{\mathbf{k}} = 2IRB\hat{\mathbf{k}}$$

To find the magnetic force on the curved part, first write an expression for the magnetic force $d\vec{\mathbf{F}}_2$ on the element $d\vec{\mathbf{s}}$ in Figure 29.19:

$$(1) \quad d\vec{\mathbf{F}}_2 = I\,d\vec{\mathbf{s}} \times \vec{\mathbf{B}} = -IB\sin\theta\,ds\,\hat{\mathbf{k}}$$

From the geometry in Figure 29.19, write an expression for ds:

$$(2) \quad ds = R\,d\theta$$

Substitute Equation (2) into Equation (1) and integrate over the angle θ from 0 to π:

$$\vec{\mathbf{F}}_2 = -\int_0^{\pi} IRB\sin\theta\,d\theta\,\hat{\mathbf{k}} = -IRB\int_0^{\pi}\sin\theta\,d\theta\,\hat{\mathbf{k}} = -IRB\left[-\cos\theta\right]_0^{\pi}\hat{\mathbf{k}}$$

$$= IRB(\cos\pi - \cos 0)\hat{\mathbf{k}} = IRB(-1-1)\hat{\mathbf{k}} = -2IRB\hat{\mathbf{k}}$$

Finalize Two very important general statements follow from this example. First, the force on the curved portion is the same in magnitude as the force on a straight wire between the same two points. In general, **the magnetic force on a curved current-carrying wire in a uniform magnetic field is equal to that on a straight wire connecting the endpoints and carrying the same current.** Furthermore, $\vec{\mathbf{F}}_1 + \vec{\mathbf{F}}_2 = 0$ is also a general result: **the net magnetic force acting on any closed current loop in a uniform magnetic field is zero.**

29.5 Torque on a Current Loop in a Uniform Magnetic Field

In Section 29.4, we showed how a magnetic force is exerted on a current-carrying conductor placed in a magnetic field. With that as a starting point, we now show that a torque is exerted on a current loop placed in a magnetic field.

Consider a rectangular loop carrying a current I in the presence of a uniform magnetic field directed parallel to the plane of the loop as shown in Figure 29.20a (page 822). No magnetic forces act on sides ① and ③ because these wires are parallel to the field; hence, $\vec{\mathbf{L}} \times \vec{\mathbf{B}} = 0$ for these sides. Magnetic forces do, however, act on sides ② and ④ because these sides are oriented perpendicular to the field. The magnitude of these forces is, from Equation 29.10,

$$F_2 = F_4 = IaB$$

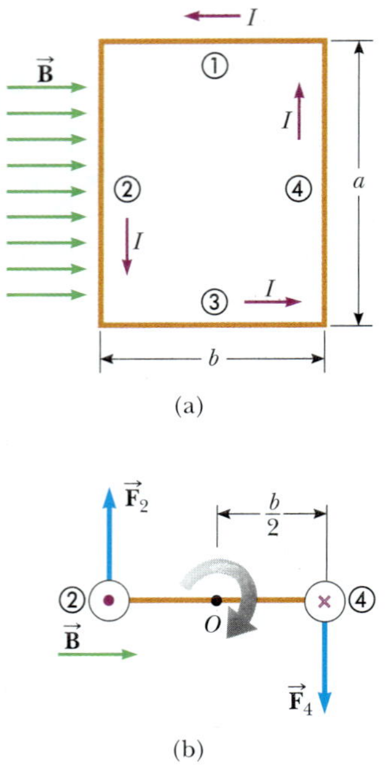

Figure 29.20 (a) Overhead view of a rectangular current loop in a uniform magnetic field. No magnetic forces are acting on sides ① and ③ because these sides are parallel to $\vec{\mathbf{B}}$. Forces are acting on sides ② and ④, however. (b) Edge view of the loop sighting down sides ② and ④ shows that the magnetic forces $\vec{\mathbf{F}}_2$ and $\vec{\mathbf{F}}_4$ exerted on these sides create a torque that tends to twist the loop clockwise. The purple dot in the left circle represents current in wire ② coming toward you; the purple cross in the right circle represents current in wire ④ moving away from you.

The direction of $\vec{\mathbf{F}}_2$, the magnetic force exerted on wire ②, is out of the page in the view shown in Figure 29.20a and that of $\vec{\mathbf{F}}_4$, the magnetic force exerted on wire ④, is into the page in the same view. If we view the loop from side ③ and sight along sides ② and ④, we see the view shown in Figure 29.20b, and the two magnetic forces $\vec{\mathbf{F}}_2$ and $\vec{\mathbf{F}}_4$ are directed as shown. Notice that the two forces point in opposite directions but are *not* directed along the same line of action. If the loop is pivoted so that it can rotate about point O, these two forces produce about O a torque that rotates the loop clockwise. The magnitude of this torque $\tau_{\max}$ is

$$\tau_{\max} = F_2\frac{b}{2} + F_4\frac{b}{2} = (IaB)\frac{b}{2} + (IaB)\frac{b}{2} = IabB$$

where the moment arm about O is $b/2$ for each force. Because the area enclosed by the loop is $A = ab$, we can express the maximum torque as

$$\tau_{\max} = IAB \tag{29.13}$$

This maximum-torque result is valid only when the magnetic field is parallel to the plane of the loop. The sense of the rotation is clockwise when viewed from side ③ as indicated in Figure 29.20b. If the current direction were reversed, the force directions would also reverse and the rotational tendency would be counterclockwise.

Now suppose the uniform magnetic field makes an angle $\theta < 90°$ with a line perpendicular to the plane of the loop as in Active Figure 29.21. For convenience, let's assume $\vec{\mathbf{B}}$ is perpendicular to sides ② and ④. In this case, the magnetic forces $\vec{\mathbf{F}}_1$ and $\vec{\mathbf{F}}_3$ exerted on sides ① and ③ cancel each other and produce no torque because they pass through a common origin. The magnetic forces $\vec{\mathbf{F}}_2$ and $\vec{\mathbf{F}}_4$ acting on sides ② and ④, however, produce a torque about *any point.* Referring to the end view shown in Active Figure 29.21, we see that the moment arm of $\vec{\mathbf{F}}_2$ about the point O is equal to $(b/2)\sin\theta$. Likewise, the moment arm of $\vec{\mathbf{F}}_4$ about O is also $(b/2)\sin\theta$. Because $F_2 = F_4 = IaB$, the magnitude of the net torque about O is

$$\begin{aligned}\tau &= F_2\frac{b}{2}\sin\theta + F_4\frac{b}{2}\sin\theta \\ &= IaB\left(\frac{b}{2}\sin\theta\right) + IaB\left(\frac{b}{2}\sin\theta\right) = IabB\sin\theta \\ &= IAB\sin\theta\end{aligned}$$

where $A = ab$ is the area of the loop. This result shows that the torque has its maximum value IAB when the field is perpendicular to the normal to the plane of the loop ($\theta = 90°$) as discussed with regard to Figure 29.20 and is zero when the field is parallel to the normal to the plane of the loop ($\theta = 0$).

A convenient expression for the torque exerted on a loop placed in a uniform magnetic field $\vec{\mathbf{B}}$ is

Torque on a current loop in a magnetic field ▶

$$\vec{\boldsymbol{\tau}} = I\vec{\mathbf{A}} \times \vec{\mathbf{B}} \tag{29.14}$$

ACTIVE FIGURE 29.21

An end view of the loop in Figure 29.20b rotated through an angle with respect to the magnetic field. If $\vec{\mathbf{B}}$ is at an angle θ with respect to vector $\vec{\mathbf{A}}$, which is perpendicular to the plane of the loop, the torque is $IAB\sin\theta$ where the magnitude of $\vec{\mathbf{A}}$ is A, the area of the loop.

Sign in at www.thomsonedu.com and go to ThomsonNOW to choose the current in the loop, the magnetic field, and the initial orientation of the loop and observe the subsequent motion.

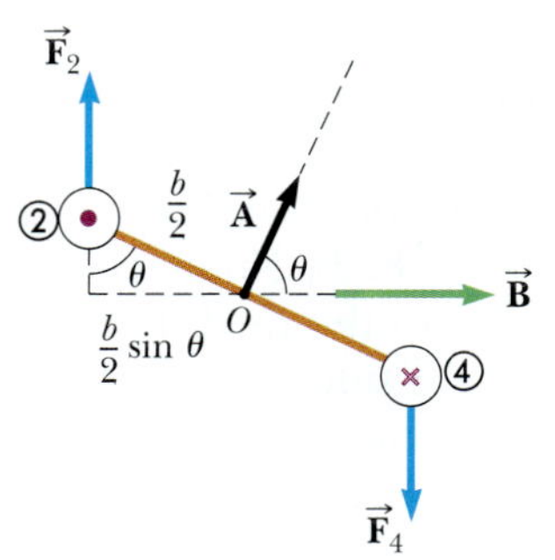

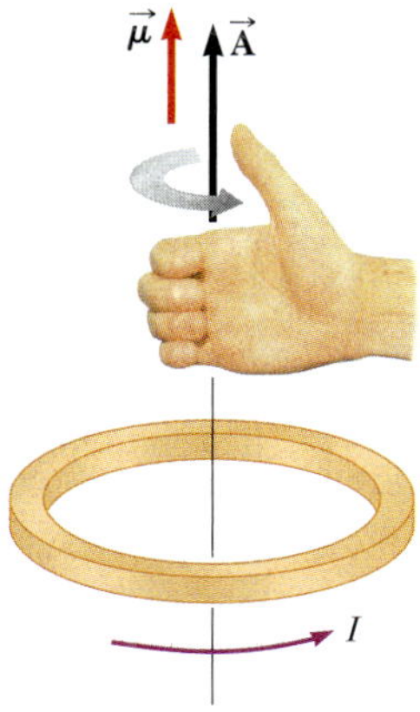

Figure 29.22 Right-hand rule for determining the direction of the vector $\vec{\mathbf{A}}$. The direction of the magnetic moment $\vec{\mu}$ is the same as the direction of $\vec{\mathbf{A}}$.

where $\vec{\mathbf{A}}$, the vector shown in Active Figure 29.21, is perpendicular to the plane of the loop and has a magnitude equal to the area of the loop. To determine the direction of $\vec{\mathbf{A}}$, use the right-hand rule described in Figure 29.22. When you curl the fingers of your right hand in the direction of the current in the loop, your thumb points in the direction of $\vec{\mathbf{A}}$. Active Figure 29.21 shows that the loop tends to rotate in the direction of decreasing values of θ (that is, such that the area vector $\vec{\mathbf{A}}$ rotates toward the direction of the magnetic field).

The product $I\vec{\mathbf{A}}$ is defined to be the **magnetic dipole moment** $\vec{\mu}$ (often simply called the "magnetic moment") of the loop:

$$\vec{\mu} \equiv I\vec{\mathbf{A}} \tag{29.15}$$

◀ Magnetic dipole moment of a current loop

The SI unit of magnetic dipole moment is the ampere-meter^2 ($A \cdot m^2$). If a coil of wire contains N loops of the same area, the magnetic moment of the coil is

$$\vec{\mu}_{\text{coil}} = NI\vec{\mathbf{A}} \tag{29.16}$$

Using Equation 29.15, we can express the torque exerted on a current-carrying loop in a magnetic field $\vec{\mathbf{B}}$ as

$$\vec{\tau} = \vec{\mu} \times \vec{\mathbf{B}} \tag{29.17}$$

◀ Torque on a magnetic moment in a magnetic field

This result is analogous to Equation 26.18, $\vec{\tau} = \vec{\mathbf{p}} \times \vec{\mathbf{E}}$, for the torque exerted on an electric dipole in the presence of an electric field $\vec{\mathbf{E}}$, where $\vec{\mathbf{p}}$ is the electric dipole moment.

Although we obtained the torque for a particular orientation of $\vec{\mathbf{B}}$ with respect to the loop, the equation $\vec{\tau} = \vec{\mu} \times \vec{\mathbf{B}}$ is valid for any orientation. Furthermore, although we derived the torque expression for a rectangular loop, the result is valid for a loop of any shape. The torque on an N-turn coil is given by Equation 29.17 by using Equation 29.16 for the magnetic moment.

In Section 26.6, we found that the potential energy of a system of an electric dipole in an electric field is given by $U = -\vec{\mathbf{p}} \cdot \vec{\mathbf{E}}$. This energy depends on the orientation of the dipole in the electric field. Likewise, the potential energy of a system of a magnetic dipole in a magnetic field depends on the orientation of the dipole in the magnetic field and is given by

$$U = -\vec{\mu} \cdot \vec{\mathbf{B}} \tag{29.18}$$

◀ Potential energy of a system of a magnetic moment in a magnetic field

This expression shows that the system has its lowest energy $U_{\text{min}} = -\mu B$ when $\vec{\mu}$ points in the same direction as $\vec{\mathbf{B}}$. The system has its highest energy $U_{\text{max}} = +\mu B$ when $\vec{\mu}$ points in the direction opposite $\vec{\mathbf{B}}$.

The torque on a current loop causes the loop to rotate; this effect is exploited practically in a **motor.** Energy enters the motor by electrical transmission, and the rotating coil can do work on some device external to the motor. For example, the motor in an car's electrical window system does work on the windows, applying a force on them and moving them up or down through some displacement. We will discuss motors in more detail in Section 31.5.

Quick Quiz 29.4 **(i)** Rank the magnitudes of the torques acting on the rectangular loops (a), (b), and (c) shown edge-on in Figure 29.23 from highest to lowest. All loops are identical and carry the same current. **(ii)** Rank the magnitudes of the net forces acting on the rectangular loops shown in Figure 29.23 from highest to lowest.

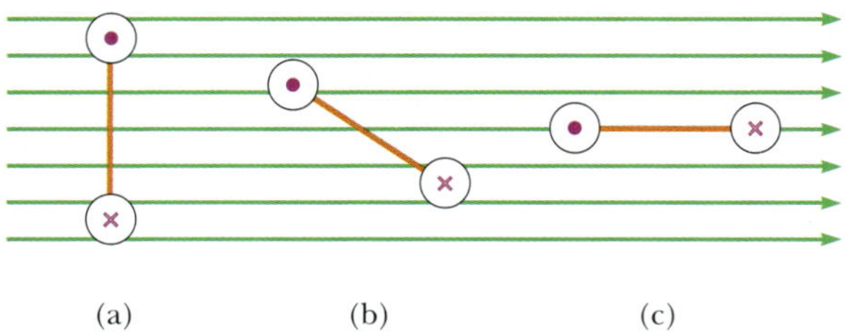

Figure 29.23 (Quick Quiz 29.4) Which current loop (seen edge-on) experiences the greatest torque, (a), (b), or (c)? Which experiences the greatest net force?

EXAMPLE 29.5 The Magnetic Dipole Moment of a Coil

A rectangular coil of dimensions 5.40 cm × 8.50 cm consists of 25 turns of wire and carries a current of 15.0 mA. A 0.350-T magnetic field is applied parallel to the plane of the coil.

(A) Calculate the magnitude of the magnetic dipole moment of the coil.

SOLUTION

Conceptualize The magnetic moment of the coil is independent of any magnetic field in which the loop resides, so it depends only on the geometry of the loop and the current it carries.

Categorize We evaluate quantities based on equations developed in this section, so we categorize this example as a substitution problem.

Use Equation 29.16 to calculate the magnetic moment:

$$\mu_{\text{coil}} = NIA = (25)(15.0 \times 10^{-3}\text{ A})(0.054\,0\text{ m})(0.085\,0\text{ m})$$
$$= \boxed{1.72 \times 10^{-3}\text{ A}\cdot\text{m}^2}$$

(B) What is the magnitude of the torque acting on the loop?

SOLUTION

Use Equation 29.17, noting that $\vec{\mathbf{B}}$ is perpendicular to $\vec{\boldsymbol{\mu}}_{\text{coil}}$:

$$\tau = \mu_{\text{coil}}B = (1.72 \times 10^{-3}\text{ A}\cdot\text{m}^2)(0.350\text{ T})$$
$$= \boxed{6.02 \times 10^{-4}\text{ N}\cdot\text{m}}$$

EXAMPLE 29.6 Rotating a Coil

Consider the loop of wire in Figure 29.24a. Imagine it is pivoted along side ④, which is parallel to the z axis and fastened so that side ④ remains fixed and the rest of the loop hangs vertically but can rotate around side ④ (Fig. 29.24b). The mass of the loop is 50.0 g, and the sides are of lengths $a = 0.200$ m and $b = 0.100$ m. The loop carries a current of 3.50 A and is immersed in a vertical uniform magnetic field of magnitude 0.010 0 T in the $+y$ direction (Fig. 29.24c). What angle does the plane of the loop make with the vertical?

SOLUTION

Conceptualize In the side view of Figure 29.24b, notice that the magnetic moment of the loop is to the left. Therefore, when the loop is in the magnetic field, the magnetic torque on the loop causes it to rotate in a clockwise direction around side ④, which we choose as the rotation axis. Imagine the loop making this clockwise rotation so that the plane of the loop is at some angle θ to the vertical as in Figure 29.24c. The gravitational force on the loop exerts a torque that would cause a rotation in the counterclockwise direction if the magnetic field were turned off.

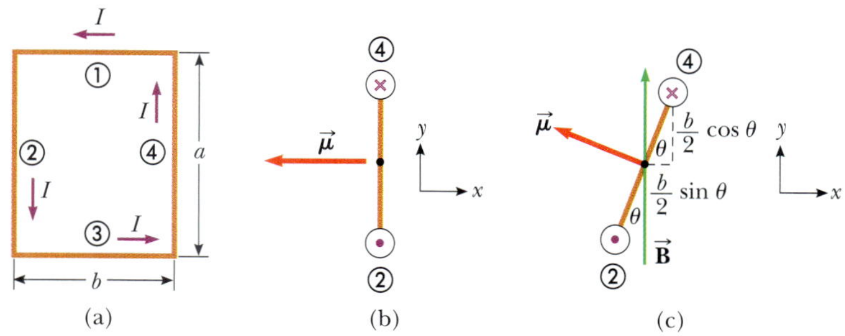

Figure 29.24 (Example 29.6) (a) Overhead view of a rectangular current loop in a uniform magnetic field. This figure is similar to the situations in Figure 29.20 and 29.21. (b) Edge view of the loop sighting down sides ② and ④. The loop hangs vertically and is pivoted so that it can rotate around side ④. (c) An end view of the loop in (b) rotated through an angle with respect to the horizontal when it is placed in a magnetic field. The magnetic torque causes the loop to rotate in a clockwise direction around side ④, whereas the gravitational torque causes a counterclockwise rotation.

Categorize At some angle of the loop, the two torques described in the *Conceptualize* step are equal in magnitude and the loop is at rest. We therefore model the loop as a rigid object in equilibrium.

Analyze Evaluate the magnetic torque on the loop from Equation 29.17:

$$\vec{\boldsymbol{\tau}}_B = \vec{\boldsymbol{\mu}} \times \vec{\mathbf{B}} = -\mu B \sin(90° - \theta)\hat{\mathbf{k}} = -IAB\cos\theta\,\hat{\mathbf{k}} = -IabB\cos\theta\,\hat{\mathbf{k}}$$

Evaluate the gravitational torque on the loop, noting that the gravitational force can be modeled to act at the center of the loop:

$$\vec{\boldsymbol{\tau}}_g = \vec{\mathbf{r}} \times m\vec{\mathbf{g}} = mg\frac{b}{2}\sin\theta\,\hat{\mathbf{k}}$$

From the rigid body in equilibrium model, add the torques and set the net torque equal to zero:

$$\sum \vec{\boldsymbol{\tau}} = -IabB\cos\theta\,\hat{\mathbf{k}} + mg\frac{b}{2}\sin\theta\,\hat{\mathbf{k}} = 0$$

Solve for $\tan\theta$:

$$IabB\cos\theta = mg\frac{b}{2}\sin\theta \quad\rightarrow\quad \tan\theta = \frac{2IaB}{mg}$$

Substitute numerical values:

$$\theta = \tan^{-1}\left(\frac{2IaB}{mg}\right)$$

$$= \tan^{-1}\left[\frac{2(3.50\text{ A})(0.200\text{ m})(0.010\,0\text{ T})}{(0.050\,0\text{ kg})(9.80\text{ m/s}^2)}\right] = \boxed{1.64°}$$

Finalize The angle is relatively small, so the loop still hangs almost vertically. If the current I or the magnetic field B is increased, however, the angle increases as the magnetic torque becomes stronger.

29.6 The Hall Effect

When a current-carrying conductor is placed in a magnetic field, a potential difference is generated in a direction perpendicular to both the current and the magnetic field. This phenomenon, first observed by Edwin Hall (1855–1938) in 1879, is known as the *Hall effect.* The arrangement for observing the Hall effect consists of a flat conductor carrying a current I in the x direction as shown in Figure 29.25 (page 826). A uniform magnetic field $\vec{\mathbf{B}}$ is applied in the y direction. If the charge

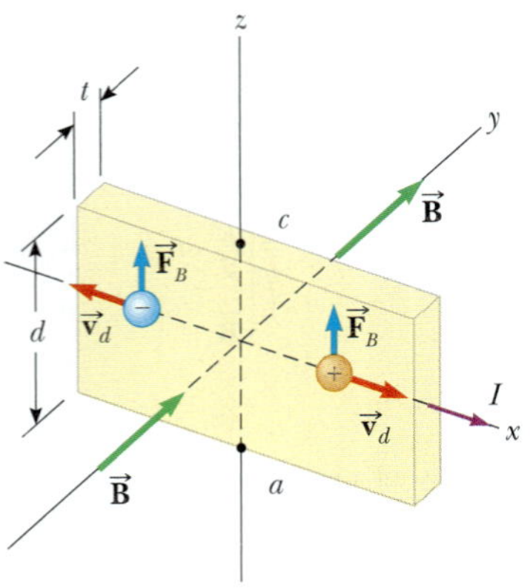

Figure 29.25 To observe the Hall effect, a magnetic field is applied to a current-carrying conductor. When I is in the x direction and $\vec{B}$ in the y direction, both positive and negative charge carriers are deflected upward in the magnetic field. The Hall voltage is measured between points a and c.

carriers are electrons moving in the negative x direction with a drift velocity $\vec{v}_d$, they experience an upward magnetic force $\vec{F}_B = q\vec{v}_d \times \vec{B}$, are deflected upward, and accumulate at the upper edge of the flat conductor, leaving an excess of positive charge at the lower edge (Fig. 29.26a). This accumulation of charge at the edges establishes an electric field in the conductor and increases until the electric force on carriers remaining in the bulk of the conductor balances the magnetic force acting on the carriers. When this equilibrium condition is reached, the electrons are no longer deflected upward. A sensitive voltmeter connected across the sample as shown in Figure 29.26 can measure the potential difference, known as the **Hall voltage** ΔV_H, generated across the conductor.

If the charge carriers are positive and hence move in the positive x direction (for rightward current) as shown in Figures 29.25 and 29.26b, they also experience an upward magnetic force $q\vec{v}_d \times \vec{B}$, which produces a buildup of positive charge on the upper edge and leaves an excess of negative charge on the lower edge. Hence, the sign of the Hall voltage generated in the sample is opposite the sign of the Hall voltage resulting from the deflection of electrons. The sign of the charge carriers can therefore be determined from measuring the polarity of the Hall voltage.

In deriving an expression for the Hall voltage, first note that the magnetic force exerted on the carriers has magnitude qv_dB. In equilibrium, this force is balanced by the electric force qE_H, where E_H is the magnitude of the electric field due to the charge separation (sometimes referred to as the *Hall field*). Therefore,

$$qv_dB = qE_H$$

$$E_H = v_dB$$

If d is the width of the conductor, the Hall voltage is

$$\Delta V_H = E_Hd = v_dBd \tag{29.19}$$

Therefore, the measured Hall voltage gives a value for the drift speed of the charge carriers if d and B are known.

We can obtain the charge carrier density n by measuring the current in the sample. From Equation 27.4, we can express the drift speed as

$$v_d = \frac{I}{nqA} \tag{29.20}$$

where A is the cross-sectional area of the conductor. Substituting Equation 29.20 into Equation 29.19 gives

$$\Delta V_H = \frac{IBd}{nqA} \tag{29.21}$$

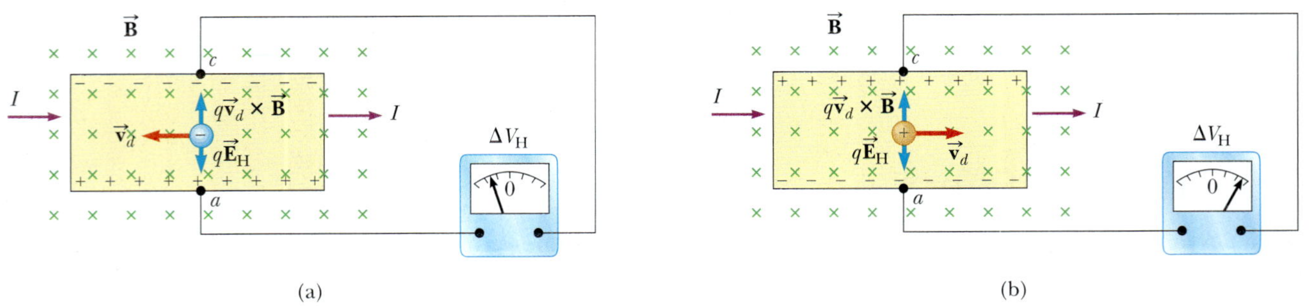

Figure 29.26 (a) When the charge carriers in a Hall-effect apparatus are negative, the upper edge of the conductor becomes negatively charged and c is at a lower electric potential than a. (b) When the charge carriers are positive, the upper edge becomes positively charged and c is at a higher potential than a. In either case, the charge carriers are no longer deflected when the edges become sufficiently charged that there is a balance on the charge carriers between the electrostatic force qE_H and the magnetic deflection force qvB.

Because $A = td$, where t is the thickness of the conductor, we can also express Equation 29.21 as

$$\Delta V_H = \frac{IB}{nqt} = \frac{R_H IB}{t} \quad (29.22)$$

◀ The Hall voltage

where $R_H = 1/nq$ is the **Hall coefficient.** This relationship shows that a properly calibrated conductor can be used to measure the magnitude of an unknown magnetic field.

Because all quantities in Equation 29.23 other than nq can be measured, a value for the Hall coefficient is readily obtainable. The sign and magnitude of R_H give the sign of the charge carriers and their number density. In most metals, the charge carriers are electrons and the charge-carrier density determined from Hall-effect measurements is in good agreement with calculated values for such metals as lithium (Li), sodium (Na), copper (Cu), and silver (Ag), whose atoms each give up one electron to act as a current carrier. In this case, n is approximately equal to the number of conducting electrons per unit volume. This classical model, however, is not valid for metals such as iron (Fe), bismuth (Bi), and cadmium (Cd) or for semiconductors. These discrepancies can be explained only by using a model based on the quantum nature of solids.

EXAMPLE 29.7 The Hall Effect for Copper

A rectangular copper strip 1.5 cm wide and 0.10 cm thick carries a current of 5.0 A. Find the Hall voltage for a 1.2-T magnetic field applied in a direction perpendicular to the strip.

SOLUTION

Conceptualize Study Figures 29.25 and 29.26 carefully and make sure you understand that a Hall voltage is developed between the top and bottom edges of the strip.

Categorize We evaluate the Hall voltage using an equation developed in this section, so we categorize this example as a substitution problem.

Assuming that one electron per atom is available for conduction, we can take the charge carrier density to be 8.46×10^{28} electrons/m^3 (see Example 27.1). Substitute this value and the given data into Equation 29.22:

$$\Delta V_H = \frac{IB}{nqt}$$

$$= \frac{(5.0\text{ A})(1.2\text{ T})}{(8.46 \times 10^{28}\text{ m}^{-3})(1.6 \times 10^{-19}\text{ C})(0.001\,0\text{ m})}$$

$$\Delta V_H = 0.44\ \mu\text{V}$$

Such an extremely small Hall voltage is expected in good conductors. (Notice that the width of the conductor is not needed in this calculation.)

What If? What if the strip has the same dimensions but is made of a semiconductor? Will the Hall voltage be smaller or larger?

Answer In semiconductors, n is much smaller than it is in metals that contribute one electron per atom to the current; hence, the Hall voltage is usually larger because it varies as the inverse of n. Currents on the order of 0.1 mA are generally used for such materials. Consider a piece of silicon that has the same dimensions as the copper strip in this example and whose value for n is 1.0×10^{20} electrons/m^3. Taking $B = 1.2$ T and $I = 0.10$ mA, we find that $\Delta V_H = 7.5$ mV. A potential difference of this magnitude is readily measured.

Summary

ThomsonNOW™ Sign in at **www.thomsonedu.com** and go to ThomsonNOW to take a practice test for this chapter.

DEFINITION

The **magnetic dipole moment** $\vec{\boldsymbol{\mu}}$ of a loop carrying a current I is

$$\vec{\boldsymbol{\mu}} \equiv I\vec{\mathbf{A}} \quad (29.5)$$

where the area vector $\vec{\mathbf{A}}$ is perpendicular to the plane of the loop and $|\vec{\mathbf{A}}|$ is equal to the area of the loop. The SI unit of $\vec{\boldsymbol{\mu}}$ is $A \cdot m^2$.

CONCEPTS AND PRINCIPLES

The magnetic force that acts on a charge q moving with a velocity $\vec{\mathbf{v}}$ in a magnetic field $\vec{\mathbf{B}}$ is

$$\vec{\mathbf{F}}_B = q\vec{\mathbf{v}} \times \vec{\mathbf{B}} \quad (29.1)$$

The direction of this magnetic force is perpendicular both to the velocity of the particle and to the magnetic field. The magnitude of this force is

$$F_B = |q|vB \sin\theta \quad (29.2)$$

where θ is the smaller angle between $\vec{\mathbf{v}}$ and $\vec{\mathbf{B}}$. The SI unit of $\vec{\mathbf{B}}$ is the **tesla** (T), where $1\ T = 1\ N/A \cdot m$.

If a charged particle moves in a uniform magnetic field so that its initial velocity is perpendicular to the field, the particle moves in a circle, the plane of which is perpendicular to the magnetic field. The radius of the circular path is

$$r = \frac{mv}{qB} \quad (29.3)$$

where m is the mass of the particle and q is its charge. The angular speed of the charged particle is

$$\omega = \frac{qB}{m} \quad (29.4)$$

If a straight conductor of length L carries a current I, the force exerted on that conductor when it is placed in a uniform magnetic field $\vec{\mathbf{B}}$ is

$$\vec{\mathbf{F}}_B = I\vec{\mathbf{L}} \times \vec{\mathbf{B}} \quad (29.10)$$

where the direction of $\vec{\mathbf{L}}$ is in the direction of the current and $|\vec{\mathbf{L}}| = L$.

If an arbitrarily shaped wire carrying a current I is placed in a magnetic field, the magnetic force exerted on a very small segment $d\vec{\mathbf{s}}$ is

$$d\vec{\mathbf{F}}_B = I\,d\vec{\mathbf{s}} \times \vec{\mathbf{B}} \quad (29.11)$$

To determine the total magnetic force on the wire, one must integrate Equation 29.11 over the wire, keeping in mind that both $\vec{\mathbf{B}}$ and $d\vec{\mathbf{s}}$ may vary at each point.

The torque $\vec{\boldsymbol{\tau}}$ on a current loop placed in a uniform magnetic field $\vec{\mathbf{B}}$ is

$$\vec{\boldsymbol{\tau}} = \vec{\boldsymbol{\mu}} \times \vec{\mathbf{B}} \quad (29.17)$$

The potential energy of the system of a magnetic dipole in a magnetic field is

$$U = -\vec{\boldsymbol{\mu}} \cdot \vec{\mathbf{B}} \quad (29.18)$$

Questions

☐ denotes answer available in *Student Solutions Manual/Study Guide;* **O** denotes objective question

Questions 2, 3, and 4 in Chapter 11 can be assigned with this chapter.

1. **O** Answer each question yes or no. Assume the motions and currents mentioned are along the x axis and fields are in the y direction. (a) Does an electric field exert a force on a stationary charged object? (b) Does a magnetic field do so? (c) Does an electric field exert a force on a moving charged object? (d) Does a magnetic field do so? (e) Does an electric field exert a force on a straight current-carrying wire? (f) Does a magnetic field do so? (g) Does an electric field exert a force on a beam of electrons? (h) Does a magnetic field do so?

2. **O** Electron A is fired horizontally with speed 1 Mm/s into a region where a vertical magnetic field exists. Electron B is fired along the same path with speed 2 Mm/s. **(i)** Which electron has a larger magnetic force exerted on it? (a) A does. (b) B does. (c) The forces have the same nonzero magnitude. (d) The forces are both zero. **(ii)** Which electron has a path that curves more sharply? (a) A does. (b) B does. (c) The particles follow the same curved path. (d) The particles continue to go straight.

3. **O** Classify each of the following as a characteristic (a) of electric forces only, (b) of magnetic forces only, (c) of both electric and magnetic forces, or (d) of neither electric nor magnetic forces. **(i)** The force is proportional to the magnitude of the field exerting it. **(ii)** The force is proportional to the magnitude of the charge of the object on which the force is exerted. **(iii)** The force exerted on a negatively charged object is opposite in direction to the force on a positive charge. **(iv)** The force exerted on a stationary charged object is zero. **(v)** The force exerted on a moving charged object is zero. **(vi)** The force exerted on a charged object is proportional to its speed. **(vii)** The force exerted on a charged object cannot alter the object's speed. **(viii)** The magnitude of the force depends on the charged object's direction of motion.

4. **O** Rank the magnitudes of the forces exerted on the following particles from the largest to the smallest. In your ranking, display any cases of equality. (a) An electron moving at 1 Mm/s perpendicular to a 1-mT magnetic field. (b) An electron moving at 1 Mm/s parallel to a 1-mT magnetic field. (c) An electron moving at 2 Mm/s perpendicular to a 1-mT magnetic field. (d) An electron moving at 1 Mm/s perpendicular to a 2-mT magnetic field. (e) A proton moving at 1 Mm/s perpendicular to a 1-mT magnetic field. (f) A proton moving at 1 Mm/s at a 45° angle to a 1-mT magnetic field.

5. **O** At a certain instant, a proton is moving in the positive x direction through a magnetic field in the negative z direction. What is the direction of the magnetic force exerted on the proton? (a) x (b) $-x$ (c) y (d) $-y$ (e) z (f) $-z$ (g) halfway between the x and $-z$ axes, at 45° to both (h) The force is zero.

6. **O** A particle with electric charge is fired into a region of space where the electric field is zero. It moves in a straight line. Can you conclude that the magnetic field in that region is zero? (a) Yes. (b) No; the field might be perpendicular to the particle's velocity. (c) No; the field might be parallel to the particle's velocity. (d) No; the particle might need to have charge of the opposite sign to have a force exerted on it. (e) No; an observation of an object with *electric* charge gives no information about a *magnetic* field.

[7.] Two charged particles are projected in the same direction into a magnetic field perpendicular to their velocities. If the particles are deflected in opposite directions, what can you say about them?

8. How can the motion of a moving charged particle be used to distinguish between a magnetic field and an electric field? Give a specific example to justify your argument.

9. **O** In the velocity selector shown in Active Figure 29.12, electrons with speed $v = E/B$ follow a straight path. Electrons moving significantly faster than this speed through the same selector will move along what kind of path? (a) a circle (b) a parabola (c) a straight line (d) a more complicated trajectory

[10.] Is it possible to orient a current loop in a uniform magnetic field such that the loop does not tend to rotate? Explain.

11. Explain why it is not possible to determine the charge and the mass of a charged particle separately by measuring accelerations produced by electric and magnetic forces on the particle.

[12.] How can a current loop be used to determine the presence of a magnetic field in a given region of space?

13. Charged particles from outer space, called cosmic rays, strike the Earth more frequently near the poles than near the equator. Why?

14. Can a constant magnetic field set into motion an electron initially at rest? Explain your answer.

Problems

WebAssign The Problems from this chapter may be assigned online in WebAssign.

ThomsonNOW Sign in at **www.thomsonedu.com** and go to ThomsonNOW to assess your understanding of this chapter's topics with additional quizzing and conceptual questions.

1, 2, 3 denotes straightforward, intermediate, challenging; ☐ denotes full solution available in *Student Solutions Manual/Study Guide;* ▲ denotes coached solution with hints available at **www.thomsonedu.com;** ▭ denotes developing symbolic reasoning; ● denotes asking for qualitative reasoning; ▪ denotes computer useful in solving problem

Section 29.1 Magnetic Fields and Forces

Problems 1, 2, 3, 4, 6, 7, and 10 in Chapter 11 can be assigned with this section.

1. ▲ Determine the initial direction of the deflection of charged particles as they enter the magnetic fields shown in Figure P29.1.

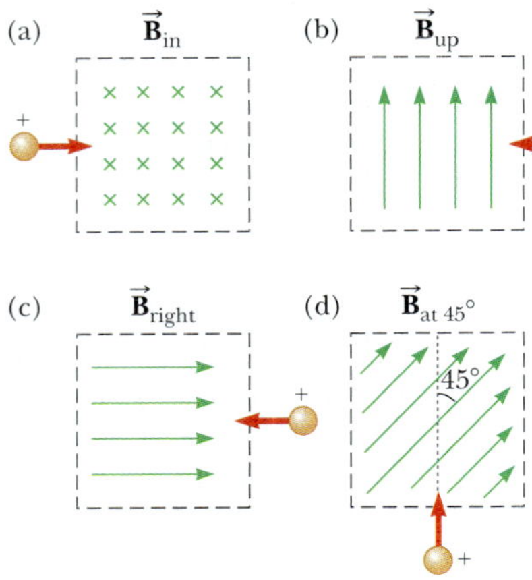

Figure P29.1

2. Consider an electron near the Earth's equator. In which direction does it tend to deflect if its velocity is (a) directed downward? (b) Directed northward? (c) Directed westward? (d) Directed southeastward?

3. A proton moves perpendicular to a uniform magnetic field $\vec{\mathbf{B}}$ at a speed of 1.00×10^7 m/s and experiences an acceleration of 2.00×10^{13} m/s² in the $+x$ direction when its velocity is in the $+z$ direction. Determine the magnitude and direction of the field.

4. A proton travels with a speed of 3.00×10^6 m/s at an angle of 37.0° with the direction of a magnetic field of 0.300 T in the $+y$ direction. What are (a) the magnitude of the magnetic force on the proton and (b) its acceleration?

5. A proton moving at 4.00×10^6 m/s through a magnetic field of magnitude 1.70 T experiences a magnetic force of magnitude 8.20×10^{-13} N. What is the angle between the proton's velocity and the field?

6. An electron is accelerated through 2 400 V from rest and then enters a uniform 1.70-T magnetic field. What are (a) the maximum and (b) the minimum values of the magnetic force this particle can experience?

7. A proton moves with a velocity of $\vec{\mathbf{v}} = (2\hat{\mathbf{i}} - 4\hat{\mathbf{j}} + \hat{\mathbf{k}})$ m/s in a region in which the magnetic field is $\vec{\mathbf{B}} = (\hat{\mathbf{i}} + 2\hat{\mathbf{j}} - 3\hat{\mathbf{k}})$ T. What is the magnitude of the magnetic force this particle experiences?

8. ● An electron in a uniform electric and magnetic field has a velocity of 1.20×10^4 m/s (in the positive x direction) and an acceleration of 2.00×10^{12} m/s² (in the positive z direction). If the electric field has a magnitude of 20.0 N/C (in the positive z direction), what can you determine about the magnetic field in the region? What can you not determine?

Section 29.2 Motion of a Charged Particle in a Uniform Magnetic Field

9. The magnetic field of the Earth at a certain location is directed vertically downward and has a magnitude of 50.0 μT. A proton is moving horizontally toward the west in this field with a speed of 6.20×10^6 m/s. (a) What are the direction and magnitude of the magnetic force the field exerts on this particle? (b) What is the radius of the circular arc followed by this proton?

10. ● An accelerating voltage of 2 500 V is applied to an electron gun, producing a beam of electrons originally traveling horizontally north in vacuum toward the center of a viewing screen 35.0 cm away. (a) What are the magnitude and direction of the deflection on the screen caused by the Earth's gravitational field? (b) What are the magnitude and direction of the deflection on the screen caused by the vertical component of the Earth's magnetic field, taken as 20.0 μT down? Does an electron in this vertical magnetic field move as a projectile, with constant vector acceleration perpendicular to a constant northward component of velocity? Is it a good approximation to assume it has this projectile motion? Explain.

11. A proton (charge $+e$, mass m_p), a deuteron (charge $+e$, mass $2m_p$), and an alpha particle (charge $+2e$, mass $4m_p$) are accelerated through a common potential difference ΔV. Each of the particles enters a uniform magnetic field $\vec{\mathbf{B}}$, with its velocity in a direction perpendicular to $\vec{\mathbf{B}}$. The proton moves in a circular path of radius r_p. Determine the radii of the circular orbits for the deuteron, r_d, and the alpha particle, r_α, in terms of r_p.

12. **Review problem.** One electron collides elastically with a second electron initially at rest. After the collision, the radii of their trajectories are 1.00 cm and 2.40 cm. The trajectories are perpendicular to a uniform magnetic field of magnitude 0.044 0 T. Determine the energy (in keV) of the incident electron.

13. **Review problem.** An electron moves in a circular path perpendicular to a constant magnetic field of magnitude 1.00 mT. The angular momentum of the electron about the center of the circle is 4.00×10^{-25} kg · m²/s. Determine (a) the radius of the circular path and (b) the speed of the electron.

14. A singly charged ion of mass m is accelerated from rest by a potential difference ΔV. It is then deflected by a uniform magnetic field (perpendicular to the ion's velocity) into a semicircle of radius R. Now a doubly charged ion of mass m' is accelerated through the same potential difference and deflected by the same magnetic field into a semicircle of radius $R' = 2R$. What is the ratio of the masses of the ions?

15. A cosmic-ray proton in interstellar space has an energy of 10.0 MeV and executes a circular orbit having a radius equal to that of Mercury's orbit around the Sun (5.80×10^{10} m). What is the magnetic field in that region of space?

16. Assume the region to the right of a certain vertical plane contains a vertical magnetic field of magnitude 1.00 mT and the field is zero in the region to the left of the plane. An electron, originally traveling perpendicular to the boundary plane, passes into the region of the field. (a) Determine the time interval required for the electron to leave the "field-filled" region, noting that its path is a semicircle. (b) Find the kinetic energy of the electron, assuming the maximum depth of penetration into the field is 2.00 cm.

Section 29.3 Applications Involving Charged Particles Moving in a Magnetic Field

17. A velocity selector consists of electric and magnetic fields described by the expressions $\vec{\mathbf{E}} = E\hat{\mathbf{k}}$ and $\vec{\mathbf{B}} = B\hat{\mathbf{j}}$, with $B = 15.0$ mT. Find the value of E such that a 750-eV electron moving along the positive x axis is undeflected.

18. ● Singly charged uranium-238 ions are accelerated through a potential difference of 2.00 kV and enter a uniform magnetic field of 1.20 T directed perpendicular to their velocities. (a) Determine the radius of their circular path. (b) Repeat for uranium-235 ions. **What If?** How does the ratio of these path radii depend on the accelerating voltage? On the magnitude of the magnetic field?

19. Consider the mass spectrometer shown schematically in Active Figure 29.13. The magnitude of the electric field between the plates of the velocity selector is 2 500 V/m, and the magnetic field in both the velocity selector and the deflection chamber has a magnitude of 0.035 0 T. Calculate the radius of the path for a singly charged ion having a mass $m = 2.18 \times 10^{-26}$ kg.

20. A cyclotron designed to accelerate protons has an outer radius of 0.350 m. The protons are emitted nearly at rest from a source at the center and are accelerated through 600 V each time they cross the gap between the dees. The dees are between the poles of an electromagnet where the field is 0.800 T. (a) Find the cyclotron frequency for the protons in this cyclotron. (b) Find the speed at which protons exit the cyclotron and (c) their maximum kinetic energy. (d) How many revolutions does a proton make in the cyclotron? (e) For what time interval does one proton accelerate?

21. A cyclotron designed to accelerate protons has a magnetic field of magnitude 0.450 T over a region of radius 1.20 m. What are (a) the cyclotron frequency and (b) the maximum speed acquired by the protons?

22. ● A particle in the cyclotron shown in Figure 29.15a gains energy $q\,\Delta V$ from the alternating power supply each time it passes from one dee to the other. The time interval for each full orbit is

$$T = \frac{2\pi}{\omega} = \frac{2\pi m}{qB}$$

so the particle's average rate of increase in energy is

$$\frac{2q\,\Delta V}{T} = \frac{q^2 B\,\Delta V}{\pi m}$$

Note that this power input is constant in time. (a) Show that the rate of increase in the radius r of its path is not constant, but is given by

$$\frac{dr}{dt} = \frac{1}{r}\frac{\Delta V}{\pi B}$$

(b) Describe how the path of the particles in Figure 29.15a could be drawn more realistically. (c) At what rate is the radial position of the protons in Problem 20 increasing immediately before the protons leave the cyclotron? (d) By how much does the radius of the protons' path increase during their last full revolution?

23. ▲ The picture tube in a television uses magnetic deflection coils rather than electric deflection plates. Suppose an electron beam is accelerated through a 50.0-kV potential difference and then through a region of uniform magnetic field 1.00 cm wide. The screen is located 10.0 cm from the center of the coils and is 50.0 cm wide. When the field is turned off, the electron beam hits the center of the screen. What field magnitude is necessary to deflect the beam to the side of the screen? Ignore relativistic corrections.

24. ● In his "discovery of the electron," J. J. Thomson showed that the same beam deflections resulted with tubes having cathodes made of *different* materials and containing *various* gases before evacuation. (a) Are these observations important? Explain your answer. (b) When he applied various potential differences to the deflection plates and turned on the magnetic coils, alone or in combination with the deflection plates, Thomson observed that the fluorescent screen continued to show a *single small* glowing patch. Argue whether his observation is important. (c) Do calculations to show that the charge-to-mass ratio Thomson obtained was huge compared to that of any macroscopic object or of any ionized atom or molecule. How can one make sense of that fact? (d) Could Thomson observe any deflection of the beam due to gravitation? Do a calculation to argue for your answer. (To obtain a visibly glowing patch on the fluorescent screen, the potential difference between the slits and the cathode must be 100 V or more.)

Section 29.4 Magnetic Force Acting on a Current-Carrying Conductor

25. ▲ A wire having a mass per unit length of 0.500 g/cm carries a 2.00-A current horizontally to the south. What are the direction and magnitude of the minimum magnetic field needed to lift this wire vertically upward?

26. A wire carries a steady current of 2.40 A. A straight section of the wire is 0.750 m long and lies along the x axis

within a uniform magnetic field, $\vec{\mathbf{B}} = 1.60\hat{\mathbf{k}}$ T. If the current is in the $+x$ direction, what is the magnetic force on the section of wire?

27. A wire 2.80 m in length carries a current of 5.00 A in a region where a uniform magnetic field has a magnitude of 0.390 T. Calculate the magnitude of the magnetic force on the wire assuming the angle between the magnetic field and the current is (a) 60.0°, (b) 90.0°, and (c) 120°.

28. Imagine a wire with linear mass density 2.40 g/m encircling the Earth at its magnetic equator, where the field is modeled as having the uniform value 28.0 μT horizontally north. What magnitude and direction of the current in the wire will keep the wire levitated immediately above the ground?

29. **Review problem.** A rod of mass 0.720 kg and radius 6.00 cm rests on two parallel rails (Fig. P29.29) that are d = 12.0 cm apart and L = 45.0 cm long. The rod carries a current of I = 48.0 A in the direction shown and rolls along the rails without slipping. A uniform magnetic field of magnitude 0.240 T is directed perpendicular to the rod and the rails. If it starts from rest, what is the speed of the rod as it leaves the rails?

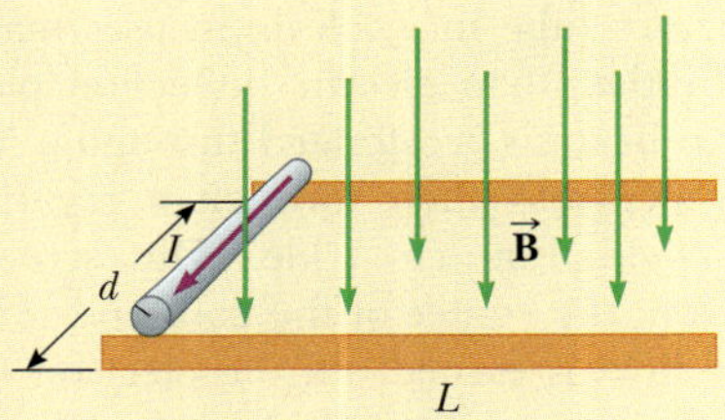

Figure P29.29 Problems 29 and 30.

30. **Review problem.** A rod of mass m and radius R rests on two parallel rails (Fig. P29.29) that are a distance d apart and have a length L. The rod carries a current I in the direction shown and rolls along the rails without slipping. A uniform magnetic field B is directed perpendicular to the rod and the rails. If it starts from rest, what is the speed of the rod as it leaves the rails?

31. ▲ *A nonuniform magnetic field exerts a net force on a magnetic dipole.* A strong magnet is placed under a horizontal conducting ring of radius r that carries current I as shown in Figure P29.31. If the magnetic field $\vec{\mathbf{B}}$ makes an angle θ with the vertical at the ring's location, what are the magnitude and direction of the resultant magnetic force on the ring?

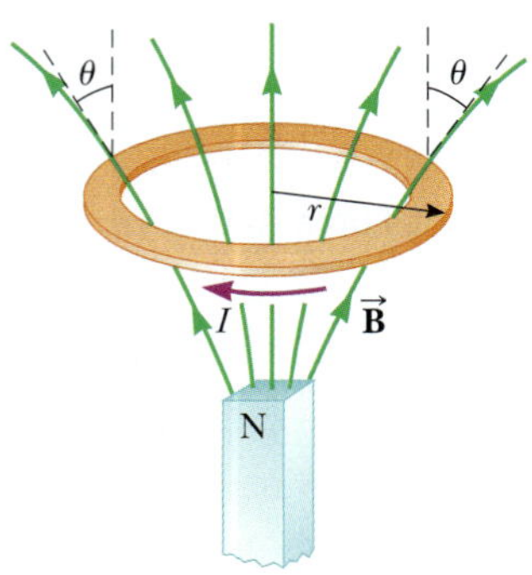

Figure P29.31

32. ● In Figure P29.32, the cube is 40.0 cm on each edge. Four straight segments of wire—ab, bc, cd, and da—form a closed loop that carries a current I = 5.00 A in the direction shown. A uniform magnetic field of magnitude B = 0.020 0 T is in the positive y direction. (a) Determine the magnitude and direction of the magnetic force on each segment. (b) Explain how you could find the force exerted on the fourth of these segments from the forces on the other three, without further calculation involving the magnetic field.

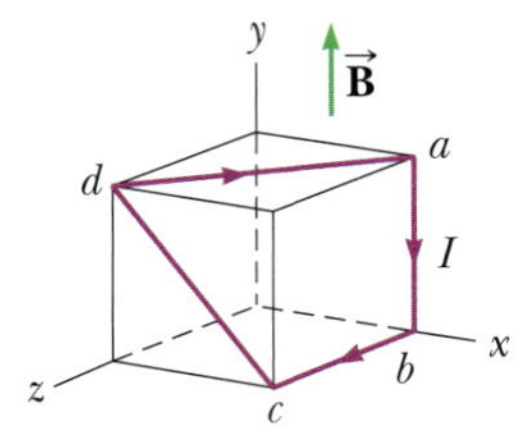

Figure P29.32

33. Assume the Earth's magnetic field is 52.0 μT northward at 60.0° below the horizontal in Atlanta, Georgia. A tube in a neon sign—situated between two diagonally opposite corners of a shop window, which lies in a north–south vertical plane—carries current 35.0 mA. The current enters the tube at the bottom south corner of the shop's window. It exits at the opposite corner, which is 1.40 m farther north and 0.850 m higher up. Between these two points, the glowing tube spells out DONUTS. Determine the total vector magnetic force on the tube. You may use the first "important statement" presented in the *Finalize* section of Example 29.4.

Section 29.5 Torque on a Current Loop in a Uniform Magnetic Field

34. A current of 17.0 mA is maintained in a single circular loop of 2.00 m circumference. A magnetic field of 0.800 T is directed parallel to the plane of the loop. (a) Calculate the magnetic moment of the loop. (b) What is the magnitude of the torque exerted by the magnetic field on the loop?

35. ▲ A rectangular coil consists of N = 100 closely wrapped turns and has dimensions a = 0.400 m and b = 0.300 m. The coil is hinged along the y axis, and its plane makes an angle θ = 30.0° with the x axis (Fig. P29.35). What is the magnitude of the torque exerted on the coil by a uniform magnetic field B = 0.800 T directed along the x axis when the current is I = 1.20 A in the direction shown? What is the expected direction of rotation of the coil?

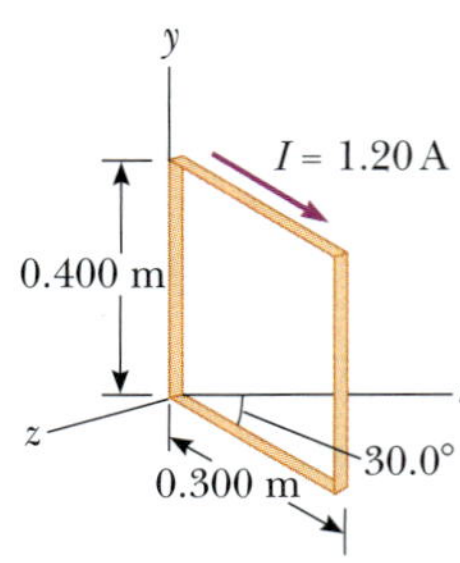

Figure P29.35

36. A current loop with magnetic dipole moment $\vec{\boldsymbol{\mu}}$ is placed in a uniform magnetic field $\vec{\mathbf{B}}$, with its moment making angle θ with the field. With the arbitrary choice of $U = 0$ for $\theta = 90°$, prove that the potential energy of the dipole–field system is $U = -\vec{\boldsymbol{\mu}} \cdot \vec{\mathbf{B}}$. You may imitate the discussion in Chapter 26 of the potential energy of an electric dipole in an electric field.

37. ● The needle of a magnetic compass has magnetic moment 9.70 mA · m^2. At its location, the Earth's magnetic field is 55.0 μT north at 48.0° below the horizontal. (a) Identify the orientations of the compass needle that represent minimum potential energy and maximum potential energy of the needle–field system. (b) How much work must be done on the needle to move it from the former to the latter orientation?

38. A wire is formed into a circle having a diameter of 10.0 cm and placed in a uniform magnetic field of 3.00 mT. The wire carries a current of 5.00 A. Find (a) the maximum torque on the wire and (b) the range of potential energies of the wire–field system for different orientations of the circle.

39. ● A wire 1.50 m long carries a current of 30.0 mA when it is connected to a battery. The whole wire can be arranged as a single loop with the shape of a circle, a square, or an equilateral triangle. The whole wire can be made into a flat, compact, circular coil with N turns. Explain how its magnetic moment compares in all these cases. In particular, can its magnetic moment go to infinity? To zero? Does its magnetic moment have a well-defined maximum value? If so, identify it. Does it have minimum value? If so, identify it.

40. The rotor in a certain electric motor is a flat, rectangular coil with 80 turns of wire and dimensions 2.50 cm by 4.00 cm. The rotor rotates in a uniform magnetic field of 0.800 T. When the plane of the rotor is perpendicular to the direction of the magnetic field, it carries a current of 10.0 mA. In this orientation, the magnetic moment of the rotor is directed opposite the magnetic field. The rotor then turns through one-half revolution. This process is repeated to cause the rotor to turn steadily at 3 600 rev/min. (a) Find the maximum torque acting on the rotor. (b) Find the peak power output of the motor. (c) Determine the amount of work performed by the magnetic field on the rotor in every full revolution. (d) What is the average power of the motor?

Section 29.6 The Hall Effect

41. In an experiment designed to measure the Earth's magnetic field using the Hall effect, a copper bar 0.500 cm thick is positioned along an east–west direction. If a current of 8.00 A in the conductor results in a Hall voltage of 5.10×10^{-12} V, what is the magnitude of the Earth's magnetic field? (Assume $n = 8.46 \times 10^{28}$ electrons/m^3 and the plane of the bar is rotated to be perpendicular to the direction of $\vec{\mathbf{B}}$.)

42. A Hall-effect probe operates with a 120-mA current. When the probe is placed in a uniform magnetic field of magnitude 0.080 0 T, it produces a Hall voltage of 0.700 μV. (a) When it is used to measure an unknown magnetic field, the Hall voltage is 0.330 μV. What is the magnitude of the unknown field? (b) The thickness of the probe in the direction of $\vec{\mathbf{B}}$ is 2.00 mm. Find the density of the charge carriers, each of which has charge of magnitude e.

Additional Problems

43. ● Heart-lung machines and artificial kidney machines employ blood pumps. A mechanical pump can mangle blood cells. Figure P29.43 represents an electromagnetic pump. The blood is confined to an electrically insulating tube, cylindrical in practice but represented as a rectangle of width w and height h. The simplicity of design makes the pump dependable. The blood is easily kept uncontaminated; the tube is simple to clean or inexpensive to replace. Two electrodes fit into the top and the bottom of the tube. The potential difference between them establishes an electric current through the blood, with current density J over a section of length L. A perpendicular magnetic field exists in the same region. (a) Explain why this arrangement produces on the liquid a force that is directed along the length of the pipe. (b) Show that the section of liquid in the magnetic field experiences a pressure increase JLB. (c) After the blood leaves the pump, is it charged? Is it current carrying? Is it magnetized? The same magnetic pump can be used for any fluid that conducts electricity, such as liquid sodium in a nuclear reactor.

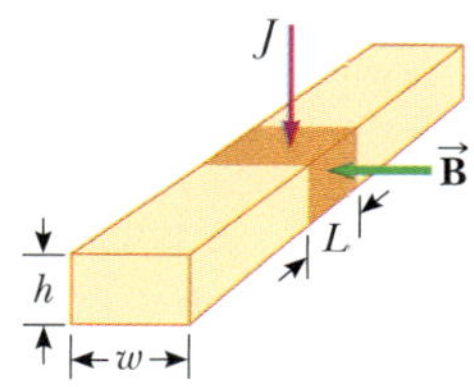

Figure P29.43

44. ● Figure 29.10 shows a charged particle traveling in a nonuniform magnetic field forming a magnetic bottle. (a) Explain why the positively charged particle in the figure must be moving clockwise. The particle travels along a helix whose radius decreases and whose pitch decreases as the particle moves into a stronger magnetic field. If the particle is moving to the right along the x axis, its velocity in this direction will be reduced to zero and it will be reflected from the right-hand side of the bottle, acting as a "magnetic mirror." The particle ends up bouncing back and forth between the ends of the bottle. (b) Explain qualitatively why the axial velocity is reduced to zero as the particle moves into the region of strong magnetic field at the end of the bottle. (c) Explain why the tangential velocity increases as the particle approaches the end of the bottle. (d) Explain why the orbiting particle has a magnetic dipole moment. (e) Sketch the magnetic moment and use the result of Problem 31 to explain again how the nonuniform magnetic field exerts a force on the orbiting particle along the x axis.

45. ● Assume in the plane of the Earth's magnetic equator the planet's field is uniform with the value 25.0 μT northward perpendicular to this plane, everywhere inside a radius of 100 Mm. Also assume the Earth's field is zero outside this circle. A cosmic-ray proton traveling at one tenth of the speed of light is heading directly toward the center of the Earth in the plane of the magnetic equator. Find the radius of curvature of the path it follows when it

enters the region of the planet's assumed field. Explain whether the proton will hit the Earth.

46. A 0.200-kg metal rod carrying a current of 10.0 A glides on two horizontal rails 0.500 m apart. What vertical magnetic field is required to keep the rod moving at a constant speed if the coefficient of kinetic friction between the rod and rails is 0.100?

47. Protons having a kinetic energy of 5.00 MeV are moving in the positive x direction and enter a magnetic field $\vec{\mathbf{B}} = 0.050\,0\hat{\mathbf{k}}$ T directed out of the plane of the page and extending from $x = 0$ to $x = 1.00$ m as shown in Figure P29.47. (a) Calculate the y component of the protons' momentum as they leave the magnetic field. (b) Find the angle α between the initial velocity vector of the proton beam and the velocity vector after the beam emerges from the field. Ignore relativistic effects and note that $1 \text{ eV} = 1.60 \times 10^{-19}$ J.

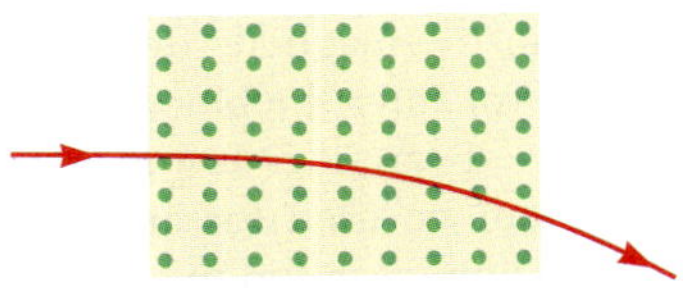

Figure P29.47

48. ● (a) A proton moving in the $+x$ direction with velocity $\vec{\mathbf{v}} = v_i\hat{\mathbf{i}}$ experiences a magnetic force $\vec{\mathbf{F}} = F_i\hat{\mathbf{j}}$ in the $+y$ direction. Explain what you can and cannot infer about $\vec{\mathbf{B}}$ from this information. (b) **What If?** In terms of F_i, what would be the force on a proton in the same field moving with velocity $\vec{\mathbf{v}} = -v_i\hat{\mathbf{i}}$? (c) What would be the force on an electron in the same field moving with velocity $\vec{\mathbf{v}} = -v_i\hat{\mathbf{i}}$?

49. A particle with positive charge $q = 3.20 \times 10^{-19}$ C moves with a velocity $\vec{\mathbf{v}} = (2\hat{\mathbf{i}} + 3\hat{\mathbf{j}} - \hat{\mathbf{k}})$ m/s through a region where both a uniform magnetic field and a uniform electric field exist. (a) Calculate the total force on the moving particle (in unit–vector notation), taking $\vec{\mathbf{B}} = (2\hat{\mathbf{i}} + 4\hat{\mathbf{j}} + \hat{\mathbf{k}})$ T and $\vec{\mathbf{E}} = (4\hat{\mathbf{i}} - \hat{\mathbf{j}} - 2\hat{\mathbf{k}})$ V/m. (b) What angle does the force vector make with the positive x axis?

50. A proton having an initial velocity of $20.0\hat{\mathbf{i}}$ Mm/s enters a uniform magnetic field of magnitude 0.300 T with a direction perpendicular to the proton's velocity. It leaves the field-filled region with velocity $-20.0\hat{\mathbf{j}}$ Mm/s. Determine (a) the direction of the magnetic field, (b) the radius of curvature of the proton's path while in the field, (c) the distance the proton traveled in the field, and (d) the time interval for which the proton is in the field.

51. **Review problem.** A wire having a linear mass density of 1.00 g/cm is placed on a horizontal surface that has a coefficient of kinetic friction of 0.200. The wire carries a current of 1.50 A toward the east and slides horizontally to the north. What are the magnitude and direction of the smallest magnetic field that enables the wire to move in this fashion?

52. **Review problem.** A proton is at rest at the plane vertical boundary of a region containing a uniform vertical magnetic field B. An alpha particle moving horizontally makes a head-on elastic collision with the proton. Immediately after the collision, both particles enter the magnetic field, moving perpendicular to the direction of the field. The radius of the proton's trajectory is R. Find the radius of the alpha particle's trajectory. The mass of the alpha particle is four times that of the proton, and its charge is twice that of the proton.

53. The circuit in Figure P29.53 consists of wires at the top and bottom and identical metal springs in the left and right sides. The upper portion of the circuit is fixed. The wire at the bottom has a mass of 10.0 g and is 5.00 cm long. The springs stretch 0.500 cm under the weight of the wire, and the circuit has a total resistance of 12.0 Ω. When a magnetic field is turned on, directed out of the page, the springs stretch an additional 0.300 cm. What is the magnitude of the magnetic field?

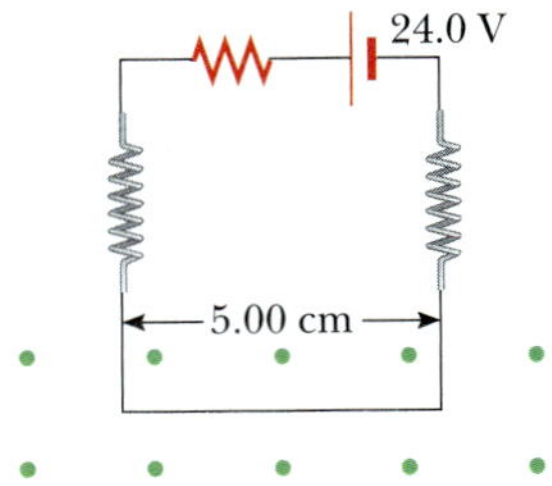

Figure P29.53

54. A handheld electric mixer contains an electric motor. Model the motor as a single flat, compact, circular coil carrying electric current in a region where a magnetic field is produced by an external permanent magnet. You need consider only one instant in the operation of the motor. (We will consider motors again in Chapter 31.) The coil moves because the magnetic field exerts torque on the coil as described in Section 29.5. Make order-of-magnitude estimates of the magnetic field, the torque on the coil, the current in it, its area, and the number of turns in the coil so that they are related according to Equation 29.17. Note that the input power to the motor is electric, given by $\mathcal{P} = I\Delta V$, and the useful output power is mechanical, $\mathcal{P} = \tau\omega$.

55. A nonconducting sphere has mass 80.0 g and radius 20.0 cm. A flat, compact coil of wire with five turns is wrapped tightly around it, with each turn concentric with the sphere. As shown in Figure P29.55, the sphere is placed on an inclined plane that slopes downward to the left, making an angle θ with the horizontal so that the coil is parallel to the inclined plane. A uniform magnetic field of 0.350 T vertically upward exists in the region of the sphere. What current in the coil will enable the sphere to rest in equilibrium on the inclined plane? Show that the result does not depend on the value of θ.

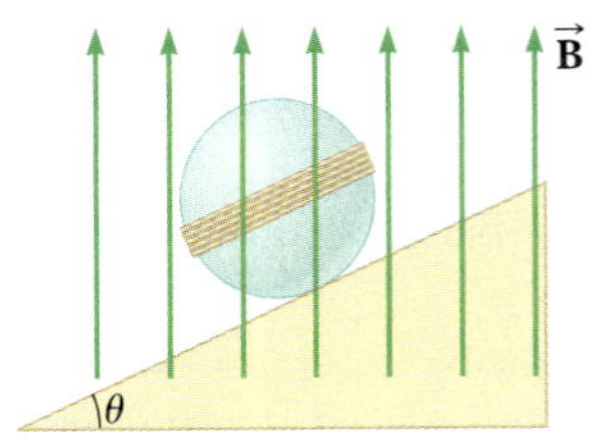

Figure P29.55

56. A metal rod having a mass per unit length λ carries a current I. The rod hangs from two vertical wires in a uniform vertical magnetic field as shown in Figure P29.56. The

2 = intermediate; 3 = challenging; □ = SSM/SG; ▲ = ThomsonNOW; ▭ = symbolic reasoning; ● = qualitative reasoning

wires make an angle θ with the vertical when in equilibrium. Determine the magnitude of the magnetic field.

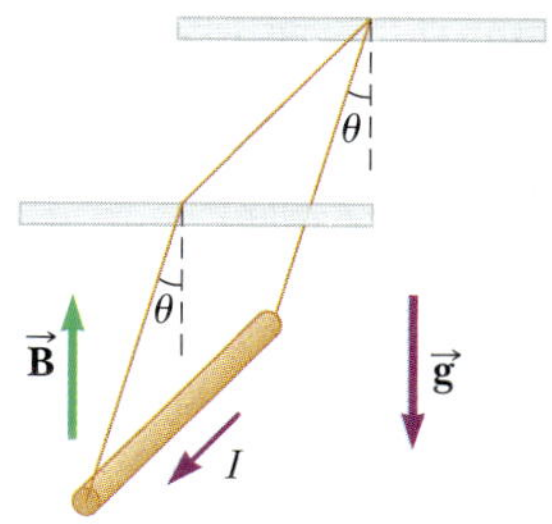

Figure P29.56

57. A cyclotron is sometimes used for carbon dating as will be described in Chapter 44. Carbon-14 and carbon-12 ions are obtained from a sample of the material to be dated and accelerated in the cyclotron. If the cyclotron has a magnetic field of magnitude 2.40 T, what is the difference in cyclotron frequencies for the two ions?

58. A uniform magnetic field of magnitude 0.150 T is directed along the positive x axis. A positron moving at 5.00×10^6 m/s enters the field along a direction that makes an angle of 85.0° with the x axis (Fig. P29.58). The motion of the particle is expected to be a helix as described in Section 29.2. Calculate (a) the pitch p and (b) the radius r of the trajectory.

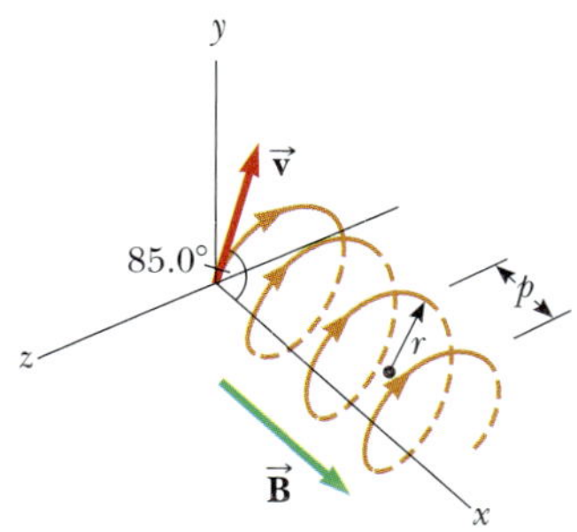

Figure P29.58

59. Consider an electron orbiting a proton and maintained in a fixed circular path of radius $R = 5.29 \times 10^{-11}$ m by the Coulomb force. Treat the orbiting particle as a current loop and calculate the resulting torque when the system is in a magnetic field of 0.400 T directed perpendicular to the magnetic moment of the electron.

60. A proton moving in the plane of the page has a kinetic energy of 6.00 MeV. A magnetic field of magnitude B = 1.00 T is directed into the page. The proton enters the magnetic field with its velocity vector at an angle θ = 45.0° to the linear boundary of the field as shown in Figure P29.60. (a) Find x, the distance from the point of entry to where the proton will leave the field. (b) Determine θ', the angle between the boundary and the proton's velocity vector as it leaves the field.

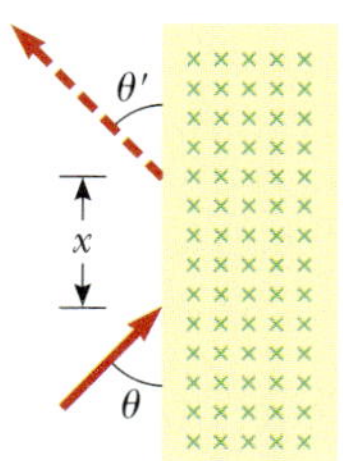

Figure P29.60

61. A heart surgeon monitors the flow rate of blood through an artery using an electromagnetic flowmeter (Fig. P29.61). Electrodes A and B make contact with the outer surface of the blood vessel, which has interior diameter 3.00 mm. (a) For a magnetic field magnitude of 0.040 0 T, an emf of 160 μV appears between the electrodes. Calculate the speed of the blood. (b) Verify that electrode A is positive as shown. Does the sign of the emf depend on whether the mobile ions in the blood are predominantly positively or negatively charged? Explain.

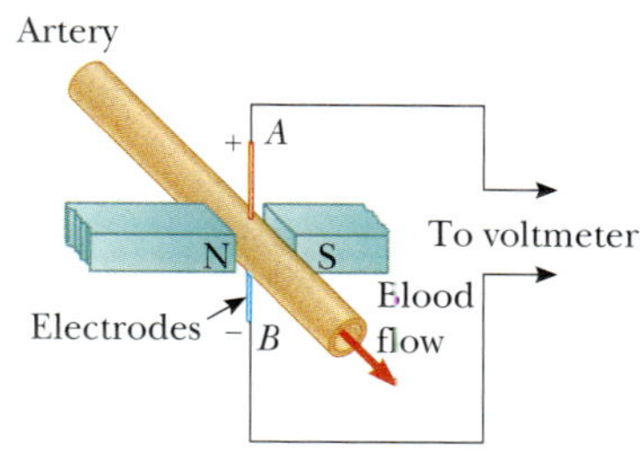

Figure P29.61

62. The following table shows measurements of a Hall voltage and corresponding magnetic field for a probe used to measure magnetic fields. (a) Plot these data and deduce a relationship between the two variables. (b) If the measurements were taken with a current of 0.200 A and the sample is made from a material having a charge-carrier density of 1.00×10^{26} kg/m³, what is the thickness of the sample?

ΔV_H (μV)	B (T)
0	0.00
11	0.10
19	0.20
28	0.30
42	0.40
50	0.50
61	0.60
68	0.70
79	0.80
90	0.90
102	1.00

63. As shown in Figure P29.63, a particle of mass m having positive charge q is initially traveling with velocity $v\hat{\mathbf{j}}$. At the origin of coordinates it enters a region between $y = 0$ and $y = h$ containing a uniform magnetic field $B\hat{\mathbf{k}}$

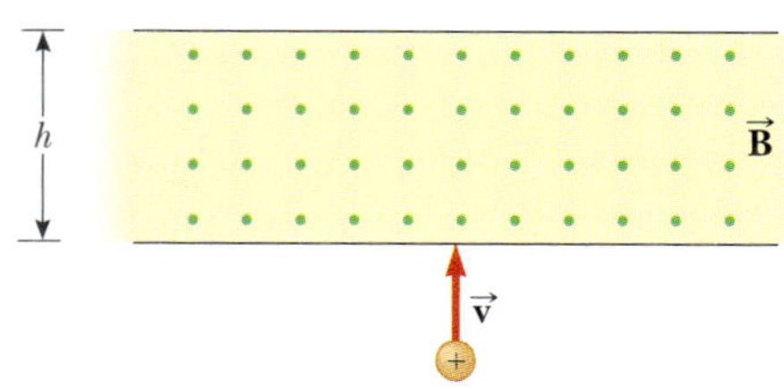

Figure P29.63

directed perpendicularly out of the page. (a) What is the critical value of v such that the particle just reaches $y = h$? Describe the path of the particle under this condition and predict its final velocity. (b) Specify the path the particle takes and its final velocity if v is less than the critical value. (c) **What If?** Specify the path the particle takes and its final velocity if v is greater than the critical value.

64. In Niels Bohr's 1913 model of the hydrogen atom, the single electron is in a circular orbit of radius 5.29×10^{-11} m and its speed is 2.19×10^{6} m/s. (a) What is the magnitude of the magnetic moment due to the electron's motion? (b) If the electron moves in a horizontal circle, counterclockwise as seen from above, what is the direction of this magnetic moment vector?

65. ● **Review problem.** Review Section 15.5 on torsional pendulums. (a) Show that a magnetic dipole in a uniform magnetic field, displaced from its equilibrium orientation and released, can oscillate as a torsional pendulum in simple harmonic motion. Is this statement true for all angular displacements, for all displacements less than 180°, or only for small angular displacements? Explain. (b) Assume the dipole is a compass needle—a light bar magnet—with a magnetic moment of magnitude μ. It has moment of inertia I about its center, where it is mounted on a frictionless vertical axle, and it is placed in a horizontal magnetic field of magnitude B. Evaluate its frequency of oscillation. (c) Explain how the compass needle can be conveniently used as an indicator of the magnitude of the external magnetic field. If its frequency is 0.680 Hz in the Earth's local field, with a horizontal component of 39.2 μT, what is the magnitude of a field in which its frequency of oscillation is 4.90 Hz?

Answers to Quick Quizzes

29.1 (e). The right-hand rule gives the direction. Be sure to account for the negative charge on the electron.

29.2 **(i),** (b). The magnetic force on the particle increases in proportion to v, but the centripetal acceleration increases according to the square of v. The result is a larger radius, as you can see from Equation 29.3. **(ii),** (a). The magnetic force on the particle increases in proportion to B. The result is a smaller radius as you can see from Equation 29.3.

29.3 (c). Use the right-hand rule to determine the direction of the magnetic field.

29.4 **(i),** (c), (b), (a). Because all loops enclose the same area and carry the same current, the magnitude of $\vec{\boldsymbol{\mu}}$ is the same for all. For (c), $\vec{\boldsymbol{\mu}}$ points upward and is perpendicular to the magnetic field and $\tau = \mu B$, the maximum torque possible. For the loop in (a), $\vec{\boldsymbol{\mu}}$ points along the direction of $\vec{\mathbf{B}}$ and the torque is zero. For (b), the torque is intermediate between zero and the maximum value. **(ii),** (a) = (b) = (c). Because the magnetic field is uniform, there is zero net force on all three loops.

A proposed method for launching future payloads into space is the use of *rail guns*, in which projectiles are accelerated by means of magnetic forces. This photo shows the firing of a projectile at a speed of over 3 km/s from an experimental rail gun at Sandia National Research Laboratories, Albuquerque, New Mexico. (Defense Threat Reduction Agency [DTRA])

30 Sources of the Magnetic Field

In Chapter 29, we discussed the magnetic force exerted on a charged particle moving in a magnetic field. To complete the description of the magnetic interaction, this chapter explores the origin of the magnetic field, moving charges. We begin by showing how to use the law of Biot and Savart to calculate the magnetic field produced at some point in space by a small current element. This formalism is then used to calculate the total magnetic field due to various current distributions. Next, we show how to determine the force between two current-carrying conductors, leading to the definition of the ampere. We also introduce Ampère's law, which is useful in calculating the magnetic field of a highly symmetric configuration carrying a steady current.

This chapter is also concerned with the complex processes that occur in magnetic materials. All magnetic effects in matter can be explained on the basis of atomic magnetic moments, which arise both from the orbital motion of electrons and from an intrinsic property of electrons known as spin.

30.1 The Biot–Savart Law

Shortly after Oersted's discovery in 1819 that a compass needle is deflected by a current-carrying conductor, Jean-Baptiste Biot (1774–1862) and Félix Savart

PITFALL PREVENTION 30.1

The Biot–Savart Law

The magnetic field described by the Biot–Savart law is the field *due to* a given current-carrying conductor. Do not confuse this field with any *external* field that may be applied to the conductor from some other source.

(1791–1841) performed quantitative experiments on the force exerted by an electric current on a nearby magnet. From their experimental results, Biot and Savart arrived at a mathematical expression that gives the magnetic field at some point in space in terms of the current that produces the field. That expression is based on the following experimental observations for the magnetic field $d\vec{\mathbf{B}}$ at a point P associated with a length element $d\vec{\mathbf{s}}$ of a wire carrying a steady current I (Fig. 30.1):

- The vector $d\vec{\mathbf{B}}$ is perpendicular both to $d\vec{\mathbf{s}}$ (which points in the direction of the current) and to the unit vector $\hat{\mathbf{r}}$ directed from $d\vec{\mathbf{s}}$ toward P.
- The magnitude of $d\vec{\mathbf{B}}$ is inversely proportional to r^2, where r is the distance from $d\vec{\mathbf{s}}$ to P.
- The magnitude of $d\vec{\mathbf{B}}$ is proportional to the current and to the magnitude ds of the length element $d\vec{\mathbf{s}}$.
- The magnitude of $d\vec{\mathbf{B}}$ is proportional to $\sin\theta$, where θ is the angle between the vectors $d\vec{\mathbf{s}}$ and $\hat{\mathbf{r}}$.

These observations are summarized in the mathematical expression known today as the **Biot–Savart law:**

Biot–Savart law ▶

$$d\vec{\mathbf{B}} = \frac{\mu_0}{4\pi}\frac{I\,d\vec{\mathbf{s}} \times \hat{\mathbf{r}}}{r^2} \tag{30.1}$$

where μ_0 is a constant called the **permeability of free space:**

Permeability of free space ▶

$$\mu_0 = 4\pi \times 10^{-7}\ \text{T}\cdot\text{m/A} \tag{30.2}$$

Notice that the field $d\vec{\mathbf{B}}$ in Equation 30.1 is the field created at a point by the current in only a small length element $d\vec{\mathbf{s}}$ of the conductor. To find the *total* magnetic field $\vec{\mathbf{B}}$ created at some point by a current of finite size, we must sum up contributions from all current elements $I\,d\vec{\mathbf{s}}$ that make up the current. That is, we must evaluate $\vec{\mathbf{B}}$ by integrating Equation 30.1:

$$\vec{\mathbf{B}} = \frac{\mu_0 I}{4\pi}\int\frac{d\vec{\mathbf{s}} \times \hat{\mathbf{r}}}{r^2} \tag{30.3}$$

where the integral is taken over the entire current distribution. This expression must be handled with special care because the integrand is a cross product and therefore a vector quantity. We shall see one case of such an integration in Example 30.1.

Although the Biot–Savart law was discussed for a current-carrying wire, it is also valid for a current consisting of charges flowing through space such as the electron beam in a television picture tube. In that case, $d\vec{\mathbf{s}}$ represents the length of a small segment of space in which the charges flow.

Interesting similarities exist between Equation 30.1 for the magnetic field due to a current element and Equation 23.9 for the electric field due to a point charge. The magnitude of the magnetic field varies as the inverse square of the distance from the source, as does the electric field due to a point charge. The directions of the two fields are quite different, however. The electric field created by a point charge is radial, but the magnetic field created by a current element is perpendicular to both the length element $d\vec{\mathbf{s}}$ and the unit vector $\hat{\mathbf{r}}$ as described by the cross product in Equation 30.1. Hence, if the conductor lies in the plane of the page as shown in Figure 30.1, $d\vec{\mathbf{B}}$ points out of the page at P and into the page at P'.

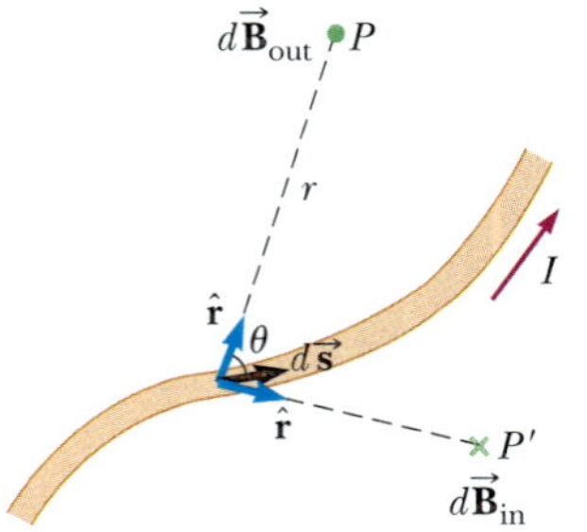

Figure 30.1 The magnetic field $d\vec{\mathbf{B}}$ at a point due to the current I through a length element $d\vec{\mathbf{s}}$ is given by the Biot–Savart law. The direction of the field is out of the page at P and into the page at P'.

Another difference between electric and magnetic fields is related to the source of the field. An electric field is established by an isolated electric charge. The Biot–Savart law gives the magnetic field of an isolated current element at some point, but such an isolated current element cannot exist the way an isolated electric charge can. A current element *must* be part of an extended current distribution because a complete circuit is needed for charges to flow. Therefore, the Biot–Savart law (Eq. 30.1) is only the first step in a calculation of a magnetic field; it must be followed by an integration over the current distribution as in Equation 30.3.

Quick Quiz 30.1 Consider the magnetic field due to the current in the length of wire shown in Figure 30.2. Rank the points *A*, *B*, and *C* in terms of magnitude of the magnetic field that is due to the current in just the length element $d\vec{\mathbf{s}}$ shown from greatest to least.

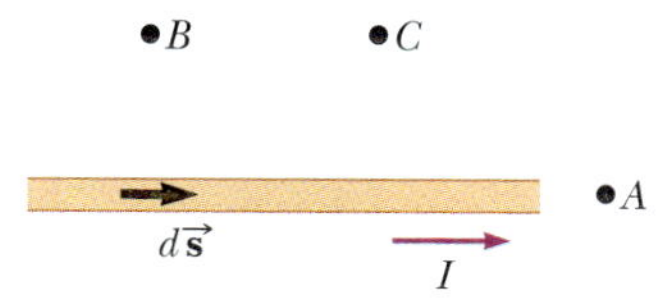

Figure 30.2 (Quick Quiz 30.1) Where is the magnetic field the greatest?

EXAMPLE 30.1 Magnetic Field Surrounding a Thin, Straight Conductor

Consider a thin, straight wire carrying a constant current I and placed along the x axis as shown in Figure 30.3. Determine the magnitude and direction of the magnetic field at point P due to this current.

SOLUTION

Conceptualize From the Biot–Savart law, we expect that the magnitude of the field is proportional to the current in the wire and decreases as the distance a from the wire to point P increases.

Categorize We are asked to find the magnetic field due to a simple current distribution, so this example is a typical problem for which the Biot–Savart law is appropriate.

Analyze Let's start by considering a length element $d\vec{\mathbf{s}}$ located a distance r from P. The direction of the magnetic field at point P due to the current in this element is out of the page because $d\vec{\mathbf{s}} \times \hat{\mathbf{r}}$ is out of the page. In fact, because *all* the current elements $I\,d\vec{\mathbf{s}}$ lie in the plane of the page, they all produce a magnetic field directed out of the page at point P. Therefore, the direction of the magnetic field at point P is out of the page and we need only find the magnitude of the field. We place the origin at O and let point P be along the positive y axis, with $\hat{\mathbf{k}}$ being a unit vector pointing out of the page.

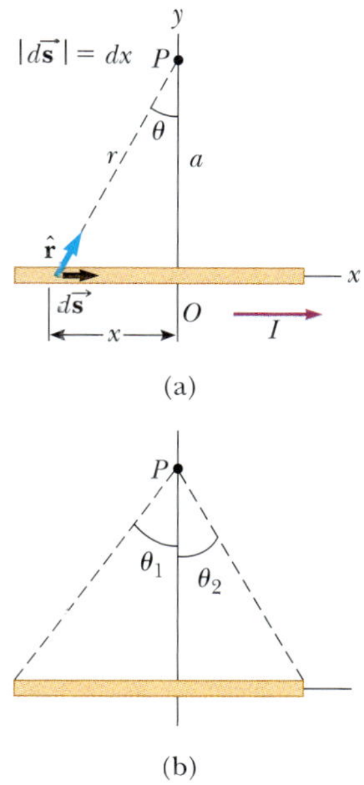

Figure 30.3 (Example 30.1) (a) A thin, straight wire carrying a current I. The magnetic field at point P due to the current in each element $d\vec{\mathbf{s}}$ of the wire is out of the page, so the net field at point P is also out of the page. (b) The angles θ_1 and θ_2 used for determining the net field.

Evaluate the cross product in the Biot–Savart law:

$$d\vec{\mathbf{s}} \times \hat{\mathbf{r}} = |d\vec{\mathbf{s}} \times \hat{\mathbf{r}}|\hat{\mathbf{k}} = \left[dx \sin\left(\frac{\pi}{2} - \theta\right)\right]\hat{\mathbf{k}} = (dx \cos\theta)\hat{\mathbf{k}}$$

Substitute into Equation 30.1:

$$(1) \quad d\vec{\mathbf{B}} = (dB)\hat{\mathbf{k}} = \frac{\mu_0 I}{4\pi}\frac{dx\cos\theta}{r^2}\hat{\mathbf{k}}$$

From the geometry in Figure 30.3a, express r in terms of θ:

$$(2) \quad r = \frac{a}{\cos\theta}$$

Notice that $\tan\theta = -x/a$ from the right triangle in Figure 30.3a (the negative sign is necessary because $d\vec{\mathbf{s}}$ is located at a negative value of x) and solve for x:

$$x = -a\tan\theta$$

Find the differential dx:

$$(3) \quad dx = -a\sec^2\theta\,d\theta = -\frac{a\,d\theta}{\cos^2\theta}$$

Substitute Equations (2) and (3) into the magnitude of the field from Equation (1):

$$(4) \quad dB = -\frac{\mu_0 I}{4\pi}\frac{(a\,d\theta)\cos\theta\cos^2\theta}{a^2\cos^2\theta} = -\frac{\mu_0 I}{4\pi a}\cos\theta\,d\theta$$

Integrate Equation (4) over all length elements on the wire, where the subtending angles range from θ_1 to θ_2 as defined in Figure 30.3b:

$$B = -\frac{\mu_0 I}{4\pi a}\int_{\theta_1}^{\theta_2} \cos\theta\, d\theta = \frac{\mu_0 I}{4\pi a}(\sin\theta_1 - \sin\theta_2) \qquad (30.4)$$

Finalize We can use this result to find the magnetic field of *any* straight current-carrying wire if we know the geometry and hence the angles θ_1 and θ_2. Consider the special case of an infinitely long, straight wire. If the wire in Figure 30.3b becomes infinitely long, we see that $\theta_1 = \pi/2$ and $\theta_2 = -\pi/2$ for length elements ranging between positions $x = -\infty$ and $x = +\infty$. Because $(\sin\theta_1 - \sin\theta_2) = (\sin\pi/2 - \sin(-\pi/2)) = 2$, Equation 30.4 becomes

$$B = \frac{\mu_0 I}{2\pi a} \qquad (30.5)$$

Equations 30.4 and 30.5 both show that the magnitude of the magnetic field is proportional to the current and decreases with increasing distance from the wire, as expected. Equation 30.5 has the same mathematical form as the expression for the magnitude of the electric field due to a long charged wire (see Eq. 24.7).

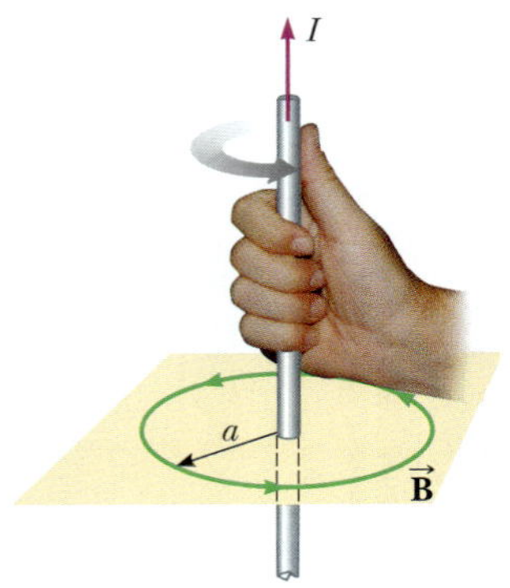

Figure 30.4 The right-hand rule for determining the direction of the magnetic field surrounding a long, straight wire carrying a current. Notice that the magnetic field lines form circles around the wire.

The result of Example 30.1 is important because a current in the form of a long, straight wire occurs often. Figure 30.4 is a perspective view of the magnetic field surrounding a long, straight, current-carrying wire. Because of the wire's symmetry, the magnetic field lines are circles concentric with the wire and lie in planes perpendicular to the wire. The magnitude of $\vec{\mathbf{B}}$ is constant on any circle of radius a and is given by Equation 30.5. A convenient rule for determining the direction of $\vec{\mathbf{B}}$ is to grasp the wire with the right hand, positioning the thumb along the direction of the current. The four fingers wrap in the direction of the magnetic field.

Figure 30.4 also shows that the magnetic field line has no beginning and no end. Rather, it forms a closed loop. That is a major difference between magnetic field lines and electric field lines, which begin on positive charges and end on negative charges. We will explore this feature of magnetic field lines further in Section 30.5.

EXAMPLE 30.2 Magnetic Field Due to a Curved Wire Segment

Calculate the magnetic field at point O for the current-carrying wire segment shown in Figure 30.5. The wire consists of two straight portions and a circular arc of radius a, which subtends an angle θ.

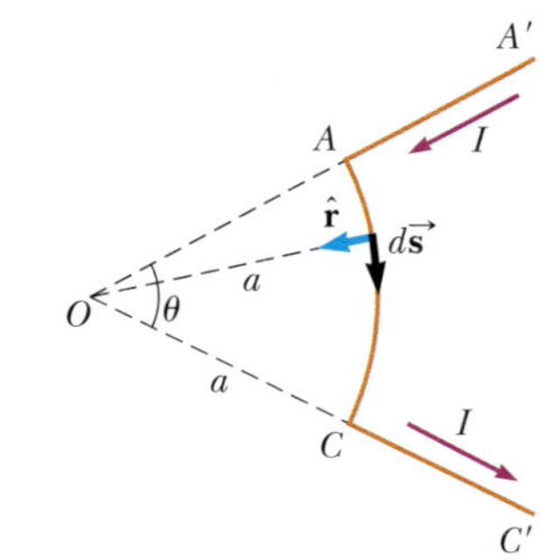

Figure 30.5 (Example 30.2) The magnetic field at O due to the current in the curved segment AC is into the page. The contribution to the field at O due to the current in the two straight segments is zero. The length of the curved segment AC is s.

SOLUTION

Conceptualize The magnetic field at O due to the current in the straight segments AA' and CC' is zero because $d\vec{\mathbf{s}}$ is parallel to $\hat{\mathbf{r}}$ along these paths, which means that $d\vec{\mathbf{s}} \times \hat{\mathbf{r}} = 0$ for these paths.

Categorize Because we can ignore segments AA' and CC', this example is categorized as an application of the Biot–Savart law to the curved wire segment AC.

Analyze Each length element $d\vec{\mathbf{s}}$ along path AC is at the same distance a from O, and the current in each contributes a field element $d\vec{\mathbf{B}}$ directed into the page at O. Furthermore, at every point on AC, $d\vec{\mathbf{s}}$ is perpendicular to $\hat{\mathbf{r}}$; hence, $|d\vec{\mathbf{s}} \times \hat{\mathbf{r}}| = ds$.

From Equation 30.1, find the magnitude of the field at O due to the current in an element of length ds:

$$dB = \frac{\mu_0}{4\pi}\frac{I\,ds}{a^2}$$

Integrate this expression over the curved path *AC*, noting that *I* and *a* are constants:

$$B = \frac{\mu_0 I}{4\pi a^2}\int ds = \frac{\mu_0 I}{4\pi a^2}s$$

From the geometry, note that $s = a\theta$ and substitute:

$$B = \frac{\mu_0 I}{4\pi a^2}(a\theta) = \frac{\mu_0 I}{4\pi a}\theta \qquad \textbf{(30.6)}$$

Finalize Equation 30.6 gives the magnitude of the magnetic field at *O*. The direction of $\vec{\mathbf{B}}$ is into the page at *O* because $d\vec{\mathbf{s}} \times \hat{\mathbf{r}}$ is into the page for every length element.

What If? What if you were asked to find the magnetic field at the center of a circular wire loop of radius *R* that carries a current *I*? Can this question be answered at this point in our understanding of the source of magnetic fields?

Answer Yes, it can. The straight wires in Figure 30.5 do not contribute to the magnetic field. The only contribution is from the curved segment. As the angle θ increases, the curved segment becomes a full circle when $\theta = 2\pi$. Therefore, you can find the magnetic field at the center of a wire loop by letting $\theta = 2\pi$ in Equation 30.6:

$$B = \frac{\mu_0 I}{4\pi a}2\pi = \frac{\mu_0 I}{2a}$$

This result is a limiting case of a more general result discussed in Example 30.3.

EXAMPLE 30.3 Magnetic Field on the Axis of a Circular Current Loop

Consider a circular wire loop of radius *a* located in the *yz* plane and carrying a steady current *I* as in Figure 30.6. Calculate the magnetic field at an axial point *P* a distance *x* from the center of the loop.

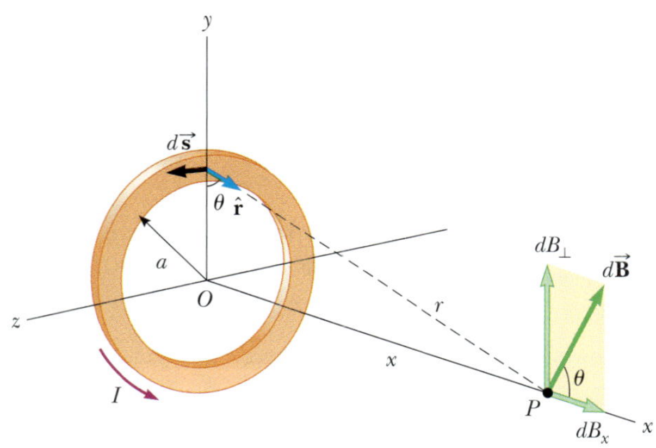

Figure 30.6 (Example 30.3) Geometry for calculating the magnetic field at a point *P* lying on the axis of a current loop. By symmetry, the total field $\vec{\mathbf{B}}$ is along this axis.

SOLUTION

Conceptualize Figure 30.6 shows the magnetic field contribution $d\vec{\mathbf{B}}$ at *P* due to a single current element at the top of the ring. This field vector can be resolved into components dB_x parallel to the axis of the ring and $dB_\perp$ perpendicular to the axis. Think about the magnetic field contributions from a current element at the bottom of the loop. Because of the symmetry of the situation, the perpendicular components of the field due to elements at the top and bottom of the ring cancel. This cancellation occurs for all pairs of segments around the ring, so we can ignore the perpendicular component of the field and focus solely on the parallel components, which simply add.

Categorize We are asked to find the magnetic field due to a simple current distribution, so this example is a typical problem for which the Biot–Savart law is appropriate.

Analyze In this situation, every length element $d\vec{\mathbf{s}}$ is perpendicular to the vector $\hat{\mathbf{r}}$ at the location of the element. Therefore, for any element, $|d\vec{\mathbf{s}} \times \hat{\mathbf{r}}| = (ds)(1)\sin 90° = ds$. Furthermore, all length elements around the loop are at the same distance *r* from *P*, where $r^2 = a^2 + x^2$.

Use Equation 30.1 to find the magnitude of $d\vec{\mathbf{B}}$ due to the current in any length element $d\vec{\mathbf{s}}$:

$$dB = \frac{\mu_0 I}{4\pi}\frac{|d\vec{\mathbf{s}} \times \hat{\mathbf{r}}|}{r^2} = \frac{\mu_0 I}{4\pi}\frac{ds}{(a^2 + x^2)}$$

Find the *x* component of the field element:

$$dB_x = \frac{\mu_0 I}{4\pi}\frac{ds}{(a^2 + x^2)}\cos\theta$$

Integrate over the entire loop:

$$B_x = \oint dB_x = \frac{\mu_0 I}{4\pi} \oint \frac{ds \cos\theta}{a^2 + x^2}$$

From the geometry, evaluate $\cos\theta$:

$$\cos\theta = \frac{a}{(a^2 + x^2)^{1/2}}$$

Substitute this expression for $\cos\theta$ into the integral and note that x, a, and θ are all constant:

$$B_x = \frac{\mu_0 I}{4\pi} \oint \frac{ds}{a^2 + x^2} \frac{a}{(a^2 + x^2)^{1/2}} = \frac{\mu_0 I}{4\pi} \frac{a}{(a^2 + x^2)^{3/2}} \oint ds$$

Integrate around the loop:

$$B_x = \frac{\mu_0 I}{4\pi} \frac{a}{(a^2 + x^2)^{3/2}} (2\pi a) = \frac{\mu_0 I a^2}{2(a^2 + x^2)^{3/2}} \qquad \textbf{(30.7)}$$

Finalize To find the magnetic field at the center of the loop, set $x = 0$ in Equation 30.7. At this special point,

$$B = \frac{\mu_0 I}{2a} \quad (\text{at } x = 0) \qquad \textbf{(30.8)}$$

which is consistent with the result of the **What If?** feature of Example 30.2.

The pattern of magnetic field lines for a circular current loop is shown in Figure 30.7a. For clarity, the lines are drawn for only the plane that contains the axis of the loop. The field-line pattern is axially symmetric and looks like the pattern around a bar magnet, which is shown in Figure 30.7c.

(a) (b) (c)

Figure 30.7 (Example 30.3) (a) Magnetic field lines surrounding a current loop. (b) Magnetic field lines surrounding a current loop, displayed with iron filings. (c) Magnetic field lines surrounding a bar magnet. Notice the similarity between this line pattern and that of a current loop.

What If? What if we consider points on the x axis very far from the loop? How does the magnetic field behave at these distant points?

Answer In this case, in which $x \gg a$, we can neglect the term a^2 in the denominator of Equation 30.7 and obtain

$$B \approx \frac{\mu_0 I a^2}{2x^3} \quad (\text{for } x \gg a) \qquad \textbf{(30.9)}$$

The magnitude of the magnetic moment μ of the loop is defined as the product of current and loop area (see Eq. 29.15): $\mu = I(\pi a^2)$ for our circular loop. We can express Equation 30.9 as

$$B \approx \frac{\mu_0}{2\pi} \frac{\mu}{x^3} \qquad \textbf{(30.10)}$$

This result is similar in form to the expression for the electric field due to an electric dipole, $E = k_e(p/y^3)$ (see Example 23.5), where $p = 2qa$ is the electric dipole moment as defined in Equation 26.16.

30.2 The Magnetic Force Between Two Parallel Conductors

In Chapter 29, we described the magnetic force that acts on a current-carrying conductor placed in an external magnetic field. Because a current in a conductor sets up its own magnetic field, it is easy to understand that two current-carrying conductors exert magnetic forces on each other. Such forces can be used as the basis for defining the ampere and the coulomb.

Consider two long, straight, parallel wires separated by a distance a and carrying currents I_1 and I_2 in the same direction as in Active Figure 30.8. Let's determine the force exerted on one wire due to the magnetic field set up by the other

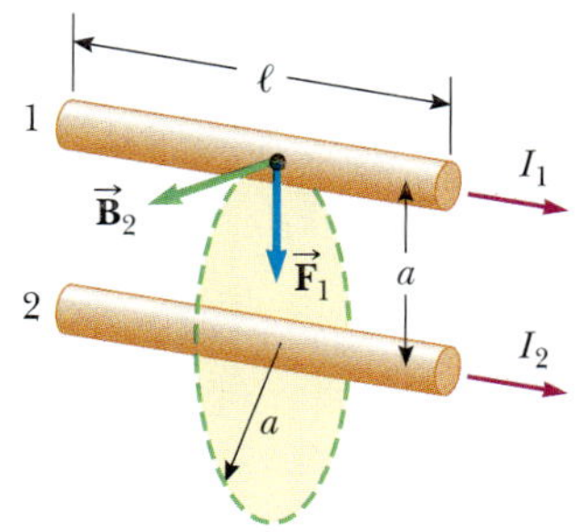

ACTIVE FIGURE 30.8

Two parallel wires that each carry a steady current exert a magnetic force on each other. The field $\vec{\mathbf{B}}_2$ due to the current in wire 2 exerts a magnetic force of magnitude $F_1 = I_1 \ell B_2$ on wire 1. The force is attractive if the currents are parallel (as shown) and repulsive if the currents are antiparallel.

Sign in at www.thomsonedu.com and go to ThomsonNOW to adjust the currents in the wires and the distance between them to see the effect on the force.

wire. Wire 2, which carries a current I_2 and is identified arbitrarily as the source wire, creates a magnetic field $\vec{\mathbf{B}}_2$ at the location of wire 1, the test wire. The direction of $\vec{\mathbf{B}}_2$ is perpendicular to wire 1 as shown in Active Figure 30.8. According to Equation 29.10, the magnetic force on a length ℓ of wire 1 is $\vec{\mathbf{F}}_1 = I_1 \vec{\boldsymbol{\ell}} \times \vec{\mathbf{B}}_2$. Because $\vec{\boldsymbol{\ell}}$ is perpendicular to $\vec{\mathbf{B}}_2$ in this situation, the magnitude of $\vec{\mathbf{F}}_1$ is $F_1 = I_1 \ell B_2$. Because the magnitude of $\vec{\mathbf{B}}_2$ is given by Equation 30.5,

$$F_1 = I_1 \ell B_2 = I_1 \ell \left(\frac{\mu_0 I_2}{2\pi a} \right) = \frac{\mu_0 I_1 I_2}{2\pi a} \ell \qquad \textbf{(30.11)}$$

The direction of $\vec{\mathbf{F}}_1$ is toward wire 2 because $\vec{\boldsymbol{\ell}} \times \vec{\mathbf{B}}_2$ is in that direction. When the field set up at wire 2 by wire 1 is calculated, the force $\vec{\mathbf{F}}_2$ acting on wire 2 is found to be equal in magnitude and opposite in direction to $\vec{\mathbf{F}}_1$, which is what we expect because Newton's third law must be obeyed. When the currents are in opposite directions (that is, when one of the currents is reversed in Active Fig. 30.8), the forces are reversed and the wires repel each other. Hence, **parallel conductors carrying currents in the same direction attract each other, and parallel conductors carrying currents in opposite directions repel each other.**

Because the magnitudes of the forces are the same on both wires, we denote the magnitude of the magnetic force between the wires as simply F_B. We can rewrite this magnitude in terms of the force per unit length:

$$\frac{F_B}{\ell} = \frac{\mu_0 I_1 I_2}{2\pi a} \qquad \textbf{(30.12)}$$

The force between two parallel wires is used to define the **ampere** as follows:

◀ Definition of the ampere

When the magnitude of the force per unit length between two long, parallel wires that carry identical currents and are separated by 1 m is 2×10^{-7} N/m, the current in each wire is defined to be 1 A.

The value 2×10^{-7} N/m is obtained from Equation 30.12 with $I_1 = I_2 = 1$ A and $a = 1$ m. Because this definition is based on a force, a mechanical measurement can be used to standardize the ampere. For instance, the National Institute of Standards and Technology uses an instrument called a *current balance* for primary current measurements. The results are then used to standardize other, more conventional instruments such as ammeters.

The SI unit of charge, the **coulomb**, is defined in terms of the ampere: When a conductor carries a steady current of 1 A, the quantity of charge that flows through a cross section of the conductor in 1 s is 1 C.

In deriving Equations 30.11 and 30.12, we assumed that both wires are long compared with their separation distance. In fact, only one wire needs to be long. The equations accurately describe the forces exerted on each other by a long wire and a straight, parallel wire of limited length ℓ.

Quick Quiz 30.2 A loose spiral spring carrying no current is hung from a ceiling. When a switch is thrown so that a current exists in the spring, do the coils (a) move closer together, (b) move farther apart, or (c) not move at all?

EXAMPLE 30.4 Suspending a Wire

Two infinitely long, parallel wires are lying on the ground 1.00 cm apart as shown in Figure 30.9a. A third wire, of length 10.0 m and mass 400 g, carries a current of $I_1 = 100$ A and is levitated above the first two wires, at a horizontal position midway between them. The infinitely long wires carry equal currents I_2 in the same direction, but in the direction opposite to that in the levitated wire. What current must the infinitely long wires carry so that the three wires form an equilateral triangle?

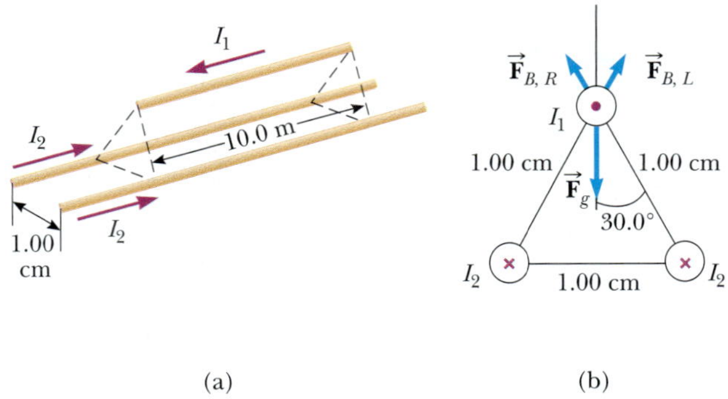

Figure 30.9 (Example 30.4) (a) Two current-carrying wires lie on the ground and suspend a third wire in the air by magnetic forces. (b) End view. In the situation described in the example, the three wires form an equilateral triangle. The two magnetic forces on the levitated wire are $\vec{F}_{B,L}$, the force due to the left-hand wire on the ground, and $\vec{F}_{B,R}$, the force due to the right-hand wire. The gravitational force $\vec{F}_g$ on the levitated wire is also shown.

SOLUTION

Conceptualize Because the current in the short wire is opposite those in the long wires, the short wire is repelled from both of the others. Imagine the currents in the long wires are increased. The repulsive force becomes stronger, and the levitated wire rises to the point at which the weight of the wire is once again levitated in equilibrium. Figure 30.9b shows the desired situation with the three wires forming an equilateral triangle.

Categorize We model the levitated wire as a particle in equilibrium.

Analyze The horizontal components of the magnetic forces on the levitated wire cancel. The vertical components are both positive and add together.

Find the total magnetic force in the upward direction on the levitated wire:

$$\vec{F}_B = 2\left(\frac{\mu_0 I_1 I_2}{2\pi a}\ell\right)\cos 30.0°\hat{k} = 0.866\frac{\mu_0 I_1 I_2}{\pi a}\ell\hat{k}$$

Find the gravitational force on the levitated wire:

$$\vec{F}_g = -mg\hat{k}$$

Apply the particle in equilibrium model by adding the forces and setting the net force equal to zero:

$$\sum \vec{F} = \vec{F}_B + \vec{F}_g = 0.866\frac{\mu_0 I_1 I_2}{\pi a}\ell\hat{k} - mg\hat{k} = 0$$

Solve for the current in the wires on the ground:

$$I_2 = \frac{mg\pi a}{0.866\mu_0 I_1 \ell}$$

Substitute numerical values:

$$I_2 = \frac{(0.400\text{ kg})(9.80\text{ m/s}^2)\pi(0.010\,0\text{ m})}{0.866(4\pi \times 10^{-7}\text{ T}\cdot\text{m/A})(100\text{ A})(10.0\text{ m})}$$

$$= 113\text{ A}$$

Finalize The currents in all wires are on the order of 10^2 A. Such large currents would require specialized equipment. Therefore, this situation would be difficult to establish in practice.

30.3 Ampère's Law

Oersted's 1819 discovery about deflected compass needles demonstrates that a current-carrying conductor produces a magnetic field. Active Figure 30.10a shows how this effect can be demonstrated in the classroom. Several compass needles are placed in a horizontal plane near a long, vertical wire. When no current is present in the wire, all the needles point in the same direction (that of the Earth's magnetic field) as expected. When the wire carries a strong, steady current, the needles all deflect in a direction tangent to the circle as in Active Figure 30.10b. These observations demonstrate that the direction of the magnetic field produced by the current in the wire is consistent with the right-hand rule described in Figure 30.4. When the current is reversed, the needles in Active Figure 30.10b also reverse.

Because the compass needles point in the direction of $\vec{\mathbf{B}}$, we conclude that the lines of $\vec{\mathbf{B}}$ form circles around the wire as discussed in Section 30.1. By symmetry, the magnitude of $\vec{\mathbf{B}}$ is the same everywhere on a circular path centered on the wire and lying in a plane perpendicular to the wire. By varying the current and distance from the wire, we find that B is proportional to the current and inversely proportional to the distance from the wire as described by Equation 30.5.

Now let's evaluate the product $\vec{\mathbf{B}} \cdot d\vec{\mathbf{s}}$ for a small length element $d\vec{\mathbf{s}}$ on the circular path defined by the compass needles and sum the products for all elements over the closed circular path.[1] Along this path, the vectors $d\vec{\mathbf{s}}$ and $\vec{\mathbf{B}}$ are parallel at each point (see Active Fig. 30.10b), so $\vec{\mathbf{B}} \cdot d\vec{\mathbf{s}} = B\,ds$. Furthermore, the magnitude of $\vec{\mathbf{B}}$ is constant on this circle and is given by Equation 30.5. Therefore, the sum of the products $B\,ds$ over the closed path, which is equivalent to the line integral of $\vec{\mathbf{B}} \cdot d\vec{\mathbf{s}}$, is

$$\oint \vec{\mathbf{B}} \cdot d\vec{\mathbf{s}} = B \oint ds = \frac{\mu_0 I}{2\pi r}(2\pi r) = \mu_0 I$$

where $\oint ds = 2\pi r$ is the circumference of the circular path. Although this result was calculated for the special case of a circular path surrounding a wire, it holds for a closed path of *any* shape (an *amperian loop*) surrounding a current that exists in an unbroken circuit. The general case, known as **Ampère's law**, can be stated as follows:

Ampère's law

The line integral of $\vec{\mathbf{B}} \cdot d\vec{\mathbf{s}}$ around any closed path equals $\mu_0 I$, where I is the total steady current passing through any surface bounded by the closed path:

$$\oint \vec{\mathbf{B}} \cdot d\vec{\mathbf{s}} = \mu_0 I \qquad \textbf{(30.13)}$$

Leonard de Selva/CORBIS

ANDRE-MARIE AMPÈRE

French Physicist (1775–1836)

Ampère is credited with the discovery of electromagnetism, which is the relationship between electric currents and magnetic fields. Ampère's genius, particularly in mathematics, became evident by the time he was 12 years old; his personal life, however, was filled with tragedy. His father, a wealthy city official, was guillotined during the French Revolution, and his wife died young, in 1803. Ampère died at the age of 61 of pneumonia. His judgment of his life is clear from the epitaph he chose for his gravestone: *Tandem Felix* (Happy at Last).

PITFALL PREVENTION 30.2

Avoiding Problems with Signs

When using Ampère's law, apply the following right-hand rule. Point your thumb in the direction of the current through the amperian loop. Your curled fingers then point in the direction that you should integrate when traversing the loop to avoid having to define the current as negative.

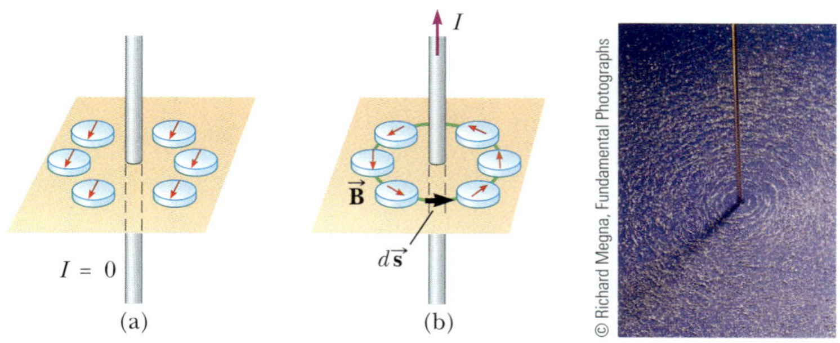

ACTIVE FIGURE 30.10

(a) When no current is present in the wire, all compass needles point in the same direction (toward the Earth's north pole). (b) When the wire carries a strong current, the compass needles deflect in a direction tangent to the circle, which is the direction of the magnetic field created by the current. (c) Circular magnetic field lines surrounding a current-carrying conductor, displayed with iron filings.

Sign in at www.thomsonedu.com and go to ThomsonNOW to change the value of the current and see the effect on the compasses.

[1] You may wonder why we would choose to evaluate this scalar product. The origin of Ampère's law is in 19th-century science, in which a "magnetic charge" (the supposed analog to an isolated electric charge) was imagined to be moved around a circular field line. The work done on the charge was related to $\vec{\mathbf{B}} \cdot d\vec{\mathbf{s}}$, just as the work done moving an electric charge in an electric field is related to $\vec{\mathbf{E}} \cdot d\vec{\mathbf{s}}$. Therefore, Ampère's law, a valid and useful principle, arose from an erroneous and abandoned work calculation!

Ampère's law describes the creation of magnetic fields by all continuous current configurations, but at our mathematical level it is useful only for calculating the magnetic field of current configurations having a high degree of symmetry. Its use is similar to that of Gauss's law in calculating electric fields for highly symmetric charge distributions.

Quick Quiz 30.3 Rank the magnitudes of $\oint \vec{\mathbf{B}} \cdot d\vec{\mathbf{s}}$ for the closed paths a through d in Figure 30.11 from least to greatest.

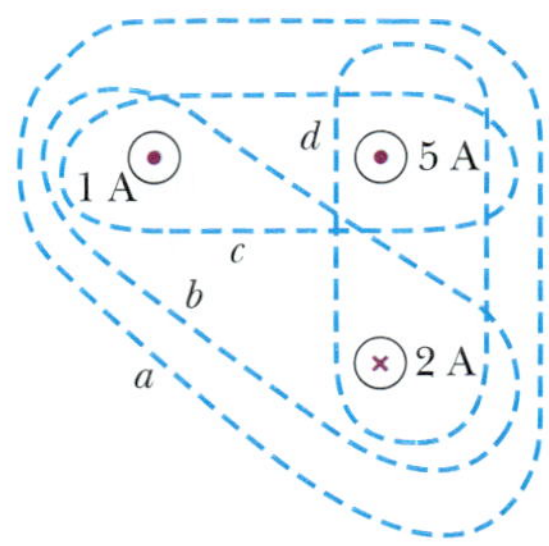

Figure 30.11 (Quick Quiz 30.3) Four closed paths around three current-carrying wires.

Quick Quiz 30.4 Rank the magnitudes of $\oint \vec{\mathbf{B}} \cdot d\vec{\mathbf{s}}$ for the closed paths a through d in Figure 30.12 from least to greatest.

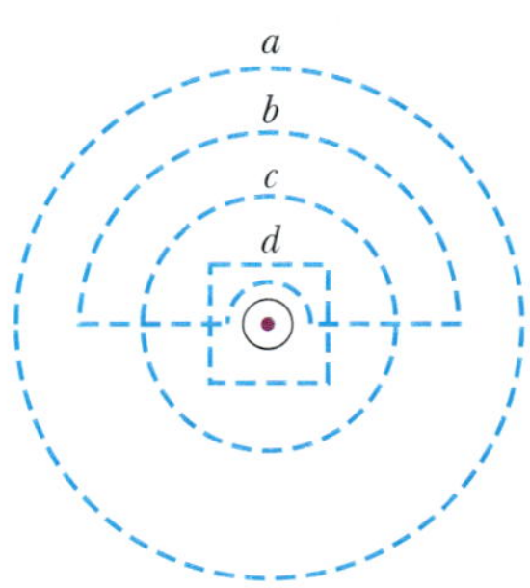

Figure 30.12 (Quick Quiz 30.4) Several closed paths near a single current-carrying wire.

EXAMPLE 30.5 The Magnetic Field Created by a Long Current-Carrying Wire

A long, straight wire of radius R carries a steady current I that is uniformly distributed through the cross section of the wire (Fig. 30.13). Calculate the magnetic field a distance r from the center of the wire in the regions $r \geq R$ and $r < R$.

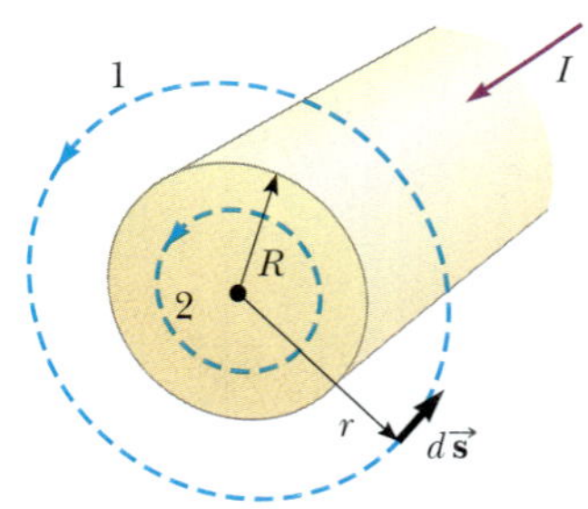

Figure 30.13 (Example 30.5) A long, straight wire of radius R carrying a steady current I uniformly distributed across the cross section of the wire. The magnetic field at any point can be calculated from Ampère's law using a circular path of radius r, concentric with the wire.

SOLUTION

Conceptualize Study Figure 30.13 to understand the structure of the wire and the current in the wire. The current creates magnetic fields everywhere, both inside and outside the wire.

Categorize Because the wire has a high degree of symmetry, we categorize this example as an Ampère's law problem. For the $r \geq R$ case, we should arrive at the same result as was obtained in Example 30.1, where we applied the Biot–Savart law to the same situation.

Analyze For the magnetic field exterior to the wire, let us choose for our path of integration circle 1 in Figure 30.13. From symmetry, $\vec{\mathbf{B}}$ must be constant in magnitude and parallel to $d\vec{\mathbf{s}}$ at every point on this circle.

Note that the total current passing through the plane of the circle is I and apply Ampère's law:

$$\oint \vec{\mathbf{B}} \cdot d\vec{\mathbf{s}} = B \oint ds = B(2\pi r) = \mu_0 I$$

Solve for B:

$$B = \frac{\mu_0 I}{2\pi r} \quad (\text{for } r \geq R) \tag{30.14}$$

Now consider the interior of the wire, where $r < R$. Here the current I' passing through the plane of circle 2 is less than the total current I.

Set the ratio of the current I' enclosed by circle 2 to the entire current I equal to the ratio of the area πr^2 enclosed by circle 2 to the cross-sectional area πR^2 of the wire:

$$\frac{I'}{I} = \frac{\pi r^2}{\pi R^2}$$

Solve for I':

$$I' = \frac{r^2}{R^2} I$$

Apply Ampère's law to circle 2:

$$\oint \vec{\mathbf{B}} \cdot d\vec{\mathbf{s}} = B(2\pi r) = \mu_0 I' = \mu_0\left(\frac{r^2}{R^2} I\right)$$

Solve for B:

$$B = \left(\frac{\mu_0 I}{2\pi R^2}\right) r \quad (\text{for } r < R) \tag{30.15}$$

Finalize The magnetic field exterior to the wire is identical in form to Equation 30.5. As is often the case in highly symmetric situations, it is much easier to use Ampère's law than the Biot–Savart law (Example 30.1). The magnetic field interior to the wire is similar in form to the expression for the electric field inside a uniformly charged sphere (see Example 24.3). The magnitude of the magnetic field versus r for this configuration is plotted in Figure 30.14. Inside the wire, $B \to 0$ as $r \to 0$. Furthermore, Equations 30.14 and 30.15 give the same value of the magnetic field at $r = R$, demonstrating that the magnetic field is continuous at the surface of the wire.

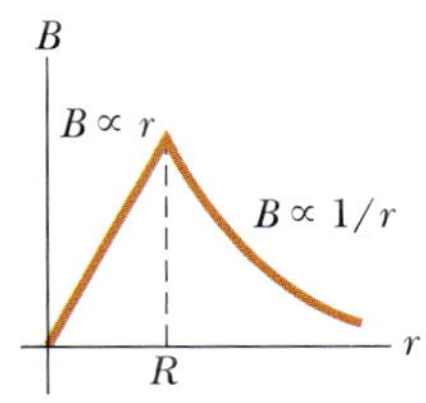

Figure 30.14 (Example 30.5) Magnitude of the magnetic field versus r for the wire shown in Figure 30.13. The field is proportional to r inside the wire and varies as $1/r$ outside the wire.

EXAMPLE 30.6 The Magnetic Field Created by a Toroid

A device called a *toroid* (Fig. 30.15) is often used to create an almost uniform magnetic field in some enclosed area. The device consists of a conducting wire wrapped around a ring (a *torus*) made of a nonconducting material. For a toroid having N closely spaced turns of wire, calculate the magnetic field in the region occupied by the torus, a distance r from the center.

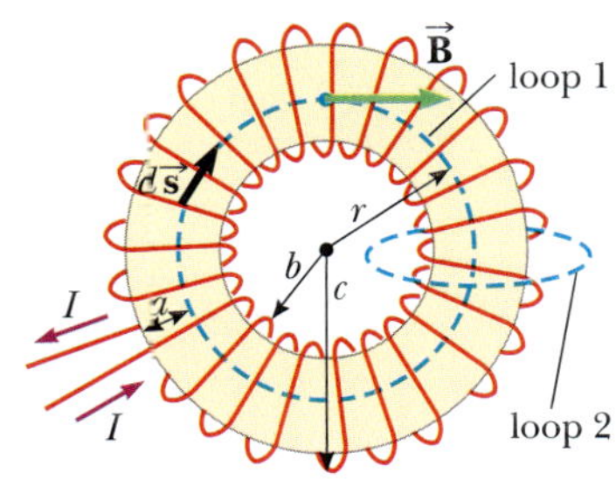

Figure 30.15 (Example 30.6) A toroid consisting of many turns of wire. If the turns are closely spaced, the magnetic field in the interior of the torus (the gold-shaded region) is tangent to the dashed circle (loop 1) and varies as $1/r$. The dimension a is the cross-sectional radius of the torus. The field outside the toroid is very small and can be described by using the amperian loop (loop 2) at the right side, perpendicular to the page.

SOLUTION

Conceptualize Study Figure 30.15 carefully to understand how the wire is wrapped around the torus. The torus could be a solid material or it could be air, with a stiff wire wrapped into the shape shown in Figure 30.15 to form an empty toroid.

Categorize Because the toroid has a high degree of symmetry, we categorize this example as an Ampère's law problem.

Analyze Consider the circular amperian loop (loop 1) of radius r in the plane of Figure 30.15. By symmetry, the magnitude of the field is constant on this circle and tangent to it, so $\vec{\mathbf{B}} \cdot d\vec{\mathbf{s}} = B\,ds$. Furthermore, the wire passes through the loop N times, so the total current through the loop is NI.

Apply Ampère's law to loop 1:

$$\oint \vec{\mathbf{B}} \cdot d\vec{\mathbf{s}} = B \oint ds = B(2\pi r) = \mu_0 NI$$

Solve for B:

$$B = \frac{\mu_0 NI}{2\pi r} \quad (30.16)$$

Finalize This result shows that B varies as $1/r$ and hence is *nonuniform* in the region occupied by the torus. If, however, r is very large compared with the cross-sectional radius a of the torus, the field is approximately uniform inside the torus.

For an ideal toroid, in which the turns are closely spaced, the external magnetic field is close to zero, but it is not exactly zero. In Figure 30.15, imagine the radius r of the amperian loop to be either smaller than b or larger than c. In either case, the loop encloses zero net current, so $\oint \vec{\mathbf{B}} \cdot d\vec{\mathbf{s}} = 0$. You might think that this result proves that $\vec{\mathbf{B}} = 0$, but it does not. Consider the amperian loop (loop 2) on the right side of the toroid in Figure 30.15. The plane of this loop is perpendicular to the page, and the toroid passes through the loop. As charges enter the toroid as indicated by the current directions in Figure 30.15, they work their way counterclockwise around the toroid. Therefore, a current passes through the perpendicular amperian loop! This current is small, but not zero. As a result, the toroid acts as a current loop and produces a weak external field of the form shown in Figure 30.7. The reason $\oint \vec{\mathbf{B}} \cdot d\vec{\mathbf{s}} = 0$ for the amperian loops of radius $r < b$ and $r > c$ in the plane of the page is that the field lines are perpendicular to $d\vec{\mathbf{s}}$, *not* because $\vec{\mathbf{B}} = 0$.

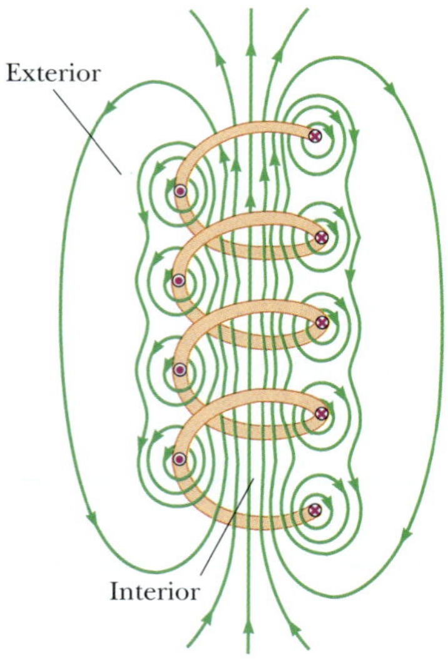

Figure 30.16 The magnetic field lines for a loosely wound solenoid.

30.4 The Magnetic Field of a Solenoid

A **solenoid** is a long wire wound in the form of a helix. With this configuration, a reasonably uniform magnetic field can be produced in the space surrounded by the turns of wire—which we shall call the *interior* of the solenoid—when the solenoid carries a current. When the turns are closely spaced, each can be approximated as a circular loop; the net magnetic field is the vector sum of the fields resulting from all the turns.

Figure 30.16 shows the magnetic field lines surrounding a loosely wound solenoid. The field lines in the interior are nearly parallel to one another, are uniformly distributed, and are close together, indicating that the field in this space is strong and almost uniform.

If the turns are closely spaced and the solenoid is of finite length, the magnetic field lines are as shown in Figure 30.17a. This field line distribution is similar to that surrounding a bar magnet (Fig. 30.17b). Hence, one end of the solenoid behaves like the north pole of a magnet and the opposite end behaves like the south pole. As the length of the solenoid increases, the interior field becomes more uniform and the exterior field becomes weaker. An *ideal solenoid* is

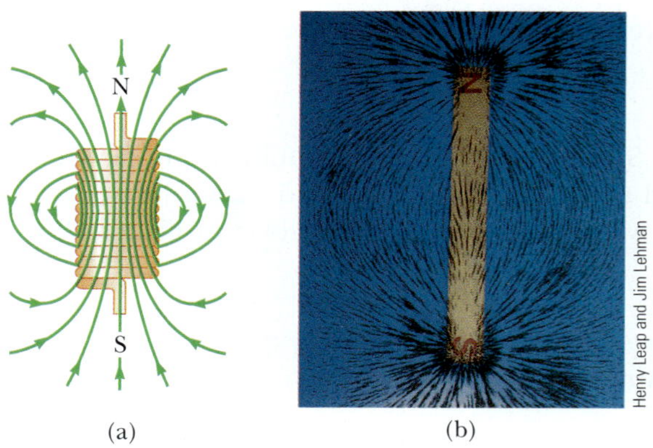

Figure 30.17 (a) Magnetic field lines for a tightly wound solenoid of finite length, carrying a steady current. The field in the interior space is strong and nearly uniform. Notice that the field lines resemble those of a bar magnet, meaning that the solenoid effectively has north and south poles. (b) The magnetic field pattern of a bar magnet, displayed with small iron filings on a sheet of paper.

approached when the turns are closely spaced and the length is much greater than the radius of the turns. Figure 30.18 shows a longitudinal cross section of part of such a solenoid carrying a current I. In this case, the external field is close to zero and the interior field is uniform over a great volume.

Consider the amperian loop (loop 1) perpendicular to the page in Figure 30.18, surrounding the ideal solenoid. This loop encloses a small current as the charges in the wire move coil by coil along the length of the solenoid. Therefore, there is a nonzero magnetic field outside the solenoid. It is a weak field, with circular field lines, like those due to a line of current as in Figure 30.4. For an ideal solenoid, this weak field is the only field external to the solenoid. We could eliminate this field in Figure 30.18 by adding a second layer of turns of wire outside the first layer, with the current carried along the axis of the solenoid in the opposite direction compared with the first layer. Then the net current along the axis is zero.

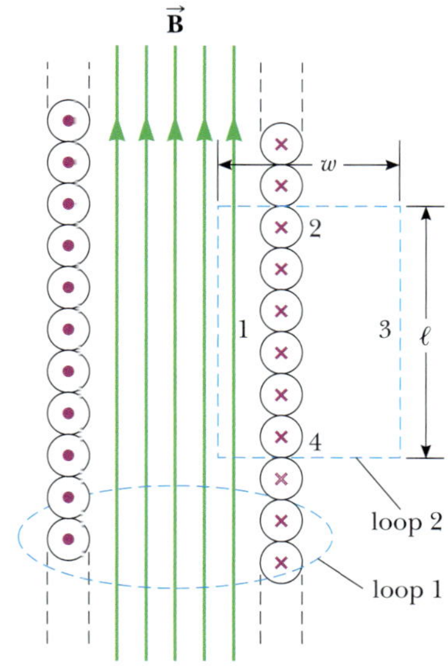

Figure 30.18 Cross-sectional view of an ideal solenoid, where the interior magnetic field is uniform and the exterior field is close to zero. Ampère's law applied to the circular path near the bottom whose plane is perpendicular to the page can be used to show that there is a weak field outside the solenoid. Ampère's law applied to the rectangular dashed path in the plane of the page can be used to calculate the magnitude of the interior field.

We can use Ampère's law to obtain a quantitative expression for the interior magnetic field in an ideal solenoid. Because the solenoid is ideal, $\vec{\mathbf{B}}$ in the interior space is uniform and parallel to the axis and the magnetic field lines in the exterior space form circles around the solenoid. The planes of these circles are perpendicular to the page. Consider the rectangular path (loop 2) of length ℓ and width w shown in Figure 30.18. Let's apply Ampère's law to this path by evaluating the integral of $\vec{\mathbf{B}} \cdot d\vec{\mathbf{s}}$ over each side of the rectangle. The contribution along side 3 is zero because the magnetic field lines are perpendicular to the path in this region. The contributions from sides 2 and 4 are both zero, again because $\vec{\mathbf{B}}$ is perpendicular to $d\vec{\mathbf{s}}$ along these paths, both inside and outside the solenoid. Side 1 gives a contribution to the integral because along this path $\vec{\mathbf{B}}$ is uniform and parallel to $d\vec{\mathbf{s}}$. The integral over the closed rectangular path is therefore

$$\oint \vec{\mathbf{B}} \cdot d\vec{\mathbf{s}} = \int_{\text{path 1}} \vec{\mathbf{B}} \cdot d\vec{\mathbf{s}} = B \int_{\text{path 1}} ds = B\ell$$

The right side of Ampère's law involves the total current I through the area bounded by the path of integration. In this case, the total current through the rectangular path equals the current through each turn multiplied by the number of turns. If N is the number of turns in the length ℓ, the total current through the rectangle is NI. Therefore, Ampère's law applied to this path gives

$$\oint \vec{\mathbf{B}} \cdot d\vec{\mathbf{s}} = B\ell = \mu_0 NI$$

$$B = \mu_0 \frac{N}{\ell} I = \mu_0 nI \qquad \textbf{(30.17)}$$

◀ Magnetic field inside a solenoid

where $n = N/\ell$ is the number of turns per unit length.

We also could obtain this result by reconsidering the magnetic field of a toroid (see Example 30.6). If the radius r of the torus in Figure 30.15 containing N turns is much greater than the toroid's cross-sectional radius a, a short section of the toroid approximates a solenoid for which $n = N/2\pi r$. In this limit, Equation 30.16 agrees with Equation 30.17.

Equation 30.17 is valid only for points near the center (that is, far from the ends) of a very long solenoid. As you might expect, the field near each end is smaller than the value given by Equation 30.17. At the very end of a long solenoid, the magnitude of the field is half the magnitude at the center (see Problem 36).

Quick Quiz 30.5 Consider a solenoid that is very long compared with its radius. Of the following choices, what is the most effective way to increase the magnetic field in the interior of the solenoid? (a) double its length, keeping the number of turns per unit length constant (b) reduce its radius by half, keeping the number of turns per unit length constant (c) overwrap the entire solenoid with an additional layer of current-carrying wire

---

30.5 Gauss's Law in Magnetism

The flux associated with a magnetic field is defined in a manner similar to that used to define electric flux (see Eq. 24.3). Consider an element of area dA on an arbitrarily shaped surface as shown in Figure 30.19. If the magnetic field at this element is $\vec{\mathbf{B}}$, the magnetic flux through the element is $\vec{\mathbf{B}} \cdot d\vec{\mathbf{A}}$, where $d\vec{\mathbf{A}}$ is a vector that is perpendicular to the surface and has a magnitude equal to the area dA. Therefore, the total magnetic flux Φ_B through the surface is

Definition of magnetic flux ▶

$$\Phi_B \equiv \int \vec{\mathbf{B}} \cdot d\vec{\mathbf{A}} \tag{30.18}$$

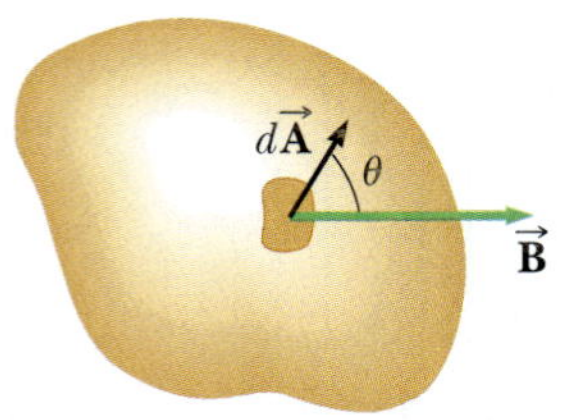

Figure 30.19 The magnetic flux through an area element dA is $\vec{\mathbf{B}} \cdot d\vec{\mathbf{A}} = B\,dA\cos\theta$, where $d\vec{\mathbf{A}}$ is a vector perpendicular to the surface.

Consider the special case of a plane of area A in a uniform field $\vec{\mathbf{B}}$ that makes an angle θ with $d\vec{\mathbf{A}}$. The magnetic flux through the plane in this case is

$$\Phi_B = BA\cos\theta \tag{30.19}$$

If the magnetic field is parallel to the plane as in Active Figure 30.20a, then $\theta = 90°$ and the flux through the plane is zero. If the field is perpendicular to the plane as in Active Figure 30.20b, then $\theta = 0$ and the flux through the plane is BA (the maximum value).

The unit of magnetic flux is $\text{T} \cdot \text{m}^2$, which is defined as a *weber* (Wb); $1 \text{ Wb} = 1 \text{ T} \cdot \text{m}^2$.

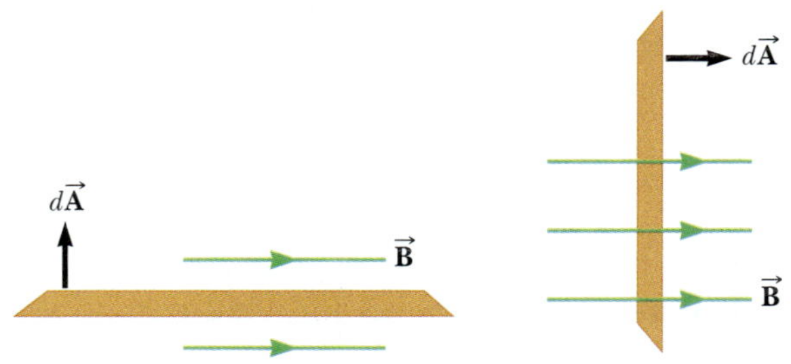

ACTIVE FIGURE 30.20

Magnetic flux through a plane lying in a magnetic field. (a) The flux through the plane is zero when the magnetic field is parallel to the plane surface. (b) The flux through the plane is a maximum when the magnetic field is perpendicular to the plane.

Sign in at www.thomsonedu.com and go to ThomsonNOW to rotate the plane and change the value of the field to see the effect on the flux.

EXAMPLE 30.7 Magnetic Flux Through a Rectangular Loop

A rectangular loop of width a and length b is located near a long wire carrying a current I (Fig. 30.21). The distance between the wire and the closest side of the loop is c. The wire is parallel to the long side of the loop. Find the total magnetic flux through the loop due to the current in the wire.

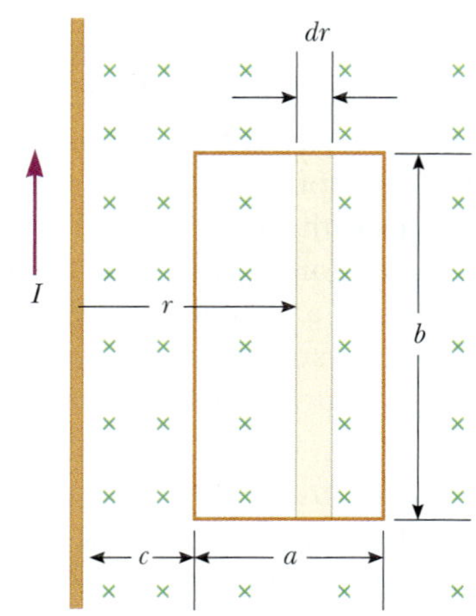

Figure 30.21 (Example 30.7) The magnetic field due to the wire carrying a current I is not uniform over the rectangular loop.

SOLUTION

Conceptualize We know that the magnetic field is a function of distance r from a long wire. Therefore, the magnetic field varies over the area of the rectangular loop.

Categorize Because the magnetic field varies over the area of the loop, we must integrate over this area to find the total flux.

Analyze Noting that $\vec{\mathbf{B}}$ is parallel to $d\vec{\mathbf{A}}$ at any point within the loop, find the magnetic flux through the rectangular area using Equation 30.18 and incorporate Equation 30.14 for the magnetic field:

$$\Phi_B = \int \vec{\mathbf{B}} \cdot d\vec{\mathbf{A}} = \int B\,dA = \int \frac{\mu_0 I}{2\pi r}\,dA$$

Express the area element (the tan strip in Fig. 30.21) as $dA = b\,dr$ and substitute:

$$\Phi_B = \int \frac{\mu_0 I}{2\pi r}\,b\,dr = \frac{\mu_0 I b}{2\pi}\int \frac{dr}{r}$$

Integrate from $r = c$ to $r = a + c$:

$$\Phi_B = \frac{\mu_0 I b}{2\pi}\int_c^{a+c}\frac{dr}{r} = \frac{\mu_0 I b}{2\pi}\ln r\bigg|_c^{a+c}$$

$$= \frac{\mu_0 I b}{2\pi}\ln\left(\frac{a+c}{c}\right) = \frac{\mu_0 I b}{2\pi}\ln\left(1 + \frac{a}{c}\right)$$

Finalize Notice how the flux depends on the size of the loop. Increasing either a or b increases the flux as expected. If c becomes large such that the loop is very far from the wire, the flux approaches zero, also as expected. If c goes to zero, the flux becomes infinite. In principle, this infinite value occurs because the field becomes infinite at $r = 0$ (assuming an infinitesimally thin wire). That will not happen in reality because the thickness of the wire prevents the left edge of the loop from reaching $r = 0$.

In Chapter 24, we found that the electric flux through a closed surface surrounding a net charge is proportional to that charge (Gauss's law). In other words, the number of electric field lines leaving the surface depends only on the net charge within it. This behavior exists because electric field lines originate and terminate on electric charges.

The situation is quite different for magnetic fields, which are continuous and form closed loops. In other words, as illustrated by the magnetic field lines of a current in Figure 30.4 and of a bar magnet in Figure 30.22, magnetic field lines do not begin or end at any point. For any closed surface such as the one outlined by the dashed line in Figure 30.22, the number of lines entering the surface equals the number leaving the surface; therefore, the net magnetic flux is zero. In contrast, for a closed surface surrounding one charge of an electric dipole (Fig. 30.23), the net electric flux is not zero.

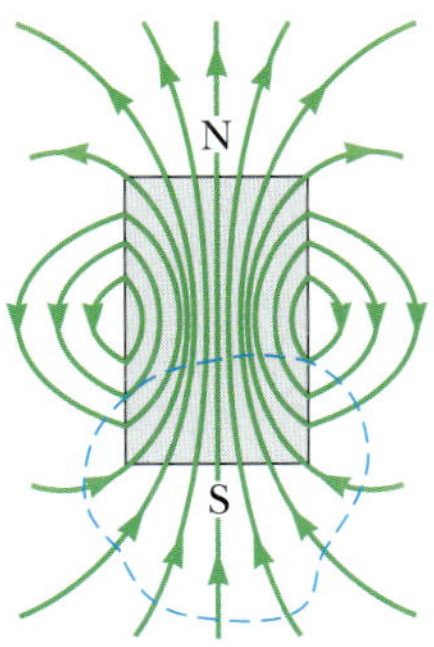

Figure 30.22 The magnetic field lines of a bar magnet form closed loops. Notice that the net magnetic flux through a closed surface surrounding one of the poles (or any other closed surface) is zero. (The dashed line represents the intersection of the surface with the page.)

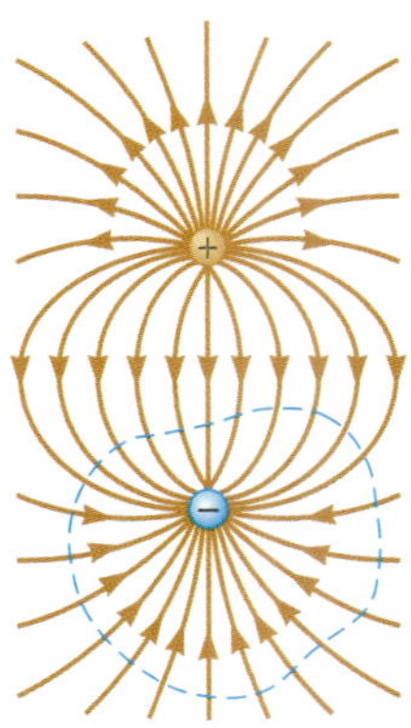

Figure 30.23 The electric field lines surrounding an electric dipole begin on the positive charge and terminate on the negative charge. The electric flux through a closed surface surrounding one of the charges is not zero.

Gauss's law in magnetism states that

Gauss's law in magnetism ▶

the net magnetic flux through any closed surface is always zero:

$$\oint \vec{\mathbf{B}} \cdot d\vec{\mathbf{A}} = 0 \qquad (30.20)$$

This statement represents that isolated magnetic poles (monopoles) have never been detected and perhaps do not exist. Nonetheless, scientists continue the search because certain theories that are otherwise successful in explaining fundamental physical behavior suggest the possible existence of magnetic monopoles.

30.6 Magnetism in Matter

The magnetic field produced by a current in a coil of wire gives us a hint as to what causes certain materials to exhibit strong magnetic properties. Earlier we found that a coil like the one shown in Figure 30.17a has a north pole and a south pole. In general, *any* current loop has a magnetic field and therefore has a magnetic dipole moment, including the atomic-level current loops described in some models of the atom.

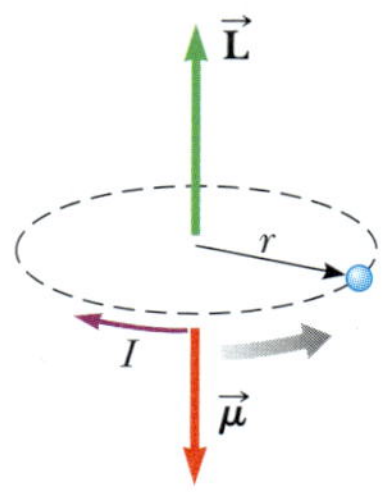

Figure 30.24 An electron moving in the direction of the gray arrow in a circular orbit of radius r has an angular momentum $\vec{\mathbf{L}}$ in one direction and a magnetic moment $\vec{\boldsymbol{\mu}}$ in the opposite direction. Because the electron carries a negative charge, the direction of the current due to its motion about the nucleus is opposite the direction of that motion.

The Magnetic Moments of Atoms

Let's begin our discussion with a classical model of the atom in which electrons move in circular orbits around the much more massive nucleus. In this model, an orbiting electron constitutes a tiny current loop (because it is a moving charge) and the magnetic moment of the electron is associated with this orbital motion. Although this model has many deficiencies, some of its predictions are in good agreement with the correct theory, which is expressed in terms of quantum physics.

In our classical model, we assume an electron moves with constant speed v in a circular orbit of radius r about the nucleus as in Figure 30.24. The current I associated with this orbiting electron is its charge e divided by its period T. Using $T = 2\pi/\omega$ and $\omega = v/r$ gives

$$I = \frac{e}{T} = \frac{e\omega}{2\pi} = \frac{ev}{2\pi r}$$

The magnitude of the magnetic moment associated with this current loop is given by $\mu = IA$, where $A = \pi r^2$ is the area enclosed by the orbit. Therefore,

$$\mu = IA = \left(\frac{ev}{2\pi r}\right)\pi r^2 = \tfrac{1}{2}evr \qquad (30.21)$$

Because the magnitude of the orbital angular momentum of the electron is given by $L = m_e vr$ (Eq. 11.12 with $\phi = 90°$), the magnetic moment can be written as

Orbital magnetic moment ▶

$$\mu = \left(\frac{e}{2m_e}\right)L \qquad (30.22)$$

This result demonstrates that **the magnetic moment of the electron is proportional to its orbital angular momentum**. Because the electron is negatively charged, the vectors $\vec{\boldsymbol{\mu}}$ and $\vec{\mathbf{L}}$ point in *opposite* directions. Both vectors are perpendicular to the plane of the orbit as indicated in Figure 30.24.

A fundamental outcome of quantum physics is that orbital angular momentum is quantized and is equal to multiples of $\hbar = h/2\pi = 1.05 \times 10^{-34}$ J·s, where h is

Planck's constant (see Chapter 40). The smallest nonzero value of the electron's magnetic moment resulting from its orbital motion is

$$\mu = \sqrt{2}\frac{e}{2m_e}\hbar \tag{30.23}$$

We shall see in Chapter 42 how expressions such as Equation 30.23 arise.

Because all substances contain electrons, you may wonder why most substances are not magnetic. The main reason is that, in most substances, the magnetic moment of one electron in an atom is canceled by that of another electron orbiting in the opposite direction. The net result is that, for most materials, **the magnetic effect produced by the orbital motion of the electrons is either zero or very small.**

In addition to its orbital magnetic moment, an electron (as well as protons, neutrons, and other particles) has an intrinsic property called **spin** that also contributes to its magnetic moment. Classically, the electron might be viewed as spinning about its axis as shown in Figure 30.25, but you should be very careful with the classical interpretation. The magnitude of the angular momentum $\vec{\mathbf{S}}$ associated with spin is on the same order of magnitude as the magnitude of the angular momentum $\vec{\mathbf{L}}$ due to the orbital motion. The magnitude of the spin angular momentum of an electron predicted by quantum theory is

$$S = \frac{\sqrt{3}}{2}\hbar$$

The magnetic moment characteristically associated with the spin of an electron has the value

$$\mu_{\text{spin}} = \frac{e\hbar}{2m_e} \tag{30.24}$$

This combination of constants is called the **Bohr magneton μ_B**:

$$\mu_B = \frac{e\hbar}{2m_e} = 9.27 \times 10^{-24}\ \text{J/T} \tag{30.25}$$

Therefore, atomic magnetic moments can be expressed as multiples of the Bohr magneton. (Note that $1\ \text{J/T} = 1\ \text{A}\cdot\text{m}^2$.)

In atoms containing many electrons, the electrons usually pair up with their spins opposite each other; therefore, the spin magnetic moments cancel. Atoms containing an odd number of electrons, however, must have at least one unpaired electron and therefore some spin magnetic moment. The total magnetic moment of an atom is the vector sum of the orbital and spin magnetic moments, and a few examples are given in Table 30.1. Notice that helium and neon have zero moments because their individual spin and orbital moments cancel.

The nucleus of an atom also has a magnetic moment associated with its constituent protons and neutrons. The magnetic moment of a proton or neutron, however, is much smaller than that of an electron and can usually be neglected. We can understand this smaller value by inspecting Equation 30.25 and replacing the mass of the electron with the mass of a proton or a neutron. Because the masses of the proton and neutron are much greater than that of the electron, their magnetic moments are on the order of 10^3 times smaller than that of the electron.

Ferromagnetism

A small number of crystalline substances exhibit strong magnetic effects called **ferromagnetism**. Some examples of ferromagnetic substances are iron, cobalt, nickel, gadolinium, and dysprosium. These substances contain permanent atomic magnetic moments that tend to align parallel to each other even in a weak external magnetic field. Once the moments are aligned, the substance remains magnetized

PITFALL PREVENTION 30.3

The Electron Does Not Spin

The electron is *not* physically spinning. It has an intrinsic angular momentum *as if it were spinning*, but the notion of rotation for a point particle is meaningless. Rotation applies only to a *rigid object*, with an extent in space, as in Chapter 10. Spin angular momentum is actually a relativistic effect.

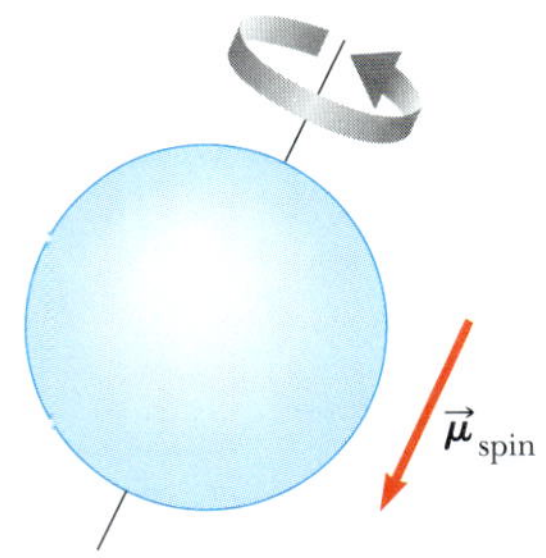

Figure 30.25 Classical model of a spinning electron. We can adopt this model to remind ourselves that electrons have an intrinsic angular momentum. The model should not be pushed too far, however; it gives an incorrect magnitude for the magnetic moment, incorrect quantum numbers, and too many degrees of freedom.

TABLE 30.1

Magnetic Moments of Some Atoms and Ions

Atom or Ion	Magnetic Moment (10^{-24} J/T)
H	9.27
He	0
Ne	0
Ce^{3+}	19.8
Yb^{3+}	37.1

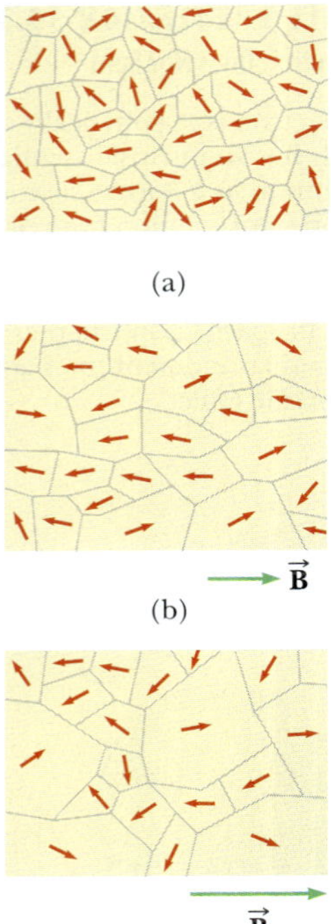

(a) (b) (c)

Figure 30.26 (a) Random orientation of atomic magnetic dipoles in the domains of an unmagnetized substance. (b) When an external field $\vec{\mathbf{B}}$ is applied, the domains with components of magnetic moment in the same direction as $\vec{\mathbf{B}}$ grow larger, giving the sample a net magnetization. (c) As the field is made even stronger, the domains with magnetic moment vectors not aligned with the external field become very small.

after the external field is removed. This permanent alignment is due to a strong coupling between neighboring moments, a coupling that can be understood only in quantum-mechanical terms.

All ferromagnetic materials are made up of microscopic regions called **domains**, regions within which all magnetic moments are aligned. These domains have volumes of about 10^{-12} to 10^{-8} m^3 and contain 10^{17} to 10^{21} atoms. The boundaries between the various domains having different orientations are called **domain walls**. In an unmagnetized sample, the magnetic moments in the domains are randomly oriented so that the net magnetic moment is zero as in Figure 30.26a. When the sample is placed in an external magnetic field $\vec{\mathbf{B}}$, the size of those domains with magnetic moments aligned with the field grows, which results in a magnetized sample as in Figure 30.26b. As the external field becomes very strong as in Figure 30.26c, the domains in which the magnetic moments are not aligned with the field become very small. When the external field is removed, the sample may retain a net magnetization in the direction of the original field. At ordinary temperatures, thermal agitation is not sufficient to disrupt this preferred orientation of magnetic moments.

Magnetic computer disks store information by alternating the direction of $\vec{\mathbf{B}}$ for portions of a thin layer of ferromagnetic material. Floppy disks have the layer on a circular sheet of plastic. Hard disks have several rigid platters with magnetic coatings on each side. Audio tapes and videotapes work the same way as floppy disks except that the ferromagnetic material is on a very long strip of plastic. Tiny coils of wire in a recording head are placed close to the magnetic material (which is moving rapidly past the head). Varying the current in the coils creates a magnetic field that magnetizes the recording material. To retrieve the information, the magnetized material is moved past a playback coil. The changing magnetism of the material induces a current in the coil as discussed in Chapter 31. This current is then amplified by audio or video equipment, or it is processed by computer circuitry.

When the temperature of a ferromagnetic substance reaches or exceeds a critical temperature called the **Curie temperature**, the substance loses its residual magnetization. Below the Curie temperature, the magnetic moments are aligned and the substance is ferromagnetic. Above the Curie temperature, the thermal agitation is great enough to cause a random orientation of the moments and the substance becomes paramagnetic. Curie temperatures for several ferromagnetic substances are given in Table 30.2.

Paramagnetism

Paramagnetic substances have a small but positive magnetism resulting from the presence of atoms (or ions) that have permanent magnetic moments. These moments interact only weakly with one another and are randomly oriented in the absence of an external magnetic field. When a paramagnetic substance is placed in an external magnetic field, its atomic moments tend to line up with the field. This alignment process, however, must compete with thermal motion, which tends to randomize the magnetic moment orientations.

TABLE 30.2

Curie Temperatures for Several Ferromagnetic Substances

Substance	T_{Curie} (K)
Iron	1 043
Cobalt	1 394
Nickel	631
Gadolinium	317
Fe_2O_3	893

Diamagnetism

When an external magnetic field is applied to a diamagnetic substance, a weak magnetic moment is induced in the direction opposite the applied field, causing diamagnetic substances to be weakly repelled by a magnet. Although diamagnetism is present in all matter, its effects are much smaller than those of paramagnetism or ferromagnetism and are evident only when those other effects do not exist.

We can attain some understanding of diamagnetism by considering a classical model of two atomic electrons orbiting the nucleus in opposite directions but with the same speed. The electrons remain in their circular orbits because of the attractive electrostatic force exerted by the positively charged nucleus. Because the magnetic moments of the two electrons are equal in magnitude and opposite in direc-

Leon Lewandowski

High Field Magnet Laboratory, University of Nijmegen, The Netherlands

(*Left*) Paramagnetism: liquid oxygen, a paramagnetic material, is attracted to the poles of a magnet. (*Right*) Diamagnetism: a frog is levitated in a 16-T magnetic field at the Nijmegen High Field Magnet Laboratory in the Netherlands. The levitation force is exerted on the diamagnetic water molecules in the frog's body. The frog suffered no ill effects from the levitation experience.

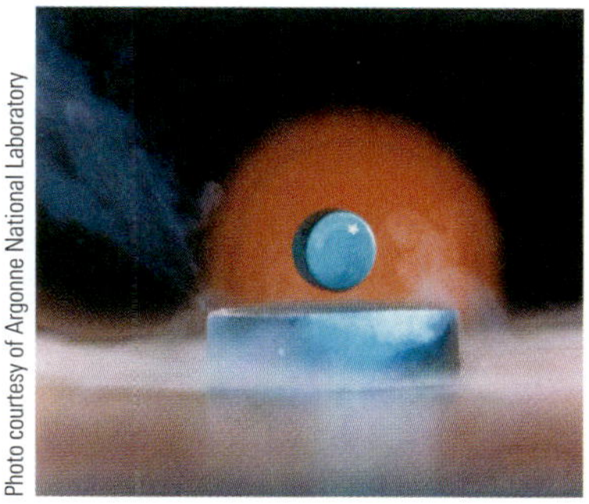

Photo courtesy of Argonne National Laboratory

Figure 30.27 An illustration of the Meissner effect, shown by this magnet suspended above a cooled ceramic superconductor disk, has become our most visual image of high-temperature superconductivity. Superconductivity is the loss of all resistance to electrical current and is a key to more-efficient energy use. In the Meissner effect, the magnet induces superconducting currents in the disk, which is cooled to $-321°F$ (77 K). The currents create a magnetic force that repels and levitates the disk.

tion, they cancel each other and the magnetic moment of the atom is zero. When an external magnetic field is applied, the electrons experience an additional magnetic force $q\vec{\mathbf{v}} \times \vec{\mathbf{B}}$. This added magnetic force combines with the electrostatic force to increase the orbital speed of the electron whose magnetic moment is antiparallel to the field and to decrease the speed of the electron whose magnetic moment is parallel to the field. As a result, the two magnetic moments of the electrons no longer cancel and the substance acquires a net magnetic moment that is opposite the applied field.

As you recall from Chapter 27, a superconductor is a substance in which the electrical resistance is zero below some critical temperature. Certain types of superconductors also exhibit perfect diamagnetism in the superconducting state. As a result, an applied magnetic field is expelled by the superconductor so that the field is zero in its interior. This phenomenon is known as the **Meissner effect**. If a permanent magnet is brought near a superconductor, the two objects repel each other. This repulsion is illustrated in Figure 30.27, which shows a small permanent magnet levitated above a superconductor maintained at 77 K.

30.7 The Magnetic Field of the Earth

When we speak of a compass magnet having a north pole and a south pole, it is more proper to say that it has a "north-seeking" pole and a "south-seeking" pole. This wording means that one pole of the magnet seeks, or points to, the north geographic pole of the Earth. Because the north pole of a magnet is attracted toward the north geographic pole of the Earth, **the Earth's south magnetic pole is located near the north geographic pole and the Earth's north magnetic pole is located near the south geographic pole**. In fact, the configuration of the Earth's magnetic field, pictured in Figure 30.28 (page 856), is very much like the one that would be achieved by burying a gigantic bar magnet deep in the interior of the Earth.

If a compass needle is suspended in bearings that allow it to rotate in the vertical plane as well as in the horizontal plane, the needle is horizontal with respect to the Earth's surface only near the equator. As the compass is moved northward, the needle rotates so that it points more and more toward the surface of the Earth. Finally, at a point near Hudson Bay in Canada, the north pole of the needle points directly downward. This site, first found in 1832, is considered to be the location of the south magnetic pole of the Earth. It is approximately 1 300 mi from the Earth's geographic North Pole, and its exact position varies slowly with time. Similarly, the north magnetic pole of the Earth is about 1 200 mi away from the Earth's geographic South Pole.

Because of this distance between the north geographic and south magnetic poles, it is only approximately correct to say that a compass needle points north.

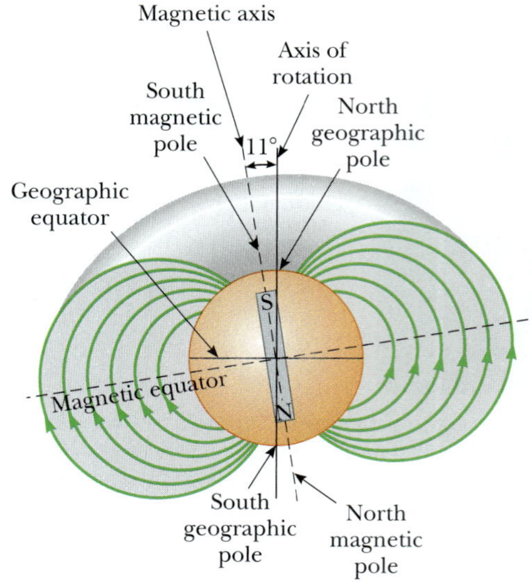

Figure 30.28 The Earth's magnetic field lines. Notice that a south magnetic pole is near the north geographic pole and a north magnetic pole is near the south geographic pole.

The difference between true north, defined as the geographic North Pole, and north indicated by a compass varies from point to point on the Earth. This difference is referred to as *magnetic declination.* For example, along a line through Florida and the Great Lakes, a compass indicates true north, whereas in the state of Washington, it aligns 25° east of true north. Figure 30.29 shows some representative values of the magnetic declination for the contiguous United States.

Although the Earth's magnetic field pattern is similar to the one that would be set up by a bar magnet deep within the Earth, it is easy to understand why the source of this magnetic field cannot be large masses of permanently magnetized material. The Earth does have large deposits of iron ore deep beneath its surface, but the high temperatures in the Earth's core prevent the iron from retaining any permanent magnetization. Scientists consider it more likely that the source of the Earth's magnetic field is convection currents in the Earth's core. Charged ions or electrons circulating in the liquid interior could produce a magnetic field just as a current loop does. There is also strong evidence that the magnitude of a planet's magnetic field is related to the planet's rate of rotation. For example, Jupiter rotates faster than the Earth, and space probes indicate that Jupiter's magnetic field is stronger than the Earth's. Venus, on the other hand, rotates more slowly than the Earth, and its magnetic field is found to be weaker. Investigation into the cause of the Earth's magnetism is ongoing.

It is interesting to point out that that the direction of the Earth's magnetic field has reversed several times during the last million years. Evidence for this reversal is provided by basalt, a type of rock that contains iron and that forms from material spewed forth by volcanic activity on the ocean floor. As the lava cools, it solidifies and retains a picture of the Earth's magnetic field direction. The rocks are dated by other means to provide a timeline for these periodic reversals of the magnetic field.

Figure 30.29 A map of the contiguous United States showing several lines of constant magnetic declination.

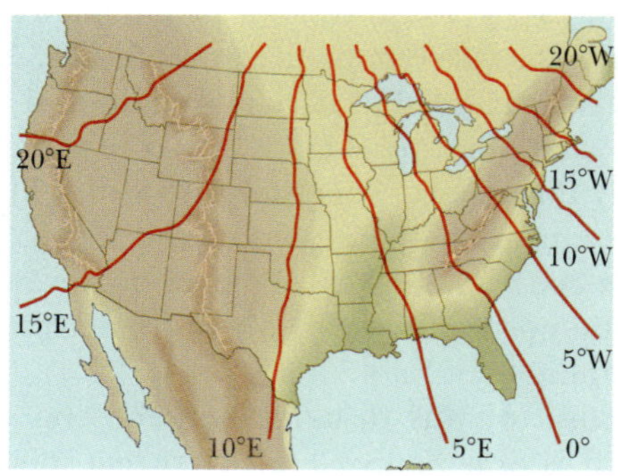

Summary

ThomsonNOW™ Sign in at **www.thomsonedu.com** and go to ThomsonNOW to take a practice test for this chapter.

DEFINITION

The **magnetic flux** Φ_B through a surface is defined by the surface integral

$$\Phi_B \equiv \int \vec{\mathbf{B}} \cdot d\vec{\mathbf{A}} \qquad \textbf{(30.18)}$$

CONCEPTS AND PRINCIPLES

The **Biot–Savart law** says that the magnetic field $d\vec{\mathbf{B}}$ at a point P due to a length element $d\vec{\mathbf{s}}$ that carries a steady current I is

$$d\vec{\mathbf{B}} = \frac{\mu_0}{4\pi}\frac{I\,d\vec{\mathbf{s}} \times \hat{\mathbf{r}}}{r^2} \qquad \textbf{(30.1)}$$

where μ_0 is the **permeability of free space**, r is the distance from the element to the point P, and $\hat{\mathbf{r}}$ is a unit vector pointing from $d\vec{\mathbf{s}}$ toward point P. We find the total field at P by integrating this expression over the entire current distribution.

The magnetic force per unit length between two parallel wires separated by a distance a and carrying currents I_1 and I_2 has a magnitude

$$\frac{F_B}{\ell} = \frac{\mu_0 I_1 I_2}{2\pi a} \qquad \textbf{(30.12)}$$

The force is attractive if the currents are in the same direction and repulsive if they are in opposite directions.

Ampère's law says that the line integral of $\vec{\mathbf{B}} \cdot d\vec{\mathbf{s}}$ around any closed path equals $\mu_0 I$, where I is the total steady current through any surface bounded by the closed path:

$$\oint \vec{\mathbf{B}} \cdot d\vec{\mathbf{s}} = \mu_0 I \qquad \textbf{(30.13)}$$

The magnitude of the magnetic field at a distance r from a long, straight wire carrying an electric current I is

$$B = \frac{\mu_0 I}{2\pi r} \qquad \textbf{(30.14)}$$

The field lines are circles concentric with the wire.

The magnitudes of the fields inside a toroid and solenoid are

$$B = \frac{\mu_0 N I}{2\pi r} \quad \text{(toroid)} \qquad \textbf{(30.16)}$$

$$B = \mu_0 \frac{N}{\ell} I = \mu_0 n I \quad \text{(solenoid)} \qquad \textbf{(30.17)}$$

where N is the total number of turns.

Gauss's law of magnetism states that the net magnetic flux through any closed surface is zero.

Substances can be classified into one of three categories that describe their magnetic behavior. **Diamagnetic** substances are those in which the magnetic moment is weak and opposite the applied magnetic field. **Paramagnetic** substances are those in which the magnetic moment is weak and in the same direction as the applied magnetic field. In **ferromagnetic** substances, interactions between atoms cause magnetic moments to align and create a strong magnetization that remains after the external field is removed.

Questions

☐ denotes answer available in *Student Solutions Manual/Study Guide;* **O** denotes objective question

1. **O** What creates a magnetic field? Choose every correct answer. (a) a stationary object with electric charge (b) a moving object with electric charge (c) a stationary conductor carrying electric current (d) a difference in electric potential (e) an electric resistor. *Note:* In Chapter 34, we will see that a changing electric field also creates a magnetic field.

2. **O** A long, vertical, metallic wire carries downward electric current. **(i)** What is the direction of the magnetic field it creates at a point 2 cm horizontally east of the center of the wire? (a) north (b) south (c) east (d) west (e) up (f) down **(ii)** What would be the direction of the field if the current consisted of positive charges moving downward, instead of electrons moving upward? Choose from the same possibilities.

3. **O** Suppose you are facing a tall makeup mirror on a vertical wall. Fluorescent tubes framing the mirror carry a clockwise electric current. **(i)** What is the direction of the magnetic field created by that current at a point slightly to the right of the center of the mirror? (a) up (b) down (c) left (d) right (e) horizontally toward you (f) away from you **(ii)** What is the direction of the field the current creates at a point on the wall outside the frame to the right? Choose from the same possibilities.

4. Explain why two parallel wires carrying currents in opposite directions repel each other.

5. **O** In Active Figure 30.8, assume $I_1 = 2$ A and $I_2 = 6$ A. What is the relationship between the magnitude F_1 of the force exerted on wire 1 and the magnitude F_2 of the force exerted on wire 2? (a) $F_1 = 6F_2$ (b) $F_1 = 3F_2$ (c) $F_1 = F_2$ (d) $F_1 = F_2/3$ (e) $F_1 = F_2/6$

6. **O** Answer each question yes or no. (a) Is it possible for each of three stationary charged particles to exert a force of attraction on the other two? (b) Is it possible for each of three stationary charged particles to repel both of the other particles? (c) Is it possible for each of three current-carrying metal wires to attract the other two? (d) Is it possible for each of three current-carrying metal wires to repel both of the other wires? André-Marie Ampère's experiments on electromagnetism are models of logical precision and included observation of the phenomena referred to in this question.

7. Is Ampère's law valid for all closed paths surrounding a conductor? Why is it not useful for calculating $\vec{\mathbf{B}}$ for all such paths?

8. Compare Ampère's law with the Biot–Savart law. Which is more generally useful for calculating $\vec{\mathbf{B}}$ for a current-carrying conductor?

9. A hollow copper tube carries a current along its length. Why is $\vec{\mathbf{B}} = 0$ inside the tube? Is $\vec{\mathbf{B}}$ nonzero outside the tube?

10. **O** **(i)** What happens to the magnitude of the magnetic field inside a long solenoid if the current is doubled? (a) It becomes 4 times larger. (b) It becomes twice as large. (c) It is unchanged. (d) It becomes one-half as large. (e) It becomes one-fourth as large. **(ii)** What happens to the field if instead the length of the solenoid is doubled, with the number of turns remaining the same? Choose from the same possibilities. **(iii)** What happens to the field if the number of turns is doubled, with the length remaining the same? Choose from the same possibilities. **(iv)** What happens to the field if the radius is doubled? Choose from the same possibilities.

11. **O** A long solenoid with closely spaced turns carries electric current. Does each turn of wire exert (a) an attractive force on the next adjacent turn, (b) a repulsive force on the next adjacent turn, (c) zero force on the next adjacent turn, or (d) either an attractive or a repulsive force on the next turn, depending on the direction of current in the solenoid?

12. **O** A uniform magnetic field is directed along the x axis. For what orientation of a flat, rectangular coil is the flux through the rectangle a maximum? (a) It is a maximum in the xy plane. (b) It is a maximum in the xz plane. (c) It is a maximum in the yz plane. (d) The flux has the same nonzero value for all these orientations. (e) The flux is zero in all cases.

13. The quantity $\oint \vec{\mathbf{B}} \cdot d\vec{\mathbf{s}}$ in Ampère's law is called *magnetic circulation.* Active Figure 30.10 and Figure 30.13 show paths around which the magnetic circulation was evaluated. Each of these paths encloses an area. What is the magnetic flux through each area? Explain your answer.

14. **O** (a) Two stationary charged particles exert forces of attraction on each other. One of the particles has negative charge. Is the other positive or negative? (b) Is the net electric field at a point halfway between the particles larger, smaller, or the same in magnitude as the field due to one charge by itself? (c) Two straight, vertical, current-carrying wires exert forces of attraction on each other. One of them carries downward current. Does the other wire carry upward or downward current? (d) Is the net magnetic field at a point halfway between the wires larger, smaller, or the same in magnitude as the field due to one wire by itself?

15. **O** Rank the magnitudes of the following magnetic fields from the largest to the smallest, noting any cases of equality. (a) the field 2 cm away from a long, straight wire carrying a current of 3 A (b) the field at the center of a flat, compact, circular coil, 2 cm in radius, with 10 turns, carrying a current of 0.3 A (c) the field at the center of a solenoid 2 cm in radius and 200 cm long, with 1 000 turns, carrying a current of 0.3 A (d) the field at the center of a long, straight metal bar, 2 cm in radius, carrying a current of 300 A (e) a field of 1 mT

16. One pole of a magnet attracts a nail. Will the other pole of the magnet attract the nail? Explain. Explain how a magnet sticks to a refrigerator door.

17. A magnet attracts a piece of iron. The iron can then attract another piece of iron. On the basis of domain alignment, explain what happens in each piece of iron.

18. Why does hitting a magnet with a hammer cause the magnetism to be reduced?

19. Which way would a compass point if you were at the north magnetic pole of the Earth?

20. Figure Q30.20 shows four permanent magnets, each having a hole through its center. Notice that the blue and yellow magnets are levitated above the red ones. (a) How does this levitation occur? (b) What purpose do the rods serve? (c) What can you say about the poles of the magnets from this observation? (d) If the upper magnet were inverted, what do you suppose would happen?

Figure Q30.20

Problems

WebAssign The Problems from this chapter may be assigned online in WebAssign.

ThomsonNOW Sign in at **www.thomsonedu.com** and go to ThomsonNOW to assess your understanding of this chapter's topics with additional quizzing and conceptual questions.

1, 2, 3 denotes straightforward, intermediate, challenging; □ denotes full solution available in *Student Solutions Manual/Study Guide;* ▲ denotes coached solution with hints available at **www.thomsonedu.com;** denotes developing symbolic reasoning; ● denotes asking for qualitative reasoning; denotes computer useful in solving problem

Section 30.1 The Biot–Savart Law

1. In Niels Bohr's 1913 model of the hydrogen atom, an electron circles the proton at a distance of 5.29×10^{-11} m with a speed of 2.19×10^6 m/s. Compute the magnitude of the magnetic field this motion produces at the location of the proton.

2. Calculate the magnitude of the magnetic field at a point 100 cm from a long, thin conductor carrying a current of 1.00 A.

3. □ (a) A conductor in the shape of a square loop of edge length $\ell = 0.400$ m carries a current $I = 10.0$ A as shown in Figure P30.3. Calculate the magnitude and direction of the magnetic field at the center of the square. (b) **What If?** If this conductor is formed into a single circular turn and carries the same current, what is the value of the magnetic field at the center?

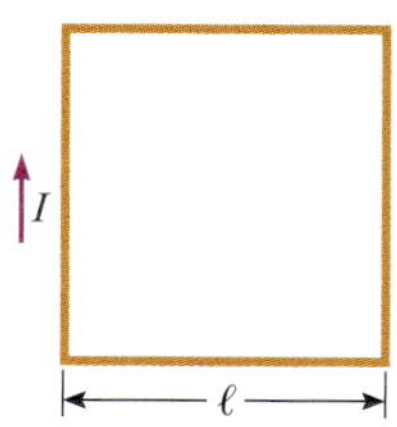

Figure P30.3

4. A conductor consists of a circular loop of radius R and two straight, long sections as shown in Figure P30.4. The wire lies in the plane of the paper and carries a current I. Find an expression for the vector magnetic field at the center of the loop.

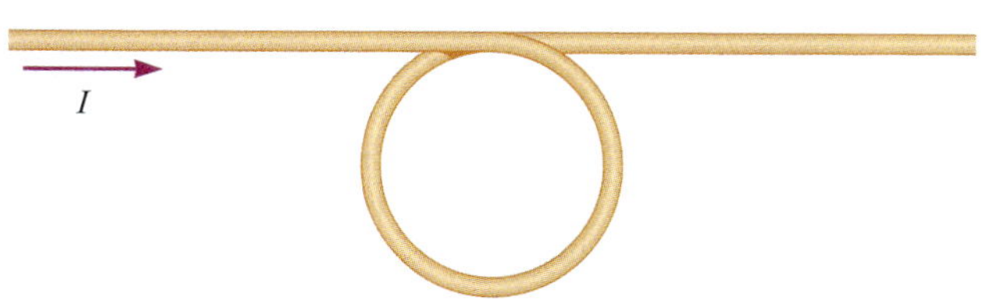

Figure P30.4

5. □ ▲ Determine the magnetic field at a point P located a distance x from the corner of an infinitely long wire bent at a right angle as shown in Figure P30.5. The wire carries a steady current I.

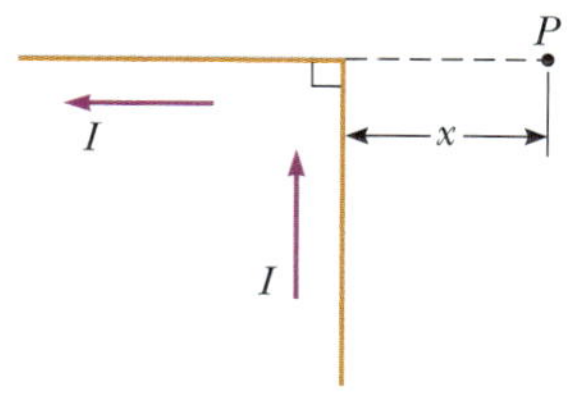

Figure P30.5

6. Consider a flat, circular current loop of radius R carrying current I. Choose the x axis to be along the axis of the loop, with the origin at the center of the loop. Plot a graph of the ratio of the magnitude of the magnetic field at coordinate x to that at the origin, for $x = 0$ to $x = 5R$. It may be useful to use a programmable calculator or a computer to solve this problem.

7. Two long, straight, parallel wires carry currents that are directed perpendicular to the page as shown in Figure P30.7. Wire 1 carries a current I_1 into the page (in the $-z$ direction) and passes through the x axis at $x = +a$. Wire 2 passes through the x axis at $x = -2a$ and carries an unknown current I_2. The total magnetic field at the origin due to the current-carrying wires has the magnitude $2\mu_0 I_1/(2\pi a)$. The current I_2 can have either of two possible values. (a) Find the value of I_2 with the smaller magnitude, stating it in terms of I_1 and giving its direction. (b) Find the other possible value of I_2.

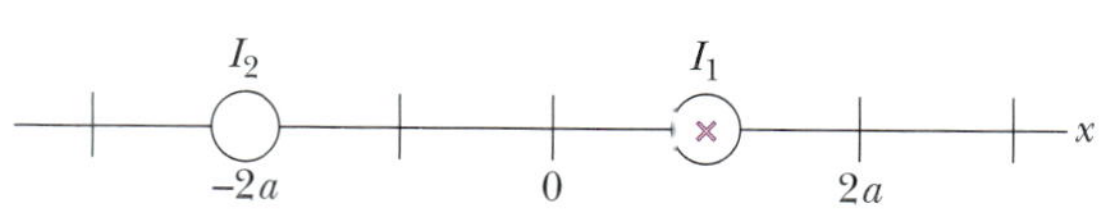

Figure P30.7

8. A long, straight wire carries current I. A right-angle bend is made in the middle of the wire. The bend forms an arc

of a circle of radius r as shown in Figure P30.8. Determine the magnetic field at the center of the arc.

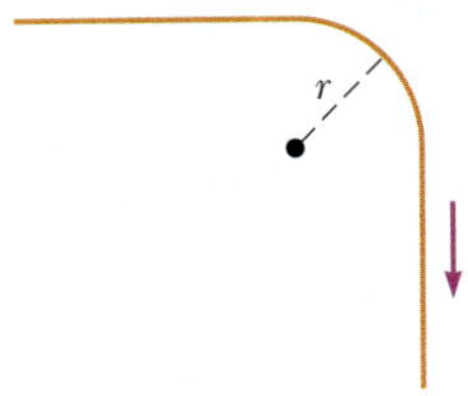

Figure P30.8

9. One long wire carries current 30.0 A to the left along the x axis. A second long wire carries current 50.0 A to the right along the line (y = 0.280 m, z = 0). (a) Where in the plane of the two wires is the total magnetic field equal to zero? (b) A particle with a charge of −2.00 μC is moving with a velocity of $150\hat{\mathbf{i}}$ Mm/s along the line (y = 0.100 m, z = 0). Calculate the vector magnetic force acting on the particle. (c) **What If?** A uniform electric field is applied to allow this particle to pass through this region undeflected. Calculate the required vector electric field.

10. A current path shaped as shown in Figure P30.10 produces a magnetic field at P, the center of the arc. If the arc subtends an angle of 30.0° and the radius of the arc is 0.600 m, what are the magnitude and direction of the field produced at P if the current is 3.00 A?

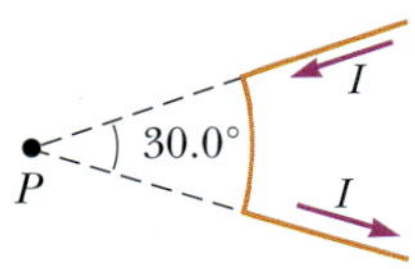

Figure P30.10

11. Three long, parallel conductors carry currents of I = 2.00 A. Figure P30.11 is an end view of the conductors, with each current coming out of the page. Taking a = 1.00 cm, determine the magnitude and direction of the magnetic field at points A, B, and C.

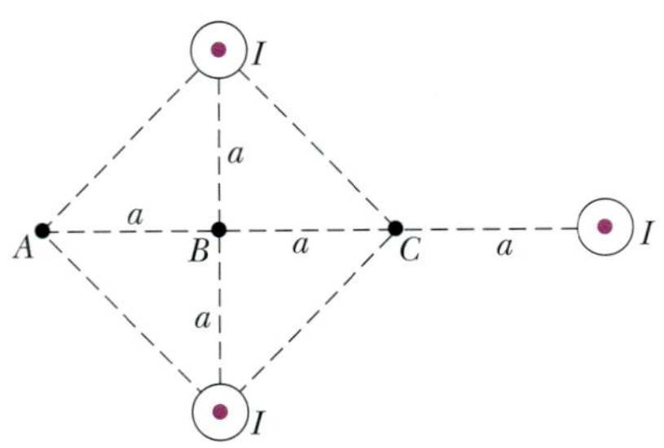

Figure P30.11

12. ● In a long, straight, vertical lightning stroke, electrons move downward and positive ions move upward, to constitute a current of magnitude 20.0 kA. At a location 50.0 m east of the middle of the stroke, a free electron drifts through the air toward the west with a speed of 300 m/s. (a) Find the vector force the lightning stroke exerts on the electron. Make a sketch showing the various vectors involved. Ignore the effect of the Earth's magnetic field. (b) Find the radius of the electron's path. Is it a good approximation to model the electron as moving in a uniform field? Explain your answer. (c) If it does not collide with any obstacles, how many revolutions will the electron complete during the 60.0-μs duration of the lightning stroke?

13. ● A wire carrying a current I is bent into the shape of an equilateral triangle of side L. (a) Find the magnitude of the magnetic field at the center of the triangle. (b) At a point halfway between the center and any vertex, is the field stronger or weaker than at the center? Give a qualitative argument for your answer.

14. Determine the magnetic field (in terms of I, a, and d) at the origin due to the current loop in Figure P30.14.

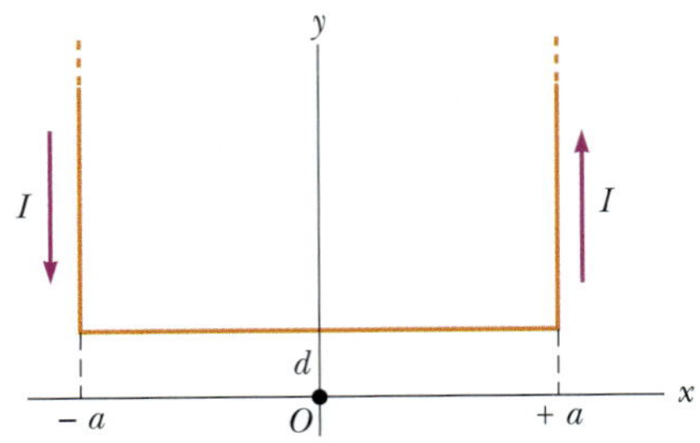

Figure P30.14

15. Two long, parallel conductors carry currents I_1 = 3.00 A and I_2 = 3.00 A, both directed into the page in Figure P30.15. Determine the magnitude and direction of the resultant magnetic field at P.

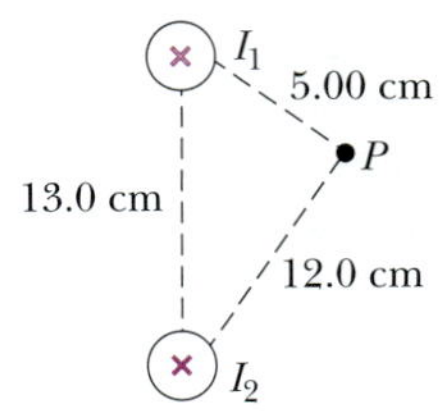

Figure P30.15

16. The idea that a magnetic field can have therapeutic value has been around for centuries. A rare-earth magnet sold to relieve joint pain is a disk 1.20 mm thick and 3.50 mm in diameter. Its circular flat faces are its north and south poles. Assume it is accurately modeled as a magnetic dipole. Also assume Equation 30.10 describes the magnetic field it produces at all points along its axis. The field is strongest, with the value 40.0 mT, at the center of each flat face. At what distance from the surface is the magnitude of the magnetic field like that of the Earth, with a value of 50.0 μT?

Section 30.2 The Magnetic Force Between Two Parallel Conductors

17. In Figure P30.17, the current in the long, straight wire is I_1 = 5.00 A and the wire lies in the plane of the rectangular loop, which carries the current I_2 = 10.0 A. The dimensions are c = 0.100 m, a = 0.150 m, and ℓ = 0.450 m. Find the magnitude and direction of the net force exerted on the loop by the magnetic field created by the wire.

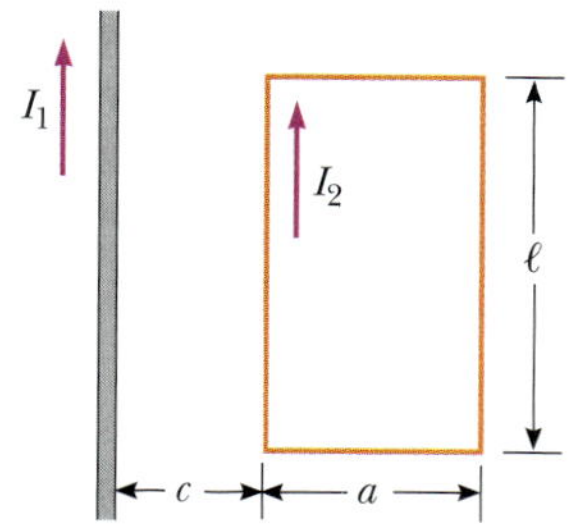

Figure P30.17

18. Two long, parallel conductors, separated by 10.0 cm, carry currents in the same direction. The first wire carries current $I_1 = 5.00$ A, and the second carries $I_2 = 8.00$ A. (a) What is the magnitude of the magnetic field created by I_1 at the location of I_2? (b) What is the force per unit length exerted by I_1 on I_2? (c) What is the magnitude of the magnetic field created by I_2 at the location of I_1? (d) What is the force per length exerted by I_2 on I_1?

19. Two long, parallel wires are attracted to each other by a force per unit length of 320 μN/m when they are separated by a vertical distance of 0.500 m. The current in the upper wire is 20.0 A to the right. Determine the location of the line in the plane of the two wires along which the total magnetic field is zero.

20. ● Three long wires (wire 1, wire 2, and wire 3) hang vertically. The distance between wire 1 and wire 2 is 20.0 cm. On the left, wire 1 carries an upward current of 1.50 A. To the right, wire 2 carries a downward current of 4.00 A. Wire 3 is to be located such that when it carries a certain current, each wire experiences no net force. (a) Is this situation possible? Is it possible in more than one way? Describe (b) the position of wire 3 and (c) the magnitude and direction of the current in wire 3.

21. ● The unit of magnetic flux is named for Wilhelm Weber. A practical-size unit of magnetic field is named for Johann Karl Friedrich Gauss. Both were scientists at Göttingen, Germany. Along with their individual accomplishments, together they built a telegraph in 1833. It consisted of a battery and switch, at one end of a transmission line 3 km long, operating an electromagnet at the other end. (André Ampère suggested electrical signaling in 1821; Samuel Morse built a telegraph line between Baltimore and Washington, D.C., in 1844.) Suppose Weber and Gauss's transmission line was as diagrammed in Figure P30.21. Two long, parallel wires, each having a mass per unit length of 40.0 g/m, are supported in a horizontal plane by strings 6.00 cm long. When both wires carry the same current I, the wires repel each other so that the angle θ between the supporting strings is 16.0°. (a) Are the currents in the same direction or in opposite directions? (b) Find the magnitude of the current. (c) If this apparatus were taken to Mars, would the current required to separate the wires by 16° be larger or smaller than on Earth? Why?

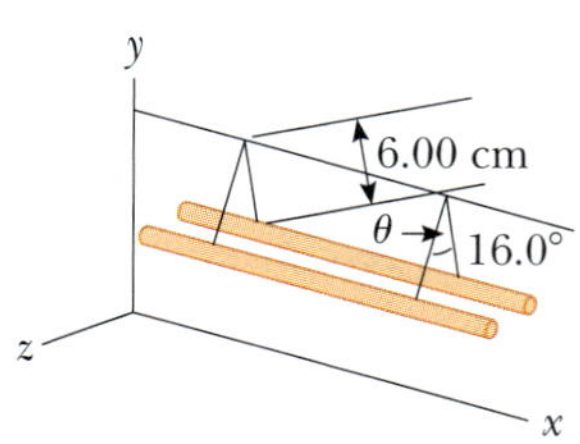

Figure P30.21

22. ● Two parallel copper conductors are each 0.500 m long. They carry 10.0-A currents in opposite directions. (a) What center-to-center separation must the conductors have if they are to repel each other with a force of 1.00 N? (b) Is this situation physically possible? Explain.

Section 30.3 Ampère's Law

23. ▲ Four long, parallel conductors carry equal currents of $I = 5.00$ A. Figure P30.23 is an end view of the conductors. The current direction is into the page at points A and B (indicated by the crosses) and out of the page at C and D (indicated by the dots). Calculate the magnitude and direction of the magnetic field at point P, located at the center of the square of edge length 0.200 m.

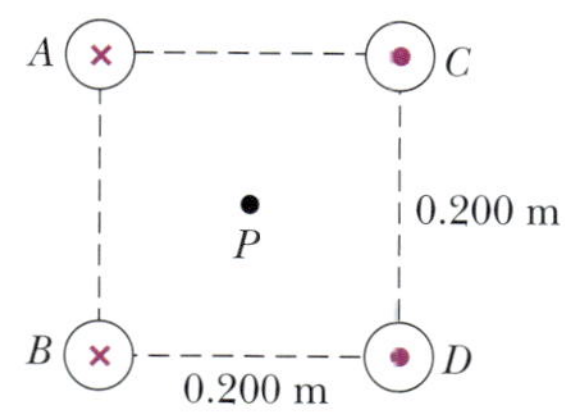

Figure P30.23

24. A long, straight wire lies on a horizontal table and carries a current of 1.20 μA. In a vacuum, a proton moves parallel to the wire (opposite the current) with a constant speed of 2.30×10^4 m/s at a distance d above the wire. Determine the value of d. You may ignore the magnetic field due to the Earth.

25. Figure P30.25 is a cross-sectional view of a coaxial cable. The center conductor is surrounded by a rubber layer, which is surrounded by an outer conductor, which is surrounded by another rubber layer. In a particular application, the current in the inner conductor is 1.00 A out of the page and the current in the outer conductor is 3.00 A into the page. Determine the magnitude and direction of the magnetic field at points a and b.

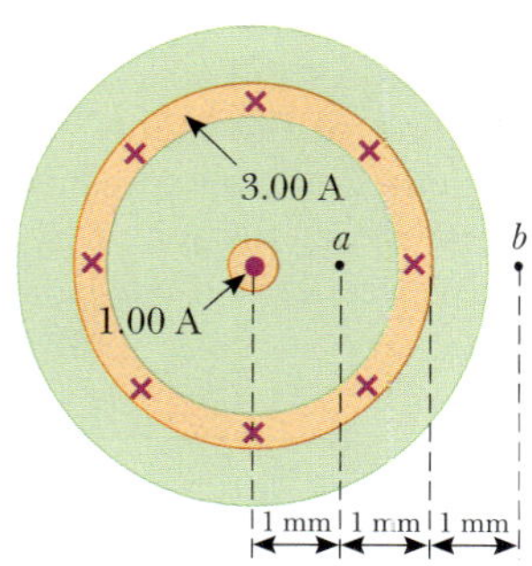

Figure P30.25

26. The magnetic field 40.0 cm away from a long,straight wire carrying current 2.00 A is 1.00 μT. (a) At what distance is it 0.100 μT? (b) **What If?** At one instant, the two conductors in a long household extension cord carry equal 2.00-A currents in opposite directions. The two wires are 3.00 mm apart. Find the magnetic field 40.0 cm away from the middle of the straight cord, in the plane of the two wires. (c) At what distance is it one-tenth as large? (d) The center wire in a coaxial cable carries current 2.00 A in one direction, and the sheath around it carries current 2.00 A in the opposite direction. What magnetic field does the cable create at points outside?

27. ● ▲ A packed bundle of 100 long, straight, insulated wires forms a cylinder of radius $R = 0.500$ cm. (a) If each wire carries 2.00 A, what are the magnitude and direction of the magnetic force per unit length acting on a wire located 0.200 cm from the center of the bundle? (b) **What If?** Would a wire on the outer edge of the bundle experience a force greater or smaller than the value calculated in part (a)? Give a qualitative argument for your answer.

28. The magnetic coils of a tokamak fusion reactor are in the shape of a toroid having an inner radius of 0.700 m and an outer radius of 1.30 m. The toroid has 900 turns of large-diameter wire, each of which carries a current of 14.0 kA. Find the magnitude of the magnetic field inside the toroid along (a) the inner radius and (b) the outer radius.

29. Consider a column of electric current passing through plasma (ionized gas). Filaments of current within the column are magnetically attracted to one another. They can crowd together to yield a very great current density and a very strong magnetic field in a small region. Sometimes the current can be cut off momentarily by this *pinch effect.* (In a metallic wire, a pinch effect is not important because the current-carrying electrons repel one another with electric forces.) The pinch effect can be demonstrated by making an empty aluminum can carry a large current parallel to its axis. Let R represent the radius of the can and I the upward current, uniformly distributed over its curved wall. Determine the magnetic field (a) just inside the wall and (b) just outside. (c) Determine the pressure on the wall.

30. Niobium metal becomes a superconductor when cooled below 9 K. Its superconductivity is destroyed when the surface magnetic field exceeds 0.100 T. Determine the maximum current a 2.00-mm-diameter niobium wire can carry and remain superconducting, in the absence of any external magnetic field.

31. A long, cylindrical conductor of radius R carries a current I as shown in Figure P30.31. The current density J, however, is not uniform over the cross section of the conductor but is a function of the radius according to $J = br$, where b is a constant. Find an expression for the magnetic field magnitude B (a) at a distance $r_1 < R$ and (b) at a distance $r_2 > R$, measured from the axis.

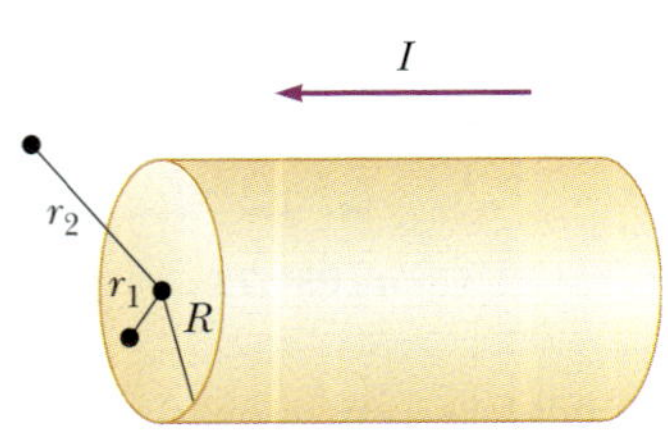

Figure P30.31

32. In Figure P30.32, both currents in the infinitely long wires are 8.00 A in the negative x direction. The wires are separated by the distance $2a = 6.00$ cm. (a) Sketch the magnetic field pattern in the yz plane. (b) What is the value of the magnetic field at the origin? At ($y = 0$, $z \rightarrow \infty$)? (c) Find the magnetic field at points along the z axis as a function of z. (d) At what distance d along the positive z axis is the magnetic field a maximum? (e) What is this maximum value?

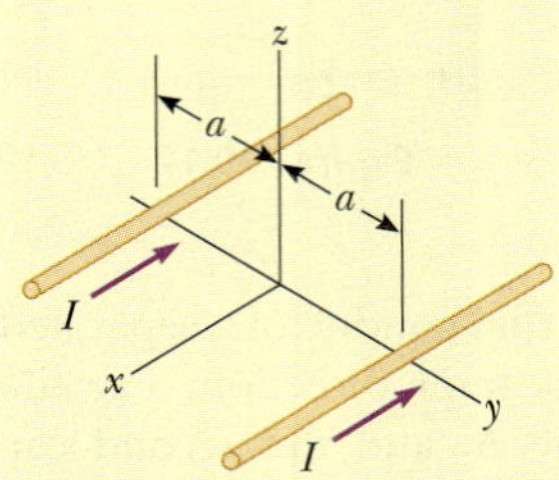

Figure P30.32

33. An infinite sheet of current lying in the yz plane carries a surface current of linear density J_s. The current is in the y direction, and J_s represents the current per unit length measured along the z axis. Figure P30.33 is an edge view of the sheet. Prove that the magnetic field near the sheet is parallel to the sheet and perpendicular to the current direction, with magnitude $\mu_0 J_s/2$. *Suggestion:* Use Ampère's law and evaluate the line integral for a rectangular path around the sheet, represented by the dashed line in Figure P30.33.

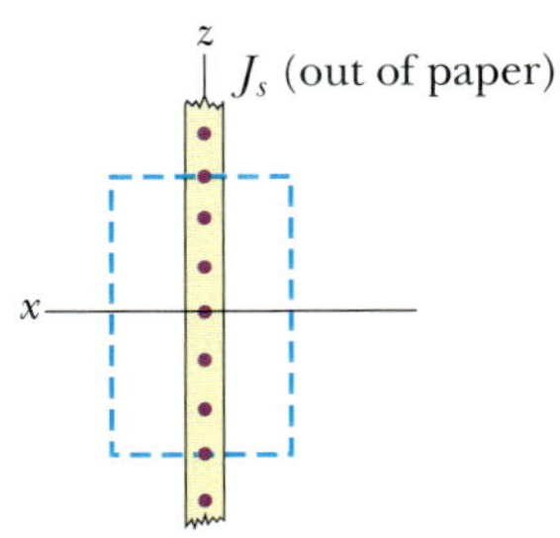

Figure P30.33

Section 30.4 The Magnetic Field of a Solenoid

34. ● You are given a certain volume of copper from which you can make copper wire. To insulate the wire, you can have as much enamel as you like. You will use the wire to make a tightly wound solenoid 20 cm long having the greatest possible magnetic field at the center and using a power supply that can deliver a current of 5 A. The solenoid can be wrapped with wire in one or more layers. (a) Should you make the wire long and thin or shorter and thick? Explain. (b) Should you make the solenoid radius small or large? Explain.

35. ▲ What current is required in the windings of a long solenoid that has 1 000 turns uniformly distributed over a length of 0.400 m to produce at the center of the solenoid a magnetic field of magnitude 1.00×10^{-4} T?

36. Consider a solenoid of length ℓ and radius R, containing N closely spaced turns and carrying a steady current I. (a) In terms of these parameters, find the magnetic field at a point along the axis as a function of distance a from the end of the solenoid. (b) Show that as ℓ becomes very long, B approaches $\mu_0 NI/2\ell$ at each end of the solenoid.

37. A single-turn square loop of wire, 2.00 cm on each edge, carries a clockwise current of 0.200 A. The loop is inside a solenoid, with the plane of the loop perpendicular to the magnetic field of the solenoid. The solenoid has 30 turns/cm and carries a clockwise current of 15.0 A. Find the force on each side of the loop and the torque acting on the loop.

38. A solenoid 10.0 cm in diameter and 75.0 cm long is made from copper wire of diameter 0.100 cm, with very thin insulation. The wire is wound onto a cardboard tube in a single layer, with adjacent turns touching each other. What power must be delivered to the solenoid if it is to produce a field of 8.00 mT at its center?

Section 30.5 Gauss's Law in Magnetism

39. A cube of edge length $\ell = 2.50$ cm is positioned as shown in Figure P30.39. A uniform magnetic field given by $\vec{\mathbf{B}} = (5\hat{\mathbf{i}} + 4\hat{\mathbf{j}} + 3\hat{\mathbf{k}})$ T exists throughout the region. (a) Calculate the magnetic flux through the shaded face. (b) What is the total flux through the six faces?

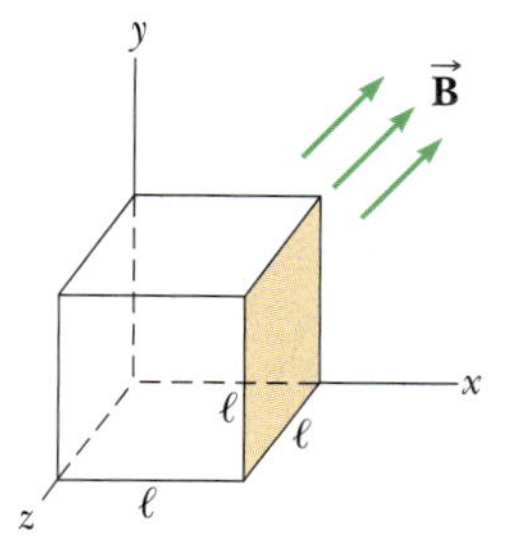

Figure P30.39

40. Consider the hemispherical closed surface in Figure P30.40. The hemisphere is in a uniform magnetic field that makes an angle θ with the vertical. Calculate the magnetic flux through (a) the flat surface S_1 and (b) the hemispherical surface S_2.

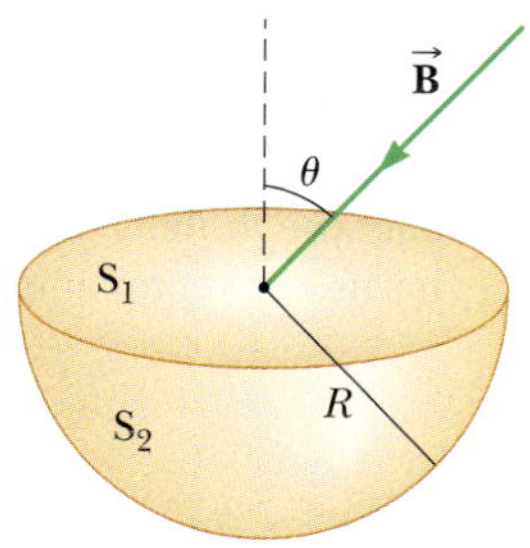

Figure P30.40

41. A solenoid 2.50 cm in diameter and 30.0 cm long has 300 turns and carries 12.0 A. (a) Calculate the flux through the surface of a disk of radius 5.00 cm that is positioned perpendicular to and centered on the axis of the solenoid, as shown in Figure P30.41a. (b) Figure P30.41b shows an enlarged end view of the same solenoid. Calculate the flux through the blue area, which is an annulus with an inner radius of 0.400 cm and an outer radius of 0.800 cm.

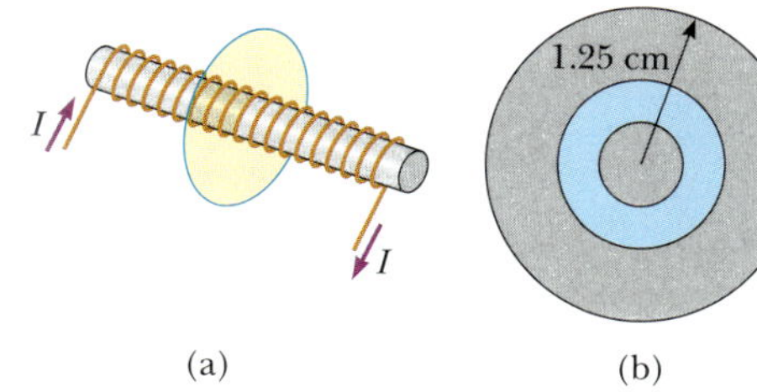

Figure P30.41

42. *Compare this problem with Problem 65 in Chapter 24.* Consider a magnetic field that is uniform in direction throughout a certain volume. Can it be uniform in magnitude? Must it be uniform in magnitude? Give evidence for your answers.

Section 30.6 Magnetism in Matter

43. At *saturation,* when nearly all the atoms have their magnetic moments aligned, the magnetic field in a sample of iron can be 2.00 T. If each electron contributes a magnetic moment of 9.27×10^{-24} A · m² (one Bohr magneton), how many electrons per atom contribute to the saturated field of iron? The number density of atoms in iron is approximately 8.50×10^{28} atoms/m³.

Section 30.7 The Magnetic Field of the Earth

44. A circular coil of 5 turns and a diameter of 30.0 cm is oriented in a vertical plane with its axis perpendicular to the horizontal component of the Earth's magnetic field. A horizontal compass placed at the center of the coil is made to deflect 45.0° from magnetic north by a current of 0.600 A in the coil. (a) What is the horizontal component of the Earth's magnetic field? (b) The current in the coil is switched off. A "dip needle" is a magnetic compass mounted so that it can rotate in a vertical north–south plane. At this location, a dip needle makes an angle of 13.0° from the vertical. What is the total magnitude of the Earth's magnetic field at this location?

45. The magnetic moment of the Earth is approximately 8.00×10^{22} A · m². (a) Imagine that the planetary magnetic field were caused by the complete magnetization of a huge iron deposit. How many unpaired electrons would participate? (b) At two unpaired electrons per iron atom, how many kilograms of iron would compose the deposit? Iron has a density of 7 900 kg/m³ and approximately 8.50×10^{28} iron atoms/m³.

46. ● A particular location on the Earth's surface is characterized by a value of gravitational field, a value of magnetic field, and a value of atmospheric pressure. (a) Which of these quantities are vectors and which are scalars? (b) Determine a value for each quantity at your current location. Include the direction of each vector quantity. State your sources. (c) Which of the quantities have separate causes from which of the others?

Additional Problems

47. A very long, thin strip of metal of width w carries a current I along its length as shown in Figure P30.47. Find the magnetic field at the point P in the diagram. The point P is in the plane of the strip at distance b away from it.

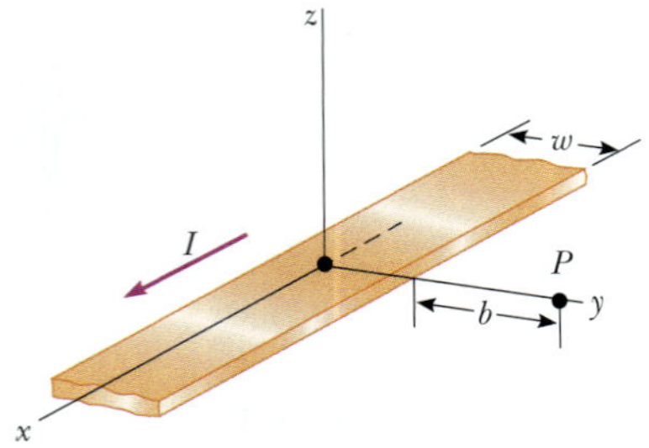

Figure P30.47

48. The magnitude of the Earth's magnetic field at either pole is approximately 7.00×10^{-5} T. Suppose the field fades away, before its next reversal. Scouts, sailors, and conservative politicians around the world join together in a program to replace the field. One plan is to use a current loop around the equator, without relying on magnetization of any materials inside the Earth. Determine the current that would generate such a field if this plan were carried out. Take the radius of the Earth as $R_E = 6.37 \times 10^6$ m.

49. A thin copper bar of length $\ell = 10.0$ cm is supported horizontally by two (nonmagnetic) contacts. The bar carries current $I_1 = 100$ A in the $-x$ direction as shown in Figure P30.49. At a distance $h = 0.500$ cm below one end of the bar, a long, straight wire carries a current $I_2 = 200$ A in the z direction. Determine the magnetic force exerted on the bar.

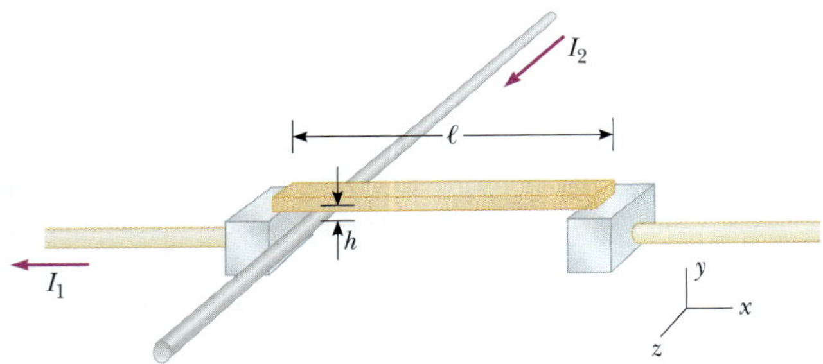

Figure P30.49

50. Suppose you install a compass on the center of the dashboard of a car. Compute an order-of-magnitude estimate for the magnetic field at this location produced by the current when you switch on the headlights. How does this estimate compare with the Earth's magnetic field? You may suppose the dashboard is made mostly of plastic.

51. ▲ A nonconducting ring of radius 10.0 cm is uniformly charged with a total positive charge 10.0 μC. The ring rotates at a constant angular speed 20.0 rad/s about an axis through its center, perpendicular to the plane of the ring. What is the magnitude of the magnetic field on the axis of the ring 5.00 cm from its center?

52. A nonconducting ring of radius R is uniformly charged with a total positive charge q. The ring rotates at a constant angular speed ω about an axis through its center, perpendicular to the plane of the ring. What is the magnitude of the magnetic field on the axis of the ring a distance $R/2$ from its center?

53. Two circular coils of radius R, each with N turns, are perpendicular to a common axis. The coil centers are a distance R apart. Each coil carries a steady current I in the same direction as shown in Figure P30.53. (a) Show that the magnetic field on the axis at a distance x from the center of one coil is

$$B = \frac{N\mu_0 IR^2}{2}\left[\frac{1}{(R^2 + x^2)^{3/2}} + \frac{1}{(2R^2 + x^2 - 2Rx)^{3/2}}\right]$$

(b) Show that dB/dx and d^2B/dx^2 are both zero at the point midway between the coils. We may then conclude that the magnetic field in the region midway between the coils is uniform. Coils in this configuration are called *Helmholtz coils.*

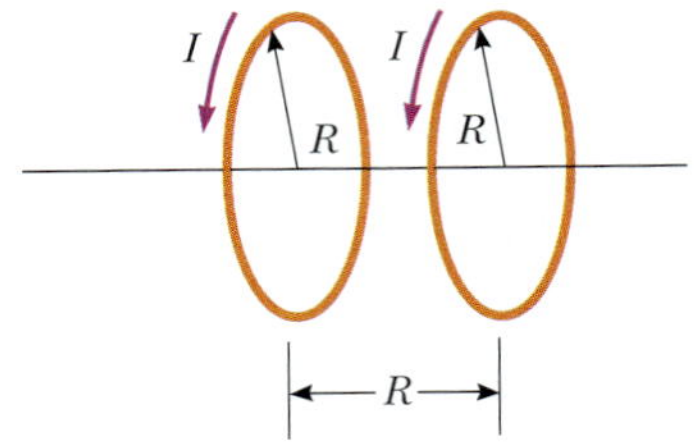

Figure P30.53 Problems 53 and 54.

54. Two identical, flat, circular coils of wire each have 100 turns and a radius of 0.500 m. The coils are arranged as a set of Helmholtz coils (see Fig. P30.53), parallel and with separation 0.500 m. Each coil carries a current of 10.0 A. Determine the magnitude of the magnetic field at a point on the common axis of the coils and halfway between them.

55. We have seen that a long solenoid produces a uniform magnetic field directed along the axis of a cylindrical region. To produce a uniform magnetic field directed parallel to a *diameter* of a cylindrical region, however, one can use the *saddle coils* illustrated in Figure P30.55. The loops are wrapped over a somewhat flattened tube. Assume the straight sections of wire are very long. The end view of the tube shows how the windings are applied. The overall current distribution is the superposition of two overlapping, circular cylinders of uniformly distributed current, one toward you and one away from you. The current density J is the same for each cylinder. The position of the axis of one cylinder is described by a position vector $\vec{\mathbf{a}}$ relative to the other cylinder. Prove that the magnetic field inside the hollow tube is $\mu_0 Ja/2$ downward. *Suggestion:* The use of vector methods simplifies the calculation.

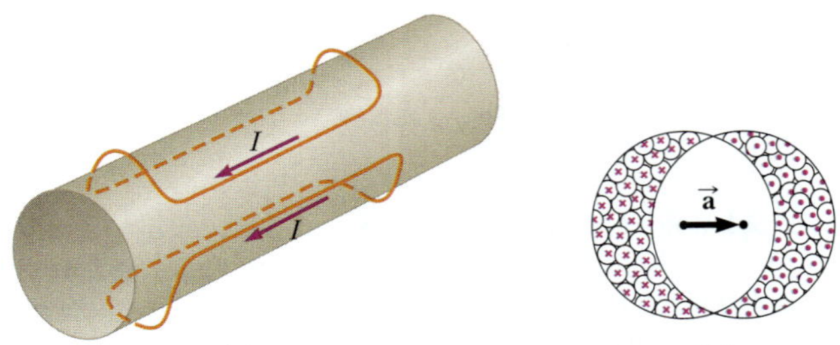

(a) (b)

Figure P30.55

56. *You may use the result of Problem 33 in solving this problem.* A very large parallel-plate capacitor carries charge with uniform charge per unit area $+\sigma$ on the upper plate and $-\sigma$ on the lower plate. The plates are horizontal and both move horizontally with speed v to the right. (a) What is the magnetic field between the plates? (b) What is the magnetic field close to the plates but outside of the capacitor? (c) What is the magnitude and direction of the magnetic force per unit area on the upper plate? (d) At what extrapolated speed v will the magnetic force on a plate balance the electric force on the plate? Calculate this speed numerically.

57. ● Two circular loops are parallel, coaxial, and almost in contact, 1.00 mm apart (Fig. P30.57). Each loop is 10.0 cm in radius. The top loop carries a clockwise current of 140 A. The bottom loop carries a counterclockwise current of 140 A. (a) Calculate the magnetic force exerted by the bottom loop on the top loop. (b) Suppose a student thinks the first step in solving part (a) is to use Equation 30.7 to find the magnetic field created by one of the loops. How would you argue for or against this idea? *Suggestion:* Think about how one loop looks to a bug perched on the other loop. (c) The upper loop has a mass of 0.021 0 kg. Calculate its acceleration, assuming the only forces acting on it are the force in part (a) and the gravitational force.

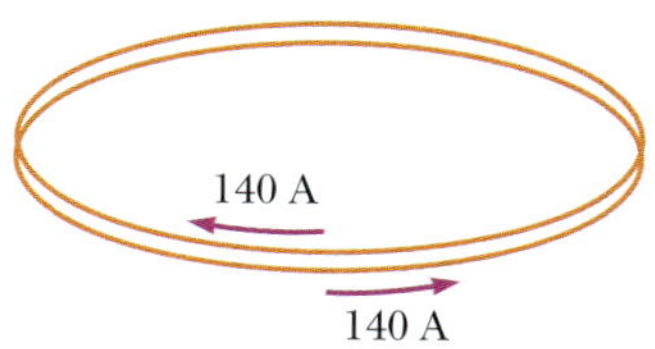

Figure P30.57

58. What objects experience a force in an electric field? Chapter 23 gives the answer: any object with electric charge, stationary or moving, other than the charged object that created the field. What creates an electric field? Any object with electric charge, stationary or moving, as you studied in Chapter 23. What objects experience a force in a magnetic field? An electric current or a moving electric charge, other than the current or charge that created the field, as discussed in Chapter 29. What creates a magnetic field? An electric current, as you studied in Section 30.1, or a moving electric charge, as shown in this problem. (a) To understand how a moving charge creates a magnetic field, consider a particle with charge q moving with velocity $\vec{\mathbf{v}}$. Define the position vector $\vec{\mathbf{r}} = r\hat{\mathbf{r}}$ leading from the particle to some location. Show that the magnetic field at that location is

$$\vec{\mathbf{B}} = \frac{\mu_0}{4\pi}\frac{q\vec{\mathbf{v}} \times \hat{\mathbf{r}}}{r^2}$$

(b) Find the magnitude of the magnetic field 1.00 mm to the side of a proton moving at 2.00×10^7 m/s. (c) Find the magnetic force on a second proton at this point, moving with the same speed in the opposite direction. (d) Find the electric force on the second proton.

59. The chapter-opening photograph shows a rail gun. Rail guns have been suggested for launching projectiles into space without chemical rockets and for ground-to-air antimissile weapons of war. A tabletop model rail gun (Fig. P30.59) consists of two long, parallel, horizontal rails 3.50 cm apart, bridged by a bar BD of mass 3.00 g. The bar is originally at rest at the midpoint of the rails and is free to slide without friction. When the switch is closed, electric current is quickly established in the circuit $ABCDEA$. The rails and bar have low electric resistance, and the current is limited to a constant 24.0 A by the power supply. (a) Find the magnitude of the magnetic field 1.75 cm from a single very long, straight wire carrying current 24.0 A. (b) Find the magnitude and direction of the magnetic field at point C in the diagram, the midpoint of the bar, immediately after the switch is closed. *Suggestion:* Consider what conclusions you can draw from the Biot–Savart law. (c) At other points along the bar BD, the field is in the same direction as at point C, but is larger in magnitude. Assume the average effective magnetic field along BD is five times larger than the field at C. With this assumption, find the magnitude and direction of the force on the bar. (d) Find the acceleration of the bar when it is in motion. (e) Does the bar move with constant acceleration? (f) Find the velocity of the bar after it has traveled 130 cm to the end of the rails.

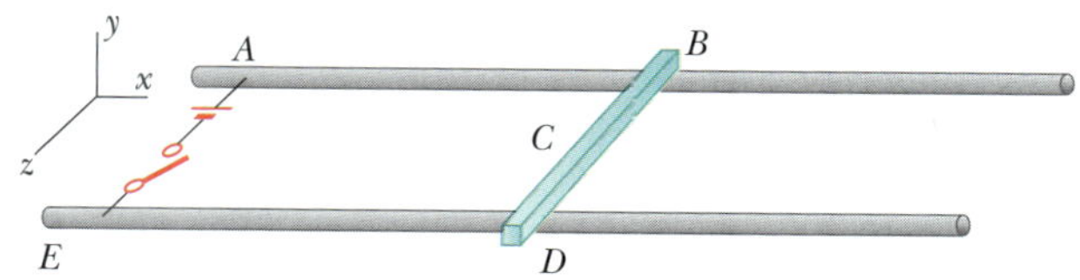

Figure P30.59

60. Fifty turns of insulated wire 0.100 cm in diameter are tightly wound to form a flat spiral. The spiral fills a disk surrounding a circle of radius 5.00 cm and extending to a radius 10.00 cm at the outer edge. Assume the wire carries current I at the center of its cross section. Approximate each turn of wire as a circle. Then a loop of current exists at radius 5.05 cm, another at 5.15 cm, and so on. Numerically calculate the magnetic field at the center of the coil.

61. An infinitely long, straight wire carrying a current I_1 is partially surrounded by a loop as shown in Figure P30.61. The loop has a length L and radius R, and it carries a current I_2. The axis of the loop coincides with the wire. Calculate the force exerted on the loop.

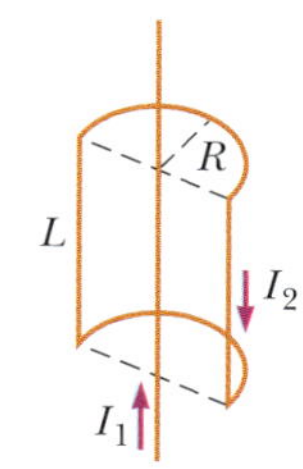

Figure P30.61

62. The magnitude of the force on a magnetic dipole $\vec{\boldsymbol{\mu}}$ aligned with a nonuniform magnetic field in the x direction is $F_x = |\vec{\boldsymbol{\mu}}|\, dB/dx$. Suppose two flat loops of wire each have radius R and carry current I. (a) The loops are arranged coaxially and separated by a variable distance x, large compared with R. Show that the magnetic force between them varies as $1/x^4$. (b) Evaluate the magnitude of this force, taking $I = 10.0$ A, $R = 0.500$ cm, and $x = 5.00$ cm.

63. A wire is formed into the shape of a square of edge length L (Fig. P30.63). Show that when the current in the loop is I, the magnetic field at point P, a distance x from the center of the square along its axis is

$$B = \frac{\mu_0 I L^2}{2\pi(x^2 + L^2/4)\sqrt{x^2 + L^2/2}}$$

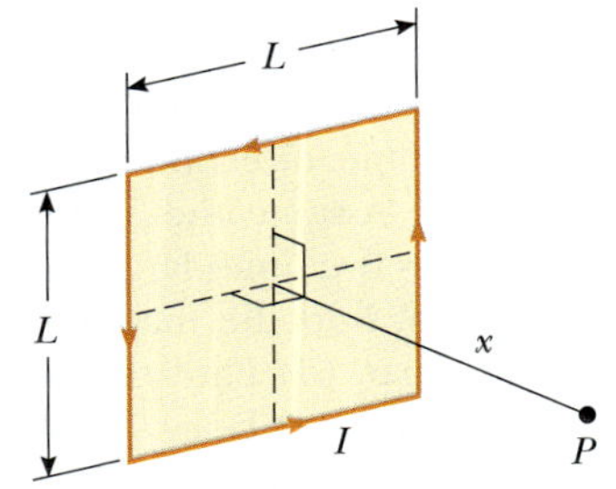

Figure P30.63

64. A wire carrying a current I is bent into the shape of an exponential spiral, $r = e^{\theta}$, from $\theta = 0$ to $\theta = 2\pi$ as suggested in Figure P30.64. To complete a loop, the ends of the spiral are connected by a straight wire along the x axis. Find the magnitude and direction of $\vec{\mathbf{B}}$ at the origin. *Suggestions:* Use the Biot–Savart law. The angle β between a radial line and its tangent line at any point on the curve $r = f(\theta)$ is related to the function as follows:

$$\tan \beta = \frac{r}{dr/d\theta}$$

Therefore, in this case $r = e^{\theta}$, $\tan \beta = 1$, and $\beta = \pi/4$, and the angle between $d\vec{\mathbf{s}}$ and $\hat{\mathbf{r}}$ is $\pi - \beta = 3\pi/4$. Also,

$$ds = \frac{dr}{\sin(\pi/4)} = \sqrt{2}\, dr$$

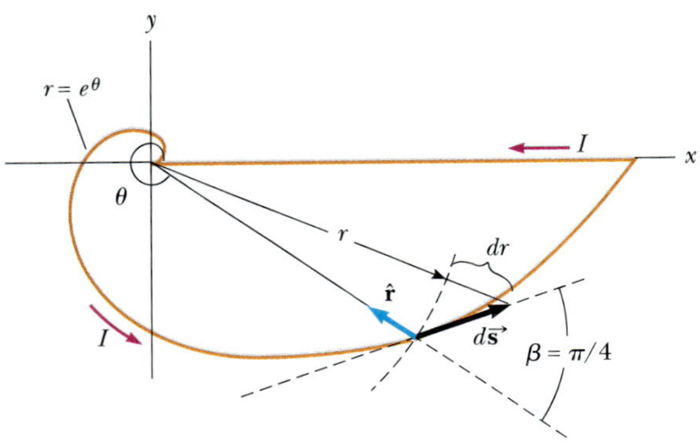

Figure P30.64

65. A sphere of radius R has a uniform volume charge density ρ. Determine the magnetic field at the center of the sphere when it rotates as a rigid object with angular speed ω about an axis through its center (Fig. P30.65).

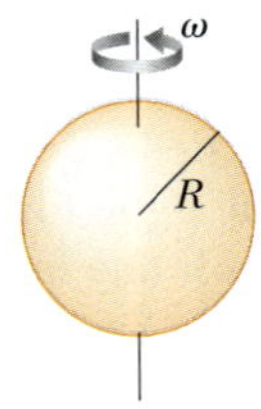

Figure P30.65 Problems 65 and 66.

66. A sphere of radius R has a uniform volume charge density ρ. Determine the magnetic dipole moment of the sphere when it rotates as a rigid body with angular speed ω about an axis through its center (Fig. P30.65).

67. A long, cylindrical conductor of radius a has two cylindrical cavities of diameter a through its entire length as shown in Figure P30.67. A current I is directed out of the page and is uniform through a cross section of the conductor. Find the magnitude and direction of the magnetic field in terms of μ_0, I, r, and a at (a) point P_1 and (b) point P_2.

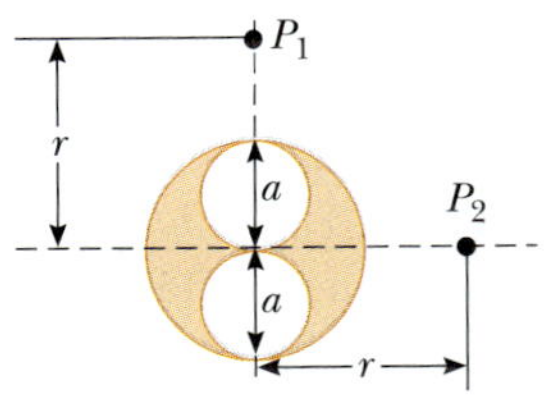

Figure P30.67

Answers to Quick Quizzes

30.1 *B, C, A*. Point *B* is closest to the current element. Point *C* is farther away, and the field is further reduced by the $\sin\theta$ factor in the cross product $d\vec{\mathbf{s}} \times \hat{\mathbf{r}}$. The field at *A* is zero because $\theta = 0$.

30.2 (a). The coils act like wires carrying parallel currents in the same direction and hence attract one another.

30.3 *b, d, a, c*. Equation 30.13 indicates that the value of the line integral depends only on the net current through each closed path. Path *b* encloses 1 A, path *d* encloses 3 A, path *a* encloses 4 A, and path *c* encloses 6 A.

30.4 *b*, then $a = c = d$. Paths *a*, *c*, and *d* all give the same nonzero value $\mu_0 I$ because the size and shape of the paths do not matter. Path *b* does not enclose the current; hence, its line integral is zero.

30.5 (c). The magnetic field in a very long solenoid is independent of its length or radius. Overwrapping with an additional layer of wire increases the number of turns per unit length.

In a commercial electric power plant, large generators transform energy that is then transferred out of the plant by electrical transmission. These generators use magnetic induction to generate a potential difference when coils of wire in the generator are rotated in a magnetic field. The source of energy to rotate the coils might be falling water, burning fossil fuels, or a nuclear reaction. (Michael Melford/Getty Images)

31 Faraday's Law

So far, our studies in electricity and magnetism have focused on the electric fields produced by stationary charges and the magnetic fields produced by moving charges. This chapter explores the effects produced by magnetic fields that vary in time.

Experiments conducted by Michael Faraday in England in 1831 and independently by Joseph Henry in the United States that same year showed that an emf can be induced in a circuit by a changing magnetic field. The results of these experiments led to a very basic and important law of electromagnetism known as *Faraday's law of induction.* An emf (and therefore a current as well) can be induced in various processes that involve a change in a magnetic flux.

By kind permission of the President and Council of the Royal Society

MICHAEL FARADAY
British Physicist and Chemist (1791–1867)
Faraday is often regarded as the greatest experimental scientist of the 1800s. His many contributions to the study of electricity include the invention of the electric motor, electric generator, and transformer as well as the discovery of electromagnetic induction and the laws of electrolysis. Greatly influenced by religion, he refused to work on the development of poison gas for the British military.

31.1 Faraday's Law of Induction

To see how an emf can be induced by a changing magnetic field, consider the experimental results obtained when a loop of wire is connected to a sensitive ammeter as illustrated in Active Figure 31.1 (page 868). When a magnet is moved toward the loop, the reading on the ammeter changes from zero in one direction, arbitrarily shown as negative in Active Figure 31.1a. When the magnet is brought to rest and held stationary relative to the loop (Active Fig. 31.1b), a reading of zero is observed. When the magnet is moved away from the loop, the reading on the ammeter changes in the opposite direction as shown in Active Figure 31.1c. Finally, when the magnet is held stationary and the loop is moved either toward or away from it, the reading changes from zero. From these observations, we conclude that the loop detects that the magnet is moving relative to it and we relate this detection to a change in magnetic field. Therefore, it seems that a relationship exists between current and changing magnetic field.

These results are quite remarkable because **a current is set up even though no batteries are present in the circuit!** We call such a current an *induced current* and say that it is produced by an *induced emf*.

Now let's describe an experiment conducted by Faraday and illustrated in Active Figure 31.2. A primary coil is wrapped around an iron ring and connected to a switch and a battery. A current in the coil produces a magnetic field when the switch is closed. A secondary coil also is wrapped around the ring and is connected to a sensitive ammeter. No battery is present in the secondary circuit, and the secondary coil is not electrically connected to the primary coil. Any current detected in the secondary circuit must be induced by some external agent.

Initially, you might guess that no current is ever detected in the secondary circuit. Something quite amazing happens when the switch in the primary circuit is either opened or thrown closed, however. At the instant the switch is closed, the ammeter reading changes from zero in one direction and then returns to zero. At the instant the switch is opened, the ammeter changes in the opposite direction and again returns to zero. Finally, the ammeter reads zero when there is either a steady current or no current in the primary circuit. To understand what happens in this experiment, note that when the switch is closed, the current in the primary circuit produces a magnetic field that penetrates the secondary circuit. Furthermore, when the switch is closed, the magnetic field produced by the current in the primary circuit changes from zero to some value over some finite time, and this changing field induces a current in the secondary circuit.

As a result of these observations, Faraday concluded that **an electric current can be induced in a loop by a changing magnetic field.** The induced current exists only while the magnetic field through the loop is changing. Once the magnetic field reaches a steady value, the current in the loop disappears. In effect, the loop behaves as though a source of emf were connected to it for a short time. It is customary to say that **an induced emf is produced in the loop by the changing magnetic field.**

The experiments shown in Active Figures 31.1 and 31.2 have one thing in common: in each case, an emf is induced in a loop when the magnetic flux through the loop changes with time. In general, this emf is directly proportional to the

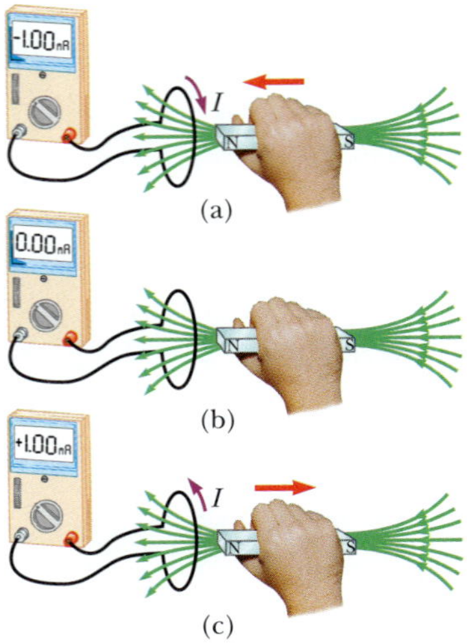

ACTIVE FIGURE 31.1

(a) When a magnet is moved toward a loop of wire connected to a sensitive ammeter, the ammeter reading changes from zero, indicating that a current is induced in the loop. (b) When the magnet is held stationary, there is no induced current in the loop, even when the magnet is inside the loop. (c) When the magnet is moved away from the loop, the ammeter reading changes in the opposite direction, indicating that the induced current is opposite that shown in (a). Changing the direction of the magnet's motion changes the direction of the current induced by that motion.

Sign in at www.thomsonedu.com and go to ThomsonNOW to move the magnet and observe the current in the ammeter.

PITFALL PREVENTION 31.1

Induced emf Requires a Change

The *existence* of a magnetic flux through an area is not sufficient to create an induced emf. The magnetic flux must *change* to induce an emf.

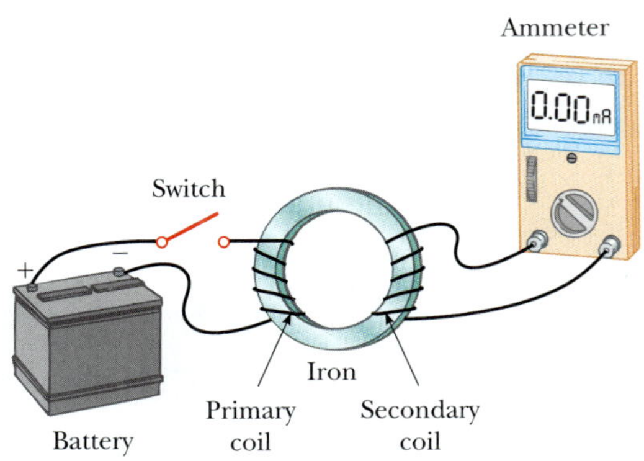

ACTIVE FIGURE 31.2

Faraday's experiment. When the switch in the primary circuit is closed, the ammeter reading in the secondary circuit changes momentarily. The emf induced in the secondary circuit is caused by the changing magnetic field through the secondary coil.

Sign in at www.thomsonedu.com and go to ThomsonNOW to open and close the switch and observe the current in the ammeter.

time rate of change of the magnetic flux through the loop. This statement can be written mathematically as **Faraday's law of induction:**

$$\mathcal{E} = -\frac{d\Phi_B}{dt} \qquad (31.1)$$

◀ Faraday's law

where $\Phi_B = \oint \vec{\mathbf{B}} \cdot d\vec{\mathbf{A}}$ is the magnetic flux through the loop. (See Section 30.5.)

If a coil consists of N loops with the same area and Φ_B is the magnetic flux through one loop, an emf is induced in every loop. The loops are in series, so their emfs add; therefore, the total induced emf in the coil is given by

$$\mathcal{E} = -N\frac{d\Phi_B}{dt} \qquad (31.2)$$

The negative sign in Equations 31.1 and 31.2 is of important physical significance as discussed in Section 31.3.

Suppose a loop enclosing an area A lies in a uniform magnetic field $\vec{\mathbf{B}}$ as in Figure 31.3. The magnetic flux through the loop is equal to $BA \cos \theta$; hence, the induced emf can be expressed as

$$\mathcal{E} = -\frac{d}{dt}(BA\cos\theta) \qquad (31.3)$$

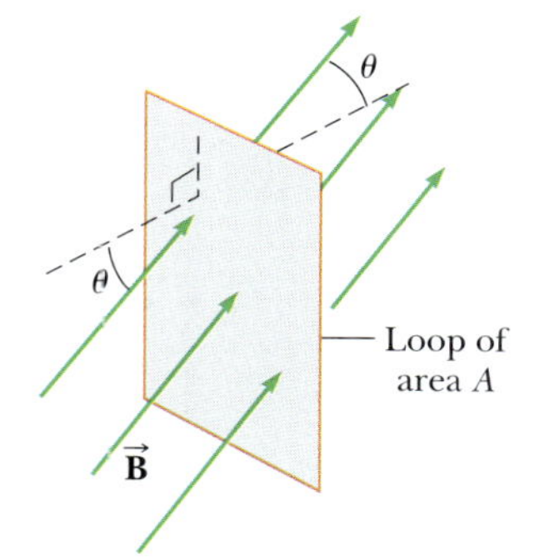

Figure 31.3 A conducting loop that encloses an area A in the presence of a uniform magnetic field $\vec{\mathbf{B}}$. The angle between $\vec{\mathbf{B}}$ and the normal to the loop is θ.

From this expression, we see that an emf can be induced in the circuit in several ways:

- The magnitude of $\vec{\mathbf{B}}$ can change with time.
- The area enclosed by the loop can change with time.
- The angle θ between $\vec{\mathbf{B}}$ and the normal to the loop can change with time.
- Any combination of the above can occur.

Quick Quiz 31.1 A circular loop of wire is held in a uniform magnetic field, with the plane of the loop perpendicular to the field lines. Which of the following will *not* cause a current to be induced in the loop? (a) crushing the loop (b) rotating the loop about an axis perpendicular to the field lines (c) keeping the orientation of the loop fixed and moving it along the field lines (d) pulling the loop out of the field

Some Applications of Faraday's Law

The ground fault interrupter (GFI) is an interesting safety device that protects users of electrical appliances against electric shock. Its operation makes use of Faraday's law. In the GFI shown in Figure 31.4, wire 1 leads from the wall outlet to the appliance to be protected and wire 2 leads from the appliance back to the wall outlet. An iron ring surrounds the two wires, and a sensing coil is wrapped around part of the ring. Because the currents in the wires are in opposite directions and of equal magnitude, there is no magnetic field surrounding the wires and the net magnetic flux through the sensing coil is zero. If the return current in wire 2 changes so that the two currents are not equal, however, circular magnetic field lines exist around the pair of wires. (This can happen if, for example, the appliance becomes wet, enabling current to leak to ground.) Therefore, the net magnetic flux through the sensing coil is no longer zero. Because household current is alternating (meaning that its direction keeps reversing), the magnetic flux through the sensing coil changes with time, inducing an emf in the coil. This induced emf is used to trigger a circuit breaker, which stops the current before it is able to reach a harmful level.

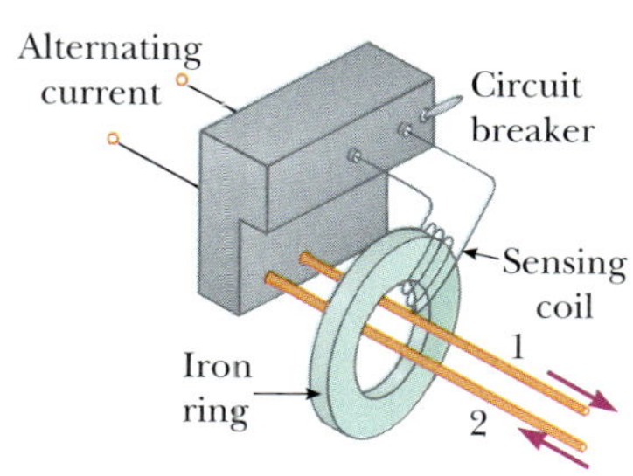

Figure 31.4 Essential components of a ground fault interrupter.

Another interesting application of Faraday's law is the production of sound in an electric guitar. The coil in this case, called the *pickup coil*, is placed near the

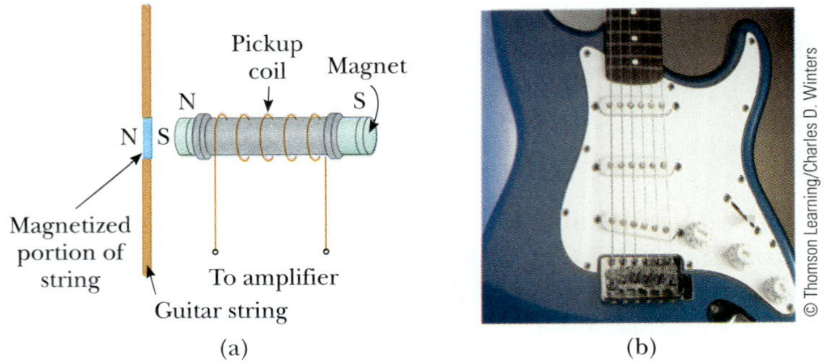

Figure 31.5 (a) In an electric guitar, a vibrating magnetized string induces an emf in a pickup coil. (b) The pickups (the circles beneath the metallic strings) of this electric guitar detect the vibrations of the strings and send this information through an amplifier and into speakers. (A switch on the guitar allows the musician to select which set of six pickups is used.)

vibrating guitar string, which is made of a metal that can be magnetized. A permanent magnet inside the coil magnetizes the portion of the string nearest the coil (Fig. 31.5a). When the string vibrates at some frequency, its magnetized segment produces a changing magnetic flux through the coil. The changing flux induces an emf in the coil that is fed to an amplifier. The output of the amplifier is sent to the loudspeakers, which produce the sound waves we hear.

EXAMPLE 31.1 Inducing an emf in a Coil

A coil consists of 200 turns of wire. Each turn is a square of side $d = 18$ cm, and a uniform magnetic field directed perpendicular to the plane of the coil is turned on. If the field changes linearly from 0 to 0.50 T in 0.80 s, what is the magnitude of the induced emf in the coil while the field is changing?

SOLUTION

Conceptualize From the description in the problem, imagine magnetic field lines passing through the coil. Because the magnetic field is changing in magnitude, an emf is induced in the coil.

Categorize We will evaluate the emf using Faraday's law from this section, so we categorize this example as a substitution problem.

Evaluate Equation 31.2 for the situation described here, noting that the magnetic field changes linearly with time:

$$|\mathcal{E}| = N\frac{\Delta\Phi_B}{\Delta t} = N\frac{\Delta(BA)}{\Delta t} = NA\frac{\Delta B}{\Delta t} = Nd^2\frac{B_f - B_i}{\Delta t}$$

Substitute numerical values:

$$|\mathcal{E}| = (200)(0.18\text{ m})^2\frac{(0.50\text{ T} - 0)}{0.80\text{ s}} = 4.0\text{ V}$$

What If? What if you were asked to find the magnitude of the induced current in the coil while the field is changing? Can you answer that question?

Answer If the ends of the coil are not connected to a circuit, the answer to this question is easy: the current is zero! (Charges move within the wire of the coil, but they cannot move into or out of the ends of the coil.) For a steady current to exist, the ends of the coil must be connected to an external circuit. Let's assume the coil is connected to a circuit and the total resistance of the coil and the circuit is 2.0 Ω. Then, the current in the coil is

$$I = \frac{\mathcal{E}}{R} = \frac{4.0\text{ V}}{2.0\ \Omega} = 2.0\text{ A}$$

EXAMPLE 31.2 An Exponentially Decaying B Field

A loop of wire enclosing an area A is placed in a region where the magnetic field is perpendicular to the plane of the loop. The magnitude of $\vec{\mathbf{B}}$ varies in time according to the expression $B = B_{max}e^{-at}$, where a is some constant. That is, at $t = 0$, the field is B_{max}, and for $t > 0$, the field decreases exponentially (Fig. 31.6). Find the induced emf in the loop as a function of time.

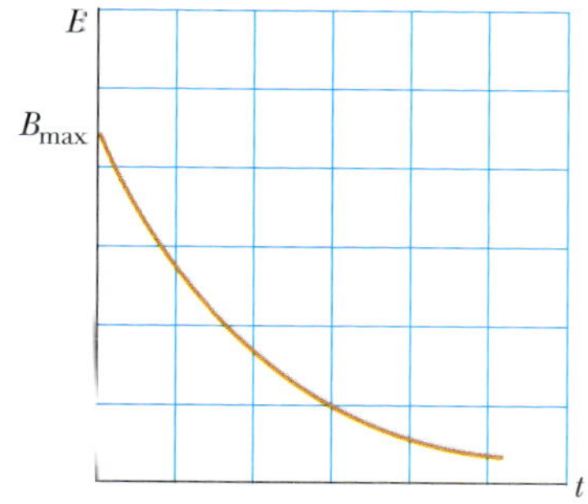

Figure 31.6 (Example 31.2) Exponential decrease in the magnitude of the magnetic field with time. The induced emf and induced current vary with time in the same way.

SOLUTION

Conceptualize The physical situation is similar to that in Example 31.1 except for two things: there is only one loop, and the field varies exponentially with time rather than linearly.

Categorize We will evaluate the emf using Faraday's law from this section, so we categorize this example as a substitution problem.

Evaluate Equation 31.1 for the situation described here:

$$\mathcal{E} = -\frac{d\Phi_B}{dt} = -\frac{d}{dt}(AB_{max}e^{-at}) = -AB_{max}\frac{d}{dt}e^{-at} = aAB_{max}e^{-at}$$

This expression indicates that the induced emf decays exponentially in time. The maximum emf occurs at $t = 0$, where $\mathcal{E}_{max} = aAB_{max}$. The plot of $\mathcal{E}$ versus t is similar to the B-versus-t curve shown in Figure 31.6.

31.2 Motional emf

In Examples 31.1 and 31.2, we considered cases in which an emf is induced in a stationary circuit placed in a magnetic field when the field changes with time. In this section, we describe **motional emf,** the emf induced in a conductor moving through a constant magnetic field.

The straight conductor of length ℓ shown in Figure 31.7 is moving through a uniform magnetic field directed into the page. For simplicity, let's assume the conductor is moving in a direction perpendicular to the field with constant velocity under the influence of some external agent. The electrons in the conductor experience a force $\vec{\mathbf{F}}_B = q\vec{\mathbf{v}} \times \vec{\mathbf{B}}$ that is directed along the length ℓ, perpendicular to both $\vec{\mathbf{v}}$ and $\vec{\mathbf{B}}$ (Eq. 29.1). Under the influence of this force, the electrons move to the lower end of the conductor and accumulate there, leaving a net positive charge at the upper end. As a result of this charge separation, an electric field $\vec{\mathbf{E}}$ is produced inside the conductor. The charges accumulate at both ends until the downward magnetic force qvB on charges remaining in the conductor is balanced by the upward electric force qE. The condition for equilibrium requires that the forces on the electrons balance:

$$qE = qvB \quad \text{or} \quad E = vB$$

The electric field produced in the conductor is related to the potential difference across the ends of the conductor according to the relationship $\Delta V = E\ell$ (Eq. 25.6). Therefore, for the equilibrium condition,

$$\Delta V = E\ell = B\ell v \qquad \textbf{(31.4)}$$

where the upper end of the conductor in Figure 31.7 is at a higher electric potential than the lower end. Therefore, **a potential difference is maintained between the ends of the conductor as long as the conductor continues to move through the uniform magnetic field.** If the direction of the motion is reversed, the polarity of the potential difference is also reversed.

A more interesting situation occurs when the moving conductor is part of a closed conducting path. This situation is particularly useful for illustrating how a

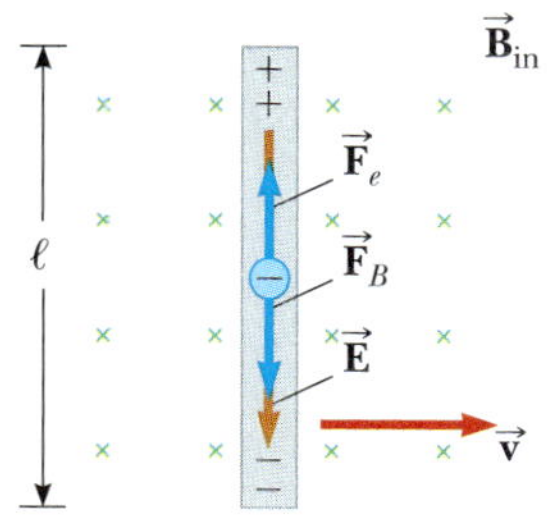

Figure 31.7 A straight electrical conductor of length ℓ moving with a velocity $\vec{\mathbf{v}}$ through a uniform magnetic field $\vec{\mathbf{B}}$ directed perpendicular to $\vec{\mathbf{v}}$. Due to the magnetic force on electrons, the ends of the conductor become oppositely charged, which establishes an electric field in the conductor. In steady state, the electric and magnetic forces on an electron in the wire are balanced.

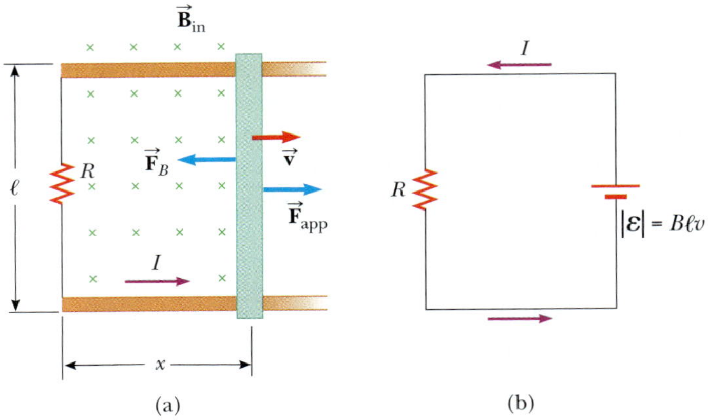

ACTIVE FIGURE 31.8

(a) A conducting bar sliding with a velocity $\vec{\mathbf{v}}$ along two conducting rails under the action of an applied force $\vec{\mathbf{F}}_{app}$. The magnetic force $\vec{\mathbf{F}}_B$ opposes the motion, and a counterclockwise current I is induced in the loop. (b) The equivalent circuit diagram for the setup shown in (a).

Sign in at www.thomsonedu.com and go to ThomsonNOW to adjust the applied force, the magnetic field, and the resistance to see the effects on the motion of the bar.

changing magnetic flux causes an induced current in a closed circuit. Consider a circuit consisting of a conducting bar of length ℓ sliding along two fixed parallel conducting rails as shown in Active Figure 31.8a. For simplicity, let's assume the bar has zero resistance and the stationary part of the circuit has a resistance R. A uniform and constant magnetic field $\vec{\mathbf{B}}$ is applied perpendicular to the plane of the circuit. As the bar is pulled to the right with a velocity $\vec{\mathbf{v}}$ under the influence of an applied force $\vec{\mathbf{F}}_{app}$, free charges in the bar experience a magnetic force directed along the length of the bar. This force sets up an induced current because the charges are free to move in the closed conducting path. In this case, the rate of change of magnetic flux through the circuit and the corresponding induced motional emf across the moving bar are proportional to the change in area of the circuit.

Because the area enclosed by the circuit at any instant is ℓx, where x is the position of the bar, the magnetic flux through that area is

$$\Phi_B = B\ell x$$

Using Faraday's law and noting that x changes with time at a rate $dx/dt = v$, we find that the induced motional emf is

$$\mathcal{E} = -\frac{d\Phi_B}{dt} = -\frac{d}{dt}(B\ell x) = -B\ell\frac{dx}{dt}$$

Motional emf ▶

$$\mathcal{E} = -B\ell v \tag{31.5}$$

Because the resistance of the circuit is R, the magnitude of the induced current is

$$I = \frac{|\mathcal{E}|}{R} = \frac{B\ell v}{R} \tag{31.6}$$

The equivalent circuit diagram for this example is shown in Active Figure 31.8b.

Let's examine the system using energy considerations. Because no battery is in the circuit, you might wonder about the origin of the induced current and the energy delivered to the resistor. We can understand the source of this current and energy by noting that the applied force does work on the conducting bar. Therefore, we model the circuit as a nonisolated system. The movement of the bar through the field causes charges to move along the bar with some average drift velocity; hence, a current is established. The change in energy in the system during some time interval must be equal to the transfer of energy into the system by work, consistent with the general principle of conservation of energy described by Equation 8.2.

Let's verify this mathematically. As the bar moves through the uniform magnetic field $\vec{\mathbf{B}}$, it experiences a magnetic force $\vec{\mathbf{F}}_B$ of magnitude $I\ell B$ (see Section 29.4). Because the bar moves with constant velocity, it is modeled as a particle in equilibrium and the magnetic force must be equal in magnitude and opposite in direction to the applied force, or to the left in Active Figure 31.8a. (If $\vec{\mathbf{F}}_B$ acted in the direction of motion, it would cause the bar to accelerate, violating the principle of conservation of energy.) Using Equation 31.6 and $F_{\text{app}} = F_B = I\ell B$, the power delivered by the applied force is

$$\mathcal{P} = F_{\text{app}}v = (I\ell B)v = \frac{B^2\ell^2v^2}{R} = \frac{\mathcal{E}^2}{R} \tag{31.7}$$

From Equation 27.21, we see that this power input is equal to the rate at which energy is delivered to the resistor.

Quick Quiz 31.2 In Active Figure 31.8a, a given applied force of magnitude F_{app} results in a constant speed v and a power input $\mathcal{P}$. Imagine that the force is increased so that the constant speed of the bar is doubled to $2v$. Under these conditions, what are the new force and the new power input? (a) $2F$ and $2\mathcal{P}$ (b) $4F$ and $2\mathcal{P}$ (c) $2F$ and $4\mathcal{P}$ (d) $4F$ and $4\mathcal{P}$

EXAMPLE 31.3 Magnetic Force Acting on a Sliding Bar

The conducting bar illustrated in Figure 31.9 moves on two frictionless, parallel rails in the presence of a uniform magnetic field directed into the page. The bar has mass m, and its length is ℓ. The bar is given an initial velocity $\vec{\mathbf{v}}_i$ to the right and is released at $t = 0$.

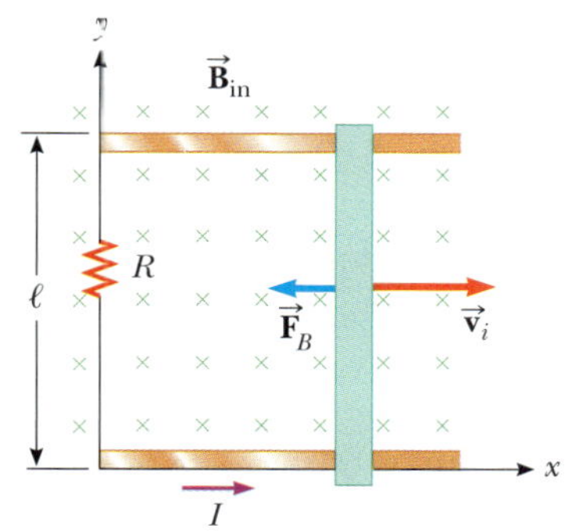

Figure 31.9 (Example 31.3) A conducting bar of length ℓ on two fixed conducting rails is given an initial velocity $\vec{\mathbf{v}}_i$ to the right.

(A) Using Newton's laws, find the velocity of the bar as a function of time.

SOLUTION

Conceptualize As the bar slides to the right in Figure 31.9, a counterclockwise current is established in the circuit consisting of the bar, the rails, and the resistor. The upward current in the bar results in a magnetic force to the left on the bar as shown in the figure. Therefore, the bar must slow down, so our mathematical solution should demonstrate that.

Categorize The text already categorizes this problem as one that uses Newton's laws. We model the bar as a particle under a net force.

Analyze From Equation 29.10, the magnetic force is $F_B = -I\ell B$, where the negative sign indicates that the force is to the left. The magnetic force is the *only* horizontal force acting on the bar.

Apply Newton's second law to the bar in the horizontal direction:

$$F_x = ma = m\frac{dv}{dt} = -I\ell B$$

Substitute $I = B\ell v/R$ from Equation 31.6:

$$m\frac{dv}{dt} = -\frac{B^2\ell^2}{R}v$$

Rearrange the equation so that all occurrences of the variable v are on the left and those of t are on the right:

$$\frac{dv}{v} = -\left(\frac{B^2\ell^2}{mR}\right)dt$$

Integrate this equation using the initial condition that $v = v_i$ at $t = 0$ and noting that $(B^2\ell^2/mR)$ is a constant:

$$\int_{v_i}^{v}\frac{dv}{v} = -\frac{B^2\ell^2}{mR}\int_0^t dt$$

$$\ln\left(\frac{v}{v_i}\right) = -\left(\frac{B^2\ell^2}{mR}\right)t$$

Define the constant $\tau = mR/B^2\ell^2$ and solve for the velocity:

$$(1) \quad v = v_i e^{-t/\tau}$$

Finalize This expression for v indicates that the velocity of the bar decreases with time under the action of the magnetic force as expected from our conceptualization of the problem.

(B) Show that the same result is found by using an energy approach.

SOLUTION

Categorize The text of this part of the problem tells us to use an energy approach for the same situation. We model the entire circuit in Figure 31.9 as an isolated system.

Finalize Consider the sliding bar as one system component possessing kinetic energy, which decreases because energy is transferring *out* of the bar by electrical transmission through the rails. The resistor is another system component possessing internal energy, which rises because energy is transferring *into* the resistor. Because energy is not leaving the system, the rate of energy transfer out of the bar equals the rate of energy transfer into the resistor.

Equate the power entering the resistor to that leaving the bar:

$$\mathcal{P}_{\text{resistor}} = -\mathcal{P}_{\text{bar}}$$

Substitute for the electrical power delivered to the resistor and the time rate of change of kinetic energy for the bar:

$$I^2R = -\frac{d}{dt}\left(\tfrac{1}{2}mv^2\right)$$

Use Equation 31.6 for the current and carry out the derivative:

$$\frac{B^2\ell^2v^2}{R} = -mv\frac{dv}{dt}$$

Rearrange terms:

$$\frac{dv}{v} = -\left(\frac{B^2\ell^2}{mR}\right)dt$$

Finalize This result is the same expression found in part (A).

What If? Suppose you wished to increase the distance through which the bar moves between the time it is initially projected and the time it essentially comes to rest. You can do so by changing one of three variables: v_i, R, or B by a factor of 2 or $\frac{1}{2}$. Which variable should you change to maximize the distance, and would you double it or halve it?

Answer Increasing v_i would make the bar move farther. Increasing R would decrease the current and therefore the magnetic force, making the bar move farther. Decreasing B would decrease the magnetic force and make the bar move farther. Which method is most effective, though?

Use Equation (1) to find the distance the bar moves by integration:

$$v = \frac{dx}{dt} = v_i e^{-t/\tau}$$

$$x = \int_0^\infty v_i e^{-t/\tau}\,dt = -v_i\tau e^{-t/\tau}\Big|_0^\infty$$

$$= -v_i\tau(0-1) = v_i\tau = v_i\left(\frac{mR}{B^2\ell^2}\right)$$

This expression shows that doubling v_i or R will double the distance. Changing B by a factor of $\frac{1}{2}$, however, causes the distance to be four times as great!

EXAMPLE 31.4 Motional emf Induced in a Rotating Bar

A conducting bar of length ℓ rotates with a constant angular speed ω about a pivot at one end. A uniform magnetic field $\vec{\mathbf{B}}$ is directed perpendicular to the plane of rotation as shown in Figure 31.10. Find the motional emf induced between the ends of the bar.

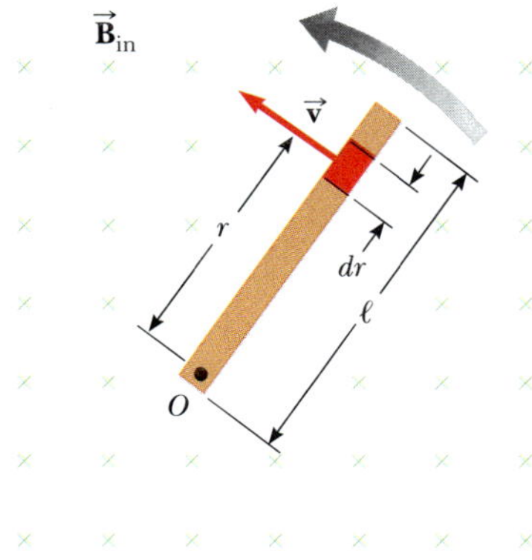

Figure 31.10 (Example 31.4) A conducting bar rotating around a pivot at one end in a uniform magnetic field that is perpendicular to the plane of rotation. A motional emf is induced across the ends of the bar.

SOLUTION

Conceptualize The rotating bar is different in nature than the sliding bar in Active Figure 31.8. Consider a small segment of the bar, however. It is a short length of conductor moving in a magnetic field and has an emf generated in it. By thinking of each small segment as a source of emf, we see that all segments are in series and the emfs add.

Categorize Based on the conceptualization of the problem, we approach this example as we did Example 31.3, with the added feature that the short segments of the bar are traveling in circular paths.

Analyze Evaluate the magnitude of the emf induced in a segment of the bar of length dr having a velocity $\vec{\mathbf{v}}$ from Equation 31.5:

$$d\mathcal{E} = Bv\,dr$$

Find the total emf between the ends of the bar by adding the emfs induced across all segments:

$$\mathcal{E} = \int Bv\,dr$$

The tangential speed v of an element is related to the angular speed ω through the relationship $v = r\omega$ (Eq. 10.10); use that fact and integrate:

$$\mathcal{E} = B\int v\,dr = B\omega\int_0^\ell r\,dr = \tfrac{1}{2}B\omega\ell^2$$

Finalize In Equation 31.5 for a sliding bar, we can increase $\mathcal{E}$ by increasing B, ℓ, or v. Increasing any one of these variables by a given factor increases $\mathcal{E}$ by the same factor. Therefore, you would choose whichever of these three variables is most convenient to increase. For the rotating rod, however, there is an advantage to increasing the length of the rod to raise the emf because ℓ is squared. Doubling the length gives four times the emf, whereas doubling the angular speed only doubles the emf.

What If? Suppose, after reading through this example, you come up with a brilliant idea. A Ferris wheel has radial metallic spokes between the hub and the circular rim. These spokes move in the magnetic field of the Earth, so each spoke acts like the bar in Figure 31.10. You plan to use the emf generated by the rotation of the Ferris wheel to power the lightbulbs on the wheel. Will this idea work?

Answer Let's estimate the emf that is generated in this situation. We know the magnitude of the magnetic field of the Earth from Table 29.1: $B = 0.5 \times 10^{-4}$ T. A typical spoke on a Ferris wheel might have a length on the order of 10 m. Suppose the period of rotation is on the order of 10 s.

Determine the angular speed of the spoke:

$$\omega = \frac{2\pi}{T} = \frac{2\pi}{10\text{ s}} = 0.63\text{ s}^{-1} \sim 1\text{ s}^{-1}$$

Assume the magnetic field lines of the Earth are horizontal at the location of the Ferris wheel and perpendicular to the spokes. Find the emf generated:

$$\mathcal{E} = \tfrac{1}{2}B\omega\ell^2 = \tfrac{1}{2}(0.5 \times 10^{-4}\text{ T})(1\text{ s}^{-1})(10\text{ m})^2$$
$$= 2.5 \times 10^{-3}\text{ V} \sim 1\text{ mV}$$

This value is a tiny emf, far smaller than that required to operate lightbulbs.

An additional difficulty is related to energy. Even assuming you could find lightbulbs that operate using a potential difference on the order of millivolts, a spoke must be part of a circuit to provide a voltage to the lightbulbs. Consequently, the spoke must carry a current. Because this current-carrying spoke is in a magnetic field, a magnetic force is exerted on the spoke in the direction opposite its direction of motion. As a result, the motor of the Ferris wheel must supply more energy to perform work against this magnetic drag force. The motor must ultimately provide the energy that is operating the lightbulbs, and you have not gained anything for free!

31.3 Lenz's Law

Faraday's law (Eq. 31.1) indicates that the induced emf and the change in flux have opposite algebraic signs. This feature has a very real physical interpretation that has come to be known as **Lenz's law:**[1]

Lenz's law ▶

> The induced current in a loop is in the direction that creates a magnetic field that opposes the change in magnetic flux through the area enclosed by the loop.

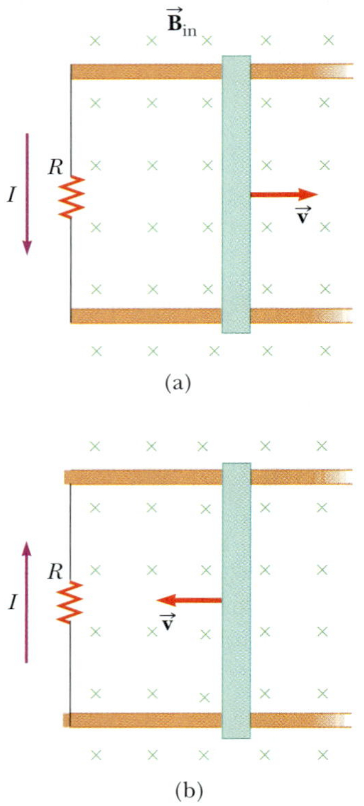

Figure 31.11 (a) As the conducting bar slides on the two fixed conducting rails, the magnetic flux due to the external magnetic field into the page through the area enclosed by the loop increases in time. By Lenz's law, the induced current must be counterclockwise to produce a counteracting magnetic field directed out of the page. (b) When the bar moves to the left, the induced current must be clockwise. Why?

That is, the induced current tends to keep the original magnetic flux through the loop from changing. We shall show that this law is a consequence of the law of conservation of energy.

To understand Lenz's law, let's return to the example of a bar moving to the right on two parallel rails in the presence of a uniform magnetic field (the *external* magnetic field; Fig. 31.11a.) As the bar moves to the right, the magnetic flux through the area enclosed by the circuit increases with time because the area increases. Lenz's law states that the induced current must be directed so that the magnetic field it produces opposes the change in the external magnetic flux. Because the magnetic flux due to an external field directed into the page is increasing, the induced current—if it is to oppose this change—must produce a field directed out of the page. Hence, the induced current must be directed counterclockwise when the bar moves to the right. (Use the right-hand rule to verify this direction.) If the bar is moving to the left as in Figure 31.11b, the external magnetic flux through the area enclosed by the loop decreases with time. Because the field is directed into the page, the direction of the induced current must be clockwise if it is to produce a field that also is directed into the page. In either case, the induced current attempts to maintain the original flux through the area enclosed by the current loop.

Let's examine this situation using energy considerations. Suppose the bar is given a slight push to the right. In the preceding analysis, we found that this motion sets up a counterclockwise current in the loop. What happens if we assume the current is clockwise such that the direction of the magnetic force exerted on the bar is to the right? This force would accelerate the rod and increase its velocity, which in turn would cause the area enclosed by the loop to increase more rapidly. The result would be an increase in the induced current, which would cause an increase in the force, which would produce an increase in the current, and so on. In effect, the system would acquire energy with no input of energy. This behavior is clearly inconsistent with all experience and violates the law of conservation of energy. Therefore, the current must be counterclockwise.

Quick Quiz 31.3 Figure 31.12 shows a circular loop of wire falling toward a wire carrying a current to the left. What is the direction of the induced current in the loop of wire? (a) clockwise (b) counterclockwise (c) zero (d) impossible to determine

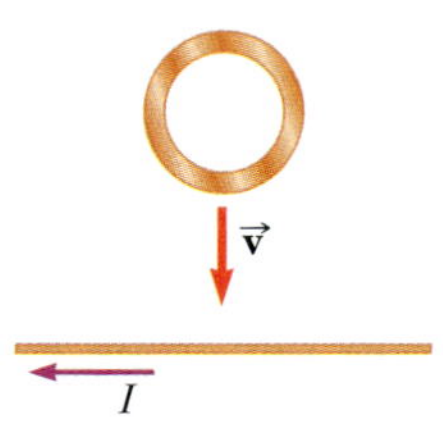

Figure 31.12 (Quick Quiz 31.3)

[1] Developed by German physicist Heinrich Lenz (1804–1865).

CONCEPTUAL EXAMPLE 31.5 Application of Lenz's Law

A magnet is placed near a metal loop as shown in Figure 31.13a.

(A) Find the direction of the induced current in the loop when the magnet is pushed toward the loop.

SOLUTION

As the magnet moves to the right toward the loop, the external magnetic flux through the loop increases with time. To counteract this increase in flux due to a field toward the right, the induced current produces its own magnetic field to the left as illustrated in Figure 31.13b; hence, the induced current is in the direction shown. Knowing that like magnetic poles repel each other, we conclude that the left face of the current loop acts like a north pole and the right face acts like a south pole.

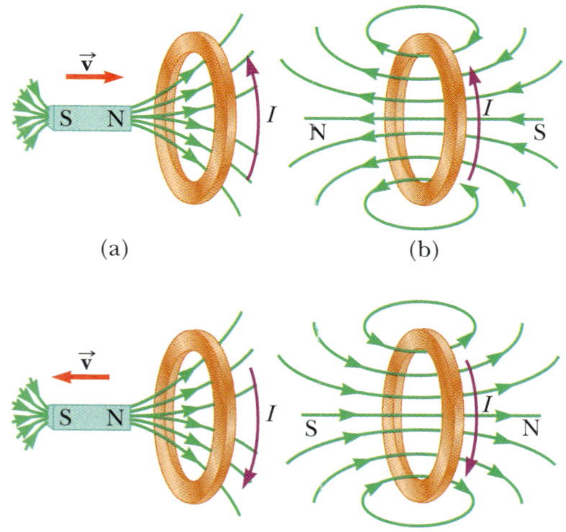

Figure 31.13 (Conceptual Example 31.5) (a) When the magnet is moved toward the stationary conducting loop, a current is induced in the direction shown. The magnetic field lines shown are those due to the bar magnet. (b) This induced current produces its own magnetic field directed to the left that counteracts the increasing external flux. The magnetic field lines shown are those due to the induced current in the ring. (c) When the magnet is moved away from the stationary conducting loop, a current is induced in the direction shown. The magnetic field lines shown are those due to the bar magnet. (d) This induced current produces a magnetic field directed to the right and so counteracts the decreasing external flux. The magnetic field lines shown are those due to the induced current in the ring.

(B) Find the direction of the induced current in the loop when the magnet is pulled away from the loop.

SOLUTION

If the magnet moves to the left as in Figure 31.13c, its flux through the area enclosed by the loop decreases in time. Now the induced current in the loop is in the direction shown in Figure 31.13d because this current direction produces a magnetic field in the same direction as the external field. In this case, the left face of the loop is a south pole and the right face is a north pole.

CONCEPTUAL EXAMPLE 31.6 A Loop Moving Through a Magnetic Field

A rectangular metallic loop of dimensions ℓ and w and resistance R moves with constant speed v to the right, as in Figure 31.14a. The loop passes through a uniform magnetic field $\vec{\mathbf{B}}$ directed into the page and extending a distance $3w$ along the x axis. Define x as the position of the right side of the loop along the x axis.

(A) Plot as a function of x the magnetic flux through the area enclosed by the loop.

SOLUTION

Figure 31.14b shows the flux through the area enclosed by the loop as a function of x. Before the loop enters the field, the flux through the loop is zero. As the loop enters the field, the flux increases linearly with position until the left edge of the loop is just inside the field. Finally, the flux through the loop decreases linearly to zero as the loop leaves the field.

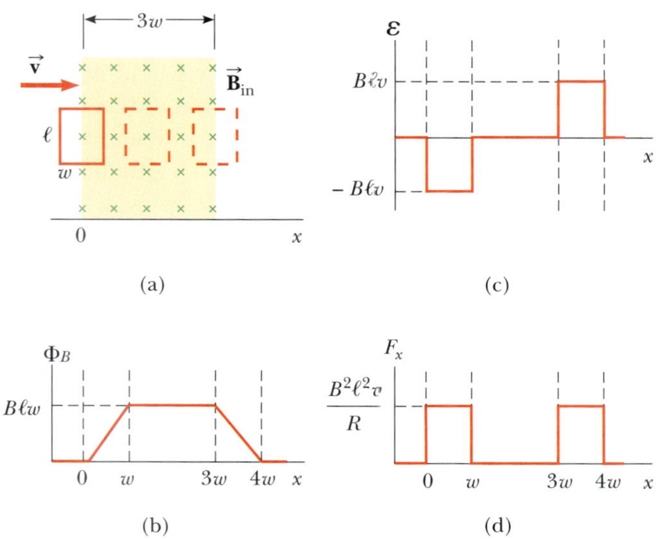

Figure 31.14 (Conceptual Example 31.6) (a) A conducting rectangular loop of width w and length ℓ moving with a velocity $\vec{\mathbf{v}}$ through a uniform magnetic field extending a distance $3w$. (b) Magnetic flux through the area enclosed by the loop as a function of loop position. (c) Induced emf as a function of loop position. (d) Applied force required for constant velocity as a function of loop position.

(B) Plot as a function of x the induced motional emf in the loop.

SOLUTION

Before the loop enters the field, no motional emf is induced in it because no field is present (Fig. 31.14c). As the right side of the loop enters the field, the magnetic flux directed into the page increases. Hence, according to Lenz's law, the induced current is counterclockwise because it must produce its own magnetic field directed out of the page. The motional emf $-B\ell v$ (from Eq. 31.5) arises from the magnetic force experienced by charges in the right side of the loop. When the loop is entirely in the field, the change in magnetic flux through the loop is zero; hence, the motional emf vanishes. That happens because once the left side of the loop enters the field, the motional emf induced in it cancels the motional emf present in the right side of the loop. As the right side of the loop leaves the field, the flux through the loop begins to decrease, a clockwise current is induced, and the induced emf is $B\ell v$. As soon as the left side leaves the field, the emf decreases to zero.

(C) Plot as a function of x the external applied force necessary to counter the magnetic force and keep v constant.

SOLUTION

The external force that must be applied to the loop to maintain this motion is plotted in Figure 31.14d. Before the loop enters the field, no magnetic force acts on it; hence, the applied force must be zero if v is constant. When the right side of the loop enters the field, the applied force necessary to maintain constant speed must be equal in magnitude and opposite in direction to the magnetic force exerted on that side. When the loop is entirely in the field, the flux through the loop is not changing with time. Hence, the net emf induced in the loop is zero and the current also is zero. Therefore, no external force is needed to maintain the motion. Finally, as the right side leaves the field, the applied force must be equal in magnitude and opposite in direction to the magnetic force acting on the left side of the loop.

From this analysis, we conclude that power is supplied only when the loop is either entering or leaving the field. Furthermore, this example shows that the motional emf induced in the loop can be zero even when there is motion through the field! A motional emf is induced *only* when the magnetic flux through the loop *changes in time.*

31.4 Induced emf and Electric Fields

We have seen that a changing magnetic flux induces an emf and a current in a conducting loop. In our study of electricity, we related a current to an electric field that applies electric forces on charged particles. In the same way, we can relate an induced current in a conducting loop to an electric field by claiming that **an electric field is created in the conductor as a result of the changing magnetic flux.**

We also noted in our study of electricity that the existence of an electric field is independent of the presence of any test charges. This independence suggests that even in the absence of a conducting loop, a changing magnetic field generates an electric field in empty space.

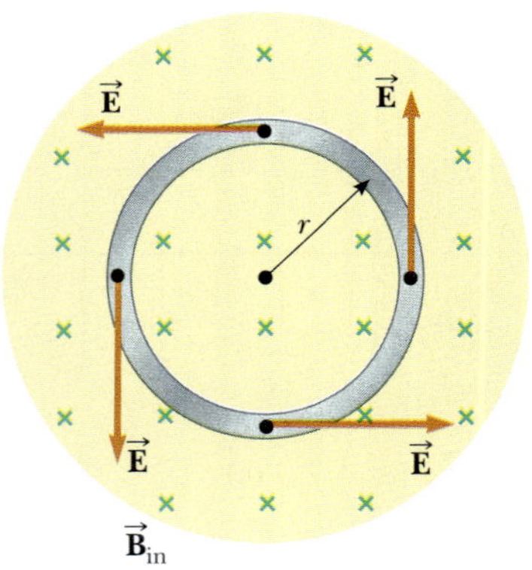

Figure 31.15 A conducting loop of radius r in a uniform magnetic field perpendicular to the plane of the loop. If $\vec{\mathbf{B}}$ changes in time, an electric field is induced in a direction tangent to the circumference of the loop.

PITFALL PREVENTION 31.2

Induced Electric Fields

The changing magnetic field does *not* need to exist at the location of the induced electric field. In Figure 31.15, even a loop outside the region of magnetic field experiences an induced electric field.

This induced electric field is *nonconservative,* unlike the electrostatic field produced by stationary charges. To illustrate this point, consider a conducting loop of radius r situated in a uniform magnetic field that is perpendicular to the plane of the loop as in Figure 31.15. If the magnetic field changes with time, an emf $\mathcal{E} = -d\Phi_B/dt$ is, according to Faraday's law (Eq. 31.1), induced in the loop. The induction of a current in the loop implies the presence of an induced electric field $\vec{\mathbf{E}}$, which must be tangent to the loop because that is the direction in which the charges in the wire move in response to the electric force. The work done by the electric field in moving a test charge q once around the loop is equal to $q\mathcal{E}$. Because the electric force acting on the charge is $q\vec{\mathbf{E}}$, the work done by the electric field in moving the charge once around the loop is $qE(2\pi r)$, where $2\pi r$ is the circumference of the loop. These two expressions for the work done must be equal; therefore,

$$q\mathcal{E} = qE(2\pi r)$$

$$E = \frac{\mathcal{E}}{2\pi r}$$

Using this result along with Equation 31.1 and that $\Phi_B = BA = B\pi r^2$ for a circular loop, the induced electric field can be expressed as

$$E = -\frac{1}{2\pi r}\frac{d\Phi_B}{dt} = -\frac{r}{2}\frac{dB}{dt} \qquad \textbf{(31.8)}$$

If the time variation of the magnetic field is specified, the induced electric field can be calculated from Equation 31.8.

The emf for any closed path can be expressed as the line integral of $\vec{\mathbf{E}} \cdot d\vec{\mathbf{s}}$ over that path: $\mathcal{E} = \oint \vec{\mathbf{E}} \cdot d\vec{\mathbf{s}}$. In more general cases, E may not be constant and the path may not be a circle. Hence, Faraday's law of induction, $\mathcal{E} = -d\Phi_B/dt$, can be written in the general form

$$\oint \vec{\mathbf{E}} \cdot d\vec{\mathbf{s}} = -\frac{d\Phi_B}{dt} \qquad (31.9)$$

◀ Faraday's law in general form

The induced electric field $\vec{\mathbf{E}}$ in Equation 31.9 is a nonconservative field that is generated by a changing magnetic field. The field $\vec{\mathbf{E}}$ that satisfies Equation 31.9 cannot possibly be an electrostatic field because were the field electrostatic and hence conservative, the line integral of $\vec{\mathbf{E}} \cdot d\vec{\mathbf{s}}$ over a closed loop would be zero (Section 25.1), which would be in contradiction to Equation 31.9.

EXAMPLE 31.7 Electric Field Induced by a Changing Magnetic Field in a Solenoid

A long solenoid of radius R has n turns of wire per unit length and carries a time-varying current that varies sinusoidally as $I = I_{max} \cos \omega t$, where I_{max} is the maximum current and ω is the angular frequency of the alternating current source (Fig. 31.16).

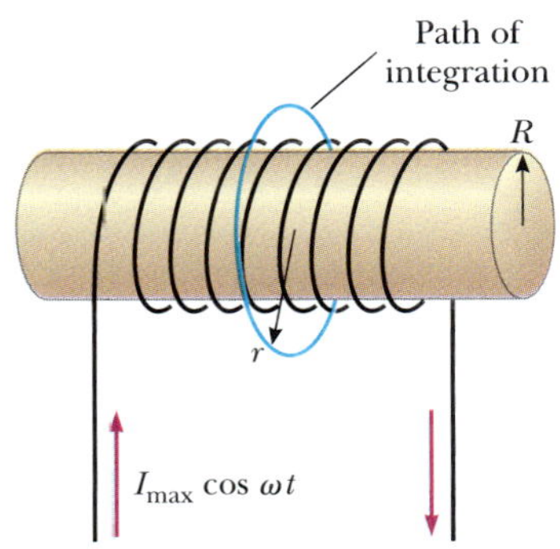

Figure 31.16 (Example 31.7) A long solenoid carrying a time-varying current given by $I = I_{max} \cos \omega t$. An electric field is induced both inside and outside the solenoid.

(A) Determine the magnitude of the induced electric field outside the solenoid at a distance $r > R$ from its long central axis.

SOLUTION

Conceptualize Figure 31.16 shows the physical situation. As the current in the coil changes, imagine a changing magnetic field at all points in space as well as an induced electric field.

Categorize Because the current varies in time, the magnetic field is changing, leading to an induced electric field as opposed to the electrostatic electric fields due to stationary electric charges.

Analyze First consider an external point and take the path for the line integral to be a circle of radius r centered on the solenoid as illustrated in Figure 31.16.

Evaluate the right side of Equation 31.9, noting that $\vec{\mathbf{B}}$ is perpendicular to the circle bounded by the path of integration and that this magnetic field exists only inside the solenoid:

$$(1) \quad -\frac{d\Phi_B}{dt} = -\frac{d}{dt}(B\pi R^2) = -\pi R^2 \frac{dB}{dt}$$

Evaluate the magnetic field in the solenoid from Equation 30.17:

$$(2) \quad B = \mu_0 n I = \mu_0 n I_{max} \cos \omega t$$

Substitute Equation (2) into Equation (1):

$$(3) \quad -\frac{d\Phi_B}{dt} = -\pi R^2 \mu_0 n I_{max} \frac{d}{dt}(\cos \omega t) = \pi R^2 \mu_0 n I_{max} \omega \sin \omega t$$

Evaluate the left side of Equation 31.9, noting that the magnitude of $\vec{\mathbf{E}}$ is constant on the path of integration and $\vec{\mathbf{E}}$ is tangent to it:

$$(4) \quad \oint \vec{\mathbf{E}} \cdot d\vec{\mathbf{s}} = E(2\pi r)$$

Substitute Equations (3) and (4) into Equation 31.9:

$$E(2\pi r) = \pi R^2 \mu_0 n I_{max} \omega \sin \omega t$$

Solve for the magnitude of the electric field:

$$E = \frac{\mu_0 n I_{max} \omega R^2}{2r} \sin \omega t \quad (\text{for } r > R)$$

Finalize This result shows that the amplitude of the electric field outside the solenoid falls off as $1/r$ and varies sinusoidally with time. As we will learn in Chapter 34, the time-varying electric field creates an additional contribution to the magnetic field. The magnetic field can be somewhat stronger than we first stated, both inside and outside the solenoid. The correction to the magnetic field is small if the angular frequency ω is small. At high frequencies, however, a new phenomenon can dominate: The electric and magnetic fields, each re-creating the other, constitute an electromagnetic wave radiated by the solenoid as we will study in Chapter 34.

(B) What is the magnitude of the induced electric field inside the solenoid, a distance r from its axis?

SOLUTION

Analyze For an interior point ($r < R$), the magnetic flux through an integration loop is given by $\Phi_B = B\pi r^2$.

Evaluate the right side of Equation 31.9:

$$(5) \quad -\frac{d\Phi_B}{dt} = -\frac{d}{dt}(B\pi r^2) = -\pi r^2 \frac{dB}{dt}$$

Substitute Equation (2) into Equation (5):

$$(6) \quad -\frac{d\Phi_B}{dt} = -\pi r^2 \mu_0 n I_{max} \frac{d}{dt}(\cos \omega t) = \pi r^2 \mu_0 n I_{max} \omega \sin \omega t$$

Substitute Equations (4) and (6) into Equation 31.9:

$$E(2\pi r) = \pi r^2 \mu_0 n I_{max} \omega \sin \omega t$$

Solve for the magnitude of the electric field:

$$E = \frac{\mu_0 n I_{max} \omega}{2} r \sin \omega t \quad (\text{for } r < R)$$

Finalize This result shows that the amplitude of the electric field induced inside the solenoid by the changing magnetic flux through the solenoid increases linearly with r and varies sinusoidally with time.

31.5 Generators and Motors

Electric generators take in energy by work and transfer it out by electrical transmission. To understand how they operate, let us consider the **alternating-current (AC) generator.** In its simplest form, it consists of a loop of wire rotated by some external means in a magnetic field (Active Fig. 31.17a).

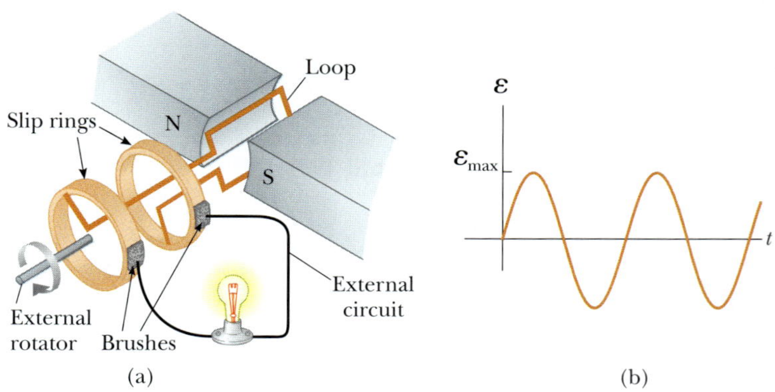

ACTIVE FIGURE 31.17

(a) Schematic diagram of an AC generator. An emf is induced in a loop that rotates in a magnetic field. (b) The alternating emf induced in the loop plotted as a function of time.

Sign in at www.thomsonedu.com and go to ThomsonNOW to adjust the speed of rotation and the strength of the field to see the effects on the emf generated.

In commercial power plants, the energy required to rotate the loop can be derived from a variety of sources. For example, in a hydroelectric plant, falling water directed against the blades of a turbine produces the rotary motion; in a coal-fired plant, the energy released by burning coal is used to convert water to steam, and this steam is directed against the turbine blades.

As a loop rotates in a magnetic field, the magnetic flux through the area enclosed by the loop changes with time, and this change induces an emf and a current in the loop according to Faraday's law. The ends of the loop are connected to slip rings that rotate with the loop. Connections from these slip rings, which act as output terminals of the generator, to the external circuit are made by stationary metallic brushes in contact with the slip rings.

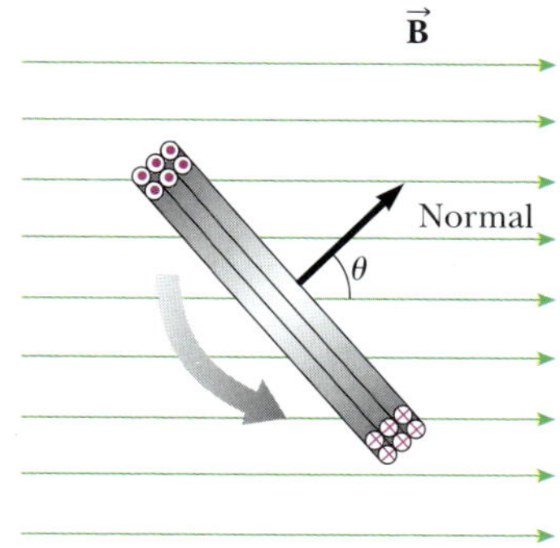

Figure 31.18 A loop enclosing an area A and containing N turns, rotating with constant angular speed ω in a magnetic field. The emf induced in the loop varies sinusoidally in time.

Instead of a single turn, suppose a coil with N turns (a more practical situation), with the same area A, rotates in a magnetic field with a constant angular speed ω. If θ is the angle between the magnetic field and the normal to the plane of the coil as in Figure 31.18, the magnetic flux through the coil at any time t is

$$\Phi_B = BA\cos\theta = BA\cos\omega t$$

where we have used the relationship $\theta = \omega t$ between angular position and angular speed (see Eq. 10.3). (We have set the clock so that $t = 0$ when $\theta = 0$.) Hence, the induced emf in the coil is

$$\mathcal{E} = -N\frac{d\Phi_B}{dt} = -NAB\frac{d}{dt}(\cos\omega t) = NAB\omega\sin\omega t \tag{31.10}$$

This result shows that the emf varies sinusoidally with time as plotted in Active Figure 31.17b. Equation 31.10 shows that the maximum emf has the value

$$\mathcal{E}_{\text{max}} = NAB\omega \tag{31.11}$$

which occurs when $\omega t = 90°$ or $270°$. In other words, $\mathcal{E} = \mathcal{E}_{\text{max}}$ when the magnetic field is in the plane of the coil and the time rate of change of flux is a maximum. Furthermore, the emf is zero when $\omega t = 0$ or $180°$, that is, when $\vec{\mathbf{B}}$ is perpendicular to the plane of the coil and the time rate of change of flux is zero.

The frequency for commercial generators in the United States and Canada is 60 Hz, whereas in some European countries it is 50 Hz. (Recall that $\omega = 2\pi f$, where f is the frequency in hertz.)

Quick Quiz 31.4 In an AC generator, a coil with N turns of wire spins in a magnetic field. Of the following choices, which does *not* cause an increase in the emf generated in the coil? (a) replacing the coil wire with one of lower resistance (b) spinning the coil faster (c) increasing the magnetic field (d) increasing the number of turns of wire on the coil

---

EXAMPLE 31.8 emf Induced in a Generator

The coil in an AC generator consists of 8 turns of wire, each of area $A = 0.090\,0\ \text{m}^2$, and the total resistance of the wire is 12.0 Ω. The coil rotates in a 0.500-T magnetic field at a constant frequency of 60.0 Hz.

(A) Find the maximum induced emf in the coil.

SOLUTION

Conceptualize Study Active Figure 31.17 to make sure you understand the operation of an AC generator.

Categorize We evaluate parameters using equations developed in this section, so we categorize this example as a substitution problem.

Use Equation 31.11 to find the maximum induced emf: $\mathcal{E}_{\text{max}} = NAB\omega = NAB(2\pi f)$

Substitute numerical values: $$\mathcal{E}_{max} = 8(0.090\,0\text{ m}^2)(0.500\text{ T})(2\pi)(60.0\text{ Hz}) = \boxed{136\text{ V}}$$

(B) What is the maximum induced current in the coil when the output terminals are connected to a low-resistance conductor?

SOLUTION

Use Equation 27.7 and the result to part (A): $$I_{max} = \frac{\mathcal{E}_{max}}{R} = \frac{136\text{ V}}{12.0\ \Omega} = \boxed{11.3\text{ A}}$$

The **direct-current (DC) generator** is illustrated in Active Figure 31.19a. Such generators are used, for instance, in older cars to charge the storage batteries. The components are essentially the same as those of the AC generator except that the contacts to the rotating coil are made using a split ring called a *commutator.*

In this configuration, the output voltage always has the same polarity and pulsates with time as shown in Active Figure 31.19b. We can understand why by noting that the contacts to the split ring reverse their roles every half cycle. At the same time, the polarity of the induced emf reverses; hence, the polarity of the split ring (which is the same as the polarity of the output voltage) remains the same.

A pulsating DC current is not suitable for most applications. To obtain a steadier DC current, commercial DC generators use many coils and commutators distributed so that the sinusoidal pulses from the various coils are out of phase. When these pulses are superimposed, the DC output is almost free of fluctuations.

A **motor** is a device into which energy is transferred by electrical transmission while energy is transferred out by work. A motor is essentially a generator operating in reverse. Instead of generating a current by rotating a coil, a current is supplied to the coil by a battery and the torque acting on the current-carrying coil (Section 29.5) causes it to rotate.

Useful mechanical work can be done by attaching the rotating coil to some external device. As the coil rotates in a magnetic field, however, the changing magnetic flux induces an emf in the coil; this induced emf always acts to reduce the current in the coil. If that were not the case, Lenz's law would be violated. The back emf increases in magnitude as the rotational speed of the coil increases. (The phrase *back emf* is used to indicate an emf that tends to reduce the supplied current.) Because the voltage available to supply current equals the difference between the supply voltage and the back emf, the current in the rotating coil is limited by the back emf.

When a motor is turned on, there is initially no back emf, and the current is very large because it is limited only by the resistance of the coil. As the coil begins

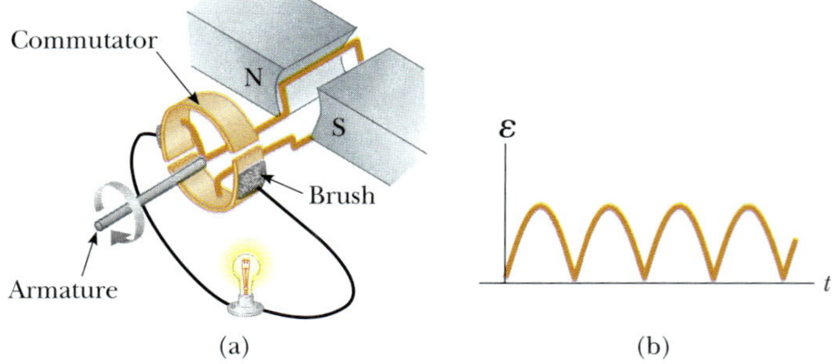

ACTIVE FIGURE 31.19

(a) Schematic diagram of a DC generator. (b) The magnitude of the emf varies in time, but the polarity never changes.

Sign in at www.thomsonedu.com and go to ThomsonNOW to adjust the speed of rotation and the strength of the field to see the effects on the emf generated.

to rotate, the induced back emf opposes the applied voltage and the current in the coil decreases. If the mechanical load increases, the motor slows down, which causes the back emf to decrease. This reduction in the back emf increases the current in the coil and therefore also increases the power needed from the external voltage source. For this reason, the power requirements for running a motor are greater for heavy loads than for light ones. If the motor is allowed to run under no mechanical load, the back emf reduces the current to a value just large enough to overcome energy losses due to internal energy and friction. If a very heavy load jams the motor so that it cannot rotate, the lack of a back emf can lead to dangerously high current in the motor's wire. This dangerous situation is explored in the **What If?** section of Example 31.9.

A modern application of motors in automobiles is seen in the development of *hybrid drive systems.* In these automobiles, a gasoline engine and an electric motor are combined to increase the fuel economy of the vehicle and reduce its emissions. Figure 31.20 shows the engine compartment of a Toyota Prius, one of a small number of hybrids available in the United States. In this automobile, power to the wheels can come from either the gasoline engine or the electric motor. In normal driving, the electric motor accelerates the vehicle from rest until it is moving at a speed of about 15 mi/h (24 km/h). During this acceleration period, the engine is not running, so gasoline is not used and there is no emission. At higher speeds, the motor and engine work together so that the engine always operates at or near its most efficient speed. The result is a significantly higher gasoline mileage than that obtained by a traditional gasoline-powered automobile. When a hybrid vehicle brakes, the motor acts as a generator and returns some of the vehicle's kinetic energy back to the battery as stored energy. In a normal vehicle, this kinetic energy is simply lost as it is transformed to internal energy in the brakes and roadway.

John W. Jewett, Jr.

Figure 31.20 The engine compartment of the Toyota Prius, a hybrid vehicle.

EXAMPLE 31.9 **The Induced Current in a Motor**

A motor contains a coil with a total resistance of 10 Ω and is supplied by a voltage of 120 V. When the motor is running at its maximum speed, the back emf is 70 V.

(A) Find the current in the coil at the instant the motor is turned on.

SOLUTION

Conceptualize Think about the motor just after it is turned on. It has not yet moved, so there is no back emf generated. As a result, the current in the motor is high. After the motor begins to turn, a back emf is generated and the current decreases.

Categorize We need to combine our new understanding about motors with the relationship between current, voltage, and resistance.

Analyze Evaluate the current in the coil from Equation 27.7 with no back emf generated:

$$I = \frac{\mathcal{E}}{R} = \frac{120\text{ V}}{10\ \Omega} = 12\text{ A}$$

(B) Find the current in the coil when the motor has reached maximum speed.

SOLUTION

Evaluate the current in the coil with the maximum back emf generated:

$$I = \frac{\mathcal{E} - \mathcal{E}_{\text{back}}}{R} = \frac{120\text{ V} - 70\text{ V}}{10\ \Omega} = \frac{50\text{ V}}{10\ \Omega} = 5.0\text{ A}$$

Finalize The current drawn by the motor when operating at its maximum speed is significantly less than that drawn before it begins to turn.

What If? Suppose this motor is in a circular saw. When you are operating the saw, the blade becomes jammed in a piece of wood and the motor cannot turn. By what percentage does the power input to the motor increase when it is jammed?

Answer You may have everyday experiences with motors becoming warm when they are prevented from turning. That is due to the increased power input to the motor. The higher rate of energy transfer results in an increase in the internal energy of the coil, an undesirable effect.

Set up the ratio of power input to the motor when jammed, which is that calculated in part (A), to that when it is not jammed, part (B):

$$\frac{\mathcal{P}_{\text{jammed}}}{\mathcal{P}_{\text{not jammed}}} = \frac{I_A{}^2 R}{I_B{}^2 R} = \frac{I_A{}^2}{I_B{}^2}$$

Substituting numerical values gives

$$\frac{\mathcal{P}_{\text{jammed}}}{\mathcal{P}_{\text{not jammed}}} = \frac{(12\text{ A})^2}{(5.0\text{ A})^2} = 5.76$$

which represents a 476% increase in the input power! Such a high power input can cause the coil to become so hot that it is damaged.

31.6 Eddy Currents

As we have seen, an emf and a current are induced in a circuit by a changing magnetic flux. In the same manner, circulating currents called **eddy currents** are induced in bulk pieces of metal moving through a magnetic field. This phenomenon can be demonstrated by allowing a flat copper or aluminum plate attached at the end of a rigid bar to swing back and forth through a magnetic field (Fig. 31.21). As the plate enters the field, the changing magnetic flux induces an emf in the plate, which in turn causes the free electrons in the plate to move, producing the swirling eddy currents. According to Lenz's law, the direction of the eddy currents is such that they create magnetic fields that oppose the change that causes the currents. For this reason, the eddy currents must produce effective magnetic poles on the plate, which are repelled by the poles of the magnet; this situation gives rise to a repulsive force that opposes the motion of the plate. (If the opposite were true, the plate would accelerate and its energy would increase after each swing, in violation of the law of conservation of energy.)

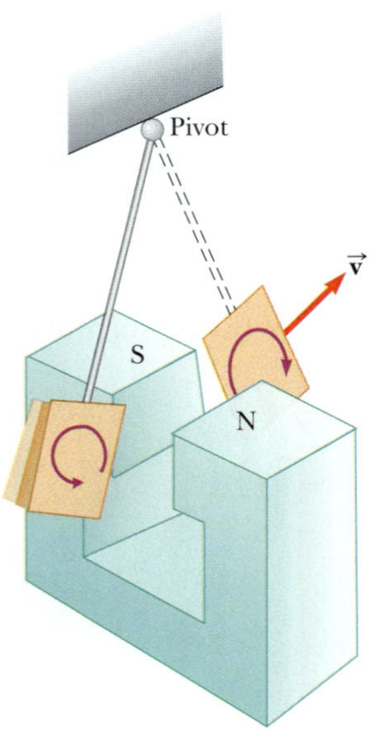

Figure 31.21 Formation of eddy currents in a conducting plate moving through a magnetic field. As the plate enters or leaves the field, the changing magnetic flux induces an emf, which causes eddy currents in the plate.

As indicated in Active Figure 31.22a, with $\vec{\mathbf{B}}$ directed into the page, the induced eddy current is counterclockwise as the swinging plate enters the field at position 1 because the flux due to the external magnetic field into the page through the plate is increasing. Hence, by Lenz's law, the induced current must provide its own magnetic field out of the page. The opposite is true as the plate leaves the field at position 2, where the current is clockwise. Because the induced eddy current always produces a magnetic retarding force $\vec{\mathbf{F}}_B$ when the plate enters or leaves the field, the swinging plate eventually comes to rest.

If slots are cut in the plate as shown in Active Figure 31.22b, the eddy currents and the corresponding retarding force are greatly reduced. We can understand this reduction in force by realizing that the cuts in the plate prevent the formation of any large current loops.

The braking systems on many subway and rapid-transit cars make use of electromagnetic induction and eddy currents. An electromagnet attached to the train is positioned near the steel rails. (An electromagnet is essentially a solenoid with an iron core.) The braking action occurs when a large current is passed through the electromagnet. The relative motion of the magnet and rails induces eddy currents in the rails, and the direction of these currents produces a drag force on the moving train. Because the eddy currents decrease steadily in magnitude as the train slows

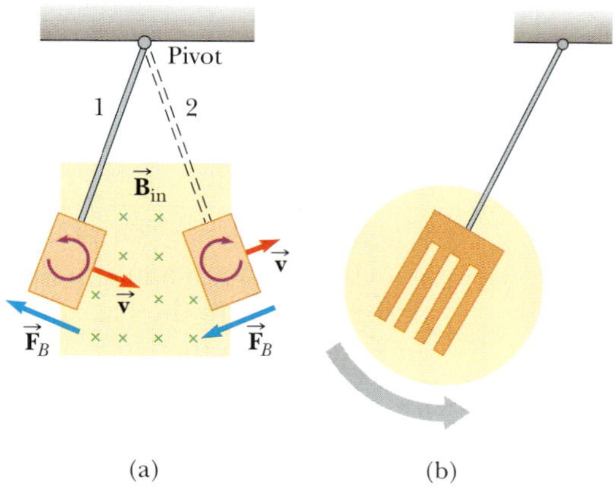

ACTIVE FIGURE 31.22

(a) As the conducting plate enters the field (position 1), the eddy currents are counterclockwise. As the plate leaves the field (position 2), the currents are clockwise. In either case, the force on the plate is opposite the velocity and eventually the plate comes to rest. (b) When slots are cut in the conducting plate, the eddy currents are reduced and the plate swings more freely through the magnetic field.

Sign in at www.thomsonedu.com and go to ThomsonNOW to choose to let a solid or a slotted plate swing through the magnetic field and observe the effect.

down, the braking effect is quite smooth. As a safety measure, some power tools use eddy currents to stop rapidly spinning blades once the device is turned off.

Eddy currents are often undesirable because they represent a transformation of mechanical energy to internal energy. To reduce this energy loss, conducting parts are often laminated; that is, they are built up in thin layers separated by a nonconducting material such as lacquer or a metal oxide. This layered structure prevents large current loops and effectively confines the currents to small loops in individual layers. Such a laminated structure is used in transformer cores (see Section 33.8) and motors to minimize eddy currents and thereby increase the efficiency of these devices.

Quick Quiz 31.5 In an equal-arm balance from the early 20th century (Fig. 31.23), an aluminum sheet hangs from one of the arms and passes between the poles of a magnet, causing the oscillations of the balance to decay rapidly. In the absence of such magnetic braking, the oscillation might continue for a long time, and the experimenter would have to wait to take a reading. Why do the oscillations decay? (a) because the aluminum sheet is attracted to the magnet (b) because currents in the aluminum sheet set up a magnetic field that opposes the oscillations (c) because aluminum is paramagnetic.

Figure 31.23 (Quick Quiz 31.5) In an old-fashioned equal-arm balance, an aluminum sheet hangs between the poles of a magnet.

Summary

ThomsonNOW™ Sign in at **www.thomsonedu.com** and go to ThomsonNOW to take a practice test for this chapter.

CONCEPTS AND PRINCIPLES

Faraday's law of induction states that the emf induced in a loop is directly proportional to the time rate of change of magnetic flux through the loop, or

$$\mathcal{E} = -\frac{d\Phi_B}{dt} \qquad (31.1)$$

where $\Phi_B = \oint \vec{\mathbf{B}} \cdot d\vec{\mathbf{A}}$ is the magnetic flux through the loop.

When a conducting bar of length ℓ moves at a velocity $\vec{\mathbf{v}}$ through a magnetic field $\vec{\mathbf{B}}$, where $\vec{\mathbf{B}}$ is perpendicular to the bar and to $\vec{\mathbf{v}}$, the **motional emf** induced in the bar is

$$\mathcal{E} = -B\ell v \qquad (31.5)$$

Lenz's law states that the induced current and induced emf in a conductor are in such a direction as to set up a magnetic field that opposes the change that produced them.

A general form of **Faraday's law of induction** is

$$\oint \vec{\mathbf{E}} \cdot d\vec{\mathbf{s}} = -\frac{d\Phi_B}{dt} \qquad (31.9)$$

where $\vec{\mathbf{E}}$ is the nonconservative electric field that is produced by the changing magnetic flux.

Questions

☐ denotes answer available in *Student Solutions Manual/Study Guide;* **O** denotes objective question

1. What is the difference between magnetic flux and magnetic field?

2. **O** Figure Q31.2 is a graph of the magnetic flux through a certain coil of wire as a function of time, during an interval while the radius of the coil is increased, the coil is rotated through 1.5 revolutions, and the external source of the magnetic field is turned off, in that order. Rank the electromotive force induced in the coil at the instants marked A through F from the largest positive value to the largest-magnitude negative value. In your ranking, note any cases of equality and also any instants when the emf is zero.

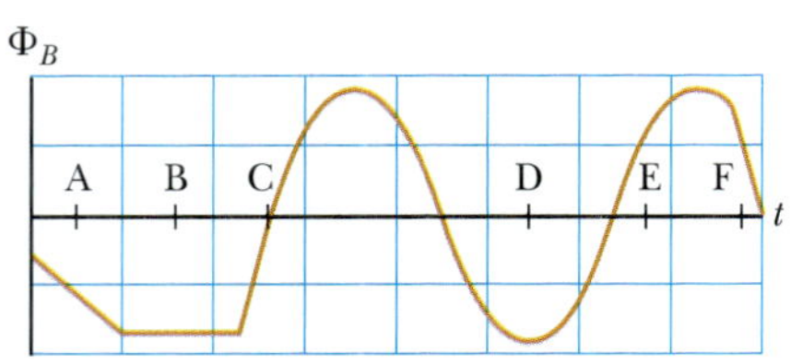

Figure Q31.2

3. **O** A flat coil of wire is placed in a uniform magnetic field that is in the y direction. **(i)** The magnetic flux through the coil is a maximum if the coil is (a) in the xy plane (b) in either the xy or the yz plane (c) in the xz plane (d) in any orientation, because it is a constant **(ii)** For what orientation is the flux zero? Choose the best answer from the same possibilities.

4. **O** A square, flat coil of wire is pulled at constant velocity through a region of uniform magnetic field directed perpendicular to the plane of the coil as shown in Figure Q31.4. **(i)** Is current induced in the coil? (a) yes, clockwise (b) yes, counterclockwise (c) no **(ii)** Does charge separation occur in the coil? (a) yes, with the top positive (b) yes, with the top negative (c) no

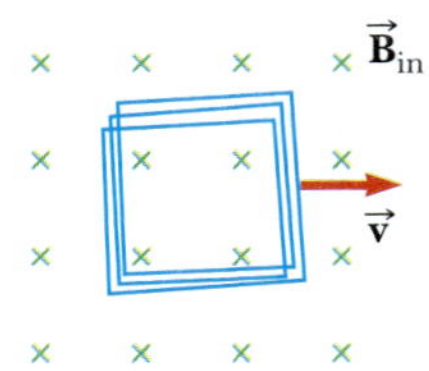

Figure Q31.4 Questions 4 and 6.

5. ☐ The bar in Figure Q31.5 moves on rails to the right with a velocity $\vec{\mathbf{v}}$, and the uniform, constant magnetic field is directed out of the page. Why is the induced current clockwise? If the bar were moving to the left, what would be the direction of the induced current?

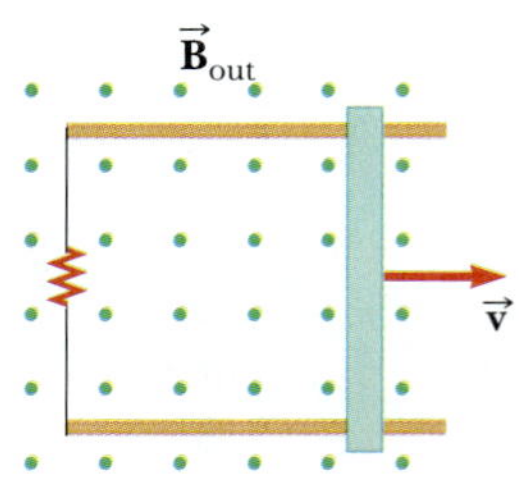

Figure Q31.5 Questions 5 and 6.

6. **O** **(i)** As the square coil of wire in Figure Q31.4 moves perpendicular to the field, is an external force required

to keep it moving with constant speed? **(ii)** Answer the same question for the bar in Figure Q31.5. **(iii)** Answer the same question for the bar in Figure Q31.6.

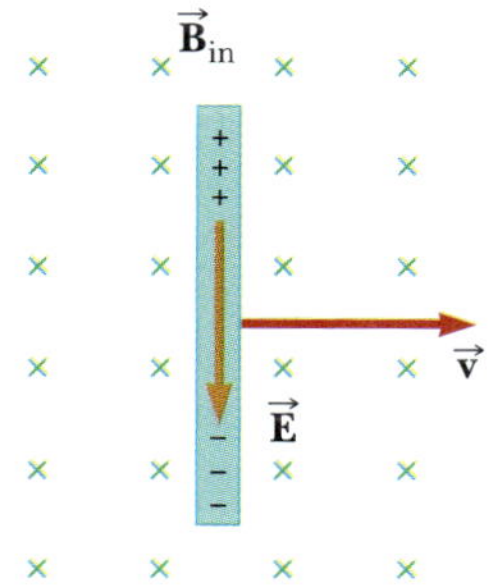

Figure Q31.6

7. In a hydroelectric dam, how is energy produced that is then transferred out by electrical transmission? That is, how is the energy of motion of the water converted to energy that is transmitted by AC electricity?

8. A piece of aluminum is dropped vertically downward between the poles of an electromagnet. Does the magnetic field affect the velocity of the aluminum?

9. **O** What happens to the amplitude of the induced emf when the rate of rotation of a generator coil is doubled? (a) It becomes 4 times larger. (b) It becomes 2 times larger. (c) It is unchanged. (d) It becomes $\frac{1}{2}$ as large. (e) It becomes $\frac{1}{4}$ as large.

10. When the switch in Figure Q31.10a is closed, a current is set up in the coil and the metal ring springs upward (Fig. Q31.10b). Explain this behavior.

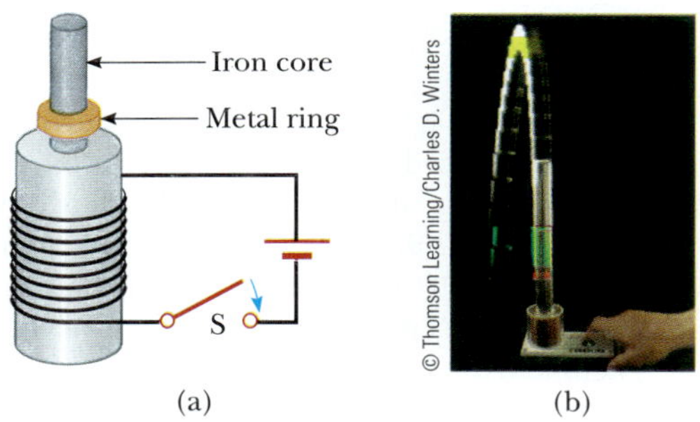

Figure Q31.10 Questions 10 and 11.

11. Assume the battery in Figure Q31.10a is replaced by an AC source and the switch is held closed. If held down, the metal ring on top of the solenoid becomes hot. Why?

12. **O** A bar magnet is held in a vertical orientation above a loop of wire that lies in a horizontal plane as shown in Figure Q31.12. The south pole of the magnet is on the bottom end, closest to the loop of wire. The magnet is dropped toward the loop. **(i)** While the magnet is falling toward the loop, what is the direction of current in the resistor? (a) to the left (b) to the right (c) there is no current (d) both to the left and to the right (e) downward **(ii)** After the magnet has passed through the loop and moves away from it, what is the direction of current in the resistor? Choose from the same possibilities. **(iii)** Now assume the magnet, producing a symmetrical field, is held in a horizontal orientation and then dropped. While it is approaching the loop, what is the direction of current in the resistor? Choose from the same possibilities.

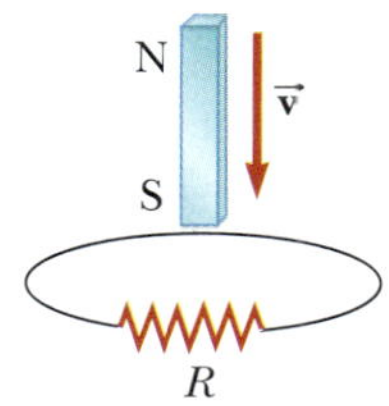

Figure Q31.12

13. **O** What is the direction of the current in the resistor in Figure Q31.13 **(i)** at an instant immediately after the switch is thrown closed, **(ii)** after the switch has been closed for several seconds, and **(iii)** at an instant after the switch has then been thrown open? Choose each answer from these possibilities: (a) left (b) right (c) both left and right (d) The current is zero.

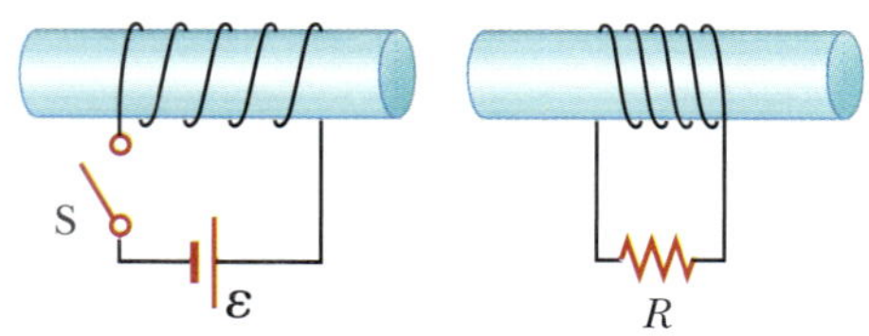

Figure Q31.13

14. In Section 7.7, we defined conservative and nonconservative forces. In Chapter 23, we stated that an electric charge creates an electric field that produces a conservative force. Argue now that induction creates an electric field that produces a nonconservative force.

Problems

WebAssign The Problems from this chapter may be assigned online in WebAssign.

ThomsonNOW Sign in at **www.thomsonedu.com** and go to ThomsonNOW to assess your understanding of this chapter's topics with additional quizzing and conceptual questions.

1, 2, 3 denotes straightforward, intermediate, challenging; ☐ denotes full solution available in *Student Solutions Manual/Study Guide;* ▲ denotes coached solution with hints available at **www.thomsonedu.com;** ■ denotes developing symbolic reasoning; ● denotes asking for qualitative reasoning; ■ denotes computer useful in solving problem

Section 31.1 Faraday's Law of Induction

Section 31.3 Lenz's Law

1. *Transcranial magnetic stimulation* is a noninvasive technique used to stimulate regions of the human brain. A small coil is placed on the scalp, and a brief burst of current in the coil produces a rapidly changing magnetic field inside the brain. The induced emf can stimulate neuronal activity. (a) One such device generates an upward magnetic field

within the brain that rises from zero to 1.50 T in 120 ms. Determine the induced emf around a horizontal circle of tissue of radius 1.60 mm. (b) **What If?** The field next changes to 0.500 T downward in 80.0 ms. How does the emf induced in this process compare with that in part (a)?

2. A flat loop of wire consisting of a single turn of cross-sectional area 8.00 cm^2 is perpendicular to a magnetic field that increases uniformly in magnitude from 0.500 T to 2.50 T in 1.00 s. What is the resulting induced current if the loop has a resistance of 2.00 Ω?

3. A 25-turn circular coil of wire has diameter 1.00 m. It is placed with its axis along the direction of the Earth's magnetic field of 50.0 μT and then in 0.200 s is flipped 180°. An average emf of what magnitude is generated in the coil?

4. ● Your physics teacher asks you to help her set up a demonstration of Faraday's law for the class. As shown in Figure P31.4, the apparatus consists of a strong, permanent magnet producing a field of 110 mT between its poles, a 12-turn coil of radius 2.10 cm cemented onto a wood frame with a handle, some flexible connecting wires, and an ammeter. The idea is to pull the coil out of the center of the magnetic field as quickly as you can and read the average current registered on the meter. The equivalent resistance of the coil, leads, and meter is 2.30 Ω. You can flip the coil out of the field in about 180 ms. The ammeter has a full-scale sensitivity of 1 000 μA. (a) Is this meter sensitive enough to show the induced current clearly? Explain your reasoning. (b) Does the meter in the diagram register a positive or a negative current? Explain your reasoning.

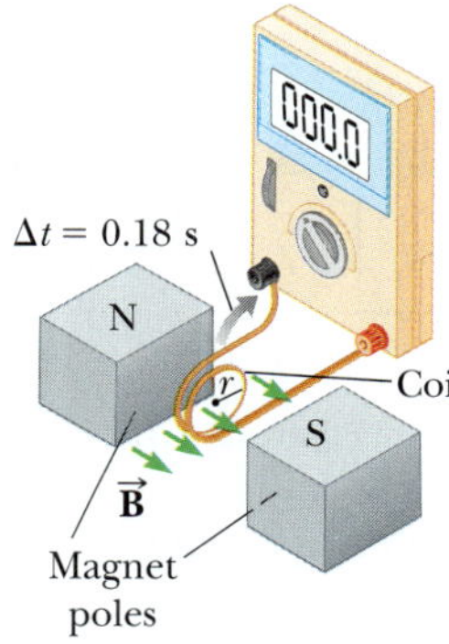

Figure P31.4

5. A rectangular loop of area A is placed in a region where the magnetic field is perpendicular to the plane of the loop. The magnitude of the field is allowed to vary in time according to $B = B_{\max}e^{-t/\tau}$, where $B_{\max}$ and τ are constants. The field has the constant value $B_{\max}$ for $t < 0$. (a) Use Faraday's law to show that the emf induced in the loop is given by

$$\mathcal{E} = \frac{AB_{\max}}{\tau}e^{-t/\tau}$$

(b) Obtain a numerical value for $\mathcal{E}$ at $t = 4.00$ s when $A = 0.160$ m^2, $B_{\max} = 0.350$ T, and $\tau = 2.00$ s. (c) For the values of A, $B_{\max}$, and τ given in part (b), what is the maximum value of $\mathcal{E}$?

6. To monitor the breathing of a hospital patient, a thin belt is girded around the patient's chest. The belt is a 200-turn coil. When the patient inhales, the area encircled by the coil increases by 39.0 cm^2. The magnitude of the Earth's magnetic field is 50.0 μT and makes an angle of 28.0° with the plane of the coil. Assuming a patient takes 1.80 s to inhale, find the average induced emf in the coil during this time interval.

7. ▲ A strong electromagnet produces a uniform magnetic field of 1.60 T over a cross-sectional area of 0.200 m^2. A coil having 200 turns and a total resistance of 20.0 Ω is placed around the electromagnet. The current in the electromagnet is then smoothly reduced until it reaches zero in 20.0 ms. What is the current induced in the coil?

8. A loop of wire in the shape of a rectangle of width w and length L and a long, straight wire carrying a current I lie on a tabletop as shown in Figure P31.8. (a) Determine the magnetic flux through the loop due to the current I. (b) Suppose the current is changing with time according to $I = a + bt$, where a and b are constants. Determine the emf that is induced in the loop if $b = 10.0$ A/s, $h = 1.00$ cm, $w = 10.0$ cm, and $L = 100$ cm. What is the direction of the induced current in the rectangle?

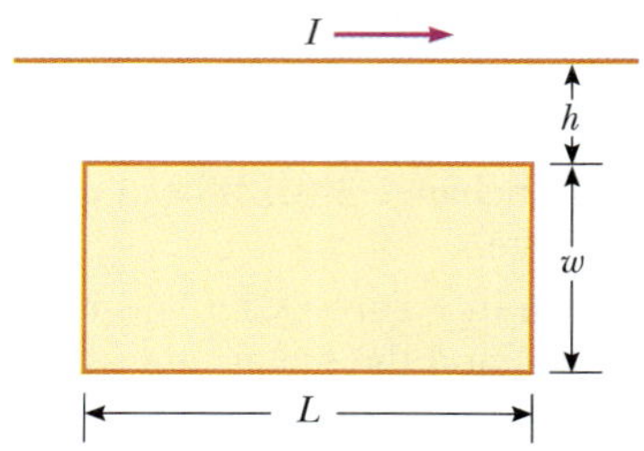

Figure P31.8 Problems 8 and 67.

9. ▲ An aluminum ring of radius 5.00 cm and resistance 3.00×10^{-4} Ω is placed around one end of a long air-core solenoid with 1 000 turns per meter and radius 3.00 cm as shown in Figure P31.9. Assume the axial component of the field produced by the solenoid is one-half as strong over the area of the end of the solenoid as at the center of the solenoid. Also assume the solenoid produces negligible field outside its cross-sectional area. The current in the solenoid is increasing at a rate of 270 A/s. (a) What is the induced current in the ring? At the center of the ring, what are (b) the magnitude and (c) the direction of the magnetic field produced by the induced current in the ring?

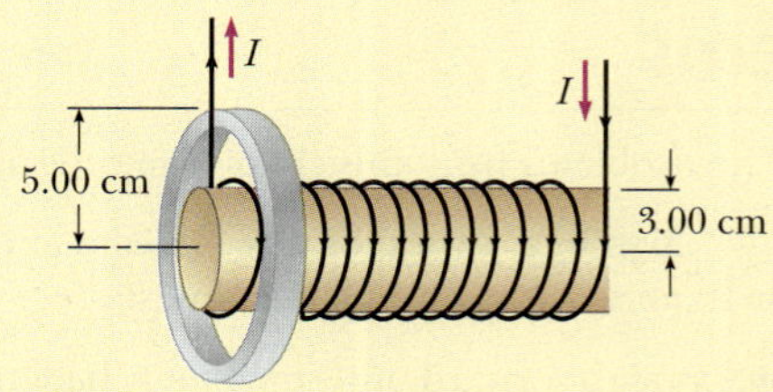

Figure P31.9 Problems 9 and 10.

10. An aluminum ring of radius r_1 and resistance R is placed around one end of a long air-core solenoid with n turns per meter and smaller radius r_2 as shown in Figure P31.9. Assume the axial component of the field produced by the solenoid over the area of the end of the solenoid is one-

half as strong as at the center of the solenoid. Also assume the solenoid produces negligible field outside its cross-sectional area. The current in the solenoid is increasing at a rate of $\Delta I/\Delta t$. (a) What is the induced current in the ring? (b) At the center of the ring, what is the magnetic field produced by the induced current in the ring? (c) What is the direction of this field?

11. A coil of 15 turns and radius 10.0 cm surrounds a long solenoid of radius 2.00 cm and 1.00×10^3 turns/meter (Fig. P31.11). The current in the solenoid changes as $I = (5.00\ \text{A})\sin(120t)$. Find the induced emf in the 15-turn coil as a function of time.

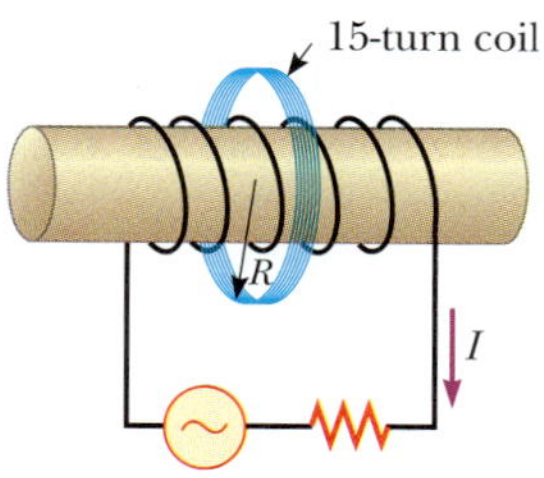

Figure P31.11

12. Two circular coils lie in the same plane. The following equation describes the emf induced in the smaller coil by a changing current in the larger coil. (a) Calculate this emf. (b) Write the statement of a problem, including data, for which the equation gives the solution.

$$\mathcal{E} = -20\frac{d}{dt}\left[\frac{130\left(4\pi \times 10^{-7}\,\dfrac{\text{T}\cdot\text{m}}{\text{A}}\right)\left(3\ \text{A} - \dfrac{(6\ \text{A})t}{13 \times 10^{-6}\ \text{s}}\right)}{2(0.40\ \text{m})}\pi(0.03\ \text{m})^2\cos 0°\right]$$

13. Find the current through section PQ of length $a = 65.0$ cm in Figure P31.13. The circuit is located in a magnetic field whose magnitude varies with time according to the expression $B = (1.00 \times 10^{-3}\ \text{T/s})t$. Assume the resistance per length of the wire is $0.100\ \Omega/\text{m}$.

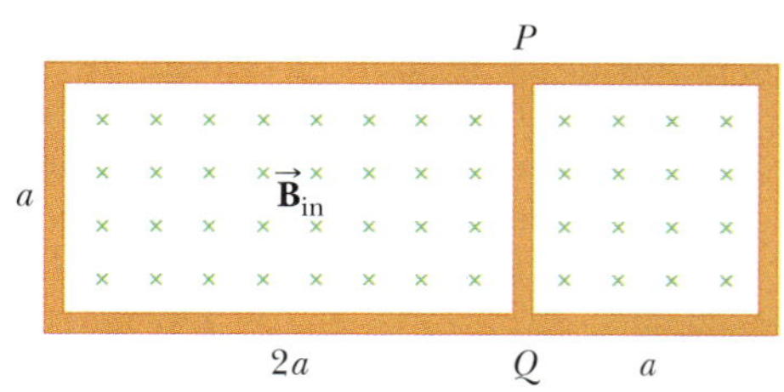

Figure P31.13

14. A 30-turn circular coil of radius 4.00 cm and resistance $1.00\ \Omega$ is placed in a magnetic field directed perpendicular to the plane of the coil. The magnitude of the magnetic field varies in time according to the expression $B = 0.010\,0t + 0.040\,0t^2$, where t is in seconds and B is in teslas. Calculate the induced emf in the coil at $t = 5.00$ s.

15. A long solenoid has $n = 400$ turns per meter and carries a current given by $I = (30.0\ \text{A})(1 - e^{-1.60t})$. Inside the solenoid and coaxial with it is a coil that has a radius of 6.00 cm and consists of a total of $N = 250$ turns of fine wire (Fig. P31.15). What emf is induced in the coil by the changing current?

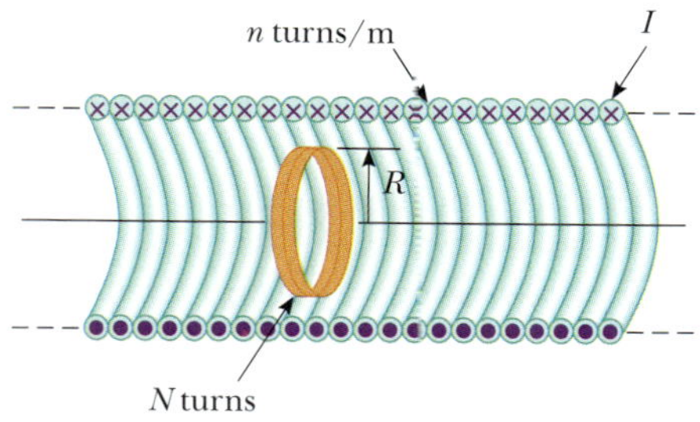

Figure P31.15

16. ● When a wire carries an AC current with a known frequency, you can use a *Rogowski coil* to determine the amplitude I_{max} of the current without disconnecting the wire to shunt the current though a meter. The Rogowski coil, shown in Figure P31.16, simply clips around the wire. It consists of a toroidal conductor wrapped around a circular return cord. Let n represent the number of turns in the toroid per unit distance along it. Let A represent the cross-sectional area of the toroid. Let $I(t) = I_{max}\sin\omega t$ represent the current to be measured. (a) Show that the amplitude of the emf induced in the Rogowski coil is $\mathcal{E}_{max} = \mu_0 nA\omega I_{max}$. (b) Explain why the wire carrying the unknown current need not be at the center of the Rogowski coil and why the coil will not respond to nearby currents that it does not enclose.

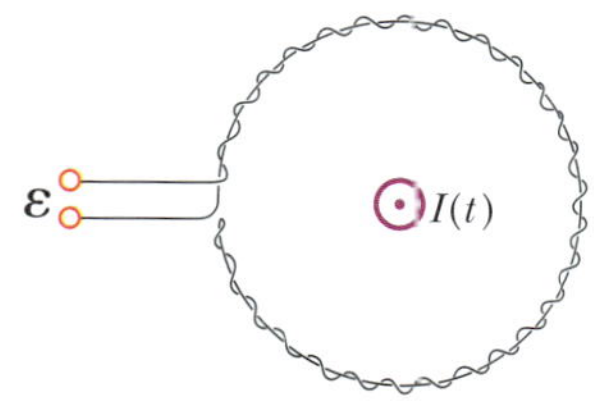

Figure P31.16

17. A coil formed by wrapping 50 turns of wire in the shape of a square is positioned in a magnetic field so that the normal to the plane of the coil makes an angle of 30.0° with the direction of the field. When the magnetic field is increased uniformly from 200 μT to 600 μT in 0.400 s, an emf of magnitude 80.0 mV is induced in the coil. What is the total length of the wire?

18. A toroid having a rectangular cross section ($a = 2.00$ cm by $b = 3.00$ cm) and inner radius $R = 4.00$ cm consists of 500 turns of wire that carries a sinusoidal current $I = I_{max}\sin\omega t$, with $I_{max} = 50.0$ A and a frequency $f = \omega/2\pi = 60.0$ Hz. A coil that consists of 20 turns of wire links with the toroid as shown in Figure P31.18. Determine the emf induced in the coil as a function of time.

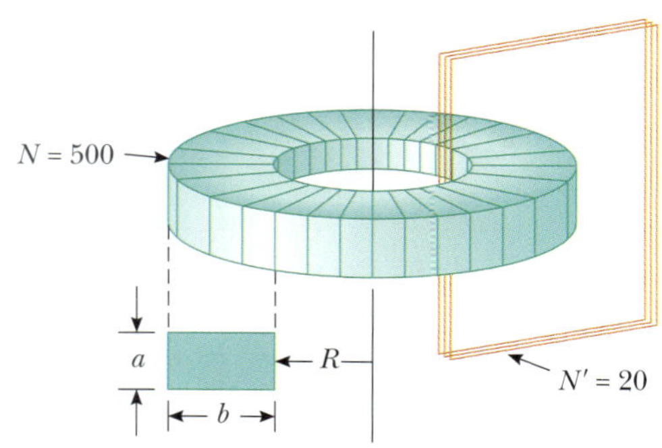

Figure P31.18

19. A piece of insulated wire is shaped into a figure 8 as shown in Figure P31.19. The radius of the upper circle is 5.00 cm and that of the lower circle is 9.00 cm. The wire has a uniform resistance per unit length of 3.00 Ω/m. A uniform magnetic field is applied perpendicular to the plane of the two circles, in the direction shown. The magnetic field is increasing at a constant rate of 2.00 T/s. Find the magnitude and direction of the induced current in the wire.

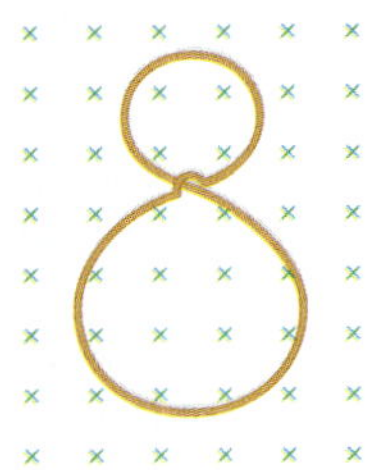

Figure P31.19

Section 31.2 Motional emf

Section 31.3 Lenz's Law

Problem 61 in Chapter 29 can be assigned with this section.

20. An automobile has a vertical radio antenna 1.20 m long. The automobile travels at 65.0 km/h on a horizontal road where the Earth's magnetic field is 50.0 μT directed toward the north and downward at an angle of 65.0° below the horizontal. (a) Specify the direction the automobile should move so as to generate the maximum motional emf in the antenna, with the top of the antenna positive relative to the bottom. (b) Calculate the magnitude of this induced emf.

21. ● A small airplane with a wingspan of 14.0 m is flying due north at a speed of 70.0 m/s over a region where the vertical component of the Earth's magnetic field is 1.20 μT downward. (a) What potential difference is developed between the wingtips? Which wingtip is at higher potential? (b) **What If?** How would the answer change if the plane turned to fly due east? (c) Can this emf be used to power a light in the passenger compartment? Explain your answer.

22. Consider the arrangement shown in Figure P31.22. Assume R = 6.00 Ω, ℓ = 1.20 m, and a uniform 2.50-T magnetic field is directed into the page. At what speed should the bar be moved to produce a current of 0.500 A in the resistor?

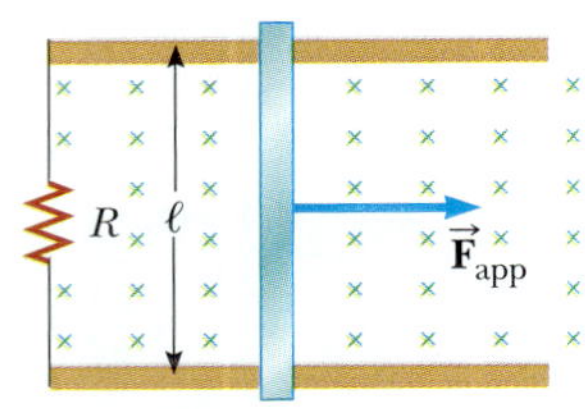

Figure P31.22 Problems 22, 23, and 24.

23. Figure P31.22 shows a top view of a bar that can slide without friction. The resistor is 6.00 Ω, and a 2.50-T magnetic field is directed perpendicularly downward, into the paper. Let ℓ = 1.20 m. (a) Calculate the applied force required to move the bar to the right at a constant speed of 2.00 m/s. (b) At what rate is energy delivered to the resistor?

24. ● A conducting rod of length ℓ moves on two horizontal, frictionless rails as shown in Figure P31.22. If a constant force of 1.00 N moves the bar at 2.00 m/s through a magnetic field $\vec{B}$ that is directed into the page, (a) what is the current in the 8.00-Ω resistor R? (b) What is the rate at which energy is delivered to the resistor? (c) What is the mechanical power delivered by the force $\vec{F}_{app}$? (d) Explain the relationship between the quantities computed in parts (b) and (c).

25. The *homopolar generator,* also called the *Faraday disk,* is a low-voltage, high-current electric generator. It consists of a rotating conducting disk with one stationary brush (a sliding electrical contact) at its axle and another at a point on its circumference as shown in Figure P31.25. A magnetic field is applied perpendicular to the plane of the disk. Assume the field is 0.900 T, the angular speed is 3 200 rev/min, and the radius of the disk is 0.400 m. Find the emf generated between the brushes. When superconducting coils are used to produce a large magnetic field, a homopolar generator can have a power output of several megawatts. Such a generator is useful, for example, in purifying metals by electrolysis. If a voltage is applied to the output terminals of the generator, it runs in reverse as a *homopolar motor* capable of providing great torque, useful in ship propulsion.

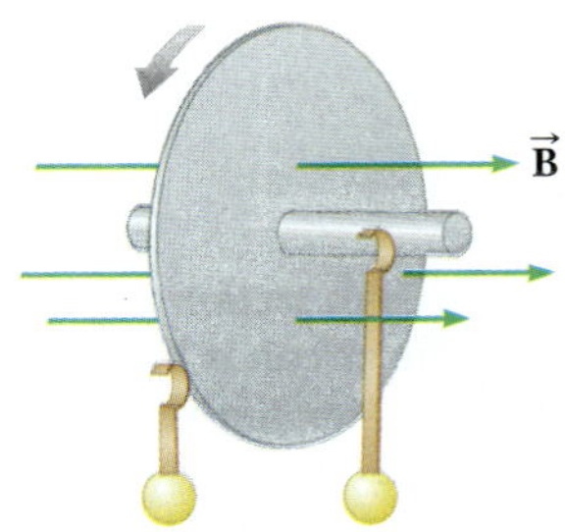

Figure P31.25

26. **Review problem.** As he starts to restring his acoustic guitar, a student attaches a single string, with linear density 3.00 × 10^{-3} kg/m, between two fixed points 64.0 cm apart, applies tension 267 N, and is distracted by a video game. His roommate attaches voltmeter leads to the ends of the metallic string and places a magnet across the string as shown in Figure P31.26. The magnet does not touch the string, but produces a uniform field of 4.50 mT

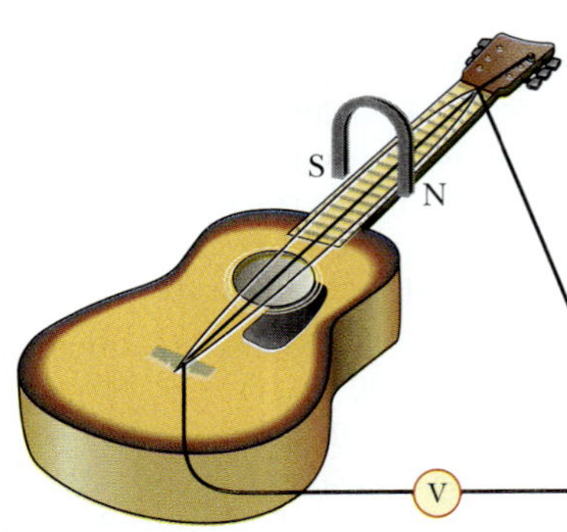

Figure P31.26

over a 2.00-cm length at the center of the string and negligible field elsewhere. Strumming the string sets it vibrating at its fundamental (lowest) frequency. The section of the string in the magnetic field moves perpendicular to the field with a uniform amplitude of 1.50 cm. Find (a) the frequency and (b) the amplitude of the electromotive force induced between the ends of the string.

27. A helicopter has blades of length 3.00 m, extending out from a central hub and rotating at 2.00 rev/s. If the vertical component of the Earth's magnetic field is 50.0 μT, what is the emf induced between the blade tip and the center hub?

28. Use Lenz's law to answer the following questions concerning the direction of induced currents. (a) What is the direction of the induced current in resistor R in Figure P31.28a when the bar magnet is moved to the left? (b) What is the direction of the current induced in the resistor R immediately after the switch S in Figure P31.28b is closed? (c) What is the direction of the induced current in R when the current I in Figure P31.28c decreases rapidly to zero? (d) A copper bar is moved to the right while its axis is maintained in a direction perpendicular to a magnetic field as shown in Figure P31.28d. If the top of the bar becomes positive relative to the bottom, what is the direction of the magnetic field?

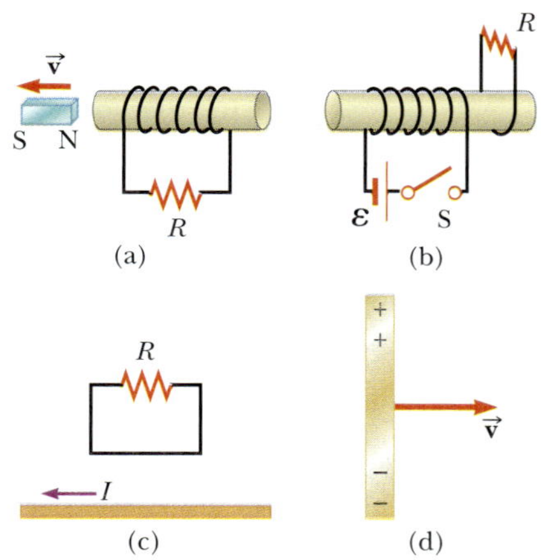

Figure P31.28

29. A rectangular coil with resistance R has N turns, each of length ℓ and width w as shown in Figure P31.29. The coil moves into a uniform magnetic field $\vec{B}$ with constant velocity $\vec{v}$. What are the magnitude and direction of the total magnetic force on the coil (a) as it enters the magnetic field, (b) as it moves within the field, and (c) as it leaves the field?

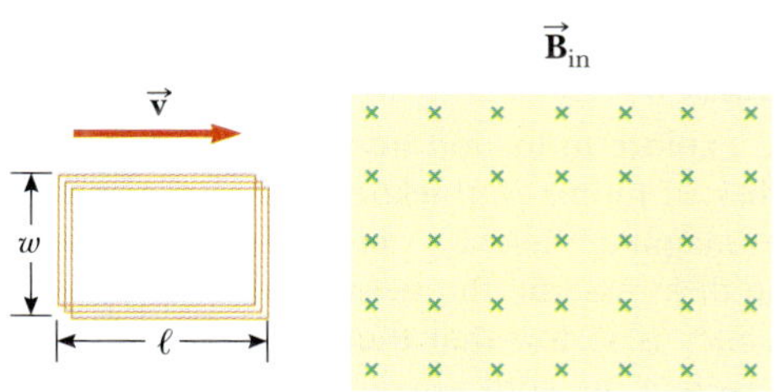

Figure P31.29

30. ● In Figure P31.30, the bar magnet is moved toward the loop. Is $V_a - V_b$ positive, negative, or zero? Explain your reasoning.

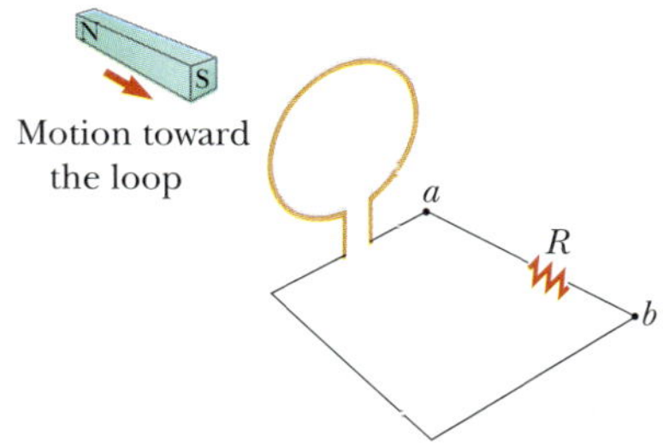

Figure P31.30

31. Two parallel rails with negligible resistance are 10.0 cm apart and are connected by a 5.00-Ω resistor. The circuit also contains two metal rods having resistances of 10.0 Ω and 15.0 Ω sliding along the rails (Fig. P31.31). The rods are pulled away from the resistor at constant speeds of 4.00 m/s and 2.00 m/s, respectively. A uniform magnetic field of magnitude 0.010 0 T is applied perpendicular to the plane of the rails. Determine the current in the 5.00-Ω resistor.

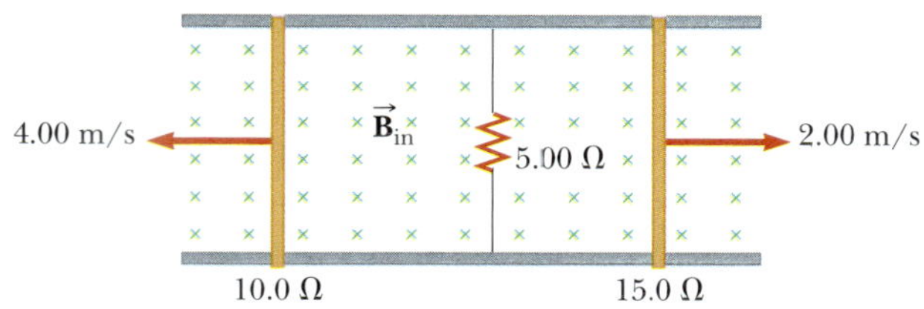

Figure P31.31

Section 31.4 Induced emf and Electric Fields

32. For the situation shown in Figure P31.32, the magnetic field changes with time according to the expression $B = (2.00t^3 - 4.00t^2 + 0.800)$ T, and $r_2 = 2R = 5.00$ cm. (a) Calculate the magnitude and direction of the force exerted on an electron located at point P_2 when $t = 2.00$ s. (b) At what instant is this force equal to zero?

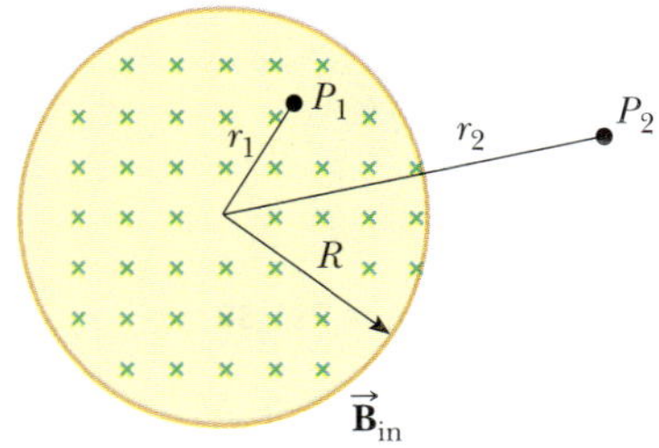

Figure P31.32 Problems 32 and 33.

33. A magnetic field directed into the page changes with time according to $B = (0.030\ 0t^2 + 1.40)$ T, where t is in seconds. The field has a circular cross section of radius $R = 2.50$ cm (Fig. P31.32). What are the magnitude and direction of the electric field at point P_1 when $t = 3.00$ s and $r_1 = 0.020\ 0$ m?

34. A long solenoid with 1 000 turns per meter and radius 2.00 cm carries an oscillating current given by $I =$ (5.00 A) sin $(100\pi t)$. What is the electric field induced at a radius $r =$ 1.00 cm from the axis of the solenoid? What is the direction of this electric field when the current is increasing counterclockwise in the coil?

Section 31.5 Generators and Motors

Problems 40 and 54 in Chapter 29 can be assigned with this section.

35. ▲ A coil of area 0.100 m^2 is rotating at 60.0 rev/s with the axis of rotation perpendicular to a 0.200-T magnetic field. (a) If the coil has 1 000 turns, what is the maximum emf generated in it? (b) What is the orientation of the coil with respect to the magnetic field when the maximum induced emf occurs?

36. In a 250-turn automobile alternator, the magnetic flux in each turn is $\Phi_B = (2.50 \times 10^{-4}$ Wb) cos ωt, where ω is the angular speed of the alternator. The alternator is geared to rotate three times for each engine revolution. When the engine is running at an angular speed of 1 000 rev/min, determine (a) the induced emf in the alternator as a function of time and (b) the maximum emf in the alternator.

37. A long solenoid, with its axis along the x axis, consists of 200 turns per meter of wire that carries a steady current of 15.0 A. A coil is formed by wrapping 30 turns of thin wire around a circular frame that has a radius of 8.00 cm. The coil is placed inside the solenoid and mounted on an axis that is a diameter of the coil and coincides with the y axis. The coil is then rotated with an angular speed of 4.00π rad/s. The plane of the coil is in the yz plane at $t = 0$. Determine the emf generated in the coil as a function of time.

38. A bar magnet is spun at constant angular speed ω around an axis as shown in Figure P31.38. A stationary, flat, rectangular conducting loop surrounds the magnet, and at $t = 0$, the magnet is oriented as shown. Make a qualitative graph of the induced current in the loop as a function of time, plotting counterclockwise currents as positive and clockwise currents as negative.

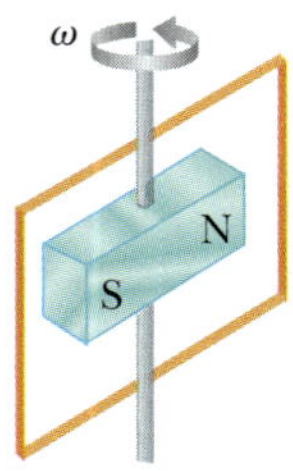

Figure P31.38

39. A motor in normal operation carries a direct current of 0.850 A when connected to a 120-V power supply. The resistance of the motor windings is 11.8 Ω. While in normal operation, (a) what is the back emf generated by the motor? (b) At what rate is internal energy produced in the windings? (c) **What If?** Suppose a malfunction stops the motor shaft from rotating. At what rate will internal energy be produced in the windings in this case? (Most motors have a thermal switch that will turn off the motor to prevent overheating when this stalling occurs.)

40. The rotating loop in an AC generator is a square 10.0 cm on each side. It is rotated at 60.0 Hz in a uniform field of 0.800 T. Calculate (a) the flux through the loop as a function of time, (b) the emf induced in the loop, (c) the current induced in the loop for a loop resistance of 1.00 Ω, (d) the power delivered to the loop, and (e) the torque that must be exerted to rotate the loop.

Section 31.6 Eddy Currents

41. ● Figure P31.41 represents an electromagnetic brake that uses eddy currents. An electromagnet hangs from a railroad car near one rail. To stop the car, a large current is sent through the coils of the electromagnet. The moving electromagnet induces eddy currents in the rails, whose fields oppose the change in the electromagnet's field. The magnetic fields of the eddy currents exert force on the current in the electromagnet, thereby slowing the car. The direction of the car's motion and the direction of the current in the electromagnet are shown correctly in the picture. Determine which of the eddy currents shown on the rails is correct. Explain your answer.

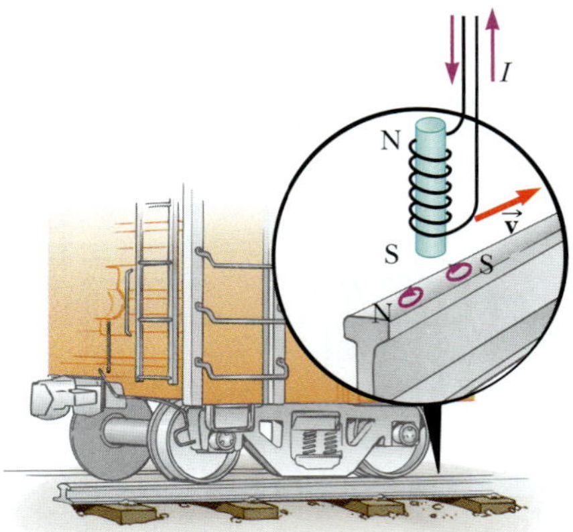

Figure P31.41

42. An *induction furnace* uses electromagnetic induction to produce eddy currents in a conductor, thereby raising the conductor's temperature. Commercial units operate at frequencies ranging from 60 Hz to about 1 MHz and deliver powers from a few watts to several megawatts. Induction heating can be used for warming a metal pan on a kitchen stove. It can be used to avoid oxidation and contamination of the metal when welding in a vacuum enclosure. At high frequencies, induced currents occur only near the surface of the conductor, in a phenomenon called the "skin effect." By creating an induced current for a short time interval at an appropriately high frequency, one can heat a sample down to a controlled depth. For example, the surface of a farm tiller can be tempered to make it hard and brittle for effective cutting while keeping the interior metal soft and ductile to resist breakage.

To explore induction heating, consider a flat conducting disk of radius R, thickness b, and resistivity ρ. A sinusoidal magnetic field B_{max} cos ωt is applied perpendicular to the disk. Assume the field is uniform in space and the frequency is so low that the skin effect is not important. Also assume the eddy currents occur in circles concentric with the disk. (a) Calculate the average power delivered to the disk. (b) **What If?** By what factor does the power change when the amplitude of the field doubles? (c) When the frequency doubles? (d) When the radius of the disk doubles?

43. ▲ ● A conducting rectangular loop of mass M, resistance R, and dimensions w by ℓ falls from rest into a magnetic field $\vec{\mathbf{B}}$ as shown in Figure P31.43. During the time interval before the top edge of the loop reaches the field, the loop approaches a terminal speed v_T. (a) Show that

$$v_T = \frac{MgR}{B^2w^2}$$

(b) Why is v_T proportional to R? (c) Why is it inversely proportional to B^2?

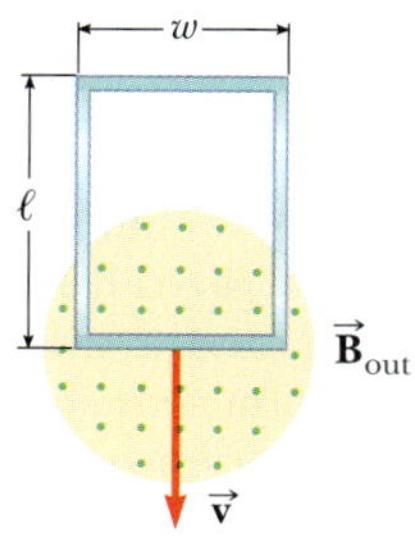

Figure P31.43

Additional Problems

44. ● Consider the apparatus shown in Figure P31.44 in which a conducting bar can be moved along two rails connected to a lightbulb. The whole system is immersed in a magnetic field of 0.400 T perpendicularly into the page. The vertical distance between the horizontal rails is 0.800 m. The resistance of the lightbulb is 48.0 Ω, assumed to be constant. The bar and rails have negligible resistance. The bar is moved toward the right by a constant force of magnitude 0.600 N. (a) What is the direction of the induced current in the circuit? (b) If the speed of the bar is 15.0 m/s at a particular instant, what is the value of the induced current? (c) Argue that the constant force causes the speed of the bar to increase and approach a certain terminal speed. Find the value of this maximum speed. (d) What power is delivered to the lightbulb when the bar is moving at its terminal speed? (e) We have assumed the resistance of the lightbulb is constant. In reality, as the power delivered to the lightbulb increases, the filament temperature increases and the resistance increases. Explain conceptually (not algebraically) whether the terminal speed found in part (c) changes if the resistance increases. If the terminal speed changes, does it increase or decrease? (f) With the assumption that the resistance of the lightbulb increases as the current increases, explain mathematically whether the power found in part (d) changes because of the increasing resistance. If it changes, is the actual power larger or smaller than the value previously found?

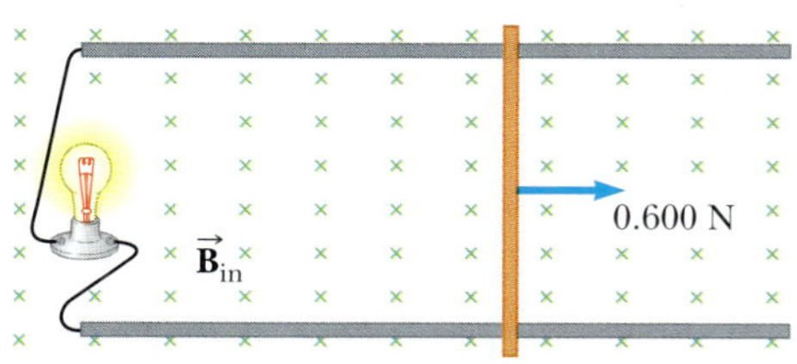

Figure P31.44

45. A guitar's steel string vibrates (Fig. 31.5a). The component of magnetic field perpendicular to the area of a pickup coil nearby is given by

$$B = 50.0 \text{ mT} + (3.20 \text{ mT}) \sin (1\,046\pi t)$$

The circular pickup coil has 30 turns and radius 2.70 mm. Find the emf induced in the coil as a function of time.

46. Strong magnetic fields are used in such medical procedures as magnetic resonance imaging, or MRI. A technician wearing a brass bracelet enclosing area 0.005 00 m^2 places her hand in a solenoid whose magnetic field is 5.00 T directed perpendicular to the plane of the bracelet. The electrical resistance around the bracelet's circumference is 0.020 0 Ω. An unexpected power failure causes the field to drop to 1.50 T in a time interval of 20.0 ms. Find (a) the current induced in the bracelet and (b) the power delivered to the bracelet. *Note:* As this problem implies, you should not wear any metal objects when working in regions of strong magnetic fields.

47. Figure P31.47 is a graph of the induced emf versus time for a coil of N turns rotating with angular speed ω in a uniform magnetic field directed perpendicular to the coil's axis of rotation. **What If?** Copy this sketch (on a larger scale) and on the same set of axes show the graph of emf versus t (a) if the number of turns in the coil is doubled, (b) if instead the angular speed is doubled, and (c) if the angular speed is doubled while the number of turns in the coil is halved.

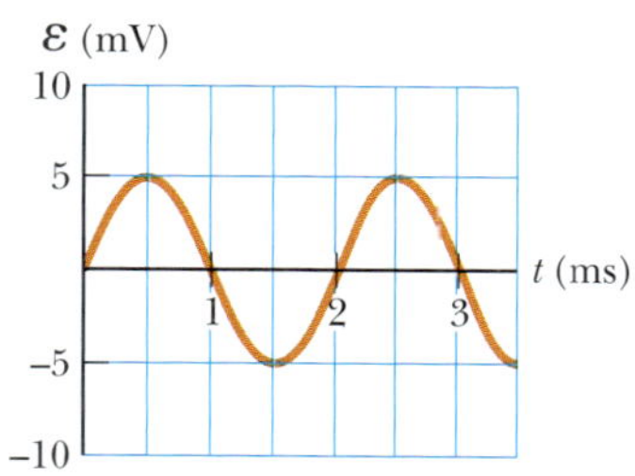

Figure P31.47

48. Two infinitely long solenoids (seen in cross section) pass through a circuit as shown in Figure P31.48. The magnitude of $\vec{\mathbf{B}}$ inside each is the same and is increasing at the rate of 100 T/s. What is the current in each resistor?

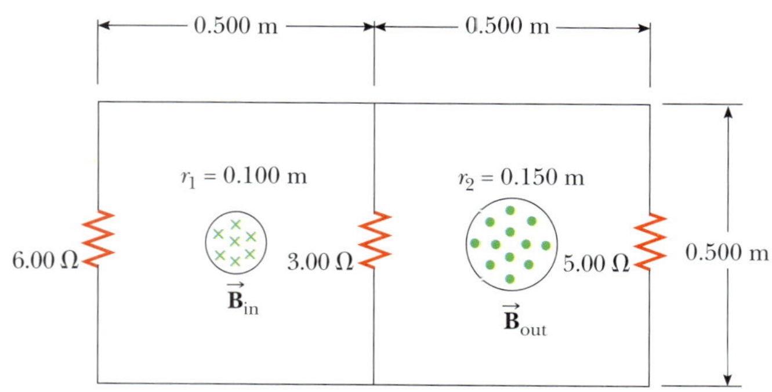

Figure P31.48

49. A conducting rod of length $\ell = 35.0$ cm is free to slide on two parallel conducting bars as shown in Figure P31.49. Two resistors $R_1 = 2.00\ \Omega$ and $R_2 = 5.00\ \Omega$ are connected across the ends of the bars to form a loop. A constant magnetic field $B = 2.50$ T is directed perpendicularly into the page. An external agent pulls the rod to the left with a constant speed of $v = 8.00$ m/s. Find (a) the currents in both resistors, (b) the total power delivered to the resistance of the circuit, and (c) the magnitude of the applied force that is needed to move the rod with this constant velocity.

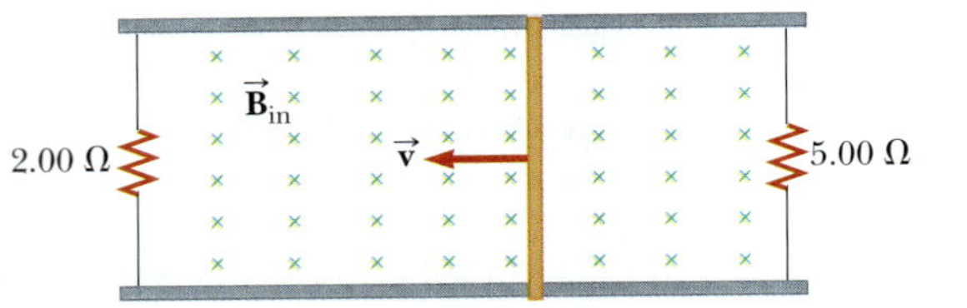

Figure P31.49

50. A bar of mass m, length d, and resistance R slides without friction in a horizontal plane, moving on parallel rails as shown in Figure P31.50. A battery that maintains a constant emf $\boldsymbol{\mathcal{E}}$ is connected between the rails, and a constant magnetic field $\vec{\mathbf{B}}$ is directed perpendicularly to the plane of the page. Assuming the bar starts from rest, show that at time t it moves with a speed

$$v = \frac{\mathcal{E}}{Bd}\left(1 - e^{-B^2d^2t/mR}\right)$$

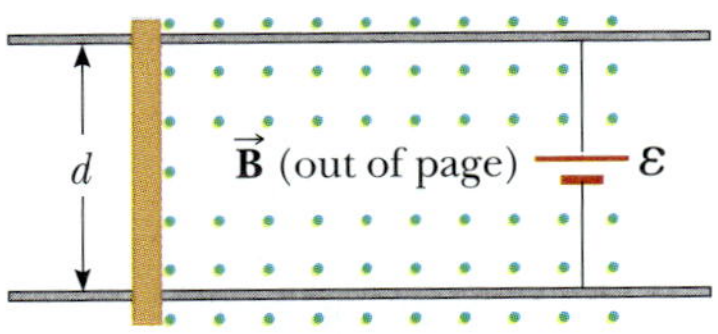

Figure P31.50

51. Suppose you wrap wire onto the core from a roll of cellophane tape to make a coil. How can you use a bar magnet to produce an induced voltage in the coil? What is the order of magnitude of the emf you generate? State the quantities you take as data and their values.

52. Magnetic field values are often determined by using a device known as a *search coil*. This technique depends on the measurement of the total charge passing through a coil in a time interval during which the magnetic flux linking the windings changes either because of the coil's motion or because of a change in the value of B. (a) Show that as the flux through the coil changes from Φ_1 to Φ_2, the charge transferred through the coil is given by $Q = N(\Phi_2 - \Phi_1)/R$, where R is the resistance of the coil and a sensitive ammeter connected across it and N is the number of turns. (b) As a specific example, calculate B when a 100-turn coil of resistance 200 Ω and cross-sectional area 40.0 cm^2 produces the following results. A total charge of 5.00×10^{-4} C passes through the coil when it is rotated in a uniform field from a position where the plane of the coil is perpendicular to the field to a position where the coil's plane is parallel to the field.

53. The plane of a square loop of wire with edge length $a = 0.200$ m is perpendicular to the Earth's magnetic field at a point where $B = 15.0\ \mu$T as shown in Figure P31.53. The total resistance of the loop and the wires connecting it to a sensitive ammeter is 0.500 Ω. If the loop is suddenly collapsed by horizontal forces as shown, what total charge passes through the ammeter?

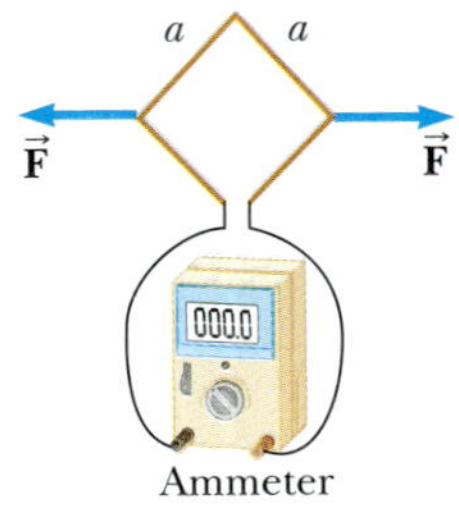

Figure P31.53

54. **Review problem.** A particle with a mass of 2.00×10^{-16} kg and a charge of 30.0 nC starts from rest, is accelerated by a strong electric field, and is fired from a small source inside a region of uniform constant magnetic field 0.600 T. The velocity of the particle is perpendicular to the field. The circular orbit of the particle encloses a magnetic flux of 15.0 μWb. (a) Calculate the speed of the particle. (b) Calculate the potential difference through which the particle accelerated inside the source.

55. ● In Figure P31.55, the rolling axle, 1.50 m long, is pushed along horizontal rails at a constant speed $v = 3.00$ m/s. A resistor $R = 0.400\ \Omega$ is connected to the rails at points a and b, directly opposite each other. The wheels make good electrical contact with the rails, so the axle, rails, and R form a closed-loop circuit. The only significant resistance in the circuit is R. A uniform magnetic field $B = 0.080\ 0$ T is vertically downward. (a) Find the induced current I in the resistor. (b) What horizontal force F is required to keep the axle rolling at constant speed? (c) Which end of the resistor, a or b, is at the higher electric potential? (d) **What If?** After the axle rolls past the resistor, does the current in R reverse direction? Explain your answer.

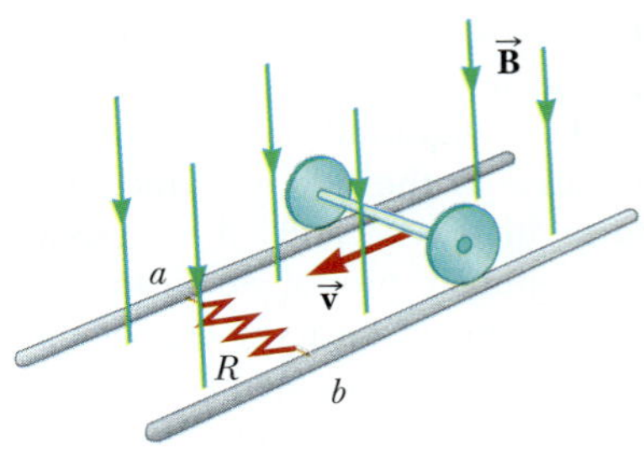

Figure P31.55

56. A conducting rod moves with a constant velocity $\vec{\mathbf{v}}$ in a direction perpendicular to a long, straight wire carrying a current I as shown in Figure P31.56. Show that the magnitude of the emf generated between the ends of the rod is

$$|\mathcal{E}| = \frac{\mu_0 v I \ell}{2\pi r}$$

In this case, note that the emf decreases with increasing r as you might expect.

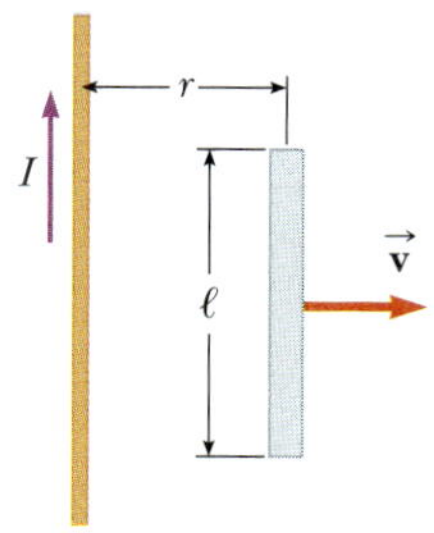

Figure P31.56

57. ● In Figure P31.57, a uniform magnetic field decreases at a constant rate $dB/dt = -K$, where K is a positive constant. A circular loop of wire of radius a containing a resistance R and a capacitance C is placed with its plane normal to the field. (a) Find the charge Q on the capacitor when it is fully charged. (b) Which plate is at the higher potential? (c) Discuss the force that causes the separation of charges.

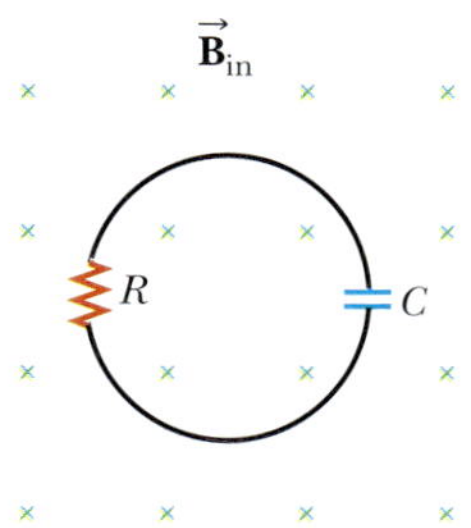

Figure P31.57

58. ● Figure P31.58 shows a compact, circular coil with 220 turns and radius 12.0 cm immersed in a uniform magnetic field parallel to the axis of the coil. The rate of change of the field has the constant magnitude 20.0 mT/s. (a) The following question cannot be answered with the information given. *Is the coil carrying clockwise or counterclockwise current?* What additional information is necessary to answer that question? (b) The coil overheats if more than 160 W of power is delivered to it. What resistance would the coil have at this critical point? To run cooler, should it have lower or higher resistance?

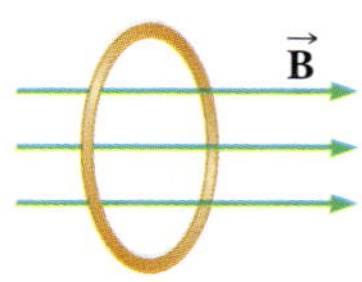

Figure P31.58

59. A rectangular coil of 60 turns, dimensions 0.100 m by 0.200 m and total resistance 10.0 Ω, rotates with angular speed 30.0 rad/s about the y axis in a region where a 1.00-T magnetic field is directed along the x axis. The rotation is initiated so that the plane of the coil is perpendicular to the direction of $\vec{\mathbf{B}}$ at $t = 0$. Calculate (a) the maximum induced emf in the coil, (b) the maximum rate of change of magnetic flux through the coil, (c) the induced emf at $t = 0.050\,0$ s, and (d) the torque exerted by the magnetic field on the coil at the instant when the emf is a maximum.

60. A small, circular washer of radius 0.500 cm is held directly below a long, straight wire carrying a current of 10.0 A. The washer is located 0.500 m above the top of a table (Fig. P31.60). (a) If the washer is dropped from rest, what is the magnitude of the average induced emf in the washer over the time interval between its release and the moment it hits the tabletop? Assume the magnetic field is nearly constant over the area of the washer and equal to the magnetic field at the center of the washer. (b) What is the direction of the induced current in the washer?

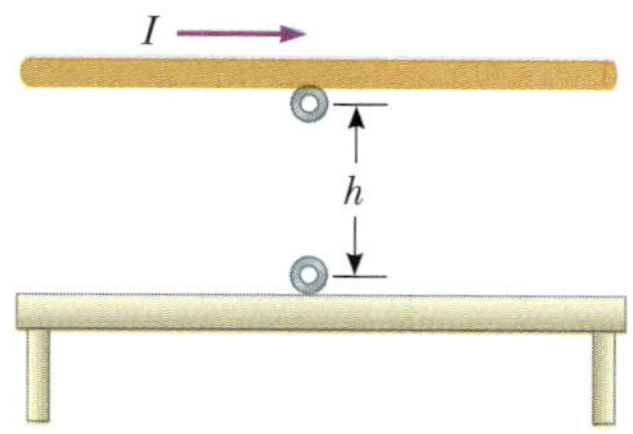

Figure P31.60

61. A conducting rod of length ℓ moves with velocity $\vec{\mathbf{v}}$ parallel to a long wire carrying a steady current I. The axis of the rod is maintained perpendicular to the wire with the near end a distance r away as shown in Figure P31.61. Show that the magnitude of the emf induced in the rod is

$$|\mathcal{E}| = \frac{\mu_0 I v}{2\pi} \ln\left(1 + \frac{\ell}{r}\right)$$

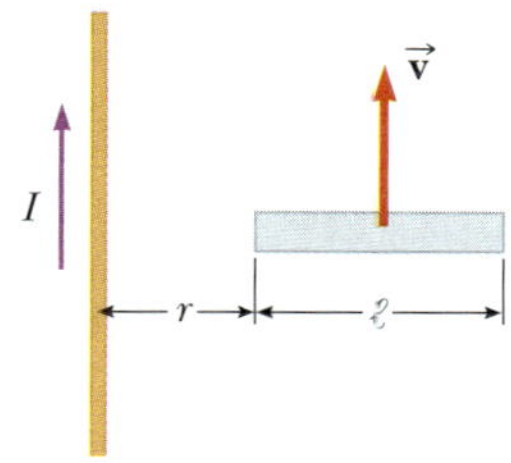

Figure P31.61

62. A rectangular loop of dimensions ℓ and w moves with a constant velocity $\vec{\mathbf{v}}$ away from a long wire that carries a current I in the plane of the loop (Fig. P31.62). The total resistance of the loop is R. Derive an expression that gives the current in the loop at the instant the near side is a distance r from the wire.

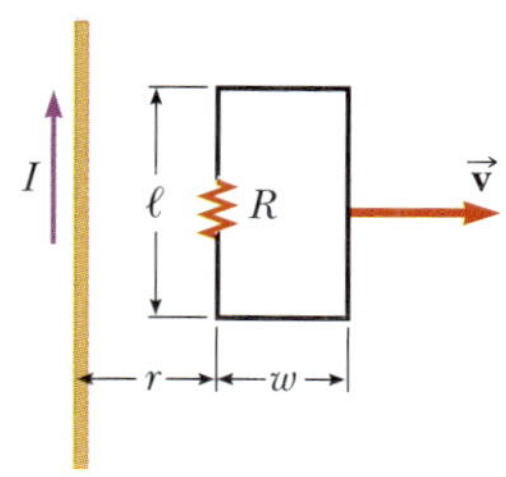

Figure P31.62

63. The magnetic flux through a metal ring varies with time t according to $\Phi_B = 3(at^3 - bt^2)$ T · m², with $a = 2.00$ s^{-3} and $b = 6.00$ s^{-2}. The resistance of the ring is 3.00 Ω. Determine the maximum current induced in the ring during the interval from $t = 0$ to $t = 2.00$ s.

64. **Review problem.** The bar of mass m in Figure P31.64 is pulled horizontally across parallel, frictionless rails by a massless string that passes over a light, frictionless pulley and is attached to a suspended object of mass M. The uniform magnetic field has a magnitude B, and the distance between the rails is ℓ. The only significant electrical resistance is the load resistor R shown connecting the rails at one end. Derive an expression that gives the horizontal speed of the bar as a function of time, assuming the suspended object is released with the bar at rest at $t = 0$.

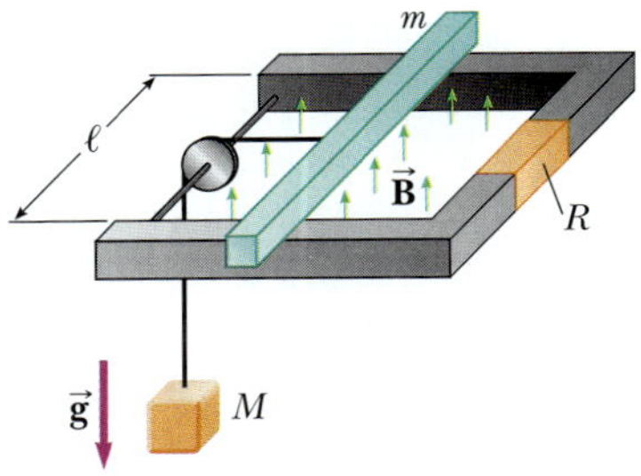

Figure P31.64

65. A *betatron* is a device that accelerates electrons to energies in the MeV range by means of electromagnetic induction. Electrons in a vacuum chamber are held in a circular orbit by a magnetic field perpendicular to the orbital plane. The magnetic field is gradually increased to induce an electric field around the orbit. (a) Show that the electric field is in the correct direction to make the electrons speed up. (b) Assume the radius of the orbit remains constant. Show that the average magnetic field over the area enclosed by the orbit must be twice as large as the magnetic field at the circle's circumference.

66. A thin wire 30.0 cm long is held parallel to and 80.0 cm above a long, thin wire carrying 200 A and resting on the horizontal floor (Fig. P31.66). The 30.0-cm wire is released at the instant $t = 0$ and falls, remaining parallel to the current-carrying wire as it falls. Assume the falling wire accelerates at 9.80 m/s². (a) Derive an equation for the emf induced in it as a function of time. (b) What is the minimum value of the emf? (c) What is the maximum value? (d) What is the induced emf 0.300 s after the wire is released?

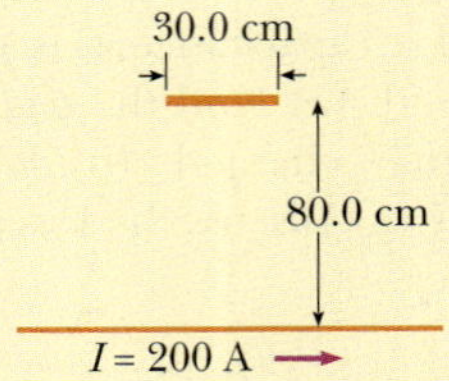

Figure P31.66

67. ▲ A long, straight wire carries a current that is given by $I = I_{max} \sin(\omega t + \phi)$. The wire lies in the plane of a rectangular coil of N turns of wire, as shown in Figure P31.8. The quantities I_{max}, ω, and ϕ are all constants. Determine the emf induced in the coil by the magnetic field created by the current in the straight wire. Assume $I_{max} = 50.0$ A, $\omega = 200\pi$ s^{-1}, $N = 100$, $h = w = 5.00$ cm, and $L = 20.0$ cm.

Answers to Quick Quizzes

31.1 (c). In all cases except this one, there is a change in the magnetic flux through the loop.

31.2 (c). The force on the wire is of magnitude $F_{app} = F_B = I\ell B$, with I given by Equation 31.6. Therefore, the force is proportional to the speed and the force doubles. Because $\mathcal{P} = F_{app}v$, the doubling of the force *and* the speed results in the power being four times as large.

31.3 (b). At the position of the loop, the magnetic field lines due to the wire point into the page. The loop is entering a region of stronger magnetic field as it drops toward the wire, so the flux is increasing. The induced current must set up a magnetic field that opposes this increase. To do so, it creates a magnetic field directed out of the page. By the right-hand rule for current loops, a counterclockwise current in the loop is required.

31.4 (a). Although reducing the resistance may increase the current the generator provides to a load, it does not alter the emf. Equation 31.11 shows that the emf depends on ω, B, and N, so all other choices increase the emf.

31.5 (b). When the aluminum sheet moves between the poles of the magnet, eddy currents are established in the aluminum. According to Lenz's law, these currents are in a direction so as to oppose the original change, which is the movement of the aluminum sheet in the magnetic field. The same principle is used in common laboratory triple-beam balances. See if you can find the magnet and the aluminum sheet the next time you use a triple-beam balance.

A treasure hunter uses a metal detector to search for buried objects at a beach. At the end of the metal detector is a coil of wire that is part of a circuit. When the coil comes near a metal object, the inductance of the coil is affected and the current in the circuit changes. This change triggers a signal in the earphones worn by the treasure hunter. We investigate inductance in this chapter. (Stone/Getty Images)

32 Inductance

In Chapter 31, we saw that an emf and a current are induced in a loop of wire when the magnetic flux through the area enclosed by the loop changes with time. This phenomenon of electromagnetic induction has some practical consequences. In this chapter, we first describe an effect known as *self-induction,* in which a time-varying current in a circuit produces an induced emf opposing the emf that initially set up the time-varying current. Self-induction is the basis of the *inductor,* an electrical circuit element. We discuss the energy stored in the magnetic field of an inductor and the energy density associated with the magnetic field.

Next, we study how an emf is induced in a coil as a result of a changing magnetic flux produced by a second coil, which is the basic principle of *mutual induction.* Finally, we examine the characteristics of circuits that contain inductors, resistors, and capacitors in various combinations.

32.1 Self-Induction and Inductance

In this chapter, we need to distinguish carefully between emfs and currents that are caused by physical sources such as batteries and those that are induced by changing magnetic fields. When we use a term (such as *emf* or *current*) without an adjective, we are describing the parameters associated with a physical source. We

North Wind Picture Archives

JOSEPH HENRY
American Physicist (1797–1878)
Henry became the first director of the Smithsonian Institution and first president of the Academy of Natural Science. He improved the design of the electromagnet and constructed one of the first motors. He also discovered the phenomenon of self-induction, but he failed to publish his findings. The unit of inductance, the henry, is named in his honor.

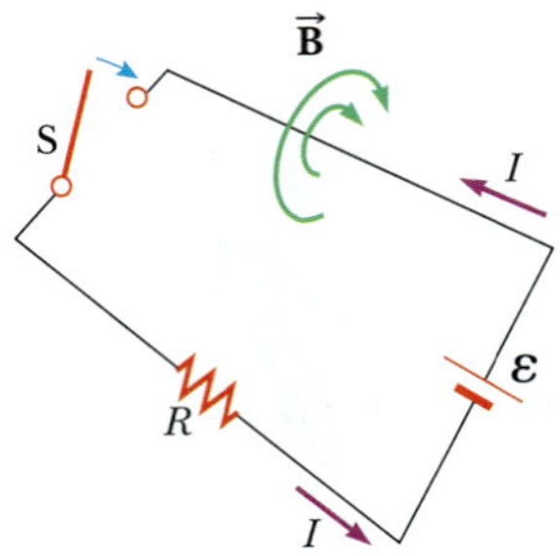

Figure 32.1 After the switch is closed, the current produces a magnetic flux through the area enclosed by the loop. As the current increases toward its equilibrium value, this magnetic flux changes in time and induces an emf in the loop.

use the adjective *induced* to describe those emfs and currents caused by a changing magnetic field.

Consider a circuit consisting of a switch, a resistor, and a source of emf as shown in Figure 32.1. The circuit diagram is represented in perspective to show the orientations of some of the magnetic field lines due to the current in the circuit. When the switch is thrown to its closed position, the current does not immediately jump from zero to its maximum value $\mathcal{E}/R$. Faraday's law of electromagnetic induction (Eq. 31.1) can be used to describe this effect as follows. As the current increases with time, the magnetic flux through the circuit loop due to this current also increases with time. This increasing flux creates an induced emf in the circuit. The direction of the induced emf is such that it would cause an induced current in the loop (if the loop did not already carry a current), which would establish a magnetic field opposing the change in the original magnetic field. Therefore, the direction of the induced emf is opposite the direction of the emf of the battery, which results in a gradual rather than instantaneous increase in the current to its final equilibrium value. Because of the direction of the induced emf, it is also called a *back emf*, similar to that in a motor as discussed in Chapter 31. This effect is called **self-induction** because the changing flux through the circuit and the resultant induced emf arise from the circuit itself. The emf $\mathcal{E}_L$ set up in this case is called a **self-induced emf.**

To obtain a quantitative description of self-induction, recall from Faraday's law that the induced emf is equal to the negative of the time rate of change of the magnetic flux. The magnetic flux is proportional to the magnetic field, which in turn is proportional to the current in the circuit. Therefore, **a self-induced emf is always proportional to the time rate of change of the current.** For any loop of wire, we can write this proportionality as

$$\mathcal{E}_L = -L\frac{dI}{dt} \tag{32.1}$$

where L is a proportionality constant—called the **inductance** of the loop—that depends on the geometry of the loop and other physical characteristics. If we consider a closely spaced coil of N turns (a toroid or an ideal solenoid) carrying a current I and containing N turns, Faraday's law tells us that $\mathcal{E}_L = -N\,d\Phi_B/dt$. Combining this expression with Equation 32.1 gives

Inductance of an N-turn coil ▶

$$L = \frac{N\Phi_B}{I} \tag{32.2}$$

where it is assumed the same magnetic flux passes through each turn and L is the inductance of the entire coil.

From Equation 32.1, we can also write the inductance as the ratio

Inductance ▶

$$L = -\frac{\mathcal{E}_L}{dI/dt} \tag{32.3}$$

Recall that resistance is a measure of the opposition to current ($R = \Delta V/I$); in comparison, Equation 32.3 shows us that inductance is a measure of the opposition to a *change* in current.

The SI unit of inductance is the **henry** (H), which as we can see from Equation 32.3 is 1 volt-second per ampere: $1\ \text{H} = 1\ \text{V}\cdot\text{s/A}$.

As shown in Example 32.1, the inductance of a coil depends on its geometry. This dependence is analogous to the capacitance of a capacitor depending on the geometry of its plates as we found in Chapter 26. Inductance calculations can be quite difficult to perform for complicated geometries, but the examples below involve simple situations for which inductances are easily evaluated.

Quick Quiz 32.1 A coil with zero resistance has its ends labeled *a* and *b*. The potential at *a* is higher than at *b*. Which of the following could be consistent with this situation? (a) The current is constant and is directed from *a* to *b*. (b) The current is constant and is directed from *b* to *a*. (c) The current is increasing and is directed from *a* to *b*. (d) The current is decreasing and is directed from *a* to *b*. (e) The current is increasing and is directed from *b* to *a*. (f) The current is decreasing and is directed from *b* to *a*.

EXAMPLE 32.1 Inductance of a Solenoid

Consider a uniformly wound solenoid having N turns and length ℓ. Assume ℓ is much longer than the radius of the windings and the core of the solenoid is air.

(A) Find the inductance of the solenoid.

SOLUTION

Conceptualize The magnetic field lines from each turn of the solenoid pass through all the turns, so an induced emf in each coil opposes changes in the current.

Categorize Because the solenoid is long, we can use the results for an ideal solenoid obtained in Chapter 30.

Analyze Find the magnetic flux through each turn of area A in the solenoid, using the expression for the magnetic field from Equation 30.17:

$$\Phi_B = BA = \mu_0 nIA = \mu_0 \frac{N}{\ell} IA$$

Substitute this expression into Equation 32.2:

$$L = \frac{N\Phi_B}{I} = \mu_0 \frac{N^2}{\ell} A \qquad \textbf{(32.4)}$$

(B) Calculate the inductance of the solenoid if it contains 300 turns, its length is 25.0 cm, and its cross-sectional area is $4.00\ \text{cm}^2$.

SOLUTION

Substitute numerical values into Equation 32.4:

$$L = (4\pi \times 10^{-7}\ \text{T}\cdot\text{m/A}) \frac{(300)^2}{25.0 \times 10^{-2}\ \text{m}} (4.00 \times 10^{-4}\ \text{m}^2)$$

$$= 1.81 \times 10^{-4}\ \text{T}\cdot\text{m}^2/\text{A} = 0.181\ \text{mH}$$

(C) Calculate the self-induced emf in the solenoid if the current it carries decreases at the rate of 50.0 A/s.

SOLUTION

Substitute $dI/dt = -50.0\ \text{A/s}$ into Equation 32.1:

$$\mathcal{E}_L = -L\frac{dI}{dt} = -(1.81 \times 10^{-4}\ \text{H})(-50.0\ \text{A/s})$$

$$= 9.05\ \text{mV}$$

Finalize The result for part (A) shows that L depends on geometry and is proportional to the square of the number of turns. Because $N = n\ell$, we can also express the result in the form

$$L = \mu_0 \frac{(n\ell)^2}{\ell} A = \mu_0 n^2 A\ell = \mu_0 n^2 V \tag{32.5}$$

where $V = A\ell$ is the interior volume of the solenoid.

32.2 *RL* Circuits

If a circuit contains a coil such as a solenoid, the inductance of the coil prevents the current in the circuit from increasing or decreasing instantaneously. A circuit element that has a large inductance is called an **inductor** and has the circuit symbol —ooo—. We always assume the inductance of the remainder of a circuit is negligible compared with that of the inductor. Keep in mind, however, that even a circuit without a coil has some inductance that can affect the circuit's behavior.

Because the inductance of an inductor results in a back emf, **an inductor in a circuit opposes changes in the current in that circuit.** The inductor attempts to keep the current the same as it was before the change occurred. If the battery voltage in the circuit is increased so that the current rises, the inductor opposes this change and the rise is not instantaneous. If the battery voltage is decreased, the inductor causes a slow drop in the current rather than an immediate drop. Therefore, the inductor causes the circuit to be "sluggish" as it reacts to changes in the voltage.

Consider the circuit shown in Active Figure 32.2, which contains a battery of negligible internal resistance. This circuit is an ***RL* circuit** because the elements connected to the battery are a resistor and an inductor. The curved lines on switch S_2 suggest this switch can never be open; it is always set to either a or b. (If the switch is connected to neither a nor b, any current in the circuit suddenly stops.) Suppose S_2 is set to a and switch S_1 is open for $t < 0$ and then thrown closed at $t = 0$. The current in the circuit begins to increase, and a back emf (Eq. 32.1) that opposes the increasing current is induced in the inductor.

With this point in mind, let's apply Kirchhoff's loop rule to this circuit, traversing the circuit in the clockwise direction:

$$\mathcal{E} - IR - L\frac{dI}{dt} = 0 \tag{32.6}$$

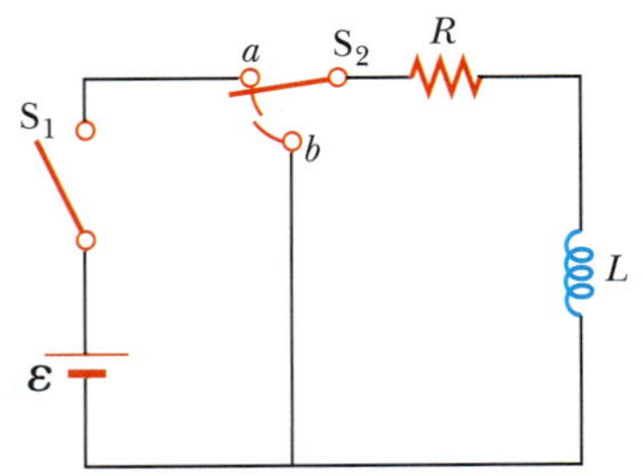

ACTIVE FIGURE 32.2

An *RL* circuit. When switch S_2 is in position a, the battery is in the circuit. When switch S_1 is thrown closed, the current increases and an emf that opposes the increasing current is induced in the inductor. When the switch is thrown to position b, the battery is no longer part of the circuit and the current decreases. The switch is designed so that it is never open, which would cause the current to stop.

Sign in at www.thomsonedu.com and go to ThomsonNOW to adjust the values of R and L and see the effect on the current. A graphical display as in Active Figure 32.3 is available.

where IR is the voltage drop across the resistor. (Kirchhoff's rules were developed for circuits with steady currents, but they can also be applied to a circuit in which the current is changing if we imagine them to represent the circuit at one *instant* of time.) Now let's find a solution to this differential equation, which is similar to that for the *RC* circuit (see Section 28.4).

A mathematical solution of Equation 32.6 represents the current in the circuit as a function of time. To find this solution, we change variables for convenience, letting $x = (\mathcal{E}/R) - I$, so $dx = -dI$. With these substitutions, Equation 32.6 becomes

$$x + \frac{L}{R}\frac{dx}{dt} = 0$$

Rearranging and integrating this last expression gives

$$\int_{x_0}^{x} \frac{dx}{x} = -\frac{R}{L}\int_0^t dt$$

$$\ln\frac{x}{x_0} = -\frac{R}{L}t$$

where x_0 is the value of x at time $t = 0$. Taking the antilogarithm of this result gives

$$x = x_0 e^{-Rt/L}$$

Because $I = 0$ at $t = 0$, note from the definition of x that $x_0 = \mathcal{E}/R$. Hence, this last expression is equivalent to

$$\frac{\mathcal{E}}{R} - I = \frac{\mathcal{E}}{R} e^{-Rt/L}$$

$$I = \frac{\mathcal{E}}{R}(1 - e^{-Rt/L})$$

This expression shows how the inductor affects the current. The current does not increase instantly to its final equilibrium value when the switch is closed, but instead increases according to an exponential function. If the inductance is removed from the circuit, which corresponds to letting L approach zero, the exponential term becomes zero and there is no time dependence of the current in this case; the current increases instantaneously to its final equilibrium value in the absence of the inductance.

We can also write this expression as

$$I = \frac{\mathcal{E}}{R}(1 - e^{-t/\tau}) \quad (32.7)$$

where the constant τ is the **time constant** of the RL circuit:

$$\tau = \frac{L}{R} \quad (32.8)$$

◀ Time constant of an RL circuit

Physically, τ is the time interval required for the current in the circuit to reach $(1 - e^{-1}) = 0.632 = 63.2\%$ of its final value $\mathcal{E}/R$. The time constant is a useful parameter for comparing the time responses of various circuits.

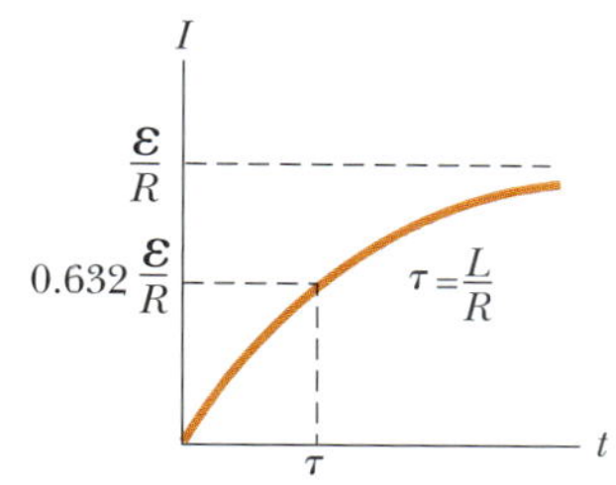

ACTIVE FIGURE 32.3

Plot of the current versus time for the RL circuit shown in Active Figure 32.2. When switch S_1 is thrown closed at $t = 0$, the current increases toward its maximum value $\mathcal{E}/R$. The time constant τ is the time interval required for I to reach 63.2% of its maximum value.

Sign in at www.thomsonedu.com and go to ThomsonNOW to observe this graph develop after switch S_1 in Active Figure 32.2 is thrown closed.

Active Figure 32.3 shows a graph of the current versus time in the RL circuit. Notice that the equilibrium value of the current, which occurs as t approaches infinity, is $\mathcal{E}/R$. That can be seen by setting dI/dt equal to zero in Equation 32.6 and solving for the current I. (At equilibrium, the change in the current is zero.) Therefore, the current initially increases very rapidly and then gradually approaches the equilibrium value $\mathcal{E}/R$ as t approaches infinity.

Let's also investigate the time rate of change of the current. Taking the first time derivative of Equation 32.7 gives

$$\frac{dI}{dt} = \frac{\mathcal{E}}{L} e^{-t/\tau} \quad (32.9)$$

This result shows that the time rate of change of the current is a maximum (equal to $\mathcal{E}/L$) at $t = 0$ and falls off exponentially to zero as t approaches infinity (Fig. 32.4).

Now consider the RL circuit in Active Figure 32.2 again. Suppose switch S_2 has been set at position a long enough (and switch S_1 remains closed) to allow the current to reach its equilibrium value $\mathcal{E}/R$. In this situation, the circuit is described by the outer loop in Active Figure 32.2. If S_2 is thrown from a to b, the circuit is now described by only the right-hand loop in Active Figure 32.2. Therefore, the battery has been eliminated from the circuit. Setting $\mathcal{E} = 0$ in Equation 32.6 gives

$$IR + L\frac{dI}{dt} = 0$$

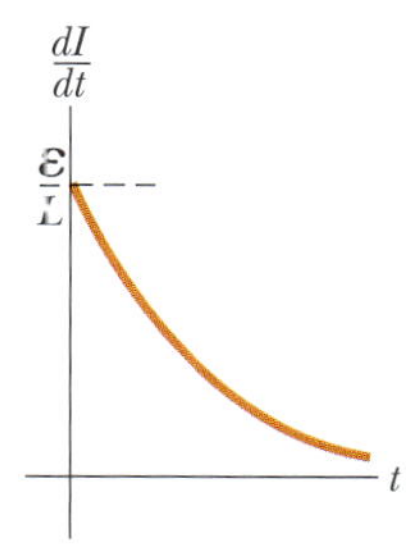

Figure 32.4 Plot of dI/dt versus time for the RL circuit shown in Active Figure 32.2. The time rate of change of current is a maximum at $t = 0$, which is the instant at which switch S_1 is thrown closed. The rate decreases exponentially with time as I increases toward its maximum value.

It is left as a problem (Problem 10) to show that the solution of this differential equation is

$$I = \frac{\mathcal{E}}{R} e^{-t/\tau} = I_i e^{-t/\tau} \quad (32.10)$$

where $\mathcal{E}$ is the emf of the battery and $I_i = \mathcal{E}/R$ is the initial current at the instant the switch is thrown to b.

If the circuit did not contain an inductor, the current would immediately decrease to zero when the battery is removed. When the inductor is present, it

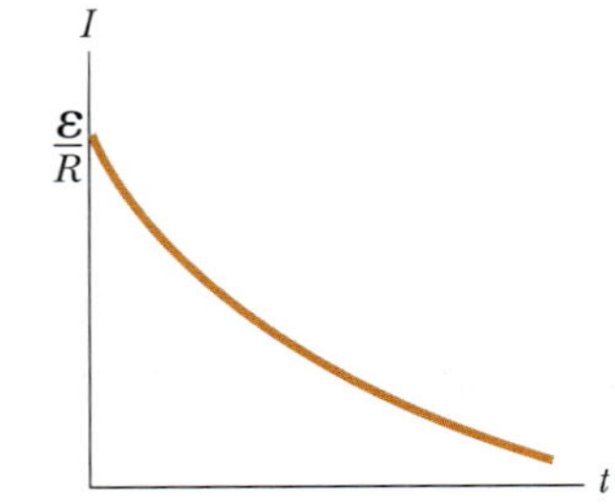

ACTIVE FIGURE 32.5

Current versus time for the right-hand loop of the circuit shown in Active Figure 32.2. For $t < 0$, switch S_2 is at position *a*. At $t = 0$, the switch is thrown to position *b* and the current has its maximum value $\mathcal{E}/R$.

Sign in at www.thomsonedu.com and go to ThomsonNOW to observe this graph develop after the switch in Active Figure 32.2 is thrown to position *b*.

opposes the decrease in the current and causes the current to decrease exponentially. A graph of the current in the circuit versus time (Active Fig. 32.5) shows that the current is continuously decreasing with time.

Quick Quiz 32.2 Consider the circuit in Active Figure 32.2 with S_1 open and S_2 at position *a*. Switch S_1 is now thrown closed. **(i)** At the instant it is closed, across which circuit element is the voltage equal to the emf of the battery? (a) the resistor (b) the inductor (c) both the inductor and resistor **(ii)** After a very long time, across which circuit element is the voltage equal to the emf of the battery? Choose from among the same answers.

EXAMPLE 32.2 Time Constant of an *RL* Circuit

Consider the circuit in Active Figure 32.2 again. Suppose the circuit elements have the following values: $\mathcal{E} = 12.0$ V, $R = 6.00\ \Omega$, and $L = 30.0$ mH.

(A) Find the time constant of the circuit.

SOLUTION

Conceptualize You should understand the behavior of this circuit from the discussion in this section.

Categorize We evaluate the results using equations developed in this section, so this example is a substitution problem.

Evaluate the time constant from Equation 32.8:

$$\tau = \frac{L}{R} = \frac{30.0 \times 10^{-3}\ \text{H}}{6.00\ \Omega} = \boxed{5.00\ \text{ms}}$$

(B) Switch S_2 is at position *a*, and switch S_1 is thrown closed at $t = 0$. Calculate the current in the circuit at $t = 2.00$ ms.

SOLUTION

Evaluate the current at $t = 2.00$ ms from Equation 32.7:

$$I = \frac{\mathcal{E}}{R}(1 - e^{-t/\tau}) = \frac{12.0\ \text{V}}{6.00\ \Omega}(1 - e^{-2.00\ \text{ms}/5.00\ \text{ms}}) = 2.00\ \text{A}\ (1 - e^{-0.400})$$

$$= \boxed{0.659\ \text{A}}$$

(C) Compare the potential difference across the resistor with that across the inductor.

SOLUTION

At the instant the switch is closed, there is no current and therefore no potential difference across the resistor. At this instant, the battery voltage appears entirely across the inductor in the form of a back emf of 12.0 V as the inductor tries to maintain the zero-current condition. (The top end of the inductor in Active Fig. 32.2 is at a higher electric potential than the bottom end.) As time passes, the emf across the inductor decreases and the current in the resistor (and hence the voltage across it) increases as shown in Figure 32.6. The sum of the two voltages at all times is 12.0 V.

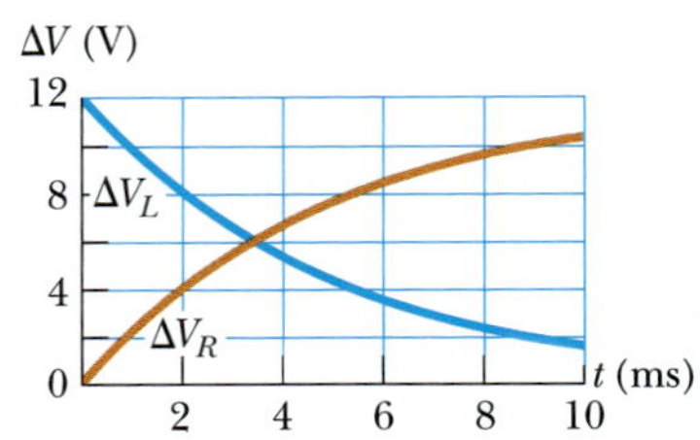

Figure 32.6 (Example 32.2) The time behavior of the voltages across the resistor and inductor in Active Figure 32.2 given the values provided in this example.

What If? In Figure 32.6, the voltages across the resistor and inductor are equal at 3.4 ms. What if you wanted to delay the condition in which the voltages are equal to some later instant, such as $t = 10.0$ ms? Which parameter, L or R, would require the least adjustment, in terms of a percentage change, to achieve that?

Answer Figure 32.6 shows that the voltages are equal when the voltage across the inductor has fallen to half its original value. Therefore, the time interval required for the voltages to become equal is the *half-life* $t_{1/2}$ of the decay. We introduced the half-life in the **What If?** section of Example 28.10 to describe the exponential decay in *RC* circuits, where $t_{1/2} = 0.693\tau$.

From the desired half-life of 10.0 ms, use the result from Example 28.10 to find the time constant of the circuit:

$$\tau = \frac{t_{1/2}}{0.693} = \frac{10.0 \text{ ms}}{0.693} = 14.4 \text{ ms}$$

Hold L fixed and find the value of R that gives this time constant:

$$\tau = \frac{L}{R} \rightarrow R = \frac{L}{\tau} = \frac{30.0 \times 10^{-3} \text{ H}}{14.4 \text{ ms}} = 2.08 \ \Omega$$

Now hold R fixed and find the appropriate value of L:

$$\tau = \frac{L}{R} \rightarrow L = \tau R = (14.4 \text{ ms})(6.00 \ \Omega) = 86.4 \times 10^{-3} \text{ H}$$

The change in R corresponds to a 65% decrease compared with the initial resistance. The change in L represents a 188% increase in inductance! Therefore, a much smaller percentage adjustment in R can achieve the desired effect than would an adjustment in L.

32.3 Energy in a Magnetic Field

A battery in a circuit containing an inductor must provide more energy than in a circuit without the inductor. Part of the energy supplied by the battery appears as internal energy in the resistance in the circuit, and the remaining energy is stored in the magnetic field of the inductor. Multiplying each term in Equation 32.6 by I and rearranging the expression gives

$$I\mathcal{E} = I^2R + LI\frac{dI}{dt} \qquad \textbf{(32.11)}$$

Recognizing $I\mathcal{E}$ as the rate at which energy is supplied by the battery and I^2R as the rate at which energy is delivered to the resistor, we see that $LI(dI/dt)$ must represent the rate at which energy is being stored in the inductor. If U is the energy stored in the inductor at any time, we can write the rate dU/dt at which energy is stored as

$$\frac{dU}{dt} = LI\frac{dI}{dt}$$

To find the total energy stored in the inductor at any instant, let's rewrite this expression as $dU = LI\,dI$ and integrate:

$$U = \int dU = \int_0^I LI\,dI = L\int_0^I I\,dI$$

$$U = \tfrac{1}{2}LI^2 \qquad \textbf{(32.12)}$$

◀ Energy stored in an inductor

where L is constant and has been removed from the integral. Equation 32.12 represents the energy stored in the magnetic field of the inductor when the current is I. It is similar in form to Equation 26.11 for the energy stored in the electric field of a capacitor, $U = \frac{1}{2}C(\Delta V)^2$. In either case, energy is required to establish a field.

PITFALL PREVENTION 32.1
Capacitors, Resistors, and Inductors Store Energy Differently

Different energy-storage mechanisms are at work in capacitors, inductors, and resistors. A charged capacitor stores energy as electrical potential energy. An inductor stores energy as what we could call magnetic potential energy when it carries current. Energy delivered to a resistor is transformed to internal energy.

We can also determine the energy density of a magnetic field. For simplicity, consider a solenoid whose inductance is given by Equation 32.5:

$$L = \mu_0 n^2 V$$

The magnetic field of a solenoid is given by Equation 30.17:

$$B = \mu_0 n I$$

Substituting the expression for L and $I = B/\mu_0 n$ into Equation 32.12 gives

$$U = \tfrac{1}{2}LI^2 = \tfrac{1}{2}\mu_0 n^2 V\left(\frac{B}{\mu_0 n}\right)^2 = \frac{B^2}{2\mu_0}V \qquad \textbf{(32.13)}$$

The magnetic energy density, or the energy stored per unit volume in the magnetic field of the inductor, is

Magnetic energy density ▶

$$u_B = \frac{U}{V} = \frac{B^2}{2\mu_0} \qquad \textbf{(32.14)}$$

Although this expression was derived for the special case of a solenoid, it is valid for any region of space in which a magnetic field exists. Equation 32.14 is similar in form to Equation 26.13 for the energy per unit volume stored in an electric field, $u_E = \frac{1}{2}\epsilon_0 E^2$. In both cases, the energy density is proportional to the square of the field magnitude.

Quick Quiz 32.3 You are performing an experiment that requires the highest-possible magnetic energy density in the interior of a very long current-carrying solenoid. Which of the following adjustments increases the energy density? (More than one choice may be correct.) (a) increasing the number of turns per unit length on the solenoid (b) increasing the cross-sectional area of the solenoid (c) increasing only the length of the solenoid while keeping the number of turns per unit length fixed (d) increasing the current in the solenoid

---

EXAMPLE 32.3 What Happens to the Energy in the Inductor?

Consider once again the *RL* circuit shown in Active Figure 32.2, with switch S_2 at position *a* and the current having reached its steady-state value. When S_2 is thrown to position *b*, the current in the right-hand loop decays exponentially with time according to the expression $I = I_i e^{-t/\tau}$, where $I_i = \mathcal{E}/R$ is the initial current in the circuit and $\tau = L/R$ is the time constant. Show that all the energy initially stored in the magnetic field of the inductor appears as internal energy in the resistor as the current decays to zero.

SOLUTION

Conceptualize Before S_2 is thrown to *b*, energy is being delivered at a constant rate to the resistor from the battery and energy is stored in the magnetic field of the inductor. After $t = 0$, when S_2 is thrown to *b*, the battery can no longer provide energy and energy is delivered to the resistor only from the inductor.

Categorize We model the right-hand loop of the circuit as an isolated system so that energy is transferred between components of the system but does not leave the system.

Analyze The energy in the magnetic field of the inductor at any time is U. The rate dU/dt at which energy leaves the inductor and is delivered to the resistor is equal to I^2R, where I is the instantaneous current.

Substitute the current given by Equation 32.10 into $dU/dt = I^2R$:

$$\frac{dU}{dt} = I^2R = (I_i e^{-Rt/L})^2 R = I_i^2 R e^{-2Rt/L}$$

Solve for dU and integrate this expression over the limits $t = 0$ to $t \to \infty$:

$$U = \int_0^\infty I_i^2 R e^{-2Rt/L}\,dt = I_i^2 R \int_0^\infty e^{-2Rt/L}\,dt$$

The value of the definite integral can be shown to be $L/2R$ (see Problem 26). Use this result to evaluate U:

$$U = I_i^2 R\left(\frac{L}{2R}\right) = \frac{1}{2}LI_i^2$$

Finalize This result is equal to the initial energy stored in the magnetic field of the inductor, given by Equation 32.12, as we set out to prove.

EXAMPLE 32.4 **The Coaxial Cable**

Coaxial cables are often used to connect electrical devices, such as your stereo system, and in receiving signals in television cable systems. Model a long coaxial cable as two thin, concentric, cylindrical conducting shells of radii a and b and length ℓ as in Figure 32.7. The conducting shells carry the same current I in opposite directions. Calculate the inductance L of this cable.

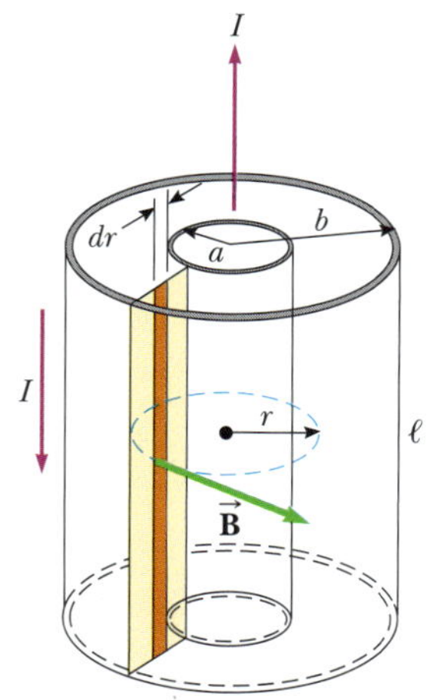

Figure 32.7 (Example 32.4) Section of a long coaxial cable. The inner and outer conductors carry equal currents in opposite directions.

SOLUTION

Conceptualize Consider Figure 32.7. Although we do not have a visible coil in this geometry, imagine a thin, radial slice of the coaxial cable such as the light gold rectangle in Figure 32.7. If the inner and outer conductors are connected at the ends of the cable (above and below the figure), this slice represents one large conducting loop. The current in the loop sets up a magnetic field between the inner and outer conductors that passes through this loop. If the current changes, the magnetic field changes and the induced emf opposes the original change in the current in the conductors.

Categorize We categorize this situation as one in which we must return to the fundamental definition of inductance, Equation 32.2.

Analyze We must find the magnetic flux through the light gold rectangle in Figure 32.7. Ampère's law (see Section 30.3) tells us that the magnetic field in the region between the shells is due to the inner conductor and that its magnitude is $B = \mu_0 I/2\pi r$, where r is measured from the common center of the shells. The magnetic field is zero outside the outer shell ($r > b$) because the net current passing through the area enclosed by a circular path surrounding the cable is zero; hence, from Ampère's law, $\oint \vec{\mathbf{B}} \cdot d\vec{\mathbf{s}} = 0$. The magnetic field is zero inside the inner shell because the shell is hollow and no current is present within a radius $r < a$.

The magnetic field is perpendicular to the light gold rectangle of length ℓ and width $b - a$, the cross section of interest. Because the magnetic field varies with radial position across this rectangle, we must use calculus to find the total magnetic flux.

Divide the light gold rectangle into strips of width dr such as the darker strip in Figure 32.7. Evaluate the magnetic flux through such a strip:

$$\Phi_B = \int B\,dA = \int B\ell\,dr$$

Substitute for the magnetic field and integrate over the entire light gold rectangle:

$$\Phi_B = \int_a^b \frac{\mu_0 I}{2\pi r}\ell\,dr = \frac{\mu_0 I\ell}{2\pi}\int_a^b \frac{dr}{r} = \frac{\mu_0 I\ell}{2\pi}\ln\left(\frac{b}{a}\right)$$

Use Equation 32.2 to find the inductance of the cable:

$$L = \frac{\Phi_B}{I} = \frac{\mu_0 \ell}{2\pi}\ln\left(\frac{b}{a}\right)$$

Finalize The inductance increases if ℓ increases, if b increases, or if a decreases. This result is consistent with our conceptualization: any of these changes increases the size of the loop represented by our radial slice and through which the magnetic field passes, increasing the inductance.

32.4 Mutual Inductance

Very often, the magnetic flux through the area enclosed by a circuit varies with time because of time-varying currents in nearby circuits. This condition induces an emf through a process known as *mutual induction,* so named because it depends on the interaction of two circuits.

Consider the two closely wound coils of wire shown in cross-sectional view in Figure 32.8. The current I_1 in coil 1, which has N_1 turns, creates a magnetic field. Some of the magnetic field lines pass through coil 2, which has N_2 turns. The magnetic flux caused by the current in coil 1 and passing through coil 2 is represented by Φ_{12}. In analogy to Equation 32.2, we can identify the **mutual inductance** M_{12} of coil 2 with respect to coil 1:

Definition of mutual inductance ▶

$$M_{12} = \frac{N_2 \Phi_{12}}{I_1} \qquad (32.15)$$

Mutual inductance depends on the geometry of both circuits and on their orientation with respect to each other. As the circuit separation distance increases, the mutual inductance decreases because the flux linking the circuits decreases.

If the current I_1 varies with time, we see from Faraday's law and Equation 32.15 that the emf induced by coil 1 in coil 2 is

$$\mathcal{E}_2 = -N_2 \frac{d\Phi_{12}}{dt} = -N_2 \frac{d}{dt}\left(\frac{M_{12} I_1}{N_2}\right) = -M_{12}\frac{dI_1}{dt} \qquad (32.16)$$

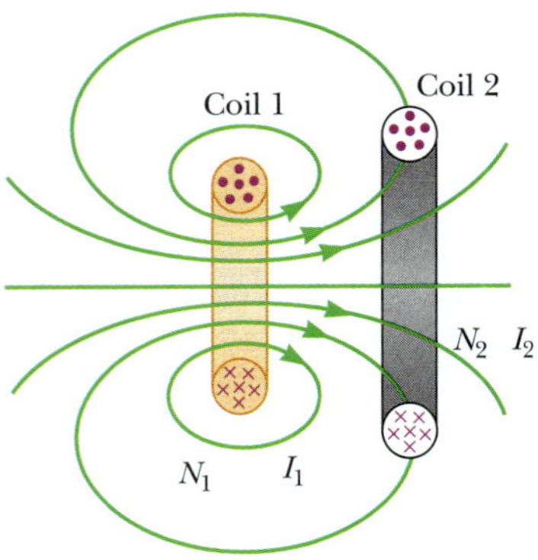

Figure 32.8 A cross-sectional view of two adjacent coils. A current in coil 1 sets up a magnetic field, and some of the magnetic field lines pass through coil 2.

In the preceding discussion, it was assumed the current is in coil 1. Let's also imagine a current I_2 in coil 2. The preceding discussion can be repeated to show that there is a mutual inductance M_{21}. If the current I_2 varies with time, the emf induced by coil 2 in coil 1 is

$$\mathcal{E}_1 = -M_{21}\frac{dI_2}{dt} \qquad (32.17)$$

In mutual induction, the emf induced in one coil is always proportional to the rate at which the current in the other coil is changing. Although the proportionality constants M_{12} and M_{21} have been treated separately, it can be shown that they are equal. Therefore, with $M_{12} = M_{21} = M$, Equations 32.16 and 32.17 become

$$\mathcal{E}_2 = -M\frac{dI_1}{dt} \quad \text{and} \quad \mathcal{E}_1 = -M\frac{dI_2}{dt}$$

These two equations are similar in form to Equation 32.1 for the self-induced emf $\mathcal{E} = -L\,(dI/dt)$. The unit of mutual inductance is the henry.

Quick Quiz 32.4 In Figure 32.8, coil 1 is moved closer to coil 2, with the orientation of both coils remaining fixed. Because of this movement, the mutual induction of the two coils (a) increases, (b) decreases, or (c) is unaffected.

EXAMPLE 32.5 "Wireless" Battery Charger

An electric toothbrush has a base designed to hold the toothbrush handle when not in use. As shown in Figure 32.9a, the handle has a cylindrical hole that fits loosely over a matching cylinder on the base. When the handle is placed on the base, a changing current in a solenoid inside the base cylinder induces a current in a coil inside the handle. This induced current charges the battery in the handle.

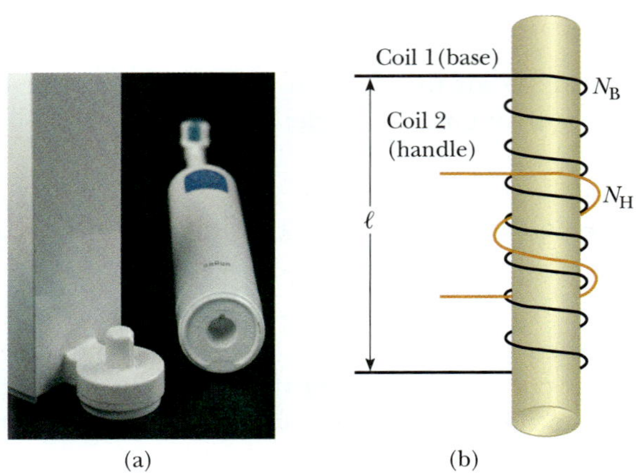

Figure 32.9 (Example 32.5) (a) This electric toothbrush uses the mutual induction of solenoids as part of its battery-charging system. (b) A coil of N_H turns wrapped around the center of a solenoid of N_B turns.

We can model the base as a solenoid of length ℓ with N_B turns (Fig. 32.9b), carrying a current I, and having a cross-sectional area A. The handle coil contains N_H turns and completely surrounds the base coil. Find the mutual inductance of the system.

SOLUTION

Conceptualize Be sure you can identify the two coils in the situation and understand that a changing current in one coil induces a current in the second coil.

Categorize We will evaluate the result using concepts discussed in this section, so we categorize this example as a substitution problem.

Use Equation 30.17 to express the magnetic field in the interior of the base solenoid:

$$B = \mu_0 \frac{N_B}{\ell} I$$

Find the mutual inductance, noting that the magnetic flux Φ_{BH} through the handle's coil caused by the magnetic field of the base coil is BA:

$$M = \frac{N_H \Phi_{BH}}{I} = \frac{N_H BA}{I} = \mu_0 \frac{N_B N_H}{\ell} A$$

Wireless charging is used in a number of other "cordless" devices. One significant example is the inductive charging used by some manufacturers of electric cars that avoids direct metal-to-metal contact between the car and the charging apparatus.

32.5 Oscillations in an *LC* Circuit

When a capacitor is connected to an inductor as illustrated in Figure 32.10, the combination is an ***LC* circuit.** If the capacitor is initially charged and the switch is then closed, both the current in the circuit and the charge on the capacitor oscillate between maximum positive and negative values. If the resistance of the circuit is zero, no energy is transformed to internal energy. In the following analysis, the resistance in the circuit is neglected. We also assume an idealized situation in which energy is not radiated away from the circuit. This radiation mechanism is discussed in Chapter 34.

Assume the capacitor has an initial charge Q_{max} (the maximum charge) and the switch is open for $t < 0$ and then closed at $t = 0$. Let's investigate what happens from an energy viewpoint.

When the capacitor is fully charged, the energy U in the circuit is stored in the capacitor's electric field and is equal to $Q_{max}^2/2C$ (Eq. 26.11). At this time, the current in the circuit is zero; therefore, no energy is stored in the inductor. After the switch is closed, the rate at which charges leave or enter the capacitor plates (which is also the rate at which the charge on the capacitor changes) is equal to the current in the circuit. After the switch is closed and the capacitor begins to discharge, the energy stored in its electric field decreases. The capacitor's discharge represents a current in the circuit, and some energy is now stored in the magnetic field of the inductor. Therefore, energy is transferred from the electric field of the capacitor to the magnetic field of the inductor. When the capacitor is fully discharged, it stores no energy. At this time, the current reaches its maximum value and all the energy in the circuit is stored in the inductor. The current continues in the same direction, decreasing in magnitude, with the capacitor eventually becoming fully charged again but with the polarity of its plates now opposite the initial polarity. This process is followed by another discharge until the circuit returns to its original state of maximum charge Q_{max} and the plate polarity shown in Figure 32.10. The energy continues to oscillate between inductor and capacitor.

The oscillations of the *LC* circuit are an electromagnetic analog to the mechanical oscillations of the block–spring system studied in Chapter 15. Much of what

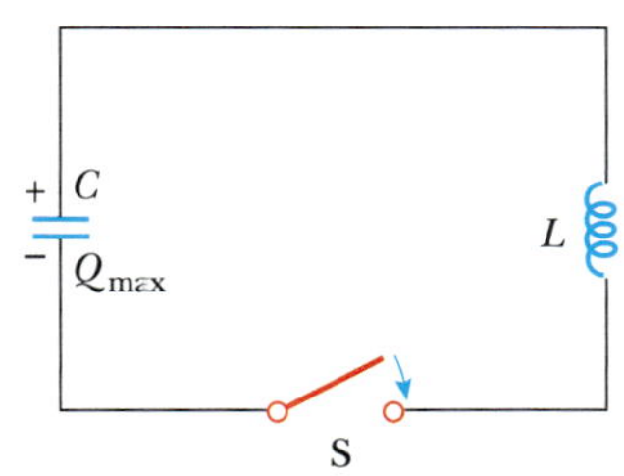

Figure 32.10 A simple *LC* circuit. The capacitor has an initial charge Q_{max}, and the switch is open for $t < 0$ and then closed at $t = 0$.

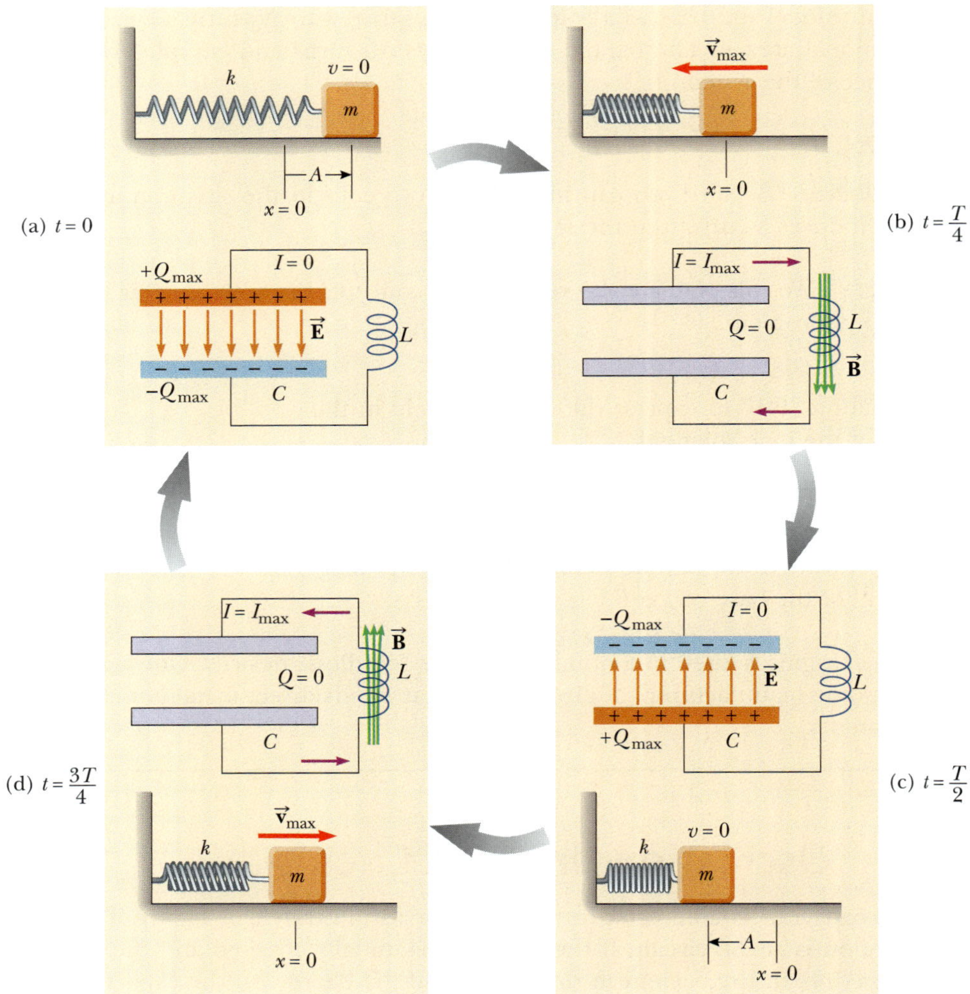

ACTIVE FIGURE 32.11

Energy transfer in a resistanceless, nonradiating LC circuit. The capacitor has a charge Q_{max} at $t = 0$, the instant at which the switch is closed. The mechanical analog of this circuit is a block–spring system.

Sign in at www.thomsonedu.com and go to ThomsonNOW to adjust the values of C and L and see the effect on the oscillating current. The block on the spring oscillates in a mechanical analog of the electrical oscillations. A graphical display as in Active Figure 32.12 is available, as is an energy bar graph.

was discussed there is applicable to LC oscillations. For example, we investigated the effect of driving a mechanical oscillator with an external force, which leads to the phenomenon of *resonance.* The same phenomenon is observed in the LC circuit. (See Section 33.7.)

A representation of the energy transfer in an LC circuit is shown in Active Figure 32.11. As mentioned, the behavior of the circuit is analogous to that of the oscillating block–spring system studied in Chapter 15. The potential energy $\frac{1}{2}kx^2$ stored in a stretched spring is analogous to the potential energy $Q_{max}^2/2C$ stored in the capacitor. The kinetic energy $\frac{1}{2}mv^2$ of the moving block is analogous to the magnetic energy $\frac{1}{2}LI^2$ stored in the inductor, which requires the presence of moving charges. In Active Figure 32.11a, all the energy is stored as electric potential energy in the capacitor at $t = 0$ (because $I = 0$), just as all the energy in a block–spring system is initially stored as potential energy in the spring if it is stretched and released at $t = 0$. In Active Figure 32.11b, all the energy is stored as magnetic energy $\frac{1}{2}LI_{max}^2$ in the inductor, where I_{max} is the maximum current. Active Figures 32.11c and 32.11d show subsequent quarter-cycle situations in which the energy is all electric or all magnetic. At intermediate points, part of the energy is electric and part is magnetic.

Consider some arbitrary time t after the switch is closed so that the capacitor has a charge $Q < Q_{max}$ and the current is $I < I_{max}$. At this time, both circuit elements store energy, but the sum of the two energies must equal the total initial energy U stored in the fully charged capacitor at $t = 0$:

Total energy stored in an LC circuit ▶

$$U = U_C + U_L = \frac{Q^2}{2C} + \frac{1}{2}LI^2 \qquad \textbf{(32.18)}$$

Because we have assumed the circuit resistance to be zero and we ignore electromagnetic radiation, no energy is transformed to internal energy and none is transferred out of the system of the circuit. Therefore, *the total energy of the system must remain constant in time.* We describe the constant energy of the system mathematically by setting $dU/dt = 0$. Therefore, by differentiating Equation 32.18 with respect to time while noting that Q and I vary with time gives

$$\frac{dU}{dt} = \frac{d}{dt}\left(\frac{Q^2}{2C} + \tfrac{1}{2}LI^2\right) = \frac{Q}{C}\frac{dQ}{dt} + LI\frac{dI}{dt} = 0 \qquad \textbf{(32.19)}$$

We can reduce this result to a differential equation in one variable by remembering that the current in the circuit is equal to the rate at which the charge on the capacitor changes: $I = dQ/dt$. It then follows that $dI/dt = d^2Q/dt^2$. Substitution of these relationships into Equation 32.19 gives

$$\frac{Q}{C} + L\frac{d^2Q}{dt^2} = 0$$

$$\frac{d^2Q}{dt^2} = -\frac{1}{LC}Q \qquad \textbf{(32.20)}$$

Let's solve for Q by noting that this expression is of the same form as the analogous Equations 15.3 and 15.5 for a block–spring system:

$$\frac{d^2x}{dt^2} = -\frac{k}{m}x = -\omega^2 x$$

where k is the spring constant, m is the mass of the block, and $\omega = \sqrt{k/m}$. The solution of this mechanical equation has the general form (Eq. 15.6):

$$x = A\cos(\omega t + \phi)$$

where A is the amplitude of the simple harmonic motion (the maximum value of x), ω is the angular frequency of this motion, and ϕ is the phase constant; the values of A and ϕ depend on the initial conditions. Because Equation 32.20 is of the same mathematical form as the differential equation of the simple harmonic oscillator, it has the solution

$$Q = Q_{\text{max}}\cos(\omega t + \phi) \qquad \textbf{(32.21)}$$

◀ Charge as a function of time for an ideal *LC* circuit

where Q_{max} is the maximum charge of the capacitor and the angular frequency ω is

$$\omega = \frac{1}{\sqrt{LC}} \qquad \textbf{(32.22)}$$

◀ Angular frequency of oscillation in an *LC* circuit

Note that the angular frequency of the oscillations depends solely on the inductance and capacitance of the circuit. Equation 32.22 gives the *natural frequency* of oscillation of the *LC* circuit.

Because Q varies sinusoidally with time, the current in the circuit also varies sinusoidally. We can show that by differentiating Equation 32.21 with respect to time:

$$I = \frac{dQ}{dt} = -\omega Q_{\text{max}}\sin(\omega t + \phi) \qquad \textbf{(32.23)}$$

◀ Current as a function of time for an ideal *LC* current

To determine the value of the phase angle ϕ, let's examine the initial conditions, which in our situation require that at $t = 0$, $I = 0$, and $Q = Q_{\text{max}}$. Setting $I = 0$ at $t = 0$ in Equation 32.23 gives

$$0 = -\omega Q_{\text{max}}\sin\phi$$

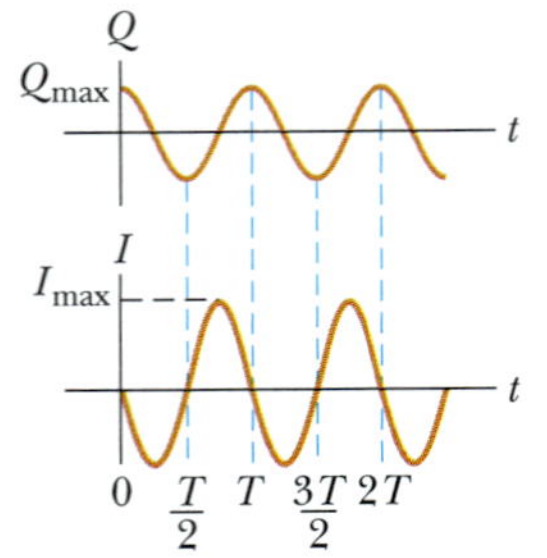

ACTIVE FIGURE 32.12

Graphs of charge versus time and current versus time for a resistanceless, nonradiating *LC* circuit. Notice that *Q* and *I* are 90° out of phase with each other.

Sign in at www.thomsonedu.com and go to ThomsonNOW to observe this graph develop for the *LC* circuit in Active Figure 32.11.

which shows that $\phi = 0$. This value for ϕ also is consistent with Equation 32.21 and the condition that $Q = Q_{max}$ at $t = 0$. Therefore, in our case, the expressions for Q and I are

$$Q = Q_{max} \cos \omega t \tag{32.24}$$

$$I = -\omega Q_{max} \sin \omega t = -I_{max} \sin \omega t \tag{32.25}$$

Graphs of Q versus t and I versus t are shown in Active Figure 32.12. The charge on the capacitor oscillates between the extreme values Q_{max} and $-Q_{max}$, and the current oscillates between I_{max} and $-I_{max}$. Furthermore, the current is 90° out of phase with the charge. That is, when the charge is a maximum, the current is zero, and when the charge is zero, the current has its maximum value.

Let's return to the energy discussion of the *LC* circuit. Substituting Equations 32.24 and 32.25 in Equation 32.18, we find that the total energy is

$$U = U_C + U_L = \frac{Q_{max}^2}{2C} \cos^2 \omega t + \tfrac{1}{2} L I_{max}^2 \sin^2 \omega t \tag{32.26}$$

This expression contains all the features described qualitatively at the beginning of this section. It shows that the energy of the *LC* circuit continuously oscillates between energy stored in the capacitor's electric field and energy stored in the inductor's magnetic field. When the energy stored in the capacitor has its maximum value $Q_{max}^2/2C$, the energy stored in the inductor is zero. When the energy stored in the inductor has its maximum value $\frac{1}{2}LI_{max}^2$, the energy stored in the capacitor is zero.

Plots of the time variations of U_C and U_L are shown in Figure 32.13. The sum $U_C + U_L$ is a constant and is equal to the total energy $Q_{max}^2/2C$, or $\frac{1}{2}LI_{max}^2$. Analytical verification is straightforward. The amplitudes of the two graphs in Figure 32.13 must be equal because the maximum energy stored in the capacitor (when $I = 0$) must equal the maximum energy stored in the inductor (when $Q = 0$). This equality is expressed mathematically as

$$\frac{Q_{max}^2}{2C} = \frac{LI_{max}^2}{2}$$

Using this expression in Equation 32.26 for the total energy gives

$$U = \frac{Q_{max}^2}{2C} (\cos^2 \omega t + \sin^2 \omega t) = \frac{Q_{max}^2}{2C} \tag{32.27}$$

because $\cos^2 \omega t + \sin^2 \omega t = 1$.

In our idealized situation, the oscillations in the circuit persist indefinitely; the total energy U of the circuit, however, remains constant only if energy transfers and transformations are neglected. In actual circuits, there is always some resistance and some energy is therefore transformed to internal energy. We mentioned at the beginning of this section that we are also ignoring radiation from the circuit. In reality, radiation is inevitable in this type of circuit, and the total energy in the circuit continuously decreases as a result of this process.

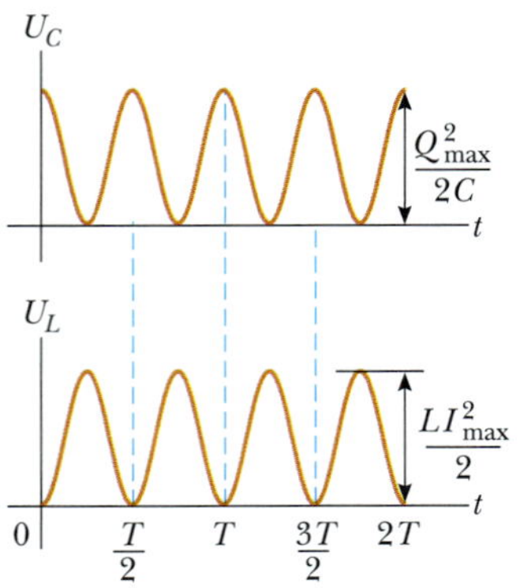

Figure 32.13 Plots of U_C versus t and U_L versus t for a resistanceless, nonradiating *LC* circuit. The sum of the two curves is a constant and is equal to the total energy stored in the circuit.

Quick Quiz 32.5 **(i)** At an instant of time during the oscillations of an *LC* circuit, the current is at its maximum value. At this instant, what happens to the voltage across the capacitor? (a) It is different from that across the inductor. (b) It is zero. (c) It has its maximum value. (d) It is impossible to determine. **(ii)** At an instant of time during the oscillations of an *LC* circuit, the current is momentarily zero. From the same choices, describe the voltage across the capacitor at this instant.

EXAMPLE 32.6 Oscillations in an *LC* Circuit

In Figure 32.14, the battery has an emf of 12.0 V, the inductance is 2.81 mH, and the capacitance is 9.00 pF. The switch has been set to position *a* for a long time so that the capacitor is charged. The switch is then thrown to position *b*, removing the battery from the circuit and connecting the capacitor directly across the inductor.

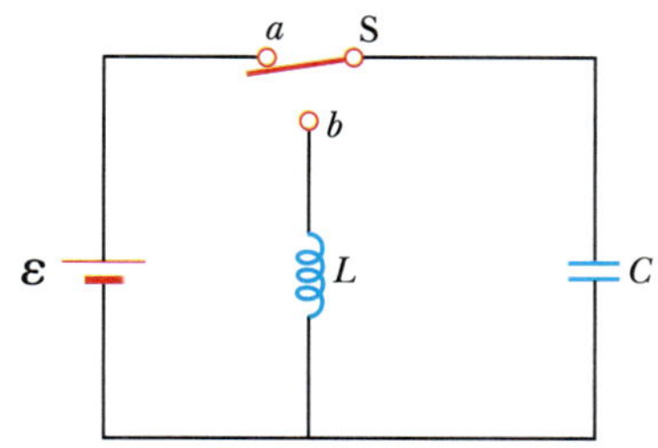

Figure 32.14 (Example 32.6) First the capacitor is fully charged with the switch set to position *a*. Then, the switch is thrown to position *b* and the battery is no longer in the circuit.

(A) Find the frequency of oscillation of the circuit.

SOLUTION

Conceptualize When the switch is thrown to position *b*, the active part of the circuit is the right-hand loop, which is an *LC* circuit.

Categorize We use equations developed in this section, so we categorize this example as a substitution problem.

Use Equation 32.22 to find the frequency:

$$f = \frac{\omega}{2\pi} = \frac{1}{2\pi\sqrt{LC}}$$

Substitute numerical values:

$$f = \frac{1}{2\pi[(2.81 \times 10^{-3}\text{ H})(9.00 \times 10^{-12}\text{ F})]^{1/2}} = 1.00 \times 10^{6}\text{ Hz}$$

(B) What are the maximum values of charge on the capacitor and current in the circuit?

SOLUTION

Find the initial charge on the capacitor, which equals the maximum charge:

$$Q_{\text{max}} = C\,\Delta V = (9.00 \times 10^{-12}\text{ F})(12.0\text{ V}) = 1.08 \times 10^{-10}\text{ C}$$

Use Equation 32.25 to find the maximum current from the maximum charge:

$$I_{\text{max}} = \omega Q_{\text{max}} = 2\pi f Q_{\text{max}} = (2\pi \times 10^{6}\text{ s}^{-1})(1.08 \times 10^{-10}\text{ C})$$
$$= 6.79 \times 10^{-4}\text{ A}$$

32.6 The *RLC* Circuit

Let's now turn our attention to a more realistic circuit consisting of a resistor, an inductor, and a capacitor connected in series as shown in Active Figure 32.15. We assume the resistance of the resistor represents all the resistance in the circuit. Suppose the switch is at position *a* so that the capacitor has an initial charge Q_{max}. The switch is now thrown to position *b*. After this instant, the total energy stored in the capacitor and inductor at any time is given by Equation 32.18. This total energy, however, is no longer constant as it was in the *LC* circuit because the resistor causes transformation to internal energy. (We continue to ignore electromagnetic radiation from the circuit in this discussion.) Because the rate of energy transformation to internal energy within a resistor is I^2R,

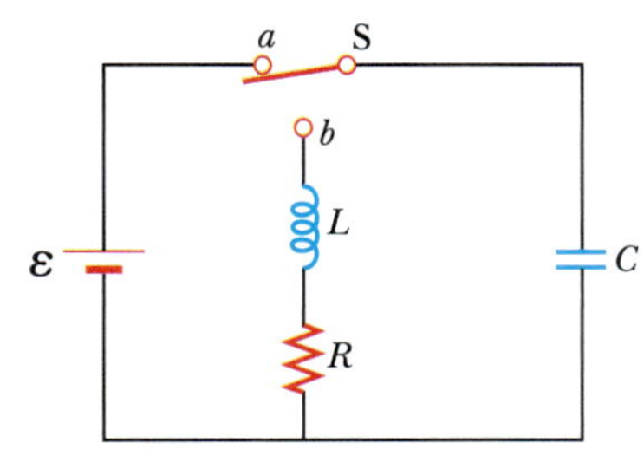

ACTIVE FIGURE 32.15

A series *RLC* circuit. The switch is set to position *a*, and the capacitor is charged. The switch is then thrown to position *b*.

Sign in at www.thomsonedu.com and go to ThomsonNOW to adjust the values of *R*, *L*, and *C* and see the effect on the decaying charge on the capacitor. A graphical display as in Active Figure 32.16a is available, as is an energy bar graph.

$$\frac{dU}{dt} = -I^2R$$

where the negative sign signifies that the energy *U* of the circuit is decreasing in time. Substituting this result into Equation 32.19 gives

$$LI\frac{dI}{dt} + \frac{Q}{C}\frac{dQ}{dt} = -I^2R \tag{32.28}$$

To convert this equation into a form that allows us to compare the electrical oscillations with their mechanical analog, we first use $I = dQ/dt$ and move all terms to the left-hand side to obtain

$$LI\frac{d^2Q}{dt^2} + I^2R + \frac{Q}{C}I = 0$$

Now divide through by I:

$$L\frac{d^2Q}{dt^2} + IR + \frac{Q}{C} = 0$$

$$L\frac{d^2Q}{dt^2} + R\frac{dQ}{dt} + \frac{Q}{C} = 0 \qquad \textbf{(32.29)}$$

The *RLC* circuit is analogous to the damped harmonic oscillator discussed in Section 15.6 and illustrated in Figure 15.20. The equation of motion for a damped block–spring system is, from Equation 15.31,

$$m\frac{d^2x}{dt^2} + b\frac{dx}{dt} + kx = 0 \qquad \textbf{(32.30)}$$

Comparing Equations 32.29 and 32.30, we see that Q corresponds to the position x of the block at any instant, L to the mass m of the block, R to the damping coefficient b, and C to $1/k$, where k is the force constant of the spring. These and other relationships are listed in Table 32.1.

TABLE 32.1

Analogies Between Electrical and Mechanical Systems

Electric Circuit		**One-Dimensional Mechanical System**
Charge	$Q \leftrightarrow x$	Position
Current	$I \leftrightarrow v_x$	Velocity
Potential difference	$\Delta V \leftrightarrow F_x$	Force
Resistance	$R \leftrightarrow b$	Viscous damping coefficient
Capacitance	$C \leftrightarrow 1/k$	(k = spring constant)
Inductance	$L \leftrightarrow m$	Mass
Current = time derivative of charge	$I = \frac{dQ}{dt} \leftrightarrow v_x = \frac{dx}{dt}$	Velocity = time derivative of position
Rate of change of current = second time derivative of charge	$\frac{dI}{dt} = \frac{d^2Q}{dt^2} \leftrightarrow a_x = \frac{dv_x}{dt} = \frac{d^2x}{dt^2}$	Acceleration = second time derivative of position
Energy in inductor	$U_L = \frac{1}{2}LI^2 \leftrightarrow K = \frac{1}{2}mv^2$	Kinetic energy of moving object
Energy in capacitor	$U_C = \frac{1}{2}\frac{Q^2}{C} \leftrightarrow U = \frac{1}{2}kx^2$	Potential energy stored in a spring
Rate of energy loss due to resistance	$I^2R \leftrightarrow bv^2$	Rate of energy loss due to friction
RLC circuit	$L\frac{d^2Q}{dt^2} + R\frac{dQ}{dt} + \frac{Q}{C} = 0 \leftrightarrow m\frac{d^2x}{dt^2} + b\frac{dx}{dt} + kx = 0$	Damped object on a spring

Because the analytical solution of Equation 32.29 is cumbersome, we give only a qualitative description of the circuit behavior. In the simplest case, when $R = 0$, Equation 32.29 reduces to that of a simple LC circuit as expected, and the charge and the current oscillate sinusoidally in time. This situation is equivalent to removing all damping in the mechanical oscillator.

When R is small, a situation that is analogous to light damping in the mechanical oscillator, the solution of Equation 32.29 is

$$Q = Q_{\max} e^{-Rt/2L} \cos \omega_d t \qquad \textbf{(32.31)}$$

where ω_d, the angular frequency at which the circuit oscillates, is given by

$$\omega_d = \left[\frac{1}{LC} - \left(\frac{R}{2L}\right)^2\right]^{1/2} \qquad \textbf{(32.32)}$$

That is, the value of the charge on the capacitor undergoes a damped harmonic oscillation in analogy with a block–spring system moving in a viscous medium. Equation 32.32 shows that when $R \ll \sqrt{4L/C}$ (so that the second term in the brackets is much smaller than the first), the frequency ω_d of the damped oscillator is close to that of the undamped oscillator, $1/\sqrt{LC}$. Because $I = dQ/dt$, it follows that the current also undergoes damped harmonic oscillation. A plot of the charge versus time for the damped oscillator is shown in Active Figure 32.16a and an oscilloscope trace for a real RLC circuit is shown in Active Figure 32.16b. The maximum value of Q decreases after each oscillation, just as the amplitude of a damped block–spring system decreases in time.

For larger values of R, the oscillations damp out more rapidly; in fact, there exists a critical resistance value $R_c = \sqrt{4L/C}$ above which no oscillations occur. A system with $R = R_c$ is said to be *critically damped*. When R exceeds R_c, the system is said to be *overdamped*.

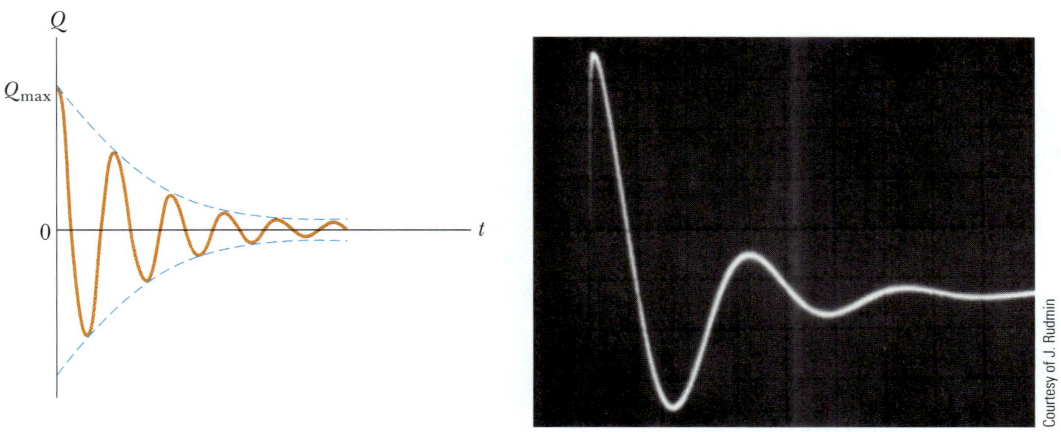

(a) (b)

ACTIVE FIGURE 32.16

(a) Charge versus time for a damped *RLC* circuit. The charge decays in this way when $R < \sqrt{4L/C}$. The Q-versus-t curve represents a plot of Equation 32.31. (b) Oscilloscope pattern showing the decay in the oscillations of an *RLC* circuit.

Sign in at www.thomsonedu.com and go to ThomsonNOW to observe this graph develop for the damped *RLC* circuit in Active Figure 32.15.

Summary

ThomsonNOW™ Sign in at **www.thomsonedu.com** and go to ThomsonNOW to take a practice test for this chapter.

CONCEPTS AND PRINCIPLES

When the current in a loop of wire changes with time, an emf is induced in the loop according to Faraday's law. The **self-induced emf** is

$$\mathcal{E}_L = -L\frac{dI}{dt} \qquad (32.1)$$

where L is the **inductance** of the loop. Inductance is a measure of how much opposition a loop offers to a change in the current in the loop. Inductance has the SI unit of **henry** (H), where 1 H = 1 V · s/A.

The inductance of any coil is

$$L = \frac{N\Phi_B}{I} \qquad (32.2)$$

where N is the total number of turns and Φ_B is the magnetic flux through the coil. The inductance of a device depends on its geometry. For example, the inductance of an air-core solenoid is

$$L = \mu_0 \frac{N^2}{\ell} A \qquad (32.4)$$

where ℓ is the length of the solenoid and A is the cross-sectional area.

If a resistor and inductor are connected in series to a battery of emf $\mathcal{E}$ at time $t = 0$, the current in the circuit varies in time according to the expression

$$I = \frac{\mathcal{E}}{R}(1 - e^{-t/\tau}) \qquad (32.7)$$

where $\tau = L/R$ is the **time constant** of the RL circuit. If we replace the battery in the circuit by a resistanceless wire, the current decays exponentially with time according to the expression

$$I = \frac{\mathcal{E}}{R} e^{-t/\tau} \qquad (32.10)$$

where $\mathcal{E}/R$ is the initial current in the circuit.

The energy stored in the magnetic field of an inductor carrying a current I is

$$U = \tfrac{1}{2}LI^2 \qquad (32.12)$$

This energy is the magnetic counterpart to the energy stored in the electric field of a charged capacitor.

The energy density at a point where the magnetic field is B is

$$u_B = \frac{B^2}{2\mu_0} \qquad (32.14)$$

The **mutual inductance** of a system of two coils is

$$M_{12} = \frac{N_2\Phi_{12}}{I_1} = M_{21} = \frac{N_1\Phi_{21}}{I_2} = M \qquad (32.15)$$

This mutual inductance allows us to relate the induced emf in a coil to the changing source current in a nearby coil using the relationships

$$\mathcal{E}_2 = -M_{12}\frac{dI_1}{dt} \quad \text{and} \quad \mathcal{E}_1 = -M_{21}\frac{dI_2}{dt} \qquad (32.16, 32.17)$$

In an LC circuit that has zero resistance and does not radiate electromagnetically (an idealization), the values of the charge on the capacitor and the current in the circuit vary sinusoidally in time at an angular frequency given by

$$\omega = \frac{1}{\sqrt{LC}} \qquad (32.22)$$

The energy in an LC circuit continuously transfers between energy stored in the capacitor and energy stored in the inductor.

In an RLC circuit with small resistance, the charge on the capacitor varies with time according to

$$Q = Q_{\max} e^{-Rt/2L} \cos \omega_d t \qquad (32.31)$$

where

$$\omega_d = \left[\frac{1}{LC} - \left(\frac{R}{2L}\right)^2\right]^{1/2} \qquad (32.32)$$

Questions

☐ denotes answer available in *Student Solutions Manual/Study Guide;* **O** denotes objective question

1. The current in a circuit containing a coil, a resistor, and a battery has reached a constant value. Does the coil have an inductance? Does the coil affect the value of the current?

2. What parameters affect the inductance of a coil? Does the inductance of a coil depend on the current in the coil?

3. O Initially, an inductor with no resistance carries a constant current. Then the current is brought to a new constant value twice as large. *After* this change, what has happened to the emf in the inductor? (a) It is larger than before the change by a factor of 4. (b) It is larger by a factor of 2. (c) It has the same nonzero value. (d) It continues to be zero. (e) It has decreased.

4. O A long, fine wire is wound into a coil with inductance 5 mH. The coil is connected across the terminals of a battery, and the current is measured a few seconds after the connection is made. The wire is unwound and wound again into a different coil with $L = 10$ mH. This second coil is connected across the same battery, and the current is measured in the same way. Compared with the current in the first coil, is the current in the second coil (a) four times as large, (b) twice as large, (c) unchanged, (d) half as large, or (e) one-fourth as large?

5. O Two solenoidal coils, A and B, are wound using equal lengths of the same kind of wire. The length of the axis of each coil is large compared with its diameter. The axial length of coil A is twice as large as that of coil B, and coil A has twice as many turns as coil B. What is the ratio of the inductance of coil A to that of coil B? (a) 8 (b) 4 (c) 2 (d) 1 (e) $\frac{1}{2}$ (f) $\frac{1}{4}$ (g) $\frac{1}{8}$

6. A switch controls the current in a circuit that has a large inductance. Is a spark (Fig. Q32.6) more likely to be produced at the switch when the switch is being closed, when it is being opened, or doesn't it matter? The electric arc can melt and oxidize the contact surfaces, resulting in high resistivity of the contacts and eventual destruction of the switch. Before electronic ignitions were invented, distributor contact points in automobiles had to be replaced regularly. Switches in power distribution networks and switches controlling large motors, generators, and electromagnets can suffer from arcing and can be very dangerous to operate.

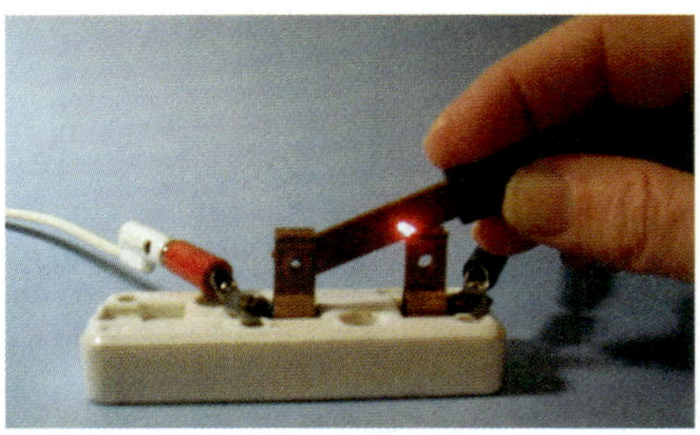

Alexandra Héder

Figure Q32.6

7. O In Figure Q32.7, the switch is left in position *a* for a long time interval and is then quickly thrown to position *b*. Rank the magnitudes of the voltages across the four circuit elements a short time thereafter from the largest to the smallest.

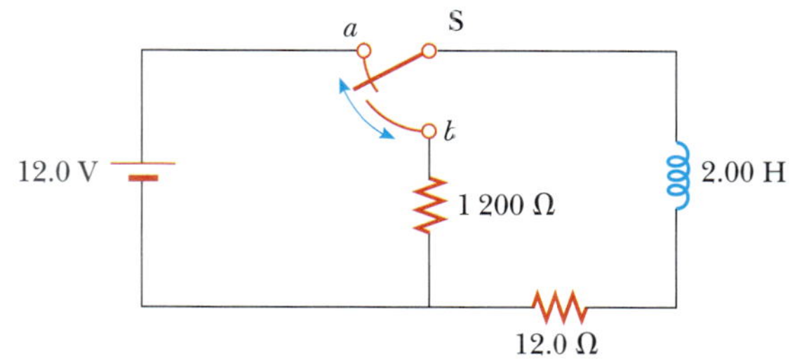

Figure Q32.7

8. Consider the four circuits shown in Figure Q32.8, each consisting of a battery, a switch, a lightbulb, a resistor, and either a capacitor or an inductor. Assume the capacitor has a large capacitance and the inductor has a large inductance but no resistance. The lightbulb has high efficiency, glowing whenever it carries electric current. **(i)** Describe what the lightbulb does in each of circuits (a), (b), (c), and (d) after the switch is thrown closed. **(ii)** Describe what the lightbulb does in each circuit after, having been closed for a long time interval, the switch is thrown open.

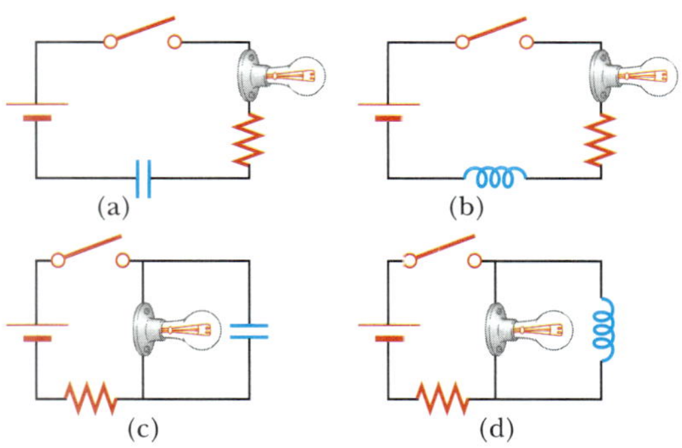

Figure Q32.8

9. O *Don't do this; it's dangerous and illegal.* Suppose a criminal wants to steal energy from the electric company by placing a flat, rectangular coil of wire close to, but not touching, one long, straight, horizontal wire in a transmission line. The long, straight wire carries a sinusoidally varying current. Which of the following statements is true? (a) The method works best if the coil is in a vertical plane surrounding the straight wire. (b) The method works best if the coil is in a vertical plane with the two long sides of the rectangle parallel to the long wire and equally far from it. (c) The method works best if the coil and the long wire are in the same horizontal plane with one long side of the rectangle close to the wire. (d) The method works for any orientation of the coil. (e) The method cannot work without contact between the coil and the long wire.

10. Consider this thesis: "Joseph Henry, America's first professional physicist, caused the most recent basic change in the human view of the Universe when he discovered self-induction during a school vacation at the Albany Academy about 1830. Before that time, one could think of the Universe as composed of only one thing: matter. The energy that temporarily maintains the current after a battery is removed from a coil, on the other hand, is not energy that belongs to any chunk of matter. It is energy in the massless magnetic field surrounding the coil. With Henry's discovery, Nature forced us to admit that the Universe consists of fields as well as matter." Argue for or against the statement. In your view, what makes up the Universe?

11. **O** If the current in an inductor is doubled, by what factor is the stored energy multiplied? (a) 4 (b) 2 (c) 1 (d) $\frac{1}{2}$ (e) $\frac{1}{4}$

12. **O** A solenoidal inductor for a printed circuit board is being redesigned. To save weight, the number of turns is reduced by one-half with the geometric dimensions kept the same. By how much must the current change if the energy stored in the inductor is to remain the same? (a) It must be four times larger (b) It must be two times larger (c) It must be larger by a factor of $\sqrt{2}$. (d) It should be left the same. (e) It should be one-half as large. (f) No change in the current can compensate for the reduction in the number of turns.

13. Discuss the similarities between the energy stored in the electric field of a charged capacitor and the energy stored in the magnetic field of a current-carrying coil.

14. The open switch in Figure Q32.14 is thrown closed at $t = 0$. Before the switch is closed, the capacitor is uncharged and all currents are zero. Determine the currents in L, C, and R and the potential differences across L, C, and R (a) at the instant after the switch is closed and (b) long after it is closed.

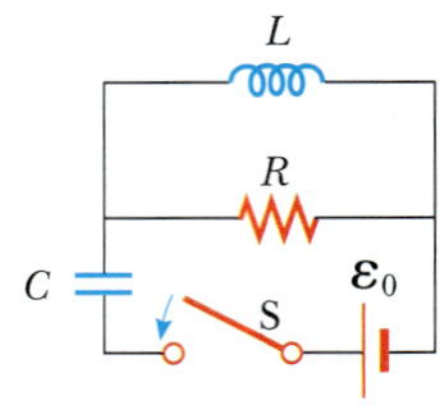

Figure Q32.14

15. **O** The centers of two circular loops are separated by a fixed distance. **(i)** For what relative orientation of the loops is their mutual inductance a maximum? (a) coaxial and lying in parallel planes (b) lying in the same plane (c) lying in perpendicular planes, with the center of one on the axis of the other (d) The orientation makes no difference. **(ii)** For what relative orientation is their mutual inductance a minimum? Choose from the same possibilities.

16. In the LC circuit shown in Figure 32.10, the charge on the capacitor is sometimes zero, but at such instants the current in the circuit is not zero. How is this behavior possible?

17. How can you tell whether an RLC circuit is overdamped or underdamped?

18. Can an object exert a force on itself? When a coil induces an emf in itself, does it exert a force on itself?

Problems

WebAssign The Problems from this chapter may be assigned online in WebAssign.

ThomsonNOW Sign in at **www.thomsonedu.com** and go to ThomsonNOW to assess your understanding of this chapter's topics with additional quizzing and conceptual questions.

1, 2, 3 denotes straightforward, intermediate, challenging; □ denotes full solution available in *Student Solutions Manual/Study Guide;* ▲ denotes coached solution with hints available at **www.thomsonedu.com;** ▭ denotes developing symbolic reasoning; ● denotes asking for qualitative reasoning; ▪ denotes computer useful in solving problem

Section 32.1 Self-Induction and Inductance

1. A 2.00-H inductor carries a steady current of 0.500 A. When the switch in the circuit is opened, the current is effectively zero after 10.0 ms. What is the average induced emf in the inductor during this time interval?

2. A coiled telephone cord forms a spiral having 70 turns, a diameter of 1.30 cm and an unstretched length of 60.0 cm. Determine the inductance of one conductor in the unstretched cord.

3. ▲ A 10.0-mH inductor carries a current $I = I_{max} \sin \omega t$, with $I_{max} = 5.00$ A and $\omega/2\pi = 60.0$ Hz. What is the self-induced emf as a function of time?

4. An emf of 24.0 mV is induced in a 500-turn coil at an instant when the current is 4.00 A and is changing at the rate of 10.0 A/s. What is the magnetic flux through each turn of the coil?

5. An inductor in the form of a solenoid contains 420 turns, is 16.0 cm in length, and has a cross-sectional area of 3.00 cm^2. What uniform rate of decrease of current through the inductor induces an emf of 175 μV?

6. The current in a 90.0-mH inductor changes with time as $I = 1.00t^2 - 6.00t$ (in SI units). Find the magnitude of the induced emf at (a) $t = 1.00$ s and (b) $t = 4.00$ s. (c) At what time is the emf zero?

7. ● A 40.0-mA current is carried by a uniformly wound air-core solenoid with 450 turns, a 15.0-mm diameter, and 12.0-cm length. Compute (a) the magnetic field inside the solenoid, (b) the magnetic flux through each turn, and (c) the inductance of the solenoid. (d) **What If?** If the current were different, which of these quantities would change?

8. A toroid has a major radius R and a minor radius r and is tightly wound with N turns of wire as shown in Figure P32.8. If $R >> r$, the magnetic field in the region enclosed by the wire of the torus, of cross-sectional area $A = \pi r^2$, is essentially the same as the magnetic field of a

solenoid that has been bent into a large circle of radius R. Modeling the field as the uniform field of a long solenoid, show that the inductance of such a toroid is approximately

$$L \approx \frac{\mu_0 N^2 A}{2\pi R}$$

(An exact expression of the inductance of a toroid with a rectangular cross section is derived in Problem 57.)

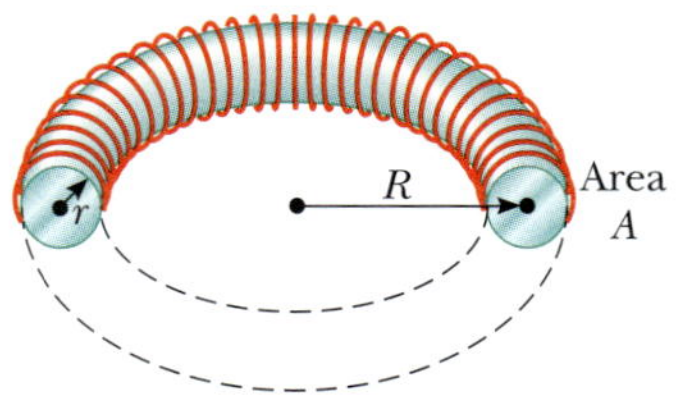

Figure P32.8

9. A self-induced emf in a solenoid of inductance L changes in time as $\boldsymbol{\mathcal{E}} = \boldsymbol{\mathcal{E}}_0 e^{-kt}$. Find the total charge that passes through the solenoid, assuming the charge is finite.

Section 32.2 *RL* Circuits

10. Show that $I = I_i e^{-t/\tau}$ is a solution of the differential equation

$$IR + L\frac{dI}{dt} = 0$$

where I_i is the current at $t = 0$ and $\tau = L/R$.

11. A 12.0-V battery is connected into a series circuit containing a 10.0-Ω resistor and a 2.00-H inductor. In what time interval will the current reach (a) 50.0% and (b) 90.0% of its final value?

12. ● In the circuit diagrammed in Figure P32.12, take $\boldsymbol{\mathcal{E}}$ = 12.0 V and R = 24.0 Ω. Assume the switch is open for $t < 0$ and is closed at $t = 0$. On a single set of axes, sketch graphs of the current in the circuit as a function of time for $t \geq 0$, assuming (a) the inductance in the circuit is essentially zero, (b) the inductance has an intermediate value, and (c) the inductance has a very large value. Label the initial and final values of the current.

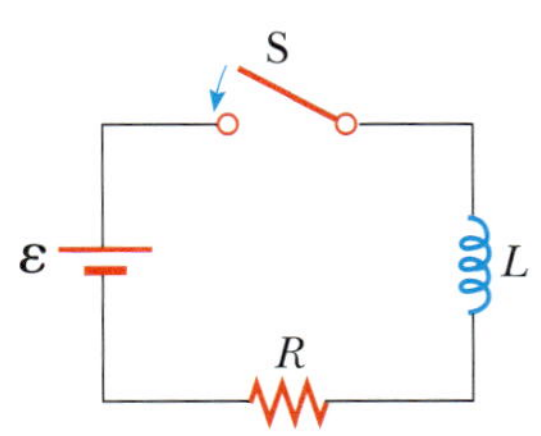

Figure P32.12 Problems 12, 13, 14, and 15.

13. Consider the circuit in Figure P32.12, taking $\boldsymbol{\mathcal{E}}$ = 6.00 V, L = 8.00 mH, and R = 4.00 Ω. (a) What is the inductive time constant of the circuit? (b) Calculate the current in the circuit 250 μs after the switch is closed. (c) What is the value of the final steady-state current? (d) After what time interval does the current reach 80.0% of its maximum value?

14. In the circuit shown in Figure P32.12, let L = 7.00 H, R = 9.00 Ω, and $\boldsymbol{\mathcal{E}}$ = 120 V. What is the self-induced emf 0.200 s after the switch is closed?

15. ▲ For the *RL* circuit shown in Figure P32.12, let the inductance be 3.00 H, the resistance 8.00 Ω, and the battery emf 36.0 V. (a) Calculate the ratio of the potential difference across the resistor to the emf across the inductor when the current is 2.00 A. (b) Calculate the emf across the inductor when the current is 4.50 A.

16. A 12.0-V battery is connected in series with a resistor and an inductor. The circuit has a time constant of 500 μs, and the maximum current is 200 mA. What is the value of the inductance of the inductor?

17. An inductor that has an inductance of 15.0 H and a resistance of 30.0 Ω is connected across a 100-V battery. What is the rate of increase of the current (a) at $t = 0$ and (b) at t = 1.50 s?

18. The switch in Figure P32.18 is open for $t < 0$ and is then thrown closed at time $t = 0$. Find the current in the inductor and the current in the switch as functions of time thereafter.

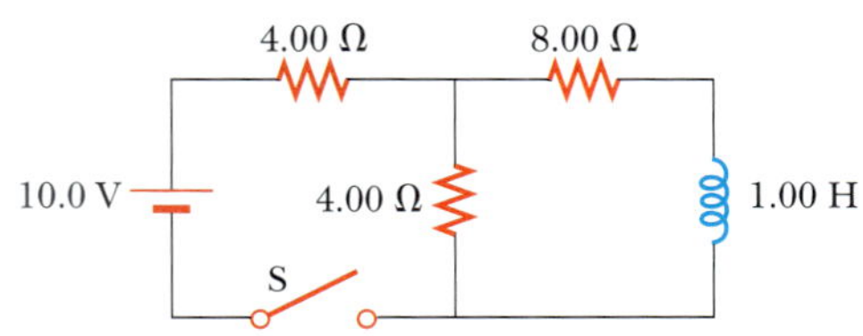

Figure P32.18 Problems 18 and 52.

19. A series *RL* circuit with L = 3.00 H and a series *RC* circuit with C = 3.00 μF have equal time constants. If the two circuits contain the same resistance R, (a) what is the value of R and (b) what is the time constant?

20. A current pulse is fed to the partial circuit shown in Figure P32.20. The current begins at zero, becomes 10.0 A between $t = 0$ and t = 200 μs, and then is zero once again. Determine the current in the inductor as a function of time.

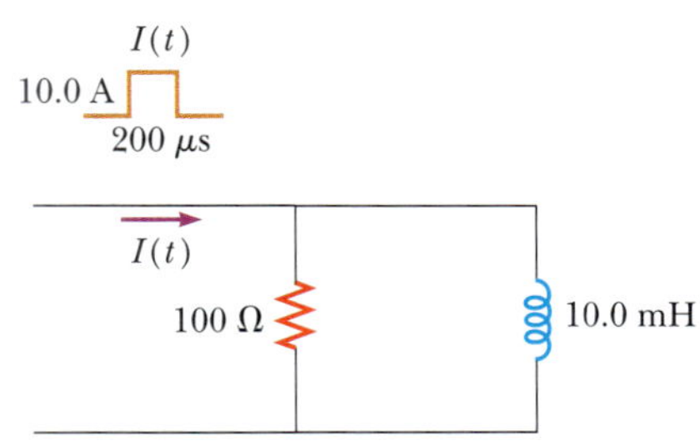

Figure P32.20

21. ▲ A 140-mH inductor and a 4.90-Ω resistor are connected with a switch to a 6.00-V battery as shown in Figure P32.21. (a) After the switch is thrown to *a* (connecting the battery), what time interval elapses before the current reaches 220 mA? (b) What is the current in the inductor 10.0 s after the switch is closed? (c) Now the switch is

2 = intermediate; 3 = challenging; □ = SSM/SG; ▲ = ThomsonNOW; = symbolic reasoning; ● = qualitative reasoning

quickly thrown from a to b. What time interval elapses before the current falls to 160 mA?

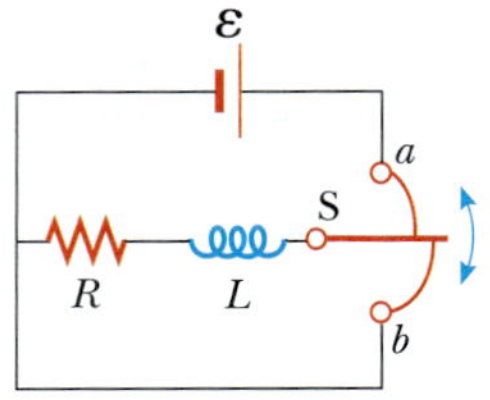

Figure P32.21

22. ● Two ideal inductors, L_1 and L_2, have *zero* internal resistance and are far apart, so their magnetic fields do not influence each other. (a) Assuming these inductors are connected in series, show that they are equivalent to a single ideal inductor having $L_{eq} = L_1 + L_2$. (b) Assuming these same two inductors are connected in parallel, show that they are equivalent to a single ideal inductor having $1/L_{eq} = 1/L_1 + 1/L_2$. (c) **What If?** Now consider two inductors L_1 and L_2 that have *nonzero* internal resistances R_1 and R_2, respectively. Assume they are still far apart so that their mutual inductance is zero. Assuming these inductors are connected in series, show that they are equivalent to a single inductor having $L_{eq} = L_1 + L_2$ and $R_{eq} = R_1 + R_2$. (d) If these same inductors are now connected in parallel, is it necessarily true that they are equivalent to a single ideal inductor having $1/L_{eq} = 1/L_1 + 1/L_2$ and $1/R_{eq} = 1/R_1 + 1/R_2$? Explain your answer.

Section 32.3 Energy in a Magnetic Field

23. An air-core solenoid with 68 turns is 8.00 cm long and has a diameter of 1.20 cm. How much energy is stored in its magnetic field when it carries a current of 0.770 A?

24. The magnetic field inside a superconducting solenoid is 4.50 T. The solenoid has an inner diameter of 6.20 cm and a length of 26.0 cm. Determine (a) the magnetic energy density in the field and (b) the energy stored in the magnetic field within the solenoid.

25. ▲ On a clear day at a certain location, a 100-V/m vertical electric field exists near the Earth's surface. At the same place, the Earth's magnetic field has a magnitude of 0.500×10^{-4} T. Compute the energy densities of the two fields.

26. Complete the calculation in Example 32.3 by proving that

$$\int_0^\infty e^{-2Rt/L}\,dt = \frac{L}{2R}$$

27. ● A flat coil of wire has an inductance of 40.0 mH and a resistance of 5.00 Ω. It is connected to a 22.0-V battery at the instant $t = 0$. Consider the moment when the current is 3.00 A. (a) At what rate is energy being delivered by the battery? (b) What is the power being delivered to the resistor? (c) At what rate is energy being stored in the magnetic field of the coil? (d) What is the relationship among these three power values? Is this relationship true at other instants as well? Explain the relationship at the moment immediately after $t = 0$ and at a moment several seconds later.

28. A 10.0-V battery, a 5.00-Ω resistor, and a 10.0-H inductor are connected in series. After the current in the circuit has reached its maximum value, calculate (a) the power being supplied by the battery, (b) the power being delivered to the resistor, (c) the power being delivered to the inductor, and (d) the energy stored in the magnetic field of the inductor.

29. Assume the magnitude of the magnetic field outside a sphere of radius R is $B = B_0(R/r)^2$, where B_0 is a constant. Determine the total energy stored in the magnetic field outside the sphere and evaluate your result for $B_0 = 5.00 \times 10^{-5}$ T and $R = 6.00 \times 10^6$ m, values appropriate for the Earth's magnetic field.

Section 32.4 Mutual Inductance

30. Two coils are close to each other. The first coil carries a current given by $I(t) = (5.00\text{ A})e^{-0.025\,0t}\sin(377t)$. At $t = 0.800$ s, the emf measured across the second coil is -3.20 V. What is the mutual inductance of the coils?

31. Two coils, held in fixed positions, have a mutual inductance of 100 μH. What is the peak emf in one coil when a sinusoidal current given by $I(t) = (10.0\text{ A})\sin(1\,000t)$ is in the other coil?

32. On a printed circuit board, a relatively long, straight conductor and a conducting rectangular loop lie in the same plane as shown in Figure P31.8 in Chapter 31. Taking $h = 0.400$ mm, $w = 1.30$ mm, and $L = 2.70$ mm, find their mutual inductance.

33. Two solenoids A and B, spaced close to each other and sharing the same cylindrical axis, have 400 and 700 turns, respectively. A current of 3.50 A in coil A produces an average flux of 300 μWb through each turn of A and a flux of 90.0 μWb through each turn of B. (a) Calculate the mutual inductance of the two solenoids. (b) What is the inductance of A? (c) What emf is induced in B when the current in A increases at the rate of 0.500 A/s?

34. ● A solenoid has N_1 turns, radius R_1, and length ℓ. It is so long that its magnetic field is uniform nearly everywhere inside it and is nearly zero outside. A second solenoid has N_2 turns, radius $R_2 < R_1$, and the same length. It lies inside the first solenoid, with their axes parallel. (a) Assume solenoid 1 carries variable current I. Compute the mutual inductance characterizing the emf induced in solenoid 2. (b) Now assume solenoid 2 carries current I. Compute the mutual inductance to which the emf in solenoid 1 is proportional. (c) State how the results of parts (a) and (b) compare with each other.

35. A large coil of radius R_1 and having N_1 turns is coaxial with a small coil of radius R_2 and having N_2 turns. The centers of the coils are separated by a distance x that is much larger than R_2. What is the mutual inductance of the coils? *Suggestion:* John von Neumann proved that the same answer must result from considering the flux through the first coil of the magnetic field produced by the second coil or from considering the flux through the second coil of the magnetic field produced by the first coil. In this problem, it is easy to calculate the flux through the small coil, but it is difficult to calculate the flux through the large coil because to do so, you would have to know the magnetic field away from the axis.

36. Two inductors having inductances L_1 and L_2 are connected in parallel as shown in Figure P32.36a. The mutual inductance between the two inductors is M. Deter-

mine the equivalent inductance L_{eq} for the system (Fig. P32.36b).

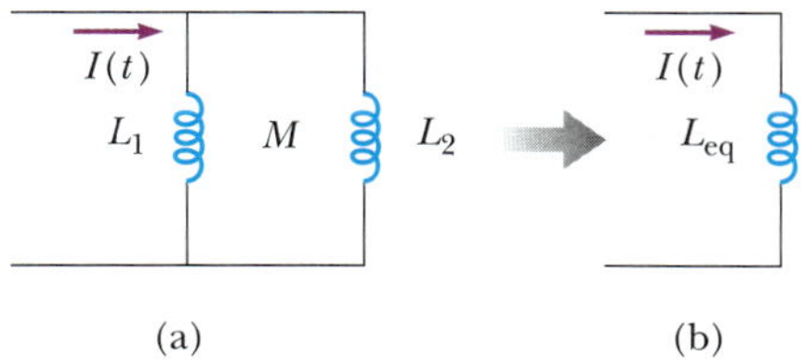

Figure P32.36

Section 32.5 Oscillations in an *LC* Circuit

37. A 1.00-μF capacitor is charged by a 40.0-V power supply. The fully charged capacitor is then discharged through a 10.0-mH inductor. Find the maximum current in the resulting oscillations.

38. An *LC* circuit consists of a 20.0-mH inductor and a 0.500-μF capacitor. If the maximum instantaneous current is 0.100 A, what is the greatest potential difference across the capacitor?

39. In the circuit of Figure P32.39, the battery emf is 50.0 V, the resistance is 250 Ω, and the capacitance is 0.500 μF. The switch S is closed for a long time interval, and zero potential difference is measured across the capacitor. After the switch is opened, the potential difference across the capacitor reaches a maximum value of 150 V. What is the value of the inductance?

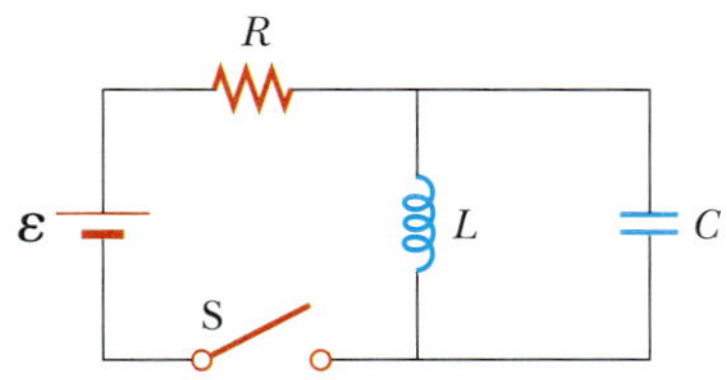

Figure P32.39

40. An *LC* circuit like the one in Figure 32.10 contains an 82.0-mH inductor and a 17.0-μF capacitor that initially carries a 180-μC charge. The switch is open for $t < 0$ and then thrown closed at $t = 0$. (a) Find the frequency (in hertz) of the resulting oscillations. At $t = 1.00$ ms, find (b) the charge on the capacitor and (c) the current in the circuit.

41. A fixed inductance $L = 1.05$ μH is used in series with a variable capacitor in the tuning section of a radiotelephone on a ship. What capacitance tunes the circuit to the signal from a transmitter broadcasting at 6.30 MHz?

42. The switch in Figure P32.42 is connected to point *a* for a long time interval. After the switch is thrown to point *b*, what are (a) the frequency of oscillation of the *LC* circuit, (b) the maximum charge that appears on the capacitor, (c) the maximum current in the inductor, and (d) the total energy the circuit possesses at $t = 3.00$ s?

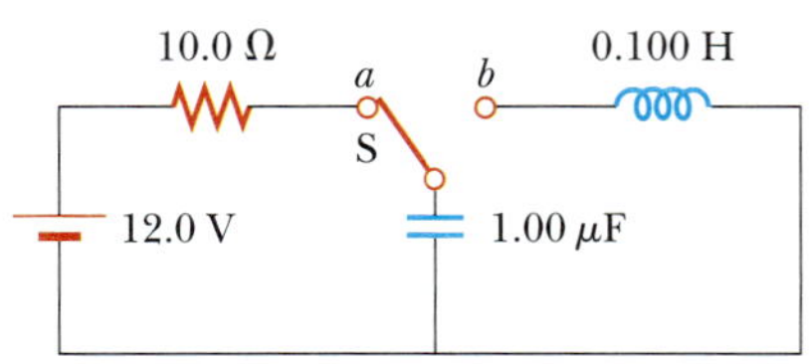

Figure P32.42

43. ▲ An *LC* circuit like that in Figure 32.10 consists of a 3.30-H inductor and an 840-pF capacitor that initially carries a 105-μC charge. The switch is open for $t < 0$ and then thrown closed at $t = 0$. Compute the following quantities at $t = 2.00$ ms: (a) the energy stored in the capacitor, (b) the energy stored in the inductor, and (c) the total energy in the circuit.

Section 32.6 The *RLC* Circuit

44. In Active Figure 32.15, let $R = 7.60$ Ω, $L = 2.20$ mH, and $C = 1.80$ μF. (a) Calculate the frequency of the damped oscillation of the circuit. (b) What is the critical resistance?

45. Consider an *LC* circuit in which $L = 500$ mH and $C = 0.100$ μF. (a) What is the resonance frequency ω_0? (b) If a resistance of 1.00 kΩ is introduced into this circuit, what is the frequency of the (damped) oscillations? (c) What is the percent difference between the two frequencies?

46. Show that Equation 32.28 in the text is Kirchhoff's loop rule as applied to the circuit in Active Figure 32.15.

47. Electrical oscillations are initiated in a series circuit containing a capacitance *C*, inductance *L*, and resistance *R*. (a) If $R \ll \sqrt{4L/C}$ (weak damping), what time interval elapses before the amplitude of the current oscillation falls to 50.0% of its initial value? (b) Over what time interval does the energy decrease to 50.0% of its initial value?

Additional Problems

48. **Review problem.** This problem extends the reasoning of Section 26.4, Problem 29 in Chapter 26, Problem 33 in Chapter 30, and Section 32.3. (a) Consider a capacitor with vacuum between its large, closely spaced, oppositely charged parallel plates. Show that the force on one plate can be accounted for by thinking of the electric field between the plates as exerting a "negative pressure" equal to the energy density of the electric field. (b) Consider two infinite plane sheets carrying electric currents in opposite directions with equal linear current densities J_s. Calculate the force per area acting on one sheet due to the magnetic field, of magnitude $\mu_0 J_s/2$, created by the other sheet. (c) Calculate the net magnetic field between the sheets and the field outside of the volume between them. (d) Calculate the energy density in the magnetic field between the sheets. (e) Show that the force on one sheet can be accounted for by thinking of the magnetic field between the sheets as exerting a positive pressure equal to its energy density. This result for magnetic pressure applies to all current configurations, not only to sheets of current.

49. A 1.00-mH inductor and a 1.00-μF capacitor are connected in series. The current in the circuit is described by $I = 20.0t$, where *t* is in seconds and *I* is in amperes. The capacitor initially has no charge. Determine (a) the voltage across the inductor as a function of time, (b) the voltage across the capacitor as a function of time, and (c) the time when the energy stored in the capacitor first exceeds that in the inductor.

50. An inductor having inductance *L* and a capacitor having capacitance *C* are connected in series. The current in the circuit increases linearly in time as described by $I = Kt$,

where K is a constant. The capacitor is initially uncharged. Determine (a) the voltage across the inductor as a function of time, (b) the voltage across the capacitor as a function of time, and (c) the time when the energy stored in the capacitor first exceeds that in the inductor.

51. A capacitor in a series LC circuit has an initial charge Q and is being discharged. Find, in terms of L and C, the flux through each of the N turns in the coil when the charge on the capacitor is $Q/2$.

52. ● In the circuit diagrammed in Figure P32.18, assume that the switch has been closed for a long time interval and is opened at $t = 0$. (a) Before the switch is opened, does the inductor behave as an open circuit, a short circuit, a resistor of some particular resistance, or none of these choices? What current does the inductor carry? (b) How much energy is stored in the inductor for $t < 0$? (c) After the switch is opened, what happens to the energy previously stored in the inductor? (d) Sketch a graph of the current in the inductor for $t \geq 0$. Label the initial and final values and the time constant.

53. ● At the moment $t = 0$, a 24.0-V battery is connected to a 5.00-mH coil and a 6.00-Ω resistor. (a) Immediately thereafter, how does the potential difference across the resistor compare to the emf across the coil? (b) Answer the same question about the circuit several seconds later. (c) Is there an instant at which these two voltages are equal in magnitude? If so, when? Is there more than one such instant? (d) After a 4.00-A current is established in the resistor and coil, the battery is suddenly replaced by a short circuit. Answer questions (a), (b), and (c) again with reference to this new circuit.

54. When the current in the portion of the circuit shown in Figure P32.54 is 2.00 A and increases at a rate of 0.500 A/s, the measured potential difference is $\Delta V_{ab} = 9.00$ V. When the current is 2.00 A and decreases at the rate of 0.500 A/s, the measured potential difference is $\Delta V_{ab} = 5.00$ V. Calculate the values of L and R.

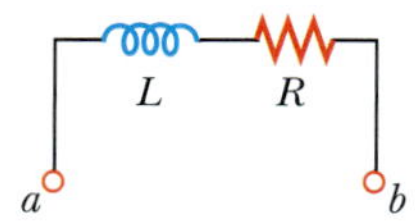

Figure P32.54

55. A time-varying current I is sent through a 50.0-mH inductor as shown in Figure P32.55. Make a graph of the potential at point b relative to the potential at point a.

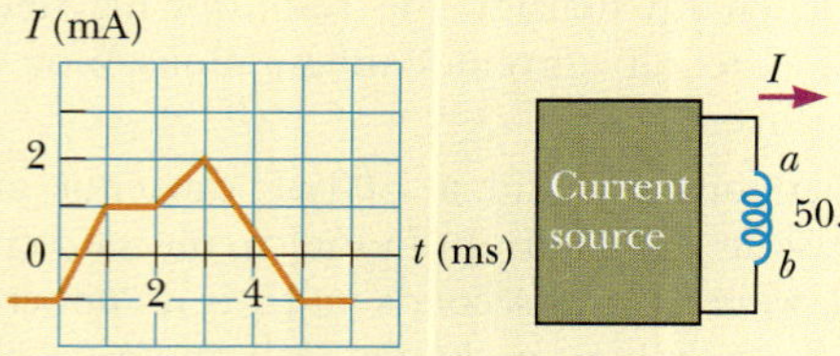

Figure P32.55

56. ● Consider a series circuit consisting of a 500-μF capacitor, a 32.0-mH inductor, and a resistor R. Explain what you can say about the angular frequency of oscillations for (a) $R = 0$, (b) $R = 4.00$ Ω, (c) $R = 15.0$ Ω, and (d) $R = 17.0$ Ω. Relate the mathematical description of the angular frequency to the experimentally measurable angular frequency.

57. ● The toroid in Figure P32.57 consists of N turns and has a rectangular cross section. Its inner and outer radii are a and b, respectively. (a) Show that the inductance of the toroid is

$$L = \frac{\mu_0 N^2 h}{2\pi} \ln \frac{b}{a}$$

(b) Using this result, compute the inductance of a 500-turn toroid for which $a = 10.0$ cm, $b = 12.0$ cm, and $h = 1.00$ cm. (c) **What If?** In Problem 8, an approximate equation for the inductance of a toroid with $R \gg r$ was derived. To get a feel for the accuracy of that result, use the expression in Problem 8 to compute the approximate inductance of the toroid described in part (b). How does that result compare with the answer to part (b)?

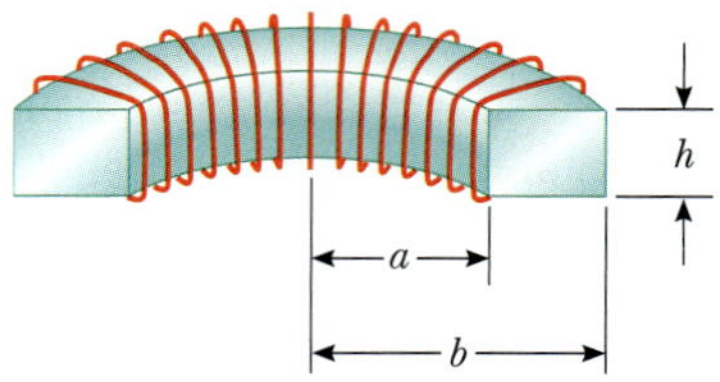

Figure P32.57

58. (a) A flat, circular coil does not actually produce a uniform magnetic field in the area it encloses. Nevertheless, estimate the inductance of a flat, compact, circular coil, with radius R and N turns, by assuming the field at its center is uniform over its area. (b) A circuit on a laboratory table consists of a 1.5-volt battery, a 270-Ω resistor, a switch, and three 30-cm-long patch cords connecting them. Suppose the circuit is arranged to be circular. Think of it as a flat coil with one turn. Compute the order of magnitude of its inductance and (c) of the time constant describing how fast the current increases when you close the switch.

59. At $t = 0$, the open switch in Figure P32.59 is thrown closed. Using Kirchhoff's rules for the instantaneous currents and voltages in this two-loop circuit, show that the current in the inductor at time $t > 0$ is

$$I(t) = \frac{\mathcal{E}}{R_1}\left[1 - e^{-(R'/L)t}\right]$$

where $R' = R_1 R_2/(R_1 + R_2)$.

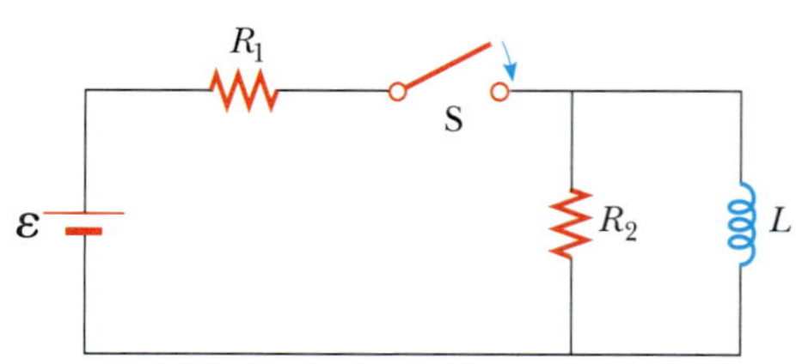

Figure P32.59

60. A wire of nonmagnetic material, with radius R, carries current uniformly distributed over its cross section. The total current carried by the wire is I. Show that the magnetic energy per unit length inside the wire is $\mu_0 I^2/16\pi$.

61. In Figure P32.61, the switch is closed for $t < 0$ and steady-state conditions are established. The switch is opened at $t = 0$. (a) Find the initial emf $\mathcal{E}_0$ across L immediately after $t = 0$. Which end of the coil, a or b, is at the higher voltage? (b) Make freehand graphs of the currents in R_1 and in R_2 as a function of time, treating the steady-state directions as positive. Show values before and after $t = 0$. (c) At what moment after $t = 0$ does the current in R_2 have the value 2.00 mA?

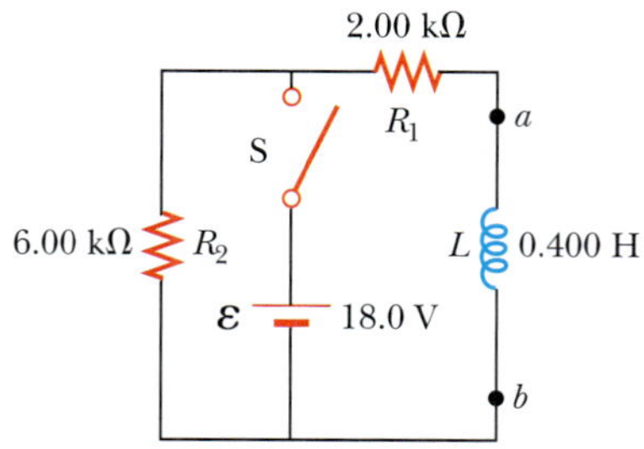

Figure P32.61

62. ● The lead-in wires from a television antenna are often constructed in the form of two parallel wires (Fig. P32.62). The two wires carry currents of equal magnitude in opposite directions. Assume the wires carry the current uniformly distributed over their surfaces and no magnetic field exists inside the wires. (a) Why does this configuration of conductors have an inductance? (b) What constitutes the flux loop for this configuration? (c) Show that the inductance of a length x of this type of lead-in is

$$L = \frac{\mu_0 x}{\pi} \ln \left(\frac{w - a}{a} \right)$$

where w is the center-to-center separation of the wires and a is their radius.

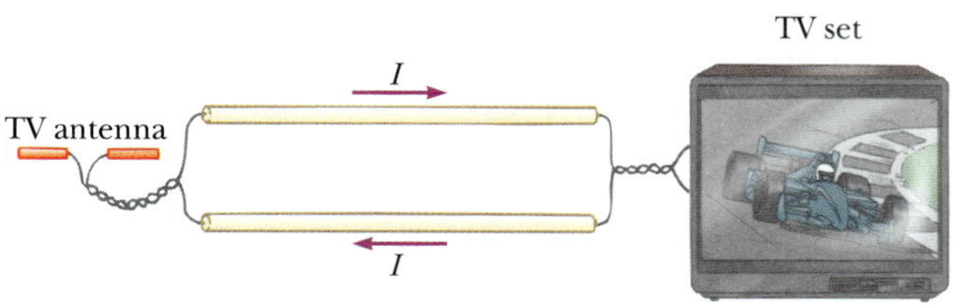

Figure P32.62

63. To prevent damage from arcing in an electric motor, a discharge resistor is sometimes placed in parallel with the armature. If the motor is suddenly unplugged while running, this resistor limits the voltage that appears across the armature coils. Consider a 12.0-V DC motor with an armature that has a resistance of 7.50 Ω and an inductance of 450 mH. Assume the magnitude of the self-induced emf in the armature coils is 10.0 V when the motor is running at normal speed. (The equivalent circuit for the armature is shown in Fig. P32.63.) Calculate the maximum resistance R that limits the voltage across the armature to 80.0 V when the motor is unplugged.

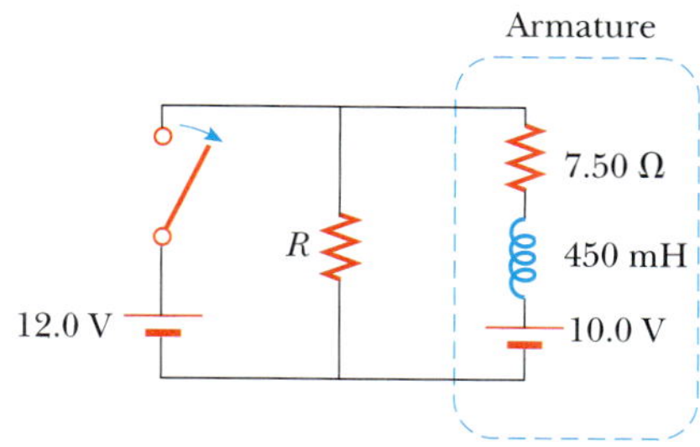

Figure P32.63

Review problems. Problems 64 through 67 apply ideas from this and earlier chapters to some properties of superconductors, which were introduced in Section 27.5.

64. *The resistance of a superconductor.* In an experiment carried out by S. C. Collins between 1955 and 1958, a current was maintained in a superconducting lead ring for 2.50 yr with no observed loss. If the inductance of the ring were 3.14×10^{-8} H and the sensitivity of the experiment were 1 part in 10^9, what was the maximum resistance of the ring? *Suggestion:* Treat the ring as an RL circuit carrying decaying current and recall that $e^{-x} \approx 1 - x$ for small x.

65. A novel method of storing energy has been proposed. A huge underground superconducting coil, 1.00 km in diameter, would be fabricated. It would carry a maximum current of 50.0 kA through each winding of a 150-turn Nb_3Sn solenoid. (a) If the inductance of this huge coil were 50.0 H, what would be the total energy stored? (b) What would be the compressive force per meter length acting between two adjacent windings 0.250 m apart?

66. *Superconducting power transmission.* The use of superconductors has been proposed for power transmission lines. A single coaxial cable (Fig. P32.66) could carry 1.00×10^3 MW (the output of a large power plant) at 200 kV, DC, over a distance of 1 000 km without loss. An inner wire of radius 2.00 cm, made from the superconductor Nb_3Sn, carries the current I in one direction. A surrounding superconducting cylinder of radius 5.00 cm would carry the return current I. In such a system, what is the magnetic field (a) at the surface of the inner conductor and (b) at the inner surface of the outer conductor? (c) How much energy would be stored in the space between the conductors in a 1 000-km superconducting line? (d) What is the pressure exerted on the outer conductor?

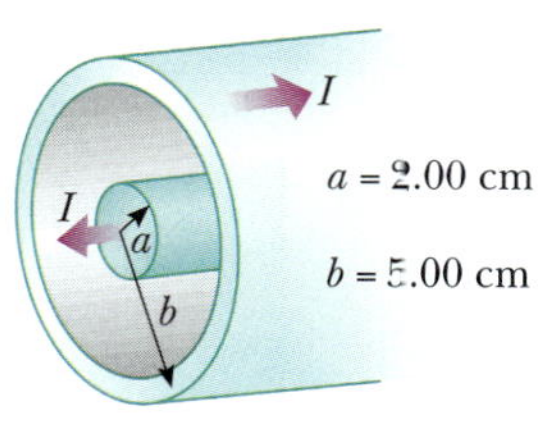

Figure P32.66

67. ● *The Meissner effect.* Compare this problem with Problem 57 in Chapter 26, pertaining to the force attracting a perfect dielectric into a strong electric field. A fundamental property of a type I superconducting material is *perfect*

diamagnetism, or demonstration of the *Meissner effect*, illustrated in Figure 30.27 in Section 30.6 and described as follows. The superconducting material has $\vec{\mathbf{B}} = 0$ everywhere inside it. If a sample of the material is placed into an externally produced magnetic field or is cooled to become superconducting while it is in a magnetic field, electric currents appear on the surface of the sample. The currents have precisely the strength and orientation required to make the total magnetic field be zero throughout the interior of the sample. This problem will help you to understand the magnetic force that can then act on the superconducting sample.

A vertical solenoid with a length of 120 cm and a diameter of 2.50 cm consists of 1 400 turns of copper wire carrying a counterclockwise current of 2.00 A as shown in Figure P32.67a. (a) Find the magnetic field in the vacuum inside the solenoid. (b) Find the energy density of the magnetic field, noting that the units J/m^3 of energy density are the same as the units N/m^2 of pressure. (c) Now a superconducting bar 2.20 cm in diameter is inserted partway into the solenoid. Its upper end is far outside the solenoid, where the magnetic field is negligible. The lower end of the bar is deep inside the solenoid. Explain how you identify the direction required for the current on the curved surface of the bar so that the total magnetic field is zero within the bar. The field created by the supercurrents is sketched in Figure P32.67b, and the total field is sketched in Figure P32.67c. (d) The field of the solenoid exerts a force on the current in the superconductor. Explain how you determine the direction of the force on the bar. (e) Calculate the magnitude of the force by multiplying the energy density of the solenoid field times the area of the bottom end of the superconducting bar.

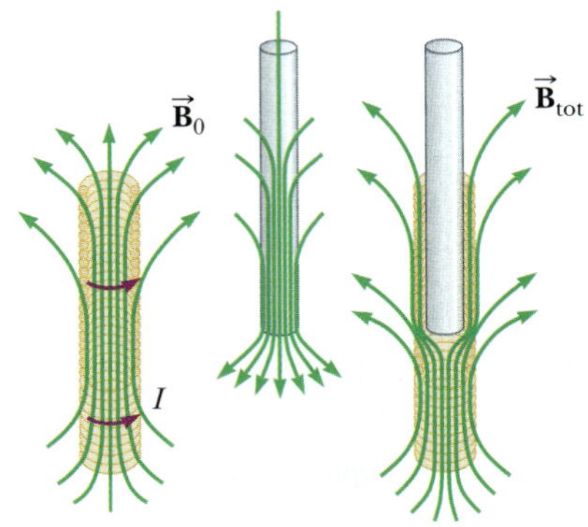

Figure P32.67

Answers to Quick Quizzes

32.1 (c), (f). For the constant current in statements (a) and (b), there is no voltage across the resistanceless inductor. In statement (c), if the current increases, the emf induced in the inductor is in the opposite direction, from b to a, making a higher in potential than b. Similarly, in statement (f), the decreasing current induces an emf in the same direction as the current, from b to a, again making the potential higher at a than at b.

32.2 **(i)**, (b). As the switch is closed, there is no current, so there is no voltage across the resistor. **(ii)**, (a). After a long time, the current has reached its final value and the inductor has no further effect on the circuit.

32.3 (a), (d). Because the energy density depends on the magnitude of the magnetic field, you must increase the magnetic field to increase the energy density. For a solenoid, $B = \mu_0 nI$, where n is the number of turns per unit length. In choice (a), increasing n increases the magnetic field. In choice (b), the change in cross-sectional area has no effect on the magnetic field. In choice (c), increasing the length but keeping n fixed has no effect on the magnetic field. Increasing the current in choice (d) increases the magnetic field in the solenoid.

32.4 (a). M increases because the magnetic flux through coil 2 increases.

32.5 **(i)**, (b). If the current is at its maximum value, the charge on the capacitor is zero. **(ii)**, (c). If the current is zero, this moment is the instant at which the capacitor is fully charged and the current is about to reverse direction.

These large transformers are used to increase the voltage at a power plant for distribution of energy by electrical transmission to the power grid. Voltages can be changed relatively easily because power is distributed by alternating current rather than direct current. (Lester Lefkowitz/Getty Images)

33 Alternating Current Circuits

In this chapter, we describe alternating-current (AC) circuits. Every time you turn on a television set, a stereo, or any of a multitude of other electrical appliances in a home, you are calling on alternating currents to provide the power to operate them. We begin our study by investigating the characteristics of simple series circuits that contain resistors, inductors, and capacitors and that are driven by a sinusoidal voltage. The primary aim of this chapter can be summarized as follows: if an AC source applies an alternating voltage to a series circuit containing resistors, inductors, and capacitors, we want to know the amplitude and time characteristics of the alternating current. We conclude this chapter with two sections concerning transformers, power transmission, and electrical filters.

33.1 AC Sources

An AC circuit consists of circuit elements and a power source that provides an alternating voltage Δv. This time-varying voltage from the source is described by

$$\Delta v = \Delta V_{\text{max}} \sin \omega t$$

where ΔV_{max} is the maximum output voltage of the source, or the **voltage amplitude.** There are various possibilities for AC sources, including generators as discussed in Section 31.5 and electrical oscillators. In a home, each electrical outlet

PITFALL PREVENTION 33.1

Time-Varying Values

We use lowercase symbols Δv and i to indicate the instantaneous values of time-varying voltages and currents. Capital letters represent fixed values of voltage and current such as ΔV_{max} and I_{max}.

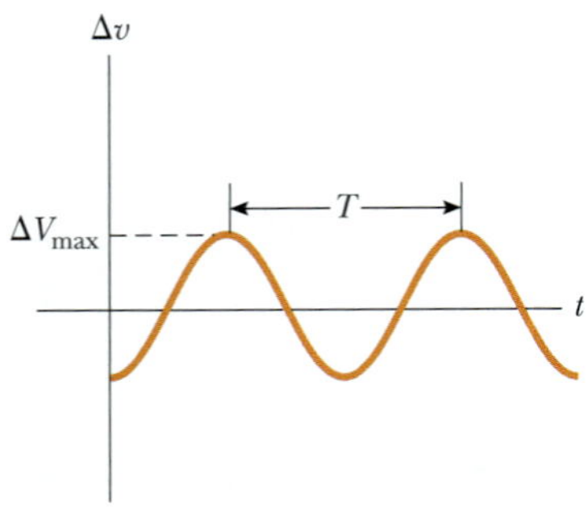

Figure 33.1 The voltage supplied by an AC source is sinusoidal with a period T.

serves as an AC source. Because the output voltage of an AC source varies sinusoidally with time, the voltage is positive during one half of the cycle and negative during the other half as in Figure 33.1. Likewise, the current in any circuit driven by an AC source is an alternating current that also varies sinusoidally with time.

From Equation 15.12, the angular frequency of the AC voltage is

$$\omega = 2\pi f = \frac{2\pi}{T}$$

where f is the frequency of the source and T is the period. The source determines the frequency of the current in any circuit connected to it. Commercial electric-power plants in the United States use a frequency of 60 Hz, which corresponds to an angular frequency of 377 rad/s.

33.2 Resistors in an AC Circuit

Consider a simple AC circuit consisting of a resistor and an AC source 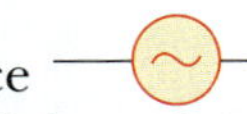 as shown in Active Figure 33.2. At any instant, the algebraic sum of the voltages around a closed loop in a circuit must be zero (Kirchhoff's loop rule). Therefore, $\Delta v + \Delta v_R = 0$ or, using Equation 27.7 for the voltage across the resistor,

$$\Delta v - i_R R = 0$$

If we rearrange this expression and substitute $\Delta V_{max} \sin \omega t$ for Δv, the instantaneous current in the resistor is

$$i_R = \frac{\Delta v}{R} = \frac{\Delta V_{max}}{R} \sin \omega t = I_{max} \sin \omega t \qquad \textbf{(33.1)}$$

where I_{max} is the maximum current:

Maximum current in a resistor ▶

$$I_{max} = \frac{\Delta V_{max}}{R} \qquad \textbf{(33.2)}$$

Equation 33.1 shows that the instantaneous voltage across the resistor is

Voltage across a resistor ▶

$$\Delta v_R = i_R R = I_{max} R \sin \omega t \qquad \textbf{(33.3)}$$

A plot of voltage and current versus time for this circuit is shown in Active Figure 33.3a. At point *a*, the current has a maximum value in one direction, arbitrarily

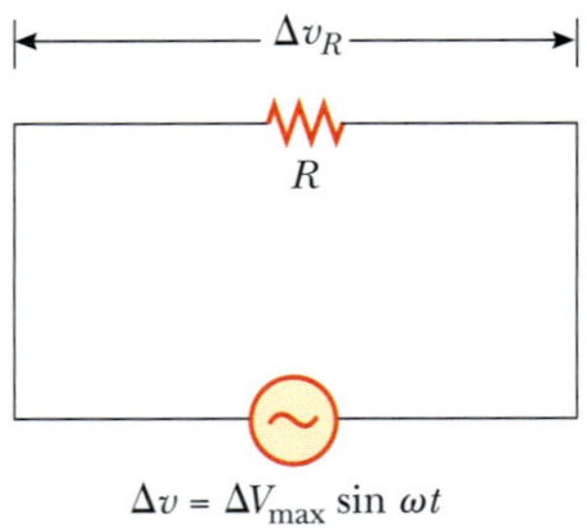

ACTIVE FIGURE 33.2

A circuit consisting of a resistor of resistance R connected to an AC source, designated by the symbol 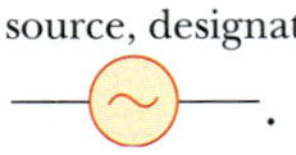.

Sign in at www.thomsonedu.com and go to ThomsonNOW to adjust the resistance, frequency, and maximum voltage. The results can be studied with the graph and the phasor diagram in Active Figure 33.3.

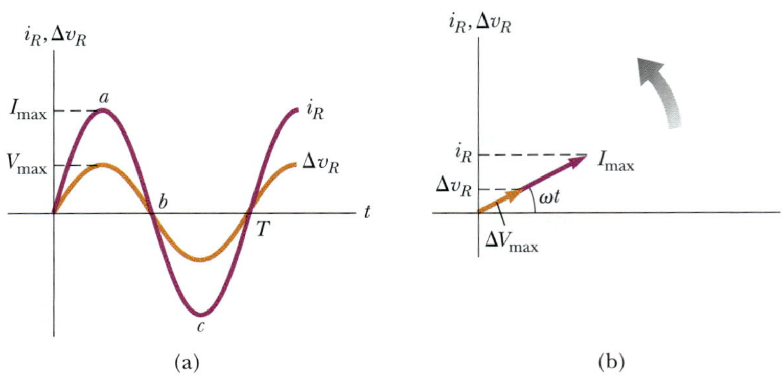

ACTIVE FIGURE 33.3

(a) Plots of the instantaneous current i_R and instantaneous voltage Δv_R across a resistor as functions of time. The current is in phase with the voltage, which means that the current is zero when the voltage is zero, maximum when the voltage is maximum, and minimum when the voltage is minimum. At time $t = T$, one cycle of the time-varying voltage and current has been completed. (b) Phasor diagram for the resistive circuit showing that the current is in phase with the voltage.

Sign in at www.thomsonedu.com and go to ThomsonNOW to adjust the resistance, frequency, and maximum voltage of the circuit in Active Figure 33.2. The results can be studied with the graph and the phasor diagram in this figure.

called the positive direction. Between points *a* and *b*, the current is decreasing in magnitude but is still in the positive direction. At point *b*, the current is momentarily zero; it then begins to increase in the negative direction between points *b* and *c*. At point *c*, the current has reached its maximum value in the negative direction.

The current and voltage are in step with each other because they vary identically with time. Because i_R and Δv_R both vary as sin ωt and reach their maximum values at the same time as shown in Active Figure 33.3a, they are said to be **in phase,** similar to the way that two waves can be in phase as discussed in our study of wave motion in Chapter 18. Therefore, **for a sinusoidal applied voltage, the current in a resistor is always in phase with the voltage across the resistor.** For resistors in AC circuits, there are no new concepts to learn. Resistors behave essentially the same way in both DC and AC circuits. That, however, is not the case for capacitors and inductors.

To simplify our analysis of circuits containing two or more elements, we use a graphical representation called a *phasor diagram.* A **phasor** is a vector whose length is proportional to the maximum value of the variable it represents (ΔV_{max} for voltage and I_{max} for current in this discussion). The phasor rotates counterclockwise at an angular speed equal to the angular frequency associated with the variable. The projection of the phasor onto the vertical axis represents the instantaneous value of the quantity it represents.

Active Figure 33.3b shows voltage and current phasors for the circuit of Active Figure 33.2 at some instant of time. The projections of the phasor arrows onto the vertical axis are determined by a sine function of the angle of the phasor with respect to the horizontal axis. For example, the projection of the current phasor in Active Figure 33.3b is I_{max} sin ωt. Notice that this expression is the same as Equation 33.1. Therefore, the projections of phasors represent current values that vary sinusoidally in time. We can do the same with time-varying voltages. The advantage of this approach is that the phase relationships among currents and voltages can be represented as vector additions of phasors using the vector addition techniques discussed in Chapter 3.

In the case of the single-loop resistive circuit of Active Figure 33.2, the current and voltage phasors lie along the same line in Active Figure 33.3b because i_R and Δv_R are in phase. The current and voltage in circuits containing capacitors and inductors have different phase relationships.

Quick Quiz 33.1 Consider the voltage phasor in Figure 33.4, shown at three instants of time. **(i)** Choose the part of the figure, (a), (b), or (c), that represents the instant of time at which the instantaneous value of the voltage has the largest magnitude. **(ii)** Choose the part of the figure that represents the instant of time at which the instantaneous value of the voltage has the smallest magnitude.

For the simple resistive circuit in Active Figure 33.2, notice that **the average value of the current over one cycle is zero.** That is, the current is maintained in the positive direction for the same amount of time and at the same magnitude as it is maintained in the negative direction. The direction of the current, however, has no effect on the behavior of the resistor. We can understand this concept by realizing that collisions between electrons and the fixed atoms of the resistor result in an increase in the resistor's temperature. Although this temperature increase depends on the magnitude of the current, it is independent of the current's direction.

We can make this discussion quantitative by recalling that the rate at which energy is delivered to a resistor is the power $\mathcal{P} = i^2R$, where i is the instantaneous current in the resistor. Because this rate is proportional to the square of the current, it makes no difference whether the current is direct or alternating, that is, whether the sign associated with the current is positive or negative. The temperature increase produced by an alternating current having a maximum value I_{max}, however, is not the same as that produced by a direct current equal to I_{max} because the alternating current has this maximum value for only an instant during each cycle (Fig. 33.5a, page 926). What is of importance in an AC circuit is an average

PITFALL PREVENTION 33.2

A Phasor Is Like a Graph

An alternating voltage can be presented in different representations. One graphical representation is shown in Figure 33.1 in which the voltage is drawn in rectangular coordinates, with voltage on the vertical axis and time on the horizontal axis. Active Figure 33.3b shows another graphical representation. The phase space in which the phasor is drawn is similar to polar coordinate graph paper. The radial coordinate represents the amplitude of the voltage. The angular coordinate is the phase angle. The vertical-axis coordinate of the tip of the phasor represents the instantaneous value of the voltage. The horizontal coordinate represents nothing at all. As shown in Active Figure 33.3b, alternating currents can also be represented by phasors.

To help with this discussion of phasors, review Section 15.4, where we represented the simple harmonic motion of a real object by the projection of an imaginary object's uniform circular motion onto a coordinate axis. A phasor is a direct analog to this representation.

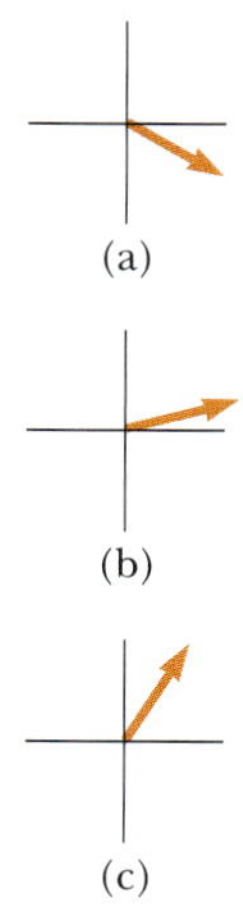

Figure 33.4 (Quick Quiz 33.1) A voltage phasor is shown at three instants of time, (a), (b), and (c).

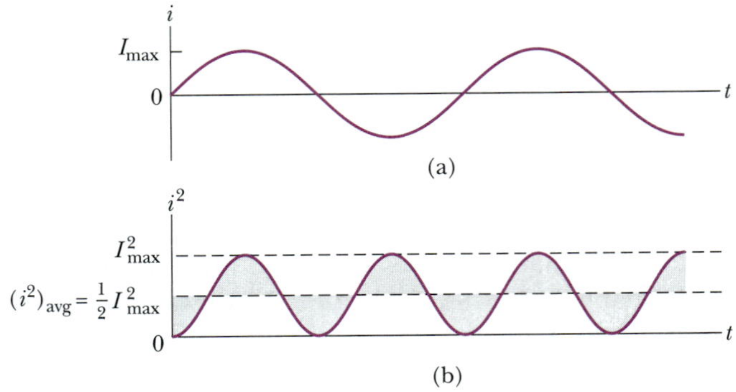

Figure 33.5 (a) Graph of the current in a resistor as a function of time. (b) Graph of the current squared in a resistor as a function of time. Notice that the gray shaded regions *under* the curve and *above* the dashed line for $\frac{1}{2}I_{\text{max}}^2$ have the same area as the gray shaded regions *above* the curve and *below* the dashed line for $\frac{1}{2}I_{\text{max}}^2$. Therefore, the average value of i^2 is $\frac{1}{2}I_{\text{max}}^2$. In general, the average value of $\sin^2 \omega t$ or $\cos^2 \omega t$ over one cycle is $\frac{1}{2}$.

value of current, referred to as the **rms current.** As we learned in Section 21.1, the notation *rms* stands for *root-mean-square,* which in this case means the square root of the mean (average) value of the square of the current: $I_{\text{rms}} = \sqrt{(i^2)_{\text{avg}}}$. Because i^2 varies as $\sin^2 \omega t$ and because the average value of i^2 is $\frac{1}{2}I_{\text{max}}^2$ (see Fig. 33.5b), the rms current is

rms current ▶

$$I_{\text{rms}} = \frac{I_{\text{max}}}{\sqrt{2}} = 0.707 I_{\text{max}} \qquad \textbf{(33.4)}$$

This equation states that an alternating current whose maximum value is 2.00 A delivers to a resistor the same power as a direct current that has a value of (0.707)(2.00 A) = 1.41 A. The average power delivered to a resistor that carries an alternating current is

Average power delivered to a resistor ▶

$$\mathcal{P}_{\text{avg}} = I_{\text{rms}}^2 R$$

Alternating voltage is also best discussed in terms of rms voltage, and the relationship is identical to that for current:

rms voltage ▶

$$\Delta V_{\text{rms}} = \frac{\Delta V_{\text{max}}}{\sqrt{2}} = 0.707\, \Delta V_{\text{max}} \qquad \textbf{(33.5)}$$

When we speak of measuring a 120-V alternating voltage from an electrical outlet, we are referring to an rms voltage of 120 V. A calculation using Equation 33.5 shows that such an alternating voltage has a maximum value of about 170 V. One reason rms values are often used when discussing alternating currents and voltages is that AC ammeters and voltmeters are designed to read rms values. Furthermore, with rms values, many of the equations we use have the same form as their direct current counterparts.

EXAMPLE 33.1 What Is the rms Current?

The voltage output of an AC source is given by the expression $\Delta v = (200 \text{ V})\sin \omega t$. Find the rms current in the circuit when this source is connected to a 100-Ω resistor.

SOLUTION

Conceptualize Active Figure 33.2 shows the physical situation for this problem.

Categorize We evaluate the current with an equation developed in this section, so we categorize this example as a substitution problem.

Comparing this expression for voltage output with the general form $\Delta v = \Delta V_{max} \sin \omega t$ shows that $\Delta V_{max} = 200$ V. Calculate the rms voltage from Equation 33.5:

$$\Delta V_{rms} = \frac{\Delta V_{max}}{\sqrt{2}} = \frac{200 \text{ V}}{\sqrt{2}} = 141 \text{ V}$$

Find the rms current:

$$I_{rms} = \frac{\Delta V_{rms}}{R} = \frac{141 \text{ V}}{100 \ \Omega} = 1.41 \text{ A}$$

33.3 Inductors in an AC Circuit

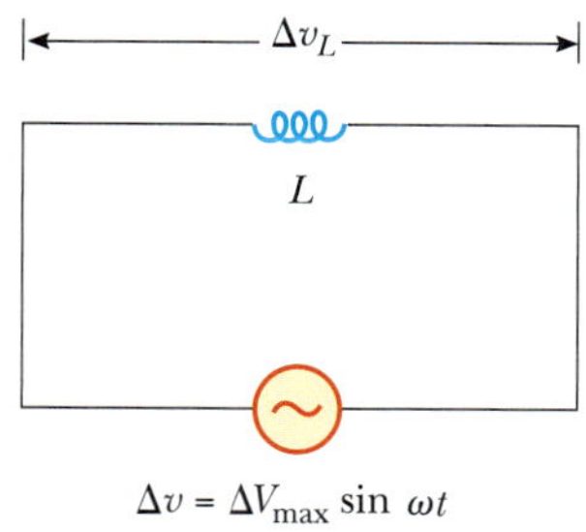

ACTIVE FIGURE 33.6

A circuit consisting of an inductor of inductance L connected to an AC source.

Sign in at www.thomsonedu.com and go to ThomsonNOW to adjust the inductance, frequency, and maximum voltage. The results can be studied with the graph and the phasor diagram in Active Figure 33.7.

Now consider an AC circuit consisting only of an inductor connected to the terminals of an AC source as shown in Active Figure 33.6. If $\Delta v_L = \mathcal{E}_L = -L(di_L/dt)$ is the self-induced instantaneous voltage across the inductor (see Eq. 32.1), Kirchhoff's loop rule applied to this circuit gives $\Delta v + \Delta v_L = 0$, or

$$\Delta v - L\frac{di_L}{dt} = 0$$

Substituting $\Delta V_{max} \sin \omega t$ for Δv and rearranging gives

$$\Delta v = L\frac{di_L}{dt} = \Delta V_{max} \sin \omega t \qquad (33.6)$$

Solving this equation for di_L gives

$$di_L = \frac{\Delta V_{max}}{L} \sin \omega t \, dt$$

Integrating this expression[1] gives the instantaneous current i_L in the inductor as a function of time:

$$i_L = \frac{\Delta V_{max}}{L}\int \sin \omega t \, dt = -\frac{\Delta V_{max}}{\omega L} \cos \omega t \qquad (33.7)$$

Using the trigonometric identity $\cos \omega t = -\sin(\omega t - \pi/2)$, we can express Equation 33.7 as

$$i_L = \frac{\Delta V_{max}}{\omega L} \sin\left(\omega t - \frac{\pi}{2}\right) \qquad (33.8)$$

◀ Current in an inductor

Comparing this result with Equation 33.6 shows that the instantaneous current i_L in the inductor and the instantaneous voltage Δv_L across the inductor are *out* of phase by $\pi/2$ rad = 90°.

A plot of voltage and current versus time is shown in Active Figure 33.7a (page 928). When the current i_L in the inductor is a maximum (point *b* in Active Fig. 33.7a), it is momentarily not changing, so the voltage across the inductor is zero (point *d*). At points such as *a* and *e*, the current is zero and the rate of change of current is at a maximum. Therefore, the voltage across the inductor is also at a maximum (points *c* and *f*). Notice that the voltage reaches its maximum value one quarter of a period before the current reaches its maximum value. Therefore, **for a sinusoidal applied voltage, the current in an inductor always lags behind the voltage across the inductor by 90° (one-quarter cycle in time).**

As with the relationship between current and voltage for a resistor, we can represent this relationship for an inductor with a phasor diagram as in Active Figure 33.7b. The phasors are at 90° to each other, representing the 90° phase difference between current and voltage.

[1] We neglect the constant of integration here because it depends on the initial conditions, which are not important for this situation.

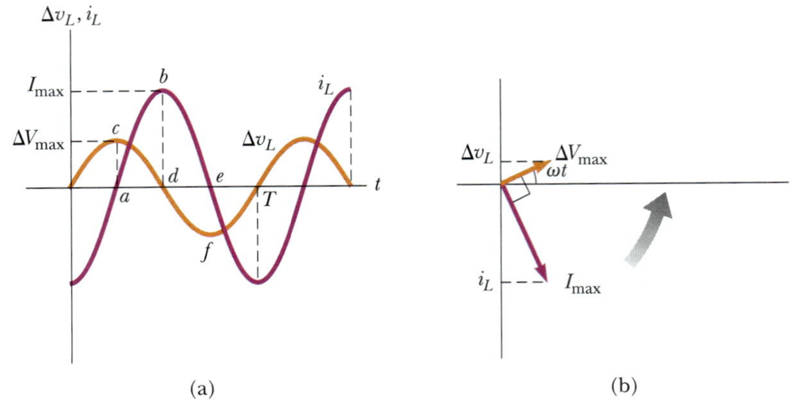

ACTIVE FIGURE 33.7

(a) Plots of the instantaneous current i_L and instantaneous voltage Δv_L across an inductor as functions of time. The current lags behind the voltage by 90°. (b) Phasor diagram for the inductive circuit, showing that the current lags behind the voltage by 90°.

Sign in at www.thomsonedu.com and go to ThomsonNOW to adjust the inductance, frequency, and maximum voltage of the circuit in Active Figure 33.6. The results can be studied with the graph and the phasor diagram in this figure.

Equation 33.7 shows that the current in an inductive circuit reaches its maximum value when $\cos \omega t = \pm 1$:

Maximum current in an inductor ▶

$$I_{max} = \frac{\Delta V_{max}}{\omega L} \quad (33.9)$$

This expression is similar to the relationship between current, voltage, and resistance in a DC circuit, $I = \Delta V/R$ (Eq. 27.7). Because I_{max} has units of amperes and ΔV_{max} has units of volts, ωL must have units of ohms. Therefore, ωL has the same units as resistance and is related to current and voltage in the same way as resistance. It must behave in a manner similar to resistance in the sense that it represents opposition to the flow of charge. Because ωL depends on the applied frequency ω, the inductor *reacts* differently, in terms of offering opposition to current, for different frequencies. For this reason, we define ωL as the **inductive reactance** X_L:

Inductive reactance ▶

$$X_L \equiv \omega L \quad (33.10)$$

Therefore, we can write Equation 33.9 as

$$I_{max} = \frac{\Delta V_{max}}{X_L} \quad (33.11)$$

The expression for the rms current in an inductor is similar to Equation 33.9, with I_{max} replaced by I_{rms} and ΔV_{max} replaced by ΔV_{rms}.

Equation 33.10 indicates that, for a given applied voltage, the inductive reactance increases as the frequency increases. This conclusion is consistent with Faraday's law: the greater the rate of change of current in the inductor, the larger the back emf. The larger back emf translates to an increase in the reactance and a decrease in the current.

Using Equations 33.6 and 33.11, we find that the instantaneous voltage across the inductor is

Voltage across an inductor ▶

$$\Delta v_L = -L\frac{di_L}{dt} = -\Delta V_{max} \sin \omega t = -I_{max} X_L \sin \omega t \quad (33.12)$$

Quick Quiz 33.2 Consider the AC circuit in Figure 33.8. The frequency of the AC source is adjusted while its voltage amplitude is held constant. When does the lightbulb glow the brightest? (a) It glows brightest at high frequencies. (b) It glows brightest at low frequencies. (c) The brightness is the same at all frequencies.

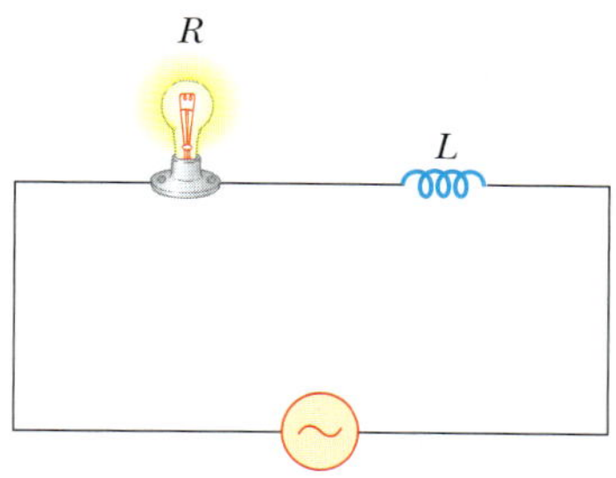

Figure 33.8 (Quick Quiz 33.2) At what frequencies does the lightbulb glow the brightest?

EXAMPLE 33.2 A Purely Inductive AC Circuit

In a purely inductive AC circuit, $L = 25.0$ mH and the rms voltage is 150 V. Calculate the inductive reactance and rms current in the circuit if the frequency is 60.0 Hz.

SOLUTION

Conceptualize Active Figure 33.6 shows the physical situation for this problem.

Categorize We evaluate the reactance and the current from equations developed in this section, so we categorize this example as a substitution problem.

Use Equation 33.10 to find the inductive reactance:

$$X_L = \omega L = 2\pi f L = 2\pi(60.0\text{ Hz})(25.0 \times 10^{-3}\text{ H})$$
$$= 9.42\ \Omega$$

Use an rms version of Equation 33.11 to find the rms current:

$$I_{\text{rms}} = \frac{\Delta V_{\text{rms}}}{X_L} = \frac{150\text{ V}}{9.42\ \Omega} = 15.9\text{ A}$$

What If? If the frequency increases to 6.00 kHz, what happens to the rms current in the circuit?

Answer If the frequency increases, the inductive reactance also increases because the current is changing at a higher rate. The increase in inductive reactance results in a lower current.

Let's calculate the new inductive reactance and the new rms current:

$$X_L = 2\pi(6.00 \times 10^3\text{ Hz})(25.0 \times 10^{-3}\text{ H}) = 942\ \Omega$$

$$I_{\text{rms}} = \frac{150\text{ V}}{942\ \Omega} = 0.159\text{ A}$$

33.4 Capacitors in an AC Circuit

Active Figure 33.9 shows an AC circuit consisting of a capacitor connected across the terminals of an AC source. Kirchhoff's loop rule applied to this circuit gives $\Delta v + \Delta v_C = 0$, or

$$\Delta v - \frac{q}{C} = 0 \tag{33.13}$$

Substituting $\Delta V_{\text{max}} \sin \omega t$ for Δv and rearranging gives

$$q = C\,\Delta V_{\text{max}} \sin \omega t \tag{33.14}$$

where q is the instantaneous charge on the capacitor. Differentiating Equation 33.14 with respect to time gives the instantaneous current in the circuit:

$$i_C = \frac{dq}{dt} = \omega C\,\Delta V_{\text{max}} \cos \omega t \tag{33.15}$$

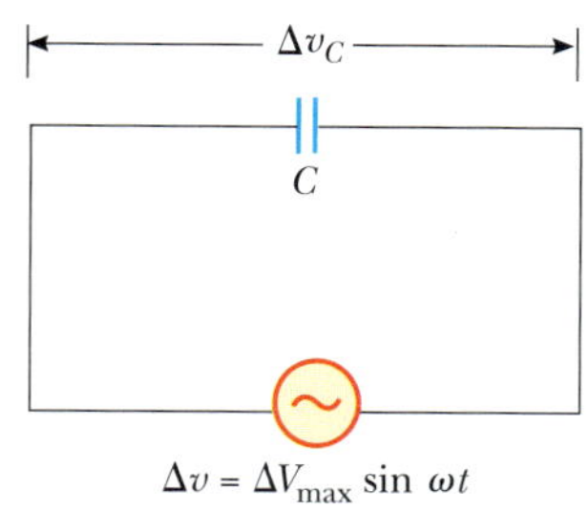

ACTIVE FIGURE 33.9

A circuit consisting of a capacitor of capacitance C connected to an AC source.

Sign in at www.thomsonedu.com and go to ThomsonNOW to adjust the capacitance, frequency, and maximum voltage. The results can be studied with the graph and the phasor diagram in Active Figure 33.10.

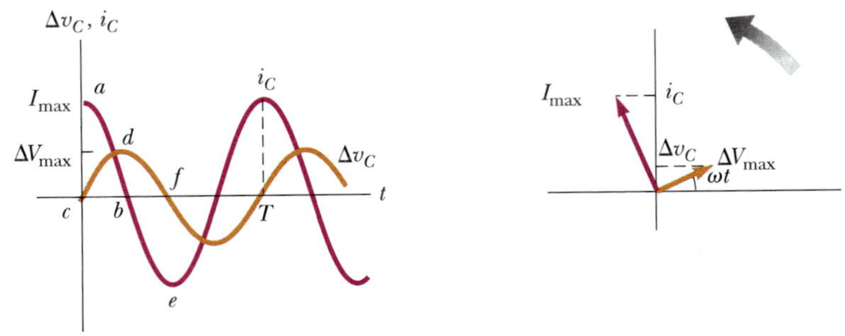

ACTIVE FIGURE 33.10

(a) Plots of the instantaneous current i_C and instantaneous voltage Δv_C across a capacitor as functions of time. The voltage lags behind the current by 90°. (b) Phasor diagram for the capacitive circuit, showing that the current leads the voltage by 90°.

Sign in at www.thomsonedu.com and go to ThomsonNOW to adjust the capacitance, frequency, and maximum voltage of the circuit in Active Figure 33.9. The results can be studied with the graph and the phasor diagram in this figure.

Using the trigonometric identity

$$\cos \omega t = \sin\left(\omega t + \frac{\pi}{2}\right)$$

we can express Equation 33.15 in the alternative form

Current in a capacitor ▶

$$i_C = \omega C \, \Delta V_{max} \sin\left(\omega t + \frac{\pi}{2}\right) \tag{33.16}$$

Comparing this expression with $\Delta v = \Delta V_{max} \sin \omega t$ shows that the current is $\pi/2$ rad $= 90°$ out of phase with the voltage across the capacitor. A plot of current and voltage versus time (Active Fig. 33.10a) shows that the current reaches its maximum value one-quarter of a cycle sooner than the voltage reaches its maximum value.

Consider a point such as *b* where the current is zero at this instant. That occurs when the capacitor reaches its maximum charge so that the voltage across the capacitor is a maximum (point *d*). At points such as *a* and *e*, the current is a maximum, which occurs at those instants when the charge on the capacitor reaches zero and the capacitor begins to recharge with the opposite polarity. When the charge is zero, the voltage across the capacitor is zero (points *c* and *f*). Therefore, the current and voltage are out of phase.

As with inductors, we can represent the current and voltage for a capacitor on a phasor diagram. The phasor diagram in Active Figure 33.10b shows that **for a sinusoidally applied voltage, the current always leads the voltage across a capacitor by 90°.**

Equation 33.15 shows that the current in the circuit reaches its maximum value when $\cos \omega t = \pm 1$:

$$I_{max} = \omega C \, \Delta V_{max} = \frac{\Delta V_{max}}{(1/\omega C)} \tag{33.17}$$

As in the case with inductors, this looks like Equation 27.7, so the denominator plays the role of resistance, with units of ohms. We give the combination $1/\omega C$ the symbol X_C, and because this function varies with frequency, we define it as the **capacitive reactance:**

Capacitive reactance ▶

$$X_C \equiv \frac{1}{\omega C} \tag{33.18}$$

We can now write Equation 33.17 as

Maximum current in a capacitor ▶

$$I_{max} = \frac{\Delta V_{max}}{X_C} \tag{33.19}$$

The rms current is given by an expression similar to Equation 33.19, with I_{max} replaced by I_{rms} and ΔV_{max} replaced by ΔV_{rms}.

Using Equation 33.19, we can express the instantaneous voltage across the capacitor as

$$\Delta v_C = \Delta V_{max} \sin \omega t = I_{max} X_C \sin \omega t \quad (33.20)$$

◀ Voltage across a capacitor

Equations 33.18 and 33.19 indicate that as the frequency of the voltage source increases, the capacitive reactance decreases and the maximum current therefore increases. The frequency of the current is determined by the frequency of the voltage source driving the circuit. As the frequency approaches zero, the capacitive reactance approaches infinity and the current therefore approaches zero. This conclusion makes sense because the circuit approaches direct current conditions as ω approaches zero and the capacitor represents an open circuit.

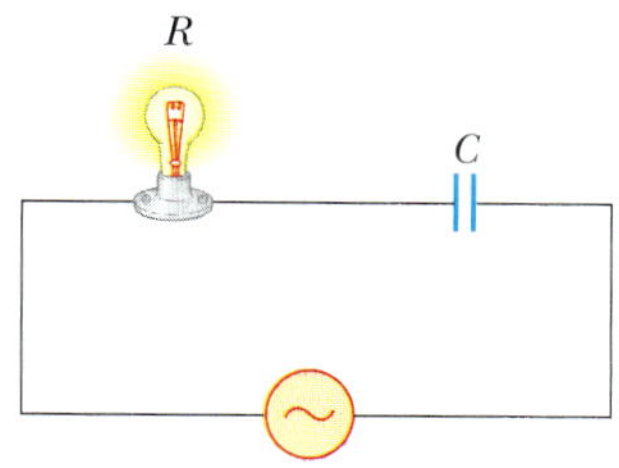

Figure 33.11 (Quick Quiz 33.3)

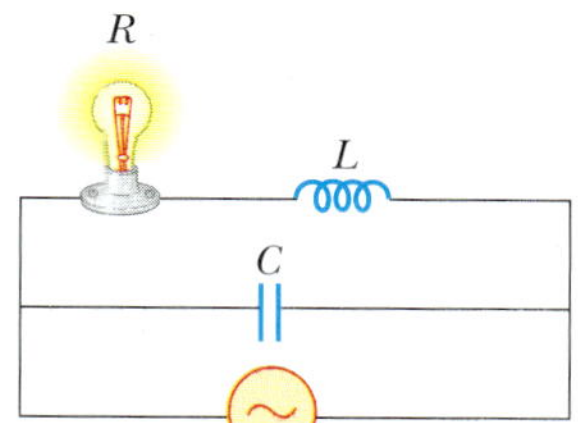

Figure 33.12 (Quick Quiz 33.4)

Quick Quiz 33.3 Consider the AC circuit in Figure 33.11. The frequency of the AC source is adjusted while its voltage amplitude is held constant. When does the lightbulb glow the brightest? (a) It glows brightest at high frequencies. (b) It glows brightest at low frequencies. (c) The brightness is the same at all frequencies.

Quick Quiz 33.4 Consider the AC circuit in Figure 33.12. The frequency of the AC source is adjusted while its voltage amplitude is held constant. When does the lightbulb glow the brightest? (a) It glows brightest at high frequencies. (b) It glows brightest at low frequencies. (c) The brightness is the same at all frequencies.

EXAMPLE 33.3 A Purely Capacitive AC Circuit

An 8.00-μF capacitor is connected to the terminals of a 60.0-Hz AC source whose rms voltage is 150 V. Find the capacitive reactance and the rms current in the circuit.

SOLUTION

Conceptualize Active Figure 33.9 shows the physical situation for this problem.

Categorize We evaluate the reactance and the current from equations developed in this section, so we categorize this example as a substitution problem.

Use Equation 33.18 to find the capacitive reactance:

$$X_C = \frac{1}{\omega C} = \frac{1}{2\pi f C} = \frac{1}{2\pi(60\ \text{Hz})(8.00 \times 10^{-6}\ \text{F})} = 332\ \Omega$$

Use an rms version of Equation 33.19 to find the rms current:

$$I_{rms} = \frac{\Delta V_{rms}}{X_C} = \frac{150\ \text{V}}{332\ \Omega} = 0.452\ \text{A}$$

What If? What if the frequency is doubled? What happens to the rms current in the circuit?

Answer If the frequency increases, the capacitive reactance decreases, which is just the opposite from the case of an inductor. The decrease in capacitive reactance results in an increase in the current.

Let's calculate the new capacitive reactance and the new rms current:

$$X_C = \frac{1}{\omega C} = \frac{1}{2\pi(120\ \text{Hz})(8.00 \times 10^{-6}\ \text{F})} = 166\ \Omega$$

$$I_{rms} = \frac{150\ \text{V}}{166\ \Omega} = 0.904\ \text{A}$$

33.5 The *RLC* Series Circuit

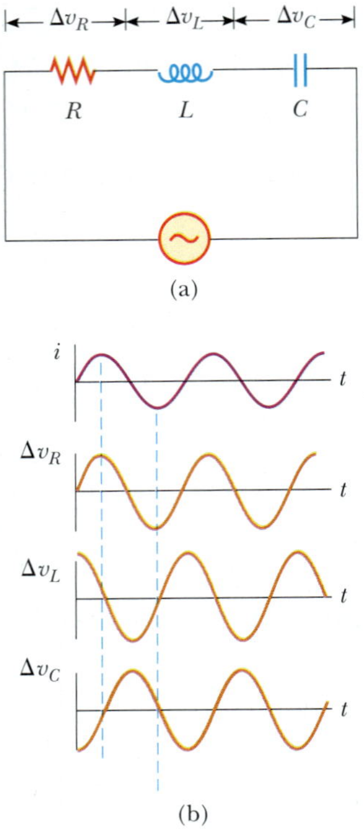

ACTIVE FIGURE 33.13

(a) A series circuit consisting of a resistor, an inductor, and a capacitor connected to an AC source. (b) Phase relationships for instantaneous voltages in the series *RLC* circuit.

Sign in at www.thomsonedu.com and go to ThomsonNOW to adjust the resistance, inductance, and capacitance. The results can be studied with the graph in this figure and the phasor diagram in Active Figure 33.15.

Active Figure 33.13a shows a circuit that contains a resistor, an inductor, and a capacitor connected in series across an alternating voltage source. If the applied voltage varies sinusoidally with time, the instantaneous applied voltage is

$$\Delta v = \Delta V_{\max} \sin \omega t$$

while the current varies as

$$i = I_{\max} \sin (\omega t - \phi)$$

where ϕ is some **phase angle** between the current and the applied voltage. Based on our discussions of phase in Sections 33.3 and 33.4, we expect that the current will generally not be in phase with the voltage in an *RLC* circuit. Our aim is to determine ϕ and $I_{\max}$. Active Figure 33.13b shows the voltage versus time across each element in the circuit and their phase relationships.

First, because the elements are in series, the current everywhere in the circuit must be the same at any instant. That is, **the current at all points in a series AC circuit has the same amplitude and phase.** Based on the preceding sections, we know that the voltage across each element has a different amplitude and phase. In particular, the voltage across the resistor is in phase with the current, the voltage across the inductor leads the current by 90°, and the voltage across the capacitor lags behind the current by 90°. Using these phase relationships, we can express the instantaneous voltages across the three circuit elements as

$$\Delta v_R = I_{\max} R \sin \omega t = \Delta V_R \sin \omega t \tag{33.21}$$

$$\Delta v_L = I_{\max} X_L \sin \left(\omega t + \frac{\pi}{2} \right) = \Delta V_L \cos \omega t \tag{33.22}$$

$$\Delta v_C = I_{\max} X_C \sin \left(\omega t - \frac{\pi}{2} \right) = -\Delta V_C \cos \omega t \tag{33.23}$$

The sum of these three voltages must equal the voltage from the AC source, but it is important to recognize that because the three voltages have different phase relationships with the current, they cannot be added directly. Figure 33.14 represents the phasors at an instant at which the current in all three elements is momentarily zero. The zero current is represented by the current phasor along the horizontal axis in each part of the figure. Next the voltage phasor is drawn at the appropriate phase angle to the current for each element.

Because phasors are rotating vectors, the voltage phasors in Figure 33.14 can be combined using vector addition as in Active Figure 33.15. In Active Figure 33.15a, the voltage phasors in Figure 33.14 are combined on the same coordinate axes. Active Figure 33.15b shows the vector addition of the voltage phasors. The voltage phasors ΔV_L and ΔV_C are in *opposite* directions along the same line, so we can construct the difference phasor $\Delta V_L - \Delta V_C$, which is perpendicular to the phasor ΔV_R. This diagram shows that the vector sum of the voltage amplitudes ΔV_R, ΔV_L, and ΔV_C equals a phasor whose length is the maximum applied voltage $\Delta V_{\max}$ and

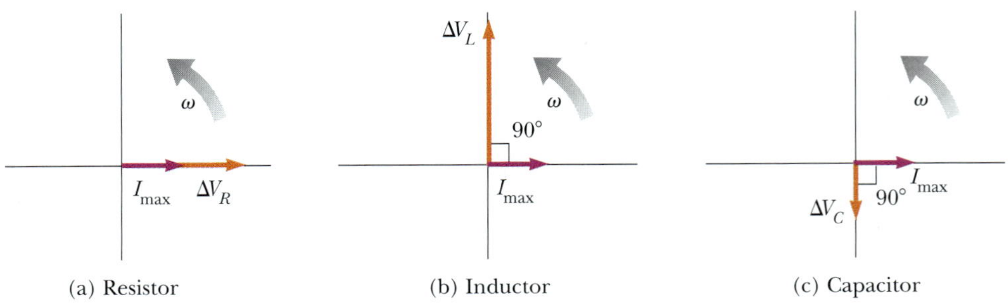

Figure 33.14 Phase relationships between the voltage and current phasors for (a) a resistor, (b) an inductor, and (c) a capacitor connected in series.

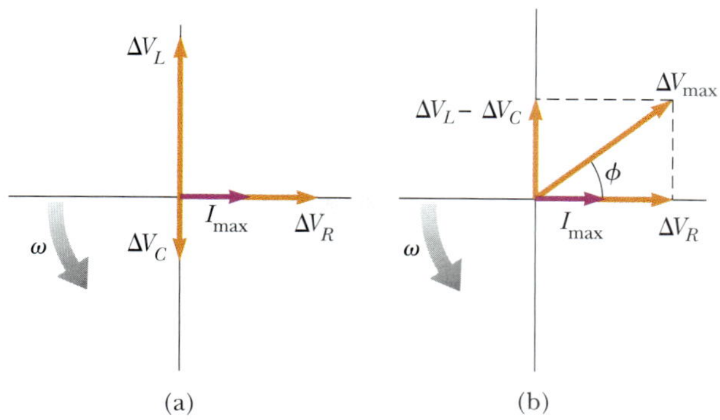

ACTIVE FIGURE 33.15

(a) Phasor diagram for the series *RLC* circuit shown in Active Figure 33.13a. The phasor ΔV_R is in phase with the current phasor I_{max}, the phasor ΔV_L leads I_{max} by 90°, and the phasor ΔV_C lags I_{max} by 90°. (b) The inductance and capacitance phasors are added together and then added vectorially to the resistance phasor. The total voltage ΔV_{max} makes an angle ϕ with I_{max}.

Sign in at www.thomsonedu.com and go to ThomsonNOW to adjust the resistance, inductance, and capacitance of the circuit in Active Figure 33.13a. The results can be studied with the graphs in Active Figure 33.13b and the phasor diagram in this figure.

which makes an angle ϕ with the current phasor I_{max}. From the right triangle in Active Figure 33.15b, we see that

$$\Delta V_{max} = \sqrt{\Delta V_R{}^2 + (\Delta V_L - \Delta V_C)^2} = \sqrt{(I_{max}R)^2 + (I_{max}X_L - I_{max}X_C)^2}$$

$$\Delta V_{max} = I_{max}\sqrt{R^2 + (X_L - X_C)^2}$$

Therefore, we can express the maximum current as

$$I_{max} = \frac{\Delta V_{max}}{\sqrt{R^2 + (X_L - X_C)^2}} \quad \textbf{(33.24)}$$

◀ Maximum current in an *RLC* circuit

Once again, this expression has the same mathematical form as Equation 27.7. The denominator of the fraction plays the role of resistance and is called the **impedance** Z of the circuit:

$$Z \equiv \sqrt{R^2 + (X_L - X_C)^2} \quad \textbf{(33.25)}$$

◀ Impedance

where impedance also has units of ohms. Therefore, Equation 33.24 can be written in the form

$$I_{max} = \frac{\Delta V_{max}}{Z} \quad \textbf{(33.26)}$$

Equation 33.26 is the AC equivalent of Equation 27.7. Note that the impedance and therefore the current in an AC circuit depend on the resistance, the inductance, the capacitance, and the frequency (because the reactances are frequency dependent).

From the right triangle in the phasor diagram in Active Figure 33.15b, the phase angle ϕ between the current and the voltage is found as follows:

$$\phi = \tan^{-1}\left(\frac{\Delta V_L - \Delta V_C}{\Delta V_R}\right) = \tan^{-1}\left(\frac{I_{max}X_L - I_{max}X_C}{I_{max}R}\right)$$

$$\phi = \tan^{-1}\left(\frac{X_L - X_C}{R}\right) \quad \textbf{(33.27)}$$

◀ Phase angle

When $X_L > X_C$ (which occurs at high frequencies), the phase angle is positive, signifying that the current lags the applied voltage as in Active Figure 33.15b. We describe this situation by saying that the circuit is *more inductive than capacitive.* When $X_L < X_C$, the phase angle is negative, signifying that the current leads the applied voltage, and the circuit is *more capacitive than inductive.* When $X_L = X_C$, the phase angle is zero and the circuit is *purely resistive.*

Quick Quiz 33.5 Label each part of Figure 33.16, (a), (b), and (c), as representing $X_L > X_C$, $X_L = X_C$, or $X_L < X_C$.

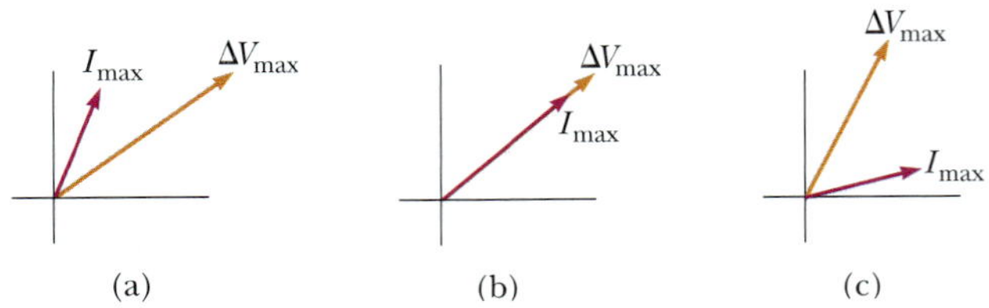

Figure 33.16 (Quick Quiz 33.5) Match the phasor diagrams to the relationships between the reactances.

EXAMPLE 33.4 Analyzing a Series *RLC* Circuit

A series *RLC* circuit has $R = 425\ \Omega$, $L = 1.25$ H, $C = 3.50\ \mu$F. It is connected to an AC source with $f = 60.0$ Hz and $\Delta V_{max} = 150$ V.

(A) Determine the inductive reactance, the capacitive reactance, and the impedance of the circuit.

SOLUTION

Conceptualize The circuit of interest in this example is shown in Active Figure 33.13a. The current in the combination of the resistor, inductor, and capacitor oscillates at a particular phase angle with respect to the applied voltage.

Categorize The circuit is a simple series *RLC* circuit, so we can use the approach discussed in this section.

Analyze Find the angular frequency:

$$\omega = 2\pi f = 2\pi(60.0\ \text{Hz}) = 377\ \text{s}^{-1}$$

Use Equation 33.10 to find the inductive reactance:

$$X_L = \omega L = (377\ \text{s}^{-1})(1.25\ \text{H}) = \boxed{471\ \Omega}$$

Use Equation 33.18 to find the capacitive reactance:

$$X_C = \frac{1}{\omega C} = \frac{1}{(377\ \text{s}^{-1})(3.50 \times 10^{-6}\ \text{F})} = \boxed{758\ \Omega}$$

Use Equation 33.25 to find the impedance:

$$Z = \sqrt{R^2 + (X_L - X_C)^2}$$

$$= \sqrt{(425\ \Omega)^2 + (471\ \Omega - 758\ \Omega)^2} = \boxed{513\ \Omega}$$

(B) Find the maximum current in the circuit.

SOLUTION

Use Equation 33.26 to find the maximum current:

$$I_{max} = \frac{\Delta V_{max}}{Z} = \frac{150\ \text{V}}{513\ \Omega} = \boxed{0.292\ \text{A}}$$

(C) Find the phase angle between the current and voltage.

SOLUTION

Use Equation 33.27 to calculate the phase angle:

$$\phi = \tan^{-1}\left(\frac{X_L - X_C}{R}\right) = \tan^{-1}\left(\frac{471\ \Omega - 758\ \Omega}{425\ \Omega}\right) = \boxed{-34.0°}$$

(D) Find the maximum voltage across each element.

SOLUTION

Use Equations 33.2, 33.11, and 33.19 to calculate the maximum voltages:

$$\Delta V_R = I_{max} R = (0.292\ \text{A})(425\ \Omega) = \boxed{124\ \text{V}}$$

$$\Delta V_L = I_{max} X_L = (0.292\ \text{A})(471\ \Omega) = \boxed{138\ \text{V}}$$

$$\Delta V_C = I_{max} X_C = (0.292\ \text{A})(758\ \Omega) = \boxed{221\ \text{V}}$$

(E) What replacement value of L should an engineer analyzing the circuit choose such that the current leads the applied voltage by 30.0°? All other values in the circuit stay the same.

SOLUTION

Solve Equation 33.27 for the inductive reactance:

$$X_L = X_C + R\tan\phi$$

Substitute Equations 33.10 and 33.18 into this expression:

$$\omega L = \frac{1}{\omega C} + R\tan\phi$$

Solve for L:

$$L = \frac{1}{\omega}\left(\frac{1}{\omega C} + R\tan\phi\right)$$

Substitute the given values:

$$L = \frac{1}{(377\ \text{s}^{-1})}\left[\frac{1}{(377\ \text{s}^{-1})(3.50\times10^{-6}\ \text{F})} + (425\ \Omega)\tan(-30.0°)\right]$$

$$L = 1.36\ \text{H}$$

Finalize Because the capacitive reactance is larger than the inductive reactance, the circuit is more capacitive than inductive. In this case, the phase angle ϕ is negative, so the current leads the applied voltage.

Using Equations 33.21, 33.22, and 33.23, the instantaneous voltages across the three elements are

$$\Delta v_R = (124\ \text{V})\sin 377t$$

$$\Delta v_L = (138\ \text{V})\cos 377t$$

$$\Delta v_C = (-221\ \text{V})\cos 377t$$

What If? What if you added up the maximum voltages across the three circuit elements? Is that a physically meaningful quantity?

Answer The sum of the maximum voltages across the elements is $\Delta V_R + \Delta V_L + \Delta V_C = 483$ V. This sum is much greater than the maximum voltage of the source, 150 V. The sum of the maximum voltages is a meaningless quantity because when sinusoidally varying quantities are added, *both their amplitudes and their phases* must be taken into account. The maximum voltages across the various elements occur at different times. Therefore, the voltages must be added in a way that takes account of the different phases as shown in Active Figure 33.15.

33.6 Power in an AC Circuit

Now let's take an energy approach to analyzing AC circuits and consider the transfer of energy from the AC source to the circuit. The power delivered by a battery to an external DC circuit is equal to the product of the current and the terminal voltage of the battery. Likewise, the instantaneous power delivered by an AC source to a circuit is the product of the current and the applied voltage. For the RLC circuit shown in Active Figure 33.13a, we can express the instantaneous power $\mathcal{P}$ as

$$\mathcal{P} = i\Delta v = I_{\text{max}}\sin(\omega t - \phi)\,\Delta V_{\text{max}}\sin\omega t$$

$$\mathcal{P} = I_{\text{max}}\Delta V_{\text{max}}\sin\omega t\sin(\omega t - \phi) \quad \textbf{(33.28)}$$

This result is a complicated function of time and is therefore not very useful from a practical viewpoint. What is generally of interest is the average power over one or more cycles. Such an average can be computed by first using the trigonometric identity $\sin(\omega t - \phi) = \sin\omega t\cos\phi - \cos\omega t\sin\phi$. Substituting this identity into Equation 33.28 gives

$$\mathcal{P} = I_{\text{max}}\Delta V_{\text{max}}\sin^2\omega t\cos\phi - I_{\text{max}}\Delta V_{\text{max}}\sin\omega t\cos\omega t\sin\phi \quad \textbf{(33.29)}$$

Let's now take the time average of $\mathcal{P}$ over one or more cycles, noting that I_{max}, ΔV_{max}, ϕ, and ω are all constants. The time average of the first term on the right of the equal sign in Equation 33.29 involves the average value of $\sin^2 \omega t$, which is $\frac{1}{2}$. The time average of the second term on the right of the equal sign is identically zero because $\sin \omega t \cos \omega t = \frac{1}{2} \sin 2\omega t$, and the average value of $\sin 2\omega t$ is zero. Therefore, we can express the **average power** $\mathcal{P}_{avg}$ as

$$\mathcal{P}_{avg} = \tfrac{1}{2} I_{max} \Delta V_{max} \cos \phi \qquad (33.30)$$

It is convenient to express the average power in terms of the rms current and rms voltage defined by Equations 33.4 and 33.5:

Average power delivered to an *RLC* circuit ▶

$$\mathcal{P}_{avg} = I_{rms} \Delta V_{rms} \cos \phi \qquad (33.31)$$

where the quantity $\cos \phi$ is called the **power factor.** Active Figure 33.15b shows that the maximum voltage across the resistor is given by $\Delta V_R = \Delta V_{max} \cos \phi = I_{max} R$. Using Equation 33.5 and $\cos \phi = I_{max} R / \Delta V_{max}$, we can express $\mathcal{P}_{avg}$ as

$$\mathcal{P}_{avg} = I_{rms} \Delta V_{rms} \cos \phi = I_{rms} \left(\frac{\Delta V_{max}}{\sqrt{2}} \right) \frac{I_{max} R}{\Delta V_{max}} = I_{rms} \frac{I_{max} R}{\sqrt{2}}$$

Substituting $I_{max} = \sqrt{2} I_{rms}$ from Equation 33.4 gives

$$\mathcal{P}_{avg} = I_{rms}^2 R \qquad (33.32)$$

In words, **the average power delivered by the source is converted to internal energy in the resistor,** just as in the case of a DC circuit. When the load is purely resistive, $\phi = 0$, $\cos \phi = 1$, and, from Equation 33.31, we see that

$$\mathcal{P}_{avg} = I_{rms} \Delta V_{rms}$$

Note that **no power losses are associated with pure capacitors and pure inductors in an AC circuit.** To see why that is true, let's first analyze the power in an AC circuit containing only a source and a capacitor. When the current begins to increase in one direction in an AC circuit, charge begins to accumulate on the capacitor and a voltage appears across it. When this voltage reaches its maximum value, the energy stored in the capacitor as electric potential energy is $\frac{1}{2} C (\Delta V_{max})^2$. This energy storage, however, is only momentary. The capacitor is charged and discharged twice during each cycle: charge is delivered to the capacitor during two quarters of the cycle and is returned to the voltage source during the remaining two quarters. Therefore, **the average power supplied by the source is zero.** In other words, **no power losses occur in a capacitor in an AC circuit.**

Now consider the case of an inductor. When the current in an inductor reaches its maximum value, the energy stored in the inductor is a maximum and is given by $\frac{1}{2} L I_{max}^2$. When the current begins to decrease in the circuit, this stored energy in the inductor returns to the source as the inductor attempts to maintain the current in the circuit.

Equation 33.31 shows that the power delivered by an AC source to any circuit depends on the phase, a result that has many interesting applications. For example, a factory that uses large motors in machines, generators, or transformers has a large inductive load (because of all the windings). To deliver greater power to such devices in the factory without using excessively high voltages, technicians introduce capacitance in the circuits to shift the phase.

Quick Quiz 33.6 An AC source drives an *RLC* circuit with a fixed voltage amplitude. If the driving frequency is ω_1, the circuit is more capacitive than inductive and the phase angle is $-10°$. If the driving frequency is ω_2, the circuit is more inductive than capacitive and the phase angle is $+10°$. At what frequency is the largest amount of power delivered to the circuit? (a) It is largest at ω_1. (b) It is largest at ω_2. (c) The same amount of power is delivered at both frequencies.

EXAMPLE 33.5 Average Power in an *RLC* Series Circuit

Calculate the average power delivered to the series *RLC* circuit described in Example 33.4.

SOLUTION

Conceptualize Consider the circuit in Active Figure 33.13a and imagine energy being delivered to the circuit by the AC source. Review Example 33.4 for other details about this circuit.

Categorize We find the result by using equations developed in this section, so we categorize this example as a substitution problem.

Use Equation 33.5 and the maximum voltage from Example 33.4 to find the rms voltage from the source:

$$\Delta V_{\rm rms} = \frac{\Delta V_{\rm max}}{\sqrt{2}} = \frac{150\ \rm V}{\sqrt{2}} = 106\ \rm V$$

Similarly, find the rms current in the circuit:

$$I_{\rm rms} = \frac{I_{\rm max}}{\sqrt{2}} = \frac{0.292\ \rm A}{\sqrt{2}} = 0.206\ \rm A$$

Use Equation 33.31 to find the power delivered by the source:

$$\mathcal{P}_{\rm avg} = I_{\rm rms} V_{\rm rms} \cos\phi = (0.206\ \rm A)(106\ \rm V)\cos(-34.0°)$$
$$= 18.1\ \rm W$$

33.7 Resonance in a Series *RLC* Circuit

A series *RLC* circuit is said to be **in resonance** when the driving frequency is such that the rms current has its maximum value. In general, the rms current can be written

$$I_{\rm rms} = \frac{\Delta V_{\rm rms}}{Z} \tag{33.33}$$

where Z is the impedance. Substituting the expression for Z from Equation 33.25 into Equation 33.33 gives

$$I_{\rm rms} = \frac{\Delta V_{\rm rms}}{\sqrt{R^2 + (X_L - X_C)^2}} \tag{33.34}$$

Because the impedance depends on the frequency of the source, the current in the *RLC* circuit also depends on the frequency. The frequency ω_0 at which $X_L - X_C = 0$ is called the **resonance frequency** of the circuit. To find ω_0, we set $X_L = X_C$, which gives $\omega_0 L = 1/\omega_0 C$, or

$$\omega_0 = \frac{1}{\sqrt{LC}} \tag{33.35}$$

◀ Resonance frequency

This frequency also corresponds to the natural frequency of oscillation of an *LC* circuit (see Section 32.5). Therefore, the rms current in a series *RLC* circuit has its maximum value when the frequency of the applied voltage matches the natural oscillator frequency, which depends only on L and C. Furthermore, at the resonance frequency, the current is in phase with the applied voltage.

Quick Quiz 33.7 What is the impedance of a series *RLC* circuit at resonance? (a) larger than R (b) less than R (c) equal to R (d) impossible to determine

A plot of rms current versus frequency for a series *RLC* circuit is shown in Active Figure 33.17a. The data assume a constant $\Delta V_{\rm rms} = 5.0$ mV, $L = 5.0\ \mu$H, and $C = 2.0$ nF. The three curves correspond to three values of R. In each case,

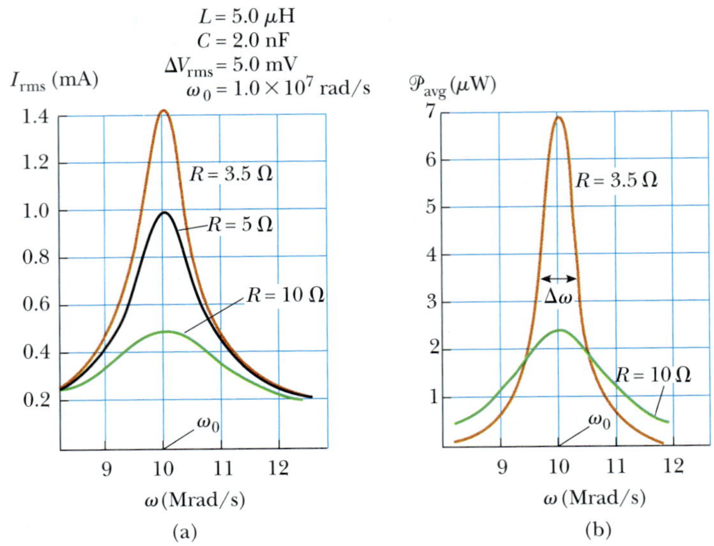

ACTIVE FIGURE 33.17

(a) The rms current versus frequency for a series *RLC* circuit for three values of *R*. The current reaches its maximum value at the resonance frequency ω_0. (b) Average power delivered to the circuit versus frequency for the series *RLC* circuit for two values of *R*.

Sign in at www.thomsonedu.com and go to ThomsonNOW to adjust the resistance, inductance, and capacitance of the circuit in Active Figure 33.13a. You can then determine the current and power for a given frequency or sweep through the frequencies to generate resonance curves as shown in this figure.

the rms current has its maximum value at the resonance frequency ω_0. Furthermore, the curves become narrower and taller as the resistance decreases.

Equation 33.34 shows that when $R = 0$, the current becomes infinite at resonance. Real circuits, however, always have some resistance, which limits the value of the current to some finite value.

We can also calculate the average power as a function of frequency for a series *RLC* circuit. Using Equations 33.32, 33.33, and 33.25 gives

$$\mathcal{P}_{\text{avg}} = I_{\text{rms}}^2 R = \frac{(\Delta V_{\text{rms}})^2}{Z^2} R = \frac{(\Delta V_{\text{rms}})^2 R}{R^2 + (X_L - X_C)^2} \tag{33.36}$$

Because $X_L = \omega L$, $X_C = 1/\omega C$, and $\omega_0^2 = 1/LC$, the term $(X_L - X_C)^2$ can be expressed as

$$(X_L - X_C)^2 = \left(\omega L - \frac{1}{\omega C}\right)^2 = \frac{L^2}{\omega^2}(\omega^2 - \omega_0^2)^2$$

Using this result in Equation 33.36 gives

Average power as a function of frequency in an *RLC* circuit ▶

$$\mathcal{P}_{\text{avg}} = \frac{(\Delta V_{\text{rms}})^2 R\omega^2}{R^2\omega^2 + L^2(\omega^2 - \omega_0^2)^2} \tag{33.37}$$

Equation 33.37 shows that **at resonance, when $\omega = \omega_0$, the average power is a maximum** and has the value $(\Delta V_{\text{rms}})^2/R$. Active Figure 33.17b is a plot of average power versus frequency for two values of *R* in a series *RLC* circuit. As the resistance is made smaller, the curve becomes sharper in the vicinity of the resonance frequency. This curve sharpness is usually described by a dimensionless parameter known as the **quality factor,**[2] denoted by *Q*:

Quality factor ▶

$$Q = \frac{\omega_0}{\Delta\omega}$$

[2] The quality factor is also defined as the ratio $2\pi E/\Delta E$, where E is the energy stored in the oscillating system and ΔE is the energy decrease per cycle of oscillation due to the resistance.

where $\Delta\omega$ is the width of the curve measured between the two values of ω for which $\mathcal{P}_{avg}$ has one-half its maximum value, called the *half-power points* (see Active Fig. 33.17b.) It is left as a problem (Problem 68) to show that the width at the half-power points has the value $\Delta\omega = R/L$ so that

$$Q = \frac{\omega_0 L}{R} \tag{33.38}$$

A radio's receiving circuit is an important application of a resonant circuit. The radio is tuned to a particular station (which transmits an electromagnetic wave or signal of a specific frequency) by varying a capacitor, which changes the receiving circuit's resonance frequency. When the circuit is driven by the electromagnetic oscillations a radio signal produces in an antenna, the tuner circuit responds with a large amplitude of electrical oscillation only for the station frequency that matches the resonance frequency. Therefore, only the signal from one radio station is passed on to the amplifier and loudspeakers even though signals from all stations are driving the circuit at the same time. Because many signals are often present over a range of frequencies, it is important to design a high-Q circuit to eliminate unwanted signals. In this manner, stations whose frequencies are near but not equal to the resonance frequency give signals at the receiver that are negligibly small relative to the signal that matches the resonance frequency.

EXAMPLE 33.6 A Resonating Series *RLC* Circuit

Consider a series *RLC* circuit for which $R = 150\ \Omega$, $L = 20.0$ mH, $\Delta V_{rms} = 20.0$ V, and $\omega = 5\ 000\ \text{s}^{-1}$. Determine the value of the capacitance for which the current is a maximum.

SOLUTION

Conceptualize Consider the circuit in Active Figure 33.13a and imagine varying the frequency of the AC source. The current in the circuit has its maximum value at the resonance frequency ω_0.

Categorize We find the result by using equations developed in this section, so we categorize this example as a substitution problem.

Use Equation 33.35 to solve for the required capacitance in terms of the resonance frequency:

$$\omega_0 = \frac{1}{\sqrt{LC}} \rightarrow C = \frac{1}{\omega_0^2 L}$$

Substitute numerical values:

$$C = \frac{1}{(5.00 \times 10^3\ \text{s}^{-1})^2 (20.0 \times 10^{-3}\ \text{H})} = 2.00\ \mu\text{F}$$

33.8 The Transformer and Power Transmission

As discussed in Section 27.6, it is economical to use a high voltage and a low current to minimize the I^2R loss in transmission lines when electric power is transmitted over great distances. Consequently, 350-kV lines are common, and in many areas, even higher-voltage (765-kV) lines are used. At the receiving end of such lines, the consumer requires power at a low voltage (for safety and for efficiency in design). In practice, the voltage is decreased to approximately 20 000 V at a distributing station, then to 4 000 V for delivery to residential areas, and finally to 120 V and 240 V at the customer's site. Therefore, a device is needed that can change the alternating voltage and current without causing appreciable changes in the power delivered. The AC transformer is that device.

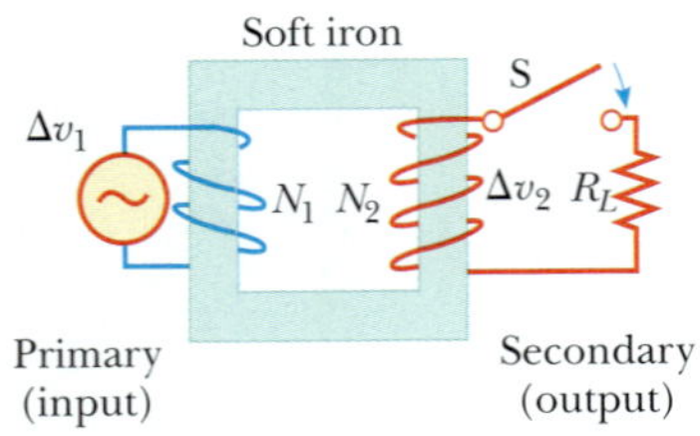

Figure 33.18 An ideal transformer consists of two coils wound on the same iron core. An alternating voltage Δv_1 is applied to the primary coil, and the output voltage Δv_2 is across the resistor of resistance R_L.

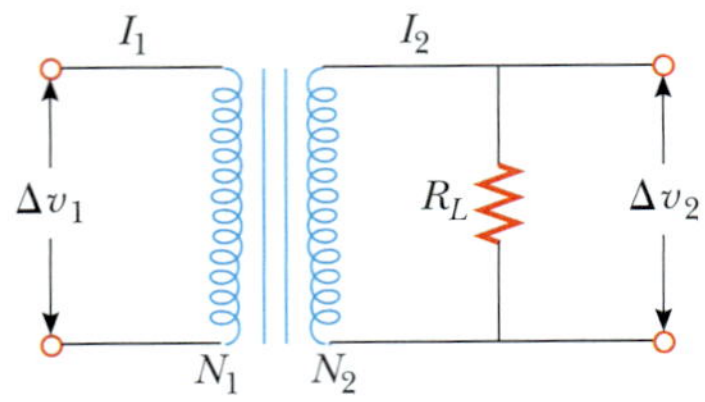

Figure 33.19 Circuit diagram for a transformer.

In its simplest form, the **AC transformer** consists of two coils of wire wound around a core of iron as illustrated in Figure 33.18. (Compare this arrangement to Faraday's experiment in Figure 31.2.) The coil on the left, which is connected to the input alternating voltage source and has N_1 turns, is called the *primary winding* (or the *primary*). The coil on the right, consisting of N_2 turns and connected to a load resistor R_L, is called the *secondary winding* (or the *secondary*). The purposes of the iron core are to increase the magnetic flux through the coil and to provide a medium in which nearly all the magnetic field lines through one coil pass through the other coil. Eddy-current losses are reduced by using a laminated core. Transformation of energy to internal energy in the finite resistance of the coil wires is usually quite small. Typical transformers have power efficiencies from 90% to 99%. In the discussion that follows, let's assume we are working with an *ideal transformer*, one in which the energy losses in the windings and core are zero.

Faraday's law states that the voltage Δv_1 across the primary is

$$\Delta v_1 = -N_1 \frac{d\Phi_B}{dt} \qquad \textbf{(33.39)}$$

where Φ_B is the magnetic flux through each turn. If we assume all magnetic field lines remain within the iron core, the flux through each turn of the primary equals the flux through each turn of the secondary. Hence, the voltage across the secondary is

$$\Delta v_2 = -N_2 \frac{d\Phi_B}{dt} \qquad \textbf{(33.40)}$$

Solving Equation 33.39 for $d\Phi_B/dt$ and substituting the result into Equation 33.40 gives

$$\Delta v_2 = \frac{N_2}{N_1}\Delta v_1 \qquad \textbf{(33.41)}$$

When $N_2 > N_1$, the output voltage Δv_2 exceeds the input voltage Δv_1. This configuration is referred to as a *step-up transformer*. When $N_2 < N_1$, the output voltage is less than the input voltage, and we have a *step-down transformer*.

When the switch in the secondary circuit is closed, a current I_2 is induced in the secondary. (In this discussion, uppercase I and ΔV refer to rms values.) If the load in the secondary circuit is a pure resistance, the induced current is in phase with the induced voltage. The power supplied to the secondary circuit must be provided by the AC source connected to the primary circuit as shown in Figure 33.19. In an ideal transformer where there are no losses, the power $I_1\,\Delta V_1$ supplied by the source is equal to the power $I_2\,\Delta V_2$ in the secondary circuit. That is,

$$I_1 \Delta V_1 = I_2 \Delta V_2 \qquad \textbf{(33.42)}$$

The value of the load resistance R_L determines the value of the secondary current because $I_2 = \Delta V_2/R_L$. Furthermore, the current in the primary is $I_1 = \Delta V_1/R_{\text{eq}}$, where

$$R_{\text{eq}} = \left(\frac{N_1}{N_2}\right)^2 R_L \qquad \textbf{(33.43)}$$

is the equivalent resistance of the load resistance when viewed from the primary side. We see from this analysis that a transformer may be used to match resistances between the primary circuit and the load. In this manner, maximum power transfer can be achieved between a given power source and the load resistance. For example, a transformer connected between the 1-kΩ output of an audio amplifier and an 8-Ω speaker ensures that as much of the audio signal as possible is transferred into the speaker. In stereo terminology, this process is called *impedance matching*.

To operate properly, many common household electronic devices require low voltages. A small transformer that plugs directly into the wall like the one illus-

NIKOLA TESLA
American Physicist (1856–1943)
Tesla was born in Croatia, but he spent most of his professional life as an inventor in the United States. He was a key figure in the development of alternating-current electricity, high-voltage transformers, and the transport of electrical power using AC transmission lines. Tesla's viewpoint was at odds with the ideas of Thomas Edison, who committed himself to the use of direct current in power transmission. Tesla's AC approach won out.

Figure 33.20 The primary winding in this transformer is directly attached to the prongs of the plug. The secondary winding is connected to the power cord on the right, which runs to an electronic device. Many of these power-supply transformers also convert alternating current to direct current.

This transformer is smaller than the one in the opening photograph of this chapter. In addition, it is a step-down transformer. It drops the voltage from 4 000 V to 240 V for delivery to a group of residences.

trated in Figure 33.20 can provide the proper voltage. The photograph shows the two windings wrapped around a common iron core that is found inside all these little "black boxes." This particular transformer converts the 120-V AC in the wall socket to 12.5-V AC. (Can you determine the ratio of the numbers of turns in the two coils?) Some black boxes also make use of diodes to convert the alternating current to direct current. (See Section 33.9.)

EXAMPLE 33.7 The Economics of AC Power

An electricity-generating station needs to deliver energy at a rate of 20 MW to a city 1.0 km away. A common voltage for commercial power generators is 22 kV, but a step-up transformer is used to boost the voltage to 230 kV before transmission.

(A) If the resistance of the wires is 2.0 Ω and the energy costs are about 10¢/kWh, estimate what it costs the utility company for the energy converted to internal energy in the wires during one day.

SOLUTION

Conceptualize The resistance of the wires is in series with the resistance representing the load (homes and businesses). Therefore, there is a voltage drop in the wires, which means that some of the transmitted energy is converted to internal energy in the wires and never reaches the load.

Categorize This problem involves finding the power delivered to a resistive load in an AC circuit. Let's ignore any capacitive or inductive characteristics of the load and set the power factor equal to 1.

Analyze Calculate I_{rms} in the wires from Equation 33.31:

$$I_{\text{rms}} = \frac{\mathcal{P}_{\text{avg}}}{\Delta V_{\text{rms}}} = \frac{20 \times 10^6 \text{ W}}{230 \times 10^3 \text{ V}} = 87 \text{ A}$$

Determine the rate at which energy is delivered to the resistance in the wires from Equation 33.32:

$$\mathcal{P}_{\text{avg}} = I_{\text{rms}}^2 R = (87 \text{ A})^2 (2.0 \ \Omega) = 15 \text{ kW}$$

Calculate the energy T_{ET} delivered to the wires over the course of a day:

$$T_{\text{ET}} = \mathcal{P}_{\text{avg}} \Delta t = (15 \text{ kW})(24 \text{ h}) = 360 \text{ kWh}$$

Find the cost of this energy at a rate of 10¢/kWh:

$$\text{Cost} = (360 \text{ kWh})(\$0.10/\text{kWh}) = \boxed{\$36}$$

(B) Repeat the calculation for the situation in which the power plant delivers the energy at its original voltage of 22 kV.

SOLUTION

Calculate I_{rms} in the wires from Equation 33.31:

$$I_{\text{rms}} = \frac{\mathcal{P}_{\text{avg}}}{\Delta V_{\text{rms}}} = \frac{20 \times 10^6 \text{ W}}{22 \times 10^3 \text{ V}} = 910 \text{ A}$$

From Equation 33.32, determine the rate at which energy is delivered to the resistance in the wires:

$$\mathcal{P}_{\text{avg}} = I_{\text{rms}}^2 R = (910 \text{ A})^2(2.0 \ \Omega) = 1.7 \times 10^3 \text{ kW}$$

Calculate the energy delivered to the wires over the course of a day:

$$T_{\text{ET}} = \mathcal{P}_{\text{avg}} \Delta t = (1.7 \times 10^3 \text{ kW})(24 \text{ h}) = 4.1 \times 10^4 \text{ kWh}$$

Find the cost of this energy at a rate of 10¢/kWh:

$$\text{Cost} = (4.1 \times 10^4 \text{ kWh})(\$0.10/\text{kWh}) = \$4.1 \times 10^3$$

Finalize Notice the tremendous savings that are possible through the use of transformers and high-voltage transmission lines. Such savings in combination with the efficiency of using alternating current to operate motors led to the universal adoption of alternating current instead of direct current for commercial power grids.

33.9 Rectifiers and Filters

Portable electronic devices such as radios and compact disc players are often powered by direct current supplied by batteries. Many devices come with AC–DC converters such as that shown in Figure 33.20. Such a converter contains a transformer that steps the voltage down from 120 V to, typically, 9 V and a circuit that converts alternating current to direct current. The AC–DC converting process is called **rectification,** and the converting device is called a **rectifier.**

The most important element in a rectifier circuit is a **diode,** a circuit element that conducts current in one direction but not the other. Most diodes used in modern electronics are semiconductor devices. The circuit symbol for a diode is —▶|—, where the arrow indicates the direction of the current in the diode. A diode has low resistance to current in one direction (the direction of the arrow) and high resistance to current in the opposite direction. To understand how a diode rectifies a current, consider Figure 33.21a, which shows a diode and a resistor connected to the secondary of a transformer. The transformer reduces the voltage from 120-V AC to the lower voltage that is needed for the device having a resistance R (the load resistance). Because the diode conducts current in only one direction, the alternating current in the load resistor is reduced to the form shown by the solid curve in Figure 33.21b. The diode conducts current only when the side of the symbol containing the arrowhead has a positive potential relative to the other side. In this situation, the diode acts as a *half-wave rectifier* because current is present in the circuit only during half of each cycle.

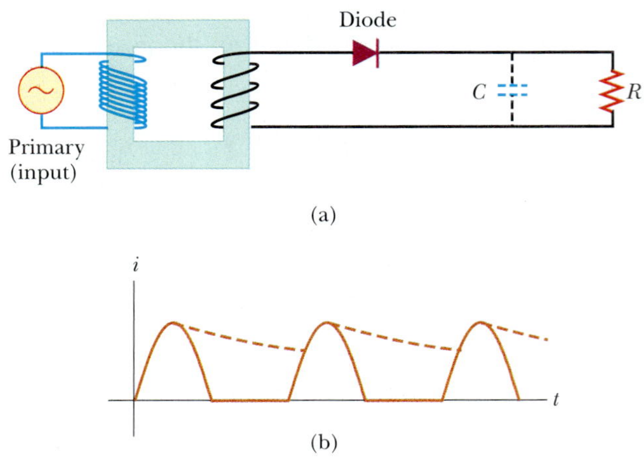

Figure 33.21 (a) A half-wave rectifier with an optional filter capacitor. (b) Current versus time in the resistor. The solid curve represents the current with no filter capacitor, and the dashed curve is the current when the circuit includes the capacitor.

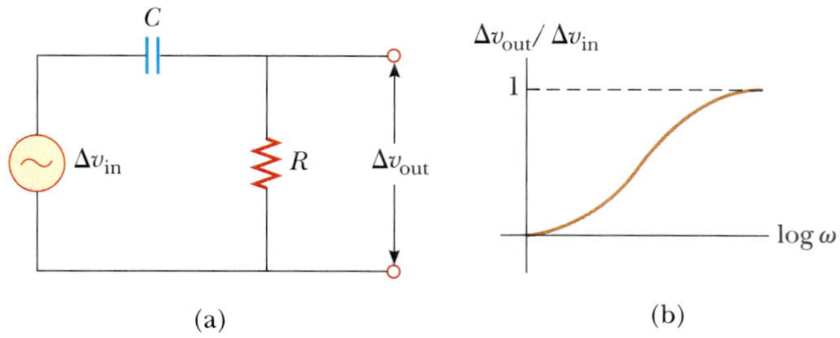

ACTIVE FIGURE 33.22

(a) A simple *RC* high-pass filter. (b) Ratio of output voltage to input voltage for an *RC* high-pass filter as a function of the angular frequency of the AC source.

Sign in at www.thomsonedu.com and go to ThomsonNOW to adjust the resistance and capacitance of the circuit in (a). You can then determine the output voltage for a given frequency or sweep through the frequencies to generate a curve like that in (b).

When a capacitor is added to the circuit as shown by the dashed lines and the capacitor symbol in Figure 33.21a, the circuit is a simple DC power supply. The time variation of the current in the load resistor (the dashed curve in Fig. 33.21b) is close to being zero, as determined by the *RC* time constant of the circuit. As the current in the circuit begins to rise at $t = 0$ in Figure 33.21b, the capacitor charges up. When the current begins to fall, however, the capacitor discharges through the resistor, so the current in the resistor does not fall as quickly as the current from the transformer.

The *RC* circuit in Figure 33.21a is one example of a **filter circuit,** which is used to smooth out or eliminate a time-varying signal. For example, radios are usually powered by a 60-Hz alternating voltage. After rectification, the voltage still contains a small AC component at 60 Hz (sometimes called *ripple*), which must be filtered. By "filtered," we mean that the 60-Hz ripple must be reduced to a value much less than that of the audio signal to be amplified because without filtering, the resulting audio signal includes an annoying hum at 60 Hz.

We can also design filters that respond differently to different frequencies. Consider the simple series *RC* circuit shown in Active Figure 33.22a. The input voltage is across the series combination of the two elements. The output is the voltage across the resistor. A plot of the ratio of the output voltage to the input voltage as a function of the logarithm of angular frequency (see Active Fig. 33.22b) shows that at low frequencies, Δv_{out} is much smaller than Δv_{in}, whereas at high frequencies, the two voltages are equal. Because the circuit preferentially passes signals of higher frequency while blocking low-frequency signals, the circuit is called an *RC* high-pass filter. (See Problem 45 for an analysis of this filter.)

Physically, a high-pass filter works because a capacitor "blocks out" direct current and AC current at low frequencies. At low frequencies, the capacitive reactance is large and much of the applied voltage appears across the capacitor rather than across the output resistor. As the frequency increases, the capacitive reactance drops and more of the applied voltage appears across the resistor.

Now consider the circuit shown in Active Figure 33.23a, where we have interchanged the resistor and capacitor and where the output voltage is taken across the capacitor. At low frequencies, the reactance of the capacitor and the voltage across the capacitor is high. As the frequency increases, the voltage across the capacitor drops. Therefore, this filter is an *RC* low-pass filter. The ratio of output voltage to input voltage (see Problem 46), plotted as a function of the logarithm of ω in Active Figure 33.23b, shows this behavior.

You may be familiar with crossover networks, which are an important part of the speaker systems for high-fidelity audio systems. These networks use low-pass filters to direct low frequencies to a special type of speaker, the "woofer," which is designed to reproduce the low notes accurately. The high frequencies are sent to the "tweeter" speaker.

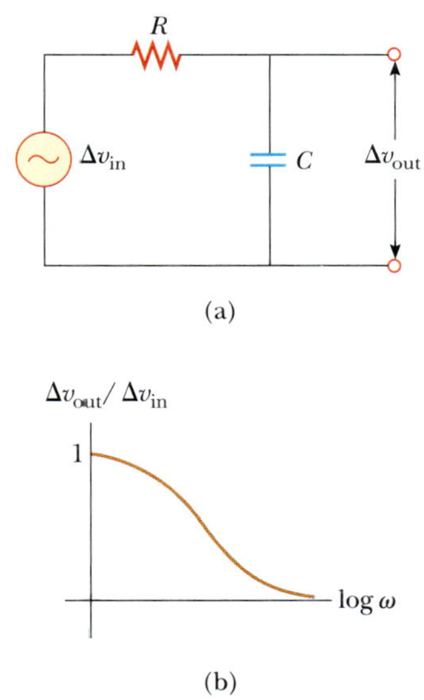

ACTIVE FIGURE 33.23

(a) A simple *RC* low-pass filter. (b) Ratio of output voltage to input voltage for an *RC* low-pass filter as a function of the angular frequency of the AC source.

Sign in at www.thomsonedu.com and go to ThomsonNOW to adjust the resistance and capacitance of the circuit in (a). You can then determine the output voltage for a given frequency or sweep through the frequencies to generate a curve like that in (b).

Summary

ThomsonNOW™ Sign in at **www.thomsonedu.com** and go to ThomsonNOW to take a practice test for this chapter.

DEFINITIONS

In AC circuits that contain inductors and capacitors, it is useful to define the **inductive reactance** X_L and the **capacitive reactance** X_C as

$$X_L \equiv \omega L \quad \textbf{(33.10)}$$

$$X_C \equiv \frac{1}{\omega C} \quad \textbf{(33.18)}$$

where ω is the angular frequency of the AC source. The SI unit of reactance is the ohm.

The **impedance** Z of an RLC series AC circuit is

$$Z \equiv \sqrt{R^2 + (X_L - X_C)^2} \quad \textbf{(33.25)}$$

This expression illustrates that we cannot simply add the resistance and reactances in a circuit. We must account for the applied voltage and current being out of phase, with the **phase angle** ϕ between the current and voltage being

$$\phi = \tan^{-1}\left(\frac{X_L - X_C}{R}\right) \quad \textbf{(33.27)}$$

The sign of ϕ can be positive or negative, depending on whether X_L is greater or less than X_C. The phase angle is zero when $X_L = X_C$.

CONCEPTS AND PRINCIPLES

The **rms current** and **rms voltage** in an AC circuit in which the voltages and current vary sinusoidally are given by

$$I_{\text{rms}} = \frac{I_{\max}}{\sqrt{2}} = 0.707 I_{\max} \quad \textbf{(33.4)}$$

$$\Delta V_{\text{rms}} = \frac{\Delta V_{\max}}{\sqrt{2}} = 0.707\,\Delta V_{\max} \quad \textbf{(33.5)}$$

where $I_{\max}$ and $\Delta V_{\max}$ are the maximum values.

If an AC circuit consists of a source and a resistor, the current is in phase with the voltage. That is, the current and voltage reach their maximum values at the same time.

If an AC circuit consists of a source and an inductor, the current lags the voltage by 90°. That is, the voltage reaches its maximum value one quarter of a period before the current reaches its maximum value.

If an AC circuit consists of a source and a capacitor, the current leads the voltage by 90°. That is, the current reaches its maximum value one quarter of a period before the voltage reaches its maximum value.

The **average power** delivered by the source in an RLC circuit is

$$\mathcal{P}_{\text{avg}} = I_{\text{rms}}\,\Delta V_{\text{rms}} \cos \phi \quad \textbf{(33.31)}$$

An equivalent expression for the average power is

$$\mathcal{P}_{\text{avg}} = I_{\text{rms}}^2 R \quad \textbf{(33.32)}$$

The average power delivered by the source results in increasing internal energy in the resistor. No power loss occurs in an ideal inductor or capacitor.

The rms current in a series RLC circuit is

$$I_{\text{rms}} = \frac{\Delta V_{\text{rms}}}{\sqrt{R^2 + (X_L - X_C)^2}} \quad \textbf{(33.34)}$$

A series RLC circuit is in resonance when the inductive reactance equals the capacitive reactance. When this condition is met, the rms current given by Equation 33.34 has its maximum value. The **resonance frequency** ω_0 of the circuit is

$$\omega_0 = \frac{1}{\sqrt{LC}} \quad \textbf{(33.35)}$$

The rms current in a series RLC circuit has its maximum value when the frequency of the source equals ω_0, that is, when the "driving" frequency matches the resonance frequency.

AC transformers allow for easy changes in alternating voltage according to

$$\Delta v_2 = \frac{N_2}{N_1}\Delta v_1 \quad \textbf{(33.41)}$$

where N_1 and N_2 are the numbers of windings on the primary and secondary coils, respectively, and Δv_1 and Δv_2 are the voltages on these coils.

Questions

☐ denotes answer available in *Student Solutions Manual/Study Guide;* **O** denotes objective question

1. O (i) What is the time average of the "square-wave" potential shown in Figure Q33.1? (a) $\sqrt{2}\,\Delta V_{\text{max}}$ (b) ΔV_{max} (c) $\Delta V_{\text{max}}/\sqrt{2}$ (d) $\Delta V_{\text{max}}/2$ (e) $\Delta V_{\text{max}}/4$ **(ii)** What is the rms voltage? Choose from the same possibilities.

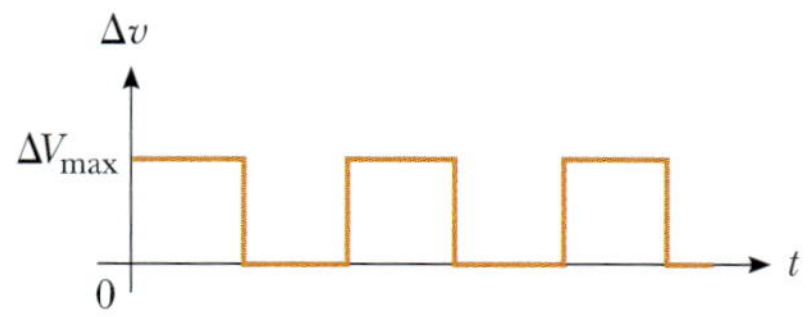

Figure Q33.1

2. O Do AC ammeters and voltmeters read (a) peak-to-valley, (b) maximum, (c) rms, or (d) average values?

3. O A sinusoidally varying potential difference has amplitude 170 V. **(i)** What is its minimum instantaneous value? (a) 240 V (b) 170 V (c) 120 V (d) 0 (e) −120 V (f) −170 V (g) −240 V **(ii)** What is its average value? **(iii)** What is its rms value? Choose from the same possibilities in each case.

4. Why does a capacitor act as a short circuit at high frequencies? Why does it act as an open circuit at low frequencies?

5. Explain how the mnemonic "ELI the ICE man" can be used to recall whether current leads voltage or voltage leads current in *RLC* circuits. Note that E represents emf $\mathcal{E}$.

6. Why is the sum of the maximum voltages across each element in a series *RLC* circuit usually greater than the maximum applied voltage? Doesn't that inequality violate Kirchhoff's loop rule?

7. Does the phase angle depend on frequency? What is the phase angle when the inductive reactance equals the capacitive reactance?

8. O (i) When a particular inductor is connected to a source of sinusoidally varying emf with constant amplitude and a frequency of 60 Hz, the rms current is 3 A. What is the rms current if the source frequency is doubled? (a) 12 A (b) 6 A (c) 4.24 A (d) 3 A (e) 2.12 A (f) 1.5 A (g) 0.75 A **(ii)** Repeat part (i) assuming the load is a capacitor instead of an inductor. **(iii)** Repeat part (i) assuming the load is a resistor instead of an inductor.

9. O What is the impedance of a series *RLC* circuit at resonance? (a) X_L (b) X_C (c) R (d) $X_L - X_C$ (e) $2X_L$ (f) $\sqrt{2}R$ (g) 0

10. O What is the phase angle in a series *RLC* circuit at resonance? (a) 180° (b) 90° (c) 0 (d) −90° (e) None of these answers is necessarily correct.

11. A certain power supply can be modeled as a source of emf in series with both a resistance of 10 Ω and an inductive reactance of 5 Ω. To obtain maximum power delivered to the load, it is found that the load should have a resistance of R_L = 10 Ω, an inductive reactance of zero, and a capacitive reactance of 5 Ω. (a) With this load, is the circuit in resonance? (b) With this load, what fraction of the average power put out by the source of emf is delivered to the load? (c) To increase the fraction of the power delivered to the load, how could the load be changed? You may wish to review Example 28.2 and Problem 4 in Chapter 28 on maximum power transfer in DC circuits.

12. As shown in Figure 7.5, a person pulls a vacuum cleaner at speed v across a horizontal floor, exerting on it a force of magnitude F directed upward at an angle θ with the horizontal. At what rate is the person doing work on the cleaner? State as completely as you can the analogy between power in this situation and in an electric circuit.

13. O A circuit containing a generator, a capacitor, an inductor, and a resistor has a high-Q resonance at 1 000 Hz. From greatest to least, rank the following contributions to the impedance of the circuit at that frequency and at lower and higher frequencies, and note any cases of equality in your ranking. (a) X_C at 500 Hz (b) X_C at 1 000 Hz (c) X_C at 1 500 Hz (d) X_L at 500 Hz (e) X_L at 1 000 Hz (f) X_L at 1 500 Hz (g) R at 500 Hz (h) R at 1 000 Hz (i) R at 1 500 Hz

14. Do some research to answer these questions: Who invented the metal detector? Why? Did it work?

15. Will a transformer operate if a battery is used for the input voltage across the primary? Explain.

16. Explain how the quality factor is related to the response characteristics of a radio receiver. Which variable most strongly influences the quality factor?

17. An ice storm breaks a transmission line and interrupts electric power to a town. A homeowner starts a gasoline-powered 120-V generator and clips its output terminals to "hot" and "ground" terminals of the electrical panel for his house. On a power pole down the block is a transformer designed to step down the voltage for household use. It has a ratio of turns N_1/N_2 of 100 to 1. A repairman climbs the pole. What voltage will he encounter on the input side of the transformer? As this question implies, safety precautions must be taken in the use of home generators and during power failures in general.

Problems

WebAssign The Problems from this chapter may be assigned online in WebAssign.

ThomsonNOW Sign in at **www.thomsonedu.com** and go to ThomsonNOW to assess your understanding of this chapter's topics with additional quizzing and conceptual questions.

1, 2, 3 denotes straightforward, intermediate, challenging; ☐ denotes full solution available in *Student Solutions Manual/Study Guide;* ▲ denotes coached solution with hints available at **www.thomsonedu.com;** ▭ denotes developing symbolic reasoning; ● denotes asking for qualitative reasoning; ▪ denotes computer useful in solving problem

Section 33.1 AC Sources

Section 33.2 Resistors in an AC Circuit

1. The rms output voltage of an AC source is 200 V and the operating frequency is 100 Hz. Write the equation giving the output voltage as a function of time.

2. (a) What is the resistance of a lightbulb that uses an average power of 75.0 W when connected to a 60.0-Hz power source having a maximum voltage of 170 V? (b) **What If?** What is the resistance of a 100-W lightbulb?

3. An AC power supply produces a maximum voltage ΔV_{max} = 100 V. This power supply is connected to a 24.0-Ω resistor, and the current and resistor voltage are measured with an ideal AC ammeter and voltmeter as shown in Figure P33.3. What does each meter read? An ideal ammeter has zero resistance and an ideal voltmeter has infinite resistance.

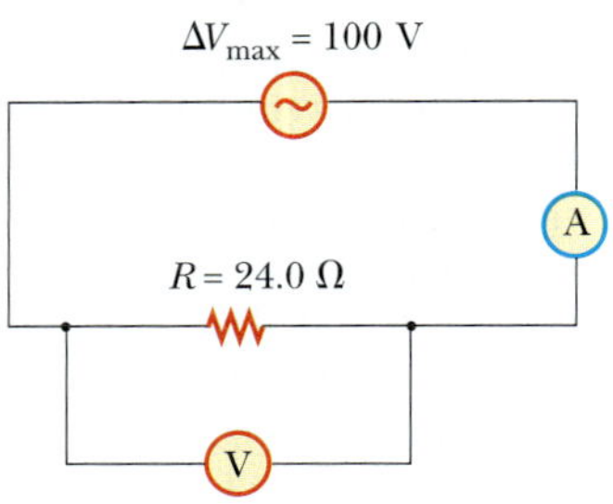

Figure P33.3

4. In the simple AC circuit shown in Active Figure 33.2, R = 70.0 Ω and $\Delta v = \Delta V_{max} \sin \omega t$. (a) If $\Delta v_R = 0.250\ \Delta V_{max}$ for the first time at t = 0.010 0 s, what is the angular frequency of the source? (b) What is the next value of t for which $\Delta v_R = 0.250\ \Delta V_{max}$?

5. The current in the circuit shown in Active Figure 33.2 equals 60.0% of the peak current at t = 7.00 ms. What is the lowest source frequency that gives this current?

6. An audio amplifier, represented by the AC source and resistor in Figure P33.6, delivers to the speaker alternating voltage at audio frequencies. If the source voltage has an amplitude of 15.0 V, R = 8.20 Ω, and the speaker is equivalent to a resistance of 10.4 Ω, what is the time-averaged power transferred to it?

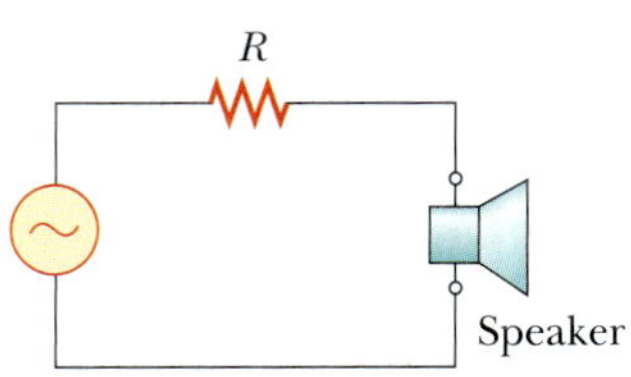

Figure P33.6

Section 33.3 Inductors in an AC Circuit

7. In a purely inductive AC circuit as shown in Active Figure 33.6, ΔV_{max} = 100 V. (a) The maximum current is 7.50 A at 50.0 Hz. Calculate the inductance L. (b) **What If?** At what angular frequency ω is the maximum current 2.50 A?

8. An inductor has a 54.0-Ω reactance at 60.0 Hz. What is the maximum current if this inductor is connected to a 50.0-Hz source that produces a 100-V rms voltage?

9. ▲ For the circuit shown in Active Figure 33.6, ΔV_{max} = 80.0 V, $\omega = 65.0\pi$ rad/s, and L = 70.0 mH. Calculate the current in the inductor at t = 15.5 ms.

10. A 20.0-mH inductor is connected to a standard electrical outlet (ΔV_{rms} = 120 V, f = 60.0 Hz). Determine the energy stored in the inductor at $t = \frac{1}{180}$ s, assuming this energy is zero at t = 0.

11. **Review problem.** Determine the maximum magnetic flux through an inductor connected to a standard electrical outlet (ΔV_{rms} = 120 V, f = 60.0 Hz).

Section 33.4 Capacitors in an AC Circuit

12. (a) For what frequencies does a 22.0-μF capacitor have a reactance below 175 Ω? (b) **What If?** What is the reactance of a 44.0-μF capacitor over this same frequency range?

13. What is the maximum current in a 2.20-μF capacitor when it is connected across (a) a North American electrical outlet having ΔV_{rms} = 120 V and f = 60.0 Hz, and (b) a European electrical outlet having ΔV_{rms} = 240 V and f = 50.0 Hz?

14. A capacitor C is connected to a power supply that operates at a frequency f and produces an rms voltage ΔV. What is the maximum charge that appears on either capacitor plate?

15. What maximum current is delivered by an AC source with ΔV_{max} = 48.0 V and f = 90.0 Hz when connected across a 3.70-μF capacitor?

16. A 1.00-mF capacitor is connected to a standard electrical outlet (ΔV_{rms} = 120 V, f = 60.0 Hz). Determine the current in the wires at $t = \frac{1}{180}$ s, assuming the energy stored in the capacitor is zero at t = 0.

Section 33.5 The *RLC* Series Circuit

17. An inductor (L = 400 mH), a capacitor (C = 4.43 μF), and a resistor (R = 500 Ω) are connected in series. A 50.0-Hz AC source produces a peak current of 250 mA in the circuit. (a) Calculate the required peak voltage ΔV_{max}. (b) Determine the phase angle by which the current leads or lags the applied voltage.

18. At what frequency does the inductive reactance of a 57.0-μH inductor equal the capacitive reactance of a 57.0-μF capacitor?

19. A series AC circuit contains the following components: a 150-Ω resistor, an inductor of 250 mH, a capacitor of 2.00 μF, and a source with ΔV_{max} = 210 V operating at 50.0 Hz. Calculate the (a) inductive reactance, (b) capacitive reactance, (c) impedance, (d) maximum current, and (e) phase angle between current and source voltage.

20. A sinusoidal voltage $\Delta v(t)$ = (40.0 V) sin (100t) is applied to a series RLC circuit with L = 160 mH, C = 99.0 μF, and R = 68.0 Ω. (a) What is the impedance of the circuit? (b) What is the maximum current? (c) Determine the numerical values for I_{max}, ω, and ϕ in the equation $i(t) = I_{max} \sin(\omega t - \phi)$.

21. ▲ An RLC circuit consists of a 150-Ω resistor, a 21.0-μF capacitor, and a 460-mH inductor connected in series with a 120-V, 60.0-Hz power supply. (a) What is the phase angle between the current and the applied voltage? (b) Which reaches its maximum earlier, the current or the voltage?

22. Four circuit elements—a capacitor, an inductor, a resistor, and an AC source—are connected together in various ways. First the capacitor is connected to the source, and the rms current is found to be 25.1 mA. The capacitor is disconnected and discharged, and then it is connected in series with the resistor and the source, making the rms current 15.7 mA. The circuit is disconnected and the capacitor discharged. The capacitor is then connected in series with the inductor and the source, making the rms current 68.2 mA. After the circuit is disconnected and the capacitor discharged, all four circuit elements are connected together in a series loop. What is the rms current in the circuit?

23. A person is working near the secondary of a transformer as shown in Figure P33.23. The primary voltage is 120 V at 60.0 Hz. The capacitance C_s, which is the stray capacitance between the hand and the secondary winding, is 20.0 pF. Assuming the person has a body resistance to ground of R_b = 50.0 kΩ, determine the rms voltage across the body. *Suggestion:* Model the secondary of the transformer as an AC source.

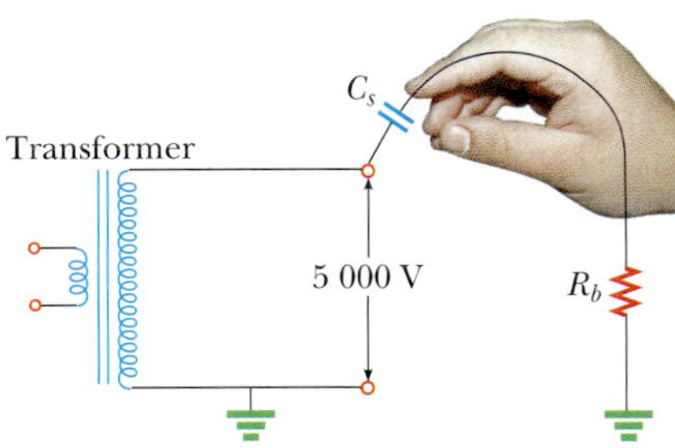

Figure P33.23

24. An AC source with ΔV_{max} = 150 V and f = 50.0 Hz is connected between points a and d in Figure P33.24. Calculate the maximum voltages between (a) points a and b, (b) points b and c, (c) points c and d, and (d) points b and d.

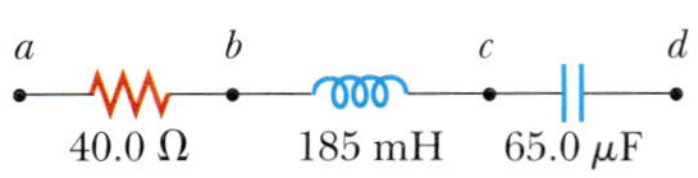

Figure P33.24 Problems 24 and 64.

25. Draw to scale a phasor diagram and determine Z, X_L, X_C, and ϕ for an AC series circuit for which R = 300 Ω, C = 11.0 μF, L = 0.200 H, and f = (500/π) Hz.

26. ● In an RLC series circuit that includes a source of alternating current operating at fixed frequency and voltage, the resistance R is equal to the inductive reactance. If the plate separation of the parallel-plate capacitor is reduced to one-half its original value, the current in the circuit doubles. Find the initial capacitive reactance in terms of R. Explain each step in your solution.

Section 33.6 Power in an AC Circuit

27. ▲ An AC voltage of the form Δv = (100 V) sin (1 000t) is applied to a series RLC circuit. Assume the resistance is 400 Ω, the capacitance is 5.00 μF, and the inductance is 0.500 H. Find the average power delivered to the circuit.

28. A series RLC circuit has a resistance of 45.0 Ω and an impedance of 75.0 Ω. What average power is delivered to this circuit when ΔV_{rms} = 210 V?

29. In a certain series RLC circuit, I_{rms} = 9.00 A, ΔV_{rms} = 180 V, and the current leads the voltage by 37.0°. (a) What is the total resistance of the circuit? (b) Calculate the reactance of the circuit ($X_L - X_C$).

30. Suppose you manage a factory that uses many electric motors. The motors create a large inductive load to the electric power line as well as a resistive load. The electric company builds an extra-heavy distribution line to supply you with a component of current that is 90° out of phase with the voltage as well as with current in phase with the voltage. The electric company charges you an extra fee for "reactive volt-amps" in addition to the amount you pay for the energy you use. You can avoid the extra fee by installing a capacitor between the power line and your factory. The following problem models this solution.

In an RL circuit, a 120-V (rms), 60.0-Hz source is in series with a 25.0-mH inductor and a 20.0-Ω resistor. What are (a) the rms current and (b) the power factor? (c) What capacitor must be added in series to make the power factor 1? (d) To what value can the supply voltage be reduced if the power supplied is to be the same as before the capacitor was installed?

31. Energy is to be transmitted at the rate of 20.0 kW with only 1.00% loss over a distance of 18.0 km at potential difference ΔV. (a) What is the diameter required for each of the two copper wires in the transmission line? Assume the current density is uniform in the conductors. (b) State how the diameter depends on ΔV. (c) Evaluate the diameter for ΔV = 1 500 V. (d) If you choose to make the diameter 3.00 mm, what potential difference is required?

32. ● A series circuit consists of an AC generator with an rms voltage of 120 V at a frequency of 60.0 Hz and a magnetic buzzer with a resistance of 100 Ω and an inductance of 100 mH. (a) Find the circuit's power factor. (b) Suppose a higher power factor is desired. Can a power factor of 1.00 be achieved by changing the inductance or any other circuit parameters? (c) Show that a power factor of 1.00 can be attained by inserting a capacitor into the original circuit, and find the value of its capacitance.

33. A diode is a device that allows current to be carried in only one direction (the direction indicated by the arrowhead in its circuit symbol). Find in terms of ΔV and R the average power delivered to the diode circuit of Figure P33.33.

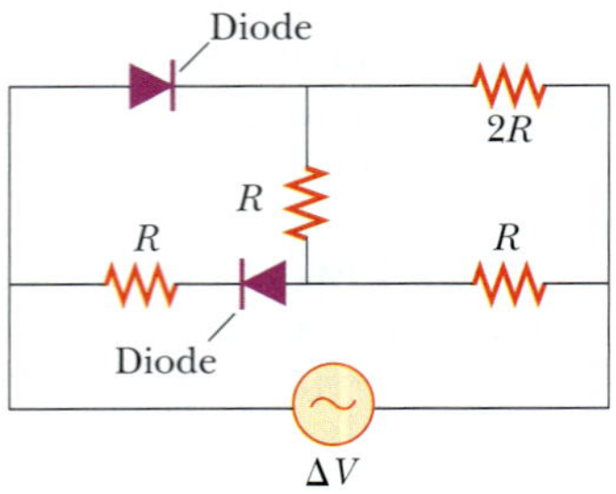

Figure P33.33

Section 33.7 Resonance in a Series *RLC* Circuit

34. A radar transmitter contains an *LC* circuit oscillating at 1.00×10^{10} Hz. (a) What capacitance resonates with a one-turn loop having an inductance of 400 pH at this frequency? (b) The capacitor has square, parallel plates separated by 1.00 mm of air. What should the edge length of the plates be? (c) What is the common reactance of the loop and capacitor at resonance?

35. An *RLC* circuit is used in a radio to tune into an FM station broadcasting at 99.7 MHz. The resistance in the circuit is 12.0 Ω, and the inductance is 1.40 μH. What capacitance should be used?

36. A series *RLC* circuit has components with the following values: $L = 20.0$ mH, $C = 100$ nF, $R = 20.0\ \Omega$, and $\Delta V_{max} = 100$ V, with $\Delta v = \Delta V_{max} \sin \omega t$. Find (a) the resonant frequency, (b) the amplitude of the current at the resonant frequency, (c) the Q of the circuit, and (d) the amplitude of the voltage across the inductor at resonance.

37. A 10.0-Ω resistor, 10.0-mH inductor, and 100-μF capacitor are connected in series to a 50.0-V (rms) source having variable frequency. Find the energy delivered to the circuit during one period if the operating frequency is twice the resonance frequency.

38. A resistor *R*, inductor *L*, and capacitor *C* are connected in series to an AC source of rms voltage ΔV and variable frequency. Find the energy delivered to the circuit during one period if the operating frequency is twice the resonance frequency.

39. Compute the quality factor for the circuits described in Problems 20 and 21. Which circuit has the sharper resonance?

Section 33.8 The Transformer and Power Transmission

40. A step-down transformer is used for recharging the batteries of portable devices such as tape players. The ratio of turns inside the transformer is 13:1, and the transformer is used with 120-V (rms) household service. If a particular ideal transformer draws 0.350 A from the house outlet, what are (a) the voltage and (b) the current supplied to a tape player from the transformer? (c) How much power is delivered?

41. A transformer has $N_1 = 350$ turns and $N_2 = 2\,000$ turns. If the input voltage is $\Delta v(t) = (170\text{ V})\cos \omega t$, what rms voltage is developed across the secondary coil?

42. A step-up transformer is designed to have an output voltage of 2 200 V (rms) when the primary is connected across a 110-V (rms) source. (a) If the primary winding has 80 turns, how many turns are required on the secondary? (b) If a load resistor across the secondary draws a current of 1.50 A, what is the current in the primary, assuming ideal conditions? (c) **What If?** If the transformer actually has an efficiency of 95.0%, what is the current in the primary when the secondary current is 1.20 A?

43. ● A transmission line that has a resistance per unit length of $4.50 \times 10^{-4}\ \Omega/\text{m}$ is to be used to transmit 5.00 MW across 400 miles (6.44×10^5 m). The output voltage of the generator is 4.50 kV. (a) What is the line loss if a transformer is used to step up the voltage to 500 kV? (b) What fraction of the input power is lost to the line under these circumstances? (c) **What If?** What difficulties would be encountered in attempting to transmit the 5.00 MW at the generator voltage of 4.50 kV?

Section 33.9 Rectifiers and Filters

44. One particular plug-in power supply for a radio looks similar to the one shown in Figure 33.20 and is marked with the following information: Input 120 V AC 8 W Output 9 V DC 300 mA. Assume these values are accurate to two digits. (a) Find the energy efficiency of the device when the radio is operating. (b) At what rate is energy wasted in the device when the radio is operating? (c) Suppose the input power to the transformer is 8.0 W when the radio is switched off and energy costs \$0.135/kWh from the electric company. Find the cost of having six such transformers around the house, each plugged in for 31 days.

45. Consider the filter circuit shown in Active Figure 33.22a. (a) Show that the ratio of the output voltage to the input voltage is

$$\frac{\Delta v_{out}}{\Delta v_{in}} = \frac{R}{\sqrt{R^2 + \left(\dfrac{1}{\omega C}\right)^2}}$$

(b) What value does this ratio approach as the frequency decreases toward zero? What value does this ratio approach as the frequency increases without limit? (c) At what frequency is the ratio equal to one-half?

46. Consider the filter circuit shown in Active Figure 33.23a. (a) Show that the ratio of the output voltage to the input voltage is

$$\frac{\Delta v_{out}}{\Delta v_{in}} = \frac{1/\omega C}{\sqrt{R^2 + \left(\dfrac{1}{\omega C}\right)^2}}$$

(b) What value does this ratio approach as the frequency decreases toward zero? What value does this ratio approach as the frequency increases without limit? (c) At what frequency is the ratio equal to one-half?

47. ▲ The *RC* high-pass filter shown in Active Figure 33.22a has a resistance $R = 0.500\ \Omega$. (a) What capacitance gives an output signal that has one-half the amplitude of a 300-Hz input signal? (b) What is the ratio $(\Delta v_{out}/\Delta v_{in})$ for a 600-Hz signal? You may use the result of Problem 45.

48. The *RC* low-pass filter shown in Active Figure 33.23a has a resistance $R = 90.0\ \Omega$ and a capacitance $C = 8.00$ nF. Calculate the ratio $(\Delta v_{out}/\Delta v_{in})$ for an input frequency of (a) 600 Hz and (b) 600 kHz. You may use the result of Problem 46.

2 = intermediate; 3 = challenging; □ = SSM/SG; ▲ = ThomsonNOW; ▭ = symbolic reasoning; ● = qualitative reasoning

49. The resistor in Figure P33.49 represents the midrange speaker in a three-speaker system. Assume its resistance to be constant at 8.00 Ω. The source represents an audio amplifier producing signals of uniform amplitude $\Delta V_{max} = 10.0$ V at all audio frequencies. The inductor and capacitor are to function as a band-pass filter with $\Delta v_{out}/\Delta v_{in} = \frac{1}{2}$ at 200 Hz and at 4 000 Hz. (a) Determine the required values of L and C. (b) Find the maximum value of the ratio $\Delta v_{out}/\Delta v_{in}$. (c) Find the frequency f_0 at which the ratio has its maximum value. (d) Find the phase shift between Δv_{in} and Δv_{out} at 200 Hz, at f_0, and at 4 000 Hz. (e) Find the average power transferred to the speaker at 200 Hz, at f_0, and at 4 000 Hz. (f) Treating the filter as a resonant circuit, find its quality factor.

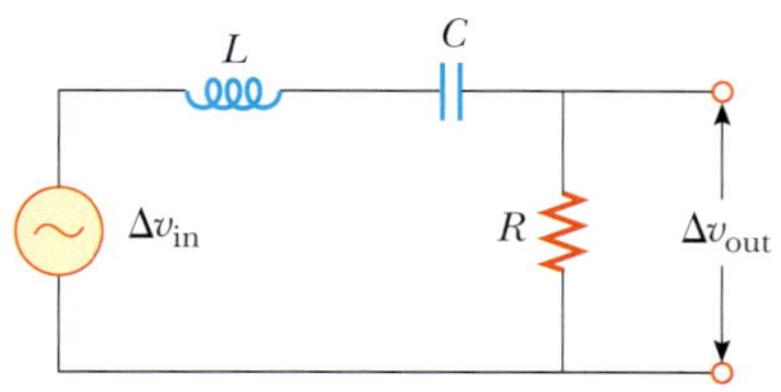

Figure P33.49

Additional Problems

50. Show that the rms value for the sawtooth voltage shown in Figure P33.50 is $\Delta V_{max}/\sqrt{3}$.

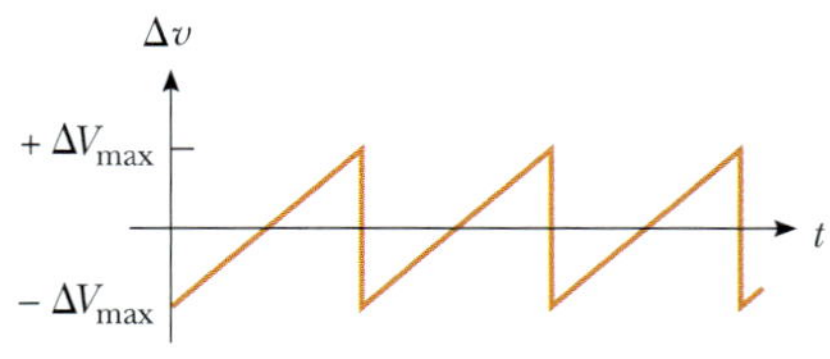

Figure P33.50

51. ● A 400-Ω resistor, an inductor, and a capacitor are in series with a generator. The reactance of the inductor is 700 Ω, and the circuit impedance is 760 Ω. (a) Explain what you can and cannot determine about the reactance of the capacitor. (b) If you find that the source power decreases as you raise the frequency, what do you know about the capacitive reactance in the original circuit? (c) Repeat part (a) assuming the resistance is 200 Ω instead of 400 Ω.

52. ● A capacitor, a coil, and two resistors of equal resistance are arranged in an AC circuit as shown in Figure P33.52. An AC generator provides an emf of 20.0 V (rms) at a frequency of 60.0 Hz. When the double-throw switch S is open as shown in the figure, the rms current is 183 mA. When the switch is closed in position 1, the rms current is 298 mA. When the switch is closed in position 2, the rms current is 137 mA. Determine the values of R, C and L. Is more than one set of values possible? Explain.

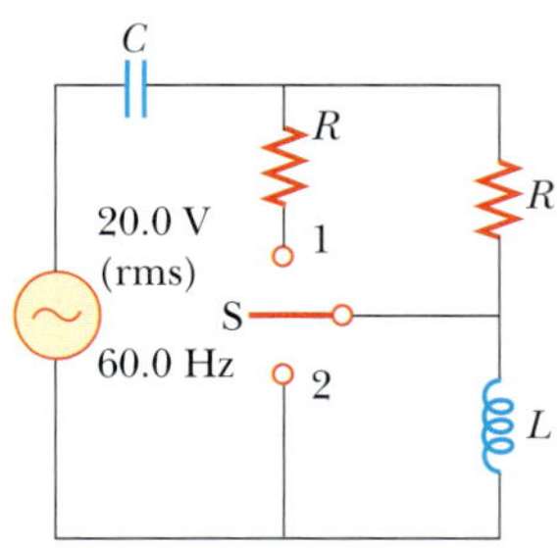

Figure P33.52

53. ▲ A series RLC circuit consists of an 8.00-Ω resistor, a 5.00-μF capacitor, and a 50.0-mH inductor. A variable frequency source applies an emf of 400 V (rms) across the combination. Determine the power delivered to the circuit when the frequency is equal to one-half the resonance frequency.

54. ● A series RLC circuit has resonance angular frequency 2 000 rad/s. When it is operating at some certain frequency, $X_L = 12.0$ Ω and $X_C = 8.00$ Ω. (a) Is this certain frequency higher than, lower than, or the same as the resonance frequency? Explain how you can tell. (b) Explain whether it is possible to determine the values of both L and C. (c) If it is possible, find L and C. If this determination is not possible, give a compact expression for the condition that L and C must satisfy.

55. ● **Review problem.** One insulated conductor from a household extension cord has a mass per length of 19.0 g/m. A section of this conductor is held under tension between two clamps. A subsection is located in a magnetic field of magnitude 15.3 mT directed perpendicular to the length of the cord. When the cord carries an AC current of 9.00 A at a frequency of 60.0 Hz, it vibrates in resonance in its simplest standing-wave vibration state. Determine the relationship that must be satisfied between the separation d of the clamps and the tension T in the cord. Determine one possible combination of values for these variables.

56. Sketch a graph of the phase angle for an RLC series circuit as a function of angular frequency from zero to a frequency much higher than the resonance frequency. Identify the value of ϕ at the resonance angular frequency ω_0. Prove that the slope of the graph of ϕ versus ω at the resonance point is $2Q/\omega_0$.

57. In Figure P33.57, find the rms current delivered by the 45.0-V (rms) power supply when (a) the frequency is very large and (b) the frequency is very small.

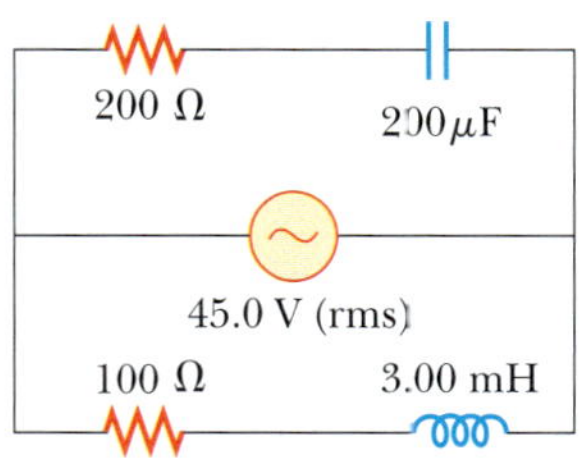

Figure P33.57

58. In the circuit shown in Figure P33.58 (page 950), assume all parameters except C are given. (a) Find the current as a function of time. (b) Find the power delivered to the circuit. (c) Find the current as a function of time after *only* switch 1 is opened. (d) After switch 2 is *also* opened, the current and voltage are in phase. Find the capacitance C. (e) Find the impedance of the circuit when both switches are open. (f) Find the maximum energy stored in the capacitor during oscillations. (g) Find the maximum energy stored in the inductor during oscillations.

(h) Now the frequency of the voltage source is doubled. Find the phase difference between the current and the voltage. (i) Find the frequency that makes the inductive reactance one-half the capacitive reactance.

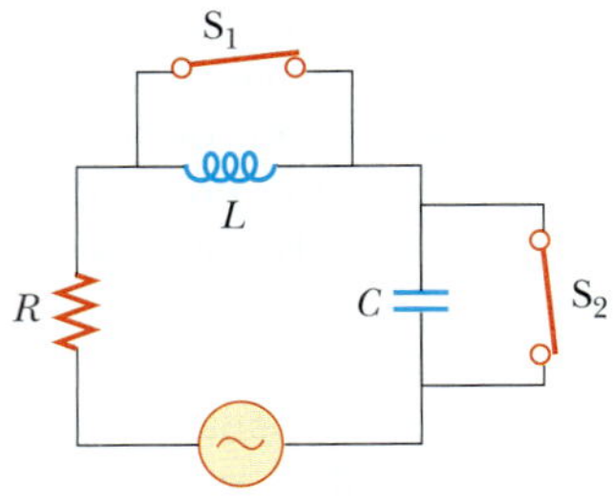

Figure P33.58

59. An 80.0-Ω resistor and a 200-mH inductor are connected in *parallel* across a 100-V (rms), 60.0-Hz source. (a) What is the rms current in the resistor? (b) By what angle does the total current lead or lag behind the voltage?

60. Make an order-of-magnitude estimate of the electric current that the electric company delivers to a town (Fig. P33.60) from a remote generating station. State the data you measure or estimate. If you wish, you may consider a suburban residential community of 20 000 people.

Eddie Hironaka/Getty Images

Figure P33.60

61. Consider a series *RLC* circuit having the following circuit parameters: $R = 200\ \Omega$, $L = 663$ mH, and $C = 26.5\ \mu$F. The applied voltage has an amplitude of 50.0 V and a frequency of 60.0 Hz. Find the following amplitudes. (a) the current I_{max} and its phase relative to the applied voltage Δv (b) the maximum voltage ΔV_R across the resistor and its phase relative to the current (c) the maximum voltage ΔV_C across the capacitor and its phase relative to the current (d) the maximum voltage ΔV_L across the inductor and its phase relative to the current

62. A voltage $\Delta v = (100\text{ V}) \sin \omega t$ is applied across a series combination of a 2.00-H inductor, a 10.0-μF capacitor, and a 10.0-Ω resistor. (a) Determine the angular frequency ω_0 at which the power delivered to the resistor is a maximum. (b) Calculate the power delivered at that frequency. (c) Determine the two angular frequencies ω_1 and ω_2 at which the power is one-half the maximum value. *Note:* The Q of the circuit is $\omega_0/(\omega_2 - \omega_1)$.

63. *Impedance matching.* Example 28.2 showed that maximum power is transferred when the internal resistance of a DC source is equal to the resistance of the load. A transformer may be used to provide maximum power transfer between two AC circuits that have different impedances Z_1 and Z_2. (a) Show that the ratio of turns N_1/N_2 needed to meet this condition is

$$\frac{N_1}{N_2} = \sqrt{\frac{Z_1}{Z_2}}$$

(b) Suppose you want to use a transformer as an impedance-matching device between an audio amplifier that has an output impedance of 8.00 kΩ and a speaker that has an input impedance of 8.00 Ω. What should your N_1/N_2 ratio be?

64. ● A power supply with $\Delta V_{rms} = 120$ V is connected between points *a* and *d* in Figure P33.24. At what frequency will it deliver a power of 250 W? Explain your answer.

65. Figure P33.65a shows a parallel *RLC* circuit, and the corresponding phasor diagram is given in Figure P33.65b. The instantaneous voltages (and rms voltages) across each of the three circuit elements are the same, and each is in phase with the current in the resistor. The currents in *C* and *L* lead or lag the current in the resistor as shown in Figure P33.65b. (a) Show that the rms current delivered by the source is

$$I_{rms} = \Delta V_{rms}\left[\frac{1}{R^2} + \left(\omega C - \frac{1}{\omega L}\right)^2\right]^{1/2}$$

(b) Show that the phase angle ϕ between ΔV_{rms} and I_{rms} is given by

$$\tan \phi = R\left(\frac{1}{X_C} - \frac{1}{X_L}\right)$$

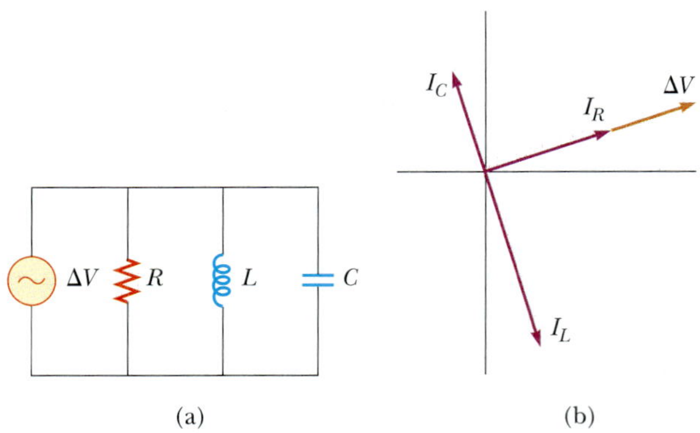

Figure P33.65

66. A certain electric circuit is described by the equations

$$\frac{200\text{ V}}{4.00\text{ A}} = \sqrt{(35.0\ \Omega)^2 + \left[\omega(205\text{ mH}) - \frac{1}{\omega C}\right]^2}$$

$$\omega = 2\pi\ (100\text{ Hz})$$

State a problem for which these equations would appear in the solution, giving the data and identifying the unknown. Evaluate the unknown quantity.

67. A series *RLC* circuit is operating at 2 000 Hz. At this frequency, $X_L = X_C = 1\,884\ \Omega$. The resistance of the circuit is 40.0 Ω. (a) Prepare a table showing the values of X_L, X_C, and Z for f = 300, 600, 800, 1 000, 1 500, 2 000, 3 000, 4 000, 6 000, and 10 000 Hz. (b) Plot on the same set of axes X_L, X_C, and Z as a function of $\ln f$.

68. A series RLC circuit with $R = 1.00\ \Omega$, $L = 1.00$ mH, and $C = 1.00$ nF is connected to an AC source delivering 1.00 V (rms). Make a precise graph of the power delivered to the circuit as a function of the frequency and verify that the full width of the resonance peak at half-maximum is $R/2\pi L$.

69. Marie Cornu, a physicist at the Polytechnic Institute in Paris, invented phasors in about 1880. This problem helps you see their general utility in representing oscillations. Two mechanical vibrations are represented by the expressions

$$y_1 = (12.0\text{ cm})\sin(4.5t)$$

and

$$y_2 = (12.0\text{ cm})\sin(4.5t + 70°)$$

Find the amplitude and phase constant of the sum of these functions (a) by using a trigonometric identity (as from Appendix B) and (b) by representing the oscillations as phasors. State the result of comparing the answers to (a) and (b). (c) Phasors make it equally easy to add traveling waves. Find the amplitude and phase constant of the sum of the three waves represented by

$$y_1 = (12.0\text{ cm})\sin(15x - 4.5t + 70°)$$

$$y_2 = (15.5\text{ cm})\sin(15x - 4.5t - 80°)$$

$$y_3 = (17.0\text{ cm})\sin(15x - 4.5t + 160°)$$

Answers to Quick Quizzes

33.1 **(i),** (c). The phasor in (c) has the largest projection onto the vertical axis. **(ii),** (b). The phasor in (b) has the smallest-magnitude projection onto the vertical axis.

33.2 (b). For low frequencies, the reactance of the inductor is small, so the current is large. Most of the voltage from the source is across the lightbulb, so the power delivered to it is large.

33.3 (a). For high frequencies, the reactance of the capacitor is small, so the current is large. Most of the voltage from the source is across the lightbulb, so the power delivered to it is large.

33.4 (b). For low frequencies, the reactance of the capacitor is large, so very little current exists in the capacitor branch. The reactance of the inductor is small, so current exists in the inductor branch and the lightbulb glows. As the frequency increases, the inductive reactance increases and the capacitive reactance decreases. At high frequencies, more current exists in the capacitor branch than the inductor branch and the lightbulb glows more dimly.

33.5 (a) $X_L < X_C$ (b) $X_L = X_C$ (c) $X_L > X_C$

33.6 (c). The cosine of $-\phi$ is the same as that of $+\phi$, so the $\cos\phi$ factor in Equation 33.31 is the same for both frequencies. The factor ΔV_{rms} is the same because the source voltage is fixed. According to Equation 33.27, changing $+\phi$ to $-\phi$ simply interchanges the values of X_L and X_C. Equation 33.25 tells us that such an interchange does not affect the impedance, so the current I_{rms} in Equation 33.31 is the same for both frequencies.

33.7 (c). At resonance, $X_L = X_C$. According to Equation 33.25, that gives us $Z = R$.

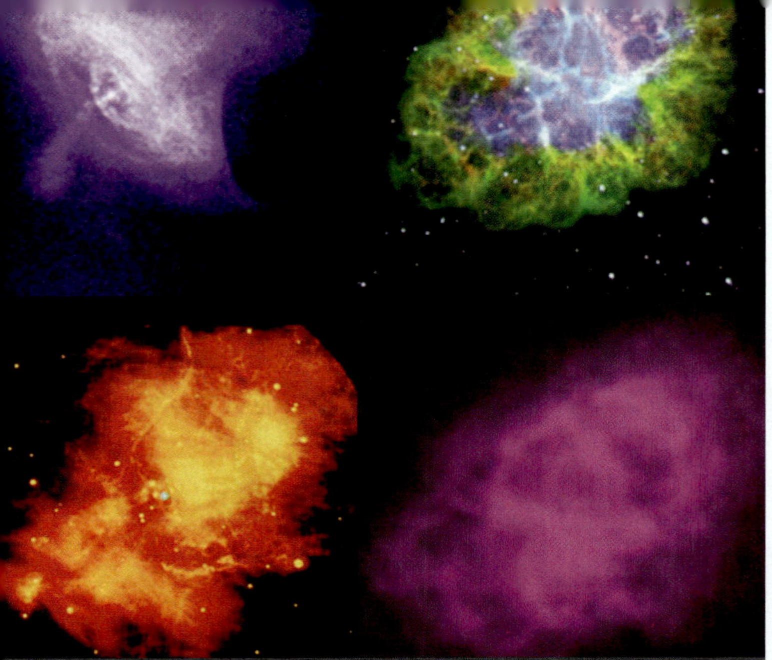

Electromagnetic waves cover a broad spectrum of wavelengths, with waves in various wavelength ranges having distinct properties. These images of the Crab Nebula show different structure for observations made with waves of various wavelengths. The images (clockwise starting from upper left) were taken with x-rays, visible light, radio waves, and infrared waves. (upper left, NASA/CXC/SAO; upper right, Palomar Observatory; lower right, VLA/NRAO; lower left, WM Keck Observatory)

34 Electromagnetic Waves

The waves described in Chapters 16, 17, and 18 are mechanical waves. By definition, the propagation of mechanical disturbances—such as sound waves, water waves, and waves on a string—requires the presence of a medium. This chapter is concerned with the properties of electromagnetic waves, which (unlike mechanical waves) can propagate through empty space.

We begin by considering Maxwell's contributions in modifying Ampère's law, which we studied in Chapter 30. We then discuss Maxwell's equations, which form the theoretical basis of all electromagnetic phenomena. These equations predict the existence of electromagnetic waves that propagate through space at the speed of light c. Heinrich Hertz confirmed Maxwell's prediction when he generated and detected electromagnetic waves in 1887. That discovery has led to many practical communication systems, including radio, television, radar, and optoelectronics.

Next, we learn how electromagnetic waves are generated by oscillating electric charges. The waves radiated from the oscillating charges can be detected at great distances. Furthermore, because electromagnetic waves carry energy and momentum, they can exert pressure on a surface. The chapter concludes with a look at many frequencies covered by electromagnetic waves.

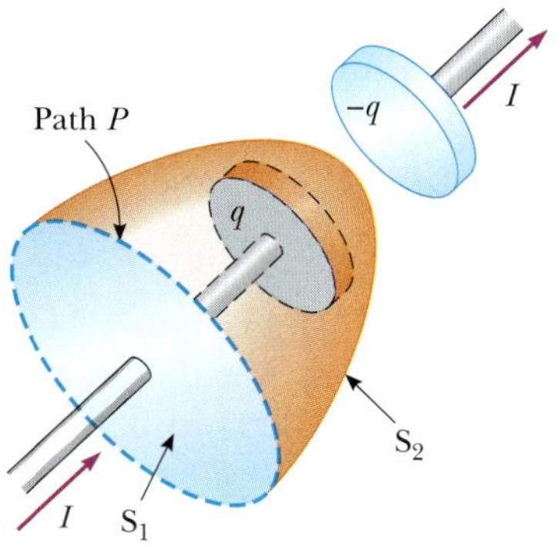

Figure 34.1 Two surfaces S_1 and S_2 near the plate of a capacitor are bounded by the same path P. The conduction current in the wire passes only through S_1, which leads to a contradiction in Ampère's law that is resolved only if one postulates a displacement current through S_2.

North Wind Picture Archives

JAMES CLERK MAXWELL
Scottish Theoretical Physicist (1831–1879)
Maxwell developed the electromagnetic theory of light and the kinetic theory of gases, and explained the nature of Saturn's rings and color vision. Maxwell's successful interpretation of the electromagnetic field resulted in the field equations that bear his name. Formidable mathematical ability combined with great insight enabled him to lead the way in the study of electromagnetism and kinetic theory. He died of cancer before he was 50.

34.1 Displacement Current and the General Form of Ampère's Law

In Chapter 30, we discussed using Ampère's law (Eq. 30.13) to analyze the magnetic fields created by currents:

$$\oint \vec{\mathbf{B}} \cdot d\vec{\mathbf{s}} = \mu_0 I$$

In this equation, the line integral is over any closed path through which conduction current passes, where conduction current is defined by the expression $I = dq/dt$. (In this section, we use the term *conduction current* to refer to the current carried by charge carriers in the wire to distinguish it from a new type of current we shall introduce shortly.) We now show that **Ampère's law in this form is valid only if any electric fields present are constant in time.** James Clerk Maxwell recognized this limitation and modified Ampère's law to include time-varying electric fields.

Consider a capacitor being charged as illustrated in Figure 34.1. When a conduction current is present, the charge on the positive plate changes but *no conduction current exists in the gap between the plates.* Now consider the two surfaces S_1 and S_2 in Figure 34.1, bounded by the same path P. Ampère's law states that $\oint \vec{\mathbf{B}} \cdot d\vec{\mathbf{s}}$ around this path must equal $\mu_0 I$, where I is the total current through *any* surface bounded by the path P.

When the path P is considered to be the boundary of S_1, $\oint \vec{\mathbf{B}} \cdot d\vec{\mathbf{s}} = \mu_0 I$ because the conduction current I passes through S_1. When the path is considered to be the boundary of S_2, however, $\oint \vec{\mathbf{B}} \cdot d\vec{\mathbf{s}} = 0$ because no conduction current passes through S_2. Therefore, we have a contradictory situation that arises from the discontinuity of the current! Maxwell solved this problem by postulating an additional term on the right side of Ampère's law, which includes a factor called the **displacement current** I_d defined as[1]

$$I_d \equiv \epsilon_0 \frac{d\Phi_E}{dt} \qquad \textbf{(34.1)}$$

◀ Displacement current

where ϵ_0 is the permittivity of free space (see Section 23.3) and $\Phi_E = \int \vec{\mathbf{E}} \cdot d\vec{\mathbf{A}}$ is the electric flux (see Eq. 24.3) through the surface bounded by the path of integration.

As the capacitor is being charged (or discharged), the changing electric field between the plates may be considered equivalent to a current that acts as a continuation of the conduction current in the wire. When the expression for the displacement current given by Equation 34.1 is added to the conduction current on the right side of Ampère's law, the difficulty represented in Figure 34.1 is resolved. No matter which surface bounded by the path P is chosen, either a conduction

[1] *Displacement* in this context does not have the meaning it does in Chapter 2. Despite the inaccurate implications, the word is historically entrenched in the language of physics, so we continue to use it.

current or a displacement current passes through it. With this new term I_d, we can express the general form of Ampère's law (sometimes called the **Ampère–Maxwell law**) as

Ampère–Maxwell law ▶

$$\oint \vec{\mathbf{B}} \cdot d\vec{\mathbf{s}} = \mu_0 (I + I_d) = \mu_0 I + \mu_0 \epsilon_0 \frac{d\Phi_E}{dt} \tag{34.2}$$

We can understand the meaning of this expression by referring to Figure 34.2. The electric flux through surface S_2 is $\Phi_E = \int \vec{\mathbf{E}} \cdot d\vec{\mathbf{A}} = EA$, where A is the area of the capacitor plates and E is the magnitude of the uniform electric field between the plates. If q is the charge on the plates at any instant, then $E = q/(\epsilon_0 A)$ (see Section 26.2). Therefore, the electric flux through S_2 is

$$\Phi_E = EA = \frac{q}{\epsilon_0}$$

Hence, the displacement current through S_2 is

$$I_d = \epsilon_0 \frac{d\Phi_E}{dt} = \frac{dq}{dt} \tag{34.3}$$

That is, the displacement current I_d through S_2 is precisely equal to the conduction current I through S_1!

By considering surface S_2, we can identify the displacement current as the source of the magnetic field on the surface boundary. The displacement current has its physical origin in the time-varying electric field. The central point of this formalism is that **magnetic fields are produced both by conduction currents and by time-varying electric fields.** This result was a remarkable example of theoretical work by Maxwell, and it contributed to major advances in the understanding of electromagnetism.

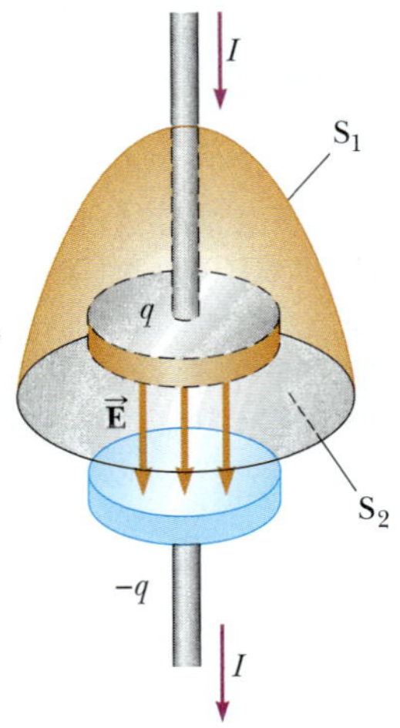

Figure 34.2 Because it exists only in the wires attached to the capacitor plates, the conduction current $I = dq/dt$ passes through S_1 but not through S_2. Only the displacement current $I_d = \epsilon_0 d\Phi_E/dt$ passes through S_2. The two currents must be equal for continuity.

Quick Quiz 34.1 In an *RC* circuit, the capacitor begins to discharge. **(i)** During the discharge in the region of space between the plates of the capacitor, is there (a) conduction current but no displacement current, (b) displacement current but no conduction current, (c) both conduction and displacement current, or (d) no current of any type? **(ii)** In the same region of space, is there (a) an electric field but no magnetic field, (b) a magnetic field but no electric field, (c) both electric and magnetic fields, or (d) no fields of any type?

EXAMPLE 34.1 Displacement Current in a Capacitor

A sinusoidally varying voltage is applied across an 8.00-μF capacitor. The frequency of the applied voltage is 3.00 kHz, and the voltage amplitude is 30.0 V. Find the displacement current in the capacitor.

SOLUTION

Conceptualize Active Figure 33.9 represents the circuit diagram for this situation. Figure 34.2 shows a close-up of the capacitor and the electric field between the plates.

Categorize We evaluate results using equations discussed in this section, so we categorize this example as a substitution problem.

Evaluate the angular frequency of the source from Equation 15.12:

$$\omega = 2\pi f = 2\pi(3.00 \times 10^3 \text{ Hz}) = 1.88 \times 10^4 \text{ s}^{-1}$$

Use Equation 33.20 to express the voltage across the capacitor as a function of time:

$$\Delta v_C = \Delta V_{\max} \sin \omega t = (30.0 \text{ V}) \sin (1.88 \times 10^4 \, t)$$

Use Equation 34.3 to find the displacement current as a function of time. Note that the charge on the capacitor is $q = C\,\Delta v_C$:

$$I_d = \frac{dq}{dt} = \frac{d}{dt}(C\,\Delta v_C) = C\frac{d}{dt}(\Delta v_C)$$

$$= (8.00 \times 10^{-6}\ \text{F})\frac{d}{dt}[(30.0\ \text{V})\sin(1.88 \times 10^4\ t)]$$

$$= (4.52\ \text{A})\cos(1.88 \times 10^4\ t)$$

34.2 Maxwell's Equations and Hertz's Discoveries

We now present four equations that are regarded as the basis of all electrical and magnetic phenomena. These equations, developed by Maxwell, are as fundamental to electromagnetic phenomena as Newton's laws are to mechanical phenomena. In fact, the theory that Maxwell developed was more far-reaching than even he imagined because it turned out to be in agreement with the special theory of relativity, as Einstein showed in 1905.

Maxwell's equations represent the laws of electricity and magnetism that we have already discussed, but they have additional important consequences. For simplicity, we present **Maxwell's equations** as applied to free space, that is, in the absence of any dielectric or magnetic material. The four equations are

$$\oint \vec{\mathbf{E}} \cdot d\vec{\mathbf{A}} = \frac{q}{\epsilon_0} \qquad \textbf{(34.4)}$$

◀ Gauss's law

$$\oint \vec{\mathbf{B}} \cdot d\vec{\mathbf{A}} = 0 \qquad \textbf{(34.5)}$$

◀ Gauss's law in magnetism

$$\oint \vec{\mathbf{E}} \cdot d\vec{\mathbf{s}} = -\frac{d\Phi_B}{dt} \qquad \textbf{(34.6)}$$

◀ Faraday's law

$$\oint \vec{\mathbf{B}} \cdot d\vec{\mathbf{s}} = \mu_0 I + \epsilon_0\mu_0\frac{d\Phi_E}{dt} \qquad \textbf{(34.7)}$$

◀ Ampère–Maxwell law

Equation 34.4 is Gauss's law: **the total electric flux through any closed surface equals the net charge inside that surface divided by** ϵ_0. This law relates an electric field to the charge distribution that creates it.

Equation 34.5 is Gauss's law in magnetism, and it states that **the net magnetic flux through a closed surface is zero.** That is, the number of magnetic field lines that enter a closed volume must equal the number that leave that volume, which implies that magnetic field lines cannot begin or end at any point. If they did, it would mean that isolated magnetic monopoles existed at those points. That isolated magnetic monopoles have not been observed in nature can be taken as a confirmation of Equation 34.5.

Equation 34.6 is Faraday's law of induction, which describes the creation of an electric field by a changing magnetic flux. This law states that **the emf, which is the line integral of the electric field around any closed path, equals the rate of change of magnetic flux through any surface bounded by that path.** One consequence of Faraday's law is the current induced in a conducting loop placed in a time-varying magnetic field.

Equation 34.7 is the Ampère–Maxwell law, and it describes the creation of a magnetic field by a changing electric field and by electric current: **the line integral of the magnetic field around any closed path is the sum of** μ_0 **times the net current through that path and** $\epsilon_0\mu_0$ **times the rate of change of electric flux through any surface bounded by that path.**

Once the electric and magnetic fields are known at some point in space, the force acting on a particle of charge q can be calculated from the expression

Lorentz force law ▶

$$\vec{\mathbf{F}} = q\vec{\mathbf{E}} + q\vec{\mathbf{v}} \times \vec{\mathbf{B}} \tag{34.8}$$

This relationship is called the **Lorentz force law.** (We saw this relationship earlier as Eq. 29.6.) Maxwell's equations, together with this force law, completely describe all classical electromagnetic interactions in a vacuum.

Notice the symmetry of Maxwell's equations. Equations 34.4 and 34.5 are symmetric, apart from the absence of the term for magnetic monopoles in Equation 34.5. Furthermore, Equations 34.6 and 34.7 are symmetric in that the line integrals of $\vec{\mathbf{E}}$ and $\vec{\mathbf{B}}$ around a closed path are related to the rate of change of magnetic flux and electric flux, respectively. Maxwell's equations are of fundamental importance not only to electromagnetism, but to all science. Hertz once wrote, "One cannot escape the feeling that these mathematical formulas have an independent existence and an intelligence of their own, that they are wiser than we are, wiser even than their discoverers, that we get more out of them than we put into them."

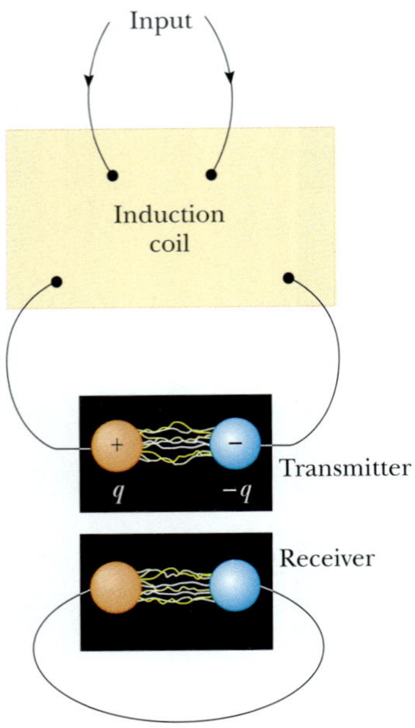

Figure 34.3 Schematic diagram of Hertz's apparatus for generating and detecting electromagnetic waves. The transmitter consists of two spherical electrodes connected to an induction coil, which provides short voltage surges to the spheres, setting up oscillations in the discharge between the electrodes. The receiver is a nearby loop of wire containing a second spark gap.

In the next section, we show that Equations 34.6 and 34.7 can be combined to obtain a wave equation for both the electric field and the magnetic field. In empty space, where $q = 0$ and $I = 0$, the solution to these two equations shows that the speed at which electromagnetic waves travel equals the measured speed of light. This result led Maxwell to predict that light waves are a form of electromagnetic radiation.

Hertz performed experiments that verified Maxwell's prediction. The experimental apparatus Hertz used to generate and detect electromagnetic waves is shown schematically in Figure 34.3. An induction coil is connected to a transmitter made up of two spherical electrodes separated by a narrow gap. The coil provides short voltage surges to the electrodes, making one positive and the other negative. A spark is generated between the spheres when the electric field near either electrode surpasses the dielectric strength for air (3×10^6 V/m; see Table 26.1). Free electrons in a strong electric field are accelerated and gain enough energy to ionize any molecules they strike. This ionization provides more electrons, which can accelerate and cause further ionizations. As the air in the gap is ionized, it becomes a much better conductor and the discharge between the electrodes exhibits an oscillatory behavior at a very high frequency. From an electric-circuit viewpoint, this experimental apparatus is equivalent to an LC circuit in which the inductance is that of the coil and the capacitance is due to the spherical electrodes.

HEINRICH RUDOLF HERTZ
German Physicist (1857–1894)
Hertz made his most important discovery of electromagnetic waves in 1887. After finding that the speed of an electromagnetic wave was the same as that of light, Hertz showed that electromagnetic waves, like light waves, could be reflected, refracted, and diffracted. Hertz died of blood poisoning at the age of 36. During his short life, he made many contributions to science. The hertz, equal to one complete vibration or cycle per second, is named after him.

Because L and C are small in Hertz's apparatus, the frequency of oscillation is high, on the order of 100 MHz. (Recall from Eq. 32.22 that $\omega = 1/\sqrt{LC}$ for an LC circuit.) Electromagnetic waves are radiated at this frequency as a result of the oscillation (and hence acceleration) of free charges in the transmitter circuit. Hertz was able to detect these waves using a single loop of wire with its own spark gap (the receiver). Such a receiver loop, placed several meters from the transmitter, has its own effective inductance, capacitance, and natural frequency of oscillation. In Hertz's experiment, sparks were induced across the gap of the receiving electrodes when the receiver's frequency was adjusted to match that of the transmitter. In this way, Hertz demonstrated that the oscillating current induced in the receiver was produced by electromagnetic waves radiated by the transmitter. His experiment is analogous to the mechanical phenomenon in which a tuning fork responds to acoustic vibrations from an identical tuning fork that is oscillating.

In addition, Hertz showed in a series of experiments that the radiation generated by his spark-gap device exhibited the wave properties of interference, diffraction, reflection, refraction, and polarization, which are all properties exhibited by light as we shall see in Part 5. Therefore, it became evident that the radio-frequency waves Hertz was generating had properties similar to those of light waves and that they differed only in frequency and wavelength. Perhaps his most convincing experiment was the measurement of the speed of this radiation. Waves of known frequency were reflected from a metal sheet and created a standing-wave interference pattern whose nodal points could be detected. The measured distance between the nodal points enabled determination of the wavelength λ. Using the relationship $v = \lambda f$ (Eq. 16.12), Hertz found that v was close to 3×10^8 m/s, the known speed c of visible light.

34.3 Plane Electromagnetic Waves

The properties of electromagnetic waves can be deduced from Maxwell's equations. One approach to deriving these properties is to solve the second-order differential equation obtained from Maxwell's third and fourth equations. A rigorous mathematical treatment of that sort is beyond the scope of this text. To circumvent this problem, let's assume the vectors for the electric field and magnetic field in an electromagnetic wave have a specific space–time behavior that is simple but consistent with Maxwell's equations.

To understand the prediction of electromagnetic waves more fully, let's focus our attention on an electromagnetic wave that travels in the x direction (the *direction of propagation*). For this wave, the electric field $\vec{\mathbf{E}}$ is in the y direction and the magnetic field $\vec{\mathbf{B}}$ is in the z direction as shown in Active Figure 34.4a. Such waves, in which the electric and magnetic fields are restricted to being parallel to a pair of perpendicular axes, are said to be **linearly polarized waves.** Furthermore, let's assume the field magnitudes E and B depend on x and t only, not on the y or z coordinate. Active Figure 34.4b shows a sinusoidal electromagnetic wave, which we discuss below.

Let's also imagine that the source of the electromagnetic waves is such that a wave radiated from *any* position in the yz plane (not only from the origin as might be suggested by Active Fig. 34.4a) propagates in the x direction and all such waves are emitted in phase. If we define a **ray** as the line along which the wave travels, all rays for these waves are parallel. This entire collection of waves is often called a **plane wave.** A surface connecting points of equal phase on all waves is a geometric plane called a **wave front,** as introduced in Chapter 17. In comparison, a point source of radiation sends waves out radially in all directions. A surface connecting points of equal phase for this situation is a sphere, so this wave is called a **spherical wave.**

PITFALL PREVENTION 34.1

What Is "a" Wave?

What do we mean by a *single* wave? The word *wave* represents both the emission from a *single point* ("wave radiated from *any* position in the yz plane" in the text) and the collection of waves from *all points* on the source ("**plane wave**" in the text). You should be able to use this term in both ways and understand its meaning from the context.

To generate the prediction of electromagnetic waves, we start with Faraday's law, Equation 34.6:

$$\oint \vec{\mathbf{E}} \cdot d\vec{\mathbf{s}} = -\frac{d\Phi_B}{dt}$$

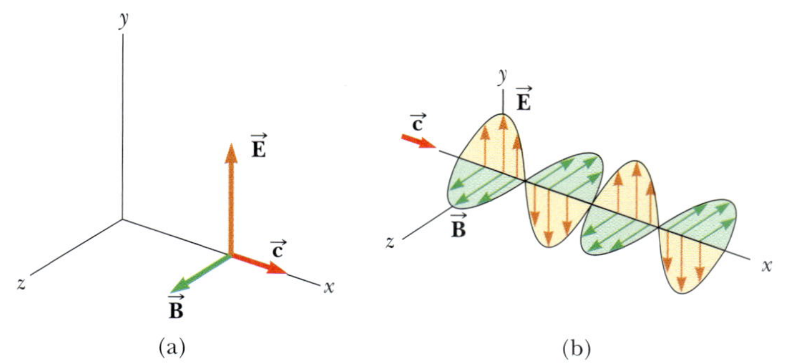

ACTIVE FIGURE 34.4

(a) An electromagnetic wave traveling at velocity $\vec{\mathbf{c}}$ in the positive x direction. The wave is shown at an instant of time at which the electric field is along the y direction and has its maximum magnitude and the magnetic field is along the z direction, also with its maximum magnitude. These fields depend only on x and t. (b) Representation of a sinusoidal electromagnetic wave moving in the positive x direction with a speed c.

Sign in at www.thomsonedu.com and go to ThomsonNOW to observe the wave in (b) and the variation of the fields in time. In addition, you can take a "snapshot" of the wave at an instant of time and investigate the electric and magnetic field vectors at that instant.

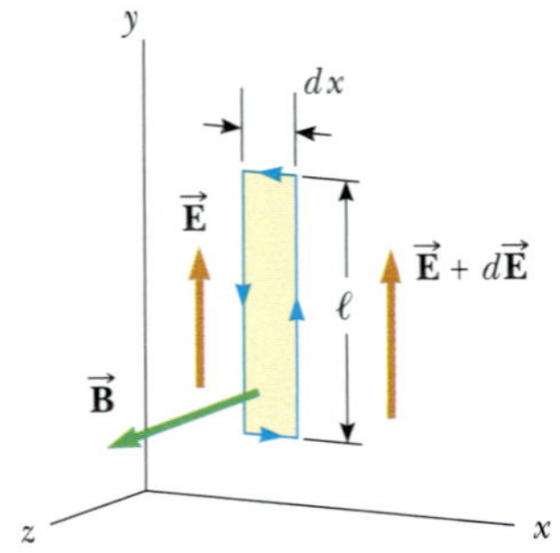

Figure 34.5 At an instant when a plane wave moving in the positive x direction passes through a rectangular path of width dx lying in the xy plane, the electric field in the y direction varies from $\vec{\mathbf{E}}$ to $\vec{\mathbf{E}} + d\vec{\mathbf{E}}$. This spatial variation in $\vec{\mathbf{E}}$ gives rise to a time-varying magnetic field along the z direction, according to Equation 34.11.

Let's again assume the electromagnetic wave is traveling in the x direction, with the electric field $\vec{\mathbf{E}}$ in the positive y direction and the magnetic field $\vec{\mathbf{B}}$ in the positive z direction.

Consider a rectangle of width dx and height ℓ lying in the xy plane as shown in Figure 34.5. To apply Equation 34.6, let's first evaluate the line integral of $\vec{\mathbf{E}} \cdot d\vec{\mathbf{s}}$ around this rectangle. The contributions from the top and bottom of the rectangle are zero because $\vec{\mathbf{E}}$ is perpendicular to $d\vec{\mathbf{s}}$ for these paths. We can express the electric field on the right side of the rectangle as

$$E(x + dx, t) \approx E(x, t) + \left.\frac{dE}{dx}\right|_{t\text{ constant}} dx = E(x, t) + \frac{\partial E}{\partial x} dx$$

where $E(x, t)$ is the field on the left side.[2] Therefore, the line integral over this rectangle is approximately

$$\oint \vec{\mathbf{E}} \cdot d\vec{\mathbf{s}} = [E(x + dx, t)]\ell - [E(x, t)]\ell \approx \ell\left(\frac{\partial E}{\partial x}\right) dx \tag{34.9}$$

Because the magnetic field is in the z direction, the magnetic flux through the rectangle of area $\ell\,dx$ is approximately $\Phi_B = B\ell\,dx$ (assuming dx is very small compared with the wavelength of the wave). Taking the time derivative of the magnetic flux gives

$$\frac{d\Phi_B}{dt} = \ell\,dx \left.\frac{dB}{dt}\right|_{x\text{ constant}} = \ell\,dx \frac{\partial B}{\partial t} \tag{34.10}$$

Substituting Equations 34.9 and 34.10 into Equation 34.6 gives

$$\ell\left(\frac{\partial E}{\partial x}\right) dx = -\ell\,dx \frac{\partial B}{\partial t}$$

$$\frac{\partial E}{\partial x} = -\frac{\partial B}{\partial t} \tag{34.11}$$

In a similar manner, we can derive a second equation by starting with Maxwell's fourth equation in empty space (Eq. 34.7). In this case, the line integral of $\vec{\mathbf{B}} \cdot d\vec{\mathbf{s}}$ is evaluated around a rectangle lying in the xz plane and having width dx and length ℓ as in Figure 34.6. Noting that the magnitude of the magnetic field changes from $B(x, t)$ to $B(x + dx, t)$ over the width dx and that the direction for taking the line integral is as shown in Figure 34.6, the line integral over this rectangle is found to be approximately

$$\oint \vec{\mathbf{B}} \cdot d\vec{\mathbf{s}} = [B(x, t)]\ell - [B(x + dx, t)]\ell \approx -\ell\left(\frac{\partial B}{\partial x}\right) dx \tag{34.12}$$

The electric flux through the rectangle is $\Phi_E = E\ell\,dx$, which, when differentiated with respect to time, gives

$$\frac{\partial \Phi_E}{\partial t} = \ell\,dx \frac{\partial E}{\partial t} \tag{34.13}$$

Substituting Equations 34.12 and 34.13 into Equation 34.7 gives

$$-\ell\left(\frac{\partial B}{\partial x}\right) dx = \mu_0\epsilon_0 \ell\,dx \left(\frac{\partial E}{\partial t}\right)$$

$$\frac{\partial B}{\partial x} = -\mu_0\epsilon_0 \frac{\partial E}{\partial t} \tag{34.14}$$

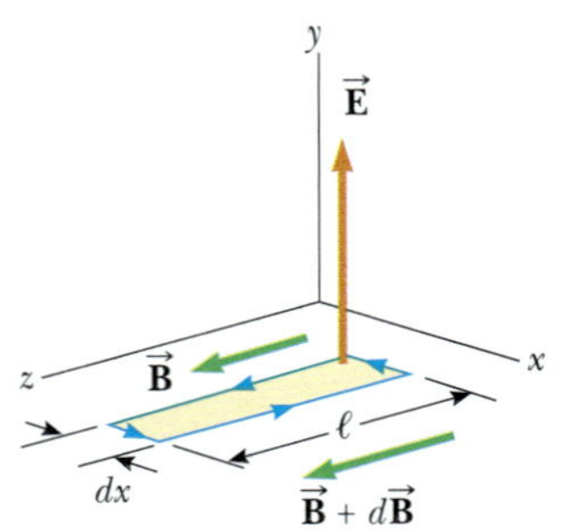

Figure 34.6 At an instant when a plane wave passes through a rectangular path of width dx lying in the xz plane, the magnetic field in the z direction varies from $\vec{\mathbf{B}}$ to $\vec{\mathbf{B}} + d\vec{\mathbf{B}}$. This spatial variation in $\vec{\mathbf{B}}$ gives rise to a time-varying electric field along the y direction, according to Equation 34.14.

[2] Because dE/dx in this equation is expressed as the change in E with x at a given instant t, dE/dx is equivalent to the partial derivative $\partial E/\partial x$. Likewise, dB/dt means the change in B with time at a particular position x; therefore, in Equation 34.10, we can replace dB/dt with $\partial B/\partial t$.

Taking the derivative of Equation 34.11 with respect to x and combining the result with Equation 34.14 gives

$$\frac{\partial^2 E}{\partial x^2} = -\frac{\partial}{\partial x}\left(\frac{\partial B}{\partial t}\right) = -\frac{\partial}{\partial t}\left(\frac{\partial B}{\partial x}\right) = -\frac{\partial}{\partial t}\left(-\mu_0\epsilon_0\frac{\partial E}{\partial t}\right)$$

$$\frac{\partial^2 E}{\partial x^2} = \mu_0\epsilon_0\frac{\partial^2 E}{\partial t^2} \qquad \textbf{(34.15)}$$

In the same manner, taking the derivative of Equation 34.14 with respect to x and combining it with Equation 34.11 gives

$$\frac{\partial^2 B}{\partial x^2} = \mu_0\epsilon_0\frac{\partial^2 B}{\partial t^2} \qquad \textbf{(34.16)}$$

Equations 34.15 and 34.16 both have the form of the general wave equation[3] with the wave speed v replaced by c, where

$$c = \frac{1}{\sqrt{\mu_0\epsilon_0}} \qquad \textbf{(34.17)}$$

◀ Speed of electromagnetic waves

Let's evaluate this speed numerically:

$$c = \frac{1}{\sqrt{(4\pi \times 10^{-7}\ \text{T}\cdot\text{m/A})(8.854\ 19 \times 10^{-12}\ \text{C}^2/\text{N}\cdot\text{m}^2)}}$$
$$= 2.997\ 92 \times 10^8\ \text{m/s}$$

Because this speed is precisely the same as the speed of light in empty space, we are led to believe (correctly) that light is an electromagnetic wave.

The simplest solution to Equations 34.15 and 34.16 is a sinusoidal wave for which the field magnitudes E and B vary with x and t according to the expressions

$$E = E_{\text{max}} \cos(kx - \omega t) \qquad \textbf{(34.18)}$$

$$B = B_{\text{max}} \cos(kx - \omega t) \qquad \textbf{(34.19)}$$

◀ Sinusoidal electric and magnetic fields

where E_{max} and B_{max} are the maximum values of the fields. The angular wave number is $k = 2\pi/\lambda$, where λ is the wavelength. The angular frequency is $\omega = 2\pi f$, where f is the wave frequency. The ratio ω/k equals the speed of an electromagnetic wave, c:

$$\frac{\omega}{k} = \frac{2\pi f}{2\pi/\lambda} = \lambda f = c$$

where we have used Equation 16.12, $v = c = \lambda f$, which relates the speed, frequency, and wavelength of any continuous wave. Therefore, for electromagnetic waves, the wavelength and frequency of these waves are related by

$$\lambda = \frac{c}{f} = \frac{3.00 \times 10^8\ \text{m/s}}{f} \qquad \textbf{(34.20)}$$

Active Figure 34.4b is a pictorial representation, at one instant, of a sinusoidal, linearly polarized plane wave moving in the positive x direction.

Taking partial derivatives of Equations 34.18 (with respect to x) and 34.19 (with respect to t) gives

$$\frac{\partial E}{\partial x} = -kE_{\text{max}} \sin(kx - \omega t)$$

$$\frac{\partial B}{\partial t} = \omega B_{\text{max}} \sin(kx - \omega t)$$

[3] The general wave equation is of the form $(\partial^2 y/\partial x^2) = (1/v^2)(\partial^2 y/\partial t^2)$, where v is the speed of the wave and y is the wave function. The general wave equation was introduced as Equation 16.27, and we suggest you review Section 16.6.

PITFALL PREVENTION 34.2

$\vec{E}$ Stronger Than $\vec{B}$?

Because the value of c is so large, some students incorrectly interpret Equation 34.21 as meaning that the electric field is much stronger than the magnetic field. Electric and magnetic fields are measured in different units, however, so they cannot be directly compared. In Section 34.4, we find that the electric and magnetic fields contribute equally to the wave's energy.

Substituting these results into Equation 34.11 shows that at any instant,

$$kE_{max} = \omega B_{max}$$

$$\frac{E_{max}}{B_{max}} = \frac{\omega}{k} = c$$

Using these results together with Equations 34.18 and 34.19 gives

$$\frac{E_{max}}{B_{max}} = \frac{E}{B} = c \tag{34.21}$$

That is, **at every instant, the ratio of the magnitude of the electric field to the magnitude of the magnetic field in an electromagnetic wave equals the speed of light.**

Finally, note that electromagnetic waves obey the superposition principle (which we discussed in Section 18.1 with respect to mechanical waves) because the differential equations involving E and B are linear equations. For example, we can add two waves with the same frequency and polarization simply by adding the magnitudes of the two electric fields algebraically.

Quick Quiz 34.2 What is the phase difference between the sinusoidal oscillations of the electric and magnetic fields in Active Figure 34.4b? (a) 180° (b) 90° (c) 0 (d) impossible to determine

EXAMPLE 34.2 An Electromagnetic Wave

A sinusoidal electromagnetic wave of frequency 40.0 MHz travels in free space in the x direction as in Figure 34.7.

(A) Determine the wavelength and period of the wave.

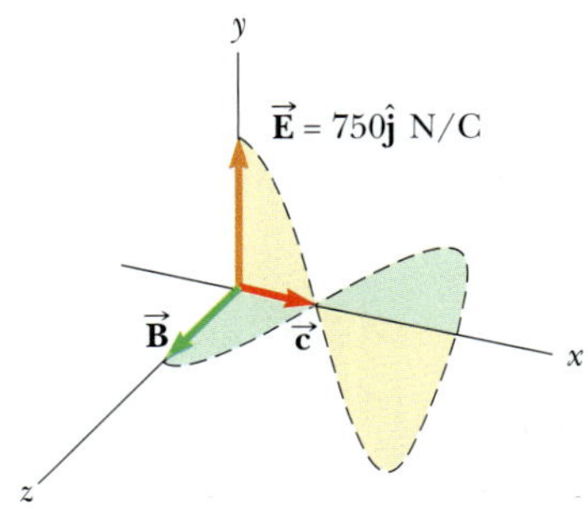

Figure 34.7 (Example 34.2) At some instant, a plane electromagnetic wave moving in the x direction has a maximum electric field of 750 N/C in the positive y direction. The corresponding magnetic field at that point has a magnitude E/c and is in the z direction.

SOLUTION

Conceptualize Imagine the wave in Figure 34.7 moving to the right along the x axis, with the electric and magnetic fields oscillating in phase.

Categorize We evaluate the results using equations developed in this section, so we categorize this example as a substitution problem.

Use Equation 34.20 to find the wavelength of the wave:

$$\lambda = \frac{c}{f} = \frac{3.00 \times 10^8 \text{ m/s}}{40.0 \times 10^6 \text{ Hz}} = 7.50 \text{ m}$$

Find the period T of the wave as the inverse of the frequency:

$$T = \frac{1}{f} = \frac{1}{40.0 \times 10^6 \text{ Hz}} = 2.50 \times 10^{-8} \text{ s}$$

(B) At some point and at some instant, the electric field has its maximum value of 750 N/C and is directed along the y axis. Calculate the magnitude and direction of the magnetic field at this position and time.

SOLUTION

Use Equation 34.21 to find the magnitude of the magnetic field:

$$B_{max} = \frac{E_{max}}{c} = \frac{750 \text{ N/C}}{3.00 \times 10^8 \text{ m/s}} = 2.50 \times 10^{-6} \text{ T}$$

Because $\vec{\mathbf{E}}$ and $\vec{\mathbf{B}}$ must be perpendicular to each other and perpendicular to the direction of wave propagation (x in this case), we conclude that $\vec{\mathbf{B}}$ is in the z direction.

34.4 Energy Carried by Electromagnetic Waves

In our discussion of the nonisolated system model in Section 8.1, we identified electromagnetic radiation as one method of energy transfer across the boundary of a system. The amount of energy transferred by electromagnetic waves is symbolized as T_{ER} in Equation 8.2. The rate of flow of energy in an electromagnetic wave is described by a vector $\vec{\mathbf{S}}$, called the **Poynting vector,** which is defined by the expression

$$\vec{\mathbf{S}} \equiv \frac{1}{\mu_0}\vec{\mathbf{E}} \times \vec{\mathbf{B}} \qquad (34.22)$$

◀ Poynting vector

The magnitude of the Poynting vector represents the rate at which energy flows through a unit surface area perpendicular to the direction of wave propagation. Therefore, the magnitude of $\vec{\mathbf{S}}$ represents *power per unit area*. The direction of the vector is along the direction of wave propagation (Fig. 34.8). The SI units of $\vec{\mathbf{S}}$ are $\text{J/s} \cdot \text{m}^2 = \text{W/m}^2$.

As an example, let's evaluate the magnitude of $\vec{\mathbf{S}}$ for a plane electromagnetic wave where $|\vec{\mathbf{E}} \times \vec{\mathbf{B}}| = EB$. In this case,

$$S = \frac{EB}{\mu_0} \qquad (34.23)$$

Because $B = E/c$, we can also express this result as

$$S = \frac{E^2}{\mu_0 c} = \frac{cB^2}{\mu_0}$$

These equations for S apply at any instant of time and represent the *instantaneous* rate at which energy is passing through a unit area.

What is of greater interest for a sinusoidal plane electromagnetic wave is the time average of S over one or more cycles, which is called the *wave intensity* I. (We discussed the intensity of sound waves in Chapter 17.) When this average is taken, we obtain an expression involving the time average of $\cos^2(kx - \omega t)$, which equals $\frac{1}{2}$. Hence, the average value of S (in other words, the intensity of the wave) is

$$I = S_{\text{avg}} = \frac{E_{\max}B_{\max}}{2\mu_0} = \frac{E_{\max}^2}{2\mu_0 c} = \frac{cB_{\max}^2}{2\mu_0} \qquad (34.24)$$

◀ Wave intensity

Recall that the energy per unit volume, which is the instantaneous energy density u_E associated with an electric field, is given by Equation 26.13:

$$u_E = \tfrac{1}{2}\epsilon_0 E^2$$

Also recall that the instantaneous energy density u_B associated with a magnetic field is given by Equation 32.14:

$$u_B = \frac{B^2}{2\mu_0}$$

Because E and B vary with time for an electromagnetic wave, the energy densities also vary with time. Using the relationships $B = E/c$ and $c = 1/\sqrt{\epsilon_0\mu_0}$, the expression for u_B becomes

$$u_B = \frac{(E/c)^2}{2\mu_0} = \frac{\epsilon_0\mu_0}{2\mu_0}E^2 = \tfrac{1}{2}\epsilon_0 E^2$$

Comparing this result with the expression for u_E, we see that

$$u_B = u_E = \tfrac{1}{2}\epsilon_0 E^2 = \frac{B^2}{2\mu_0}$$

PITFALL PREVENTION 34.3
An Instantaneous Value

The Poynting vector given by Equation 34.22 is time dependent. Its magnitude varies in time, reaching a maximum value at the same instant as the magnitudes of $\vec{\mathbf{E}}$ and $\vec{\mathbf{B}}$ do. The *average* rate of energy transfer is given by Equation 34.24.

PITFALL PREVENTION 34.4
Irradiance

In this discussion, intensity is defined in the same way as in Chapter 17 (as power per unit area). In the optics industry, however, power per unit area is called the *irradiance*. Radiant intensity is defined as the power in watts per solid angle (measured in steradians).

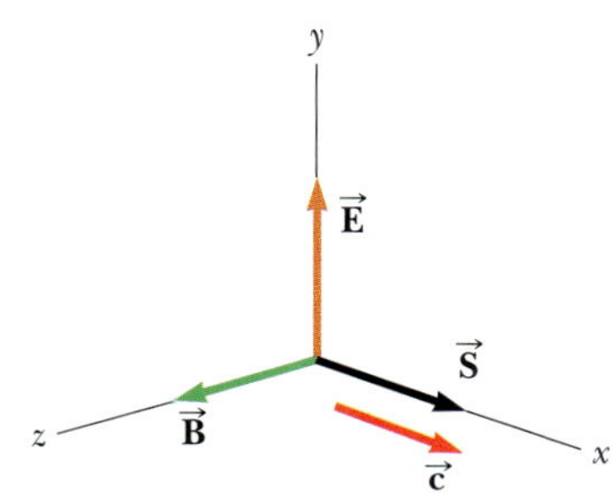

Figure 34.8 The Poynting vector $\vec{\mathbf{S}}$ for a plane electromagnetic wave is along the direction of wave propagation.

That is, **the instantaneous energy density associated with the magnetic field of an electromagnetic wave equals the instantaneous energy density associated with the electric field.** Hence, in a given volume, the energy is equally shared by the two fields.

The **total instantaneous energy density** u is equal to the sum of the energy densities associated with the electric and magnetic fields:

Total instantaneous energy density of an electromagnetic wave ▶

$$u = u_E + u_B = \epsilon_0 E^2 = \frac{B^2}{\mu_0}$$

When this total instantaneous energy density is averaged over one or more cycles of an electromagnetic wave, we again obtain a factor of $\frac{1}{2}$. Hence, for any electromagnetic wave, the total average energy per unit volume is

Average energy density of an electromagnetic wave ▶

$$u_{\text{avg}} = \epsilon_0 (E^2)_{\text{avg}} = \tfrac{1}{2}\epsilon_0 E_{\text{max}}^2 = \frac{B_{\text{max}}^2}{2\mu_0} \qquad \textbf{(34.25)}$$

Comparing this result with Equation 34.24 for the average value of S, we see that

$$I = S_{\text{avg}} = c u_{\text{avg}} \qquad \textbf{(34.26)}$$

In other words, **the intensity of an electromagnetic wave equals the average energy density multiplied by the speed of light.**

The Sun delivers about 10^3 W/m^2 of energy to the Earth's surface via electromagnetic radiation. Let's calculate the total power that is incident on the roof of a home. The roof's dimensions are 8.00 m × 20.0 m. We assume the average magnitude of the Poynting vector for solar radiation at the surface of the Earth is $S_{\text{avg}} =$ 1 000 W/m^2. This average value represents the power per unit area, or the light intensity. Assuming the radiation is incident normal to the roof, we obtain

$$\mathcal{P}_{\text{avg}} = S_{\text{avg}} A = (1\,000 \text{ W/m}^2)(8.00 \text{ m} \times 20.0 \text{ m}) = 1.60 \times 10^5 \text{ W}$$

This power is large compared to the power requirements of a typical home. If this power were maintained for 24 hours per day and the energy could be absorbed and made available to electrical devices, it would provide more than enough energy for the average home. Solar energy is not easily harnessed, however, and the prospects for large-scale conversion are not as bright as may appear from this calculation. For example, the efficiency of conversion from solar energy is typically 10% for photovoltaic cells, reducing the available power by an order of magnitude. Other considerations reduce the power even further. Depending on location, the radiation is most likely not incident normal to the roof and, even if it is (in locations near the equator), this situation exists for only a short time near the middle of the day. No energy is available for about half of each day during the nighttime hours, and cloudy days further reduce the available energy. Finally, while energy is arriving at a large rate during the middle of the day, some of it must be stored for later use, requiring batteries or other storage devices. All in all, complete solar operation of homes is not currently cost effective for most homes.

Quick Quiz 34.3 An electromagnetic wave propagates in the $-y$ direction. The electric field at a point in space is momentarily oriented in the $+x$ direction. In which direction is the magnetic field at that point momentarily oriented? (a) the $-x$ direction (b) the $+y$ direction (c) the $+z$ direction (d) the $-z$ direction

EXAMPLE 34.3 Fields on the Page

Estimate the maximum magnitudes of the electric and magnetic fields of the light that is incident on this page because of the visible light coming from your desk lamp. Treat the lightbulb as a point source of electromagnetic radiation that is 5% efficient at transforming energy coming in by electrical transmission to energy leaving by visible light.

SOLUTION

Conceptualize The filament in your lightbulb emits electromagnetic radiation. The brighter the light, the larger the magnitudes of the electric and magnetic fields.

Categorize Because the lightbulb is to be treated as a point source, it emits equally in all directions, so the outgoing electromagnetic radiation can be modeled as a spherical wave.

Analyze Recall from Equation 17.7 that the wave intensity I a distance r from a point source is $I = \mathcal{P}_{avg}/4\pi r^2$, where $\mathcal{P}_{avg}$ is the average power output of the source and $4\pi r^2$ is the area of a sphere of radius r centered on the source.

Set this expression for I equal to the intensity of an electromagnetic wave given by Equation 34.24:

$$I = \frac{\mathcal{P}_{avg}}{4\pi r^2} = \frac{E_{max}^2}{2\mu_0 c}$$

Solve for the electric field magnitude:

$$E_{max} = \sqrt{\frac{\mu_0 c \mathcal{P}_{avg}}{2\pi r^2}}$$

Let's make some assumptions about numbers to enter in this equation. The visible light output of a 60-W lightbulb operating at 5% efficiency is approximately 3.0 W by visible light. (The remaining energy transfers out of the lightbulb by conduction and invisible radiation.) A reasonable distance from the lightbulb to the page might be 0.30 m.

Substitute these values:

$$E_{max} = \sqrt{\frac{(4\pi \times 10^{-7}\ \text{T}\cdot\text{m/A})(3.00 \times 10^8\ \text{m/s})(3.0\ \text{W})}{2\pi(0.30\ \text{m})^2}}$$

$$= 45\ \text{V/m}$$

Use Equation 34.21 to find the magnetic field magnitude:

$$B_{max} = \frac{E_{max}}{c} = \frac{45\ \text{V/m}}{3.00 \times 10^8\ \text{m/s}} = 1.5 \times 10^{-7}\ \text{T}$$

Finalize This value of the magnetic field magnitude is two orders of magnitude smaller than the Earth's magnetic field.

34.5 Momentum and Radiation Pressure

Electromagnetic waves transport linear momentum as well as energy. As this momentum is absorbed by some surface, pressure is exerted on the surface. In this discussion, let's assume the electromagnetic wave strikes the surface at normal incidence and transports a total energy T_{ER} to the surface in a time interval Δt. Maxwell showed that if the surface absorbs all the incident energy T_{ER} in this time interval (as does a black body, introduced in Section 20.7), the total momentum $\vec{\mathbf{p}}$ transported to the surface has a magnitude

$$p = \frac{T_{ER}}{c} \quad \text{(complete absorption)} \qquad \textbf{(34.27)}$$

◀ Momentum transported to a perfectly absorbing surface

The pressure exerted on the surface is defined as force per unit area F/A, which when combined with Newton's second law gives

$$P = \frac{F}{A} = \frac{1}{A}\frac{dp}{dt}$$

Substituting Equation 34.27 into this expression for P gives

$$P = \frac{1}{A}\frac{dp}{dt} = \frac{1}{A}\frac{d}{dt}\left(\frac{T_{ER}}{c}\right) = \frac{1}{c}\frac{(dT_{ER}/dt)}{A}$$

PITFALL PREVENTION 34.5

So Many p's

We have p for momentum and P for pressure, and they are both related to $\mathcal{P}$ for power! Be sure to keep all these symbols straight.

We recognize $(dT_{ER}/dt)/A$ as the rate at which energy is arriving at the surface per unit area, which is the magnitude of the Poynting vector. Therefore, the radiation pressure P exerted on the perfectly absorbing surface is

Radiation pressure exerted on a perfectly absorbing surface ▶

$$P = \frac{S}{c} \tag{34.28}$$

If the surface is a perfect reflector (such as a mirror) and incidence is normal, the momentum transported to the surface in a time interval Δt is twice that given by Equation 34.27. That is, the momentum transferred to the surface by the incoming light is $p = T_{ER}/c$ and that transferred by the reflected light also is $p = T_{ER}/c$. Therefore,

$$p = \frac{2T_{ER}}{c} \quad \text{(complete reflection)} \tag{34.29}$$

The radiation pressure exerted on a perfectly reflecting surface for normal incidence of the wave is

Radiation pressure exerted on a perfectly reflecting surface ▶

$$P = \frac{2S}{c} \tag{34.30}$$

The pressure on a surface having a reflectivity somewhere between these two extremes has a value between S/c and $2S/c$, depending on the properties of the surface.

Although radiation pressures are very small (about 5×10^{-6} N/m^2 for direct sunlight), NASA is exploring the possibility of *solar sailing* as a low-cost means of sending spacecraft to the planets. Large sheets would experience radiation pressure from sunlight and would be used in much the way canvas sheets are used on earthbound sailboats. In 1973, NASA engineers took advantage of the momentum of the sunlight striking the solar panels of *Mariner 10* to make small course corrections when the spacecraft's fuel supply was running low. (This procedure was carried out when the spacecraft was in the vicinity of Mercury. Would it have worked as well near Neptune?)

Quick Quiz 34.4 To maximize the radiation pressure on the sails of a spacecraft using solar sailing, should the sheets be (a) very black to absorb as much sunlight as possible or (b) very shiny to reflect as much sunlight as possible?

CONCEPTUAL EXAMPLE 34.4 Sweeping the Solar System

A great amount of dust exists in interplanetary space. Although in theory these dust particles can vary in size from molecular size to a much larger size, very little of the dust in our solar system is smaller than about 0.2 μm. Why?

SOLUTION

The dust particles are subject to two significant forces: the gravitational force that draws them toward the Sun and the radiation-pressure force that pushes them away from the Sun. The gravitational force is proportional to the cube of the radius of a spherical dust particle because it is proportional to the mass and therefore to the volume $4\pi r^3/3$ of the particle. The radiation pressure is proportional to the square of the radius because it depends on the planar cross section of the particle. For large particles, the gravitational force is greater than the force from radiation pressure. For particles having radii less than about 0.2 μm, the radiation-pressure force is greater than the gravitational force. As a result, these particles are swept out of our solar system by sunlight.

EXAMPLE 34.5 Pressure from a Laser Pointer

When giving presentations, many people use a laser pointer to direct the attention of the audience to information on a screen. If a 3.0-mW pointer creates a spot on a screen that is 2.0 mm in diameter, determine the radiation pressure on a screen that reflects 70% of the light that strikes it. The power 3.0 mW is a time-averaged value.

SOLUTION

Conceptualize The pressure should not be very large.

Categorize This problem involves a calculation of radiation pressure using an approach like that leading to Equation 34.28 or Equation 34.30, but it is complicated by the 70% reflection.

Analyze We begin by determining the magnitude of the beam's Poynting vector.

Divide the time-averaged power delivered via the electromagnetic wave by the cross-sectional area of the beam:

$$S_{avg} = \frac{\mathcal{P}_{avg}}{A} = \frac{\mathcal{P}_{avg}}{\pi r^2} = \frac{3.0 \times 10^{-3}\text{ W}}{\pi\left(\dfrac{2.0 \times 10^{-3}\text{ m}}{2}\right)^2} = 955\text{ W/m}^2$$

Now let's determine the radiation pressure from the laser beam. Equation 34.30 indicates that a completely reflected beam would apply an average pressure of $P_{avg} = 2S_{avg}/c$. We can model the actual reflection as follows. Imagine that the surface absorbs the beam, resulting in pressure $P_{avg} = S_{avg}/c$. Then the surface emits the beam, resulting in additional pressure $P_{avg} = S_{avg}/c$. If the surface emits only a fraction f of the beam (so that f is the amount of the incident beam reflected), the pressure due to the emitted beam is $P_{avg} = fS_{avg}/c$.

Use this model to find the total pressure on the surface due to absorption and re-emission (reflection):

$$P_{avg} = \frac{S_{avg}}{c} + f\frac{S_{avg}}{c} = (1 + f)\frac{S_{avg}}{c}$$

Evaluate this pressure for a beam that is 70% reflected:

$$P_{avg} = (1 + 0.70)\frac{955\text{ W/m}^2}{3.0 \times 10^8\text{ m/s}} = \boxed{5.4 \times 10^{-6}\text{ N/m}^2}$$

Finalize Consider the magnitude of the Poynting vector, $S_{avg} = 955\text{ W/m}^2$. It is about the same as the intensity of sunlight at the Earth's surface. For this reason, it is not safe to shine the beam of a laser pointer into a person's eyes, which may be more dangerous than looking directly at the Sun. The pressure has an extremely small value, as expected. (Recall from Section 14.2 that atmospheric pressure is approximately 10^5 N/m^2.)

What If? What if the laser pointer is moved twice as far away from the screen? Does that affect the radiation pressure on the screen?

Answer Because a laser beam is popularly represented as a beam of light with constant cross section, you might think that the intensity of radiation, and therefore the radiation pressure, is independent of distance from the screen. A laser beam, however, does not have a constant cross section at all distances from the source; rather, there is a small but measurable divergence of the beam. If the laser is moved farther away from the screen, the area of illumination on the screen increases, decreasing the intensity. In turn, the radiation pressure is reduced.

In addition, the doubled distance from the screen results in more loss of energy from the beam due to scattering from air molecules and dust particles as the light travels from the laser to the screen. This energy loss further reduces the radiation pressure on the screen.

34.6 Production of Electromagnetic Waves by an Antenna

Stationary charges and steady currents cannot produce electromagnetic waves. Whenever the current in a wire changes with time, however, the wire emits electromagnetic radiation. **The fundamental mechanism responsible for this radiation is the acceleration of a charged particle. Whenever a charged particle accelerates, it radiates energy.**

Let's consider the production of electromagnetic waves by a *half-wave antenna*. In this arrangement, two conducting rods are connected to a source of alternating voltage (such as an LC oscillator) as shown in Figure 34.9 (page 966). The length

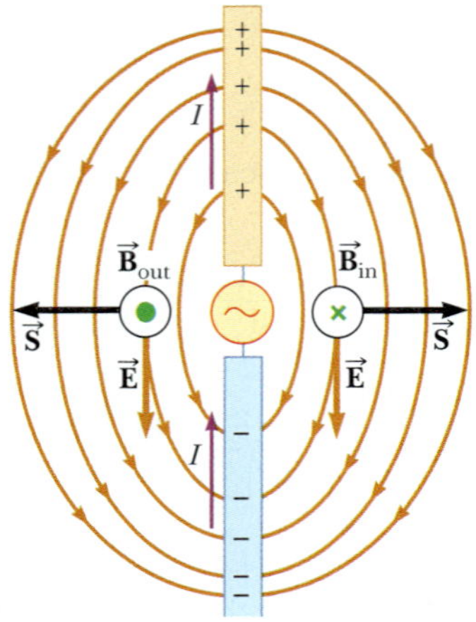

Figure 34.9 A half-wave antenna consists of two metal rods connected to an alternating voltage source. This diagram shows $\vec{E}$ and $\vec{B}$ at an arbitrary instant when the current is upward. Notice that the electric field lines resemble those of a dipole (shown in Fig. 23.20).

of each rod is equal to one quarter of the wavelength of the radiation emitted when the oscillator operates at frequency f. The oscillator forces charges to accelerate back and forth between the two rods. Figure 34.9 shows the configuration of the electric and magnetic fields at some instant when the current is upward. The separation of charges in the upper and lower portions of the antenna make the electric field lines resemble those of an electric dipole. (As a result, this type of antenna is sometimes called a *dipole antenna.*) Because these charges are continuously oscillating between the two rods, the antenna can be approximated by an oscillating electric dipole. The current representing the movement of charges between the ends of the antenna produces magnetic field lines forming concentric circles around the antenna that are perpendicular to the electric field lines at all points. The magnetic field is zero at all points along the axis of the antenna. Furthermore, $\vec{E}$ and $\vec{B}$ are 90° out of phase in time; for example, the current is zero when the charges at the outer ends of the rods are at a maximum.

At the two points where the magnetic field is shown in Figure 34.9, the Poynting vector $\vec{S}$ is directed radially outward, indicating that energy is flowing away from the antenna at this instant. At later times, the fields and the Poynting vector reverse direction as the current alternates. Because $\vec{E}$ and $\vec{B}$ are 90° out of phase at points near the dipole, the net energy flow is zero. From this fact, you might conclude (incorrectly) that no energy is radiated by the dipole.

Energy is indeed radiated, however. Because the dipole fields fall off as $1/r^3$ (as shown in Example 23.5 for the electric field of a static dipole), they are negligible at great distances from the antenna. At these great distances, something else causes a type of radiation different from that close to the antenna. The source of this radiation is the continuous induction of an electric field by the time-varying magnetic field and the induction of a magnetic field by the time-varying electric field, predicted by Equations 34.6 and 34.7. The electric and magnetic fields produced in this manner are in phase with each other and vary as $1/r$. The result is an outward flow of energy at all times.

The angular dependence of the radiation intensity produced by a dipole antenna is shown in Figure 34.10. Notice that the intensity and the power radiated are a maximum in a plane that is perpendicular to the antenna and passing through its midpoint. Furthermore, the power radiated is zero along the antenna's axis. A mathematical solution to Maxwell's equations for the dipole antenna shows that the intensity of the radiation varies as $(\sin^2\theta)/r^2$, where θ is measured from the axis of the antenna.

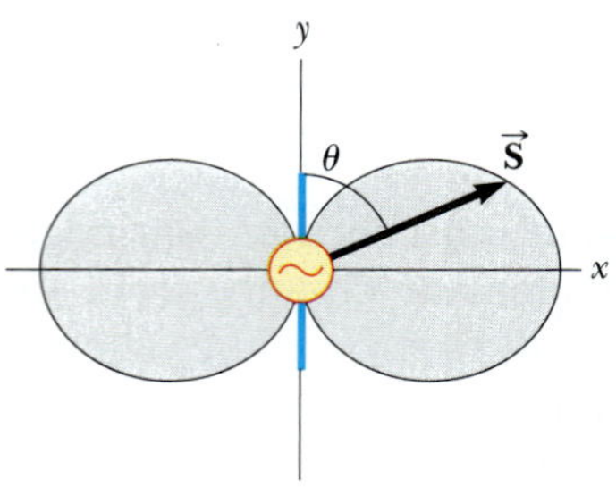

Figure 34.10 Angular dependence of the intensity of radiation produced by an oscillating electric dipole. The distance from the origin to a point on the edge of the gray shape is proportional to the intensity of radiation.

Electromagnetic waves can also induce currents in a receiving antenna. The response of a dipole receiving antenna at a given position is a maximum when the antenna axis is parallel to the electric field at that point and zero when the axis is perpendicular to the electric field.

Quick Quiz 34.5 If the antenna in Figure 34.9 represents the source of a distant radio station, what would be the best orientation for your portable radio antenna located to the right of the figure? (a) up–down along the page (b) left–right along the page (c) perpendicular to the page

34.7 The Spectrum of Electromagnetic Waves

The various types of electromagnetic waves are listed in Figure 34.11, which shows the **electromagnetic spectrum.** Notice the wide ranges of frequencies and wavelengths. No sharp dividing point exists between one type of wave and the next. Remember that **all forms of the various types of radiation are produced by the same phenomenon, accelerating charges.** The names given to the types of waves are simply a convenient way to describe the region of the spectrum in which they lie.

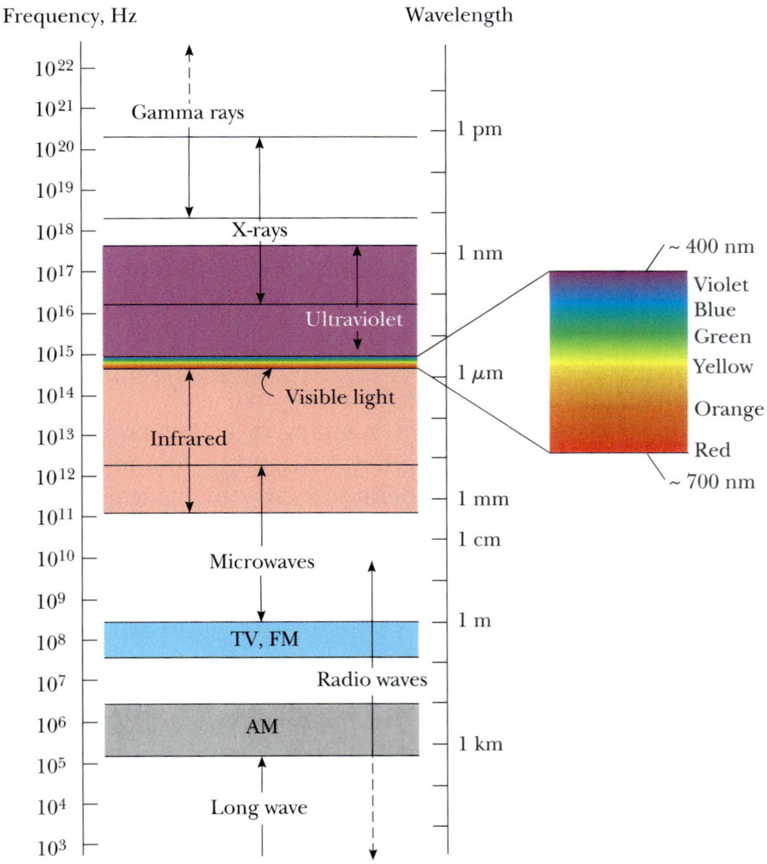

Figure 34.11 The electromagnetic spectrum. Notice the overlap between adjacent wave types. The expanded view to the right shows details of the visible spectrum.

Radio waves, whose wavelengths range from more than 10^4 m to about 0.1 m, are the result of charges accelerating through conducting wires. They are generated by such electronic devices as *LC* oscillators and are used in radio and television communication systems.

Microwaves have wavelengths ranging from approximately 0.3 m to 10^{-4} m and are also generated by electronic devices. Because of their short wavelengths, they are well suited for radar systems and for studying the atomic and molecular properties of matter. Microwave ovens are an interesting domestic application of these waves. It has been suggested that solar energy could be harnessed by beaming microwaves to the Earth from a solar collector in space.

Infrared waves have wavelengths ranging from approximately 10^{-3} m to the longest wavelength of visible light, 7×10^{-7} m. These waves, produced by molecules and room-temperature objects, are readily absorbed by most materials. The infrared (IR) energy absorbed by a substance appears as internal energy because the energy agitates the object's atoms, increasing their vibrational or translational motion, which results in a temperature increase. Infrared radiation has practical and scientific applications in many areas, including physical therapy, IR photography, and vibrational spectroscopy.

Visible light, the most familiar form of electromagnetic waves, is the part of the electromagnetic spectrum the human eye can detect. Light is produced by the rearrangement of electrons in atoms and molecules. The various wavelengths of visible light, which correspond to different colors, range from red ($\lambda \approx 7 \times 10^{-7}$ m) to violet ($\lambda \approx 4 \times 10^{-7}$ m). The sensitivity of the human eye is a function of wavelength, being a maximum at a wavelength of about 5.5×10^{-7} m. With that in mind, why do you suppose tennis balls often have a yellow-green color? Table 34.1 (page 968) provides approximate correspondences between the wavelength of

PITFALL PREVENTION 34.6
"Heat Rays"

Infrared rays are often called "heat rays," but this terminology is a misnomer. Although infrared radiation is used to raise or maintain temperature as in the case of keeping food warm with "heat lamps" at a fast-food restaurant, all wavelengths of electromagnetic radiation carry energy that can cause the temperature of a system to increase. As an example, consider a potato baking in your microwave oven.

TABLE 34.1

Approximate Correspondence Between Wavelengths of Visible Light and Color

Wavelength Range (nm)	Color Description
400–430	Violet
430–485	Blue
485–560	Green
560–590	Yellow
590–625	Orange
625–700	Red

Note: The wavelength ranges here are approximate. Different people will describe colors differently.

Ron Chapple/Getty Images

Wearing sunglasses that do not block ultraviolet (UV) light is worse for your eyes than wearing no sunglasses at all. The lenses of any sunglasses absorb some visible light, thereby causing the wearer's pupils to dilate. If the glasses do not also block UV light, more damage may be done to the lens of the eye because of the dilated pupils. If you wear no sunglasses at all, your pupils are contracted, you squint, and much less UV light enters your eyes. High-quality sunglasses block nearly all the eye-damaging UV light.

visible light and the color assigned to it by humans. Light is the basis of the science of optics and optical instruments, to be discussed in Chapters 35 through 38.

Ultraviolet waves cover wavelengths ranging from approximately 4×10^{-7} m to 6×10^{-10} m. The Sun is an important source of ultraviolet (UV) light, which is the main cause of sunburn. Sunscreen lotions are transparent to visible light but absorb most UV light. The higher a sunscreen's solar protection factor, or SPF, the greater the percentage of UV light absorbed. Ultraviolet rays have also been implicated in the formation of cataracts, a clouding of the lens inside the eye.

Most of the UV light from the Sun is absorbed by ozone (O_3) molecules in the Earth's upper atmosphere, in a layer called the stratosphere. This ozone shield converts lethal high-energy UV radiation to IR radiation, which in turn warms the stratosphere.

X-rays have wavelengths in the range from approximately 10^{-8} m to 10^{-12} m. The most common source of x-rays is the stopping of high-energy electrons upon bombarding a metal target. X-rays are used as a diagnostic tool in medicine and as a treatment for certain forms of cancer. Because x-rays can damage or destroy living tissues and organisms, care must be taken to avoid unnecessary exposure or overexposure. X-rays are also used in the study of crystal structure because x-ray wavelengths are comparable to the atomic separation distances in solids (about 0.1 nm).

Gamma rays are electromagnetic waves emitted by radioactive nuclei (such as ^{60}Co and ^{137}Cs) and during certain nuclear reactions. High-energy gamma rays are a component of cosmic rays that enter the Earth's atmosphere from space. They have wavelengths ranging from approximately 10^{-10} m to less than 10^{-14} m. Gamma rays are highly penetrating and produce serious damage when absorbed by living tissues. Consequently, those working near such dangerous radiation must be protected with heavily absorbing materials such as thick layers of lead.

Quick Quiz 34.6 In many kitchens, a microwave oven is used to cook food. The frequency of the microwaves is on the order of 10^{10} Hz. Are the wavelengths of these microwaves on the order of (a) kilometers, (b) meters, (c) centimeters, or (d) micrometers?

Quick Quiz 34.7 A radio wave of frequency on the order of 10^5 Hz is used to carry a sound wave with a frequency on the order of 10^3 Hz. Is the wavelength of this radio wave on the order of (a) kilometers, (b) meters, (c) centimeters, or (d) micrometers?

Summary

ThomsonNOW™ Sign in at **www.thomsonedu.com** and go to ThomsonNOW to take a practice test for this chapter.

DEFINITIONS

In a region of space in which there is a changing electric field, there is a **displacement current** defined as

$$I_d \equiv \epsilon_0 \frac{d\Phi_E}{dt} \qquad \textbf{(34.1)}$$

where ϵ_0 is the permittivity of free space (see Section 23.3) and $\Phi_E = \int \vec{\mathbf{E}} \cdot d\vec{\mathbf{A}}$ is the electric flux.

The rate of flow of energy crossing a unit area by electromagnetic radiation is described by the **Poynting vector** $\vec{\mathbf{S}}$, where

$$\vec{\mathbf{S}} \equiv \frac{1}{\mu_0} \vec{\mathbf{E}} \times \vec{\mathbf{B}} \qquad \textbf{(34.22)}$$

(continued)

CONCEPTS AND PRINCIPLES

When used with the **Lorentz force law, $\vec{\mathbf{F}} = q\vec{\mathbf{E}} + q\vec{\mathbf{v}} \times \vec{\mathbf{B}}$, Maxwell's equations** describe all electromagnetic phenomena:

$$\oint \vec{\mathbf{E}} \cdot d\vec{\mathbf{A}} = \frac{q}{\epsilon_0} \quad \textbf{(34.4)}$$

$$\oint \vec{\mathbf{B}} \cdot d\vec{\mathbf{A}} = 0 \quad \textbf{(34.5)}$$

$$\oint \vec{\mathbf{E}} \cdot d\vec{\mathbf{s}} = -\frac{d\Phi_B}{dt} \quad \textbf{(34.6)}$$

$$\oint \vec{\mathbf{B}} \cdot d\vec{\mathbf{s}} = \mu_0 I + \epsilon_0 \mu_0 \frac{d\Phi_E}{dt} \quad \textbf{(34.7)}$$

Electromagnetic waves, which are predicted by Maxwell's equations, have the following properties:

- The electric field and the magnetic field each satisfy a wave equation. These two wave equations, which can be obtained from Maxwell's third and fourth equations, are

$$\frac{\partial^2 E}{\partial x^2} = \mu_0 \epsilon_0 \frac{\partial^2 E}{\partial t^2} \quad \textbf{(34.15)}$$

$$\frac{\partial^2 B}{\partial x^2} = \mu_0 \epsilon_0 \frac{\partial^2 B}{\partial t^2} \quad \textbf{(34.16)}$$

- The waves travel through a vacuum with the speed of light c, where

$$c = \frac{1}{\sqrt{\mu_0 \epsilon_0}} \quad \textbf{(34.17)}$$

- Numerically, the speed of electromagnetic waves in a vacuum is 3.00×10^8 m/s.
- The electric and magnetic fields are perpendicular to each other and perpendicular to the direction of wave propagation.
- The instantaneous magnitudes of $\vec{\mathbf{E}}$ and $\vec{\mathbf{B}}$ in an electromagnetic wave are related by the expression

$$\frac{E}{B} = c \quad \textbf{(34.21)}$$

- Electromagnetic waves carry energy.
- Electromagnetic waves carry momentum.

Because electromagnetic waves carry momentum, they exert pressure on surfaces. If an electromagnetic wave whose Poynting vector is $\vec{\mathbf{S}}$ is completely absorbed by a surface upon which it is normally incident, the radiation pressure on that surface is

$$P = \frac{S}{c} \quad \text{(complete absorption)} \quad \textbf{(34.28)}$$

If the surface totally reflects a normally incident wave, the pressure is doubled.

The electric and magnetic fields of a sinusoidal plane electromagnetic wave propagating in the positive x direction can be written as

$$E = E_{\max} \cos (kx - \omega t) \quad \textbf{(34.18)}$$

$$B = B_{\max} \cos (kx - \omega t) \quad \textbf{(34.19)}$$

where ω is the angular frequency of the wave and k is the angular wave number. These equations represent special solutions to the wave equations for E and B. The wavelength and frequency of electromagnetic waves are related by

$$\lambda = \frac{c}{f} = \frac{3.00 \times 10^8 \text{ m/s}}{f} \quad \textbf{(34.20)}$$

The average value of the Poynting vector for a plane electromagnetic wave has a magnitude

$$S_{\text{avg}} = \frac{E_{\max} B_{\max}}{2\mu_0} = \frac{E_{\max}^2}{2\mu_0 c} = \frac{c B_{\max}^2}{2\mu_0} \quad \textbf{(34.24)}$$

The intensity of a sinusoidal plane electromagnetic wave equals the average value of the Poynting vector taken over one or more cycles.

The electromagnetic spectrum includes waves covering a broad range of wavelengths, from long radio waves at more than 10^4 m to gamma rays at less than 10^{-14} m.

Questions

☐ denotes answer available in *Student Solutions Manual/Study Guide;* **O** denotes objective question

1. What new concept did Maxwell's generalized form of Ampère's law include?

2. Do Maxwell's equations allow for the existence of magnetic monopoles? Explain.

3. Radio stations often advertise "instant news." If that means you can hear the news the instant the radio announcer speaks it, is the claim true? What approximate time interval is required for a message to travel from Maine to California by radio waves? Assume the waves can be detected at this range.

4. When light (or other electromagnetic radiation) travels across a given region, what is it that oscillates? What is it that is transported?

5. If a high-frequency current exists in a solenoid containing a metallic core, the core becomes warm due to induction. Explain why the material rises in temperature in this situation.

6. **O** A student working with transmitting apparatus like Heinrich Hertz's wishes to adjust the electrodes to generate electromagnetic waves with a frequency half as large as before. **(i)** How large should she make the effective capacitance of the pair of electrodes? (a) 4 times larger than before (b) 2 times larger than before (c) $\frac{1}{2}$ as large as before (d) $\frac{1}{4}$ as large as before (e) None of these answers would have the desired effect. **(ii)** After she makes the required adjustment, what will the wavelength of the transmitted wave be? (a) 4 times larger than before (b) 2 times larger than before (c) the same as before (d) $\frac{1}{2}$ as large as before (e) $\frac{1}{4}$ as large as before (f) None of these answers is necessarily true.

7. **O** Assume you charge a comb by running it through your hair and then hold the comb next to a bar magnet. Do the electric and magnetic fields produced constitute an electromagnetic wave? (a) Yes they do, necessarily. (b) Yes they do because charged particles are moving inside the bar magnet. (c) They can, but only if the electric field of the comb and the magnetic field of the magnet are perpendicular. (d) They can, but only if both the comb and the magnet are moving. (e) They can, if either the comb or the magnet or both are accelerating.

8. **O** A small source radiates an electromagnetic wave with a single frequency into vacuum, equally in all directions. **(i)** As the wave moves, does its frequency (a) increase, (b) decrease, or (c) stay constant? Answer the same question about **(ii)** its wavelength, **(iii)** its speed, **(iv)** its intensity, and **(v)** the amplitude of its electric field.

9. **O** A plane electromagnetic wave with a single frequency moves in vacuum in the $+x$ direction. Its amplitude is uniform over the yz plane. **(i)** As the wave moves, does its frequency (a) increase, (b) decrease, or (c) stay constant? Answer the same question about **(ii)** its wavelength, **(iii)** its speed, **(iv)** its intensity, and **(v)** the amplitude of its magnetic field.

10. List as many similarities and differences between sound waves and light waves as you can.

11. Describe the physical significance of the Poynting vector.

12. **O** Assume the amplitude of the electric field in a plane electromagnetic wave is E_1 and the amplitude of the magnetic field is B_1. The source of the wave is then adjusted so that the amplitude of the electric field doubles to become $2E_1$. **(i)** What happens to the amplitude of the magnetic field in this process? (a) It becomes 4 times larger. (b) It becomes 2 times larger. (c) It can stay constant. (d) It becomes $\frac{1}{2}$ as large. (e) It becomes $\frac{1}{4}$ as large. (f) None of these answers is necessarily true. **(ii)** What happens to the intensity of the wave? Choose from the same possibilities.

13. **O** A spherical interplanetary dust grain of radius 0.2 μm is at distance r_1 from the Sun. The gravitational force exerted by the Sun on the grain just balances the force due to radiation pressure from the Sun's light. **(i)** Assume the grain is moved to a distance $2r_1$ from the Sun and released. At this location, what is the net force exerted on the grain? (a) toward the Sun (b) away from the Sun (c) zero (d) impossible to determine without knowing the mass of the grain **(ii)** Now assume the grain is moved back to its original location at r_1, compressed so that it crystallizes into a sphere with significantly higher density, and released. In this situation, what is the net force exerted on the grain? (a) toward the Sun (b) away from the Sun (c) zero (d) impossible to determine without knowing the mass of the grain

14. For a given incident energy of an electromagnetic wave, why is the radiation pressure on a perfectly reflecting surface twice as great as that on a perfect absorbing surface?

15. Some television viewers use "rabbit ears" atop their sets (Fig. Q34.15) instead of purchasing cable television service or satellite dishes. Certain orientations of the receiving antenna on a television set give better reception than others. Furthermore, the best orientation varies from station to station. Explain.

© Thomson Learning/George Semple

Figure Q34.15 Question 15 and Problem 63.

16. What does a radio wave do to the charges in the receiving antenna to provide a signal for your car radio?

17. **O** **(i)** Rank the following kinds of waves according to their wavelength ranges from those with the smallest typical or average wavelength to the largest, noting any cases of equality: (a) gamma rays (b) infrared light (c) microwaves (d) radio waves (e) ultraviolet light (f) visible light (g) x-rays **(ii)** Rank the kinds of waves according to their frequencies from lowest to highest. **(iii)** Rank the kinds of waves according to their speeds in a vacuum from slowest to fastest.

18. An empty plastic or glass dish being removed from a microwave oven can be cool to the touch, even when food on an adjoining dish is hot. How is this phenomenon possible?

19. Why should an infrared photograph of a person look different from a photograph taken with visible light?

20. Suppose a creature from another planet has eyes that are sensitive to infrared radiation. Describe what the alien would see if it looked around the room you are now in. In particular, what would be bright and what would be dim?

21. A home microwave oven uses electromagnetic waves with a wavelength of about 12.2 cm. Some 2.4-GHz cordless telephones suffer noisy interference when a microwave oven is used nearby. Locate the waves used by both devices on the electromagnetic spectrum. Do you expect them to interfere with each other?

Problems

WebAssign The Problems from this chapter may be assigned online in WebAssign.

ThomsonNOW Sign in at **www.thomsonedu.com** and go to ThomsonNOW to assess your understanding of this chapter's topics with additional quizzing and conceptual questions.

1, 2, 3 denotes straightforward, intermediate, challenging; □ denotes full solution available in *Student Solutions Manual/Study Guide;* ▲ denotes coached solution with hints available at **www.thomsonedu.com;** ▭ denotes developing symbolic reasoning; ● denotes asking for qualitative reasoning; 🖥 denotes computer useful in solving problem

Section 34.1 Displacement Current and the General Form of Ampère's Law

1. A 0.100-A current is charging a capacitor that has square plates 5.00 cm on each side. The plate separation is 4.00 mm. Find (a) the time rate of change of electric flux between the plates and (b) the displacement current between the plates.

2. A 0.200-A current is charging a capacitor that has circular plates 10.0 cm in radius. If the plate separation is 4.00 mm, (a) what is the time rate of increase of electric field between the plates? (b) What is the magnetic field between the plates 5.00 cm from the center?

3. Consider the situation shown in Figure P34.3. An electric field of 300 V/m is confined to a circular area 10.0 cm in diameter and directed outward perpendicular to the plane of the figure. If the field is increasing at a rate of 20.0 V/m · s, what are the direction and magnitude of the magnetic field at the point P, 15.0 cm from the center of the circle?

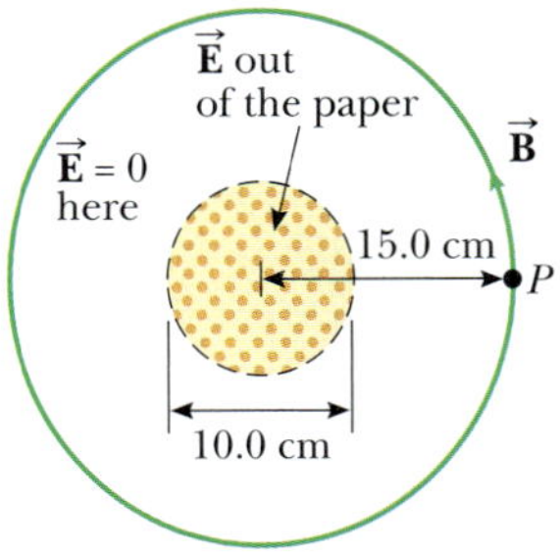

Figure P34.3

Section 34.2 Maxwell's Equations and Hertz's Discoveries

4. A very long, thin rod carries electric charge with the linear density 35.0 nC/m. It lies along the x axis and moves in the x direction at a speed of 15.0 Mm/s. (a) Find the electric field the rod creates at the point ($x = 0$, $y = 20.0$ cm, $z = 0$). (b) Find the magnetic field it creates at the same point. (c) Find the force exerted on an electron at this point, moving with a velocity of $(240\hat{\mathbf{i}})$ Mm/s.

5. A proton moves through a uniform electric field given by $\vec{\mathbf{E}} = 50.0\hat{\mathbf{j}}$ V/m and a uniform magnetic field $\vec{\mathbf{B}} = (0.200\hat{\mathbf{i}} + 0.300\hat{\mathbf{j}} + 0.400\hat{\mathbf{k}})$ T. Determine the acceleration of the proton when it has a velocity $\vec{\mathbf{v}} = 200\hat{\mathbf{i}}$ m/s.

6. An electron moves through a uniform electric field $\vec{\mathbf{E}} = (2.50\hat{\mathbf{i}} + 5.00\hat{\mathbf{j}})$ V/m and a uniform magnetic field $\vec{\mathbf{B}} = (0.400\hat{\mathbf{k}})$ T. Determine the acceleration of the electron when it has a velocity $\vec{\mathbf{v}} = 10.0\hat{\mathbf{i}}$ m/s.

Section 34.3 Plane Electromagnetic Waves

Note: Assume the medium is vacuum unless specified otherwise.

7. (a) The distance to the North Star, Polaris, is approximately 6.44×10^{18} m. If Polaris were to burn out today, in what year would we see it disappear? (b) What time interval is required for sunlight to reach the Earth? (c) What is the transit time for a microwave radar signal traveling from the Earth to the Moon and back? (d) In what time interval does a radio wave travel once around the Earth in a great circle, close to the planet's surface? (e) How long does it take for light to reach you from a lightning stroke 10.0 km away?

8. The speed of an electromagnetic wave traveling in a transparent nonmagnetic substance is $v = 1/\sqrt{\kappa\mu_0\epsilon_0}$, where κ is the dielectric constant of the substance. Determine the speed of light in water, which has a dielectric constant at optical frequencies of 1.78.

9. ▲ Active Figure 34.4b shows a plane electromagnetic sinusoidal wave propagating in the x direction. Suppose the wavelength is 50.0 m and the electric field vibrates in the xy plane with an amplitude of 22.0 V/m. Calculate (a) the frequency of the wave and (b) the magnitude and direction of $\vec{\mathbf{B}}$ when the electric field has its maximum value in the negative y direction. (c) Write an expression for $\vec{\mathbf{B}}$ with the correct unit vector, with numerical values for B_{max}, k, and ω, and with its magnitude in the form

$$B = B_{max} \cos (kx - \omega t)$$

10. An electromagnetic wave in vacuum has an electric field amplitude of 220 V/m. Calculate the amplitude of the corresponding magnetic field.

11. In SI units, the electric field in an electromagnetic wave is described by

$$E_y = 100 \sin (1.00 \times 10^7 x - \omega t)$$

Find (a) the amplitude of the corresponding magnetic field oscillations, (b) the wavelength λ, and (c) the frequency f.

12. Verify by substitution that the following equations are solutions to Equations 34.15 and 34.16, respectively:

$$E = E_{\max} \cos (kx - \omega t)$$

$$B = B_{\max} \cos (kx - \omega t)$$

13. **Review problem.** A standing-wave pattern is set up by radio waves between two metal sheets 2.00 m apart, which is the shortest distance between the plates that produces a standing-wave pattern. What is the frequency of the radio waves?

14. A microwave oven is powered by an electron tube, called a magnetron, that generates electromagnetic waves of frequency 2.45 GHz. The microwaves enter the oven and are reflected by the walls. The standing-wave pattern produced in the oven can cook food unevenly, with hot spots in the food at antinodes and cool spots at nodes, so a turntable is often used to rotate the food and distribute the energy. If a microwave oven intended for use with a turntable is instead used with a cooking dish in a fixed position, the antinodes can appear as burn marks on foods such as carrot strips or cheese. The separation distance between the burns is measured to be 6 cm $\pm$ 5%. From these data, calculate the speed of the microwaves.

Section 34.4 Energy Carried by Electromagnetic Waves

15. How much electromagnetic energy per cubic meter is contained in sunlight if the intensity of sunlight at the Earth's surface under a fairly clear sky is 1 000 W/m^2?

16. An AM radio station broadcasts isotropically (equally in all directions) with an average power of 4.00 kW. A dipole receiving antenna 65.0 cm long is at a location 4.00 mi from the transmitter. Compute the amplitude of the emf that is induced by this signal between the ends of the receiving antenna.

17. What is the average magnitude of the Poynting vector 5.00 mi from a radio transmitter broadcasting isotropically (equally in all directions) with an average power of 250 kW?

18. ● The power of sunlight reaching each square meter of the Earth's surface on a clear day in the tropics is close to 1 000 W. On a winter day in Manitoba, the power concentration of sunlight can be 100 W/m^2. Many human activities are described by a power-per-footprint-area on the order of 10^2 W/m^2 or less. (a) Consider, for example, a family of four paying \$80 to the electric company every 30 days for 600 kWh of energy carried by electrical transmission to their house, which has floor dimensions of 13.0 m by 9.50 m. Compute the power-per-area measure of this energy use. (b) Consider a car 2.10 m wide and 4.90 m long traveling at 55.0 mi/h using gasoline having "heat of combustion" 44.0 MJ/kg with fuel economy 25.0 mi/gal. One gallon of gasoline has a mass of 2.54 kg. Find the power-per-area measure of the car's energy use. It can be similar to that of a steel mill in which rocks are melted in blast furnaces. (c) Explain why direct use of solar energy is not practical for a conventional automobile. What are some uses of solar energy that are on their face more practical?

19. ▲ A community plans to build a facility to convert solar radiation to electrical power. The community requires 1.00 MW of power, and the system to be installed has an efficiency of 30.0% (that is, 30.0% of the solar energy incident on the surface is converted to useful energy that can power the community). What must be the effective area of a perfectly absorbing surface used in such an installation, assuming sunlight has a constant intensity of 1 000 W/m^2?

20. ● Assuming the antenna of a 10.0-kW radio station radiates spherical electromagnetic waves, compute the maximum value of the magnetic field 5.00 km from the antenna and state how this value compares with the surface magnetic field of the Earth.

21. ▲ The filament of an incandescent lamp has a 150-Ω resistance and carries a direct current of 1.00 A. The filament is 8.00 cm long and 0.900 mm in radius. (a) Calculate the Poynting vector at the surface of the filament, associated with the static electric field producing the current and the current's static magnetic field. (b) Find the magnitudes of the static electric and magnetic fields at the surface of the filament.

22. One of the weapons being considered for the "Star Wars" antimissile system is a laser that could destroy ballistic missiles. When a high-power laser is used in the Earth's atmosphere, the electric field can ionize the air, turning it into a conducting plasma that reflects the laser light. In dry air at 0°C and 1 atm, electric breakdown occurs for fields with amplitudes above about 3.00 MV/m. (a) What laser beam intensity will produce such a field? (b) At this maximum intensity, what power can be delivered in a cylindrical beam of diameter 5.00 mm?

23. In a region of free space, the electric field at an instant of time is $\vec{\mathbf{E}} = (80.0\hat{\mathbf{i}} + 32.0\hat{\mathbf{j}} - 64.0\hat{\mathbf{k}})$ N/C and the magnetic field is $\vec{\mathbf{B}} = (0.200\hat{\mathbf{i}} + 0.080\,0\hat{\mathbf{j}} + 0.290\hat{\mathbf{k}})$ μT. (a) Show that the two fields are perpendicular to each other. (b) Determine the Poynting vector for these fields.

24. Model the electromagnetic wave in a microwave oven as a plane traveling wave moving to the left, with an intensity of 25.0 kW/m^2. An oven contains two cubical containers of small mass, each full of water. One has an edge length of 6.00 cm, and the other, 12.0 cm. Energy falls perpendicularly on one face of each container. The water in the smaller container absorbs 70.0% of the energy that falls on it. The water in the larger container absorbs 91.0%. That is, the fraction 0.3 of the incoming microwave energy passes through a 6-cm thickness of water, and the fraction $(0.3)(0.3) = 0.09$ passes through a 12-cm thickness. Find the temperature change of the water in each container over a time interval of 480 s. Assume a negligible amount of energy leaves either container by heat.

25. High-power lasers in factories are used to cut through cloth and metal (Fig. P34.25). One such laser has a beam diameter of 1.00 mm and generates an electric field having an amplitude of 0.700 MV/m at the target. Find (a) the amplitude of the magnetic field produced, (b) the intensity of the laser, and (c) the power delivered by the laser.

26. At one location on the Earth, the rms value of the magnetic field caused by solar radiation is 1.80 μT. From this value, calculate (a) the rms electric field due to solar radiation, (b) the average energy density of the solar component of electromagnetic radiation at this location, and (c) the average magnitude of the Poynting vector for the Sun's radiation.

Philippe Plailly/SPL/Photo Researchers

Figure P34.25

27. Consider a bright star in our night sky. Assume its distance from Earth is 20.0 ly and its power output is 4.00×10^{28} W, about 100 times that of the Sun. (a) Find the intensity of the starlight at the Earth. (b) Find the power of the starlight the Earth intercepts.

Section 34.5 Momentum and Radiation Pressure

28. A possible means of space flight is to place a perfectly reflecting aluminized sheet into orbit around the Earth and then use the light from the Sun to push this "solar sail." Suppose a sail of area 6.00×10^5 m^2 and mass 6 000 kg is placed in orbit facing the Sun. (a) What force is exerted on the sail? (b) What is the sail's acceleration? (c) What time interval is required for the sail to reach the Moon, 3.84×10^8 m away? Ignore all gravitational effects, assume the acceleration calculated in part (b) remains constant, and assume a solar intensity of 1 370 W/m^2.

29. A radio wave transmits 25.0 W/m^2 of power per unit area. A flat surface of area A is perpendicular to the direction of propagation of the wave. Calculate the radiation pressure on it, assuming the surface is a perfect absorber.

30. ● A plane electromagnetic wave of intensity 6.00 W/m^2, moving in the x direction, strikes a small pocket mirror, of area 40.0 cm^2, held in the yz plane. (a) What momentum does the wave transfer to the mirror each second? (b) Find the force that the wave exerts on the mirror. (c) Explain the relationship between the answers to parts (a) and (b).

31. ▲ A 15.0-mW helium–neon laser (λ = 632.8 nm) emits a beam of circular cross section with a diameter of 2.00 mm. (a) Find the maximum electric field in the beam. (b) What total energy is contained in a 1.00-m length of the beam? (c) Find the momentum carried by a 1.00-m length of the beam.

32. ● The intensity of sunlight at the Earth's distance from the Sun is 1 370 W/m^2. (a) Assume the Earth absorbs all the sunlight incident upon it. Find the total force the Sun exerts on the Earth due to radiation pressure. (b) Explain how this force compares with the Sun's gravitational attraction.

33. ● Assume the intensity of solar radiation incident on the upper atmosphere of the Earth is 1 370 W/m^2 and use data from Table 13.2 as necessary. Determine (a) the intensity of solar radiation incident on Mars, (b) the total power incident on Mars, and (c) the radiation force that acts on that planet if it absorbs nearly all the light. (d) State how this force compares with the gravitational attraction exerted by the Sun on Mars. (e) Compare the ratio of the gravitational force to the light-pressure force exerted on Earth, found in Problem 32, and the ratio of these forces exerted on Mars, found in part (d).

34. A uniform circular disk of mass 24.0 g and radius 40.0 cm hangs vertically from a fixed, frictionless, horizontal hinge at a point on its circumference. A beam of electromagnetic radiation with intensity 10.0 MW/m^2 is incident on the disk in a direction perpendicular to its surface. The disk is perfectly absorbing, and the resulting radiation pressure makes the disk rotate. Find the angle through which the disk rotates as it reaches its new equilibrium position. Assume the radiation is *always* perpendicular to the surface of the disk.

Section 34.6 Production of Electromagnetic Waves by an Antenna

35. Figure 34.9 shows a Hertz antenna (also known as a half-wave antenna because its length is $\lambda/2$). The antenna is located far enough from the ground that reflections do not significantly affect its radiation pattern. Most AM radio stations, however, use a Marconi antenna, which consists of the top half of a Hertz antenna. The lower end of this (quarter-wave) antenna is connected to Earth ground, and the ground itself serves as the missing lower half. What are the heights of the Marconi antennas for radio stations broadcasting at (a) 560 kHz and (b) 1 600 kHz? In the United States, these stations are at 560 and 1 600 on the AM dial.

36. Two handheld radio transceivers with dipole antennas are separated by a large, fixed distance. If the transmitting antenna is vertical, what fraction of the maximum received power will appear in the receiving antenna when it is inclined from the vertical by (a) 15.0°? (b) 45.0°? (c) 90.0°?

37. Two vertical radio-transmitting antennas are separated by half the broadcast wavelength and are driven in phase with each other. In what horizontal directions are (a) the strongest and (b) the weakest signals radiated?

38. **Review problem.** Accelerating charges radiate electromagnetic waves. Calculate the wavelength of radiation produced by a proton moving in a circle of radius R perpendicular to a magnetic field of magnitude B.

39. A large, flat sheet carries a uniformly distributed electric current with current per unit width J_s. Problem 33 in Chapter 30 demonstrated that the current creates a magnetic field on both sides of the sheet, parallel to the sheet and perpendicular to the current, with magnitude $B = \frac{1}{2}\mu_0 J_s$. If the current is in the y direction and oscillates in time according to

$$J_{\max}(\cos \omega t)\,\hat{\mathbf{j}} = J_{\max}[\cos(-\omega t)]\,\hat{\mathbf{j}}$$

the sheet radiates an electromagnetic wave as shown in Figure P34.39. The magnetic field of the wave is described

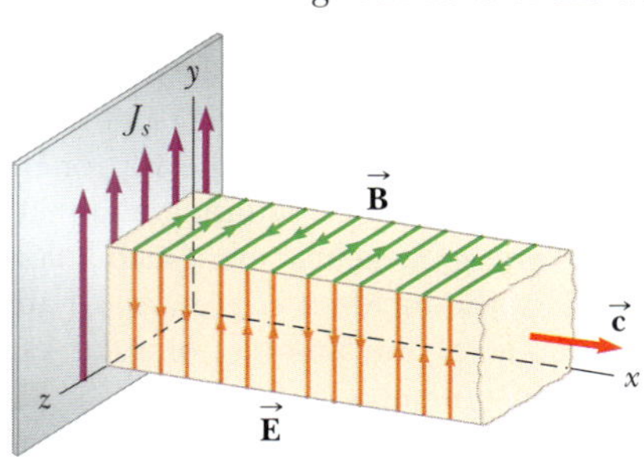

Figure P34.39

by the wave function $\vec{\mathbf{B}} = \frac{1}{2}\mu_0 J_{max}[\cos(kx - \omega t)]\hat{\mathbf{k}}$. (a) Find the wave function for the electric field in the wave. (b) Find the Poynting vector as a function of x and t. (c) Find the intensity of the wave. (d) **What If?** If the sheet is to emit radiation in each direction (normal to the plane of the sheet) with intensity 570 W/m^2, what maximum value of sinusoidal current density is required?

Section 34.7 The Spectrum of Electromagnetic Waves

40. Classify waves with frequencies of 2 Hz, 2 kHz, 2 MHz, 2 GHz, 2 THz, 2 PHz, 2 EHz, 2 ZHz, and 2 YHz on the electromagnetic spectrum. Classify waves with wavelengths of 2 km, 2 m, 2 mm, 2 μm, 2 nm, 2 pm, 2 fm, and 2 am.

41. What are the wavelengths of electromagnetic waves in free space that have frequencies of (a) 5.00×10^{19} Hz and (b) 4.00×10^{9} Hz?

42. Compute an order-of-magnitude estimate for the frequency of an electromagnetic wave with wavelength equal to (a) your height and (b) the thickness of a sheet of paper. How is each wave classified on the electromagnetic spectrum?

43. Twelve VHF television channels (channels 2 through 13) lie in the range of frequencies between 54.0 MHz and 216 MHz. Each channel is assigned a width of 6.0 MHz, with the two ranges 72.0–76.0 MHz and 88.0–174 MHz reserved for non-TV purposes. (Channel 2, for example, lies between 54.0 and 60.0 MHz.) Calculate the broadcast wavelength range for (a) channel 4, (b) channel 6, and (c) channel 8.

44. ● *This just in!* An important news announcement is transmitted by radio waves to people sitting next to their radios 100 km from the station and by sound waves to people sitting across the newsroom 3.00 m from the newscaster. Who receives the news first? Explain. Take the speed of sound in air to be 343 m/s.

45. ● The United States Navy has long proposed the construction of extremely low-frequency (ELF) communication systems. Such waves could penetrate the oceans to reach distant submarines. Calculate the length of a quarter-wavelength antenna for a transmitter generating ELF waves of frequency 75.0 Hz. How practical is this plan?

Additional Problems

46. Write expressions for the electric and magnetic fields of a sinusoidal plane electromagnetic wave having a frequency of 3.00 GHz and traveling in the positive x direction. The amplitude of the electric field is 300 V/m.

47. Assume the intensity of solar radiation incident on the cloud tops of the Earth is 1 370 W/m^2. (a) Calculate the total power radiated by the Sun, taking the average Earth–Sun separation to be 1.496×10^{11} m. (b) Determine the maximum values of the electric and magnetic fields in the sunlight at the Earth's location.

48. The intensity of solar radiation at the top of the Earth's atmosphere is 1 370 W/m^2. Assuming that 60% of the incoming solar energy reaches the Earth's surface and you absorb 50% of the incident energy, make an order-of-magnitude estimate of the amount of solar energy you absorb in a 60-min sunbath.

49. ● You may wish to review Sections 16.5 and 17.3 on the transport of energy by string waves and sound. Active Figure 34.4b is a graphical representation of an electromagnetic wave moving in the x direction. A mathematical representation is given by the two equations $E = E_{max}\cos(kx - \omega t)$ and $B = B_{max}\cos(kx - \omega t)$, where E_{max} is the amplitude of the electric field and B_{max} is the amplitude of the magnetic field. (a) Sketch a graph of the electric field in this wave at the instant $t = 0$, letting your flat paper represent the xy plane. (b) Compute the energy density u_E in the electric field as a function of x at the instant $t = 0$. Compute (c) the energy density in the magnetic field u_B and (d) the total energy density u as functions of x. Add a curve to your graph representing $u = u_E + u_B$. The energy in a rectangular box of length dx along the x direction and area A parallel to the yz plane is $uA\,dx$. The energy in a "shoebox" of length λ and frontal area A is $E_\lambda = \int_0^\lambda uA\,dx$. (The symbol E_λ for energy in a wavelength imitates the notation of Sections 16.5 and 17.3.) (e) Perform the integration to compute the amount of this energy in terms of A, λ, E_{max}, and universal constants. We may think of the energy transport by the whole wave as a series of these shoeboxes going past as if carried on a conveyor belt. Each cycle passes in a time interval defined as the period $T = 1/f$ of the wave. The power the wave carries through area A is then $\mathcal{P} = T_{ER}/\Delta t = E_\lambda/T$. The intensity of the wave is $I = \mathcal{P}/A = E_\lambda/AT$. (f) Compute this intensity in terms of E_{max} and universal constants. Explain how your result compares with that given in Equation 34.24.

50. Consider a small, spherical particle of radius r located in space a distance R from the Sun. (a) Show that the ratio F_{rad}/F_{grav} is proportional to $1/r$, where F_{rad} is the force exerted by solar radiation and F_{grav} is the force of gravitational attraction. (b) The result of part (a) means that, for a sufficiently small value of r, the force exerted on the particle by solar radiation exceeds the force of gravitational attraction. Calculate the value of r for which the particle is in equilibrium under the two forces. Assume the particle has a perfectly absorbing surface and a mass density of 1.50 g/cm^3. Let the particle be located 3.75×10^{11} m from the Sun and use 214 W/m^2 as the value of the solar intensity at that point.

51. A dish antenna having a diameter of 20.0 m receives (at normal incidence) a radio signal from a distant source as shown in Figure P34.51. The radio signal is a continuous sinusoidal wave with amplitude $E_{max} = 0.200\ \mu$V/m. Assume the antenna absorbs all the radiation that falls on the dish. (a) What is the amplitude of the magnetic field in this wave? (b) What is the intensity of the radiation received by this antenna? (c) What is the power received by the antenna? (d) What force is exerted by the radio waves on the antenna?

Figure P34.51

52. ● (a) A stationary charged particle at the origin creates an electric flux of 487 $N \cdot m^2/C$. Find the electric field it creates in the empty space around it as a function of radial distance r away from the particle. (b) A small source at the origin emits an electromagnetic wave with a single frequency into vacuum, equally in all directions, with power 25.0 W. Find the electric field amplitude as a function of radial distance away from the source. Show explicitly how the units in your expression work out. (c) At what distance is the amplitude of the electric field in the wave equal to 3.00 MV/m, representing the dielectric strength of air? (d) As the distance doubles, what happens to the field amplitude? State how this behavior compares with the behavior of the field in part (a).

53. In 1965, Arno Penzias and Robert Wilson discovered the cosmic microwave radiation left over from the Big Bang expansion of the Universe. Suppose the energy density of this background radiation is 4.00×10^{-14} J/m^3. Determine the corresponding electric field amplitude.

54. ● A handheld cellular telephone operates in the 860- to 900-MHz band and has a power output of 0.600 W from an antenna 10.0 cm long (Fig. P34.54). (a) Find the average magnitude of the Poynting vector 4.00 cm from the antenna at the location of a typical person's head. Assume the antenna emits energy with cylindrical wave fronts. (The actual radiation from antennas follows a more complicated pattern.) (b) The ANSI/IEEE C95.1-1991 maximum exposure standard is 0.57 mW/cm^2 for persons living near cellular telephone base stations, who would be continuously exposed to the radiation. State how the answer to part (a) compares with this standard.

Amos Morgan/Getty Images

Figure P34.54

55. A linearly polarized microwave of wavelength 1.50 cm is directed along the positive x axis. The electric field vector has a maximum value of 175 V/m and vibrates in the xy plane. (a) Assuming the magnetic field component of the wave can be written in the form $B = B_{max} \sin(kx - \omega t)$, give values for B_{max}, k, and ω. Also, determine in which plane the magnetic field vector vibrates. (b) Calculate the average value of the Poynting vector for this wave. (c) What radiation pressure would this wave exert if it were directed at normal incidence onto a perfectly reflecting sheet? (d) What acceleration would be imparted to a 500-g sheet (perfectly reflecting and at normal incidence) with dimensions of 1.00 m × 0.750 m?

56. ● The Earth reflects approximately 38.0% of the incident sunlight from its clouds and surface. (a) Given that the intensity of solar radiation is 1 370 W/m^2, find the radiation pressure on the Earth, in pascals, at the location where the Sun is straight overhead. (b) State how this quantity compares with normal atmospheric pressure at the Earth's surface, which is 101 kPa.

57. An astronaut, stranded in space 10.0 m from her spacecraft and at rest relative to it, has a mass (including equipment) of 110 kg. Because she has a 100-W light source that forms a directed beam, she considers using the beam as a photon rocket to propel herself continuously toward the spacecraft. (a) Calculate the time interval required for her to reach the spacecraft by this method. (b) **What If?** Suppose she throws the light source away in the direction away from the spacecraft instead. The mass of the light source is 3.00 kg, and, after being thrown, it moves at 12.0 m/s relative to the recoiling astronaut. After what time interval will the astronaut reach the spacecraft?

58. **Review problem.** A 1.00-m-diameter mirror focuses the Sun's rays onto an absorbing plate 2.00 cm in radius, which holds a can containing 1.00 L of water at 20.0°C. (a) If the solar intensity is 1.00 kW/m^2, what is the intensity on the absorbing plate? (b) What are the maximum magnitudes of the fields $\vec{\mathbf{E}}$ and $\vec{\mathbf{B}}$? (c) If 40.0% of the energy is absorbed, what time interval is required to bring the water to its boiling point?

59. Lasers have been used to suspend spherical glass beads in the Earth's gravitational field. (a) A black bead has a mass of 1.00 μg and a density of 0.200 g/cm^3. Determine the radiation intensity needed to support the bead. (b) If the bead has a radius of 0.200 cm, what power is required for this laser?

60. Lasers have been used to suspend spherical glass beads in the Earth's gravitational field. (a) A black bead has mass m and density ρ. Determine the radiation intensity needed to support the bead. (b) If the bead has radius r, what power is required for this laser?

61. The electromagnetic power radiated by a nonrelativistic particle with charge q moving with acceleration a is

$$\mathcal{P} = \frac{q^2 a^2}{6\pi\epsilon_0 c^3}$$

where ϵ_0 is the permittivity of free space (also called the permittivity of vacuum) and c is the speed of light in vacuum. (a) Show that the right side of this equation has units of watts. (b) An electron is placed in a constant electric field of magnitude 100 N/C. Determine the acceleration of the electron and the electromagnetic power radiated by this electron. (c) **What If?** If a proton is placed in a cyclotron with a radius of 0.500 m and a magnetic field of magnitude 0.350 T, what electromagnetic power does this proton radiate?

62. A plane electromagnetic wave varies sinusoidally at 90.0 MHz as it travels along the $+x$ direction. The peak value of the electric field is 2.00 mV/m, and it is directed along the $\pm y$ direction. (a) Find the wavelength, the period, and the maximum value of the magnetic field. (b) Write expressions in SI units for the space and time variations of the electric field and of the magnetic field. Include both numerical values and subscripts to indicate coordinate directions. (c) Find the average power per unit area that this wave carries through space. (d) Find the average energy density in the radiation (in joules per cubic meter). (e) What radiation pressure would this wave exert upon a perfectly reflecting surface at normal incidence?

63. ▲ **Review problem.** In the absence of cable input or a satellite dish, a television set can use a dipole-receiving antenna for VHF channels and a loop antenna for UHF

channels. In Figure Q34.15, the "rabbit ears" form the VHF antenna and the smaller loop of wire is the UHF antenna. The UHF antenna produces an emf from the changing magnetic flux through the loop. The television station broadcasts a signal with a frequency f, and the signal has an electric-field amplitude E_{max} and a magnetic-field amplitude B_{max} at the location of the receiving antenna. (a) Using Faraday's law, derive an expression for the amplitude of the emf that appears in a single-turn, circular loop antenna with a radius r that is small compared with the wavelength of the wave. (b) If the electric field in the signal points vertically, what orientation of the loop gives the best reception?

> *Note:* Section 8.1 introduced electromagnetic radiation as a mode of energy transfer, which was discussed in more detail in Section 20.7. The next three problems use ideas introduced there and in this chapter.

64. ● **Review problem.** Eliza is a black cat with four black kittens. Eliza's mass is 5.50 kg, and each kitten has mass 0.800 kg. One cool night, all five sleep snuggled together on a mat, with their bodies forming one hemisphere. (a) Assuming the purring heap has a uniform density of 990 kg/m^3, find the radius of the hemisphere. (b) Find the area of its curved surface. (c) Assume the surface temperature is 31.0°C and the emissivity is 0.970. Find the intensity of radiation emitted by the cats at their curved surface and (d) the radiated power from this surface. (e) You may think of the emitted electromagnetic wave as having a single predominant frequency (of 31.2 THz). Find the amplitude of the electric field just outside the surface of the cozy pile and (f) the amplitude of the magnetic field. (g) Are the sleeping cats charged? Are they current carrying? Are they magnetic? Are they a radiation source? Do they glow in the dark? Give an explanation for your answers so that they do not seem contradictory. (h) **What If?** The next night, the kittens all sleep alone, curling up into separate hemispheres like their mother. Find the total radiated power of the family. (For simplicity, ignore the cats' absorption of radiation from the environment.)

65. **Review problem.** (a) An elderly couple has a solar water heater installed on the roof of their house (Fig. P34.65). The heater is a flat, closed box with excellent thermal insulation. Its interior is painted black, and its front face is made of insulating glass. Its emissivity for visible light is 0.900, and its emissivity for infrared light is 0.700. Light from the noon Sun is incident perpendicular to the glass with an intensity of 1 000 W/m^2, and no water enters or leaves the box. Find the steady-state temperature of the box's interior. (b) **What If?** The couple builds an identical box with no water tubes. It lies flat on the ground in front of the house. They use it as a cold frame, where they plant seeds in early spring. Assuming the same noon Sun is at an elevation angle of 50.0°, find the steady-state temperature of the interior of the box when its ventilation slots are tightly closed.

Figure P34.65

66. ● **Review problem.** The study of Creation suggests a Creator with an inordinate fondness for beetles and for small, red stars. A small, red star radiates electromagnetic waves with power 6.00×10^{23} W, which is only 0.159% of the luminosity of the Sun. Consider a spherical planet in a circular orbit around this star. Assume the emissivity of the planet is equal for infrared and for visible light and the planet has a uniform surface temperature. Identify the projected area over which the planet absorbs starlight and the radiating area of the planet. If beetles thrive at a temperature of 310 K, what should the radius of the planet's orbit be?

Answers to Quick Quizzes

34.1 **(i),** (b). There can be no conduction current because there is no conductor between the plates. There is a time-varying electric field because of the decreasing charge on the plates, and the time-varying electric flux represents a displacement current. **(ii),** (c). There is a time-varying electric field because of the decreasing charge on the plates. This time-varying electric field produces a magnetic field.

34.2 (c). Active Figure 34.4b shows that the $\vec{\mathbf{B}}$ and $\vec{\mathbf{E}}$ vectors reach their maximum positive values at the same time.

34.3 (c). The $\vec{\mathbf{B}}$ field must be in the $+z$ direction in order that the Poynting vector be directed along the $-y$ direction.

34.4 (b). To maximize the pressure on the sails, they should be perfectly reflective, so the pressure is given by Equation 34.30.

34.5 (a). The best orientation is parallel to the transmitting antenna because that is the orientation of the electric field. The electric field moves electrons in the receiving antenna, thereby inducing a current that is detected and amplified.

34.6 (c). Either Equation 34.20 or Figure 34.11 can be used to find the order of magnitude of the wavelengths.

34.7 (a). Either Equation 34.20 or Figure 34.11 can be used to find the order of magnitude of the wavelength.

PART 5

Light and Optics

Light is basic to almost all life on the Earth. For example, plants convert the energy transferred by sunlight to chemical energy through photosynthesis. In addition, light is the principal means by which we are able to transmit and receive information to and from objects around us and throughout the Universe. Light is a form of electromagnetic radiation and represents energy transfer from the source to the observer.

Many phenomena in our everyday life depend on the properties of light. When you watch a color television or view photos on a computer monitor, you are seeing millions of colors formed from combinations of only three colors that are physically on the screen: red, blue, and green. The blue color of the daytime sky is a result of the optical phenomenon of *scattering* of light by air molecules, as are the red and orange colors of sunrises and sunsets. You see your image in your bathroom mirror in the morning or the images of other cars in your car's rearview mirror when you are driving. These images result from *reflection* of light. If you wear glasses or contact lenses, you are depending on *refraction* of light for clear vision. The colors of a rainbow result from *dispersion* of light as it passes through raindrops hovering in the sky after a rainstorm. If you have ever seen the colored circles of the glory surrounding the shadow of your airplane on clouds as you fly above them, you are seeing an effect that results from *interference* of light. The phenomena mentioned here have been studied by scientists and are well understood.

In the introduction to Chapter 35, we discuss the dual nature of light. In some cases, it is best to model light as a stream of particles; in others, a wave model works better. Chapters 35 through 38 concentrate on those aspects of light that are best understood through the wave model of light. In Part 6, we will investigate the particle nature of light.

The Grand Tetons in western Wyoming are reflected in a smooth lake at sunset. The optical principles we study in this part of the book will explain the nature of the reflected image of the mountains and why the sky appears red. (David Muench/CORBIS)

This photograph of a rainbow shows a distinct secondary rainbow with the colors reversed. The appearance of the rainbow depends on three optical phenomena discussed in this chapter: reflection, refraction, and dispersion. (Mark D. Phillips/Photo Researchers, Inc.)

35 The Nature of Light and the Laws of Geometric Optics

Courtesy of Rijksmuseum voor de Geschiedenis der Natuurwetenschappen and Niels Bohr Library

CHRISTIAN HUYGENS
Dutch Physicist and Astronomer (1629–1695)
Huygens is best known for his contributions to the fields of optics and dynamics. To Huygens, light was a type of vibratory motion, spreading out and producing the sensation of light when impinging on the eye. On the basis of this theory, he deduced the laws of reflection and refraction and explained the phenomenon of double refraction.

This first chapter on optics begins by introducing two historical models for light and discussing early methods for measuring the speed of light. Next we study the fundamental phenomena of geometric optics: reflection of light from a surface and refraction as the light crosses the boundary between two media. We will also study the dispersion of light as it refracts into materials, resulting in visual displays such as the rainbow. Finally, we investigate the phenomenon of total internal reflection, which is the basis for the operation of optical fibers and the burgeoning technology of fiber optics.

35.1 The Nature of Light

Before the beginning of the nineteenth century, light was considered to be a stream of particles that either was emitted by the object being viewed or emanated from the eyes of the viewer. Newton, the chief architect of the particle model of light, held that particles were emitted from a light source and that these particles stimulated the sense of sight upon entering the eye. Using this idea, he was able to explain reflection and refraction.

Most scientists accepted Newton's particle model. During Newton's lifetime, however, another model was proposed, one that argued that light might be some sort of wave motion. In 1678, Dutch physicist and astronomer Christian Huygens showed that a wave model of light could also explain reflection and refraction.

In 1801, Thomas Young (1773–1829) provided the first clear experimental demonstration of the wave nature of light. Young showed that under appropriate conditions light rays interfere with one another. Such behavior could not be explained at that time by a particle model because there was no conceivable way in which two or more particles could come together and cancel one another. Additional developments during the nineteenth century led to the general acceptance of the wave model of light, the most important resulting from the work of Maxwell, who in 1873 asserted that light was a form of high-frequency electromagnetic wave. As discussed in Chapter 34, Hertz provided experimental confirmation of Maxwell's theory in 1887 by producing and detecting electromagnetic waves.

Although the wave model and the classical theory of electricity and magnetism were able to explain most known properties of light, they could not explain some subsequent experiments. The most striking phenomenon is the photoelectric effect, also discovered by Hertz: when light strikes a metal surface, electrons are sometimes ejected from the surface. As one example of the difficulties that arose, experiments showed that the kinetic energy of an ejected electron is independent of the light intensity. This finding contradicted the wave model, which held that a more intense beam of light should add more energy to the electron. Einstein proposed an explanation of the photoelectric effect in 1905 using a model based on the concept of quantization developed by Max Planck (1858–1947) in 1900. The quantization model assumes the energy of a light wave is present in particles called *photons;* hence, the energy is said to be quantized. According to Einstein's theory, the energy of a photon is proportional to the frequency of the electromagnetic wave:

$$E = hf \qquad \textbf{(35.1)}$$

◀ Energy of a photon

where the constant of proportionality $h = 6.63 \times 10^{-34}\ \text{J}\cdot\text{s}$ is called *Planck's constant.* We study this theory in Chapter 40.

In view of these developments, light must be regarded as having a dual nature. **Light exhibits the characteristics of a wave in some situations and the characteristics of a particle in other situations.** Light is light, to be sure. The question "Is light a wave or a particle?" is inappropriate, however. Sometimes light acts like a wave, and other times it acts like a particle. In the next few chapters, we investigate the wave nature of light.

35.2 Measurements of the Speed of Light

Light travels at such a high speed (to three digits, $c = 3.00 \times 10^8$ m/s) that early attempts to measure its speed were unsuccessful. Galileo attempted to measure the speed of light by positioning two observers in towers separated by approximately 10 km. Each observer carried a shuttered lantern. One observer would open his lantern first, and then the other would open his lantern at the moment he saw the light from the first lantern. Galileo reasoned that by knowing the transit time of the light beams from one lantern to the other and the distance between the two lanterns, he could obtain the speed. His results were inconclusive. Today, we realize (as Galileo concluded) that it is impossible to measure the speed of light in this manner because the transit time for the light is so much less than the reaction time of the observers.

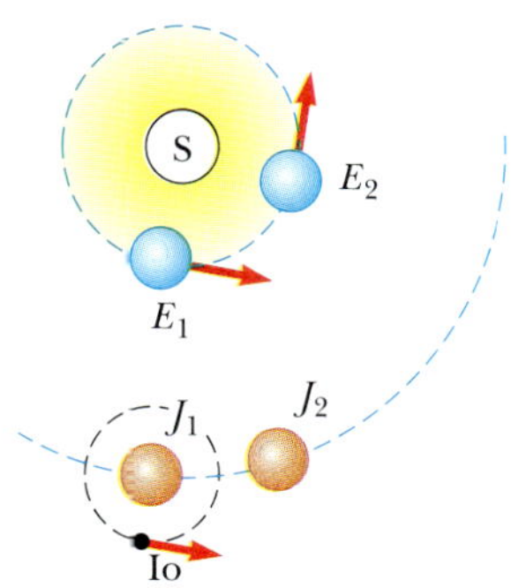

Figure 35.1 Roemer's method for measuring the speed of light. In the time interval during which the Earth travels 90° around the Sun (three months), Jupiter travels only about 7.5° (drawing not to scale).

Roemer's Method

In 1675, Danish astronomer Ole Roemer (1644–1710) made the first successful estimate of the speed of light. Roemer's technique involved astronomical observations of Io, one of the moons of Jupiter. Io has a period of revolution around Jupiter of approximately 42.5 h. The period of revolution of Jupiter around the Sun is about 12 yr; therefore, as the Earth moves through 90° around the Sun, Jupiter revolves through only $(\frac{1}{12})90° = 7.5°$ (Fig. 35.1).

An observer using the orbital motion of Io as a clock would expect the orbit to have a constant period. After collecting data for more than a year, however, Roemer observed a systematic variation in Io's period. He found that the periods were longer than average when the Earth was receding from Jupiter and shorter than average when the Earth was approaching Jupiter. Roemer attributed this variation in period to the distance between the Earth and Jupiter changing from one observation to the next.

Using Roemer's data, Huygens estimated the lower limit for the speed of light to be approximately 2.3×10^8 m/s. This experiment is important historically because it demonstrated that light does have a finite speed and gave an estimate of this speed.

Fizeau's Method

The first successful method for measuring the speed of light by means of purely terrestrial techniques was developed in 1849 by French physicist Armand H. L. Fizeau (1819–1896). Figure 35.2 represents a simplified diagram of Fizeau's apparatus. The basic procedure is to measure the total time interval during which light travels from some point to a distant mirror and back. If d is the distance between the light source (considered to be at the location of the wheel) and the mirror and if the time interval for one round trip is Δt, the speed of light is $c = 2d/\Delta t$.

To measure the transit time, Fizeau used a rotating toothed wheel, which converts a continuous beam of light into a series of light pulses. The rotation of such a wheel controls what an observer at the light source sees. For example, if the pulse traveling toward the mirror and passing the opening at point A in Figure 35.2 should return to the wheel at the instant tooth B had rotated into position to cover the return path, the pulse would not reach the observer. At a greater rate of rotation, the opening at point C could move into position to allow the reflected pulse to reach the observer. Knowing the distance d, the number of teeth in the wheel, and the angular speed of the wheel, Fizeau arrived at a value of 3.1×10^8 m/s. Similar measurements made by subsequent investigators yielded more precise values for c, which led to the currently accepted value of $2.997\,9 \times 10^8$ m/s.

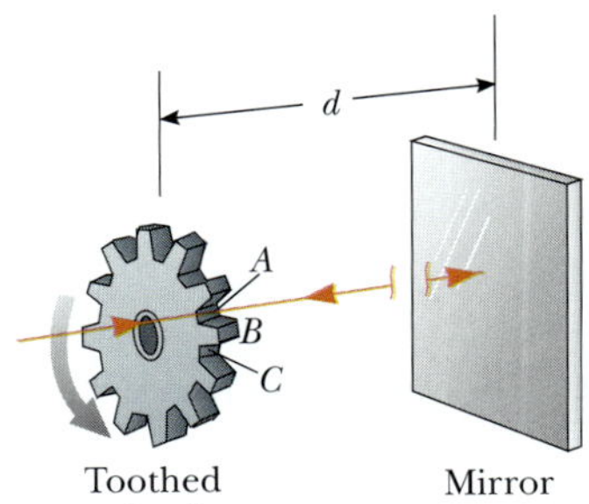

Figure 35.2 Fizeau's method for measuring the speed of light using a rotating toothed wheel. The light source is considered to be at the location of the wheel; therefore, the distance d is known.

EXAMPLE 35.1 Measuring the Speed of Light with Fizeau's Wheel

Assume Fizeau's wheel has 360 teeth and rotates at 27.5 rev/s when a pulse of light passing through opening A in Figure 35.2 is blocked by tooth B on its return. If the distance to the mirror is 7 500 m, what is the speed of light?

SOLUTION

Conceptualize Imagine a pulse of light passing through opening A in Figure 35.2 and reflecting from the mirror. By the time the pulse arrives back at the wheel, tooth B has rotated into the position previously occupied by opening A.

Categorize We model the wheel as a rigid object under constant angular speed and the pulse of light as a particle under constant speed.

Analyze The wheel has 360 teeth, so it must have 360 openings. Therefore, because the light passes through opening A but is blocked by the tooth immediately adjacent to A, the wheel must rotate through an angular displacement of $\frac{1}{720}$ rev in the time interval during which the light pulse makes its round trip.

Use the rigid object under constant angular speed model to find the time interval for the pulse's round trip:

$$\Delta t = \frac{\Delta\theta}{\omega} = \frac{\frac{1}{720}\text{ rev}}{27.5\text{ rev/s}} = 5.05 \times 10^{-5}\text{ s}$$

From the particle under constant speed model, find the speed of the pulse of light:

$$c = \frac{2d}{\Delta t} = \frac{2(7\,500\text{ m})}{5.05 \times 10^{-5}\text{ s}} = 2.97 \times 10^8\text{ m/s}$$

Finalize This result is very close to the actual value of the speed of light.

35.3 The Ray Approximation in Geometric Optics

The field of **geometric optics** involves the study of the propagation of light. Geometric optics assumes light travels in a fixed direction in a straight line as it passes through a uniform medium and changes its direction when it meets the surface of a different medium or if the optical properties of the medium are nonuniform in either space or time. In our study of geometric optics here and in Chapter 36, we use what is called the **ray approximation.** To understand this approximation, first notice that the rays of a given wave are straight lines perpendicular to the wave fronts as illustrated in Figure 35.3 for a plane wave. In the ray approximation, a wave moving through a medium travels in a straight line in the direction of its rays.

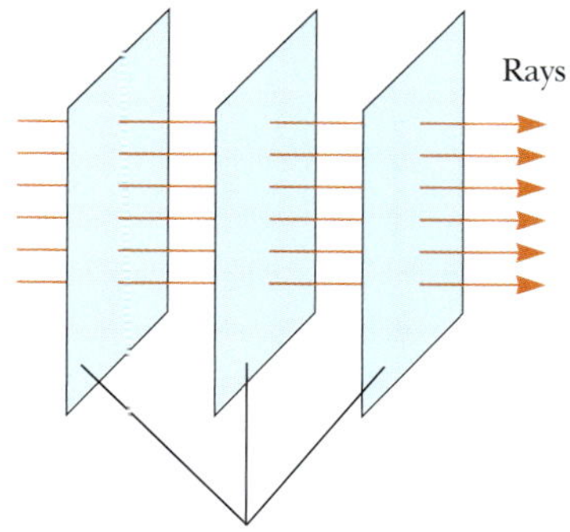

Figure 35.3 A plane wave propagating to the right. Notice that the rays, which always point in the direction of the wave propagation, are straight lines perpendicular to the wave fronts.

If the wave meets a barrier in which there is a circular opening whose diameter is much larger than the wavelength as in Active Figure 35.4a, the wave emerging from the opening continues to move in a straight line (apart from some small edge effects); hence, the ray approximation is valid. If the diameter of the opening is on the order of the wavelength as in Active Figure 35.4b, the waves spread out from the opening in all directions. This effect, called *diffraction,* will be studied in Chapter 37. Finally, if the opening is much smaller than the wavelength, the opening can be approximated as a point source of waves as shown in Active Fig. 35.4c.

Similar effects are seen when waves encounter an opaque object of dimension d. In that case, when $\lambda \ll d$, the object casts a sharp shadow.

The ray approximation and the assumption that $\lambda \ll d$ are used in this chapter and in Chapter 36, both of which deal with geometric optics. This approximation is very good for the study of mirrors, lenses, prisms, and associated optical instruments such as telescopes, cameras, and eyeglasses.

35.4 The Wave Under Reflection

We introduced the concept of reflection of waves in a discussion of waves on strings in Section 16.4. As with waves on strings, when a light ray traveling in one medium encounters a boundary with another medium, part of the incident light is reflected. For waves on a one-dimensional string, the reflected wave must necessarily be restricted to a direction along the string. For light waves traveling in three-dimensional space, no such restriction applies and the reflected light waves can be

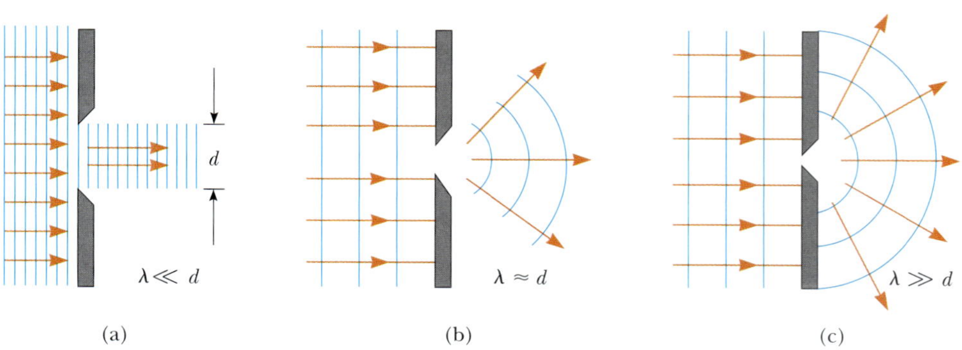

ACTIVE FIGURE 35.4

A plane wave of wavelength λ is incident on a barrier in which there is an opening of diameter d. (a) When $\lambda \ll d$, the rays continue in a straight-line path and the ray approximation remains valid. (b) When $\lambda \approx d$, the rays spread out after passing through the opening. (c) When $\lambda \gg d$, the opening behaves as a point source emitting spherical waves.

Sign in at www.thomsonedu.com and go to ThomsonNOW to adjust the size of the opening and observe the effect on the waves passing through.

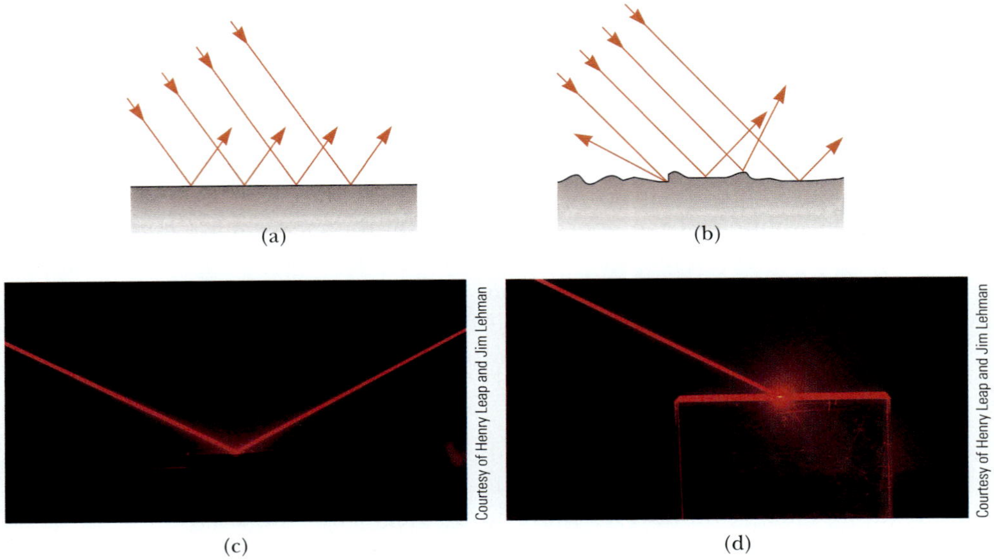

Figure 35.5 Schematic representation of (a) specular reflection, where the reflected rays are all parallel to each other, and (b) diffuse reflection, where the reflected rays travel in random directions. (c) and (d) Photographs of specular and diffuse reflection using laser light.

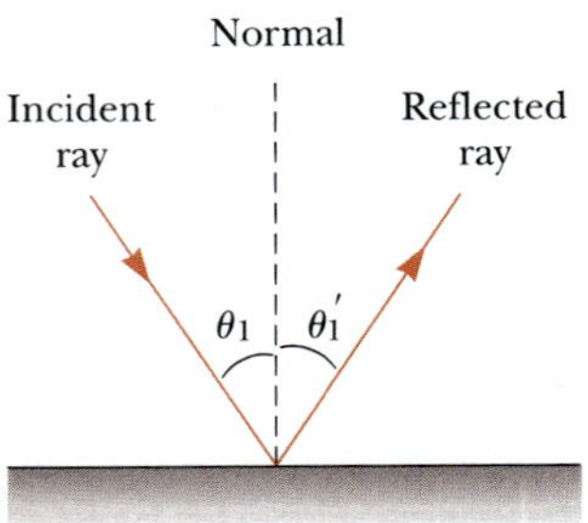

ACTIVE FIGURE 35.6

According to the wave under reflection model, $\theta_1' = \theta_1$. The incident ray, the reflected ray, and the normal all lie in the same plane.

Sign in at www.thomsonedu.com and go to ThomsonNOW to vary the incident angle and see the effect on the reflected ray.

in directions different from the direction of the incident waves. Figure 35.5a shows several rays of a beam of light incident on a smooth, mirror-like, reflecting surface. The reflected rays are parallel to one another as indicated in the figure. The direction of a reflected ray is in the plane perpendicular to the reflecting surface that contains the incident ray. Reflection of light from such a smooth surface is called **specular reflection.** If the reflecting surface is rough as in Figure 35.5b, the surface reflects the rays not as a parallel set but in various directions. Reflection from any rough surface is known as **diffuse reflection.** A surface behaves as a smooth surface as long as the surface variations are much smaller than the wavelength of the incident light.

The difference between these two kinds of reflection explains why it is more difficult to see while driving on a rainy night than on a dry, sunny day. If the road is wet, the smooth surface of the water specularly reflects most of your headlight beams away from your car (and perhaps into the eyes of oncoming drivers). When the road is dry, its rough surface diffusely reflects part of your headlight beam back toward you, allowing you to see the road more clearly. In this book, we restrict our study to specular reflection and use the term *reflection* to mean specular reflection.

Consider a light ray traveling in air and incident at an angle on a flat, smooth surface as shown in Active Figure 35.6. The incident and reflected rays make angles θ_1 and θ_1', respectively, where the angles are measured between the normal and the rays. (The normal is a line drawn perpendicular to the surface at the point where the incident ray strikes the surface.) Experiments and theory show that **the angle of reflection equals the angle of incidence:**

Law of reflection ▶

$$\theta_1' = \theta_1 \tag{35.2}$$

This relationship is called the **law of reflection.** Because reflection of waves from an interface between two media is a common phenomenon, we identify an analysis model for this situation: the **wave under reflection.** Equation 35.2 is the mathematical representation of this model.

PITFALL PREVENTION 35.1

Subscript Notation

The subscript 1 refers to parameters for the light in the initial medium. When light travels from one medium to another, we use the subscript 2 for the parameters associated with the light in the new medium. In this discussion, the light stays in the same medium, so we only have to use the subscript 1.

Quick Quiz 35.1 In the movies, you sometimes see an actor looking in a mirror and you can see his face in the mirror. During the filming of such a scene, what does the actor see in the mirror? (a) his face (b) your face (c) the director's face (d) the movie camera (e) impossible to determine

---

EXAMPLE 35.2 The Double-Reflected Light Ray

Two mirrors make an angle of 120° with each other as illustrated in Figure 35.7a. A ray is incident on mirror M_1 at an angle of 65° to the normal. Find the direction of the ray after it is reflected from mirror M_2.

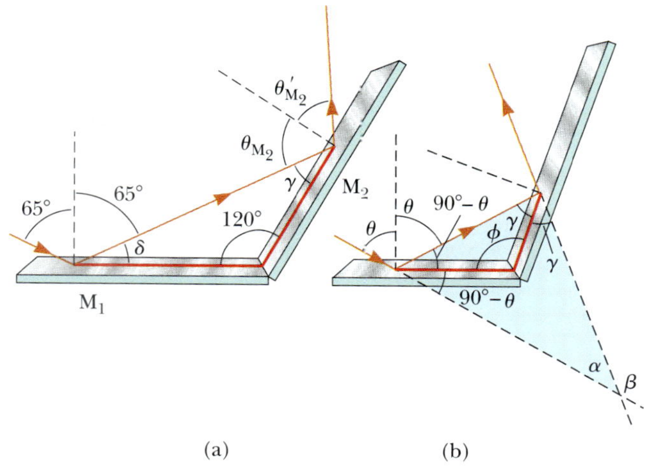

Figure 35.7 (Example 35.2) (a) Mirrors M_1 and M_2 make an angle of 120° with each other. (b) The geometry for an arbitrary mirror angle.

SOLUTION

Conceptualize Figure 35.7a helps conceptualize this situation. The incoming ray reflects from the first mirror, and the reflected ray is directed toward the second mirror. Therefore, there is a second reflection from the second mirror.

Categorize Because the interactions with both mirrors are simple reflections, we apply the wave under reflection model and some geometry.

Analyze From the law of reflection, the first reflected ray makes an angle of 65° with the normal.

Find the angle the first reflected ray makes with the horizontal:

$$\delta = 90° - 65° = 25°$$

From the triangle made by the first reflected ray and the two mirrors, find the angle the reflected ray makes with M_2:

$$\gamma = 180° - 25° - 120° = 35°$$

Find the angle the first reflected ray makes with the normal to M_2:

$$\theta_{M_2} = 90° - 35° = 55°$$

From the law of reflection, find the angle the second reflected ray makes with the normal to M_2:

$$\theta'_{M_2} = \theta_{M_2} = 55°$$

Finalize Let's explore variations in the angle between the mirrors as follows.

What If? If the incoming and outgoing rays in Figure 35.7a are extended behind the mirror, they cross at an angle of 60° and the overall change in direction of the light ray is 120°. This angle is the same as that between the mirrors. What if the angle between the mirrors is changed? Is the overall change in the direction of the light ray always equal to the angle between the mirrors?

Answer Making a general statement based on one data point or one observation is always a dangerous practice! Let's investigate the change in direction for a general situation. Figure 35.7b shows the mirrors at an arbitrary angle ϕ and the incoming light ray striking the mirror at an arbitrary angle θ with respect to the normal to the mirror surface. In accordance with the law of reflection and the sum of the interior angles of a triangle, the angle γ is given by $180° - (90° - \theta) - \phi = 90° + \theta - \phi$.

Consider the triangle highlighted in blue in Figure 35.7b and determine α:

$$\alpha + 2\gamma + 2(90° - \theta) = 180° \quad \rightarrow \quad \alpha = 2(\theta - \gamma)$$

Notice from Figure 35.7b that the change in direction of the light ray is angle β. Use the geometry in the figure to solve for β:

$$\beta = 180° - \alpha = 180° - 2(\theta - \gamma)$$
$$= 180° - 2[\theta - (90° + \theta - \phi)] = 360° - 2\phi$$

Notice that β is not equal to ϕ. For $\phi = 120°$, we obtain $\beta = 120°$, which happens to be the same as the mirror angle; that is true only for this special angle between the mirrors, however. For example, if $\phi = 90°$, we obtain $\beta = 180°$. In that case, the light is reflected straight back to its origin.

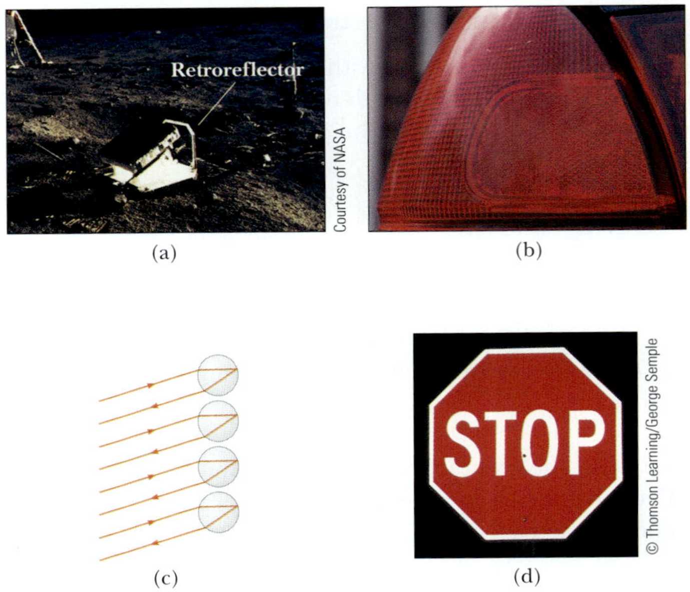

Figure 35.8 Applications of retroreflection. (a) This panel on the Moon reflects a laser beam directly back to its source on the Earth. (b) An automobile taillight has small retroreflectors to ensure that headlight beams are reflected back toward the car that sent them. (c) A light ray hitting a transparent sphere at the proper position is retroreflected. (d) This stop sign appears to glow in headlight beams because its surface is covered with a layer of many tiny retroreflecting spheres. What would you see if the sign had a mirror-like surface?

If the angle between two mirrors is 90°, the reflected beam returns to the source parallel to its original path as discussed in the **What If?** section of the preceding example. This phenomenon, called *retroreflection,* has many practical applications. If a third mirror is placed perpendicular to the first two so that the three form the corner of a cube, retroreflection works in three dimensions. In 1969, a panel of many small reflectors was placed on the Moon by the *Apollo 11* astronauts (Fig. 35.8a). A laser beam from the Earth is reflected directly back on itself, and its transit time is measured. This information is used to determine the distance to the Moon with an uncertainty of 15 cm. (Imagine how difficult it would be to align a regular flat mirror so that the reflected laser beam would hit a particular location on the Earth!) A more everyday application is found in automobile taillights. Part of the plastic making up the taillight is formed into many tiny cube corners (Fig. 35.8b) so that headlight beams from cars approaching from the rear are reflected back to the drivers. Instead of cube corners, small spherical bumps are sometimes used (Fig. 35.8c). Tiny clear spheres are used in a coating material found on many road signs. Due to retroreflection from these spheres, the stop sign in Figure 35.8d appears much brighter than it would if it were simply a flat, shiny surface. Retroreflectors are also used for reflective panels on running shoes and running clothing to allow joggers to be seen at night.

Another practical application of the law of reflection is the digital projection of movies, television shows, and computer presentations. A digital projector uses an optical semiconductor chip called a *digital micromirror device.* This device contains an array of tiny mirrors (Fig. 35.9a) that can be individually tilted by means of signals to an address electrode underneath the edge of the mirror. Each mirror corresponds to a pixel in the projected image. When the pixel corresponding to a given mirror is to be bright, the mirror is in the "on" position and is oriented so as to reflect light from a source illuminating the array to the screen (Fig. 35.9b). When the pixel for this mirror is to be dark, the mirror is "off" and is tilted so that the light is reflected away from the screen. The brightness of the pixel is determined by the total time interval during which the mirror is in the "on" position during the display of one image.

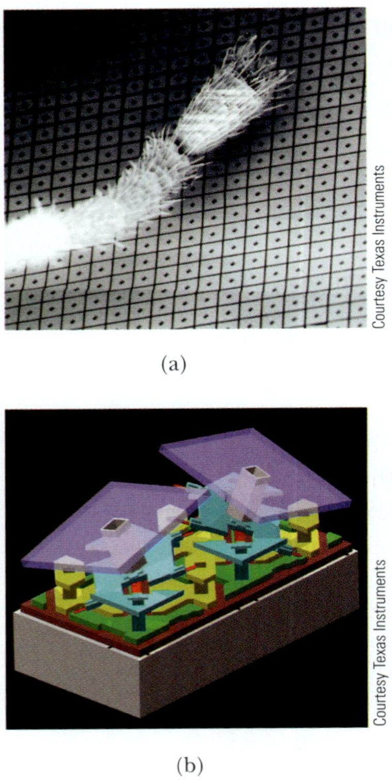

Figure 35.9 (a) An array of mirrors on the surface of a digital micromirror device. Each mirror has an area of approximately 16 μm^2. To provide a sense of scale, the leg of an ant appears in the photograph. (b) A close-up view of two single micromirrors. The mirror on the left is "on," and the one on the right is "off."

Digital movie projectors use three micromirror devices, one for each of the primary colors red, blue, and green, so that movies can be displayed with up to 35 trillion colors. Because information is stored as binary data, a digital movie does not degrade with time as does film. Furthermore, because the movie is entirely in the form of computer software, it can be delivered to theaters by means of satellites, optical discs, or optical fiber networks.

Several movies have been projected digitally to audiences, and polls show that 85% of viewers describe the image quality as "excellent." The first all-digital movie, from cinematography to postproduction to projection, was *Star Wars Episode II: Attack of the Clones* in 2002.

35.5 The Wave Under Refraction

In addition to the phenomenon of reflection discussed for waves on strings in Section 16.4, we also found that some of the energy of the incident wave transmits into the new medium. Similarly, when a ray of light traveling through a transparent medium encounters a boundary leading into another transparent medium as shown in Active Figure 35.10, part of the energy is reflected and part enters the second medium. As with reflection, the direction of the transmitted wave exhibits an interesting behavior because of the three-dimensional nature of the light waves. The ray that enters the second medium is bent at the boundary and is said to be **refracted.** The incident ray, the reflected ray, and the refracted ray all lie in the same plane. The **angle of refraction,** θ_2 in Active Figure 35.10a, depends on the properties of the two media and on the angle of incidence θ_1 through the relationship

$$\frac{\sin\theta_2}{\sin\theta_1} = \frac{v_2}{v_1} \tag{35.3}$$

where v_1 is the speed of light in the first medium and v_2 is the speed of light in the second medium.

The path of a light ray through a refracting surface is reversible. For example, the ray shown in Active Figure 35.10a travels from point A to point B. If the ray originated at B, it would travel along line BA to reach point A and the reflected ray would point downward and to the left in the glass.

Quick Quiz 35.2 If beam ① is the incoming beam in Active Figure 35.10b, which of the other four red lines are reflected beams and which are refracted beams?

From Equation 35.3, we can infer that when light moves from a material in which its speed is high to a material in which its speed is lower as shown in Active

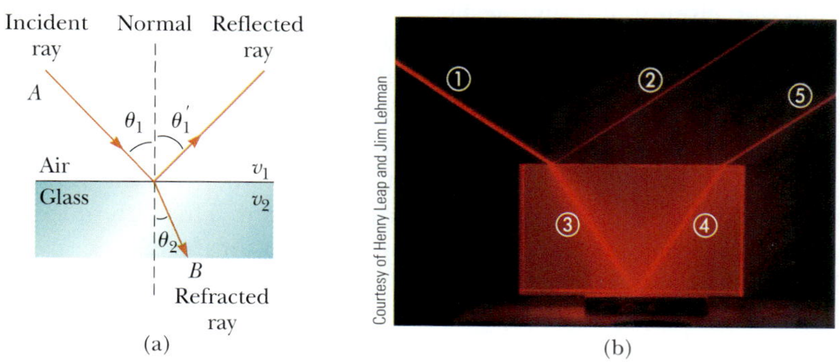

ACTIVE FIGURE 35.10

(a) A ray obliquely incident on an air–glass interface behaves according to the wave under refraction model. The refracted ray is bent toward the normal because $v_2 < v_1$. All rays and the normal lie in the same plane. (b) Light incident on the Lucite block refracts both when it enters the block and when it leaves the block.

Sign in at www.thomsonedu.com and go to ThomsonNOW to vary the incident angle and see the effect on the reflected and refracted rays.

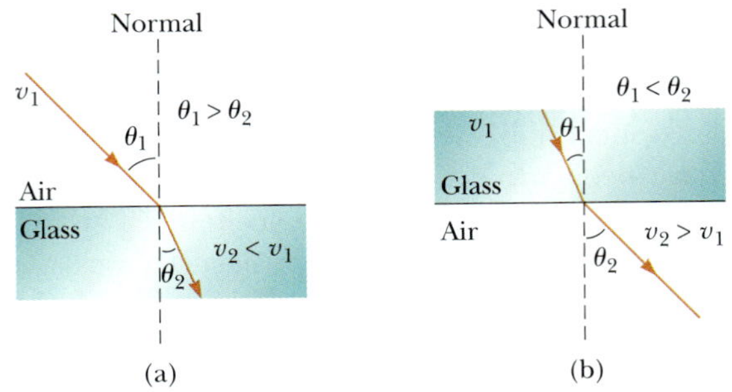

ACTIVE FIGURE 35.11

(a) When the light beam moves from air into glass, the light slows down upon entering the glass and its path is bent toward the normal. (b) When the beam moves from glass into air, the light speeds up upon entering the air and its path is bent away from the normal.

Sign in at www.thomsonedu.com and go to ThomsonNOW to observe light passing through three layers of material. You can vary the incident angle and see the effect on the refracted rays for a variety of values of the index of refraction (defined in Equation 35.4) of the three materials.

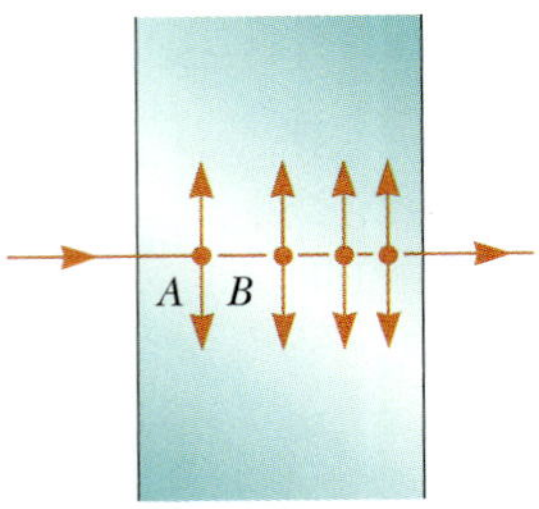

Figure 35.12 Light passing from one atom to another in a medium. The dots are electrons, and the vertical arrows represent their oscillations.

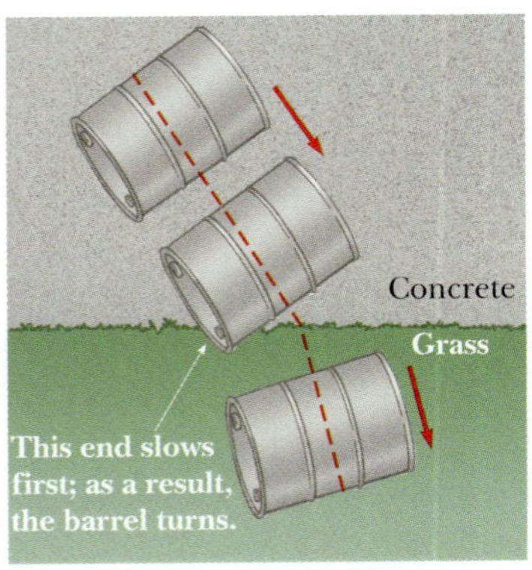

Figure 35.13 Overhead view of a barrel rolling from concrete onto grass.

Figure 35.11a, the angle of refraction θ_2 is less than the angle of incidence θ_1 and the ray is bent *toward* the normal. If the ray moves from a material in which light moves slowly to a material in which it moves more rapidly as illustrated in Active Figure 35.11b, θ_2 is greater than θ_1 and the ray is bent *away* from the normal.

The behavior of light as it passes from air into another substance and then re-emerges into air is often a source of confusion to students. When light travels in air, its speed is 3.00×10^8 m/s, but this speed is reduced to approximately 2×10^8 m/s when the light enters a block of glass. When the light re-emerges into air, its speed instantaneously increases to its original value of 3.00×10^8 m/s. This effect is far different from what happens, for example, when a bullet is fired through a block of wood. In that case, the speed of the bullet decreases as it moves through the wood because some of its original energy is used to tear apart the wood fibers. When the bullet enters the air once again, it emerges at a speed lower than it had when it entered the wood.

To see why light behaves as it does, consider Figure 35.12, which represents a beam of light entering a piece of glass from the left. Once inside the glass, the light may encounter an electron bound to an atom, indicated as point *A*. Let's assume light is absorbed by the atom, which causes the electron to oscillate (a detail represented by the double-headed vertical arrows). The oscillating electron then acts as an antenna and radiates the beam of light toward an atom at *B*, where the light is again absorbed. The details of these absorptions and radiations are best explained in terms of quantum mechanics (Chapter 42). For now, it is sufficient to think of light passing from one atom to another through the glass. Although light travels from one atom to another at 3.00×10^8 m/s, the absorption and radiation that take place cause the *average* light speed through the material to fall to about 2×10^8 m/s. Once the light emerges into the air, absorption and radiation cease and the light travels at a constant speed of 3.00×10^8 m/s.

A mechanical analog of refraction is shown in Figure 35.13. When the left end of the rolling barrel reaches the grass, it slows down, whereas the right end remains on the concrete and moves at its original speed. This difference in speeds causes the barrel to pivot, which changes the direction of travel.

Index of Refraction

In general, the speed of light in any material is *less* than its speed in vacuum. In fact, *light travels at its maximum speed c in vacuum.* It is convenient to define the **index of refraction** n of a medium to be the ratio

Index of refraction ▶

$$n \equiv \frac{\text{speed of light in vacuum}}{\text{speed of light in a medium}} \equiv \frac{c}{v} \tag{35.4}$$

TABLE 35.1

Indices of Refraction

Substance	Index of Refraction	Substance	Index of Refraction
Solids at 20°C		*Liquids at 20°C*	
Cubic zirconia	2.20	Benzene	1.501
Diamond (C)	2.419	Carbon disulfide	1.628
Fluorite (CaF_2)	1.434	Carbon tetrachloride	1.461
Fused quartz (SiO_2)	1.458	Ethyl alcohol	1.361
Gallium phosphide	3.50	Glycerin	1.473
Glass, crown	1.52	Water	1.333
Glass, flint	1.66		
Ice (H_2O)	1.309	*Gases at 0°C, 1 atm*	
Polystyrene	1.49	Air	1.000 293
Sodium chloride (NaCl)	1.544	Carbon dioxide	1.000 45

Note: All values are for light having a wavelength of 589 nm in vacuum.

PITFALL PREVENTION 35.2

n Is Not an Integer Here

The symbol n has been used several times as an integer, such as in Chapter 18 to indicate the standing wave mode on a string or in an air column. The index of refraction n is *not* an integer.

This definition shows that the index of refraction is a dimensionless number greater than unity because v is always less than c. Furthermore, n is equal to unity for vacuum. The indices of refraction for various substances are listed in Table 35.1.

As light travels from one medium to another, its frequency does not change but its wavelength does. To see why that is true, consider Figure 35.14. Waves pass an observer at point A in medium 1 with a certain frequency and are incident on the boundary between medium 1 and medium 2. The frequency with which the waves pass an observer at point B in medium 2 must equal the frequency at which they pass point A. If that were not the case, energy would be piling up or disappearing at the boundary. Because there is no mechanism for that to occur, the frequency must be a constant as a light ray passes from one medium into another. Therefore, because the relationship $v = \lambda f$ (Eq. 16.12) must be valid in both media and because $f_1 = f_2 = f$, we see that

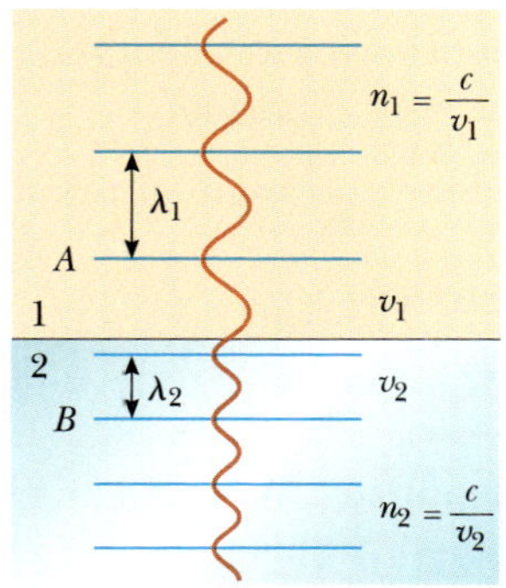

Figure 35.14 As a wave moves from medium 1 to medium 2, its wavelength changes but its frequency remains constant.

$$v_1 = \lambda_1 f \quad \text{and} \quad v_2 = \lambda_2 f \tag{35.5}$$

Because $v_1 \neq v_2$, it follows that $\lambda_1 \neq \lambda_2$ as shown in Figure 35.14.

We can obtain a relationship between index of refraction and wavelength by dividing the first Equation 35.5 by the second and then using Equation 35.4:

$$\frac{\lambda_1}{\lambda_2} = \frac{v_1}{v_2} = \frac{c/n_1}{c/n_2} = \frac{n_2}{n_1} \tag{35.6}$$

This expression gives

$$\lambda_1 n_1 = \lambda_2 n_2$$

If medium 1 is vacuum or, for all practical purposes, air, then $n_1 = 1$. Hence, it follows from Equation 35.6 that the index of refraction of any medium can be expressed as the ratio

$$n = \frac{\lambda}{\lambda_n} \tag{35.7}$$

where λ is the wavelength of light in vacuum and λ_n is the wavelength of light in the medium whose index of refraction is n. From Equation 35.7, we see that because $n > 1$, $\lambda_n < \lambda$.

PITFALL PREVENTION 35.3

An Inverse Relationship

The index of refraction is *inversely* proportional to the wave speed. As the wave speed v decreases, the index of refraction n increases. Therefore, the higher the index of refraction of a material, the more it *slows down* light from its speed in vacuum. The more the light slows down, the more θ_2 differs from θ_1 in Equation 35.8.

We are now in a position to express Equation 35.3 in an alternative form. Replacing the v_2/v_1 term in Equation 35.3 with n_1/n_2 from Equation 35.6 gives

$$n_1 \sin\theta_1 = n_2 \sin\theta_2 \tag{35.8}$$

◀ Snell's law of refraction

The experimental discovery of this relationship is usually credited to Willebrord Snell (1591–1626) and it is therefore known as **Snell's law of refraction.** We shall examine this equation further in Section 35.6. Refraction of waves at an interface between two media is a common phenomenon, so we identify an analysis model for this situation: the **wave under refraction.** Equation 35.8 is the mathematical representation of this model for electromagnetic radiation. Other waves, such as seismic waves and sound waves, also exhibit refraction according to this model, and the mathematical representation for these waves is Equation 35.3.

Quick Quiz 35.3 Light passes from a material with index of refraction 1.3 into one with index of refraction 1.2. Compared to the incident ray, what happens to the refracted ray? (a) It bends toward the normal. (b) It is undeflected. (c) It bends away from the normal.

EXAMPLE 35.3 Angle of Refraction for Glass

A light ray of wavelength 589 nm traveling through air is incident on a smooth, flat slab of crown glass at an angle of 30.0° to the normal.

(A) Find the angle of refraction.

SOLUTION

Conceptualize Study Active Figure 35.11a, which illustrates the refraction process occurring in this problem.

Categorize We evaluate results by using equations developed in this section, so we categorize this example as a substitution problem.

Rearrange Snell's law of refraction to find $\sin\theta_2$:

$$\sin\theta_2 = \frac{n_1}{n_2}\sin\theta_1$$

Substitute the incident angle and, from Table 35.1, $n_1 = 1.00$ for air and $n_2 = 1.52$ for crown glass:

$$\sin\theta_2 = \left(\frac{1.00}{1.52}\right)\sin 30.0° = 0.329$$

Solve for θ_2:

$$\theta_2 = \sin^{-1}(0.329)$$
$$= 19.2°$$

(B) Find the speed of this light once it enters the glass.

SOLUTION

Solve Equation 35.4 for the speed of light in the glass:

$$v = \frac{c}{n}$$

Substitute numerical values:

$$v = \frac{3.00 \times 10^8\ \text{m/s}}{1.52} = 1.97 \times 10^8\ \text{m/s}$$

(C) What is the wavelength of this light in the glass?

SOLUTION

Use Equation 35.7 to find the wavelength in the glass:

$$\lambda_n = \frac{\lambda}{n} = \frac{589\ \text{nm}}{1.52} = 388\ \text{nm}$$

EXAMPLE 35.4 Light Passing Through a Slab

A light beam passes from medium 1 to medium 2, with the latter medium being a thick slab of material whose index of refraction is n_2 (Fig. 35.15). Show that the beam emerging into medium 1 from the other side is parallel to the incident beam.

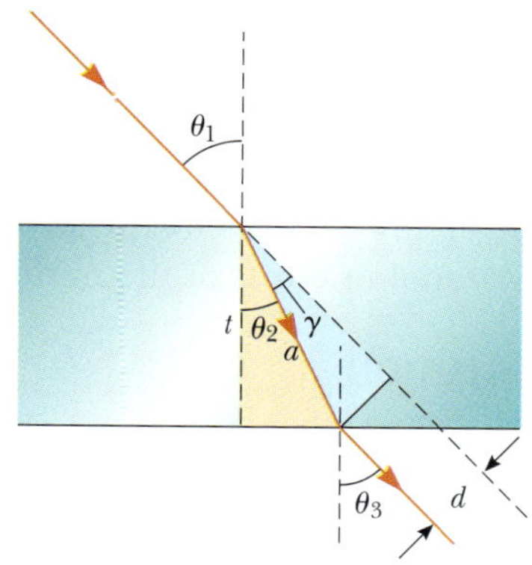

Figure 35.15 (Example 35.4) When light passes through a flat slab of material, the emerging beam is parallel to the incident beam; therefore, $\theta_1 = \theta_3$. The dashed line drawn parallel to the ray coming out the bottom of the slab represents the path the light would take were the slab not there.

SOLUTION

Conceptualize Follow the path of the light beam as it enters and exits the slab of material in Figure 35.15. The ray bends toward the normal upon entering and away from the normal upon leaving.

Categorize We evaluate results by using equations developed in this section, so we categorize this example as a substitution problem.

Apply Snell's law of refraction to the upper surface:

$$(1)\quad \sin\theta_2 = \frac{n_1}{n_2}\sin\theta_1$$

Apply Snell's law to the lower surface:

$$(2)\quad \sin\theta_3 = \frac{n_2}{n_1}\sin\theta_2$$

Substitute Equation (1) into Equation (2):

$$\sin\theta_3 = \frac{n_2}{n_1}\left(\frac{n_1}{n_2}\sin\theta_1\right) = \sin\theta_1$$

Therefore, $\theta_3 = \theta_1$ and the slab does not alter the direction of the beam. It does, however, offset the beam parallel to itself by the distance d shown in Figure 35.15.

What If? What if the thickness t of the slab is doubled? Does the offset distance d also double?

Answer Consider the region of the light path within the slab in Figure 35.15. The distance a is the hypotenuse of two right triangles.

Find an expression for a from the gold triangle:

$$a = \frac{t}{\cos\theta_2}$$

Find an expression for d from the blue triangle:

$$d = a\sin\gamma = a\sin(\theta_1 - \theta_2)$$

Combine these equations:

$$d = \frac{t}{\cos\theta_2}\sin(\theta_1 - \theta_2)$$

For a given incident angle θ_1, the refracted angle θ_2 is determined solely by the index of refraction, so the offset distance d is proportional to t. If the thickness doubles, so does the offset distance.

In Example 35.4, the light passes through a slab of material with parallel sides. What happens when light strikes a prism with nonparallel sides as shown in Figure 35.16? In this case, the outgoing ray does not propagate in the same direction as the incoming ray. A ray of single-wavelength light incident on the prism from the left emerges at angle δ from its original direction of travel. This angle δ is called the **angle of deviation.** The **apex angle** Φ of the prism, shown in the figure, is defined as the angle between the surface at which the light enters the prism and the second surface that the light encounters.

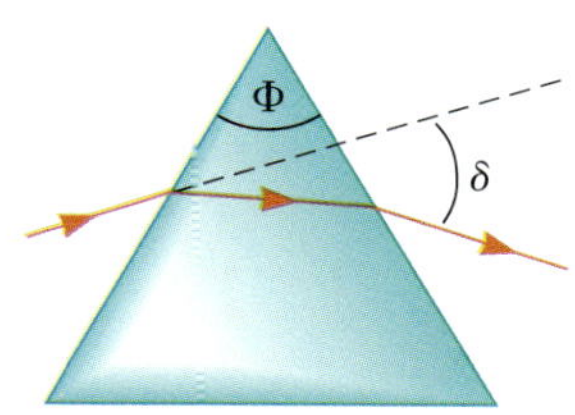

Figure 35.16 A prism refracts a single-wavelength light ray through an angle of deviation δ.

EXAMPLE 35.5 Measuring *n* Using a Prism

Although we do not prove it here, the minimum angle of deviation δ_{min} for a prism occurs when the angle of incidence θ_1 is such that the refracted ray inside the prism makes the same angle with the normal to the two prism faces[1] as shown in Figure 35.17. Obtain an expression for the index of refraction of the prism material in terms of the minimum angle of deviation and the apex angle Φ.

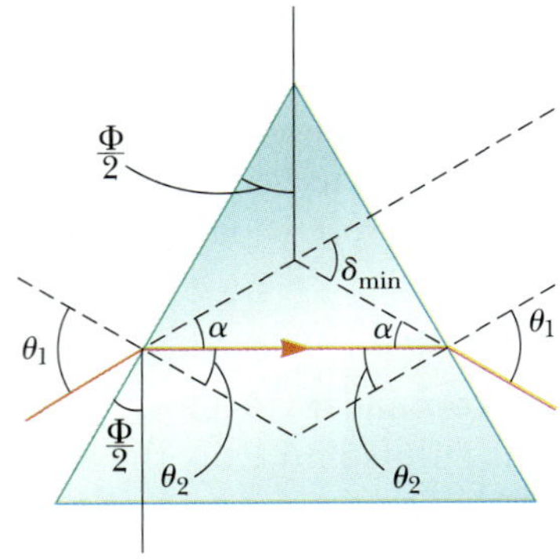

Figure 35.17 (Example 35.5) A light ray passing through a prism at the minimum angle of deviation δ_{min}.

SOLUTION

Conceptualize Study Figure 35.17 carefully and be sure you understand why the light ray comes out of the prism traveling in a different direction.

Categorize In this example, light enters a material through one surface and leaves the material at another surface. Let's apply the wave under refraction model at each surface.

Analyze Consider the geometry in Figure 35.17. The reproduction of the angle $\Phi/2$ at the location of the incoming light ray shows that $\theta_2 = \Phi/2$. The theorem that an exterior angle of any triangle equals the sum of the two opposite interior angles shows that $\delta_{min} = 2\alpha$. The geometry also shows that $\theta_1 = \theta_2 + \alpha$.

Combine these three geometric results:

$$\theta_1 = \theta_2 + \alpha = \frac{\Phi}{2} + \frac{\delta_{min}}{2} = \frac{\Phi + \delta_{min}}{2}$$

Apply the wave under refraction model at the left surface and solve for n:

$$(1.00)\sin\theta_1 = n\sin\theta_2 \quad\rightarrow\quad n = \frac{\sin\theta_1}{\sin\theta_2}$$

Substitute for the incident and refracted angles:

$$n = \frac{\sin\left(\dfrac{\Phi + \delta_{min}}{2}\right)}{\sin(\Phi/2)} \qquad (35.9)$$

Finalize Knowing the apex angle Φ of the prism and measuring δ_{min}, you can calculate the index of refraction of the prism material. Furthermore, a hollow prism can be used to determine the values of n for various liquids filling the prism.

35.6 Huygens's Principle

PITFALL PREVENTION 35.4
Of What Use Is Huygens's Principle?

At this point, the importance of Huygens's principle may not be evident. Predicting the position of a future wave front may not seem to be very critical. We will use Huygens's principle in later chapters to explain additional wave phenomena for light, however.

In this section, we develop the laws of reflection and refraction by using a geometric method proposed by Huygens in 1678. **Huygens's principle** is a geometric construction for using knowledge of an earlier wave front to determine the position of a new wave front at some instant. In Huygens's construction, **all points on a given wave front are taken as point sources for the production of spherical secondary waves, called wavelets, that propagate outward through a medium with speeds characteristic of waves in that medium. After some time interval has passed, the new position of the wave front is the surface tangent to the wavelets.**

First, consider a plane wave moving through free space as shown in Figure 35.18a. At $t = 0$, the wave front is indicated by the plane labeled AA'. In Huygens's construction, each point on this wave front is considered a point source. For clarity, only three points on AA' are shown. With these points as sources for the wavelets, we draw circles, each of radius $c\,\Delta t$, where c is the speed of light in vacuum and Δt is some time interval during which the wave propagates. The surface drawn tangent to these wavelets is the plane BB', which is the wave front at a later

[1] The details of this proof are available in texts on optics.

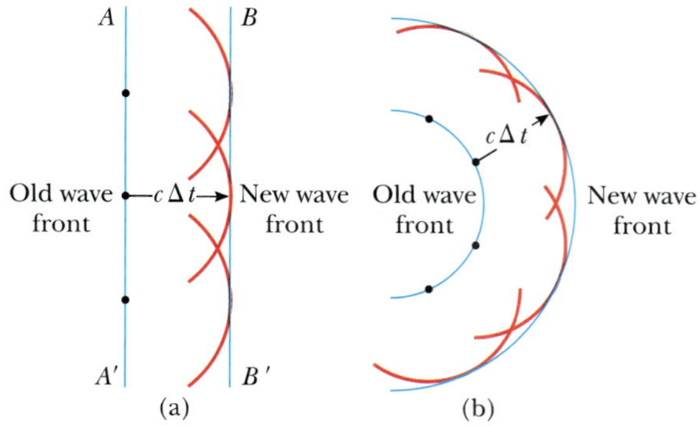

Figure 35.18 Huygens's construction for (a) a plane wave propagating to the right and (b) a spherical wave propagating to the right.

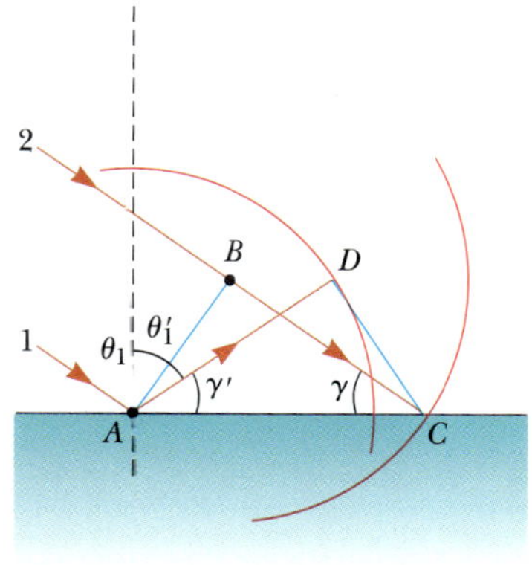

Figure 35.19 Huygens's construction for proving the law of reflection. The instant ray 1 strikes the surface, it sends out a Huygens wavelet from A and ray 2 sends out a Huygens wavelet from B. We choose a radius of the wavelet to be $c\,\Delta t$, where Δt is the time interval for ray 2 to travel from B to C. Triangle ADC is congruent to triangle ABC.

time, and is parallel to AA'. In a similar manner, Figure 35.18b shows Huygens's construction for a spherical wave.

Huygens's Principle Applied to Reflection and Refraction

The laws of reflection and refraction were stated earlier in this chapter without proof. We now derive these laws, using Huygens's principle.

For the law of reflection, refer to Figure 35.19. The line AB represents a plane wave front of the incident light just as ray 1 strikes the surface. At this instant, the wave at A sends out a Huygens wavelet (the red circular arc centered on A). The reflected light propagates toward D. At the same time, the wave at B emits a Huygens wavelet (the red circular arc centered on B) with the light propagating toward C. Figure 35.19 shows these wavelets after a time interval Δt, after which ray 2 strikes the surface. Because both rays 1 and 2 move with the same speed, we must have $AD = BC = c\,\Delta t$.

The remainder of our analysis depends on geometry. Notice that the two triangles ABC and ADC are congruent because they have the same hypotenuse AC and because $AD = BC$. Figure 35.19 shows that

$$\cos\gamma = \frac{BC}{AC} \quad \text{and} \quad \cos\gamma' = \frac{AD}{AC}$$

where $\gamma = 90° - \theta_1$ and $\gamma' = 90° - \theta_1'$. Because $AD = BC$,

$$\cos\gamma = \cos\gamma'$$

Therefore,

$$\gamma = \gamma'$$

$$90° - \theta_1 = 90° - \theta_1'$$

and

$$\theta_1 = \theta_1'$$

which is the law of reflection.

Now let's use Huygens's principle and Figure 35.20 to derive Snell's law of refraction. We focus our attention on the instant ray 1 strikes the surface and the subsequent time interval until ray 2 strikes the surface. During this time interval, the wave at A sends out a Huygens wavelet (the red arc centered on A) and the light refracts toward D. In the same time interval, the wave at B sends out a Huygens wavelet (the red arc centered on B) and the light continues to propagate toward C. Because these two wavelets travel through different media, the radii of the wavelets are different. The radius of the wavelet from A is $AD = v_2\,\Delta t$, where v_2 is the wave speed in the second medium. The radius of the wavelet from B is $BC = v_1\,\Delta t$, where v_1 is the wave speed in the original medium.

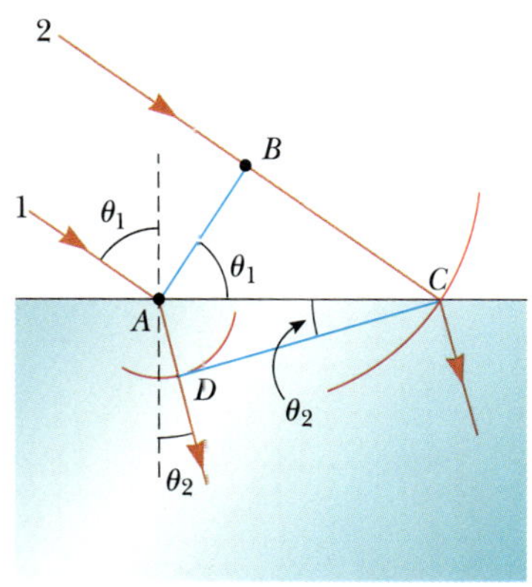

Figure 35.20 Huygens's construction for proving Snell's law of refraction. The instant ray 1 strikes the surface, it sends out a Huygens wavelet from A and ray 2 sends out a Huygens wavelet from B. The two wavelets have different radii because they travel in different media.

From triangles *ABC* and *ADC*, we find that

$$\sin\theta_1 = \frac{BC}{AC} = \frac{v_1\,\Delta t}{AC} \quad \text{and} \quad \sin\theta_2 = \frac{AD}{AC} = \frac{v_2\,\Delta t}{AC}$$

Dividing the first equation by the second gives

$$\frac{\sin\theta_1}{\sin\theta_2} = \frac{v_1}{v_2}$$

From Equation 35.4, however, we know that $v_1 = c/n_1$ and $v_2 = c/n_2$. Therefore,

$$\frac{\sin\theta_1}{\sin\theta_2} = \frac{c/n_1}{c/n_2} = \frac{n_2}{n_1}$$

and

$$n_1 \sin\theta_1 = n_2 \sin\theta_2$$

which is Snell's law of refraction.

35.7 Dispersion

An important property of the index of refraction n is that, for a given material, the index varies with the wavelength of the light passing through the material as Figure 35.21 shows. This behavior is called **dispersion.** Because n is a function of wavelength, Snell's law of refraction indicates that light of different wavelengths is refracted at different angles when incident on a material.

Figure 35.21 shows that the index of refraction generally decreases with increasing wavelength. For example, violet light refracts more than red light does when passing into a material.

Now suppose a beam of *white light* (a combination of all visible wavelengths) is incident on a prism as illustrated in Figure 35.22. Clearly, the angle of deviation δ depends on wavelength. The rays that emerge spread out in a series of colors known as the **visible spectrum.** These colors, in order of decreasing wavelength, are red, orange, yellow, green, blue, and violet. Newton showed that each color has a particular angle of deviation and that the colors can be recombined to form the original white light.

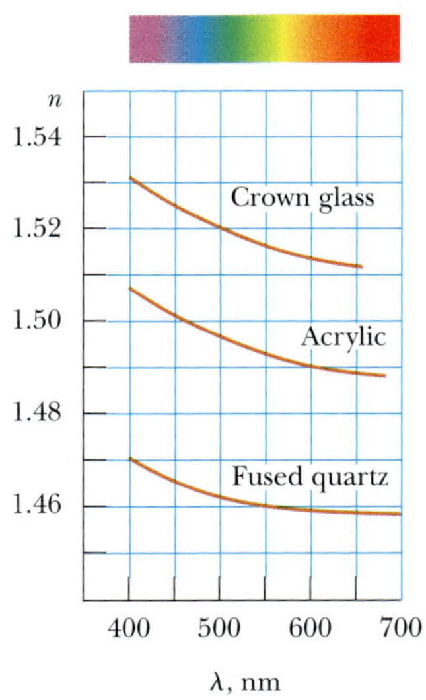

Figure 35.21 Variation of index of refraction with vacuum wavelength for three materials.

David Parker/Science Photo Library/Photo Researchers, Inc.

Figure 35.22 White light enters a glass prism at the upper left. A reflected beam of light comes out of the prism below the incoming beam. The beam moving toward the lower right shows distinct colors. Different colors are refracted at different angles because the index of refraction of the glass depends on wavelength. Violet light deviates the most; red light deviates the least.

The dispersion of light into a spectrum is demonstrated most vividly in nature by the formation of a rainbow, which is often seen by an observer positioned between the Sun and a rain shower. To understand how a rainbow is formed, consider Active Figure 35.23. A ray of sunlight (which is white light) passing overhead strikes a drop of water in the atmosphere and is refracted and reflected as follows. It is first refracted at the front surface of the drop, with the violet light deviating the most and the red light the least. At the back surface of the drop, the light is reflected and returns to the front surface, where it again undergoes refraction as it moves from water into air. The rays leave the drop such that the angle between the incident white light and the most intense returning violet ray is 40° and the angle between the incident white light and the most intense returning red ray is 42°. This small angular difference between the returning rays causes us to see a colored bow.

Now suppose an observer is viewing a rainbow as shown in Figure 35.24. If a raindrop high in the sky is being observed, the most intense red light returning from the drop reaches the observer because it is deviated the most; the most intense violet light, however, passes over the observer because it is deviated the least. Hence, the observer sees red light coming from this drop. Similarly, a drop lower in the sky directs the most intense violet light toward the observer and appears violet to the observer. (The most intense red light from this drop passes below the observer's eye and is not seen.) The most intense light from other colors of the spectrum reaches the observer from raindrops lying between these two extreme positions.

The opening photograph for this chapter shows a *double rainbow.* The secondary rainbow is fainter than the primary rainbow, and the colors are reversed. The secondary rainbow arises from light that makes two reflections from the interior surface before exiting the raindrop. In the laboratory, rainbows have been observed in which the light makes more than 30 reflections before exiting the water drop. Because each reflection involves some loss of light due to refraction out of the water drop, the intensity of these higher-order rainbows is small compared with that of the primary rainbow.

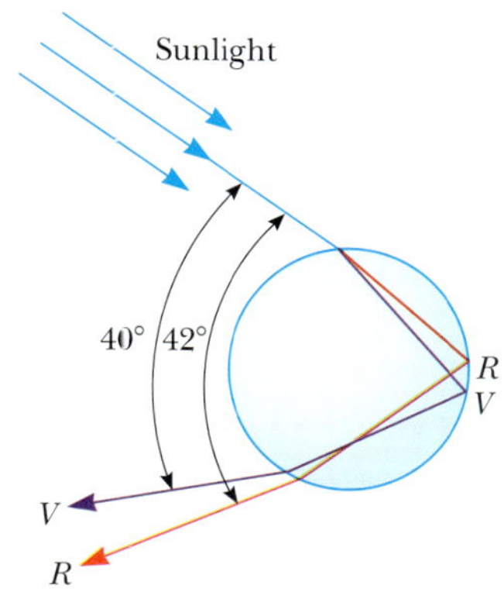

ACTIVE FIGURE 35.23

Path of sunlight through a spherical raindrop. Light following this path contributes to the visible rainbow.

Sign in at www.thomsonedu.com and go to ThomsonNOW to vary the point at which the sunlight enters the raindrop and verify that the angles shown are the maximum angles.

PITFALL PREVENTION 35.5

A Rainbow of Many Light Rays

Pictorial representations such as Active Figure 35.23 are subject to misinterpretation. The figure shows one ray of light entering the raindrop and undergoing reflection and refraction, exiting the raindrop in a range of 40° to 42° from the entering ray. This illustration might be interpreted incorrectly as meaning that *all* light entering the raindrop exits in this small range of angles. In reality, light exits the raindrop over a much larger range of angles, from 0° to 42°. A careful analysis of the reflection and refraction from the spherical raindrop shows that the range of 40° to 42° is where the *highest-intensity light* exits the raindrop.

Quick Quiz 35.4 In film photography, lenses in a camera use refraction to form an image on a film. Ideally, you want all the colors in the light from the object being photographed to be refracted by the same amount. Of the materials shown in Figure 35.21, which would you choose for a single-element camera lens? (a) crown glass (b) acrylic (c) fused quartz (d) impossible to determine

35.8 Total Internal Reflection

An interesting effect called **total internal reflection** can occur when light is directed from a medium having a given index of refraction toward one having a lower index of refraction. Consider Active Figure 35.25a (page 994), in which a light ray travels in medium 1 and meets the boundary between medium 1 and medium 2, where n_1 is greater than n_2. In the figure, labels 1 through 5 indicate various possible directions of the ray consistent with the wave under refraction model. The refracted rays are bent away from the normal because n_1 is greater than n_2. At some particular angle of incidence θ_c, called the **critical angle,** the refracted light ray moves parallel to the boundary so that $\theta_2 = 90°$ (Active Fig. 35.25b). For angles of incidence greater than θ_c, the ray is entirely reflected at the boundary as shown by ray 5 in Active Figure 35.25a.

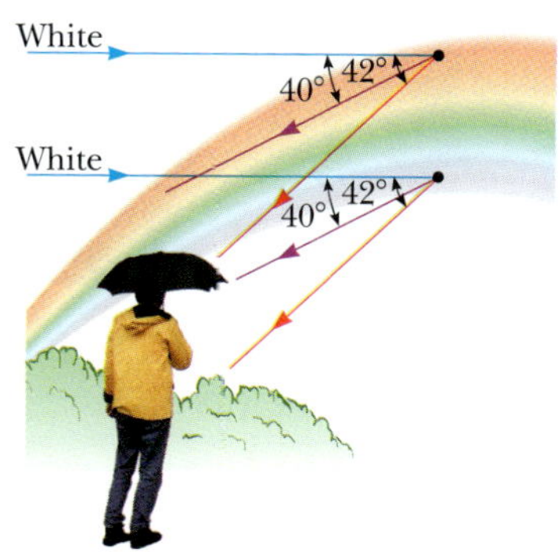

Figure 35.24 The formation of a rainbow seen by an observer standing with the Sun behind his back.

We can use Snell's law of refraction to find the critical angle. When $\theta_1 = \theta_c$, $\theta_2 = 90°$ and Equation 35.8 gives

$$n_1 \sin \theta_c = n_2 \sin 90° = n_2$$

$$\sin \theta_c = \frac{n_2}{n_1} \quad (\text{for } n_1 > n_2) \qquad (35.10)$$

◀ Critical angle for total internal reflection

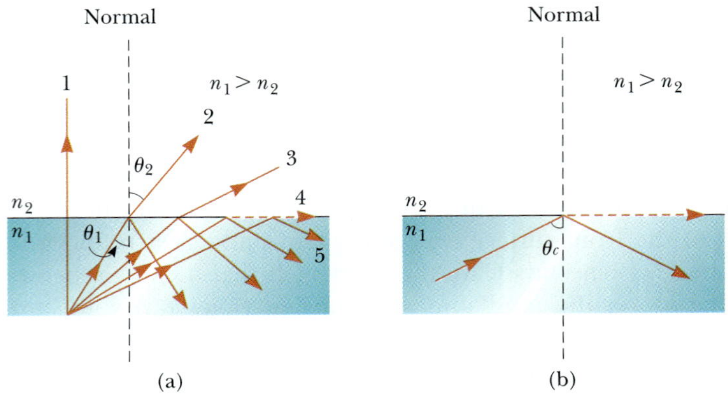

ACTIVE FIGURE 35.25

(a) Rays travel from a medium of index of refraction n_1 into a medium of index of refraction n_2, where $n_2 < n_1$. As the angle of incidence θ_1 increases, the angle of refraction θ_2 increases until θ_2 is 90° (ray 4). The dashed line indicates that no energy actually propagates in this direction. For even larger angles of incidence, total internal reflection occurs (ray 5). (b) The angle of incidence producing an angle of refraction equal to 90° is the critical angle θ_c. At this angle of incidence, all the energy of the incident light is reflected.

Sign in at www.thomsonedu.com and go to ThomsonNOW to vary the incident angle and see the effect on the refracted ray and the distribution of incident energy between the reflected and refracted rays.

This equation can be used only when n_1 is greater than n_2. That is, **total internal reflection occurs only when light is directed from a medium of a given index of refraction toward a medium of lower index of refraction.** If n_1 were less than n_2, Equation 35.10 would give $\sin \theta_c > 1$, which is a meaningless result because the sine of an angle can never be greater than unity.

The critical angle for total internal reflection is small when n_1 is considerably greater than n_2. For example, the critical angle for a diamond in air is 24°. Any ray inside the diamond that approaches the surface at an angle greater than 24° is completely reflected back into the crystal. This property, combined with proper faceting, causes diamonds to sparkle. The angles of the facets are cut so that light is "caught" inside the crystal through multiple internal reflections. These multiple reflections give the light a long path through the medium, and substantial dispersion of colors occurs. By the time the light exits through the top surface of the crystal, the rays associated with different colors have been fairly widely separated from one another.

Cubic zirconia also has a high index of refraction and can be made to sparkle very much like a diamond. If a suspect jewel is immersed in corn syrup, the difference in n for the cubic zirconia and that for the corn syrup is small and the critical angle is therefore great. Hence, more rays escape sooner; as a result, the sparkle completely disappears. A real diamond does not lose all its sparkle when placed in corn syrup.

Courtesy of Henry Leap and Jim Lehman

Figure 35.26 (Quick Quiz 35.5) Five nonparallel light rays enter a glass prism from the left.

Quick Quiz 35.5 In Figure 35.26, five light rays enter a glass prism from the left. **(i)** How many of these rays undergo total internal reflection at the slanted surface of the prism? (a) 1 (b) 2 (c) 3 (d) 4 (e) 5 **(ii)** Suppose the prism in Figure 35.26 can be rotated in the plane of the paper. For *all five* rays to experience total internal reflection from the slanted surface, should the prism be rotated (a) clockwise or (b) counterclockwise?

EXAMPLE 35.6 A View from the Fish's Eye

Find the critical angle for an air–water boundary. (The index of refraction of water is 1.33.)

SOLUTION

Conceptualize Study Active Figure 35.25 to understand the concept of total internal reflection and the significance of the critical angle.

Categorize We use concepts developed in this section, so we categorize this example as a substitution problem.

Apply Equation 35.10 to the air–water interface:

$$\sin\theta_c = \frac{n_2}{n_1} = \frac{1.00}{1.33} = 0.752$$

$$\theta_c = 48.8°$$

What If? What if a fish in a still pond looks upward toward the water's surface at different angles relative to the surface as in Figure 35.27? What does it see?

Answer Because the path of a light ray is reversible, light traveling from medium 2 into medium 1 in Active Figure 35.25a follows the paths shown, but in the *opposite* direction. A fish looking upward toward the water surface as in Figure 35.27 can see out of the water if it looks toward the surface at an angle less than the critical angle. Therefore, when the fish's line of vision makes an angle of $\theta = 40°$ with the normal to the surface, for example, light from above the water reaches the fish's eye. At $\theta = 48.8°$, the critical angle for water, the light has to skim along the water's surface before being refracted to the fish's eye; at this angle, the fish can, in principle, see the entire shore of the pond. At angles greater than the critical angle, the light reaching the fish comes by means of total internal reflection at the surface. Therefore, at $\theta = 60°$, the fish sees a reflection of the bottom of the pond.

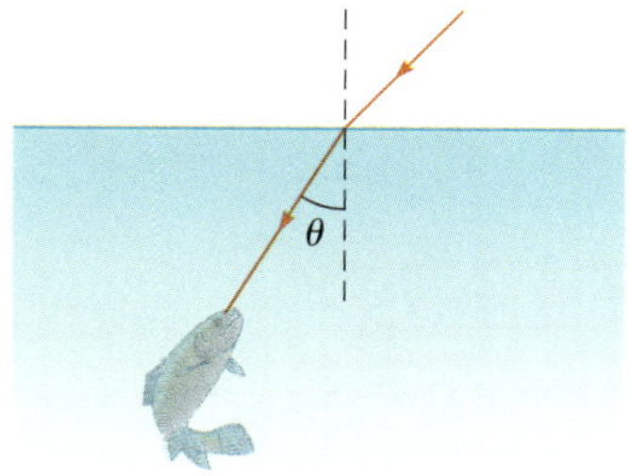

Figure 35.27 (Example 35.6) **What If?** A fish looks upward toward the water surface.

Optical Fibers

Another interesting application of total internal reflection is the use of glass or transparent plastic rods to "pipe" light from one place to another. As indicated in Figure 35.28, light is confined to traveling within a rod, even around curves, as the result of successive total internal reflections. Such a light pipe is flexible if thin fibers are used rather than thick rods. A flexible light pipe is called an **optical fiber.** If a bundle of parallel fibers is used to construct an optical transmission line, images can be transferred from one point to another. This technique is used in a sizable industry known as *fiber optics.*

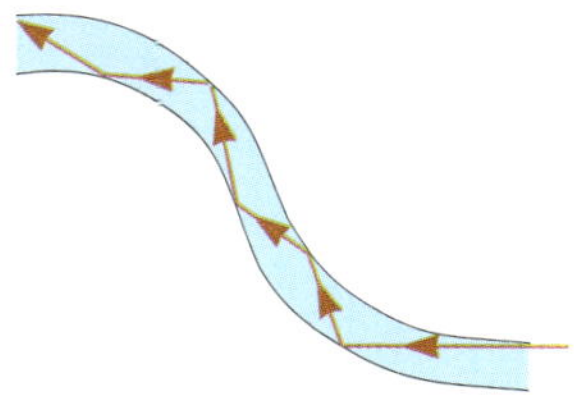

Figure 35.28 Light travels in a curved transparent rod by multiple internal reflections.

A practical optical fiber consists of a transparent core surrounded by a *cladding,* a material that has a lower index of refraction than the core. The combination may be surrounded by a plastic *jacket* to prevent mechanical damage. Figure 35.29 shows a cutaway view of this construction. Because the index of refraction of the cladding is less than that of the core, light traveling in the core experiences total internal reflection if it arrives at the interface between the core and the cladding at an angle

Dennis O'Clair/Tony Stone Images

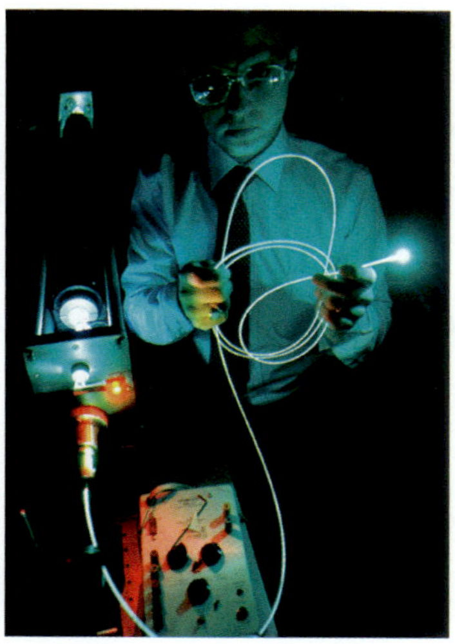

Hank Morgan/Photo Researchers, Inc.

(Left) Strands of glass optical fibers are used to carry voice, video, and data signals in telecommunication networks. *(Right)* A bundle of optical fibers is illuminated by a laser.

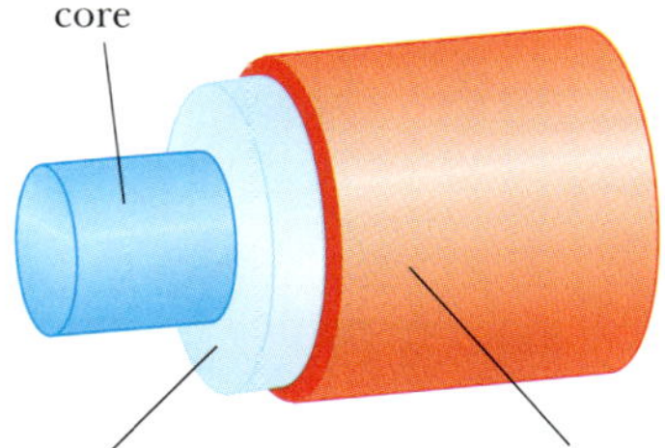

Figure 35.29 The construction of an optical fiber. Light travels in the core, which is surrounded by a cladding and a protective jacket.

of incidence that exceeds the critical angle. In this case, light "bounces" along the core of the optical fiber, losing very little of its intensity as it travels.

Any loss in intensity in an optical fiber is essentially due to reflections from the two ends and absorption by the fiber material. Optical fiber devices are particularly useful for viewing an object at an inaccessible location. For example, physicians often use such devices to examine internal organs of the body or to perform surgery without making large incisions. Optical fiber cables are replacing copper wiring and coaxial cables for telecommunications because the fibers can carry a much greater volume of telephone calls or other forms of communication than electrical wires can.

Summary

ThomsonNOW™ Sign in at **www.thomsonedu.com** and go to ThomsonNOW to take a practice test for this chapter.

DEFINITION

The **index of refraction** n of a medium is defined by the ratio

$$n \equiv \frac{c}{v} \tag{35.4}$$

where c is the speed of light in a vacuum and v is the speed of light in the medium.

CONCEPTS AND PRINCIPLES

In geometric optics, we use the **ray approximation,** in which a wave travels through a uniform medium in straight lines in the direction of the rays.

Total internal reflection occurs when light travels from a medium of high index of refraction to one of lower index of refraction. The **critical angle** θ_c for which total internal reflection occurs at an interface is given by

$$\sin\theta_c = \frac{n_2}{n_1} \quad (\text{for } n_1 > n_2) \tag{35.10}$$

ANALYSIS MODELS FOR PROBLEM SOLVING

Wave Under Reflection. The **law of reflection** states that for a light ray (or other type of wave) incident on a smooth surface, the angle of reflection θ_1' equals the angle of incidence θ_1:

$$\theta_1' = \theta_1 \tag{35.2}$$

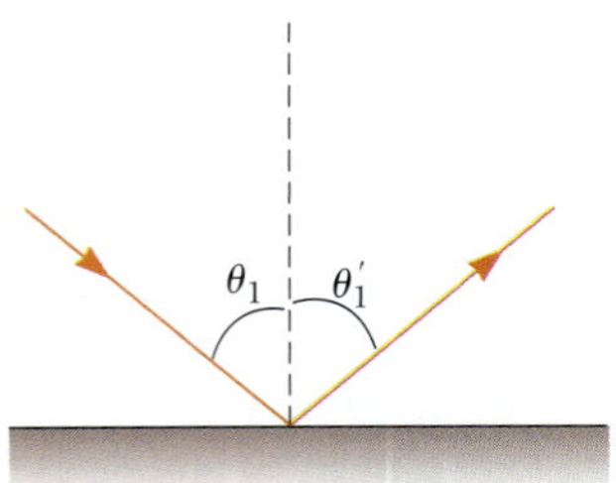

Wave Under Refraction. A wave crossing a boundary as it travels from medium 1 to medium 2 is **refracted,** or bent. The angle of refraction θ_2 is related to the incident angle θ_1 by the relationship

$$\frac{\sin\theta_2}{\sin\theta_1} = \frac{v_2}{v_1} \tag{35.3}$$

where v_1 and v_2 are the speeds of the wave in medium 1 and medium 2, respectively. The incident ray, the reflected ray, the refracted ray, and the normal to the surface all lie in the same plane.

For light waves, **Snell's law of refraction** states that

$$n_1 \sin\theta_1 = n_2 \sin\theta_2 \tag{35.8}$$

where n_1 and n_2 are the indices of refraction in the two media.

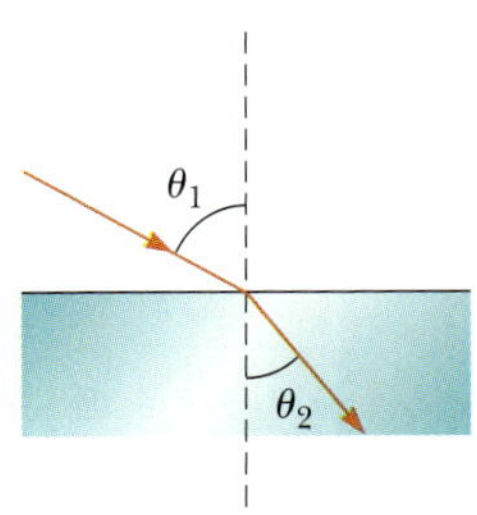

Questions

☐ denotes answer available in *Student Solutions Manual/Study Guide;* **O** denotes objective question

1. Why do astronomers looking at distant galaxies talk about looking backward in time?

2. O What is the order of magnitude of the time interval required for light to travel 10 km as in Galileo's attempt to measure the speed of light? (a) several seconds (b) several milliseconds (c) several microseconds (d) several nanoseconds

3. O In each of the following situations, a wave passes through an opening in an absorbing wall. Rank the situations in order from the one in which the wave is best described by the ray approximation to the one in which the wave coming through the opening spreads out most nearly equally in all directions in the hemisphere beyond the wall. (a) The sound of a low whistle at 1 kHz passes through a doorway 1 m wide. (b) Red light passes through the pupil of your eye. (c) Blue light passes through the pupil of your eye. (d) The wave broadcast by an AM radio station passes through a doorway 1 m wide. (e) An x-ray passes through the space between bones in your elbow joint.

4. The display windows of some department stores are slanted slightly inward at the bottom. This tilt is to decrease the glare from streetlights and the Sun, which would make it difficult for shoppers to see the display inside. Sketch a light ray reflecting from such a window to show how this design works.

5. You take a child for walks around the neighborhood. She loves to listen to echoes from houses when she shouts or when you clap loudly. A house with a large, flat front wall can produce an echo if you stand straight in front of it and reasonably far away. Draw a bird's-eye view of the situation to explain the production of the echo. Shade the area where you can stand to hear the echo. **What If?** The child helps you discover that a house with an L-shaped floor plan can produce echoes if you are standing in a wider range of locations. You can be standing at any reasonably distant location from which you can see the inside corner. Explain the echo in this case and draw another diagram for comparison. **What If?** What if the two wings of the house are not perpendicular? Will you and the child, standing close together, hear echoes? **What If?** What if a rectangular house and its garage have perpendicular walls that would form an inside corner but have a breezeway between them so that the walls do not meet? Will this structure produce strong echoes for people in a wide range of locations? Explain your answers with diagrams.

6. The F-117A stealth fighter (Figure Q35.6) is specifically designed to be a *non*retroreflector of radar. What aspects of its design help accomplish this purpose? *Suggestion:* Answer the previous question as preparation for this one. Notice that the bottom of the plane is flat and that all the flat exterior panels meet at odd angles.

Courtesy of U.S. Air Force, Langley Air Force Base

Figure Q35.6

7. O A light wave moves between medium 1 and medium 2. Which of the following are correct statements relating its speed, frequency, and wavelength in the two media, the indices of refraction of the media, and the angles of incidence and refraction? Choose all correct statements. (a) $v_1/\sin\theta_1 = v_2/\sin\theta_2$ (b) $\csc\theta_1/n_1 = \csc\theta_2/n_2$ (c) $\lambda_1/\sin\theta_1 = \lambda_2/\sin\theta_2$ (d) $f_1/\sin\theta_1 = f_2/\sin\theta_2$ (e) $n_1/\cos\theta_1 = n_2/\cos\theta_2$

8. Sound waves have much in common with light waves, including the properties of reflection and refraction. Give examples of these phenomena for sound waves.

9. O Consider light traveling from one medium into another with a different index of refraction. (a) Does its wavelength change? (b) Does its frequency change? (c) Does its speed change? (d) Does its direction always change?

10. A laser beam passing through a nonhomogeneous sugar solution follows a curved path. Explain.

11. **O** (a) Can light undergo total internal reflection at a smooth interface between air and water? If so, in which medium must it be traveling originally? (b) Can sound undergo total internal reflection at a smooth interface between air and water? If so, in which medium must it be traveling originally?

12. Explain why a diamond sparkles more than a glass crystal of the same shape and size.

13. Total internal reflection is applied in the periscope of a submarine to let the user "see around corners." In this device, two prisms are arranged as shown in Figure Q35.13 so that an incident beam of light follows the path shown. Parallel tilted, silvered mirrors could be used, but glass prisms with no silvered surfaces give higher light throughput. Propose a reason for the higher efficiency.

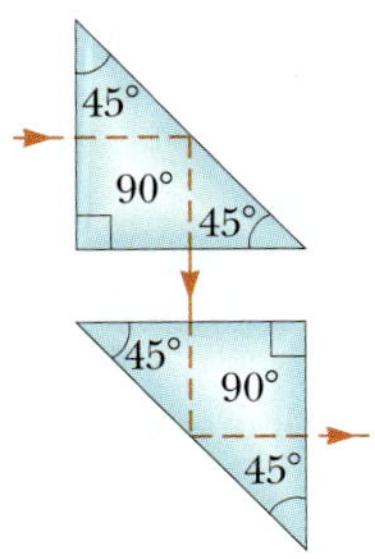

Figure Q35.13

14. **O** Suppose you find experimentally that two colors of light, A and B, originally traveling in the same direction in air, are sent through a glass prism, and A changes direction more than B. Which travels more slowly in the prism, A or B? Alternatively, is there insufficient information to determine which moves more slowly?

15. Retroreflection by transparent spheres, mentioned in Section 35.4, can be observed with dewdrops. To do so, look at the shadow of your head where it falls on dewy grass. Compare your observations to the reactions of two other people: Renaissance artist Benvenuto Cellini described the phenomenon and his reaction in his *Autobiography,* at the end of Part One, and American philosopher Henry David Thoreau did the same in *Walden,* "Baker Farm," second paragraph. The optical display around the shadow of your head is called *heiligenschein,* which is German for *holy light.* Try to find a person you know who has seen the heiligenschein. What did that person think about it?

16. How is it possible that a complete circle of a rainbow can sometimes be seen from an airplane? With a stepladder, a lawn sprinkler, and a sunny day, how can you show the complete circle to children?

17. At one restaurant, a worker uses colored chalk to write the daily specials on a blackboard illuminated with a spotlight. At another restaurant, a worker writes with colored grease pencils onto a flat, smooth sheet of transparent acrylic plastic with index of refraction 1.55. The panel hangs in front of a piece of black felt. Small, bright electric lights are installed all along the edges of the sheet, inside an opaque channel. Figure Q35.17 shows a cutaway view of the sign. Explain why viewers at both restaurants see the letters shining against a black background. Explain why the sign at the second restaurant may use less energy from the electric company than the illuminated blackboard at the first restaurant. What would be a good choice for the index of refraction of the material in the grease pencils?

Figure Q35.17

18. **O** The core of an optical fiber transmits light with minimal loss if it is surrounded by what? (a) water (b) diamond (c) air (d) glass (e) fused quartz.

19. Under what conditions is a mirage formed? On a hot day, what are we seeing when we observe "water on the road"?

Problems

WebAssign The Problems from this chapter may be assigned online in WebAssign.

ThomsonNOW Sign in at **www.thomsonedu.com** and go to ThomsonNOW to assess your understanding of this chapter's topics with additional quizzing and conceptual questions.

1, 2, 3 denotes straightforward, intermediate, challenging; □ denotes full solution available in *Student Solutions Manual/Study Guide;* ▲ denotes coached solution with hints available at **www.thomsonedu.com;** ▭ denotes developing symbolic reasoning; ● denotes asking for qualitative reasoning; ▪ denotes computer useful in solving problem

Section 35.1 The Nature of Light

Section 35.2 Measurements of the Speed of Light

1. ● The *Apollo 11* astronauts set up a panel of efficient corner-cube retroreflectors on the Moon's surface (Fig. 35.8a). The speed of light can be found by measuring the time interval required for a laser beam to travel from the Earth, reflect from the panel, and return to the Earth. Assume this interval is measured to be 2.51 s at a station where the Moon is at the zenith. What is the measured speed of light? Take the center-to-center distance from the Earth to the Moon to be 3.84×10^8 m. Explain whether it is necessary to consider the sizes of the Earth and the Moon in your calculation.

2. As a result of his observations, Roemer concluded that eclipses of Io by Jupiter were delayed by 22 min during a six-month period as the Earth moved from the point in its orbit where it is closest to Jupiter to the diametrically opposite point where it is farthest from Jupiter. Using 1.50×10^8 km as the average radius of the Earth's orbit around the Sun, calculate the speed of light from these data.

3. □ In an experiment to measure the speed of light using the apparatus of Fizeau (see Fig. 35.2), the distance between light source and mirror was 11.45 km and the wheel had 720 notches. The experimentally determined value of c was 2.998×10^8 m/s. Calculate the minimum angular speed of the wheel for this experiment.

Section 35.3 The Ray Approximation in Geometric Optics

Section 35.4 The Wave Under Reflection

Section 35.5 The Wave Under Refraction

Note: You may look up indices of refraction in Table 35.1.

4. A dance hall is built without pillars and with a horizontal ceiling 7.20 m above the floor. A mirror is fastened flat against one section of the ceiling. Following an earthquake, the mirror is in place and unbroken. An engineer makes a quick check of whether the ceiling is sagging by directing a vertical beam of laser light up at the mirror and observing its reflection on the floor. (a) Show that if the mirror has rotated to make an angle ϕ with the horizontal, the normal to the mirror makes an angle ϕ with the vertical. (b) Show that the reflected laser light makes an angle 2ϕ with the vertical. (c) Assume the reflected laser light makes a spot on the floor 1.40 cm away from the point vertically below the laser. Find the angle ϕ.

5. The two mirrors illustrated in Figure P35.5 meet at a right angle. The beam of light in the vertical plane P strikes mirror 1 as shown. (a) Determine the distance the reflected light beam travels before striking mirror 2. (b) In what direction does the light beam travel after being reflected from mirror 2?

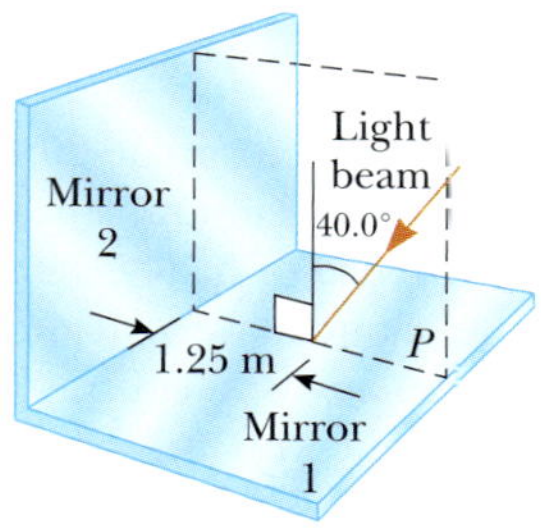

Figure P35.5

6. Two flat, rectangular mirrors, both perpendicular to a horizontal sheet of paper, are set edge to edge with their reflecting surfaces perpendicular to each other. (a) A light ray in the plane of the paper strikes one of the mirrors at an arbitrary angle of incidence θ_1. Prove that the final direction of the ray, after reflection from both mirrors, is opposite its initial direction. In a clothing store, such a pair of mirrors shows you an image of yourself as others see you, with no apparent right–left reversal. (b) **What If?** Now assume the paper is replaced with a third flat mirror, touching edges with the other two and perpendicular to both. The set of three mirrors is called a *corner-cube reflector.* A ray of light is incident from any direction within the octant of space bounded by the reflecting surfaces. Argue that the ray will reflect once from each mirror and that its final direction will be opposite to its original direction. The *Apollo 11* astronauts placed a panel of corner-cube retroreflectors on the Moon. Analysis of timing data taken with it reveals that the radius of the Moon's orbit is increasing at the rate of 3.8 cm/yr as it loses kinetic energy because of tidal friction.

7. The distance of a lightbulb from a large plane mirror is twice the distance of a person from the plane mirror. Light from the lightbulb reaches the person by two paths. It travels to the mirror at an angle of incidence θ and reflects from the mirror to the person. It also travels directly to the person without reflecting off the mirror. The total distance traveled by the light in the first case is twice the distance traveled by the light in the second case. Find the value of the angle θ.

8. Two light pulses are emitted simultaneously from a source. Both pulses travel to a detector, but mirrors shunt one pulse along a path that carries it through 6.20 m of ice along the way. Determine the difference in the pulses' times of arrival at the detector.

9. A narrow beam of sodium yellow light, with wavelength 589 nm in vacuum, is incident from air onto a smooth water surface at an angle of incidence of 35.0°. Determine the angle of refraction and the wavelength of the light in water.

10. ● A plane sound wave in air at 20°C, with wavelength 589 mm, is incident on a smooth surface of water at 25°C at an angle of incidence of 3.50°. Determine the angle of refraction for the sound wave and the wavelength of the sound in water. Compare and contrast the behavior of the sound in this problem with the behavior of the light in Problem 9.

11. An underwater scuba diver sees the Sun at an apparent angle of 45.0° above the horizontal. What is the actual elevation angle of the Sun above the horizontal?

12. The wavelength of red helium–neon laser light in air is 632.8 nm. (a) What is its frequency? (b) What is its wavelength in glass that has an index of refraction of 1.50? (c) What is its speed in the glass?

13. A ray of light is incident on a flat surface of a block of crown glass that is surrounded by water. The angle of refraction is 19.6°. Find the angle of reflection.

14. A laser beam with vacuum wavelength 632.8 nm is incident from air onto a block of Lucite as shown in Active Figure 35.10b. The line of sight of the photograph is perpendicular to the plane in which the light moves. Find (a) the speed, (b) the frequency, and (c) the wavelength of the light in the Lucite. *Suggestion:* Use a protractor.

15. Find the speed of light in (a) flint glass, (b) water, and (c) cubic zirconia.

16. A narrow beam of ultrasonic waves reflects off the liver tumor illustrated in Figure P35.16. The speed of the wave is 10.0% less in the liver than in the surrounding medium. Determine the depth of the tumor.

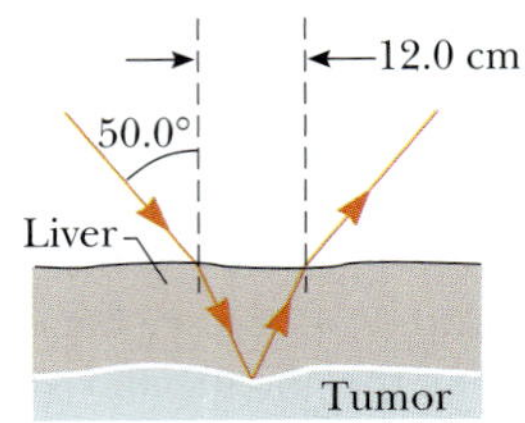

Figure P35.16

17. ▲ A ray of light strikes a flat block of glass ($n = 1.50$) of thickness 2.00 cm at an angle of 30.0° with the normal. Trace the light beam through the glass and find the angles of incidence and refraction at each surface.

18. An opaque cylindrical tank with an open top has a diameter of 3.00 m and is completely filled with water. When the afternoon Sun reaches an angle of 28.0° above the horizon, sunlight ceases to illuminate any part of the bottom of the tank. How deep is the tank?

19. When the light illustrated in Figure P35.19 passes through the glass block, it is shifted laterally by the distance d. Taking $n = 1.50$, find the value of d.

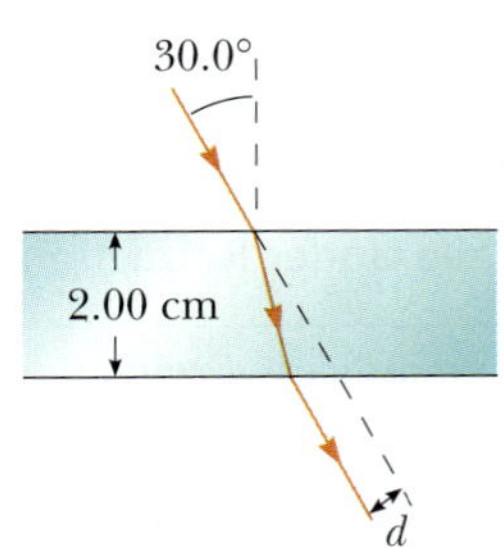

Figure P35.19

20. Find the time interval required for the light to pass through the glass block described in Problem 19.

21. The light beam shown in Figure P35.21 makes an angle of 20.0° with the normal line NN' in the linseed oil. Determine the angles θ and θ'. (The index of refraction of linseed oil is 1.48.)

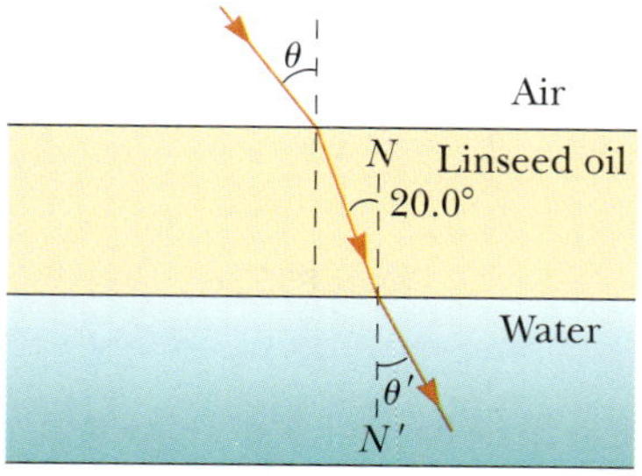

Figure P35.21

22. Three sheets of plastic have unknown indices of refraction. Sheet 1 is placed on top of sheet 2, and a laser beam is directed onto the sheets from above so that it strikes the interface at an angle of 26.5° with the normal. The refracted beam in sheet 2 makes an angle of 31.7° with the normal. The experiment is repeated with sheet 3 on top of sheet 2, and, with the same angle of incidence, the refracted beam makes an angle of 36.7° with the normal. If the experiment is repeated again with sheet 1 on top of sheet 3, what is the expected angle of refraction in sheet 3? Assume the same angle of incidence.

23. ● Light passes from air into flint glass. (a) Is it possible for the component of its velocity perpendicular to the interface to remain constant? Explain your answer. (b) **What If?** Can the component of velocity parallel to the interface remain constant during refraction? Explain your answer.

24. When you look through a window, by what time interval is the light you see delayed by having to go through glass instead of air? Make an order-of-magnitude estimate on the basis of data you specify. By how many wavelengths is it delayed?

25. A prism that has an apex angle of 50.0° is made of cubic zirconia, with $n = 2.20$. What is its angle of minimum deviation?

26. Light of wavelength 700 nm is incident on the face of a fused quartz prism at an angle of 75.0° (with respect to the normal to the surface). The apex angle of the prism is 60.0°. Use the value of n from Figure 35.21 and calculate the angle (a) of refraction at the first surface, (b) of incidence at the second surface, (c) of refraction at the second surface, and (d) between the incident and emerging rays.

27. A triangular glass prism with apex angle $\Phi = 60.0°$ has an index of refraction $n = 1.50$ (Fig. P35.27). What is the smallest angle of incidence θ_1 for which a light ray can emerge from the other side?

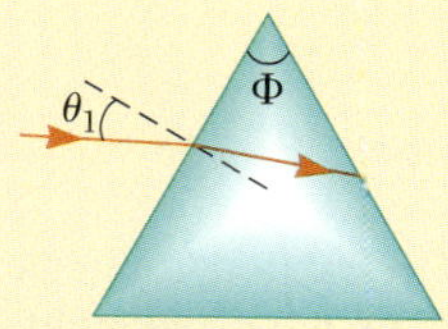

Figure P35.27 Problems 27 and 28.

28. A triangular glass prism with apex angle Φ has index of refraction n. (See Fig. P35.27.) What is the smallest angle of incidence θ_1 for which a light ray can emerge from the other side?

29. A triangular glass prism with apex angle 60.0° has an index of refraction of 1.50. (a) Show that if its angle of incidence on the first surface is $\theta_1 = 48.6°$, light will pass symmetrically through the prism as shown in Figure 35.17. (b) Find the angle of deviation δ_{min} for $\theta_1 = 48.6°$. (c) **What If?** Find the angle of deviation if the angle of incidence on the first surface is 45.6°. (d) Find the angle of deviation if $\theta_1 = 51.6°$.

Section 35.6 Huygens's Principle

30. The speed of a water wave is described by $v = \sqrt{gd}$, where d is the water depth, assumed to be small compared to the wavelength. Because their speed changes, water waves refract when moving into a region of different depth. Sketch a map of an ocean beach on the eastern side of a landmass. Show contour lines of constant depth under water, assuming reasonably uniform slope. (a) Suppose waves approach the coast from a storm far away to the north–northeast. Demonstrate that the waves move nearly perpendicular to the shoreline when they reach the beach. (b) Sketch a map of a coastline with alternating bays and headlands as suggested in Figure P35.30. Again

Ray Atkeson/Image Archive

Figure P35.30

make a reasonable guess about the shape of contour lines of constant depth. Suppose waves approach the coast, carrying energy with uniform density along originally straight wave fronts. Show that the energy reaching the coast is concentrated at the headlands and has lower intensity in the bays.

Section 35.7 Dispersion

31. ▲ The index of refraction for violet light in silica flint glass is 1.66 and that for red light is 1.62. What is the angular spread of visible light passing through a prism of apex angle 60.0° if the angle of incidence is 50.0°? See Figure P35.31.

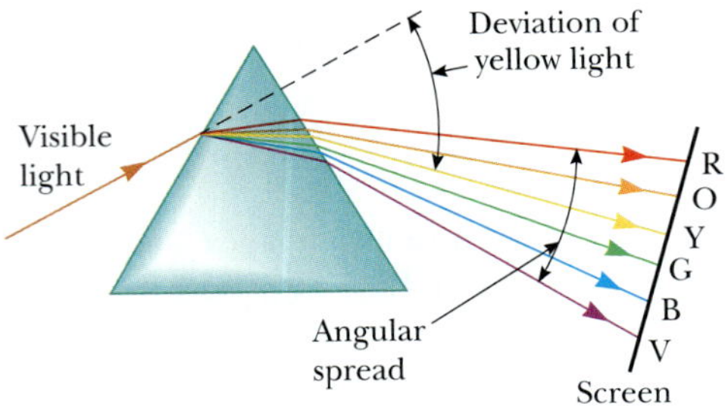

Figure P35.31

32. A narrow, white light beam is incident on a block of fused quartz at an angle of 30.0°. Find the angular spread of the light beam inside the quartz due to dispersion.

Section 35.8 Total Internal Reflection

33. For 589-nm light, calculate the critical angle for the following materials surrounded by air. (a) diamond (b) flint glass (c) ice

34. A glass fiber ($n = 1.50$) is submerged in water ($n = 1.33$). What is the critical angle for light to stay inside the optical fiber?

35. Consider a common mirage formed by superheated air immediately above a roadway. A truck driver whose eyes are 2.00 m above the road, where $n = 1.000\,3$, looks forward. She perceives the illusion of a patch of water ahead on the road, where her line of sight makes an angle of 1.20° below the horizontal. Find the index of refraction of the air immediately above the road surface. *Suggestion:* Treat this problem as one about total internal reflection.

36. Determine the maximum angle θ for which the light rays incident on the end of the pipe in Figure P35.36 are subject to total internal reflection along the walls of the pipe. Assume the pipe has an index of refraction of 1.36 and the outside medium is air. Your answer defines the size of the *cone of acceptance* for the light pipe.

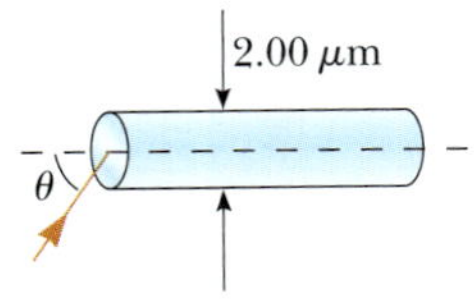

Figure P35.36

37. ● An optical fiber has index of refraction n and diameter d. It is surrounded by air. Light is sent into the fiber along its axis as shown in Figure P35.37. (a) Find the smallest outside radius R permitted for a bend in the fiber if no light is to escape. (b) **What If?** Does the result for part (a) predict reasonable behavior as d approaches zero? As n increases? As n approaches 1? (c) Evaluate R assuming the fiber diameter is 100 μm and its index of refraction is 1.40.

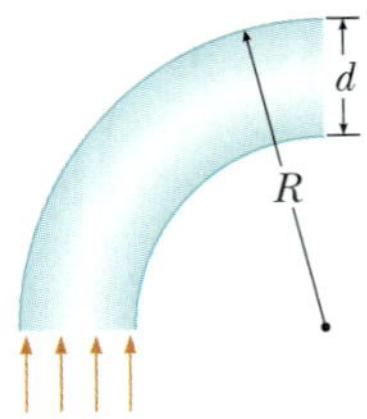

Figure P35.37

38. ● A room contains air in which the speed of sound is 343 m/s. The walls of the room are made of concrete in which the speed of sound is 1 850 m/s. (a) Find the critical angle for total internal reflection of sound at the concrete–air boundary. (b) In which medium must the sound be traveling if it is undergo total internal reflection? (c) "A bare concrete wall is a highly efficient mirror for sound." Give evidence for or against this statement.

39. ● Around 1965, engineers at the Toro Company invented a gasoline gauge for small engines diagrammed in Figure P35.39. The gauge has no moving parts. It consists of a flat slab of transparent plastic fitting vertically into a slot in the cap on the gas tank. None of the plastic has a reflective coating. The plastic projects from the horizontal top down nearly to the bottom of the opaque tank. Its lower edge is cut with facets making angles of 45° with the horizontal. A lawn mower operator looks down from above and sees a boundary between bright and dark on the gauge. The location of the boundary, across the width of the plastic, indicates the quantity of gasoline in the tank. Explain how the gauge works. Explain the design requirements, if any, for the index of refraction of the plastic.

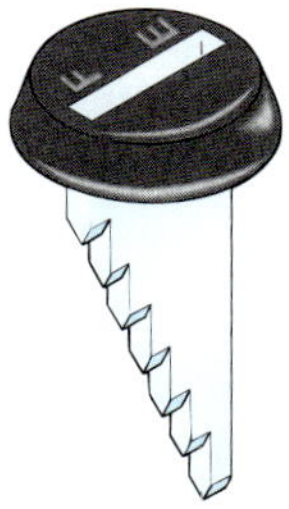

Figure P35.39

41. ● Figure P35.41a shows a desk ornament globe containing a photograph. The flat photograph is in air, inside a vertical slot located behind a water-filled compartment having the shape of one half of a cylinder. Suppose you are looking at the center of the photograph and then rotate the globe about a vertical axis. You find that the center of the photograph disappears when you rotate the globe beyond a certain maximum angle (Fig. P35.41b). Account for this phenomenon and calculate the maximum angle. Describe what you see when you turn the globe beyond this angle.

(a) (b)

Figure P35.41

Additional Problems

40. A digital videodisc records information in a spiral track approximately 1 μm wide. The track consists of a series of pits in the information layer (Fig. P35.40a) that scatter light from a laser beam sharply focused on them. The laser shines in through transparent plastic of thickness t = 1.20 mm and index of refraction 1.55 (Fig. P35.40b). Assume the width of the laser beam at the information layer must be a = 1.00 μm to read from only one track and not from its neighbors. Assume the width of the beam as it enters the transparent plastic from below is w = 0.700 mm. A lens makes the beam converge into a cone with an apex angle $2\theta_1$ before it enters the videodisc. Find the incidence angle θ_1 of the light at the edge of the conical beam. This design is relatively immune to small dust particles degrading the video quality. Particles on the plastic surface would have to be as large as 0.7 mm to obscure the beam.

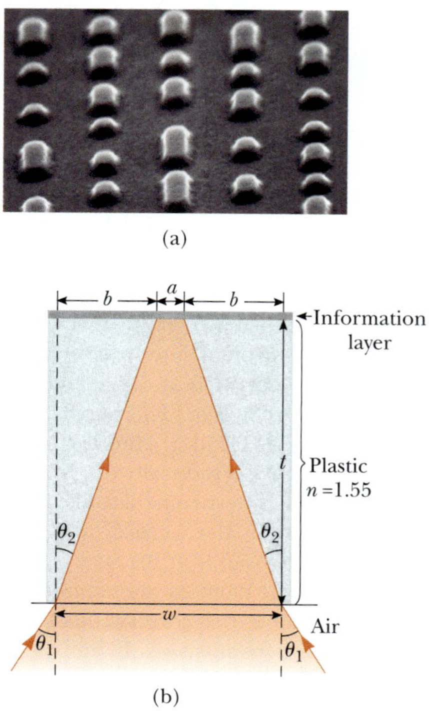

Figure P35.40

42. ● A light ray enters the atmosphere of a planet and descends vertically to the surface a distance h below. The index of refraction where the light enters the atmosphere is 1.000, and it increases linearly with distance to have the value n at the planet surface. (a) Over what time interval does the light traverse this path? (b) State how this travel time compares with the time interval required in the absence of an atmosphere.

43. A narrow beam of light is incident from air onto the surface of glass with index of refraction 1.56. Find the angle of incidence for which the corresponding angle of refraction is half the angle of incidence. *Suggestion:* You might want to use the trigonometric identity $\sin 2\theta = 2 \sin\theta \cos\theta$.

44. (a) Consider a horizontal interface between air above and glass of index 1.55 below. Draw a light ray incident from the air at angle of incidence 30.0°. Determine the angles of the reflected and refracted rays and show them on the diagram. (b) **What If?** Now suppose the light ray is incident from the glass at angle of incidence 30.0°. Determine the angles of the reflected and refracted rays and show all three rays on a new diagram. (c) For rays incident from the air onto the air–glass surface, determine and tabulate the angles of reflection and refraction for all the angles of incidence at 10.0° intervals from 0° to 90.0°.

(d) Do the same for light rays coming up to the interface through the glass.

45. ▲ A small light fixture on the bottom of a swimming pool is 1.00 m below the surface. The light emerging from the still water forms a circle on the water surface. What is the diameter of this circle?

46. The walls of a prison cell are perpendicular to the four cardinal compass directions. On the first day of spring, light from the rising Sun enters a rectangular window in the eastern wall. The light traverses 2.37 m horizontally to shine perpendicularly on the wall opposite the window. A young prisoner observes the patch of light moving across this western wall and for the first time forms his own understanding of the rotation of the Earth. (a) With what speed does the illuminated rectangle move? (b) The prisoner holds a small, square mirror flat against the wall at one corner of the rectangle of light. The mirror reflects light back to a spot on the eastern wall close beside the window. With what speed does the smaller square of light move across that wall? (c) Seen from a latitude of 40.0° north, the rising Sun moves through the sky along a line making a 50.0° angle with the southeastern horizon. In what direction does the rectangular patch of light on the western wall of the prisoner's cell move? (d) In what direction does the smaller square of light on the eastern wall move?

47. A hiker stands on an isolated mountain peak near sunset and observes a rainbow caused by water droplets in the air at a distance of 8.00 km along her line of sight. The valley is 2.00 km below the mountain peak and entirely flat. What fraction of the complete circular arc of the rainbow is visible to the hiker? (See Fig. 35.24.)

48. Figure P35.48 shows a top view of a square enclosure. The inner surfaces are plane mirrors. A ray of light enters a small hole in the center of one mirror. (a) At what angle θ must the ray enter if it exits through the hole after being reflected once by each of the other three mirrors? (b) **What If?** Are there other values of θ for which the ray can exit after multiple reflections? If so, sketch one of the ray's paths.

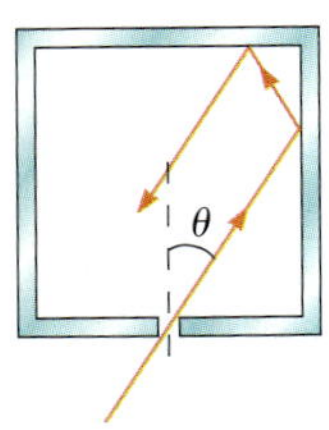

Figure P35.48

49. ▲ A laser beam strikes one end of a slab of material as shown in Figure P35.49. The index of refraction of the slab is 1.48. Determine the number of internal reflections of the beam before it emerges from the opposite end of the slab.

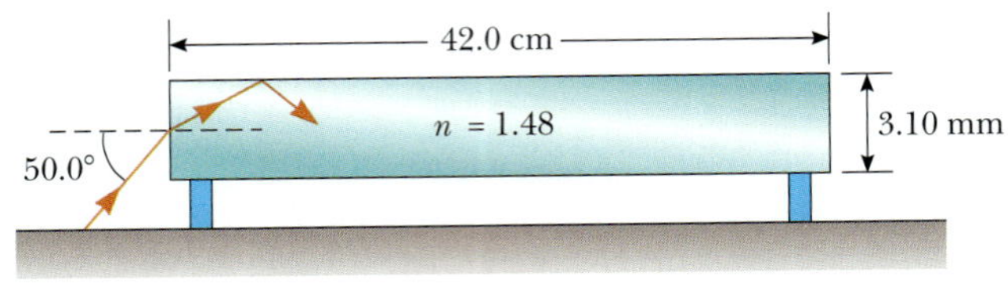

Figure P35.49

50. A 4.00-m-long pole stands vertically in a lake having a depth of 2.00 m. The Sun is 40.0° above the horizontal. Determine the length of the pole's shadow on the bottom of the lake. Take the index of refraction for water to be 1.33.

51. The light beam in Figure P35.51 strikes surface 2 at the critical angle. Determine the angle of incidence θ_1.

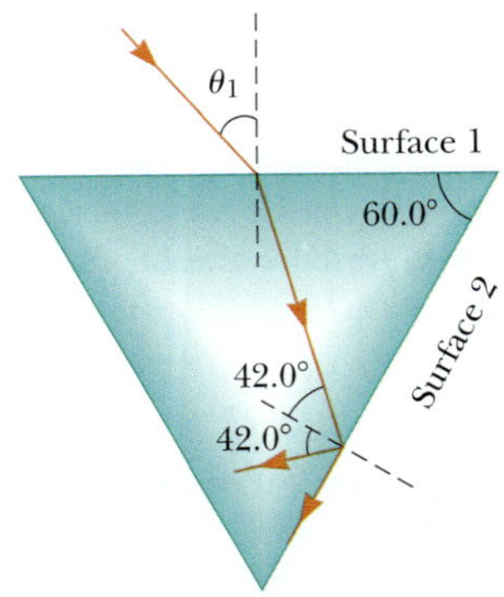

Figure P35.51

52. Builders use a leveling instrument in which the beam from a fixed helium–neon laser reflects in a horizontal plane from a small, flat mirror mounted on a vertical rotating shaft. The light is sufficiently bright and the rotation rate is sufficiently high that the reflected light appears as a horizontal line, wherever it falls on a wall. (a) Assume the mirror is at the center of a circular grain elevator of radius 3.00 m. The mirror spins with constant angular velocity 35.0 rad/s. Find the speed of the spot of laser light on the curved wall. (b) Now assume the spinning mirror is at a perpendicular distance of 3.00 m from point O on a long, flat, vertical wall. When the spot of laser light on the wall is at distance x from point O, what is its speed? (c) What is the minimum value for the speed? What value of x corresponds to it? How does the minimum speed compare with the speed you found in part (a)? (d) What is the maximum speed of the spot on the flat wall? (e) In what time interval does the spot change from its minimum to its maximum speed?

53. ▲ ● A light ray of wavelength 589 nm is incident at an angle θ on the top surface of a block of polystyrene as shown in Figure P35.53. (a) Find the maximum value of θ for which the refracted ray undergoes total internal reflection at the left vertical face of the block. **What If?** Repeat the calculation for the case in which the polystyrene block is immersed in (b) water and (c) carbon disulfide. You will need to explain your answers.

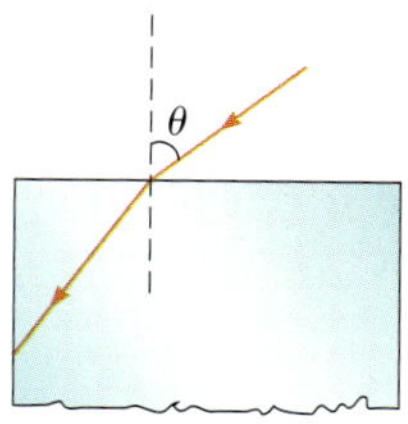

Figure P35.53

54. ● As sunlight enters the Earth's atmosphere, it changes direction due to the small difference between the speeds of light in vacuum and in air. The duration of an *optical* day is defined as the time interval between the instant when the top of the rising Sun is just visible above the horizon and the instant when the top of the Sun just disappears below the horizontal plane. The duration of the *geometric* day is defined as the time interval between the instant when a mathematically straight line between an observer and the top of the Sun just clears the horizon and the instant at which this line just dips below the horizon. (a) Explain which is longer, an optical day or a geometric day. (b) Find the difference between these two time intervals. Model the Earth's atmosphere as uniform, with index of refraction 1.000 293, a sharply defined upper surface, and depth 8 614 m. Assume the observer is at the Earth's equator so that the apparent path of the rising and setting Sun is perpendicular to the horizon.

55. A shallow glass dish is 4.00 cm wide at the bottom as shown in Figure P35.55. When an observer's eye is located as shown, the observer sees the edge of the bottom of the empty dish. When this dish is filled with water, the observer sees the center of the bottom of the dish. Find the height of the dish.

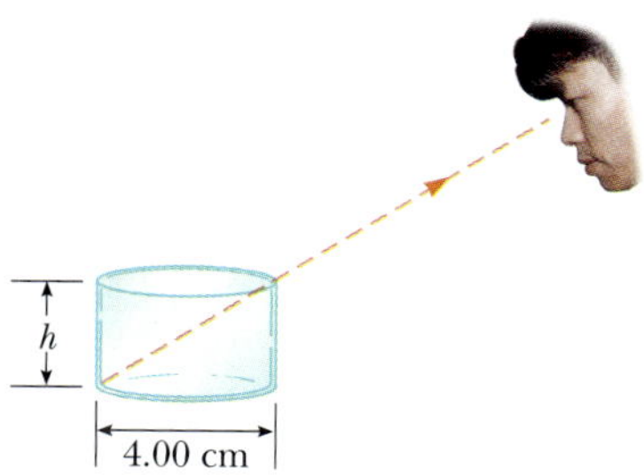

Figure P35.55

56. A ray of light passes from air into water. For its deviation angle $\delta = |\theta_1 - \theta_2|$ to be 10.0°, what must its angle of incidence be?

57. A material having an index of refraction n is surrounded by a vacuum and is in the shape of a quarter circle of radius R (Fig. P35.57). A light ray parallel to the base of the material is incident from the left at a distance L above the base and emerges from the material at the angle θ. Determine an expression for θ.

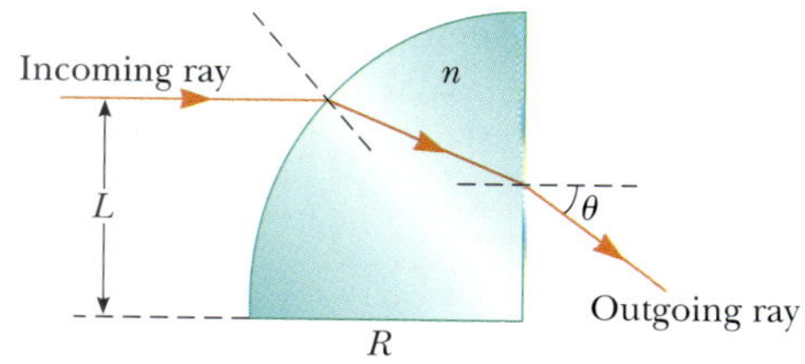

Figure P35.57

58. *Fermat's principle.* Pierre de Fermat (1601–1665) showed that whenever light travels from one point to another, its actual path is the path that requires the smallest time interval. The simplest example is for light propagating in a homogeneous medium. It moves in a straight line because a straight line is the shortest distance between two points. Derive Snell's law of refraction from Fermat's principle. Proceed as follows. In Figure P35.58, a light ray travels from point P in medium 1 to point Q in medium 2. The two points are respectively at perpendicular distances a and b from the interface. The displacement from P to Q has the component d parallel to the interface, and we let x represent the coordinate of the point where the ray enters the second medium. Let $t = 0$ be the instant at which the light starts from P. (a) Show that the time at which the light arrives at Q is

$$t = \frac{r_1}{v_1} + \frac{r_2}{v_2} = \frac{n_1\sqrt{a^2 + x^2}}{c} + \frac{n_2\sqrt{b^2 + (d - x)^2}}{c}$$

(b) To obtain the value of x for which t has its minimum value, differentiate t with respect to x and set the derivative equal to zero. Show that the result implies

$$\frac{n_1 x}{\sqrt{a^2 + x^2}} = \frac{n_2(d - x)}{\sqrt{b^2 + (d - x)^2}}$$

(c) Show that this expression in turn gives Snell's law

$$n_1 \sin\theta_1 = n_2 \sin\theta_2$$

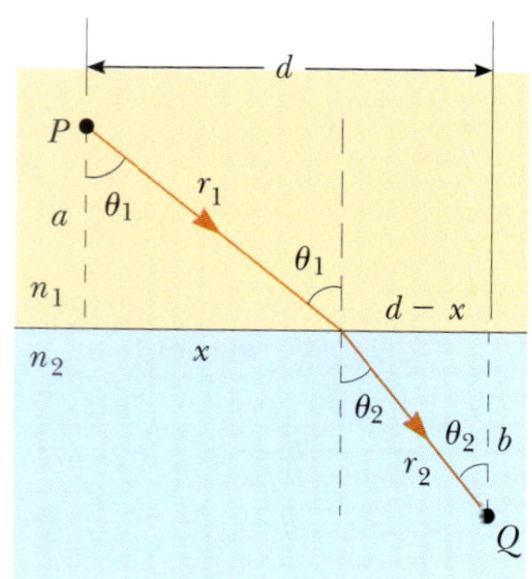

Figure P35.58

59. Refer to Problem 58 for the statement of Fermat's principle of least time. Derive the law of reflection (Eq. 35.2) from Fermat's principle.

60. A transparent cylinder of radius $R = 2.00$ m has a mirrored surface on its right half as shown in Figure P35.60. A light ray traveling in air is incident on the left side of the cylinder. The incident light ray and exiting light ray are parallel, and $d = 2.00$ m. Determine the index of refraction of the material.

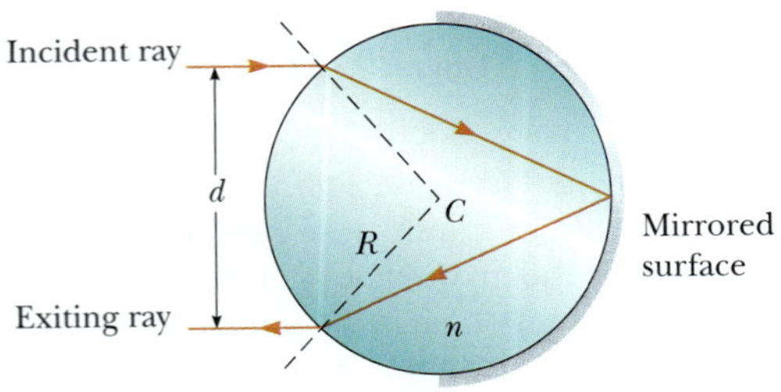

Figure P35.60

61. Suppose a luminous sphere of radius R_1 (such as the Sun) is surrounded by a uniform atmosphere of radius R_2 and index of refraction n. When the sphere is viewed from a location far away in vacuum, what is its apparent radius? You will need to distinguish between the two cases (a) $R_2 > nR_1$ and (b) $R_2 < nR_1$.

62. ● A. H. Pfund's method for measuring the index of refraction of glass is illustrated in Figure P35.62. One face of a slab of thickness t is painted white, and a small hole scraped clear at point P serves as a source of diverging rays when the slab is illuminated from below. Ray PBB' strikes the clear surface at the critical angle and is totally reflected as are rays such as PCC'. Rays such as PAA' emerge from the clear surface. On the painted surface, there appears a dark circle of diameter d surrounded by an illuminated region, or halo. (a) Derive an equation for n in terms of the measured quantities d and t. (b) What is the diameter of the dark circle if $n = 1.52$ for a slab 0.600 cm thick? (c) If white light is used, dispersion causes the critical angle to depend on color. Is the inner edge of the white halo tinged with red light or with violet light? Explain.

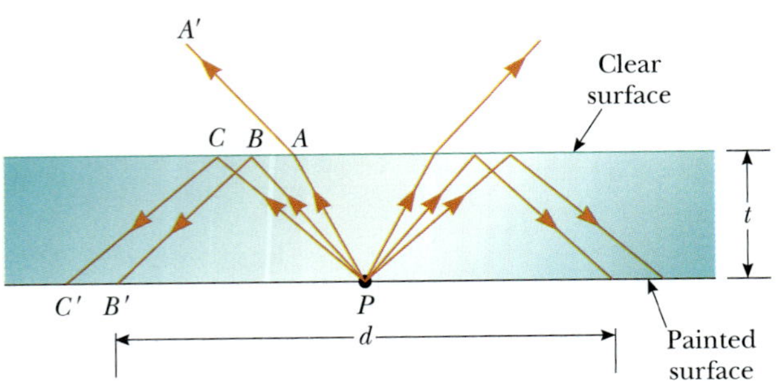

Figure P35.62

63. A light ray enters a rectangular block of plastic at an angle $\theta_1 = 45.0°$ and emerges at an angle $\theta_2 = 76.0°$ as shown in Figure P35.63. (a) Determine the index of refraction of the plastic. (b) If the light ray enters the plastic at a point $L = 50.0$ cm from the bottom edge, what time interval is required for the light ray to travel through the plastic?

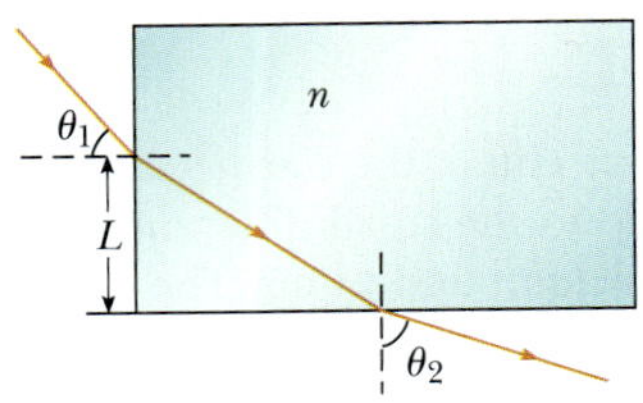

Figure P35.63

64. ● Students allow a narrow beam of laser light to strike a water surface. They measure the angle of refraction for selected angles of incidence and record the data shown in the accompanying table. Use the data to verify Snell's law of refraction by plotting the sine of the angle of incidence versus the sine of the angle of refraction. Explain what the shape of the graph demonstrates. Use the resulting plot to deduce the index of refraction of water, explaining how you do so.

Angle of Incidence (degrees)	Angle of Refraction (degrees)
10.0	7.5
20.0	15.1
30.0	22.3
40.0	28.7
50.0	35.2
60.0	40.3
70.0	45.3
80.0	47.7

65. **Review problem.** A mirror is often "silvered" with aluminum. By adjusting the thickness of the metallic film, one can make a sheet of glass into a mirror that reflects anything between, say, 3% and 98% of the incident light, transmitting the rest. Prove that it is impossible to construct a "one-way mirror" that would reflect 90% of the electromagnetic waves incident from one side and reflect 10% of those incident from the other side. *Suggestion:* Use Clausius's statement of the second law of thermodynamics.

Answers to Quick Quizzes

35.1 (d). The light rays from the actor's face must reflect from the mirror and into the camera. If these light rays are reversed, light from the camera reflects from the mirror into the eyes of the actor.

35.2 Beams ② and ④ are reflected; beams ③ and ⑤ are refracted.

35.3 (c). Because the light is entering a material in which the index of refraction is lower, the speed of light is higher and the light bends away from the normal.

35.4 (c). An ideal camera lens would have an index of refraction that does not vary with wavelength so that all colors would be bent through the same angle by the lens. Of the three choices, fused quartz has the least variation in n across the visible spectrum. A lens designer can do even better by stacking two lenses of different materials together to make an *achromatic doublet.*

35.5 **(i),** (b). The two bright rays exiting the bottom of the prism on the right in Figure 35.26 result from total internal reflection at the right face of the prism. Notice that there is no refracted light exiting the slanted side for these rays. The light from the other three rays is divided into reflected and refracted parts. **(ii),** (b). Counterclockwise rotation of the prism will cause the rays to strike the slanted side of the prism at a larger angle. When the five rays strike at an angle larger than the critical angle, they all undergo total internal reflection.

The light rays coming from the leaves in the background of this scene did not form a focused image on the film of the camera that took this photograph. Consequently, the background appears very blurry. Light rays passing though the raindrop, however, have been altered so as to form a focused image of the background leaves on the film. In this chapter, we investigate the formation of images as light rays reflect from mirrors and refract through lenses. (Don Hammond/CORBIS)

36 Image Formation

This chapter is concerned with the images that result when light rays encounter flat and curved surfaces. Images can be formed by either reflection or refraction, and we can design mirrors and lenses to form images with desired characteristics. We continue to use the ray approximation and assume light travels in straight lines. These two steps lead to valid predictions in the field called *geometric optics.* Subsequent chapters cover interference and diffraction effects, which are the objects of study in the field of *wave optics.*

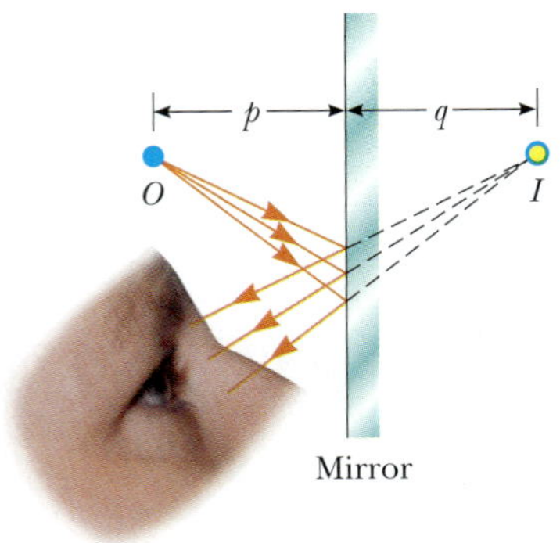

Figure 36.1 An image formed by reflection from a flat mirror. The image point I is located behind the mirror a perpendicular distance q from the mirror (the image distance). The image distance has the same magnitude as the object distance p.

36.1 Images Formed by Flat Mirrors

Image formation by mirrors can be understood through the analysis of light rays following the wave under reflection model. We begin by considering the simplest possible mirror, the flat mirror. Consider a point source of light placed at O in Figure 36.1, a distance p in front of a flat mirror. The distance p is called the **object distance.** Diverging light rays leave the source and are reflected from the mirror. Upon reflection, the rays continue to diverge. The dashed lines in Figure 36.1 are extensions of the diverging rays back to a point of intersection at I. The diverging rays appear to the viewer to originate at the point I behind the mirror. Point I, which is a distance q behind the mirror, is called the **image** of the object at O. The distance q is called the **image distance.** Regardless of the system under study, images can always be located by extending diverging rays back to a point at which they intersect. **Images are located either at a point from which rays of light *actually* diverge or at a point from which they *appear* to diverge.**

Images are classified as **real** or **virtual. A real image is formed when light rays pass through and diverge from the image point; a virtual image is formed when the light rays do not pass through the image point but only appear to diverge from that point.** The image formed by the mirror in Figure 36.1 is virtual. The image of an object seen in a flat mirror is *always* virtual. Real images can be displayed on a screen (as at a movie theater), but virtual images cannot be displayed on a screen. We shall see an example of a real image in Section 36.2.

We can use the simple geometry in Active Figure 36.2 to examine the properties of the images of extended objects formed by flat mirrors. Even though there are an infinite number of choices of direction in which light rays could leave each point on the object (represented by a blue arrow), we need to choose only two rays to determine where an image is formed. One of those rays starts at P, follows a path perpendicular to the mirror, and reflects back on itself. The second ray follows the oblique path PR and reflects as shown in Active Figure 36.2 according to the law of reflection. An observer in front of the mirror would extend the two reflected rays back to the point at which they appear to have originated, which is point P' behind the mirror. A continuation of this process for points other than P on the object would result in a virtual image (represented by a yellow arrow) of the entire object behind the mirror. Because triangles PQR and $P'QR$ are congruent, $PQ = P'Q$, so that $|p| = |q|$. Therefore, **the image formed of an object placed in front of a flat mirror is as far behind the mirror as the object is in front of the mirror.**

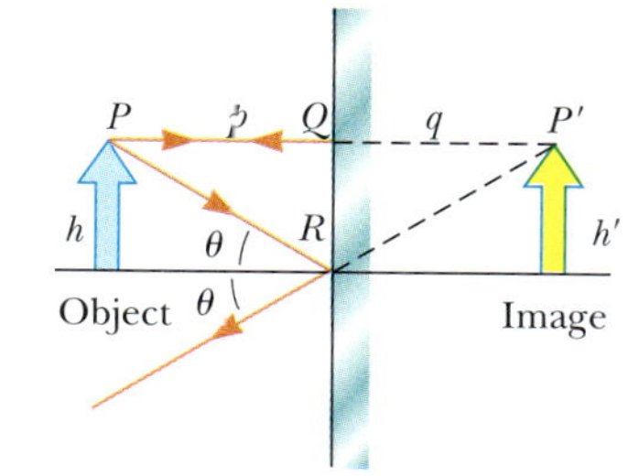

ACTIVE FIGURE 36.2

A geometric construction that is used to locate the image of an object placed in front of a flat mirror. Because the triangles PQR and $P'QR$ are congruent, $|p| = |q|$ and $h = h'$.

Sign in at www.thomsonedu.com and go to ThomsonNOW to move the object and see the effect on the image.

The geometry in Active Figure 36.2 also reveals that the object height h equals the image height h'. Let us define **lateral magnification** M of an image as follows:

$$M \equiv \frac{\text{image height}}{\text{object height}} = \frac{h'}{h} \qquad (36.1)$$

◀ Lateral magnification

This general definition of the lateral magnification for an image from any type of mirror is also valid for images formed by lenses, which we study in Section 36.4. For a flat mirror, $M = +1$ for any image because $h' = h$. The positive value of the magnification signifies that the image is upright. (By upright we mean that if the object arrow points upward as in Active Figure 36.2, so does the image arrow.)

PITFALL PREVENTION 36.1

Magnification Does Not Necessarily Imply Enlargement

For optical elements other than flat mirrors, the magnification defined in Equation 36.1 can result in a number with a magnitude larger *or* smaller than 1. Therefore, despite the cultural usage of the word *magnification* to mean *enlargement*, the image could be smaller than the object.

A flat mirror produces an image that has an *apparent* left–right reversal. You can see this reversal by standing in front of a mirror and raising your right hand as shown in Figure 36.3. The image you see raises its left hand. Likewise, your hair appears to be parted on the side opposite your real part, and a mole on your right cheek appears to be on your left cheek.

This reversal is not *actually* a left–right reversal. Imagine, for example, lying on your left side on the floor with your body parallel to the mirror surface. Now your head is on the left and your feet are on the right. If you shake your feet, the image does not shake its head! If you raise your right hand, however, the image again raises its left hand. Therefore, the mirror again appears to produce a left–right reversal but in the up–down direction!

The reversal is actually a *front–back reversal,* caused by the light rays going forward toward the mirror and then reflecting back from it. An interesting exercise is to stand in front of a mirror while holding an overhead transparency in front of you so that you can read the writing on the transparency. You will also be able to read the writing on the image of the transparency. You may have had a similar experience if you have attached a transparent decal with words on it to the rear window of your car. If the decal can be read from outside the car, you can also read it when looking into your rearview mirror from inside the car.

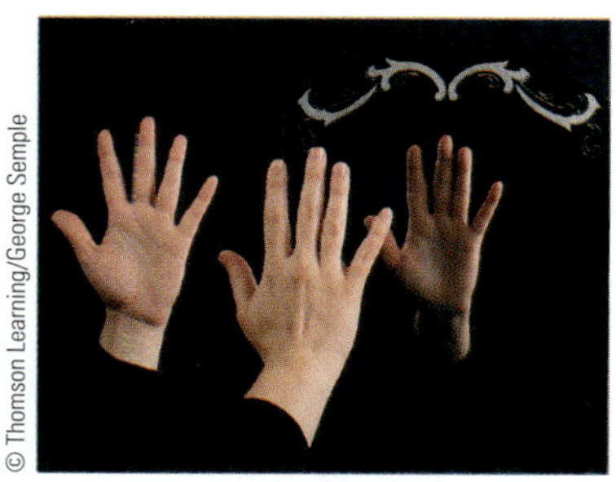

Figure 36.3 The image in the mirror of a person's right hand is reversed front to back, which makes the right hand appear to be a left hand. Notice that the thumb is on the left side of both real hands and on the left side of the image. That the thumb is not on the right side of the image indicates that there is no left-to-right reversal.

Quick Quiz 36.1 You are standing approximately 2 m away from a mirror. The mirror has water spots on its surface. True or False: It is possible for you to see the water spots and your image both in focus at the same time.

CONCEPTUAL EXAMPLE 36.1 Multiple Images Formed by Two Mirrors

Two flat mirrors are perpendicular to each other as in Figure 36.4, and an object is placed at point *O*. In this situation, multiple images are formed. Locate the positions of these images.

SOLUTION

The image of the object is at I_1 in mirror 1 (violet rays) and at I_2 in mirror 2 (blue rays). In addition, a third image is formed at I_3 (brown rays). This third image is the image of I_1 in mirror 2 or, equivalently, the image of I_2 in mirror 1. That is, the image at I_1 (or I_2) serves as the object for I_3. To form this image at I_3, the rays reflect twice after leaving the object at *O*.

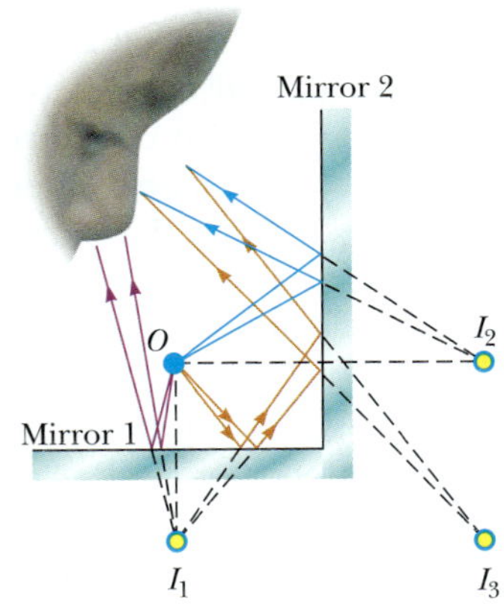

Figure 36.4 (Conceptual Example 36.1) When an object is placed in front of two mutually perpendicular mirrors as shown, three images are formed. Follow the different-colored light rays to understand the formation of each image.

CONCEPTUAL EXAMPLE 36.2 The Tilting Rearview Mirror

Most rearview mirrors in cars have a day setting and a night setting. The night setting greatly diminishes the intensity of the image so that lights from trailing vehicles do not temporarily blind the driver. How does such a mirror work?

SOLUTION

Figure 36.5 shows a cross-sectional view of a rearview mirror for each setting. The unit consists of a reflective coating on the back of a wedge of glass. In the day setting (Fig. 36.5a), the light from an object behind the car strikes the glass wedge at point 1. Most of the light enters the wedge, refracting as it crosses the front surface, and reflects from the back surface to return to the front surface, where it is refracted again as it re-enters the air as ray *B* (for *bright*). In addition, a small portion of the light is reflected at the front surface of the glass as indicated by ray *D* (for *dim*).

This dim reflected light is responsible for the image observed when the mirror is in the night setting (Fig. 36.5b). In that case, the wedge is rotated so that the path followed by the bright light (ray *B*) does not lead to the eye. Instead, the dim light reflected from the front surface of the wedge travels to the eye, and the brightness of trailing headlights does not become a hazard.

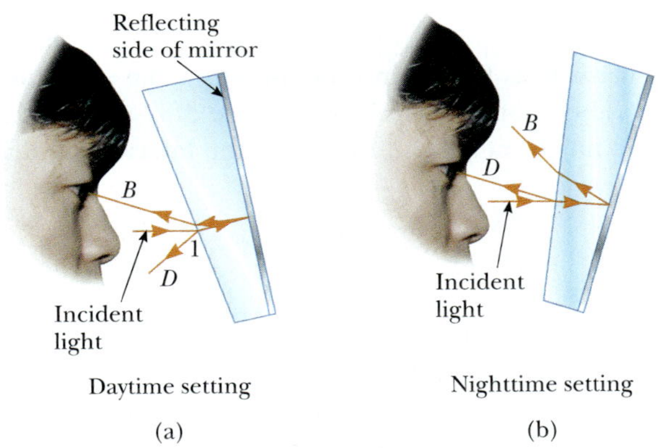

Figure 36.5 (Conceptual Example 36.2) Cross-sectional views of a rearview mirror. (a) With the day setting, the silvered back surface of the mirror reflects a bright ray *B* into the driver's eyes. (b) With the night setting, the glass of the unsilvered front surface of the mirror reflects a dim ray *D* into the driver's eyes.

36.2 Images Formed by Spherical Mirrors

In the preceding section, we considered images formed from flat mirrors. Now we study images formed by curved mirrors. Although a variety of curvatures are possible, we will restrict our investigation to spherical mirrors. As its name implies, a **spherical mirror** has the shape of a section of a sphere.

Concave Mirrors

We first consider reflection of light from the inner, concave surface of a spherical mirror as shown in Figure 36.6. This type of reflecting surface is called a **concave mirror.** Figure 36.6a shows that the mirror has a radius of curvature *R*, and its center of curvature is point *C*. Point *V* is the center of the spherical section, and a line through *C* and *V* is called the **principal axis** of the mirror. Figure 36.6a shows a

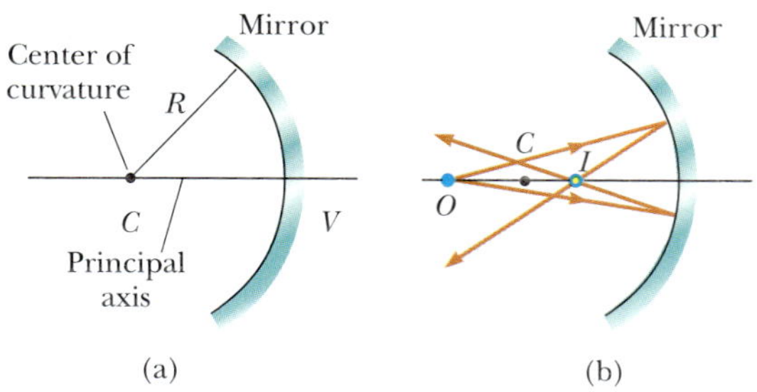

Figure 36.6 (a) A concave mirror of radius R. The center of curvature C is located on the principal axis. (b) A point object placed at O in front of a concave spherical mirror of radius R, where O is any point on the principal axis farther than R from the mirror surface, forms a real image at I. If the rays diverge from O at small angles, they all reflect through the same image point.

Ken Kay/Fundamental Photographs

Figure 36.7 Red, blue, and green light rays are reflected by a curved mirror. Notice that the three colored beams meet at a point.

cross section of a spherical mirror, with its surface represented by the solid, curved black line. (The blue band represents the structural support for the mirrored surface, such as a curved piece of glass on which the silvered surface is deposited.) This type of mirror focuses incoming parallel rays to a point as demonstrated by the colored light rays in Figure 36.7.

Now consider a point source of light placed at point O in Figure 36.6b, where O is any point on the principal axis to the left of C. Two diverging light rays that originate at O are shown. After reflecting from the mirror, these rays converge and cross at the image point I. They then continue to diverge from I as if an object were there. As a result, the image at point I is real.

In this section, we shall consider only rays that diverge from the object and make a small angle with the principal axis. Such rays are called **paraxial rays.** All paraxial rays reflect through the image point as shown in Figure 36.6b. Rays that are far from the principal axis such as those shown in Figure 36.8 converge to other points on the principal axis, producing a blurred image. This effect, called *spherical aberration,* is present to some extent for any spherical mirror and is discussed in Section 36.5.

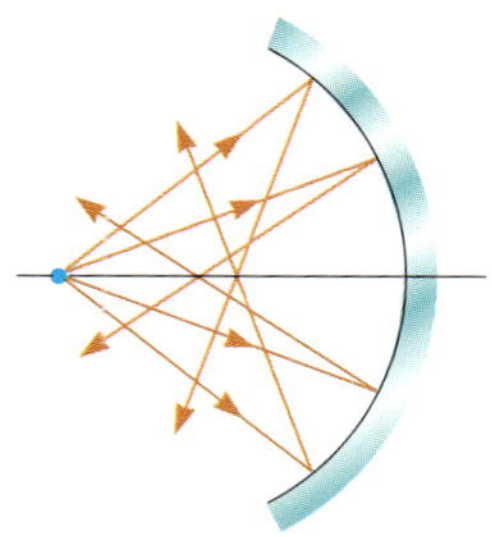

Figure 36.8 Rays diverging from the object at large angles from the principal axis reflect from a spherical concave mirror to intersect the principal axis at different points, resulting in a blurred image. This condition is called *spherical aberration.*

If the object distance p and radius of curvature R are known, we can use Figure 36.9 to calculate the image distance q. By convention, these distances are measured from point V. Figure 36.9 shows two rays leaving the tip of the object. One of these rays passes through the center of curvature C of the mirror, hitting the mirror perpendicular to the mirror surface and reflecting back on itself. The second ray strikes the mirror at its center (point V) and reflects as shown, obeying the law of reflection. The image of the tip of the arrow is located at the point where these two rays intersect. From the large, gold right triangle in Figure 36.9, we see that $\tan\theta = h/p$, and from the blue right triangle, we see that $\tan\theta = -h'/q$. The negative sign is introduced because the image is inverted, so h' is taken to be negative. Therefore, from Equation 36.1 and these results, we find that the magnification of the image is

$$M = \frac{h'}{h} = -\frac{q}{p} \tag{36.2}$$

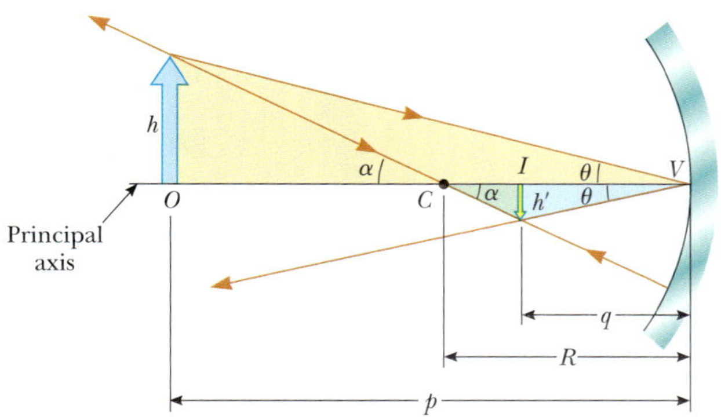

Figure 36.9 The image formed by a spherical concave mirror when the object O lies outside the center of curvature C. This geometric construction is used to derive Equation 36.4.

Courtesy of Thomson Consumer Electronics

A satellite-dish antenna is a concave reflector for television signals from a satellite in orbit around the Earth. Because the satellite is so far away, the signals are carried by microwaves that are parallel when they arrive at the dish. These waves reflect from the dish and are focused on the receiver.

Also notice from the green right triangle in Figure 36.9 and the smaller gold right triangle that

$$\tan \alpha = \frac{-h'}{R-q} \quad \text{and} \quad \tan \alpha = \frac{h}{p-R}$$

from which it follows that

$$\frac{h'}{h} = -\frac{R-q}{p-R} \tag{36.3}$$

Comparing Equations 36.2 and 36.3 gives

$$\frac{R-q}{p-R} = \frac{q}{p}$$

Simple algebra reduces this expression to

Mirror equation in terms of radius of curvature ▶

$$\frac{1}{p} + \frac{1}{q} = \frac{2}{R} \tag{36.4}$$

which is called the *mirror equation.* We present a modified version of this equation shortly.

If the object is very far from the mirror—that is, if p is so much greater than R that p can be said to approach infinity—then $1/p \approx 0$, and Equation 36.4 shows that $q \approx R/2$. That is, when the object is very far from the mirror, the image point is halfway between the center of curvature and the center point on the mirror as shown in Figure 36.10a. The incoming rays from the object are essentially parallel in this figure because the source is assumed to be very far from the mirror. The image point in this special case is called the **focal point** F, and the image distance the **focal length** f, where

Focal length ▶

$$f = \frac{R}{2} \tag{36.5}$$

In Figure 36.7, the colored beams are traveling parallel to the principal axis and the mirror reflects all three beams to the focal point. Notice that the point at which the three beams intersect and the colors add is white.

Because the focal length is a parameter particular to a given mirror, it can be used to compare one mirror with another. Combining Equations 36.4 and 36.5, the **mirror equation** can be expressed in terms of the focal length:

Mirror equation in terms of focal length ▶

$$\frac{1}{p} + \frac{1}{q} = \frac{1}{f} \tag{36.6}$$

Notice that the focal length of a mirror depends only on the curvature of the mirror and not on the material from which the mirror is made because the formation of the image results from rays reflected from the surface of the material. The situ-

PITFALL PREVENTION 36.2

The *Focal* Point Is Not the *Focus* Point

The focal point *is usually not* the point at which the light rays focus to form an image. The focal point is determined solely by the curvature of the mirror; it does not depend on the location of the object. In general, an image forms at a point different from the focal point of a mirror (or a lens). The *only* exception is when the object is located infinitely far away from the mirror.

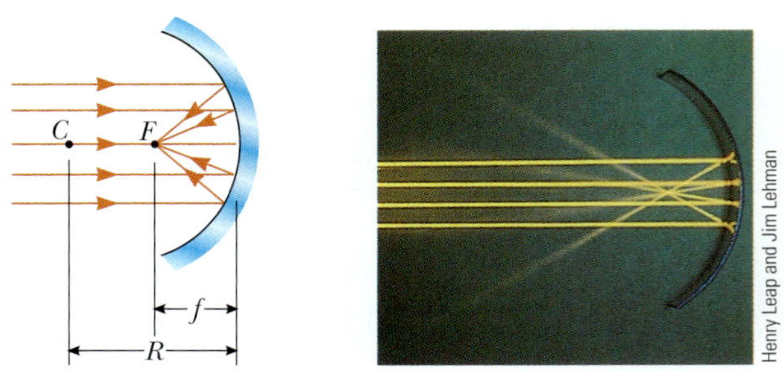

(a) (b)

Figure 36.10 (a) Light rays from a distant object ($p \rightarrow \infty$) reflect from a concave mirror through the focal point F. In this case, the image distance $q \approx R/2 = f$, where f is the focal length of the mirror. (b) Reflection of parallel rays from a concave mirror.

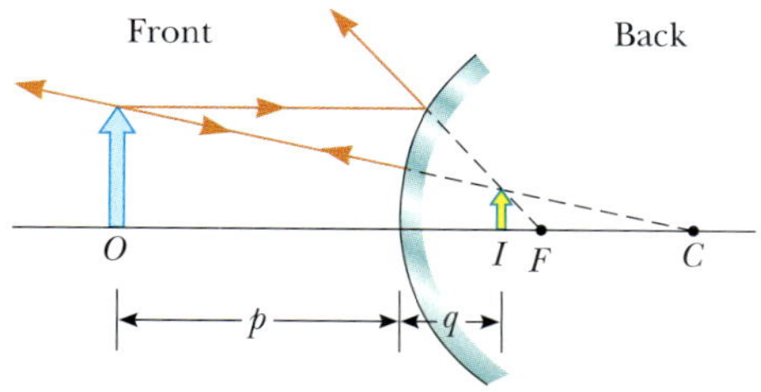

Figure 36.11 Formation of an image by a spherical convex mirror. The image formed by the object is virtual and upright.

ation is different for lenses; in that case, the light actually passes through the material and the focal length depends on the type of material from which the lens is made. (See Section 36.4.)

Convex Mirrors

Figure 36.11 shows the formation of an image by a **convex mirror**, that is, one silvered so that light is reflected from the outer, convex surface. It is sometimes called a **diverging mirror** because the rays from any point on an object diverge after reflection as though they were coming from some point behind the mirror. The image in Figure 36.11 is virtual because the reflected rays only appear to originate at the image point as indicated by the dashed lines. Furthermore, the image is always upright and smaller than the object. This type of mirror is often used in stores to foil shoplifters. A single mirror can be used to survey a large field of view because it forms a smaller image of the interior of the store.

We do not derive any equations for convex spherical mirrors because Equations 36.2, 36.4, and 36.6 can be used for either concave or convex mirrors if we adhere to the following procedure. We will refer to the region in which light rays originate and move toward the mirror as the *front side* of the mirror and the other side as the *back side*. For example, in Figures 36.9 and 36.11, the side to the left of the mirrors is the front side and the side to the right of the mirrors is the back side. Figure 36.12 states the sign conventions for object and image distances, and Table 36.1 summarizes the sign conventions for all quantities. One entry in the table, a *virtual object*, is formally introduced in Section 36.4.

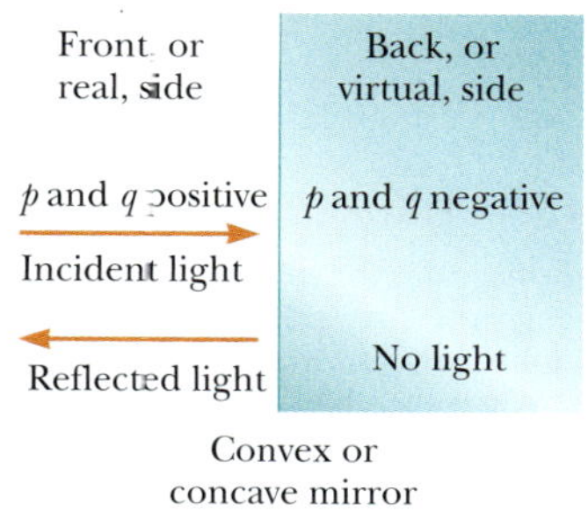

Figure 36.12 Signs of p and q for convex and concave mirrors.

Ray Diagrams for Mirrors

The positions and sizes of images formed by mirrors can be conveniently determined with *ray diagrams*. These pictorial representations reveal the nature of the image and can be used to check results calculated from the mathematical representation using the mirror and magnification equations. To draw a ray diagram, you must know the position of the object and the locations of the mirror's focal point and center of curvature. You then draw three rays to locate the image as

TABLE 36.1

Sign Conventions for Mirrors

Quantity	Positive When . . .	Negative When . . .
Object location (p)	object is in front of mirror (real object).	object is in back of mirror (virtual object).
Image location (q)	image is in front of mirror (real image).	image is in back of mirror (virtual image).
Image height (h')	image is upright.	image is inverted.
Focal length (f) and radius (R)	mirror is concave.	mirror is convex.
Magnification (M)	image is upright.	image is inverted.

PITFALL PREVENTION 36.3

Watch Your Signs

Success in working mirror problems (as well as problems involving refracting surfaces and thin lenses) is largely determined by proper sign choices when substituting into the equations. The best way to success is to work a multitude of problems on your own.

PITFALL PREVENTION 36.4

Choose a Small Number of Rays

A *huge* number of light rays leave each point on an object (and pass through each point on an image). In a ray diagram, which displays the characteristics of the image, we choose only a few rays that follow simply stated rules. Locating the image by calculation complements the diagram.

shown by the examples in Active Figure 36.13. These rays all start from the same object point and are drawn as follows. You may choose any point on the object; here, let's choose the top of the object for simplicity. For concave mirrors (see Active Figs. 36.13a and 36.13b), draw the following three rays:

- Ray 1 is drawn from the top of the object parallel to the principal axis and is reflected through the focal point F.
- Ray 2 is drawn from the top of the object through the focal point (or as if coming from the focal point if $p < f$) and is reflected parallel to the principal axis.
- Ray 3 is drawn from the top of the object through the center of curvature C and is reflected back on itself.

The intersection of any two of these rays locates the image. The third ray serves as a check of the construction. The image point obtained in this fashion must always agree with the value of q calculated from the mirror equation. With concave mirrors, notice what happens as the object is moved closer to the mirror. The real, inverted image in Active Figure 36.13a moves to the left and becomes larger as the object approaches the focal point. When the object is at the focal point, the image is infinitely far to the left. When the object lies between the focal point and the mirror surface as shown in Active Figure 36.13b, however, the image is to the right, behind the object, and virtual, upright, and enlarged. This latter situation applies when you use a shaving mirror or a makeup mirror, both of which are concave. Your face is closer to the mirror than the focal point, and you see an upright, enlarged image of your face.

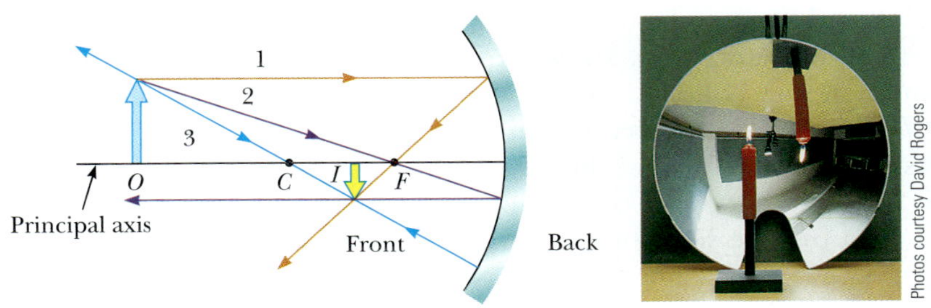

(a)

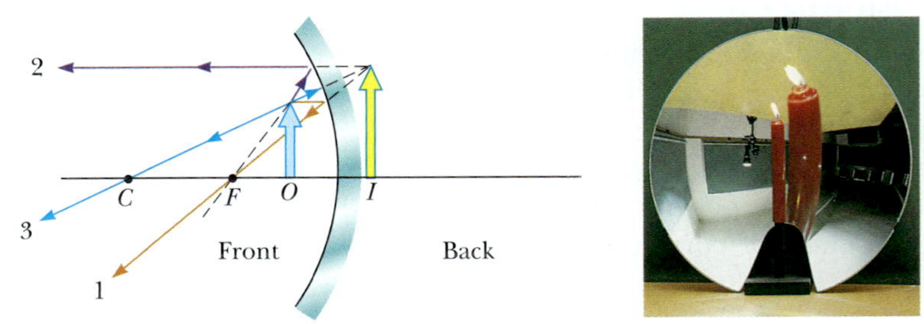

(b)

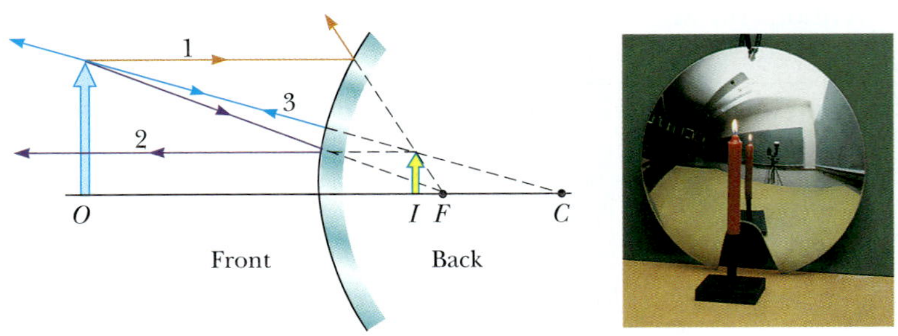

(c)

ACTIVE FIGURE 36.13

Ray diagrams for spherical mirrors along with corresponding photographs of the images of candles. (a) When the object is located so that the center of curvature lies between the object and a concave mirror surface, the image is real, inverted, and reduced in size. (b) When the object is located between the focal point and a concave mirror surface, the image is virtual, upright, and enlarged. (c) When the object is in front of a convex mirror, the image is virtual, upright, and reduced in size.

Sign in at www.thomsonedu.com and go to ThomsonNOW to move the objects and change the focal length of the mirrors to see the effect on the images.

For convex mirrors (see Active Fig. 36.13c), draw the following three rays:

- Ray 1 is drawn from the top of the object parallel to the principal axis and is reflected *away from* the focal point *F*.
- Ray 2 is drawn from the top of the object toward the focal point on the back side of the mirror and is reflected parallel to the principal axis.
- Ray 3 is drawn from the top of the object toward the center of curvature *C* on the back side of the mirror and is reflected back on itself.

In a convex mirror, the image of an object is always virtual, upright, and reduced in size as shown in Active Figure 36.13c. In this case, as the object distance decreases, the virtual image increases in size and moves away from the focal point toward the mirror as the object approaches the mirror. You should construct other diagrams to verify how image position varies with object position.

Quick Quiz 36.2 You wish to start a fire by reflecting sunlight from a mirror onto some paper under a pile of wood. Which would be the best choice for the type of mirror? (a) flat (b) concave (c) convex

Quick Quiz 36.3 Consider the image in the mirror in Figure 36.14. Based on the appearance of this image, would you conclude that (a) the mirror is concave and the image is real, (b) the mirror is concave and the image is virtual, (c) the mirror is convex and the image is real, or (d) the mirror is convex and the image is virtual?

NASA

Figure 36.14 (Quick Quiz 36.3) What type of mirror is shown here?

EXAMPLE 36.3 The Image Formed by a Concave Mirror

A spherical mirror has a focal length of +10.0 cm.

(A) Locate and describe the image for an object distance of 25.0 cm.

SOLUTION

Conceptualize Because the focal length of the mirror is positive, it is a concave mirror (see Table 36.1). We expect the possibilities of both real and virtual images.

Categorize Because the object distance in this part of the problem is larger than the focal length, we expect the image to be real. This situation is analogous to that in Active Figure 36.13a.

Analyze Find the image distance by using Equation 36.6:

$$\frac{1}{q} = \frac{1}{f} - \frac{1}{p}$$

$$\frac{1}{q} = \frac{1}{10.0 \text{ cm}} - \frac{1}{25.0 \text{ cm}}$$

$$q = 16.7 \text{ cm}$$

Find the magnification of the image from Equation 36.2:

$$M = -\frac{q}{p} = -\frac{16.7 \text{ cm}}{25.0 \text{ cm}} = -0.668$$

Finalize The absolute value of M is less than unity, so the image is smaller than the object, and the negative sign for M tells us that the image is inverted. Because q is positive, the image is located on the front side of the mirror and is real. Look into the bowl of a shiny spoon or stand far away from a shaving mirror to see this image.

(B) Locate and describe the image for an object distance of 10.0 cm.

SOLUTION

Categorize Because the object is at the focal point, we expect the image to be infinitely far away.

Analyze Find the image distance by using Equation 36.6:

$$\frac{1}{q} = \frac{1}{f} - \frac{1}{p}$$

$$\frac{1}{q} = \frac{1}{10.0\text{ cm}} - \frac{1}{10.0\text{ cm}}$$

$$q = \infty$$

Finalize This result means that rays originating from an object positioned at the focal point of a mirror are reflected so that the image is formed at an infinite distance from the mirror; that is, the rays travel parallel to one another after reflection. Such is the situation in a flashlight or an automobile headlight, where the bulb filament is placed at the focal point of a reflector, producing a parallel beam of light.

(C) Locate and describe the image for an object distance of 5.00 cm.

SOLUTION

Categorize Because the object distance is smaller than the focal length, we expect the image to be virtual. This situation is analogous to that in Active Figure 36.13b.

Analyze Find the image distance by using Equation 36.6:

$$\frac{1}{q} = \frac{1}{f} - \frac{1}{p}$$

$$\frac{1}{q} = \frac{1}{10.0\text{ cm}} - \frac{1}{5.00\text{ cm}}$$

$$q = -10.0\text{ cm}$$

Find the magnification of the image from Equation 36.2:

$$M = -\frac{q}{p} = -\left(\frac{-10.0\text{ cm}}{5.00\text{ cm}}\right) = +2.00$$

Finalize The image is twice as large as the object, and the positive sign for M indicates that the image is upright (see Active Fig. 36.13b). The negative value of the image distance tells us that the image is virtual, as expected. Put your face close to a shaving mirror to see this type of image.

What If? Suppose you set up the candle and mirror apparatus illustrated in Active Figure 36.13a and described here in part (A). While adjusting the apparatus, you accidentally bump the candle and it begins to slide toward the mirror at velocity v_p. How fast does the image of the candle move?

Answer Solve the mirror equation, Equation 36.6, for q:

$$q = \frac{fp}{p - f}$$

Differentiate this equation with respect to time to find the velocity of the image:

$$(1)\quad v_q = \frac{dq}{dt} = \frac{d}{dt}\left(\frac{fp}{p - f}\right) = -\frac{f^2}{(p - f)^2}\frac{dp}{dt} = -\frac{f^2 v_p}{(p - f)^2}$$

Substitute numerical values from part (A):

$$v_q = -\frac{(10.0\text{ cm})^2\, v_p}{(25.0\text{ cm} - 10.0\text{ cm})^2} = -0.444 v_p$$

Therefore, the speed of the image is less than that of the object in this case.

We can see two interesting behaviors of the function for v_q in Equation (1). First, the velocity is negative regardless of the value of p or f. Therefore, if the object moves toward the mirror, the image moves toward the left in Active Figure 36.13 without regard for the side of the focal point at which the object is located or whether the mirror is concave or convex. Second, in the limit of $p \to 0$, the velocity v_q approaches $-v_p$. As the object moves very close to the mirror, the mirror looks like a plane mirror, the image is as far behind the mirror as the object is in front, and both the object and the image move with the same speed.

EXAMPLE 36.4 **The Image Formed by a Convex Mirror**

An automobile rearview mirror as shown in Figure 36.15 shows an image of a truck located 10.0 m from the mirror. The focal length of the mirror is -0.60 m.

(A) Find the position of the image of the truck.

© Bo Zaunders/CORBIS

Figure 36.15 (Example 36.4) An approaching truck is seen in a convex mirror on the right side of an automobile. Because the image is reduced in size, the truck appears to be farther away than it actually is. Notice also that the image of the truck is in focus, but the frame of the mirror is not, which demonstrates that the image is not at the same location as the mirror surface.

SOLUTION

Conceptualize This situation is depicted in Active Figure 36.13c.

Categorize Because the mirror is convex, we expect it to form an upright, reduced, virtual image for any object position.

Analyze Find the image distance by using Equation 36.6:

$$\frac{1}{q} = \frac{1}{f} - \frac{1}{p}$$

$$\frac{1}{q} = \frac{1}{-0.60 \text{ m}} - \frac{1}{10.0 \text{ m}}$$

$$q = -0.57 \text{ m}$$

(B) Find the magnification of the image.

SOLUTION

Analyze Use Equation 36.2:

$$M = -\frac{q}{p} = -\left(\frac{-0.57 \text{ m}}{10.0 \text{ m}}\right) = +0.057$$

Finalize The negative value of q in part (A) indicates that the image is virtual, or behind the mirror, as shown in Active Figure 36.13c. The magnification in part (B) indicates that the image is much smaller than the truck and is upright because M is positive. Because of the image's small size, these mirrors carry the inscription, "Objects in this mirror are closer than they appear." Look into your rearview mirror or the back side of a shiny spoon to see an image of this type.

36.3 Images Formed by Refraction

In this section, we describe how images are formed when light rays follow the wave under refraction model at the boundary between two transparent materials. Consider two transparent media having indices of refraction n_1 and n_2, where the boundary between the two media is a spherical surface of radius R (Fig. 36.16). We assume the object at O is in the medium for which the index of refraction is n_1. Let's consider the paraxial rays leaving O. As we shall see, all such rays are refracted at the spherical surface and focus at a single point I, the image point.

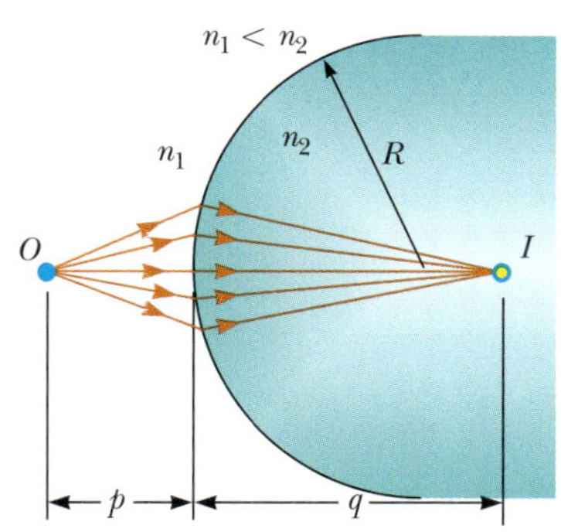

Figure 36.16 An image formed by refraction at a spherical surface. Rays making small angles with the principal axis diverge from a point object at O and are refracted through the image point I.

Figure 36.17 (page 1018) shows a single ray leaving point O and refracting to point I. Snell's law of refraction applied to this ray gives

$$n_1 \sin \theta_1 = n_2 \sin \theta_2$$

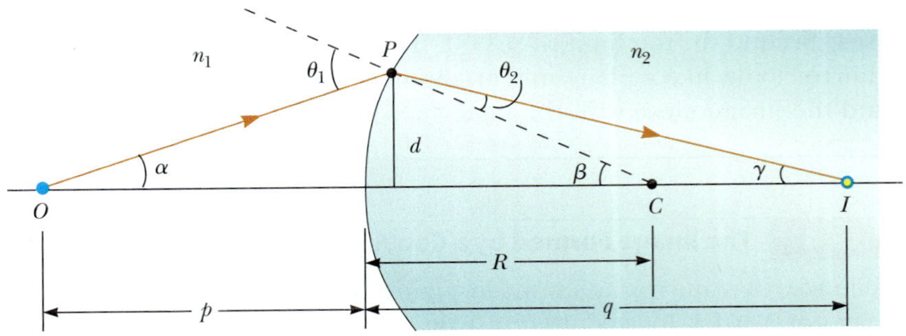

Figure 36.17 Geometry used to derive Equation 36.8, assuming that $n_1 < n_2$.

Because θ_1 and θ_2 are assumed to be small, we can use the small-angle approximation $\sin\theta \approx \theta$ (with angles in radians) and write Snell's law as

$$n_1\theta_1 = n_2\theta_2$$

We know that an exterior angle of any triangle equals the sum of the two opposite interior angles, so applying this rule to triangles *OPC* and *PIC* in Figure 36.17 gives

$$\theta_1 = \alpha + \beta$$

$$\beta = \theta_2 + \gamma$$

Combining all three expressions and eliminating θ_1 and θ_2 gives

$$n_1\alpha + n_2\gamma = (n_2 - n_1)\beta \tag{36.7}$$

Figure 36.17 shows three right triangles that have a common vertical leg of length *d*. For paraxial rays (unlike the relatively large-angle ray shown in Fig. 36.17), the horizontal legs of these triangles are approximately p for the triangle containing angle α, R for the triangle containing angle β, and q for the triangle containing angle γ. In the small-angle approximation, $\tan\theta \approx \theta$, so we can write the approximate relationships from these triangles as follows:

$$\tan\alpha \approx \alpha \approx \frac{d}{p} \qquad \tan\beta \approx \beta \approx \frac{d}{R} \qquad \tan\gamma \approx \gamma \approx \frac{d}{q}$$

Substituting these expressions into Equation 36.7 and dividing through by *d* gives

Relation between object and image distance for a refracting surface ▶

$$\frac{n_1}{p} + \frac{n_2}{q} = \frac{n_2 - n_1}{R} \tag{36.8}$$

For a fixed object distance p, the image distance q is independent of the angle the ray makes with the axis. This result tells us that all paraxial rays focus at the same point *I*.

As with mirrors, we must use a sign convention to apply Equation 36.8 to a variety of cases. We define the side of the surface in which light rays originate as the front side. The other side is called the back side. In contrast with mirrors, where real images are formed in front of the reflecting surface, real images are formed by refraction of light rays to the back of the surface. Because of the difference in location of real images, the refraction sign conventions for q and R are opposite the reflection sign conventions. For example, q and R are both positive in Figure 36.17. The sign conventions for spherical refracting surfaces are summarized in Table 36.2.

We derived Equation 36.8 from an assumption that $n_1 < n_2$ in Figure 36.17. This assumption is not necessary, however. Equation 36.8 is valid regardless of which index of refraction is greater.

TABLE 36.2

Sign Conventions for Refracting Surfaces

Quantity	Positive When . . .	Negative When . . .
Object location (p)	object is in front of surface (real object).	object is in back of surface (virtual object).
Image location (q)	image is in back of surface (real image).	image is in front of surface (virtual image).
Image height (h')	image is upright.	image is inverted.
Radius (R)	center of curvature is in back of surface.	center of curvature is in front of surface.

Flat Refracting Surfaces

If a refracting surface is flat, then R is infinite and Equation 36.8 reduces to

$$\frac{n_1}{p} = -\frac{n_2}{q}$$

$$q = -\frac{n_2}{n_1}p \qquad \textbf{(36.9)}$$

From this expression, we see that the sign of q is opposite that of p. Therefore, according to Table 36.2, **the image formed by a flat refracting surface is on the same side of the surface as the object** as illustrated in Active Figure 36.18 for the situation in which the object is in the medium of index n_1 and n_1 is greater than n_2. In this case, a virtual image is formed between the object and the surface. If n_1 is less than n_2, the rays on the back side diverge from one another at lesser angles than those in Active Figure 36.18. As a result, the virtual image is formed to the left of the object.

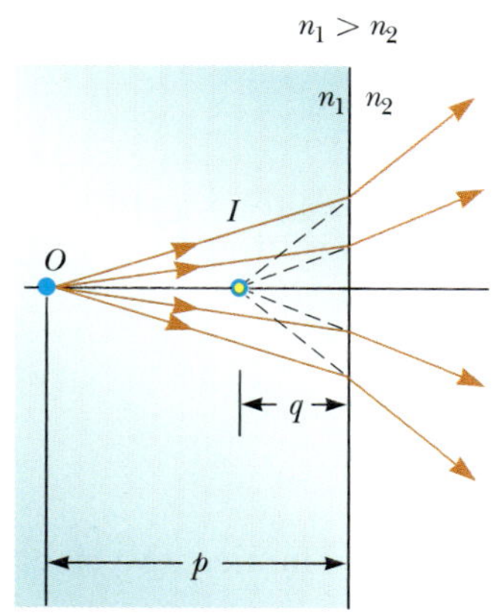

ACTIVE FIGURE 36.18

The image formed by a flat refracting surface is virtual and on the same side of the surface as the object. All rays are assumed to be paraxial.

Sign in at www.thomsonedu.com and go to ThomsonNOW to move the object and see the effect on the location of the image.

Quick Quiz 36.4 In Figure 36.16, what happens to the image point I as the object point O is moved to the right from very far away to very close to the refracting surface? (a) It is always to the right of the surface. (b) It is always to the left of the surface. (c) It starts off to the left, and at some position of O, I moves to the right of the surface. (d) It starts off to the right, and at some position of O, I moves to the left of the surface.

Quick Quiz 36.5 In Active Figure 36.18, what happens to the image point I as the object point O moves toward the right-hand surface of the material of index of refraction n_1? (a) It always remains between O and the surface, arriving at the surface just as O does. (b) It moves toward the surface more slowly than O so that eventually O passes I. (c) It approaches the surface and then moves to the right of the surface.

CONCEPTUAL EXAMPLE 36.5 **Let's Go Scuba Diving!**

Objects viewed under water with the naked eye appear blurred and out of focus. A scuba diver using a mask, however, has a clear view of underwater objects. Explain how that works, using the information that the indices of refraction of the cornea, water, and air are 1.376, 1.333, and 1.000 29, respectively.

SOLUTION

Because the cornea and water have almost identical indices of refraction, very little refraction occurs when a person under water views objects with the naked eye. In this case, light rays from an object focus behind the retina, resulting in a blurred image. When a mask is used, however, the air space between the eye and the mask surface provides the normal amount of refraction at the eye–air interface; consequently, the light from the object focuses on the retina.

EXAMPLE 36.6 Gaze into the Crystal Ball

A set of coins is embedded in a spherical plastic paperweight having a radius of 3.0 cm. The index of refraction of the plastic is $n_1 = 1.50$. One coin is located 2.0 cm from the edge of the sphere (Fig. 36.19). Find the position of the image of the coin.

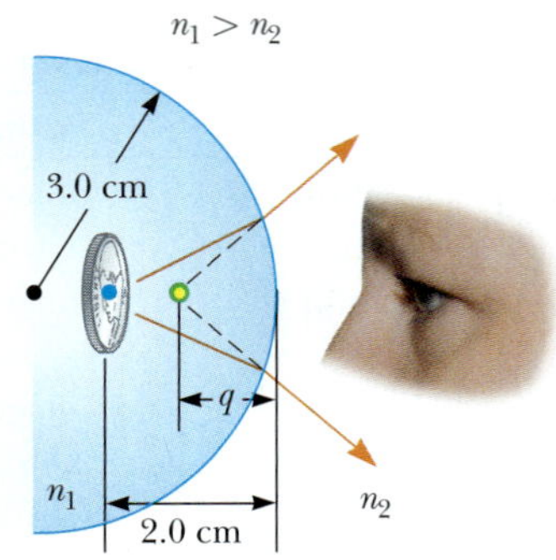

Figure 36.19 (Example 36.6) Light rays from a coin embedded in a plastic sphere form a virtual image between the surface of the object and the sphere surface. Because the object is inside the sphere, the front of the refracting surface is the *interior* of the sphere.

SOLUTION

Conceptualize Because $n_1 > n_2$, where $n_2 = 1.00$ is the index of refraction for air, the rays originating from the coin in Figure 36.19 are refracted away from the normal at the surface and diverge outward.

Categorize Because the light rays originate in one material and then pass through a curved surface into another material, this example involves an image formed by refraction.

Analyze Apply Equation 36.8, noting from Table 36.2 that R is negative:

$$\frac{n_2}{q} = \frac{n_2 - n_1}{R} - \frac{n_1}{p}$$

$$\frac{1}{q} = \frac{1.00 - 1.50}{-3.0 \text{ cm}} - \frac{1.50}{2.0 \text{ cm}}$$

$$q = -1.7 \text{ cm}$$

Finalize The negative sign for q indicates that the image is in front of the surface; in other words, it is in the same medium as the object as shown in Figure 36.19. Therefore, the image must be virtual. (See Table 36.2.) The coin appears to be closer to the paperweight surface than it actually is.

EXAMPLE 36.7 The One That Got Away

A small fish is swimming at a depth d below the surface of a pond (Fig. 36.20).

(A) What is the apparent depth of the fish as viewed from directly overhead?

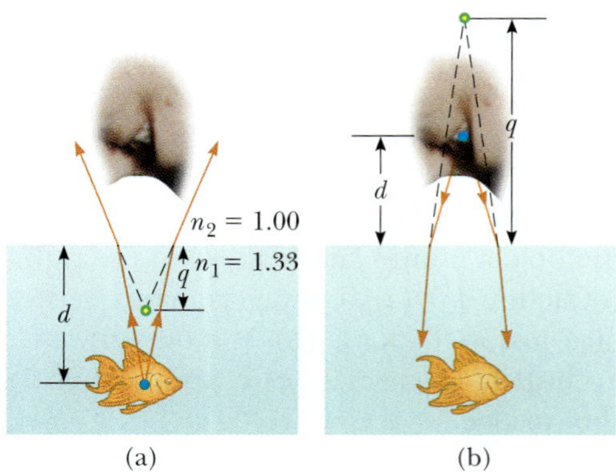

Figure 36.20 (Example 36.7) (a) The apparent depth q of the fish is less than the true depth d. All rays are assumed to be paraxial. (b) Your face appears to the fish to be higher above the surface than it is.

SOLUTION

Conceptualize Because $n_1 > n_2$, where $n_2 = 1.00$ is the index of refraction for air, the rays originating from the fish in Figure 36.20a are refracted away from the normal at the surface and diverge outward.

Categorize Because the refracting surface is flat, R is infinite. Hence, we can use Equation 36.9 to determine the location of the image with $p = d$.

Analyze Use the indices of refraction given in Figure 36.20a in Equation 36.9:

$$q = -\frac{n_2}{n_1}p = -\frac{1.00}{1.33}d = -0.752d$$

Finalize Because q is negative, the image is virtual as indicated by the dashed lines in Figure 36.20a. The apparent depth is approximately three-fourths the actual depth.

(B) If your face is a distance d above the water surface, at what apparent distance above the surface does the fish see your face?

SOLUTION

The light rays from your face are shown in Figure 36.20b.

Conceptualize Because the rays refract toward the normal, your face appears higher above the surface than it actually is.

Categorize Because the refracting surface is flat, R is infinite. Hence, we can use Equation 36.9 to determine the location of the image with $p = d$.

Analyze Use Equation 36.9 to find the image distance:

$$q = -\frac{n_2}{n_1}p = -\frac{1.33}{1.00}d = -1.33d$$

Finalize The negative sign for q indicates that the image is in the medium from which the light originated, which is the air above the water.

What If? What if you look more carefully at the fish and measure its apparent *height* from its upper fin to its lower fin? Is the apparent height h' of the fish different from the actual height h?

Answer Because all points on the fish appear to be fractionally closer to the observer, we expect the height to be smaller. Let the distance d in Figure 36.20a be measured to the top fin, and let the distance to the bottom fin be $d + h$. Then the images of the top and bottom of the fish are located at

$$q_{\text{top}} = -0.752d$$

$$q_{\text{bottom}} = -0.752(d + h)$$

The apparent height h' of the fish is

$$h' = q_{\text{top}} - q_{\text{bottom}} = -0.752d - [-0.752(d + h)] = 0.752h$$

Hence, the fish appears to be approximately three-fourths its actual height.

36.4 Thin Lenses

Lenses are commonly used to form images by refraction in optical instruments such as cameras, telescopes, and microscopes. Let's use what we just learned about images formed by refracting surfaces to help locate the image formed by a lens. Light passing through a lens experiences refraction at two surfaces. The development we shall follow is based on the notion that **the image formed by one refracting surface serves as the object for the second surface.** We shall analyze a thick lens first and then let the thickness of the lens be approximately zero.

Consider a lens having an index of refraction n and two spherical surfaces with radii of curvature R_1 and R_2 as in Figure 36.21 (page 1022). (Notice that R_1 is the radius of curvature of the lens surface the light from the object reaches first and R_2 is the radius of curvature of the other surface of the lens.) An object is placed at point O at a distance p_1 in front of surface 1.

Let's begin with the image formed by surface 1. Using Equation 36.8 and assuming $n_1 = 1$ because the lens is surrounded by air, we find that the image I_1 formed by surface 1 satisfies the equation

$$\frac{1}{p_1} + \frac{n}{q_1} = \frac{n - 1}{R_1} \qquad \textbf{(36.10)}$$

where q_1 is the position of the image formed by surface 1. If the image formed by surface 1 is virtual (Fig. 36.21a), q_1 is negative; it is positive if the image is real (Fig. 36.21b).

Now let's apply Equation 36.8 to surface 2, taking $n_1 = n$ and $n_2 = 1$. (We make this switch in index because the light rays approaching surface 2 are *in the material*

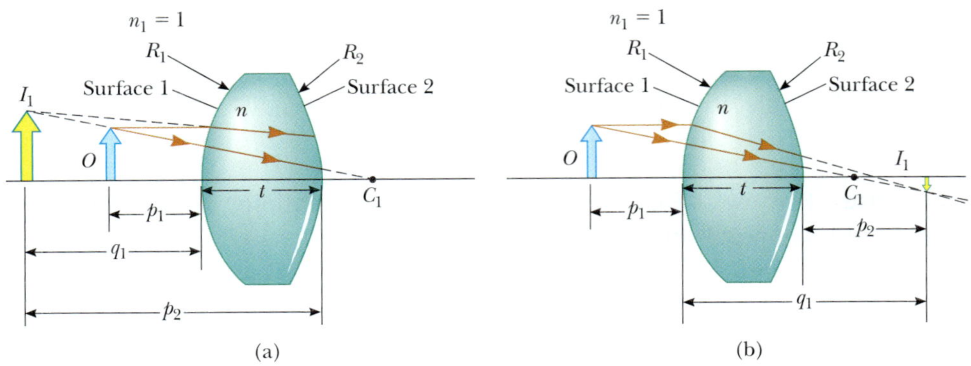

Figure 36.21 To locate the image formed by a lens, we use the virtual image at I_1 formed by surface 1 as the object for the image formed by surface 2. The point C_1 is the center of curvature of surface 1. (a) The image due to surface 1 is virtual, so I_1 is to the left of the surface. (b) The image due to surface 1 is real, so I_1 is to the right of the surface.

of the lens, and this material has index n.) Taking p_2 as the object distance for surface 2 and q_2 as the image distance gives

$$\frac{n}{p_2} + \frac{1}{q_2} = \frac{1 - n}{R_2} \qquad \textbf{(36.11)}$$

We now introduce mathematically that the image formed by the first surface acts as the object for the second surface. If the image from surface 1 is virtual as in Figure 36.21a, we see that p_2, measured from surface 2, is related to q_1 as $p_2 = -q_1 + t$, where t is the thickness of the lens. Because q_1 is negative, p_2 is a positive number. Figure 36.21b shows the case of the image from surface 1 being real. In this situation, q_1 is positive and $p_2 = -q_1 + t$, where the image from surface 1 acts as a virtual object, so p_2 is negative. Regardless of the type of image from surface 1, the same equation describes the location of the object for surface 2 based on our sign convention. For a *thin* lens (one whose thickness is small compared with the radii of curvature), we can neglect t. In this approximation, $p_2 = -q_1$ for either type of image from surface 1. Hence, Equation 36.11 becomes

$$-\frac{n}{q_1} + \frac{1}{q_2} = \frac{1 - n}{R_2} \qquad \textbf{(36.12)}$$

Adding Equations 36.10 and 36.12 gives

$$\frac{1}{p_1} + \frac{1}{q_2} = (n - 1)\left(\frac{1}{R_1} - \frac{1}{R_2}\right) \qquad \textbf{(36.13)}$$

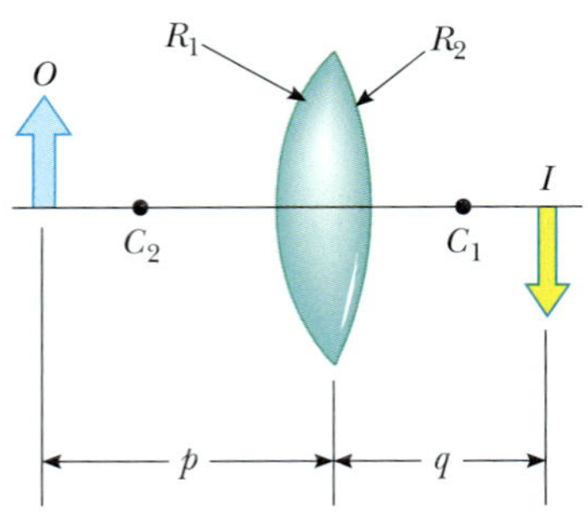

Figure 36.22 Simplified geometry for a thin lens.

For a thin lens, we can omit the subscripts on p_1 and q_2 in Equation 36.13 and call the object distance p and the image distance q, as in Figure 36.22. Hence, we can write Equation 36.13 as

$$\frac{1}{p} + \frac{1}{q} = (n - 1)\left(\frac{1}{R_1} - \frac{1}{R_2}\right) \qquad \textbf{(36.14)}$$

This expression relates the image distance q of the image formed by a thin lens to the object distance p and to the lens properties (index of refraction and radii of curvature). It is valid only for paraxial rays and only when the lens thickness is much less than R_1 and R_2.

The **focal length** f of a thin lens is the image distance that corresponds to an infinite object distance, just as with mirrors. Letting p approach ∞ and q approach f in Equation 36.14, we see that the inverse of the focal length for a thin lens is

Lens-makers' equation ▶

$$\frac{1}{f} = (n - 1)\left(\frac{1}{R_1} - \frac{1}{R_2}\right) \qquad \textbf{(36.15)}$$

This relationship is called the **lens-makers' equation** because it can be used to determine the values of R_1 and R_2 needed for a given index of refraction and a desired focal length f. Conversely, if the index of refraction and the radii of curvature of a lens are given, this equation can be used to find the focal length. If the

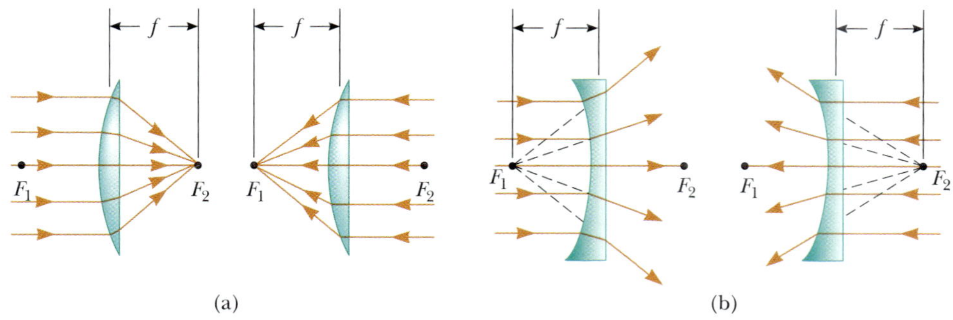

Figure 36.23 Parallel light rays pass through (a) a converging lens and (b) a diverging lens. The focal length is the same for light rays passing through a given lens in either direction. Both focal points F_1 and F_2 are the same distance from the lens.

lens is immersed in something other than air, this same equation can be used, with n interpreted as the *ratio* of the index of refraction of the lens material to that of the surrounding fluid.

Using Equation 36.15, we can write Equation 36.14 in a form identical to Equation 36.6 for mirrors:

$$\frac{1}{p} + \frac{1}{q} = \frac{1}{f} \qquad \textbf{(36.16)}$$

This equation, called the **thin lens equation,** can be used to relate the image distance and object distance for a thin lens.

Because light can travel in either direction through a lens, each lens has two focal points, one for light rays passing through in one direction and one for rays passing through in the other direction. These two focal points are illustrated in Figure 36.23 for a plano-convex lens (a converging lens) and a plano-concave lens (a diverging lens).

Figure 36.24 is useful for obtaining the signs of p and q, and Table 36.3 gives the sign conventions for thin lenses. These sign conventions are the *same* as those for refracting surfaces (see Table 36.2).

Various lens shapes are shown in Figure 36.25. Notice that a converging lens is thicker at the center than at the edge, whereas a diverging lens is thinner at the center than at the edge.

PITFALL PREVENTION 36.5

A Lens Has Two Focal Points but Only One Focal Length

A lens has a focal point on each side, front and back. There is only one focal length, however; each of the two focal points is located the same distance from the lens (Fig. 36.23). As a result, the lens forms an image of an object at the same point if it is turned around. In practice, that might not happen because real lenses are not infinitesimally thin.

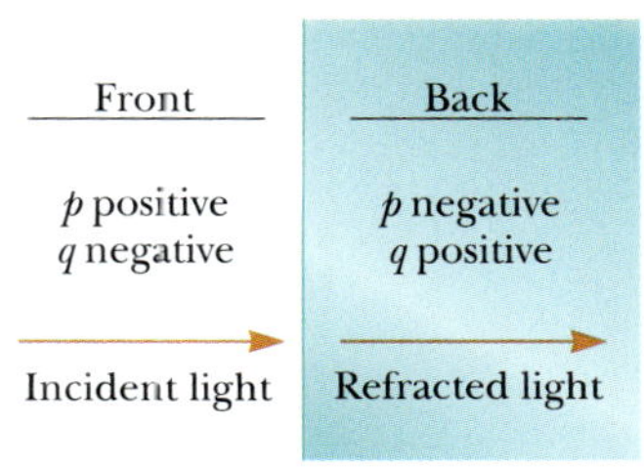

Figure 36.24 A diagram for obtaining the signs of p and q for a thin lens. (This diagram also applies to a refracting surface.)

Magnification of Images

Consider a thin lens through which light rays from an object pass. As with mirrors (Eq. 36.2), a geometric construction shows that the lateral magnification of the image is

$$M = \frac{h'}{h} = -\frac{q}{p} \qquad \textbf{(36.17)}$$

From this expression, it follows that when M is positive, the image is upright and on the same side of the lens as the object. When M is negative, the image is inverted and on the side of the lens opposite the object.

TABLE 36.3

Sign Conventions for Thin Lenses

Quantity	Positive When . . .	Negative When . . .
Object location (p)	object is in front of lens (real object).	object is in back of lens (virtual object).
Image location (q)	image is in back of lens (real image).	image is in front of lens (virtual image).
Image height (h')	image is upright.	image is inverted.
R_1 and R_2	center of curvature is in back of lens.	center of curvature is in front of lens.
Focal length (f)	a converging lens.	a diverging lens.

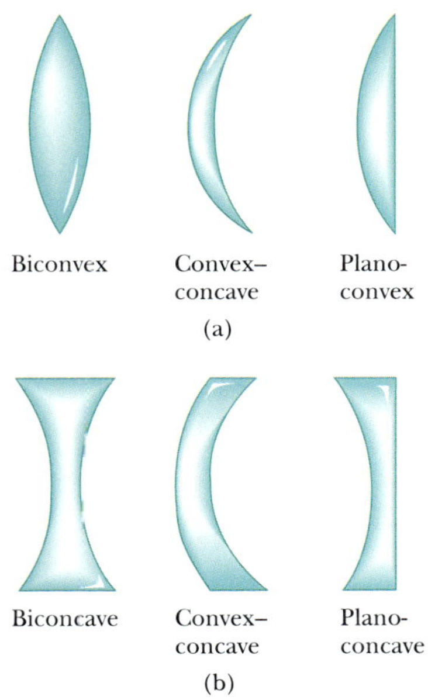

Figure 36.25 Various lens shapes. (a) Converging lenses have a positive focal length and are thickest at the middle. (b) Diverging lenses have a negative focal length and are thickest at the edges.

Ray Diagrams for Thin Lenses

Ray diagrams are convenient for locating the images formed by thin lenses or systems of lenses. They also help clarify our sign conventions. Active Figure 36.26 shows such diagrams for three single-lens situations.

To locate the image of a *converging* lens (Active Fig. 36.26a and b), the following three rays are drawn from the top of the object:

- Ray 1 is drawn parallel to the principal axis. After being refracted by the lens, this ray passes through the focal point on the back side of the lens.
- Ray 2 is drawn through the center of the lens and continues in a straight line.
- Ray 3 is drawn through the focal point on the front side of the lens (or as if coming from the focal point if $p < f$) and emerges from the lens parallel to the principal axis.

To locate the image of a *diverging* lens (Active Fig. 36.26c), the following three rays are drawn from the top of the object:

- Ray 1 is drawn parallel to the principal axis. After being refracted by the lens, this ray emerges directed away from the focal point on the front side of the lens.
- Ray 2 is drawn through the center of the lens and continues in a straight line.
- Ray 3 is drawn in the direction toward the focal point on the back side of the lens and emerges from the lens parallel to the principal axis.

For the converging lens in Active Figure 36.26a, where the object is to the left of the focal point ($p > f$), the image is real and inverted. When the object is between the focal point and the lens ($p < f$) as in Active Figure 36.26b, the image is virtual and upright. In that case, the lens acts as a magnifying glass, which we study in more detail in Section 36.8. For a diverging lens (Active Fig. 36.26c), the image is always virtual and upright, regardless of where the object is placed. These geometric constructions are reasonably accurate only if the distance between the rays and the principal axis is much less than the radii of the lens surfaces.

Notice that refraction occurs only at the surfaces of the lens. A certain lens design takes advantage of this behavior to produce the *Fresnel lens,* a powerful lens

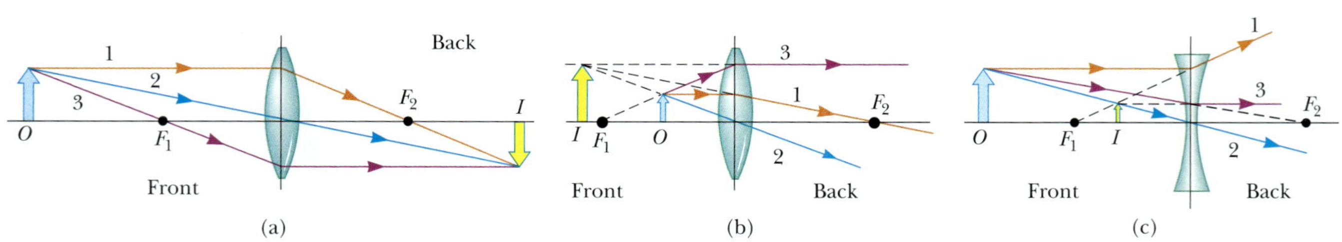

ACTIVE FIGURE 36.26

Ray diagrams for locating the image formed by a thin lens. (a) When the object is in front of and outside the focal point of a converging lens, the image is real, inverted, and on the back side of the lens. (b) When the object is between the focal point and a converging lens, the image is virtual, upright, larger than the object, and on the front side of the lens. (c) When an object is anywhere in front of a diverging lens, the image is virtual, upright, smaller than the object, and on the front side of the lens.

Sign in at www.thomsonedu.com and go to ThomsonNOW to move the objects and change the focal length of the lenses to see the effect on the images.

without great thickness. Because only the surface curvature is important in the refracting qualities of the lens, material in the middle of a Fresnel lens is removed as shown in the cross sections of lenses in Figure 36.27. Because the edges of the curved segments cause some distortion, Fresnel lenses are generally used only in situations in which image quality is less important than reduction of weight. A classroom overhead projector often uses a Fresnel lens; the circular edges between segments of the lens can be seen by looking closely at the light projected onto a screen.

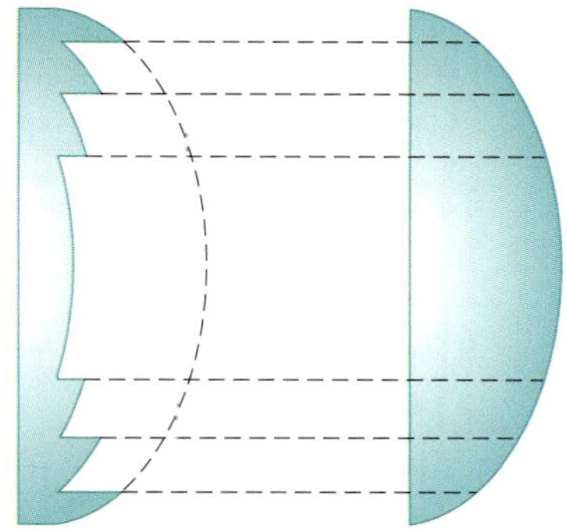

Figure 36.27 The Fresnel lens on the left has the same focal length as the thick lens on the right but is made of much less glass.

Quick Quiz 36.6 What is the focal length of a pane of window glass? (a) zero (b) infinity (c) the thickness of the glass (d) impossible to determine

EXAMPLE 36.8 Images Formed by a Converging Lens

A converging lens has a focal length of 10.0 cm.

(A) An object is placed 30.0 cm from the lens. Construct a ray diagram, find the image distance, and describe the image.

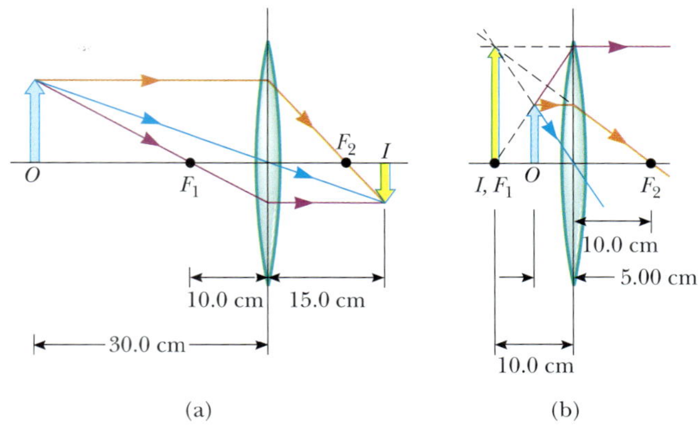

Figure 36.28 (Example 36.8) An image is formed by a converging lens. (a) The object is farther from the lens than the focal point. (b) The object is closer to the lens than the focal point.

SOLUTION

Conceptualize Because the lens is converging, the focal length is positive (see Table 36.3). We expect the possibilities of both real and virtual images.

Categorize Because the object distance is larger than the focal length, we expect the image to be real. The ray diagram for this situation is shown in Figure 36.28a.

Analyze Find the image distance by using Equation 36.16:

$$\frac{1}{q} = \frac{1}{f} - \frac{1}{p}$$

$$\frac{1}{q} = \frac{1}{10.0 \text{ cm}} - \frac{1}{30.0 \text{ cm}}$$

$$q = +15.0 \text{ cm}$$

Find the magnification of the image from Equation 36.17:

$$M = -\frac{q}{p} = -\frac{15.0 \text{ cm}}{30.0 \text{ cm}} = -0.500$$

Finalize The positive sign for the image distance tells us that the image is indeed real and on the back side of the lens. The magnification of the image tells us that the image is reduced in height by one half, and the negative sign for M tells us that the image is inverted.

(B) An object is placed 10.0 cm from the lens. Find the image distance and describe the image.

SOLUTION

Categorize Because the object is at the focal point, we expect the image to be infinitely far away.

Analyze Find the image distance by using Equation 36.16:

$$\frac{1}{q} = \frac{1}{f} - \frac{1}{p}$$

$$\frac{1}{q} = \frac{1}{10.0 \text{ cm}} - \frac{1}{10.0 \text{ cm}}$$

$$q = \infty$$

Finalize This result means that rays originating from an object positioned at the focal point of a lens are refracted so that the image is formed at an infinite distance from the lens; that is, the rays travel parallel to one another after refraction.

(C) An object is placed 5.00 cm from the lens. Construct a ray diagram, find the image distance, and describe the image.

SOLUTION

Categorize Because the object distance is smaller than the focal length, we expect the image to be virtual. The ray diagram for this situation is shown in Figure 36.28b.

Analyze Find the image distance by using Equation 36.16:

$$\frac{1}{q} = \frac{1}{f} - \frac{1}{p}$$

$$\frac{1}{q} = \frac{1}{10.0 \text{ cm}} - \frac{1}{5.00 \text{ cm}}$$

$$q = -10.0 \text{ cm}$$

Find the magnification of the image from Equation 36.17:

$$M = -\frac{q}{p} = -\left(\frac{-10.0 \text{ cm}}{5.00 \text{ cm}}\right) = +2.00$$

Finalize The negative image distance tells us that the image is virtual and formed on the side of the lens from which the light is incident, the front side. The image is enlarged, and the positive sign for M tells us that the image is upright.

What If? What if the object moves right up to the lens surface, so that $p \to 0$? Where is the image?

Answer In this case, because $p \ll R$, where R is either of the radii of the surfaces of the lens, the curvature of the lens can be ignored. The lens should appear to have the same effect as a flat piece of material, which suggests that the image is just on the front side of the lens, at $q = 0$. This conclusion can be verified mathematically by rearranging the thin lens equation:

$$\frac{1}{q} = \frac{1}{f} - \frac{1}{p}$$

If we let $p \to 0$, the second term on the right becomes very large compared with the first and we can neglect $1/f$. The equation becomes

$$\frac{1}{q} = -\frac{1}{p} \quad \to \quad q = -p = 0$$

Therefore, q is on the front side of the lens (because it has the opposite sign as p) and right at the lens surface.

EXAMPLE 36.9 **Images Formed by a Diverging Lens**

A diverging lens has a focal length of 10.0 cm.

(A) An object is placed 30.0 cm from the lens. Construct a ray diagram, find the image distance, and describe the image.

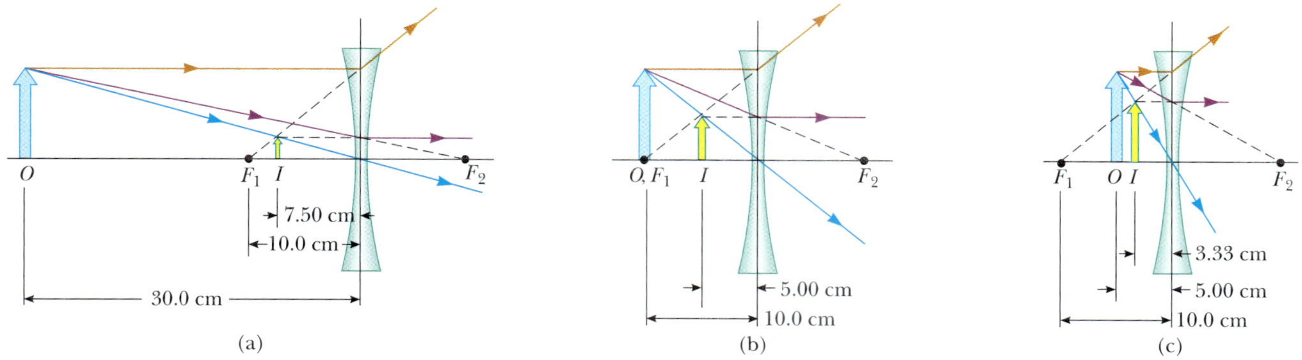

Figure 36.29 (Example 36.9) An image is formed by a diverging lens. (a) The object is farther from the lens than the focal point. (b) The object is at the focal point. (c) The object is closer to the lens than the focal point.

SOLUTION

Conceptualize Because the lens is diverging, the focal length is negative (see Table 36.3). The ray diagram for this situation is shown in Figure 36.29a.

Categorize Because the lens is diverging, we expect it to form an upright, reduced, virtual image for any object position.

Analyze Find the image distance by using Equation 36.16:

$$\frac{1}{q} = \frac{1}{f} - \frac{1}{p}$$

$$\frac{1}{q} = \frac{1}{-10.0\text{ cm}} - \frac{1}{30.0\text{ cm}}$$

$$q = -7.50\text{ cm}$$

Find the magnification of the image from Equation 36.17:

$$M = -\frac{q}{p} = -\left(\frac{-7.50\text{ cm}}{30.0\text{ cm}}\right) = +0.250$$

Finalize This result confirms that the image is virtual, smaller than the object, and upright. Look through the diverging lens in a door peephole to see this type of image.

(B) An object is placed 10.0 cm from the lens. Construct a ray diagram, find the image distance, and describe the image.

SOLUTION

The ray diagram for this situation is shown in Figure 36.29b.

Analyze Find the image distance by using Equation 36.16:

$$\frac{1}{q} = \frac{1}{f} - \frac{1}{p}$$

$$\frac{1}{q} = \frac{1}{-10.0\text{ cm}} - \frac{1}{10.0\text{ cm}}$$

$$q = -5.00\text{ cm}$$

Find the magnification of the image from Equation 36.17:

$$M = -\frac{q}{p} = -\left(\frac{-5.00\text{ cm}}{10.0\text{ cm}}\right) = +0.500$$

Finalize Notice the difference between this situation and that for a converging lens. For a diverging lens, an object at the focal point does not produce an image infinitely far away.

(C) An object is placed 5.00 cm from the lens. Construct a ray diagram, find the image distance, and describe the image.

SOLUTION

The ray diagram for this situation is shown in Figure 36.29c.

Analyze Find the image distance by using Equation 36.16:

$$\frac{1}{q} = \frac{1}{f} - \frac{1}{p}$$

$$\frac{1}{q} = \frac{1}{-10.0\text{ cm}} - \frac{1}{5.00\text{ cm}}$$

$$q = -3.33\text{ cm}$$

Find the magnification of the image from Equation 36.17:

$$M = -\left(\frac{-3.33\text{ cm}}{5.00\text{ cm}}\right) = +0.667$$

Finalize For all three object positions, the image position is negative and the magnification is a positive number smaller than 1, which confirms that the image is virtual, smaller than the object, and upright.

Combination of Thin Lenses

If two thin lenses are used to form an image, the system can be treated in the following manner. First, the image formed by the first lens is located as if the second lens were not present. Then a ray diagram is drawn for the second lens, with the image formed by the first lens now serving as the object for the second lens. The second image formed is the final image of the system. If the image formed by the first lens lies on the back side of the second lens, that image is treated as a **virtual object** for the second lens (that is, in the thin lens equation, p is negative). The same procedure can be extended to a system of three or more lenses. Because the magnification due to the second lens is performed on the magnified image due to the first lens, the **overall magnification of the image due to the combination of lenses is the product of the individual magnifications:**

$$M = M_1 M_2 \tag{36.18}$$

This equation can be used for combinations of any optical elements such as a lens and a mirror. For more than two optical elements, the magnifications due to all elements are multiplied together.

Let's consider the special case of a system of two lenses of focal lengths f_1 and f_2 in contact with each other. If $p_1 = p$ is the object distance for the combination, application of the thin lens equation (Eq. 36.16) to the first lens gives

$$\frac{1}{p} + \frac{1}{q_1} = \frac{1}{f_1}$$

where q_1 is the image distance for the first lens. Treating this image as the object for the second lens, we see that the object distance for the second lens must be $p_2 = -q_1$. (The distances are the same because the lenses are in contact and assumed to be infinitesimally thin. The object distance is negative because the object is virtual.) Therefore, for the second lens,

$$\frac{1}{p_2} + \frac{1}{q_2} = \frac{1}{f_2} \quad \rightarrow \quad -\frac{1}{q_1} + \frac{1}{q} = \frac{1}{f_2}$$

where $q = q_2$ is the final image distance from the second lens, which is the image distance for the combination. Adding the equations for the two lenses eliminates q_1 and gives

$$\frac{1}{p} + \frac{1}{q} = \frac{1}{f_1} + \frac{1}{f_2}$$

If the combination is replaced with a single lens that forms an image at the same location, its focal length must be related to the individual focal lengths by the expression

$$\frac{1}{f} = \frac{1}{f_1} + \frac{1}{f_2} \tag{36.19}$$

◄ Focal length for a combination of two thin lenses in contact

Therefore, **two thin lenses in contact with each other are equivalent to a single thin lens having a focal length given by Equation 36.19.**

EXAMPLE 36.10 Where Is the Final Image?

Two thin converging lenses of focal lengths $f_1 = 10.0$ cm and $f_2 = 20.0$ cm are separated by 20.0 cm as illustrated in Figure 36.30. An object is placed 30.0 cm to the left of lens 1. Find the position and the magnification of the final image.

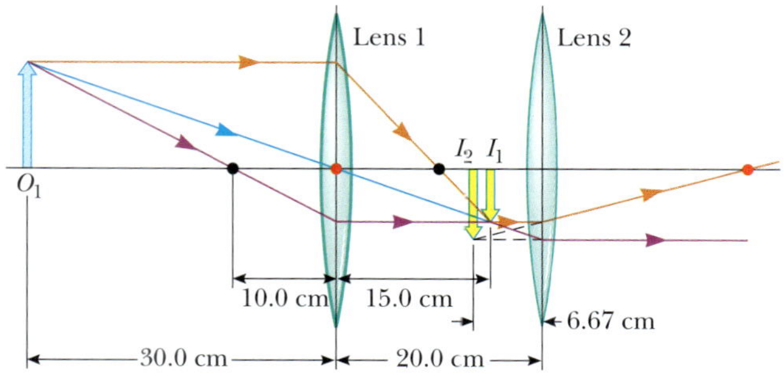

Figure 36.30 (Example 36.10) A combination of two converging lenses. The ray diagram shows the location of the final image due to the combination of lenses. The black dots are the focal points of lens 1 and the red dots are the focal points of lens 2.

SOLUTION

Conceptualize Imagine light rays passing through the first lens and forming a real image (because $p > f$) in the absence of a second lens. Figure 36.30 shows these light rays forming the inverted image I_1. Once the light rays converge to the image point, they do not stop. They continue through the image point and interact with the second lens. The rays leaving the image point behave in the same way as the rays leaving an object. Therefore, the image of the first lens serves as the object of the second lens.

Categorize We categorize this problem as one in which the thin lens equation is applied in a stepwise fashion to the two lenses.

Analyze Find the location of the image formed by lens 1 from the thin lens equation:

$$\frac{1}{q_1} = \frac{1}{f} - \frac{1}{p_1}$$

$$\frac{1}{q_1} = \frac{1}{10.0 \text{ cm}} - \frac{1}{30.0 \text{ cm}}$$

$$q_1 = +15.0 \text{ cm}$$

Find the magnification of the image from Equation 36.17:

$$M_1 = -\frac{q_1}{p_1} = -\frac{15.0 \text{ cm}}{30.0 \text{ cm}} = -0.500$$

The image formed by this lens acts as the object for the second lens. Therefore, the object distance for the second lens is 20.0 cm − 15.0 cm = 5.00 cm.

Find the location of the image formed by lens 2 from the thin lens equation:

$$\frac{1}{q_2} = \frac{1}{20.0 \text{ cm}} - \frac{1}{5.00 \text{ cm}}$$

$$q_2 = -6.67 \text{ cm}$$

Find the magnification of the image from Equation 36.17:

$$M_2 = -\frac{q_2}{p_2} = -\frac{(-6.67 \text{ cm})}{5.00 \text{ cm}} = +1.33$$

Find the overall magnification of the system from Equation 36.18:

$$M = M_1 M_2 = (-0.500)(1.33) = -0.667$$

Finalize The negative sign on the overall magnification indicates that the final image is inverted with respect to the initial object. Because the absolute value of the magnification is less than 1, the final image is smaller than the object. Because q_2 is negative, the final image is on the front, or left, side of lens 2. These conclusions are consistent with the ray diagram in Figure 36.30.

What If? Suppose you want to create an upright image with this system of two lenses. How must the second lens be moved?

Answer Because the object is farther from the first lens than the focal length of that lens, the first image is inverted. Consequently, the second lens must invert the image once again so that the final image is upright. An inverted image is only formed by a converging lens if the object is outside the focal point. Therefore, the image formed by the first lens must be to the left of the focal point of the second lens in Figure 36.30. To make that happen, you must move the second lens at least as far away from the first lens as the sum $q_1 + f_2 = 15.0 \text{ cm} + 20.0 \text{ cm} = 35.0 \text{ cm}$.

36.5 Lens Aberrations

Our analysis of mirrors and lenses assumes rays make small angles with the principal axis and the lenses are thin. In this simple model, all rays leaving a point source focus at a single point, producing a sharp image. Clearly, that is not always true. When the approximations used in this analysis do not hold, imperfect images are formed.

A precise analysis of image formation requires tracing each ray, using Snell's law at each refracting surface and the law of reflection at each reflecting surface. This procedure shows that the rays from a point object do not focus at a single point, with the result that the image is blurred. The departures of actual images from the ideal predicted by our simplified model are called **aberrations.**

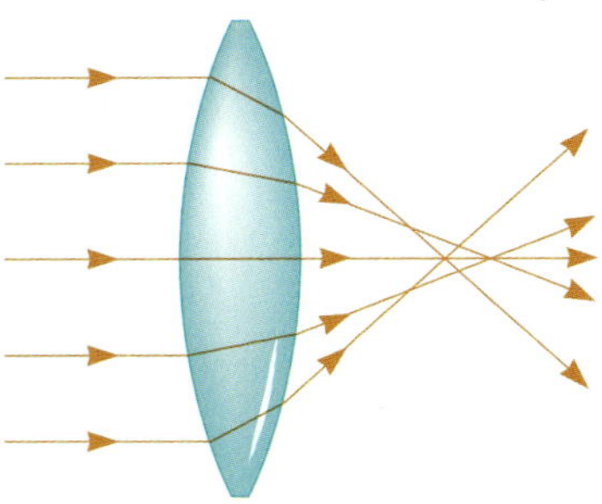

Figure 36.31 Spherical aberration caused by a converging lens. Does a diverging lens cause spherical aberration?

Spherical Aberration

Spherical aberration occurs because the focal points of rays far from the principal axis of a spherical lens (or mirror) are different from the focal points of rays of the same wavelength passing near the axis. Figure 36.31 illustrates spherical aberration for parallel rays passing through a converging lens. Rays passing through points near the center of the lens are imaged farther from the lens than rays passing through points near the edges. Figure 36.8 earlier in the chapter showed a similar situation for a spherical mirror.

Many cameras have an adjustable aperture to control light intensity and reduce spherical aberration. (An aperture is an opening that controls the amount of light passing through the lens.) Sharper images are produced as the aperture size is reduced; with a small aperture, only the central portion of the lens is exposed to the light and therefore a greater percentage of the rays are paraxial. At the same time, however, less light passes through the lens. To compensate for this lower light intensity, a longer exposure time is used.

In the case of mirrors, spherical aberration can be minimized through the use of a parabolic reflecting surface rather than a spherical surface. Parabolic surfaces are not used often, however, because those with high-quality optics are very expensive to make. Parallel light rays incident on a parabolic surface focus at a common point, regardless of their distance from the principal axis. Parabolic reflecting surfaces are used in many astronomical telescopes to enhance image quality.

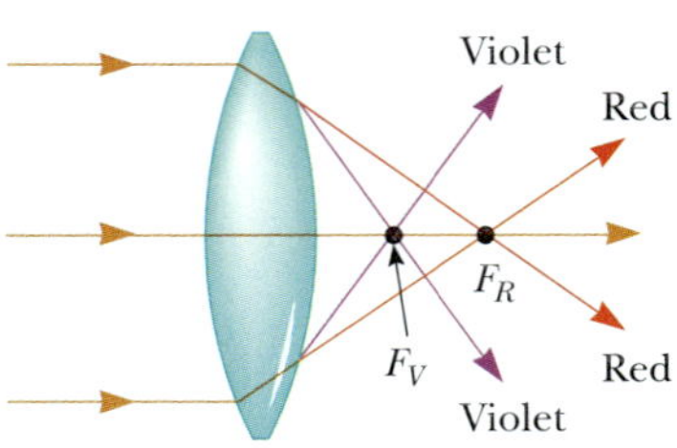

Figure 36.32 Chromatic aberration caused by a converging lens. Rays of different wavelengths focus at different points.

Chromatic Aberration

In Chapter 35, we described dispersion, whereby a material's index of refraction varies with wavelength. Because of this phenomenon, violet rays are refracted more than red rays when white light passes through a lens (Fig. 36.32). The figure shows that the focal length of a lens is greater for red light than for violet light. Other wavelengths (not shown in Fig. 36.32) have focal points intermediate between those of red and violet, which causes a blurred image and is called **chromatic aberration.**

Chromatic aberration for a diverging lens also results in a shorter focal length for violet light than for red light, but on the front side of the lens. Chromatic aberration can be greatly reduced by combining a converging lens made of one type of glass and a diverging lens made of another type of glass.

36.6 The Camera

The photographic **camera** is a simple optical instrument whose essential features are shown in Figure 36.33. It consists of a lighttight chamber, a converging lens that produces a real image, and a film behind the lens to receive the image.

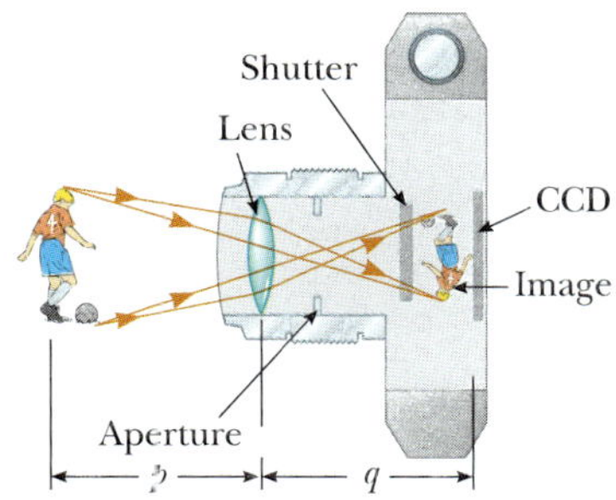

Figure 36.33 Cross-sectional view of a simple digital camera. The CCD is the light-sensitive component of the camera. In a nondigital camera, the light from the lens falls onto photographic film. In reality, $p \gg q$.

Digital cameras are similar to film cameras except that the light does not form an image on photographic film. The image in a digital camera is formed on a *charge-coupled device* (CCD), which digitizes the image, turning it into binary code as we discussed for sound in Section 17.5. (A CCD is described in Section 40.2.) The digital information is then stored on a memory chip for playback on the camera's display screen, or it can be downloaded to a computer. In the discussion that follows, we assume the camera is digital.

A camera is focused by varying the distance between the lens and the CCD. For proper focusing—which is necessary for the formation of sharp images—the lens-to-CCD distance depends on the object distance as well as the focal length of the lens.

The shutter, positioned behind the lens, is a mechanical device that is opened for selected time intervals, called *exposure times.* You can photograph moving objects by using short exposure times or photograph dark scenes (with low light levels) by using long exposure times. If this adjustment were not available, it would be impossible to take stop-action photographs. For example, a rapidly moving vehicle could move enough in the time interval during which the shutter is open to produce a blurred image. Another major cause of blurred images is the movement of the camera while the shutter is open. To prevent such movement, either short exposure times or a tripod should be used, even for stationary objects. Typical shutter speeds (that is, exposure times) are $\frac{1}{30}$ s, $\frac{1}{60}$ s, $\frac{1}{125}$ s, and $\frac{1}{250}$ s. In practice, stationary objects are normally shot with an intermediate shutter speed of $\frac{1}{60}$ s.

The intensity I of the light reaching the CCD is proportional to the area of the lens. Because this area is proportional to the square of the diameter D, it follows that I is also proportional to D^2. Light intensity is a measure of the rate at which energy is received by the CCD per unit area of the image. Because the area of the image is proportional to q^2 and $q \approx f$ (when $p \gg f$, so p can be approximated as infinite), we conclude that the intensity is also proportional to $1/f^2$ and therefore that $I \propto D^2/f^2$.

The ratio f/D is called the ***f*-number** of a lens:

$$f\text{-number} \equiv \frac{f}{D} \tag{36.20}$$

Hence, the intensity of light incident on the CCD varies according to the following proportionality:

$$I \propto \frac{1}{(f/D)^2} \propto \frac{1}{(f\text{-number})^2} \tag{36.21}$$

The *f*-number is often given as a description of the lens's "speed." The lower the *f*-number, the wider the aperture and the higher the rate at which energy from the light exposes the CCD; therefore, a lens with a low *f*-number is a "fast" lens. The conventional notation for an *f*-number is "*f*/" followed by the actual number. For example, "$f/4$" means an f-number of 4; it *does not* mean to divide f by 4! Extremely fast lenses, which have f-numbers as low as approximately $f/1.2$, are expensive because it is very difficult to keep aberrations acceptably small with light rays passing through a large area of the lens. Camera lens systems (that is, combinations of lenses with adjustable apertures) are often marked with multiple f-numbers, usually $f/2.8$, $f/4$, $f/5.6$, $f/8$, $f/11$, and $f/16$. Any one of these settings can be selected by adjusting the aperture, which changes the value of D. Increasing the setting from one f-number to the next higher value (for example, from $f/2.8$ to $f/4$) decreases the area of the aperture by a factor of 2. The lowest f-number setting on a camera lens corresponds to a wide-open aperture and the use of the maximum possible lens area.

Simple cameras usually have a fixed focal length and a fixed aperture size, with an f-number of about $f/11$. This high value for the f-number allows for a large **depth of field,** meaning that objects at a wide range of distances from the lens form reasonably sharp images on the CCD. In other words, the camera does not have to be focused.

Quick Quiz 36.7 A camera can be modeled as a simple converging lens that focuses an image on the CCD, acting as the screen. A camera is initially focused on a distant object. To focus the image of an object close to the camera, must the lens be (a) moved away from the CCD, (b) left where it is, or (c) moved toward the CCD?

EXAMPLE 36.11 Finding the Correct Exposure Time

The lens of a digital camera has a focal length of 55 mm and a speed (an f-number) of $f/1.8$. The correct exposure time for this speed under certain conditions is known to be $\frac{1}{500}$ s.

(A) Determine the diameter of the lens.

SOLUTION

Conceptualize Remember that the f-number for a lens relates its focal length to its diameter.

Categorize We evaluate results using equations developed in this section, so we categorize this example as a substitution problem.

Solve Equation 36.20 for D and substitute numerical values:

$$D = \frac{f}{f\text{-number}} = \frac{55 \text{ mm}}{1.8} = 31 \text{ mm}$$

(B) Calculate the correct exposure time if the f-number is changed to $f/4$ under the same lighting conditions.

SOLUTION

The total light energy hitting the CCD is proportional to the product of the intensity and the exposure time. If I is the light intensity reaching the CCD, the energy per unit area received by the CCD in a time interval Δt is proportional to $I\,\Delta t$. Comparing the two situations, we require that $I_1\,\Delta t_1 = I_2\,\Delta t_2$, where Δt_1 is the correct exposure time for $f/1.8$ and Δt_2 is the correct exposure time for $f/4$.

Use this result and substitute for I from Equation 36.21:

$$I_1\,\Delta t_1 = I_2\,\Delta t_2 \quad \rightarrow \quad \frac{\Delta t_1}{(f_1\text{-number})^2} = \frac{\Delta t_2}{(f_2\text{-number})^2}$$

Solve for Δt_2 and substitute numerical values:

$$\Delta t_2 = \left(\frac{f_2\text{-number}}{f_1\text{-number}}\right)^2 \Delta t_1 = \left(\frac{4}{1.8}\right)^2 \left(\tfrac{1}{500}\text{ s}\right) \approx \tfrac{1}{100}\text{ s}$$

As the aperture size is reduced, the exposure time must increase.

36.7 The Eye

Like a camera, a normal eye focuses light and produces a sharp image. The mechanisms by which the eye controls the amount of light admitted and adjusts to produce correctly focused images, however, are far more complex, intricate, and effective than those in even the most sophisticated camera. In all respects, the eye is a physiological wonder.

Figure 36.34 shows the basic parts of the human eye. Light entering the eye passes through a transparent structure called the *cornea* (Fig. 36.35), behind which are a clear liquid (the *aqueous humor*), a variable aperture (the *pupil,* which is an opening in the *iris*), and the *crystalline lens.* Most of the refraction occurs at the outer surface of the eye, where the cornea is covered with a film of tears. Relatively little refraction occurs in the crystalline lens because the aqueous humor in contact with the lens has an average index of refraction close to that of the lens. The iris, which is the colored portion of the eye, is a muscular diaphragm that controls pupil size. The iris regulates the amount of light entering the eye by dilating, or opening, the pupil in low-light conditions and contracting, or closing, the pupil in high-light conditions. The f-number range of the human eye is approximately $f/2.8$ to $f/16$.

The cornea–lens system focuses light onto the back surface of the eye, the *retina,* which consists of millions of sensitive receptors called *rods* and *cones.* When stimulated by light, these receptors send impulses via the optic nerve to the brain, where an image is perceived. By this process, a distinct image of an object is observed when the image falls on the retina.

The eye focuses on an object by varying the shape of the pliable crystalline lens through a process called **accommodation.** The lens adjustments take place so swiftly that we are not even aware of the change. Accommodation is limited in that objects very close to the eye produce blurred images. The **near point** is the closest distance for which the lens can accommodate to focus light on the retina. This distance usually increases with age and has an average value of 25 cm. At age 10, the near point of the eye is typically approximately 18 cm. It increases to approximately 25 cm at age 20, to 50 cm at age 40, and to 500 cm or greater at age 60. The **far point** of the eye represents the greatest distance for which the lens of the relaxed eye can focus light on the retina. A person with normal vision can see very distant objects and therefore has a far point that can be approximated as infinity.

Recall that the light leaving the mirror in Figure 36.7 becomes white where it comes together but then diverges into separate colors again. Because nothing but air exists at the point where the rays cross (and hence nothing exists to cause the colors to separate again), seeing white light as a result of a combination of colors must be a visual illusion. In fact, that is the case. Only three types of color-sensitive

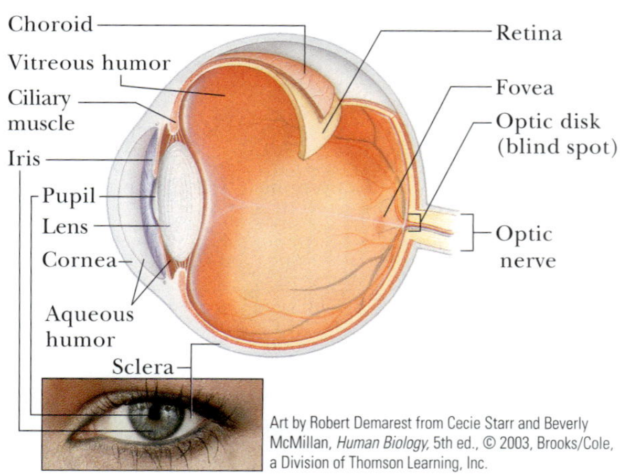

Art by Robert Demarest from Cecie Starr and Beverly McMillan, *Human Biology,* 5th ed., © 2003, Brooks/Cole, a Division of Thomson Learning, Inc.

Figure 36.34 Important parts of the eye.

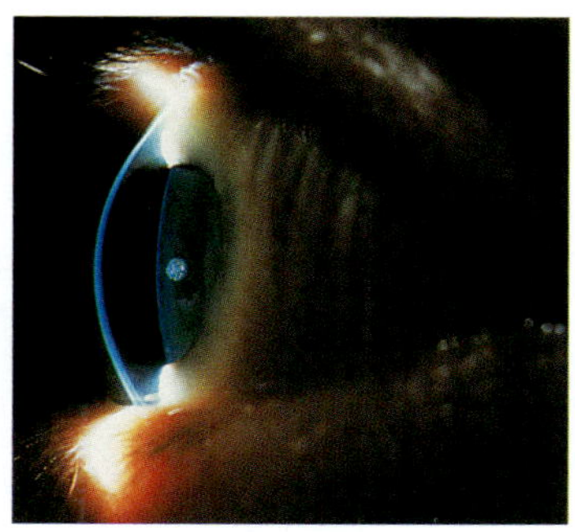

From Lennart Nilsson, in collaboration with Jan Lindberg, *Behold Man: A Photographic Journey of Discovery Inside the Body,* Boston, Little, Brown & Co., 1974.

Figure 36.35 Close-up photograph of the cornea of the human eye.

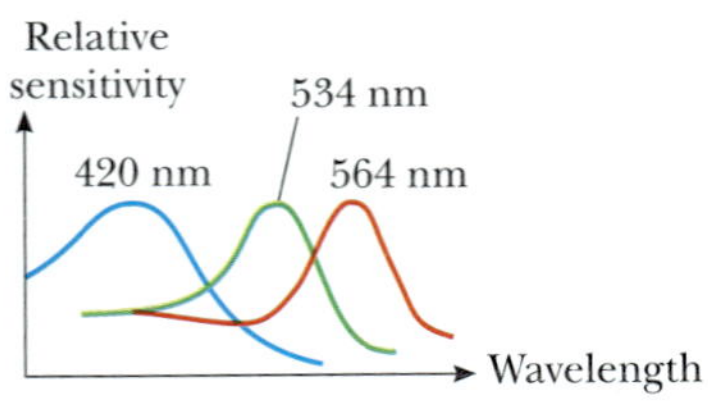

Figure 36.36 Approximate color sensitivity of the three types of cones in the retina.

cells are present in the retina. They are called red, green, and blue cones because of the peaks of the color ranges to which they respond (Fig. 36.36). If the red and green cones are stimulated simultaneously (as would be the case if yellow light were shining on them), the brain interprets what is seen as yellow. If all three types of cones are stimulated by the separate colors red, blue, and green as in Figure 36.7, white light is seen. If all three types of cones are stimulated by light that contains *all* colors, such as sunlight, again white light is seen.

Color televisions take advantage of this visual illusion by having only red, green, and blue dots on the screen. With specific combinations of brightness in these three primary colors, our eyes can be made to see any color in the rainbow. Therefore, the yellow lemon you see in a television commercial is not actually yellow, it is red and green! The paper on which this page is printed is made of tiny, matted, translucent fibers that scatter light in all directions, and the resultant mixture of colors appears white to the eye. Snow, clouds, and white hair are not actually white. In fact, there is no such thing as a white pigment. The appearance of these things is a consequence of the scattering of light containing all colors, which we interpret as white.

Conditions of the Eye

When the eye suffers a mismatch between the focusing range of the lens–cornea system and the length of the eye, with the result that light rays from a near object reach the retina before they converge to form an image as shown in Figure 36.37a, the condition is known as **farsightedness** (or *hyperopia*). A farsighted person can usually see faraway objects clearly but not nearby objects. Although the near point of a normal eye is approximately 25 cm, the near point of a farsighted person is much farther away. The refracting power in the cornea and lens is insufficient to focus the light from all but distant objects satisfactorily. The condition can be corrected by placing a converging lens in front of the eye as shown in Figure 36.37b. The lens refracts the incoming rays more toward the principal axis before entering the eye, allowing them to converge and focus on the retina.

A person with **nearsightedness** (or *myopia*), another mismatch condition, can focus on nearby objects but not on faraway objects. The far point of the nearsighted eye is not infinity and may be less than 1 m. The maximum focal length of the nearsighted eye is insufficient to produce a sharp image on the retina, and rays from a distant object converge to a focus in front of the retina. They then continue past that point, diverging before they finally reach the retina and causing blurred vision (Fig. 36.38a). Nearsightedness can be corrected with a diverging lens as shown in Figure 36.38b. The lens refracts the rays away from the principal axis before they enter the eye, allowing them to focus on the retina.

Beginning in middle age, most people lose some of their accommodation ability as their visual muscles weaken and the lens hardens. Unlike farsightedness, which is a mismatch between focusing power and eye length, **presbyopia** (literally, "old-age vision") is due to a reduction in accommodation ability. The cornea and

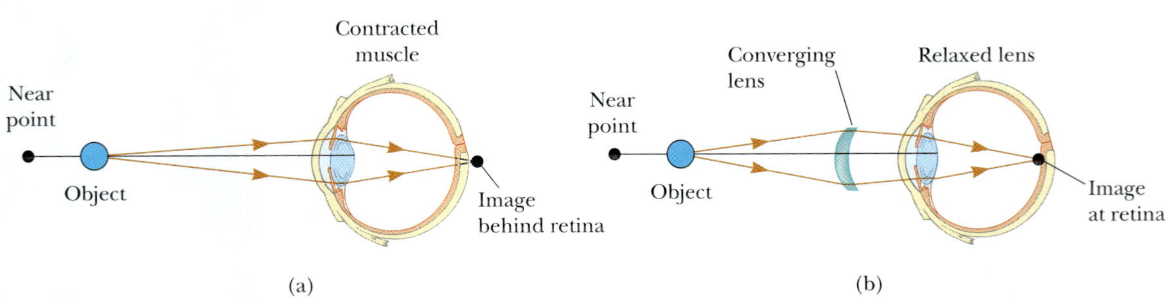

Figure 36.37 (a) When a farsighted eye looks at an object located between the near point and the eye, the image point is behind the retina, resulting in blurred vision. The eye muscle contracts to try to bring the object into focus. (b) Farsightedness is corrected with a converging lens.

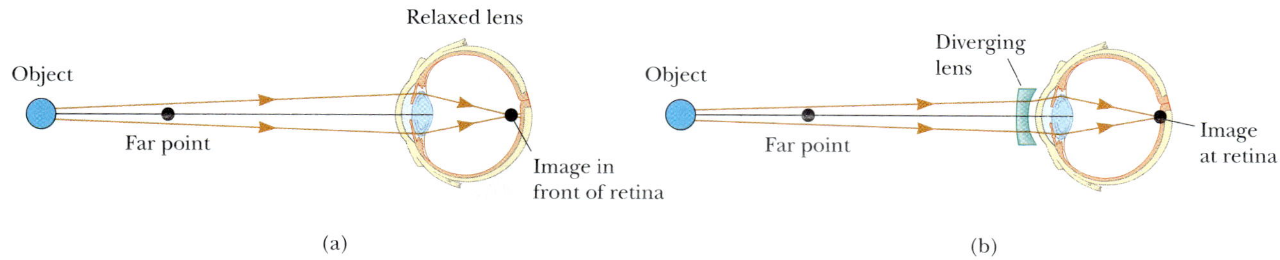

Figure 36.38 (a) When a nearsighted eye looks at an object that lies beyond the eye's far point, the image is formed in front of the retina, resulting in blurred vision. (b) Nearsightedness is corrected with a diverging lens.

lens do not have sufficient focusing power to bring nearby objects into focus on the retina. The symptoms are the same as those of farsightedness, and the condition can be corrected with converging lenses.

In eyes having a defect known as **astigmatism,** light from a point source produces a line image on the retina. This condition arises when either the cornea, the lens, or both are not perfectly symmetric. Astigmatism can be corrected with lenses that have different curvatures in two mutually perpendicular directions.

Optometrists and ophthalmologists usually prescribe lenses[1] measured in **diopters:** the **power** P of a lens in diopters equals the inverse of the focal length in meters: $P = 1/f$. For example, a converging lens of focal length +20 cm has a power of +5.0 diopters, and a diverging lens of focal length −40 cm has a power of −2.5 diopters.

Quick Quiz 36.8 Two campers wish to start a fire during the day. One camper is nearsighted, and one is farsighted. Whose glasses should be used to focus the Sun's rays onto some paper to start the fire? (a) either camper (b) the nearsighted camper (c) the farsighted camper

36.8 The Simple Magnifier

The simple magnifier, or magnifying glass, consists of a single converging lens. This device increases the apparent size of an object.

Suppose an object is viewed at some distance p from the eye as illustrated in Figure 36.39. The size of the image formed at the retina depends on the angle θ subtended by the object at the eye. As the object moves closer to the eye, θ increases and a larger image is observed. An average normal human eye, however, cannot focus on an object closer than about 25 cm, the near point (Fig. 36.40a, page 1036). Therefore, θ is maximum at the near point.

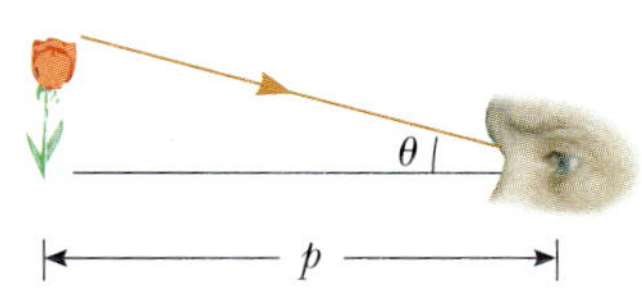

Figure 36.39 The size of the image formed on the retina depends on the angle θ subtended at the eye.

To further increase the apparent angular size of an object, a converging lens can be placed in front of the eye as in Figure 36.40b, with the object located at point O, immediately inside the focal point of the lens. At this location, the lens forms a virtual, upright, enlarged image. We define **angular magnification** m as the ratio of the angle subtended by an object with a lens in use (angle θ in Fig. 36.40b) to the angle subtended by the object placed at the near point with no lens in use (angle θ_0 in Fig. 36.40a):

$$m \equiv \frac{\theta}{\theta_0} \qquad \textbf{(36.22)}$$

[1] The word *lens* comes from *lentil,* the name of an Italian legume. (You may have eaten lentil soup.) Early eyeglasses were called "glass lentils" because the biconvex shape of their lenses resembled the shape of a lentil. The first lenses for farsightedness and presbyopia appeared around 1280; concave eyeglasses for correcting nearsightedness did not appear until more than 100 years later.

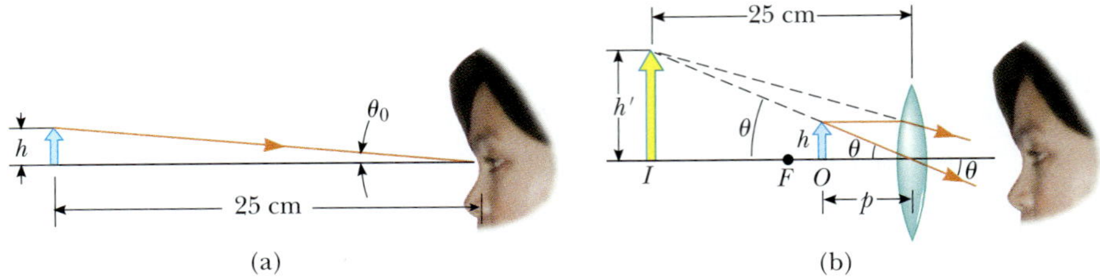

Figure 36.40 (a) An object placed at the near point of the eye (p = 25 cm) subtends an angle $\theta_0 \approx h/25$ at the eye. (b) An object placed near the focal point of a converging lens produces a magnified image that subtends an angle $\theta \approx h'/25$ at the eye.

The angular magnification is a maximum when the image is at the near point of the eye, that is, when $q = -25$ cm. The object distance corresponding to this image distance can be calculated from the thin lens equation:

$$\frac{1}{p} + \frac{1}{-25 \text{ cm}} = \frac{1}{f} \quad \rightarrow \quad p = \frac{25f}{25 + f}$$

where f is the focal length of the magnifier in centimeters. If we make the small-angle approximations

$$\tan \theta_0 \approx \theta_0 \approx \frac{h}{25} \quad \text{and} \quad \tan \theta \approx \theta \approx \frac{h}{p} \qquad \textbf{(36.23)}$$

Equation 36.22 becomes

$$m_{\max} = \frac{\theta}{\theta_0} = \frac{h/p}{h/25} = \frac{25}{p} = \frac{25}{25f/(25 + f)}$$

$$m_{\max} = 1 + \frac{25 \text{ cm}}{f} \qquad \textbf{(36.24)}$$

Although the eye can focus on an image formed anywhere between the near point and infinity, it is most relaxed when the image is at infinity. For the image formed by the magnifying lens to appear at infinity, the object has to be at the focal point of the lens. In this case, Equations 36.23 become

$$\theta_0 \approx \frac{h}{25} \quad \text{and} \quad \theta \approx \frac{h}{f}$$

and the magnification is

$$m_{\min} = \frac{\theta}{\theta_0} = \frac{25 \text{ cm}}{f} \qquad \textbf{(36.25)}$$

With a single lens, it is possible to obtain angular magnifications up to about 4 without serious aberrations. Magnifications up to about 20 can be achieved by using one or two additional lenses to correct for aberrations.

George Semple

A simple magnifier, also called a magnifying glass, is used to view an enlarged image of a portion of a map.

EXAMPLE 36.12 Magnification of a Lens

What is the maximum magnification that is possible with a lens having a focal length of 10 cm, and what is the magnification of this lens when the eye is relaxed?

SOLUTION

Conceptualize Study Figure 36.40b for the situation in which a magnifying glass forms an enlarged image of an object placed inside the focal point. The maximum magnification occurs when the image is located at the near point of the eye. When the eye is relaxed, the image is at infinity.

Categorize We evaluate results using equations developed in this section, so we categorize this example as a substitution problem.

Evaluate the maximum magnification from Equation 36.24:

$$m_{\max} = 1 + \frac{25\text{ cm}}{f} = 1 + \frac{25\text{ cm}}{10\text{ cm}} = 3.5$$

Evaluate the minimum magnification, when the eye is relaxed, from Equation 36.25:

$$m_{\min} = \frac{25\text{ cm}}{f} = \frac{25\text{ cm}}{10\text{ cm}} = 2.5$$

36.9 The Compound Microscope

A simple magnifier provides only limited assistance in inspecting minute details of an object. Greater magnification can be achieved by combining two lenses in a device called a **compound microscope** shown in Active Figure 36.41a. It consists of one lens, the *objective,* that has a very short focal length $f_o < 1$ cm and a second lens, the *eyepiece,* that has a focal length f_e of a few centimeters. The two lenses are separated by a distance L that is much greater than either f_o or f_e. The object, which is placed just outside the focal point of the objective, forms a real, inverted image at I_1, and this image is located at or close to the focal point of the eyepiece. The eyepiece, which serves as a simple magnifier, produces at I_2 a virtual, enlarged image of I_1. The lateral magnification M_1 of the first image is $-q_1/p_1$. Notice from Active Figure 36.41a that q_1 is approximately equal to L and that the object is very close to the focal point of the objective: $p_1 \approx f_o$. Therefore, the lateral magnification by the objective is

$$M_o \approx -\frac{L}{f_o}$$

The angular magnification by the eyepiece for an object (corresponding to the image at I_1) placed at the focal point of the eyepiece is, from Equation 36.25,

$$m_e = \frac{25\text{ cm}}{f_e}$$

The overall magnification of the image formed by a compound microscope is defined as the product of the lateral and angular magnifications:

$$M = M_o m_e = -\frac{L}{f_o}\left(\frac{25\text{ cm}}{f_e}\right) \qquad (36.26)$$

The negative sign indicates that the image is inverted.

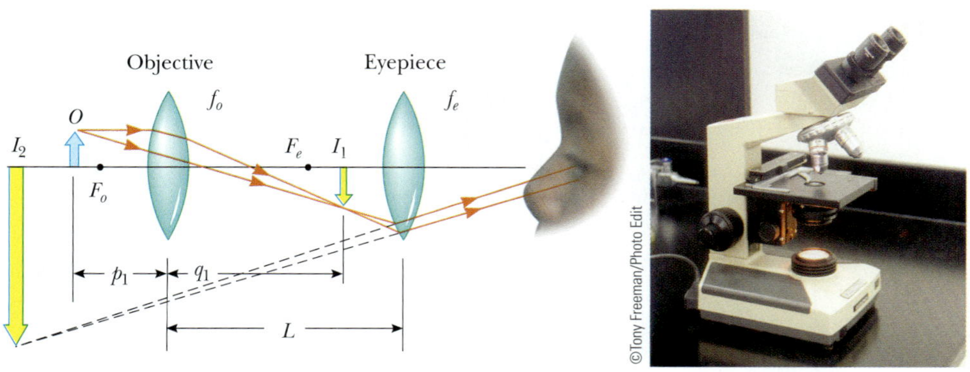

(a) (b)

ACTIVE FIGURE 36.41

(a) Diagram of a compound microscope, which consists of an objective lens and an eyepiece lens. (b) A compound microscope. The three-objective turret allows the user to choose from several powers of magnification. Combinations of eyepieces with different focal lengths and different objectives can produce a wide range of magnifications.

Sign in at www.thomsonedu.com and go to ThomsonNOW to adjust the focal lengths of the objective and eyepiece lenses and see the effect on the final image.

The microscope has extended human vision to the point where we can view previously unknown details of incredibly small objects. The capabilities of this instrument have steadily increased with improved techniques for precision grinding of lenses. A question often asked about microscopes is, "If one were extremely patient and careful, would it be possible to construct a microscope that would enable the human eye to see an atom?" The answer is no, as long as light is used to illuminate the object. For an object under an optical microscope (one that uses visible light) to be seen, the object must be at least as large as a wavelength of light. Because the diameter of any atom is many times smaller than the wavelengths of visible light, the mysteries of the atom must be probed using other types of "microscopes."

36.10 The Telescope

Two fundamentally different types of **telescopes** exist; both are designed to aid in viewing distant objects, such as the planets in our solar system. The **refracting telescope** uses a combination of lenses to form an image, and the **reflecting telescope** uses a curved mirror and a lens.

Like the compound microscope, the refracting telescope shown in Active Figure 36.42a has an objective and an eyepiece. The two lenses are arranged so that the objective forms a real, inverted image of a distant object very near the focal point of the eyepiece. Because the object is essentially at infinity, this point at which I_1 forms is the focal point of the objective. The eyepiece then forms, at I_2, an enlarged, inverted image of the image at I_1. To provide the largest possible magnification, the image distance for the eyepiece is infinite. The light rays exit the eyepiece lens parallel to the principal axis, and the image due to the objective lens must form at the focal point of the eyepiece. Hence, the two lenses are separated by a distance $f_o + f_e$, which corresponds to the length of the telescope tube.

The angular magnification of the telescope is given by θ/θ_o, where θ_o is the angle subtended by the object at the objective and θ is the angle subtended by the final image at the viewer's eye. Consider Active Figure 36.42a, in which the object is a very great distance to the left of the figure. The angle θ_o (to the *left* of the objective) subtended by the object at the objective is the same as the angle (to the *right* of the objective) subtended by the first image at the objective. Therefore,

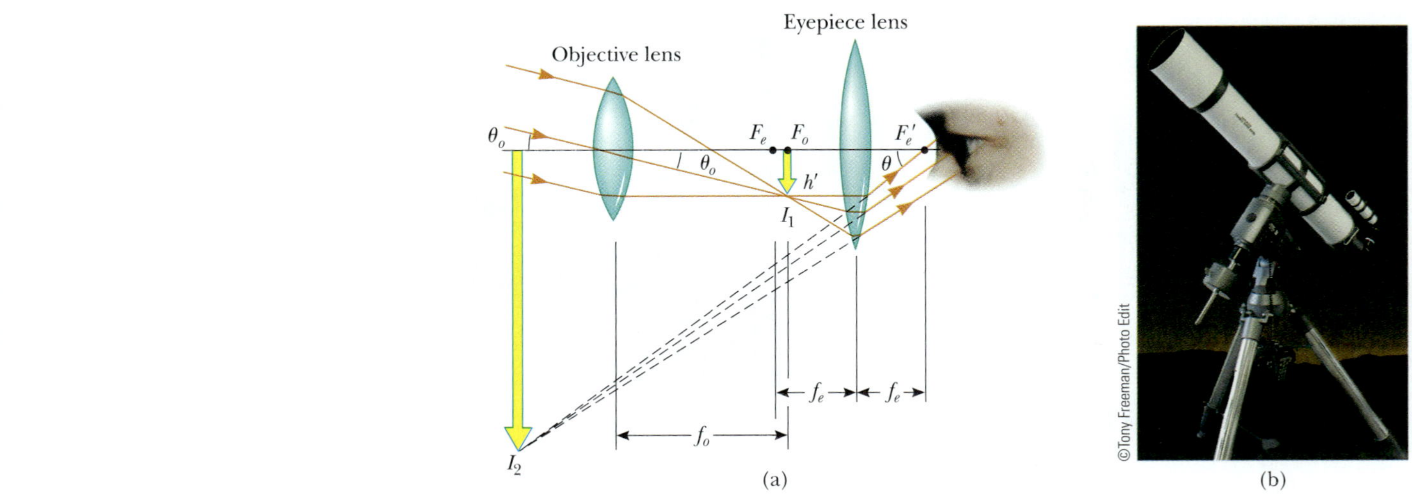

ACTIVE FIGURE 36.42

(a) Lens arrangement in a refracting telescope, with the object at infinity. (b) A refracting telescope.

Sign in at www.thomsonedu.com and go to ThomsonNOW to adjust the focal lengths of the objective and eyepiece lenses and see the effect on the final image.

$$\tan \theta_o \approx \theta_o \approx -\frac{h'}{f_o}$$

where the negative sign indicates that the image is inverted.

The angle θ subtended by the final image at the eye is the same as the angle that a ray coming from the tip of I_1 and traveling parallel to the principal axis makes with the principal axis after it passes through the lens. Therefore,

$$\tan \theta \approx \theta \approx \frac{h'}{f_e}$$

We have not used a negative sign in this equation because the final image is not inverted; the object creating this final image I_2 is I_1, and both it and I_2 point in the same direction. Therefore, the angular magnification of the telescope can be expressed as

$$m = \frac{\theta}{\theta_o} = \frac{h'/f_e}{-h'/f_o} = -\frac{f_o}{f_e} \tag{36.27}$$

This result shows that the angular magnification of a telescope equals the ratio of the objective focal length to the eyepiece focal length. The negative sign indicates that the image is inverted.

When you look through a telescope at such relatively nearby objects as the Moon and the planets, magnification is important. Individual stars in our galaxy, however, are so far away that they always appear as small points of light no matter how great the magnification. To gather as much light as possible, large research telescopes used to study very distant objects must have a large diameter. It is difficult and expensive to manufacture large lenses for refracting telescopes. Another difficulty with large lenses is that their weight leads to sagging, which is an additional source of aberration.

These problems associated with large lenses can be partially overcome by replacing the objective with a concave mirror, which results in a reflecting telescope. Because light is reflected from the mirror and does not pass through a lens, the mirror can have rigid supports on the back side. Such supports eliminate the problem of sagging.

Figure 36.43a shows the design for a typical reflecting telescope. The incoming light rays are reflected by a parabolic mirror at the base. These reflected rays converge toward point A in the figure, where an image would be formed. Before this image is formed, however, a small, flat mirror M reflects the light toward an opening in the tube's side and it passes into an eyepiece. This particular design is said to have a Newtonian focus because Newton developed it. Figure 36.43b shows such a telescope. Notice that the light never passes through glass (except through the

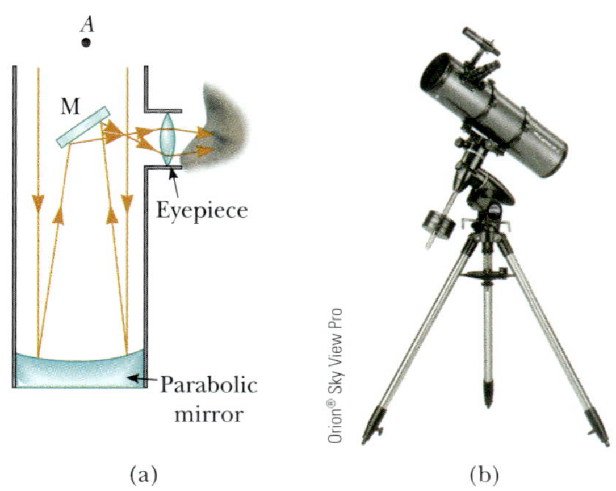

Figure 36.43 (a) A Newtonian-focus reflecting telescope. (b) A reflecting telescope. This type of telescope is shorter than that in Figure 36.42b.

small eyepiece) in the reflecting telescope. As a result, problems associated with chromatic aberration are virtually eliminated. The reflecting telescope can be made even shorter by orienting the flat mirror so that it reflects the light back toward the objective mirror and the light enters an eyepiece in a hole in the middle of the mirror.

The largest reflecting telescopes in the world are at the Keck Observatory on Mauna Kea, Hawaii. The site includes two telescopes with diameters of 10 m, each containing 36 hexagonally shaped, computer-controlled mirrors that work together to form a large reflecting surface. Discussions and plans have been initiated for telescopes with different mirrors working together, as at the Keck Observatory, resulting in an effective diameter up to 21 m. In contrast, the largest refracting telescope in the world, at the Yerkes Observatory in Williams Bay, Wisconsin, has a diameter of only 1 m.

Summary

ThomsonNOW™ Sign in at **www.thomsonedu.com** and go to ThomsonNOW to take a practice test for this chapter.

DEFINITIONS

The **lateral magnification** M of the image due to a mirror or lens is defined as the ratio of the image height h' to the object height h. It is equal to the negative of the ratio of the image distance q to the object distance p:

$$M \equiv \frac{\text{image height}}{\text{object height}} = \frac{h'}{h} = -\frac{q}{p} \qquad \textbf{(36.1, 36.2, 36.17)}$$

The **angular magnification** m is the ratio of the angle subtended by an object with a lens in use (angle θ in Fig. 36.40b) to the angle subtended by the object placed at the near point with no lens in use (angle θ_0 in Fig. 36.40a):

$$m \equiv \frac{\theta}{\theta_0} \qquad \textbf{(36.22)}$$

The ratio of the focal length of a camera lens to the diameter of the lens is called the ***f*-number** of the lens:

$$f\text{-number} \equiv \frac{f}{D} \qquad \textbf{(36.20)}$$

CONCEPTS AND PRINCIPLES

In the paraxial ray approximation, the object distance p and image distance q for a spherical mirror of radius R are related by the **mirror equation:**

$$\frac{1}{p} + \frac{1}{q} = \frac{2}{R} = \frac{1}{f} \qquad \textbf{(36.4, 36.6)}$$

where $f = R/2$ is the **focal length** of the mirror.

An image can be formed by refraction from a spherical surface of radius R. The object and image distances for refraction from such a surface are related by

$$\frac{n_1}{p} + \frac{n_2}{q} = \frac{n_2 - n_1}{R} \qquad \textbf{(36.8)}$$

where the light is incident in the medium for which the index of refraction is n_1 and is refracted in the medium for which the index of refraction is n_2.

(*continued*)

The inverse of the **focal length** f of a thin lens surrounded by air is given by the **lens-makers' equation:**

$$\frac{1}{f} = (n - 1)\left(\frac{1}{R_1} - \frac{1}{R_2}\right) \quad \textbf{(36.15)}$$

Converging lenses have positive focal lengths, and **diverging lenses** have negative focal lengths.

For a thin lens, and in the paraxial ray approximation, the object and image distances are related by the **thin lens equation:**

$$\frac{1}{p} + \frac{1}{q} = \frac{1}{f} \quad \textbf{(36.16)}$$

The maximum magnification of a single lens of focal length f used as a simple magnifier is

$$m_{\max} = 1 + \frac{25\text{ cm}}{f} \quad \textbf{(36.24)}$$

The overall magnification of the image formed by a compound microscope is:

$$M = -\frac{L}{f_o}\left(\frac{25\text{ cm}}{f_e}\right) \quad \textbf{(36.26)}$$

where f_o and f_e are the focal lengths of the objective and eyepiece lenses, respectively, and L is the distance between the lenses.

The angular magnification of a refracting telescope can be expressed as

$$m = -\frac{f_o}{f_e} \quad \textbf{(36.27)}$$

where f_o and f_e are the focal lengths of the objective and eyepiece lenses, respectively. The angular magnification of a reflecting telescope is given by the same expression where f_o is the focal length of the objective mirror.

Questions

☐ denotes answer available in *Student Solutions Manual/Study Guide;* **O** denotes objective question

1. Consider a concave spherical mirror with a real object. Is the image always inverted? Is the image always real? Give conditions for your answers.

2. Repeat Question 1 for a convex spherical mirror.

3. **O** **(i)** What is the focal length of a plane mirror? (a) 0 (b) 1 (c) −1 (d) ∞ (e) equal to the mirror height (f) Neither the focal length nor its reciprocal can be defined. **(ii)** What magnification does a plane mirror produce? (a) 0 (b) 1 (c) −1 (d) ∞ (e) Neither the magnification nor its reciprocal can be defined.

4. Do the equations $1/p + 1/q = 1/f$ and $M = -q/p$ apply to the image formed by a flat mirror? Explain your answer.

5. **O** Lulu looks at her image in a makeup mirror. It is enlarged when she is close to the mirror. As she backs away, the image becomes larger, then impossible to identify when she is 30 cm from the mirror, then upside down when she is beyond 30 cm, and finally small, clear, and upside down when she is much farther from the mirror. **(i)** Is the mirror (a) convex, (b) plane, or (c) concave? **(ii)** What is the magnitude of its focal length? (a) 0 (b) 15 cm (c) 30 cm (d) 60 cm (e) ∞

6. Consider a spherical concave mirror with the object located to the left of the mirror beyond the focal point. Using ray diagrams, show that the image moves to the left as the object approaches the focal point.

7. **O** **(i)** Consider the mirror in Figure 36.11. What are the signs of the following? (a) the object distance (b) the image distance (c) the mirror radius (d) the focal length (e) the object height (f) the image height (g) the magnification **(ii)** Consider the objective lens in Active Figure 36.41a. What are the signs of the following? (a) the object distance (b) the image distance (c) the focal length (d) the object height (e) the image height (f) the magnification **(iii)** Answer the same questions (a) through (f) as in part (ii) for the eyepiece in Active Figure 36.41a.

8. **O** A person spearfishing from a boat sees a stationary fish a few meters away in a direction about 30° below the horizontal. To spear the fish, should the person (a) aim above where he sees the fish, (b) aim precisely at the fish, or (c) aim below the fish? Assume the dense spear does not change direction when it enters the water.

9. **O** A single converging lens can be used to constitute a scale model of each of the following devices in use simply by changing the distance from the lens to a candle representing the object. Rank the cases according to the distance from the object to the lens from the largest to the smallest. (a) a movie projector (b) Batman's signal, used to project an image on clouds high above Gotham City (c) a magnifying glass (d) a burning glass, used to make a sharp image of the Sun on tinder (e) an astronomical refracting telescope, used to make a sharp image of stars on an electronic detector (f) a searchlight, used to produce a beam of parallel rays from a point source. (g) a camera lens, used to photograph a soccer game.

10. In Active Figure 36.26a, assume the blue object arrow is replaced by one that is much taller than the lens. How many rays from the top of the object will strike the lens? How many principal rays can be drawn in a ray diagram?

11. **O** A converging lens in a vertical plane receives light from an object and forms an inverted image on a screen. An opaque card is then placed next to the lens, covering only the upper half of the lens. What happens to the image on the screen? (a) The upper half of the image disappears. (b) The lower half of the image disappears. (c) The entire image disappears. (d) The entire image is still visible, but is dimmer. (e) Half of the image disappears and the rest is dimmer. (f) No change in the image occurs.

12. **O** A converging lens of focal length 8 cm forms a sharp image of an object on a screen. What is the smallest possible distance between the object and the screen? (a) 0 (b) 4 cm (c) 8 cm (d) 16 cm (e) 32 cm (f) ∞

13. Explain this statement: "The focal point of a lens is the location of the image of a point object at infinity." Discuss the notion of infinity in real terms as it applies to object distances. Based on this statement, can you think of a simple method for determining the focal length of a converging lens?

14. Discuss the proper position of a photographic slide relative to the lens in a slide projector. What type of lens must the slide projector have?

15. **O** In this chapter's opening photograph, a water drop functions as a biconvex lens with radii of curvature of small magnitude. What is the location of the image photographed? (a) inside the water drop (b) on the back surface of the drop, farthest from the camera (c) somewhat beyond the back surface of the drop (d) on the front surface of the drop, closest to the camera (e) somewhat closer to the camera than the front surface of the drop

16. Explain why a mirror cannot give rise to chromatic aberration.

17. Can a converging lens be made to diverge light if it is placed into a liquid? **What If?** What about a converging mirror?

18. Explain why a fish in a spherical goldfish bowl appears larger than it really is.

19. Why do some emergency vehicles have the symbol ƎƆИA⅃UꓭMA written on the front?

20. Lenses used in eyeglasses, whether converging or diverging, are always designed so that the middle of the lens curves away from the eye like the center lenses of Figures 36.25a and 36.25b. Why?

21. **O** The faceplate of a diving mask can be a corrective lens for a diver who does not have perfect vision and who needs essentially the same prescription for both eyes. Then the diver does not have to wear glasses or contact lenses. The proper design allows the person to see clearly both under water and in the air. Normal eyeglasses have lenses with both the front and back surfaces curved. Should the lens of a diving mask be curved (a) on the outer surface only, (b) on the inner surface only, or (c) on both surfaces?

22. In Figures Q36.22a and Q36.22b, which glasses correct nearsightedness and which correct farsightedness?

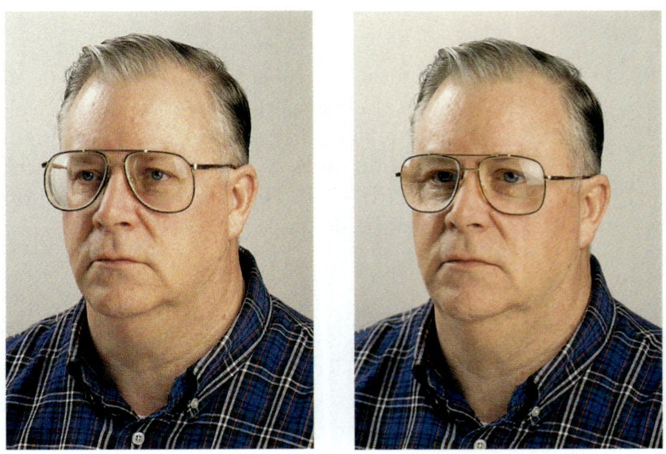

(a) (b)

Figure Q36.22 Questions 22 and 23.

23. A child tries on either his hyperopic grandfather's or his myopic brother's glasses and complains, "Everything looks blurry." Why do the eyes of a person wearing glasses not look blurry? (See Figure Q36.22.)

24. In a Jules Verne novel, a piece of ice is shaped to form a magnifying lens to focus sunlight to start a fire. Is that possible?

25. A solar furnace can be constructed by using a concave mirror to reflect and focus sunlight into a furnace enclosure. What factors in the design of the reflecting mirror would guarantee very high temperatures?

26. Figure Q36.26 shows a lithograph by M. C. Escher titled *Hand with Reflection Sphere (Self-Portrait in Spherical Mirror)*. Escher said about the work:

> The picture shows a spherical mirror, resting on a left hand. But as a print is the reverse of the original drawing on stone, it was my right hand that you see depicted. (Being left-handed, I needed my left hand to make the drawing.) Such a globe reflection collects almost one's whole surroundings in one disk-shaped image. The whole room, four walls, the floor, and the ceiling, everything, albeit distorted, is compressed into that one small circle. Your own head, or more exactly the point between your eyes, is the absolute center. No matter how you turn or twist yourself, you can't get out of that central point. You are immovably the focus, the unshakable core, of your world.

Comment on the accuracy of Escher's description.

Figure Q36.26

27. A converging lens of short focal length can take light diverging from a small source and refract it into a beam of parallel rays. A Fresnel lens as shown in Figure 36.27 is used in a lighthouse for this purpose. A concave mirror can take light diverging from a small source and reflect it into a beam of parallel rays. Is it possible to make a Fresnel mirror? Is this idea original, or has it already been done? *Suggestion:* Look at the walls and ceiling of an auditorium.

Problems

WebAssign The Problems from this chapter may be assigned online in WebAssign.

ThomsonNOW Sign in at **www.thomsonedu.com** and go to ThomsonNOW to assess your understanding of this chapter's topics with additional quizzing and conceptual questions.

1, 2, 3 denotes straightforward, intermediate, challenging; □ denotes full solution available in *Student Solutions Manual/Study Guide;* ▲ denotes coached solution with hints available at **www.thomsonedu.com;** ▭ denotes developing symbolic reasoning; ● denotes asking for qualitative reasoning; ▪ denotes computer useful in solving problem

Section 36.1 Images Formed by Flat Mirrors

1. Does your bathroom mirror show you older or younger than you actually are? Compute an order-of-magnitude estimate for the age difference based on data you specify.

2. In a church choir loft, two parallel walls are 5.30 m apart. The singers stand against the north wall. The organist faces the south wall, sitting 0.800 m away from it. To enable her to see the choir, a flat mirror 0.600 m wide is mounted on the south wall, straight in front of her. What width of the north wall can the organist see? *Suggestion:* Draw a top-view diagram to justify your answer.

3. Determine the minimum height of a vertical flat mirror in which a person 5 ft 10 in. in height can see his or her full image. (A ray diagram would be helpful.)

4. A person walks into a room that has two flat mirrors on opposite walls. The mirrors produce multiple images of the person. When the person is 5.00 ft from the mirror on the left wall and 10.0 ft from the mirror on the right wall, find the distance from the person to the first three images seen in the mirror on the left.

5. A periscope (Fig. P36.5) is useful for viewing objects that cannot be seen directly. It can be used in submarines and when watching golf matches or parades from behind a crowd of people. Suppose the object is a distance p_1 from the upper mirror and the two flat mirrors are separated by a distance h. (a) What is the distance of the final image from the lower mirror? (b) Is the final image real or virtual? (c) Is it upright or inverted? (d) What is its magnification? (e) Does it appear to be left–right reversed?

Figure P36.5

Section 36.2 Images Formed by Spherical Mirrors

6. A concave spherical mirror has a radius of curvature of 20.0 cm. Find the location of the image for object distances of (a) 40.0 cm, (b) 20.0 cm, and (c) 10.0 cm. For each case, state whether the image is real or virtual and upright or inverted. Find the magnification in each case.

7. ▲ A spherical convex mirror has a radius of curvature with a magnitude of 40.0 cm. Determine the position of the virtual image and the magnification for object distances of (a) 30.0 cm and (b) 60.0 cm. (c) Are the images upright or inverted?

8. At an intersection of hospital hallways, a convex mirror is mounted high on a wall to help people avoid collisions. The magnitude of the mirror's radius of curvature is 0.550 m. Locate and describe the image of a patient 10.0 m from the mirror. Determine the magnification of the image.

9. A concave mirror has a radius of curvature of 60.0 cm. Calculate the image position and magnification of an object placed in front of the mirror at distances of (a) 90.0 cm and (b) 20.0 cm. (c) Draw ray diagrams to obtain the image characteristics in each case.

10. A large church has a niche in one wall. On the floor plan, the niche appears as a semicircular indentation of radius 2.50 m. A worshiper stands on the centerline of the niche, 2.00 m out from its deepest point, and whispers a prayer. Where is the sound concentrated after reflection from the back wall of the niche?

11. A dentist uses a mirror to examine a tooth. The tooth is 1.00 cm in front of the mirror, and the image is formed 10.0 cm behind the mirror. Determine (a) the mirror's radius of curvature and (b) the magnification of the image.

12. A certain Christmas tree ornament is a silver sphere having a diameter of 8.50 cm. Determine an object location for which the size of the reflected image is three-fourths the object's size. Use a principal-ray diagram to arrive at a description of the image.

13. (a) A concave mirror forms an inverted image four times larger than the object. Find the focal length of the mirror, assuming the distance between object and image is 0.600 m. (b) A convex mirror forms a virtual image half the size of the object. Assuming the distance between image and object is 20.0 cm, determine the radius of curvature of the mirror.

14. To fit a contact lens to a patient's eye, a *keratometer* can be used to measure the curvature of the eye's front surface, the cornea. This instrument places an illuminated object of known size at a known distance p from the cornea. The cornea reflects some light from the object, forming an image of the object. The magnification M of the image is measured by using a small viewing telescope that allows comparison of the image formed by the cornea with a second calibrated image projected into the field of view by a prism arrangement. Determine the radius of curvature of the cornea for the case $p = 30.0$ cm and $M = 0.013\,0$.

15. An object 10.0 cm tall is placed at the zero mark of a meterstick. A spherical mirror located at some point on the meterstick creates an image of the object that is upright, 4.00 cm tall, and located at the 42.0-cm mark of the meterstick. (a) Is the mirror convex or concave? (b) Where is the mirror? (c) What is the mirror's focal length?

16. A dedicated sports car enthusiast polishes the inside and outside surfaces of a hubcap that is a section of a sphere. When she looks into one side of the hubcap, she sees an image of her face 30.0 cm in back of the hubcap. She then flips the hubcap over and sees another image of her face 10.0 cm in back of the hubcap. (a) How far is her face from the hubcap? (b) What is the radius of curvature of the hubcap?

17. A spherical mirror is to be used to form, on a screen located 5.00 m from the object, an image five times the size of the object. (a) Describe the type of mirror

required. (b) Where should the mirror be positioned relative to the object?

18. You unconsciously estimate the distance to an object from the angle it subtends in your field of view. This angle θ in radians is related to the linear height of the object h and to the distance d by $\theta = h/d$. Assume you are driving a car and another car, 1.50 m high, is 24.0 m behind you. (a) Suppose your car has a flat passenger-side rearview mirror, 1.55 m from your eyes. How far from your eyes is the image of the car following you? (b) What angle does the image subtend in your field of view? (c) **What If?** Now suppose your car has a convex rearview mirror with a radius of curvature of magnitude 2.00 m (as suggested in Fig. 36.15). How far from your eyes is the image of the car behind you? (d) What angle does the image subtend at your eyes? (e) Based on its angular size, how far away does the following car appear to be?

19. ● **Review problem.** A ball is dropped at $t = 0$ from rest 3.00 m directly above the vertex of a concave mirror that has a radius of curvature of 1.00 m and lies in a horizontal plane. (a) Describe the motion of the ball's image in the mirror. (b) At what instant or instants do the ball and its image coincide?

Section 36.3 Images Formed by Refraction

20. A flint glass plate ($n = 1.66$) rests on the bottom of an aquarium tank. The plate is 8.00 cm thick (vertical dimension) and is covered with a layer of water ($n = 1.33$) 12.0 cm deep. Calculate the apparent thickness of the plate as viewed from straight above the water.

21. A cubical block of ice 50.0 cm on a side is placed over a speck of dust on a level floor. Find the location of the image of the speck as viewed from above. The index of refraction of ice is 1.309.

22. One end of a long glass rod ($n = 1.50$) is formed into a convex surface with a radius of curvature of 6.00 cm. An object is located in air along the axis of the rod. Find the image positions corresponding to object distances of (a) 20.0 cm, (b) 10.0 cm, and (c) 3.00 cm from the end of the rod.

23. A glass sphere ($n = 1.50$) with a radius of 15.0 cm has a tiny air bubble 5.00 cm above its center. The sphere is viewed looking down along the extended radius containing the bubble. What is the apparent depth of the bubble below the surface of the sphere?

24. Figure P36.24 shows a curved surface separating a material with index of refraction n_1 from a material with index n_2. The surface forms an image I of object O. The ray shown in blue passes through the surface along a radial line. Its angles of incidence and refraction are both zero, so its direction does not change at the surface. For the ray shown in brown, the direction changes according to $n_1 \sin\theta_1 = n_2 \sin\theta_2$. For paraxial rays, we assume θ_1 and θ_2 are small, so we may write $n_1 \tan\theta_1 = n_2 \tan\theta_2$. The magnification is defined as $M = h'/h$. Prove that the magnification is given by $M = -n_1 q/n_2 p$.

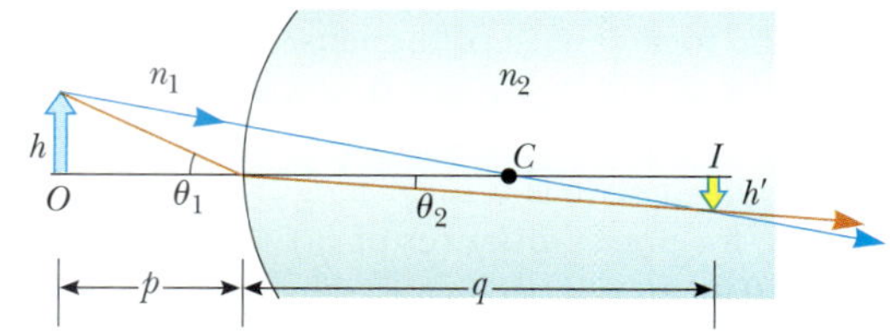

Figure P36.24

25. ● As shown in Figure P36.25, a water tank containing lobsters has a curved front made of plastic with uniform thickness and a radius of curvature of magnitude 80.0 cm. Locate and describe the images of lobsters (a) 30.0 cm and (b) 90.0 cm from the base of the front wall. (c) Find the magnification of each image. You may use the result of Problem 24. (d) The lobsters are both 9.00 cm in height. Find the height of each image. (e) Explain why you do not need to know the index of refraction of the plastic to solve this problem.

Alexandra Héder

Figure P36.25

26. A goldfish is swimming at 2.00 cm/s toward the front wall of a rectangular aquarium. What is the apparent speed of the fish measured by an observer looking in from outside the front wall of the tank? The index of refraction of water is 1.33.

Section 36.4 Thin Lenses

27. ▲ The left face of a biconvex lens has a radius of curvature of magnitude 12.0 cm, and the right face has a radius of curvature of magnitude 18.0 cm. The index of refraction of the glass is 1.44. (a) Calculate the focal length of the lens. (b) **What If?** Calculate the focal length the lens has after it is turned around to interchange the radii of curvature of the two faces.

28. A contact lens is made of plastic with an index of refraction of 1.50. The lens has an outer radius of curvature of +2.00 cm and an inner radius of curvature of +2.50 cm. What is the focal length of the lens?

29. A converging lens has a focal length of 20.0 cm. Locate the image for object distances of (a) 40.0 cm, (b) 20.0 cm, and (c) 10.0 cm. For each case, state whether the image is real or virtual and upright or inverted. Find the magnification in each case.

30. An object located 32.0 cm in front of a lens forms an image on a screen 8.00 cm behind the lens. (a) Find the focal length of the lens. (b) Determine the magnification. (c) Is the lens converging or diverging?

31. ▲ The nickel's image in Figure P36.31 has twice the diameter of the nickel and is 2.84 cm from the lens. Determine the focal length of the lens.

Figure P36.31

32. Suppose an object has thickness dp so that it extends from object distance p to $p + dp$. Prove that the thickness dq of its image is given by $(-q^2/p^2)\,dp$. Then the longitudinal magnification is $dq/dp = -M^2$, where M is the lateral magnification.

33. An object is located 20.0 cm to the left of a diverging lens having a focal length $f = -32.0$ cm. Determine (a) the location and (b) the magnification of the image. (c) Construct a ray diagram for this arrangement.

34. The projection lens in a certain slide projector is a single thin lens. A slide 24.0 mm high is to be projected so that its image fills a screen 1.80 m high. The slide-to-screen distance is 3.00 m. (a) Determine the focal length of the projection lens. (b) How far from the slide should the lens of the projector be placed so as to form the image on the screen?

35. The use of a lens in a certain situation is described by the equation

$$\frac{1}{p} + \frac{1}{-3.50p} = \frac{1}{7.50\text{ cm}}$$

Determine (a) the object distance and (b) the image distance. (c) Use a ray diagram to obtain a description of the image. (d) Identify a practical device described by the given equation and write the statement of a problem for which the equation appears in the solution.

36. An antelope is at a distance of 20.0 m from a converging lens of focal length 30.0 cm. The lens forms an image of the animal. If the antelope runs away from the lens at a speed of 5.00 m/s, how fast does the image move? Does the image move toward or away from the lens?

37. ● An object is at a distance d to the left of a flat screen. A converging lens with focal length $f < d/4$ is placed between object and screen. (a) Show that two lens positions exist that form an image on the screen and determine how far these positions are from the object. (b) How do the two images differ from each other?

38. ● In Figure P36.38, a thin converging lens of focal length 14.0 cm forms an image of the square *abcd*, which is $h_c = h_b = 10.0$ cm high and lies between distances of $p_d = 20.0$ cm and $p_a = 30.0$ cm from the lens. (a) Let a', b', c', and d' represent the respective corners of the image. Let q_a represent the image distance for points a' and b', q_d represent the image distance for points c' and d', h'_b represent the distance from point b' to the axis, and h'_c represent the height of c'. Evaluate each of these quantities. Make a sketch of the image. (b) The area of the object is 100 cm². By carrying out the following steps, you will evaluate the area of the image. Let q represent the image distance of any point between a' and d', for which the object distance is p. Let h' represent the distance from the axis to the point at the edge of the image between b' and c' at image distance q. Demonstrate that

$$|h'| = (10\text{ cm})q\left(\frac{1}{14\text{ cm}} - \frac{1}{q}\right)$$

(c) Explain why the geometric area of the image is given by

$$\int_{q_a}^{q_d} |h'|\,dq$$

Carry out the integration to find the area of the image.

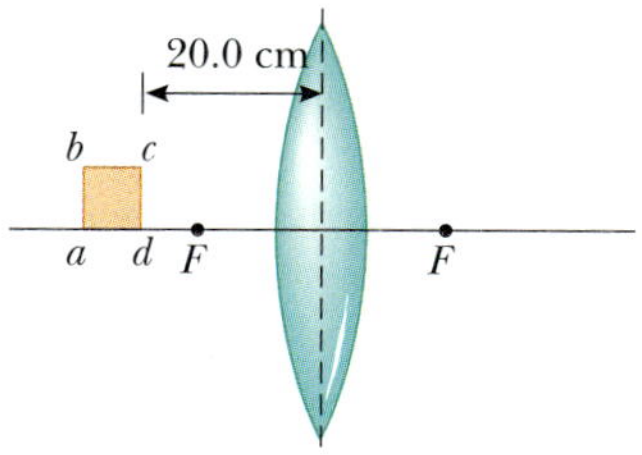

Figure P36.38

39. Figure 36.33 diagrams a cross section of a camera. It has a single lens of focal length 65.0 mm that is to form an image on the CCD at the back of the camera. Suppose the position of the lens has been adjusted to focus the image of a distant object. How far and in what direction must the lens be moved to form a sharp image of an object that is 2.00 m away?

Section 36.5 Lens Aberrations

40. The magnitudes of the radii of curvature are 32.5 cm and 42.5 cm for the two faces of a biconcave lens. The glass has index of refraction 1.53 for violet light and 1.51 for red light. For a very distant object, locate and describe (a) the image formed by violet light and (b) the image formed by red light.

41. Two rays traveling parallel to the principal axis strike a large plano-convex lens having an index of refraction of 1.60 (Fig. P36.41). If the convex face is spherical, a ray near the edge does not pass through the focal point (spherical aberration occurs). Assume this face has a radius of curvature of 20.0 cm and the two rays are at distances $h_1 = 0.500$ cm and $h_2 = 12.0$ cm from the principal axis. Find the difference Δx in the positions where each crosses the principal axis.

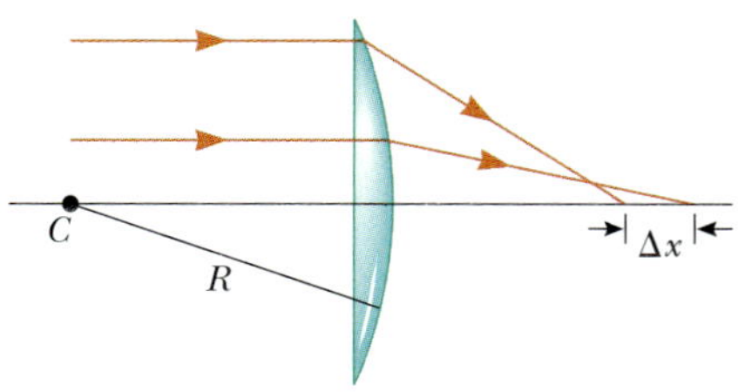

Figure P36.41

Section 36.6 The Camera

42. A camera is being used with a correct exposure at $f/4$ and a shutter speed of $\frac{1}{16}$ s. To photograph a rapidly moving subject, the shutter speed is changed to $\frac{1}{128}$ s. Find the new f-number setting needed to maintain satisfactory exposure.

Section 36.7 The Eye

43. A nearsighted person cannot see objects clearly beyond 25.0 cm (her far point). If she has no astigmatism and contact lenses are prescribed for her, what power and type of lens are required to correct her vision?

44. The accommodation limits for nearsighted Nick's eyes are 18.0 cm and 80.0 cm. When he wears his glasses, he can see faraway objects clearly. At what minimum distance is he able to see objects clearly?

Section 36.8 The Simple Magnifier

Section 36.9 The Compound Microscope

Section 36.10 The Telescope

45. A lens that has a focal length of 5.00 cm is used as a magnifying glass. (a) To obtain maximum magnification, where should the object be placed? (b) What is the magnification?

46. The distance between eyepiece and objective lens in a certain compound microscope is 23.0 cm. The focal length of the eyepiece is 2.50 cm and that of the objective is 0.400 cm. What is the overall magnification of the microscope?

47. The refracting telescope at the Yerkes Observatory has a 1.00-m diameter objective lens of focal length 20.0 m. Assume it is used with an eyepiece of focal length 2.50 cm. (a) Determine the magnification of Mars as seen through this telescope. (b) Are the Martian polar caps right side up or upside down?

48. ● Astronomers often take photographs with the objective lens or mirror of a telescope alone, without an eyepiece. (a) Show that the image size h' for such a telescope is given by $h' = fh/(f - p)$, where h is the object size, f is the objective focal length, and p is the object distance. (b) **What If?** Simplify the expression in part (a) for the case in which the object distance is much greater than objective focal length. (c) The "wingspan" of the International Space Station is 108.6 m, the overall width of its solar panel configuration. Find the width of the image formed by a telescope objective of focal length 4.00 m when the station is orbiting at an altitude of 407 km.

49. A certain telescope has an objective mirror with an aperture diameter of 200 mm and a focal length of 2 000 mm. It captures the image of a nebula on photographic film at its prime focus with an exposure time of 1.50 min. To produce the same light energy per unit area on the film, what is the required exposure time to photograph the same nebula with a smaller telescope that has an objective with a diameter of 60.0 mm and a focal length of 900 mm?

Additional Problems

50. A *zoom lens* system is a combination of lenses that produces a variable magnification of a fixed object as it maintains a fixed image position. The magnification is varied by moving one or more lenses along the axis. Multiple lenses are used in practice to obtain high-quality images, but the effect of zooming in on an object can be demonstrated with a simple two-lens system. An object, two converging lenses, and a screen are mounted on an optical bench. The first lens, which is to the right of the object, has a focal length of 5.00 cm, and the second lens, which is to the right of the first lens, has a focal length of 10.0 cm. The screen is to the right of the second lens. Initially, an object is situated at a distance of 7.50 cm to the left of the first lens, and the image formed on the screen has a magnification of +1.00. (a) Find the distance between the object and the screen. (b) Both lenses are now moved along their common axis, while the object and the screen maintain fixed positions, until the image formed on the screen has a magnification of +3.00. Find the displacement of each lens from its initial position in part (a). Can the lenses be displaced in more than one way?

51. The distance between an object and its upright image is 20.0 cm. If the magnification is 0.500, what is the focal length of the lens being used to form the image?

52. The distance between an object and its upright image is d. If the magnification is M, what is the focal length of the lens being used to form the image?

53. A real object is located at the zero end of a meterstick. A large concave mirror at the 100-cm end of the meterstick forms an image of the object at the 70.0-cm position. A small convex mirror placed at the 20.0-cm position forms a final image at the 10.0-cm point. What is the radius of curvature of the convex mirror?

54. The lens and mirror in Figure P36.54 have focal lengths of +80.0 cm and −50.0 cm, respectively. An object is placed 1.00 m to the left of the lens as shown. Locate the final image, formed by light that has gone through the lens twice. State whether the image is upright or inverted and determine the overall magnification.

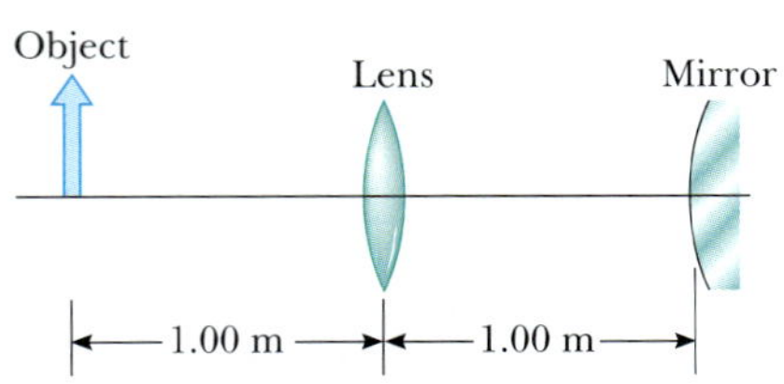

Figure P36.54

55. ● An object is originally at the $x_i = 0$ cm position of a meterstick located on the x axis. A converging lens of focal length 26.0 cm is fixed at the position 32.0 cm. Then we gradually slide the object to the position $x_f =$ 12.0 cm. Find the location x' of the object's image as a function of the object position x. Describe the pattern of the motion of the image with reference to a graph or a table of values. As the object moves 12 cm to the right, how far does the image move? In what direction or directions?

56. The object in Figure P36.56 is midway between the lens and the mirror. The mirror's radius of curvature is 20.0 cm, and the lens has a focal length of −16.7 cm. Considering only the light that leaves the object and travels first toward the mirror, locate the final image formed by this system. Is this image real or virtual? Is it upright or inverted? What is the overall magnification?

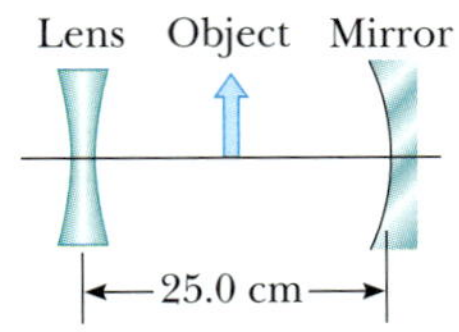

Figure P36.56

57. ● In many applications, it is necessary to expand or decrease the diameter of a beam of parallel rays of light. This change can be made by using a converging lens and a diverging lens in combination. Suppose you have a converging lens of focal length 21.0 cm and a diverging lens of focal length −12.0 cm. How can you arrange these lenses to increase the diameter of a beam of parallel rays? By what factor will the diameter increase?

58. The lens-makers' equation applies to a lens immersed in a liquid if n in the equation is replaced by n_2/n_1. Here n_2 refers to the index of refraction of the lens material and n_1 is that of the medium surrounding the lens. (a) A certain lens has focal length 79.0 cm in air and index of refraction 1.55. Find its focal length in water. (b) A certain mirror has focal length 79.0 cm in air. Find its focal length in water.

59. ▲ A parallel beam of light enters a glass hemisphere perpendicular to the flat face as shown in Figure P36.59. The

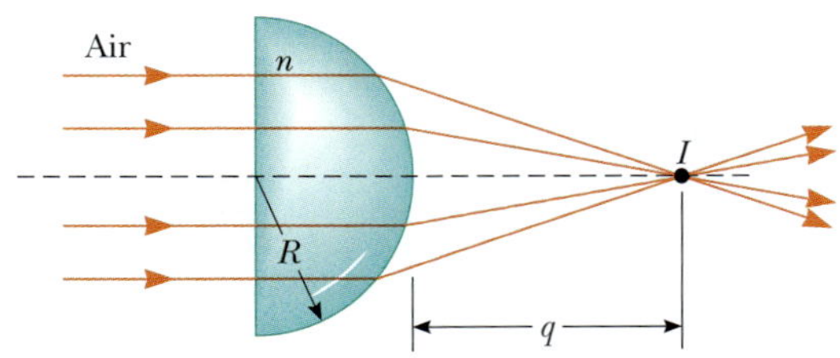

Figure P36.59

magnitude of the radius is 6.00 cm, and the index of refraction is 1.560. Determine the point at which the beam is focused. (Assume paraxial rays.)

60. **Review problem.** A spherical lightbulb of diameter 3.20 cm radiates light equally in all directions, with power 4.50 W. (a) Find the light intensity at the surface of the lightbulb. (b) Find the light intensity 7.20 m away from the center of the lightbulb. (c) At this 7.20-m distance, a lens is set up with its axis pointing toward the lightbulb. The lens has a circular face with a diameter 15.0 cm and has a focal length of 35.0 cm. Find the diameter of the image of the lightbulb. (d) Find the light intensity at the image.

61. An object is placed 12.0 cm to the left of a diverging lens of focal length −6.00 cm. A converging lens of focal length 12.0 cm is placed a distance d to the right of the diverging lens. Find the distance d so that the final image is at infinity. Draw a ray diagram for this case.

62. Assume the intensity of sunlight is 1.00 kW/m^2 at a particular location. A highly reflecting concave mirror is to be pointed toward the Sun to produce a power of at least 350 W at the image. (a) Find the required radius R_a of the circular face area of the mirror. (b) Now suppose the light intensity is to be at least 120 kW/m^2 at the image. Find the required relationship between R_a and the radius of curvature R of the mirror. The disk of the Sun subtends an angle of 0.533° at the Earth.

63. ▲ The disk of the Sun subtends an angle of 0.533° at the Earth. What are the position and diameter of the solar image formed by a concave spherical mirror with a radius of curvature of 3.00 m?

64. ● Figure P36.64 shows a thin converging lens for which the radii of curvature are $R_1 = 9.00$ cm and $R_2 = -11.0$ cm. The lens is in front of a concave spherical mirror with the radius of curvature $R = 8.00$ cm. (a) Assume its focal points F_1 and F_2 are 5.00 cm from the center of the lens. Determine its index of refraction. (b) The lens and mirror are 20.0 cm apart, and an object is placed 8.00 cm to the left of the lens. Determine the position of the final image and its magnification as seen by the eye in the figure. (c) Is the final image inverted or upright? Explain.

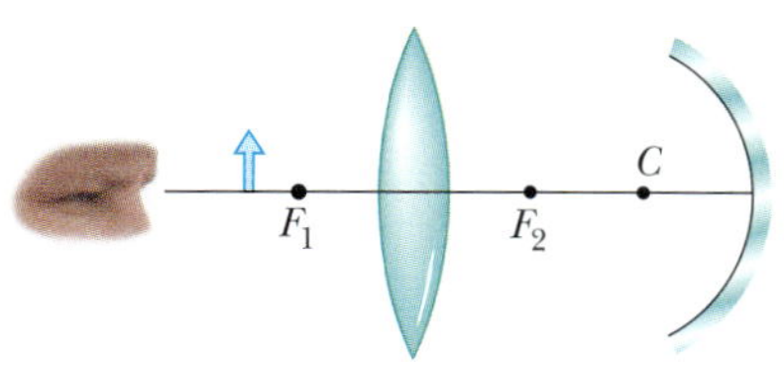

Figure P36.64

65. In a darkened room, a burning candle is placed 1.50 m from a white wall. A lens is placed between candle and wall at a location that causes a larger, inverted image to form on the wall. When the lens is moved 90.0 cm toward the wall, another image of the candle is formed. Find (a) the two object distances that produce the specified images and (b) the focal length of the lens. (c) Characterize the second image.

66. ● A floating strawberry illusion is achieved with two parabolic mirrors, each having a focal length 7.50 cm, facing each other so that their centers are 7.50 cm apart (Fig. P36.66). If a strawberry is placed on the lower mirror, an image of the strawberry is formed at the small opening at the center of the top mirror. Show that the final image is formed at that location and describe its characteristics. *Note:* A very startling effect is to shine a flashlight beam on this image. Even at a glancing angle, the incoming light beam is seemingly reflected from the image! Do you understand why?

Figure P36.66

67. An object 2.00 cm high is placed 40.0 cm to the left of a converging lens having a focal length of 30.0 cm. A diverging lens with a focal length of −20.0 cm is placed 110 cm to the right of the converging lens. (a) Determine the position and magnification of the final image. (b) Is the image upright or inverted? (c) **What If?** Repeat parts (a) and (b) for the case in which the second lens is a converging lens having a focal length of +20.0 cm.

68. Two lenses made of kinds of glass having different indices of refraction n_1 and n_2 are cemented together to form an *optical doublet.* Optical doublets are often used to correct chromatic aberrations in optical devices. The first lens of a certain doublet has one flat side and one concave side with a radius of curvature of magnitude R. The second lens has two convex sides with radii of curvature also of magnitude R. Show that the doublet can be modeled as a single thin lens with a focal length described by

$$\frac{1}{f} = \frac{2n_2 - n_1 - 1}{R}$$

Answers to Quick Quizzes

36.1 False. The water spots are 2 m away from you, and your image is 4 m away. You cannot focus your eyes on both at the same time.

36.2 (b). A concave mirror focuses the light from a large area of the mirror onto a small area of the paper, resulting in a very high power input to the paper.

36.3 (b). A convex mirror always forms an image with a magnification less than 1, so the mirror must be concave. In a concave mirror, only virtual images are upright. This particular photograph is of the Hubble Space Telescope primary mirror. The scientists acting as the object for the image are to the left of the photograph and not visible to us.

36.4 (d). When O is far away, the rays refract into the material of index n_2 and converge to form a real image as in Figure 36.16. For certain combinations of R and n_2 as O moves very close to the refracting surface, the incident angle of the rays increases so much that rays are no longer refracted back toward the principal axis. The result is a virtual image as shown in the next column.

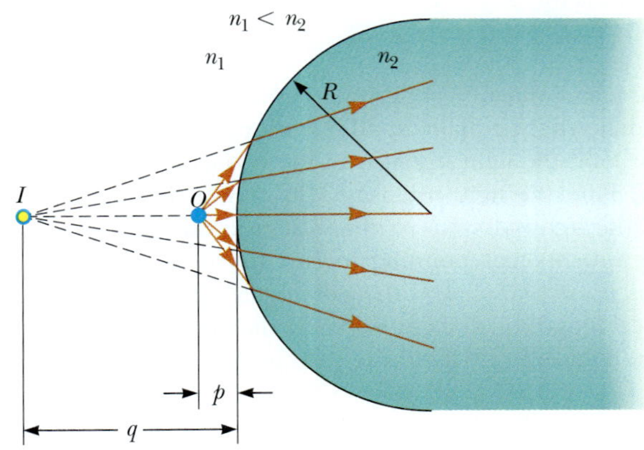

36.5 (a). No matter where O is, the rays refract into the air away from the normal and form a virtual image between O and the surface.

36.6 (b). Because the flat surfaces of the plane have infinite radii of curvature, Equation 36.15 indicates that the focal length is also infinite. Parallel rays striking the plane focus at infinity, which means that they remain parallel after passing through the glass.

36.7 (a). If the object is brought closer to the lens, the image moves farther away from the lens, behind the plane of the CCD. To bring the image back up to the CCD, the lens is moved toward the object and away from the CCD.

36.8 (c). The Sun's rays must converge onto the paper. A far-sighted person wears converging lenses.

The colors in many of a hummingbird's feathers are not due to pigment. The *iridescence* that makes the brilliant colors that often appear on the bird's throat and belly is due to an interference effect caused by structures in the feathers. The colors will vary with the viewing angle. (RO-MA/Index Stock Imagery)

37 Interference of Light Waves

In Chapter 36, we studied light rays passing through a lens or reflecting from a mirror to describe the formation of images. This discussion completed our study of *geometric optics.* In this chapter and in Chapter 38, we are concerned with *wave optics* or *physical optics,* the study of interference, diffraction, and polarization of light. These phenomena cannot be adequately explained with the ray optics used in Chapters 35 and 36. We now learn how treating light as waves rather than as rays leads to a satisfying description of such phenomena.

37.1 Conditions for Interference

In Chapter 18, we studied the waves in interference model and found that the superposition of two mechanical waves can be constructive or destructive. In constructive interference, the amplitude of the resultant wave is greater than that of either individual wave, whereas in destructive interference, the resultant amplitude is less than that of the larger wave. Light waves also interfere with one another. Fundamentally, all interference associated with light waves arises when the electromagnetic fields that constitute the individual waves combine.

If two lightbulbs are placed side by side so that light from both bulbs combines, no interference effects are observed because the light waves from one bulb are emitted independently of those from the other bulb. The emissions from the two lightbulbs do not maintain a constant phase relationship with each other over time. Light waves from an ordinary source such as a lightbulb undergo random phase changes in time intervals of less than a nanosecond. Therefore, the conditions for constructive interference, destructive interference, or some intermediate state are maintained only for such short time intervals. Because the eye cannot follow such rapid changes, no interference effects are observed. Such light sources are said to be **incoherent.**

To observe interference of waves from two sources, the following conditions must be met:

Conditions for interference ▶

- The sources must be **coherent**; that is, **they must maintain a constant phase with respect to each other.**
- The sources should be **monochromatic**; that is, they should be of a single wavelength.

As an example, single-frequency sound waves emitted by two side-by-side loudspeakers driven by a single amplifier can interfere with each other because the two speakers are coherent. In other words, they respond to the amplifier in the same way at the same time.

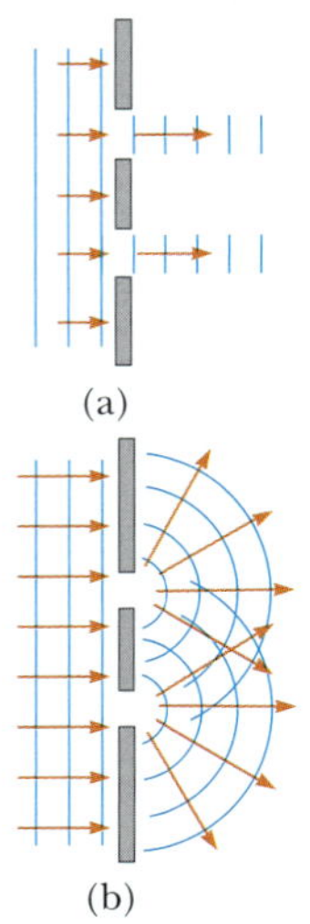

Figure 37.1 (a) If light waves did not spread out after passing through the slits, no interference would occur. (b) The light waves from the two slits overlap as they spread out, filling what we expect to be shadowed regions with light and producing interference fringes on a screen placed to the right of the slits.

37.2 Young's Double-Slit Experiment

A common method for producing two coherent light sources is to use a monochromatic source to illuminate a barrier containing two small openings, usually in the shape of slits. The light emerging from the two slits is coherent because a single source produces the original light beam and the two slits serve only to separate the original beam into two parts (which, after all, is what is done to the sound signal from two side-by-side loudspeakers). Any random change in the light emitted by the source occurs in both beams at the same time. As a result, interference effects can be observed when the light from the two slits arrives at a viewing screen.

If the light traveled only in its original direction after passing through the slits as shown in Figure 37.1a, the waves would not overlap and no interference pattern would be seen. Instead, as we have discussed in our treatment of Huygens's principle (Section 35.6), the waves spread out from the slits as shown in Figure 37.1b. In other words, the light deviates from a straight-line path and enters the region that

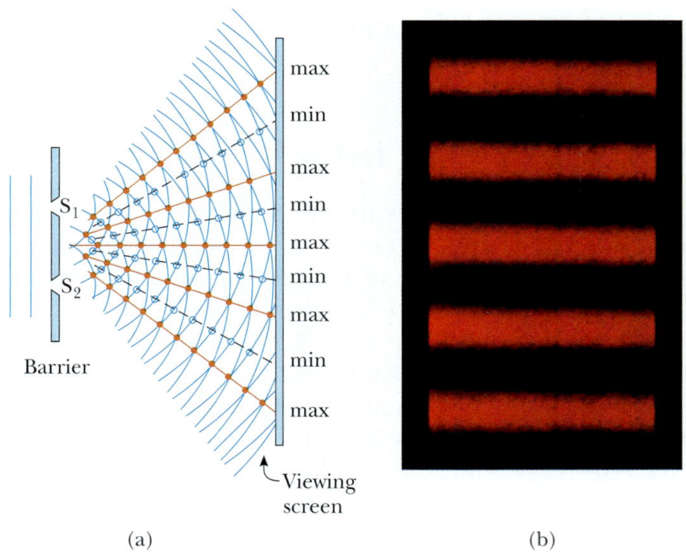

M. Cagnet, M. Françon, J. C. Thrierr, *Atlas of Optical Phenomena*, Berlin, Springer-Verlag, 1962

ACTIVE FIGURE 37.2

(a) Schematic diagram of Young's double-slit experiment. Slits S_1 and S_2 behave as coherent sources of light waves that produce an interference pattern on the viewing screen (drawing not to scale). (b) An enlargement of the center of a fringe pattern formed on the viewing screen.

Sign in at www.thomsonedu.com and go to ThomsonNOW to adjust the slit separation and the wavelength of the light and see the effect on the interference pattern.

Figure 37.3 An interference pattern involving water waves is produced by two vibrating sources at the water's surface. The pattern is analogous to that observed in Young's double-slit experiment. Notice the regions of constructive (*A*) and destructive (*B*) interference.

would otherwise be shadowed. As noted in Section 35.3, this divergence of light from its initial line of travel is called **diffraction.**

Interference in light waves from two sources was first demonstrated by Thomas Young in 1801. A schematic diagram of the apparatus Young used is shown in Active Figure 37.2a. Plane light waves arrive at a barrier that contains two parallel slits S_1 and S_2. The light from S_1 and S_2 produces on a viewing screen a visible pattern of bright and dark parallel bands called **fringes** (Active Fig. 37.2b). When the light from S_1 and that from S_2 both arrive at a point on the screen such that constructive interference occurs at that location, a bright fringe appears. When the light from the two slits combines destructively at any location on the screen, a dark fringe results. Figure 37.3 is a photograph of an interference pattern produced by two coherent vibrating sources in a water tank.

Figure 37.4 shows some of the ways in which two waves can combine at the screen. In Figure 37.4a, the two waves, which leave the two slits in phase, strike the screen at the central point *O*. Because both waves travel the same distance, they arrive at *O* in phase. As a result, constructive interference occurs at this location and a bright fringe is observed. In Figure 37.4b, the two waves also start in phase, but here the lower wave has to travel one wavelength farther than the upper wave to reach point *P*. Because the lower wave falls behind the upper one by exactly one wavelength, they still arrive in phase at *P* and a second bright fringe appears at this location. At point *R* in Figure 37.4c, however, between points *O* and *P*, the lower wave has fallen half a wavelength behind the upper wave and a trough of the upper wave overlaps a crest of the lower wave, giving rise to destructive interference at point *R*. A dark fringe is therefore observed at this location.

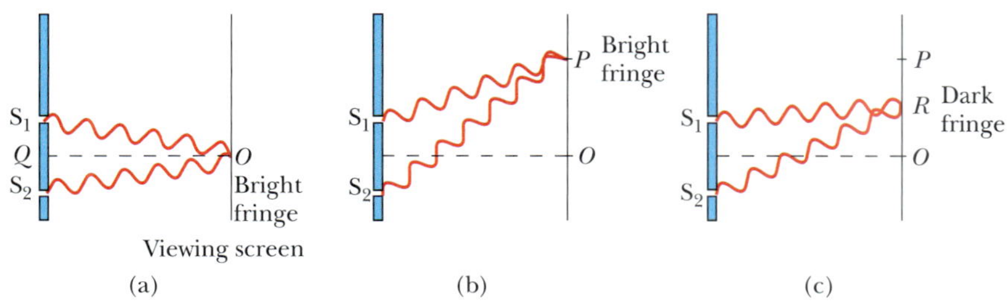

Figure 37.4 (a) Constructive interference occurs at point *O* when the waves combine. (b) Constructive interference also occurs at point *P*. (c) Destructive interference occurs at point *R* when the two waves combine because the lower wave falls one-half a wavelength behind the upper wave. (All figures not to scale.)

37.3 Light Waves in Interference

We can describe Young's experiment quantitatively with the help of Figure 37.5. The viewing screen is located a perpendicular distance L from the barrier containing two slits, S_1 and S_2 (Fig. 37.5a). These slits are separated by a distance d, and the source is monochromatic. To reach any arbitrary point P in the upper half of the screen, a wave from the lower slit must travel farther than a wave from the upper slit by a distance $d \sin \theta$ (Fig. 37.5b). This distance is called the **path difference** δ (Greek letter delta). If we assume the rays labeled r_1 and r_2 are parallel, which is approximately true if L is much greater than d, then δ is given by

Path difference ▶

$$\delta = r_2 - r_1 = d \sin \theta \tag{37.1}$$

The value of δ determines whether the two waves are in phase when they arrive at point P. If δ is either zero or some integer multiple of the wavelength, the two waves are in phase at point P and constructive interference results. Therefore, the condition for bright fringes, or **constructive interference,** at point P is

Conditions for constructive interference ▶

$$d \sin \theta_{\text{bright}} = m\lambda \quad (m = 0, \pm 1, \pm 2, \cdots) \tag{37.2}$$

The number m is called the **order number.** For constructive interference, the order number is the same as the number of wavelengths that represents the path difference between the waves from the two slits. The central bright fringe at $\theta_{\text{bright}} = 0$ is called the *zeroth-order maximum*. The first maximum on either side, where $m = \pm 1$, is called the *first-order maximum*, and so forth.

When δ is an odd multiple of $\lambda/2$, the two waves arriving at point P are 180° out of phase and give rise to destructive interference. Therefore, the condition for dark fringes, or **destructive interference,** at point P is

Conditions for destructive interference ▶

$$d \sin \theta_{\text{dark}} = (m + \tfrac{1}{2})\lambda \quad (m = 0, \pm 1, \pm 2, \cdots) \tag{37.3}$$

These equations provide the *angular* positions of the fringes. It is also useful to obtain expressions for the *linear* positions measured along the screen from O to P. From the triangle OPQ in Figure 37.5a, we see that

$$\tan \theta = \frac{y}{L} \tag{37.4}$$

Using this result, the linear positions of bright and dark fringes are given by

$$y_{\text{bright}} = L \tan \theta_{\text{bright}} \tag{37.5}$$

$$y_{\text{dark}} = L \tan \theta_{\text{dark}} \tag{37.6}$$

where θ_{bright} and θ_{dark} are given by Equations 37.2 and 37.3.

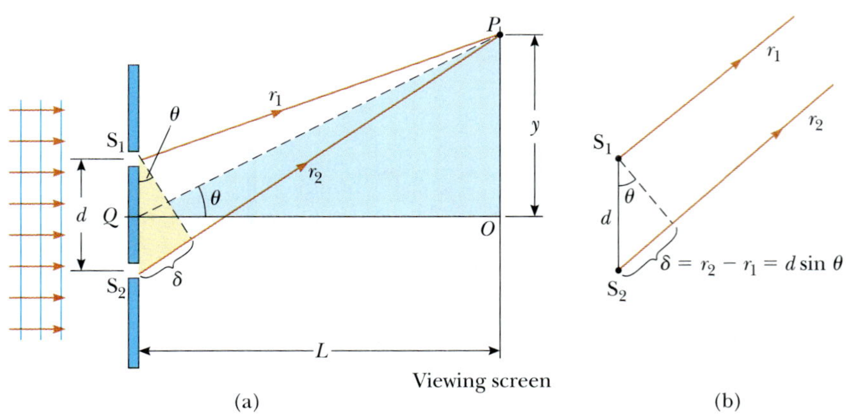

Figure 37.5 (a) Geometric construction for describing Young's double-slit experiment (not to scale). (b) When we assume r_1 is parallel to r_2, the path difference between the two rays is $r_2 - r_1 = d \sin \theta$. For this approximation to be valid, it is essential that $L \gg d$.

When the angles to the fringes are small, the positions of the fringes are linear near the center of the pattern. This can be verified by noting that for small angles, $\tan\theta \approx \sin\theta$, so Equation 37.5 gives the positions of the bright fringes as $y_{\text{bright}} = L\sin\theta_{\text{bright}}$. Incorporating Equation 37.2 gives

$$y_{\text{bright}} = L\left(\frac{m\lambda}{d}\right) \quad \text{(small angles)} \tag{37.7}$$

This result shows that y_{bright} is linear in the order number m, so the fringes are equally spaced.

As demonstrated in Example 37.1, Young's double-slit experiment provides a method for measuring the wavelength of light. In fact, Young used this technique to do precisely that. In addition, his experiment gave the wave model of light a great deal of credibility. It was inconceivable that particles of light coming through the slits could cancel one another in a way that would explain the dark fringes.

The principles discussed in this section are the basis of the **waves in interference** analysis model. This model was applied to mechanical waves in one dimension in Chapter 18. Here we see the details of applying this model in three dimensions to light.

John W. Jewett Jr.

The faint pastel-colored bows beneath the main rainbow are called *supernumerary bows.* They are formed by interference between rays of light leaving raindrops at angles slightly smaller than the angle of maximum intensity. (See Section 35.7 for a discussion of the rainbow.)

Quick Quiz 37.1 Which of the following causes the fringes in a two-slit interference pattern to move farther apart? (a) decreasing the wavelength of the light (b) decreasing the screen distance L (c) decreasing the slit spacing d (d) immersing the entire apparatus in water

EXAMPLE 37.1 Measuring the Wavelength of a Light Source

A viewing screen is separated from a double slit by 1.2 m. The distance between the two slits is 0.030 mm. Monochromatic light is directed toward the double slit and forms an interference pattern on the screen. The second-order bright fringe ($m = 2$) is 4.5 cm from the center line on the screen.

(A) Determine the wavelength of the light.

SOLUTION

Conceptualize Study Figure 37.5 to be sure you understand the phenomenon of interference of light waves.

Categorize We evaluate results using equations developed in this section, so we categorize this example as a substitution problem.

Solve Equation 37.7 for the wavelength and substitute numerical values:

$$\lambda = \frac{y_{\text{bright}}d}{mL} = \frac{(4.5\times10^{-2}\text{ m})(3.0\times10^{-5}\text{ m})}{2(1.2\text{ m})}$$

$$= 5.6\times10^{-7}\text{ m} = 560\text{ nm}$$

(B) Calculate the distance between adjacent bright fringes.

SOLUTION

Find the distance between adjacent bright fringes from Equation 37.7 and the results of part (A):

$$y_{m+1} - y_m = L\frac{(m+1)\lambda}{d} - L\left(\frac{m\lambda}{d}\right)$$

$$= L\left(\frac{\lambda}{d}\right) = 1.2\text{ m}\left(\frac{5.6\times10^{-7}\text{ m}}{3.0\times10^{-5}\text{ m}}\right)$$

$$= 2.2\times10^{-2}\text{ m} = 2.2\text{ cm}$$

EXAMPLE 37.2 Separating Double-Slit Fringes of Two Wavelengths

A light source emits visible light of two wavelengths: $\lambda = 430$ nm and $\lambda' = 510$ nm. The source is used in a double-slit interference experiment in which $L = 1.50$ m and $d = 0.025\,0$ mm. Find the separation distance between the third-order bright fringes for the two wavelengths.

SOLUTION

Conceptualize In Figure 37.5a, imagine light of two wavelengths incident on the slits and forming two interference patterns on the screen.

Categorize We evaluate results using equations developed in this section, so we categorize this example as a substitution problem.

Use Equation 37.7, with $m = 3$, to find the fringe positions corresponding to these two wavelengths:

$$y_{\text{bright}} = L\left(\frac{m\lambda}{d}\right) = L\left(\frac{3\lambda}{d}\right) = 1.50 \text{ m}\left[\frac{3(430 \times 10^{-9} \text{ m})}{0.025\,0 \times 10^{-3} \text{ m}}\right]$$

$$= 7.74 \times 10^{-2} \text{ m}$$

$$y'_{\text{bright}} = L\left(\frac{m\lambda'}{d}\right) = L\left(\frac{3\lambda'}{d}\right) = 1.50 \text{ m}\left[\frac{3(510 \times 10^{-9} \text{ m})}{0.025\,0 \times 10^{-3} \text{ m}}\right]$$

$$= 9.18 \times 10^{-2} \text{ m}$$

Evaluate the separation distance between the two fringes:

$$\Delta y = 9.18 \times 10^{-2} \text{ m} - 7.74 \times 10^{-2} \text{ m}$$

$$= 1.44 \times 10^{-2} \text{ m} = 1.44 \text{ cm}$$

What If? What if we examine the entire interference pattern due to the two wavelengths and look for overlapping fringes? Are there any locations on the screen where the bright fringes from the two wavelengths overlap exactly?

Answer Find such a location by setting the location of any bright fringe due to λ equal to one due to λ', using Equation 37.7:

$$L\left(\frac{m\lambda}{d}\right) = L\left(\frac{m'\lambda'}{d}\right) \rightarrow \frac{m'}{m} = \frac{\lambda}{\lambda'}$$

Substitute the wavelengths:

$$\frac{m'}{m} = \frac{430 \text{ nm}}{510 \text{ nm}} = \frac{43}{51}$$

Use Equation 37.7 to find the value of y for these fringes:

$$y = 1.50 \text{ m}\left[\frac{51(430 \times 10^{-9} \text{ m})}{0.025\,0 \times 10^{-3} \text{ m}}\right] = 1.32 \text{ m}$$

This value of y is comparable to L, so the small-angle approximation used for Equation 37.7 is *not* valid. This conclusion suggests we should not expect Equation 37.7 to give us the correct result. If you use Equation 37.5, you can show that the bright fringes do indeed overlap when the same condition, $m'/m = \lambda/\lambda'$, is met (see Problem 38). Therefore, the 51st fringe of the 430-nm light does overlap with the 43rd fringe of the 510-nm light, but not at the location of 1.32 m. You are asked to find the correct location as part of Problem 38.

37.4 Intensity Distribution of the Double-Slit Interference Pattern

Notice that the edges of the bright fringes in Active Figure 37.2b are not sharp; rather, there is a gradual change from bright to dark. So far, we have discussed the locations of only the centers of the bright and dark fringes on a distant screen. Let's now direct our attention to the intensity of the light at other points between the positions of maximum constructive and destructive interference. In other words, we now calculate the distribution of light intensity associated with the double-slit interference pattern.

Again, suppose the two slits represent coherent sources of sinusoidal waves such that the two waves from the slits have the same angular frequency ω and are in phase. The total magnitude of the electric field at point P on the screen in Figure 37.6 is the superposition of the two waves. Assuming that the two waves have the same amplitude E_0, we can write the magnitude of the electric field at point P due to each wave separately as

$$E_1 = E_0 \sin \omega t \quad \text{and} \quad E_2 = E_0 \sin (\omega t + \phi) \tag{37.8}$$

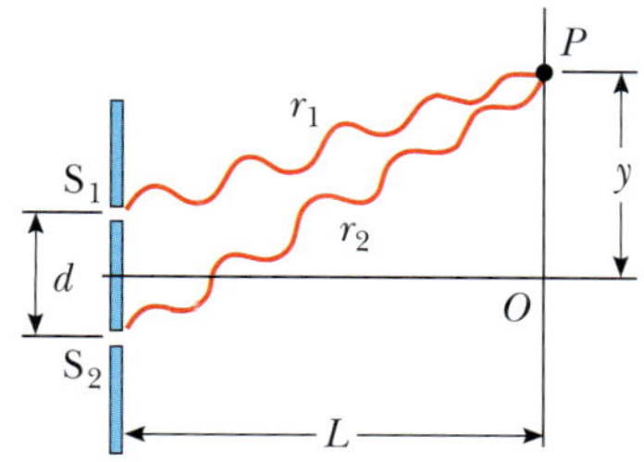

Figure 37.6 Construction for analyzing the double-slit interference pattern. A bright fringe, or intensity maximum, is observed at O.

Although the waves are in phase at the slits, *their phase difference ϕ at P depends on the path difference* $\delta = r_2 - r_1 = d \sin \theta$. A path difference of λ (for constructive interference) corresponds to a phase difference of 2π rad. A path difference of δ is the same fraction of λ as the phase difference ϕ is of 2π. We can describe this fraction mathematically with the ratio

$$\frac{\delta}{\lambda} = \frac{\phi}{2\pi}$$

which gives

$$\phi = \frac{2\pi}{\lambda}\delta = \frac{2\pi}{\lambda} d \sin \theta \tag{37.9}$$

◀ Phase difference

This equation shows how the phase difference ϕ depends on the angle θ in Figure 37.5.

Using the superposition principle and Equation 37.8, we obtain the following expression for the magnitude of the resultant electric field at point P:

$$E_P = E_1 + E_2 = E_0[\sin \omega t + \sin (\omega t + \phi)] \tag{37.10}$$

We can simplify this expression by using the trigonometric identity

$$\sin A + \sin B = 2 \sin \left(\frac{A + B}{2}\right) \cos \left(\frac{A - B}{2}\right)$$

Taking $A = \omega t + \phi$ and $B = \omega t$, Equation 37.10 becomes

$$E_P = 2E_0 \cos \left(\frac{\phi}{2}\right) \sin \left(\omega t + \frac{\phi}{2}\right) \tag{37.11}$$

This result indicates that the electric field at point P has the same frequency ω as the light at the slits but that the amplitude of the field is multiplied by the factor $2 \cos (\phi/2)$. To check the consistency of this result, note that if $\phi = 0, 2\pi, 4\pi, \ldots,$ the magnitude of the electric field at point P is $2E_0$, corresponding to the condition for maximum constructive interference. These values of ϕ are consistent with Equation 37.2 for constructive interference. Likewise, if $\phi = \pi, 3\pi, 5\pi, \ldots,$ the magnitude of the electric field at point P is zero, which is consistent with Equation 37.3 for total destructive interference.

Finally, to obtain an expression for the light intensity at point P, recall from Section 34.4 that *the intensity of a wave is proportional to the square of the resultant electric field magnitude at that point* (Eq. 34.24). Using Equation 37.11, we can therefore express the light intensity at point P as

$$I \propto E_P{}^2 = 4E_0{}^2 \cos^2 \left(\frac{\phi}{2}\right) \sin^2 \left(\omega t + \frac{\phi}{2}\right)$$

Most light-detecting instruments measure time-averaged light intensity, and the time-averaged value of $\sin^2 (\omega t + \phi/2)$ over one cycle is $\frac{1}{2}$. (See Fig. 33.5.) Therefore, we can write the average light intensity at point P as

$$I = I_{\text{max}} \cos^2 \left(\frac{\phi}{2}\right) \tag{37.12}$$

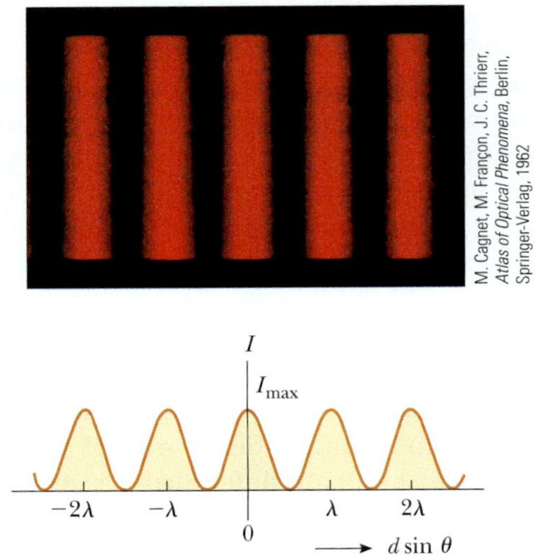

Figure 37.7 Light intensity versus $d \sin \theta$ for a double-slit interference pattern when the screen is far from the two slits ($L \gg d$).

where I_{max} is the maximum intensity on the screen and the expression represents the time average. Substituting the value for ϕ given by Equation 37.9 into this expression gives

$$I = I_{max} \cos^2\left(\frac{\pi d \sin \theta}{\lambda}\right) \quad \textbf{(37.13)}$$

Alternatively, because $\sin \theta \approx y/L$ for small values of θ in Figure 37.5, we can write Equation 37.13 in the form

$$I = I_{max} \cos^2\left(\frac{\pi d}{\lambda L} y\right) \quad \textbf{(37.14)}$$

Constructive interference, which produces light intensity maxima, occurs when the quantity $\pi\, dy/\lambda L$ is an integral multiple of π, corresponding to $y = (\lambda L/d)\, m$. This result is consistent with Equation 37.7.

A plot of light intensity versus $d \sin \theta$ is given in Figure 37.7. The interference pattern consists of equally spaced fringes of equal intensity. Remember, however, that this result is valid only if the slit-to-screen distance L is much greater than the slit separation and only for small values of θ.

Figure 37.8 shows similar plots of light intensity versus $d \sin \theta$ for light passing through multiple slits. For more than two slits, the pattern contains primary and secondary maxima. For three slits, notice that the primary maxima are nine times more intense than the secondary maxima as measured by the height of the curve because the intensity varies as E^2. For N slits, the intensity of the primary maxima is N^2 times greater than that due to a single slit. As the number of slits increases, the primary maxima increase in intensity and become narrower, while the secondary maxima decrease in intensity relative to the primary maxima. Figure 37.8 also shows that as the number of slits increases, the number of secondary maxima also increases. In fact, the number of secondary maxima is always $N - 2$, where N is the number of slits. In Section 38.4, we shall investigate the pattern for a very large number of slits in a device called a *diffraction grating.*

Quick Quiz 37.2 Using Figure 37.8 as a model, sketch the interference pattern from six slits.

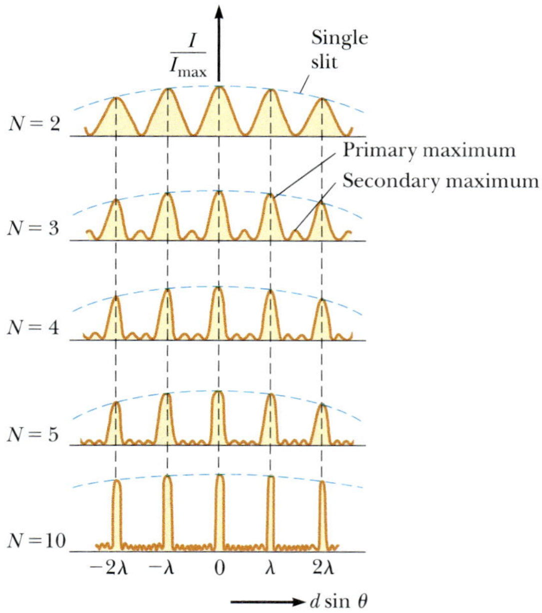

Figure 37.8 Multiple-slit interference patterns. As *N*, the number of slits, is increased, the primary maxima (the tallest peaks in each graph) become narrower but remain fixed in position and the number of secondary maxima increases. For any value of *N*, the decrease in intensity in maxima to the left and right of the central maximum, indicated by the blue dashed arcs, is due to *diffraction patterns* from the individual slits, which are discussed in Chapter 38.

37.5 Change of Phase Due to Reflection

Young's method for producing two coherent light sources involves illuminating a pair of slits with a single source. Another simple, yet ingenious, arrangement for producing an interference pattern with a single light source is known as *Lloyd's mirror*[1] (Fig. 37.9). A point light source S is placed close to a mirror, and a viewing screen is positioned some distance away and perpendicular to the mirror. Light waves can reach point P on the screen either directly from S to P or by the path involving reflection from the mirror. The reflected ray can be treated as a ray originating from a virtual source S′. As a result, we can think of this arrangement as a double-slit source with the distance between sources S and S′ comparable to length d in Figure 37.5. Hence, at observation points far from the source ($L \gg d$), we expect waves from S and S′ to form an interference pattern exactly like the one formed by two real coherent sources. An interference pattern is indeed observed. The positions of the dark and bright fringes, however, are reversed relative to the pattern created by two real coherent sources (Young's experiment). Such a reversal can only occur if the coherent sources S and S′ differ in phase by 180°.

To illustrate further, consider point P', the point where the mirror intersects the screen. This point is equidistant from sources S and S′. If path difference alone were responsible for the phase difference, we would see a bright fringe at P' (because the path difference is zero for this point), corresponding to the central bright fringe of the two-slit interference pattern. Instead, a dark fringe is observed at P'. We therefore conclude that a 180° phase change must be produced by reflection from the mirror. In general, **an electromagnetic wave undergoes a phase change of 180° upon reflection from a medium that has a higher index of refraction than the one in which the wave is traveling.**

It is useful to draw an analogy between reflected light waves and the reflections of a transverse wave pulse on a stretched string (Section 16.4). The reflected pulse on a string undergoes a phase change of 180° when reflected from the boundary

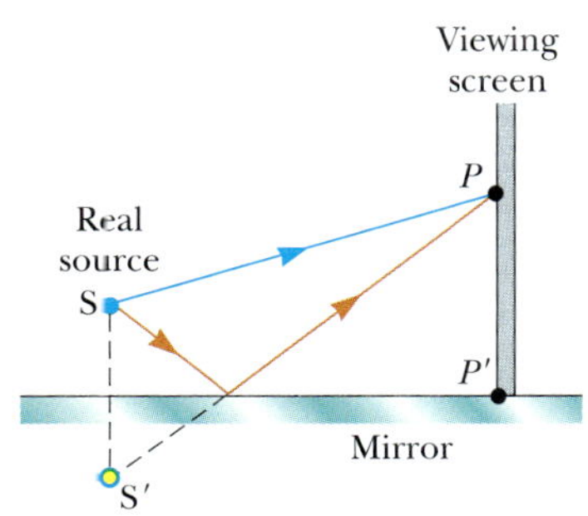

Figure 37.9 Lloyd's mirror. An interference pattern is produced at point P on the screen as a result of the combination of the direct ray (blue) and the reflected ray (brown). The reflected ray undergoes a phase change of 180°.

[1] Developed in 1834 by Humphrey Lloyd (1800–1881), Professor of Natural and Experimental Philosophy, Trinity College, Dublin.

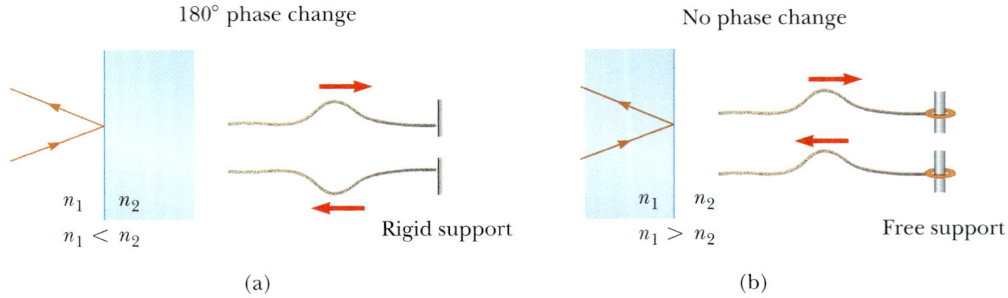

Figure 37.10 (a) For $n_1 < n_2$, a light ray traveling in medium 1 when reflected from the surface of medium 2 undergoes a 180° phase change. The same thing happens with a reflected pulse traveling along a string fixed at one end. (b) For $n_1 > n_2$, a light ray traveling in medium 1 undergoes no phase change when reflected from the surface of medium 2. The same is true of a reflected wave pulse on a string whose supported end is free to move.

of a denser medium, but no phase change occurs when the pulse is reflected from the boundary of a less dense medium. Similarly, an electromagnetic wave undergoes a 180° phase change when reflected from a boundary leading to an optically denser medium (defined as a medium with a higher index of refraction), but no phase change occurs when the wave is reflected from a boundary leading to a less dense medium. These rules, summarized in Figure 37.10, can be deduced from Maxwell's equations, but the treatment is beyond the scope of this text.

37.6 Interference in Thin Films

Interference effects are commonly observed in thin films, such as thin layers of oil on water or the thin surface of a soap bubble. The varied colors observed when white light is incident on such films result from the interference of waves reflected from the two surfaces of the film.

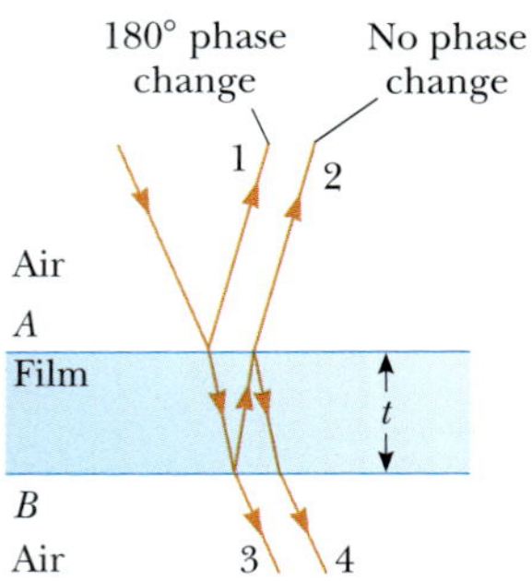

Figure 37.11 Interference in light reflected from a thin film is due to a combination of rays 1 and 2 reflected from the upper and lower surfaces of the film. Rays 3 and 4 lead to interference effects for light transmitted through the film.

Consider a film of uniform thickness t and index of refraction n as shown in Figure 37.11. Let's assume the light rays traveling in air are nearly normal to the two surfaces of the film. The wavelength of light λ_n in the film (see Section 35.5) is

$$\lambda_n = \frac{\lambda}{n}$$

where λ is the wavelength of the light in free space and n is the index of refraction of the film material.

Reflected ray 1, which is reflected from the upper surface (A) in Figure 37.11, undergoes a phase change of 180° with respect to the incident wave. Reflected ray 2, which is reflected from the lower film surface (B), undergoes no phase change because it is reflected from a medium (air) that has a lower index of refraction. Therefore, ray 1 is 180° out of phase with ray 2, which is equivalent to a path difference of $\lambda_n/2$. We must also consider, however, that ray 2 travels an extra distance $2t$ before the waves recombine in the air above surface A. (Remember that we are considering light rays that are close to normal to the surface. If the rays are not close to normal, the path difference is larger than $2t$.) If $2t = \lambda_n/2$, rays 1 and 2 recombine in phase and the result is constructive interference. In general, the condition for *constructive* interference in thin films is[2]

$$2t = (m + \tfrac{1}{2})\lambda_n \quad (m = 0, 1, 2, \ldots) \qquad \textbf{(37.15)}$$

This condition takes into account two factors: (1) the difference in path length for the two rays (the term $m\lambda_n$) and (2) the 180° phase change upon reflection (the term $\frac{1}{2}\lambda_n$). Because $\lambda_n = \lambda/n$, we can write Equation 37.15 as

Conditions for constructive interference in thin films ▶

$$2nt = (m + \tfrac{1}{2})\lambda \quad (m = 0, 1, 2, \ldots) \qquad \textbf{(37.16)}$$

[2] The full interference effect in a thin film requires an analysis of an infinite number of reflections back and forth between the top and bottom surfaces of the film. We focus here only on a single reflection from the bottom of the film, which provides the largest contribution to the interference effect.

If the extra distance $2t$ traveled by ray 2 corresponds to a multiple of λ_n, the two waves combine out of phase and the result is destructive interference. The general equation for *destructive* interference in thin films is

$$2nt = m\lambda \quad (m = 0, 1, 2, \dots) \qquad \textbf{(37.17)}$$

◀ Conditions for destructive interference in thin films

The foregoing conditions for constructive and destructive interference are valid when the medium above the top surface of the film is the same as the medium below the bottom surface or, if there are different media above and below the film, the index of refraction of both is less than n. If the film is placed between two different media, one with $n < n_{\text{film}}$ and the other with $n > n_{\text{film}}$, the conditions for constructive and destructive interference are reversed. In that case, either there is a phase change of 180° for both ray 1 reflecting from surface A and ray 2 reflecting from surface B or there is no phase change for either ray; hence, the net change in relative phase due to the reflections is zero.

Rays 3 and 4 in Figure 37.11 lead to interference effects in the light transmitted through the thin film. The analysis of these effects is similar to that of the reflected light. You are asked to explore the transmitted light in Problems 23, 29, and 30.

PITFALL PREVENTION 37.1
Be Careful with Thin Films

Be sure to include *both* effects—path length and phase change—when analyzing an interference pattern resulting from a thin film. The possible phase change is a new feature we did not need to consider for double-slit interference. Also think carefully about the material on either side of the film. You may have a situation in which there is a 180° phase change at *both* surfaces or at *neither* surface, if there are different materials on either side of the film.

Quick Quiz 37.3 One microscope slide is placed on top of another with their left edges in contact and a human hair under the right edge of the upper slide. As a result, a wedge of air exists between the slides. An interference pattern results when monochromatic light is incident on the wedge. What is at the left edges of the slides? (a) a dark fringe (b) a bright fringe (c) impossible to determine

Newton's Rings

Another method for observing interference in light waves is to place a plano-convex lens on top of a flat glass surface as shown in Figure 37.12a. With this arrangement, the air film between the glass surfaces varies in thickness from zero at the point of contact to some value t at point P. If the radius of curvature R of the lens is much greater than the distance r and the system is viewed from above, a pattern of light and dark rings is observed as shown in Figure 37.12b. These circular fringes, discovered by Newton, are called **Newton's rings.**

The interference effect is due to the combination of ray 1, reflected from the flat plate, with ray 2, reflected from the curved surface of the lens. Ray 1 undergoes a phase change of 180° upon reflection (because it is reflected from a medium of higher index of refraction), whereas ray 2 undergoes no phase change (because it is reflected from a medium of lower index of refraction). Hence, the conditions for constructive and destructive interference are given by Equations 37.16 and 37.17, respectively, with $n = 1$ because the film is air. Because there is

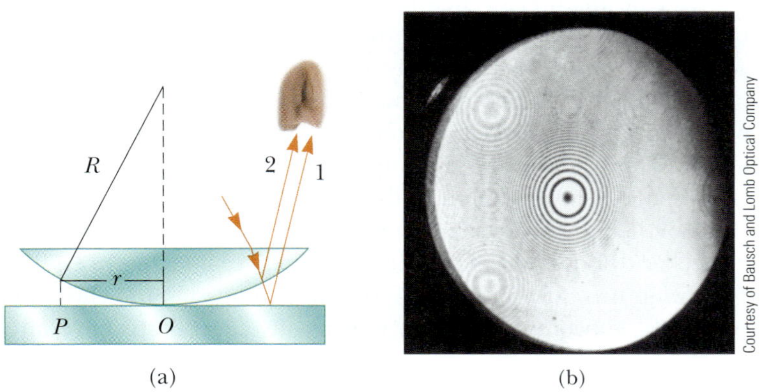

Figure 37.12 (a) The combination of rays reflected from the flat plate and the curved lens surface gives rise to an interference pattern known as Newton's rings. (b) Photograph of Newton's rings.

(a)

(b)

(a) A thin film of oil floating on water displays interference, shown by the pattern of colors when white light is incident on the film. Variations in film thickness produce the interesting color pattern. The razor blade gives you an idea of the size of the colored bands. (b) Interference in soap bubbles. The colors are due to interference between light rays reflected from the front and back surfaces of the thin film of soap making up the bubble. The color depends on the thickness of the film, ranging from black, where the film is thinnest, to magenta, where it is thickest.

no path difference and the total phase change is due only to the 180° phase change upon reflection, the contact point at O is dark as seen in Figure 37.12b.

Using the geometry shown in Figure 37.12a, we can obtain expressions for the radii of the bright and dark bands in terms of the radius of curvature R and wavelength λ. For example, the dark rings have radii given by the expression $r \approx \sqrt{m\lambda R/n}$. The details are left as a problem (see Problem 60). We can obtain the wavelength of the light causing the interference pattern by measuring the radii of the rings, provided R is known. Conversely, we can use a known wavelength to obtain R.

One important use of Newton's rings is in the testing of optical lenses. A circular pattern like that pictured in Figure 37.12b is obtained only when the lens is ground to a perfectly symmetric curvature. Variations from such symmetry produce a pattern with fringes that vary from a smooth, circular shape. These variations indicate how the lens must be reground and repolished to remove imperfections.

PROBLEM-SOLVING STRATEGY **Thin-Film Interference**

The following features should be kept in mind when working thin-film interference problems.

1. *Conceptualize.* Think about what is going on physically in the problem. Identify the light source and the location of the observer.

2. *Categorize.* Confirm that you should use the techniques for thin-film interference by identifying the thin film causing the interference.

3. *Analyze.* The type of interference that occurs is determined by the phase relationship between the portion of the wave reflected at the upper surface of the film and the portion reflected at the lower surface. Phase differences between the two portions of the wave have two causes: differences in the distances traveled by the two portions and phase changes occurring on reflection. *Both* causes must be considered when determining which type of interference occurs. If the media above and below the film both have index of refraction larger than that of the film or if both indices are smaller, use Equation 37.16 for constructive interference and Equation 37.17 for destructive interference. If the film is located between two different media, one with $n < n_{\text{film}}$ and the other with $n > n_{\text{film}}$, reverse these two equations for constructive and destructive interference.

4. *Finalize.* Inspect your final results to see if they make sense physically and are of an appropriate size.

EXAMPLE 37.3 Interference in a Soap Film

Calculate the minimum thickness of a soap-bubble film that results in constructive interference in the reflected light if the film is illuminated with light whose wavelength in free space is $\lambda = 600$ nm. The index of refraction of the soap film is 1.33.

SOLUTION

Conceptualize Imagine that the film in Figure 37.11 is soap, with air on both sides.

Categorize We evaluate the result using an equation from this section, so we categorize this example as a substitution problem.

The minimum film thickness for constructive interference in the reflected light corresponds to $m = 0$ in Equation 37.16. Solve this equation for t and substitute numerical values:

$$t = \frac{(0 + \frac{1}{2})\lambda}{2n} = \frac{\lambda}{4n} = \frac{(600 \text{ nm})}{4(1.33)} = 113 \text{ nm}$$

What If? What if the film is twice as thick? Does this situation produce constructive interference?

Answer Using Equation 37.16, we can solve for the thicknesses at which constructive interference occurs:

$$t = (m + \tfrac{1}{2})\frac{\lambda}{2n} = (2m + 1)\frac{\lambda}{4n} \qquad (m = 0, 1, 2, \ldots)$$

The allowed values of m show that constructive interference occurs for *odd* multiples of the thickness corresponding to $m = 0$, $t = 113$ nm. Therefore, constructive interference does *not* occur for a film that is twice as thick.

EXAMPLE 37.4 Nonreflective Coatings for Solar Cells

Solar cells—devices that generate electricity when exposed to sunlight—are often coated with a transparent, thin film of silicon monoxide (SiO, n = 1.45) to minimize reflective losses from the surface. Suppose a silicon solar cell ($n = 3.5$) is coated with a thin film of silicon monoxide for this purpose (Fig. 37.13a). Determine the minimum film thickness that produces the least reflection at a wavelength of 550 nm, near the center of the visible spectrum.

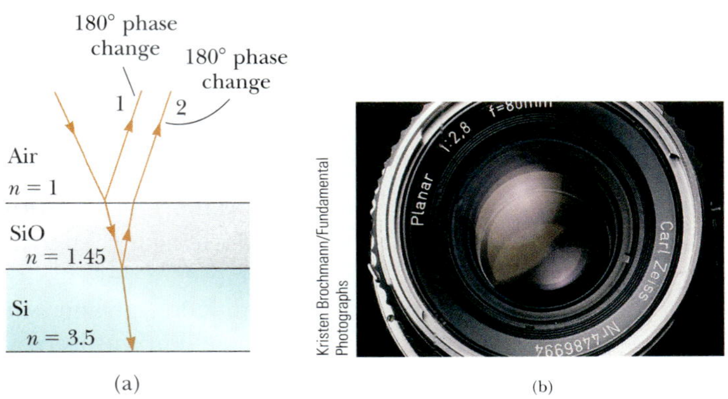

Figure 37.13 (Example 37.4) (a) Reflective losses from a silicon solar cell are minimized by coating the surface of the cell with a thin film of silicon monoxide. (b) The reflected light from a coated camera lens often has a reddish-violet appearance.

SOLUTION

Conceptualize Figure 37.13a helps us visualize the path of the rays in the SiO film that result in interference in the reflected light.

Categorize Based on the geometry of the SiO layer, we categorize this example as a thin-film interference problem.

Analyze The reflected light is a minimum when rays 1 and 2 in Figure 37.13a meet the condition of destructive interference. In this situation, *both* rays undergo a 180° phase change upon reflection: ray 1 from the upper SiO surface, and ray 2 from the lower SiO surface. The net change in phase due to reflection is therefore zero, and the condition for a reflection minimum requires a path difference of $\lambda_n/2$, where λ_n is the wavelength of the light in SiO. Hence, $2nt = \lambda/2$, where λ is the wavelength in air and n is the index of refraction of SiO.

Solve the equation $2nt = \lambda/2$ for t and substitute numerical values:

$$t = \frac{\lambda}{4n} = \frac{550 \text{ nm}}{4(1.45)} = 94.8 \text{ nm}$$

Finalize A typical uncoated solar cell has reflective losses as high as 30%, but a coating of SiO can reduce this value to about 10%. This significant decrease in reflective losses increases the cell's efficiency because less reflection means that more sunlight enters the silicon to create charge carriers in the cell. No coating can ever be made perfectly nonreflecting because the required thickness is wavelength-dependent and the incident light covers a wide range of wavelengths.

Glass lenses used in cameras and other optical instruments are usually coated with a transparent thin film to reduce or eliminate unwanted reflection and to enhance the transmission of light through the lenses. The camera lens in Figure 37.13b has several coatings (of different thicknesses) to minimize reflection of light waves having wavelengths near the center of the visible spectrum. As a result, the small amount of light that is reflected by the lens has a greater proportion of the far ends of the spectrum and often appears reddish violet.

37.7 The Michelson Interferometer

The **interferometer,** invented by American physicist A. A. Michelson (1852–1931), splits a light beam into two parts and then recombines the parts to form an interference pattern. The device can be used to measure wavelengths or other lengths with great precision because a large and precisely measurable displacement of one of the mirrors is related to an exactly countable number of wavelengths of light.

A schematic diagram of the interferometer is shown in Active Figure 37.14. A ray of light from a monochromatic source is split into two rays by mirror M_0, which is inclined at 45° to the incident light beam. Mirror M_0, called a *beam splitter,* transmits half the light incident on it and reflects the rest. One ray is reflected from M_0 vertically upward toward mirror M_1, and the second ray is transmitted horizontally through M_0 toward mirror M_2. Hence, the two rays travel separate paths L_1 and L_2. After reflecting from M_1 and M_2, the two rays eventually recombine at M_0 to produce an interference pattern, which can be viewed through a telescope.

The interference condition for the two rays is determined by the difference in their path length. When the two mirrors are exactly perpendicular to each other, the interference pattern is a target pattern of bright and dark circular fringes, similar to Newton's rings. As M_1 is moved, the fringe pattern collapses or expands, depending on the direction in which M_1 is moved. For example, if a dark circle appears at the center of the target pattern (corresponding to destructive interference) and M_1 is then moved a distance $\lambda/4$ toward M_0, the path difference changes by $\lambda/2$. What was a dark circle at the center now becomes a bright circle. As M_1 is moved an additional distance $\lambda/4$ toward M_0, the bright circle becomes a dark circle again. Therefore, the fringe pattern shifts by one-half fringe each time M_1 is moved a distance $\lambda/4$. The wavelength of light is then measured by counting the number of fringe shifts for a given displacement of M_1. If the wavelength is accurately known, mirror displacements can be measured to within a fraction of the wavelength.

We will see an important historical use of the Michelson interferometer in our discussion of relativity in Chapter 39. Modern uses include the following two applications, Fourier transform infrared spectroscopy and the laser interferometer gravitational-wave observatory.

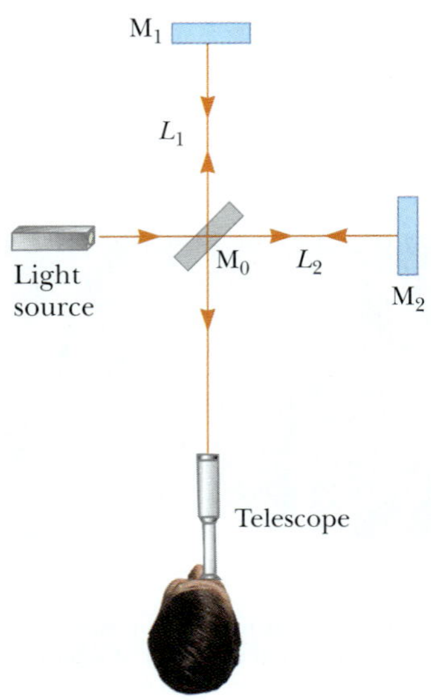

ACTIVE FIGURE 37.14

Diagram of the Michelson interferometer. A single ray of light is split into two rays by mirror M_0, which is called a beam splitter. The path difference between the two rays is varied with the adjustable mirror M_1. As M_1 is moved, an interference pattern changes in the field of view.

Sign in at www.thomsonedu.com and go to ThomsonNOW to move the mirror and see the effect on the interference pattern and use the interferometer to measure the wavelength of light.

Fourier Transform Infrared Spectroscopy

Spectroscopy is the study of the wavelength distribution of radiation from a sample that can be used to identify the characteristics of atoms or molecules in the sample. Infrared spectroscopy is particularly important to organic chemists when analyzing organic molecules. Traditional spectroscopy involves the use of an optical element, such as a prism (Section 35.5) or a diffraction grating (Section 38.4), which spreads out various wavelengths in a complex optical signal from the sample into different angles. In this way, the various wavelengths of radiation and their intensities in the signal can be determined. These types of devices are limited in their resolution and effectiveness because they must be scanned through the various angular deviations of the radiation.

The technique of *Fourier transform infrared* (FTIR) *spectroscopy* is used to create a higher-resolution spectrum in a time interval of 1 second that may have required 30 minutes with a standard spectrometer. In this technique, the radiation from a sample enters a Michelson interferometer. The movable mirror is swept through the zero-path-difference condition, and the intensity of radiation at the viewing position is recorded. The result is a complex set of data relating light intensity as a function of mirror position, called an *interferogram.* Because there is a relationship between mirror position and light intensity for a given wavelength, the interferogram contains information about all wavelengths in the signal.

In Section 18.8, we discussed Fourier analysis of a waveform. The waveform is a function that contains information about all the individual frequency components that make up the waveform.[3] Equation 18.13 shows how the waveform is generated from the individual frequency components. Similarly, the interferogram can be analyzed by computer, in a process called a *Fourier transform,* to provide all of the wavelength components. This information is the same as that generated by traditional spectroscopy, but the resolution of FTIR spectroscopy is much higher.

Laser Interferometer Gravitational-Wave Observatory

Einstein's general theory of relativity (Section 39.10) predicts the existence of *gravitational waves.* These waves propagate from the site of any gravitational disturbance, which could be periodic and predictable, such as the rotation of a double star around a center of mass, or unpredictable, such as the supernova explosion of a massive star.

In Einstein's theory, gravitation is equivalent to a distortion of space. Therefore, a gravitational disturbance causes an additional distortion that propagates through space in a manner similar to mechanical or electromagnetic waves. When gravitational waves from a disturbance pass by the Earth, they create a distortion of the local space. The laser interferometer gravitational-wave observatory (LIGO) apparatus is designed to detect this distortion. The apparatus employs a Michelson interferometer that uses laser beams with an effective path length of several kilometers. At the end of an arm of the interferometer, a mirror is mounted on a massive pendulum. When a gravitational wave passes by, the pendulum and the attached mirror move and the interference pattern due to the laser beams from the two arms changes.

Two sites for interferometers have been developed in the United States—in Richland, Washington, and in Livingston, Louisiana—to allow coincidence studies of gravitational waves. Figure 37.15 shows the Washington site. The two arms of

LIGO Hanford Observatory

Figure 37.15 The Laser Interferometer Gravitational-Wave Observatory (LIGO) near Richland, Washington. Notice the two perpendicular arms of the Michelson interferometer.

[3] In acoustics, it is common to talk about the components of a complex signal in terms of frequency. In optics, it is more common to identify the components by wavelength.

the Michelson interferometer are evident in the photograph. Five data runs have been performed as of 2007. These runs have been coordinated with other gravitational wave detectors, such as GEO in Hannover, Germany, TAMA in Mitaka, Japan, and VIRGO in Cascina, Italy. So far, gravitational waves have not yet been detected, but the data runs have provided critical information for modifications and design features for the next generation of detectors.

Summary

ThomsonNOW™ Sign in at **www.thomsonedu.com** and go to ThomsonNOW to take a practice test for this chapter.

CONCEPTS AND PRINCIPLES

Interference in light waves occurs whenever two or more waves overlap at a given point. An interference pattern is observed if (1) the sources are coherent and (2) the sources have identical wavelengths.

The **intensity** at a point in a double-slit interference pattern is

$$I = I_{\max} \cos^2\left(\frac{\pi d \sin\theta}{\lambda}\right) \quad \textbf{(37.13)}$$

where $I_{\max}$ is the maximum intensity on the screen and the expression represents the time average.

A wave traveling from a medium of index of refraction n_1 toward a medium of index of refraction n_2 undergoes a 180° phase change upon reflection when $n_2 > n_1$ and undergoes no phase change when $n_2 < n_1$.

The condition for constructive interference in a film of thickness t and index of refraction n surrounded by air is

$$2nt = (m + \tfrac{1}{2})\lambda \quad (m = 0, 1, 2, \ldots) \quad \textbf{(37.16)}$$

where λ is the wavelength of the light in free space.

Similarly, the condition for destructive interference in a thin film surrounded by air is

$$2nt = m\lambda \quad (m = 0, 1, 2, \ldots) \quad \textbf{(37.17)}$$

ANALYSIS MODEL FOR PROBLEM SOLVING

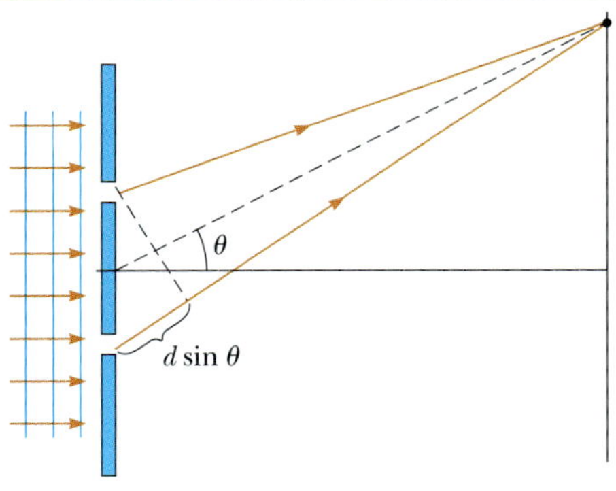

Waves in Interference. Young's double-slit experiment serves as a prototype for interference phenomena involving electromagnetic radiation. In this experiment, two slits separated by a distance d are illuminated by a single-wavelength light source. The condition for bright fringes **(constructive interference)** is

$$d \sin\theta_{\text{bright}} = m\lambda \quad (m = 0, \pm 1, \pm 2, \ldots) \quad \textbf{(37.2)}$$

The condition for dark fringes **(destructive interference)** is

$$d \sin\theta_{\text{dark}} = (m + \tfrac{1}{2})\lambda \quad (m = 0, \pm 1, \pm 2, \ldots) \quad \textbf{(37.3)}$$

The number m is called the **order number** of the fringe.

Questions

☐ denotes answer available in *Student Solutions Manual/Study Guide;* **O** denotes objective question

Question 4 in Chapter 18 may be assigned with this chapter.

☐**1.** What is the necessary condition on the path length difference between two waves that interfere (a) constructively and (b) destructively?

2. Explain why two flashlights held close together do not produce an interference pattern on a distant screen.

3. **O** Four trials of Young's double-slit experiment are conducted. (a) In the first trial, blue light passes through two fine slits 400 μm apart and forms an interference pattern on a screen 4 m away. (b) In a second trial, red light passes through the same slits and falls on the same screen. (c) A third trial is performed with red light and the same screen, but with slits 800 μm apart. (d) A final trial is performed with red light, slits 800 μm apart, and a screen 8 m away. **(i)** Rank the trials (a) through (d) from largest to smallest value of the angle between the central maximum and the first-order side maximum. In your ranking, note any cases of equality. **(ii)** Rank the same trials according to the distance between the central maximum and the first-order side maximum on the screen.

4. Suppose you blow smoke into the space between the barrier and the viewing screen in Young's double-slit experiment, shown in Active Figure 37.2a. Would the smoke show evidence of interference within this space? Explain your answer.

5. **O** Suppose Young's double-slit experiment is performed in air using red light and then the apparatus is immersed in water. What happens to the interference pattern on the screen? (a) It disappears. (b) The bright and dark fringes stay in the same locations, but the contrast is reduced. (c) The bright fringes are closer together. (d) The color shifts toward blue. (e) The bright fringes are farther apart. (f) The bright fringes are in continuous motion. (g) No change happens in the interference pattern.

☐**6.** In Young's double-slit experiment, why do we use monochromatic light? If white light is used, how would the pattern change?

7. **O** Suppose you perform Young's double-slit experiment with the slit separation slightly smaller than the wavelength of the light. As a screen, you use a large half-cylinder with its axis along the midline between the slits. What interference pattern will you see on the interior surface of the cylinder? (a) bright and dark fringes so closely spaced as to be indistinguishable (b) one central bright fringe and two dark fringes only (c) a completely bright screen with no dark fringes (d) one central dark fringe and two bright fringes only (e) a completely dark screen with no bright fringes

8. As a soap bubble evaporates, it appears black immediately before it breaks, as at the top of the circular film shown in Figure Q37.8. Explain this phenomenon in terms of the phase changes that occur on reflection from the two surfaces of the soap film.

Figure Q37.8 Question 8 and Problem 63.

9. **O** A film of oil on a puddle in a parking lot shows a variety of bright colors in swirled patches. What can you say about the thickness of the oil film? (a) It is much less than the wavelength of visible light. (b) It is of the same order of magnitude as the wavelength of visible light. (c) It is much greater than the wavelength of visible light. (d) It might have any relationship to the wavelength of visible light.

10. **O** Assume the index of refraction of flint glass is 1.66 and the index of refraction of crown glass is 1.52. **(i)** A film formed by one drop of sassafras oil, on a horizontal surface of a flint glass block, is viewed by reflected light. The film appears brightest at its outer margin, where it is thinnest. A film of the same oil on crown glass appears dark at its outer margin. What can you say about the index of refraction of the oil? (a) It must be less than 1.52. (b) It must be between 1.52 and 1.66. (c) It must be greater than 1.66. (d) None of statements (a) through (c) is necessarily true. **(ii)** Could a very thin film of some other liquid appear bright by reflected light on both of the glass blocks? **(iii)** Could it appear dark on both?

(iv) Could it appear dark on crown glass and bright on flint glass? Experiments described by Thomas Young suggested this question.

11. A lens with outer radius of curvature R and index of refraction n rests on a flat glass plate. The combination is illuminated with white light from above and observed from above. Is there a dark spot or a light spot at the center of the lens? What does it mean if the observed rings are noncircular?

12. Why is the lens on a good-quality camera coated with a thin film?

13. **O** Green light has a wavelength of 500 nm in air. **(i)** Assume green light is reflected from a mirror with angle of incidence 0°. The incident and reflected waves together constitute a standing wave with what distance from one node to the next antinode? (a) 1 000 nm (b) 500 nm (c) 250 nm (d) 125 nm (e) 62.5 nm **(ii)** The green light is sent into a Michelson interferometer that is adjusted to produce a central bright circle. How far must the interferometer's moving mirror be shifted to change the center of the pattern into a dark circle? Choose from the same possibilities. **(iii)** The light is reflected perpendicularly from a thin film of a plastic with index of refraction 2.00. The film appears bright in the reflected light. How much additional thickness would make the film appear dark?

14. **O** Using a Michelson interferometer, shown in Active Figure 37.14, you are viewing a dark circle at the center of the interference pattern. As you gradually move the light source toward the central mirror M_0, through a distance $\lambda/2$, what do you see? (a) There is no change in the pattern. (b) The dark circle changes into a bright circle. (c) The dark circle changes into a bright circle and then back into a dark circle. (d) The dark circle changes into a bright circle, then into a dark circle, and then into a bright circle.

Problems

WebAssign The Problems from this chapter may be assigned online in WebAssign.

ThomsonNOW Sign in at **www.thomsonedu.com** and go to ThomsonNOW to assess your understanding of this chapter's topics with additional quizzing and conceptual questions.

1, 2, 3 denotes straightforward, intermediate, challenging; ☐ denotes full solution available in *Student Solutions Manual/Study Guide;* ▲ denotes coached solution with hints available at **www.thomsonedu.com;** ▭ denotes developing symbolic reasoning; ● denotes asking for qualitative reasoning; 🖥 denotes computer useful in solving problem

Section 37.1 Conditions for Interference

Section 37.2 Young's Double-Slit Experiment

Section 37.3 Light Waves in Interference

Note: Problems 4, 5, 6, 7, 8, and 10 in Chapter 18 can be assigned with this section.

1. A laser beam (λ = 632.8 nm) is incident on two slits 0.200 mm apart. How far apart are the bright interference fringes on a screen 5.00 m away from the double slits?

2. A Young's interference experiment is performed with monochromatic light. The separation between the slits is 0.500 mm, and the interference pattern on a screen 3.30 m away shows the first side maximum 3.40 mm from the center of the pattern. What is the wavelength?

3. ▲ Two radio antennas separated by 300 m as shown in Figure P37.3 simultaneously broadcast identical signals at the same wavelength. A radio in a car traveling due north receives the signals. (a) If the car is at the position of the second maximum, what is the wavelength of the signals? (b) How much farther must the car travel to encounter the next minimum in reception? *Note:* Do not use the small-angle approximation in this problem.

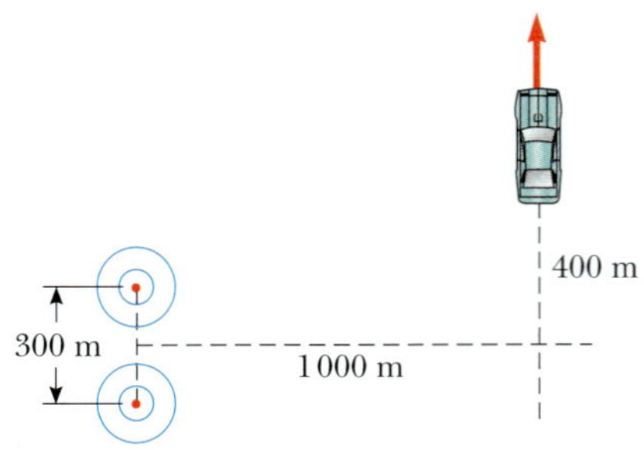

Figure P37.3

4. In a location where the speed of sound is 354 m/s, a 2 000-Hz sound wave impinges on two slits 30.0 cm apart. (a) At what angle is the first maximum located? (b) **What If?** If the sound wave is replaced by 3.00-cm microwaves, what slit separation gives the same angle for the first maximum? (c) **What If?** If the slit separation is 1.00 μm, what frequency of light gives the same first maximum angle?

5. ▲ Young's double-slit experiment is performed with 589-nm light and a distance of 2.00 m between the slits and the screen. The tenth interference minimum is observed 7.26 mm from the central maximum. Determine the spacing of the slits.

6. ● Write the statement of a problem, including data, for which the following equations appear in the solution.

$$\lambda = \frac{343 \text{ m/s}}{1\,620/\text{s}} \qquad (35.0 \text{ cm}) \sin \theta_0 = 0\lambda$$

$$(35.0 \text{ cm}) \sin \theta_{1 \text{ soft}} = 0.5\lambda \qquad (35.0 \text{ cm}) \sin \theta_{1 \text{ loud}} = 1\lambda$$

$$(35.0 \text{ cm}) \sin \theta_{2 \text{ soft}} = 1.5\lambda \qquad (35.0 \text{ cm}) \sin \theta_{2 \text{ loud}} = 2\lambda$$

State the solution to the problem, including values for each quantity that appears as an unknown. State what you can conclude from the last of the set of six equations. Does this equation describe an angle $\theta_{2 \text{ loud}}$ that is larger than 90°?

7. Two narrow, parallel slits separated by 0.250 mm are illuminated by green light (λ = 546.1 nm). The interference pattern is observed on a screen 1.20 m away from the plane of the slits. Calculate the distance (a) from the central maximum to the first bright region on either side of the central maximum and (b) between the first and second dark bands.

8. A riverside warehouse has two open doors as shown in Figure P37.8. Its walls are lined with sound-absorbing material. A boat on the river sounds its horn. To person A, the sound is loud and clear. To person B, the sound is barely audible. The principal wavelength of the sound waves is 3.00 m. Assuming person B is at the position of the first minimum, determine the distance between the doors, center to center.

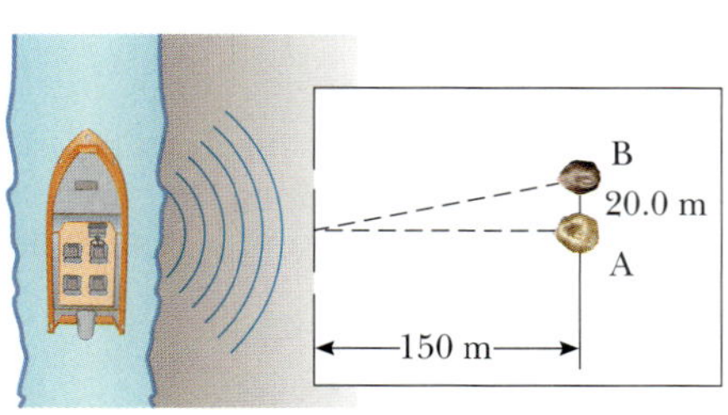

Figure P37.8

9. Light with wavelength 442 nm passes through a double-slit system that has a slit separation d = 0.400 mm. Determine how far away a screen must be placed so that dark fringes appear directly opposite both slits, with only one bright fringe between them.

10. Two slits are separated by 0.320 mm. A beam of 500-nm light strikes the slits, producing an interference pattern. Determine the number of maxima observed in the angular range $-30.0° < \theta < 30.0°$.

11. ● Young's double-slit experiment underlies the *instrument landing system* used to guide aircraft to safe landings when the visibility is poor. Although real systems are more complicated than the example described here, they operate on the same principles. A pilot is trying to align her plane with a runway as suggested in Figure P37.11a. Two radio antennas A_1 and A_2 are positioned adjacent to the runway, separated by 40.0 m. The antennas broadcast unmodulated coherent radio waves at 30.0 MHz. (a) Find the wavelength of the waves. The pilot "locks onto" the strong signal radiated along an interference maximum, and steers the plane to keep the received signal strong. If she has found the central maximum, the plane will have precisely the right heading to land when it reaches the runway. (b) **What If?** Suppose the plane is flying along the first side maximum instead (Fig. P37.11b). How far to the side of the runway centerline will the plane be when it is 2.00 km from the antennas? (c) It is possible to tell the pilot that she is on the wrong maximum by sending out two signals from each antenna and equipping the aircraft with a two-channel receiver. The ratio of the two frequencies must not be the ratio of small integers (such as $\frac{3}{4}$). Explain how this two-frequency system would work and

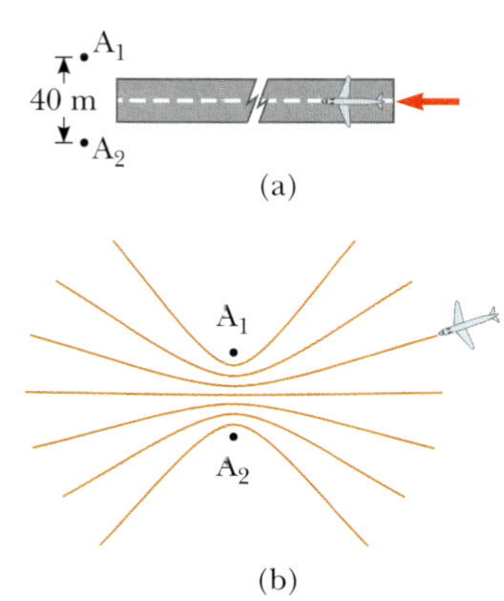

Figure P37.11

why it would not necessarily work if the frequencies were related by an integer ratio.

12. A student holds a laser that emits light of wavelength 633 nm. The beam passes though a pair of slits separated by 0.300 mm, in a glass plate attached to the front of the laser. The beam then falls perpendicularly on a screen, creating an interference pattern on it. The student begins to walk directly toward the screen at 3.00 m/s. The central maximum on the screen is stationary. Find the speed of the first-order maxima on the screen.

13. In Figure 37.5, let L = 1.20 m and d = 0.120 mm and assume the slit system is illuminated with monochromatic 500-nm light. Calculate the phase difference between the two wave fronts arriving at P when (a) $\theta = 0.500°$ and (b) y = 5.00 mm. (c) What is the value of θ for which the phase difference is 0.333 rad? (d) What is the value of θ for which the path difference is $\lambda/4$?

14. Coherent light rays of wavelength λ strike a pair of slits separated by distance d at an angle θ_1 as shown in Figure P37.14. Assume an interference maximum is formed at an angle θ_2 a great distance from the slits. Show that $d(\sin \theta_2 - \sin \theta_1) = m\lambda$, where m is an integer.

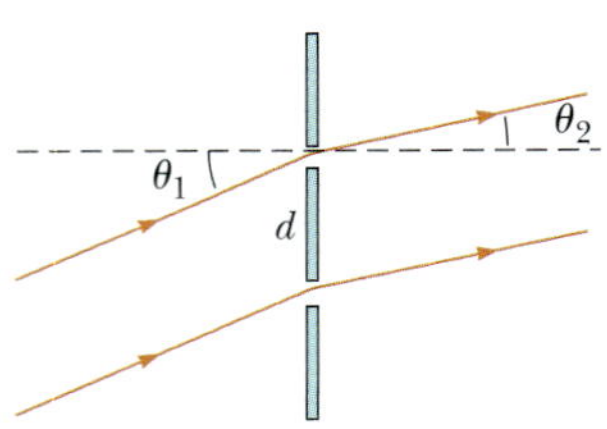

Figure P37.14

15. ● In a double-slit arrangement of Figure 37.5, d = 0.150 mm, L = 140 cm, λ = 643 nm, and y = 1.80 cm. (a) What is the path difference δ for the rays from the two slits arriving at P? (b) Express this path difference in terms of λ. (c) Does P correspond to a maximum, a minimum, or an intermediate condition? Give evidence for your answer.

Section 37.4 Intensity Distribution of the Double-Slit Interference Pattern

16. The intensity on the screen at a certain point in a double-slit interference pattern is 64.0% of the maximum value. (a) What minimum phase difference (in radians) between sources produces this result? (b) Express this phase difference as a path difference for 486.1-nm light.

17. ▲ In Figure 37.5, let L = 120 cm and d = 0.250 cm. The slits are illuminated with coherent 600-nm light. Calculate the distance y above the central maximum for which the average intensity on the screen is 75.0% of the maximum.

18. Two slits are separated by 0.180 mm. An interference pattern is formed on a screen 80.0 cm away by 656.3-nm light. Calculate the fraction of the maximum intensity 0.600 cm above the central maximum.

19. ▲ Show that the two waves with wave functions E_1 = $6.00 \sin (100\pi t)$ and $E_2 = 8.00 \sin (100\pi t + \pi/2)$ add to give a wave with the wave function $E_R \sin (100\pi t + \phi)$. Find the required values for E_R and ϕ.

20. Make a graph of I/I_{max} as a function of θ for the interference pattern produced by the arrangement described in Problem 7. Let θ range over the interval from $-0.3°$ to $+0.3°$.

21. Two narrow, parallel slits separated by 0.850 mm are illuminated by 600-nm light, and the viewing screen is 2.80 m away from the slits. (a) What is the phase difference between the two interfering waves on a screen at a point 2.50 mm from the central bright fringe? (b) What is the ratio of the intensity at this point to the intensity at the center of a bright fringe?

22. ● Monochromatic coherent light of amplitude E_0 and angular frequency ω passes through three parallel slits each separated by a distance d from its neighbor. (a) Show that the time-averaged intensity as a function of the angle θ is

$$I(\theta) = I_{max}\left[1 + 2\cos\left(\frac{2\pi d \sin\theta}{\lambda}\right)\right]^2$$

(b) Explain how this expression describes both the primary and the secondary maxima. Determine the ratio of the intensities of the primary and secondary maxima.

Section 37.5 Change of Phase Due to Reflection

Section 37.6 Interference in Thin Films

23. ● An oil film (n = 1.45) floating on water is illuminated by white light at normal incidence. The film is 280 nm

thick. Find (a) the color of the light in the visible spectrum most strongly reflected and (b) the color of the light in the spectrum most strongly transmitted. Explain your reasoning.

24. A soap bubble ($n = 1.33$) is floating in air. If the thickness of the bubble wall is 115 nm, what is the wavelength of the light that is most strongly reflected?

25. A thin film of oil ($n = 1.25$) is located on smooth, wet pavement. When viewed perpendicular to the pavement, the film reflects most strongly red light at 640 nm and reflects no green light at 512 nm. How thick is the oil film?

26. A possible means for making an airplane invisible to radar is to coat the plane with an antireflective polymer. If radar waves have a wavelength of 3.00 cm and the index of refraction of the polymer is $n = 1.50$, how thick would you make the coating?

27. A material having an index of refraction of 1.30 is used as an antireflective coating on a piece of glass ($n = 1.50$). What should the minimum thickness of this film be to minimize reflection of 500-nm light?

28. A film of MgF_2 ($n = 1.38$) having thickness 1.00×10^{-5} cm is used to coat a camera lens. Are any wavelengths in the visible spectrum intensified in the reflected light?

29. ● Astronomers observe the chromosphere of the Sun with a filter that passes the red hydrogen spectral line of wavelength 656.3 nm, called the H_α line. The filter consists of a transparent dielectric of thickness d held between two partially aluminized glass plates. The filter is held at a constant temperature. (a) Find the minimum value of d that produces maximum transmission of perpendicular H_α light if the dielectric has an index of refraction of 1.378. (b) **What If?** If the temperature of the filter increases above the normal value, what happens to the transmitted wavelength? (The index of refraction of the filter does not change significantly.) (c) The dielectric will also pass what near-visible wavelength? One of the glass plates is colored red to absorb this light.

30. A beam of 580-nm light passes through two closely spaced glass plates as shown in Figure P37.30. For what minimum nonzero value of the plate separation d is the transmitted light bright?

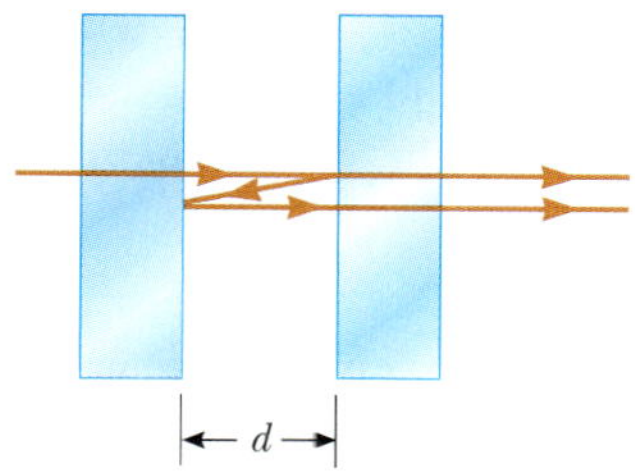

Figure P37.30

31. ▲ An air wedge is formed between two glass plates separated at one edge by a very fine wire as shown in Figure P37.31. When the wedge is illuminated from above by 600-nm light and viewed from above, 30 dark fringes are observed. Calculate the radius of the wire.

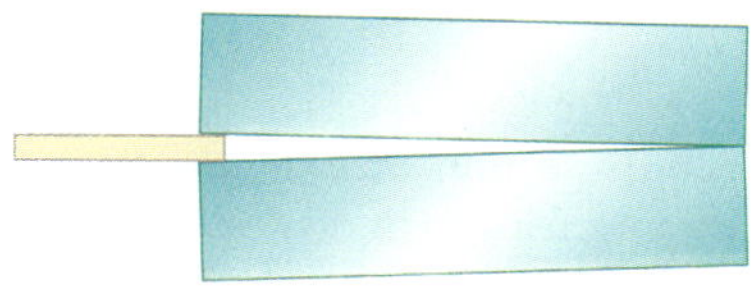

Figure P37.31 Problems 31 and 33.

32. When a liquid is introduced into the air space between the lens and the plate in a Newton's-rings apparatus, the diameter of the tenth ring changes from 1.50 to 1.31 cm. Find the index of refraction of the liquid.

33. Two glass plates 10.0 cm long are in contact at one end and separated at the other end by a thread 0.050 0 mm in diameter (Fig. P37.31). Light containing the two wavelengths 400 nm and 600 nm is incident perpendicularly and viewed by reflection. At what distance from the contact point is the next dark fringe?

Section 37.7 The Michelson Interferometer

34. Monochromatic light is beamed into a Michelson interferometer. The movable mirror is displaced 0.382 mm, causing the interferometer pattern to reproduce itself 1 700 times. Determine the wavelength of the light. What color is it?

35. Mirror M_1 in Active Figure 37.14 is moved through a displacement ΔL. During this displacement, 250 fringe reversals (formation of successive dark or bright bands) are counted. The light being used has a wavelength of 632.8 nm. Calculate the displacement ΔL.

36. One leg of a Michelson interferometer contains an evacuated cylinder of length L, having glass plates on each end. A gas is slowly leaked into the cylinder until a pressure of 1 atm is reached. If N bright fringes pass on the screen during this process when light of wavelength λ is used, what is the index of refraction of the gas?

Additional Problems

37. In an experiment similar to that of Example 37.1, green light with wavelength 560 nm, sent through a pair of slits 30.0 μm apart, produces bright fringes 2.24 cm apart on a screen 1.20 m away. Calculate the fringe separation for this same arrangement assuming that the apparatus is submerged in a tank containing a sugar solution with index of refraction 1.38.

38. In the **What If?** section of Example 37.2, it was claimed that overlapping fringes in a two-slit interference pattern for two different wavelengths obey the following relationship even for large values of the angle θ:

$$\frac{m'}{m} = \frac{\lambda}{\lambda'}$$

(a) Prove this assertion. (b) Using the data in Example 37.2, find the nonzero value of y on the screen at which the fringes from the two wavelengths first coincide.

39. One radio transmitter A operating at 60.0 MHz is 10.0 m from another similar transmitter B that is 180° out of phase with A. How far must an observer move from A toward B along the line connecting the two transmitters to reach the nearest point where the two beams are in phase?

40. **Review problem.** This problem extends the result of Problem 10 in Chapter 18. Figure P37.40 shows two adjacent vibrating balls dipping into a pan of water. At distant points, they produce an interference pattern of water waves as shown in Figure 37.3. Let λ represent the wavelength of the ripples. Show that the two sources produce a standing wave along the line segment, of length d, between them. In terms of λ and d, find the number of nodes and the number of antinodes in the standing wave. Find the number of zones of constructive and of destructive interference in the interference pattern far away from the sources. Each line of destructive interference springs from a node in the standing wave, and each line of constructive interference springs from an antinode.

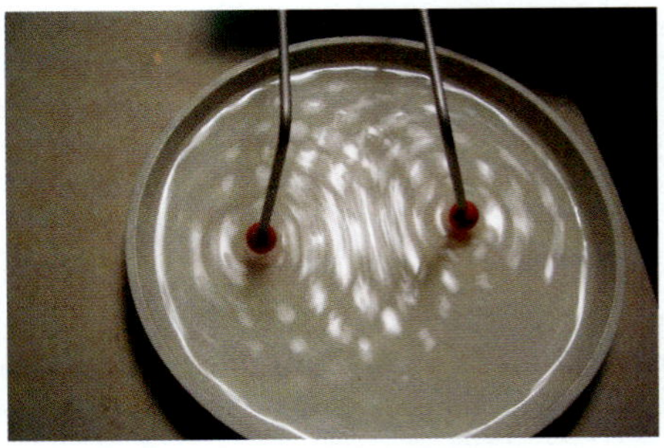

Alexandra Héder

Figure P37.40

41. Raise your hand and hold it flat. Think of the space between your index finger and your middle finger as one slit and think of the space between middle finger and ring finger as a second slit. (a) Consider the interference resulting from sending coherent visible light perpendicularly through this pair of openings. Compute an order-of-magnitude estimate for the angle between adjacent zones of constructive interference. (b) To make the angles in the interference pattern easy to measure with a plastic protractor, you should use an electromagnetic wave with frequency of what order of magnitude? How is this wave classified on the electromagnetic spectrum?

42. Two coherent waves, coming from sources at different locations, move along the x axis. Their wave functions are

$$E_1 = (860 \text{ V/m}) \sin\left[\frac{2\pi x_1}{650 \text{ nm}} - 2\pi(462 \text{ THz})t + \frac{\pi}{6}\right]$$

and

$$E_2 = (860 \text{ V/m}) \sin\left[\frac{2\pi x_2}{650 \text{ nm}} - 2\pi(462 \text{ THz})t + \frac{\pi}{8}\right]$$

Determine the relationship between x_1 and x_2 that produces constructive interference when the two waves are superposed.

43. In a Young's interference experiment, the two slits are separated by 0.150 mm and the incident light includes two wavelengths: λ_1 = 540 nm (green) and λ_2 = 450 nm (blue). The overlapping interference patterns are observed on a screen 1.40 m from the slits. Calculate the minimum distance from the center of the screen to a point where a bright fringe of the green light coincides with a bright fringe of the blue light.

44. In a Young's double-slit experiment using light of wavelength λ, a thin piece of Plexiglas having index of refraction n covers one of the slits. If the center point on the screen is a dark spot instead of a bright spot, what is the minimum thickness of the Plexiglas?

45. **Review problem.** A flat piece of glass is held stationary and horizontal above the flat top end of a 10.0-cm-long vertical metal rod that has its lower end rigidly fixed. The thin film of air between the rod and glass is observed to be bright by reflected light when it is illuminated by light of wavelength 500 nm. As the temperature is slowly increased by 25.0°C, the film changes from bright to dark and back to bright 200 times. What is the coefficient of linear expansion of the metal?

46. A certain crude oil has an index of refraction of 1.25. A ship dumps 1.00 m^3 of this oil into the ocean, and the oil spreads into a thin uniform slick. If the film produces a first-order maximum of light of wavelength 500 nm normally incident on it, how much surface area of the ocean does the oil slick cover? Assume the index of refraction of the ocean water is 1.34.

47. Astronomers observe a 60.0-MHz radio source both directly and by reflection from the sea. If the receiving dish is 20.0 m above sea level, what is the angle of the radio source above the horizon at first maximum?

48. Interference effects are produced at point P on a screen as a result of direct rays from a 500-nm source and reflected rays from the mirror as shown in Figure P37.48. Assume the source is 100 m to the left of the screen and 1.00 cm above the mirror. Find the distance y to the first dark band above the mirror.

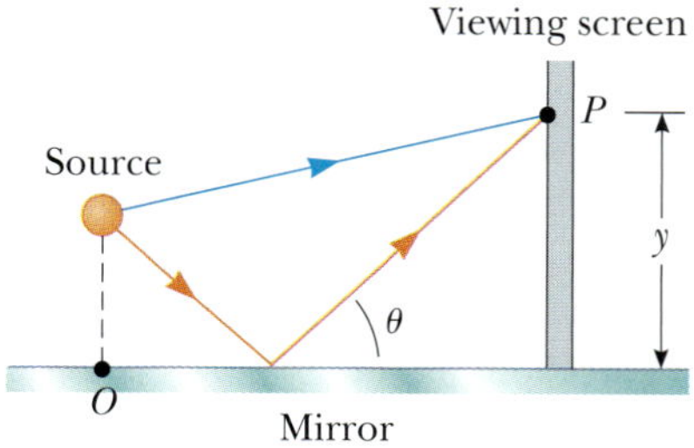

Figure P37.48

49. The waves from a radio station can reach a home receiver by two paths. One is a straight-line path from transmitter to home, a distance of 30.0 km. The second path is by reflection from the ionosphere (a layer of ionized air molecules high in the atmosphere). Assume this reflection takes place at a point midway between receiver and transmitter and the wavelength broadcast by the radio station is 350 m. Find the minimum height of the ionospheric layer that could produce destructive interference between the direct and reflected beams. Assume no phase change occurs on reflection.

50. Many cells are transparent and colorless. Structures of great interest in biology and medicine can be practically invisible to ordinary microscopy. To indicate the size and shape of cell structures, an *interference microscope* reveals a difference in index of refraction as a shift in interference fringes. The idea is exemplified in the following problem. An air wedge is formed between two glass plates in contact along one edge and slightly separated at the opposite edge. When the plates are illuminated with monochromatic light from above, the reflected light has 85 dark fringes. Calculate the number of dark fringes that appear if water (n = 1.33) replaces the air between the plates.

51. Measurements are made of the intensity distribution within the central bright fringe in a Young's interference pattern (see Fig. 37.7). At a particular value of y, it is found that I/I_{max} = 0.810 when 600-nm light is used. What wavelength of light should be used to reduce the relative intensity at the same location to 64.0% of the maximum intensity?

52. Our discussion of the techniques for determining constructive and destructive interference by reflection from a thin film in air has been confined to rays striking the film at nearly normal incidence. **What If?** Assume a ray is incident at an angle of 30.0° (relative to the normal) on a film with index of refraction 1.38. Calculate the minimum thickness for constructive interference of sodium light with a wavelength of 590 nm.

53. The condition for constructive interference by reflection from a thin film in air as developed in Section 37.6 assumes nearly normal incidence. **What If?** Show that if the light is incident on the film at a nonzero angle ϕ_1 (relative to the normal), the condition for constructive interference is $2nt \cos \theta_2 = (m + \frac{1}{2})\lambda$, where θ_2 is the angle of refraction.

54. ● The quantity δ in Equation 37.1 is called the path difference. Its size in comparison to the wavelength controls the character of the interference between two beams in vacuum by controlling the phase difference between the beams. The analogous quantity nt in Equations 37.16 and 37.17 is called the *optical path length* corresponding to the geometrical distance t. The optical

path length is proportional to n because a larger index of refraction shortens the wavelength, so more cycles of a wave fit into a particular geometrical distance. (a) Assume a mixture of corn syrup and water is prepared in a tank, with its index of refraction n increasing uniformly from 1.33 at $y = 20.0$ cm at the top to 1.90 at $y = 0$. Write the index of refraction $n(y)$ as a function of y. (b) Compute the optical path length corresponding to the 20-cm height of the tank by calculating

$$\int_0^{20\text{ cm}} n(y)\,dy$$

(c) Suppose a narrow beam of light is directed into the mixture with its original direction between horizontal and vertically upward. Qualitatively describe its path.

55. (a) Both sides of a uniform film that has index of refraction n and thickness d are in contact with air. For normal incidence of light, an intensity minimum is observed in the reflected light at λ_2 and an intensity maximum is observed at λ_1, where $\lambda_1 > \lambda_2$. Assuming that no intensity minima are observed between λ_1 and λ_2, show that the integer m in Equations 37.16 and 37.17 is given by $m = \lambda_1/2(\lambda_1 - \lambda_2)$. (b) Determine the thickness of the film, assuming $n = 1.40$, $\lambda_1 = 500$ nm, and $\lambda_2 = 370$ nm.

56. Figure P37.56 shows a radio-wave transmitter and a receiver separated by a distance d and both a distance h above the ground. The receiver can receive signals both directly from the transmitter and indirectly from signals that reflect from the ground. Assume the ground is level between the transmitter and receiver and a 180° phase shift occurs upon reflection. Determine the longest wavelengths that interfere (a) constructively and (b) destructively.

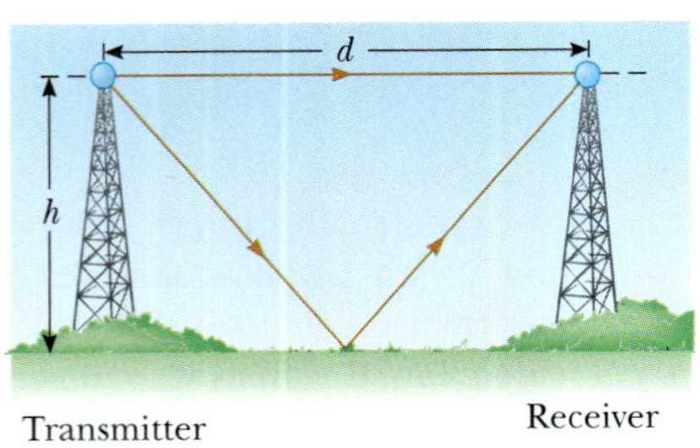

Figure P37.56

57. Consider the double-slit arrangement shown in Figure P37.57, where the slit separation is d and the distance from the slit to the screen is L. A sheet of transparent plastic having an index of refraction n and thickness t is placed over the upper slit. As a result, the central maximum of the interference pattern moves upward a distance y'. Find y'.

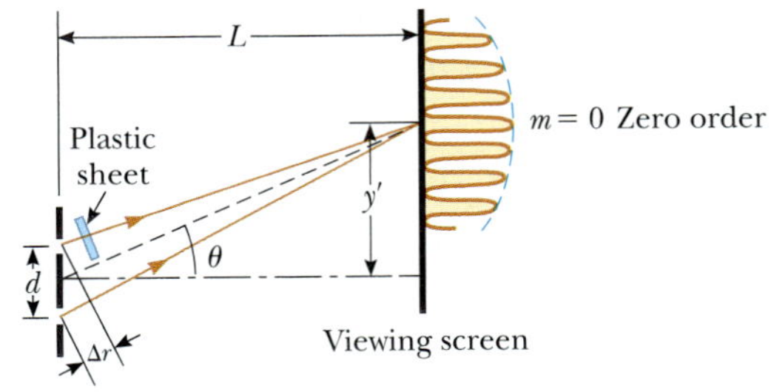

Figure P37.57

58. A piece of transparent material having an index of refraction n is cut into the shape of a wedge as shown in Figure P37.58. The angle of the wedge is small. Monochromatic light of wavelength λ is normally incident from above and is viewed from above. Let h represent the height of the wedge and ℓ its width. Show that bright fringes occur at the positions $x = \lambda\ell(m + \frac{1}{2})/2hn$ and dark fringes occur at the positions $x = \lambda\ell m/2hn$, where $m = 0, 1, 2, \ldots$ and x is measured as shown.

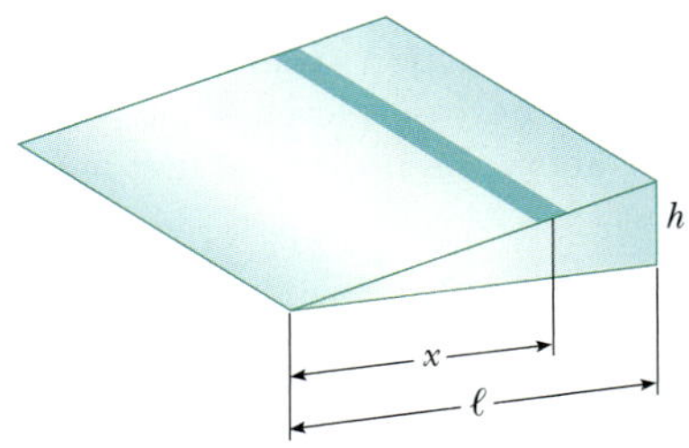

Figure P37.58

59. In a Newton's-rings experiment, a plano-convex glass ($n = 1.52$) lens having diameter 10.0 cm is placed on a flat plate as shown in Figure 37.12a. When 650-nm light is incident normally, 55 bright rings are observed, with the last one precisely on the edge of the lens. (a) What is the radius of curvature of the convex surface of the lens? (b) What is the focal length of the lens?

60. A plano-convex lens has index of refraction n. The curved side of the lens has radius of curvature R and rests on a flat glass surface of the same index of refraction, with a film of index n_{film} between them, as shown in Fig. 37.12a. The lens is illuminated from above by light of wavelength λ. Show that the dark Newton's rings have radii given approximately by

$$r \approx \sqrt{\frac{m\lambda R}{n_{\text{film}}}}$$

where m is an integer and r is much less than R.

61. A plano-concave lens having index of refraction 1.50 is placed on a flat glass plate as shown in Figure P37.61. Its curved surface, with radius of curvature 8.00 m, is on the bottom. The lens is illuminated from above with yellow sodium light of wavelength 589 nm, and a series of concentric bright and dark rings is observed by reflection. The interference pattern has a dark spot at the center, surrounded by 50 dark rings, the largest of which is at the outer edge of the lens. (a) What is the thickness of the air layer at the center of the interference pattern? (b) Calculate the radius of the outermost dark ring. (c) Find the focal length of the lens.

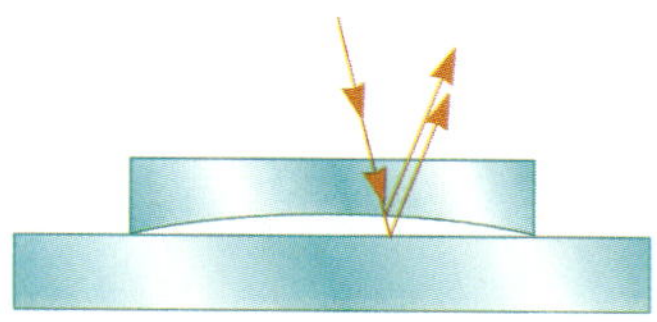

Figure P37.61

62. A plano-convex lens having a radius of curvature of $r = 4.00$ m is placed on a concave glass surface whose radius of curvature is $R = 12.0$ m as shown in Figure P37.62. Determine the radius of the 100th bright ring, assuming 500-nm light is incident normal to the flat surface of the lens.

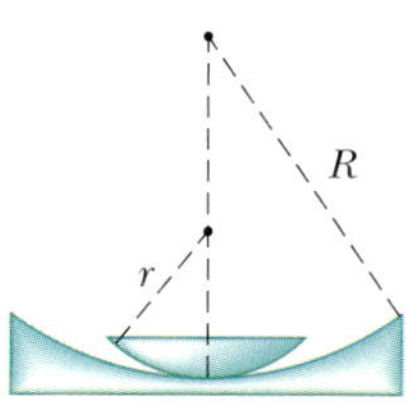

Figure P37.62

63. Figure Q37.8 shows an unbroken soap film in a circular frame. The film thickness increases from top to bottom, slowly at first and then rapidly. As a simpler model, consider a soap film ($n = 1.33$) contained within a rectangular wire frame. The frame is held vertically so that the film drains downward and forms a wedge with flat faces. The thickness of the film at the top is essentially zero. The film is viewed in reflected white light with near-normal incidence, and the first violet ($\lambda = 420$ nm) interference band is observed 3.00 cm from the top edge of the film. (a) Locate the first red ($\lambda = 680$ nm) interference band. (b) Determine the film thickness at the positions of the violet and red bands. (c) What is the wedge angle of the film?

64. Compact disc (CD) and digital videodisc (DVD) players use interference to generate strong signals from tiny bumps, shown in Figure P35.40. A pit's depth is chosen to be one-quarter of the wavelength of the laser light used to read the disc. Then light reflected from the pit and light reflected from the adjoining flat surface differ in path length traveled by one-half wavelength, interfering destructively at the detector. As the disc rotates, the light intensity drops significantly every time light is reflected from near a pit edge. The space between the leading and trailing edges of a pit determines the time interval between the fluctuations. The series of time intervals is decoded into a series of zeros and ones that carries the stored information. Assume infrared light with a wavelength of 780 nm in vacuum is used in a CD player. The disc is coated with plastic having an index of refraction of 1.50. What should the depth of each pit be? A DVD player uses light of a shorter wavelength, and the pit dimensions are correspondingly smaller. This reduction is one factor resulting in a DVD's greater storage capacity compared with that of a CD.

65. Interference fringes are produced using Lloyd's mirror and a 606-nm source as shown in Figure 37.9. Fringes 1.20 mm apart are formed on a screen 2.00 m from the real source S. Find the vertical distance h of the source above the reflecting surface.

66. Slit 1 of a double slit is wider than slit 2 so that the light from slit 1 has an amplitude 3.00 times that of the light from slit 2. Show that Equation 37.12 is replaced by the equation $I = (4I_{max}/9)(1 + 3\cos^2\phi/2)$ for this situation.

67. Monochromatic light of wavelength 620 nm passes through a very narrow slit S and then strikes a screen in which are two parallel slits, S_1 and S_2, as shown in Figure P37.67. Slit S_1 is directly in line with S and at a distance of $L = 1.20$ m away from S, whereas S_2 is displaced a distance d to one side. The light is detected at point P on a second screen, equidistant from S_1 and S_2. When either slit S_1 or S_2 is open, equal light intensities are measured at point P. When both slits are open, the intensity is three times larger. Find the minimum possible value for the slit separation d.

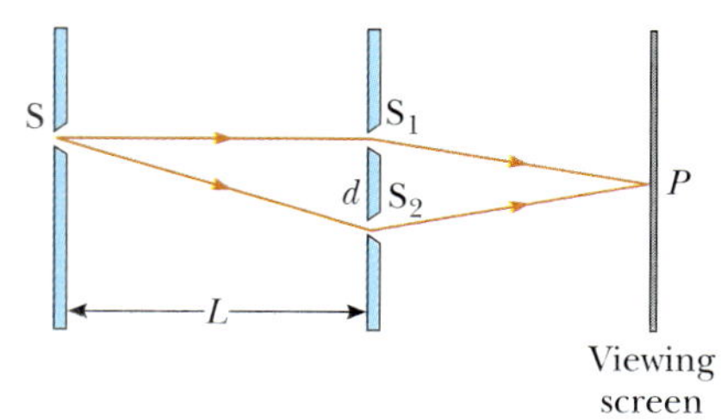

Figure P37.67

Answers to Quick Quizzes

37.1 (c). Equation 37.7 shows that decreasing λ or L will bring the fringes closer together. Immersing the apparatus in water decreases the wavelength so that the fringes move closer together.

37.2 The graph is shown below. The width of the primary maxima is slightly narrower than the $N = 5$ primary width but wider than the $N = 10$ primary width. Because $N = 6$, the secondary maxima are $\frac{1}{36}$ as intense as the primary maxima.

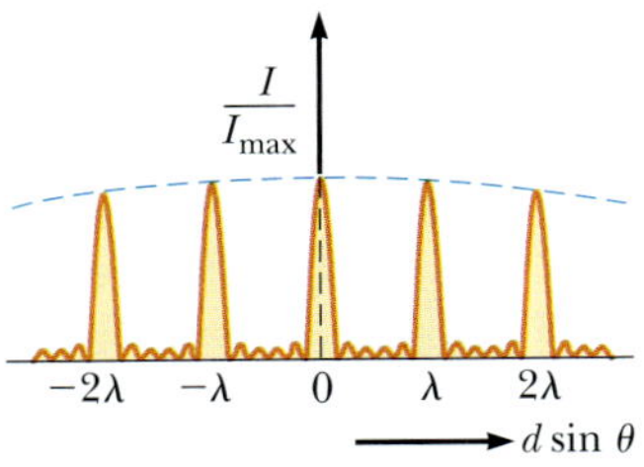

37.3 (a). At the left edge, the air wedge has zero thickness and the only contribution to the interference is the 180° phase shift as the light reflects from the upper surface of the glass slide.

The Hubble Space Telescope does its viewing above the atmosphere and does not suffer from the atmospheric blurring, caused by air turbulence, that plagues ground-based telescopes. Despite this advantage, it does have limitations due to diffraction effects. In this chapter, we show how the wave nature of light limits the ability of any optical system to distinguish between closely spaced objects. (© Denis Scott/CORBIS)

38 Diffraction Patterns and Polarization

When plane light waves pass through a small aperture in an opaque barrier, the aperture acts as if it were a point source of light, with waves entering the shadow region behind the barrier. This phenomenon, known as diffraction, can be described only with a wave model for light as discussed in Section 35.3. In this chapter, we investigate the features of the *diffraction pattern* that occurs when the light from the aperture is allowed to fall upon a screen.

In Chapter 34, we learned that electromagnetic waves are transverse. That is, the electric and magnetic field vectors associated with electromagnetic waves are perpendicular to the direction of wave propagation. In this chapter, we show that under certain conditions these transverse waves with electric field vectors in all possible transverse directions can be *polarized* in various ways. In other words, only certain directions of the electric field vectors are present in the polarized wave.

38.1 Introduction to Diffraction Patterns

In Section 35.3, we discussed that light of wavelength comparable to or larger than the width of a slit spreads out in all forward directions upon passing through the slit. This phenomenon is called *diffraction.* When light passes through a narrow slit, it spreads beyond the narrow path defined by the slit into regions that would be in shadow if light traveled in straight lines. Other waves, such as sound waves and

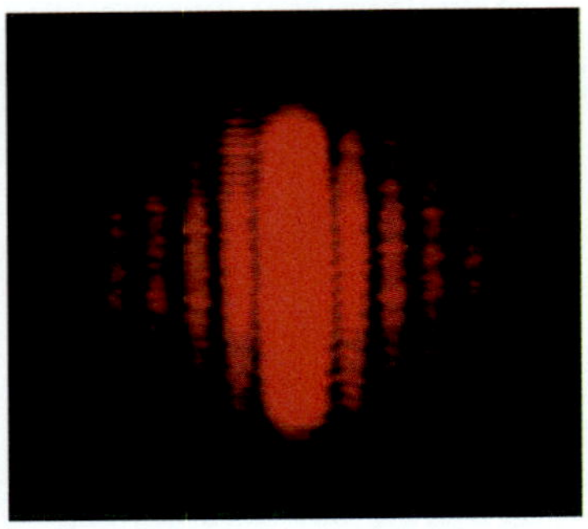

Douglas C. Johnson/California State Polytechnic University, Pomona

Figure 38.1 The diffraction pattern that appears on a screen when light passes through a narrow vertical slit. The pattern consists of a broad central fringe and a series of less intense and narrower side fringes.

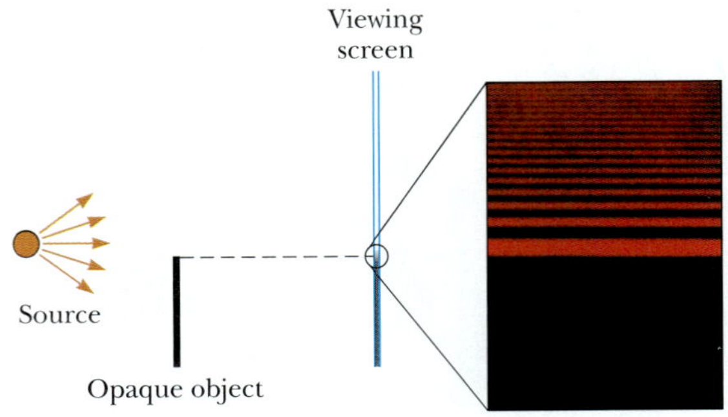

From M. Cagnet, M. Françon, and J. C. Thrierr, *Atlas of Optical Phenomena*, Berlin, Springer-Verlag, 1962, plate 32

Figure 38.2 Light from a small source passes by the edge of an opaque object and continues on to a screen. A diffraction pattern consisting of bright and dark fringes appears on the screen in the region above the edge of the object.

water waves, also have this property of spreading when passing through apertures or by sharp edges.

You might expect that the light passing through a small opening would simply result in a broad region of light on a screen due to the spreading of the light as it passes through the opening. We find something more interesting, however. A **diffraction pattern** consisting of light and dark areas is observed, somewhat similar to the interference patterns discussed earlier. For example, when a narrow slit is placed between a distant light source (or a laser beam) and a screen, the light produces a diffraction pattern like that shown in Figure 38.1. The pattern consists of a broad, intense central band (called the **central maximum**) flanked by a series of narrower, less intense additional bands (called **side maxima** or **secondary maxima**) and a series of intervening dark bands (or **minima**). Figure 38.2 shows a diffraction pattern associated with light passing by the edge of an object. Again we see bright and dark fringes, which is reminiscent of an interference pattern.

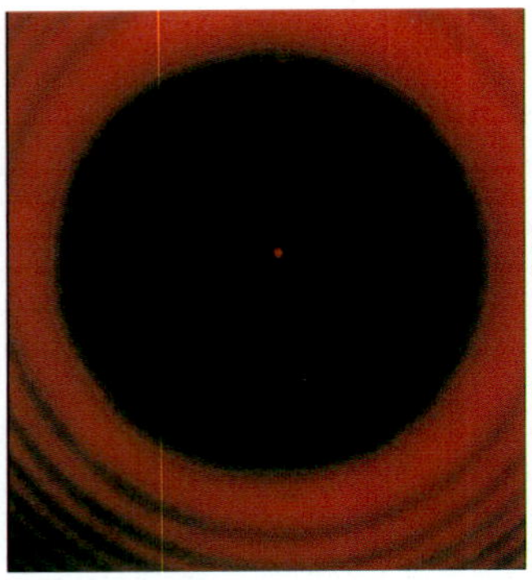

P. M. Rinard, *Am. J. Phys.* 44: 70 1976

Figure 38.3 Diffraction pattern created by the illumination of a penny, with the penny positioned midway between screen and light source. Note the bright spot at the center.

Figure 38.3 shows a diffraction pattern associated with the shadow of a penny. A bright spot occurs at the center, and circular fringes extend outward from the shadow's edge. We can explain the central bright spot by using the wave theory of light, which predicts constructive interference at this point. From the viewpoint of geometric optics (in which light is viewed as rays traveling in straight lines), we expect the center of the shadow to be dark because that part of the viewing screen is completely shielded by the penny.

Shortly before the central bright spot was first observed, one of the supporters of geometric optics, Simeon Poisson, argued that if Augustin Fresnel's wave theory of light were valid, a central bright spot should be observed in the shadow of a circular object illuminated by a point source of light. To Poisson's astonishment, the spot was observed by Dominique Arago shortly thereafter. Therefore, Poisson's prediction reinforced the wave theory rather than disproving it.

PITFALL PREVENTION 38.1

Diffraction Versus Diffraction Pattern

Diffraction refers to the general behavior of waves spreading out as they pass through a slit. We used diffraction in explaining the existence of an interference pattern in Chapter 37. A *diffraction pattern* is actually a misnomer, but is deeply entrenched in the language of physics. The diffraction pattern seen on a screen when a single slit is illuminated is actually another interference pattern. The interference is between parts of the incident light illuminating different regions of the slit.

38.2 Diffraction Patterns from Narrow Slits

Let's consider a common situation, that of light passing through a narrow opening modeled as a slit and projected onto a screen. To simplify our analysis, we assume the observing screen is far from the slit and the rays reaching the screen are approximately parallel. (This situation can also be achieved experimentally by using a converging lens to focus the parallel rays on a nearby screen.) In this model, the pattern on the screen is called a **Fraunhofer diffraction pattern.**[1]

Active Figure 38.4a shows light entering a single slit from the left and diffracting as it propagates toward a screen. Active Figure 38.4b is a photograph of a single-slit Fraunhofer diffraction pattern. A bright fringe is observed along the axis at

[1] If the screen is brought close to the slit (and no lens is used), the pattern is a *Fresnel* diffraction pattern. The Fresnel pattern is more difficult to analyze, so we shall restrict our discussion to Fraunhofer diffraction.

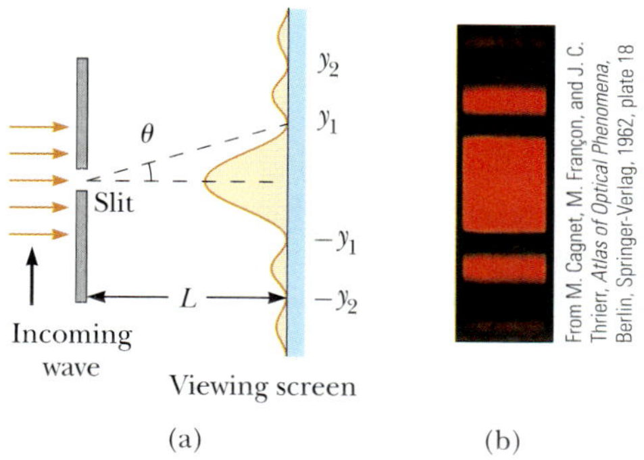

ACTIVE FIGURE 38.4

(a) Fraunhofer diffraction pattern of a single slit. The pattern consists of a central bright fringe flanked by much weaker maxima alternating with dark fringes. (Drawing not to scale.) The labels y indicate positions of dark fringes on the screen. (b) Photograph of a single-slit Fraunhofer diffraction pattern.

Sign in at www.thomsonedu.com and go to ThomsonNOW to adjust the slit width and the wavelength of the light to see the effect on the diffraction pattern.

$\theta = 0$, with alternating dark and bright fringes on each side of the central bright fringe.

Until now, we have assumed slits are point sources of light. In this section, we abandon that assumption and see how the finite width of slits is the basis for understanding Fraunhofer diffraction. We can explain some important features of this phenomenon by examining waves coming from various portions of the slit as shown in Figure 38.5. According to Huygens's principle, **each portion of the slit acts as a source of light waves.** Hence, light from one portion of the slit can interfere with light from another portion, and the resultant light intensity on a viewing screen depends on the direction θ. Based on this analysis, we recognize that a diffraction pattern is actually an interference pattern in which the different sources of light are different portions of the single slit!

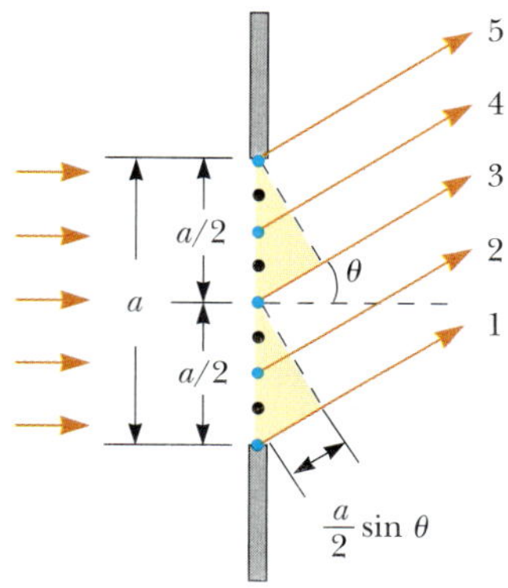

Figure 38.5 Paths of light rays that encounter a narrow slit of width a and diffract toward a screen in the direction described by angle θ. Each portion of the slit acts as a point source of light waves. The path difference between rays 1 and 3, rays 2 and 4, or rays 3 and 5 is $(a/2)\sin\theta$. (Drawing not to scale.)

To analyze the diffraction pattern, let's divide the slit into two halves as shown in Figure 38.5. Keeping in mind that all the waves are in phase as they leave the slit, consider rays 1 and 3. As these two rays travel toward a viewing screen far to the right of the figure, ray 1 travels farther than ray 3 by an amount equal to the path difference $(a/2)\sin\theta$, where a is the width of the slit. Similarly, the path difference between rays 2 and 4 is also $(a/2)\sin\theta$, as is that between rays 3 and 5. If this path difference is exactly half a wavelength (corresponding to a phase difference of 180°), the pairs of waves cancel each other and destructive interference results. This cancellation occurs for any two rays that originate at points separated by half the slit width because the phase difference between two such points is 180°. Therefore, waves from the upper half of the slit interfere destructively with waves from the lower half when

$$\frac{a}{2}\sin\theta = \pm\frac{\lambda}{2}$$

or when

$$\sin\theta = \pm\frac{\lambda}{a}$$

Dividing the slit into four equal parts and using similar reasoning, we find that the viewing screen is also dark when

$$\sin\theta = \pm 2\frac{\lambda}{a}$$

Likewise, dividing the slit into six equal parts shows that darkness occurs on the screen when

$$\sin\theta = \pm 3\frac{\lambda}{a}$$

Therefore, the general condition for destructive interference is

Condition for destructive interference for a single slit ▶

$$\sin\theta_{\text{dark}} = m\frac{\lambda}{a} \qquad m = \pm 1, \pm 2, \pm 3, \ldots \tag{38.1}$$

This equation gives the values of θ_{dark} for which the diffraction pattern has zero light intensity, that is, when a dark fringe is formed. It tells us nothing, however, about the variation in light intensity along the screen. The general features of the intensity distribution are shown in Active Figure 38.4. A broad, central bright fringe is observed; this fringe is flanked by much weaker bright fringes alternating with dark fringes. The various dark fringes occur at the values of θ_{dark} that satisfy Equation 38.1. Each bright-fringe peak lies approximately halfway between its bordering dark-fringe minima. Notice that the central bright maximum is twice as wide as the secondary maxima.

PITFALL PREVENTION 38.2
Similar Equation Warning!

Equation 38.1 has exactly the same form as Equation 37.2, with d, the slit separation, used in Equation 37.2 and a, the slit width, used in Equation 38.1. Equation 37.2, however, describes the *bright* regions in a two-slit interference pattern, whereas Equation 38.1 describes the *dark* regions in a single-slit diffraction pattern. Furthermore, $m = 0$ does not represent a dark fringe in the diffraction pattern.

Quick Quiz 38.1 Suppose the slit width in Active Figure 38.4 is made half as wide. Does the central bright fringe (a) become wider, (b) remain the same, or (c) become narrower?

EXAMPLE 38.1 Where Are the Dark Fringes?

Light of wavelength 580 nm is incident on a slit having a width of 0.300 mm. The viewing screen is 2.00 m from the slit. Find the positions of the first dark fringes and the width of the central bright fringe.

SOLUTION

Conceptualize Based on the problem statement, we imagine a single-slit diffraction pattern similar to that in Active Figure 38.4.

Categorize We categorize this example as a straightforward application of our discussion of single-slit diffraction patterns.

Analyze Evaluate Equation 38.1 for the two dark fringes that flank the central bright fringe, which correspond to $m = \pm 1$:

$$\sin\theta_{\text{dark}} = \pm\frac{\lambda}{a} = \pm\frac{5.80 \times 10^{-7}\text{ m}}{0.300 \times 10^{-3}\text{ m}} = \pm 1.933 \times 10^{-3}$$

From the triangle in Active Figure 38.4a, notice that $\tan\theta_{\text{dark}} = y_1/L$. Because θ_{dark} is very small, we can use the approximation $\sin\theta_{\text{dark}} \approx \tan\theta_{\text{dark}}$; therefore, $\sin\theta_{\text{dark}} \approx y_1/L$.

Use this result to find the positions of the first minima measured from the central axis:

$$y_1 \approx L\sin\theta_{\text{dark}} = (2.00\text{ m})(\pm 1.933 \times 10^{-3}) = \pm 3.866 \times 10^{-3}\text{ m}$$

Find the width of the central bright fringe:

$$2|y_1| = 7.73 \times 10^{-3}\text{ m} = 7.73\text{ mm}$$

Finalize Notice that this value is much greater than the width of the slit. Let's explore what happens if we change the slit width.

What If? What if the slit width is increased by an order of magnitude to 3.00 mm? What happens to the diffraction pattern?

Answer Based on Equation 38.1, we expect that the angles at which the dark bands appear will decrease as a increases. Therefore, the diffraction pattern narrows.

From Equation 38.1, find the sines of the angles θ_{dark} for the $m = \pm 1$ dark fringes:

$$\sin\theta_{\text{dark}} = \pm\frac{\lambda}{a} = \pm\frac{5.80 \times 10^{-7}\text{ m}}{3.00 \times 10^{-3}\text{ m}} = \pm 1.933 \times 10^{-4}$$

Use this result to find the positions of the first minima measured from the central axis:

$$y_1 \approx L \sin \theta_{\text{dark}} = (2.00 \text{ m})(\pm 1.933 \times 10^{-4})$$
$$= \pm 3.866 \times 10^{-4} \text{ m}$$

Find the width of the central bright fringe:

$$2|y_1| = 7.73 \times 10^{-4} \text{ m} = 0.773 \text{ mm}$$

Notice that this result is *smaller* than the width of the slit. In general, for large values of a, the various maxima and minima are so closely spaced that only a large, central bright area resembling the geometric image of the slit is observed. This concept is very important in the performance of optical instruments such as telescopes.

Intensity of Single-Slit Diffraction Patterns

Analysis of the intensity variation in a diffraction pattern from a single slit of width a shows that the intensity is given by

$$I = I_{\text{max}}\left[\frac{\sin(\pi a \sin\theta/\lambda)}{\pi a \sin\theta/\lambda}\right]^2 \quad \textbf{(38.2)}$$

◀ Intensity of a single-slit Fraunhofer diffraction pattern

where I_{max} is the intensity at $\theta = 0$ (the central maximum) and λ is the wavelength of light used to illuminate the slit. This expression shows that *minima* occur when

$$\frac{\pi a \sin\theta_{\text{dark}}}{\lambda} = m\pi$$

or

$$\sin\theta_{\text{dark}} = m\frac{\lambda}{a} \qquad m = \pm 1, \pm 2, \pm 3, \ldots$$

◀ Condition for intensity minima for a single slit

in agreement with Equation 38.1.

Figure 38.6a represents a plot of Equation 38.2, and Figure 38.6b is a photograph of a single-slit Fraunhofer diffraction pattern. Notice that most of the light intensity is concentrated in the central bright fringe.

Intensity of Two-Slit Diffraction Patterns

When more than one slit is present, we must consider not only diffraction patterns due to the individual slits but also the interference patterns due to the waves coming from different slits. Notice the curved dashed lines in Figure 37.8, which indicate a decrease in intensity of the interference maxima as θ increases. This decrease is due to a diffraction pattern. To determine the effects of both two-slit

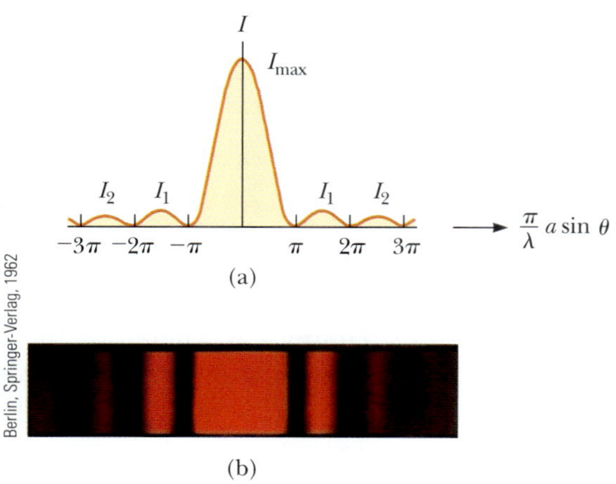

From M. Cagnet, M. Françon, and J. C. Thrierr, *Atlas of Optical Phenomena*, Berlin, Springer-Verlag, 1962

Figure 38.6 (a) A plot of light intensity I versus $(\pi/\lambda)a \sin\theta$ for the single-slit Fraunhofer diffraction pattern. (b) Photograph of a single-slit Fraunhofer diffraction pattern.

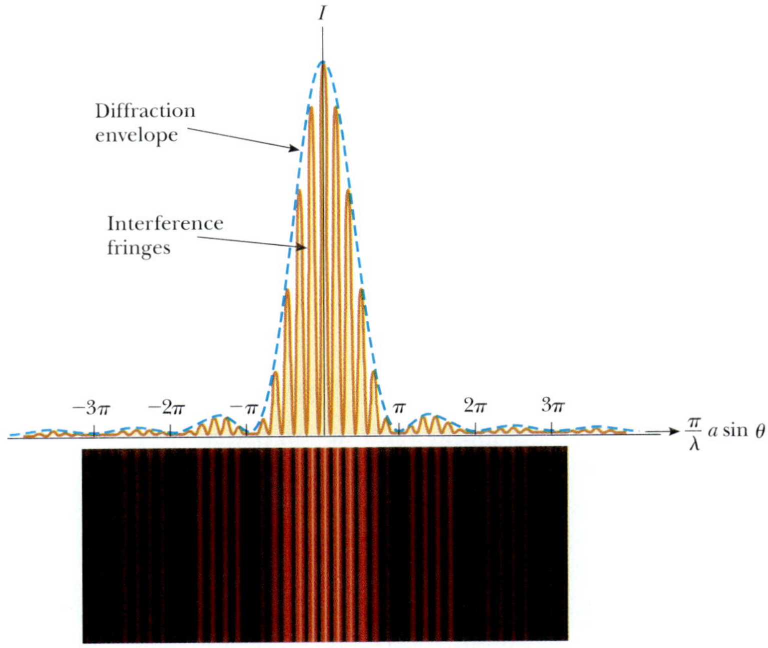

ACTIVE FIGURE 38.7

The combined effects of two-slit and single-slit interference. This pattern is produced when 650-nm light waves pass through two 3.0-μm slits that are 18 μm apart. Notice how the diffraction pattern acts as an "envelope" and controls the intensity of the regularly spaced interference maxima.

Sign in at www.thomsonedu.com and go to ThomsonNOW to adjust the slit width, slit separation, and the wavelength of the light to see the effect on the interference pattern.

interference and a single-slit diffraction pattern from each slit, we combine Equations 37.12 and 38.2:

$$I = I_{\max} \cos^2\left(\frac{\pi d \sin\theta}{\lambda}\right)\left[\frac{\sin(\pi a \sin\theta/\lambda)}{\pi a \sin\theta/\lambda}\right]^2 \qquad \textbf{(38.3)}$$

Although this expression looks complicated, it merely represents the single-slit diffraction pattern (the factor in square brackets) acting as an "envelope" for a two-slit interference pattern (the cosine-squared factor) as shown in Active Figure 38.7. The broken blue curve in Active Figure 38.7 represents the factor in square brackets in Equation 38.3. The cosine-squared factor by itself would give a series of peaks all with the same height as the highest peak of the brown curve in Active Figure 38.7. Because of the effect of the square-bracket factor, however, these peaks vary in height as shown.

Equation 37.2 indicates the conditions for interference maxima as $d \sin\theta = m\lambda$, where d is the distance between the two slits. Equation 38.1 specifies that the first diffraction minimum occurs when $a \sin\theta = \lambda$, where a is the slit width. Dividing Equation 37.2 by Equation 38.1 (with $m = 1$) allows us to determine which interference maximum coincides with the first diffraction minimum:

$$\frac{d \sin\theta}{a \sin\theta} = \frac{m\lambda}{\lambda}$$

$$\frac{d}{a} = m \qquad \textbf{(38.4)}$$

In Active Figure 38.7, $d/a = 18\ \mu\text{m}/3.0\ \mu\text{m} = 6$. Therefore, the sixth interference maximum (if we count the central maximum as $m = 0$) is aligned with the first diffraction minimum and cannot be seen.

Quick Quiz 38.2 Consider the central peak in the diffraction envelope in Active Figure 38.7. Suppose the wavelength of the light is changed to 450 nm. What happens to this central peak? (a) The width of the peak decreases, and the number of interference fringes it encloses decreases. (b) The width of the peak decreases, and the number of interference fringes it encloses increases. (c) The width of the peak decreases, and the number of interference fringes it encloses remains the same. (d) The width of the peak increases, and the number of interference fringes it encloses decreases. (e) The width of the peak increases, and the number of interference fringes it encloses increases. (f) The width of the peak increases, and the number of interference fringes it encloses remains the same. (g) The width of the peak remains the same and the number of interference fringes it encloses decreases. (h) The width of the peak remains the same and the number of interference fringes it encloses increases. (i) The width of the peak remains the same and the number of interference fringes it encloses remains the same.

38.3 Resolution of Single-Slit and Circular Apertures

The ability of optical systems to distinguish between closely spaced objects is limited because of the wave nature of light. To understand this limitation, consider Figure 38.8, which shows two light sources far from a narrow slit of width a. The sources can be two noncoherent point sources S_1 and S_2; for example, they could be two distant stars. If no interference occurred between light passing through different parts of the slit, two distinct bright spots (or images) would be observed on the viewing screen. Because of such interference, however, each source is imaged as a bright central region flanked by weaker bright and dark fringes, a diffraction pattern. What is observed on the screen is the sum of two diffraction patterns: one from S_1 and the other from S_2.

If the two sources are far enough apart to keep their central maxima from overlapping as in Figure 38.8a, their images can be distinguished and are said to be *resolved.* If the sources are close together as in Figure 38.8b, however, the two central maxima overlap and the images are not resolved. To determine whether two images are resolved, the following condition is often used:

When the central maximum of one image falls on the first minimum of another image, the images are said to be just resolved. This limiting condition of resolution is known as **Rayleigh's criterion.**

From Rayleigh's criterion, we can determine the minimum angular separation $\theta_{\min}$ subtended by the sources at the slit in Figure 38.8 for which the images are just resolved. Equation 38.1 indicates that the first minimum in a single-slit diffraction pattern occurs at the angle for which

$$\sin\theta = \frac{\lambda}{a}$$

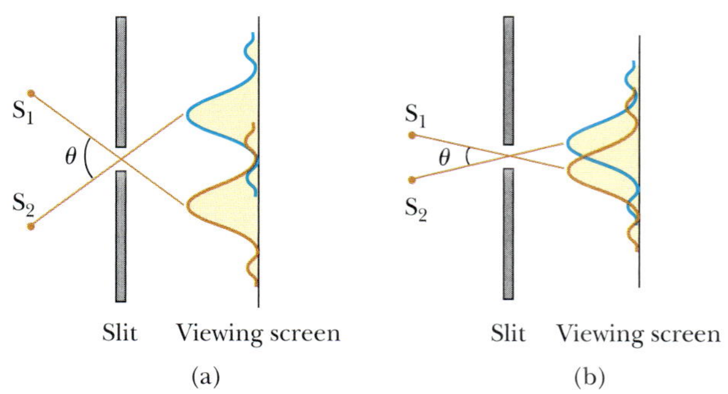

Figure 38.8 Two point sources far from a narrow slit each produce a diffraction pattern. (a) The angle subtended by the sources at the slit is large enough for the diffraction patterns to be distinguishable. (b) The angle subtended by the sources is so small that their diffraction patterns overlap, and the images are not well resolved. (Notice that the angles are greatly exaggerated. The drawing is not to scale.)

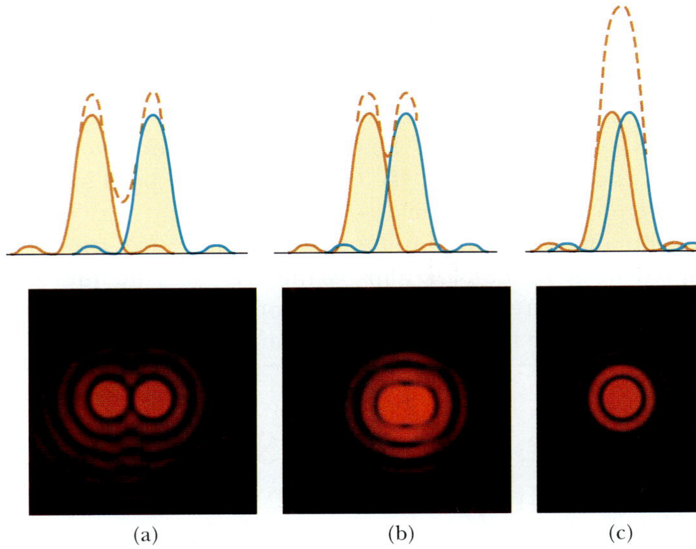

From M. Cagnet, M. Françon, and J. C. Thrierr, *Atlas of Optical Phenomena*, Berlin, Springer-Verlag, 1962, plate 16

Figure 38.9 Individual diffraction patterns of two point sources (solid curves) and the resultant patterns (dashed curves) for various angular separations of the sources as the light passes through a circular aperture. In each case, the dashed curve is the sum of the two solid curves. (a) The sources are far apart, and the patterns are well resolved. (b) The sources are closer together such that the angular separation just satisfies Rayleigh's criterion, and the patterns are just resolved. (c) The sources are so close together that the patterns are not resolved.

where a is the width of the slit. According to Rayleigh's criterion, this expression gives the smallest angular separation for which the two images are resolved. Because $\lambda \ll a$ in most situations, $\sin\theta$ is small and we can use the approximation $\sin\theta \approx \theta$. Therefore, the limiting angle of resolution for a slit of width a is

$$\theta_{\text{min}} = \frac{\lambda}{a} \tag{38.5}$$

where θ_{min} is expressed in radians. Hence, the angle subtended by the two sources at the slit must be greater than λ/a if the images are to be resolved.

Many optical systems use circular apertures rather than slits. The diffraction pattern of a circular aperture as shown in the photographs of Figure 38.9 consists of a central circular bright disk surrounded by progressively fainter bright and dark rings. Figure 38.9 shows diffraction patterns for three situations in which light from two point sources passes through a circular aperture. When the sources are far apart, their images are well resolved (Fig. 38.9a). When the angular separation of the sources satisfies Rayleigh's criterion, the images are just resolved (Fig. 38.9b). Finally, when the sources are close together, the images are said to be unresolved (Fig. 38.9c).

Analysis shows that the limiting angle of resolution of the circular aperture is

Limiting angle of resolution for a circular aperture ▶

$$\theta_{\text{min}} = 1.22\,\frac{\lambda}{D} \tag{38.6}$$

where D is the diameter of the aperture. This expression is similar to Equation 38.5 except for the factor 1.22, which arises from a mathematical analysis of diffraction from the circular aperture.

Quick Quiz 38.3 Cat's eyes have pupils that can be modeled as vertical slits. At night, would cats be more successful in resolving (a) headlights on a distant car or (b) vertically separated lights on the mast of a distant boat?

Quick Quiz 38.4 Suppose you are observing a binary star with a telescope and are having difficulty resolving the two stars. You decide to use a colored filter to maximize the resolution. (A filter of a given color transmits only that color of light.) What color filter should you choose? (a) blue (b) green (c) yellow (d) red

---

EXAMPLE 38.2 Resolution of the Eye

Light of wavelength 500 nm, near the center of the visible spectrum, enters a human eye. Although pupil diameter varies from person to person, estimate a daytime diameter of 2 mm.

(A) Estimate the limiting angle of resolution for this eye, assuming its resolution is limited only by diffraction.

SOLUTION

Conceptualize In Figure 38.9, identify the aperture through which the light travels as the pupil of the eye. Light passing through this small aperture causes diffraction patterns to occur on the retina.

Categorize We evaluate the result using equations developed in this section, so we categorize this example as a substitution problem.

Use Equation 38.6, taking $\lambda = 500$ nm and $D = 2$ mm:

$$\theta_{\min} = 1.22\frac{\lambda}{D} = 1.22\left(\frac{5.00 \times 10^{-7}\text{ m}}{2 \times 10^{-3}\text{ m}}\right)$$

$$\approx 3 \times 10^{-4}\text{ rad} \approx 1\text{ min of arc}$$

(B) Determine the minimum separation distance d between two point sources that the eye can distinguish if the point sources are a distance $L = 25$ cm from the observer (Fig. 38.10).

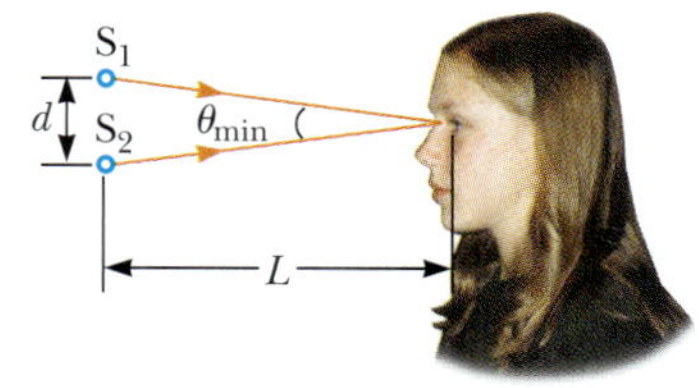

Figure 38.10 (Example 38.2) Two point sources separated by a distance d as observed by the eye.

SOLUTION

Noting that $\theta_{\min}$ is small, find d:

$$\sin\theta_{\min} \approx \theta_{\min} \approx \frac{d}{L} \rightarrow d = L\theta_{\min}$$

Substitute numerical values:

$$d = (25\text{ cm})(3 \times 10^{-4}\text{ rad}) = 8 \times 10^{-3}\text{ cm}$$

This result is approximately equal to the thickness of a human hair.

EXAMPLE 38.3 Resolution of a Telescope

The Keck telescope at Mauna Kea, Hawaii, has an effective diameter of 10 m. What is its limiting angle of resolution for 600-nm light?

SOLUTION

Conceptualize In Figure 38.9, identify the aperture through which the light travels as the opening of the telescope. Light passing through this aperture causes diffraction patterns to occur in the final image.

Categorize We evaluate the result using equations developed in this section, so we categorize this example as a substitution problem.

Use Equation 38.6, taking $\lambda = 6.00 \times 10^{-7}$ m and $D =$ 10 m:

$$\theta_{\min} = 1.22\frac{\lambda}{D} = 1.22\left(\frac{6.00 \times 10^{-7}\text{ m}}{10\text{ m}}\right)$$

$$= 7.3 \times 10^{-8}\text{ rad} \approx 0.015\text{ s of arc}$$

Any two stars that subtend an angle greater than or equal to this value are resolved (if atmospheric conditions are ideal).

What If? What if we consider radio telescopes? They are much larger in diameter than optical telescopes, but do they have better angular resolutions than optical telescopes? For example, the radio telescope at Arecibo, Puerto Rico, has a diameter of 305 m and is designed to detect radio waves of 0.75-m wavelength. How does its resolution compare with that of the Keck telescope?

Answer The increase in diameter might suggest that radio telescopes would have better resolution than the Keck telescope, but Equation 38.6 shows that $\theta_{\min}$ depends on *both* diameter and wavelength. Calculating the minimum angle of resolution for the radio telescope, we find

$$\theta_{\min} = 1.22\frac{\lambda}{D} = 1.22\left(\frac{0.75\text{ m}}{305\text{ m}}\right)$$

$$= 3.0 \times 10^{-3}\text{ rad} \approx 10\text{ min of arc}$$

This limiting angle of resolution is measured in *minutes* of arc rather than the *seconds* of arc for the optical telescope. Therefore, the change in wavelength more than compensates for the increase in diameter. The limiting angle of resolution for the Arecibo radio telescope is more than 40 000 times larger (that is, *worse*) than the Keck minimum.

The Keck telescope discussed in Example 38.3 can never reach its diffraction limit because the limiting angle of resolution is always set by atmospheric blurring at optical wavelengths. This seeing limit is usually about 1 s of arc and is never smaller than about 0.1 s of arc. The atmospheric blurring is caused by variations in index of refraction with temperature variations in the air. This blurring is one reason for the superiority of photographs from the Hubble Space Telescope, which views celestial objects from an orbital position above the atmosphere.

As an example of the effects of atmospheric blurring, consider telescopic images of Pluto and its moon, Charon. Figure 38.11a, an image taken in 1978, represents the discovery of Charon. In this photograph, taken from an Earth-based telescope, atmospheric turbulence causes the image of Charon to appear only as a bump on the edge of Pluto. In comparison, Figure 38.11b shows a photograph taken with the Hubble Space Telescope. Without the problems of atmospheric turbulence, Pluto and its moon are clearly resolved.

38.4 The Diffraction Grating

The **diffraction grating,** a useful device for analyzing light sources, consists of a large number of equally spaced parallel slits. A *transmission grating* can be made by cutting parallel grooves on a glass plate with a precision ruling machine. The spaces between the grooves are transparent to the light and hence act as separate slits. A *reflection grating* can be made by cutting parallel grooves on the surface of a

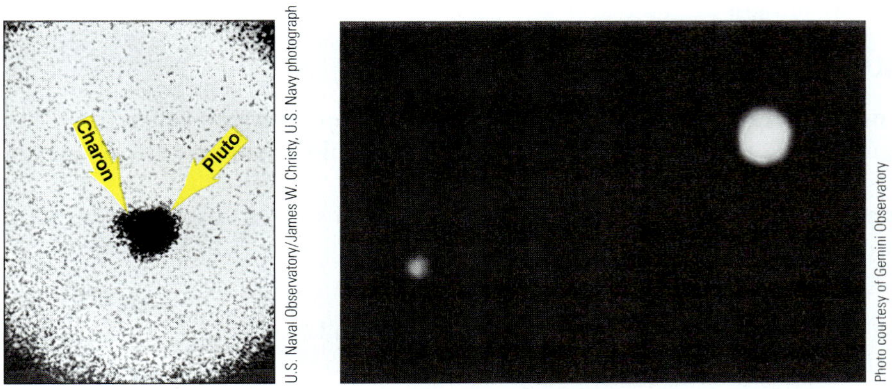

(a) (b)

Figure 38.11 (a) The photograph on which Charon, the moon of Pluto, was discovered in 1978. From an Earth-based telescope, atmospheric blurring results in Charon appearing only as a subtle bump on the edge of Pluto. (b) A Hubble Space Telescope photo of Pluto and Charon, clearly resolving the two objects.

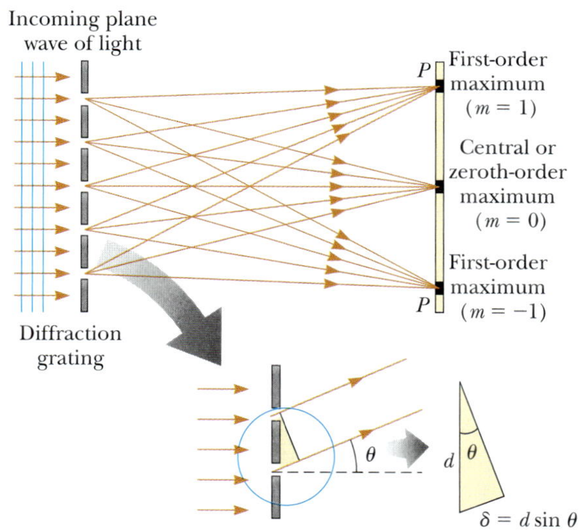

Figure 38.12 Side view of a diffraction grating. The slit separation is d, and the path difference between adjacent slits is $d \sin \theta$.

reflective material. The reflection of light from the spaces between the grooves is specular, and the reflection from the grooves cut into the material is diffuse. Therefore, the spaces between the grooves act as parallel sources of reflected light like the slits in a transmission grating. Current technology can produce gratings that have very small slit spacings. For example, a typical grating ruled with 5 000 grooves/cm has a slit spacing $d = (1/5\,000)$ cm $= 2.00 \times 10^{-4}$ cm.

A section of a diffraction grating is illustrated in Figure 38.12. A plane wave is incident from the left, normal to the plane of the grating. The pattern observed on the screen far to the right of the grating is the result of the combined effects of interference and diffraction. Each slit produces diffraction, and the diffracted beams interfere with one another to produce the final pattern.

The waves from all slits are in phase as they leave the slits. For an arbitrary direction θ measured from the horizontal, however, the waves must travel different path lengths before reaching the screen. Notice in Figure 38.12 that the path difference δ between rays from any two adjacent slits is equal to $d \sin \theta$. If this path difference equals one wavelength or some integral multiple of a wavelength, waves from all slits are in phase at the screen and a bright fringe is observed. Therefore, the condition for *maxima* in the interference pattern at the angle θ_{bright} is

$$d \sin \theta_{\text{bright}} = m\lambda \quad m = 0, \pm 1, \pm 2, \pm 3, \ldots \qquad (38.7)$$

◀ Condition for interference maxima for a grating

We can use this expression to calculate the wavelength if we know the grating spacing d and the angle θ_{bright}. If the incident radiation contains several wavelengths, the mth-order maximum for each wavelength occurs at a specific angle. All wavelengths are seen at $\theta = 0$, corresponding to $m = 0$, the zeroth-order maximum. The first-order maximum ($m = 1$) is observed at an angle that satisfies the relationship $\sin \theta_{\text{bright}} = \lambda/d$, the second-order maximum ($m = 2$) is observed at a larger angle θ_{bright}, and so on. For the small values of d typical in a diffraction grating, the angles θ_{bright} are large, as we see in Example 38.5.

The intensity distribution for a diffraction grating obtained with the use of a monochromatic source is shown in Active Figure 38.13. Notice the sharpness of the principal maxima and the broadness of the dark areas compared with the broad bright fringes characteristic of the two-slit interference pattern (see Fig. 37.7). You should also review Figure 37.8, which shows that the width of the intensity maxima decreases as the number of slits increases. Because the principal maxima are so sharp, they are much brighter than two-slit interference maxima.

Quick Quiz 38.5 Ultraviolet light of wavelength 350 nm is incident on a diffraction grating with slit spacing d and forms an interference pattern on a screen

PITFALL PREVENTION 38.3

A Diffraction Grating Is an Interference Grating

As with *diffraction pattern*, *diffraction grating* is a misnomer, but is deeply entrenched in the language of physics. The diffraction grating depends on diffraction in the same way as the double slit, spreading the light so that light from different slits can interfere. It would be more correct to call it an *interference grating*, but *diffraction grating* is the name in use.

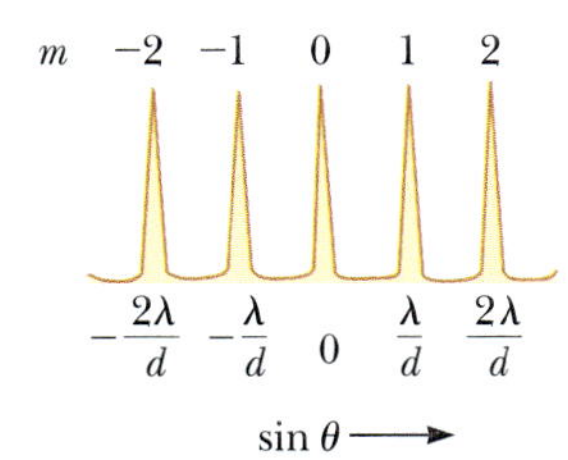

ACTIVE FIGURE 38.13

Intensity versus $\sin \theta$ for a diffraction grating. The zeroth-, first-, and second-order maxima are shown.

Sign in at www.thomsonedu.com and go to ThomsonNOW to choose the number of slits to be illuminated and see the effect on the interference pattern.

a distance L away. The angular positions θ_{bright} of the interference maxima are large. The locations of the bright fringes are marked on the screen. Now red light of wavelength 700 nm is used with a diffraction grating to form another diffraction pattern on the screen. Will the bright fringes of this pattern be located at the marks on the screen if (a) the screen is moved to a distance $2L$ from the grating, (b) the screen is moved to a distance $L/2$ from the grating, (c) the grating is replaced with one of slit spacing $2d$, (d) the grating is replaced with one of slit spacing $d/2$, or (e) nothing is changed?

CONCEPTUAL EXAMPLE 38.4 A Compact Disc Is a Diffraction Grating

Light reflected from the surface of a compact disc is multicolored, as shown in Figure 38.14. The colors and their intensities depend on the orientation of the CD relative to the eye and relative to the light source. Explain how this works.

SOLUTION

The surface of a CD has a spiral grooved track (with adjacent grooves having a separation on the order of 1 μm). Therefore, the surface acts as a reflection grating. The light reflecting from the regions between these closely spaced grooves interferes constructively only in certain directions that depend on the wavelength and the direction of the incident light. Any section of the CD serves as a diffraction grating for white light, sending different colors in different directions. The different colors you see upon viewing one section change when the light source, the CD, or you change position. This change in position causes the angle of incidence or the angle of the diffracted light to be altered.

Figure 38.14 (Conceptual Example 38.4) A compact disc observed under white light. The colors observed in the reflected light and their intensities depend on the orientation of the CD relative to the eye and relative to the light source.

EXAMPLE 38.5 The Orders of a Diffraction Grating

Monochromatic light from a helium–neon laser (λ = 632.8 nm) is incident normally on a diffraction grating containing 6 000 grooves per centimeter. Find the angles at which the first- and second-order maxima are observed.

SOLUTION

Conceptualize Study Figure 38.12 and imagine that the light coming from the left originates from the helium–neon laser.

Categorize We evaluate results using equations developed in this section, so we categorize this example as a substitution problem.

Calculate the slit separation as the inverse of the number of grooves per centimeter:

$$d = \frac{1}{6\,000}\text{ cm} = 1.667 \times 10^{-4}\text{ cm} = 1\,667\text{ nm}$$

Solve Equation 38.7 for $\sin\theta$ and substitute numerical values for the first-order maximum ($m = 1$) to find θ_1:

$$\sin\theta_1 = \frac{(1)\lambda}{d} = \frac{632.8\text{ nm}}{1\,667\text{ nm}} = 0.379\,7$$

$$\theta_1 = 22.31°$$

Repeat for the second-order maximum ($m = 2$):

$$\sin\theta_2 = \frac{(2)\lambda}{d} = \frac{2(632.8\text{ nm})}{1\,667\text{ nm}} = 0.759\,4$$

$$\theta_2 = 49.41°$$

What If? What if you looked for the third-order maximum? Would you find it?

Answer For $m = 3$, we find $\sin\theta_3 = 1.139$. Because $\sin\theta$ cannot exceed unity, this result does not represent a realistic solution. Hence, only zeroth-, first-, and second-order maxima can be observed for this situation.

Applications of Diffraction Gratings

A schematic drawing of a simple apparatus used to measure angles in a diffraction pattern is shown in Active Figure 38.15. This apparatus is a *diffraction grating spectrometer.* The light to be analyzed passes through a slit, and a collimated beam of light is incident on the grating. The diffracted light leaves the grating at angles that satisfy Equation 38.7, and a telescope is used to view the image of the slit. The wavelength can be determined by measuring the precise angles at which the images of the slit appear for the various orders.

The spectrometer is a useful tool in *atomic spectroscopy,* in which the light from an atom is analyzed to find the wavelength components. These wavelength components can be used to identify the atom. We shall investigate atomic spectra in Chapter 42 of the extended version of this text.

Another application of diffraction gratings is the *grating light valve* (GLV), which may compete in the near future in video projection with the digital micromirror devices (DMDs) discussed in Section 35.4. A GLV is a silicon microchip fitted with an array of parallel silicon nitride ribbons coated with a thin layer of aluminum (Fig. 38.16). Each ribbon is approximately 20 μm long and 5 μm wide and is separated from the silicon substrate by an air gap on the order of 100 nm. With no voltage applied, all ribbons are at the same level. In this situation, the array of ribbons acts as a flat surface, specularly reflecting incident light.

When a voltage is applied between a ribbon and the electrode on the silicon substrate, an electric force pulls the ribbon downward, closer to the substrate. Alternate ribbons can be pulled down, while those in between remain in an elevated configuration. As a result, the array of ribbons acts as a diffraction grating such that the constructive interference for a particular wavelength of light can be directed toward a screen or other optical display system. By using three such devices—one each for red, blue, and green light—full-color display is possible.

A GLV tends to be simpler to fabricate and higher in resolution than comparable DMDs. On the other hand, DMDs have already made an entry into the market. It will be interesting to watch this technology competition in future years.

Another interesting application of diffraction gratings is **holography,** the production of three-dimensional images of objects. The physics of holography was developed by Dennis Gabor (1900–1979) in 1948 and resulted in the Nobel Prize

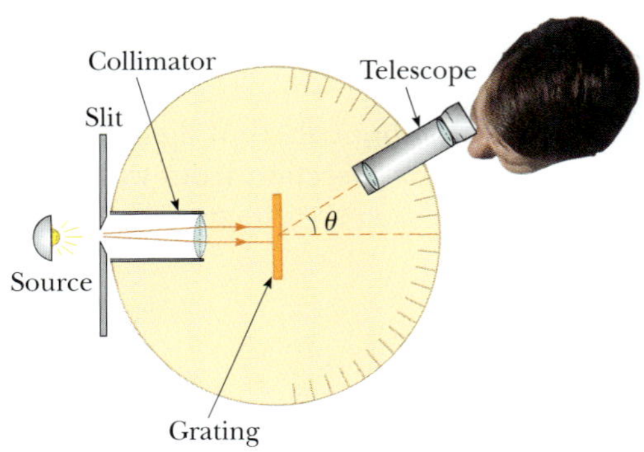

ACTIVE FIGURE 38.15

Diagram of a diffraction grating spectrometer. The collimated beam incident on the grating is spread into its various wavelength components with constructive interference for a particular wavelength occurring at the angles θ_{bright} that satisfy the equation $d\sin\theta_{\text{bright}} = m\lambda$, where $m = 0, \pm1, \pm2, \ldots$.

Sign in at www.thomsonedu.com and go to ThomsonNOW to use the spectrometer and observe constructive interference for various wavelengths.

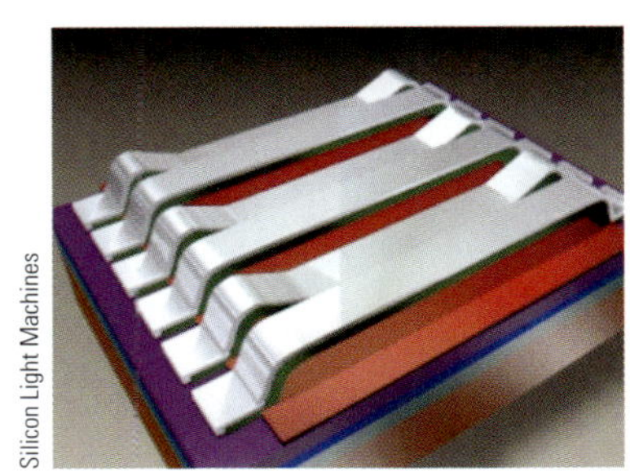

Figure 38.16 A small portion of a grating light valve. The alternating reflective ribbons at different levels act as a diffraction grating, offering very high-speed control of the direction of light toward a digital display device.

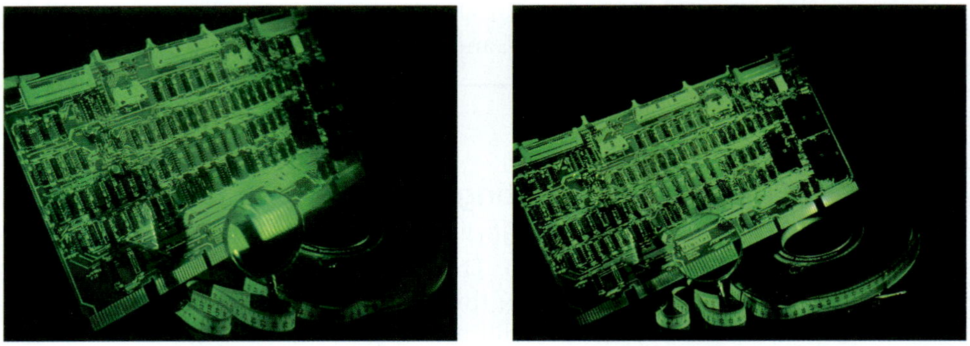

(a) (b)

Figure 38.17 In this hologram, a circuit board is shown from two different views. Notice the difference in the appearance of the measuring tape and the view through the magnifying lens in (a) and (b).

in Physics for Gabor in 1971. The requirement of coherent light for holography delayed the realization of holographic images from Gabor's work until the development of lasers in the 1960s. Figure 38.17 shows a hologram and the three-dimensional character of its image. Notice in particular the difference in the view through the magnifying glass in Figures 38.17a and 38.17b.

Figure 38.18 shows how a hologram is made. Light from the laser is split into two parts by a half-silvered mirror at *B*. One part of the beam reflects off the object to be photographed and strikes an ordinary photographic film. The other half of the beam is diverged by lens L_2, reflects from mirrors M_1 and M_2, and finally strikes the film. The two beams overlap to form an extremely complicated interference pattern on the film. Such an interference pattern can be produced only if the phase relationship of the two waves is constant throughout the exposure of the film. This condition is met by illuminating the scene with light coming through a pinhole or with coherent laser radiation. The hologram records not only the intensity of the light scattered from the object (as in a conventional photograph), but also the phase difference between the reference beam and the beam scattered from the object. Because of this phase difference, an interference pattern is formed that produces an image in which all three-dimensional information available from the perspective of any point on the hologram is preserved.

In a normal photographic image, a lens is used to focus the image so that each point on the object corresponds to a single point on the film. Notice that there is no lens used in Figure 38.18 to focus the light onto the film. Therefore, light from each point on the object reaches *all* points on the film. As a result, each region of the photographic film on which the hologram is recorded contains information about all illuminated points on the object, which leads to a remarkable result: if a small section of the hologram is cut from the film, the complete image can be formed from the small piece! (The quality of the image is reduced, but the entire image is present.)

A hologram is best viewed by allowing coherent light to pass through the developed film as one looks back along the direction from which the beam comes. The interference pattern on the film acts as a diffraction grating. Figure 38.19 shows

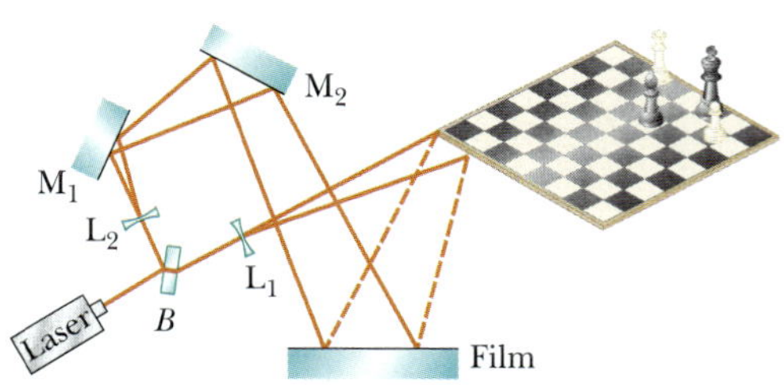

Figure 38.18 Experimental arrangement for producing a hologram.

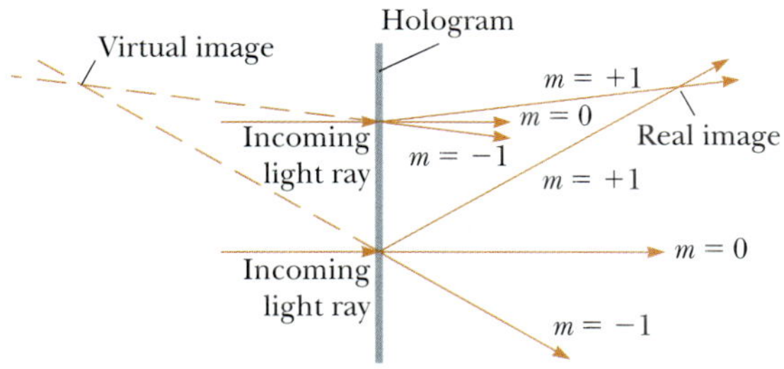

Figure 38.19 Two light rays strike a hologram at normal incidence. For each ray, outgoing rays corresponding to $m = 0$ and $m = \pm 1$ are shown. If the $m = -1$ rays are extended backward, a virtual image of the object photographed in the hologram exists on the front side of the hologram.

two rays of light striking and passing through the film. For each ray, the $m = 0$ and $m = \pm 1$ rays in the diffraction pattern are shown emerging from the right side of the film. The $m = +1$ rays converge to form a real image of the scene, which is not the image that is normally viewed. By extending the light rays corresponding to $m = -1$ behind the film, we see that there is a virtual image located there, with light coming from it in exactly the same way that light came from the actual object when the film was exposed. This image is what one sees when looking through the holographic film.

Holograms are finding a number of applications. You may have a hologram on your credit card. This special type of hologram is called a *rainbow hologram* and is designed to be viewed in reflected white light.

38.5 Diffraction of X-Rays by Crystals

In principle, the wavelength of any electromagnetic wave can be determined if a grating of the proper spacing (on the order of λ) is available. X-rays, discovered by Wilhelm Roentgen (1845–1923) in 1895, are electromagnetic waves of very short wavelength (on the order of 0.1 nm). It would be impossible to construct a grating having such a small spacing by the cutting process described at the beginning of Section 38.4. The atomic spacing in a solid is known to be about 0.1 nm, however. In 1913, Max von Laue (1879–1960) suggested that the regular array of atoms in a crystal could act as a three-dimensional diffraction grating for x-rays. Subsequent experiments confirmed this prediction. The diffraction patterns from crystals are complex because of the three-dimensional nature of the crystal structure. Nevertheless, x-ray diffraction has proved to be an invaluable technique for elucidating these structures and for understanding the structure of matter.

Figure 38.20 shows one experimental arrangement for observing x-ray diffraction from a crystal. A collimated beam of monochromatic x-rays is incident on a crystal. The diffracted beams are very intense in certain directions, corresponding

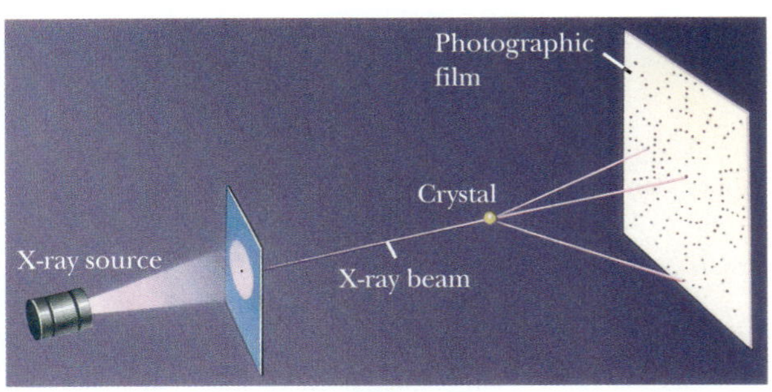

Figure 38.20 Schematic diagram of the technique used to observe the diffraction of x-rays by a crystal. The array of spots formed on the film is called a Laue pattern.

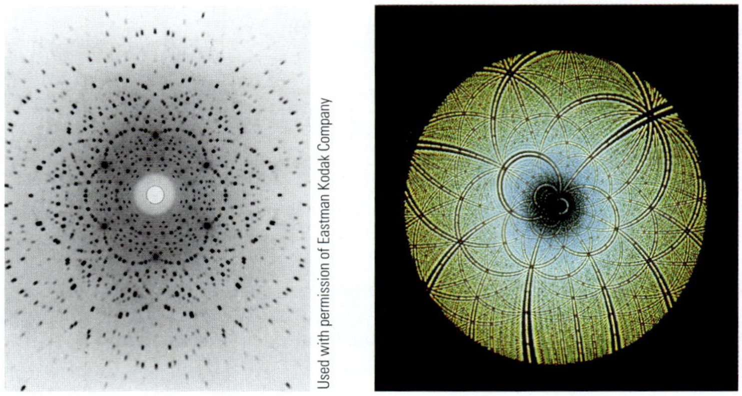

Used with permission of Eastman Kodak Company

© I. Andersson Oxford Molecular Biophysics Laboratory/Photo Researchers, Inc.

(a) (b)

Figure 38.21 (a) A Laue pattern of a single crystal of the mineral beryl (beryllium aluminum silicate). Each dot represents a point of constructive interference. (b) A Laue pattern of the enzyme Rubisco, produced with a wideband x-ray spectrum. This enzyme is present in plants and takes part in the process of photosynthesis. The Laue pattern is used to determine the crystal structure of Rubisco.

to constructive interference from waves reflected from layers of atoms in the crystal. The diffracted beams, which can be detected by a photographic film, form an array of spots known as a *Laue pattern* as in Figure 38.21a. One can deduce the crystalline structure by analyzing the positions and intensities of the various spots in the pattern. Figure 38.21b shows a Laue pattern from a crystalline enzyme, using a wide range of wavelengths so that a swirling pattern results.

PITFALL PREVENTION 38.4

Different Angles

Notice in Figure 38.23 that the angle θ is measured from the reflecting surface rather than from the normal as in the case of the law of reflection in Chapter 35. With slits and diffraction gratings, we also measured the angle θ from the normal to the array of slits. Because of historical tradition, the angle is measured differently in Bragg diffraction, so interpret Equation 38.8 with care.

The arrangement of atoms in a crystal of sodium chloride (NaCl) is shown in Figure 38.22. Each unit cell (the geometric solid that repeats throughout the crystal) is a cube having an edge length a. A careful examination of the NaCl structure shows that the ions lie in discrete planes (the shaded areas in Fig. 38.22). Now suppose an incident x-ray beam makes an angle θ with one of the planes as in Figure 38.23. The beam can be reflected from both the upper plane and the lower one, but the beam reflected from the lower plane travels farther than the beam reflected from the upper plane. The effective path difference is $2d \sin\theta$. The two beams reinforce each other (constructive interference) when this path difference equals some integer multiple of λ. The same is true for reflection from the entire family of parallel planes. Hence, the condition for *constructive* interference (maxima in the reflected beam) is

Bragg's law ▶

$$2d \sin\theta = m\lambda \quad m = 1, 2, 3, \ldots \tag{38.8}$$

This condition is known as **Bragg's law,** after W. L. Bragg (1890–1971), who first derived the relationship. If the wavelength and diffraction angle are measured, Equation 38.8 can be used to calculate the spacing between atomic planes.

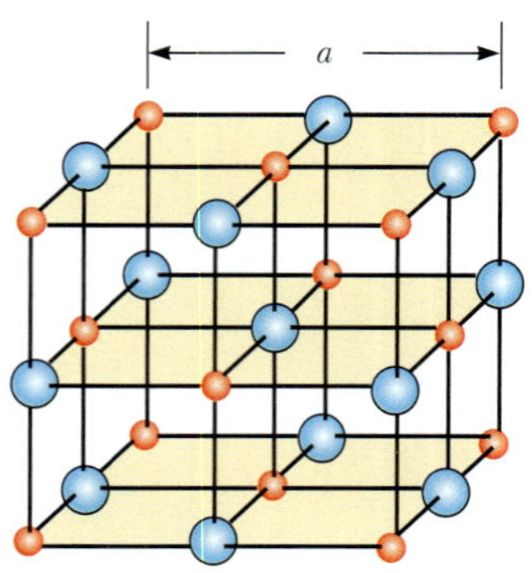

Figure 38.22 Crystalline structure of sodium chloride (NaCl). The blue spheres represent Cl^- ions, and the red spheres represent Na^+ ions. The length of the cube edge is $a = 0.562\ 737$ nm.

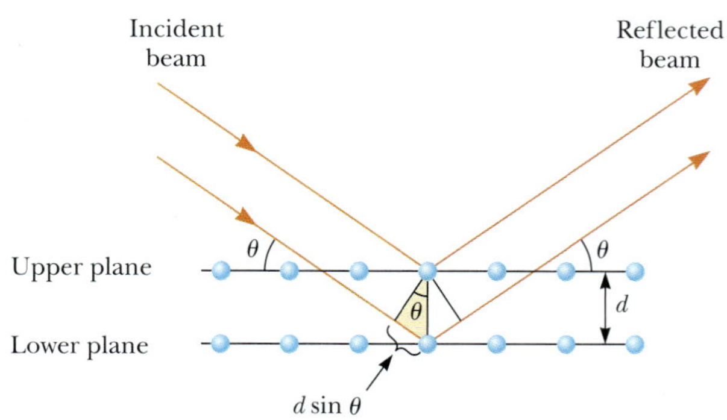

Figure 38.23 A two-dimensional description of the reflection of an x-ray beam from two parallel crystalline planes separated by a distance d. The beam reflected from the lower plane travels farther than the beam reflected from the upper plane by a distance $2d \sin\theta$.

38.6 Polarization of Light Waves

In Chapter 34, we described the transverse nature of light and all other electromagnetic waves. Polarization, discussed in this section, is firm evidence of this transverse nature.

An ordinary beam of light consists of a large number of waves emitted by the atoms of the light source. Each atom produces a wave having some particular orientation of the electric field vector $\vec{\mathbf{E}}$, corresponding to the direction of atomic vibration. The *direction of polarization* of each individual wave is defined to be the direction in which the electric field is vibrating. In Figure 38.24, this direction happens to lie along the y axis. An individual electromagnetic wave, however, could have its $\vec{\mathbf{E}}$ vector in the yz plane, making any possible angle with the y axis. Because all directions of vibration from a wave source are possible, the resultant electromagnetic wave is a superposition of waves vibrating in many different directions. The result is an **unpolarized** light beam, represented in Figure 38.25a. The direction of wave propagation in this figure is perpendicular to the page. The arrows show a few possible directions of the electric field vectors for the individual waves making up the resultant beam. At any given point and at some instant of time, all these individual electric field vectors add to give one resultant electric field vector.

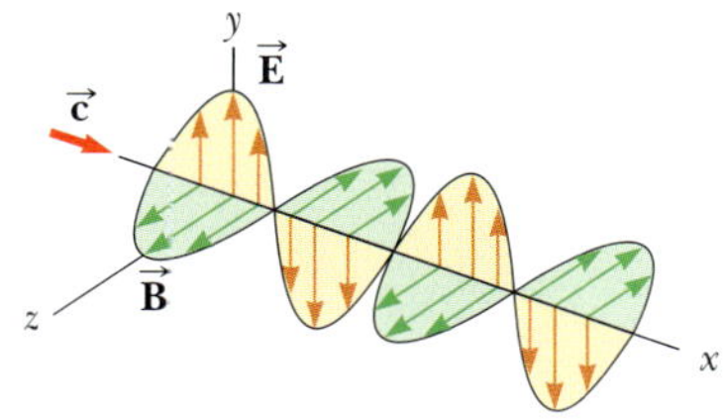

Figure 38.24 Schematic diagram of an electromagnetic wave propagating at velocity $\vec{\mathbf{c}}$ in the x direction. The electric field vibrates in the xy plane, and the magnetic field vibrates in the xz plane.

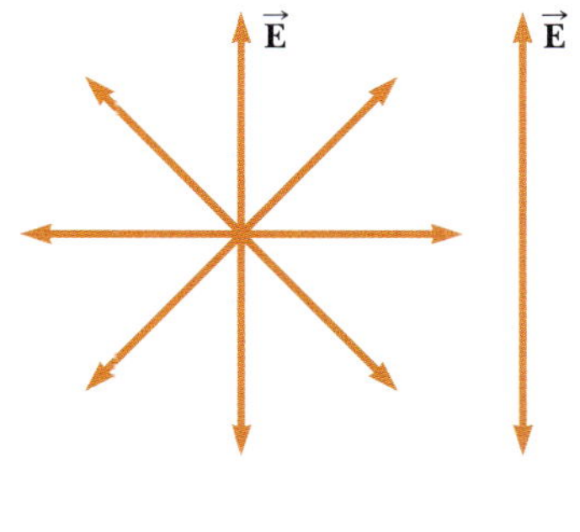

Figure 38.25 (a) A representation of an unpolarized light beam viewed along the direction of propagation (perpendicular to the page). The transverse electric field can vibrate in any direction in the plane of the page with equal probability. (b) A linearly polarized light beam with the electric field vibrating in the vertical direction.

As noted in Section 34.3, a wave is said to be **linearly polarized** if the resultant electric field $\vec{\mathbf{E}}$ vibrates in the same direction *at all times* at a particular point as shown in Figure 38.25b. (Sometimes, such a wave is described as *plane-polarized,* or simply *polarized.*) The plane formed by $\vec{\mathbf{E}}$ and the direction of propagation is called the *plane of polarization* of the wave. If the wave in Figure 38.24 represents the resultant of all individual waves, the plane of polarization is the xy plane.

A linearly polarized beam can be obtained from an unpolarized beam by removing all waves from the beam except those whose electric field vectors oscillate in a single plane. We now discuss four processes for producing polarized light from unpolarized light.

Polarization by Selective Absorption

The most common technique for producing polarized light is to use a material that transmits waves whose electric fields vibrate in a plane parallel to a certain direction and that absorbs waves whose electric fields vibrate in all other directions.

In 1938, E. H. Land (1909–1991) discovered a material, which he called *Polaroid,* that polarizes light through selective absorption. This material is fabricated in thin sheets of long-chain hydrocarbons. The sheets are stretched during manufacture so that the long-chain molecules align. After a sheet is dipped into a solution containing iodine, the molecules become good electrical conductors. Conduction takes place primarily along the hydrocarbon chains because electrons can move easily only along the chains. If light whose electric field vector is parallel to the chains is incident on the material, the electric field accelerates electrons along the chains and energy is absorbed from the radiation. Therefore, the light does not pass through the material. Light whose electric field vector is perpendicular to the chains passes through the material because electrons cannot move from one molecule to the next. As a result, when unpolarized light is incident on the material, the exiting light is polarized perpendicular to the molecular chains.

It is common to refer to the direction perpendicular to the molecular chains as the *transmission axis.* In an ideal polarizer, all light with $\vec{\mathbf{E}}$ parallel to the transmission axis is transmitted and all light with $\vec{\mathbf{E}}$ perpendicular to the transmission axis is absorbed.

Active Figure 38.26 (page 1094) represents an unpolarized light beam incident on a first polarizing sheet, called the *polarizer.* Because the transmission axis is oriented vertically in the figure, the light transmitted through this sheet is polarized vertically. A second polarizing sheet, called the *analyzer,* intercepts the beam. In

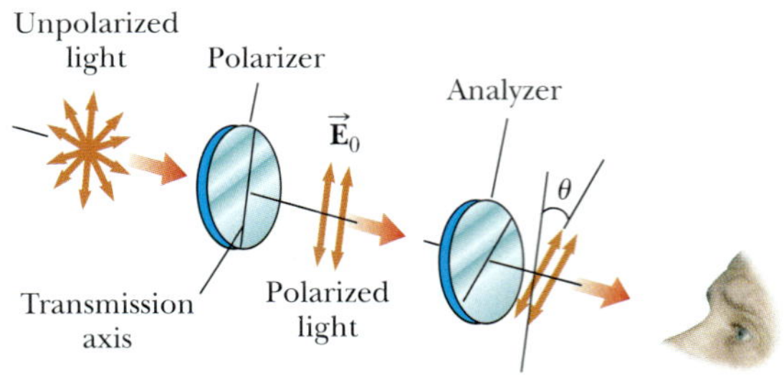

ACTIVE FIGURE 38.26

Two polarizing sheets whose transmission axes make an angle θ with each other. Only a fraction of the polarized light incident on the analyzer is transmitted through it.

Sign in at www.thomsonedu.com and go to ThomsonNOW to rotate the analyzer and see the effect on the transmitted light.

Active Figure 38.26, the analyzer transmission axis is set at an angle θ to the polarizer axis. We call the electric field vector of the first transmitted beam $\vec{\mathbf{E}}_0$. The component of $\vec{\mathbf{E}}_0$ perpendicular to the analyzer axis is completely absorbed. The component of $\vec{\mathbf{E}}_0$ parallel to the analyzer axis, which is transmitted through the analyzer, is $E_0 \cos\theta$. Because the intensity of the transmitted beam varies as the square of its magnitude, we conclude that the intensity I of the (polarized) beam transmitted through the analyzer varies as

Malus's law ▶

$$I = I_{\max} \cos^2\theta \tag{38.9}$$

where $I_{\max}$ is the intensity of the polarized beam incident on the analyzer. This expression, known as **Malus's law,**[2] applies to any two polarizing materials whose transmission axes are at an angle θ to each other. This expression shows that the intensity of the transmitted beam is maximum when the transmission axes are parallel ($\theta = 0$ or $180°$) and is zero (complete absorption by the analyzer) when the transmission axes are perpendicular to each other. This variation in transmitted intensity through a pair of polarizing sheets is illustrated in Figure 38.27.

Polarization by Reflection

When an unpolarized light beam is reflected from a surface, the reflected light may be completely polarized, partially polarized, or unpolarized, depending on the angle of incidence. If the angle of incidence is $0°$, the reflected beam is unpolarized. For other angles of incidence, the reflected light is polarized to some extent, and for one particular angle of incidence, the reflected light is completely polarized. Let's now investigate reflection at that special angle.

Suppose an unpolarized light beam is incident on a surface as in Figure 38.28a. Each individual electric field vector can be resolved into two components: one parallel to the surface (and perpendicular to the page in Fig. 38.28, represented by

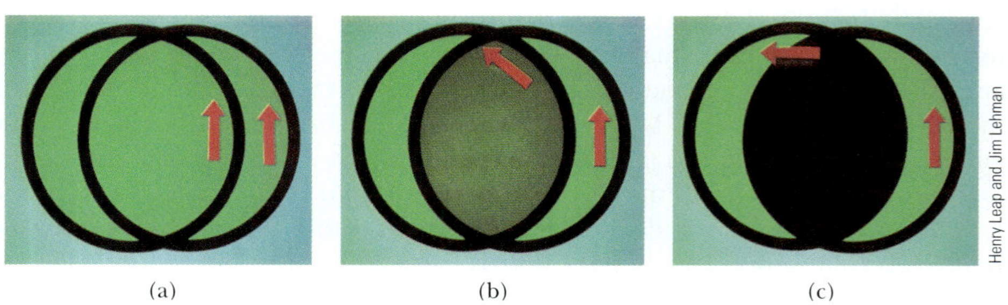

Figure 38.27 The intensity of light transmitted through two polarizers depends on the relative orientation of their transmission axes. (a) The transmitted light has maximum intensity when the transmission axes are aligned with each other. (b) The transmitted light has lesser intensity when the transmission axes are at an angle of 45° with each other. (c) The transmitted light intensity is a minimum when the transmission axes are perpendicular to each other.

[2] Named after its discoverer, E. L. Malus (1775–1812). Malus discovered that reflected light was polarized by viewing it through a calcite ($CaCO_3$) crystal.

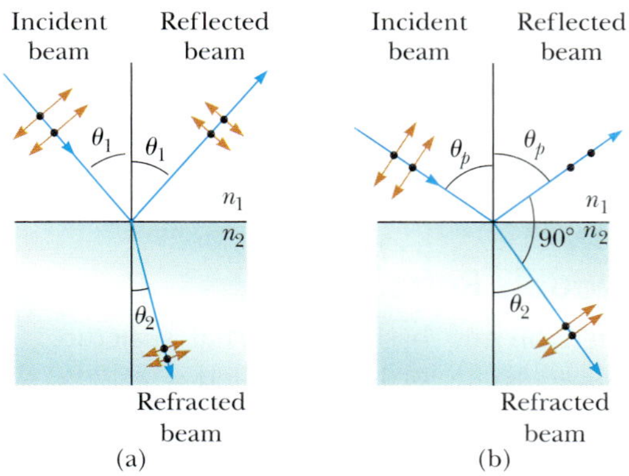

Figure 38.28 (a) When unpolarized light is incident on a reflecting surface, the reflected and refracted beams are partially polarized. (b) The reflected beam is completely polarized when the angle of incidence equals the polarizing angle θ_p, which satisfies the equation $n = \tan\theta_p$. At this incident angle, the reflected and refracted rays are perpendicular to each other.

the dots) and the other (represented by the brown arrows) perpendicular both to the first component and to the direction of propagation. Therefore, the polarization of the entire beam can be described by two electric field components in these directions. It is found that the parallel component reflects more strongly than the perpendicular component, resulting in a partially polarized reflected beam. Furthermore, the refracted beam is also partially polarized.

Now suppose the angle of incidence θ_1 is varied until the angle between the reflected and refracted beams is 90° as in Figure 38.28b. At this particular angle of incidence, the reflected beam is completely polarized (with its electric field vector parallel to the surface) and the refracted beam is still only partially polarized. The angle of incidence at which this polarization occurs is called the **polarizing angle** θ_p.

We can obtain an expression relating the polarizing angle to the index of refraction of the reflecting substance by using Figure 38.28b. From this figure, we see that $\theta_p + 90° + \theta_2 = 180°$; therefore, $\theta_2 = 90° - \theta_p$. Using Snell's law of refraction (Eq. 35.8) gives

$$\frac{n_2}{n_1} = \frac{\sin\theta_1}{\sin\theta_2} = \frac{\sin\theta_p}{\sin\theta_2}$$

Because $\sin\theta_2 = \sin(90° - \theta_p) = \cos\theta_p$, we can write this expression as $n_2/n_1 = \sin\theta_p/\cos\theta_p$, which means that

$$\tan\theta_p = \frac{n_2}{n_1} \qquad \textbf{(38.10)}$$

◀ Brewster's law

This expression is called **Brewster's law,** and the polarizing angle θ_p is sometimes called **Brewster's angle,** after its discoverer, David Brewster (1781–1868). Because n varies with wavelength for a given substance, Brewster's angle is also a function of wavelength.

We can understand polarization by reflection by imagining that the electric field in the incident light sets electrons at the surface of the material in Figure 38.28b into oscillation. The component directions of oscillation are (1) parallel to the arrows shown on the refracted beam of light and (2) perpendicular to the page. The oscillating electrons act as antennas radiating light with a polarization parallel to the direction of oscillation. For the oscillations in direction 1, there is no radiation in the perpendicular direction, which is along the reflected ray (see the $\theta = 90°$ direction in Fig. 34.10). For oscillations in direction 2, the electrons radiate light with a polarization perpendicular to the page (the $\theta = 0$ direction in Fig. 34.10). Therefore, the light reflected from the surface at this angle is completely polarized parallel to the surface.

Polarization by reflection is a common phenomenon. Sunlight reflected from water, glass, and snow is partially polarized. If the surface is horizontal, the electric field vector of the reflected light has a strong horizontal component. Sunglasses made of polarizing material reduce the glare of reflected light. The transmission axes of such lenses are oriented vertically so that they absorb the strong horizontal component of the reflected light. If you rotate sunglasses through 90°, they are not as effective at blocking the glare from shiny horizontal surfaces.

Polarization by Double Refraction

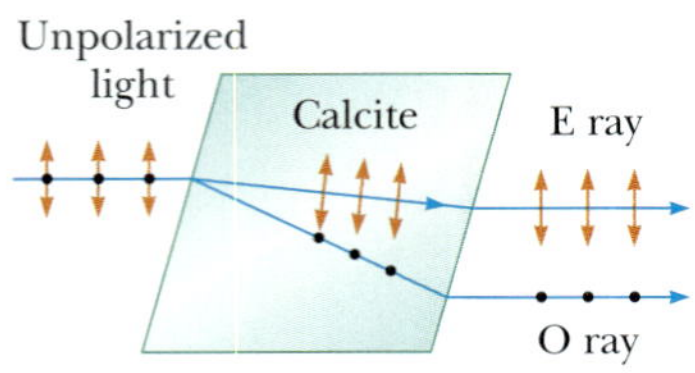

Figure 38.29 Unpolarized light incident at an angle to the optic axis in a calcite crystal splits into an ordinary (O) ray and an extraordinary (E) ray. These two rays are polarized in mutually perpendicular directions. (Drawing not to scale.)

Solids can be classified on the basis of internal structure. Those in which the atoms are arranged in a specific order are called *crystalline;* the NaCl structure of Figure 38.22 is one example of a crystalline solid. Those solids in which the atoms are distributed randomly are called *amorphous.* When light travels through an amorphous material such as glass, it travels with a speed that is the same in all directions. That is, glass has a single index of refraction. In certain crystalline materials such as calcite and quartz, however, the speed of light is not the same in all directions. In these materials, the speed of light depends on the direction of propagation *and* on the plane of polarization of the light. Such materials are characterized by two indices of refraction. Hence, they are often referred to as **double-refracting** or **birefringent** materials.

When unpolarized light enters a birefringent material, it may split into an **ordinary (O) ray** and an **extraordinary (E) ray.** These two rays have mutually perpendicular polarizations and travel at different speeds through the material. The two speeds correspond to two indices of refraction, n_O for the ordinary ray and n_E for the extraordinary ray.

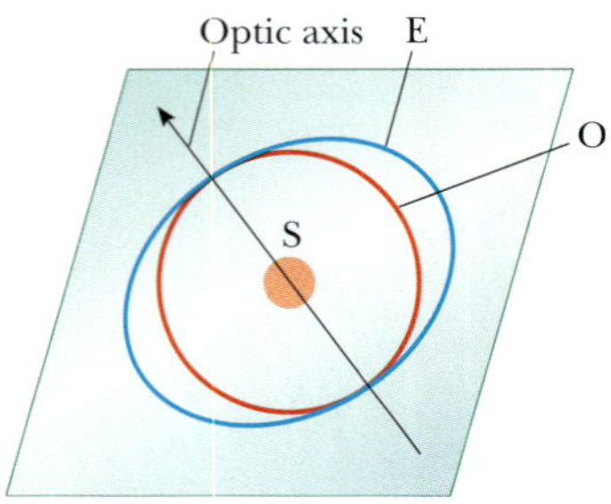

Figure 38.30 A point source S inside a double-refracting crystal produces a spherical wave front corresponding to the ordinary (O) ray and an elliptical wave front corresponding to the extraordinary (E) ray. The two waves propagate with the same velocity along the optic axis.

There is one direction, called the **optic axis,** along which the ordinary and extraordinary rays have the same speed. If light enters a birefringent material at an angle to the optic axis, however, the different indices of refraction will cause the two polarized rays to split and travel in different directions as shown in Figure 38.29.

The index of refraction n_O for the ordinary ray is the same in all directions. If one could place a point source of light inside the crystal as in Figure 38.30, the ordinary waves would spread out from the source as spheres. The index of refraction n_E varies with the direction of propagation. A point source sends out an extraordinary wave having wave fronts that are elliptical in cross section. The difference in speed for the two rays is a maximum in the direction perpendicular to the optic axis. For example, in calcite, $n_O = 1.658$ at a wavelength of 589.3 nm, and n_E varies from 1.658 along the optic axis to 1.486 perpendicular to the optic axis. Values for n_O and the extreme value of n_E for various double-refracting crystals are given in Table 38.1.

If you place a calcite crystal on a sheet of paper and then look through the crystal at any writing on the paper, you would see two images as shown in Figure 38.31. As can be seen from Figure 38.29, these two images correspond to one formed by the ordinary ray and one formed by the extraordinary ray. If the two images are viewed through a sheet of rotating polarizing glass, they alternately appear and dis-

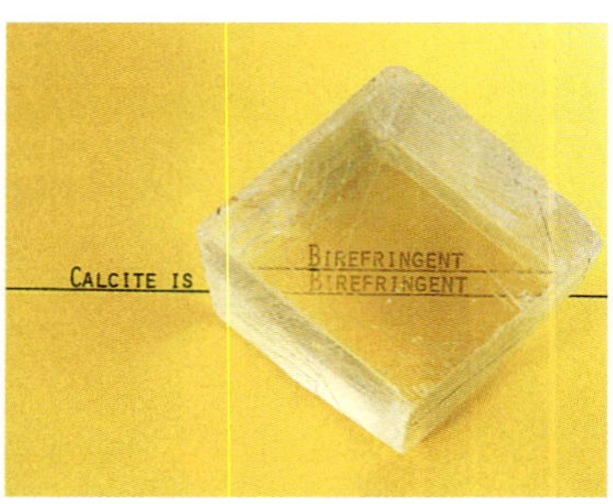

Henry Leap and Jim Lehman

Figure 38.31 A calcite crystal produces a double image because it is a birefringent (double-refracting) material.

TABLE 38.1

Indices of Refraction for Some Double-Refracting Crystals at a Wavelength of 589.3 nm

Crystal	n_O	n_E	n_O/n_E
Calcite ($CaCO_3$)	1.658	1.486	1.116
Quartz (SiO_2)	1.544	1.553	0.994
Sodium nitrate ($NaNO_3$)	1.587	1.336	1.188
Sodium sulfite ($NaSO_3$)	1.565	1.515	1.033
Zinc chloride ($ZnCl_2$)	1.687	1.713	0.985
Zinc sulfide (ZnS)	2.356	2.378	0.991

Sepp Seitz 1981 (a) Peter Aprahamian/Science Photo Library (b)

Figure 38.32 (a) Strain distribution in a plastic model of a hip replacement used in a medical research laboratory. The pattern is produced when the plastic model is viewed between a polarizer and analyzer oriented perpendicular to each other. (b) A plastic model of an arch structure under load conditions observed between perpendicular polarizers. Such patterns are useful in the optimal design of architectural components.

appear because the ordinary and extraordinary rays are plane-polarized along mutually perpendicular directions.

Some materials such as glass and plastic become birefringent when stressed. Suppose an unstressed piece of plastic is placed between a polarizer and an analyzer so that light passes from polarizer to plastic to analyzer. When the plastic is unstressed and the analyzer axis is perpendicular to the polarizer axis, none of the polarized light passes through the analyzer. In other words, the unstressed plastic has no effect on the light passing through it. If the plastic is stressed, however, regions of greatest stress become birefringent and the polarization of the light passing through the plastic changes. Hence, a series of bright and dark bands is observed in the transmitted light, with the bright bands corresponding to regions of greatest stress.

Engineers often use this technique, called *optical stress analysis,* in designing structures ranging from bridges to small tools. They build a plastic model and analyze it under different load conditions to determine regions of potential weakness and failure under stress. Two examples of plastic models under stress are shown in Figure 38.32.

Polarization by Scattering

When light is incident on any material, the electrons in the material can absorb and reradiate part of the light. Such absorption and reradiation of light by electrons in the gas molecules that make up air is what causes sunlight reaching an observer on the Earth to be partially polarized. You can observe this effect—called **scattering**—by looking directly up at the sky through a pair of sunglasses whose lenses are made of polarizing material. Less light passes through at certain orientations of the lenses than at others.

Figure 38.33 illustrates how sunlight becomes polarized when it is scattered. The phenomenon is similar to that creating completely polarized light upon reflection from a surface at Brewster's angle. An unpolarized beam of sunlight traveling in the horizontal direction (parallel to the ground) strikes a molecule of one of the gases that make up air, setting the electrons of the molecule into vibration. These vibrating charges act like the vibrating charges in an antenna. The horizontal component of the electric field vector in the incident wave results in a horizontal component of the vibration of the charges, and the vertical component of the vector results in a vertical component of vibration. If the observer in Figure 38.33 is looking straight up (perpendicular to the original direction of propagation of the light), the vertical oscillations of the charges send no radiation toward the observer. Therefore, the observer sees light that is completely polarized in the horizontal direction as indicated by the brown arrows. If the observer looks in other directions, the light is partially polarized in the horizontal direction.

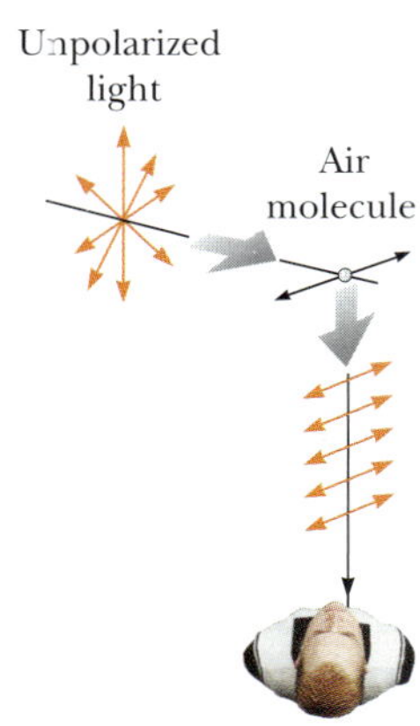

Figure 38.33 The scattering of unpolarized sunlight by air molecules. The scattered light traveling perpendicular to the incident light is plane-polarized because the vertical vibrations of the charges in the air molecule send no light in this direction.

Variations in the color of scattered light in the atmosphere can be understood as follows. When light of various wavelengths λ is incident on gas molecules of diameter d, where $d \ll \lambda$, the relative intensity of the scattered light varies as $1/\lambda^4$. The condition $d \ll \lambda$ is satisfied for scattering from oxygen (O_2) and nitrogen (N_2) molecules in the atmosphere, whose diameters are about 0.2 nm. Hence,

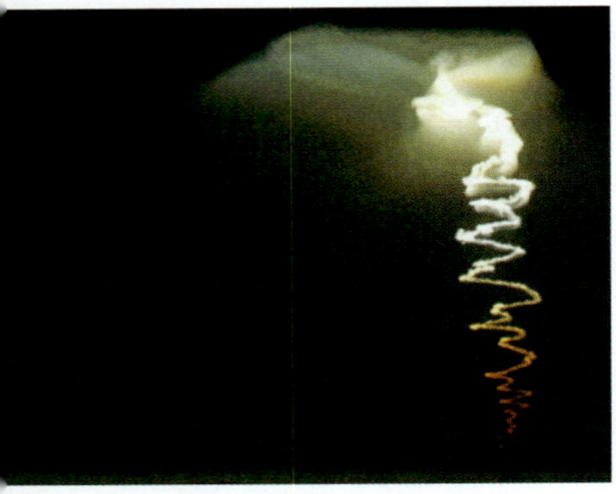

This photograph is a view of a rocket launch from Vandenburg Air Force Base in California. The trail left by the rocket shows the effects of scattering of light by air molecules. The lower portion of the trail appears red due to the scattering of wavelengths at the violet end of the spectrum as the light from the Sun travels through a large portion of the atmosphere to light up the trail. The upper portion of the trail is illuminated by light that has traveled through much less atmosphere and appears white.

short wavelengths (violet light) are scattered more efficiently than long wavelengths (red light). Therefore, when sunlight is scattered by gas molecules in the air, the short-wavelength radiation (violet) is scattered more intensely than the long-wavelength radiation (red).

When you look up into the sky in a direction that is not toward the Sun, you see the scattered light, which is predominantly violet. Your eyes, however, are not very sensitive to violet light. Light of the next color in the spectrum, blue, is scattered with less intensity than violet, but your eyes are far more sensitive to blue light than to violet light. Hence, you see a blue sky. If you look toward the west at sunset (or toward the east at sunrise), you are looking in a direction toward the Sun and are seeing light that has passed through a large distance of air. Most of the blue light has been scattered by the air between you and the Sun. The light that survives this trip through the air to you has had much of its blue component scattered and is therefore heavily weighted toward the red end of the spectrum; as a result, you see the red and orange colors of sunset.

Optical Activity

Many important applications of polarized light involve materials that display **optical activity.** A material is said to be optically active if it rotates the plane of polarization of any light transmitted through the material. The angle through which the light is rotated by a specific material depends on the length of the path through the material and on concentration if the material is in solution. One optically active material is a solution of the common sugar dextrose. A standard method for determining the concentration of sugar solutions is to measure the rotation produced by a fixed length of the solution.

Molecular asymmetry determines whether a material is optically active. For example, some proteins are optically active because of their spiral shape.

The liquid crystal displays found in most calculators have their optical activity changed by the application of electric potential across different parts of the display. Try using a pair of polarizing sunglasses to investigate the polarization used in the display of your calculator.

Quick Quiz 38.6 A polarizer for microwaves can be made as a grid of parallel metal wires approximately 1 cm apart. Is the electric field vector for microwaves transmitted through this polarizer (a) parallel or (b) perpendicular to the metal wires?

Quick Quiz 38.7 You are walking down a long hallway that has many light fixtures in the ceiling and a very shiny, newly waxed floor. When looking at the floor, you see reflections of every light fixture. Now you put on sunglasses that are polarized. Some of the reflections of the light fixtures can no longer be seen. (Try it!) Are the reflections that disappear those (a) nearest to you, (b) farthest from you, or (c) at an intermediate distance from you?

Summary

ThomsonNOW™ Sign in at **www.thomsonedu.com** and go to ThomsonNOW to take a practice test for this chapter.

CONCEPTS AND PRINCIPLES

Diffraction is the deviation of light from a straight-line path when the light passes through an aperture or around an obstacle. Diffraction is due to the wave nature of light.

The **Fraunhofer diffraction pattern** produced by a single slit of width a on a distant screen consists of a central bright fringe and alternating bright and dark fringes of much lower intensities. The angles θ_{dark} at which the diffraction pattern has zero intensity, corresponding to destructive interference, are given by

$$\sin\theta_{\text{dark}} = m\frac{\lambda}{a} \qquad m = \pm 1, \pm 2, \pm 3, \ldots \qquad \textbf{(38.1)}$$

Rayleigh's criterion, which is a limiting condition of resolution, states that two images formed by an aperture are just distinguishable if the central maximum of the diffraction pattern for one image falls on the first minimum of the diffraction pattern for the other image. The limiting angle of resolution for a slit of width a is $\theta_{\text{min}} = \lambda/a$, and the limiting angle of resolution for a circular aperture of diameter D is given by $\theta_{\text{min}} = 1.22\lambda/D$.

A **diffraction grating** consists of a large number of equally spaced, identical slits. The condition for intensity maxima in the interference pattern of a diffraction grating for normal incidence is

$$d\sin\theta_{\text{bright}} = m\lambda \qquad m = 0, \pm 1, \pm 2, \pm 3, \ldots \qquad \textbf{(38.7)}$$

where d is the spacing between adjacent slits and m is the order number of the intensity maximum.

When polarized light of intensity I_{max} is emitted by a polarizer and then is incident on an analyzer, the light transmitted through the analyzer has an intensity equal to $I_{\text{max}}\cos^2\theta$, where θ is the angle between the polarizer and analyzer transmission axes.

In general, reflected light is partially polarized. Reflected light, however, is completely polarized when the angle of incidence is such that the angle between the reflected and refracted beams is 90°. This angle of incidence, called the **polarizing angle** θ_p, satisfies **Brewster's law:**

$$\tan\theta_p = \frac{n_2}{n_1} \qquad \textbf{(38.10)}$$

where n_1 is the index of refraction of the medium in which the light initially travels and n_2 is the index of refraction of the reflecting medium.

Questions

☐ denotes answer available in *Student Solutions Manual/Study Guide;* **O** denotes objective question

1. ☐ Why can you hear around corners, but not see around corners?

2. Holding your hand at arm's length, you can readily block sunlight from reaching your eyes. Why can you not block sound from reaching your ears this way?

3. ☐ **O** Consider a wave passing through a single slit. What happens to the width of the central maximum of its diffraction pattern as the slit is made half as wide? (a) The central maximum becomes one-fourth as wide. (b) It becomes one-half as wide. (c) Its width does not change. (d) It becomes two times wider. (e) It becomes four times wider.

4. **O** Assume Figure 38.1 was photographed with red light of a single wavelength λ_0. The light passed through a single slit of width a and traveled distance L to the screen where the photograph was made. Consider the width of the central bright fringe, measured between the dark fringes on both sides of it. Rank from largest to smallest the widths of the central fringe in the following situations and note any cases of equality. (a) The experiment is performed as photographed. (b) The experiment is performed with

light whose frequency is increased by 50%. (c) The experiment is performed with light whose wavelength is increased by 50%. (Its wavelength is $3\lambda_0/2$.) (d) The experiment is performed with the original light and with a slit of width $2a$. (e) The experiment is performed with the original light and slit, and with distance $2L$ to the screen. (f) The experiment is performed with light of twice the original intensity.

5. **O** In Active Figure 38.4, assume the slit is in a barrier that is opaque to x-rays as well as to visible light. The photograph in Active Figure 38.4b shows the diffraction pattern produced with visible light. What will happen if the experiment is repeated with x-rays as the incoming wave and with no other changes? (a) The diffraction pattern is similar. (b) There is no noticeable diffraction pattern but rather a projected shadow of high intensity on the screen, having the same width as the slit. (c) The central maximum is much wider and the minima occur at larger angles than with visible light. (d) No x-rays reach the screen.

6. **O** Off in the distance, you see the headlights of a car, but they are indistinguishable from the single headlight of a motorcycle. Assume the car's headlights are now switched from low beam to high beam so that the light intensity you receive becomes three times greater. What then happens to your ability to resolve the two light sources? (a) It increases by a factor of 9. (b) It increases by a factor of 3. (c) It remains the same. (d) It becomes one-third as good. (e) It becomes one-ninth as good.

7. A laser beam is incident at a shallow angle on a horizontal machinist's ruler that has a finely calibrated scale. The engraved rulings on the scale give rise to a diffraction pattern on a vertical screen. Discuss how you can use this technique to obtain a measure of the wavelength of the laser light.

8. **O** When you receive a chest x-ray at a hospital, the x-rays pass through a set of parallel ribs in your chest. Do your ribs act as a diffraction grating for x-rays? (a) Yes. They produce diffracted beams that can be observed separately. (b) Not to a measurable extent. The ribs are too far apart. (c) Essentially not. The ribs are too close together. (d) Essentially not. The ribs are too few in number. (e) Absolutely not. X-rays cannot diffract.

9. **O** Certain sunglasses use a polarizing material to reduce the intensity of light reflected as glare from water or automobiles. What orientation should the polarizing filters have to be most effective? (a) The polarizers should absorb light with its electric field horizontal. (b) The polarizers should absorb light with its electric field vertical. (c) The polarizers should absorb both horizontal and vertical electric fields. (d) The polarizers should not absorb either horizontal or vertical electric fields.

10. Is light from the sky polarized? Why is it that clouds seen through Polaroid glasses stand out in bold contrast to the sky?

11. **O** When unpolarized light passes straight through a diffraction grating, does it become polarized? (a) No, it does not. (b) Yes, it does, with the transmission axis parallel to the slits or grooves in the grating. (c) Yes, it does, with the transmission axis perpendicular to the slits or grooves in the grating. (d) It possibly does because an electric field above some threshold is blocked out by the grating if the field is perpendicular to the slits.

12. If a coin is glued to a glass sheet and this arrangement is held in front of a laser beam, the projected shadow has diffraction rings around its edge and a bright spot in the center. How are these effects possible?

13. How could the index of refraction of a flat piece of opaque volcanic glass be determined?

14. A laser produces a beam a few millimeters wide, with uniform intensity across its width. A hair is stretched vertically across the front of the laser to cross the beam. How is the diffraction pattern it produces on a distant screen related to that of a vertical slit equal in width to the hair? How could you determine the width of the hair from measurements of its diffraction pattern?

15. A radio station serves listeners in a city to the northeast of its broadcast site. It broadcasts from three adjacent towers on a mountain ridge, along a line running east and west. Show that by introducing time delays among the signals the individual towers radiate, the station can maximize net intensity in the direction toward the city (and in the opposite direction) and minimize the signal transmitted in other directions. The towers together are said to form a *phased array*.

16. John William Strutt, Lord Rayleigh (1842–1919), is known as the last person to understand all physics and all mathematics. He invented an improved foghorn. To warn ships of a coastline, a foghorn should radiate sound in a wide horizontal sheet over the ocean's surface. It should not waste energy by broadcasting sound upward. It should not emit sound downward because the water in front of the foghorn would reflect that sound upward. Rayleigh's foghorn trumpet is shown in Figure Q38.16. Is it installed in the correct orientation? Decide whether the long dimension of the rectangular opening should be horizontal or vertical, and argue for your decision.

Figure Q38.16

Problems

WebAssign The Problems from this chapter may be assigned online in WebAssign.

ThomsonNOW Sign in at **www.thomsonedu.com** and go to ThomsonNOW to assess your understanding of this chapter's topics with additional quizzing and conceptual questions.

1, 2, 3 denotes straightforward, intermediate, challenging; □ denotes full solution available in *Student Solutions Manual/Study Guide;* ▲ denotes coached solution with hints available at **www.thomsonedu.com;** ▭ denotes developing symbolic reasoning; ● denotes asking for qualitative reasoning; 🖥 denotes computer useful in solving problem

Section 38.2 Diffraction Patterns from Narrow Slits

1. Helium–neon laser light (λ = 632.8 nm) is sent through a 0.300-mm-wide single slit. What is the width of the central maximum on a screen 1.00 m from the slit?

2. A beam of monochromatic green light is diffracted by a slit of width 0.550 mm. The diffraction pattern forms on a wall 2.06 m beyond the slit. The distance between the positions of zero intensity on both sides of the central bright fringe is 4.10 mm. Calculate the wavelength of the light.

3. ▲ A screen is placed 50.0 cm from a single slit, which is illuminated with 690-nm light. If the distance between the first and third minima in the diffraction pattern is 3.00 mm, what is the width of the slit?

4. Coherent microwaves of wavelength 5.00 cm enter a tall, narrow window in a building otherwise essentially opaque to the microwaves. If the window is 36.0 cm wide, what is the distance from the central maximum to the first-order minimum along a wall 6.50 m from the window?

5. Sound with a frequency 650 Hz from a distant source passes through a doorway 1.10 m wide in a sound-absorbing wall. Find the number and approximate directions of the diffraction-maximum beams radiated into the space beyond.

6. ● A horizontal laser beam of wavelength 632.8 nm has a circular cross section 2.00 mm in diameter. A rectangular aperture is to be placed in the center of the beam so that when the light falls perpendicularly on a wall 4.50 m away, the central maximum fills a rectangle 110 mm wide and 6.00 mm high. The dimensions are measured between the minima bracketing the central maximum. (a) Find the required width and height of the aperture. (b) Is the longer dimension of the central bright patch in the diffraction pattern horizontal or vertical? Is the longer dimension of the aperture horizontal or vertical? Explain the relationship between these two rectangles, using a diagram.

7. A diffraction pattern is formed on a screen 120 cm away from a 0.400-mm-wide slit. Monochromatic 546.1-nm light is used. Calculate the fractional intensity I/I_{max} at a point on the screen 4.10 mm from the center of the principal maximum.

8. **What If?** Assume the light in Figure 38.5 strikes the single slit at an angle β from the perpendicular direction. Show that Equation 38.1, the condition for destructive interference, must be modified to read

$$\sin\theta_{dark} = m\left(\frac{\lambda}{a}\right) - \sin\beta$$

9. Assume light with a wavelength of 650 nm passes through two slits 3.00 μm wide, with their centers 9.00 μm apart. Make a sketch of the combined diffraction and interference pattern in the form of a graph of intensity versus $\phi = (\pi a \sin \theta)/\lambda$. You may use Active Figure 38.7 as a starting point.

10. Coherent light of wavelength 501.5 nm is sent through two parallel slits in a large, flat wall. Each slit is 0.700 μm wide. Their centers are 2.80 μm apart. The light then falls on a semicylindrical screen, with its axis at the midline between the slits. (a) Predict the direction of each interference maximum on the screen as an angle away from the bisector of the line joining the slits. (b) Describe the pattern of light on the screen, specifying the number of bright fringes and the location of each. (c) Find the intensity of light on the screen at the center of each bright fringe, expressed as a fraction of the light intensity I_{max} at the center of the pattern.

Section 38.3 Resolution of Single-Slit and Circular Apertures

11. The pupil of a cat's eye narrows to a vertical slit of width 0.500 mm in daylight. What is the angular resolution for horizontally separated mice? Assume the average wavelength of the light is 500 nm.

12. ● Yellow light of wavelength 589 nm is used to view an object under a microscope. The objective diameter is 9.00 mm. (a) What is the limiting angle of resolution? (b) Suppose it is possible to use visible light of any wavelength. What color should you choose to give the smallest possible angle of resolution, and what is this angle? (c) Suppose water fills the space between the object and the objective. What effect does this change have on the resolving power when 589-nm light is used?

13. ▲ A helium–neon laser emits light that has a wavelength of 632.8 nm. The circular aperture through which the beam emerges has a diameter of 0.500 cm. Estimate the diameter of the beam 10.0 km from the laser.

14. ● Narrow, parallel, glowing gas-filled tubes in a variety of colors form block letters to spell out the name of a nightclub. Adjacent tubes are all 2.80 cm apart. The tubes forming one letter are filled with neon and radiate predominantly red light with a wavelength of 640 nm. For another letter, the tubes emit predominantly blue light at 440 nm. The pupil of a dark-adapted viewer's eye is 5.20 mm in diameter. If she is in a certain range of distances away, the viewer can resolve the separate tubes of one color but not the other. Which color is easier to resolve? State how you decide. The viewer's distance must be in what range for her to resolve the tubes of only one of these two colors?

15. Impressionist painter Georges Seurat created paintings with an enormous number of dots of pure pigment, each of which was approximately 2.00 mm in diameter. The idea was to have colors such as red and green next to each other to form a scintillating canvas (Fig. P38.15). Outside what distance would one be unable to discern individual dots on the canvas? Assume λ = 500 nm and a pupil diameter of 4.00 mm.

Figure P38.15

16. What are the approximate dimensions of the smallest object on the Earth that astronauts can resolve by eye when they are orbiting 250 km above the Earth? Assume λ = 500 nm and a pupil diameter of 5.00 mm.

17. A spy satellite can consist of a large-diameter concave mirror forming an image on a digital-camera detector and sending the picture to a ground receiver by radio waves. In effect, it is an astronomical telescope in orbit, looking down instead of up. Can a spy satellite read a license plate? Can it read the date on a dime? Argue for your answers by making an order-of-magnitude calculation, specifying the data you estimate.

18. A circular radar antenna on a Coast Guard ship has a diameter of 2.10 m and radiates at a frequency of 15.0 GHz. Two small boats are located 9.00 km away from the ship. How close together could the boats be and still be detected as two objects?

Section 38.4 The Diffraction Grating

Note: In the following problems, assume the light is incident normally on the gratings.

19. White light is spread out into its spectral components by a diffraction grating. If the grating has 2 000 grooves per centimeter, at what angle does red light of wavelength 640 nm appear in first order?

20. Light from an argon laser strikes a diffraction grating that has 5 310 grooves per centimeter. The central and first-order principal maxima are separated by 0.488 m on a wall 1.72 m from the grating. Determine the wavelength of the laser light.

21. ▲ The hydrogen spectrum includes a red line at 656 nm and a blue-violet line at 434 nm. What are the angular separations between these two spectral lines obtained with a diffraction grating that has 4 500 grooves/cm?

22. A helium–neon laser (λ = 632.8 nm) is used to calibrate a diffraction grating. If the first-order maximum occurs at 20.5°, what is the spacing between adjacent grooves in the grating?

23. Three discrete spectral lines occur at angles of 10.09°, 13.71°, and 14.77° in the first-order spectrum of a grating spectrometer. (a) If the grating has 3 660 slits/cm, what are the wavelengths of the light? (b) At what angles are these lines found in the second-order spectrum?

24. Show that whenever white light is passed through a diffraction grating of any spacing size, the violet end of the continuous visible spectrum in third order always overlaps with red light at the other end of the second-order spectrum.

25. A refrigerator shelf is an array of parallel wires with uniform spacing of 1.30 cm between centers. In air at 20°C, ultrasound with a frequency of 37.2 kHz from a distant source falls perpendicularly on the shelf. Find the number of diffracted beams leaving the other side of the shelf. Find the direction of each beam.

26. The laser in a CD player must precisely follow the spiral track, along which the distance between one loop of the spiral and the next is only about 1.25 μm. A feedback mechanism lets the player know if the laser drifts off the track so that the player can steer it back again. Figure P38.26 shows how a diffraction grating is used to provide information to keep the beam on track. The laser light passes through a diffraction grating before it reaches the disk. The strong central maximum of the diffraction pattern is used to read the information in the track of pits. The two first-order side maxima are used for steering. The grating is designed so that the first-order maxima fall on the flat surfaces on both sides of the information track. Both side beams are reflected into their own detectors. As long as both beams are reflecting from smooth nonpitted surfaces, they are detected with constant high intensity. If the main beam wanders off the track, however, one of the side beams begins to strike pits on the information track and the reflected light diminishes. This change is used with an electronic circuit to guide the beam back to the desired location. Assume the laser light has a wavelength of 780 nm and the diffraction grating is positioned 6.90 μm from the disk. Assume the first-order beams are to fall on the disk 0.400 μm on either side of the information track. What should be the number of grooves per millimeter in the grating?

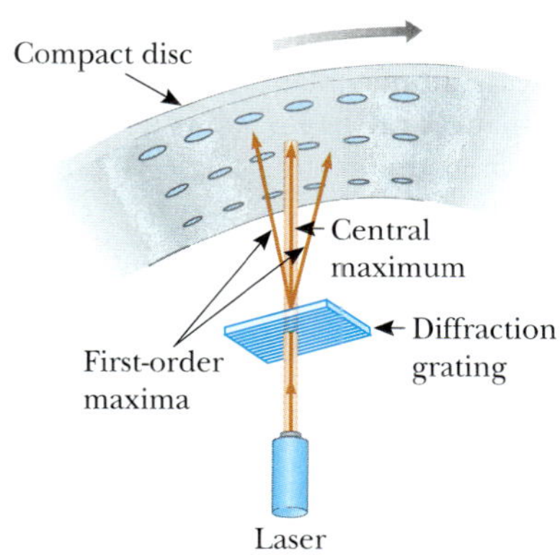

Figure P38.26

27. A grating with 250 grooves/mm is used with an incandescent light source. Assume the visible spectrum to range in

wavelength from 400 to 700 nm. In how many orders can one see (a) the entire visible spectrum and (b) the short-wavelength region?

28. ● A wide beam of laser light with a wavelength of 632.8 nm is directed through several narrow parallel slits, separated by 1.20 mm, and falls on a sheet of photographic film 1.40 m away. The exposure time is chosen so that the film stays unexposed everywhere except at the central region of each bright fringe. (a) Find the distance between these interference maxima. The film is printed as a transparency; it is opaque everywhere except at the exposed lines. Next, the same beam of laser light is directed through the transparency and allowed to fall on a screen 1.40 m beyond. (b) Argue that several narrow parallel bright regions, separated by 1.20 mm, appear on the screen as real images of the original slits. If the screen is removed, light diverges from the images of the original slits with the same reconstructed wave fronts as the original slits produced. *Suggestion:* You may find it useful to draw diagrams similar to Figure 38.12. A similar train of thought, at a soccer game, led Dennis Gabor to invent holography.

29. A diffraction grating has 4 200 rulings/cm. On a screen 2.00 m from the grating, it is found that for a particular order m, the maxima corresponding to two closely spaced wavelengths of sodium (589.0 nm and 589.6 nm) are separated by 1.59 mm. Determine the value of m.

Section 38.5 Diffraction of X-Rays by Crystals

30. Potassium iodide (KI) has the same crystalline structure as NaCl, with atomic planes separated by 0.353 nm. A monochromatic x-ray beam shows a first-order diffraction maximum when the grazing angle is 7.60°. Calculate the x-ray wavelength.

31. ▲ If the interplanar spacing of NaCl is 0.281 nm, what is the predicted angle at which 0.140-nm x-rays are diffracted in a first-order maximum?

32. In water of uniform depth, a wide pier is supported on pilings in several parallel rows 2.80 m apart. Ocean waves of uniform wavelength roll in, moving in a direction that makes an angle of 80.0° with the rows of posts. Find the three longest wavelengths of waves that are strongly reflected by the pilings.

33. ● The atoms in a crystal lie in planes separated by a few tenths of a nanometer. Can they produce a diffraction pattern for visible light as they do for x-rays? Explain your answer with reference to Bragg's law.

Section 38.6 Polarization of Light Waves

Problem 36 in Chapter 34 can be assigned with this section.

34. Unpolarized light passes through two ideal Polaroid sheets. The axis of the first is vertical and that of the second is at 30.0° to the vertical. What fraction of the incident light is transmitted?

35. Plane-polarized light is incident on a single polarizing disk with the direction of $\vec{\mathbf{E}}_0$ parallel to the direction of the transmission axis. Through what angle should the disk be rotated so that the intensity in the transmitted beam is reduced by a factor of (a) 3.00, (b) 5.00, and (c) 10.0?

36. The angle of incidence of a light beam onto a reflecting surface is continuously variable. The reflected ray in air is completely polarized when the angle of incidence is 48.0°. What is the index of refraction of the reflecting material?

37. The critical angle for total internal reflection for sapphire surrounded by air is 34.4°. Calculate the polarizing angle for sapphire.

38. For a particular transparent medium surrounded by air, show that the critical angle for total internal reflection and the polarizing angle are related by $\cot \theta_p = \sin \theta_c$.

39. ● **Review problem.** (a) A transparent plate with index of refraction n_2 is immersed in a medium with index n_1. Light traveling in the surrounding medium strikes the top surface of the plate at Brewster's angle. Show that if and only if the surfaces of the plate are parallel, the refracted light strikes the bottom surface of the plate at Brewster's angle for that interface. (b) **What If?** Instead of a plate, consider a prism of index of refraction n_2 separating

media of different refractive indices n_1 and n_3. The light propagates in a plane, containing also the apex angle of the prism. Is there one particular apex angle between the surfaces of the prism for which light can fall on both its surfaces at Brewster's angle as it passes through the prism? If so, determine it.

40. In Figure P38.40, suppose the transmission axes of the left and right polarizing disks are perpendicular to each other. Also, let the center disk be rotated on the common axis with an angular speed ω. Show that if unpolarized light is incident on the left disk with an intensity I_{max}, the intensity of the beam emerging from the right disk is

$$I = \frac{1}{16} I_{max}(1 - \cos 4\omega t)$$

This result means that the intensity of the emerging beam is modulated at a rate four times the rate of rotation of the center disk. *Suggestion:* Use the trigonometric identities $\cos^2 \theta = (1 + \cos 2\theta)/2$ and $\sin^2 \theta = (1 - \cos 2\theta)/2$, and recall that $\theta = \omega t$.

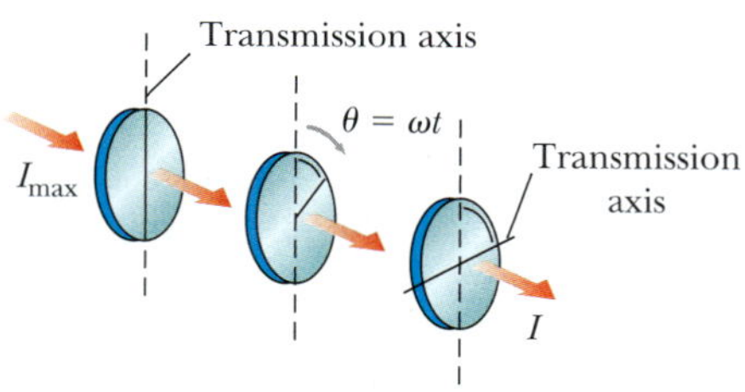

Figure P38.40

41. ● An unpolarized beam of light is incident on a stack of ideal polarizing filters. The axis of the first filter is perpendicular to the axis of the last filter in the stack. Find the fraction by which the transmitted beam's intensity is reduced in the following three cases. (a) Three filters are in the stack, each with its transmission axis at 45.0° relative to the preceding filter. (b) Four filters are in the stack, each with its transmission axis at 30.0° relative to the preceding filter. (c) Seven filters are in the stack, each with its axis at 15.0° relative to the preceding filter. (d) Comment on comparing the answers to parts (a), (b), and (c).

Additional Problems

42. ● Laser light with a wavelength of 632.8 nm is directed through one slit or two slits and allowed to fall on a screen 2.60 m beyond. Figure P38.42 shows the pattern on the screen, with a centimeter ruler below it. Did the light pass through one slit or two slits? Explain how you can tell. If one slit, find its width. If two slits, find the distance between their centers.

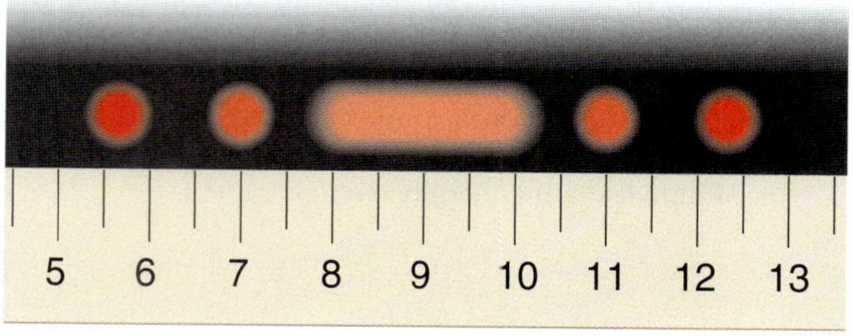

Figure P38.42

43. You use a sequence of ideal polarizing filters, each with its axis making the same angle with the axis of the previous filter, to rotate the plane of polarization of a polarized light beam by a total of 45.0°. You wish to have an intensity reduction no larger than 10.0%. (a) How many polarizers do you need to achieve your goal? (b) What is the angle between adjacent polarizers?

44. ● Figure P38.44 shows a megaphone in use. Construct a theoretical description of how a megaphone works. You may assume the sound of your voice radiates just through the opening of your mouth. Most of the information in speech is carried not in a signal at the fundamental frequency, but in noises and in harmonics, with frequencies of a few thousand hertz. Does your theory allow any prediction that is simple to test?

© Doug Pensigner/Getty Images

Figure P38.44

45. In a single-slit diffraction pattern, (a) find the ratio of the intensity of the first-order side maximum to the intensity of the central maximum. (b) Find the ratio of the intensity of the second-order side maximum to the intensity of the central maximum. You may assume each side maximum is halfway between the adjacent minima.

46. Consider a light wave passing through a slit. Give a mathematical argument that more than 90% of the transmitted energy is in the central maximum of the diffraction pattern. *Suggestions:* Think of the energy as represented by the shaded area in Figure 38.6a. Solve Problem 45 as preparation for this one. You are not expected to calculate the precise percentage, but explain the steps of your reasoning. You may use the identification

$$\frac{1}{1^2}+\frac{1}{3^2}+\frac{1}{5^2}+\cdots=\frac{\pi^2}{8}$$

47. Light from a helium–neon laser ($\lambda = 632.8$ nm) is incident on a single slit. What is the maximum width of the slit for which no diffraction minima are observed?

48. ● Two motorcycles separated laterally by 2.30 m are approaching an observer who is holding a "snooper scope" sensitive to infrared light of wavelength 885 nm. What aperture diameter is required if the motorcycles' headlights are to be resolved at a distance of 12.0 km? Assume the light propagates through perfectly steady and uniform air. Comment on how realistic this assumption is.

49. **Review problem.** A beam of 541-nm light is incident on a diffraction grating that has 400 grooves/mm. (a) Determine the angle of the second-order ray. (b) **What If?** If the entire apparatus is immersed in water, what is the new second-order angle of diffraction? (c) Show that the two diffracted rays of parts (a) and (b) are related through the law of refraction.

50. The *Very Large Array* (VLA) is a set of 27 radio telescope dishes in Catron and Socorro counties, New Mexico (Fig. P38.50). The antennas can be moved apart on railroad tracks, and their combined signals give the resolving power of a synthetic aperture 36.0 km in diameter. (a) If the detectors are tuned to a frequency of 1.40 GHz, what is the angular resolution of the VLA? (b) Clouds of hydrogen radiate at this frequency. What must be the separation distance of two clouds at the center of the galaxy, 26 000 light-years away, if they are to be resolved? (c) **What If?** As the telescope looks up, a circling hawk looks down. Find the angular resolution of the hawk's eye. Assume the hawk is most sensitive to green light having a wavelength of 500 nm and it has a pupil of diameter 12.0 mm. (d) A mouse is on the ground 30.0 m below. By what distance must the mouse's whiskers be separated if the hawk can resolve them?

Riccardo Giovanelli and Martha Haynes, Cornell University

Figure P38.50

51. A 750-nm light beam hits the flat surface of a certain liquid, and the beam is split into a reflected ray and a refracted ray. If the reflected ray is completely polarized at 36.0°, what is the wavelength of the refracted ray?

52. ● Iridescent peacock feathers are shown in Figure P38.52a. The surface of one microscopic barbule is composed of transparent keratin that supports rods of dark brown melanin in a regular lattice, represented in Figure P38.52b. (Your fingernails are made of keratin, and melanin is the dark pigment giving color to human skin.) In a portion of the feather that can appear turquoise

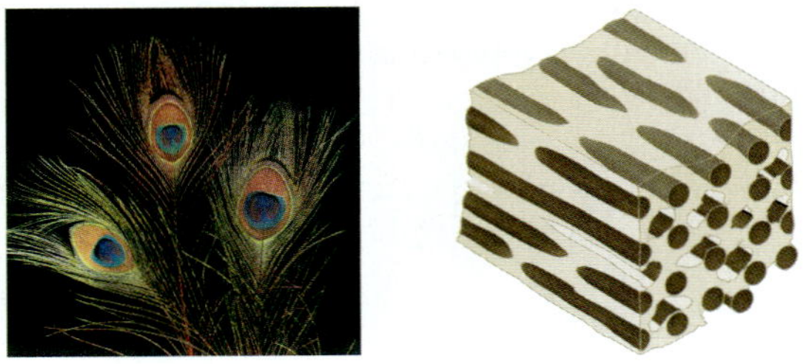

© Diane Schiumo 1988, Fundamental Photographs

(a) (b)

Figure P38.52

(blue-green), assume the melanin rods are uniformly separated by 0.25 μm, with air between them. (a) Explain how this structure can appear blue-green when it contains no blue or green pigment. (b) Explain how it can also appear violet if light falls on it in a different direction. (c) Explain how it can present different colors to your two eyes simultaneously, which is a characteristic of iridescence. (d) A compact disc can appear to be any color of the rainbow. Explain why this portion of the feather cannot appear yellow or red. (e) What could be different about the array of melanin rods in a portion of the feather that does appear to be red?

53. Light of wavelength 500 nm is incident normally on a diffraction grating. If the third-order maximum of the diffraction pattern is observed at 32.0°, (a) what is the number of rulings per centimeter for the grating? (b) Determine the total number of primary maxima that can be observed in this situation.

54. ● Light in air strikes a water surface at the polarizing angle. The part of the beam refracted into the water strikes a submerged slab of material with refractive index n as shown in Figure P38.54. The light reflected from the upper surface of the slab is completely polarized. (a) Find the angle θ between the water surface and the surface of the slab as a function of n. (b) Identify the maximum imaginable value of θ and describe the physical situation to which it corresponds. (c) Identify the minimum imaginable value of θ and describe the physical situation to which it corresponds.

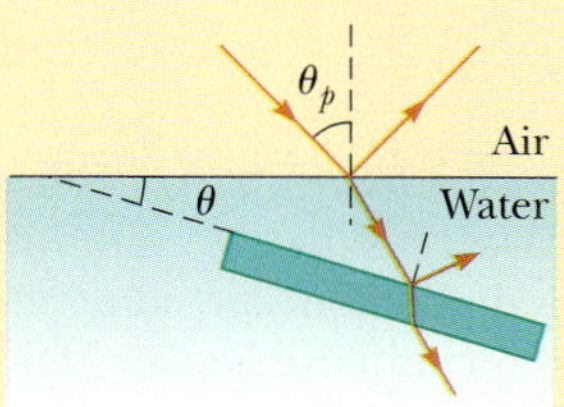

Figure P38.54

55. A beam of bright red light of wavelength 654 nm passes through a diffraction grating. Enclosing the space beyond the grating is a large screen forming one half of a cylinder centered on the grating, with its axis parallel to the slits in the grating. Fifteen bright spots appear on the screen. Find the maximum and minimum possible values for the slit separation in the diffraction grating.

56. A *pinhole camera* has a small circular aperture of diameter D. Light from distant objects passes through the aperture into an otherwise dark box, falling on a screen located a distance L away. If D is too large, the display on the screen will be fuzzy because a bright point in the field of view will send light onto a circle of diameter slightly larger than D. On the other hand, if D is too small, diffraction will blur the display on the screen. The screen shows a reasonably sharp image if the diameter of the central disk of the diffraction pattern, specified by Equation 38.6, is equal to D at the screen. (a) Show that for monochromatic light with plane wave fronts and $L \gg D$, the condition for a sharp view is fulfilled if $D^2 = 2.44\lambda L$. (b) Find the optimum pinhole diameter for 500-nm light projected onto a screen 15.0 cm away.

57. An American standard television picture is composed of approximately 485 horizontal lines of varying light intensity. Assume your ability to resolve the lines is limited only by the Rayleigh criterion and the pupils of your eyes are 5.00 mm in diameter. Calculate the ratio of minimum viewing distance to the vertical dimension of the picture such that you will not be able to resolve the lines. Assume the average wavelength of the light coming from the screen is 550 nm.

58. (a) Light traveling in a medium of index of refraction n_1 is incident at an angle θ on the surface of a medium of index n_2. The angle between reflected and refracted rays is β. Show that

$$\tan \theta = \frac{n_2 \sin \beta}{n_1 - n_2 \cos \beta}$$

Suggestion: Use the identity $\sin (A + B) = \sin A \cos B + \cos A \sin B$. (b) **What If?** Show that this expression for $\tan \theta$ reduces to Brewster's law when $\beta = 90°$, $n_1 = 1$, and $n_2 = n$.

59. Suppose the single slit in Active Figure 38.4 is 6.00 cm wide and in front of a microwave source operating at 7.50 GHz. (a) Calculate the angle subtended by the first minimum in the diffraction pattern. (b) What is the relative intensity I/I_{max} at $\theta = 15.0°$? (c) Assume two such sources, separated laterally by 20.0 cm, are behind the slit. What must the maximum distance between the plane of the sources and the slit be if the diffraction patterns are to be resolved? In this case, the approximation $\sin \theta \approx \tan \theta$ is not valid because of the relatively small value of a/λ.

60. ● (a) Two polarizing sheets are placed together with their transmission axes crossed so that no light is transmitted. A third sheet is inserted between them with its

transmission axis at an angle of 45.0° with respect to each of the other axes. Find the fraction of incident unpolarized light intensity transmitted by the three-sheet combination. Assume each polarizing sheet is ideal. (b) A parent is searching an Internet collection of recipes for easy main dishes. She uses a computer to eliminate all the recipes containing meat and then to eliminate all the remaining recipes containing cheese. No recipes from the original database remain after these two sorting processes. With the same database, she next tries a sequence of three selection rounds: eliminating all recipes containing meat, eliminating remaining recipes containing more than a little meat or cheese, and eliminating remaining recipes containing any cheese. Will any recipes remain in the list after these three sorting processes? Compare and contrast the results of the recipe-sorting experiment with the results of the polarization experiment in part (a).

61. The scale of a map is a number of kilometers per centimeter, specifying the distance on the ground that any distance on the map represents. The scale of a spectrum is its *dispersion,* a number of nanometers per centimeter, specifying the change in wavelength that a distance across the spectrum represents. You must know the dispersion if you want to compare one spectrum with another or to make a measurement of, for example, a Doppler shift. Let y represent the position relative to the center of a diffraction pattern projected onto a flat screen at distance L by a diffraction grating with slit spacing d. The dispersion is $d\lambda/dy$. (a) Prove that the dispersion is given by

$$\frac{d\lambda}{dy} = \frac{L^2 d}{m(L^2 + y^2)^{3/2}}$$

(b) Calculate the dispersion in first order for light with a mean wavelength of 550 nm, analyzed with a grating having 8 000 rulings/cm, and projected onto a screen 2.40 m away.

62. Two closely spaced wavelengths of light are incident on a diffraction grating. (a) Starting with Equation 38.7, show that the angular dispersion of the grating is given by

$$\frac{d\theta}{d\lambda} = \frac{m}{d\cos\theta}$$

(b) A square grating 2.00 cm on each side containing 8 000 equally spaced slits is used to analyze the spectrum of mercury. Two closely spaced lines emitted by this element have wavelengths of 579.065 nm and 576.959 nm. What is the angular separation of these two wavelengths in the second-order spectrum?

63. Figure P38.63a is a three-dimensional sketch of a birefringent crystal. The dotted lines illustrate how a thin, parallel-faced slab of material could be cut from the larger specimen with the crystal's optic axis parallel to the faces of the plate. A section cut from the crystal in this manner is known as a *retardation plate.* When a beam of light is incident on the plate perpendicular to the direction of the optic axis as shown in Figure P38.63b, the O ray and the E ray travel along a single straight line, but with different speeds. (a) Let the thickness of the plate be d. Show that the phase difference between the O ray and the E ray is

$$\theta = \frac{2\pi d}{\lambda}|n_O - n_E|$$

where λ is the wavelength in air. (b) In a particular case, the incident light has a wavelength of 550 nm. Find the minimum value of d for a quartz plate for which $\theta = \pi/2$. Such a plate is called a *quarter-wave plate.* Use values of n_O and n_E from Table 38.1.

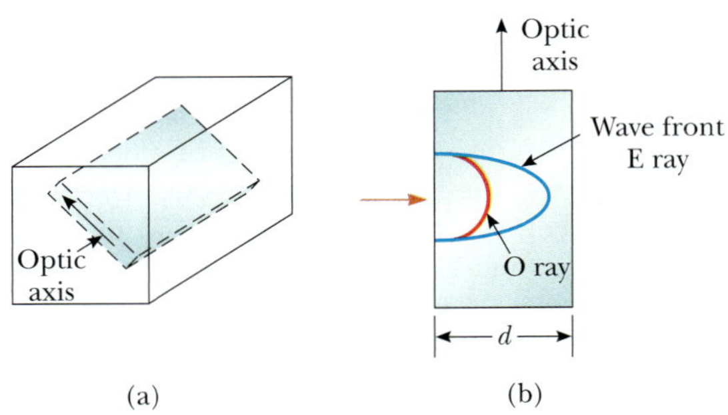

Figure P38.63

64. How much diffraction spreading does a light beam undergo? One quantitative answer is the *full width at half maximum* of the central maximum of the single-slit Fraunhofer diffraction pattern. You can evaluate this angle of spreading in this problem and the next. (a) In Equation 38.2, define $\pi a \sin\theta/\lambda = \phi$ and show that, at the point where $I = 0.5I_{max}$, we must have $\sin\phi = \phi/\sqrt{2}$. (b) Let $y_1 = \sin\phi$ and $y_2 = \phi/\sqrt{2}$. Plot y_1 and y_2 on the same set of axes over a range from $\phi = 1$ rad to $\phi = \pi/2$ rad. Determine ϕ from the point of intersection of the two curves. (c) Then show that if the fraction λ/a is not large, the angular full width at half maximum of the central diffraction maximum is $\Delta\theta = 0.886\lambda/a$.

65. Another method to solve the transcendental equation $\phi = \sqrt{2}\sin\phi$ in Problem 64 is to guess a first value of ϕ, use a computer or calculator to see how nearly it fits, and

continue to update your estimate until the equation balances. How many steps (iterations) does this process take?

horizontal tangent at maxima and also at minima. You will need to solve a transcendental equation.

66. The diffraction pattern of a single slit is described by the equation

$$I_\theta = I_{max}\frac{\sin^2\phi}{\phi^2}$$

where $\phi = (\pi a \sin\theta)/\lambda$. The central maximum is at $\phi = 0$, and the side maxima are *approximately* at $\phi = (m + \frac{1}{2})\pi$ for $m = 1, 2, 3, \ldots$. Determine more precisely (a) the location of the first side maximum, where $m = 1$, and (b) the location of the second side maximum. Observe in Figure 38.6a that the graph of intensity versus ϕ has a horizontal tangent at maxima and also at minima. You will need to solve a transcendental equation.

67. Light of wavelength 632.8 nm illuminates a single slit, and a diffraction pattern is formed on a screen 1.00 m from the slit. Using the data in the following table, plot relative intensity versus position. Choose an appropriate value for the slit width a and, on the same graph used for the experimental data, plot the theoretical expression for the relative intensity

$$\frac{I_\theta}{I_{max}} = \frac{\sin^2\phi}{\phi^2}$$

where $\phi = (\pi a \sin\theta)/\lambda$. What value of a gives the best fit of theory and experiment?

Relative Intensity	Position Relative to Central Maximum (mm)	Relative Intensity	Position Relative to Central Maximum (mm)
1.00	0	0.029	10.5
0.95	0.8	0.013	11.3
0.80	1.6	0.002	12.1
0.60	2.4	0.000 3	12.9
0.39	3.2	0.005	13.7
0.21	4.0	0.012	14.5
0.079	4.8	0.016	15.3
0.014	5.6	0.015	16.1
0.003	6.5	0.010	16.9
0.015	7.3	0.004 4	17.7
0.036	8.1	0.000 6	18.5
0.047	8.9	0.000 3	19.3
0.043	9.7	0.003	20.2

Answers to Quick Quizzes

38.1 (a). Equation 38.1 shows that a decrease in a results in an increase in the angles at which the dark fringes appear.

38.2 (i). In Equation 38.4, the ratio d/a is independent of wavelength, so the number of interference fringes in the central diffraction pattern peak remains the same. Equation 38.1 tells us that a decrease in wavelength causes a decrease in the width of the central peak if I is graphed against y as in Active Figure 38.4. If I is graphed against (π/λ) $a \sin\theta$ as in Active Figure 38.7, however, the peak width is independent of λ.

38.3 (b). The effective slit width in the vertical direction of the cat's eye is larger than that in the horizontal direction. Therefore, the cat's eye has more resolving power for lights separated in the vertical direction and would be more effective at resolving the mast lights on the boat.

38.4 (a). We would like to reduce the minimum angular separation for two objects below the angle subtended by the two stars in the binary system. That can be done by reducing the wavelength of the light, which in essence makes the aperture larger, relative to the light wavelength, increasing the resolving power. Therefore, we should choose a blue filter.

38.5 (c). Doubling the wavelength makes the pattern wider. Choices (a) and (d) make the pattern even wider. From Equation 38.10, we see that choice (b) causes $\sin\theta$ to be twice as large. Because the small angle approximation cannot be used, however, doubling $\sin\theta$ is not the same as doubling θ, which would translate to a doubling of the position of a maximum along the screen. If we only consider small-angle maxima, choice (b) would work, but it does not work in the large-angle case.

38.6 (b). Electric field vectors parallel to the metal wires cause electrons in the metal to oscillate parallel to the wires. Therefore, the energy from the waves with these electric field vectors is transferred to the metal by accelerating these electrons and is eventually transformed to internal energy through the resistance of the metal. Waves with electric field vectors perpendicular to the metal wires pass through because they are not able to accelerate electrons in the wires.

38.7 (c). At some intermediate distance, the light rays from the fixtures will strike the floor at Brewster's angle and reflect to your eyes. Because this light is polarized horizontally, it will not pass through your polarized sunglasses. Tilting your head to the side will cause the reflections to reappear.

Modern Physics

At the end of the 19th century, many scientists believed they had learned most of what there was to know about physics. Newton's laws of motion and theory of universal gravitation, Maxwell's theoretical work in unifying electricity and magnetism, the laws of thermodynamics and kinetic theory, and the principles of optics were highly successful in explaining a variety of phenomena.

At the turn of the 20th century, however, a major revolution shook the world of physics. In 1900, Max Planck provided the basic ideas that led to the formulation of the quantum theory, and in 1905, Albert Einstein formulated his special theory of relativity. The excitement of the times is captured in Einstein's own words: "It was a marvelous time to be alive." Both theories were to have a profound effect on our understanding of nature. Within a few decades, they inspired new developments in the fields of atomic physics, nuclear physics, and condensed-matter physics.

In Chapter 39, we shall introduce the special theory of relativity. The theory provides us with a new and deeper view of physical laws. Although the predictions of this theory often violate our common sense, the theory correctly describes the results of experiments involving speeds near the speed of light. The extended version of this textbook, *Physics for Scientists and Engineers with Modern Physics,* covers the basic concepts of quantum mechanics and their application to atomic and molecular physics, and we introduce solid-state physics, nuclear physics, particle physics, and cosmology.

Even though the physics that was developed during the 20th century has led to a multitude of important technological achievements, the story is still incomplete. Discoveries will continue to evolve during our lifetimes, and many of these discoveries will deepen or refine our understanding of nature and the Universe around us. It is still a "marvelous time to be alive."

A portion of the accelerator tunnel at Fermilab, near Chicago, Illinois. The tunnel is circular and 1.9 km in diameter. Using electric and magnetic fields, protons and antiprotons are accelerated to speeds close to that of light and then allowed to collide head-on, so that the production of new particles can be investigated. (Fermilab Photo)

Standing on the shoulders of a giant. David Serway, son of one of the authors, watches over two of his children, Nathan and Kaitlyn, as they frolic in the arms of Albert Einstein's statue at the Einstein memorial in Washington, D.C. It is well known that Einstein, the principal architect of relativity, was very fond of children. (Emily Serway)

39 Relativity

Our everyday experiences and observations involve objects that move at speeds much less than the speed of light. Newtonian mechanics was formulated by observing and describing the motion of such objects, and this formalism is very successful in describing a wide range of phenomena that occur at low speeds. Nonetheless, it fails to describe properly the motion of objects whose speeds approach that of light.

Experimentally, the predictions of Newtonian theory can be tested at high speeds by accelerating electrons or other charged particles through a large electric potential difference. For example, it is possible to accelerate an electron to a speed of $0.99c$ (where c is the speed of light) by using a potential difference of several million volts. According to Newtonian mechanics, if the potential difference is increased by a factor of 4, the electron's kinetic energy is four times greater and its speed should double to $1.98c$. Experiments show, however, that the speed of the electron—as well as the speed of any other object in the Universe—always remains less than the speed of light, regardless of the size of the accelerating voltage. Because it places no upper limit on speed, Newtonian mechanics is contrary to modern experimental results and is clearly a limited theory.

In 1905, at the age of only 26, Einstein published his special theory of relativity. Regarding the theory, Einstein wrote:

> The relativity theory arose from necessity, from serious and deep contradictions in the old theory from which there seemed no escape. The strength of the new theory lies in the consistency and simplicity with which it solves all these difficulties.[1]

Although Einstein made many other important contributions to science, the special theory of relativity alone represents one of the greatest intellectual achievements of all time. With this theory, experimental observations can be correctly predicted over the range of speeds from $v = 0$ to speeds approaching the speed of light. At low speeds, Einstein's theory reduces to Newtonian mechanics as a limiting situation. It is important to recognize that Einstein was working on electromagnetism when he developed the special theory of relativity. He was convinced that Maxwell's equations were correct, and to reconcile them with one of his postulates, he was forced into the revolutionary notion of assuming that space and time are not absolute.

This chapter gives an introduction to the special theory of relativity, with emphasis on some of its predictions. In addition to its well-known and essential role in theoretical physics, the special theory of relativity has practical applications, including the design of nuclear power plants and modern global positioning system (GPS) units. These devices do not work if designed in accordance with nonrelativistic principles.

39.1 The Principle of Galilean Relativity

To describe a physical event, we must establish a frame of reference. You should recall from Chapter 5 that an inertial frame of reference is one in which an object is observed to have no acceleration when no forces act on it. Furthermore, any frame moving with constant velocity with respect to an inertial frame must also be an inertial frame.

There is no absolute inertial reference frame. Therefore, the results of an experiment performed in a vehicle moving with uniform velocity must be identical to the results of the same experiment performed in a stationary vehicle. The formal statement of this result is called the **principle of galilean relativity:**

The laws of mechanics must be the same in all inertial frames of reference.

◀ Principle of Galilean relativity

Let's consider an observation that illustrates the equivalence of the laws of mechanics in different inertial frames. A pickup truck moves with a constant velocity as shown in Figure 39.1a. If a passenger in the truck throws a ball straight up

Figure 39.1 (a) The observer in the truck sees the ball move in a vertical path when thrown upward. (b) The Earth-based observer sees the ball's path as a parabola.

[1] A. Einstein and L. Infeld, *The Evolution of Physics* (New York: Simon and Schuster, 1961).

and if air effects are neglected, the passenger observes that the ball moves in a vertical path. The motion of the ball appears to be precisely the same as if the ball were thrown by a person at rest on the Earth. The law of universal gravitation and the equations of motion under constant acceleration are obeyed whether the truck is at rest or in uniform motion.

Both observers agree on the laws of physics: they each throw a ball straight up, and it rises and falls back into their own hand. Do the observers agree on the path of the ball thrown by the observer in the truck? The observer on the ground sees the path of the ball as a parabola as illustrated in Figure 39.1b, while, as mentioned earlier, the observer in the truck sees the ball move in a vertical path. Furthermore, according to the observer on the ground, the ball has a horizontal component of velocity equal to the velocity of the truck. Although the two observers disagree on certain aspects of the situation, they agree on the validity of Newton's laws and on such classical principles as conservation of energy and conservation of linear momentum. This agreement implies that no mechanical experiment can detect any difference between the two inertial frames. The only thing that can be detected is the relative motion of one frame with respect to the other.

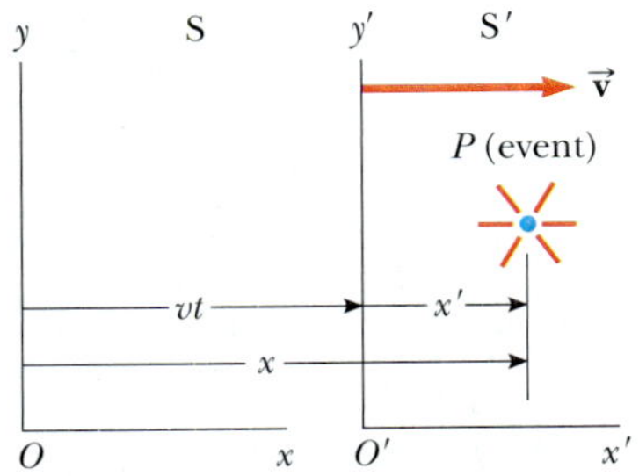

Figure 39.2 An event occurs at a point *P*. The event is seen by two observers in inertial frames S and S′, where S′ moves with a velocity $\vec{\mathbf{v}}$ relative to S.

Quick Quiz 39.1 Which observer in Figure 39.1 sees the ball's *correct* path? (a) the observer in the truck (b) the observer on the ground (c) both observers

---

Suppose some physical phenomenon, which we call an *event,* occurs and is observed by an observer at rest in an inertial reference frame. The wording "in a frame" means that the observer is at rest with respect to the origin of that frame. The event's location and time of occurrence can be specified by the four coordinates (x, y, z, t). We would like to be able to transform these coordinates from those of an observer in one inertial frame to those of another observer in a frame moving with uniform relative velocity compared with the first frame.

Consider two inertial frames S and S′ (Fig. 39.2). The S′ frame moves with a constant velocity $\vec{\mathbf{v}}$ along the common x and x' axes, where $\vec{\mathbf{v}}$ is measured relative to S. We assume the origins of S and S′ coincide at $t = 0$ and an event occurs at point P in space at some instant of time. An observer in S describes the event with space–time coordinates (x, y, z, t), whereas an observer in S′ uses the coordinates (x', y', z', t') to describe the same event. As we see from the geometry in Figure 39.2, the relationships among these various coordinates can be written

Galilean transformation equations ▶

$$x' = x - vt \qquad y' = y \qquad z' = z \qquad t' = t \tag{39.1}$$

These equations are the **Galilean space–time transformation equations.** Note that time is assumed to be the same in both inertial frames. That is, within the framework of classical mechanics, all clocks run at the same rate, regardless of their velocity, so the time at which an event occurs for an observer in S is the same as the time for the same event in S′. Consequently, the time interval between two successive events should be the same for both observers. Although this assumption may seem obvious, it turns out to be incorrect in situations where v is comparable to the speed of light.

PITFALL PREVENTION 39.1
The Relationship Between the S and S′ Frames

Many of the mathematical representations in this chapter are true *only* for the specified relationship between the S and S′ frames. The x and x' axes coincide, except their origins are different. The y and y' axes (and the z and z' axes) are parallel, but they do not coincide due to the displacement of the origin of S′ with respect to that of S. We choose the time $t = 0$ to be the instant at which the origins of the two coordinate systems coincide. If the S′ frame is moving in the positive x direction relative to S, then v is positive; otherwise, it is negative.

Now suppose a particle moves through a displacement of magnitude dx along the x axis in a time interval dt as measured by an observer in S. It follows from Equations 39.1 that the corresponding displacement dx' measured by an observer in S′ is $dx' = dx - v\,dt$, where frame S′ is moving with speed v in the x direction relative to frame S. Because $dt = dt'$, we find that

$$\frac{dx'}{dt'} = \frac{dx}{dt} - v$$

or

$$u'_x = u_x - v \tag{39.2}$$

where u_x and u'_x are the x components of the velocity of the particle measured by observers in S and S′, respectively. (We use the symbol $\vec{\mathbf{u}}$ rather than $\vec{\mathbf{v}}$ for particle velocity because $\vec{\mathbf{v}}$ is already used for the relative velocity of two reference frames.) Equation 39.2 is the **Galilean velocity transformation equation.** It is consistent with our intuitive notion of time and space as well as with our discussions in Section 4.6. As we shall soon see, however, it leads to serious contradictions when applied to electromagnetic waves.

Quick Quiz 39.2 A baseball pitcher with a 90-mi/h fastball throws a ball while standing on a railroad flatcar moving at 110 mi/h. The ball is thrown in the same direction as that of the velocity of the train. If you apply the Galilean velocity transformation equation to this situation, is the speed of the ball relative to the Earth (a) 90 mi/h, (b) 110 mi/h, (c) 20 mi/h, (d) 200 mi/h, or (e) impossible to determine?

The Speed of Light

It is quite natural to ask whether the principle of Galilean relativity also applies to electricity, magnetism, and optics. Experiments indicate that the answer is no. Recall from Chapter 34 that Maxwell showed that the speed of light in free space is $c = 3.00 \times 10^8$ m/s. Physicists of the late 1800s thought light waves moved through a medium called the *ether* and the speed of light was c only in a special, absolute frame at rest with respect to the ether. The Galilean velocity transformation equation was expected to hold for observations of light made by an observer in any frame moving at speed v relative to the absolute ether frame. That is, if light travels along the x axis and an observer moves with velocity $\vec{\mathbf{v}}$ along the x axis, the observer measures the light to have speed $c \pm v$, depending on the directions of travel of the observer and the light.

Because the existence of a preferred, absolute ether frame would show that light is similar to other classical waves and that Newtonian ideas of an absolute frame are true, considerable importance was attached to establishing the existence of the ether frame. Prior to the late 1800s, experiments involving light traveling in media moving at the highest laboratory speeds attainable at that time were not capable of detecting differences as small as that between c and $c \pm v$. Starting in about 1880, scientists decided to use the Earth as the moving frame in an attempt to improve their chances of detecting these small changes in the speed of light.

Observers fixed on the Earth can take the view that they are stationary and that the absolute ether frame containing the medium for light propagation moves past them with speed v. Determining the speed of light under these circumstances is similar to determining the speed of an aircraft traveling in a moving air current, or wind; consequently, we speak of an "ether wind" blowing through our apparatus fixed to the Earth.

A direct method for detecting an ether wind would use an apparatus fixed to the Earth to measure the ether wind's influence on the speed of light. If v is the speed of the ether relative to the Earth, light should have its maximum speed $c + v$ when propagating downwind as in Figure 39.3a. Likewise, the speed of light should have its minimum value $c - v$ when the light is propagating upwind as in Figure 39.3b and an intermediate value $(c^2 - v^2)^{1/2}$ when the light is directed such that it travels perpendicular to the ether wind as in Figure 39.3c. If the Sun is assumed to be at rest in the ether, the velocity of the ether wind would be equal to the orbital velocity of the Earth around the Sun, which has a magnitude of approximately 30 km/s or 3×10^4 m/s. Because $c = 3 \times 10^8$ m/s, it is necessary to detect a change in speed of approximately 1 part in 10^4 for measurements in the upwind or downwind directions. Although such a change is experimentally measurable, all attempts to detect such changes and establish the existence of the ether wind (and hence the absolute frame) proved futile! We shall discuss the classic experimental search for the ether in Section 39.2.

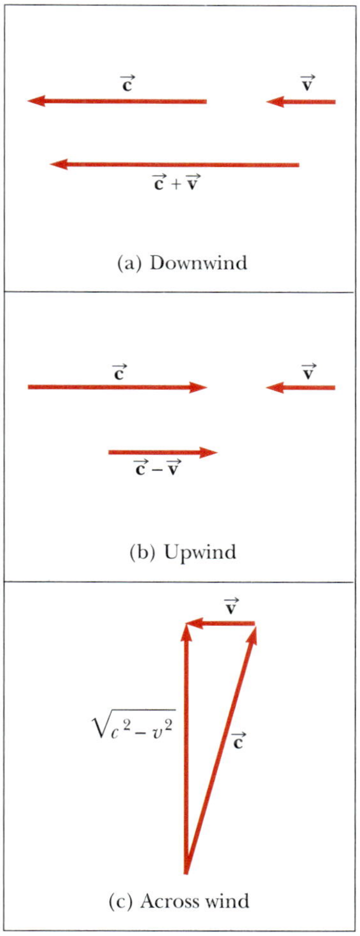

Figure 39.3 If the velocity of the ether wind relative to the Earth is $\vec{\mathbf{v}}$ and the velocity of light relative to the ether is $\vec{\mathbf{c}}$, the speed of light relative to the Earth is (a) $c + v$ in the downwind direction, (b) $c - v$ in the upwind direction, and (c) $(c^2 - v^2)^{1/2}$ in the direction perpendicular to the wind.

The principle of Galilean relativity refers only to the laws of mechanics. If it is assumed the laws of electricity and magnetism are the same in all inertial frames, a paradox concerning the speed of light immediately arises. That can be understood by recognizing that Maxwell's equations imply that the speed of light always has the fixed value 3.00×10^8 m/s in all inertial frames, a result in direct contradiction to what is expected based on the Galilean velocity transformation equation. According to Galilean relativity, the speed of light should *not* be the same in all inertial frames.

To resolve this contradiction in theories, we must conclude that either (1) the laws of electricity and magnetism are not the same in all inertial frames or (2) the Galilean velocity transformation equation is incorrect. If we assume the first alternative, a preferred reference frame in which the speed of light has the value c must exist and the measured speed must be greater or less than this value in any other reference frame, in accordance with the Galilean velocity transformation equation. If we assume the second alternative, we must abandon the notions of absolute time and absolute length that form the basis of the Galilean space–time transformation equations.

39.2 The Michelson–Morley Experiment

The most famous experiment designed to detect small changes in the speed of light was first performed in 1881 by Albert A. Michelson (see Section 37.7) and later repeated under various conditions by Michelson and Edward W. Morley (1838–1923). As we shall see, the outcome of the experiment contradicted the ether hypothesis.

The experiment was designed to determine the velocity of the Earth relative to that of the hypothetical ether. The experimental tool used was the Michelson interferometer, which was discussed in Section 37.7 and is shown again in Active Figure 39.4. Arm 2 is aligned along the direction of the Earth's motion through space. The Earth moving through the ether at speed v is equivalent to the ether flowing past the Earth in the opposite direction with speed v. This ether wind blowing in the direction opposite the direction of the Earth's motion should cause the speed of light measured in the Earth frame to be $c - v$ as the light approaches mirror M_2 and $c + v$ after reflection, where c is the speed of light in the ether frame.

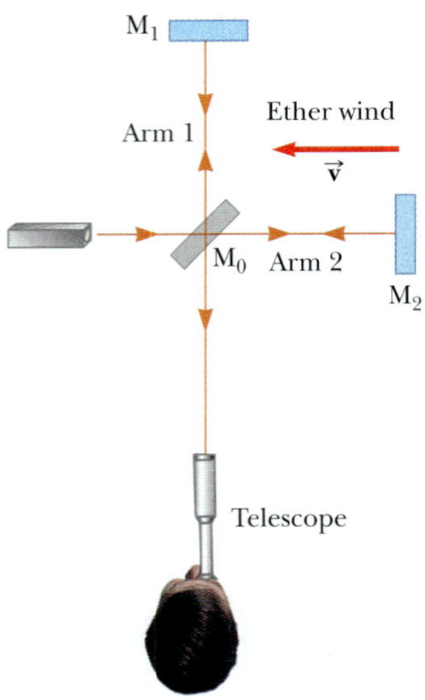

ACTIVE FIGURE 39.4

According to the ether wind theory, the speed of light should be $c - v$ as the beam approaches mirror M_2 and $c + v$ after reflection.

Sign in at www.thomsonedu.com and go to ThomsonNOW to adjust the speed of the ether wind and see the effect on the light beams if there were an ether.

The two light beams reflect from M_1 and M_2 and recombine, and an interference pattern is formed as discussed in Section 37.7. The interference pattern is observed while the interferometer is rotated through an angle of 90°. This rotation interchanges the speed of the ether wind between the arms of the interferometer. The rotation should cause the fringe pattern to shift slightly but measurably. Measurements failed, however, to show any change in the interference pattern! The Michelson–Morley experiment was repeated at different times of the year when the ether wind was expected to change direction and magnitude, but the results were always the same: **no fringe shift of the magnitude required was ever observed.**[2]

The negative results of the Michelson–Morley experiment not only contradicted the ether hypothesis, but also showed that it is impossible to measure the absolute velocity of the Earth with respect to the ether frame. Einstein, however, offered a postulate for his special theory of relativity that places quite a different interpretation on these null results. In later years, when more was known about the nature of light, the idea of an ether that permeates all of space was abandoned. **Light is now understood to be an electromagnetic wave, which requires no medium for its propagation.** As a result, the idea of an ether in which these waves travel became unnecessary.

[2] From an Earth-based observer's point of view, changes in the Earth's speed and direction of motion in the course of a year are viewed as ether wind shifts. Even if the speed of the Earth with respect to the ether were zero at some time, six months later the speed of the Earth would be 60 km/s with respect to the ether and as a result a fringe shift should be noticed. No shift has ever been observed, however.

Details of the Michelson–Morley Experiment

To understand the outcome of the Michelson–Morley experiment, let's assume the two arms of the interferometer in Active Figure 39.4 are of equal length L. We shall analyze the situation as if there were an ether wind because that is what Michelson and Morley expected to find. As noted above, the speed of the light beam along arm 2 should be $c - v$ as the beam approaches M_2 and $c + v$ after the beam is reflected. We model a pulse of light as a particle under constant speed. Therefore, the time interval for travel to the right for the pulse is $\Delta t = L/(c - v)$, and the time interval for travel to the left is $\Delta t = L/(c + v)$. The total time interval for the round trip along arm 2 is

$$\Delta t_{\text{arm 2}} = \frac{L}{c+v} + \frac{L}{c-v} = \frac{2Lc}{c^2 - v^2} = \frac{2L}{c}\left(1 - \frac{v^2}{c^2}\right)^{-1}$$

Now consider the light beam traveling along arm 1, perpendicular to the ether wind. Because the speed of the beam relative to the Earth is $(c^2 - v^2)^{1/2}$ in this case (see Fig. 39.3c), the time interval for travel for each half of the trip is $\Delta t = L/(c^2 - v^2)^{1/2}$ and the total time interval for the round trip is

$$\Delta t_{\text{arm 1}} = \frac{2L}{(c^2 - v^2)^{1/2}} = \frac{2L}{c}\left(1 - \frac{v^2}{c^2}\right)^{-1/2}$$

The time difference Δt between the horizontal round trip (arm 2) and the vertical round trip (arm 1) is

$$\Delta t = \Delta t_{\text{arm 2}} - \Delta t_{\text{arm 1}} = \frac{2L}{c}\left[\left(1 - \frac{v^2}{c^2}\right)^{-1} - \left(1 - \frac{v^2}{c^2}\right)^{-1/2}\right]$$

Because $v^2/c^2 \ll 1$, we can simplify this expression by using the following binomial expansion after dropping all terms higher than second order:

$$(1 - x)^n \approx 1 - nx \quad (\text{for } x \ll 1)$$

In our case, $x = v^2/c^2$, and we find that

$$\Delta t = \Delta t_{\text{arm 2}} - \Delta t_{\text{arm1}} \approx \frac{Lv^2}{c^3} \qquad \textbf{(39.3)}$$

This time difference between the two instants at which the reflected beams arrive at the viewing telescope gives rise to a phase difference between the beams, producing an interference pattern when they combine at the position of the telescope. A shift in the interference pattern should be detected when the interferometer is rotated through 90° in a horizontal plane so that the two beams exchange roles. This rotation results in a time difference twice that given by Equation 39.3. Therefore, the path difference that corresponds to this time difference is

$$\Delta d = c(2\,\Delta t) = \frac{2Lv^2}{c^2}$$

Because a change in path length of one wavelength corresponds to a shift of one fringe, the corresponding fringe shift is equal to this path difference divided by the wavelength of the light:

$$\text{Shift} = \frac{2Lv^2}{\lambda c^2} \qquad \textbf{(39.4)}$$

In the experiments by Michelson and Morley, each light beam was reflected by mirrors many times to give an effective path length L of approximately 11 m. Using this value, taking v to be equal to 3.0×10^4 m/s (the speed of the Earth around the Sun), and using 500 nm for the wavelength of the light, we expect a fringe shift of

$$\text{Shift} = \frac{2(11\text{ m})(3.0 \times 10^4\text{ m/s})^2}{(5.0 \times 10^{-7}\text{ m})(3.0 \times 10^8\text{ m/s})^2} = 0.44$$

The instrument used by Michelson and Morley could detect shifts as small as 0.01 fringe, but **it detected no shift whatsoever in the fringe pattern!** The experiment has been repeated many times since by different scientists under a wide variety of conditions, and no fringe shift has ever been detected. Therefore, it was concluded that the motion of the Earth with respect to the postulated ether cannot be detected.

Many efforts were made to explain the null results of the Michelson–Morley experiment and to save the ether frame concept and the Galilean velocity transformation equation for light. All proposals resulting from these efforts have been shown to be wrong. No experiment in the history of physics received such valiant efforts to explain the absence of an expected result as did the Michelson–Morley experiment. The stage was set for Einstein, who solved the problem in 1905 with his special theory of relativity.

39.3 Einstein's Principle of Relativity

AIP Niels Bohr Library

ALBERT EINSTEIN
German-American Physicist (1879–1955)
Einstein, one of the greatest physicists of all times, was born in Ulm, Germany. In 1905, at age 26, he published four scientific papers that revolutionized physics. Two of these papers were concerned with what is now considered his most important contribution: the special theory of relativity.

In 1916, Einstein published his work on the general theory of relativity. The most dramatic prediction of this theory is the degree to which light is deflected by a gravitational field. Measurements made by astronomers on bright stars in the vicinity of the eclipsed Sun in 1919 confirmed Einstein's prediction, and Einstein became a world celebrity as a result. Einstein was deeply disturbed by the development of quantum mechanics in the 1920s despite his own role as a scientific revolutionary. In particular, he could never accept the probabilistic view of events in nature that is a central feature of quantum theory. The last few decades of his life were devoted to an unsuccessful search for a unified theory that would combine gravitation and electromagnetism.

In the previous section, we noted the impossibility of measuring the speed of the ether with respect to the Earth and the failure of the Galilean velocity transformation equation in the case of light. Einstein proposed a theory that boldly removed these difficulties and at the same time completely altered our notion of space and time.[3] He based his special theory of relativity on two postulates:

1. **The principle of relativity:** The laws of physics must be the same in all inertial reference frames.
2. **The constancy of the speed of light:** The speed of light in vacuum has the same value, $c = 3.00 \times 10^8$ m/s, in all inertial frames, regardless of the velocity of the observer or the velocity of the source emitting the light.

The first postulate asserts that *all* the laws of physics—those dealing with mechanics, electricity and magnetism, optics, thermodynamics, and so on—are the same in all reference frames moving with constant velocity relative to one another. This postulate is a generalization of the principle of Galilean relativity, which refers only to the laws of mechanics. From an experimental point of view, Einstein's principle of relativity means that any kind of experiment (measuring the speed of light, for example) performed in a laboratory at rest must give the same result when performed in a laboratory moving at a constant velocity with respect to the first one. Hence, no preferred inertial reference frame exists, and it is impossible to detect absolute motion.

Note that postulate 2 is required by postulate 1: if the speed of light were not the same in all inertial frames, measurements of different speeds would make it possible to distinguish between inertial frames. As a result, a preferred, absolute frame could be identified, in contradiction to postulate 1.

Although the Michelson–Morley experiment was performed before Einstein published his work on relativity, it is not clear whether or not Einstein was aware of the details of the experiment. Nonetheless, the null result of the experiment can be readily understood within the framework of Einstein's theory. According to his principle of relativity, the premises of the Michelson–Morley experiment were incorrect. In the process of trying to explain the expected results, we stated that when light traveled against the ether wind, its speed was $c - v$, in accordance with the Galilean velocity transformation equation. If the state of motion of the observer or of the source has no influence on the value found for the speed of

[3] A. Einstein, "On the Electrodynamics of Moving Bodies," *Ann. Physik* **17:**891, 1905. For an English translation of this article and other publications by Einstein, see the book by H. Lorentz, A. Einstein, H. Minkowski, and H. Weyl, *The Principle of Relativity* (New York: Dover, 1958).

light, however, one always measures the value to be c. Likewise, the light makes the return trip after reflection from the mirror at speed c, not at speed $c + v$. Therefore, the motion of the Earth does not influence the fringe pattern observed in the Michelson–Morley experiment, and a null result should be expected.

If we accept Einstein's theory of relativity, we must conclude that relative motion is unimportant when measuring the speed of light. At the same time, we must alter our commonsense notion of space and time and be prepared for some surprising consequences. As you read the pages ahead, keep in mind that our commonsense ideas are based on a lifetime of everyday experiences and not on observations of objects moving at hundreds of thousands of kilometers per second. Therefore, these results may seem strange, but that is only because we have no experience with them.

39.4 Consequences of the Special Theory of Relativity

As we examine some of the consequences of relativity in this section, we restrict our discussion to the concepts of simultaneity, time intervals, and lengths, all three of which are quite different in relativistic mechanics from what they are in Newtonian mechanics. In relativistic mechanics, for example, the distance between two points and the time interval between two events depend on the frame of reference in which they are measured.

Simultaneity and the Relativity of Time

A basic premise of Newtonian mechanics is that a universal time scale exists that is the same for all observers. Newton and his followers took simultaneity for granted. In his special theory of relativity, Einstein abandoned this assumption.

Einstein devised the following thought experiment to illustrate this point. A boxcar moves with uniform velocity, and two bolts of lightning strike its ends as illustrated in Figure 39.5a, leaving marks on the boxcar and on the ground. The marks on the boxcar are labeled A' and B', and those on the ground are labeled A and B. An observer O' moving with the boxcar is midway between A' and B', and a ground observer O is midway between A and B. The events recorded by the observers are the striking of the boxcar by the two lightning bolts.

The light signals emitted from A and B at the instant at which the two bolts strike later reach observer O at the same time as indicated in Figure 39.5b. This observer realizes that the signals traveled at the same speed over equal distances and so concludes that the events at A and B occurred simultaneously. Now consider the same events as viewed by observer O'. By the time the signals have reached observer O, observer O' has moved as indicated in Figure 39.5b. Therefore, the signal from B' has already swept past O', but the signal from A' has not yet reached O'. In other words, O' sees the signal from B' before seeing the signal from A'. According to Einstein, *the two observers must find that light travels at the same*

PITFALL PREVENTION 39.2
Who's Right?

You might wonder which observer in Figure 39.5 is correct concerning the two lightning strikes. *Both are correct* because the principle of relativity states that *there is no preferred inertial frame of reference.* Although the two observers reach different conclusions, both are correct in their own reference frame because the concept of simultaneity is not absolute. That, in fact, is the central point of relativity: any uniformly moving frame of reference can be used to describe events and do physics.

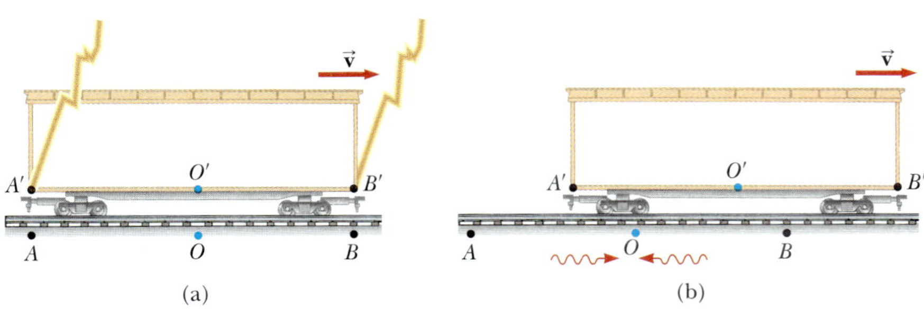

Figure 39.5 (a) Two lightning bolts strike the ends of a moving boxcar. (b) The events appear to be simultaneous to the stationary observer O who is standing midway between A and B. The events do not appear to be simultaneous to observer O', who claims that the front of the car is struck before the rear. Notice in (b) that the leftward-traveling light signal has already passed O', but the rightward-traveling signal has not yet reached O'.

speed. Therefore, observer O' concludes that one lightning bolt strikes the front of the boxcar *before* the other one strikes the back.

This thought experiment clearly demonstrates that the two events that appear to be simultaneous to observer O do *not* appear to be simultaneous to observer O'. In other words,

> two events that are simultaneous in one reference frame are in general not simultaneous in a second frame moving relative to the first.

Simultaneity is not an absolute concept but rather one that depends on the state of motion of the observer. Einstein's thought experiment demonstrates that two observers can disagree on the simultaneity of two events. **This disagreement, however, depends on the transit time of light to the observers and therefore does *not* demonstrate the deeper meaning of relativity.** In relativistic analyses of high-speed situations, simultaneity is relative even when the transit time is subtracted out. In fact, in all the relativistic effects that we discuss, we ignore differences caused by the transit time of light to the observers.

Time Dilation

To illustrate that observers in different inertial frames can measure different time intervals between a pair of events, consider a vehicle moving to the right with a speed v such as the boxcar shown in Active Figure 39.6a. A mirror is fixed to the ceiling of the vehicle, and observer O' at rest in the frame attached to the vehicle holds a flashlight a distance d below the mirror. At some instant, the flashlight emits a pulse of light directed toward the mirror (event 1), and at some later time after reflecting from the mirror, the pulse arrives back at the flashlight (event 2). Observer O' carries a clock and uses it to measure the time interval Δt_p between these two events. (The subscript p stands for *proper*, as we shall see in a moment.) We model the pulse of light as a particle under constant speed. Because the light pulse has a speed c, the time interval required for the pulse to travel from O' to the mirror and back is

$$\Delta t_p = \frac{\text{distance traveled}}{\text{speed}} = \frac{2d}{c} \qquad \textbf{(39.5)}$$

Now consider the same pair of events as viewed by observer O in a second frame as shown in Active Figure 39.6b. According to this observer, the mirror and the flashlight are moving to the right with a speed v, and as a result, the sequence of

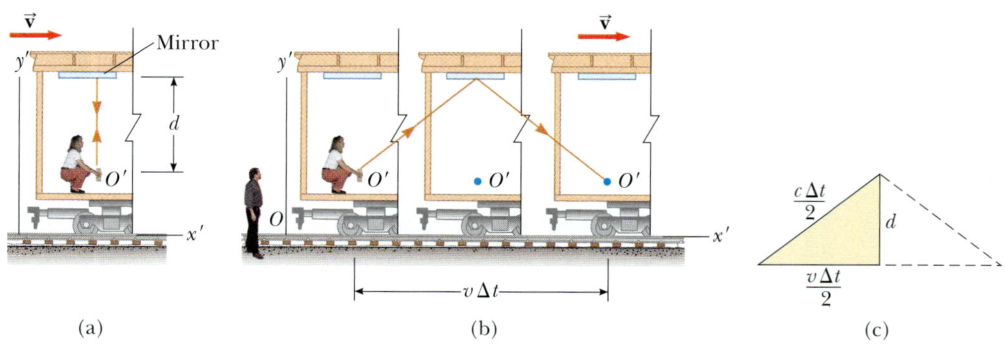

ACTIVE FIGURE 39.6

(a) A mirror is fixed to a moving vehicle, and a light pulse is sent out by observer O' at rest in the vehicle. (b) Relative to a stationary observer O standing alongside the vehicle, the mirror and O' move with a speed v. Notice that what observer O measures for the distance the pulse travels is greater than $2d$. (c) The right triangle for calculating the relationship between Δt and Δt_p.

Sign in at www.thomsonedu.com and go to ThomsonNOW to observe the bouncing of the light pulse for various speeds of the train.

events appears entirely different. By the time the light from the flashlight reaches the mirror, the mirror has moved to the right a distance $v\,\Delta t/2$, where Δt is the time interval required for the light to travel from O' to the mirror and back to O' as measured by O. Observer O concludes that because of the motion of the vehicle, if the light is to hit the mirror, it must leave the flashlight at an angle with respect to the vertical direction. Comparing Active Figure 39.6a with Active Figure 39.6b, we see that the light must travel farther in part (b) than in part (a). (Notice that neither observer "knows" that he or she is moving. Each is at rest in his or her own inertial frame.)

According to the second postulate of the special theory of relativity, both observers must measure c for the speed of light. Because the light travels farther according to O, the time interval Δt measured by O is longer than the time interval Δt_p measured by O'. To obtain a relationship between these two time intervals, let's use the right triangle shown in Active Figure 39.6c. The Pythagorean theorem gives

$$\left(\frac{c\,\Delta t}{2}\right)^2 = \left(\frac{v\,\Delta t}{2}\right)^2 + d^2$$

Solving for Δt gives

$$\Delta t = \frac{2d}{\sqrt{c^2 - v^2}} = \frac{2d}{c\sqrt{1 - \frac{v^2}{c^2}}} \qquad \textbf{(39.6)}$$

Because $\Delta t_p = 2d/c$, we can express this result as

$$\Delta t = \frac{\Delta t_p}{\sqrt{1 - \frac{v^2}{c^2}}} = \gamma \Delta t_p \qquad \textbf{(39.7)}$$

◀ Time dilation

where

$$\gamma = \frac{1}{\sqrt{1 - \frac{v^2}{c^2}}} \qquad \textbf{(39.8)}$$

Because γ is always greater than unity, Equation 39.7 shows that **the time interval Δt measured by an observer moving with respect to a clock is longer than the time interval Δt_p measured by an observer at rest with respect to the clock. This effect is known as time dilation.**

Time dilation is not observed in our everyday lives, which can be understood by considering the factor γ. This factor deviates significantly from a value of 1 only for very high speeds as shown in Figure 39.7 and Table 39.1. For example, for a speed of $0.1c$, the value of γ is 1.005. Therefore, there is a time dilation of only 0.5% at one-tenth the speed of light. Speeds encountered on an everyday basis are far slower than $0.1c$, so we do not experience time dilation in normal situations.

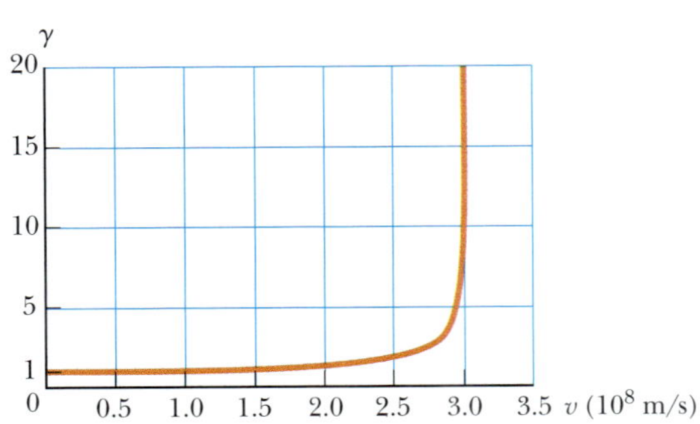

Figure 39.7 Graph of γ versus v. As the speed approaches that of light, γ increases rapidly.

TABLE 39.1

Approximate Values for γ at Various Speeds

v/c	γ
0.001 0	1.000 000 5
0.010	1.000 05
0.10	1.005
0.20	1.021
0.30	1.048
0.40	1.091
0.50	1.155
0.60	1.250
0.70	1.400
0.80	1.667
0.90	2.294
0.92	2.552
0.94	2.931
0.96	3.571
0.98	5.025
0.99	7.089
0.995	10.01
0.999	22.37

PITFALL PREVENTION 39.3
The Proper Time Interval

It is *very* important in relativistic calculations to correctly identify the observer who measures the proper time interval. The proper time interval between two events is always the time interval measured by an observer for whom the two events take place at the same position.

The time interval Δt_p in Equations 39.5 and 39.7 is called the **proper time interval.** (Einstein used the German term *Eigenzeit,* which means "own-time.") In general, **the proper time interval is the time interval between two events measured by an observer who sees the events occur at the same point in space.**

If a clock is moving with respect to you, the time interval between ticks of the moving clock is observed to be longer than the time interval between ticks of an identical clock in your reference frame. Therefore, it is often said that a moving clock is measured to run more slowly than a clock in your reference frame by a factor γ. We can generalize this result by stating that all physical processes, including mechanical, chemical, and biological ones, are measured to slow down when those processes occur in a frame moving with respect to the observer. For example, the heartbeat of an astronaut moving through space keeps time with a clock inside the spacecraft. Both the astronaut's clock and heartbeat are measured to slow down relative to a clock back on the Earth (although the astronaut would have no sensation of life slowing down in the spacecraft).

Quick Quiz 39.3 Suppose the observer O' on the train in Active Figure 39.6 aims her flashlight at the far wall of the boxcar and turns it on and off, sending a pulse of light toward the far wall. Both O' and O measure the time interval between when the pulse leaves the flashlight and when it hits the far wall. Which observer measures the proper time interval between these two events? (a) O' (b) O (c) both observers (d) neither observer

Quick Quiz 39.4 A crew on a spacecraft watches a movie that is two hours long. The spacecraft is moving at high speed through space. Does an Earth-based observer watching the movie screen on the spacecraft through a powerful telescope measure the duration of the movie to be (a) longer than, (b) shorter than, or (c) equal to two hours?

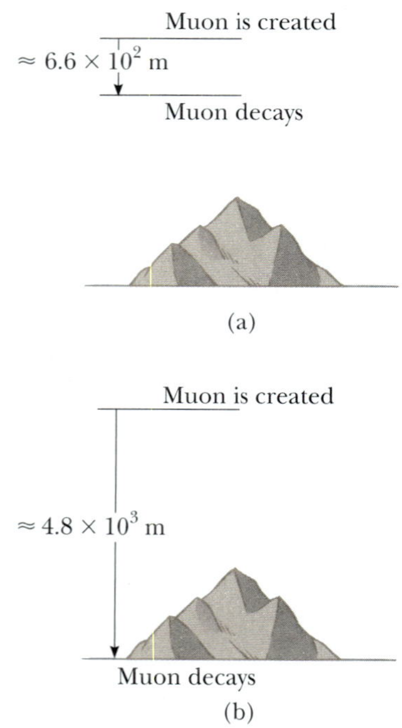

Figure 39.8 Travel of muons according to an Earth-based observer. (a) Without relativistic considerations, muons created in the atmosphere and traveling downward with a speed of $0.99c$ travel only about 6.6×10^2 m before decaying with an average lifetime of 2.2 μs. Therefore, very few muons reach the surface of the Earth. (b) With relativistic considerations, the muon's lifetime is dilated according to an observer on the Earth. Hence, according to this observer, the muon can travel about 4.8×10^3 m before decaying. The result is many of them arriving at the surface.

Time dilation is a very real phenomenon that has been verified by various experiments involving natural clocks. One experiment reported by J. C. Hafele and R. E. Keating provided direct evidence of time dilation.[4] Time intervals measured with four cesium atomic clocks in jet flight were compared with time intervals measured by Earth-based reference atomic clocks. To compare these results with theory, many factors had to be considered, including periods of speeding up and slowing down relative to the Earth, variations in direction of travel, and the weaker gravitational field experienced by the flying clocks than that experienced by the Earth-based clock. The results were in good agreement with the predictions of the special theory of relativity and were explained in terms of the relative motion between the Earth and the jet aircraft. In their paper, Hafele and Keating stated that "relative to the atomic time scale of the U.S. Naval Observatory, the flying clocks lost 59 ± 10 ns during the eastward trip and gained 273 ± 7 ns during the westward trip."

Another interesting example of time dilation involves the observation of *muons,* unstable elementary particles that have a charge equal to that of the electron and a mass 207 times that of the electron. Muons can be produced by the collision of cosmic radiation with atoms high in the atmosphere. Slow-moving muons in the laboratory have a lifetime that is measured to be the proper time interval $\Delta t_p = 2.2\ \mu$s. If we take 2.2 μs as the average lifetime of a muon and assume their speed is close to the speed of light, we find that these particles can travel a distance of approximately $(3.0 \times 10^8\ \text{m/s})(2.2 \times 10^{-6}\ \text{s}) \approx 6.6 \times 10^2$ m before they decay (Fig. 39.8a). Hence, they are unlikely to reach the surface of the Earth from high in the atmosphere where they are produced. Experiments show, however, that a large number of muons *do* reach the surface. The phenomenon of time dilation explains this effect. As measured by an observer on the Earth, the muons have a dilated lifetime

[4] J. C. Hafele and R. E. Keating, "Around the World Atomic Clocks: Relativistic Time Gains Observed," *Science* **177:**168, 1972.

equal to $\gamma \Delta t_p$. For example, for $v = 0.99c$, $\gamma \approx 7.1$, and $\gamma \Delta t_p \approx 16\ \mu s$. Hence, the average distance traveled by the muons in this time interval as measured by an observer on the Earth is approximately $(0.99)(3.0 \times 10^8 \text{ m/s})(16 \times 10^{-6} \text{ s}) \approx 4.8 \times 10^3$ m as indicated in Figure 39.8b.

In 1976, at the laboratory of the European Council for Nuclear Research in Geneva, muons injected into a large storage ring reached speeds of approximately $0.999\,4c$. Electrons produced by the decaying muons were detected by counters around the ring, enabling scientists to measure the decay rate and hence the muon lifetime. The lifetime of the moving muons was measured to be approximately 30 times as long as that of the stationary muon, in agreement with the prediction of relativity to within two parts in a thousand.

EXAMPLE 39.1 What Is the Period of the Pendulum?

The period of a pendulum is measured to be 3.00 s in the reference frame of the pendulum. What is the period when measured by an observer moving at a speed of $0.960c$ relative to the pendulum?

SOLUTION

Conceptualize Let's change frames of reference. Instead of the observer moving at $0.960c$, we can take the equivalent point of view that the observer is at rest and the pendulum is moving at $0.960c$ past the stationary observer. Hence, the pendulum is an example of a clock moving at high speed with respect to an observer.

Categorize Based on the Conceptualize step, we can categorize this problem as one involving time dilation.

Analyze The proper time interval, measured in the rest frame of the pendulum, is $\Delta t_p = 3.00$ s.

Use Equation 39.7 to find the dilated time interval:

$$\Delta t = \gamma\,\Delta t_p = \frac{1}{\sqrt{1 - \dfrac{(0.960c)^2}{c^2}}}\,\Delta t_p = \frac{1}{\sqrt{1 - 0.921\,6}}\,\Delta t_p$$

$$= 3.57(3.00 \text{ s}) = 10.7 \text{ s}$$

Finalize This result shows that a moving pendulum is indeed measured to take longer to complete a period than a pendulum at rest does. The period increases by a factor of $\gamma = 3.57$.

What If? What if the speed of the observer increases by 4.00%? Does the dilated time interval increase by 4.00%?

Answer Based on the highly nonlinear behavior of γ as a function of v in Figure 39.7, we would guess that the increase in Δt would be different from 4.00%.

Find the new speed if it increases by 4.00%:

$$v_{\text{new}} = (1.040\,0)(0.960c) = 0.998\,4c$$

Perform the time dilation calculation again:

$$\Delta t = \gamma\,\Delta t_p = \frac{1}{\sqrt{1 - \dfrac{(0.998\,4c)^2}{c^2}}}\,\Delta t_p = \frac{1}{\sqrt{1 - 0.996\,8}}\,\Delta t_p$$

$$= 17.68(3.00 \text{ s}) = 53.1 \text{ s}$$

Therefore, the 4.00% increase in speed results in almost a 400% increase in the dilated time!

EXAMPLE 39.2 **How Long Was Your Trip?**

Suppose you are driving your car on a business trip and are traveling at 30 m/s. Your boss, who is waiting at your destination, expects the trip to take 5.0 h. When you arrive late, your excuse is that clock in your car registered the passage of 5.0 h but that you were driving fast and so your clock ran more slowly than the clock in your boss's office. If your car clock actually did indicate a 5.0-h trip, how much time passed on your boss's clock, which was at rest on the Earth?

SOLUTION

Conceptualize The observer is your boss standing stationary on the Earth. The clock is in your car, moving at 30 m/s with respect to your boss.

Categorize The speed of 30 m/s suggests we might categorize this problem as one in which we use classical concepts and equations. Based on the problem statement that the moving clock runs more slowly than a stationary clock however, we categorize this problem as one involving time dilation.

Analyze The proper time interval, measured in the rest frame of the car, is $\Delta t_p = 5.0$ h.

Use Equation 39.8 to evaluate γ:

$$\gamma = \frac{1}{\sqrt{1 - \frac{v^2}{c^2}}} = \frac{1}{\sqrt{1 - \frac{(3.0 \times 10^1 \text{ m/s})^2}{(3.0 \times 10^8 \text{ m/s})^2}}} = \frac{1}{\sqrt{1 - 10^{-14}}}$$

If you try to determine this value on your calculator, you will probably obtain $\gamma = 1$. Instead, perform a binomial expansion:

$$\gamma = (1 - 10^{-14})^{-1/2} \approx 1 + \tfrac{1}{2}(10^{-14}) = 1 + 5.0 \times 10^{-15}$$

Use Equation 39.7 to find the dilated time interval measured by your boss:

$$\Delta t = \gamma \, \Delta t_p = (1 + 5.0 \times 10^{-15})(5.0 \text{ h})$$

$$= 5.0 \text{ h} + 2.5 \times 10^{-14} \text{ h} = 5.0 \text{ h} + 0.090 \text{ ns}$$

Finalize Your boss's clock would be only 0.090 ns ahead of your car clock. You might want to think of another excuse!

The Twin Paradox

An intriguing consequence of time dilation is the *twin paradox* (Fig. 39.9). Consider an experiment involving a set of twins named Speedo and Goslo. When they are 20 years old, Speedo, the more adventuresome of the two, sets out on an epic journey from the Earth to Planet X, located 20 lightyears away. One lightyear (ly) is the distance light travels through free space in 1 year. Furthermore, Speedo's spacecraft is capable of reaching a speed of $0.95c$ relative to the inertial frame of his twin brother back home on the Earth. After reaching Planet X, Speedo becomes homesick and immediately returns to the Earth at the same speed $0.95c$.

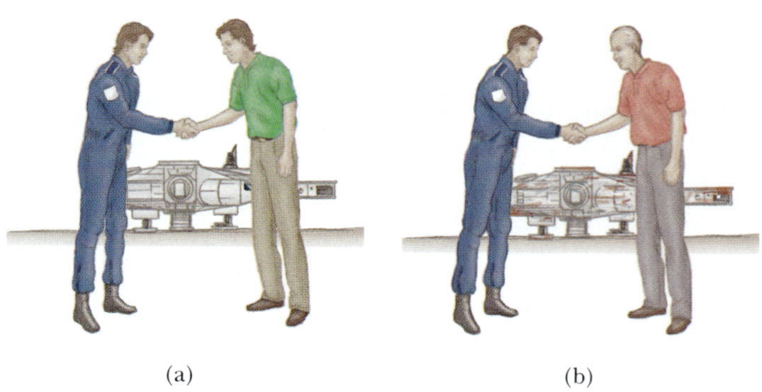

Figure 39.9 (a) As one twin leaves his brother on the Earth, both are the same age. (b) When Speedo returns from his journey to Planet X, he is younger than his twin Goslo.

Upon his return, Speedo is shocked to discover that Goslo has aged 42 years and is now 62 years old. Speedo, on the other hand, has aged only 13 years.

The paradox is *not* that the twins have aged at different rates. Here is the paradox. From Goslo's frame of reference, he was at rest while his brother traveled at a high speed away from him and then came back. According to Speedo, however, he himself remained stationary while Goslo and the Earth raced away from him and then headed back. Therefore, we might expect Speedo to claim that Goslo ages more slowly than himself. The situation appears to be symmetrical from either twin's point of view. Which twin *actually* ages more slowly?

The situation is actually not symmetrical. Consider a third observer moving at a constant speed relative to Goslo. According to the third observer, Goslo never changes inertial frames. Goslo's speed relative to the third observer is always the same. The third observer notes, however, that Speedo accelerates during his journey when he slows down and starts moving back toward the Earth, *changing reference frames in the process.* From the third observer's perspective, there is something very different about the motion of Goslo when compared to Speedo. Therefore, there is no paradox: only Goslo, who is always in a single inertial frame, can make correct predictions based on special relativity. Goslo finds that instead of aging 42 years, Speedo ages only $(1 - v^2/c^2)^{1/2}(42 \text{ years}) = 13$ years. Of these 13 years, Speedo spends 6.5 years traveling to Planet X and 6.5 years returning.

Quick Quiz 39.5 Suppose astronauts are paid according to the amount of time they spend traveling in space. After a long voyage traveling at a speed approaching c, would a crew rather be paid according to (a) an Earth-based clock, (b) their spacecraft's clock, or (c) either clock?

Length Contraction

The measured distance between two points in space also depends on the frame of reference of the observer. **The proper length L_p of an object is the length measured by someone at rest relative to the object.** The length of an object measured by someone in a reference frame that is moving with respect to the object is always less than the proper length. This effect is known as **length contraction.**

PITFALL PREVENTION 39.4

The Proper Length

As with the proper time interval, it is *very* important in relativistic calculations to correctly identify the observer who measures the proper length. The proper length between two points in space is always the length measured by an observer at rest with respect to the points. Often, the proper time interval and the proper length are *not* measured by the same observer.

To understand length contraction, consider a spacecraft traveling with a speed v from one star to another. There are two observers: one on the Earth and the other in the spacecraft. The observer at rest on the Earth (and also assumed to be at rest with respect to the two stars) measures the distance between the stars to be the proper length L_p. According to this observer, the time interval required for the spacecraft to complete the voyage is $\Delta t = L_p/v$. The passages of the two stars by the spacecraft occur at the same position for the space traveler. Therefore, the space traveler measures the proper time interval Δt_p. Because of time dilation, the proper time interval is related to the Earth-measured time interval by $\Delta t_p = \Delta t/\gamma$. Because the space traveler reaches the second star in the time Δt_p, he or she concludes that the distance L between the stars is

$$L = v\,\Delta t_p = v\frac{\Delta t}{\gamma}$$

Because the proper length is $L_p = v\,\Delta t$, we see that

$$L = \frac{L_p}{\gamma} = L_p\sqrt{1 - \frac{v^2}{c^2}} \qquad \textbf{(39.9)}$$

◀ Length contraction

where $\sqrt{1 - v^2/c^2}$ is a factor less than unity. **If an object has a proper length L_p when it is measured by an observer at rest with respect to the object, its length L when it moves with speed v in a direction parallel to its length is measured to be shorter according to $L = L_p\sqrt{1 - v^2/c^2} = L_p/\gamma$.**

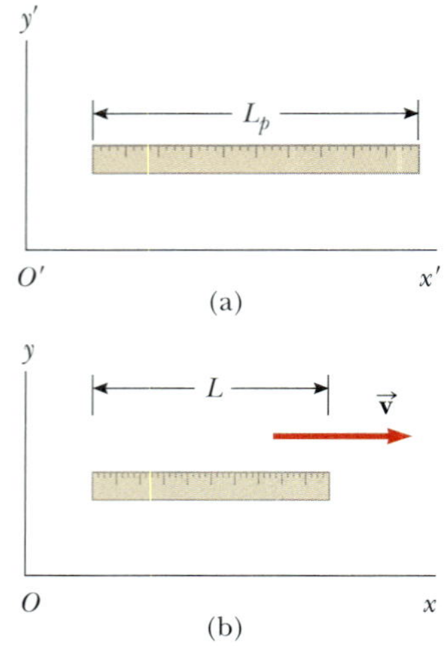

ACTIVE FIGURE 39.10

(a) A meterstick measured by an observer in a frame attached to the stick (that is, both have the same velocity) has its proper length L_p. (b) The meterstick measured by an observer in a frame in which the stick has a velocity $\vec{v}$ relative to the frame is measured to be shorter than its proper length L_p by a factor $(1 - v^2/c^2)^{1/2}$.

Sign in at www.thomsonedu.com and go to ThomsonNOW to view the meterstick from the points of view of two observers and compare the measured length of the stick.

For example, suppose a meterstick moves past a stationary Earth-based observer with speed v as in Active Figure 39.10. The length of the meterstick as measured by an observer in a frame attached to the stick is the proper length L_p shown in Active Figure 39.10a. The length of the stick L measured by the Earth observer is shorter than L_p by the factor $(1 - v^2/c^2)^{1/2}$ as suggested in Active Figure 39.10b. Notice that **length contraction takes place only along the direction of motion.**

The proper length and the proper time interval are defined differently. The proper length is measured by an observer for whom the end points of the length remain fixed in space. The proper time interval is measured by someone for whom the two events take place at the same position in space. As an example of this point, let's return to the decaying muons moving at speeds close to the speed of light. An observer in the muon's reference frame measures the proper lifetime, whereas an Earth-based observer measures the proper length (the distance between the creation point and the decay point in Fig. 39.8b). In the muon's reference frame, there is no time dilation, but the distance of travel to the surface is shorter when measured in this frame. Likewise, in the Earth observer's reference frame, there is time dilation, but the distance of travel is measured to be the proper length. Therefore, when calculations on the muon are performed in both frames, the outcome of the experiment in one frame is the same as the outcome in the other frame: more muons reach the surface than would be predicted without relativistic effects.

Quick Quiz 39.6 You are packing for a trip to another star. During the journey, you will be traveling at 0.99c. You are trying to decide whether you should buy smaller sizes of your clothing, because you will be thinner on your trip due to length contraction. You also plan to save money by reserving a smaller cabin to sleep in because you will be shorter when you lie down. Should you (a) buy smaller sizes of clothing, (b) reserve a smaller cabin, (c) do neither of these things, or (d) do both of these things?

Quick Quiz 39.7 You are observing a spacecraft moving away from you. You measure it to be shorter than when it was at rest on the ground next to you. You also see a clock through the spacecraft window, and you observe that the passage of time on the clock is measured to be slower than that of the watch on your wrist. Compared with when the spacecraft was on the ground, what do you measure if the spacecraft turns around and comes *toward* you at the same speed? (a) The spacecraft is measured to be longer, and the clock runs faster. (b) The spacecraft is measured to be longer, and the clock runs slower. (c) The spacecraft is measured to be shorter, and the clock runs faster. (d) The spacecraft is measured to be shorter, and the clock runs slower.

Space–Time Graphs

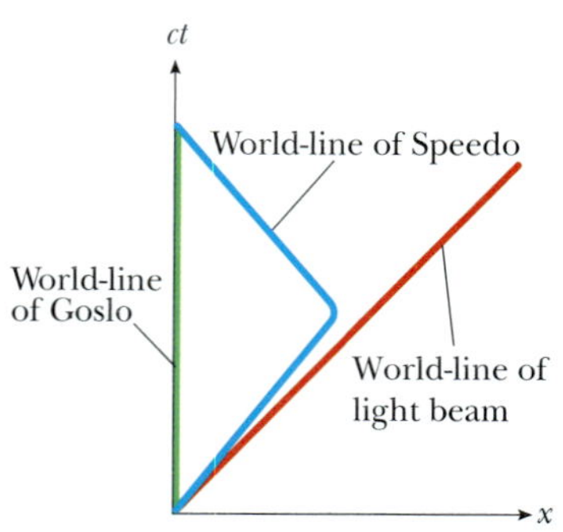

Figure 39.11 The twin paradox on a space–time graph. The twin who stays on the Earth has a world-line along the ct axis (green). The path of the traveling twin through space–time is represented by a world-line that changes direction (blue).

It is sometimes helpful to represent a physical situation with a *space–time graph,* in which ct is the ordinate and position x is the abscissa. The twin paradox is displayed in such a graph in Figure 39.11 from Goslo's point of view. A path through space–time is called a **world-line.** At the origin, the world-lines of Speedo (blue) and Goslo (green) coincide because the twins are in the same location at the same time. After Speedo leaves on his trip, his world-line diverges from that of his brother. Goslo's world-line is vertical because he remains fixed in location. At Goslo and Speedo's reunion, the two world-lines again come together. It would be impossible for Speedo to have a world-line that crossed the path of a light beam that left the Earth when he did. To do so would require him to have a speed greater than c (which, as shown in Sections 39.6 and 39.7, is not possible).

World-lines for light beams are diagonal lines on space–time graphs, typically drawn at 45° to the right or left of vertical (assuming that the x and ct axes have the same scales), depending on whether the light beam is traveling in the direction of increasing or decreasing x. These two world-lines mean that all possible future events for Goslo and Speedo lie within two 45° lines extending from the origin. Either twin's presence at an event outside this "light cone" would require that twin to move at a speed greater than c, which we have said is not possible. Also, the only past events that Goslo and Speedo could have experienced occur between two similar 45° world-lines that approach the origin from below the x axis.

EXAMPLE 39.3 A Voyage to Sirius

An astronaut takes a trip to Sirius, which is located a distance of 8 lightyears from the Earth. The astronaut measures the time of the one-way journey to be 6 years. If the spaceship moves at a constant speed of $0.8c$, how can the 8-ly distance be reconciled with the 6-year trip time measured by the astronaut?

SOLUTION

Conceptualize An observer on the Earth measures light to require 8 years to travel from Earth to Sirius. The astronaut measures a time interval of only 6 years. Is the astronaut traveling faster than light?

Categorize Because the astronaut is measuring a length of space between Earth and Sirius that is in motion with respect to her, we categorize this example as a length contraction problem.

Analyze The distance of 8 ly represents the proper length from the Earth to Sirius measured by an observer on the Earth seeing both objects nearly at rest.

Calculate the contracted length measured by the astronaut using Equation 39.9:

$$L = \frac{8\text{ ly}}{\gamma} = (8\text{ ly})\sqrt{1 - \frac{v^2}{c^2}} = (8\text{ ly})\sqrt{1 - \frac{(0.8c)^2}{c^2}} = 5\text{ ly}$$

Use the particle under constant speed model to find the travel time measured on the astronaut's clock:

$$\Delta t = \frac{L}{v} = \frac{5\text{ ly}}{0.8c} = \frac{5\text{ ly}}{0.8(1\text{ ly/yr})} = 6\text{ yr}$$

Finalize Note that we have used the value for the speed of light as $c = 1$ ly/yr. The trip takes a time interval shorter than 8 years for the astronaut because, to her, the distance between the Earth and Sirius is measured to be shorter.

What If? What if this trip is observed with a very powerful telescope by a technician in Mission Control on the Earth? At what time will this technician *see* that the astronaut has arrived at Sirius?

Answer The time interval the technician measures for the astronaut to arrive is

$$\Delta t = \frac{L_p}{v} = \frac{8\text{ ly}}{0.8c} = 10\text{ yr}$$

For the technician to *see* the arrival, the light from the scene of the arrival must travel back to the Earth and enter the telescope. This travel requires a time interval of

$$\Delta t = \frac{L_p}{v} = \frac{8\text{ ly}}{c} = 8\text{ yr}$$

Therefore, the technician sees the arrival after 10 yr + 8 yr = 18 yr. If the astronaut immediately turns around and comes back home, she arrives, according to the technician, 20 years after leaving, only 2 years *after the technician saw her arrive!* In addition, the astronaut would have aged by only 12 years.

EXAMPLE 39.4 The Pole-in-the-Barn Paradox

The twin paradox, discussed earlier, is a classic "paradox" in relativity. Another classic "paradox" is as follows. Suppose a runner moving at $0.75c$ carries a horizontal pole 15 m long toward a barn that is 10 m long. The barn has front and rear doors that are initially open. An observer on the ground can instantly and simultaneously close and open the two doors by remote control. When the runner and the pole are inside the barn, the ground observer closes and then opens both doors so that the runner and pole are momentarily captured inside the barn and then proceed to exit the barn from the back door. Do both the runner and the ground observer agree that the runner makes it safely through the barn?

SOLUTION

Conceptualize From your everyday experience, you would be surprised to see a 15-m pole fit inside a 10-m barn.

Categorize The pole is in motion with respect to the ground observer so that the observer measures its length to be contracted, whereas the stationary barn has a proper length of 10 m. We categorize this example as a length contraction problem.

Analyze Use Equation 39.9 to find the contracted length of the pole according to the ground observer:

$$L_{\text{pole}} = L_p\sqrt{1-\frac{v^2}{c^2}} = (15\text{ m})\sqrt{1-(0.75)^2} = 9.9\text{ m}$$

Therefore, the ground observer measures the pole to be slightly shorter than the barn and there is no problem with momentarily capturing the pole inside it. The "paradox" arises when we consider the runner's point of view.

Use Equation 39.9 to find the contracted length of the barn according to the running observer:

$$L_{\text{barn}} = L_p\sqrt{1-\frac{v^2}{c^2}} = (10\text{ m})\sqrt{1-(0.75)^2} = 6.6\text{ m}$$

Because the pole is in the rest frame of the runner, the runner measures it to have its proper length of 15 m. How can a 15-m pole fit inside a 6.6-m barn? Although this question is the classic one that is often asked, it is not the question we have asked because it is not the important one. We asked, *"Does the runner make it safely through the barn?"*

The resolution of the "paradox" lies in the relativity of simultaneity. The closing of the two doors is measured to be simultaneous by the ground observer. Because the doors are at different positions, however, they do not close simultaneously as measured by the runner. The rear door closes and then opens first, allowing the leading end of the pole to exit. The front door of the barn does not close until the trailing end of the pole passes by.

We can analyze this "paradox" using a space–time graph. Figure 39.12a is a space–time graph from the ground observer's point of view. We choose $x = 0$ as the position of the front door of the barn and $t = 0$ as the instant at which the leading end of the pole is located at the front door of the barn. The world-lines for the two doors of the barn are separated by 10 m and are vertical because the barn is not moving relative to this observer. For the pole, we follow two tilted world-lines, one for each end of the moving pole. These world-lines are 9.9 m apart horizontally, which is the con-

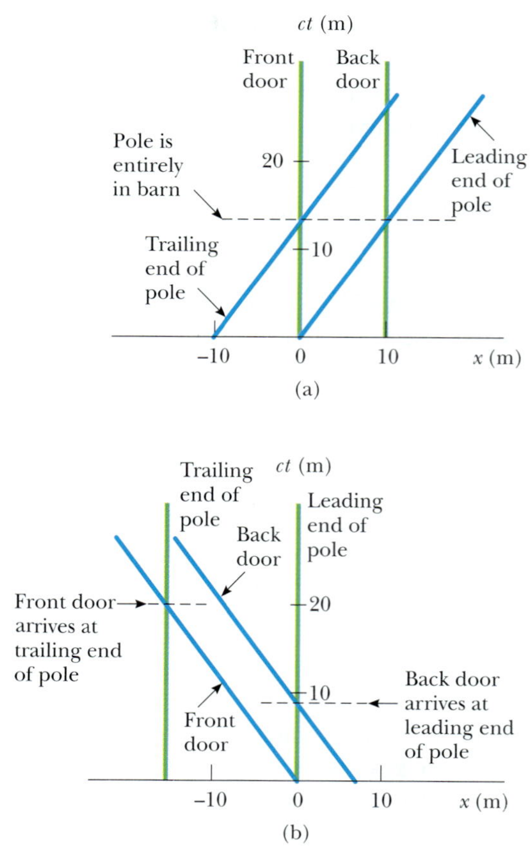

Figure 39.12 (Example 39.4) Space–time graphs for the pole-in-the-barn paradox. (a) From the ground observer's point of view, the world-lines for the front and back doors of the barn are vertical lines. The world-lines for the ends of the pole are tilted and are 9.9 m apart horizontally. The front door of the barn is at $x = 0$, and the leading end of the pole enters the front door at $t = 0$. The entire pole is inside the barn at the time indicated by the dashed line. (b) From the runner's point of view, the world-lines for the ends of the pole are vertical. The barn is moving in the negative direction, so the world-lines for the front and back doors are tilted to the left. The leading end of the pole exits the back door before the trailing end arrives at the front door.

tracted length seen by the ground observer. As seen in Figure 39.12a, the pole is entirely within the barn at one instant.

Figure 39.12b shows the space–time graph according to the runner. Here, the world-lines for the pole are separated by 15 m and are vertical because the pole is at rest in the runner's frame of reference. The barn is hurtling *toward* the runner, so the world-lines for the front and rear doors of the barn are tilted to the left. The world-lines for the barn are separated by 6.6 m, the contracted length as seen by the runner. The leading end of the pole leaves the rear door of the barn long before the trailing end of the pole enters the barn. Therefore, the opening of the rear door occurs before the closing of the front door.

From the ground observer's point of view, use the particle under constant velocity model to find the time after $t = 0$ at which the trailing end of the pole enters the barn:

$$(1) \quad t = \frac{\Delta x}{v} = \frac{9.9 \text{ m}}{0.75c} = \frac{13.2 \text{ m}}{c}$$

From the runner's point of view, use the particle under constant velocity model to find the time at which the leading end of the pole leaves the barn:

$$(2) \quad t = \frac{\Delta x}{v} = \frac{6.6 \text{ m}}{0.75c} = \frac{8.8 \text{ m}}{c}$$

Find the time at which the trailing end of the pole enters the front door of the barn:

$$(3) \quad t = \frac{\Delta x}{v} = \frac{15 \text{ m}}{0.75c} = \frac{20 \text{ m}}{c}$$

Finalize From Equation (1), the pole should be completely inside the barn at a time corresponding to $ct = 13.2$ m. This situation is consistent with the point on the ct axis in Figure 39.12a where the pole is inside the barn. From Equation (2), the leading end of the pole leaves the barn at $ct = 8.8$ m. This situation is consistent with the point on the ct axis in Figure 39.12b where the back door of the barn arrives at the leading end of the pole. Equation (3) gives $ct = 20$ m, which agrees with the instant shown in Figure 39.12b at which the front door of the barn arrives at the trailing end of the pole.

The Relativistic Doppler Effect

Another important consequence of time dilation is the shift in frequency observed for light emitted by atoms in motion as opposed to light emitted by atoms at rest. This phenomenon, known as the Doppler effect, was introduced in Chapter 17 as it pertains to sound waves. In the case of sound, the motion of the source with respect to the medium of propagation can be distinguished from the motion of the observer with respect to the medium. Light waves must be analyzed differently, however, because they require no medium of propagation and no method exists for distinguishing the motion of a light source from the motion of the observer.

If a light source and an observer approach each other with a relative speed v, the frequency f_{obs} measured by the observer is

$$f_{\text{obs}} = \frac{\sqrt{1 + v/c}}{\sqrt{1 - v/c}} f_{\text{source}} \qquad \textbf{(39.10)}$$

where f_{source} is the frequency of the source measured in its rest frame. This relativistic Doppler shift equation, unlike the Doppler shift equation for sound, depends only on the relative speed v of the source and observer and holds for relative speeds as great as c. As you might expect, the equation predicts that $f_{\text{obs}} > f_{\text{source}}$ when the source and observer approach each other. We obtain the expression for the case in which the source and observer recede from each other by substituting negative values for v in Equation 39.10.

The most spectacular and dramatic use of the relativistic Doppler effect is the measurement of shifts in the frequency of light emitted by a moving astronomical object such as a galaxy. Light emitted by atoms and normally found in the extreme violet region of the spectrum is shifted toward the red end of the spectrum for atoms in other galaxies, indicating that these galaxies are *receding* from us. American astronomer Edwin Hubble (1889–1953) performed extensive measurements of this *red shift* to confirm that most galaxies are moving away from us, indicating that the Universe is expanding.

39.5 The Lorentz Transformation Equations

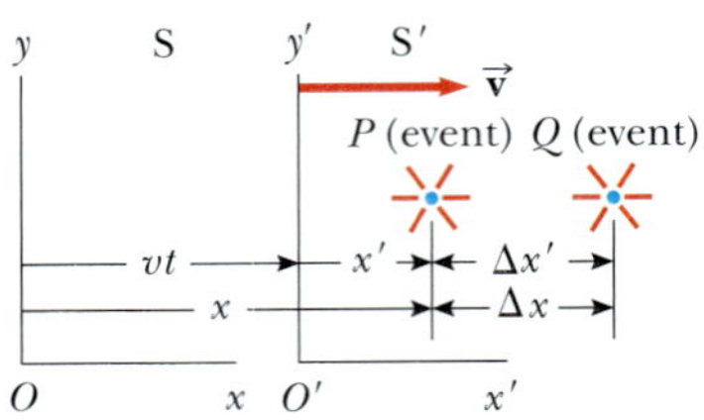

Figure 39.13 Events occur at points P and Q and are observed by an observer at rest in the S frame and another in the S′ frame, which is moving to the right with a speed v.

Suppose two events occur at points P and Q and are reported by two observers, one at rest in a frame S and another in a frame S′ that is moving to the right with speed v as in Figure 39.13. The observer in S reports the events with space–time coordinates (x, y, z, t) and the observer in S′ reports the same events using the coordinates (x', y', z', t'). Equation 39.1 predicts that the distance between the two points in space at which the events occur does not depend on motion of the observer: $\Delta x = \Delta x'$. Because this prediction is contradictory to the notion of length contraction, the Galilean transformation is not valid when v approaches the speed of light. In this section, we present the correct transformation equations that apply for all speeds in the range $0 < v < c$.

The equations that are valid for all speeds and that enable us to transform coordinates from S to S′ are the **Lorentz transformation equations:**

Lorentz transformation for S → S′ ▶

$$x' = \gamma(x - vt) \qquad y' = y \qquad z' = z \qquad t' = \gamma\left(t - \frac{v}{c^2}x\right) \tag{39.11}$$

These transformation equations were developed by Hendrik A. Lorentz (1853–1928) in 1890 in connection with electromagnetism. It was Einstein, however, who recognized their physical significance and took the bold step of interpreting them within the framework of the special theory of relativity.

Notice the difference between the Galilean and Lorentz time equations. In the Galilean case, $t = t'$. In the Lorentz case, however, the value for t' assigned to an event by an observer O' in the S′ frame in Figure 39.13 depends both on the time t and on the coordinate x as measured by an observer O in the S frame, which is consistent with the notion that an event is characterized by four space–time coordinates (x, y, z, t). In other words, in relativity, space and time are *not* separate concepts but rather are closely interwoven with each other.

If you wish to transform coordinates in the S′ frame to coordinates in the S frame, simply replace v by $-v$ and interchange the primed and unprimed coordinates in Equations 39.11:

Inverse Lorentz transformation for S′ → S ▶

$$x = \gamma(x' + vt') \qquad y = y' \qquad z = z' \qquad t = \gamma\left(t' + \frac{v}{c^2}x'\right) \tag{39.12}$$

When $v \ll c$, the Lorentz transformation equations should reduce to the Galilean equations. As v approaches zero, $v/c \ll 1$; therefore, $\gamma \rightarrow 1$ and Equations 39.11 indeed reduce to the Galilean space–time transformation equations in Equation 39.1.

In many situations, we would like to know the difference in coordinates between two events or the time interval between two events as seen by observers O and O'. From Equations 39.11 and 39.12, we can express the differences between the four variables x, x', t, and t' in the form

$$\left.\begin{aligned} \Delta x' &= \gamma(\Delta x - v\,\Delta t) \\ \Delta t' &= \gamma\left(\Delta t - \frac{v}{c^2}\Delta x\right) \end{aligned}\right\} \text{S} \rightarrow \text{S}' \tag{39.13}$$

$$\left.\begin{aligned} \Delta x &= \gamma(\Delta x' + v\,\Delta t') \\ \Delta t &= \gamma\left(\Delta t' + \frac{v}{c^2}\Delta x'\right) \end{aligned}\right\} \text{S}' \rightarrow \text{S} \tag{39.14}$$

where $\Delta x' = x_2' - x_1'$ and $\Delta t' = t_2' - t_1'$ are the differences measured by observer O' and $\Delta x = x_2 - x_1$ and $\Delta t = t_2 - t_1$ are the differences measured by observer O. (We have not included the expressions for relating the y and z coordinates because they are unaffected by motion along the x direction.[5])

EXAMPLE 39.5 Simultaneity and Time Dilation Revisited

(A) Use the Lorentz transformation equations in difference form to show that simultaneity is not an absolute concept.

SOLUTION

Conceptualize Imagine two events that are simultaneous and separated in space such that $\Delta t' = 0$ and $\Delta x' \neq 0$ according to an observer O' who is moving with speed v relative to O.

Categorize The statement of the problem tells us to categorize this example as one involving the Lorentz transformation.

Analyze From the expression for Δt given in Equation 39.14, find the time interval Δt measured by observer O:

$$\Delta t = \gamma\left(\Delta t' + \frac{v}{c^2}\Delta x'\right) = \gamma\left(0 + \frac{v}{c^2}\Delta x'\right) = \gamma\frac{v}{c^2}\Delta x'$$

Finalize The time interval for the same two events as measured by O is nonzero, so the events do not appear to be simultaneous to O.

(B) Use the Lorentz transformation equations in difference form to show that a moving clock is measured to run more slowly than a clock that is at rest with respect to an observer.

SOLUTION

Conceptualize Imagine that observer O' carries a clock that he uses to measure a time interval $\Delta t'$. He finds that two events occur at the same place in his reference frame ($\Delta x' = 0$) but at different times ($\Delta t' \neq 0$). Observer O' is moving with speed v relative to O.

Categorize The statement of the problem tells us to categorize this example as one involving the Lorentz transformation.

Analyze From the expression for Δt given in Equation 39.14, find the time interval Δt measured by observer O:

$$\Delta t = \gamma\left(\Delta t' + \frac{v}{c^2}\Delta x'\right) = \gamma\left(\Delta t' + \frac{v}{c^2}(0)\right) = \gamma\Delta t'$$

Finalize This result is the equation for time dilation found earlier (Eq. 39.7), where $\Delta t' = \Delta t_p$ is the proper time interval measured by the clock carried by observer O'. Therefore, O measures the moving clock to run slow.

39.6 The Lorentz Velocity Transformation Equations

Suppose two observers in relative motion with respect to each other are both observing an object's motion. Previously, we defined an event as occurring at an instant of time. Now let's interpret the "event" as the object's motion. We know that the Galilean velocity transformation (Eq. 39.2) is valid for low speeds. How do the observers' measurements of the velocity of the object relate to each other if the speed of the object is close to that of light? Once again, S′ is our frame moving

[5] Although relative motion of the two frames along the x axis does not change the y and z coordinates of an object, it does change the y and z velocity components of an object moving in either frame as noted in Section 39.6.

at a speed v relative to S. Suppose an object has a velocity component u'_x measured in the S′ frame, where

$$u'_x = \frac{dx'}{dt'} \tag{39.15}$$

Using Equation 39.11, we have

$$dx' = \gamma(dx - v\,dt)$$

$$dt' = \gamma\left(dt - \frac{v}{c^2}\,dx\right)$$

Substituting these values into Equation 39.15 gives

$$u'_x = \frac{dx - v\,dt}{dt - \dfrac{v}{c^2}\,dx} = \frac{\dfrac{dx}{dt} - v}{1 - \dfrac{v}{c^2}\dfrac{dx}{dt}}$$

The term dx/dt, however, is simply the velocity component u_x of the object measured by an observer in S, so this expression becomes

Lorentz velocity transformation for S → S′

$$u'_x = \frac{u_x - v}{1 - \dfrac{u_x v}{c^2}} \tag{39.16}$$

If the object has velocity components along the y and z axes, the components as measured by an observer in S′ are

$$u'_y = \frac{u_y}{\gamma\left(1 - \dfrac{u_x v}{c^2}\right)} \quad \text{and} \quad u'_z = \frac{u_z}{\gamma\left(1 - \dfrac{u_x v}{c^2}\right)} \tag{39.17}$$

Notice that u'_y and u'_z do not contain the parameter v in the numerator because the relative velocity is along the x axis.

When v is much smaller than c (the nonrelativistic case), the denominator of Equation 39.16 approaches unity and so $u'_x \approx u_x - v$, which is the Galilean velocity transformation equation. In another extreme, when $u_x = c$, Equation 39.16 becomes

$$u'_x = \frac{c - v}{1 - \dfrac{cv}{c^2}} = \frac{c\left(1 - \dfrac{v}{c}\right)}{1 - \dfrac{v}{c}} = c$$

PITFALL PREVENTION 39.5

What Can the Observers Agree On?

We have seen several measurements that the two observers O and O' do *not* agree on: (1) the time interval between events that take place in the same position in one of the frames, (2) the distance between two points that remain fixed in one of their frames, (3) the velocity components of a moving particle, and (4) whether two events occurring at different locations in both frames are simultaneous or not. The two observers *can* agree on (1) their relative speed of motion v with respect to each other, (2) the speed c of any ray of light, and (3) the simultaneity of two events which take place at the same position *and* time in some frame.

This result shows that a speed measured as c by an observer in S is also measured as c by an observer in S′, independent of the relative motion of S and S′. This conclusion is consistent with Einstein's second postulate: the speed of light must be c relative to all inertial reference frames. Furthermore, we find that the speed of an object can never be measured as larger than c. That is, the speed of light is the ultimate speed. We shall return to this point later.

To obtain u_x in terms of u'_x, we replace v by $-v$ in Equation 39.16 and interchange the roles of u_x and u'_x:

$$u_x = \frac{u'_x + v}{1 + \dfrac{u'_x v}{c^2}} \tag{39.18}$$

Quick Quiz 39.8 You are driving on a freeway at a relativistic speed. **(i)** Straight ahead of you, a technician standing on the ground turns on a searchlight and a beam of light moves exactly vertically upward as seen by the technician. As you observe the beam of light, do you measure the magnitude of the vertical component of its velocity as (a) equal to c, (b) greater than c, or (c) less than c? **(ii)** If the technician aims the searchlight directly at you instead of upward, do you measure the magnitude of the horizontal component of its velocity as (a) equal to c, (b) greater than c, or (c) less than c?

EXAMPLE 39.6 Relative Velocity of Two Spacecraft

Two spacecraft A and B are moving in opposite directions as shown in Figure 39.14. An observer on the Earth measures the speed of spacecraft A to be $0.750c$ and the speed of spacecraft B to be $0.850c$. Find the velocity of spacecraft B as observed by the crew on spacecraft A.

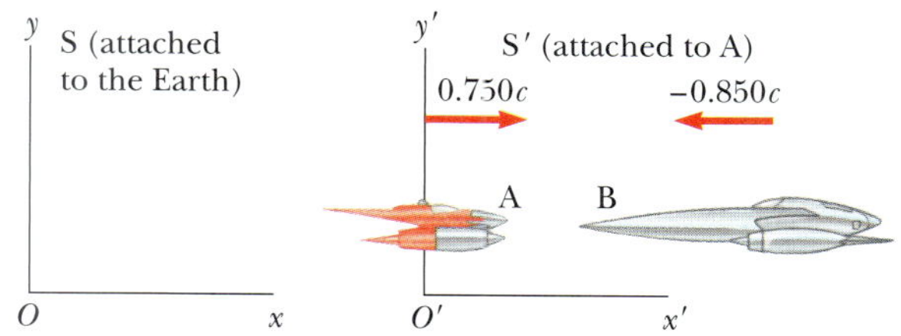

Figure 39.14 (Example 39.6) Two spacecraft A and B move in opposite directions. The speed of spacecraft B relative to spacecraft A is *less* than c and is obtained from the relativistic velocity transformation equation.

SOLUTION

Conceptualize There are two observers, one on the Earth and one on spacecraft A. The event is the motion of spacecraft B.

Categorize Because the problem asks to find an observed velocity, we categorize this example as one requiring the Lorentz velocity transformation.

Analyze The Earth-based observer at rest in the S frame makes two measurements, one of each spacecraft. We want to find the velocity of spacecraft B as measured by the crew on spacecraft A. Therefore, $u_x = -0.850c$. The velocity of spacecraft A is also the velocity of the observer at rest in spacecraft A (the S′ frame) relative to the observer at rest on the Earth. Therefore, $v = 0.750c$.

Obtain the velocity u'_x of spacecraft B relative to spacecraft A using Equation 39.16:

$$u'_x = \frac{u_x - v}{1 - \dfrac{u_x v}{c^2}} = \frac{-0.850c - 0.750c}{1 - \dfrac{(-0.850c)(0.750c)}{c^2}} = -0.977c$$

Finalize The negative sign indicates that spacecraft B is moving in the negative x direction as observed by the crew on spacecraft A. Is that consistent with your expectation from Figure 39.14? Notice that the speed is less than c. That is, an object whose speed is less than c in one frame of reference must have a speed less than c in any other frame. (Had you used the Galilean velocity transformation equation in this example, you would have found that $u'_x = u_x - v = -0.850c - 0.750c = -1.60c$, which is impossible. The Galilean transformation equation does not work in relativistic situations.)

What If? What if the two spacecraft pass each other? What is their relative speed now?

Answer The calculation using Equation 39.16 involves only the velocities of the two spacecraft and does not depend on their locations. After they pass each other, they have the same velocities, so the velocity of spacecraft B as observed by the crew on spacecraft A is the same, $-0.977c$. The only difference after they pass is that spacecraft B is receding from spacecraft A, whereas it was approaching spacecraft A before it passed.

EXAMPLE 39.7 Relativistic Leaders of the Pack

Two motorcycle pack leaders named David and Emily are racing at relativistic speeds along perpendicular paths as shown in Figure 39.15. How fast does Emily recede as seen by David over his right shoulder?

Figure 39.15 (Example 39.7) David moves east with a speed $0.75c$ relative to the police officer, and Emily travels south at a speed $0.90c$ relative to the officer.

SOLUTION

Conceptualize The two observers are David and the police officer in Figure 39.15. The event is the motion of Emily. Figure 39.15 represents the situation as seen by the police officer at rest in frame S. Frame S′ moves along with David.

Categorize Because the problem asks to find an observed velocity, we categorize this problem as one requiring the Lorentz velocity transformation. The motion takes place in two dimensions.

Analyze Identify the velocity components for David and Emily according to the police officer:

$$\text{David: } v_x = v = 0.75c \qquad v_y = 0$$

$$\text{Emily: } u_x = 0 \qquad u_y = -0.90c$$

Using Equations 39.16 and 39.17, calculate u'_x and u'_y for Emily as measured by David:

$$u'_x = \frac{u_x - v}{1 - \dfrac{u_x v}{c^2}} = \frac{0 - 0.75c}{1 - \dfrac{(0)(0.75c)}{c^2}} = -0.75c$$

$$u'_y = \frac{u_y}{\gamma\left(1 - \dfrac{u_x v}{c^2}\right)} = \frac{\sqrt{1 - \dfrac{(0.75c)^2}{c^2}}\,(-0.90c)}{\left(1 - \dfrac{(0)(0.75c)}{c^2}\right)} = -0.60c$$

Using the Pythagorean theorem, find the speed of Emily as measured by David:

$$u' = \sqrt{(u'_x)^2 + (u'_y)^2} = \sqrt{(-0.75c)^2 + (-0.60c)^2} = 0.96c$$

Finalize This speed is less than c, as required by the special theory of relativity.

PITFALL PREVENTION 39.6
Watch Out for "Relativistic Mass"

Some older treatments of relativity maintained the conservation of momentum principle at high speeds by using a model in which a particle's mass increases with speed. You might still encounter this notion of "relativistic mass" in your outside reading, especially in older books. Be aware that this notion is no longer widely accepted; today, mass is considered as *invariant*, independent of speed. The mass of an object in all frames is considered to be the mass as measured by an observer at rest with respect to the object.

39.7 Relativistic Linear Momentum

To describe the motion of particles within the framework of the special theory of relativity properly, you must replace the Galilean transformation equations by the Lorentz transformation equations. Because the laws of physics must remain unchanged under the Lorentz transformation, we must generalize Newton's laws and the definitions of linear momentum and energy to conform to the Lorentz transformation equations and the principle of relativity. These generalized definitions should reduce to the classical (nonrelativistic) definitions for $v \ll c$.

First, recall from the isolated system model that when two particles (or objects that can be modeled as particles) collide, the total momentum of the isolated system of the two particles remains constant. Suppose we observe this collision in a reference frame S and confirm that the momentum of the system is conserved. Now imagine that the momenta of the particles are measured by an observer in a second reference frame S′ moving with velocity $\vec{\mathbf{v}}$ relative to the first frame. Using the Lorentz velocity transformation equation and the classical definition of linear momentum, $\vec{\mathbf{p}} = m\vec{\mathbf{u}}$ (where $\vec{\mathbf{u}}$ is the velocity of a particle), we find that linear momentum is *not* measured to be conserved by the observer in S′. Because the laws of physics are the same in all inertial frames, however, linear momentum of the system must be conserved in all frames. We have a contradiction. In view of this contradiction and assuming the Lorentz velocity transformation equation is correct, we must modify the definition of linear momentum so that the momentum of an isolated system is conserved for all observers. For any particle, the correct relativistic equation for linear momentum that satisfies this condition is

Definition of relativistic linear momentum ▶

$$\vec{\mathbf{p}} \equiv \frac{m\vec{\mathbf{u}}}{\sqrt{1 - \dfrac{u^2}{c^2}}} = \gamma m\vec{\mathbf{u}} \qquad (39.19)$$

where m is the mass of the particle and $\vec{\mathbf{u}}$ is the velocity of the particle. When u is much less than c, $\gamma = (1 - u^2/c^2)^{-1/2}$ approaches unity and $\vec{\mathbf{p}}$ approaches $m\vec{\mathbf{u}}$. Therefore, the relativistic equation for $\vec{\mathbf{p}}$ reduces to the classical expression when u is much smaller than c, as it should.

The relativistic force $\vec{\mathbf{F}}$ acting on a particle whose linear momentum is $\vec{\mathbf{p}}$ is defined as

$$\vec{\mathbf{F}} \equiv \frac{d\vec{\mathbf{p}}}{dt} \tag{39.20}$$

where $\vec{\mathbf{p}}$ is given by Equation 39.19. This expression, which is the relativistic form of Newton's second law, is reasonable because it preserves classical mechanics in the limit of low velocities and is consistent with conservation of linear momentum for an isolated system ($\vec{\mathbf{F}}_{\text{ext}} = 0$) both relativistically and classically.

It is left as an end-of-chapter problem (Problem 61) to show that under relativistic conditions, the acceleration $\vec{\mathbf{a}}$ of a particle decreases under the action of a constant force, in which case $a \propto (1 - u^2/c^2)^{3/2}$. This proportionality shows that as the particle's speed approaches c, the acceleration caused by any finite force approaches zero. Hence, it is impossible to accelerate a particle from rest to a speed $u \geq c$. This argument reinforces that the speed of light is the ultimate speed, the speed limit of the Universe. It is the maximum possible speed for energy transfer and for information transfer. Any object with mass must move at a lower speed.

EXAMPLE 39.8 Linear Momentum of an Electron

An electron, which has a mass of 9.11×10^{-31} kg, moves with a speed of $0.750c$. Find the magnitude of its relativistic momentum and compare this value with the momentum calculated from the classical expression.

SOLUTION

Conceptualize Imagine an electron moving with high speed. The electron carries momentum, but the magnitude of its momentum is not given by $p = mu$ because the speed is relativistic.

Categorize We categorize this example as a substitution problem involving a relativistic equation.

Use Equation 39.19 with $u = 0.750c$ to find the momentum:

$$p = \frac{m_e u}{\sqrt{1 - \frac{u^2}{c^2}}}$$

$$p = \frac{(9.11 \times 10^{-31}\text{ kg})(0.750)(3.00 \times 10^8\text{ m/s})}{\sqrt{1 - \frac{(0.750c)^2}{c^2}}}$$

$$= 3.10 \times 10^{-22}\text{ kg} \cdot \text{m/s}$$

The classical expression (used incorrectly here) gives $p_{\text{classical}} = m_e u = 2.05 \times 10^{-22}$ kg · m/s. Hence, the correct relativistic result is 50% greater than the classical result!

39.8 Relativistic Energy

We have seen that the definition of linear momentum requires generalization to make it compatible with Einstein's postulates. This conclusion implies that the definition of kinetic energy must most likely be modified also.

To derive the relativistic form of the work–kinetic energy theorem, imagine a particle moving in one dimension along the x axis. A force in the x direction causes the momentum of the particle to change according to Equation 39.20. In what follows, we assume the particle is accelerated from rest to some final speed u. The work done by the force F on the particle is

$$W = \int_{x_1}^{x_2} F\,dx = \int_{x_1}^{x_2} \frac{dp}{dt}\,dx \tag{39.21}$$

To perform this integration and find the work done on the particle and the relativistic kinetic energy as a function of u, we first evaluate dp/dt:

$$\frac{dp}{dt} = \frac{d}{dt}\frac{mu}{\sqrt{1-\dfrac{u^2}{c^2}}} = \frac{m}{\left(1-\dfrac{u^2}{c^2}\right)^{3/2}}\frac{du}{dt}$$

Substituting this expression for dp/dt and $dx = u\,dt$ into Equation 39.21 gives

$$W = \int_0^t \frac{m}{\left(1-\dfrac{u^2}{c^2}\right)^{3/2}}\frac{du}{dt}(u\,dt) = m\int_0^u \frac{u}{\left(1-\dfrac{u^2}{c^2}\right)^{3/2}}\,du$$

where we use the limits 0 and u in the integral because the integration variable has been changed from t to u. Evaluating the integral gives

$$W = \frac{mc^2}{\sqrt{1-\dfrac{u^2}{c^2}}} - mc^2 \tag{39.22}$$

Recall from Chapter 7 that the work done by a force acting on a system consisting of a single particle equals the change in kinetic energy of the particle. Because we assumed the initial speed of the particle is zero, its initial kinetic energy is zero. Therefore, the work W in Equation 39.22 is equivalent to the relativistic kinetic energy K:

Relativistic kinetic energy ►

$$K = \frac{mc^2}{\sqrt{1-\dfrac{u^2}{c^2}}} - mc^2 = \gamma mc^2 - mc^2 = (\gamma - 1)mc^2 \tag{39.23}$$

This equation is routinely confirmed by experiments using high-energy particle accelerators.

At low speeds, where $u/c \ll 1$, Equation 39.23 should reduce to the classical expression $K = \frac{1}{2}mu^2$. We can check that by using the binomial expansion $(1-\beta^2)^{-1/2} \approx 1 + \frac{1}{2}\beta^2 + \cdots$ for $\beta \ll 1$, where the higher-order powers of β are neglected in the expansion. (In treatments of relativity, β is a common symbol used to represent u/c or v/c.) In our case, $\beta = u/c$, so

$$\gamma = \frac{1}{\sqrt{1-\dfrac{u^2}{c^2}}} = \left(1-\frac{u^2}{c^2}\right)^{-1/2} \approx 1 + \frac{1}{2}\frac{u^2}{c^2}$$

Substituting this result into Equation 39.23 gives

$$K \approx \left[\left(1 + \frac{1}{2}\frac{u^2}{c^2}\right) - 1\right]mc^2 = \frac{1}{2}mu^2 \quad (\text{for } u/c \ll 1)$$

which is the classical expression for kinetic energy. A graph comparing the relativistic and nonrelativistic expressions is given in Figure 39.16. In the relativistic case, the particle speed never exceeds c, regardless of the kinetic energy. The two curves are in good agreement when $u \ll c$.

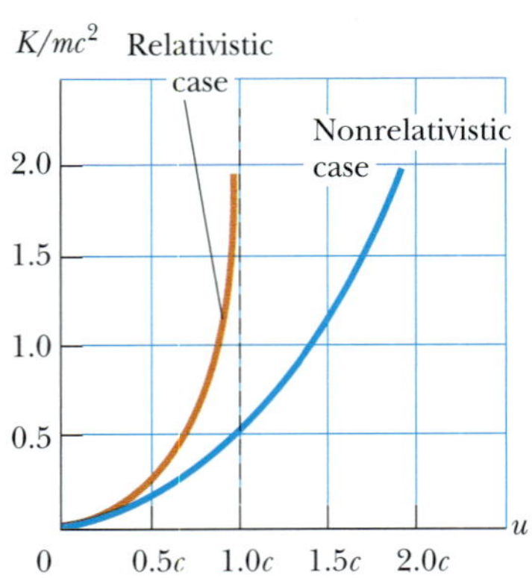

Figure 39.16 A graph comparing relativistic and nonrelativistic kinetic energy of a moving particle. The energies are plotted as a function of particle speed u. In the relativistic case, u is always less than c.

The constant term mc^2 in Equation 39.23, which is independent of the speed of the particle, is called the **rest energy** E_R of the particle:

$$E_R = mc^2 \tag{39.24}$$

◀ Rest energy

Equation 39.24 shows that **mass is a form of energy,** where c^2 is simply a constant conversion factor. This expression also shows that a small mass corresponds to an enormous amount of energy, a concept fundamental to nuclear and elementary-particle physics.

The term γmc^2 in Equation 39.23, which depends on the particle speed, is the sum of the kinetic and rest energies. It is called the **total energy** E:

$$\text{Total energy} = \text{kinetic energy} + \text{rest energy}$$

$$E = K + mc^2 \tag{39.25}$$

or

$$E = \frac{mc^2}{\sqrt{1 - \dfrac{u^2}{c^2}}} = \gamma mc^2 \tag{39.26}$$

◀ Total energy of a relativistic particle

In many situations, the linear momentum or energy of a particle rather than its speed is measured. It is therefore useful to have an expression relating the total energy E to the relativistic linear momentum p, which is accomplished by using the expressions $E = \gamma mc^2$ and $p = \gamma mu$. By squaring these equations and subtracting, we can eliminate u (Problem 37). The result, after some algebra, is[6]

$$E^2 = p^2c^2 + (mc^2)^2 \tag{39.27}$$

◀ Energy–momentum relationship for a relativistic particle

When the particle is at rest, $p = 0$, so $E = E_R = mc^2$.

In Section 35.1, we introduced the concept of a particle of light, called a **photon.** For particles that have zero mass, such as photons, we set $m = 0$ in Equation 39.27 and find that

$$E = pc \tag{39.28}$$

This equation is an exact expression relating total energy and linear momentum for photons, which always travel at the speed of light (in vacuum).

Finally, because the mass m of a particle is independent of its motion, m must have the same value in all reference frames. For this reason, m is often called the **invariant mass.** On the other hand, because the total energy and linear momentum of a particle both depend on velocity, these quantities depend on the reference frame in which they are measured.

When dealing with subatomic particles, it is convenient to express their energy in electron volts (Section 25.1) because the particles are usually given this energy by acceleration through a potential difference. The conversion factor, as you recall from Equation 25.5, is

$$1 \text{ eV} = 1.60 \times 10^{-19} \text{ J}$$

For example, the mass of an electron is 9.11×10^{-31} kg. Hence, the rest energy of the electron is

$$\begin{aligned} m_ec^2 &= (9.109 \times 10^{-31}\text{ kg})(2.998 \times 10^8 \text{ m/s})^2 = 8.187 \times 10^{-14}\text{ J} \\ &= (8.187 \times 10^{-14}\text{ J})(1 \text{ eV}/1.602 \times 10^{-19}\text{ J}) = 0.511 \text{ MeV} \end{aligned}$$

Quick Quiz 39.9 The following *pairs* of energies—particle 1: E, $2E$; particle 2: E, $3E$; particle 3: $2E$, $4E$—represent the rest energy and total energy of three different particles. Rank the particles from greatest to least according to their (a) mass, (b) kinetic energy, and (c) speed.

[6] One way to remember this relationship is to draw a right triangle having a hypotenuse of length E and legs of lengths pc and mc^2.

EXAMPLE 39.9 The Energy of a Speedy Proton

(A) Find the rest energy of a proton in units of electron volts.

SOLUTION

Conceptualize Even if the proton is not moving, it has energy associated with its mass. If it moves, the proton possesses more energy, with the total energy being the sum of its rest energy and its kinetic energy.

Categorize The phrase "rest energy" suggests we must take a relativistic rather than a classical approach to this problem.

Analyze Use Equation 39.24 to find the rest energy:

$$E_R = m_pc^2 = (1.673 \times 10^{-27}\text{ kg})(2.998 \times 10^8\text{ m/s})^2$$

$$= (1.504 \times 10^{-10}\text{ J})\left(\frac{1.00\text{ eV}}{1.602 \times 10^{-19}\text{ J}}\right) = \boxed{938\text{ MeV}}$$

(B) If the total energy of a proton is three times its rest energy, what is the speed of the proton?

SOLUTION

Use Equation 39.26 to relate the total energy of the proton to the rest energy:

$$E = 3m_pc^2 = \frac{m_pc^2}{\sqrt{1 - \frac{u^2}{c^2}}} \rightarrow 3 = \frac{1}{\sqrt{1 - \frac{u^2}{c^2}}}$$

Solve for u:

$$\left(1 - \frac{u^2}{c^2}\right) = \tfrac{1}{9} \rightarrow \frac{u^2}{c^2} = \tfrac{8}{9}$$

$$u = \frac{\sqrt{8}}{3}c = 0.943c = \boxed{2.83 \times 10^8\text{ m/s}}$$

(C) Determine the kinetic energy of the proton in units of electron volts.

SOLUTION

Use Equation 39.25 to find the kinetic energy of the proton:

$$K = E - m_pc^2 = 3m_pc^2 - m_pc^2 = 2m_pc^2$$

$$= 2(938\text{ MeV}) = \boxed{1.88 \times 10^3\text{ MeV}}$$

(D) What is the proton's momentum?

SOLUTION

Use Equation 39.27 to calculate the momentum:

$$E^2 = p^2c^2 + (m_pc^2)^2 = (3m_pc^2)^2$$

$$p^2c^2 = 9(m_pc^2)^2 - (m_pc^2)^2 = 8(m_pc^2)^2$$

$$p = \sqrt{8}\,\frac{m_pc^2}{c} = \sqrt{8}\,\frac{(938\text{ MeV})}{c} = \boxed{2.65 \times 10^3\text{ MeV}/c}$$

Finalize The unit of momentum in part (D) is written MeV/c, which is a common unit in particle physics. For comparison, you might want to solve this example using classical equations.

What If? In classical physics, if the momentum of a particle doubles, the kinetic energy increases by a factor of 4. What happens to the kinetic energy of the proton in this example if its momentum doubles?

Answer Based on what we have seen so far in relativity, it is likely you would predict that its kinetic energy does not increase by a factor of 4.

Find the new doubled momentum:

$$p_{\text{new}} = 2\left(\sqrt{8}\,\frac{m_p c^2}{c}\right) = 4\sqrt{2}\,\frac{m_p c^2}{c}$$

Use this result in Equation 39.27 to find the new total energy:

$$E_{\text{new}}^2 = p_{\text{new}}^2 c^2 + (m_p c^2)^2$$

$$E_{\text{new}}^2 = \left(4\sqrt{2}\,\frac{m_p c^2}{c}\right)^2 c^2 + (m_p c^2)^2 = 33(m_p c^2)^2$$

$$E_{\text{new}} = \sqrt{33}(m_p c^2) = 5.7 m_p c^2$$

Use Equation 39.25 to find the new kinetic energy:

$$K_{\text{new}} = E_{\text{new}} - m_p c^2 = 5.7 m_p c^2 - m_p c^2 = 4.7 m_p c^2$$

This value is a little more than twice the kinetic energy found in part (C), not four times. In general, the factor by which the kinetic energy increases if the momentum doubles depends on the initial momentum, but it approaches 4 as the momentum approaches zero. In this latter situation, classical physics correctly describes the situation.

39.9 Mass and Energy

Equation 39.26, $E = \gamma mc^2$, represents the total energy of a particle. This important equation suggests that even when a particle is at rest ($\gamma = 1$), it still possesses enormous energy through its mass. The clearest experimental proof of the equivalence of mass and energy occurs in nuclear and elementary-particle interactions in which the conversion of mass into kinetic energy takes place. Consequently, we cannot use the principle of conservation of energy in relativistic situations as it was outlined in Chapter 8. We must modify the principle by including rest energy as another form of energy storage.

This concept is important in atomic and nuclear processes, in which the change in mass is a relatively large fraction of the initial mass. In a conventional nuclear reactor, for example, the uranium nucleus undergoes *fission,* a reaction that results in several lighter fragments having considerable kinetic energy. In the case of ^{235}U, which is used as fuel in nuclear power plants, the fragments are two lighter nuclei and a few neutrons. The total mass of the fragments is less than that of the ^{235}U by an amount Δm. The corresponding energy Δmc^2 associated with this mass difference is exactly equal to the total kinetic energy of the fragments. The kinetic energy is absorbed as the fragments move through water, raising the internal energy of the water. This internal energy is used to produce steam for the generation of electricity.

Next, consider a basic *fusion* reaction in which two deuterium atoms combine to form one helium atom. The decrease in mass that results from the creation of one helium atom from two deuterium atoms is $\Delta m = 4.25 \times 10^{-29}$ kg. Hence, the corresponding energy that results from one fusion reaction is $\Delta mc^2 = 3.83 \times 10^{-12}$ J = 23.9 MeV. To appreciate the magnitude of this result, consider that if only 1 g of deuterium were converted to helium, the energy released would be on the order of 10^{12} J! In 2007's cost of electrical energy, this energy would be worth approximately \$30 000. We shall present more details of these nuclear processes in Chapter 45 of the extended version of this textbook.

EXAMPLE 39.10 Mass Change in a Radioactive Decay

The ^{216}Po nucleus is unstable and exhibits radioactivity (Chapter 44). It decays to ^{212}Pb by emitting an alpha particle, which is a helium nucleus, ^{4}He. The relevant masses are $m_i = m(^{216}\text{Po}) = 216.001\,905$ u, and $m_f = m(^{212}\text{Pb}) + m(^4\text{He}) = 211.991\,888\text{ u} + 4.002\,603$ u.

(A) Find the mass change of the system in this decay.

SOLUTION

Conceptualize The initial system is the ^{216}Po nucleus. Imagine the mass of the system decreasing during the decay and transforming to kinetic energy of the alpha particle and the ^{212}Pb nucleus after the decay.

Categorize We use concepts discussed in this section, so we categorize this example as a substitution problem.

Calculate the mass change:

$$\Delta m = 216.001\,905 \text{ u} - (211.991\,888 \text{ u} + 4.002\,603 \text{ u})$$
$$= 0.007\,414 \text{ u} = \boxed{1.23 \times 10^{-29} \text{ kg}}$$

(B) Find the energy this mass change represents.

SOLUTION

Use Equation 39.24 to find the energy associated with this mass change:

$$E = \Delta mc^2 = (1.23 \times 10^{-29} \text{ kg})(3.00 \times 10^8 \text{ m/s})^2$$
$$= 1.11 \times 10^{-12} \text{ J} = \boxed{6.92 \text{ MeV}}$$

39.10 The General Theory of Relativity

Up to this point, we have sidestepped a curious puzzle. Mass has two seemingly different properties: a *gravitational attraction* for other masses and an *inertial* property that represents a resistance to acceleration. To designate these two attributes, we use the subscripts g and i and write

$$\text{Gravitational property:} \quad F_g = m_g g$$

$$\text{Inertial property:} \quad \sum F = m_i a$$

The value for the gravitational constant G was chosen to make the magnitudes of m_g and m_i numerically equal. Regardless of how G is chosen, however, the strict proportionality of m_g and m_i has been established experimentally to an extremely high degree: a few parts in 10^{12}. Therefore, it appears that gravitational mass and inertial mass may indeed be exactly proportional.

Why, though? They seem to involve two entirely different concepts: a force of mutual gravitational attraction between two masses and the resistance of a single mass to being accelerated. This question, which puzzled Newton and many other physicists over the years, was answered by Einstein in 1916 when he published his theory of gravitation, known as the *general theory of relativity*. Because it is a mathematically complex theory, we offer merely a hint of its elegance and insight.

In Einstein's view, the dual behavior of mass was evidence for a very intimate and basic connection between the two behaviors. He pointed out that no mechanical experiment (such as dropping an object) could distinguish between the two situations illustrated in Figures 39.17a and 39.17b. In Figure 39.17a, a person standing in an elevator on the surface of a planet feels pressed into the floor due to the gravitational force. If he releases his briefcase, he observes it moving toward the floor with acceleration $\vec{\mathbf{g}} = -g\hat{\mathbf{j}}$. In Figure 39.17b, the person is in an elevator in empty space accelerating upward with $\vec{\mathbf{a}}_{el} = +g\hat{\mathbf{j}}$. The person feels pressed into the floor with the same force as in Figure 39.17a. If he releases his briefcase, he observes it moving toward the floor with acceleration g, exactly as in the previous situation. In each situation, an object released by the observer undergoes a downward acceleration of magnitude g relative to the floor. In Figure 39.17a, the person is at rest in an inertial frame in a gravitational field due to the planet. In Figure 39.17b, the person is in a noninertial frame accelerating in gravity-free space. Einstein's claim is that these two situations are completely equivalent.

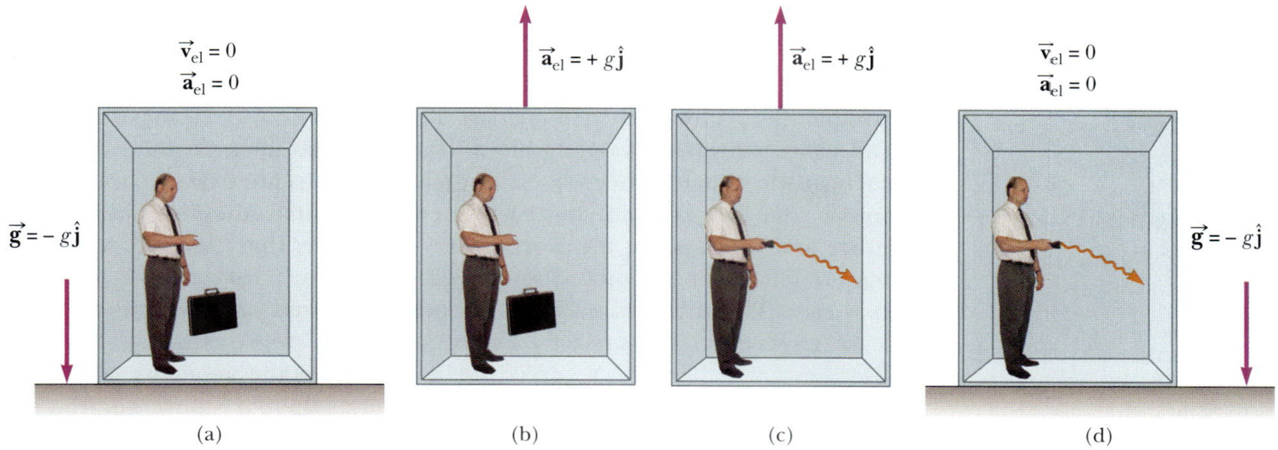

Figure 39.17 (a) The observer is at rest in an elevator in a uniform gravitational field $\vec{g} = -g\hat{j}$, directed downward. The observer drops his briefcase, which moves downward with acceleration g. (b) The observer is in a region where gravity is negligible, but the elevator moves upward with an acceleration $\vec{a}_{el} = +g\hat{j}$. The observer releases his briefcase, which moves downward (according to the observer) with acceleration g relative to the floor of the elevator. According to Einstein, the frames of reference in (a) and (b) are equivalent in every way. No local experiment can distinguish any difference between the two frames. (c) In the accelerating frame, a ray of light would appear to bend downward due to the acceleration. (d) If (a) and (b) are truly equivalent, as Einstein proposed, (c) suggests that a ray of light would bend downward in a gravitational field.

Einstein carried this idea further and proposed that *no* experiment, mechanical or otherwise, could distinguish between the two situations. This extension to include all phenomena (not just mechanical ones) has interesting consequences. For example, suppose a light pulse is sent horizontally across the elevator as in Figure 39.17c, in which the elevator is accelerating upward in empty space. From the point of view of an observer in an inertial frame outside the elevator, the light travels in a straight line while the floor of the elevator accelerates upward. According to the observer on the elevator, however, the trajectory of the light pulse bends downward as the floor of the elevator (and the observer) accelerates upward. Therefore, based on the equality of parts (a) and (b) of the figure, Einstein proposed that **a beam of light should also be bent downward by a gravitational field** as in Figure 39.17d. Experiments have verified the effect, although the bending is small. A laser aimed at the horizon falls less than 1 cm after traveling 6 000 km. (No such bending is predicted in Newton's theory of gravitation.)

Einstein's **general theory of relativity** has two postulates:

◀ Postulates of the general theory of relativity

- All the laws of nature have the same form for observers in any frame of reference, whether accelerated or not.
- In the vicinity of any point, a gravitational field is equivalent to an accelerated frame of reference in gravity-free space (the **principle of equivalence**).

One interesting effect predicted by the general theory is that time is altered by gravity. A clock in the presence of gravity runs slower than one located where gravity is negligible. Consequently, the frequencies of radiation emitted by atoms in the presence of a strong gravitational field are *redshifted* to lower frequencies when compared with the same emissions in the presence of a weak field. This gravitational redshift has been detected in spectral lines emitted by atoms in massive stars. It has also been verified on the Earth by comparing the frequencies of gamma rays emitted from nuclei separated vertically by about 20 m.

The second postulate suggests a gravitational field may be "transformed away" at any point if we choose an appropriate accelerated frame of reference, a freely falling one. Einstein developed an ingenious method of describing the acceleration necessary to make the gravitational field "disappear." He specified a concept, the *curvature of space–time*, that describes the gravitational effect at every point. In fact, the curvature of space–time completely replaces Newton's gravitational theory. According to Einstein, there is no such thing as a gravitational force. Rather, the presence of a mass causes a curvature of space–time in the vicinity of the mass,

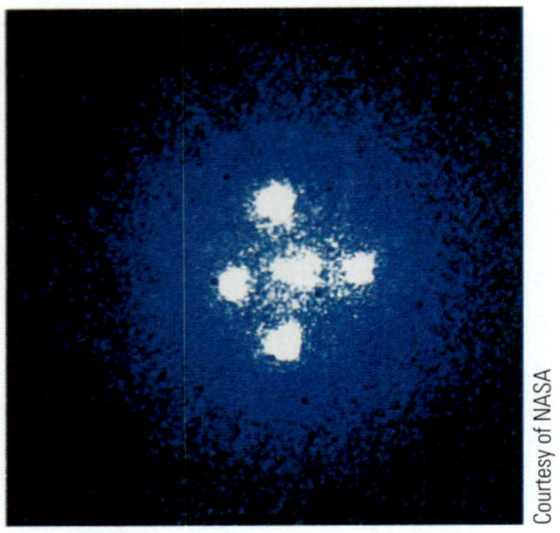

Einstein's cross. The four bright spots are images of the same galaxy that have been bent around a massive object located between the galaxy and the Earth. The massive object acts like a lens, causing the rays of light that were diverging from the distant galaxy to converge on the Earth. (If the intervening massive object had a uniform mass distribution, we would see a bright ring instead of four spots.)

and this curvature dictates the space–time path that all freely moving objects must follow.

As an example of the effects of curved space–time, imagine two travelers moving on parallel paths a few meters apart on the surface of the Earth and maintaining an exact northward heading along two longitude lines. As they observe each other near the equator, they will claim that their paths are exactly parallel. As they approach the North Pole, however, they notice that they are moving closer together and will meet at the North Pole. Therefore, they claim that they moved along parallel paths, but moved toward each other, *as if there were an attractive force between them.* The travelers make this conclusion based on their everyday experience of moving on flat surfaces. From our mental representation, however, we realize they are walking on a curved surface, and it is the geometry of the curved surface, rather than an attractive force, that causes them to converge. In a similar way, general relativity replaces the notion of forces with the movement of objects through curved space–time.

One prediction of the general theory of relativity is that a light ray passing near the Sun should be deflected in the curved space–time created by the Sun's mass. This prediction was confirmed when astronomers detected the bending of starlight near the Sun during a total solar eclipse that occurred shortly after World War I (Fig. 39.18). When this discovery was announced, Einstein became an international celebrity.

If the concentration of mass becomes very great as is believed to occur when a large star exhausts its nuclear fuel and collapses to a very small volume, a **black hole** may form. Here, the curvature of space–time is so extreme that within a certain distance from the center of the black hole all matter and light become trapped as discussed in Section 13.6.

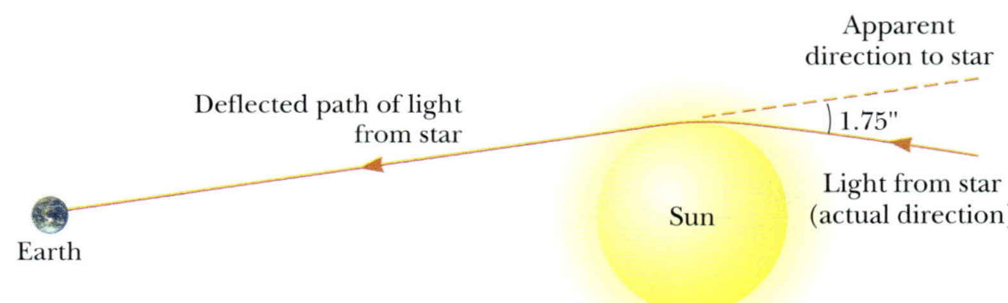

Figure 39.18 Deflection of starlight passing near the Sun. Because of this effect, the Sun or some other remote object can act as a *gravitational lens.* In his general theory of relativity, Einstein calculated that starlight just grazing the Sun's surface should be deflected by an angle of 1.75 s of arc.

Summary

ThomsonNOW™ Sign in at **www.thomsonedu.com** and go to ThomsonNOW to take a practice test for this chapter.

DEFINITIONS

The relativistic expression for the **linear momentum** of a particle moving with a velocity $\vec{\mathbf{u}}$ is

$$\vec{\mathbf{p}} \equiv \frac{m\vec{\mathbf{u}}}{\sqrt{1-\frac{u^2}{c^2}}} = \gamma m\vec{\mathbf{u}} \quad (39.19)$$

The relativistic force $\vec{\mathbf{F}}$ acting on a particle whose linear momentum is $\vec{\mathbf{p}}$ is defined as

$$\vec{\mathbf{F}} \equiv \frac{d\vec{\mathbf{p}}}{dt} \quad (39.20)$$

(continued)

CONCEPTS AND PRINCIPLES

The two basic postulates of the special theory of relativity are as follows:

- The laws of physics must be the same in all inertial reference frames.
- The speed of light in vacuum has the same value, $c = 3.00 \times 10^8$ m/s, in all inertial frames, regardless of the velocity of the observer or the velocity of the source emitting the light.

Three consequences of the special theory of relativity are as follows:

- Events that are measured to be simultaneous for one observer are not necessarily measured to be simultaneous for another observer who is in motion relative to the first.
- Clocks in motion relative to an observer are measured to run slower by a factor $\gamma = (1 - v^2/c^2)^{-1/2}$. This phenomenon is known as **time dilation.**
- The length of objects in motion are measured to be contracted in the direction of motion by a factor $1/\gamma = (1 - v^2/c^2)^{1/2}$. This phenomenon is known as **length contraction.**

To satisfy the postulates of special relativity, the Galilean transformation equations must be replaced by the **Lorentz transformation equations:**

$$x' = \gamma(x - vt) \qquad y' = y \qquad z' = z \qquad t' = \gamma\left(t - \frac{v}{c^2}x\right) \tag{39.11}$$

where $\gamma = (1 - v^2/c^2)^{-1/2}$ and the S′ frame moves in the x direction relative to the S frame.

The relativistic form of the **Lorentz velocity transformation equation** is

$$u'_x = \frac{u_x - v}{1 - \dfrac{u_x v}{c^2}} \tag{39.16}$$

where u'_x is the x component of the velocity of an object as measured in the S′ frame and u_x is its component as measured in the S frame.

The relativistic expression for the **kinetic energy** of a particle is

$$K = \frac{mc^2}{\sqrt{1 - \dfrac{u^2}{c^2}}} - mc^2 = (\gamma - 1)mc^2 \tag{39.23}$$

The constant term mc^2 in Equation 39.23 is called the **rest energy** E_R of the particle:

$$E_R = mc^2 \tag{39.24}$$

The total energy E of a particle is given by

$$E = \frac{mc^2}{\sqrt{1 - \dfrac{u^2}{c^2}}} = \gamma mc^2 \tag{39.26}$$

The relativistic linear momentum of a particle is related to its total energy through the equation

$$E^2 = p^2c^2 + (mc^2)^2 \tag{39.27}$$

Questions

□ denotes answer available in *Student Solutions Manual/Study Guide;* **O** denotes objective question

1. The speed of light in water is 230 Mm/s. Suppose an electron is moving through water at 250 Mm/s. Does that motion violate the principle of relativity?

2. **O** You measure the volume of a cube at rest to be V_0. You then measure the volume of the same cube as it passes you in a direction parallel to one side of the cube. The speed of the cube is $0.98c$, so $\gamma \approx 5$. Is the volume you measure close to (a) $V_0/125$, (b) $V_0/25$, (c) $V_0/5$, (d) V_0, (e) $5V_0$, (f) $25V_0$, or (g) $125V_0$?

3. **O** A spacecraft built in the shape of a sphere moves past an observer on the Earth with a speed of $0.5c$. What shape does the observer measure for the spacecraft as it goes by? (a) a sphere (b) a cigar shape, elongated along the direction of motion (c) a round pillow shape, flattened along the direction of motion (d) a conical shape, pointing in the direction of motion

4. **O** A spacecraft zooms past the Earth with a constant velocity. An observer on the Earth measures that an undamaged clock on the spacecraft is ticking at one-third the rate of an identical clock on the Earth. What does an observer on the spacecraft measure about the Earth clock's ticking rate? (a) It runs more than three times faster than his own clock. (b) It runs three times faster than his own. (c) It runs at the same rate as his own. (d) It runs at approximately half the rate of his own. (e) It runs at one-third the rate of his own. (f) It runs at less than one-third the rate of his own.

5. Explain why, when defining the length of a rod, it is necessary to specify that the positions of the ends of the rod are to be measured simultaneously.

6. **O** Two identical clocks are set side by side and synchronized. One remains on the Earth. The other is put into orbit around the Earth moving toward the east. **(i)** As measured by an observer on the Earth, while in rapid motion does the orbiting clock (a) run faster than the Earth-based clock, (b) run at the same rate, or (c) run slower? **(ii)** The orbiting clock is returned to its original location and brought to rest relative to the Earth. Thereafter, (a) its reading lags farther and farther behind the Earth-based clock, (b) it lags behind the Earth-based clock by a constant amount, (c) it is synchronized with the Earth-based clock, (d) it is ahead of the Earth-based clock by a constant amount, or (e) it gets farther and farther ahead of the Earth-based clock.

7. A train is approaching you at very high speed as you stand next to the tracks. Just as an observer on the train passes you, you both begin to play the same Beethoven symphony on portable CD players. (a) According to you, whose CD player finishes the symphony first? (b) **What If?** According to the observer on the train, whose CD player finishes the symphony first? (c) Whose CD player actually finishes the symphony first?

8. List some ways our day-to-day lives would change if the speed of light were only 50 m/s.

9. How is acceleration indicated on a space–time graph?

10. Explain how the Doppler effect with microwaves is used to determine the speed of an automobile.

11. In several cases, a nearby star has been found to have a large planet orbiting about it, although light from the planet could not be seen separately from the starlight. Using the ideas of a system rotating about its center of mass and of the Doppler shift for light, explain how an astronomer could determine the presence of the invisible planet.

12. A particle is moving at a speed less than $c/2$. If the speed of the particle is doubled, what happens to its momentum?

13. **O** Rank the following particles according to the magnitudes of their momentum from the largest to the smallest. If any have equal amounts of momentum, or zero momentum, display that fact in your ranking. (a) a 1-MeV photon (b) a proton with kinetic energy $K = 1$ MeV (c) an electron with $K = 1$ MeV (d) a grain of dust with $K = 1$ MeV $= 160$ fJ

14. Give a physical argument that shows it is impossible to accelerate an object of mass m to the speed of light, even with a continuous force acting on it.

15. **O** **(i)** Does the speed of an electron have an upper limit? (a) yes, the speed of light c (b) yes, with another value (c) no **(ii)** Does the magnitude of an electron's momentum have an upper limit? (a) yes, m_ec (b) yes, with another value (c) no **(iii)** Does the electron's kinetic energy have an upper limit? (a) yes, m_ec^2 (b) yes, $\frac{1}{2}m_ec^2$ (c) yes, with another value (d) no

16. **O** A distant astronomical object (a quasar) is moving away from us at half the speed of light. What is the speed of the light we receive from this quasar? (a) greater than c (b) c (c) between $c/2$ and c (d) $c/2$ (e) between 0 and $c/2$ (f) 0

17. "Newtonian mechanics correctly describes objects moving at ordinary speeds and relativistic mechanics correctly describes objects moving very fast." "Relativistic mechanics must make a smooth transition as it reduces to Newtonian mechanics in a case in which the speed of an object becomes small compared with the speed of light." Argue for or against each of these two statements.

18. Two cards have straight edges. Suppose the top edge of one card crosses the bottom edge of another card at a small angle as shown in Figure Q39.18a. A person slides the cards together at a moderately high speed. In what direction does the intersection point of the edges move? Show that the intersection point can move at a speed greater than the speed of light.

A small flashlight is suspended in a horizontal plane and set into rapid rotation. Show that the spot of light it produces on a distant screen can move across the screen at a speed greater than the speed of light. (If you use a laser pointer as shown in Figure Q39.18b, make sure the direct laser light cannot enter a person's eyes.) Argue that these experiments do not invalidate the principle that no material, no energy, and no information can move faster than light moves in a vacuum.

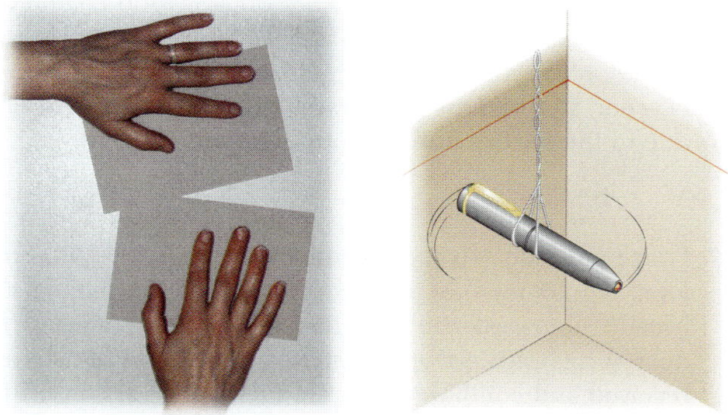

(a) (b)

Figure Q39.18

19. With regard to reference frames, how does general relativity differ from special relativity?

20. Two identical clocks are in the same house, one upstairs in a bedroom and the other downstairs in the kitchen. Which clock runs slower? Explain.

Problems

WebAssign The Problems from this chapter may be assigned online in WebAssign.

ThomsonNOW Sign in at **www.thomsonedu.com** and go to ThomsonNOW to assess your understanding of this chapter's topics with additional quizzing and conceptual questions.

1, 2, 3 denotes straightforward, intermediate, challenging; ☐ denotes full solution available in *Student Solutions Manual/Study Guide;* ▲ denotes coached solution with hints available at **www.thomsonedu.com;** ▭ denotes developing symbolic reasoning; ● denotes asking for qualitative reasoning; 🖥 denotes computer useful in solving problem

Section 39.1 The Principle of Galilean Relativity

☐ **1.** In a laboratory frame of reference, an observer notes that Newton's second law is valid. Show that it is also valid for an observer moving at a constant speed, small compared with the speed of light, relative to the laboratory frame.

2. Show that Newton's second law is *not* valid in a reference frame moving past the laboratory frame of Problem 1 with a constant acceleration.

3. A 2 000-kg car moving at 20.0 m/s collides and locks together with a 1 500-kg car at rest at a stop sign. Show that momentum is conserved in a reference frame moving at 10.0 m/s in the direction of the moving car.

Section 39.2 The Michelson–Morley Experiment

Section 39.3 Einstein's Principle of Relativity

Section 39.4 Consequences of the Special Theory of Relativity

Problem 37 in Chapter 4 can be assigned with this section.

4. How fast must a meterstick be moving if its length is measured to shrink to 0.500 m?

5. At what speed does a clock move if it is measured to run at a rate one-half the rate of a clock at rest with respect to an observer?

6. An astronaut is traveling in a space vehicle moving at $0.500c$ relative to the Earth. The astronaut measures her pulse rate at 75.0 beats per minute. Signals generated by the astronaut's pulse are radioed to the Earth when the vehicle is moving in a direction perpendicular to the line that connects the vehicle with an observer on the Earth. (a) What pulse rate does the Earth-based observer measure? (b) **What If?** What would be the pulse rate if the speed of the space vehicle were increased to $0.990c$?

☐ **7.** An atomic clock moves at 1 000 km/h for 1.00 h as measured by an identical clock on the Earth. At the end of the 1.00-h interval, how many nanoseconds slow will the moving clock be compared with the Earth clock?

8. A muon formed high in the Earth's atmosphere travels at speed $v = 0.990c$ for a distance of 4.60 km before it decays into an electron, a neutrino, and an antineutrino ($\mu^- \rightarrow e^- + \nu + \bar{\nu}$). (a) For what time interval does the muon live as measured in its reference frame? (b) How far does the Earth travel as measured in the frame of the muon?

9. ▲ A spacecraft with a proper length of 300 m takes 0.750 μs to pass an Earth-based observer. Determine the speed of the spacecraft as measured by the Earth observer.

10. (a) An object of proper length L_p takes a time interval Δt to pass an Earth-based observer. Determine the speed of the object as measured by the Earth observer. (b) A column of tanks, 300 m long, takes 75.0 s to pass a child waiting at a street corner on her way to school. Determine the speed of the armored vehicles. (c) Show that the answer to part (a) includes the answer to Problem 9 as a special case and includes the answer to part (b) as another special case.

11. ● **Review problem.** In 1963, astronaut Gordon Cooper orbited the Earth 22 times. The press stated that for each orbit, he aged 2 millionths of a second less than he would have had he remained on the Earth. (a) Assuming that he was 160 km above the Earth in a circular orbit, determine the difference in elapsed time between someone on the Earth and the orbiting astronaut for the 22 orbits. You may use the approximation

$$\frac{1}{\sqrt{1-x}} \approx 1 + \frac{x}{2}$$

for small x. (b) Did the press report accurate information? Explain.

12. For what value of v does $\gamma = 1.010\,0$? Observe that for speeds lower than this value, time dilation and length contraction are effects amounting to less than 1%.

13. A friend passes by you in a spacecraft traveling at a high speed. He tells you that his spacecraft is 20.0 m long and that the identically constructed spacecraft you are sitting in is 19.0 m long. According to your observations, (a) how long is your spacecraft, (b) how long is your friend's spacecraft, and (c) what is the speed of your friend's spacecraft?

14. The identical twins Speedo and Goslo join a migration from the Earth to Planet X. It is 20.0 ly away in a reference frame in which both planets are at rest. The twins, of the same age, depart at the same moment on different spacecraft. Speedo's spacecraft travels steadily at $0.950c$ and Goslo's at $0.750c$. Calculate the age difference between the twins after Goslo's spacecraft lands on Planet X. Which twin is older?

15. **Review problem.** An alien civilization occupies a brown dwarf, nearly stationary relative to the Sun, several lightyears away. The extraterrestrials have come to love original broadcasts of *I Love Lucy,* on television channel 2, at carrier frequency 57.0 MHz. Their line of sight to us is in the plane of the Earth's orbit. Find the difference between the highest and lowest frequencies they receive due to the Earth's orbital motion around the Sun.

16. Police radar detects the speed of a car (Fig. P39.16) as follows. Microwaves of a precisely known frequency are broadcast toward the car. The moving car reflects the microwaves with a Doppler shift. The reflected waves are received and combined with an attenuated version of the transmitted wave. Beats occur between the two microwave signals. The beat frequency is measured. (a) For an electromagnetic wave reflected back to its source from a mirror approaching at speed v, show that the reflected wave has frequency

$$f = f_{\text{source}}\frac{c+v}{c-v}$$

where f_{source} is the source frequency. (b) When v is much less than c, the beat frequency is much smaller than the transmitted frequency. In this case, use the approximation $f + f_{\text{source}} \approx 2f_{\text{source}}$ and show that the beat frequency can be written as $f_{\text{beat}} = 2v/\lambda$. (c) What beat frequency is measured for a car speed of 30.0 m/s if the microwaves have frequency 10.0 GHz? (d) If the beat frequency measurement is accurate to ± 5 Hz, how accurate is the speed measurement?

Trent Steffler/David R. Frazier Photolibrary

Figure P39.16

17. *The redshift.* A light source recedes from an observer with a speed v_{source} that is small compared with c. (a) Show that the fractional shift in the measured wavelength is given by the approximate expression

$$\frac{\Delta\lambda}{\lambda} \approx \frac{v_{\text{source}}}{c}$$

This phenomenon is known as the redshift because the visible light is shifted toward the red. (b) Spectroscopic measurements of light at $\lambda = 397$ nm coming from a galaxy in Ursa Major reveal a redshift of 20.0 nm. What is the recessional speed of the galaxy?

2 = intermediate; 3 = challenging; □ = SSM/SG; ▲ = ThomsonNOW; ▭ = symbolic reasoning; ● = qualitative reasoning

18. A physicist drives through a stop light. When he is pulled over, he tells the police officer that the Doppler shift made the red light of wavelength 650 nm appear green to him, with a wavelength of 520 nm. The police officer writes out a traffic citation for speeding. How fast was the physicist traveling, according to his own testimony?

Section 39.5 The Lorentz Transformation Equations

19. Suzanne observes two light pulses to be emitted from the same location, but separated in time by 3.00 μs. Mark observes the emission of the same two pulses to be separated in time by 9.00 μs. (a) How fast is Mark moving relative to Suzanne? (b) According to Mark, what is the separation in space of the two pulses?

20. A moving rod is observed to have a length of 2.00 m and to be oriented at an angle of 30.0° with respect to the direction of motion as shown in Figure P39.20. The rod has a speed of 0.995c. (a) What is the proper length of the rod? (b) What is the orientation angle in the proper frame?

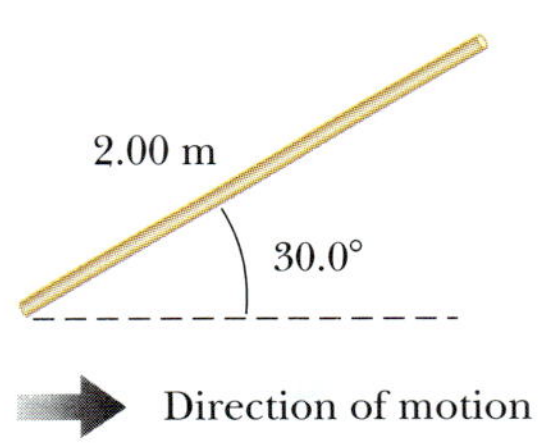

Figure P39.20

21. An observer in reference frame S measures two events to be simultaneous. Event A occurs at the point (50.0 m, 0, 0) at the instant 9:00:00 Universal time on January 15, 2008. Event B occurs at the point (150 m, 0, 0) at the same moment. A second observer, moving past with a velocity of $0.800c\hat{\mathbf{i}}$, also observes the two events. In her reference frame S′, which event occurred first and what time interval elapsed between the events?

22. A red light flashes at position $x_R = 3.00$ m and time $t_R = 1.00 \times 10^{-9}$ s, and a blue light flashes at $x_B = 5.00$ m and $t_B = 9.00 \times 10^{-9}$ s, all measured in the S reference frame. Reference frame S′ moves uniformly to the right and has its origin at the same point as S at $t = t' = 0$. Both flashes are observed to occur at the same place in S′. (a) Find the relative speed between S and S′. (b) Find the location of the two flashes in frame S′. (c) At what time does the red flash occur in the S′ frame?

Section 39.6 The Lorentz Velocity Transformation Equations

23. ▲ Two jets of material from the center of a radio galaxy are ejected in opposite directions. Both jets move at 0.750c relative to the galaxy. Determine the speed of one jet relative to the other.

24. A Klingon spacecraft moves away from the Earth at a speed of 0.800c (Fig. P39.24). The starship *Enterprise* pursues at a speed of 0.900c relative to the Earth. Observers on the Earth measure the *Enterprise* to be overtaking the Klingon craft at a relative speed of 0.100c. With what speed is the *Enterprise* overtaking the Klingon craft as measured by the crew of the *Enterprise*?

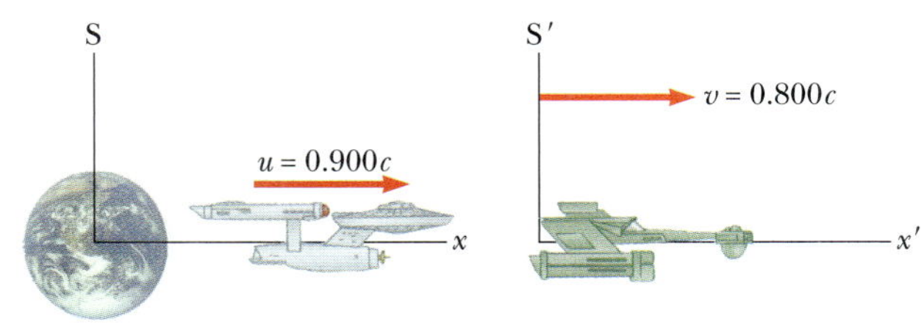

Figure P39.24

Section 39.7 Relativistic Linear Momentum

25. Calculate the momentum of an electron moving with a speed of (a) 0.010 0c, (b) 0.500c, and (c) 0.900c.

26. The nonrelativistic expression for the momentum of a particle, $p = mu$, agrees with experiment if $u \ll c$. For what speed does the use of this equation give an error in the momentum of (a) 1.00% and (b) 10.0%?

27. A golf ball travels with a speed of 90.0 m/s. By what fraction does its relativistic momentum magnitude p differ from its classical value mu? That is, calculate the ratio $(p - mu)/mu$.

28. The speed limit on a certain roadway is 90.0 km/h. Suppose speeding fines are made proportional to the amount by which a vehicle's momentum exceeds the momentum it would have when traveling at the speed limit. The fine for driving at 190 km/h (that is, 100 km/h over the speed limit) is $80.0. What then will be the fine for traveling (a) at 1 090 km/h? (b) At 1 000 000 090 km/h?

29. ▲ An unstable particle at rest spontaneously breaks into two fragments of unequal mass. The mass of the first fragment is 2.50×10^{-28} kg, and that of the other is 1.67×10^{-27} kg. If the lighter fragment has a speed of $0.893c$ after the breakup, what is the speed of the heavier fragment?

Section 39.8 Relativistic Energy

30. An electron has a kinetic energy five times greater than its rest energy. Find its (a) total energy and (b) speed.

31. A proton in a high-energy accelerator moves with a speed of $c/2$. Use the work–kinetic energy theorem to find the work required to increase its speed to (a) $0.750c$ and (b) $0.995c$.

32. Show that for any object moving at less than one-tenth the speed of light, the relativistic kinetic energy agrees with the result of the classical equation $K = \frac{1}{2}mu^2$ to within less than 1%. Therefore, for most purposes, the classical equation is good enough to describe these objects.

33. Find the momentum of a proton in MeV/c units assuming its total energy is twice its rest energy.

34. ● (a) Find the kinetic energy of a 78.0-kg spacecraft launched out of the solar system with speed 106 km/s by using the classical equation $K = \frac{1}{2}mu^2$. (b) **What If?** Calculate its kinetic energy using the relativistic equation. (c) Explain the result of comparing the results of parts (a) and (b).

35. ▲ A proton moves at $0.950c$. Calculate its (a) rest energy, (b) total energy, and (c) kinetic energy.

36. An unstable particle with a mass of 3.34×10^{-27} kg is initially at rest. The particle decays into two fragments that fly off along the x axis with velocity components $0.987c$ and $-0.868c$. Find the masses of the fragments. *Suggestion:* Use conservation of both energy and momentum.

37. Show that the energy–momentum relationship $E^2 = p^2c^2 + (mc^2)^2$ follows from the expressions $E = \gamma mc^2$ and $p = \gamma mu$.

38. In a typical color television picture tube, the electrons are accelerated from rest through a potential difference of 25 000 V. (a) What speed do the electrons have when they strike the screen? (b) What is their kinetic energy in joules?

39. The rest energy of an electron is 0.511 MeV. The rest energy of a proton is 938 MeV. Assume both particles have kinetic energies of 2.00 MeV. Find the speed of (a) the electron and (b) the proton. (c) By how much does the speed of the electron exceed that of the proton? (d) Repeat the calculations assuming both particles have kinetic energies of 2 000 MeV.

40. Consider electrons accelerated to an energy of 20.0 GeV in the 3.00-km-long Stanford Linear Accelerator. (a) What is the γ factor for the electrons? (b) What is the electrons' speed? (c) How long does the accelerator appear to the electrons?

41. A pion at rest ($m_\pi = 273m_e$) decays to a muon ($m_\mu = 207m_e$) and an antineutrino ($m_{\bar{\nu}} \approx 0$). The reaction is written $\pi^- \rightarrow \mu^- + \bar{\nu}$. Find the kinetic energy of the muon and the energy of the antineutrino in electron volts. *Suggestion:* Use conservation of both energy and momentum for the decay process.

42. Consider a car moving at highway speed u. Is its actual kinetic energy larger or smaller than $\frac{1}{2}mu^2$? Make an order-of-magnitude estimate of the amount by which its actual kinetic energy differs from $\frac{1}{2}mu^2$. In your solution, state the quantities you take as data and the values you measure or estimate for them. You may find Appendix B.5 useful.

Section 39.9 Mass and Energy

43. ● When 1.00 g of hydrogen combines with 8.00 g of oxygen, 9.00 g of water is formed. During this chemical reaction, 2.86×10^5 J of energy is released. Is the mass of the water larger or smaller than the mass of the reactants? What is the difference in mass? Explain whether the change in mass is likely to be detectable.

44. In a nuclear power plant, the fuel rods last 3 yr before they are replaced. If a plant with rated thermal power 1.00 GW operates at 80.0% capacity for 3.00 yr, what is the loss of mass of the fuel?

45. The power output of the Sun is 3.85×10^{26} W. How much mass is converted to energy in the Sun each second?

46. A gamma ray (a high-energy photon) can produce an electron (e^-) and a positron (e^+) when it enters the electric field of a heavy nucleus: $\gamma \rightarrow e^+ + e^-$. What minimum gamma-ray energy is required to accomplish this task? *Note:* The masses of the electron and the positron are equal.

Section 39.10 The General Theory of Relativity

47. An Earth satellite used in the global positioning system (GPS) moves in a circular orbit with period 11 h 58 min. (a) Determine the radius of its orbit. (b) Determine its speed. (c) The satellite contains an oscillator producing the principal nonmilitary GPS signal. Its frequency is 1 575.42 MHz in the reference frame of the satellite. When it is received on the Earth's surface, what is the fractional change in this frequency due to time dilation as described by special relativity? (d) The gravitational "blueshift" of the frequency according to general relativity is a separate effect. It is called a blueshift to indicate a change to a higher frequency. The magnitude of that fractional change is given by

$$\frac{\Delta f}{f} = \frac{\Delta U_g}{mc^2}$$

where ΔU_g is the change in gravitational potential energy of an object–Earth system when the object of mass m is moved between the two points where the signal is observed. Calculate this fractional change in frequency. (e) What is the overall fractional change in frequency? Superposed on both of these relativistic effects is a Doppler shift that is generally much larger. It can be a redshift or a blueshift, depending on the motion of a particular satellite relative to a GPS receiver (Fig. P39.47).

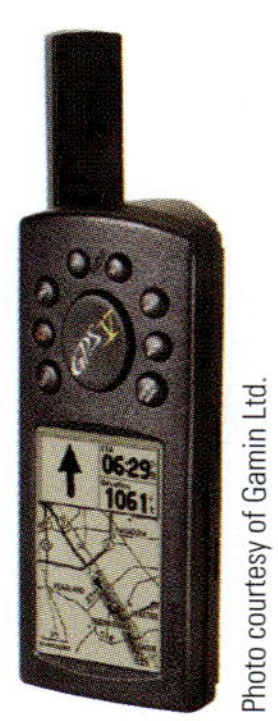

Figure P39.47

Additional Problems

48. *Houston, we've got a problem.* An astronaut wishes to visit the Andromeda galaxy, making a one-way trip that will take 30.0 yr in the spacecraft's frame of reference. Assume the galaxy is 2.00×10^6 ly away and the astronaut's speed is constant. (a) How fast must he travel relative to the Earth? (b) What will be the kinetic energy of his 1 000-metric-ton spacecraft? (c) What is the cost of this energy if it is purchased at a typical consumer price for electric energy of \$0.130/kWh?

49. ▲ The cosmic rays of highest energy are protons that have kinetic energy on the order of 10^{13} MeV. (a) How long would it take a proton of this energy to travel across the Milky Way galaxy, having a diameter $\sim 10^5$ ly, as measured in the proton's frame? (b) From the point of view of the proton, how many kilometers across is the galaxy?

50. An electron has a speed of $0.750c$. (a) Find the speed of a proton that has the same kinetic energy as the electron. (b) **What If?** Find the speed of a proton that has the same momentum as the electron.

51. ● The equation

$$K = \left(\frac{1}{\sqrt{1 - u^2/c^2}} - 1\right) mc^2$$

gives the kinetic energy of a particle moving at speed u. (a) Solve the equation for u. (b) From the equation for u, identify the minimum possible value of speed and the corresponding kinetic energy. (c) Identify the maximum possible speed and the corresponding kinetic energy. (d) Differentiate the equation for u with respect to time to obtain an equation describing the acceleration of a particle as a function of its kinetic energy and the power input to the particle. (e) Observe that for a nonrelativistic particle we have $u = (2K/m)^{1/2}$ and that differentiating this equation with respect to time gives $a = \mathcal{P}/(2mK)^{1/2}$. State the limiting form of the expression in part (d) at low energy. State how it compares with the nonrelativistic expression. (f) State the limiting form of the expression in part (d) at high energy. (g) Consider a particle with constant input power. Explain how the answer to part (f) helps account for the answer to part (c).

52. Ted and Mary are playing a game of catch in frame S′, which is moving at $0.600c$ with respect to frame S, while Jim, at rest in frame S, watches the action (Fig. P39.52). Ted throws the ball to Mary at $0.800c$ (according to Ted), and their separation (measured in S′) is 1.80×10^{12} m. (a) According to Mary, how fast is the ball moving? (b) According to Mary, what time interval is required for the ball to reach her? (c) According to Jim, how far apart are Ted and Mary and how fast is the ball moving? (d) According to Jim, what time interval is required for the ball to reach Mary?

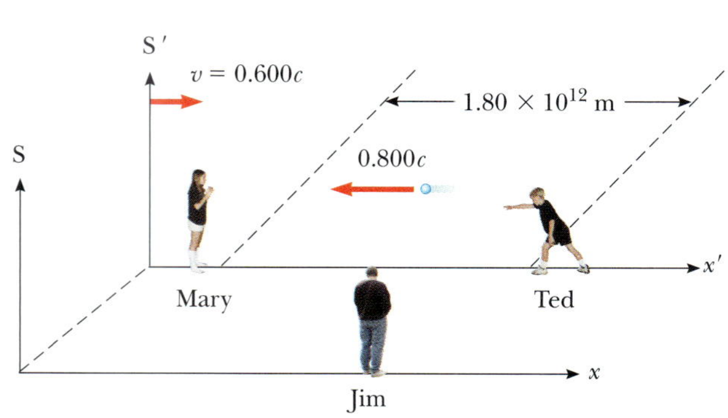

Figure P39.52

53. The net nuclear fusion reaction inside the Sun can be written as $4\,^1\text{H} \rightarrow {}^4\text{He} + E$. The rest energy of each hydrogen atom is 938.78 MeV, and the rest energy of the helium-4 atom is 3 728.4 MeV. Calculate the percentage of the starting mass that is transformed to other forms of energy.

54. An object disintegrates into two fragments. One fragment has mass $1.00\ \text{MeV}/c^2$ and momentum $1.75\ \text{MeV}/c$ in the positive x direction. The other fragment has mass $1.50\ \text{MeV}/c^2$ and momentum $2.00\ \text{MeV}/c$ in the positive y direction. Find the (a) mass and (b) speed of the original object.

55. Spacecraft I, containing students taking a physics exam, approaches the Earth with a speed of $0.600c$ (relative to the Earth), while spacecraft II, containing professors proctoring the exam, moves at $0.280c$ (relative to the Earth) directly toward the students. If the professors stop the exam after 50.0 min have passed on their clock, for what time interval does the exam last as measured by (a) the students and (b) an observer on the Earth?

56. **Review problem.** An electron is traveling through water at a speed 10.0% faster than the speed of light in water. Determine the electron's (a) total energy, (b) kinetic energy, and (c) momentum. The electron gives off Cerenkov radiation, the electromagnetic equivalent of a bow wave or a sonic boom. (d) Find the angle between the shock wave and the electron's direction of motion. Around the core of a nuclear reactor shielded by a large pool of water, Cerenkov radiation appears as a blue glow.

57. An alien spaceship traveling at $0.600c$ toward the Earth launches a landing craft with an advance guard of purchasing agents and environmental educators. The landing craft travels in the same direction with a speed of $0.800c$ relative to the mother ship. As observed on the Earth, the spaceship is 0.200 ly from the Earth when the landing craft is launched. (a) What speed do the Earth-based observers measure for the approaching landing craft? (b) What is the distance to the Earth at the moment of the landing craft's launch as observed by the aliens? (c) What travel time is required for the landing craft to reach the Earth as observed by the aliens on the mother ship? (d) If the landing craft has a mass of 4.00×10^5 kg, what is its kinetic energy as observed in the Earth reference frame?

58. ● *Speed of light in a moving medium.* The motion of a transparent medium influences the speed of light. This effect was first observed by Fizeau in 1851. Consider a light beam in water that moves with speed v in a horizontal pipe. Assume the light travels in the same direction as the water. The speed of light with respect to the water is c/n, where $n = 1.33$ is the index of refraction of water. (a) Use the velocity transformation equation to show that the speed of the light measured in the laboratory frame is

$$u = \frac{c}{n}\left(\frac{1 + nv/c}{1 + v/nc}\right)$$

(b) Show that for $v \ll c$, the expression from part (a) becomes, to a good approximation,

$$u \approx \frac{c}{n} + v - \frac{v}{n^2}$$

Argue for or against the view that we should expect the result to be $u = (c/n) + v$ according to the Galilean transformation and that the presence of the term $-v/n^2$ represents a relativistic effect appearing even at "nonrelativistic" speeds. (c) Evaluate u in the limit as the speed of the water approaches c.

59. A supertrain (proper length 100 m) travels at a speed of $0.950c$ as it passes through a tunnel (proper length 50.0 m). As seen by a trackside observer, is the train ever completely within the tunnel? If so, how much space is there to spare?

60. Imagine that the entire Sun collapses to a sphere of radius R_g such that the work required to remove a small mass m from the surface would be equal to its rest energy mc^2. This radius is called the *gravitational radius* for the Sun. Find R_g. The ultimate fate of very massive stars is thought to be collapsing beyond their gravitational radii into black holes.

61. ● A particle with electric charge q moves along a straight line in a uniform electric field $\vec{\mathbf{E}}$ with a speed of u. The electric force exerted on the charge is $q\vec{\mathbf{E}}$. The motion and the electric field are both in the x direction. (a) Show that the acceleration of the particle in the x direction is given by

$$a = \frac{du}{dt} = \frac{qE}{m}\left(1 - \frac{u^2}{c^2}\right)^{3/2}$$

(b) Discuss the significance of the dependence of the acceleration on the speed. (c) **What If?** If the particle starts from rest at $x = 0$ at $t = 0$, how would you proceed to find the speed of the particle and its position at time t?

62. An observer in a coasting spacecraft moves toward a mirror at speed v relative to the reference frame labeled by S in Figure P39.62. The mirror is stationary with respect to S. A light pulse emitted by the spacecraft travels toward the mirror and is reflected back to the spacecraft. The front of the spacecraft is a distance d from the mirror (as measured by observers in S) at the moment the light pulse leaves the spacecraft. What is the total travel time of the pulse as measured by observers in (a) the S frame and (b) the front of the spacecraft?

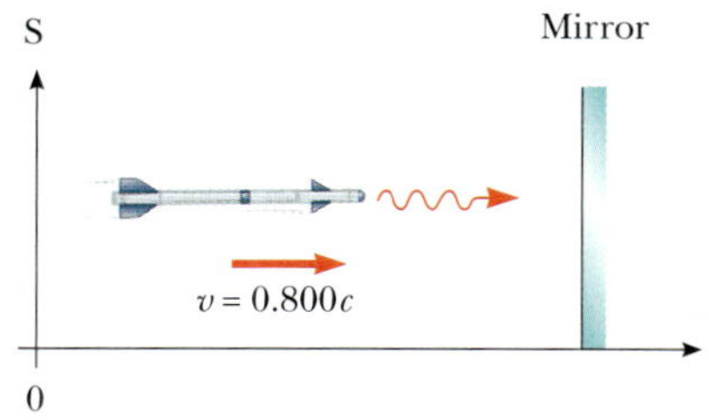

Figure P39.62

63. Massive stars ending their lives in supernova explosions produce the nuclei of all the atoms in the bottom half of the periodic table, by fusion of smaller nuclei. This problem roughly models that process. A particle of mass m moving along the x axis with a velocity component $+u$ collides head-on and sticks to a particle of mass $m/3$ moving along the x axis with the velocity component $-u$. (a) What is the mass M of the resulting particle? (b) Evaluate the expression from part (a) in the limit $u \to 0$. Explain whether the result agrees with what you should expect from nonrelativistic physics.

64. The creation and study of new elementary particles is an important part of contemporary physics. Especially interesting is the discovery of a very massive particle. To create a particle of mass M requires an energy Mc^2. With enough energy, an exotic particle can be created by allowing a fast-moving particle of ordinary matter, such as a proton, to collide with a similar target particle. Consider a perfectly inelastic collision between two protons: an incident proton with mass m_p, kinetic energy K, and momentum magnitude p joins with an originally stationary target proton to form a single product particle of mass M. You might think that the creation of a new product particle, nine times more massive than in a previous experiment, would require only nine times more energy for the incident proton. Unfortunately, not all the kinetic energy of the incoming proton is available to create the product particle because conservation of momentum requires that the system as a whole still must have some kinetic energy after the collision. Only a fraction of the energy of the incident particle is therefore available to create a new particle. In this problem, you must determine how the energy available for particle creation depends on the energy of the moving proton. Show that the energy available to create a product particle is given by

$$Mc^2 = 2m_pc^2\sqrt{1 + \frac{K}{2m_pc^2}}$$

This result shows that when the kinetic energy K of the incident proton is large compared with its rest energy m_pc^2, then M approaches $(2m_pK)^{1/2}/c$. Therefore, if the energy of the incoming proton is increased by a factor of 9, the mass you can create increases only by a factor of 3. This disappointing result is the main reason that most modern accelerators such as those at CERN (in Europe), at Fermilab (near Chicago), at SLAC (at Stanford), and at DESY (in Germany) use *colliding beams*. Here the total momentum of a pair of interacting particles can be zero. The center of mass can be at rest after the collision, so, in principle, all the initial kinetic energy can be used for particle creation, according to

$$Mc^2 = 2mc^2 + K = 2mc^2\left(1 + \frac{K}{2mc^2}\right)$$

where K is the total kinetic energy of two identical colliding particles. Here if $K \gg mc^2$, we have M directly proportional to K as we would desire. These machines are difficult to build and to operate, but they open new vistas in physics.

65. Suppose our Sun is about to explode. In an effort to escape, we depart in a spacecraft at $v = 0.800c$ and head toward the star Tau Ceti, 12.0 ly away. When we reach the midpoint of our journey from the Earth, we see our Sun explode, and, unfortunately, at the same instant we see Tau Ceti explode as well. (a) In the spacecraft's frame of reference, should we conclude that the two explosions occurred simultaneously? If not, which occurred first? (b) **What If?** In a frame of reference in which the Sun and Tau Ceti are at rest, did they explode simultaneously? If not, which exploded first?

66. Prepare a graph of the relativistic kinetic energy and the classical kinetic energy, both as a function of speed, for an object with a mass of your choice. At what speed does the classical kinetic energy underestimate the experimental value by 1%? By 5%? By 50%?

67. A ^{57}Fe nucleus at rest emits a 14.0-keV photon. Use conservation of energy and momentum to deduce the kinetic energy of the recoiling nucleus in electron volts. Use $Mc^2 = 8.60 \times 10^{-9}$ J for the final state of the ^{57}Fe nucleus.

Answers to Quick Quizzes

39.1 (c). Although the observers' measurements differ, both are correct.

39.2 (d). The Galilean velocity transformation gives us $u_x = u'_x + v = 90$ mi/h + 110 mi/h = 200 mi/h.

39.3 (d). The two events (the pulse leaving the flashlight and the pulse hitting the far wall) take place at different locations for both observers, so neither measures the proper time interval.

39.4 (a). The two events are the beginning and the end of the movie, both of which take place at rest with respect to the spacecraft crew. Therefore, the crew measures the proper time interval of 2 h. Any observer in motion with respect to the spacecraft, which includes the observer on Earth, will measure a longer time interval due to time dilation.

39.5 (a). If their on-duty time is based on clocks that remain on the Earth, the astronauts will have larger paychecks. A shorter time interval will have passed for the astronauts in their frame of reference than for their employer back on the Earth.

39.6 (c). Both your body and your sleeping cabin are at rest in your reference frame; therefore, they will have their proper length according to you. There will be no change in measured lengths of objects, including yourself, within your spacecraft.

39.7 (d). Time dilation and length contraction depend only on the relative speed of one observer relative to another, not on whether the observers are receding or approaching each other.

39.8 **(i),** (c). Because of your motion toward the source of the light, the light beam has a horizontal component of velocity as measured by you. The magnitude of the vector sum of the horizontal and vertical component vectors must be equal to c, so the magnitude of the vertical component must be smaller than c. **(ii),** (a). In this case, there is only a horizontal component of the velocity of the light and you must measure a speed of c.

39.9 (a) $m_3 > m_2 = m_1$; the rest energy of particle 3 is $2E$, whereas it is E for particles 1 and 2. (b) $K_3 = K_2 > K_1$; the kinetic energy is the difference between the total energy and the rest energy. The kinetic energy is $4E - 2E = 2E$ for particle 3, $3E - E = 2E$ for particle 2, and $2E - E = E$ for particle 1. (c) $u_2 > u_3 = u_1$; from Equation 39.26, $E = \gamma E_R$. Solving for the square of the particle speed u, we find that $u^2 = c^2(1 - (E_R/E)^2)$. Therefore, the particle with the smallest ratio of rest energy to total energy will have the largest speed. Particles 1 and 3 have the same ratio as each other, and the ratio of particle 2 is smaller.

TABLE A.1

Conversion Factors

Length

	m	cm	km	in.	ft	mi
1 meter	1	10^2	10^{-3}	39.37	3.281	6.214×10^{-4}
1 centimeter	10^{-2}	1	10^{-5}	0.393 7	3.281×10^{-2}	6.214×10^{-6}
1 kilometer	10^3	10^5	1	3.937×10^4	3.281×10^3	0.621 4
1 inch	2.540×10^{-2}	2.540	2.540×10^{-5}	1	8.333×10^{-2}	1.578×10^{-5}
1 foot	0.304 8	30.48	3.048×10^{-4}	12	1	1.894×10^{-4}
1 mile	1 609	1.609×10^5	1.609	6.336×10^4	5 280	1

Mass

	kg	g	slug	u
1 kilogram	1	10^3	6.852×10^{-2}	6.024×10^{26}
1 gram	10^{-3}	1	6.852×10^{-5}	6.024×10^{23}
1 slug	14.59	1.459×10^4	1	8.789×10^{27}
1 atomic mass unit	1.660×10^{-27}	1.660×10^{-24}	1.137×10^{-28}	1

Note: 1 metric ton = 1 000 kg.

Time

	s	min	h	day	yr
1 second	1	1.667×10^{-2}	2.778×10^{-4}	1.157×10^{-5}	3.169×10^{-8}
1 minute	60	1	1.667×10^{-2}	6.994×10^{-4}	1.901×10^{-6}
1 hour	3 600	60	1	4.167×10^{-2}	1.141×10^{-4}
1 day	8.640×10^4	1 440	24	1	2.738×10^{-5}
1 year	3.156×10^7	5.259×10^5	8.766×10^3	365.2	1

Speed

	m/s	cm/s	ft/s	mi/h
1 meter per second	1	10^2	3.281	2.237
1 centimeter per second	10^{-2}	1	3.281×10^{-2}	2.237×10^{-2}
1 foot per second	0.304 8	30.48	1	0.681 8
1 mile per hour	0.447 0	44.70	1.467	1

Note: 1 mi/min = 60 mi/h = 88 ft/s.

Force

	N	lb
1 newton	1	0.224 8
1 pound	4.448	1

(*Continued*)

TABLE A.1

Conversion Factors *(Continued)*

Energy, Energy Transfer

	J	ft · lb	eV
1 joule	1	0.737 6	6.242×10^{18}
1 foot-pound	1.356	1	8.464×10^{18}
1 electron volt	1.602×10^{-19}	1.182×10^{-19}	1
1 calorie	4.186	3.087	2.613×10^{19}
1 British thermal unit	1.055×10^{3}	7.779×10^{2}	6.585×10^{21}
1 kilowatt-hour	3.600×10^{6}	2.655×10^{6}	2.247×10^{25}

	cal	Btu	kWh
1 joule	0.238 9	9.481×10^{-4}	2.778×10^{-7}
1 foot-pound	0.323 9	1.285×10^{-3}	3.766×10^{-7}
1 electron volt	3.827×10^{-20}	1.519×10^{-22}	4.450×10^{-26}
1 calorie	1	3.968×10^{-3}	1.163×10^{-6}
1 British thermal unit	2.520×10^{2}	1	2.930×10^{-4}
1 kilowatt-hour	8.601×10^{5}	3.413×10^{2}	1

Pressure

	Pa	atm
1 pascal	1	9.869×10^{-6}
1 atmosphere	1.013×10^{5}	1
1 centimeter mercury[a]	1.333×10^{3}	1.316×10^{-2}
1 pound per square inch	6.895×10^{3}	6.805×10^{-2}
1 pound per square foot	47.88	4.725×10^{-4}

	cm Hg	lb/in.^{2}	lb/ft^2
1 pascal	7.501×10^{-4}	1.450×10^{-4}	2.089×10^{-2}
1 atmosphere	76	14.70	2.116×10^{3}
1 centimeter mercury[a]	1	0.194 3	27.85
1 pound per square inch	5.171	1	144
1 pound per square foot	3.591×10^{-2}	6.944×10^{-3}	1

[a]At 0°C and at a location where the free-fall acceleration has its "standard" value, 9.806 65 m/s^2.

TABLE A.2

Symbols, Dimensions, and Units of Physical Quantities

Quantity	Common Symbol	Unit[a]	Dimensions[b]	Unit in Terms of Base SI Units
Acceleration	$\vec{\mathbf{a}}$	m/s^2	L/T^2	m/s^2
Amount of substance	n	MOLE		mol
Angle	θ, ϕ	radian (rad)	1	
Angular acceleration	$\vec{\boldsymbol{\alpha}}$	rad/s^2	T^{-2}	s^{-2}
Angular frequency	ω	rad/s	T^{-1}	s^{-1}
Angular momentum	$\vec{\mathbf{L}}$	kg · m^2/s	ML^2/T	kg · m^2/s
Angular velocity	$\vec{\boldsymbol{\omega}}$	rad/s	T^{-1}	s^{-1}
Area	A	m^2	L^2	m^2
Atomic number	Z			
Capacitance	C	farad (F)	Q^{2}T^2/ML^2	A^2 · s^4/kg · m^2
Charge	q, Q, e	coulomb (C)	Q	A · s

(Continued)

TABLE A.2

Symbols, Dimensions, and Units of Physical Quantities *(Continued)*

Charge density				
Line	λ	C/m	Q/L	A · s/m
Surface	σ	C/m^2	Q/L^2	$A \cdot s/m^2$
Volume	ρ	C/m^3	Q/L^3	$A \cdot s/m^3$
Conductivity	σ	$1/\Omega \cdot m$	Q^2T/ML^3	$A^2 \cdot s^3/kg \cdot m^3$
Current	I	AMPERE	Q/T	A
Current density	J	A/m^2	Q/TL^2	A/m^2
Density	ρ	kg/m^3	M/L^3	kg/m^3
Dielectric constant	κ			
Electric dipole moment	$\vec{\mathbf{p}}$	C · m	QL	A · s · m
Electric field	$\vec{\mathbf{E}}$	V/m	ML/QT^2	$kg \cdot m/A \cdot s^3$
Electric flux	Φ_E	V · m	ML^3/QT^2	$kg \cdot m^3/A \cdot s^3$
Electromotive force	$\mathcal{E}$	volt (V)	ML^2/QT^2	$kg \cdot m^2/A \cdot s^3$
Energy	E, U, K	joule (J)	ML^2/T^2	$kg \cdot m^2/s^2$
Entropy	S	J/K	ML^2/T^2K	$kg \cdot m^2/s^2 \cdot K$
Force	$\vec{\mathbf{F}}$	newton (N)	ML/T^2	$kg \cdot m/s^2$
Frequency	f	hertz (Hz)	T^{-1}	s^{-1}
Heat	Q	joule (J)	ML^2/T^2	$kg \cdot m^2/s^2$
Inductance	L	henry (H)	ML^2/Q^2	$kg \cdot m^2/A^2 \cdot s^2$
Length	ℓ, L	METER	L	m
Displacement	$\Delta x, \Delta\vec{\mathbf{r}}$			
Distance	d, h			
Position	$x, y, z, \vec{\mathbf{r}}$			
Magnetic dipole moment	$\vec{\boldsymbol{\mu}}$	N · m/T	QL^2/T	$A \cdot m^2$
Magnetic field	$\vec{\mathbf{B}}$	tesla (T) (= Wb/m^2)	M/QT	$kg/A \cdot s^2$
Magnetic flux	Φ_B	weber (Wb)	ML^2/QT	$kg \cdot m^2/A \cdot s^2$
Mass	m, M	KILOGRAM	M	kg
Molar specific heat	C	J/mol · K		$kg \cdot m^2/s^2 \cdot mol \cdot K$
Moment of inertia	I	$kg \cdot m^2$	ML^2	$kg \cdot m^2$
Momentum	$\vec{\mathbf{p}}$	kg · m/s	ML/T	kg · m/s
Period	T	s	T	s
Permeability of free space	μ_0	N/A^2 (= H/m)	ML/Q^2	$kg \cdot m/A^2 \cdot s^2$
Permittivity of free space	ϵ_0	$C^2/N \cdot m^2$ (= F/m)	Q^2T^2/ML^3	$A^2 \cdot s^4/kg \cdot m^3$
Potential	V	volt (V) (= J/C)	ML^2/QT^2	$kg \cdot m^2/A \cdot s^3$
Power	$\mathcal{P}$	watt (W) (= J/s)	ML^2/T^3	$kg \cdot m^2/s^3$
Pressure	P	pascal (Pa) (= N/m^2)	M/LT^2	$kg/m \cdot s^2$
Resistance	R	ohm (Ω) (= V/A)	ML^2/Q^2T	$kg \cdot m^2/A^2 \cdot s^3$
Specific heat	c	J/kg · K	L^2/T^2K	$m^2/s^2 \cdot K$
Speed	v	m/s	L/T	m/s
Temperature	T	KELVIN	K	K
Time	t	SECOND	T	s
Torque	$\vec{\boldsymbol{\tau}}$	N · m	ML^2/T^2	$kg \cdot m^2/s^2$
Velocity	$\vec{\mathbf{v}}$	m/s	L/T	m/s
Volume	V	m^3	L^3	m^3
Wavelength	λ	m	L	m
Work	W	joule (J) (= N · m)	ML^2/T^2	$kg \cdot m^2/s^2$

[a]The base SI units are given in uppercase letters.

[b]The symbols M, L, T, K, and Q denote mass, length, time, temperature, and charge, respectively.

This appendix in mathematics is intended as a brief review of operations and methods. Early in this course, you should be totally familiar with basic algebraic techniques, analytic geometry, and trigonometry. The sections on differential and integral calculus are more detailed and are intended for students who have difficulty applying calculus concepts to physical situations.

B.1 Scientific Notation

Many quantities used by scientists often have very large or very small values. The speed of light, for example, is about 300 000 000 m/s, and the ink required to make the dot over an *i* in this textbook has a mass of about 0.000 000 001 kg. Obviously, it is very cumbersome to read, write, and keep track of such numbers. We avoid this problem by using a method incorporating powers of the number 10:

$$10^0 = 1$$

$$10^1 = 10$$

$$10^2 = 10 \times 10 = 100$$

$$10^3 = 10 \times 10 \times 10 = 1\ 000$$

$$10^4 = 10 \times 10 \times 10 \times 10 = 10\ 000$$

$$10^5 = 10 \times 10 \times 10 \times 10 \times 10 = 100\ 000$$

and so on. The number of zeros corresponds to the power to which ten is raised, called the **exponent** of ten. For example, the speed of light, 300 000 000 m/s, can be expressed as 3.00×10^8 m/s.

In this method, some representative numbers smaller than unity are the following:

$$10^{-1} = \frac{1}{10} = 0.1$$

$$10^{-2} = \frac{1}{10 \times 10} = 0.01$$

$$10^{-3} = \frac{1}{10 \times 10 \times 10} = 0.001$$

$$10^{-4} = \frac{1}{10 \times 10 \times 10 \times 10} = 0.000\ 1$$

$$10^{-5} = \frac{1}{10 \times 10 \times 10 \times 10 \times 10} = 0.000\ 01$$

In these cases, the number of places the decimal point is to the left of the digit 1 equals the value of the (negative) exponent. Numbers expressed as some power of ten multiplied by another number between one and ten are said to be in **scientific notation.** For example, the scientific notation for 5 943 000 000 is 5.943×10^9 and that for 0.000 083 2 is 8.32×10^{-5}.

When numbers expressed in scientific notation are being multiplied, the following general rule is very useful:

$$10^n \times 10^m = 10^{n+m} \qquad \textbf{(B.1)}$$

where n and m can be *any* numbers (not necessarily integers). For example, $10^2 \times 10^5 = 10^7$. The rule also applies if one of the exponents is negative: $10^3 \times 10^{-8} = 10^{-5}$.

When dividing numbers expressed in scientific notation, note that

$$\frac{10^n}{10^m} = 10^n \times 10^{-m} = 10^{n-m} \tag{B.2}$$

Exercises

With help from the preceding rules, verify the answers to the following equations:

1. $86\,400 = 8.64 \times 10^4$
2. $9\,816\,762.5 = 9.816\,762\,5 \times 10^6$
3. $0.000\,000\,039\,8 = 3.98 \times 10^{-8}$
4. $(4.0 \times 10^8)(9.0 \times 10^9) = 3.6 \times 10^{18}$
5. $(3.0 \times 10^7)(6.0 \times 10^{-12}) = 1.8 \times 10^{-4}$
6. $\dfrac{75 \times 10^{-11}}{5.0 \times 10^{-3}} = 1.5 \times 10^{-7}$
7. $\dfrac{(3 \times 10^6)(8 \times 10^{-2})}{(2 \times 10^{17})(6 \times 10^5)} = 2 \times 10^{-18}$

B.2 Algebra

Some Basic Rules

When algebraic operations are performed, the laws of arithmetic apply. Symbols such as x, y, and z are usually used to represent unspecified quantities, called the **unknowns.**

First, consider the equation

$$8x = 32$$

If we wish to solve for x, we can divide (or multiply) each side of the equation by the same factor without destroying the equality. In this case, if we divide both sides by 8, we have

$$\frac{8x}{8} = \frac{32}{8}$$

$$x = 4$$

Next consider the equation

$$x + 2 = 8$$

In this type of expression, we can add or subtract the same quantity from each side. If we subtract 2 from each side, we have

$$x + 2 - 2 = 8 - 2$$

$$x = 6$$

In general, if $x + a = b$, then $x = b - a$.

Now consider the equation

$$\frac{x}{5} = 9$$

If we multiply each side by 5, we are left with x on the left by itself and 45 on the right:

$$\left(\frac{x}{5}\right)(5) = 9 \times 5$$

$$x = 45$$

In all cases, *whatever operation is performed on the left side of the equality must also be performed on the right side.*

The following rules for multiplying, dividing, adding, and subtracting fractions should be recalled, where a, b, c, and d are four numbers:

	Rule	Example
Multiplying	$\left(\frac{a}{b}\right)\left(\frac{c}{d}\right) = \frac{ac}{bd}$	$\left(\frac{2}{3}\right)\left(\frac{4}{5}\right) = \frac{8}{15}$
Dividing	$\frac{(a/b)}{(c/d)} = \frac{ad}{bc}$	$\frac{2/3}{4/5} = \frac{(2)(5)}{(4)(3)} = \frac{10}{12}$
Adding	$\frac{a}{b} \pm \frac{c}{d} = \frac{ad \pm bc}{bd}$	$\frac{2}{3} - \frac{4}{5} = \frac{(2)(5) - (4)(3)}{(3)(5)} = -\frac{2}{15}$

Exercises

In the following exercises, solve for x.

	Answers
1. $a = \frac{1}{1 + x}$	$x = \frac{1 - a}{a}$
2. $3x - 5 = 13$	$x = 6$
3. $ax - 5 = bx + 2$	$x = \frac{7}{a - b}$
4. $\frac{5}{2x + 6} = \frac{3}{4x + 8}$	$x = -\frac{11}{7}$

Powers

When powers of a given quantity x are multiplied, the following rule applies:

$$x^n x^m = x^{n+m} \quad \textbf{(B.3)}$$

For example, $x^2x^4 = x^{2+4} = x^6$.

When dividing the powers of a given quantity, the rule is

$$\frac{x^n}{x^m} = x^{n-m} \quad \textbf{(B.4)}$$

For example, $x^8/x^2 = x^{8-2} = x^6$.

A power that is a fraction, such as $\frac{1}{3}$, corresponds to a root as follows:

$$x^{1/n} = \sqrt[n]{x} \quad \textbf{(B.5)}$$

For example, $4^{1/3} = \sqrt[3]{4} = 1.587\,4$. (A scientific calculator is useful for such calculations.)

Finally, any quantity x^n raised to the mth power is

$$(x^n)^m = x^{nm} \quad \textbf{(B.6)}$$

Table B.1 summarizes the rules of exponents.

TABLE B.1

Rules of Exponents

$x^0 = 1$

$x^1 = x$

$x^nx^m = x^{n+m}$

$x^n/x^m = x^{n-m}$

$x^{1/n} = \sqrt[n]{x}$

$(x^n)^m = x^{nm}$

Exercises

Verify the following equations:

1. $3^2 \times 3^3 = 243$
2. $x^5x^{-8} = x^{-3}$
3. $x^{10}/x^{-5} = x^{15}$
4. $5^{1/3} = 1.709\,976$ (Use your calculator.)
5. $60^{1/4} = 2.783\,158$ (Use your calculator.)
6. $(x^4)^3 = x^{12}$

Factoring

Some useful formulas for factoring an equation are the following:

$ax + ay + az = a(x + y + z)$ common factor

$a^2 + 2ab + b^2 = (a + b)^2$ perfect square

$a^2 - b^2 = (a + b)(a - b)$ differences of squares

Quadratic Equations

The general form of a quadratic equation is

$$ax^2 + bx + c = 0 \quad \textbf{(B.7)}$$

where x is the unknown quantity and a, b, and c are numerical factors referred to as **coefficients** of the equation. This equation has two roots, given by

$$x = \frac{-b \pm \sqrt{b^2 - 4ac}}{2a} \quad \textbf{(B.8)}$$

If $b^2 \geq 4ac$, the roots are real.

EXAMPLE B.1

The equation $x^2 + 5x + 4 = 0$ has the following roots corresponding to the two signs of the square-root term:

$$x = \frac{-5 \pm \sqrt{5^2 - (4)(1)(4)}}{2(1)} = \frac{-5 \pm \sqrt{9}}{2} = \frac{-5 \pm 3}{2}$$

$$x_+ = \frac{-5 + 3}{2} = -1 \qquad x_- = \frac{-5 - 3}{2} = -4$$

where x_+ refers to the root corresponding to the positive sign and x_- refers to the root corresponding to the negative sign.

Exercises

Solve the following quadratic equations:

		Answers	
1.	$x^2 + 2x - 3 = 0$	$x_+ = 1$	$x_- = -3$
2.	$2x^2 - 5x + 2 = 0$	$x_+ = 2$	$x_- = \frac{1}{2}$
3.	$2x^2 - 4x - 9 = 0$	$x_+ = 1 + \sqrt{22}/2$	$x_- = 1 - \sqrt{22}/2$

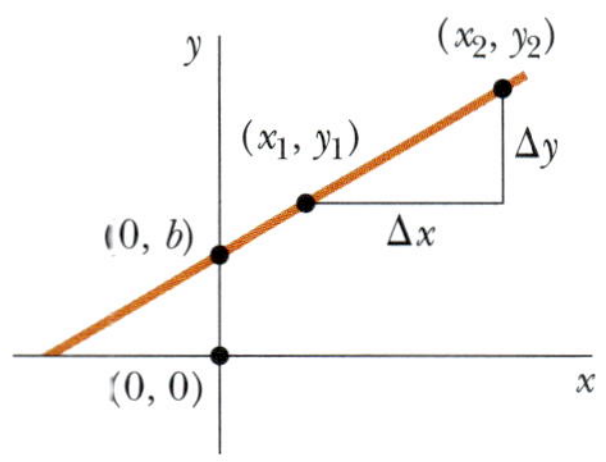

Figure B.1 A straight line graphed on an xy coordinate system. The slope of the line is the ratio of Δy to Δx.

Linear Equations

A linear equation has the general form

$$y = mx + b \quad \textbf{(B.9)}$$

where m and b are constants. This equation is referred to as linear because the graph of y versus x is a straight line as shown in Figure B.1. The constant b, called the **y-intercept,** represents the value of y at which the straight line intersects the y axis. The constant m is equal to the **slope** of the straight line. If any two points on the straight line are specified by the coordinates (x_1, y_1) and (x_2, y_2) as in Figure B.1, the slope of the straight line can be expressed as

$$\text{Slope} = \frac{y_2 - y_1}{x_2 - x_1} = \frac{\Delta y}{\Delta x} \quad \textbf{(B.10)}$$

Note that m and b can have either positive or negative values. If $m > 0$, the straight line has a *positive* slope as in Figure B.1. If $m < 0$, the straight line has a *negative* slope. In Figure B.1, both m and b are positive. Three other possible situations are shown in Figure B.2.

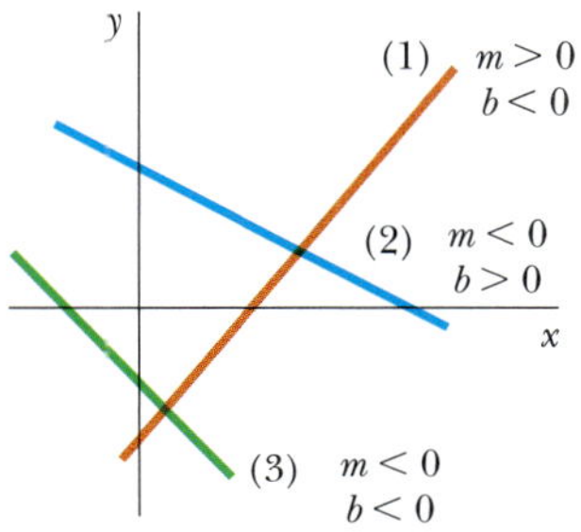

Figure B.2 The brown line has a positive slope and a negative y-intercept. The blue line has a negative slope and a positive y-intercept. The green line has a negative slope and a negative y-intercept.

Exercises

1. Draw graphs of the following straight lines: (a) $y = 5x + 3$ (b) $y = -2x + 4$ (c) $y = -3x - 6$
2. Find the slopes of the straight lines described in Exercise 1.

Answers (a) 5 (b) -2 (c) -3

3. Find the slopes of the straight lines that pass through the following sets of points: (a) (0, −4) and (4, 2) (b) (0, 0) and (2, −5) (c) (−5, 2) and (4, −2)

Answers (a) $\frac{3}{2}$ (b) $-\frac{5}{2}$ (c) $-\frac{4}{9}$

Solving Simultaneous Linear Equations

Consider the equation $3x + 5y = 15$, which has two unknowns, x and y. Such an equation does not have a unique solution. For example, ($x = 0$, $y = 3$), ($x = 5$, $y = 0$), and ($x = 2$, $y = \frac{9}{5}$) are all solutions to this equation.

If a problem has two unknowns, a unique solution is possible only if we have *two* equations. In general, if a problem has n unknowns, its solution requires n equations. To solve two simultaneous equations involving two unknowns, x and y, we solve one of the equations for x in terms of y and substitute this expression into the other equation.

EXAMPLE B.2

Solve the two simultaneous equations

$$(1) \qquad 5x + y = -8$$

$$(2) \qquad 2x - 2y = 4$$

Solution From Equation (2), $x = y + 2$. Substitution of this equation into Equation (1) gives

$$5(y + 2) + y = -8$$

$$6y = -18$$

$$y = -3$$

$$x = y + 2 = -1$$

Alternative Solution Multiply each term in Equation (1) by the factor 2 and add the result to Equation (2):

$$10x + 2y = -16$$

$$\underline{2x - 2y = 4}$$

$$12x \qquad = -12$$

$$x = -1$$

$$y = x - 2 = -3$$

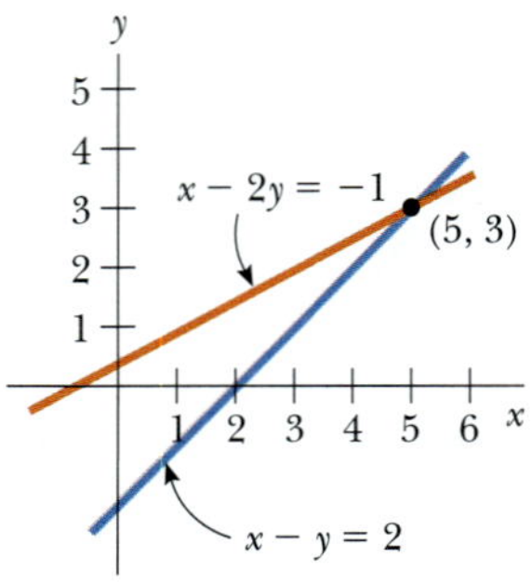

Figure B.3 A graphical solution for two linear equations.

Two linear equations containing two unknowns can also be solved by a graphical method. If the straight lines corresponding to the two equations are plotted in a conventional coordinate system, the intersection of the two lines represents the solution. For example, consider the two equations

$$x - y = 2$$

$$x - 2y = -1$$

These equations are plotted in Figure B.3. The intersection of the two lines has the coordinates $x = 5$ and $y = 3$, which represents the solution to the equations. You should check this solution by the analytical technique discussed earlier.

Exercises

Solve the following pairs of simultaneous equations involving two unknowns:

		Answers
1.	$x + y = 8$ $x - y = 2$	$x = 5,\ y = 3$
2.	$98 - T = 10a$ $T - 49 = 5a$	$T = 65,\ a = 3.27$
3.	$6x + 2y = 6$ $8x - 4y = 28$	$x = 2,\ y = -3$

Logarithms

Suppose a quantity x is expressed as a power of some quantity a:

$$x = a^y \tag{B.11}$$

The number a is called the **base** number. The **logarithm** of x with respect to the base a is equal to the exponent to which the base must be raised to satisfy the expression $x = a^y$:

$$y = \log_a x \tag{B.12}$$

Conversely, the **antilogarithm** of y is the number x:

$$x = \text{antilog}_a\, y \tag{B.13}$$

In practice, the two bases most often used are base 10, called the *common* logarithm base, and base $e = 2.718\,282$, called Euler's constant or the *natural* logarithm base. When common logarithms are used,

$$y = \log_{10} x \quad (\text{or } x = 10^y) \tag{B.14}$$

When natural logarithms are used,

$$y = \ln x \quad (\text{or } x = e^y) \tag{B.15}$$

For example, $\log_{10} 52 = 1.716$, so $\text{antilog}_{10}\, 1.716 = 10^{1.716} = 52$. Likewise, $\ln 52 = 3.951$, so $\text{antiln}\, 3.951 = e^{3.951} = 52$.

In general, note you can convert between base 10 and base e with the equality

$$\ln x = (2.302\,585) \log_{10} x \tag{B.16}$$

Finally, some useful properties of logarithms are the following:

$$\left.\begin{aligned} \log(ab) &= \log a + \log b \\ \log(a/b) &= \log a - \log b \\ \log(a^n) &= n \log a \end{aligned}\right\} \text{any base}$$

$$\ln e = 1$$

$$\ln e^a = a$$

$$\ln\left(\frac{1}{a}\right) = -\ln a$$

B.3 Geometry

The **distance** d between two points having coordinates (x_1, y_1) and (x_2, y_2) is

$$d = \sqrt{(x_2 - x_1)^2 + (y_2 - y_1)^2} \tag{B.17}$$

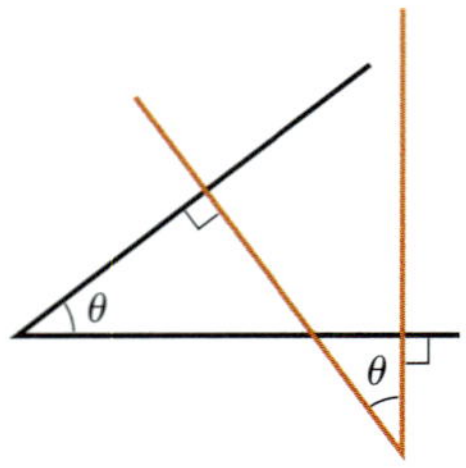

Figure B.4 The angles are equal because their sides are perpendicular.

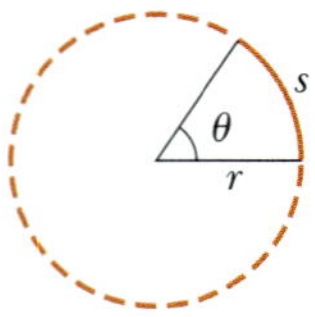

Figure B.5 The angle θ in radians is the ratio of the arc length s to the radius r of the circle.

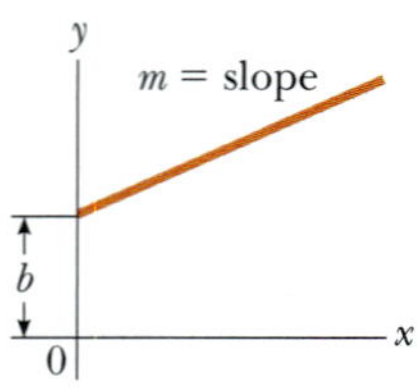

Figure B.6 A straight line with a slope of m and a y-intercept of b.

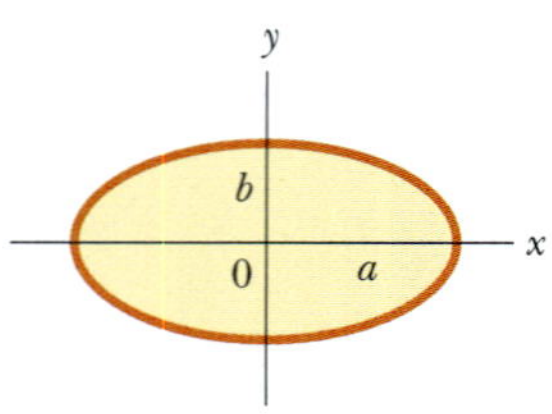

Figure B.7 An ellipse with semimajor axis a and semiminor axis b.

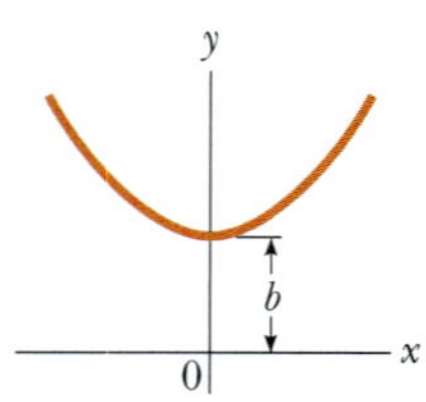

Figure B.8 A parabola with its vertex at $y = b$.

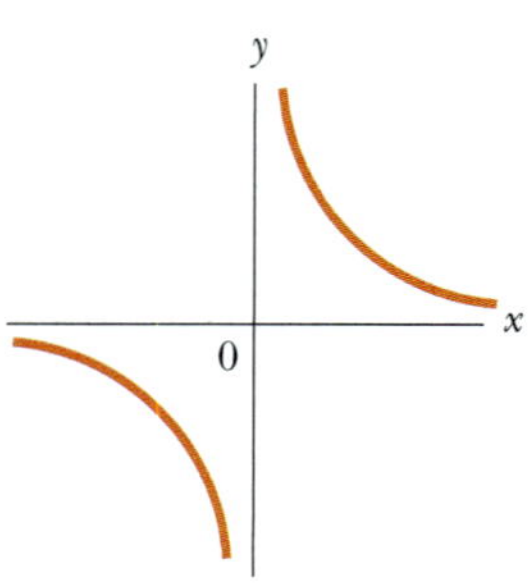

Figure B.9 A hyperbola.

TABLE B.2

Useful Information for Geometry

Shape	Area or Volume	Shape	Area or Volume
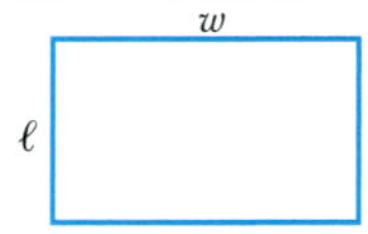 Rectangle	Area $= \ell w$	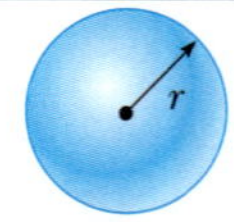 Sphere	Surface area $= 4\pi r^2$ Volume $= \frac{4\pi r^3}{3}$
 Circle	Area $= \pi r^2$ Circumference $= 2\pi r$	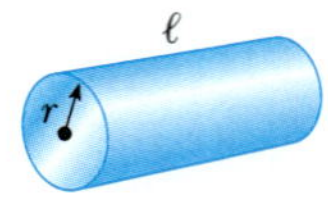 Cylinder	Lateral surface area $= 2\pi r\ell$ Volume $= \pi r^2 \ell$
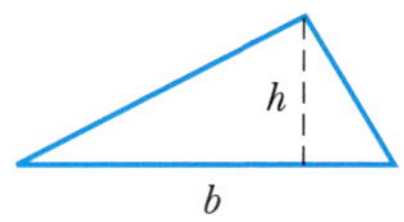 Triangle	Area $= \frac{1}{2}bh$	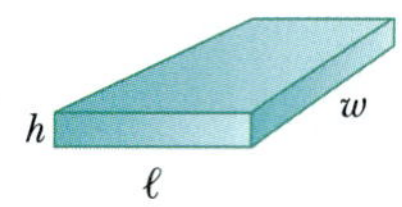 Rectangular box	Surface area $= 2(\ell h + \ell w + hw)$ Volume $= \ell wh$

Two angles are equal if their sides are perpendicular, right side to right side and left side to left side. For example, the two angles marked θ in Figure B.4 are the same because of the perpendicularity of the sides of the angles. To distinguish the left and right sides of an angle, imagine standing at the angle's apex and facing into the angle.

Radian measure: The arc length s of a circular arc (Fig. B.5) is proportional to the radius r for a fixed value of θ (in radians):

$$s = r\theta \qquad \theta = \frac{s}{r} \tag{B.18}$$

Table B.2 gives the **areas** and **volumes** for several geometric shapes used throughout this text.

The equation of a **straight line** (Fig. B.6) is

$$y = mx + b \tag{B.19}$$

where b is the y-intercept and m is the slope of the line.

The equation of a **circle** of radius R centered at the origin is

$$x^2 + y^2 = R^2 \tag{B.20}$$

The equation of an **ellipse** having the origin at its center (Fig. B.7) is

$$\frac{x^2}{a^2} + \frac{y^2}{b^2} = 1 \tag{B.21}$$

where a is the length of the semimajor axis (the longer one) and b is the length of the semiminor axis (the shorter one).

The equation of a **parabola** the vertex of which is at $y = b$ (Fig. B.8) is

$$y = ax^2 + b \tag{B.22}$$

The equation of a **rectangular hyperbola** (Fig. B.9) is

$$xy = \text{constant} \tag{B.23}$$

B.4 Trigonometry

That portion of mathematics based on the special properties of the right triangle is called trigonometry. By definition, a right triangle is a triangle containing a 90° angle. Consider the right triangle shown in Figure B.10, where side a is opposite the angle θ, side b is adjacent to the angle θ, and side c is the hypotenuse of the triangle. The three basic trigonometric functions defined by such a triangle are the sine (sin), cosine (cos), and tangent (tan). In terms of the angle θ, these functions are defined as follows:

$$\sin\theta = \frac{\text{side opposite } \theta}{\text{hypotenuse}} = \frac{a}{c} \quad \textbf{(B.24)}$$

$$\cos\theta = \frac{\text{side adjacent to } \theta}{\text{hypotenuse}} = \frac{b}{c} \quad \textbf{(B.25)}$$

$$\tan\theta = \frac{\text{side opposite } \theta}{\text{side adjacent to } \theta} = \frac{a}{b} \quad \textbf{(B.26)}$$

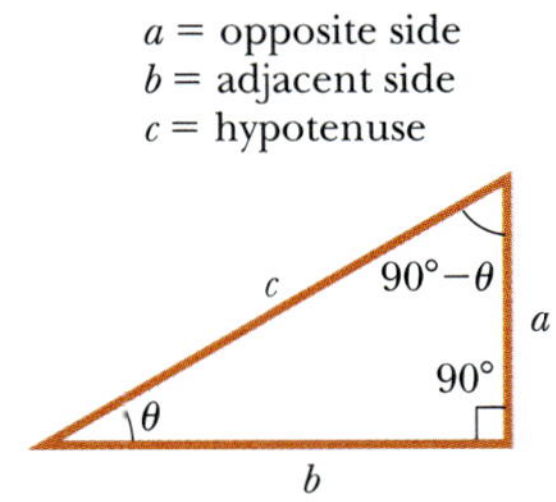

Figure B.10 A right triangle, used to define the basic functions of trigonometry.

The Pythagorean theorem provides the following relationship among the sides of a right triangle:

$$c^2 = a^2 + b^2 \quad \textbf{(B.27)}$$

From the preceding definitions and the Pythagorean theorem, it follows that

$$\sin^2\theta + \cos^2\theta = 1$$

$$\tan\theta = \frac{\sin\theta}{\cos\theta}$$

The cosecant, secant, and cotangent functions are defined by

$$\csc\theta = \frac{1}{\sin\theta} \qquad \sec\theta = \frac{1}{\cos\theta} \qquad \cot\theta = \frac{1}{\tan\theta}$$

The following relationships are derived directly from the right triangle shown in Figure B.10:

$$\sin\theta = \cos(90° - \theta)$$

$$\cos\theta = \sin(90° - \theta)$$

$$\cot\theta = \tan(90° - \theta)$$

Some properties of trigonometric functions are the following:

$$\sin(-\theta) = -\sin\theta$$

$$\cos(-\theta) = \cos\theta$$

$$\tan(-\theta) = -\tan\theta$$

The following relationships apply to *any* triangle as shown in Figure B.11:

$$\alpha + \beta + \gamma = 180°$$

$$\text{Law of cosines} \begin{cases} a^2 = b^2 + c^2 - 2bc\cos\alpha \\ b^2 = a^2 + c^2 - 2ac\cos\beta \\ c^2 = a^2 + b^2 - 2ab\cos\gamma \end{cases}$$

$$\text{Law of sines} \qquad \frac{a}{\sin\alpha} = \frac{b}{\sin\beta} = \frac{c}{\sin\gamma}$$

Table B.3 (page A-12) lists a number of useful trigonometric identities.

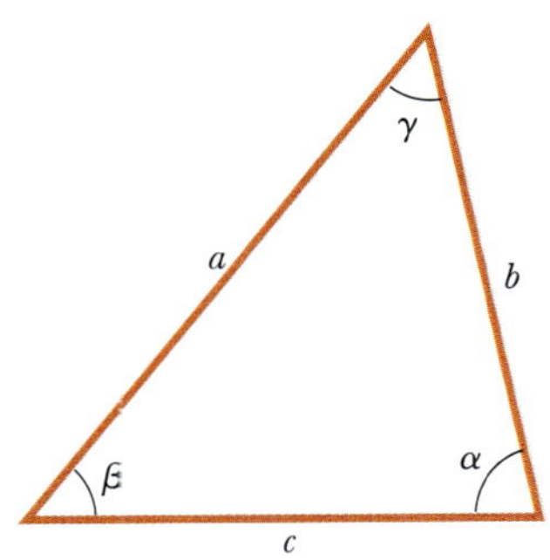

Figure B.11 An arbitrary, nonright triangle.

EXAMPLE B.3

Consider the right triangle in Figure B.12 in which $a = 2.00$, $b = 5.00$, and c is unknown. From the Pythagorean theorem, we have

$$c^2 = a^2 + b^2 = 2.00^2 + 5.00^2 = 4.00 + 25.0 = 29.0$$

$$c = \sqrt{29.0} = 5.39$$

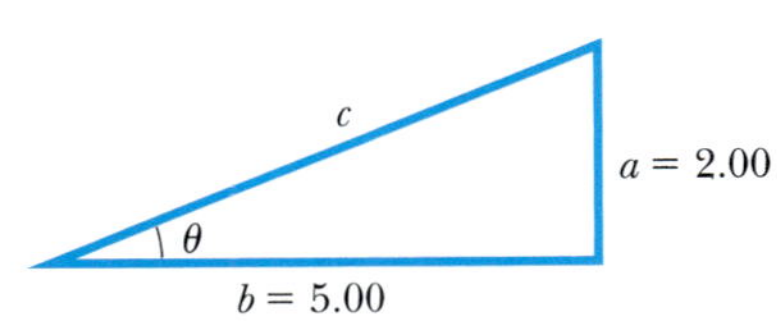

Figure B.12 (Example B.3)

To find the angle θ, note that

$$\tan \theta = \frac{a}{b} = \frac{2.00}{5.00} = 0.400$$

Using a calculator, we find that

$$\theta = \tan^{-1}(0.400) = 21.8°$$

where $\tan^{-1}(0.400)$ is the notation for "angle whose tangent is 0.400," sometimes written as arctan (0.400).

TABLE B.3

Some Trigonometric Identities

$\sin^2\theta + \cos^2\theta = 1$	$\csc^2\theta = 1 + \cot^2\theta$
$\sec^2\theta = 1 + \tan^2\theta$	$\sin^2\frac{\theta}{2} = \frac{1}{2}(1 - \cos\theta)$
$\sin 2\theta = 2\sin\theta\cos\theta$	$\cos^2\frac{\theta}{2} = \frac{1}{2}(1 + \cos\theta)$
$\cos 2\theta = \cos^2\theta - \sin^2\theta$	$1 - \cos\theta = 2\sin^2\frac{\theta}{2}$
$\tan 2\theta = \frac{2\tan\theta}{1 - \tan^2\theta}$	$\tan\frac{\theta}{2} = \sqrt{\frac{1 - \cos\theta}{1 + \cos\theta}}$

$$\sin(A \pm B) = \sin A\cos B \pm \cos A\sin B$$
$$\cos(A \pm B) = \cos A\cos B \mp \sin A\sin B$$
$$\sin A \pm \sin B = 2\sin\left[\tfrac{1}{2}(A \pm B)\right]\cos\left[\tfrac{1}{2}(A \mp B)\right]$$
$$\cos A + \cos B = 2\cos\left[\tfrac{1}{2}(A + B)\right]\cos\left[\tfrac{1}{2}(A - B)\right]$$
$$\cos A - \cos B = 2\sin\left[\tfrac{1}{2}(A + B)\right]\sin\left[\tfrac{1}{2}(B - A)\right]$$

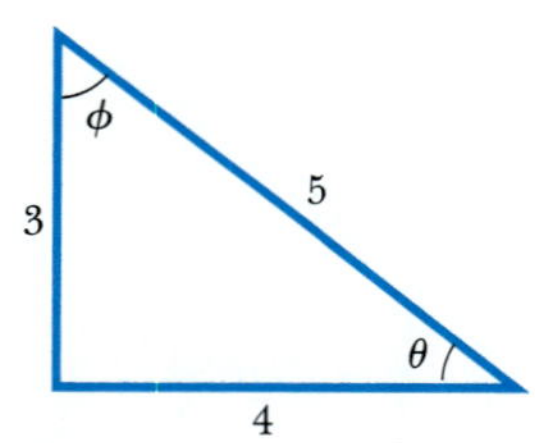

Figure B.13 (Exercise 1)

Exercises

1. In Figure B.13, identify (a) the side opposite θ (b) the side adjacent to ϕ and then find (c) $\cos\theta$, (d) $\sin\phi$, and (e) $\tan\phi$.

Answers (a) 3 (b) 3 (c) $\frac{4}{5}$ (d) $\frac{4}{5}$ (e) $\frac{4}{3}$

2. In a certain right triangle, the two sides that are perpendicular to each other are 5.00 m and 7.00 m long. What is the length of the third side?

Answer 8.60 m

3. A right triangle has a hypotenuse of length 3.0 m, and one of its angles is 30°. (a) What is the length of the side opposite the 30° angle? (b) What is the side adjacent to the 30° angle?

Answers (a) 1.5 m (b) 2.6 m

B.5 Series Expansions

$$(a + b)^n = a^n + \frac{n}{1!}a^{n-1}b + \frac{n(n-1)}{2!}a^{n-2}b^2 + \cdots$$

$$(1 + x)^n = 1 + nx + \frac{n(n-1)}{2!}x^2 + \cdots$$

$$e^x = 1 + x + \frac{x^2}{2!} + \frac{x^3}{3!} + \cdots$$

$$\ln(1 \pm x) = \pm x - \tfrac{1}{2}x^2 \pm \tfrac{1}{3}x^3 - \cdots$$

$$\left.\begin{aligned} \sin x &= x - \frac{x^3}{3!} + \frac{x^5}{5!} - \cdots \\ \cos x &= 1 - \frac{x^2}{2!} + \frac{x^4}{4!} - \cdots \\ \tan x &= x + \frac{x^3}{3} + \frac{2x^5}{15} + \cdots \quad |x| < \frac{\pi}{2} \end{aligned}\right\} x \text{ in radians}$$

For $x << 1$, the following approximations can be used:[1]

$$(1 + x)^n \approx 1 + nx \qquad \sin x \approx x$$

$$e^x \approx 1 + x \qquad \cos x \approx 1$$

$$\ln(1 \pm x) \approx \pm x \qquad \tan x \approx x$$

B.6 Differential Calculus

In various branches of science, it is sometimes necessary to use the basic tools of calculus, invented by Newton, to describe physical phenomena. The use of calculus is fundamental in the treatment of various problems in Newtonian mechanics, electricity, and magnetism. In this section, we simply state some basic properties and "rules of thumb" that should be a useful review to the student.

First, a **function** must be specified that relates one variable to another (e.g., a coordinate as a function of time). Suppose one of the variables is called y (the dependent variable), and the other x (the independent variable). We might have a function relationship such as

$$y(x) = ax^3 + bx^2 + cx + d$$

If a, b, c, and d are specified constants, y can be calculated for any value of x. We usually deal with continuous functions, that is, those for which y varies "smoothly" with x.

The **derivative** of y with respect to x is defined as the limit as Δx approaches zero of the slopes of chords drawn between two points on the y versus x curve. Mathematically, we write this definition as

$$\frac{dy}{dx} = \lim_{\Delta x \to 0} \frac{\Delta y}{\Delta x} = \lim_{\Delta x \to 0} \frac{y(x + \Delta x) - y(x)}{\Delta x} \tag{B.28}$$

where Δy and Δx are defined as $\Delta x = x_2 - x_1$ and $\Delta y = y_2 - y_1$ (Fig. B.14). Note that dy/dx *does not* mean dy divided by dx, but rather is simply a notation of the limiting process of the derivative as defined by Equation B.28.

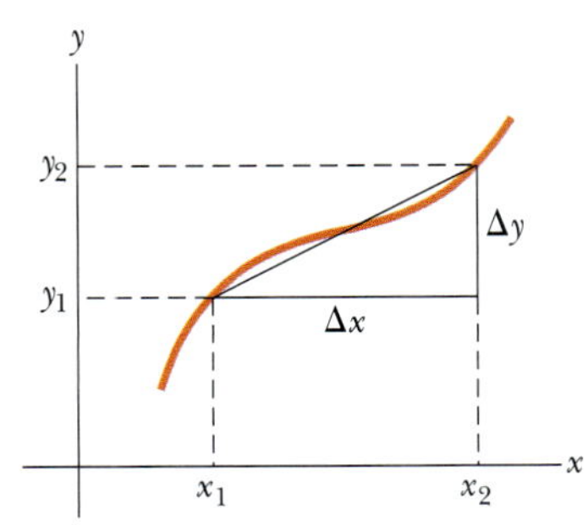

Figure B.14 The lengths Δx and Δy are used to define the derivative of this function at a point.

A useful expression to remember when $y(x) = ax^n$, where a is a *constant* and n is *any* positive or negative number (integer or fraction), is

$$\frac{dy}{dx} = nax^{n-1} \tag{B.29}$$

If $y(x)$ is a polynomial or algebraic function of x, we apply Equation B.29 to *each* term in the polynomial and take $d[\text{constant}]/dx = 0$. In Examples B.4 through B.7, we evaluate the derivatives of several functions.

[1]The approximations for the functions sin x, cos x, and tan x are for $x \leq 0.1$ rad.

TABLE B.4

Derivative for Several Functions

$$\frac{d}{dx}(a) = 0$$

$$\frac{d}{dx}(ax^n) = nax^{n-1}$$

$$\frac{d}{dx}(e^{ax}) = ae^{ax}$$

$$\frac{d}{dx}(\sin ax) = a\cos ax$$

$$\frac{d}{dx}(\cos ax) = -a\sin ax$$

$$\frac{d}{dx}(\tan ax) = a\sec^2 ax$$

$$\frac{d}{dx}(\cot ax) = -a\csc^2 ax$$

$$\frac{d}{dx}(\sec x) = \tan x\sec x$$

$$\frac{d}{dx}(\csc x) = -\cot x\csc x$$

$$\frac{d}{dx}(\ln ax) = \frac{1}{x}$$

$$\frac{d}{dx}(\sin^{-1} ax) = \frac{a}{\sqrt{1-a^2x^2}}$$

$$\frac{d}{dx}(\cos^{-1} ax) = \frac{-a}{\sqrt{1-a^2x^2}}$$

$$\frac{d}{dx}(\tan^{-1} ax) = \frac{a}{1+a^2x^2}$$

Note: The symbols a and n represent constants.

Special Properties of the Derivative

A. Derivative of the product of two functions If a function $f(x)$ is given by the product of two functions—say, $g(x)$ and $h(x)$—the derivative of $f(x)$ is defined as

$$\frac{d}{dx}f(x) = \frac{d}{dx}[g(x)h(x)] = g\frac{dh}{dx} + h\frac{dg}{dx} \qquad \text{(B.30)}$$

B. Derivative of the sum of two functions If a function $f(x)$ is equal to the sum of two functions, the derivative of the sum is equal to the sum of the derivatives:

$$\frac{d}{dx}f(x) = \frac{d}{dx}[g(x) + h(x)] = \frac{dg}{dx} + \frac{dh}{dx} \qquad \text{(B.31)}$$

C. Chain rule of differential calculus If $y = f(x)$ and $x = g(z)$, then dy/dz can be written as the product of two derivatives:

$$\frac{dy}{dz} = \frac{dy}{dx}\frac{dx}{dz} \qquad \text{(B.32)}$$

D. The second derivative The second derivative of y with respect to x is defined as the derivative of the function dy/dx (the derivative of the derivative). It is usually written as

$$\frac{d^2y}{dx^2} = \frac{d}{dx}\left(\frac{dy}{dx}\right) \qquad \text{(B.33)}$$

Some of the more commonly used derivatives of functions are listed in Table B.4.

EXAMPLE B.4

Suppose $y(x)$ (that is, y as a function of x) is given by

$$y(x) = ax^3 + bx + c$$

where a and b are constants. It follows that

$$y(x + \Delta x) = a(x + \Delta x)^3 + b(x + \Delta x) + c$$
$$= a(x^3 + 3x^2\,\Delta x + 3x\,\Delta x^2 + \Delta x^3) + b(x + \Delta x) + c$$

so

$$\Delta y = y(x + \Delta x) - y(x) = a(3x^2\,\Delta x + 3x\,\Delta x^2 + \Delta x^3) + b\,\Delta x$$

Substituting this into Equation B.28 gives

$$\frac{dy}{dx} = \lim_{\Delta x\to 0}\frac{\Delta y}{\Delta x} = \lim_{\Delta x\to 0}[3ax^2 + 3ax\,\Delta x + a\,\Delta x^2] + b$$

$$\frac{dy}{dx} = 3ax^2 + b$$

EXAMPLE B.5

Find the derivative of

$$y(x) = 8x^5 + 4x^3 + 2x + 7$$

Solution Applying Equation B.29 to each term independently and remembering that d/dx (constant) $= 0$, we have

$$\frac{dy}{dx} = 8(5)x^4 + 4(3)x^2 + 2(1)x^0 + 0$$

$$\frac{dy}{dx} = 40x^4 + 12x^2 + 2$$

EXAMPLE B.6

Find the derivative of $y(x) = x^3/(x+1)^2$ with respect to x.

Solution We can rewrite this function as $y(x) = x^3(x+1)^{-2}$ and apply Equation B.30:

$$\frac{dy}{dx} = (x+1)^{-2}\frac{d}{dx}(x^3) + x^3\frac{d}{dx}(x+1)^{-2}$$

$$= (x+1)^{-2}3x^2 + x^3(-2)(x+1)^{-3}$$

$$\frac{dy}{dx} = \frac{3x^2}{(x+1)^2} - \frac{2x^3}{(x+1)^3}$$

EXAMPLE B.7

A useful formula that follows from Equation B.30 is the derivative of the quotient of two functions. Show that

$$\frac{d}{dx}\left[\frac{g(x)}{h(x)}\right] = \frac{h\dfrac{dg}{dx} - g\dfrac{dh}{dx}}{h^2}$$

Solution We can write the quotient as gh^{-1} and then apply Equations B.29 and B.30:

$$\frac{d}{dx}\left(\frac{g}{h}\right) = \frac{d}{dx}(gh^{-1}) = g\frac{d}{dx}(h^{-1}) + h^{-1}\frac{d}{dx}(g)$$

$$= -gh^{-2}\frac{dh}{dx} + h^{-1}\frac{dg}{dx}$$

$$= \frac{h\dfrac{dg}{dx} - g\dfrac{dh}{dx}}{h^2}$$

B.7 Integral Calculus

We think of integration as the inverse of differentiation. As an example, consider the expression

$$f(x) = \frac{dy}{dx} = 3ax^2 + b \tag{B.34}$$

which was the result of differentiating the function

$$y(x) = ax^3 + bx + c$$

in Example B.4. We can write Equation B.34 as $dy = f(x)\ dx = (3ax^2 + b)\ dx$ and obtain $y(x)$ by "summing" over all values of x. Mathematically, we write this inverse operation as

$$y(x) = \int f(x)\ dx$$

For the function $f(x)$ given by Equation B.34, we have

$$y(x) = \int (3ax^2 + b)\ dx = ax^3 + bx + c$$

where c is a constant of the integration. This type of integral is called an *indefinite integral* because its value depends on the choice of c.

A general **indefinite integral** $I(x)$ is defined as

$$I(x) = \int f(x)\ dx \tag{B.35}$$

where $f(x)$ is called the *integrand* and $f(x) = dI(x)/dx$.

For a *general continuous* function $f(x)$, the integral can be described as the area under the curve bounded by $f(x)$ and the x axis, between two specified values of x, say, x_1 and x_2, as in Figure B.15.

The area of the blue element in Figure B.15 is approximately $f(x_i)\ \Delta x_i$. If we sum all these area elements between x_1 and x_2 and take the limit of this sum as $\Delta x_i \rightarrow 0$, we obtain the *true* area under the curve bounded by $f(x)$ and the x axis, between the limits x_1 and x_2:

$$\text{Area} = \lim_{\Delta x_i \rightarrow 0} \sum_i f(x_i)\Delta x_i = \int_{x_1}^{x_2} f(x)\ dx \tag{B.36}$$

Integrals of the type defined by Equation B.36 are called **definite integrals.**

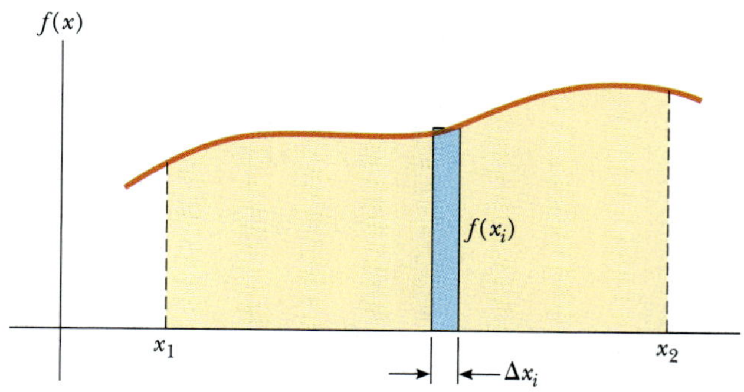

Figure B.15 The definite integral of a function is the area under the curve of the function between the limits x_1 and x_2.

One common integral that arises in practical situations has the form

$$\int x^n\, dx = \frac{x^{n+1}}{n+1} + c \quad (n \neq -1) \tag{B.37}$$

This result is obvious, being that differentiation of the right-hand side with respect to x gives $f(x) = x^n$ directly. If the limits of the integration are known, this integral becomes a *definite integral* and is written

$$\int_{x_1}^{x_2} x^n\, dx = \frac{x^{n+1}}{n+1}\bigg|_{x_1}^{x_2} = \frac{x_2^{\,n+1} - x_1^{\,n+1}}{n+1} \quad (n \neq -1) \tag{B.38}$$

EXAMPLES

1. $\displaystyle\int_0^a x^2\, dx = \frac{x^3}{3}\bigg]_0^a = \frac{a^3}{3}$

2. $\displaystyle\int_0^b x^{3/2} dx = \frac{x^{5/2}}{5/2}\bigg]_0^b = \tfrac{2}{5}b^{5/2}$

3. $\displaystyle\int_3^5 x\, dx = \frac{x^2}{2}\bigg]_3^5 = \frac{5^2 - 3^2}{2} = 8$

Partial Integration

Sometimes it is useful to apply the method of *partial integration* (also called "integrating by parts") to evaluate certain integrals. This method uses the property

$$\int u\, dv = uv - \int v\, du \tag{B.39}$$

where u and v are *carefully* chosen so as to reduce a complex integral to a simpler one. In many cases, several reductions have to be made. Consider the function

$$I(x) = \int x^2 e^x\, dx$$

which can be evaluated by integrating by parts twice. First, if we choose $u = x^2$, $v = e^x$, we obtain

$$\int x^2 e^x\, dx = \int x^2\, d(e^x) = x^2 e^x - 2\int e^x x\, dx + c_1$$

Now, in the second term, choose $u = x$, $v = e^x$, which gives

$$\int x^2 e^x\, dx = x^2 e^x - 2x\, e^x + 2\int e^x\, dx + c_1$$

or

$$\int x^2 e^x\, dx = x^2 e^x - 2xe^x + 2e^x + c_2$$

TABLE B.5

Some Indefinite Integrals (An arbitrary constant should be added to each of these integrals.)

$$\int x^n\,dx = \frac{x^{n+1}}{n+1} \quad (\text{provided } n \neq 1)$$

$$\int \frac{dx}{x} = \int x^{-1}\,dx = \ln x$$

$$\int \frac{dx}{a+bx} = \frac{1}{b}\ln(a+bx)$$

$$\int \frac{x\,dx}{a+bx} = \frac{x}{b} - \frac{a}{b^2}\ln(a+bx)$$

$$\int \frac{dx}{x(x+a)} = -\frac{1}{a}\ln\frac{x+a}{x}$$

$$\int \frac{dx}{(a+bx)^2} = -\frac{1}{b(a+bx)}$$

$$\int \frac{dx}{a^2+x^2} = \frac{1}{a}\tan^{-1}\frac{x}{a}$$

$$\int \frac{dx}{a^2-x^2} = \frac{1}{2a}\ln\frac{a+x}{a-x} \quad (a^2 - x^2 > 0)$$

$$\int \frac{dx}{x^2-a^2} = \frac{1}{2a}\ln\frac{x-a}{x+a} \quad (x^2 - a^2 > 0)$$

$$\int \frac{x\,dx}{a^2 \pm x^2} = \pm\tfrac{1}{2}\ln(a^2 \pm x^2)$$

$$\int \frac{dx}{\sqrt{a^2-x^2}} = \sin^{-1}\frac{x}{a} = -\cos^{-1}\frac{x}{a} \quad (a^2 - x^2 > 0)$$

$$\int \frac{dx}{\sqrt{x^2+a^2}} = \ln(x + \sqrt{x^2 \pm a^2})$$

$$\int \frac{x\,dx}{\sqrt{a^2-x^2}} = -\sqrt{a^2-x^2}$$

$$\int \frac{x\,dx}{\sqrt{x^2 \pm a^2}} = \sqrt{x^2 \pm a^2}$$

$$\int \sqrt{a^2-x^2}\,dx = \tfrac{1}{2}\left(x\sqrt{a^2-x^2} + a^2\sin^{-1}\frac{x}{a}\right)$$

$$\int x\sqrt{a^2-x^2}\,dx = -\tfrac{1}{3}(a^2-x^2)^{3/2}$$

$$\int \sqrt{x^2 \pm a^2}\,dx = \tfrac{1}{2}\left[x\sqrt{x^2 \pm a^2} \pm a^2\ln(x + \sqrt{x^2 \pm a^2})\right]$$

$$\int x(\sqrt{x^2 \pm a^2})\,dx = \tfrac{1}{3}(x^2 \pm a^2)^{3/2}$$

$$\int e^{ax}\,dx = \frac{1}{a}e^{ax}$$

$$\int \ln ax\,dx = (x\ln ax) - x$$

$$\int xe^{ax}\,dx = \frac{e^{ax}}{a^2}(ax - 1)$$

$$\int \frac{dx}{a+be^{cx}} = \frac{x}{a} - \frac{1}{ac}\ln(a + be^{cx})$$

$$\int \sin ax\,dx = \frac{1}{a}\cos ax$$

$$\int \cos ax\,dx = \frac{1}{a}\sin ax$$

$$\int \tan ax\,dx = -\frac{1}{a}\ln(\cos ax) = \frac{1}{a}\ln(\sec ax)$$

$$\int \cot ax\,dx = \frac{1}{a}\ln(\sin ax)$$

$$\int \sec ax\,dx = \frac{1}{a}\ln(\sec ax + \tan ax) = \frac{1}{a}\ln\left[\tan\left(\frac{ax}{2} + \frac{\pi}{4}\right)\right]$$

$$\int \csc ax\,dx = \frac{1}{a}\ln(\csc ax - \cot ax) = \frac{1}{a}\ln\left(\tan\frac{ax}{2}\right).$$

$$\int \sin^2 ax\,dx = \frac{x}{2} + \frac{\sin 2ax}{4a}$$

$$\int \cos^2 ax\,dx = \frac{x}{2} + \frac{\sin 2ax}{4a}$$

$$\int \frac{dx}{\sin^2 ax} = -\frac{1}{a}\cot ax$$

$$\int \frac{dx}{\cos^2 ax} = \frac{1}{a}\tan ax$$

$$\int \tan^2 ax\,dx = \frac{1}{a}(\tan ax) - x$$

$$\int \cot^2 ax\,dx = -\frac{1}{a}(\cot ax) - x$$

$$\int \sin^{-1} ax\,dx = x(\sin^{-1} ax) + \frac{\sqrt{1-a^2x^2}}{a}$$

$$\int \cos^{-1} ax\,dx = x(\cos^{-1} ax) - \frac{\sqrt{1-a^2x^2}}{a}$$

$$\int \frac{dx}{(x^2+a^2)^{3/2}} = \frac{x}{a^2\sqrt{x^2+a^2}}$$

$$\int \frac{x\,dx}{(x^2+a^2)^{3/2}} = -\frac{1}{\sqrt{x^2+a^2}}$$

TABLE B.6

Gauss's Probability Integral and Other Definite Integrals

$$\int_0^\infty x^n e^{-ax}\,dx = \frac{n!}{a^{n+1}}$$

$$I_0 = \int_0^\infty e^{-ax^2}\,dx = \frac{1}{2}\sqrt{\frac{\pi}{a}} \qquad \text{(Gauss's probability integral)}$$

$$I_1 = \int_0^\infty xe^{-ax^2}\,dx = \frac{1}{2a}$$

$$I_2 = \int_0^\infty x^2 e^{-ax^2}\,dx = -\frac{dI_0}{da} = \frac{1}{4}\sqrt{\frac{\pi}{a^3}}$$

$$I_3 = \int_0^\infty x^3 e^{-ax^2}\,dx = -\frac{dI_1}{da} = \frac{1}{2a^2}$$

$$I_4 = \int_0^\infty x^4 e^{-ax^2}\,dx = \frac{d^2 I_0}{da^2} = \frac{3}{8}\sqrt{\frac{\pi}{a^5}}$$

$$I_5 = \int_0^\infty x^5 e^{-ax^2}\,dx = \frac{d^2 I_1}{da^2} = \frac{1}{a^3}$$

$$\vdots$$

$$I_{2n} = (-1)^n \frac{d^n}{da^n} I_0$$

$$I_{2n+1} = (-1)^n \frac{d^n}{da^n} I_1$$

The Perfect Differential

Another useful method to remember is that of the *perfect differential,* in which we look for a change of variable such that the differential of the function is the differential of the independent variable appearing in the integrand. For example, consider the integral

$$I(x) = \int \cos^2 x \sin x\, dx$$

This integral becomes easy to evaluate if we rewrite the differential as $d(\cos x) = -\sin x\, dx$. The integral then becomes

$$\int \cos^2 x \sin x\, dx = -\int \cos^2 x\, d(\cos x)$$

If we now change variables, letting $y = \cos x$, we obtain

$$\int \cos^2 x \sin x\, dx = -\int y^2\, dy = -\frac{y^3}{3} + c = -\frac{\cos^3 x}{3} + c$$

Table B.5 (page A-18) lists some useful indefinite integrals. Table B.6 gives Gauss's probability integral and other definite integrals. A more complete list can be found in various handbooks, such as *The Handbook of Chemistry and Physics* (Boca Raton, FL: CRC Press, published annually).

B.8 Propagation of Uncertainty

In laboratory experiments, a common activity is to take measurements that act as raw data. These measurements are of several types—length, time interval, temperature, voltage, and so on—and are taken by a variety of instruments. Regardless of the measurement and the quality of the instrumentation, **there is always uncertainty associated with a physical measurement.** This uncertainty is a combination of that associated with the instrument and that related to the system being measured. An example of the former is the inability to exactly determine the position of a length measurement between the lines on a meterstick. An example of uncertainty related to the system being measured is the variation of temperature within a sample of water so that a single temperature for the sample is difficult to determine.

Uncertainties can be expressed in two ways. **Absolute uncertainty** refers to an uncertainty expressed in the same units as the measurement. Therefore, the length of a computer disk label might be expressed as (5.5 ± 0.1) cm. The uncertainty of ± 0.1 cm by itself is not descriptive enough for some purposes, however. This uncertainty is large if the measurement is 1.0 cm, but it is small if the measurement is 100 m. To give a more descriptive account of the uncertainty, **fractional uncertainty** or **percent uncertainty** is used. In this type of description, the uncertainty is divided by the actual measurement. Therefore, the length of the computer disk label could be expressed as

$$\ell = 5.5 \text{ cm} \pm \frac{0.1 \text{ cm}}{5.5 \text{ cm}} = 5.5 \text{ cm} \pm 0.018 \quad \text{(fractional uncertainty)}$$

or as

$$\ell = 5.5 \text{ cm} \pm 1.8\% \quad \text{(percent uncertainty)}$$

When combining measurements in a calculation, the percent uncertainty in the final result is generally larger than the uncertainty in the individual measurements. This is called **propagation of uncertainty** and is one of the challenges of experimental physics.

Some simple rules can provide a reasonable estimate of the uncertainty in a calculated result:

Multiplication and division: When measurements with uncertainties are multiplied or divided, add the *percent uncertainties* to obtain the percent uncertainty in the result.

Example: The Area of a Rectangular Plate

$$A = \ell w = (5.5 \text{ cm} \pm 1.8\%) \times (6.4 \text{ cm} \pm 1.6\%) = 35 \text{ cm}^2 \pm 3.4\%$$
$$= (35 \pm 1) \text{ cm}^2$$

Addition and subtraction: When measurements with uncertainties are added or subtracted, add the *absolute uncertainties* to obtain the absolute uncertainty in the result.

Example: A Change in Temperature

$$\Delta T = T_2 - T_1 = (99.2 \pm 1.5)°\text{C} - (27.6 \pm 1.5)°\text{C} = (71.6 \pm 3.0)°\text{C}$$
$$= 71.6°\text{C} \pm 4.2\%$$

Powers: If a measurement is taken to a power, the percent uncertainty is multiplied by that power to obtain the percent uncertainty in the result.

Example: The Volume of a Sphere

$$V = \tfrac{4}{3}\pi r^3 = \tfrac{4}{3}\pi(6.20 \text{ cm} \pm 2.0\%)^3 = 998 \text{ cm}^3 \pm 6.0\%$$
$$= (998 \pm 60) \text{ cm}^3$$

For complicated calculations, many uncertainties are added together, which can cause the uncertainty in the final result to be undesirably large. Experiments should be designed such that calculations are as simple as possible.

Notice that uncertainties in a calculation always add. As a result, an experiment involving a subtraction should be avoided if possible, especially if the measurements being subtracted are close together. The result of such a calculation is a small difference in the measurements and uncertainties that add together. It is possible that the uncertainty in the result could be larger than the result itself!

Symbol — Ca 20 — Atomic number
Atomic mass† — 40.078
$4s^2$ — Electron configuration

Group I	Group II	Transition elements						
H 1 1.007 9 $1s$								
Li 3 6.941 $2s^1$	**Be** 4 9.0122 $2s^2$							
Na 11 22.990 $3s^1$	**Mg** 12 24.305 $3s^2$							
K 19 39.098 $4s^1$	**Ca** 20 40.078 $4s^2$	**Sc** 21 44.956 $3d^14s^2$	**Ti** 22 47.867 $3d^24s^2$	**V** 23 50.942 $3d^34s^2$	**Cr** 24 51.996 $3d^54s^1$	**Mn** 25 54.938 $3d^54s^2$	**Fe** 26 55.845 $3d^64s^2$	**Co** 27 58.933 $3d^74s^2$
Rb 37 85.468 $5s^1$	**Sr** 38 87.62 $5s^2$	**Y** 39 88.906 $4d^15s^2$	**Zr** 40 91.224 $4d^25s^2$	**Nb** 41 92.906 $4d^45s^1$	**Mo** 42 95.94 $4d^55s^1$	**Tc** 43 (98) $4d^55s^2$	**Ru** 44 101.07 $4d^75s^1$	**Rh** 45 102.91 $4d^85s^1$
Cs 55 132.91 $6s^1$	**Ba** 56 137.33 $6s^2$	57–71*	**Hf** 72 178.49 $5d^26s^2$	**Ta** 73 180.95 $5d^36s^2$	**W** 74 183.84 $5d^46s^2$	**Re** 75 186.21 $5d^56s^2$	**Os** 76 190.23 $5d^66s^2$	**Ir** 77 192.2 $5d^76s^2$
Fr 87 (223) $7s^1$	**Ra** 88 (226) $7s^2$	89–103**	**Rf** 104 (261) $6d^27s^2$	**Db** 105 (262) $6d^37s^2$	**Sg** 106 (266)	**Bh** 107 (264)	**Hs** 108 (277)	**Mt** 109 (268)

*Lanthanide series	**La** 57 138.91 $5d^16s^2$	**Ce** 58 140.12 $5d^14f^16s^2$	**Pr** 59 140.91 $4f^36s^2$	**Nd** 60 144.24 $4f^46s^2$	**Pm** 61 (145) $4f^56s^2$	**Sm** 62 150.36 $4f^66s^2$
**Actinide series	**Ac** 89 (227) $6d^17s^2$	**Th** 90 232.04 $6d^27s^2$	**Pa** 91 231.04 $5f^26d^17s^2$	**U** 92 238.03 $5f^36d^17s^2$	**Np** 93 (237) $5f^46d^17s^2$	**Pu** 94 (244) $5f^66d^07s^2$

Note: Atomic mass values given are averaged over isotopes in the percentages in which they exist in nature.

†For an unstable element, mass number of the most stable known isotope is given in parentheses.

††Elements 112 and 114 have not yet been named.

†††For a description of the atomic data, visit *physics.nist.gov/PhysRefData/Elements/per_text.html*

			Group III	Group IV	Group V	Group VI	Group VII	Group 0
							H 1 1.007 9 $1s^1$	**He** 2 4.002 6 $1s^2$
			B 5 10.811 $2p^1$	**C** 6 12.011 $2p^2$	**N** 7 14.007 $2p^3$	**O** 8 15.999 $2p^4$	**F** 9 18.998 $2p^5$	**Ne** 10 20.180 $2p^6$
			Al 13 26.982 $3p^1$	**Si** 14 28.086 $3p^2$	**P** 15 30.974 $3p^3$	**S** 16 32.066 $3p^4$	**Cl** 17 35.453 $3p^5$	**Ar** 18 39.948 $3p^6$
Ni 28 58.693 $3d^84s^2$	**Cu** 29 63.546 $3d^{10}4s^1$	**Zn** 30 65.41 $3d^{10}4s^2$	**Ga** 31 69.723 $4p^1$	**Ge** 32 72.64 $4p^2$	**As** 33 74.922 $4p^3$	**Se** 34 78.96 $4p^4$	**Br** 35 79.904 $4p^5$	**Kr** 36 83.80 $4p^6$
Pd 46 106.42 $4d^{10}$	**Ag** 47 107.87 $4d^{10}5s^1$	**Cd** 48 112.41 $4d^{10}5s^2$	**In** 49 114.82 $5p^1$	**Sn** 50 118.71 $5p^2$	**Sb** 51 121.76 $5p^3$	**Te** 52 127.60 $5p^4$	**I** 53 126.90 $5p^5$	**Xe** 54 131.29 $5p^6$
Pt 78 195.08 $5d^96s^1$	**Au** 79 196.97 $5d^{10}6s^1$	**Hg** 80 200.59 $5d^{10}6s^2$	**Tl** 81 204.38 $6p^1$	**Pb** 82 207.2 $6p^2$	**Bi** 83 208.98 $6p^3$	**Po** 84 (209) $6p^4$	**At** 85 (210) $6p^5$	**Rn** 86 (222) $6p^6$
Ds 110 (271)	**Rg** 111 (272)	112†† (285)		114†† (289)				

Eu 63 151.96 $4f^76s^2$	**Gd** 64 157.25 $4f^75d^16s^2$	**Tb** 65 158.93 $4f^85d^16s^2$	**Dy** 66 162.50 $4f^{10}6s^2$	**Ho** 67 164.93 $4f^{11}6s^2$	**Er** 68 167.26 $4f^{12}6s^2$	**Tm** 69 168.93 $4f^{13}6s^2$	**Yb** 70 173.04 $4f^{14}6s^2$	**Lu** 71 174.97 $4f^{14}5d^16s^2$
Am 95 (243) $5f^77s^2$	**Cm** 96 (247) $5f^76d^17s^2$	**Bk** 97 (247) $5f^86d^17s^2$	**Cf** 98 (251) $5f^{10}7s^2$	**Es** 99 (252) $5f^{11}7s^2$	**Fm** 100 (257) $5f^{12}7s^2$	**Md** 101 (258) $5f^{13}7s^2$	**No** 102 (259) $5f^{14}7s^2$	**Lr** 103 (262) $6d^15f^{14}7s^2$

TABLE D.1

SI Units

Base Quantity	SI Base Unit Name	SI Base Unit Symbol
Length	meter	m
Mass	kilogram	kg
Time	second	s
Electric current	ampere	A
Temperature	kelvin	K
Amount of substance	mole	mol
Luminous intensity	candela	cd

TABLE D.2

Some Derived SI Units

Quantity	Name	Symbol	Expression in Terms of Base Units	Expression in Terms of Other SI Units
Plane angle	radian	rad	m/m	
Frequency	hertz	Hz	s^{-1}	
Force	newton	N	$kg \cdot m/s^2$	J/m
Pressure	pascal	Pa	$kg/m \cdot s^2$	N/m^2
Energy	joule	J	$kg \cdot m^2/s^2$	$N \cdot m$
Power	watt	W	$kg \cdot m^2/s^3$	J/s
Electric charge	coulomb	C	$A \cdot s$	
Electric potential	volt	V	$kg \cdot m^2/A \cdot s^3$	W/A
Capacitance	farad	F	$A^2 \cdot s^4/kg \cdot m^2$	C/V
Electric resistance	ohm	Ω	$kg \cdot m^2/A^2 \cdot s^3$	V/A
Magnetic flux	weber	Wb	$kg \cdot m^2/A \cdot s^2$	$V \cdot s$
Magnetic field	tesla	T	$kg/A \cdot s^2$	
Inductance	henry	H	$kg \cdot m^2/A^2 \cdot s^2$	$T \cdot m^2/A$

CHAPTER 1

1. 5.52×10^3 kg/m^3, between the density of aluminum and iron and greater than the densities of typical surface rocks
3. 23.0 kg
5. 7.69 cm
7. (b) only
9. The units of G are m^3/kg · s^2.
11. 1.39×10^3 m^2
13. Not with the pages from Volume 1, but yes with the pages from the full version. Each page has area 0.059 m^2. The room has wall area 37 m^2, requiring 630 sheets, which would be counted as 1 260 pages.
15. 11.4×10^3 kg/m^3
17. (a) 250 yr (b) 3.09×10^4 times
19. 1.00×10^{10} lb
21. 151 μm
23. 2.86 cm
25. $\sim 10^6$ balls
27. $\sim 10^2$ kg; $\sim 10^3$ kg
29. $\sim 10^2$ tuners
31. (a) 3 (b) 4 (c) 3 (d) 2
33. (a) 797 (b) 1.1 (c) 17.66
35. 8.80%
37. 9
39. 63
41. 108° and 288°
43. 48.6 kg
45. (a) smaller by nine times (b) Δt is inversely proportional to d^2. (c) Plot Δt on the vertical axis and $1/d^2$ on the horizontal axis. (d) $4QL/[k\pi(T_h - T_c)]$
47. (a) m = 346 g − (14.5 g/cm^3)a^3 (b) $a = 0$ (c) 346 g (d) yes (e) a = 2.60 cm (f) 90.6 g (g) yes (h) 218 g (i) No; 218 g is not equal to 314 g. (j) Parts (b), (c), and (d) describe a uniform solid sphere with ρ = 4.70 g/cm^3 as a approaches zero. Parts (e), (f), and (g) describe a uniform liquid drop with ρ = 1.23 g/cm^3 as a approaches 2.60 cm. The function $m(a)$ is not a linear function, so a halfway between 0 and 2.60 cm does not give a value for m halfway between the minimum and maximum values. The graph of m versus a starts at $a = 0$ with a horizontal tangent. Then it curves down more and more steeply as a increases. The liquid drop of radius 1.30 cm has only one eighth the volume of the whole sphere, so its presence brings down the mass by only a small amount, from 346 g to 314 g. (k) The answer would not change as long as the wall of the shell is unbroken.
49. 5.0 m
51. $0.579t$ ft^3/s + $(1.19 \times 10^{-9})t^2$ ft^3/s^2
53. 3.41 m
55. 0.449%
57. (a) 0.529 cm/s (b) 11.5 cm/s
59. 1×10^{10} gal/yr

CHAPTER 2

1. (a) 5 m/s (b) 1.2 m/s (c) −2.5 m/s (d) −3.3 m/s (e) 0
3. (a) 3.75 m/s (b) 0
5. (a) −2.4 m/s (b) −3.8 m/s (c) 4.0 s
7. (a) and (c)

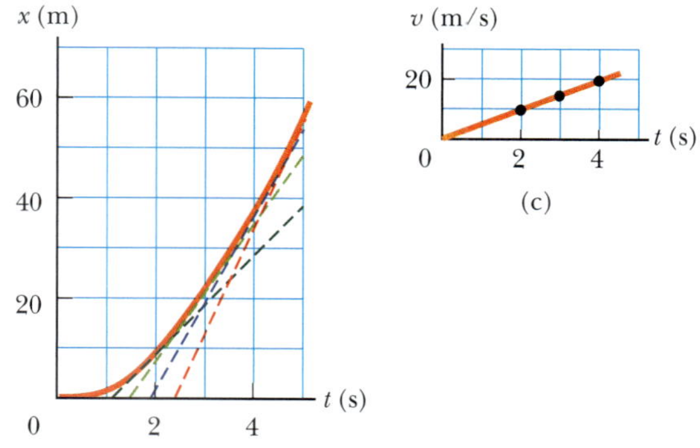

(a) (c)

(b) $v_{t=5.0\text{ s}}$ = 23 m/s, $v_{t=4.0\text{ s}}$ = 18 m/s, $v_{t=3.0\text{ s}}$ = 14 m/s, $v_{t=2.0\text{ s}}$ = 9.0 m/s (c) 4.6 m/s^2 (d) 0
9. 5.00 m
11. (a) 20.0 m/s, 5.00 m/s (b) 262 m
13. (a) 2.00 m (b) −3.00 m/s (c) −2.00 m/s^2
15. (a) 13.0 m/s (b) 10.0 m/s, 16.0 m/s (c) 6.00 m/s^2 (d) 6.00 m/s^2
17.

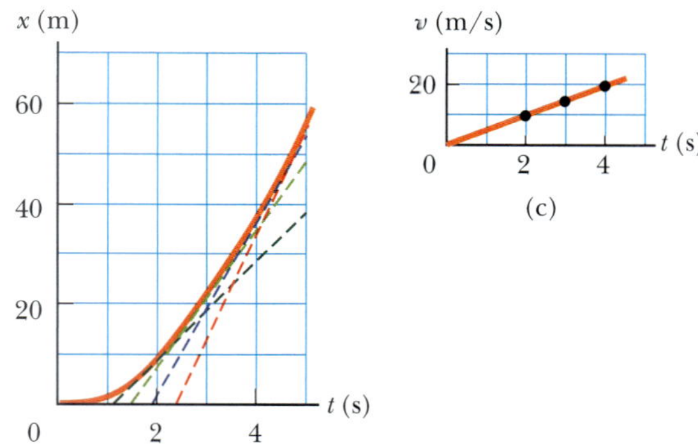

(a) (c)

19. (a) 9.00 m/s (b) 5.00 m/s (c) 3.00 m/s (d) −3.00 m/s (e) 17.0 m/s (f) The graph of velocity versus time is a straight line passing through 13 m/s at 10:05 a.m. and sloping downward, decreasing by 4 m/s for each second thereafter. (g) If and only if we know the object's velocity at one instant of time, knowing its acceleration tells us its velocity at every other moment, as long as the acceleration is constant.
21. −16.0 cm/s^2
23. (a) 20.0 s (b) It cannot; it would need a longer runway.
25. 3.10 m/s
27. (a) −202 m/s^2 (b) 198 m
29. (a) 4.98×10^{-9} s (b) 1.20×10^{15} m/s^2
31. (a) False unless the acceleration is zero. We define constant acceleration to mean that the velocity is changing steadily in time. Then the velocity cannot be changing steadily in space. (b) True. Because the velocity is changing steadily in time, the velocity halfway through an interval is equal to the average of its initial and final values.
33. (a) 3.45 s (b) 10.0 ft
35. (a) 19.7 cm/s (b) 4.70 cm/s^2 (c) The time interval required for the speed to change between Ⓐ and Ⓑ is

sufficient to find the acceleration, more directly than we could find it from the distance between the points.

37. We ignore air resistance. We assume the worker's flight time, "a mile," and "a dollar" were measured to three-digit precision. We have interpreted "up in the sky" as referring to free-fall time, not to the launch and landing times. Therefore, the wage was \$99.3/h.

39. (a) 10.0 m/s up (b) 4.68 m/s down

41. (a) 29.4 m/s (b) 44.1 m

43. (a) 7.82 m (b) 0.782 s

45. 38.2 m

47. (a) $a_x(t) = a_{xi} + Jt$, $v_x(t) = v_{xi} + a_{xi}t + \frac{1}{2}Jt^2$, $x(t) = x_i + v_{xi}t + \frac{1}{2}a_{xi}t^2 + \frac{1}{6}Jt^3$

49. (a) 0 (b) 6.0 m/s^2 (c) −3.6 m/s^2 (d) 6 s and 18 s (e) 18 s (f) 84 m (g) 204 m

51. (a) 41.0 s (b) 1.73 km (c) −184 m/s

53. (a) 5.43 m/s^2 and 3.83 m/s^2 (b) 10.9 m/s and 11.5 m/s (c) Maggie by 2.62 m

55. 155 s, 129 s

57. (a) 3.00 s (b) −15.3 m/s (c) 31.4 m/s down and 34.8 m/s down

59. (a) 5.46 s (b) 73.0 m (c) $v_{\text{Stan}} = 22.6$ m/s, $v_{\text{Kathy}} = 26.7$ m/s

61. (a) yes, to two significant digits (b) 0.742 s (c) Yes; the braking distance is proportional to the square of the original speed. (d) −19.7 ft/s^2 = −6.01 m/s^2

63. $0.577v$

CHAPTER 3

1. (−2.75, −4.76) m

3. (a) 2.24 m (b) 2.24 m at 26.6°

5. (a) r, 180° − θ (b) $2r$, 180° + θ (c) $3r$, $-\theta$

7. 70.0 m

9. (a) 10.0 m (b) 15.7 m (c) 0

11. (a) 5.2 m at 60° (b) 3.0 m at 330° (c) 3.0 m at 150° (d) 5.2 m at 300°

13. approximately 420 ft at −3°

15. 47.2 units at 122°

17. Yes. The speed of the camper should be 28.3 m/s or greater.

19. (a) $(-11.1\hat{\mathbf{i}} + 6.40\hat{\mathbf{j}})$ m (b) $(1.65\hat{\mathbf{i}} + 2.86\hat{\mathbf{j}})$ cm (c) $(-18.0\hat{\mathbf{i}} - 12.6\hat{\mathbf{j}})$ in.

21. 358 m at 2.00° S of E

23. 196 cm at 345°

25. (a) $2.00\hat{\mathbf{i}} - 6.00\hat{\mathbf{j}}$ (b) $4.00\hat{\mathbf{i}} + 2.00\hat{\mathbf{j}}$ (c) 6.32 (d) 4.47 (e) 288°, 26.6°

27. 9.48 m at 166°

29. 4.64 m at 78.6° N of E

31. (a) 185 N at 77.8° from the $+x$ axis (b) $(-39.3\hat{\mathbf{i}} - 181\hat{\mathbf{j}})$ N

33. $|\vec{\mathbf{B}}| = 7.81$, $\theta_x = 59.2°$, $\theta_y = 39.8°$, $\theta_z = 67.4°$

35. (a) 5.92 m is the magnitude of $(5.00\hat{\mathbf{i}} - 1.00\hat{\mathbf{j}} - 3.00\hat{\mathbf{k}})$ m. (b) 19.0 m is the magnitude of $(4.00\hat{\mathbf{i}} - 11.0\hat{\mathbf{j}} - 15.0\hat{\mathbf{k}})$ m.

37. (a) $8.00\hat{\mathbf{i}} + 12.0\hat{\mathbf{j}} - 4.00\hat{\mathbf{k}}$ (b) $2.00\hat{\mathbf{i}} + 3.00\hat{\mathbf{j}} - 1.00\hat{\mathbf{k}}$ (c) $-24.0\hat{\mathbf{i}} - 36.0\hat{\mathbf{j}} + 12.0\hat{\mathbf{k}}$

39. (a) $(3.12\hat{\mathbf{i}} + 5.02\hat{\mathbf{j}} - 2.20\hat{\mathbf{k}})$ km (b) 6.31 km

41. (a) $-3.00\hat{\mathbf{i}} + 2.00\hat{\mathbf{j}}$ (b) 3.61 at 146° (c) $3.00\hat{\mathbf{i}} - 6.00\hat{\mathbf{j}}$

43. (a) $49.5\hat{\mathbf{i}} + 27.1\hat{\mathbf{j}}$ (b) 56.4 units at 28.7°

45. (a) $[(5 + 11f)\hat{\mathbf{i}} + (3 + 9f)\hat{\mathbf{j}}]$ m (b) $(5\hat{\mathbf{i}} + 3\hat{\mathbf{j}})$ m is reasonable because it is the starting point. (c) $(16\hat{\mathbf{i}} + 12\hat{\mathbf{j}})$ m is reasonable because it is the endpoint.

47. 1.15°

49. 2.29 km

51. (a) 7.17 km (b) 6.15 km

53. 390 mi/h at 7.37° N of E

55. $(0.456\hat{\mathbf{i}} - 0.708\hat{\mathbf{j}})$ m

57. 240 m at 237°

59. (a) (10.0 m, 16.0 m) (b) You will arrive at the treasure if you take the trees in any order. The directions take you to the average position of the trees.

61. 106°

CHAPTER 4

1. (a) 4.87 km at 209° from E (b) 23.3 m/s (c) 13.5 m/s at 209°

3. 2.50 m/s

5. (a) $(0.800\hat{\mathbf{i}} - 0.300\hat{\mathbf{j}})$ m/s^2 (b) 339° (c) $(360\hat{\mathbf{i}} - 72.7\hat{\mathbf{j}})$ m, −15.2°

7. (a) $\vec{\mathbf{v}} = 5\hat{\mathbf{i}} + 4t^{3/2}\hat{\mathbf{j}}$ (b) $\vec{\mathbf{r}} = 5t\hat{\mathbf{i}} + 1.6t^{5/2}\hat{\mathbf{j}}$

9. (a) $3.34\hat{\mathbf{i}}$ m/s (b) −50.9°

11. (7.23×10^3 m, 1.68×10^3 m)

13. 53.1°

15. (a) 22.6 m (b) 52.3 m (c) 1.18 s

17. (a) The ball clears by 0.889 m. (b) while descending

19. (a) 18.1 m/s (b) 1.13 m (c) 2.79 m

21. 9.91 m/s

23. $\tan^{-1}[(2gh)^{1/2}/v]$

25. 377 m/s^2

27. (a) 6.00 rev/s (b) 1.52 km/s^2 (c) 1.28 km/s^2

29. 1.48 m/s^2 inward and 29.9° backward

31. (a) 13.0 m/s^2 (b) 5.70 m/s (c) 7.50 m/s^2

33. (a) 57.7 km/h at 60.0° W of vertical (b) 28.9 km/h downward

35. 2.02×10^3 s; 21.0% longer

37. $t_{\text{Alan}} = \dfrac{2L/c}{1 - v^2/c^2}$, $t_{\text{Beth}} = \dfrac{2L/c}{\sqrt{1 - v^2/c^2}}$. Beth returns first.

39. 15.3 m

41. 27.7° E of N

43. (a) 9.80 m/s^2 down (b) 3.72 m

45. (a) 41.7 m/s (b) 3.81 s (c) $(34.1\hat{\mathbf{i}} - 13.4\hat{\mathbf{j}})$ m/s; 36.7 m/s

47. (a) 25.0 m/s^2; 9.80 m/s^2
(b)

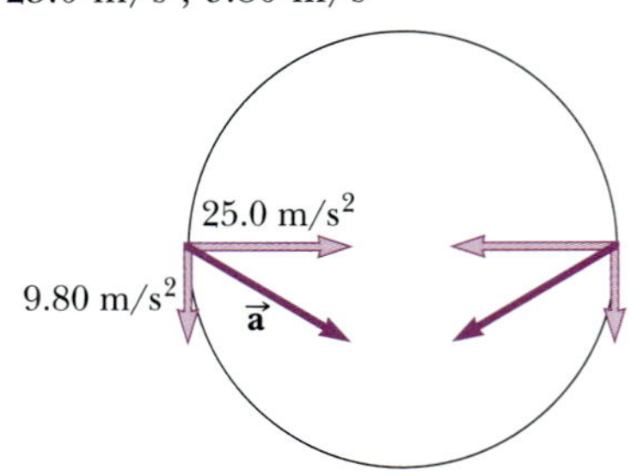

(c) 26.8 m/s^2 inward at 21.4° below the horizontal

49. (a)

t (s)	0	1	2	3	4	5
r (m)	0	45.7	82.0	109	127	136

t (s)	6	7	8	9	10
r (m)	138	133	124	117	120

(b) The vector $\vec{\mathbf{v}}$ tells how $\vec{\mathbf{r}}$ is changing. If $\vec{\mathbf{v}}$ at a particular point has a component along $\vec{\mathbf{r}}$, then $\vec{\mathbf{r}}$ will be increasing in magnitude (if $\vec{\mathbf{v}}$ is at an angle less than 90° from $\vec{\mathbf{r}}$) or decreasing (if the angle between $\vec{\mathbf{v}}$ and $\vec{\mathbf{r}}$ is more than 90°). To be at a maximum, the distance from the origin must be momentarily staying constant, and the only way that can happen is if the angle between velocity and position is a right angle. Then $\vec{\mathbf{r}}$ will be changing in direction at that point, but not in magnitude. (c) The requirement for perpendicularity can be defined as equality between the tangent of the angle between $\vec{\mathbf{v}}$ and the *x* direction and the tangent of the angle between $\vec{\mathbf{r}}$ and the *y* direction. In symbols, this equality can be written $(9.8t - 49)/12 = 12t/(49t - 4.9t^2)$, which has the solution $t = 5.70$ s, giving, in turn, $r = 138$ m. Alternatively, we can require $dr^2/dt = 0 = (d/dt)[(12t)^2 + (49t - 4.9t^2)^2]$, which results in the same equation with the same solution.

51. (a) 26.6° (b) 0.949
53. (a) 6.80 km (b) 3.00 km vertically above the impact point (c) 66.2°
55. (a) 46.5 m/s (b) −77.6° (c) 6.34 s
57. (a) 20.0 m/s, 5.00 s (b) $(16.0\hat{\mathbf{i}} - 27.1\hat{\mathbf{j}})$ m/s (c) 6.53 s (d) $24.5\hat{\mathbf{i}}$ m
59. (a) 43.2 m (b) $(9.66\hat{\mathbf{i}} - 25.5\hat{\mathbf{j}})$ m/s. Air resistance would ordinarily make the jump distance smaller and the final horizontal and vertical velocity components both somewhat smaller. When the skilled jumper makes his body into an airfoil, he deflects downward the air through which he passes so that it deflects him upward, giving him more time in the air and a longer jump.
61. Safe distances are less than 270 m or greater than 3.48×10^3 m from the western shore.

CHAPTER 5

1. $(6.00\hat{\mathbf{i}} + 15.0\hat{\mathbf{j}})$ N; 16.2 N
3. (a) $(2.50\hat{\mathbf{i}} + 5.00\hat{\mathbf{j}})$ N (b) 5.59 N
5. (a) 3.64×10^{-18} N (b) 8.93×10^{-30} N is 408 billion times smaller
7. 2.55 N for an 88.7-kg person
9. (a) 5.00 m/s² at 36.9° (b) 6.08 m/s² at 25.3°
11. (a) $\sim 10^{-22}$ m/s² (b) $\sim 10^{-23}$ m
13. (a) 15.0 lb up (b) 5.00 lb up (c) 0
15. (a) 3.43 kN (b) 0.967 m/s horizontally forward
17.

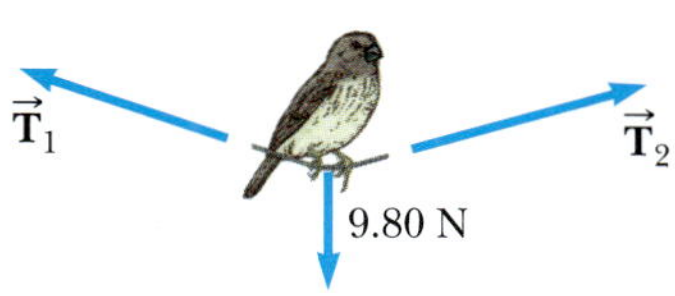

613 N

19. (a) $P \cos 40° - n = 0$ and $P \sin 40° - 220\text{ N} = 0$; $P = 342$ N and $n = 262$ N (b) $P - n \cos 40° - (220\text{ N}) \sin 40° = 0$ and $n \sin 40 - (220\text{ N}) \cos 40° = 0$; $n = 262$ N and $P = 342$ N (c) The results agree. The methods have a similar level of difficulty. Each involves one equation in one unknown and one equation in two unknowns. If we are interested in finding *n* without finding *P*, method (b) is simpler.
23. (a) 49.0 N (b) 49.0 N (c) 98.0 N (d) 24.5 N
25. 8.66 N east
27. (a) 646 N up (b) 646 N up (c) 627 N up (d) 589 N up
29. 3.73 m
31. (a) $F_x > 19.6$ N (b) $F_x \leq -78.4$ N (c)

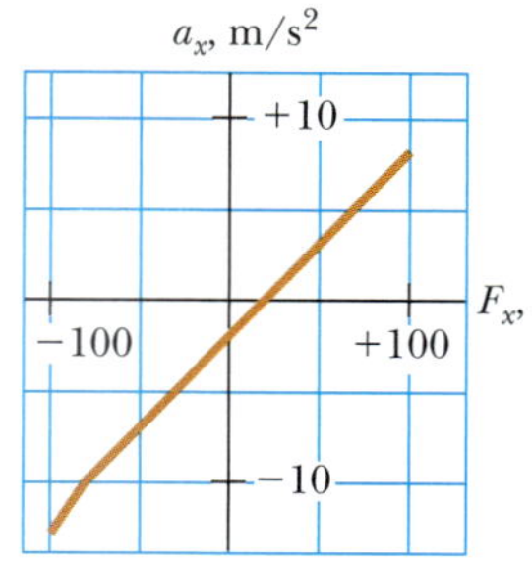

33. (a) 706 N (b) 814 N (c) 706 N (d) 648 N
35. (a) 256 m (b) 42.7 m
37. (a) no (b) 16.9 N backwards + 37.2 N upward = 40.9 N upward and backward at 65.6° with the horizontal
39. (a) 1.78 m/s² (b) 0.368 (c) 9.37 N (d) 2.67 m/s
41. 37.8 N
43. (a)

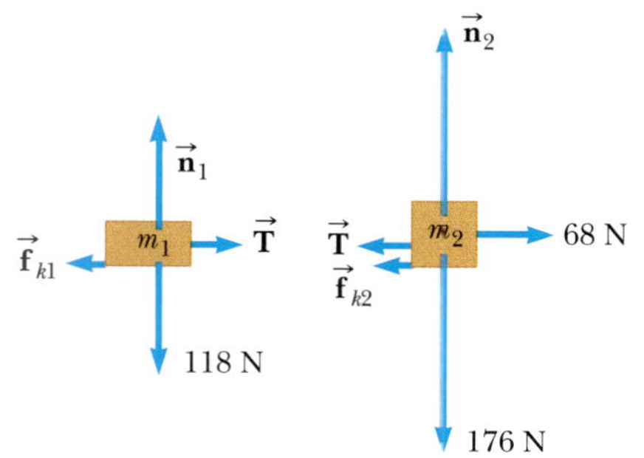

(b) 27.2 N, 1.29 m/s²

45. (a) $a = 0$ if $P < 8.11$ N; $a = -3.33$ m/s² + (1.41/kg) *P* to the right if $P > 8.11$ N (b) $a = 0$; 3.99 N horizontally backward (c) 10.8 m/s² to the right; 3.45 N to the left (d) The acceleration is zero for all values of *P* less than 8.11 N. When *P* passes this threshold, the acceleration jumps to its minimum nonzero value of 8.14 m/s². From there it increases linearly with *P* toward arbitrarily high values.
47. 72.0 N
49. (a) 2.94 m/s² forward (b) 2.45 m/s² forward (c) 1.19 m/s² up the incline (d) 0.711 m/s² up the incline (e) 16.7° (f) The mass makes no difference. Mathematically, the mass divides out in determinations of acceleration. If several packages of dishes were placed in the truck, they would all slide together, whether they were tied to one another or not.
51. (a)

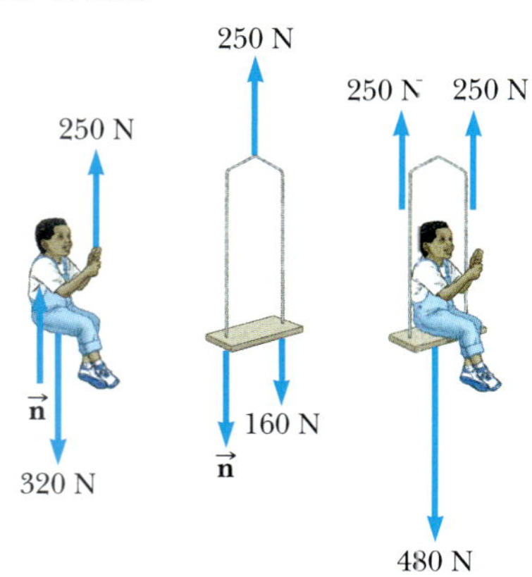

(b) 0.408 m/s² (c) 83.3 N

53. (a) 3.00 s (b) 20.1 m (c) $(18.0\hat{\mathbf{i}} - 9.00\hat{\mathbf{j}})$ m

55. (a) $a = 12\text{ N}/(4\text{ kg} + m_1)$ forward (b) $12\text{ N}/(1 + m_1/4\text{ kg})$ forward (c) 2.50 m/s² forward and 10.0 N forward (d) The force approaches zero (e) The force approaches 12.0 N (f) The tension in a cord of negligible mass is constant along its length.

57. (a) $Mg/2$, $Mg/2$, $Mg/2$, $3Mg/2$, Mg (b) $Mg/2$

59. (a) Both are equal respectively. (b) 1.61×10^4 N (c) 2.95×10^4 N (d) 0 N; 3.51 m/s upward. The first 3.50 m/s of the speed of 3.51 m/s needs no dynamic cause; the motion of the cable continues on its own, as described by the law of "inertia" or "pigheadedness." The increase from 3.50 m/s to 3.51 m/s must be caused by some total upward force on the section of cable. Because its mass is very small compared to a thousand kilograms, however, the force is very small compared to 1.61×10^4 N, the nearly uniform tension of this section of cable.

61. (b)

$\boldsymbol{\theta}$	0	15°	30°	45°	60°
P (N)	40.0	46.4	60.1	94.3	260

63. (a) The net force on the cushion is in a fixed direction, downward and forward making angle $\tan^{-1}(F/mg)$ with the vertical. Starting from rest, it will move along this line with (b) increasing speed. Its velocity changes in magnitude. (c) 1.63 m (d) It will move along a parabola. The axis of the parabola is parallel to the dashed line in the problem figure. If the cushion is thrown in a direction above the dashed line, its path will be concave downward, making its velocity become more and more nearly parallel to the dashed line over time. If the cushion is thrown down more steeply, its path will be concave upward, again making its velocity turn toward the fixed direction of its acceleration.

65. (a) 19.3° (b) 4.21 N

67. $(M + m_1 + m_2)(m_2 g/m_1)$

69. (a) 30.7° (b) 0.843 N

71. (a) $T_1 = \dfrac{2mg}{\sin\theta_1}$, $T_2 = \dfrac{mg}{\sin\theta_2} = \dfrac{mg}{\sin\left[\tan^{-1}\left(\frac{1}{2}\tan\theta_1\right)\right]}$,

$$T_3 = \frac{2\,mg}{\tan\theta_1}$$

(b) $\theta_2 = \tan^{-1}\left(\dfrac{\tan\theta_1}{2}\right)$

CHAPTER 6

1. Any speed up to 8.08 m/s

3. (a) 8.32×10^{-8} N toward the nucleus (b) 9.13×10^{22} m/s² inward

5. (a) static friction (b) 0.085 0

7. 2.14 rev/min

9. $v \le 14.3$ m/s

11. (a) 108 N (b) 56.2 N

13. (a) 4.81 m/s (b) 700 N up

15. No. Tarzan needs a vine of tensile strength 1.38 kN.

17. 3.13 m/s

19. (a) 3.60 m/s² (b) zero (c) An observer in the car (a noninertial frame) claims an 18.0-N force toward the left and an 18.0-N force toward the right. An inertial observer (outside the car) claims only an 18.0-N force toward the right.

21. (a) 17.0° (b) 5.12 N

23. (a) 491 N (b) 50.1 kg (c) 2.00 m/s

25. 0.092 8°

27. (a) 32.7 s⁻¹ (b) 9.80 m/s² down (c) 4.90 m/s² down

29. 3.01 N up

31. (a) 1.47 N·s/m (b) 2.04×10^{-3} s (c) 2.94×10^{-2} N

33. (a) 78.3 m/s (b) 11.1 s (c) 121 m

35. (a) $x = k^{-1}\ln(1 + kv_0 t)$ (b) $v = v_0 e^{-kx}$

37. (a) 0.034 7 s⁻¹ (b) 2.50 m/s (c) $a = -cv$

39. $v = v_0 e^{-bt/m}$

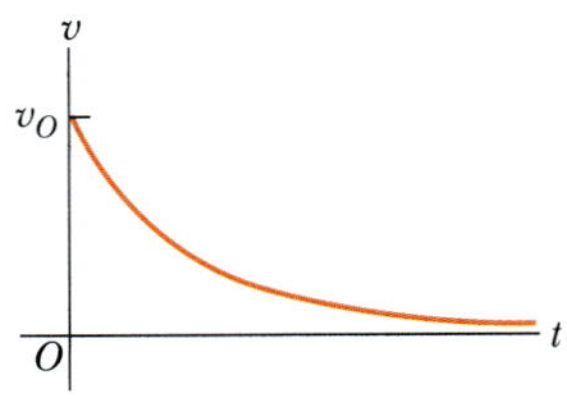

In this model, the object keeps moving forever. It travels a finite distance in an infinite time interval.

41. (a) 106 N up the incline (b) 0.396

43. (a) 11.5 kN (b) 14.1 m/s

45. (a) 0.016 2 kg/m (b) $\frac{1}{2}D\rho A$ (c) 0.778 (d) 1.5% (e) For stacked coffee filters falling in air at terminal speed, the graph of air resistance force as a function of the square of speed demonstrates that the force is proportional to the speed squared, within the experimental uncertainty estimated as 2%. This proportionality agrees with the theoretical model of air resistance at high speeds. The drag coefficient of a coffee filter is $D = 0.78 \pm 2\%$.

47. $g(\cos\phi\tan\theta - \sin\phi)$

49. (b) 732 N down at the equator and 735 N down at the poles

51. (a) The only horizontal force on the car is the force of friction, with a maximum value determined by the surface roughness (described by the coefficient of static friction) and the normal force (here equal to the gravitational force on the car). (b) 34.3 m (c) 68.6 m (d) Braking is better. You should not turn the wheel. If you used any of the available friction force to change the direction of the car, it would be unavailable to slow the car and the stopping distance would be longer. (e) The conclusion is true in general. The radius of the curve you can barely make is twice your minimum stopping distance.

53. (a) 5.19 m/s (b) $T = 555$ N

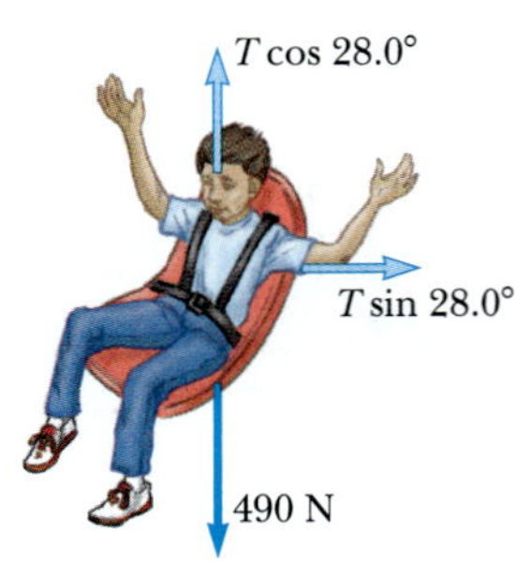

55. (b) 2.54 s; 23.6 rev/min (c) The gravitational and friction forces remain constant. The normal force increases. The person remains in motion with the wall. (d) The gravitational force remains constant. The normal and friction forces decrease. The person slides relative to the wall and downward into the pit.

57. (a) $v_{\min} = \sqrt{\dfrac{Rg(\tan\theta - \mu_s)}{1 + \mu_s\tan\theta}}$, $v_{\max} = \sqrt{\dfrac{Rg(\tan\theta + \mu_s)}{1 - \mu_s\tan\theta}}$

(b) $\mu_s = \tan\theta$ (c) $8.57\text{ m/s} \le v \le 16.6\text{ m/s}$

59. (a) 0.013 2 m/s (b) 1.03 m/s (c) 6.87 m/s

61. 12.8 N

CHAPTER 7

1. (a) 31.9 J (b) 0 (c) 0 (d) 31.9 J
3. −4.70 kJ
7. (a) 16.0 J (b) 36.9°
9. (a) 11.3° (b) 156° (c) 82.3°
11. $\vec{\mathbf{A}}$ = 7.05 m at 28.4°
13. (a) 24.0 J (b) −3.00 J (c) 21.0 J
15. (a) 7.50 J (b) 15.0 J (c) 7.50 J (d) 30.0 J
17. (a) 0.938 cm (b) 1.25 J
19. 7.37 N/m
21. 0.299 m/s
23. (a) 0.020 4 m (b) 720 N/m
25. (b) mgR
27. (a) 0.600 J (b) −0.600 J (c) 1.50 J
29. (a) 1.20 J (b) 5.00 m/s (c) 6.30 J
31. (a) 60.0 J (b) 60.0 J
33. 878 kN up
35. (a) 4.56 kJ (b) 6.34 kN (c) 422 km/s^2 (d) 6.34 kN (e) The forces are the same. The two theories agree.
37. (a) 259 kJ, 0, −259 kJ (b) 0, −259 kJ, −259 kJ
39. (a) −196 J (b) −196 J (c) −196 J. The force is conservative.
41. (a) 125 J (b) 50.0 J (c) 66.7 J (d) The force is nonconservative. The results differ.
43. (a) 40.0 J (b) −40.0 J (c) 62.5 J
45. (A/r^2) away from the other particle
47. (a) + at Ⓑ, − at Ⓓ, 0 at Ⓐ, Ⓒ, and Ⓔ (b) Ⓒ stable; Ⓐ and Ⓔ unstable
(c)

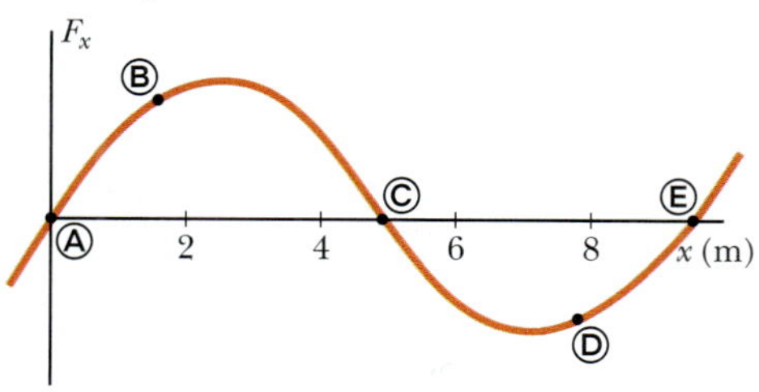

49. (c)

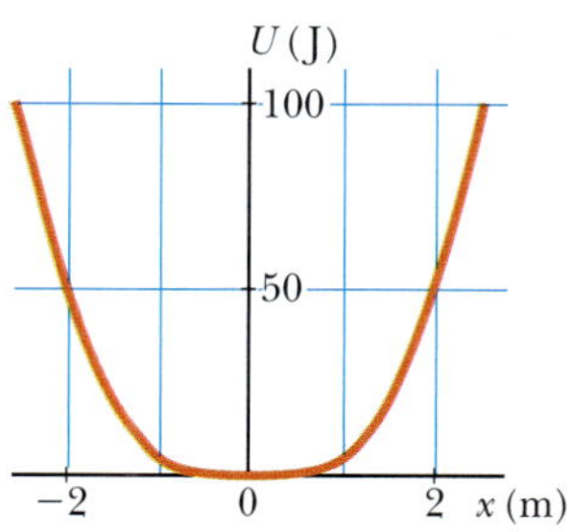

Equilibrium at $x = 0$

(d) 0.823 m/s
51. 90.0 J
53. (a) $x = (3.62\ m)/(4.30 - 23.4m)$ where x is in meters and m is in kilograms (b) 0.095 1 m (c) 0.492 m (d) 6.85 m (e) The situation is impossible. (f) The extension is directly proportional to m when m is only a few grams. Then it grows faster and faster, diverging to infinity for m = 0.184 kg.
55. $U(x) = 1 + 4e^{-2x}$. The force must be conservative because the work the force does on the object on which it acts depends only on the original and final positions of the object, not on the path between them.
57. 1.68 m/s
59. 0.799 J

CHAPTER 8

1. (a) $\Delta E_{\text{int}} = Q + T_{\text{ET}} + T_{\text{ER}}$ (b) $\Delta K + \Delta U + \Delta E_{\text{int}} = W + Q + T_{\text{MW}} + T_{\text{MT}}$ (c) $\Delta U = Q + T_{\text{MT}}$ (d) $0 = Q + T_{\text{MT}} + T_{\text{ET}} + T_{\text{ER}}$
3. (a) $v = (3gR)^{1/2}$ (b) 0.098 0 N down
5. 10.2 m
7. (a) 4.43 m/s (b) 5.00 m
9. 5.49 m/s
11. (a) 25.8 m (b) 27.1 m/s^2
13. (a) 650 J (b) 588 J (c) 0 (d) 0 (e) 62.0 J (f) 1.76 m/s
15. (a) −168 J (b) 184 J (c) 500 J (d) 148 J (e) 5.65 m/s
17. 2.04 m
19. 3.74 m/s
21. (a) −160 J (b) 73.5 J (c) 28.8 N (d) 0.679
23. (a) 1.40 m/s (b) 4.60 cm after release (c) 1.79 m/s
25. (a) 0.381 m (b) 0.143 m (c) 0.371 m
27. (a) $a_x = -\mu_k gx/L$ (b) $v = (\mu_k gL)^{1/2}$
29. 875 W
31. ~ 10^4 W
33. $46.2
35. (a) 10.2 kW (b) 10.6 kW (c) 5.82 MJ
37. (a) 11.1 m/s (b) 19.6 m/s^2 upward (c) 2.23×10^3 N upward (d) 1.01×10^3 J (e) 5.14 m/s (f) 1.35 m (g) 1.39 s
39. (a) $(2 + 24t^2 + 72t^4)$ J (b) $12t$ m/s^2; $48t$ N (c) $(48t + 288t^3)$ W (d) 1 250 J
41. (a) 1.38×10^4 J (b) 3.02×10^4 W
43. (a) 4.12 m (b) 3.35 m
45. (a) 2.17 kW (b) 58.6 kW
47. (a) x = −4.0 mm (b) −1.0 cm
49. 33.4 kW
51. (a) 0.225 J (b) ΔE_{mech} = −0.363 J (c) No. The normal force changes in a complicated way.
53. (a) 100 J (b) 0.410 m (c) 2.84 m/s (d) −9.80 mm (e) 2.85 m/s
55. 0.328
57. 1.24 m/s
59. (a) 0.400 m (b) 4.10 m/s (c) The block stays on the track.
61. $2m$
65. (a) 14.1 m/s (b) −7.90 kJ (c) 800 N (d) 771 N (e) 1.57 kN up

CHAPTER 9

1. (a) $(9.00\hat{\mathbf{i}} - 12.0\hat{\mathbf{j}})$ kg · m/s (b) 15.0 kg · m/s at 307°
3. ~ 10^{-23} m/s
5. (b) $p = \sqrt{2mK}$
7. (a) 13.5 N · s (b) 9.00 kN (c) 18.0 kN
9. 260 N normal to the wall
11. (a) $12.0\hat{\mathbf{i}}$ N · s (b) $4.80\hat{\mathbf{i}}$ m/s (c) $2.80\hat{\mathbf{i}}$ m/s (d) $2.40\hat{\mathbf{i}}$ N
13. (b) small (d) large (e) no difference
15. 301 m/s
17. (a) 2.50 m/s (b) 37.5 kJ (c) Each process is the time-reversal of the other. The same momentum conservation equation describes both.
19. 0.556 m
21. (a) $\vec{\mathbf{v}}_g = 1.15\hat{\mathbf{i}}$ m/s (b) $\vec{\mathbf{v}}_p = -0.346\hat{\mathbf{i}}$ m/s
23. (a) 0.284 (b) 115 fJ and 45.4 fJ

25. 91.2 m/s
27. 2.50 m/s at −60.0°
29. $v_{\text{orange}} = 3.99$ m/s, $v_{\text{yellow}} = 3.01$ m/s
31. $(3.00\hat{\mathbf{i}} - 1.20\hat{\mathbf{j}})$ m/s
33. (a) $(-9.33\hat{\mathbf{i}} - 8.33\hat{\mathbf{j}})$ Mm/s (b) 439 fJ
35. $\vec{\mathbf{r}}_{\text{CM}} = (0\hat{\mathbf{i}} + 1.00\hat{\mathbf{j}})$ m
37. $\vec{\mathbf{r}}_{\text{CM}} = (11.7\hat{\mathbf{i}} + 13.3\hat{\mathbf{j}})$ cm
39. (a) 15.9 g (b) 0.153 m
41. (a) $(1.40\hat{\mathbf{i}} + 2.40\hat{\mathbf{j}})$ m/s (b) $(7.00\hat{\mathbf{i}} + 12.0\hat{\mathbf{j}})$ kg · m/s
43. 0.700 m
45. (a) Yes. $18.0\hat{\mathbf{i}}$ kg · m/s. (b) No. The floor does zero work. (c) Yes. We could say that the final momentum of the cart came from the floor or from the Earth through the floor. (d) No. The kinetic energy came from the original gravitational energy of the elevated load, in amount 27.0 J. (e) Yes. The acceleration is caused by the static friction force exerted by the floor that prevents the caterpillar tracks from slipping backward.
47. (b) 2.06 m/s (c) Yes. The bumper continues to exert a force to the left until the particle has swung down to its lowest point.
49. (a) 3.75 kg · m/s² to the right (b) 3.75 N to the right (c) 3.75 N (d) 2.81 J (e) 1.41 J (f) Friction between sand and belt converts half of the input work into extra internal energy.
51. (a) 39.0 MN (b) 3.20 m/s² up
53. (a) 442 metric tons (b) 19.2 metric tons. This amount is much less than the value suggested. Mathematically, the logarithm in the rocket propulsion equation is not a linear function. Physically, a higher exhaust speed has an extra-large cumulative effect on the rocket frame's final speed, by counting again and again in the speed the frame attains second after second during its burn.
55. 240 s
57. $\left(\dfrac{M+m}{m}\right)\sqrt{\dfrac{gd^2}{2h}}$
59. (a) 0; inelastic
(b) $(-0.250\hat{\mathbf{i}} + 0.750\hat{\mathbf{j}} - 2.00\hat{\mathbf{k}})$ m/s; perfectly inelastic
(c) either $a = -6.74$ with $\vec{\mathbf{v}} = -0.419\,\hat{\mathbf{k}}$ m/s or $a = 2.74$ with $\vec{\mathbf{v}} = -3.58\,\hat{\mathbf{k}}$ m/s
61. (a) $m/M = 0.403$ (b) no changes; no difference
63. (b) 0.042 9 (c) 1.00 (d) Energy is an entirely different thing from momentum. A comparison: When children eat their soup, they do not eat the tablecloth. Another comparison: When a photographer's single-use flashbulb flashes, a magnesium filament oxidizes. Chemical energy disappears. (Internal energy appears and light carries some energy away.) The measured mass of the flashbulb is the same before and after. It can be the same despite the 100% energy conversion because energy and mass are totally different things in classical physics. In the ballistic pendulum, conversion of energy from mechanical into internal does not upset conservation of mass or conservation of momentum.
65. (a) $-0.256\hat{\mathbf{i}}$ m/s and $0.128\hat{\mathbf{i}}$ m/s
(b) $-0.064\,2\hat{\mathbf{i}}$ m/s and 0 (c) 0 and 0
67. (a) 100 m/s (b) 374 J
69. $(3Mgx/L)\hat{\mathbf{j}}$

CHAPTER 10

1. (a) 5.00 rad, 10.0 rad/s, 4.00 rad/s² (b) 53.0 rad, 22.0 rad/s, 4.00 rad/s²
3. (a) 4.00 rad/s² (b) 18.0 rad
5. (a) 5.24 s (b) 27.4 rad
7. (a) 7.27×10^{-5} rad/s (b) 2.57×10^{4} s = 428 min
9. 50.0 rev
11. $\sim 10^{7}$ rev
13. (a) 8.00 rad/s (b) 8.00 m/s, $a_r = -64.0$ m/s², $a_t = 4.00$ m/s² (c) 9.00 rad
15. (a) $(-2.73\hat{\mathbf{i}} + 1.24\hat{\mathbf{j}})$ m (b) in the second quadrant, at 156° (c) $(-1.85\hat{\mathbf{i}} - 4.10\hat{\mathbf{j}})$ m/s (d) toward the third quadrant, at 246°

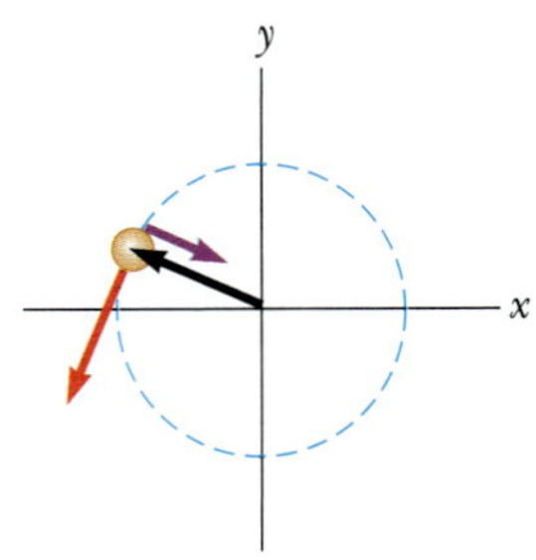

(e) $(6.15\hat{\mathbf{i}} - 2.78\hat{\mathbf{j}})$ m/s² (f) $(24.6\hat{\mathbf{i}} - 11.1\hat{\mathbf{j}})$ N
17. (a) 126 rad/s (b) 3.77 m/s (c) 1.26 km/s² (d) 20.1 m
19. 0.572
21. (a) 143 kg · m² (b) 2.57 kJ
25. (a) 24.5 m/s (b) no; no; no; no; yes
27. 1.28 kg · m²
29. $\sim 10^{0}$ kg · m²
33. −3.55 N · m
35. (a) 24.0 N · m (b) 0.035 6 rad/s² (c) 1.07 m/s²
37. (a) 0.309 m/s² (b) 7.67 N and 9.22 N
39. 21.5 N
41. 24.5 km
43. 149 rad/s
45. (a) 1.59 m/s (b) 53.1 rad/s
47. (a) 11.4 N, 7.57 m/s², 9.53 m/s down (b) 9.53 m/s
51. (a) $2(Rg/3)^{1/2}$ (b) $4(Rg/3)^{1/2}$ (c) $(Rg)^{1/2}$
53. (a) 500 J (b) 250 J (c) 750 J
55. (a) $\frac{2}{3}g\sin\theta$ for the disk, larger than $\frac{1}{2}g\sin\theta$ for the hoop (b) $\frac{1}{3}\tan\theta$
57. 1.21×10^{-4} kg · m²; height is unnecessary
59. $\frac{1}{3}\ell$
61. (a) 4.00 J (b) 1.60 s (c) yes
63. (a) $\omega = 3F\ell/b$ (b) $\alpha = 3F\ell/mL^2$ (c) and (d) Both larger. A component of the thrust force, exerted by the water about to spray from the ends of the arms, causes a forward torque on the rotor. Notice also that the rotor with bent arms has a slightly smaller moment of inertia than it would if the same metal tubes were straight.
65. (a) $(3g/L)^{1/2}$ (b) $3g/2L$ (c) $-\frac{3}{2}g\hat{\mathbf{i}} - \frac{3}{4}g\hat{\mathbf{j}}$ (d) $-\frac{3}{2}Mg\hat{\mathbf{i}} + \frac{1}{4}Mg\hat{\mathbf{j}}$
67. −0.322 rad/s²
71. (a) 118 N and 156 N (b) 1.17 kg · m²
73. (a) $\alpha = -0.176$ rad/s² (b) 1.29 rev (c) 9.26 rev
75. (a) $\omega(2h^3/g)^{1/2}$ (b) 0.011 6 m (c) Yes; the deflection is only 0.02% of the original height.
79. (a) $2.70R$ (b) $\Sigma F_x = -20mg/7$, $\Sigma F_y = -5mg/7$
81. (a) $(3gh/4)^{1/2}$ (b) $(3gh/4)^{1/2}$
83. (c) $(8Fd/3M)^{1/2}$
85. to the left

CHAPTER 11

1. $-7.00\hat{\mathbf{i}} + 16.0\hat{\mathbf{j}} - 10.0\hat{\mathbf{k}}$
3. (a) $-17.0\hat{\mathbf{k}}$ (b) 70.6°
5. 0.343 N · m horizontally north
7. 45.0°
9. $F_3 = F_1 + F_2$; no
11. $17.5\hat{\mathbf{k}}$ kg · m²/s
13. $(60.0\hat{\mathbf{k}})$ kg · m²/s
15. $mvR[\cos(vt/R) + 1]\hat{\mathbf{k}}$
17. (a) zero (b) $(-mv_i^3 \sin^2\theta \cos\theta/2g)\hat{\mathbf{k}}$ (c) $(-2mv_i^3 \sin^2\theta\cos\theta/g)\hat{\mathbf{k}}$ (d) The downward gravitational force exerts a torque in the $-z$ direction.
19. (a) $-m\ell gt\cos\theta\hat{\mathbf{k}}$ (b) The planet exerts a gravitational torque on the ball. (c) $-mg\ell\cos\theta\hat{\mathbf{k}}$
23. (a) 0.360 kg · m²/s (b) 0.540 kg · m²/s
25. (a) 0.433 kg · m²/s (b) 1.73 kg · m²/s
27. (a) 1.57×10^8 kg · m²/s (b) 6.26×10^3 s = 1.74 h
29. (a) $\omega_f = \omega_i I_1/(I_1 + I_2)$ (b) $I_1/(I_1 + I_2)$
31. (a) 11.1 rad/s counterclockwise (b) No. 507 J is transformed into internal energy. (c) No. The turntable bearing promptly imparts impulse 44.9 kg · m/s north into the turntable-clay system and thereafter keeps changing the system momentum.
33. 7.14 rev/min
35. (a) Mechanical energy is not conserved; some chemical energy is converted into mechanical energy. Momentum is not conserved. The turntable bearing exerts an external northward force on the axle. Angular momentum is conserved. (b) 0.360 rad/s counterclockwise (c) 99.9 J
37. (a) $mv\ell$ down (b) $M/(M + m)$
39. (a) $\omega = 2mv_i d/[M + 2m]R^2$ (b) No; some mechanical energy changes into internal energy. (c) Momentum is not conserved. The axle exerts a backward force on the cylinder.
41. $\sim 10^{-13}$ rad/s
43. 5.45×10^{22} N · m
45. (a) $1.67\hat{\mathbf{i}}$ m/s (b) 0.033 5 = 3.35% (c) $1.67\hat{\mathbf{i}}$ m/s (d) 15.8 rad/s (e) 1.00 = 100%
47. (a) $7md^2/3$ (b) $mgd\hat{\mathbf{k}}$ (c) $3g/7d$ counterclockwise (d) $2g/7$ upward (e) mgd (f) $\sqrt{6g/7d}$ (g) $m\sqrt{14gd^3}/3$ (h) $\sqrt{2gd/21}$
49. 0.910 km/s
51. (a) $v_i r_i/r$ (b) $T = (mv_i^2 r_i^2)r^{-3}$ (c) $\frac{1}{2}mv_i^2(r_i^2/r^2 - 1)$ (d) 4.50 m/s, 10.1 N, 0.450 J
53. (a) 3 750 kg · m²/s (b) 1.88 kJ (c) 3 750 kg · m²/s (d) 10.0 m/s (e) 7.50 kJ (f) 5.62 kJ
55. (a) $2mv_0$ (b) $2v_0/3$ (c) $4m\ell v_0/3$ (d) $4v_0/9\ell$ (e) mv_0^2 (f) $26mv_0^2/27$ (g) No horizontal forces act on the bola from outside after release, so the horizontal momentum stays constant. Its center of mass moves steadily with the horizontal velocity it had at release. No torques about its axis of rotation act on the bola, so its spin angular momentum stays constant. Internal forces cannot affect momentum conservation and angular momentum conservation, but they can affect mechanical energy. Energy $mv_0^2/27$ changes from mechanical energy into internal energy as the bola takes its stable configuration.
57. An increase of 0.550 s. It is not a significant change.

CHAPTER 12

1. $[(m_1 + m_b)d + m_1\ell/2]/m_2$
3. (3.85 cm, 6.85 cm)
5. (−1.50 m, −1.50 m)
7. (2.54 m, 4.75 m)
9. 177 kg
11. (a) $f_s = 268$ N, $n = 1\,300$ N (b) 0.324
13. 2.94 kN on each rear wheel and 4.41 kN on each front wheel
15. (a) 29.9 N (b) 22.2 N
17. (a) 1.73 rad/s² (b) 1.56 rad/s (c) $(-4.72\hat{\mathbf{i}} + 6.62\hat{\mathbf{j}})$ kN (d) $38.9\hat{\mathbf{j}}$ kN
19. 2.82 m
21. 88.2 N and 58.8 N
23. 4.90 mm
25. 23.8 μm
27. (a) 3.14×10^4 N (b) 6.28×10^4 N
29. 1.65×10^8 N/m²
31. 0.860 mm
33. $n_A = 5.98 \times 10^5$ N, $n_B = 4.80 \times 10^5$ N
35. 9.00 ft
37. (a)

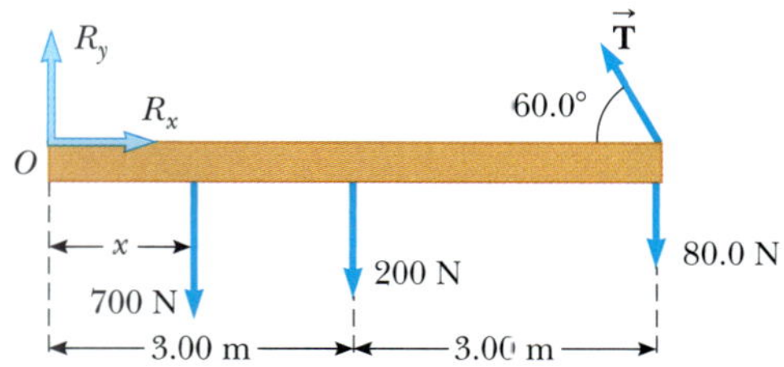

(b) $T = 343$ N, $R_x = 171$ N to the right, $R_y = 683$ N up (c) 5.13 m
39. (a) $T = F_g(L + d)/[\sin\theta\,(2L + d)]$ (b) $R_x = F_g(L + d)\cot\theta/(2L + d)$, $R_y = F_gL/(2L + d)$
41. $\vec{\mathbf{F}}_A = (-6.47 \times 10^5\hat{\mathbf{i}} + 1.27 \times 10^5\hat{\mathbf{j}})$ N, $\vec{\mathbf{F}}_B = 6.47 \times 10^5\hat{\mathbf{i}}$ N
43. 5.08 kN, $R_x = 4.77$ kN, $R_y = 8.26$ kN
45. (a) 20.1 cm to the left of the front edge; $\mu_k = 0.571$ (b) 0.501 m
47. (a) $M = (m/2)(2\mu_s\sin\theta - \cos\theta)(\cos\theta - \mu_s\sin\theta)^{-1}$ (b) $R = (m + M)g(1 + \mu_s^2)^{1/2}$, $F = g[M^2 + \mu_s^2(m + M)^2]^{1/2}$
49. (b) AB compression 732 N, AC tension 634 N, BC compression 897 N
51. (a) 133 N (b) $n_A = 429$ N and $n_B = 257$ N (c) $R_x = 133$ N and $R_y = -257$ N
55. 1.09 m
57. (a) 4 500 N (b) 4.50×10^6 N/m² (c) The board will break.
59. (a) $P_y = (F_g/L)(d - ah/g)$ (b) 0.306 m (c) $(-306\hat{\mathbf{i}} + 553\hat{\mathbf{j}})$ N

CHAPTER 13

1. $\sim 10^{-7}$ N toward you
3. (a) 2.50×10^{-5} N toward the 500-kg object (b) between the objects and 0.245 m from the 500-kg object
5. $(-100\hat{\mathbf{i}} + 59.3\hat{\mathbf{j}})$ pN
7. 7.41×10^{-10} N
9. 0.613 m/s² toward the Earth
11. $\rho_{\text{Moon}}/\rho_{\text{Earth}} = \frac{2}{3}$
13. 1.26×10^{32} kg
15. 1.90×10^{27} kg
17. 8.92×10^7 m
19. After 3.93 yr, Mercury would be farther from the Sun than Pluto.

21. $\vec{\mathbf{g}} = \frac{Gm}{\ell^2}\left(\frac{1}{2} + \sqrt{2}\right)$ toward the opposite corner

23. (a) $\vec{\mathbf{g}} = 2MGr(r^2 + a^2)^{-3/2}$ toward the center of mass (b) At $r = 0$, the fields of the two objects are equal in magnitude and opposite in direction, to add to zero. (d) When r is much greater than a, the fact that the two masses are separate is unimportant. They create a total field like that of a single object of mass $2M$.

25. (a) 1.84×10^9 kg/m^3 (b) 3.27×10^6 m/s^2 (c) -2.08×10^{13} J

27. (a) -1.67×10^{-14} J (b) Each object will slowly accelerate toward the center of the triangle, where the three will simultaneously collide.

29. (b) 340 s

31. 1.66×10^4 m/s

35. (a) 5.30×10^3 s (b) 7.79 km/s (c) 6.43×10^9 J

37. (b) 1.00×10^7 m (c) 1.00×10^4 m/s

39. (a) 0.980 (b) 127 yr (c) -2.13×10^{17} J

43. (b) $2[Gm^3(1/2r - 1/R)]^{1/2}$

45. (a) -7.04×10^4 J (b) -1.57×10^5 J (c) 13.2 m/s

47. 7.79×10^{14} kg

49. $\omega = 0.057\,2$ rad/s or 1 rev in 110 s

51. (a) $m_2(2G/d)^{1/2}(m_1 + m_2)^{-1/2}$ and $m_1(2G/d)^{1/2}(m_1 + m_2)^{-1/2}$; relative speed $(2G/d)^{1/2}(m_1 + m_2)^{1/2}$ (b) 1.07×10^{32} J and 2.67×10^{31} J

53. (a) 200 Myr (b) $\sim 10^{41}$ kg; $\sim 10^{11}$ stars

55. $(GM_E/4R_E)^{1/2}$

59. $(800 + 1.73 \times 10^{-4})\hat{\mathbf{i}}$ m/s and $(800 - 1.73 \times 10^{-4})\hat{\mathbf{i}}$ m/s

61. 18.2 ms

CHAPTER 14

1. 0.111 kg

3. 6.24 MPa

5. 1.62 m

7. 7.74×10^{-3} m^2

9. 271 kN horizontally backward

11. 5.88×10^6 N down; 196 kN outward; 588 kN outward

13. 0.722 mm

15. 10.5 m; no because some alcohol and water evaporate

17. 98.6 kPa

19. (a) 1.57 Pa, 1.55×10^{-2} atm, 11.8 mm Hg (b) The fluid level in the tap should rise. (c) blockage of flow of the cerebrospinal fluid

21. 0.258 N down

23. (a) $1.017\,9 \times 10^3$ N down, $1.029\,7 \times 10^3$ N up (b) 86.2 N (c) By either method of evaluation, the buoyant force is 11.8 N up.

25. (a) 1.20×10^3 N/s (b) 0

27. (a) 7.00 cm (b) 2.80 kg

31. 1 430 m^3

33. 1 250 kg/m^3 and 500 kg/m^3

35. (a) 17.7 m/s (b) 1.73 mm

37. 31.6 m/s

39. 0.247 cm

41. (a) 2.28 N toward Holland (b) 1.74×10^6 s

43. (a) 1 atm + 15.0 MPa (b) 2.95 m/s (c) 4.34 kPa

45. 2.51×10^{-3} m^3/s

47. (a) 4.43 m/s (b) The siphon can be no higher than 10.3 m.

49. 12.6 m/s

51. 1.91 m

55. 0.604 m

57. If the helicopter could create the air it expels downward, the mass flow rate of the air would have to be at least 233 kg/s. In reality, the rotor takes in air from above, which is moving over a larger area with lower speed, and blows it downward at higher speed. The amount of this air has to be at least a few times larger than 233 kg every second.

61. 17.3 N and 31.7 N

63. 90.04%

65. 758 Pa

67. 4.43 m/s

69. (a) 1.25 cm (b) 13.8 m/s

71. (c) 1.70 m^2

CHAPTER 15

1. (a) The motion repeats precisely. (b) 1.81 s (c) No, the force is not in the form of Hooke's law

3. (a) 1.50 Hz, 0.667 s (b) 4.00 m (c) π rad (d) 2.83 m

5. (b) 18.8 cm/s, 0.333 s (c) 178 cm/s^2, 0.500 s (d) 12.0 cm

7. 40.9 N/m

9. 18.8 m/s, 7.11 km/s^2

11. (a) 40.0 cm/s, 160 cm/s^2 (b) 32.0 cm/s, -96.0 cm/s^2 (c) 0.232 s

13. 0.628 m/s

15. 2.23 m/s

17. (a) 28.0 mJ (b) 1.02 m/s (c) 12.2 mJ (d) 15.8 mJ

19. 2.60 cm and -2.60 cm

21. (a) at 0.218 s and at 1.09 s (b) 0.014 6 W

23. (b) 0.628 s

25. Assuming simple harmonic motion, (a) 0.820 m/s, (b) 2.57 rad/s^2, and (c) 0.641 N. More precisely, (a) 0.817 m/s, (b) 2.54 rad/s^2, and (c) 0.634 N. The answers agree to two digits. The answers computed from conservation of energy and from Newton's second law are more precisely correct. With this amplitude, the motion of the pendulum is approximately simple harmonic.

29. 0.944 kg · m^2

33. (a) 5.00×10^{-7} kg · m^2 (b) 3.16×10^{-4} N · m/rad

35. 1.00×10^{-3} s^{-1}

37. (a) 7.00 Hz (b) 2.00% (c) 10.6 s

39. (a) 1.00 s (b) 5.09 cm

41. 318 N

43. 1.74 Hz

45. (a) 2.09 s (b) 0.477 Hz (c) 36.0 cm/s (d) $(0.064\,8\ \text{m}^2/\text{s}^2)m$ (e) $(9.00/\text{s}^2)m$ (f) Period, frequency, and maximum speed are all independent of mass in this situation. The energy and the force constant are directly proportional to mass.

47. (a) $2Mg$, $Mg(1 + y/L)$ (b) $T = (4\pi/3)(2L/g)^{1/2}$, 2.68 s

49. 6.62 cm

51. 9.19×10^{13} Hz

53. (a)

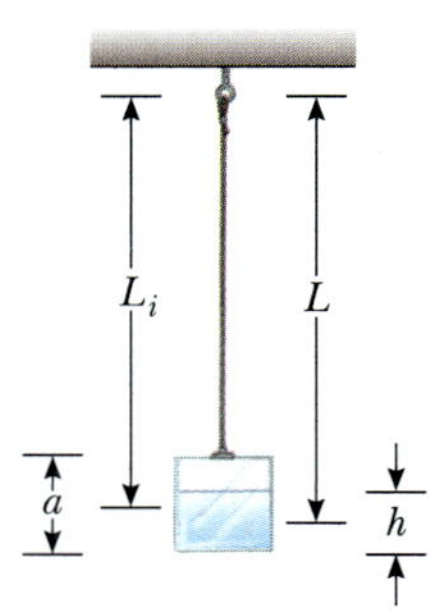

(b) $\dfrac{dT}{dt} = \dfrac{\pi\, dM/dt}{2\rho a^2 g^{1/2}[L_i + (dM/dt)t/2\rho a^2]^{1/2}}$

(c) $T = 2\pi g^{-1/2}\left[L_i + \left(\dfrac{dM}{dt}\right)\left(\dfrac{t}{2\rho a^2}\right)\right]^{1/2}$

55. $f = (2\pi L)^{-1}\left(gL + \dfrac{kh^2}{M}\right)^{1/2}$

57. (b) 1.23 Hz

59. (a) 3.00 s (b) 14.3 J (c) 25.5°

61. If the cyclist goes over washboard bumps at one certain speed, they can excite a resonance vibration of the bike, so large in amplitude as to make the rider lose control. $\sim 10^1$ m

69. (b) after 42.2 minutes

CHAPTER 16

1. $y = 6\,[(x - 4.5t)^2 + 3]^{-1}$

3. (a) the P wave (b) 665 s

5. (a) $(3.33\hat{\mathbf{i}})$ m/s (b) −5.48 cm (c) 0.667 m, 5.00 Hz (d) 11.0 m/s

7. 0.319 m

9. 2.00 cm, 2.98 m, 0.576 Hz, 1.72 m/s

11. (a) 31.4 rad/s (b) 1.57 rad/m (c) $y = (0.120 \text{ m}) \sin(1.57x - 31.4t)$ where x is in meters and t is in seconds (d) 3.77 m/s (e) 118 m/s²

13. (a) 0.250 m (b) 40.0 rad/s (c) 0.300 rad/m (d) 20.9 m (e) 133 m/s (f) $+x$

15. (a) $y = (8.00 \text{ cm}) \sin(7.85x + 6\pi t)$ (b) $y = (8.00 \text{ cm}) \sin(7.85x + 6\pi t - 0.785)$

17. (a) −1.51 m/s, 0 (b) 16.0 m, 0.500 s, 32.0 m/s

19. (a) 0.500 Hz, 3.14 rad/s (b) 3.14 rad/m (c) $(0.100 \text{ m}) \sin(3.14\, x/\text{m} - 3.14\, t/\text{s})$ (d) $(0.100 \text{ m}) \sin(-3.14\, t/\text{s})$ (e) $(0.100 \text{ m}) \sin(4.71 \text{ rad} - 3.14\, t/\text{s})$ (f) 0.314 m/s

21. 80.0 N

23. 520 m/s

25. 1.64 m/s²

27. 13.5 N

29. 185 m/s

31. 0.329 s

35. 55.1 Hz

37. (a) 62.5 m/s (b) 7.85 m (c) 7.96 Hz (d) 21.1 W

39. $\sqrt{2}\,\mathcal{P}_0$

41. (a) $A = 40$ (b) $A = 7.00$, $B = 0$, $C = 3.00$. One can take the dot product of the given equation with each one of $\hat{\mathbf{i}}, \hat{\mathbf{j}}$, and $\hat{\mathbf{k}}$. (c) $A = 0$, $B = 7.00$ mm, $C = 3.00$/m, $D = 4.00$/s, $E = 2.00$. Consider the average value of both sides of the given equation to find A. Then consider the maximum value of both sides to find B. You can evaluate the partial derivative of both sides of the given equation with respect to x and separately with respect to t to obtain equations yielding C and D upon chosen substitutions for x and t. Then substitute $x = 0$ and $t = 0$ to obtain E.

45. ~ 1 min

47. 0.456 m/s

49. (a) 39.2 N (b) 0.892 m (c) 83.6 m/s

51. (a) The energy a wave crest carries is constant in the absence of absorption. Then the rate at which energy moves beyond a fixed distance from the source, which is the power of the wave, is constant. The power is proportional to the square of the amplitude and to the wave speed. The speed decreases as the wave moves into shallower water near shore, so the amplitude must increase. (b) 8.31 m (c) As the water depth goes to zero, our model would predict zero speed and infinite amplitude. The amplitude must be finite as the wave comes ashore. As the speed decreases, the wavelength also decreases. When it becomes comparable to the water depth, or smaller, the expression $v = \sqrt{gd}$ no longer applies.

53. (a) $\mathcal{P} = (0.050\,0 \text{ kg/s})v_{y,\max}^2$ (b) The power is proportional to the square of the maximum element speed. (c) $(7.5 \times 10^{-4} \text{ kg})v_{y,\max}^2 = \frac{1}{2}m_3 v_{y,\max}^2$ (d) $(0.300 \text{ kg})v_{y,\max}^2$

55. 0.084 3 rad

59. (a) $(0.707)2(L/g)^{1/2}$ (b) $L/4$

61. 3.86×10^{-4}

63. (a) $\dfrac{\mu\omega^3}{2k}A_0^{\,2}e^{-2bx}$ (b) $\dfrac{\mu\omega^3}{2k}A_0^{\,2}$ (c) e^{-2bx}

65. (a) $\mu_0 + (\mu_L - \mu_0)x/L$

CHAPTER 17

1. 5.56 km. As long as the speed of light is much greater than the speed of sound, its actual value does not matter.

3. 0.196 s

5. 7.82 m

7. (a) 826 m (b) 1.47 s

9. (a) 0.625 mm (b) 1.50 mm to 75.0 μm

11. (a) 2.00 μm, 40.0 cm, 54.6 m/s (b) −0.433 μm (c) 1.72 mm/s

13. $\Delta P = (0.200 \text{ N/m}^2)\sin(62.8x/\text{m} - 2.16 \times 10^4 t/\text{s})$

15. 5.81 m

17. 66.0 dB

19. (a) 3.75 W/m² (b) 0.600 W/m²

21. (a) 2.34 m and 0.390 m (b) 0.161 N/m² for both notes (c) 4.25×10^{-7} m and 7.09×10^{-8} m (d) The wavelengths and displacement amplitudes would be larger by a factor of 1.09. The answer to part (b) would be unchanged.

23. (a) 1.32×10^{-4} W/m² (b) 81.2 dB

25. (a) 0.691 m (b) 691 km

27. 65.6 dB

29. (a) 30.0 m (b) 9.49×10^5 m

31. (a) 332 J (b) 46.4 dB

33. (a) 3.04 kHz (b) 2.08 kHz (c) 2.62 kHz, 2.40 kHz

35. 26.4 m/s

37. 19.3 m

39. (a) 56.3 s (b) 56.6 km farther along

41. 2.82×10^8 m/s

43. It is unreasonable, implying a sound level of 123 dB. Nearly all the missing mechanical energy becomes internal energy in the latch.

45. (a) f is a few hundred hertz. λ ~ 1 m, duration ~ 0.1 s. (b) Yes. The frequency can be close to 1 000 Hz. If the person clapping his or her hands is at the base of the pyramid, the echo can drop somewhat in frequency and in loudness as sound returns, with the later cycles coming from the smaller and more distant upper risers. The sound could imitate some particular bird and could in fact be a recording of the call.

49. (a) 0.515/min (b) 0.614/min

51. (a) 55.8 m/s (b) 2 500 Hz

53. 1 204.2 Hz

55. (a) 0.642 W (b) 0.004 28 = 0.428%

57. (a) The sound through the metal arrives first. (b) (365 m/s) Δt (c) 46.3 m (d) The answer becomes

$$\ell = \frac{\Delta t}{\dfrac{1}{331\text{ m/s}} - \dfrac{1}{v_r}}$$

where v_r is the speed of sound in the rod. As v_r goes to infinity, the travel time in the rod becomes negligible. The answer approaches (331 m/s) Δt, which is the distance the sound travels in air during the delay time.

59. (a) 0.948° (b) 4.40°

61. 1.34×10^4 N

63. (a) 6.45 (b) 0

CHAPTER 18

1. (a) −1.65 cm (b) −6.02 cm (c) 1.15 cm

3. (a) $+x$, $-x$ (b) 0.750 s (c) 1.00 m

5. (a) 9.24 m (b) 600 Hz

7. (a) 2 (b) 9.28 m and 1.99 m

9. (a) 156° (b) 0.058 4 cm

11. 15.7 m, 31.8 Hz, 500 m/s

13. At 0.089 1 m, 0.303 m, 0.518 m, 0.732 m, 0.947 m, 1.16 m from one speaker

15. (a) 4.24 cm (b) 6.00 cm (c) 6.00 cm (d) 0.500 cm, 1.50 cm, 2.50 cm

17. 0.786 Hz, 1.57 Hz, 2.36 Hz, 3.14 Hz

19. (a) 350 Hz (b) 400 kg

21. (a) 163 N (b) 660 Hz

23. $\dfrac{Mg}{4Lf^2 \tan\theta}$

25. (a) 3 loops (b) 16.7 Hz (c) 1 loop

27. (a) 3.66 m/s (b) 0.200 Hz

29. (a) 0.357 m (b) 0.715 m

31. 0.656 m and 1.64 m

33. n(206 Hz) for $n = 1$ to 9 and n(84.5 Hz) for $n = 2$ to 23

35. 50.0 Hz, 1.70 m

37. (a) 350 m/s (b) 1.14 m

39. (21.5 ± 0.1) m. The data suggest 0.6-Hz uncertainty in the frequency measurements, which is only a little more than 1%.

41. (a) 1.59 kHz (b) odd-numbered harmonics (c) 1.11 kHz

43. 5.64 beats/s

45. (a) 1.99 beats/s (b) 3.38 m/s

47. The second harmonic of E is close to the third harmonic of A, and the fourth harmonic of C# is close to the fifth harmonic of A.

49. (a) The yo-yo's downward speed is $dL/dt = (0.8\text{ m/s}^2)(1.2\text{ s}) = 0.960$ m/s. The instantaneous wavelength of the fundamental string wave is given by $d_{NN} = \lambda/2 = L$, so $\lambda = 2L$ and $d\lambda/dt = 2\,dL/dt = 2(0.96\text{ m/s}) = 1.92$ m/s. (b) For the second harmonic, the wavelength is equal to the length of the string. Then the rate of change of wavelength is equal to $dL/dt = 0.960$ m/s, half as much as for the first harmonic. (c) A yo-yo of different mass will hold the string under different tension to make each string wave vibrate with a different frequency, but the geometrical argument given in parts (a) and (b) still applies to the wavelength. The answers are unchanged: $d\lambda_1/dt = 1.92$ m/s and $d\lambda_2/dt = 0.960$ m/s.

51. (a) 34.8 m/s (b) 0.977 m

53. 3.85 m/s away from the station or 3.77 m/s toward the station

55. (a) 59.9 Hz (b) 20.0 cm

57. (a) $\frac{1}{2}$ (b) $[n/(n+1)]^2T$ (c) $\frac{9}{16}$

59. $y_1 + y_2 = 11.2 \sin(2.00x - 10.0t + 63.4°)$

61. (a) 78.9 N (b) 211 Hz

CHAPTER 19

1. (a) −274°C (b) 1.27 atm (c) 1.74 atm

3. (a) −320°F (b) 77.3 K

5. 3.27 cm

7. (a) 0.176 mm (b) 8.78 μm (c) 0.093 0 cm^3

9. (a) −179°C is attainable. (b) −376°C is below 0 K and unattainable.

11. (a) 99.8 mL (b) about 6% of the volume change of the acetone

13. (a) 99.4 cm^3 (b) 0.943 cm

15. 5 336 images

17. (a) 400 kPa (b) 449 kPa

19. 1.50×10^{29} molecules

21. 472 K

23. (a) 41.6 mol (b) 1.20 kg, nearly in agreement with the tabulated density

25. (a) 1.17 g (b) 11.5 mN (c) 1.01 kN (d) The molecules must be moving very fast.

27. 4.39 kg

29. (a) 7.13 m (b) The open end of the tube should be at the bottom after the bird surfaces so that the water can drain out. There is no other requirement. Air does not tend to bubble out of a narrow tube.

31. (a) 94.97 cm (b) 95.03 cm

33. 3.55 cm

35. It falls by 0.094 3 Hz.

37. (a) Expansion makes density drop. (b) $5 \times 10^{-5}(°\text{C})^{-1}$

39. (a) $h = nRT/(mg + P_0A)$ (b) 0.661 m

41. We assume $\alpha\,\Delta T$ is much less than 1.

43. Yes, as long as the coefficients of expansion remain constant. The lengths L_C and L_S at 0°C need to satisfy $17L_C = 11L_S$. Then the steel rod must be longer. With $L_S - L_C = 5.00$ cm, the only possibility is $L_S = 14.2$ cm and $L_C = 9.17$ cm.

45. (a) 0.340% (b) 0.480%

47. 2.74 m

49. (b) 1.33 kg/m^3

53. No. Steel would need to be 2.30 times stronger.

55. (a) $L_f = L_i e^{\alpha\Delta T}$ (b) 2.00×10^{-4}%; 59.4%

57. (a) 6.17×10^{-3} kg/m (b) 632 N (c) 580 N; 192 Hz

59. 4.54 m

CHAPTER 20

1. (10.0 + 0.117)°C

3. 0.234 kJ/kg · °C

5. 1.78×10^4 kg

7. 29.6°C

9. (a) 0.435 cal/g · °C (b) We cannot make a definite identification. The material might be an unknown alloy or a material not listed in the table. It might be beryllium.

11. 23.6°C

13. 1.22×10^5 J

15. 0.294 g

17. 0.414 kg

19. (a) 0°C (b) 114 g

21. −1.18 MJ

23. -466 J

25. (a) $-4P_iV_i$ (b) It is proportional to the square of the volume, according to $T = (P_i/nRV_i)V^2$.

27. $Q = -720$ J

29.

	Q	W	ΔE_{int}
BC	$-$	0	$-$
CA	$-$	+	$-$
AB	+	$-$	+

31. (a) 7.50 kJ (b) 900 K

33. -3.10 kJ, 37.6 kJ

35. (a) 0.041 0 m^3 (b) +5.48 kJ (c) -5.48 kJ

37. 10.0 kW

39. 51.2°C

41. 74.8 kJ

43. (a) 0.964 kg or more (b) The test samples and the inner surface of the insulation can be prewarmed to 37.0°C as the box is assembled. Then nothing changes in temperature during the test period, and the masses of the test samples and insulation make no difference.

45. 3.49×10^3 K

47. Intensity is defined as power per area perpendicular to the direction of energy flow. The direction of sunlight is along the line from the Sun to the object. The perpendicular area is the projected flat, circular area enclosed by the *terminator*, the line that separates day and night on the object. The object radiates infrared light outward in all directions. The area perpendicular to this energy flow is its spherical surface area. The steady-state surface temperature is 279 K = 6°C. We find this temperature to be chilly, well below comfortable room temperatures.

49. 2.27 km

51. (a) 16.8 L (b) 0.351 L/s

53. $c = \mathcal{P}/\rho R\,\Delta T$

55. 5.87×10^4°C

57. 5.31 h

59. 1.44 kg

61. 38.6 m^3/d

63. 9.32 kW

65. (a) The equation $dT/dr = \mathcal{P}/4\pi kr^2$ represents the law of thermal conduction, incorporating the definition of thermal conductivity, applied to a spherical surface within the shell. The rate of energy transfer $\mathcal{P}$ must be the same for all radii so that each bit of material stays at a temperature that is constant in time. (b) We separate the variables T and r in the thermal conduction equation and integrate the equation between points on the interior and exterior surfaces. (c) 18.5 W (d) With $\mathcal{P}$ now known, we separate the variables again and integrate between a point on the interior surface and any point within the shell. (e) $T = 5°\text{C} + 184\ \text{cm}\cdot°\text{C}\ [1/(3\ \text{cm}) - 1/r]$ (f) 29.5°C

CHAPTER 21

1. (a) 4.00 u = 6.64×10^{-24} g (b) 55.9 u = 9.28×10^{-23} g (c) 207 u = 3.44×10^{-22} g

3. 0.943 N, 1.57 Pa

5. 3.21×10^{12} molecules

7. 3.32 mol

9. (a) 3.54×10^{23} atoms (b) 6.07×10^{-21} J (c) 1.35 km/s

11. (a) 8.76×10^{-21} J for both (b) 1.62 km/s for helium and 514 m/s for argon

13. (a) 3.46 kJ (b) 2.45 kJ (c) -1.01 kJ

15. Between 10^{-2}°C and 10^{-3}°C

17. $13.5PV$

19. (a) 1.39 atm (b) 366 K, 253 K (c) 0, -4.66 kJ, -4.66 kJ

21. 227 K

23. (a)

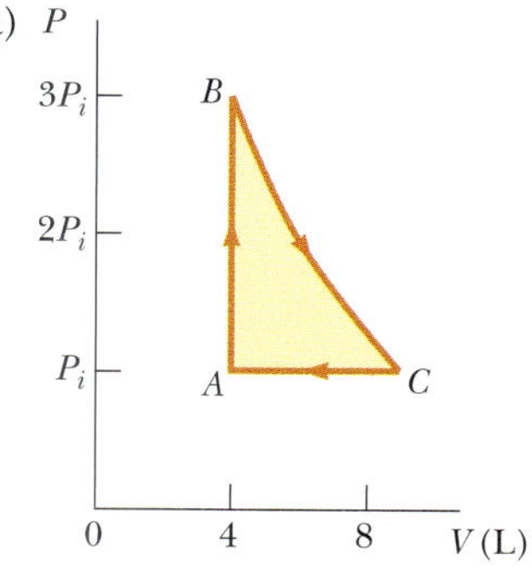

(b) 8.77 L (c) 900 K (d) 300 K (e) -336 J

25. (a) 28.0 kJ (b) 46.0 kJ (c) isothermal process: $P_f = 10.0$ atm; adiabatic process: $P_f = 25.1$ atm

27. (a) 9.95 cal/K, 13.9 cal/K (b) 13.9 cal/K, 17.9 cal/K

29. Sulfur dioxide is the gas in Table 21.2 with the greatest molecular mass. If the effective spring constants for various chemical bonds are comparable, SO_2 can then be expected to have low frequencies of atomic vibration. Vibration can be excited at lower temperature for sulfur dioxide than for the other gases. Some vibration may be going on at 300 K.

31. (a) 6.80 m/s (b) 7.41 m/s (c) 7.00 m/s

35. (a) 2.37×10^4 K (b) 1.06×10^3 K

37. (b) 0.278

39. (a) 100 kPa, 66.5 L, 400 K, 5.82 kJ, 7.48 kJ, -1.66 kJ
(b) 133 kPa, 49.9 L, 400 K, 5.82 kJ, 5.82 kJ, 0
(c) 120 kPa, 41.6 L, 300 K, 0, -909 J, +909 J
(d) 120 kPa, 43.3 L, 312 K, 722 J, 0, +722 J

41. (b) 447 J/kg·°C agrees with the tabulated value within 0.3%. (c) 127 J/kg·°C agrees with the tabulated value within 2%.

43. (b) The expressions are equal because $PV = nRT$ and $\gamma = (C_V + R)/C_V = 1 + R/C_V$ give $R = (\gamma - 1)C_V$, so $PV = n(\gamma - 1)C_VT$ and $PV/(\gamma - 1) = nC_VT$

45. 510 K and 290 K

47. 0.623

49. (a) Pressure increases as volume decreases.
(d) 0.500 atm^{-1}, 0.300 atm^{-1}

51. (a) 7.27×10^{-20} J (b) 2.20 km/s (c) 3 510 K. The evaporating molecules are exceptional, at the high-speed tail of the distribution of molecular speeds. The average speed of molecules in the liquid and in the vapor is appropriate only to room temperature.

53. (a) 0.514 m^3 (b) 2.06 m^3 (c) 2.38×10^3 K (d) -480 kJ (e) 2.28 MJ

55. 1.09×10^{-3}, 2.69×10^{-2}, 0.529, 1.00, 0.199, 1.01×10^{-41}, $1.25 \times 10^{-1\,082}$

59. (a) 0.203 mol (b) $T_B = T_C = 900$ K, $V_C = 15.0$ L

(c, d)	P, atm	V, L	T, K	E_{int}, kJ
A	1.00	5.00	300	0.760
B	3.00	5.00	900	2.28
C	1.00	15.0	900	2.28
A	1.00	5.00	300	0.760

(e) Lock the piston in place and put the cylinder into an oven at 900 K. Keep the gas in the oven while gradually letting the gas expand to lift a load on the piston as far as it can. Move the cylinder from the oven back to the 300-K room and let the gas cool and contract.

(f, g)	Q, kJ	W, kJ	ΔE_{int}, kJ
AB	1.52	0	1.52
BC	1.67	−1.67	0
CA	−2.53	+1.01	−1.52
$ABCA$	0.656	−0.656	0

61. (b) 1.60×10^4 K

CHAPTER 22

1. (a) 6.94% (b) 335 J
3. (a) 10.7 kJ (b) 0.533 s
5. 55.4%
7. 77.8 W
9. (a) 67.2% (b) 58.8 kW
11. The actual efficiency of 0.069 8 is less than four-tenths of the Carnot efficiency of 0.177.
13. (a) 741 J (b) 459 J
15. (a) 564 K (b) 212 kW (c) 47.5%
17. (b) $1 - T_c/T_h$, the same as for a single reversible engine (c) $(T_c + T_h)/2$ (d) $(T_hT_c)^{1/2}$
19. 9.00
23. 72.2 J
25. 23.1 mW
27. (a) 244 kPa (b) 192 J
29. (a) 51.2% (b) 36.2%
33. 195 J/K
35. 1.02 kJ/K
37. ~ 10^0 W/K from metabolism; much more if you are using high-power electric appliances or an automobile
39. 5.76 J/K; the temperature is constant if the gas is ideal.
41. (a) 1 (b) 6
43. (a)

Result	Number of ways to draw
All R	1
2 R, 1 G	3
1 R, 2 G	3
All G	1

(b)

Result	Number of ways to draw
All R	1
4R, 1G	5
3R, 2G	10
2R, 3G	10
1R, 4G	5
All G	1

45. (a) 214 J, 64.3 J (b) −35.7 J, −35.7 J. The net effect would be the transport of energy by heat from the cold to the hot reservoir without expenditure of external work. (c) 333 J, 233 J (d) 83.3 J, 83.3 J, 0. The net effect would be converting energy, taken in by heat, entirely into energy output by work in a cyclic process. (e) −0.111 J/K. The entropy of the Universe would have decreased.
47. (a) 5.00 kW (b) 763 W
49. (a) $2nRT_i \ln 2$ (b) 0.273
51. 5.97×10^4 kg/s
53. (a) 8.48 kW (b) 1.52 kW (c) 1.09×10^4 J/K (d) The COP drops by 20.0%.
55. (a) $10.5nRT_i$ (b) $8.50nRT_i$ (c) 0.190 (d) This efficiency is much less than the 0.833 for a Carnot engine operating between the temperatures used here.
57. (a) $nC_P \ln 3$ (b) Both ask for the change in entropy between the same two states of the same system. Entropy is a state variable. The change in entropy does not depend on path, but only on original and final states.
61. (a) 20.0°C (c) $\Delta S = +4.88$ J/K (d) The mixing is irreversible. It is clear that warm water and cool water do not come unmixed, and the entropy change is positive.

CHAPTER 23

1. (a) +160 zC, 1.01 u (b) +160 zC, 23.0 u (c) −160 zC, 35.5 u (d) +320 zC, 40.1 u (e) −480 zC, 14.0 u (f) +640 zC, 14.0 u (g) +1.12 aC, 14.0 u (h) −160 zC, 18.0 u
3. The force is ~ 10^{26} N.
5. (a) 1.59 nN away from the other (b) 1.24×10^{36} times larger (c) 8.61×10^{-11} C/kg
7. 0.872 N at 330°
9. (a) 2.16×10^{-5} N toward the other (b) 8.99×10^{-7} N away from the other
11. (a) 82.2 nN toward the other particle (b) 2.19 Mm/s
13. (a) 55.8 pN/C down (b) 102 nN/C up
15. The field at the origin can be to the right if the unknown charge is $-9Q$, or the field can be to the left if and only if the unknown charge is $+27Q$.
17. (a) $5.91k_eq/a^2$ at 58.8° (b) $5.91k_eq^2/a^2$ at 58.8°
19. (a) $k_eQx\hat{\mathbf{i}}/(R^2 + x^2)^{3/2}$ (b) As long as the charge is symmetrically placed, the number of charges does not matter. A continuous ring corresponds to n becoming larger without limit.
21. 1.59×10^6 N/C toward the rod
23. (a) $6.64\hat{\mathbf{i}}$ MN/C (b) $24.1\hat{\mathbf{i}}$ MN/C (c) $6.40\hat{\mathbf{i}}$ MN/C (d) $0.664\hat{\mathbf{i}}$ MN/C, taking the axis of the ring as the x axis
25. (a) 93.6 MN/C; the near-field approximation is 104 MN/C, about 11% high. (b) 0.516 MN/C; the charged-particle approximation is 0.519 MN/C, about 0.6% high.
27. $-21.6\hat{\mathbf{i}}$ MN/C
31. (a) 86.4 pC for each (b) 324 pC, 459 pC, 459 pC, 432 pC (c) 57.6 pC, 106 pC, 154 pC, 96.0 pC
33.

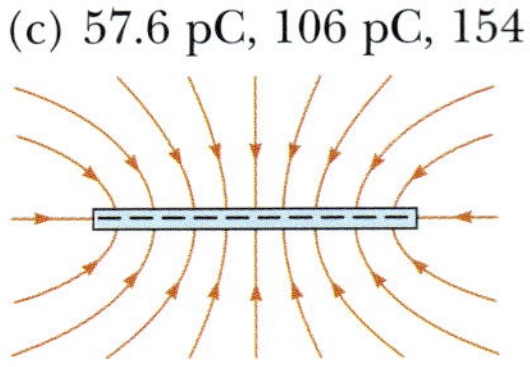

35. (a)

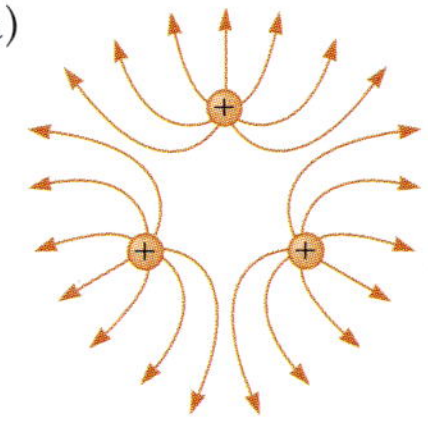

The field is zero at the center of the triangle. (b) $1.73\, k_eq\hat{\mathbf{j}}/a^2$
37. (a) 61.3 Gm/s² (b) 19.5 μs (c) 11.7 m (d) 1.20 fJ
39. K/ed in the direction of motion
41. (a) 111 ns (b) 5.68 mm (c) $(450\hat{\mathbf{i}} + 102\hat{\mathbf{j}})$ km/s

43. (a) 21.8 μm (b) 2.43 cm
45. (a) 10.9 nC (b) 5.44 mN
47. 40.9 N at 263°
49. $Q = 2L\sqrt{\frac{k(L - L_i)}{k_e}}$
53. $-707\hat{\mathbf{j}}$ mN
55. (a) $\theta_1 = \theta_2$
57. (a) 0.307 s (b) Yes. Ignoring gravity makes a difference of 2.28%.
59. (a) $\vec{\mathbf{F}} = 1.90(k_e q^2/s^2)(\hat{\mathbf{i}} + \hat{\mathbf{j}} + \hat{\mathbf{k}})$ (b) $\vec{\mathbf{F}} = 3.29(k_e q^2/s^2)$ in the direction away from the diagonally opposite vertex
65. $\frac{k_e \lambda_0}{2x_0}(-\hat{\mathbf{i}})$

CHAPTER 24

1. 4.14 MN/C
3. (a) aA (b) bA (c) 0
5. 1.87 kN $\cdot$ m^2/C
7. (a) -6.89 MN $\cdot$ m^2/C (b) The number of lines entering exceeds the number leaving by 2.91 times or more.
9. $-Q/\epsilon_0$ for S_1; 0 for S_2; $-2Q/\epsilon_0$ for S_3; 0 for S_4
11. (a) $+Q/2\epsilon_0$ (b) $-Q/2\epsilon_0$
13. -18.8 kN $\cdot$ m^2/C
15. 0 if $R \le d$; $(2\lambda/\epsilon_0)\sqrt{R^2 - d^2}$ if $R > d$
17. (a) 3.20 MN $\cdot$ m^2/C (b) 19.2 MN $\cdot$ m^2/C (c) The answer to part (a) could change, but the answer to part (b) would stay the same.
19. 2.33×10^{21} N/C
21. 508 kN/C up
23. -2.48 μC/m^2
25. 5.94×10^5 m/s
27. $\vec{\mathbf{E}} = \rho r/2\epsilon_0$ away from the axis
29. (a) 0 (b) 7.19 MN/C away from the center
31. (a) 51.4 kN/C outward (b) 646 N $\cdot$ m^2/C
33. (a) 0 (b) 5 400 N/C outward (c) 540 N/C outward
35. (a) $+708$ nC/m^2 and -708 nC/m^2 (b) $+177$ nC and -177 nC
37. 2.00 N
39. (a) $-\lambda, +3\lambda$ (b) $3\lambda/2\pi\epsilon_0 r$ radially outward
41. (a) 80.0 nC/m^2 on each face (b) $9.04\hat{\mathbf{k}}$ kN/C (c) $-9.04\hat{\mathbf{k}}$ kN/C
43. (b) $Q/2\epsilon_0$ (c) Q/ϵ_0
45. (a) The charge on the exterior surface is -55.7 nC distributed uniformly. (b) The charge on the interior surface is $+55.7$ nC. It might have any distribution. (c) The charge within the shell is -55.7 nC. It might have any distribution.
47. (a) $\rho r/3\epsilon_0$, $Q/4\pi\epsilon_0 r^2$, 0, $Q/4\pi\epsilon_0 r^2$, all radially outward (b) $-Q/4\pi b^2$ and $+Q/4\pi c^2$
49. $\theta = \tan^{-1}[qQ/(2\pi\epsilon_0 dmv^2)]$
51. (a) σ/ϵ_0 away from both plates (b) 0 (c) σ/ϵ_0 away from both plates
53. $\sigma/2\epsilon_0$ radially outward
57. $\vec{\mathbf{E}} = a/2\epsilon_0$ radially outward
61. (b) $\vec{\mathbf{g}} = GM_E r/R_E^{\ 3}$ radially inward
63. (a) -4.00 nC (b) $+9.56$ nC (c) $+4.00$ nC and $+5.56$ nC
65. (a) If the volume charge density is nonzero, the field cannot be uniform in magnitude. (b) The field must be uniform in magnitude along any line in the direction of the field. The field magnitude can vary between points in a plane perpendicular to the field lines.

CHAPTER 25

1. (a) 152 km/s (b) 6.49 Mm/s
3. 1.67 MN/C
5. 38.9 V; the origin
7. (a) $2QE/k$ (b) QE/k (c) $2\pi\sqrt{m/k}$ (d) $2(QE - \mu_k mg)/k$
9. (a) 0.400 m/s (b) It is the same. Each bit of the rod feels a force of the same size as before.
11. (a) 1.44×10^{-7} V (b) -7.19×10^{-8} V (c) -1.44×10^{-7} V, $+7.19 \times 10^{-8}$ V
13. (a) 6.00 m (b) -2.00 μC
15. -11.0 MV
17. 8.95 J
21. (a) no point at a finite distance from the particles (b) $2k_e q/a$
23. (a) 10.8 m/s and 1.55 m/s (b) Greater. The conducting spheres will polarize each other, with most of the positive charge of one and of the negative charge of the other on their inside faces. Immediately before they collide, their centers of charge will be closer than their geometric centers, so they will have less electric potential energy and more kinetic energy.
25. $5k_e q^2/9d$
27. $\left[\left(1 + \sqrt{\tfrac{1}{8}}\right)\frac{k_e q^2}{mL}\right]^{1/2}$
29. (a) 10.0 V, -11.0 V, -32.0 V (b) 7.00 N/C in the $+x$ direction
31. $\vec{\mathbf{E}} = (-5 + 6xy)\hat{\mathbf{i}} + (3x^2 - 2z^2)\hat{\mathbf{j}} - 4yz\hat{\mathbf{k}}$; 7.07 N/C
33. $E_y = \frac{k_e Q}{y\sqrt{\ell^2 + y^2}}$
35. (a) C/m^2 (b) $k_e\alpha[L - d\ln(1 + L/d)]$
37. -1.51 MV
39. (a) 0, 1.67 MV (b) 5.84 MN/C away, 1.17 MV (c) 11.9 MN/C away, 1.67 MV
41. (a) 248 nC/m^2 (b) 496 nC/m^2
43. (a) 450 kV (b) 7.51 μC
45. (a) 1.42 mm (b) 9.20 kV/m
47. 253 MeV
49. (a) -27.2 eV (b) -6.80 eV (c) 0
51. (a) Yes. The inverse proportionality of potential to radius is sufficient to show that $200R = 150(R + 10\text{ cm})$, so $R = 30.0$ cm. Then $Q = 6.67$ nC. (b) Almost but not quite. Two possibilities exist: $R = 29.1$ cm with $Q = 6.79$ nC and $R = 3.44$ cm with $Q = 804$ pC.
53. 4.00 nC at (-1.00 m, 0) and -5.01 nC at (0, 2.00 m)
55. $k_e Q^2/2R$
57. $V_2 - V_1 = (-\lambda/2\pi\epsilon_0)\ln(r_2/r_1)$
61. (b) $E_r = 2k_e p\cos\theta/r^3$; $E_\theta = k_e p\sin\theta/r^3$; yes; no (c) $V = k_e py(x^2 + y^2)^{-3/2}$; $\vec{\mathbf{E}} = 3k_e pxy(x^2 + y^2)^{-5/2}\hat{\mathbf{i}} + k_e p(2y^2 - x^2)(x^2 + y^2)^{-5/2}\hat{\mathbf{j}}$
63. $V = \pi k_e C\left[R\sqrt{x^2 + R^2} + x^2\ln\left(\frac{x}{R + \sqrt{x^2 + R^2}}\right)\right]$
65. (a) 488 V (b) 78.1 aJ (c) 306 km/s (d) 390 Gm/s^2 toward the negative plate (e) 651 aN toward the negative plate (f) 4.07 kN/C
67. Outside the sphere, $E_x = 3E_0a^3xz(x^2 + y^2 + z^2)^{-5/2}$, $E_y = 3E_0a^3yz(x^2 + y^2 + z^2)^{-5/2}$, and $E_z = E_0 + E_0a^3(2z^2 - x^2 - y^2)(x^2 + y^2 + z^2)^{-5/2}$. Inside the sphere, $E_x = E_y = E_z = 0$.

CHAPTER 26

1. (a) 48.0 μC (b) 6.00 μC
3. (a) 1.33 μC/m^2 (b) 13.3 pF
5. (a) 11.1 kV/m toward the negative plate (b) 98.3 nC/m^2 (c) 3.74 pF (d) 74.7 pC
7. 4.42 μm
9. (a) 2.68 nF (b) 3.02 kV
11. (a) 15.6 pF (b) 256 kV
13. (a) 3.53 μF (b) 6.35 V and 2.65 V (c) 31.8 μC on each
15. 6.00 pF and 3.00 pF
17. (a) 5.96 μF (b) 89.5 μC on 20 μF, 63.2 μC on 6 μF, 26.3 μC on 15 μF and on 3 μF
19. 120 μC; 80.0 μC and 40.0 μC
21. ten
23. 6.04 μF
25. 12.9 μF
27. (a) 216 μJ (b) 54.0 μJ
31. (a) 1.50 μC (b) 1.83 kV
35. 9.79 kg
37. (a) 81.3 pF (b) 2.40 kV
39. 1.04 m
41. 22.5 V
43. (b) -8.78×10^6 N/C·m; $-5.53 \times 10^{-2}\hat{\mathbf{i}}$ N
45. 19.0 kV
47. (a) 11.2 pF (b) 134 pC (c) 16.7 pF (d) 66.9 pC
49. (a) 40.0 μJ (b) 500 V
51. 0.188 m^2
55. Gasoline has 194 times the specific energy content of the battery and 727 000 times that of the capacitor.
57. (a) $Q_0^2 d(\ell - x)/(2\ell^3\epsilon_0)$ (b) $Q_0^2 d/(2\ell^3\epsilon_0)$ to the right (c) $Q_0^2/(2\ell^4\epsilon_0)$ (d) $Q_0^2/(2\ell^4\epsilon_0)$; they are precisely the same.
59. 4.29 μF
61. (a) The additional energy comes from work done by the electric field in the wires as it forces more charge onto the already-charged plates. (b) The charge increases according to $Q/Q_0 = \kappa$.
63. 750 μC on C_1 and 250 μC on C_2
65. $\frac{4}{3}C$

CHAPTER 27

1. 7.50×10^{15} electrons
3. (a) $0.632I_0\tau$ (b) $0.999\,95I_0\tau$ (c) $I_0\tau$
5. (a) 17.0 A (b) 85.0 kA/m^2
7. (a) 2.55 A/m^2 (b) 5.31×10^{10} m^{-3} (c) 1.20×10^{10} s
9. (a) 221 nm (b) No. The deuterons are so far apart that one does not produce a significant potential at the location of the next.
11. 6.43 A
13. (a) 1.82 m (b) 280 μm
15. $6.00 \times 10^{-15}/\Omega \cdot$m
17. 0.180 V/m
19. (a) 31.5 n$\Omega \cdot$m (b) 6.35 MA/m^2 (c) 49.9 mA (d) 659 μm/s (e) 0.400 V
21. 0.125
23. 5.00 A, 24.0 Ω
25. 5.49 Ω
27. 36.1%
29. (a) 3.17 m (b) 340 W
31. (a) 0.660 kWh (b) \$0.039 6
33. \$0.232
35. \$0.269/day
37. (a) 184 W (b) 461°C
39. ~ \$1
41. Any diameter d and length ℓ related by $d^2 = (4.77 \times 10^{-8}\text{ m})\ell$, such as length 0.900 m and diameter 0.207 mm. Yes.
45. Experimental resistivity = 1.47 $\mu\Omega \cdot$m ±4%, in agreement with 1.50 $\mu\Omega \cdot$m
47. (a) 8.00 V/m in the x direction (b) 0.637 Ω (c) 6.28 A (d) 200 MA/m^2 in the x direction
49. (a) 667 A (b) 50.0 km
51.

Material	$\alpha' = \alpha/(1 - 20\alpha)$
Silver	$4.1 \times 10^{-3}/°C$
Copper	$4.2 \times 10^{-3}/°C$
Gold	$3.6 \times 10^{-3}/°C$
Aluminum	$4.2 \times 10^{-3}/°C$
Tungsten	$4.9 \times 10^{-3}/°C$
Iron	$5.6 \times 10^{-3}/°C$
Platinum	$4.25 \times 10^{-3}/°C$
Lead	$4.2 \times 10^{-3}/°C$
Nichrome	$0.4 \times 10^{-3}/°C$
Carbon	$-0.5 \times 10^{-3}/°C$
Germanium	$-24 \times 10^{-3}/°C$
Silicon	$-30 \times 10^{-3}/°C$

53. It is exact. The resistance can be written $R = \rho L^2/V$ and the stretched length as $L = L_i(1 + \delta)$. Then the result follows directly.
55. (b) Charge is conducted by current in the direction of decreasing potential. Energy is conducted by heat in the direction of decreasing temperature.
59. Coat the surfaces of entry and exit with a material of much higher conductivity than the bulk material of the object. The electric potential will be essentially uniform over each of these electrodes. Current will be distributed over the whole area where each electrode is in contact with the resistive object.
61. (a) $\dfrac{\epsilon_0 \ell}{2d}(\ell + 2x + \kappa\ell - 2\kappa x)$
(b) $\dfrac{\epsilon_0 \ell v\,\Delta V(\kappa - 1)}{d}$ clockwise
63. 2.71 MΩ
65. 2 020°C

CHAPTER 28

1. (a) 6.73 Ω (b) 1.97 Ω
3. (a) 12.4 V (b) 9.65 V
5. (a) 17.1 Ω (b) 1.99 A for 4 Ω and 9 Ω, 1.17 A for 7 Ω, 0.818 A for 10 Ω
7. (a) 227 mA (b) 5.68 V
9. (a) 75.0 V (b) 25.0 W, 6.25 W, and 6.25 W; 37.5 W
11. $R_1 = 1.00$ kΩ, $R_2 = 2.00$ kΩ, $R_3 = 3.00$ kΩ
13. It decreases. Closing the switch opens a new path with resistance of only 20 Ω. $R = 14.0\ \Omega$
15. 14.2 W to 2 Ω, 28.4 W to 4 Ω, 1.33 W to 3 Ω, 4.00 W to 1 Ω
17. 846 mA down in the 8-Ω resistor; 462 mA down in the middle branch; 1.31 A up in the right-hand branch
19. (a) −222 J and 1.88 kJ (b) 687 J, 128 J, 25.6 J, 616 J, 205 J (c) 1.66 kJ of chemical energy is transformed into internal energy.

21. 0.395 A and 1.50 V

23. 1.00 A up in 200 Ω, 4.00 A up in 70 Ω, 3.00 A up in 80 Ω, 8.00 A down in 20 Ω, 200 V

25. (a) 909 mA (b) $-1.82\text{ V} = V_b - V_a$

27. (a) 5.00 s (b) 150 μC (c) 4.06 μA

29. (a) −61.6 mA (b) 0.235 μC (c) 1.96 A

31. (a) 6.00 V (b) 8.29 μs

33. 0.302 Ω

35. 16.6 kΩ

37. (a) 12.5 A, 6.25 A, 8.33 A (b) No. Together they would require 27.1 A.

39. (a) 1.02 A down (b) 0.364 A down (c) 1.38 A up (d) 0 (e) 66.0 μC

41. 2.22 h

43. a is 4.00 V higher

45. 87.3 %

47. 6.00 Ω, 3.00 Ω

49. (a) $I_1 = \dfrac{IR_2}{R_1 + R_2}$, $I_2 = \dfrac{IR_1}{R_1 + R_2}$

51. (a) $R \leq 1\,050\ \Omega$ (b) $R \geq 10.0\ \Omega$

53. (a) 9.93 μC (b) 33.7 nA (c) 334 nW (d) 337 nW

55. (a) 40.0 W (b) 80.0 V, 40.0 V, 40.0 V

57. (a) 9.30 V, 2.51 Ω (b) 186 V and 3.70 A (c) 1.09 A (d) 143 W (e) 0.162 Ω (f) 3.00 mW (g) 2.21 W (h) The power output of the emf depends on the resistance connected to it. A question about "the rest of the power" is not meaningful when it compares circuits with different currents. The net emf produces more current in the circuit where the copper wire is used. The net emf delivers more power when the copper wire is used, 687 W rather than 203 W without the wire. Nearly all this power results in extra internal energy in the internal resistance of the batteries, which rapidly rise to a high temperature. The circuit with the copper wire is unsafe because the batteries overheat. The circuit without the copper wire is unsafe because it delivers an electric shock to the experimenter.

61. (a) 0 in 3 kΩ and 333 μA in 12 kΩ and 15 kΩ (b) 50.0 μC (c) $(278\ \mu\text{A})e^{-t/180\text{ ms}}$ (d) 290 ms

63. (a) $R_x = R_2 - R_1/4$ (b) $R_x = 2.75\ \Omega$. The station is inadequately grounded.

65. (a) $2\Delta t/3$ (b) $3\Delta t$

CHAPTER 29

1. (a) up (b) toward you, out of the plane of the paper (c) no deflection (d) into the plane of the paper

3. $(-20.9\hat{\mathbf{j}})$ mT

5. 48.9° or 131°

7. 2.34 aN

9. (a) 49.6 aN south (b) 1.29 km

11. $r_\alpha = r_d = \sqrt{2}r_p$

13. (a) 5.00 cm (b) 8.78×10^6 m/s

15. 7.88 pT

17. 244 kV/m

19. 0.278 m

21. (a) 4.31×10^7 rad/s (b) 51.7 Mm/s

23. 70.1 mT

25. 0.245 T east

27. (a) 4.73 N (b) 5.46 N (c) 4.73 N

29. 1.07 m/s

31. $2\pi rIB \sin\theta$ up

33. 2.98 μN west

35. 9.98 N · m clockwise as seen looking down from above

37. (a) Minimum: pointing north at 48.0° below the horizontal; maximum: pointing south at 48.0° above the horizontal. (b) 1.07 μJ

39. The magnetic moment cannot go to infinity. Its maximum value is 5.37 mA · m² for a single-turn circle. Smaller by 21% and by 40% are the magnetic moments for the single-turn square and triangle. Circular coils with several turns have magnetic moments inversely proportional to the number of turns, approaching zero as the number of turns goes to infinity.

41. 43.1 μT

43. (a) The electric current experiences a magnetic force. (c) no, no, no

45. 12.5 km. It will not hit the Earth, but it will perform a hairpin turn and go back parallel to its original direction.

47. (a) -8.00×10^{-21} kg · m/s (b) 8.90°

49. (a) $(3.52\hat{\mathbf{i}} - 1.60\hat{\mathbf{j}})$ aN (b) 24.4°

51. 128 mT north at an angle of 78.7° below the horizontal

53. 0.588 T

55. 0.713 A counterclockwise as seen from above

57. 2.75 Mrad/s

59. 3.70×10^{-24} N · m

61. (a) 1.33 m/s (b) Positive ions moving toward you in magnetic field to the right feel upward magnetic force and migrate upward in the blood vessel. Negative ions moving toward you feel downward magnetic force and accumulate at the bottom of this section of vessel. Therefore, both species can participate in the generation of the emf.

63. (a) $v = qBh/m$. If its speed is slightly less than the critical value, the particle moves in a semicircle of radius h and leaves the field with velocity $-v\hat{\mathbf{j}}$. If its speed is incrementally greater, the particle moves in a quarter circle of the same radius and moves along the boundary outside the field with velocity $v\hat{\mathbf{i}}$. (b) The particle moves in a smaller semicircle of radius mv/qB, attaining final velocity $-v\hat{\mathbf{j}}$. (c) The particle moves in a circular arc of radius $r = mv/qB$, leaving the field with velocity $v\sin\theta\,\hat{\mathbf{i}} + v\cos\theta\,\hat{\mathbf{j}}$, where $\theta = \sin^{-1}(h/r)$.

65. (a) For small angular displacements, the torque on the dipole is equal to a negative constant times the displacement.

(b) $f = \dfrac{1}{2\pi}\sqrt{\dfrac{\mu B}{I}}$

(c) The equilibrium orientation of the needle shows the direction of the field. In a stronger field, the frequency is higher. The frequency is easy to measure precisely over a wide range of values. 2.04 mT.

CHAPTER 30

1. 12.5 T

3. (a) 28.3 μT into the paper (b) 24.7 μT into the paper

5. $\dfrac{\mu_0 I}{4\pi x}$ into the paper

7. (a) $2I_1$ out of the page (b) $6I_1$ into the page

9. (a) along the line ($y = -0.420$ m, $z = 0$) (b) $(-34.7\hat{\mathbf{j}})$ mN (c) $(17.3\hat{\mathbf{j}})$ kN/C

11. at A, 53.3 μT toward the bottom of the page; at B, 20.0 μT toward the bottom of the page; at C, zero.

13. (a) $4.5\dfrac{\mu_0 I}{\pi L}$

(b) Stronger. Each of the two sides meeting at the nearby vertex contributes more than twice as much to the net field at the new point.

15. $(-13.0\hat{\mathbf{j}})\ \mu\text{T}$

17. $(-27.0\hat{\mathbf{i}})\ \mu\text{N}$

19. parallel to the wires and 0.167 m below the upper wire

21. (a) opposite directions (b) 67.8 A (c) Smaller. A smaller gravitational force would be pulling down on the wires, therefore tending to reduce the angle.

23. 20.0 μT toward the bottom of the page

25. at a, 200 μT toward the top of the page; at b, 133 μT toward the bottom of the page

27. (a) 6.34 mN/m inward (b) Greater. The magnetic field increases toward the outside of the bundle, where more net current lies inside a particular radius. The larger field exerts a stronger force on the strand we choose to monitor.

29. (a) 0

(b) $\dfrac{\mu_0 I}{2\pi R}$ tangent to the wall in a counterclockwise sense

(c) $\dfrac{\mu_0 I^2}{(2\pi R)^2}$ inward

31. (a) $\mu_0 b r_1{}^2/3$ (b) $\mu_0 b R^3/3r_2$

35. 31.8 mA

37. 226 μN away from the center of the loop, 0

39. (a) 3.13 mWb (b) 0

41. (a) 7.40 μWb (b) 2.27 μWb

43. 2.02

45. (a) 8.63×10^{45} electrons (b) 4.01×10^{20} kg

47. $\dfrac{\mu_0 I}{2\pi w} \ln\left(1 + \dfrac{w}{b}\right)\hat{\mathbf{k}}$

49. $(-12.0\hat{\mathbf{k}})$ mN

51. 143 pT

57. (a) 2.46 N upward (b) The magnetic field at the center of the loop or on its axis is much weaker than the magnetic field immediately outside the wire. The wire has negligible curvature on the scale of 1 mm, so we model the lower loop as a long, straight wire to find the field it creates at the location of the upper wire. (c) 107 m/s^2 upward

59. (a) 274 μT (b) $(-274\hat{\mathbf{j}})\ \mu$T (c) $(1.15\hat{\mathbf{i}})$ mN (d) $(0.384\hat{\mathbf{i}})$ m/s^2 (e) acceleration is constant (f) $(0.999\hat{\mathbf{i}})$ m/s

61. $\dfrac{\mu_0 I_1 I_2 L}{\pi R}$ to the right

65. $\frac{1}{3}\rho\mu_0\omega R^2$

67. (a) $\dfrac{\mu_0 I(2r^2 - a^2)}{\pi r(4r^2 - a^2)}$ to the left

(b) $\dfrac{\mu_0 I(2r^2 + a^2)}{\pi r(4r^2 + a^2)}$ toward the top of the page

CHAPTER 31

1. (a) 101 μV tending to produce clockwise current as seen from above (b) It is twice as large in magnitude and in the opposite sense.

3. 9.82 mV

5. (b) 3.79 mV (c) 28.0 mV

7. 160 A

9. (a) 1.60 A counterclockwise (b) 20.1 μT (c) left

11. $-(14.2\text{ mV})\cos(120t)$

13. 283 μA upward

15. $(68.2\text{ mV})e^{-1.6t}$, tending to produce counterclockwise current

17. 272 m

19. 13.3 mA counterclockwise in the lower loop and clockwise in the upper loop

21. (a) 1.18 mV. The wingtip on the pilot's left is positive. (b) no change (c) No. If we try to connect the wings into a circuit with the lightbulb, we run an extra insulated wire along the wing. In a uniform field, the total emf generated in the one-turn coil is zero.

23. (a) 3.00 N to the right (b) 6.00 W

25. 24.1 V with the outer contact positive

27. 2.83 mV

29. (a) $F = N^2B^2w^2v/R$ to the left (b) 0 (c) $F = N^2B^2w^2v/R$ to the left

31. 145 μA upward in the picture

33. 1.80 mN/C upward and to the left, perpendicular to r_1

35. (a) 7.54 kV (b) The plane of the loop is parallel to $\vec{\mathbf{B}}$.

37. $(28.6\text{ mV})\sin(4\pi t)$

39. (a) 110 V (b) 8.53 W (c) 1.22 kW

41. Both are correct. The current in the magnet creates an upward magnetic field 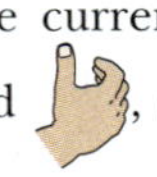, so the N and S poles on the solenoid core are shown correctly. On the rail in front of the brake, the upward magnetic flux increases as the coil approaches, so a current is induced here to create downward magnetic field . This current is clockwise, so the S pole on the rail is shown correctly. On the rail behind the brake, the upward magnetic flux is decreasing. The induced current in the rail will produce upward magnetic field by being counterclockwise  as the picture correctly shows.

43. (b) Larger R makes current smaller, so the loop must travel faster to maintain equality of magnetic force and weight. (c) The magnetic force is proportional to the product of field and current, while the current is itself proportional to field. If B becomes two times smaller, the speed must become four times larger to compensate.

45. $-(7.22\text{ mV})\cos(2\pi\ 523t/\text{s})$

47.

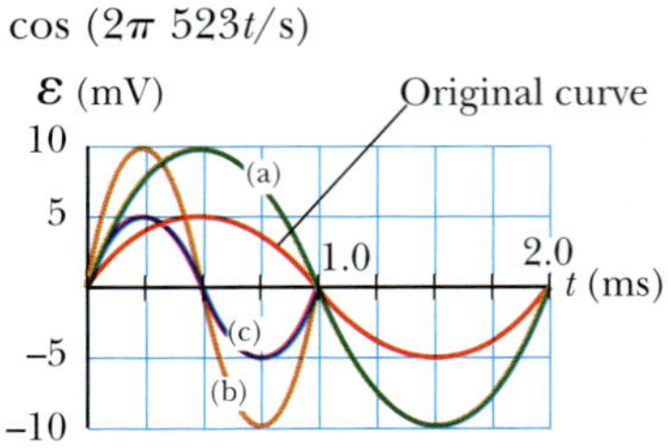

(a) Doubling N doubles amplitude. (b) Doubling ω doubles the amplitude and halves the period. (c) Doubling ω and halving N leaves the amplitude the same and cuts the period in half.

49. (a) 3.50 A up in 2 Ω, and 1.40 A up in 5 Ω (b) 34.3 W (c) 4.29 N

51. $\sim 10^{-4}$ V, by reversing a 20-turn coil of diameter 3 cm in 0.1 s in a field of 10^{-3} T

53. 1.20 μC

55. (a) 0.900 A from b toward a (b) 0.108 N (c) b (d) No. Instead of decreasing downward magnetic flux to induce clockwise current, the new loop will see increasing downward flux to cause counterclockwise current, but the current in the resistor is still from b to a.
57. (a) $C\pi a^2 K$ (b) the upper plate (c) The changing magnetic field within the loop induces an electric field around the circumference, which pushes on charged particles in the wire.
59. (a) 36.0 V (b) 600 mWb/s (c) 35.9 V (d) 4.32 N · m
63. 6.00 A
67. $(-87.1 \text{ mV}) \cos (200\pi t + \phi)$

CHAPTER 32

1. 100 V
3. $-(18.8 \text{ V}) \cos (377t)$
5. -0.421 A/s
7. (a) 188 μT (b) 33.3 nT · m² (c) 0.375 mH (d) B and Φ_B are proportional to current; L is independent of current
9. $\mathcal{E}_0/k^2L$
11. (a) 0.139 s (b) 0.461 s
13. (a) 2.00 ms (b) 0.176 A (c) 1.50 A (d) 3.22 ms
15. (a) 0.800 (b) 0
17. (a) 6.67 A/s (b) 0.332 A/s
19. (a) 1.00 kΩ (b) 3.00 ms
21. (a) 5.66 ms (b) 1.22 A (c) 58.1 ms
23. 2.44 μJ
25. 44.2 nJ/m³ for the $\vec{\mathbf{E}}$ field and 995 μJ/m³ for the $\vec{\mathbf{B}}$ field
27. (a) 66.0 W (b) 45.0 W (c) 21.0 W (d) At all instants after the connection is made, the battery power is equal to the sum of the power delivered to the resistor and the power delivered to the magnetic field. Immediately after $t = 0$, the resistor power is nearly zero and nearly all the battery power is going into the magnetic field. Long after the connection is made, the magnetic field is absorbing no more power and the battery power is going into the resistor.
29. $\dfrac{2\pi B_0^2 R^3}{\mu_0} = 2.70 \times 10^{18}$ J
31. 1.00 V
33. (a) 18.0 mH (b) 34.3 mH (c) -9.00 mV
35. $M = \dfrac{N_1 N_2 \pi \mu_0 R_1^2 R_2^2}{2(x^2 + R_1^2)^{3/2}}$
37. 400 mA
39. 281 mH
41. 608 pF
43. (a) 6.03 J (b) 0.529 J (c) 6.56 J
45. (a) 4.47 krad/s (b) 4.36 krad/s (c) 2.53%
47. (a) $0.693(2L/R)$ (b) $0.347(2L/R)$
49. (a) -20.0 mV (b) $-(10.0 \text{ MV/s}^2)t^2$ (c) 63.2 μs
51. $(Q/2N)(3L/C)^{1/2}$
53. (a) Immediately after the circuit is connected, the potential difference across the resistor is zero and the emf across the coil is 24.0 V. (b) After several seconds, the potential difference across the resistor is 24.0 V and that across the coil is 0. (c) The two voltages are equal to each other, both being 12.0 V, only once, at 0.578 ms after the circuit is connected. (d) As the current decays, the potential difference across the resistor is always equal to the emf across the coil.

55.

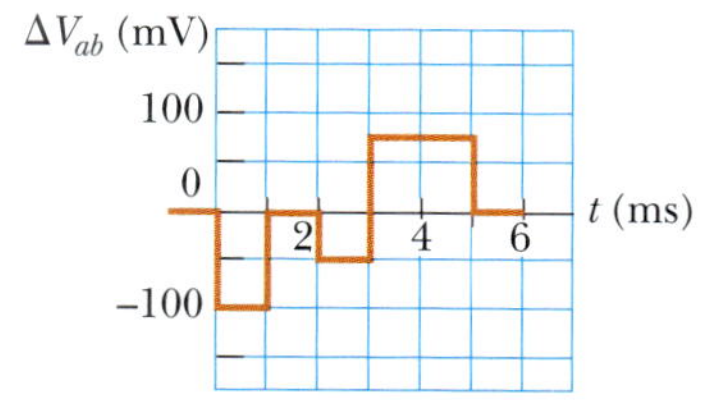

57. (b) 91.2 μH (c) 90.9 μH is only 0.3% smaller
61. (a) 72.0 V; b
(b)

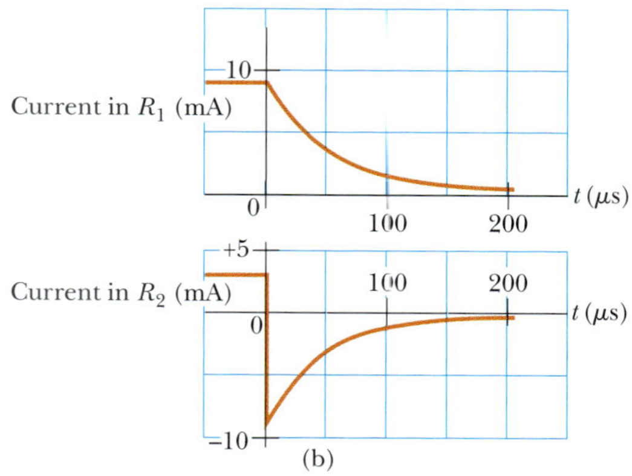

(c) 75.2 μs
63. 300 Ω
65. (a) 62.5 GJ (b) 2 000 N
67. (a) 2.93 mT up (b) 3.42 Pa (c) The supercurrents must be clockwise to produce a downward magnetic field that cancels the upward field of the current in the windings. (d) The field of the windings is upward and radially outward around the top of the solenoid. It exerts a force radially inward and upward on each bit of the clockwise supercurrent. The total force on the supercurrents in the bar is upward. (e) 1.30 mN

CHAPTER 33

1. $\Delta v(t) = (283 \text{ V}) \sin (628t)$
3. 2.95 A, 70.7 V
5. 14.6 Hz
7. (a) 42.4 mH (b) 942 rad/s
9. 5.60 A
11. 0.450 Wb
13. (a) 141 mA (b) 235 mA
15. 100 mA
17. (a) 194 V (b) current leads by 49.9°
19. (a) 78.5 Ω (b) 1.59 kΩ (c) 1.52 kΩ (d) 138 mA (e) $-84.3°$
21. (a) 17.4° (b) The voltage leads the current.
23. 1.88 V
25.

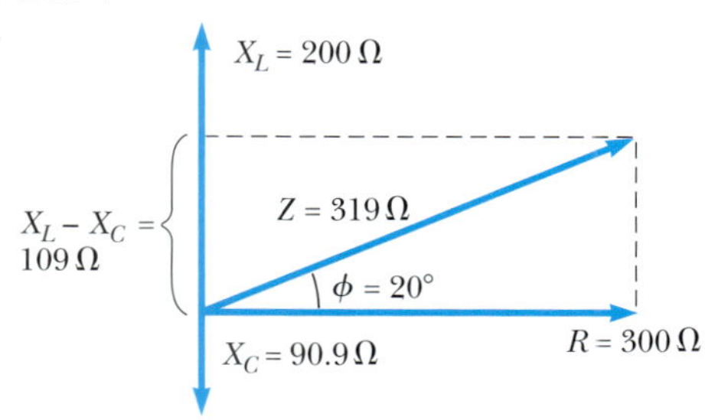

27. 8.00 W
29. (a) 16.0 Ω (b) -12.0 Ω
31. (a) 39.5 V · m/ΔV (b) The diameter is inversely proportional to the potential difference. (c) 26.3 mm (d) 13.2 kV

33. $11(\Delta V_{rms})^2/14R$
35. 1.82 pF
37. 242 mJ
39. 0.591 and 0.987; the circuit in Problem 21
41. 687 V
43. (a) 29.0 kW (b) 5.80×10^{-3} (c) If the generator were limited to 4 500 V, no more than 17.5 kW could be delivered to the load, never 5 000 kW.
45. (b) 0; 1 (c) $f_h = (10.88RC)^{-1}$
47. (a) 613 μF (b) 0.756
49. (a) 580 μH and 54.6 μF (b) 1 (c) 894 Hz (d) Δv_{out} leads Δv_{in} by 60.0° at 200 Hz. Δv_{out} and Δv_{in} are in phase at 894 Hz. Δv_{out} lags Δv_{in} by 60.0° at 4 000 Hz. (e) 1.56 W, 6.25 W, 1.56 W (f) 0.408
51. (a) X_C could be 53.8 Ω or it could be 1.35 kΩ. (b) X_C must be 53.8 Ω. (c) X_C must be 1.43 kΩ.
53. 56.7 W
55. Tension T and separation d must be related by $T = (274\ \text{N/m}^2)d^2$. One possibility is $T = 10.9$ N and $d = 0.200$ m.
57. (a) 225 mA (b) 450 mA
59. (a) 1.25 A (b) The current lags the voltage by 46.7°.
61. (a) 200 mA; voltage leads by 36.8° (b) 40.0 V; $\phi = 0°$ (c) 20.0 V; $\phi = -90.0°$ (d) 50.0 V; $\phi = +90.0°$
63. (b) 31.6
67. (a)

f (Hz)	X_L (Ω)	X_C (Ω)	Z (Ω)
300	283	12 600	12 300
600	565	6 280	5 720
800	754	4 710	3 960
1 000	942	3 770	2 830
1 500	1 410	2 510	1 100
2 000	1 880	1 880	40.0
3 000	2 830	1 260	1 570
4 000	3 770	942	2 830
6 000	5 650	628	5 020
10 000	9 420	377	9 040

(b)
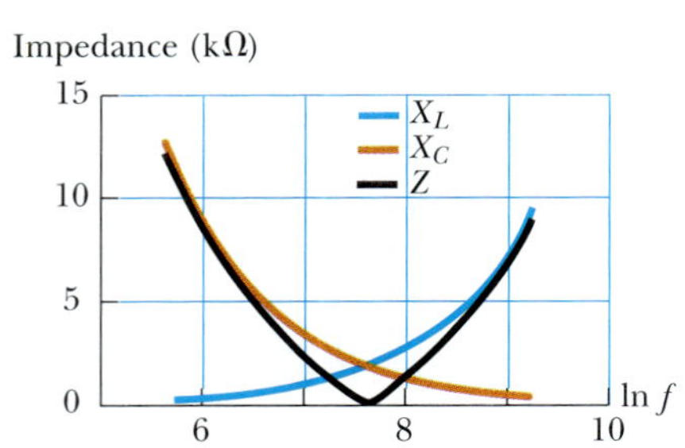

69. (a) and (b) 19.7 cm at 35.0°. The answers are identical. (c) 9.36 cm at 169°

CHAPTER 34

1. (a) 11.3 GV · m/s (b) 0.100 A
3. 1.85 aT up
5. $(-2.87\hat{\mathbf{j}} + 5.75\hat{\mathbf{k}})$ Gm/s²
7. (a) the year 2.69×10^3 (b) 499 s (c) 2.56 s (d) 0.133 s (e) 33.3 μs
9. (a) 6.00 MHz (b) $(-73.3\hat{\mathbf{k}})$ nT (c) $\vec{\mathbf{B}} = [(-73.3\hat{\mathbf{k}})\text{ nT}]\cos(0.126x - 3.77 \times 10^7 t)$
11. (a) 0.333 μT (b) 0.628 μm (c) 477 THz
13. 75.0 MHz
15. 3.33 μJ/m³
17. 307 μW/m²
19. 3.33×10^3 m²
21. (a) 332 kW/m² radially inward (b) 1.88 kV/m and 222 μT
23. (a) $\vec{\mathbf{E}} \cdot \vec{\mathbf{B}} = 0$ (b) $(11.5\hat{\mathbf{i}} - 28.6\hat{\mathbf{j}})$ W/m²
25. (a) 2.33 mT (b) 650 MW/m² (c) 510 W
27. (a) 88.8 nW/m² (b) 11.3 MW
29. 83.3 nPa
31. (a) 1.90 kN/C (b) 50.0 pJ (c) 1.67×10^{-19} kg · m/s
33. (a) 590 W/m² (b) 2.10×10^{16} W (c) 70.1 MN (d) The gravitational force is $\sim 10^{13}$ times stronger and in the opposite direction. (e) On the Earth, the Sun's gravitational force is also $\sim 10^{13}$ times stronger than the light-pressure force and in the opposite direction.
35. (a) 134 m (b) 46.9 m
37. (a) away along the perpendicular bisector of the line segment joining the antennas (b) along the extensions of the line segment joining the antennas
39. (a) $\vec{\mathbf{E}} = \frac{1}{2}\mu_0 c J_{max}[\cos(kx - \omega t)]\hat{\mathbf{j}}$ (b) $\vec{\mathbf{S}} = \frac{1}{4}\mu_0 c J_{max}^2[\cos^2(kx - \omega t)]\hat{\mathbf{i}}$ (c) $I = \dfrac{\mu_0 c J_{max}^2}{8}$ (d) 3.48 A/m
41. (a) 6.00 pm (b) 7.50 cm
43. (a) 4.17 m to 4.55 m (b) 3.41 m to 3.66 m (c) 1.61 m to 1.67 m
45. 1.00 Mm = 621 mi; not very practical
47. (a) 3.85×10^{26} W (b) 1.02 kV/m and 3.39 μT
49. (a)
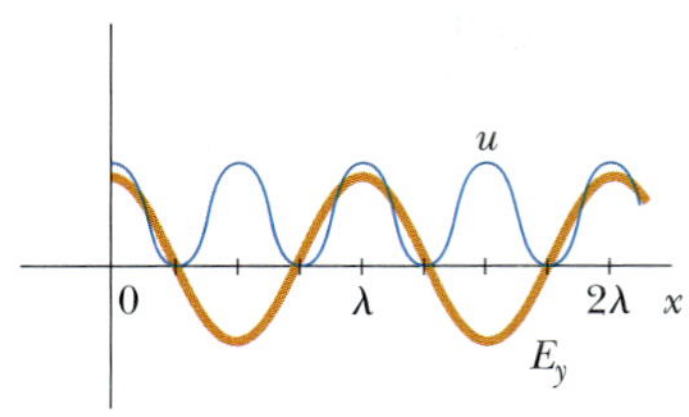

(b), (c) $u_E = u_B = \frac{1}{2}\epsilon_0 E_{max}^2 \cos^2(kx)$
(d) $u = \epsilon_0 E_{max}^2 \cos^2(kx)$ (e) $E_\lambda = \frac{1}{2}A\lambda\epsilon_0 E_{max}^2$
(f) $I = \frac{1}{2}c\epsilon_0 E_{max}^2 = \frac{1}{2}\sqrt{\dfrac{\epsilon_0}{\mu_0}}E_{max}^2$. This result agrees with $I = \dfrac{E_{max}^2}{2\mu_0 c}$ in Equation 34.24.
51. (a) 6.67×10^{-16} T (b) 5.31×10^{-17} W/m² (c) 1.67×10^{-14} W (d) 5.56×10^{-23} N
53. 95.1 mV/m
55. (a) $B_{max} = 583$ nT, $k = 419$ rad/m, $\omega = 126$ Grad/s; $\vec{\mathbf{B}}$ vibrates in xz plane (b) $\vec{\mathbf{S}}_{avg} = (40.6\hat{\mathbf{i}})$ W/m² (c) 271 nPa (d) $(406\hat{\mathbf{i}})$ nm/s²
57. (a) 22.6 h (b) 30.6 s
59. (a) 8.32×10^7 W/m² (b) 1.05 kW
61. (b) 17.6 Tm/s², 1.75×10^{-27} W (c) 1.80×10^{-24} W
63. (a) $2\pi^2 r^2 f B_{max}\cos\theta$, where θ is the angle between the magnetic field and the normal to the loop (b) The loop

should be in the vertical plane containing the line of sight to the transmitter.

65. (a) 388 K (b) 363 K

CHAPTER 35

1. 300 Mm/s. The sizes of the objects need to be accounted for; otherwise, the answer would be too large by 2%.

3. 114 rad/s

5. (a) 1.94 m (b) 50.0° above the horizontal

7. 23.3°

9. 25.5°, 442 nm

11. 19.5° above the horizon

13. 22.5°

15. (a) 181 Mm/s (b) 225 Mm/s (c) 136 Mm/s

17. 30.0° and 19.5° at entry; 19.5° and 30.0° at exit

19. 3.88 mm

21. 30.4° and 22.3°

23. (a) yes, if the angle of incidence is 58.9° (b) No. Both the reduction in speed and the bending toward the normal reduce the component of velocity parallel to the interface. This component cannot remain constant unless the angle of incidence is 0°.

25. 86.8°

27. 27.9°

29. (b) 37.2° (c) 37.3° (d) 37.3°

31. 4.61°

33. (a) 24.4° (b) 37.0° (c) 49.8°

35. 1.000 08

37. (a) $dn/(n-1)$ (b) yes; yes; yes (c) 350 μm

39. Skylight incident from above travels down the plastic. If the index of refraction of the plastic is greater than 1.41, the rays close in direction to the vertical are totally reflected from the side walls of the slab and from both facets at the bottom of the plastic, where it is not immersed in gasoline. This light returns up inside the plastic and makes it look bright. Unless the index of refraction of the plastic is unrealistically large (greater than about 2.1), total internal reflection is frustrated where the plastic is immersed in gasoline. There the downward-propagating light passes from the plastic out into the gasoline. Little light is reflected up, and the gauge looks dark.

41. Scattered light leaving the photograph in all forward horizontal directions in air is gathered by refraction into a fan in the water of half-angle 48.6°. At larger angles, you see things on the other side of the globe, reflected by total internal reflection at the back surface of the cylinder.

43. 77.5°

45. 2.27 m

47. 62.2%

49. 82 reflections

51. 27.5°

53. (a) Total internal reflection occurs for all values of θ, or the maximum angle is 90°. (b) 30.3° (c) Total internal reflection never occurs as the light moves from lower-index polystyrene into higher-index carbon disulfide.

55. 2.36 cm

57. $\theta = \sin^{-1}\left[\frac{L}{R^2}\left(\sqrt{n^2R^2 - L^2} - \sqrt{R^2 - L^2}\right)\right]$

61. (a) nR_1 (b) R_2

63. (a) 1.20 (b) 3.40 ns

CHAPTER 36

1. ~ 10^{-9} s younger

3. 35.0 in.

5. (a) $-(p_1 + h)$ (b) virtual (c) upright (d) 1.00 (e) no

7. (a) −12.0 cm; 0.400 (b) −15.0 cm; 0.250 (c) upright

9. (a) $q = 45.0$ cm; $M = -0.500$
(b) $q = -60.0$ cm; $M = 3.00$
(c) Image (a) is real, inverted, and diminished. Image (b) is virtual, upright, and enlarged.

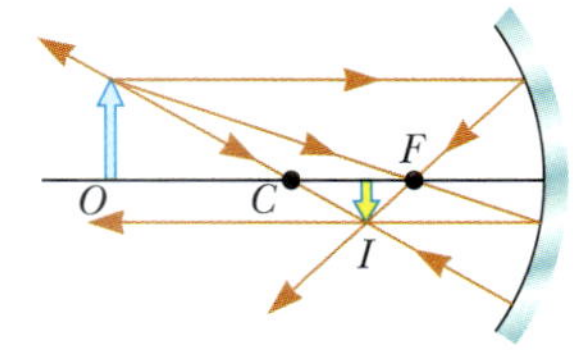

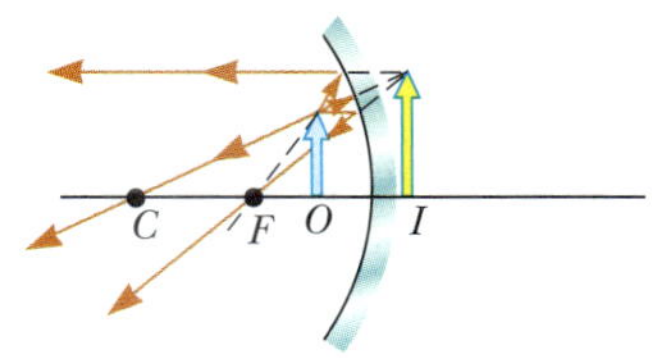

11. (a) 2.22 cm (b) 10.0

13. (a) 160 mm (b) $R = -267$ mm

15. (a) convex (b) at the 30.0 cm mark (c) −20.0 cm

17. (a) a concave mirror with radius of curvature 2.08 m
(b) 1.25 m from the object

19. (a) The image starts 60.0 cm above the mirror and moves up faster and faster, running out to an infinite distance above the mirror. At that moment, the image rays are parallel and the image is equally well described as infinitely far below the mirror. From there, the image moves up, slowing down as it moves, to reach the mirror vertex.
(b) at 0.639 s and 0.782 s

21. 38.2 cm below the top surface of the ice

23. 8.57 cm

25. (a) inside the tank, 24.9 cm behind the front wall; virtual, right side up, enlarged (b) inside the tank, 93.9 cm behind the front wall; virtual, right side up, enlarged (c) 1.10 and 1.39 (d) 9.92 cm and 12.5 cm (e) The plastic has uniform thickness, so the surfaces of entry and exit for any particular ray are very nearly parallel. The ray is slightly displaced, but it would not be changed in direction by going through the plastic wall with air on both sides. Only the difference between the air and water is responsible for the refraction of the light.

27. (a) 16.4 cm (b) 16.4 cm

29. (a) $q = 40.0$ cm, real and inverted, actual size $M = -1.00$
(b) $q = \infty$, $M = \infty$, no image is formed (c) $q = -20.0$ cm, upright, virtual, enlarged $M = +2.00$

31. 2.84 cm

33. (a) −12.3 cm, to the left of the lens (b) 0.615
(c)

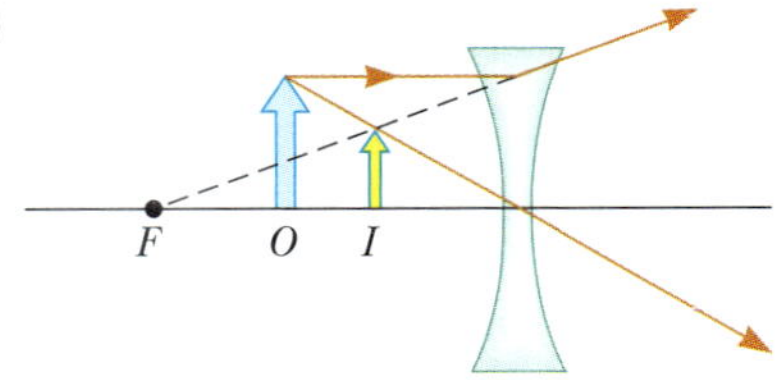

35. (a) 5.36 cm (b) −18.8 cm (c) virtual, right side up, enlarged (d) A magnifying glass with focal length 7.50 cm is used to form an image of a stamp, enlarged 3.50 times. Find the object distance. Locate and describe the image.

37. (a) $p = \frac{d}{2} \pm \sqrt{\frac{d^2}{4} - fd}$
(b) Both images are real and inverted. One is enlarged, the other diminished.
39. 2.18 mm away from the film plane
41. 21.3 cm
43. −4.00 diopters, a diverging lens
45. (a) at 4.17 cm (b) 6.00
47. (a) −800 (b) image is inverted
49. 3.38 min
51. −40.0 cm
53. −25.0 cm
55. $x' = (1\,024 \text{ cm} - 58x)\text{ cm}/(6 \text{ cm} - x)$. The image starts at the position $x_i' = 171$ cm and moves in the positive x direction, faster and faster, until it is out at infinity when the object is at the position $x = 6$ cm. At this instant, the rays from the top of the object are parallel as they leave the lens. Their intersection point can be described as at $x' = \infty$ to the right or equally well as at $x' = -\infty$ on the left. From $x' = -\infty$, the image continues moving to the right, now slowing down. It reaches, for example, −280 cm when the object is at 8 cm and −55 cm when the object is finally at 12 cm. The image has traveled always to the right, to infinity and beyond.
57. Align the lenses on the same axis and 9.00 cm apart. Let the light pass first through the diverging lens and then through the converging lens. The diameter increases by a factor of 1.75.
59. 0.107 m to the right of the vertex of the hemispherical face
61. 8.00 cm. Ray diagram:

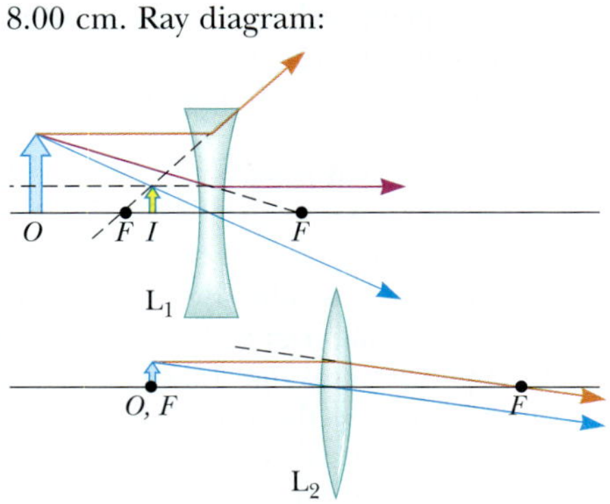

63. 1.50 m in front of the mirror; 1.40 cm (inverted)
65. (a) 30.0 cm and 120 cm (b) 24.0 cm (c) real, inverted, diminished with $M = -0.250$
67. (a) 20.0 cm to the right of the second lens, −6.00
(b) inverted (c) 6.67 cm to the right of the second lens, −2.00, inverted

CHAPTER 37

1. 1.58 cm
3. (a) 55.7 m (b) 124 m
5. 1.54 mm
7. (a) 2.62 mm (b) 2.62 mm
9. 36.2 cm
11. (a) 10.0 m (b) 516 m (c) Only the runway centerline is a maximum for the interference patterns for both frequencies. If the frequencies were related by a ratio of small integers k/ℓ, the plane could by mistake fly along the kth side maximum of one signal, where it coincides with the ℓth side maximum of the other. The plane cannot make a sharp turn at the end of the runway, so it would then be headed in the wrong direction to land.
13. (a) 13.2 rad (b) 6.28 rad (c) 0.012 7 degree
(d) 0.059 7 degree
15. (a) 1.93 μm (b) 3.00λ (c) It corresponds to a maximum. The path difference is an integer multiple of the wavelength.
17. 48.0 μm
19. $E_1 + E_2 = 10.0 \sin(100\pi t + 0.927)$
21. (a) 7.95 rad (b) 0.453
23. (a) green (b) violet
25. 512 nm
27. 96.2 nm
29. (a) 238 nm (b) The wavelength increases because of thermal expansion of the filter material. (c) 328 nm
31. 4.35 μm
33. 1.20 mm
35. 39.6 μm
37. 1.62 cm
39. 1.25 m
41. (a) ~ 10^{-3} degree (b) ~ 10^{11} Hz, microwave
43. 2.52 cm
45. $20.0 \times 10^{-6}\ ^\circ\text{C}^{-1}$
47. 3.58°
49. 1.62 km
51. 421 nm
55. (b) 266 nm
57. $y' = (n - 1)tL/d$
59. (a) 70.6 m (b) 136 m
61. (a) 14.7 μm (b) 1.53 cm (c) −16.0 m
63. (a) 4.86 cm from the top (b) 78.9 nm and 128 nm
(c) 2.63×10^{-6} rad
65. 0.505 mm
67. 0.498 mm

CHAPTER 38

1. 4.22 mm
3. 0.230 mm
5. three maxima, at 0° and near 46° on both sides
7. 0.016 2
9.

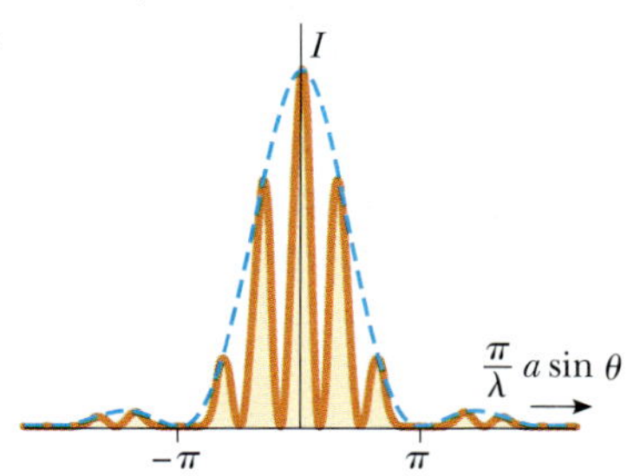

11. 1.00 mrad
13. 3.09 m
15. 13.1 m
17. Neither. It can resolve no objects closer than several centimeters apart.
19. 7.35°
21. 5.91° in first order, 13.2° in second order, 26.5° in third order
23. (a) 478.7 nm, 647.6 nm, and 696.6 nm (b) 20.51°, 28.30°, and 30.66°
25. three, at 0° and at 45.2° to the right and left.

27. (a) five orders (b) ten orders in the short-wavelength region
29. 2
31. 14.4°
33. The crystal cannot produce diffracted beams of visible light. Bragg's law cannot be satisfied for a wavelength much larger than the distance between atomic planes in the crystal.
35. (a) 54.7° (b) 63.4° (c) 71.6°
37. 60.5°
39. (b) For light confined to a plane, yes.

$$\left|\tan^{-1}\left(\frac{n_3}{n_2}\right) - \tan^{-1}\left(\frac{n_1}{n_2}\right)\right|$$

41. (a) 0.875 (b) 0.789 (c) 0.670 (d) You can get more and more of the incident light through the stack of ideal filters, approaching 50%, by reducing the angle between the axes of each one and the next.
43. (a) 6 (b) 7.50°
45. (a) 0.045 0 (b) 0.016 2
47. 632.8 nm
49. (a) 25.6° (b) 19.0°
51. 545 nm
53. (a) 3.53×10^3 grooves/cm (c) 11 maxima
55. 4.58 μm $< d <$ 5.23 μm
57. 15.4
59. (a) 41.8° (b) 0.593 (c) 0.262 m
61. (b) 3.77 nm/cm
63. (b) 15.3 μm
65. ϕ = 1.391 557 4 after 17 steps or fewer
67. a = 99.5 μm ± 1%

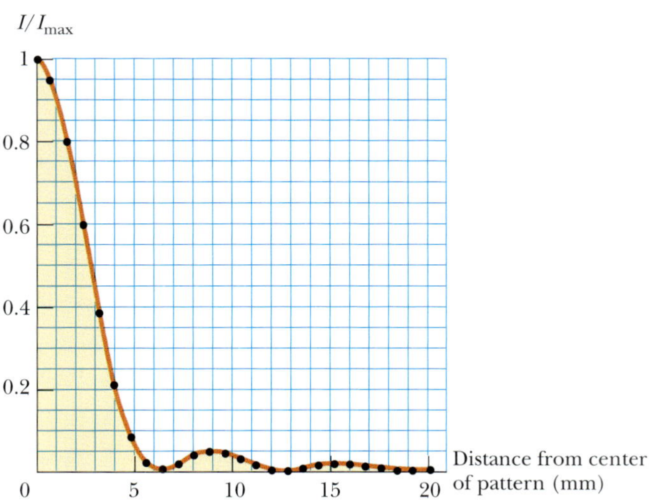

CHAPTER 39

5. 0.866c
7. 1.54 ns
9. 0.800c
11. (a) 39.2 μs (b) It was accurate to one digit. Cooper aged 1.78 μs less in each single orbit.
13. (a) 20.0 m (b) 19.0 m (c) 0.312c
15. 11.3 kHz
17. (b) 0.050 4c
19. (a) 0.943c (b) 2.55 km
21. B occurred 4.44×10^{-7} s before A
23. 0.960c
25. (a) 2.73×10^{-24} kg · m/s (b) 1.58×10^{-22} kg · m/s (c) 5.64×10^{-22} kg · m/s
27. 4.50×10^{-14}
29. 0.285c
31. (a) 5.37×10^{-11} J (b) 1.33×10^{-9} J
33. 1.63×10^3 MeV/c
35. (a) 938 MeV (b) 3.00 GeV (c) 2.07 GeV
39. (a) 0.979c (b) 0.065 2c (c) 0.914c = 274 Mm/s (d) 0.999 999 97c; 0.948c; 0.052 3c = 15.7 Mm/s
41. 4.08 MeV and 29.6 MeV
43. smaller by 3.18×10^{-12} kg, which is too small a fraction of 9 g to be measured
45. 4.28×10^9 kg/s
47. (a) 26.6 Mm (b) 3.87 km/s (c) -8.34×10^{-11} (d) 5.29×10^{-10} (e) $+4.46 \times 10^{-10}$
49. (a) a few hundred seconds (b) ~ 10^8 km
51. (a) $u = c\left(\dfrac{2H + H^2}{1 + 2H + H^2}\right)^{1/2}$, where $H = K/mc^2$

(b) u goes to 0 as K goes to 0. (c) u approaches c as K increases without limit.

(d) $a = \dfrac{\mathcal{P}}{mcH^{1/2}(2 + H)^{1/2}(1 + H)^2}$

(e) $a = \dfrac{\mathcal{P}}{mc\,(2H)^{1/2}} = \dfrac{\mathcal{P}}{(2mK)^{1/2}}$, in agreement with the nonrelativistic case.

(f) a approaches $\mathcal{P}/mcH^3 = \mathcal{P}m^2c^5/K^3$ (g) As energy is steadily imparted to the particle, the particle's acceleration decreases. It decreases steeply, proportionally to $1/K^3$ at high energy. In this way, the particle's speed cannot reach or surpass a certain upper limit, which is the speed of light in vacuum.
53. 0.712%
55. (a) 76.0 min (b) 52.1 min
57. (a) 0.946c (b) 0.160 ly (c) 0.114 yr (d) 7.50×10^{22} J
59. yes, with 18.8 m to spare
61. (b) For u small compared with c, the relativistic expression agrees with the classical expression As u approaches c, the acceleration approaches zero, so the object can never reach or surpass the speed of light.
(c) Perform the operation $\int(1 - u^2/c^2)^{-3/2}du = (qE/m)\int dt$ to obtain $u = qEct(m^2c^2 + q^2E^2t^2)^{-1/2}$ and then $\int dx = \int qEct(m^2c^2 + q^2E^2t^2)^{-1/2}dt$ to obtain $x = (c/qE)[(m^2c^2 + q^2E^2t^2)^{1/2} - mc]$
63. (a) $M = \dfrac{2m\sqrt{4 - u^2/c^2}}{3\sqrt{1 - u^2/c^2}}$

(b) $M = 4m/3$. The result agrees with the arithmetic sum of the masses of the two colliding particles.
65. (a) The refugees conclude that Tau Ceti exploded 16.0 yr before the Sun. (b) A stationary observer at the midpoint concludes that Tau Ceti and the Sun exploded simultaneously.
67. 1.82×10^{-3} eV

Locator note: **boldface** indicates a definition; *italics* indicates a figure; *t* indicates a table.

Standard Abbreviations and Symbols for Units

Symbol	Unit	Symbol	Unit
A	ampere	K	kelvin
u	atomic mass unit	kg	kilogram
atm	atmosphere	kmol	kilomole
Btu	British thermal unit	L	liter
C	coulomb	lb	pound
°C	degree Celsius	ly	lightyear
cal	calorie	m	meter
d	day	min	minute
eV	electron volt	mol	mole
°F	degree Fahrenheit	N	newton
F	farad	Pa	pascal
ft	foot	rad	radian
G	gauss	rev	revolution
g	gram	s	second
H	henry	T	tesla
h	hour	V	volt
hp	horsepower	W	watt
Hz	hertz	Wb	weber
in.	inch	yr	year
J	joule	Ω	ohm

Mathematical Symbols Used in the Text and Their Meaning

Symbol	Meaning
$=$	is equal to
$\equiv$	is defined as
$\neq$	is not equal to
$\propto$	is proportional to
$\sim$	is on the order of
$>$	is greater than
$<$	is less than
$\gg(\ll)$	is much greater (less) than
$\approx$	is approximately equal to
Δx	the change in x
$\sum_{i=1}^{N} x_i$	the sum of all quantities x_i from $i = 1$ to $i = N$
$\lvert x \rvert$	the magnitude of x (always a nonnegative quantity)
$\Delta x \rightarrow 0$	Δx approaches zero
$\frac{dx}{dt}$	the derivative of x with respect to t
$\frac{\partial x}{\partial t}$	the partial derivative of x with respect to t
$\int$	integral